The Greenland Entomofauna

Fauna Entomologica Scandinavica

Editors

N.P. Kristensen†
V. Michelsen

VOLUME 44

The titles published in this series are listed at *brill.com/fes*

The Greenland Entomofauna

An Identification Manual of Insects, Spiders and Their Allies

Edited by

Jens Böcher, Niels P. Kristensen†, Thomas Pape and Lars Vilhelmsen

BRILL

LEIDEN | BOSTON

Cover illustration: Upper: *Dolichopus groenlandicus* on leaf of *Dryopteris* sp.; Middle: *Nysius groenlandicus*, pair in copulation on fruit head of *Dryas* sp.; Lower: *Pardosa groenlandica*, male.

Library of Congress Cataloging-in-Publication Data

The Greenland entomofauna : an identification manual of insects, spiders and their allies / edited by Jens Böcher, Niels P. Kristensen, Thomas Pape and Lars Vilhelmsen.
pages cm. -- (Fauna entomologica Scandinavica ; volume 44)
Includes bibliographical references and index.
ISBN 978-90-04-25640-8 (hardback : alk. paper) -- ISBN 978-90-04-26105-1 (e-book) 1. Insects--Greenland--Identification. 2. Spiders--Greenland--Identification. 3. Myriapoda--Greenland--Identification. I. Böcher, Jens, editor. II. Kristensen, Niels P., editor. III. Pape, Thomas, 1960- editor. IV. Vilhelmsen, Lars, editor. V. Series: Fauna entomologica Scandinavica ; v. 44.

QL482.G83G73 2015
595.70998'2--dc23

2015015717

This publication has been typeset in the multilingual "Brill" typeface. With over 5,100 characters covering Latin, IPA, Greek, and Cyrillic, this typeface is especially suitable for use in the humanities. For more information, please see www.brill.com/brill-typeface.

ISSN 0106-8377
ISBN 978-90-04-25640-8 (hardback)
ISBN 978-90-04-26105-1 (e-book)

This book is printed on acid-free paper.

Contents

Foreword

While the Greenland 'entomofauna' certainly is depauperate by global standards, it is sufficiently large (now known to comprise close to 1.200 species) that the project of preparing the present comprehensive identification manual proved to be a long and laborious task. Its completion has been entirely dependent on the fact that our invitations to contribute have been met with favourable, indeed often enthusiastic, acceptance from a large international group of distinguished specialists.

Undoubtedly, an important reason for this response is that the while the Greenland fauna is a young one with little (if any? – time will show) endemism, it is indeed of special interest from a number of perspectives: It is *arctic*, hence composed of taxa whose adaptations to extreme habitats pose challenging questions to autecologists and comparative physiologists. It is *isolated*, but located at the *gateway between North America and Europe*; hence the origins, ages and invasion modes of its different elements have been intensely discussed for many years, especially by Scandinavian and British biogeographers. Last but not least, it is a general expectation that the biota at high latitudes will prove particularly sensitive to those *changes in global climate* which have come to figure so prominently on the public agenda during the last decades; hence monitoring of shifts in the Greenland distribution ranges of individual species now appears a matter of some urgency.

Hence, while Greenland does not belong to the geographic area covered in ordinary volumes in the *Fauna Entomologica Scandinavica* series (namely 'Fennoscandia and Denmark'), there is a rationale for publishing this book as a special volume in that series. Most likely many workers involved in the study of NW Palaearctic and/or NE Nearctic faunas will find it of considerable interest to learn which members of their chosen groups have been able to colonize Greenland – and will similarly find the overall picture of the Greenland entomofauna fascinating in its own right.

This account of the Greenland entomofauna is arguably long overdue. The latest comprehensive review of the 'insect' (including Collembola) fauna is that by Kai L. Henriksen more than 70 years ago. In total, he enumerated 595 species, a figure which through subsequent collecting efforts and critical re-examination of available material has now been almost doubled. Suitable identification tools have been available for only a small proportion of the taxa. A richly illustrated semipopular Danish-language guide to the Greenland terrestrial and limnic invertebrates (Jens Böcher: *Insekter og andre smådyr – i Grønlands field og ferskvand*, Atuagkat, Copenhagen 2001) is an early outcome of the preparations for the present work; it is comprehensive in the case of the beetles, moths & butterflies and some smaller insect groups, but only serves as a first introduction to the more species-rich taxa.

The initiation of the present manual project was made possible by a grant from the Danish *Environmental Protection Agency* through its DANCEA (**DAN**ish **C**ooperation for **E**nvironment in the **A**rctic) programme (project 50108) which funded three years of salary for Jens Böcher as managing editor. The Agency also provided substantial economic support to the production of this book at the completion/closure of the project. During the first project years it was realized that a southwestern area of Greenland, renowned for its floristic richness, was practically *terra incognita* in terms of entomology, and filling this knowledge gap appeared a mandatory requirement for a meaningful completion of the present work. A targeted entomological inventory of the area in question was, then, made by a specialist team (Jens Böcher, Arne Fjellberg, Ole Karsholt, Søren Langemark, Ole Martin, Sofie Tind Nielsen, Lars Vilhelmsen) constituting the 'Kissavik-expedition' in 2003, and we are greatly indebted to *Aase og Jørgen Münters Fond*, *Knud Højgaards Fond* and, in particular, the *Carlsberg Foundation* for the generous funding of that expedition. The *Carlsberg Foundation* also supported the project through the crucially important funding of Olga Makarova's preparatory work (including two stays in Copenhagen) on the large mite chapter. Financial support of the project from *Niels Bohr Legatet* (Royal Danish Academy of Sciences and Letters) and *Bøje Benzons Støttefond* (Natural History Museum of Denmark/Zoological Museum) is similarly much appreciated. J. Roháček's study on Greenland Sphaeroceridae was supported by grant No. IGS201103 of the Silesian Museum, Opava (Czech Republic).

Several colleagues in the Entomology Department, Natural History Museum of Denmark, helped with innumerable scientific and practical problems related to the preparation of this work. We are particularly grateful to David K.B. Cheung for his extensive and many-sided assistance with illustrations. Our sincere thanks for support and advice of many kinds are also due to Ole Karsholt, Ole Martin, Jan Pedersen, Jesper Julskov Schlie and Verner Michelsen. Geert Brovad, formerly of the Museum's photography unit, prepared and arranged numerous photographs. We are most grateful for the scanning of a multitude of diapositives by Henrik Havemann. We are also most grateful to Michiel Thijssen, Sabine Steenbeek,

Marjolein Schaake and Hubert Turner of the E.J. Brill publishing house for enjoyable collaboration on several issues in the production/publication process.

The publication of this work at a price that makes it available to a wide range of users was rendered possible by generous additional support from *Aage V. Jensens Fond*; for this we are immensely grateful.

Above all, of course, we are indebted to all the contributing taxonomist authors, without whose efforts this work would not have come into existence.

Copenhagen December 2014
The editors

List of Contributors

Achterberg, Cornelius van
Department of Entomology, Nationaal Natuurhistorisch Museum, Leiden, Netherlands. E-mail: achterberg @naturalis.nnm.nl

Ahola, Matti
Reisjärvi, Finland. E-mail mj.ahola@kotinet.com

Barták, Miroslav
Czech University of Life Sciences, Department of Zoology and Fisheries, Praha, Czech Republic. E-mail: bartak@af .czu.cz

Baur, Hannes
Department of Invertebrates, Natural History Museum, Bern, Switzerland. E-mail: hannes.baur@nmbe.unibe.ch

Behan-Pelletier, Valerie
Invertebrate Biodiversity Program, Agriculture and Agri-Food Canada, Ottawa, Ontario, Canada. E-mail: behanpv@agr.ge.ca or behanpv@gmail.com

Bird, Jeremy M
School of Natural Sciences and Psychology, Liverpool John Moores University, Liverpool, UK. E-mail: j.m.bird@ljmu .ac.uk

Böcher, Jens
Natural History Museum of Denmark, University of Copenhagen, Denmark. E-mail: jjbocher@snm .ku.dk

Bøg, Katrine
Institut for Miljø, Samfund og Rumlig Forandring (ENSPAC), Roskilde Universitetscenter, Roskilde, Denmark. E-mail: gkb@ruc.dk

Brodo, Fenja
Canadian Museum of Nature, Ottawa, Ontario, Canada. E-mail: fbrodo@sympatico.ca

Buhl, Peter N
c/o Natural History Museum of Denmark, University of Copenhagen, Denmark. E-mail: pnbuhl@hotmail.com

Dahl, Christine
(Formerly) Department of Systematic Zoology, Uppsala University, Sweden. E-mail: christine.dahl@telia.com

Disney, R Henry L
Department of Zoology, University of Cambridge, Cambridge, UK. E-mail: rhld2@hermes. cam.ac.uk

Dittmar, Katharina
State University of New York at Buffalo, Buffalo, New York, USA. E-mail: katharinad@gmail.com

Fjellberg, Arne
Tjörne, Norway. E-mail: arnecoll@gmail.com

Gammelmo, Øivind
Natural History Museum, University of Oslo, Oslo, Norway

Forshage, Mattias
Swedish Museum of Natural History, Stockholm, Sweden. E-mail: Mattias.Forshage@nrm.se

Gerecke, Reinhard
Tübingen, Germany. E-mail: reinhard-gerecke@uni .tuebingen.de

Gertsson, Carl-Axel
Lund, Sweden. E-mail: carl-axel.gertsson@mailbox .swipnet.se

Hastriter, Michael M L
Bean Life Science Museum, Brigham Young University, Provo, UT, USA. E-mail: michaelhastriter@ comcast.net

Haenni, Jean-Paul
Museum d'histoire naturelle, Neuchâtel, Switzerland. E-mail: Jean-Paul.Haenni@unine.ch

Heie, Ole E
Holte, Denmark. E-mail: o.e.heie@hotmail.com

Heraty, John M
Department of Entomology, University of California, CA, USA. E-mail: john.heraty@ucr.edu

Hodgson, Chris
Department of Biodiversity & Biological Systematics, National Museum of Wales, Cardiff, UK. E-mail: HodgsonCJ@Cardiff.ac.uk

Hodkinson, Ian D
Biological and Earth Sciences, Liverpool John Moores University, Liverpool, UK. E-mail: i.d.hodkinson@ljmu.ac.uk

Horsfield, David
National Museums Collection Centre, Edinburgh, UK. E-mail: d.horsfield@nms.ac.uk

Huber, John T
Natural Resources Canada, Ottawa, Canada. E-mail: huberjh@agr.gc.ca

Jaschhof, Mathias
Senckenberg German Entomological Institute Müncheberg, Germany. E-mail: mathias.jaschhof@senckenberg.de

Jensen, Frank
(Formerly) Naturhistorisk Museum, Århus, Denmark. E-mail: fj@nathist.aau.dk

Johanson, Kjell Arne
Swedish Museum of Natural History, Stockholm, Sweden. E-mail: kjell.arne.johanson@nrm.se

Jussila, Reijo
Zoological Museum, Department of Biology, University of Turku, Finland. E-mail: reijus@utu.fi

Karsholt, Ole
Natural History Museum of Denmark, University of Copenhagen, Denmark. E-mail: okarsholt@snm.ku.dk

Kristensen, Niels P
Deceased. Formerly Natural History Museum of Denmark, University of Copenhagen, Denmark.

Krzemińska, Ewa
Department of Invertebrate Zoology, Polish Academy of Sciences, Krakow, Poland. E-mail: krzeminska@isez.pan.Krakow.pl

Lantsov, Vladimir I
Tembotov Institute of Ecology of Mountain Territories, Kabardino Balkarian Research Center, Russian Academy of Sciences, Nalchic, Russia. E-mail: lantsov@megalog.ru

Láska, Pavel
Department of Zoology, Palacký University, Olomouc, Czech Republic. E-mail: laskap@seznam.cz

Lindegaard, Claus
(Formerly) Freshwater Biology Institute, University of Copenhagen, Denmark. E-mail: clindegaard@mail.dk

Lyneborg, Leif
Deceased. Formerly Natural History Museum of Denmark, University of Copenhagen, Denmark

Makarova, Olga
Institute of Ecology and Evolution, Russian Academy of Sciences, Moscow, Russia. E-mail: ol_makarova@mail.ru

Marusik, Yura M
Institute of Biological Problems of the North, Magadan, Russia. E-mail: yurmar@mail.ru

Mathis, Wayne N
(Formerly) National Museum of Natural History, Smithsonian Institution, Washington DC, USA. E-mail: mathisw@si.edu

Mazánek, Libor
Department of Zoology, Palacký University, Olomouc, Czech Republic. E-mail: syrphidae@centrum.cz

Michelsen, Verner
Natural History Museum of Denmark, University of Copenhagen, Denmark. E-mail: vmichelsen@snm.ku.dk

Munk, Thorkild
Deceased. Formerly Natural History Museum Aarhus, Aarhus, Denmark.

Murphy, William L
(Formerly) National Museum of Natural History, Smithsonian Institution, Washington DC, USA. E-mail: billmurphy8@sbcglobal.net

Nielsen, Søren A
Institut for Miljø, Samfund og Rumlig Forandring (ENSPAC), Roskilde Universitetscenter, Roskilde, Denmark. E-mail: san@ruc.dk

Nielsen, Tore R
Sandnes, Norway. E-mail: Tore@Nielsen.cc

Noyes, John S
Natural History Museum, London, UK. E-mail: jsn@nhm.ac.uk

Oosterbroek, Pjotr
NCB-Naturalis, Leiden, Netherlands. E-mail: p.oosterbroek@chello.nl

Ozerov, Andrey L
Zoological Museum, Moscow State University, Russia. E-mail: ozerov2455@rambler.ru

Pape, Thomas
Natural History Museum of Denmark, University of Copenhagen, Denmark. E-mail: Tpape@snm.ku.dk

Pinto, John D
Department of Entomology, University of California, Riverside CA, USA. E-mail: john.pinto@ucr.edu

Pollet, Marc
Royal Belgian Institute of Natural Sciences, Brussels, Belgium. E-mail: mpollet.doli@gmail.com

Rindal, Eirik
Natural History Museum, University of Oslo, Oslo, Norway. E-mail: eirikrindal@nhm.uio.no

Rohácek, Jindrich
Silesian Museum, Opava, Czeck Republic. E-mail: rohacek@szmo.cz

Simonsen, Thomas J
Natural History Museum, London, UK. E-mail: t.simonsen@nhm.ac.uk

Smith, Vincent S
Natural History Museum, London, UK. E-mail: v.smith@nhm.ac.uk

Söli, Geir
Natural History Museum, University of Oslo, Oslo, Norway. E-mail: g.e.e.soli@nhm.uio.no

Starý, Jaroslav
Silesian Museum, Opava, Czech Republic. E-mail: stary.cranefly@gmail.com

Strassen, Richard zur
Deceased. Formerly Forschungsinstitut Senckenberg, Frankfurt am Main, Germany.

Svensson, Bo W
c/o Department of Biology. University of Lund, Sweden. E-mail: Bo-W.Svensson@ecb.uu.se

Vilhelmsen, Lars
Natural History Museum of Denmark, University of Copenhagen, Denmark. E-mail: lbvilhelmsen@snm.ku.dk

Vilkamaa, Pekka
Finnish Museum of Natural History, Helsinki, Finland. E-mail: pekka.vilkamaa@helsinki.fi

Wilson, Michael
National Museum & Galleries of Wales, Cardiff, UK. E-mail: Mike.Wilson@museumwales.ac.uk

Zatwarnicki, Tadeusz
Department of Biosystematics, University of Opole, Poland. E-mail: zatwar@uni.opole.pl

CHAPTER 1

Introduction

Jens Böcher and Niels P. Kristensen†

The 'entomofauna' in the conventional sense of the term comprises the arthropods in terrestrial and limnic habitats, except for the Crustacea. Hence the present work is an identification manual to the Greenland members of the lineages conservatively grouped into the taxa Arachnida, Myriapoda and Hexapoda.

By means of this work it should be possible to identify the great majority of the Greenland representatives of these animal groups. In certain cases, however, the current taxonomic state does not permit consistent identification below generic level; this is, for instance, true for some dipteran families and Acari.

To which degree does the present treatment live up to contemporary expectations for "a good identification handbook"? Obviously the requirements spelled out by Hulcr (2009), and with (almost all of) which we agree in principle, are met to somewhat variable degrees in the following taxon accounts. It will, however, undoubtedly be generally appreciated that contributions to a collaborative work like the present, with its inevitable constraints in terms of time, space, material availability, etc., cannot be judged by the standards appropriate for individual studies devoid of such straight-jackets. Similarly there will undoubtedly be a general appreciation of the fact that when it comes to recruiting taxonomic specialists to revise any at all sizable material, the current situation is 'sellers' market': demand being far greater than supply. We have greatly appreciated every collaborator's acceptance to contribute, and we have accepted that constraints on the resources (including time!) of individual workers have necessitated a certain unevenness in the coverage of their contributions. So has, as already mentioned, the very uneven current 'taxonomic maturity' of individual groups.

While some editorial modifications have been made to enhance uniformity in presentation, certain details which are considered unlikely to cause confusion (for instance use of American and British spelling, or of Arabic or Roman numerals for segment numbers) have been left un-harmonized between chapters.

The chapter division adopted primarily serves a practical purpose, and chapters may have quite different coverages: an order, an assemblage of orders or a subset of a single order. In each chapter the illustrations are consecutively numbered, and the relevant references are placed at the end.

The choice of key characters has been made with the purpose of rendering the keys as simple/user-friendly as possible. Since the Greenland representation of most taxa is very depauperate, the keys can emphatically not be expected to be applicable in northern continental territories, let alone on a wider geographical basis.

In the individual species accounts synonyms are, as a general rule, given only where another name has been used in the preceding comprehensive treatment of the Greenland taxa in question. In most cases this will be Henriksen (1939), but for Acari-Oribatida and Collembola Hammer (1944), for spiders Larsen & Scharff (2003) and for Lepidoptera Wolff (1964). The descriptive sections, when present, mostly focus on characters useful for distinction between similar Greenland species, hence supplementing information presented in the keys. The sections on 'biology' contain information about habitat, food and phenology. The Greenland distribution and the general distribution of individual species is presented in a condensed way. For the purpose of simplicity, Greenland has been divided into a number of 'zoogeographic districts' (S, SW, NW, N, NE, SE). These are based on distributional patterns found among both arthropods and other organism and are characterized in chapter 4 based on Böcher (1988, p. 70). When a species is found in all districts, this pan-Greenlandic distribution is denoted by G.

For the general distribution, the following terms are used: nearctic (North America), palaearctic (Europe, northern Africa and temperate/subtropical Asia), circumpolar (all arctic areas) and holarctic (nearctic + palaearctic, sometimes with major gaps). From north to south the climatic distribution is termed arctic (including high arctic and low arctic), temperate (including subarctic, boreal = northern temperate, and southern temperate), and subtropical. A marked alpine occurrence may also be mentioned. In rare cases, even remote occurrences, *e.g.* in neotropical areas, are also mentioned.

The preparation of the present work led to the discovery of a number of undescribed species, many, if not all, of which will undoubtedly prove to occur also in high-latitude biotas elsewhere on the northern hemisphere. It was early decided that descriptions of new species would be misplaced in the present work, and hence these descriptions have appeared in separate taxonomic publications (Achterberg 2006, Gertsson 2004, Gertsson & Hodgson 2005, Hodgson 2005). In addition, a number of groups have elsewhere been treated in more detail, or from different perspectives (Makarova & Böcher 2009, Marusik *et al.* 2006, Michelsen 2006, Oosterbroek *et al.* 2007, Pollet & Böcher 2005).

References

Achterberg, C. van (2006): The Braconidae of Greenland. – *Zoologische Mededelingen* 80: 13–62.

Böcher, J. (1988): The Coleoptera of Greenland. – *Meddelelser om Grønland Bioscience* 26: 100 pp.

Gertsson, C.-A. (2004): The scale insects (Hemiptera: Coccoidea) of Greenland. – *Proceedings of the X International Symposium on Scale Insect Studies 19th–23rd April 2004*: 331–337.

Gertsson, C.-A. & Hodgson, C. (2005): Four new species of mealybug (Pseudococcidae) and a new species of soft scale (Coccidae: Coccoidea: Hemiptera) from Greenland and redescriptions of a mealybug and a soft scale from Arctic Canada. – *Zootaxa* 897: 1–24.

Hammer, M. (1944): Studies on the oribatids and collemboles of Greenland. – *Meddelelser om Grønland* 141 (3): 210 pp.

Henriksen, K.L. (1939): A revised index of the insects of Grønland. – *Meddelelser om Grønland* 119 (10): 112 pp.

Hulcr, J. (2009): What makes a good insect identification handbook? – *Systematic Entomology* 34: 402–405.

Hodgson, C. (2005): The adult males of Coccoidea (Hemiptera: Sternorrhyncha) known from Greenland. – *Zootaxa* 907: 1–48.

Larsen, S. & Scharff, N. (2003): The spiders of Greenland – a checklist (Arachnida: Araneae). – *Entomologiske Meddelelser* 71(1): 53–61.

Makarova, O.L. & Böcher, J. (2009): Diversity and geographical ranges of Greenland mites (Acari: Oribatida and Mesostigmata). *In*: Golovatch, S.I., Babenko, A.B., Makarova, O.L. & Penev, L. (eds) *Species and Communities in Extreme Environments.* Festschrift towards the 75th Birthday and a Laudatio in Honour of Academician Yuri Ivanovich Chernov, pp. 165–186.

Michelsen, V. (2006): Annotated catalogue of the Anthomyiidae, Fanniidae, Muscidae and Scathophagidae (Diptera: Muscoidea) of Greenland. – *Steenstrupia* 29(2): 105–126.

Oosterbroek, P., Brodo, F., Lantsov, V. & Stary, J. (2007): The Tipulidae and Limoniidae of Greenland (Diptera, Nematocera, Craneflies). – *Entomologiske Meddelelser* 75: 3–33.

Pollet, M. & Böcher, J. (2005): The Dolichopodidae (Diptera) of Greenland. – *Entomologiske Meddelelser* 73: 3–15.

Wolff, N. (1964): The Lepidoptera of Greenland. – *Meddelelser om Grønland* 159 (11): 74 pp.

CHAPTER 2

Arthropod structure and systematics

Niels P. Kristensen† and Jens Böcher

Discussions about the composition and interrelationships of the principal arthropod taxa have been as vigorous in the 21st century as ever in zoological history, but while these subjects are a significant scientific challenge in their own right, they are of little consequence for the purpose of an identification manual. Suffice it here to note that while there has long been near-universal agreement that the Arachnida are a 'natural group', in as much as they are considered to be monophyletic in the strict modern sense (*i.e.*, the group comprises all known descendants of its members' last common ancestor), the monophyly of the Myriapoda has been continuously debated, and the monophyly of the Hexapoda has recently become so too.

A comprehensive account of arthropod structure, function and phylogeny is evidently outside the scope of the present work; reference must be made to a text of invertebrate zoology such as Ruppert *et al.* (2004) or Brusca & Brusca (2003), as well as to the recent information-rich treatise edited by Minelli *et al.* (2013).

Arthropod body design

Few zoologists will contest the notion that the extant arthropod groups descended from an ancestor with many overall similar segments, most if not all of which bore one pair of segmented limbs. The arthropod body wall comprises a single cell layer ('epidermis') which secretes an outer covering cuticle mainly composed of chitin in a protein matrix. It is the shape and pigmentation of the cuticle that primarily determines the external appearance of an arthropod. The cuticle forms an outer skeleton ('exoskeleton') on which most of the animal's muscles are attached (other muscles surround inner organs such as the gut and genital ducts). Most of the muscle attachments are located on little-flexible parts of the body wall ('sclerites'), where the cuticle is hardened and usually darkened; the intervening flexible (and often unpigmented) body wall areas are referred to as 'membranes'. A generalized arthropod segment is equipped with a dorsal and ventral sclerite ('tergum' and 'sternum', respectively) with the segmental limbs arising from the lateral soft-bodied 'pleural' areas, which may bear sclerotizations between the limbs and the tergum.

The patterns of 'tagmosis' (or 'tagmatization'), *i.e.*, the integration of adjacent segments into functional units, 'tagmata' are diagnostic of the principal arthropod lineages. The most anterior tagma bears the mouth opening and a series of adjacent mouth appendages which because of their role in food uptake are more or less profoundly modified relative to the following locomotory limbs. This anterior tagma similarly bears the visual organs, and in myriapods and hexapods moreover a pair of antennae, on which numerous sensory organs (chemoreceptors in particular) are located. The opening of the gut is morphologically located on the hind end of the animal, but can topographically appear shifted more or less extensively forward. The location of the genital opening is also diagnostic of major arthropod lineages.

Small protuberances of the exoskeleton come in many forms and may contribute significantly to the appearance of more or less extensive body regions. Of special note are the 'sensilla', comprising, together with the visual organs, the principal receptors of external stimuli. A basic sensillum type is the socketed hair or 'seta', which arises from a flexible socket. Setae may be innervated and respond to mechanical or chemical stimuli, but chemoreceptors can also have other shapes (cones, plates etc.), and they may be elevated or sunk in depressions; their surfaces have single (larger) or numerous (minute) pores permitting access for molecules of external substances. Additional sensillum types exist, but it is mostly chemoreceptors on antennae and (particularly) the larger mechanoreceptive setae on certain trunk and limb regions that have taxonomic importance (with respect to number, size, position etc.) in many groups. Broad and flat ('lamellar') setae are referred to as 'scales'. They are particularly well known from the wings of moths and butterflies, but also form dense covers on some springtails, and occur in scattered members of other groups as well (*e.g.*, along wing veins of mosquitoes). Most scales are not innervated and this also applies to some 'ordinary' (hair-like) setae, particularly in cases where they form dense coverings.

The ground plan configuration of the paired segmental appendages has been much debated (for a recent review see Boxhall 2013). A subunit in an appendage is referred to as a 'podomere[1]', or as a 'segment' (like the units in

1 The termination -mere (from Greek *méros*, part, portion) is also encountered in terms such as 'antennomeres' (=subunits of the antenna), 'flagellomeres' (=subunits of the antennal flagellum of Insecta *s.str.*) and 'tarsomeres' (=subunits of the tarsus of the leg).

the trunk); it is typically moved by antagonistic muscles arising in the podomere proximad from it. The number of immediately identifiable podomeres varies across the principal arthropod lineages.

For obvious reasons most internal organ systems play little role for practical identification of arachnids and hexapods. Suffice it to note that the external openings ('spiracles') of the invaginated, branching tracheal (respiratory) systems of many taxa are often easily discernible and their location, structure and/or the modification of adjacent body regions is occasionally taxonomically informative. Similarly, the regions surrounding (or immediately inside) the openings of the genital systems in both sexes are often characteristically modified.

Systematization

Mainstream zoological systematics endeavours to classify animals according to their phylogeny, i.e., into a hierarchy reflecting their relationships in terms of relative recency of common ancestry. The 'maturity' of contemporary taxonomy varies considerably across the arthropods, and it is practically certain that some of the current concepts of higher taxa (*e.g.*, families) adopted in the present work will eventually prove to refer to entities that are not monophyletic in the strict sense, i.e., not including *all* descendants of their members' last common ancestor. While reference is here made, conservatively, to 'orders' of arthropods, it must be emphasized that these like all other supraspecific taxa are not in any sense 'equivalent' evolutionary entities, and no formal ranks are here ascribed to other higher taxa, which are above family-group rank and hence not covered by naming conventions required in the *International Code of Zoological Nomenclature*.

Arachnida

Arachnids are immediately distinguishable from other extant terrestrial/limnic arthropods by the absence of antennae and by presence of four pairs of locomotory limbs (early-instar mite juveniles excepted, they have just three). Tagmosis is also characteristic, with the body typically divided into an anterior prosoma and a posterior abdomen, alternatively called opisthosoma (Fig. **18.1**–1).[2] Again the mites are aberrant in lacking this kind of body division; their tagmosis patterns come in different forms (Figs **18.3.1**–1&2).

2 **Emboldened** numerals refer to other chapters.

In most arachnids the prosoma is covered by a solid dorsal carapace, while its ventral surface has one or more sternal plates or is covered by the coxae of the appendages. Appendages are borne on the prosoma (and its mite homologues). They comprise an anterior pair called 'chelicerae' followed by the 'pedipalps' (or just 'palps') and the locomotory legs. Most arachnids are carnivores that capture and kill the prey (usually small arthropods) with the chelicerae. The pedipalps perform a variety of functions related to food uptake and in some taxa (male spiders being an important example) to sperm transfer. Seven podomeres are usually recognizable in arachnid locomotory legs; they are here referred to (from base to tip) as coxa, trochanter, femur, patella/genu (mite workers' term), tibia, tarsus and pretarsus/metatarsus.

While compound eyes as found in insects (and crustaceans) are entirely absent from arachnids, simple eyes may occur in variable numbers and arrangements. Invaginations to cuticle-lined respiratory cavities (book lungs and/or tracheae) are variably located. Some mites lack special respiratory organs, and gas exchange then takes place through the general body wall. The genital orifice of both sexes is found on ventral side of the second abdominal segment except in some mites. Indirect sperm transmission is characteristic of most arachnids and in many cases the male produces a spermatophore; frequently a more or less complex courtship behaviour precedes copulation.

Arachnid juveniles are overall similar to adults (leg numbers in 1st instar mites excepted), but characters necessary for species identification are often undeveloped before sexual maturation.

Currently about ten 'orders' are recognized in the Arachnida; just three of these are represented in Greenland: Araneae (**18.1**), Opiliones (**18.2**) and Acari (**18.3**).

The myriapods

The myriapods are an assemblage (perhaps indeed monophyletic) of arthropods which all have an elongate trunk comprising numerous leg-bearing segments, hence they are 'many-legged', as the scientific name implies. Like the crustaceans and hexapods they are 'mandibulate arthropods' in as much as their most anterior paired mouth appendages are biting/chewing mandibles. They moreover resemble hexapods in having a single pair of antennae and the segment between the latter and the mandibles being devoid of appendages. Most myriapod visual organs are aggregations of simple eyes. The segmental composition of the locomotory limbs varies across the principal myriapod lineages.

Myriapods are represented in Greenland only by a single (probably) native species of Chilopoda (**19**); another chilopod has been accidentally introduced.

The hexapods

Until recently there was a near-consensus that the hexapod, *i.e.*, 'six-legged', arthropods comprising the Insecta *s.str*. and the Collembola (plus the similarly 'entognathan' Diplura and Protura, unrepresented in Greenland) constitute a 'natural'/monophyletic group, and 'Hexapoda' was therefore unreservedly used as a taxon name in a phylogenetic systematization of the arthropods. Hexapod monophyly was seriously challenged first by Nardi *et al.* (2003) whose analysis based on mithochondrial DNA data placed the Collembola outside a putative monophylum comprising the remaining hexapods plus the Crustacea. Subsequent analyses of other data sets have variously recovered a conventional monophyletic Hexapoda, or a phylogeny in which the Diplura + Protura are outside a monophylum comprising (Insecta + Collembola) + Crustacea. Recently the pendulum is swinging back to hexapod monophyly being favoured; for further information reference can be made to reviews such as Giribet & Edgecombe (2013).

In any case, however, a suite of characteristic features are shared by the hexapods whether due to convergence or immediate common ancestry. The head tagma is always of apparently similar basic construction. It consists of a cranial capsule which bears visual organs, a pair of antennae (with numerous sensilla, particularly chemoreceptors, but also mechanoreceptors etc.), and the mouthparts. The ground pattern of the latter comprises an anterior median labrum (a plate-like 'upper lip') and three pairs of mouth appendages, viz., the mandibles, maxillae and labium. The hexapod mandible is usually a solid 'one-piece' appendage and the animal's principal biting/chewing instrument. The maxillae and labium bear segmented appendages, palps, the segment numbers, length, shape and vestiture (including sensillum complement) of which furnish many taxonomic characters; the same is true for two inner lobes borne mediad to the palps and termed, respectively, lacinia and galea (on the maxilla), and glossa and paraglossa (on the labium). The right and left labium components are medially fused, so that they form a median functional 'lower lip'.

Behind the head follows the thorax, formed by three segments, pro-, meso- and metathorax. The anterior part of the prothorax is largely membranous and allows a considerable mobility of the head relative to the trunk. In the ground pattern each thoracic segment bears a pair of walking legs above the sternal sclerotization, and more or less extensive (very large in the case of adult winged insects) pleural sclerites are located in the lateral body wall, partly encircling the leg base (and often considered to be derived from a basal limb segment, a 'subcoxa'). Hexapod leg segments (Fig. **6**–1) are referred to as coxa, trochanter, femur, tibia, tarsus and pretarsus, with the principal 'knee' being the femur-tibia joint; an obvious counterpart of the arachnid patella is lacking. Fusion of segments may occur; thus the tibia and tarsus are fused in Collembola (Fig. **6**–1) and larvae of polyphagan beetles (Fig. **12**–2B).

The posterior tagma, the abdomen, bears no walking legs; its configuration is otherwise very different in Collembola and Insecta. It always bears the genital opening near the posterior end.

Collembola

This high-rank hexapod taxon, immediately distinctive by the low number (six) of abdominal segments is well represented in Greenland (**6**).

Insecta

Insects are usually immediately identifiable as three-legged arthropods with an abdomen comprising 11 segments, of which at least nine are easily recognizable (segmental boundaries in the posteriormost region are often difficult, and occasionally impossible, to trace with any certainty in postembryonic stages). The most generalized 'orders' of 'true' insects, without native representatives in Greenland, are wingless throughout their life. In contrast, the ground pattern of the vast majority of insects, constituting the monophylum Pterygota ('winged insects'), is characterized by the presence of paired wings on the meso- and metathorax in the sexually mature life stage, the adult or 'imago'. In the extant fauna only the Ephemeroptera (**7**) have two winged stages.

The basic type of insect head bears a pair of large compound eyes (each individual unit of which is externally recognizable by an arched cuticular lens) and three simple eyes (ocelli), a pair and one unpaired. The insect antenna is uniform in basic structure, but highly variable in size and shape; it comprises two basal segments (scapus and pedicellus) followed by a variably segmented flagellum. Eyes and antennae may become reduced in taxa /life stages with life styles that have lessened dependency on certain sensory cues. Loss of ocelli (sometimes just the medial one) is particularly widespread, but more extensive eye reduction and even total blindness also occurs and is expectedly characteristic of endophytic, endoparasitic and certain soil-dwelling taxa. Ocelli are always absent in (Greenland) endopterygote larvae which also have the compound eyes replaced by simple larval eyes, 'stemmata'. Excessive antenna reduction is less frequent; larvae of Trichoptera (**14**) and many Diptera (**17**) are examples.

In the thorax the two wing-bearing segments (often referred to collectively as the pterothorax) are often markedly larger than the prothorax since they must provide space for the muscular 'flight motor'; they are characterized also by the large size of the pleural sclerotizations. The presence of a 'pleural ridge', formed by a linear thickening/infolding of the cuticle is a very conspicuous feature; it extends from the leg articulation upwards (sometimes pronouncedly forwards-slanting) to a process on the upper pleural margin which is a pivot for wing movement. The pleural regions in front of and behind the ridge are referred to, respectively, as the episternum and epimeron. The tergum of a thoracic segment is often referred to as the 'notum' (hence one speaks about pro-, meso-, and metanotum). The pterothoracic nota may be transversely divided (by furrows/strengthening ridges) into an anterior 'prescutum' followed by a 'scutum' and a posterior 'scutellum'; particularly the two last-mentioned sections may be clearly demarcated from each other.

An insect wing is a flat fold in the body wall between the notum and the pleuron, hence it has cuticle-bearing upper and lower walls, which are closely appressed against each other. It is strengthened primarily by a system of dichotomously branching longitudinal veins, which are convexities in the wing wall (upper, lower or both) that provide space for nerves, tracheae and blood flow; small sclerites at their bases articulate with processes on the notal and pleural margins to which movements of the thoracic 'box' are transferred. The longitudinal veins may be connected by a more or less rich complement of transverse veins. The configuration of the wing veins ('venation') provide identification characters of considerable importance in most winged insect orders, and while the ancestral venation pattern remains debatable, the one depicted in Fig. 1 is probably not far off the mark. Deviations from the depicted scheme mostly come in the form of reductions, basal fusions and distal anastomoses of veins, but some cases of particularly rich peripheral branching may also be derived. The interpretation and naming of veins differ even among workers on the same insect 'order', hence descriptions and figures in individual taxonomic treatments must always be checked to make sure one understands the meaning in the treatment at hand; *this also applies to contributions to the present work.*

Wing morphology varies immensely. All pterygotes apart from Ephemeroptera (7) and the extralimital dragonflies have the capacity to fold the wings backwards with their lower surface facing the body (or they descend from ancestors with this capacity). In this position the mesothoracic fore wings cover/protect the metathoracic hind wings, and hence they are often more rigid and have a stronger pigmentation pattern than the latter; the 'elytra' of Coleoptera (**13**) and 'hemelytra' of Hemiptera-Heteroptera (**10.6**) are extreme examples. The two wing

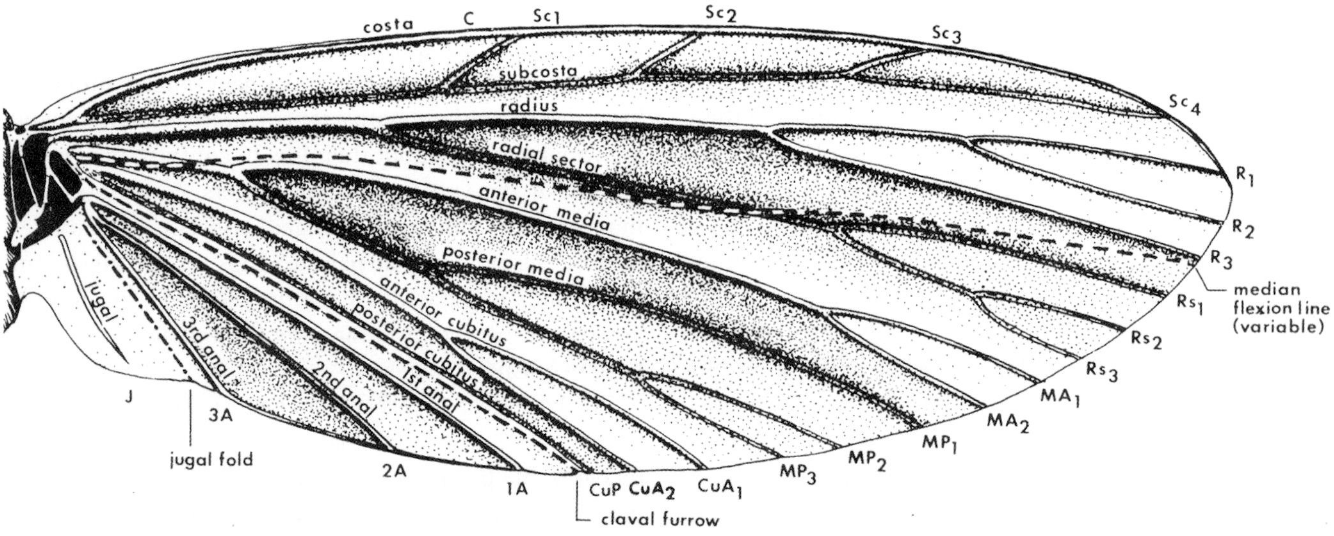

FIGURE 1 *Generalized insect wing (as reconstructed by Wootton,* Systematic Entomology *4: 92, 1979, with permission from Wiley, publishers) showing names of principal wing veins (with standard abbreviations) and folds. Many authors do not distinguish between radius and radial sector, and between anterior and posterior media, but refer to all veins in the two systems as 'R' and 'M' veins respectively.*

pairs may be overall uniform and contribute significantly and independently to flight, or the two become linked together and are moved as an entity; such linking has happened independently on several occasions and by a variety of structural devices. Alternatively, one wing pair may have come to play a dominant role in flight, in some cases to the extent that the other pair is completely lost and/or functionally transformed (Diptera, **17**; winged male Coccoidea, **10.4**). In insects with certain life-style strategies selection has favoured abandoning the costly resource investment in wings and their muscular motor. 'Secondary winglessness' is therefore a very widespread phenomenon within most pterygote insect orders, and it even characterizes all members of sizable taxa (parasitic lice, **8.2**; fleas, **16**); sometimes wing development is polymorphic within a single species. In the keys and descriptions in this book the abbreviations 'FW' (fore wings) and 'HW' (hind wings) are often used.

The ground pattern of the pterygote leg has the tarsus subdivided into five tarsomeres; reduction of this number (through fusion and perhaps sometimes loss) has occurred innumerable times. The pretarsus usually bears paired claws and an additional complement of tiny paired and/or unpaired processes which may be taxonomically useful and to which largely order-specific names have been applied. Legs may become reduced or entirely lost, particularly in taxa/stages with concealed life-style.

In the abdomen of winged insects the first segment is always more or less complexly united with the metathorax and hence modified. The following 'pregenital segments' (up to VII or VIII) are often overall uniform with simple terga and sterna and ample lateral pleural membranes. In both sexes the posterior abdominal segments are more or less extensively modified to serve sperm transfer (which in all Greenland pterygotes is a genuine gonopore-to-gonopore copulation) and, in females, oviposition. An 'ovipositor' comprising paired ventral appendages on segments VIII and IX pertains to the ground pattern of female insects, but in the Greenland fauna it is well developed only in some representatives of the Hemiptera (**10**), Thysanoptera (**9**) and in the Hymenoptera (**11**). The sometimes truly elaborate male genital/postgenital segments (which in many cases possess 'gonopods', appendages that are likely derivatives of the segmental limbs on IX) are often even more taxonomically informative at species level than their female counterparts. In the insect ground pattern the limbs on segment XI are long, annulated and thread-like appendages ('cerci'), while the dorsal part of the segment is produced into a somewhat similar median 'caudal (or 'terminal') filament. In the native Greenland insect fauna 'tails' of this kind are, however, retained only in the Ephemeroptera (**7**), and the median one is even lacking in the adult of the single species present. A formidable, and in many cases order-specific, nomenclature has been applied to the parts of the postabdominal segments in both sexes, and the statements above (concerning wing venation) about the necessity of checking the meaning of terms used in any given taxonomic treatment must be re-emphasized here.

Insects respire by means of a tracheal system which opens on a maximum of 10 pairs of spiracles situated laterally on the thorax (near the pro-meso- and meso-metathoracic boundaries) and the first eight abdominal segments (often in the pleural membranes). Reduction in number of functional spiracles is widespread, and particularly pronounced in many aquatic forms; in extreme cases such as, *e.g.*, trichopteran larvae (**14**), all spiracles are closed (except during moulting, when they open to permit the withdrawal of the old tracheae).

The primitive developmental mode in insects is via a series of juvenile stages, 'nymphs', which are overall similar to the adults (though sometimes with special structures which are obliterated at the final moult), but the genital-segment specializations are undeveloped, at least in the early instars, and so are the wings, which (when present) only are small dorsolateral lobes on the pterothoracic segments. Members of this evolutionary 'grade' within the winged insects are referred to as being 'exopterygote'. However, the vast majority of the insects belong to a likely monophyletic entity, the Endopterygota, which are characterized by a so-called 'complete metamorphosis'. Here the first juvenile stage (which comprises several instars), the 'larva' (note that the term 'larva' is by some authors also applied to nymphs of non-endopterygotes), differ markedly from the adults. Like young nymphs they have undeveloped genitalia, but moreover their legs (when retained) and antennae are relatively shorter than the adult counterparts and their body sclerotization is almost always much weaker and less extensive. Their eyes are special larval structures, which are broken down at metamorphosis when the adult's compound eyes develop, and they are present just as small groups of simple units; no Greenland endopterygote larva has externally facetted compound eyes. The wing rudiments are never external buds, but develop in epidermal pockets below the cuticle. The last immature stage, the 'pupa', is structurally comparable to a non-endopterygote nymph, with sizable external wing buds. Pupae do not feed and usually have little mobility, but in some cases the adult insect can move actively while still enclosed in the pupal skin (the 'pharate adult'). The most primitive pupal type has movable mandibles ('decticous' type) and has free antennae, wing buds and legs ('exarate'), but specialized types can have the mandibles (when present) immovable ('adecticous') and/or have the wings buds and appendages glued to the general body surface ('obtect'). Many pupae rest in a special shelter (cocoon) prepared by the last-instar larva just before pupation; silken threads, secreted by labial glands or Malpighian tubules, are often an integral component of such shelters.

Only four of 14 recognized exopterygote orders have indigenous representatives in Greenland: Ephemeroptera (7), Psocodea (8), Thysanoptera (9), Hemiptera (10). In contrast, seven of the *c.* 10 endopterygote orders are represented: Hymenoptera (11), Neuroptera (12), Coleoptera (13), Trichoptera (14), Lepidoptera (15), Siphonaptera (16), Diptera (17). The last-mentioned order accounts for close to one-third of the species richness in the currently known Greenland entomofauna.

Key to the 'classes' and 'orders' (and some major subordinal taxa) of Greenland arachnids and hexapods (the single chilopod is unmistakable with its elongate, near-homonomous, multilegged trunk, Fig. 19–1)

1. With usable wings (FW sometimes of' 'hemelytra' (*e.g.,* Figs **10.6**–2C–D) or 'elytra'-type (*e.g.,* Figs **13**–1A & 18–25) .... (hexapods, part) 2
– Without wings, or wings reduced to tiny stumps, useless for flight 14
2. Two pairs of wings (usually differently developed and/or uneven in size) 3
– Only FW developed 12
3. FW, at least their basal part, developed as tough elytra, more or less extensively concealing the abdomen and membranous hind wings (when present) 11
– FW not developed as 'elytra' (but sometimes more pigmented and slightly more firm than HW), their venation always clearly discernible 4
4. Wing surfaces with extensive covers of hairs or scales 5
– Wings with hairs/scales on margins and veins only, or completely naked 6
5. Wings (at least FW) with more or less open cover of hairs .... Trichoptera (caddis-flies, **14**)
– Wings completely covered by coloured scales .... Lepidoptera (moths and butterflies, **15**)
6. Small (≤5 mm long) and slender insects, wings narrow, near-linear, with marginal fringes of very long hairs, facies as Fig. **9**–1) .... Thysanoptera (thrips, **9**).
– Wings not near-linear, sizes and body proportions very variable 7
7. Hind end of abdomen with 2 long thread-like "tails" .... Ephemeroptera (mayflies, **7**)
– Hind end of abdomen without long "tails" 8
8. FW and HW of almost similar size and shape, with dense venational network; several veins with small apical forks (Fig. **12**–1A–B) .... Neuroptera (lacewings, **12**)
– Wings markedly dissimilar in size and/or shape, with much more open venation; never several veins with small apical forks 9
9. Antennae very short, flagellum a delicate bristle-like 'arista'. FW markedly more firm and pigmented than HW. Facies as Fig. **10.5**–1 .... Hemiptera-Cicadiformia (leafhoppers, **10.5**)
– Antennal flagellum long and not particularly delicate. FW not more firm, and rarely with more pigment than HW 10
10. FW usually with one or more cells closed by 'internal' veins (venation may be highly reduced in miniaturized taxa); HW on front margin with several hooks linking it to FW. Tarsi usually with 5, and at least with 3 tarsomeres. Mandibles 'normal', easily observable ....Hymenoptera (bees, wasps, **11**)
– FW with closed cells at margins only; HW on front margin at most with a single hook linking it to FW. Tarsi with 1–2 tarsomeres. Mandibles (like maxillae) stylet-like, concealed in labial sheath of piercing rostrum ....Hemiptera-Sternorrhyncha (part: psyllids, **10.2** & aphids, **10.3**)
11. Only basal part of FW developed as elytron, distal part membranous, with discernible veins. Mouthparts developed as a sucking rostrum. Facies as Figs **10.6**–2C–D .... Hemiptera-Heteroptera (true bugs)
– Entire FW developed as elytron. Mouthparts generalized, biting. Facies as Figs **13**–18–25)............... ....Coleoptera (beetles)
12. Tiny (ca. 1 mm) insects. Wings with 2 longitudinal veins. Tarsi 1-segmented, with single pretarsal claw.... .... Hemiptera-Sternorrhyncha (male Coccoidea, **10.4**)
– Size highly variable. Wing venation rarely (Cecidomyiidae, **17.12**) reduced to 2 longitudinal veins. Tarsi 5-segmented, with paired pretarsal claws .... Diptera (true flies, **17**)
13. With legs 14
– Without legs 24
14. With 3 pairs of legs and one pair of antennae ....(Insecta, part) 15
– With 4 pairs of legs and no antennae .... (Arachnida) 25
15. Abdomen with only 6 segments. With a ventral tube on abdominal segment I and usually a fork-like 'jumping organ' on the fourth segment (Fig. **6**–1).... ....Collembola (springtails, **6**)

– Abdomen with 11 segments (segments IX, X and XI sometimes not easily discernible or distinguishable), without ventral tube on abdominal segment I ... 16

16. Ectoparasites on birds and mammals 17

– Not such ectoparasites 18

17. Dorsoventrally flattened, body sclerotization variable. Tarsi with 1–2 tarsomeres. Facies as Figs **8.2**-2 .... Psocodea (part, **8.2**)

– Laterally compressed, body strongly sclerotized. Tarsi with 5 tarsomeres. Facies as Figs **16**-1A) .... Siphonaptera (fleas, **16**)

18. With compound eyes 19

– With 1–6 simple eyes, or eyeless ....(endopterygote insect larvae) 20

19. Aquatic. Abdomen with 3 'tails' (Fig. 7–1B) .... Ephemeroptera (mayfly nymphs, 7)

– Terrestrial, abdomen without such 'tails' ...A diverse assemblage of adult insects belonging to otherwise fully winged taxa will key out here: Psocodea: *Trogium pulsatorium* (**8.1**); Hemiptera-Sternorrhyncha: parthenogenetic females of Aphidoidea (**10.3**), female Coccoidea (**10.4**); Hemiptera-Heteroptera (Fig. **10.6**): *Nabis flavomarginatus, Cimex lectularius*; some Thysanoptera (**9**); some Hymenoptera-Encyrtidae (**11.5**); Hymenoptera-Scelionidae (**11.11**): female *Baeus*; Hymenoptera-Ichneumonidae (**11.15**): *Gelis nigritulus*; some female Hymenoptera-Dryinidae (**11.16**); Lepidoptera (**15**): female *Operophthera bruceata*.

20. With prolegs on some abdominal segments, almost exclusively herbivores 21

– Without abdominal prolegs, life-styles highly variable 22

21. More than 5 pairs of prolegs, from abdominal segment II backwards; prolegs without cuticular hooks. Only a single pair of (large) simple eyes (Fig. **11.2**–2) ....Hymenoptera (part, sawfly larvae, **11**),

– At most 5 pairs of prolegs, on abdominal segment X and II–VI, *e.g.,* Fig. **15**–9H–I (or V–VI, Fig. **15**–9C–D, or just VI, Fig. **15**–6A–B); prolegs with apical complement of cuticular hooks. Six pairs of simple eyes (Fig. **15**–2B–C) ....Lepidoptera (moth and butterfly larvae, **15**)

22. Maxillae stylet-like, united with long and sickle-shaped mandibles ....Neuroptera (lacewing larvae, **12**)

– Maxillae not stylet-like, discrete 23

23. Aquatic, abdomen with tracheal gills, in cases of foreign bodies .... Trichoptera (caddis-fly larvae, **14**)

– Mostly terrestrial, if aquatic not case-makers, and (apart from *Gyrinus*, Fig. **13**–8) devoid of tracheal gills .... Coleoptera (beetle larvae, **13**)

24. Trunk with segmental whorls of prominent setae. Head capsule well developed. In terrestrial locations inhabited by bird or mammal hosts .... Siphonaptera (flea larvae, **16**)

– Trunk segments without whorls of prominent setae A diverse assemblage of legless endopterygote larvae will key out here. The vast majority of those encountered will be Diptera (**17**; including many aquatic taxa and also all the characteristic cyclorrhaphan 'maggots' with the head capsule completely desclerotized, fragmented and concealed within the thorax). Furthermore Hymenoptera-Apocrita (**11.3**–**18**; mostly endoparasitoids in other insects, vespids and apids in characteristic nests) and Coleoptera-Curculionoidea (**13**; head capsule strongly sclerotized, trunk wall with prominent longitudinal and transverse folds; endophytic or in soil).

25. Body markedly constricted between cephalothorax and abdomen. Size highly variable, several taxa >5 mm .... Araneae (spiders, **18.1**)

– Body without such constriction 26

26. Body outline near-elliptical. Longest legs >2× body length, which in adults of the single Greenland species is about 5 mm .... Opiliones (harvestmen, **18.2**).

– Body outline variable, often anteriorly tapering. Length always < (usually<<) 4 mm, with the notable exception of *Ixodes uriae* (Fig. **18.3**–3). Legs < twice body length .... Acari (mites, **18.3**)

References

Boxhall, G.A. (2013): Arthropod limbs and their development. *In*: Minelli, A., Boxshall, G. & Fusco, G. (eds): *Arthropod Biology and Evolution. Molecules, Development, Morphology*, pp. 241–267. – Springer, Heidelberg/New York/Dordrecht/London.

Brusca, R.C. & Brusca, G.J. (2002): *Invertebrates* (2nd edition). – Sinauer Associates, Sunderland.

Giribet, G. & Edgecombe, G.D. (2013): The Arthropoda: a phylogenetic framework. *In*: Minelli, A., Boxshall, G. & Fusco, G. (eds): *Arthropod Biology and Evolution. Molecules, Development, Morphology*, pp. 17–40. – Springer, Heidelberg/New York/Dordrecht/London.

Minelli, A., Boxshall, G. & Fusco, G. (eds) (2013): *Arthropod Biology and Evolution. Molecules, Development, Morphology.* – Springer, Heidelberg/New York/Dordrecht/London.

Nardi, F., Spinsanti, G., Boore, J.L., Carapelli, A., Dallai, R. & Frati, F. (2003): Hexapod Origins: Monophyletic or Paraphyletic? – *Science* 299: 1887–1889.

Ruppert, E.E., Fox, R.S. & Barnes, R.D. (2004): *Invertebrate Zoology* (7th edition). – Brooks/Cole, Belmont.

CHAPTER 3

The Greenland entomofauna in its arctic context

Jens Böcher

The present account is a succinct outline. For more detailed information with abundant references to the primary literature see Bale *et al.* 1997; Born & Böcher (eds) 2001; Böcher 1988; Böcher & Petersen 1997; Chernov 1995; Corbet 1972; Danks 1981, 1986, 1990, 1991, 2006a; Downes 1962, 1964, 1965; Hodkinson 2005; Kevan & Danks 1986; Lee & Denlinger (eds) 1991; MacLean 1975; Oliver 1968; Storey & Storey 1988; Strathdee & Bale 1998; Sømme & Block 1991; Woodin & Marquiss (eds) 1997. The classical reviews by Downes can be particularly recommended, as can the very comprehensive and useful handbook by Danks (1981). See also the "CAFF 2013. Arctic Biodiversity Assessment" (Meltofte 2013).

FIGURE 1 *Subarctic low forest dominated by* Betula pubescens *and* Salix glauca. *Narsarsuaq, S.*

Environment

Greenland is almost totally positioned within the Arctic, which is defined as those areas of the Northern Hemisphere where the average temperature at sea level does not exceed 10°C during the warmest month of the year. The terrestrial Arctic is essentially identical to the land masses north of the tree line, so that proper forest does not exist in the Arctic.

The Arctic biome is characterized in general by low temperatures and a short growth season. That means long, cold winters including a period with continuous darkness, and short, cool summers with a period of continuous light. Typically there is snow cover for up to 8 months of the year. In a high arctic environment at 78°N, the mean monthly temperature is above 0°C (but below 5°C) for at most three months of the year, June-August, when the sun is constantly above the horizon. The short growth period and the changing light regime during the year separates the Arctic from montane regions further south, where the plants may grow throughout the year, but where the difference between day and night may be comparable to those between summer and winter in the Arctic.

The Arctic is geographically (climatically) subdivided into the Low Arctic and the High Arctic by the 5°C isotherm in warmest month. This subdivision coincides with some important biological boundaries, *e.g.,* true willow copses do not occur in the High Arctic.

The inland areas along the fiords of Southwest and South Greenland are actually not parts of the Arctic, but are referred to the Subarctic, the northernmost part of the temperate vegetation zone, because average temperatures in the warmest month exceed 10°C and the growth season lasts 4–5 months. Low birch forest and copse grow at protected sites (Fig. 1).

Some Danish botanists now divide the Greenland Arctic into three zones: Low-, Middle- and High Arctic, which are further subdivided into oceanic and continental parts. The northernmost, oceanic part of the High Arctic is also termed Arctic Desert (Bay 1998, T.W. Böcher 2000). Russian scientists generally subdivide the Arctic into three subzones (from south to north): Typical tundra, Arctic Tundra, and Arctic Desert, of which the two last correspond to the High Arctic. South of these are found the subarctic Forest Tundra and Northern Taiga (Aleksandrova 1980, Chernov 2002, Makarova 2000).

The biological significance of the light arctic "night" during summer (with midnight sun north of the Arctic Circle) is a positive one. The long daily period of light/sun enables the plants to carry out photosynthesis and the poikilothermic animals to be active most of the time, provided the temperature is sufficient. This partly compensates for the short growing season.

Habitats

In temperate and tropical regions terrestrial arthropods occur practically everywhere. The diversity of biotopes available decreases as one travels northwards and being

FIGURE 2 *Fell field, oligotrophic lakes and westward view towards the large nunatak area, Dronning Louise Land. Dove Bugt, NE.*

FIGURE 3 *Desert-like sandy area along a large melt water river. Most vegetation is made up by* Chamerion latifolium (*the Greenland national plant*). *Easternmost Søndre Strømfjord/Kangerlussuaq, nSW.*

FIGURE 4 *Steppe vegetation dominated by* Carex supina, Calamagrostis purpurascens, Poa glauca, Cerastium alpinum, Campanula gieseckiana. *Here the Hemiptera* Nysius groenlandicus, Chlamydatus pullus *and* Psammotettix lividellus *exhibit huge populations. Paradisdalen west of Søndre Strømfjord/Kangerlussuaq, nSW.*

FIGURE 5 *Tidal meadow in the high arctic North East. The vegetation is almost confined to* Stellaria humifusa *and* Carex subspathacea. *This is the habitat of the high arctic staphylinid beetle* Gnypeta cavicollis. *Zackenberg, NE.*

lowest in the extreme High Arctic (Fig. 4–4A), where only the sparsely occurring fresh water biotopes and vegetation-covered soils are inhabitable. The larvae of flies and midges – here the dominant insect groups – usually take several years to develop in such biotopes. The adults leave the water or soil to take up a brief aerial existence, sometimes lasting only for a few days or even hours.

The cold, unfettered winds that blow over the wide open spaces (Figs 2–5) of the Arctic cause considerable problems of chilling, desiccation, and displacement. The latter factor is especially important under arctic conditions, where appropriate life conditions often occur as small 'islands' in an otherwise huge, barren area. Close to the ground the microclimate offers reduced wind speeds, higher temperatures, especially in sunshine, and a reasonable humidity. This benign layer is of crucial importance to the small terrestrial animals of the Arctic, but its vertical extent is limited to a few cm. Extended arctic areas are characterized by patterned ground, polygons (Fig. 6), hummocks etc. Even small differences in the micro-topographical relief (Fig. 7) are responsible for significant differences in the microenvironment and thereby for a marked variety in the microclimate, making it suitable for a number of different arthropods. In general, exposed south-facing slopes (Fig. 8) offer the warmest conditions.

FIGURE 6 *Patterned ground in Northeast Greenland. The pattern is caused by the underlying permafrost. Vegetation mostly* Salix arctica, *mosses and lichens. NE.*

FIGURE 7 *Small differences in altitude give rise to great differences in plant community and insect associations. Uppermost (at left) a steppe-like community with* Carex *spp.and fruiting* Dryas *sp., at a little lower level heath of* Vaccinium uliginosum *(red autumn leaves) and further down (about 1.5 metres) tussocky marsh with many mosses and* Eriophorum scheuchzeri. Nysius groenlandicus *(Heteroptera) was exclusively found in the two uppermost plant communities. Zackenberg, NE.*

FIGURE 8 *Lush meadow vegetation at a hot spring close to Godhavn/Qeqertarsuaq (the orchids* Leuchorchis albida *and* Platanthera hyperborea *in addition to* Alchemilla glomerulans, Potentilla crantzii, Bartsia alpina *etc.). The carabid* Bembidion grapii *and the coccid* Arctorthecia cataphracta *are both common here. nSW.*
(PHOTO: M. PHILIPP)

The most coherent vegetation types in the Arctic are different kinds of heath (Figs 9–10) which cover vast areas.

The snow layer during winter is an extremely important ecological factor. Snow insulates the soil from the low air temperatures, and below a thick layer of snow (at least 20 cm) the conditions are relatively stable throughout the winter, providing moderately low temperatures, high humidity, and windless conditions. Many arctic arthropods are dependent on the protection of snow during winter, especially those that are not particularly frost resistant.

The disadvantage of being dependent on snow-protection is that a thick layer of snow prolongs the winter in that particular place. In the more extreme cases snow-patches do not disappear until late summer. In dry areas, especially in the High Arctic, snow-patches (Fig. 9) play an important role by supplying melt water to lower lying areas.

The body of warm-blooded animals, birds and mammals, is a special and well-protected biotope which in the Arctic harbours many ectoparasitic insects, including Psocodea and Siphonaptera.

In general it must be stated that the more high arctic a biotope, the more importance the abiotic factors have so that in the most extreme situations the biotic interrelationship probably only plays a minor role.

FIGURE 9 *Low arctic heath dominated by* Salix glauca, Cassiope tetragona *and* Vaccinium uliginosum. *Notice the extended snow areas in July. Western Disko/Qeqertarsuaq, nSW.*

FIGURE 10 *High arctic heath in autumn dominated by* Dryas octopetala, Cassiope tetragona *and* Vaccinium uliginosum (*red*). *Fligely Fjord, NE.*

General adaptations to arctic conditions

The structural, biological, and behavioural characteristics of arctic arthropods are mainly adaptations to the special arctic conditions: the general low temperature and short growing season. Of course, it is of crucial importance for the small ectothermic (poikilothermic) animals to be able to withstand cold, especially in the hibernating stages).

In species that occur from, *e.g.*, the Subarctic to the High Arctic, individuals often become smaller, darker and more hairy the further north they live. The reduction in size is probably due to poorer growing conditions, whereas the dark colouration and the hairiness optimize the animal's exploitation of solar radiation. In a number of cases it has been shown that the physiological rates of arctic insects are elevated compared to insects from other climatic regions (*e.g.*, Strømme *et al.* 1986).

Direct sunshine plays a tremendous role in the ability of the arctic arthropods to carry out their activities. In coastal areas, where there are comparatively few hours of sunshine, the number of species is likewise small (Born & Böcher 2001). Butterflies and other insects bask on the ground, with their body and wings positioned perpendicularly to the sun's rays (Kevan & Shorthouse 1970). They take flight when their body temperature is sufficiently elevated. An insect often flies just above the ground, following the undulations of the landscape, until its temperature has dropped so much that it is forced to land and once again heat up within the microclimate. As soon as the sun disappears behind a cloud or a mountain, insect activity is often almost halted, at least regarding flying Generally, the importance for arctic arthropods of behavioural thermoregulation can not be underestimated (Kevan *et al.* 1982; Kukal *et al.* 1988).

It has been established that a number of arctic arthropods have very low thresholds of activity and development. Some insects are able to fly at low temperatures, *e.g.*, mosquitoes at 1.7°C, and spiders can be active at subzero temperatures down to -8°C (Aitchison 1984, Hodkinson 2005). Most arctic soil microarthropods are still active at 0°C, and locomotory activity is not terminated until -5°C, which is 5–10°C lower than in related temperate species (Bale *et al.* 1997). On the other hand, the common and widespread Greenland seed bug (*Nysius groenlandicus*) has a threshold of activity around 10°C and a preferred temperature of about 30°C (Böcher & Nachman 2001).

The large, densely haired arctic bumblebees (*Bombus*) are exceptional in being to a certain degree able to produce their own heat by activating the flight muscles before flying. This allows them to take to the wing even in cloudy weather at low temperatures (down to 0°C). This behaviour also allows the bumblebees to maintain nest temperatures in excess of 30°C, to the benefit of the larvae and pupae (Pape 1983).

In the Arctic the daily variations in light intensity during summer is much smaller than at temperate latitudes, and the illumination is continuously at levels that further south inhibit the activity of nocturnal Lepidoptera. In spite of this, arctic noctuid moths (Noctuidae) show a distinct daily periodicity with the activity occurring at about the same time of day as in the same or related species from lower latitudes. Obviously the moths have adapted to the high light intensities during the arctic summer (Dreisig 1981).

On the other hand, the slight variations in light regime during the high arctic summer means that the clue found

FIGURE 11 *Autumn colours (mostly* Salix glauca, Vaccinium uliginosum, Betula nana) *at the hot spring area Østerlien, immediately north of the Arctic Station, University of Copenhagen. The herb vegetation is similar to that in Fig. 8. Godhavn/Qeqertarsuaq, nSW.*

further south of a decreasing day length, telling that winter is approaching, does not exist in the High Arctic.

Physiological adaptation to low temperatures. Disregarding the parasites of warm-blooded animals, there are only two biotopes at which the animals of the Arctic can avoid frost throughout the year: near the bottom of a lake and surroundings of hot springs (Fig. 11).

Thus the arctic arthropods must be able to withstand frost in their wintering stage. There are two fundamentally different ways of accomplishing this: by *super cooling* of the body fluids (freeze avoidance), or by exhibiting *freeze tolerance* (*e.g.,* Ring & Tesar 1981; Lee & Denlinger 1991; Danks, Kukal & Ring 1994; Danks 2006a).

Even pure water can be super cooled to a few degrees below zero if it is contained within a small compartment, like for example the body of an insect. The addition of certain soluble compounds (salts, sugars) will further lower the freezing point. Most terrestrial invertebrates in temperate regions are adapted to super cooling and will die if ice is formed in their body fluids. These animals prepare for the winter by producing a number of cryo-protective substances such as glycerol and various sugars (sorbitol, mannitoil, trehalose, glucose, and fructose). These compounds lower the freezing point of the hemolymph, in extreme cases to -50°C or more, when freezing occurs with lethal effect. In preparation for hibernation the animal rids itself of superfluous free water, empties its digestive tract, and finds a dry place. If ice crystals form around a grain of sand or a small amount of water in the intestine, they will grow and eventually cause the body fluids to freeze.

However, many arctic insects and other arthropods do tolerate freezing, but only of extracellular fluids, not the formation of intracellular ice crystals. Most insects and spiders that tolerate freezing produce special proteins in the hemolymph that actually facilitate the formation of ice crystals at temperatures slightly below 0°C in order to counteract super cooling. This keeps the production of ice crystals under control in such a manner that the dehydration of their cells takes place gradually. The formation of ice leads to an increased concentration of extracellular dissolved substances, and the elevated osmotic pressure draws water out of the cells. The above-mentioned cryo-protective sugars also facilitate the slow, controlled growth of a network of small ice crystals. This avoids the principal danger, which is the rapid formation and growth of large ice crystals that penetrate and destroy the tissues.

Tolerance to freezing is less hazardous than super cooling, which is based on an unstable equilibrium; if the equilibrium is disturbed, the result is certain death. This is why freeze tolerance is so common among arctic terrestrial invertebrates. According to Sformo *et al.* (2010) a third method of avoiding freezing, namely *vitrification*, should be added to the list of insect overwintering strategies.

Frost resistant species often spend the winter in completely unprotected locations, *e.g.,* on the ground where the snow blows away. Compared to species dependent on snow protection, they have the advantage of being able to immediately make use of the sun's warmth in spring, to be active and commence growth in the favourable microclimate on the soil surface.

All arctic soil mites and collembolans hitherto tested show some ability to survive anoxia. Overwintering soil animals can easily be encased in ice during freezing of soil water, and many species show a remarkable resistance by surviving several months exposed to anoxia in ice, *e.g.* the cryptostigmatic mite *Camisia anomia* (Hodkinson 2005).

Structural modifications. Adaptations of insects to arctic conditions also include reductions of various structures (Danks 1981) in cases where scarcity of food resources is counterbalanced by minimizing activity and by remaining within a protective microclimate close to the ground. Like mountain insects, many arctic insects have lost the ability to fly, and in some cases the wings as well. This considerably reduces the capacity to disperse, but also minimizes the risk of being transported to unsuitable locations by the wind. Thus many Chironomidae, Simuliidae, Tipulidae and other insects have reduced wings and sense organs and show other extreme adaptations to the extreme situations to which flying insects are exposed in the Arctic. In many cases mating takes place on the ground, and the large antennae and specialized eyes by which many male nematocerans identify females in an aerial swarm are usually reduced in the ground mating species (Downes 1962; Danks 1981).

Modifications of developmental and reproductive strategies. Reduction in the number of generations is common in arctic arthropods, partly caused by a time-constraint because of the shortness of the active season. For instance, in temperate regions aphids (Aphidoidea) go through a great number (up to 20 parthenogenetic and a single sexual; Dixon 1998) of generations each year. In the Arctic only two generations are normally present, and the usual shift from a woody winter host to an herbaceous summer host has often been deleted. The latter serves as host throughout the year (Lambers 1952; Strathdee *et al.* 1993; Bale *et al.* 1997). None the less, even during the short Spitsbergen summer, the endemic aphid *Acyrthosiphon svalbardicum*, monophagous on *Dryas octopetala*, in favourable years may produce a third annual generation: Some of the winged founder individuals (1st generation), which normally produce apterous, egg-laying males and females, sometimes produce another generation of winged individuals (Strathdee *et al.* 1993, Hodkinson *et al.* 2002b). However, the production of a second parthenogenetic generation could not be confirmed by Hullé & Bonhomme (2008).

Parthenogenesis may be regarded as a reduction of the male sex. Parthenogenesis is fairly common among arctic insects, found primarily in Diptera (Tipulidae, Chironomidae, Simuliidae etc.) and Coleoptera (Staphylinidae, Curculionidae), but also in Ephemeroptera, Trichoptera, Heteroptera, Coccoidea (and as a normal feature of the life cycle in Aphidoidea), and Hymenoptera. Parthenogenesis facilitates the dispersal of a species, because a single female is able to colonize a new territory. On the other hand, it results in reduced genetic variability of the species involved (Böcher 1971; Böcher & Enghoff 1984; Böcher & Nachman 2010, 2011).

Perenniality. Most temperate and tropical arthropods complete their life cycle from egg to reproducing adult within a year or less. A few arctic insects also have an obligatory life cycle that must be completed within a year, *e.g.,* the Greenlandic heteropterans *Nysius groenlandicus* and *Chlamydatus pullus* and generally the psyllids (MacLean 1983; Böcher 1971, 1975, 1976; Böcher & Nachman 2001; Hodkinson 1997). However, most species are "opportunists" that grow and carry out their activities as soon and as often as conditions (primarily temperatures) allow. Otherwise, they remain inactive or in torpor. Surveys are given by Danks (1999, 2006b).

The species often have a perennial development of indefinite length, dependent on whether the conditions for growth are favourable or not. Thus the larval life of midges (Chironomidae) may last from 2 to 7 years. The species *Tanytarsus gracilentus* completes two annual generations in Iceland but its larval development in the High Arctic is perennial (Danks 1981).

One of the most studied, well known, highly adapted arctic insects is the arctic woolly bear *Gynaephora groenlandica* (Wocke) (Erebidae-Lymantriinae). This species has a life cycle of at least 7 years. More than 6 years are spent as hardy, hairy larvae, whereas pupal, adult and egg stages are limited to a few weeks during a single summer (*e.g.* Kukal 1991, 1995; Kukal & Kevan 1987; Kukal *et al.* 1988a, 1988b; Morewood & Ring 1998).

In collembolans and mites growth and development depend directly on temperature, and life cycles may extend over several years without any apparent synchrony with the season. These free-running life cycles result in mixed-aged populations (Bale *et al.* 1997).

Trophic groups. The predominant trophic niche occupied by the terrestrial and limnic invertebrates of the Arctic is the *saprophagous* one. Many species living at the bottom of streams, ponds and lakes or in moist soil and feed on dead organic matter. Most collembolans and oribatid mites and many types of insect larvae, *e.g.,* most Diptera, belong to this trophic group. Several of them are actually omnivorous because along with the dead organic matter they ingest large amounts of bacteria, fungal hyphae, algal cells, and protozoans, probably the most important component of their food.

Compared to the saprophages, the *herbivores* are less prevalent in the Arctic and even rare in the High Arctic (Danks 1986). Few species live directly on the green parts of plants, *e.g.,* caterpillars (Lepidoptera), sawflies (Tenthredinoidea), aphids (Aphidomorpha) and jumping plant lice (Psylloidea). However, sometimes violent outbreaks of masses of lepidopterous larvae occur, when the birch shrubs and heaths are defoliated and the grass vegetation devastated over fairly large areas. This in particular happens in the southwestern part of Greenland and is caused by a limited number of geometrid and noctuid moths (Tenow 1990). A few insects feed on seeds (*e.g.,* the lygaeid bug *Nysius groenlandicus*; Böcher 1972) and some eat roots, *e.g.,* fly maggots and beetle larvae.

It thus appears that a number of herbivore niches are "empty", or under-exploited in the Arctic, the more the further north one gets (*e.g.,* Hodkinson 1997). The vast quantities of mosses in the Arctic are hardly exploited at all as a food item. Exceptions are the pill beetles

(Byrrhidae), which are specialists on mosses, and some species of water bears (Tardigrada). To some extent, the same applies to lichens, whereas fungi are eaten by, *e.g.,* fungus gnat larvae (Mycetophilidae).

Flowers do, however, play an important role for adult insects that eat pollen and suck nectar. Nectar provides fuel for their very energy-demanding flight activities – especially for flies, butterflies, and bumblebees. The protein-rich pollen is important in the egg production of many flies and, together with nectar, for the raising of bumblebee larvae. (*e.g.,* Kevan 1972; Philipp *et al.* 1990). Especially the flowers of *Dryas* spp. play a significant role for Arctic insects – and the vegetative parts are essential for a number of insects (*e.g.,* members of Aphididae, Tortricidae, Noctuidae, Nymphalidae, Agromyzidae).

The omnivorous saprophages of the soil are partly to be regarded as *carnivores.* True carnivores are found among the insects and mites. The spiders are exclusively carnivores that catch their prey in a number of different ways.

The term *parasitoid* is used of insects that lay their eggs in or on another arthropod host from which the larvae feed and finally kill. This applies to a large number of Hymenoptera, of which the Ichneumonidae has the greatest number of arctic representatives. Some of the larger species live on caterpillars. Tachinid flies (Tachinidae) have a similar life style.

True *parasites* comprise *ectoparasites*, which parasitize on the surface of other animals, and *endoparasites* that live inside other animals. Among arthropods many forms are ectoparasites in the Arctic, *e.g.,* Mallophaga (bird lice), Anoplura (sucking lice) and Siphonaptera (fleas), which are stationary parasites. The blood-sucking Diptera, (Culicidae, Simuliidae, Ceratopogonidae etc.) are temporary ectoparasites on mammals and birds and are very prominent and ecologically important in the Arctic (*e.g.,* Corbet & Danks 1973). Certain mites (*Leptus*, *Erythraeus* etc.) live as ectoparasites on insects (flies, bumblebees) and other arachnids, but very often the association of mites with their hosts is phoretic (a means of transportation). A couple of species of botflies (Oestridae) are harmful endoparasites in arctic reindeer and caribou.

Hodkinson *et al.* (2002a, 2004) studied the invertebrate communities along proglacial chronosequences in the High Arctic (Svalbard) and found that – contrary to common belief – heterotrophic organisms and omnivorous carnivors (cryptostigmatic mites, surface-active Collembola, spiders) are the first to colonize newly exposed land, before vascular plants arrived and soils developed.

Population density and production. The microarthropods, Collembola and oribatid mites, that occur in large numbers in arctic soils, play a significant role in the turnover of dead and living organic matter, primarily by breaking it up into smaller particles, allowing bacteria and fungi to complete the decomposition process. The grazing of microarthropods on *e.g.* algae and fungi stimulates the growth and reproduction of these producers. Among the insects, the midge family (Chironomidae) is without comparison the most successful and important group, regarding both production and number of species. The larvae constitute an important food item for waders and arctic charr. Ducks and other birds consume large amounts of pupae when they come up to the water surface to hatch; and small passerine birds primarily feed their young with adult midges.

Few quantitative studies have been carried out on the productivity of arctic arthropods. The density of collembolans can be staggering; 10.000–40.000 per m^2, up to around 270.000 per m^2 have been reported. Mite populations are somewhat smaller, up to about 80.000 per m^2, but about 250.000 have been found in manured grassland in Spitsbergen (Bengtson *et al.* 1974). The local variations may be enormous. In polygonal soils (Fig. 6) of Alaska the density of collembolans in the soggy, cold centres of the polygons was measured as about 24.000 per m^2, whereas along the well-drained edges of the polygons, where vegetation is denser and the microclimate drier and warmer, their number was about 180.000 per m^2 (Fjellberg 1986).

The density of midge larvae in an eutrophic arctic pond like that shown in Fig. 12 is generally about 6.000–9.000 or even up to 27.000 per m^2. A square metre of fertile soil may contain about 50 fly maggots, but more than 2.000 on wet moss tundra in Spitsbergen (Bengtson *et al.* 1974). One m^2 of Greenlandic steppe may house 50–100 individuals of the seed bug *Nysius groenlandicus* (Böcher 1976). However, the density of most arctic insects is much lower, *e.g.* for larvae of the tussock moth *Gynaephora groenlandica* (Erebidae-Lymantriinae) never more than 0.1 per m^2 (Danks 1981).

It is as yet not possible to quantify the role of terrestrial and limnic arthropods in the productivity of arctic ecosystems. Our knowledge of the lifecycles, food preferences, and population densities of individual species is too inadequate. It must, however, be assumed that the great number of saprophagous insect larvae and microarthropods play a very important part in the turnover of nutrients in decomposer food webs.

Nowhere in the Arctic are insects likely to consume more than 2% of the primary production of higher plants (Danks 1981). The fraction of total herbivore production that is eventually consumed by parasitoids and carnivores is hardly any greater.

FIGURE 12 *Pond in interior West Greenland with luxuriant vegetation of* Eriophorum scheuchzeri, Ranunculus confervoides, Hippuris vulgaris *etc. Generally such ponds also have a very rich fauna with* Colymbetes dolabratus, Hydroporus morio, Gyrinus opacus, *larvae of Diptera, especially Culicidae, larvae of Trichoptera; the snail* Lymnaea vahlii, *Cladocera, Copepoda, Ostracoda. Kløftsøerne, eastern Søndre Strømfjord/Kangerlussuaq, nSW.*

Insects and spiders are important food items for birds, especially for the large populations of waders that breed in the High Arctic. Three fourths of the midge production in a lake in arctic Canada was consumed by arctic charr; the availability of larvae of midges and mosquitoes is crucial to this species. Insects also play an important role in the diet of the arctic fox where remains of bumblebees may constitute almost an entire fox excrement.

Recently Roslin *et al.* (2013) and Várkonyi & Roslin (2013) have questioned the general belief that arctic ecosystem are "simple". In a careful investigation of the *Dryas-Sympistis zetterstedti* (as *S. nigrita*)-Hymenoptera parasitoids trophic system in a high arctic environment (Zackenberg, Northeast Greenland) they stressed the complexity of food webs and give quantifications of the trophic interactions.

Diversity

The arctic fauna is overall empoverished compared to that of other regions. There are, however, a few taxa such as Tardigrada (water bears) that have the highest number of species in the Arctic and other extreme environments (R.M. Kristensen, pers. comm.). As in most of the world, soils in the Arctic harbour a comparatively rich fauna, both in terms of species numbers and abundance of individuals; among soil arthropods, springtails (Collembola) and mites (Acari, in particular Oribatida) are prominent. Some mesostigmate mites (genus *Arctoseius*) seem confined to the High Arctic, and the same applies to many Araneae (*e.g. Alopecosa exasperans*), Diptera (*e.g. Gonarcticus arcticus, Spilogona* spp., Chironomidae spp.), Hymenoptera (Ichneumonidae spp., Braconidae spp.), Lepidoptera (*e.g. Gynaephora groenlandica, Psychophora sabini*), and the rove beetle (Staphylinidae) *Gnypeta cavicollis*. On the other hand, a number of important taxa further south are absent in the Arctic. Examples are Odonata (dragonflies), Orthoptera (grasshoppers) and Formicidae (ants), although a few species of ants do occur just north of the tree line in Canada. Other species-rich taxa, such as Hemiptera and Coleoptera, are relatively very poorly represented. In the Russian Arctic a gradual decline of the fauna of Carabidae from south to north is found throughout the tundra (Chernov *et al.* 2000).

There are several reasons for this state of affairs. One is that many ecological niches do not occur north of the tree line. Another is that following the most recent glaciation (Weichselian/Wisconsin), most of the present Arctic Region had to be recolonized as recent as since only *c.* 10.000 years ago (Böcher 2012). A third reason is that prospective inhabitants have to possess or develop specialized adaptations in order to survive under arctic conditions. The ability of different groups to cope with these requirements apparently varies strongly (Chernov 2002). The microarthropods, *e.g.,* springtails and mites, are highly successful arctic inhabitants. As a whole, the Diptera (flies) are successful and have a relatively high number of arctic representatives, including about 340 species in Greenland. Beetles (Coleoptera), on the other hand, the largest animal order, have not done well in the Arctic, with only 37 species known from Greenland – compared to 3.770 from Denmark (Jørum *et al.* 2006) and around 360.000 named beetle species worldwide (Foottit & Adler 2009). As mentioned in chapter 4 the poor representation of *e.g.* Coleoptera is undoubtedly partly caused by the isolated position of Greenland.

References

Aitchison, C.W. (1984): Low temperature feeding by winteractive spiders. – *Journal of Arachnology* 12: 297–305.

Aleksandrova, V.D. (1980): *The Arctic and Antarctic: their division into geobotanical areas*. Cambridge University Press. 247 pp.

Bale, J.S., Hodkinson, I.D., Block, W., Webbs, N.R., Coulson, S.C. & Strathdee, A.T. (1997): Life strategies of arctic terrestrial arthropods. *In*: Woodin, S.J. & Marquiss, M. (eds) *Ecology of Arctic Environments*, pp. 137–165. British Entomological Society/Blackwell Science, Oxford.

Bay, C. (1998): *Vegetation mapping of Zackenberg Valley, Northeast Greenland*. Danish Polar Center and Botanical Museum, University of Copenhagen. Copenhagen, Denmark.

Bengtson, S.-A., Fjellberg, A. & Solhöy, T. (1974): Abundance of Tundra Arthropods in Spitsbergen. – *Entomologica scandinavica* 5: 137–142.

Born, E.W. & Böcher, J. (eds) (2001): *The Ecology of Greenland*. Atuakkiorfik Education, Nuuk.

Böcher, J. (1971): Preliminary studies on the biology and ecology of *Chlamydatus pullus* (Reuter) (Heteroptera: Miridae) in Greenland. – *Meddelelser om Grønland* 191 (3): 29 pp.

Böcher, J. (1972): Feeding biology of *Nysius groenlandicus* (Heteroptera: Miridae) in Greenland. With a note on oviposition in relation to food-source and dispersal of the species. – *Meddelelser om Grønland* 191 (4): 41 pp.

Böcher, J. (1975): Notes on the reproductive biology and egg-diapause in *Nysius groenlandicus* (Zett.) (Heteroptera: Lygaeidae). – *Videnskabelige Meddelelser fra Dansk Naturhistorisk Forening* 138: 21–38.

Böcher, J. (1976): Population studies on *Nysius groenlandicus* (Zett.) (Heteroptera: Lygaeidae) in Greenland, with particular reference to climatic factors, especially the snow cover. – *Videnskabelige Meddelelser fra Dansk Naturhistorisk Forening* 139: 61–89.

Böcher, J. (1988): The Coleoptera of Greenland. – *Meddelelser om Grønland. Bioscience* 26: 100 pp.

Böcher, J. (2012): Interglacial insects and their possible survival in Greenland during the last glacial stage. – *Boreas* 41: 644–659.

Böcher, J. & Enghoff, H. (1984): A centipede in Greenland: *Lamyctes fulvicornis* Meinert, 1868 (Chilopoda, Lithobiomorpha, Henicopidae). – *Entomologiske Meddelelser* 52: 49–50.

Böcher, J. & Petersen, P.M. (1997): Greenland. Chapter 20 *in* Wielgolaski, F.E. (ed.). *Ecosystems of the World 3. Polar and Alpine Tundra*, pp. 685–720.

Böcher, J. & Nachman, G. (2001): Temperature and humidity responses of the arctic-alpine seed bug *Nysius groenlandicus*. – *Entomologia Experimentalis et Applicata* 99: 319–330.

Böcher, J. & Nachman, G. (2010): Are environmental factors responsible for the sex ratio of the Greenlandic seed-bug, *Nysius groenlandicus*? – *Entomologia Experimentalis et Applicata* 134: 122–130.

Böcher, J. & Nachman, G. (2011): Coexistence of bisexual and unisexual populations of *Nysius groenlandicus* in the Zackenberg Valley, Northeast Greenland. – *Entomologia Experimentalis et Applicata* 140: 196–206.

Böcher, T.W. (2000) (Fredskild, B. & Böcher, J. eds): *Det grønne Grønland*. Rhodos, 432 pp. København.

Chernov, Y.I. (1995): Diversity of the Arctic Terrestrial Fauna. *In*: Chapin, F.S. & Körner, C. (eds) *Arctic and Alpine Biodiversity. Patterns, Causes and Ecosystem Consequences*: 81–95. Springer-Verlag, Berlin etc.

Chernov, Y.I. (2002): Arctic biota: Taxonomic diversity. – *Entomological Review* 82. Suppl. 1: S1–S23.

Chernov, Y.I., Makarov, K.V. & Eremon, P.K. (2000): Family of ground beetles (Coleoptera, Carabidae) in the Arctic fauna: Communication 1. – *Entomological Review* 80 (8): 1009–1021. Translated from *Zoologicheskii Zhurnal* 79 (12). 2000.

Corbet, P.S. (1972). The microclimate of arctic plants and animals, on land and in fresh water. – *Acta Arctica* 18: 43 pp.

Corbet, P.S. & Danks, H.V. (1973): Seasonal emergence and activity of mosquitoes (Diptera: Culicidae) in a high-arctic locality. – *The Canadian Entomologist* 105: 837–872.

Danks, H.V. (1981): *Arctic Arthropods. A review of systematics and ecology with particular reference to the North American fauna*. Entomological Society of Canada, Ottawa. 608 pp.

Danks, H.V. (1986): Insect plant interactions in arctic regions. – *Revue d´entomologie du Quebec* 31(1–2): 52–75.

Danks, H.V. (1990): Arctic insects: instructive diversity. *In*: Harington, C.R. (ed.) *Canada´s missing dimension: Science and history in the Canadian arctic islands*. – Canadian Museum of Nature, Ottawa. 855 pp.

Danks, H.V. (1991): Winter Habitats and Ecological Adaptations for Winter Survival. *In*: Lee, R.E. & Denlinger, D.L. (eds). *Insects at Low Temperature*, pp. 231–259. – Chapman and Hall, New York & London.

Danks, H.V. (1999): Life cycles in polar arthropods – flexible or programmed? – *European Journal of Entomology* 96: 83–102.

Danks, H.V. (2006a): Insect adaptations to cold and changing environments. – *Canadian Entomologist* 138: 1–23.

Danks, H.V. (2006b): Canadian perspectives: Life-cycle types in the Arctic. – *Newsletter of the Biological Survey of Canada (Terrestrial Arthropods)* 25 (2): 37–69

Danks, H.V., Kukal, O. & Ring, R.A. (1994): Insect coldhardiness: Insights from the Arctic. – *Arctic* 47(5): 391–404.

Dixon, A.F.G. (1998): *Aphid Ecology*. Springer Verlag. 300 pp.

Downes, J.A. (1962): What is an Arctic Insect? – *Canadian Entomologist* 94: 143–162.

Downes, J.A. (1964): Arctic insects and their environment. – *Canadian Entomologist* 96: 279–307.

Downes, J.A. (1965): Adaptations of insects in the Arctic. – *Annual Review of Entomology* 10: 257–274.

Dreisig, H. (1981): Daily flight activity of moths in the continuous daylight of the arctic summer. – *Holarctic Ecology* 4: 36–42.

Fjellberg, A. (1986): Collembola from the Canadian high arctic. Review and additional records. – *Canadian Journal of Zoology* 64: 2386–2390.

Foottit, R.G. & Adler, P.H. (2009) (eds): Insect Biodiversity: Science and Society. – Blackwell Publishing.

Hodkinson, I.D. (1997): Progressive restriction of host plant exploitation along a climatic gradient: the willow psyllid *Cacopsylla groenlandica* in Greenland. – *Ecological Entomology* 22: 47–54.

Hodkinson, I.D. (2005): Adaptations of invertebrates of terrestrial Arctic environments. – *Det Kongelige Norske Videnskabers Selskab Skrifter* 2005 (2). 44 pp.

Hodkinson, I.D., Webb, N.R. & Coulson, S.J. (2002a): Primary community assembly on land – the missing stages: why are the heterotrophic organisms always the first? – *Journal of Ecology* 90: 569–577.

Hodkinson, I.D., Coulson, S.J., Bird, J. & Webb, N.R. (2002b): Discovery of the rare alate morph of *Acyrthosiphon svalbardicum* Heikinheimo (Homoptera; Aphididae): descriptions and implications for species ecology. – *Norwegian Journal of Entomology* 49: 77–80.

Hodkinson, I.D., Coulson, S.J. & Webb, N.R. (2004): Invertebrate community assembly along proglacial chronosequences in the high Arctic. – *Journal of Animal Ecology* 73: 556–568.

Hullé, M. & Bonhomme, J. (2008): Is the cycle of high arctic aphids adapted to climate change? – *Polar Biology* 31: 1037–1042.

Jørum, P., Mahler, V. & Pedersen, J. (2006): Fund af biller i Danmark 2005 (Coleoptera). – *Entomologiske Meddelelser* 74 (2): 107–134.

Kevan, P.G. (1972): Insect pollination of high arctic flowers. – *Journal of Ecology* 60(3): 831–847.

Kevan, P.G. & Shorthouse, J.D. (1970): Behavioural thermoregulation by high arctic butterflie. – *Arctic* 23(4): 268–279.

Kevan, P.G. & Danks, H.V. (1986): Arctic insects. *In*: Sage, B. (ed.) *The arctic and its wildlife*, pp. 72–77. Croom Helm, Beckenham.

Kevan, P., Jensen, T.S. & Shorthouse, J.D. (1982): Body temperatures and behavioral thermoregulation of high arctic woollybear caterpillars and pupae (*Gynaephora rossii*, Lymantriidae: Lepidoptera) and the importance of sunshine. – *Arctic and Alpine Research* 14(2): 125–136.

Kukal, O. (1991): Behavioral and Physiological Adaptations to Cold in a Freeze-Tolerant Arctic Insect. *In*: Lee, R.E. Jr. & Denlinger, D.L. (eds) *Insects at Low Temperature*, pp. 276–300. Chapman and Hall, New York and London.

Kukal, O. (1995): Winter mortality and the function of larval hibernacula during the 14-year life cycle of an arctic moth, *Gynaephora groenlandica*. – *Canadian Journal of Zoology* 73(4): 657–662.

Kukal, O. & Kevan, P.G. (1987): The influence of parasitism on the life history of a high arctic insect, *Gynaephora groenlandica* (Wöcke) (Lepidoptera: Lymantriidae). – *Canadian Journal of Zoology* 65: 156–163.

Kukal, O., Heinrich, B. & Duman, J.G. (1988a): Behavioural thermoregulation in the freeze-tolerant arctic caterpillar, *Gynaephora groenlandica*. – *Journal of experimental Biology* 138: 181–193.

Kukal, O., Duman, J.G. & Serianni, A.S. (1988b): Cold-induced mitochondrial degradation and cryoprotectant synthesis in freeze-tolerant arctic caterpillars. – *Journal of Comparative Physiology* B 158: 661–671.

Lambers, D.H.R. (1952): The Aphid Fauna of Greenland. – *Meddelelser om Grønland* 136(1): 33 pp.

Lee, R.E. & Denlinger, D.L. (eds) (1991): *Insects at Low Temperature*. – Chapman & Hall, New York & London. 513 pp.

MacLean, S.F. (1975): Ecological adaptations of tundra invertebrates. *In*: Vernberg, F.J. (ed.) *Physiological Adaptations to the Environment*, pp. 269–300. – Intext Educational Publishers, New York.

MacLean, S.F. (1983): Life cycles and the distribution of psyllids (Homoptera: Psylloidea) in arctic and subarctic Alaska. – *Oikos* 40: 445–451.

Makarova, O.L. (2000): To a study of mites of the genus *Arctoseius* (Parasitiformes, Ascidae) from the far north: 3. Ranges and ecological preferences of species. – *Entomological Review* 80, Suppl. 1: S143–S150. Translated from *Zoologizheskii Zhurnal* 79 (9).

Meltofte, H. (ed.) (2013): Arctic Biodiversity Assessment. Status and trends in Arctic biodiversity. Conservation of Arctic Flora and Fauna, Akureyri.

Morewood, D. & Ring, R. (1998): Revision of the life history of the High Arctic moth *Gynaephora groenlandica* (Wocke) (Lepidoptera:Lymantriidae). – *Canadian Journal of Zoology* 76: 1371–1381.

Oliver, D.R. (1968): Adaptations of Arctic Chironomidae. – *Annales Zoologici Fennici* 5: 111–118.

Pape, T. (1983): Observations on nests of *Bombus polaris* Curtis usurped by *B. hyperboreus* Schönherr in Greenland. – *Entomologiske Meddelelser* 50: 145–150.

Philipp, M., Böcher, J., Mattson, O. & Woodell, S.R.J. (1990): A quantitative approach to the sexual reproductive biology and population structure in some arctic flowering plants: *Dryas integrifolia*, *Silene acaulis* and *Ranunculus nivalis*. – *Meddelelser om Grønland. Bioscience* 34: 60 pp.

Ring, R.A. & Tesar, D. (1981): Adaptations to Cold in Canadian Arctic Insects. – *Cryobiology* 18: 199–211.

Roslin, T., Wirta, H., Hopkins, T., Hardwick, B. & Várkonyi, G. (2013): Insect interactions in the High Arctic. – *PLOS ONE* 8(6): e67367. dol: 10.1371/journal.pone.0067367.

Sformo, T., Walters, K., Jeannet, K., Wowk, B., Fahy, G.M., Barnes, B.M. & Duman, J.G. (2010): Deep supercooling, vitrification and limited survival to -100°C in the Alaskan beetle *Cucujus clavipes poniceus* (Coleoptera: Cucyjidae) larvae. – *The Journal of Experimental Biology* 2010 dol: 10. 1242/jeb.035758.

Storey, K.B. & Storey, J.M. (1988): Freeze Tolerance in Animals. – *Physiological Reviews* 68(1): 27–84.

Strathdee, A.T., Bale, J.S., Block, W.C., Webb, N.R., Hodkinson, I.D. & Coulson, S.J. (1993): Extreme adaptive life-cycle in a high arctic aphid, *Acyrthosiphon svalbardicum*. – *Ecological Entomology* 18: 254–258.

Strathdee, A.T. & Bale, J.S. (1998): Life on the Edge: Insect Ecology in Arctic Environments. – *Annual Review of Entomology* 43: 85–106.

Sømme, L. & Block, W. (1991): Adaptations to Alpine and Polar Environments in Insects and Other Terrestrial Arthropods. *In*: Lee, R.E. & Denlinger, D.L. (eds). *Insects at Low Temperature*, pp 318–359. Chapman and Hall, New York & London.

Strømme, J.A., Ngari, T.W. & Zachariassen, KE. (1986): Physiological adaptations in Coleoptera on Spitsbergen. – *Polar Research* 4.n.s.: 199–204.

Tenow, O. (1990): Altitudinal and withincrown distribution of insects grazing on Mountain birch (*Betula pubescens* coll.) in SW Greenland and NW Norway. – *Meddelelser om Grønland. Bioscience* 33: 35–42.

Várkonyi, G. & Roslin, T. (2013): Freezing cold and yet diverse: dissecting a high-Arctic parasitoid community associated with Lepidoptera hosts. – *Canadian Entomologist* 145: 193–218.

Woodin, S.J. & Marquiss, M. (eds) (1997): *Ecology of Arctic Environments*. British Entomological Society/Blackwell Science, Oxford. 286 p.

CHAPTER 4

The Greenland entomofauna: zoogeography and history

Jens Böcher

Greenland is almost exclusively located within the *Arctic climatic zone*, geographically defined as the northern global area in which average temperatures during the warmest month of the year (usually July) are below 10°C. This temperature boundary is more or less coincident with the timberline. The arctic climatic zone is subdivided into a *high arctic zone*, in which average temperatures during the warmest month lie below 5°C, and a *low arctic zone*, where average temperatures during the warmest month lie between 5°C and 10°C. Between the temperate, coniferous zone (in which forestry is economically feasible) and the arctic climatic zone lies a narrow transitional region termed the *subarctic climatic zone*. Here, average temperatures during the warmest month lie slightly above 10°C and the growth season is just long enough to allow the development of low, scattered forest. In Greenland the subarctic zone is only found in the interior parts of the southern fiords, and the tree involved is generally *Betula pubescens* (Born & Böcher 2001).

Distribution types

For the purpose of summarizing distribution patterns of Greenland animals the country has been divided (J. Böcher 1988) into a number of "faunal districts" as shown in Fig. 1. These are admittedly fairly crude, but are in agreement with those proposed by Salomonsen (1981) and partly with currently accepted floristic provinces and districts (T.W. Böcher *et al.* 1968). The districts may be characterized as follows: S comprises the climatically most favourable, subarctic part of Greenland where *Betula pubescens*, *Sorbus groenlandica* and *Salix glauca* form small woodlands (Fig. 3–1) in inland areas. SW is the low arctic West Greenland with a relatively mild, oceanic climate along the coast and a continental climate with fairly warm, dry summers in the interior. SE is the low arctic East Greenland with a cool, humid oceanic climate. NW and NE are relatively dry, high arctic areas. N is the extreme High Arctic, also with a dry climate and comprising arctic desert. The northern boundary of SW and SE coincides approximately with the northern limits of real willow coppice and herb slope communities (T.W. Böcher 1979, 1981; Salomonsen 1972, 1979). SW is subdivided into sSW and nSW, especially according to the presently known distribution of a number of insects.

Concerning *general distribution* the following terms are used: *Circumpolar* (all arctic areas), *Holarctic* (both nearctic and palaearctic areas, possibly with essential absences in an otherwise circumpolar range), *Nearctic* (North America), *Palaearctic* (Europe + northern Asia + northern Africa).

Greenland is traditionally included in the Nearctic Region, implying that the occurrence of a species in Greenland automatically renders it 'nearctic'. **In this book Greenland is considered a zoogeographical unit of its own** (Böcher 1988, Bergersen 1955). A palaearctic species occurring in Greenland, but not in North America, is accordingly *not* termed holarctic.

The *climatic zones* outside Greenland are given as follows: *Northern temperate* or *boreal* (with coniferous forests), *southern temperate* (with deciduous forests), and *Mediterranean* (primarily with maquis).

Biodiversity

There is a general tendency towards decreasing biological diversity with increasing latitude (Downes 1964, Danks 1981), and Greenland is almost entirely within the Arctic Biome (Born & Böcher 2001). However, this decline is very variable between different taxa, but unexpectedly high in most groups of insects when compared with other areas at similar latitude. For instance, the number of Greenland butterflies (Papilionoidea) amounts to only 9% of the butterfly fauna of arctic North America (Danks 1981), and Coope (1986) showed that Greenland has only 17% of the expected number of Coleoptera. The ground beetles (Carabidae), generally the most succesful beetle group in the Arctic, have only four Greenland representatives – against at least 50 in the arctic part of Fennoscandia (Lindroth 1985). On the other hand, the Diptera are comparatively well represented, especially Chironomidae and Muscidae, which are still more dominating the more extreme and high arctic the biotope. Also the soil fauna, collembolans and mites, is very well represented even in the High Arctic, often with immense numbers per square metre. A number of globally important taxa are totally absent from the native fauna of Greenland (*e.g.*, Diplopoda,

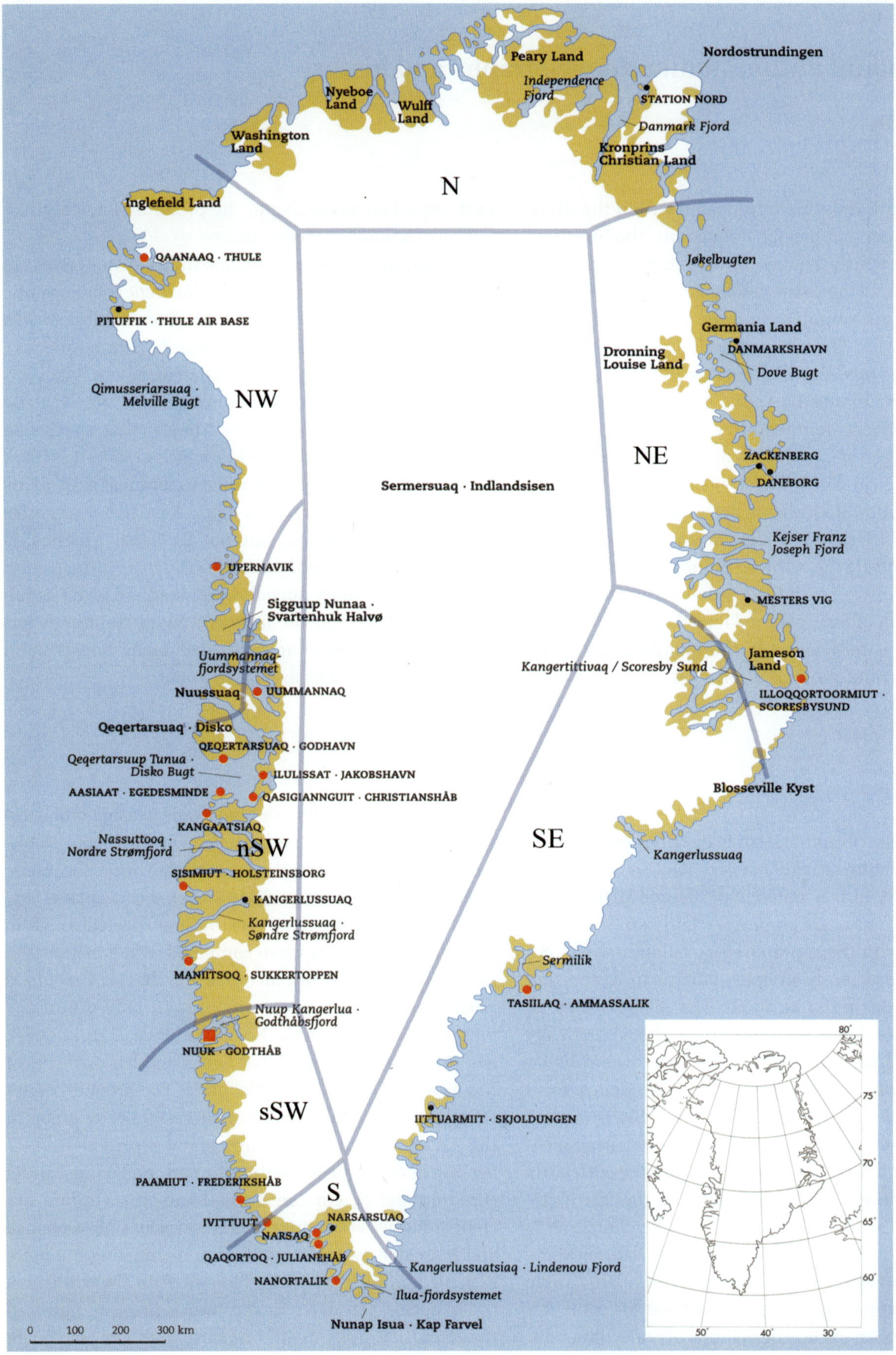

FIGURE 1 *Map of Greenland with the geographical names used and the zoogeographical zones indicated. Inset: outline map showing principal long/lat coordinates, and position relative to adjacent land-masses (Canada, Iceland, Spitsbergen).*

Odonata, Orthoptera, Hymenoptera-Formicidae) and some are represented by only very few species (*e.g.*, Chilopoda, Ephemeroptera, Heteroptera, Cicadomorpha, Hymenoptera-Apoidea).

Zoogeographic insights emerging from the present work

This work presents a total of ≈1190 species of insects, centipedes and arachnids living in Greenland, a considerable augmentation since the last surveys (*c.* 680 species, according to Henriksen 1939, Henriksen & Lundbeck 1917, Larsen & Scharff 2003).

It is by no means easy to make general statements about the zoogeographical results emerging from the studies made in preparation for the present work. Every species has its own distribution reflecting both ecological requirements and dispersal powers together with a unique history. Another difficulty is the still highly variable state of knowledge of different groups. For example, it is very difficult to compare Chironomidae with Carabidae – or a family of prostigmatic mites with the well known lycosid spiders. None the less, a number of striking tendencies may be enumerated.

Some groups represent extremes regarding their affinity to either the Nearctic or the Palaearctic Region. Thus all of the 5 butterflies (Papilionoidea) are nearctic and arctic (Downes 1966). Trichoptera (with a total of 8 species) are also markedly nearctic, with 6 North American, 1 palaearctic and 1 circumpolar species. Another predominantly nearctic taxon is Araneae with 75 species of which 20+ are nearctic, close to 10 palaearctic and the rest circumpolar/Beringian/holarctic. The same is the case with Muscidae (altogether 37 species of which 10 are nearctic, the others holarctic). Anthomyidae are similar, but not so marked, with 32 species of which 3 nearctic and no palaearctic representatives; the many remainders are holarctic or circumpolar.

The other extreme is represented by the Coleoptera, of which the zoogeographical situation and immigration history has been debated for more than a century (see below, and Böcher 1988). Of the 36 naturally occurring species just 1 is nearctic and the remaining almost equally divided between palaearctic and holarctic/circumpolar ones. It is noteworthy, that findings from the last interglacial in Jameson Land are similar. The beetle fauna (19 species) was also then composed of palaearctic and circumpolar species, without any nearctic elements (see below, and Böcher 1997, Bennike & Böcher 1994). Another highly palaearctic dominated group is Collembola: while 10+ species are classified as cosmopolitans, just a single is nearctic, and the bulk has almost equal representation of palaearctic and circumpolar/holarctic species. Aphidoidea (21 species) have no nearctic species, but 6 palaearctic, 5 c/h, while no less than 9 are currently considered endemic.

Great reservation must be made with respect to 'endemism' of Greenland species. It is possible, even likely, that most or at least many of these species in fact occur elsewhere and will appear when material from other geographic areas are scrutinized by specialists. For instance, possibly the Eurasian fauna of Collembola is better studied than that of North America. Nonetheless, taking into account how near Greenland is to North America, it is astonishing that the fauna is so highly marked by palaearctic species. The taxa with a nearctic dominance are generally those which are fairly easily dispersed, either by the wind or because they are strong fliers (Diptera, Araneae, see Brændegaard 1938) whereas it is remarkable that most of the more earthbound taxa (mites, Collembola, Coleoptera) are markedly palaearctic. This was also the conclusions by Omodeo (1957) and Nurminen (1973), studying the Greenland earthworms and Enchytraeidae, respectively. It is still not possible to give a full and adequate explanation of this strange state of things.

Postglacial immigration

Obvious reasons for the taxonomical poverty of the Greenland entomofauna are the isolated position of Greenland and the short time that has enabled species to become established. Following the almost total covering of the land mass by ice during the latest Pleistocene glaciation, the ice retreated so that an ice free fringe of land was ready to be repopulated about 10.000 years BP (Bennike 1999). But from where? In North America and Eurasia the biota could survive the glacial intervals in suitable areas south of the ice sheets and simply follow the retreating ice northwards. But Greenland has no such retreat area in the south. Insect life was probably all but wiped out during the latest glacial interval, by extensive glacier advancement and the adverse climatic conditions that accompanied them (see below). Immigration had to take place across large stretches of ocean, by the wind or with ocean currents, or via the Canadian Arctic Archipelago which is separated from Greenland by only comparatively narrow straits. But this last route is rendered difficult by the hostile, high arctic conditions of the area, filtering out all but the most well adapted arctic forms (see also Dolichopodidae).

Nonetheless, a number of species succeeded in colonizing Greenland. According to Downes (1966, 1988) the Greenland insect fauna is made up by mainly two elements: high arctic species and northern boreal species. The first group is likely to have immigrated through the northwestern gateway from the Canadian archipelago. In fact, the lepidopteran fauna of North Greenland is almost identical to that of Ellesmere Island. The other group is heterogeneous and of adventitious origin. It probably immigrated by passive transportation from both North America and Europe, in a few cases (certain Noctuidae and Nymphalidae) possibly by active flight (Elton 1925, Downes 1988).

This different origin is evident when the fauna of northern Greenland is compared with that of southern Greenland. All the five species of butterflies are high arctic, probably originating in arctic Canada where they are all common. In southern Greenland none of the low arctic butterflies which 'ought to be there' are present; on the other hand two of the high arctic species, *Clossiana chariclea* and *Colias hecla* are found, but generally not below a certain altitude in the mountains. Most of the Greenland 'entomofauna' has different kinds of southern distributions which rarely extend into the High Arctic. A very special and puzzling situation is found among the Coleoptera, because the European element is considerable and moreover the fauna is largely made up by flightless species with a fairly southern general distribution; only one species (*Arctobyrrhus subcanus*) has a clear North American origin. Accordingly, their presence in Greenland is quite enigmatic and has been widely discussed for more than a century (Böcher 1988, 1997, 2012).

In the beginning of the twentieth century it was widely believed that a large fraction of the flora and fauna of Greenland had been introduced by the Norse people, *i.e.* after 982 AD (*e.g.*, Ostenfeld 1926, Jensen 1928). However, findings of postulated introductions a considerable distance away from the former Norse settlements in South- and Southwest Greenland (*e.g.*, of *Bembidion grapii* and *Otiorrhynchus arcticus* at Wiedemanns Fjord, Blosseville Kyst; Degerbøl 1937) and subsequent findings of such taxa in archaeological excavations of Inuit layers predating the Norse settlements (Buckland 1988, Böcher 1997, Böcher & Fredskild 1993) rendered this hypothesis improbable. While a number of plants and invertebrates may have entered Greenland together with the Norsemen (among arthropods the chilopod *Lamyctes emarginatus*, the bug *Nabis flavomarginatus*, and the beetles *Omalium excavatum* and *Nephus redtenbacheri*), but all of these might as well be native (Böcher 1988). A number of clearly synanthropic beetle species were indeed transported to Greenland with the Norse people, but nearly all died out with the disappearance of human settlements (Böcher 1988; Buckland *et al.* 1983; McGovern *et al.* 1983).

Other workers strongly opposed the 'man-mediated-introduction' explanations and favoured immigration along hypothetical late glacial or interglacial land bridges connecting Greenland and Europe. However, there is no geological evidence of such land bridges later than the Middle Tertiary (Matthews 1980, Downes 1988).

Coope (1969, 1979b, 1986; Buckland 1988; Buckland *et al.* 1986) broadened the discussion by adding precise knowledge of the postglacial insect history of northern Europe from the fossil record, and also from information on past sea surface circulation patterns in the North Atlantic, derived from deep sea cores. Coope suggested an elegant solution to the 'European beetle problem', namely an invasion from northern Europe to southern Greenland by rafting over a brackish North Atlantic Ocean on a deflected Gulf Stream – a situation which seemed to occur during the Younger Dryas-Holocene transition *c.* 10.000 B.P. (Ruddiman & McIntyre 1981). However, newer studies of the past circulation pattern in the North Atlantic do not seem to support Coope's ideas (Hald & Aspeli 1997).

Bennike (1999) and Bennike *et al.* (2000) stressed the possible importance of migratory birds as vectors of plants and invertebrates from Europe to Greenland. Concerning especially the enigmatic palaearctic beetles, this mode of dispersal appears unrealistic, but it may be significant for other organisms, especially plants.

According to yet a third hypothesis a considerable part of the biota survived in refuge areas and on nunataks (isolated mountains surrounded by ice) *inside* Greenland – at least during the last glacial stage, and possibly since the Tertiary (*e.g.*, T.W. Böcher 1956, 1963; Lindroth 1970, 1979; Böcher 2012). This was, however, contradicted by studies of ice cores from the Greenland inland ice which showed that the yearly minimum temperature during the last glacial maximum at c. 25,000 years B.P., on a yearly basis was around 23°C lower than at present (Dahl-Jensen *et al.* 1998), rendering it difficult to imagine survival of any but the most hardy organisms in terrestrial Greenland (Bennike *et al.* 1999, 2000). But this temperature lowering may not be decisive, as at present the yearly average temperature in certain areas of north eastern Siberia – with taiga vegetation and undoubtedly a rich arthropod life – is around -18°C (Böcher 2012).

Studies of both terrestrial erratics (Håkansson *et al.* 2008) and marine geophysics (Evans *et al.* 2008) indicated almost total covering of Greenland by ice during the height of the last glaciation. However, this evidence is not convincing (Funder *et al.* 2011), and it has been shown that large areas in coastal East Greenland (*e.g.*, parts of Jameson Land, Hochstetter Forland etc.) were *not* glaciated during the last glaciation (Funder 1978, 1979) in which case the possibility of survival in Greenland since the last interglacial (Eemian) must have existed.

In this connection it is relevant to mention the earthbound fauna (mites, collembolans, earthworms) which is remarkably homogeneous over the entire northern circumpolar area and for which a preglacial origin in Greenland has been argued (Haarløv 1942; Hammer 1944, 1955; Omodeo 1957; Nurminen 1973). Extremely well adapted arctic species, like the moth *Gynaephora groenlandica* and the staphylinid beetle *Gnypeta cavicollis*, are also obvious candidates of glacial surviving.

Dispersal within Greenland

Arriving in Greenland is one thing. Another is to survive, become established and disperse. It has been argued against the ice-rafting hypothesis that an establishment in Greenland of the 'ice passengers' is unlikely, because at that time only a small fringe of land was ice free. However, vicinity to the Ice Cap by itself is not important, if only the climatic conditions are favourable. In Greenland (and Iceland: Lindroth 1931, 1965) luxuriant plant communities exist close to the present ice margin. In many cases ballooning spiders and winged insects have no serious problems with dispersal in Greenland during the Holocene. However, flightless species encounter a lot of difficulties. Two heteropteran species, *Nysius groenlandicus* (Lygaeidae) and *Chlamydatus pullus* (Miridae), demonstrate some of the difficulties in understanding the modern Greenlandic range of the species in a historical perspective. Both have wide distributions, *Chlamydatus* being circumpolar, *Nysius* probably palaearctic with westernmost occurrence in West Greenland (the existence of *N. groenlandicus* in subarctic Canada has yet to be confirmed; Böcher 1976, Danks 1981). Both species might like some of the beetles have been transported to Greenland from Europe during the Holocene. The seed-feeding *N. groenlandicus* is almost ubiquitous in Greenland, absent only from the north coast; *C. pullus* is also widely distributed, extending into the High Arctic in East Greenland. Taking into account the very limited dispersal powers of these species, it is difficult to understand how such an extensive distribution has been accomplished during the last 10.000 years. Both species are long-winged, but *N. groenlandicus* is able to fly only when the surrounding temperature is above 30°C, and at most for about 50 cm (Böcher 1975). In Greenland *C. pullus* has never been observed using its wings, not even as an aid in leaping (Böcher 1971). Passive transportation with the wind is a possibility, but considering the large inhospitable stretches of fiords, tall mountains and glaciers along the coasts and the marked thermophily and xerophily of both species (Böcher 1971, 1975, 1976; Böcher & Nachman 2001) this method would appear to be of minor importance.

However, in these bugs parthenogenetic reproduction may offer at least a partial explanation: No male *C. pullus* has been found in Greenland, and in some inland Northeast Greenland populations (Jameson Land, Zackenberg, Bessel Fjord) of *Nysius groenlandicus* males are rare or missing altogether (Böcher 1990; Böcher & Nachman 2010, 2011). This strategy enhances dispersal potential relative to bisexual reproduction, because a single individual is sufficient to colonise a new area. Also some Coleoptera, *Otiorhynchus nodosus* (Curculionidae) and four species of *Atheta* (Staphylinidae), occur in Greenland in parthenogenetic forms (Böcher 1988).

While feeding, *N. groenlandicus* frequently deposits the eggs on fruits and seeds equipped with a parachute device for wind dispersal, thereby 'hitchhiking' on the plants' means of dispersal. This may constitute part of the background for the seed bug's efficient colonization of Greenland (Böcher 1972, 1975).

Arthropods connected with Eskimo and Norse sites

Vibe (1953) pointed out that during the entire period after the start of Danish colonization of Greenland in 1721 not a single instance of an introduced insect species established outdoors in Greenland has been reported. Nonetheless, archaeological investigations in the Nuuk/Godthåb area, the Norse Vesterbygd, have revealed a multitude of insect remains, partly from markedly synanthropic species (Buckland 1988, Buckland *et al.* 1983, McGovern *et al.* 1983). Some of these were here recorded for the first and only time in Greenland and they probably succumbed when the Norse disappeared. Others (*Quedius mesomelinus* and *Xylodromus concinnus*, Staphylinidae) were collected frequently during the first decades of the preceding century, but are today very rare or extinct in Greenland. Possibly these species survived the disappearance of the

Norse by switching over to a life associated with the Eskimo turf huts, which have now almost vanished (Böcher 1988).

In addition to the synanthropic and undoubtedly anthropochorous fauna, Buckland and collaborators naturally also found remains of the modern arthropods. Some of these were collected from below the Norse culture layers and thereby represent the indigenous fauna, including both widespread, circumpolar species and also, *e.g.*, the weevil *Otiorhynchus arcticus* which is otherwise confined to northwestern and alpine Europe and thus constitutes a certain non-anthropochorous palaearctic species.

A number of non-synanthropic species nowadays confined to the former Norse habitations might also be human introductions. It must, however, be stressed that these areas at the same time are the climatically most favoured in Greenland, so that this may be the cause of the limited distribution. The single exclusively Nearctic beetle species, *Arctobyrrhus subcanus,* has its only known Greenland occurrences in the Norse Østerbygd (Eastern Settlement).

A fairly detailed picture of the insect fauna associated with the first appearance of man in Greenland has been obtained during excavations of paleo-Eskimo sites dating from *c.* 4.000–3.000 B.P. (Haarløv 1967, Böcher & Fredskild 1993, Meldgaard 2004). In Qeqertasussuk, Disko Bugt, the samples studied were surprisingly rich in arthropod remains, well preserved by the permafrost (Fig. 2). For instance, of the 21 beetle species presently known from this faunal district (Böcher 1988), 12 were found, and some of them north of their present northern limit, indicating a warmer climate than today – in accordance with that the deposition of the Qeqertasussuk culture layers took place during the last part of the postglacial hypsithermal (in Greenland *c.* 8.000–4.000 B.P; Fredskild 1973, Böcher & Fredskild 1993, Christiansen *et al.* 2002).

Excavations from the palaeo-Eskimo site Asummiut at Sisimiut/Holsteinsborg, dated to around 3.000 B.P. among other species yielded the remarkable find of the tiny rove beetle *Euaestethus laeviusculus* which is unknown in modern Greenland. However, the small size and concealed habits of the species makes it possible that it is still living in Greenland but hitherto has escaped discovery (Böcher 1997, 1998, 1999).

Early Holocene insects

Evidence from different sources now throw a fascinating light on the 'postglacial warming', the Holocene Climatic Optimum, in early Holocene, which appears to have been much more marked in Northeast Greenland than hitherto thought (Christiansen *et al.* 2002). From an entomological point of view the most remarkable is the remains of *Phratora* cf. *polaris* (Coleoptera-Chrysomelidae) which have turned up in about 7.000 years old layers from two widely separated localities: Jameson Land (Lollands Elv, Coloradodal) and a coastal cliff at Zackenberg, Wollaston Forland (Böcher & Bennike 1996; Bennike *et al.* 1999, 2000; Christiansen *et al.* 2002). No leaf beetle species are found in modern Greenland; the known distribution of *P. polaris* covers Iceland, Fennoscandia and northern Russia, where it feeds on *Salix herbacea* in heath communities and in scrubs along water courses and outskirts of forests (Larsson & Giggja 1959). The wide distribution in early Holocene of this subarctic species in high arctic Northeast Greenland is really puzzling, and indeed a strong indication of climatic conditions much differing from the present. Like a number of other palaearctic beetle species, *P. polaris* probably has colonized Greenland from northern Europe in early Holocene, but it is indeed amazing to detect it so early and so widespread at such northern latitudes. The species may later have succumbed during the climatic deterioration. Also the widespread *Nysius groenlandicus* was found in the *c.* 7.000 years old deposits from both Jameson Land and Zackenberg (Böcher & Bennike 1991, 1996; Christiansen *et al.* 2002).

In many cases lake corings for pollen analysis have yielded fossil arthropods (Eisner *et al.* 1995, Bennike 2000, Böcher *et al.* 2012). Until recently *Hydroporus morio* (Dytiscidae) was the oldest known Holocene Greenlandic insect, as its remains were found almost throughout the stratigraphy of a lake sediment core in South Greenland (Qassiarsuk), dating back to 8.600 B.P. (Fredskild *et al.* 1975). However, investigations in Jameson Land, East Greenland, showed that the other Greenland dytiscid beetle, *Colymbetes dolabratus*, was present even at 10.800 B.P. (Björck *et al.* 1994) – which is about the time when the ice of the last glacial interval retreated from the area.

Interglacial fauna

Since 1986 an increasing number of sites with terrogene deposits from the last interglacial (Eemian: Isotope Stage 5e, 100.000–130.000 B.P.) have been discovered, first in the Thule area, Northwest Greenland, later in several localities along the south coast of Jameson Land, East Greenland (Böcher 1989b; Böcher & Bennike 1991; Bennike & Böcher

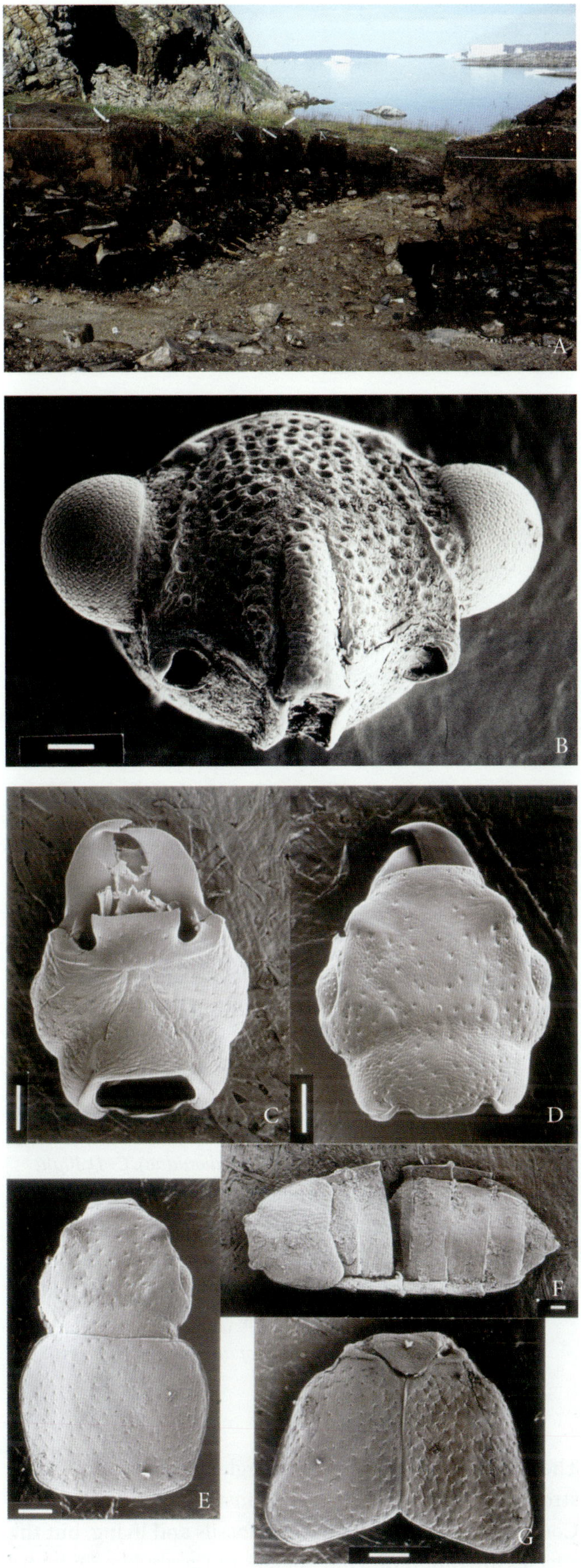

FIGURE 2
The palaeo-Eskimo site on the island Qeqertasussuk in Qeqertarsuup Tunua/Disko Bugt, West Greenland and some of its insect fossils. A. The excavation site (in 1987) looking NW, when the overlying organic material from the culture layers had been removed, exposing the sandy beach on which the palaeo-Eskimo colonization originally took place. B. Head of Nysius groenlandicus *(Heteroptera). Scale: 100 µm. C–G. Heads and articulated fragments of* Micralymma brevilingue *(Coleoptera, Staphylinidae). Scales: 100 µm.*

1992, 1994; Böcher 2012; Figs 3A–D). The present beetle fauna of Jameson Land consists of only three species, *Colymbetes dolabratus*, *Gnypeta cavicollis* and *Coccinella transversoguttata*, but *Micralymma brevilingue* and *Byrrhus fasciatus* are probably to be found too. The interglacial fauna of Coleoptera was surprisingly rich and, in contrast to the flora of vascular plants, comprised several species which are absent from the modern fauna of Greenland (Table 1). Also the seed-bug *Nysius groenlandicus* was found in both Northwest and East Greenland interglacial sediments (Bennike & Böcher 1992, 1994). In Jameson Land the fossil flora (*e.g.*, *Betula pubescens* and *Alnus crispa*) as well as the insect fauna clearly indicate subarctic conditions, implying a mean summer temperature at least 5°C higher than today and, compared to the present situation, a displacement of the subarctic bioclimatic zone from southernmost Greenland at least 1.000 km northwards. A similar shift is seen in some of the Coleoptera, as most of the interglacial species are primarily boreal. These conclusions are in agreement with the results from ice coring in East Greenland (Johnsen *et. al.* 1992, Andersen *et al.* 2004).

During the last interglacial the arctic/alpine carabid beetle *Amara alpina* (Paykull, 1790) was present in both the Thule area and in Jameson Land, indicating a wide range in Greenland at that time. The species is absent from Greenland today, but has otherwise a wide circumpolar distribution, *e.g.*, found as close to Northwest Greenland as on Devon Island in high arctic Canada (Lindroth 1957, 1968; Allan V. Morgan, personal communication). It is accordingly difficult to understand why the species is missing in modern Greenland. One explanation could be that the Holocene Climatic Optimum was just not warm enough to allow the species to enter through the northeastern gateway (Böcher 1989b). The occurrence of another arctic/holarctic species, *Isochnus arcticus* (Korotyaev, 1976) (Curculionidae) in sediments from both Thule and Jameson Land is also in favour of this explanation (Bennike & Böcher 1992).

The most remarkable zoogeographical characteristic of the interglacial beetle fauna is the total absence of exclusively nearctic taxa of today, whereas palaearctic and circumpolar species account for 67 and 33%, respectively (Table 1). This is the same situation as found in modern Greenland, but in an even more extreme version. Accordingly, the fossil beetle fauna demonstrates that even at the beginning of the previous interglacial a similar situation (ice-rafting?) might have occurred, capable of transporting European taxa over the North Atlantic to Greenland (Bennike & Böcher 1994, Böcher 2012).

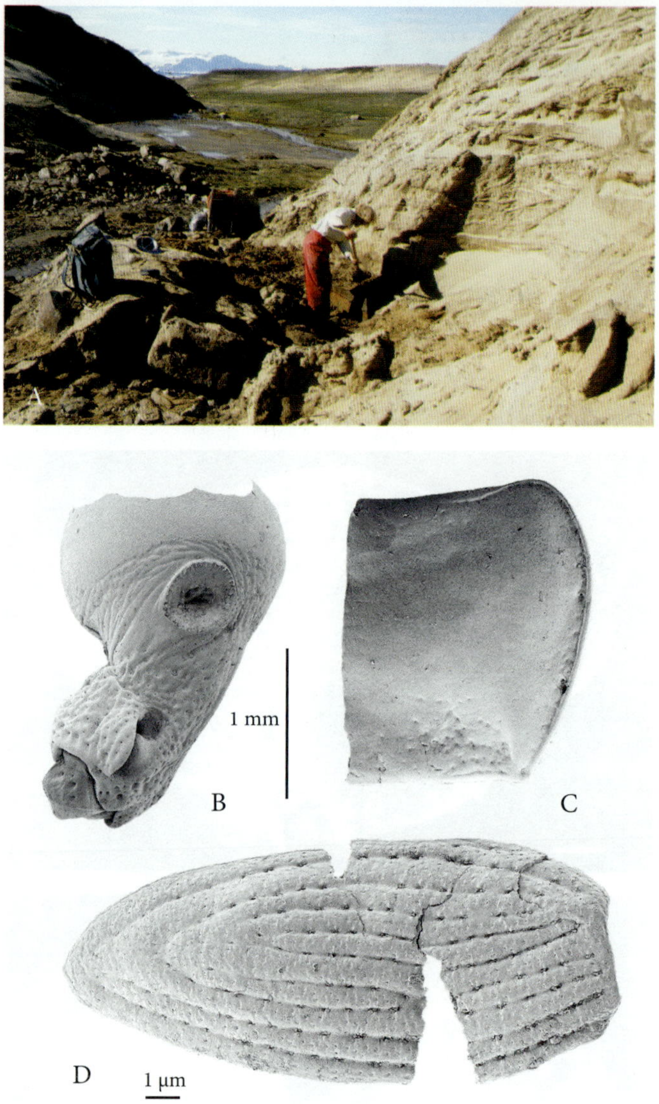

FIGURE 3 *Sediments from the last interglacial and some of their insect fossils. A. Excavation of a coastal cliff with sediments from the last interglacial in southernmost Jameson Land, middle East Greenland, looking south. B. Head of* Otiorrhyncus nodosus *(Coleoptera, Curculionidae). C–D. Right half of pronotum of (C)* Amara alpina *(Paykull, 1790) (Carabidae) and (D) right elytron of* Isochnus arcticus *(Korotyaev, 1976) (Curculionidae). Both species are absent from present day Greenland; the fragment of* A. alpina *(from the Thule-area) was the first indication of an interglacial insect fauna in Greenland.*

The composition of the interglacial fauna deviated strongly from the modern. For instance, two species of Coccinellidae are found both as fossils and living, but the species are different. Four species of Carabidae are found

TABLE 1 *Survey of last interglacial Greenland Coleoptera.* Ecological range, *a: aquatic, e: eurytopic, h: hygrophilic, p: phytophagous, r: riparian, s: forest, t: tundra, x: xerophilic.* Climatic range, *HA: high arctic, LA: low arctic, SA: subarctic (northern boreal), B: boreal (northern temperate), ST: temperate (southern temperate), TR: tropical.*

	Ecological range	Climatic range					
		HA	LA	SA	B	ST	TR
Elaphrus tuberculatus	r		xxxxxxxxxxxxxxxxx				
Amara alpina	t	xxxxxxxxxxxxxxxxxxxxxxxxxxxxxxxx					
Cymindis vaporariorum	x			xxxxxxxxxxxxxx			
Hydroporus morio	a	xxxxxxxxxxxxxxxxxxxxxxxxxxxxxxxxxxxxxxxxxxxxxxxxxxxx					
Hydroporus cf. *planus*	a				xxxxxxxxxxxxxxxxxxxxxxxxxxxxxxxxx		
Agabus bipustulatus	a				xxxxxxxxxxxxxxxxxxxxxxxxxxxxxxxxx		
Colymbetes dolabratus	a		xxxxxxxxxxxxxxxxxxxxxxxxxxxxxxxxxxxxxx				
Eucnecosum tenue/ brachypterum	h			xxxxxxxxxxxxxxxxxxxxxx			
Olophrum consimile	h			xxxxxxxxxxxxxxxxxxxxxx			
Stenus carbonarius	s-h			xxxxxxxxxxxxxxxxxxxxxxxxxxxxxx			
Atheta sp.							
Simplocaria metallica	r		xxxxxxxxxxxxxxxxxxxxxxxxxxxxx				
Byrrhus fasciatus	r		xxxxxxxxxxxxxxxxxxxxxxxxxxxxxxxxxxxxxxxxxxxxxx				
Nephus cf. *bisignatus*	x				xxxxxxxxxxxxxxxxxx		
Coccinella cf. *hieroglyphica*	s-x				xxxxxxxxxxxxxxxxxx		
Chrysomelidae sp.	(p)						
Otiorhynchus arcticus	e(x)		xxxxxxxxxxxxxxxxxxxxxxxxxxxxx				
Otiorhynchus dubius	e(x)		xxxxxxxxxxxxxxxxxxxxxxxxxxxxx				
Lepyrus cf. *canadensis*	p		xxxxxxxxxxxxxxxxxxxxxxxxxxxxx				
Isochnus arcticus	p	xxxxxxxxxxxxxxxxxxxxxxxxxxxxx					

today, but three different species are recorded as interglacial fossils. This striking dissimilarity presumably reflects the marked element of chance in successful dispersal from Europe.

The Tertiary-Quaternary transition, and earlier periods

The remainder of the Pleistocene – more than two million years – left no known trace of terrestrial life in Greenland. However, on the east coast of Peary Land deposits dating back 2–2.5 Ma were discovered in 1979. The Kap København Formation has now yielded detailed information concerning the palaeoenvironment at the Tertiary-Quaternary transition (Funder & Hjort 1980; Fredskild & Røen 1982; Funder *et al.* 1984, 1985; Røen 1988; Böcher 1989a, 1995; Bennike 1990; Bennike & Böcher 1990; Feyling-Hanssen 1990; Brouwers *et al.* 1991; Heie 1995; Símonarson *et al.* 1998, Funder *et al.* 2001).

At present, Peary Land, the northernmost landmass on earth, is an extreme high arctic environment (Fig. 4A), partly a polar desert (Bay 1992) with a very poor insect fauna mainly including species of Diptera (Böcher 1988, 1995, 2003). However, two species of Coleoptera (*Hydroporus morio*, *Gnypeta cavicollis*), and *Nysius groenlandicus* have been reported from interior Peary Land (Anthony Downes, personal communication 1990).

In striking contrast, an abundance of plant and insect fossils from the Kap København Formation show that at the very beginning of the Pleistocene a rich and varied biota existed here. Forest environments comprised temperate trees such as *Thuja* and *Taxus* in addition to *Picea* and *Larix*, which are all absent from Greenland today (Bennike 1990). The fossil insect fauna is impressive,

FIGURE 4 *Sediments from the Plio-Pleistocene Kap København Formation, Peary Land, northernmost Greenland, and some of their insect fossils. A. Northern part of the Kap København Formation with the river Ladegårdsåen. B. Part of a rich sample from the Kap København Formation. Species of* Bembidion, Pterostichus (Cryobius) *and Dytiscidae dominate. In the centre the metallic green pronotum of* Chrysomela populi *L. (Coleoptera, Chrysomelidae), immediately to the left of this elytra of* Hydroporus *sp. and* Elaphrus tuberculatus. *C. Left elytron of* Elaphrus tuberculatus *Mäklin, 1877 (Coleoptera, Carabidae) – a common species in the Kap København Formation, but absent from present day Greenland.*
(PHOTO: G. BROVAD (B-C)).

comprising 155 named species (130 genera) of which 142 species (96 genera) are Coleoptera, mostly Carabidae, Staphylinidae and Curculionidae (Figs 4B–C). Also the diversity of other groups is remarkable. For instance, four genera of Formicidae are represented, among them *Formica* and *Camponotus* (ants are absent from modern Greenland), and the Trichoptera include 9 named species and 12 genera, indicating the existence of a great variety of freshwater biotopes.

Except in a few cases – the carabid *Diacheila matthewsi* and a couple of undescribed staphylinids which probably represent extinct species – all the named taxa have been referred to living species. Thus once more the longevity of insect species is demonstrated (Coope 1970, 1978, 1979a; Buckland & Coope 1991; Böcher 1995). Nine of the beetles from the Kap København Formation are found in Greenland today, most of them widespread Holarctic, low arctic-subarctic species like the boreo-alpine carabid *Nebria rufescens*.

The insect fauna of the Kap København Formation shows a high degree of similarity with that from the 0.5–1 million years older Beaufort Formation on Meighen Island, Arctic Canadian Archipelago (Elias & Matthews 2002; Elias *et al.* 2006; Matthews 1977, 1979a, 1979b, unpublished). For instance, in both sites the genus *Bembidion* and the subgenus *Cryobius* of *Pterostichus* (Carabidae) are predominant. There are, however, important differences as well. The genus *Carabus* is represented by several species on Meighen Island but not found in the Kap København sediments. These, in return, contain a species of *Cicindela,* a genus not represented in the Beaufort Formation. A strange absence in the Kap København Formation is that of *Amara alpina,* so common both in last interglacial sites from Greenland and in the Beaufort Formation (Elias & Matthews 2002, Matthews unpublished).

The insect fauna of the Kap København Formation allows detailed palaeoecological reconstruction of a highly diverse environment. Freshwater, freshwater shores, humid terrestrial biotopes, forests and alpine biotopes each contributes about equal numbers of species, with somewhat fewer taxa from dry environments, including steppes and saline ponds (Böcher 1995).

Biogeographically the Kap København fossil flora was dominated by taxa with a modern circumpolar or Nearctic range (Bennike 1990), whereas the insect fauna had an equal share of what are today nearctic and palaearctic elements (Böcher 1995). The greater part of the insects shows temperate conditions, but heterogeneous compositions of some of the insect assemblages indicate either subarctic or temperate summer temperatures (Böcher 1995).

From several points of view the Kap København Formation is very important. It is so well dated that it may be regarded a key site of the Early Pleistocene of the Northern Hemisphere (Elias & Matthews 2002). Stratigraphically it can be placed just after the first major cooling event with widespread glaciations on the Northern Hemisphere (*c.* 2.5 Ma; Funder *et al.* 2001; Jansen *et al.* 1990). This cold period is possibly represented in the lower member (A) of the formation which does not contain terrestrial remains. Member A is correlated with the Hvitland Beds on Ellesmere Island with a sparse arctic biota (Repenning & Brouwers 1992; Elias & Matthews 2002; Matthews unpublished).

The fossils from the younger member (B) of the Kap København Formation show that the biota recovered and a rich temperate flora and fauna developed. It existed at an extreme northern latitude, merely 800 km from the North Pole – just as today with five months without sunshine during winter and five months with continual day during summer, since a couple of million years ago Greenland was situated almost at the present position in relation to the North Pole (Rowley & Lottes 1988). The excellent preservation of most Kap København fossils indicates that the warm period came to a sudden end. Since then, and apparently uninterrupted by interglacial warmings, the Kap København sediments have been kept in the natural freezer of the arctic permafrost.

Without doubt, the Kap København-fauna represents the last remains of the rich Tertiary fauna which inhabited Greenland for millions of years (Heer 1883, Böcher 1997, Elias *et al.* 2006). This fauna probably died out in Greenland with the end of the climatic warming *c.* 2.5 million years ago, and the faunas during the many Quaternary interglacials had to re-immigrate every time. And every time Greenland presented special problems like those in the Holocene, and the arthropods involved had to possess some adaptations to cold – they had to be preadapted to a cool environment.

The origin of this pre-adapted fauna has not been so widely discussed (*e.g.*, Bale *et al.* 1997, Chernov 1985, Chernov *et al.* 2000). The most natural origin seems to be the alpine faunas living in the large Russian, Siberian and North American mountains, which in most cases extend from (present) temperate regions to the Arctic. According to Olga Makarova (personal communication 2007) the most diverse set of several arctic taxa is found around the Chukotka Peninsula in northeastern Siberia (*e.g.*, the gamasid mite genus *Arctoseius*; Makarova 2000). This area ('Eoarctic') is highly varied ecologically and contains very different landscapes and biotopes. But, in contrast to, *e.g.*, Greenland, the area has for most of the Quaternary been free from glaciations and the biological diversity is exceedingly high. This area might have served a 'store' of arctic taxa during the glaciations.

Very little has become known about Greenland insects older than the Pliocene. Heer (1883) illustrated a number of insects from Paleocene coal beds in West Greenland, Henriksen (1917) reported a cast of a weevil-elytron from Eocene-Oligocene layers at Kap Dalton, East Greenland, and Birket-Smith (1981) described the cast of a wing of an alleged phasmatodean from the Cretaceous of Nuussuaq, West Greenland.

Current climate changes and their effects

It is not possible to conclude this chapter without addressing the issue of seemingly marked (and likely at least partly man-made) climate change with which we are currently confronted. According to all predictions the change will be especially pronounced in high latitudes, and not least in Grenland (IPCC–reports), where many species attain their northern limit of existence. Because arthropods like most other organisms are ectothermic, receiving almost all their body heat from the surroundings, a rising temperature must have a profound effect on their life cycle and distribution. A general trend may be a movement towards higher latitudes and altitudes of species with a general northern distribution (*e.g.*, Böcher 2009). On the other hand, some arctic adapted forms may disappear from their southern range, in particular if the climate change involves enhanced humidity and diminishing sunshine. Such effects are probably already in force, but remain undocumented pending investigations specifically targeted at tracing northern and altitudinal limits of certain common species in the Arctic. Naturally, these questions have concerned several arctic entomologists (*e.g.*, Bale *et al.* 1997; Danks 1992, 2007; Hodkinson & Bird 1998; Høye & Forchammer 2008; Høye & Sikes 2012; Meltofte 2013). It is also essential here to mention the multitude of examples on rapid shifts of insect faunas recently found in palaeo-entomological studies (*e.g.*, Böcher 1995, 1997; Coope 1970, 1979a, 1995; Coope & Lemdahl 1995; Matthews 1977, 1979a, 1989; Elias 1991; Elias *et al.* 2006), These all indicate that insect fauna compositions almost instantly reflect climate change, and often on an amazing geographic scale.

References

Andersen, K.K. *et al.* (49 authors) (2004): High-resolution record of Northern Hemisphere climate extending into the last interglacial period. – *Nature* 43: 147–151.

Bale, J.S., Hodkinson, I.D., Block, W., Webb, N.R., Coulson, S.C. & Strathdee, A.T. (1997): *Life strategies of arctic terrestrial arthropods. In*: Woodin, S.J. & Marquiss, M. (eds). Ecology of Arctic Environment, pp. 137–165. Blackwell Science, U.K.

Bay, C. (1992): A phytogeographical study of the vascular plants of northern Greenland – north of 74° northern latitude. – *Meddelelser om Grønland. Bioscience* 36: 102 pp.

Bennike, O. (1990): The Kap København Formation: stratigraphy and palaeobotany of a Plio-Pleistocene sequence in Peary Land, North Greenland. – *Meddelelser om Grønland. Geoscience* 23: 85 pp.

Bennike, O. (1999): Colonisation of Greenland by plants and animals after the last ice age: a review. – *Polar Record* 35: 323–336.

Bennike, O. (2000): Palaeoecological studies of Holocene lake sediments from West Greenland. – *Palaeoecology, Palaeoclimatology, Palaeogeography* 155: 285–304.

Bennike, O. & Böcher, J. (1990): Forest-tundra neighbouring the North Pole: plant and insect remains from the Plio-Pleistocene Kap København Formation, North Greenland. – *Arctic* 43 (4): 331–338.

Bennike, O. & Böcher, J. (1992): Early Weichselian interstadial land biotas at Thule, Northwest Greenland. – *Boreas* 21: 111–117.

Bennike, O. & Böcher, J. (1994): Land biotas of the last interglacial/glacial cycle on Jameson Land, East Greenland. – *Boreas* 23: 479–487.

Bennike, O., Björck, S., Böcher, J., Hansen, L., Heinemeier, J. & Wohlfarth, B. (1999): Early Holocene plant and animal remains from North-east Greenland. – *Journal of Biogeography* 26: 667–677.

Bennike, O., Björck, S., Böcher, J. & Walker, I.R. (2000): The Quaternary arthropod fauna of Greenland: a review with new data. – *Bulletin of the Geological Society of Denmark* 47: 111–134.

Bergersen, R. (1995:) Is Greenland a zoogeographical unit of its own? – *Journal of Biogeography* 22: 1–6.

Birket-Smith, S.J.R. (1981): A wing of Phasmida from the Cretaceous Period (Insecta). – *Entomologica Scandinavica* 12: 245–249.

Björck, S., Bennike, O., Ingólfsson, O., Barnekow, L. & Penney, D.N. (1994): Lake Boksehandsken's earliest postglacial sediments and their palaeoenvironmental implications, Jameson Land, East Greenland. – *Boreas* 23: 459–472.

Böcher, J. (1971): Preliminary studies on the biology and ecology of *Chlamydatus pullus* (Reuter) (Heteroptera: Miridae) in Greenland. – *Meddelelser om Grønland* 191 (3): 29 pp.

Böcher, J. (1972): Feeding biology of *Nysius groenlandicus* (Heteroptera: Miridae) in Greenland. With a note on oviposition in relation to food-source and dispersal of the species. – *Meddelelser om Grønland* 191 (4): 41 pp.

Böcher, J. (1975): Spredningen af den grønlandske frøtæge, *Nysius groenlandicus* (Zett.) (Heteroptera: Lygaeidae). (With a summary: Dispersal of *Nysius groenlandicus* (Zett.) in Greenland). – *Entomologiske Meddelelser* 43: 105–109.

Böcher, J. (1976): Population studies on *Nysius groenlandicus* (Zett.) (Heteroptera: Lygaeidae) in Greenland, with particular reference to climatic factors, especially the snow cover. – *Videnskabelige Meddelelser fra Dansk Naturhistorisk Forening* 139: 61–89.

Böcher, J. (1988): The Coleoptera of Greenland. – *Meddelelser om Grønland. Bioscience* 26: 100 pp.

Böcher, J. (1989a): Boreal insects in northernmost Greenland: palaeo-entomological evidence from the Kap København Formation (Plio-Pleistocene), Peary Land. – *Fauna norvegica* Ser B 36: 37–43.

Böcher, J. (1989b): First record of an interstadial insect from Greenland: *Amara alpina* (Paykull, 1790) (Coleoptera: Carabidae). – *Boreas* 18: 1–4.

Böcher, J. (1990): Biologisk-arkæologisk kortlægning af Grønlands østkyst mellem 75°N og 79°30'N. Del 4: Entomologiske undersøgelser: Feltarbejde i 1989 og foreløbige resultater. – *Grønlands Hjemmestyre, Miljø- og Naturforvaltning. Teknisk rapport* 17, maj 1990. 28 pp.

Böcher, J. (1995): Palaeoentomology of the Kap København Formation, a Plio-Pleistocene sequence in Peary Land, North Greenland. – *Meddelelser om Grønland. Geoscience* 33: 81 pp.

Böcher, J. (1997): History of the Greenland insect fauna with emphasis on living and fossil beetles. – *Quaternary Proceedings* 5. *In*: Ashworth, A.C., Buckland, P.C. & Sadler, J.P. (eds): *Studies in Quaternary Entomology. An Inordinate Fondness for Insects*, pp. 35–47.

Böcher, J. (1998): *Insect remains from Asummiut. In*: Arneborg, J. & Gulløv, H.C. (eds) Man, culture and environment in ancient Greenland, pp.: 133–134. Danish Polar Center, Copenhagen.

Böcher, J. (1999): Mystisk opdukken af biller i Grønland. – *Dyr i natur og museum* 1999: 22–24.

Böcher, J. (2003): Landdyr i Peary Land. *In:* Martens, G., Jensen, J.F., Meldgaard, M. & Meltofte, H. (eds.): Peary Land, pp 269–272. Forlaget Atuagkat, Nuuk.

Böcher, J. (2012): Interglacial insects and their possible survival in Greenland during the last glacial stage. – *Boreas* DOI 10.1111/j.1502–3885.2012.00251.x.

Böcher, J. (2009): Fund af den grønlandske mariehøne (*Coccinella transversoguttata* Falderman, 1835) i Zackenbergdalen, Nordøstgrønland). – *Entomologiske Meddelelser* 77: 115–116. (With an English summary).

Böcher, J. & Bennike, O. (1991): Interglacial land biotas of Jameson Land, East Greenland. – *LUNDQUA Report* 33: 129–136.

Böcher, J. & Bennike, O. (1996): Early Holocene insect and plant remains from Jameson Land, East Greenland. – *Boreas* 25: 187–193.

Böcher, J. & Enghoff, H. (1984): A centipede in Greenland: *Lamyctes fulvicornis* Meinert, 1868 (Chilopoda, Lithobiomorpha, Henicopidae). – *Entomologiske Meddelelser* 52: 49–50.

Böcher, J. & Fredskild, B. (1993): Plant and arthropod remains from the palaeo-Eskimo site on Qeqertasussuk, West Greenland. – *Meddelelser om Grønland. Geoscience* 30: 35 pp.

Böcher, J. & Nachman, G. (2001): Temperature and humidity responses of the arctic-alpine seed bug *Nysius groenlandicus*. – *Entomologia Experimentalis et Applicata* 99: 319–330.

Böcher, J. & Nachman, G. (2010): Are environmental factors responsible for geographic variation in the sex ratio of the Greenlandic seed-bug *Nysius groenlandicus*. – *Entomologia Experimentalis et Applicata* 134: 122–130.

Böcher, J. & Nachman, G. (2011): Coexistence of bisexual and unisexual populations of *Nysius groenlandicus* in the Zackenberg Valley, Northeast Greenland. – *Entomologia Experimentalis et Applicata* 140: 196–206.

Böcher, J., Bennike, O. & Wagner, B. (2012): Holocene insect remains from south-western Greenland. – *Polar Research* 2012, 31, 18367 http: 11dx.doi.org/10.3402/polar.v31i0.18367.

Böcher, T.W. (1956): Area-limits and isolations of plants in relation to the physiography of the southern parts of Greenland. – *Meddelelser om Grønland* 124 (8): 40 pp.

Böcher, T.W. (1963): *Phytogeography of Greenland in the light of recent investigations*. – *In*: Löve, A. & Löve, D. (eds) North Atlantic Biota and Their History, pp. 285–295. Pergamon Press, Oxford.

Böcher, T.W. (1979): Birch woodland and tree growth in southern Greenland. – *Holarctic Ecology* 2: 218–221.

Böcher, T.W. (1981): *Hovedtræk af Grønlands plantegeografi*. – *In*: Nørrevang, A. & Lundø, J. (eds). Danmarks Natur 11, Grønland, pp. 350–364. Politikens Forlag, København.

Böcher, T.W., Holmen, K. & Jakobsen, K. (1968): *The Flora of Greenland*. P. Haase & Søn, Copenhagen. 312 pp.

Born, E.W. & Böcher, J. (eds) (2001): *The Ecology of Greenland*. Atuakkiorfik Education. Ilinniusiorfik, Nuuk. 429 pp.

Brændegaard, J. (1938): Aeronautic spiders in the Arctic. – *Meddelelser om Grønland* 119 (5): 9 pp.

Brouwers, E.M., Jørgensen, N.O. & Cronin, T.M. (1991): Climatic significance of the ostracode fauna from the Pliocene Kap København Formation, North Greenland. – *Micropaleontology* 37 (3): 245–267.

Buckland, P.C. (1988): North Atlantic faunal connections – introduction or endemics? – *Entomologica Scandinavica Supplementum* 32: 7–29.

Buckland, P.C. & Coope, G.R. (1991): *A Bibliography and Literature Review of Quaternary Entomology*. J.R. Collis Publications, Department of Archaeology & Prehistory, University of Sheffield. 85 pp.

Buckland, P.C., Sveinbjarnardóttir, G., McGovern, T.H., Skidmore, P. & Andreasen, C. (1983): Norsemen at Nipáitsoq, Greenland: A palaeoecological investigation. – *Norwegian Archaeological Review* 16: 86–98.

Buckland, P.C., Perry, D.W., Gislason, G.M. & Dugmore, A.J. (1986): The pre-Landnám fauna of Iceland: a palaeontological contribution. – *Boreas* 15: 173–184.

Chernov, Y.I. (1985): *The Living Tundra*. Cambridge University Press. Cambridge. 213 pp.

Chernov, Y.I., Makarov, K.V. & Eremon, P.K. (2000): Family of ground beetles (Coleoptera, Carabidae) in the Arctic fauna: Communication 1. – *Entomological Review* 80 (8): 1009–1021. Translated from *Zoologicheskii Zhurnal* 79 (12). 2000.

Christiansen, H.H., Bennike, O., Böcher, J., Elberling, B., Humlum, O. & Jakobsen, B.H. (2002): Holocene environmental reconstruction from deltaic deposits in northeast Greenland. – *Journal of Quaternary Science* 17(2): 145–160.

Coope, G.R. (1969): The contribution that the Coleoptera of Glacial Britain could have made to subsequent colonisation of Scandinavia. – *Opuscula Entomologica* 34: 95–108.

Coope, G.R. (1970): Interpretations of Quaternary insect fossils. – *Annual Review of Entomology* 15: 97–120.

Coope, G.R. (1978): *Constancy of insect species versus inconstancy of Quaternary environments*. – *In*: Mound, L.A. & Waloff, N. (eds.) Diversity of Insect Faunas, pp. 176–187. (Symposia of the Royal Entomological Society of London 9). Blackwell, Oxford.

Coope, G.R. (1979a): Late Cenozoic fossil Coleoptera: evolution, biogeography and ecology. – *Annual Review of Ecology and Systematics* 10: 247–257.

Coope, G.R.(1979b): *The Carabidae of the Glacial Refuge in the British Isles and their Contribution to the Post Glacial Colonization of Scandinavia and the North Atlantic Islands*. – *In*: Erwin, T.L., Ball, G.E. & Whitehead, D.R. (eds). Carabid Beetles, their Evolution, Natural History, and Classification, pp. 407–424. Dr. W. Junk Publishers, the Hague.

Coope, G.R. (1986): The invasion and colonization of the North Atlantic Islands: a palaeoecological solution to a biogeographic problem. – *Philosophical Transactions of the Royal Society of London* B 314: 619–635.

Coope, G.R. (1995): *The effects of Quaternary climate changes on insect populations: lessons from the past*. *In*: Harrington, R. & Stork, N.E. (eds). Insects in a changing environment. Academic Press, London.

Coope, G.R. & Lemdahl, G. (1995): Regional differencies in the lateglacial climate of northern Europe based on coleopteran analysis. – *Journal of Quaternary Science* 10: 391–395.

Coope, G.R. & Lemdahl, G. (1996): Validations for the use of beetle remains as reliable indicators of Quaternary climates:

a reply to criticism by Johan Andersen. – *Journal of Biogeography* 23: 115–120.

Dahl-Jensen, D., Mosegaard, K., Gundestrup, N., Clow, G.D., Johnsen, S.J., Hansen, A.W. & Balling, B. (1998): Past temperatures directly from the Greenland ice sheet. – *Science* 282: 268–271.

Danks, H.V. (1981): *Arctic Arthropods. A review of systematics and ecology with particular reference to the North American fauna.* Entomological Society of Canada, Ottawa. 608 pp.

Danks, H.V. (1992): Arctic insects as indicators of environmental change. – *Arctic* 45: 159–166.

Danks, H.V. (2007): Canadian Perspectives: Climate change impacts on insects in northern Canada. – *Newsletter of the Biological Survey of Canada (Terrestrial Arthropods)* 26 (1): 15–18.

Degerbøl, M. (1937): A contribution to the investigation of the fauna of the Blosseville Kyst, East Greenland, with special reference to zoogeography. – *Meddelelser om Grønland* 104 (19): 36 pp.

Downes, J.A. (1964): Arctic insects and their environment. – *Canadian Entomologist* 96: 279–307.

Downes, J.A. (1966): The Lepidoptera of Greenland; some geographic considerations. – *Canadian Entomologist* 98: 1135–1144.

Downes, J.A. (1988): The Post-glacial colonization of the North Atlantic Islands. – *Memoirs of the Entomological Society of Canada* 144: 55–92.

Eisner, W.R., Törnqvist, T.E., Koster, E.A., Bennike, O. & van Leeuwen, J.F.N. (1995): Paleoecological studies of a Holocene lacustrine record from the Kangerlussuaq (Søndre Strømfjord) region of West Greenland. – *Quaternary Research* 43: 55–66.

Elias, S.A. (1991): Insects and climate change. Fossil evidence from the Rocky Mountains. – *BioScience* 41 (8): 552–559.

Elias, S.A. & Matthews, J.V. Jr, (2002): Arctic North American seasonal temperatures in the Pliocene and Early Pleistocene, based on mutual climatic range analysis of fossil beetle assemblages. – *Canadian Journal of Earth Sciences* 39: 911–920.

Elias, S.A., Kuzmina, S. & Kiselyov, S. (2006): Late Tertiary origins of the Arctic beetle fauna. – *Palaeogeography, Palaeoclimatology, Palaeoecology* 241: 373–392.

Elton, C.S. (1925): The dispersal of insects to Spitzbergen. – *Transactions of the Entomological Society of London* 1925, 1–2: 289–299.

Evans, J., Cofaigh, C.O., Dowdeswell, J.A. & Wadhams, P. (2008): Marine geophysical evidence for former expansion and flow of the Greenland Ice Sheet across the north-east Greenland continental shelf. – *Journal of Quaternary Science* DOI: 10.1002/jqs.1231. ISSN 0267–8179. 15 pp.

Feyling-Hanssen, R.W. (1990): Foraminiferal stratigraphy in the Plio-Pleistocene Kap København Formation, North Greenland. – *Meddelelser om Grønland. Geoscience* 24. 32 pp.

Fredskild, B. (1973): Studies in the vegetational history of Greenland. Palaeobotanical investigations of some holocene lake and bog deposits. – *Meddelelser om Grønland* 198 (4): 245 pp.

Fredskild, B. & Røen, U. (1982): Macrofossils in an interglacial peat deposit at Kap København, North Greenland. – *Boreas* 11: 181–185.

Fredskild, B., Jacobsen, N. & Røen, U. (1975): Remains of mosses and freshwater animals in some Holocene lake and bog sediments from Greenland. – *Meddelelser om Grønland* 198 (5): 44 pp.

Funder, S. (1978): Holocene stratigraphy and vegetational history in the Scoresby Sund area, East Greenland. – *Grønlands Geologiske Undersøgelse, Bulletin* 129: 66 pp.

Funder, S. (1979): Ice-age plant refugia in East Greenland. – *Palaeogeography, Palaeoclimatology, Palaeoecology* 28: 279–295.

Funder, S. & Hjort, C. (1980): A reconnaissance of the Quaternary geology of eastern North Greenland. – *Grønlands Geologiske Undersøgelse. Rapport* 99: 99–105.

Funder, S., Bennike, O., Mogensen, G.S., Nygaard, B., Pedersen, S.A.S. & Petersen, K.S. (1984): The Kap København Formation, a late Cainozoic sedimentary sequence in North Greenland. – *Grønlands Geologiske Undersøgelse. Rapport* 120: 9–18.

Funder, S., Abrahamsen, N., Bennike, O. & Feyling-Hanssen, R.W. (1985): Forested Arctic: Evidence from North Greenland. – *Geology* 13: 542–546.

Funder, S., Bennike, O., Böcher, J., Israelson, C., Petersen, K.S. & Símonarson, L.A. (2001): Late Pliocene Greenland – The Kap København Formation in North Greenland. – *Bulletin of the Geological Society of Denmark* 48: 117–134.

Funder, S., Kjellerup, K.K., Kjær, K.H. & Ó Cofaigh, C. (2011): The Greenland ice sheet, the last 300,000 years: a review. *In*: Ehlers, F. & Gibbard, P. (eds) *Quarternary Glaciations – Extent and Chronology. Part IV: A closer look*, pp. 699–713. Elsevier.

Haarløv, N. (1942): A morphologic-systematic-ecological investigation of Acarina. – *Meddelelser om Grønland* 128 (1): 71 pp.

Haarløv, N. (1967): Arthropoda (Acarina, Diptera) from subfossil layers in West Greenland. – *Meddelelser om Grønland* 184 (3): 17 pp.

Håkansson, L., Alexanderson, H., Hjort, C., Möller, P., Briner, J.P., Aldahan, A. & Possnert, G. (2008): Late Pleistocene glacial history of Jameson Land, central East Greenland, derived from cosmogenic ^{10}Be and ^{26}Al exposure dating. – *Boreas* 10.1111/j.1502–3885.2008.00064.x.ISSN 0300–9483. 17 pp.

Hald, M. & Aspeli, R. (1997): Rapid climatic shifts of the Norvegian Sea during the last glaciation and the Holocene. – *Boreas* 26: 15–28.

Hammer, M. (1944): Studies on the oribatids and collembolans of Greenland. – *Meddelelser om Grønland* 141: 210 pp.

Hammer, M. (1955): Some aspects of the distribution of microfauna in the Arctic. – *Arctic* 8 (2): 115–126.

Heer, O. (1883): Oversigt over Grønlands fossile Flora. (With a summary: Recherches sur fossiles appartenant aux formation crétacée et miocéne, sur la côte occidentale du Grønland. Communiqué par M.F. Johnstrup). – *Meddelelser om Grønland* 5 (3): 81–199. (With a supplement volume containing 109 plates).

Heie, O.E. (1995): An aphid from the Plio-Pleistocene Kap København Formation, North Greenland (Homoptera, Aphidoidea). – *Entomologiske Meddelelser* 63: 17–18.

Henriksen, K.L. (1917): En Bemærkning om Tertiæret ved Kap Dalton i Øst-Grønland. – *Meddelelser om Grønland* 56 (5): 205–206.

Henriksen, K.L. (1939): A revised index of the insects of Grønland. – *Meddelelser om Grønland* 119 (10): 112 pp.

Henriksen, K.L. & Lundbeck, W. (1917): Grønlands Landarthropoder (Insecta et Arachnida Groenlandicæ. – *Meddelelser om Grønland* 22: 483–821.

Hodkinson, I. & Bird, J. (1998): Host-specifik insect herbivores as sensors of climate change in arctic and alpine environments. – *Arctic and Alpine Research* 30 (1): 78–83.

Høye, T.T. & Forchammer, M.C. (2008): Phenology of High-Arctic Arthropods: Effect of Climate on Spatial, Seasonal, and Inter-Annual Variation. *In*: Meltofte, H., Christensen, T.R., Elberling, B., Forchammer, M.C. & Rasch, M. High-Arctic Ecosystem Dynamics in a Changing Climate. Ten years of monitoring and research at Zackenberg Research Station, Northeast Greenland. – *Advances in Ecological Research* 40: 299–324. Elsevier, Amsterdam etc.

Høye, T.T. & Sikes, D.S. (2013): Arctic entomology in the 21st century. – *Canadian Entomologist* 145: 125–130.

IPPC–reports (The Intergovernmental Panel on Climate Change). United Nations.

Jansen, E., Sjøholm, J., Bleil, U. & Erichsen, J.A. (1990): *Neogene and Pleistocene glaciations in the Northern Hemisphere and Late Miocene-Pliocene global ice volume fluctuations: evidence from the Norwegian Sea.* – *In*: Bleil, U. & Thiede, J. (eds). Geological History of the Polar Ocean; Arctic Versus Antarctic, pp. 677–705. Kluwer, Amsterdam.

Jensen, A.S. (1928): *Grønlands Fauna, et Forsøg på en Oversigt.* Festskrift udgivet af Københavns Universitet i Anledning af Hans Majestæt Kongens Fødselsdag 26. September 1928. København. 88 pp.

Johnsen, S.J., Clausen, H.B., Dansgaard, W., Gundestrup, N.S., Hansson, M., Jonsson, P., Steffensen, J.P. & Sveinbjørnsdottir, A.E. (1992): A "deep" ice core from East Greenland. – *Meddelelser om Grønland. Geoscience* 29: 22 pp.

Larsen, S. & Scharff, N. (2003): The spiders of Greenland – a checklist (Arachnida: Araneae). – *Entomologiske Meddelelser* 71 (1): 53–61.

Larsson, S.G. & Gigja, G. (1959): *Coleoptera 1. Synopsis.* The Zoology of Iceland 3, 46a. Ejnar Munksgaard, Copenhagen. 218 pp.

Lindroth, C.H. (1931): *Die Insektenfauna Islands und ihre Probleme.* Uppsala. 589 pp.

Lindroth, C.H. (1957): *The Faunal Connection between Europe and North America.* Almquist & Wiksell, Stockholm. 344 pp.

Lindroth, C.H. (1965): Skaftafell, Iceland, a living glacial refugium. – *Oikos Supplementum* 6. 142 pp.

Lindroth, C.H. (1968): Distribution and distributional centers of North Atlantic insects. – *Bulletin of the Entomological Society of America* 14 (2): 91–95.

Lindroth, C.H. (1970): Survival of animals and plants on ice-free refugia during the Pleistocene glaciations. – *Endeavour* 29 (108); 129–134.

Lindroth, C.H. (1979): *The Theory of Glacial Refugia. In*: Erwin, T.L., Ball, G.E. & Whitehead, D.R. (eds). Carabid Beetles, their Evolution, Natural History, and Classification, pp. 385–394. Dr. W. Junk Publishers, the Hague.

Lindroth, C.H. (1985–1986): *The Carabidae (Coleoptera) of Fennoscandia and Denmark.* Fauna Entomologica Scandinavica 15, 1–2. 497 pp.

Makarova, O.L. (2000): To a study of the genus *Arctoseius* (Parasitiformes, Ascidae) from the far north: 3. Ranges and ecological preferences of species. – *Entomological Review* 80 (Suppl. 1): S145–S150.

Matthews, J.V. Jr. (1977): Tertiary Coleoptera fossils from the North American Arctic. – *The Coleopterists Bulletin* 31 (2): 297–308.

Matthews, J.V. Jr. (1979a): *Late Tertiary Carabid Fossils from Alaska and the Canadian Archipelago. In*: Erwin, T.L., Ball, G.E. & Whitehead, D.R. (eds). Carabid Beetles, their Evolution, Natural History and Classification, pp. 425–445. Junk, the Hague.

Matthews, J.V. Jr. (1979b): *Fossil Beetles and the Late Cenozoic History of the Tundra Environment. In*: Gray, J. & Boucot, A.J. (eds). Historical Biogeography, Plate Tectonics, and the Changing Environmen, pp. 371–376. Oregon State University Press.

Matthews, J.V. Jr. (1980): Tertiary land bridges and their climate: Backdrop for development of the present Canadian insect fauna. – *Canadian Entomologist* 112: 1089–1103.

Matthews, J.V. Jr. (1989): Late Tertiary Environments: A Vision of the Future? – *Geos* 18 (3): 14–18.

Matthews, J.V. Jr. (unpublished). *Late Cenozoic plant and insect macrofossils from Alaska and Arctic/Subarctic Canada: new data, intercontinental comparisons and correlations.*

McGovern, T.H., Buckland, P.C., Savory, D., Sveinbjarnardóttir, G., Andreasen, C. & Skidmore, P. (1983): A study of the faunal and floral remains from two Norse farms in the Western Settlement, Greenland. – *Arctic Anthropology* 20: 93–120.

Meldgaard, M. (2004): Ancient Harp Seal Hunters of Disko Bay. Subsistence and Settlement at the Saqqaq Site Qeqertasussuk (2100–1400 BC), West Greenland. – *Meddelelser om Grønland. Man & Society* 30: 189 pp.

Meltofte, H. (2013) (ed.) Arctic Biodiversity Assessment. Status and trends in Arctic biodiversity. – Conservation of Arctic Flora and Fauna, Akureyri.

Nurminen, M. (1973): Distributions of northern Enchytraeids (Oligochaeta). – *Annales Zoologici Fennici* 10: 483–486.

Omodeo, P. (1957): Lumbricidae and Lumbriculidae of Greenland. – *Meddelelser om Grønland* 124 (6): 27 pp.

Ostenfeld, C.H. (1926): The flora of Greenland and its origin. – *Det Kongelige Danske Videnskabernes Selskab, Biologiske Meddelelser* 6 (3): 71 pp.

Repenning, C.A. & Brouwers, E.M. (1992): Late Pliocene-Early Pleistocene ecologic changes in the Arctic Ocean borderland. – *U.S. Geological Survey Bulletin* 2036: 37 pp.

Røen, U. (1988): Remains of early glacial Cladocera from North Greenland. – *Verhandlungen der Internationalen Vereinigung für Limnologie* 23: 845–847.

Rowley, D.B. & Lottes, A.L. (1988): Plate-kinematic reconstructions of the North Atlantic and Arctic: Late Jurassic to Present. – *Tectonophysics* 155: 73–120.

Ruddiman, W.F. & McIntyre, A. (1981): The North Atlantic Ocean during the last deglaciation. – *Palaeogeography, Palaeoclimatology, Palaeoecology* 35: 145–214.

Salomonsen, F. (1972): *Zoogeographical and ecological problems in arctic birds*. – *In*: Proceedings of the XVth International Ornithological Congress, The Hague, September 1970, pp. 25–77. E.J. Brill, Leiden.

Salomonsen, F. (1979): Ornithological and ecological studies in S.W. Greenland (59°46'-62°27' N. Lat.). – *Meddelelser om Grønland* 204 (6): 1–214.

Símonarson, L.A., Petersen, K.S. & Funder, S. (1998): Molluscan palaeontology of the Pliocene-Pleistocene Kap København Formation, North Greenland. – *Meddelelser om Grønland. Geoscienc* 36: 103 pp.

Vibe, C. (1953): *The zoogeography of Greenland*. *In*: Proceedings of the XIV International Congress of Zoology, Copenhagen 5.-12. August 1953, pp. 126–127.

CHAPTER 5

The inventory of the Greenland entomofauna

Jens Böcher

The present account is primarily based on Henriksen (1921–37), Henriksen & Lundbeck (1917) and Henriksen (1939). Principal sources of supplementary information are Wolff (1964) and Danks (1981).

The initial Danish inventory of the Greenland fauna is totally dominated by a single figure, namely Otto (or Otho) Fabricius (1744–1822), "the founder of Greenlandic zoology" (Fig. 1); he was a distant relative of the contemporary famous entomologist, Johann C. Fabricius. Otto Fabricius was a missionary in Paamiut/Frederikshåb, South Greenland, during five years (1768–73) and his 1780 book 'Fauna Groenlandica' – is *the* classical work on arctic animal life. Because of the wealth of accurate biological and faunistic information, this work has retained its value in modern time. It described about 470 animal species, including 62 insects and 19 arachnids. More than a fourth are "*novae species*", and with few exceptions all the species are so carefully described that they are today fairly easily identified.

FIGURE 1 *Otto Fabricius (Contemporary painting by unknown artist).*

For 77 years 'Fauna Groenlandica' was the only written source on Greenland zoology, although additional species were discovered. C. P. Holbøll (1795–1856) collected 'naturalia' (including several insects) for the 'Kongelige Naturhistoriske Museum' (the Royal Museum of Natural History, one of the forerunners of Copenhagen Zoological Museum and hence of the present 'Natural History Museum of Denmark'). J. L. M. Vahl (1796–1854) in 1828–36 travelled along the west coast of Greenland in order to collect plants and animals including insects, and later he – together with H. Krøyer (renowned ichthyologist and carcinologist) – joined a French expedition (led by P. Gaimard) to Nordkap (Norway) and Svalbard. While Vahl's botanical achievements were enormous, his entomological contributions were modest; nonetheless, this new Greenland material forms the basis of Stæger's overview (1845) of the Greenland Diptera.

Also a number of foreign expeditions added knowledge about the Greenland fauna of insects and arachnids. From the British expedition to East Greenland 1822 led by W. Scoresby, Jameson & Wilson (in Scoresby 1823) reported the finds of the first insect species collected in East Greenland, and Lefebvre (1836a, 1836b) described a few Lepidoptera.

In 'Insecta Lapponica' Zetterstedt (1840) enumerated 61 species of insects from Greenland, while Stæger (1845) increased the number of known Diptera to 55, of which eight were new to science.

A systematic list of all the known Greenland insects and arachnids by Schiødte (1857) was for a long time *the* reference work on the Greenland fauna of land arthropods. Subsequent developments in the knowledge about this fauna was in no small measure due to Danish laymen resident in West Greenland, including C.G.F. Pfaff and C.J.S.I. Thaarup (medical practitioners in Ilulissat/Jakobshavn and Ivittuut, respectively) and the civil servant E. C. Fencker. One of the most important private collections of Greenland insects was procured by G. A. Meldorf, an officer of health in the Qaqortoq/Julianehåb district 1897–1903. J. No Nygaard, assistant 1906–12 to the botanist Morten Pedersen Porsild at the 'Den Danske Arktiske Station på Disko' (now 'The Arctic Station,

University of Copenhagen', Fig. 2) at Qeqertarsuaq/Disko; he procured a large material of Lepidoptera, especially from the Ameralik Fjord south of Nuuk/Godthåb. The collections formed by all of these are (entirely, or for the greater parts) now in the Natural History Museum of Denmark.

During the last half of the nineteenth century the exploration of Greenland was to a high degree carried out by expeditions and scientists from countries other than Denmark. Holmgren (1872) treated and published on the insect material collected during A.E. Nordenskiöld's 1870 expedition to Northwest Greenland (Aasiat/Egedesminde, Qeqertarsuaq/Disko). During A.E. Nordenskiöld's expedition in 1883 to Indlandsisen (the Ice Cap), G. Kolthoff collected insects in the Arfersiorfik fiord (*ca.* 68°N). The results were published by Aurivillius (1890), Sahlberg (1901), and Forsslund (1932).

'Grönland-Expedition der Gesellschaft für Erdkunde in Berlin' (1891–93) under the leadership of Erich von Drygalski operated in the innermost part of the Uummannaq fiord-complex (70°30'N), where E. Vanhöffen collected insects and arachnids, published in 1897 by Kramer, Lenz, Rübsaamen, and Vanhöffen.

The Norwegian '2den 'Fram' Ekspedition' 1898–1902 with O. Sverdrup as leader and E. Bay and H. G. Simmons as zoological collectors worked in both middle West Greenland (Qeqertarsuup Tunua/Disko Bugt) and on Ellesmere Island. Strand (1905) published the entomological outcome.

The insects collected during 'Die zweite deutsche Nordpolarfahrt' 1869 & 1870, led by Koldewey and Payer, onboard the 'Germania' to Northeast Greenland between 73° and 77° N were treated by Gerstäcker (1874) and v. Homeyer (1875), who thereby added considerably to the scanty knowledge of the insect fauna of that remote region. The relatively meagre entomological outcome of the American 'Polaris Expedition' in 1872 (led by C.F. Hall) to Ellesmere Land, Northwest- and North Greenland (78°–82° N) was published Osten-Sacken (1878), Packard (1877) and Thorell (1878), while McLachlan (1878) reported on the findings by a contemporary English expedion to the same area.

In 1878 'Kommissionen for Ledelsen af de geologiske og geografiske Undersøgelser i Grønland' was established (now 'The Commision for Scientific Research in Greenland'). This meant that the scientific exploration of Greenland now became coordinated and centrally organized. The results of the activities were (and still are to a great extent) published in the series 'Meddelelser om Grønland' and its successors (including 'Greenland Bioscience'). Some of the scientists sent to Greenland by the Commision besides their main task collected insects for the Zoological Museum, in particular two geologists, A. Kornerup (1857–1881) and the later professor K.J.V. Steenstrup (1842–1913). Kornerup took part in three expeditions, *e.g.*, together with J.A.D. Jensen to a nunatak-area in the Paamiut/Frederikshåb district which were later named after Jensen. Kornerup (1879) described the biota on one of the nunataks.

FIGURE 2 *The Arctic Station, University of Copenhagen, in March 1968 when the old main building (right) was still white; following a thorough renovation (1980) it was painted red. The station is positioned 1 km east of Qeqertarsuaq/Godhavn on the south coast of the island Qeqertarsuaq/Disko. It was erected in 1906 by the botanist Morten P. Porsild, and is the oldest research station in the Arctic. The main building comprises guest facilities and the home of the station manager. To the left the resident scientist's private home, in middle the laboratory and library building. In the background the mountain Skarvefjeld.*

In the years 1883–85 the Danish 'Konebåd Ekspeditionen' ('the Umiak Expedition') sailed along the southeastern coast from Nunap Isua/Kap Farvel to Tasiilaq/Ammassalik. The biologist member of the expedition, P. Eberlin, in addition to making large collections of plants and minerals collected a few insects. Much more important for the exploration of the Greenland insect and arachnid fauna was 'Skibsekspeditionen til Grønlands Østkyst 1891–1892', which explored the huge Scoresby Sund fiord complex onboard the 'Hekla'. The expedition overwintered in 'Hekla Havn' on Danmarks Ø. Insects were collected partly by the zoologist and geologist E. Bay, but to a higher degree by the physician of the expedition, Henrik Deichmann. He not only collected a large material, but also carried out interesting phenological studies on insect life during winter and spring. The material was published by Deichmann & Lundbeck (1895), for which

Deichmann prepared the general biological account and Lundbeck (see below) the fauna list.

During 1898–1899 and 1900 two Swedish expedition – led by the botanist G. Nathorst onboard the 'Antarctic' and by G. Kolthoff onboard the 'Frithjof' – explored the large East Greenland fiords, Franz Joseph Fjord, Kong Oscar Fjord, and Scoresby Sund. I. Arwidsson collected insects and arachnids on Nathorst's expedition. The entomological material has been treated by Aurivillius (1900), Wahlgren (1900), and Mjöberg (1910).

In 1900 followed the third greater Danish expedition to East Greenland onboard the 'Antarctic' (acquired after Nathorst's expedition). The very rich insect material brought home by this expedition is due to the afore-mentioned H. Deichmann and the zoologist S. Jensen. On the basis of their contribution together with earlier material, J.C. Nielsen was able to compile 'The Insects of East Greenland' (1907).

During 1906–08 followed the great 'Danmark-Expeditionen til Grønlands Nordøstkyst' onboard the 'Danmark'. The ship anchored in 'Danmarks Havn' (now Danmarkshavn) on Germania Land (76°48'N) and functioned as the expedition base; Fritz Johansen (1882–1957) was the zoologist member of the expedition. The entomological yield of the 'Danmark Expedition' was worked up by Johansen & Nielsen (1910); here Johansen wrote the general part and Nielsen the systematic part. The arachnids were treated by Trägårdh (1914) and Emerton (1918).

In 1889–1890 the Commission sent out a combined entomological/botanical expedition to West Greenland, with William Lundbeck (1863–1941), as entomologist and N. Hartz as botanist. They travelled by umiak from Paamiut/Frederikshåb to Qaqortoq/Julianehåb on the southern stretch of the coast (1889) and next summer from Egedesminde/Aasiat to northern Disko Bugt/Qeqertarsuup Tunua. Lundbeck (Fig. 3) collected by far the most important entomological material so far procured in Greenland, and based on this he published a number of papers (1891, 1891–1892, 1896a, 1896b, 1898) including a comprehensive survey of the insects known from East Greenland (1895). Lundbeck's material similarly formed the basis for treatments of Greenland arthropods by other workers (Bang-Haas 1896, Meinert 1896, Rostrup 1900, Sørensen 1898, Trägårdh 1909).

An annotated list of the insects and arachnids of Greenland in the series *'Conspectus Faunæ Groenlandicæ'* (1917) was prepared jointly by Lundbeck and K. L. Henriksen. In addition to Lundbeck's own material it treated the above-mentioned collections from East Greenland (Nielsen 1907, 1910), from the 'Danmark Ekspedition' 1906–08 (Johansen & Nielsen 1910, Johansen 1911) and Swedish collections published by Holmgren (1872), Aurivillius (1890, 1900), Wahlgren (1900), Roman (1916), and some other minor contributions. *'Conspectus'* includes 437 species of insects and 124 species of arachnids. Henriksen later (1939) revised this list of Greenland insects, now enumerating 592 species. The list also includes material published in some works overlooked by Henriksen & Lundbeck (1917), *e.g.* the collections by the 2nd Norwegian Arctic Expedition in the 'Fram' 1898–1902 (Strand 1905).

FIGURE 3 *William Lundbeck (Photograph in Natural History Museum of Denmark).*

The time between *'Conspectus'* and Henriksen's revised list, 1917–1939 was a period of high exploratory activity in Greenland, but from Danish side almost exclusively in East- and Northeast Greenland, where a number of expeditions had biologist members who – in addition to other activities – made valuable collections of insects and arachnids. The Swedish botanist Thorild Wulff, participating in the '2nd Thule Expedition' along the North Coast of Greenland (during which he tragically succumbed 1916) in addition to his botanical work also collected insects.

The ornithologist Alwin Pedersen during both 1924–25 and 1927–29 mainly studied mammals and birds in

Kangertittivaq/Scoresby Sund (Jameson Land and Liverpool Land), but also collected a sizable material of insects.

On the geological expeditions (under leadership of Lauge Koch) to the Kejser Franz Joseph Fjord-area (c.73° 30'N) 1929–30 onboard the 'Godthaab', a number of insects were collected by Bernt Løppenthin (ornithologist) and Gunnar Seidenfaden (botanist) (see Graversen 1931).

The '6th and 7th Thule Expeditions' (1932–33, under the leadership of Knud Rasmussen) investigated the southern stretch of the east coast, between *ca.* 60° and 68° N. The entomological material was collected by Marie Hammer (née Jørgensen) and Richard Bøgvad (geologist). At the same time (1932) 'The Scoresby Sund Committee's 2nd East Greenland Expedition' led by Ejnar Mikkelsen (1933) explored Kong Christian IX Land, the Blosseville Kyst between Kap Dalton and Kangerlussuaq (*ca.* 68°-69°30'N), with the mammalogist Magnus Degerbøl as zoologist collector, while the 'Treårsekspeditionen til Christian X Land' 1931–34 led by Lauge Koch included a number of zoologists (Frode Søgaard Andersen, Jens Brændegaard, Holger Madsen, Alwin Pedersen, Ragnar Spärck, Gunnar Thorson) who collected between *ca.* 69°30' and 76°N. G. Thorson and in particular M. Jørgensen/Hammer primarily sampled micro-arthropods by means of Berlese-funnels.

On the 'Thor-expedition' (1935) to southernmost East Greenland, E. Berthelsen (ichthyologist) collected insects, mainly in Kangerlussuatsiaq/Lindenow Fjord (c. 60°30'N).

Thus, together these expeditions covered the entire east coast north to c. 76°N and provided a large material of insects and arachnids from hitherto unexplored parts of Greenland. Papers dealing with the results from the expeditions were published by Andersen (1937), Brændegaard, Henriksen & Spärck (1935), Degerbøl (1937), Jørgensen/Hammer (1934a, 1934b, 1937, 1938) and Madsen (1936). However, the bulk of the insect collections was treated by Henriksen (1935, 1937).

Also the 'Norwegian Government Expeditions to East Greenland' onboard the 'Veslekari' 1929 (leadership: Anders K. Orwin) and 1930 (leadership Adolf Hoel) carried out important entomological collections between 71° and 75°N, the collector being Nils Knaben. The material has been dealt with by Barca (1937), Friese (1935), Goetghebuer (1935), Knaben (1937), Lindberg (1935), Linnaniemi (1935) Malaise (1933), Ringdahl (1933), and Roman (1933).

A number of foreign expeditions collected insects in the Kangertittivaq/Scoresby Sund-area. On the French 'Pourqoi Pas?'-expedition in 1926 (led by J.B. Charcot), Paul Remy collected around Hurry Inlet, *ca.* 70° 25'N (Remy 1928a, 1928b, 1928c). On Alfred Wegener's 'Deutsche Grönland Expedition' 1929 and 1930–31, H.B. Peters worked near the 'East Station' (*ca.* 71°N). The 'Cambridge Expedition to Scoresby Sound' (1933), on which David Lack collected insects, also worked mainly around Hurry Inlet (*ca.* 70°40'N). Finally, the English Wilkins Expedition (1933) went to Lake Fjord on the Blosseville Coast (66°17'N), where F. Spencer Chapman collected insects (Jackson 1934, Meyrick 1934, Collin 1937, Edwards 1935, Lack 1934, Roman 1934).

In the same period, investigations of West Greenland were due mainly to three English expeditions from the Oxford University: In 1928 R.W.G. Hingston collected insects in the Nuup Kangerlua/Godthåbsfjord at Qussuk and Majuala at 64°45'N. In 1934 J.M. Wordie on his way to Ellesmere Land collected insects between Upernavik (72°45'N) and Amdrup Ø (74°43'N), and at Etah (78°20'N). In 1936 H.G. Vevers worked in the Sisimiut/Holsteinsborg area at *ca.* 67°N at Utoqqaat in Amerloq and Camp Lloyd (Strømfjordshavn) at the head of Kangerlussuaq/Søndre Strømfjord. A number of papers deal with the material collected by the Oxford University Greenland Expeditions (Blair 1930, Brown 1937, Carpenter 1937, 1938, China 1934, Collin 1930, 1931, Cooper 1931, Edwards 1931, Jackson 1930, Longstaff 1936, Mosely 1929, Richards 1931, Roman 1930).

Christian Vibe in 1939 made entomological collections during the 'Thule and Ellesmere Land Expedition' (1939–40). In 1948 he was appointed leader of the so called 'Greenland Zootopographical Investigations', established at the then Copenhagen Zoological Museum, and during the following years he made collections and studies of the fauna in many parts of Greenland. While Vibe personally published little on insects (1950), he initiated collaboration with a number of specialists whose publications in a sub-series of 'Meddelelser om Grønland' ('The Danish Zoogeographical Investigations in Greenland') are mainly based on his material. In this way a number of taxa were revised with many additions to the Greenland species list, *e.g.*, Aphidoidea by Lambers (1952, 1960), Lepidoptera by Wolff (1964), Agromyzidae by Griffiths (1966). In 1981 Vibe wrote a popular survey of Greenland terrestrial invertebrates.

The material collected by Vibe from Qaqortoq/Julianehåb 1953 unfortunately has been been mixed up with his collections from Mestersvig (NE) the same year. This means that *e.g. Wesmaelius nervosus* probably does not occur in East Greenland; and *Nabis flavomarginatus* is certainly restricted to localities in S and SW.

After the Second World War various expeditions and collecting trips by single persons augmented the knowledge and added to the list of species.

The zoologist Palle Johnsen and the glaciologist Børge Fristrup took part in the 'Danish Peary Land Expedition' (1947–1950) and collected insects and arachnids in the (by then entomologically entirely unknown) Heilprinland-area, about 82°N. Later, from the '2nd Danish Peary Land Expedition' (1963), the botanist Bent Fredskild brought home a small collection of Lepidoptera from the surroundings of Jørgen Brønlund Fjord (82°10'–82°15'N).

During the 2nd French Polar Expedition ('Missions Paul Emil Victor') in 1949, H. de Lesse collected insects at Eqe close to the Ice Cap (69°45'N) on the west coast (de Lesse 1950 etc.).

In 1955 the Icelandic entomologist Hálfdan Björnsson collected insects at Mestersvig in East Greenland (72°13'N). The same year T. Feddersen collected Lepidoptera during his service as a medical officer in the Danish Navy at Kangilinnguit/Grønnedal in southwestern Greenland.

N.L. Wolf in 1962 collected Lepidoptera in three places in West Greenland (Narsarsuaq, Kangerlussuaq/Søndre Strømfjord, Qeqertarsuaq/Disko) and later revised the Lepidoptera fauna of Greenland.

Holm (1967) published on spiders collected in 1962 mainly at Qeqertarsuaq/ Godhavn (69°15'N) and Nuuk/ Godthåb (64°10'N).

From the 1930s onwards entomological research in Greenland has increasingly focussed on ecological/biological aspects rather than the purely faunistic exploration. The first real ecological work on Greenland invertebrates, which includes information on terrestrial arthropods, is Madsen's treatment (1936) of the shore fauna of East Greenland, in which the main emphasis is on the zonation and the different living conditions of the animals along the marine-terrestrial gradient. Andersen (1937, 1946) treated populations of Chironomidae from a biological/ecological viewpoint. Haarløv (1942) and Hammer (1944), studying the Greenland soil micro-arthropods, introduced ecological approaches. Hammer identified associations of collembolans and oribatid mites and discussed their relationship to plant communities, microclimate and the humidity of the soil. Haarløv especially considered the phenology of the soil arthropods, in addition to the occurrence of soil-microfauna associations in relation to physical parameters and plant communities. Brændegaard (1946) in his work on the spider fauna of East Greenland grouped the species in a number of 'climatic-ecological fauna elements' and discussed the 'ecology of distribution' and the immigration of the different climatic/zoogeographic elements in the Greenland fauna of Araneae and other animal groups. He also (1938) stressed the importance of aerial dispersal of spiders in the Arctic. B.L. Madsen (1959) made observations on the biology of the water beetles in Kangilinnguit/Grønnedal. E.T. Nielsen and H.T. Nielsen (1966) studied the swarming behaviour of mosquitoes in relation to the daily light regime at different latitudes in Greenland (Narsarsuaq, Qeqertarsuaq/Godhavn). Similarly, Dreisig (1981) at Qeqertarsuaq/Godhavn investigated the daily flight activity of moths in the continuous daylight of the arctic summer. U. Røen (*e.g.* 1963, 1966, 1981) published information on freshwater invertebrates partly based on his own collections. Peter Nielsen extensively sampled grassland and soil fauna in South Greenland 1983–85.

J. Böcher in 1958 joined a botanical expedition (led by T.W. Böcher) along the west coast of Greenland (between Maniitsoq/Sukkertoppen and northern Qeqertarsuup Tunua/Disko Bugt), during which he did some arthropod collecting. During his tenure (1967–70) of the position as resident scientific head of the Arctic Station, University of Copenhagen, situated in Qeqertarsuaq/Godhavn Böcher made biological/ecological studies on the two common species of Heteroptera (*Nysius groenlandicus*, *Chlamydatus pullus*) and moreover collected insects and arachnids in the surroundings of Qeqertarsuaq/Godhavn and especially in the hitherto entomologically almost unexplored Uummannaq district (c. 70°20'–72°20'N). In 1970 Böcher was a member of the 'Kap Farvel Expedition' (led by limnologist Ulrik Røen) and here made general entomological collections and biological studies on *Nysius groenlandicus* (Böcher 1971, 1972, 1975a, 1975b, 1976, Böcher & Nachman 2001). Based on material in the Zoological Museum, his continued personal collecting and P. Nielsen's pitfall trap samples, Böcher revised the Coleoptera fauna of Greenland (1988), including distributional and biological/ecological information, and also discussed colonization of Greenland by the beetles. He continued faunistic collections and field studies during a number of subsequent stays in different parts of Greenland, collaborated with botanists on pollination studies (Böcher 1990, Philipp *et al.* 1990, 1996); and eventually coauthored a textbook on Greenland ecology (Born & Böcher 2001). For a decade Böcher concentrated on palaeo-entomological studies with special emphasis on the Kap København Formation, Peary Land, a Plio-Pleistocene sequence (Böcher 1989, 1991, 1992–93, 1995, 1996, 1997a, 1997b, 1998, 1999, 2000; Böcher & Bennike 1991, 1996; Böcher & Fredskild 1993; Bennike & Böcher 1990, 1992, 1994, 1996; Bennike *et al.* 1999, 2000; Christiansen *et al.* 2002).

Biological/ecological student projects made as part of course work at the above-mentioned Arctic Station in Qeqertarsuaq/ Godhavn include Kristensen & Vestergaard

(1975, 24-hour rhythm during arctic summer conditions in the collembolan *Sminthurides malmgreni* Tullberg (1975) and Pape (1983, observations on nests of *Bombus polaris* usurped by *B. hyperboreus* and measurements of the almost constant temperatures inside a bumblebee nest). Pape later revised the Scathophagidae (Diptera) of Greenland (1984) and described the insects, mites and tardigrades associated with a bumblebee nest (1986).

A number of people with different backgrounds made valuable contributions to the biology of Greenland arthropods. Jeppesen (1986) gave an account on the biology of the dytiscid beetles. Koponen (1960, 1979) and Tenow (1990) studied the arthropod fauna associated with tree birch (*Betula pubescens*) in South Greenland. Johannesen (1990) during his term as a teacher collected and studied Lepidoptera in the Narsaq area. Hodkinson (1997) made a thorough study on the progressive restriction of host plants (*Salix* spp.) of the psyllid *Cacopsylla groenlandica* along the south–north gradient.

The establishment in 1995 of the permanent Danish/Greenlandic research station at Zackenberg (Fig. 4) in high arctic N.E.Greenland (74°28'N, 20°34'W) offers new possibilities to entomological work in Greenland. The framework of the station, 'ZERO' ('Zackenberg Ecological Research Operations') includes a biological long-time monitoring programme ('BioBasis'), and as part of this faunistic and phenological investigations of arthropods are now carried out by means of routine collections using window traps and yellow pitfall traps situated in different plant communities (Böcher 1997c).

FIGURE 5.4 *The Zackenberg Research Station on Wollaston Forland, Northeast Greenland (74°30'N, 21°00'W), officially opened in 1997. The station is mainly an ecosystems monitoring facility with 10 buildings holding scientific equipment and laboratories, accommodations and a conference room. It can host up to 25 persons at the same time.*

A need for an intensive inventory of the botanically rich, but entomology-wise strangely little-explored coastal stretch north of the area so thoroughly investigated by Lundbeck became obvious during the preparation of the present work. A special collecting expedition with seven entomologist participants was launched to fill this knowledge gap; based on former naval cutter 'Kissavik' it worked along the west coast from Paamiut/Frederikshåb in south to Sisimiut/ Holsteinsborg in north (Böcher 2004).

References

Andersen, F.S. (1937): Über die Metamorphose der Ceratopogoniden und Chironomiden Nordost-Grönlands. – *Meddelelser om Grønland* 116 (1): 95 pp.

Andersen, F.S. (1946): East Greenland lakes as habitats for chironomid larvae. Studies of the systematics and biology of Chironomidae, II. – *Meddelelser om Grønland* 100 (10): 65 pp.

Aurivillius, C. (1890): Grönlands insektfauna I. Lepidoptera, Hymenoptera. – *Bihang til kongliga Svenska Vetenskaps-Akademiens Handlingar* 15, 4: 34 pp.

Aurivillius, C. (1900): Lepidoptera och Coleoptera insamlade under professor A.G. Nathorst's arktiska expeditioner 1898 och 1899, under den svenska expeditionen till Beeren Eiland 1899 och under konservator G. Koltoff's expedition till Grönland 1900. – *Öfversigt av Kongliga Vetenskaps-Akademiens Förhandlingar* 1900, 10: 1135–1144.

Bang-Haas, A. (1896): Lepidoptera Groenlandica. – *Videnskabelige Meddelelser fra den naturhistoriske Forening i Kjøbenhavn* 1896: 178–195.

Barca, E. (1937): Mikrolepidopteren aus Nordostgrönland. Zoological results of the Norwegian scientific expedition to East Greenland VI. – *Skrifter om Svalbard og Ishavet* 71: 31–37.

Bennike, O. & Böcher, J. (1990): Forest-tundra neighbouring the North Pole: plant and insect remains from the Plio-Pleistocene Kap København Formation, North Greenland. – *Arctic* 43 (4): 331–338.

Bennike, O. & Böcher, J. (1992): Early Weichselian interstadial land biotas at Thule, Northwest Greenland. – *Boreas* 21: 111–117.

Bennike, O. & Böcher, J. (1994): Land biotas of the last interglacial/glacial cycle on Jameson Land, East Greenland. – *Boreas* 23: 479–487.

Bennike, O. & Böcher, J. (1996): Et mildt og frodigt Østgrønland. – *Forskning i Grønland/ Tusaat* 1/96: 10–17.

Bennike, O., Björck, S., Böcher, J., Hansen, L., Heinemeier, J. & Wohlfarth, B. (1999): Early Holocene plant and animal remains from North-east Greenland. – *Journal of Biogeography* 26: 667–677.

Bennike, O., Björck, S., Böcher, J. & Walker, I.R. (2000): The Quaternary arthropod fauna of Greenland: a review with new data. – *Bulletin of the Geological Society of Denmark* 47: 111–134.

Blair, K.G. (1930). Coleoptera from Greenland. Oxford University Greenland Expedition, 1928. – *The Annals and Magazine of Natural History*, Tenth Series, 5: 394–400.

Böcher, J. (1971): Preliminary studies on the biology and ecology of *Chlamydatus pullus* (Reuter) (Heteroptera: Miridae) in Greenland. – *Meddelelser om Grønland* 191 (3): 29 pp.

Böcher, J. (1972): Feeding biology of *Nysius groenlandicus* (Heteroptera: Miridae) in Greenland. With a note on oviposition in relation to food-source and dispersal of the species. – *Meddelelser om Grønland* 191 (4): 41 pp.

Böcher, J. (1975a): Notes on the reproductive biology and egg-diapause in *Nysius groenlandicus* (Zett.) (Heteroptera: Lygaeidae). – *Videnskabelige Meddelelser fra Dansk Naturhistorisk Forening* 138: 21–38.

Böcher, J. (1975b): Spredningen af den grønlandske frøtæge, *Nysius groenlandicus* (Zett.) (Heteroptera: Lygaeidae). (With a summary: Dispersal of *Nysius groenlandicus* (Zett.) in Greenland). – *Entomologiske Meddelelser* 43: 105–109.

Böcher, J. (1976): Population studies on *Nysius groenlandicus* (Zett.) (Heteroptera: Lygaeidae) in Greenland, with particular reference to climatic factors, especially the snow cover. – *Videnskabelige Meddelelser fra Dansk Naturhistorisk Forening* 139: 61–89.

Böcher, J. (1988): The Coleoptera of Greenland. – *Meddelelser om Grønland. Bioscience* 26: 100 pp.

Böcher, J. (1989): First record of an interstadial insect from Greenland: *Amara alpina* (Paykull, 1790) (Coleoptera: Carabidae). – *Boreas* 18: 1–4.

Böcher, J. (1990): Biologisk-arkæologisk kortlægning af Grønlands østkyst mellem 75°N og 79°30'N. Del 4: Entomologiske undersøgelser: Feltarbejde i 1989 og foreløbige resultater. – *Grønlands Hjemmestyre, Miljø- og Naturforvaltning.* Teknisk rapport nr. 17 – maj 1990. 28 pp.

Böcher, J. (1991): Biller og fluer – Qeqertasussuk set med insektøjne. – *Tidsskriftet Grønland* 4–7, 1991: 159–171.

Böcher, J. (1992–93): Den grønlandske insektverdens historie, I–II (Kalaallit Nunaanni sullinikkut oqaluttuassartaat, I–II). – *Forskning i Grønland/Tusaat* 4/92: 12–18, 1/94: 29–40.

Böcher, J. (1995): Palaeoentomology of the Kap København Formation, a Plio-Pleistocene sequence in Peary Land, North Greenland. – *Meddelelser om Grønland, Geoscience* 33: 81 pp.

Böcher, J. (1996): Fossil Insects as Indicators of Ancient Environments as Illustrated by Examples from Greenland. *In*: Grønnow, B. (ed.). *The Paleo-Eskimo Cultures of Greenland. New Perspectives in Greenlandic Archaeology* pp. 253–256. Danish Polar Center, Copenhagen.

Böcher, J. (1997a): *Studier af den grønlandske insektfaunas historie – med særligt henblik på billerne* (*The history of the Greenland insect fauna – with emphasis on living and fossil beetles*). – Zoologisk Museum, Københavns Universitet. 20 pp.

Böcher, J. (1997b): History of the Greenland insect fauna with emphasis on living and fossil beetles. – *Quaternary Proceedings* 5. *In*: Ashworth, A.C., Buckland, P.C. & Sadler, J.P. (eds). *Studies in Quaternary Entomology. An Inordinate Fondness for Insects*, pp. 35–47.

Böcher, J. (1997c): Feature locality: The Zackenberg Ecological Research Station, Greenland. – *Arctic Insect News* 8: 4–9.

Böcher, J. (1998): Insect remains from Asummiut. *In*: Arneborg, J. & Gulløv, H.C. (eds). *Man, culture and environment in ancient Greenland* pp. 133–134. – Danish Polar Center, Copenhagen.

Böcher, J. (1999): Mystisk opdukken af biller i Grønland. – *Dyr i natur og museum* 1999: 22–24.

Böcher, J. (2000): Biller og Grønlands forhistorie. *In*: Jakobsen, B.H. *et al.* (eds). *Topografisk Atlas Grønland*, pp. 134–139. – Det Kongelige Danske Geografiske Selskab & Kort- og Matrikelstyrelsen/C.A. Reitzels Forlag, København.

Böcher, J, (2004): På jagt efter Grønlands små landdyr. – *Dyr i natur og museum* 2004,1: 14–17.

Böcher, J. & Bennike, O. (1991): Interglacial land biotas of Jameson Land, East Greenland. – *Lundqua Report* 33: 129–136.

Böcher, J. & Bennike, O. (1996): Early Holocene insect and plant remains from Jameson Land, East Greenland. – *Boreas* 25: 187–193.

Böcher, J. & Fredskild, B. (1993): Plant and arthropod remains from the palaeo-Eskimo site on Qeqertasussuk, West Greenland. – *Meddelelser om Grønland. Geoscience* 30: 35 pp.

Böcher, J. & Nachman, G. (2001): Temperature and humidity responses of the arctic-alpine seed bug *Nysius groenlandicus*. – *Entomologia Experimentalis et Applicata* 99: 319–330.

Born, E.W. & Böcher, J. (eds) 2001: *The Ecology of Greenland.* Atuakkiorfik Education, Nuuk. 429 pp.

Brændegaard, J. (1938): Aeronautic spiders in the Arctic. – *Meddelelser om Grønland* 119 (5): 9 pp.

Brændegaard, J. (1946): The spiders (Araneina) of East Greenland. A faunistic and zoogeographical investigation. – *Meddelelser om Grønland* 121 (15): 128 pp.

Brændegaard, J., Henriksen, K.L. & Spärck, R. (1935): Insects and Arachnids. – *Meddelelser om Grønland* 104 (16): 1–18.

Brown, W.J. (1937): The Coleoptera of Canada's eastern arctic. – *Canadian Entomologist* 69: 106–111.

Carpenter, G.D.H. (1937): Notes on insects collected in Northwest Greenland by the Oxford University Ellesmereland Expedition 1934–35. With description of a new species of *Rhamphomyia* (Dipt.) by J.E. Collin. – *The Annals and Magazine of Natural History*, Tenth Series, 20: 401–409.

Carpenter, G.D.H. (1938): Notes on insects collected in West Greenland by the Oxford University Greenland Expedition,

1936. With descriptions of a new species of *Angitia* (Hymenoptera: Ichneumonidae) by A. Roman, and of *Fannia* (Diptera, Anthomyidae) by J.E. Collin. – *The Annals and Magazine of Natural History*, Eleventh Series, 1: 529–553.

China, W.E. (1934): Hemiptera collected by the Oxford University Expedition to West Greenland, 1928. – *The Annals and Magazine of Natural History*, Tenth Series, 13: 330–333.

Christiansen, H.H., Bennike, O., Böcher, J., Elberling, B., Humlum, O. & Jakobsen, B.H. (2002): Holocene environmental reconstruction from deltaic deposits in northeast Greenland. – *Journal of Quaternary Science* 17(2): 145–160.

Collin, J.E. (1930): A revision of the Greenland species of the anthomyid genus *Limnophora* sens. lat. (Diptera), with figures of the male genitalia of these and many other palaearctic species. – *Transactions of the entomological Sociesty of London* 78: 255–281.

Collin, J.E. (1931): The Oxford University Expedition to Greenland, 1928. Diptera (Orthorrhapha Brachycera and Cyclorrhapha) from Greenland. – *The Annals and Magazine of Natural History*, Tenth Series, 7: 67–91.

Collin, J.E. (1937): On a collection of Diptera from East Greenland. – *The Annals and Magazine of Natural History*, Tenth Series, 19: 594–595.

Cooper, J.O. (1931): *Gyrinus opacus* var. *lecontei* O.-C. (nec Fall. 1921) = *G. opacus* var. *blairi* nom. nov. – *Entomological monthly Magazine* 67: 239.

Danks, H.V. (1981): *Arctic Arthropods. A review of systematics and ecology with particular reference to the North American fauna.* Entomological Society of Canada, Ottawa. 608 pp.

Degerbøl, M. (1937): A contribution to the investigation of the fauna of the Blosseville Kyst, East Greenland, with special reference to zoogeography. – *Meddelelser om Grønland* 104 (19): 36 pp.

Deichmann, H. & Lundbeck, W. (1895): III. Østgrønlandske Insekter. *In*: Den østgrønlandske Expedition, udført i Aarene 1891–92 under Ledelse af C. Ryder. Tredie Del. – *Meddelelser om Grønland* 19: 97–120.

de Lesse, H. (1950): Notes zoologiques et botaniques sur l'Ouest de Groenland. Expeditions polaires francaises (Missions P.-E. Victor). – *La Terre et la Vie* 1950 (4): 175–201.

Dreisig, H. (1981): Daily flight activity of moths in the continuous daylight of the arctic summer. – *Holarctc Ecology* 4: 36–42.

Edwards, F.W. (1931): Oxford University Greenland Expedition, 1928. Diptera Nematocera. – *The Annals and Magazine of Natural History*, Tenth Series, 8: 617–618.

Edwards, F.W. (1935): Diptera Nematocera from East Greenland. – *The Annals and Magazine of Natural History*, Tenth Series, 15: 467–473.

Emerton, J.H. (1918): Notes on the spiders collected by Frits Johansen of the 'Danmark Expedition' 1906–08 to Northeastern Greenland, Latitude 76½° to 77° North. – *Videnskabelige Meddelelser fra Dansk naturhistorisk Forening i København* 70: 143–145.

Fabricius, O. (1780): *Fauna Groenlandica.* – Hafniae et Lipsiae. 452 pp.

Forsslund, K.-H. (1932): Zur Kenntnis der Trichopteren Grönlands. – *Entomologisk Tidskrift* 53: 56–59.

Friese, H. (1935). Apiden aus Nordost-Grönland. Zoological results of the Norwegian Expedition to East-Greenland IV. – *Skrifter om Svalbard og Ishavet* 65: 3–10.

Gerstäcker, A. (1874): Hymenopteren und Dipteren. *Die zweite deutsche Nordpolarfahrt 1869 und 1870*, II: 404–406.

Goetghebuer, M. (1935): Chironomides du Groenland oriental, du Svalbard et de la Terre de Francois Joseph. Zoological results of the Norwegian Expedition to East-Greenland II. – *Skrifter om Svalbard og Ishavet* 53: 19–31.

Graversen, C.P. (1931): Notizen über grönländische Oribatiden. – *Meddelelser om Grønland* 91 (2). 18 pp.

Griffiths, G.C.D. (1966): The Agromyzidae (Diptera) of Greenland. – *Meddelelser om Grønland* 170 (4): 32 pp.

Haarløv, N. (1942): A morphologic-systematic-ecological investigation of Acarina and other representatives of the microfauna of the soil around Mørkefjord, Northeast Greenland. – *Meddelelser om Grønland* 128 (1): 71 pp.

Hammer, M. (1937): A quantitative and qualitative investigation of the microfauna communities of the soil at Angmagssalik and in Mikis Fjord. 6. og 7. Thule Expedition til Sydøstgrønland 1931–33. – *Meddelelser om Grønland* 108 (2). 53 pp.

Hammer, M. (1938): The Zoology of East Greenland. Collemboles. – *Meddelelser om Grønland* 121 (2): 42 pp.

Hammer, M. (1944): Studies on the oribatids and collemboles of Greenland. – *Meddelelser om Grønland* 141: 210 pp.

Henriksen, K.L. (1921–1937): Oversigt over Dansk Entomologis Historie. – *Entomologiske Meddelelser* 15. 578 pp.

Henriksen, K.L. (1935): Insects (exclusive of Diptera) and Acarina. The Scoresby Sound Committee's 2nd East Greenland Expedition in 1932 to King Christion IX's Land. – *Meddelelser om Grønland* 104 (16): 1–9.

Henriksen, K.L. (1937): Insects. – *Reports of the fifth Thule Expedition 1921–24*, vol. II No. 8. København. 34 pp.

Henriksen, K.L. (1939): A revised index of the insects of Grønland. – *Meddelelser om Grønland* 119 (10): 112 pp.

Henriksen, K.L. & Lundbeck, W. (1917): Landarthropoder (Insecta et Arachnida). Conspectus Faunæ Groenlandicæ, Pars secunda. – *Meddelelser om Grønland* 22: 481–823.

Hodkinson, I.D. (1997): Progressive restriction of host plant exploitation along a climatic gradient: the willow psyllid *Cacopsylla groenlandica* in Greenland. – *Ecological Entomology* 22: 47–54.

Holm, Å. (1967): Spiders (Araneae) from West Greenland. – *Meddelelser om Grønland* 184 (1): 99 pp.

Holmgren, A.E. (1872): Insekter från Nordgrönland, samlade av Prof. A.E. Nordenskiöld år 1870. – *Öfversigt af Kongliga Vetenskaps-Akademiens Förhandlingar* 29 (6): 97–105.

v. Homeyer, A. (1874): Lepidopteren. – *Die zweite deutsche Nordpolarfahrt 1869 und 1870*, II: 407–410.

Jackson, A.R. (1930): Araneæ and Opiliones collected by Major R.W.G. Hingston, with some notes on Icelandic spiders. – *The Annals and Magazine of Natural History*, Tenth Series, 6: 639–656.

Jackson, A.R. (1934): Notes on Arctic spiders obtained in 1933. – *The Annals and Magazine of Natural History*, Tenth Series, 14: 611–620.

Jeppesen, P.C. (1986): Dytiscid beetles in Greenland, with description of the three larval stages of *Hydroporus melanocephalus* (Marsham, 1802). – *Entomologica Basilensis* 11: 67–69.

Johannesen, P. (1990): Observationer af sommerfugle i Syd- og Sydvestgrønland (Lepidoptera). – *Entomologiske Meddelelser* 58: 1–8.

Johansen, F. & Nielsen, I.C. (1910a): The insects of the 'Danmark' Expedition. Danmark-Ekspeditionen til Grønlands Nordøstkyst 1906–1908. III (2). – *Meddelelser om Grønland* 43: 1–36.

Johansen, F. (1911): Freshwater life in North-East Greenland. Danmark-Ekspeditionen til Grønlands Nordøstkyst 1906–1908. V(6). – *Meddelelser om Grønland* 45: 321–327.

Johansen, F. & Nielsen, I.C.(1910b): The insects of the 'Danmark' Expedition. Danmark-Ekspeditionen til Grønlands Nordøstkyst 1906–1908, 3(2). – *Meddelelser om Grønland* 43: 35–68.

Jørgensen, M. (1934a): A quantitative investigation of the microfauna communities of the soil in East Greenland (Preliminary Report). Treårsekspeditionen til Christian den X's Land 1931–34. – *Meddelelser om Grønland* 100 (9): 39 pp.

Jørgensen, M. (1934b): Ein revidiertes Verzeichnis über grönländische Milben. – *Zoologische Anzeiger* 107: 40–47.

Knaben, N. (1937): Makrolepidopteren aus Nordostgrönland. Zoological Results of the Norwegian Scientific Expedition to East-Greenland. – *Skrifter om Svalbard og Ishavet* 71: 3–30.

Koponen, R. (1960): Notes on herbivorous insects of the birch in southern Greenland. – *Reports from the Kevo Subarctic Research Station* 14: 13–17. (*Annales Universitatis Turkuensis* A 2 (60)).

Koponen, S. (1979): Herbivora insekter på björk i nordliga regioner. [Herbivorous insects of the birch in northern regions] – *Entomologisk Tidskrift* 100: 231–233.

Kornerup, A. (1879): Det organiske Liv på den østligste Nunatak.- *Meddelelser om Grønland* 1879: 150–152.

Kramer, P. (1897): Grönländische Milben. Zoologische Ergebnisse der von der Gesellschaft für Erdkunde zu Berlin unter Leitung Dr. von Drygalski aus gesandten Grönlandexpedition nach Dr. Vanhöffen's Sammlung bearbeitet, VI. – *Bibliotheca Zoologica* 8 (=Heft 20): 77–83.

Kristensen, R.M. & Vestergaard, K. (1975): Døgnaktivitet under arktiske sommerbetingelser hos springhalen *Sminthurides malmgreni* Tullberg (Collembola). – *Entomologiske Meddelelser* 43: 21–32.

Lack, D. (1934): Some insects from the Scoresby Sound Region, East Greenland, with an account of the fauna of a nunatak. – *The Annals and Magazine of Natural History*, Tenth Series, 14: 599–606.

Lambers, D.H.R. (1952): The aphid fauna of Greenland. – *Meddelelser om Grønland* 136 (1): 33 pp.

Lambers, D.H.R. (1960): Additions to the aphid fauna of Greenland. – *Meddelelser om Grønland* 159 (5): 18 pp.

Lefebvre, A. (1836a): Description d'une nouvelle espèce de Coliade. – *Annales de la Société Entomologiques de France* 5: 383–387.

Lefebvre, A. (1836b): Description de quelques Lépidoptères nocturnes hyperboréens. – *Annales de la Société Entomologiques de France* 5: 389–399.

Lenz, H. (1897): Grönländische Spinnen. Zoologische Ergebnisse der von der Gesellschaft für Erdkunde zu Berlin unter Leitung Dr. von Drygalski aus gesandten Grönlandexpedition nach Dr. Vanhöffen's Sammlung bearbeitet, VI. – *Bibliotheca Zoologica* 8 (=Heft 20): 73–76.

Lindberg, H. (1935): Hemiptera aus Nordost-Grönland. Zoological Results of the Norwegian Scientific Expedition to East-Greenland IV. – *Skrifter om Svalbard og Ishavet* 65: 11–17.

Linnaniemi, W.M. (1935): Collembolen aus Nordost-Grönland. Zoological Results of the Norwegian Scientific Expedition to East-Greenland IV. – *Skrifter om Svalbard og Ishavet* 65: 19–25.

Longstaff, T.G. (1936): Entomological note on Mr. J. M. Wordie's expedition to N.W. Greenland and E. Baffinland in 1934. – *The Annals and Magazine of Natural History*, Tenth Series, 18: 531–533.

Lundbeck, W. (1891a): Entomologiske Undersøgelser i Vest-Grønland 1889 og 1890. – *Meddelelser om Grønland* 7 (4): 107–144.

Lundbeck, W. (1891b): Notitser om Grønlands entomologiske Fauna. – *Entomologiske Meddelelser* 3: 45–52.

Lundbeck, W. (1895): Østgrønlandske Insekter. Fortegnelse over de indsamlede Insekter. *In*: Den østgrønlandske Expedition, udført i Aarene 1891–92 under Ledelse af C. Ryder. Tredie Del. – *Meddelelser om Grønland* 19: 105–120.

Lundbeck, W. (1896a): Coleoptera Groenlandica. – *Videnskabelige Meddelelser fra den naturhistoriske Forening i Kjøbenhavn* 1896: 196–219.

Lundbeck, W. (1896b): Hymenoptera Groenlandica. – *Videnskabelige Meddelelser fra den naturhistoriske Forening i Kjøbenhavn* 1896: 220–251.

Lundbeck, W. (1898 & 1900): Diptera Groenlandica I–II. – *Videnskabelige Meddelelser fra den naturhistoriske Forening i Kjøbenhavn* 1898: 236–314, 1900: 281–316.

Madsen, B.L. (1959): Limnologiske undersøgelser fra Sydvestgrønland. – *Flora og Fauna* 65 (3): 41–48.

Madsen, H. (1936): Investigations on the shore fauna of East Greenland, with a survey of the shores of the other arctic regions. Treårsexspeditionen til Christian den X's Land 1931–34. – *Meddelelser om Grønland* 100 (8): 1–79.

Malaise, R. (1935): Eine neue Blattwespe aus Ost-Grönland. Zoological Results of the Norwegian Scientific Expedition to East-Greenland II. – *Skrifter om Svalbard og Ishavet* 53: 3–4.

McLachlan, R. (1878): Report on the Insecta (including Arachnida) collected by Captain Feilden and Mr. Hart between the Parallels of 78° and 83° North Latitude, during the recent Arctic Expedition. – *Journal of the Linnean Society of London* 14: 98–122.

Meinert, F. (1896): Neuroptera, Pseudoneuroptera, Thysanopoda, Mallophaga, Collembola, Suctoria, Siphunculata Groenlandica. – *Videnskabelige Meddelelser fra den naturhistoriske Forening i Kjøbenhavn* 1896: 154–177.

Meyrich, E. (1934): A new Micro Lepidopteron (Lepidoptera Tineina) from Greenland. – *Entomologist* 67: 59.

Mjöberg, E. (1910): Studien über Mallophagen und Anopluren. – *Arkiv för Zoologi* 6 (13): 296 pp.

Mosely, M.E. (1929): Oxford University Greenland Expedition, 1928. Trichoptera and Ephemeroptera of Greenland: Additional records made by the Oxford University Expedition to Kugssuk, Godthåb Fjord, W.Greenland. – *The Annals and Magazine of Natural History*, Tenth Series, 4: 501–508.

Nielsen, E.T. & Nielsen, H. T. (1966): Observations on mosquitoes in Greenland. – *Meddelelser om Grønland* 170 (3). 27 pp.

Nielsen, I.C. (1907): The insects of East-Greenland. – *Meddelelser om Grønland* 29 (10): 365–414.

Nielsen, I.C. (1910): A catalogue of the insects of Northeast Greenland with descriptions of some larvae. – *Meddelelser om Grønland* 43: 55–68.

Osten-Sacken, C.R. (1878): Report on the Diptera brought home by Dr. Bessels from the arctic voyage of the 'Polaris' in 1872. – *Proceedings of the Boston Society for Natural History* 19: 41–43.

Packard, A.S. (1877): Explorations of the Polaris expedition to the North Pole. – *American Naturalist* 11: 51–53 & *Entomological monthly Magazine* 13: 228–229.

Pape, T. (1983): Observations on nests of *Bombus polaris* Curtis usurped by *B. hyperboreus* Schönherr in Greenland. – *Entomologiske Meddelelser* 50: 145–150.

Pape, T. (1984): Grønlands møgfluer (Diptera: Scathophagidae). – *Entomologiske Meddelelser* 52: 44–48.

Pape, T. (1986): Bjørnedyr, mider og insekter fra en humlebirede i Grønland. – *Entomologiske Meddelelser* 53: 75–81.

Philipp, M., Böcher, J., Mattson, O. & Woodell, S.R.J. (1990): A quantitative approach to the sexual reproductive biology and population structure in some arctic flowering plants: *Dryas integrifolia*, *Silene acaulis* and *Ranunculus nivalis*. – *Meddelelser om Grønland. Bioscience* 34: 60 pp.

Philipp, M., Böcher, J., S.R.J. Woodell & O. Mattson (1996): Reproductive biology of four species of *Pedicularis* (Schrophulariaceae) in West Greenland. – *Arctic and Alpine Research* 28 (4): 403–413.

Remy, P. (1928a): Quelques Arthropodes récoltés dans la région du Scoresby-Sund de la croisière arctique du 'Pourquoi-Pas' en 1926. – *Meddelelser om Grønland* 74: 84–95.

Remy, P. (1928b): Arthropodes Terrestres récoltés au Groenland en cours de la croisière du 'Pourquoi-Pas' en 1926. – *Bulletin de la Societé Linnéenne* 7: 7.

Remy, P. (1928c): Les Collemboles du Groenland. – *Meddelelser om Grønland* 74: 57–70.

Richards, O.W. (1931): Some notes on the bumble-bees allied to *Bombus alpinus* L. – *Tromsø Museums Årshefter* 50 (1927), No. 6. 32 pp.

Ringdahl, O. (1933): Tachiniden und Musciden aus Nordost-Grönland. Zoological Results of the Norwegian Scientific Expedition to East-Greenland II. – *Skrifter om Svalbard og Ishavet* 53: 15–18.

Rink, H. (1857). (*See*: Schiødte, J.C. (1857)).

Røen, U. (1963): Nogle udbredelsestyper i den grønlandske ferskvandsfauna. – *Tidsskriftet Grønland*, oktober 1963: 361–374.

Røen, U. (1966): Pearylands ferske vande og deres dyreliv. – *Tidsskriftet Grønland*, marts 1966: 90–101.

Røen, U. (1981): *Ferskvandsfaunaen*. *In*: Nørrevang, A. & Lundø, J. (eds). Danmarks Natur 11 pp. 459–473. Politikens Forlag, København.

Roman, A. (1916): Ichneumoniden aus West-Grönland. – *Arkiv för Zoologie* 10 (22): 12 pp.

Roman, A. (1930): Oxford University Greenland expedition, 1928. Ichneumonidae collected by Major R.W.G. Hingston. – *The Annals and Magazine of Natural History*, Tenth Series, 5: 281–288.

Roman, A. (1933): Schlupfwespen aus Ost-Grönland. Zoological Results of the Norwegian Scientific Expedition to East-Greenland II. – *Skrifter om Svalbard og Ishavet* 53: 5–13.

Roman, A. (1934): East Greenland ichneumonids. – *The Annals and Magazine of Natural History*, Tenth Series, 14: 606–611.

Rostrup, S. (1900): Grønlands Phytoptider. – *Videnskabelige Meddelelser fra den naturhistoriske Forening i Kjøbenhavn* 1900: 241–249.

Rübsaamen, E.H. (1897): Grönländische Mycetophiliden, Sciariden, Cecidomyiden, Psylliden, Aphiden und Gallen. Zoologische Ergebnisse der von der Gesellschaft für Erdkunde zu Berlin unter Leitung Dr. von Drygalski aus gesandten Grönlandexpedition nach Dr. Vanhöffen's Sammlung bearbeitet,VI. – *Bibliotheca Zoologica* 8 (=Heft 20): 103–108.

Sahlberg, J. (1901): Aleocharider insamlade i Polarregionerna af svenska expeditionerna 1883 och 1899. – *Entomologisk Tidskrift* 1901: 167–169.

Schiødte, J.C. (1845): Om Slægten *Micralymma*. – *Naturhistorisk Tidsskrift* 2 (1): 370–389.

Schiødte, J.C. (1857): Tillæg Nr. 3. Udsigt over Grønlands Land-, Ferskvands- og Strandbreds-Arthropoder, pp. 50–74 *In*: Reinhardt, I., Schiødte, J.C., Mørch, O.M.L., Lütken, C.F., Lange, J. & Rink, H. (1857): *Naturhistoriske Tillæg til en geografisk og statistisk Beskrivelse af Grønland. Særskilt Aftryk af Tillæggene til Rink, H. Grønland geografisk og statistisk beskrevet*, pp. 50–74. A.F. Høst, København.

Scoresby, W. (1823): Appendix III: List of animals met with on the Eastern coast of West Greenland with notes and memoranda. V. Insecta. Pp. 423–428 in *Journal of a voyage to the northern Whale-Fishery, including researches and discoveries on the eastern coast of West Greenland, made in the summer of 1822, in the Ship Baffin of Liverpool*. Constable, Edinburgh.

Sørensen, W. (1898): Arachnida Groenlandica. – *Videnskabelige Meddelelser fra den naturhistoriske Forening i Kjøbenhavn* 1898: 176–235.

Stæger, R.C. (1845): Grønlands Antliater. – *Naturhistorisk Tidsskrift* 2: 346–369.

Strand, E. (1905): Coleoptera, Hymenoptera, Lepidoptera und Araneae. – *Report of the Second Norwegian Arctic Expedition in the 'Fram' 1898–1902*, I (3): 30 pp.

Tenow, O. (1990): Altitudinal and within-crown distribution of insects grazing on Mountain birch (*Betula pubescens* coll.) in SW Greenland and NW Norway. – *Meddelelser om Grønland. Bioscience* 33: 35–42.

Thorell, T. (1872): Om några Arachnider från Grønland. – *Öfversigt av Kongliga Vetenskaps-Akademiens Förhandlingar* 29: 147–166.

Thorell, T. (1878): Notice of the spiders of the Polaris expedition. – *American Naturalist* 12: 393–395.

Trägårdh, I. (1909): Monographie der arktischen Acariden. – *Fauna Arctica* IV: 1–78. Jena.

Trägårdh, I. (1914): Acari. The insects of the 'Danmark' Expedition. Danmark-Ekspeditionen til Grønlands Nordøstkyst 1906–1908. III (14). – *Meddelelser om Grønland* 43: 415–427.

Vanhöffen, E. (1897): *Die Fauna und Flora Grönlands*. Grönland-Expedition der Gesellschaft für Erdkunde zu Berlin 1891–1893 unter Leitung von Erich von Drygalski II (2). 383 pp.

Vibe, C. (1950): Some insects new to Greenland. – *Entomologiske Meddelelser* 25 (6): 419–420.

Vibe, V. (1981): *Lavere dyr i Grønland*. *In*: Nørrevang, A. & Lundø, L. (eds). Danmarks Natur 11, Grønland, pp 542–550. Politikens Forlag, København.

Wahlgren, E. (1900): Collembola während der schwedischen Grönlandsexpedition 1899 auf Jan Mayen und Ost-Grönland eingesammelt. – *Öfversigt av Kongliga Vetenskaps-Akademiens Förhandlingar* 57 (3): 353–375.

Wolff, N.L. (1964): The Lepidoptera of Greenland. – *Meddelelser om Grønland* 159 (11): 74 pp.

Zetterstedt, J.W. (1838–40): *Insecta Lapponica*. Lipsiae, Sumtibus Leopoldi Voss. 1140 pp.

CHAPTER 6

Collembola (springtails, collembolans)

Arne Fjellberg

The Collembola (springtails) are one of the dominant animal groups in most soil ecosystems including those of the arctic regions. Their unique body design with the low number of abdominal segments render them immediately recognizable among the antennate hexapod arthropods. Their exact phylogenetic relationships to other arthropods are currently a subject of much discussion, see chapter 2.

Morphology

Collembola are enthognathous hexapod arthropods, *i.e.*, their mandibles and maxillae are enclosed in the head capsule. Most species are 0.5–1.5 mm long, with some species reaching more than 3 mm.The antennae are 4–segmented, the eye-field has 8 socalled 'ocelli' (dispersed compound eye ommatidia, hence non-homologous with insect ocelli) which often become reduced. Each of the three thoracic segments bears a leg consisting of claw, tibiotarsus, femur, trochanter, coxa and subcoxa. Some of the six abdominal segments have structures derived from paired appendages: *Ventral tube* on the first segment, *retinaculum* on the third, and *furca* (the jumping organ) on the fourth. Genital organs open on the fifth and anus on the sixth (terminal) segment. The genital aperture of the female is a transverse slit with an anterior and a posterior lip, the male aperture is a roundish papilla with an incision from behind Fig.1 shows the general morphology of the Greenland species *Hypogastrura concolor*. For a more detailed description Fjellberg (1998) may be consulted.

Abbreviations used in the text (see also Fig.1). a: Anterior; A: Apical setae of tibiotarsus; abd.: Abdomen (abd.1: First abdominal segment); ant: Antenna (ant.1: First antennal segment); cl: Claw; m: Median; M: Macrochaeta; ms: Spine-like microsensillum; p: Posterior; PAO: Post antennal organ; s: Sensillum; T: Accessorial seta of the apical

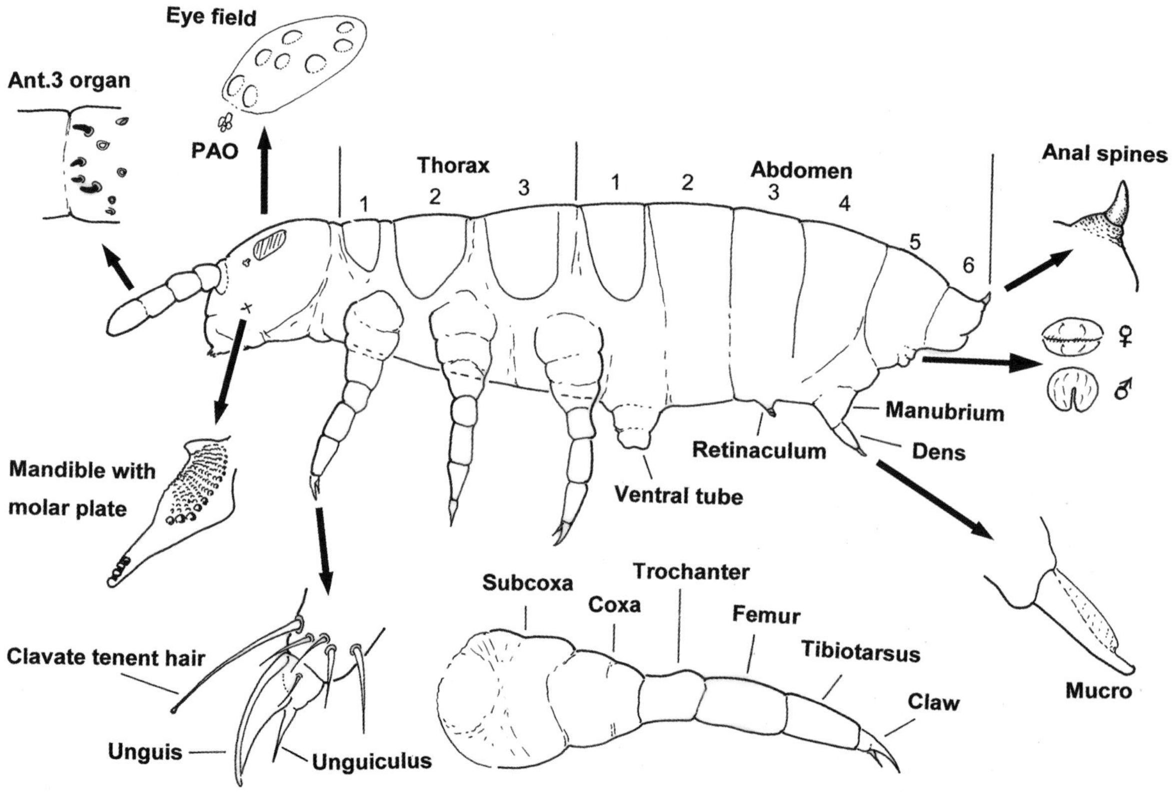

FIGURE 1 *General morphology,* Hypogastrura concolor.

whorl on tibiotarsus; th: Thorax (th.1: First thoracic segment); tib: Tibiotarsus

Reproduction and development

Both sexual and parthenogenetic reproduction are common in Collembola. Some species may switch between the modes, depending on environmental conditions. The juvenile hatch from the egg with all body segments developed. Several instars separated by moults are passed before the adult appears. Under arctic conditions the life span may cover several years and maturity is not reached until at least two years age in some species. Sperm transfer is indirect, with males setting off spermatophores on the substrate. These are picked up by the females, somtimes after a complex courtship (Goloshchapova *et al.* 2005).

Collecting and identification

Collembola inhabit a most diverse assemblage of terrestrial biotopes from coastal seashores to inland nunataks. Mass occurrences may be observed in littoral wrack beds or other accumulations of rich organic substrate, at shore of lakes or on melt water at snow thaw in spring. Specimens are easily collected by hand into a vial with 80% ethanol, but more efficient is extraction of soil/litter samples placed on a fine wire mesh in a funnel which is heated from above by an electrical light bulb. Animals will move away from the heat and finally fall through the mesh into a collecting vial placed under the funnel. Samples of specimens collected directly in alcohol should be heated on a water bath until boiling, to ensure proper preservation of the material.

With some training a number of species will be recognised at low magnification while floating in alcohol, or even by eye in the field. Smaller species always need to be mounted in slides and examined by a microscope. Specimens may be cleared for microscopy by heating in a mixture of glycerol and lactic acid (concentrated) in proportions 1:3. Small juveniles may not always be safely identified by the following keys.

Key to families

1. Body elongate, clearly segmented (suborder Arthropleona, Fig. 1) 2
– Body more or less globular, segments of thorax and abdomen fused (suborders Neelipleona and Symphypleona, Figs 19, 20, 21) 6
2. Th.1 large, with some dorsal setae. Mostly short and thick species with short extremities. Many species completely white, these have no furca but usually a pair of anal spines present on tip of abdomen (except *Anurida/Micranurida*; section Poduromorpha) 3
– Th.1 reduced, dorsal setae absent. Generally well pigmented, elongate forms with long extremities. The few white species have furca present (except *Pseudanurophorus*), but anal spines absent (section Entomobryomorpha) 5
3. Body pigmented or white. None of the white species have pseudocelli on the body or finger-like papillae in the ant.3 organ 4
– All species white with pseudocelli on the body. The larger species (>1.0 mm) also with finger-like papillae in the ant.3 organ (Fig. 2A) .... Onychiuridae, p. 57
4. Mandibles strong, with distinct molar plate (Fig. 1). Tip of abdomen with 2 anal spines (absent in *Willemia denisi*) .... Hypogastruridae, p. 50
– Mandibles reduced, no molar plate. Anal spines either absent or 3 or 5 (Figs 7A, B) .... Neanuridae, p. 55
5. Body with an open cover of simple hairs only. PAO present (except in the white, blind *Isotomiella minor*) .... Isotomidae, p. 63
– Body with scales or a dense cover of ciliated, apically expanded macrochaetae (Figs 2B, C). PAO absent, eyes and pigment present in the two Greenland species .... Entomobryidae, p. 76
6. Larger species, eyes present, antennae longer than head (suborder Symphypleona) 7
– Very small (0.5 mm), blind species with antennae shorter than head (suborder Neelipleona, Fig. 19A) .... Neelidae, p. 76
7. Head with 8+8 ocelli, darker species 8
– Head with 1+1 ocelli, pigmentation weak or absent (Fig. 19B) .... Arrhopalitidae, p. 77
8. Males with simple antennae, females with a pair of modified setae below anus (Fig. 2H). Tibiotarsal organ absent 9
– Males with modified antennae (Fig. 2D), females without modified setae below anus. Tibiotarsal organ present (Fig. 19G) ...... Sminthurididae, p. 77
9. Ant.4 subdivided (Fig. 2G) 10
– Ant.4 simple .... Katiannidae, p. 78
10. Tibiotarsi with simple apical hairs (Fig. 2F) .... Sminthuridae, p. 80
– Tibiotarsi with clavate apical hairs (Fig. 2E) .... Bourletiellidae, p. 79

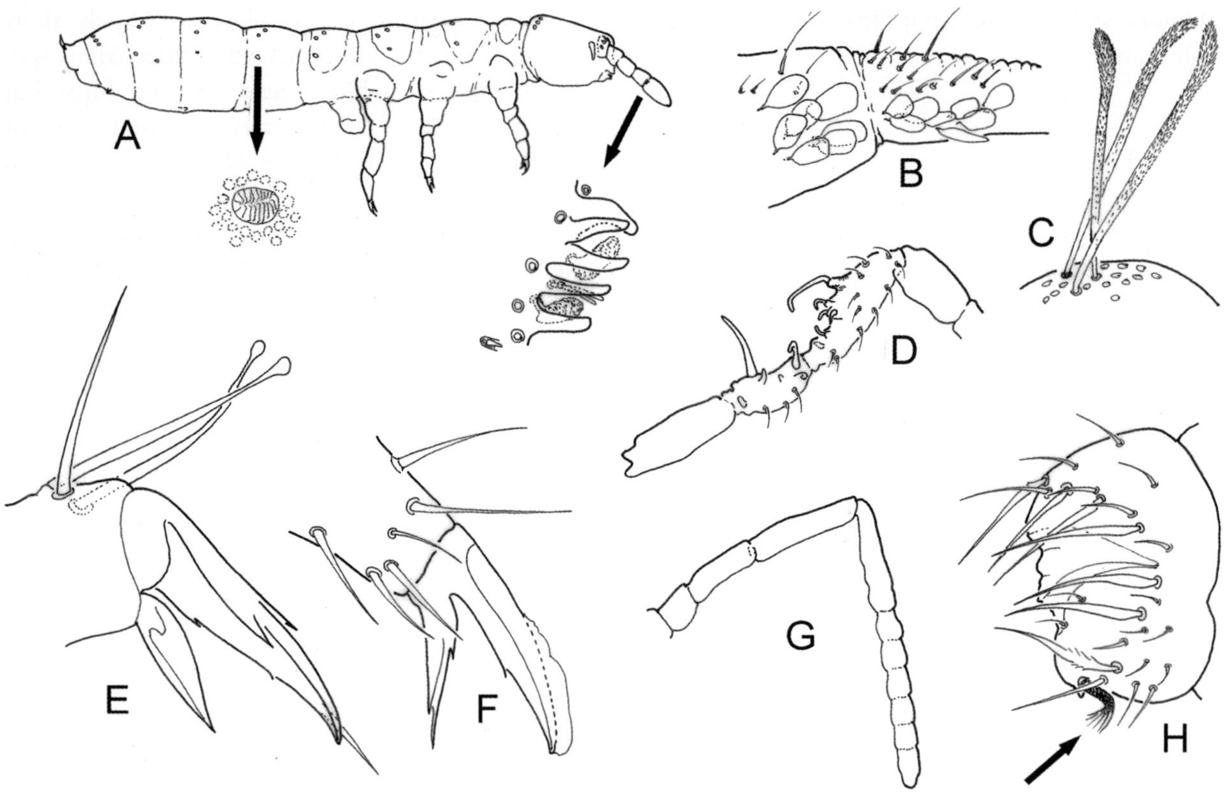

FIGURE 2 *Morphology of Collembola species. A.* Protaphorura subuliginata *with pseudocelli and antennal organ (arrows); B.* Lepidocyrtus violaceus, *scales on manubrium/dens; C.* Entomobrya nivalis, *ciliate macrochaetae from anterior edge of mesothorax; D.* Sminthurides schoetti, *male clasping organ on ant.3–4; E.* Bourletiella hortensis, *claw and tip of tib. 3; F.* Sminthurus nigromaculatus, *ditto; G.* Bourletiella hortensis, *antenna; H.* Sminthurinus aureus, *abd.6 of female with modified subanal seta (arrow).*

HYPOGASTRURIDAE

Key to genera

1. Pigmented species, eyes and furca present..........2
– White, blind species without furca .... *Willemia*
2. Dens with more than 2 setae, ocelli 8+8, PAO present 3
– Dens with 2 setae, ocelli 5+5, PAO absent ....*Xenylla*
3. Ant.3–4 with a ventral protrusible sac (Fig. 4G). Anal spines as long as or longer than claws (Fig. 4F). Retinaculum with 4+4 teeth. Macrochaetae strongly differentiated from ordinary setae ....*Ceratophysella*
– Ant.3–4 without protrusible sac. Anal spines shorter than claws. Retinaculum with 3+3 teeth. Macrochaetae poorly differentiated ....*Hypogastrura*

***Hypogastrura* Bourlet, 1839**

This genus, with 3 Greenlandic species, differs from other genera of the family by a set of mostly plesiomorphic characters: PAO small with 4 subequal lobes (Fig. 3E), body hairs slightly differentiated, claws with unguiculus present, furca strong with 6 setae on dens, anal spines shorter than claws (Fig. 1), body granulation moderate, retinaculum with 3+3 teeth.

Key to species

1. Tibiotarsi with 2–3–3 clavate tenent hairs, all apical (Fig. 3J). Colour bluish black.........................2
– Tibiotarsi with 2–3–2 clavate tenent hairs, on tib.2–3 only one in apical position (Figs 3A, B). Colour more brownish ....*purpurescens*
2. Body hairs long, macrochaetae clearly developed (Fig. 3F). Mucro with broad, angular lamella (Fig. 3G) ....*viatica*
– Body hairs short, spine-like (Fig. 3K). Mucro with narrow lamella (Fig. 1) ....*concolor*

***Hypogastrura purpurescens* (Lubbock, 1867)**

Figs 3A–3E

Greyish-blue to reddish-brown, 2.0 mm. Body hairs moderately long, macrochaetae only differentiated from ordinary setae at base of legs and laterally on abd. 4–6.

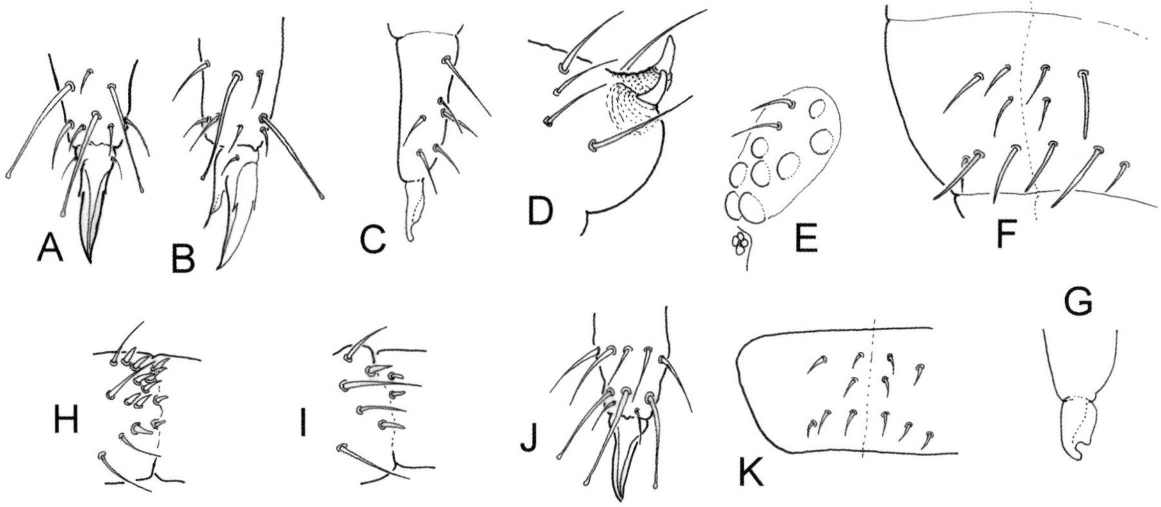

FIGURE 3 *Morphology of* Hypogastrura *species. A–E.* Hypogastrura purpurescens. *Left tib.2 (A), left tib.3 (B), dens and mucro (C), anal spines (D), Ocelli and PAO (E); F–G.* H. viatica. *Median part of abd.4 (F), mucro (G); H.* H. tullbergi, *ant.3 organ; I–K.* H. concolor. *Ant.3 organ (I), left tib.2 (J), median part of abd.4 (K).*

Tibiotarsi with 2–3–2 clavate tenent hairs. On tib.2–3 only one clavate tenent hair in apical position, others subapical (Figs 3A, 3B). Mucro elongate, with a low lateral lamella (Fig. 3C).

The slender mucro, unique arrangement of tibiotarsal tenent hairs and pale colour readily identify this species.

BIOLOGY. Mostly in organic debris in human habitations and along shores of sea and lakes. In the inner parts of Kangerluarsunguaq/Buksefjord and Kangerlussuatsiaq/Evighedsfjord in litter under *Alnus*.

DISTRIBUTION. S, SW, NW, NE. Scattered records both on the west coast (Qaqortoq/Julianehåb north to Upernavik) and east coast (Illoqqortoormiut/Scoresbysund north to North Fjord).

Cosmopolitan.

***Hypogastrura viatica* (Tullberg, 1872) Figs 3F, 3G**
Henriksen 1939: *Hypogastrura viatica* Tullb.

Bluish black, 1.9 mm. Body hairs differentiated, with stiff, straight macrochaetae and curved mesochaetae (Fig. 3F). Tergites of Abd.1–3 often with roundish lateral lobes. Tibiotarsi with 2–3–3 clavate tenent hairs, on tib.2–3 all in apical position (as Fig. 3J). Mucro compact, with a strong, angular lateral lamella (Fig. 3G).

Easily identified by the strong mucronal lamella and long body hairs. None of the other *Hypogastrura* species will occur in masses along seashores.

BIOLOGY. Very common in the marine littoral, often in masses in deposits of rotting seaweeds. Also on shores of inland lakes and in other biotopes with abundance of organic matter.

DISTRIBUTION. S, SW, NW, NE. Many records on the west coast (Qaqortoq/Julianehåb north to Upernavik). On the east coast only reported from Ella Ø (Hammer 1944).

Cosmopolitan.

***Hypogastrura concolor* (Carpenter, 1900) Figs 1, 3I–3K**
Henriksen 1939: *Hypogastrura tullbergi* Schäff.
Hammer 1954: *Neogastrura tullbergi* (Schäffer)

Bluish black, 1.5 mm. General habitus as Fig. 1. Body hairs short, undifferentiated, in large specimens almost spine-like on tergites (Fig. 3K). Tibiotarsi with 2–3–3 apical clavate tenent hairs (Fig. 3J). Mucro slender, with a variable lateral lamella, often sligtly angular (Fig. 1).

Identified by the short, undifferentiated body hairs and the slender mucro. Unlike *viatica* it prefers biotopes with an arctic character and is never found in masses along seashores. The similar species *H. tullbergi* (Schäffer, 1900) may show up in Greenland, but no verified records so far. It is almost identical to *concolor*, but differs by having more than 4 spines in the ant.3 organ (Figs 3H, 3I). Like *concolor* it is circumpolar, but apparently absent in arctic Canada.

BIOLOGY. In various habitats and vegetation types, often in dry tundra sites.

DISTRIBUTION. SW, NW, N, NE. Common on the east coast from Illoqqortoormiut/Scoresbysund north to Peary Land. Probably also common in northern part of the west coast (Pituffik/Thule), uncommon in the southern part (lake Aasivit at inner Itilleq, *c.* 66°N).

Circumpolar, arctic.

Ceratophysella Börner, 1932

The 3 Greenland species of this genus are recognised by their long anal spines (Fig. 4F), coarse body granulation, long body hairs with strong macrochaetae, broad-tipped mucro (Fig. 4K) and presence of a protrusible sac ventrally between antennal segments 3–4 (Fig. 4G). PAO with 4 unequal lobes (Fig. 4E). Retinaculum with 4+4 teeth.

Key to species

1. Dens with 7 dorsal setae (Fig. 4D). Th.2 with seta a_3 much longer than a_2 (Fig. 4B). Axial setae on abd.4 strongly diverging (Fig. 4C) 2
– Dens with 6 dorsal setae (Fig. 4K). Th.2 with a_2 and a_3 subequal (Fig. 4I). Axial setae on abd.4 nearly in parallel rows (Fig. 4J) ....*succinea*
2. Abd.5 with 3+3 a-setae between sensilla p_3 (Fig. 4A) ....*denticulata*
– Abd.5 with 2+2 a-setae between sensilla p_3 (Fig. 4H) ....*engadinensis*

Ceratophysella denticulata (Bagnall, 1941) Figs 4A–4G

Henriksen 1939: *Hypogastrura armata* Nic.
Hammer 1952: *Ceratophysella armata* (Nic.)

Greyish blue, pigmentation often spotted. 1.8 mm. Th.2 with setae a_3 much longer than a_2 (Fig. 4B). Axial setae (a_1–m_1–p_1) on abd.4 strongly diverging (Fig. 4C). Chaetotaxy of abd.5 as Fig. 4A, with 3+3 a-setae between the sensilla p_3. Dens with 7 dorsal setae, inner 2–3 flattened, expanded in the middle (Fig. 4D). Papilla C present on the labial palp (cf. Fig. 23 in Fjellberg 1998).

Early authors on Greenland Collembola usually reported this species as *Hypogastrura armata* (Nicolet, 1841), which is not present in Greenland. The only difference between *denticulata* and *engadinensis* seems to be the loss of seta a_2' on abd.5 in the latter (Figs 4A, 4H).

BIOLOGY. Often abundant in organic litter, in particular from manured sites near human dwellings.

DISTRIBUTION. S, SW, NE. Common on both west coast (Qaqortoq/Julianehåb north to Qeqertarsuaq/Disko) and east coast (Illoqqortoormiut/Scoresbysund north to Zackenberg).

Cosmopolitan.

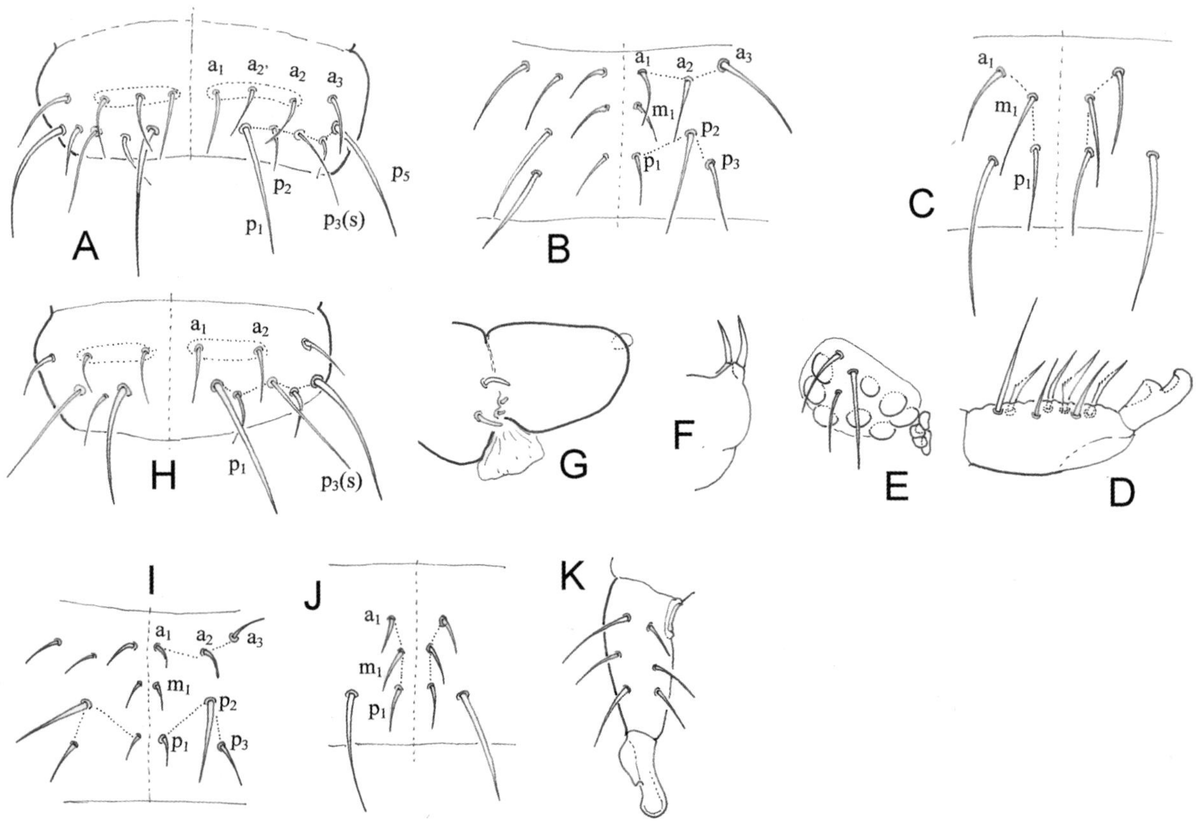

FIGURE 4 *Morphology of* Ceratophysella *species. A–F.* C. denticulata. *Abd.5 (A), median part of th.2 (B), median part of abd.4 (C), dens and mucro (D), ocelli and PAO (E), (F), ant.3–4 with expanded protrusible sac (G); H.* C. engadinensis, *abd.5; I–K.* C. succinea. *Median part of th.2 (I), median part of abd.4 (J), dens and mucro (K).*

***Ceratophysella engadinensis* (Gisin, 1949) Fig. 4H**
Identical to *denticulata* apart from absence of seta a_2' on abd.5 (Fig. 4H).

BIOLOGY. Found in organic debris.

DISTRIBUTION. SW. Only found once from a bird cliff (Innasuaq, Søndre Isortoq). New species to Greenland.

Probably cosmopolitan.

***Ceratophysella succinea* Gisin, 1949 Figs 4I–4K**
Brownish, 1.8 mm. Abd.5 without seta a_2' (as in Fig. 4H). Th.2 with seta a_2 and a_3 subequal. Axial setae on abd.4 in almost parallell lines (Fig. 4J). Dens with 6 dorsal setae, all slender (Fig. 4K). Labial palp without papilla C (Fig. 23 in Fjellberg 1998).

Pigmentation usually darker than in previous species. The presence of only 6 setae on dens and the chaetotaxy characters described above, readily separates this species from *denticulata*/*engadinensis*. Loss of papilla C on the labial palp is an unique character for the species, identifying also small juveniles.

BIOLOGY. Usually found in plant litter along shores of lakes and streams.

DISTRIBUTION. SW, NE. Only two Greenland records (lake Aasivit at Itilleq, *c.* 66°N; Illoqqortoormiut/Scoresbysund). New species to Greenland

Cosmopolitan.

***Xenylla* Tullberg, 1869**
Dark coloured species, differing from other genera of the family by absence of PAO. Ocelli 5+5. Anal spines present, very small. Tibiotarsi with 2 clavate tenent hairs in the apical row, claws without unguiculus (Fig. 5A). Furca well developed. My be confused with *Hypogastrura*, but dens only with 2 setae. Only two species in Greenland.

Key to species

1. Mucro long, clearly set off from dens (Fig. 5C). Head with 2+2 postlabial setae along ventral line (Fig. 5B). Dark, large species .... *humicola*
– Mucro short, fused to dens (Fig. 5D). Head with 3+3 postlabial setae. Small, pale species ...... *grisea*

***Xenylla humicola* (O. Fabricius, 1780) Figs 5A–5C**
Violet blue to almost black, 2.0 mm. Body hairs short and spiny, slightly serrated. Anal spines small, straight. Furca strong, mucro long and pointed, clearly set off from dens (Fig. 5A). Last pair of legs with a fleshy subcoxal lobe. The dark colour, large size and strong furca separate this species from *grisea*.

BIOLOGY. Usually found in salt meadows and organic debris along seashores, often in large numbers.

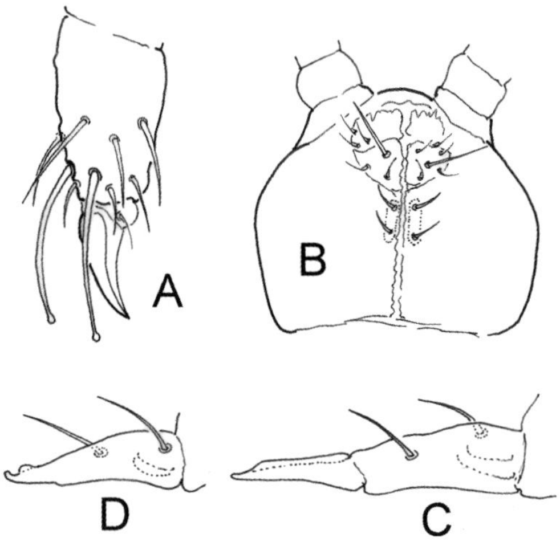

FIGURE 5 *Morphology of* Xenylla *species. A–C.* X. humicola. *Right tib.3 (A), ventral side of head (B), dens and mucro (C); D.* X. grisea, *dens and mucro.*

DISTRIBUTION. SW, NW, SE, NE. Common both on west coast (Kangerluarsusuaq/Grædefjord north to Upernavik) and east coast (Saqqisikuik/Skjoldungen north to Illoqortoormiut/Scoresbysund).

Cosmopolitan.

***Xenylla grisea* Axelson, 1900 Fig. 5D**
Bluish grey, rather pale, 1.0 mm. Mucro short, not clearly set off from dens (Fig. 5D).

BIOLOGY. Lives in soils with rich organic content.

DISTRIBUTION. S. Only found once in a moist meadow at Narsarsuaq. New species to Greenland.

Cosmopolitan.

***Willemia* Börner, 1901**
Among the Greenland Hypogastruridae, species of *Willemia* are recognised by small size (less than 1 mm) and absence of eyes, body pigment and furca. PAO present, with 4 or more lobes in a flower-like arrangement. A pair of small anal spines are either present or absent.

Key to species

1. Anal spines present (Fig. 6A). Ant.4 sensilla slender (Fig. 6F). Head without subcutane network 2
– Anal spines absent. Ant.4 sensilla drop-shaped (Fig. 6E). Central part of head with a subcutane network (Fig. 6D)................................ *denisi*
2. Abd.4 with 3+3 axial setae (m_1 present, Fig. 6A) 3
– Abd.4 with 2+2 axial setae (m_1 absent, Fig. 6J) .... *scandinavica*

3. PAO with 5 (4–6) spacy lobes (Fig. 6I). Abd.1–3 with seta m_4 present, m_3 absent (Fig. 6B). Ant.3 organ narrow, small sensilla partly covered (Fig. 4G) .... *anophthalma*

– PAO with 6–9 packed lobes (Fig. 6H). Abd.1–3 with m_4 absent, m_3 present (Fig. 6C). Ant.3 organ wide, small sensilla free (Fig. 6F) .... *similis*

Willemia anophthalma **Börner, 1901 Figs 6A, 6B, 6G, 6I**

White, 0.7 mm. PAO normally with 4 lobes sitting on a protruding cone (Fig. 6I). Unlike other species the ant.3 organ is narrow (small distance between the lateral guards), and the 2 small central sensilla are partly covered by an integumentary fold (Fig. 6G). Prelabral setae 4.

The characteristic PAO and ant.3 organ makes the species easy to identify. Also presence om seta m_4 on abd.1–2 is an unique character.

BIOLOGY. Frequent in soil/litter in lush meadows and in thickets of *Salix* and *Alnus*. Also in dry heathland with *Empetrum* and *Rhododendron*.

DISTRIBUTION. S, SW, NW, NE. Widely distributed (Kangerlussuatsiaq/Lindenow Fjord north to Upernavik), but records from the east coast north of Lindenow Fjord need verification. In Hammer's material of *anophthalma* from the east coast (Hammer 1944) both *similis* and *scandinavica* are present.

Holarctic. Mainly in forested regions, but penetrates into alpine and arctic environments.

Willemia similis **Mills, 1934 Figs 6C, 6F, 6H**

White, 0.7 mm. Prelabral setae 2. Similar to previous species in presence of anal spines and 3+3 axial setae on abd.4, but readily identified by different PAO (Fig. 6H) and ant.3 organ (Fig. 6F). The m_3 seta on abd.1–3 sometimes absent on one or more tergites.

BIOLOGY. Found in soil/litter from various vegetation, both shrubs and meadows. In Kangerlussuaq/Søndre Strømfjord common in dry heathland.

DISTRIBUTION. SW, NW, NE (many sites from Kangerluarsunguaq/Buksefjord north to Pituffik/Thule; Ella Ø, Zackenberg). New species to Greenland.

Holarctic, with preference for arctic environments (in Europe only Iceland and Svalbard).

Willemia scandinavica **Stach, 1949 Fig. 6J**

White, 0.7 mm. Prelabral setae 4. Abd.1–3 without m_3/m_4. The only *Willemia* with 2+2 axial setae on abd.4 (Fig. 6J) combined with presence of anal spines.

BIOLOGY. Mostly found in dry meadows and heaths, avoiding lush places.

DISTRIBUTION. S, SW, NE. Probably the most common *Willemia* in Greenland, with many records both

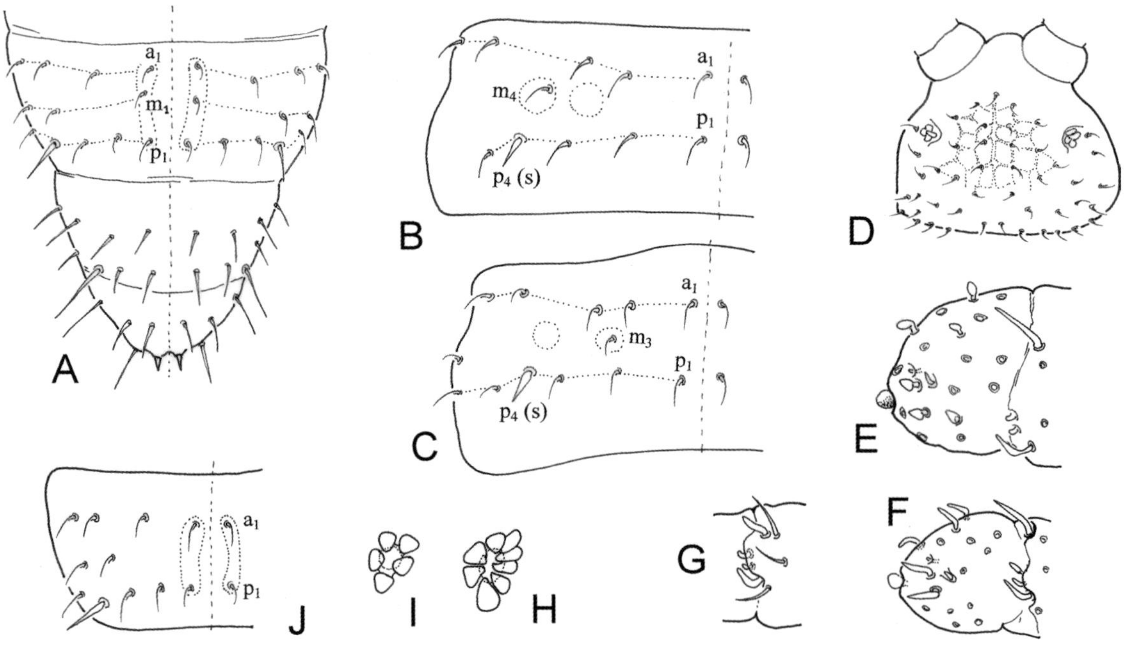

FIGURE 6 *Morphology of* Willemia *species. A.* W. anophthalma, *abd.4–6; B–C. Abd.1 in* W. anophthalma *(B) and* W. similis *(C); D.* W. denisi, *head with subcutane network; E–F. Ant.3–4 in* W. denisi *(E) and* W. similis *(F); G.* W. anophthalma, *ant.3 organ; H–I. PAO in* W. similis *(H) and* W. anophthalma *(I); J.* W. scandinavica, *abd.4.*

on the west coast (Narsarsuaq north to Kangerdluarssukasik) and the east coast (Kangerlussuatsiaq/Lindenow Fjord north to Zackenberg). New species to Greenland.

Holarctic.

***Willemia denisi* Mills, 1932 Figs 6D, 6E**

White, 1.0 mm. Prelabral setae 4. Setae m_3/m_4 absent on abd.1–3. Abd.4 with 2+2 axial setae (as Fig. 6J). Easy to identify by the absence of anal spines. In mixed samples it is recognised by broader body shape and slightly larger size.

BIOLOGY. Few Greenland records from wet biotopes (bogs, snowbed, spring, litter of *Betula nana/Salix*).

DISTRIBUTION. SW, NE (Kuanersoq/Kvanefjord north to Ikamiut Kangerluarsuat; Ella Ø, Zackenberg). New species to Greenland.

Holarctic.

NEANURIDAE

Key to genera

1. Abd.6 without anal spines 2
– Abd.6 with 3 or 5 anal spines (Figs 7A, B) .... *Friesea*
2. White or faintly pigmented species with short body hairs and smooth skin 3
– Bluish grey, body warted with long, spiny hairs (Fig. 9) .... *Neanura*
3. Very small (0.5 mm), ocelli 2+2. Traces of body pigmentation usually present .... *Micranurida*
– Larger (above 1.0 mm), ocelli and pigment absent .... *Anurida*

***Friesea* Dalla Torre, 1895**

Unlike other Greenland species, members of this genus have either 3 or 5 spines on tip of abdomen. The sickle shaped maxilla (Fig. 7I) is a unique character for the genus. PAO absent. Claws without unguiculus.

Key to species

1. Anal spines 3, ocelli 8+8 (Figs 7A, G) 2
– Anal spines 5, ocelli 5+5 (Figs 7B, H) .... *quinquespinosa*
2. Abdominal macrochaetae pointed 3
– Macrochaetae at tip of abdomen clavate (Fig. 7A) .... *claviseta*
3. A small mucro present (Fig. 7D). .... Inner side of tibiotarsi with seta A4 present (Fig. 7E) .. .. *mirabilis*
– Mucro absent (Fig. 7C), tibiotarsi without A_4 (Fig. 7F) .... *truncata*

***Friesea quinquespinosa* Wahlgren, 1900 Figs 7B, 7H**

Bluish grey, 1.5 mm. Furca short, with 3 setae, mucro absent. Tibiotarsi with both A_4–A_5 present (as Fig. 7E). One of the anal spines may frequently be absent or irregularly placed.

The presence of 5 anal spines immediately identifies *quinquespinosa* from all other Greenland Collembola.

BIOLOGY. Common in organic debris in wet localities (seashores, springs, seabird colonies), such as the banks of the duck pond in the city of Godthåb/Nuuk.

DISTRIBUTION. SW, NE, SE. Both on the west coast (Kangerluarsussuaq/Grædefjord north to Kangerlussuatsiaq/Evighedsfjord) and the east coast (Atingat, Ammassalik, Ella Ø).

Circumpolar, only in arctic sites.

***Friesea mirabilis* (Tullberg, 1871) Figs 7D, 7E, 7G, 7I**

Bluish grey of variable intensity, 1.9 mm. Among the 3-spined species characterized by pointed abdominal macrochaetae and presence of a full set of apical setae on the tibiotarsi (Fig. 7E).

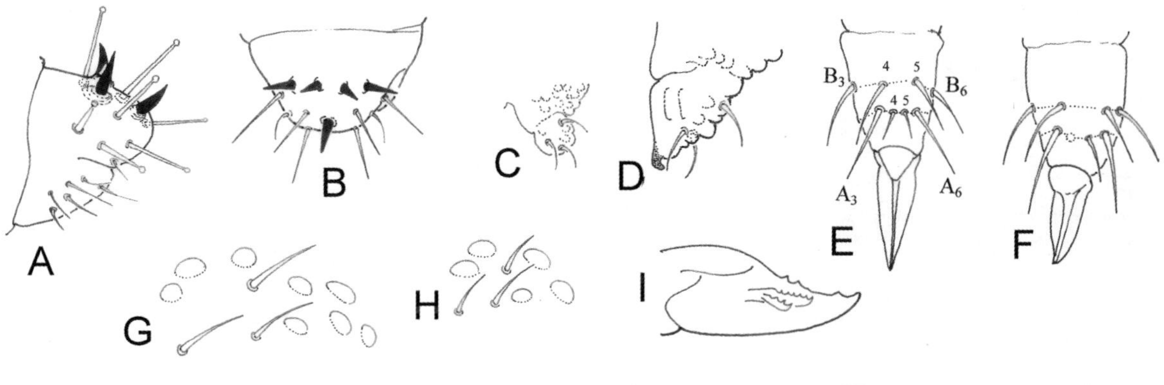

FIGURE 7 *Morphology of* Friesea species. *A–B. Abd.6 in* F. claviseta (*A*) *and* F. quinquespinosa (*B*); *C–D. Furca in* F. truncata (*C*) *and* F. mirabilis (*D*); *E–F. Inner side of left tib.2 in* F. mirabilis (*E*) *and* F. truncata (*F*); *G–H. Ocellar field in* F. mirabilis (*G*) *and* F. quinquespinosa (*H*); *I.* F. mirabilis, *maxilla.*

BIOLOGY. The Greenland records are only from seashore wrackbeds and salt meadows.

DISTRIBUTION. SW (Kvanefjord north to Kangerlussuaq/Søndre Strømfjord). Cosmopolitan.

Friesea truncata Cassagnau, 1958 Figs 7C, 7F

Bluish grey, 1.0 mm. Apart from reduced mucro and absence of A_4 on tibiotarsi (Figs 7C, 7F), inseparable from *mirabilis*.

BIOLOGY. Found in soil/litter in various vegetation types.

DISTRIBUTION. S. A few samples from Narsarsuaq only. New species to Greenland.

Transpalaearctic, but distribution incompletely known due to confusion with *mirabilis*.

Friesea claviseta Axelson, 1900 Fig. 7A

Bluish grey, 1.0 mm. Seta A_3/A_4 both present on tibiotarsi. Mucro present on furca. Tibiotarsal clavate tenent hairs longer than in other species. The clavate abdominal hairs readily identifies this species.

BIOLOGY. Found in dry places (moss/lichens on stones and on tree stems).

DISTRIBUTION. SW. Only found once in moss/lichens on a large boulder at Tasersuaq, Ikamiut Kangerluarssuat.

Cosmopolitan.

Micranurida pygmaea Börner, 1901 Figs 8A, 8B

Hammer 1952: *Micranurida pygmaea* Börner, *Micranurida hasai* Ksen.

White or with diffuse greyish blue pigment, 0.5 mm. Ocelli 2+2, as dark spots. PAO with 6–8 lobes in a circular arrangement (Fig. 8B). Maxilla styliform. Ant.4 with 5 thick, hammer-shaped or egg-shaped sensilla (Fig. 8A). Body hairs smooth and short. Anal spines and furca absent. Claws without unguiculus.

The very small size, weak or absent body pigmentation and 2+2 pigmented ocelli identify this species among other Greenland Collembola.

BIOLOGY. Common in various vegetation types (litter of *Alnus*/*Betula*/*Salix*, snowbeds, meadows, heathland).

DISTRIBUTION. S, SW, NE (Narsarsuaq north to Kangerlussuaq/Søndre Strømfjord; Ella Ø, Zackenberg).

Cosmopolitan.

Anurida Laboulbène, 1865

Greenlandic species unpigmented, without ocelli and furca. Maxilla short, with toothed capitulum and 3 serrated lamella.

Key to species

1. Sensilla (p_3) on Th.2–3 on line with setae p_2–p_4. Abd.4 with a_3 and p_3 present (Fig. 8G). Slender species 2
– Sensilla on Th.2–3 set in front of p_2–p_4 (Fig. 8H). Abd.1–4 without a_3 and p_3 (Fig. 8I). Thick and stout species .... *granaria*
2. Th.1 with more than 2+2 setae between lateral macrochaetae (Fig. 8F). PAO with regular lobes (Fig. 8D) .... *thalassophila*
– Th.1 with 2+2 (sometimes 2+3) setae between lateral macrochatae. PAO lobes irregular (Fig. 8C) ... *polaris*

Anurida granaria (Nicolet, 1847) Figs 8H, 8I

White or slightly yellowish, 2.0 mm. PAO with up to 20 regular lobes in circular arrangement. Body granulation uniform, coarse. Ant.4 with 6 thick sensilla. Th.1 normally with 3+3 setae between the lateral macrochaetae. The chaetotaxy characters readily separates *granaria* from the other Greenland species.

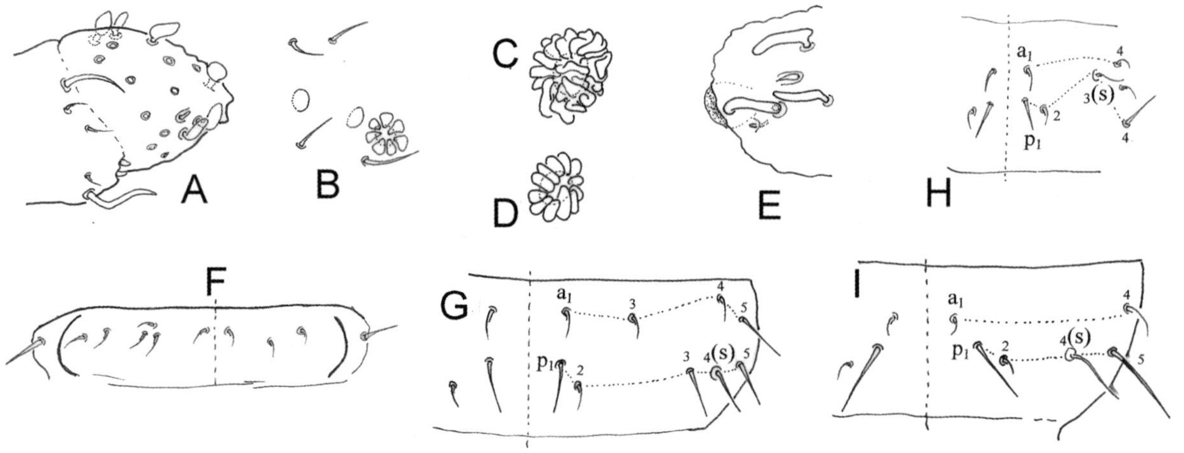

FIGURE 8 *Morphology of* Micranurida *(A–B) and* Anurida *(C–I) species. A–B.* M. pygmaea. *Ant.3–4 (A), ocelli and PAO (B); C–D. PAO in* A.polaris *(C) and* A. thalassophila *(D); E–G.* A. thalassophila. *Sensilla on tip of ant.4 (E), th.1 (F), abd.4 (G); H–I.* A. granaria. *Meadian part of th.2 (H), abd.4 (I).*

BIOLOGY. Usually found in salt meadows and in rich organic soils near seashores. The species is capable of producing light (bioluminescens).

DISTRIBUTION. S, SW, NW, NE (Kuanersoq/Kvanefjord north to Upernavik; Kangerlussuatsiaq/Lindenow Fjord north to Illoqqortoormiut/Scoresbysund).

Palaearctic, but distribution unclear due to confusion with other species.

***Anurida thalassophila* (Bagnall, 1939) Figs 8D–8G**
White, 1.5 mm. PAO with 10–15 regular lobes in circular arrangement (Fig. 8D). Body granulation uniform, finer than in previous species. Ant. 4 with 7 thick, partly hammer-shaped sensilla (Fig. 8E).

Very similar to *polaris*, but normally separated by the key characters. Small juveniles with incomplete chatotaxy and immature PAO may be unseparable.

BIOLOGY. Only found in salt meadows and other seashore habitats rich in organic matter.

DISTRIBUTION. SW. Many records along the west coast (Allumersat/Bjørnesund north to Sisimiut/Holsteinsborg). Probably also on the east coast, but it may have been confused with *granaria* by earlier authors. New species to Greenland.

Western Palaearctic.

***Anurida polaris* (Hammer, 1954) Fig. 8C**
Hammer 1954: *Micranurida polaris* n. sp.

White, 1.5 mm. Identical to previous species apart from the key characters above. It may be a parthenogenetic arcto-alpine form of the marine littoral *thalassophila*. The question calls for modern DNA studies.

BIOLOGY. Lives in cold arctic environments, usually in damp moss/grass along melt water streams.

DISTRIBUTION. N, NE. Originally described by Hammer (1954) from Peary Land. Recent records from Zackenberg. No records from the west coast, but probably present in alpine and northern parts.

Holarctic. In Europe only Svalbard and Scandinavian high mountains.

***Neanura muscorum* (Templeton, 1835) Fig. 9**
Henriksen 1939: *Achorutes muscorum* Templ.

Bluish grey, 2–3 mm. Body with large, wart-like granulation bearing long spine-like setae (Fig. 9). Ocelli 3+3, PAO absent. Maxilla styliform. Ant. 4 with 8 finger-like sensilla. Furca absent. This large, spiny and sluggish species is easily identified by the naked eye in the field.

BIOLOGY. In litter of *Alnus*/*Salix* and in other moist biotopes rich in organic matter. Often spotted under stones or sitting on the underside of boards on seashores.

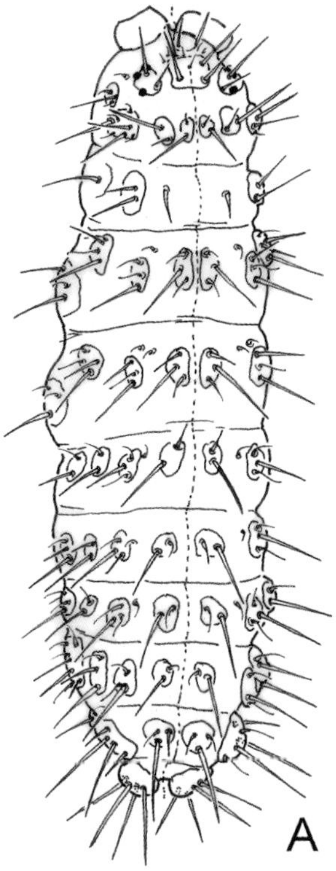

FIGURE 9 Neanura muscorum, *dorsal chaetotaxy.*

DISTRIBUTION. S, SW, SE. On the west coast from Narsarsuaq north to Kangerlussuaq/Søndre Strømfjord. On the east coast north to Ammassalik.

Holarctic.

ONYCHIURIDAE

Key to genera

1. Ant.3 organ with slender, finger-like papillae covering the sensillary elements (Fig. 2A). Larger species (>1.0 mm) (subfamily Onychiurinae) 2
– Ant.3 organ without papillae, or these are short and blunt, not finger-like. Smaller species (<1.0 mm) (subfamily Tullbergiinae) 6
2. PAO large, with many vesicles (Figs 11J–L) 3
– PAO small, with a single lobed vesicle (Fig. 11A) .... *Oligaphorura*
3. Head with pseudocelli present along posterior edge of head. Antennal base with more than 1 ps.oc. (Fig. 2A) 4
– Head without posterior pseudocelli, antennal base with 1 ps.oc. (Fig. 11P) .... *Hymenaphorura*
4. Ventral side of abd.4 without traces of furca 5
– Abd.4 with a reduced flap-like furca (Fig. 11N) .... *Protaphorura*

5. Tibiotarsi with 9 setae in the apical whorl, T_2–T_3 absent (Fig. 11G). Smaller, white species (<1.5 mm) .. .. *Thalassaphorura*
– Tibiotarsi with 11 apical setae, all 4 T–setae present (Fig. 11M). Large species, up to 3.5 mm. Live specimens yellowish .... *Megaphorura*
6. Abd.6 with a pair of crescentic ridges in anterior part (Fig. 12N). Smaller species with short anal spines 7
– Abd.6 without crescentic ridges (Fig. 12O). Larger species with longer anal spines .... *Chaetaphorura*
7. PAO with many simple vesicles in 2 rows (Fig. 12R) . ... *Mesaphorura*
– PAO with 6 elongate vesicles in star-shaped arrangement (Fig. 12S) .... *Karlstejnia*

***Oligaphorura* Bagnall, 1949**

The reduced PAO, with an almost invisible vesicle (Fig. 11A), characterize the 2 Greenland species of this genus. Tibiotarsi with all T–setae present.

Key to species

1. Lateral microsensillum present on th.2–3 (Fig. 11B). Labium with terminal sensilla A and C blunt-tipped (Fig. 11D) .... *groenlandica*
– Lateral microsensillum present on th.2, absent on th.3. Labium with only one blunt-tipped terminal sensillum (Fig. 11E) .... *ursi*

***Oligaphorura groenlandica* (Tullberg, 1876) Figs 11A–11D**

Henriksen 1939: *Onychiurus groenlandicus* Tullb.

White or slightly yellowish, 1.6 mm. Body shape slightly pear-shaped, with abdomen widening towards tip. Dorsal pseudocelli distributed as 32/133/33343. Abd.6 with fine granulation, not coarser than on abd.5. Abd.5 usually with seta m_1 erect, longer than a_1 and p_1 (Fig. 11C).

BIOLOGY. Usually found in damp places.

DISTRIBUTION. SW, NE. Probably widely distributed, but confused with the following species. Verified specimens on the west coast from Kangerluarssuaq/Grædefjord north to Itilleq (*c.* 66°35'N). On the east coast specimens are known from Charcot Land, Ella Ø and Zackenberg.

Holarctic.

***Oligaphorura ursi* (Fjellberg, 1984) Figs 11E, 11F**

White, 1.3 mm. Similar to *groenlandica*, but always distinguished by absence of lateral microsensillum on metathorax. Also the body is smaller, more cylindric, not widened towards tip. Abd.6 has enlarged skin granules, clearly larger than on abd.5. On abd.5 seta m_1 usually curved, shorter than a_1 and p_1 (Fig. 11F).

BIOLOGY. Usually found in lush biotopes, in meadows and litter under *Alnus* and *Salix*.

DISTRIBUTION. SW, NE (Kuanersoq/Kvanefjord north to Kangerlussuatsiaq/Evighedsfjord; Ella Ø, Zackenberg). New species to Greenland.

Palaearctic.

***Thalassaphorura* Bagnall, 1949**

PAO ovale, with many simple lobes (Fig. 11L). Ventral side of abd.4 without traces of furca, only 4 small setulae indicate the position (Fig. 11I). Th.1 with pseudocelli present. Tibiotarsi with reduced chaetotaxy, setae T_2 and T_3 absent in the apical whorl (Fig. 11G).

Key to species

1. Th.1 with 2+2 pseudocelli (Fig.11H) .... *duplopunctata*
– Th.1 with 2+2 pseudocelli (lateral missing)........ .... *debilis*

***Thalassaphorura duplopunctata* (Strenzke, 1954) Figs 11G–11I, 11L**

White, 1.4 mm. Pseudocelli on dorsal side as 32/233/33343. On ventral side of abd.2–4 usually 1+1 pseudocelli.

The presence of 2+2 ps.oc. on th.1 will separate this species from *debilis*.

BIOLOGY. In various biotopes along seashores (marine littoral and supralittoral).

DISTRIBUTION. SW, NE (Kuanersoq/Kvanefjord north to Kangerluarsunguaq/Buksefjord; Ella Ø, Zackenberg). New species to Greenland.

Holarctic.

***Thalassaphorura debilis* (Moniez, 1890)**

White, 1.4 mm.

Differs from the previous species by loss of the lateral pseudocelli on th.1.

Also ventral side of abd.2–4 usually without ps.oc. The status of the two taxa is not settled. Possibly there is only one variable species with different dominance of the two forms in a north–south gradient.

BIOLOGY. Lives in the marine littoral zone, sometimes also along shores of inland lakes.

DISTRIBUTION. SW, NE. (Eqaluit Paarliit/Præstefjord, Zackenberg). New species to Greenland.

Palaearctic.

***Megaphorura* Fjellberg, 1998**

PAO with many simple vesicles. Th.1 without pseudocelli. Tibiotarsal chaetotaxy complete, all 4 T-setae present (Fig. 11M). No traces of furca.

***Megaphorura arctica* (Tullberg, 1876) Fig. 11M**

Henriksen 1939: (?) *Onychiurus armatus* Tullb. var. *arcticus* Tullb.

Live specimens pale yellow, becoming white in alcohol. Up to 3.5 mm. Dorsal pseudocelli as 32/023/33354

The large size and yellow colour makes this species unmistakable, even in the field.

BIOLOGY. Often in masses under stones and vegetation along seashores and in bird cliffs. Once found in thousands in thick *Alnus/Salix* litter close to the sea (Kangiusaq in Kangerlussuatsiaq/Evighedsfjord).

DISTRIBUTION. SW. Kuanersoq/Kvanefjord north to Kangerlussuatsiaq/Evighedsfjord).

Palacarctic.

***Hymenaphorura* Bagnall, 1948**

Members of this genus differ from other Onychiurinae by absence of pseudocelli along posterior edge of the head (Fig. 11P) and the PAO which has rather few vesicles which are arranged obliquely in relation to the length axis of the organ (Fig. 11K). Body more flattened than in the cylindrous *Protaphorura*. Tibiotarsi with complete chaetotaxy (apical whorl with 11 setae, T_1–T_4 present). Only one species in Greenland:

***Hymenaphorura* sp. near *anatolii* Pomorski, 2001 Figs 11K, 11P, 11Q**

Henriksen 1939: *Onychiurus sibiricus* Tullb.

White, 1.8 mm. Dorsal pseudocelli as 10/011/11112. Furca absent, furcal field marked as 2 small finely granulated patches. Abd. 5–6 as Fig. 11Q, p_1 setae usually absent on abd.5. Claws with small lateral teeth in basal 1/3.

After modern consepts of the genus *Hymenaphorura*, the old Greenland "*Onychiurus sibiricus*" comes close to *H. anatolii* which was described from N Russia, extending into the arctic tundra. Greenland specimens differ from the original description by presence of lateral teeth on the claws.

BIOLOGY. According to Hammer (1938, 1944) the species is common in various types of biotopes, but most abundant in wet or humid sites.

DISTRIBUTION. NE (Ella Ø, Ymer Ø, Suess Land, Nordfjord, Kempe Fjord – all in the Kejser Franz Joseph Fjord-Kong Oscar Fjord area). Apparently absent from the west coast.

Palaearctic.

FIGURE 10 *Megaphorura arctica.* (PHOTO: J. BÖCHER)

***Protaphorura* Absolon, 1901**

Members of this genus all have PAO with many regular vesicles arranged at right angle to the length axis of the organ (Fig. 11J). Tibiotarsal chaetotaxy complete (T_1–T_4 present). Furca present as a small flap ventrally on abd.4 (Fig. 11N). The Greenland species all belong to the *armata*-group, in which the species separation is still not settled. Earlier Greenland records of this complex were collectively called "*Onychiurus armatus* Tullb." (Henriksen 1939). All species are white, up to 2.0 mm. Due to frequent individual variation in pseudocellar number, several specimens from each sample should be examined.

Key to species

1. Th.3 with 3+3 pseudocelli 2
– Th.3 with 2+2 pseudocelli .... *procampata*
2. Antennal base and abd.1–3 with 3+3 ps.oc 3
– Antennal base and abd.1–3 with 4+4 ps.oc .... *macfadyeni*
3. Ps.oc. as 34/023/33353. Ventral tube with frontal seta (Fig. 11O) .... *pseudovanderdrifti*
– Ps.oc. as 33/023/33343. Ventral tube without frontal setae .... *subuliginata*

***Protaphorura macfadyeni* (Gisin, 1953) Fig. 11J**

The pseudocellar formula (43/023/4453) normally identifies this species, though the ps.oc. number on abd.1–3 is rather unstable with 3–5 pseudocelli on each side.

BIOLOGY. All Greenland records are from marine seashores (wrackbeds, salt meadows, old whalebones in turf).

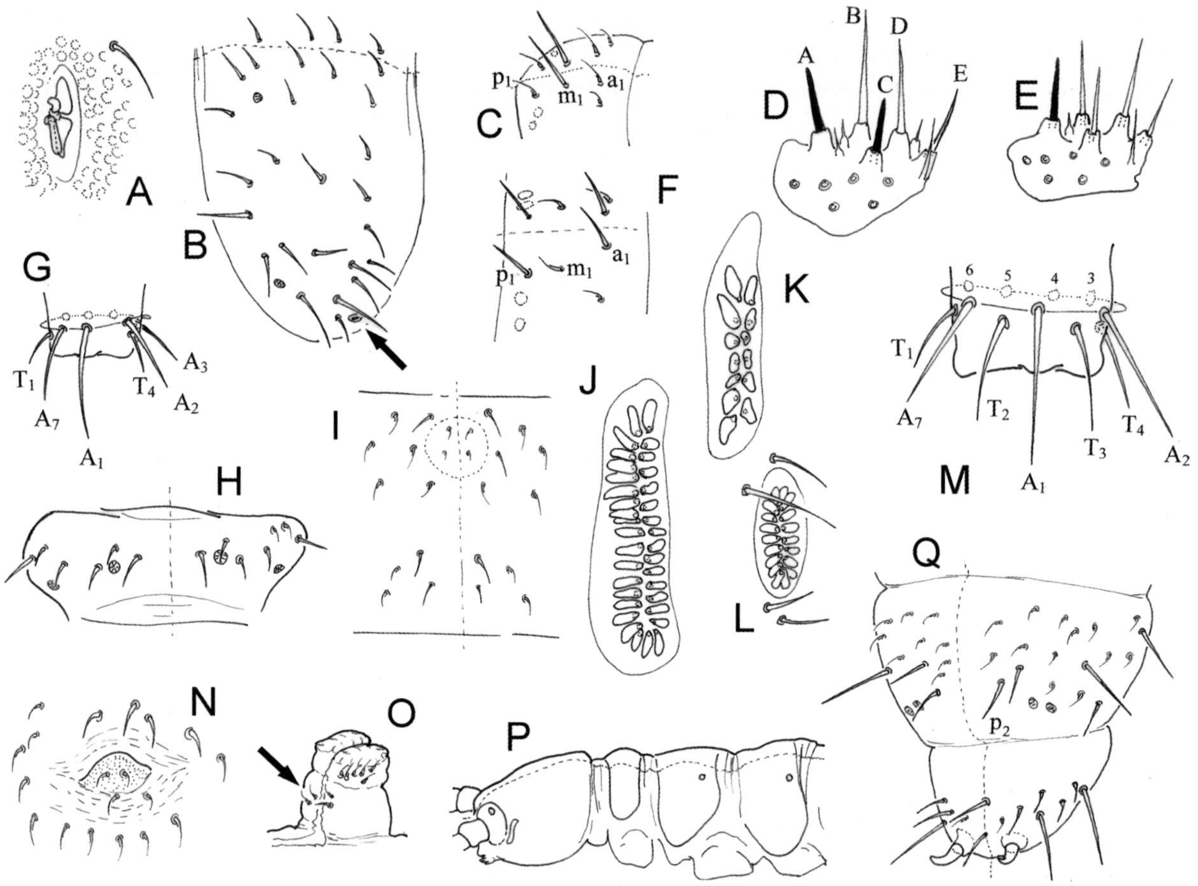

FIGURE 11 *Morphology of Onychiurinae. A–D.* Oligaphorura groenlandica. *PAO (A), right side of th.2 with lateral sensillum marked (B), median part of abd.5 (C), labial palp (D); E–F.* Oligaphorura ursi. *Labial palp (E), median part of abd.5 (F); G–I.* Thalassaphorura duplopunctata. *Apical setae on tib.2 (G), th.1 (H), midventral part of abd.3 with furcal field encircled (I), J–L. PAO in* Protaphorura macfadyeni *(J),* Hymenaphorura *sp. near* anatolii *(K),* Thalassaphorura duplopunctata *(L); M.* Megaphorura arctica, *apical setae on right tib.2; N–O.* Protaphorura pseudovanderdrifti. *Ventral side of abd.4 with furcal flap (N), ventral tube with anterior setae marked (O); P–Q.* Hymenaphorura anatolii. *Distribution of pseudocelli on head-metathorax (P), abd.5–6 (Q).*

DISTRIBUTION. SW (Allumersat/Bjørnesund north to Tasersuaq, Ikamiut Kangerluarsuat). New species to Greenland.

Palaearctic.

Protaphorura pseudovanderdrifti (Gisin, 1957) Figs 11N–11O

The presence of 4+4 pseudocelli at posterior edge of the head and 5+5 on abd.5 usually identify this species. An additional unique character is the presence of 1–2 anterior setae on each side of the ventral tube (Fig. 11O).

BIOLOGY. Few records, in dry plant communities (*Alchemilla*, *Thymus* in gravel slopes).

DISTRIBUTION. S, SW (Upernaviarsuk, Qeqertarsuatsiaat/Fiskenæsset). New species to Greenland.

Palaearctic.

Protaphorura subuliginata (Gisin, 1956) Fig. 2A

Pseudocellar formula will normally identify this species.

BIOLOGY. Usually in rich localities (lush meadows, bird cliffs, seashores).

DISTRIBUTION. SW (Allumersat/Bjørnesund north to Sisimiut/Holsteinsborg). New species to Greenland.

Palaearctic.

Protaphorura procampata (Gisin, 1956)

The presence of 2+2 pseudocelli on th.3 is a good diagnostic character, but abnormal specimens of *subuliginata* with reduced number on one or both sides should be considered.

BIOLOGY. Only few Grenland records from various localities (moss on whale bones, rich fern slope, *Betula* forest).

DISTRIBUTION. S, SW (Narsarsuaq, Qeqertarsuatsiaat/Fiskenæsset). New species to Greenland.

Palaearctic.

***Mesaphorura* Börner, 1901**

All *Mesaphorura* species are white, slender and very small (0.5–0.7 mm). PAO with many small lobes in 2 parallel rows (Fig. 12R). Ant.3 organ with 2 large sensory clubs (Fig. 12M). Tergites of abd.2–3 without pseudocelli, abd.1 and abd.4–5 each with 1+1 pseudocelli. Abd.6 with crescentic ridges (Fig. 12N). Earlier authors on Greenland Collembola used to call all *Mesaphorura* species "*Tullbergia krausbaueri*". Nine species have been recorded in recent years.

Key to species

1. Th.3 with 1+1 ps.oc. Abd.4 without unpaired axial seta 2
– Th.3 with 2+2 ps.oc (Fig. 12A). Abd.4 with unpaired axial seta present (Fig. 12G) .... *tenuisensillata*
2. Abd.5 with 3+3 short a-setae between the long a_4 (Fig. 12B) 3
– Abd.5 with 2+2 short a setae between the long a_4 (Fig. 12C) 7
3. Abd.4 with p_1–p_1 distance shorter than p_2–p_2 (Fig. 12D) 4
– Abd.4 with p_1–p_1 longer than p_2–p_2 (Fig. 12F) .... *italica*
4. Th.3 with seta a_2 present (Fig. 12H) 5
– Th.3 with a_2 absent (Fig. 12A) .... *petterdassi*
5. Anal lobe with seta l_2' present (Fig. 12J) 6
– Anal lobe with l_2' absent (Fig. 12I) .... *krausbaueri*
6. Abd.4 with macrochaeta m_5 present (as Fig. 12D) .... *macrochaeta*
– Abd.4 with m_5 absent .... *arbeai*
7. Th.3 without seta a_2 (as Fig. 12A). Abd.4 with distance p_1–p_1 shorter than p_2–p_2 (as Fig. 12D) 8
– Th.3 with a_2 present. Abd.4 with p_1–p_1 longer than p_2–p_2 .... *critica*
8. Th.2 with lateral ps.oc. present (as Fig. 12A) .... *hylophila*
– Th.2 with lateral ps.oc. absent (Fig. 12K) .... *jirii*

***Mesaphorura tenuisensillata* Rusek, 1974 Figs 12A, 12G, 12P**

Th.3 without seta a_2. Abd.4 with distance p_1–p_1 shorter than p_2–p_2. Pseudocelli on thorax in position m_5–p_5 (th.2) and m_5–p_5/p_3–p_4 (Fig. 12A). Abd.6 with l_2' present. Th.3 with a_2 absent. Abd.5 with 3+3 short a-setae in central field.

The key characters are unique for the species and makes it easy to identify. Another unique character is tibiotarsal chaetotaxy where seta B_1 is absent, B_2 and B_7 present (Fig. 12P).

BIOLOGY. Common in soil in rich meadows and *Alnus*/*Salix* thickets, but also in dry shrub heath (*Empetrum*, *Rhododendron*).

DISTRIBUTION. SW. Kuanersoq/Kvanefjord north to Kangerlussuaq/Søndre Strømfjord). Probably also on the east coast. New species to Greenland.

Palaearctic.

***Mesaphorura krausbaueri* Börner, 1901**
Figs 12D, 12I, 12N, 12R

Henriksen 1939: *Tullbergia krausbaueri* Börner

Th.2–3 with ps.oc. in median position (p_3–p_4). Seta a_2 present on Th.3. Macrochaeta m_5 present of abd.4 (as Fig. 12D). Abd.6 with anal seta l_2' absent (Fig. 12I).

Apart from absence of l_2' identical to *macrochaeta*.

BIOLOGY. Only 3 verified Greenland records, two of them from very dry biotopes (*Salix*-litter in warm slope at Kangerlussuaq/Søndre Strømfjord, *Carex*/*Artemisia* on top of a sand dune in Ameralik).

DISTRIBUTION. S (Narsarsuaq), SW.
Cosmopolitan.

***Mesaphorura macrochaeta* Rusek, 1976 Figs 12B, 12H, 12J, 12Q**

The presence of the anal seta l_2' (Fig. 12J) separates it from *krausbaueri*.

BIOLOGY. Common in dry meadows, especially sandy seashores.

DISTRIBUTION. SW. Many records Kuanersoq/Kvanefjord north to Kangerlussuatsiaq/Evighedsfjord. New species to Greenland.

Cosmopolitan.

***Mesaphorura arbeai* Simon *et al.*, 1994**

Identical to *macrochaeta* apart from absence of macrochaeta m_5 on abd.4 (as Fig. 12D).

BIOLOGY. Found only once in soil on a stony river bank (with *Artemisia*, *Papaver*, *Festuca*, moss).

DISTRIBUTION. SW (Allumersat/Bjørnesund). New species to Greenland.

Palaearctic, but little known. Recently found in Iceland and Norway.

***Mesaphorura petterdassi* (Fjellberg, 1988) Fig. 12E**

Seta a_2 absent on th.3. Other body marks are as in *macrochaeta*, but body hairs are generally shorter, body granulation is coarser and uniform and on abd.4 seta p_1 is moved forwards, almost on line with p_2–p_3 (Fig. 12E).

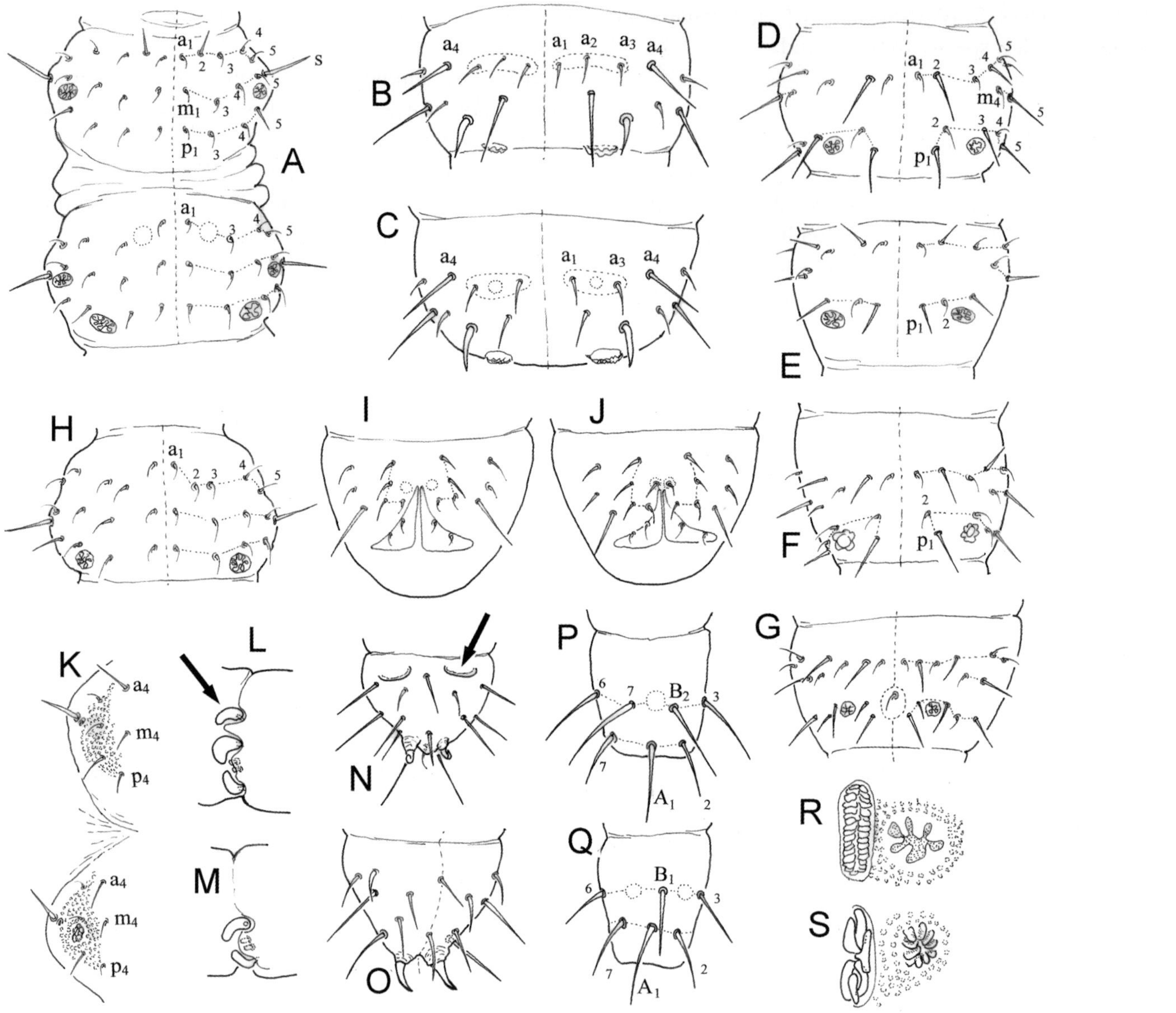

FIGURE 12 *Morphology of Tullbergiinae. A.* Mesaphorura tenuisensillata, *th.2–3; B–C. Abd.5 in* M. macrochaeta (*B*) *and* M. jirii (*C*)*; D–G. Abd.4 in* M. krausbaueri (*D*), M. petterdassi (*E*), M. italica (*F*), M. tenuisensillata (*G*)*; H.* M. macrochaeta, *th.3; I–J. Anal fields of* M. krausbaueri (*I*) *and* M. macrochaeta (*J*)*; K.* M. jirii, *left side of th.2–3; L–M. Ant.3 organ in* Chaetaphorura bella (*L*) *and* M. italica (*M*)*; N–O. Abd.6 in* M. krausbaueri (*N*) *and* Chaetaphorura bella (*O*)*; P–Q. Outer setae on right tib.2 in* M. tenuisensillata (*P*) *and* M. macrochaeta (*Q*)*; R–S. PAO in* M. krausbaueri (*R*) *and* Karlstejnia norvegica (*S*).

BIOLOGY. A characteristic species living on sandy seashores (sand dunes, wrack beds, salt meadows).

DISTRIBUTION. SW. Some records from Kuanersoq/Kvanefjord north to Kangerlussuatsiaq/Evighedsfjord. New species to Greenland.

Palaearctic (Norway, Iceland).

***Mesaphorura italica* (Rusek, 1971) Fig. 12F**

Differs from other species with 3+3 a-setae on abd.5, by the lateral position of pseudocelli on th.2–3, between setae m_5 and p_5 (see Fig. 12A), and the long distance between p_1–p_1 setae on abd.4 (Fig. 12F). On abd.6 the anal seta l_2' is absent (as in *krausbaueri*, cf. Fig. 12I).

BIOLOGY. Found in a few samples form lush slopes (grass and tall herbs, *Archangelica*, *Hieracium*).

DISTRIBUTION. S, SW. Scattered occurrences from Narsarsuaq north to Tasersuaq, Ikamiut Kangerlluarsuat. New species to Greenland.

Palaearctic.

***Mesaphorura critica* Ellis, 1976**
The combination of 2+2 short a-setae on abd.5 (as Fig. 12C) and presence of seta a_2 on th.3 (as Fig.12H) easily identify this species.

BIOLOGY. Found in dry, sandy localities.

DISTRIBUTION. SW. Only one record from inner part of Itilleq (*c.* 66°35'N). New species to Greenland.

Palaearctic.

***Mesaphorura hylophila* Rusek, 1982**
Differs from the previous species by absence of seta a_2 on th.3 and short distance between p_1–p_1 on abd.4 (as Fig. 85).

BIOLOGY. Found in dry, sandy localities.

DISTRIBUTION. SW. Only found once on plant roots (*Elymus*, etc.) in a coastal sand dune at inner part of Itilleq (*c.* 66°35'N). New species to Greenland.

Palaearctic.

***Mesaphorura jirii* Rusek, 1982 Figs 12C, 12K**
Similar to *hylophila* apart from absence of pseudocelli on mesothorax (Fig. 12K), which is an unique condition among the Greenland members of the genus.

BIOLOGY. Lives in various soil types, often in soil/litter under *Alnus, Salix, Betula.*

DISTRIBUTION. S, SW. Several records from Narsarsuaq north to Kangerlussuaq/Søndre Strømfjord. New species to Greenland.

Palaearctic.

***Karlstejnia* Rusek, 1974**
The single Greenland species of this genus is very similar to a *Mesaphorura*, but is easily recognised by the peculiar PAO bearing 6 elongate lobes in a characteristic arrangement (Fig. 12S).

***Karlstejnia norvegica* Fjellberg, 1974 Fig. 12S**
White, 0.5 mm. Distribution of pseudocelli on head-abd.5 as 11/022/11111. Seta a_2 present on th.3. Abd.4 with short distance between setae p_1–p_1, unpaired axial seta present (as Fig. 12G). Abd.5 with 3+3 short a-setae between the a_4. Anal seta l_2' present.

BIOLOGY. Found in dry heathland with sand or gravel (*Empetrum, Louisleurea, Silene, Diapensia*).

DISTRIBUTION. SW. Kuanersoq/Kvanefjord north to Kangerlussuatsiaq/Evighedsfjord.

Palaearctic. New species to Greenland.

***Chaetaphorura* Rusek, 1976**
The single Greenland species looks like a large *Mesaphorura*, but is rcognised by absence of crescentic ridges on abd.6 and longer anal spines (Fig. 12O). In ant.3 organ there is one additional sensory club (Fig. 12L) and mesothorax has 2+2 pseudocelli (1+1 or 0 in *Mesaphorura*). PAO of the *Mesaphorura*-type (as Fig. 12R), but larger, with more lobes.

***Chaetaphorura bella* (Fjellberg, 1984) Figs 12L, 12O**
Hammer 1952: *Tullbergia collis* Bacon (?)

White, 0.8 mm. Distribution of pseudocelli on head–abd.5 as 11/122/11121.

The absence of crescentic ridges on abd.6 easily identifies the species among Greenland Tullbergiinae (Fig. 12O). A similar species, *C. arctica* (Wahlgren, 1900), may show up in Greenland. It differs by having 2+2 pseudocelli on the tergites of abd.1–3 and 5 setae in the basolateral field of labium (4 in *bella*).

BIOLOGY. On sandy seashores (*Elymus*, sand dunes).

DISTRIBUTION. SW (only from Søndre Isortoq and Ikamiut Kangerluarsuat). New species to Greenland.

Holarctic.

ISOTOMIDAE

Key to genera

1. Tip of abdomen with spines 2
– Tip of abdomen with simple setae only 3
2. Abd.5–6 with 4 large spines (Fig. 13A) .... *Tetracanthella*
– Abd.5–6 with 8 small spines (Fig. 16E) .... *Weberacantha*
3. At most 2 last abdominal segments fused 4
– 3 last abdominal segments fused (Fig. 14N) .... *Folsomia*
4. Furca absent 5
– Furca present 6
5. Ocelli at least 6+6. Dark species. Clavate tibiotarsal hairs present. Larger species (>1.0 mm) .... *Anurophorus*
– Ocelli at most 1+1. Pigment absent or present as indistinct granules. Clavate hairs absent on tiobiotarsi. Smaller species (0.5 mm) .... *Pseudanurophorus*
6. Ocelli and PAO present. Mostly dark species 7
– Ocelli and PAO absent. White species .... *Isotomiella*
7. Manubrium with anterior setae 8
– Manubrium without anterior setae .... *Archisotoma*
8. Manubrium with more than 1+1 anterior setae 9
– Manubrium with 1+1 anterior setae .... *Cryptopygus*
9. Tib.2 with 2 expanded, feather-like outer setae (Fig. 18T). Ventral tube with 2+2 laterodistal

setae. Strictly marine littoral species .... *Halisotoma*

– Tib.2 only with normal setae. Ventral tube always with more than 2+2 laterodistal setae. No species restricted to the marine littoral zone 10

10. Subapical seta on dens not passing mucro, claws without tunica 11

– Subapical seta on dens long, passing tip of mucro (Fig. 17D). Claws with a basal membraneous tunica (Fig. 17C) .... *Agrenia*

11. Tibiotarsi with clavate apical hairs (Figs 18U, X) 12

– Tibiotarsal apical hairs pointed 13

12. Abd.5–6 fused. Tibiotarsi without T_2–T_3 setae (Fig. 18U) .... *Pseudisotoma*

– Abd.5–6 separate. Tibiotarsi with all T setae present (Fig. 18X) .... *Vertagopus*

13. Ocelli 8+8 14

– Ocelli at most 4+4 (Fig. 17G) .... *Parisotoma*

14. Manubrium with simple ventroapical setae (Fig. 17P). Mucro mostly with 4 teeth (3 in one species) .... *Desoria*

– Manubrium with thick, spine-like ventroapical setae (Fig. 18R). Mucro with 3 teeth .... *Isotoma*

Tetracanthella Schött, 1891

The characteristic 4–spined abdomen readily distinguish the single Greenland species belonging to this genus (Fig. 13A).

Tetracanthella sibirica Deharveng, 1987 Figs 13A–13C

Henriksen 1939: *Tetracanthella wahlgreni* Linnan.

Violet black, 1.5 mm. Ocelli 8+8. Furca as Fig. 13C, dens with 2 posterior and 1 anterior seta. Mucro absent. Retinaculum with 2+2 teeth and 1 seta. Abd.1–3 with 2–3–3 long macrochaetae on each side. Abd.4 with p_1 short, usually as long as or shorter than distance between their bases. Head with 3–5 pp–setae on each side between midline an p_3 seta (Fig. 13B). Abd.4 with a variable smooth field (devoid of integumental pattern) in posterior part between setae a_1–p_1.

Greenland specimens belong to the taxon *sibirica* as defined by Potapov (2001), which is not sharply set off from *arctica* Cassagnau, 1959.

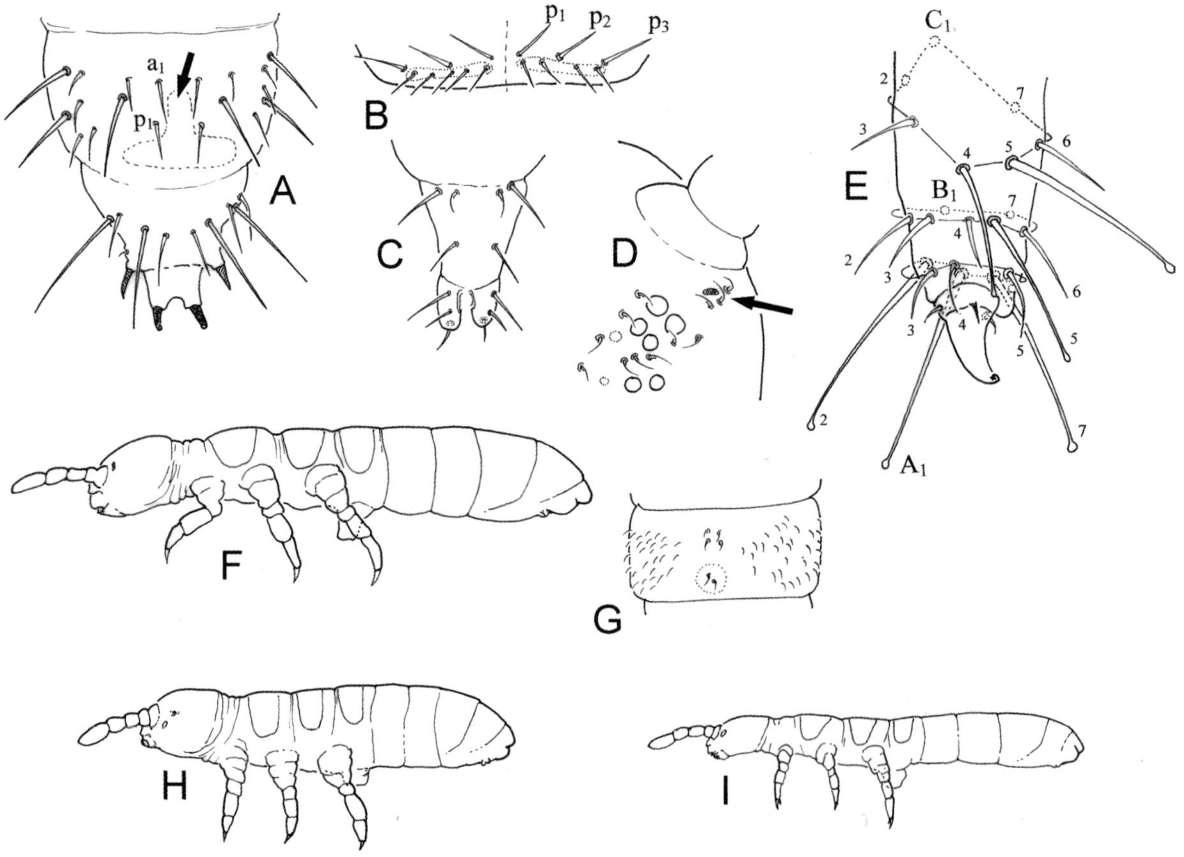

FIGURE 13 *Morphology of Isotomidae. A–C.* Tetracanthella sibirica. *Abd.4–6, smooth field indicated (A), posterior edge of head (B), furca (C); D–G.* Anurophorus fulvus. *Ocellar field and PAO, marked (D), inner side of right tib.2 (E), habitus (F), ventral side of abd.3 with retinacular field encircled (G); H.* Pseudanurophorus binoculatus, *habitus; I.* P. alticolus, *habitus.*

BIOLOGY. Common in a variety of localities, including lush meadows, dry heaths, moss, salt littoral meadows, bogs.

DISTRIBUTION. S, SW, NE (Narsarsuaq north to Qeqertarsuaq/Disko; Illoqqortoormiut/Scoresbysund, Ammassalik, Kangerlussuatsiaq/Lindenow Fjord).

Holarctic.

***Anurophorus* Nicolet, 1842**

The cylindrous body, absence of furca and distinct pigmentation identify the single Greenland species of this genus. General morpholgy as Fig. 13F.

***Anurophorus fulvus* Fjellberg, 1988 Figs 13D–13G**

Yellowish brown to violet, 1.5 mm. Head with 8+8 ocelli, 2 smaller. PAO small, narrow (Fig. 13D). Maxillary outer lobe with 2 sublobal hairs. Tibiotarsi with 2–3–3 long, clavate tenent hairs on outer side, 3–3–1 on inner side (Fig. 13E). Longest macrochaetae on tip of abdomen blunt-tipped. Retinacular field with less than 4 setae (Fig. 13G).

A similar species, *A. laricis* Nicolet, 1842, may occur in Greenland. It differs in having only one subobal hair on the maxillary outer lobe and more than 6 setae in the retinacular field.

BIOLOGY. The typical habitat is in lichens on top of large boulders which have been manured by resting birds. Also in moss/grass in bird cliffs. Only 3 Greenland records.

DISTRIBUTION. SW (Kangerlussuatsiaq/Evighedsfjord, Kangerluarssuk Tulleq). New species to Greenland.

Palaearctic.

***Pseudanurophorus* Stach, 1922**

Similar to previous genus by complete absence of furca, differing by smaller size, reduced eye-number (0 or 1+1) and absent or weak pigmentation.

Key to species

1. Ocelli 1+1, weak pigmentation present .... *binoculatus*
– Ocelli absent, no pigment .... *alticolus*

***Pseudanurophorus binoculatus* Kseneman, 1934 Fig. 13H**

White, with scattered grains of pigmentation, eye-spot dark. 0.5 mm. Body shape rather short and plump (Fig. 13H).

BIOLOGY. Both found in dry heathland (*Empetrum, Dryas, Juniperus*) and in lush forests/meadows (*Betula, Salix*, ferns).

DISTRIBUTION. SW. Many records from Kuanersoq/Kvanefjord north to Kangerlussuaq/Søndre Strømfjord. New species to Greenland.

Holarctic.

***Pseudanurophorus alticola* Bagnall, 1949 Fig. 13I**

White, 0.5 mm. No eyes or pigment. Body very elongate and slender (Fig. 13I).

No other Greenland isotomid is blind, without furca and pigment.

BIOLOGY. Only one record from a rocky area.

DISTRIBUTION. NE. One specimen found in a sample from Ymer Ø (Søgaard Andersen, 18.IX.1933). New species to Greenland.

Palaearctic.

***Folsomia* Willem, 1902**

The fusion of the 3 last abdominal segments separates this genus from the rest of the family. Most species have a cylindrous body shape with a rather short furca (Fig. 14N).

Key to species

1. Manubrium with 1+1 anterior setae (Fig. 14B) 2
– Manubrium with more than 1+1 anterior setae 3
2. Ocelli 2+2 .... *quadrioculata*
– Ocelli 1+1 .... *palaearctica*
3. Ocelli absent 7
– Ocelli present 4
4. Ocelli 1+1 5
– Ocelli 2+2 .... sp. near *penicula*
– Ocelli 3+3.. 6
– Ocelli 8+8 .... *variabilis*
5. Manubrium with 2+2 anterior setae .... *agrelli*
– Manubrium with more than 2+2 anterior setae .... *regularis*
6. Dorsomedian macrochaeta on abd.5 more than 3 times as long as mucro. Pale species .... *thalassophila*
– Dorsomedian macrochaeta less than 2.5 as long as mucro, dark species .... *sexoculata*
7. Abd.5 on each side with a pair of broad sensilla. Ventral setae present on th.3 (Fig. 15D) 10
– Abd.5 only with hair-like sensilla. Th.3 without ventral setae 8
8. Manubrium with 2+2 anterior setae 9
– Manubrium with more than 2+2 anterior setae (Fig. 14F) .... *nivalis*
9. Microsensilla present on abd.1 (Fig. 15F, see also Fig. 15A). Manubrium without seta ml_1 (Fig. 14I) .... *bisetosa*
– Microsensilla absent on abd.1. Manubrial seta ml_1 present (Fig. 14J) .... *bisetosella*
10. Dens with 2 outer setae in mid-section (Fig. 14H). Manubrium with more than 3+3 anterior setae. Body pigment usually present .... *coeruleogrisea*

– Dens with 1 outer seta in mid-section (Fig. 14G). Manubrium usually with 2+2 anterior setae. Never pigmented .... *sensibilis*

Folsomia quadrioculata **(Tullberg, 1871) Figs 14B, 14K, 14N, 14O**

Spotted greyish, 1.5 mm, sometimes larger. 2+2 ocelli. Rarely white, without pigment. Body shape characteristic (Fig. 14N). Macrochaetae on abdomen long (Fig. 14K). Dens with 3 posterior setae in basal half, anterior side with 7–8 setae (Fig. 14B). Sensillum at hind-corner of th.2 set in front of the p-row (Fig. 14O).

Presence of 2+2 ocelli in combination with 1+1 anterior manubrial setae identify this species. A related species, *F. manolachei* Bagnall, 1939, may show up in southern Greenland. It differs from *quadrioculata* by having shorter body hairs and hind corner sensillum on th.2 set within the p-row.

BIOLOGY. Common in many different biotopes, most abundant in humid or moist conditions (meadows, forest, shrubs, snowbeds, beachfront, bird cliffs).

DISTRIBUTION. G. Common on both the west and the east coast, north to Pituffik/Thule and Peary Land.

Holarctic.

Folsomia palaearctica **Potapov & Babenko, 2000**

Henriksen 1939: *Folsomia diplophthalma* Linnan.

Hammer 1952: *Folsomia diplophthalma* (Axels.)

Pale greyish, eye-spots darker, 0.7 mm. 1+1 ocelli. A small, slender species with short macrochaetae.

Presence of 1+1 ocelli and 1+1 manubrial anterior setae identify the species.

BIOLOGY. In dry, shrubby biotopes (*Dryas*, *Rhododendron*). Numerous in *Salix*-litter in extremely dry slopes in Kangerlussuaq/Søndre Strømfjord. Hammer (1938) also reports it from wet moss.

DISTRIBUTION. SW, SE, NE, N. On the west coast from Søndre Isortoq north to Kangerlussuaq/Søndre Strømfjord, on the east coast from Mikis Fjord (Blosseville Kyst) north to Peary Land.

Holarctic.

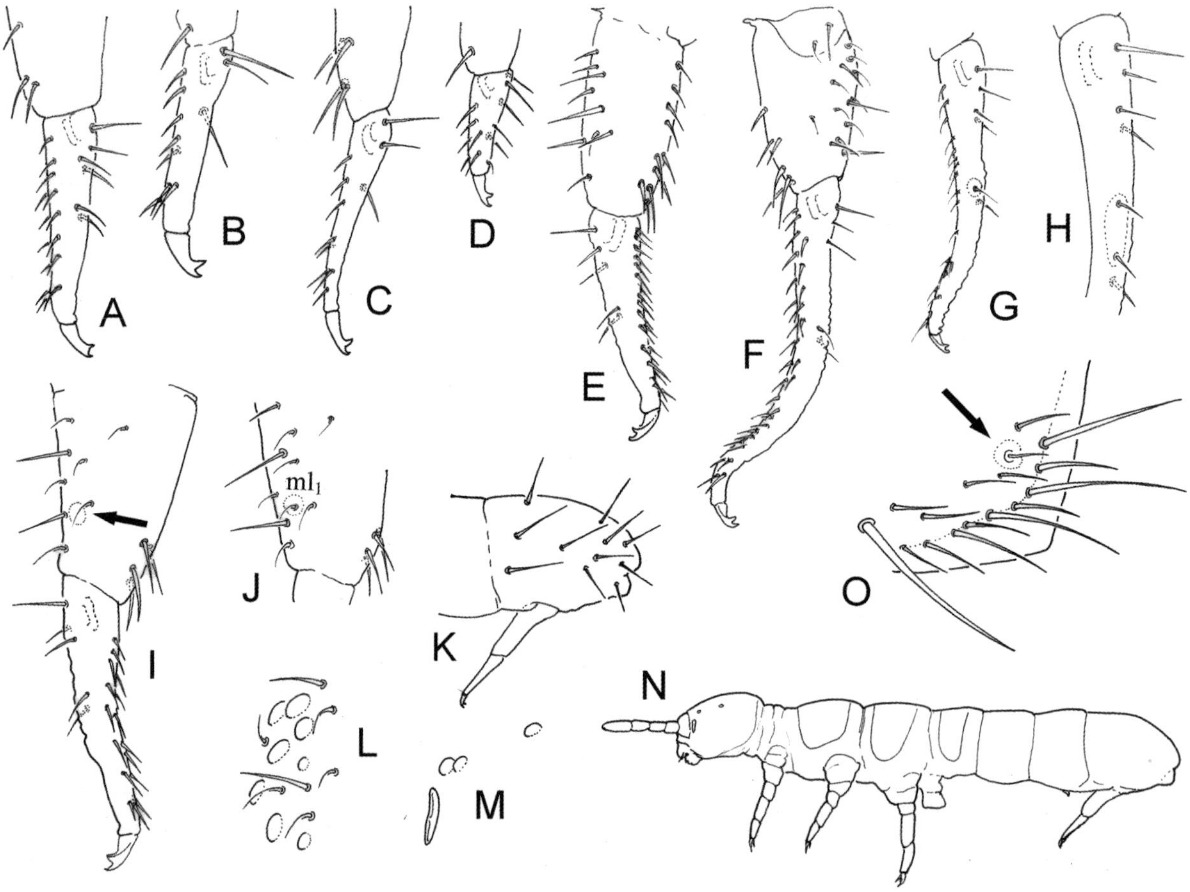

FIGURE 14 *Morphology of* Folsomia *species. Furca of* F. sexoculata (*A*), F. quadrioculata (*B*). F. agrelli (*C*), F. variabilis (*D*), *F. sp. near* penicula (*E*), F. nivalis (*F*), F. sensibilis, *dens and mucro* (*G*), F. coeruleogrisea, *dens* (*H*), F. bisetosa (*I*), F. bisetosella, *manubrium* (*J*); *K.* F. quadrioculata, *macrochaetae on abd.4–6; L.* F. variabilis, *left eye-field; M.* F. sexoculata, *PAO and ocelli; N–O.* F. quadrioculata. *Habitus* (*N*), *hind corner of mesothorax with sensillum marked* (*O*).

***Folsomia sexoculata* (Tullberg, 1871) Figs 14A, 14M**
Greyish black, 2.0 mm. 3+3 ocelli (Fig. 14M). Body with a dense cover of short setae. Macrochaetae only distinct on last abdominal segments. Manubrium with 2+2 anteroapical setae, with 1–3 additionals higher up. Dens with 4+2 posterior setae, anterior side with more than 10 setae (Fig. 14A).

The only other Greenland *Folsomia* with 3+3 ocelli is *thalassophila*, which differs by paler colour and longer body hairs.

BIOLOGY. Only found in the marine littoral zone, occassionally at the shore of inland lakes (Lake Aussivit, Itivdleq).

DISTRIBUTION. SW, NE, SE (Kuanersoq/Kvanefjord north to Sisimiut/Holsteinsborg; Ammassalik north to Ella Ø).

Holarctic.

***Folsomia thalassophila* Bagnall, 1940**
(*Folsomia norvegica* Altner, 1963)

With spotted dark pigmentation, sometimes almost white, 1.5 mm.

BIOLOGY. A marine littoral species.

DISTRIBUTION. SW (Allumersat/Bjørnesund north to Kangerlussuatsiaq/ Evighedsfjord). New species to Greenland.

Palaearctic.

***Folsomia agrelli* Gisin, 1941 Fig. 14C**
White with dark eyespots, 1.2 mm. 1+1 ocelli. Body with long macrochaetae like *quadrioculata*. Manubrium with 2+2 anterior setae, dens with 3 posterior setae in basal part and 7–8 anterior setae (Fig. 14C).

The long body hairs, 2+2 manubrial setae and 1+1 ocelli identify this species.

BIOLOGY. Found only once at the shore of a lake. In Scandinavia a characteristic high alpine species.

DISTRIBUTION. SW (Tasersuaq, Ikamiut Kangerluarsuat). New species to Greenland.

Palaearctic.

***Folsomia regularis* Hammer, 1953 Fig. 15G**
White, 1.6 mm. Th.3 with an erect dorsomedian macrochaeta (Fig. 15G).

Very similar to previous species, but body hairs generally shorter and manubrium with 3–5 anterior setae on each side of the midline.

BIOLOGY. Lives in damp high arctic sites.

DISTRIBUTION. N. A few specimens present in the material from Peary Land collected in 1949. Originally described from Ellesmere Island.

Holarctic (in Eurasia only on Wrangel Island).

***Folsomia variabilis* Fjellberg, 1984 Figs 14D, 14L**
Greyish black of variable intensity, 1.0 mm. Body shape rather short and stout with dens shorter than manubrium. Ocelli 8+8, 2 smaller (Fig. 14L). PAO narrow elongate, shorter than width of ant.1. Body hairs short, longest macrochaetae on tip of abdomen less than twice as long as inner edge of last claw. Upper sensilla on th.3–abd.3 set in front of the p-row. Dens with 6 posterior and 6–7 anterior setae (Fig. 14D). Retinaculum with 4+4 teeth and 1–3 setae.

The only Greenland species of *Folsomia* with 8+8 ocelli. The species was described from alpine communities in Colorado and Washington. Greenland specimens have been compared with material from Mt. Rainier (Washington), and no differences were observed.

BIOLOGY. Only 3 records (grass/moss at edge of the duck pond in the city of Nuuk/Godthåb, moss on an old reindeer skull 400 m asl. in Kangerluarsusuaq/Grædefjord, moss/grass at edge of a muddy glacier stream at Tasersuaq, Ikamiut Kangerluarsuat).

DISTRIBUTION. SW (details above). New species to Greenland.

Nearctic.

***Folsomia bisetosa* Gisin, 1953 Figs 14I, 15E, 15F**
White, 0.8 mm. Ocelli absent. A slender, elongate species with short macrochaetae. Upper sensilla of abd.2–3 set within the p-row (Fig. 15E). Furca as Fig. 14I. Dens with 5 posterior and about 15–20 anterior setae. PAO elongate, about as long as width of ant.1.

Very similar to *bisetosella*, but in addition to key characters, *bisetosa* has 4 prelabral setae while *bisetosella* usually has 3.

BIOLOGY. Common in a variety of biotopes (meadows, shrublands, heaths). Often in dry situations.

DISTRIBUTION. SW, N, NE (Kuanersoq/Kvanefjord north to Pituffik/Thule; Peary Land; Ella Ø).

Holarctic (almost circumpolar).

***Folsomia bisetosella* Fjellberg, 2005 Fig. 14J**
White, 0.6 mm.

Very similar to the previous species, but generally smaller. In particular the absence of microsensilla on abd.1 is significant.

BIOLOGY. Common in different vegetation types, with a preference for dry, sandy habitats along seashores and streambanks.

DISTRIBUTION. S, SW, NE (Narsarsuaq north to Kangerlussuaq/Søndre Strømfjord; Zackenberg; Ella Ø).

Palaearctic. In Scandinavia it has a more southern distribution than *bisetosa*.

***Folsomia nivalis* (Packard, 1873) Fig. 14F**
White, 1.0 mm. A slender species with slightly pear-shaped abdomen having long macrochaetae which are finely

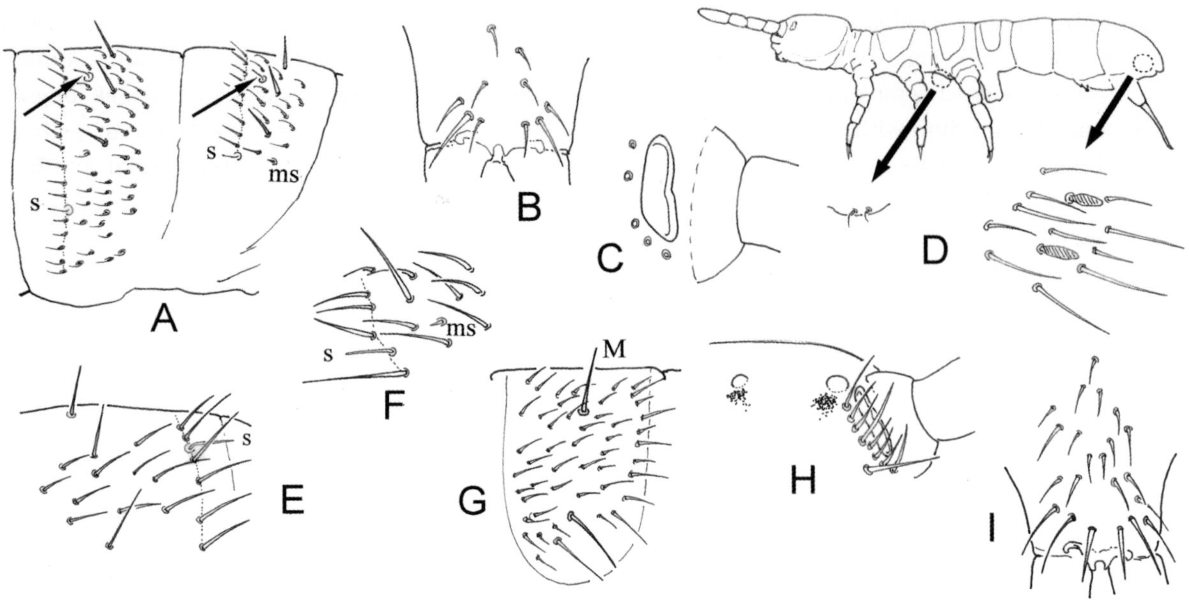

FIGURE 15 *Morphology of* Folsomia *species. A–B.* F.coeruleogrisea. *Right side of abd.1 with position of upper sensilla marked (A), anterior side of manubrium (B); C–D.* F. sensibilis. *PAO and antennal base (C), habitus with details of ventral setae on th.3 and broad sensilla on abd.5 (D); E–F.* F. bisetosa. *Left side of abd.3 with sensillum (s) in the p-row (E), right side of abd.1 (F); G.* F. regularis, *left side of th.2, M: Dorsomedian macrochaeta. Specimen from Pary Land; H–I.* F. *sp. near* penicula. *Ocelli and PAO (H), anterior side of manubrium (I).*

serrated towards tip. Longest abdominal macrochaetae 2.5–3.0 as long as inner edge of last claw. Upper sensilla on abd.1–3 set in the p-row. Anterolateral microsensillum present on abd.1 (as Fig. 15F). Furca strong, dens longer than manubrium. Anterior setae on manubrium usually arranged as 3+2+1 on each side of the midline (Fig. 14F).

Absence of pigment and ocelli in combination with increased numer of anterior setae on manubrium will identify this species.

BIOLOGY. Found in organic litter in various places.

DISTRIBUTION. S (Narsarsuaq, Kiattuut). Older published records as *Folsomia fimetaria* (Linnaeus, 1758) from Mikis Fjord (Blosseville Kyst), Illoqqortoormiut/ Scoresbysund, Peary Land and Upernavik (Lack 1934, Hammer 1938, 1944, 1954) are unreliable and may include *nivalis* as well as *bisetosa/bisetosella*.

New species to Greenland. Nearctic.

***Folsomia sensibilis* Kseneman, 1936 Figs 14G, 15C, 15D**

White, 0.6 mm. Ocelli absent. PAO characteristic, broad, slightly sickle-shaped (Fig. 15C). Body slender, cylindrous, macrochaetae moderately long (about twice as long as inner length of last claw). Upper sensilla on abd.1–3 set well in front of the p-row. Furca long, dens longer than manubrium. Posterior side of dens with 4 setae in basal part, 2 in middle (Fig. 14G).

Easily identified by presence of broad sensilla on abd.5 and 1+1 midventral setae on th.3. For separation from *coeruleogrisea*, see below.

BIOLOGY. Often in lush biotopes like warm slopes with ferns and litter of *Alnus*, *Betula* and *Salix*, but also in dry slopes and shrub heaths.

DISTRIBUTION. S, SW, NE (Narsarsuaq north to Kangerlussuatsiaq/Evighedsfjord; Ella Ø). New species to Greenland.

Palaearctic.

***Folsomia coeruleogrisea* (Hammer, 1938) Figs 14H, 15A, 15B**

Hammer 1952: *Isotomina coeruleogrisea* (Hammer)

White, usually with diffuse bluish grey pigmentation, 1.2 mm. Upper sensillum on abd.1–3 set well in front of the p-row (Fig. 15A).

Similar to *sensibilis* by presence of midventral setae on th.3 and 2+2 broad abd.5 sensilla, differing by larger size, greyish colour, more anterior setae on manubrium and 2 outer setae in middle part of dens (Figs 14G, 14H).

BIOLOGY. Common in a variety of biotopes, often lush meadows and snowbeds, but also in drier sites (shrub heaths, sandy beaches).

DISTRIBUTION. S, SE, NE (Narsarsuaq north to Itivleq; Kangerlussuatsiaq/ Lindenow Fjord north to Ella Ø).

Holarctic (circumpolar).

***Folsomia* sp. near *penicula* Bagnall, 1939 Figs 14E, 15H, 15I**

Spotted greyish, 1.6 mm. A longhaired species of same shape as *quadrioculata*. Ocelli 2+2. PAO narrow elongate, almost as long as width on ant.1 (Fig. 15H). Manubrium with more than 20 anterior setae in large specimens (Fig. 15I). Dens with 3+2 posterior setae (Fig. 14E).

The species comes close to *penicula* and *taigicola* Stebaeva & Potapov, 1997. From the former it differs by having 3 setae in the basal group on dens (not 2), from the latter it differs by having 2+2 ocelli (not 1+1). The Greenland sample may represent an undescribed species.

BIOLOGY. Several specimens were found in a sandy moss/lichen tundra at Nuk in Søndre Isortoq.

DISTRIBUTION. SW (only found in Søndre Isortoq).

***Cryptopygus* Willem, 1901**

The single Greenland species of this genus is easily recognised by a combination of fused abd.5–6, 1+1 manubrial setae and 2-toothed mucro, large body size and dark colour.

***Cryptopygus roberti* (Fjellberg, 1991) Figs 16A–16D**

Henriksen 1939: *Proisotoma tenella* Reuter

Dark bluish grey, 1.4 mm. Body shape as Fig. 15A. Two last abdominal segments fused. Body with short, uniform cover of hairs, macrochaetae hardly developed. Ocelli 8+8. PAO elongate (Fig. 16D). Maxillary palp bifurcate, 3 sublobal hairs. Furca as Fig. 16B, manubrium with 1+1 anterior setae. Dens with 7 posterior setae, anterior side with 12–15 setae. Mucro with 2 teeth. Claws elongate, tibiotarsi with a blunt-tipped or clavate spur hair (Fig. 16C).

BIOLOGY. In wet sites (moss in snowbeds, shores of lakes, wet meadows). Originally described from samples taken at the bottom of a stream at Qaqortoq/Julianehåb (Fjellberg 1991). Like the similar species *C. clavatus* (Schött, 1893), it probably has a wetable cuticle which allows it to move under water.

DISTRIBUTION. S, SW, SE (Qaqortoq/Julianehåb north to Maniitsoq/Sukkertoppen; Ammassalik). So far no records outside Greenland.

***Weberacantha* Christiansen, 1951**

A unique generic character is the presence of 8 short spines on abd.5–6 (Fig. 16E). Body cylindrous with a full set of eyes and a strong furca. Dark coloured species with distincly reticulate integument.

***Weberacantha octa* Christiansen, 1951 Figs 16E, 16F**

Bluish black, 1.3 mm. Ocelli 8+8. PAO narrow, elongate. Body with a cover of short, spiny hairs. Abd.1–4 with 2–2–2–3 erect macrochaetae on each side. Abd.5–6 as Fig. 16E. Furca as Fig. 16F, dens shorter than manubrium, with 4 posterior and 1 anteroapical seta. Manubrium with 1+1 anteroapical setae and many posterior. Mucro strong, with 3 teeth. Apical setae of tibiotarsi pointed, not prolonged.

The species is unmistakable by the 8 abdominal spines.

BIOLOGY. Found in dry sites with sparse vegetation.

DISTRIBUTION. NE (Illoqqortoormiut/Scoresbysund area, reported by Hammer (1953) and recollected by a German expedition in 1999 at Gåseelv, 4 km N of Constable Pynt, Scoresbysund).

Holarctic (in Eurasia only in Chukotka).

***Archisotoma* Linnaniemi, 1912**

The Greenland members of this genus are distingushed from other isotomids by an unspined abominal tip, dark body colour and a strong furca without anteroapical setae on manubrium. The mouthparts (labium and maxilla) are strongly modified. Abd.5–6 laterally fused, upper part with 3+3 erect sensorial hairs. All species of *Archisotoma* live in the marine littoral zone. Three species present in Greenland.

Key to species

1. Tib.3 with a dorsoapical spine-like seta (Fig. 16G). Frontoclypeal field with at most 2–3 setae 2
– Tib.3 with all apical setae slender, not spine-like (Fig. 16L). Frontoclypeal field with many setae (Fig. 16M) .... *megalops*
2. Small species (0.5 mm), dens with few setae (4 anterior, 6 posterior, Fig. 16H) .... *theae*
– Larger species (>0.6 mm), dens with numerous setae (Fig. 16I) .... *besselsi*

***Archisotoma megalops* (Bagnall, 1939) Figs 16J–16M**

Greyish brown, 1.5 mm. Ocelli large, 8+8. PAO narrow, elongate (Fig.16K). Furca strong, ventral side of dens with hairs almost to base (Fig. 16J).

May be confused with *besselsi*, but the key characters will normally separate the species. *A. megalops* has a set of plesiomorphic characters which are not seen in other species of the genus (dorsal sensilla present on tergites of th.3–abd.3, hypostomal setae unmodified – not hook-like, maxilla, labral setae and sublobal hairs less modified).

BIOLOGY. A marine littoral species.

DISTRIBUTION. SW (Qeqertarsuatsiaat/Fiskenæsset). In Greenland one record only, but probably widely distributed. New species to Greenland.

Palaearctic.

***Archisotoma besselsi* (Packard, 1877) Figs 16G, 16I**

Henriksen 1939: *Archisotoma besselsi* Pack.

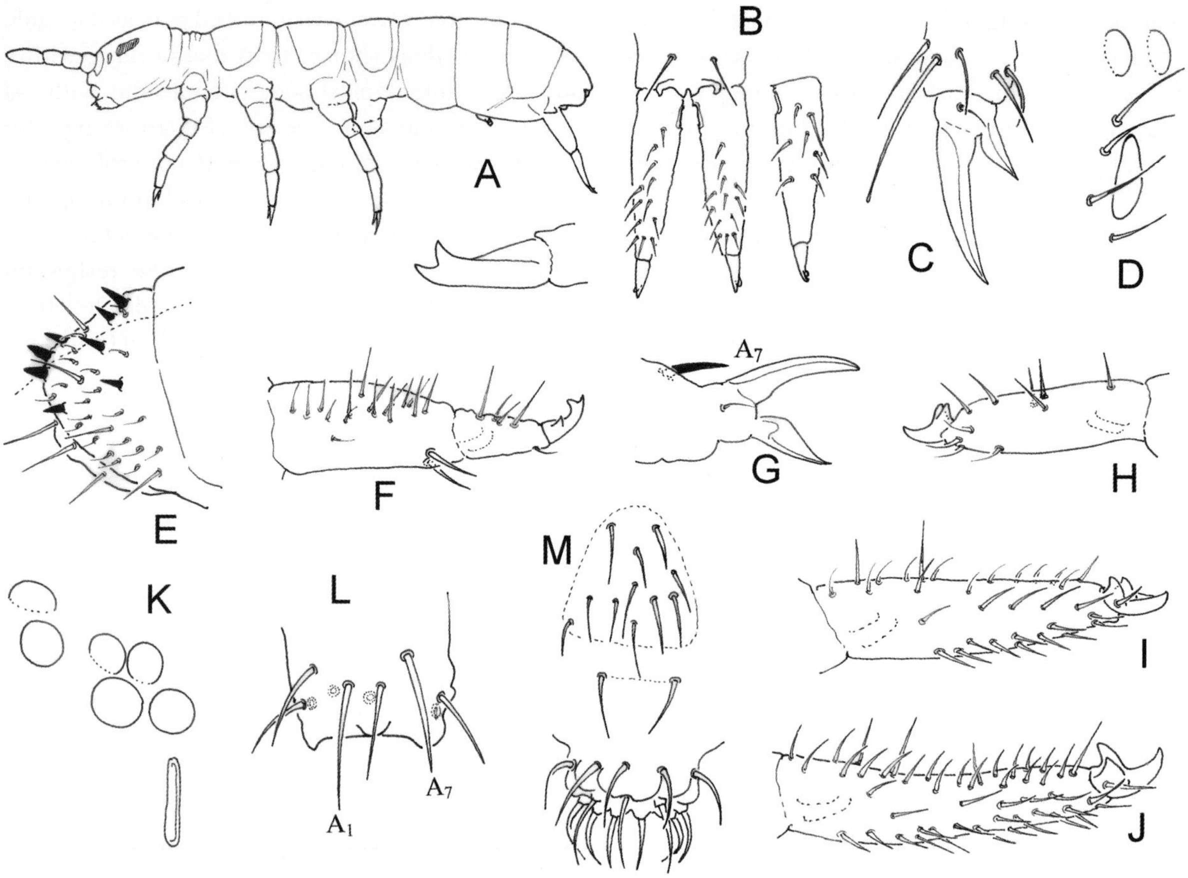

FIGURE 16 *Morphology of Isotomidae. A–D.* Proisotoma roberti. *Habitus with detail of mucro (A), anterior side of dens and manubrium, dorsal side of dens to the right (B), claw and apical part of tib.1 (C), PAO and nearest ocelli (D); E–F.* Weberacantha octa. *Spines on tip of abdomen (E), furca (F); G.* Archisotoma besselsi, *claw and enlarged spur hair on tib.3; H–J. Dens and mucro in* A. theae *(H),* A. besselsi *(I),* A. megalops *(J); K–M.* A. megalops. *PAO and ocelli (K), apical setae on left tib.3 (L), frontoclypeal field (encircled) and labrum (M).*

Bluish black, often very dark, 1.5 mm (usually smaller).

The key characters distinguish the species. It also differs from *megalops* in having anterior side of dens free of hairs in the basal 1/3, while *megalops* is haired almost to base (Figs 16I, 16J).

BIOLOGY. Marine littoral species. In littoral sand, under stones and in organic debris. Also at shores of inland lakes with slightly saline water.

DISTRIBUTION. SW, N, NE (Kuanersoq/Kvanefjord north to Itilleq: *c.* 66°35'N); Ella Ø; Zackenberg; Polaris Bugt at Polaris Forland: the original type locality of the species.

Holarctic.

Archisotoma theae **Fjellberg, 1980 Fig. 16H**

Greyish brown, eyes darker, 0.5 mm.

The small size (adults) and sparsely haired dens easily identify the species.

BIOLOGY. Found on a marine, sandy shore.

DISTRIBUTION. SW (inner part of Kangia). One record only, but probably widely distributed. New species to Greenland.

Palaearctic.

NOTE. A fourth species, *Archisotoma polaris* Fjellberg & Poinsot, 1975, probably occurs in northern Greenland as it is recorded from the northeastern coast of Ellesmere Island and Svalbard. It is a large species (1.8 mm), greyish brown, with tib.3 apical spur as in *besselsi*, differing from that species in having 4–8 setae in the frontoclypeal field and a different type of maxilla.

Isotomiella **Bagnall, 1939**

The single Greenland species is easy to identify by white colour, absence of eyes and PAO, long furca and fused abd.5–6 (Fig. 17B).

Isotomiella minor **(Schäffer, 1896) Figs 17A, 17B**
White, 1.0 mm. PAO and ocelli absent. Ant.4 with a dorsoapical group of enlarged sensilla (Fig. 17A). Labrum with 4 slender, back-curved apical papillae. Abd.5 with long, serrated macrochaetae and a large, curved sensillum (Fig. 17B). Manubrium with 5–6 anterior setae on each side. Dens with 6 posterior setae. Mucro 3–toothed.

The only white, blind Isotomidae without PAO. The many unique details of this species makes it easy to identify.

BIOLOGY. A characteristic species in lush humid biotopes like rich slope meadows, snow bed meadows, litter of *Betula, Alnus, Salix*.

DISTRIBUTION. S, SW, SE. Northwards to Kangerlussuatsiaq/Evighedsfjord and Mikis Fjord (Blosseville Kyst).

Holarctic.

Agrenia **Börner, 1906**
One large, dark species with long furca characterised by a very long subapical seta on dens (Fig. 17D), a sheet-like tunica at base of claws (Fig. 17C), and absence of sublobal hairs on the maxillary outer lobe (all these characters unique among Greenland Isotomidae).

Agrenia bidenticulata **(Tullberg, 1876). Figs 17C, 17D**
Olive brown of variable intensity, 1.9 mm. An unmistakable species by the generic characters described above.

BIOLOGY. In wet sites along snowfields, streambanks and shores of lakes.

DISTRIBUTION. N, NE (Peary Land, Ile de France).

Holarctic.

Parisotoma **Bagnall, 1940**
The reduced number of eyes (1+1 or 4+4 ocelli), long furca with many anterior setae on dens and 3–toothed mucro will identify the two Greenlandic species from other Isotomidae. Tibiotarsi with 7 apical setae in Greenland species (T_1 absent).

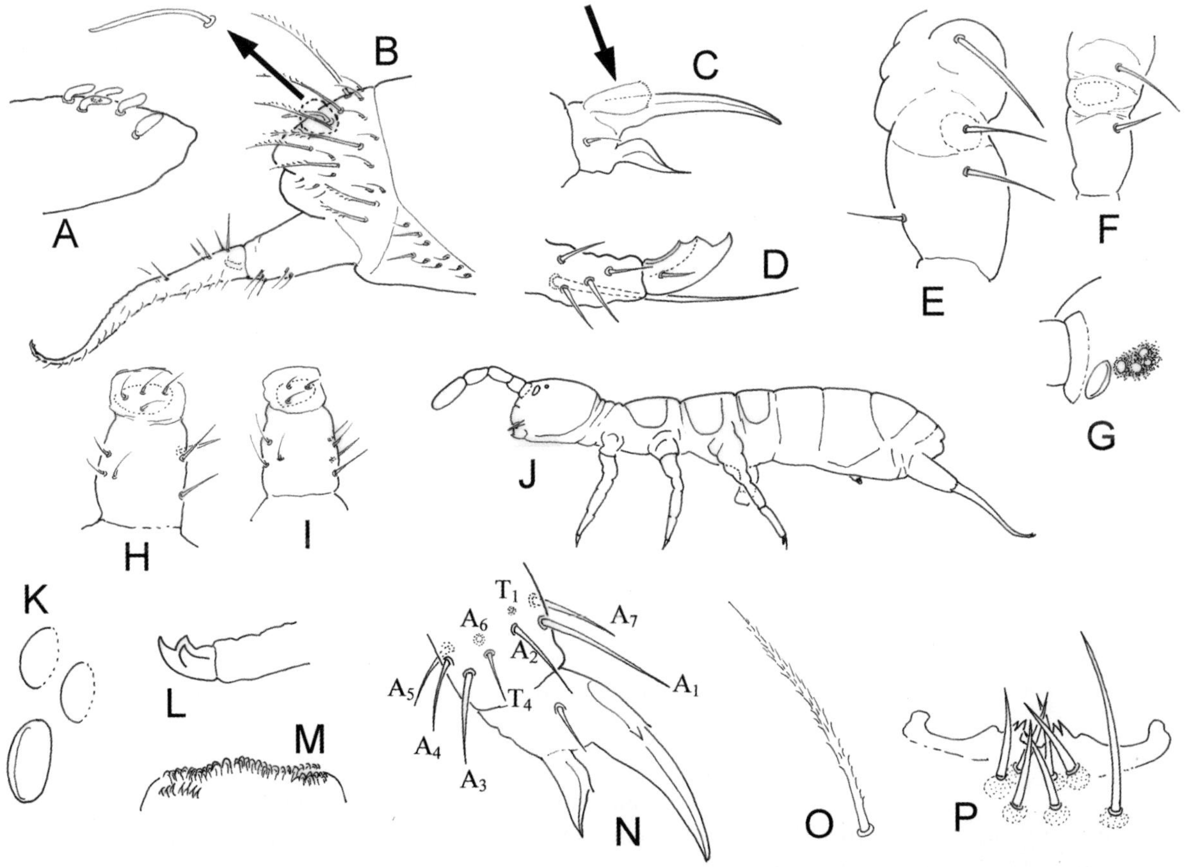

FIGURE 17 *Morphology of Isotomidae. A–B.* Isotomiella minor. *Ant. 4 sensilla (A), tip of abdomen with enlarged sensillum marked (B); C–D.* Agrenia bidenticulata. *Claw with tunica, marked (C), mucro and apical part of dens (D); E–F. Basal part of first leg in* Parisotoma agrelli *(E) and* P. notabilis *(F); G.* P. notabilis, *PAO and nearest ocelli; H–I. Ventral tube in* P. ekmani *(H) and* P. notabilis *(I); J.* P. ekmani, *habitus; K–P.* Desoria multisetis. *PAO and nearest ocelli (K), mucro (L), ventroapical ciliation of labrum (M), claw and apical setae of left tib.3 (N), ciliate abdominal macrochaeta (O), setae and ventroapical thickening of manubrium (P).*

Key to species

1. Ocelli 4+4 (Fig. 17G). Body with greyish pigment. Ventral tube with 3+3 laterodistal setae (Fig. 17I). Dens with 2 inner setae on posterior side (mid-section) .... *notabilis*
– Ocelli 1+1. Body white. Ventral tube with 4+4 laterodistal setae (Fig. 17H). Dens with 3 setae on inner side .... *ekmani*

***Parisotoma notabilis* (Schäffer, 1896) Figs 17F, 17G, 17I**

Greyish brown of variable intensity, 1.0 mm. Ocelli 4+4, sometimes difficult to see. PAO large (Fig. 15G). Ventral tube with 3+3 laterodistal setae (Fig. 17I). Lower subcoxa on first leg without seta (Fig. 17F). Dens normally with 7 posterior setae, 2 on the inner side.

Adults will easily be identified by the key characters. Small juveniles may be problematic, but traces of body pigment will normally separate them from *ekmani*.

BIOLOGY. Usually found in lush slopes with ferns and tall herbs (*Archangelica*) and in litter under *Alnus, Betula, Salix*.

DISTRIBUTION. S, SW (Narsarsuaq north to Kangerlussuaq/Søndre Strømfjord).

Cosmopolitan.

***Parisotoma ekmani* (Fjellberg, 1977) Figs 17H, 17J**

Henriksen 1939: *Isotoma bipunctata* Linnan.

Hammer 1952, 1954: *Isotoma notabilis* Schäffer var. *pallida* Agrell

White, apart from a small dark eye-spot, 0.7 mm.

The presence of only 1+1 ocelli set on a small, dark eye-spot normally separates this species from *notabilis*. Also the 4+4 laterodistal setae on the ventral tube is a good character. Dens normally has 8 posterior setae, 3 on the inner side. The maxilla of the two species are very different (Potapov 2001).

BIOLOGY. Both in dry and wet biotopes, often in bogs, snowbeds and shrub heaths.

DISTRIBUTION. SW, NW, N, NE, SE (no records in S Greenland). On the west coast from Ameralik north to Upernavik, on the east coast from Atingat north to Peary Land.

Holarctic.

NOTE. A third species of *Parisotoma*, *P. agrelli* (Delamare Deboutteville, 1950), may be found in S Greenland. It lives on marine seashores (salt meadows etc.) and is present in Iceland. Like *ekmani* it has only 1+1 ocelli, but is larger (1.0 mm) and has spotted dark pigment all over the body. Lower subcoxa on first leg has an outer seta (Fig. 17E) which is absent in *ekmani*. Maxilla also differs (Potapov 2001).

***Desoria* Nicolet, 1841**

Large, well pigmented isotomids with 8+8 ocelli and a strong furca. Tibiotarsi with 11 setae in the apical whorl (T_1–4 present), except in one species (*multisetis*). Six species have been found in Greenland.

Key to species

1. Mucro with 4 teeth. Body macrochaetae smooth. Tibiotarsi with 11 apical setae. Body hairs with simple bases... 2
– Mucro with 3 teeth (Fig. 17L). Abdominal macrochaetae densely ciliate (Fig. 17O). Tibiotarsi with 8–9 apical setae (Fig. 17N). Body hairs with glandular bases (Fig. 17P) .... *multisetis*
2. Manubrial teeth not free 3
– Manubrial teeth free (Fig. 18H)............ *neglecta*
3. Mucro without lateral seta. Maxillary palp simple or bifurcate (Figs 18A, 18B). Ventral tube with 4 setae in the posterior apical row (Fig. 18K) 4
– Mucro with lateral seta (Fig. 18G). Maxillary palp trifurcate (Fig. 18C). Ventral tube with 2 setae in the posterior apical row 5
4. Edge of labrum simple or weakly undulating. PAO elongate (Fig. 18I). Mucro with very large apical tooth (Fig. 18E). Maxillary palp simple (Fig. 18A) ....*tigrina*
– Labral edge with 4 sharp folds (Fig. 18D). PAO shorter (Fig. 18J). Apical tooth of mucro smaller (Fig. 18F). Maxillary palp bifurcate (Fig. 18B) .... *tolya*
5. Median macrochaetae on abd.5 are 0.5–0.6 as long as tergite, 1.1–1.2 as long as inner edge of last claw. Ant.4 pin-seta simple (Fig. 18L) .... *olivacea*
– Abd.5 macrochaetae 0.7–1.0 as long as tergite, 1.7–2.5 as long as inner edge of last claw. Ant.4 pin-seta bifurcate (Fig. 18M)...................... *tshernovi*

***Desoria multisetis* (Carpenter & Phillips, 1922) Figs 17K–17P**

Dark blue, head paler on dorsal (lateral and posterior) and ventral sides, 1.6 mm. PAO ovale, about 1.5 as long as diameter of nearest ocellus (Fig. 17K). Ant.4 with a small spherical subapical organite, pin-seta shortly bifurcate. Maxillary lamellae short, unmodified. Labrum with 4 sharp apical folds and a simple ventroapical ciliation (Fig. 17M). Maxillary outer lobe bifurcate, 4 sublobal hairs. Labium with 3 proximal setae, lateral guard e_7 present, basomedian field with 6–7 setae. Abdominal macrochaetae long, median macrochaetae on abd.5 about 3 times as long as inner edge of last claw. Ventral tube with about 10+10 lateroapical setae, posterior side with 4 setae in the apical transverse row. Retinaculum with 4+4 teeth and 9–11 setae. Manubrium with split ventroapical teeth (Fig. 17P). Tib.1–3 with 8–8–9 apical setae (T_1 present on tib.1–2, T_1 and T_4 present on tib.3, Fig. 17N).

Easily identified by the dark colour, 3–toothed mucro and multiciliate macrochaetae. The species takes an isolated systematic position. Glandular setal bases, reduced tibiotarsal chaetotaxy and 3–toothed mucro points to the genus *Isotoma* s.str.

BIOLOGY. A few specimens appeared in a *Cassiope* heath at Gåseelv (at Scoresbysund) in July 1999 (C. Albrect leg.).

DISTRIBUTION. SW, NE (Qeqertarsuaq/Disko; Dove Bugt, Ella Ø, Gåseelv: 4 km north of Constable Pynt at Scoresbysund). New species to Greenland.

Holarctic (circumpolar), but probably a complex of several species (Potapov 2001). Originally described from Bear Island (Norway).

***Desoria neglecta* (Schäffer, 1900) Fig. 18H**

Hammer 1952: *Isotoma oliv.* var. *neglecta* Schäff.

Colour variable, usually greyish brown, sometimes partly green or red, 1.8 mm. Maxillary palp bifurcate. Abd.5 macrochaetae about 2.5 as long as inner edge of last claw. Ventral tube with 7–13 laterodistal setae. Mucro with apical tooth slightly larger than subapical tooth (in winter apical tooth becomes smaller).

The free manubrial teeth in combination with a 4-toothed mucro identify this species.

BIOLOGY. Hammer (1952) reports specimens from *Salix herbacea* in a snowbed at Kangaamiut (entrance to Kangerlussuatsiaq/Evighedsfjord). In Scandinavia typically in bogs and other wet biotopes.

DISTRIBUTION. SW, NE (Kangaamiut; Dove Bugt).

Holarctic.

***Desoria tigrina* Nicolet, 1842 Figs 18A, 18E, 18I**

Greyish brown, 2.0 mm. PAO elongate (Fig. 18I). Maxillary palp simple (Fig. 18A). Ant.1 with a group of thickened spine-like sensilla on ventral side. Median macrochaetae on abd.5 short, 1.4–1.6 as long as inner length of last claw.

Unlike other Greenland *Desoria*, it has strongly developed maxillary lamellae which extends beyond the teeth of capitulum. Also abd.1 has no anterolateral spine-like microsensillum (present in *tolya*).

BIOLOGY. Usually found in organic debris near human dwellings (compost, manure, etc.), but also in wrack along seashores.

DISTRIBUTION. SW (Kuanersoq/Kvanefjord, Maniitsoq/Sukkertoppen). New species to Greenland.

Holarctic.

NOTE. The few Greenland specimens have shorter body hairs than specimens from continental Europe. The *tigrina* complex is in need of revision, and the Greenland samples may represent an undescribed species.

***Desoria tolya* (Fjellberg, 2007) Figs 18B, 18D, 18J, 18K**

Henriksen 1939: *Isotoma violacea* Tullb.

Dark blue, extremities and parts of head whitish, 2.0 mm. PAO ovale, about 1.5 as long as diameter of nearest ocellus (Fig. 18J). Median macrochaetae on abd.5 moderately long, 2.5–3.0 as long as inner edge of last claw. Claws with strong lateral teeth. Ventroapical cilation of labrum simple.

Isotoma violacea was originally described from Siberian materials. The populations in North Europe proved to be an unnamed species which was formally described by Fjellberg (2007: 120).

BIOLOGY. Common in a variety of biotopes like, alpine heaths, lush meadows, forest litter and both wet and dry seashores. .

DISTRIBUTION. S, SW, NW, SE, NE. North to Upernavik and (in NE) Ella Ø.

Holarctic.

***Desoria olivacea* (Tullberg, 1871) Figs 18C, 18G, 18L**

Henriksen 1939: *Isotoma olivacea* Tullb.

Pale yellowish brown to olive green, of variable intensity, 1.8 mm. Body hairs short. Abdominal macrochaetae not clearly differentiated from ground cover of hairs. Labrum with 4 sharp apical folds, ventroapical ciliation composite, like a dense brush. Abd.1 without anterolateral microsensillum.

The presence of a mucronal seta (Fig. 18G) readily separates this species from other *Desoria* spp. except *tshernovi* (see below).

BIOLOGY. Usually in damp mossy biotopes in snowbeds and along streambanks and springs. Also in wet meadows and damp litter in *Alnus*/*Salix* thickets.

DISTRIBUTION. S, SW, SE, NE. North to Kangerlussuaq/Søndre Strømfjord and (in NE) Ella Ø.

Holarctic.

***Desoria tshernovi* (Martynova, 1974) Fig. 18M**

Yellowish or greyish green, 1.5 mm. Body hairs generally longer than in *olivacea*, abdominal macrochaetae clearly differentiated. Winter animals have shorter apical tooth on mucro, shorter sensilla on tergites and a rod-shaped (not globular) subapical organite on ant.4.

Very similar to *olivacea*, but separable by longer abdominal macrochaetae and bifurcate ant.4 pin-seta (Fig. 18M).

BIOLOGY. Lives in damp tundra.

DISTRIBUTION. SW, NE (Disko; Illoqqortoormiut/Scoresbysund; Dove Bugt).

Holarctic. More northern than *olivacea*.

***Halisotoma* Bagnall, 1949**
The single Greenland species is separated from other isotomid genera by the peculiar feather-like setae on outer side of tib.2 (Fig. 18T). In small juveniles these are poorly developed.

***Halisotoma maritima* (Tullberg, 1871) Figs 18S, 18T**
Greyish of variable intensity, 2.0 mm. Ocelli 8+8. PAO large, ovale, 2–3 times as long as diameter of nearest ocellus. Maxillary palp trifurcate. Maxillae and labial palps strongly modified. Body macrochaetae moderately long, clearly differentiated from ground cover of hairs. Dens with a long subapical seta, mucro 3–toothed, with large apical tooth (Fig. 18S).

BIOLOGY. Found in the marine littoral under stones and in organic debris.

DISTRIBUTION. SW (Kuanersoq/Kvanefjord north to Kangerlussuatsiaq/Evighedsfjord). New species to Greenland.

Palaearctic (world distribution unclear, the species complex needs revision).

***Isotoma* Bourlet, 1839**
Large isotomids with strong furca and long, serrated abdominal macrochaetae. The spine-like anteroapical manubrial setae (Fig. 18R) identify the genus. Maxilla with short, unmodified lamellae. Ventral tube with numerous setae in anterior, laterodistal and posterior positions. Mucro with 3 teeth. Claws with 2 teeth on the inner edge. Setal bases glandular (cf. Fig. 17P). Northern species were recently revised by Fjellberg (2003). The two Greenland species were earlier collectively refered to as *Isotoma viridis* Bourl. (Henriksen 1939). The true *viridis* is not seen in material from Greenland.

Key to species

1. Manubrial macrochaeta A > B. Dens with 3 posterior basal macrochaetae (Fig. 18P) .... *anglicana*
– Manubrial macrochaeta A < B. Dens with 2 posterior macrochaetae in basal part (Fig. 18Q) ..... *caerulea*

***Isotoma anglicana* Lubbock, 1873 Figs 18N–18P**
Uniformly blue or violet of variable intensity, up to 3 mm. Abdominal macrochaetae long, curved, clearly serrated (Fig. 18O). Manubrial teeth bispinose (as Fig. 18R). PAO roundish, smaller than an ocellus (Fig. 18N). Maxillary palp bifurcate.

One of the largest Greenland springtails.Very similar to *caerulea*, but the manubrial chaetotaxy character is always safe. First instar juveniles of *anglicana* have no anteroapical setae on manubrium, while the first instar of *caerulea* has 1+1.

BIOLOGY. Common in moist habitats with moss and grass, often in cultivated fields and near human dwellings.

DISTRIBUTION. S, SW, SE. North to Qeqertarsuaq/Disko and Saqqisikuik/Skjoldungen.

Palaearctic.

***Isotoma caerulea* Bourlet, 1839 Figs 18Q, 18R**
Colour variable. Juveniles pale, adults bluish or brownish red. Up to 3 mm.

BIOLOGY. In same biotopes as previous species, sometimes in drier conditions.

DISTRIBUTION. S, SW (Narsarsuaq north to Qeqertarsuaq/Disko).

Holarctic.

***Pseudisotoma* Handschin, 1924**
The single Greenland species is recognised by the fused abd.5–6 segments and the long, clavate tibotarsal hairs (Fig. 18U). Maxillae with short, unmodified lamellae. Setal bases glandular.

***Pseudisotoma sensibilis* (Tullberg, 1876) Fig. 18U**
Dark blue, 1.7 mm. Ocelli 8+8. PAO small, ovale, slightly longer than diameter of nearest ocellus. Maxillary palp trifurcate, 4 sublobals. Macrochaetae short in anterior part of the body, long and distinct on abd.4–6, finely serrated. Tibiotarsi with 2–3–3 clavate tenent hairs. T_1 present on tib.1, T_1 and T_4 present on tib.2–3 (Fig. 18U). Mucro with 3 subequal teeth.

BIOLOGY. Common in meadows and mossy habitats, but also in alpine tundra and dry heathland.

DISTRIBUTION. SW, NE (Kangilinnguit/Grønnedal north to Sisimiut/Holsteinsborg; Jameson Land; Dove Bugt).

Holarctic.

***Vertagopus* Bagnall, 1939**
Species of this genus have a 4–toothed mucro, separate abd.5–6 and clavate tibiotarsal hairs. Unlike *Pseudisotoma* the apical whorl on tibiotarsi have seta T_2–T_3 present (Fig. 18X) and setal bases are non-glandular.

Key to species

1. Th.3 with 1–2 seta on each side of the ventral line. Macrosensilla on th.2–abd. 3 as 44/333 (Fig. 18Y). Ventral tube with 2 posterior setae in the apical transverse row .... *arcticus*
– Th.3 with 3–4 (adults) ventral seta on each side. Macrosensilla of th.2–abd.3 as 55/444 (Fig. 18Z). Ventral tube with 4 posterior setae in the apical transverse row .... *cinereus*

***Vertagopus arcticus* (Martynova, 1969) Figs 18X, 18Y, 18W, 18Ø**
Henriksen 1939: *Isotoma cinerea* Nic. (?)

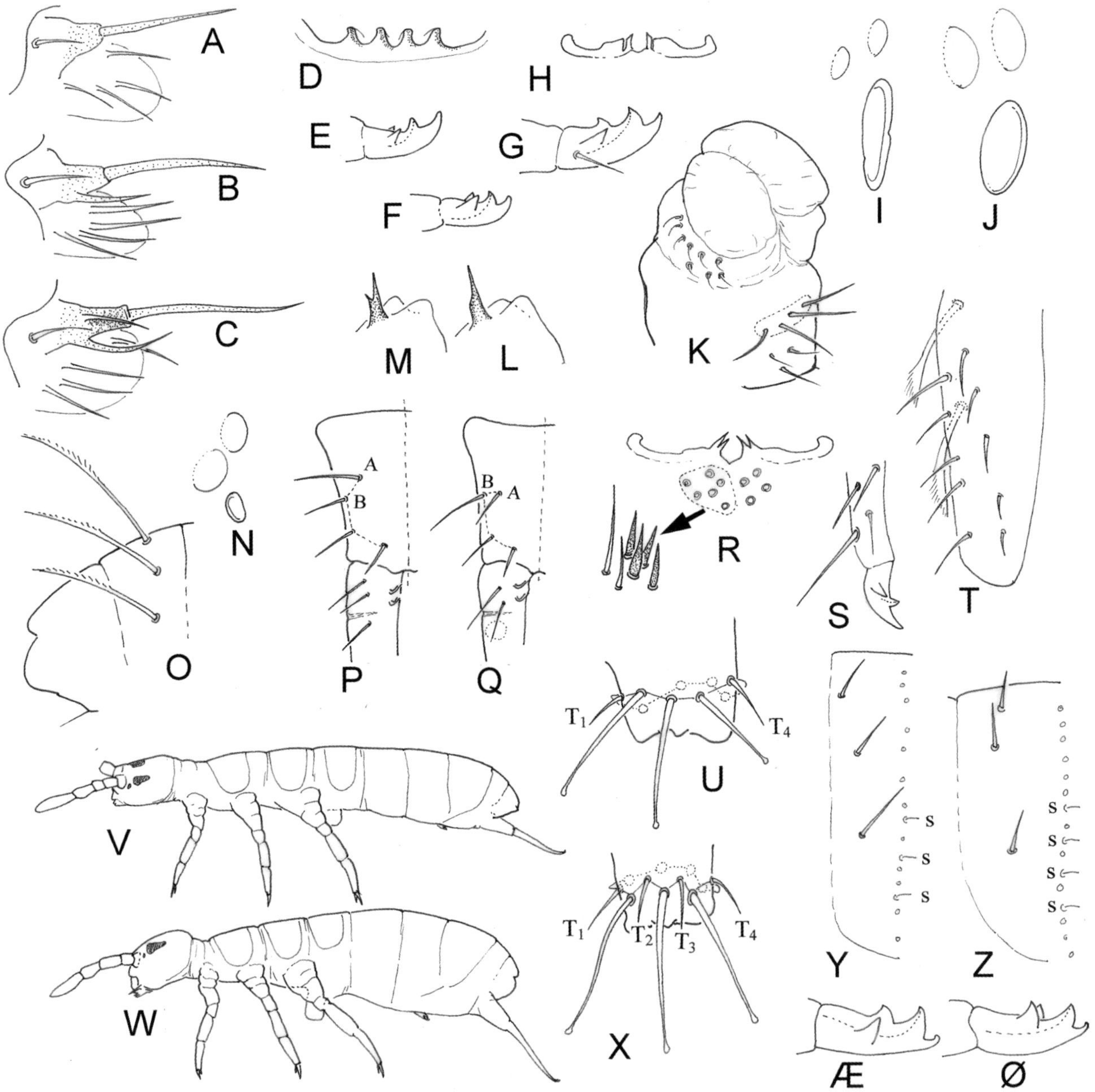

FIGURE 18 *Morphology of Isotomidae. A–M.* Desoria *spp. A–C. Maxillary palp in* D. tigrina, *simple (A),* D. violacea, *bifurcate (B),* D. olivacea, *trifurcate (C); D.* D.violacea, *apical folds of labrum; E–G. Mucro in* D. tigrina *(E),* D. violacea *(F),* D. olivacea *(G); H.* D. neglecta, *ventroapical manubrial thickening with free spines; I–J. PAO and nearest ocelli in* D. tigrina *(I) and* D. violacea *(J); K.* D. violacea, *ventral tube with posterior apical row encircled; L–M. Subapical pin-seta of ant.4 in* D. olivacea *(L) and* D. tshernovi *(M); N–O.* Isotoma anglicana. *PAO and nearest ocelli (N), serrate macrochaetae on abd.5 (O); P–Q. Posterior macrochaetae on manubrium and proximal part of dens in* I. anglicana *(P) and* I. caerulea *(Q); R.* I. caerulea, *ventroapical manubrial thickening and enlargement of adjoining setae; S–T.* Halisotoma maritima. *Mucro and apical part of dens (S), feathered setae on tib.2 (T); U.* Pseudisotoma sensibilis, *apical setae on right tib.2; V–W. Habitus of* Vertagopus cinereus *(V) and* V. arcticus *(W); X.* V. arcticus, *apical setae of right tib.3; Y–Z. Macrochaetae and sensilla on left side of abd.1 in* V. arcticus *(Y) and* V. cinereus *(Z); Æ–Ø. Mucro in* V. cinereus *(Æ) and* V. arcticus *(Ø).*

Hammer 1954: *Vertagopus arborea* (L.) (?)

Violet blue with posterior part of head and extremities paler, 1.5 mm. PAO ovale, 1.5–2.0 as long as diameter of nearest ocellus. Maxillary palp bifurcate. Tibiotarsi with 2–3–3 clavate tenent hairs, all 4 T–setae present (Fig. 18X).

Similar to *cinereus*, but body more pear-shaped with relatively longer furca (Fig. 18W). The 4–toothed mucro is symmetric (Fig. 18Ø), in *cinereus* the outer basal tooth is laterally displaced (Fig. 18Æ). The differences in sensillary chaetotaxy are desicive characters.

BIOLOGY. The habitat is arctic tundra, usually in damp and cold sites.

DISTRIBUTION. N? (Peary Land), NE (Illoqqortoormiut/ Scoresbysund: Kraemers Bugt), SE? (Kangerlussuaq). A tube with some unidentified specimens marked "Scoresbysund 1933, B.T. 21" were found in Zoological Museum, Copenhagen. They probably were collected by the Danish 3–year Expedition and are clearly *V. arcticus*. One record from SE refers to Brown's "*Isotoma arborea*" (1935) and is unverified, so is Hammer's (1954) collections from Peary Land.
Holarctic, arctic.

***Vertagopus cinereus* (Nicolet, 1841) Figs 18V, 18Z, 18Æ**
Greyish brown, 1.8 mm.

The Greenland specimens (two samples from pitfall traps in moist meadows at Narsarsuaq, Aug. 1991, J. Böcher leg.) have been compared with Norwegian *cinereus*, and no differences were found except that Greenland specimens have posterior part of abdomen bluish, distinctly darker than anterior part, while Norwegian specimens are usually uniformly greyish brown. Also the habitat is unusual, as *cinereus* is normally a corticole species. Lack's (1934) record of *cinereus* from a nunatak (1.000 m asl.) at Illoqqortoormiut/ Scoresbysund is probably *arcticus*.

BIOLOGY. Two samples from moist meadows.

DISTRIBUTION. S (Narsarsuaq).

Palaearctic.

ENTOMOBRYIDAE

Key to genera

1. Body with scales, uniformly metallic blue (Figs 2B, 19A) .... *Lepidocyrtus*
– Body without scales, bluish colour pattern (Fig. 19D) .... *Entomobrya*

***Entomobrya* Rondani, 1861**
Body shape and colour pattern of the single Greenland species is characteristic (Fig. 19D). Body with hairs only, no scales. Body macrochaetae long, expanded towards apex (Fig. 2C).

***Entomobrya nivalis* (Linnaeus, 1758) Figs 2C, 19D**
Ground colour pale yellow or greenish white with dark blue patterns as Fig. 19D. Characteristic is the V-shaped figure on abd.4. Intensity of pigmentation variable. Up to 2.0 mm. Antennae about 2/3 as long as body. Unmistakable by colour pattern.

BIOLOGY. Collected by pitfall traps in grasslands, several records.

DISTRIBUTION. S (Narsarsuaq). New species to Greenland.

Cosmopolitan.

***Lepidocyrtus* Bourlet, 1839**
Body shape charateristic (Fig. 19A), uniformly blue. Body with scales giving metallic reflections. Macrochaetae ciliated, but not apically expanded as in previous genus.

***Lepidocyrtus violaceus* Lubbock, 1873 Figs 2B, 19A–C**
Henriksen 1939: *Lepidocyrtus cyaneus* Tullb.

Uniformly blue, live specimens appear metallic. 1.5 mm. Body with scales, including feet, furca and basal segments of antennae (Fig. 2B). Basomedian field of labium with 2 smooth and 4 ciliate setae (Fig. 19B). Mucro with 2 teeth and 1 spine (Fig. 19C). Antennae about 1/3 as long as body. Easy to identify by the scaled, blue body.

BIOLOGY. Common in rich organic litter in thickets of *Betula*/*Alnus*/*Salix*, also along streambanks and seashores and in meadows near human dwellings.

DISTRIBUTION. S, SW, SE (Narsarsuaq north to Qeqertarsuaq/Disko; Saqqisikuik/Skjoldungen).

Palaearctic.

NEELIDAE

***Megalothorax* Willem, 1900**
The single Greenland species is recognised by minute size and absence of eyes and pigment. Body shape characteristic (Fig. 20A).

***Megalothorax minimus* Willem, 1900 Fig. 20A**
White, 0.4 mm. Ocelli absent, antennae very short. Body sparsely haired, with peculiar sensory fields on head (eyes), at base of 2. and 3. leg and on dorsum of thorax and abdomen. Dens with a secondary joint in the middle, mucro pointed in apical 1/3. Several large digestive balls usually visible inside the body (Fig. 20A). Easily identified by the minute size.

BIOLOGY. Found in various substrates (*Alnus* litter, bogs, meadows), but easily overlooked due to the small size.

DISTRIBUTION. SW, NE, N (Kangerlussuatsiaq/ Evighedsfjord, Ella Ø, Zackenberg, Peary Land).

Cosmopolitan.

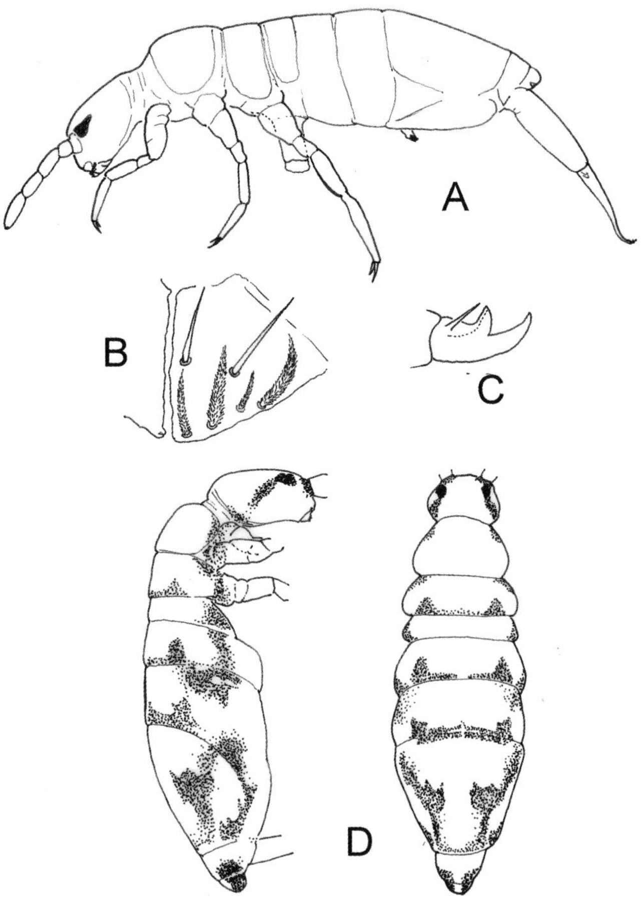

FIGURE 19 *Morphology of Entomobryidae. A–C.* Lepidocyrtus violaceus. *Habitus (A), setae in basomedian field of labium (B), mucro (C); D.* Entomobrya nivalis, *pattern of pigmentation, lateral and dorsal.*

ARRHOPALITIDAE

***Arrhopalites* Börner, 1906**

The single Greenland species is characteristic by having only 1+1 ocelli and only traces of brown or pink body pigment.

***Arrhopalites principalis* Stach, 1945 Figs 20B, 20C**

Henriksen 1939: *Arrhopalites pygmaeus* Wankel

Pale brownish red or almost white, 1.0 mm. Head with 1+1 ocelli, unpigmented. Ant.4 divided into 6 subsegments. Adults with 6+6 spines on top of head (Fig. 20C). Females with feather-shaped anal appendages.

Unmistakable by body shape (Fig. 20B) and only 1+1 ocelli. Earlier authors on Greenland Collembola (Hammer 1937, 1938) misidentified the species and called it *A. pygmaeus*. Several juveniles were checked in a sample from Ella Ø, collected in 1934 (Søgaard Andersen).

BIOLOGY. Usually in moss in wet sites (bogs, streambanks, shores of lakes).

DISTRIBUTION. S, SE, NE (Kangerlussuatsiaq/Lindenow Fjord, Ammassalik, Illoqqortoormiut/Scoresbysund, Ella Ø).

Holarctic.

SMINTHURIDIDAE

***Sminthurides* Börner, 1900**

Species of this genus are recognised by a full set of eyes, small body size, specialised mucro, absence of anal appendages in females and presence of antennal clasping organ in males.

Key to species

1. Mucro elongate, with narrow lamellae and tubular apex (Fig. 20D). Females with ant.4 subdivided into 4 segments (Fig. 20G) .... *schoetti*
– Mucro shorter, with broad lamellae, apex not tubular. Females with simple ant.4 2
2. Mucro about twice as long as high (Fig. 20F). Large species (up to 1.0 mm).......................... .... *aquaticus*
– Mucro about 3 times as long as high (Fig. 20E). Smaller (0.6 mm) .... *malmgreni*

***Sminthurides schoetti* (Axelson, 1903) Figs 20D, 20G, 20H**

Pale yellow with diffuse blue pigment, sometimes in 2 dorsal bands. 0.5 mm. Male clasping organ as Fig. 20H.

The characteristic mucro and the subdividet 4th. antennal segment in females identify the species.

BIOLOGY. In wet sites (bogs, etc.).

DISTRIBUTION. NE (Ella Ø, Zackenberg, Dove Bugt). New species to Greenland.

Palaearctic.

***Sminthurides malmgreni* (Tullberg, 1876) Fig. 20E**

Yellowish with blue dorsal and lateral bands which merge in dark individuals. 0.6 mm. Inner side of of tib.3 with a tibiotarsal organ having 2 small knobs and a lamellate long seta (as Fig. 20G).

Mucro shape and the 3 dark bands on the body are characteristic.

BIOLOGY. Common in various wetlands like small ponds, shores of lakes, snowbeds. Specialised for jumping on water surfaces.

DISTRIBUTION. G.

Holarctic.

***Sminthurides aquaticus* (Bourlet, 1843) Fig. 20F**

Pale reddish yellow, somewhat darkened dorsally. 1.0 mm. Mucro as Fig. 20F.

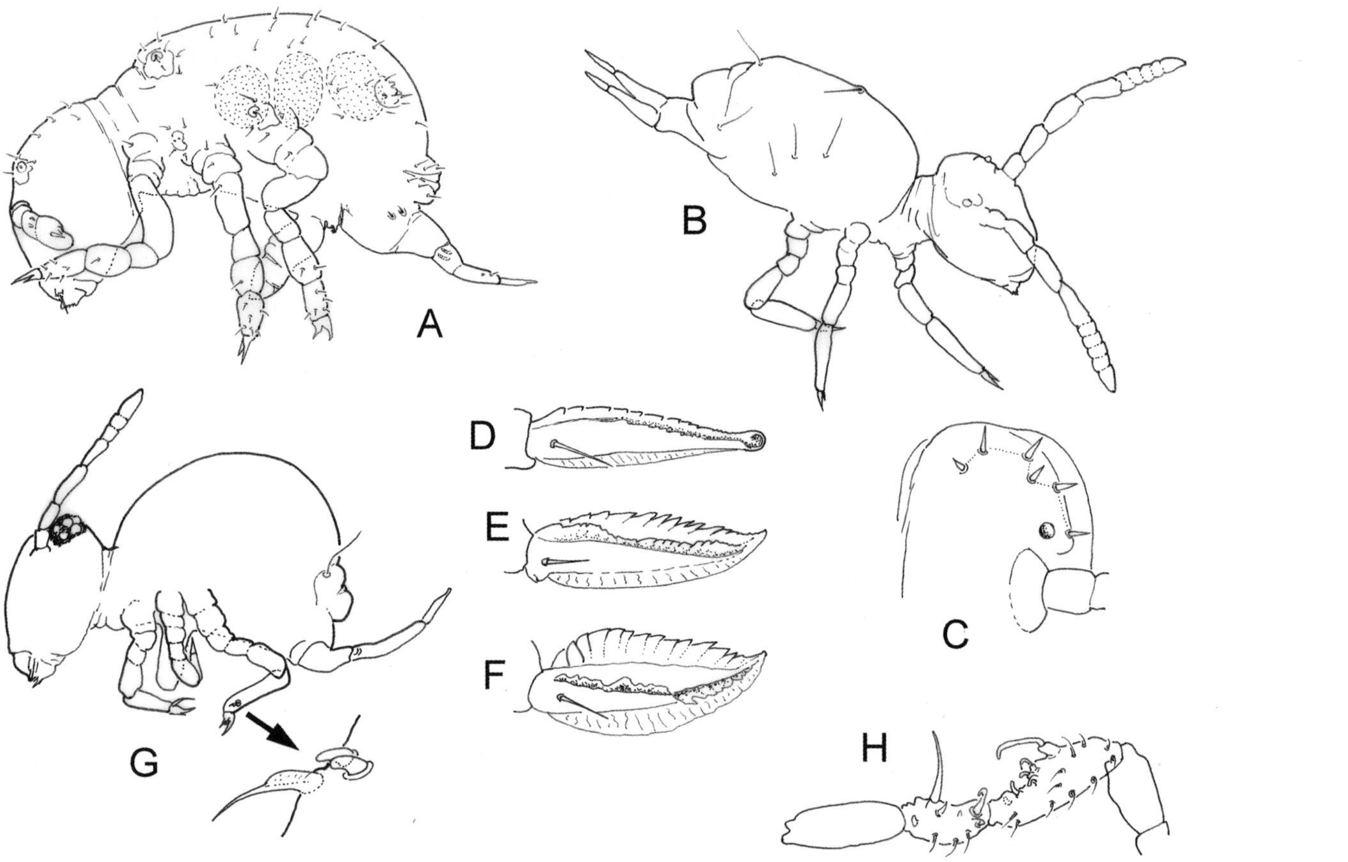

FIGURE 20 *Morphology of Neelidae (A), Arrhopalitidae (B–C), Sminthurididae (D–H). A.* Megalothorax minimus, *habitus; B–C.* Arrhopalites principalis. *Habitus, juv. (B), cephalic spines on right side of head (C); D–F. Mucro in* Sminthurides schoetti *(D),* S. malmgreni *(E),* S. aquaticus *(F); G–H.* Sminthurides schoetti. *Habitus and enlargement of tibiotarsal organ (G), male clasping organ on ant.2–3 (H).*

Larger than previous species, mucro shorter. Long seta of tibiotarsal organ simple, without basal lamella (cf. Fig. 20G).

BIOLOGY. Specialised for living on water surfaces of ponds and lakes. Only one Greenland record.

DISTRIBUTION. SW (Kangerluarsuk Tulleq, lake at inner part of the fiord). New species to Greenland.

Holarctic.

KATIANNIDAE

***Sminthurinus* Börner, 1901**

Full-eyed species with simple ant.4 and modified subanal appendages in females. The two Greenland species have a characteristic habitus (Fig. 21A).

Key to species

1. Abd.5 set off from abd.4 (Fig. 21E). Dens with 5 basal and 2 subapical ventral setae. Both inner and outer edge of mucro serrated (Fig. 21D) .... *concolor*
– Abd.5 not set off from abd.4 (Fig. 21A). Dens with 3 basal and no subapical ventral setae. Inner mucro edge serrated, outer simple (Fig. 21C) .... *aureus*

***Sminthurinus concolor* (Meiner, 1896) Figs 21D, 21E**

Entirely bluish black, 0.8 mm. Circumanal setae constricted at base, subanal appendage feather-shaped as in *aureus* (cf. Fig. 21B).

The differences in the fusion of the last abdominal segments readily separated the two *Sminthurinus* species (Figs 21A, 21E). Also the chaetotaxy of dens offers sharp diagnostic characters (Figs 21C, 21D).

BIOLOGY. Found in various types of meadows and tundra vegetation.

DISTRIBUTION. SW, NW, SE (Paamiut/Frederikshåb north to Qeqertarsuaq/Disko; Ammasalik; Zackenberg). Some of the old records may be wrong, but verified specimens are seen from Illoqqortoormiut/Scoresbysund and Zackenberg.

Holarctic.

***Sminthurinus aureus* (Lubbock, 1862) Figs 21A–21C**

Colour variable from bluish black to pale greyish yellow. 0.9 mm. Body shape as Fig. 21A. Abd.6 as Fig. 21B.

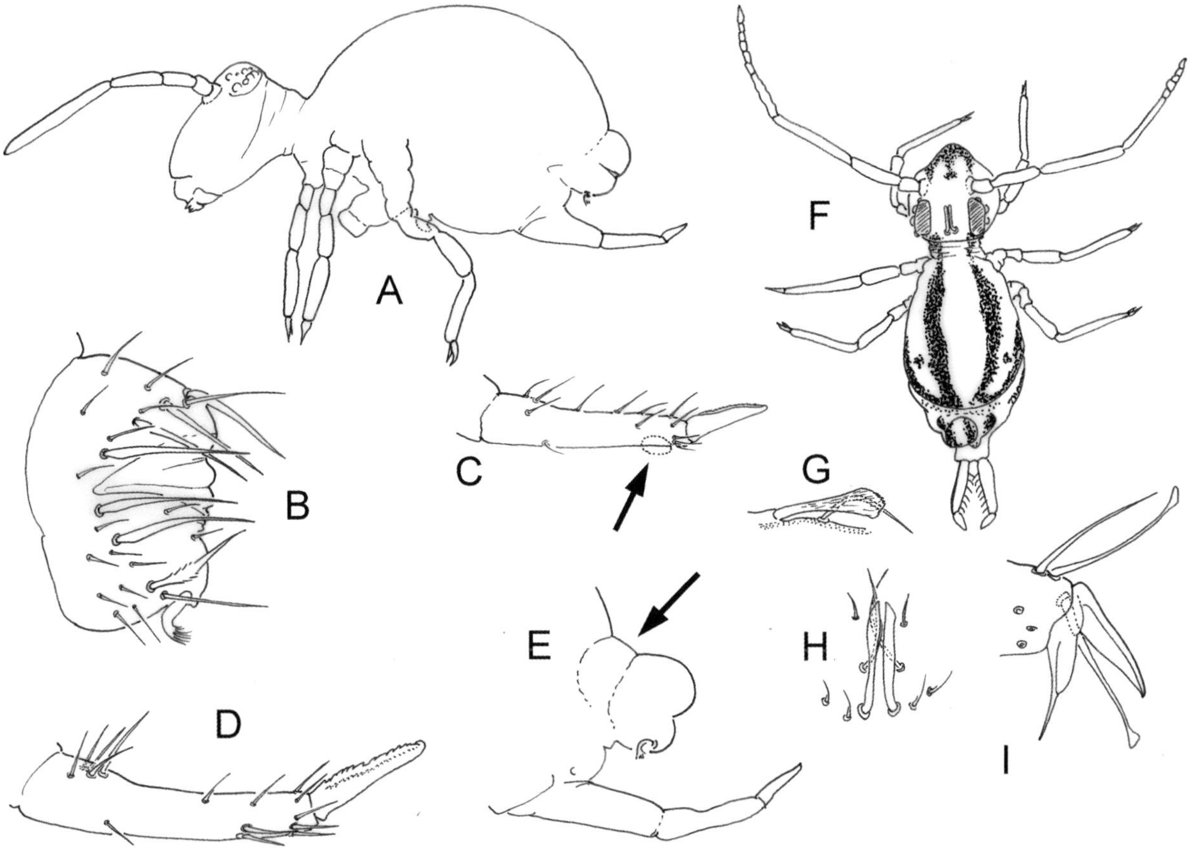

FIGURE 21 *Morphology of Katiannidae (A–E) and Bourletiellidae (F–I). A–C.* Sminthurinus aureus. *Habitus (A), left side of abd.6 in female (B), mucro and dens with absence of subapical setae (C); D–E.* Sminthurinus concolor. *Dens and mucro (D), abd.5–6 with abd.5 set off from abd.4 (E); F–.* Heterosminthurus claviger. *Habitus, male (F), male cephalic setae, lateral (G), ditto, dorsal (H), claw and apical setae of right tib.3 (I).*

BIOLOGY. Found in various meadow and heath vegetation. Once from grass fields at the duck pond in the city of Nuuk/Godthåb.

DISTRIBUTION. SW (Qeqertarsuatsiaat/Fiskenæsset north to Nuuk/Godthåb). New species to Greenland.

Palaearctic.

BOURLETIELLIDAE

Key to genera

1. Unguiculus on tib.2–3 with broad basal lamella (Fig. 21I). Males without dorsal organ on abd.6 .... *Heterosminthurus*
– Unguiculus gradually tapering (Fig. 22C). Abd.6 in males with a special dorsal organ (Fig. 22A) .... *Bourletiella*

Bourletiella **Banks, 1899**

One Greenland species which is recogised by the subdivided ant.4 (Fig. 22B), claw structures and the clavate tenent hairs on tibiotarsi (Fig. 22C).

Bourletiella hortensis **(Fitch, 1863) Figs 22A–22C**

Bluish black with a yellowish field between eyes, 1.3 mm. Males with a special dorsal organ on abd.6 consisting of cuticular ridges, curved spines and a long middorsal curled setaceous filament (Fig. 22A).

BIOLOGY. Several samples from pitfall traps in grass fields at Saqqisikuik/Skjoldungen (Juli 1992, S. Andersen).

DISTRIBUTION. SE (Saqqisikuik/Skjoldungen). New species to Greenland.

Holarctic.

Heterosminthurus **Stach, 1955**

The single Greenland species is recognised by the long antennae with subdivided segment 4 and the colouration of the body.

Heterosminthurus claviger **Gisin, 1958 Figs 21F–21I**

Body yellowish with bluish longitudinal patterns as in Fig. 21F. 0.9 mm. Males with prolonged antennae and 2 thick, clavate setae on top of head (Figs 21G, 21H).

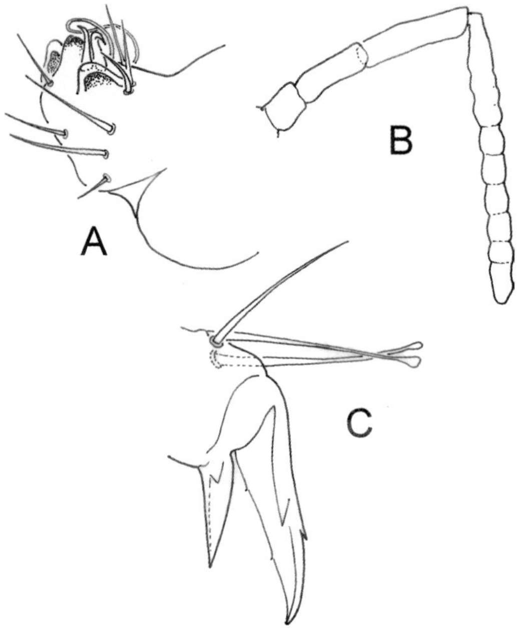

FIGURE 22 Bourletiella hortensis. *Dorsal organ of abd. 6, male (A), antenna (B), claw and apical setae on tib.3 (C).*

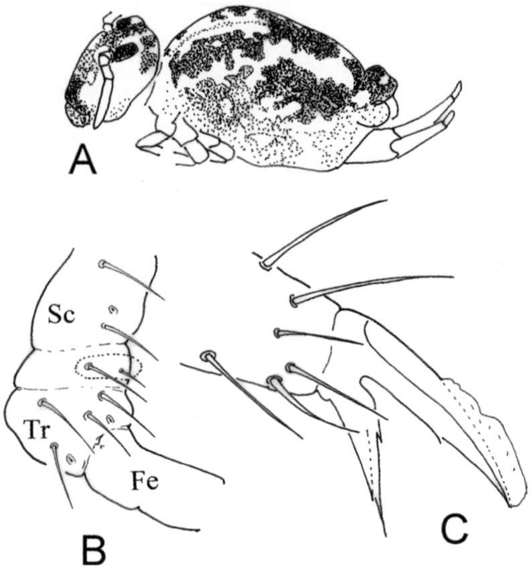

FIGURE 23 Sminthurus nigromaculatus. *Pattern of pigmentation (A), basal parts of 2. leg. Sc: subcoxa, Tr: trochanter, Fe: femur (B), claw and apical part of tib.3 (C).*

The colour pattern and specialised male setae on head identify the species.

BIOLOGY. Common in various types of meadows and heathland, both wet and dry.

DISTRIBUTION. S, SW (Qaqortoq/Julianehåb north to Qeqertarsuaq/Disko). New species to Greenland.

Palaearctic.

SMINTHURIDAE

Sminthurus Latreille, 1804

Recognised by large size, simple tibiotarsal apical hairs and colour pattern.

***Sminthurus nigromaculatus* (Tullberg, 1872)**
Figs 23A–23C

Henriksen 1939: *Sminthurus viridis* L.

Ground colour pale with spotted bluish grey patterns as in Fig. 23A. Abd.6 more or less bluish grey, with a pair of pale dorsal spots near base. 1.5–2.0 mm, females largest.

Earlier authors on Greenland collembola list this spcies as *S. viridis*. All checked specimens prove to be *nigromaculatus*, and *viridis* is probably not present. *S. viridis* differs from *nigromaculatus* in having only one short seta in lower part of subcoxa 3 (2 in *nigromaculatus*, Fig. 23B).

BIOLOGY. Found in meadows, grass fields and snow beds with rich herb vegetation.

DISTRIBUTION. S, NE (Upernaviarssuk, Jameson Land).

Holarctic.

DUBIOUS GREENLAND RECORDS

Hypogastrura manubrialis (Tullberg, 1869) (Henriksen 1939: *Hypogastrura manubrialis* Tullb.). Records from NE Greenland date back to Wahlgren (1900). There are no recent reports of the species and the identification was probably wrong.

Ceratophysella longispina (Tullberg, 1876) (Henriksen 1939: *Hypogastrura ? longispina* Tullb.). Hammer (1938b: 7) reports a dubious specimen from Illoqqortoormiut/Scoresbysund and shows that some of the older records of "*longispina*" are in reality *denticulata*. Although the true *longispina* lives in the Norwegian arctic islands (Fjellberg 1994), its presence in Greenland must be verified.

Protaphorura pseudarmata (Folsom, 1917) (Henriksen 1939: *Onychiurus pseudarmatus* Folsom). Presence of this North American species (described from St. Paul Island, Alaska) refers to a single specimen collected at Constable Pynt, Scoresby Sound, published by Lack (1934). The identity of Folsom's species is unclear, and the Greenland record needs verification.

Tullbergia collis Bacon, 1914 (Hammer 1952: *Tullbergia collis* Bacon). Hammer (1952) reports several samples in the Kangerlussuaq/Søndre Strømfjord area. None of her specimens are seen, but they may belong to *Chaetaphorura bella*.

Proisotoma tenella Reuter, 1895 (Henriksen 1939: *Proisotoma tenella* Reuter). Hammer (1944: 198) refers to a series of specimens collected at Ammassalik during July – Sept. 1933. Agrell (1939: 165) reports the species from Greenland, and in the Lund collection (Zool. Mus.)

there is a *tenella*-slide marked "Arnmagssalik, 12/9–33" containing *Proisotoma roberti*. Hammer's specimens might be a mixture of forms, but *tenella* is dubious.

Vertagopus arboreus (Linnaeus, 1758). (Henriksen 1939: *Isotoma* (*Vertagopus*) *arborea* L. No specimens are seen, and the published records from Kraemers Bugt, Kangerlussuaq (Brown 1937) and Peary Land (Hammer 1954) are dubious and probably refer to *arcticus*.

Isotoma finitima Tscherbakow, 1899 (Henriksen 1939: *Isotoma finitima* Stscherb.). Published by Wahlgren (1900). His 0.5 mm white isotomid from Kong Oscar Fjord is probably *Parisotoma ekmani* or a *Folsomia*.

Lepidocyrtus lanuginosus (Gmelin, 1788). (Henriksen 1939: *Lepidocyrtus lanuginosus* Gmel.). Recorded from NE Greenland (Ella Ø) by Jørgensen (1934). No further records. Probably it was a pale *violaceus*.

Lepidocyrtus cyaneus Tullberg, 1871 (Henriksen 1939: *Lepidocyrtus cyaneus* Tullb.). All *Lepidocyrtus* from Greenland which have been checked are *violaceus* and the presence of *cyaneus* seems doubtful.

Sminthurinus niger (Lubbock, 1876) (Henriksen 1939: *Sminthurinus niger* Lubb.). Hammer (1938b) referes to Lack's (1934) original record from Illoqqortoormiut/ Scoresbysund, identified by Carpenter. No further records exist, and a misidentification of either *concolor* or *aureus* seems likely.

Sminthurus viridis (Linnaeus, 1758) (Henriksen 1939: *Sminthurus viridis* L.). All Greenland specimens of *Sminthurus* which have been checked are *nigromaculatus*. Presence of *viridis* seems doubtful.

References

Agrell, I. (1939): Zur Kenntnis der schwedischen Collembolen mit Beschreibung von vier neuen Arten und einigen Varietäten. – *Opuscula Entomologica 1939*: 159–168.

Brown, J.M. (1937): On some Collembola from Iceland and Greenland. – *Annals and Magazine of Natural History*, Ser.10. Vol. 20: 514–520.

Carpenter, G.D.H. & Holm, A. (1939): Insecta and Araneae collected in the Kangerdlugsuak region of East Greenland by the British Expedition, 1935–1936. – *Annals and Magazine of Natural History*, Ser.11. Vol. 3: 60–80.

Fabricius, O. (1780): *Fauna Groenlandica*. Hafniae et Lipsiae. 211 pp.

Fjellberg, A. (1991): Proisotoma roberti n.sp. from Greenland, and redescription of P. ripicola Linnaniemi, 1912 (Collembola, Isotomidae). – *Entomologiske Meddelelser* 59: 81–83.

Fjellberg, A. (1994): The Collembola of the Norwegian Arctic Islands. – *Meddelelser* 133: 1–57. Norsk Polarinstitutt, Oslo.

Fjellberg, A. (1998): The Collembola of Fennoscandia and Denmark. Part 1: Poduromorpha. – *Fauna Entomologica Scandinavica* 35: 1–183.

Fjellberg, A. (2003): Revision of six northern species of the Isotoma viridis Bourlet, 1839 complex (Collembola, Isotomidae). – *Norwegian Journal of Entomology* 50: 91–98.

Fjellberg, A. (2007): The Collembola of Fennoscandia and Denmark. Part II: Entomobryomorpha and Symphypleona. – *Fauna Entomologica Scandinavica* 42: 1–264.

Goloshchapova, N.P., Potapov, M.B. & Chernova, N.M. (2005): Sexual behaviour of three species of the family Isotomidae (Collembola). – *Entomological Review* 85: 89–94.

Greenwood, J.J.D. (1977): Activity patterns and habitat distribution of tundra invertebrates. *In* Wood, G.H. & Greenwood, J.J.D. (eds.) *Report of Joint Biological Expedition to N.E. Greenland, 1974*, pp. 168–182. Dundee University, Dundee.

Hammer, M. (1937): A quantitative and qualitative investigation of the microfauna communities of the soil at Angmagssalik and in Mikis Fjord. – *Meddelelser om Grønland* 108 (2): 1–53.

Hammer, M. (1938a): Eine neue *Isotoma*-Art aus Ostgrönland. – *Zoologischer Anzeiger* 121: 43–45.

Hammer, M. (1938b): The zoology of East Greenland. Collemboles. – *Meddelelser om Grønland* 121 (1–2): 1–42.

Hammer, M. (1944): Studies on the Oribatids and Collemboles of Greenland. *Meddelelser om Grønland* 122 (1): 1–39.

Hammer, M. (1952): The Oribatid and Collembola fauna in some soil samples from Søndre Strømfjord. – *Entomologiske Meddelelser* 26: 404–414.

Hammer, M. (1953a): Investigations on the microfauna of Northern Canada. II. Collembola. – *Acta Arctica* 6: 1–108.

Hammer, M. (1953b): Collemboles and oribatids from Thule district (North West Greenland) and Ellesmere Island (Canada). – *Meddelelser om Grønland* 136 (5): 1–16.

Hammer, M. (1954): Collemboles and oribatids from Peary Land (North Greenland) – *Meddelelser om Grønland* 137 (5): 1–28.

Hammer, M. (1955): Some aspects of the distribution of microfauna in the Arctic. – *Arctic* 8: 115–126.

Henriksen, K.L. (1937): Insects collected on the fifth Thule Expedition. – *Report of the fifth Thule expedition, 1921–1924* 2 (8): 1–34.

Henriksen, K.L. (1939): A revised index of the insects of Grønland. – *Meddelelser om Grønland* 119 (10): 1–112.

Henriksen, K.L. & Lundbeck, W. (1917): Grønlands landarthropoder. Collembola. – *Meddelelser om Grønland* 22 (2): 736–743.

Haarløv, N. (1942): A morphologic-systematic-ecological investigation of Acarina and other representatives of the microfauna of the soil around Mørkefjord, Northwest Greenland. – *Meddelelser om Grønland* 128 (1): 1–71.

Jørgensen, M. (1934): A quantitative investigation of the microfauna communities of the soil in East Greenland. Preliminary report. – *Meddelelser om Grønland* 100 (9): 55–70.

Kristensen, R.M. & Vestergaard, K. (1975): Døgnaktivitet under arktiske sommerbetingelser hos springhalen *Sminthurides malmgreni* Tullberg (Collembola). – *Entomologiske Meddelelser* 43: 21–32.

Lack, D. (1934): Some insects from the Scoresby Sound region, East Greenland, with an account of the fauna of a nunatak (Collembola). – *Annals and Magazine of Natural History*. Ser.10. Vol. 14: 599–606.

Linnaniemi, W.M. (1935): Collembolen aus Nordost-Grönland. Zoological Researches of Norwegian Scientific Expedition to East-Greenland. – *Skrifter om Svalbard og Ishavet* 65: 19–25.

Madsen, H. (1936): Investigation on the Shore fauna of East Greenland with a Survey of the Shores of the other arctic regions. – *Meddelelser om Grønland* 100 (8): 1–79.

Meinert, F. (1897): Neuroptera, Pseudoneuroptera, Thysanoptera, Mallophaga, Collembola, Suctoria, Siphunculata Groenlandica. – *Videnskabelige Meddelelser fra den naturhistoriske Forening i Kjøbenhavn*: 154–177.

Müller, O.Fr. (1776): *Zoologiae Danicae prodromus, seu animalium Daniae et Norwegiae indigenarum characters nomina et synonyma imprimis popularium.* Havniae XXXII Copenhagen, Hallageriis. 282 pp.

Nardi, F., Spinsanti, G., Boore, J.L., Carapelli, A., Dallai, R. & Frati, F. (2003): Hexapod Origins: Monophyletic or Paraphyletic? – *Science* 299: 1887–1889.

Nielsen, J.C. (1907): The insects of East Greenland. – *Meddelelser om Grønland* 29: 363–409.

Peters, H.B. (1934): Anthropologie und Zoologie. *Wissenschaftliche Ergebnisse der Deutschen Grönland-Expedition Alfred Wegener 1929 und 1930–1931.* Bd. 6. Leipzig.

Remy, P. (1928): Arthropodes terrestres récoltés au Groenland au cours de la croisière du "Porquoi-Pas?" en 1926. – *Bulletin de la Société Linnéenne* 7. Lyon.

Remy, P. (1930): Les Collemboles du Groenland. – *Meddelelser om Grønland* 74: 57–70.

Schäffer, C. (1900): Die arktischen und subarktischen Collembola. – *Fauna Arctica* 1: 237–258.

Vanhöffen, E. (1897): Die fauna und flora Grönlands. – *Grönland-Expedition der Gesellschaft für Erdkunde zu Berlin 1891–1893 under Leitung von Erich von Drygalski* 2: 1–383.

Wahlgren, E. (1900): Collembola während der schwedischen Grönland-expedition 1899 auf Jan Mayen und Ost-Grönland eingesammelt. – *Öfversigt af Kongliga Vetenskaps-Akademiens Förhandlingar* 57: 353–375.

CHAPTER 7

Ephemeroptera (Mayflies)

Jens Böcher and Niels P. Kristensen†

Adults are fragile and weakly flying insects with short antennae and vestigial mouthparts. The first pair of wings is much larger than the second pair, which in some taxa is entirely missing. The abdomen terminates in two long cerci (and in many extralimital taxa there is a similar, median 'caudal filament'). The nymphs are aquatic, they consistently have a terminal filament in addition to long cerci, and and they have dorsolateral tracheal gills on a number of abdominal segments. The mature terminal stage is preceded by another winged stage, the subimago.

Morphology and behaviour: adults

Ocelli and compound eyes are well developed, especially the upper ommatidia in the males may be considerably enlarged. The mouthparts are of the biting type, but vestigial in the adults which are non-feeding. The complex wing venation includes a number of 'intercalary' veins (short peripheral veins between the main longitudinal vein branches) and cross veins. At rest the wings are held upwards over the abdomen, dorsal surfaces facing each other. The legs are delicately built and little used for walking; however, male forelegs are usually elongated and used for gripping the female (from below) during copulation in flight. The configuration of the genital segments/ appendages is of crucial importance for species identification in faunas, where several congeneric species coexist.

The *subimago* stage is unique among extant pterygote insects. The two winged stages are very similar, but the subimago generally has a duller colour and shorter abdominal appendages. After a brief flight of the subimago the last moult takes place, and the definitive adult stage is also very short-lived, a few days or even hours. The males of some species aggregate in mating swarms which attract the females.

Juvenile stages

Nymphs are found in both stagnant and running fresh water, on stones and plants, exposed on surface of the bottom or digging in mud. In accordance with the varied nymph biology, the morphology is correspondingly varied. The mouthparts are biting and variable in details depending on principal feeding habits, but many taxa are opportunists with broad potential diet spectra. Generally there are seven pairs of movable tracheal gills of highly varying form along the abdomen, and always two cerci and a median filament. The number of ecdyses is exceptionally high, up to 27 have been recorded.

FIGURE 1 *North European* Baetis *specimens morphologically identical with Nearctic* B. bundyae. *A, female subimago (inset: male head at higher magnification showing 'turban eyes', eye pigmentation distinct only in lower eye sectors); B, nymph (drawing by E. Engblom).*

Greenland fauna

In Greenland only one species occurs; it belongs to the Baetidae, one of the largest families, which includes a number of species in northern and arctic areas. Almost all genera, including *Baetis* itself, show striking sexual dimorphism in the development of the compound eyes of the winged stages: those of the males are markedly enlarged, with the upwards directed parts ('turban eyes') separated from the lateroventral parts.

Baetis bundyae Lehmkuhl, 1973 Fig. 1

Henriksen 1939: *Baetis tenax* Eaton.

Adult *c.* 5 mm without caudal filament. Hind wings very small.

Nymphs have pronouncedly elongate gills (gill 3 length ≥ 2.6× width), with no externally identifiable tracheation).

BIOLOGY. *Baetis* nymphs are detritus or periphyton feeders (Bauernfeind & Soldán 2012). The nymphs of the Greenland species have always been found on boulders and rocks in rivers, which are in some cases draining fairly eutrophic lakes. In localities close to Narssaq, SW, the species occurred abundantly in the beginning of September, both nymphs, subimagos, and adults. Of *c.* 100 adults collected all were female, indicating parthenogenesis (J. Skriver, pers. comm.). Most adults have been collected around August 1 (Upernavik: 7 July).

DISTRIBUTION. S, SW, [NE]. Possibly fairly rare, but widespread along the west coast (north to Upernavik). [NE: only recorded at Mestersvig by C. Vibe, but see note in Chapter 5].

Circumpolar (?), arctic-subarctic, alpine in temperate regions.

NOTE: The relation between the taxa *bundyae* (justified emendation of *bundyi* Lehmkuhl, 1973), described from Canada, and *macani* Kimmins, 1957, described from N. Finland, cannot be considered conclusively settled; the two have alternatively been considered distinctive species, or the former has been considered a junior synonym, a geographical subspecies or an ecomorph of the latter. Recent molecular evidence moreover indicates the possible existence of an additional undescribed cryptic species in this complex; see Bauernfeind & Soldan (2012) and references therein.

Acknowledgements

Field notes from Jens Skriver, and comments on taxonomy as well as loan of material and illustrations from Eva Engblom are gratefully acknowledged.

References

Bauernfeind, E. & Soldan, T. (2012): *The Mayflies of Europe (Ephemeroptera)*. Apollo Books, Ollerup 781 pp.

Brittain, J.E. (1975): The life cycle of *Baëtis macani* Kimmins (Ephemerida) in a Norwegian mountain biotope. – *Entomologica scandinavica* 6: 47–51.

Harper, P.P. & Harper, F. (1997): Mayflies (Ephemeroptera) of the Yukon. *In:* Danks, H.V. & Downes, J.A. (eds): Insects of the Yukon. – *Biological Survey of Canada Monograph series* 2: 152–167. Biological Survey of Canada (Terrestrial Arthropods).

Lemkuhl, D.M. (1973): A new species of *Baetis* (Ephemeroptera) from ponds in the Canadian Arctic, with biological notes. – *Canadian Entomologist* 105: 343–346.

Macan, T.T. (1957): A description of the nymph of *Baëtis macani* Kimmins. – *Notulae Entomologicae* 37: 58–60.

Morihara, D.K. & McCafferty, W.P. (1979): Subspecies of the transatlantic species, *Baetis macani* (Ephemeroptera: Baetidae). – *Proceedings of the Entomological Society of Washington* 81: 34–37.

CHAPTER 8

Psocodea
(Bark/dust/book lice and their parasitic relatives)

8.1. Generalities and non-parasitic taxa

Bo. W. Svensson

The Psocodea are an assemblage of small insects comprising free-living taxa (bark lice, dust lice, book lice) as well as taxa (chewing and sucking lice) which are ectoparasites on birds and mammals. Conventionally the free-living and parasitic lineages have been grouped under the names 'Psocoptera'/'psocids' and 'Phthiraptera' respectively. It has long been recognized that the parasitic lice are phylogenetically subordinate in the free-living psocodeans, and recent molecular and morphological analyses (Johnson *et al.* 2004, Yoshizawa & Johnson 2006) indicate that the parasitic lifestyle has evolved independently in two lineages, viz., in the Amblycera and in the Ischnocera + Anoplura (+ the extralimital Rhynchophthirina). The most overall generalized psocodeans are winged, but wings have been lost on a number of occasions within the free-living lineages and they are consistently absent in the parasitic lice. Free-living taxa feed on fungus hyphae and other organic material; they are occasionally indoor pests. It is easy to propose plausible scenarios of taxa living on debris in mammal dens and birds' nests giving rise to the ectoparasitic lineages. The principal diagnostic features of the Psocodea are found in mouthpart features (hypopharyngeal sclerites in particular) which are not visible without dissection. Like the Thysanoptera and Hemiptera the Psocodea have few (three or fewer) tarsal segments and lack cerci. The few Greenland free-living psocodeans are indoor synanthropists and are immediately recognizable by the prominently bulging anterior part of the head (the clypeus).

'Psocids' comprising the dust lice (Trogiomorpha), book lice (Troctomorpha) and bark lice (Psocomorpha), include 4.400 described species (Lienhard & Smithers 2002), of which 244 are recorded in Europe (Lienhard 1998) and 82 species in the Nordic countries (Svensson 2010).

Adults

Adult 'psocids' are of superficially varying general appearance, but share a number of characters. Soft-bodied and small (body length 1–6 mm). Orthognathous or prognathous, head mobile with a characteristic large and bulging postclypeus. Mouthparts chewing with robust and slightly asymmetric mandibles. Lacinia typically rod like, shape not rarely species specific. Eyes from small to usually large; ocelli normally three, or absent (apterous taxa or morphs). Antennae usually long, 13 up to 50 segments. Basal flagellomeres with a weakened cuticle resulting in the 'antennal rupturing mechanism', which like the specialized hypopharyngeal sclerites also occur in the true lice. These sclerites are involved in the ability to absorb moisture from atmospheric vapour very effectively, a unique ability among insects in the adult stage. Prothorax is reduced, mesothorax enlarged. Macropterous, brachypterous, micropterous or apterous.

Most taxa of the most diverse group, Psocomorpha, are macropterous (forewings extended beyond the tip of abdomen), but in some only males are long-winged. Normally two pairs of membraneous wings, with relatively few cross-veins. The presence and shape of the characteristic cell 'areola postica' has considerable taxonomic relevance. The wings are usually held steeply roof-like, but in Ectopsocidae they are hold flat over the abdomen (especially when moving). Legs slender, tarsi two- or three-segmented, pretarsal claws usally with basal appendage, pulvillus, and in many families an apical tooth. Hind coxa of many species with a Pearman's organ, a supposed stridulatory mechanism. Abdomen membraneous except for the last two or three segments which are usually well sclerotized. In some species the sclerotized terga 8–10 or 9–10 form a clunium, sometimes of taxonomic importance.

Psocids occupy a wide range of habitats, from living as indoor synanthropists (dust lice, book lice) to outdoor living in ground litter, but most are represented on bark or foliage of dead or living trees and shrubs (bark lice). Arboreal psocids usually have a broad range of tree species specificity, but an ecological grouping seems evident. Thus, some taxa are clearly predominant on conifers and others on deciduous trees, whereas many appear indifferent. Some species, however, show clear preferences within these two major habitat categories (Svensson 2010). Rather few taxa are mainly found on low vegetation, *e.g.* grass and herbs, and may then be difficult to separate ecologically from those of ground litter and from shrubs or small trees. Many species of Trogiomorpha and Troctomorpha predominantly occur indoor, often in stored products, *e.g.* grain and flour (not rarely as pests in warehouses in the tropics), old furnitures and books etc. Several species, including some Psocomorpha, have also been documented from nests of insects (*e.g.* bees and ants), birds and mammals.

FIGURE 1
Habitus of Trogium pulsatorium (*A*) *and* Lachesilla pedicularia (*B*). (PHOTO: K. HALL, AFTER SVENSSON (2010))

Psocids are generally assumed to be grazing herbivores or detritivores. The range of foods of the microflora includes the green alga *Pleurococcus*, fungal spores and hyphae, lichens (especially the algal components) and to some extent pollen grains. The food type does usually not differ between species in a certain habitat, as for nymphs and adults.

Sex ratio differs between species, but in most females are predominating as found from various types of sampling. In fact, parthenogenesis in psocids is wide spread.

Immature stages

Eggs are laid singly or in batches. Many foliage psocids cover the batches with a thick layer of labial silk. Bark-frequenting species, *e.g.* among Psocomorpha, often cover the eggs with a hard encrustation from the rectum, and sometimes with additional fragments of bark or grains which make the nymphs almost invisible on bark surfaces. The number of nymphal instars is usually six, but can be reduced to five or four in a few taxa. Some psocids are univoltine, but rather many temperate species produce more than one generation per year (usually two or three). Most species of the Nordic fauna pass the winter as egg and only a handful overwinter as larva.

Greenland fauna

A number of free-living psocodean species pertaining to different families are near-cosmopolitan synanthropists. So far representatives of two such 'dust-lice' families have been encountered in Greenland.

Key to species

1. Antenna with at least 20 segments. Forewing reduced to flat scale (0.2–0.25 mm) (Trogiidae)....2
– Antenna with 13 segments. Macropterous, forewing length 1.5–2.1 mm (females sometimes brachypterous) (Lachesillidae) .... *Lachesilla pedicularia*
2. Terminal segment of maxillary palps ca three times as long as broad, with a forked sensillum at outer mid-margin (high magnification needed)......... .... *Lepinotus patruelis*
– Terminal segment of maxillary palps ca two times as long as broad, without forked sensillum .... *Trogium pulsatorium*

TROGIIDAE (DUST LICE)

Maxillary palps with conical sensillum (short and stout) on inner border of second segment.

Tarsi 3-segmented, tarsal claw without apical tooth.

Lepinotus patruelis Pearman, 1931

Differs from other Holarctic *Lepinotus* species by no presence of 1–2 preapical spines along inner margin of hind tibia: no forewing reticulation. Usually easily separated by its dark brown stripe between the eye to the antennal base.

BIOLOGY. Mainly a domestic species, but also to be found in stables and birds' nests. Can produce a ticking sound which differs between sexes.

DISTRIBUTION. Two records from Greenland. Nuuk/Godthaab, April 1943. Dansk Pearyland Exp. Heilprinland, December 1948 ["in a box in house of the expedition"], P. Johnsen leg.

Probably cosmopolitan (Lienhard 1998).

Trogium pulsatorium (Linnaeus, 1758) Fig. 1A

Easily recognized by the rows of reddish-brown triangular spots on abdomen. Body creme yellow, head with median longtitudinal reddish brown stripe.

BIOLOGY. In Greenland a domestic and possibly an occasional species. In Europe sometimes found outdoor on grass, palms (the Mediterranean) and nests of insects and birds. Well known for the female's habit of tapping the end of the abdomen rapidly on the substrate to produce an audible ticking ("the death watch" New 2005).

DISTRIBUTION. Two records from Greenland. See Henriksen (1939), no data; also Qaqortoq/Julianehåb.

Probably cosmopolitan.

"*Troctes divinatorius* Müll." is recorded from Greenland by Henriksen (1939), but the identity of the taxon thereby referred to remains uncertain, and no voucher material has been retrieved.

LACHESILLIDAE (BARK LICE)

Usually macropterous, sometimes brachypterous or rarely micropterous. Wings glabrous. Tarsi 2-segmented, tarsal claws with a subapical tooth and a row of short hairs inside.

Lachesilla pedicularia (Linnaeus, 1758) Fig. 1B

Especially females easily distinguished from close relatives on broader and shorter pterostigma, males especially characterized by a curved hook on the epiproct.

BIOLOGY. Found in various habitats, but especially near ground in grass etc and hay in barns, often in colonies. Sometimes observed migrating in huge swarms in late summer. Has been found higher up in the sky than any other psocid, i.e. 3000 m (Glick 1939), and far from land (Harrell & Holzapfel 1966). One of the few species found at the highest suction trap (93 m) in a clear-cutting in a mixed coniferous woodland (Svensson & Solbreck 2008).

DISTRIBUTION. Found in Ivittuut/Ivigtut ["in stable"], August 1950 (C. Vibe).

Cosmopolitan.

References

Glick, P.A. (1939): The distribution of insects, spiders, and mites in the air. – *U.S. Department of Agriculture Technical Bulletin* 673: 1–150.

Harrell, J.C. & Holzapfel, E. (1966): Trapping air-borne insects on ships in the Pacific, Part 6. – *Pacific Insects* 8: 33–42.

Johnson, K.P., Yoshizawa, K. & Smith, V.S. (2004): Multiple origins of parasitism in lice. – *Proceedings of the Royal Society of London (B)* 271: 1771–1776.

Lienhard, C. (1998): Psocoptères Euro-Méditerranéens. – *Faune de France* 83: xx + 517 pp.

Lienhard, C. & Smithers, C.N. (2002): Psocoptera. World Catalogue and Bibliography. – *Instrumenta Biodiversitatis* 5: xli + 745 pp. Muséum d'histoire naturelle, Genève.

New, T.R. (2005): Psocoptera. – *Handbooks for the Identification of British Insects* 1(7): iv, 146 pp. Royal Entomological Society, London.

Svensson, B.W. (2010): *Stövsländor Psocoptera*. Nationalnyckeln till Sveriges flora och fauna. ArtDatabanken, SLU, Uppsala.

Svensson, B.W. & Solbreck, C. (2008): Stövsländor på väg i lufthavet. Fångster i sugfällor på en TV-mast i Värmland. [Psocids in suction traps on a TV-tower in a mixed coniferous forest in South Central Sweden]. Insecta: Psocodea: "Psocoptera". – *Entomologisk Tidskrift* 129: 29–39.

Yoshizawa, K. & Johnson, K.P. (2006): Morphology of male genitalia in lice and their relatives and phylogenetic implications. – Systematic Entomology 31: 350–361.

8.2. Lice: the parasitic Psocodea

Vincent S. Smith

Wingless, permanent obligate ectoparasites of mammals and birds. Highly host specific. Body dorso-ventrally flattened, adults 0.35–11 mm long. Eyes variable, antennae 3–5 segmented, mouthparts mandibulate or highly modified for sucking and/or piercing. Pronotum usually free. Meso- and metanotum complete or partially fused. Legs usually short/stout bearing single or paired pretarsal claws. Abdomen comprising 8–11 visible segments variably sclerotized, cerci absent. Colouration usually cryptic matching host pelage or plumage. Three nymphal instars.

Adult

Significant morphological features important for identification appear in Fig. 1. Head dorso-ventrally compressed and more or less prognathous with reduced or greatly modified mouthparts. Annular antennae short, small, or rarely absent, comprising a scape, pedicel, and three terminal flagellomeres, the latter two of which bear sensilla. Terminal flagellomeres may be fused and are often sexually dimorphic. Variable compound eyes (may be absent) and no ocelli. Tentorium reduced or absent. Clypeus variably extended, displacing the labrum to the ventral surface. This forms a conspicuous hyaline pad-like protrusion (the pulvinus) which sits anterior to the mandibular mouthparts of most chewing lice (i.e. Amblycera, Ischnocera, and Rhynchophthirina) but is absent in sucking lice (Anoplura). When present the pulvinus and mandibles assist in anchoring the louse to its host. Maxillae greatly reduced in chewing lice but highly modified as stylet guides in sucking lice. Labium and hypopharynx distinct in Amblycera and Ischnocera but highly modified in Rhynchophthirina (not present in Greenland) and Anoplura.

Pronotum reduced but usually distinct except in Anoplura. Meso- and metanotum usually indivisibly fused except in some Amblycera. Leg articulations with the thorax pleuro- (i.e. lateral) or sternocoxal (i.e. ventral), usually both. Legs variously modified for locomotion and attachment to host. Tarsus subdivided into two tarsomeres that are variably fused. Pretarsus bears two claws for lice parasitizing birds, or one claw for lice parasitizing mammals (no exceptions in Greenland).

In the abdomen one or more of the eleven segments are always partly or wholly suppressed. Typically nine segments are visible, the first comprising segment I and II (true segment number indicated by Roman numerals) and the ninth comprising fused segments IX, X, and XI. Tergal, sternal and tergopleural sclerotized plates are variably distributed over the abdomen. Six abdominal spiracles are normally borne from segments III–VIII, although reduction has occurred in many mammal-infesting species – usually sequentially from segment VIII. The posterior margin of sternum VII forms the ventral margin of the vulva. Gonapophyses may be present on segment VIII and a female genital lobe of uncertain homology on segment IX. The male genital opening is posterior to sternum IX. Male external genitalia of Phthiraptera are highly variable.

Immature stages

Three nymphal instars successively smaller than the adult stage can be recognised in a series of specimens. Often but not always these have non-overlapping size classes. Ontogenetic transformations in morphology from first instar to adult range from minor to significant and include changes in the chaetotaxy (distribution of setae) and degree of sclerotization. These can provide a means of stage determination independent of specimen size, although the instar morphology of most phthirapteran taxa remains undocumented. With the exception of the human body louse, lice attach their eggs to the pelage/plumage of their host. Usually eggs are cemented close to the host's skin, and remain firmly attached (even after hatching) as the pelage/plumage grows outward. A distinct cap (operculum) is present at the free end of the egg, bearing pores, tubercles, and/or other ornamentations that are taxon specific. Eggs usually hatch within a week of oviposition, although the incubation period is temperature and taxon dependent. Maturity from oviposition to adult typically takes 15–30 days. After hatching the empty egg cases often take on a shiny appearance and are referred to as a 'nits'.

Greenland fauna

As permanent obligate ectoparasites, louse distribution essentially mirrors that of the mammal and bird species they parasitize. Consequently, it is the diversity of these hosts rather than any geographic or ecological correlate that is the best predictor of louse diversity. Whilst Greenland is home to just 13 species of terrestrial/semi-terrestrial mammals, the rich avifauna of the country, comprising an unusual mix of Arctic species combined with European and North American vagrants, provides a varied habitat for the many species of lice that infest these birds.

For the size of the country, the lice of Greenland are probably the least known of any country in the world. The following species lists are only possible because extensive world checklists of louse associations have been compiled by other authors, and because of the host specificity exhibited by parasitic louse taxa. Price, Palma, and Hellenthal (2003) provide the definitive world checklist of chewing lice, whilst Durden and Musser (1994) detail known associations for the worlds sucking lice. From these publications the associations are extracted for those hosts known to breed in Greenland (with the exception of domesticated animals). Some of these taxa (*e.g.*, the musk ox, *Ovibos moschatus*) may not have been properly investigated. The data presented here must thus be regarded as a prediction of the likely louse diversity for Greenland and in no way substitutes the need for collection of specimens. Scant collecting opportunities combined with limited material in existing natural history collections have persuaded the author that a 'predictive' checklist of this type is of some use. Nonetheless, the reader should treat these data with caution, because real distributions of lice within a hosts' geographic range are known to occur for some louse species. It is the author's hope that this checklist will enable amateur and professional entomologists alike to better assess Greenland's louse diversity.

Key to the louse families (applicable to adults of Greenland species)

1. Third antennal segment pedunculate (wine-glass shaped with a distinct stalk); maxillary palps present. Post-spiracular seta of at least one abdominal segment with two minute associated setae Amblycera. 2
– Third antennal segment not pedunculate; maxillary palps absent. Post-spiracular setae lacking two minute associated setae.............................4
2. Labial palps absent; labrum with conspicuous hyaline protrusions (pulvinus) either side of the head. Mesonotum, metanotum, and tergum I fused together................................... Ricinidae
– Labial palps four segmented; pulvinus absent or undivided. Mesonotum, metanotum, and tergum I not all fused..3
3. Temples with area of sculpturing and outer rows of peg-like projections. Venter of 3rd femur and some sternites with distinctive patches of microtrichia (microsetae); meso- and metanotum fused. .... Laemobothriidae
– Temples without sculpturing or rows of peg-like projections. Venter of 3rd femur and sternites without patches of microtrichia; meso- and metanotum not fused................................. Menoponidae
4. Mandibulate mouthparts. Pronotum reduced or fully developed. Ischnocera..........................5
– Piercing mouthparts (sac containing 3 stylets). Pronotum not apparent. Anoplura......................6
5. Two pretarsal claws. Restricted to birds. .... Philopteridae
– One pretarsal claw. Restricted to mammals......... ....Trichodectidae
6. Head with distinct eyes or subacute ocular points on lateral margins posterior to antennae. Parasitizing humans...7
– Head without external evidence of eyes or prominent ocular points. Not parasitizing humans 8
7. Compact body less than twice as long as wide; thorax very wide with short abdomen; prominent tubercular lateral lobes. Forelegs slender, mid-, and hindlegs stout with a well developed stout pretarsal claw.......................................Pthiridae
– Slender body more than twice as long as wide; thorax not wide; abdomen slender, longer than thorax; paratergites as sclerotized caps/lobes. Fore-, mid-, and hindlegs slender without stout pretarsal claws. .... Pediculidae
8. Head and thorax thickly beset with setae. Abdomen covered with distinct spiniform, scale and/or peg-like setae; spiracular atrium tubular. Restricted to aquatic carnivores ....Echinophthiriidae
– Head and thorax with a few setae. Abdomen without scales; spiracular atrium bulbous. Restricted to terrestrial mammals...................................9
9. Abdomen without distinct paratergites or at most small tubercles posterior to each spiracle. Forecoxae widely separated from each other.....Linognathidae
– Abdomen with distinct paratergites apically free from the body. Forecoxae mesally close together...... 10
10. Sternites of abdominal segment II extended laterally on each side to articulate with the corresponding paratergal plates. Hindlegs largest of the three pairs and stout with a stout pretarsal claw................ .... Hoplopleuridae
– Sternites of abdominal segment II narrow, not extending laterally. Midlegs usually subequal to hindlegs in shape and size.............Polyplacidae

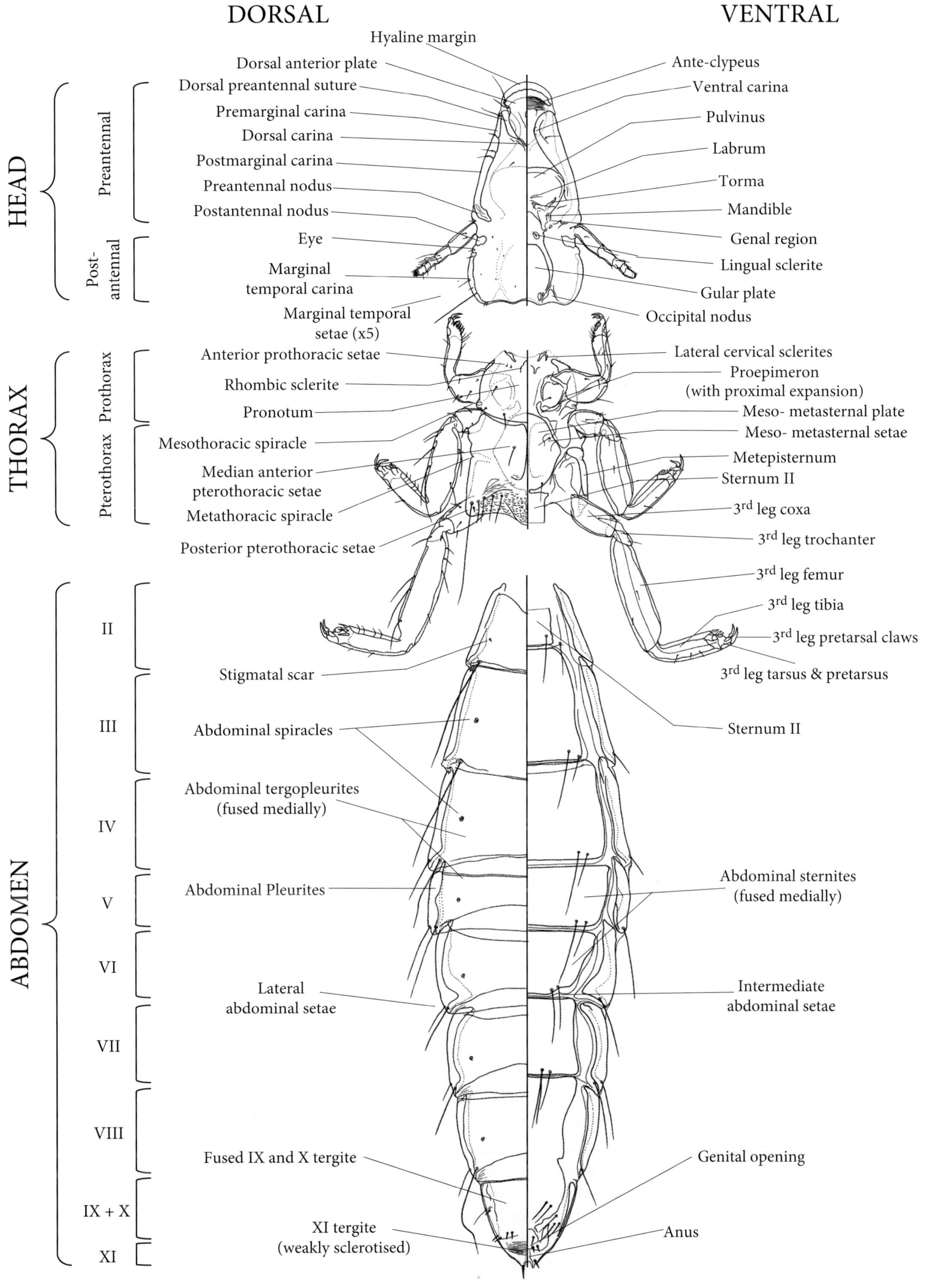

FIGURE 1 *Male* Philoceanus *sp. (Phthiraptera: Ischnocera), illustrating the major characters and nomenclature used throughout the treatment of the parasitic lice. Modified from Smith (2001). For comparable illustrations of the other principal louse groups consult Marshall (2003) for Amblycera and Kim, Pratt, and Stojanovich (1986) for Anoplura.*

AMBLYCERA

1.0–11 mm long. Head large and developed laterally, often projecting posteriorly over prothorax. Tentorium complete with exception of dorsal arms. Cranial plates and sutures often poorly developed. Antennae capitate, four- or usually five-segmented with the third segment (first flagellomere) pedunculate. Second and third flagellomeres bear sensilla. Eyes variable, may be large, reduced, or absent. Maxillary palps filiform and well developed. Mandibular mouthparts (usually asymmetric) that articulate horizontally with the head. Thoracic segments large, well developed, and ellipsoid. Pronotum always separate. Meso- and metanotum variably fused or separate. Single pair of 'mesothoracic' spiracles present on prothorax. Legs large and variable, particularly with respect to the tarsal segments and their angle of connection with the thorax. Paired pretarsal claws present except in the family Gyropidae that is not known from Greenland. Abdomen shape large and variable, typically broad and 'fat'. Eight pregenital segments, although segments I and II are sometimes fused. In adults the ninth visible segment bears the genital apparatus. Segments range from highly sclerotized to membranous. Five or usually six pairs of abdominal spiracles. The post-spiracular seta of at least one abdominal segment with two minute associated setae. External genitalia usually well developed in both sexes, often with distinct gonapophyses arising as lobes on some females. Potential hosts in Greenland: Birds.

RICINIDAE

Fig. 2A. Medium Amblycera (up to 5 mm). Head elongated, almost conical with a blunt, rounded apex. Antennae four- or five-segmented with a single plate-like (coeloconica) sensilla on the second and third flagellomere, or adjacent if these segments are fused. Labial palps absent. Maxillary palps four segmented but not bearing subterminal setae on the terminal segment. Mandibles somewhat reduced and symmetrical, modified for piercing, not chewing, with pointed needle-like tips in some taxa. Probably blood pool feeders (telmophages). Labrum with conspicuous hyaline protrusions (pulvinus) present either side of the mouthparts. Comb-like subocular setal row absent. Alveolus of temporal submarginal setae 26 and 27 (*sensu* Clay, 1970) not contiguous and head chaetotaxy atypical for Amblycera. Pronotum separate but mesonotum, metanotum, and tergum I of the first abdominal segment fused with a continuous lateral buttress of internal tergal thickening each side. Abdomen somewhat elongated as in Laemobothriidae and well sclerotized. Six spiracles present on segments III–VIII. Male external genitalia well developed. Female anal corona of setae absent. Potential hosts in Greenland: Passeriformes.

Potential Greenland species: ***Ricinus serratus*** (Durant, 1906) [*Eremophila alpestris*]; ***R. calcarii*** Nelson, 1972 [*Calcarius lapponicus*]; ***R. diffusus*** (Helloogg, 1896) [*Plectrophenax nivalis*]; ***R. elongatus*** (Olfers, 1816) [*Sturnus vulgaris, Turdus iliacus, Turdus pilaris*]; ***R. fringillae*** De Geer, 1778 [*Plectrophenax nivalis, Carduelis flammea, Anthus pratensis, Anthus rubescens, Motacilla alba*]; ***R. major*** Fresca, 1924 [*Plectrophenax nivalis*]; ***R. thoracicus*** (Packard, 1870) [*Plectrophenax nivalis*].

LAEMOBOTHRIIDAE

Fig. 2B. Very large Amblycera superficially similar to the smaller Ricinidae. Head elongated, somewhat conical with a blunt rounded apex. Antennae four- or five-segmented with three (not two) plate-like (coeloconica) sensilla on the second and third flagellomere. Labial palps present bearing five distal setae. Maxillary palps four-segmented and well developed, bearing two subterminal setae on the terminal segment, one of which is usually peg-like. Alveolus of temporal submarginal setae 26 and 27 (*sensu* Clay, 1970) usually contiguous but otherwise head chaetotaxy somewhat atypical for Amblycera. Mentum (*sensu* Symmons, 1952, fig. 18) with a bladder-like lobe conspicuous in unmounted specimens. Distinctive sculpturing on the temples with an outer row of peg-like projections. Pronotum separate but meso- and metanotum fused. Venter of third femur and some sternites with patches of characteristic microtrichia (microsetae). Distal end of tibiae II and III with dorsal patch of microtrichia and an anterior patch of smaller microtrichia. Male external genitalia well developed. Female anal setal corona absent. Potential hosts in Greenland: Falconiformes (Accipitridae and Falconidae) and Gruiformes (Rallidae).

Potential Greenland species: ***Laemobothrion (Laemobothrion) vulturis*** (Fabricius [I.C.], 1775) [*Haliaeetus albicilla*]; ***L. (L.) tinnunculi*** (Linnaeus, 1758) [*Falco peregrinus*].

MENOPONIDAE

Fig. 2C. Small to large Amblycera. Head typically broadly triangular, expanded behind the eyes. Antennae four- or five-segmented with a single plate-like (coeloconica) sensilla on the second and third flagellomere, or adjacent if these segments are fused. Labial palps present bearing five distal setae. Maxillary palps four-segmented and well developed, bearing two subterminal setae on the terminal segment, one of which is usually peg-like. Comb-like subocular setal row present. Alveolus of temporal submarginal setae 26 and 27 (*sensu* Clay, 1970) contiguous.

Pro-, meso-, and metanotum separate. Setal-bearing protuberances of the mesonotum absent. Foreleg coxa elongated. Mid- and hindlegs with two pretarsal claws. Tergite of abdominal segment I developed, especially in males and not fused to metanotum. Abdomen well sclerotized bearing six spiracles of similar size on segments III–VIII. Shape variable. Post-spiracular setal complex present at both ends of central tergites. Male external genitalia variable, often though not always with a long thin basal apodeme. Female anal corona of setae present. Potential hosts in Greenland: Anseriformes, Apodiformes, Charadriiformes, Ciconiiformes, Columbiformes, Cuculiformes, Falconiformes, Gruiformes, Passeriformes, Pelecaniformes, Piciformes, Podicipediformes, Procellariiformes, and Strigiformes.

Potential Greenland species: ***Actornithophilus bicolor*** (Piaget, 1880) [*Arenaria interpres*]; ***A. canuti*** Price & Leibovitz, 1969 [*Calidris canutus*]; ***A. ocellatus*** (Rudow, 1869) [*Numenius phaeopus*]; ***A. ochraceus*** (Nitzsch, 1818) [*Charadrius hiaticula, Pluvialis apricaria, Pluvialis dominica, Pluvialis squatarola*]; ***A. pediculoides*** (Mjöberg, 1910) [*Arenaria interpres*]; ***A. piceus lari*** (Packard, 1870) [*Larus canus, Larus argentatus, Larus fuscus, Larus glaucoides, Larus hyperboreus, Larus marinus, Larus ridibundus, Pagophila eburnea, Rhodostethia rosea, Rissa tridactyla, Xema sabini*]; ***A. piceus piceus*** (Denny, 1842) [*Sterna paradisea*]; ***A. umbrinus*** (Burmeister, 1838) [*Calidris alba, Calidris alpina, Calidris bairdii, Calidris cornutus, Calidris maritima, Phalaropus fulicaria, Phalaropus umbrinus*]; ***Ancistrona vagelli*** (Fabricius [J.C.], 1787) [*Fulmarus glacialis, Puffinus gravis*]; ***Austromenopon aegialitidis*** (Durrant, 1906) [*Charadrius hiaticula*]; ***A. alpinum*** Timmermann, 1959 [*Calidris alpina*]; ***A. atrofulvum*** (Piaget, 1880) [*Sterna paradisea*]; ***A. brevifimbriatum*** (Piaget, 1880) [*Fulmarus glacialis*]; ***A. durisetosum*** (Blagoveshtchensky, 1948) [*Gallinago gallinago*]; ***A. erilis*** Złotorzycka, 1968 [*Calidris maritima*]; ***A. fuscofasciatum*** (Piaget, 1880) [*Stercorarius longicaudus, Stercorarius parasiticus, Stercorarius pomarinus*]; ***A. lutescens*** (Burmeister, 1838) [*Arenaria interpres, Calidris canutus*]; ***A. nigropleurum*** (Denny, 1842) [*Alca torda, Alle alle, Fratercula arctica, Uria aalge, Uria lomvia*]; ***A. paululum*** (Kelogg & Chapman, 1902) [*Puffinus gravis*]; ***A. pelagicum*** Timmermann, 1963 [*Oceanodroma leucorhoa*]; ***A. phaeopodis*** (Schrank, 1802) [*Numenius phaeopus*]; ***A. spenceri*** Timmermann, 1956 [*Phalaropus lobatus*]; ***A. squatarolae*** Timmermann, 1954 [*Pluvialis squatarola*]; ***A. transversum*** (Denny, 1842) [*Larus canus, Larus argentatus, Larus fuscus, Larus glaucoides, Larus hyperboreus, Larus marinus, Larus ridibundus, Pagophila eburnea, Rhodostethia rosea, Rissa tridactyla, Xema sabini*]; ***Ciconiphilus cygni*** Price & Beer, 1965 [*Cygnus cygnus*]; ***C. pectiniventris*** (Harrison, 1916) [*Anser albifrons, Anser brachyrhynchus, Anser caerulescens, Branta bernicla, Branta canadensis, Branta leucopsis*]; ***Colcocephalum brachysomum*** Kelogg & Cgapman, 1902 [*Asio flammeus*]; ***C. flavescens*** (De Haan, 1829) [*Haliaeetus albicilla*]; ***C. fregili*** Denny, 1842 [*Corvus corax*]; ***C. nanum*** Piaget, 1890 [*Larus canus*]; ***C. subzerafae*** Tendeiro, 1988 [*Falco pereginus*]; ***C. zerafae*** Ansari, 1955 [*Falco peregrinus*]; ***Eidmaniella pelucida*** (Rudow, 1869) [*Phalacrocorax carbo*]; ***E. postulosa*** (Nitzsch, [in Giebel] 1866) [*Morus bassanus*]; ***Holomenopon acutae*** Price, 1971 [*Anas acuta*]; ***H. clypeilargum*** Eichler, 1943 [*Anas acuta, Anas crecca, Anas penelope*]; ***H. leucoxanthum*** (Burmeister, 1838) [*Anas acuta, Anas crecca, Anas penelope, Anas platyrhynchos, Branta canadensis, Clangula hyemalis*]; ***H. loomisii*** (Kelogg, 1896) [*Mergus serrato, Somateria mollissima,* Somateria spectabilis]; ***H. setigerum*** (Blagoveshtchensky, 1948) [*Anas acuta, Anas crecca*]; ***H. tadornae*** (Gervais, 1844) [*Branta bernicla*]; ***Kurodaia* (*Conciella*) *flammei*** Price & Beer, 1963 [*Asio flammeus*]; ***Menacanthus affinis*** Fedorenko [& Belskaya], 1977 [*Oenanthe oenanthe*]; ***M. alaudae*** (Schrank, 1776) [*Eremophila alpestris, Plectrophenax nivalis, Carduelis flammea*]; ***M. eurysternus*** (Burmeister, 1838) [*Calcarius lapponicus, Carduelis flammea, Sturnus vulgaris, Turdus pilaris*]; ***M. exilis*** (Nitzsch, 1866) [*Oenanthe oenanthe*]; ***M. malleus*** (Burmeister, 1838) [*Hirundo rustica*]; ***M. gonophaeus*** (Burmeister, 1838) [*Corvus corax*]; ***M. nogoma*** Uchida, 1926 [*Motacilla alba*]; ***M. pusillus*** (Nitzsch, 1866) [*Anthus pratensis, Motacilla alba*]; ***Myrsidea cucullaris*** (Nitzsch, 1818) [*Sturnus vulgaris*]; ***M. iliaci*** (Denny, 1842) [*Turdus iliacus*]; ***M. major*** (Piaget, 1880) [*Plectrophenax nivalis*]; ***M. rustica*** (Giebel, 1874) [*Hirundo rustica*]; ***Nosopon lucidum*** (Rudow, 1869) [*Falco peregrinus*]; ***Trinoton anserinum*** (Fabricius, [J.C.] 1805) [*Anser albifrons, Anser caerulescens, Branta canadensis, Cygnus cygnus*]; ***T. querquedulae*** (Linnaeus, 1758) [*Anas acuta, Anas crecca, Anas penelope, Anas platyrhynchos, Bucephala islandica, Clangula hyemalis, Histrionicus histrionicus, Mergus serrator, Somateria mollissima, Somateria spectabilis*].

ISCHNOCERA

0.6 – 9.5 mm long. Head large and often developed laterally, may project posteriorly over prothorax. Tentorium greatly reduced to a delicate ligament. Cranial plates and sutures well developed. Antennae filiform in females, males often sexually dimorphic. Antennae always exposed, three- or usually five-segmented. Penultimate and terminal flagellomeres bear sensilla. Eyes variable. Maxillary palps absent. Mandibular mouthparts (usually asymmetrical) that articulate vertically with the head. Thoracic segments large, well developed. Pronotum

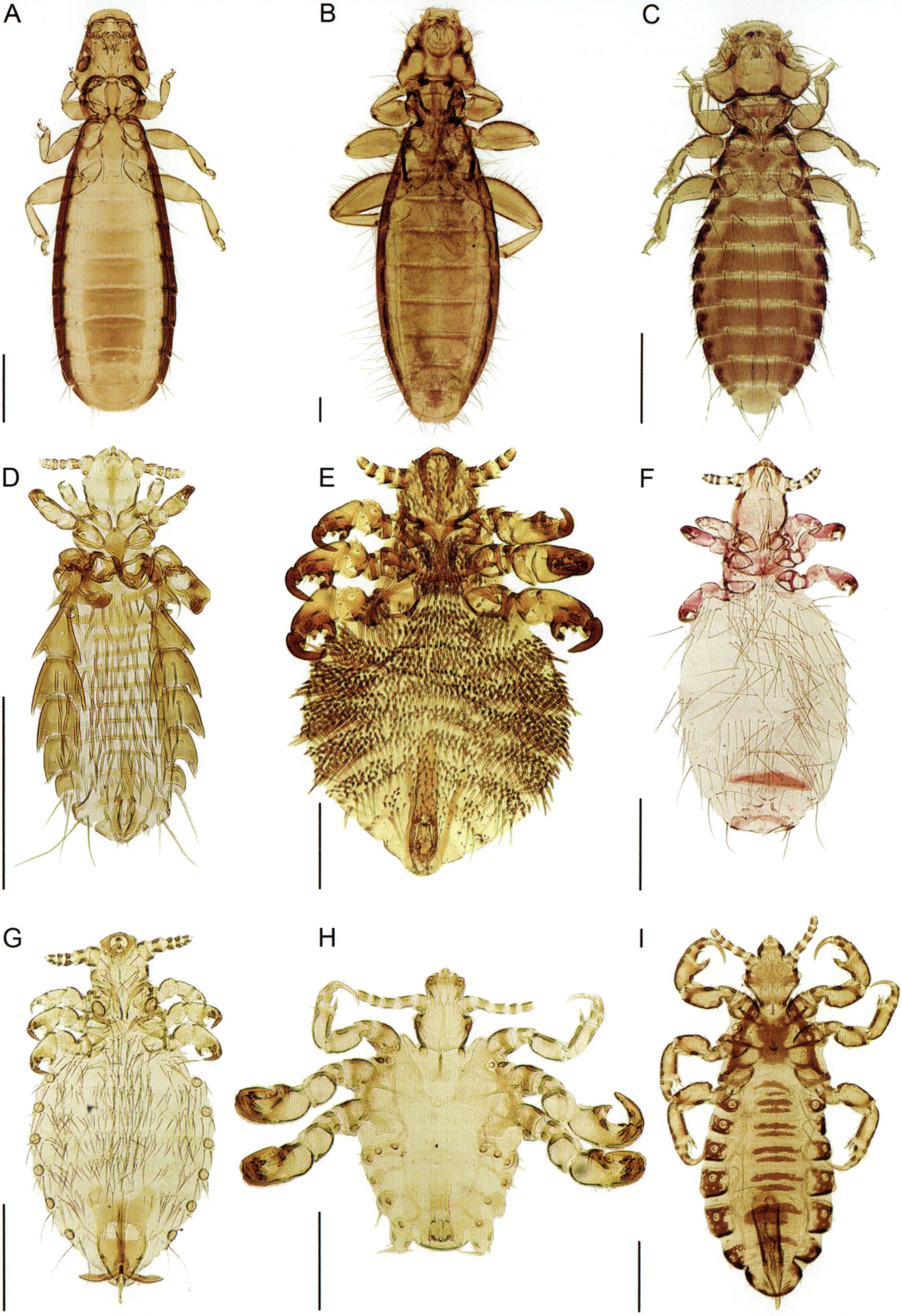

FIGURE 2 *Exemplars of the amblyceran and anopluran louse families represented in Greenland. Family names in parentheses. All scale bars 0.5 mm. A–C, Amblycera: A,* Ricinus fringillae *(Ricinidae); B,* Laemobothrion maximum *(Laemobothriidae); C,* Actornithophilus uniseriatus *(Menoponidae). D–I, Anoplura: D,* Hoplopleura acanthopus *(Hoplopleuridae); E,* Echinophthirius horridus *(Echinophthiriidae); F,* Haemodipsus lyriocephalus *(Polyplacidae); G,* Linognathus setosus *(Linognathidae); H,* Pthirus pubis *(Pthiridae); I,* Pediculus humanus *(Pediculidae).*

always separate. Meso- and metanotum completely fused (except in Heptapsogasteridae, not present in Greenland), often trapezoid in shape. 'Mesothoracic' spiracles present on prothorax. Legs variable, often well developed. Paired pretarsal claws present in philopterid taxa infesting birds, single claw present for those infesting mammals (Trichodectidae). Abdomen large and variable, ranging from broad and fat to elongated and thin. Shape corresponds to the host microhabitat occupied by the louse. Eight pregenital segments. In adults the ninth visible segment bears the genital apparatus. Segments usually highly sclerotized, although the abdomen of some mammalian infesting taxa may be membranous. Always six pairs of abdominal spiracles except in some Trichodectidae. External genitalia usually well developed in both sexes. Potential hosts in Greenland: Mammals and birds.

PHILOPTERIDAE

Fig. 1. A large and heterogeneous group of dubious monophyly. Head form variable ('circumfasciate' or 'non-circumfasciate' sensu, Symmons, 1952) with the clypeal margin not divided by the pulvinus. Distinctive plates and sutures present. Antenna five segmented, always filiform in females and sexual dimorphism common. Mandibles large and asymmetric, especially in forms occupying the head/neck and rump of the host. Head and thoracic chaetotaxy distinctive. Vestigial spiracular-like structures present on the metathorax in some taxa infesting seabirds. Distal end of tarsus usually bearing three stout setae and a pair of pretarsal claws. Abdominal form variable, usually highly sclerotized. Tergum I never present or indistinguishable from Tergum II. Female gonapophyses absent, except in some species of *Goniodes*. Restricted to birds. Potential hosts in Greenland: Anseriformes, Caprimulgiformes, Charadriiformes, Ciconiiformes, Columbiformes, Coraciiformes, Cuculiformes, Falconiformes, Galliformes, Gaviiformes, Gruiformes, Passeriformes, Pelecaniformes, Pelecaniformes, Piciformes , Podicipediformes, Procellariiformes, and Strigiformes.

Potential Greenland species: ***Anaticola anseris*** (Linnaeus, 1758) [*Anser albifrons, Anser caerulescens, Branta canadensis, Branta leucopsis*]; ***A. branderi*** Eichler & Hackman, 1973 [*Clangula hyemalis*]; ***A. clangulae*** (Fabricius [O.], 1780) [*Bucephala islandica*]; ***A. crassicornis*** (Scopoli, 1763) [*Anas acuta, Anas crecca, Anas penelope, Anas platyrhynchos, Cygnus cygnus, Somateria spectabilis*]; ***A. mergiserrati*** (De Geer, 1778) [*Mergus serrator, Somateria mollissima*]; ***A. rheinwaldi*** Fischer & Vasjukova, 1980 [*Branta bernicla*]; ***A. tamarae*** Eichler, 1981 [*Histrionicus histrionicus*]; ***A. tordae*** (Muller, 1927) [*Alca torda*]; ***A. cygni*** (Deny, 1842) [*Cygnus cygnus*]; ***A. dentatus*** (Scopoli, 1763) [*Anas acuta, Anas crecca, Anas penelope, Anas platyrhynchos, Anser albifrons, Anser caerulescens, Branta bernicla, Branta canadebsis, Branta leucopsis, Bucephala islandica, Clangula hyemalis, Mergus serrator, Somateria mollissima, Somateria spectabilis*]; ***A. icterodes*** (Nitzsch, 1818) [*Anas acuta, Anas crecca, Anas penelope, Anas platyrhynchos, Anser albifrons, Branta bernicla, Branta canadensis, Branta leucopsi, Bucephala islandicas, Clangula hyemalis, Mergus serrator, Somateria mollissima*]; ***Brueelia antimarginalis*** Eichler, 1951 [*Turdus pilaris*]; ***B. biocellata*** (Piaget, 1880) [*Corvus corax*]; ***B. corydalla*** Timmermann, 1950 [*Anthus pratensis*]; ***B. haftorni*** (Balat, 1981) [*Turdus iliacus*]; ***B. iliaci*** (Denny, 1842) [*Turdus iliacus*]; ***B. infrequens*** (Carriker, 1902) [*Calcarius lapponicus*]; ***B. marginata*** (Burmeister, 1838) [*Turdus pilaris*]; ***B. nebulosa*** (Burmeister, 1838) [*Sturnus vulgaris*]; ***B. nivalis*** (Giebel, 1874) [*Plectrophenax nivalis*]; ***Carduiceps meinertzhageni*** Timmermann, 1954 [*Calidris alpina, Calidris maritima*]; ***C. subscalaris*** (Piaget, 1880) [*Phalaropus lobatus*]; ***C. zonarius*** (Notzsch [in Giebel], 1866) [*Calidris alba, Calidris bairdii, Calidris canutus*]; ***Craspedonirmus colymbianus*** (Denny, 1842) [*Gavia stellata*]; ***C. immer*** Emerson, 1955 [*Gavia immer*]; ***Craspedorrhynchus macrocephalus*** (Nitzsch [in Giebel], 1874) [*Haliaeetus albicilla*]; ***Cummingsiella ambigua*** (Burmeister, 1938) [*Gallinago gallinago*]; ***Degeeriella discocephalus*** (Burmeister, 1838) [*Haliaeetus albicilla*]; ***D. rufa*** (Burmeister, 1838) [*Falco peregrinus, Falco rusticolus*]; ***Falcoliperus sulcifrons*** (Denny, 1842) [*Haliaeetus albicilla*]; ***Goniodes lagopi*** (Linnaeus, 1758) [*Lagopus mutus*]; ***Haffneria grandis*** (Piaget, 1880) [*Stercorarius longicaudus, Stercorarius parasiticus, Stercorarius pomarinus, Stercorarius skua*]; ***Halipeurus*** (***Halipeurus***) ***gravis*** Timmermann, 1961 [*Puffinus gravis*]; ***H.*** (***Synnautes***) ***pelagicus*** (Denny, 1842) [*Oceanodroma leucorhoa*]; ***Lagopoecus affinis*** (Children, 1836) [*Lagopus mutus*]; ***Lunaceps actophilus*** (Kelogg & Chapman, 1899) [*Calidris alba, Calidris alpina*]; ***L. drosti*** Timmermann, 1954 [*Calidris canutus*], ***L. incoensis*** (Kelogg & Chapman, 1899) [*Pluvialis squatarola*]; ***L. nereis*** Timmermann, 1954 [*Calidris maritima*]; ***Ornithobius cygni*** (Linnaeus, 1758) [*Cygnus cygnus*]; ***O. goniopleurus*** *Denny*, 1842 [*Branta canadensis*]; ***O. hexophthalmus*** (Nitzsch [in Giebel], 1861) [*Branta leucopsis*]; ***O. mathisi*** (Neumann, 1912) [*Anser albifrons*]; ***Pectinopygus bassani*** (Fabricius [O.], 1780) [*Morus bassanus*]; ***P. gyricornis*** (Denny, 1842) [*Phalacrocorax carbo*], ***Perineus nigrolimbatus*** (Giebel, 1874) [*Fulmarus glacialis*]; ***Philoceanus amadoni*** Timmermann, 1961 [*Oceanodroma leucorhoa*], ***Philopterus bischoffi*** (Eichler, 1951) [*Turdus pilaris*]; ***P. corvae*** (Linnaeus, 1758) [*Corvus corax*], ***P. hamatus*** (Packard, 1870)

[*Plectrophenax nivalis*]; ***P. kekilovae*** Fedorenko, 1984 [*Eremophila alpestris*]; ***P. linariae*** (Piaget, 1885) [*Carduelis flammea*]; ***P. microsomaticus*** Tandan, 1955 [*Hirundo rustica*]; ***P. oenanthe*** Fedorenko, 1983 [*Oenanthe oenanthe*]; ***P. osborni*** Edwards, 1952 [*Corvus corax*]; ***P. passerinus*** (Denny, 1842) [*Motacilla alba*], ***Quadraceps aethereus*** (Giebel, 1874) [*Alle alle*]; ***Q. alcae*** (Denny, 1842) [*Alca torda, Cepphus grylle*]; ***Q. bryki*** Timmermann, 1952 [*Rhodostethia rosa*]; ***Q. connexus*** (Kelogg & Mann) [*Phalaropus lobatus*]; ***Q. charadrii*** (Linnaeus, 1758) [*Pluvialis apricaria*]; ***Q. fissus*** (Burmeister, 1838) [*Charadrius hiaticula*]; ***Q. hiaticolae*** (Fabricius [O.], 1780) [*Charadrius hiaticola*]; ***Q. helgovauki*** (Timmermann, 1974) [*Fratercula arctica*]; ***Q. hospes*** (Nitzsch [in Giebel], 1866) [*Pluvialis squatarola*]; ***Q. houri*** Hopkins, 1949 [*Sterna paradisaea*]; ***Q. lineatus*** (Piaget, 1880) [*Xema sabini*]; ***Q. normifer*** (Grube, 1851) [*Stercorarius parasiticus*]; ***Q. obliquus*** (Mjöberg, 1910) [*Uria aalge, Uria lomvia*]; ***Q. ornatus lineolatus*** (Nitzsch [in Giebel], 1866) [*Rissa tridactyla*], ***Q. ornatus ornatus*** (Grube, 1851) [*Larus canus*]; ***Q. ornatus paulschulzei*** (Timmermann, 1949) [*Rissa tridactyla*]; ***Q. ornatus striolatus*** (Nitzsch [in Giebel], 1866) [*Larus argentatus, Larus glaucoides, Larus hyperboreus, Larus marinus*], ***Q. parvopallidus*** (Eichler, 1951) [*Stercorarius longicaudus*]; ***Q. phalaropi*** (Benny, 1842) [*Phalaropus fulicari*]; ***Q. punctatus punctatus*** (Burmeister, 1838) [*Larus ridibundus*]; ***Q. punctatus regressus*** Timmermann, 1952 [*Larus canus, Larus fuscus*]; ***Q. stellaepolaris*** Timmermann, 1952 [*Stercorarius pomarinus, Stercorarius skua*]; ***Q. strepsilaris*** (Denny, 1842) [*Arenaria interpres*]; ***Rhynonirmus scolopacis*** (Denny, 1842) [*Gallinago gallinago*]; ***Saemundssonia*** (***Saemundssonia***) ***alaskensis*** (Kelogg & Kuwana, 1901) [*Rhodostethia rosea*]; ***S.*** (***S.***) ***calva*** (Kelogg, 1896) [*Uria aalge, Uria lomvia*]; ***S.*** (***S.***) ***celidoxa*** (Burmeister, 1838) [*Alca torda*]; ***S.*** (***S.***) ***cephalus*** (Denny, 1842) [*Stercorarius parasiticus*]; ***S.*** (***S.***) ***conica conica*** (Denny, 1842) [*Pluvialis apricaria*]; ***S.*** (***S.***) ***conica naumanni*** (Giebel, 1874) [*Pluvialis squatarola*]; ***S.*** (***S.***) ***euryrhyncha*** (Giebel, 1874) [*Stercorarius pomarinus, Stercorarius skua*], ***S.*** (***S.***) ***fraterculae*** (Overgaard, 1942) [*Fratercula arctica*]; ***S.*** (***S.***) ***grylle*** (Fabricius [O.], 1780) [*Cepphus grylle*]; ***S.*** (***S.***) ***incisa*** Timmermann, 1950 [*Oceanodroma leucorhoa*]; ***S.*** (***S.***) ***inexpectata*** Timmermann, 1951 [*Stercorarius longicaudus*]; ***S.*** (***S.***) ***kratochvili*** Balat, 1950 [*Gallinago gallinago*]; ***S.*** (***S.***) ***lari*** (Fabricius [O.], 1780) [*Larus canus, Larus argentatus, Larus fuscus, Larus glaucoides, Larus hyperboreus, Larus marinus, Larus ridibundus, Pagophila eburnea, Rissa tridactyl, Xema sabinia*]; ***S.*** (***S.***) ***lockleyi*** Clay, 1949 [*Sterna paradisaea*]; ***S.*** (***S.***) ***merguli*** (Denny, 1842) [*Alle alle*]; ***S.*** (***S.***) ***occidentalis*** (Kelogg, 1896) [*Fulmarus glacialis*]; ***S.*** (***S.***) ***platygaster islandica*** Timmermann, 1951 [*Calidris canutus*]; ***S.*** (***S.***) ***platygaster jadwigiae*** Timmermann, 1969 [*Calidris alba*]; ***S.*** (***S.***) ***platygaster platygaster*** (Denny, 1842) [*Charadrius hiaticola*]; ***S.*** (***S.***) ***platygaster stenrami*** Timmermann, 1969 [*Arenaria interpres*]; ***S.*** (***S.***) ***scolopacisphaeopodis scolopacisphaeopodis*** (Schrank, 1803) [*Numenius phaeopus*]; ***S.*** (***S.***) ***tringae*** (Fabricius [O.], 1780) [*Arenaria interpres, Calidris alba, Calidris alpina, Calidris bairdii, Calidris canutus, Calidris maritima, Phalaropus lobatus*]; ***Strigiphilus ceblebrachys*** (Denny, 1842) [*Nyctea scandiaca*]; ***S. cursor*** (Burmeister, 1838) [*Asio flammeus*], ***Sturnidoecus aeneas*** (Piaget, 1885) [*Motacilla alba*]; ***S. sturni*** (Schrank, 1776) [*Sturnus vulgaris*].

TRICHODECTIDAE

Head form distinctive, with the clypeal margin usually divided by the pulvinus ('trichodectid' form sensu, Symmons, 1952) never 'non-circumfasciate'. Cranial plates and sutures less dominant than in Philopteridae. Antenna three segmented in all males (except *Eurytrichophilus*, not present in Greenland) and most females. Antennal sexual dimorphism common, although the segments are filiform in females. Mandibles asymmetric, usually ridged with three apical teeth on the right side and two on the left. The right mandible is usually largest. Distal end of tarsus bearing a single stout apically rounded seta. Additional stout subterminal setae may be present on the mid- and hindlegs. Abdomen sclerotized but occasionally membranous. Tergum I usually present and distinct. One to six pairs of spiracles present between segments III–VIII, or rarely completely absent. Female gonapophyses present and usually well developed. Restricted to mammals. Potential hosts in Greenland: Artiodactyla (Bovidae and Cervidae) and Carnivora (Canidae and Mustelidae).

Potential Greenland species: ***Trichodectes canis*** (De Geer, 1778) [*Canis lupus*]; ***Stachiella ermineae*** Hopkins, 1941 [*Mustela erminea*].

ANOPLURA (SUCKING LICE)

0.35–8 mm long. Head usually conical, often pointed and lacking a tentorial structure. Cranial plates and sutures occasionally developed in nymphs but usually obliterated in the adult instar. Antennae short, filiform, three or usually five segmented. Sensilla present on the second and third flagellomere. Compound eyes absent or reduced. Mouthparts highly modified comprising three protrusible flexible stylets formed from the fused maxillae, normally withdrawn inside a haustella sac. When extruded these pierce the hosts skin to locate suitable blood vessels. True solenophages (vessel feeders). Thorax usually small and completely fused bearing a single pair of spiracles. Legs often well developed with a modified tibia and one segmented tarsus that bears a single pretarsal claw. Abdomen comprises eight pregenital segments, although the most anterior pair are usually fused. The ninth visible segment bears the genital apparatus. Tergal and sternal plates across the abdomen may be present and are usually weakly sclerotized, although the paratergites are often well developed. Six, three, two or a single pair of spiracles borne on segments III–VIII. External genitalia usually well developed in both sexes. Potential hosts in Greenland: Eutherian mammals.

HOPLOPLEURIDAE

Fig. 2D. Small to medium Anoplura. Eyes absent. Antennae five, occasionally 4 segmented. Thoracic notal pit absent and sternal plate well developed with its posterior apex free. Forelegs small bearing a single acuminate pretarsal claw. Midlegs intermediate in size between the forelegs and large hindlegs, the latter bearing a stout pretarsal claw. Tibial thumbs well developed. Abdominal segments bearing distinct tergal and sternal plates with dorsal and ventral transverse rows of setae. Sternal plate of segment II prolonged laterally on each side bearing a ventral pair of thickened setae. This plate articulates with the corresponding paratergal plate of segment II. Paratergites well developed with their apex free from the body wall and overlapping the succeeding segment. Male external genitalia well developed. Potential hosts in Greenland: Rodentia (Muridae).

Potential Greenland species: ***Bovicola* (*Bovicola*) *ovis*** (Schrank, 1781) [*Ovis aries*]; ***B.* (*B.*) *tarandi*** (Mjöberg, 1910) [*Rangifer tarandus*]; ***Hoplopleura acanthopus*** (Burmeister, 183 [*Dicrostonyx groenlandicus*].

ECHINOPHTHIRIIDAE

Fig. 2E. Medium to large Anoplura. Body usually beset with distinct spiniform, scale and/or peg-like setae. Posterior margin of head rimmed by long setae. Thorax with mesothoracic phragmata connected across the dorsum, usually enclosing a notal pit. Sternal plate absent. Meso- and metathoracic legs well developed and of similar size bearing a distinct, blunt pretarsal claw. Tibial thumb elaborate. Abdomen membranous bearing six small spiracles of similar size with a distinct tubular atrium. Male external genitalia large and well developed with a distinct basal apodeme and parameres. Male and female subgenital plate weakly developed. Potential hosts in Greenland: Carnivora (Odobenidae and Phocidae).

Potential Greenland species: ***Antarctophtthirius tricheci*** (Boheman, 1865) [*Odobenus rosmarus*]; ***Echinophthirius horridus*** (van Olfers, 1816) [*Cystophora cristata, Erignathus barbatus, Phoca groenlandica, Phoca hispida, Phoca vitulina*].

POLYPLACIDAE

Fig. 2F. Small to medium Anoplura. Head without eyes. Antennae five segmented and usually sexually dimorphic. Thoracic notal pit absent and sternal plate well developed, although occasionally absent. When present, sternal plate posterior apex may be free or fused. Forelegs always small and slender bearing a single acuminate pretarsal claw. Mid- and hindlegs subequal in size and shape. Abdominal sclerites usually well developed although absent in Greenland's fauna whose sole representative of this family bears a membranous abdomen. When present, sternal plate of segment II not extending laterally to articulate with corresponding paratergites. Male external genitalia well developed. Potential hosts in Greenland: Lagomorpha (Leporidae).

Potential Greenland species: ***Haemodipsus lyriocephalus*** (Burmeister, 1839) [*Lepus arcticus*].

LINOGNATHIDAE

Fig. 2G. Anoplura of medium size. Head usually elongate without eyes. Antennae five segmented and not sexually dimorphic. Thoracic notal pit usually distinct but sternal plate absent or if present with posterior apex not free from body. Forelegs small and slender with acuminate pretarsal claw. Mid- and hindlegs subequal with distinct stout pretarsal claw and larger than forelegs. Tibial thumbs well developed with a distinct spiniform seta. Abdomen membranous with no trace of tergal, sternal or paratergal plates except in genital segments. Occasionally paratergites may be represented by small tubercles anterior to each spiracle on segments III–VIII. Spiracles are usually distinct with prominent internal ridges. Basal apodeme of male genitalia slender. Parameres well developed and free at their apex. Male subgenital plate large and well developed. Female subgenital plate variously shaped, sometimes lacking. Potential hosts in Greenland: Artiodactyla (Bovidae and Cervidae) and Carnivora (Carnidae).

Potential Greenland species: ***Linognathus africanus*** Kelogg & Paine, 1911 [*Ovis aries*]; ***L. ovillus*** (Neumann, 1907) [*Ovis aries*]; ***L. pedalis*** (Osborn, 1896) [*Ovis aries*]; ***L. setosus*** (van Olfers, 1816) [*Alopex lagopus, Canis lupus*]; ***Solenopotes tarandi*** (Mjöberg, 1915) [*Rangifer tarandus*].

PTHIRIDAE

Fig. 2H. Medium Anoplura. Compact body with a short head much narrower than thorax. Eyes distinct, present on well developed ocular nodi. Antennae five segmented and not sexually dimorphic. Thorax very short and very wide, completely fused with the anterior abdominal segments. Notal pit and sternal plate absent. Legs articulate with the extreme lateral margin of the thorax. Abdomen membranous and short, less than the width of the anterior abdominal segment. Tergites and sternites absent except for the genital segments. Distinct lateral lobes on segments V–VIII well developed and heavily sclerotized. Six pairs of abdominal spiracles, the anterior three of which are displaced toward the anterior portion of the abdomen. Segmental setae short and arranged in transverse rows. Male external genitalia well developed with a large basal apodeme and thick short parameres. Male subgenital plate present though weakly sclerotized. Female subgenital plate wide and distinct. Hosts in Greenland: Primates (Hominidae – man).

Potential Greenland species: ***Pthirus pubis*** (Linnaeus, 1758) [*Homo sapiens*].

PEDICULIDAE

Fig. 2I. Medium Anoplura. Short head with distinct eyes. Antennae five-segmented and not sexually dimorphic. Thorax with well developed phragmata (internal ridges), thoracic sternal plate weakly developed or absent. Legs all subequal in size and shape. Abdomen long and largely membranous, bearing lobe-like paratergites enclosing six pairs of spiracles. Short abdominal setae arranged transversely with multiple rows on each segment. Male external genitalia well developed with large basal apodeme and very short parameres. Male subgenital plate absent or not sclerotized but female subgenital plate large and distinct. Hosts in Greenland: Primates (Hominidae – man).

Potential Greenland species: ***Pediculus humanus*** Linnaeus, 1758 [*Homo sapiens*].

References

Clay, T. (1970): The Amblycera (Phthiraptera: Insecta). – *Bulletin of the British Museum (Natural History) Entomology* 25: 75–98.

Durden, L.A. & Musser, G.G. (1994): The sucking lice (Insecta, Anoplura) of the world: a taxonomic checklist with records of mammalian hosts and geographic distributions. – *Bulletin of the American Museum of Natural History* 218: 1–90.

Howard, R. & Moore, A. (1991): *A complete checklist of birds of the world*, 2nd edn. – Academic Press, London.

Howard, R. & Moore, A. (2003): *A complete checklist of birds of the world*, 3rd edn. – Christopher Helm, London.

Kim, K.C., Pratt, H.D. & Stojanovich, C.J. (1986): *The Sucking Lice of North America*. – Pennsylvania State University, University Park.

Marshall, I.K. (2003): A morphological phylogeny for four families of amblyceran lice (Phthiraptera: Amblycera: Menoponidae, Boopiidae, Laemobothriidae, Ricinidae). – *Zoological Journal of the Linnean Society* 138: 39–82.

Price, R.D., Hellenthal, R.A. & Palma, R.L. (2003): World checklist of chewing lice with host associations and keys to families and genera. *In*: Price, R.D., Hellenthal, R.A., Palma, R.L., Johnson, K.P. & Clayton, D.H. (eds) *The chewing lice: world checklist and biological overview*, pp. 1–448. – Illinois Natural History Survey, Champaign-Urbana, Illinois.

Sage, B.L. (1986): *The Arctic and its wildlife*. – Facts On File Inc., New York.

Smith, V.S. (2001): Avian louse phylogeny (Phthiraptera: Ischnocera): a cladistic study based morphology. – *Zoological Journal of the Linnean Society* 132: 81–144.

Symmons, S. (1952): Comparative anatomy of the Mallophagan head. – *Transactions of the Zoological Society of London* 27: 349–436.

Wilson, D.E. & Reeder, D.M. (1993): *Mammal species of the world: a taxonomic and geographic reference*, 2nd edn. – Smithsonian Institution Press, Washington and London.

Appendix: List of species probably occurring in Greenland, arranged by hosts

Avian host classification follows Howard and Moore (1991, 2003). The English names of each bird taxon is followed by their current status as listed by Birdlife International (http://www.birdlife.net/).

Chewing louse classification follows Price, Palma and Hellenthal (2003) whilst sucking louse classification is derived from Durden and Musser (1994). The list is arranged alphabetically by host order, family and species respectively. Louse taxa are suffixed by their abbreviated subordinal names (Amblycera, Amb.; Ischnocera, Isch; Anoplura, Anop.) and familial names (Echinophthiriidae, Echino.; Hoplopleuridae, Hoplo.; Laemobothriidae, Laemob.; Linognathidae, Linog.; Menoponidae, Menop.; Philopteridae, Philop.; Polyplacidae; Poly.; Trichodectidae, Tricho.).

AVES

Anseriformes–Anatidae

Anas acuta Linnaeus (Northern Pintail): *Anaticola crassicornis* (Scopoli 1763) [Isch.: Philop.]; *Anatoecus dentatus* (Scopoli 1763) [Isch.: Philop.]; *Anatoecus icterodes* (Nitzsch 1818) [Isch.: Philop.]; *Holomenopon acutae* Price 1971 [Amb.: Menop.]; *Holomenopon clypeilargum* Eichler 1943 [Amb.: Menop.]; *Holomenopon leucoxanthum* (Burmeister 1838) [Amb.: Menop.]; *Holomenopon setigerum* (Blagoveshtchensky 1948) [Amb.: Menop.]; *Trinoton querquedulae* (Linnaeus 1758) [Amb.: Menop.].

Anas crecca Linnaeus (Common Teal) – Rare/ Accidental: *Anaticola crassicornis* (Scopoli 1763) [Isch.; Philop.]; *Anatoecus dentatus* (Scopoli 1763) [Isch.: Philop.]; *Anatoecus icterodes* (Nitzsch 1818) [Isch.: Philop.]; *Holomenopon clypeilargum* Eichler 1943 [Amb.: Menop.]; *Holomenopon leucoxanthum* (Burmeister 1838) [Amb.: Menop.]; *Holomenopon setigerum* (Blagoveshtchensky 1948) [Amb.: Menop.]; *Trinoton querquedulae* (Linnaeus 1758) [Amb.: Menop.].

Anas penelope Linnaeus (Eurasian Wigeon) – Rare/ Accidental: *Anaticola crassicornis* (Scopoli 1763) [Isch.: Philop.]; *Anatoecus dentatus* (Scopoli 1763) [Isch.: Philop.]; *Anatoecus icterodes* (Nitzsch 1818) [Isch.: Philop.]; *Holomenopon clypeilargum* Eichler 1943 [Amb.: Menop.]; *Holomenopon leucoxanthum* (Burmeister 1838) [Amb.: Menop.]; *Trinoton querquedulae* (Linnaeus 1758) [Amb.: Menop.].

Anas platyrhynchos Linnaeus (Mallard): *Anaticola crassicornis* (Scopoli 1763) [Isch.: Philop.]; *Anatoecus dentatus* (Scopoli 1763) [Isch.: Philop.]; *Anatoecus icterodes* (Nitzsch 1818) [Isch.: Philop.]; *Holomenopon leucoxanthum* (Burmeister 1838) [Amb.: Menop.]; *Trinoton querquedulae* (Linnaeus 1758) [Amb.: Menop.].

Anser albifrons (Scopoli) (Greater White-fronted Goose): *Anaticola anseris* (Linnaeus 1758) [Isch.: Philop.]; *Anatoecus dentatus* (Scopoli 1763) [Isch.: Philop.]; *Anatoecus icterodes* (Nitzsch 1818) [Isch.: Philop.]; *Ciconiphilus pectiniventris* (Harrison 1916) [Amb.: Menop.]; *Trinoton anserinum* (Fabricius [J.C.] 1805) [Amb.: Menop.]; *Ornithobius mathisi* (Neumann 1912) [Isch.: Philop.]; *Trinoton anserinum* (Fabricius [J.C.] 1805) [Amb.: Menop.].

Anser brachyrhynchus Baillon (Pink-footed Goose): *Anaticola anseris* (Linnaeus 1758) [Isch.: Philop.]; *Ciconiphilus pectiniventris* (Harrison 1916) [Amb.: Menop.].

Anser caerulescens (Linnaeus) (Snow Goose): *Anaticola anseris* (Linnaeus 1758) [Isch.: Philop.]; *Anatoecus dentatus* (Scopoli 1763) [Isch.: Philop.]; *Anatoecus icterodes* (Nitzsch 1818) [Isch.: Philop.]; *Ciconiphilus pectiniventris* (Harrison 1916) [Amb.: Menop.]; *Trinoton anserinum* (Fabricius [J.C.] 1805) [Amb.: Menop.].

Branta bernicla (Linnaeus) (Brent Goose): *Anaticola rheinwaldi* Eichler & Vasjukova 1980 [Isch.: Philop.]; *Anatoecus dentatus* (Scopoli 1763) [Isch.: Philop.]; *Anatoecus icterodes* (Nitzsch 1818) [Isch.: Philop.]; *Ciconiphilus pectiniventris* (Harrison 1916) [Amb.: Menop.]; *Holomenopon tadornae* (Gervais 1844) [Amb.: Menop.].

Branta canadensis (Linnaeus) (Canada Goose): *Anaticola anseris* (Linnaeus 1758) [Isch.: Philop.]; *Anatoecus dentatus* (Scopoli 1763) [Isch.: Philop.]; *Anatoecus icterodes* (Nitzsch 1818) [Isch.: Philop.]; *Ciconiphilus pectiniventris* (Harrison 1916) [Amb.: Menop.]; *Holomenopon leucoxanthum* (Burmeister 1838) [Amb.: Menop.]; *Ornithobius goniopleurus* Denny 1842 [Isch.: Philop.]; *Trinoton anserinum* (Fabricius [J.C.] 1805) [Amb.: Menop.].

Branta leucopsis (Bechstein) (Barnacle Goose): *Anaticola anseris* (Linnaeus 1758) [Isch.: Philop.]; *Anatoecus dentatus* (Scopoli 1763) [Isch.: Philop.]; *Anatoecus icterodes* (Nitzsch 1818) [Isch.: Philop.]; *Ciconiphilus pectiniventris* (Harrison 1916) [Amb.: Menop.]; *Ornithobius hexophthalmus* (Nitzsch [In Giebel] 1861) [Isch.: Philop.].

Bucephala islandica (Gmelin) (Barrow's Goldeneye): *Anaticola clangulae* (Fabricius [O.] 1780) [Isch.: Philop.]; *Anatoecus dentatus* (Scopoli 1763) [Isch.: Philop.]; *Anatoecus icterodes* (Nitzsch 1818) [Isch.: Philop.]; *Trinoton querquedulae* (Linnaeus 1758) [Amb.: Menop.].

Clangula hyemalis (Linnaeus) (Long-tailed Duck): *Anaticola branderi* Eichler & Hackman 1973 [Isch.: Philop.]; *Anatoecus dentatus* (Scopoli 1763) [Isch.: Philop.]; *Anatoecus icterodes* (Nitzsch 1818) [Isch.: Philop.]; *Holomenopon leucoxanthum* (Burmeister 1838) [Amb.: Menop.]; *Trinoton querquedulae* (Linnaeus 1758) [Amb.: Menop.].

Cygnus cygnus (Linnaeus) (Whooper Swan) – Rare/Accidental: *Anaticola crassicornis* (Scopoli 1763) [Isch.: Philop.]; *Anatoecus cygni* (Denny 1842) [Isch.: Philop.]; *Ciconiphilus cygni* Price & Beer 1965 [Amb.: Menop.]; *Ornithobius cygni* (Linnaeus 1758) [Isch.: Philop.]; *Trinoton anserinum* (Fabricius [J.C.] 1805) [Amb.: Menop.].

Histrionicus histrionicus (Linnaeus) (Harlequin Duck): *Anaticola tamarae* Eichler 1981 [Isch.: Philop.]; *Trinoton querquedulae* (Linnaeus 1758) [Amb.: Menop.].

Mergus serrator Linnaeus (Red-breasted Merganser): *Anaticola mergiserrati* (De Geer 1778) [Isch.: Philop.]; *Anatoecus dentatus* (Scopoli 1763) [Isch.: Philop.]; *Anatoecus icterodes* (Nitzsch 1818) [Isch.: Philop.]; *Holomenopon loomisii* (Kellogg 1896) [Amb.: Menop.]; *Trinoton querquedulae* (Linnaeus 1758) [Amb.: Menop.].

Somateria mollissima (Linnaeus) (Common Eider): *Anaticola mergiserrati* (De Geer 1778) [Isch.: Philop.]; *Anatoecus dentatus* (Scopoli 1763) [Isch.: Philop.]; *Anatoecus icterodes* (Nitzsch 1818) [Isch.: Philop.]; *Holomenopon loomisii* (Kellogg 1896) [Amb.: Menop.]; *Trinoton querquedulae* (Linnaeus 1758) [Amb.: Menop.].

Somateria spectabilis (Linnaeus) (King Eider): *Anaticola crassicornis* (Scopoli 1763) [Isch.: Philop.]; *Anatoecus dentatus* (Scopoli 1763) [Isch.: Philop.]; *Holomenopon loomisii* (Kellogg 1896) [Amb.: Menop.]; *Trinoton querquedulae* (Linnaeus 1758) [Amb.: Menop.].

Charadriiformes–Alcidae

Alca torda (Razorbill): *Anaticola tordae* (Muller 1927) [Isch.: Philop.]; *Austromenopon nigropleurum* (Denny 1842) [Amb.: Menop.]; *Quadraceps alcae* (Denny 1842) [Isch.: Philop.]; *Saemundssonia (Saemundssonia) celidoxa* (Burmeister 1838) [Isch.: Philop.].

Alle alle (Linnaeus) (Little Auk): *Austromenopon nigropleurum* (Denny 1842) [Amb.: Menop.]; *Quadraceps aethereus* (Giebel 1874) [Isch.: Philop.]; *Saemundssonia (Saemundssonia) merguli* (Denny 1842) [Isch.: Philop.].

Cepphus grylle (Linnaeus) (Black Guillemot): *Quadraceps alcae* (Denny 1842) [Isch.: Philop.]; *Saemundssonia (Saemundssonia) grylle* (Fabricius [O.] 1780) [Isch.: Philop.].

Fratercula arctica (Linnaeus) (Atlantic Puffin): *Austromenopon nigropleurum* (Denny 1842) [Amb.: Menop.]; *Quadraceps helgovauki* (Timmermann 1974) [Isch.: Philop.]; *Saemundssonia (Saemundssonia) fraterculae* (Overgaard 1942) [Isch.: Philop.].

Uria aalge (Pontoppidan) (Guillemot): *Austromenopon nigropleurum* (Denny 1842) [Amb.: Menop.]; *Quadraceps obliquus* (Mjöberg 1910) [Isch.: Philop.]; *Saemundssonia (Saemundssonia) calva* (Kellogg 1896) [Isch.: Philop.].

Uria lomvia (Linnaeus) (Thick-billed Murre): *Austromenopon nigropleurum* (Denny 1842) [Amb.: Menop.]; *Quadraceps obliquus* (Mjöberg 1910) [Isch.: Philop.]; *Saemundssonia (Saemundssonia) calva* (Kellogg 1896) [Isch.: Philop.].

Charadriiformes–Charadriidae

Charadrius hiaticula Linnaeus (Common Ringed Plover): *Actornithophilus ochraceus* (Nitzsch 1818) [Amb.: Menop.]; *Austromenopon aegialitidis* (Durrant 1906) [Amb.: Menop.]; *Quadraceps fissus* (Burmeister 1838) [Isch.: Philop.]; *Quadraceps hiaticulae* (Fabricius [O.] 1780) [Isch.: Philop.]; *Saemundssonia (Saemundssonia) platygaster platygaster* (Denny 1842) [Isch.: Philop.].

Pluvialis apricaria A C Meinertzhagen (Eurasian Golden-Plover) – Rare/Accidental: *Actornithophilus ochraceus* (Nitzsch 1818) [Amb.: Menop.]; *Quadraceps charadrii* (Linnaeus 1758) [Isch.: Philop.]; *Saemundssonia (Saemundssonia) conica conica* (Denny 1842) [Isch.: Philop.].

Pluvialis dominica (P L S Muller) (American Golden-Plover) – Rare/Accidental: *Actornithophilus ochraceus* (Nitzsch 1818) [Amb.: Menop.]; *Quadraceps orarius* (Kellogg 1896) [Isch.: Philop.]; *Saemundssonia (Saemundssonia) conica conica* (Denny 1842) [Isch.: Philop.].

Pluvialis squatarola (Linnaeus) (Grey Plover) – Rare/Accidental: *Actornithophilus ochraceus* (Nitzsch 1818) [Amb.: Menop.]; *Austromenopon squatarolae* Timmermann 1954 [Amb.: Menop.]; *Lunaceps incoenis* (Kellogg & Chapman 1899) [Isch.: Philop.]; *Quadraceps hospes* (Nitzsch [In Giebel] 1866) [Isch.: Philop.]; *Saemundssonia (Saemundssonia) conica naumanni* (Giebel 1874) [Isch.: Philop.].

Charadriiformes–Laridae

Larus canus Linnaeus (Mew Gull): *Actornithophilus piceus lari* (Packard 1870) [Amb.: Menop.]; *Austromenopon transversum* (Denny 1842) [Amb.: Menop.]; *Colpocephalum nanum* Piaget 1890 [Amb.: Menop.]; *Quadraceps ornatus ornatus* (Grube 1851) [Isch.: Philop.]; *Quadraceps punctatus regressus* Timmermann 1952 [Isch.: Philop.]; *Saemundssonia (Saemundssonia) lari* (Fabricius [O.] 1780) [Isch.: Philop.].

Larus argentatus Pontoppidan (Herring Gull) – Rare/Accidental: *Actornithophilus piceus lari* (Packard 1870)

[Amb.: Menop.]; *Austromenopon transversum* (Denny 1842) [Amb.: Menop.]; *Quadraceps ornatus striolatus* (Nitzsch [In Giebel] 1866) [Isch.: Philop.]; *Saemundssonia* (*Saemundssonia*) *lari* (Fabricius [O.] 1780) [Isch.: Philop.].

Larus fuscus Linnaeus (Lesser Black-backed Gull) – Rare/Accidental: *Actornithophilus piceus lari* (Packard 1870) [Amb.: Menop.]; *Austromenopon transversum* (Denny 1842) [Amb.: Menop.]; *Quadraceps punctatus regressus* Timmermann 1952 [Isch.: Philop.]; *Saemundssonia* (*Saemundssonia*) *lari* (Fabricius [O.] 1780) [Isch.: Philop.].

Larus glaucoides Meyer (Iceland Gull): *Actornithophilus piceus lari* (Packard 1870) [Amb.: Menop.]; *Austromenopon transversum* (Denny 1842) [Amb.: Menop.]; *Quadraceps ornatus striolatus* (Nitzsch [In Giebel] 1866) [Isch.: Philop.]; *Saemundssonia* (*Saemundssonia*) *lari* (Fabricius [O.] 1780) [Isch.: Philop.].

Larus hyperboreus Gunnerus (Glaucous Gull): *Actornithophilus piceus lari* (Packard 1870) [Amb.: Menop.]; *Austromenopon transversum* (Denny 1842) [Amb.: Menop.]; *Quadraceps ornatus striolatus* (Nitzsch [In Giebel] 1866) [Isch.: Philop.]; *Saemundssonia* (*Saemundssonia*) *lari* (Fabricius [O.] 1780) [Isch.: Philop.].

Larus marinus Linnaeus (Great-Black-backed Gull): *Actornithophilus piceus lari* (Packard 1870) [Amb.: Menop.]; *Austromenopon transversum* (Denny 1842) [Amb.: Menop.]; *Quadraceps ornatus striolatus* (Nitzsch [In Giebel] 1866) [Isch.: Philop.]; *Saemundssonia* (*Saemundssonia*) *lari* (Fabricius [O.] 1780) [Isch.: Philop.].

Larus ridibundus Linnaeus (Common Black-headed Gull) – Rare/Accidental: *Actornithophilus piceus lari* (Packard 1870) [Amb.: Menop.]; *Austromenopon transversum* (Denny 1842) [Amb.: Menop.]; *Quadraceps punctatus punctatus* (Burmeister 1838) [Isch.: Philop.]; *Saemundssonia* (*Saemundssonia*) *lari* (Fabricius [O.] 1780) [Isch.: Philop.].

Pagophila eburnea (Phipps) (Ivory Gull): *Actornithophilus piceus lari* (Packard 1870) [Amb.: Menop.]; *Austromenopon transversum* (Denny 1842) [Amb.: Menop.]; *Saemundssonia* (*Saemundssonia*) *lari* (Fabricius [O.] 1780) [Isch.: Philop.].

Rhodostethia rosea (Macgillivray) (Ross's Gull): *Actornithophilus piceus lari* (Packard 1870) [Amb.: Menop.]; *Austromenopon transversum* (Denny 1842) [Amb.: Menop.]; *Quadraceps bryki* Timmermann 1952 [Isch.: Philop.]; *Saemundssonia* (*Saemundssonia*) *alaskensis* (Kellogg & Kuwana 1901) [Isch.: Philop.].

Rissa tridactyla (Linnaeus) (Black-legged Kittiwake): *Actornithophilus piceus lari* (Packard 1870) [Amb.: Menop.]; *Austromenopon transversum* (Denny 1842) [Amb.: Menop.]; *Quadraceps ornatus lineolatus* (Nitzsch [In Giebel] 1866) [Isch.: Philop.]; *Quadraceps ornatus paulschulzei* (Timmermann 1949) [Isch.: Philop.]; *Saemundssonia* (*Saemundssonia*) *lari* (Fabricius [O.] 1780) [Isch.: Philop.].

Xema sabini (J Sabine) (Sabine's Gull): *Actornithophilus piceus lari* (Packard 1870) [Amb.: Menop.]; *Austromenopon transversum* (Denny 1842) [Amb.: Menop.]; *Quadraceps lineatus* (Piaget 1880) [Isch.: Philop.]; *Saemundssonia* (*Saemundssonia*) *lari* (Fabricius [O.] 1780) [Isch.: Philop.].

Sterna paradisaea Pontoppidan (Arctic Tern): *Actornithophilus piceus piceus* (Denny 1842) [Amb.: Menop.]; *Austromenopon atrofulvum* (Piaget 1880) [Amb.: Menop.]; *Quadraceps houri* Hopkins 1949 [Isch.: Philop.]; *Saemundssonia* (*Saemundssonia*) *lockleyi* Clay 1949 [Isch.: Philop.].

Charadriiformes–Scolopacidae

Arenaria interpres (Linnaeus) (Ruddy Turnstone): *Actornithophilus bicolor* (Piaget 1880) [Amb.: Menop.]; *Actornithophilus pediculoides* (Mjöberg 1910) [Amb.: Menop.]; *Austromenopon lutescens* (Burmeister 1838) [Amb.: Menop.]; *Quadraceps strepsilaris* (Denny 1842) [Isch.: Philop.]; *Saemundssonia* (*Saemundssonia*) *platygaster stenrami* Timmermann 1969 [Isch.: Philop.]; *Saemundssonia* (*Saemundssonia*) *tringae* (Fabricius [O.] 1780) [Isch.: Philop.].

Calidris alba (Pallas) (Sanderling): *Actornithophilus umbrinus* (Burmeister 1838) [Amb.: Menop.]; *Carduiceps zonarius* (Nitzsch [In Giebel] 1866) [Isch.: Philop.]; *Lunaceps actophilus* (Kellogg & Chapman 1899) [Isch.: Philop.]; *Saemundssonia* (*Saemundssonia*) *platygaster jadwigae* Timmermann 1969 [Isch.: Philop.]; *Saemundssonia* (*Saemundssonia*) *tringae* (Fabricius [O.] 1780) [Isch.: Philop.].

Calidris alpina (Linnaeus) (Dunlin): *Actornithophilus umbrinus* (Burmeister 1838) [Amb.: Menop.]; *Austromenopon alpinum* Timmermann 1954 [Amb.: Menop.]; *Carduiceps meinertzhageni* Timmermann 1954 [Isch.: Philop.]; *Lunaceps actophilus* (Kellogg & Chapman 1899) [Isch.: Philop.]; *Saemundssonia* (*Saemundssonia*) *tringae* (Fabricius [O.] 1780) [Isch.: Philop.].

Calidris bairdii (Coues) (Baird's Sandpiper): *Actornithophilus umbrinus* (Burmeister 1838) [Amb.: Menop.]; *Carduiceps zonarius* (Nitzsch [In Giebel] 1866) [Isch.: Philop.]; *Saemundssonia* (*Saemundssonia*) *tringae* (Fabricius [O.] 1780) [Isch.: Philop.].

Calidris canutus (Linnaeus) (Red Knot): *Actornithophilus canuti* Price & Leibovitz 1969 [Amb.: Menop.]; *Actornithophilus umbrinus* (Burmeister 1838) [Amb.: Menop.]; *Austromenopon lutescens* (Burmeister 1838) [Amb.: Menop.]; *Carduiceps zonarius* (Nitzsch [In

Giebel] 1866) [Isch.: Philop.]; *Lunaceps drosti* Timmermann 1954 [Isch.: Philop.]; *Saemundssonia (Saemundssonia) platygaster islandica* Timmermann 1951 [Isch.: Philop.]; *Saemundssonia (Saemundssonia) tringae* (Fabricius [O.] 1780) [Isch.: Philop.].

Calidris maritima (Brunnich) (Purple Sandpiper): *Actornithophilus umbrinus* (Burmeister 1838) [Amb.: Menop.]; *Austromenopon erilis* Zlotorzycka 1968 [Amb.: Menop.]; *Carduiceps meinertzhageni* Timmermann 1954 [Isch.: Philop.]; *Lunaceps nereis* Timmermann 1954 [Isch.: Philop.]; *Saemundssonia (Saemundssonia) tringae* (Fabricius [O.] 1780) [Isch.: Philop.].

Gallinago gallinago (Linnaeus) (Common Snipe) – Rare/Accidental: *Austromenopon durisetosum* (Blagoveshtchensky 1948) [Amb.: Menop.]; *Cummingsiella ambigua* (Burmeister 1838) [Isch.: Philop.]; *Rhynonirmus scolopacis* (Denny 1842) [Isch.: Philop.]; *Saemundssonia (Saemundssonia) kratochvili* Balat 1950 [Isch.: Philop.].

Numenius phaeopus (Linnaeus) (Whimbrel) – Rare/Accidental: *Actornithophilus ocellatus* (Rudow 1869) [Amb.: Menop.]; *Austromenopon phaeopodis* (Schrank 1802) [Amb.: Menop.]; *Saemundssonia (Saemundssonia) scolopacisphaeopodis scolopacisphaeopodis* (Schrank 1803) [Isch.: Philop.].

Phalaropus fulicaria (Linnaeus) (Red Phalarope): *Actornithophilus umbrinus* (Burmeister 1838) [Amb.: Menop.]; *Austromenopon corporosum* (Kellogg & Kuwana 1901) [Amb.: Menop.]; *Quadraceps phalaropi* (Denny 1842) [Isch.: Philop.].

Phalaropus lobatus (Linnaeus) (Red-necked Phalarope): *Actornithophilus umbrinus* (Burmeister 1838) [Amb.: Menop.]; *Austromenopon spenceri* Timmermann 1956 [Amb.: Menop.]; *Carduiceps subscalaris* (Piaget 1880) [Isch.: Philop.]; *Quadraceps connexus* (Kellogg & Mann 1912) [Isch.: Philop.]; *Saemundssonia (Saemundssonia) tringae* (Fabricius [O.] 1780) [Isch.: Philop.].

Charadriiformes–Stercorariidae

Stercorarius longicaudus Vieillot (Long-tailed Jaeger): *Austromenopon fuscofasciatum* (Piaget 1880) [Amb.: Menop.]; *Haffneria grandis* (Piaget 1880) [Isch.: Philop.]; *Quadraceps parvopallidus* (Eichler 1951) [Isch.: Philop.]; *Saemundssonia (Saemundssonia) inexspectata* Timmermann 1951 [Isch.: Philop.].

Stercorarius parasiticus (Linnaeus) (Arctic Skua): *Austromenopon fuscofasciatum* (Piaget 1880) [Amb.: Menop.]; *Haffneria grandis* (Piaget 1880) [Isch.: Philop.]; *Quadraceps normifer* (Grube 1851) [Isch.: Philop.]; *Saemundssonia (Saemundssonia) cephalus* (Denny 1842) [Isch.: Philop.].

Stercorarius pomarinus (Temminck) (Pomarine Skua): *Austromenopon fuscofasciatum* (Piaget 1880) [Amb.: Menop.]; *Haffneria grandis* (Piaget 1880) [Isch.: Philop.]; *Quadraceps stellaepolaris* Timmermann 1952 [Isch.: Philop.]; *Saemundssonia (Saemundssonia) euryrhyncha* (Giebel 1874) [Isch.: Philop.].

Stercorarius skua Brunnich (Great Skua): *Haffneria grandis* (Piaget 1880) [Isch.: Philop.]; *Quadraceps stellaepolaris* Timmermann 1952 [Isch.: Philop.]; *Saemundssonia (Saemundssonia) euryrhyncha* (Giebel 1874) [Isch.: Philop.].

Falconiformes–Accipitridae

Haliaeetus albicilla (Linnaeus) (White-tailed Sea Eagle) – Near-threatened: *Colpocephalum flavescens* (De Haan 1829) [Amb.: Menop.]; *Craspedorrhynchus macrocephalus* (Nitzsch [In Giebel] 1874) [Isch.: Philop.]; *Degeeriella discocephalus* (Burmeister 1838) [Isch.: Philop.]; *Falcolipeurus sulcifrons* (Denny 1842) [Isch.: Philop.]; *Laemobothrion (Laemobothrion) vulturis* (Fabricius [J.C.] 1775) [Amb.: Laemob.].

Falconiformes–Falconidae

Falco peregrinus Tunstall (Peregrine Falcon): *Colpocephalum subzerafae* Tendeiro 1988 [Amb.: Menop.]; *Colpocephalum zerafae* Ansari 1955 [Amb.: Menop.]; *Degeeriella rufa* (Burmeister 1838) [Isch.: Philop.]; *Laemobothrion (Laemobothrion) tinnunculi* (Linnaeus 1758) [Amb.: Laemob.]; *Nosopon lucidum* (Rudow 1869) [Amb.: Menop.].

Falco rusticolus Linnaeus (Gyrfalcon): *Degeeriella rufa* (Burmeister 1838) [Isch.: Philop.].

Galliformes–Phasianidae

Lagopus mutus (Montin) (Rock Ptarmigan): *Goniodes lagopi* (Linnaeus 1758) [Isch.: Philop.]; *Lagopoecus affinis* (Children 1836) [Isch.: Philop.].

GaviiformesGaviidae

Gavia immer (Brunnich) (Great Northern Diver): *Craspedonirmus immer* Emerson 1955 [Isch.: Philop.]; ***Gavia stellata*** (Pontoppidan) (Red-throated Diver); *Craspedonirmus colymbinus* (Denny 1842) [Isch.: Philop.].

Passeriformes–Alaudidae

Eremophila alpestris (Linnaeus) (Horned Lark) – Rare/Accidental: *Menacanthus alaudae* (Schrank 1776) [Amb.: Menop.]; *Philopterus kekilovae* Fedorenko 1984 [Isch.: Philop.]: *Ricinus serratus* (Durrant 1906) [Amb.: Ricinidae].

Passeriformes–Corvidae

Corvus corax Linnaeus (Common Raven): *Brueelia biocellata* (Piaget 1880) [Isch.: Philop.]; *Colpocephalum fregili*

Denny 1842 [Amb.: Menop.]; *Menacanthus gonophaeus* (Burmeister 1838) [Amb.: Menop.]; *Philopterus corvi* (Linnaeus 1758) [Isch.: Philop.]; *Philopterus osborni* Edwards 1952 [Isch.: Philop.].

Passeriformes–Emberizidae

Calcarius lapponicus (Linnaeus) (Lapland Longspur): *Brueelia infrequens* (Carriker 1902) [Isch.: Philop.]; *Menacanthus eurysternus* (Burmeister 1838) [Amb.: Menop.]; *Ricinus calcarii* Nelson 1972 [Amb.: Ricinidae];.

Plectrophenax nivalis (Linnaeus) (Snow Bunting): *Brueelia nivalis* (Giebel 1874) [Isch.: Philop.]; *Menacanthus alaudae* (Schrank 1776) [Amb.: Menop.]; *Myrsidea major* (Piaget 1880) [Amb.: Menop.]; *Philopterus hamatus* (Packard 1870) [Isch.: Philop.]; *Ricinus diffusus* (Kellogg 1896) [Amb.: Ricinidae]; *Ricinus fringillae* De Geer 1778 [Amb.: Ricinidae]; *Ricinus major* Fresca 1924 [Amb.: Ricinidae]; *Ricinus thoracicus* (Packard 1870) [Amb.: Ricinidae].

Passeriformes–Fringillidae

Carduelis flammea (Linnaeus) (Common Redpoll): *Menacanthus alaudae* (Schrank 1776) [Amb.: Menop.]; *Menacanthus eurysternus* (Burmeister 1838) [Amb.: Menop.]; *Philopterus linariae* (Piaget 1885) [Isch.: Philop.]; *Ricinus fringillae* De Geer 1778 [Amb.: Ricinidae].

Passeriformes–Hirundinidae

Hirundo rustica Linnaeus (Barn Swallow) – Rare/Accidental: *Machaerilaemus malleus* (Burmeister 1838) [Amb.: Menop.]; *Myrsidea rustica* (Giebel 1874) [Amb.: Menop.]; *Philopterus microsomaticus* Tandan 1955 [Isch.: Philop.].

Passeriformes–Motacillidae

Anthus pratensis (Linnaeus) (Meadow Pipit): *Brueelia corydalla* Timmermann 1950 [Isch.: Philop.]; *Menacanthus pusillus* (Nitzsch 1866) [Amb.: Menop.]; *Ricinus fringillae* De Geer 1778 [Amb.: Ricinidae].

Anthus rubescens (Tunstall) (Buff-bellied Pipit) (According to Salomonsen (1967): *A. spinoletta rubescens* (Turnstall)): *Ricinus fringillae* De Geer 1778 [Amb.: Ricinidae].

Motacilla alba Linnaeus (White Wagtail): *Menacanthus nogoma* Uchida 1926 [Amb.: Menop.]; *Menacanthus pusillus* (Nitzsch 1866) [Amb.: Menop.]; *Philopterus passerinus* (Denny 1842) [Isch.: Philop.]; *Ricinus fringillae* De Geer 1778 [Amb.: Ricinidae]; *Sturnidoecus aeneas* (Piaget 1885) [Isch.: Philop.].

Passeriformes–Muscicapidae

Oenanthe oenanthe (Linnaeus) (Northern Wheatear): *Menacanthus affinis* Fedorenko [& Belskaya] 1977 [Amb.: Menop.]; *Menacanthus exilis* (Nitzsch 1866) [Amb.: Menop.]; *Philopterus oenanthe* Fedorenko 1983 [Isch.: Philop.].

Passeriformes–Sturnidae

Sturnus vulgaris Linnaeus (European Starling) – Rare/Accidental: *Brueelia nebulosa* (Burmeister 1838) [Isch.: Philop.]; *Menacanthus eurysternus* (Burmeister 1838) [Amb.: Menop.]; *Myrsidea cucullaris* (Nitzsch 1818) [Amb.: Menop.]; *Ricinus elongatus* (Olfers 1816) [Amb.: Ricinidae]; *Sturnidoecus sturni* (Schrank 1776) [Isch.: Philop.].

Passeriformes–Turdidae

Turdus iliacus Linnaeus (Redwing) – Rare/Accidental: *Brueelia haftorni* (Balat 1981) [Isch.: Philop.];.

Brueelia iliaci (Denny 1842) [Isch.: Philop.]; *Myrsidea iliaci* Eichler 1951 [Amb.: Menop.]; *Ricinus elongatus* (Olfers 1816) [Amb.: Ricinidae].

Turdus pilaris Linnaeus (Fieldfare): *Brueelia antimarginalis* Eichler 1951 [Isch.: Philop.]; *Brueelia marginata* (Burmeister 1838) [Isch.: Philop.]; *Menacanthus eurysternus* (Burmeister 1838) [Amb.: Menop.]; *Philopterus bischoffi* (Eichler 1951) [Isch.: Philop.]; *Ricinus elongatus* (Olfers 1816) [Amb.: Ricinidae].

Pelecaniformes–Sulidae

Morus bassanus (Linnaeus) (Northern Gannett) – Rare/Accidental: *Eidmanniella pustulosa* (Nitzsch [In Giebel] 1866) [Amb.: Menop.]; *Pectinopygus bassani* (Fabricius [O.] 1780) [Isch.: Philop.].

Pelecaniformes–Phalacrocoracidae

Phalacrocorax carbo (Linnaeus) (Great Cormorant): *Eidmanniella pellucida* (Rudow 1869) [Amb.: Menop.]; *Pectinopygus gyricornis* (Denny 1842) [Isch.: Philop.].

Procellariiformes–Hydrobatidae

Oceanodroma leucorhoa (Vieillot) (Leach's Storm Petrel) – Rare/Accidental: *Austromenopon pelagicum* Timmermann 1963 [Amb.: Menop.]; *Halipeurus (Synnautes) pelagicus* (Denny 1842) [Isch.: Philop.]; *Philoceanus amadoni* Timmermann 1961 [Isch.: Philop.]; *Saemundssonia (Saemundssonia) incisa* Timmermann 1950 [Isch.: Philop.].

Procellariiformes–Procellariidae

Fulmarus glacialis (Linnaeus) (Northern Fulmar): *Ancistrona vagelli* (Fabricius [J.C.] 1787) [Amb.: Menop.]; *Austromenopon brevifimbriatum* (Piaget 1880) [Amb.: Menop.]; *Perineus nigrolimbatus* (Giebel 1874) [Isch.: Philop.]; *Saemundssonia (Saemundssonia) occidentalis* (Kellogg 1896) [Isch.: Philop.].

Puffinus gravis (O'Reilly) (Great Shearwater) – Rare/ Accidental: *Ancistrona vagelli* (Fabricius [J.C.] 1787) [Amb.: Menop.]; *Austromenopon paululum* (Kellogg & Chapman 1899) [Amb.: Menop.]; *Halipeurus* (*Halipeurus*) *gravis* Timmermann 1961 [Isch.: Philop.].

Strigiformes–Strigidae

Asio flammeus (Pontoppidan) (Short-eared Owl) – Rare/ Accidental: *Colpocephalum brachysomum* Kellogg & Chapman 1902 [Amb.: Menop.]; *Kurodaia* (*Conciella*) *flammei* Price & Beer 1963 [Amb.: Menop.]; *Strigiphilus cursor* (Burmeister 1838) [Isch.: Philop.].

Nyctea scandiaca (Linnaeus) (Snowy Owl): *Strigiphilus ceblebrachys* (Denny 1842) [Isch.: Philop.].

MAMMALIA

Artiodactyla–Bovidae

Ovis aries Linnaeus (Sheep): *Bovicola* (*Bovicola*) *ovis* (Schrank 1781) [Isch.: Tricho.]; *Linognathus africanus* Kellogg and Paine 1911 [Anop.: Linog.]; *Linognathus ovillus* (Neumann 1907) [Anop.: Linog.]; *Linognathus pedalis* (Osborn 1896) [Anop.: Linog.].

Artiodactyla–Cervidae

Rangifer tarandus (Linnaeus) (Caribou/Reindeer): *Solenopotes tarandi* (Mjöberg 1915) [Anop.: Linog.]; *Bovicola* (*Bovicola*) *tarandi* (Mjöberg 1910) [Isch.: Tricho.].

Carnivora–Canidae

Alopex lagopus Linnaeus (Arctic fox): *Linognathus setosus* (von Olfers 1816) [Anop.: Linog.].

Canis lupus Linnaeus (Gray wolf): *Linognathus setosus* (von Olfers 1816) [Anop.: Linog.]; *Linognathus setosus* (von Olfers 1816) [Anop.: Linog.]; *Trichodectes canis* (De Geer 1778) [Isch.: Tricho.].

Carnivora–Mustelidae

Mustela erminea Barrett-Hamilton (Ermine/Stoat): *Stachiella ermineae* Hopkins 1941 [Isch.: Tricho.].

Carnivora–Odobenidae

Odobenus rosmarus (Linnaeus) (Walrus): *Antarctophthirus trichechi* (Bohemann 1865) [Anop.: Echino.].

Carnivora–Phocidae

Cystophora cristata Erxleben (Hooded seal): *Echinophthirius horridus* (von Olfers 1816) [Anop.: Echino.].

Erignathus barbatus Erxleben (Bearded seal): *Echinophthirius horridus* (von Olfers 1816) [Anop.: Echino.].

Phoca groenlandica (Erxleben) (Harp seal): *Echinophthirius horridus* (von Olfers 1816) [Anop.: Echino.].

Phoca hispida (Schreber) (Ringed seal): *Echinophthirius horridus* (von Olfers 1816) [Anop.: Echino.].

Phoca vitulina Linnaeus (Harbor seal): *Echinophthirius horridus* (von Olfers 1816) [Anop.: Echino.].

Lagomorpha–Leporidae

Lepus arcticus Ross (Arctic hare): *Haemodipsus lyriocephalus* (Burmeister 1839) [Anop.: Poly.].

Primates–Hominidae

Homo sapiens Linnaeus (Human): *Pediculus humanus* Linnaeus 1758 [Anop.: Pediculidae]; *Pthirus pubis* (Linnaeus 1758) [Anop.: Pthiridae.].

Rodentia–Muridae

Dicrostonyx groenlandicus Traill (Northern collared lemming): *Hoplopleura acanthopus* (Burmeister 1839) [Anop.: Hoplo.].

CHAPTER 9

Thysanoptera (Thrips)

Richard zur Strassen†

Thrips are minute insects occurring all over the world from low plains up to high alpine mountainous regions. The average body size is 1.2–2.5 mm. The majority of the species are macropterous, having two pairs of very slender wings. Other species are brachypterous with short wing pads, or are even apterous, having no wings at all. The scientific name of the whole order (Thysanoptera) refers to the dense hair fringes along their wing margins. Important morphological peculiarities in thrips are that the hypognathous mouth parts are shifted ventrally under the prothorax, and the right mandible is absent. Thrips feed by piercing host tissues in order to imbibe the cell contents. Most of the worldwide 5500 species are fungiphagous or phytophagous, roughly 300 species are zoophagous (zur Strassen 1995).

The thrips fauna of Greenland has so far not been explored with a similar intensity as is the case with most of the European countries and eastern North America. The first notice of a record of a thrips from Greenland was published by Meinert (1896: 163) who mentioned but one species, *i.e.*, *Physopus vulgatissimus* (Haliday, 1836). No additional taxa were cited by either Nielsen (1907: 408) or Henriksen & Lundbeck (1917: 708). Five species were listed by Richter (1928: 850) and the same species were included by Henriksen (1939). Only the family Thripidae is involved.

FIGURE 1 *Female habitus, dorsal view. A,* Thrips vulgatissimus*; B,* Apterothrips secticornis.

Key to the Greenland species of Thripidae (females)

1. Specimens apterous, wings lacking, no ocelli between compound eyes 2
– Specimens macropterous, wings fully developed, ocelli present between compound eyes 4
2. Body stout, head slightly broader than long. Posterior margin of tergites with a craspedon (Fig. 2C). Seta S2 on tergite IX short and somewhat thickened .... *Apterothrips secticornis*
– Body slender, head somewhat longer than broad. Posterior margin of tergites without a craspedon. Seta S2 on tergite IX long and thin.................. 3
3. Antenna 8-segmented. Setae S1 on tergite IX (placed between the two posterior campaniform sensilla) easy to see, 43–67 µm long, clearly longer than their

distance from each other (Fig. 2E)................... .... *Aptinothrips stylifer*

– Antenna 6-segmented. Setae S1 on tergite IX (placed between the two posterior campaniform sensilla) minute, barely visible, much shorter than their distance from each other (Fig. 2G) ....*Aptinothrips rufus*

4. Tergites VII and VIII laterally near the spiraculum with a well developed ctenidium (Fig. 2H). Pronotum with two postero-angular setae at each side (Fig. 2J). Dorsal seta S1 on abdominal segment X thin and long. Sense cones on antennal segments III and IV forked...5

– Tergites VII and VIII laterally without ctenidium. Pronotum with one postero-angular seta at each side. Dorsal seta S1 on abdominal segment X rather short, thorn-like (Fig. 2I). Sense cone on antennal segments III and IV simple, unforked (Fig. 2K) .... *Limothrips cerealium*

5. Anterior margin of pronotum with two pairs of long setae. Anterior vein of forewing with a complete row of equally distant setae ....*Frankliniella intonsa*

– Anterior margin of pronotum with at most one pair of elongated setae. Anterior vein of forewing in its distal half generally with three (1 + 2) setae, rarely with four or five.....................................6

6. Antenna 8-segmented. Metanotum with two campaniform sensilla (Fig. 3E). Hind margin of tergite VIII with a complete comb of microtrichia. Sternites with discal setae (Fig. 3G).......*Thrips vulgatissimus*

– Antenna 7-segmented. Metanotum without campaniform sensilla (Fig. 3F). Hind margin of tergite VIII medially without a comb of microtrichia, being developed only laterally. Sternites without discal setae (Fig. 3H)........................... *Thrips major*

***Apterothrips secticornis* (Trybom, 1896) Figs 1B, 2A, C**

Hemianaphothrips postumus Richter, 1928 [**n. syn.**]

Henriksen (1939): *Anaphothrips secticornis* Tryb. & *Anaphothrips* (*Hemianaphothrips*) *postumus* Richter.

Seems to be the second-most abundant thrips in Greenland (after *Thrips vulgatissimus*), making up about 9% of the specimens studied.

DISTRIBUTION. S, SW, NW, SE

Holarctic, in cold temperate latitudes, particularly in mountainous and alpine areas.

Unplaced name: ***Anaphothrips lundbecki* Richter, 1928**

As explicitly noted in the original description (Richter, 1928: 850) the unique type specimen is "filled with disc-shaped, so far not interpretable bodies of 0,014 mm diameter" [translated from German]. The formations in question are spores of a pathogenic fungus, likely of the genus *Neozygites* (J. Eilenberg, pers. comm), which is known from other thysanopteran taxa (Carris & Humber 1998). The specimen is, then, most probably a teratological *A. secticornis* whose allegedly diagnostic structural details (wide separation of middle coxae) are effects of the infestation, but no formal synonymization is made here.

***Aptinothrips rufus* Haliday, 1836 Figs 2F, G**

Henriksen (1939): *Aptinothrips nitidulus groenlandicus* Hal.

Seems to be quite rare in Greenland.

DISTRIBUTION. SW (recorded from four localities from Kuanersoq/Kvanefjord in the south to Ukkusissat Fjord: Pangnertoq in the north), NE (Eskimonæs).

Cosmopolitan.

***Aptinothrips stylifer* Trybom, 1894 Figs 2D, E**

DISTRIBUTION. nSW (1 specimen from Kangerlussuaq/Søndre Strømfjord, 1 from Evighedsfjord: Kangiussaq).

Holarctic.

***Frankliniella intonsa* (Trybom, 1895) Fig. 2J**

DISTRIBUTION. nSW (1 doubtful specimen from Arnangarnup Qoorua/Paradisdal at middle Kangerlussuaq/Søndre Strømfjord; also 1 specimen from 1889 without locality).

Palaearctic (except northern Africa), northern part of the Oriental region, Canada.

***Limothrips cerealium* Haliday, 1836 Figs 2B, I, K**

DISTRIBUTION. N (1 specimen from Heilprin Land).

Almost cosmopolitan.

***Thrips* (?) *major* Uzel, 1895 Figs 3F, H**

DISTRIBUTION. nSW (1 specimen from Kangerlussuaq/Søndre Strømfjord and 1 from Kangiussap Imaa, Svartenhuk).

Palaearctic.

***Thrips vulgatissimus* Haliday, 1836 Figs 1A, 3E, G**

Henriksen (1939): *Taeniothrips inconsequens* Uzel [likely misidentification].

By far the most abundant thrips in Greenland, making up about 90% of the material studied.

DISTRIBUTION. S, SW, NW, NE, SE (not in N).

Circumpolar, arctic-temperate, also New Zealand, Australia.

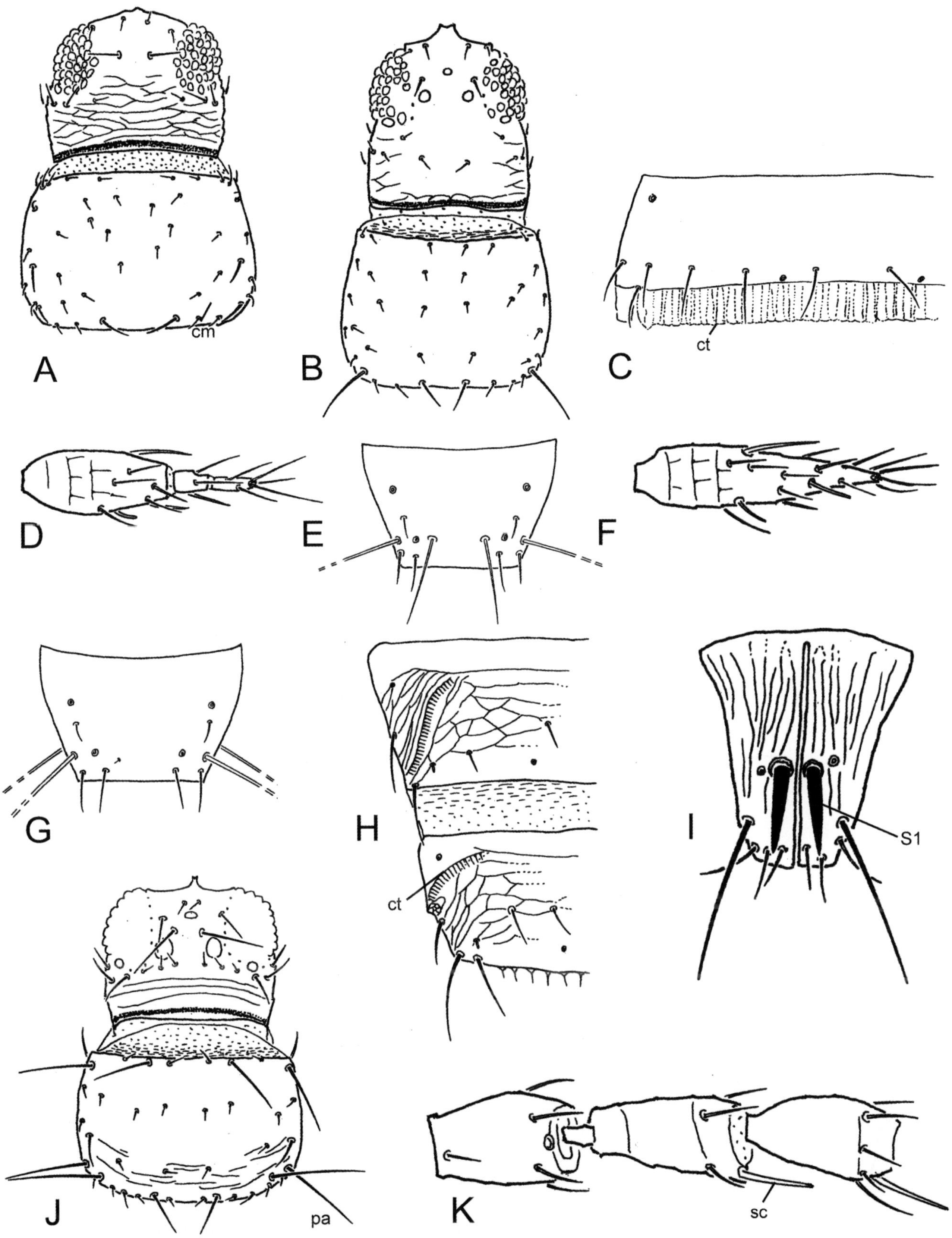

FIGURE 2 *Structural details. A,* Apterothrips secticornis, *female head and pronotum. dorsally; B, same of* Limothrips cerealium*; C,* Apterothrips secticornis, *female, tergite IV (left hand half) showing the craspedon (cr); D,* Aptinothrips stylifer, *female, right antenna, segments VI–VIII; E, same, female, tergite IX (sculpture omitted); F,* Aptinothrips rufus, *female, segment VI of right antenna; G, same, female, tergite IX; H,* Frankliniella intonsa, *female, tergites VII and VIII (left hand half) showing the ctenidium (ct); I,* Limothrips cerealium, *female, tergite X showing the strong seta S1; J,* Frankliniella intonsa, *female, head and pronotum dorsally showing the postero-angular setae (pa); K,* Limothrips cerealium *Haliday, female, right antenna, segments II–IV, sc = sense cone. A–E, G–J from zur Strassen (2003; A–B, D slightly altered).*

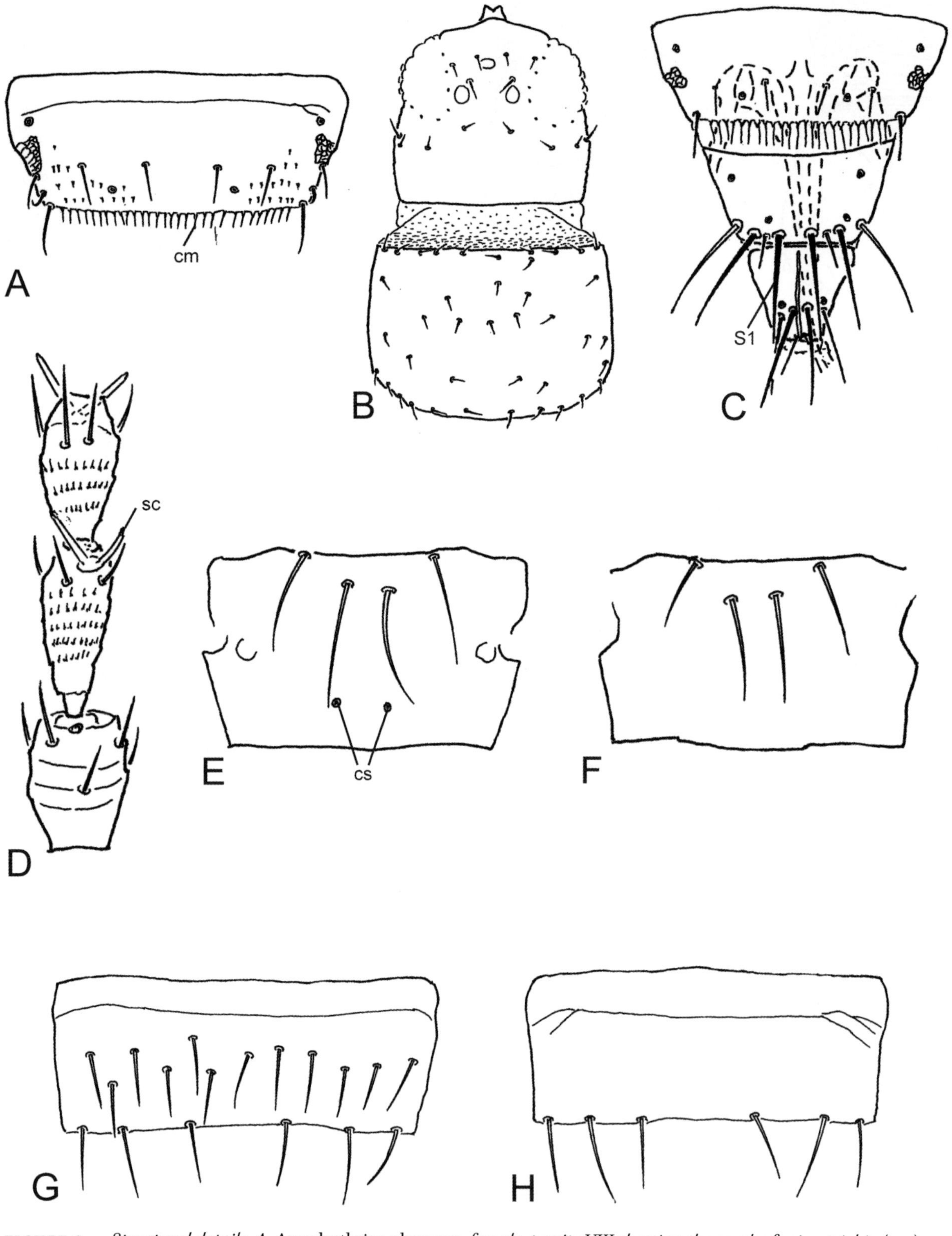

FIGURE 3 *Structural details. A,* Anaphothrips obscurus, *female, tergite VIII showing the comb of microtrichia (cm); B,* Anaphothrips obscurus, *female, head and pronotum dorsally (ommatidia and sculpture omitted); C,* Anaphothrips obscurus, *female, abdominal segments VIII–X dorsally with seta S1 on IX (sculpture omitted); D,* Anaphothrips obscurus, *female, right antenna, segments II–IV, sc = sense cone; E,* Thrips vulgatissimus, *female, metanotum showing the campaniform sensillae (cs) (sculpture omitted); F,* Thrips major, *female, metanotum (sculpture omitted); G,* Thrips vulgatissimus, *female, sternite VI (sculpture omitted); H,* Thrips major, *female, sternite VI (sculpture omitted). A–C from zur Strassen (2003).*

NOTES. It is noteworthy that not a single specimen of the true *Taeniothrips inconsequens* (Uzel ,1895) was found. The occurrence of this species in southern Greenland is not unlikely. The species was mentioned by Richter (1928: 850) and Henriksen (1939: 19). The present author assumes that Richter was misled by the number of setae in the distal half of the anterior vein of the forewing of those specimens he regarded as *inconsequens.* This species has 4–8 such setae, but in *Thrips vulgatissimus* usually only three distal setae are present. Very rarely, however, four or even five of these distal setae may occur in the latter species. It therefore seems possible that the record of *inconsequens* from Greenland is based on a misidentification of *vulgatissimus*. The sternites in *vulgatissimus* have a transverse row of discal (accessory) setae which are often absent in *inconsequens*.

Acknowledgement

Prof. J. Eilenberg (Department of Agriculture and Ecology, University of Copenhagen) is thanked for providing an expert opinion on the contents of body cavity in the type specimen of *A. lundbecki*.

References

Carris, L.M. & Humber, R.A. (1998): *Neozygites parvispora*, a pathogen of *Limothrips* sp. associated with *Lolium multiflorum* in Oregon. – *Mycologia* 90: 565–568.

Henriksen, K.L. (1939): A revised index of the insects of Grønland. – *Meddelelser om Grønland* 119 (10): 1–112.

Henriksen, K.L. & Lundbeck, W. (1917): Landarthropoder (Insecta et Arachnida). Conspectus Faunæ Groenlandicæ, pars 2. – *Meddelelser om Grønland* 22: 481–823.

Meinert, F. (1896): Neuroptera, Pseudoneuroptera, Thysanopoda, Mallophaga, Collembola, Suctoria, Siphunculata Groenlandica. – *Videnskabelige Meddelelser fra den naturhistoriske Forening i Kjøbenhavn* 48: 154–177.

Nielsen, J.C. (1907): The insects of East-Greenland. – *Meddelelser om Grønland* 29: 363–414.

Richter, W. (1928): Die Thysanopteren des arktischen Gebietes. – *Fauna arctica* 5: 837–850.

zur Strassen, R. (1995): Bionomial data of some predacious thrips. 325–328 *In*: Parker, B.L., Skinner, M. & Lewis, T. (eds.): *Thrips, biology and management,* NATO Science Series (A): 325–328.

zur Strassen, R. (2003): Die Terebranten Thysanopteren Europas und das Mittelmeer-Gebietes. *Die Tierwelt Deutschlands* 74, 277 pp.

CHAPTER 10

Hemiptera
(*Plant bugs, true bugs, and relatives*)

10.1. Generalities and key to principal groups

Jens Böcher and Niels P. Kristensen†

The order Hemiptera is the dominant group of exopterygote insects. Its members are extremely diverse and do not share any immediately observable characteristics. However, they all possess a highly specialised configuration of the mouthparts, which are adapted for piercing and sucking liquid contents of plants and animals. This mouth apparatus (rostrum or proboscis, Fig. 1) comprises paired mandibular and maxillary stylets which are enclosed within a 'sheath' formed by the grooved labium and proximally they are covered by the labrum. The rostrum tip is equipped with sensory organs which facilitate selection of a suitable feeding site. When in position, the proboscis 'sheath' is bent backwards and the stylets are protruded and inserted into the substrate. Each maxillary stylet has on the inner surface two grooves which, when the stylets are apposed together in the mid-line, constitute two fine, longitudinal canals; food is sucked up through the anterior one and saliva pumped down the posterior. No remains of palps (maxillary or labial) can be traced. Hemipteran wing structure is very diverse. The fore wings may be be more thickened and pigmented than the hind wings, which are always membraneous; in the Cicadomorpha and (particularly) the Heteroptera the fore wings have the apical portion less thickened than the proximal part. Devices for coupling the fore- and hindwings have evolved independently on several occasions.

The Hemiptera comprise just over 100.000 described species worldwide. several are terrestrial and phytophagous; only the Heteroptera include families whose members are all or predominantly aquatic and/or predacious. While the time-honoured collective group 'Homoptera' comprising the non-heteropteran hemiptera has long been recognized as being a likely non-monophyletic assemblage, the monophyly of the taxon Sternorrhyncha and its four constituent high-rank taxa (Psylloidea, Aleyrodoidea, Aphidomorpha,

FIGURE 1 *Head of* Nabis flavomarginatus, *illustrating the hemipteran rostrum.*

Coccoidea) receives continued support. Most principal hemipteran groups are represented in the country, the exceptions being Aleyrodoidea (which may, however, eventually become introduced with indoors cultivated plants) and Fulgoromorpha. Compared to the situation in high-latitude continental Europe the Sternorrhyncha constitute an unusually high proportion of the Greenland hemipteran fauna.

Key to the principal groups of Greenland Hemiptera

Male Coccoidea which have completely reduced mouthparts and are not immediately identifiable as Hemiptera are omitted from the key. They resemble small nematoceran Diptera, but their tiny hind wing vestiges are slender appendages linked to the forewing bases rather than discrete club-like halteres, and their tarsi are 1-segmented (almost always 5-segmented in Diptera). Older instars of many female Coccoidea ('scale insects') are also not immediately identifiable as Hemiptera, or even as insects.

1. Long axis of head forwards pointing, hence proboscis apparently originating from front end of head. Fore wings (when present) held horizontally in repose, with sharply delimited apical portion more membranous than the proximal part. .... Heteroptera (True bugs), 10.6
– Long axis of head reflexed, hence proboscis originating behind front end of head, sometimes close to front leg bases. Fore wings (when present) held rooflike in repose; homogeneously membranous throughout, or thicker proximal region not sharply delimited against less thickened/pigmented apical area 2
2. Antennae very short, with terminal arista. Rostrum clearly arising in front of fore legs. Tarsi 3-segmented. .... Cicadomorpha, 10.5
– Antennae longer and without terminal arista. Rostrum (if present) appears to arise between fore legs. Tarsi with 1–2 segments .... 3 (Sternorrhyncha)
3. Wings well developed 4
– Wings represented by small rudiments/buds, or absent 6
4. Fore wings whitish due to delicate wax covering, venation highly reduced.. .... [Aleyrodoidea],
– Fore wings transparent, venation distinct 5
5. Fore wing with veins branching off from a central common stem, forming a tree-like figure (as Figs 10.2–3B). ....Psylloidea, 10.2
– Fore wings immediately behind front margin with a strong (composite) longitudinal vein from which a few obliquely-transverse veins branch off in a ± comb-like fashion. Abdomen often with paired tubular processes on 5th or 6th segment. ....Aphidomorpha, 10.3
6. Distal antennal segment with apical part ('processus terminalis') distinctly set off from wider proximal part. Abdomen often with paired tubular processes on 5th or 6th segment. Legs always well developed, tarsi 2–segmented and with paired claws ....Aphidomorpha, 10.3
– Distal antennal segment with apical part not distinctly set off from proximal part. Abdomen never with paired tubular processes on 5th or 6th segment. Legs differently developed, sometimes strongly reduced 7
7. Tarsal claws paired. Body sometimes with waxy filaments, but no solid covers .... Psylloidea (nymphs), 10.2
– Tarsal claws unpaired. Body with ± solid cover of wax plates and/or other secretions and exuviae of earlier instars ....Coccoidea (females and nymphs), 10.4

10.2. Psyllidae (Jumping plant-lice, psyllids)

Ian D. Hodkinson and Jerry M. Bird

Adult members of this family (Fig. 2) are reminiscent of tiny (2–4 mm) cicadas, but the venation of the fore wings (Figs 3B, 4B, 5B) and the long antennae of the Psylloidea make them readily identifiable. Psyllids are poor fliers but they are excellent jumpers using their enlarged hind legs. Most species are highly host-plant specific, occurring on one or two closely related plant species, often woody plants. Several species form galls on their host. The flat, relatively immobile nymphs often produce fluffy wax threads (Fig. 1) to which droplets of honey-dew frequently remain attached. When psyllids are abundant, the copious honey-dew excreted by the nymphs may encourage fungal infection of leaves.

Psyllids, like aphids, are plant phloem sap-sucking insects that depend on their host plant for food and shelter. Close phenological synchrony with the growth cycle of their host plant is necessary for the psyllids successfully to complete their life cycle, particularly during the short Arctic growing season. Typically this cycle comprises of an egg, five nymphal instars and an adult stage. All the Greenland species are univoltine, but the overwintering stage tends to differ among genera. Nymphs are, to a greater or lesser extent, dorso-ventrally flattened and are usually found associated with the newly growing shoot tips, expanding leaves or catkins of their host plant. Adults are often long-lived and change body colour considerably over time, from pale colours, such as yellow or pale orange of newly emerged individuals to reddish or dirty white background colour with strong brown or black markings typical of the heavily-sclerotised older individuals. Body colour is thus an unreliable character for identification and emphasis should be placed on stable morphological features.

FIGURE 1 Psylla betulaenanae *nymphs producing copious amounts of white flocculent wax within which the liquid, sugary faeces are trapped.* (PHOTO: A. FJELLBERG)

Morphology: adults

Characterised by the strongly developed hind legs adapted for leaping. The large, curved hind coxae are fused to the thorax and each bears a short, curved, backwardly directed tapering process, the meracanthus. The hind femur and tibia are elongate; the latter, together with the first tarsal segment, bears a variable number of black apical saltatorial spurs. Adult psyllids normally have 10–segmented antennae and in many taxa the genae are developed into a pair of conspicuous forward or downward directed processes referred to as genal cones (Figs 3A, 4A, 5A). Major forewing characters that distinguish taxa include the pattern of wing venation, the size and shape of cells, the size and shape of pterostigma and the presence/absence of an associated costal break (Figs 3B, 4B, 5B). The distribution pattern of surface microtrichiae on the membrane (referred to as spinules) is often used to distinguish species.

Male and female terminalia are also of high taxonomic significance. Male terminalia (Figs 3D, 4D, 5D) comprise a subgenital plate within which are attached the bases of a pair of claspers, the parameres (Figs 3E, 4E, 5E), and the fold-away two-segmented aedeagus (Figs 3F, 4F, 5F). The male proctiger, which bears the anus terminally, often takes the form of a simple tube attached basally to the subgenital plate (Figs 3D, 4D, 5D). Female terminalia (Figs 3C, 4C, 5C) comprise two elongate sclerotised outer valves. The upper valve, the proctiger, which bears basally the anus with its circumanal pore field, often tapers to a slender drawn out apex. The lower valve, the subgenital plate, is usually shorter than the proctiger, somewhat triangular in shape and often with the apex narrowly acute. The proctiger and the subgenital plate together enclose the inner palpi, the valvulae ventrales and the ovipositor.

Immature stages

Nymphal stages are recognized on the number of segments in the antennae, the numbers of setae and the number of tarsal segments. Final instar nymphs (Figs 3G, 4G, 5G) of different species are primarily separated by general shape of the body and appendages, size, shape, position and number of setae, and by position and configuration of the anal pore fields (Figs 3H, 4H, I, 5H, I).

Greenland fauna

The known psyllid fauna of Greenland is depauperate, comprising just three species. An understanding of their geographical origin depends on knowledge of the history of their host plants in Greenland. The circumboreal dwarf shrub *Betula nana*, for example, is thought initially to have colonised the east coast of Greenland from Iceland by around 8000 years B.P., crossing the Icecap to reach the west coast at Nuup Kangerlua/Godthåbsfjord by around 6500 years B.P. and then subsequently spreading southwards (Hultén 1968, T.W. Böcher *et al.* 1978, Fredskild 1996). It is still absent from the extreme south of Greenland. The North American *Betula glandulosa* by contrast is first recorded from the SW Greenland floral province around 5700 B.P., subsequently extending its range to S Greenland (Hultén 1968, Fredskild 1996). *Betula pubescens* is similarly confined to SW and S Greenland. *Alnus crispa*, a species distributed throughout eastern Siberia and northern parts of Canada and U.S.A., arrived in SW Greenland significantly later, by around 4000 years B.P. and retains this restricted distribution. By contrast *Salix glauca*, the primary host of *Cacopsylla groenlandica* (Šulc), was probably the earliest psyllid host plant to arrive, being known from the Paamiut/Frederikshåb area of SW Greenland by around 9000 years B.P. This species is now widely established with a continuous range north to the low arctic-high arctic transition. By contrast, two other important *Salix* hosts, with their present distribution centres in boreal North America, *Salix arctophila* and *Salix uva-ursi*, are restricted to SW Greenland.

The evidence thus suggests strongly that the primary psyllid host plants have colonised Greenland from North America. The psyllid data appear to support this contention (Hodkinson 1978b). *Cacopsylla groenlandica* appears confined to the west coast of Greenland, despite the wider distribution of its host plant. Its nearest relative is the Alaskan/Beringean *Salix*-feeding species *Cacopsylla toolikensis* (Hodkinson) (Hodkinson 1978a) from which it differs primarily in small details of the male terminalia. Earlier records of *C. groenlandica* for Europe (*e.g.* Šulc 1913) are incorrect (Klimaszewski 1973). *Cacopsylla septentrionalis* (Šulc), described from Greenland, appears to be a form of *C. groenlandica*. Similarly, the closest relative of *Psylla borealis* (Horvath) is the Alaskan form of the North American *Psylla viridescens* (Provancher) (=*Psylla galeaformis* Patch), which likewise feeds on *Alnus crispa* but from which it can again be distinguished by features of the male terminalia. A resemblance to the European *Psylla alni* (L.) led previous European authors (Šulc 1913, Henriksen 1939) incorrectly to regard this species as a small dark subspecies of the former. The origins of *Psylla betulaenanae* Ossianilsson are a little more problematical and one can postulate initial colonization occurring on either the east or the west coast of Greenland or perhaps more interestingly, given the disjunct distribution pattern of the insect and its hosts, an independent colonization of either coast (Hodkinson 1988, Ossiannilsson 1970, 1973). This, and other relationships of the Greenland psyllids to related extant species elsewhere, will only be resolved by detailed genetic analyses.

Key to species

1. Antennae (Fig. 2C) short, less than 1.8 times maximum head breadth (including eyes). Forewing (Fig. 3B) with relatively broad pterostigma. Male paramere in lateral view (Figs 3D, E) tapering to a narrow drawn out apex; extreme apex when viewed from above appearing as a dark slender, forward-facing hook. Aedeagus (Fig. 3F) with strongly hooked apex. Female terminalia (Fig. 3C) shorter, more wedge-shaped. Head in dorsal view as in Fig. 3A. Nymph (Fig. 3G) strongly dorso-ventrally flattened, with anal pore ventral, surrounded by simple narrow pore fields (Fig. 3H). On *Salix* spp .... *Cacopsylla groenlandica*

– Antennae (Figs 2A, B) long, more than twice maximum head breadth (including eyes). Forewing (Figs 4B, 5B) with a narrow, gradually tapering pterostigma. Male paramere in lateral view (Figs 4D, E, 5D, E) club-shaped, with apex somewhat expanded; extreme apex when seen from above comprising two inwardly pointing denticles. Aedeagus (Figs 4F, 5F) with apex at most weakly hooked. Female terminalia in lateral view (Figs 4C, 5C) long and slender. Nymph (Figs 4G, 5G) less flattened; anus terminal, with a large broad pore field that spreads onto both dorsal and ventral surfaces of abdomen (Figs 3H, I, 4H, I). On *Alnus* or *Betula* 2

2. Forewing membrane (Figs 2B, 4B) uniformly light orange-brown; veins concolorous. Female terminalia (Fig. 4C) with proctiger (dorsal valve) in profile with dorsal margin more or less straight. Male paramere in lateral view as in Figs 4D, E. Aedeagus (Fig. 4F) with apical portion oblong-oval. Head in dorsal view (Fig. 5A) with apex of genal processes broadly rounded, processes slightly shorter than length of vertex. General ground colour of body in mature specimens reddish orange, with strong dark brown to black markings (Fig. 2B). Nymph with shape of dorsal and ventral anal pore fields as in Figs 4H, I. On *Betula* spp. .... *Psylla betulaenanae*

– Forewing membrane (Figs 2A, 5B) clear, not light orange-brown, sometimes with suffuse greyish brown clouding in the cells; veins dark brown to black. Female terminalia (Fig. 5C) with dorsal margin of proctiger in profile appearing shallowly concave. Male paramere as in Figs 5D, E. Aedeagus (Fig. 5F) with apical portion relatively shorter, somewhat kidney-shaped. Head in dorsal view (Fig. 5A) with genal processes having a more narrowly rounded apex, processes slightly longer than vertex. General ground colour of body in mature specimens (Fig. 2A) dirty white to dirty yellow with strong dark brown to black markings. Nymph (Fig. 5G) with shape of the dorsal and ventral anal pore fields as in Figs 5H, I. On *Alnus crispa*. ....*Psylla borealis*

***Cacopsylla groenlandica* (Šulc, 1913) Figs 2C, 3**

Henriksen 1939: *Psylla ?ambigua* Frst.

BIOLOGY. Confirmed host plants are *Salix glauca*, *Salix herbacea*, *Salix arctophila* and *S. uva-ursi*. The life cycle of *C. groenlandica* was described in detail by Hodkinson (1997). It overwinters as a mature adult in protected situations within the leaf litter beneath its host plant *Salix*. Active adults are found on stems of the host plant during the very early growing season, often when snow is still on the ground, where they mate, often before the willows break buds. Mature mated females are thus ready to oviposit the moment the buds break and the newly growing *Salix* tissues become available. Egg laying and nymphal development take place mainly on the female catkin, although in the slightly longer and thus phenologically less demanding growing season of southern Greenland development can also take place on growing shoots. The number of *Salix* host species, available for exploitation, diminishes northwards as some willows such as *S. uva-ursi* and *S. arctophila* reach their northern range limits. *C. groenlandica* thus shows increasing restriction in the range of host-plants and tissues exploited with increasing latitude. At higher latitudes, where phenological synchrony with the host is most critical, development is confined to female catkins of *S. glauca*. Furthermore, successful completion of life cycle depends on nymphs completing their development before maturing catkins become unsuitable through drying out and dehiscing. The cycle is thus completed during the early growing

FIGURE 2
Greenland psyllid species in lateral view. A. Psylla borealis, *female (left) and male (right), B.* Psylla betulaenanae, *female (left) and male (right), C.* Cacopsylla groenlandica, *female (left) and male (right).*

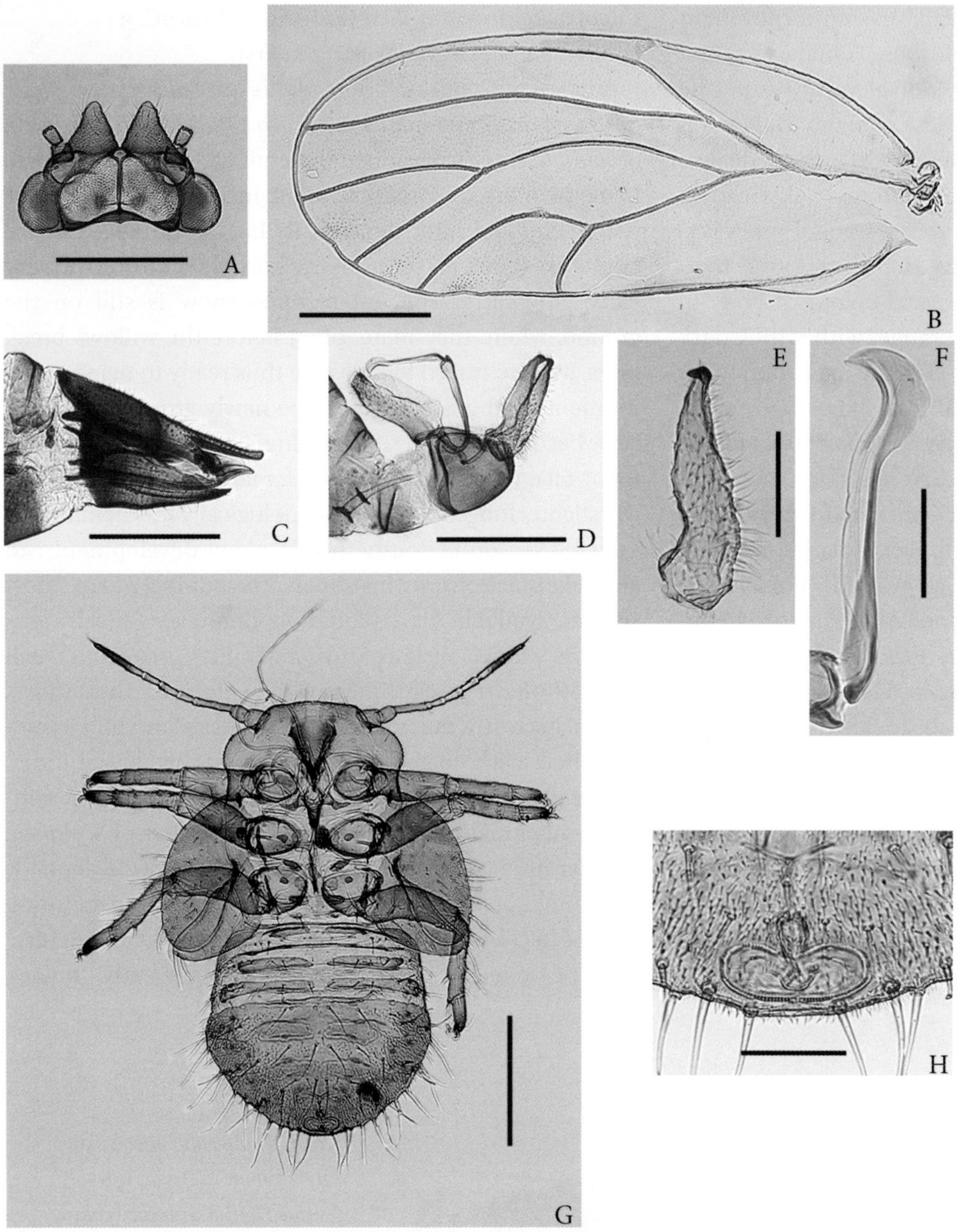

FIGURE 3
Cacopsylla groenlandica.
A. Head, dorsal view showing the genal processes, B. Forewing, C. Female terminalia in lateral view, D. Male terminalia in lateral view, showing the subgenital plate, the proctiger, paired parameres and the aedeagus, E. Male paramere enlarged, outer view, F. Aedeagus enlarged, G. Final instar nymph, H. Anus with ventral circumanal pore field. The scale line for head, forewing, male and female terminalia and nymph is 0.5 mm, for enlarged paramere is 0.2 mm and for the male aedeagus and nymphal circum-anal pore field is 0.1 mm.

season and adults spend the rest of the season feeding on their host before moving downwards to hibernation sites later in the year, coincident with leaf fall.

DISTRIBUTION. S, SW, NW (north to the Upernavik area). There are no reliable records for East Greenland. Possibly endemic to Greenland.

***Psylla betulaenanae* Ossiannilsson, 1970**
Figs 1, 2B, 4
Henriksen 1939: not recorded.

BIOLOGY. *P. betulaenanae* is known to complete its development on *Betula nana* and *Betula glandulosa*, although the free hybridisation between birch species sometimes makes the exact provenance of some host plants difficult to ascertain. *Betula pubescens* is another potential host plant. By contrast with *C. groenlandica*, *P. betulaenanae* overwinters as eggs laid the previous summer on the shoots of *Betula* spp. close to the overwintering buds. Eggs hatch following bud burst the subsequent year and nymphs move onto the growing shoots. Early instar nymphs tend to be found in the leaf clusters at shoot apex but older nymphs are often associated with more mature foliage. Nymphs are easily found as they produce copious amounts of white flocculent wax within which the liquid sugary faeces become trapped. These wax-covered nymphs are conspicuous against the darker greens and browns of the plant (Fig. 1). Nymphs complete their development in the early growing season, generally

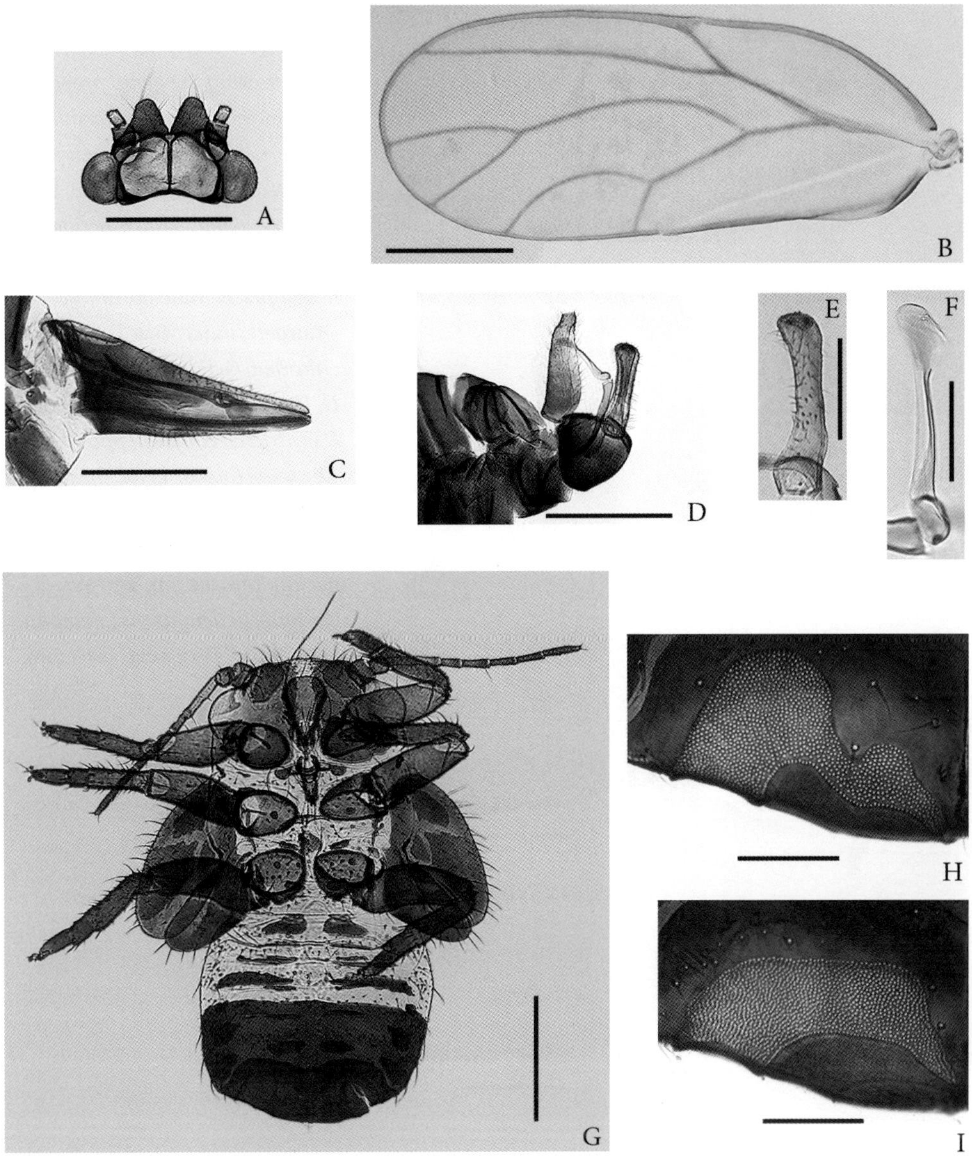

FIGURE 4
Psylla betulaenanae. *A. Head, dorsal view showing the genal processes, B. Forewing, C. Female terminalia in lateral view, D. Male terminalia in lateral view, showing the subgenital plate, the proctiger, paired parameres and the aedeagus. E. Male paramere enlarged, outer view, F. Aedeagus enlarged, G. Final instar nymph, H. Dorsal circumanal pore field, I. Ventral circumanal pore field. The scale line for head, forewing, male and female terminalia and nymph is 0.5 mm, for enlarged paramere is 0.2 mm and for the male aedeagus and nymphal circumanal pore fields is 0.1 mm.*

before *Betula* leaves reach maturity. Emerging adults then persist throughout summer, with females ovipositing until the leaves begin to senesce and fall.

DISTRIBUTION. S, SW, SE (a disjunct distribution, north to Kangerlussuaq/Søndre Strømfjord and at localities around Ammasalik).

Circumpolar, low arctic-alpine.

***Psylla borealis* (Horvath, 1908) Figs 2A, 5**

Henriksen 1939: *Psylla alni* L.

BIOLOGY. Known only from *Alnus crispa*.

The detailed biology of *Psylla borealis* is unknown but almost certainly mirrors that of *P. betulaenanae* and other *Alnus*-feeding psyllids. It almost undoubtedly overwinters as eggs laid near the dormant buds. Mature nymphs, again characterised by their conspicuous covering of loose flocculent wax, are found on the growing shoots early in the growing season as the *Alnus* leaves approach maturity. Emerging adult females mate and continue to oviposit throughout the remaining season.

DISTRIBUTION. S, SW. Confined to sites where its host plant is found along the inner fjords of S and SW Greenland. Possibly endemic to Greenland.

References

Böcher, T.W., Fredskild, B., Holmen, K. & Jakobsen, K. (1978): *Grønlands Flora*. 3rd edition. P. Haase & Søns Forlag, København.

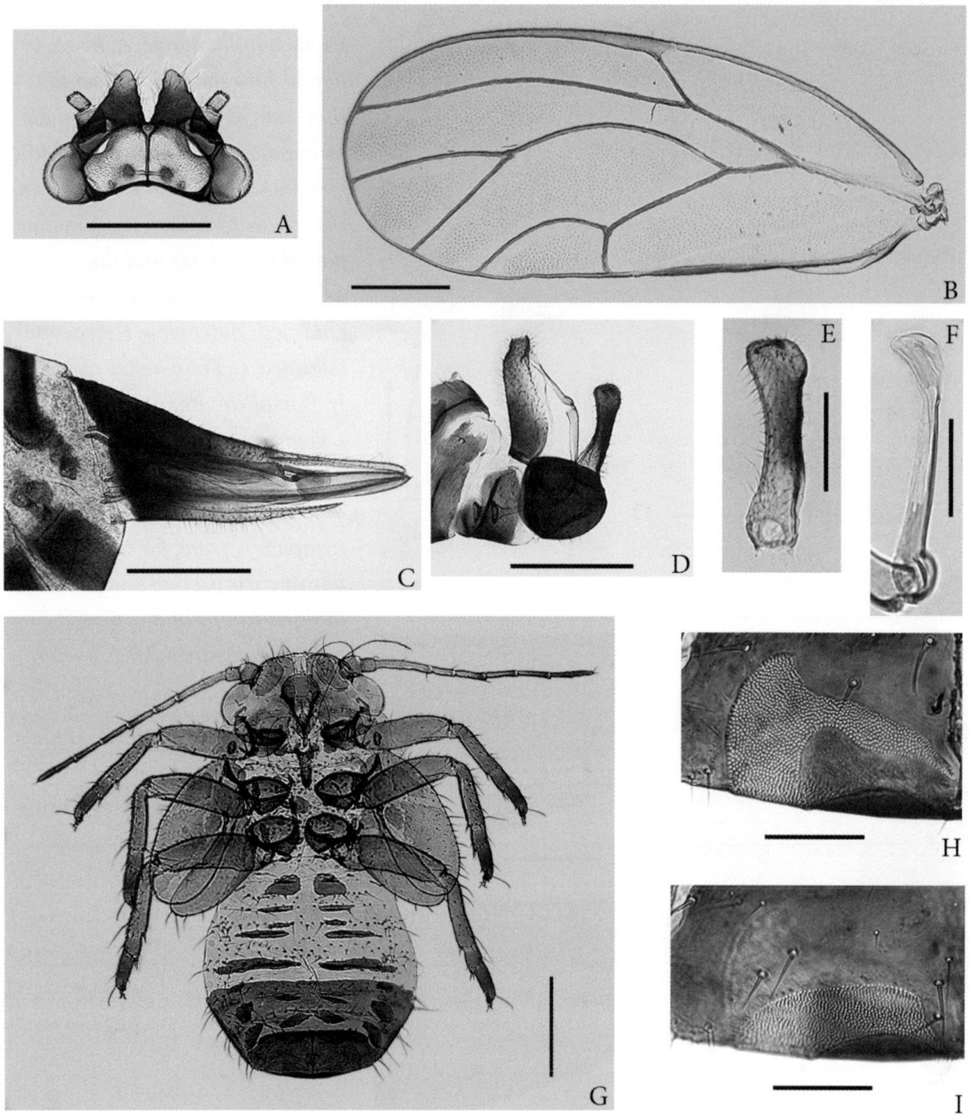

FIGURE 5

Psylla borealis. *A. Head, dorsal view showing the genal processes, B. Forewing, C. Female terminalia in lateral view, D. Male terminalia in lateral view, showing the subgenital plate, the proctiger, paired parameres and the aedeagus. E. Male paramere enlarged, outer view, F. Aedeagus enlarged, G. Final instar nymph, H. Dorsal circumanal pore field, I. Ventral circumanal pore field. The scale line for head, forewing, male and female terminalia and nymph is 0.5 mm, for enlarged paramere is 0.2 mm and for the male aedeagus and nymphal circumanal pore fields is 0.1 mm.*

Fredskild, B. (1996): A phytogeographical study of the vascular plants of West Greenland. – *Meddelelser om Grønland, Bioscience* 45: 1–157.

Henriksen, K.L. (1939): A revised index of the insects of Grønland. – *Meddelelser om Grønland* 119: 1–111.

Hodkinson, I.D. (1978a): The psyllids (Homoptera: Psylloidea) of Alaska. – *Systematic Entomology* 3: 333–360.

Hodkinson, I.D. (1978b): Present-day distribution patterns of the holarctic Psylloidea with particular reference to the origins of the Nearctic fauna. – *Journal of Biogeography* 7: 127–146.

Hodkinson, I.D. (1988): The nearctic Psylloidea (Insecta: Homoptera): an annotated check list. – *Journal of Natural History* 22: 1179–1243.

Hodkinson, I.D. (1987): Progressive restriction of host plant exploitation along a climatic gradient: the willow psyllid *Cacopsylla groenlandica* in Greenland. – *Ecological Entomology* 22: 47–54.

Hodkinson, I.D. (1993): The identities of psyllids (Homoptera: Psylloidea) in the Provancher collection with the designation of lectotypes. – *Canadian Entomologist* 125: 1141–1142.

Horvath, GH. & (1980): Rémarques sur quelques Hémiptères de l'Amerique du Nord. – *Annales Musei Nationalis Hungarici* 6: 555–569.

Hultén, E. (1968): *Flora of Alaska and Neighbouring Territories.* – Stanford University Press, Stanford.

Klimaszewski, S.M. (1973): The jumping plant-lice or psyllids of the Palaearctic. – *Annales Zoologici, Warszawa* 30: 188–198.

Ossiannilsson, F. (1970): Contributions to the knowledge of Swedish psyllids (Hemiptera: Psylloidea). – *Entomologica Scandinavica* 12: 22–26.

Ossiannilsson, F. (1992): The Psylloidea (Homoptera) of Fennoscandia and Denmark. – *Fauna Entomologica Scandinavica* 26: 1–347.

Patch, E.M. (1911): Psyllidae. – *Bulletin of the Maine Agricultural Experimental Station* 187: 10–20.

Šulc, K. (1913): Zur kenntnis einiger Psylla-arten aus dem Ungarischen National Museum in Budapest. – *Annales Historico-Naturales Musei Nationalis Hungarici* 11: 409–435.

Šulc, K. (1939): Description of a new psyllid from Greenland (Homoptera). – *Annals and Magazine of Natural History* (11) 4: 78–81.

Tuthill, L.D. (1943): The psyllids of America north of Mexico (*Psyllidae: Homoptera*). – *Iowa State College Journal of Science* 17: 443–660.

[Aleyrodoidea (Whiteflies)

Jens Böcher

Minute insects (wing span about 3 mm) with highly reduced venation of the fore wings which are covered by a fine waxy powder. Honey-dew is excreted in large quantities in all stages, particularly by the nymphs, from a specialized anal opening. Members of this superfamily has not yet been recorded from Greenland, but will undoubtedly sooner or later be found feeding on the underside of leaves of cultivated plants in greenhouses and inside houses. Most species are tropical, but a number of serious pests are present in greenhouses in the temperate region.

The life history is specialized. The flat, oval first-instar nymph is active, but after the first moult antennae and legs degenerate and the following three instars are immobile, often surrounded/protected by a waxy secretion. Feeding continues into the first part of the fourth instar, but stops after the final apolysis. The pharate adult is referred to as a 'puparium' (by analogy with conditions in cyclorrhaphan flies).

Parthenogenesis is common. Eggs are generally laid in an arc and provided with a stalk by which they are attached to leaves etc.]

10.3. Aphidomorpha (Aphids)

Ole E. Heie

A large group of Hemiptera (about 4500 species worldwide). Many aphids occur in immense number, and as most of them are parthenogenetic females, they can reproduce very quickly. All are small, the majority *c.* 2–3 mm. The body is usually pear-shaped with a small head and bulbous abdomen. The dominant colours are shades of green and brown. Wings are clear and membranous with reduced and very constant venation, but wingless forms are very common in most species. A specialization present in many groups is a pair of tubular processes (*siphunculi*) on the rear of abdomen which exude a defensive fluid when the aphid is attacked by a predator. Siphunculi are openings of modified wax glands. Many aphids also produce a powdery wax in dermal glands.

Specialized life cycles and types of polymorphism are typical. In most cases aphids pass through several generations a year, and the annual cycle involves a sequence of different forms (morphs).

Morphology

The wing venation is very characteristic and nearly identical in all species. The fore wing (Fig. 2A) has a narrow pterostigma and a single (composite) longitudinal vein consisting of radius, media and the basal part of cubitus. From this vein runs the radial sector near the wing apex, then the free part of media, which usually has two forks, and finally close to the wing base the free part of cubitus, which consists of two branches. The hind wing has only a longitudinal vein and two oblique veins.

Antennae (Fig. 1A) normally consisting of 6 segments, but some species have less. The ultimate segment has a basal part with the same width as the penultimate segment and a thinner distal part, *processus terminalis*. The antennae carry two kinds of rhinaria. *Primary rhinaria*, present in all morphs, are placed on the two ultimate segments, one on the anterior part of the penultimate segment, one at the end of the thick basal part of the ultimate segment. S*econdary rhinaria* may be absent, but most often present on antennal segment III. There may be many of them, especially in males. In several species the antennae are located on lateral frontal processes (tubercles).

Eyes (Figs 1B–C) are very characteristic. Generally large compound eyes are present, and behind each of them is a small tubercle, so-called *triommatidium*, which is a small eye with only three lenses. This is the only eye in nymphs and apterous morphs of some aphids, *e.g.* belonging to the family Eriosomatidae.

Rostrum consists of four segments of which the apical one is sometimes subdivided. Tarsi are 2–segmented: segment I very short, segment II much longer and carries 2 claws.

The siphunculi may be long or short, sometimes only made up by pores, cylindrical or swollen, ending in an aperture normally surrounded by a flange (the collar around the apical opening). In some cases siphunculi are entirely missing. When an aphid is attacked, the siphunculi produce alarm pheromones. The most posterior body segment is the *cauda* (tail). It covers the anus and may be long and finger-shaped, or shorter, triangular, tongue-shaped, often constricted at base and then called knobbed, but may also be rounded, occasionally so short as nearly invisible. The cauda may be used to remove the sticky excrements, "honeydew", from the body and is short only in species which can remove excrement droplets in other ways, *e.g.* with the hind legs or by means of ants (highly attracted to honeydew). The cauda is only finally developed in adults.

The ventral side of the segment between the genital aperture and anus is termed the *anal plate*; in front of the genital aperture lies the *genital plate* (Fig. 1F).

The cuticle may be membranous or sclerotised, in latter case pigmented or not. Pigmentation is usually present on sclerites. Most sclerites are segmental and placed along the midline of the body (spinal sclerites). Marginal sclerites are found along the body margins with pleural sclerites between them. Also intersegmental pleural sclerites may occur. Segmental sclerites may be fused into cross bars, which may furthermore be fused into a dorsal patch.

Life cycle

Aphids have polymorphism, and in some cases the morphs of one species are more different from each other than the same morph of two different species are

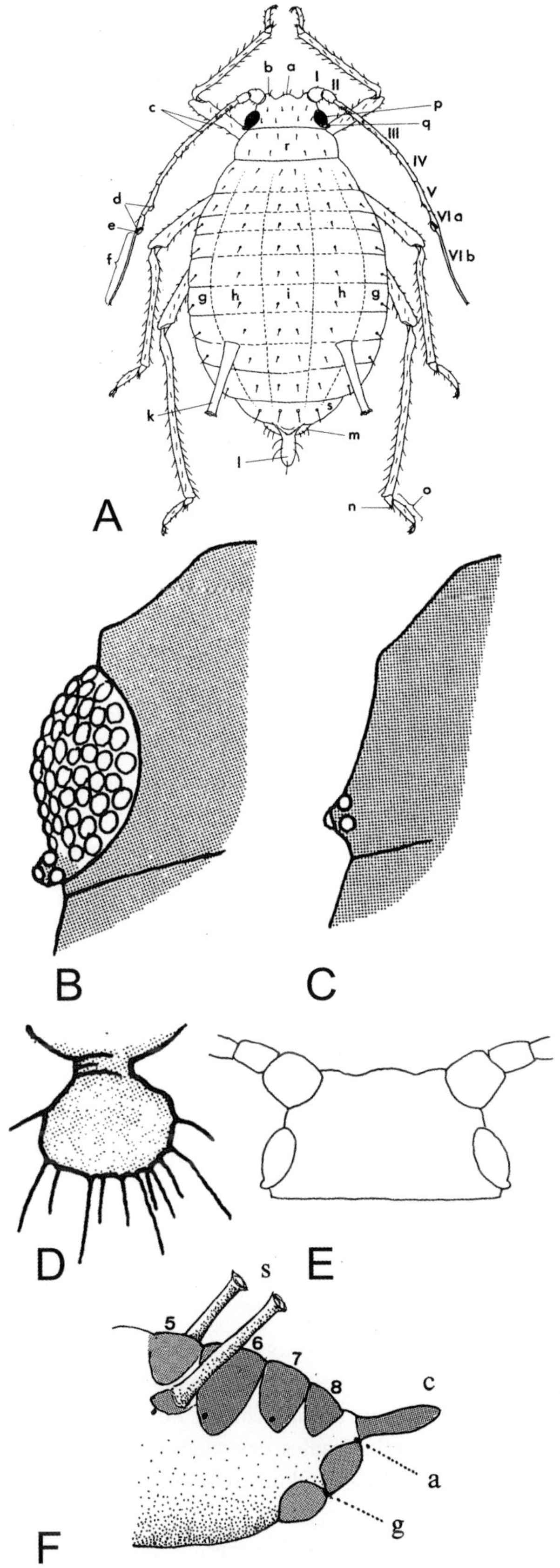

FIGURE 1

Structural details of importance for aphid identification. A, apterous viviparous female; a, median frontal tubercle; b, lateral frontal tubercle; c, secondary rhinaria: d, primary rhinaria; e, accessory rhinaria (close to primary rhinarium of ant. segm. VI); f, processus terminalis; g, marginal areas of dorsum; h, pleural areas of dorsum; i, spinal area of dorsum; k, siphunculus; l, cauda; m, anal plate; n, hairs on first tarsal segment; o, second tarsal segment; p, compound eye; q, ocular tubercle (triommatidion); r, pronotum; s, abd. segm. VIII; I–V, ant. segm. I–V; Via, basal part of ultimate ant. segm.; VIb, antennal processus terminalis; B–C, two types of lateral eyes: compound eye with ocular tubercle/triommatidion (B) and triommatidion only (C); D, 'knobbed' cauda; E, little developed frontal tubercles; F, abdomen of viviparous female, left lateral view. 5-8: abdominal segments V-VIII; c: cauda; a: anus; g: genital aperture; s: siphunculus; the plate between a and g is the anal plate; the plate in front of g is the genital plate. (After Heie 1980).

different. Species able to produce males, oviparous females and fertilized eggs are *holocyclic*, while *anholocyclic* species have lost this ability, having parthenogenetic reproduction only. The morph which in holocyclic species appears from an egg in the beginning of the year, is the *fundatrix*, generally an apterous, parthenogenetic, viviparous and fairly thick stem mother, able to produce a lot of nymphs which grow up to parthenogenetic females belonging to two different morphs, the *apterous viviparous females* and the *alate viviparous females*. The former are usually represented by most specimens and are the best known, and in rarely found species frequently the only morph which has been described. There are at least two more morphs of holocyclic aphids, namely the *sexuales*, which reproduce by eggs after copulation and include the *oviparous females*, which normally are apterous and attract the males by means of scent plaques on the hind tibia, and the *males*, which sense the females by means of a large number of rhinaria on their antennae. In most aphids sexual reproduction takes place only once a year, and diapausing eggs are laid by the oviparous females after copulation. Males may be alate or apterous and do not occur in any other generation.

Some species have host alternation. Their *primary host* is the plant, usually a tree, on which eggs are laid. The *secondary host* is a herbaceous plant or the roots of a woody plant on which only the parthenogenetic generations feeds. In Greenland no species is host-alternating. This is probably partly caused by the short warm season, partly because trees and bushes are relatively rare.

The aphids feed on phloem sap of many kinds of plants, not only seed plants, but also some spore plants like mosses. Usually each species is specialised on only one plant genus or even species.

Greenland fauna

The Aphidoidea comprise 11 families (by some authors ranked as subfamilies). Only four of them are represented in Greenland, with altogether 21 species. One species is found as fossil, viz. *Chaitophorus salijaponicus* subsp. *niger* Mordvilko in deposits at Kap København, Peary Land, from the Plio-Pleistocene, *c.* 2.4 million years old (Heie 1995a).

A treatment of the Greenland aphids was published by Hille Ris Lambers (1952, 1960). The following is based on these papers and Heie (1980–1995).

No less than 8 of the species are regarded as endemic to Greenland; at least they have not been recorded from other parts of the world. The remainder have also been found in Europe. Most of the species are undoubtedly of palaearctic origin, even if they have also been recorded from North America. The probably endemic *Paducia aterrima* has its closest relatives in North America and may be of nearctic origin. The origin of *Myzodium modestum* is uncertain.

Aphids have been found so far north as Pituffik/Thule, but they are not common in Greenland, and the number of arctic species is low compared with the situation in temperate climates.

At least three species are anholocyclic, while the majority probably is holocyclic and definitely not host-alternating. The number of annual generations is low, probably three, perhaps four, because of the short summer. The holocyclic species pass the winter as diapausing eggs, while the overwintering of anholocyclic species must be as parthenogenetic females, either nymphs or adults, the only representatives of anholocyclic species. The two *Pemphigus* species are subterranean, and *Myzodium modestum* lives among mosses—in both cases in biotopes undoubtedly with winter temperatures higher than in the air.

It is interesting that the sexuales appear under long-day conditions in Greenland, in August or even July, because short days together with low temperature are the normal conditions for the appearance of sexuales in regions at lower latitudes.

Betula is host for 3 species, *Salix* for another 3 species. 2 species feed on grasses, 2 on *Carex*, 1 on mosses, and 1 on each of the genera *Juniperus, Epilobium, Potentilla, Dryas, Cerastium, Saxifraga,* and 1 on an unknown genus (perhaps *Sedum*). The most common species seem to be *Euceraphis punctipenni* which feeds on *Betula pubescens* in S Greenland, and *Myzodium modestum* which feeds on mosses. None of the aphids occurring in Greenland can be regarded pests of cultivated plants.

It is a problem that so relatively few of the specimens caught in Greenland are taken *on* plants, because the relationship to a host plant in most cases is a decisive factor for a definitive identification.

Key to species

The key includes *only apterous parthenogentic females*. In *Euceraphis punctipennis* all parthenogenetic females are alate, hence this species is not included. Keys to fundatrices, alate parthenogenetic females, or sexuales cannot be given, because these morphs are unknown for some of the species.

1. Siphunculi pore-shaped or missing 2
– Siphunculi well developed, low and truncate or longer 8
2. Head with large compound eyes with triommatidia (Fig. 1B) 3

– Head without large compound eyes, only with triommatidia (Fig. 1C) 6
3. Siphuncular pores placed on black sclerites or cones. Cauda short and rounded. Body rather broad, spider-like. On twigs of *Juniperus* .... *Cinara juniperi*
– Siphuncular pores not placed on black sclerites or cones. Cauda knobbed (Fig.1D). Body elongated. On *Carex* 4
4. Anal plate distinctly bilobed .... *Thripsaphis sensoriata*
– Anal plate with a nearly straight hind margin. .... *Thripsaphis* (*Trichocallis*) *vibei,* 5
5. Antennae half as long as body or shorter .... *Thripsaphis* (*Trichocallis*) *vibei* subsp. *vibei*
– Antennae slightly longer than half of the body .... *Thripsaphis* (*Trichocallis*) *vibei* subsp. *arctica*
6. Processus terminalis up to 1/4 of the length of the basal part of ultimate antennal segment. On roots of grasses ....7 (*Pemphigus groenlandicus*)
– Processus terminalis up to 1/3 of the length of the ultimate antennal segment. On roots of willow .... *Pemphigus saliciradicis*
7. Processus terminalis 0.20–0.25× the length of the ultimate antennal segment. Body oval. The pigmented part of the genital plate emarginated on the posterior margin only .... *Pemphigus groenlandicus* subsp. *groenlandicus*
– Processus terminalis 0.16–0.20× the length of the basal part of ultimate antennal segment. Body elongated. The pigmented part of the genital plate emarginated both on the anterior and the posterior margin .... *Pemphigus groenlandicus* subsp. *crassicornis*
8. Siphunculi short, truncate, about as long as wide, shorter than 0.07× body length 9
– Siphunculi at least 0.07× body length 11
9. Cauda rounded. Antenna about 0.3× body length. Not on birch .... *Brachycaudus* (*Thuleaphis*) *acaudatus*
– Cauda well developed, not rounded. Antenna longer than 0.4× body length. On birch (*Betula nana*) 10
10. Cauda conical. Antenna about 0.5× body length. Processus terminalis shorter than the basal part of the ultimate antennal segment ..... *Betulaphis pelei*
– Cauda slightly constricted at base. Antenna at least 1.4× body length. Processus terminalis about twice as long as the base of the ultimate antennal segment .... *Calaphis arctica*
11. A process present on abdominal segment VII just above the cauda: the supracaudal process. Cauda tongue-shaped. On leaves of *Salix* .... *Cavariella borealis*
– A supracaudal process absent. On roots or twigs of *Salix*, and if so then cauda rounded, or not on *Salix* ... .12
12. Siphunculi swollen 13
– Siphunculi cylindrical 16
13. Cauda rounded. Lateral frontal tubercles absent. On *Salix* 14
– Cauda well developed, longer than wide. Distinct lateral frontal tubercles present. Not on *Salix* 15
14. Antenna 4–segmented. Processus terminalis 2.5–2.9× the length of the basal part of the ultimate antennal segment. On roots of *Salix* .... *Paducia aterrima*
– Antenna 6–segmented. Processus terminalis 1.3–1.7× the length of the basal part of the ultimate antennal segment. On twigs of *Salix* .... *Pterocomma groenlandicum*
15. Inner sides of lateral frontal tubercles converging. Antennal segment III without secondary rhinaria. On *Cerastium* .... *Myzus polaris*
– Inner sides of lateral frontal tubercles parallel, angular. Antennal segment III with 1–2 secondary rhinaria. On grasses .... *Utamphorophora vibei*
16. Body black. With diverging inner sides of the lateral frontal tubercles. Siphunculi not slightly S-curved. On *Saxifraga* .... *Nasonovia* (*Kakimia*) *saxifragae*
– Body not totally black, or—if with dark pigmentation—lateral frontal tubercles very low and siphunculi slightly S-curved and then on mosses. Not on *Saxifraga* 17
17. Siphunculi 0.10× body length or shorter. On *Epilobium* .... *Aphis nivalis*
– Siphunculi longer than 0.10× body length. Not on *Epilobium* 18
18. Lateral frontal tubercles very low (as in Fig. 1E). Siphunculi black and slightly S-curved. Dorsal cuticle sclerotic and dark pigmented. On mosses .... *Myzodium modestum*
– Lateral frontal tubercles well developed, with diverging inner sides. Siphunculi pale and not curved. Dorsum not pigmented. Not on mosses 19
19. Siphunculi 0.21–0.30× body length. On *Potentilla* .... *Acyrthosiphon boreale*
– Siphunculi 0.14–0.18× body length. Not on *Potentilla* 20
20. Processus terminalis 3.3× the basal part of the ultimate antennal segment or longer. Apical segment of rostrum 0.9× the length of segment II of hind tarsus. Antenna 0.9–1.3× body length. On *Vaccinium* .... *Acyrthosiphon brachysiphon*

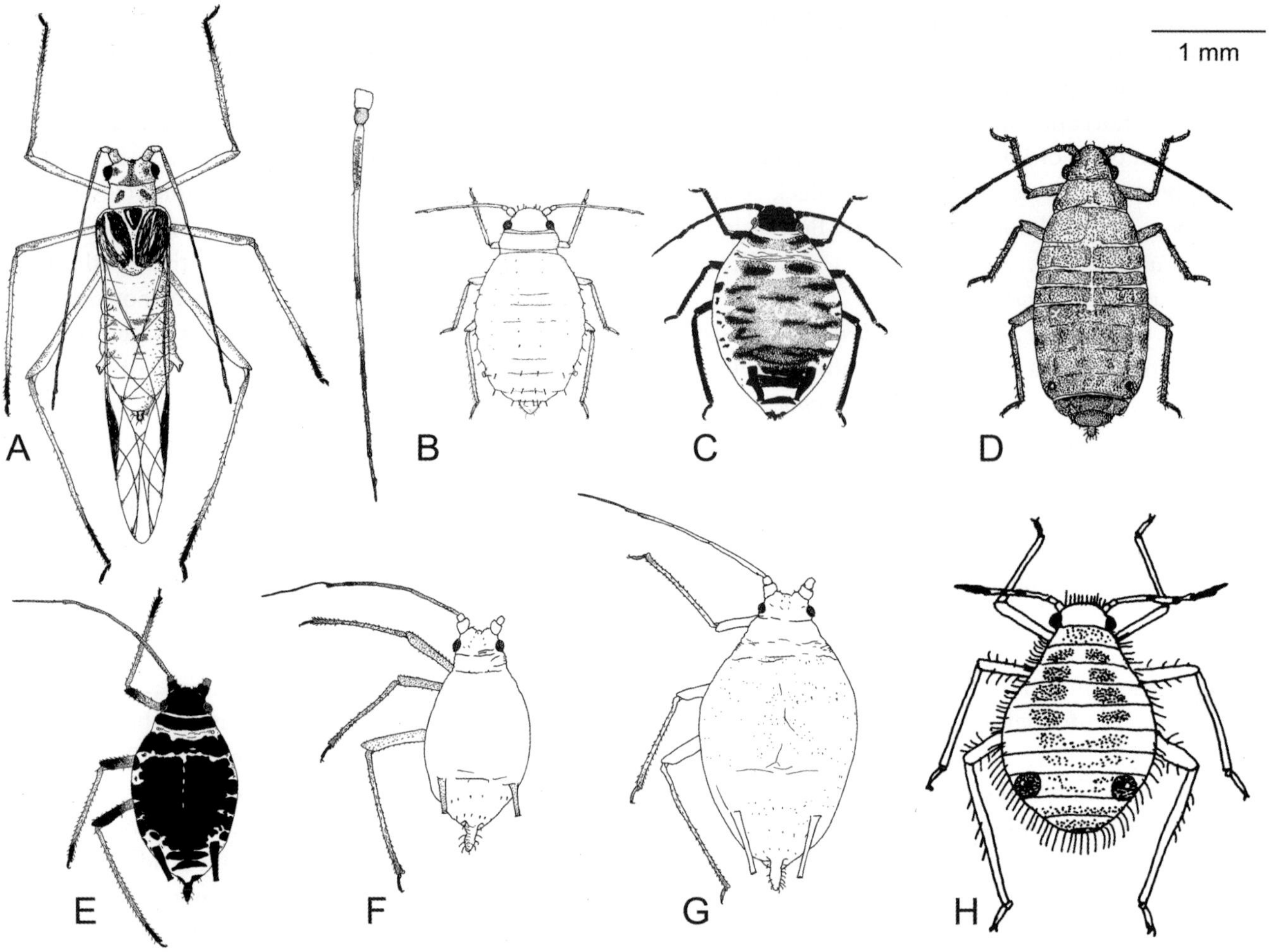

FIGURE 2 *Habitus of Greenland aphids. A, alate viviparous female* Euceraphis punctipennis (*right:* Euceraphis *antenna*); *B, apterous parthenogenetic female* Betulaphis pelei; *C, apterous parthenogenetic female* Myzodium modestum; *D, apterous parthenogenetic female* Thripsaphis vibei arctica; *E, apterous parthenogenetic female* Nasonovia saxifragae; *F, apterous parthenogenetic female* Acyrthosiphon brachysiphon; *G, apterous parthenogenetic female* Acyrthosiphon boreale; *H, apterous parthenogenetic female* Cinara juniperi. *After Heie (1982: A–B, D; 1992: C; 1994: E–G; 1995: H).*

– Processus terminalis 2.5× the basal part of the ultimate antennal segment or shorter. Apical segment of rostrum about as long as segment II of hind tarsus. Antenna 0.8× body length or shorter. On *Dryas*. .... *Acyrthosiphon brevicorne*

Eriosomatidae

Characterized by small, apterous sexuales without mouthparts. They are born by parthenogenetic sexuparae. All Greenland species of Eriosomatidae are probably anholocyclic, and sexuales have never been found here. The apterous parthenogenetic morph has no compound eyes, only triommatidia. Siphunculi pore-shaped or missing. Cauda short and rounded.

***Pemphigus groenlandicus* (Rübsaamen, 1898)**
Henriksen 1939: not recorded.

Anholocyclic, feeding on grass roots. Other species of *Pemphigus* holocyclic with *Populus* as primary host. Sexuales lay eggs on this tree, while parthenogenetic females occur on herbaceous plants, the secondary hosts. The species has been divided into two subspecies:

***Pemphigus groenlandicus* subspecies *groenlandicus* (Rübsaamen, 1898)**
Henriksen 1939: not recorded.

Apterous parthenogenetic female broadly oval-almost round, 1.6–2.1 mm. Wax glands present on abdomen, 4 on each of tergites III–VI, 2 on tergite VII, usually also 2 glands on the head. Antennae with 5 or, less frequently, 6 segments, 0.25× body length. Processus terminalis about

0.20–0.25× length of basal part of ultimate antennal segment. Secondary rhinaria and siphunculi absent. Cauda not well developed. Only posterior margin of pigmented part of genital plate emarginated. Anterior part of genital plate carries 6–10 hairs, hardly stouter than hairs along posterior margin.

Alate parthenogenetic female with black head and thorax and pale abdomen. Antennae consisting of 6 segments, a little shorter than half body length. Antennal segment III carries 4–6 secondary rhinaria, segment IV with 2. Siphunculi absent as in the apterous morph.

DISTRIBUTION. SW (only known from Kapisillit and Qooqqut in Nuup Kangerlua/ Godthåbsfjord). Possibly endemic.

***Pemphigus groenlandicus* (Rübsaamen, 1898) subspecies *crassicornis* Hille Ris Lambers, 1952**
Henriksen 1939: not recorded.

Apterous parthenogenetic female very similar to *P. groenlandicus groenlandicus*, but normally more elongated, 1.7–2.0 mm. Antennae consisting of 5–6 segments; only 0.18–0.25× body length; in specimens with 5 antennal segments number IV nearly globular. Processus terminalis 0.16–0.20× the length of basal part of ultimate antennal segment. In most specimens pigmented part of genital plate emarginated at both anterior and posterior margin. Anterior part of genital plate with 4–6 stout hairs.

Alate parthenogenetic female similar to nominal subspecies, but basal part of antennal segment VI considerably swollen near primary rhinarium, here more than twice as wide as at base. Processus terminalis 0.15–0.22× length of basal part of antennal segment VI. Antennal segment III carries 3–6 secondary rhinaria, segment IV with 2.

BIOLOGY. Known from roots of grasses.

DISTRIBUTION. SW (only known from Kapisillit and Aasiaat/Egedesminde). Possibly endemic.

***Pemphigus saliciradicis* (Börner, 1950)**
Syn. *Pemphigus salicicola* Hille Ris Lambers 1952
Henriksen 1939: not recorded.

Apterous parthenogenetic female broadly oval and pale. Head, thorax and legs frequently rather dark. 1.6–2.4 mm. Antennae consisting of 5 or 6 segments, only 0.2–0.3× body length. Processus terminalis up to one-third of length of basal part of ultimate antennal segment. Secondary rhinaria absent. Oval wax gland plates on abdomen and occasionally also on head.

BIOLOGY. Found on roots of *Salix herbacea* at several localities.

DISTRIBUTION. S, SW, NW.

Europe.

DREPANOSIPHIDAE

Sexuales very different from those of Eriosomatidae. Not very small, and oviparous female always apterous with thick hind tibiae, while male usually alate. Greenland representatives have pore-shaped or short, truncate siphunculi. Cauda knobbed, broadly rounded or short triangular. Secondary rhinaria frequently narrow and transverse, in some cases circular. None of the species have host alternation, but are holocyclic, either on woody or herbaceous plants (frequently grasses or *Carex* spp.). All species on woody hosts in Greenland feed on *Betula* spp.

***Euceraphis punctipennis* (Zetterstedt, 1828) Fig. 2A**
Henriksen 1939: *Aphis punctipennis* Zett.
Viviparous, parthenogenetic female winged (Fig. 2A), also fundatrix, which is apterous in most other aphid species. Parthenogenetic female pale green, frequently wax powdered with tufts of wax threads on antennae and legs. Head with dark dorsal spot or longitudinal stripe, prothorax with two brown spots. Meso- and metathorax brown. In some cases a short, black cross bar on abdomen. 3.0–4.8 mm. Antennae (Fig. 2A, inset) as long as the body, with rather short processus terminalis. The latter shorter than basal part of antennal segment VI, 0,29–0.43 mm. Antennal segment III carries 14–26 narrow, transverse secondary rhinaria. Apical segment of rostrum about 0,7× length of segment II of hind tarsus. Siphunculi shorter than cauda, truncate, about 0.03× body length. Cauda knobbed.

Oviparous female the only apterous morph. Yellowish brown or dark brown. Secondary rhinaria absent from antennae. Posterior part of abdomen prolonged, conical, functioning as an ovipositor. Hind tibia slightly thickened and carries numerous scent plaques.

Male similar to parthenogenetic female, but more slender and always with several dark cross bars on abdomen. More secondary rhinaria, 65–80 on antennal segment III, 0–2 on segment IV and 0–19 on segment V.

BIOLOGY. Feeds on leaves of *Betula pubescens*, especially close to the shoot ends. Very common in subarctic Greenland.

DISTRIBUTION. S.

Widespread in Europe, introduced into North America.

***Betulaphis pelei* Hille Ris Lambers, 1952 Fig. 2B**
Henriksen 1939: not recorded.

Apterous parthenogenetic female oval, somewhat depressed and rather pale. 1.4–1.7 mm. Antennae about half as long as body, siphunculi very short and truncate. Processus terminalis a little shorter than basal part of antennal segment VI. Frons with three pairs of capitate hairs, each of them placed on a conical tubercle. Abdomen with very short, rather thick, blunt hairs on dorsum and somewhat longer hairs along margins. Cauda short and conical.

Alate parthenogenetic female has dark marginal sclerites and some small spots. Dorsal hairs rather long and fine. Antennae about 0.7× body length; segment III carries 7–12 secondary rhinaria on basal two-thirds.

BIOLOGY. Seems to be common in Greenland, where *Betula nana* is the host plant.

DISTRIBUTION. SW. Originally described from Saqqaq and also found at Qooqqut in Nuup Kangerlua/Godthåbsfjord.

Possibly circumpolar (northern Scandinavia, Khibiny Mountains in northern Russia, northern Canada).

Calaphis arctica Hille Ris Lambers, 1952

Henriksen 1939: not recorded.

Apterous parthenogenetic female pale with short dorsal hairs situated on small, dark spots. 1.3–2.0 mm. Antennae 0.9–1.4× body length, slightly darker than head and with darker apices of segments than the rest of them. Processus terminalis about twice as long as basal part of segment VI. Antennal segment III carries about 2–6 large, flat secondary rhinaria. Siphunculi short and truncate. Apical segment of rostrum 0.8–1.0× segment II of hind tarsus. Cauda constricted at base.

Alate parthenogenetic female pale without pigmented abdomen. Antennal segment III carries 6–8 secondary rhinaria. Fore wings with slightly shadowed veins, radial sector frequently hardly visible. Abdominal margins have semiglobular processes.

Oviparous female similar to parthenogenetic viviparous female. Antennal segment III may carry up to 11 secondary rhinaria. Hind tibia with about 35 scent plaques on basal half.

Male apterous and slender, antennae longer than body and carrying 7–13 secondary rhinaria on segment III.

BIOLOGY. *Betula nana* is host plant. The species feeds on undersides of leaves, singly or a few specimens together.

DISTRIBUTION. S (only known from Kangilinnguit/Grønnedal).

Palaearctic, low arctic-boreal and alpine (Finland, Central Europe).

Thripsaphis (*Thripsaphis*) *sensoriata* Hille Ris Lambers, 1952

Henriksen 1939: not recorded.

Only alate parthenogenetic female known. Body elongated, head and thorax black, abdomen with broad, dark cross bars and large marginal sclerites. About 2 mm. Antennae dark and half as long as body. Processus terminalis very short, about 0.4× length of basal part of antennal segment VI. Antennal segment III carries 6–8 secondary rhinaria, segment IV 2, segment V 1. Siphunculi pore-shaped. Cauda strongly knobbed, anal plate deeply bilobed.

BIOLOGY. Found on "a woody plant" at Saqqaq, but the real host is probably a *Carex* sp.

DISTRIBUTION. SW. Only known from Saqqaq. Possibly endemic.

Thripsaphis (*Trichocallis*) *vibei* Hille Ris Lambers, 1952 Fig. 2D

The species comprises two subspecies, both occurring in Greenland and feeding on *Carex*.

T. (*Trichocallis*) *vibei* subspecies *vibei* Hille Ris Lambers, 1952

Henriksen 1939: not recorded.

The apterous parthenogenetic female has a mottled, dark dorsum with a paler longitudinal stripe from mesonotum to the anterior margin of abdominal segment III. 2.3–2.8 mm. Dorsal cuticle covered with very low, rounded nodules and small wax gland pores. Antennae blackish and half as long as body or shorter. Processus terminalis slightly shorter than basal part of antennal segment VI. Antennal segment III carries 0–4 inconspicuous secondary rhinaria placed on its distal half. Siphunculi mere pores on middle of tergite VI. Posterior margin of abdominal tergite VIII straight. Cauda strongly constricted at base.

Alate parthenogenetic female black on head and thorax. Abdomen with dark marginal sclerites and transverse cross bars, those on tergites III–VI broad and partly fused. Antennal segment III has 7–10 secondary rhinaria, usually with 2 placed on a pale part. Wings narrow, veins of fore wings faintly shadowed. Basal oblique vein of hind wing frequently bifurcated or absent.

DISTRIBUTION. S, SW (known from Uunartoq and Paamiut/Frederikshaab). Possibly endemic.

T. (*Trichocallis*) *vibei* subspecies *arctica* Hille Ris Lambers, 1960

Henriksen 1939: not recorded.

Apterous parthenogenetic female with same colouration as other subspecies, and abdominal tergites III–VI also fused. 2.3–2.7 mm. Wax pores in groups around spinal, pleural and marginal hairs and also on tergites VII and VIII. Antennae slightly longer than half the body length. Siphunculi pore-shaped. Processus terminalis 0.8–1.0× basal part of antennal segment VI. Antennal segment III carries 0–3 secondary rhinaria on distal half.

Alate parthenogenetic female similar to *T. vibei vibei*. Antennal segment III carries 9–10 secondary rhinaria.

Oviparous female smaller and more slender than apterous parthenogenetic female, and abdominal tergites III–VI rarely completely fused. On underside of body, below siphunculi very large wax gland plates present. Basal half or slightly more of hind tibia indistinctly swollen and carries about 30–50 scent plaques.

Male apterous and sclerotised as oviparous female. Antennal segment III carries along one side 8–10 small, secondary rhinaria , segment IV 3–5, segment V about 7, the basal part of segment VI 3–4.

BIOLOGY. In Greenland sexuales are produced already at the end of July.

DISTRIBUTION. S (only found at Narsarsuaq).

Palaearctic, low arctic (Iceland, northern Sweden).

APHIDIDAE

The aphidoid family which is richest in species. Sexuales rather similar to parthenogenetic females. As in Drepanosiphidae, oviparous female apterous with thick hind tibiae, male may be apterous or alate. If apterous, thorax similar to that of an alate male. Siphunculi normally well developed, short or long, cylindrical or swollen. Cauda in most species finger- or tongue-shaped, never knobbed. Secondary rhinaria more or less circular, sometimes missing in apterous morphs.

***Pterocomma groenlandicum* Hille Ris Lambers, 1952**

Henriksen 1939: not recorded.

Apterous parthenogenetic female broadly oval, 2.8–3.7 mm and long-haired. Head, broad cross bands on thorax, and small sclerites on abdomen dark. Several abdominal segments carry dark marginal tubercles. Antennae about half as long as body. Processus terminalis 1.3–1.7× length of basal part of antennal segment VI. Secondary rhinaria absent. Siphunculi 0.08–0.10× body length, slightly swollen with a constriction at apices. Cauda rounded, 0.6–0.7× the length of siphunculus. In some cases hind tibia carries a few scent plaques.

Alate parthenogenetic female with rather large marginal sclerites and broad cross bars on abdomen. Antennae slightly longer than half body length; segment III carries 14–18 secondary rhinaria. Siphunculi more distinctly swollen than in apterous morph, cauda more elongated and dark. Wing veins pale brownish.

Male apterous and similar to alate female, but smaller and more slender. Antennae 0.7–0.9× body length; segment III carries 40–58 secondary rhinaria, segment IV 13–21, segment V 12–24, basal part of segment VI 0–1.

BIOLOGY. On twigs of *Salix*. In Greenland males have been found in August; the number of generations per year must be 3 or 4.

DISTRIBUTION. SW, NW (north to Pituffik/Thule). Perhaps endemic to Greenland, if *Pterocomma baicalense* Ivanovskaja, 1976 from Siberia is not a synonym.

***Paducia aterrima* Hille Ris Lambers, 1952**

Henriksen 1939: not recorded.

Apterous parthenogenetic female with a strongly swollen, black body, 2.1–2.2 mm. Almost entirely sclerotised dorsum from mesonotum to abdominal tergite VII covered by large blackish patch bearing long, fine hairs. Flat marginal tubercles present on most body segments. Antennae consisting of only 4 segments, slightly shorter than half body length and dark except close to primary rhinarium on segment IV. A curved processus terminalis 2.5–2.9× the length of basal part of antennal segment IV. Secondary rhinaria absent. Siphunculi blackish, swollen, thick-walled, and imbricate, nearly reticulated below the wide, swollen flange. Cauda small, narrow, at base a little wider than the base of siphunculi, consisting of a basal conical and a distal oval part.

BIOLOGY. Has been found on roots of *Salix* sp.

DISTRIBUTION. SW (only found at Kapisillit), possibly endemic to Greenland. The genus only known from North America.

***Aphis nivalis* Hille Ris Lambers, 1960**

Henriksen 1939: not recorded.

Apterous parthenogenetic female with dark, rather large marginal sclerites on thorax, smaller ones on abdomen, and narrow cross bars on abdominal tergites VII and VIII. Very small marginal tubercles on pronotum and abdominal segments I and VII.1.8–2.0 mm. Frons W-shaped. Antennae slightly longer than half body length. Processus terminalis 1.3–1.7× basal part of antennal segment VI, shorter than in most other species of *Aphis*. Secondary rhinaria absent. Siphunculi dark, rather thick or conical, 08–0.09× body length, imbricated, with a small flange. Cauda paler and slightly longer than siphunculi, with 7–11 hairs.

BIOLOGY. Host *Epilobium* sp.

DISTRIBUTION. nSW. Only found at Kangerlussuaq/ Søndre Strømfjord Airport. Possibly endemic.

***Brachycaudus* (*Thuleaphis*) *acaudatus* (Hille Ris Lambers, 1960)**

Henriksen 1939: not recorded.

Apterous parthenogenetic female broadly oval, slightly depressed, 1.5–1.7 mm. Head dark with a pale area around each compound eye. Body segments with pairs of dark, rectangular, broad, spinal sclerites, fused into cross bars on abdominal segments (IV–)V–VIII and small marginal and pleural sclerites. Flat marginal tubercles on pronotum and abdominal segments II–V. Slightly larger spinal tubercles on abdominal segments VII and VIII, occasionally also on head. Antennae with 5 or 6 segments, about one-third of body length. Processus terminalis about 1.5× basal part of ultimate antennal segment. Secondary rhinaria absent. Siphunculi very short, as long as their basal width or shorter, conical, dark and smooth, with a very distinctly developed flange. Cauda rounded, much wider than long (*acaudatus*), with about 8 hairs.

Alate parthenogenetic female with dark head, thorax and a large dorsal patch on abdominal segments III–VI, and cross bars on segments VII and VIII. Antennae 6-segmented and about two-thirds of body length. Antennal segment III carries 10–16 transverse secondary rhinaria of different sizes, segment IV 0–3. Cauda much narrower than in apterous morph, at least twice as long as siphunculi. Wing veins pale brown; media of fore wing frequently with only one fork instead of the normal two.

Oviparous female similar to viviparous female, but only head and cross bars on abdominal segments VI–VIII dark. Hind tibia swollen all along and carries numerous scent plaques.

Male apterous and only 1.1–1.3 mm. Antennae with 6 segments (or 5 if III and IV are fused), about as long as the body. Antennal segment III (or in 6–segmented antennae III+IV) carries 48–52 rather small secondary rhinaria, penultimate segment 5–7, basal part of ultimate segment 5–7.

BIOLOGY. Has been found on grasses, but the true host is unknown. A close relative in NW Europe, *B.* (*T.*) *sedi* Jacob, feeds on *Sedum roseum*.

DISTRIBUTION. NW (only found at Pituffik/Thule). Possibly endemic.

Cavariella borealis **Hille Ris Lambers, 1952**
Henriksen 1939: not recorded.

Apterous parthenogenetic female more or less pigmented, occasionally with a large, dark dorsal patch. 0.7–2.3 mm. Antennae consist of 5 or 6 segments and are 0.3–0.4× body length. Processus terminalis about length of basal part of ultimate antennal segment. Secondary rhinaria absent. Siphunculi 0.14–0.17× body length, scaly and cylindrical or slightly swollen on the distal part, with a small flange. Cauda thick, blunt, tongue-shaped, about 0.55–0.67× length of siphunculi, with 4–5 laterally arranged hairs. Supracaudal, cylindrical or conical process present, like in other species of the genus.

Alate parthenogenetic female with black head, thorax and broad, transverse, more or less fused sclerites on abdominal segments III–VII and also marginal sclerites, occasionally with tubercles. Antennae with 6 segments and 0.55–0.63× body length. Processus terminalis usually longer than basal part of antennal segment VI. Antennal segment III carries 12–24 secondary rhinaria, segment IV 3–7, segment V 0–3. Siphunculi 0.11–0.13× body length, dusky, darker towards apices and imbricate. Cauda about 0.7× length of siphunculi, conical and rather dark. Supracaudal process only a tubercle with two hairs. Alate female may be brachypterous.

Oviparous female very similar to viviparous female, but much less pigmented. Cauda has 7–11 hairs. Hind tibia very thick, more or less cylindrical and carries a large number of scent plaques on inner side.

Male either apterous or alate. Apterous male about 1.3 mm and with dark cross bars on abdominal segments III–VIII, and marginal sclerites. Thorax similar to that of alate female. Antennae about 0.8× the body length and dark. Antennal segment III carries 31–46 secondary rhinaria, segment IV 10–19, segment V 9–16, basal part of segment VI 1–4. Alate male similar, but with well developed wings, about 1.5 mm. Antennal segment III carries 41–49 secondary rhinaria, segment IV 13–18, segment V 9–15, basal part of segment VI 2.

BIOLOGY. Host plants are *Salix* spp. (*arctica* or *herbacea*). The species lives on the underside of leaves.

DISTRIBUTION. (S), NW (has been found at Inglefield Bredning and Pituffik/Thule, perhaps also at Kangilinnguit/Grønnedal). Possibly endemic.

Myzus* (*Nectarosiphon*) *polaris **Hille Ris Lamber, 1952**
Henriksen 1939: not recorded.

Apterous parthenogenetic female 1.8–2.1 mm. Dorsal cuticle slightly corrugated with pale brown pleural intersegmental sclerites. Well developed frontal tubercles with converging inner margins. Antennae slightly shorter than body, gradually darker towards apices, with scabrous inner side of first segment. Processus terminalis 2.1–2.8× the length of basal part of segment VI. Secondary rhinaria absent. Siphunculi about 0.20× body length, distinctly swollen and coarsely imbricate, with a small flange. Cauda thick, rather blunt, half as long as siphunculi or shorter and pigmented like bases of siphunculi, with 6–8 hairs.

Abdomen of alate parthenogenetic female with marginal sclerites and a large central sclerite or patch and cross bars on segments VII and VIII. Antennae nearly as long as body and black. Antennal segment III carries 12–18 slightly elevated secondary rhinaria along one side, segment IV 1–4. Siphunculi shorter than in the apterous female, swollen and evenly pigmented. Cauda thinner than in apterous female, triangular and almost pointed.

Oviparous female similar to viviparous female, but with shorter antennae. Hind tibia only slightly swollen, with about 25 scent plaques on basal half.

Male apterous and slender. Antennal segment III carries 42–45 small secondary rhinaria, which are not placed along one side; segment IV 18–19, segment V 16–18.

BIOLOGY. The host plant is *Cerastium alpinum*.

DISTRIBUTION. SW, NW, NE (found at Kangerlussuaq/Søndre Strømfjord, Saqqaq, Pituffik/Thule, and Zackenberg). Palaearctic, low arctic (also known from Iceland).

***Myzodium modestum* (Hottes, 1926) Fig. 2C**
Henriksen 1939: not recorded.

Apterous parthenogenetic female shining reddish brown to dark brown or olive brown with posterior end of body occasionally greenish. 1.5–1.9 mm. Cuticle sclerotised, wrinkled or granulate, especially on head and body margins. Antennae 0.6–0.7× body length. Processus terminalis pointed and 2.4–2.5(–3.0)× length of basal part of antennal segment VI. Siphunculi slightly S-curved, not swollen, strongly squamous and 0.18–0.25× body length, 3–4× length of cauda, with a couple of transverse stripes or traces of reticulation and a constriction below the very large flange. Cauda with broad basal and narrower distal part, partly covered by abdominal tergite VIII.

Alate parthenogenetic female with large, dark dorsal patch on abdomen. Antennal segment III carries 21–45 secondary rhinaria, segment IV 7–13, segment V 0–4.

BIOLOGY. Lives on mosses, *e.g. Polytrichum* spp., probably anholocyclic

DISTRIBUTION. S, SW, NE (Pamialluk in Kap Farvel-area, Narsarsuaq, Kapisillit, Jameson Land).

Holarctic (not recorded from Siberia, perhaps originally nearctic), arctic-temperate.

***Nasonovia* (*Kakimia*) *saxifragae* (Doncaster & Stroyan, 1952) Fig. 2E**
Henriksen 1939: not recorded.

Apterous parthenogenetic female 2.0–2.1 mm, blackish from metanotum to abdominal tergites V or VI, and has dorsal sclerites on abdominal segments VI–VIII. Head, antennae, legs, siphunculi and cauda dark. Marginal tubercles on abdominal segments II–V (or absent from one or both sides of one of these segments). Antennae 0.7–1.0× body length. Processus terminalis 2.5–5.0 length of basal part of antennal segment VI. Antennal segment III carries 4–25 secondary rhinaria, segment IV 0–2. Siphunculi 0.14–0.20× body length, 1.2–1.6× length of cauda, a little imbricated. Cauda finger-shaped and slightly constricted, carries 5 or rarely 6 hairs.

Alate parthenogenetic female 2.4–2.7 mm, with dark marginal sclerites and dorsal markings consisting of a central, occasionally subdivided patch. Also pleural and spinal sclerites on the other segments, and postsiphuncular and intersegmental sclerites. Processus terminalis 4.0–5.2× length of basal part of antennal segment VI. Antennal segment III carries 32–46 secondary rhinaria, segment IV 11–22, segment V 0–2.

Oviparous female with small abdominal sclerites. Processus terminalis 3.0–3.8× length of basal part of antennal segment VI. Antennal segment III carries 3–22 secondary rhinaria. Hind tibia is more or less distinctly swollen, occasionally dark and carrying up to about 50 scent plaques.

Male apterous and rather similar to the alate female. Antennal segment III carries 20–39 secondary rhinaria, segment IV 7–21, segment V 5–8.

BIOLOGY. Lives on various species of *Saxifraga*. In Greenland sexuales have been found in September.

DISTRIBUTION. NE (only found at Skeldal, northern Scoresby Land).

Palaearctic, arctic-temperate (Iceland, Jan Mayen, Norway, Britain).

***Utamphorophora vibei* (Hille Ris Lambers, 1952)**
(*Sitomyzus vibei* Hille Ris Lambers, 1952)
Henriksen 1939: not recorded.

Apterous parthenogenetic female sclerotised and coarsely corrugated, but not pigmented. 2.6–2.8 mm. Lateral frontal tubercles well developed and scabrous, with rectangular, parallel inner sides. Spinal and marginal tubercles completely absent, opposite the situation in all related species. Antennae 0.6–0.8× body length, rather thin, imbricate and gradually darker towards apices. Processus terminalis about 3.5× length of basal part of antennal segment VI, about as long as antennal segment III, which carries 1–2 slightly protruding, small secondary rhinaria. Siphunculi about 0.17–0.18× body length, about 1.7× length of cauda, swollen on distal half, slightly imbricate and brownish, with a small flange. Cauda thick, carries 2 pairs of lateral hairs and 1–2 dorsal hairs.

Alate parthenogenetic female with brownish head, thorax and sharply bordered, pleural intersegmental sclerites on abdomen and also small, pale brown marginal sclerites. Antennae blackish, imbricate and slightly shorter than body. Processus terminalis 3.5–3.9× length of basal part of antennal segment VI. Antennal segment III carries 4–9 elevated secondary rhinaria, segment IV 0–1. Siphunculi nearly smooth.

Oviparous female very similar to viviparous female; however, dorsal cuticle not sclerotised but wrinkled. Antennal segment III carries 0–1 secondary rhinarium. Hind tibia hardly swollen and carries 25–35 scent plaques on basal half.

BIOLOGY. Found on an unidentified species of grass. Oviparous females collected in the beginning of August. Complete life cycle seems to consist of only 3–4 generations.

DISTRIBUTION. S, SW (found at Ivittuut and Kapisillit). Possibly endemic.

***Acyrthosiphon boreale* Hille Ris Lambers, 1952 Fig. 2G**
Henriksen 1939: not recorded.

Apterous parthenogenetic female green with anterior body yellowish or yellowish green. Dorsal cuticle slightly sclerotised and wrinkled, but not pigmented. Lateral frontal tubercles well developed with diverging, rough inner sides. Median frontal tubercle conspicuous. 2.7–3.2 mm. Antennae 0.7–1.1× body length. Processus terminalis very long, 4.6–6.0× length of basal part of antennal segment VI. Antennal segment III carries 2–9 slightly elevated secondary rhinaria, irregularly placed in a row, mainly on basal half. Siphunculi 0.21–0.30× body length, 1.9–2.3× length of cauda, cylindrical and imbricated, with a rather well developed flange. Cauda rather thick, not constricted or only slightly so, carries 7–11 hairs.

Fundatrix similar, but processus terminalis only about 2.7× basal part of ultimate antennal segment. Male apterous (Eastop, 1971).

BIOLOGY. Host plants are *Potentilla* spp. on which the species may be present in huge numbers.

DISTRIBUTION. S, SW (found at Qaqortoq/Julianehåb, Igaliku, and Qeqertarsuup Tunua/Disko Bugt: Kangersuneq).

Possibly circumpolar, boreo-alpine (Iceland, Finland, northern Sweden, Hungary, Switzerland, Kazachstan, Quebec).

Acyrthosiphon brachysiphon **Hille Ris Lambers, 1952 Fig. 2F**

Henriksen 1939: not recorded.

Apterous parthenogenetic female 1.8–2.8 mm, green and wax-powdered. Dorsal cuticle membranous or slightly sclerotised. Lateral frontal tubercles low with diverging, rough inner sides. Antennae 0.9–1.3× body length, gradually darker towards apices. Processus terminalis 3.3–4.8× length of basal part of antennal segment VI. Antennal segment III carries 0–9 secondary rhinaria. Siphunculi 0.14–0.17× body length, 1.0–1.5× length of cauda, cylindrical, imbricate with a few flat hexagonal cells below the very small flange. Cauda tongue-shaped and carries 6–8 hairs.

Fundatrix similar but with less developed frontal tubercles and shorter antennae, which may consist of only 5 segments. Processus terminalis only 2.1–2.9× length of basal part of ultimate antennal segment. Antennal segment III carries 0–3 small rhinaria.

Alate parthenogenetic female with large marginal sclerites, which are partly fused with large pleural intersegmental sclerites. Cross bars on abdominal segments VII and VIII. Antennal segment III carries 5–17 rhinaria. Siphunculi and cauda dark.

Oviparous female pale with slightly darker pleural intersegmental sclerites. Hind tibia somewhat swollen and carries 70–90 scent plaques. Cauda thicker than in viviparous female.

Male apterous and slender. Abdomen with marginal sclerites, pleural intersegmental sclerites and more or less fused dorsal cross bars. Antennae up to 1.4× body length; segment III carries 27–68 secondary rhinaria, segment IV 14–38, segment V 22–25.

BIOLOGY. Feeds on underside of leaves and on young shoots of *Vaccinium uliginosum*. In Greenland only three generations per year: fundatrices in July, one further generation of viviparous females, and sexuales in August (Hille Ris Lambers, 1952).

DISTRIBUTION. S, SW (north to Saqqaq), NE (Dove Bugt: Lindhard Ø).

Holarctic (not recorded from Siberia), low arctic-boreal and alpine.

Acyrthosiphon brevicorne **Hille Ris Lambers, 1960**

Henriksen 1939: not recorded.

Apterous parthenogenetic female 1.8–2.3 mm with dorsal cuticle membranous or slightly sclerotised. Lateral frontal tubercles low and nearly smooth, with diverging inner sides. Median frontal tubercle very low. Antennae consisting of 5 or 6 segments, about 0.7–0.8× body length. Processus terminalis 1.6–2.4× length of basal part of ultimate antennal segment. Both antennae and processus terminalis short compared with those of other species of *Acyrthosiphon*. Antennal segment III carries 0–1 secondary rhinarium near base. Hind tibia occasionally with a few scent plaques. Siphunculi 0.15–0.18× body length, 1.1–1.4× length of cauda, cylindrical, imbricated on the apical three-fourths and usually constricted below the small flange. Cauda oblong conical, rather thick, not constricted or only slightly so and carrying 6–8 hairs.

Fundatrix thicker and has always only 5 antennal segments. Antennae only about 0.5× body length. Processus terminalis only 1.5× basal part of antennal segment V. Siphunculi about 0.13× body length and rather thick.

Alate parthenogenetic female with small marginal sclerites and slightly paler, small pleural intersegmental sclerites. Antennae 0.8–0.9× body length. Processus terminalis 2.5–3.2× basal part of antennal segment VI. Antennal segment III carries 4–6 secondary rhinaria on basal half or two-thirds. Siphunculi thinner than in apterous female, cauda more slender and pointed.

Oviparous female green and like viviparous female, but with relatively longer processus terminalis (2.3–2.8× basal part of ultimate antennal segment). Basal half of hind tibia considerably swollen and dark, carries 64–80 scent plaques.

Male only 1.3–1.7 mm, slender and apterous, regarding pigmentation similar to alate female. Antennae relatively longer, 1.0–1.3× body length. Antennal segment III carries 17–40 secondary rhinaria, segment IV 12–23, segment V 8–17.

BIOLOGY. Lives on *Dryas* sp. Fundatrix found in the beginning of July, sexuales already in last part of July (at Pituffik/Thule; Hille Ris Lambers 1960). Accordingly the number of generations per year must be three or perhaps four.

DISTRIBUTION. S, SW, NW (north to Pituffik/Thule).

Possibly circumpolar, arctic-boreal, alpine (northern Sweden, Switzerland, northern Canada).

LACHNIDAE

Apical segment of rostrum distinctly subdivided into two segments. Most species somewhat spider-like and haired. Frontal and marginal tubercles always absent. Antennae 6–segmented and rather short. Processus terminalis shorter than basal part of antennal segment VI. Secondary rhinaria circular. Siphunculi pore-shaped and normally placed on dark sclerites or low cones. Cauda rounded.

Cinara juniperi **(De Geer, 1773) Fig. 2H**

Henriksen 1939: not recorded.

Apterous parthenogenetic female 2.6–3.1 mm, somewhat spider-like, light brown or pinkish brown, occasionally blackish grey, wax-powdered, with two longitudinal dorsal stripes without wax. Abdomen with intersegmental sclerites. Spinal or pleurospinal and marginal sclerites present on mesonotum, metanotum and abdominal tergites I, II(–V) and VII, and on VIII as a frequently broken cross bar. Antennae about half as long as body. Processus terminalis carries 3 subapical hairs. Antennal segment III without rhinaria, segment IV 1–3, segment V 0–1. The siphunculi are pores on flat, black, hairy cones or sclerites, 0.27–0.59 mm broad. Cauda very short and rounded.

Alate parthenogenetic female similar to apterous female, but has smaller siphuncular sclerites or cones. Antennal segment III carries 4–8 secondary rhinaria, segment IV 2–3, segment V 0–2.

FIGURE 3 *Very often nymphal aphids appear below a lifted stone in Greenland. These are hardly identified if not found on well defined plant parts.* (PHOTO: A. FJELLBERG)

BIOLOGY. Host plant *Juniperus communis*, where the species feeds on the needle-carrying twigs.

DISTRIBUTION. S, SW (known from Tasiusaq and Kapisillit).

Almost cosmopolitan, low arctic-subtropical.

References

Eastop, V.F. (1971): Keys for the identification of *Acyrthosiphon* (Homoptera; Aphididae). – *Bulletin of the British Museum (Natural History)*, Entomology Series 26: 1–115.

Edwards, J.S. (1966): Aphids of the King Oscars Fjord area, North East Greenland. – *Entomologiske Meddelelser* 34: 301–346.

Heie, O.E., (1980): The Aphidoidea (Hemiptera) of Fennoscandia and Denmark I. General part. The families Mindaridae, Hormaphididae, Thelaxidae, Anoeciidae, and Pemphigidae. – *Fauna Entomologica Scandinavica* 9: 1–236.

Heie, O.E. (1982): The Aphidoidea (Hemiptera) of Fennoscandia and Denmark II. The family Drepanosiphidae. – *Fauna Entomologica Scandinavica* 11: 11–176.

Heie, O.E. (1986): The Aphidoidea (Hemiptera) of Fennoscandia and Denmark III. Family Aphididae: subfamily Pterocommatinae & tribe Aphidini of subfamily Aphidinae. – *Fauna Entomologica Scandinavica* 17: 1–314, E.J.Brill/Scandinavian Science Press Ltd., Leiden/Copenhagen.

Heie, O.E. (1992): The Aphidoidea (Hemiptera) of Fennoscandia and Denmark. IV. Aphididae, Aphidinae: Part 1 of tribe Macrosiphini. – *Fauna Entomologica Scandinavica* 25: 1–188.

Heie, O.E. (1994): The Aphidoidea (Hemiptera) of Fennoscandia and Denmark. V. Aphididae, Aphidinae: Part 2 of tribe Macrosiphini. – *Fauna Entomologica Scandinavica* 28: 1–239.

Heie, O.E. (1995a): The Aphidoidea (Hemiptera) of Fennoscandia and Denmark. VI. Part 3 of Macrosiphini and family Lachnidae. – *Fauna Entomologica Scandinavica* 31: 1–217.

Heie, O.E. (1964): Aphids collected in Iceland in August 1961 (Homoptera: Aphididae). – *Entomologiske Meddelelser* 32: 220–235.

Heie, O.E. (1995a): An aphid from the Plio-Pleistocene København Formation, North Greenland (Homoptera, Aphidoidea). – *Entomologiske Meddelelser* 63: 17–18.

Hille Ris Lambers, D. (1952): The aphid fauna of Greenland. – *Meddelelser om Grønland* 136 (1): 1–33.

Hille Ris Lambers, D. (1960): Additions to the aphid fauna of Greenland. – *Meddelelser om Grønland* 159 (5): 1–18.

Holman, J. (1996): *Symydobius nanae* sp. N. (Homoptera, Aphidoidea) and other aphids living on *Betula nana* in the Sumava National Park, Czech Republic. – *European Journal of Entomology* 93: 239–248.

10.4. Coccoidea (Scale insects)

Carl-Axel Gertsson and Chris Hodgson[1]

The approximately 20 families of scale insects are grouped into two major assemblages, the archaeococcids and the neococcids. Adult females of the former group possess abdominal spiracles, and adult males have compound eyes. The neococcids, which include the bulk of the remaining families, lack abdominal spiracles at all stages and the adult males have simple eyes. Examples of both groups occur in Greenland.

The reproductive systems of the Coccoidea include hermaphroditism (the only known insect example) and seven kinds of parthenogenesis; they also have six major types of chromosome systems. Scale insects have the greatest diversity of sperm structure and sex determining systems of any known group of organisms.

Scale insects owe their vernacular name to the very diverse coverings which protect juveniles and adult females (Fig. 1). These coverings are formed by waxy dermal secretions, which may be mixed with secretions from the gut; often exuviae of early instars are also built into the 'scale'.

More than 7300 species of scale insects are known. The Diaspididae (armoured scales), Pseudococcidae (mealybugs) and Coccidae (soft scales or coccids) are the largest families. The Diaspididae are rare north of 60° latitude (Arctic) but both the pseudococcids and coccids are represented. Information on the scale insects of the world is available in a database called ScaleNet (http://www.sel.barc.usda.gov/scalenet/scalenet.htm).

FIGURE 1 *Female* Arctorthezia cataphracta. (PHOTO: J. BÖCHER)

1 The accounts of females and males are mainly the responsibility of C–EG and CH, respectively.

Morphology: adults

Scale insects are mostly small (generally less than 5 mm long) and are markedly sexually dimorphic. Females are neotenic, passing through 2 or 3 nymphal instars before becoming adult. Structurally, the separation between head, thorax and abdomen is ill-defined and they have no wings and often have reduced legs (indeed, legs are entirely absent on many non-Greenland species). Usually they have well-developed mouthparts, with feeding stylets of greatly variable length (probably depending on host-plant) and the adult female may live for many weeks or months.

Adult males resemble very small flies, with well-defined head, thorax and abdomen; they generally have a pair of forewings but the hind wings are either absent or reduced to a pair of vestigial *hamulohalteres*. Males pass through 4 immature stages (including a so-called 'prepupal' and a 'pupal' stage) before becoming adult; they lack mouthparts and are short lived (a few days at the most). Relatively few adult male Coccoidea have ever been described in detail, mainly because they are very short lived (generally only 1–3 days), are mostly extremely small and (as here) are almost impossible to identify unless collected with the associated females. They are also relatively difficult to mount satisfactorily and their structure is complex. However, males are now known from most families and are proving important in phylogenetics.

For information on the morphological terms used, see the figure legends.

Life cycle

The first-instar nymphs or 'crawlers' are mobile and it is this stage which ensures the dispersal of the species, either by walking or as aerial plankton. This stage is normally elongate elliptical or oval, generally very small (often 0.3 mm or less) and always have well-developed legs, antennae and mouthparts. Male and female crawlers are generally morphologically identical. Once settled, they feed and then moult into 2nd-stage nymphs. This is a major feeding stage for both the males and females and these nymphs are sexually dimorphic. In some species, perhaps depending on environmental conditions, a 3rd female nymphal stage may occur. Diapause during the winter can occur at any stage, depending on species. Development of the male is similar to the female up to the 2nd moult, going through two feeding nymphal stages. However, it never feeds again, changing first into a prepupa, then into a pupa and finally into the adult male.

Greenland fauna

Only 4 coccoid families, with a total of 11 species, have been collected from Greenland to date: Ortheziidae, Pseudococcidae, Eriococcidae and Coccidae. The adult males of 9 of these species have been collected: six of mealybug species (Pseudococcidae), possibly two of felt scale species (Eriococcidae) and one soft scale (Coccidae). Despite having a circumpolar distribution, the male of *A. cataphracta*, the only ortheziid, has never been collected and it is assumed that this species is parthenogenetic.

However, with male Coccoidea, there is always some doubt about their identity! Species of Coccoidea are almost always described based on female structure and most males are so poorly known that they are unidentifiable. Even when a male is bred from a pupa associated with a female of a known species, there is a slight doubt about the association as more than one species of scale insect is frequently found on a given plant; and often male nymphs (crawlers) are known to settle some way from where the female nymphs (crawlers) settle. However, this is less likely to be a problem in Greenland because few species are known and because only one or two species have been collected from most sites in any one year. Although many of the males were collected in pit-fall traps, luckily most have also been collected at least once associated with a female and so the identity suggested here is thought to be reasonably likely.

The males of none of these 9 species have been previously described, although males of *Spinococcus* (*Peliococcus*) *callunetti* (Lindinger), which is thought not to have been collected in this study, were briefly described and illustrated by Schmutterer (1952). His specimens were from Germany. Unfortunately neither the description nor the illustration were adequate to be included here. Four of the mealybug species and the only species of Coccidae were new to science (Hodgson 2005, Gertsson & Hodgson, 2005).

Family key to adult females known from Greenland

1. Abdominal spiracles present, although not very conspicuous. Wax on body (of unmounted insect) in distinctly separate plates (Fig. 1). Ovisac carried around by wandering adult female ..... Ortheziidae
– Abdominal spiracles absent. Wax on body (of unmounted insect), if present, not in distinct plates. Ovisac fixed to substrate, not carried around by adult females 2
2. Body of adult females broadly oval and not very convex. Cottony ovisac secreted from beneath abdomen. Anal opening covered dorsally by a pair of sclerotised triangular anal plates .... *Pulvinaria glacialis* (Coccidae)
– Body of adult female more or less round in cross-section. Cottony ovisac, when present, secreted from both surfaces. Anal opening not covered by a pair of anal plates 3
3. Ostioles, circuli, up to 18 pairs of cerarii, trilocular pores and oral collar ducts all present. Anal lobes wider than long and with conical cerarian setae .... Pseudococcidae
– Ostioles, circuli, cerarii, trilocular pores and oral collar ducts all absent. Dorsal surface and margins generally with strongly spinose setae. Anal lobes as long as or longer than wide, with spinose setae along the margins .... Eriococcidae

Key to adult males (all families) known from Greenland

1. Penial sheath at least 4× longer than wide; scutum with a large quadrangular membranous area medially; ocular sclerites and genae covered in polygonal reticulations; loculate pores entirely absent (apart from in glandular pouches); body with frequent fleshy setae, even on abdomen .... *Pulvinaria glacialis* (Coccidae)
– Penial sheath never more than 2× longer than broad; scutum without a large quadrangular area medially; polygonal reticulations absent from genae and restricted to around each simple eye on ocular sclerites; loculate pores usually present, at least on head but also generally on abdomen; body without fleshy setae on abdomen 2
2. Postoccipital ridge strong; scutum with two small membranous areas medially; scutellum with a well developed scutellar ridge; claw digitules capitate and longer than claw; loculate pores absent apart from on head; apical antennal segment almost as wide as long; cranial apophysis well developed and trilobate; ostioles absent .... 8 (Eriococcidae)
– Postoccipital ridge weak or absent; scutum without two small membranous areas medially; scutellum

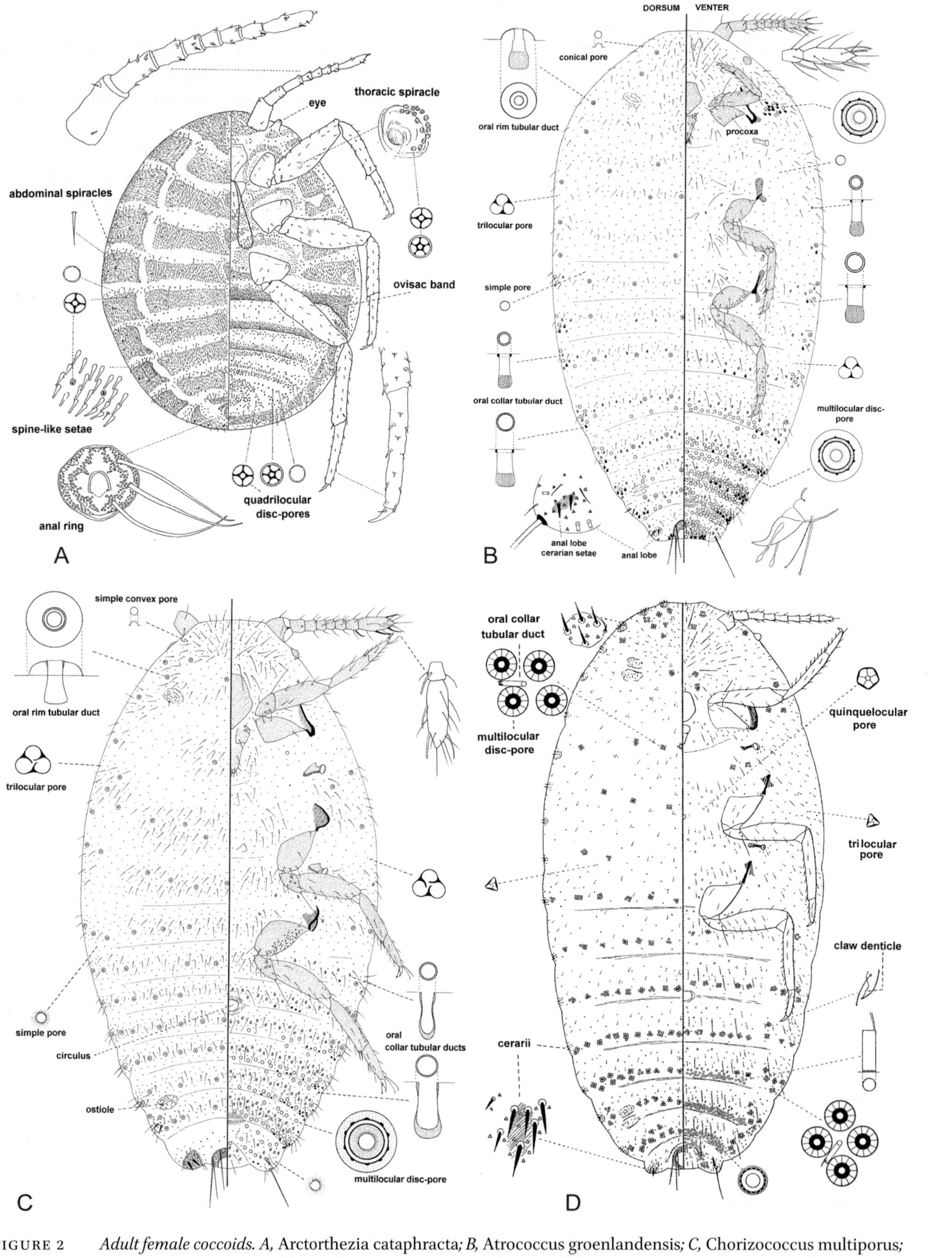

FIGURE 2 *Adult female coccoids. A,* Arctorthezia cataphracta*; B,* Atrococcus groenlandensis*; C,* Chorizococcus multiporus*; D,* Peliococcus balteatus *(A modified after Kozár, 2004; D modified after Williams, 1962).*

without a scutellar ridge; claw digitules setose and much shorter than claw; loculate pores often frequent and nearly always present in pleural region of abdomen; apical antennal segment at least twice as long as broad; cranial apophysis poorly developed or absent; ostioles present ....3 (Pseudococcidae)

3. Apterous; mesothorax mainly membranous; postoccipital ridge entirely absent .... *Chorizococcus multiporus*
– Macropterous; mesothorax heavily sclerotised; postoccipital ridge present 4
4. With two pairs of glandular pouches, one pair on abdominal segment VII and the other on segment VIII; fleshy setae on antennae setose and at least twice as long as width of flagellar antennal segments .... *Peliococcus balteatus*
– With only one pair of glandular pouches, on abdominal segment VIII; fleshy setae on antennae thick, with a blunt apex, clearly different from hair-like setae and each about as long as width of flagellar antennal segments 5
5. Several postmesospiracular setae present anterior to basisternum; fleshy setae clearly differentiated from hair-like setae on head; each gena with 4+ genal setae, some often fleshy; small convex pores on ventral epicranium of head 6
– Postmesospiracular setae very few or absent anterior to basisternum; fleshy setae not differentiated from hair-like setae on head; each gena with 3 or fewer genal setae, none fleshy; pores on head, when present, on dorsal epicranium 7
6. Fleshy setae on head quite short; 3rd, 4th and 5th antennal segments all equal to or shorter than 2nd segment; postmesospiracular disc-pores and setae few immediately anterior to basisternum; antennae fleshy setae short, about width of antennal segments .... *Trionymus heterelymus*
– Fleshy setae on head almost as long as hair-like setae; 3rd antennal segment longer than 2nd segment and 3rd and 4th segments subequal in length; postmesospiracular disc-pores few or absent but setae quite abundant immediately anterior to basisternum; antennae fleshy setae long, longer than width of antennal segments ..... *Trionymus bocheri*
7. With 0 or 1 loculate pore on dorsal surface of head; pleural region of each abdominal segment with 0 or 1 loculate pore; abdominal segment VIII with a pair of transverse sclerotisations mediolaterally on sternite; loculate pores absent from dorsal and ventral surfaces of abdomen; with several metatergal setae .... *Atrococcus groenlandensis*
– With several loculate pores on dorsal surface of head; pleural region of each abdominal segment with 2–5+ loculate pores; abdominal segment VIII without a pair of transverse sclerotisations mediolaterally on sternite; loculate pores present on dorsal and ventral surfaces of abdomen; with only 1 pair of metatergal setae .... *Trionymus thulensis*
8. Length of fleshy setae on antennae equal to or less than width of flagellar segments of antennae; fleshy setae with a blunt apex; some pores on head quite large .... *Eriococcus pseudoinsignis*
– Length of fleshy setae on antennae clearly greater than width of flagellar segments of antennae; fleshy setae with sharply rounded apices; pores on head all quite small .... *Eriococcus munroi*

NOTE. Males of *Arctorthezia cataphracta* (Olafsen) are apparently unknown (Kosztarab & Kozar, 1988), while males of *Spinococcus* (*Peliococcus*) *callunetti* (Lindinger) are not represented in the available Greenland material. However, the small illustration by Schmutterer (1952) clearly shows that *P. callunetti* differs from *P. balteatus* in having only a single pair of glandular pouches.

ORTHEZIIDAE (ENSIGN SCALES)

The Ortheziidae are considered to be one of the most primitive groups of scale insects. The name 'ensign scale' refers to the flag-like ovisac carried by the females. The dorsal surface of the female is covered with a series of white waxy plates. The antennae are 7 or 8 segmented, with a short, stout seta on the tip of the apical segment. Eyes on short stalks. Legs very long. Body covered in spine-like setae.

Arctorthezia cataphracta (Olafsen, 1772)

Figs 1, 2A

Henriksen (1939): *Orthecia cataphracta* Olafs.

Female (unmounted, Fig. 1) with tube-like ovisac up to 5 mm long. Adult covered with dorsal white wax plates as follows: with 9 pairs in each marginal row, 8 in each median row and with 3 triangular plates in centre of thorax, each of latter as long as wide. Antennae 7 or 8 segmented. Mounted specimens (Fig. 2A): Dorsum: Body 2.5–3.5 mm long. Abdomen with 8 pairs of small, inconspicuous spiracles. Quadrilocular disc pores sparsely present. Short thick spines present in bands and clusters. Venter: Spines forming a typical ovisac band on ventral surface of abdomen. Within this ovisac band 3 rows of spines.

BIOLOGY. Very common in moist places, such as under stones and in or under moss and lichens. In Greenland it

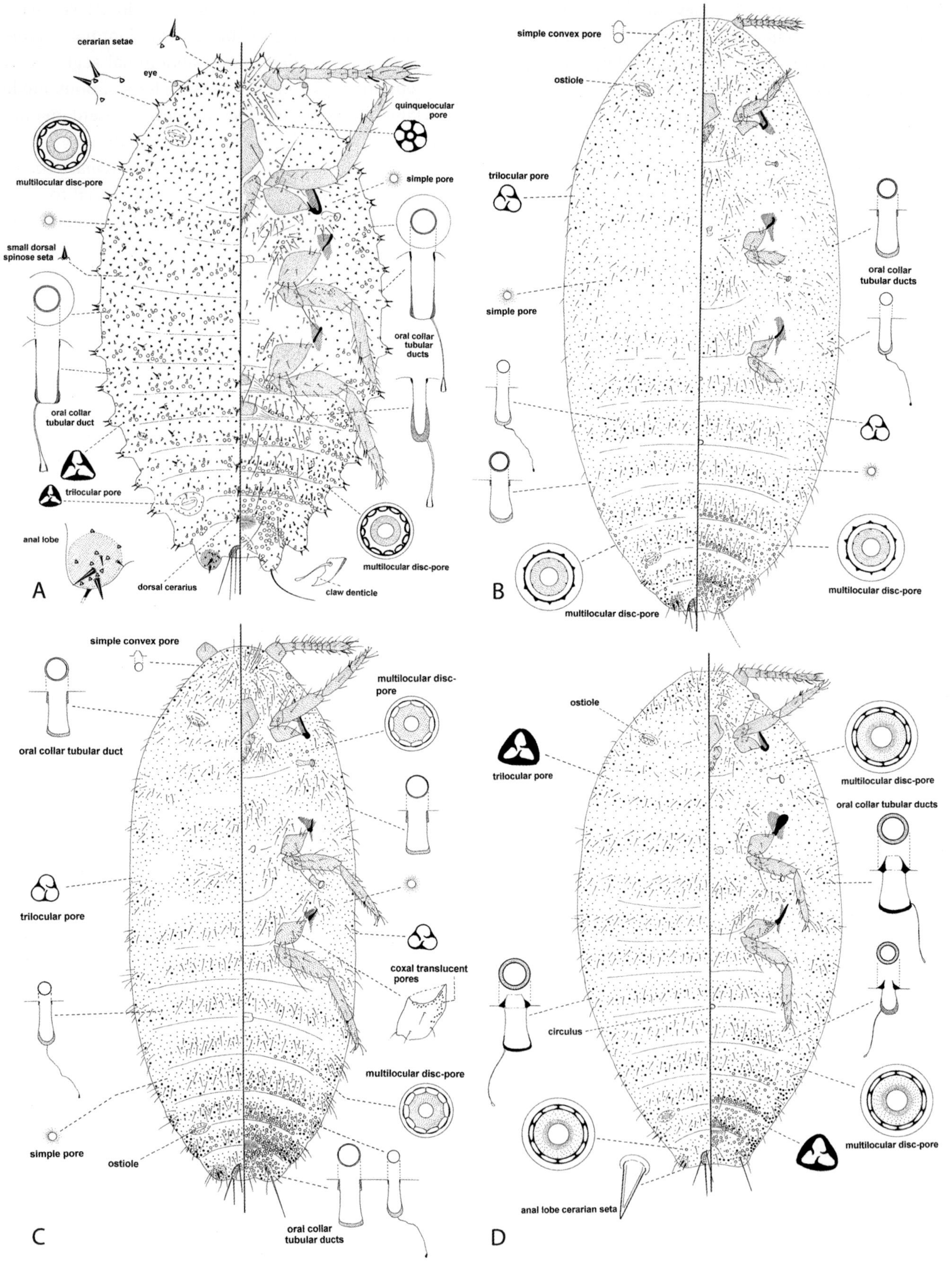

FIGURE 3 *Adult female coccoids. A* Peliococcus (Spinococcus) calluneti, *Greenland form; B,* Trionymus bocheri*; C,* Trionymus heterelymus*; D,* Trionymus thulensis, *Greenland form.*

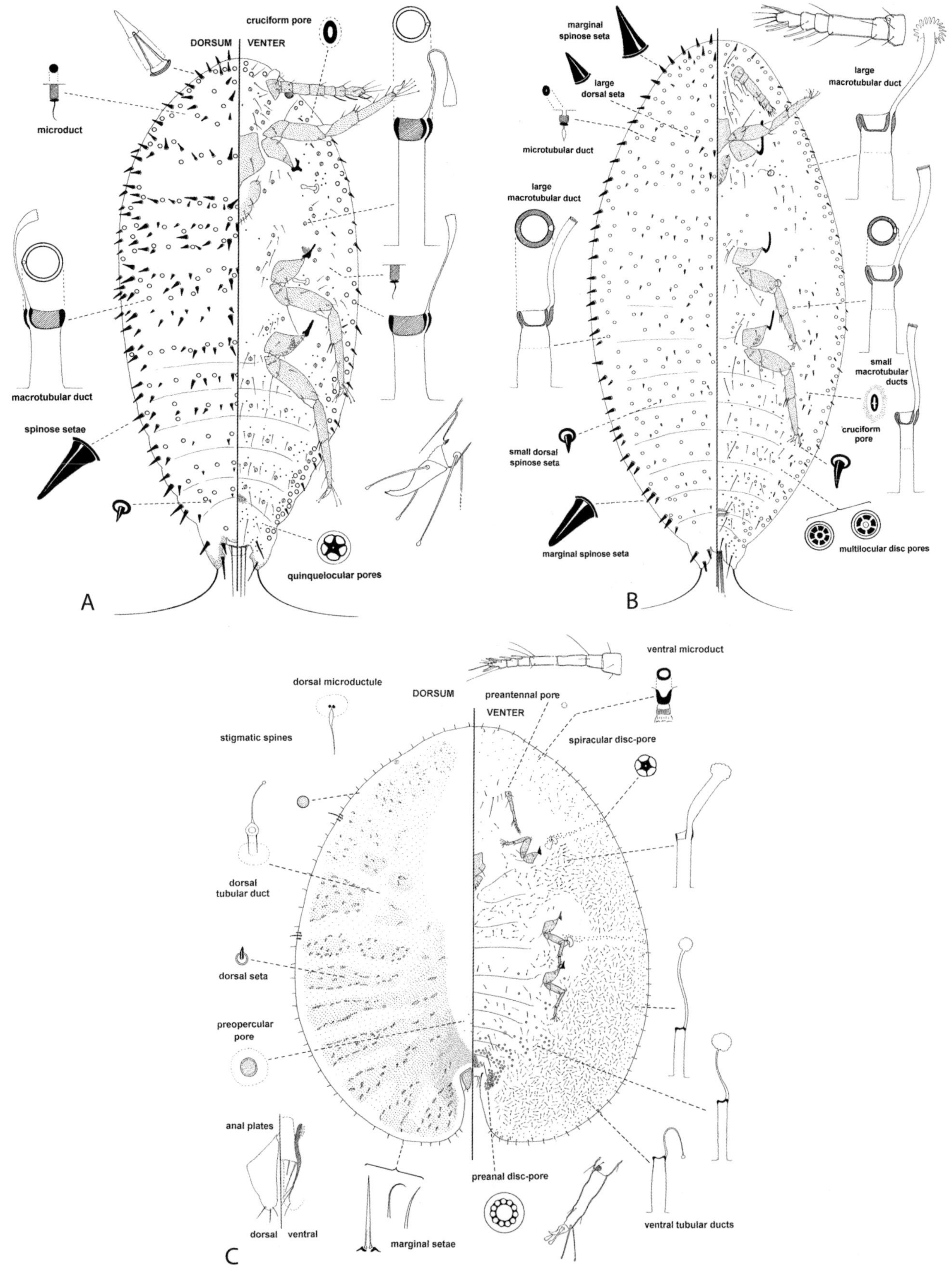

FIGURE 4 *Adult female coccoids. A,* Eriococcus munroi, *Greenland form; B,* Eriococcus pseudinsignis; *C,* Pulvinaria glacialis *(B modified after Williams, 1985).*

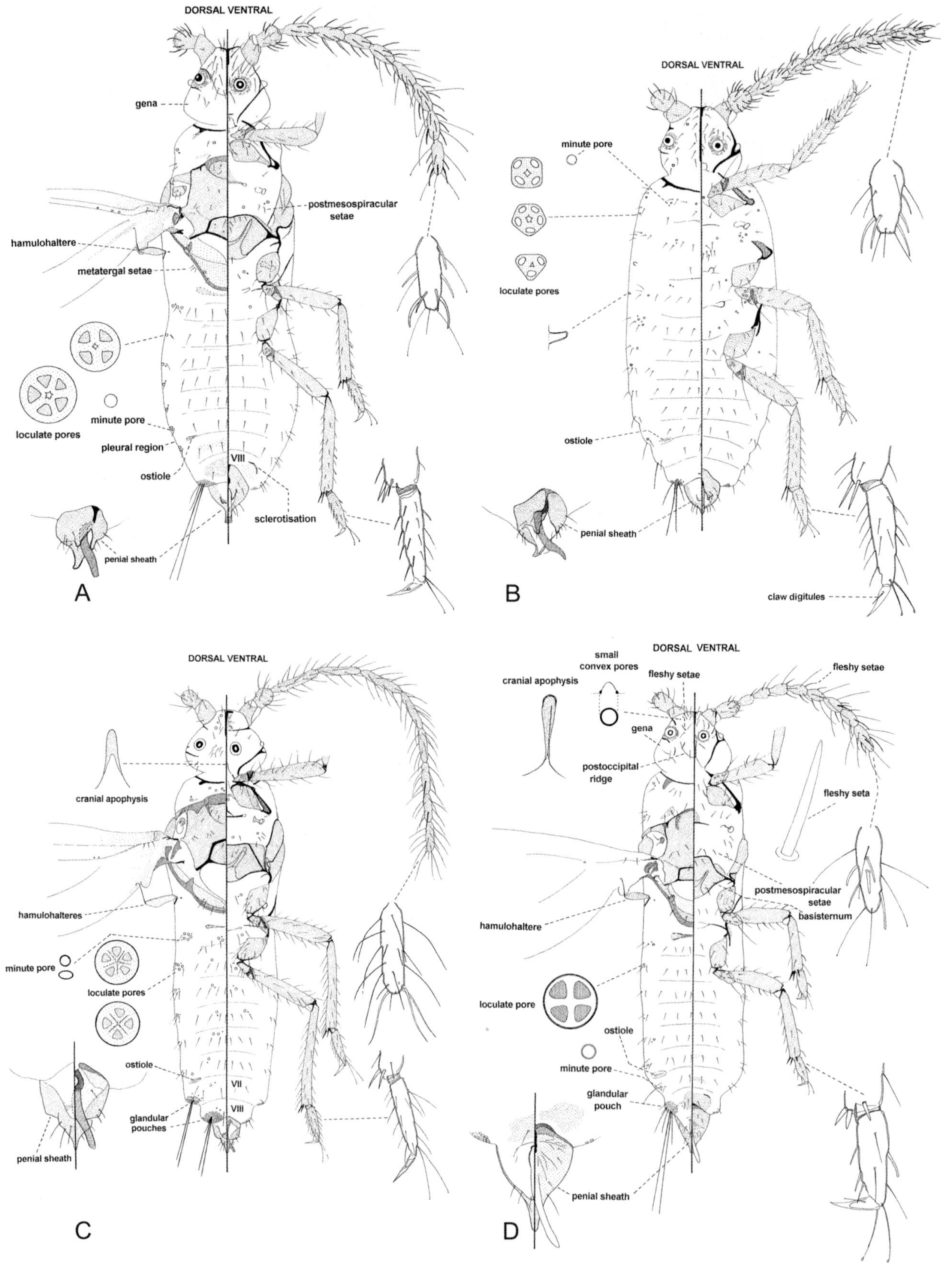

FIGURE 5 *Coccoid males. A,* Atrococcus groenlandensis*; B,* Corizococcus multiporus*; C,* Peliococcus balteatus*; D,* Trionymus bocheri.

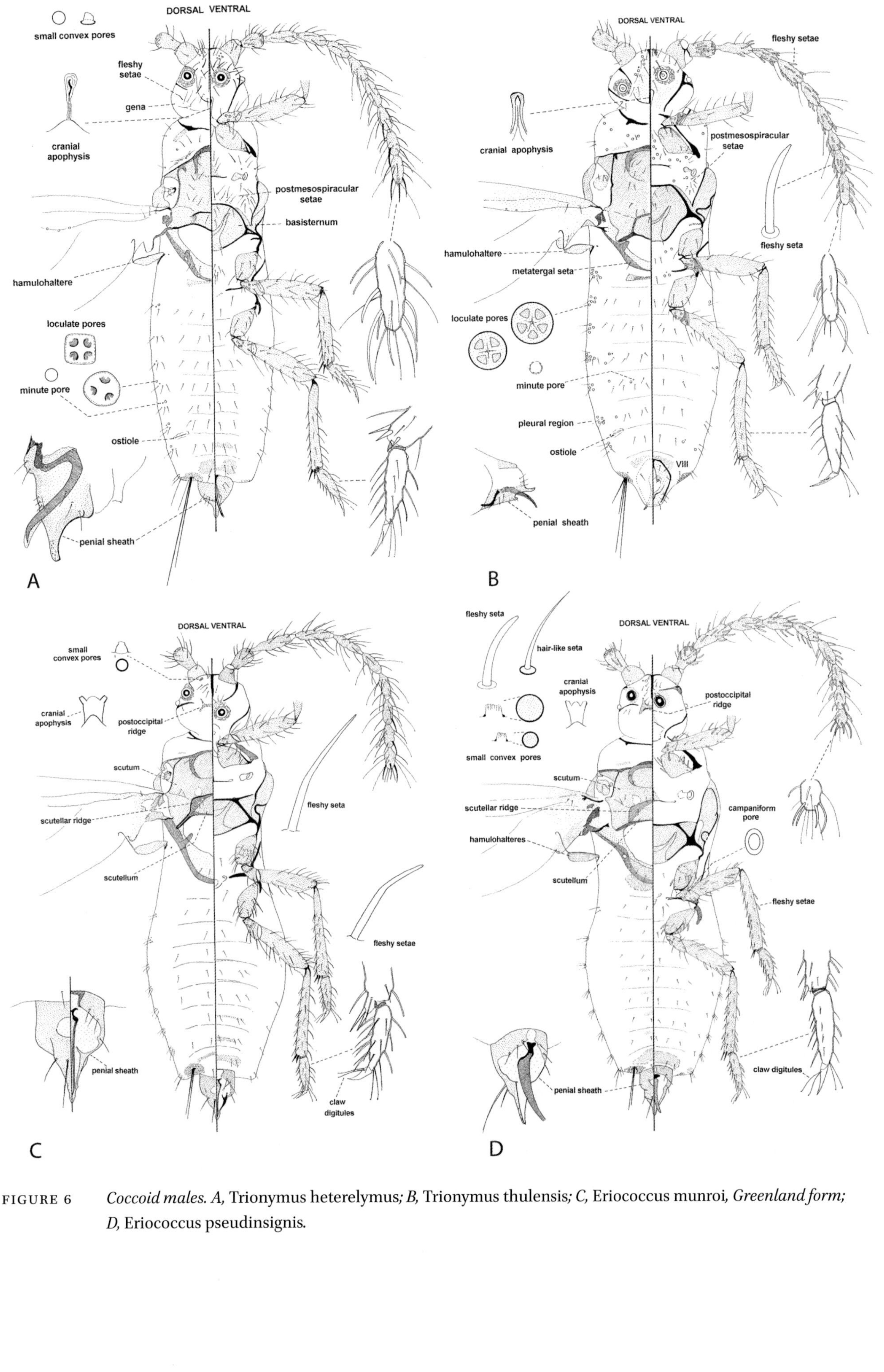

FIGURE 6 *Coccoid males. A,* Trionymus heterelymus*; B,* Trionymus thulensis*; C,* Eriococcus munroi*, Greenland form; D,* Eriococcus pseudinsignis.

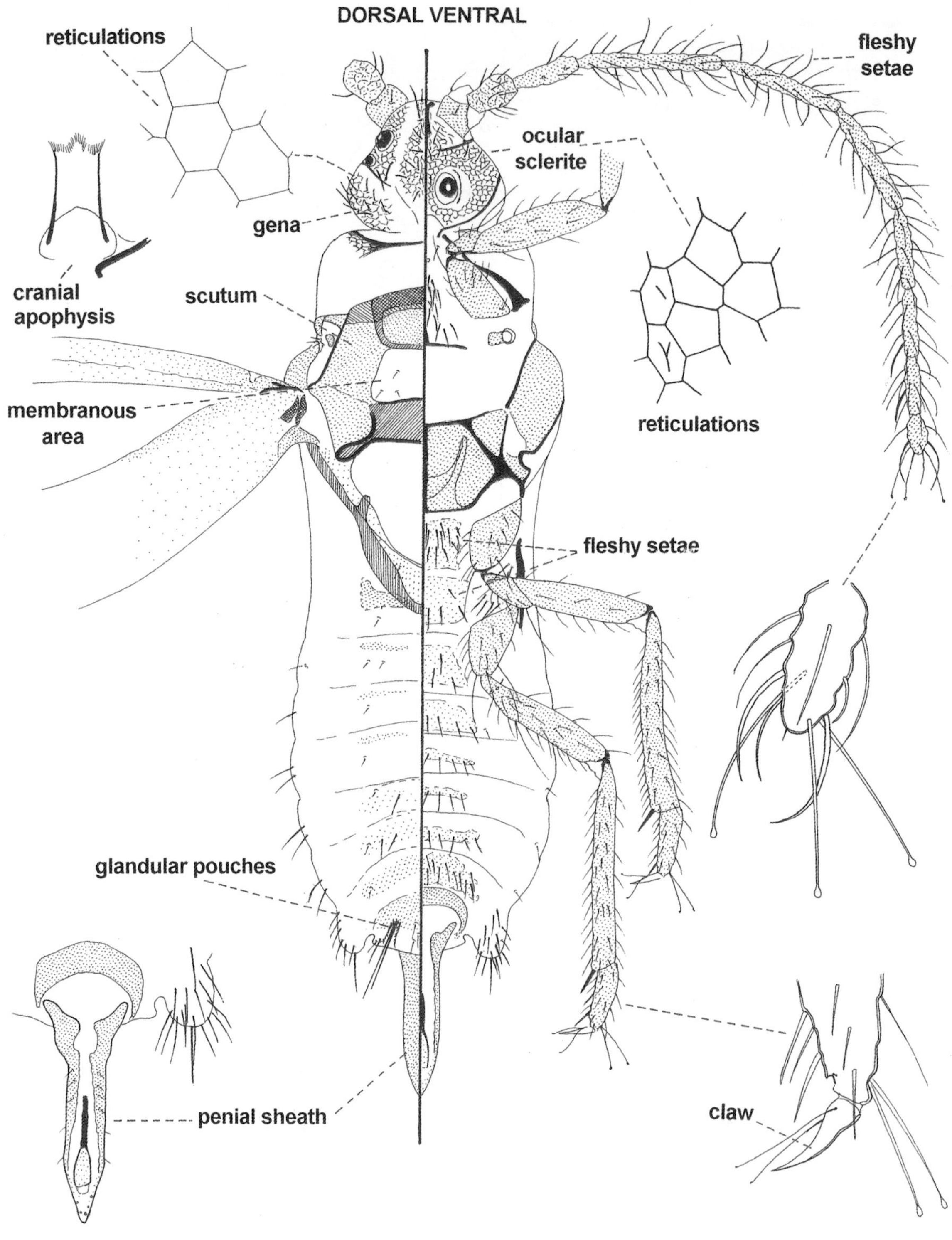

FIGURE 7 *Male* Pulvinaria glacialis.

is especially numerous in typical, luxuriant herb slope vegetation. Overwinters in various life stages.

DISTRIBUTION. S, SW, SE (approximately north to the Low Arctic-High Arctic transition).
Circumpolar, low arctic-boreal and alpine.

PSEUDOCOCCIDAE (MEALYBUGS)

Unmounted adult female mealybugs covered in a white powdery or mealy wax, although this may be reduced on species living in concealed locations, such as under grass sheaths. The margin of the body often has a series of white, lateral wax protrusions, arranged segmentally, generally most prominent posteriorly. Body elongate or oval, the eight abdominal segments generally visible dorsally. On mounted specimens, the circulus (when present) located medially between abdominal segments III and IV. Mealybugs have several different types of wax-secreting pores, all of taxonomic significance, *e.g.*, multilocular disc-pores, quinquelocular pores and trilocular pores—the latter unique to this family. Two types of tubular ducts generally present: oral rim ducts and oral collar ducts. Antennae 6–8 segmented. Legs well developed on Greenland species. The largest family in Greenland.

Key to adult female Pseudococcidae

1. 18 pairs of cerarii, several with more than 2 cerarian setae; quinquelocular pores present throughout venter; claws with a denticle 2
– Less than 5 cerarii, each with 2 cerarian setae; quinquelocular pores absent from venter; claws without a denticle 3
2. Multilocular disc-pores tending to be in distinct groups of 2–4, each group with a small oral collar tubular duct in centre; dorsal cerarii absent .... *Peliococcus balteatus*
– Multilocular disc-pores not tending to be in distinct groups but generally close to an oral collar tubular duct; dorsal cerarii present along mid-line of all segments .... *Peliococcus* (*Spinococcus*) *calluneti*
3. Circulus absent; with a group of multilocular disc-pores and large oral collar ducts laterad to each anterior coxa .... *Atrococcus groenlandensis*
– Circulus present; without a distinct group of multilocular disc-pores laterad to each anterior coxa 4
4. With many oral rim ducts present throughout dorsum; oral collar ducts absent from dorsum .... *Chorizococcus multiporus*
– Oral rim ducts absent from dorsum; oral collar ducts present on dorsum, often fairly abundant 5
5. Circulus large, membranous and approximately rectangular .... *Trionymus elymus*
– Circulus small, oval to almost circular and sometimes slightly sclerotised 6
6. Dorsally, multilocular disc-pores very few on posterior 3 abdominal segments or absent; multilocular disc-pores present ventrally on head anterior to clypeolabral shield .... *Trionymus thulensis*
– Dorsally, multilocular disc-pores present on posterior 5 abdominal segments; multilocular disc-pores absent ventrally from head .... *Trionymus bocheri*

Atrococcus groenlandensis Gertsson & Hodgson, 2005 Figs 2B, 5A

Henriksen 1939: not recorded.
Adult female (mounted material, Fig. 2B) oval in outline, 1.0–2.6 mm and 0.5–1.0 mm wide; anal lobes moderately developed; antennae 8 segmented: legs normally developed; claw without a denticle; circulus absent; with 2 pairs of distinct cerarii plus a third which may be poorly differentiated.

Dorsum: Oral rim ducts distributed as follows: VIII 0; VII–III 1–6; II 0. Oral collar ducts of two sizes, larger ducts restricted to a small group marginally on abdominal segments; smaller ducts in transverse lines across abdominal segments. Multilocular disc-pores present on abdominal segments I–VII, mainly near margin. Venter: Oral rim ducts similar to those on dorsum, with 0 or 1 marginally on abdominal segments I–V and on meta- and mesothorax. Oral collar ducts of two sizes as on dorsum, with larger ducts along posterior margin of each segment; quite frequent on posterior abdominal segments. Multilocular disc-pores present across abdominal segments VIII–IV and laterally on III–I.

The adult female of *A. groenlandensis* is very similar to that of *A. cracens* Williams (a Trans-Palaearctic species), differing mainly in having many fewer multilocular disc-pores, with none present medially anterior to segment IV (on *A. cracens*, multilocular disc-pores are present as far forward as the head). Additionally, *A. cracens* has a group of 15–27 multilocular disc-pores laterad to each procoxa, equal to or exceeding the number of oral collar ducts, whilst *A. groenlandensis* has only 0–6 disc-pores, significantly fewer than the number of oral collar ducts in the group (9–20). The cerarii on abdominal segment VI on *A. groenlandensis* also seem to be hardly differentiated, whereas they are more obvious on *A. cracens*, which may also have cerarii on segment V. For further details see Gertsson & Hodgson (2005). Note: during preparation in potassium hydroxide, the body contents of *Atrococcus* species become black.

Adult male (Fig. 5A). While easily separable from the other species discussed here using the key characters, it must be emphasized that males of apparently closely related *A. cracens* are currently unknown.

BIOLOGY. Polyphagous, found feeding on *Campanula gieseckiana, Melandrium affine/triflorum, Viscaria alpina, Artemisia borealis, Hieracium rigorosum, Arctostaphylos uva-ursi, Vaccinium uliginosum, Poa glauca, Thymus drucei, Chamerion angustifolium, Rumex acetosella, Coptis trifoliata, Dryas integrifolia, Potentilla tridentata, Salix arctophila, Salix glauca, Saxifraga tricuspidata.* Males have been collected from several sites, quite often in pitfall traps and with adult females on several occasions.

DISTRIBUTION. S, SW (several places between Paamiut/Frederikshåb and Sisimiut/Holsteinsborg), NE (only Jameson Land: Muslingeelvbugt).

Currently only known from Greenland.

Chorizococcus multiporus **Gertsson & Hodgson, 2005 Figs 2C, 5B**

Henriksen 1939: not recorded.

Adult female (mounted material, Fig. 2C). Oval in outline, 1.2–1.3 mm long and 0.6–0.7 mm wide; anal lobes moderately developed; antennae 7 or 8 segmented; legs well developed; circulus quite large; with 3 distinct cerarii; oral rim ducts abundant on dorsum. Dorsum: Cerarii clearly present on posterior 3 abdominal segments, each with two stout conical setae. Oral rim ducts present as follows: VIII 0; I–VII 5–14; meso- and metathorax 4–14; prothorax 2–3; head 1–7 (total range 75–115). Oral collar ducts absent. Multilocular disc-pores absent. Venter: Oral rim ducts present, generally with 1 or 2 (occasionally 0) on lateral margins of most segments. Oral collar ducts of two sizes, larger ducts located along posterior margin of each abdominal segment and near margin of abdominal segments II–VIII; small ducts in a transverse line medially across all abdominal segments; large ducts absent more anteriorly to abdomen but small ducts present on at least meso- and metathorax. Multilocular disc-pores quite large, present on abdomen along anterior and posterior margins of segments V–VIII, and along posterior margin only of segment IV; absent more anteriorly apart from occasional disc-pores between procoxa and anterior spiracle. Translucent pores quite abundant on lateral half of each metacoxa and also with a few on femur and tibia.

The adult females of *C. multiporus* from Greenland are superficially similar to *C. altoarcticus* Richards (Richards, 1964) from Ellesmere Island, arctic Canada, but differ significantly in detail, particularly in the following (character-states on Canadian material in brackets): (i) oral rim ducts abundant on dorsum, with a total of 75–115 (total of 10–30), and (ii) oral collar ducts absent on dorsum (a few present on all abdominal segments). The limbs of *C. multiporus* also tend to be slightly larger. For further details, see Gertsson & Hodgson (2005).

Adult males (Fig. 5B) are immediately separable from all other coccoid males known from Greenland in being apterous. The adult males of *C. multiporus* were compared with males from the type series of *C. altoarcticus* Richards from Ellesmere Island, Canada, and appear to be nearly identical.

BIOLOGY. Unknown. Males were collected from only three sites, although quite abundant where found, particularly in Zackenberg; all were collected in pit-fall traps. Along with *Pulvinaria glacialis* the most northerly occurring coccid species.

DISTRIBUTION. SW (only known from Qeqertarsuup Kangerlua/Disko Bugt: Qeqertasussuk, and Kangerlussuaq/Søndre Strømfjord), NW (Pituffik/Thule), NE (several sites, males only at Zackenberg, Jameson Land: Muslingeelvbugt), SE (Ammasalik).

Currently only known from Greenland. Was referred to as *C. altoarcticus* Richards by Gertsson (2004).

Peliococcus balteatus **(Green, 1928) Figs 2D, 5C**

Henriksen (1939): *Phenacoccus venustus* Green.

Adult female (mounted material, Fig. 2D). Elongate oval, up to 2.7 mm long. Antennae of Greenland specimens with 8 segments. Legs rather large, claw with a denticle. Derm with groups of multilocular disc-pores on both body surfaces, in clusters of mostly 2–4, forming transverse rows. Isolated groups of multilocular disc-pores present on venter close to antennae, proboscis and each coxa. Small tubular ducts present associated with multilocular disc-pore clusters. With 18 pairs of cerarii. Anal lobe cerarii with 4 spinose cerarian setae and 1 or 2 smaller conical setae. Cerarii 1–3 (on head) with three to five spinose setae. Circulus absent. Dorsum: Multilocular pores in clusters of 2 or 3, each cluster associated with a small tubular duct; clusters in transverse rows on abdominal segments I–VII. Vente: Anal lobe setae longer than setae of anal ring. Multilocular disc-pore of two sizes: larger disc-pores in clusters on head and in transverse rows on abdominal segments III–V; smaller disc-pores in transverse rows across posterior abdominal segments. Quinquelocular disc-pores present associated with labium and near hind coxae.

This species was mentioned by Henriksen (1939) (as *Phenacoccus venustus* (Green), synonymised with *P. balteatus* by Danzig (2001)) in his Index of insects from Greenland. It differs from the illustrations and descriptions of both Williams (1962) and Danzig (2001) in having no circulus. In addition, the lateral cerarii are indistinct and the antennae are 8 segmented. For further descriptions, see Williams (1962) and Danzig (1986, 2001).

Adult male (Fig. 5C). The Greenland male specimens identified (based on sex co-occurrence, see below) as *P. balteatus* are immediately separable from the other mealybug males known from th country in having two pairs of glandular pouches, one pair each on abdominal segments VII and VIII.

BIOLOGY. Oligophagous on grasses, where it lives on leaves and in leaf sheaths (Danzig 2001). Males referred to this species were collected from two sites only, all in pit-fall traps. Since only *P. balteatus* females were collected (also in pit-fall traps) from these sites on the dates when the male specimens were collected, is assumed that the latter belong to this species.

DISTRIBUTION. S (males only from Kap Farvel-area: Anorliuitsoq on Pamialluk island, and Narsarsuaq), SW, SE.

Holarctic, low arctic-temperate (known from USA, Sweden, England, Germany and eastern Europe (Ben-Dov, 1994).

Peliococcus calluneti (Lindinger, 1912) Fig. 3A
Henriksen 1939: not recorded.

Adult female (mounted material, Fig. 3A). Body length up to 1.6 mm. Antennae 7 or 8 segmented. Legs of normal size; claws with a denticle. With 18 pairs of cerarii, most on distinct lateral protrusions. Anal lobe cerarii on a sclerotised plate with 2 conical cerarian setae plus 1 or 2 small seta. Dorsal cerarii present medially and with cerarian setae also present elsewhere on dorsum. Multilocular disc-pores more or less in segmental rows on both body surfaces. Greenland specimens with 2 circuli.

The female material from Greenland agrees fairly closely with the description and figure in Danzig (2001), except that: (i) the association of the multilocular disc-pores and oral collar tubular ducts in groups is not as obvious; (ii) the spinose setae on the dorsum are much smaller submedially and submarginally; and (iii) the oral collar ducts medially on venter of abdomen are much more abundant. Williams (1962) also redescribed *P. calluneti*, but the Greenland material differs significantly from his figure, as follows: (i) the marginal cerarii are all on distinct, large lobular protrusions; (ii) both circuli are large; (iii) the setae of dorsal cerarii medially on the abdomen are much larger and more distinct; (iv) other dorsal spinose setae are much more abundant on the head; (v) oral collar ducts much more abundant posteriorly on venter of abdomen, and (vi) rather fewer multilocular disc-pores present laterally on thorax and head. Despite these differences, it is considered that this material is referrable to *P. calluneti*. For further details, see Williams (1962) and Danzig (1986, 2001).

BIOLOGY. In Greenland collected on *Arctostaphylos alpina*. Elsewhere, mainly known off the roots of the families Ericaceae, Empetraceae and Pyrolaceae (Danzig 2001). In Europe overwintering as an egg (Kosztarab & Kozár, 1988).

DISTRIBUTION. SW, NE. Only collected in two places: Kangerluarsuk Tulleq at Sisimiut/Holsteinsborg (SW) and Rypefjord at Scoresby Sund (NE, on *Arctostaphylos alpina*). Palaearctic, temperate. Known from several places in Europe (Ben-Dov, 1994).

Trionymus bocheri Gertsson & Hodgson, 2005 Figs 3B, 5D
Henriksen 1939: not recorded.

Adult female (mounted material, Fig. 3B). Elongate oval in outline, 3.6 mm long and 1.6 mm wide; anal lobes rounded and moderately sclerotised; antennae 8 segmented; legs relatively quite short; claw without denticle; circulus small and round; with 2 distinct cerarii; oral rim ducts absent. Dorsum: Cerarii clearly present on abdominal segments VII and VIII, each with 2 stout conical setae. Anal lobe distinctly sclerotised, with about 20–22 trilocular pores on each sclerotised area. Oral collar ducts of two sizes: larger ducts frequent throughout; smaller ducts very scarce, present on abdominal segments only. Multilocular disc-pores present on abdominal segments IV–VII.

Venter: Oral collar ducts of two sizes as on dorsum, both sizes present throughout but larger most frequent on posterior abdominal segments, otherwise mainly along posterior margin of posterior abdominal segments; smaller ducts generally frequent, in a transverse line medially across abdominal segments II–IV and on each anal lobe; very scarce on thorax and head. Multilocular disc-pores present on anterior and posterior margins of abdominal segments V–VIII, and on posterior margin only of segment IV; also one medially on segment II; absent more anteriorly apart from 0–2 more or less laterad to posterior spiracles.

Only a single female specimen of this species has been collected. It is very similar to those of *T. perrisii* (Signoret), but differs (character-states on *T. perrisii* in brackets) mainly in the complete absence of multilocular disc-pores on the dorsum anterior to abdominal segment IV (present sparsely throughout dorsum) and almost complete absence ventrally anterior to abdomen (present sparsely on thorax and head).

T. bocheri differs most obviously from *T. elymus* in having a small, almost round, slightly sclerotised circulus, without an intersegmental membrane. Small circuli are also found on *T. thulensis* Green, from which *T. bocheri* differs (character-states on *T. thulensis* in brackets) in having some multilocular disc-pores on the abdominal dorsum (absent) and in having oral collar ducts of 2 sizes on dorsum (of one size only). For further details, see Gertsson & Hodgson (2005).

Adult male (Fig. 5D). Very close to *Trionymus heterelymus* which also has the fleshy setae on the head fairly well defined. For other differences see under *Trionymus heterelymus* below.

BIOLOGY. Not known. Males (along with a single female) have only been collected in pit-fall traps at Narsarsuaq in 1991. Male described below in next section.

DISTRIBUTION. S (only Narsarsuaq). Currently only known from Greenland.

***Trionymus heterelymus* Gertsson & Hodgson, 2005**
Figs 3C, 6A
Trionymus elymus Gertsson & Hodgson, 2005: homonym of *T. elymus* Borchsenius (1937)
Henriksen 1939: not recorded.

Adult female (mounted material, Fig. 3C). Elongate oval, 1.6–2.6 mm long and 0.7–1.0 mm wide. Anal lobes moderately developed and lightly sclerotised. Antennae 8 segmented. Legs normally developed; claw without denticle. Large oval circulus present, with a distinct intersegmental fold. With 2 distinct cerarii. Oral rim ducts absent.

Dorsum: Cerarii clearly present on abdominal segments VII and VIII, each with 2 stout spinose setae. Oral collar ducts of two sizes present: larger ducts frequent throughout, but most abundant on segment VII, smaller ducts in transverse lines across abdominal segments III–VII. Multilocular disc-pores sparse on abdominal segments V–VII. Venter: Oral collar ducts of two sizes, as on dorsum; larger ducts present throughout but most frequent on posterior abdominal segments, particularly laterally, otherwise mainly along posterior margin of posterior abdominal segments and apparently randomly elsewhere; smaller ducts in a transverse line medially across abdominal segments III–VII and randomly on anal lobe; very scarce on thorax and head. Multilocular disc-pores present on abdominal segments IV–VIII, extremely abundant posteriorly and apparently randomly distributed; absent more anteriorly apart from 1–3 between each procoxa and anterior spiracle. Translucent pores quite abundant on lateral 1/3 of each metacoxa but absent from femur and tibia.

The adult female of *T. heterelymus* is very similar to that of *T. tomlini* Green, but differs (character-states on *T. tomlini* in brackets) mainly in the complete absence of multilocular disc-pores on the dorsum anterior to abdominal segment V (present on all abdominal segments) and their almost complete absence ventrally anterior to abdominal segment IV (frequent throughout venter). In addition, the degree of sclerotisation of the anal lobe appears to be less, and the smaller oral collar ducts are also absent anterior to abdominal segment III (present on all abdominal and thoracic segments).

T. heterelymus differs most obviously from the other *Trionymus* species known from Greenland (*T. bocheri* and *T. thulensis* Green) in having a large oval to almost square circulus, with an intersegmental membrane, whereas those on *T. bocheri* and *T. thulensis* are small and round and lack any intersegmental membrane. For further details, see Gertsson & Hodgson (2005).

Adult male (Fig. 6A). This species, along with *Trionymus bocheri*, has the fleshy setae on the head fairly well defined. In addition (character-states on other species in brackets): the males of *T. heterelymus* have (i) the 2nd and 3rd antennal segments shorter than or subequal to the length of the pedicel (segments clearly longer on *T. bocheri*); (ii) rather few postmesospiracular setae, particularly medially (abundant, particularly medially), and (iii) fleshy setae on antennae about as long as width of antennal segments (longer than width).

BIOLOGY. On *Elymus mollis*. At Kangerlussuatsiaq/Evighedsfjorden (Kangiusaq) both sexes were collected on this plant; at Narsarsuaq males were collected in pitfall traps.

DISTRIBUTION. S, SW. Kangerlussuatsiaq/Evighedsfjorden (Kangiusaq, Tasiusaq [females only]), Sisimiut/Holsteinsborg district (head of Itilleq Fjord, females only), Narsarsuaq (males only). Currently only known from Greenland.

***Trionymus thulensis* Green, 1931 Figs 3D, 6B**
Henriksen 1939: not recorded.

Adult female (mounted material, Fig. 3D). Elongate oval, up to 4 mm long and 1.2–2 mm wide. Antennae 7 or 8 segmented. Legs normally developed; hind coxae only with a few translucent pores. Circulus small. With 2 pairs of cerarii. Anal lobe cerarii each with 2 conical setae and about 7 or 8 trilocular pores. Dorsum: On Greenland material, a few multilocular disc-pores present on posterior abdominal segments. Oral collar ducts of one size only, present throughout, mainly in transverse rows. Venter: Anal ring setae twice as long as diameter of anal ring. Anal lobe setae longer than anal ring setae. Multilocular disc-pores in transverse bands on segments VIII–IV, along both anterior and posterior margins of V–VIII. Oral collar ducts of two sizes present: larger ducts mixed with smaller ducts in transverse rows on abdominal segments.

The female of this species has been very well described by Williams (1962). However, the material from Greenland, whilst considered to represent this species, does differ slightly from Williams' description and so it is here illustrated. Specifically, it differs in having (character-states on non-Greenland material in brackets): (i) a few multilocular disc-pores on the dorsum of the abdomen, although these are very few and may sometimes be absent (entirely absent), and (ii) there is no sclerotisation on the anal lobes (present between the two cerarian setae). The distribution and frequency of the other structures appear to be nearly identical to Williams' description.

Adult male (Fig. 6B). All setae on the head appear to be hair-like, whereas adult male *Trionymus heterelymus* and *T. bocheri* have setae on the head which are obviously fleshy. In addition, *T. thulensis* differs from the other two *Trionymus* species (character-state on other two species in brackets) in having: (i) several disc-pores on the pleural region of each abdominal segment (0–2); (ii) no obvious fleshy setae on the legs (present); (iii) multilocular disc-pores on the dorsal epicranium (absent), and (iv) fleshy setae on the antennae very obviously short, much less than antennal width, and stout (longer, thinner and more pointed).

BIOLOGY. On *Poa glauca*. Males were collected at two sites only: Qeqertarsuaq/Godhavn and Narsarsuaq. All specimens were collected in pit-fall traps but some adult females were also caught at the same time at Narsarsuaq.

DISTRIBUTION. S, SW (only found in few places: Narsarsuaq, Kangerluarsuk Tulleq, Qeqertarsuaq/Godhavn).

Palaearctic, low arctic-temperate (known from Iceland, Sweden, England and Wales, France, Hungary, Poland, China (Ben-Dov, 1994).

ERIOCOCCIDAE (FELT SCALES)

The name refers to the felt-like sac that encloses the adult females and their eggs. The body is oval or elongate, narrowing towards the posterior end. Spinose setae often numerous, in transverse bands or rows on dorsal surface and along body margin. Each anal lobe has a long apical seta and 3 short conical setae on dorsal surface. Ventral setae mainly hair-like. Antennae 6 or 7 segmented. Legs normally developed; claw with a denticle. Macrotubular ducts with a cup-shaped internal invagination. Trilocular pores, cerarii and circuli all absent.

Key for separation of adult female Eriococcidae known from Greenland

1. Larger spinose setae restricted to body margins; smaller spinose setae found throughout dorsum .... *Eriococcus pseudinsignis*
– Large spinose setae abundant over most of dorsum; smaller spinose setae restricted to posterior 3 or 4 abdominal segments .... *Eriococcus munroi*

Eriococcus munroi (Boratynski, 1962)

Figs 4A, 6C

Henriksen 1939: not recorded.

Adult female (mounted material, Fig. 4A). Elongate-oval, derm slightly nodulose; length 1.3 mm. Antennae 7 segmented. Legs well developed. Dorsum: spinose setae of various sizes more or less in segmental rows: present throughout but those on abdominal segments V–VII very small (length about equal to width of basal socket). Macrotubular ducts of one size, frequent throughout. Venter: with hair-like setae. A few spinose setae along margins and with about 4 or 5 small spinose setae submarginally on thorax. Macrotubular ducts of 2 sizes, larger ducts (similar to those on dorsum) located marginally, smaller ducts sparse throughout venter. Cruciform pores extremely few, in submargin only.

The female material from Greenland agrees quite well with Williams' (1985) description, except that the larger spinose setae on the dorsum are absent on abdominal segments IV–VII, being replaced by small to minute spinose setae. On the material seen by Williams (1962), Danzig (1986) and Kosztarab & Kozár (1988), the larger spinose setae are present on segment IV, only becoming small on segment V. Nonetheless, the Greenland material is here considered to be referable to this species.

Adult male (Fig. 6C), Very similar to that of the next species. For differences, see the key.

BIOLOGY. In Greenland found on *Dryas integrifolia* and *Hieracium rigorosum*. Elsewhere polyphagous on stems, leaves and root crown area of 13 plant families (Miller & Gimpel, 2000). Associated with mixed grasses according to Danzig (1986). In England two generations a year. Eggs of 2nd generation overwinter (Boratynski, 1962).

DISTRIBUTION. S (Nunap Isua/Kap Farvel-area, Narsarsuaq), SW (Qeqertarsuatsiat Kangerluat/Fiskenæsfjord, Kangerlussuatsiaq/Evighedsfjord: Tasiusaq).

Palaearctic, temperate (found only in some parts of Europe and northwestern Asia (Miller & Gimpel, 2000).

Eriococcus pseudinsignis Green, 1921

Figs 4B, 6D

Henriksen 1939: not recorded.

Adult female (mounted material, Fig. 4B). Elongate-oval, 1.5–2.0 mm long. Anal lobes wide. Each lobe dorsally with 1 outer and 2 inner enlarged setae. Antennae 6 segmented.

Dorsum: Largest spinose setae along margin, with 3 large setae on each side of each abdominal segment; smaller setae scattered over rest of dorsum. On the posterior abdominal segments these setae are more rounded, blunt, pointed toward head. Macrotubular ducts of one size. Venter: Multilocular disc-pores with 5 or 7 loculi. Macrotubular ducts of 2 sizes, larger type on both body surfaces. The larger type around margins only. For further details, see Williams (1985) and Kosztarab & Kozár (1988).

Adult male (Fig. 6D). Very similar to *E. munroi*; for differences, see the key.

BIOLOGY. Primarily on grasses, infesting the leaves (Kosztarab & Kozár, 1988). In Europe one annual generation, overwintering as eggs. Adults appear during the first part of June, laying 29–78 eggs per female from the end of June (Miller & Gimpel, 2000). Males have been collected (in pit-fall traps) at only 2 sites: Anorlliuitsoq on Pamialluk island in Nunap Isua/Kap Farvel-area, and Narsarsuaq. Adult females of *E. pseudinsignis* have also been collected at Narsarsuaq.

DISTRIBUTION. S (Nunap Isua/Kap Farvel-area: Pamialluk (Anorlliuitsoq), Narsarsuaq, Qassiarsuk).

Palaearctic, temperate, especially found in Europe east to Ukraine (Miller & Gimpel, 2000).

COCCIDAE (SOFT SCALES)

The adult females in this family are more or less convex and often occur in groups. Older adult females become heavily sclerotised. The only represented genus in Greenland is *Pulvinaria*. This genus produces a white waxen ovisac from under the abdomen. Adult female Coccidae have a distinct anal cleft at posterior end of the body, with a pair of anal plates covering the anus at the anterior end of the cleft. Each spiracular furrow has 3 spinose setae on the margin, differentiated from the marginal setae. Species of *Pulvinaria* are also characterised by having numerous ventral tubular ducts of 2–4 types, distributed throughout most of the venter.

Pulvinaria glacialis Gertsson & Hodgson, 2005
Figs 4C, 7

Henriksen 1939: not recorded.

Adult female (mounted material, Fig. 4C). 1.8–3.4 mm long and 1.3–3.3 mm wide, more or less oval but slightly more pointed at anterior end. Legs well developed, claw broad; claw denticle very indistinct or absent. Stigmatic clefts absent. Antennae 7 segmented. Dorsum: Derm of older females lightly sclerotised at maturity, with a series of small, slightly more heavily sclerotised, areas of irregular shape, arranged more or less segmentally, most abundant near margin, becoming less frequent medially. Dorsal setae extremely scarce or absent; when present with a blunt apex and short, only slightly longer than width of basal socket. Dorsal tubular ducts present. Preopercular pores few. Anal plates each approximately triangular, with 3 setae near apex. Margin: Marginal setae stoutly setose, sharply pointed, sometimes slightly to strongly bent towards apex. Stigmatic clefts absent, each stigmatic area with 3 stigmatic spines, median spine longest. Venter: Preanal disc-pores abundant around vulva but becoming less frequent on anterior abdominal segments; 1 or 2 present on metathorax. Spiracular disc-pores in moderately broad bands between spiracle and margin. Ventral microducts quite large, frequent in a marginal band which broadens on head and prothorax. Ventral tubular ducts of 3 types, mainly in a broad submarginal band but absent anterior to antennae; also present medially on head, thorax and abdomen. For further details, see Gertsson & Hodgson (2005).

The adult females of *P. glacialis* are superficially very similar to those of *P. ellesmerensis* Richards (Richards, 1964) from Ellesmere Island, arctic Canada, also collected off *Salix arctica*. However, they differ quite significantly in a number of features (character-states for *P. ellesmerensis* in brackets): (i) dorsal setae extremely few, possibly even absent on most specimens (frequent, though sparsely distributed); (ii) dorsal setae (when present) blunt and extremely short, about as long as width of basal socket (at least 2× width of basal socket and setose); (iii) antennae 7 segmented (antennae 8 segmented); (iv) marginal setae significantly more abundant than ventral submarginal setae (marginal setae as frequent or less frequent than ventral submarginal setae), and (v) preopercular pores very few, less than 10 (quite abundant, more than 40).

Adult male (Fig. 7). See key characters. As male specimens of only a single species of Coccidae have been collected in Greenland, it seems reasonable to assume that these belong to *P. glacialis*, particularly since they show close similarities to the males of *P. ?betulae* (L.) and *P. acericola* (Walsh & Riles), described by Giliomee (1967). The male of the closely related *P. ellesmerensis* Richards (Richards, 1964) from arctic Canada has not been described.

BIOLOGY. Only known off *Salix arctica*. Males have only been collected in, and SW (head of Itilleq, *c.* 66°32'N).

DISTRIBUTION. S (only males, Nunap Isua/Kap Farvel-area: Anorlliuitsoq on Pamialluk island), SW (several places, including the head of Kangerlussuaq/Søndre Strømfjord.), NW (Nuussuaq, Olrik Fjord), NE (several places). Currently only known from Greenland.—Together with *C. multiporus*, the most northerly coccid species in Greenland. Was referred to as *P. ellesmerensis* Richards in Gertsson (2004).

References

Ben-Dov, Y. (1994): *A Systematic Catalogue of the Mealybugs of the World (Insecta: Homoptera: Coccoidea: Pseudococcidae and Putoidae), with data on Geographical Distribution, Host Plants, Biology and Economic Importance*. – Intercept. 686 pp.

Boratynski, K.L. (1962): A new species of *Acanthococcus* Signoret, 1875 (Eriococcidae, Coccoidea, Homoptera) from Britain. – *Proceedings of the Royal Entomological Society of London* (B) 31: 55–60.

Danzig, E.M. (1986): *Coccids of the Far-Eastern USSR. Phylogenetic analysis of Coccids of the World Fauna.* Translated from 1980 Russian edition. – Amerind Publishing, India. 450 pp.

Danzig, E.M. (2001): Mealybugs of the genera *Peliococcus* and *Peliococcopsis* from Russia and neighbouring countries (Homoptera: Coccinea: Pseudococcidae). – *Zoosystematica Rossica* 9: 123–154.

Gertsson, C.-A. (2004): The scale insects (Hemiptera: Coccoidea) of Greenland. – *ISSIS (Proceedings of the X International Symposium on Scale Insect Studies) 19th-23rd April 2004*: 331–337.

Gertsson, C–A. & Hodgson, C.J. (2005): Four new species of mealybug (Pseudococcidae) and a new species of soft scale (Coccidae: Coccoidea: Hemiptera) from Greenland and redescriptions of a mealybug and a soft scale from arctic Canada. – *Zootaxa* 897: 1–24.

Giliomee, J.H. (1967): Morphology and Taxonomy of adult males of the family Coccidae (Homoptera: Coccoidea). – *Bulletin of the British Museum (Natural History), Entomology,* Supplement 7: 1–168.

Henriksen, K.L. (1939): A revised index of the insects of Grønland. – *Meddelelser om Grønland* 119 (10): 1–111.

Hodgson, C. (2005): The adult males of Coccoidea (Hemiptera: Sternorrhyncha) known from Greenland. – *Zootaxa* 907: 1–48.

Kosztarab, M. & Kozár, F. (1988): *Scale Insects of Central Europe.* Junk. 456 pp.

Kozár, F. 2004. *Ortheziidae of the World.* Plant Protection Institute, Hungarian Academy of Sciences. 525 pp.

Miller, D.R. & Gimpel, M.E. (2000): *A Systematic Catalogue of the Eriococcidae (Felt Scales) (Hemiptera: Coccoidea) of the World.* Intercept. 589 pp.

Richards, W.R. (1964): The scale insects of the Canadian arctic (Homoptera: Coccoidea). – *The Canadian Entomologist* 96: 1457–1462.

Schmutterer, H. (1952): Die Ökologie der Cocciden (Homoptera: Coccoidea) Frankens. – *Zeitschrift für Angewandte Entomologie* 33: 369–420.

Williams, D.J. (1962): The British Pseudococcidae. – *Bulletin of the British Museum (Natural History)*, Entomology Series 12 (1): 1–79.

Williams, D.J. (1885): The British and some European Eriococcidae (Homoptera: Coccoidea). – *Bulletin of the British Museum (Natural History)*, Entomology Series 51 (4): 348–393.

10.5. Cicadomorpha

Michael Wilson and Jens Böcher

Cicadomorphans share with the extralimital Fulgoromorpha a specialized antennal gross morphology and abdominal sound producing organs. The two have conventionally been united in a taxon 'Auchenorrhyncha', the monophyly of which has, however, in recent decades been repeatedly questioned on both morphological and molecular grounds.

Morphology

Most families (including the single one represented in Greenland) are characterized by their very short antennae, with the flagellum (*i.e.*, the segments beyond the two basal ones) reduced to form a bristle-like 'arista'. The other characteristic trait, viz., the sound-producing organ is located in the anterior part of abdomen. It consists of an arched part of the lateral body-wall (*tymbal*) on which a strong muscle is attached; apodemes accommodating insertions of these muscles sometimes furnish useful taxonomic characters. Sounds of Greenland leaf hoppers cannot be heard by humans; they are perceived by the insects by means of chordotonal organs in the tibiae.

Greenland fauna

Just two 'auchenorrhynchan' species are present in Greenland. It is perhaps surprising that no fulgoromorphan planthoppers of the family Delphacidae have been found especially since members of the genus *Javesella* are known from other northern latitudes such as Iceland (Remane 2005).

CICADELLIDAE (LEAF HOPPERS)

A large family of worldwide occurrence. They often occur in large numbers especially in grassland biotopes. The family is distinguished by the very broad hind coxae and one or more rows of small spines on hind tibiae. The membranous hind wings have a 'peripheral rib' along most of the hind margin. Many members of the family jump extremely well by means of the hind legs. The ovipositor is adapted for piercing plant tissues, where the eggs are deposited either singly or in rows. The family contains a number of serious pests which damage cultivated plants by sap-sucking and also by the transmission of phytopathogenic viruses and phytoplasmas.

Two easily distinguishable species occur in Greenland, *Psammotettix lividellus* and *Macrosteles laevis*. It is not surprising that species from these genera are found. Both are widely distributed in the Holarctic and contain a large number of species, some of which are very common in grassland (Ossiannilsson 1983, Nickel 2003, Remane 2005).

***Psammotettix lividellus* (Zetterstedt, 1838) Figs 1A, 1B**
Henriksen 1939: *Deltocephalus lividellus* Zett.

Easily distinguished from *Macrosteles laevis* by the almost uniformly coloured, yellow-grey surface without black spots on head or pronotum, and by relatively smaller spines on hind tibiae (length of largest about the width of tibia). Fore wings with distinct lighter longitudinal markings (veins) and the front rim is yellow.

BIOLOGY. Found in dry and warm, generally south-exposed slopes, very often together with the Heteroptera, *Nysius groenlandicus* and *Chlamydatus pullus*. In grassy areas, especially dry steppes, often in immense numbers. Presumably associated with grasses (Poaceae) on which it sucks the sap. In Greenland one yearly generation and hibernation as egg. The species jumps readily but seems unable to fly.

DISTRIBUTION. S, SW (north to Qeqertarsuup Tunua/ Disko Bugt).

Nearctic, low arctic-boreal. The widest distributed of the Canadian leafhoppers, and the only species of the suborder found in Arctic surroundings (Greene 1971).

***Macrosteles laevis* (Ribaut, 1927) Fig. 1C**
Henriksen 1939: not recorded.

General colouration yellow to yellowish green. Easily separated from *P. lividellus* by dark spots on both head and pronotum. Generally 6 spots in 3 pairs on head, but their size are highly varying and may be more or less confluent into blots. In most cases 2 distinct spots on scutellum with a central dark streak further back. Hind tibiae with strong spines, some of which are longer than width of tibia. Fore wings without distinct veins.

BIOLOGY. Occurs in large number in grassy areas, especially in the old Norse fields and modern cultivated fields, but not in dry localities. In Greenland one yearly generation, in temperate Europe 2–3. Here sometimes outbreaks and serious damage of cereal crops. *M. laevis* belongs to the group of species which may have been introduced to Greenland by the Norsemen. It is a pioneer species of sparsely vegetated mud banks, fertilised meadows, ruderal

FIGURE 1 *A–C: Cicadellid anterior body parts, dorsal view. A,* Psammotettix lividellus*; C,* Macrosteles laevis; *B,* Psammotettix lividellus.

sites etc. Polyphagous, living on grasses, sedges and also dicotyledonous herbs.

DISTRIBUTION. S, SW (north to Nuup Kangerlua/ Godthåbsfjord).

Circumpolar (Eurasia, Alaska), low arctic-temperate. Very widely distributed and recorded in almost all countries of the Palaearctic.

References

Greene, J.F. (1971): A revision of the Nearctic species of the genus *Psammotettix*. – *Smithsonian Contributions to Zoology* 74: 1–40.

Nickel, H. (2003): *The leafhoppers and planthoppers of Germany (Hemiptera, Auchenorrhyncha): Patterns and strategies in a highly diverse group of phytophagous insects*. Pensoft & Goecke & Evers. 460 pp.

Ossiannilsson, F. (1983): The Auchenorrhyncha (Homoptera) of Fennoscandia and Denmark. Part 3: The Family Cicadellidae: Deltocephalinae, Catalogue, Literature and Index. – *Scandinavian Science Press*, Copenhagen, pp. 591–979.

Remane, R. (2005): Zur Zikaden-Fauna Islands, einer nur geologisch "alten" Vulkaninsel: Analyses eines 2001 von Dr M. v. Tschirnhaus gesammelten Materials (Insecta Rhynchota Auchenorrhyncha). – *Marburger Entomologische Publikationen* 3: 111–123.

10.6. Heteroptera (True bugs)

Jens Böcher

Morphology: adults

Body in most cases flattened and wings folded flat when at rest. Head generally prognathous, the rostrum can be directed forwards. Compound eyes always present, generally also 2 ocelli (absent *e.g.* in Miridae). Antennae never with more than 5 segments, but may be long. Mouthparts fairly uniform, labial sheath with 3 or 4 segments. When not in use the proboscis (labium and stylets) is bent backwards under the body. Prothorax always large and conspicuous, occupying most of the thorax viewed from above. Mesonotum mostly (apterous taxa excepted) concealed between pronotum and fore wings, but prominent triangular scutellum region visible between wing bases; metathorax totally hidden in dorsal view. The fore wings are termed *hemelytra*; they are composed of a proximal hardened portion and a distal membranous part, the *membrane*. The hardened portion is divided into two or more regions along the claval furrow separating an anterior *corium* from a posterior *clavus* (adjacent to scutellum when wings are folded). The venation is also highly modified and reduced in the hind wings. Both pairs of wings are used in flight. There are at most 3 tarsal segments and two claws.

Immature stages

The eggs often have a delicate sculpturing and a detachable lid. They are frequently attached to plants or placed inside plant tissues. There are generally 5 nymphal stages, in some cases fewer. First instar nymphs may look quite unlike the parents (fewer antennal and tarsal segments, other colours), but at each moult the nymphs gradually get more like the adults, attaining functional wings and sexual organs at last moult. Wing buds are usually obvious from fourth instar (Fig. 1B). Generally nymphs ingest the same food as adults and are in most cases found together with them.

Greenland fauna

Only 3 species, representing 3 families, are known from outdoor habitats in Greenland, and one of them (*Nabis flavomarginatus*) may have been introduced by the Norse people about 1000 years ago. *Nysius groenlandicus* and *Chlamydatu pullus* are widespread, and the former in particular is an important member component of the ecosystems almost everywhere in the country; with respect to behaviour and ecology it has been studied in greater depth than any other Greenland terrestrial arthropod.

Key to the Greenland families of Heteroptera

1. Body flat, brown and wingless .... Cimicidae
– Not as above 2
2. Body (adult) more than 5 mm. Rostrum curved, not pressed against body at rest. Fore legs somewhat raptorial .... Nabidae
– Body less than 5 mm. Rostrum pressed against body at rest. Fore legs not raptorial 3
3. Ocelli present. Membrane with 4 veins ... Lygaeidae
– Ocelli absent. Membrane veins forming a crescent cell .... Miridae

LYGAEIDAE (SEED BUGS)

Four antennal segments, antennae inserted well down on sides of head. Two ocelli and not more than 5 veins in the membrane. Colours mostly rather dull, in some species bright red and black (warning colouration). A large family of world-wide occurrence. Most species are polyphagous seed-eaters and live in warm, xeric environments.

***Nysius groenlandicus* (Zetterstedt, 1838)**
(Greenland seed bug) Figs 1, 2C
Henriksen 1939: *Nysius groenlandicus* Zett.

One of the most characteristic and most widely distributed Greenland insects. *C.* 3–4 mm. Quite variable in size and colour which may be from pale grey shadings to very dark, almost black, especially in the males. Populations entirely or almost consisting of females have been discovered in Northeast Greenland. While having functional wings, the species flies only as an escape reaction, not for more than ½ m and only if ambient temperature is above *c.* 30°C.

BIOLOGY. The species is found almost everywhere in Greenland where conditions permit. The life cycle is restricted to a single season, and accordingly the species is found only where temperatures are high enough to produce a sufficient number of day-degrees. In general, inland

FIGURE 1 *A–B:* Nysius groenlandicus. *A, copulation in fruit head of* Dryas *sp.; B, V–instar nymph.*

populations are denser and more eurytopic than coastal populations, and the species is usually absent from humid and north-facing places. The highest local densities (>100 per m^2) are found in xeric, well-drained sites, such as south-facing herb fields and steppe-like communities. This pattern indicates thermophily (preferred temperature ca. 30°C) and xerophily which is also confirmed experimentally (Böcher 1972, 1975a, 1975b, 1976, 1990, 1997; Böcher & Nachman, 2001, 2010, 2011).

N. groenlandicus is a polyphagous seed-eater which may be found on a great variety of plants and on the soil surface, sucking fallen seeds. The limited means of dispersal is surprising in view of the almost universal occurrence in Greenland, which (as indicated by the fossil record) must have existed almost since the last glaciation (Böcher & Bennike 1996, Böcher 1997). A possible explanation is the tendency of females to deposit eggs while feeding in the capitulae of plants producing seeds and fruit with a parachute device for seed dispersal. Wind-dispersed seeds with eggs may be observed in the field, and probably this 'dispersal parasitism' can explain the ubiquity of the species in Greenland (Böcher 1972, 1975a).

In spite of some laboratory results indicating the existence of a partial second generation (Böcher 1975b), the life cycle in nature appears to be strictly univoltine, and the winter is passed in the egg-stage. First nymphal instars appear just after snow-melt, the other stages are passed during summer. The nymphs are quite conspicuous due to the reddish colour of their abdomen whereas the dull-coloured adults are cryptic. Adults appear in late summer, mostly August, to mate and lay eggs. They pass away during the first days of winter with severe night frost.

DISTRIBUTION. G. Apart from partly unexplored areas especially in NW and N, *N. groenlandicus* seems to inhabit the whole of Greenland, even small islands and skerries far from the main land. There are unpublished records from Peary Land (J.A. Downes, personal communication 1990).

?Palaearctic, arctic-alpine.

NOTE. The distribution in Asia is incompletely known, but probably the species is found in most mountain ranges, even up to 5200 m a.s. in the Himalayas (Mani 1962, 1968, as "*Nysius ericae obscuratus/alticola*"). Makarova & Makarov (2006) report the species from Dolgii Island in the Barents Sea. The species is also recorded from Alaska and many places in Canada (Danks 1981, Henry & Froeschner 1988). However, the occurrence in Canada is still questionable due to identification problems and because all findings are from areas south of the Arctic. Indeed the taxonomic status of the populations currently referred to as *N. groenlandicus* needs detailed reinvestigation.

NABIDAE (DAMSEL BUGS)

Antennae very thin, with 4 or 5 segments; rostrum 4-segmented. Fore legs somewhat raptorial: front femora shorter and thicker than middle and hind femora, on the posterior side with 2 rows of short and strong spines. Scent-gland openings conspicuous on metapleura.

All nabids are active predators occurring on herbage where they attack smaller arthropods. The bug is attracted by scent and movements of the prey, and usually it is seized by the fore legs whereupon the rostrum is inserted. Nabid females lay eggs in plant stems. Males of some species have a stridulatory organ on the apex of abdomen.

Nabis flavomarginatus Scholtz, 1847 Figs 2A, 2B

Henriksen 1939: *Nabis flavomarginatus* Schltx.

Unmistakable among Greenland insects. Front wings nearly always shortened, generally covering less than half of the abdomen. In Greenland the long-winged form is extremely rare.

BIOLOGY. The species is found in grassy vegetation and may be abundant (*e.g.* inner Ameralik). In some cases stridulation (by males) may be observed, but it cannot be heard by humans. In Europe overwintering as nymph is reported (Southwood & Leston 1959); this is possible in Greenland as well, but not proven. There is one annual generation, and in the second half of July most specimens are adult.

DISTRIBUTION. S, SW (north to Nuup Kangerlua/Godthåbsfjord). [Also reported by C. Vibe from Mestersvig, NE, but this is erroneous; see note in Chapter 5.] The species appears to be restricted to former Norse habitations and is thus one of the candidates of a human introduction.

Circumpolar, low arctic-temperate.

MIRIDAE (PLANT BUGS)

Usually green or brownish, delicate insects. Ocelli absent, rostrum 4-segmented. FW with *cuneus*, a furrow-delimited area along front margin proximad from membrane; membrane with a distinctive, reduced vein pattern. The family is highly diverse ecologically and represented in most biotopes. By far the majority are plant feeders, some are omnivorous, and a few carnivorous. The family is cosmopolitan and the largest in Heteroptera.

Chlamydatus pullus (Reuter, 1870) Fig. 2D

Henriksen 1939: *Chlamydatus pulicarius* Fall.

In Greenland easily recognized by the dark colour of adults, while the nymphs are different shades of green. Legs yellow with black spines rising from black dots. When disturbed the bug leaps 10–15 cm by means of the strong hind legs, but it seems unable to fly in spite of normally developed wings (Böcher 1971).

BIOLOGY. In Europe two yearly generations and wintering in the egg-stage (Kullenberg 1944, Southwood & Leston 1959). In Greenland, too, the eggs hibernate, but there is only one annual generation. The first nymphs appear in June–July, and the adults are found as late in the season that the sun no longer melts the snow in the habitat. All adults known from Greenland are females, indicating that the species here reproduces parthenogenetically (a very rare phenomenon among Heteroptera). In all other parts of its range a normal sex-ratio has been found. In Greenland a maximum of 13 ripe eggs have been counted per female (Böcher 1971).

The habitat is open, dry, sandy and sunny places with low vegetation, where the species very often occurs together with *Nysius groenlandicus* (Böcher 1971), but generally it is active at lower temperatures than the seed bug. In Swedish Lapland it is often found in subarctic birchwood (Lindberg 1935).

C. pullus is mainly phytophagous and highly polyphagous. In Greenland it sucks stems, leaves, flowers and fruits of mainly *Potentilla* spp., *Sibbaldia procumbens* and *Polygonum viviparum*. In some cases it predates other small insects, *e.g.* Psyllidae (Böcher 1971). When attacked, the species sometimes enters a cataleptic state for several minutes, in some cases combined with loosing one of the hind legs by autotomi (Böcher 1971).

DISTRIBUTION. S, SW (north to Ukkusissat Fjord), NE, SE. The Greenland distribution appears to be remarkably incoherent .

Circumpolar, low arctic-temperate.

CIMICIDAE (BED BUGS)

Flat and broad parasites of mammals and birds. Ocelli missing. The single Greenland species has highly reduced hemelytra.

Cimex lectularius Linnaeus, 1758 (Bed bug) Fig. 2E

C. 4–5 mm. Micropterous parasite, entirely anthropocorous and anthropophilous.

BIOLOGY. Always feeding primarily on man, probably introduced to Greenland by both Palaeoeskimos and Europeans. The species is strictly nocturnal, and during daytime hides (often in crowds) in crevices in walls or furniture. For development at least one blood-meal is necessary in each instar, and females must feed before ovipositing. There is no evidence that bed bugs transmit human diseases. *C. lectularius* is polytypic, and some identifiable morphotypes may prove to be distinct species (*e.g.* Usinger 1966).

In heated rooms breeding goes on throughout the year. The threshold for hatching, nymphal development and adult activity is 13–15°C, although the species can tolerate extremes as low as -15°C for brief periods (Usinger 1966, Benoit *et al.* 2009).

DISTRIBUTION. Common in towns and settlements during first half of the twentieth century, when at a time 23 of 71 houses in Sisimiut/Holsteinsborg were infested (Henriksen 1939). Later only minor, scattered outbreaks have been reported.

Almost cosmopolitan.

References

Benoit, J.B., Lopez-Martinez, G., Teets, N.M., Phillips, S.A. & Denlinger, D.L. (2009): Responses of the bed bug, *Cimex*

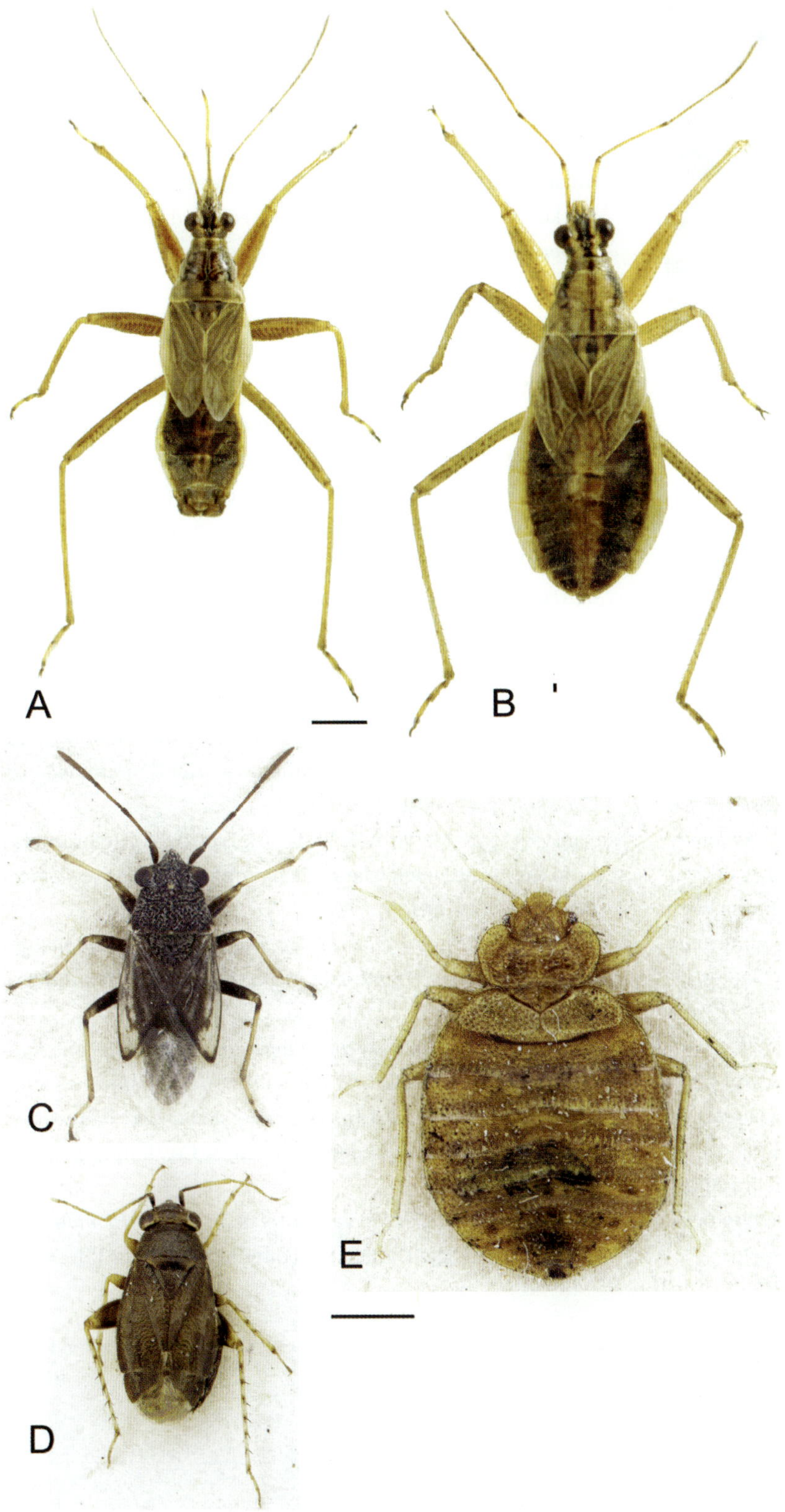

FIGURE 2

A–B, Nabis flavomarginatus ♂ *(A)* & ♀ *(B); C,* Nysius groenlandicus, *male; D,* Chlamydatus pullus. *E,* Cimex lectularius. *Scale bars 1 mm, for A–B and C–E.*

lectularius, to temperature extremes and dehydration: levels of tolerance, rapid cold hardening and expression of heat shock proteins. – *Medical and Veterinary Entomology* 23: 418–425.

Böcher, J. (1971): Preliminary studies on the biology and ecology of *Chlamydatus pullus* (Reuter) (Heteroptera: Miridae) in Greenland. – *Meddelelser om Grønland* 191 (3): 29 pp.

Böcher, J. (1972): Feeding biology of *Nysius groenlandicus* (Heteroptera: Miridae) in Greenland. With a note on oviposition in relation to food-source and dispersal of the species. – *Meddelelser om Grønland* 191 (4): 41 pp.

Böcher, J. (1975a): Notes on the reproductive biology and egg-diapause in *Nysius groenlandicus* (Zett.) (Heteroptera:

FIGURE 3 Chlamydatus pullus *in flower head of* Taraxacum croceumens.
(PHOTO: J. BÖCHER)

Lygaeidae). – *Videnskabelige Meddelelser fra Dansk Naturhistorisk Forening* 138: 21–38.

Böcher, J. (1975b): Spredningen af den grønlandske frøtæge, *Nysius groenlandicus* (Zett.) (Heteroptera: Lygaeidae). (With a summary: Dispersal of *Nysius groenlandicus* (Zett.) in Greenland). – *Entomologiske Meddelelser* 43: 105–109.

Böcher, J. (1976): Population studies on *Nysius groenlandicus* (Zett.) (Heteroptera: Lygaeidae) in Greenland, with particular reference to climatic factors, especially the snow cover. – *Videnskabelige Meddelelser fra Dansk Naturhistorisk Forening* 139: 61–89.

Böcher, J. (1997): History of the Greenland insect fauna with emphasis on living and fossil beetles. – Quaternary Proceedings 5. *In*: Ashworth, A.C., Buckland, P.C. & Sadler, J.P. (eds). *Studies in Quaternary Entomology. An Inordinate Fondness for Insects*: 35–47.

Böcher, J. & Bennike, O. (1996): Early Holocene insect and plant remains from Jameson Land, East Greenland. – *Boreas* 25: 187–193.

Böcher, J. & Nachman, G. (2001): Temperature and humidity responses of the arctic-alpine seed bug *Nysius groenlandicus*. – *Entomologia Experimentalis et Applicata* 99: 319–330.

Böcher, J. & Nachman, G. (2010): Are environmental factors responsible for geographic variation in the sex ratio of the Greenlandic seed bug *Nysius groenlandicus*? – *Entomologia Experimentalis et Applicata* 124: 122–130.

Böcher, J. & Nachman, G. (2011): Coexistence of bisexual and unisexual populations of *Nysius groenlandicus* in the Zackenberg Valley, Northeast Greenland. – *Entomologia Experimentalis et Applicata* 140: 196–206.

Böcher, J. & Petersen, P.M. (1997): Greenland. Chapter 20 *in* Wielgolaski, F.E. (ed.). *Ecosystems of the World 3. Polar and Alpine Tundra*: 685–720. – Elsevier Amsterdam etc.

Danks, H.V. (1981): Arctic Arthropods. A review of systematics and ecology with particular reference to the North American fauna. – *Entomological Society of Canada*, Ottawa. 608 pp.

Henriksen, K.L. (1939): A revised index of the insects of Grønland. – *Meddelelser om Grønland* 119 (10): 1–111.

Henry, J. & Froeschner, R.C. (eds) (1988): *Catalog of the Heteroptera, or True Bugs, of Canada and the Continental United States*. – E.J. Brill, Leiden-New York-København-Köln. 958 pp.

Kullenberg, B. (1944): Studien über die Biologie der Capsiden. – *Zoologiska Bidrag från Uppsala*. 522 pp.

Lindberg, H. (1935): Über einige arktische und subarktische Hemipteren aus Fennoskandien. – *Norsk Entomologisk Tidsskrift* 3: 382–394.

Makarova, O.L. & Makarov, K.V. (2006): True bugs (Heteroptera) from the arctic Dolgii Island, the Barents Sea. – *Entomological Review* 86: 423–432.

Mani, M.S. (1962): *Introduction to High Altitude Entomology. Insect Life Above Timber-line in the North-west Himalaya*. – Methuen & Co Ltd., London. 302 pp.

Mani, M.S. (1968): *Ecology and Biogeography of High Altitude Insects*. – Dr. W. Junk N.V. Publishers, The Hague. 527 pp.

Southwood, T.R.E. & Leston, D. (1959): *Land and Water Bugs of the British Isles*. – Frederick Warne & Co., London & New York. 436 pp.

Usinger, R.L. (1966): *Monograph of Cimicidae (Hemiptera – Heteroptera)*. – The Thomas Say Foundation, Volume VII. 585 pp.

CHAPTER 11

Hymenoptera (*Wasps*)

11.1. Generalities and family key

Lars Vilhelmsen

The Hymenoptera is one of the four megadiverse insect orders, comprising more than 125,000 described species. Traditionally, it has been divided into two suborders, the almost exclusively herbivorous 'Symphyta' (sawflies and woodwasps) and the predominantly parasitoid Apocrita (wasp-waisted wasps). However, this classification has been rendered obsolete by recent advances in hymenopteran phylogenetics, which has demonstrated the paraphyly of the 'Symphyta'. Current concepts in the classification the Hymenoptera are focussed on the superfamilies as the operative units. Nevertheless, 'Symphyta' and Apocrita can be retained as informal terms useful when making summary statements about hymenopteran diversity.

Hymenoptera are well represented in Greenland with about 200 species, being second only to the Diptera with regard to species diversity. Most Hymenoptera are parasitic wasps, attacking a broad range of other insect's larva. The two main parasitoid lineages are the Ichneumonoidea and the Chalcidoidea; the former comprise more than half the species of Hymenoptera known from Greenland. Other important lifestyles are herbivory (sawflies = Tenthredinoidea, gallwasps = Cynipidae, bees = Apoidea – Apiformes) or a predatory lifestyle combined with nest construction and provisioning of the offspring (digger wasps = Apoidea – 'Spheciformes', hornets and yellowjackets = Vespoidea). Eusociality has been developed to its highest complexity within the Hymenoptera, having evolved several times: in Apidae, Formicidae, and Vespidae. The Apidae, with two species of *Bombus*, are the only naturally occurring social species represented in Greenland.

Adult anatomy

HEAD. The head capsule is well sclerotised, usually hypognathous. Three ocelli are almost always present. The antennae are usually filiform, but very variable with regard to the number of antennomeres. Biting/chewing mandibles are almost universally present, but may be combined with a proboscis formed from the labiomaxillary complex; a proboscis has evolved on numerous occasions within the order. In most Hymenoptera, the tips of the mandibles are displaced anteriorly of the reduced labrum, allowing them to serve as tools for purposes other than food ingestion.

THORAX. The lateral cervical sclerites are fused with the propleura, together with the prosternum and fore legs forming the propectus, a functional unit connecting the head with the remainder of the thorax. The fore leg has an antenna cleaner apparatus developed to varying degrees, involving the apical tibial spur and the basitarsus. The pronotum is more or less fused to the mesothorax, which is in turn closely integrated with the metathorax. The mesothorax contains the main flight musculature and is by far the largest segment in the hymenopteran body. The meso – and metathoracic leg bases are situated very close together.

WINGS. The fore wings are much the largest, being coupled to the hind wings by several groups of tiny hooks, the hamuli, situated on the anterior margin of the hind wing and gripping the posterior margin of the fore wing. This effectively makes the fore and hind wing one functional unit during flight. The wings are usually placed above the body when at rest, at most being folded longitudinally. The wing venation is characteristic of the order, but highly variable within; it is usually reduced to a single supporting strut anteriorly in the wing in most of the smaller parasitoid wasps.

ABDOMEN. The first abdominal tergum is always closely associated with the metathorax. In the Apocrita, comprising the majority of the Hymenoptera, it has become totally integrated in the thorax as the *propodeum*. The propodeum and thorax together form the mesosoma, the remainder of the abdomen the metasoma; meso- and metasoma are separated by a constriction, the wasp-waist, the metasoma being highly manoeuvrable in many taxa. This is made possible by the variously developed articulations between the first and second abdominal terga. The first abdominal sternum is always absent. The abdominal terga are extended laterally to accommodate the spiracles (when present). Extensive fusion of the abdominal segments often occurs. A fully developed, functional ovipositor is present, but it is often concealed within the abdomen. The male genitalia consist of an unpaired aedoeagus as well as paired claspers, the latimeres/gonoforceps. Another pair of claspers, the volsellae, is situated ventromedially of the latimeres.

LARVA. Two main types of hymenopteran larvae can be recognised. The externally feeding sawfly larva (most Tenthredinoidea) resembles a caterpillar, with a well sclerotised head capsule and well developed eyes (always with only a single common cornea for each eye), antennae, and mouthparts of the chewing type, with well developed maxillary and labial endites and palps. Segmented thoracic legs with claws as well as abdominal prolegs are also present. In contrast, the mostly carnivorous apocritan larvae are grub-like with eyes, thoracic legs and abdominal prolegs absent, and the antennae and mouthparts very reduced. These differences reflect their divergent lifestyles.

PUPA. The pupa is adecticous and usually exarate. A cocoon may or may not be spun, if spun it is usually done from the labial glands. Pupation takes place in soil or on host plants, in parasitoid species usually in or close to the remains of the host. Nest constructing species pupate in their nest cells.

Key to adults of superfamilies/families of Hymenoptera occurring in Greenland

1. Wings reduced or absent 2
– Wings fully developed 5
2. Very small (body length < 1.3 mm); antenna with scapus elongate (longer than combined length of two following antennomeres), distal most antennomeres thickened (Fig. 1G) 3
– Larger (body length > 2 mm); antenna with scapus shorter than combined length of two following antennomeres, distalmost antennomeres not thickened 4
3. Body dark; head hypognathous; first gastral tergite very long, occupying most of gaster .... Platygastridae, Platygastroidea (female *Baeus*), p. 207
– Body pale; head sub-prognathous; first gastral tergite much shorter, occupying at most one third of gaster.... .... Encyrtidae, Chalcidoidea (female *Rhopus Tetracyclos*), pp. 172, 173 [??]
4. Antenna inserted low on frons, close to clypeus, with ten segments; female with conspicuous chela on foretarsus (Fig. 1K) .. Dryinidae, Chrysidoidea, p. 251
– Antenna inserted higher on frons, between eyes, with more than ten segments; female without chela on foretarsus .... Ichneumonidae, Ichneumonoidea (*Gelis*), p. 244
5. No constriction (wasp-waist) between first and second abdominal segments............................ .... Tenthredinidae, Tenthredinoidea, p. 162
– Constriction between between first and second abdominal segments................... 6 (Apocrita)
6. Wing venation well developed (at least three closed cells in forewing, usually more than one in hindwing; Figs 2A, H–K) 7
– Wing venation reduced (at most two closed cells present in forewing, at most one in hindwing; Figs 2B–G)..10
7. Fore margin of forewing proximally of pterostigma with two parallel veins (costa and radius) distinctly separated by elongate, narrow costal cell (Fig. 2A); females with ovipositor always concealed; large species (body length at least 10 mm........ 8 (Aculeata)
– Fore margin of forewing proximally of pterostigma with costa and radius fused or abutting, costal cell obliterated (Figs 2H–I); females with ovipositor exposed; body size variable .... 9 (Ichneumonoidea)
8. Body conspicuously hairy, obscuring large parts of the underlying integument; hairs branched; eye without incurvation in inner margin opposite antennal base (*e.g.,* Figs 1H, J); wings not folded longitudinally in resting position.............................. .... Apidae, Apoidea (Fig. 3D), p. 254
– Body with only scattered hairs, integument visible between them throughout; hairs unbranched; eye with distinct incurvation opposite antennal base (Fig. 1I); wings folded longitudinally in resting position ....Vespidae, Vespoidea (Fig. 3E), p. 252
9. Forewing usually with vein 1Rs + M present and vein 2 m-cu absent (Fig. 2H); hindwing with vein 1rm basal to separation of R1 and Rs (Fig. 2J) ....Braconidae, Ichneumonoidea, p. 212
– Forewing usually with 1Rs + m absent and vein 2 m-cu present (Fig. 2I); hindwing with vein 1rm distal to separation of R1 and Rs (Fig. 2K).......................... Ichneumonidae, Ichneumonoidea, (Figs 3B–C), p. 229
10. Forewing with two parallel veins (costa and radius) distinctly separated by elongate, closed costal cell (Fig. 2B) 11
– Forewing fore margin with at most one vein (Fig. 2F), costal cell not closed...............................12
11. Antennae inserted on projection between eyes (Fig. 1H); forewing pterostigma small, hardly broader than costal cell (Fig. 2B) ....Diapriidae, Proctotrupoidea, p. 208
– Antennae not inserted on projection; forewing pterostigma larger, triangular, broader than costal cell (Fig. 2G)...... Proctotrupidae, Proctotrupoidea, p. 211
12. Pterostigma well developed (Fig. 2F); foretibia with two apical spurs..................................... .... Megaspilidae, Ceraphronoidea, p. 165
– Pterostigma absent (Figs 2C – E); foretibia with only one apical spur....................................13
13. Pronotum does not reach tegula (Fig. 1D); fore – and hindwing without any closed cells (Fig. 2C); body

often green or blue green with metallic lustre .... 14 (Chalcidoidea)
– Pronotum reaches tegula laterally (Fig. 1E); closed cells sometimes present in forewing; body brownish or yellowish, always without metallic lustre 19
14. Metasoma without apparent petiole, hence appearing broadly attached to mesosoma (Fig. 1B)........15
– Metasoma petiolate, hence distincly constricted anteriorly (Fig. 1C) 17
15. Tarsi 3-segmented; forewings with distinct radiating setal lines (see 11.9, Fig. 1C)..Trichogrammatidae, p. 198
– Tarsi 5-segmented; forewings without radiating setal lines 16
16. Forewing with marginal vein very long, up to 10 or more times as long as stigmal vein (see 11.4, Fig. 1B); antenna with only 6 segments (11.4, Figs 1C, F); mesoscutum with notauli distinct (11.4, Fig. 1A) .... Aphelinidae, p. 168
– Forewing with marginal vein very short, at most slightly longer than stigmal vein (11.5, Figs 1C, J); antenna most often with more than 6 segments (11.5, Figs 1A, 2A); mesoscutum with notauli indistinct .... Encyrtidae, p. 170
17. Forewing with strongly reduced venation, marginal vein short and both postmarginal and stigmal vein indistinct (see 11.7, Figs 1A, C, 2A, C); hindwing with membrane not extending to base of wing; wings with long to very long marginal setae; antennal toruli closer to inner eye margin than to each other; body without metallic lustre, minute to small (0.3–1.0 mm)........................... Mymaridae, p. 184
– Forewing with normally developed venation, marginal vein moderately long and at least stigmal vein distinct (11.6, Figs 1C, 3B; 11.8, Figs 1F, 4E); hindwing with membrane extending to base of wing; wings with short marginal setae; antennal toruli closer to each other than to inner eye margin (11.8, Figs 1C, 4B); body mostly with moderate to strong metallic lustre and more than 1 mm...............................18
18. Tarsi 4-segmented; protibial spur straight; marginal vein usually more than 3 times as long as stigmal vein (see 11.6, Fig. 1C)............. Eulophidae, p. 177
– Tarsi 5-segmented; protibial spur bent at apex; marginal vein usually less than 3 times as long as stigmal vein (11.8, Fig. 1F) .... Pteromalidae, p. 186
19. Antennae situated between eyes (*e.g.*, Fig. 1H); forewing with cross veins as well as longitudinal veins, closed cells might also be present (Fig. 2D) .... Figitidae, Cynipoidea, p. 200
– Antennae situated below eyes (Fig. 1J); forewing with at most one distally bifid longitudinal vein (Figs 2E) .... Platygastridae, Platygastroidea, p. 204.

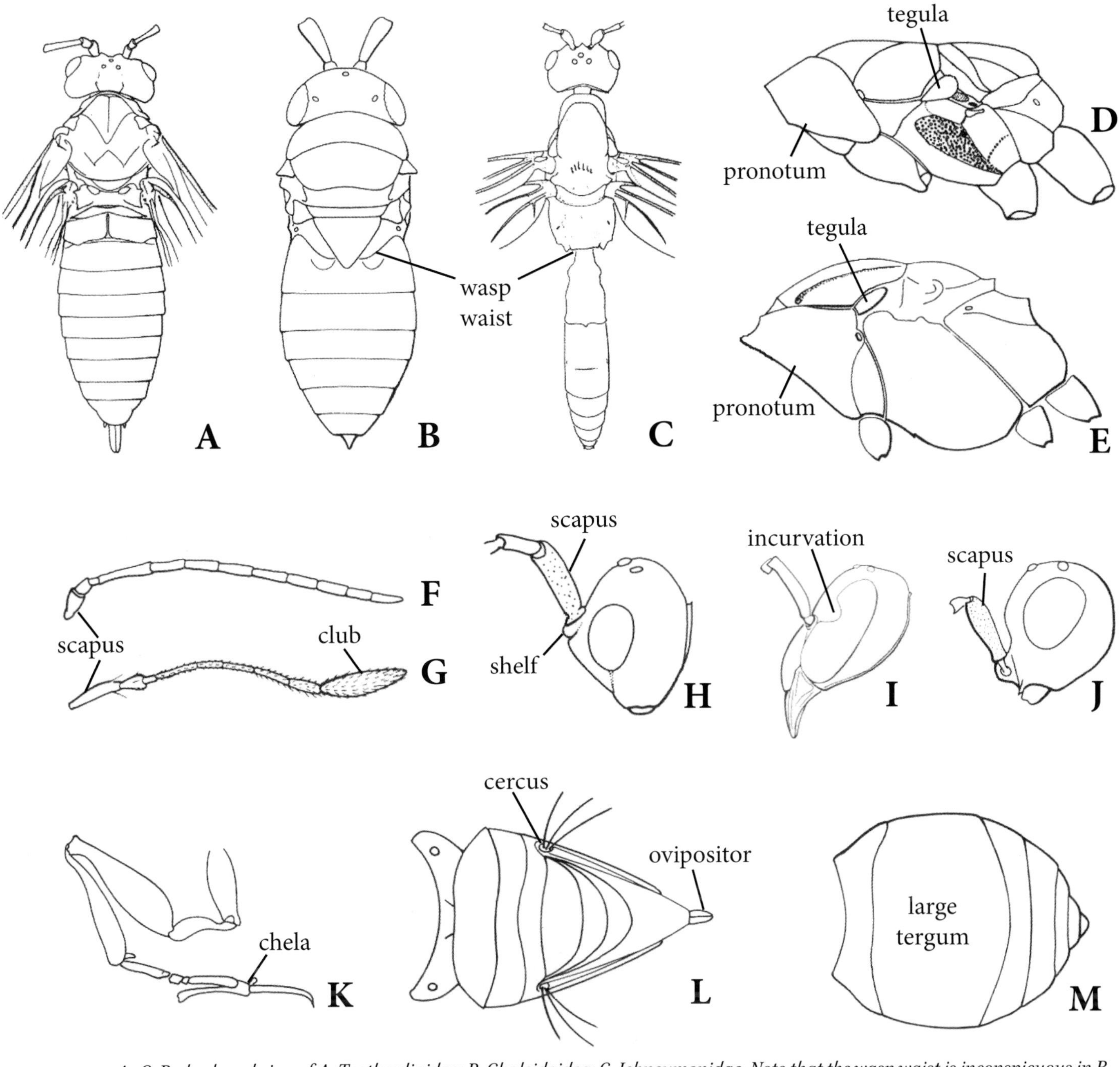

FIGURE 1 *A–C. Body, dorsal view of A. Tenthredinidae; B. Chalcidoidea; C. Ichneumonidae. Note that the wasp waist is inconspicuous in B. D–E. Mesosoma, lateral view of d. Chalcidoidea; E. Proctotrupoidea; F–G. Antenna of F. Aculeata; G. Chalcidoidea. H–J. Head, lateral view of H. Diapriidae; I. Vespidae; J. Platygastroidea. K. Foreleg with chela of Dryinidae. L–M. Metasoma, dorsal view of L. Encyrtidae (Chalcidoidea); M. Platygastroidea.*
MODIFIED FROM GOULET & HUBER 1993.

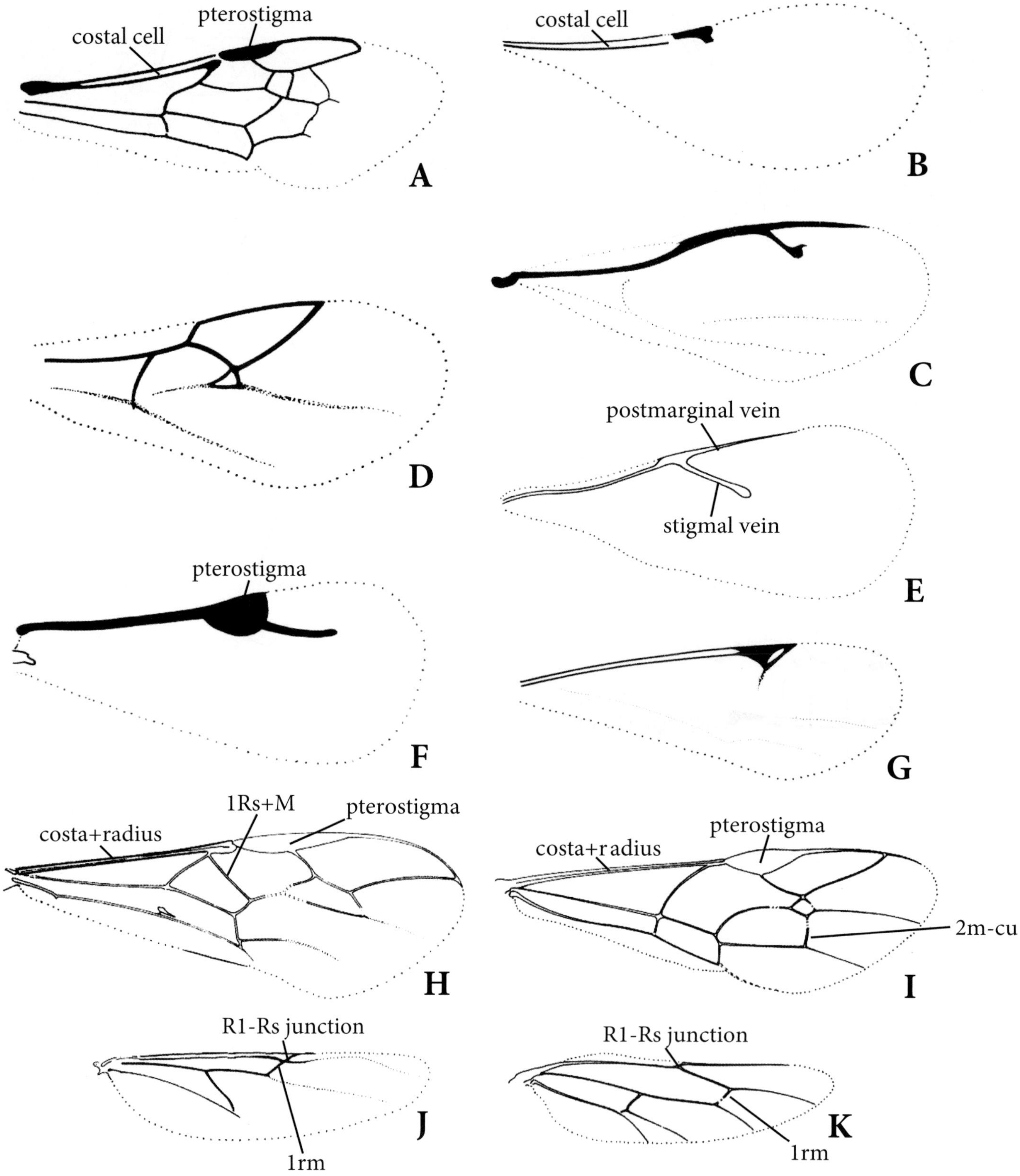

FIGURE 2 *Forewing of A. Aculeata; B. Diapriidae; C. Chalcidoidea; D. Figitidae; E. Platygastridae; F. Megaspilidae; G. Proctotrupidae; H. Braconidae; I. Ichneumonidae. J–K. Hindwing of J. Braconidae; K. Ichneumonidae.*
MODIFIED FROM GOULET & HUBER 1993.

IGURE 3 *Life pictures of Hymenoptera. A.* Gelis *sp.* (*Ichneumonidae*)*; B.* Buathra *sp.* (*Ichneumonidae*)*; C.* Pimpla sodalis (*Ichneumonidae*)*; D.* Bombus polaris (*Apidae*)*; E.* Ancistrocerus waldenii (*Vespidae*)*, building nest chamber. F. Nest of* A. waldenii*; chambers broken open, displaying caterpillars collected by the adult* (*left and middle*)*, eggs* (*middle*)*, and mature vespid larva* (*right*)*.*
(PHOTOS: J. BÖCHER (A & D); G. VÁRKONYI (B-C); T. VALENTIN (E); A. FJELLBERG (F)).

11.2. Tenthredinidae (Tenthredinoidea) (True sawflies)

Lars Vilhelmsen

Antenna of both female and male usually with nine segments; wasp-waist absent; both fore- and hindwing with a number of closed cells, including at least one closed anal cell in forewing; females with short, serrated ovipositor; male genitalia rotated 180 degrees around longitudinal axis even in resting position. Larva usually with well developed eyes, antennae, mouthparts, thoracic legs, and 7–8 pairs of abdominal prolegs.

Most species are external feeders on leaves; a few are leafminers, leafrollers, or gallers. The egg is placed in the host plant in a cavity cut with the ovipositor. The larva finishes feeding in a few weeks; it lies dormant until next spring, usually in the inactive prepupal stage, often in an earthen cell, but sometimes within plant tissue. Pupation takes place in spring. The adults emerge a couple of weeks later. Some species have more than one generation a year, but this is not known to occur in Greenland.

Tenthredinidae comprises some 6,000 species worldwide, placed in seven subfamilies. Most species occur in the Holarctic. It is the only family of non-apocritan wasps represented in Greenland, with two subfamilies (Allantinae and Nematinae).

Key to genera

ADULT

1. Forewing vein 2r present, two anal cells present and abutting, only separated by short anal cross vein (Fig. 1A); hindwing cells Rs and M absent (Fig. 1C) *Ametastegia* ....(Allantinae)
– Forewing vein 2r absent, only one anal cells present (Fig. 1B); hindwing cells Rs and M present (Fig. 1D) .... 2 (Nematinae)
2. Head with labiomaxillary complex elongate, especially cardo + stipes, maxillary palp comparatively short, segments subequal in length (Fig. 1E); tarsal claws bifid at apex, proximal tooth well developed, elongate, close to distal tooth....... *Amauronematus*
– Labiomaxillary complex short, maxillary palp comparatively elongate, third segment is longest (Fig. 1F). Tarsal claws with only minute proximal tooth, well separated from distal tooth........... *Pachynematus*

LARVA

1. Antenna erect, with five segments arranged concentrically (Fig. 2A); abdominal segments 2–8 and 10 with prolegs; ventral eversible glands on abdominal segments 1–8 absent ...... *Ametastegia* (Allantinae)
– Antenna flattened, with four segments in a row (Fig. 2B); abdominal segments 2–7 and 10 with prolegs; ventral eversible glands on abdominal segments 1–8 present..................... 2 (Nematinae)
2. Upper margin of left mandible with several teeth ....*Amauronematus*
– Upper margin of left mandible with one continuous cutting edge .... *Pachynematus*

ALLANTINAE

***Ametastegia* sp.**

ADULT (according to Smith 1979). Clypeus shallowly emarginated; forewing with veins M and Rs + M meeting in the same point on Sc + R, vein 2r present, veins 1 m-cu and 2 m-cu joining vein M in different cells (Fig. 1A); hindwing without cells Rs and M (Fig. 1B).

LARVA. Right mandible with mesal ridge without teeth, left mandible with 3 ventral teeth of uneven size, one seta on each mandible only; thoracic leg with trochanter shorter than femur; abdominal segments with setae on annulets 2 and 4 only; head reddish-brown, lighter below eyes, without dark spots; body darker dorsally, without dark spots.

BIOLOGY. Host plant in Greenland unknown, elsewhere known to feed on *Rumex*, *Polygonum*, *Viola*, and occasionally *Salix*. External feeder. Pupation might take place inside fruits, berry-canes, or under bark, not necessarily on the host plant. Most species occurring outside Greenland have several generations a year, the condition in Greenland is unknown.

DISTRIBUTION. S. Only known from a few larvae collected in Upernaviarsuk and Narsarsuaq. Adults have not been collected in Greenland so far.

Holarctic and Oriental regions.

NEMATINAE

Most Tenthredinidae found in arctic and alpine environments belong in this subfamily. A feature associated with high altitudes and latitudes is the scarcity of males (none were present in the material examined for the present publication). These species are likely to produce females asexually (thelytokious parthenogenesis).

***Amauronematus* Konow, 1890. Fig. 3A**

Key to species (adults only)

1. Sawsheath erect, at an angle relative to the longitudinal axis of the abdomen, expanded distally, with three distinct lobes distally (Figs 1G, H) .... *amentorum*
– Sawsheath subparallel with longitudinal axis of abdomen, not expanded distally and with evenly rounded apex (Fig. 1I). 2
2. Body colour with distinct light brown markings around eyes, laterally on pronotum, on mesonotum along notauli and laterally and on mesoscutellum, and on upper part of mesopleura; legs with femora dark brown, distal parts light brown....*groenlandicus*
– Body uniformly black or darkly brown, without extensive light brown markings ....*nitidipleuris*

Amauronematus amentorum **(Förster, 1854)**
C. 5.0 mm. Sawsheath erect, expanded and three-lobed distally; body colour almost entirely dark brown to black, except for anterior part of abdomen and distal parts of appendages, which are lighter brown. Darker than specimens collected outside Greenland, which have white markings on lower part of frons, pronotum, tegulae, legs and underside of abdomen.

BIOLOGY. The larva feeds on developing seeds in catkins of *Salix*.

DISTRIBUTION. SW, N, NE.

Cicumboreal, known from large parts of the Holarctic.

Amauronematus groenlandicus **Malaise, 1933**
C. 7.0 mm. Abdomen broad; body colouration predominantly dark brown to black, but with extensive light brown markings (see key to species).

DISTRIBUTION. NE, SE.

Circumboreal.

Amauronematus nitidipleuris **Malaise, 1931**
Henriksen 1939: not recorded.

C. 7.0 mm. Sawsheath subparallel with abdomen, not expanded distally, evenly rounded; body colour predominantly black, clypeus, labrum, tibiae and tarsi dark brown.

DISTRIBUTION. NW, NE, SE.

Circumpolar.

Pachynematus **Konow, 1890 . Fig. 3B**
Key to species (adults only)

Pterostigma yellow; abdominal sterna 2–6 predominantly dark brown, at most with light posterior margins .... *excisus*
Pterostigma dark brown to black; abdominal sterna entirely whitish yellow.....................*obductus*

Pachynematus excisus **(Thomson, 1862)**
Henriksen 1939: not recorded.

4.5–5.5 mm. Head with reddish brown spot posterodorsally of the eye; labrum yellow; pterostigma yellow; hindtibia entirely whitish-yellow; inner hindtibial apical spur less than half the length of hindbasitarsus; abdominal sterna 2–6 predominantly dark brown, at most with light posterior margins; sawsheath yellow to light brown.

DISTRIBUTION. S, SW, NW, SE.

Northern Europe.

Pachynematus obductus **(Hartig, 1837)**
5.0–6.0 mm. Head uniformly dark brown to black posteriorly of eye; labrum dark brown-black; pterostigma dark brown to black; hind tibia partly dark brown to black distally; inner hind tibial apical spur about half the length of hindbasitarsus; abdominal sterna 2–6 entirely whitish-yellow; sawsheath dark brown to black.

BIOLOGY. Larvae are external feeders on Poaceae.

DISTRIBUTION. S, SE.

Circumpolar, known throughout the northern Holarctic.

NOTE. Additional species of Tenthredinidae listed by Henriksen (1939) are *Trichiocampus viminalis* (Fallén, 1808), *Amauronematus viduatus* (Zetterstedt, 1838), and *Amauronematus borealis* (Marlatt, 1892). Currently, no material from Greenland can be associated with these names, even though it is supposed to be located in the ZMUC collection.

Reference

Smith, D.R. (1979): Nearctic sawflies. IV: Allantinae: Adults and larvae (Hymenoptera: Tenthredinidae). – *Technical Bulletin United States Department of Agriculture* 1595: 1–196.

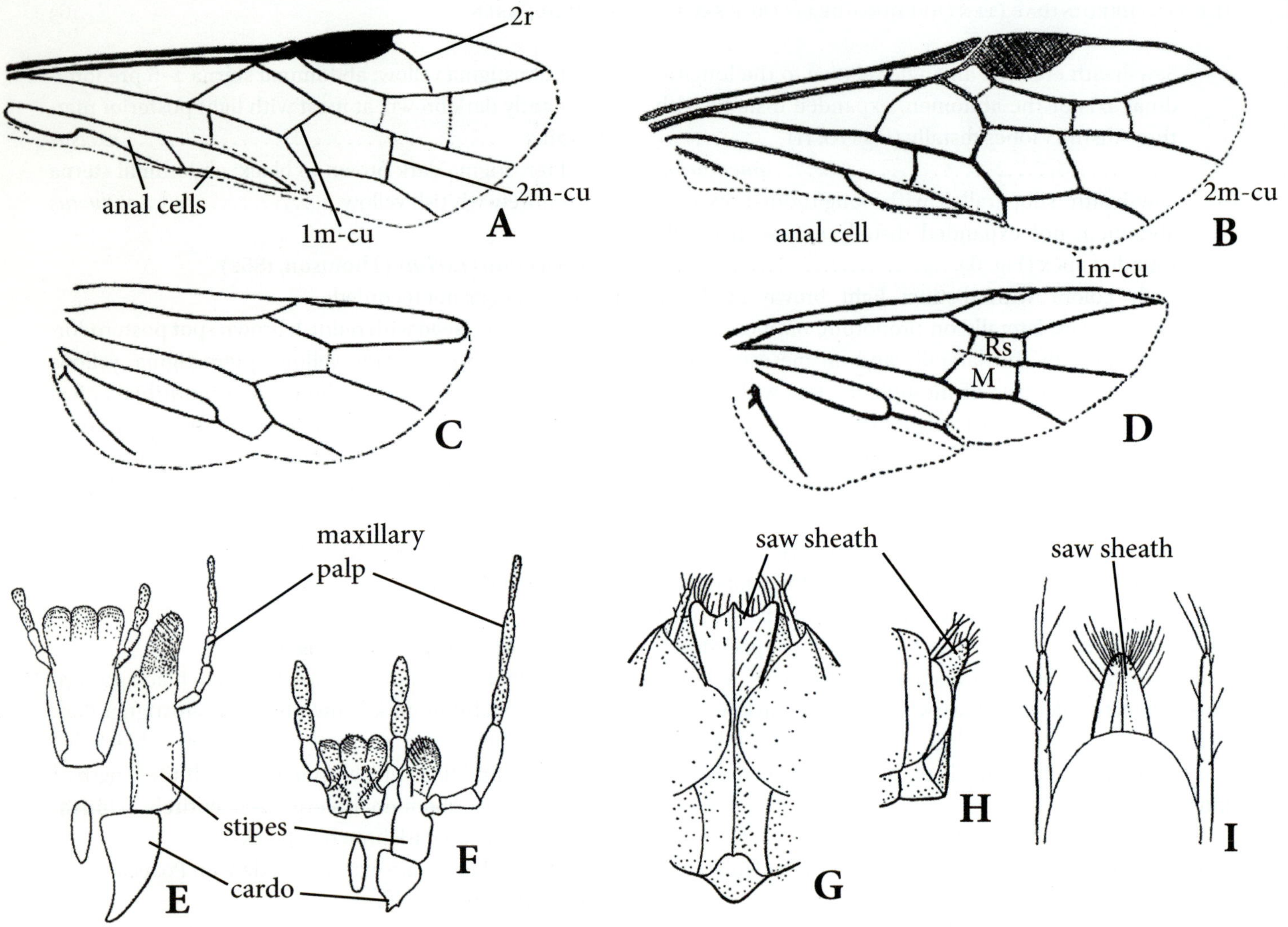

FIGURE 1 *Forewing of A. Allantinae; B. Nematinae. Hindwing of C. Allantinae; D. Nematinae. Labiomaxillary complex, ventral view, right maxilla removed, of E.* Amauronematus; *F.* Pachynematus. *G. Sawsheath of* Amauronematus amentorum, *ventral view. H. Sawsheath of* A. amentorum, *lateral view. I. Sawsheath of* A. nitidipleuris, *dorsal view.*
MODIFIED FROM BENSON (1952, 1958)

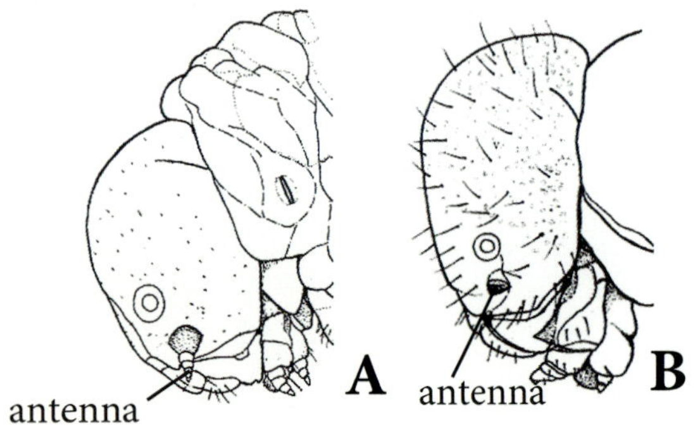

FIGURE 2 *Larval head, lateral view of A. Allantinae; B. Nematinae.*
MODIFIED FROM YUASA (1922)

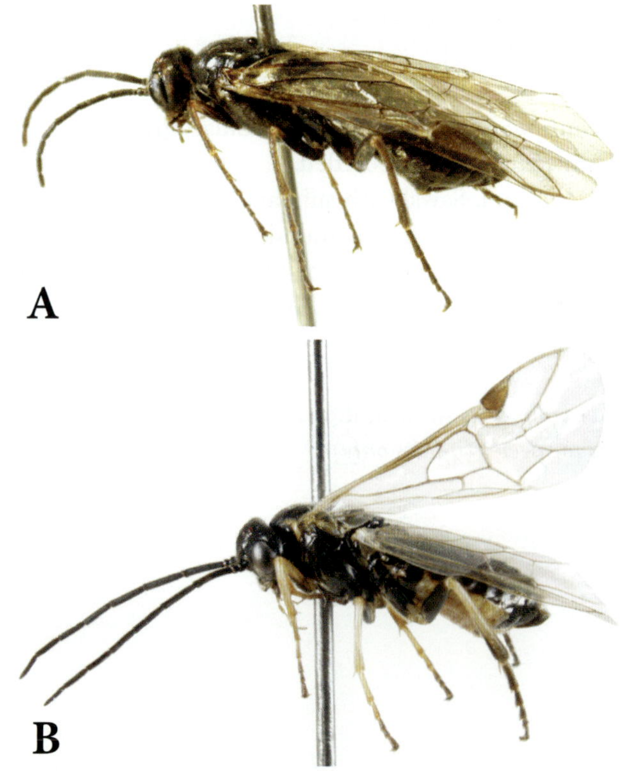

FIGURE 3 *Habitus of A.* Amauronematus *sp.; B.* Pachynematus obductus.

11.3. Megaspilidae (Ceraphronoidea)

Peter N. Buhl

Small (about 1–3 mm) dark species; antennae of both sexes with 11 segments, scape long, flagellar segments often modified in male (in *Dendrocerus* segments dilated and with long hairs); mesoscutum with median sulcus as well as notauli (Fig. 1C); wings with large pterostigma (see 11.1, Fig. 2F) and incomplete radial vein, costal cell absent; 1st tergum of metasoma by far the largest, also forming entire anterior constriction (petiole), 1st sternum much reduced.

Cosmopolitan family with about 450 described species.

Dendrocerus aphidum **(Rondani, 1877)**
Henriksen 1939: not recorded.

1.2–1.5 mm. Female antenna (Fig. 1A) with segment 2–3 about equal, 4 distinctly shorter than 3, preapical segments about 1.5 times as long as wide. First flagellar segment of male twice as long as wide, longer than following segments which are weakly triangular, each about 1.5 times as long as wide; preapical segments cylindrical, fully twice as long as wide (Fig. 1B). Pterostigma about 1.6 times as long as wide. Habitus of male (Fig. 1C). Male genitalia cf. Dessart (1972).

BIOLOGY. Hyperparasitoid on Aphidiinae (Braconidae). Host records from outside Greenland: *Hyalopterus pruni* Geoffroy (Aphidiidae; primary host), *Praon volucre* Haliday, *Aphidius picipes* Nees, *Diaeretiella rapae* M'ltosh (all Aphidiinae; secondary host) (all host records from Alekseev 1978).

DISTRIBUTION. NV (Zackenberg, Jameson Land: Coloradodal).

Europe, temperate, widespread.

Dendrocerus bifoveatus **(Kieffer, 1907) Fig. 2**
1.8–2.3 mm. Female antennal segments 2–4 of almost equal length, preapical antennal segments almost twice as long as wide (Fig. 1D); flagellar segments of male cylindrical, 2.3–3.3× as long as wide (Fig. 1E); pterostigma 2.0–2.2× as long as wide (Fig. 1F). Male genitalia cf. Dessart (1972).

BIOLOGY. Hyperparasitoid on Aphidiinae. Outside Greenland reared from *Aphis* sp. (Aphidiidae) (Alekseev 1978).

DISTRIBUTION. S (Kap Farvel area: Pamialluk, Nanortalik), NE (Zackenberg).
Europe, low arctic-temperate (Iceland, Sweden, France, Russia).

References

Alekseev, V.N. (1978): Ceraphronoidea. – *Opredeliteli Fauna SSSR* 120: 665–691.

Dessart, P. (1972): Révision des espèces européennes du genre *Dendrocerus* Ratzeburg, 1852 (Hymenoptera Ceraphronoidea). – *Mémoires de la Societé Royale Belge d'Entomologie* 32: 1–310.

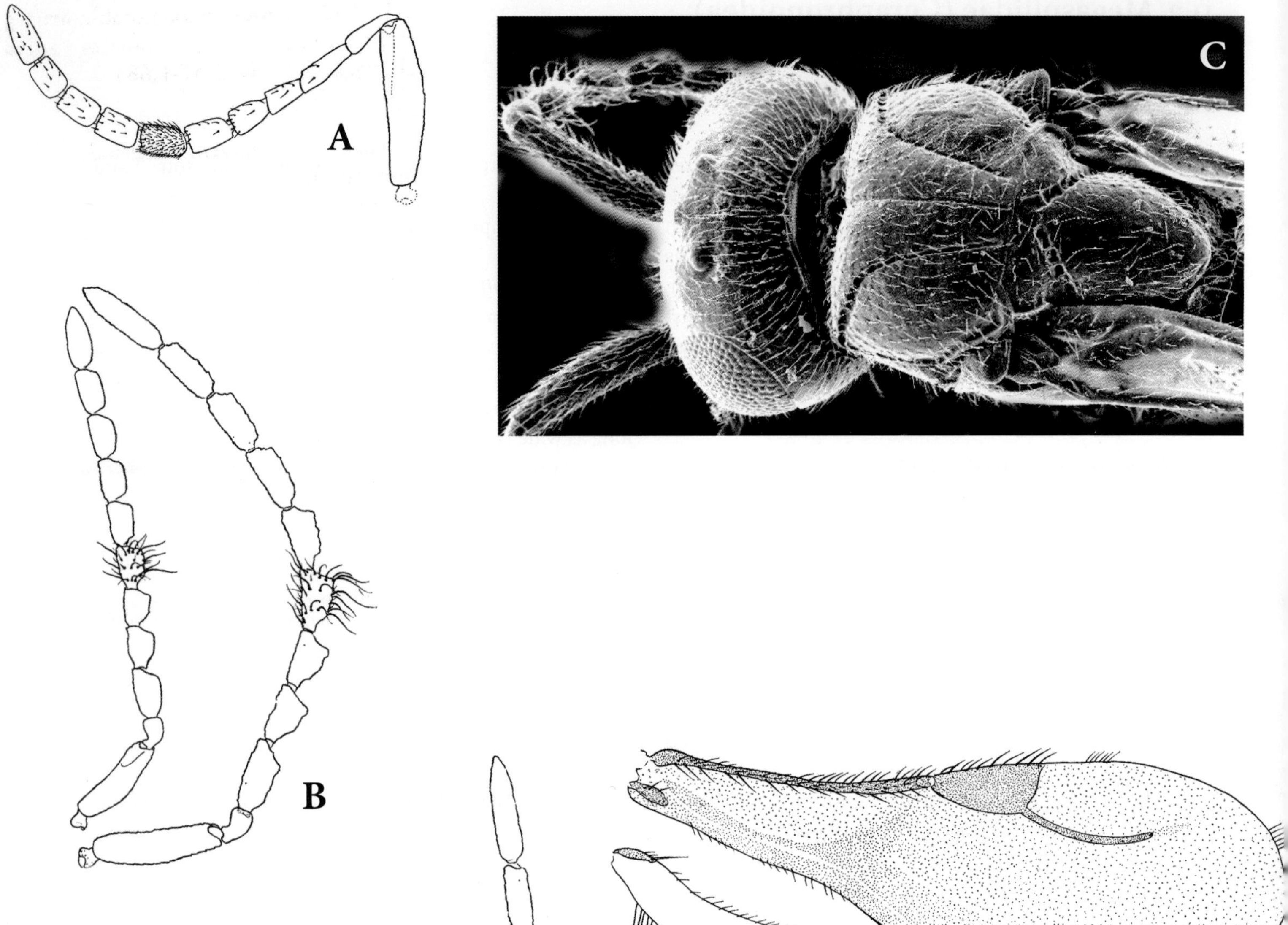

FIGURE 1 *A.* Dendrocerus aphidum, *female antenna; B.* D. aphidum, *antenna of small and large male. C.* D. aphidum, *head and mesosoma in dorsal view; D.* Dendrocerus bifoveatus, *female antenna; E.* D. bifoveatus, *male antenna; F.* D. bifoveatus, *fore- and hindwing.*
FIGS A, B, D-F FROM DESSART 1972.

FIGURE 2 Dendrocerus bifoveatus, *habitus.*

11.4. Aphelinidae (Chalcidoidea)

John M. Heraty

Aphelinidae are minute wasps (0.5–2 mm) that are parasites of whiteflies, aphids and armored scale insects (Hayat 1998). The family is cosmopolitan in distribution and represented by more than 1168 species in 33 genera (Noyes 2003). Only one species of *Aphelinus* is known from Greenland. The subfamily Aphelininae is recognized by having a reduced number of antennal segments (1 1 3 1 antennal formula, Figs 1C, F), the fore wing usually with a linea calva (lc, Fig. 1B), five tarsomeres, and a prominent curved and bifid calcar (Hayat 1998). The subfamily includes 299 species in 11 genera (Noyes 2003). The genus *Aphelinus* is recognized by the presence of a prominent hypopygium in females, and a prominent patch of porous sensillae on the ventral aspect of the male antennal scape. Currently 86 species of *Aphelinus* are recognized (Noyes 2003).

***Aphelinus* (*Mesidia*) sp. (unidentified to species)**
2 mm. Body pale yellow, wings hyaline. Mesosoma with scattered small setae on midlobe of mesoscutum and two pairs of prominent setae on scutellum, with anterior pair slightly medial to posterior pair (Fig. 1A). Fore wing with long marginal vein, short truncate stigmal vein and linea calva (lc, Fig. 1B). Antenna with 6 segments (Figs 1C, F); funicular segments (between pedicel and terminal segment) of female slightly longer than broad and increasing in length (Fig. 1C), funicular segments of male each more than twice longer than broad and about equal in length, scape of male with ventral patch of prominent porous sensillae. Ovipositor (Fig. 1D) small in proportion to midtibia (Fig. 1E), hypopygium prominent and triangular.

NOTE. Hayat (1998) divided the subgenus into three species groups (*argiope*, *subflavescens* and *automatus*). The species from Greenland is included within the *argiope* group on the basis of having all funicular segments at least slightly longer than broad (Fig. 1C), and much longer than broad in the male (Fig. 1F).

BIOLOGY. Internal parasitoids of aphids (Hemiptera: Aphididoidea) (Hayat 1998).

DISTRIBUTION. S (Kap Farvel-area: Pamialluk). At least 12 species of *Aphelinus* may be included in the subgenus *Mesidia*, which is found mostly in the Nearctic and Palaearctic regions (including India), but with a single species known from the Ethiopian region (Noyes 2003).

References

Hayat, M. (1998): Aphelinidae of India (Hymenoptera: Chalcidoidea): A taxonomic revision. *Memoirs on Entomology International* 13: 416 pp.

Noyes, J.S. (2003): *Universal Chalcidoidea Database*. – World Wide Web electronic publication. www.nhm.ac.uk/entomology/chalcidoids/index.html [accessed 22/04/2005].

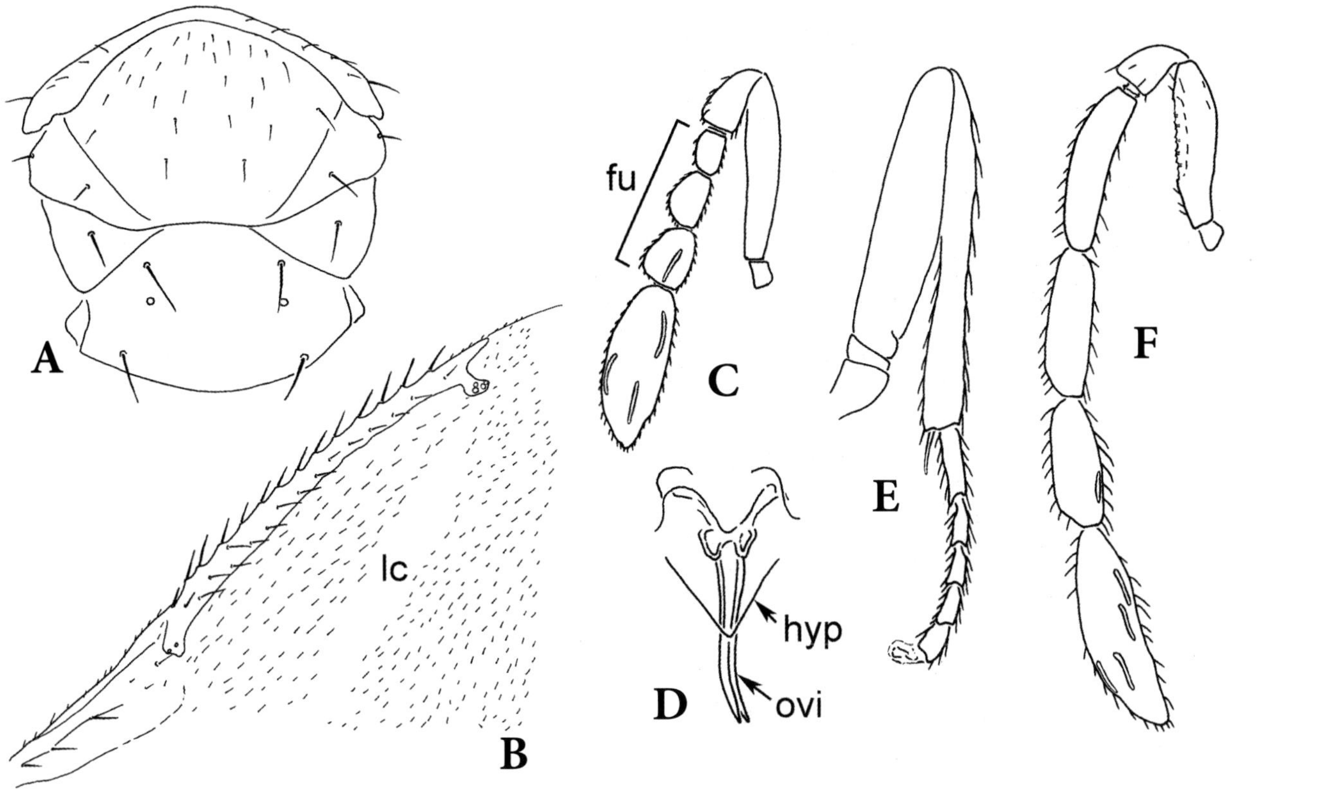

FIGURE 1 Aphelinus *sp.: A, mesosoma in dorsal view; B, fore wing with linea calva (lc); C, antenna of female; D, metasoma apex, hypopyium (hyp) cleared to show ovipositor (ovi); E, midleg; F, antenna of male.*

11.5. Encyrtidae (Chalcidoidea)

John S. Noyes

Size *c.* 0.5 to 2.0 mm, robust, mostly fully winged, occasionally brachypterous. Antennal flagellum in female 5–9-segmented, in male 5–8-segmented, apical segment larger than preceding segments, or apical 2 or 3 segments fused forming a distinct clava. Pronotum strongly transverse; mesoscutum normally without notaular lines, but if present extremely shallow; anterior margins of axillae forming a straight line; scutellum shield-shaped; prepectus moderately large and free, not forming shoulders in dorsal view; mesopleuron convex, undivided, occupying more than half the mesosoma in side view, often touching the basal segment of the gaster. Fore wing with a bare line (linea calva) extending obliquely across wing from stigmal vein; marginal vein generally very short. Mid coxa, in profile, level with middle of mesopleuron; mid tibial spur relatively long and stout, usually subequal in length to mid basitarsus. Gaster most often broadly attached to mesosoma; tergites VII and VIII fused; cercal plates usually well advanced, sometimes in anterior part of gaster; ovipositor hidden or well exserted; outer plates of ovipositor separated from tergite VIII, at most connected by a weak membrane or an elongate sclerotized strip. Most species of Encyrtidae are primary endoparasitoids of other arthropods, whilst a few are hyperparasitoids via other Encyrtidae, or Aphelinidae, Pteromalidae, Braconidae, Dryinidae, etc. A few species of *Microterys*, however, have been recorded as predators of the eggs of scale insects (Coccidae). Most encyrtids parasitize Hemiptera, with about one third of the species being recorded from mealybugs (Pseudococcidae) and another third from soft scales (Coccidae). Nearly all species belonging to the subfamily Tetracneminae are parasitoids of Pseudococcidae. Host associations of Encyrtinae are more diverse, with many attacking immature stages of holometabolous insects. Some encyrtids (*e.g. Copidosoma* spp.) are polyembryonic egg-larval parasitoids of Lepidoptera. A large number of species belonging to the family Encyrtidae have been associated with classical biological control programmes.

Nine species of Encyrtidae are known from Greenland. The family includes about 4700 described species worldwide and is most diverse in tropical and subtropical areas.

The following regional reviews may prove useful: Trjapitzin, 1989 (Palaearctic); Noyes, 1980 (Neotropical); Noyes & Hayat, (Indo-Pacific, including Australia); Noyes, Woolley & Zolnerowich, 1997 (Nearctic); Prinsloo & Annecke, 1979 (Afrotropical). The various subfamily and tribal classifications have been summarised by Trjapitzin (1973a, b), Trjapitzin (1989) and Noyes & Hayat (1994). Terminology follows Noyes (1980, 2005).

Key to species

1. Antenna with fewer than 6 segments in funicle, all segments very short and transverse (Figs 2L, M), female clava 1- or 2-segmented 2
– Antenna with 6 segments in funicle, segments often much longer than broad (Figs 1A, B, D, F, I, K, 2A, F, J, N, P), female clava always 3-segmented 3
2. Funicle 4-segmented (Fig. 2M); mesoscutum and scutellum both very short and strap-like, together a little more than 2× as broad as long ...... *Tetracyclos*
– Funicle 5-segmented (Fig. 2L); mesoscutum and scutellum longer, together about 1.5× as broad as long .. .. *Stemmatosteres*
3. Clava 3-segmented (Figs 1A, D, I, 2A, F, N) (females) 4
– Clava solid (Figs 1B, F, K, 2C, J, P), sometimes appearing 2-segmented (males)........................... 10
4. Tegula entirely dark brown; thorax dorsally generally dark brown with a moderate to strong metallic green sheen 5
– Tegula white or yellow, at least in basal half; thorax mostly yellow or orange, although occasionally dark brown with an almost imperceptible sheen 7
5. Frontovertex more than one-third head width (*i.e.* wider than eye in frontal or dorsal view); ovipositor exserted, the exserted part longer than mid tibial spur.................................. *Pseudencyrtus*
– Frontovertex less than one-third head width (i.e. narrower than eye in frontal or dorsal view); ovipositor not exserted or hardly so, the exserted part conspicuously shorter than mid tibial spur 6
6. Fore wing hyaline to weakly infuscate with area distad of venation hyaline to apex (Fig. 2O); scutellum convex and with deeper, coarser sculpture than on mesoscutum.......................... *Trichomasthus*
– Fore wing distinctly infuscate with a hyaline band distad of apex of venation, apical area also infuscate (Fig. 1J); scutellum flat and with sculpture similar to that on mesoscutum .... *Microterys*

7. Thorax dorsally orange, mixed dark brown to completely dark brown; hypopygium not reaching more than 4/5 along gaster (*Metaphycus*) 8
– Thorax almost entirely yellow to dusky pale orange, sometimes weakly infuscate; hypopygium reaching apex of gaster 9
8. Posterior margin of pronotum dark brown .... *Metaphycus ater*
– Posterior margin of pronotum translucent white to pale yellow .... *Metaphycus groenlandicus*
9. Clava white contrasting with dark brown funicle segments (Fig. 2F); posterior margin of pronotum hardly concave so that it is clearly visible medially.. .... *Pseudococcobius*
– Clava brown contrasting with yellow to pale brown funicle segments; posterior margin of pronotum strongly concave and V-shaped so that median part is hardly visible behind head ....*Rhopus*
10. Tegula entirely dark brown 11
– Tegula at least partially white or yellow...........12
11. Frontovertex at least 2× as broad as long; funicle clothed in long setae, each of which is conspicuously longer than diameter of segments (Fig. 2C) ....*Pseudencyrtus*
– Frontovertex hardly broader than long; funicle clothed in setae that are each about as long as diameter of segments (Fig. 2P)..........*Trichomasthus*
12. Dorsum of thorax generally dark brown 13
– Dorsum of thorax generally yellow to dusky yellow...15
13. Frontovertex and dorsum of thorax with a metallic green lustre............................... *Microterys*
– Frontovertex and dorsum of thorax without metallic green lustre (*Metaphycus*) 14
14. Posterior margin of pronotum dark brown ....*Metaphycus ater*
– Posterior margin of pronotum translucent white to pale yellow .... *Metaphycus groenlandicus*
15. Antenna with funicle segments broader than long, clava longer than funicle (similar to Fig. 2F) ....*Pseudococcobius*
– Antenna with funicle segments longer than broad, clava less than half as long as funicle (Fig. 2J)... *Rhopus*

ENCYRTINAE-ECHTHROPLEXIELLINI

***Metaphycus ater* (Mercet, 1925). Figs 1A–C**

Henriksen 1939: not recorded.

Female about 1.1 mm. Body generally dark brown; antenna (Fig. 1A) pale brown, F5 and F6 yellow; legs yellow with hind femur brown; tegula white in basal half, apex dark brown; wings hyaline. Head hypognathous; scape subcylindrical, about 3.5× as long as broad; funicle segments all transverse; clava 3-segmented, nearly as long as funicle. Thorax with visible part of pronotum about 0.2× as long as mesoscutum; mesoscutum slightly shorter than scutellum, about 2× as broad as long, notaular lines absent; wings fully developed, fore wing (Fig. 1C) reaching past apex of gaster [elsewhere brachypterous forms known]. Gaster with hypopygium reaching about 2/3 along gaster; ovipositor not exserted, or hardly so. Male, length about 0.9–1.1 mm. Very similar to female, sexes very difficult to separate. Antenna (Fig. 1B) like that of female, but clava solid.

BIOLOGY. Recorded as a parasitoid of *Eriococcus greeni* Newstead and *E. inermis* (Green) (Hemiptera: Eriococcidae) feeding on grasses (Poaceae).

DISTRIBUTION. S (Nunap Isua/Kap Farvel-area, Narsarsuaq). Throughout Europe. New species to Greenland.

***Metaphycus groenlandicus* Buhl, 1997. Figs 1D–H**

Henriksen 1939: not recorded.

Female about 1.0–1.2 mm. Head mostly pale yellow with some dark brown markings; scape (Fig. 1D) dark brown with dorsal margin pale yellow; proximal funicle segments dark brown, at least F5 and F6 yellow; clava dark brown; dorsum of thorax mostly dark brown, side and venter dusky pale yellow; legs dusky pale yellow with a pair of rather diffuse brown bands on each tibia; tegula basally white, apex dark brown; wings hyaline. Head hypognathous; scape (Fig. 1D) broadened and flattened, about 2.5× as long as broad; funicle segments transverse or quadrate; clava 3-segmented. Pronotum about 0.2× as long as mesoscutum, mesoscutum slightly shorter than scutellum, notaular lines visible in anterior half; wings fully developed, fore wing (Fig. 1E) exceeding apex of gaster. Gaster with hypopygium reaching about 2/3 along gaster, ovipositor not exserted, or hardly so. Male, length about 0.9–1 mm. Generally very similar to female but for antenna. Scape slightly more than 3× as broad; all funicle segments longer than broad (Fig. 1F); clava solid and about one-third length of funicle; antennal torulus with associated pores along inner margin (Fig. 1G).

BIOLOGY. Host unknown but almost certainly a scale insect (Hemiptera: Coccidae) on *Salix*.

DISTRIBUTION. NW (Pituffik/Thule), NE (Dove Bugt, Zackenberg, Mestersvig, inner Scoresby Sund).

Recorded from Greenland only, probably endemic.

***Stemmatosteres ?apterus* Timberlake, 1918. Fig. 2L**

Henriksen 1939: not recorded.

Female (excluding ovipositor) about 0.6–0.7 mm. Body generally pale yellow to pale orange. Head sub-prognathous; ocelli minute, virtually absent; antenna (Fig. 2L) with funicle 5-segmented, all segmented broader than long; clava 2-segmented. Thorax with pronotum clearly visible in dorsal view; mesoscutum and scutellum both very transverse, subequal in length and together about 1.5× as broad as long; apex of scutellum truncate; wings absent. Gaster with ovipositor exserted, the exserted part about 1.5× as long as mid tibial spur. Male. Unknown, but probably very similar to female in appearance.

BIOLOGY. Unknown, but species of the genus have been reared elsewhere from mealybugs (Hemiptera: Pseudococcidae) feeding on grasses (Poaceae).

DISTRIBUTION. S (only collected in Narsarsuaq). New species to Greenland. *Stemmatosteres apterus* is found in North America.

Tetracyclos boreios **Kryger, 1942. Fig. 2M**
Henriksen 1939: not recorded.

Female about 0.5–0.7 mm. Body generally pale yellow to pale orange. Head sub-prognathous; eye small, only about 0.7× as long as frontovertex width; ocelli minute, virtually absent; antenna (Fig. 2M) with funicle 4-segmented, all segments much broader than long; clava solid and more than 2× as long as funicle. Thorax with pronotum clearly visible in dorsal view, visible part at least 1.5× as long as mesoscutum, posterior margin almost straight; apex of scutellum broadly truncate; mesoscutum and scutellum both very transverse, subequal in length and together a little more than 2× as broad as long; wings absent. Gaster with cercal plates near apex; hypopygium reaching apex; ovipositor not exserted, or hardly so, exserted part not longer than mid tibial spur. Male. Unknown, but would probably very similar to female in appearance.

BIOLOGY. Host unknown but probably mealybugs (Hemiptera: Pseudococcidae) feeding on grasses (Poaceae).

DISTRIBUTION. NE (Zackenberg, Jameson Land: Muslingeelvbugt, Lollands Elv, Constable Pynt). Endemic. Recorded from Greenland only.

Pseudococcobius obenbergeri **(Novicky, 1926). Figs 2F, G**
Henriksen 1939: not recorded.

Female (excluding ovipositor) about 0.6–1.3 mm. Body generally pale dusky orange; antenna (Fig. 2F) with funicle dark brown and clava white; sclerites of thoracic dorsum margined pale brown; gaster with a transverse, median brown band on most tergites. Head hypognathous; funicle 6-segmented, clava 3-segmented. Visible part of pronotum slightly shorter than mesoscutum; mesoscutum about 2× as broad as long, with notaular lines visible anteriorly; scutellum with apex rounded; fore wing shorter than tegula, hind wing minute [fully winged forms not known in Greenland, but where they occur elsewhere the fore wing is almost hyaline but has a very faint, broad infuscate band below marginal vein (Fig. 2G)]. Gaster with ovipositor exserted, the exserted part about 2× as long as mid tibial spur. Male, length about 0.7 mm. Generally very similar to female but clava solid and fore wing reaches past base of gaster. In only specimen available from Greenland the funicle appears to be 5-segmented, but it is likely that F1 is small and obscured by pedicel.

BIOLOGY. Host unknown, but recorded elsewhere as a parasitoid of *Trionymus multivorus* (Kiritshenko) and *T. perrisii* (Signoret) feeding on *Festuca* (Poaceae).

DISTRIBUTION. S, SW (only known from Narsarsuaq and Kangerlussuaq/Søndre Strømfjord). New species to Greenland.

Northern and central Europe.

ENCYRTINAE-DISCODINI

Microterys curio **Trjapitzin, 1966. Figs 1I–L**
Henriksen 1939: not recorded.

Female about 1.4–1.7 mm. Head dark orange, occasionally mixed dark brown with metallic green areas; antenna (Fig. 1I) with scape, pedicel and clava orange brown, funicle a little paler; thorax brown, darker dorsally and with a distinct metallic blue or blue-green sheen, slightly coppery on pronotum; tegula dark brown; fore wing (Fig. 1J) infuscate with hyaline areas; legs generally brown; gaster brown. Head hypognathous; scape (Fig. 1I) about 3× as long as broad; clava 3-segmented. Thorax without notaular lines; scutellum flat, shield-shaped; sculpture on mesoscutum and scutellum similar; wings fully developed or occasionally slightly shortened so that fore wing only just reaches apex of gaster. Gaster with ovipositor not exserted, or hardly so. Male, length about 1.2–1.6 mm. Head mostly metallic blue-green to green; scape yellow; pedicel brown; flagellum pale yellow-brown; dorsum of thorax mostly metallic green to blue-green; wings hyaline (Fig. 1L); coxae brown; fore and mid legs otherwise pale orange, mid tibia mixed brown; gaster brown. Antenna (Fig. 1K) with scape broadened and slightly flattened, about 2× as long as broad; funicle with all funicle longer than broad; clava entire; setae on flagellum clearly shorter than diameter of segments.

BIOLOGY. Recorded elsewhere as a parasitoid of an unidentified scale insect (Hemiptera: Coccoidea, probably Coccidae) on *Salix*.

DISTRIBUTION. S, SW, NW (north to the Pituffik/Thule-area). Appears to be very common in western Greenland. New species to Greenland.

Northern Europe (Finland, northern Russia).

Trichomasthus ?marsus **(Walker, 1837). Figs 2N–Q**
Henriksen 1939: not recorded.

Female about 1.3–1.8 mm. Head virtually black with a metallic sheen; antenna (Fig. 2N) dark brown; thorax brown with a distinct metallic sheen; tegula dark brown; fore wing (Fig. 2O) hyaline or weakly infuscate to apex of venation; legs generally brown; gaster brown. Head hypognathous; antenna (Fig. 2N) with scape about 4× as long as broad; funicle 6-segmented; clava 3-segmented. Thorax with visible part of pronotum about one-third as long as mesoscutum; mesoscutum with notaular lines absent; scutellum shield-shaped with apex rounded; sculpture on scutellum mostly a little deeper than that on mesoscutum; wings fully developed reaching past apex of gaster. Gaster with ovipositor not exserted, or hardly so. Male, length about 0.8–1.5 mm. Similar to female but pedicel brown; flagellum (Fig. 2P) pale brown; wings virtually hyaline; fore wing very weakly infuscate below marginal vein (Fig. 2Q); antenna (Fig. 2P) with scape slightly less than 3× as long as broad; funicle filiform, all segments longer than broad; clava entire; longest setae on flagellum very slightly longer than diameter of segments; wings slightly reduced only just reaching apex of gaster to fully developed and clearly reaching past apex of gaster.

BIOLOGY. Host unknown, but other species of the genus have been recorded as parasitoids of various scale insects (Hemiptera, especially Coccidae and Eriococcidae) feeding on a variety of plants, especially grasses (Poaceae), Asteraceae and Fabaceae.

DISTRIBUTION. S (Narsarsuaq), SW (north to Upernavik Laksefjord). Widespread in SW Greenland. New species to Greenland.

Trichomasthus marsus (Walker) was previously recorded from the British Isles only.

ENCYRTINAE-BOTHRIOTHORACINI

***Pseudencyrtus* sp. nov. Figs 2A–E**

Female (excluding ovipositor) about 1.0–1.5 mm. Body generally dark brown with weak metallic reflections on head and dorsum of thorax; antenna (Fig. 2A) dark brown; tegula dark brown; wings hyaline, but fore wing (Fig. 2B) with a small, infuscate area immediately below marginal and stigmal veins; legs generally dark brown. Head hypognathous; scape about 4.5× as long as broad; all funicle segments very slightly longer than broader; clava 3-segmented. Thorax with visible part of pronotum about one-fifth as long as mesoscutum; mesoscutum with notaular lines absent; scutellum shield-shaped; sculpture on scutellum similar to that on mesoscutum; wings slightly reduced, fore wing (Fig. 2B) not quite reaching apex of gaster. Gaster with ovipositor exserted, the exserted part longer than mid tibial spur. Male, length about 1.0–1.8 mm. Colour as female. Antenna (Fig. 2C) with scape broadened and slightly flattened, about 4× as long as broad; funicle with all segments longer than broad; clava entire; longest setae on flagellum about 2× as long as diameter of segments. Wings varying from fully developed (Fig. 2E) to strongly reduced (Fig. 2D) and reaching only about half way along gaster.

BIOLOGY. Host unknown, but probably a parasitoid of Cecidomyiidae (Diptera) on *Salix*.

DISTRIBUTION. NW (Pituffik/Thule), NE (Zackenberg, Clavering Ø, Jameson Land: Børglum Elvdal). Only known from Greenland. New species to Greenland and almost certainly endemic.

TETRACNEMINAE-ANAGYRINI

***Rhopus sulphureus* (Westwood, 1837). Figs 2H–K**

Henriksen 1939: not recorded.

Female about 1.0–1.2 mm. Body generally yellow; F1–F4 pale brown; clava dark brown (Fig. 2H). Body dorso-ventrally flattened; head sub-prognathous; antenna (Fig. 2H) with scape about 2.5× as long as broad; all funicle segments slightly transverse; clava 3-segmented. Thorax with pronotum very short medially and not quite visible behind head, longitudinally divided; mesoscutum about 1.5× as broad as long, notaular lines absent; scutellum shield-shaped, with apex more or less rounded; wings reduced, fore wing not reaching base of gaster [in extralimital areas fully winged females are known (similar to Fig. 2I)]. Ovipositor not exserted. Male, length about 0.6–1.0 mm. Colour as female, but dorsum of thorax often mixed brownish. Structure generally as for female, but antenna (Fig. 2J) filiform with all funicle segments about 1.5–2× as long as broad and clothed with long setae, the longest of which is about 1.75× as long as diameter of segments, F6 with a line of 6 or 7 small scale-like structures ventrally (Fig. 2K); wings fully developed (Fig. 2I) [in extralimital areas brachypterous males are known].

BIOLOGY. Recorded elsewhere as a parasitoid of mealybugs (Hemiptera: Pseudococcidae) feeding on grasses (Poaceae).

DISTRIBUTION. S (only collected from Narsarsuaq, Qassiarsuk, Upernaviarsuk). New species to Greenland. Northern and central Europe.

References

Noyes, J.S. (1980): A review of the genera of Neotropical Encyrtidae (Hymenoptera: Chalcidoidea). – *Bulletin of the British Museum (Natural History) (Entomology)* 41:107–253.

Noyes, J.S. (2005): *Universal Chalcidoidea Database*. Worldwide web electronic publication: www.nhm.ac.uk/research-curation/projects/chalcidoids/.

Noyes, J.S. & Hayat, M. (1984): A review of the genera of Indo-Pacific Encyrtidae (Hymenoptera: Chalcidoidea). – *Bulletin of the British Museum (Natural History) (Entomology)* 48: 131–395.

Noyes, J.S. & Hayat, M. (1994): *Oriental mealybug parasitoids of the Anagyrini (Hymenoptera: Encyrtidae)*: viii+554pp CAB International, Wallingford, UK.

Noyes, J.S., Woolley, J.B. & Zolnerowich, G. (1997): *Chapter 8. Encyrtidae. In* Gibson, G.A.P., Huber, J.T. & Woolley, J.B. (eds.) Annotated keys to the genera of Nearctic Chalcidoidea (Hymenoptera), pp. 170–320. – NRC Research Press, Ottawa.

Prinsloo, G.L. & Annecke, D.P. (1979): A key to the genera of Encyrtidae from the Ethiopian region, with descripitions of three new genera (Hymenoptera: Chalcidoidea). *Journal of the Entomological Society of Southern Africa* 42 (2): 349–382.

Trjapitzin, V.A. (1973a): The classification of the family Encyrtidae (Hymenoptera, Chalcidoidea). Part 1. Survey of the systems of classification. The subfamily Tetracneminae Howard, 1892. – *Entomologicheskoe Obozrenie* 52 (1): 163–175.

Trjapitzin, V.A. (1973b): Classification of the parasitic Hymenoptera of the family Encyrtidae (Chalcidoidea). Part II. Subfamily Encyrtinae Walker, 1837. – *Entomologicheskoe Obozrenie* 52 (2): 416–429.

Trjapitzin, V.A. (1989): Parasitic Hymenoptera of the Fam. Encyrtidae of Palaearctics. *Opredeliteli po Faune* SSSR 158: 1–489 Zoologicheskim Institutom Akademii Nauk SSR, Leningrad.

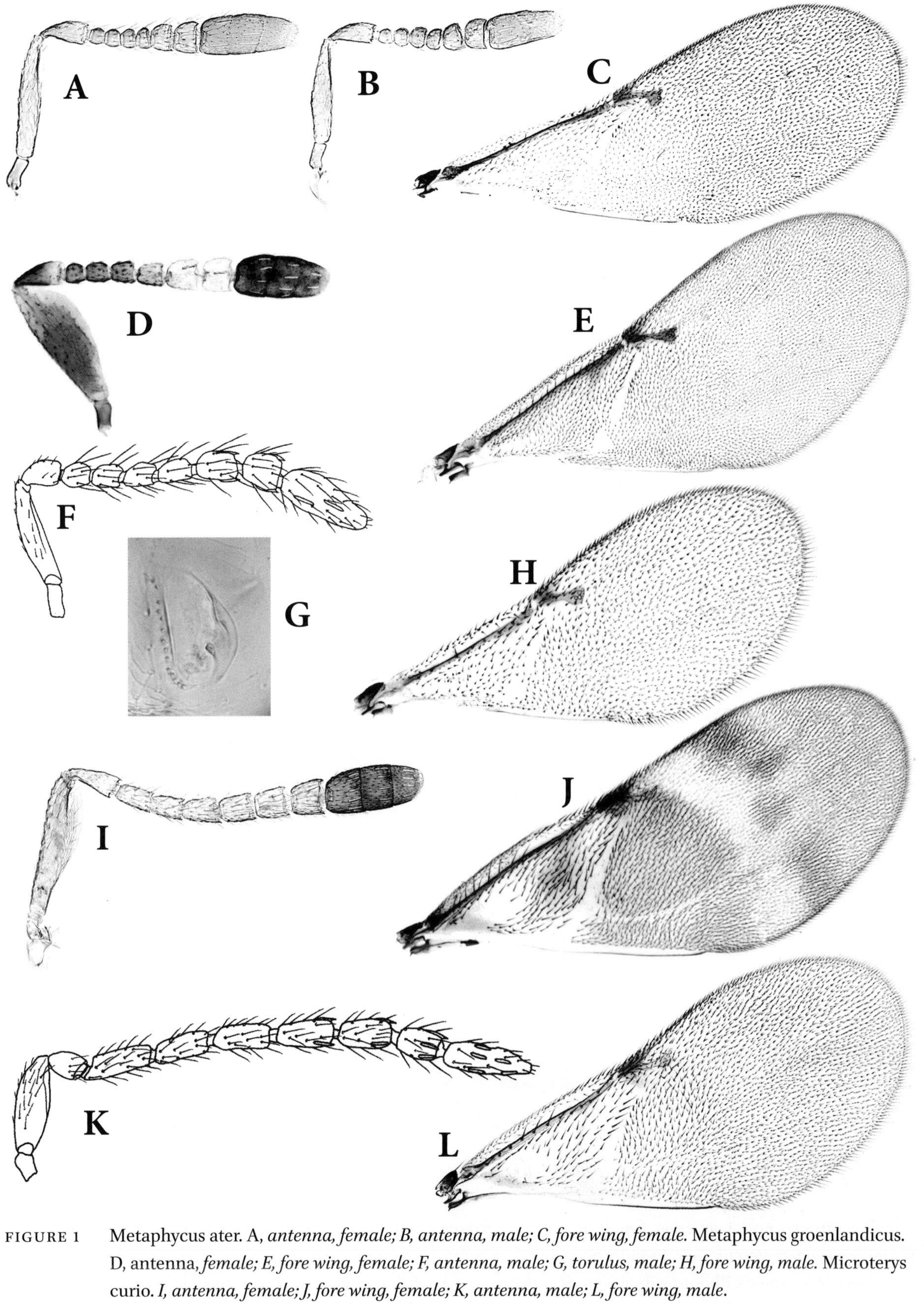

FIGURE 1 Metaphycus ater. A, *antenna, female; B, antenna, male; C, fore wing, female.* Metaphycus groenlandicus. D, antenna, *female; E, fore wing, female; F, antenna, male; G, torulus, male; H, fore wing, male.* Microterys curio. *I, antenna, female; J, fore wing, female; K, antenna, male; L, fore wing, male.*

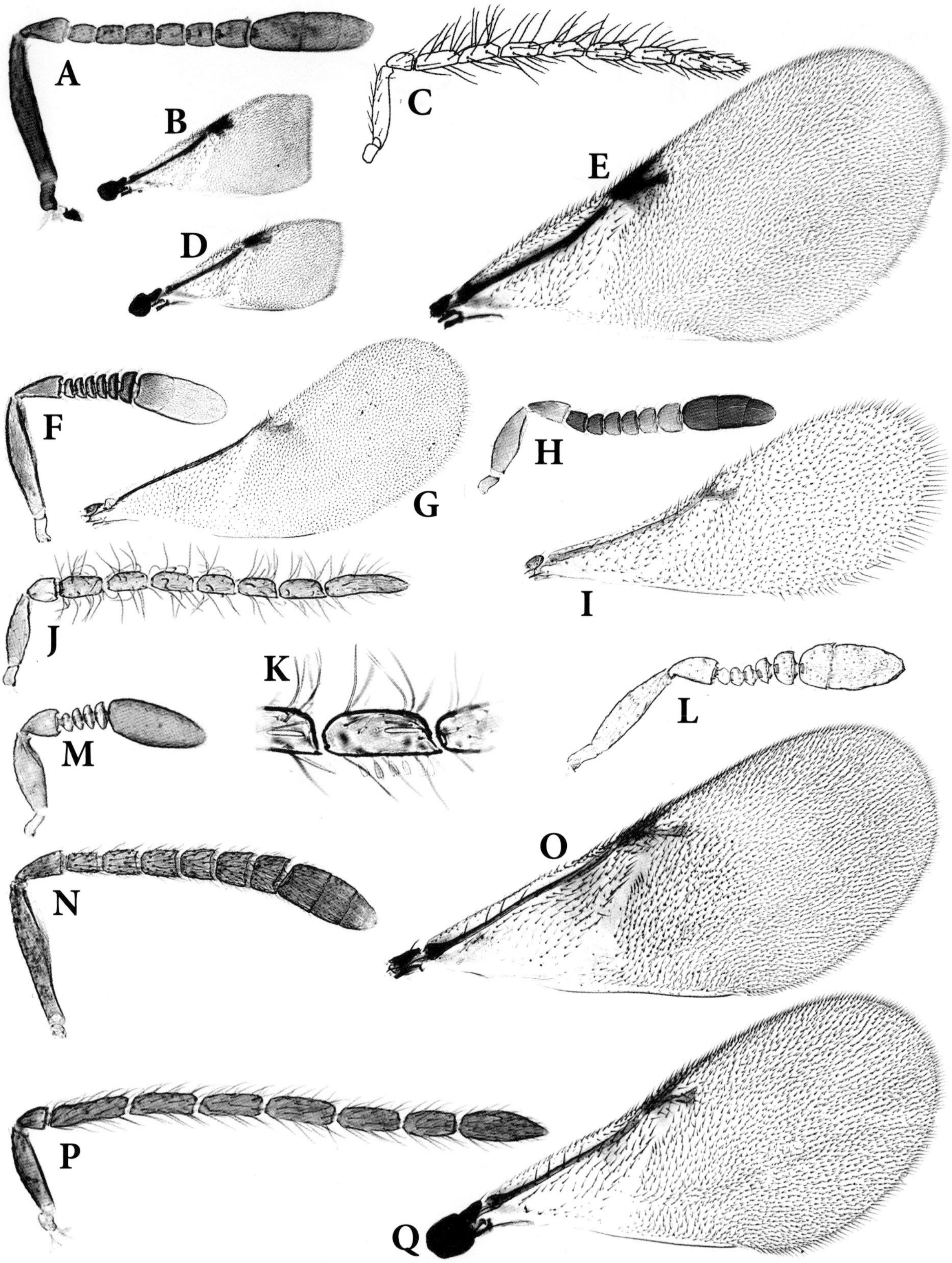

FIGURE 2 Pseudencyrtus sp. nov. *A, antenna, female; B, fore wing, female; C, antenna, male; D, fore wing, brachypterous, male; E, fore wing, macropterous male.* Pseudococcobius obenbergeri. *F, antenna, female; G, fully developed fore wing, female (extralimital specimen).* Rhopus sulphureus. *H, antenna, female; I, fore wing, male; J, antenna, male; K, F6, showing scale-like structures, male.* Stemmatosteres ? apterus. *L, antenna, female.* Tetracyclos boreios. *M, antenna, female.* Trichomasthus ? spatie marsus. *N, antenna, female; O, fore wing, female; P, antenna, male; Q, fore wing, male.*

11.6. Eulophidae (Chalcidoidea)

Hannes Baur

Body minute to small (0.4–6 mm), varying from yellowish to greenish or black, often with ± strong metallic tinge; antenna with 7–12 segments, usually with only 2–4 funicular segments; protibial spur short and straight; tarsi usually 4-segmented; fore wing with marginal vein long, usually more than 3 times as long as short stigmal or postmarginal veins; gaster constricted at base, hence narrowly attached to the mesosoma; petiole mostly transverse, occasionally elongate.

Eulophids are very diverse in biology and their range of hosts is probably wider than in any other family of parasitic wasps. Most species are primary or secondary parasitoids of Lepidoptera, Coleoptera, Diptera, Hymenoptera and few other insect orders, but they also parasitize other arthropods, such as gall mites (Acari), eggs of spiders (Araneae), and nematodes (Nematoda). Most eulophids attack concealed hosts, for instance leaf-miners, leaf-rollers, woodborers, or gall-formers. All host stages, from egg, larva, pupa to adults, are attacked as either ectoparasitoid or – less commonly – endoparasitoid. Phytophagy occurs in the subfamily Tetrastichinae, including gall-formers, seed predators and inquilines. Eulophids are economically important and many species are used for biological pest control.

The Eulophidae are worldwide distributed, comprising some 5200 species in 334 genera (Noyes 2014). The Greenland fauna has been reviewed by Baur (2005), who recognized 6 species and genera in the 3 subfamilies Entedoninae, Eulophinae and Terastichinae. Eulophids are generally abundant and belong to the most commonly collected chalcid waps, also in Greenland.

Terminology and morphology follows Gibson (1997). Information concerning biology and general distribution originates from Baur (2005) and Noyes (2014).

Key to species of Greenland Eulophidae

1. Scutellum with a single pair of setae placed slightly behind middle of sclerite (Fig. 2A); submarginal vein with two setae (Fig. 2D) (Entedoninae) 2
– Scutellum with two or more pairs of setae (Figs 1B, 3E, 4C); submarginal vein with more than two setae (Figs 1D, 4D) 4
2. Propodeum with median carina diverging posteriorly, joining complete plica laterally (Fig. 4F); petiole conspicuous, with subparallel lateral margins, from almost as long as broad (female) to distinctly longer than broad (male), dorsal surface strongly reticulate (Fig. 4F) .... *Pediobius alaspharus*
– Propodeum with indistinct median carina and plica indicated at most in posterior half (*e.g.* Figs 1B, 2E, 3C, 4C); petiole inconspicuous and transverse, lateral margins converging anteriorly and dorsal surface ± smooth (Fig. 2E)..............................3
3. Antenna with 3 funicular segments (Figs 1E, F); mesoscutum and scutellum strongly sculptured, reticulation consists of rather large and high meshes (Fig. 2A); postmarginal vein about 1.6–1.8 times as long as stigmal vein (Figs 2B, C) .... *Chrysocharis pubicornis*
– Antenna with 2 funicular segments (Fig. 2F); mesoscutum and scutellum feebly sculptured, reticulation consists of minute and very superficial meshes (Fig. 3A); postmarginal vein about half as long as stigmal vein (Fig. 3B)......*Neochrysocharis diastatae*
4. Notaulus complete and very deep, groove like (Fig. 1B); postmarginal vein a short stub (Fig. 1C). Antenna in females with 3 funicular segments, in males (Fig. 1A) with 4 funicular segments bearing subbasal whorls of long setae which are much longer than length of funicular segment which bears them (Tetrastichinae) .... *Aprostocetus meltoftei*
– Notaulus either incomplete (Fig. 3E) or only moderately deep, suture like (Fig. 4C); postmarginal vein at least as long as stigmal vein. Antenna in both sexes with either 2 (Fig. 3D) or 4 funicular segments, in males without subbasal whorls of long setae (Eulophinae) 5
5. Antenna with 2 funicular segments (Fig. 3D); mid lobe of mesoscutum with only two pairs of long setae, notaulus incomplete (Fig. 3E); axilla very distinctly produced anteriorly (Fig. 3E); fore wing without speculum (Fig. 3F); postmarginal vein about as long as stigmal vein (Figs 4A, B); petiole inconspicuous, broader than long and smooth on upper side .... *Diglyphus isaea*
– Antenna with 4 funicular segments; mid lobe of mesoscutum with numerous scattered short setae and a single pair of longer setae at posterior margin of the sclerite, notaulus complete (Fig. 4C); axilla only slightly produced anteriorly (Fig. 4C); fore wing with relatively large speculum (Fig. 4D); postmarginal

vein 1.6–1.8 times as long as stigmal vein (Fig. 4E); petiole conspicuous, about as long as broad and finely reticulate on upper side .... *Elachertus fenestratus*

TETRASTICHINAE

***Aprostocetus meltoftei* Buhl, 1997 Figs 1A–D**

Henriksen 1939: not recorded.

0.9–1.2 mm. Antenna in females with 3 funicular segments, in males with 4 funicular segments (Fig. 1A) bearing subbasal whorls of long setae which are much longer than length of funicular segment which bears them. Mesosoma (Fig. 1B) with notaulus complete and very deep, groove like; mesoscutum and scutellum with very fine, superficial reticulation; mid lobe of mesoscutum with a row of 4–5 setae along each notaulus; scutellum with two pairs of setae, the first placed in about middle, the second at hind margin of sclerite; axilla very distinctly produced anteriorly; fore wing with small speculum (Fig. 1D); submarginal vein with (3–)4–5(–6) setae (Fig. 1D); postmarginal vein hardly one seventh as long as stigmal vein, i.e. only a short stub (Fig. 1C); propodeum (Fig. 1B) with complete, slightly raised median carina but without plica. Metasoma with petiole transverse, lateral margins converging anteriorly and dorsal surface smooth.

DISTRIBUTION. NW (Pituffik/ Thule), NE (Zackenberg, Jameson Land: Langelandselv).

ENTEDONINAE

***Chrysocharis pubicornis* (Zetterstedt, 1838). Figs 1E, F, 2A–E**

Henriksen 1939: not recorded.

1.0–1.9 mm. Antenna (Figs 1E, F) with 3 funicular segments; male funicular segments with numerous setae but proximally without compact whorls of long setae. Mesosoma (Fig. 2A) with notaulus complete, widening in posterior third; mesoscutum and scutellum strongly sculptured, reticulation consists of rather large and high meshes; mid lobe of mesoscutum with two pairs of long setae; scutellum with a single pair of setae placed slightly behind middle of sclerite; axilla only slightly produced anteriorly; fore wing (Fig. 2B) with moderately large speculum (Fig. 2D); submarginal vein with two setae (Fig. 2D); postmarginal vein about 1.6–1.8 times as long as stigmal vein (Fig. 2C); propodeum (Fig. 2E) without median carina but anteriorly with a raised anchor- or y-shaped carina, plica indicated in posterior third to half. Metasoma with petiole (Fig. 2E) transverse, lateral margins converging anteriorly and dorsal surface weakly reticulate in posterior part.

BIOLOGY. Mainly a primary parasitoid of various leaf-miner flies (Diptera: Agromyzidae) and moths (Lepidoptera: Gracillariidae, Nepticulidae; Tischeriidae, among others).

DISTRIBUTION. S, SW.

Holarctic, widely distributed.

***Neochrysocharis diastatae* (Howard, 1881). Figs 2F, 3A–C**

Henriksen 1939: not recorded.

0.6–1.4 mm. Antenna (Fig. 2F) with 2 funicular segments; male funicular segments with numerous setae but without compact whorls of long setae. Mesosoma (Fig. 3A) with notaulus complete, widening in posterior third; mesoscutum and scutellum feebly sculptured, reticulation consists of minute and very superficial meshes; mid lobe of mesoscutum with two pairs of long setae; scutellum with a single pair of setae placed slightly behind middle of sclerite; axilla distinctly produced anteriorly; fore wing (Fig. 3B) with moderately large speculum; submarginal vein with two setae; postmarginal vein about half as long as stigmal vein; propodeum (Fig. 3C) without median carina and plica. Metasoma with petiole inconspicuous and transverse, lateral margins converging anteriorly and dorsal surface smooth.

BIOLOGY. Mainly a primary parasitoid of various leaf-miner flies (Diptera: Agromyzidae), but also of a few wood wasps (Hymenoptera: Sympyta) and moths (Microlepidoptera) with concealed larvae.

DISTRIBUTION. S (Ilua: Tupaasat, Narsarsuaq).

Holarctic (widely distributed in North and Central America and China).

***Pediobius alaspharus* (Walker, 1839). Fig. 4F**

Henriksen 1939: not recorded.

1.8–2.5 mm. Antenna with 3 funicular segments in female and 4 in male; male funicular segments with numerous setae but proximally without compact whorls of long setae. Mesosoma with notaulus complete, widening in posterior third; mesoscutum and scutellum strongly sculptured, reticulation consists of moderately large to large meshes; scutellum with a single pair of setae placed in posterior third; axilla only slightly produced anteriorly; mid lobe of mesoscutum with two pairs of long setae; fore wing without speculum; submarginal vein with 2 setae; postmarginal vein about 1.4–1.5 times as long as stigmal vein; propodeum (Fig. 4F) with median carina diverging posteriorly, joining complete plica laterally. Metasoma (Fig. 4F) with petiole conspicuous, with subparallel lateral margins, from almost as long as broad (female) to distinctly longer than broad (male), dorsal surface strongly reticulate.

BIOLOGY. Primary parasitoid of *Tetramesa cornuta* (Walker) (Hymenoptera: Eurytomidae), a phytophagus chalcidoid forming galls in stems of the Quack grass *Agropyron repens* (Poaceae).

DISTRIBUTION. S (recorded only from Ivittuut).

Northern and Central Europe.

EULOPHINAE

Diglyphus isaea (Walker, 1838). Figs 3D–F, 4A, B

1.0–1.6 mm. Antenna (Fig. 3D) with 2 funicular segments; male funicular segments with numerous setae but proximally without compact whorls of long setae. Mesosoma (Fig. 3E) with notaulus incomplete; reticulation of mesoscutum consists of small and slightly raised meshes, which are distinctly smaller and engraved on scutellum; mid lobe of mesoscutum with two pairs of long setae; scutellum with a single pair of setae placed slightly behind middle of sclerite; axilla distinctly produced anteriorly; fore wing without speculum (Fig. 3F); submarginal vein with about 4–7 setae (Fig. 3F); postmarginal vein about as long as stigmal vein (Figs 4A, B); propodeum (Fig. 3E) with indistinct median carina and plica indicated only in posterior fifth. Metasoma with petiole transverse, lateral margins converging anteriorly and dorsal smooth.

BIOLOGY. Mainly a primary parasitoid of various leafminer flies (Diptera: Agromyzidae), but also of a few leaf-miner moths (Lepidoptera: Gracillariidae, Lyonetiidae, Nepticulidae).

DISTRIBUTION. S, SW.

Palaearctic, widely distributed.

Elachertus fenestratus Nees, 1834. Figs 4C–E

1.3–2.2 mm. Antenna with 4 funicular segments; male funicular segments with numerous setae but proximally without compact whorls of long setae. Mesosoma (Fig. 4C) with notaulus complete and narrowly impressed throughout; mesoscutum and scutellum with mostly engraved reticulation, on scutellum with distinctly smaller meshes; mid lobe of mesoscutum with numerous scattered short setae and a single pair of longer setae at posterior margin of the sclerite; scutellum with two pairs of setae, the first placed slightly before middle, the second in posterior quarter; axilla only slightly produced anteriorly; fore wing with moderately large speculum (Figs 4D, E); submarginal vein with about 4–7 setae; postmarginal vein 1.6–1.8 times as long as stigmal vein (Fig. 4E); propodeum with sharp and straight median carina but without plica. Metasoma with petiole conspicuous, about as long as broad and finely reticulate on upper side.

BIOLOGY. Mainly a primary parasitoid of moths of various microlepidopteran families with concealed larvae (Coleophoridae, Gelechiidae, Gracillariidae, Tortricidae, among others).

DISTRIBUTION. N (few localities).

Circumpolar (widely distributed in Palaearctis and North and Central America).

References

Baur, H. (2005): A review of the Eulophidae and Pteromalidae (Hymenoptera: Chalcidoidea) of Greenland. – *Acta Societatis Zoologicae Bohemicae* 69: 23–34.

Gibson, G.A.P. (1997): Morphology and terminology. Pp. 16–44. *In* Gibson, G.A.P., Huber, J.T. & Woolley, J.B. (eds): Annotated keys to the genera of Nearctic Chalcidoidea (Hymenoptera). – Ottawa: NRC Research Press. xi + 794 pp.

Noyes, J.S. (2014): *Universal Chalcidoidea Database*. World Wide Web electronic publication. http://www.nhm.ac.uk/chalcidoids [accessed 2-June-2014].

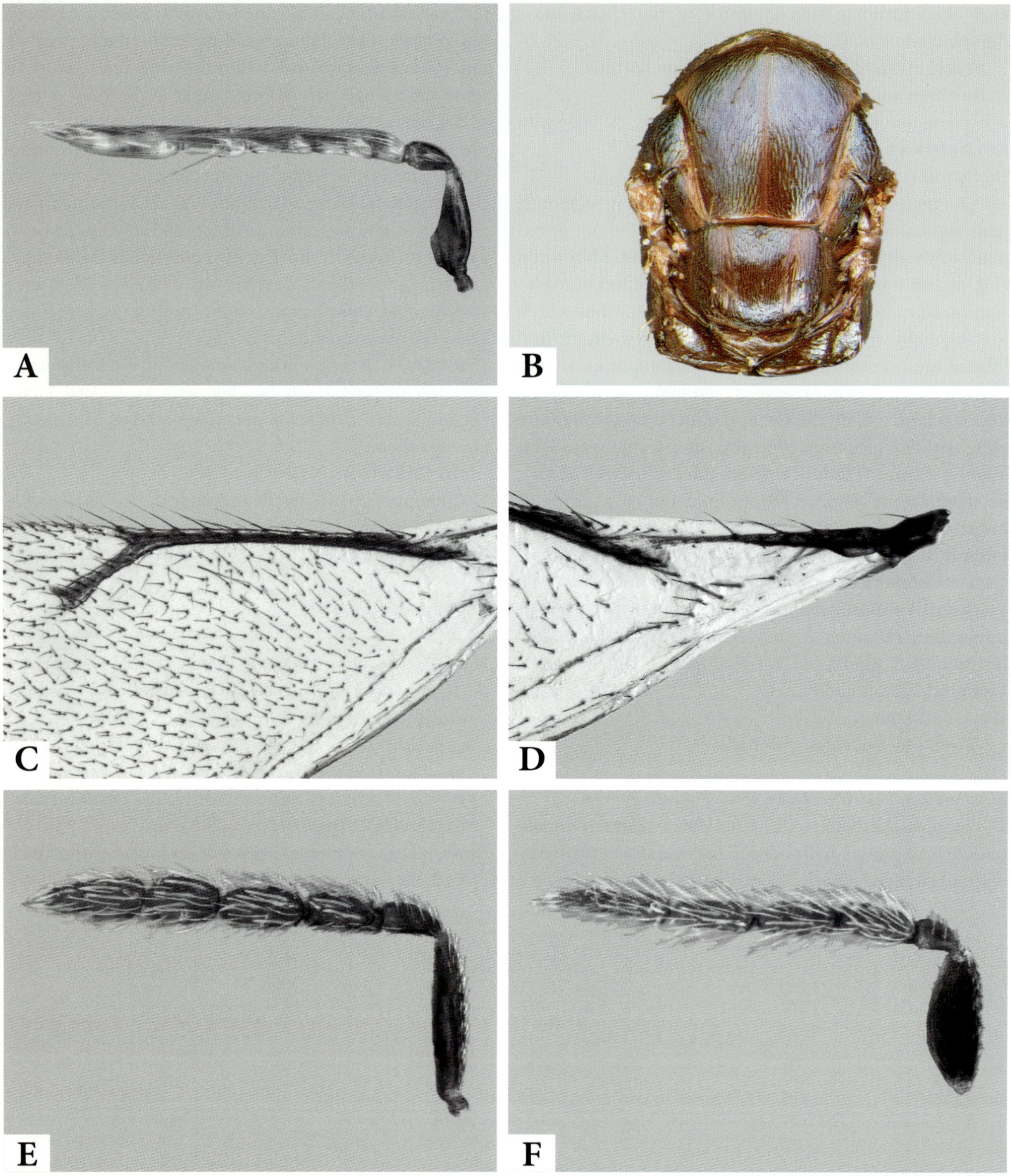

FIGURE 1 *A–D,* Aprostocetus meltoftei *Buhl: antenna male (A), mesosoma female B), marginal and stigmal vein female (C), speculum and submarginal vein female (D). E–F,* Chrysocharis pubescens *(Zetterstedt): antenna female (E), antenna male (F).*

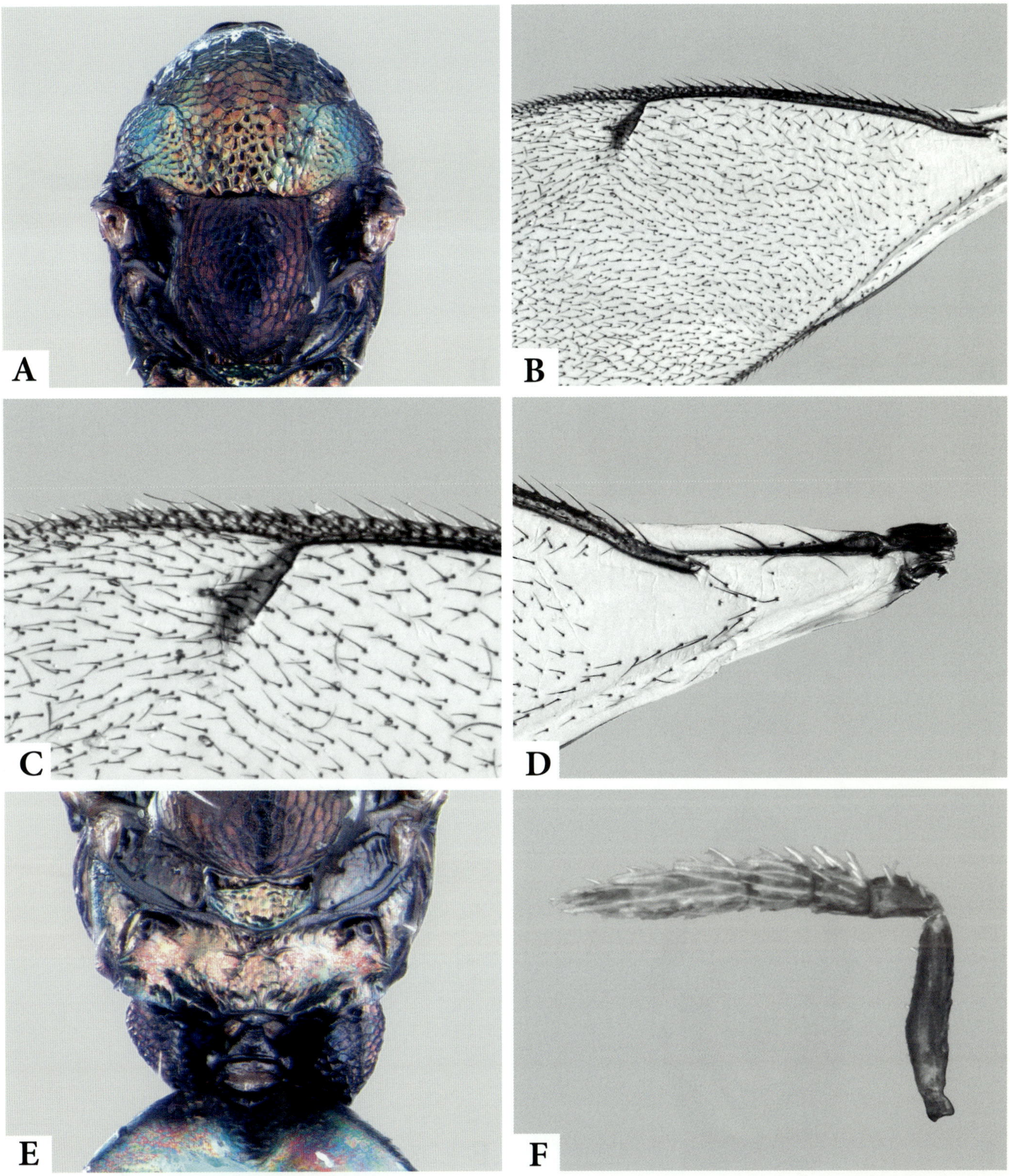

FIGURE 2 *A–E,* Chrysocharis pubescens *(Zetterstedt): mesosoma female (A), fore wing female (B), stigmal and postmarginal vein female (C), speculum and submarginal vein female (D), propodeum and petiole female (E). F.* Neochrysocharis diastatae *(Howard): antenna female.*

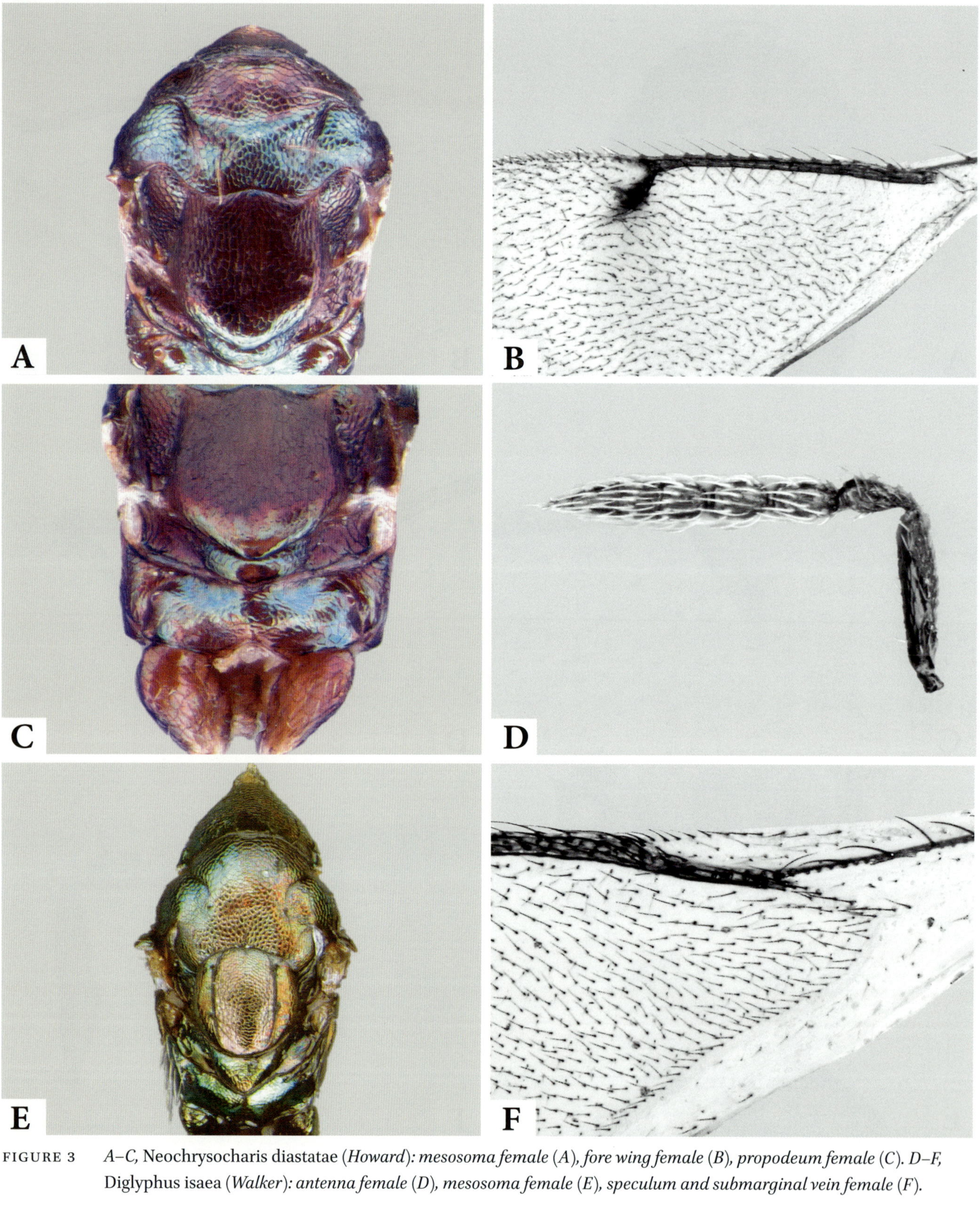

FIGURE 3 *A–C,* Neochrysocharis diastatae (*Howard*)*: mesosoma female (A), fore wing female (B), propodeum female (C). D–F,* Diglyphus isaea (*Walker*)*: antenna female (D), mesosoma female (E), speculum and submarginal vein female (F).*

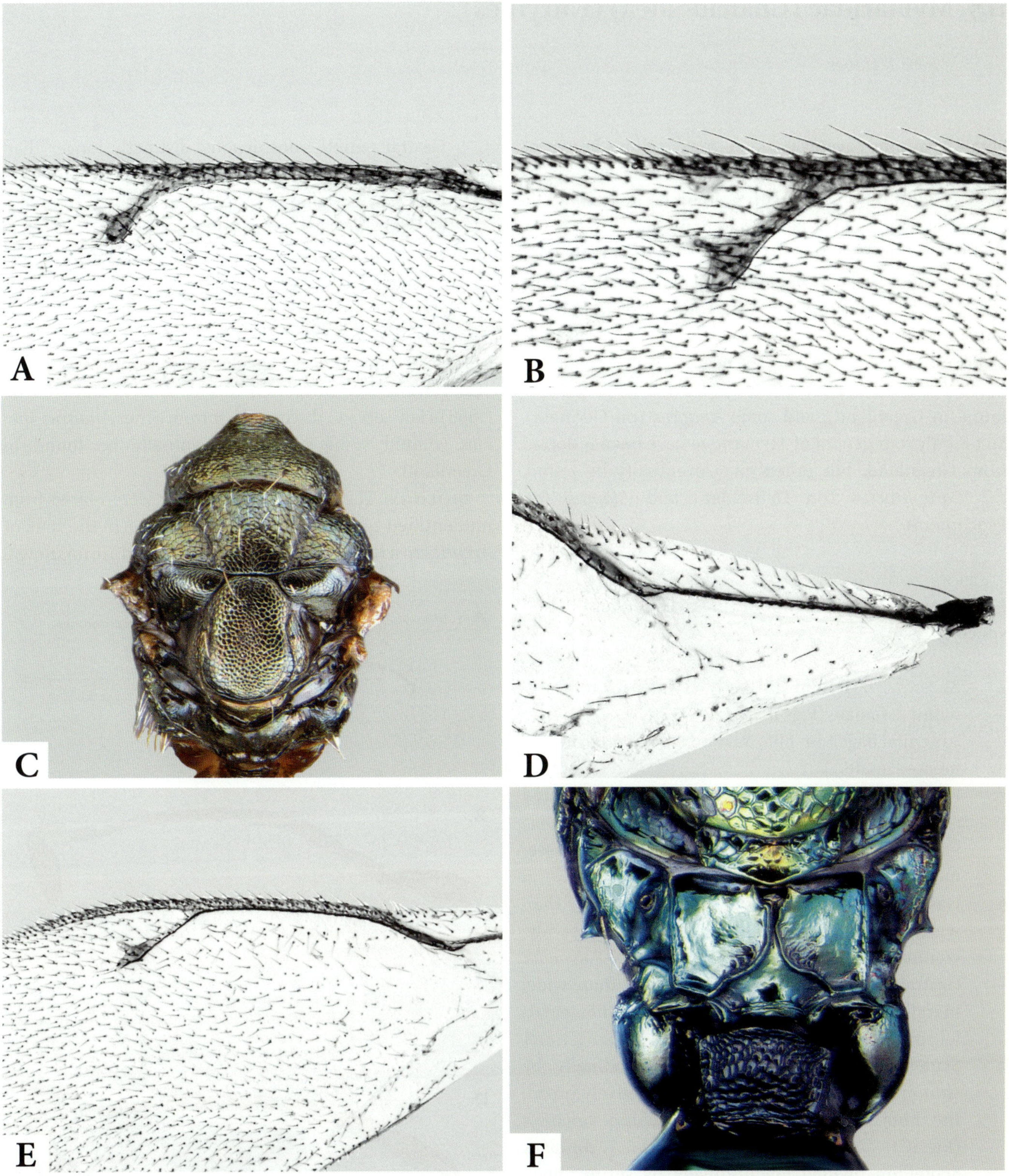

FIGURE 4 *A–B,* Diglyphus isaea *(Walker): marginal, postmarginal and stigmal vein female (A), stigmal and postmarginal vein female (B). C–E,* Elachertus fenestratus *(Nees): mesosoma female (C), speculum and submarginal vein female (D), marginal, postmarginal and stigmal vein female (E). F,* Pediobius alaspharus *(Walker): propodeum and petiole female.*

11.7. Mymaridae (Chalcidoidea) (Fairyflies)

John T. Huber

Small to minute (average: 0.5–1.0 mm), yellow, brown or black, non-metallic chalcids with antennae at least as long as body, hind wings with membrane not extending to base of wing and with long marginal fringes, and vertex separated from face and eyes by an H-shaped pattern of sutures and associated bars of thickened cuticle. Almost all species are parasitoids in eggs of other insects, particularly Auchenorrhyncha (Hemiptera) but no hosts have yet been reported the species occurring in Greenland. Most species are terrestrial, except *Caraphractus* (not yet found in Greenland), and some *Anagrus* (on Odonata). So far only four genera of Mymaridae have been reported from Greenland, but others may eventually be found, *e.g. Caraphractus* (on Dytiscidae) and *Alaptus* (on Psocoptera).

Key to mymarid genera (applicable to Greenland species)

1. Antennal flagellum distinctly clavate, up to 9-segmented; female (Figs 1B, 1D, 2B, D) 2
– Antennal flagellum filiform, 11- or, apparently, 10-segmented; male..5
2. Funicle 8-segmented (Fig. 2B); tarsi 5-segmented ....*Gonatocerus* (*Lymaenon*)
– Funicle 6-segmented (Figs 1B, 1D, 2D); tarsi 4-segmented 3
3. Gastral petiole conspicuous, distinctly longer than wide; marginal vein about four times as long as wide, comma-shaped (Fig. 2C).................. *Polynema*
– Gastral petiole inconspicuous, shorter than wide; marginal vein about 10 times as long as wide, parallel-sided (Figs 1A, C) 4
4. Scutellum entire, not divided longitudinally by suture; body dark brown; forewing relatively broad, the hind margin without lobe behind venation (Fig. 1C).................................... *Anaphes*
– Scutellum divided longitudinally by suture; body pale, yellow or light brown; forewing relatively narrow, the hind margin with distinct lobe behind venation (Fig. 1A)...5
5. Tarsi 5-segmented.........*Gonatocerus* (*Lymaenon*)
– Tarsi 4-segmented..................................6
6. Flagellum apparently 10-segmented........ *Anaphes*
– Flagellum 11-segmented.............................7
7. Gastral petiole conspicuous, distinctly longer than wide; marginal + stigmal veins short, about 4 times as long as wide, comma-shaped (Fig. 2C) ..... *Polynema*
– Gastral petiole inconspicuous, shorter than wide; marginal + stigmal veins at least 9× as long as wide, parallel sided................................*Anagrus*

***Anagrus nigriceps* (Smits van Burgst, 1914). Figs 1A, B**
This is the only *Anagrus* species known from Greenland. The generic key presented above, and Chiapini *et al.* (1996)'s key serve to distinguish it from other *Anagrus* species, should some of them eventually be found in Greenland.

BIOLOGY. The types of *A. nigriceps* were reared from unidentified eggs, probably of Cicadellidae, in the stalk of *Avena sativa* (oats). The host(s) in Greenland are unknown.

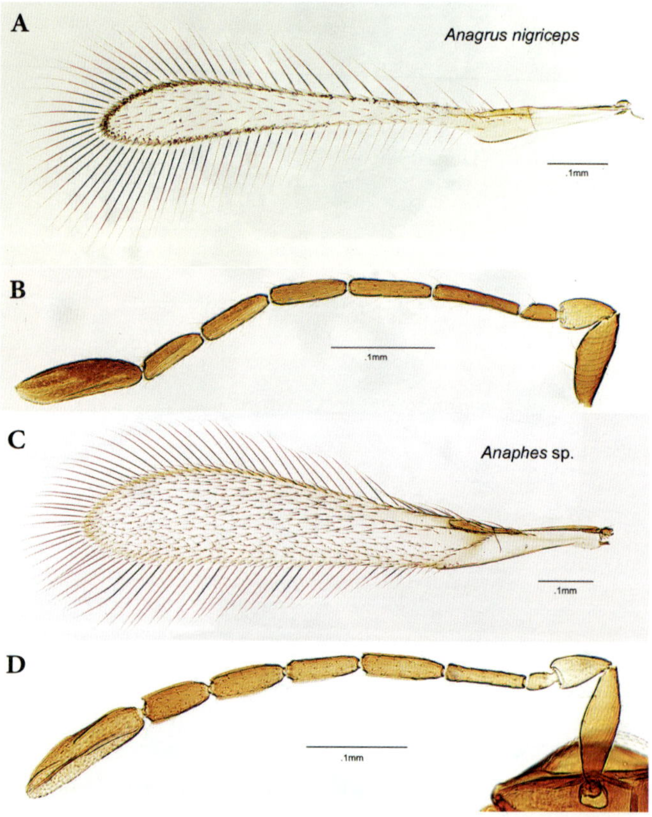

FIGURE 1 *A, B.* Anagrus nigriceps. *A, male forewing (specimen from Qagssiarssuk, 26.vi-21.vii.1984, P. Nielsen); B, female antenna (specimen from Upernaviarssuk, 9–11.viii.1982, P. Nielsen. C, D.* Anaphes *sp., female (specimen from Kussuag, 19–23.viii.1982, P. Nielsen). C, forewing; D, antenna.*

DISTRIBUTION. S (Upernaviarsuk, Qassiarsuk). Europe (Holland, possibly, England).

***Anaphes* sp. Figs 1C, D**
Henriksen 1939: not recorded.

There appears to be only one *Anaphes* species known from Greenland, where it is widespread. It belongs to the *fuscicornis* species group, recognized by the 1-segmented clava (2-segmented in the *crassicornis* group; Huber 1992), but most critical point dried specimens appear to have a faint and incomplete suture. Cleared slide mounts show no trace of a suture, however. The genus is currently under review for Europe and for North America. Once these are completed, it may be possible to identify the Greenland specimens confidently to species.

BIOLOGY. Species of *Anaphes* parasitize Coleoptera (Curculionidae, Chrysomelidae, Byrrhidae), Diptera (Ephydridae, Tipulidae) and, less frequently, Symphyta (Tenthredinidae) and Hemiptera (Miridae). Eggs of these hosts should be collected for possible host records.

DISTRIBUTION. S, SW, SE (Saqqisikuik/Skjoldungen).

***Gonatocerus*, probably *litoralis* (Haliday, 1833). Figs 2A, B**
Henriksen 1939: not recorded.

The only species found belongs to the subgenus *Lymaenon*. It keys out best to *G. litoralis* (Haliday) in Matthews (1986). The ovipositor/middle tibia ratio is about 1.0, which corresponds to the measurements provided by Mathews in his redescription (not 1.6, as stated in his key). The forewing appears to be more truncate than his illustration suggests but the forewing length/breadth ratio of 3.1–3.3 falls within the range he provided (2.75–5.24).

BIOLOGY. Species of *Gonatocerus* (*Lymaenon*) mainly parasitize Hemiptera (Cicadellidae). Matthews (1986) reported *Cicadula sexnotata* (Fallen) and *Acocephalus* sp. as hosts, but the hosts for Greenland specimens are unknown.

DISTRIBUTION. S, SW (only Saqqaq).

***Polynema* sp. Figs 2C, D**
Henriksen 1939: not recorded.

At least one species has been found in Greenland, but until the European and North American species are revised it cannot be identified confidently to species.

BIOLOGY. Species of *Polynema* mainly parasitize Hemiptera. Eggs of these hosts should be collected for the probable host species.

DISTRIBUTION. S.

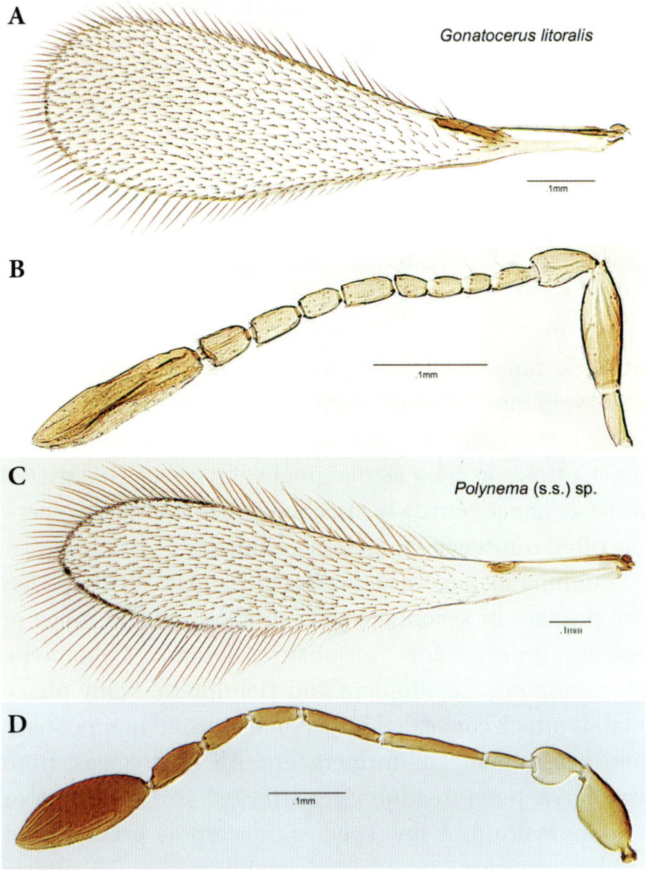

FIGURE 2 *A, B.* Gonatocerus litoralis (*specimen from Kap Farvel, Pamiagadluk, Anordliutsoq, 29.viii.1970, J. Bocher. A, forewing; B, antenna. C, D.* Polynema (s.s.) sp. (*specimen from Kussuag, 19–23.viii.1982, P. Nielsen*). *C, forewing; D, antenna.*

References

Chiappini, E., Triapitsyn, S.V. & Donev, A. (1996): Key to the Holarctic species of *Anagrus* Haliday (Hymenoptera: Mymaridae) with a review of the Nearctic and Palaearctic (other than European) species and descscriptions of new taxa. – *Journal of Natural History* 30: 551–595.

Huber, J.T. (1992): The subgenera, species groups, and synonyms of *Anaphes* (Hymenoptera: Mymaridae) with a review of the described Nearctic species of the *fuscipennis* group of *Anaphes s.s.* and the described species of *Anaphes* (*Yungaburra*). – *Proceedings of the Entomological Society of Ontario* 123: 23–110.

Matthews, M.J. (1986): The British species of *Gonatocerus* Nees (Hymenoptera: Mymaridae), egg parasitoids of Homoptera. – *Systematic Entomology* 11: 213–229.

11.8. Pteromalidae (Chalcidoidea)

Hannes Baur

Holarctic species small to moderately large, 1.1–7(–18) mm, varying from bright to dark green or black, often with strong metallic lustre; antenna with 8–13 segments, usually with 5–7 funicular segments and a 2–3 segmented clava; protibial spur relatively long, curved and bifurcate at apex; tarsi usually 5-segmented; fore wing with marginal vein most often distinctly less than 3 times as long as moderately long stigmal vein, postmarginal vein mostly 0.5–1.3 times as long as marginal vein; gaster constricted at base, hence narrowly attached to the mesosoma; petiole often transverse, sometimes elongate.

Pteromalids are very diverse in biology. Most species are primary or secondary parasitoids of a wide range of insect orders, for instance Diptera, Coleoptera, Hymenoptera, Lepidoptera and Hemiptera. Many pteromalids attack concealed hosts, such as wood borers, stem- and leaf-miners, gall-formers, etc. All host stages, from egg, larva, pupa to adults, are attacked as either ecto- or endoparasitoid. A few species develop as predators of spiders or insect eggs. Phytophagy is very rare. The pteromalids are economically less important than other chalcidoid wasps (*e.g.* Aphelinidae, Encyrtidae, Eulophidae), though a number of species are successfully used for biological pest control, *e.g.* against synanthropic flies.

Pteromalidae are distributed worldwide, comprising about 3700 species in 636 genera (Noyes 2014). The Greenland fauna has been reviewed by Baur (2005), who recognized 9 species in 8 genera and 3 subfamilies (Asaphinae, Miscogasterinae and Ptermalinae). Pteromalids are usually slightly less abundant than eulophids, but not in Greenland, where they are apparently more commonly collected.

Terminology and morphology follows Gibson (1997). Information concerning biology and general distribution originates from Baur (2005) and Noyes (2014).

Key to species of Pteromalidae

1. Notaulus complete, extending to hind margin of mesoscutum (Figs 1E, 3C, 7F); but *if* indistinct posteriorly (Figs 4D, 5D) *then* anterior margin of clypeus with 2–3 asymetric teeth (Figs 1B, 4A, 5A) . . 2
– Notaulus incomplete, not extending to hind margin of mesoscutum (Fig. 7C) and anterior margin of clypeus slightly emarginate, without asymetric teeth (Figs 6A, F, 8C) 7

2. Head with strongly arched occipital carina (Fig. 2C); gena bordered by a strong carina (Fig. 2D); pronotum collar large, subrectangular, anteriorly broadly rounded (Fig. 3C); anterior margin of clypeus truncate, without teeth (Fig. 2E); metapleuron with at least 3 but usually 10 or more setae in lower half (Fig. 3D) . . . . *Asaphes hirsutus*
– Head without occipital carina (Fig. 1A); gena without carina, rounded (Fig. 3 F); pronotum collar either without collar (Figs 5C, D, 7E, F) or collar margined anteriorly (Figs 1D, E) or with distinct edge (Figs 4C, D); anterior margin of clypeus with 2–3 teeth (Figs 1B, 4A, 5A); metapleuron bare (Figs 2A, 4C, 5C, 7E) 3

3. Pronotum sloping from hind margin, almost without differentiation of a collar (Figs 5C, D, 7E, F) 4
– Pronotum with a subhorizontal collar which is either margined anteriorly (Figs 1D, E) or at least delimited by a distinct edge (Figs 4C, D) 6

4. Stigma of fore wing large, subcircular (Fig. 5E); notaulus superficial posteriorly (Fig. 5D); petiole about 1.5 times as long as broad (Fig. 5F) . . . . *Glyphognathus laevigatus*
– Stigma of fore wing smaller, elongate (Fig. 8A); notaulus distinctly impressed throughout (Fig. 7F); petiole broader than long 5

5. Mesosoma strongly arched dorsally, in lateral view scutellum appearing distinctly convex (cf. *G. laevigatus*, Fig. 5C) . . . . *Seladerma* cf. *breve*
– Mesosoma weakly arched dorsally, in lateral view scutellum appearing almost flat (Fig. 7E) . . . . *Seladerma geniculatum*

6. Speculum of fore wing reduced to a small bare area below parastigma, separated by 2–4 rows of setae from basal and cubital fold (Fig. 1F); pronotal collar sharply margined anteriorly (Figs 1D, E); notaulus distinct throughout (Fig. 1E); petiole 1.1–1.3 times as long as broad (Fig. 2B) . . . . *Ardilea convexa*
– Speculum of fore wing large (Fig. 4E), extending to basal fold and open below; pronotal collar not margined but with a distinct edge anteriorly (Figs 4C, D); notaulus very superficial in posterior part (Fig. 4D); petiole 1.9–2.0 times as long as broad (Fig. 4F) . . . . *Collentis suecicus*

7. Marginal vein widening distally, much wider than stigma and at most slightly longer than stigmal vein

(Fig. 6D); gena slightly compressed near mouth margin (Fig. 6A, B) .... *Pachyneuron groenlandicum*
– Marginal vein not widening distally, at most as wide as stigma and much longer than stigmal vein; gena slightly rounded near mouth margin 8
8. Occiput without carina (Fig. 6E). Antenna inserted distinctly above level of lower ocular line (Fig. 7A); female antenna almost filiform, all funicular segments longer than wide (Fig. 7B) .... *Pteromalus* nr. *semotus*
– Occiput with fine carina (Fig. 8B). Antenna inserted at level of lower ocular line (Fig. 8D); female antenna distinctly clavate, distal funicular segments transverse (Fig. 8E) .... *Trichomalopsis fucicola*

ASAPHINAE

***Asaphes hirsutus* Gibson & Vikberg, 1998. Figs 2C–F, 3A–E**

Henriksen 1939: not recorded.

1.3–2.5 mm. Head with strongly arched occipital carina (Fig. 2C); gena bordered by a strong carina (Fig. 2D); anterior margin of clypeus truncate, without teeth (Fig. 2E); antenna inserted slightly below level of lower ocular line (Fig. 2F), with 2 anelli and 6 funicular segments (Fig. 3A); second anellus large, 0.8–1.1 times as long as broad and sometimes provided with 1–2 longitudinal sensilla (Fig. 3B); female flagellum very distinctly clavate, last funicular segment slightly transverse (Fig. 3A). Mesosoma arched dorsally, in profile scutellum appearing distinctly convex; pronotal collar large, subrectangular, anteriorly rounded (Fig. 3C); notaulus complete (Fig. 3C); fore wing with speculum not extending to basal fold, reduced to a narrow strip below parastigma; marginal vein slightly widened distally, 1.1–1.4 times as long as stigmal vein, stigma small; metapleuron with at least 3 setae in lower half, but most often with a batch of 10 or more setae, otherwise smooth (Fig. 3D); median area of propodeum with strong sculpture. Metasoma with petiole conspicuous, 0.9–1.3 times as long as broad, lateral margins subparallel, upper surface strongly sculptured and often with some longitudinal carinae (Fig. 3E).

BIOLOGY. Hyperparasitoid of Braconidae and Megaspilidae (Hymenoptera) with aphids (Hemiptera: Aphididae) as the primary host.

DISTRIBUTION. S, SW. Widely distributed.

Holarctic, mostly alpine.

MISCOGASTRINAE

***Ardilea convexa* (Walker, 1833). Figs 1, 2A, B**

Henriksen 1939: not recorded.

1.3–2.5 mm. Head without occipital carina (Fig. 1A); gena without carina, more or less rounded; anterior margin of clypeus with 3 strong, slightly asymmetric teeth (Fig. 1B); antenna inserted slightly above lower ocular line (Fig. 1C), with 2 anelli and 6 funicular segments; female flagellum distinctly clavate, last funicular segment almost quadrate. Mesosoma arched dorsally, in profile scutellum appearing distinctly convex (Fig. 1D); pronotal collar sharply margined (Fig. 1E); notaulus complete (Fig. 1E); fore wing with speculum reduced to a small bare area below parastigma (Fig. 1F), separated by 2–4 rows of setae from basal and cubital fold; marginal vein not widened, 1.7–2.1 times as long as stigmal vein, stigma small (Fig. 1F); metapleuron bare, reticulate (Fig. 2A). Metasoma with petiole conspicuous, 1.1–1.3 times as long as broad, lateral margins subparallel, upper surface finely reticulate (Fig. 2B).

BIOLOGY. Hyperparasitoid of an unidentified aphid species (Hemiptera: Aphididae).

DISTRIBUTION. S, SW. Widely distributed.

Northern Europe.

***Collentis suecicus* (Graham, 1969). Figs 3 F, 4**

Henriksen 1939: not recorded.

1.5–2.4 mm. Head without occipital carina; gena without carina, more or less rounded (Fig. 3 F); anterior margin of clypeus with 3 asymmetric teeth (Fig. 4A); antenna inserted slightly above lower ocular line (Fig. 4B), with 2 anelli and 6 funicular segments; female flagellum very distinctly clavate, last funicular segment transverse. Mesosoma arched dorsally, in profile scutellum appearing distinctly convex (Fig. 4C); pronotal collar not margined but delimited by a distinct edge anteriorly (Figs 4C, D); notaulus complete but very superficial in posterior third (Fig. 4D); fore wing with speculum open below, extending on upper side to beginning of marginal vein (Fig. 4E); marginal vein not widened, 1.6–1.8 times as long as stigmal vein, stigma small (Fig. 4E); metapleuron bare, reticulate. Metasoma with petiole conspicuous, about 1.9–2.0 times as long as broad, almost as long as median length of propodeum, lateral margins converging towards base, upper surface finely reticulate (Fig. 4F).

DISTRIBUTION. G. Widely distributed.

Northern and Central Europe.

***Glyphognathus laevigatus* (Delucchi, 1953). Fig. 5**

Henriksen 1939: not recorded.

1.8–2.2 mm. Head without occipital carina; gena without carina, more or less rounded; anterior margin of clypeus with 2 slightly asymetric teeth (Fig. 5A); antenna inserted distinctly above lower ocular line (Fig. 5B), with 2 anelli and 6 funicular segments; female flagellum very slightly clavate, last funicular segment slightly longer than broad. Mesosoma arched dorsally, in profile scutellum appearing strongly convex (Fig. 5C); pronotum sloping from hind margin, almost without differentiation of a collar (Figs 5C, D); notaulus complete but superficial in posterior third (Fig. 5D); fore wing with speculum closed below, extending on upper side to beginning of marginal vein; marginal vein not widened, 1.1–1.35 times as long as stigmal vein, stigma large and subcircular, separated by 1.25–1.35 times its height from postmarginal vein (Fig. 5E); metapleuron bare, feebly reticulate (Fig. 5C). Metasoma with petiole conspicuous, about 1.5 times as long as broad, lateral margins converging towards base, upper surface very feebly reticulate (Fig. 5F).

BIOLOGY. Primary parasitoid of the leaf-miner fly *Liriomyza demeijerei* Hering (Diptera: Agromyzidae).

DISTRIBUTION. S, SW (known from only few localities). Northern and Central Europe, widely distributed.

***Seladerma* cf. *breve* (Walker, 1834)**

Henriksen 1939: not recorded.

1.7–1.9 mm. Head without occipital carina; gena without carina, more or less rounded; anterior margin of clypeus with 3 slightly asymmetric teeth, middle tooth stronger than lateral teeth; antenna inserted distinctly above lower ocular line, with 2 anelli and 6 funicular segments; female flagellum slightly clavate, last funicular segment about quadrate. Mesosoma arched dorsally, in profile scutellum appearing distinctly convex; pronotum sloping from hind margin, almost without differentiation of a collar; notaulus complete and distinct throughout; fore wing with speculum closed below, extending on upper side to beginning of marginal vein; marginal vein not widened, about 1.4–1.5 times as long as stigmal vein, stigma moderately large and subrectangular, separated by about 2.2–2.3 times its height from postmarginal vein; metapleuron bare, feebly reticulate. Metasoma with petiole inconspicuous, broader than long.

BIOLOGY. *S. breve* was recorded as a primary parasitoid of the dung fly *Norellia* (*Norellisoma*) *spinimanum* (Fallén) (Diptera: Scathophagidae).

DISTRIBUTION. NE (only recorded from Mørkefjord). Europe (known from Germany and England).

***Seladerma geniculatum* (Zetterstedt, 1838). Figs 7E, F, 8A**

Henriksen 1939: not recorded

1.6–2 mm. Head without occipital carina; gena without carina, more or less rounded; anterior margin of clypeus with 3 slightly asymmetric teeth; antenna inserted distinctly above lower ocular line, with 2 anelli and 6 funicular segments; female flagellum slightly clavate, last funicular segment transverse. Mesosoma flattened dorsally, in profile scutellum appearing virtually flat (Fig. 7E); pronotum sloping from hind margin, almost without differentiation of a collar (Figs 7E, F); notaulus complete and distinct throughout (Fig. 7F); fore wing with speculum closed below, extending on upper side to beginning of marginal vein; marginal vein not widened, about 1.2–1.4 times as long as stigmal vein, stigma moderately large and subrectangular, separated by about 1.7–2.5 times its height from postmarginal vein (Fig. 8A); metapleuron bare, feebly reticulate (Fig. 7E). Metasoma with petiole inconspicuous, broader than long.

BIOLOGY. Primary parasitoid of the leaf-miner fly *Phytomyza varipes* Macquart (Diptera: Agromyzidae).

DISTRIBUTION. S (only known from Eqaluit). Northern and Central Europe, widely distributed.

PTEROMALINAE

***Pachyneuron groenlandicum* (Holmgren, 1872). Figs 6A–D**

1.5–2.2 mm. Head without occipital carina; gena without carina, but slightly compressed near mouth margin (Fig. 6B); anterior margin of clypeus slightly protruding and medially weakly emarginate (Fig. 6A); antenna inserted above lower ocular line (Fig. 6C), with 2 anelli and 6 funicular segments; female flagellum filiform, last funicular segment about quadrate. Mesosoma arched dorsally, in profile scutellum appearing distinctly convex; pronotal collar finely margined; notaulus incomplete; fore wing with speculum open below, extending on upper side to distal end of marginal vein; basal fold mostly with a complete row of setae; marginal vein conspicuously widening distally, at least 1.5 times as broad as height of small stigma and 1.0–1.1 times as long as stigmal vein (Fig. 6D); metapleuron bare, reticulate. Metasoma with petiole conspicuous, usually about 0.9–1.2 times as long as broad, lateral margins converging towards base, upper surface feebly reticulate.

BIOLOGY. Mainly a primary parasitoid of various hoverflies (Diptera: Syrphidae) and the frit fly *Oscinella frit* (Linnaeus) (Diptera: Chloropidae); also known as a hyperparasitoid of aphids and soft scales (Hemiptera: Aphididae, Coccidae).

DISTRIBUTION. S, SW, SE.

Palaearctic, widely distributed.

***Pteromalus* sp. near *semotus* (Walker, 1834). Figs 6E, F, 7A–D**

1.8–3.3 mm. Head without occipital carina (Fig. 6E); gena without carina, more or less rounded; anterior margin of clypeus shallowly and broadly emarginate (Fig. 6F); antenna inserted distinctly above lower ocular line (Fig. 7A), with 2 anelli and 6 funicular segments

(Fig. 7B); female flagellum almost filiform, last funicular segment slightly longer than broad (Fig. 7B). Mesosoma arched dorsally, in profile scutellum appearing distinctly convex; pronotal collar not margined but with a rather sharp edge anteriorly (Fig. 7C); notaulus incomplete (Fig. 7C); fore wing with speculum open below, extending on upper side to middle of marginal vein; marginal vein not widened, 1.4–1.6 times as long as stigmal vein, stigma small; metapleuron bare, reticulate. Metasoma with petiole inconspicuous, broader than long (Fig. 7D).

DISTRIBUTION. Recorded only from a few localities in western and southern Greenland.

Trichomalopsis fucicola **(Walker, 1835). Figs 8B–F**
Henriksen 1939: not recorded.

1.9–2.4 mm. Head with fine occipital carina (Fig. 8B); gena without carina, more or less rounded; anterior margin of clypeus very shallowly emarginate (Fig. 8C); antenna inserted hardly above lower ocular line (Fig. 8D), with 2 anelli and 6 funicular segments (Fig. 8E); female flagellum distinctly clavate, last funicular segment transverse (Fig. 8E). Mesosoma arched dorsally, in profile scutellum appearing distinctly convex (Fig. 8F); pronotal collar not margined but with a rounded edge; notaulus incomplete; fore wing with speculum open below, extending on upper side to middle of marginal vein; marginal vein not widened, 1.5–1.8 times as long as stigmal vein, stigma small; in males fore wings usually considerably shortened; metapleuron bare, reticulate (Fig. 8F). Metasoma with petiole inconspicuous, broader than long.

BIOLOGY. Primary pupal parasitoid of flies (Diptera: Anthomyiidae, Sepsidae, Sphaeroceridae) associated with seaweed.

DISTRIBUTION. SW (only recorded from Kapisillit).

Europe (known from Iceland, Germany, Ireland, United Kingdom).

References

Baur, H. (2005): A review of the Eulophidae and Pteromalidae (Hymenoptera: Chalcidoidea) of Greenland. – *Acta Societatis Zoologicae Bohemicae* 69: 23–34.

Gibson, G.A.P. (1997): Morphology and terminology. Pp. 16–44. *In*: Gibson, G.A.P., Huber, J.T. & Woolley, J.B. (eds): Annotated keys to the genera of Nearctic Chalcidoidea (Hymenoptera). – Ottawa: NRC Research Press. xi+794 pp.

Noyes, J.S. (2014): *Universal Chalcidoidea Database*. World Wide Web electronic publication. http://www.nhm.ac.uk/chalcidoids [accessed 2-June-2014].

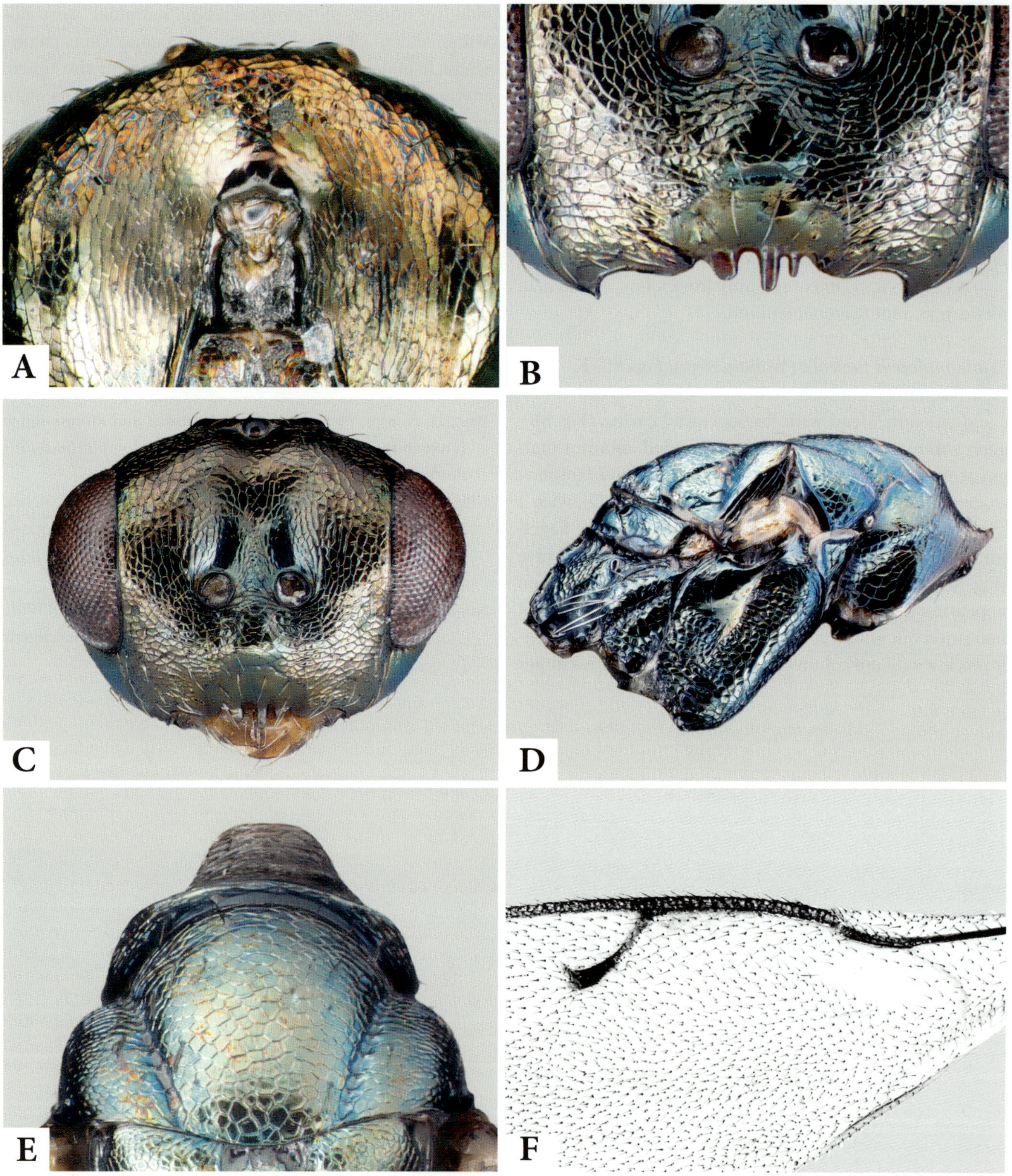

FIGURE 1 Ardilea convexa (*Walker*), *female: head, posterior view* (*A*), *clypeus* (*B*), *head, frontal view* (*C*), *mesosoma, lateral view* (*D*), *pronotum and mesoscutum, dorsal view* (*E*), *speculum, marginal and stigmal vein of forewing* (*F*).

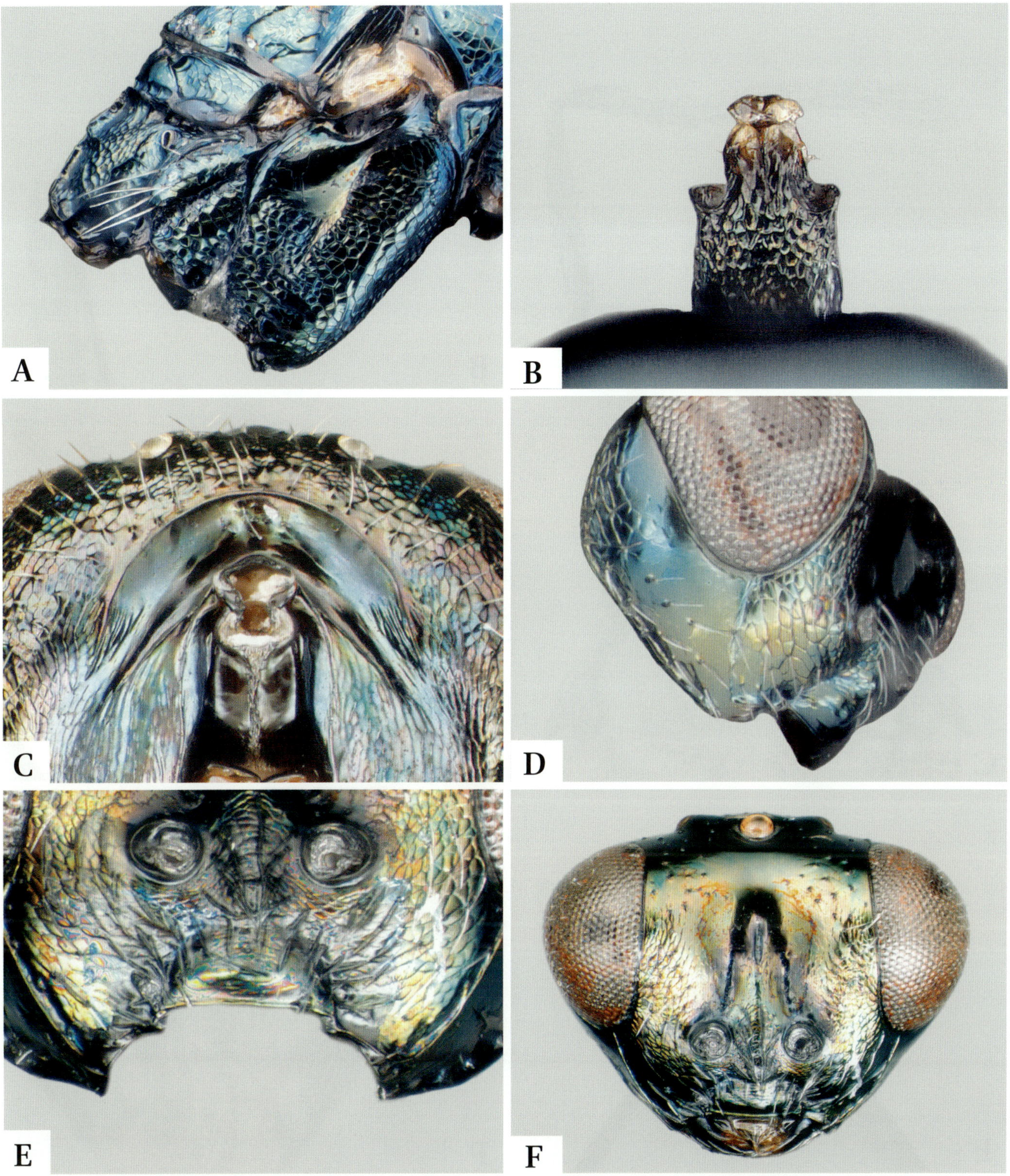

FIGURE 2 *A–B,* Ardilea convexa *(Walker), female: meso- and metapleuron (A), petiole attached to gaster, dorsal view (B). C–F,* Asaphes hirsutus *Gibson & Vikberg: head, posterior view (C), gena, lateral view and slightly from below (D), clypeus slightly from below (E), head, frontal view (F).*

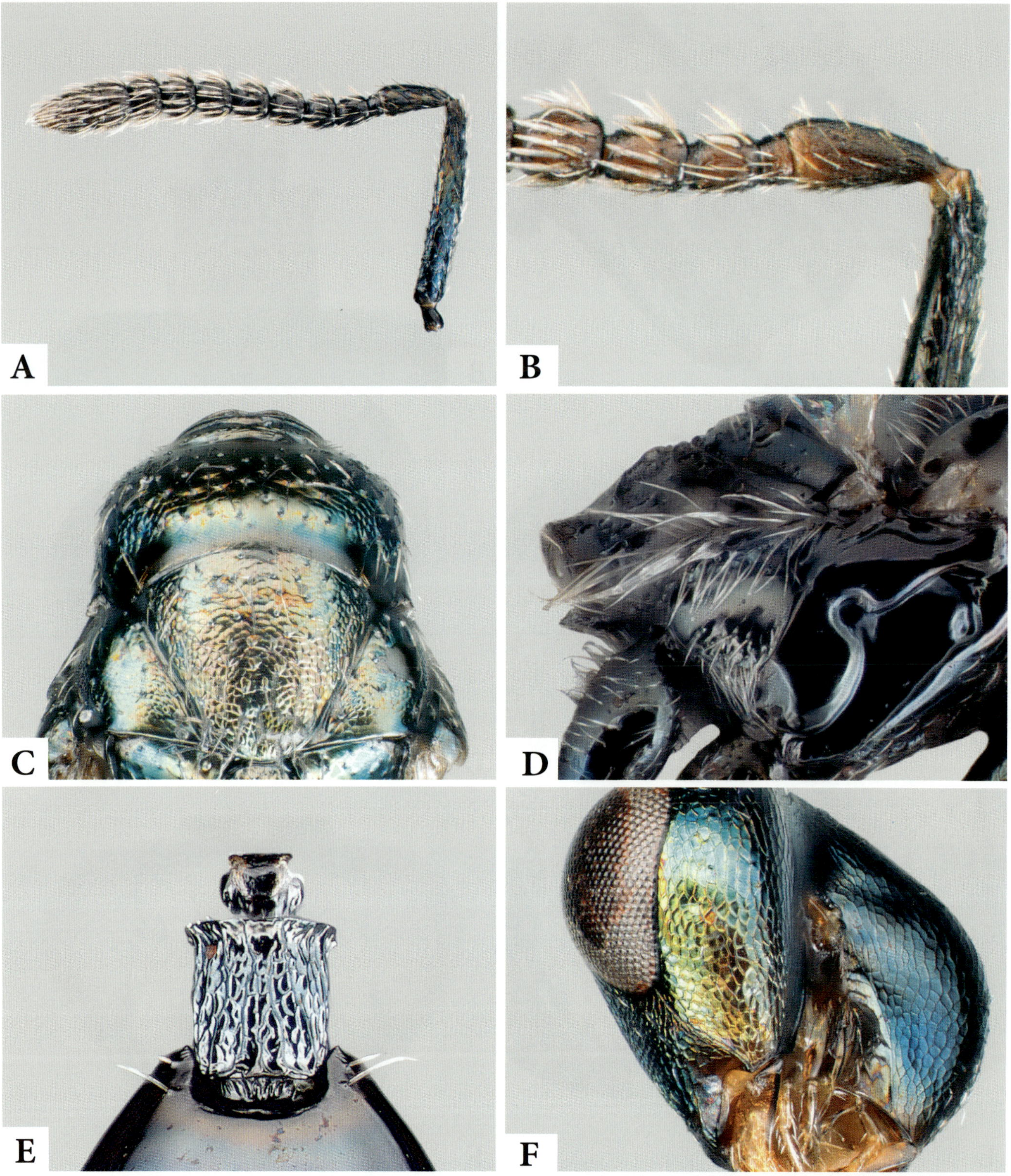

FIGURE 3 *A–E,* Asaphes hirsutus *Gibson & Vikberg, female: antenna (A), pedicel and basal flagellar segments (B), pronotum and mesoscutum, dorsal view (C), meso- and metapleuron (D), petiole attached to gaster, dorsal view. (E). F.* Collentis suecicus *(Graham), female: gena, posterior view and slightly from below.*

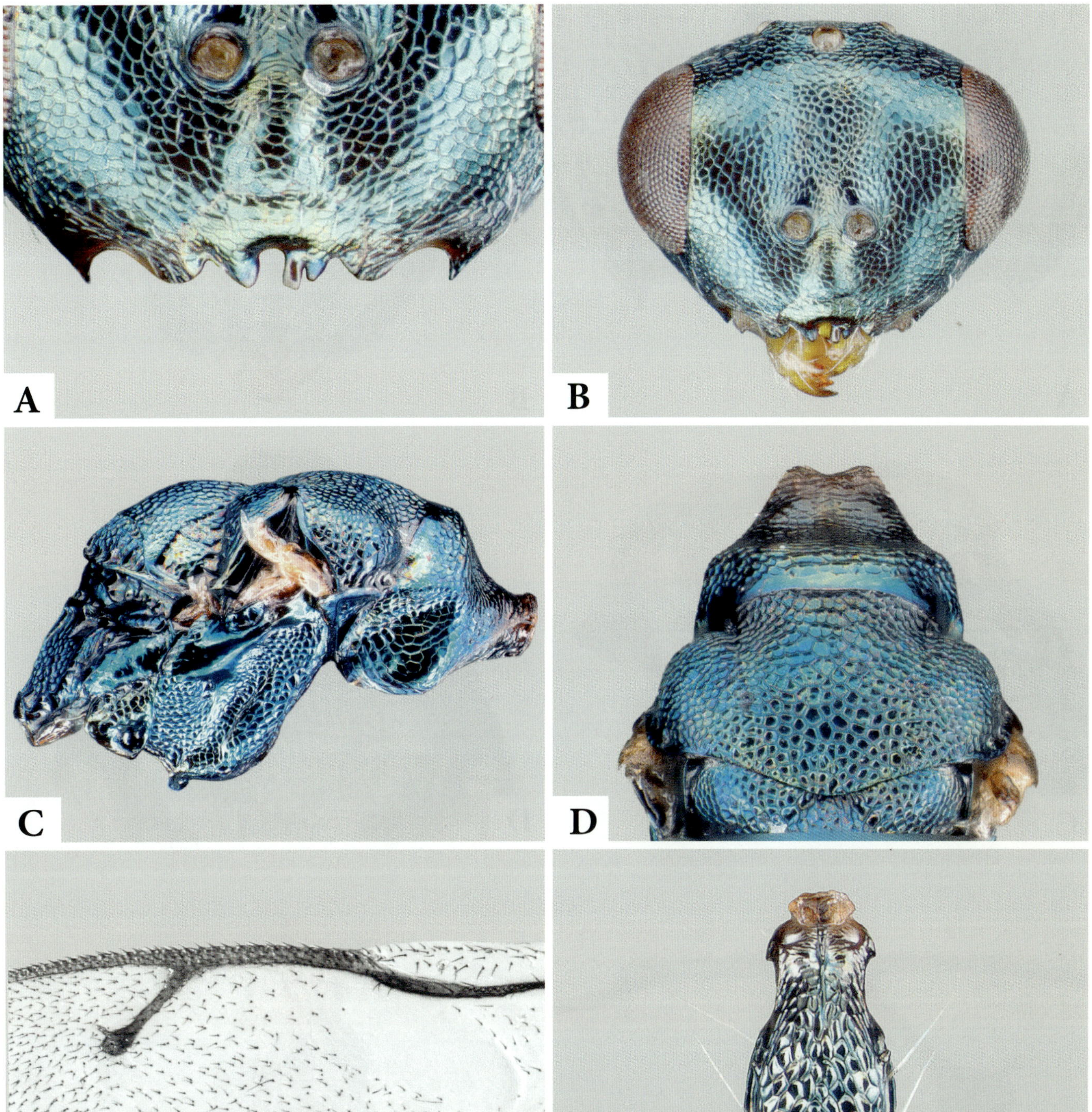

FIGURE 4 Collentis suecicus (*Graham*), *female: clypeus* (*A*), *head, frontal view* (*B*), *mesosoma, lateral view* (*C*), *pronotum and mesoscutum, dorsal view* (*D*), *speculum, marginal and stigmal vein of forewing* (*E*), *petiole attached to gaster, dorsal view* (*F*).

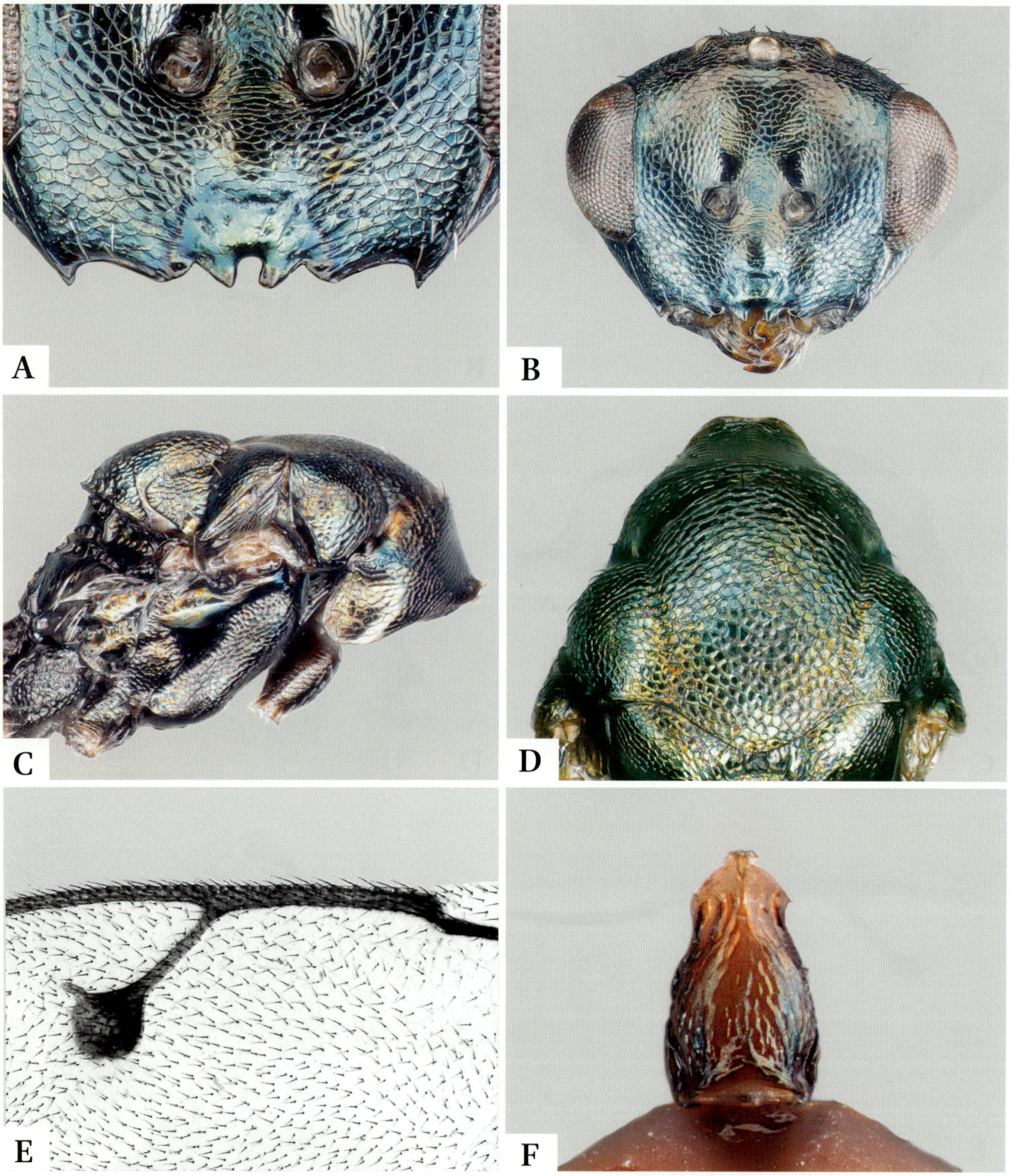

FIGURE 5 Glyphognathus laevigatus (*Delucchi*), *female: clypeus (A), head, frontal view (B), mesosoma, lateral view (C), pronotum and mesoscutum, dorsal view (D), marginal and stigmal vein of forewing (E), petiole attached to gaster, dorsal view (F).*

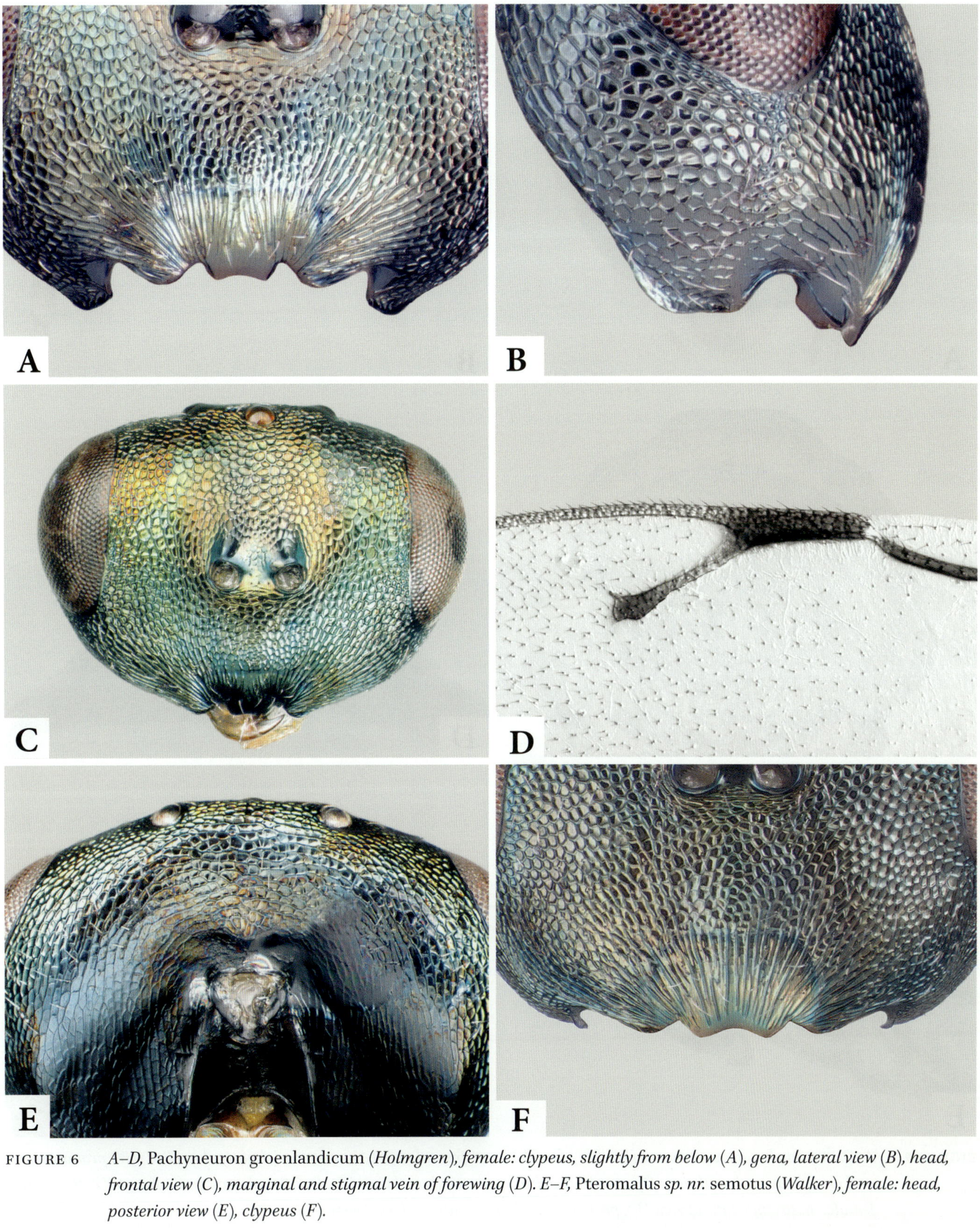

FIGURE 6 *A–D,* Pachyneuron groenlandicum *(Holmgren), female: clypeus, slightly from below (A), gena, lateral view (B), head, frontal view (C), marginal and stigmal vein of forewing (D). E–F,* Pteromalus *sp. nr.* semotus *(Walker), female: head, posterior view (E), clypeus (F).*

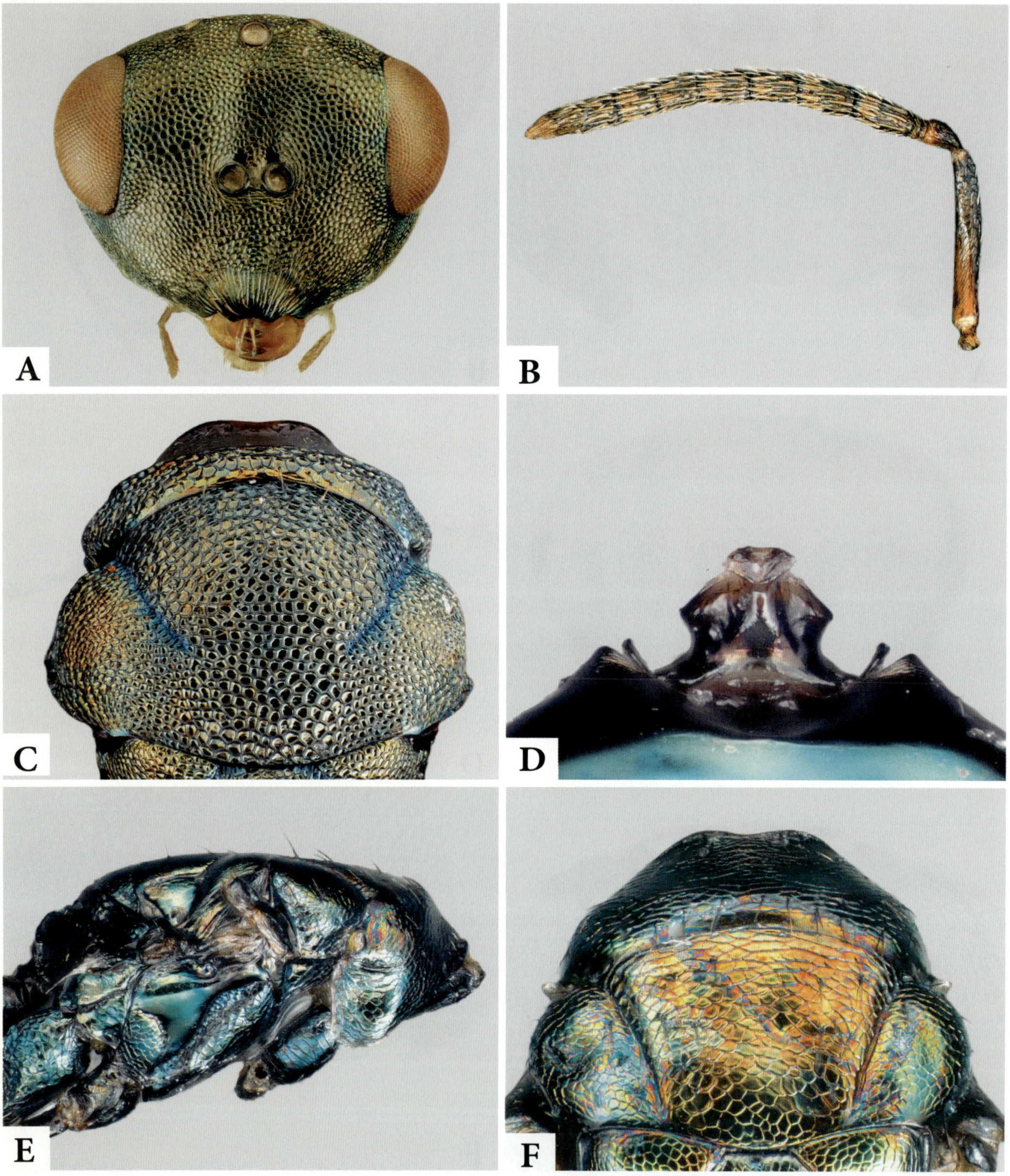

FIGURE 7 *A–D,* Pteromalus *sp. nr.* semotus *(Walker), female: head, frontal view (A), antenna (slightly shrivelled) (B), pronotum and mesoscutum, dorsal view (C), petiole attached to gaster, dorsal view (D). E–F,* Seladerma geniculatum *(Zetterstedt), female: mesosoma, lateral view (E), pronotum and mesoscutum, dorsal view (F).*

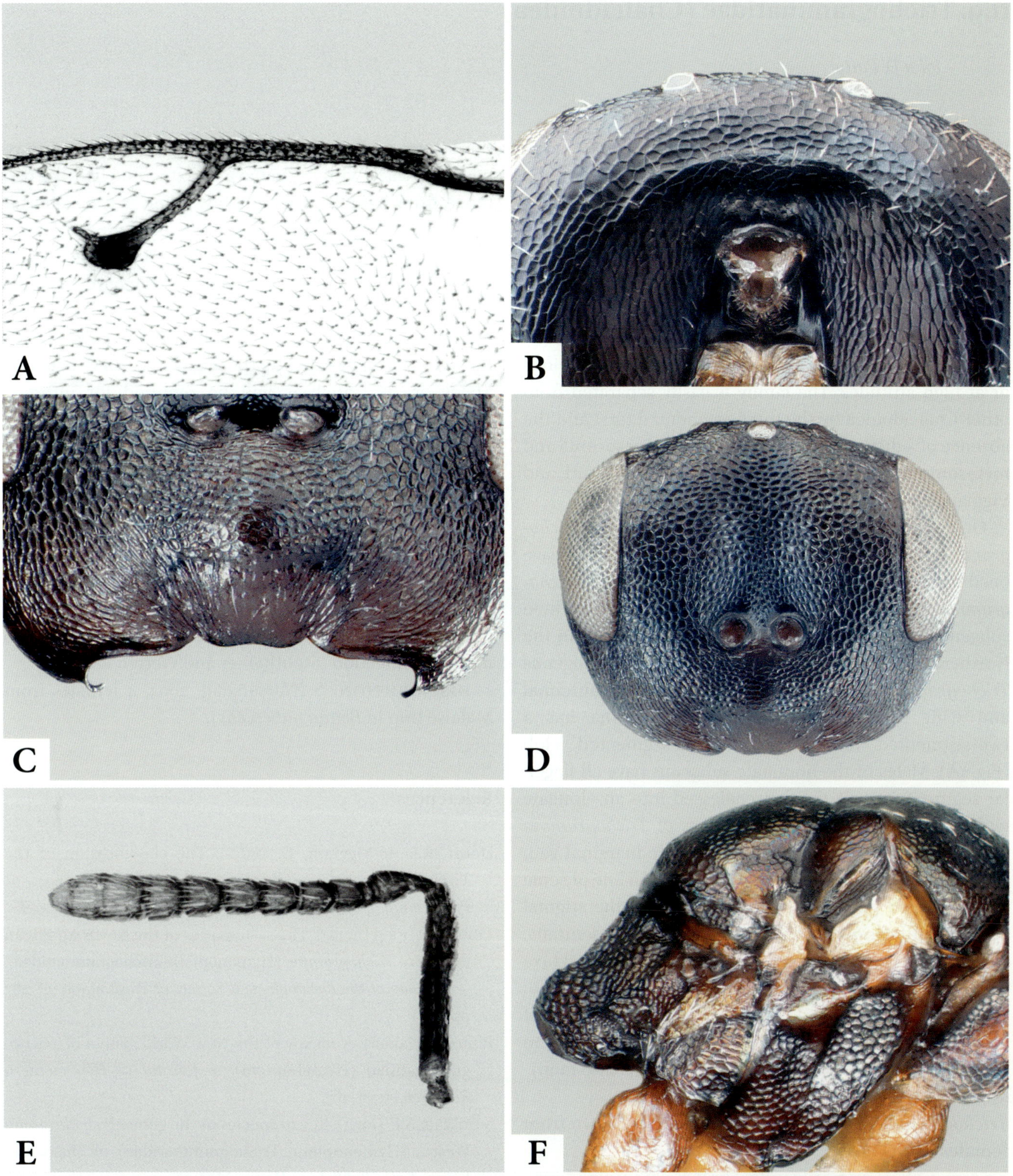

FIGURE 8 *A,* Seladerma geniculatum *(Zetterstedt), female: marginal and stigmal vein of forewing. B–F,* Trichomalopsis fucicola *(Walker), female: head, posterior view (B), clypeus, slightly from below (C), head, frontal view (D), antenna (E), meso- and metapleuron (F).*

11.9. Trichogrammatidae (Chalcidoidea)

John D. Pinto

Wasps belonging to the Trichogrammatidae are among the smallest of insects, ranging in size from 0.2–1.5 mm. Virtually all species are insect egg parasitoids. They attack the eggs of a large number of insect groups, but are most commonly associated with Lepidoptera, Hemiptera and Coleoptera (Pinto 2006). The family is cosmopolitan with over 700 species assigned to 88 genera. Although widespread, trichogrammatids are most diverse in arid, semiarid and tropical areas. Few genera occur in boreal regions or at high altitudes. The family is easily separated from other Chalcidoidea by the three-segmented tarsi and the absence of a distinct constriction between mesosoma and metasoma. Keys to genera include those by Doutt and Viggiani (1968, world) and Pinto (2006, New World).

Only one species, a member of the genus *Trichogramma*, has been found in Greenland. Only females have been collected. With *c.* 230 species described, *Trichogramma* is the most speciose genus in the family. Three subgenera are recognized (Pinto 1999). All species in the Northern Hemisphere belong to the nominate subgenus *Trichogramma*. The genus is easily identified by antennal and wing structure. In females the antenna has a two-segmented funicle and a one-segmented club (Fig. 1A). Males of the nominate subgenus have all flagellar segments except the first coalesced into an elongate club segment (Fig. 1B). The forewings of *Trichogramma* (Fig. 1C) have a sinuate venation with the marginal vein gradually curving onto the stigmal vein, and a line of setae (RS1) directed posteriorly from the apex of the stigmal vein. Identification to species requires male genitalia, although DNA sequences can be used to verify several taxa if specimens are properly stored after collection (Stouthamer *et al.* 1999). A key to the world species of *Trichogramma* was published by Sorokina (1993); a key to the North American species also is available (Pinto 1999).

Trichogramma sp. (unidentified to species; known from females only; Fig. 1)

0.5 mm. Colour primarily yellow brown with propodeum, extreme base of metasoma, antennae and legs beyond coxae lighter, yellowish; wings hyaline with base of forewings slightly fumose. Antenna (Fig. 1A) with funicle segments as long as wide. Forewing (Fig. 1C) *c.* half as wide as long, fringe setae at posteroapical margin *c.* 0.15 times maximum wing width; membrane moderately densely setose, with distinct linear setal tracks; RS1 track distinct, composed of 5 setae in material examined. Hindwing with three setal tracks; middle track complete to wing apex, posterior track with *c.* 7 setae and extending half the distance from hamuli to wing apex, anterior track shorter than posterior track. Scutellum with anterior pair of setae minute, much shorter than posterior pair. Ovipositor not extending beyond apex of metasoma, its length subequal to that of hind tibia.

NOTE. Females of several *Trichogramma* species have the same characteristics as those collected in Greenland. Males or DNA analysis will be required for ultimate identification.

BIOLOGY. The host of the Greenland species is unknown. *Trichogramma* parasitizes eggs of at least seven orders of insects. Lepidoptera eggs are the most frequent hosts; certain species in boreal areas are also known from eggs of sawflies (Hymenoptera) (Pinto 1999). *Trichogramma* species generally are bisexual. However, at least one common Holarctic species, *T. cacaeciae* Marchal, is genetically unisexual. Others may become unisexual when subject to microbial infection (Stouthamer and Werren 1993).

DISTRIBUTION. S (Narsarsuaq, 2004, 4 females from Malaise trap in *Betula pubescens*).

References

Doutt, R.L. & Viggiani, G. (1968): The classification of the Trichogrammatidae (Hymenoptera: Chalcidoidea). – *Proceedings of the California Academy of Sciences* 35: 477–586.

Pinto, J.D. (1999 (1998)): The systematics of the North American species of *Trichogramma* (Hymenoptera: Trichogrammatidae). – *Memoirs of the Entomological Society of Washington* 22. 287 pp.

Pinto, J.D. (2006): A review of the New World genera of Trichogrammatidae (Hymenoptera). – *Journal of Hymenoptera Research* 15: 38–163.

Sorokina, A.P. (1993): Key to species of the genus *Trichogramma* Westw. (Hymenoptera, Trichogrammatidae) of the world fauna- *Kolos Publishing House*, Moscow, 77 pp. [In Russian.]

Stouthamer, R. & Werren, J.H. (1993): Microbes associated with parthenogenesis in wasps of the genus *Trichogramma*. – *Journal of Invertebrate Pathology* 61: 6–9.

Stouthamer, R., Hu, J., van Kan, F., Platner, G.R. & Pinto, J.D. (1999): The utility of internally transcribed spacer 2 DNA sequences of the nuclear ribosomal gene for distinguishing sibling species of *Trichogramma* (Hymenoptera: Trichogrammatidae). – *BioControl* 43: 421–440.

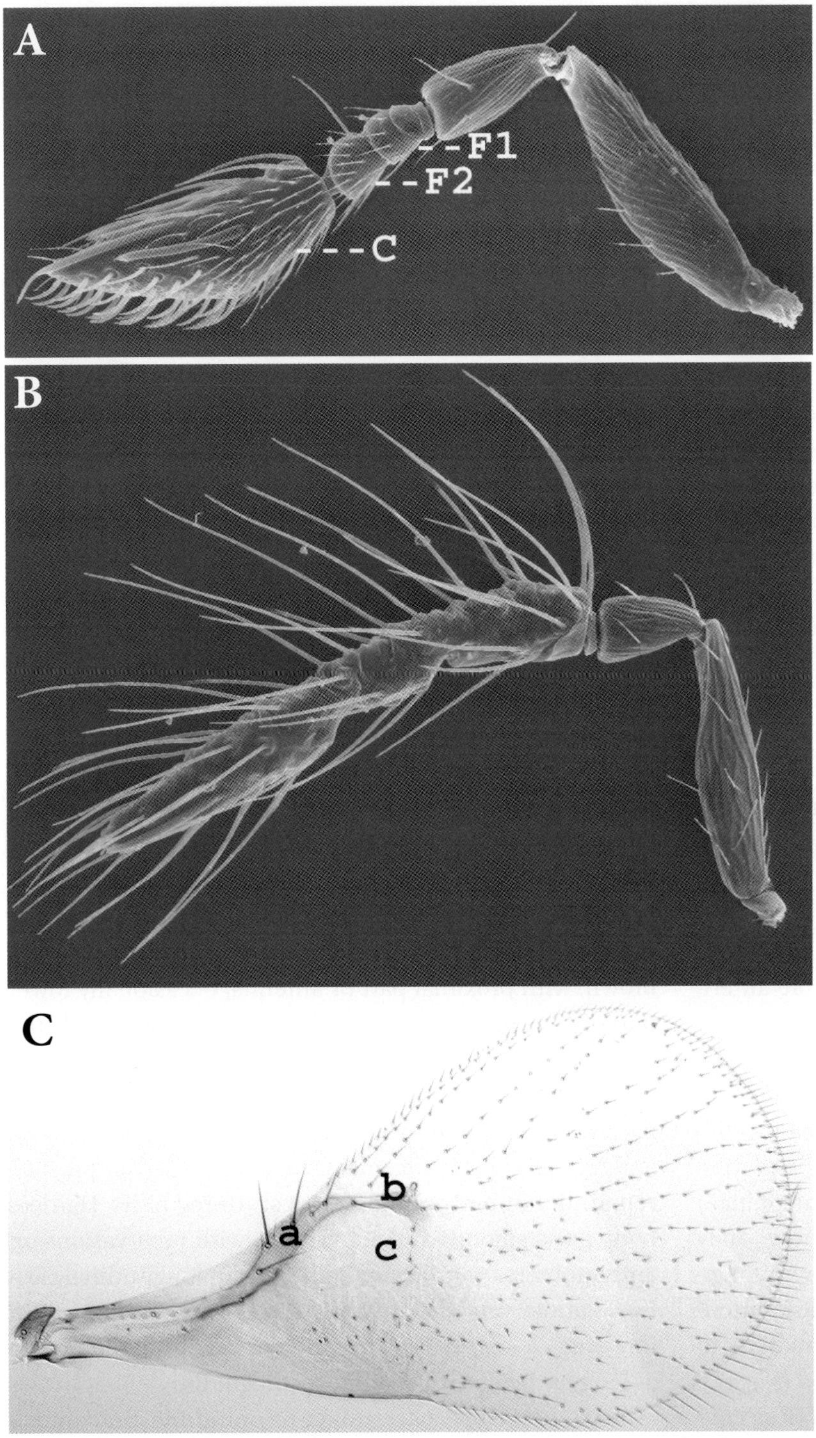

FIGURE 1 A–C, Trichogramma (Trichogramma): *female antenna (A; F1, F2 = funicular segments; C = club), male antenna (B), forewing (C; '**a**' at marginal vein, '**b**' at stigmal vein, '**c**' to left of RS1 setal track). Species illustrated is* T. pretiosum *Riley, a common New World species. The female antenna and forewing illustrated (A and C) are virtually identical to those of the Greenland species. The male antenna (B) is intended for subgeneric characterization only (male unknown for Greenland species).*

11.10. Figitidae (Cynipoidea)

Lars Vilhelmsen and Mattias Forshage

Thorax compact, mesoscutellum extending posteriorly of vertically oriented metanotum; fore wing vein Rs+M proximally issuing from posterior part of basal vein; female T9 subdivided by flexion points opposite proximal ends of 3rd valvulae.

The Figitidae (formerly subdivided into the families Charipidae, Eucoilidae, and Figitidae s. str., see Ronquist 1995) is the largest family (approx. 1500 species; Ronquist 1999) within the Cynipoidea and the only one of the superfamily represented in Greenland. All members of the family are parasitoids, unlike its putative sistergroup, the Cynipidae (the true gallwasps). The diversity of the Greenland fauna is low, only four genera and very few species recorded so far.

Key to genera of Figitidae in Greenland

1. Metasoma with distinct petiole and gaster, petiole longer than hind coxa (Fig. 1A); metasoma without hairy ring surrounding petiole posteriorly .... *Anacharis*

– Metasoma with petiole inconspicuous, much shorter than hind coxa; hairy ring surrounds petiole anteriorly on metasoma (Fig. 1B)..........................2

2. Scutellum without median raised area, evenly rounded in lateral view (cf. Fig. 1C) .... *Alloxysta*

– Scutellum with distinct raised median area (scutellar plate), in lateral view with distinct projection posterodorsally (Fig. 1D) ....3 (Eucoilinae)

3. Median raised area on scutellum narrow, only slightly expanded posteriorly; scutellum with longitudinal ridges (Fig. 1D). In female, distalmost antennomeres (usually three) swollen compared to remaining; male third antennomere curved (Fig. 1E). Forewing lateral margin with small notch (less conspicuous in male, wing appears truncated laterally) (Fig. 1F) .... *Kleidotoma*

– Median raised area on scutellum expanded posteriorly, tear-shaped; scutellum with coarse punctures, but no ridges. Distal antennomeres not conspicuously swollen in females; male third antennomere short and straight. Forewing lateral margin rounded................................*Trybliographa*

***Alloxysta arcuata* Kieffer, 1902**
Henriksen 1939: not recorded.

1–1.2 mm. Dark brown, with legs and proximal parts of antennae paler. Propodeum and anterolateral parts of metasoma with dense pilosity, scutellum and pronotum with rather dense setae, other parts of mesosoma as well as head with sparse setae. Smooth and shiny. Male with second antennomere slightly curved. Propodeum with longitudinal carinae. Radial cell in forewing closed, rather short and deep (not more than twice as long as deep) with distinctly curved Rs.

BIOLOGY: Hyperparasitoids of aphidiine Braconidae, probably also Aphelinidae, with aphids as the primary host. Eggs are laid in the larva of the primary parasitoid within the aphid host, but do not hatch until the aphid has been completely devoured. Pupation takes place in the cocoon spun by the primary parasitoid.

DISTRIBUTION: SW (Kangerlussuaq).

Almost cosmopolitan (Europe, South America, Africa, Asia).

***Alloxysta victrix* (Westwood, 1833)**
Henriksen 1939: not recorded.

1.1–1.5 mm. Variable in color, often predominantly dark brown, with proximal part of antenna, occasionally lower part of face, mouthparts, distal parts of legs (except distalmost tarsomeres) paler, sometimes entire head and legs yellow. Pronotum, dorsal part of propleuron, anterodorsal part of mesopleuron, metapleuron, propodeum, and anterolateral parts of metasoma with dense pilosity, remainder of body at most with scattered hairs. Hairless body parts smooth and shiny. Male with incurvations on antennomeres 3–5. Propodeum without longitudinal carinae. Petiole very short, hardly discernible. Radial cell in forewing closed, rather long and narrow, more than twice as long as deep, with an almost straight Rs.

BIOLOGY. Hyperparasitoids of aphidiine Braconidae with aphids as the primary host. Eggs are laid in the larva of the primary parasitoid within the aphid host, but does not hatch until the aphid has been completely devoured. Pupation takes place in the cocoon spun by the primary parasitoid. A wide host range has been recorded for this common species, with several genera of Aphidinae and Macrosiphinae as secondary hosts, and several genera of Braconidae Aphidiinae and Aphelinidae as primary hosts (Ferrer-Suay *et al.* 2012).

DISTRIBUTION. S, SW, NW (widely distributed), NE (Zackenberg).

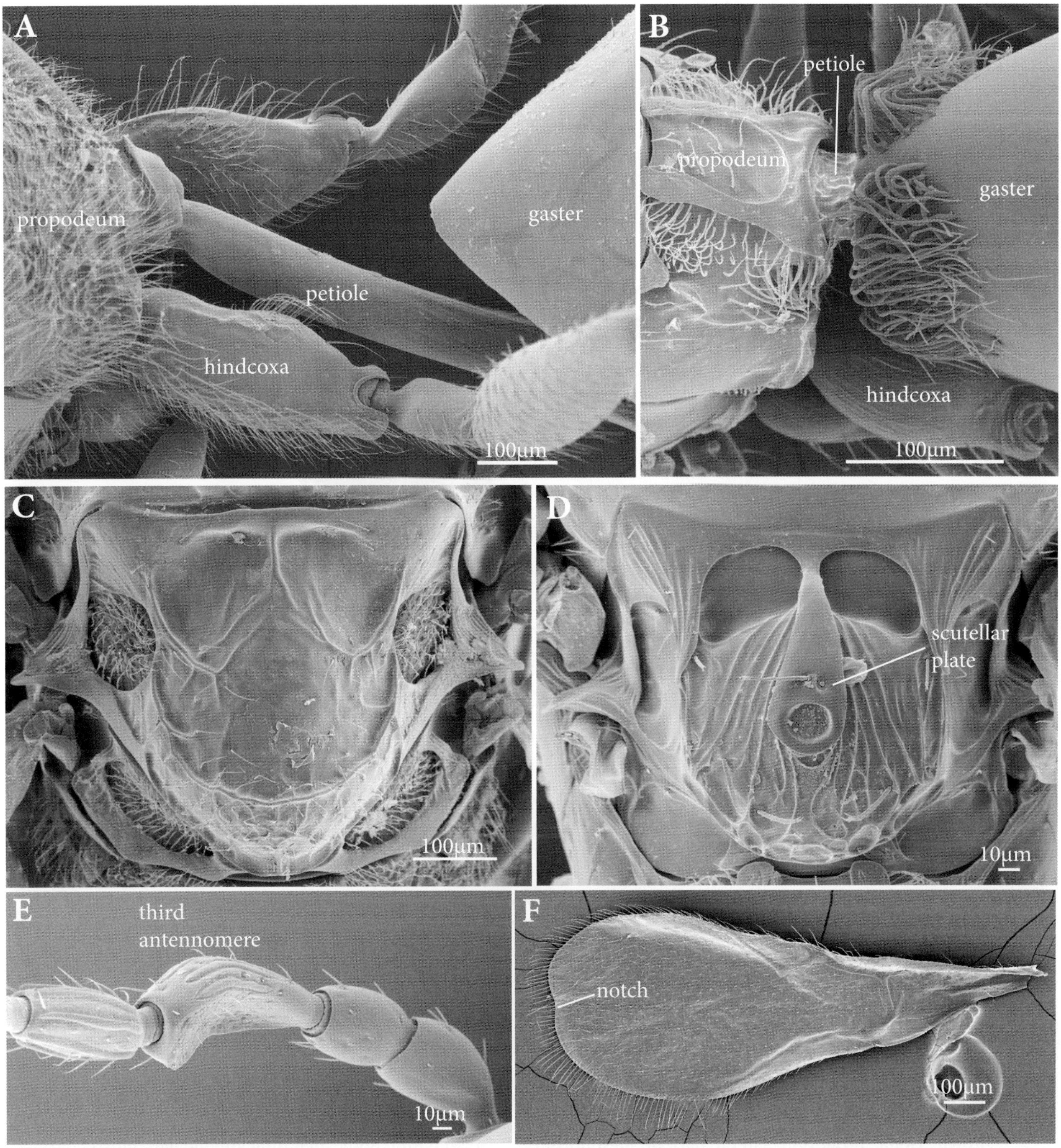

FIGURE 1 *A–B. Mesosoma-metasoma connection of A.* Anacharis eucharioides, *B.* Kleidotoma psiloides. *C–D. Mesoscutellum of C.* A. eucharioides, *D.* K. psiloides. *E. Proximal antennomeres of male* K. psiloides. *F. Forewing of female* K. psiloides.

FIGURE 2 *Habitus images of A.* Anacharis eucharioides, *B.* Alloxysta *sp., C.* Kleidotoma psiloides. *Not to scale.*

Almost cosmopolitan (Europe, North America, South America, Australia).

NOTE. Several specimens of *Alloxysta spp.* (Fig. 2B) from Greenland are not yet identified to the species level and could represent additional species.

Anacharis eucharioides **(Dalman, 1818). Figs 1A, C, 2A**
3.0–3.5mm. Body, including antenna and coxae, dark brown to black, mouthparts and remainder of legs light brown. Gena, pronotum, lateral part of mesonotum, dorsal part of mesopleuron, metapleuron, and propodeum with dense pilosity, remainder of body at most with only scattered hairs. Petiole conspicuously elongate (Fig. 1A).

BIOLOGY. Parasitoid of larvae of Neuroptera: Hemerobiidae of the genera *Hemerobius* and *Wesmaelius* (Fergusson 1986). The larva develops internally until the host spins a cocoon; the parasitoid larva then emerges from the host's body and finishes feeding externally. It then pupates inside the host's cocoon.

DISTRIBUTION. S, SW, NW (widely distributed). Palaearctic (northern and central Europe, Siberia).

Kleidotoma psiloides **Westwood, 1833. Figs 1B, D–F, 2C**
Henriksen 1939: not recorded.

1.2–1.6 mm. Body, including appendages, mostly dark brown to black, legs getting paler distally, metasoma sometimes lighter brown than mesosoma. Female antenna with three distalmost antennomeres distinctly swollen; male antenna with third segment distinctly curved (Fig. 1E). Pronotum without conspicuous fringe of hairs laterally; hair fringe surrounding metasomal base posteriorly present (Fig. 1B). Median raised area on mesoscutellum much narrower than long (Fig. 1D). Forewing with distinct notch laterally, radial cell triangular and wide open (Fig. 1F).

BIOLOGY. Parasitoid of Drosophilidae and Sphaeroceridae (Diptera).

DISTRIBUTION. S, SW, SE, W.

Northwestern Europe.

NOTE. Some specimens of *Kleidotoma spp.* from Greenland are still unidentified to the species level, and represent additional species, at least one in the difficult group with female antennae with four distinctly swollen antennomeres.

***Trybliographa simulatrix* (Ruthe, 1859)**

Henriksen 1939: not recorded.

1.7–2.5 mm. Body, antennal base, and basal parts of legs dark brown to black, mouthparts, distal parts of antennae and legs light brown. Female antenna not distinctly swollen distally, male antenna with third antennomere not curved. Pronotum anterolaterally with narrow fringe of hairs; fringe also present around posterior part of petiole. Median raised area on mesoscutellum broad, tear-shaped. Fore wing rounded laterally, without notch in lateral margin.

BIOLOGY. Parasitoid of Anthomyiidae (Forshage & Nordlander 2008).

DISTRIBUTION. S, SW (only Kapisillit). Apparently Holarctic (northern and Central Europe, North America including Alaska).

References

Fergusson, N.D.M. (1986): Charipidae, Ibaliidae & Figitidae. – *Handbooks for the identification of British insects* 8 (1c): 55 pp.

Ferrer-Suay, M., Paretas-Martínez, J., Selfa, J. & Pujade-Villar, J. (2012): Taxonomic and synonymic catalogue of the Charipinae and notes about this subfamily (Hymenoptera: Cynipoidea: Figitidae). – *Zootaxa* 3376: 1–92.

Forshage, M. & Nordlander, G. (2008): Identification key to European genera of Eucoilinae (Hymenoptera, Cynipoidea, Figitidae). – *Insect Systematics & Evolution* 39: 341–359.

Ronquist, F. (1995): Phylogeny and early evolution of the Cynipoidea (Hymenoptera). – *Systematic Entomology* 20: 309–335.

Ronquist, F. (1999): Phylogeny, classification and evolution of the Cynipoidea. – *Zoologica Scripta* 28: 139–164.

11.11. Platygastridae (Platygastroidea)

Peter N. Buhl

Small (0.3–7.5 mm), mostly dark, unmetallic species with venation reduced, in Platygastrinae venation nearly or entirely absent. Antenna with 6–12 segments, composed of long scape, a pedicel and a flagellum most often ending in a club (females) or filiform (males). Metasoma laterally sharply angled, except in Telenominae. 2nd metasomal tergum often by far the longest in Platygastrinae and Telenominae, not the longest in Scelioninae and Teleasinae.

All species of the subfamily Platygastrinae with the large cosmopolitan genera *Platygaster*, *Synopeas* and *Leptacis* are presumed to be egg-larval parasitoids of cecidomyiids (Diptera). All members of Scelioninae, Teleasinae and Telenominae are egg parasitoids of spiders or many insect orders. Almost 5.000 described species known worldwide (only a small fraction of the total).

Key to Greenland species of Platygastridae

1. Body length distinctly less than 1 mm, body very compact; antenna of male 11–segmented, in female 7–segmented (Fig. 2B) .... *Baeus seminulum*
– If length slightly less than 1 mm then body more elongate; male antenna with 10 or 12 segments, in female with 10, 11 or 12 segments.................... 2
2. Antennae of both males and females 10–segmented 3
– Antennae of males with 12, of females with 11 or 12 segments 4
3. Fore wing with a distinct submarginal vein ending in a knob; tergum 1 of female with a long horn above mesosoma.................... *Inostemma* cf. *walkeri*
– Fore wing entirely without veins; tergum 1 of female without a horn .... *Platygaster opaca*
4. Metasomal tergum 2 much longer than other terga; metasoma not sharply margined laterally; fore wing with stigmalis longer than marginalis (Fig. 1G); antenna of male 12–segmented, of female 11–segmented 5
– Metasomal tergum 3 much longer than tergum 2; metasoma sharply margined laterally; fore wing (when developed) with stigmalis shorter than marginalis; antenna of males and females with 12 segments... 6
5. Head at most twice as wide as long; antennal segment 3 of female 1.5 times as long as wide; female flagellar segments virtually bare; male antenna with only a slight difference between segments 3–5 and preapical antennal segments, all segments slightly longer than wide .... *Telenomus kolbei*
– Head at least 2.5 times as wide as long; antennal segment 3 of female about twice as long as wide; female flagellar segments distinctly pubescent; male antenna with contrast between segments 3–5 which are about 1.75 times as long as wide and preapical antennal segments which are not longer than wide . ... *Telenomus nitidulus*
6. Antennal segment 3 of female almost twice as long as wide; male with antennal segments 3–12 about 3 times as long as wide; tergum 3 dull, distinctly sculptured; body length about 1.5 mm .... *Trimorus punctulator*
– Antennal segment 3 of female hardly longer than wide; male with antennal segments 3–12 1.5–2 times as long as wide; tergum 3 almost smooth; body length about 1.0 mm .... *Trimorus orbiculatus*

Inostemma cf. _walkeri_ Kieffer, 1914
Henriksen 1939: not recorded.

1.1–2.2 mm. A single female collected by Thorkild Munk at Narsarsuaq in July 2009 could not be determined with certainty, but in the available keys to this very characteristic genus it comes closest to *I. walkeri*.

I. walkeri has been recorded from the British Isles and Denmark.

Platygaster opaca Ruthe, 1859
Henriksen 1939: not recorded.

1.5–2.5 mm. Head (Fig. 1A) broadly rounded behind eyes, evenly reticulate-coriaceous, occiput without striae; preapical antennal segments of female about as long as wide; mesoscutum sculptured as head, notauli complete; 2nd tergum striated to about three-fourths of its length.

NOTE. I doubt if this species is distinct from NW European *P. otanes* Walker, 1835, cf. Vlug (1985) (and *P. opaca* compared with Danish specimens of *P. otanes*).

DISTRIBUTION. S, SW (only recorded from Saqqaq). Also recorded from Iceland.

Telenomus kolbei Mayr, 1879. Fig. 2D
Henriksen 1939: not recorded.

0.7–0.8 mm. Head (Figs 1B–C, 2D) at most twice as wide as long, without hyperoccipital carina; female with 2nd antennal segment 1.4 times as long as 3rd segment, 3rd

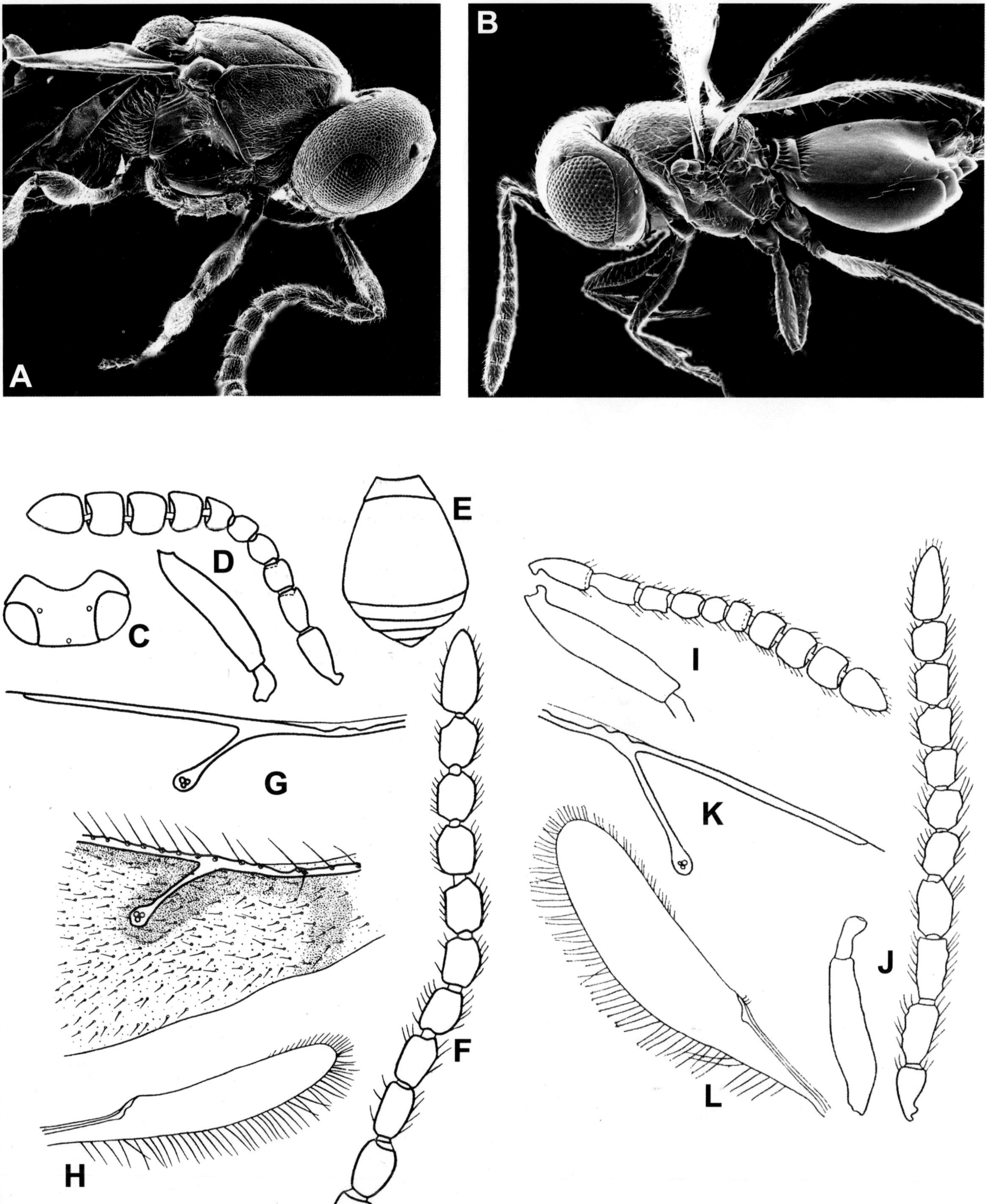

FIGURE 1 *A.* Platygaster opaca, *female head and mesosoma in lateral view. B–H.* Telenomus kolbei. *B. Female in lateral view. C. Head in dorsal view. D. Female antenna. E. Female metasoma in dorsal view. F. Male antenna. Fig. G. Forewing venation and basal part of forewing. H. Hindwing. I–L.* Telenomus nitidulus *I. Female antenna. J. Male antenna. K. Forewing venation. L. Hindwing.* (*C–L from* Kozlov & Kononova 1983).

FIGURE 2 *A–B.* Baeus seminulum. *A. female habitus and B. antenna. C.* Trimorus punctulator, *habitus. D.* Telenomus kolbei *habitus. Not to scale.*

segment 1.5 times as long as wide, and 1.5 times as long as following segment (Fig. 1D); mesoscutum without notauli, scutellum smooth; second metasomal tergum not longer than wide (Fig. 1E), at base with very short striation; legs brown. Male with antennal segment 4 slightly longer than segments 3 and 5, twice as long as wide; preapical segments 1.3 times as long as wide (Fig. 1F). Fore- and hindwing (Figs 1G, H).

NOTE. Males from Greenland seem to have the antennae slightly more slender than in European specimens, cf. Kozlov & Kononova (1983).

BIOLOGY. Probably parasitoid of Lepidoptera eggs.

DISTRIBUTION. S, SW (north to Qeqertarsuup Tunua/ Disko Bugt).

Europe (Denmark, Austria, Crimea).

***Telenomus nitidulus* (Thomson, 1860)**

Henriksen 1939: not recorded.

0.8–1.0 mm. Head 2.5–3 times as wide as long, without hyperoccipital carina; female with 2nd antennal segment at most 1.2 times as long as 3rd segment which is 1.8 times as long as following segment (Fig. 1I); mesoscutum without notauli, scutellum smooth with very few punctures; second tergum wider than long, at base with weak striation which is hardly as long as first tergum; legs brown. Male with antennal segment 2 about 0.75 times as long as segment 3 (Fig. 1J). Fore- and hindwing (Figs 1K, L).

BIOLOGY. Known as a parasitoid of eggs of several species of Lepidoptera, viz. *Leucoma salicis* (L.), *Agrotis segetum* Schiff., and some species of *Bombyx* and *Smerinthus*.

DISTRIBUTION. S (only known from Narsarsuaq, 1 female and 1 male). First record from Greenland (Malaise trap and swept, together with *T. kolbei*).

Palaearctic, low arctic-southern temperate (Iceland, western Europe, Moldavia, Armenia, and Altai territory).

Baeus seminulum **Haliday, 1833. Fig. 2A, B**

Henriksen 1939: not recorded.

Female: 0.3–0.6 mm. Antenna with large 1–segmented club (Fig. 2B); wings absent; scutellum shaped as narrow strip (Fig. 2A). Male: Fore wings long but narrow, with distinct basal vein and with marginal cilia only slightly shorter than width of wing.

BIOLOGY. In Denmark reared from spider eggs (*Theridion* sp.; Kryger 1910).

DISTRIBUTION. S, SW (Qeqertasussuk in Qeqertarsuup Tunua/Disko Bugt).

Europe, low arctic-temperate (including Iceland, Moldavia, Russia).

Trimorus orbiculatus **(Thomson, 1859)**

Henriksen 1939: not recorded.

0.8–1.1 mm. Head almost smooth, tooth of postscutellum very small, wings of female reaching end of propodeum or middle of metasoma, 3rd metasomal tergum without furrows; female metasoma almost circular, less elongate than in *T. punctulator*. Male most often fully winged.

BIOLOGY. As most members of subfamily Teleasinae probably a parasitoid on eggs of some beetle (Gauld & Bolton 1988).

DISTRIBUTION. S (widespread).

Also known from Scandinavia and Russia.

Trimorus punctulator **(Ruthe, 1859). Fig. 2C**

Henriksen 1939: not recorded.

1.1–1.6 mm. Head and terga posterior to petiole roughly reticulate-coriaceous; mesoscutum without notauli, postscutellum without tooth, wings reaching posterior margin of propodeum, 3rd tergum without furrows, legs dark. Female metasoma relatively elongate (Fig. 2C). See also key. The male is described by Petersen (1956).

BIOLOGY. Cf. *T. orbiculatus* above.

DISTRIBUTION. S (Narsarsuaq, Upernaviarsuk)

Also known from Iceland, Moldavia and Russia.

References

Gauld I. & Bolton, B. (eds) (1988): *The Hymenoptera*. 332 pp. Oxford.

Kozlov, M.A. & Kononova, S.V. (1983): *Telenominae of the SSSR*. – Opredeliteli Fauna SSSR 136: 336 pp.

Kryger, J.P, (1910): Snyltere i Edderkoppeæg. – *Entomologiske Meddelelser* 2 (3): 257–285.

Petersen, B. (1956): Hymenoptera. – *The Zoology of Iceland* III (49–50): 1–176. Copenhagen and Reykjavik.

Vlug, H.J. (1985): The types of Platygastridae (Hymenoptera, Scelionoidea) described by Haliday and Walker and preserved in the National Museum of Ireland and in the British Museum (Natural History). 2. Keys to species, redescriptions, synonymy. – *Tijdschrift voor Entomologie* 127: 179–224.

11.12. Diapriidae (Proctotrupoidea)

Peter N. Buhl

Body length 2–5 mm; antennae inserted on shelf, with 12–15 segments, scape long, antennae markedly sexually dimorphic, 1st or 2nd flagellar segment in male usually with an emargination and/or tooth; fore wings without stigma but with an elongate marginal vein along wing margin, a short stigmal vein, and a long radial vein parallel to wing margin more or less closing a narrow triangular radial cell; metasoma with petiole, following tergum 2 the largest; between tergum 2 and the two apical terga usually with two (Diapriinae) or four (Belytinae) short ring-shaped terga.

The biology of very few species is known, but most species probably are endoparasitoids on prepupae or pupae of Diptera, a few on immature Coleoptera, Hymenoptera and Lepidoptera. About 2.300 species are known worldwide.

Key to species of Diapriidae in Greenland

1. Antenna with 14 or 15 segments in female; male with 3rd antennal segment modified; fore wing with long, open radial cell; between large tergum and the two apical terga with four short ring-shaped terga; body length at least about 2 mm (Belytinae)............. 2
– Antenna with 12 segments in female; male with 4th antennal segment modified; fore wing without indication of radialis; between large tergum and the two apical terga with two short ring-shaped terga; body length at most about 2 mm (Diapriinae) 4
2. Antenna of female with 14 segments, scape at apex with two angular projections; mandibles somewhat beak-like prolonged................. *Polypeza ciliata*
– Antenna of female with 15 segments, scape without projections at apex; mandibles short and curved.... 3
3. Tergite 2 without distinct punctures; female antenna with segment 3 hardly twice as long as wide......... .... *Zygota ruficornis*
– Tergite 2 with distinct punctures (Fig. 2E); female antenna with segment 3 fully 2.5 times as long as wide (Fig. 2F)................... *Zygota groenlandica*
4. Female antenna with sharply differentiated club (Fig. 1B); male macropterous, with basal vein .... *Basalys bifoveata*
– Female antenna with relatively weak club (Fig. 1C); male micropterous, wings hardly reaching end of thorax....................... *Trichopria nigricornis*

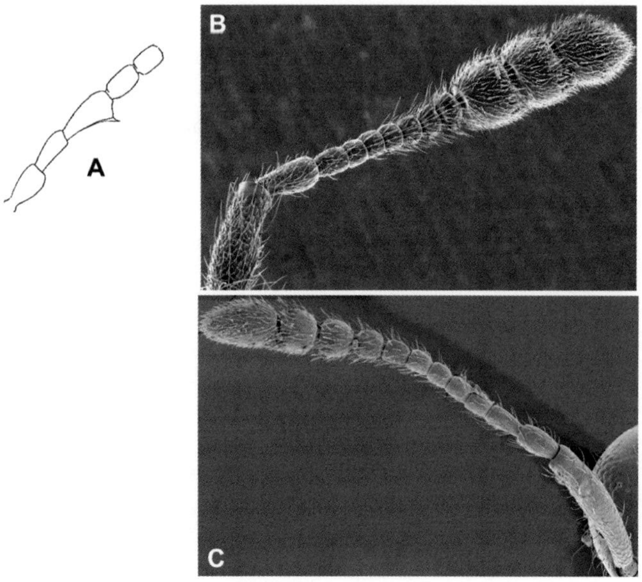

FIGURE 1 *A–B.* Basalys bifoveata. *A. Male antennal segments 2–6. (From Nixon 1980). B. Female antenna. Fig. C.* Trichopria nigricornis, *female antenna.*

***Basalys bifoveata* (Kieffer, 1911)**

1.0–1.8 mm. Male: Antenna with scape shorter than following three segments combined, segment 2 elongate, segment 4 large (Fig. 1A), preapical segment one and a third times as long as wide; head from above cubical; base of second metasomal tergite without pubescence. Female (not known from Greenland): Antenna (Fig. 1B). Fore wings reaching at most middle of 2nd metasomal tergum.

NOTE. In Greenland only known from an incomplete male specimen. The existing body parts agree closely with those of Danish specimens of *B. bifoveata*. The female is probably identical to *B. parva* Thomson, 1859. Probably the species recorded here is also identical to the species recorded from the Faroes and Iceland by Petersen (1956) as *Loxotropa suecica* Kieffer, 1911.

DISTRIBUTION. nSW (only recorded from Kangerlussuaq/Søndre Strømfjord).

Europe, temperate (variable, common and widely distributed).

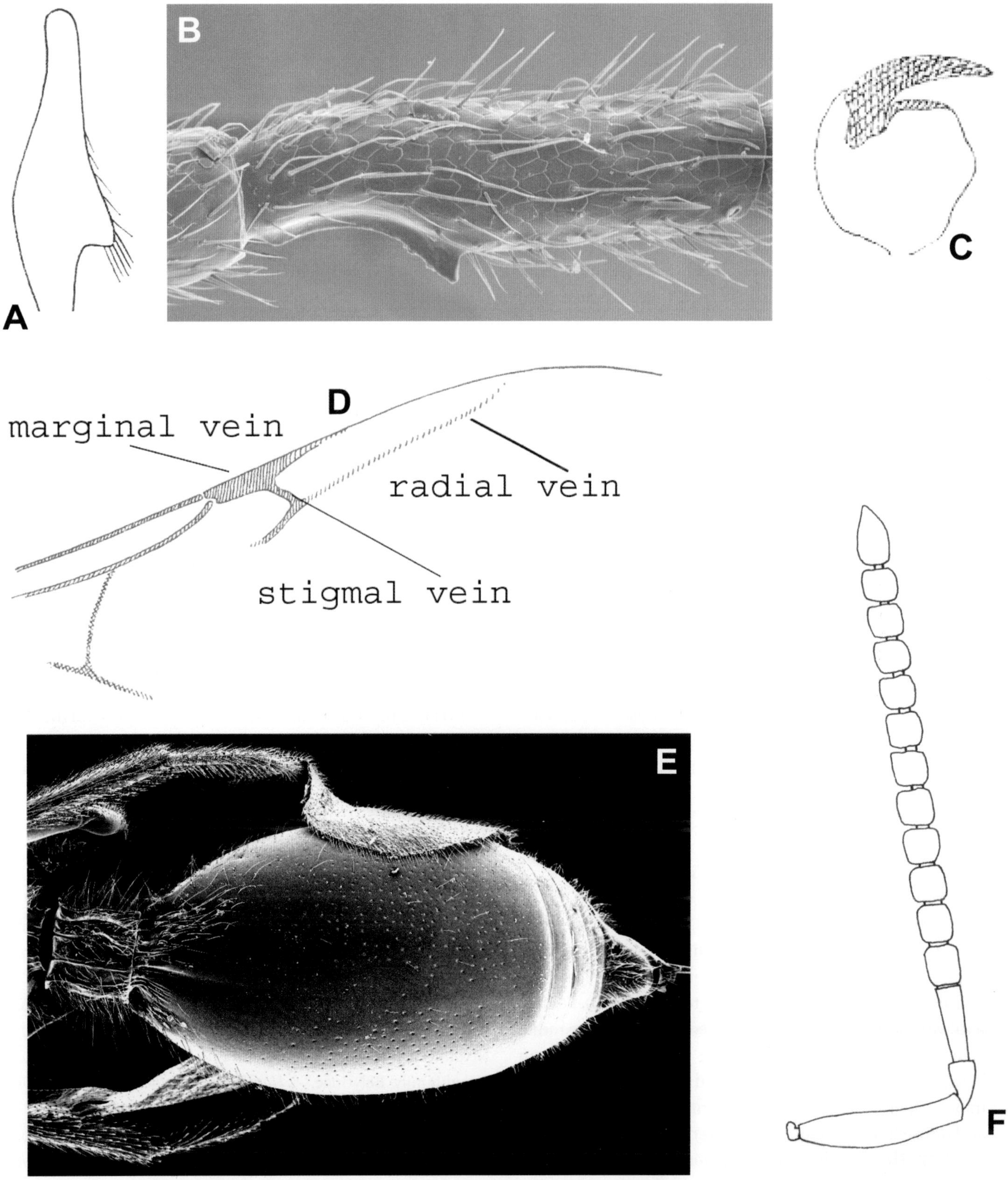

FIGURE 2 *A–D.* Zygota dentatipes *A. Fore tibia of male. B. Male antennal segment 3. C. Digitus of male genitalia. D. Forewing venation. (Figs A, C–D from* Nixon 1957*). E–F.* Zygota groenlandica. *E. Female metasoma in dorsal view. F. Female antenna. (From Buhl* 1995*).*

***Trichopria nigricornis* (Marshall, 1868)**
Henriksen 1939: not recorded.

1.4–1.5 mm. Wings shortened, in female reaching hind margin of metasomal tergum 2, shorter in male (hitherto not found in Greenland). Head subcircular in dorsal view, 2nd flagellar segment of female hardly longer than wide (Fig. 1C), scutellum without keel, cf. Nixon (1980).

NOTE. I could find no differences between the Greenlandic specimens and Danish specimens in the ZMUC. The four Greenlandic females available were all collected 2 August on different localities, one of them in *Ledum decumbens* at the shore of a lake. Nixon (1980) states that it is "probably a common coastal species".

DISTRIBUTION. nSW (only recorded from Kangerlussuaq/Søndre Strømfjord). First record from Greenland.

Northwestern Europe.

***Polypeza ciliata* (Thomson, 1859)**
Henriksen 1939: not recorded.

1.6–2.5 mm. Easily recognised by the structure of the female antenna. First segment of flagellum about three times as long as wide, the following segments hardly longer than wide. The male is recognised by the following characters: Scape as long as the following two segments combined, first flagellar segment twice as long as next segment, this and the following segments each 2.5–3 times as long as wide. A generally dark, blackish species.

NOTE. Described as *Atelopsilus borealis* by Petersen (1956) from Iceland, synonymized by Macek (1993).

DISTRIBUTION. S (Narsarsuaq). First record from Greenland.

Rather common in NW Europe.

***Zygota dentatipes* (Kieffer, 1908). Fig. 3**
Henriksen 1939: not recorded.

2.5–3.0 mm. Male with distinct triangular projection medially on fore tibia, having distinct notch distally (Fig. 2A), 1st flagellar segment with emargination (Fig. 2B), digitus of genitalia with single large tooth (Fig. 2C). Female with very long radial cell as indicated by incomplete radial vein (Fig. 2D), fully twice as long as distance between basal vein and marginal vein, and open at apex (as in male), preapical antennal segments slightly transverse, and dense hairs at sides of apical terga.

NOTE. The female runs to *Z. norvegica* (Kieffer, 1912) in Nixon's (1957) key though antennal structure more indicates *Z. ruficornis* (Curtis, 1831) (only females of these species known). *Z. dentatipes* is probably a junior synonym of one of these two species, but for the time being I use this name, as the males collected in Greenland are no doubt identical to *dentatipes*. The male of this species can be separated from the closely allied *Z. excisipes* (Kieffer, 1908) by the presence of several smaller teeth on the digitus of the genitalia in the latter.

FIGURE 3 Zygota dentatipes, *habitus*.

DISTRIBUTION. S, SW (north to Ukkusissat Sulluat). Very common and widespread.

Northwestern Europe, widespread.

***Zygota groenlandica* Buhl, 1995**
Henriksen 1939: not recorded.

2.8–3.9 mm. Differs from *Z. dentatipes* by characters mentioned in key.

NOTE. Only the female known, the species might be parthenogenetic.

DISTRIBUTION. S, SW, SE (only recorded from Saqqisikuik/Skjoldungen). Endemic to Greenland.

References

Buhl, P.N. (1995): The proctotrupoid wasps of Greenland (Hymenoptera, Proctotrupoidea s.l.). – *Entomologiske Meddelelser* 63: 7–10.

Macek, J. (1993): Revision of Holarctic *Polypeza* (Hymenoptera, Diapriidae). – *Folia Heyrovskyana* 1: 19–24.

Nixon, G.E.J. (1957): Hymenoptera Proctotrupoidea. Diapriidae subfamily Belytinae. – *Handbooks for the identification of British Insects* VIII, 3 (dii): 107 pp.

Nixon, G.E.J. (1980): Diapriidae (Diapriinae). Hymenoptera, Proctotrupoidea. – *Handbooks for the identification of British Insects* VIII, 3 (di): 55pp.

Petersen, B. (1956): Hymenoptera. – *The Zoology of Iceland* III (49–50): 1–176. Copenhagen and Reykjavik.

11.13. Proctotrupidae (Proctotrupoidea)

Peter N. Buhl

Mostly black species, generally 3–6 mm, with filiform, 13-segmented antennae. The forewing has a distinct stigma, with relatively wide closed costal cell and a generally short, closed radial cell, rest of veins hardly indicated. The ovipositor is projecting from apex of the metasoma and has heavily sclerotized sheaths with characteristic shape and sculpture important for identification of species. The majority of the species are endoparasitoids of Coleoptera larvae. About 320 described species worldwide.

Exallonyx cf. _microcerus_ Kieffer, 1908

2.5–3.5 mm. Preapical antennal segments of female hardly longer than wide. A single, rather worn female collected by Thorkild Munk at Narsarsuaq in July 2009 could not be determined with certainty, but in the available keys it comes closest to *E. microcerus* which is a common NW European species known as a parasitoid of larvae of Staphylinidae.

11.14. Braconidae (Ichneumonoidea) (Shield wasps or braconids)

Cornelis van Achterberg

A very large family of usually medium-sized to small insects in the temperate regions, usually with length of fore wing between 1–5 mm (and similar body length). Larger species (up to 15 mm) occur but are far less common and even larger species are found in the tropics (with wings up to 23 mm). Typical for Braconidae are the joined second and third metasomal tergites (forming a kind of shield: "Shield wasps"), the comparatively basally situated and reclivous vein 1r-m of the hind wing and the absence of vein m-cu of fore wing (except in one subfamily from Chile). The hind wing is smaller than the fore wing and has less venation. The venation of the fore wing in this family is extremely variable and an important character for identification. Nearly all Braconidae are primary parasitoids (some Euphorinae are hyperparasitoids of Ichneumonidae) and mostly of larvae. An important part of the Euphorinae (including the tribe Neoneurini on adult ants) and the Aphidiinae are koinobiont endoparasitoids of adult insects. Several species are koinobiont ovo-larval endoparasitoids: the egg is deposited in the egg of the host, remain nearly dormant (only manipulate the host to avoid encapsulation in the larval stage; in the egg stage there is no risk of encapsulation) and develops when the larva reaches its third or fourth instar. The members of the microgastroid group (main members are the subfamilies Microgastrinae and Cheloninae) use a specialised Braco-virus (PDV or PolyDNAVirus) to avoid encapsulation. The "Cyclostomes" (=one of the main clades of Braconidae) consists partly of idiobiont ectoparasitoids (mainly the majority of the subfamilies Doryctinae and Braconinae), the remainder of the Braconidae are nearly all koinobiont endoparasitoids.

The larvae are simple and legless; the main visible structure consists of the mouthparts. In Greenland the Microgastrinae and the Aphidiinae are the most dominant subfamilies. Microgastrinae are parasitoids of lepidopterous larvae and may be gregarious; Aphidiinae are parasitoids of aphids. One subfamily is obviously undercollected: the Alysiinae, which are endoparasitoids of cyclorrhaphous Diptera. There are many more species of cyclorrhaphous Diptera known from Greenland than of Alysiinae. In a recent paper van Achterberg (2006) listed 16 genera and 30 species for Greenland, of which 9 (= 56%) and 17 (56%), respectively, were new to the fauna. Seven species were considered new to science (= 23%). However, the material of one species included in van Achterberg (2006), p. 37; *Protapanteles immunis* (Haliday, 1834)) is actually from Iceland, so this species has no Greenland record.

The terminology follows van Achterberg (1988). For additional information (synonyms, distribution, biology, etc.) of the species described before 2005, see the Taxapad database (Yu *et al.*, 2005).

Key to the Braconidae of Greenland

1. Mandibles straight or curved outwards, their tips not touching when closed (Fig. 3E), and with 3 or 4 teeth or lobes; Figs 2C, H, 3I, K: "exodont braconids"; .... 2 (subfamily Alysiinae)
– Mandibles curved inwards, their tips touching when closed (Figs 4B, G), and with 2 teeth (Fig. 17C)...... 5
2. Vein r-m of fore wing absent (Figs 3A, J) 3
– Vein r-m of fore wing present (Figs 2A, F).......... 4
3. Setae of metapleuron partly directed dorsally and anteriorly; mandible with a more or less developed minute fourth tooth, if absent or obsolescent then central tooth enlarged (Fig. 3K) and hind coxa often with tufts of setae dorsally .... *Chorebus* sp. near *cytherea*
– Setae of metapleuron directed posteriorly; mandible with 3 teeth (Fig. 3I); hind coxa without tufts of setae dorsally .... *Dacnusa groenlandica*
4. Third antennal segment slightly longer than fourth segment (Fig. 2E); vein 1–SR+M of fore wing present (Fig. 2F) .... *Alysia alticola*
– Third antennal segment shorter than fourth segment (Fig. 2D); vein 1–SR+M of fore wing absent (Fig. 2A) .... *Aphaereta minuta*
5. Hypoclypeal depression deep and wide, and middle of apparent ventral margin of clypeus distinctly above upper level of mandibular bases (Figs 4B, G, 5F); bottom of hypoclypeal depression consists of the concave labrum and usually a depressed part of clypeus ("cyclostome braconids") 6
– Hypoclypeal depression absent (Figs 14H, 15H, 17E), medio-ventral margin of clypeus close to upper level of mandibular bases (Fig. 14H); labrum flat and ventral part of clypeus not part of a hypoclypeal depression... 9
6. Fore tibia with row of stout pegs or spines, which are at most as long as about 6 times their width (Fig. 4F);

first metasomal tergite elongate (Fig. 4H); posterior flange of propleuron largely dorsally situated (Fig. 4I); subfamily Doryctinae .... *Spathius exarator*

– Fore tibia without row of pegs or spines (Fig. 13G), length of bristles at least about 8 times their width; first tergite robust (Figs 4D, 5H); posterior flange of propleuron situated mainly subapically (Fig. 4E) or obsolescent... 7

7. Second and third metasomal tergites largely membranous dorsally (Fig. 4D), nearly always less sclerotised than their epipleura; pedicellus nearly as long as scapus; inner sides of eyes hardly or not emarginate (Fig. 4B); subfamily Hormiinae s. str. .... *Hormius moniliatus*

– Second and third tergites similarly or more strongly sclerotised than their epipleura; pedicellus distinctly shorter than scapus (Figs 5B, L); inner side of eyes distinctly emarginate (Fig. 5F); subfamily Rogadinae .... 8 (*Aleiodes*)

8. Clypeus strongly convex (Fig. 5L); stemmaticum distinctly protruding above level of vertex (Fig. 5L); head roundly narrowed below eyes in anterior view; area of precoxal sulcus distinctly rugose; tarsal claws narrowly brownish pectinate; third antennal segment robust (Fig. 5L)...... *Aleiodes* sp. near *signatus*

– Clypeus weakly convex (Fig. 5B); stemmaticum weakly protruding above level of vertex (Fig. 5B); head linearly narrowed below eyes in anterior view (Fig. 5F); area of precoxal sulcus granulate or rugulose; tarsal claws setose; third antennal segment normal (Fig. 5B); antenna of female with 32–33 segments, of male with 38–39 segments .... *Aleiodes borealis*

9. Vein SR1 of fore wing partly or completely unsclerotised, resulting in a distally open marginal cell distally (Figs 6A, 14A, 17G); metasoma often hardly longer than mesosoma............................. 10

– Vein SR1 of fore wing completely sclerotised, tubular, reaching margin of wing, resulting in a closed marginal cell distally (Figs 18H, 19G, 20A); metasoma usually distinctly longer than mesosoma 26

10. Vein SR1 of fore wing completely unsclerotised (Figs 6A, 8A, H, 10A); vein cu-a of hind wing present; scutellar sulcus with a median carina or crenulate; antenna with 18 segments .... 11 (subfamily Microgastrinae)

– Vein SR1 of fore wing basally sclerotised (Figs 14A, 15A, 16F, 17G); vein cu-a of hind wing absent (Fig. 14A); scutellar sulcus smooth; number of antennal segments variable.... 22 (subfamily Aphidiinae)

11. Hind coxa not surpassing level of apex of first metasomal tergite and less than twice as long as middle coxa; veins r-m and 3–CU1 of fore wing present (Figs 6A, J); precoxal sulcus at least medially present.................................... 12 (*Microplitis*)

– Hind coxa distinctly surpassing level of apex of first metasomal tergite and at least about twice as long as middle coxa; veins r-m and 3–CU1 of fore wing absent (Figs 10A, 11A, I); precoxal sulcus usually absent medially 16

12. Hind femur of both sexes distinctly inflated and short (Fig. 6B); antenna of female very short, submedial segments about as wide as long (Fig. 6E); face densely punctate; first tergite shiny and largely smooth; fore tarsal segments both sexes short (Fig. 6C); [cocoons white, without longitudinal ribs] .... *Microplitis coactus*

– Hind femur of both sexes normal (Figs 6F, 7E); antenna of female medium-sized, submedial segments distinctly longer than wide (Figs 6I, 7C, H); face densely and finely rugulose; first tergite usually less shiny and more or less microsculptured (except apically); fore tarsal segments of both sexes normal to rather robust (Figs 6H, 7B, G).................... 13

13. Basal half of antenna (after pedicellus) largely yellowish-brown; hind femur brownish-yellow; apical border of third and following tergite ivory laterally; hind tarsus brownish-yellow or infuscate; fore basitarsus slender (Fig. 6H); [first tergite comparatively robust, apical third of first metasomal tergite weakly convex medially and tergite weakly or not narrowed apically; cocoons brown and with distinct longitudinal ribs] .... *Microplitis mandibularis*

– Basal half of antenna (after pedicellus) largely dark brown; hind femur dark brown; apical border of third and following tergite orange or/and dark brown, respectively; hind tarsus largely dark brown; fore basitarsus comparatively robust (Figs 7B, G).... 14

14. Penultimate antennal segments of female hardly longer than wide; first tergite comparatively slender (Fig. 8E), apical third of tergite strongly convex medially and rather narrowed apically; colour of basal third of pterostigma variable 15

– Penultimate antennal segments of female nearly twice longer than wide (Fig. 7C); first tergite comparatively robust, apical third of tergite weakly convex medially and tergite weakly or not narrowed apically; basal 0.3–0.4 of pterostigma yellow, contrasting with dark brown remainder of pterostigma (Fig. 7A) .... *Microplitis sofron*

15. First subdiscal cell of fore wing comparatively wide compared to first discal cell (Fig. 8A); pterostigma largely or completely dark brown basally (Fig. 8A); vein r of fore wing strongly inclivous, forming an acute angle with anterior margin of pterostigma (Fig. 8A); vein r-m of fore wing comparatively long, resulting in a more or less quadrangular second submarginal cell (Fig. 8A); hind femur comparatively robust (Fig. 7I); vein 2–SR of fore wing less oblique; vein m-cu about as long as vein 2–SR+M; clypeus normally convex .... *Microplitis lugubris*

– First subdiscal cell of fore wing normal compared to first discal cell (Fig. 8H); basal third of pterostigma distinctly yellowish basally (Fig. 8H); vein r of fore wing subvertical or weakly oblique, forming a rectangular angle with anterior margin of pterostigma or nearly so (Fig. 8H); vein r-m of fore wing comparatively short, resulting in a more or less triangular second submarginal cell; hind femur less robust (Fig. 8H); vein 2–SR of fore wing more oblique; vein m-cu somewhat longer than vein 2–SR+M; clypeus strongly convex .... *Microplitis lugubroides*

16. Vein 1–SR of fore wing pointed to vein 1–CU1 or near vein cu-a (Figs 9A, 10A); first metasomal tergite wide posteriorly, subparallel-sided and stout (Fig. 9G); if to vein 2–CU1 (Fig. 11I) then vein 1–SR hardly longer than wide; Figs 11A, I) *and* propodeum usually strongly to coarsely reticulate and/or rugose; first discal cell of fore wing comparatively wide (Fig. 11I; 0.7–0.8 times higher than wide); scutellar sulcus distinctly crenulate medially; scutellum postero-laterally without distinct pit ....17 (*Cotesia*)

– Vein 1–SR of fore wing usually pointed to vein 2–CU1, if pointed to vein 1–CU1 or to vein cu-a (Fig. 12C: *P. pallipes*) then first tergite slender and narrowed posteriorly (Fig. 12A); vein 1–SR slender and comparatively long and propodeum largely smooth or finely sculptured; first discal cell of fore wing usually narrower; scutellar sulcus usually weakly crenulate or with obsolescent sculpture medially; scutellum postero-laterally with pit ....21 (*Protapanteles*)

17. Hind femur of female about 4.3 times as long as wide (Fig. 9C) and usually basally yellowish-brown, and more or less dark brown apically; penultimate segments of antenna of female 1.6–1.8 times as long as wide (Fig. 9B); vein r of fore wing subvertical; [vein 1–SR of fore wing distinctly longer than wide and pointing to vein cu-a; lateral lobe of mesoscutum densely punctate; third tergite largely smooth basally] .... *Cotesia yakutatensis*

– Hind femur of female 3.0–4.0 times as long as wide and at least partly blackish or dark brown basally (Fig. 9D); penultimate segments of antenna of female 1.0–1.7 times as long as wide (Figs 10E, 11B, G); vein r of fore wing pointed outwards (Figs 11A, I) or subvertical (Fig. 9J) 18

18. First discal cell of fore wing comparatively high (Fig. 9J); vein r of fore wing subvertical (Fig. 9J); outer side of hind femur with longitudinal yellowish streak; mesopleuron anteriorly and hind coxa smooth .... *Cotesia fascifemorata*

– First discal cell of fore wing comparatively transverse (Figs 10A, 11A, I); vein r of fore wing pointing outwards (Figs 11A, I); outer side of hind femur completely blackish; sculpture of mesopleuron anteriorly and hind coxa variable.........................19

19. Frons smooth posteriorly; fore and middle femora of female moderately slender (Figs 11C, D, J); upper margin of hind femur of female nearly straight (Fig. 11E); segments of fore tarsus of female comparatively slender (Figs 11F, H); hind coxa usually matt dorsally and more or less rugulose; inner hind spur about as long as outer spur; mesopleuron usually distinctly sculptured anteriorly; sculpture of third tergite variable 20

– Frons densely pimply (or rather punctate) posteriorly, very rarely reduced; fore and middle femora of female robust (Figs 10B, C); upper margin of hind femur of female distinctly convex (Fig. 10D); segments of fore tarsus of female robust (Fig. 10F); hind coxa shiny dorsally and more or less punctate; inner hind spur 1.2 times long as outer spur; mesopleuron usually sparsely sculptured anteriorly; third tergite more or less sculptured medially.................... ....*Cotesia crassifemorata*

20. Inner hind tibial spur 0.5 times as long as hind basitarsus; third metasomal tergite shiny and largely smooth; vein cu-a of fore wing distinctly shorter than vein 1–CU1 of fore wing (Fig. 11A); [outer side of hind coxa largely rugulose or densely punctate; vein 1–SR of fore wing distinctly longer than wide and pointing to vein cu-a in specimens from Greenland, but in Continental specimens hardly longer than wide and pointing to vein 2–CU1; one female from Greenland has 1–CU1 as long as vein 2–CU1, normally about 0.7 times as long] .... *Cotesia hallii*

– Inner hind tibial spur 0.6 times as long as hind basitarsus; third tergite matt and often distinctly sculptured; vein cu-a of fore wing often about as long as vein 1–CU1 of fore wing (Fig. 11I) ....*Cotesia eliniae*

21. Posterior half of mesoscutum strongly shiny and more or less punctate; median carina of propodeum irregular and usually distinct anteriorly; vein 1–SR of fore wing pointing to vein 1–CU1 (Fig. 12C); scutellum sparsely setose and rather narrow; hind femur less slender (Fig. 13C) .... *Protapanteles pallipes*
– Posterior half of mesoscutum with satin sheen and usually mainly smooth (usually only laterally with some punctures); median carina of propodeum largely absent and weakly developed, if intermediate then regular; vein 1–SR of fore wing pointing to vein 2–CU1 (Fig. 13K); scutellum densely setose and comparatively wide; hind femur slender (Fig. 13H) .. .. *Protapanteles fulvipes*
22. Hind wing with a closed cell (Fig. 14A); vein M+CU of hind wing distinctly sclerotised (Fig. 14A); vein 1–SR+M of fore wing present, at least as a weakly pigmented trace (Fig. 14A); notauli complete .... *Praon brevistigma*
– Hind wing without closed cells; vein M+CU of hind wing unsclerotised, only pigmented; vein 1–SR+M of fore wing absent (Figs 15A, G, 16F); notauli usually absent on mesoscutal disc 23
23. Veins m-cu and r-m of fore wing absent (Fig. 15A); genitalia of female with prongs (Fig. 15D); vein r+3–SR+SR1 of fore wing comparatively long and strongly curved (Fig. 15A) and comparatively far removed from level of apex of pterostigma (Fig. 15A)......... .... *Trioxys compressicornis*
– Veins m-cu and r-m of fore wing present (Figs 15G, 17G); genitalia of female without prongs (Fig. 15E); vein r+3–SR+SR1 of fore wing medium-sized and slightly curved, near level of apex of pterostigma (Figs 15G, 16F, 17G) .... 24 (*Aphidius*)
24. Head gradually narrowed posteriorly in dorsal view (Fig. 17B); face 1.1–1.7 times as wide as high (Figs 15H, 17A); fifth tergite of female sparsely setose 25
– Head directly narrowed posteriorly (Fig. 17F, sometimes less so in male); face about twice as wide as high (Fig. 17E); fifth tergite of female rather extensively setose .... *Aphidius cingulatus*
25. Fore tarsus comparatively robust (Fig. 15F); antenna of female with about 15 segments (of male 17 or 18); apical quarter of first tergite rather matt and largely finely sculptured sublaterally; first tergite baso-laterally very finely and densely aciculate, interspaces almost absent (Fig. 16E); first subdiscal cell of fore wing narrow (Figs 15G, I); vein 1–R1 of fore wing (metacarp) about as long as pterostigma (Fig. 15G).. .... *Aphidius tarsalis*
– Fore tarsus slender (Fig. 16L); antenna of female with 17–18 segments (of male 20–21 segments); apical quarter of first tergite strongly shiny and largely smooth; first tergite baso-laterally with 2–8 well separated sinuate carinae (Fig. 16I); first subdiscal cell of fore wing medium-sized (Fig. 16F); vein 1–R1 of fore wing distinctly shorter than pterostigma (Fig. 16F) .. .. *Aphidius avenae*
26. First metasomal tergite distinctly petiolate (Figs 19A, J, 20B); spiracle of first tergite usually medially or behind middle of tergite (Figs 19A, 20B); anterior subalar depression with at least one carina......... .... 27 (subfamily Euphorinae)
– First tergite sessile (Fig. 18D); spiracle of first tergite situated before middle of tergite (Fig. 18D); anterior subalar depression smooth .... *Blacus groenlandicus* (subfamily Blacinae)
27. Vein r-m of fore wing present (Fig. 19G); vein SR1 of fore wing straight (Fig. 19G); scapus comparatively short (Fig. 19I) .... 28 (*Meteorus*)
– Vein r-m of fore wing absent (Fig. 20A); vein SR1 of fore wing curved (Fig. 20A); scapus comparatively long (Fig. 20C).............. *Dinocampus coccinellae*
28. Setae of face nearly as long as setae of clypeus (Fig. 20D); first metasomal tergite slightly more robust and usually more abruptly narrowed submedially (but in front of spiracles largely parallel-sided; Figs 19A, B, D) .... *Meteorus arcticus*
– Setae of face much shorter than setae of clypeus (Fig. 19F); first tergite comparatively slender and gradually narrowed submedially (Fig. 19J) .... *Meteorus rubens*

***Alysia alticola* (Ashmead, 1890). Figs 1A, 2E–H**

Henriksen 1939: *Alysia manducator* Fabr.

3.5–7.0 mm. Mainly black. One of the larger and more conspicuous species of Braconidae from Greenland. As in many other braconid wasps size differences of 100% are normal. Very similar to *Alysia manducator* (Panzer, 1799), which has a similar biology. The comparatively dense and short setosity of the tenth and its surrounding antennal segments (about half as long as width of segment; Fig. 2E) show that the Greenland specimens belong to *A. alticola* (Ashmead). *A. manducator* (Panzer, 1799), which is widely distributed in the Holarctic region but has not yet been found in Greenland, has a more slender third antennal segment and the setae of the tenth and neighbouring antennal segments are 0.7–0.8 times as wide as the

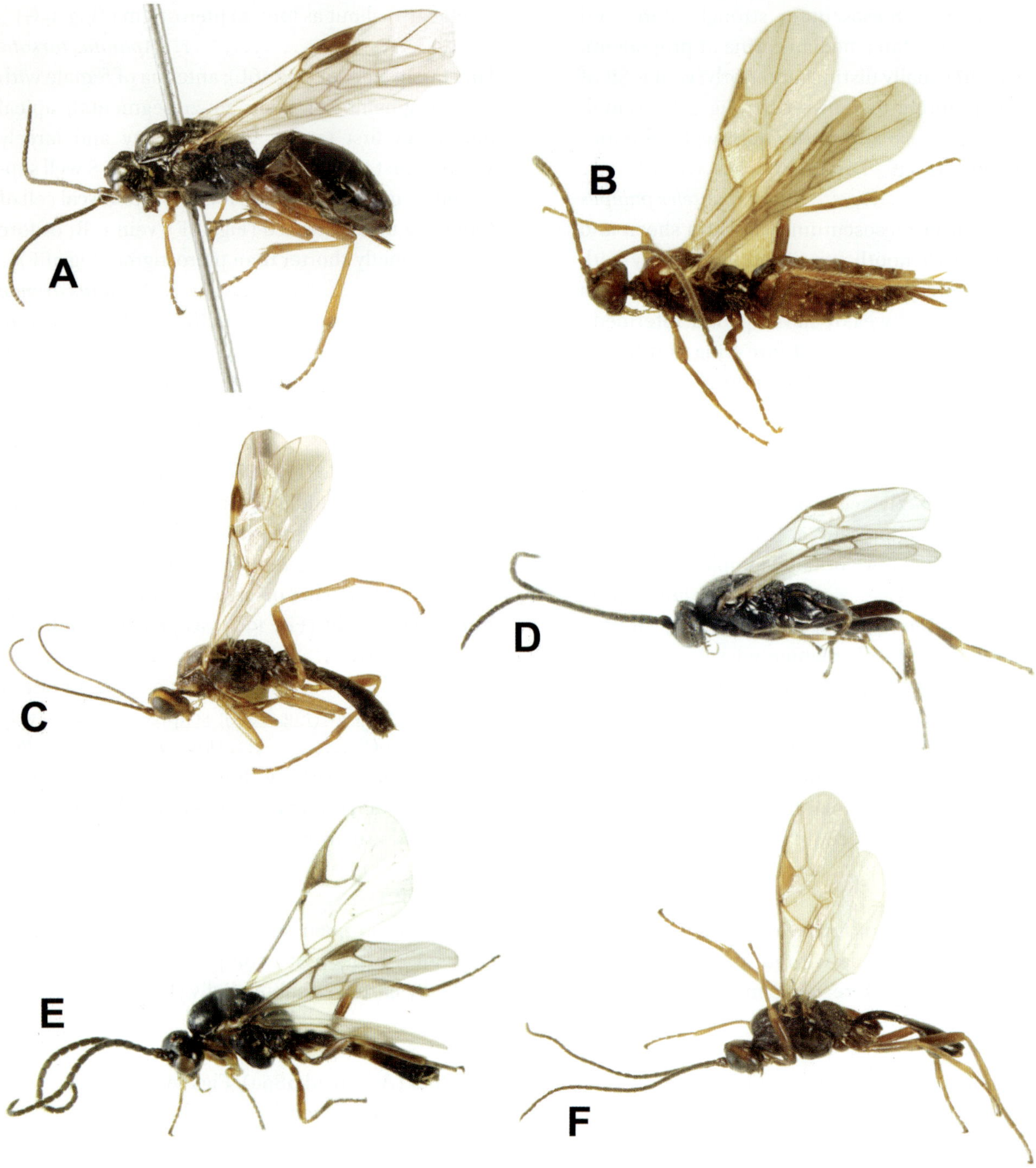

FIGURE 1 *Habitus images of selected Braconidae. A.* Alysia alticola*; B.* Hormius moniliatus*; C.* Aleiodes borealis*; D.* Cotesia crassifemorata*; E.* Aphidius (Euaphidius) cingulatus*; F.* Meteorus rubens*. Not to scale.*

segments and arranged mainly in two rows: a subbasal and a subapical one.

BIOLOGY. Koinobiont endoparasitoid of *Calliphora* spp. (Calliphoridae) especially in carcasses.

DISTRIBUTION. S, SW (north to Kangerlussuaq/Søndre Strømfjord).
Circumpolar, temperate, widely distributed.

Aphaereta minuta **(Nees, 1811). Figs 2A–D**
Henriksen 1939: not recorded.

C. 2–3 mm. Hind basitarsus comparatively robust (Fig. 2B), the base of the hind femur and of the hind tarsus are more or less darkened and the precoxal sulcus is often rather widely sculptured. This seems to fall within the variation present in this species in Europe.

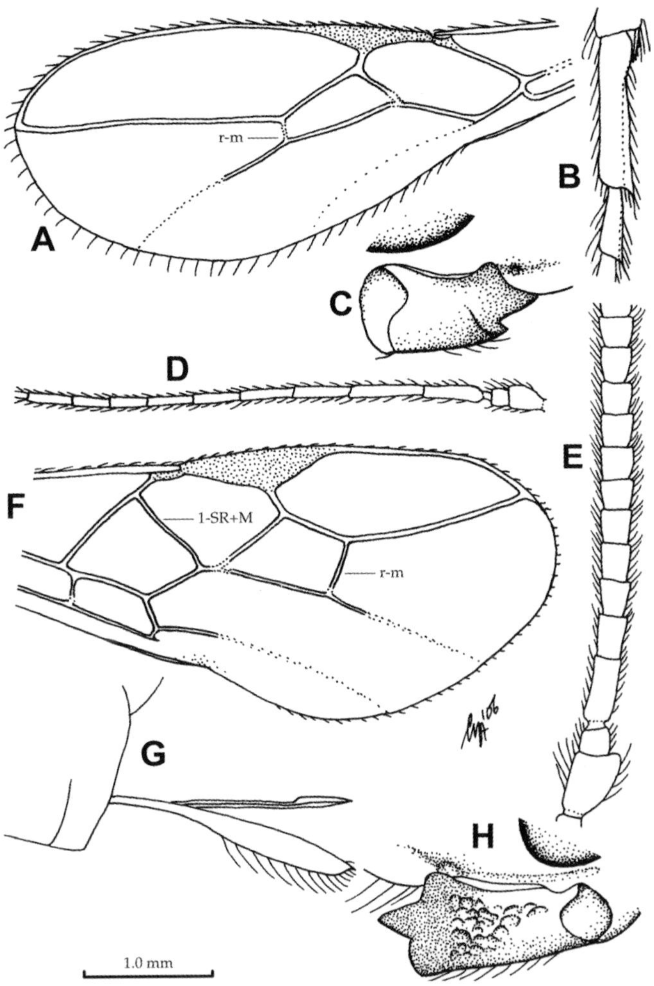

FIGURE 2 *A–D,* Aphaereta minuta *(Nees), female, Greenland, Evighedsfjorden;* Figs 2E–H, Alysia alticola *(Ashmead), female, S Greenland, Narsarsuaq. A, F, fore wing; B, hind basitarsus, lateral aspect; C, H, mandible; D, E, base of antenna; G, ovipositor and its sheath, lateral aspect. A, D: 2.4× scale-line; B, C: 3.8×; E, G, H: 2.2×; F: 1.0×.*

BIOLOGY. Koinobiont endoparasitoid of *Scatophaga* species (Scathophagidae) in rotting organic matter (*e.g.* seaweed on the beach, human excrements).

DISTRIBUTION. SW (north to Kangerlussuatsiaq/ Evighedsfjorden).
Palaearctic, low arctic-subtropical.

***Chorebus* sp. near *cytherea* (Nixon, 1939). Figs 3J–K**
Henriksen 1939: not recorded.

Similar to *C. cytherea* (Nixon, 1939) but differs by the more robust marginal cell of the fore wing (Fig. 3J), the slender mandible (Fig. 3K), the setae of first tergite directed posteriorly, the precoxal sulcus narrow (and completely smooth), and the completely dark brown legs and palpi. The only known specimen has incomplete antenna and may be one of the only 15 reported species of the genus from the Nearctic region (241 species are known from Europe) or belongs to an undescribed species. The genus has never been revised for the Nearctic area and no compiled key is available.

BIOLOGY. Not reared but related species are koinobiont endoparasitoid of mining Diptera.

DISTRIBUTION. NW (only known from Pituffik/Thule).
? Canada.

***Dacnusa groenlandica* van Achterberg, 2006. Figs 3A–I**
Henriksen 1939: not recorded.

1.8 mm. Similar to *D. arctica* Griffiths, 1984, from Lapland (N. Sweden) and Alaska, however *D. groenlandica* differs as follows: the middle mesoscutal lobe completely setose (only near the notauli and posteriorly in *D. arctica*), vein m-cu of the fore wing distinctly antefurcal (subinterstitial), the pterostigma of both sexes comparatively wide subapically (distinctly narrowed subapically), the mesosoma 1.3–1.4 times as long as high (1.1–1.2 times) and the antenna with 22–26 segments (20–24 segments).

BIOLOGY. Not reared but related species are koinobiont endoparasitoid of mining Diptera.

DISTRIBUTION. N (only recorded from Peary Land: Kap København).
? Canada.

[*Spathius exarator* (Linnaeus, 1758). Figs 4F–I
Henriksen 1939: not recorded.

The only specimen examined of this rather large species (usually 4–6 mm) is an artificially flattened specimen, which was possible imported with timber. In view of the paucity of records, this conspicuous species, which parasitizes small wood-boring Coleoptera, is unlikely to be permanently resident in Greenland, but might be regularly imported.

BIOLOGY. Idiobiont ectoparasitoid of larvae of Anobiidae, but also reported from Cerambycidae and Curculionidae living under the bark of dead branches or trees.

DISTRIBUTION. N (only found once, in Peary Land: Kap København; probably introduced with expedition goods).

Palaearctic (from Finland to Spain and Bulgaria eastward to far eastern Russia, Korea, Japan). Also found in New Zealand.]

***Hormius moniliatus* (Nees, 1811). Figs 2B, 4A–E**
C. 2–3 mm. May be recognised by the sculpture of the head and by the venation and the colour of the fore wing.

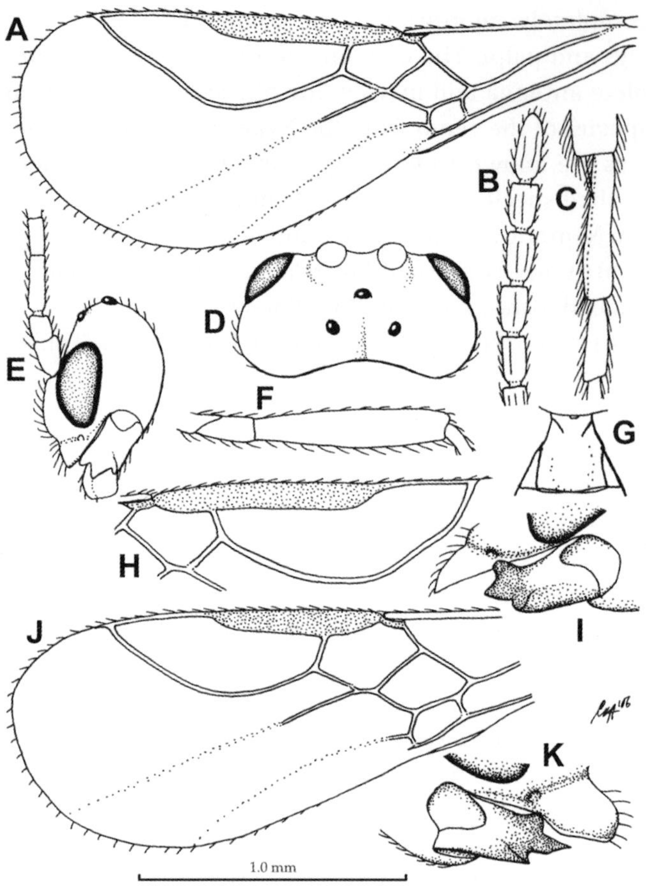

FIGURE 3 *A–I,* Dacnusa groenlandica *spec. nov., female, holotype, but H of male, paratype; J–K,* Chorebus *spec. near* cytherea *(Nixon), female, N Greenland, Thule. A, J, fore wing; B, apex of antenna; C, hind basitarsus, lateral aspect; D, head, dorsal aspect; E, head, lateral aspect; F, hind femur, lateral aspect; G, first metasomal tergite, dorsal aspect; H, detail of pterostigma; I, K, mandible, full view on first tooth. A, H, J: 1.0 × scale-line; B, C, I, K: 2.3 ×; D–F: 1.5 ×; G: 1.2 ×.*

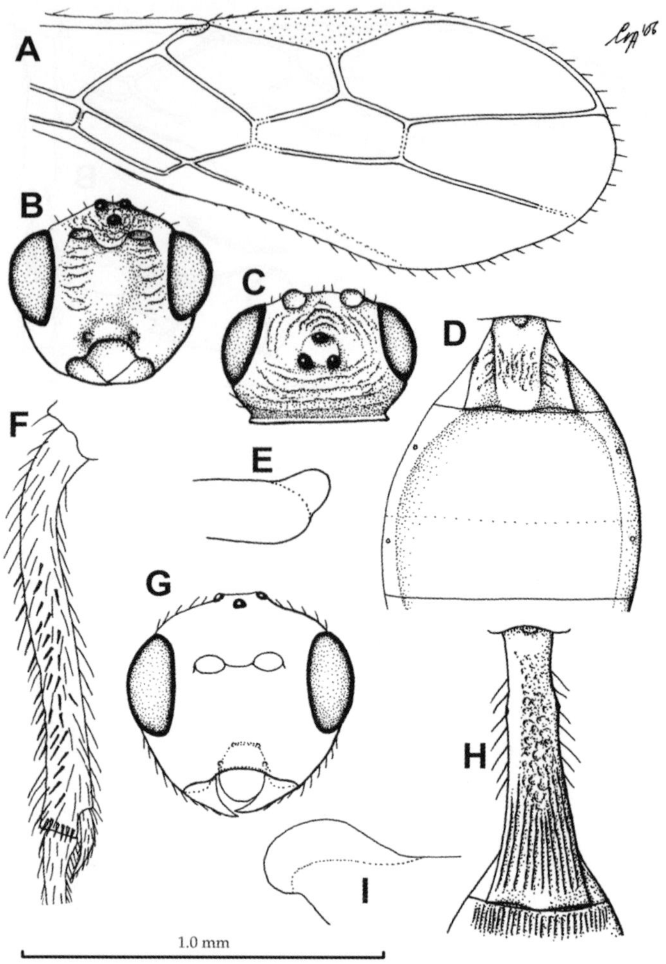

FIGURE 4 *A–E,* Hormius moniliatus *(Nees), female, neotype, but E of female, Greenland, Mestersvig; Figs F–I,* Spathius exarator *(Linnaeus), female, Netherlands, Melissant. A, fore wing; B, G, head, anterior aspect; C, head, dorsal aspect; D, first-third metasomal tergites, dorsal aspect; E, I, propleural flange; F, fore tibia, anterior aspect; H, first metasomal tergite, dorsal aspect. A–D, F, I: 1.0 × scale-line; E: 2.1 ×; G: 1.6 ×; H: 0.9 ×.*

BIOLOGY. Idiobiont ectoparasitoid of larvae of Pyralidae, Depressariidae, Gelechiidae and Tortricidae.

DISTRIBUTION. S (Narsarsuaq), SW (Maniitsoq/Sukkertoppen), NE (Scoresby Sund: Hekla Havn; Jameson Land: Mestersvig).

Palaearctic, Oriental, low arctic-subtropical.

***Aleiodes borealis* (Thomson, 1892) sensu lato. Figs 2C, 5A–H**

Henriksen 1939: not recorded.

Rather large (4–5 mm) and mainly blackish.

BIOLOGY. Koinobiont endoparasitoid of larvae of Noctuidae.

DISTRIBUTION. SW (only found at Nuup Kangerlua: Kapisillit and at Kangerlussuaq/Søndre Strømfjord airport).

Palaearctic, low arctic.

***Aleiodes* sp. near *signatus* (Nees, 1811). Figs 5I–L**

Henriksen 1939: not recorded.

BIOLOGY. Unknown, but related species are koinobiont endoparasitoids of larvae of especially Noctuidae and Geometridae.

DISTRIBUTION. N (only found in Wulff Land).

Nearctic (? Canada).

***Microplitis coactus* (Lundbeck, 1896). Figs 6A–E**

2.5–4.0 mm. Most commonly collected species of the genus *Microplitis* in Greenland.

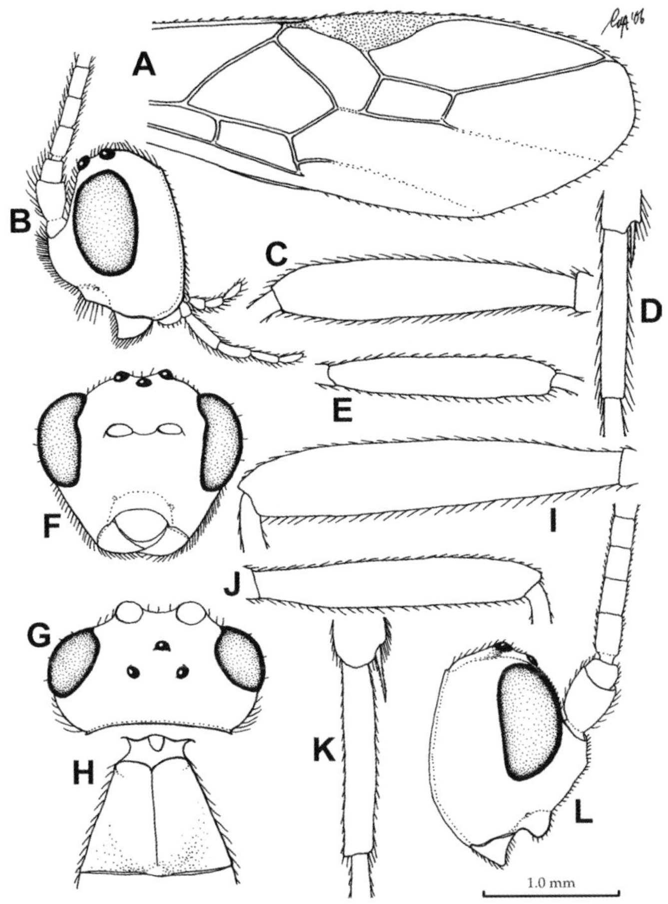

FIGURE 5 *A–H,* Aleiodes borealis *(Thomson) s. lato, female, S Greenland, Narsarsuaq; I–L,* A. *spec. near* signatus *(Nees), female, N Greenland, Wulff Land Peninsula. A, fore wing; B, L, head, lateral aspect; C, I, hind femur, lateral aspect; D, K, hind basitarsus, lateral aspect; E, J, fore femur, lateral aspect; F, head, anterior aspect; G, head, dorsal aspect; H, first metasomal tergite, dorsal aspect. A: 1.0× scale-line; B, F–H: 1.2×; C–E: 2.0×; I, J, L: 1.5×; K: 1.8×.*

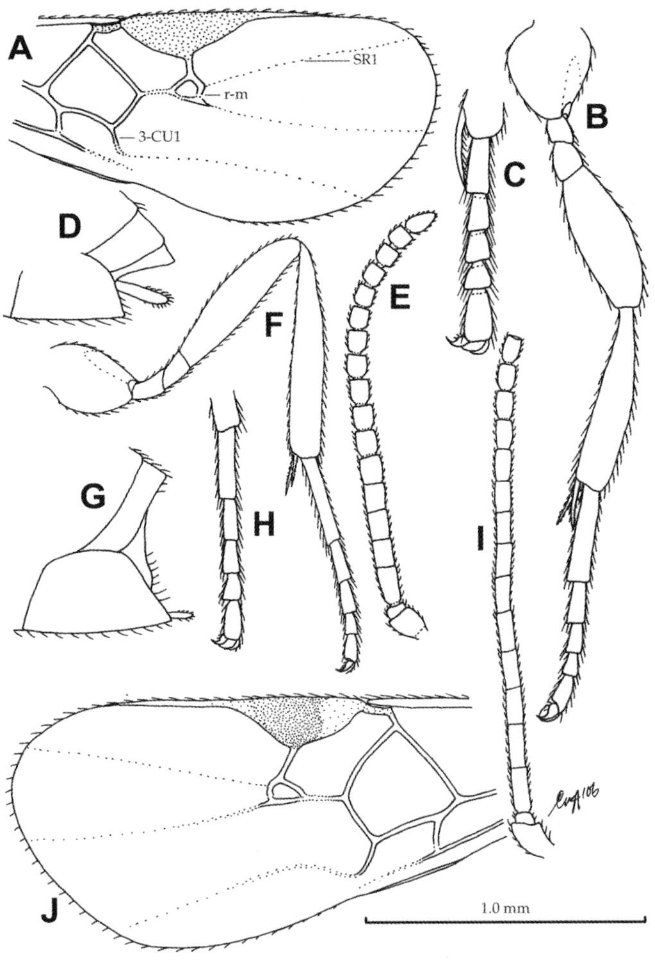

FIGURE 6 *A–E,* Microplitis coactus *(Lundbeck), female, S Greenland, Narssarssuaq; F–J,* M. mandibularis *(Thomson), female, Greenland. A, J, fore wing; B, F, hind leg; C, fore tarsus, lateral aspect; D, G, ovipositor sheath; E, I, antenna; H, fore tarsus, dorsal aspect. A–E, G, H: 1.0 × scale-line; F, 49: 0.6 ×; I: 0.9 ×.*

BIOLOGY. Reported as a parasitoid of a *Noctua* caterpillar (Noctuidae). Cocoons white without longitudinal ribs.

DISTRIBUTION. S, SW (north to Uummannaq), SE (Saqqisikuik/Skjoldungen).
Holarctic (only known from Iceland and possibly Canada).

Microplitis lugubris **(Ruthe, 1860). Figs 7G–I, 8A**
Henriksen 1939: not recorded.

2.5–3.0 mm. Recognisable by the combination of comparatively wide first subdiscal cell of the fore wing and the dark pterostigma.

BIOLOGY. Solitary koinobiont endoparasitoid of caterpillars belonging to Noctuidae and Arctiidae.

DISTRIBUTION. NW (Zackenberg, Ella Ø, Mestersvig). Palaearctic (Finland to Turkey, Mongolia, Central Asia).

Microplitis lugubroides **van Achterberg, 2006. Figs 8B–H**
Henriksen 1939: not recorded.

2–3 mm. Similar to *M. lugubris* but has a narrower first subdiscal cell of the fore wing and with pale basal patch of the pterostigma.

BIOLOGY. Unknown, but related species are koinobiont endoparasitoid of caterpillars belonging to Noctuidae and Arctiidae.

DISTRIBUTION. SW (north to Ukkusissat Sulluat: Pangnertooq). Only recorded from Greenland.

Microplitis mandibularis **(Thomson, 1895). Figs 6F–J**
Henriksen 1939: *Microplitis mediator* Hal.

2.5–3.5 mm. Easily recognisable among the other *Microplitis* species, because of its colouration. Lundbeck (1897) reported *M. mediator* (Haliday, 1834), but without any additional information; likely it concerns a misidentified *M. mandibularis* (Thomson). The female specimen seen of *M. mandibularis*, might have been the one reported by Lundbeck (1897).

BIOLOGY. Gregarious koinobiont endoparasitoid of caterpillars belonging to the Noctuidae. The only examined specimen has a typically ribbed and brownish cocoon connected to the pin.

DISTRIBUTION. Only one specimen known from Greenland, no precise locality.
Palaearctic (Finland to Spain and Turkey, Mongolia).

***Microplitis sofron* Nixon, 1970. Figs 7A–F**
Henriksen 1939: not recorded.

About 3 mm. Once reported by Papp (1984) but no additional information and no specimens available.

BIOLOGY. Koinobiont endoparasitoid of caterpillars belonging to the Noctuidae.

DISTRIBUTION. No precise locality given in Greenland.
Palaearctic (Norway to Turkey, Far East Russia, Central Asia).

***Cotesia crassifemorata* van Achterberg, 2006. Figs 2D, 10**
Henriksen 1939: not recorded.

1.8–2.4 mm. Easily recognisable because of the very robust femora (Figs 10B–D). Similar to *C. setebis* (Nixon, 1974), but this species has a more slender hind femur (with the dorsal border straight) and fore tarsus, the antennal segments somewhat more slender and the face smooth, with a satin lustre.

BIOLOGY. Unknown, but related species are koinobiont endoparasitoid of caterpillars belonging to Geometridae.

DISTRIBUTION. S, SW (north to Uummannaq), SE (Saqqisikuik/Skjoldungen). Only known from Greenland.

***Cotesia eliniae* Papp, 1989. Figs 11G–K**
Henriksen 1939: not recorded.

About 3 mm. Very close to *C. hallii* (Packard), and weakly sculptured specimens may easily be confused with this species. Hind femur hardly convex dorsally and 3.6–4.0 times as long as wide.

BIOLOGY. Unknown, but related species are koinobiont endoparasitoid of caterpillars.

DISTRIBUTION. nSW (Qeqertarsuaq/Godhavn, Nuussuaq: Marraat), NW (Pituffik/Thule), NE (widespread from Holm Land: Eskimonæs south to Illoqqortoormiut/Scoresby Sund), SE (Saqqisikuit/Skjoldungen). Only known from Greenland.

***Cotesia fascifemorata* van Achterberg, 2006 Figs 9D–J**
Henriksen 1939: not recorded.

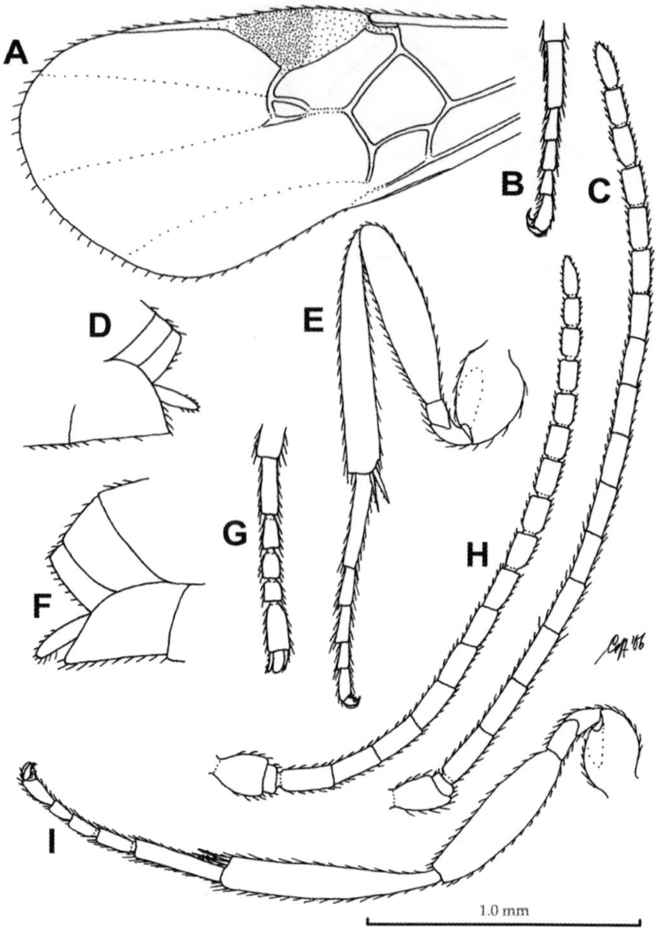

FIGURE 7 *A–F,* Microplitis sofron *Nixon, female, Spain, Briviesca; G–I,* M. lugubris *(Ruthe), female, Greenland, Neqé. A, fore wing; B, fore tarsus, lateral aspect; C, H, antenna; D, F, ovipositor sheath; E, I, hind leg; G, fore tarsus , dorsal aspect. A, C, E, H, I: 1.0 × scale-line; B, D, F, G: 1.1 ×.*

1.8–2.4 mm. Easily recognisable because of the moniliform apical half the antenna of the female. Similar to *C. kazak* (Telenga, 1949), but this species has the apical third of the female antenna not moniliform, veins 1– and 2–CU1 of the fore wing without any pigmentation (as the surrounding veins), the outer side of the hind femur evenly dark brown, the propodeum (except narrowly anteriorly) strongly declivous and a long vein r of fore wing. *C. brevicornis* (Wesmael, 1837) has a similar antenna, but has the apical third of the first metasomal tergite distinctly narrowed, veins 1–R1 and C+SC+R of the fore wing distinctly paler than the pterostigma and the fore tarsus more slender.

BIOLOGY. Unknown, but related species are koinobiont endoparasitoid of caterpillars.

DISTRIBUTION. SW (found only in Allumersat/Bjørnesund). Only known from Greenland.

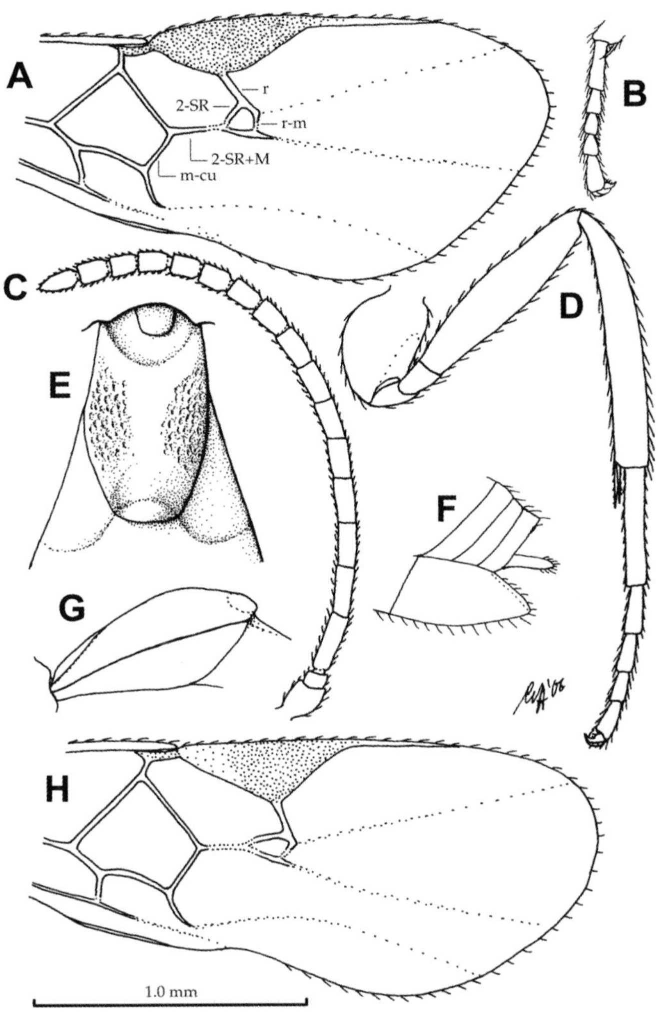

FIGURE 8 *A,* Microplitis lugubris *(Ruthe), female, Greenland, Neqé; B–H,* M. lugubroides *spec. nov., female, holotype. A, H, fore wing; B, fore tarsus, lateral aspect; C, antenna; D, hind leg; E, first metasomal tergite; F, ovipositor sheath, lateral aspect; G, first tergite, lateral aspect. A: 1.0 × scale-line; B–D, F, H: 1.2 ×; E, G: 1.6 ×.*

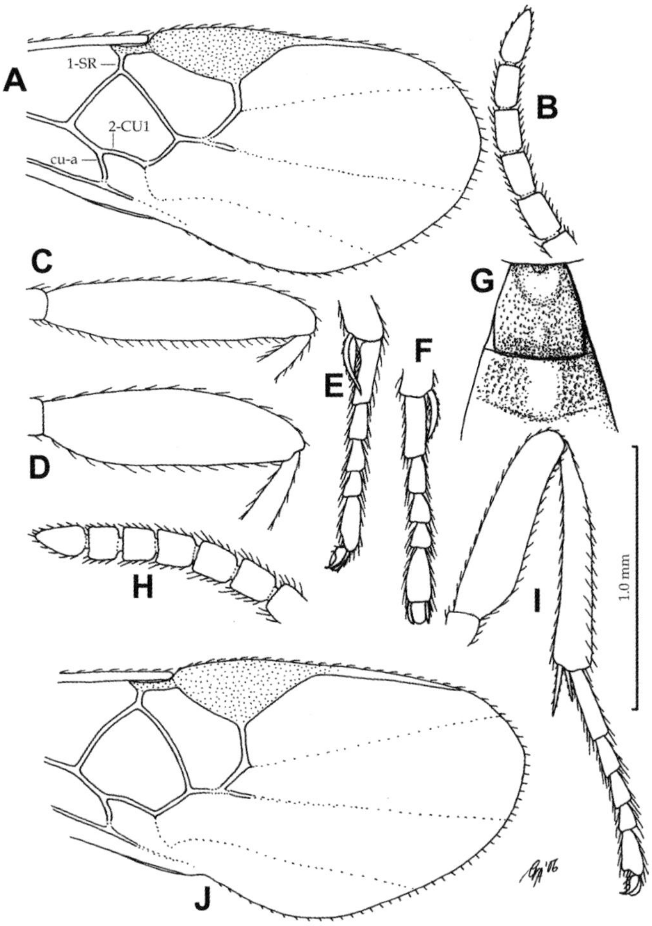

FIGURE 9 *A–C,* Cotesia yakutatensis *(Ashmead), female, Canada, Quebec, Gt. Whale R.; Figs D–J,* C. fascifemorata *spec. nov., female, holotype. A, J, fore wing; B, H, apex of antenna; C, D, hind femur, lateral aspect; E, fore tarsus, lateral aspect; F, fore tarsus, dorsal aspect; G, first and second metasomal tergites, dorsal aspect; I, middle leg, lateral aspect. A, J: 1.0 × scale-line; B, D, i: 1.5 ×; C: 1.4 ×; E, f: 1.7 ×; G: 1.1 ×; H: 1.9 ×.*

***Cotesia hallii* (Packard, 1877). Figs 11A–F**
Henriksen 1939: *Apanteles halli* Pack.

About 3 mm. Close to *C. eliniae* Papp, but has the inner hind tibial spur 0.5 times as long as the hind basitarsus, the third metasomal tergite shiny and largely smooth, vein cu-a of the fore wing distinctly shorter than vein 1–CU1 of the fore wing (Fig. 11A) and the outer side of the hind coxa largely rugulose or densely punctate. Vein 1–SR of fore wing is distinctly longer than wide and pointing to vein cu-a in specimens from Greenland, but in Canadian specimens hardly longer than wide and pointing to vein 2–CU1; one female from Greenland has vein 1–CU1 as long as vein 2–CU1, about vein 1–CU1 is 0.7 times as long as 2–CU1.

BIOLOGY. Unknown, but related species are koinobiont endoparasitoid of caterpillars.

DISTRIBUTION. nSW (Qeqertarsuaq/Godhavn, Saqqaq), NW (Pituffik/Thule), NE (Daneborg, Jameson Land: Coloradodal, Muslingeelvdal; Scoresby Sund: Gåseland, Hekla Havn).

Nearctic (Canada: Northwest Territories).

***Cotesia yakutatensis* (Ashmead, 1902). Figs 9A–C**
Henriksen 1939: not recorded.

2.5–3.0 mm. This blackish species has the hind femur of the female usually yellowish-brown basally, and more or less dark brown apically, the penultimate segments of antenna of the female are 1.6–1.8 times as long as wide

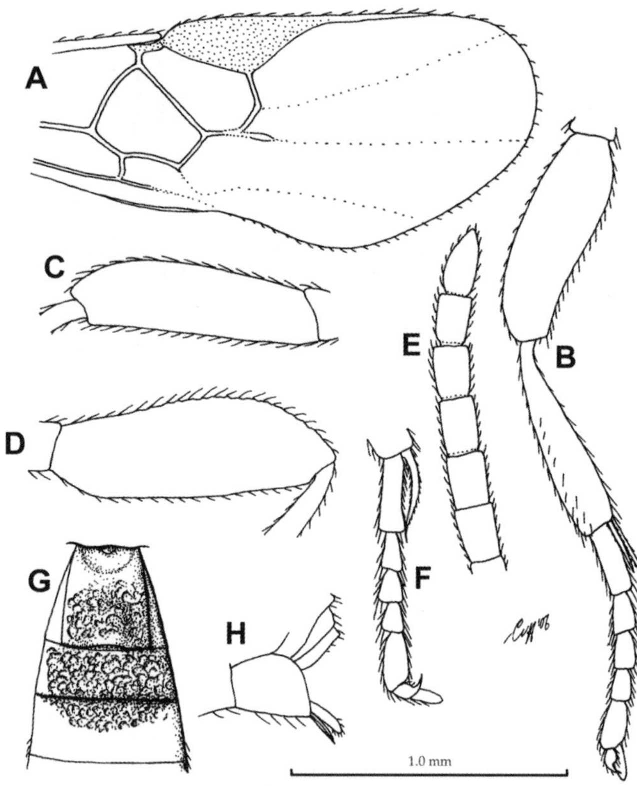

FIGURE 10 Cotesia crassifemorata *Achterberg, female, holotype. A, fore wing; B, middle leg, lateral aspect; C, fore femur, lateral aspect; D, hind femur, lateral aspect; E, apex of antenna; F, fore tarsus, lateral aspect; G, first-third metasomal tergites, dorsal aspect; H, ovipositor sheath. A, G: 1.0 × scale-line; B–D: 1.5 ×; E, F, H: 1.9 ×.*

(Fig. 9B), vein r of the fore wing is subvertical (Fig. 9A), vein 1–SR of fore wing distinctly longer than wide and pointing to vein cu-a (Fig. 9A); the lateral lobes of the mesoscutum are densely punctate and the third metasomal tergite is largely smooth basally.

BIOLOGY. Gregarious koinobiont endoparasitoid of caterpillars of Noctuidae.

DISTRIBUTION. No specimens seen from Greenland, but reported from Greenland by Papp (1989).

Nearctic (western U.S.A, Canada: Quebec).

***Protapanteles fulvipes* (Haliday, 1834). Figs 13F–L**
Henriksen 1939: *Apanteles fulvipes* Hal.

2.5–4.0 mm. The propodeum is largely regularly and more or less obliquely striate in Greenland specimens; in European specimens the propodeum is usually more irregularly rugose or rugulose. Rather variable in colour, dark specimens may run to the Palaearctic *P. nivalis* (Papp, 1983), but *P. fulvipes* has the hind femur 5 times as long as wide and the area of the second metasomal tergite comparatively narrow.

BIOLOGY. Gregarious koinobiont endoparasitoid of caterpillars of mainly Noctuidae, Geometridae, Lasiocampidae, Notodontidae and Lymantriidae.

DISTRIBUTION. S, SW, NW (north to Pituffik/Thule), NE (widespread from Île de France south to southern Scoresby Sund), SE (Saqqisikuik/Skjoldungen).

Holarctic (widespread in Palaearctis, eastern U.S.A.).

***Protapanteles pallipes* (Reinhard, 1880). Figs 12, 13A–E**
Henriksen 1939: not recorded.

1.8–3.0 mm. Dark with posterior half of mesoscutum strongly shiny and more or less punctate, the median carina of propodeum irregular and usually distinct anteriorly, vein 1–SR of the fore wing pointing to vein 1–CU1 (Fig. 12C) and second tergite distinctly sculptured medially (Fig. 12A).

BIOLOGY. Gregarious koinobiont endoparasitoid of caterpillars of mainly Noctuidae.

DISTRIBUTION. Recently reported from Greenland by Papp (1989) but no specimens found in the ZMUC collection.

Holarctic and Oriental, very widespread.

***Praon brevistigma* van Achterberg, 2006. Fig. 14**
Henriksen 1939: *Praon abjectus* Hal.

1.2–2.1 mm. Small, dark brown species. Similar to the Palaearctic *P. abjectum* (Haliday, 1834) but differs by the wider pterostigma, the slightly lower number of antennal segments (female: 14–15, rarely 13) and the short vein 1–R1 of fore wing (about 1.5 times width of pterostigma). *P. abjectum* is reported also from the Nearctic region (Yu *et al.*, 2005) but not recorded in the revision by Johnson (1987); this may indicate that the occurrence in the Nearctic region is based on misidentifications. Runs in the key to the Nearctic *Praon* spp. by Johnson (1987) to *P. unicum* Smith, 1944, but this species differs by the shape of the pterostigma (length of pterostigma 4.5 times its width (about 3 times in Greenland specimens)), length of vein 1–R1 of fore wing (2.5 times longer than width of pterostigma (about 1.5 times)), ovipositor sheath slightly concave dorsally and rather wide apically (straight dorsally and comparatively narrow apically), first and second metasomal tergites yellowish-brown (dark brown) and vein m-cu of fore wing completely sclerotised (more or less reduced). The shape of the pterostigma and the colour of the body are reminiscent of the Nearctic *P. artemisaphis* Smith, 1944, but this species has vein 1–R1 about as long as width of the pterostigma, the antenna of female with 17 segments, the ovipositor sheath less narrowed apically, the

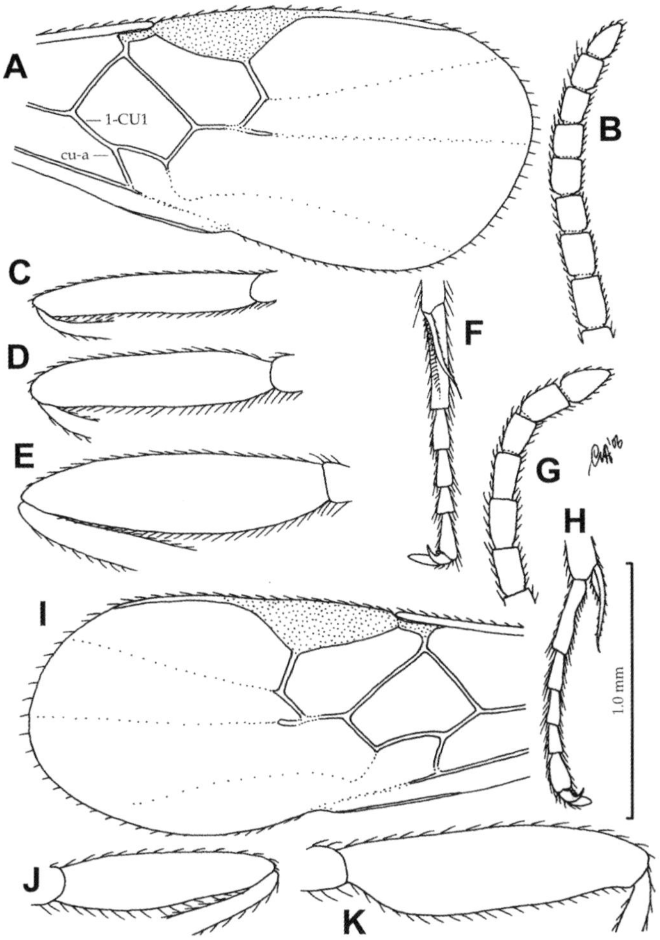

FIGURE 11 *A–F,* Cotesia hallii *(Packard), female, N Greenland, Thule; G–K,* C. eliniae *(Papp), female, Greenland, Eskimonæs. A, I, fore wing; B, G, apex of antenna; C, J, fore femur, lateral aspect; D, middle femur, lateral aspect; E, K, hind femur, lateral aspect; F, H, fore tarsus, lateral aspect. A: 1.0 × scale-line; B–F: 1.5 ×; G, H, J, K: 1.7 ×; I : 1.1 ×.*

lateral mesoscutal lobes nearly glabrous and the legs darker.

Praon peregrinum Ruthe, 1859, was described from Iceland and is one of the northern species with a dark third antennal segment. It does not fit the new species; the length of the antenna is about as long as the body, and consists of 16–18 segments (female) or 18–20 segments (male). Moreover, it is considered to be a synonym of *P. abjectum* (Haliday) (Yu *et al.*, 2005).

BIOLOGY. Unknown, but related species are koinobiont endoparasitoids of aphids (Aphididae).

DISTRIBUTION. S, SW (north to Uummannaq district: Ukkusissat Sulluat), NE (Zackenberg; Jameson Land: Coloradodal, Muslingeelvbugt). Only found in Greenland.

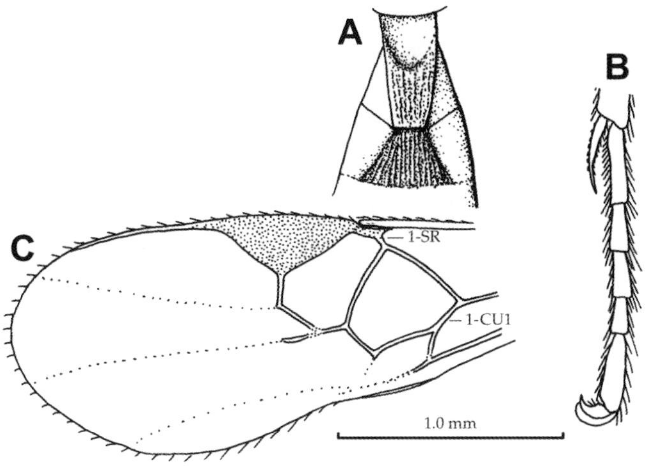

FIGURE 12 Protapanteles pallipes *(Reinhard), female, Netherlands, Lelystad. A, first and second metasomal tergites, dorsal aspect; B, fore tarsus, lateral aspect; C fore wing. A: 1.5 × scale-line; B: 2.3 × ; C: 1.3 ×.*

***Trioxys compressicornis* Ruthe, 1859. Figs 15A–D**
Henriksen 1939: not recorded.

2.3–2.5 mm. The only Aphidiine species in Greenland with a largely yellow head, and the female has very conspicuous prongs (Fig. 15D).

BIOLOGY. Koinobiont endoparasitoid of Callaphidid aphids on birch (*Betula* spp.).

DISTRIBUTION. S (only found in Narsarsuaq).

Holarctic, low arctic-temperate (Iceland and Finland to Andorra, Canada: Quebec and Yukon).

***Aphidius (Aphidius) avenae* Haliday, 1834. Figs 16F–L, 17A–C**
Henriksen 1939: *Aphidius picipes* Nees.

1.5–3.5 mm. Black or dark brown. Specimens with comparatively fine baso-lateral carinae at the first tergite are similar to the Holarctic *A. urticae* Haliday, 1834, but differ by the distinctly convex clypeus, the less flattened face and the shorter vein 1–R1 (but this last character state is rather variable). Completely black males occur with even the clypeus blackish; normally the clypeus is brownish-yellow.

BIOLOGY. Koinobiont endoparasitoids of many species of aphids (Aphididae).

DISTRIBUTION. S, SW (north to Upernavik Laksefjord). Widespread..

Palaearctic, Oriental, northern temperate-subtropical, widely distributed. The species has been introduced in Africa, South America and U.S.A.

***Aphidius (Euaphidius) cingulatus* (Ruthe, 1859). Figs 2E, 17D–I, 18A, B, E**
Henriksen 1939: not recorded.

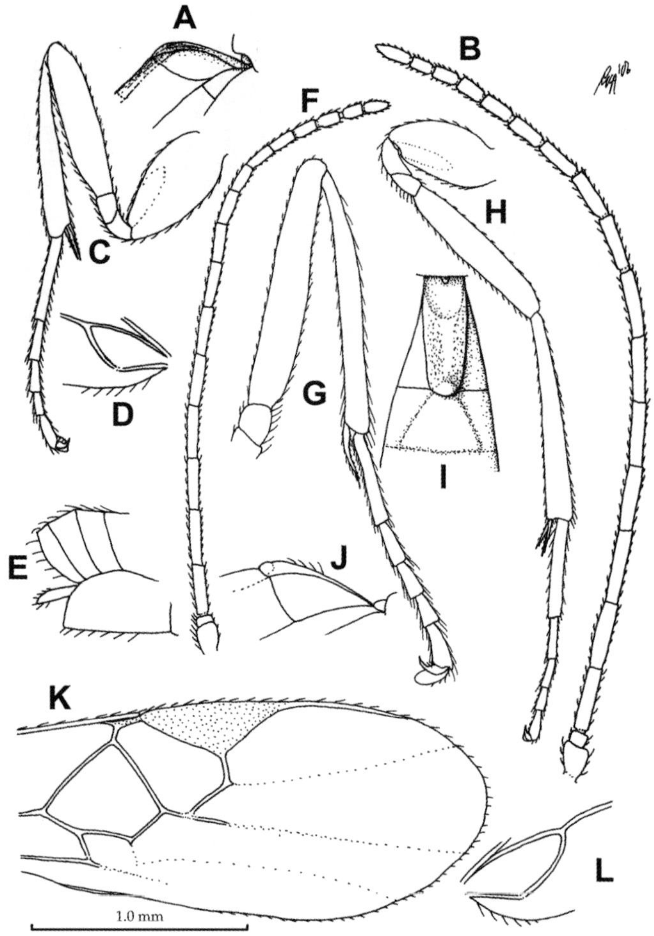

FIGURE 13 *A–E,* Protapanteles pallipes *(Reinhard) female, Netherlands, Lelystad; F–L,* P. fulvipes *(Haliday), female, W Greenland, Søndrestrømfjord Air Base. A, J, first metasomal tergite, lateral aspect; B, F, antenna; C, H, hind leg, lateral aspect; D, L, detail of vein cu-a of hind wing; E, ovipositor sheath, lateral aspect; G, fore leg, lateral aspect; I, first and second metasomal tergites, dorsal aspect; K, fore wing. A: 1.3 × scale-line; B, D–F, I, J, L: 1.2 ×; C, K: 1.0 ×; G: 1.6 ×; H: 0.9 ×.*

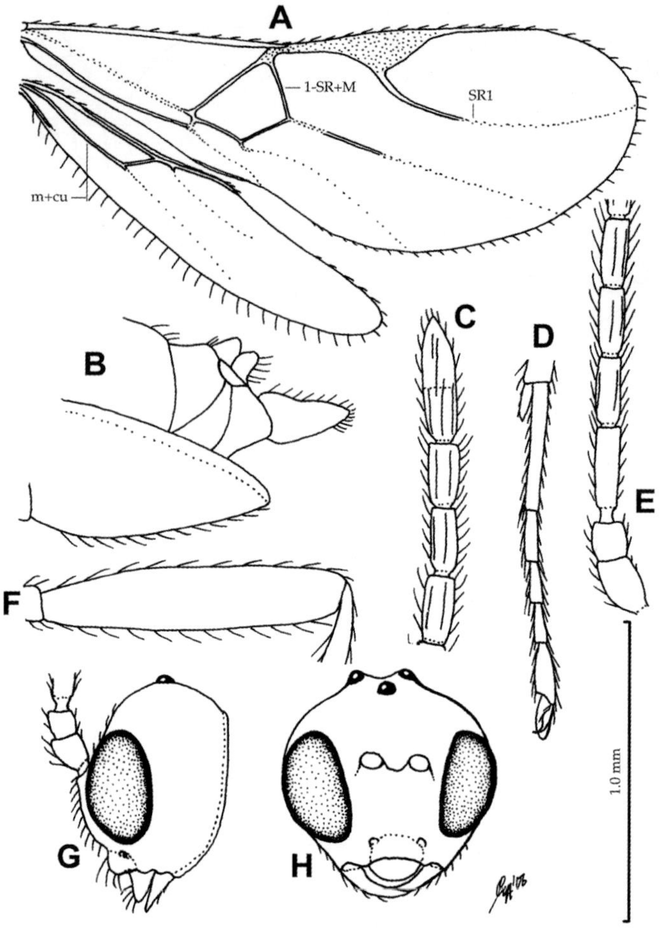

FIGURE 14 Praon brevistigmus *Achterberg, female, holotype. A, wings; B, ovipositor sheath, lateral aspect; C, apex of antenna; D, fore basitarsus, lateral aspect; E, base of antenna; F, hind femur, lateral aspect, G, head, lateral aspect; H, head, anterior aspect. A: 1.0 × scale-line; B–F: 2.3 ×; G, H: 1.7 ×.*

2–4 mm. Mainly dark brown species. The mesosoma (except for the pronotum ventrally and the mesopleuron antero-dorsally) is dark brown as in other boreal specimens and the temple slightly longer (length of eye about 1.4 times temple in dorsal view; 1.6–1.8 times in European specimens). Males (as in the subgenus *Aphidius* Nees) have the third and fourth antennal segments much more robust than the females and the baso-lateral striae of first tergite are less developed. *A. cingulatus* has the clypeus rather wide and comparatively weakly convex (Figs 17D; but sometimes in males moderately convex), the ventral half of pronotal side yellowish-brown or pale brown (exceptionally dark brown), the pterostigma dark brown and the palpal segments robust. The temples are less directly narrowed in males than in females.

BIOLOGY. Koinobiont endoparasitoids of many species of aphids (Aphididae: Pterocommatinae) on *Populus* and *Salix* spp.

DISTRIBUTION. S, SW, NW (north to Pituffik/Thule), NE (Scoresby Sund: Charcot Land, Gåseland: Faxe Sø; Mestersvig). Seems to be common in Greenland.

Holarctic, Oriental, northern temperate-subtropical, widely distributed.

***Aphidius* (*Aphidius*) *tarsalis* van Achterberg, 2006.**
Figs 15E–J, 16A–E

Henriksen 1939: not recorded.

1.3–1.9 mm. Dark, small species. Similar to *A. matricariae* Haliday, 1834, because of the rather robust fore tarsus, the narrow first subdiscal cell of fore wing and the comparatively

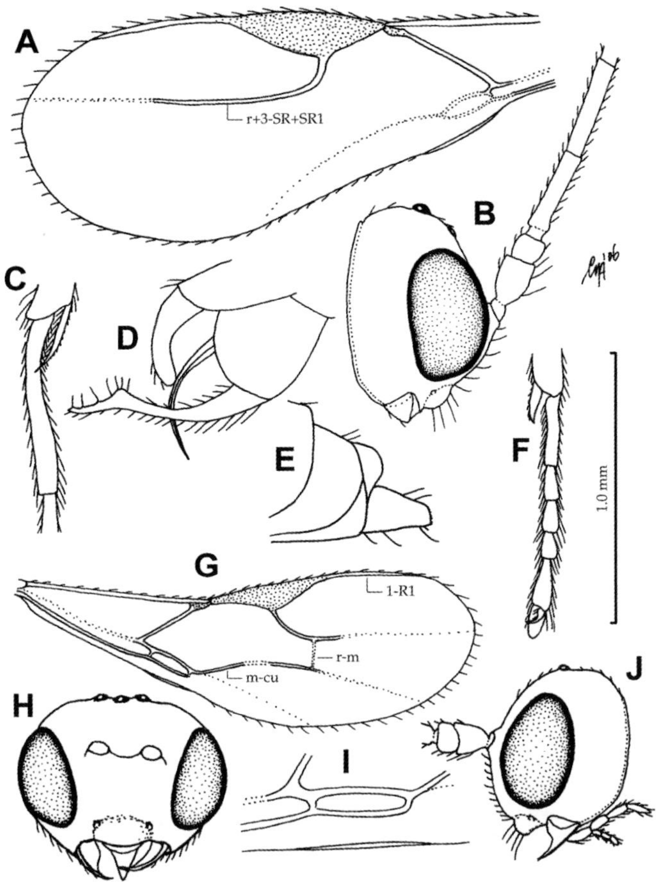

FIGURE 15 *A–D,* Trioxys compressicornis *Ruthe, female, S Greenland, Narsarsuaq; E–J,* Aphidius tarsalis *Achterberg, female, holotype. A, G, fore wing; B, J, head, lateral aspect; C, fore basitarsus, lateral aspect; D, genitalia, lateral aspect; E, ovipositor sheath, lateral aspect; F, fore tarsus, lateral aspect; H, head, anterior aspect; I, detail of first subdiscal cell of fore wing. A: 1.0 × scale-line; B–D: 1.5 ×; E: 2.8 ×; F, I: 2.3 × G: 1.2 ×; H, J: 1.9 ×.*

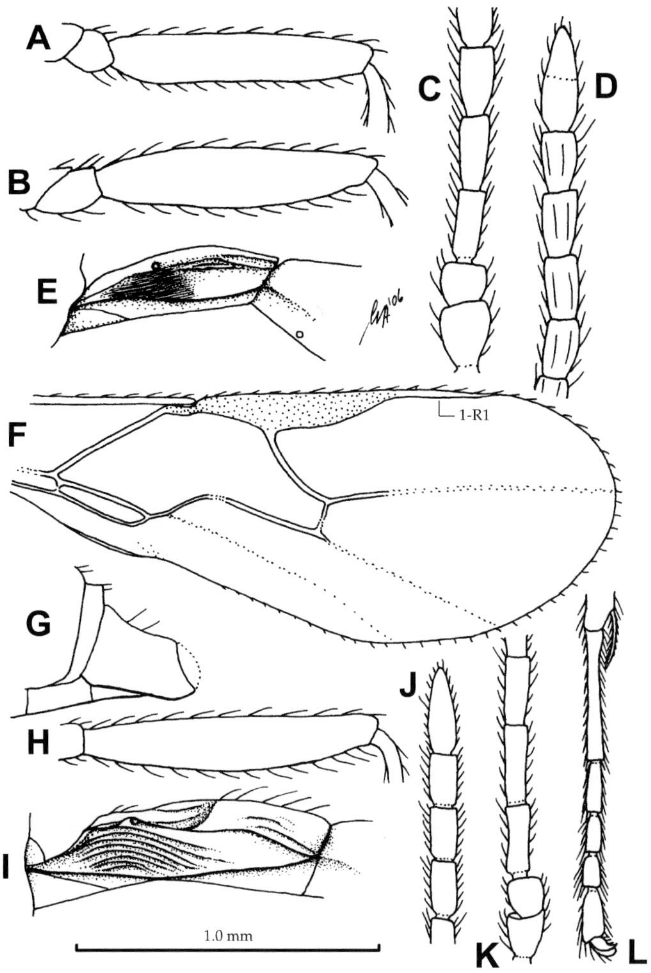

FIGURE 16 *A–E,* Aphidius tarsalis *Achterberg, female, holotype; F–L,* A. avenae *Haliday, female, W Greenland, Upernavik. A, H, fore femur, lateral aspect; B, hind femur, lateral aspect; C, K, base of antenna; D, J, apex of antenna; E, I, first metasomal tergite, lateral aspect; F, fore wing; G, ovipositor sheath, lateral aspect; L, fore tarsus, lateral aspect. A–E: 1.1 × scale-line; F: 1.0 ×; G, I: 2.3 ×; H, J–L: 1.5 ×.*

narrow fore wing, but *A. tarsalis* has normal 4–segmented maxillar palpi. Among the Nearctic species it resembles *A. berberidus* Smith, 1944. However, this species has a widely truncate apex of the ovipositor sheath, the face paler than the vertex, vein 1–R1 of the fore wing distinctly shorter than the pterostigma, the first tergite smooth and its males have the antenna 16– or 17–segmented.

BIOLOGY. Unknown, but related species are koinobiont endoparasitoids of aphids (Aphididae).

DISTRIBUTION. S. (Nunap Isua/Kap Farvel-area: Pamialluk; Narsarsuaq). Only recorded from Greenland.

***Blacus* (*Blacus*) *groenlandicus* van Achterberg, 2006**
Figs 18C, D, F–H
Henriksen 1939: not recorded.

About 2 mm. Black. It does not fit in the key to the Nearctic species of the genus *Blacus* (subgenus *Blacus*) by van Achterberg (1988). In the key to the Palaearctic species it runs either to *B. modestus* Haeselbarth, 1973, known from Denmark to Bulgaria, or to *B. nivalis* van Achterberg, 1988, from the Canary Islands. It differs from *B. modestus* by having the length of malar space about equal to basal width of mandible, the length of the ovipositor sheath about 0.35 times as long as the fore wing, the first discal cell of the fore wing of female comparatively widely truncate anteriorly and parastigma rather long (Fig. 315 in van Achterberg, 1988). *B. nivalis* has vein cu-a of the fore wing postfurcal by about its length, vein 3–SR+SR1 of fore wing ending near apex of wing, first discal cell of fore wing of female acute anteriorly, and parastigma short (Fig. 303 in van Achterberg, 1988).

BIOLOGY. Unknown, but related species are koinobiont endoparasitoids of beetle larvae belonging to the families Nitidulidae, Curculionidae, Anobiidae and Cryptophagidae.

DISTRIBUTION. S (the holotype is from Ivittuut; only known from Greenland).

Dinocampus coccinellae **(Schrank, 1802). Figs 20A–C**
Henriksen 1939: not recorded.

2.0-4.5 mm. Mainly blackish with some parts of the body dark reddish.

BIOLOGY. A parthenogenetic (but rarely males occur) and koinobiont endoparasitoid of adult Coccinellidae. It is able to re-parasitize its host beetle.

DISTRIBUTION. SW (only found in Itilleq, eastern end).

Nearly cosmopolitan, but not reported from the Afrotropical region.

Meteorus arcticus **Papp, 1989. Figs 19A–E, 20D**
Henriksen 1939: not recorded.

4–5 mm. Slightly larger than the common *M. rubens* (Nees) described from Scoresby Sund (E Greenland). Collected in yellow pan traps placed in vegetation with *Vaccinium uliginosum, Empetrum nigrum, Polygonum viviparum, Betula nana, Dryas octopetala, Rhododendron lapponicum* and *Salix glauca*.

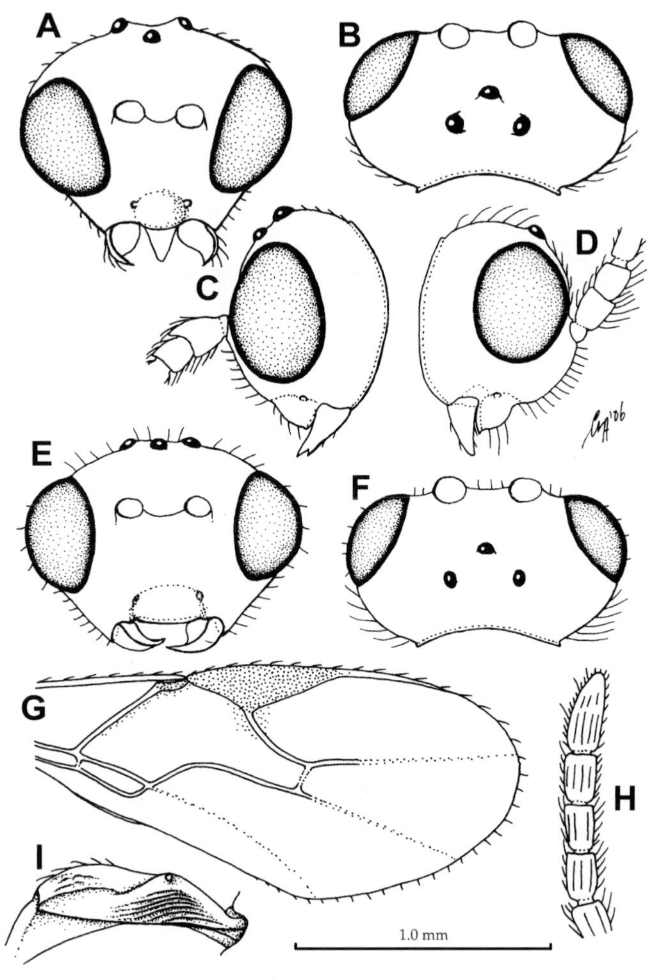

FIGURE 17 *A–C,* Aphidius avenae *Haliday, female, W Greenland, Upernavik; D–I,* A. cingulatus (*Ruthe*), *female, W Greenland, Søndre Ivortoq. A, E, head, anterior aspect; B, F, head, dorsal aspect; C, D, head, lateral aspect; G, fore wing; H, apex of antenna; I, first metasomal tergite, lateral aspect. A–C, H, I: 1.8 × scale-line; D–F: 1.6 ×; G: 1.0 ×.*

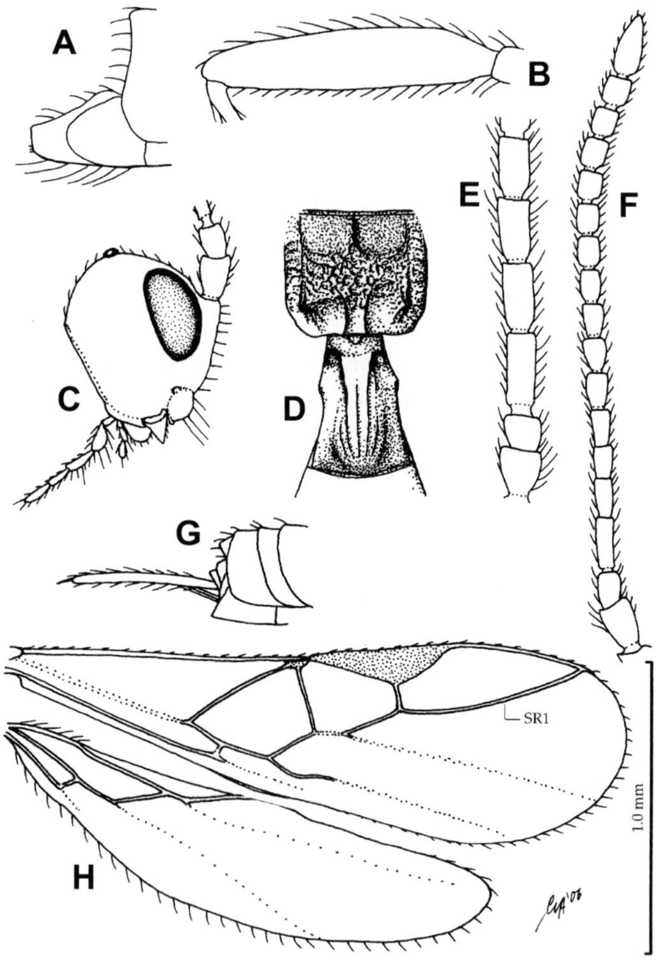

FIGURE 18 *A, B, E,* Aphidius cingulatus (*Ruthe*)*, female, W Greenland, Søndre Ivortoq; C, D, F–H,* Blacus groenlandicus *Achterberg., female, holotype. A, G, ovipositor sheath, lateral aspect; B, hind femur, lateral aspect; C, head, lateral aspect; D, propodeum and first metasomal tergite, dorsal aspect; E, base of antenna; F, antenna; H, wings. A: 1.4 × scale-line; B–F: 1.5 ×; G, H: 1.0 ×.*

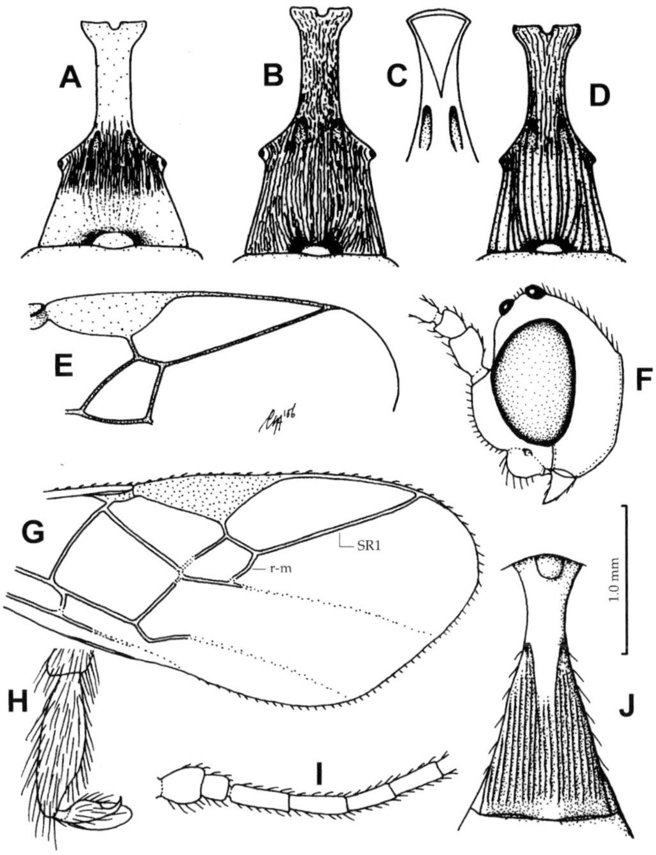

FIGURE 19 *A–E,* Meteorus arcticus *Papp, after* Papp (1989); *F–J,* M. rubens *(Nees), female, W Greenland, Buksefjorden. A, B, D, J, first metasomal tergite, dorsal aspect; C, the same, but ventro-basal aspect; E, detail of pterostigma and second submarginal cell of fore wing; F, head, lateral aspect; G, fore wing; H, outer hind claw; I, base of antenna. F, H–J: 2.0 × scale-line; G: 1.0 ×.*

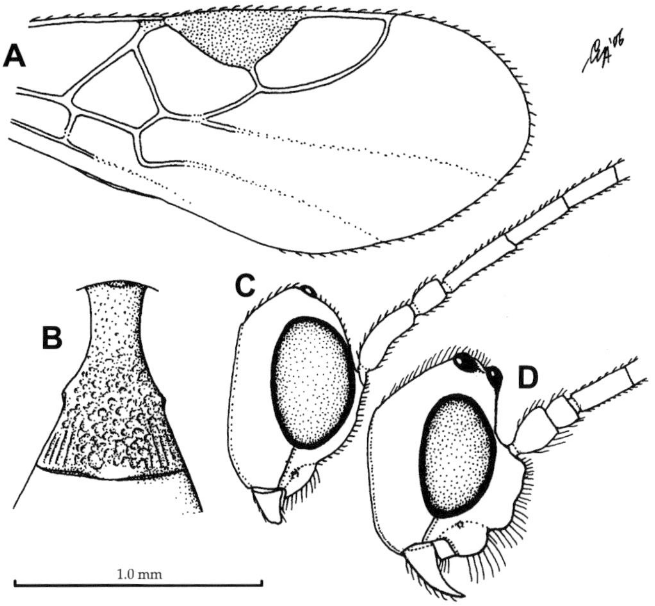

FIGURE 20 *A–C,* Dinocampus coccinellae *(Schrank), female, Greenland, Itiveq;. D,* Meteorus arcticus *Papp, female, Greenland, Godhavn. A, fore wing; B, first metasomal tergite, dorsal aspect; C, D, head, lateral aspect. A: 1.0 × scale-line; B–D: 1.2 ×.*

An uncommon species very closely related to *M. rubens* (Nees). It may have the dorsope of the first tergite weakly impressed (but absent in examined specimens; the dorsope may be indistinctly developed also in *M. rubens*), the tergite slightly more abruptly narrowed submedially (but in front of the spiracles largely parallel-sided; Figs 19A, B, D), the second submarginal cell of fore wing is widened posteriorly (Fig. 19E, but this is both in *M. rubens* and in this species variable); and the tarsal claws more slender subbasally. The few character states which separate both species are given in the key.

BIOLOGY. Unknown, but similar species are koinobiont endoparasitoids of lepidopterous larvae.

DISTRIBUTION. SW (Kapisillit, Nassuttooq/Nordre Strømfjord: Nuersorfiaraq, Qeqertarsuaq/Godhavn), NE (Zackenberg). Only known from Greenland.

Meteorus rubens **(Nees, 1811). Figs 2F, 19F–J**

Henriksen 1939: *Meteorus islandicus* Ruthe.

3–6 mm. Large and conspicuous. Very variable in colour (normally largely yellowish but nearly completely blackish specimens occur) and sculpture (the face and clypeus are normally distinctly sculptured but face and clypeus may be largely smooth).

BIOLOGY. Koinobiont endoparasitoid of larvae mainly of the family Noctuidae. Has been reared from Greenland *Rhyacia quadrangula* (Zetterstedt).

DISTRIBUTION. S, SW, NW (north to Sigguup Nunaa/Svartenhuk: Kangiussap Qinngua), NE (Scoresby Sund: Charcot Land, Rypefjord; Jameson Land: Lollandselv), SE (Saqqisikuik/Skjoldungen). Seems to be very common.

Holarctic, Oriental, Neotropical, low arctic-tropical, widespread.

References

Achterberg, C. van, (1988): Revision of the subfamily Blacinae Foerster (Hymenoptera, Braconidae). – *Zoologische Verhandelingen Leiden* 249: 1–324, Figs 2–1250.

Achterberg, C. van. (2006): The Braconidae (Hymenoptera) of Greenland. – *Zoologische Mededelingen Leiden* 80 (1): 13–62.

Henriksen, K.L. (1939): A revised index of the insects of Grønland, containing a supplement of the insect list in Kai L Henriksen & Will. Lundbeck: Grønlands Landarthropoder. – *Meddelelser om Grønland* 119 (10): 112 pp.

Lundbeck, W. (1897): Hymenoptera groenlandica. – *Videnskabelige Meddelelser fra den Naturhistoriske Forening i Kjøbenhavn* 1896: 220–251.

Papp, J. (1984): Palaearctic species of *Microgaster* Latreille (=*Microplitis* Förster) with description of seven new species (Hymenoptera, Braconidae, Microgastrinae). – *Entomologische Abhandlungen* 47: 95–140.

Papp, J. (1989): Contribution to the braconid wasps of Greenland, Denmark (Hymenoptera: Braconidae). – *Folia Entomologica Hungarica* 100: 95–104, Figs 2–26.

11.15. Ichneumonidae (Ichneumonoidea)

Reijo Jussila

Ichneumonidae is probably the largest insect family. The members have a long and thin body with a narrow waist. The antennae have at least 12 flagellomeres. The cross-vein 2m-cu of the forewing is almost always present. The middle and hind legs have distinct trochantelli. The metasoma normally attaches near the bottom of the propodeum, and tergites 1–7 have spiracles.

Ichneumonidae are important ecto- or endoparasitoids of other insects and arachnids. Ectoparasitoids are usually also idiobionts, terminating the development of the host by venom injected at oviposition and the parasitoid larva initiating its development immediately. Endoparasitism is usually correlated with the koinobiont lifestyle, where the host will continue its development for some time after being parasitized, the parasitoid only developing when the host has reached a certain developmental stage. Mostly immature stages, *e.g.*, larvae or pupae, are attacked. The female usually locates the food plant of the host and searches with her antennae for a suitable host. The ovipositor of some species is longer than their body. A long ovipositor allows the female to inject her eggs into hidden hosts such as leaf-rolling or stem-boring caterpillars. However, most ichneumonids have distinctly shorter ovipositors. Some species lay their eggs externally and the larvae attack the host from the outside. The larvae either pupate inside the dead host or form an outside cocoon. Most adults are active in daytime and feed on flowers.

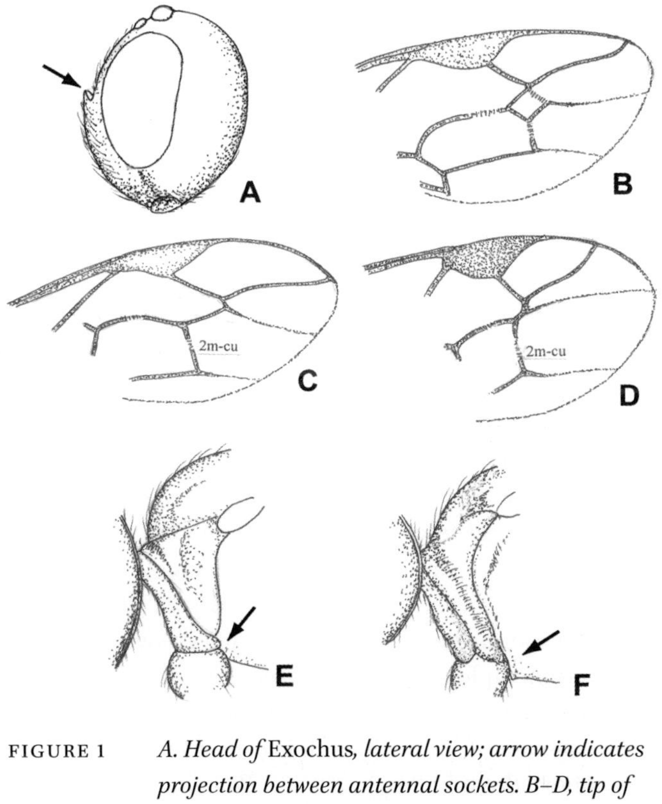

FIGURE 1 *A. Head of* Exochus, *lateral view; arrow indicates projection between antennal sockets. B–D, tip of right front wing of B.* Mesochorus, *C.* Ophion, *D.* Barycnemis. *E–F. Front part or thorax, lateral view of E.* Campoletis, *F.* Barycnemis. *Arrows indicate ventroposterior corner of propleuron.*

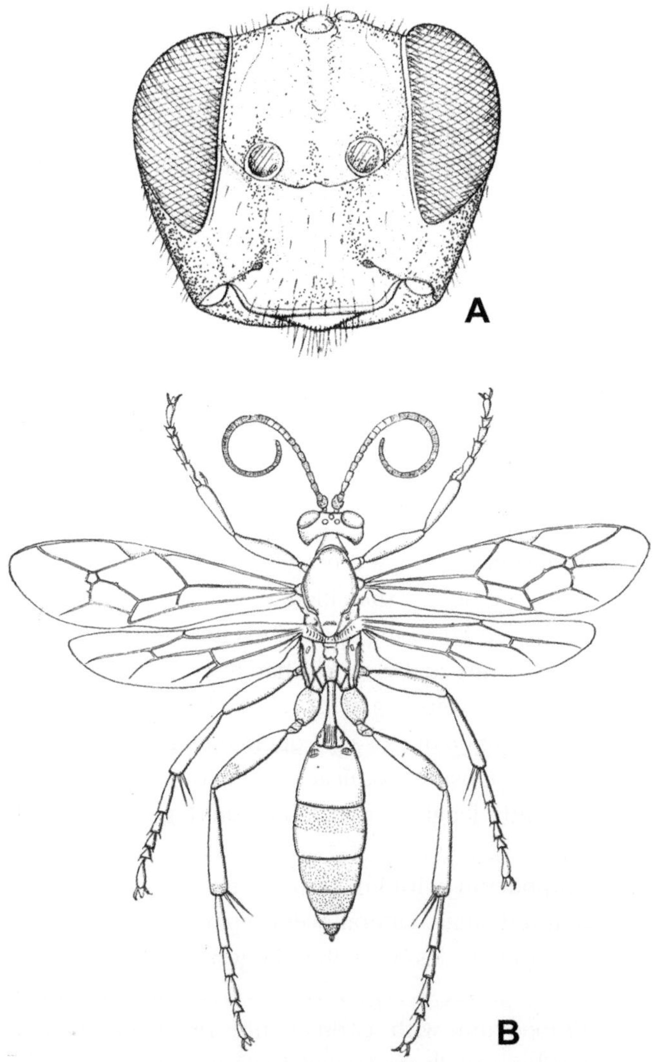

FIGURE 2 Ichneumon sarcitorius. *A. Face; B. Female habitus (according to* Constantineanu 1959).

Key to subfamilies of Ichneumonidae found in Greenland

(For terminology, see Richards 1956 and Fitton *et al.* 1988)

1. Wings absent....................... *Gelis* (Cryptinae)
– Wings normal or vestigial............................2
2. Face protuberant, forming an almost continuous strongly convex bulge in lateral view, clypeus not separated from face (Fig. 1A) 3
– Face more or less flat.................................4
3. Upper edge of face with a projection over between antennal sockets (Fig. 1A). Scape 1.2–1.7 as long as wide .... Metopiinae
– Upper edge of face without projection over between antennal sockets. Scape subcylindric, 1.8–4 as long as wide ...... *Orthocentrus*, *Stenomacrus* and *Plectiscus* (Orthocentrinae)
4. Areolet of fore wing large and more or less rhombic, always complete (Fig. 1B). Clypeus not separated from face by distinct groove. Spiracle of metasomal segment 1 near the middle. Ovipositor rather long but never as long as metasoma, with no subapical notch. Male clasper long and spine-like............. .... Mesochorinae
– Areolet various, but usually not rhombic. Clypeus usually separated from face by distinct groove. Male clasper never long and spine-like...................5
5. Spiracle of metasomal segment 1 posterior to middle...6
– Spiracle of metasomal segment 1 at or anterior to middle... 13
6. Metasoma compressed, its segments 3 and 4 deeper than wide...7
– Metasoma depressed or cylindrical, its segments 3 and 4 not deeper than high 11
7. Areolet of fore wing open, the remaining vein apical to vein 2m-cu by more than two-fifths length of remaining vein (Fig. 1C)................Ophioninae
– Areolet open or closed; if open, then remaining vein basal, opposite, or apical to vein 2m-cu by less than two-fifths length of remaining vein (Figs 1B, 5A, B) 8
8. Propodeum without carinae, its sculpture coarse and reticulate. Lateral ocelli separated from occipital carina by less than their largest diameter......... ....Anomaloninae
– Propodeum with carinae, not reticulate. Lateral ocellus usually separated from occipital carina by more than its largest diameter.......................9
9. Ventroposterior corner of propleuron strongly produced, the more or less angulated lobe touching or overlapping pronotum (Fig. 1E). Postpectal carina of mesothorax complete. Clypeus not separated from face by distinct groove 10
– Ventroposterior corner of propleuron not developed as a distinct lobe, not angulated (Fig. 1F). Postpectal carina incomplete. Clypeus separated from face. Apical margin of clypeus with a fringe of long parallel setae................................ Tersilochinae
10. Clypeus separated from face by a groove. Stigma of fore wing relatively wide (Fig. 1D)......Cremastinae
– Clypeus weakly or not separated from face. Stigma relatively slender (cf. Fig. 1C) ....Campopleginae
11. Inner margins of eyes strongly convergent ventrally ....*Symplecis* (Orthocentrinae)
– Inner margins of eyes not convergent ventrally....12
12. Clypeus more or less wide and flattened, its apical margin as if truncate (Fig. 2A). Vein M+Cu of hind wing straight (Fig. 2B). Postpetiole more or less transverse and tergum 2 with well developed thyridia and transverse gastrocoeli (Fig. 2B). Ovipositor rarely extending conspicuously beyond metasomal apex .... Ichneumoninae
– Clypeus moderately to strongly convex, its apical margin usually convex and impressed (Figs. 3A, 9C). Vein M+Cu often strongly arched. Postpetiole elongate, tergum 2 without gastrocoeli and thyridia usually small or absent (Fig. 3C). Ovipositor often extending conspicuously posterior to metasomal apex ....Cryptinae
13. Dorsal tooth of mandible wide, more ore less distinctly divided into upper and lower teeth. Tergum 1 rectangular (Fig. 3D). Ovipositor not extending beyond metasomal apex .... Diplazontinae
– Dorsal tooth of mandible not divided into two teeth 14
14. Front tibia with an apical tooth ...Ctenopelmatinae
– Front tibia without apical tooth....................15
15. Metasomal terga 2–4 with submedian pair of deep oblique grooves. Ovipositor rather long............. .... *Glypta* (Banchinae)
– Metasomal terga 2–4 without such grooves 16
16. Submetapleural carina raised anteriorly (Fig. 3E). Metasoma compressed. Ovipositor short. Vein Cu+cu-a of hind wing (cf. Fig. 4E, F) intercepted far above the middle............*Banchus* in Banchinae
– Submetapleural carina not raised anteriorly...... 17
17. Apical margin of clypeus with a median notch (appearing bilobed) (Fig. 3F). Terga 3–5 of metasoma

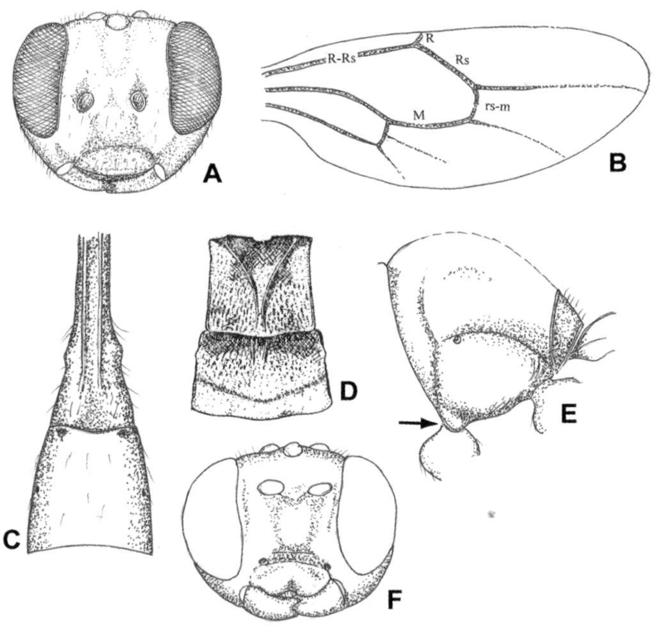

FIGURE 3 *A. Head of* Atractodes gilvipes. *B. Right hind wing of* Atractodes. *C–D. Metasomal tergites 1–2, dorsal view of D.* Atractodes *D.* Diplazon. *E. Propodeum of* Banchus, *lateral view; arrow indicates submetapleural carina. F. Head of* Pimpla.

with a large, ill-defined tubercle on each side of middle. Tarsal claws often with a tooth or basal lobe, especially in females. Metasomal segment 1 with glymma. . . . . Pimplinae
– Clypeus, metasomal terga and tarsal claws not as above 18
18. Clypeus small and strongly convex, without parallel setae. Epomia of pronotum not noticeable. Small and delicate specimens (length 3–5 mm) . . . . several species in Orthocentrinae
– Clypeus wide and not strongly convex, its apical margin with fringe of parallel setae. Epomia strong. Eggs often attached to ovipositor by stalk. More robust species (length 5–7 mm). . . . . Tryphoninae

PIMPLINAE

Small to large species. Clypeus separated from face by a groove, its apical margin with a median notch giving it a bilobed appearance. Mesoscutum smooth; propodeum often with reduced carinae. Metasoma cylindrical or dorso-ventrally flattened; its segment 1 usually short and wide, usually with glymma and with spiracle before its middle; terga 2–4 usually with surface impressions and swellings. Ovipositor short to long; its apex without dorsal notch, but often with ventral ridges or teeth.

Most Pimplinae are idiobiont parasitoids of prepupae or pupae of Lepidoptera.

Key to genera

1. Mesopleural furrow angled opposite episternal scrobe (Fig. 4A) 2
– Mesopleural furrow not angled opposite episternal scrobe (Fig. 4B). 3
2. Forewing with vein 3rs-m present (as in Fig. 5B). . . . . *Scambus* (*Epiurus*)
– Forewing with vein 3rs-m absent (as in Fig. 5A) . . . . *Tromatobia*
3. Segment 2 of hind tarsus distinctly shorter than segment 5 and inner margin of eye strongly indented just above level of antennal socket (Fig. 4C) . . . .*Itoplectis*
– Segment 2 of hind tarsus not distinctly shorter than segment 5 and inner margin of eye only weakly indented just above level of antennal socket (Fig. 4D) . . . . *Pimpla*

***Scambus brevicornis* (Gravenhorst, 1829)**

Henriksen 1939: not recorded.

BIOLOGY. Endoparasitoid of Lepidoptera pupae.

DISTRIBUTION. SW (Allumersat/Bjørnesund, Ameralik, Nuup Kangerlua/Godthåbsfjord: Kapisillit; Kangerlussuaq/Søndre Strømfjord). New species to Greenland.

Holarctic.

***Tromatobia forsiusi* (Hellén, 1915)**

Henriksen 1939: not recorded.

BIOLOGY. All *Tromatobia* species are gregarious predators in spider egg sacs.

DISTIRIBUTION: SW (one male specimen from Søndre Isortoq).

Hitherto only Palaearctic.

***Itoplectis quadricingulata* (Provancher, 1880)**

Henriksen 1939: *Itoplectis alternans* Grav.

BIOLOGY. A primary parasitoid of weakly concealed or cocooned but easily accessible smaller Lepidoptera pupae.

DISTRIBUTION. S, SW (north to Uummannaq district). Seems to be common in Greenland.

Holarctic.

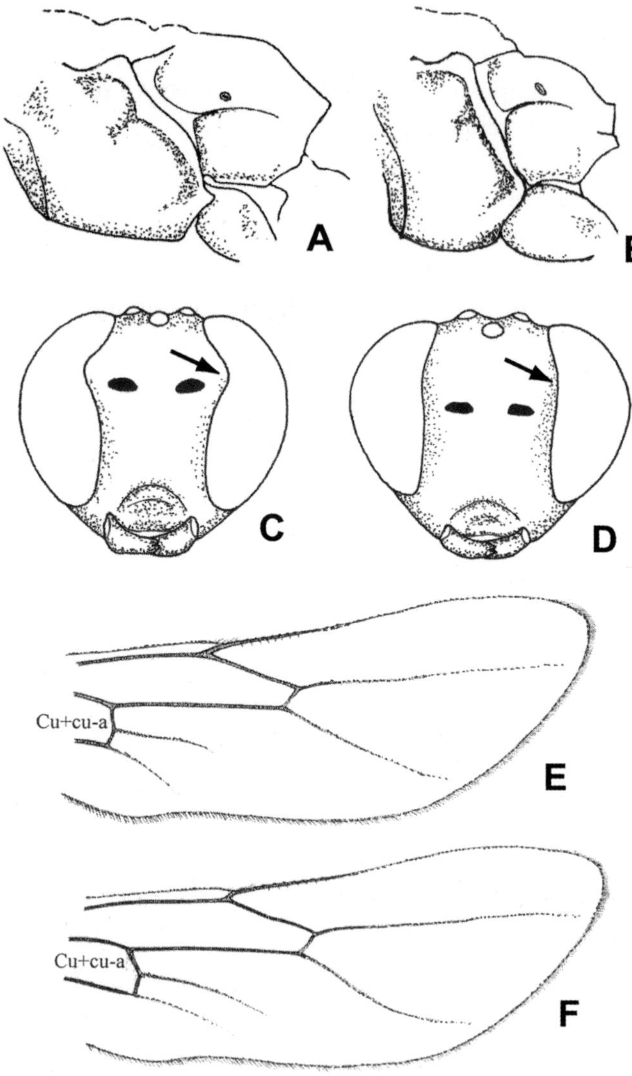

FIGURE 4 *A–B. Mesosoma, lateral view of A.* Scambus, *B.* Pimpla. *C–D. Head of C.* Itoplectis; *D.* Pimpla. *Arrows indicate inner eye margins. E–F. Right hind wing of E.* Mesoleius; *F.* Campodorus ultimus.

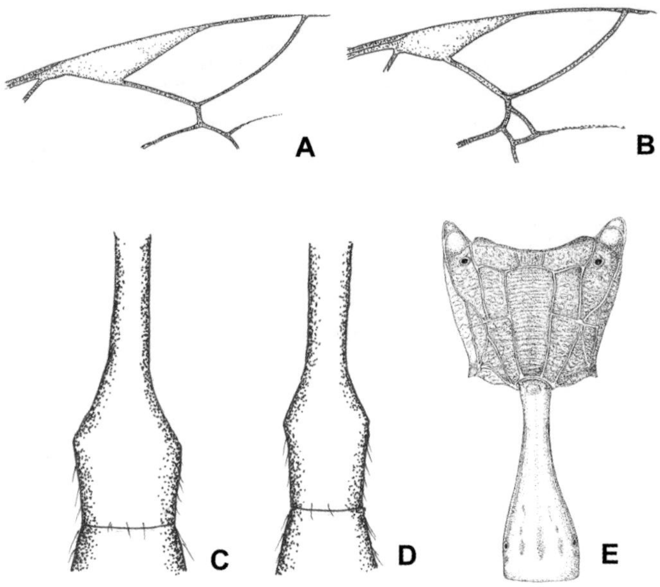

FIGURE 5 *A–B. Part of right front wing in A.* Enytus; *B.* Campoletis. *C–D. Metasomal tergite 1, dorsal view in C.* Hyposoter; *D.* Olesicampe. *E. Propodeum and metasomal tergite 1 of* Atractodes, *dorsal view (according to* Jussila 2001).

***Pimpla sodalis* Ruthe, 1859. Figs 11.1–3C, 10A**

BIOLOGY. Attacks the geometrid moth *Epirrita autumnata* in the Fennoscandian Lapland (Nuorteva & Jussila 1967, 1969, Jussila & Nuorteva 1968). All *Pimpla* species are endoparasitoids of Lepidoptera pupae.

DISTRIBUTION. S, SW (very common north to Uummannaq district: Ukkusissat Sulluat), NE (widespread in Scoresby Sund), SE (Saqqisikuik/Skjoldungen). Holarctic.

NOTES. *Itoplectis alternans* (Gravenhorst, 1829) has been confused with this species (Henriksen 1939). Townes (1960) clarified the species differences. *Pimpla turionellae* (Linnaeus, 1758) (*freyi* Hellén, 1940) is an incorrect information in Yu 1999. *Theronia atalantae* (Poda, 1761) ssp. *fulvescens* (Cresson, 1865) is an incorrect information in Henriksen (1937) and Yu (1999).

TRYPHONINAE

Small to large species. Clypeus convex, separated from face by a groove; its apical margin with a fringe of long parallel setae. Ovipositor mostly short, not longer than metasomal height at apex, without dorsal subapical notch.

***Ctenochira* Förster, 1855**

Vein 2m-cu without a strong zig-zag. Propodeum with distinct carinae. Ovipositor decurved and its sheath broad. The members of this genus are koinobiont ectoparasitoids of Tenthredinidae sawflies.

***Ctenochira rufipes* (Gravenhorst, 1829)**

Henriksen 1939: not recorded.

DISTRIBUTION. NW (only one specimen has been found, from Pituffik/Thule). New species to Greenland. Hitherto only Palaearctic.

BANCHINAE

Small to large species. Clypeus convex, separated from face by a groove. Anterior part of submetapleural carina usually produced as a long lobe. Female hypopygium large and triangular in lateral view, not extending beyond metasomal apex, the apex with a median notch; ovipositor more or less long with a dorsal subapical notch. Banchinae are koinobiont endoparasitoids of Lepidoptera larvae.

Key to genera

See subfamily key, couplets 15–16.

Glypta arctica **Dasch, 1988**

Henriksen 1939: not recorded.

BIOLOGY. The hosts of *Glypta* species live in leaf rolls or tunnels in herbaceous stems, twigs or fruits.

DISTRIBUTION. N (only Peary Land: Kap København; 6 male specimens). New species to Greenland.

Nearctic (only known from Canada: Quebec and Northwest Territories).

Banchus **Fabricius, 1798**

The hosts are primarily Noctuidae.

Key to species

1. Segment 5 of maxillary palp less than 0.2 in ♀ and 0.5 in ♂ as long as segment 4 (Figs. 6F, G). Metasoma usually entirely black .... *palpalis*
– Segment 5 of maxillary palp more than 0.4 in ♀ and 0.7 in ♂ as long as segment 4 (Figs. 6H, I) Metasomal terga 1, 2 and 3 of ♀ often with ferruginous areas, and all terga of ♂ broadly yellow on posterior areas...... .... *volutatorius*

Banchus palpalis **Ruthe, 1859. Fig. 10B**

= *groenlandicus* Aurivillius, 1890

Henriksen 1939: *Banchus monileatus* Grav. [*partim*]

DISTRIBUTION. S, SW (common north to Uummannaq Fjord), SE (Saqqisikuik/Skjoldungen).

Holarctic.

Banchus volutatorius **(Linnaeus, 1758)**

Henriksen 1939: *Banchus monileatus* Grav.

BIOLOGY. According to Henriksen (1939) the species was common in 1932 when at the same time the noctuid *Eurois occulta* was abundant. The latter is probably the host.

DISTRIBUTION. S, SW (north to Kangerlussuaq/Søndre Strømfjord). Seems to be rare.

Holarctic. Widespread.

CTENOPELMATINAE

Small to large specimens. Clypeus fairly flat, usually wide and short, and separated from face by groove; its apical margin often blunt or rounded. Apex of front tibia with a tooth on its dorsal margin. Ovipositor short, barely extending beyond metasomal apex; the dorsal subapical notch usually present. All Greenland Ctenopelmatinae are koinobiont endoparasitoids of Symphyta.

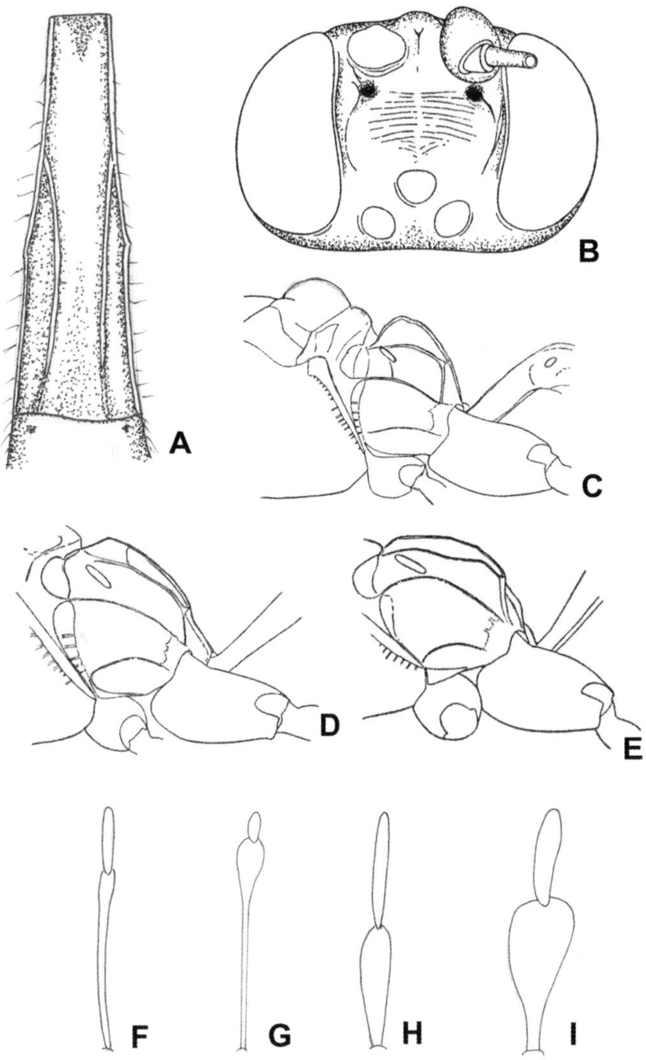

FIGURE 6 *A. Metasomal tergite 1 of* Bathythrix, *dorsal view. B. Head of* Buathra, *dorsal view. C–E. Propodeum, lateral view (according to Perkins 1959). C.* Groenlabus thulensis *D.* Coelichneumonops. *E.* Ichneumon. *F–G. Segments 4 and 5 of maxillary palp of* Banchus palpalis. *F. Female; G. Male. H–I. Segments 4 and 5 of maxillary palp of* B. volutatorius. *H. Female; I. Male.*

Key to genera

1. Glymma of metasomal segment 1 present 2
– Glymma of metasomal segment 1 absent 5
2. Dorsal profile of tergite 1 rather strongly convex. Ovipositor sheath wide and short, broadly elliptic. Apical half of female metasoma strongly compressed .... *Saotis*

– Dorsal profile of tergite 1 less convex. Ovipositor sheath not wide and short. Apical half of female metasoma not strongly compressed................3
3. Punctures on the lower half of more and less polished mesopleuron coarse ....*Otlophorus*
– Punctures on the lower half of more and less mat mesopleuron small to medium sized...............4
4. Vein Cu+cu-a of hind wing subvertical (Fig. 4E). Apical margin of clypeus sharp at the midline. Notaulus absent or not reaching to front edge of mesoscutum.............................. *Mesoleius*
– Vein Cu+cu-a inclivous (Fig. 4F). Apical margin of clypeus usually blunt at midline or much swollen near the middle of apical margin. Notaulus usually reaching front edge of mesoscutum.. *Campodorus*
5. Tarsal claws distinctly pectinate........... *Occapes*
– Tarsal claws apparently simple ....*Hypamblys*

***Otlophorus vibei* Jussila, 2006**
Henriksen 1939: not recorded.

DISTRIBUTION. NW. The holotype male is from Pituffik/Thule (Jussila 2006).

Possibly endemic to Greenland.

***Campodorus* Förster, 1869**
Key to species
1. Tegula whitish.......................................2
– Tegula black...3
2. Mesopleuron mat and rough. Metasomal terga black, only apical margin of tergum 2 somewhat light apically...*ultimus*
– Mesopleuron wrinkled, ventrally more polished with shallow punctures. Middle terga ferruginous ... *elini*
3. Metasoma from tergum 2 ferruginous.......*lituratus*
– Metasomal terga black..............................4
4. Metasomal sterna (plica) black...............*polaris*
– Sterna (plica) light ....*suspicax*

***Campodorus elini* Jussila, 1996**
Henriksen 1939: not recorded.

DISTRIBUTION. N (Peary Land: Heilprin Land), NE (Scoresby Sund: Hekla Havn).

Possibly endemic to Greenland.

***Campodorus lituratus* (Holmgren, 1857). Fig. 10C**
Henriksen 1939: not recorded.

DISTRIBUTION. SW (Itilleq Fjord, Kangerluarsuq Tulleq), NE (Zackenberg). New species to Greenland.

Holarctic (northern Europe, Romania, northern parts of Russia, Canada: Newfoundland).

***Campodorus polaris* Jussila, 1996**
Henriksen 1939: not recorded.

The male closely resembles the female but its antennae are thicker (length-to-thickness ratio of male: 1st flagellomere about 3.5, 2nd 2.1 and 7th about 1.8; of female: 1st flagellomere about 4.4, 2nd 2.9 and 7th about 2.0) (Jussila 1996).

DISTRIBUTION. NE (Scoresby Sund: Hekla Havn). Only two males and two females found. The male sex hitherto unknown. Possibly endemic.

***Campodorus suspicax* (Holmgren, 1876)**
Henriksen 1939: *Mesoleius* cf. *suspicax* Holmgr.

DISTRIBUTION. NW (only one male found at Pituffik/Thule). New species to Greenland.

Hitherto only Palaearctic (known from Finland, Sweden, Germany).

***Campodorus ultimus* Jussila, 2006**
Henriksen 1939: not recorded.

The male specimen closely resembles the holotype female but its antennae are wholly black and the metasomal tergites without yellowish brown colour.

DISTRIBUTION. SW (Allumersat/Bjørnesund; only one male specimen, the male sex was hitherto unknown.), N (the female holotype is from Peary Land: Børglum Elvdal; Jussila 2006). Possibly endemic to Greenland.

***Mesoleius groenlandicus* Roman, 1930**
DISTRIBUTION. S (Nunap Isua/Kap Farvel-area: Pamialluk), SW (Nuup Kangerlua/Godthåbsfjord: Qallunaatsigarfik/Majuola, Kapisillit), NE (Ella Ø, Hekla Havn).

Endemic to Greenland.

***Saotis hoeli* Roman, 1933**
DISTRIBUTION. NE (only one female from Ymers Ø, mouth of Sofia Sund).

Possibly endemic to Greenland.

***Hypamblys albopictus* (Gravenhorst, 1829)**
DISTRIBUTION. S, SW (north to Kangerlussuaq/Søndre Strømfjord).

Holarctic (Europe, North America: Alaska, Washington).

***Occapes hinzi* Jussila, 1996**
Henriksen 1939: not recorded.

The female closely resembles the male but its face is wholly black (the male has a yellow blotch in the middle) and its flagellum has less and thinner flagellomeres (the female has 23, the male 25 flagellomeres; length-to-thickness ratio of 1st flagellomere about 3.2, 2nd 3.0, 7th 2.0 and penultimate about 1.8 in the female, and that of 1st flagellomere about 2.4, 2nd 2.3, 7th 2.0 and penultimate about 1.8 in the male) (Jussila 1996).

DISTRIBUTION. N (Inglefield Land: 1 female; Kap Humboldt: 1 female), NE (Mestersvig: 1 male; Illoqqortoormiut/Scoresbysund: 2 males). The female sex was hitherto unknown.

Possibly endemic to Greenland.

CREMASTINAE

Small to medium-sized species. Clypeus usually convex, separated from face by a groove. Postpectal carina complete; propodeal carinae complete or almost so. Stigma of front wing often wide and triangular. Meso- and metatibial spurs separated from tarsomere 1 by a sclerotized bridge. Metasoma compressed laterally; its tergum 1 elongate; ovipositor long with a dorsal subapical notch.

Cremastinae are koinobiont endoparasitoids of Lepidoptera and Coleoptera larvae in tunnels, leaf rolls, buds, galls and other concealed situations.

***Cremastus* Gravenhorst, 1829**

Cremastus incompletus (Provancher, 1875). Fig. 10D.
Henriksen 1939: not recorded.

DISTRIBUTION. NW (one male and one female from Pituffik/Thule), NE (Ymers Ø). New species to Greenland.

Nearctic (U.S.A., Canada).

TERSILOCHINAE

Small to medium-sized species. Clypeus wide, separated from face by a groove; its apical margin with a fringe of long parallel setae. Postpectal carina of mesosoma incomplete. Fore wing with areolet open, vein 2rs-m very short; stigma large and triangular. Metasoma laterally compressed, laterotergites of terga wide and not separated from median tergites by crease; segment 1 slender with its spiracle near the apex; ovipositor with a dorsal subapical notch.

Most species are koinobiont parasitoids of Coleoptera larvae, especially those of Curculionidae and Chrysomelidae.

***Barycnemis* Förster, 1869**
= *Porizon* auct.

Key to species

1. Hind tarsus with segment 1 shorter than tibia. Ovipositor not extending beyond metasomal apex.. .... *claviventris*
– Hind tarsus with segment 1 as long as tibia. Ovipositor extending distinctly beyond metasomal apex.... .... *bellator*

***Barycnemis bellator* (Müller, 1776). Fig. 10E**
= *laeviceps* (Thomson, 1889)
Henriksen 1939: *Porizon claviventris* Grav.

DISTRIBUTION. S, SW (common, north to Uummannaq Fjord), SE (Saqqisikuik/Skjoldungen).

Holarctic.

***Barycnemis claviventris* (Gravenhorst, 1829)**
Henriksen 1939: *Porizon claviventris* Grav.

DISTRIBUTION. SW (one female from Nuup Kangerlua/Godthåbsfjord: Qallunaatsigarfik/ Majuola; Roman 1930).

Holarctic.

OPHIONINAE

Medium to large species. Clypeus separated from face by a distinct groove; ocelli always large, with lateral ocelli separated from eye by less than their diameter; antenna long and slender. Areolet of fore wing open, the remaining vein apical to vein 2m-cu by more than two-fifths length of the remaining vein. Metasoma strongly compressed laterally; its segment 1 long, without glymma, and with spiracle near its apex; ovipositor short, equal to the metasomal height at apex, the dorsal subapical notch present. Body more or less yellowish.

The species are koinobiont endoparasitoids of Lepidoptera.

***Ophion* Fabricius, 1798**
Hosts are mostly Noctuidae, but various other medium-sized or moderately large Lepidoptera are attacked.

***Ophion luteus* (Linnaeus, 1758). 10F**

DISTRIBUTION. S (few localities: Narsarsuaq, Qaqortoq/Julianehåb, Upernaviarsuk).

All continents except South America.

ANOMALONINAE

Medium-sized to large species. Clypeus convex and often not separated from face by a groove. Propodeum without regular carinae and coarsely reticulate in the species of Greenland, with its apex projecting between hind coxae. Areolet of fore wing open. Metasoma strongly compressed laterally; its segment 1 long and slender, without glymma, and with spiracle near its apex; ovipositor short and the dorsal subapical notch present.

Anomaloninae are koinobiont endoparasitoids of Lepidoptera and Coleoptera larvae, with emergence always from the host pupa.

***Erigorgus* Förster, 1869**
The hosts are larvae of moderately large moths.

Key to species

1. Clypeus rather evenly rounded with a weak median tooth (Fig. 7A). Hind trochanter and trochantellus usually black .... *melanobatus*
– Clypeus pointed, with strong median apical tooth (Fig. 7B). Hind trochanter black and hind trochantellus usually yellow .... *cerinops*

Erigorgus cerinops **(Gravenhorst, 1829). Fig. 11A**
Henriksen 1939: not recorded.

DISTRIBUTION. nSW (only recorded from Kangerlussuaq/Søndre Strømfjord and Qeqertarsuaq/ Godhavn).

Greenland and Palaearctic Region.

Erigorgus melanobatus **(Gravenhorst, 1829)**
= *ferrugator* (Gravenhorst, 1829)
Lundbeck 1897: *Anomalon ferrugator* Grav.
= *pubescens* (Zetterstedt, 1838)
Henriksen 1939: *Anomalon pubescens* Zett.

DISTRIBUTION. W (without loc.)

Holarctic.

CAMPOPLEGINAE

Small to medium-sized species. Clypeus usually not separated from face. Postpectal carina complete and propodeum with a fairly complete set of carinae in the species of Greenland. Areolet of fore wing closed or open. Metasoma more or less compressed laterally; its segment 1 long and slender, widened apically, and with spiracle near its apex; ovipositor long or short, dorsal subapical notch almost always present.

The species are koinobiont endoparasitoids mainly of Lepidoptera or Symphyta.

Key to genera

1. Areolet of forewing absent (Fig. 5A) .... *Enytus*
– Areolet present (Fig. 5B) 2
2. Glymma of first metasomal segment absent.........
.... *Bathyplectes*

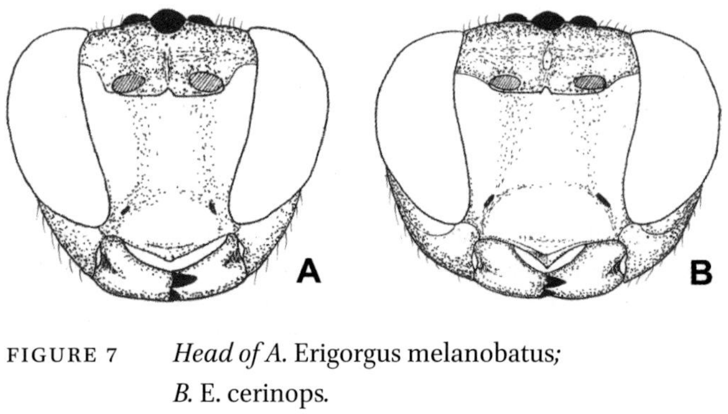

FIGURE 7 *Head of A.* Erigorgus melanobatus; *B.* E. cerinops.

– Glymma present 3
3. Areolet receiving 2m–cu vein somewhat basad of the middle (Fig. 5B). Apical margin of clypeus with a more or less distinct tooth (Figs 8A–C)... *Campoletis*
– Areolet receiving 2m–cu vein distad of the middle. Apical margin of clypeus without a trace of median tooth... 4
4. Basal part of lower margin of mandible bearing a moderately wide flange. Apical margin of narrower clypeus convex. Postpetiole distinctly widened (Fig. 5C) .... *Hyposoter*
– Basal part of lower margin of mandible bearing a narrow flange. Apical margin of moderately wide clypeus not convex. Postpetiole not so distinctly widened (Fig. 5D) 5
5. Clypeus not wide, its apical margin more or less convex. Ovipositor as long as or somewhat longer than metatibia .... *Diadegma*
– Clypeus wide, its apex more or less truncate. Ovipositor extending only a little beyond metasomal apex.
.... *Olesicampe*

Bathyplectes glacialis **Jussila, 2006**
Henriksen 1939: not recorded.

BIOLOGY. The species is probably a parasitoid of Curculionidae (Coleoptera).

DISTRIBUTION. S. The holotype female is from Narsarsuaq (Jussila 2006). Possibly endemic to Greenland.

Campoletis **Förster, 1869**
The usual hosts are larvae of Noctuidae.

Key to species

1. Malar space about 1.0× width of mandible base (Fig. 8C). Carination of propodeum almost complete, areola as long as broad (Fig. 8F)......... *horstmanni*
– Malar space about 1.3× width of mandible base (Figs. 8A, B). Carination of propodeum incomplete (Figs. 8D, E)... 2
2. Head with long and dense hairs (Fig. 8A). Areola of propodeum about as long as broad (Fig. 8D).........
.... *longiceps*
– Head with shorter hairs (Fig. 8B). Areola longer than broad (Fig. 8E) .... *rostrata*

Campoletis horstmanni **Jussila, 1996**
Henriksen 1939: not recorded.

The male resembles exactly the female.

DISTRIBUTION. S, SW, N, NE. The male sex hitherto unknown (five in the material). Possibly endemic to Greenland.

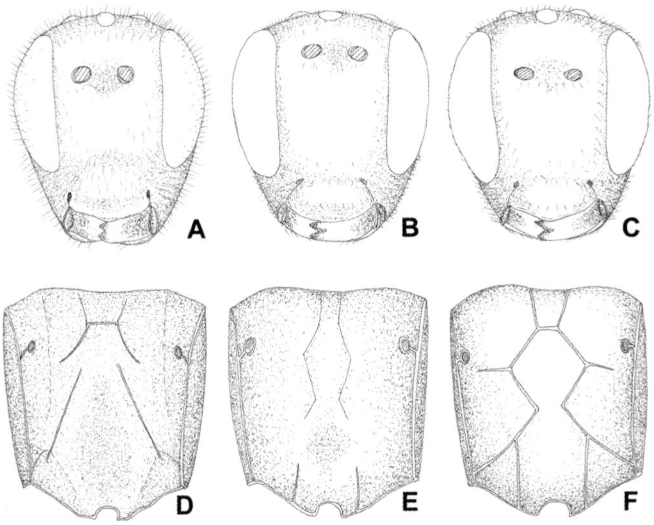

FIGURE 8 *A–C. Head (according to Jussila 1996) of A.* Campoletis longiceps; *B.* C. rostrata; *C.* C. horstmanni. *D–F Propodeum, dorsal view (according to* Jussila 1996) *of D.* C. longiceps; *E.* C. rostrata; *F.* C. horstmanni.

***Campoletis longiceps* (Roman, 1926)**
Henriksen 1939: *Ecphoropsis longiceps* Roman.

DISTRIBUTION. nSW, NW, N, NE (Ymers Ø). Seemingly rather common, but only north of Qeqertarsuup Tunua/Disko Bugt and Scoresby Sund.

Holarctic (including northern Russia).

***Campoletis rostrata* Jussila, 1996**
Henriksen 1939: not recorded.

DISTRIBUTION. SW (north to Uummannaq Fjord: Qarassap Nunataa), NE (Île de France, Scoresby Sund: Charcot Land).

Possibly endemic to Greenland.

***Enytus* Cameron, 1905**
The hosts are various small or medium-sized Lepidoptera.

Key to species

1. Area superomedia of propodeum open posteriorly... *apostata*
– Area superomedia divided from area posteromedia with posterior transverse carina........... *montanus*

***Enytus apostata* (Gravenhorst, 1829)**
= *exareolatus* (Ratzeburg, 1852)

Henriksen 1939: *Angitia exareolata* Ratz.

DISTRIBUTION. S, SW (Itilleq).

Holarctic, also known from India and South Africa.

***Enytus montanus* (Ashmead, 1890) (Fig. 11B)**
Henriksen 1939: *Angitia laticeps* Roman, 1938.

DISTRIBUTION. S, SW (common, north to Qeqertarsuup Tunua/Disko Bugt: Kangersuneq). Holarctic (Europe, North America).

***Diadegma* Förster, 1869**
The hosts are various small or medium-sized Lepidoptera.

Key to species

1. Hind tibia ferruginous with somewhat darker colour on apex and beyond base. Ovipositor shorter than metatibia............................... *combinatum*
– Hind tibia with yellowish or whitish colour. Ovipositor as long as or longer than metatibia............ 2
2. Hind tibia with whitish medium and base, and dark brownish on apex and near base. Metasomal segment 2 in ♀ as long as wide and in ♂ about 1.3 longer than wide............................. *groenlandicum*
– Hind tibia with yellowish medium and base, and brownish on apex and near base. Metasomal segment 2 in ♀ longer than wide and in ♂ about 1.5 longer than wide........................... *majale*

***Diadegma combinatum* (Holmgren, 1860)**
Henriksen 1939: *Angitia combinata* Holmgr.

DISTRIBUTION. S, SW, NW (Pituffik/Thule), N (Inglefield Land: Marshall Bugt), NE (Hekla Havn). Widely distributed but sparse.

Holarctic.

***Diadegma groenlandicum* (Roman, 1916)**
Henriksen 1939: *Angitia groenlandica* Roman

BIOLOGY: One male specimen reared from *Gnorimoschema vibei* (Lepidoptera: Gelechiidae).

DISTRIBUTION. S (Quingua, Narsarsuaq), SW (many places but sparse, north to Qeqertarsuaq/Godhavn), NE (Charcot Land, Jameson Land).

Possibly endemic to Greenland.

***Diadegma majale* (Gravenhorst, 1829)**
Henriksen 1939: *Angitia* cf. *claripennis* Thoms.

DISTRIBUTION. Found in western Greenland (Carpenter 1938) and in East Greenland (Kap Herschel, Wollaston Forland; Roman 1933, Henriksen 1939).

Holarctic.

***Olesicampe* Förster, 1869**
= *Holocremna* Thomson, 1887

The hosts are sawflies, with oviposition into the sawfly larva and emergence from the cocoon.

Key to species

1. Clypeus with a dull tooth. Hind tarsus with whitish medium and base, and dark brownish on apex and near base .... *dentata*
– Clypeus without apical tooth. Hind tarsus with yellowish medium and base, and brownish on apex and near base .... *extrema*

Olesicampe dentata **(Provancher, 1874)**
Henriksen 1939: not recorded.

DISTRIBUTION. nSW (only one male from Kangerlussuaq/Søndre Strømfjord). New species to Greenland.

Nearctic (hitherto only Canada: Newfoundland).

Olesicampe extrema **(Holmgren, 1872)**
Henriksen 1939: *Holocremna extrema* Holmgr.

DISTRIBUTION. S (only Kangilinnguit/Grønnedal), SW (Qeqertarsuaq/Godhavn; Qeqertarsuup Tunua/Disko Bugt: Kangersuneq; Nuussuaq: Marraat), SE (Saqqisikuik/Skjoldungen: Dronning Maries Dal).

Hitherto only Palaearctic (recorded from Sweden).

Hyposoter **Förster, 1869**
The hosts are lepidopterous larvae.

Key to species

1. Tibia of hind leg with whitish medium and base, and dark brownish on apex and near base....... *frigidus*
– Tibia of hind leg evenly ferruginous ..... *deichmanni*

Hyposoter deichmanni **(Nielsen, 1907)**
Henriksen 1939: *Anilasta pectinata* Thoms.

Regarded as a junior synonym of *H. pectinatus* (Thomson, 1887) (Yu *et al.* 2005) but is more strongly pubescent, face distinctly flatter in lateral view, and malar space distinctly longer than in *pectinatus* (about 1.1 and 0.7 the basal width of mandible).

BIOLOGY. A specimen from Hekla Havn hatched from a larva of *Gynaephora groenlandica*.

DISTRIBUTION. N (Hall Land), NE (Germania Land; Scoresby Sund: Hekla Havn; Traill Ø).

Possibly endemic to Greenland.

Hyposoter frigidus **(Lundbeck, 1897) (Fig. 11C).**
Henriksen 1939: *Anilasta frigida* Lundb.

DISTRIBUTION. Possibly G. S (several localities), SW (few records, north to Nuussuaq: Marraat), N (Inglefield Land: Marshall Bugt; Peary Land: Brønlund Fjord), NE (several localities), SE (Saqqisikuik/Skjoldungen). Appears to be common. Possibly endemic to Greenland.

MESOCHORINAE

Small to moderately large species. Clypeus not distinctly separated from face by groove. Areolet of fore wing large and usually rhombic. Metasoma usually weakly compressed, at least on its apical half; its segment 1 slender, with glymma large and deep, and with spiracle near or behind middle; hypopygium of female large and triangular in lateral view; ovipositor without dorsal subapical notch; clasper of male ending into a long, narrow rod.

Almost all species are koinobiont hyperparasitoids of Ichneumonoidea.

Key to genera

1. Lower face with transverse groove from edge of clypeus to eye................................ *Mesochorus*
– Lower face lacking transverse groove from edge of clypeus to eye .... *Cidaphus*

Cidaphus atricilla **(Haliday, 1838) (Fig. 11D)**
Henriksen 1939: not recorded.

BIOLOGY. A secondary parasitoid ovipositing into large-sized ichneumonoid hosts.

DISTRIBUTION. S (Nunap Isua/Kap Farvel-area and Narsarsuaq). New species to Greenland.

Hitherto only Palaearctic.

Mesochorus agilis **Cresson, 1865**
= *nigriceps* Thomson, 1886.

Henriksen 1939: *Mesochorus nigripes* Ths.

BIOLOGY. Secondary parasitoid ovipositing into ichneumonoid hosts.

DISTRIBUTION. S (widespread), SW (several localities), NW (Pittufik/Thule), NE (Illoqqoortormiut/Scoresby Sund).

Nearctic, common.

METOPIINAE

Small to medium-sized species. Upper edge of face with a triangular projection over between antennal sockets. In the species of Greenland clypeus not separated from face by a groove, together forming an evenly convex surface. Metasoma dorso-ventrally depressed; its tergum 1 short and stout, with glymma, and with spiracle before middle; ovipositor short.

The species are koinobiont endoparasitoids of Lepidoptera. The oviposition is into the larva and the emergence is from the host pupa.

Exochus **Gravenhorst, 1829**

Exochus pullatus **Townes, 1959**
Henriksen 1939: not recorded.

DISTRIBUTION. NE (only one female found at Illoqqortoormiut/Scoresby Sund; Jussila 1996).

Holarctic, common.

ORTHOCENTRINAE

Small species. Clypeus either small and separated from face by a groove or face protuberant forming in lateral view an almost continuous strongly convex bulge, hence clypeus not separated from face. Segment 1 of metasoma slender to stout, with spiracle near or in front of middle except in *Symplecis*, where spiracle is posterior to midline, cf. subfamily key, and with or without glymma; ovipositor fairly long to very short, the dorsal subapical notch present or absent.

Probably all species are koinobiont endoparasitoids of Mycetophilidae and Sciaridae (Diptera).

Key to genera

1. Face protuberant, forming in lateral view an almost continuous strongly convex bulge, clypeus not separated from face (as in Fig. 1A) 2
– Face more or less flat 4
2. Lower edge of clypeus weakly convex, adjacent to mandibles; labrum not exposed below clypeus. Vein Cu+cu-a of hind wing vertical to inclivous .... *Orthocentrus*
– Lower edge of clypeus truncate or subtruncate, at midline at a short distance from mandibles; labrum exposed below clypeus. Vein Cu+cu-a vertical to reclivous 3
3. Prepectal carina of pronotum present. Wings often vestigial .... *Stenomacrus*
– Prepectal carina entirely lacking. Wings normal .... *Plectiscus*
4. Inner margins of eyes strongly convergent ventrally .... *Symplecis*
– Inner margins of eyes not convergent ventrally 5
5. Glymma of metasomal tergite 1 present. Apex of metasomal sternite 1 in front of the middle of its tergite .... *Pantisarthrus*
– Glymma absent. Apex of sternite 1 at or behind the middle of its tergite .... *Plectiscidea*

Pantisarthrus Förster, 1871

Key to species

1. Hind wing with section of vein Rs from rs–m to apex longer than section Rs between R and rs–m (as in Fig. 3B) .... *luridus*
– Hind wing with section of vein Rs from rs–m to apex equal to or slightly shorter than section Rs between R and rs–m .... *lubricus*

Pantisarthrus lubricus (Förster, 1871)

= *inaequalis* (Förster, 1871).

Henriksen 1939: *Aniseres lubricus* Först.

DISTRIBUTION. SW (several localities, north to Kangerluarsuk Tulleq).

Holarctic (Europe, North America).

Pantisarthrus luridus Förster, 1871

Henriksen 1939: *Plectiscus luridus* Först.

DISTRIBUTION. S, SW (north to Uummannaq district: Ukkusissat Sulluat: Pannertooq).

Holarctic.

Plectiscidea Viereck, 1914

= *Plectiscus* auct.

Key to species

1. Females 2
– Males 7
2. Ovipositor sheath 0.1–0.15 of length of front wing ... . *bistriata*
– Ovipositor sheath longer than 0.25 of length of front wing 3
3. Ovipositor sheath 0.4–0.5 of length of front wing 4
– Ovipositor sheath about 0.3 of length of front wing... 5
4. Metasomal segment 1 about 2.6 its apical width. Propleuron yellowish .... *collaris*
– Metasomal segment 1 about 2.0 its apical width. Propleuron blackish .... *monticola*
5. Metasomal tergite 2 distinctly pustulate basally .... *prolata*
– Metasomal tergite 2 not distinctly pustulate basally. 6
6. Metasoma entirely black .... *hyperborea*
– Metasoma centrally yellowish brown .... *arctica*
7. Metasoma entirely black .... *hyperborea*
– Metasoma not entirely black 8
8. Hind leg dark .... *prolata*
– Hind leg light 9
9. Length of metasomal segment 1 about 2.6 the apical width. Propleuron yellowish .... *collaris*
– Metasomal segment 1 shorter. Propleuron blackish . ... *arctica, bistriata, monticola*

Plectiscidea arctica (Roman, 1930)

Henriksen 1939: *Plectiscus arcticus* Först.

DISTRIBUTION. SW (Allumersat/Bjørnesund, Nuup Kangerlua/Godthåbsfjord: Majuala; north to Qeqertarsuup Tunua/Disko Bugt: Saqqaq, Torsukattak). Possibly endemic to Greenland.

***Plectiscidea bistriata* (Thomson, 1888)**
Henriksen 1939: not recorded.

DISTRIBUTION. nSW (not found south of Kangerlussuaq/Søndre Strømfjord; north to Uummannaq district: Ukkusissat Sulluat: Pannertooq).

Palaearctic (earlier only recorded from Europe).

***Plectiscidea collaris* (Gravenhorst, 1829)**
Henriksen: *Plectiscus collaris* Grav.

DISTRIBUTION. S, SW (several localities, north to Uummannaq district: Ukkusissat Sulluat: Pannertooq), SE (Saqqisikuik/Skjoldungen).

Holarctic, common.

***Plectiscidea hyperborea* (Holmgren, 1869)**
Henriksen 1939: *Plectiscus hyperboreus* Holmgr.

DISTRIBUTION. S, SW (several localities), NW (Pituffik/Thule), NE (several localities), SE (Ammassalik). Widespread and common in Greenland.

Western Palaearctic and Nearctic (Greenland).

***Plectiscidea monticola* (Förster, 1871). Fig. 11E**
Henriksen 1939: not recorded.

DISTRIBUTION. S, SW (several localities, north to Qeqertarsuup Tunua: Saqqaq), SE (Saqqisikuik/Skjoldungen). New species to Greenland.

Hitherto only Palaearctic.

***Plectiscidea prolata* Dasch, 1992**
Henriksen 1939: not recorded.

DISTRIBUTION. S, SW (several localities, north to Uummannaq district: Ukkusissat Sulluat: Pannertooq).

Nearctic, common.

***Symplecis* Förster, 1869**
= *Blapticus* Förster, 1869

Key to species

1. Face entirely black .... *bicingulata*
– Facial orbit yellow.......................... *clipeator*

***Symplecis bicingulata* (Gravenhorst, 1829)**
DISTRIBUTION. S (1 male from Narsarsuaq), SW (1 male from Allumersat/Bjørnesund) New species to Greenland.

Holarctic.

***Symplecis clipeator* (Lundbeck, 1897)**
Henriksen 1939: *Hemiteles clipeator* Lundb.

DISTRIBUTION. SW (few finds from Nuuk/Godthåb and Qeqertarsuup Tunua/Disko Bugt).

Nearctic, common.

***Orthocentrus* Gravenhorst, 1829**
Key to species

1. Face with transverse pustule rows.......... *frontator*
– Pustules not organized in transverse rows 2
2. Females..3
– Males 4
3. Flagellomeres with tyloids dorsally...... *monilicornis*
– Flagellomeres without tyloids .... *winnertzii*
4. Pterostigma of front wing broad (cf. Fig. 1C) .... *winnertzii*
– Pterostigma of front wing narrower (cf. Fig. 5B) .... *monilicornis*

***Orthocentrus frontator* (Zetterstedt, 1838)**
= *repentinus* Holmgren, 1858.

Henriksen 1939: *Orthocentrus repentinus* Holmgr.

DISTRIBUTION. S (Nunap Isua/Kap Farvel-area: Kangikitsoq), SW (Uummannaq district: Ukkusissat Sulluat: Pannertooq), NE (Scoresby Sund: Charcot Land). New species to Greenland.

***Orthocentrus monilicornis* Holmgren, 1858**
DISTRIBUTION. nSW (Qeqertarsuup Tunua/Disko Bugt: Kangersuneq), N (Peary Land: Heilprin Land). New species to Greenland.
Hitherto only Palaearctic, common.

***Orthocentrus winnertzii* Förster, 1850 ssp. *groenlandicus* Jussila, 1996**
= *Orthocentrus stigmaticus* Holmgren, 1858 ssp. *groenlandicus* Jussila, 1996

DISTRIBUTION. SW (1 specimen from Kangerlussuatsiaq/Evighedsfjorden), NE (only Illoqqortoormiut/Scoresby Sund; Jussila 1996). The subspecies is endemic to Greenland.

Holarctic.

***Stenomacrus* Förster, 1869**
Key to species

1. Females..2
– Males 7
2. Wings short, useless for flight........................3
– Wings of normal length.............................4
3. Metasomal tergite 2 rugose and longitudinally striate. Face more pustulate; head not narrowed behind eyes .... *micropennis*
– Metasomal tergite 2 smooth and shining. Face somewhat pustulate; head narrowed behind eyes......... .... *groenlandicus*
4. Front wing with vein 3rs-m existing. Flagellomere 1 cubic.. *laticollis*

– Vein 3rs-m missing. Flagellomere 1 longer than wide 5
5. Ovipositor sheath somewhat longer than hind tibia. ....*terebrator*
– Ovipositor sheath about 0.4 as long as hind tibia....6
6. Metasomal tergite 2 rugose and longitudinally striate. Face pustulate; head not narrowed behind eyes. ....*difficilis*
– Metasomal tergite 2 smooth and shining. Face only somewhat pustulate at most on its dorsal part; head narrowed behind eyes ....*groenlandicus*
7. Face yellowish...............................*laticollis*
– Face dark..8
8. Hind leg 1.15–1.25 longer than mesosoma + metasoma. Face granulate and mat..............*longipes*
– Hind leg shorter. Face at most somewhat pustulate, not mat..9
9. Metasomal tergite 2 wholly smooth and shining. Face more or less pustulate; head narrowed behind eyes...................................*groenlandicus*
– Base of metasomal tergite 2 striate. Face more or less smooth and shining; head not narrowed behind eyes..10
10. Length of metasomal segment 1 about 1.2 its apical width.....................................*terebrator*
– Length of metasomal segment 1 about 1.5 its apical width..11
11. Femora black.................................*atratus*
– Femora light.......................................12
12. Face smooth and shining with small punctures .... *micropennis*
– Face with weak striae and punctures........*difficilis*

***Stenomacrus atratus* (Holmgren, 1858)**
Henriksen 1939: *Stenomacrus ? atratus* Holmgr.

This species belongs to "species incertae sedis" according to Aubert (1981) but is valid according to Yu *et al.* (2005).

European and Greenlandic.

***Stenomacrus difficilis* Jussila, 1996**
Henriksen 1939: not recorded.

The male distinctly resembles the female but its vertex is somewhat narrowing behind the eyes and its antenna has 20–21 flagellomeres (in the female about 17). Length-to-thickness ratio of 1st flagellomere is 1.2–1.5, 2nd 1.2–1.4, 10th and penultimate about 1.1–1.2 in the male, and of 1st and 2nd flagellomeres about 1.5, 10th and penultimate about 2.0 in the female. In fore wing vein 3rs-m either lacking or it is weakly developed. Metasoma seen from above more or less compressed from segment 3 to the apex. The male sex hitherto unknown.

DISTRIBUTION. S, NW (Pituffik/Thule), NE (Daneborg: Sandodden). Seems to be rare. Possibly endemic to Greenland.

***Stenomacrus groenlandicus* Jussila, 1996. Fig. 11F**
Henriksen 1939: not recorded.

DISTRIBUTION. G (widespread and common in Greenland). Also recorded from Svalbard (Coulson *et al.* 2003).

***Stenomacrus laticollis* (Holmgren, 1883)**
DISTRIBUTION. SW (Kangerluarsussuaq/Grædefjord, Nuup Kangerlua/Godthåbsfjord, Nuussuaq: Paatuut), NE (Scoresby Sund: Amdrup Havn).

Hitherto only Palaearctic.

***Stenomacrus longipes* Jussila, 1996**
Henriksen 1939: not recorded.

DISTRIBUTION. NE (only 4 male specimens from Illoqqortoormiut/ Scoresby Sund; Jussila 1996).

Possibly endemic to Greenland.

***Stenomacrus micropennis* Jussila, 2006**
= *brevipennis* Jussila, 1996: 152.

NOTE. The name *brevipennis* is a junior secondary homonym of *Stenomacrus brevipennis* (Ashmead, 1902) [*Synopsis brevipennis* Ashmead 1902: Wash. Acad. Sci., Proc. 4: 226, ♀].

Henriksen 1939: not recorded.

DISTRIBUTION. S (Ikerasassuaq/Prins Christian Sund, Tasiusaq, Tissaluk), NE (only caught at Île de France, several specimens).

Possibly endemic to Greenland.

***Stenomacrus terebrator* Roman, 1934**
DISTRIBUTION. N (only 1 female from Peary Land: Jørgen Brønlund Fjord, 300 m a.s.l.).

Possibly endemic to Greenland.

***Plectiscus minutus* (Holmgren, 1858)**
DISTRIBUTION. SW (Kangerlussuatsiaq/Evighedsfjorden, Itilleq Fjord, Qeqertarsuaq/Godhavn, Uummannaq Fjord), NE (Illoqqortoormiut /Scoresby Sund). New species to Greenland.

Palaearctic (known from Fennoscandia and Lithuania).

DIPLAZONTINAE

Small to medium-sized species. Clypeus small and separated from face by a groove, the apical margin usually concave and notched; dorsal tooth of mandible wide, divided into upper and lower teeth. Metasoma of all species of Greenland dorso-ventrally depressed, segment 1 short, wide at base and only slightly wider at apex, glymma small and shallow, and spiracle in front of middle; ovipositor short, dorsal notch present at about its middle.

The species are koinobiont endoparasitoids of Syrphidae (Diptera), with oviposition into the larva (or sometimes into the egg); the emergence is from the puparium.

Key to genera

1. Notaulus of mesoscutum distinct. Areolet of front wing absent 2
– Notaulus absent....................................3
2. Metasomal tergites 1–3 each with a postmedian transverse groove .... *Diplazon*
– Tergites 1–3 without postmedian transverse groove . ... *Tymmophorus*
3. Hind tibia white at base, blackened on rest of length.................................... *Enizemum*
– Hind tibia not white at base .... *Syrphoctonus*

Diplazon **Nees von Esenbeck, 1918**

Key to species

1. Mesoscutum and mesopleuron smooth and polished. Transverse impressions on metasomal tergites 1–3 strong................................. *annulatus*
– Mesoscutum and mesopleuron rugose with small punctures. Transverse impressions on metasomal tergites 1–3 very weak................. *hyperboreus*

Diplazon annulatus **(Gravenhorst, 1829)**

DISTRIBUTION. NE, SE (Roman 1933, 1934; Yu *et al.* 2005).

Holarctis and India.

Diplazon hyperboreus **(Marshall, 1877)**

= *algidus* Dasch, 1964.

Henriksen 1939: *Diplazon annulatus* Gr.

DISTRIBUTION. N (Peary Land: Heilprin Land: Børglum Elvdal), NE (Kap Herschell, Dickson Fjord, Mestersvig).

Holarctic.

Tymmophorus gelidus **Dasch, 1964**

Henriksen 1939: not recorded.

DISTRIBUTION. NW (Pituffik/Thule), NE (Clavering Ø, Hold With Hope).

Nearctic (Canada: North West Territories).

Syrphoctonus **Förster, 1869**

= *Homotropus* Förster, 1869.

Key to species

1. Hind tibia dark at base and apex, whitish banded medially...2
– Hind tibia coloured otherwise......................3
2. Coxae black .... *melanogaster*
– Coxae ferruginous................... .*nigriclypealis*
3. Mesoscutum and mesopleuron smooth and polished..................................... *nigritarsus*
– Mesoscutum and mesopleuron rugose, polished only anteriorly and laterally on mesoscutum .... *aquilonius*

Syrphoctonus aquilonius **(Dasch, 1964)**

DISTRIBUTION. SW (Kangerlussuatsiaq/Evighedsfjorden, Qeqertarsuup Tunua/Disko Bugt: Saqqaq; Uummannaq district: Ukkusissat Sulluat: Pannertooq), NE (Nordfjord). Possibly endemic to Greenland.

Syrphoctonus melanogaster **(Holmgren, 1872). Fig. 12A**

Henriksen 1939: *Homotropus melanogaster* Holmgr.

DISTRIBUTION. S, SW (several localities, north to Uummannaq district: Ukkusissat Sulluat: Pannertooq), NE (only one record from Scoresby Sund: Nordfjord).

Nearctic, northern.

Syrphoctonus nigriclypealis **(Dasch, 1964)**

Henriksen 1939: not recorded.

DISTRIBUTION. S (Tassiusak, Tissaluk; Dasch 1964), SW (several localities).

Nearctic (Canada).

Syrphoctonus nigritarsus **(Gravenhorst, 1829) ssp.** ***groenlandicus*** **(Holmgren, 1872)**

Henriksen 1939: *Homotropus elegans* Grav. var. *nigritarsus* Roman.

DISTRIBUTION. S, SW (common, several localities north to Uummannaq district: Ukkusissat Sulluat: Pannertooq), NE (Scoresby Sund: Gåseland), SE (Saqqisikuik/Skjoldungen). Very common in southern West Greenland. The subspecies is endemic to Greenland.

Holarctic, widespread.

Enizemum ornatum **(Gravenhorst, 1829)**

Henriksen 1939: *Homotropus ornatus* Grav.

DISTRIBUTION. S, SW (several localities, north to Uummannaq Fjord).

Holarctic and Oriental, common.

CRYPTINAE

Small to large species. Clypeus usually convex, separated from face by a groove. Carination of propodeum varies from complete to missing. Areolet of fore wing pentagonal when closed. Metasoma dorso-ventrally depressed; its segment 1 more or less long with a more or less widened postpetiole, without glymma, spiracle usually behind its middle; ovipositor short to long, without a dorsal subapical notch.

Most species are idiobiont ectoparasitoids of pupae and prepupae. Many species can be secondary parasitoids.

Key to genera

1. Wings absent .... *Gelis*
– Wings fully developed 2
2. Vein 2m–cu of front wing with two bullae (but sometimes with one), meeting posteriorly vein Cu at more or less oblique angle (as in Fig. 1C) 3
– Vein 2m–cu with one bulla, meeting posteriorly vein Cu at right angle (as in Figs 1B, D) 15
3. Dorsal side of pronotum, just behind collar, with a strong but short median longitudinal carina crossing the transverse sulcus .... *Acrolyta*
– Dorsal side of pronotum without strong median longitudinal carina 4
4. Area petiolaris of propodeum separated from area superomedia by a transverse carina (as in Fig. 2B) 5
– Area superomedia and petiolaris forming a median area without transverse carina (Fig. 5E) 14
5. Lateral edge of metasomal tergite 2 joining its epipleuron without separation by a crease and curving over to the ventral side without an interruption 6
– Lateral edge of metasomal tergite 2 joining its epipleuron with a distinct crease 7
6. Vein Cu+cu–a of hind wing intercepted (as in Figs 4E, F) .... *Xiphulcus*
– Vein Cu+cu–a not intercepted .... *Aclastus*
7. Outer face of mandible with a strong subbasal swelling .... *Gelis*
– Outer face of mandible without subbasal swelling 8
8. Vein Cu+cu–a of hind wing more or less vertical (as in Fig. 4E) 9
– Vein Cu+cu–a distinctly inclivous (as in Fig. 4F) .. 12
9. First metasomal tergite very slender, its spiracle near midlength (Fig. 6A) .... *Bathythrix*
– First metasomal tergite quite stout, its spiracle more or less behind midlength 10
10. Notaulus sharp, reaching beyond centre of mesoscutum .... *Sulcarius*
– Notaulus not reaching to centre of mesoscutum 11
11. Apical truncation of antennal scape about 60 degrees from the transverse. Ovipositor about 0.45 as long as front wing .... *Theroscopus*
– Apical truncation of antennal scape 20–40 degrees from the transverse. Ovipositor shorter .... *Phygadeuon*
12. Prescutellar transverse groove of mesosoma with a strong median longitudinal ridge that divides the groove into two halves .... *Endasys*
– Prescutellar transverse edge without a median longitudinal ridge 13
13. Body moderately stout (length of 1st metasomal tergite about 1.2 its width) .... *Brachypimpla*
– Body slender (length of 1st metasomal tergite 2.1–3.0 its width) .... *Charitopes*
14. Metasoma more or less depressed, in females its tergite 2 without lateral crease separating epipleuron .... *Stilpnus*
– Metasoma more or less compressed, in females its tergite 2 with lateral crease extending from base at least as far as stigma .... *Atractodes*
15. Propodeum with longitudinal and transverse carinae .... *Oresbius*
– Propodeum without longitudinal carinae 16
16. Frons with distinct dorsal tentorial pits (Fig. 6B) .... *Buathra*
– Dorsal tentorial pits not apparent .... *Cryptus*

***Acrolyta glacialis* Jussila, 1996.**
Henriksen 1939: not recorded.

DISTRIBUTION. NE (only found at Illoqqortoormiut/Scoresby Sund; Jussila 1996).

Possibly endemic to Greenland.

***Aclastus* Förster, 1869**
Key to species

1. Face quadrate; cheek as long as width of mandible base .... *borealis*
– Length of face less than its width; cheek distinctly shorter than mandible base .... *gracilis*

***Aclastus borealis* (Boheman, 1866)**
= *septentrionalis* (Holmgren, 1869).

Henriksen 1939: not recorded.

DISTRIBUTION. S (only found in Narsarsuaq and Kuunnaat). New species to Greenland.

Hitherto only Palaearctic.

***Aclastus gracilis* (Thomson, 1884)**
DISTRIBUTION. S, SW (few records, *e.g.* Søndre Isortoq).

Hitherto only Palaearctic.

Xiphulcus plectiscinus (Roman, 1916)
Henriksen 1939: *Phygadeuon plectiscinus* Roman
DISTRIBUTION. S, SW (north to Uummannaq district: Ukkusissat Sulluat: Pannertooq). Appears to be common. Possibly endemic to Greenland.

***Gelis* Thunberg, 1827. Fig. 11.1–3A**
= *Pezomachus* Gravenhorst, 1829.
Hosts reside in a large variety of small cocoons or silken bags: Psychidae and Coleophoridae, spiders, Chrysopidae (Neuroptera), small Lepidoptera, Ichneumonidae and Braconidae (Hymenoptera).

Key to species

1. Wings absent 2
– Wings fully developed 3
2. Transverse carina of propodeum absent .... *nigritulus* (females and part of males)
– Transverse carina of propodeum present .... *curvicauda*
3. Mesonotum and mesopleuron polished .... *glacialis*
– Mesonotum and mesopleuron more or less mat..... 4
4. Mesonotum and mesopleuron rough .... *maesticolor*
– Mesonotum and mesopleuron densely punctulate and somewhat shining............................... .... *nigritulus* (part of males)

***Gelis curvicauda* Horstmann, 1993**
Henriksen 1939: not recorded.
DISTRIBUTION. S (only one female found on Pamialluk Island, Kap Farvel area).
Hitherto only Palaearctic (found in Sweden, United Kingdom, Germany, Austria, Italy).

***Gelis glacialis* (Holmgren, 1869)**
Henriksen 1939: not recorded.
DISTRIBUTION. NE (only found at Illoqqortoormiut/Scoresby Sund; Jussila 1996).
Hitherto only Palaearctic (found in Finland, Sweden, Norway, Russia, Azerbaijan).

***Gelis maesticolor* (Roman, 1933)**
Henriksen 1939: *Astomaspis maesticolor* Roman; Jussila 1996: *Gelis melanogaster* (Thomson, 1884).
DISTRIBUTION. SW (Ameralik, Kangerlussuaq/Søndre Strømfjord, Qeqertarsuaq/Godhavn), NE (according to Henriksen 1939: Clavering Ø, Kap Humboldt). Possibly endemic to Greenland.

***Gelis nigritulus* (Zetterstedt, 1838). Fig. 12B**
= *terebrator* (Ratzeburg, 1848).
Henriksen 1939: *Gelis terebrator* Ratz.
BIOLOGY. Hosts are Braconidae and small Lepidoptera.
DISTRIBUTION. S, SW (several localities, north to Uummannaq district: Ukkusissat Sulluat: Pannertooq).
Hitherto only Palaearctic.

***Endasys bicolor* (Lundbeck, 1897)**
Henriksen 1939: *Stylocryptus bicolor* Lundbeck.
BIOLOGY: Hosts are probably parasitoids of sawflies.
DISTRIBUTION. S (Tasermiut fjord), SW (Nuup Kangerlua: Qooqqut; Qeqertarsuaq/Godhavn, Uummannaq district: Nuugaatsiaq). Seems to be rare.
Nearctic (Canada) and Western Palaearctic.

***Brachypimpla brachyura* Strobl, 1902**
Henriksen 1939: not recorded.
DISTRIBUTION. nSW (one male specimen recorded from Uummannaq Fjord: Kangerluarssuk). New species to Greenland.
Hitherto only Palaearctic.

***Charitopes* Förster, 1869**
The species are parasitoids of Hemerobiidae.

Key to species

1. Mesoscutum entirely mat .... *albilabris*
– Mesoscutum polished on lateral lobe .... *piceiventris*

***Charitopes albilabris* Townes, 1983**
Henriksen 1939: not recorded.
DISTRIBUTION. S (Narsarsuaq), SW (several localities, north to Itilleq Fjord). New species to Greenland.
Nearctic (Canada, U.S.A.).

***Charitopes piceiventris* (Harrington, 1894)**
Henriksen 1939: not recorded.
DISTRIBUTION. SW (Eqaluit Paarliit/Præstefjord, Itilleq Fjord, Uummannaq Fjord). New species to Greenland.
Nearctic (Canada, U.S.A.).

***Bathythrix longiceps* Townes, 1983**
Henriksen 1939: not recorded.
BIOLOGY. Hosts could be various small cocoons.
DISTRIBUTION. NW (only two males and one female recorded at Pituffik/Thule). New species to Greenland.
Nearctic.

***Sulcarius* Townes, 1970**
The hosts are Limnephilidae (Trichoptera) (see Yu *et al.* 2005).

Key to species

1. Lower front part of mesopleuron with a rough area . ... *nigricornis*
– Lower front area of mesopleuron smooth.. *biannulatus*

Sulcarius biannulatus **(Gravenhorst, 1829)**

DISTRIBUTION. SW (Kuanersoq/Kvanefjord, Itilleq Fjord, Qeqertarsuup Tunua: Saqqaq). New species to Greenland.

Hitherto only Palaearctic.

Sulcarius nigricornis **(Thomson, 1884)**

Henriksen 1939: not recorded.

DISTRIBUTION. S (only one female found at Narsarsuaq). New species to Greenland.

Holarctic.

Theroscopus arcticus **Jussila, 1996**

Henriksen 1939: not recorded.

DISTRIBUTION. NE (only one female found at Illoqqortoormiut/Scoresby Sund). New species to Greenland. Probably endemic to Greenland.

Phygadeuon **Gravenhorst, 1829**

The species are parasitoids of cyclorrhaphous Diptera.

Key to species

1. Thorax cylindrical, its length in female about 2.3 and in male about 2.0 its height. Pterostigma of front wing narrow............................ *cylindraceus*
– Thorax normal, pterostigma broadly triangular.... 2
2. Metasoma wholly black. Head not narrowed behind eyes.. *solidus*
– Metasoma medially reddish brown. Head somewhat narrowed behind eyes .... *fumator*

Phygadeuon cylindraceus **Ruthe, 1859**

DISTRIBUTION. S, SW (several localities, north to Qeqertarsuaq/Godhavn), NE (Hekla Havn).

Hitherto only Palaearctic (common).

Phygadeuon fumator **(Gravenhorst, 1829) ssp. *nivalis* Holmgren, 1883**

Henriksen 1939: not recorded.

DISTRIBUTION. S, SW, NW (Upernavik), NE (Sabine Ø). New species to Greenland.

The species is Holarctic.

Phygadeuon solidus **Lundbeck, 1897**

DISTRIBUTION. S, SW (several localities, north to Uummannaq district: Ukkusissat Sulluat: Pannertooq), NW (Pituffik/Thule), NE (Scoresby Sund: Hekla Havn).

Possibly endemic to Greenland.

Stilpnus tenebricosus **(Gravenhorst, 1829)**

Henriksen 1939: *Atractodes tenebricosus* Grav.

BIOLOGY. The species is a parasitoid of cyclorrhaphous Diptera.

DISTRIBUTION. S, SW (several localities, north to Qeqertarsuup Tunua: Saqqaq), NE (only Zackenberg).

Holarctic (Palaearctis, Alaska).

Atractodes **Gravenhorst, 1829**

The species are parasitoids of cyclorrhaphous Diptera.

Key to species

1. Females 2
– Males.. 10
2. Mesopleuron strongly rugose. Eye with at most sparse and short hairs..............................3
– Mesopleuron not rugose. Eye sometimes with dense and long hairs......................................4
3. Metasoma and legs blackish............... *aterrimus*
– Metasoma (except for tergite 1) and legs light .... *alpinus*
4. Tarsal claws longer than arolium, narrow and apically as if stretched (Fig. 9A). Apical flagellomeres more or less spherical...............................5
– Tarsal claws as long as arolium, thicker (Fig. 9B). Apical flagellomeres not spherical 9
5. Metasomal tergite 1 rugose and streaked, mat .... *alpestris*
– Metasomal tergite 1 smooth and shining, at most very obscurely rugose...............................6
6. Length of flagellomere 2 at least 4.4 its thickness. Eyes with long setae.......................... *gilvipes*
– Length of flagellomere 2 at most 3.9 its thickness 7
7. Hypopygium punctured. Vein 3rs-m of front wing present .... *bicolor*
– Hypopygium not punctured.........................8
8. Eye bare or with some short hairs.......... *obsoletor*
– Eye with long hairs........................... *pusillus*
9. Legs dark .... *picipes*
– Legs light .... *exilis*
10. Mesopleuron strongly rugose....................... 11
– Mesopleuron not rugose............................13
11. Surface of eye with elongate conspicuous hairs. Legs wholly black (or tibiae lighter), also metasoma black.. *picipes*
– Surface of face at most with sparse pubescence. Legs lighter 12
12. Metasoma and legs blackish............... *aterrimus*
– Metasoma and legs partly light.............. *alpinus*

13. Tarsal claws longer than arolium, narrow and apically as if stretched (cf. Fig. 9A; it should be noted that in many males claws are not as long as in females) 14
– Tarsal claws as long as arolium, thicker (Fig. 9B)..... .... *exilis*
14. Tegula and hind corner of pronotum more or less light... 15
– Tegula dark, pronotum wholly black 16
15. Clypeus higher (width at most 2.5 its height), upper edge distinctly more convex than lower edge, rather polished with large punctures on its upper part (Fig. 9C). Length 5.0–6.5 mm .... *bicolor*
– Clypeus more transverse (width at least 2.5 its height), upper edge as convex as lower edge, more rugose with smaller punctures on its upper part (Fig. 3A). Length 4.0–5.0 mm .... *gilvipes*
16. Metasomal tergite 1 rugose and streaked, mat; postpetiole subquadrate .... *alpestris*
– Metasomal tergite 1 smooth and shining, length of postpetiole about 1.5 its width..................... 17
17. Hind coxa ventrally mat. Mesopleuron around sternaulus more or less rough .... *obsoletor*
– Hind coxa ventrally coarsely punctuated and shining. Mesopleuron around sternaulus at most slightly coarsely sculptured .... *pusillus*

Atractodes alpestris **Roman, 1918**
Henriksen 1939: not recorded.

DISTRIBUTION. SW (Kuanersoq/Kvanefjord, Kangerluarsussuaq/Grædefjord, Nuup Kangerlua: Kapisillit; Qeqertarsuup Tunua/Disko Bugt: Saqqaq; Nuussuaq: Marraat), N (Brønlund Fjord, Børglum Elvdal), NE (Clavering Ø; Scoresby Sund: Hekla Havn; Jameson Land).

Palaearctic and Nearctic (Greenland).

Atractodes alpinus **Förster, 1876**
Henriksen 1939: not recorded.

DISTRIBUTION. S, SW (several localities), NW (Pituffik/Thule), NE (Clavering Ø; Scoresby Sund: Hekla Havn). New species to Greenland.

Hitherto only Palaearctic.

Atractodes aterrimus **Holmgren, 1872 (Fig. 12C)**
DISTRIBUTION. S, SW (several localities), NW (Pituffik/Thule). Seems to be very common, especially at Qeqertarsuaq/Godhavn.

Palaearctic and Nearctic (Greenland).

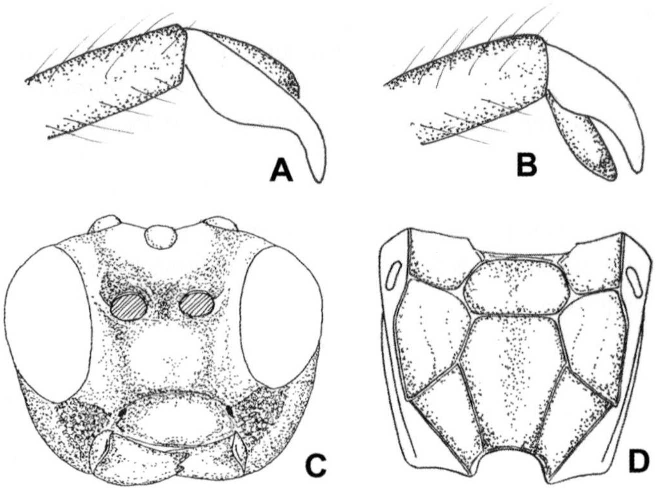

FIGURE 9 *A–B. Apex of hind tarsus, lateral view of B.* Atractodes gilvipes *female; B.* A. exilis *female. C. Head of* A. bicolor. *D. Propodeum of* Homotherus erythromelas, *dorsal view.*

Atractodes bicolor **Gravenhorst, 1829 ssp. *arcticus* Holmgren, 1872**
DISTRIBUTION. SW (only north of Nuuk/Godthåb, north to Qeqertarsuup Tunua/Disko Bugt), NE (Nathorst Land).

This subspecies is Holarctic, arctic (Yu *et al.* 2005).

Atractodes exilis **Haliday, 1838**
Henriksen 1939: not recorded.

DISTRIBUTION. NE (Nordfjord).

Palaearctic and Nearctic (Greenland).

Atractodes gilvipes **Holmgren, 1860**
Henriksen 1939: not recorded.

DISTRIBUTION. NW (only recorded from Pituffik/Thule). New species to Greenland.

Hitherto only Palaearctic.

Atractodes obsoletor **(Zetterstedt, 1838)**
Henriksen 1939: not recorded.

DISTRIBUTION. S (Tissaluk, Eqaluit, Kuunnaat), SW (Kuanersoq/Kvanefjord, Allumersat/Bjørnefjord, Ikkamiut Kangerluarssuat, Kangerlussuatsiaq/Evighedsfjorden). New species to Greenland.

Hitherto only Palaearctic (Europe).

Atractodes picipes **Holmgren, 1860**
Henriksen 1939: not recorded.
= *nigripes* Förster, 1876

DISTRIBUTION. nSW (only recorded from Qeqertarsuaq/Godhavn).

Holarctic, arctic.

***Atractodes pusillus* Förster, 1876**
Henriksen 1939: not recorded.

DISTRIBUTION. S, SW (several localities, north to Uummannaq Fjord), N (Peary Land: Heilprin Land), NE (Zackenberg; Scoresby Sund: Hekla Havn). Widespread and common in Greenland. New species to Greenland?

Palaearctic and Nearctic (Greenland).

***Oresbius* sp.**
BIOLOGY. Hosts are probably tenthredinid cocoons.

DISTRIBUTION. NW (one incomplete specimen from Pituffik/Thule).

***Buathra laborator* (Thunberg, 1824). Figs 11.1–3B, 12D, E**
Henriksen 1939: *Cryptus laborator* Thbg.
= *fabricii* (Schiødte, 1839)

BIOLOGY. The usual hosts are pupae or prepupae of Lepidoptera.

DISTRIBUTION. S, SW (north to Qeqertarsuup Tunua/Disko Bugt: Saqqaq), NE (Kejser Franz Josephs Fjord, Jameson Land), SE (Saqqisikuik/Skjoldungen)

Holarctic (common).

***Cryptus* Fabricius, 1804**
Host records include a variety of Lepidoptera.

Key to species

1. Head, thorax and metasoma with long erect hairs. Metasomal terga ferruginous except for black tergum 1; femora, tibiae and tarsi light, too .... *arcticus*
– Head, thorax and metasoma with moderately short hairs. Metasoma and legs blackish........... *leechi*

***Cryptus arcticus* Schiødte, 1857. Fig. 12F**
DISTRIBUTION. SW (south to Nuuk/Godthåb), N (Peary Land: Nedre Midsommersø), NE (several localities).

Holarctic (mainly Nearctic, but also found in Austria).

***Cryptus leechi* Mason, 1968**
Henriksen 1939: not recorded.

DISTRIBUTION. N (Peary Land: Nedre Midsommersø), NE (Loch Fyne, Jameson Land).

Nearctic, arctic.

ICHNEUMONINAE

Medium-sized to large species in Greenland. Clypeus wide and flat and separated from face by a weak groove, the apex widely truncate. Postpectal carina usually incomplete; carinae of propodeum more or less complete. Areolet of fore wing pentagonal or sub-triangular, almost always closed; vein M+Cu of hind wing almost straight. Metasoma dorso-ventrally depressed; its segment 1 slender anteriorly, widened posteriorly, without glymma, and spiracle near apex; tergum 2 usually with deep gastrocoeli; ovipositor short, without dorsal subapical notch. Many species are sexually dichromatic.

The species are either koinobiont or idiobiont endoparasitoids of Lepidoptera; the emergence is always from the pupa.

Key to genera

1. Propodeum with lateral profile an evenly curved line from base of areola to metasomal attachment, second lateral area long and reaching far down so that its apex is closer to metasomal attachment than costula (Fig. 6D) .... *Coelichneumonops*
– Propodeum with lateral profile having distinguishable dorsal and posterodorsal faces meeting at more or less distinct angle and thus giving lateral profile an angulate outline, second lateral area closer to costula than metasomal attachment (Figs 6C, E) 2
2. Thyridia each wider than their interval ...... *Aoplus*
– Thyridia not wider than their interval.............. 3
3. Postpetiole distinctly punctate .... *Crypteffigies, Homotherus*
– Postpetiolus not punctate, striate to smooth 4
4. Carina closing apical side of second lateral area (lateral section of posterior transverse carina) approximately straight, not distinctly angled at its juncture with apical section of median longitudinal carina, receiving longitudinal carina at or close to lower lateral corner of areola................. *Ichneumon*
– Carina closing apical side of second lateral area angled downward at its juncture with apical section of median longitudinal carina, receiving longitudinal carina a little distance below lower lateral corner of areola 5
5. Gastrocoeli deep, adjacent to base of tergite 2; thyridia separated from base of tergite 2 by less than its width. Median field of postpetiolus and base of tergite 2 with weak longitudinal striae .... *Groenlabus*
– Gastrocoeli only slightly excavated, inconspicuous, sometimes superficial or absent; thyridia separated from base of tergite 2 by 2× its width or greater. Median field of postpetiolus smooth, base of tergite 2 without longitudinal striae .... *Cratichneumon*

***Groenlabus thulensis* Jussila, 2006**
Henriksen 1939: not recorded.

BIOLOGY. Hosts are probably medium-sized Lepidoptera.

DISTRIBUTION. NW. The holotype male from Pituffik/Thule (Jussila 2006).

Possibly endemic to Greenland.

***Crypteffigies* Heinrich, 1961 and *Homotherus* Förster, 1869**

Hosts are medium-sized Lepidoptera.

Key to species

1. Areola of propodeum and metasomal segment 1 noticeable wide (Fig. 9D) .... *Homotherus erythromelas*
– Areola of propodeum and metasomal segment 1 normal..2
2. Mesosoma black and metasoma blackish........... ....*Homotherus magus*
– More or less ferruginous colour on mesosoma and metasoma 3
3. Females..4
– Males 5
4. Ovipositor more than 2× apical depth of metasoma. ....*Crypteffigies confusus*
– Ovipositor shorter..........*Homotherus porcelariae*
5. Hind coxa ferruginous ....*Crypteffigies confusus*
– Hind coxa dark............ *Homotherus porcelariae*

***Crypteffigies confusus* (Ashmead, 1902)**

Henriksen 1939: not recorded.

BIOLOGY. Hosts are medium-sized Lepidoptera.

DISTRIBUTION. S (Kap Farvel area, Ivittuut), sSW (north to Kangerluarsunnguaq/Buksefjord). New species to Greenland.

Nearctic (Canada, Alaska).

***Homotherus erythromelas* (MacLachlan, 1878)**

Henriksen 1939: *Cratichneumon ? erythromelas* McLachl.

DISTRIBUTION. S (Nanortalik), SW (Nuup Kangerlua/Godthåbsfjord: Qussuk),

N (Peary Land).

Nearctic, arctic.

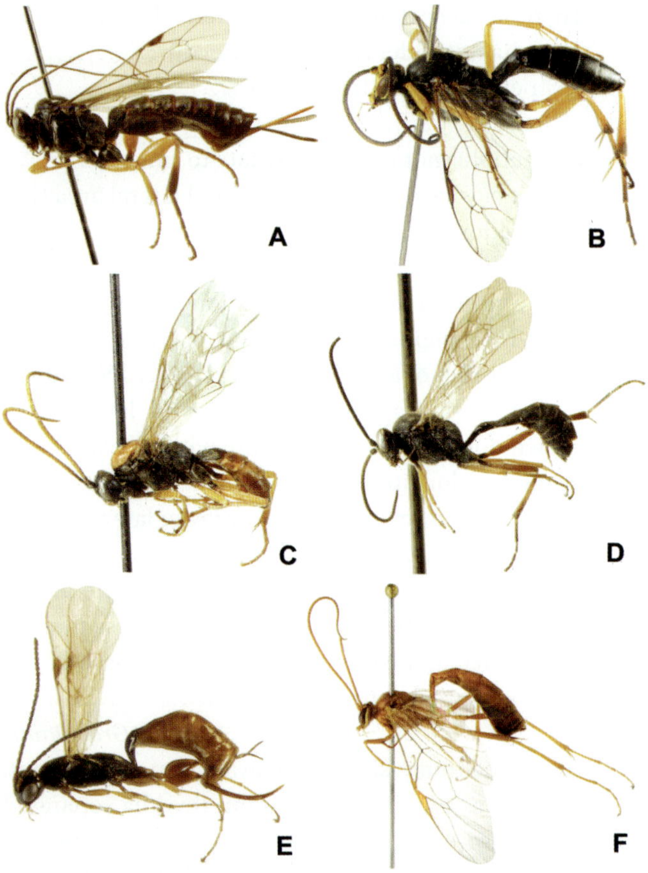

FIGURE 10 *Habitus images of A.* Pimpla sodalis*; B.* Banchus palpalis*; C.* Campodorus lituratus*; D.* Hyposoter *sp.; E.* Barycnemis bellator*; F.* Ophion luteus.

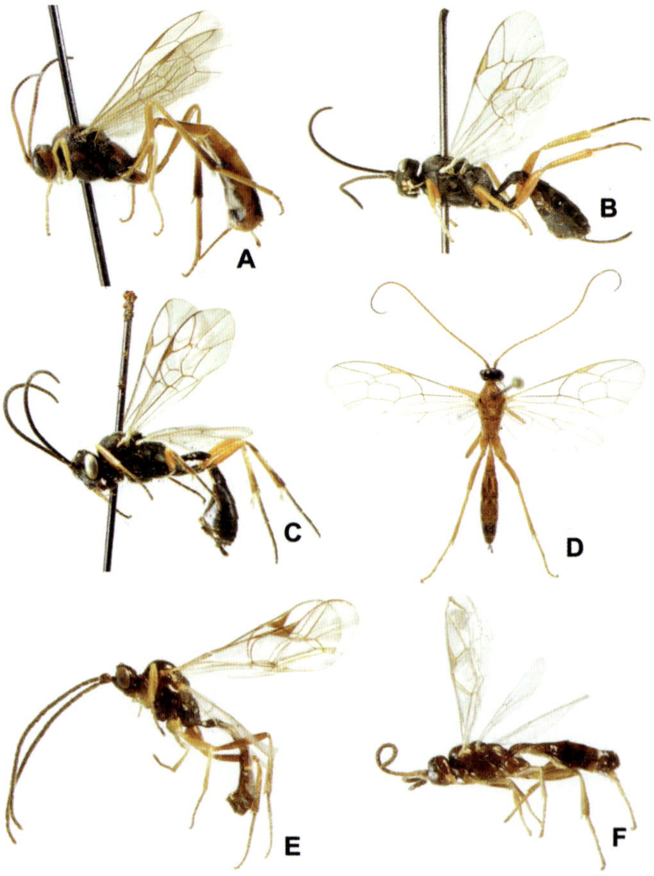

FIGURE 11 *Habitus images of A.* Erigorgus cerinops*; B.* Enytus montanus*; C.* Hyposoter frigidus*; D.* Cidaphus atricilla *E.* Plectiscidea monticola*; F.* Stenomacrus groenlandicus.

***Homotherus magus* (Wesmael, 1855)**
= *nitidus* (Bridgman, 1886)

Henriksen 1939: *Cratichneumon magus* Wesm.

DISTRIBUTION. S, SW (north to Kangerlussuaq/Søndre Strømfjord).

Holarctic (Palaearctis, Canada).

***Homotherus porcelariae* Heinrich, 1961**
Henriksen 1939: not recorded.

DISTRIBUTION. S (Nanortalik), SW (Kangerluarsunnguaq/Buksefjord, Eqaluit Paarliit/Præstefjord, Illutalik/Fiskefjord, Itilleq Fjord). New species to Greenland.

Nearctic (Canada).

***Cratichneumon vescus* (Provancher, 1877)**
Henriksen 1939: not recorded.

BIOLOGY. Hosts are medium-sized Lepidoptera.

DISTRIBUTION. S, SW (common, several localities north to Kangerlussuaq/Søndre Strømfjord; only males found). New species to Greenland.

Nearctic.

***Aoplus groenlandicus* (Lundbeck, 1897)**
Henriksen 1939: *Ichneumon groenlandicus* Lundb.

BIOLOGY. The species is a parasitoid of Geometridae (Lepidoptera).

DISTRIBUTION. NE (two unlabelled females, one female found in Scoresby Sund: Charcot Land).

Possibly endemic to Greenland.

***Ichneumon* Linnaeus, 1758**
Hosts are medium-sized Lepidoptera.

Key to species

1. Females 2
– Males 6
2. Mesoscutum and scutellum ferruginous 3
– Mesoscutum and scutellum black 4
3. Hind coxa without scopa. The punctures of metasomal tergites 2 and 3 separated by about two times their diameters .... *discoensis*
– Hind coxa with an indistinct scopa. The punctures of metasomal tergites 2 and 3 separated by about their own diameters .... *lariae* (part of specimens)
4. Hind femur wholly black. Metasomal terga 2 and 3 wholly ferruginous .... *bucculentus*
– Hind femur not wholly black. Metasomal terga 2 and 3 with more or less black colour 5
5. Apical half of hind femur and parts of terga 2 and 3 black .... *sarcitorius*
– Femur ferruginous. Metasoma blackish .... *lariae* (part of specimens)
6. Metasoma black .... *lariae*
– Metasoma not wholly black 7
7. Metasoma with terga 2 and 3 black with apical margins broadly yellow .... *sarcitorius*
– Metasomal terga 2 and 3 wholly ferruginous 8
8. Basal part of hind tibia yellowish .... *bucculentus*
– Hind tibia wholly ferruginous .... *discoensis*

***Ichneumon bucculentus* Wesmael, 1845**
Listed with ? by Henriksen (1939), who did not provide locality information. Perhaps misidentified.

Holarctic.

***Ichneumon lariae* Curtis, 1835 ssp. *aurivillii* (Roman, 1916). Fig. 12G**
Henriksen 1939: *Ichneumon aurivillii* Roman.

DISTRIBUTION. S, SW(several localities), NW, N (Inglefield Land), NE (Zackenberg, Hochstetter Forland, Ella Ø, Hudson Land: Moskusoksefjord; Jameson Land). The subspecies is endemic to Greenland.

The species is Holarctic (Russia, North America).

FIGURE 12 *Habitus images of A.* Syrphoctonus melanogaster; *B.* Gelis nigritulus; *C.* Atractodes aterrimus; *D, E.* Buathra laborator; *F.* Cryptus arcticus; *G.* Ichneumon lariae.

***Ichneumon sarcitorius* Linnaeus, 1758**
Henriksen 1939: not recorded.

DISTRIBUTION. S (only recorded from Narsaq, labelled *ex pupa* "*R. quadricingulata*"). New species to Greenland.

Holarctic (including northern Ethiopia).

***Ichneumon discoensis* Fox, 1892**
= *scoresbysundensis* Jussila, 1996, **syn. nov.**

DISTIBUTION: nSW, NE (Zackenberg, Scoresby Sund, Mestersvig). Possibly endemic to Greenland.

***Coelichneumonops occidentalis* (Roman, 1934)**
Henriksen 1939: *Coelichneumon solutus occidentalis* Roman

BIOLOGY: Hosts are perhaps medium-sized Lepidoptera.

DISTRIBUTION. NW (only one female from Pituffik/Thule).

Nearctic (Canada: Newfoundland) and European (Iceland).

References

Aubert, J.-F. (1981): Révision des Ichneumonides *Stenomacrus* sensu lato. – *Mitteilungen der Münchner Entomologischen Gesellschaft* 71: 139–159.

Carpenter, G.D.H. (1938): Notes on insects collected in West Greenland by the Oxford University.

Greenland expedition 1936. With descriptions of a new species of *Angitia* by A. Roman and of *Fannia* by J. E. Collin. – *Anual Magazine of natural History* (11) 1: 529–553.

Constantineanu, M.J. (1959): *Familia Ichneumonidae. Subfamilia Ichneumoninae Stenopneusticae.* – Fauna R. P. R., Insecta 9, Fasc. 4, Bucureşti: 1249 pp.

Coulson, S.J., Hodkinson, I.D., Jussila, R., Webb, N.R. & Harrison, J.A. (2003): *Atractodes pusillus* Förster, 1876 and *Stenomacrus groenlandicus* Jussila, 1996 (Hymenoptera, Ichneumonidae), new to Svalbard. – *Norwegian Journal of Entomology* 48: 328.

Fitton, M.G., Shaw, M.R. & Gauld, I.D. (1988): Pimpline Ichneumon-flies. Hymenoptera, Ichneumonidae (Pimplinae). – *Handbooks for the identification of British insects* 7 (1): 110 pp.

Henriksen, K.L. (1937): *Insects collected on the fifth Thule Expedition.* – Report 5. – Thule Expedition 1921–1924, København 2: 1–34.

Henriksen, K.L. (1939): A revised index of the insects of Grønland, containing a supplement of the insect list in Kai L. Henriksen & Will. Lundbeck: Grønlands Landarthropoder. – *Meddelelser om Grønland* 119 (10): 112 pp.

Jussila, R. (1965): The Ichneumonidae of the Kevojoki area in Inari Lapland (Finland). – Reports from the *Kevo Subarctic Research Station* 2: 1–186.

Jussila, R. (1996): Ichneumonidae (Hymenoptera) of Greenland found in Scoresbysund (Ittoggortoormiit). – *Entomologica Fennica* 7: 145–156.

Jussila, R. (2001): Additions to the revision of the genus *Atractodes* (Hymenoptera: Ichneumonidae) of the Palaearctic Region. III. – *Entomologica Fennica* 12: 193–216.

Jussila, R.(2006): A new genus, four new species and a new name in Ichneumonidae (Hymenoptera) from Greenland. – *Entomologiske Meddelelser* 74: 73–79.

Perkins, J. F. (1959): Hymenoptera, Ichneumonidea, Ichneumonidae, key to subfamilies and Ichneumoninae. I. – *Handbooks for the identification of British insects* 7 (2): 116 pp.

Richards, O.W. (1956): Hymenoptera. Introduction and keys to families. – *Handbooks for the identification of British insects* 6 (1): 94 pp.

Roman, A. (1916): Ichneumoniden aus West-Grönland. – *Arkiv för Zoologi* 10: 1–12.

Roman, A. (1930): Oxford University Greenland expedition, 1928. Ichneumonidae collected by Major G. Hingston on the Oxford University Expedition to Greenland, 1928. – *Annual Magazine of Natural History* 5: 281–288.

Roman, A. (1933): Schlupfwespen aus Ost-Grönland. – *Skrifter om Svalbard og Ishavet* 53: 5–13.

Townes, H. (1960): Ichneumon-flies of America north of Mexico: 2. *Subfamilies Ephialtinae, Xoridinae, Acaenitinae. – United. States Natural Museum* 216 (2): 676 pp.

Wahl, D.B. (1993): *Family Ichneumonidae. In*: Goulet, H. & Huber, J.T. (eds.), Hymenoptera of the world. An identification guide to families. – Ottawa, 668 pp.

Yu, D.S. (1999): *Interactive catalogue of World Ichneumonidae* 1998. – CD–ROM, Taxapad, Bentall Centre, Vancouver, Canada.

Yu, D.S., Achterberg, K. van & Horstmann, K. (2005): *Biological and taxonomical information: Ichneumonoidea 2004.*– Taxapad Interactive Catalogue, Vancouver.

11.16. Dryinidae (Chrysidoidea)

Lars Vilhelmsen

Antenna with 10 segments, inserting near clypeus; frons flat, without prominence accommodating the antennal insertions; sexual dimorphism displayed to various degrees; males usually macropterous, rarely brachypterous, without modified foretarsus; females macropterous to apterous, usually with foretarsus having distinct chela formed by the enlarged fifth tarsomere and one claw; remaining claw reduced or absent.

The larvae parasitise nymphs and adults of Homoptera: Auchenorrhyncha. The host is grasped by the female wasp with the foretarsal chelae (when present) and temporarily paralysed. The egg is placed between two of the host's abdominal sclerites, and the larva spends most of its development in a sac formed by the exuviae of its previous instars, feeding from the host's tissues, most of which will be consumed eventually. The sac is visible externally, projecting from the host's abdomen. Pupation takes place in the soil or on the host plant of the host.

The family is worldwide in distribution, comprising *c.* 1100 species in 11 subfamilies and 58 genera.

GONATOPODINAE

***Gonatopus brooksi* Olmi, 1984. Fig. 1**

3.2–4.6 mm. Female apterous; foretarsus with chela (illustrated by Olmi 1984 and Andersen & Nielsen 1987); fifth tarsomere having 14–19 lamellae in a single row ventrally. Colour predominantly dark brown to black, with part of the head capsule, antennal base, legs except forefemur, and sometimes abdomen lighter brown. Male not known.

Andersen & Nielsen (1987) provided the first record of this species and the Dryinidae as a whole. One additional specimen has been located in the ZMUC ethanol collection, matching fairly well with the descriptions of Olmi (1984) and Andersen & Nielsen (1987).

FIGURE 1 Gonatopus brooksi, *female habitus.*

BIOLOGY. The ethanol specimen was collected in a pitfall trap together with specimens of *Psammotettix lividellus* (Zetterstedt) (Cicadellidae). One of these specimens had a dryinid larva attached to it, making *P. lividellus* a putative host of *G. brooksi* (Andersen & Nielsen 1987).

DISTRIBUTION. S (Qaqortoq/Julianehåb-area), SW (Nuuk/Godthåb-area).

Nearctic (Canada: Saskatchewan).

References

Andersen, M. & Nielsen, P. (1987): *Gonatopus brooksi* Olmi, 1984 found in Greenland (Hymenoptera, Dryinidae). – *Entomologiske Meddelelser* 55: 21–22.

Olmi, M. (1984): Revision of the Dryinidae. – *Memoirs of the American Entomological Institute* 37: 1–1913.

11.17. Vespidae (Vespoidea)

Lars Vilhelmsen

Inner margin of eye usually with distinct incurvation opposite antennal base; female antenna with 12 segments, male antenna with 13; wings longitudinally folded at rest; sexual dimorphism slight, both sexes macropterous, with several closed cells in both fore- and hindwings; species often strikingly aposematically coloured, usually in black and yellow patterns.

Most Vespidae are predators, provisioning their larvae with insect prey; a few (Masarinae) provide nectar and pollen instead. The members of the subfamilies Vespinae and Polistinae are eusocial, feeding their larvae continuously on masticated insects. In the Eumeninae, the only subfamily registered so far in Greenland (except for occasional introduced individuals of *Vespula*), the majority of species are solitary, constructing either exposed mud nests or nests in burrows in soil or wood. Mud is often utilised at least partly in the construction of the nest. Single nest cells are built for each egg; after the egg is laid, the cell is provisioned with a number of paralyzed insect larvae, usually Coleoptera and/or Lepidoptera, but occasionally Hymenoptera (Symphyta). The cell is sealed after being provisioned, and the larva completes its entire development within it.

The Vespidae are worldwide in distribution, comprising *c.* 4.000 species in six subfamilies in about 200 genera. The Eumeninae is by far the largest of the subfamilies, comprising three quarters of the species and genera of the Vespidae.

EUMENINAE

***Ancistrocerus waldenii* (Vierick, 1906). Figs 11.1–3E, F, 1**
Henriksen 1939: not recorded.

10–12 mm. Predominantly black. Yellow markings present in both sexes between antennal bases, posteriorly of dorsal part of eye (small), anterior margin of pronotum (transversely elongate, swollen in the middle), anterior/lateral margin of the tegula (variable; tegula entirely black in some males), no marks on mesopleuron. Some females have additional yellow markings on lateral part of mesoscutellum (small); all females have most of femora black and remainder of legs reddish brown rather than yellow; on metasoma, females have yellow markings on posterior margin of T2–6 and S3, medially on T7 and laterally on

FIGURE 1 Ancistrocerus waldeni, *female habitus.*

S4–(5). Males have additional yellow markings on entire clypeus, labrum and most of mandibles, lower part of the scapus, anterior part of midcoxa, distal half of femora and remainder of legs; metasoma in males with yellow markings on entire posterior margin of T2–5 and S3 and laterally on S4, occasionally median posterior margin of T6 and small spots laterally on S5. Specimens collected outside Greenland are more extensively marked with yellow (see Becquaert 1925).

BIOLOGY. The species seems to be especially associated with inland localities with moderate precipitation. It builds exposed nests on rocks. Nests constructed of mud and provisioned with numerous larvae of Tortricidae (Figs 1.1-E–F) have been observed in Greenland.

DISTRIBUTION. SW (known from the northeastern Godthåbsfjord/Nuup Kangerlua and Kangerlussuaq/Søndre Strømfjord areas). The species was recently reported from Qeqertarsuaq on Disko (Wahlberg *et al.* 2014), expanding the northern limit of this species substantially.

Nearctic.

References

Bequaert, J. (1925): The genus *Ancistrocerus* (Hymenoptera, Vespidae) in North America, with a partial key to species. – *Transactions of the American Entomological Society* 51: 57–117.

Wahlberg, E., Baião, G.C., Häggqvist, S., Martinsson, S., Pistone, D. & Pape, T. (2014): *Ancistrocerus waldenii waldenii* (Viereck 1906) (Hymenoptera: Vespidae, Eumeninae), a new addition to the fauna of Greenland. *Zootaxa* 3838: 143–150.

11.18. Formicidae (Vespoidea) (Ants)

Lars Vilhelmsen

Body with at least two, sometimes three constrictions on abdomen, separating petiole (and sometimes postpetiole) from the rest of the metasoma. Females (queens and workers) with elongate scapus, making antenna elbowed. Queens and males winged, workers wingless, with much reduced thorax.

All ants are eusocial or social parasites. The size of the colonies varies from a handful of individuals to billions. In some Neotropical rainforests, ants may constitute about one quarter of the total animal biomass. Nests are often underground, but may be in or on vegetation; some ants have no permanent nests. Most ants are predators, but the biology across the family is extremely diverse, including scavengers, foragers, fungus growers, seed collectors, slave drivers etc. Ants rely on pheromone mediated recruitment of nest mates to efficiently exploit food sources they discover. Other interactions within the colony are also controlled to a large extent by pheromones and may be highly complex. Many ants have close associations with other insects, in particular with plantsucking Homoptera (*e.g.,* aphids), from which they collect surplus carbohydrates (honeydew) to supplement their diet. Many other insects also lives associated with ants as nest commensals or parasites. Ants have huge ecological impact and introduction of ant pests (*e.g.,* argentine ants, fire ants) have caused major problems in large parts of the world.

Monomorium pharaonis **Linnaeus, 1758**

Henriksen 1939: not recorded.

Queens 3.5–4.5 mm, males ca. 3 mm, workers 2–2.5 mm. Body light reddish-brown, posterior part of metasoma darker. Antennae slender, with elongate scapus, clubbed towards the tip. Mesosoma with marked transverse groove but without propodeal spines. Abdomen with both petiole and postpetiole.

BIOLOGY. The presence of this small ant is easily overlooked until it reaches large population size, by which time it may become a serious and resilient pest. Pharao ants have a very broad diet, scavenging on whatever potential food sources they come across. They have a complex foraging system involving several pheromones. Each colony contains many queens, making it possible to create daughter colonies by budding. Colonies may also be started by single, mated founder queens.

DISTRIBUTION. nSW (only recorded from Kangerlussuaq/Søndre Strømfjord).

Found worldwide, associated with buildings in colder climates. Area of origin unknown, probably in the tropical region.

NOTES. Introduced species. A single male was caught in a Malaise trap in the summer of 2013 a couple of km from Kangerlussuag airport. It is highly likely that the species has been introduced through the airport and established a population in the buildings around it. If this is the case, it can be expected to spread to the other major population centres in Greenland.

11.19. Apidae (Apoidea) (Bees)

Lars Vilhelmsen

Body often densely hairy, at least some hairs branched; labiomaxillary complex often elongate, forming tongue; hind tibia in female usually flattened and fringed by hairs, forming corbicula (pollen basket); hind basitarsus usually flattened and enlarged.

The Apoidea is subdivided into the Spheciformes and the Apiformes. The Spheciformes comprise an assemblage of predatory wasps (digger wasps, sand wasps etc.) provisioning their larvae with insect prey; it is not a natural group, but paraphyletic with respect to the Apiformes. No members of the Spheciformes have been recorded from Greenland. The Apiformes comprises the bees, most of which are nectar and pollen collectors. The pollen is collected by the females, easily getting caught in the body hairs. Most bees transfer the pollen to areas on the body which are specialised for transportation, usually on the hindlegs. The nectar is imbibed and stored in the crop, from which it can later be regurgitated. Nests are constructed in a variety of biotopes and with various materials, the cells being provisioned with nectar and pollen. A few groups utilize other food sources, *e.g.*, floral oils and carrion. A number of bee taxa have specialised in parasitizing other bees, either by laying eggs in their nests (cleptoparasitism) or by taking over the nest of another species (social parasitism); these parasites usually do not collect pollen themselves. The degree of care for the brood differs. Sociality has developed to various degrees several times within the bees; some of the highly eusocial species have thousands of individuals and provision their larvae continuously.

Bees are found throughout the world, being especially abundant in dry habitats. About 16.000 species have been described. Of the 7 families of bees currently recognized, only the Apidae are represented in Greenland with 2 species of bumblebees (*Bombus*) and the introduced honeybee (*Apis mellifera*).

Key to species of Apidae in Greenland

1. Body slender; thorax dorsally with uniform coverage of yellow/brown hairs................. *Apis mellifera*
– Body stocky; thorax dorsally with anterior and posterior fringe of white/yellow hairs separated by area with black/dark brown hairs ....2 (*Bombus* spp.)
2. Malar space length/mandibular base width approx. 1:1 .... *Bombus polaris*
– Malar space length/mandibular base width approx. 1,5:1 .... *Bombus hyperboreus*

[*Apis mellifera* Linnaeus, 1758.]
Henriksen 1939: not recorded.

15–16 mm (reproductives), 12–13 mm (workers). Males recognized by enlarged eyes (meeting on top of the head), queens by larger size, especially of abdomen. Body covered with yellow/brown hairs of various density, striping pattern created by alternating densely and sparsely setose areas.

BIOLOGY. Eusocial, hive size 20.000–200.000 individuals, mostly workers (unreproductive females), but governed through pheromone control by the reproductive queen. Virgin queens and males (drones) are produced at intervals. Workers typically live for a month during summer, established queens live for several years, males are shortlived, die after mating. Workers are active collecting pollen and nectar throughout summer. The hive is able to survive winter by living on stored pollen and honey. The eggs are placed in cells constructed of wax. The brood is provisioned continuously by the workers prior to pupation. The caste (queen/worker) of the offspring is decided by the amount and quality of the food it receives during growth.

DISTRIBUTION. S (only in the Qaqortoq/Julianehåb district).

Native to Africa, Europe and Asia, introduced to the Americas and Australia.

NOTE. Introduced species. The stock introduced belongs to the subspecies *A. m. mellifera* (the Brown bee), acquired from the island Læsø, Denmark. *A. mellifera* is entirely dependent on the attendance of humans for its survival in Greenland. The first abortive attempt at introducing *A. mellifera* to Greenland was made in the 1950'ies. It foundered due to lack of interest in beekeeping. The present attempt, initiated in 1998, is apparently more successful, the produced honey being marketable for its exotic value. Beekeeping may prove a valuable supplementary source of income to farmers in southern Greenland.

Bombus hyperboreus **Schönherr, 1809. Figs 1A, B**
Henriksen 1939: *Alpinibombus hyperboreus* Schönh.

18–22 mm (queens), 13–18 mm (males). With dense patches of elongate hairs; hairs yellow-orange on pronotum, yellowish white on mesoscutellum and anterior part

FIGURE 1 *Habitus images of A, B.* Bombus hyperboreus, *female and male; C–E.* Bombus polaris, *female (queen), worker and male.*

of metasoma, remaining hairs dark brown. Very similar to *B. polaris*, difference: see key.

BIOLOGY. Social parasite of *B. polaris* in Greenland. The queens of *B. hyperboreus* emerge later than those of *B. polaris*, after the latter have constructed their nests. The nest is usurped by the queen of *B. hyperboreus*, who kills the *B. polaris* queen. The brood of the *B. hyperboreus* queen is then raised by the *B. polaris* workers. Workers of *B. hyperboreus* do not occur in Greenland; however, the queens of *B. hyperboreus* apparently collect pollen (Pape 1983).

DISTRIBUTION. G (possibly absent from some of the northernmost areas).

Circumpolar, arctic-alpine.

Bombus polaris **Curtis, 1835. Figs 11.1–3D, 1C–E**
Henriksen 1939: *Alpinibombus arcticus* Kby.

17–19 mm (queens), 11–15 mm (males), 9–11 mm (workers). With dense patches of elongate hairs; hairs yellow-orange on pronotum, yellowish white on mesoscutellum and anterior part of metasoma (occasionally also on tip of metasoma), remaining hairs dark brown. Very similar to *B. hyperboreus*, difference: see key.

BIOLOGY. Nests are placed on the ground or in low vegetation, preferably on a sunny spot. The nest is constructed from plant parts, especially mosses, shaped into a loose sphere. The colony is founded by a single queen, workers eventually taking over most of the brood care and foraging. Typically 20 workers will be present in the colony when it is fully developed, fewer in the northernmost latitudes, more in the south. The brood is raised on pollen, being incubated in the process; by shivering, the adult(s) maintain a fairly constant temperature of 25–30° C (Pape 1983). Regurgitated nectar/honey is stored in honey pots for use in bad weather and during incubation.

DISTRIBUTION. G. Widespread and common throughout Greenland; in the southern parts most frequent at higher altitudes.

Circumpolar, arctic-alpine.

References

Pape, T. (1983): Observations on nests of *Bombus polaris* Curtis usurped by *B. hyperboreus* Schönherr in Greenland (Hymenoptera: Apidae). – *Entomologiske Meddelelser* 50: 145–150.

CHAPTER 12

Neuroptera or Planipennia[1] (Lacewings)

N. P. Kristensen

A medium-sized order with ca 6.000 described species worldwide. Adult lacewings are generally slenderly built insects with modest flight capacity; they are often predacious, feeding on little-mobile soft-bodied insects (*e.g.*, aphids), as are practically all larvae.

Morphology: adults

Neuroptera are not characterized by striking specializations in adult morphology, but in Greenland adult lacewings can be distinguished from other winged insects by their two pairs of broad, unscaled and overall similar wings with a dense venational network, notable for secondary (peripheral) branching of the principal longitudinal veins and abundant cross-veins. The wings are held steeply roof-like in resting position.

Immature stages

The principal diagnostic specialization of the order Neuroptera is the configuration of the piercing/sucking mouth parts of the larvae (Fig. 1C). Their elongate mandibles are slender (usually curved) and pointed, and the concave lower/inner wall of each is coapted with a somewhat similarly shaped maxillary stylet; the two stylets enclose between them the canal through which fluid food is sucked up after extraoral digestion. A venomous (and/or digestive?) secretion is injected into the prey through a canal inside the cuticle of the maxillary stylet. No other terrestrial insect larvae in Greenland have sickle-shaped mandibles which project far beyond the head capsule and clearly exceed the latter in length. The long antennae (also exceeding head-capsule length) are similarly unique among terrestrial insect larvae in this country. There are usually three larval instars.

The pupae are exarate and have movable mandibles; they are enclosed in a cocoon formed by a secretion from the larval malpighian tubules.

1 The names Neuroptera and Plannipennia are both in current use for the taxon characterized by the specialized larval mouth apparatus. Previously the name Neuroptera has been used in a wider sense for a taxon comprising also the Raphidioptera (snake flies) and Megaloptera (alderflies, dobson flies), both of which are unrepresented in Greenland. In this sense 'Neuroptera' equals what is currently referred to as the superordinal taxon Neuropterida.

Greenland fauna

Only one family and two species have so far been recorded from Greenland.

Hemerobiidae (Brown lacewings)

In the northern Holarctic this is the most species rich family of Neuroptera. Most species are brownish/greyish with delicate markings and have a wing span of 1 (in Greenland species ≈1.4)–2 cm. Diagnostic features are the forked cross-veins in the costal field and pectinate arrangement of Rs branches on vein R. While examination of characters in the genital segments is often essential for species identification in richer faunas, the taxa in Greenland can be separated on easily observable external characters. Note, however, that the pigmentation pattern of hemerobiid wings is variably developed and in some specimens it may be almost indiscernible; this variability is at least partly dependent on post-eclosion age, the pattern being weakest in newly emerged individuals. According to Veenstra *et al.* (1990) the larvae of the Greenland taxa are separable by differences in the relative lengths of their mandibles.

It has not been possible to confirm the presence in Greenland of a third species, listed as "*Hemerobius* ? *perelegans* Steph." by Henriksen (1939).

***Wesmaelius nervosus* (Fabricius, 1793) Figs 1B, 1C**

Henriksen 1939: *Boriomya betulina* Ström.

FW with basal cross-vein present between posterior Rs branch and anterior M branch (Fig. 1B, arrow); it may be weakly pigmented, particularly in its posterior part, and is then much more distinctly seen in transmitted than in incident light. In FW pigmented spot around middle M–Cu cross-vein not markedly stronger/more extensive than spot around distal cross-vein.

Larva (Fig. 1C) with mandibles markedly (ca 1.5×) longer than width of head capsule.

BIOLOGY. In Greenland found in different heath habitats, especially inland. Collection data indicate that there is one annual generation in Greenland, with hibernation in the egg stage.

DISTRIBUTION. S, SW (north to northern Uummannaq district: Qallatoq in Ukkusissat Fjord), [NE: only found by C. Vibe at Mestersvig; but see note in Chapter 5], NE (Jameson Land).
Palaearctic, arctic-subtropical, also alpine.

FIGURE 1 Hemerobius simulans *(A) and fore wing of* Wesmaelius nervosus *(B), arrows indicate diagnostic features (see text); C,* Wesmaelius nervosus *larva (painting by Denys Ovenden).*

FIGURE 2 Hemerobius simulans *in resting position on young leaf of* Salix glauca. (PHOTO: J. BÖCHER)

***Hemerobius simulans* Walker, 1853 Fig. 1A, 2**

FW without basal cross-vein between posterior Rs branch and anterior M branch. In FW pigmented spot around middle M–Cu cross-vein markedly stronger/more extensive than spot around distal cross-vein (Fig. 1A, arrow).

Larval mandibles not longer than width of head capsule.

BIOLOGY. In Greenland not known to differ from that of *W. nervosus*. In Europe this species is almost exclusively found on conifers, especially *Larix*.

DISTRIBUTION. S, SW (north to Qeqertarsuup Tunua/ Disko Bugt: Kangersuneq). Seems to have a much more restricted distribution in Greenland than *W. nervosus*.

Circumpolar, arctic-temperate, alpine.

[**CRYSOPIDAE (GREEN LACEWINGS)**

This family has not yet been recorded from Greenland, but will probably prove to be represented here. The species are overall larger than hemerobiids and have more elongate wings; body and wing veins are generally shades of green, and the compound eyes are brilliantly metallic (golden); unlike hemerobiids, chrysopids have the usual state of unforked costal cross-veins and just one Rs stem arising from R. Members of the *Chrysoperla carnea* species complex (including the most abundant northern European lacewings) hibernate as adults and often enter houses in the autumn; they would appear prone to casual introduction, but may not thrive outdoors in Greenland (One specimen of *C. carnea* was in fact caught in *Nuuk* in 2014). Chrysopid eggs are borne on stalks of hardened gland secretion. In some taxa the larva obtains camouflage by covering itself with emptied skins of prey insects.]

Acknowledgments

Prof. U. Aspöck and Dr. A. Popov kindly checked the identity of the bulk of the Greenland Neuroptera material available.

Reference

Veenstra, C.; Feichter, F. & Gepp, J. (1990): Larval diagnosis of the European genera of Hemerobiidae (Insecta: Neuroptera). *In*: Mansell, M.W. & Aspöck, H. (eds): *Advances in neuropterology. Proceedings of the 3rd International Symposium on Neuropterology.* Department of Agricultural Development. Pretoria, pp. 211–213.

CHAPTER 13

Coleoptera (Beetles)

Jens Böcher

The Coleoptera comprise considerably more than quarter of a million described species, and in terms of currently known species numbers they may be considered the most successful extant animal group. Their species richness reflects a considerable diversity in feeding strategies and habitat requirements. However, beetles are not doing particularly well north of the tree line, and only very few species have their main distribution within the Arctic (and accordingly are truly 'Arctic adapted'). High arctic Coleoptera have been treated by Chernov & Makarova (2008), and beetles in polar deserts of the Severnaya Zemlya Archipelago, Russia, are reported by Makarova *et al.* (2007). The number of extant indigenous beetle species now known from Greenland is no more than 36. This number is matched by that of evidently introduced species; these are briefly listed at the end of this chapter.

Adult beetles are usually (and always in the Greenland fauna) immediately recognizable by the modification of the fore wings into tough *elytra*, which meet in the mid line and cover and protect the membranous hind wings (when present) and, to a variable extent, the abdomen. The prothorax is large and conspicuous.

Morphological features most important for identification appear from Fig. 1. Generalized antennae have 11 segments and are filiform, while those in specialized lineages may be highly varied in shape. Two ocelli are sometimes present, but in most cases ocelli are missing. The prothorax is the largest thoracic segments and usually freely movable, meso- and metathorax are immovably fused. The mesoscutellum is visible from above as a triangular plate between the bases of the elytra. The latter are generally characteristically sculptured and equipped with longitudinal ribs; during flight they are almost always (no exceptions in Greenland) lifted. The hind wings are membranous with highly modified venation. Because the hind wings are longer than the elytra, they are folded at least once in the resting position. In many taxa the hind wings are more or less reduced or entirely missing. The legs are normally adapted for walking and running, increasing in size from front to rear. In some families one or more pairs may be modified for burrowing in soil, tunnelling in wood, swimming or jumping. The tibiae are generally equipped with strong spines and spurs. The tarsi are 5–, 4–, or 3–segmented.

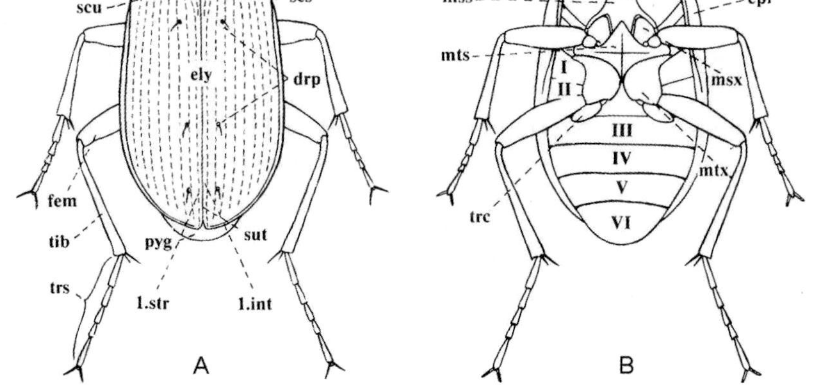

FIGURE 1
General structure of a carabid beetle in dorsal (A) and ventral (B) view.
ant:** antenna; **bsf:** basal fovea of prothorax; **cly:** clypeus;* **drp: dorsal punctures; **ely:** elytra; **epl:** epipleura of elytra;* eye*: compound eye; **fem:** femur; **frf:** frontal furrow; **fro:** frons;* **gul***: gula; **hum:** humerus; **lbp:** labial palp; **lbr:** labrum; **max:** maxilla; **mnd:** mandible; **mnt:** mentum; **mss:** mesosternum; **msx:** meso-coxa; **mts:** metasternum; **mtx:** meta-coxa; **mxp:** maxillary palp; **ppe:** setae of prothorax; **pre:** pro-episterna; **prs:** prosternum; **prt:** prothorax; **prx:** pro-coxa; **pyg:** pygidium; **scx:** scutellar stria; **scu:** scutellum; **sut:** suture of elytra; **tib:** tibia; **trc:** trochanter; **trs:** tarsus; **1.int:** first elytral interval; **1.str:** first elytral stria; **I–VI:** abdominal sternites. From Lindroth (1983).*

In the abdomen of most beetles the first 5 or 6 terga are lightly sclerotised, but the hindmost may be hardened when they are exposed beyond the elytra.

In most cases beetle eggs are simple and ovoid, but a sculptured chorion is found in some groups whose members oviposit on surfaces of plants etc.

Beetle larvae are highly diverse and may be apodous. They have a well developed and sclerotised head capsule, antennae with 4 or fewer segments, 6 or fewer stemmata. Mandibles are of the chewing type. Abdomen has never prolegs, but a pair of appendages, 'urogomphi', may be borne on the 9th tergum; their morphological nature is debated. The basic number of leg segments is 6 or 5 (including tarsus and pretarsus with 1 or 2 claws), but reduction of legs has occurred in many groups. Normally beetle larvae pass through 3 instars.

The diversity of biological types is especially evident in the larvae which are morphologically highly diverse and adapted to different modes of life. Most Greenland beetle species have free living larvae (Carabidae, Dytiscidae, Staphylinidae, Coccinellidae). The moss-eating Byrrhidae have 'scarabaeoid' larvae of crescentic shape, with head and legs well developed, and with soft, inflated abdomen. The subterraneous or plant boring larvae of Curculionidae are maggot-like, without appendages (apodous) and with highly reduced eyes and antennae.

The pupae are adecticous and nearly always exarate. Most are soft and pale, found in the soil or in food plants, generally enclosed in a cocoon made by a secretion from the malpighian tubes. The Coccinellidae, however, have freely exposed, obtect and coloured pupae.

The classification of families follows Crowson (1981). There is a comprehensive treatment of the Greenland Coleoptera by Böcher (1988) to which the reader is referred for further details and literature references. The extensive information on the species' biology in Iceland given by Larsson & Gigja (1959) is mostly not repeated here.

The peculiarities of the Greenland beetle fauna are discussed in chapter 4. The similarity to the coleopterous fauna of Iceland is evident, although that of Greenland is much less diverse (*c.* 160 indigenous species in Iceland; Larsson & Gigja 1959). Of particular interest is a comparison with the beetles of the High Arctic Spitsbergen (Norway) situated between about 74°N and 81°N. The fauna here is astonishingly rich for the latitude, comprising 10 non-synanthropic species (possibly 12), but only 3 are shared with Greenland, viz., *Micralymma marinum*, *Simplocaria metallica*, *Enicmus minutus* (Bengtson *et al.* 1975, Coulson 2000, Coulson *et al.* 2014, Fjellberg 1983, Hågvar 1969, Kangas 1967, 1973, Prestrud *et al.* 2004, Sømme 1979, Strand 1942).

Key to the families of Greenland beetles (adults). Families which in Greenland only comprise introduced species are included [in square brackets]

1. Antennae simple, filiform or moniliform. Tarsi always with 5 segments. Hind coxae immovably fused with metasternum and completely dividing 1st visible abdominal sternum (Fig. 1B). .... 2 (suborder Adephaga)
– Antennae ± clubbed, serrate or geniculate, if filiform hind coxae usually movably articulated to metasternum and not dividing 1st visible abdominal sternum. Tarsi in some cases with fewer than 5 segments ..... 4 (suborder Polyphaga)
2. Terrestrial insects with projecting sensory bristles on various parts of the body. Hind coxae do not extend laterally to meet the elytra.................... Carabidae
– Streamlined aquatic insects without projecting sensory bristles. Hind coxae extend laterally to meet the elytra 3
3. Compound eyes completely divided into dorsal and ventral parts. Antennae short and thick. Mid and hind legs modified into short and broad paddles ....Gyrinidae
– Compound eyes not divided. Mid and hind legs not modified................................. Dytiscidae
4. Head more or less distinctly produced into a rostrum. Antennae clubbed and elbowed, 1st segment can be concealed in a groove in the rostrum......... Curculionidae
– Head not prolonged into a rostrum. Antennae not clubbed and elbowed 5
5. Elytra more or less truncate so that at least 3 terminal abdominal segments are exposed .... Staphylinidae
– Almost entire abdomen covered by the elytra 6
6. Maxillary palps longer than antennae................ Hydrophilidae
– Antennae much longer than maxillary palps 7
7. Legs and antennae can be withdrawn into grooves on underside of body. Middle coxae widely separated from each other...................... Byrrhidae
– No grooves on underside of body to receive legs and antennae. Middle coxae close to each other 8
8. Head more or less covered by a 'hood' formed by pronotum 9
– No such hood formed by pronotum...............10
9. Head only partly covered by pronotal 'hood'. Antennae more or less serrate............ Anobiidae

– Head completely covered by pronotal 'hood' Species less than 1 cm, with long antennae and legs......... .... Ptinidae
10. Antennae more or less distinctly clubbed.......... 11
– Antennae not clubbed. Tarsi apparently 4-segmented, with 3rd segment bilobed.......................... 17
11. Antennae shorter than head width. Rounded species with head partly concealed by pronotum viewed from above .... Coccinellidae
– Antennae longer than head width................. 12
12. More or less hairy species with no conspicuous sculpturing of pronotum or elytra................. 13
– Generally smooth species, in most cases with conspicuous sculpture 14
13. More or less rounded species. Pronotum posteriorly much widened, hind angles more or less acute. Most species with a single large ocellus on top of head.... .... Dermestidae
– Elongate species without ocelli...................... .... Cryptophagidae
14. Pronotum longer than wide; single Greenland representative a synanthropic species, *c.* 3.5 mm. long with, strongly toothed lateral pronotal margins .... [Cucujiidae]
– Pronotum wider than long......................... 15
15. Species less than 5 mm with sculptured elytra....... .... [Latridiidae]
– Species of different size, generally not with conspicuous sculpture 16
16. Carabid-like body design, anterior pronotal corners prominent, acute. All legs with 5 tarsal segments.... .... [Trogossitidae]
– Near-cylindrical bodies, anterior pronotal corners not prominently produced, Hind legs with only 4 tarsal segments....................... Tenebrionidae
17. Elongate species with antennae at least two-thirds as long as whole body. All tibiae with two spurs........ .... [Cerambycidae]
– Generally rounded, bare and shining species. Antennae less than two-thirds of body length. Not all tibiae with two spurs .... [Chrysomelidae]

Key to the larvae of the principal native beetle families in Greenland

1. Legs with discrete, one-segmented tarsus and two claws (Fig. 2A) Adephaga 2
– Legs with composite tibiotarsus and a single claw (Fig. 2B), or legs absent Polyphaga 4

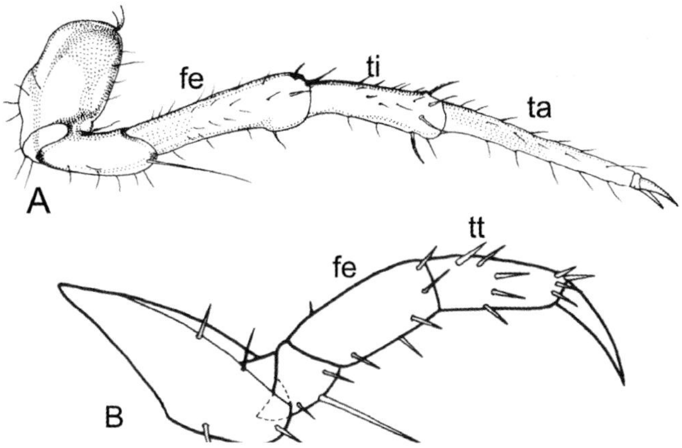

FIGURE 2 *Larval leg of adephagan (A) and polyphagan (B) beetle. fe, femur; ta, tarsus; ti, tibia; tt, fused tibiotarsus. A from Luff (1993), B from M. Hansen* (Biologiske Skrifter *40, 1991*).

2. Abdomen soft without distinct terga and with one pair of tracheal gills on each segment............... .... Limnic Gyrinidae
– Abdomen with distinct terga and without tracheal gills 3
3. Mandibles with a tooth (retinaculum) on the inner side .... Terrestrial Carabidae
– Mandibles with a longitudinal furrow, but without a retinaculum .... Limnic Dytiscidae
4. Urogomphi present, trunk near-cylindrical......... .... Staphylinidae
– Urogomphi absent 5
5. Trunk strongly coloured ('warning colouration'), blackish blue and orange .... Coccinellidae (part, *Coccinella*)
– Trunk pale, or with brownish sclerites 6
6. Thoracic legs absent .... Curculionidae
– Thoracic legs present 7
7. Sluggish moss-feeders, with crescentic shape and sizable sclerotized terga .... Byrrhidae
– Active, free-living predators with only minute trunk sclerotizations......... Coccinellidae (part, *Nephus*)

CARABIDAE (GROUND BEETLES)

Terrestrial, generally elongate, dark coloured, glabrous and in many cases metallic beetles with projecting sensory bristles on various parts of the body. Hind coxae not extending laterally to meet elytra. Good runners with long legs and antennae. Many have the elytra firmly soldered and the hind wings atrophied; others (*e.g.*, *Bembidion grapii*) are wing polymorphic. Larvae campodeoid and generally active, fast-running predators. Abdominal segment 9 (the 'anal tube') in most cases bears long urogomphi equipped with sensory setae; the urogomphi may be secondarily segmented. The larvae have been thoroughly treated by Luff (1993).

Key to the Greenland Carabidae

IMAGINES

1. > 8 mm in length 2
– < 6 mm in length 3
2. Pronotum much wider than long .... *Nebria rufescens*
– Pronotum approximately as wide as long .... *Patrobus septentrionus*
3. Side margin of pronotum S-shaped, hind corners approximately rectangular .... *Bembidion grapii*
– Side margin of pronotum convex, hind corners smoothly rounded .... *Trichocellus cognatus*

LARVAE

1. Head with a marked 'neck', roundish with pattern of dark markings. Nasale with 4 marked teeth. Third instar c. 20 mm (Figs C-D) .... *Nebria rufescens*
– Head without 'neck', almost quadratic 2
2. 2. Nasale with (about) 5 inconspicuous teeth. Third instar c. 20 mm (Figs A-B) .... *Patrobus septentrionis*
– Nasale without teeth or with 8 small teeths. Third instar 4–5 mm 3
3. Nasale without teeth. Mandible slender with retinaculum longer than broad, its tip pointing backwards (Figs 4A, B) .... *Bembidion grapii*
– Nasale with c. 8 teeth. Mandibles relatively short and stout, retinaculum not longer than its basal width (Figs 4C, D) .... *Trichocellus cognatus*

Nebria rufescens **(Ström, 1768) Figs 3C-D 5, 18A**
Henriksen 1939: *Nebria gyllenhali* Schönh.

Largest of the Greenland Carabidae, *c.* 9–12 mm. Furrows (striae) of elytra deep. Pronotum much wider than long. Legs long and strong. Hind wings always fully developed.

The larva has a conspicuous 'neck'.

BIOLOGY. Markedly hygrophilous and generally in cool habitats. In Greenland mainly found under stones on sandy beaches and river banks at low altitude. Frequently fairly numerous along brooks and small rivers surrounded by vigorous vegetation, in marshy areas with *Eriophorum*, and in luxuriant grass vegetation. The species is probably nocturnal, even in Greenland; it has

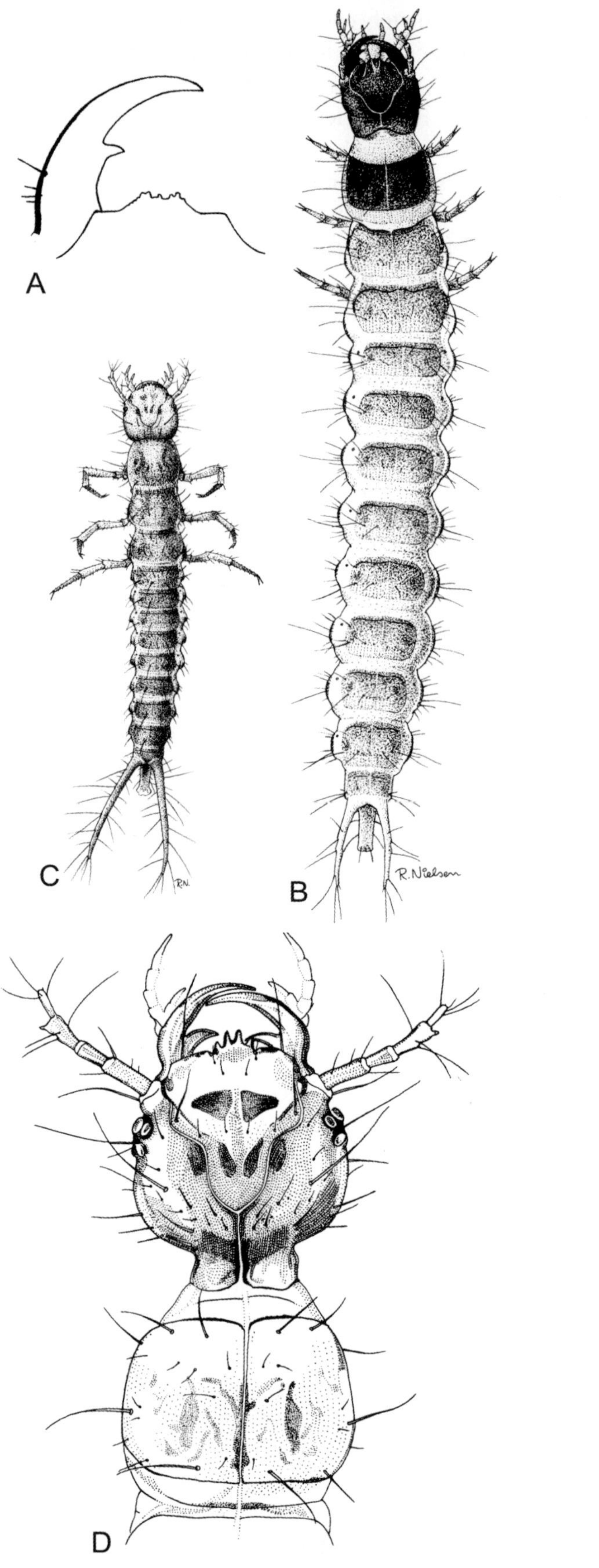

FIGURE 3 *A–B,* Patrobus septentrionus; *left larval mandible (A) and larva (B). C–D,* Nebria rufescens, *second instar larva (C), larval head and pronotum. (D). B, C from Böcher (1988), A, D from Luff (1993).*

never been observed flying. Greenland collection data indicate a two-year life cycle with both larval and adult hibernation. Further south the cycle appears to be annual.

DISTRIBUTION. S, SW (north to Maniitsoq/Sukkertoppen). Local and fairly rare, quite common only in the Qaqortoq/Julianehåb district and in the eastern Nuup Kangerlua/Godthåbsfjord area.

Circumpolar, low arctic-subarctic, in Europe boreo-alpine.

NOTE. A polymorphic species with geographically variation in frequency of colour morphs. In the Palaearctic the black-legged nominate form coexists with a red-legged variety (var. *balbi*) (Bengtson *et al.* 1983). In Eurasia the latter becomes more frequent towards higher altitudes and latitudes and with drier conditions. In Greenland *balbi* predominates, but also intermediate, brown-legged forms are found. In North America *balbi* appears entirely absent.

Patrobus septentrionis Dejean, 1821 Figs 3A-B, 18B

Similar to *Nebria rufescens* in general shape and colour, but smaller (*c.* 8–10 mm). Pronotum about as long as wide, striae of elytra narrow and shallow, legs relatively shorter. Elytra often more or less rufous, especially in alpine environments. Hind wings always fully developed.

BIOLOGY. Generally found in humid situations in highly diverse biotopes. May be present in large numbers on beaches, lake shores and along water courses (in company with *Nebria rufescens*), as well as in lush shrubs, herb slopes, grasslands and heaths (up to 500 m a.s.). It has a low thermal threshold of activity and may be active during milder spells of winter. In Scandinavia strictly nocturnal, whereas in Greenland observed in activity during daytime; spontaneous flight has been observed, but not in Greenland. Greenland data definitely show that the species here, like in other parts of its range, is biennial with hibernation of both larvae and adults.

DISTRIBUTION. S, SW, SE (north to Sisimiut/Holsteinsborg and Kangerlussuatsiaq/Lindenow Fjord). Common and widespread in SW Greenland.

Circumpolar, low arctic-boreal, in Europe boreo-alpine,

Bembidion grapii Gyllenhal, 1827 Figs 4A, B, 18C

Henriksen 1939: *Bembidion* (*Peryphus*) *grapei* Gyll.

C. 4–4.5 mm. Head, pronotum and elytra shining black, usually metallic. Side margins of pronotum S-shaped, hind angles approximately rectangular.

Polymorphic with respect to hind wing development; they may vary from being fully functional to rudimentary, scale-like. Four main wing types may be distinguished,

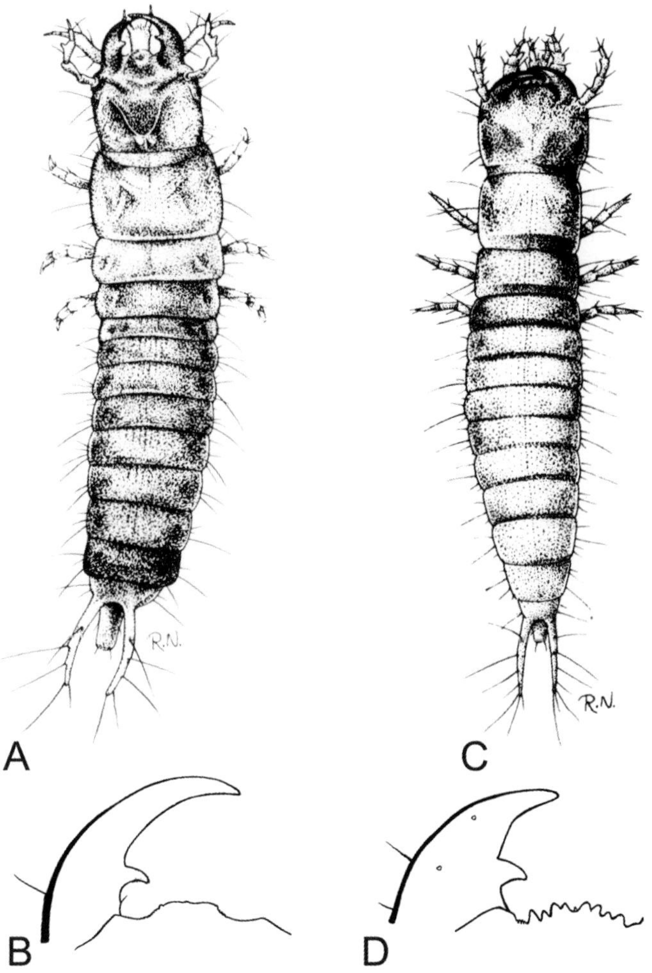

FIGURE 4 *A–B,* Bembidion grapii*; larva (A) and larval left mandible (B). C–D,* Trichocellus cognatus*; larva (C) and larval left mandible (D). A, C from Böcher (1988); B, D from Luff (1993).*

but at least in Greenland there are intermediates between two brachypterous forms. In Greenland brachyptery predominates, but macropterous individuals have been found everywhere.

Larval mandibles slender, no teeth on nasale between mandibles.

BIOLOGY. Mainly found under stones, sometimes adults and larvae together, in fairly warm and dry situations, *e.g.*among low and sparse steppe-like vegetation or in birch coppices on south-facing slopes, but most often close to more luxuriant vegetation (*e.g.* bordering small streams); never on barren fell-fields or really dry ground, and frequently in company with *Trichocellus cognatus*. The highest locality recorded (580 m, at Narsarsuaq) was in a heath rich in mosses and lichens. It is not certain whether the macropterous forms are able to fly, as observations are lacking. In captivity the species was observed eating collembolans and oribatid mites. Greenland data indicate an annual cycle. Adults hibernate to lay eggs in spring, the larval stages are passed during summer.

DISTRIBUTION. S, SW (north to Ukkusissat Fjord), SE (north to Blosseville Kyst). *B. grapii* reaches farthest north of the Greenland carabids, almost to the limits of the High Arctic. A single reported find from NE Greenland (Mørkefjord; Haarløv 1942) must be considered very doubtful (Böcher 1988, Böcher & Bennike 1996).

Circumpolar, low arctic-subarctic and alpine.

Trichocellus cognatus **(Gyllenhal, 1827) Figs 4C, D, 18D**
Size (*c*. 4–5 mm) and general shape similar to *Bembidion grapii*, but readily distinguished by the light brownish-yellow margins of pronotum and elytra. Hind angles of pronotum rounded, side margins convex. Hind wings always well developed.

Larva with broad and stout mandibles, teeth on nasale easily visible between mandibles.

BIOLOGY. The habitat is not easily described in Greenland, though similar to that of *B. grapii*. The species is generally found under stones in both humid and dry, but fairly warm surroundings. Very often it occurs together with *B. grapii*, but is arkedly less frequent and widespread. Flight not observed in Greenland, but frequently in Europe. The scanty evidence indicates a biennial cycle in Greenland, with hibernation of both larvae and adults.

FIGURE 5 Nebria rufescens.

DISTRIBUTION. S, SW (north to inner Kangerlussuaq/Søndre Strømfjord), SE (Saqqisikuik/Skjoldungen). Like *B. grapii* it seems to disappear from the coast towards north in Greenland.

Circumpolar and very widespread, low arctic-temperate.

DYTISCIDAE (PREDACIOUS WATER BEETLES)
Streamlined aquatic insects without projecting sensory bristles. Hind coxae extend laterally to meet elytra. Hind legs function as swimming organs: flattened, widely separated and fringed with long hairs. Adults can without problems exist on land. In the water they carry under the elytra a reservoir of air, which is renewed by penetrating the water surface with the abdomen tip. Most Dytiscidae are excellent, nocturnal fliers. In males of certain genera the first three segments of the fore tarsi are dilated to form adhesive pads provided beneath with suckers. These enable them to get a firm hold of females during copulation.

Larvae swim by both rowing with legs and movements of total body. Like the adults they are carnivores and attack and pierce their victim with the long, sickle-shaped mandibles, which are traversed by a fine, almost closed groove. A secretion from the gut is injected into the prey through these channels, and following an external digestion the dissolved interior of the prey is taken up the same way. The larvae get air from the atmosphere through the posterior spiracles, while pendent from the surface film. Pupation generally takes place in small crevices on the shore close to the water, *e.g.*, under stones or in moss.

Hydroporus morio **Aubé, 1838 Figs 6D, E, 19B**
Henriksen 1939: *Hydroporus melanocephalus* Marsh.

Unmistakable among Greenland water beetles due to the oval, somewhat flattened shape and the small size (*c*. 3–3.5 mm). Always fully winged.

The larva has a characteristic triangular head (Fig. 6E).

BIOLOGY. *H. morio* is able to fly and inhabits any kind of small ponds and pools, frequently temporary, and also marshes. According to B. L. Madsen (pers. comm.) ponds with low water and muddy bottom are preferred. When found in lakes, the habitat proper is a sheltered cove or a small, secluded portion. The principal food is possibly chironomid larvae, small crustaceans and rotifers (Jeppesen 1986). Madsen (1959) observed adults attacking and eating chironomid larvae, frequently a number of beetles devouring the same larva.

Adults have been collected from early May throughout summer (until October; Madsen 1959). Females collected at Kangilinnguit/Grønnedal produced eggs on 4 May, and lar-

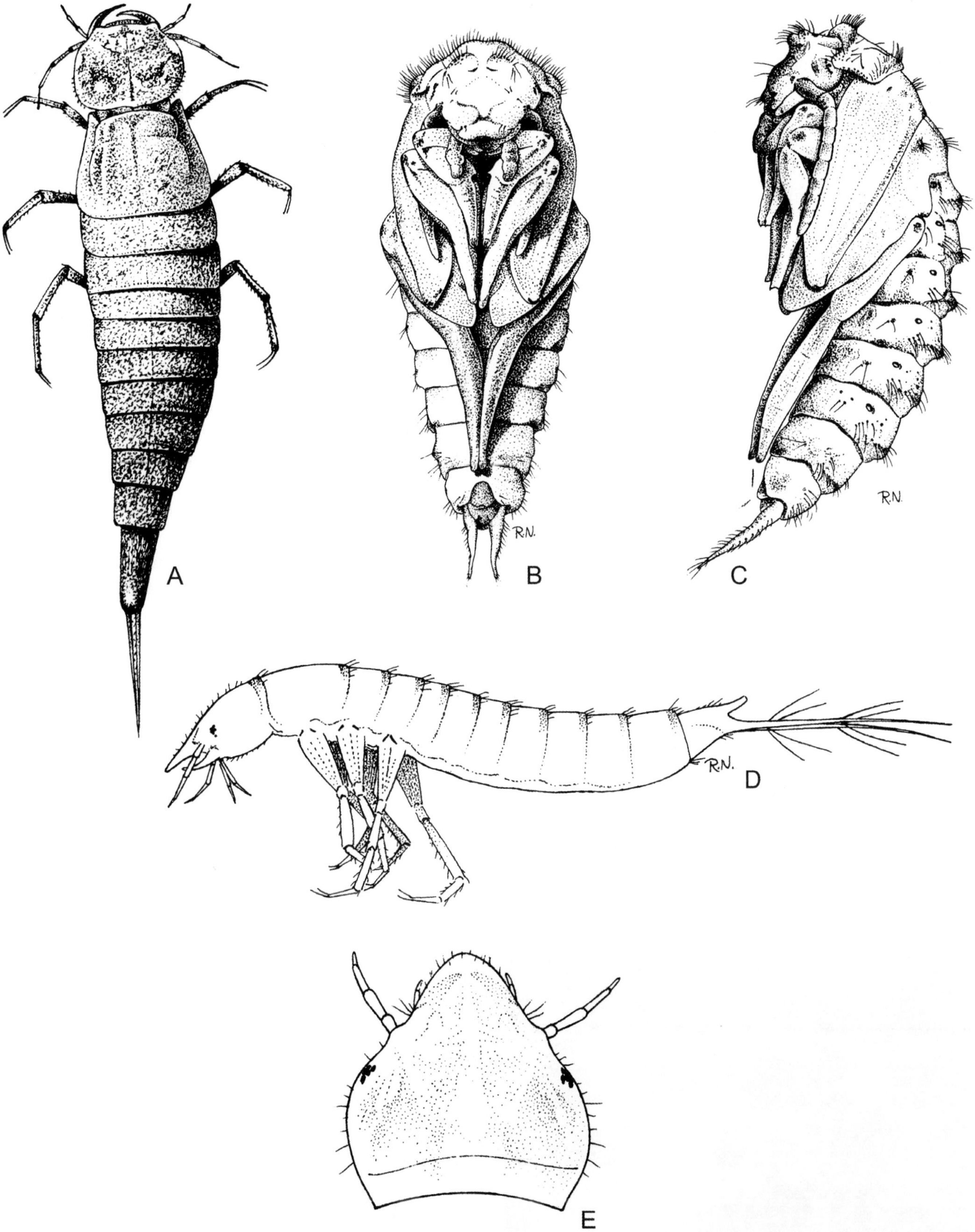

FIGURE 6 *A–C,* Colymbetes dolabratus; *larva, dorsal view (A), pupa in ventral (B) and lateral (C) view. D–E,* Hydroporus morio; *larva, lateral view (D) and larval head, dorsal view (E). From Böcher (1988).*

vae appeared 3 weeks later (Madsen 1959). Jeppesen (1986) found the three larval stages successively during summer. Imaginal hibernation is unquestionable, but possibly a certain number of larvae also survive the winter to pupate in early summer. Larvae about to pupate and newly pupated individuals were observed on August 16–23 in moist gravel under stones at the bank of a pond, 10–20 cm from the water edge. At the same time no larvae were seen in the water (at Tutttilik, SE Greenland, B. L. Madsen, pers. comm.). The species must be able to tolerate freezing during winter, possibly mostly in water mosses and in the bottom sediment. During September 2006 the species was searched for in vain in many suitable biotopes in the Kangilinnguit/Grønnedal area (S). In the few places where it did occur it was highly aggregated, always in submerged moss at the pond margins. In a small pond about 50 adult individuals were in a walnut-sized lump of moss. Their sluggish behaviour may indicate that they were in a state of dormancy, preparing for winter (B. L. Madsen, pers. comm.).

DISTRIBUTION. S, SW, NW, N(?), SE. Very common and widely distributed in western Greenland, but not found in NE between Peary Land and Ammassalik (the record from Peary Land has to be confirmed, A. Downes, pers. comm. 1990).

Circumpolar, high arctic-boreal, in Europe boreo-alpine.

NOTE. *H. morio* is member of an intricate complex of species, and it is not excluded that other species of the genus occur in Greenland. An extensive treatment is found in Nilsson & Holmen (1995). Jeppesen (1986) described and keyed the larval instars.

Colymbetes dolabratus **(Paykull, 1798) Figs 6A–C, 7, 19A**

Largest beetle in Greenland, *c.* 13–17 mm, and quite unmistakable due to rounded, flat shape, the flattened, hairy middle- and (especially) hind legs, and the habitat. Hind wings always fully developed and functional. The larval stages and the pupa have been figured by Galewski (1967, 1968). A careful treatment and a well illustrated key to the Scandinavian species is found in Nilsson & Holmen (1995).

FIGURE 7 Colymbetes dolabratus. (PHOTO: B.L. MADSEN)

BIOLOGY. The species typically lives in ponds with a rich vegetation, but is highly eurytopic and may be present in almost any kind of stagnant water; it has frequently been observed flying. The feeding biology has not been studied, but in Greenland undoubtedly larval nematocerans and crustaceans play a significant part in the diet.

In summer larvae and adults are found together. According to Røen (1963, 1981) in late summer imagines migrate to hibernate in lakes, which do not freeze to the bottom, *i.e.* in Greenland generally are deeper than 3 m. Migration of adult *C. dolabratus* between breeding localities before and after winter was confirmed through observations in 1994 and 1995 by Svensson (2005) on the situation in a small oval lake (0,15 ha, max depth 1.8 m) in West Greenland (67°27'N, 50°30'W). In June – July larvae and adults were abundant, but the population dramatically decreased in August, and no larvae found late in that month. In August 1994 one individual was observed flying away from the lake in the evening. On 9 May 1995 a noteworthy series of records were made of immigrating specimens: During one hour, from about 11 a.m., 33 individuals were heard when landing on the lake surface, and about one third of these were actually observed descending from a considerable height (sky clear, air temp. 10°C, no wind). The beetles immediately dived into the lake, of which about half was still ice-covered. Similar mass-migrations are characteristic also of other dytiscids (*e.g. Dytiscus* spp.), as can in central Sweden be observed on the first warm and sunny days in May (B.W. Svensson, pers. comm.).

In S Greenland (Ivittut) imagines were observed March 9 (1959) in a small lake under 105 cm of ice (B.L. Madsen, pers. comm.). In ponds, at low altitude only, at Kangilinnguit/Grønnedal (S) adults were observed on May 3 (2006) when the lakes were still covered with ice. Mature larvae were seen during July through late August, and metamorphosis was observed in mid- August (B.L. Madsen, pers. comm.).

During winter adults must depend on diffusion of oxygen into the air bubble stored under the elytra, but it decreases in size when not replenished. The northern range limit of the species in Greenland (around 74°N in both East- and West Greenland) likely reflects this problem, because north of *c.* 74°N lakes are ice-covered for at least 10 months, which is the maximum duration of the functional respiratory air-bubble (Røen 1963, 1981). South of *c.* 67°N larval and pupal stages are passed during summer, so that only adults hibernate to lay eggs the following spring. Further north fully grown larvae have been caught in spring, presumably representing individuals which take two years to complete the life cycle. Pupae are found during mid-summer.

At Kummiut and Tuttilik (Ammassalik area) larvae were never observed in lakes and ponds at higher altitude. It is suggested that here duration of the ice cover is too long for adults to overwinter, so that the winter is passed in the egg-stage. The species is thus restricted to ponds which melt early in the season and soon reach a high temperature (B. L. Madsen, pers. comm.).

DISTRIBUTION. S, SW, NW, NE, SE. The range has a marked northern limit at c. 74°N in both West- and East Greenland. One of the most commnon and generally known, most widely distributed and most thoroughly studied of the Greenland insects.

Circumpolar, high arctic-boreal.

GYRINIDAE (WHIRLIGIG BEETLES)

Smooth and shiny, streamlined aquatic beetles, which like dytiscids have no projecting sensory bristles and whose hind coxae extend laterally to meet elytra. Unmistakable due to the short, paddle-like middle and hind legs, very short and thick antennae, and specialized compound eyes divided into an upper and a lower part, adapted for vision in air and water, respectively. They are gregarious predators, which skate on the surface of stagnant and slowly flowing water, hunting for insects caught in the surface film. The prey is seized by the long fore legs. The bottom-dwelling larvae are likewise carnivorous and easily recognizable due to the pale, elongated body with long tracheal gills along sides of abdomen; their tracheal system is closed.

Gyrinus opacus Sahlberg, 1819 Figs 8, 19C

Unmistakable among Greenland insects. Length *c.* 5.5–6 mm.

BIOLOGY. Found in different stagnant waters, most often ponds with a rich vegetation, but also lakes. Svensson 2005) studied a particularly large (up to 2500 individuals) population in one small lake (0,15 ha, max depth 1.8 m) among some 30 inhabited by the species near its northern range limit in West Greenland (67°27'N, 50°30'W). The adults hibernate on the bottom, as was directly demonstrated by the finding of two individually marked beetles after the 6–7 cm thick ice was sawn up in early October (1996). Some individuals overwintered twice and also reproduced two years in this lake. At the onset of hibernation the beetles had lower fat contents than individuals in a coastal rock pool population in southern central Sweden, but the proportional loss of fat during hibernation did not

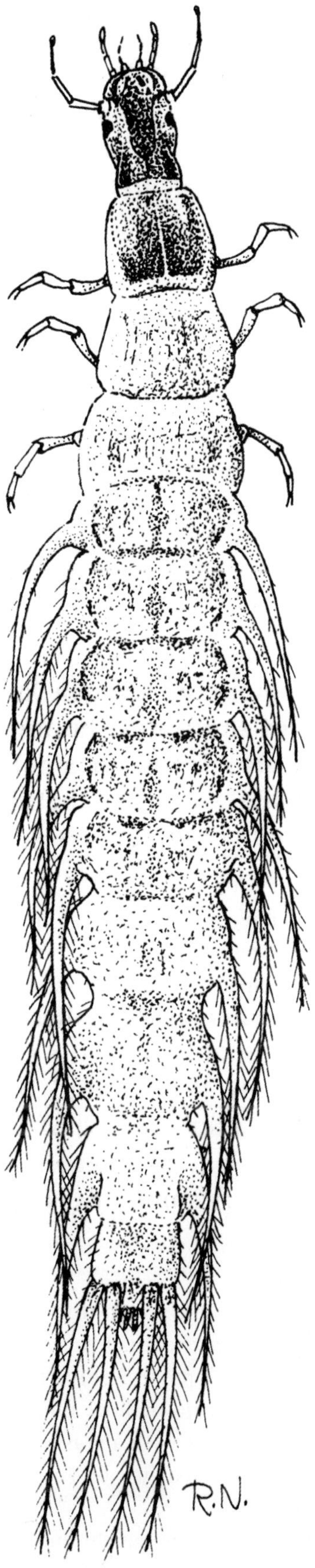

FIGURE 8 Gyrinus opacus, *larva. From Böcher (1988).*

differ. Reproductive females continued to lose fat in spring and reached lower mean values than Swedish females, being close to levels experimentally found to be lethal to the latter. During the period of fat reserve build-up from late August to late September/early October, examination of gut contents revealed no insect prey. Instead floating cladocerans (*Eurycercus* sp.) were regularly consumed, and the beetles also intensively grazed epiphytes on submerged *Potamogeton filiformis,* on mollusc shells and cases of caddisfly larvae. The accumulation of energy reserves of adults by consuming resources not normally utilized by boreal populations may be seen as an important strategy for surviving the severe early reproduction period the following spring in the Arctic.

DISTRIBUTION. S, SW (north to Kangerluarsuk Tulleq at Sisimiut/Holsteinsborg). Fairly common especially in southern part of the Greenland range, northward more often scattered occurrences and mainly inland.

Circumpolar, low arctic-boreal.

HYDROPHILIDAE (WATER SCAVENGER BEETLES)

Antennae small, less prominent than maxillary palps and terminating in a short club. Most species are aquatic with a superficial resemblance to the Dytiscidae, but without flattened, oar-like legs; the species generally crawl on water plants. Adults are in most cases vegetarians, feeding on dacaying plants, but many larvae are carnivorous. The underside of adults is covered with fine hairs that hold a layer of air, and another air bubble is located under elytra. Both air supplies are replenished when an antenna is pushed through the water surface. The larvae have different shapes, and may in some cases be similar to those of Carabidae.

Helophorus brevipalpis Bedel, 1881 Fig. 19D

C. 3 mm. Easily distinguished from other Greenland water beetles by bronze-metallic glow of head and pronotum, both rugged and with furrows, and the brownish yellow elytra with marked striae. Antennae of same length as maxillary palps. Hind wings always fully developed; the species is an active flier.

BIOLOGY. Unknown in Greenland, since the only record is from a pitfall trap placed in a heath of *Vaccinium uliginosum* on a dry, south facing slope. In Europe fairly eurytopic in stagnant water, preferring shallow, more or less open, frequently temporary pools with a grassy bottom, but also found in the lower reaches of running waters.

The single Greenland specimen was caught June 13 (1984). In Europe the eggs are normally laid in spring (sometimes late autumn) in cocoons placed in mud at the waterline. The larvae are fully grown in two or three weeks and pupate in the soil on land. The new generation returns to the water to hibernate there (Hansen 1987).

DISTRIBUTION. nSW. The single find was from Eqalummiut Nunaat at the head of Kuup Akua/Nordre Strømfjord (Stroud 1984, pers. comm.).

Palaearctic, temperate.

STAPHYLINIDAE (ROVE BEETLES)

A very large family of generally dark coloured beetles of highly differing size. Elytra more or less truncate, at least 3 terminal abdominal segments exposed. These are very movable and have sclerotized terga. Despite the small elytral size, hind wings are large, and as a rule the species are good fliers. Most rove beetles are carnivorous. Larvae are typically campodeiform and often closely resemble those of the Carabidae. The larvae are often found together with adults and are likewise in many cases known to be predacious. In Greenland by far the dominant beetle family. A number of species are only known as subfossils from Norse farms. These are mentioned in chapter 4.

Key to the Greenland Staphylinidae

ADULTS

1. Less than 2 mm. Antennae distinctly clubbed .... *Euaestethus laeviusculus*
– More than 2 mm. Antennae less distinctly clubbed 2
2. Body broad and somewhat flattened. Two ocelli between compound eyes 3
– Body elongate and not flattened. No ocelli.......... 6
3. Head and pronotum each with two longitudinal grooves. Elytra black, at least of same length as the exposed part of abdomen .... *Omalium excavatum*
– Head and pronotum without longitudinal grooves. Elytra distinctly shorter than the visible part of abdomen 4
4. Elytra light brown, a little shorter than the exposed part of abdomen .... *Xylodromus concinnus*
– Elytra dark, much shorter than the exposed part of abdomen 5
5. Dull, black species. No antennal segment distinctly shorter than long. Lives in the tidal zone at the beach .... *Micralymma marinum*
– Shiny black, more slender. Antennal segments 8–10 distinctly shorter than long.......................... .... *Micralymma brevilingue*
6. Antennal bases in front of the eyes. Elytra with side margins.. 7

– Antennal bases between the eyes. Elytra without side margins..10
7. Head more or less constricted posteriorly, to form a 'neck'. Body not spindle-shaped (narrowing both anteriorly and posteriorly)...........................8
– Head without a 'neck', relatively small and partly concealed by the prothorax. Body spindle-shaped .. .. *Mycetoporus nigrans*
8. Head posteriorly strongly constricted. Prothorax subrectangular, longer than broad ....*Lathrobium fulvipenne*
– Head posteriorly less constricted. Pronotum rounded 9
9. Length c. 10 mm. Width of eyes less than half of head width ....*Quedius mesomelinus*
– Length c. 6 mm. Width of eyes c. 2/3 head width .... *Quedius fellmanni*
10. Head posteriorly markedly constricted. Length of antennae about the same length as head + prothorax ....*Gnypeta cavicollis*
– Head posteriorly only slightly constricted. Antennae longer than head + prothorax ....11 (*Atheta*)
11. Head with scattered, large but shallow punctures. Pronotum about same width as head, side margins backwards slightly concave. Length c. 3–4 mm ....*Atheta* (*Thinobaena*) *vestita*
– Head with fine punctures. Pronotum distinctly wider than head, side margins not concave................ 12
12. Midline hairs of pronotum directed backwards; inwards-bent lateral parts of pronotum in lateral view barely visible; side margins slightly and evenly curved. Length c. 2–2.5 mm......................... ....*Atheta* (*Mocyta*) *groenlandica*
– Midline hairs of pronotum directed forwards; inwards-bent lateral parts of pronotum in lateral view visible; anterior part of side margins curved down towards front angles.........................13
13. Pronotum with shallow, wide depression in middle of posterior half. Inner margins of elytra longer than pronotum. In female hind margin of 6th sternum distinctly concave in the middle and with fringe of long hairs. Body length c. 3.5–4 mm.............. .... *Atheta* (*Boreophila*) *hyperborea*
– Pronotum without or with only a slight, narrow depression along midline. Inner margins of elytra shorter than pronotum. In female hind margin of 6th sternum evenly curved and with fringe of short hairs. Body length c. 3–3.5 mm .... *Atheta* (*Boreophila*) *islandica*

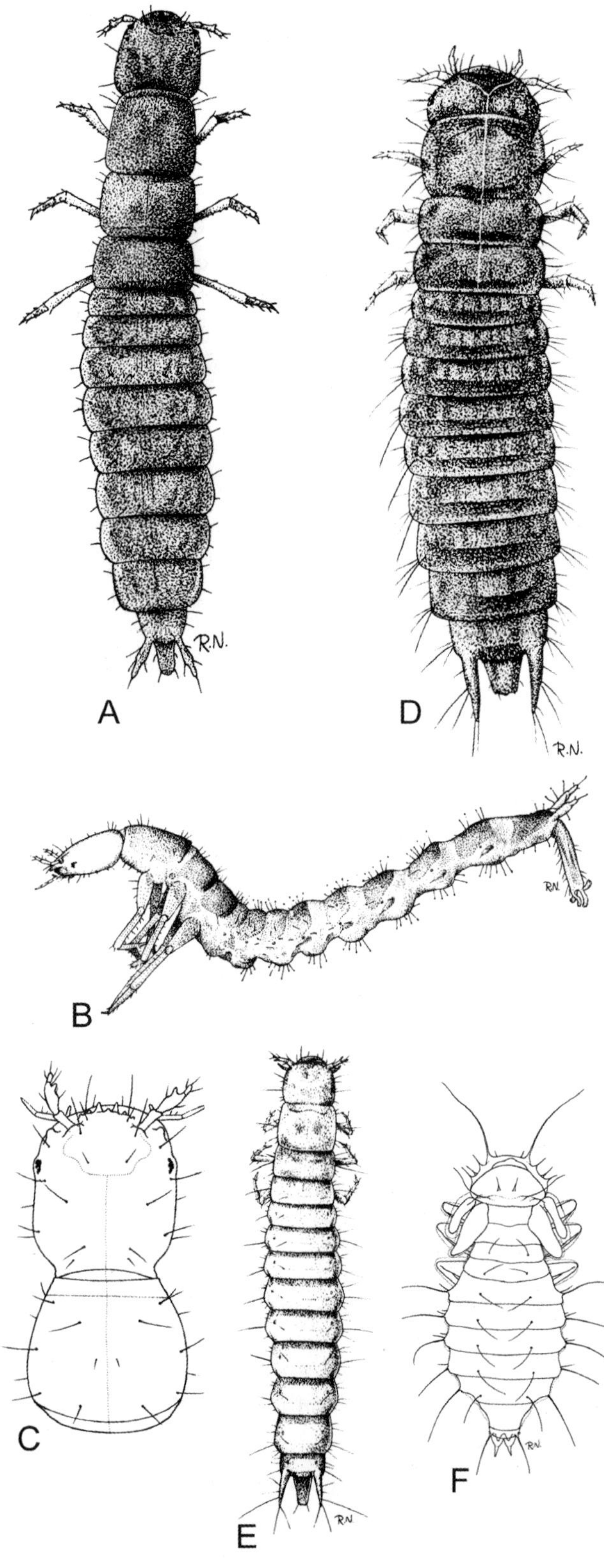

FIGURE 9 *A,* Quedius mesomelinus, *larva. B–C,* Quedius fellmanni, *larva, lateral view* (*B*), *larval head and prothorax, dorsal view* (*C*). *D,* Micralymma marinum, *larva, dorsal view. E–F,* Micralymma brevilingue, *larva* (*E*) *and pupa* (*F*), *dorsal view. From Böcher* (*1988*).

LARVAE (TO SUBFAMILY)

1. Mandible with preapical tooth. Antenna with 3 segments 2
– Mandible without preapical tooth. Antenna with 3 or 4 segments.................................... 3
2. One stemma. Antenna short, the second segment bearing a sensory appendage. Urogomphus with 1 or 2 segments. Abdominal segment 8 with a prominent dorsal (defensive) gland........................ .... Aleocharin (*Amischa, Atheta, Gnypeta, Ocalea*)
– Six stemmata. Urogomphus with 2 segments ....Tachyporinae (*Mycetoporus*)
3. 6 stemmata, antenna with 4 segments 4
– Four or five stemmata, antenna with 3 or 4 segments 5
4. Urogomphus with 2 segments. Antenna slender. Anterior margin of labrum with 2 pairs of large teeth and a median pair of small teeth.............. .... Paederinae (*Lathrobium*)
– Urogomphus with 1 segment. Antennae stout .... Euaestethinae (*Euaestethus*)
5. Five stemmata. Antenna with 3 segments. Urogomphus with 1 segment. Labrum with unsclerotized area at base........... Omaliinae (*Anthobium, Eusphalerum, . Micralymma, Omalium, Xylodromus*)
– Four stemmata. Antenna with 4 segments. Urogomphus with 2 segments. Anterior margin of labrum with 9 blunt teeth. Anterior tibia with a brush of bifid spines. All terga with 2 pairs of very long setae. ....Staphylininae (*Othius, Philonthus, Quedius*)

Omalium excavatum **Stephens, 1834 Fig. 21A**

C. 3–4 mm. Distinguished by the broad, black elytra which are at least as long as the visible, dorsal part of abdomen. Pronotum with two elongate depressions in the middle.

Larva undescribed and has not been found in Greenland.

BIOLOGY. In Greenland distinctly synanthropic and only found close to cultivated land in the southernmost part of the country. In Scandinavia partly synanthropic and present in highly different, mostly humid biotopes away from human habitations. In Iceland and Scandinavia the adults overwinter.

DISTRIBUTION. S. Rare and always singly in or close to cultivated land in the Narsaq- and Qaqortoq/Julianehåb districts. *O. excavatum* is thus one of the few likely candidates for having been introduced to Greenland by the Norsemen.

Palaearctic, subarctic-subtropical.

Xylodromus concinnus **(Marsham, 1802) Fig. 21B**

C. 3.5–4 mm. Elytra light brown and relatively long, but shorter than the uncovered portion of abdomen.

BIOLOGY. In Greenland markedly synanthropic, found in the original turf houses and never outdoors. Like *Quedius mesomelinus* collected almost exclusively during 1876–1899, with only one later record (1945). The explanation must be the same in both cases, namely that the transition from turf houses to wooden buildings destroyed the habitat. Now presumably extinct in Greenland. The larva is described by Larsson & Gigja (1959) and Steel (1970).

DISTRIBUTION. (S, SW). Earlier found in a number of towns in West Greenland.

Probably almost cosmopolitan.

[*Euaestethus laeviusculus* **Mannerheim, 1844 Fig. 21D**

C. 1.8 mm. Fragments have recently been discovered in archaeological samples from excavation of the palaeo-eskimo settlement Asummiut at Sisimiut/Holsteinsborg (*c.* 3000 years old) and also from a Norse farm in eastern Nuup Kangerlua/Godthåbsfjord area (*c.* 1000 years old). The occurrence so relatively late in history makes it probable that the species is still living in Greenland, but because of its hidden habitat – below rotting plant debris – the tiny beetle may easily escape discovery (Böcher 1998, 1999).

DISTRIBUTION. Palaearctic, subarctic-temperate (Europe, not including Iceland and the Faeroes).]

Lathrobium fulvipenne **Gravenhorst, 1806 Fig. 20D**

C. 8 mm. Head posteriorly strongly constricted, forming a "neck". Prothorax longer than broad, subrectangular, constricted behind to form a "waist". Extensions on inner side of foreleg femora and tibia. Hind wings rudimentary, but a macropterous form is known from central Europe.

The larva is undescribed and has not been found in Greenland.

BIOLOGY. Nothing is known about the single Greenland finding. In Iceland and the Faeroes most often encountered under stones in grassland. In the rest of Europe found in more humid places, reaching the alpine zone in Scandinavia. In Iceland the species hibernates as adult and passes through the larval stages during summer.

DISTRIBUTION. S. In Greenland only found at Ivittuut (1883) and might thus be an accidental introduction. However, species of *Lathrobium* are never synanthropic and not easily transported by man, and regarding the general distribution it might well be indigenous in Greenland (Lundbeck 1896; Lindroth 1931, 1957).

Palaearctic, subarctic-southern temperate.

***Quedius mesomelinus* (Marsham, 1802) Figs 9A, 20A**
Largest Greenland rove beetle, *c.* 6.5–11 mm. Distinguished from *Q. fellmanni* by the size and the relatively much smaller eyes (width about one-third of head length). There is considerable variation in size, and sex-related variations in general shape. Hind wings slightly reduced and possibly nonfunctional.

BIOLOGY. More or less synanthropic, especially in the northern part of its range. In Greenland formerly principally found in the turf houses, mainly in the walls, but also under stones close to the houses. Apparently free-living individuals have been taken only a few times in southernmost Greenland. Most specimens were collected before 1902; since then the species has rarely been found, suggesting that its abundance has decreased drastically. The decline took place during the period when the original turf houses were replaced by wooden houses. Because the walls of the turf houses seemingly represented the most important habitat of the species, this is probably the main cause of its decrease/disappearance (*cf.* Chapter 4). During a warm spell in the summer 2003 adults were suddenly quite common in the city of Paamiut/Frederikshåb. They probably left their concealed living places in the town in order to mate and swarm (B. Knudsen 2003, pers. comm.).

Larvae have only been collected in late summer, adults all the year round, even in mid winter. Possibly the larval and pupal development is completed during one summer, so that only adults hibernate.

DISTRIBUTION. (S, SW). The finds are concentrated in a southern area (Narsaq district) and a northern area (southern Qeqertarsuup Tunua/Disko Bugt) with Nuuk/Godthåb and Sisimiut/Holsteinsborg in between.

Almost cosmopolitan, in most places synanthropic and probably introduced.

***Quedius fellmanni* (Zetterstedt, 1838) Figs 9B, C, 20B**
Henriksen 1939: *Quedius boops* Grav.

Similar to *Q. mesomelinus*, but smaller, *c.* 5–6 mm, and with large eyes (width more than two-thirds of head length). Hind wings reduced and not fit for flight.

BIOLOGY. In Greenland an inland species preferring fairly warm places, but otherwise its habitat spectrum appears to be extremely wide and to include a great variety of plant communities in both dry and humid situations. Collection data indicate adult overwintering.

DISTRIBUTION. S, SW. In Greenland fairly rare with a typical southwestern distribution, north to Nuup Kangerlua/Godthåbsfjord.

Circumpolar, low arctic-boreal.

***Mycetoporus nigrans* Mäklin, 1853 Fig. 20C**
Among Greenland beetles easily recognized by the spindle-shaped body. *C.* 7 mm.

BIOLOGY. Found among plant debris in *Salix glauca* shrubs, under stones and washed-up seaweeds and at margins of melting snow patches.

DISTRIBUTION. SWs. Recently discovered in Greenland, but seems to be fairly widespread in the Nuup Kangerlua/Godthåbsfjord area.

Circumpolar, low arctic-subarctic and alpine.

***Micralymma marinum* (Ström, 1783) Figs 9D, 10, 21C**
Micralymma (with two Greenland species) are characterized, i.a., by very short elytra, which are posteriorly distinctly widened. Hind wings entirely reduced. Broad and rather flat species, widest immediately behind mid- abdomen (Poppius 1908). *M. marinum* is most easily distinguished from *M. brevilingue* (below) by the dull, black colour, by the antennae, of which no segment is markedly shorter than long, and by larger compound eyes with larger facets, separated by less than their diameter. Length *c.* 3 mm.

The larva is described by Larsson & Gigja (1959) and Steel (1970).

BIOLOGY. Exceptional by being a true coastal insect, probably spending its entire life cycle in the tidal zone and being regularly submerged during high tide. Structural and biological features have been studied in some detail byThayer (1985), Elliot *et al.* (1983) and King *et al.* (1980). Adults and larvae are sometimes found together on vertical or lower sides of stones and rocks. Probably predacious (mainly feeding on mites and springtails?), but direct observations are lacking. It apparently has a two-year cycle, with overwintering both as larvae and adults.

DISTRIBUTION. S, SW, SE (north to Qeqertarsuup Tunua/Disko Bugt and Timmiarmiut). Probably present almost everywhere along Greenland rocky coasts, at least as northerly as indicated above.

Amphiatlantic, arctic-temperate. Probably distributed along all the rocky coasts of the northern Atlantic Ocean which are influenced by the Gulf Stream; even found on Svalbard and Bear Island (Lindroth 1931, 1957).

***Micralymma brevilingue* Schiödte, 1845 Figs 9E, F, 21E**
C. 3 mm, dark brown to black, somewhat shining. Antennae, legs and palpi dark brown (black in *M. marinum*). Antennal segments 8–10 clearly wider than long. Eyes smaller than in *M. marinum*, their facets often separated by spaces as large as, or larger than their diameter (Schiödte 1845).

FIGURE 10 Micralymma marinum *among barnacles on a rock at low tide.*

BIOLOGY. Mainly coastal, frequently found under washed-up seaweed or stones close to the high water mark, but also in fairly humid places at some distance from the sea, *e.g.* in herb fields and snow beds. The habitat thus appears to be very wide, from the beach to damp situations inland. In a couple of cases *M. brevilingue* and *M. marinum* have been collected together on seashores. According to Chernov & Makarova (2008) and Makarova *et al.* (2007), *M. brevilingue* is the only beetle species which is common in a wide range of habitats in the polar deserts of Severnaya Zemlya archipelago (Siberia), and is assumed to be the most cold-tolerant Coleoptera species of the Northern Hemisphere. In the laboratory *M. brevilingue* consumed collembolans (*Folsomia* spp.; Makarova *et al.* 2007). Collection data indicate imaginal and probably also larval hibernation.

DISTRIBUTION. S, SW, NW, NE, SE. One of the most widespread beetles in Greenland. The scanty finds from East Greenland is probably mainly a reflection of the limited collecting intensity there. Like a few other beetle species, it seems to be absent in the southernmost part of the High Arctic.

Circumpolar, arctic.

Gnypeta cavicollis Sahlberg, 1880 Figs 11, 21F

C. 3–3.5 mm. Fairly similar to an *Atheta* (see below), but head posteriorly distinctly constricted, forming a "neck", and antennae relatively shorter than in any Greenland *Atheta* species (antennal length approximately equal to length of head + pronotum). First segment of posterior tarsus longer than second (shorter or equal to the second in *Atheta*). Black, dull on head and pronotum, shining on abdomen. Head with a conspicuous depression in the middle, pronotum with a shallow depression along the midline.

The larva is undescribed.

BIOLOGY. In most instances found in High Arctic salt marshes and sand flats close to the sea (Cotton 1979), with sparse vegetation of, *e.g.*, *Carex subspathacea*, *Puccinellia phryganodes* and *Stellaria humifusa*. However, it has also been reported inland in Peary Land (in a meadow and at the lakes Midsommersøerne). Life cycle unknown, but larvae have been reported to be present during summer (Cotton 1979).

DISTRIBUTION. N, NE. Among Greenland beetles, as well as beetles in general, *G. cavicollis* is exceptional in being exclusively High Arctic, apparently with a southern limit at the Low Arctic-High Arctic transition.

Outside Greenland only known from Novaja Zemlja and two localities in the arctic part of the Yenissei Valley, Siberia.

NOTE. Klimaszewski *et al.* (2008) published a revision of *Gnypeta* from Canada, Alaska and Greenland. This revision does not mention *G. cavicollis* and does not include material from the Copenhagen Zoological Museum, nor does it cite Danish literature. Presumably the taxon *G. groenlandica* Lohse, 1989 of which the material studied originated from Canada (Canadian Museum of Nature) and collected at Nedre Midsommersø, Peary Land, is identical with *G. cavicollis*, as the description agrees with this species in key characters. An inquiry into this potential synonymy is now in progress.

Atheta Thomson, 1858

A large genus of small, dark, elongate staphylinids with head and pronotum of about equal width and head not or only

FIGURE 11 Gnypeta cavicollis *on a mud flat at low tide, Northeast Greenland.*

slightly constricted posteriorly. Hind wings well developed and functional. For identification of the Greenland species, see the key. Distinction from *Gnypeta cavicollis*, see above.

No larvae of the genus has hitherto been collected in Greenland, and larvae of the Greenland species are undescribed. Larvae of some *Atheta s.l.* are described and illustrated by Paulian (1941) and Topp (1975). In most cases collecting data indicate adult hibernation in Greenland.

***Atheta* (*Mocyta*) *groenlandica* Mahler, 1988 Fig. 22C**

Henriksen 1939: *Acrotona fungi* Grav.

C. 2 mm. Only females have been collected, and probably the species reproduces parthenogenetically.

BIOLOGY. Apperently fairly hygrophilous, occurring under stones and leaf litter in different vigorous plant communities – willow shrubs, grasslands, marshes and lake shores.

DISTRIBUTION. S (several places), SE (Saqqisikuik/Skjoldungen). Only known from Greenland.

***Atheta* (*Boreophila*) *islandica* (Kraatz, 1856) Fig. 22A**

C. 2.5 mm. Only females known from Greenland, whereas in Iceland and Scandinavia the frequency of the two sexes is similar.

BIOLOGY. Collected in different, generally more or less humid places, *e.g.*, grasslands, marshes and willow shrubs. Outside Greenland a highly eurytopic species.

DISTRIBUTION. S, SW (north to Aasiat/Egedesminde), SE (Timmiarmiut). Probably widely distributed in southern Greenland, but the finds are very scattered.

Palaearctic, low arctic-subarctic, in Scandinavia mainly alpine.

***Atheta* (*Boreophila*) *hyperborea* Brundin, 1940 Fig. 22B**

C. 4 mm. Only females have been collected in Greenland.

BIOLOGY. A specimen from Qassiarsuk, S Greenland, was collected in a typical snow-patch vegetation dominated by *Salix herbacea*.

DISTRIBUTION. S, SW. Presumably rare in Greenland. Six specimens have been collected from three widely separated areas (subarctic S Greenland, Eqaluit Paarliit/Præstefjord, head of Nassuttooq/Nordre Strømfjord).

Palaearctic, low arctic-boreal (only known from northern Fennoscandia).

***Atheta* (*Thinobaena*) *vestita* (Gravenhorst, 1806) Fig. 22D**

C. 3 mm.

BIOLOGY. Found on sandy and gravelly seashores, chiefly under washed-up seaweeds. In Iceland also reported from grassland and wasteland close to the beach.

DISTRIBUTION. S, sSW. Undoubtedly rare in Greenland, northernmost in Ameralik (*c.* 64°N).

Palaearctic, low arctic-temperate. Along the European North Atlantic beaches.

BYRRHIDAE (PILL BEETLES)

A small family of rounded, dark beetles. Hind coxae widely separated. Legs and antennae can be more or less withdrawn into grooves on the underside of the body, the insects feigning dead in this manner. Generally moss-feeders. The c-shaped larvae have a large, well-developed hypognathous head with 6 stemmata. The 5-segmented legs are well-developed, and in many cases the tip of abdomen is swollen.

Key to Greenland species

1. Dorsal surface dull and very hairy, more or less patterned in black and brown with lighter hairs. If present, longitudinal furrows of elytra are weak 2
– Dorsal surface less hairy, black and somewhat shining, with marked longitudinal elytral furrows....... 3
2. Dorsal surface patterned in brown and black with light transverse bands. All tibiae strongly flattened, and with grooves to receive the feet during withdrawal. C. 6–8 mm .... *Byrrhus fasciatus*
– Dorsal surface dark brown with faint, dark markings and a few light hairs. Only front tibiae somewhat flattened on outer side. C. 4 mm...................... ....*Arctobyrrhus* (*Tylicus*) *subcanus*
3. C. 2.5–3 mm. Pronotum c. twice as wide as long..... ....*Simplocaria metallica*
– C. 3.5 mm. Pronotum c. 1.75x as wide as long .... *Simplocaria elongata*

***Simplocaria metallica* (Sturm, 1807) Figs 12C, 23C**

Dorsal surface black, shining, with light hairs. Tibiae flattened on outer side (to receive feet during withdrawal). *C.* 2.5 – 3 mm.

The larva is not described, but here illustrated on the basis of material which most likely is referable to this species (Fig. 12C).

BIOLOGY. Most often found under stones and in moss, but otherwise it is difficult to characterize the habitat, which is as diverse as that of *Byrrhus fasciatus*. *S. metallica* appears to be somewhat dependent on the presence of water, as it has been found most abundantly on fjord beaches, river banks, along brooks (also at a melt-water brook on a barren moraine, interior Qeqertarsuaq/Disko). The habitat also comprises more xeric inland plant communities, *e.g.*, mixed heaths with an undercover of mosses and coppices of *Salix glauca* on south facing slopes. Presumably the presence of the appropriate species of mosses, on which *S. metallica* like other byrrhids is assumed to feed, is the most important requirement of the species. Collection data suggest that both larvae and adults hibernate, and possibly the reproductive cycle is not fixed.

DISTRIBUTION. S, SW (north to Qeqertarsuaq/Disko), SE (Saqqisikuik/Skjoldungen). Found rather scattered, but is fairly common in the subarctic northeastern part of the Qaqortoq/Julianehåb Bugt area. Recent finds of fossil remains show that the species was present in middle West Greenland already 9920 cal. years B.P. (Böcher *et al.* 2012).

Holarctic (Europe and North America), low arctic-boreal, in Europe boreo-alpine.

Simplocaria elongata **Sahlberg, 1903 Fig. 23D**
Distinction from *S. metallica*, see the key. The elytral striae are also more strongly impressed.

BIOLOGY. Collected from mosses in humid places. In Canada adults have been found throughout summer, possibly indicating an indetermined reproductive cycle.

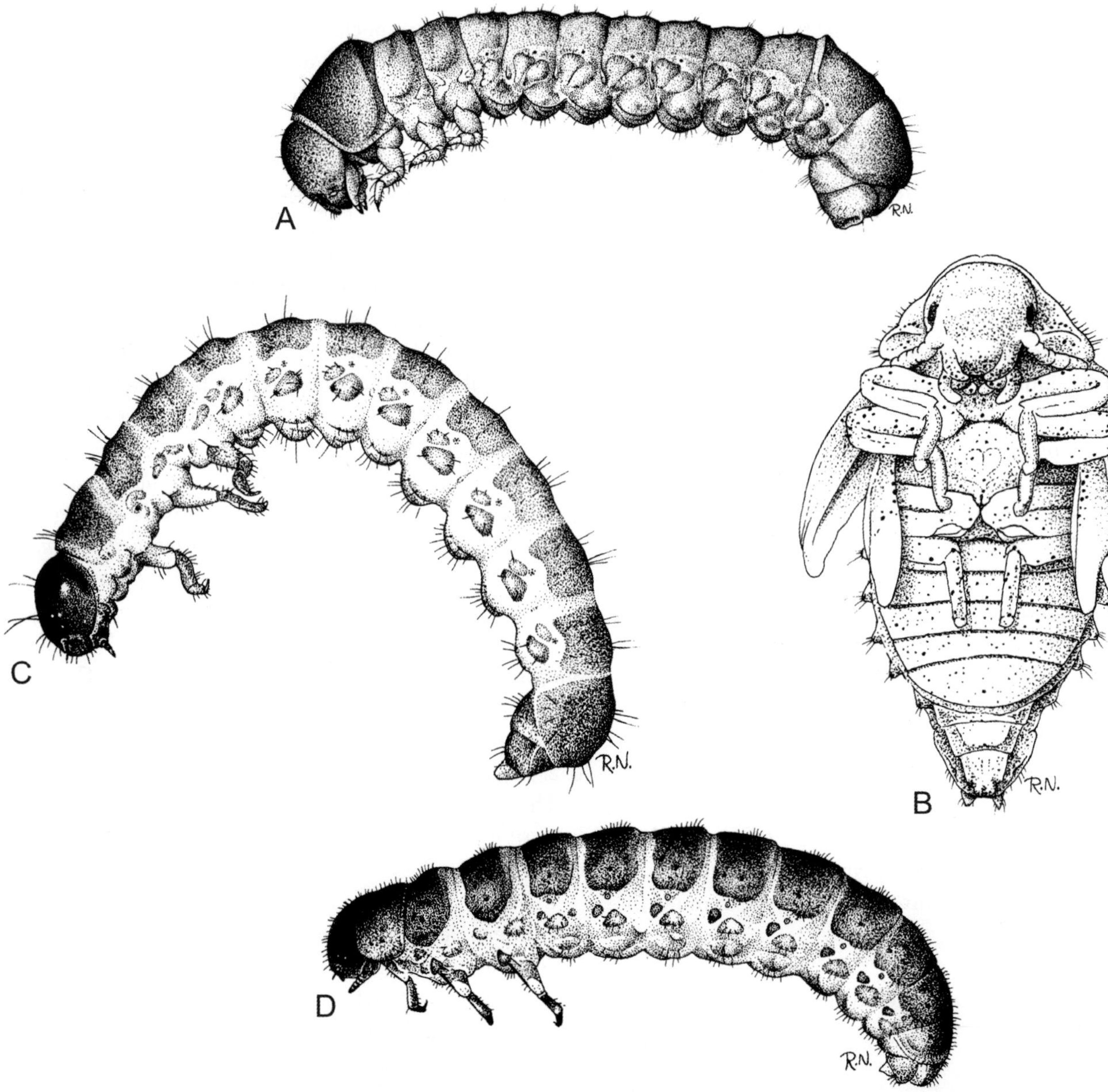

FIGURE 12 *A–B:* Byrrhus fasciatus; *larva, lateral view (A) and pupa, ventral view (B). C,* Simplocaria *cfr.* metallica *larva, lateral view. D,* Arctobyrrhus *cfr.* subcanus *larva, lateral view. From Böcher (1988).*

DISTRIBUTION. sSW. Only one specimen has hitherto been collected, from Kilaarsarfik/Sandnæs at Amarella fjord east of Nuuk/Godthåb (P.S. Buckland, pers. comm.).

Circumpolar, arctic-subarctic.

***Arctobyrrhus* (*Tylicus*) *subcanus* (LeConte, 1878) Figs 12D, 23B**

Looks like a small *Byrrhus*. Dorsal surface dark and dull with faint, black longitudinal markings and light hairs. Elytra smooth. At least in the Greenland specimens the hind wings are so short that flight is impossible. Only front tibiae somewhat flattened on outer side. *C.* 4 mm.

The larva has not been described, but is illustrated here on the basis of material which most likely is referable to this species (Fig. 12D).

BIOLOGY. Found in a dry heath rich in lichens and mosses (Narsarsuaq) and on a lake shore (Tasersuaq at Tasermiut). In North America it occurs in dry places in mountains. The larva is active on the surface of mosses and does not burrow except to pupate. In North America univoltine with larval hibernation.

DISTRIBUTION. S. Hitherto only found in two localities in S Greenland (Narsarsuaq, Tasermiut).

Nearctic, low arctic-subarctic.

***Byrrhus fasciatus* Forster, 1771 Figs 12A, B, 23A**

Dorsal surface dull and very hairy, dark brown with a variable pattern of longitudinal, black stripes and spots and light transverse bands. Tibiae very flattened and with grooves to receive feet during withdrawal. *C.* 6–8 mm. A highly variable species, both regarding colour and general shape and size. In Greenland the variation is less pronounced, the colour is dark brown with more or less distinct black, brown and grey markings produced by differently coloured hairs. Hind wings fully developed.

BIOLOGY. Commonly found under stones, under vegetation and in moss, but otherwise in highly different, both humid and dry situations: along water courses, in snow beds, coppices, herb slopes, grasslands, humid and dry heaths.

Adults and larvae feed on mosses and are often found together, adults regularly fly in sunny weather. Both adults and larvae (in all stages) hibernate, with pupation taking place during summer. The reproductive cycle thus seems to be fairly indefinite, with possibly more than one hibernation.

DISTRIBUTION. S, SW, NW, NE, SE. Presumably present almost universally in the southern half of Greenland with a northern limit at the Low Arctic/High Arctic transition. Very common especially along the southern part of the west coast.

Circumpolar-circumtemperate, low arctic-southern temperate.

CRYPTOPHAGIDAE

Small, elongate species, generally less than 3 mm. Antennae more or less clubbed, longer than the head is wide. In most cases scavengers or fungus feeders. Some of the larvae are illustrated by Hinton (1945).

***Caenoscelis ferruginea* (Sahlberg, 1820) Fig. 25A**

Elongate, oval, yellow-red, hairy and fairly shining. Antennae very strong and clubbed. A fine groove along the side margins of pronotum delimits a narrow rim. *C.* 2 mm.

BIOLOGY. The Greenland specimens were caught by sweeping a dry, grassy south-facing slope. In Scandinavia the species is mostly connected with trees, found, *e.g.*, under bark and decaying stumps, but also in hay and by sweeping meadows. It presumably feeds on moulds.

DISTRIBUTION. sSW. Only found once (by C. Vibe), at Kapisillit in Nuup Kangerlua/Godthåbsfjord.

Circumpolar-circumtemperate, low arctic-southern temperate.

COCCINELLIDAE (LADYBIRDS)

'Typical ladybirds' are smooth, brightly coloured, rounded beetles with head partly concealed by pronotum when viewed from above. Antennae shorter than head width. Like the adults the larvae are active predators on small insects, especially aphids, psyllids and scale insects. The bright yellow/black or red/black colours of both adults and some larvae are interpreted as a warning, announcing inedibility.

***Nephus redtenbacheri* (Mulsant, 1846) Figs 14C, 24B**

Henriksen 1939: *Scymnus* (*Nephus*) *limonii* Donis.

A small, hairy ladybird. Length of antennae less than half the width of head. Black and orange-brownish pattern on elytra characteristic. *C.* 2 mm.

BIOLOGY. Found in a number of plant communities, including birch coppice, hygrophilous vegetation along water courses, grasslands, and dry heaths. The species is often found under stones and is regularly associated with different species of scale insects (Coccoidea). In Greenland it has been found together with *Arctorthecia cataphracta*, on which it probably feeds. The scanty evidence suggests a reproductive cycle similar to that of *Coccinella transversoguttata*: adult hibernation and larval development during summer.

DISTRIBUTION. S, sSW. In Greenland a rare species, however sometimes occurring fairly abundantly. North to Kuup Kangerlua/Godthåbsfjord.

Palaearctic, subarctic-subtropical (Europe, northern Africa).

Coccinella transversoguttata **Falderman, 1835 Figs 13, 14A, B, 24A**

Unmistakable among Greenland beetles, but somewhat variable in size and shape of the black spots on elytra. In addition to the median and apical spots a sublateral spot between the scutellar transverse band and the median spot is sometimes found. The size of this spot is highly variable, from minute to half the size of the median spot, sometimes confluent with the scutellar transverse band. Rarely the median and apical spots are also confluent. Also the reddish colour of the elytra varies from yellow to dark brownish red. *C.* 4–6 mm.

BIOLOGY. Fairly eurytopic, found in almost any kind of vegetation. It is most abundant in dry heaths with *Betula nana* and *Salix glauca*, where adults and larvae together hunt for psyllids and aphids. In Greenland it overwinters as adult and reproduces in spring. The new generation reaches adult stage well before winter, thus spending the warmest part of the summer as larvae.

DISTRIBUTION. S, SW, NW, NE, SE. The ladybird is the terrestrial beetle with the widest Greenland distribution, reaching the southern High Arctic in both NW and NE. It is most ubiquitous inland. In recent years the species has expanded its distribution northwards in NE, now reaching Zackenbergdalen at 74°30'N (Böcher 2009).

Holarctic (absent from Europe), low arctic-southern temperate.

FIGURE 13 Coccinella transversoguttata *on leaf of* Betula nana.

LATRIDIIDAE (MINUTE BROWN SCAVENGER BEETLES)

Small and minute species with clubbed antennae and sculptured elytra. Most are associated with decaying vegetable matter and feed on moulds.

Latridius minutus **(Linnaeus, 1767) Fig. 25B**

Henriksen 1939: *Enicmus minutus* L.

The small size (*c.* 1.5 – 2.0 mm) and the shape of pronotum render this species characteristic among Greenland beetles. It is, however, fairly similar to *L.* (*Enicmus*) *transversus* (below), but generally smaller and the pronotum is clearly broadest anteriorly and with slightly concave sides.

Neither larvae (described by Hinton [1945]) nor pupae are known from Greenland.

BIOLOGY. Generally considered synanthropic, and in Greenland most specimens have been collected inside buildings. It has, however, also been caught far from human habitations; it is a good flier that readily takes to

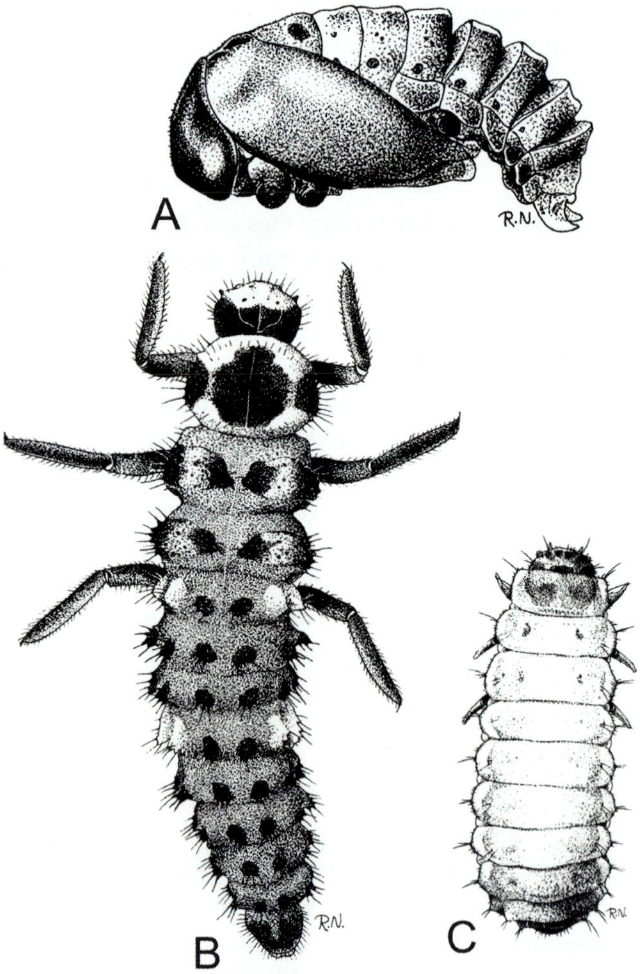

FIGURE 14 *A–B,* Coccinella transversoguttata*; pupa, lateral view (A) and larva, dorsal view (B). D,* Nephus redtenbacheri *larva, dorsal view. From Böcher (1988).*

the wings. The immense global distribution may be caused by flight in combination with wind dispersal. In Scandinavia frequently caught in the wild. Associated with rotting and mouldy vegetable matter and feeds exclusively on moulds. Inside buildings it probably, like other synanthropes, reproduces almost continuously.

DISTRIBUTION. S, SW, NE, SE. Highly scattered occurrences in the southern half of Greenland, recently also found at the Zackenberg Research Station (NE).

Almost cosmopolitan.

***Latridius* (*Enicmus*) *transversus* (Olivier, 1790) Fig. 25D**

C. 2 mm. Similar to *Latridius minutus*, but on average larger. Pronotum only slightly broadened anteriorly, with rounded, slightly convex sides.

BIOLOGY. Like the other members of the family found in places where moulds occur. The circumstances under which it was found in Greenland seem highly unusual: 3 specimens were caught in pitfall traps in a very dry steppe-like environment about 3 km east of the airfield in Kangerlussuaq/Søndre Strømfjord. Most probably the species has been introduced to Greenland, but here caught in the wild. In Denmark also generally found outdoors.

DISTRIBUTION. nSW (only found at the head of Kangerlussuaq/Søndre Strømfjord).

Palaearctic, temperate (widely distributed).

***Corticaria rubripes* Mannerheim, 1844 Fig. 25C**

Most readily distinguished by rounded pronotum and elongate shape. From the occasionally introduced *Corticaria serrata* separable by the small, blunt spines along side margins of pronotum (the spines are pointed and very prominent in *C. serrata*).

The larva is undescribed, but immature stages of a congener (*C. fulva*) are described and figured by Hinton (1945).

BIOLOGY. Caught in pitfalls on a south facing slope along a stream bordered by fairly luxuriant vegetation, including *Salix glauca* coppice. Also collected in moss in grassland on a silty moraine, and in pitfalls in a dry steppe locality. In Scandinavia reported from highly varied biotopes, also synanthropically in barns and manure, feeding on moulds.

DISTRIBUTION. nSW. Hitherto found in only three Greenland localities, all inland and close to Kangerlussuaq/Søndre Strømfjord: Arnangarnup Qoorua, Camp Lloyd/Strømfjordshavn (close to the head of the fjord), and close to the airport (in pitfalls together with *Latridius transversus*).

Palaearctic, low arctic-southern temperate.

[CHRYSOMELIDAE (LEAF BEETLES)

Middle sized, generally rounded, smooth and shining species, many with bright, metallic colours. Length of antennae less than two-thirds of body length. Tibiae without or with single (weak) apical spur. Almost all species are leaf-feeders, both adult and larva, which are commonly found together on the food plants. A very large family represented in Greenland by a single species which is so far only known from Holocene fossils.

Chrysomelidae, especially the genus *Chrysolina* (Danks 1991, Chernov & Makarova 2008), play an important part in other parts of the Arctic biome. Even in the high arctic (a polar desert), at Severnaya Zemlya archipelago, Siberia, two species have been found (Makarova *et al.* 2007).

***Phratora polaris* (Sparre Schneider, 1886) Fig. 24C**

Well preserved fragments have been found in *c.* 7000 years old sediments from NE Greenland (northern Jameson Land, Zackenberg; Böcher & Bennike 1996, Christiansen *et al.* 2002), but also in much more recent (*c.* 1000 years old) deposits from Norse habitations in Nuup Kangerlua/Godthåbsfjord (P.C. Buckland, pers. comm.). The last mentioned finding indicates that the species might still be extant in Greenland. *P. polaris* feeds on *Salix* spp., especially *S. herbacea* and *S. polaris*. It is a subarctic, mostly montane species distributed in Iceland, Fennoscandia, and northern Russia.]

CURCULIONIDAE (WEEVILS)

Head more or less prolonged into a snout (rostrum). Antennae clubbed and elbowed, 1st segment (scape) can be concealed in a groove along the side of the beak. All species are plant feeders. The larvae are pale and grub-like, living a concealed life boring in plants or in the soil. A huge family with a small representation in Greenland.

Key to Greenland species

ADULTS

1. Body covered with hairs and scales .... *Hypera diversipunctata*
– Body without scales.................................2
2. Snout short, not longer than its width in front of eyes...3
– Snout markedly longer than its width in front of eyes...4
3. Shining black and fairly glabrous.................... ....*Otiorhynchus arcticus*
– Dull black and unevenly hairy. Pronotum coarsely warty-puncturate. Elytra with marked, wrinkled striae.......................... *Otiorhynchus nodosus*
4. Length c. 2 mm. Abdomen (with elytra) almost spherical....................... *Rutidosoma globulus*

– Length c. 3.5–4 mm. Abdomen (with elytra) about twice as long as broad.........*Dorytomus imbecillus*

LARVAE (to subfamily)

1. Without stemmata.................................... ....Otiorhynchinae (*Othiorhynchus*)
– One pair of stemmata................................ ....Curculioninae (*Dorytomus, Rutidosoma*)
– Two pairs of large stemmata.... Hyperinae (*Hypera*)

Otiorhynchus arcticus **(O. Fabricius, 1780) Figs 15A, C, 16, 26B**

Black, glabrous and shining. Pronotum with fine, dense punctures. Elytral striae vary from being almost invisible to fairly marked grooves. Hind wings always rudimentary. In Greenland a considerable variation in size, *c.* 5.5 – 8.5 mm.

Larvae of the two Greenland species of *Otiorhynchus* are undescribed and probably impossible to separate on external characters (Böcher 1988). There is a tremendous variation in size and shape of the larval mandibles (Böcher 1988).

BIOLOGY. *O. arcticus* has usually been collected under stones, where it hides during day, and where numerous exoskeletons from dead individuals may accumulate. Highly eurytopic in Greenland, occurring in almost any plant community. Probably the most typical biotope is fairly dry, mixed heath with *Empetrum* and *Vaccinium uliginosum*. In most situations *O. arcticus* is found together with *O. nodosus* and by far outnumbers that species, but in more humid places *O. arcticus* may be less frequent or absent. In contrast, as one moves towards more dry, exposed localities with sparse vegetation, *O. nodosus* is found to disappear before *O. arcticus*. *O. arcticus* is mainly nocturnal. During night and in dark weather adults climb different herbaceous plants on which they feed.

The larva has four instars. It lives in the soil as a root-feeder and is presumably, like the adult, highly polyphagous. It has been collected under stones, in moss and under dead leaves in a coppice of *Betula pubescens*. Adults have been found throughout the season, and probably adults as well as larvae in different stages hibernate. Larvae may overwinter more than once, and possibly there is no fixed reproductive cycle.

DISTRIBUTION. S, SW (north to Itilleq), SE (north to the Blosseville Kyst; Degerbøl 1937). In southern Greenland *O. arcticus* is the most widespread and abundant of all beetles. It is peculiar by extending further north on the harsh east coast than on the climatically more favoured west coast.

Palaearctic, low arctic-boreal, in Europe boreo-alpine.

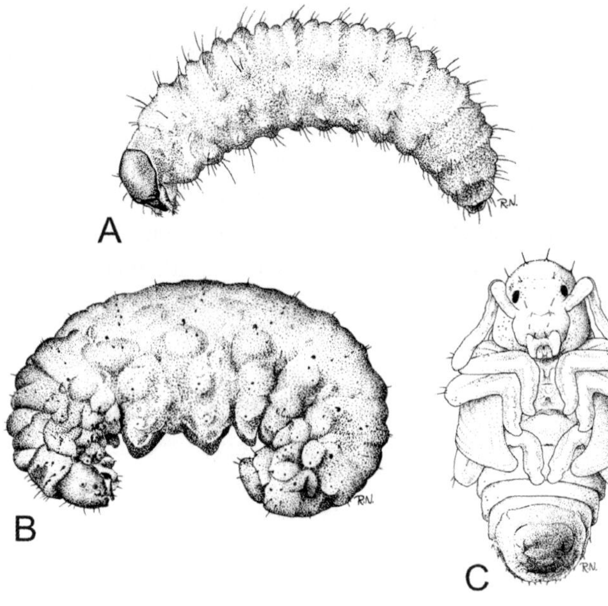

FIGURE 15 *A, C,* Otiorrhynchus arcticus; *larva, lateral view (A) and pupa, ventral view (C). B,* Hypera diversipunctata *larva, lateral view. From Böcher (1988).*

Otiorhynchus nodosus **(Müller, 1764) Fig. 26A**

Black, unevenly hairy and fairly dull. Pronotum coarsely warty-punctured. Elytra with marked, wrinkled striae. *C.* 7–9 mm, hence the size-variation in Greenland specimens is not as pronounced as in *O. arcticus*. *O. nodosus* is parthenogenetic in Greenland and throughout its North European range, where only females have been found.

A single larva with circular antennal socket (lentil-shaped in all other Greenland curculionid larvae studied) might be ascribed to this species (see Böcher 1988).

BIOLOGY. In Greenland *O. nodosus* is highly eurytopic. Usually found in different kinds of heath, where deposits of the resistant exoskeletons from both species of *Otiorrhynchus* are common under stones. Collected in almost all kinds of vegetation and up to about 600 m a.s. As a rule it is found together with *O. arcticus*, but clearly prefering more humid biotopes. *O. nodosus* is scarce or absent in fairly dry localities, where *O. arcticus* may be abundant.

In continental Europe and Iceland adult *O. nodosus* are nocturnal, climbing the vegetation to feed on a variety of plants during night or in dark weather, hiding under stones and in moss during day. In Greenland there are observations of activity in clear weather during daytime. The larva is presumably a polyphagous root-feeder living permanently in the soil. Phenological evidence indicates both larval and adult hibernation and probably the absence of a fixed reproductive cycle.

DISTRIBUTION. S, SW (north to Sisimiut/Holsteinsborg), SE (north to Kangerlussuatsiaq/Lindenow Fjord).

FIGURE 16 Otiorrhynchus arcticus.

Widely distributed and almost ubiquitous in southern Greenland; its range extends only a little along the east coast, but further north than that of *O. arcticus* on the west coast.

Palaearctic (Europe), low arctic-boreal, boreo-alpine.

***Hypera diversipunctata* (Schrank, 1798) Figs 15B, 26C**
Henriksen 1939: *Phytonomus elongatus* Payk.

Covered by hairs and bifurcate scales in colours grading from pale grey to reddish brown. In Greenland hind wings highly reduced. *C.* 5.5 – 6.5 mm.

The larva (Fig. 15B) is somewhat caterpillar-like with two pairs of large stemmata, proleg-like ventral protuberances and a terminal pygopod.

BIOLOGY. The species has been found in an *Empetrum*-heath, under stones on a fjord-beach (vegetation of *Festuca, Potentilla tridentata, Rumex acetosella*), in a luxuriant south-facing slope (dominated by *Anthoxanthum odoratum* and *Salix glauca*), and in a former settlement. Both adult and larva are probably polyphagous; outside Greenland the adult has been found associated with Fabaceae, the larva with Caryophyllaceae (Böcher 1988).

DISTRIBUTION. SW (north to Arfersiorfik fjord). Presumably rare in Greenland, findings are remarkably scattered, often singletons.

Circumpolar, low arctic-boreal.

***Dorytomus imbecillus* Faust, 1882 Figs 17, 26D**
Head and pronotum almost black, densely punctured. Elytra reddish brown with darker markings, somewhat shining, with marked punctured striae. Hind wings of normal size and probably usable. *C.* 3.5–4 mm (Korotyaev 1976).

BIOLOGY. Appears to be confined to *Salix* spp., mostly *S. arctophila*. Both larvae and imagines are found on leaves and catkins, where adults are freely exposed, and the larvae bore in the central part of the catkins (Fig. 17). Pupae are present in litter and the soil under low *Salix*-bushes from mid-July. Adults have been caught from June to September; newly hatched adults from the middle of July, pointing to adult hibernation. The species is rare in the inner part of fjords (Fjellberg & Böcher 2006).

DISTRIBUTION. S, SW (at least north to Kangerlussuatsiaq/Evighedsfjorden).

Holarctic (absent from Europe), low arctic-boreal.

***Rutidosoma globulus* (Herbst, 1795) Fig. 26E**
Henriksen 1939: *Rhytidosomus globulus* Hbst.

C. 2 mm. Unmistakable among Greenland beetles due to the small size and the near-spherical elytra. The hind wings are rudimentary.

BIOLOGY. Taken by sweeping in a small valley with snow-patch vegetation (*Salix herbacea, S. arctophila, S. glauca, Polygonum viviparum* and *Carex* spp.). In Europe the food plant is *Populus tremulus. Populus* is absent from Greenland, and accordingly the food plant here may be one or more species of *Salix*.

DISTRIBUTION. nSW. On the "Kissavik Expedition" (2003) the species was re-discovered by one specimen in the Søndre Isortoq, where one specimen was also found in 1885 (leg. S. Hansen). A third allegedly Greenland specimen is without locality data. In 2013 several specimens were found in Isua north of Nuup Kangerlua/Godthåbsfjord. Here the habitats were a south-facing slope with mixed vegetation as well as a NE facing slope with meagre snow bed vegetation (R. R. Hansen, pers. comm.). Examination of Holocene lake sediment material (5840–3300 cal. years before present) close to Sisimiut/Holsteinsborg has revealed a number of skeletal parts of the species (Böcher *et al.* 2012).

Palaearctic (Europe), low arctic-southern temperate.

Introduced species of Coleoptera (from Böcher [1988] with later additions)

A comparatively large number of beetle species, which are undoubtedly introduced by man, have been collected in Greenland. Some of them have been established indoors and are probably able to reproduce here. Others are unable to survive or reproduce in Greenland.

HYDROPHILIDAE (WATER SCAVENGER BEETLES)
***Cercyon obsoletus* (Gyllenhal, 1808) Fig. 19E**
C. 3.5–4 mm. Easily distinguished from *Helophorus brevipalpis* by dark, glabrous, evenly rounded body without

FIGURE 17 *Larvae of* Dorytomus imbecillus *in fruithead (opened) of* Salix arctica. (PHOTO: A. FJELLBERG)

constriction between pronotum and elytra. Elytra with very finely punctured striae. Particularly in droppings of various mammals (cow, deer), but also found in other kinds of decaying organic matter (compost, old mushrooms, carrion, etc.). Only one Greenlandic finding, from Kilaarsarfik/Sandnæs in Amarella Fjord (P.C. Buckland, pers. comm.). Considering the general distribution of the species, the occurrence in Greenland is provisionally regarded an accidental introduction. – Palaearctic, temperate-subtropical (in Scandinavia only southern part).

STAPHYLINIDAE (ROVE BEETLES)

A number of species have only been found as fossils from excavations of Norse farms, in most cases from Kilaarsarfik/Sandnæs in Amarella Fjord east of Nuuk/Godthåb. The species are more or less synanthropic and have undoubtedly been introduced by the Norsemen, only to become extinct with their disappearance.

Eusphalerum sorbi **(Gyllenhal, 1810)**
(Henriksen 1939: *Anthobium sorbi* Gyll.)
Undoubtedly an accidental introduction. Generally found in flowers. Also found as fossil (Kilaarsarfik/Sandnæs; P.C. Buckland, pers. comm.).

Ocalea **cf. *picata* (Stephens, 1832)**
(Kilaarsarfik/Sandnæs; P.C. Buckland, pers. comm.).

Othius angustus **Stephens, 1833**
(Kilaarsarfik/Sandnæs; P.C. Buckland, pers. comm.).

Philonthus **cf. *cephalotes* (Gravenhorst, 1802)**
(Kilaarsarfik/Sandnæs; P.C. Buckland, pers. comm.).

Philonthus politus **(Linnaeus, 1758)**
Only known from the Norse farm at Nipaatsoq, Amarella fjord east of Nuuk/Godthåb (Buckland *et al.* 1983).

BUPRESTIDAE (JEWEL BEETLES)

Mainly shiny, metallic beetles. Most larvae are wood-borers, tunelling just under the bark.

Melanophila acuminata **(De Geer, 1774)**

DERMESTIDAE (SKIN BEETLES)

Sombre species covered with downy hairs or scales. Most species with a large, single ocellus on top of the head. Antennae distinctly clubbed, and together with the legs they can be concealed on the underside of the body. Many species are of considerable economic importance since the larvae may damage furs, fabrics and stored food products. Most larvae are covered with long hairs. No species have so far been recorded from skin-stores in Greenland. Generally cosmopolitans.

Anthrenus museorum **(Linnaeus, 1758)**
Attagenus pellio **(Linnaeus, 1758)**
Dermestes lardarius **(Linnaeus, 1758)**
Reesa vespulae **(Milliron, 1939)**

LYCTIDAE (POWDER-POST BEETLES)

Pronotum covering much of the head. The wood-boring larvae produce a fine dust, and some are serious pests of timber.

Lyctus brunneus **(Stephens, 1830)**

ANOBIIDAE (WOOD BORERS)

Antennae originate well separated and have no distinct club. The family contains species in which the larvae tunnel into dead wood and may destroy furniture, timber etc.

Ernobius mollis **(Linnaeus, 1758)**
Anobium punctatum **(De Geer, 1774)**

PTINIDAE (SPIDER BEETLES)

Strongly rounded beetles closedly related to the Anobiidae, but their antennae originate close together. A number of species are scavengers and pests of stored vegetable products.

Ptinus fur **(Linnaeus, 1758)**
Ptinus tectus **Boieldieu, 1856**
Ptinus raptor **Sturm, 1837**

TROGOSSITIDAE

Tenebroides mauritanicus **(Linnaeus, 1758)**
A predator found in food stores where it hunts for pest insects.

FIGURE 18 *A,* Nebria rufescens; *B,* Patrobus septentrionis; *C,* Bembidion grapii; *D,* Trichocellus cognatus.

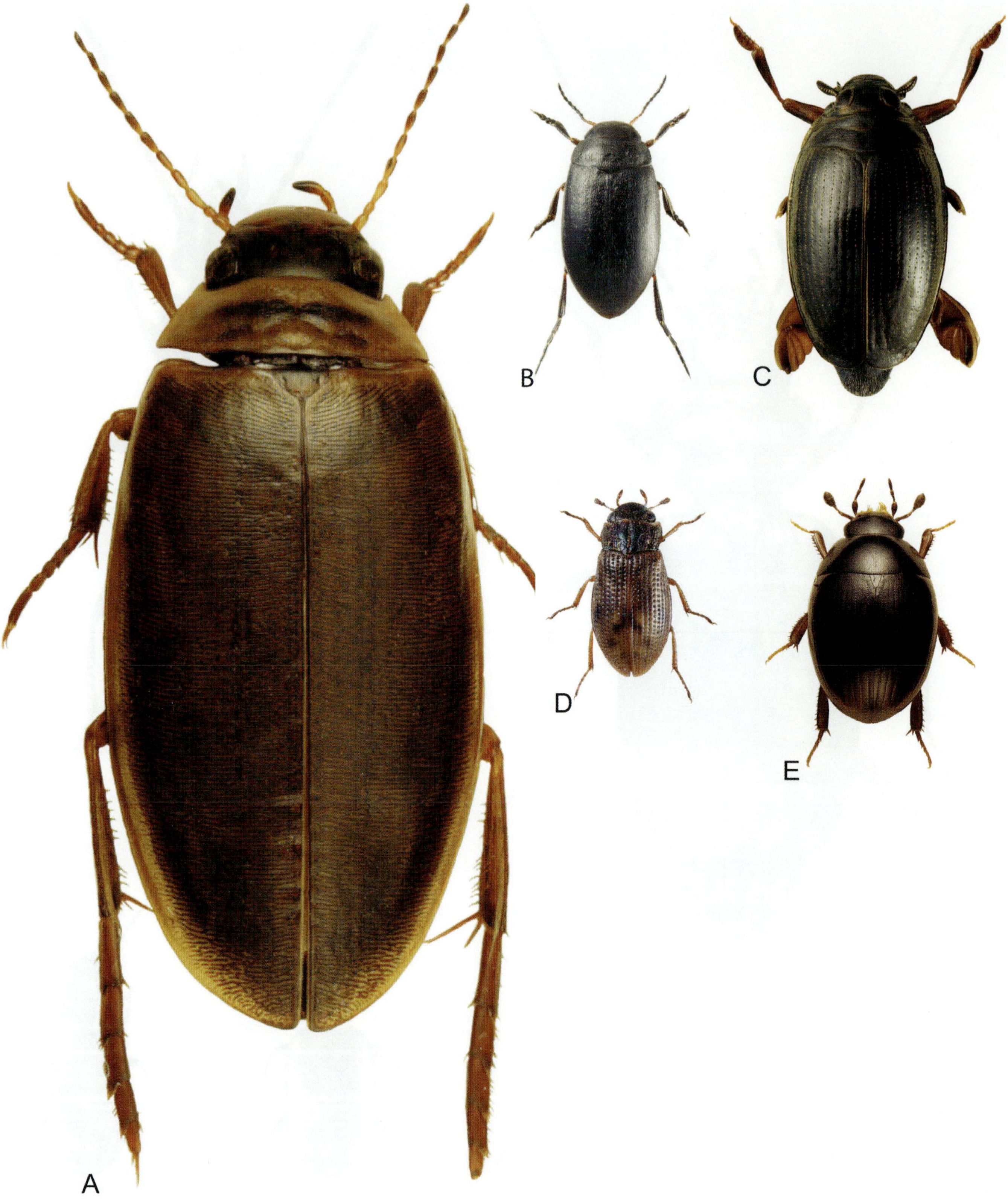

FIGURE 19 *A,* Colymbetes dolabratus*; B,* Hydroporus morio*; C,* Gyrinus opacus*; D,* Helophorus brevipalpis*; E,* Cercyon obsoletus.

FIGURE 20 *A*, Quedius mesomelinus; *B*, Quedius fellmanni; *C*, Mycetoporus nigrans; *D*, Lathrobium fulvipenne.

FIGURE 21 *A,* Omalium excavatum*; B,* Xylodromus concinnus*: C,* Micralymma marinum*; D,* Euaestethus laeviusculus*; E,* Micralymma brevilingue; *F,* Gnypeta cavicollis.

FIGURE 22 *A,* Atheta islandica*; B,* Atheta hyperborea*; C,* Atheta groenlandica*; D,* Atheta vestita.

FIGURE 23 *A,* Byrrhus fasciatus*; B,* Arctobyrrhus subcanus*; C,* Simplocaria metallica*; D,* Simplocaria elongata.

FIGURE 24 *A,* Coccinella transversoguttata*; B,* Nephus redtenbacheri*; C,* Phratora polaris.

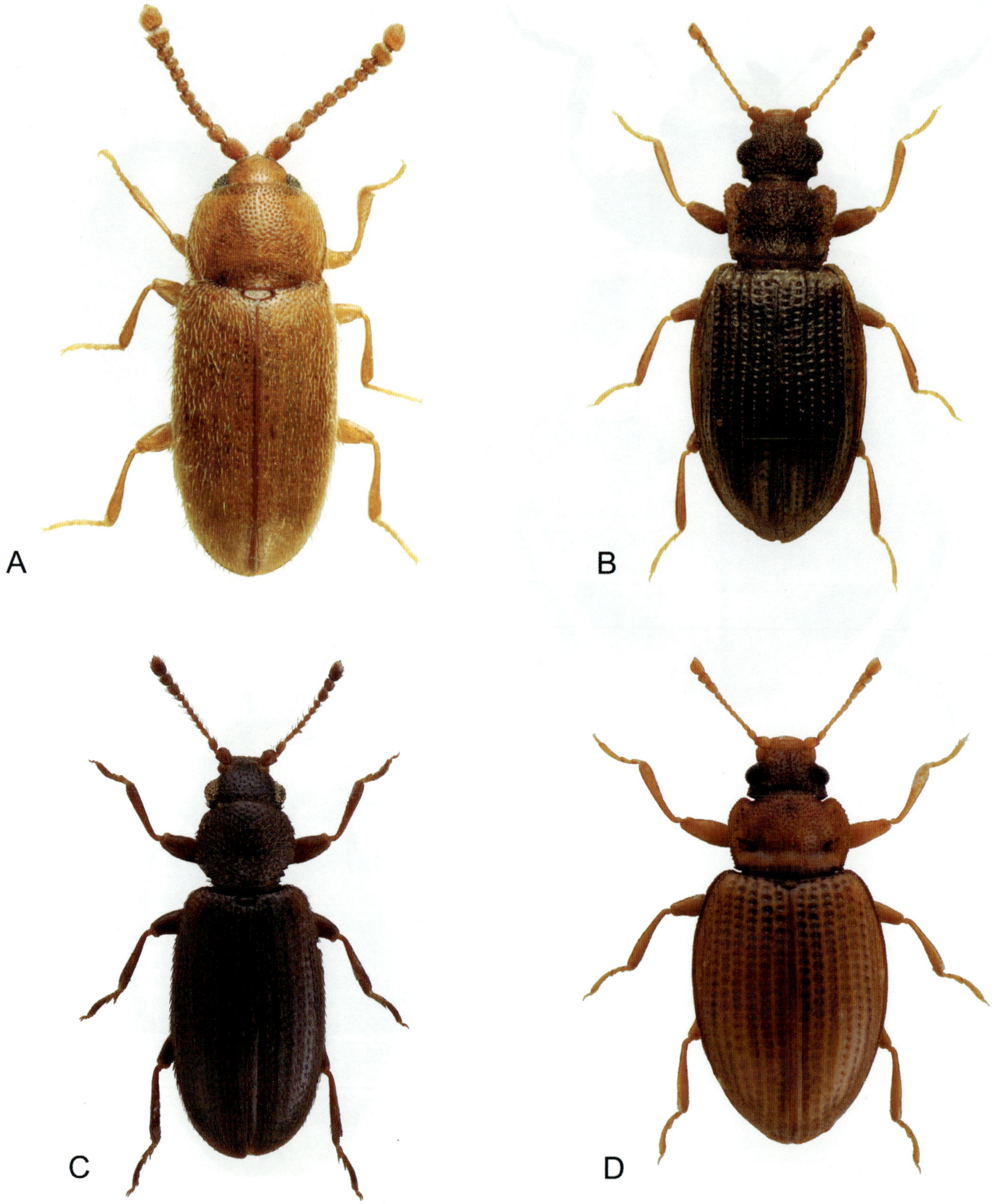

FIGURE 25 *A:* Caenoscelis ferruginea*; B,* Latridius minutus*; C,* Corticaria rubripes*; D,* Latridius transversus.

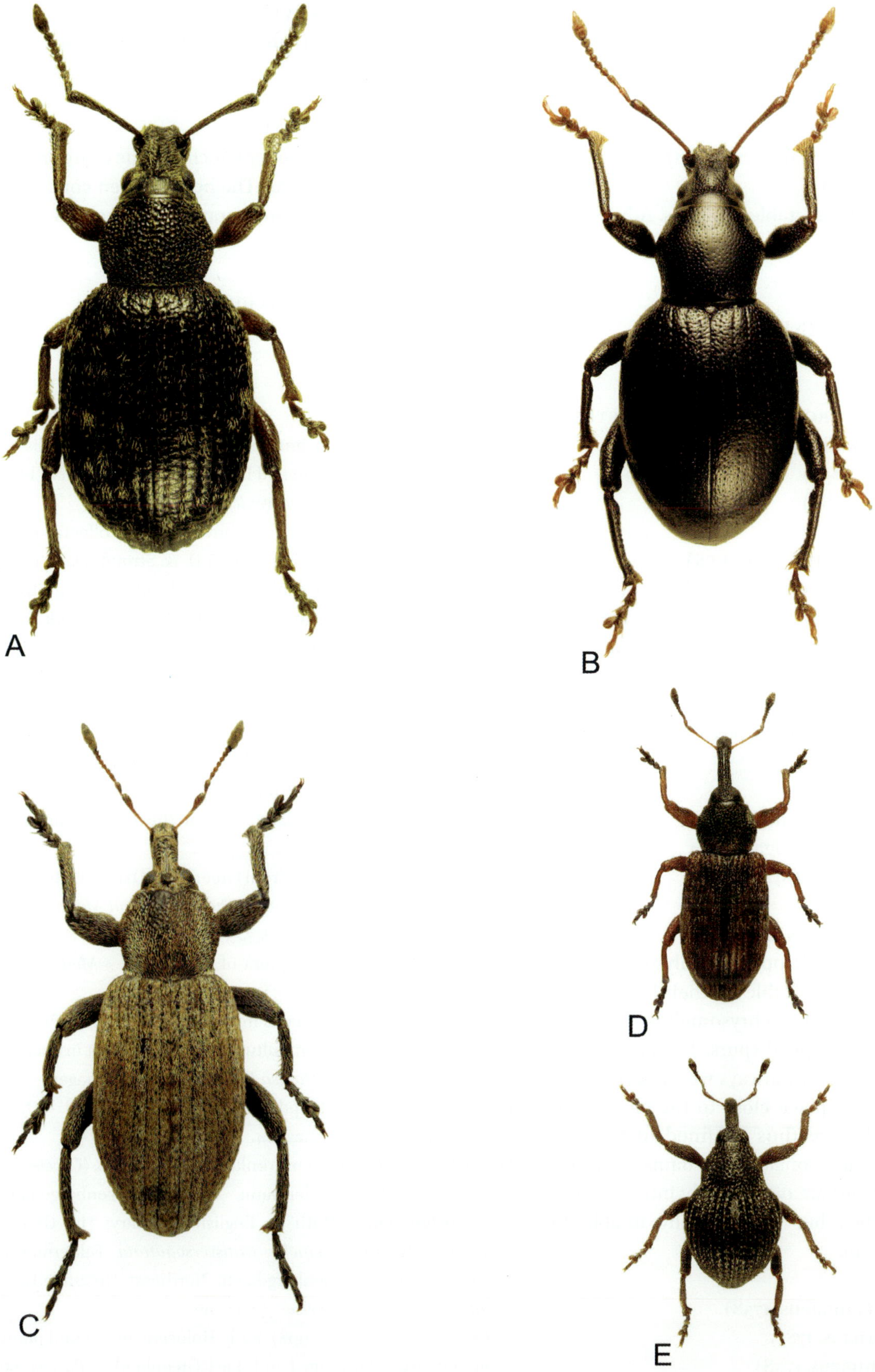

FIGURE 26 *A,* Otiorhynchus arcticus*; B,* Otiorhynchus nodosus*; C,* Hypera diversipunctata*; D,* Dorytomus imbecillus*; E,* Rutidosoma globulus.

MALACHIIDAE
Malachius aeneus **(Linnaeus, 1758)**
A predator generally found in flowers.

CUCUJIDAE (FLAT BARK BEETLES)
Oryzaephilus surinamensis **(Linnaeus, 1758) (Saw-toothed grain beetle)**
The larva feeds on stored vegetable products. Adults also feed on animal substances.

CRYPTOPHAGIDAE
Cryptophagus acutangulus **Gyllenhal, 1827**
Cryptophagus lapponicus **Gyllenhal, 1827**

LATRIDIIDAE (MINUTE BROWN SCAVENGER BEETLES)
Corticaria serrata **(Paykull, 1798)**
Dienerella filum **(Aubé, 1850)**
Thes bergrothi **(Reitter, 1880)**

TENEBRIONIDAE (DARKLING BEETLES)
Species encountered in Greenland are dark (brown to black), near-cylindrical beetles with clearly 'heteromerous' tarsi (i.e., the number of tarsal segments on the fore-, mid- and hindlegs are 5, 5 and 4, respectively). The larvae are overall uniform: elongate, cylindrical, glabrous and evenly sclerotized. Some species are pests in granaries and stored cereal products.

Tribolium confusum **Jacquelin du Val, 1863**
Tribolium destructor **Uyttenboogart, 1934**
Tenebrio obscurus **Fabricius, 1792**

CERAMBYCIDAE (LONGHORN BEETLES)
A very large, predominantly tropical family of elongate beetles, many of which are notable for their long antennae They are closely related to chrysomelids, but their tibiae consistently have 2 apical spurs. The pale, fleshy and apodous larvae are almost always wood-borers, and pupation generally takes place close to the bark, or in the bark. Most species are thus confined to forested regions where some cause considerable damage to timber. A number of species are occasionally introduced to Greenland with timber, but are naturally unable to become established here.

Callidium violaceum **(Linnaeus, 1758)**
Gracilia minuta **(Fabricius, 1781)**
Molorchus minor **(Linnaeus, 1758)**
Phymatodes testaceus **(Linnaeus, 1758)**
Pogonocherus fasciculatus **(De Geer, 1775)**
Tetropium castaneum **(Linnaeus, 1758)**
Xylotrechus colonus **(Fabricius, 1775)**

CURCULIONIDAE (WEEVILS)
Otiorhynchus sulcatus **(Fabricius, 1775) (Black vine weevil)**

SCOLYTIDAE (BARK BEETLES)
Small, dark, cylindrical beetles which include a large number of bark tunnelling pests. The head is often concealed from above by the large prothorax, and the elytra often scooped out at the back.

Pityogenes chalcographus **(Linnaus, 1761)**

References

Andersen, J. (1982): Contribution to the knowledge of the distribution, habitat selection and life history of the riparian beetles in Norway. – *Fauna norvegica* (B) 29: 62–68.

Bell, J.E. (1981): An annotated list of invertebrates collected from Eqalungmiut Nunaat. – *In*: Fox, A.D. & Stroud, D.A. (eds). Report of the 1979 Greenland White-fronted Goose Study Expedition to Eqalungmiut Nunaat, West Greenland: 227–229. *Greenland White-fronted Goose Study, Aberystwyth.*

Bengtson, S.-A., Fjellberg, A. & Solhøy, T. (1975): *Amara quenseli* Schn. (Coleoptera, Carabidae) new to Svalbard. – *Norwegian Journal of Entomology* 22: 81–82.

Bengtson, S.-A., Enckell, P.H., Erikstad, K.E. & Solhöy, T. (1983): Geography of leg-colour dimorphism in the carabid beetle *Nebria gyllenhali* in Iceland and on the Faroe Islands. – *Entomologica scandinavica* 14: 57–66.

Blair, K.G. (1930): Coleoptera from Greenland. Oxford University Greenland Expedition, 1928. – *The Annals and Magazine of Natural History*, Tenth Series, 5: 394–400.

Böcher, J. (1988): The Coleoptera of Greenland. – *Meddelelser om Grønland Bioscience* 26: 100 pp.

Böcher, J. (1998): Insect remains from Asummiut. *In*: Arneborg, J. & Gulløv, H.C. (eds) Man, culture and environment in ancient Greenland: 133–134. – *Danish Polar Center*, Copenhagen.

Böcher, J. (1999): Mystisk opdukken af biller i Grønland. – *Dyr i natur og museum* 1999: 22–24.

Böcher, J. (2009): Fund af den grønlandske mariehøne (*Coccinella transversoguttata* Faldermann, 1835) i Zackenbergdalen, Nordøstgrønland. (With an English summary: The Greenlandic ladybird *Coccinella transversoguttata* Faldermann, 1835, found in Zackenbergdalen, Northeast Greenland). – *Entomologiske Meddelelser* 77: 115–116.

Böcher, J. & Bennike, O. (1996): Early Holocene insect and plant remains from Jameson Land, East Greenland. – *Boreas* 25: 187–193.

Böcher, J., Bennike, O. & Wagner, B. (2012) Holocene insect remains from south-western Greenland. – *Polar Research* 2012, 31, 18367, http://dx.doiorg/10.3402/polar.v31i0.18367.

Brinck, P. (1940): Beitrag zur Kenntnis der Wasserkäfer Grönlands. – *Opuscula Entomologica* 5: 37–41.

Buckland, P.C., Sveinbjarnardóttir, G., Savory, D., McGovern, T.H., Skidmore, P. & Andreasen, C. (1983): Norsemen at Nipaitsoq, Greenland: A palaeoecological investigation. – *Norwegian Archaeological Review* 16 (2): 86–98.

Chernov, Yu.I. & Makarova, O.L. (2008): Beetles (Coleoptera) in High Arctic. – *In*: Penev, L., Erwin, T. & Assmann, T. (eds). Back to the Roots and Back to the Future? Towards a New Synthesis between Taxonomic, Ecological and Biogeographical Approaches in Carabidology. – *Proceedings of the XIII European Carabidologists Meeting, Blagoevgrad*, August 20–24, 2007: 213–246. Pensoft Publishers, Sofia-Moscow.

Christiansen, H.H., Bennike, O., Böcher, J., Elberling, B., Humlum, O. & Jakobsen, B.H. (2002): Holocene environmental reconstruction from deltaic deposits in northeast Greenland. – *Journal of Quaternary Science* 17(2): 145–160.

Cotton, M.J. (1979): Records of insects from North-East Greenland. – *The Entomologist's monthly Magazine* 113: 213–217.

Crowson, R.A. (1981): *The Biology of the Coleoptera.* – Academic Press, London etc. 802 pp.

Coulson, S.J. (2000): A review of the terrestrial and freshwater invertebrate fauna of the High Arctic archipelago of Svalbard. – *Norwegian Journal of Entomology* 47: 41–63.

Coulson *et al.* [38 authors] (2014): The terrestrial and freshwater invertebrate biodiversity of the archipelagos of the Barents Sea; Svalbard, Franz Josef Land and Novaya Zemlya. – *Soil Biology & Biochemistry* 68: 440–470.

Danks, H.V. (1981): *Arctic Arthropods. A review of systematics and ecology with particular reference to the North American fauna.* – Biological Survey Project, Entomological Society of Canada, 608 pp.

Danks, H.V. & Downes, J.A. (eds) (1997): *Insects of the Yukon. Biological Survey of Canada Monograph series No. 2.* – Biological Survey of Canada (Terrestrial Arthropods). 1034 pp.

Degerbøl, M. (1937): A contribution to the investigation of the fauna of the Blosseville Kyst, East Greenland, with special reference to zoogeography. – *Meddelelser om Grønland* 104 (19): 36 pp.

Elliott, P., King, P.E. & Fordy, M.R. (1983): Observations on staphylinid beetles living on rocky shores. – *Journal of Natural History* 17: 575–581.

Fabricius, O. (1780): *Fauna Groenlandica.* – Hafniæ et Lipsiæ, 452 pp.

Fjellberg, A. (1983): Three species of staphylinid Coleoptera new to Spitsbergen. – *Fauna norvegica* Ser. B. 30: 110.

Fjellberg, A. & Böcher, J. (2006): Biological observations on *Dorytomus imbecillus* Faust, 1882, in West Greenland (Coleoptera, Curculionidae). – *Entomologiske Meddelelser* 74: 81–86.

Galewski, K. (1967): The description of pupae of *Colymbetes paykulli* Er, and *C. dolabratus* (Payk.) with a key to the identification of pupae of the European species of *Colymbetes* Clairv. (Coleoptera, Dytiscidae). – *Annales Zoologici* 24: 367–374.

Galewski, K. (1968): The descriptions of pupae of *Colymbetes dolabratus* (Payk.) with keys to the identifications of larvae of the European species of *Colymbetes* Clairv. (Coleoptera, Dytiscidae). – *Annales Zoologici* 26: 227–238.

Hågvar, S. (1970): Some observations on Coleoptera, Hymenoptera and Siphonaptera in Svalbard 1968. – *Årsbok 1969, Norsk Polarinstitutt*, Oslo 1970: 101–106.

Haarløv, N. (1942): A morphologic-systematic-ecological investigation of Acarina. – *Meddelelser om Grønland* 128 (1): 71 pp.

Hansen, M. (1987): The Hydrophiloidea (Coleoptera) of Fennoscandia and Denmark. – *Fauna Entomologica Scandinavica* 18. E.J. Brill/Scandinavian Science Press Ltd. Leiden-Copenhagen, 254 pp.

Hansen, M. (1991): The Hydrophiloid Beetles. Classification and a Revision of the Genera (Coleoptera, Hydrophiloidea). – *Biologiske Skrifter* 40, Det Kgl. Danske Videnskabernes Selskab. 368 pp.

Henriksen, K.E. (1939): A revised index of the insects of Grønland. – *Meddelelser om Grønland* 119 (10): 112 pp.

Henriksen, K.L. & Lundbeck, W. (1917): Landarthropoder (Insecta et Arachnida). Conspectus Faunæ Groenlandicæ, Pars secunda. – *Meddelelser om Grønland* 22: 481–823.

Hinton, H.E. (1945): *A Monograph of the Beetles associated with Stored Products 1.* – British Museum (Natural History), London, 443 pp.

Jeppesen, P.C. (1986): Dytiscid beetles in Greenland, with description of the three larval stages of *Hydrophorus melanocephalus* (Marsham, 1802). – *Entomologica Basilensis* 11: 67–79.

Kangas, E. (1967): Identification of the Coleoptera collected by the Finnish Spitsbergen expeditions. – *Annales entomologici fennici* 33: 41–43.

Kangas, E. (1973): Über die Coleopterenfauna in Spitsbergen. – *Annales entomologici fennici* 39: 68–70.

King, P.E., Fordy, M. & Al-Khalifa, M.S. (1980): Observations on the intertidal *Micralymma marinum* (Strøm) (Col., Staphylinidae). – *The Entomologist's monthly Magazine* 115: 133–136.

Klimaszewski, J., Savard, K., Pelletier, G. & Webster, R. (2008): Species review of the genus *Gnypeta* Thomson from Canada, Alaska and Greenland (Coleoptera, Staphylinidae, Aleocharinae): systematics, bionomics and distribution. – *ZooKeys* 2: 11–84.

Korotyaev, B.A. (1976): A review of weevils of the genus *Dorytomus* Germ. (Coleoptera, Curculionidae) of North-eastern Asia. – *Revue d'Entomologie de l'URSS* 55: 124–136.

Larsson, S.G. & Gigja, G. (1959): Coleoptera 1. Synopsis. – *In*: Frídriksson, Á. & Tuxen, S.L. (eds). *The Zoology of Iceland* 3, Part 46a. Ejnar Munksgaard, Copenhagen & Reykjavik, 218 pp.

Lindroth, C.H. (1931): *Die Insektenfauna Islands und ihre Probleme.* – Uppsala, 589 pp.

Lindroth, C.H. (1957): *The Faunal Connection between Europe and North America.* – Almqvist & Wiksell, Stockholm, 344 pp.

Lindroth, C.H. (1983): The Carabidae (Coleoptera) of Fennoscandia and Denmark. – *Fauna Entomologica Scandinavica* 15, 1. E.J. Brill/Scandinavian Science Press, Leiden-Copenhagen, 126 pp.

Luff, M.L. (1993): The Carabidae (Coleoptera) larvae of Fennoscandia and Denmark. – *Fauna Entomologica Scandinavica* 27. E.J. Brill, Leiden, New York, Köln, 186 pp.

Lundbeck, W. (1896): Coleoptera Groenlandica. – *Videnskabelige Meddelelser fra dansk naturhistorisk Forening* 1896: 196–219.

Mahler, V. (1988): A new species of *Atheta* Thomson, 1858, from Greenland (Coleoptera, Staphylinidae, Aleocharinae). – *Steenstrupia* 14: 93–97.

Makarova, O.L., Bienkowski, A.O., Bulavintsev, V.I. & Sokolov, A.V. (2007): Beetles (Coleoptera) in polar deserts of the Severnaya Zemlya Archipelago. – *Entomological Reviews* 87(9): 1142–1154.

Nilsson, N. & Holmen, M. (1995): The aquatic Adephaga (Coleoptera) of Fennoscandia and Denmark. II. Dytiscidae. – *Fauna Entomologica Scandinavica* 32. 192 pp. – E.J. Brill, Leiden, New York, Köln, 188 pp.

Paulian, R. (1941): Les premiers état des Staphylinoidea. – *Mémoires du Muséum National d'Histoire Naturelle*, Nouvelle Série (1). 361 pp.

Poppius, B. (1908): Zur Kenntnis der Staphyliden-Gattung *Micralymma* Westw. – *Meddelanden från Societatis Fauna et Flora Fennica* 35: 96–99.

Poppius, B.(1910): Die Coleopteren des arktischen Gebietes. – *Fauna Arctica* 5: 289–447.

Prestrud, P., Strøm, H. & Goldman, H.V. (eds) (2004): *Skrifter 201. A catalogue of the terrestrial and marine animals of Svalbard.* – Norwegian Polar Institute, Tromsø, 137 pp.

Røen, U. (1963): Nogle udbredelsestyper i den grønlandske ferskvandsfauna. – *Tidsskriftet Grønland*, oktober 1963: 361–374.

Røen, U. (1981): Ferskvandsfaunaen. *In*: Nørrevang, A. & Lundø, J. (eds). *Danmarks Natur 11, Grønland.* Politikens Forlag, København, pp. 459–473.

Schiødte, J.C. (1845): Om Slægten *Micralymma*. – *Naturhistorisk Tidsskrift* 2: 370–389.

Schiødte, J.C. (1857): Tillæg Nr. 3. Udsigt over Grønlands Land-, Ferskvands- og Strandbreds-Arthropoder, *In*: Reinhardt, I., Schiødte, J.C., Mørch, O.M.L., Lütken, C.F., Lange, J. & Rink, H. (eds): *Naturhistoriske Tillæg til en geografisk og statistisk Beskrivelse af Grønland.* Særskilt Aftryk af Tillæggene til Rink, H. (1852–1857): Grønland geografisk og statistisk beskrevet: 50–74. A.F. Høst, København.

Steel, W.O. (1970): The larvae of the genera of the Omaliinae (Coleoptera: Staphylinae) with particular reference to the British fauna. – *Transactions of the Royal entomological Society of London* 122: 1–47.

Strand, A. (1943): Die Käferfauna von Svalbard. – *Norsk entomologisk Tidsskrift* B 6: 53–69.

Strand, A. (1946): *Nord-Norges Coleoptera.* – *Tromsø Museums Årshefter* 67, 629 pp.

Svensson, B.W. (2005): Seasonal fat content fluctuations in a gyrinid beetle: comparison between an arctic and a temperate zone population (Coleoptera: Gyrinidae). – *Aquatic Insects* 27: 11–19.

Sømme, L. (1979): Insektliv på Svalbard. – *Fauna* 32: 137–144.

Thayer, M.K. (1985): *Micralymma marinum* (Stroem) in North America: biological notes and new ditributional records (Coleoptera: Staphylinidae). – *Psyche* 92: 49–55.

Topp, W. (1975): Zur Larvalmorphologie der Athetae (Col., Staphylinidae). – *Stuttgarter Beiträge zur Naturkunde*, Ser. A, 268: 1–23.

CHAPTER 14

Trichoptera (Caddisflies)

Kjell Arne Johanson

Trichoptera are holometabolous insects closely related to Lepidoptera (butterflies and moths). They are moderately large insects, the adults being moth-like with seta-covered wings, the larvae aquatic and near-cylindrical. So far ca. 13,600 species are described (Morse 2010). With about 1600 species from western Palaearctic, 1180 species from eastern Palaearctic and 2220 species from the Nearctic, the northern hemisphere fauna holds about 40% of the totally described species of the order. However, the tropical faunas are to a substantial degree unexplored, and the total richness might be several times higher.

Morphology: adults

The adults are characteristic in having wings covered by long setae and a haustellum (feeding apparatus) formed by fusion of prelabium and hypopharynx. The maxillary and labial palps are well developed, mandibles reduced. The forewings are longer than the hindwings, and the wing venation is characteristic, involving looping forewing anal veins (as in Lepidoptera) and a strong M-stem (unlike in Lepidoptera). Caddisflies have a characteristic resting posture, with the antennae pointing forwards (Fig. 6).

Immature stages

In Greenland the larvae are easily separated from those of other orders by being fully aquatic and by building portable cases of plant or mineral particles glued together by silk produced in the labial gland (Fig. 5). They have a sclerotized pronotal tergum and each thoracic segment bears one pair of walking legs. Abdomen with a pair of anal prolegs on segment X, but lacking cerci. Spiracles non-functional, scale-like gills absent.

The larva pupates within the case. The pharate adult escapes from the case and can swim about in the water before shedding the pupal cuticle in the terrestrial environment. The larvae of a few species (none in Greenland) in the families Limnephilidae and Leptoceridae have secondarily developed a terrestrial life style.

Most species have a one-year life cycle, but at high altitudes and latitudes the development lasts longer. The larvae are diverse in feeding behaviour, involving predators, filterers, shredders of organic material, or grazers on macrophytes. A few families of the 'suborder' Spicipalpia have species with free-living larvae (*e.g.* Rhyacophilidae) or different types of cases. Most species of the suborder Annulipalpia, which are often confined to running water, have larvae which build nets or tunnels. All Greenland taxa belong to the suborder Integripalpia in which larvae produce portable cases, probably as a protection against predation. Larvae are ecologically diverse and many species are available as biological indicators of water quality due to different tolerance for pollution. The adults of some species are capable of imbibing water or of feeding on energy-rich fluids like nectar. Most adults are mainly active during night, but some species are swarming also at dusk and even at daytime. Adults of most species hatch in the warmest periods of summer or early autumn and may be active for several weeks.

Greenland fauna

Only 3 of totally 47 families are present in Greenland: Phryganeidae (one species), Apataniidae (one species), and Limnephilidae (six species). The larva has been described for only three species: *Apatania zonella*, *Limnephilus griseus*, and *Grammotaulius interrogationis*. *Agrypnia* larvae probably belonging to *A. glacialis*, were, however, collected from various Greenland localities.

Key to Greenland families

ADULTS

1. Antennal scape about as long as broad. Each male maxillary palp with four segments (female always with five-segmented maxillay palps). Each foreleg tibia with 2 tibial spurs. Each midleg tibia with 4 tibial spurs (Phryganeidae) .... *Agrypnia glacialis*

– Antennal scape longer than broad. Male maxillary palps with three segments (female always with five-segmented maxillay palps). Each foreleg tibia with 1 spur. Each midleg tibia with 2 or 3 spurs 2

2. Each midleg tibia with 2 spurs.................... .... (Apataniidae) *Apatania zonella*

– Each midleg tibia with 3 spurs.................... .... Limnephilidae

LARVAE

The knowledge of Greenland Limnephilidae larvae is incomplete, and the key is restricted to family level.

1. Dorsum of mesothorax without large sclerites; setal arrangement on mesothorax similar to that

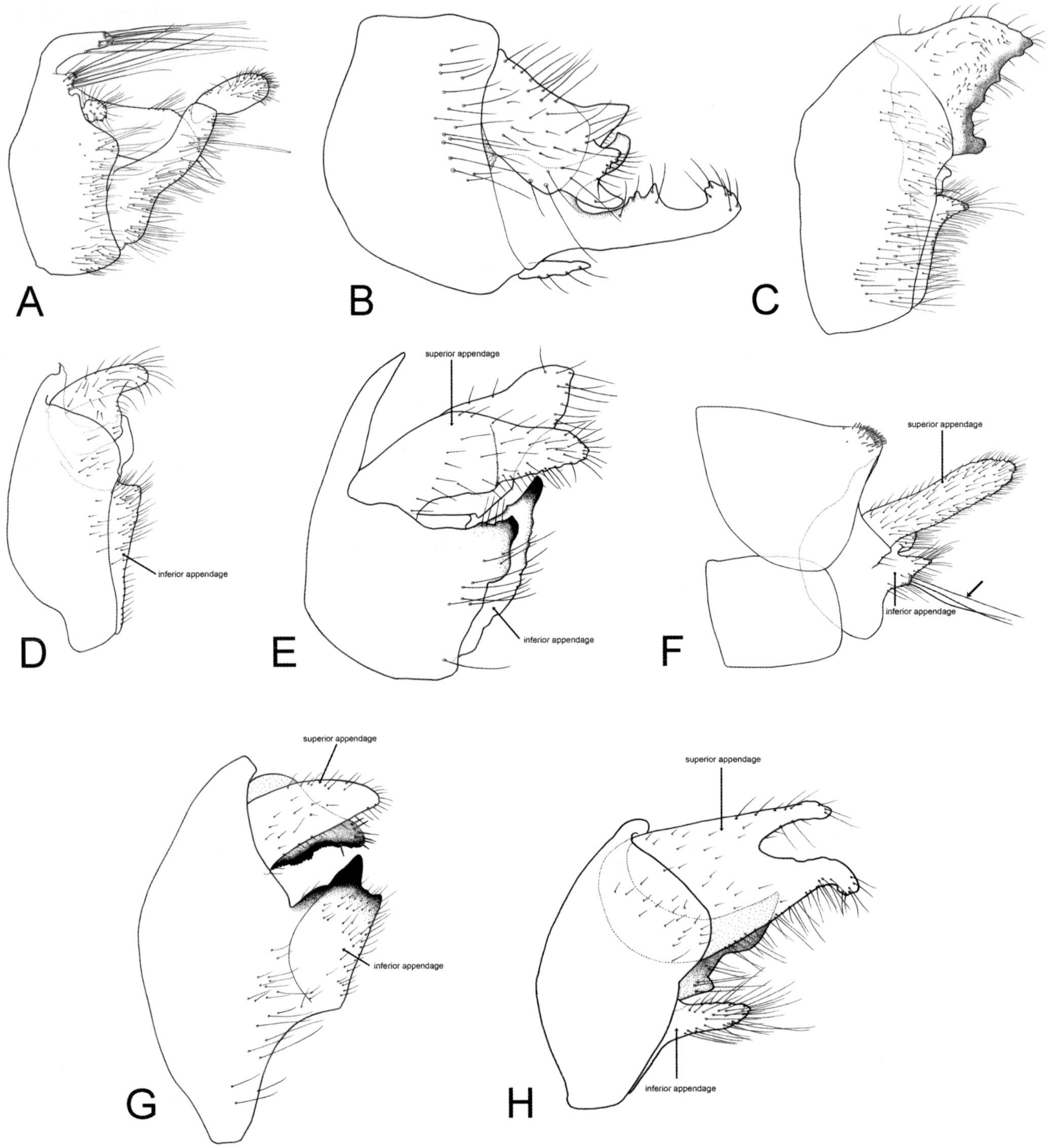

FIGURE 1 *Male genitalia of Greenland Trichoptera species, lateral view. A.* Agrypnia glacialis*; B.* Apatania zonella*; C.* Grammotaulius interrogationis*; D.* Limnephilus tarsalis*; E.* Limnephilus kennicotti*; F.* Limnephilus moestus*; G.* Limnephilus griseus*; H.* Limnephilus ornatus.

in the metathorax .... (Phryganeidae) *Agrypnia glacialis*

– Dorsum of mesothorax with large sclerites; setal arrangement on mesothorax different from that in the metathorax 2

2. Dorsum of metathorax without anteromedian sclerites .... (Apataniidae) *Apatania zonella*

– Dorsum of metathorax with one or more anteromedian sclerites .... Limnephilidae

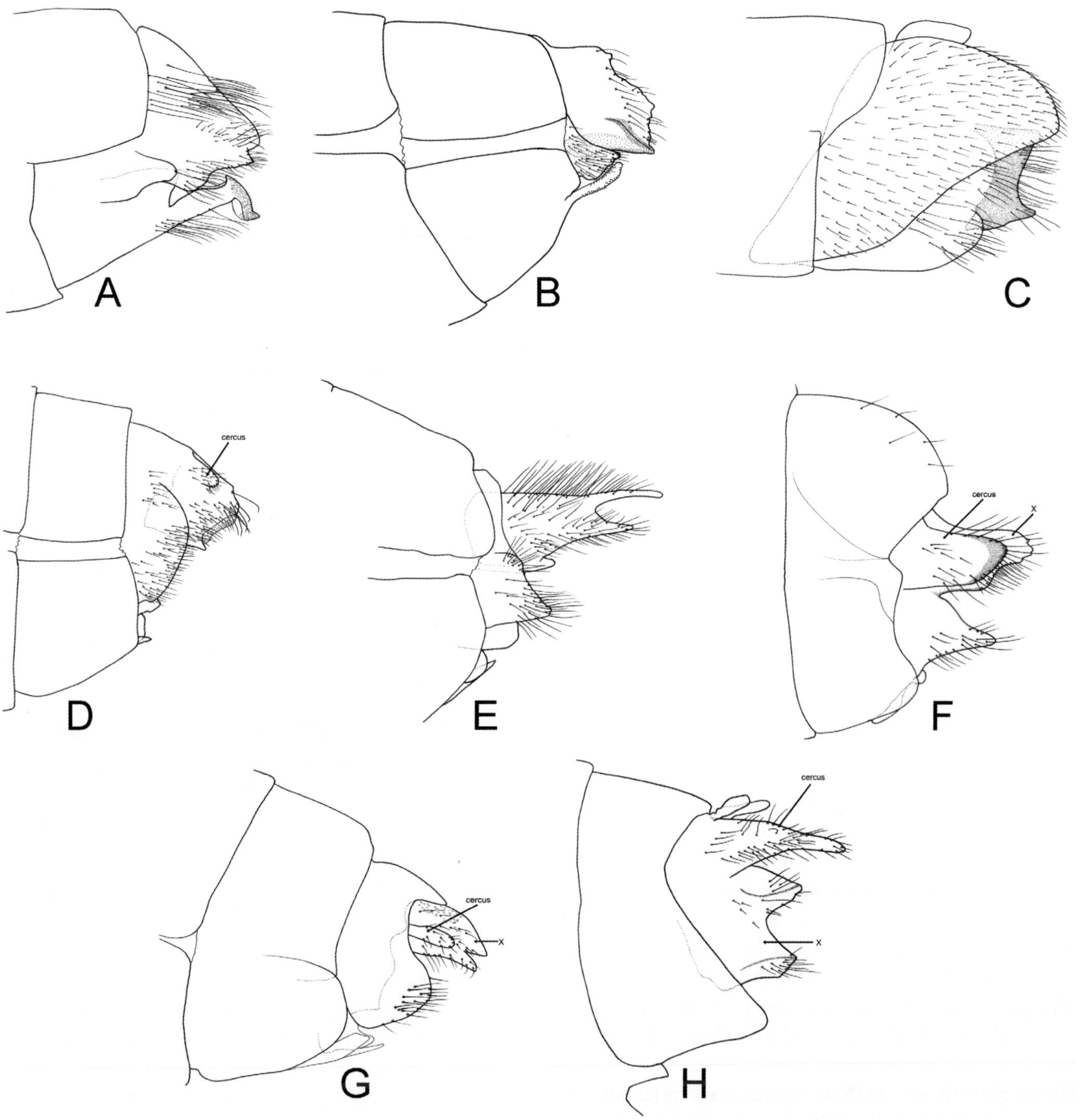

FIGURE 2 *Female genitalia of Greenland Trichoptera species, lateral view. A.* Agrypnia glacialis; *B.* Apatania zonella; *C.* Grammotaulius interrogationis; *D.* Limnephilus tarsalis; *E.* Limnephilus kennicotti; *F.* Limnephilus moestus; *G.* Limnephilus griseus; *H.* Limnephilus ornatus.

PHRYGANEIDAE (LARGE CADDISFLIES)

Adults robust compared to those in other families, antennae always stout and about as long as forewings. Three large ocelli. Male maxillary palps four-segmented, five-segmented in female. Tibial spurs 2, 4, 4. One genus known from Greenland, *Agrypnia* Curtis, 1835.

***Agrypnia glacialis* Hagen, 1873 Figs 1A, 2A, 3A, 4A**

Separable from other Greenland species by the characters given for the family. Female slightly larger than male. Male forewing length 9.5–10.3 mm (N=3). Female forewing length 11.0–11.2 mm (N=2). Wing spots absent, but contrasting wing pattern appears due to pale brownish wing membrane and darker brown veins (Fig. 4A). Genitalia illustrated in Figs 1A (male) and 2A (female).

BIOLOGY. Larvae of other species of the genus build cases from plant pieces cut from dead plants and arranged in a characteristic spiral pattern (Wiggins 1998).

DISTRIBUTION. S–sSW (Fiskefjord, Qaqortoq/Julianehaab, Kangerlussuaq/Søndre Strømfjord).

Nearctic, temperate (Canada, USA south to northern California).

APATANIIDAE

The small ocelli are characteristic of the family. Male maxillary palps three-segmented, while five-segmented in the female. Antennae thin and shorter than forewings. Tibial spurs 1, 2, 4. Also characteristic by the forewing subcosta which ends in a crossvein instead of at the wing margin. The family is separated from Phryganeidae and Limnephilidae also by absence of a discoidal cell in hindwing. In Greenland one genus, *Apatania* Kolenati, 1848.

***Apatania zonella* (Zetterstedt, 1840)**
Figs 1B, 2B, 3B, 4B, 5

Henriksen 1939: *Apatelia zonella* Zett.

The species has a complex taxonomic history and appears in the literature under many synonym names. Only the synonym applied to Greenland material is listed above. Larvae of *Halesus ?radiatus* Curtis were reported from Greenland by Henriksen & Lundbeck (1917) and Henriksen (1939), probably due to a misidentification and according to Stoltze (unpublished) belong to *A. zonella*. The species is easily separated from other Greenland species by the family characters. The forewings (Fig. 4B) of both male and female are uniformly grayish-brown. Male forewing length: 8.8 mm (N=1). Female forewing length: 6.5–10.1 (N=89). The genitalia are illustrated in Figs 1B and 2B.

BIOLOGY. In Iceland, larvae of *A. zonella* inhabit stony streams, rivers and shores of lakes (Gislason 1979). Nielsen (1969) reported that *Ranunculus glacialis* probably is dependent on this caddisfly for pollination due to the overlapping distribution of the two species. In northern Sweden, eggs of *A. zonella* hatch in the autumn and the

FIGURE 3 *Larvae of Greenland Trichoptera species, dorsal view. A.* Agrypnia glacialis, *head and thorax; B.* Apatania zonella, *thorax; C. Limnephilidae indet., head and thorax. Arrows in 3C indicate position of anteromedian sclerites of metathorax.*

FIGURE 4 *Adults of Greenland Trichoptera species, dorsal view. A.* Agrypnia glacialis*; B.* Apatania zonella*; C.* Grammotaulius interrogationis, *arrows show position of forewing R5 and hindwing R5; D.* Limnephilus tarsalis, *arrows show position of forewing R5 and hindwing R5; E.* Limnephilus kennicotti*; F.* Limnephilus moestus*; G.* Limnephilus griseus*; H.* Limnephilus ornatus.

young larvae overwinter (Otto & Svensson 1980). From Greenland larvae have been sampled from Qeqertarsuaq/ Disko in a 1.8–3 m wide, vegetation rich and groundwater-fed streams with pH 7.1–7.3 (Friberg *et al.* 2001). The species is well known for having the potential of parthenogenesis, which is also illustrated by the overrepresentation of females in various samples.

DISTRIBUTION. G.

FIGURE 5 *Larvae of* Apatania zonella *in their case, in lateral (above) and ventral (below) views.* (PHOTO: J. BÖCHER)

Circumpolar, arctic-alpine (Iceland, Svalbard, Bear Island, Novaya Zemlya, a band from Scandinavia to Kamtchatka (including Mongolia and Turkestan), and Canada.

LIMNEPHILIDAE (NORTHERN CADDISFLIES)

A diverse family with modifications both in wing venation and spur formula. Ocelli well developed. Maxillary palps three-segmented in males, five-segmented in females. Common spur formula for all Greenland species is 1, 3, 4. Subcosta vein of forewings ends in the wing margin. Hindwings have a closed discoidal cell.

Key to species

1. R5 of both fore- and hindwings more darkly pigmented than other apical veins (arrows in Fig. 4C). In hindwing, fork 1 petiolate (1 in Fig. 4C) .... *Grammotaulius interrogationis*
– R5 of both fore- and hindwing equally strongly pigmented as other apical veins (arrows in Fig. 4D). In hindwing, fork 1 sessile (1 in Fig. 4D) .... 2 (*Limnephilus*)
2. In male genitalia, each inferior appendage form a vertical plate well separated from segment IX in lateral view (Fig. 1D). In female genitalia, cercus minute (Fig. 2D) .... *Limnephilus tarsalis*
– In male genitalia, each inferior appendage forms a horizontal plate more (Figs 1E, H) or less (Figs 1F, G) separated from segment IX in lateral view. In female genitalia, cercus absent (Fig. 2E) or well developed (Figs 2F–H) 3
3. In male genitalia, superior appendage slightly angled posteriorly at midway (Fig. 1E). In female genitalia, cercus absent (Figs 2E) ...... *Limnephilus kennicotti*
– In male genitalia, superior appendage not angled along its length (Figs 1F–H). In female genitalia, cercus well developed (Figs 2F–H)................... 4
4. In male genitalia, each superior appendage straight, about four times longer than wide (Fig. 1F); inferior appendages fused with segment IX in lateral view, and with long, dark and thick setae along its ventral margin (arrow, Fig. 1F). In female genitalia, cercus broad, nearly as long as segment X (Fig. 2F) .... *Limnephilus moestus*
– In male genitalia, superior appendage shorter than three times its width (Figs 1G, H); inferior appendages fused with or separate from segment IX in lateral view, without thick, long, dark setae along its ventral margin (Figs 1G, H). In female genitalia, each cercus broad, much shorter (Fig. 2G) or longer (Fig. 2H) than segment X........................ 5
5. In male genitalia, each superior appendage simple (Fig. 1G); inferior appendages bifurcate, fused with segment IX in lateral view (Fig. 1G). In female genitalia, each cerci much shorter than segment X and club shaped, located laterally on segment X (Fig. 2G) ... . *Limnephilus griseus*
– In male genitalia, each superior appendage deeply bifurcate (Fig. 1H); inferior appendages simple, separated from segment IX in lateral view (Fig. 1H). In female genitalia, each cerci much longer than segment X and tapering along its length, located dorsally on segment X (Fig. 2H).... *Limnephilus ornatus*

***Grammotaulius interrogationis* (Zetterstedt, 1840)**
Figs 1C, 2C, 4C

With a forewing length of *c.* 17 mm in males, and 14–20 mm in females this is the largest species of Trichoptera in Greenland. Forewings pale brownish with small, brownish spots mainly within the marginal cells, and a darkly pigmented R5 of both forewing and hindwing (arrows in Fig. 4C). Male genitalia include a pair of prominent superior appendages having darkly pigmented posterior margin, and a small inferior appendage (Fig. 1C). Female genitalia with a pair of very large cerci more or less covering dorsal part of segment X in lateral view (Fig. 2C).

BIOLOGY. According to Lepneva (1971) European species of the genus *Grammotaulius* are found in small, lentic waters with detritus. In Canada larvae of *G. interrogationis* inhabit detritus rich, slow-running waters (Clifford 1969).

DISTRIBUTION. S, SW (north to the Maniitsoq/ Sukkertoppen area).

Nearctic, temperate (southern Canada, northern USA excluding Alaska).

Limnephilus tarsalis **(Banks, 1920)**
Figs 1D, 2D, 4D
With a forewing length of 6.0–9.0 mm in males among the smallest species in Greenland. Forewings brownish with small, pale spots scattered over wing surface (Fig. 4D). Setae on central anterior part of forewing often lost. Male genitalia have short superior appendages with dorsal part being slightly produced posteriorly (Fig. 1D). Inferior appendages very short, each forming a vertical plate with about right angled dorsal corner. Female genitalia with pair of minute cerci on dorsolateral surface of segment X (Fig. 2D).

DISTRIBUTION. SW (north to Saqqaq).

Nearctic (USA, Canada, Alaska), low arctic-boreal and alpine.

Limnephilus kennicotti **Banks, 1920**
Figs 1E, 2E, 4E, 6
Henriksen 1939: *Limnophilus fenestratus* Zett.

Forewings grayish brown with many pale spots, most frequently in all apical cells (Fig. 4E). A characteristic rhomboid, pale spot centrally in the forewing thyridium cell. Male forewing length 6.7–9.7 mm, female forewing length 7.4–10.2 mm. Male genitalia (Fig. 1E) with large, somewhat posteriorly bent superior appendages and a prominent segment X. Inferior appendages fused with segment IX, bifurcated and with dark, dorsally oriented apices. Female genitalia (Fig. 2E) without cerci. Segment X long, narrowing, deeply incised; dorsal branches about twice the length of ventral branchs.

DISTRIBUTION. S, SW (north to Saqqaq), SE (Danell Fjord).

Nearctic (Canada), low arctic-subarctic.

NOTE. Records of *Limnephilus fenestratus* (Zetterstedt, 1840) from Greenland by several early authors (Forsslund 1932, Fristrup 1942, Henriksen 1939, Ulmer 1932) are due to misidentification.

Limnephilus moestus **Banks, 1908 Figs 1F, 2F, 4F**
Moderately large species with male forewing length 9.9–11.8 mm, female forewing length 7.4–12.0 mm. Forewing brown to pale brown, with contrasting pale spots as in Fig. 4F. Dark brown area present in thyridium cell and between M2 and M3+4. Posterior part of abdominal tergum VIII with group of short spines (Fig. 1F). Male genitalia (Fig. 1F) with long and slender superior appendages.

FIGURE 6 *Adult of* Limnephilus kennicotti. (PHOTO: J. BÖCHER)

Each inferior appendage short, posteriorly pointed, with long, dark, stout setae on posterior margin. Female genitalia (Fig. 2F) with broad cerci nearly as long as segment X.

BIOLOGY. Cobb *et al.* (1993) reported *L. moestus* from NW Ontario, Canada, emerging from peat land (poor fen), a semi aquatic larval habitat.

DISTRIBUTION. S, SW (north to Maniitsoq/ Sukkertoppen), SE (Saqqisikuik/Skjoldungen).

Nearctic, low arctic-temperate.

Limnephilus griseus **(Linnaeus, 1758)**
Figs 1G, 2G, 4G
Forewing brownish with contrasting pale spots (Fig. 4G). Pterostigma large and somewhat darker than ground colour of wing. A large pale, rhomboid spot diagonally at mid part of wing between R1 and Cu1. Hindwing apex brownish pigmented. Forewing length 9.8–12.8 mm in male, in female 9.4–11.5 mm. Male genitalia (Fig. 1G) with wide and long, slightly curving superior appendage. Segment X tapering distally into rounded apex; ventral margin blackish. Inferior appendages fused with segment

IX, dorsally pointed, apex blackish. Female genitalia (Fig. 2G) with slender, club shaped cerci. Segment X slightly curving ventrally, incised at apex.

BIOLOGY. In Iceland Gislason (1979) reported larvae from small pools, often temporary, in spring-fed streams and lake outlets, and stony shores and swamps. Also temporary acidic pools are inhabited by larvae (Hiley 1976). Otto & Svensson (1980) reports that eggs hatch in autumn and the young larvae produce cases of organic particles, while old cases are made from mineral particles.

DISTRIBUTION. S, SW (north to Nuuk/Godthåb).

Palaearctic, low arctic-temperate (widely distributed from northwestern Europe to Siberia).

Limnephilus ornatus **Banks, 1897 Figs 1H, 2H, 4H**

Easily recognized by large forewings and colour pattern consisting of contrasting whitish and brownish longitudinal spots on a pale brownish ground (Fig. 4H). Forewing length in male 13.4–14.8 mm (N=4), in female 14.0–15.8 mm (N=2). Male genitalia (Fig. 1H) unique in having deeply incised superior appendage, and tuboid, posteriorly oriented, interior appendage. Female genitalia (Fig. 2H) include a long, tapering, cercus exceeding segment X.

BIOLOGY. Little is known, but Cobb *et al.* (1993) reported adults from NW Ontario, Canada, emerging from peat land (poor fen), a semi aquatic larval habitat.

DISTRIBUTION. S, SW (north to Sisimiut/Holsteinsborg).

Nearctic, low arctic-temperate.

References

Banks, N. (1897): New North American neuropteroid insects. – *Transactions of the American Entomological Society* 24: 21–31.

Banks, N. (1908): Some Trichoptera, and allied insects, from Newfoundland. – *Psyche* 15: 61–67.

Banks, N. (1920): New neuropteroid insects. – *Bulletin of the Museum of ComparativeZoology* 64: 297–362.

Clifford, H.F. (1969): Limnological features of a northern brownwater stream, with special reference to the life histories of the aquatic insects. – *American Midland Naturalist* 82: 578–597.

Cobb, D.G., Flannagan, J.F., Rosenberg, D.M. & Wiens, A.P. (1993): Emergence of Trichoptera from aquatic habitats in the Experimental Lakes Area, northwestern Ontario, Canada. *In*: Otto, C. (ed.) *Proceedings of the Seventh International Symposium on Trichoptera*, pp. 219–222. Backhuys Publishers, Leiden, the Netherlands.

Curtis, J. (1835): *British Entomology – being illustrations and descriptions of the genera of insects found in Great Britain and Ireland; containing coloured figures from nature of the most rare and beautiful species, and in many instances of the plants upon which they are found.* Plate 540. – Richard Taylor, London.

Forsslund, K.H. (1932): Zur Kenntnis der Trichopteren Grönlands. – *Entomologisk Tidskrift* 53: 56–59.

Friberg, N., Milner, A.M., Svendsen, L.M., Lindegaard, C. & Larsen, S.E. (2001): Macroinvertebrate stream communities along regional and physicochemical gradients in Western Greenland. – *Freshwater Biology* 46: 1753–1764.

Fristrup, B. 1942. Neuroptera and Trichoptera. – *Zoology of Iceland* 3: 3–23.

Gislason, G.M. (1979): Identification of Icelandic caddis larvae, with description of *Limnephilus fenestratus* (Zett.) and *L. picturatus* McL. (Trichoptera: Limnephilidae, Phryganeidae). – *Entomologica scandinavica* 10: 161–176.

Hagen, H.A. (1873): North American Phryganeidae. – *Proceedings of the Boston Society of Natural History* 15: 384–385.

Henriksen, K.L. & Lundbeck, W. (1917): Landarthropoder (Insecta et Arachnida). – *Meddelelser om Grønland* 22: 483–823.

Henriksen, K.L. (1939): A revised index of the insects of Greenland. – *Meddelelser om Grønland* 119(10): 1–112.

Hiley, P.H. (1976): The identification of British limnephilid larvae (Trichoptera). – *Systematic Entomology* 1: 147–167.

Kolenati, F.A. (1848): Genera et species Trichopterorum. Pars prior. – *Acta Regiae Bohemoslovenicae Societatis Scientiarum. Prague* 6: 1–108.

Lepneva, S.G. (1971): Fauna SSSR, Rucheinik, Vol. II, No. 2, Lichinki i kukolki podotryada tsel'noshchupikovykh. *Zoologicheskii Institut Akademii Nauk SSSR, 95.* [Translated into English as *Fauna of the USSR, Trichoptera*, Vol. II, No. 2. Larvae and pupae of Integripalpia. Israel Program for Scientific Translations, Jerusalem].

Linnaeus, C. (1758): *Systema Naturae per Regna tria Naturae, Secundum Classes, Ordines, Genera, Species, cum Characteribus, Differentiis, Synonymis, Locis.* 10th Edition, Volume, 1: Regnum Animalia. Holmiae [Stockholm], Laurentii Salvii.

Martynov, A.V. (1924): Rucheiniki (caddisflies [Trichoptera])[in Russian]. *In* Bogdanova-Kat'kova, N.N. (ed.) pp. 1–384. – *Prakticheskaya entomologiya* 5.

Morse, J.C. (2010): *Trichoptera World Checklist*, Clemson University web-page. Available from: http://entweb.clemson.edu/database/trichopt/index.htm (accessed 19 October 2010).

Mosely, M.E. (1929): Trichoptera and Ephemeroptera of Greenland. Additional records made by the Oxford University Expedition to Kugssuk, Godthaab Fjord, W. Greenland, 1928. – *Annals and Magazine of Natural History* 10 (10): 501–509.

Nielsen, A. (1969): On the subspecies of *Apatania muliebris* McL. (Trichoptera). – *Entomologiske Meddelelser* 37: 313–318.

Otto, C. & Svensson, B.S. (1980): The significance of case material selection for the survival of caddis larvae. – *Journal of Animal Ecology* 49: 855–865.

Ross, H.H. (1938): Lectotypes of North American caddis flies in the Museum of Comparative Zoology. – *Psyche* 45: 1–61.

Ross, H.H. (1944): The caddis flies, or Trichoptera, of Illinois. – *Bulletin of the Illinois Natural History Survey* 23: 1–326.

Stoltze, M. (unpublished): *Grønlandske vårfluers udbredelse og indvandringshistorie*. Unpublished undergraduate project work, Zoologisk Museum, Universitety of Copenhagen. 15 pp.

Ulmer, G. (1932): *Die Trichopteren, Ephemeropteren und Plecopteren des arktischen Gebietes*. *In* Römer, F. & Schaudinn, F. (eds) Fauna Arctica. Eine Zusammenstellung der arktischen Tierformen VI, pp. 208–218. Gustav Fischer Verlag, Jena.

Walker, F. (1852): *Catalogue of the specimens of neuropterous insects in the collection of the British Museum, Part I: Phryganides-Perlides*. 192 pp. British Museum, London.

Wiggins, G.B. (1998): *The caddisfly family Phryganeidae (Trichoptera)*. University of Toronto Press. Toronto.

Zetterstedt, J.W. (1840): *Insecta Lapponica, sectio quinta Neuroptera*, pp 1025–1074. Leipzig.

CHAPTER 15

Lepidoptera (Moths and butterflies)

Ole Karsholt, Niels P. Kristensen†, Thomas J. Simonsen and Matti Ahola[1]

In terms of currently described species (150.000+, Nieukerken *et al.* 2011) the order Lepidoptera is the third-largest of the 4 'megadiverse' insect orders[2]; so it is in Greenland, but its representation here is very modest. Adult Lepidoptera are immediately recognizable by the dense scale-covering of the wings, which confers on the latter a more or less elaborate colour pattern. The body and legs also have a dense scale and/or hair vestiture, which has no counterpart in other Greenland insects. Almost all adult Lepidoptera (all in Greenland) feed on fluid substances, particularly nectar and honeydew (the sugar-rich fluid excrements of sternorrhynchan Hemiptera), or they do not feed at all, and hence are short-living. The vast majority of lepidopteran larvae ('caterpillars') are herbivores, but some are fungivores or feed on other organic material; in Greenland they are all terrestrial. They can almost always be immediately distinguished from other insect larvae by their complement of *crochet-bearing* abdominal prolegs (see below). The feeding style and ± pronounced host specificity of several Lepidoptera larvae facilitates species identification.

Morphology: adults

While members of the most basal extant families in this order are quite generalized endopterygote insects, the bulk of the lepidopterans, including all Greenland representatives, belong to the subordinate taxon Ditrysia, which is characterized by a suite of remarkable specializations evolved on top of those pertaining to the lepidopteran ground pattern. Only this advanced morphology type is referred to in the following succinct account focusing on structures useful in identification. Not all specialized terms used in keys and descriptions are explained here; reference should be made to the morphology glossary.

On the head the compound eyes account for a major part of the surface area; the median ocellus is consistently absent, while lateral ocelli are of variable occurrence. Antennae are always well developed, and in several moths the flagellum (sometimes in males only) is *pectinate/bipectinate*, while butterflies are unusual by having the flagellum apex more or less abruptly thickened. A 'chaetosema' is a useful diagnostic feature of some higher taxa; it is a ± convex cranial area located behind the antennae and bearing a distinctive seta vestiture (Fig. 1B). The mandibles are non-functional, small to minute protuberances, and the adult insects can feed on fluid substances exclusively. The feeding apparatus is a remarkable type of proboscis ('tongue') formed by the two maxillary galeae which are linked together by cuticular processes; the inner surface of each galea is concave, hence the coapted galae enclose a canal between them. In repose the proboscis is spirally coiled (Fig. 1C); it can be extended by blood pressure, while recoil is due to wall elasticity combined with contractions of intrinsic muscle fibres. Some moths do not feed at all in the adult stage, and in these the proboscis may be highly reduced. Among ditrysian Lepidoptera only some Tineoidea (in Greenland *Monopis*) have retained prominent maxillary palps; in all other they are small, sometimes barely identifiable. In contrast, the labial palps in Greenland Lepidoptera are almost always well developed and often very prominent (Fig. 1A, note that the term 'palp' in most Lepidoptera literature will apply to these only); their structure and vestiture often provide useful taxonomic characters.

The prothorax is considerably smaller than the pterothoracic segments, and of the latter the mesothorax is the dominant one. This is related to the fact that two wing pairs in most ditrysian Lepidoptera are effectively coupled and the mesothoracic muscles moving the forewings are the principal flight motor. The wing coupling device is usually sexually dimorphic: In males a hook-like cuticular lobe, the 'retinaculum', arising from the lower surface of the anterior

1 OK prepared the first draft of the taxonomic treatment. TJS & MA contributed drafts of the keys to (and some descriptive notes on) the larvae; the habitus figures of the latter are mostly from Scandinavian specimens supplied by K. Silvonen. NPK drafted most of the introductory sections and some keys to adults, amalgamated the other authors' contributions and revised them more or less extensively in the process. The definitive version is due to the collaborative efforts of NPK and OK; unpublished observations (largely made on live specimens during the 'Kissavik' expedition by the latter are demarcated by (OK).

2 It may, however, be safely assumed that on a global scale it will eventually prove to be *the* least species-rich of the four, as is known to be the case in the best inventoried part of the world, viz., NW Europe.

part of the forewing base grips a stout bristle, 'frenulum' (actually a composite structure formed through fusion of several adjacent bristles), arising from an elevated socket on the basal front margin of the hindwing. In females the frenulum typically comprises a number of discrete bristles, while the retinaculum analogue is a group of more or less stiff, forwards-directed scales or hairs on the mid/posterior part of the forewing underside base. Examination of frenulum structure (males single, female multiple) hence often permits a most convenient rapid sexing of ditrysian moths whose genital segments may be strongly retracted in dead specimens, but atypical cases do occur; for instance, the phycitine pyralids have but a single female frenulum bristle. The butterflies lack a special wing-coupling device but still move the (basally considerably overlapping) two wing pairs as an entity. The scales along the apical and posterior wing margins (and in many smaller moths also on the greater part of the front margin) are often long and slender, hair-like; they are then referred to as 'fringes' or 'cilia'. In resting moths the forewings usually hide the hindwings, and only on the former is an elaborate pattern developed; in contrast, butterflies in repose fold their wings, uppersides-against-uppersides, over the body, and their wing undersides (of the hindwings in particular) are also elaborately patterned. Wing patterns are very diverse within the order. Very often there are more or less prominent transverse fasciae, and a 'median' spot located at the outer margin of the discal cell is similarly recurrent. In most noctuids and also in some pyraloids this spot is more or less pronouncedly kidney-shaped, hence referred to as the 'reniform stigma'; in the same taxa two more proximal markings are termed, respectively, the 'orbicular stigma' (inside the cell) and the 'claviform stigma' (below the cell). The legs are overall unspecialized except that the fore tibia usually carries, near mid-length, a small appendage, the 'epiphysis', used for grooming the antenna. Leg vestiture and spinosity may furnish useful identification characters.

In an extended abdomen of both sexes 9 segments are immediately recognizable in dorsal view, whereas on the ventral side of segment I (hidden above the backwards extending hind coxae) is much shortened and devoid of any sclerotization; in repose the terminal abdominal segments are often partly/largely withdrawn into segment VIII. While examination of the genital segment is of paramount importance for identification of Lepidoptera species in many parts of the world, the overall depauperate Greenland fauna comprises very few species groups whose members, when well preserved, cannot easily be distinguished on external features alone.

The most conspicuous parts of the male genital apparatus pertain to segment IX; this apparatus is generally ± retracted into the preceding segment. It comprises a ventral sclerotization, the vinculum, which posteriorly articulates with sizable paired clasping 'valves' (likely derivatives of the segmental appendages), while its upper ends articulate with/are synscleritous with the dorsal sclerotization of the segment, the tegumen. The latter in turn articulates with, or is synscleritous with, a variably shaped process, the uncus, which is formed by tergum X. Below the uncus lies the (partly membranous) anal cone, which morphologically is the most posterior part of the body. The intromittent organ (phallus) is a sclerotized tube enclosing an eversible 'vesica', which often bears characteristic spines ('cornuti'); it is situated between the valves and can be ± extensively protruded during copulation (Figs 10A-B).

The tip of the female abdomen (Figs 10C–D, 11C & F) bears 2 lateral, ± hairy and sclerotized lateral lobes, the 'anal

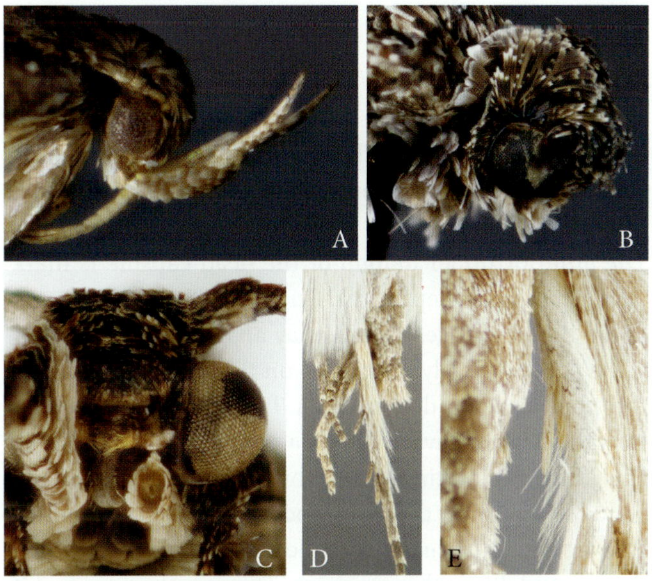

FIGURE 1 *Taxonomically informative structures of adult Lepidoptera. A, gelechioid head* (Bryotropha), *lateral view, showing large, upcurved labial palps and scale-covered basal part of proboscis. B, tortricid head* (Acleris), *lateral view, showing ocellus adjacent to upper margin of compound eye, and chaetosema* (*with delicate hair-scales, here bluish*) *immediately behind. C, tortricid head* (Acleris), *anteroventral view* (*one labial palp removed beyond basal segment*), *showing coiled proboscis and 'tortricid-type' head scaling* (*scales on lower median part of head-capsule pointing inwards-upwards*). *D, gelechioid hind tibia* (Agonopterix), *showing dorsal vestiture of long, raised hairs. E, pyraloid hind tibia* (Gesneria), *showing dorsal smooth vestiture of appressed scales.*

papillae' which enclose the anus and the egg-exit-opening ('ovipore') between them. A variably long skeletal rod, the 'posterior apophysis' arises on or near the anterior margin of each papilla; papillae plus apophyses belong to the complexly united segments IX and X. The Ditrysia (to which, as mentioned, all Greenland Lepidoptera pertain) are remarkable for having a female genitalia configuration in which the copulation takes place through a special opening, separate from the ovipore. This opening, the 'ostium (bursae)' is located on the ventral surface of segment VIII; it is often surrounded by specialized sclerotizations; the duct leading from the opening into the copulatory sac (bursa copulatrix; here the sperm is deposited on copulation) as well as the bursa itself may similarly bear characteristic sclerotizations. In most Lepidoptera paired 'apophyses anteriores', sclerotized rods reminiscent of the posterior pair, arise from the anterior margin of segment VIII.

Immature stages

Larvae (Figs 2, 3, 5, 6, 9). The larval head capsule is more or less sclerotised, and though some internal feeders are prognathous most larvae are more or less hypognathous. The anterior/upper margin of the head capsule is markedly indented ('epicranial notch'). V- or Y-shaped strengthening ridges (which may appear as grooves and/or be conspicuously darkened) on the anterior surface of the head capsule demarcate a 'facial triangle', enclosed by the clear 'cleavage lines' along which the capsule slits during moulting. The short antennae are situated ventro-laterally on the head, above/behind the mandible. The stemmata (usually 6, sometimes fewer) are located dorso-caudally to the antennae. The mouthparts are biting, with generalized labrum and mandibles. Each maxilla bears a single inner lobe (galea or fused galea + lacinia) and a small maxillary palp. The labium (with minute palps) is fused with the hypopharynx; the composite lobe bears an unpaired spinneret (a small process with the silk gland opening on the apex), and its anterior/dorsal surface (forming the floor of the mouth cavity) has a complement of spines which occasionally furnish characters useful for identification.

All 3 thoracic segments bear a pair of legs, and the first (T1) usually has a more or less sclerotised dorsal thoracic shield. Ten segments (AI–AX) are easily identifiable in the abdomen. 9 pairs of lateral spiracles are usually prominent; the first pair topographically on T1, the following on AI to AVIII. Paired abdominal prolegs are generally present on AIII to AVI, but may be absent on all but AVI. The caudal segment AX also has a pair of prolegs ('claspers'). The prolegs comprise a proleg base and a terminal part called the planta. The planta never bears setae, but has a complement of short cuticular hooks, the so-called 'crochets', the arrangement of which often is crucial for identification (Figs 2F–G). AX often has a dorso-caudal 'anal shield'.

'Chaetotaxy', the study of the occurrence, positions and individual properties of head and body setae, has longstanding application in caterpillar taxonomy. 3 categories of setae are recognized: primary, subprimary and secondary setae. Primary setae are present on the first instar, subprimary setae from the second instar throughout the development, whereas secondary setae are new setae added in later instars. The relationships and positions of the primary and subprimary setae are well understood in Lepidoptera larvae and the positions of the individual setae are crucial for identifying larvae; general seta maps for the head and different body segments are shown in Figs 2A–D. A flat sclerotised plate bearing a seta (sometimes more than 1 seta) is a *pinaculum*, an elevated pinaculum with a single seta is a *chalaza*. A very large chalaza-like structure, sometimes bearing spines and/or several setae is a *scolus*, while a non-sclerotised, fleshy lobe with multiple setae is a *verricule*.

Pupae. The pupae of ditrysian Lepidoptera are obtect. They are rarely sampled in faunistic/ecological surveys, and an identification guide to taxa in this stage is outside the scope of the present account. Pupae of many species also occurring in Central Europe are treated (with illustrations of diagnostic traits) by Patočka & Turčan (2005).

Greenland fauna

A mere 58 Lepidoptera species have been recorded from Greenland. Of these 7 are (more or less regularly occurring) introduced synanthropes, another 7 are casual immigrants and just 44 are believed to be 'natural' residents. While the inventory of moths and butterflies in Greenland is largely covered in the general survey given in chapter 5 of the inventory of the entomofauna of the country, some special features pertinent to the exploration of this particular order should be highlighted here. Due to the delicateness of the wings and (in particular) their scale vestiture these insects are particularly easily damaged, and their identification hence much impeded, by non-targeted collecting and handling. Even standard sweep-net samples may be problematical, and the problems are exacerbated with material obtained by automated collecting methods such as interception and

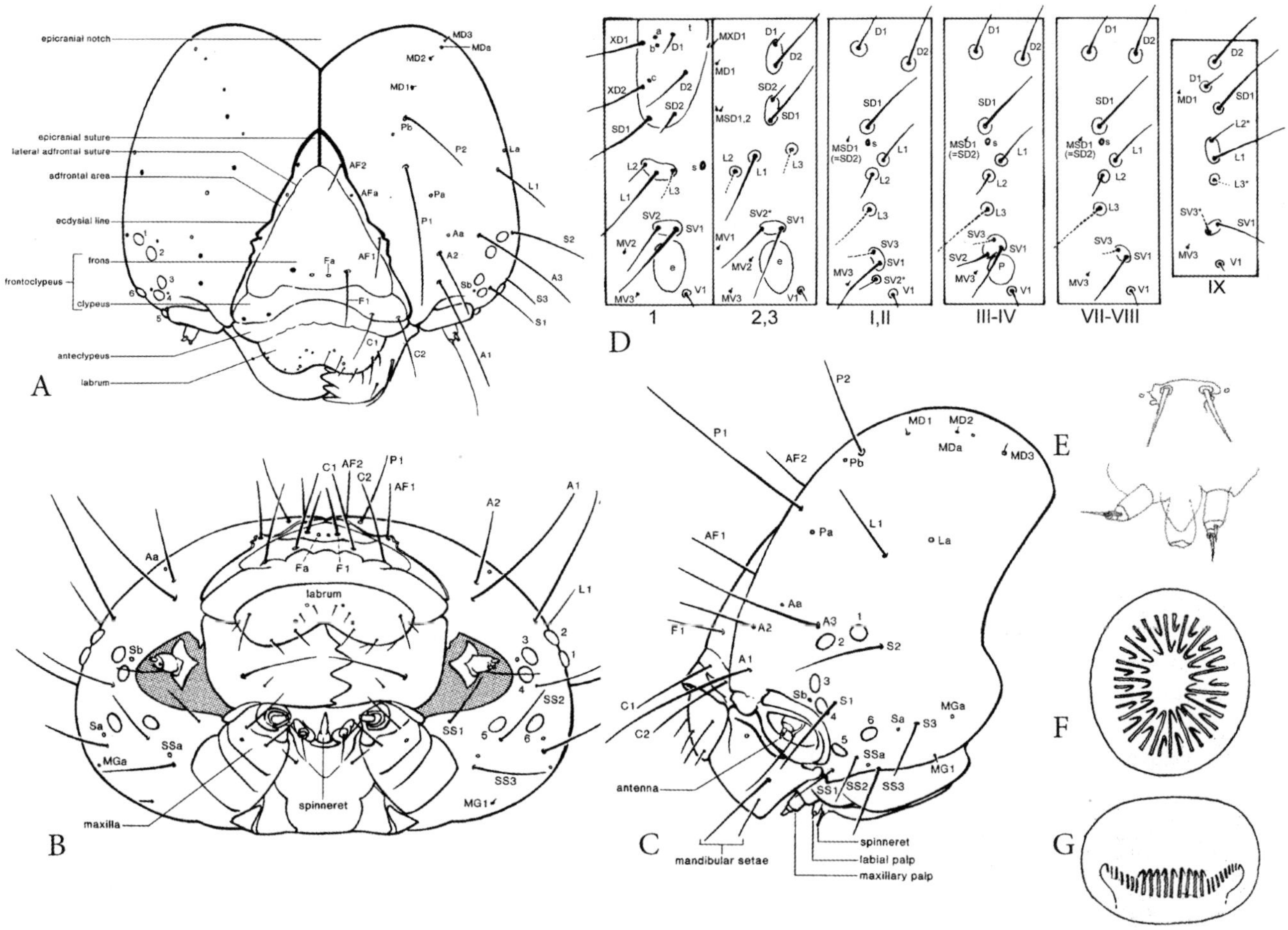

FIGURE 2 *Diagrams of larval structures. A, seta map of the head in antero-dorsal view. B, same in ventral view. C, same in lateral view. D, 'seta map' of trunk segments. E, ventral view of labium of* Sympistis lapponica *with very long premental setae. F, diagrammatic surface view of tineid proleg planta showing ring-shaped 'micromoth crochet arrangement'. G, same of noctuoid, showing 'macromoth crochet arrangement' with crochets on medial planta edge only. A–C, F–G after* Stehr (*1987*).

Malaise trapping. Lepidoptera (small-sized taxa in particular) therefore tend to be under-represented in insect samples obtained by non-lepidopterist naturalists. In contrast, the faunistic inventory of Greenland Lepidoptera more than that of any other insect order has benefited from the attention/activity of experienced and knowledgeable amateur collectors. In addition to publishing workers mentioned in Chapter 5, Roland Jacobsen, Brian Martinsen (both resident in Greenland for a number of years), Alex Madsen, Knud Bech and Erkki Laasonen deserve special mention. Several other Danish lepidopterists have contributed less extensive, but still valuable, information about findings of Greenland moths and butterflies. Hence Greenland distribution patterns of Lepidoptera may now be considered reasonably well documented.

In most parts of the world nocturnally active adult moths (which constitute the vast majority of the Lepidoptera) are most efficiently sampled by UV light trapping, and it has evidently been of considerable consequence for the inventory of Greenland moths that such techniques are of little or no use at the ambient light regimes prevalent even at mid-night at these high latitudes. It is only in the southernmost part of the country that light trapping (operated by Peter Nielsen) has proved productive.

Wolff's comprehensive 1964-treatment of Greenland Lepidoptera (with a supplement, 1971a) rendered this order the then best known larger terrestrial invertebrate group in the country. Subsequently acquired knowledge, including changing concepts of taxonomic status/nomenclature of individual taxa, is therefore here documented in greater detail for this order than has been done (and indeed would have been appropriate) for most other orders in the present work. Only taxon names or combinations which differ from the names/combinations used in these comprehensive accounts are listed below; species subsequently recorded from Greenland are similarly flagged.

A few of the Greenland lepidopterans were first described on the basis of material from this country; these are marked with an asterisk (*) in the systematic account.

Identification of Greenland Lepidoptera

Greenland Lepidoptera can almost always be immediately referred to one of three long-recognized assemblages; this is a convenient starting point for species identification. Comparisons with the habitus illustrations (Figs 12–17) will in most cases suffice for a safe identification to species or at least a small group of superficially similar species. Descriptive remarks focus on traits not observable (or not immediately obvious) in the plate figures, as well as on variation in the wing patterns. The scale covering, and hence the wing pattern, may be more or less extensively lost in older specimens, or in specimens that have become wet; such loss can be particularly pronounced in material collected in traps with liquid preservatives. However, in the overall depauperate Greenland fauna even such specimens can often be identified on the basis of size, wing shape and structural characters referred to below.

Micro-moths, the 'grade' comprising the most overall generalized lepidopteran lineages; in Greenland represented by the superfamilies Tineoidea, Gracillarioidea, Yponomeutoidea, Gelechioidea, Tortricoidea, Pterophoroidea and Pyraloidea. These are small to medium-sized insects (FWL 2.5–15 mm), with facies as illustrated in Figs 12–14M. All Lepidoptera with a scale covering on the basal part of the proboscis are referable to this assemblage, and so are all small, pronouncedly narrow-winged moths.

Prolegs of micro-moth larvae have crochets (Fig. 2F) arrranged in a circle or ellipse (when partly open referred to as a 'penellipse').

Butterflies (Papilionoidea *s.lat.*) a derived lepidopteran lineage whose medium-sized to large, broad-winged and often colourful day-flying members (Figs 5D, 15A–O) are among the few insects that are familiar even to non-entomologists; their apically widened/clubbed antennae have no counterparts in other Greenland Lepidoptera.

Greenland butterfly larvae have secondary setae. Their proleg crochets are arranged in a 'mesoseries': a longitudinal row on the medial side of the planta (Fig. 2G). This arrangement (which facilitates grasping of plant stems/twigs and is generally associated with an external-feeder lifestyle) is parallelled in macro-moth larvae.

Macro-moths (Macroheterocera), an apparently monophyletic group of overall derived lepidopterans which in Greenland is represented by the superfamilies Geometroidea and Noctuoidea. Together with the butterflies it constitutes the time-honoured assemblage 'Macrolepidoptera', which has recently proved not to be a 'natural' (monophyletic in the strict sense) entity. All larger and/or more robust Greenland moths belong here, their facies are illustrated in Figs **14**N–17. Size-wise some of the smaller Geometridae and Noctuidae overlap with the larger micro-moths in wing span, but as will be seen from the habitus illustrations their wing shapes are usually distinctive. Moreover they differ from the pyraloids and larger gelechioids in consistently lacking scales on the proboscis, and from larger tortricoids in that their lower facial scales are directed downwards (as indeed in most Lepidoptera) rather than inwards-upwards (Fig. 1C).

The larvae of Greenland macro-moths have no secondary setae (except for the uniquely 'furry' *Gynaephora* caterpillar, Fig. 9B); as in butterflies their proleg crochets are arranged in a mesoseries.

A single Greenland lepidopteran is **brachypterous**, with wings reduced to small stumps and hence useless for flight; this is the female of the geometrid *Operophtera bruceata*, Fig. **15**Z. The scaly body vestiture of this insect suffices to ensure its ordinal assignment.

Guide to identification of Greenland Lepidoptera larvae based on bionomics

Some lepidopteran larvae attract more attention in the larval stage than as adults, and it is almost consistently the larvae that play a significant role in ecosystems (and hence as pests!). Hosts plants and feeding styles provide useful hints on identification, and together with a few, readily observable structural characters permit immediate placement of most Greenland caterpillars in family/genus group. It must be noted, however, that some lepidopteran larvae are regularly or occasionally polyphagous, and particularly in the case of mass-occurrence can unusual food plants be adopted. Also, few lepidopteran immatures have been closely observed in Greenland, and it is likely that their host plant spectrum here may deviate from that in northeastern America and/or northwestern Europe; the same is true for the life cycle: it would be unsurprising if more species than currently known here prove to take more than one year to complete their development, or hibernate in another stage, than at lower latitudes. For this reason, and also because geographical variation in larval colour patterns remains little studied, the geographical source of information on immatures is given in the species accounts when the information is not always based on observations on Greenland material.

Indoors synanthropes. *Monopis, Tineola, Hoffmanophila, Endrosis, *Plodia, *Ephestia*

Living in or just above ground level, feeding on roots or mosses. *Bryotropha, Phiaris, *Tehama, *Gesneria, *#Euxoa, *#Agrotis*

Case-bearing feeders on *Juncus* and *Luzula* seeds: *Coleophora*

Leaf miner on *Ledum*: *Phyllonorycter*

On leaves of herbaceous plants, feeding openly or in a web
Polyphagous: **Nomophila, *#Agriades, *#†Psychophora; *#††Autographa, *#Mniotype, *#Polia, *#Lasionycta, *#Rhyacia, *#Spaelotis*
Poaceae: **Tehama, *#Apamea*
***Polygonum*:** **#Boloria chariclea*
***Oxyria, Rumex*:** **#Lycaena, *#†Operophtera*
***Saxifraga*:** *Stenoptilia*
***Chamerion*:** *Scythris*
***Dryas*:** *Argyroploce, *#Sympistis*
***Urtica*:** [*Vanessa*]
Brassicaceae: *Plutella, Rhigognostis*
***Pedicularis*:** *Phiaris*
***Antennaria, Gnaphalium*:** *Agonopterix*
***Erigeron*:** *Gnorimoschema valesiella*

On leaves of shrubs, feeding openly or in a web
Polyphagous: *Pyla, *#†Operophtera, *#†Eupithecia gelidata, *#Sympistis lapponica, *#Mniotype, *#Polia, *#Spaelotis, *#Eurois*
***Juniperus*:** **#†Eupithecia pusillata*
Poaceae: **#Apamea*
***Salix*:** *Gnorimoschema vibei, Acleris arcticana, *#Boloria chariclea, *#Entephria, *#††Syngrapha parilis*
***Betula*:** *Gnorimoschema vibei, Acleris caryosphena, Epinotia*
***Calluna, Erica*:** **Pyla*
***Vaccinium*:** **#††Syngrapha u-aureum*

Symbols: *2 (not 3) prespiracular L–setae on T1 (Fig. 2D); #macrolepidopteran proleg (Fig. 2G, crochets in single row, not in circle/ellipse/'penellipse'); † geometrid proleg complement (Fig. 6, on VI and X only); †† 'plusiine' proleg complement (Figs 9C–D, on V, VI and X).

Key to micro-moth families/superfamilies

ADULTS
(Abbreviations: FW, forewing; HW, hindwing; L, length; B, breadth)

1. FW elongate, outer 2/5 subdivided into 2 lobes, HW 3–lobed (Fig. **14A**) .... (Pterophoridae) *Stenoptilia mengeli*
– Wings not subdivided...............................2
2. Very small species (FWL 2.5–3.5 mm); FW bright orange-brown with white markings................. ....(Gracillariidae) *Phyllonorycter ledella*
– Larger species (FWL > 5 mm).......................3
3. Basal part of proboscis with scales (Fig. 1A) 4
– Basal part of proboscis without scales 5
4. Hind tibia with smooth dorsal vestiture of appressed scales (Fig. 1E). In HW vein Sc+R (the vein closest to the front margin) anastomosing with the vein behind. Labial palp usually porrect, never markedly sickle-shaped. Overall larger species (FWL 7–15 mm and FWB ≥ 2 mm), facies as Figs 14B–M .... Pyraloidea
– In HW vein Sc+R not anastomosing with the vein behind. If FWL ≥ 7 mm labial palp markedly sickle-shaped (Fig. 1A) *or* FWB ≤ 2 mm. Overall smaller species (FWL 4.5–12 mm), facies as Figs 12K–T .... Gelechioidea
5. Segment 2 of labial palp with lateral bristles, ocelli absent .... Tineidae
– Segment 2 of labial palp without lateral bristles, ocelli present.......................................6
6. Well-developed chaetosema present behind ocellus (Fig. 1B). Upper facial scales downwards directed, lower facial scales mediad-upwards directed (Fig. 1C). Rather broad-winged moths with facies as in Fig. **13** ....Tortricidae
– No such chaetosema present. Scales on entire anterior head surface ('face') downwards directed. Narrow-winged moths with facies as Figs 12G–J ....Plutellidae

LARVAE
See Figs 2A–D for seta terminology

1. Prothorax with 3 L–setae 2
– Prothorax with 2 L–setae .... Pyraloidea
2. Larva with secondary setae on body 3
– Larva with subprimary or primary setae only......4
3. Stemmata 3–5 (see Fig. 2C) close together forming a near-vertical line, stemmata 1–3 separated by the approximate width of a stemmata, stemma 6 clearly

separated caudally from other stemmata .... (Gelechioidea) *Scythris*
– Stemmata not arranged as above..................... .... (Pterophoridae) *Stenoptilia mengeli*
4. Abdomen with 3 pairs of pre-terminal prolegs on abdomen, mining on underside of *Ledum* leaves .... (Gracillaridae) *Phyllonorycter ledella*
– Abdomen with 4 pairs of pre-terminal prolegs on abdominal segments, not leaf mining..............5
5. Setae L1 and L2 well separated on body 6
– Setae L1 and L2 close together, often on same wart.. 7
6. Prolegs short, crochets form an ellipse ....Tineidae
– Prolegs long, crochets form a circle ....Plutellidae
7. D2–setae well separated on abdominal segment IX .... Gelechioidea (part)
– D2–setae on abdominal segment IX on same wart, or at least close together ....Tortricidae

TINEIDAE

Adult tineids are the only Greenland moths with a vestiture of erect piliform scales not only on the dorsal but also on the anterior head surface; the labial palps are characteristic by having long erect bristles projecting from the scale covering of segment 2. The presence of 5-segmented maxillary palps, which may be as long (*Niditinea*) or longer (*Monopis*) as the labial palps are similarly unique, but in *Tineola* the regression of these palps (4-segmented, and much shorter than the labial palps) has paralelled that of higher moths.

Tineids are unusual Lepidoptera in that the vast majority have fungus- and/or detritus-feeding rather than herbivorous larvae. The few Greenland taxa all belong to the subfamily Tineinae in which the larvae feed on keratin or arthropod remains, a niche which is likely to have been invaded *via* fungivory (see, *e.g.*, Robinson & Nielsen 1993). They can be immediately identified by the combination of their feeding situation, trisetose prothoracic L group and well separated trunk setae L1 and L2.

Key to species

ADULTS

1. FW unicolourous, pale ochreous .... *Tineola bisseliella*
– FW blackish brown mottled with yellowish scales; in the middle a round, hyaline spot ....*Monopis laevigella*

LARVAE

1. L–group bisetose on abdominal segment IX; SV–setae near-horizontal on thoracic segments 2–3..... ....*Tineola biselliella*
– L–group trisetose abdominal segment IX; SV–setae more vertical on thoracic segments 2–3 2
2. SV–group unisetose on abdominal segment VIII; L1 and L2 more vertical and position of L1–seta behind spiracles on abdominal segments I–VI ....*Monopis laevigella*
– SV–group bisetose on abdominal segment VIII; L1 and L2 more horizontal and position of seta L1 ventrally from spiracle on abdominal segments I–VI .... [*Niditinea fuscella*]

[*Trichophaga tapetzella* (Linnaeus, 1758)
Syn.: *greenlandica* Anonymous, 1836

The description of *Tinea greenlandica* was based on specimens bred from a skin of the polar bear from Greenland kept in the Royal Society of Dublin, Ireland. We agree with Gaedike (2009: 86) in that the infestation probably occurred in the British Isles and not in Greenland as *T. tapetzella* have not been reported from that country].

***Tineola bisselliella* (Hummel, 1823) Fig. 12F**
FWL 5–7 mm, FWB 1.5–2 mm. FWs unicolourous except for darker brown costal base.

Head and FW colour varying from lighter to darker yellowish brown.

Larva white with pale brown head with darker margins; prothoracic plate hardly darker than rest of integument (UK: Pelham-Clinton 1985: 193).

BIOLOGY. Larva in a flimsy white silken gallery to which particles of food material adhere, producing much additional webbing. The species is particularly known as a pest on wool products ('The Common Clothes Moth'), but many other kinds of material are eaten, including fish and flour. Indoors there is a continuous succession of generations (Pelham-Clinton 1985: 193). In Greenland adults have been found in July and early August.

DISTRIBUTION. S, NW. Only known from Qaqortoq/Julianehåb (60°43'N) "in immense numbers" (Wolff 1964: 52) and 1 specimen from Upernavik (72°45'N). Probably not a permanent resident in Greenland.

Cosmopolitan. In temperate climates only as synanthropic.

[*Niditinea fuscella* (Linnaeus, 1758)
Wolff (1964: 52): *Niditinea* (*Tinea*) *fuscipunctella* Haw.

This cosmopolitan indoors pest is known from one specimen labelled "Øst Grønland Deichmann", probably collected around 1900 (Wolff 1964: 52). As it has not been found again, it seems that indoors populations in Greenland cannot be sustained, hence it is not here considered as pertaining to "the Greenland fauna"].

Monopis laevigella **(Denis & Schiffermüller, 1775)**
Fig. 12E
Wolff 1964: *Monopis rusticella* Hb.

FWL 6–10 mm, FWB 1.5–2.5 mm; the considerable size variation being at least partly dependent of the larval food source. FW with variable suffusion of yellowish scales and intensity of violet tinge. Some specimens with small, yellowish white tornal spot.

Larva whitish; head brownish with darker margins; T1 plate yellowish, divided medially (UK: Pelham-Clinton 1985: 187).

BIOLOGY. The larva is known to make silken tunnels and webbing on a great variety of animal material, including debris in birds nest, dead birds and animals; it also feeds on various stored products (Pelham-Clinton 1985: 187). In Europe it is mostly found outdoors or in unheated buildings. Three of the four specimens from Greenland were found dead indoors in empty buildings (Wolff 1964: 7–8, 52). Adults have been found in August.

DISTRIBUTION. S. Only known from one specimen from Sermilik/Bredefjord (60°55'N), 1948 and the above mentioned three specimens from Narsarsuaq (61°10'N), 1962. Probably not a permanent resident in Greenland.

Holarctic. Widespread in eastern Canada (J.-F. Landry, *in litt.*), less common in the USA (D.R. Davis, *in litt.*). Common in Europe, with more scattered distribution in Asia.

GRACILLARIIDAE

The most species rich family of leaf mining Lepidoptera. The adult moths have pronouncedly narrow wings (hindwings extreme) with long cilia, a configuration which among Greenland moths is otherwise most closely approached in (clearly larger) members of Gelechioidea-Coleophoridae and Scythrididae. The single Greenland representative belongs in the subfamily Lithocolletinae which have sap-feeding early larval instars (as other gracillariids) with markedly modified mouthparts and lacking legs and prolegs; later instars remain miners, but are tissue feeding and have a more generalized caterpillar structure, though prolegs are absent from AVI. Pupation takes place in the mine.

Phyllonorycter ledella **(Walsingham, 1889)**
Figs 12A–B
Wolff 1964: not recorded.
Wolff 1971a: *Lithocolletis junoniella* Zeller.

By far the smallest Greenland moth. FWL 2.5–3.5 mm, FWB 1 mm. FW colour variying from lighter to darker bright orange brown, worn specimens becoming lighter yellow-brown. Small specimens tend to have white FW markings reduced.

Early-instars larvae undescribed, adult larval head light brown, trunk yellowish, translucent (dark gut contents visible) (OK).

BIOLOGY. Larva upper-surface miner in *Ledum groenlandicum*. Mine somewhat inflated, in smaller leaves occupying entire leaf; light yellowish green, turning brown after the larva has pupated (OK). The larva presumably hibernates in the non-deciduous leaves. Johannesen suggested (1990: 3) that *P. ledella* may have a two-year cycle in Greenland. Adults fly from late July to early August and are best found by sweeping the host plant. The species seems to prefer sheltered places at low altitudes (OK; Johannesen 1990). The examined material from Greenland contains more than three times as many females as males.

DISTRIBUTION. S, SW. Widely distributed and locally common in southwestern Greenland from Tasiusaq Alleq (60°08'N) to Itilleq (63°33'N).

Nearctic. Eastern Canada and north-eastern and western USA (D.R. Davis & J.-F. Landry, *in litt.*).

NOTE. The Greenland *Phyllonorycter* was recorded (Wolff 1971a: 72, 78; Johannesen 1990: 3–4) as *P. junoniella* (Zeller, 1846), which is a Palaearctic species considered monophageous on *Vaccinium vitis-idaea* where it is a lower-surface miner. Targeted search by OK on *Vaccinium vitis-idaea* failed to disclose any mines on this plant, even where it occurs abundantly in the same SW Greenland areas where *P. ledella* feeds on *Ledum groenlandicum*. *P. ledella* is a Nearctic species (described from California), known to make upper-surface mines on *Ledum glandulosum* (Walsingham 1889: 80). It is overall similar to *P. junoniella* in FW markings and genitalia of both sexes, but slight differences in the male genitalia suggest that the two taxa are indeed not conspecific. According to a sketch of a paratype male genitalia made by D.R. Davis (*in litt.*) it differs slightly from the Greenland *Phyllonorycter*, but the latter differs even more from *P. junoniella*. Further study may prove the Greenland (and East Canadian) *Phyllonorycter* species to be distinct from *P. ledella*, but on the basis of present knowledge we consider a tentative assignment to the latter justified.

PLUTELLIDAE

The Greenland plutellids are narrow-winged moths with prominent ocelli. A scale tuft on the lower surface of labial palp segment 2 is prominent; it is paralleled in some non-Greenlandic gelechioids. The larvae have conspicuously long and slender abdominal prolegs on which the crochets form an unserial circle. Both Greenland species feed on Brassicaceae.

Key to species

ADULTS

1. Smaller species (FWL 6–7 mm). Scale tuft on segment 2 of labial palp almost as long as segment 3 ... .*Plutella xylostella*
– Larger species (FWL 8–11 mm). Scale tuft on segment 2 of labial palp markedly shorter than segment 3 ....*Rhigognostis senilella*

LARVAE

1. Distance between setae P1–P2 on head > P1–A2, distance D2–D2 > D2–SD1 on anal shield, number of SV–setae on abdominal segments VII–IX: 2–1–1 ....*Plutella xylostella*
– Setae not arranged as above ....*Rhigognostis senilella*

Plutella xylostella **(Linnaeus, 1758) Figs 12G–H**
Wolff 1964: *Plutella maculipennis* Curt.

FWL 6–7 mm, FWB 1.5 mm. Labial palp segment 2 with tuft of erect scales almost as long as segment 3, but the tuft may be lost in worn specimens. Head and thorax vestiture white to ochreous. FW pattern variable: costal 2/3 of forewing usually brown (darkest towards dorsum), dorsal third ochreous to white, delimited by angled line (hence the name 'diamond-back moth', referring to the appearance of the light dorsal wing areas when the two forewing margins are held against each others in repose); markings sometimes quite indistinct, especially in females.

Larva green with darker dorsal line and black setae; head pale brown speckled dark brown; prothoracic and anal plates whitish speckled brown (*e.g.*, Agassiz 1996: 99).

BIOLOGY. The larva feeds on *Brassica* and several other Brassicaceae (Robinson & Sattler 2001: 17). It has been bred from *Brassica oleracea* in South Greenland (both cultivated and imported) (Johannesen 1990: 3; specimen label in ZMUC). The larva feeds openly, usually on the underside of a leaf (occasionally on florets), sometimes in a slight web, making translucent patches on the leaf which gradual shrivels and turns brown (Agassiz 1996: 99). Adults have been found in nature (mostly in light traps) in second half of June into August; bred Greenland specimens emerged in August and September.

DISRIBUTION. **S, SW**, NE. Most specimens are from the southwesternmost part, but there are singletons from Jameson Land (70°40'N) and Ravns Fjord (68°30'N). During automatical light trapping in SW: Upernaviarsuk in the years 1983–91 the number of specimens varied from 0 in 1990 to 138 in 1983 (P. Nielsen, pers. comm.). The occurrence of *P. xylostella* in Greenland is most likely the result of migration, and only in S and SW it can perhaps sometimes survive the Greenland winter.

Cosmopolitan, including Iceland and Svalbard. One of the most widely distributed species of Lepidoptera.

Rhigognostis senilella **(Zetterstedt, 1839)* Figs 12I–J**
Wolff 1964: *Plutella senilella* Zett.

FWL 8–11 mm, FWB 2.2.5 mm, hence considerably larger than the preceding. Labial palp segment 2 with tuft of erect scales markedly shorter than segment 3. FW colour varying from light grey-brown to dark grey-brown, the latter type mostly found in East Greenland. Hibernated specimens often become more or less unicolorous grey-brown.

Larva 12–13 mm, broadest in middle, dull green; head yellowish green spreckled brownish; prothoracic shield with black dots which, together with the dark hairs, form a Gothic B (Staudinger 1857: 274–275).

BIOLOGY. The larva has been observed in Greenland (Sisimiut 28th July) crawling on a cliff between *Arabis alpina* and *Cochlearia officinalis*, both of which are eaten in captivity (OK); it has been bred from cultivated *Brassica oleracea* in SW Greenland. Several genera of Brassicaceae are recorded as hosts elsewhere (Agassiz 1996), in Iceland it has been found on *Arabis* sp., *Cardamine pratensis* and *Draba* sp. (Wolff 1971b: 22). It feeds in a flimsy web, usually under the leaves. Pupation takes place in a characteristic coarse-meshed cocoon. Adults fly from the end of July, hibernating until June next year, and can thus be found in all months of the year.

DISTRIBUTION. S, SW, NW, NE, SE. Found between Narsaq Qujalleq/Frederiksdal (60°N) and Eqi (69°45'N) in west and between Ammassalik (65°35'N) and Ella Ø (72°50'N) in east.

Holarctic. Known from coastal localities around the North Atlantic, mountains of northern, central and southern Europe, the Caucasus and Middle Asia to Japan (Kyrki 1989: 445); also reported from Alaska (J.-F. Landry, pers. comm.).

THE GELECHIOID FAMILIES

The following families (Elachistidae – Gelechiidae) belong to a vast and diverse micro-moth assemblage whose members like the pyraloids have prominent scale coverings on the basal part of the proboscis. They have long been considered to constitute a superfamily, Gelechioidea, the monophyly of which still remains supported.

The family-level classification follows that adopted in Nieukerken *et al.* (2011), which reflects advances due to large-scale analyses (see Kaila *et al.* 2011) since the publication of the *Handbook of Zoology* treatment (Hodges 1998).

Key to species

ADULTS

1. HW termen below apex concave .... 2 (Gelechiidae)
– HW termen below apex not concave 4
2. Scape of antenna with single bristle (high magnification needed!). Facies as Fig. 12Q .... *Bryotropha similis*
– Scape of antenna without bristle. Facies as Figs 12P, R–T 3
3. Costal and tornal spots whitish yellow (known only from West Greenland).. *Gnorimoschema vibei*
– Costal and tornal spots indistinct or missing (known only from East Greenland)... *Gnorimoschema valesiella*
4. Labial palps 2–3× eye diameter, porrect or slightly recurved..5
– Labial palps > 3× eye diameter, markedly recurved.. 7
5. Labial palps about 2× eye diameter. Antenna unicolorous black .... *Scythris noricella* (Scythrididae)
– Labial palps about 3× eye diameter. Antenna ringed brownish and white.. 6 (Coleophoridae: *Coleophora*)
6. Antennae ringed brown and white in basal half only. Male genitalia with tip of sacculus tapering. Female genitalia with sclerotisation of ductus bursae about as long as segment VIII........ *Coleophora alticolella*
– Antennae ringed brown and white for most of their length. Male genitalia with tip of sacculus awl-shaped. Female genitalia with sclerotisation of ductus bursae more than double the length of segment VIII .... *Coleophora glaucicolella*
7. Head and thorax white. Apex of HW pointed....... .... *Endrosis sarcitrella* (Oecophoridae)
– Head and thorax greyish or brownish. Apex of HW rounded 8
8. Base and basal part of costa lighter than rest of FW. In middle of wing 1–2 white dots, surrounded with black...................... *Agonopterix antennariella* (Elachistidae–Depressariinae)
– Base and basal part of costa not lighter than rest of FW. In middle of wing 3–4 black dots .... *Hofmannophila pseudospretella* (Oecophoridae)

LARVAE

1. Larva case-bearer .... 2 (*Coleophora*)
– Larva not case-bearer 3
2. Metathorax with 1 pair of brown, dorsal sclerites.... .... *Coleophora glaucicolella*
– Metathorax without dorsal sclerites.................. .... *Coleophora alticolella*
3. Larva with secondary setae .... *Scythris noricella*
– Larva without secondary setae 4
4. Pinaculum with setae L1 and L2 located anterior to spiracle on abdominal segments I–VI, setae vertically on pinaculum .... 5 (Oecophoridae)
– Pinaculum with setae L1 and L2 located ventrally of spiracle on abdominal segments I–VI (Fig. 2D), setae situated more horizontally on pinaculum..........6
5. With 4 or 5 stemmata per side....................... .... *Hoffmanophila pseudospretella*
– With only two stemmata per side .... *Endrosis sarcitrella*
6. Position of seta D1 anterior of line D2–SD1 on 9th abdominal segment.......................... .... *Agonopterix antennariella*
– Setae D1, D2 and SD1 vertically in line on 9. abdominal segment .... 7 (Gelechiidae)
7. Head and prothoracic shield dark-brown, trunk brown. Moss-feeder .... *Bryotropha similis*
– Larva pale with darker pinaculae. Miners in low leaves, silken webbings extending from ground to mine entrances.....................................8
8. On Asteraceae (probably *Erigeron*) .... *Gnorimoschema valesiella*
– On *Salix* and (occasionally) *Betula*.................. .... *Gnorimoschema vibei*

OECOPHORIDAE

Ocelli absent. Labial palp long, upcurved. Only represented in Greenland by 2 widespread synanthropic species; observations on their immatures in the country have not been recorded.

Endrosis sarcitrella **(Linnaeus, 1758) Fig. 12K**
FWL 6–9 mm, FWB 1.5–2 mm. White-coloured head and thorax (with tegulae) very characteristic, labial palp with brown bands. FW brownish-greyish, mottled with black and white scales. Specimens vary from light to dark depending on the abundance of blackish grey scales in the FW.

Larval head and T1 plate brown. Trunk whitish with pinacula and anal plate coloured like adjacent skin (Harper *et al.* 2002: 89).

BIOLOGY. The larva feeds, in a silken gallery mixed with frass, on a variety of dead animal and vegetable matter including cereals, fruits, dried skin, flesh and wool; it has been found in birds' nests (North America, Europe: Hodges 1974: 132; Harper *et al.* 2002: 89). Adults are found from June to September.

DISTRIBUTION. S, SW, NW. Known from a number of western Greenland localities between 60°43'N (Qaqortoq/Julianehåb) and 69°45'N (Eqi) (Wolff 1964: 50). There are no records since 1950.

Cosmopolitan, in cooler areas only indoors.

NOTE. Wolf (1964: 50) (cited by Hodges 1974: 132) gave the time of occurrence as "June-October", but was probably misled by some specimens bearing an extra label "Meldorf / 6/10.1912", which is the date they were accessed in the ZMUC.

Hofmannophila pseudospretella **(Stainton, 1849) Fig. 12N**
FWL 8–12 mm, FWB 2.5–3.5 mm. Head, thorax and tegulae brown. FW brown with many black scales; 3–4 black dots in middle of wing.

Larva greyish cream with reddish brown head and ochreous thoracic plate (Harper *et al.* 2002: 88).

BIOLOGY. The larva feeds on a wide variety of dead animal and plant material including stored plants, seeds, grain, dried skins, dead insects, decaying mammalian carcasses and refuse in birds' nests (N. America, Europe: Hodges 1974: 129; Harper *et al.* 2002: 88). In Greenland adults have been found from September to November, probably always indoors.

DISTRIBUTION. S. Known from 3 specimens found in Qaqortoq/Julianehåb 1898–1899 and more specimens (dead and alive) from Arsuk (61°13'N) 1969 by Johannesen (1990: 3), who stated that the species seemed to be able to upheld an indoors population in Greenland.

Cosmopolitan, in cooler areas only breeding indoors.

NOTE. Lack (1934: 604) reported a specimen found onboard a vessel in NE: Scoresby Sund, stating that it "was clearly introduced", which is probably correct.

ELACHISTIDAE–DEPRESSARIINAE
Ocelli present. Labial palp long, recurved. *Agonopterix* and its relatives were formerly included in the Oecophoridae and subsequently in a family-group taxon of their own (Depressariidae). In the arrangement followed here the latter has status as a subfamily in the Elachistidae (centred on the species-rich, mainly grass-/sedge-feeding *Elachista* group of genera, of which no Greenland representatives have so far been found), but this placement may not prove tenable; it is not supported in the recent analysis by Kaila *et al.* (2011).

[*Agonopterix heracliana* (Linnaeus, 1758)
A specimens of this common Europaean moth was found indoors during Christmas 1980 in Nanortalik (Johannesen 1990: 3). It is undoubtedly introduced, probably from Denmark. The adult hibernates and is easily transported with goods, but its larva does not feed on stored products. *A. heracliana* superficially resemble *A. antennariella* (see below) and may be mistaken for that species.]

Agonopterix antennariella **Clarke, 1941 Figs 12L–M**
Wolff 1964: *Agonopteryx victori* de Lesse & Viette.*

FWL 7–10 mm, FWB 2.5–3 mm. Labial palp recurved. Outer margin of FW rounded. Extent of greyish white FW scaling variable, specimens with few of such scales look dark reddish grey-brown. Hibernated specimens are normally much lighter.

The larval head is brown; thoracic shield blackish brown, darkest in posterior part, with a light stripe in the middle; trunk dirty green with scattered, black hairs (OK).

BIOLOGY. Larvae feed from end of June and in July on *Antennaria canescens* and *Gnaphalium norvegicum*. First they spin a chamber by folding a leaf together along the mid-rib; later more leaves are included, or a new dwelling constructed. Only the green parts of the leaves are eaten; the whitish grey felt remains, often mixed with excrements, thereby revealing the presence of the larva. Sometimes leaves are also tied together along the stem. Pupation takes place in the dwelling or, more often, in a cocoon in the ground. *A. antennariella* is most often found in dry, sheltered places among rocks (OK). De Lesse & Viette (1949: 89) recorded the larva from *Antennaria alpina*. Adults emerge in August, hibernating until next spring or early next summer.

DISTRIBUTION. S, SW, NW. Widely distributed between 60°N (Narsaq) and 69°45'N (Eqi), but normally in low numbers.

Nearctic. Only reported from a few places in British Columbia and Washington (Hodges 1974: 40).

NOTE. *Agonopteryx victori* de Lesse & Viette, 1949 was described on material collected in West Greenland. It was synonymised with *A. antennariella* by Wolff (1971a: 78).

COLEOPHORIDAE

These pronouncedly slender-winged moths, when alive, are (like in the Plutellidae) recognizable by the posture of the antennae which are held close together and forwards pointing; preserved moths are distinguishable by having paired scale-free patches of stout spiny setae on most abdominal terga. The labial palps, about 3 times eye diameter, are slender, porrect or weakly upcurving.

The first larval instars of the Greenland species feed inside the seeds of *Juncus*; later instars are case bearers and easily recognised as such. The portable case is made from plant material and/or silk and may incorporate frass or other material as well. The 2 resident Greenland species can be separated on the basis of thoracic sclerites (see key).

Both Greenland coleophorids have only been recorded from the country in recent decades, and their distribution is quite restricted. This begets the question whether their presence in Greenland is due to man-mediated introduction, an assumption which appears particularly straightforward in the case of *C. alticolella* which is only recorded from the immediate vicinity of the Narsarsuak airport.

Coleophora glaucicolella Wood, 1892 Fig. 12C

Wolff 1964: not recorded.

FWL 5–7 mm, FWB 0.5–1 mm. Antenna ringed dark brown and white along entire length. FW pale buff to light greyish brown, with some suffusion of white scales particularly along veins; fresh specimens are light brownish, worn specimens greyish. Tip of abdomen with yellowish grey hairs. The size variation in the available material perhaps just reflects differences between wild-caught and bred specimens.

Larva similar to that of *C. alticolella* (see below), but differs as follows: metathorax with 1 pair of brown dorsal sclerites and with lateral sclerites nearly as large as those on prothorax; body without black spot on posterior border of segment IX. The larva constructs a case of whitish silk. When fully formed it is 5–6 mm, subcylindrical, pale buff or greyish, with darker granular material (UK: Emmet *et al.* 1996: 324; Wood 1892).

BIOLOGY. The larva feeds from its case on seeds of *Luzula spicata* (Johannesen 1990: 5) and pupates in the case. Outside Greenland *C. glaucicolela* seems to prefer *Juncus* species as host plants.

DISTRIBUTION. S. Only found in the Narsaq/Narsarsuak area (60°57' to 61°25'N), first finding in 1976 [the specimen in question, deposited in ZMUC, is mislabelled "18/7–1977"] (Koponen 1981: 28).

Holarctic. Scattered localities in North America and Asia. Widely distributed and common in Europe (Koponen 1981: 28).

Coleophora alticolella Zeller, 1849 Fig. 12D

Wolff 1964: not recorded.

FWL 4.5–5.5 mm, FWB 0.5 mm. Basal half of antenna ringed brown and white. Hairs on abdominal tip lighter yellow than in *C. glaucicolella*. However, even newly emerged specimens of *C. alticolella* are sometimes hardly distinguishable from *C. glaucicolella* on external features. There are distinct differences in the genitalia between the two species, see the key.

Larval head and prothoracic plate blackish brown; mesothorax with 2 pairs of dark brown sclerites, anterior pair lateral to the posterior; metathorax without dorsal sclerites; all thoracic segments with dark brown lateral sclerites, metathoracic only one-quarter size of mesothoracic; anal plate chestnut-brown; sometimes a small, dull blackish spot dorsally on posterior margin of segment IX. The larva constructs a case which is indistinguishable from that of *C. glaucicolella* (UK: Emmet *et al.* 1996: 326).

BIOLOGY. The larva feeds from its case on seeds of *Juncus arcticus* (Johannesen 1990: 5) and pupates in the case.

DISTRIBUTION. S. Only found near the airport of Narsarsuak (61°10'N), and first found in 1983 (Johannesen 1990: 5).

Holarctic. Canada, USA, Iceland, most of Europe, Iran, Siberia (Emmet *et al.* 1996: 327; Sinev 2008: 76; Wolff 1971b: 27).

SCYTHRIDIDAE

Head with appressed scales. Larvae unusual in having secondary setae, especially on the lateral, subventral and ventral trunk areas, and are moreover characterized by having stemmata 3 and 4 near-contiguous (Stehr 1987). Setae SD1 on AI–VII is surrounded by an elaborate sclerotised ring. Prolegs somewhat telescopic, with crochets in a circle. *Scythris* is devoid of the external traits that immediately characterize other slender-winged Greenland gelechioid moths.

Scythris noricella (Zeller, 1843) Figs 12O–P

Syn.: *latifoliella* Wolff, 1964.*

FWL 6–8 mm, FWB 1.5 mm, males are on average larger than females. Labial palp short, about 2 times length of eye diameter, slender, porrect or weakly upcurved. Head dark grey. FW with 3–4 black spots or streaks. Moths may look black or whitish grey, depending on the abundance of white scales in the FWs; worn specimens tend to be pale grey. Male genital segment unusually long.

Larval head black with 2 oblique, light, longitudinal lines and 2 light brown patches; thoracic shield black with anterior margin and medial stripe bone-white; abdomen reddish brown, dorsally lighter, ventral side and dorsal stripe light yellow; each side of abdomen with 4 rows with black, hairy pinacula (OK; de Lesse & Viette 1949: 86: abdominal chaetotaxy).

BIOLOGY. The larva feeds until the middle of July on *Chamerion latifolium*, eating leaves and flowers from a loose web (OK). The habitat is stony places, especially along streams, where the host plant grows. The adult flies from middle of July into September.

DISTRIBUTION. S, SW, NW, SE. Found between Pamialluk (60°05'N) and Kangiusap Qinngua, Svartenhuk (71°46'N), and at Saqqisikuik/Skjoldungen: Qoornoq (62°30'N).

Holarctic. Northwestern part of North America, with single localities in Oregon, Colorado and Québec; Kamchatka and northern Siberia; northern Scandinavia and mountains of central Europe (Landry 1991: 336 (map); Sinev 2008: 64).

NOTE. The Greenland population was described as a distinct subspecies *latifoliella* Wolff, 1964; it differs mainly from the nominotypical European subspecies *noricella* Zeller, 1843 in being smaller.

GELECHIIDAE

Gelechiids are slender-winged moths with a characteristic hindwing outline: the termen is markedly concave and the apical area correspondingly narrowed. The palps are recurved.

The larvae of Greenland Gelechiidae are most similar to those of Oecophoridae and *Agonopterix*, but differ in the positions of dorsal and subdorsal setae on abdomen IX (Stehr 1987); they feed either on mosses (*Bryotropha*) or on vascular plants (*Gnorimoschema*).

Bryotropha similis (Stainton, 1854) Fig. 12Q

FWL 3.5–5.5 mm, FWB 1–1.5 mm. Antennal scape with single bristle (pecten). FW glossy, varying from rather light with 2–4 distinct dark spots to blackish without distinctive markings, sometimes mottled with white scales; costal and tornal patches vary from white to ochreous and can be rather prominent, often fused to an outwards convex fascia; subapical area with many black scales.

Larval head and T1 shield dark brown, trunk brown (UK: Meyrick 1928).

BIOLOGY. Larva on mosses, has been found (as have pupae) under moss on walls as well as on boulders and stones on the ground (Karsholt & Rutten 2005: 142). Adults fly from late June to first part of August.

DISTRIBUTION. S, SW (widely distributed and common north to 70°N), NE (Mestersvig, 72°10'N), SE (Saqqisikuik/Skjoldungen, 62°30'N. Note: the dot in N in the map by Rutten & Karsholt (2004: 41) is misplaced.

Holarctic. Widely distributed throughout large parts of the Nearctic and Palaearctic Regions (Karsholt & Rutten 2005: map fig. 422; Rutten & Karsholt 2004: map fig. 77). Also in Iceland (Wolff 1971b).

NOTE. Greenland *B. similis* are on average smaller and tend to have the FW more mottled with white scales than specimens from other parts of the distribution range.

Gnorimoschema valesiella (Staudinger, 1877) Fig. 12R

Wolff 1964: *Phthorimaea valesiella* Staud.

FWL 7–8 mm, FWB 2 mm. Antennal scape without even a single bristle. FW blackish brown with 2–4 indistinct black spots, surrounded by some yellow-orange scales and more or less distinct white costal and tornal spots.

Larva undescribed.

BIOLOGY. The host plant in Greenland is unknown. In Finland the larva feeds on *Solidago virgaurea* and *Erigeron* (only the latter occurs in Greenland). It mines the lower leaves from a silken gallery extending from the lower leaf surface to below ground level (Kaitila 1996: 103; Huemer & Karsholt 2010: 51). Adults have been found in July and August.

DISTRIBUTION. NE. Known from a dozen of specimens collected between 69°24'N (Kap Dalton) and 72°10'N (Mestersvig).

Palaearctic (see below). Iceland, northern Europe, mountains of central and southwest Europe, Caucasus, Siberia (Povolný 2002: 26; Wolff 1971b).

NOTE. Povolný (1967, 1998) recorded an Alaskan *Gnorimoschema* species-group taxon *alaskense* (Povolný, 1967) as a subspecies of *valesiella*. However, based on characters in the genitalia these taxa are here considered to be separate species, and the Greenland moths to be conspecific with the Palaearctic *valesiella* rather than with *Gnorimoschema alaskense* (Huemer & Karsholt 2010: 51).

***Gnoriomoschema vibei* (Wolff, 1964)* Figs 3, 12S–T**
Wolff 1964: *Phthorimaea vibei* n.sp.

FWL 5–6.5 mm, FWB 1–1.5 mm. Antennal scape as in *G. valesiella* without bristles. FW colour varying from almost black to light grey, depending on the suffusion of whitish grey scales; occasionally these light scales have a yellowish lustre, and such specimens have a light yellowish brown appearance. Dark central spots surrounded by yellow-orange scales; light costal and tornal spots may be confluent, forming a transverse band.

Larva (Fig. 3A) with head light brown. Thoracic shield light greenish brown, with darker, waved markings along anterior and posterior margins. Trunk light green, dorsally darker, and with small black pinacula; anal shield light greenish brown (OK).

FIGURE 3 Gnoriomoschema vibei. *A, sand-encrusted silken tube extending from mined* Salix *leaf, and larva removed from tube. B, several larval tubes (almost certainly made by one single individual) extending from mined leaf.*
(PHOTOS: A. FJELLBERG)

BIOLOGY. Details recorded here for the first time (OK). The larva feeds on *Salix glauca* (exceptionally also on *Betula nana*) growing in white sand dunes; it mines leaves which lie on the sand, starting at the leaf base and forming a blotch-mine. From the mine entrance one or more silken tubes lead down into the sand (Fig. 3), permitting rapid escape. The larva changes leaf several times. Larvae of different age found between July 22th and 29th pupated in Denmark in early August; unexpectedly adults emerged already from mid-August to early September. The few (five) previously known specimens (all in bad condition) were collected as adults in late June and early July. The species probably hibernates as an adult.

DISTRIBUTION. SW. Known only from a few localities in western Greenland around 67°N, but should be searched for in other places with white sand dunes.

Until recently known only from Greenland, but some *Gnorimoschema* specimens collected in N Quebec (ca 55° N) have very recently proved to belong to this species (Adamski *et al.* 2014)

[*Phthorimaea operculella* (Zeller, 1873)
An adult specimen of the Potato Tuberworm was found indoors in Nanortalik in August 1981 (Johannesen 1990: 5). It has undoubtedly been introduced, most likely with potatoes from Europe.]

TORTRICOIDEA-TORTRICIDAE
One of the most species rich lepidopteran families in the Northern Hemisphere, but represented in Greenland by only 5 native species. Well preserved adult tortricids are perhaps most easily recognized by the characteristic facial scale vestiture, with the lower scales being directed upwards (Fig. 1C) contrary to the upper scales that are directed downwards (as are all facial scales in most Lepidoptera). The moths have chaetosemata and ocelli (Fig. 1B), but no proboscis scales, and in the labial palp segment 3 is markedly smaller than segment 2; there are no abdominal tympanic organs or tergal spines.

The larvae of Greenland Tortricidae are easily recognised by having the D2 setae on AIX placed close together, often on the same pinaculum.

Besides the species listed below, Johannesen (1990: 6) reported a specimen of *Acleris laterana* (Fabricius, 1794) from Paamiut/Frederikshåb 1970 "in leaves of imported apples". No exact date was given, nor a statement whether the finding was of a larva, pupa or adult. The specimen could not be located in the ZMUC where Johannesen's collection of Lepidoptera from Greenland is kept. If correctly identified this species is likely introduced from Europe, but since its identity cannot be verified no formal treatment is given here.

Key to species

ADULTS

1. FW markedly curved near base, hence wing appearing near-rectangular (Figs 13A–J). In HW basal part of vein CuA (posterior frame of 'cell') without row of close-set long hair-scales. Antennae with 2 scale whorls per flagellum segment (Fig. 4A). Hibernating as adults .... 2 (Tortricinae)

– FW more triangular (Figs 13K–Q). In HW basal part of vein CuA with row of close-set long hair-scales (Fig. 4C). Antennae with a single scale whorl per flagellum segment (Fig. 4B). Hibernating as immatures .... 3 (Olethreutinae)

2. Smaller species (FWL 5–8 mm). Termen of FW slightly emarginate below apex. FW without small spot of erect scales at 1/5 from base near dorsum.... .... *Acleris caryosphena*

– Larger species (FWL 7–11 mm). Apex of FW appears rounded, termen not emarginate. FW with small spot of erect scales at 1/5 from base near dorsum.... .... *Acleris arcticana*

3. Larger species (FWL 8–14 mm), facies as Figs 13K–L. FW ground colour orange/reddish-brown .... *Phiaris inquietana*

– Smaller species (FWL 7–8 mm). FW ground colour not reddish-brown 4

4. FW in male with costal fold. Only known from southern/western Greenland .... *Epinotia indecorana*

– FW in male without costal fold. Only known from northern/eastern Greenland......................... .... *Argyroploce mengelana*

LARVAE

1. Feeding on herbaceous plants....................... 2

– Feeding on shrubs (*Betula, Salix*).................... 3

2. On *Pedicularis*.................... *Phiaris inquietana*

– On *Dryas* .... *Argyroploce mengelana*

3. On *Salix* .... *Acleris arcticana*

– On *Betula*... 4

4. Pinacula of same colour as adjacent greenish trunk surface......................... *Acleris caryosphena*

– Pinacula dark.................. *Epinotia indecorana*

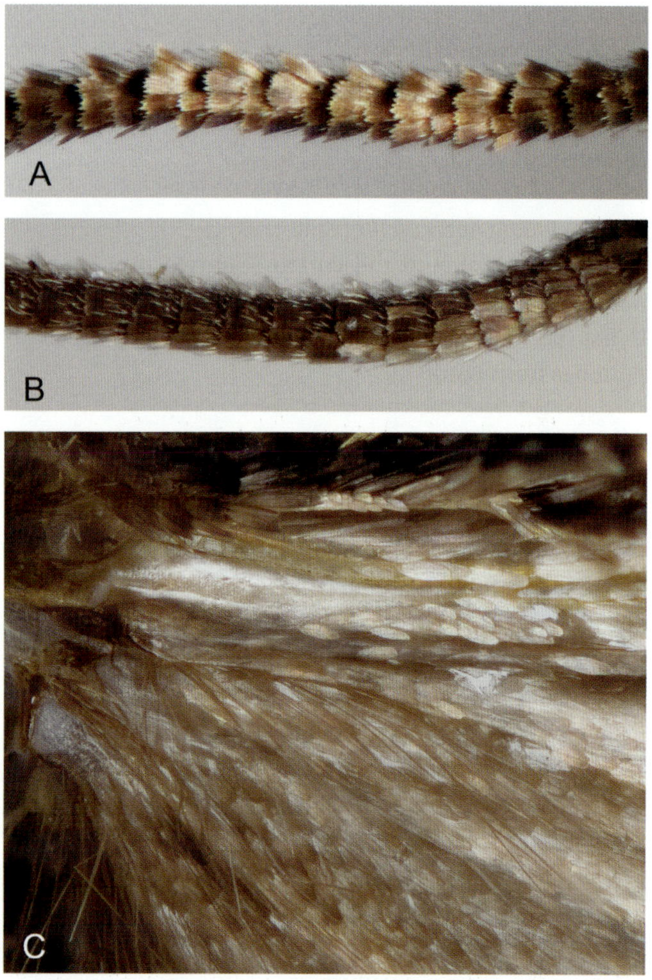

FIGURE 4 *Tortricid characters. A, detail of tortricine antenna showing two scale whorls per flagellomere. B, same of olethreutine antenna, showing single scale whorl per flagellomere. C, olethreutine hindwing base showing 'cubital pecten'.*

TORTRICINAE

Moths characterized by having two scale whorls per antennal flagellum segment and no hindwing 'cubital pecten'. The two Greenland representatives both belong to the genus *Acleris* and like many congeners they have almost rectangular forewings with remarkably variable patterns.

Acleris caryosphena **(Meyrick, 1937)* Figs 13A–D**

FWL 5–8 mm, FWB 2–3 mm. FW with slightly pointed apex and slightly emarginate termen; it may be unicolorous grey or brown, or may have a very variable pattern of brown or black markings. A particularly characteristic form has light grey FW with a reddish-brown tapering fascia extending dorsad from costa. Worn and hibernated specimens become more greyish or blackish brown.

A. caryosphena is superficially similar to small specimens of *A. arcticana*, but the FW of the latter has a more rounded apex and only rarely a dark triangular fascia/patch from costa. *A. caryosphena* has no erect scales on FW at 1/5 from base near dorsum.

Larval head and prothoracic plate black, the latter divided in middle. Thoracic legs black. Abdomen green, darkest on dorsal side and with a darker dorsal line; pinacula of same colour as adjacent integument, with hairs as long as half the body diameter; anal plate light green with darker spots (OK).

BIOLOGY. The larva feeds in June and first half of July on *Betula nana*, spinning leaves together two by two and fastening them to the twigs; leaves are skeletonized, ribs left untouched. The larva pupates in its dwelling (OK). Attacked leaves become brown; during mass occurrence large stands of *Betula nana* become reddish brown as in autumn (OK; Wolff 1964: 40). In S Greenland the species has also been bred from *Betula glandulosa* (Johannesen 1990: 5). It hibernates in the adult stage, flying August-September and again in May.

DISTRIBUTION. S, SW. Found between Ilua Fjord (60°10'N) and Qeqertarsuup Tunua/Disko Bugt: Kangersuneq (68°50'N).

General distribution uncertain due to unsettled taxonomic status (see below).

NOTES. *A. caryosphaena* is a close relative of the Palaearctic *A. notana* (Donovan, 1806) (=*ferrugana sensu* Treitschke, 1835, *nec* Denis & Schiffermüller, 1775), mainly differing in female genitalia characters (Wolff 1964). Razowski (1984: 265–266) concluded that *caryosphena* is a subspecies of *A. tripunctana* (=*A. notana*). Wolff (1964: 39) suggested that the *Acleris* recorded from North America under the name of *ferrugana* is in fact *caryosphena*. Obraztsov (1963: 220–223) treated the species in question under the praeoccupied name *A. tripunctana* (Hübner); his description of (external features of) North American specimens supports Wolff's suggestion. We have been unable to obtain North American material for examination, but like Wolff we find it likely that the North American population is referable to *caryosphena* rather than to *notana* (=*tripunctana*), and we follow his arguments for ascribing these taxa species-level distinctness. *A. caryosphaena* was recently erroneously listed as a synonym of *A. arcticana* (Guenée) (Brown *et al.* 2005: 40).

While the species description is based on two specimens collected among *Betula nana* and *Salix* at Søndre Strømfjord, Camp Lloyd, Meyrick (1937: 157) wrote: "Probably allied to *hastiana* and feeding on *Salix*". This likely led de Lesse & Viette (1949: 90–92) to the erroneous assumption that the *Peronea* (now *Acleris*) they bred from *Salix glauca* was *caryosphena* (Wolff 1964: 39). They added to the confusion by figuring the male genitalia of *Eucosma petalonota* Meyrick (now *Epinotia indecorana* (Zetterstedt)) as belonging to *A. caryosphena*.

Meyrick stated one of the two original specimens to be "broken in transit", hence Wolff considered the other to be the holotype. The statement that his illustration of the phallus of the latter (Wolff 1964, pl. 17, fig. 3) shows the "Holotype" constitutes a valid indirect lectotype designation of *Peronea caryosphena* Meyrick (ICZN 1999: art. 74.5).

Acleris arcticana (Guenée, 1845) Figs 13E–J
Wolff 1964: *Acleris boreana* n.sp.*

FWL 7–11 mm, FWB 2.5–4 mm. FW almost rectangular with slightly rounded apex; a small group of erect scales almost always present in the middle and near dorsum 1/5 from base. FW occasionally unicolorous grey, but often patterned with dark transverse or longitudinal fasciae on ground colours varying from light greyish to warm orange. It has only rarely a central triangular reddish patch/fascia extending from costa. Worn and hibernated specimens become more greyish or blackish brown. Females are generally smaller than males. See preceding species for discrimination.

Larva head light brown. Thoracic plate light brownish green with darker admixture. Abdomen light green with darker dorsal line; anal plate is concolorous with adjacent trunk surface (OK).

BIOLOGY. The larva feeds on *Salix glauca*, spinning leaves together; in late July it pupates in the dwelling (OK). *A. arcticana* hibernates in the adult stage, flying August-September and again from April to mid-June. The moth can be found flying among *Salix glauca* in the sunshine, and is also attracted to light. Hibernating adults can been found during mild winter days under bushes of *Betula* and *Salix* growing among rocks (Johannesen 1990: 6).

DISTRIBUTION. S, SW. Found between Ilua (59°55'N) and Ingia Fjord (71°52'N).

Holarctic. Canada (Razowski 1984: 239); Northern Europe (Fennoscandia); apparently not yet recorded from Asia.

NOTE. Razowski (1984: 238–239) stated that compared to 'typical' specimens, those from southern Greenland ("Fjord Ata, 69°45'N", which is nSW) have the FW unusually dark, and in females particularly narrow, acuminate from costal bending to apex, and with outer margin more emarginate; males are rather small, while females have relatively large abdomens. It was not stated whether other Greenland material was studied, but the description applies well to many of the >200 specimens we have examined from various Greenland localities.

[***Epiphyas postvittana*** **(Walker, 1863)**

An adult specimen of the Light Brown Apple Moth was found indoors in Thule in October 1974 (leg. H. Engelund). It has undoubtedly been introduced.]

OLETHREUTINAE

Moths characterized by having only a single scale whorl per antennal flagellum segment (Fig. 4B) and by the presence of a 'cubital pecten', *i.e.*, a row of close-set hair-scales on the basal part of the vein (CuA) that forms the posterior frame of the hindwing 'cell' (Fig. 4C).

Phiaris inquietana **(Walker, 1863) Figs 13K–L**

Wolff 1964: *Argyroploce inquietana* Wlk.

FWL 8–14 mm, FWB 4–6 mm. Stands out for the large size. The deep orange/reddish-brown colouration of the FWs varies in brilliance; in some specimens it is rather dull and the whole wing may be suffused with black scales so that the light markings become obscure. Rarely only the orange-brown areas have become dark, and in such specimens the white markings are very clear. Cilia sometimes (but not consistently) 'chequered', *i.e.*, groups of white hair-scales regularly alternate with fuscous ones. Size somewhat variable, females being normally smaller than males.

Larval head reddish brown to dark brown. Thoracic and anal shields, as well as thoracic legs brown; pinacula brownish; trunk colour undescribed (Canada: Mackay & Downes 1969: chaetotaxy).

BIOLOGY. The larva feeds on *Pedicularis capitata* and *P. hirsuta*. It has been found burrowing into the root (only in *P. hirsuta*) or in webs on ground bases of flowering shoots (as *P. capitata* is too small to be tunnelled by the larva). In Ellesmere Island it was found mining stems and roots also of *P. arctica* and *P. lanata*. Larvae of *P. inquietana* normally have a life cycle of 2 years (Mackay & Downes 1969). Flight season from late May to mid-August.

DISTRIBUTION. NW, N, NE. From Dundas (76°33'N) through Inglefield Land (79°N) to Kap Dalton (69°24'N).

Holarctic. Alaska, northern part of Canada, south to New England; northern Asia east of the Urals (Lafontaine & Wood 1997: 773; Wolff 1964: 40).

NOTE. The generic placement of *inquietana* is disputed. Here we follow Aarvik (2011) in placing it in the genus *Phiaris* Hübner; this is also in accordance with Wolff (1964). Brown *et al.* (2005: 438) included *Phiaris* in a larger concept of *Olethreutes* Hübner, 1822.

Argyroploce mengelana **(Fernald, 1894) Figs 13M–N**

Wolff 1964: *Olethreutes* (*Argyroploce*) *mengelana* Fern.

FWL 7–8 mm, FWB 2.5–3 mm. Forewing black, mottled with brown and lead-grey, and with 2 broad, white, irregular, transverse fasciae. Inner white fascia sometimes visible only near costa. Pale specimens sometimes with several white, costal marks.

Larval head yellowish or reddish brown with blackish ocellar area, lateral bar on cheek and most of frontal area; thoracic shield brown; thoracic legs and anal shield pale brown; pinacula brown on prothorax, paler than body colour on other segments; spinules dense, with dark bases, giving the body a dark appearance; body colour unrecorded; maximum length 8 mm (Mackay & Downes 1969: chaetotaxy of Canadian larvae; identity of studied material not confirmed by rearing, Miller & Jalava 2000: 49).

BIOLOGY. The host plant in Canada is *Dryas integrifolia*. Larvae of *A. mengelana* probably have a life cycle of 2 years (Mackay & Downes 1969). Adults fly from early June to early August. They fly in dry, sunny places with *Cassiope* vegetation (Nielsen 1910: 67).

DISTRIBUTION. NW, N, NE. From Dundas (76°35'N) through Børglum Elv (82°34'N) in Peary Land to Scoresby Sund: Hekla Havn (70°27'N).

Holarctic. Arctic Canada, Alaska (Miller & Jalava 2000: 49). Taymyr Peninsula (Kozlov *et al.* 2006: 140) and Novaya Zemlya (Aarvik 2011) in northernmost Russia.

NOTE. The generic placement of *mengelana* is disputed. Here we follow Aarvik (2011) in placing it in the genus *Argyroploce* Hübner; this is also in accordance with Wolff (1964). Brown *et al.* (2005: 442) included *mengelana* in *Olethreutes* Hübner, 1822 and misspelt the synonym *groenlandicana* Bang-Haas, 1896* as *groenlandica*.

Epinotia indecorana **(Zetterstedt, 1839)* Figs 13O–Q**

FWL 6–8 mm, FWB 2–3 mm. FW in male with costal fold. A variable species, with females generally more variegated than males. Worn specimens in particular may be almost uniformly grey-brown, whereas newly emerged specimens may be very contrast-rich, with black borders between the light greyish/whitish and dark brown areas. When exposed to light for some time the colour becomes light brown (Wolff 1964: 43).

Larva dark green with dark head, shields, thoracic legs and pinacula (Scandinavia: Benander 1950: 138).

BIOLOGY. Larva in June and July between spun leaves of *Betula nana* (OK) and *B. glandulosa* (Johannesen 1990: 6). Flight season from late July to early September.

DISTRIBUTION. S, SW, NW. Found between Pamialluk (60°05'N) and Eqi (69°45'N).

Holarctic. Canada: British Columbia, Northwest Territories, Yukon Territory (R. Brown, *in litt.*) and Quebec (J. Kullberg, *in litt.*). Scandinavia through northern part of Russia to southern Siberia and Mongolia (Kuznetzov 1989: 696).

NOTE. This species was described as *Teras indecorana* Zetterstedt, 1839, based on one specimen from Norway and one from Greenland. Wolff (1964: 43) designated the latter as lectotype, thereby fixing Greenland as type locality of *indecorana*, and he synonymised *Eucosma petalonota* Meyrick, 1937* with it. *E. indecorana* has genitalia similar to those of *E. trigonella* (Linnaeus, 1758), though R. Brown (*in litt.*) reported small differences, and is often regarded a synonym/northern subspecies of the latter, whose nominotypical form occurs throughout the Palaearctic Region and in North America (Kuznetzov 1989: 693). An inquiry into the conspecificity or otherwise of *indecorana* and *trigonella* is outside the scope of the present treatment.

PTEROPHORIDAE (PLUME MOTHS)

The vast majority of the pterophorids including the single Greenland species are characterized by having the apical parts of the narrow and elongate wings divided: the FW into 2 lobes, and the hindwing into 3 lobes (longer and more slender than the FW lobes).

The larvae of Pterophoridae have secondary setae on the body and are most easily confused with the larvae of Pyralidae or Scythrididae. They are separable from Pyralidae by the presence of 3 L–setae on the prothorax and from Scythrididae by a different stemmata arrangement (Stehr 1987).

***Stenoptilia mengeli* Fernald, 1898 Fig. 14A**

Wolff (1964: 36): *Stenoptilia islandica* Staud.

FWL 8–10 mm, FWB 2–2.5 mm. This species cannot be confused with any other Greenland moth. Variation is slight: in some specimens the black streak at the lower FW lobe is missing.

Larva undescribed.

BIOLOGY. Host plant unknown, but suggested to be *Saxifraga* spp. (E. Laasonen, *in litt.*). Flight season from July to first part of August, moths being especially active in the evening sun (E. Laasonen, *in litt.*). Specimens have been collected from sea level up to 1280 m (at Celsiusberget) in NE Greenland (Barca 1937: 33).

DISTRIBUTION. NW, N, NE. Found from Foulke Fjord (78°15'N) in north-west through North Heilprin Land (82°N) in north to Sabine Ø (74°40'N) and Scoresby Sund: Gåseland (70°15'N) in north-east. Apparently a rare species.

Nearctic. Artic Canada (Arenberger 2005: 82; Wolff 1971a: 76).

NOTE. The *Stenoptilia* species occurring in Greenland was in earlier literature referred to as *S. islandicus* (Staudinger, 1857), a species occurring in Iceland and cooler parts of Europe. Wolff (1971a) proved that the Greenland species is the Nearctic *S. mengeli*, which differs from *S. islandicus* in details of the male and female genitalia. Arenberger's recent record (2005) of both of the said *Stenoptilia* species from Greenland is most likely due to a misunderstanding. The material listed and the specimens used for illustrating *S. mengeli* by Arenberger came from ZMUC, but no illustrations (adults or genitalia) of *S. islandicus* from Greenland were given, and the Greenland localities (Landingsdalen, Celsiusberget and Holmvika) mentioned are those published by Barca (1937: 33) as referring to *S. islandicus* before Wolff's clarification of the species identity.

THE PYRALOID FAMILIES

The two following families together clearly constitute a monophylum, and there is no universal agreement on the appropriateness of their separation (adopted here in accordance with Munroe & Solis 1998). The pyraloid moths which constitute one of the largest (if not *the* largest) family-group taxa in the Lepidoptera are primarily characterized by the presence of a distinctive type of tympanic (hearing) organ, which like that of the Geometridae is situated in the lower region of abdominal segments I+II, and hence does not lend itself easily to observation (particularly not in dried collection specimens). Pyraloids like gelechioids have scales on the basal part of the proboscis, but they never have narrow hindwings with 'fringe' length exceeding wing breadth, or prominently upturned, sickle-shaped labial palps

Pyraloid larvae are generally near-cylindrical and concealed feeders, but occupy very diverse niches. They are easily identifiable by the combined presence of prolegs with 'micro-moth-type' crochets (in circle/modified circle) and of only 2 L–setae on the prothorax.

Key to adults

1. Smaller species, FW length < 9 mm. Indoors synanthropes 2
– Larger species, FW length ≥ 10 mm. Synanthropes, or occurring in natural habitats 3
2. FW with basal half pale, distal half orange-brown .... *Plodia interpunctella*
– FW ground colour greyish-brownish throughout .... *Ephestia elutella*
3. In HW basal part of vein CuA (posterior frame of 'cell') with row of close-set long hair-scales 4
– In HW basal part of vein CuA without row of close-set long hair-scales 6

4. FW without transverse fasciae and distinct markings. Female frenulum of 2–more bristles.. *Tehama bonifatella*
– FW with transverse fasciae and distinct markings. Female frenulum a single bristle.................. 5
5. HW dark brownish grey, with veins CuA1, M3 and M2 forming a 'double Y'; male antenna with basal segments excavated, forming a 'sinus' covered by blackish brown scale tufts; in outdoors habitats .... *Pyla fusca*
– HW whitish, vein M2 absent; male antenna simple; indoors synanthrope............ *Ephestia kuehniella*
6. Large species with narrow FW (FW ≥ 14 mm, L/W ratio ≈ 2.9). Facies as Fig. **14M** .... *Nomophila nearctica*
– Usually smaller; if FW > 12 mm its L/W ratio < 2.7 7
7. FW markedly triangular. HW light grey, underside with indistinct 'central spot'. Head without 'chaetosema'.............................. *Udea torvalis*
– FW less markedly triangular. HW dark grey, occasionally lighter towards base, underside with distinct 'central spot'. Head with 'chaetosema'.. *Gesneria centuriella*

Larvae, family distinction:
Pyralidae: Abdominal segment VIII with sclerotized ring around base of seta SD1. All indoor synanthropes belong here.
Crambidae*:* Abdominal segment VIII without sclerotized ring around base of seta SD1.

PYRALIDAE

Devoid of a scaled integumental fold in front of the tympanic organs. Female frenulum consisting of a single bristle as the male counterpart.

The larvae of the indoors pyralid synanthropes are well-known cosmopolitan pests on a wide variety of stored plant products (Hasenfuss 1960; Neunzig 1987).

Key to larvae

1. Pinacula of dorsal and lateral setae on abdominal segments I–IX darkened, distinct (usually most so on posterior segments) 2
– Pinacula of dorsal and lateral setae on abdominal segments I–IX unpigmented, not easily identifiable. .... *Plodia interpunctella*
2. Spiracle on abdominal segment VIII at least as large as area enclosed by the ring around seta D1......... .... *Ephestia kuehniella*
– Spiracle on abdominal segment VIII distinctly smaller than area enclosed by the ring around seta D1.................................. *Ephestia elutella*

Pyla fusca **(Haworth, 1811) Figs 14B–C**

FWL 10–14 mm, FWB 3.5–5 mm. Labial palp 2–3 times eye diameter, slightly upturned. Antenna of male with basal segments modified forming a mesial sinus covering blackish brown scale tufts; antenna in female unmodified. FW narrow with blunt rounded apex, varying from having contrasting light grey and black markings to being completely black. Specimens from East Greenland are generally lighter than those from West Greenland.

Larval head, T1 shield and dorsal line black; trunk blackish brown, more brownish between segments (northern Europe: Palm 1986: 58).

BIOLOGY. Neunzig's statement (2003: 148) that the larva probably feeds on *Vaccinium* in Greenland is conjectural, while the statement that it is unknown in North America is contradicted by Wilterding & Balogh's record (2002: 498) of it having been bred from "blueberry". In Europe it feeds in a silken web on shoots of *Erica*, *Calluna*, *Vaccinium myrtillus* and probably also *Salix* and *Betula* (Palm 1986; Wolff 1971b). Flight season from mid-June to late August. Adults have been observed up to 1280 m altitude in NE Greenland (Barca 1937: 31).

DISTRIBUTION. S, SW, NW, NE, SE. Distributed between Pamialluk (60°05'N) and Upernavik (72°45'N) in west and between Skjoldungen (62°30'N) and Godfred Hansens Ø (76°45'N) in east.

Holarctic. North America apart from northernmost Canada and southern USA, Palaearctic region from Iceland and the British Isles to Japan (Neunzig 2003, Wilterding & Balogh 2002: 497, Wolff 1971b). *P. fusca* is one of the few Lepidoptera species occurring in Svalbard (Coulson *et al.* 2003).

Plodia interpunctella **(Hübner, 1813) Fig. 14D**

FWL 7 mm, FWB 2 mm. Male antenna unmodified. Apically rounded FW appears 'bicoloured': basal half of FW pale, distal half orange-brown with fasciae and markings silvery grey.

Larval head light reddish brown; facial triangle high, extending 2/3 (or more) distance to epicranial notch. T1 shield brownish yellow. Trunk greyish white or yellowish white to pink (Hasenfuss 1960; Neunzig 1987, 1990).

BIOLOGY. A pest of stored products such as seeds, or milled seeds of grasses, dried fruits, chocolate, and many other materials (Neunzig 1990).

DISTRIBUTION. S, SW, NW. Only known from three specimens found in Nanortalik (1981), Aasiaat/ Egedesminde (1955), and Qullissat (1957), and three specimens from Thule (1976–79); also reported from larvae in walnut in Christianshåb, 1967.

Cosmopolitan. In temperate climates only as synantropes.

***Ephestia kuehniella* Zeller, 1879**

Wolff 1964: *Anagasta* (*Ephestia*) *kühniella* Zell.

FWL 11 mm, FWB 3.5 mm. Male antenna simple. FW slender, light grey mottled with black and light brown. HW shining white without yellow scale tufts near base. In HW veins CuA1 and M3 arising separately from cell.

Larval head and T1 shield yellowish brown to brown, the former with dark streaks posterolaterally; facial triangle varible, usually low as in following species. Trunk white or yellowish white to pink (Neunzig 1987, 1990).

BIOLOGY. An important pest on seeds and flour of cereals, and the larva also attacks other materials, notably nuts. The larva spins silken threads through the flour, making it look greyish (Neunzig 1990: 107; Palm 1986: 100).

DISTRIBUTION. S, SW. Only known from very few Greenland records. One specimen was found in a kitchen in Aasiaat/Egedesminde.

Cosmopolitan, probably originating from USA (Neunzig 1990: 109). In temperate climates only as synantropic.

NOTE. Henriksen (1939: 36) recorded the species from Qaqortoq/Julianehåb in S, but Wolff (1964: 34) showed that it was based on misidentification of *E. elutella*. A subsequent record by Vestergaard (1956: 6), also from Qaqortoq/Julianehåb could not be confirmed as no voucher specimens apparently have been kept.

***Ephestia elutella* (Hübner, 1796) Fig. 14E**

FWL 7–8 mm, FWB 2.3 mm. Somewhat similar to the preceding, but smaller. In HW veins CuA1 and M3 arising with common 'stalk' from cell. FW ground colour more brownish, mottled with black white-tipped scales; there are often also some brick red scales, and some specimens have the dorsal part of the wing brick red. Male hindwing grey with two yellow scale tufts near base.

Larval head pale brownish yellow to brown, slightly darker edged; facial triangle low, little more than half distance to epicranial notch. T1 shield brownish yellow. Body white or yellowish white to pink (Neunzig 1987, 1990).

BIOLOGY. Polyphageous on all kinds of stored materials, notably cacao and tobacco, but also on stored cereals; it also breeds in barns and ricks where it evidently feeds on dry grasses and seeds. Whereas it in Europe also lives out of doors that is not the case in North America (Goater 1986: 135; Neunzig 1990: 107).

DISTRIBUTION. S, SW. Only two specimens collected (Qaqortoq/Julianehåb, in house 1938, and Ameralik/ Lysefjord, 1950).

Cosmopolitan. In temperate climates partly synantropic.

CRAMBIDAE

Scaled integumental fold ('precinctorium') present in front of the tympanic organs. Female frenulum consisting of 2–more bristles as in most ditrysian moths.

Little is known about the Greenland crambid larvae, none of which have been observed in the country. *Tehama bonifatella* is a grass-feeder. *Gesneria centuriella* is very probably a moss-feeder. *Nomophila nearctica* (not resident in Greenland) feeds on a wide variety of lower plants, and the same likely applies to *Udea torvalis* whose congeners feed on a broad spectrum of flowering plants (both herbs and shrubs), often in a ± loose web or between tied leaves. *Tehama* and *Gesneria* will expectedly (judging from crambid subfamily differences established on the basis of C. European taxa; Hasenfuss 1960) differ from *Nomophila* and *Udea* by having the ventral setae on the two sides of AIX more separate than on AX. A moth larva (found together with a pupa under a stone) described and figured in a preliminary way by Nielsen (1910: 65) is likely a crambid.

***Tehama bonifatella* (Hulst, 1887) Figs 14F–G**

FWL 8–10 mm, FWB 3–3.5 mm. FW unusually narrow, with weakly pointed apex; veins (especially in distal third) with white scales; abundance of dark and black scales variable, but blackish brown specimens are not known from Greenland. Worn specimens tend to be light yellowish grey.

Larval head capsule and T1 shield very variably darkened (from light brown to jet black), colour of pinacula also variable; trunk surface light reddish to light greenish grey-brown.

BIOLOGY. This is the North American Western 'Lawn Moth' which is an occasional turfgrass pest, feeding on several grass species. Larvae are nocturnal, 1st instars being leaf skeletonizers, later instars bringing leaves/leaf pieces to the silk-lined galleries in the soil where they spend the day; see *e.g.*, http://www.ipm.ucdavis.edu/ PMG/r785301011.html#DESCRIPTION. Adults fly in July to early August.

DISTRIBUTION. S, SW. Found between Tasiusaq (60°11'N) and Qoornoq (64°30'N). Local, but often common where it occurs (Johannesen 1990: 6).

Nearctic. Western part from Alaska to New Mexico and in eastern part recorded from Nova Scotia, Labrador and Newfoundland (Wolff 1964: 34, 64).

NOTE. The synonymy of *Crambus inornatellus* Walker, 1863 with *Spermatophthora bonifatella* Hulst, 1887 was first published by Błeszyński (1965). He did, however, not use the senior synonym, but stated it to be a *nomen oblitum* according to the nomenclatorial code then in use (ICZN 1961: article 23b). Even though Błeszyński did not fulfil the demand of that code (*op cit.* article 23(a)i) the name *T. bonifatella* stayed in use, with *T. inornatellus* as a synonym. This is in accordance with the present code (ICZN 1999: article 23.9) and is followed here, even though it is left for future revisionary work to fulfil the demands of article 23.9.1 and 23.9.2 of that code.

***Gesneria centuriella* (Denis & Schiffermüller, 1775) Figs 14H–J**

Syn.: *borealis* Duponchel, 1836*; *hybridalis* Zetterstedt, 1839*; *caecalis* Walker, 1858

Wolff 1964: *Scoparona* (*Scoparia*) *centuriella* Den. & Schiff.

FWL 10–14 mm, FWB 4–6 mm. Head with chaetosema. FW slender subtriangular. Very variable with respect to size, colour and distinctness of markings. Females are normally darker than males, with the FW more or less extensively suffused with grey-brown scales. HW dark grey, occasionally lighter towards base. On HW underside 'central spot' (at end of the 'cell') distinct.

Larva pale, with head and thoracic shield yellowish brown, the latter bordered brown posteriorly; thoracic legs brownish; pinacula large, flat, brownish and distinct; anal shield brownish (Canada: Mackay 1972a: 10, 33: chaetotaxy figures).

BIOLOGY. The host plant (probably a moss) and feeding style seem to be unrecorded (Munroe 1972: 19; Nuss 2005: 178). The biotope in Greenland is especially moist, stony places like riverbeds, often close to glaciers. Adults have been found among *Chamenerion latifolium* both in SW Greenland (OK) and in 650 m altitude in NE Greenland (Barca 1937: 32). Flight season from end of June to mid-August.

DISTRIBUTION. S, SW, NE, SE. Found between Lichtenau Fjord (60°30'N) and Nuussuaq: Marraat (70°40'N) in the west and between Saqqisikuik/Skjoldungen (62°30'N) (65°35'N) and Clavering Ø (74°15'N) in the east.

Holarctic. Widely distributed in the northern parts of the Nearctic and Palaearctic Regions, also alpine to the south (Munroe 1972; Nuss 2005).

NOTE. Munroe (1972) grouped the Nearctic populations of *centuriella* into 4 subspecies, restricting ssp. *borealis* (Duponchel, 1836) to those from Greenland. He apparently saw only little Greenland material: The diagnosis and figures given (Munroe 1972: 19, pl. 1, figs 1–2) reflect only a limited range of the variability we have observed in 140+ specimens from the country, which is actually better reflected in his figures (loc cit., figs 3–10) of the northern N. American *G. centuriella* ssp. *caecalis* (Walker, 1858). The two are here formally placed in synonymy (**syn. n.**).

***Udea torvalis* (Möschler, 1864) Fig 14K–L**

FWL 9–12 mm, FWB 3.5–5 mm. Head without chaetosema. FW more pronouncedly triangular than in other Greenland crambids, varying from light grey to slate-black depending on the abundance of whitish-grey scales; these are easily lost when specimens becomes worn. HW light grey, underside with 'central spot' indistinct or lacking.

A ca 8–9 mm long 'spindle-shaped' moth larva (found together with a pupa under a stone) superficially described (p. 65) and illustrated by Nielsen (1910) likely represents this species.

BIOLOGY. The larva is likely polyphagous on herbs, as are many of its congeners. Adults fly in June to August.

DISTRIBUTION. S, SW, NW, N, NE, SE. Widely distributed, but in South-East Greenland only found at Ammagssalik (65°35'N).

Holarctic. From Yukon through arctic Canada to Labrador (Lafontaine & Wood 1997: 768). Taimyr Peninsula in northern Sibirea (Kozlov *et al.* 2006: 141, 150).

***Nomophila nearctica* Munroe, 1973 Fig 14M**

Wolff (1964: 35): *Nomophila noctuella* Den. & Schiff.

Wolff (1975: 129ff): *Nomophila nearctica* Munroe.

FWL 14–15 mm, FWB 5.5.5 mm. This large pyraloid with narrow forewings bears no close resemblance to other Greenland moths.

Larval head yellowish brown, with darker brown pattern; thoracic shield brownish, thoracic legs pale; anal shield brown and body colour pale (Canada, as *N. noctuella*: Mackay 1972c).

BIOLOGY. The larva feeds on a wide range of lower plants (Munroe 1973: 192). The species most probably cannot survive the winter in Greenland.

DISTRIBUTION. S. Only known from 3 specimens found at Kuunnaat (61°13'N) and 1 specimen at Tissaluuk (61°20'N) (all 1954).

Nearctic and Neotropical. From central America through North America to Labrador (Munroe 1973).

NOTES. Based on studies of the meteorological conditions in the period of capture of the Greenland specimens Mikkola (1968: 509–510) concluded that at least the Kuunaat specimens found on 8th July originated from the south-eastern part of USA.

It is likely that also the Palaearctic *N. noctuella* may turn up in Greenland, since this species has repeatedly been found in Iceland (Wolff 1971b: 45; Olafsson, *in litt.*). The two species are overall similar, but can be separated by differences in the genitalia.

PAPILIONOIDEA (BUTTERFLIES)

Three butterfly families are represented in Greenland. Altogether 4 species are known with certainty to have native populations in the country, while another two occur only as rare casual immigrants and the status of one (*Lycaena*) is uncertain. With the exception of the two fritillary butterflies (*Boloria*) they can all be immediately identified by comparison with the plate figures. In contrast, separation of the two fritillaries is occasionally quite difficult; the characters listed in the species treatment below seem to offer the most reliable guide.

While most butterflies (not members of that pierid lineage to which *Colias* belongs) may rest temporarily with the wings ± openly spread, they are unusual in holding the wings vertically above the body (upper wing surfaces facing each other) in the definitive resting position (Fig. 5D). The larvae have proleg crochets in the 'macrolepidopteran' configuration, *i.e.,* arranged in a single median row (a 'mesoseries'). They moreover have (short) secondary setae on the head and trunk, whereby they differ from all other 'macrolepidopteran' caterpillars in Greenland except for the unmistakable long-haired *Gynaephora* larva (Fig. 9B).

Larvae of resident Greenland butterflies

Pieridae, *Colias hecla*: Trunk near-cylindrical ('ordinary caterpillar type'), without conspicuous processes (Fig. 5A).
Lycaenidae, *Lycaeana phlaeas* (Fig. 5B), *Agriades glandon*: Trunk woodlouse-like with long neck, head retracted into prothorax at rest.
Nymphalidae, *Boloria*: Trunk near-cylindrical, with conspicuous spined processes ('chalazae') (Fig. 5C).

PIERIDAE

In many regions members of this medium-sized family are among the most numerically dominant butterflies. The single Greenland representative belongs to the predominantly Holarctic genus *Colias*, which comprises several species in high altitude and high latitude biota.

***Colias hecla* Lefèbvre, 1836 Figs 5A, 15A–B**

FWL 19–25 mm, FWB 12–15 mm.Unmistakeable, wing pattern in the two sexes clearly different. There is some variation in the colour of the wings, from yellowish to more or less warm orange. Rarely females have the wings almost covered with blackish grey. A form with the orange colour of the wings replaced by cream with pinkish tint is described from West Greenland (Skinner & Mengel 1892: 156; Skinner 1892: pl. 2, fig. 4). The variability of *C. hecla* in NE Greenland was discussed in detail by Knaben (1937: 12–13).

Larva (Fig. 5A) green with narrow, yellow sub-dorsal lines and a dark, latero-posterior spot on each segment just ventrally to the sub-dorsal line (Canada, northern Europe: Layberry *et al.* 1998, Henriksen & Kreutzer 1982). Knaben (1937: pl. 2, fig. 10) figured a pupa found on a blade of grass.

BIOLOGY. Perhaps surprisingly, the Greenland host plant of this spectacular and familiar butterfly is unknown. In Canada (Layberry *et al.* 1998: 109) the larva feeds on "a number of northern legumes", especially *Astragalus alpinus*; the latter is also the butterfly's host in northern Scandinavia, but is not native in Greenland. Flight season from late June to mid-August.

FIGURE 5 *Butterflies. A–C, larvae of* Colias hecla (*A*), Lycaena phlaeas (*B*) *and* Boloria polaris (*C*). *D,* Boloria polaris *in typical butterfly resting posture.*
(PHOTOS: K. SILVONEN (A–C), G. VÁRKONYI (D))

DISTRIBUTION. G. A common species in northern and central parts of Greenland, though less frequent in the south where it seems to be more or less restricted to higher elevations. Almost absent from SE, only found on the Blosseville Kyst northwards from Søkongen Bugt (68°45'N).

Holarctic. Throughout the northern part of Canada and Alaska (distribution map by Ferris 1982: 17) and Northern Eurasia.

NOTE. *C. hecla* is described on the basis of Greenland material (Karsholt & Kristensen unpublished). It has sometimes been considered specifically distinct from the Palaearctic taxon *sulitelma* Aurivilius, 1890 (see, *e.g.*, Grieshuber *et al.* 2001: 171), but recent studies do not support this (Grieshuber, *in litt.*).

LYCAENIDAE

Small butterflies. The larvae are short and stocky with a long neck; head retracted into the prothorax at rest.

***Lycaena phlaeas* (Linnaeus, 1761) Figs 15C–D**
Syn.: *feildeni* M'Lachlan, 1878
Wolff 1964: *Lycaena* (*Chrysophanus*) *phlaeas* L.

FWL 13–15 mm, FWB 9–10 mm. The copper-orange FW ground colour and HW margin immediately distinguishes the species from the other Greenland lycaenid.

Larva green with a reddish mid-dorsal line. In Canada it is stated to vary from red to yellowish green, with varying red to white stripes, and is covered with downy hairs (Layberry *et al.* 1998: 12).

BIOLOGY. The host plant in Greenland is unknown. In arctic Canada the larva feeds on *Oxyria digyna* (Layberry *et al.* 1998: 121). Throughout its range it also feeds on different species of *Rumex*. Flight season July-August.

DISTRIBUTION. NW. Only known from 2 specimens captured at Qaanaaq at MacCormick Fjord, in 1941 (Wolff 1964: 14) and at Inersusat near Pituffik/Thule Air Base, 1973 (Kristiansen 1974: 213–214). The two localities are situated within a distance of 25 km (Boertmann 2007: 133). Also Rydell *et al.* (2001: 134–135) recorded a specimen from the Thule Air Force Base in 1999.

Holarctic. Widely distributed throughout the temperate parts of the Nearctic and Palaearctic Regions. In Canada it is found in Ellesmere Island close to Greenland (Layberry *et al.* 1998: 121; Wolff 1964: 14).

NOTE. The Greenland specimens have been ascribed to ssp. *feildeni* (M'Lachlan, 1878), a Nearctic race characterized by the FW ground colour and HW margin being somewhat pale orange, whereas the nominotypical European *phlaeas* Linnaeus usually are more red, as are also some races in other parts of North America.

***Agriades glandon* (de Prunner, 1798) Figs 15E–F**
Syn.: *aquilo* Boisduval, 1832; *franklinii* Curtis, 1835
Wolff 1964: *Plebeius* (*Lycaena*) *glandon* Prunn.

FWL 10–13 mm, FWB 7–9 mm. Sexually dimorphic. Wing uppersides are greyish blue in males, brownish grey in females. Wing undersides are similar in both sexes, ground colour light grey.

Larva pale green with reddish marks on the dorsal surface, covered with long, fine hairs (Canada: Layberry *et al.* 1998: 162).

BIOLOGY. The host plant in Greenland is unknown. In Canada the larva feeds on various plants belonging to Diapensiaceae, Fabaceae, Primulaceae and Saxifragaceae. In Ellesmere Island *A. glandon* was reared from *Saxifraga oppositifolia* (Layberry *et al.* 1998: 162). The species prefers dry, sandy places (Knaben 1937: 18). Flight season from early July to late August.

DISTRIBUTION. NW, N, NE. Only found in the Greenland High Arctic, from Dundas (76° 33') in NW to Jørgen Brønlund Fjord (82°10') in the north; in the east not south of Illoqqortoormiut/Scoresby Sund (70°29'). Rarely abundant.

Holarctic. Widely distributed in North America; troughout Asia to Europe, mainly in mountains or arctic areas.

NOTE. The Greenland specimens have been ascribed to ssp. *franklinii* Curtis, 1835, which occurs in northern Canada and Alaska (Layberry *et al.* 1998: 162). The taxonomy of the *Plebejus glandon* complex has been debated and is probably not yet definitively resolved. The species has been listed from Greenland as *P. aquilio* (Henriksen 1939: 26) and *P. franklinii* (Böcher 2001: 134).

NYMPHALIDAE

The Greenland nymphalids belong to lineages in which the larvae possess large scoli with secondary setae on each segment. *Vanessa atalanta* and *V. cardui* are migrants and their larvae are not likely to be found in Greenland.

***Vanessa atalanta* (Linnaeus, 1758) Fig. 15G**
Wolff 1964: not recorded.

FWL 28–35 mm, FWB 19–23 mm. Unmistakeable; sexes similar.

BIOLOGY. The larva feeds on *Urtica* spp. (in Canada also on *Laportea canadensis*) (Layberry *et al.* 1998: 207). As noted, breeding is unlikely in Greenland, and it is unthinkable that the species can survive the winter here.

DISTRIBUTION. Only known from 1 specimen, probably a migrant, from Aasiaat/Egedesminde, early September 1967 (Langer 1967: 104).

Holarctic. Strongly migratory.

NOTE. The single Greenland specimen was recorded as a migrant, probably from North America (Langer

1967: 104). Wolff (1971a: 75) believed that the specimen more likely had been introduced from Denmark, but it is noteworthy that in October 2005 *V. atalanta* turned up first in the Faroe Islands and a few days later in Iceland (J.-K. Jensen & E. Olafsson, *in litt*). Hence it is well possible that specimens could also reach Greenland, either from Europe or from North America (see remarks on *Autographa gamma* L.).

Vanessa cardui **(Linnaeus, 1758) Fig. 15H**
Wolff 1964: not recorded.

FWL 27–31 mm, FWB 18–20 mm. Unmistakeable; sexes similar.

BIOLOGY. The larva feeds on a wide range of plants (Scott 1986: 281–282), especially *Carduus* and *Cirsius* spp. As with the preceding species breeding is unlikely and hibernation unthinkable in Greenland.

DISTRIBUTION. S, SW. First recorded from the country ('N Greenland', but without any details, Scott 1986: 281), ocurrence subsequently documented by the capture of two specimens in Qaqortoq/Julianehåb and Upernaviarsuk, 20th July 1991 and observations of several specimens in Greenland between 7th July and 4th August 1991; more specimens were also observed in S. Greenland in 1992 (P. Nielsen, pers. comm.). The Greenland specimens may have immigrated either from Europe or North America.

Cosmopolitan. The most widely distributed butterfly species (Shields 1992: 235).

[*Aglais urticae* (Linnaeus, 1758)
This well known European butterfly has been observed twice in Greenland. 1 specimen was found dead in a store in Christianshåb in 1967, and another was seen in Nanortalik in 1981 (Johannesen 1990: 6). Both are considered as introduced from Denmark.

A single specimen from Canada is also regarded as having been imported from Europe (Layberry *et al.* 1998: 249).]

Boloria **(Fritillary butterflies)**
While the two fritillary butterfly species occurring in Greenland are not each other's closest relatives (Simonsen & Wahlberg 2010) their separation is occasionally beset with difficulties because of their considerable intraspecific variability in colour pattern; both are also quite variable in size. Principal distinctive features may be summarised as follows (see Figs 15I–O.):

Boloria polaris (Figs 15I–K): HW underside at 2/3 from base with a row of distinct black spots proximally edged with white; elongate whitish spot above lower 'cell' margin transversely divided; white marginal spots usually prominent, T-, arrow- or mushroom-shaped

Boloria chariclea (Figs 15L–O): HW underside at 2/3 from base with spots that are less distinct and not edged with white; elongate whitish spot above lower cell margin not transversely divided; white marginal spots simple and flat, semilunar or rectangular.

Boloria polaris **(Boisduval, 1828) Figs 5D, 15I–K**
Syn.: *americana* Strand, 1905.
Wolff 1964: *Clossiana* (*Argynnis*) *polaris* Bdv.
FWL 17–22 mm, FWB 10–14 mm. FW ground colour usually dull orange-brown. The intensity of black at the upper side of the wing is variable, and specimens may ocasionally be as bright orange brown as *B. chariclea*. The underside is less variable than the upper side. *B. polaris* is also rather variable in size.

Larva (Fig. 5C) probably generally differing from that of *B. chariclea* by having light greyish rather than orange scoli.

BIOLOGY. The host plant in Greenland is unknown. In Canada the larva feeds on *Vaccinium uliginosum* and *Dryas* spp., having a two years life cycle (Layberry *et al.* 1998: 180).

Collecting dates range from June to September 3, with most specimens found in July.

DISTRIBUTION. N, NE. According to Kristiansen (1974: 212–214) also in NW: Thule (77°35'N). Less common than *B. chariclea*, but in the High Arctic found in the same localities.

Holarctic. Throughout the arctic areas of Eurasia (Layberry *et al.* 1998: 180).

NOTE: According to Knaben (1937: 17) specimens from Greenland differ from Norwegian specimens in that the black patches on the inner patch-row in the FW are larger and almost square (smaller and almost circular in Norwegian specimens). We have refrained from ascribing the N Nearctic/ Greenland populations subspecies status, hence the name *americana* Strand, 1905 is left in synonymy.

Boloria chariclea **(Schneider, 1794) Figs 15L–O**
Wolff 1964: *Clossiana* (*Argynnis*) *chariclea* Schn.

FWL 16–22 mm, FWB 10–14 mm. Ground colour variable, usually brighter orange-brown than in *B. polaris*, but may occasionally be as dull/dark orange brown as in the latter; females are on average darker than males. The extensive intraspecific variation was discussed in detail by Knaben (1937: 13–16), who named a few infrasubspecific forms from East Greenland.

Larva darkish grey with black dorsal and lateral stripes and orange spines (Canada: Layberry *et al.* 1998, Bird *et al.* 1995). The head is black (Aurivillius 1900: 1036).

BIOLOGY. The host plant in Greenland is unknown. In Canada the larva feeds on *Polygonum bistortoides* and *Salix*, using two years to develop in colder areas (Layberry *et al.* 1998: 183). Adults fly from late June to August in places with a rich vegetation of flowering plants. Knaben (1937: pl. 2, fig. **15**) figured a pupa from Greenland.

DISTRIBUTION. G. A common species in most parts of Greenland, though less frequent in the south where it seems to be almost restricted to higher elevations.

Holarctic. Widespread in northern North America (Layberry *et al.* 1998: 180) and throughout northern Asia to northern Scandinavia (Lukhtanov & Lukhtanov 1994: 200).

NOTE. The Greenland population has been ascribed to ssp. *arctica* (Zetterstedt, 1839) (Layberry *et al.* 1998: 183). Knaben (1937: 14) compared a large material of *A. chariclea* from East Greenland with series from Norway and found it impossible consistently to separate specimens of ssp. *arctica* from specimens of the nominotypical race.

[***Oeneis bore*** **(Schneider, 1792)**
This species has been listed from Greenland in various lists (*e.g.*, Zetterstedt 1839: 902; Henriksen & Lundbeck 1917: 700), based on a record by Boisduval (1832–41: 193) [not seen]. In ZMUC is a specimen labeled "Grønland, Coll. Dr. H. C. Nissen, O. bore taygete". Nissen surely purchased the specimen from a dealer, and it is most likely mislabelled. The species is Holarctic and common in subarctic Canada (Layberry *et al.* 1998: 232), but there are no later, and reliable, records from Greenland. In the same work Boisduval (*loc. cit.*: 189) also erroneously recorded '*Oeneis* (*Chionbas*) *balder* Boisd.' (a synonym of *O. jutta* (Hübner)) from Greenland. Only few decades ago Danks (1981: 501) listed another *Oeneis*, viz. *O. taygete* (Geyer) from Greenland. There is no evidence that any *Oeneis* species has been found in Greenland.]

'MACRO-MOTHS'

Macro-moths are represented in Greenland by the Geometridae and the Noctuoidea. Both are among the most species-rich family-group taxa within the Lepidoptera, and worldwide they comprise many members that cannot from a superficial examination be easily placed in either of the two. However, in the depauperate Greenland macro-moth fauna where ennomine geometrids (of which many have quite stout bodies) are absent, and 'erebid' noctuoids (of which many have quite slender bodies and broad forewings) are represented only by the distinctive *Gynaephora*, immediate family assignment of adult moths on 'general appearance' is unproblematical. Also, the reduced proleg complement of Greenland geometrid larvae permits their immediate separation from those of noctuoids.

Family-group distinction

Geometridae: Medium-sized moths with slender bodies and relatively broad wings, see figures **14**N–R, 15P–Y. Ocelli absent; chaetosema present behind antenna. Tympanal (hearing) organs in lower part of abdominal base. Larvae with only 2 pairs of prolegs (on VI, and the 'anal claspers' on X, Fig. 6).

Noctuoidea: Medium-sized to large, stout-bodied moths with forewings relatively narrow, see figures **16**–17. Ocelli present (*Gynaephora* excepted), situated behind the antennae; chaetosema absent. Tympanal organs in metathorax (below/behind hindwing base). Larvae with 3, or all 5 pairs of prolegs.

GEOMETRIDAE

All Greenland representatives of this family belong to the subfamily Larentiinae, which is the dominant geometrid lineage at high latitudes.

Key to species

ADULTS

1. Forewings broadly rounded, light greyish brown. Flight-season from mid-September to mid-November .... *Operophtera bruceata* (male)
– Forewing apices more acute, hence wings more nearly triangular. Flight-season early June to mid-September..2
2. Larger (FWL 12–16 mm, FWB 7–9 mm), broad-winged species, with greyish forewings .... *Entephria punctipes*
– Smaller (FWL 8–13 mm, FWB 4–8 mm), more slender-winged species 3
3. Larger species (FWL 11–13 mm), antenna in male bipectinate; not in S............. *Psychophora sabini*
– Smaller species (FWL 8–12 mm), male antenna simple, lower (unscaled) surface shortly 'ciliate', not in N...4
4. Relatively more slender-winged, FW L:B ratio ≥ 2.4. FW light grey to blackish grey; scale vestiture on vertex appears grey. Flight-season from early June to early August..................... *Eupithecia gelidata*
– Relatively more broad-winged, FW L:B ratio ≤ 2.3. FW brownish grey; scale vestiture on vertex appears

FIGURE 6 *Geometrid larvae. A,* Psychophora sabini. *B–C,* Entephria punctipes (*C camouflaged among* Dryas *foliage*). *D,* Operophtera bruceata. *E,* Eupithecia gelidata. *F,* Eupithecia pusillata. (PHOTOS: K. SILVONEN (A–B, E–F), J. BÖCHER (D), G. VÁRKONYI (C))

whitish. Flight-season from late July to early September .... *Eupithecia pusillata*

LARVAE

1. Secondary setae on side of ventral prolegs number 6 or less 2
– Secondary setae on side of ventral prolegs number 8 or more 4
2. Setae D1 on anal shield anterior to line joining SD-setae, trunk cuticle rough 3
– Setae D1 on anal shield posterior to line joining SD-setae, trunk cuticle smooth .... *Operophtera bruceata*
3. Trunk green or light brown with broad, reddish brown middorsal line, which is often broken into flecks; anal shield pale with darker mid-dorsum. On *Juniperus* .... *Eupithecia pusillata*
– Trunk yellowish green with narrow mid-dorsal line, not broken into flecks; anal shield shades of red. On *Ledum* (more rarely on *Salix*, *Betula* and *Vaccinium*) .... *Eupithecia gelidata*
4. Abdomen dorsally with segmentally arranges heart-shaped flecks, which have red centres and blackish margins; ground colour green or reddish; setae not on pinacula .... *Entephria punctipes*
– Dorsal markings on abdomen different; larva dark with pale flecks between segments; setae on conspicuous, large and pale pinacula .... *Psychophora sabini*

Psychophora sabini **Kirby, 1824 Figs 6A, 15P–R**

Wolff 1964: *Psychophora* (*Cidaria*) *sabinii* Kirby.

FWL 11–13 mm, FWB 6–8 mm. Antenna in male pectinate. FW with pointed apex and usually with darker central (and basal) area, other transverse fascia at most faintly indicated. Fringes unicolourous to slightly chequered light grey and brownish grey. Wing undersides almost unmarked, whitish grey. Specimens from the western part of North Greenland are, on average, smaller and have more unicolorous (often lighter) wings, with no or only faint markings.

Larva (Fig. 6A) easily recognizable, darkish with densely granulated cuticle. Seta bases encircled by pale spots. There are previous descriptions of the larva from Greenland by Packard (1876) and Strand (1905).

BIOLOGY. The larva is polyphagous on herbs.

DISTRIBUTION. NW, N, NE. From Thule (77°35'N) though Midsommer Sø (82°15'N) south to Scoresby Sund (70°15'N).

Holarctic. Northern North America. In the Palaearctic known from northern Fennoscandia and northern Sibiria (J. Viidalepp, pers. comm.).

NOTE. There is an old record of *sabini* from a much more southern locality in North-West Greenland, based on two specimens from Upernavik (72°45'N) (M'Lachlan 1878) [not seen]. According to Langer (1968: 169–170) these specimens are paler and completely without markings. Such specimens have earlier been considered as belonging to a separate species, *P. sabiniaria* (Packard) (syn. *P. immaculata* (Skinner)) from western North Greenland and were discussed and figured by Downes (1964: 297, pl. 2, figs 18, 19) and stated to occur on the Ellef Ringnes Island, Arctic Canada. The material studied by M'Lachlan included insects collected both in NW Greenland and in northeastern arctic Canada. Since *sabini* has not been re-found in that area of NW Greenland it cannot be excluded that M'Lachlan's record is based on mislabelled specimens.

Entephria punctipes **(Curtis, 1835) Figs 6B–C, 15S–U**
Wolff 1964: *Entephria* (*Cidaria*) *polata* Dup.

FWL 12–16 mm, FWB 7–9 mm. Male antenna very shortly ciliate. FW relatively broader than in preceding species, with more or less distinct darker basal and central areas and light-and-dark transverse zigzag-fascia; specimens with almost unmarked dark grey wings occasionally occur. Fringes more or less distinctly chequered grey and white. Wing undersides grey, with darker lines and spots.

Larval head dark brown, with diffuse darker small spots. Abdominal segments each with two black oblique stripes forming an aqute angle with each other; area within these stripes with indistinct brownish longitudinal stripes; last segment with broad, black median line; the two last thoracic segments posteriory with faint markings like those of the abdomen; lower surface and legs greyish brown (Nielsen 1910: 67).

BIOLOGY. The larva has been found on *Dryas* in North-East Greenland (Nielsen 1910: 67). In northern Europe *E. punctipes* has been reared in captivity on *Salix aurita* (J. Viidalepp, pers. comm.). Flight season from end of June to middle of August. Adults occur both in lowland and in the mountains (Knaben 1937: 26).

DISTRIBUTION. G. Widespread and locally common.

Holarctic. Throughout the northern Nearctic Region, and in high altitudes of the Rocky Mountains. In the Palaearctic know from northern Fennoscandia and northern Sibiria (Troubridge 1997: 135–136).

NOTES. The Greenland *Entephria* was formerly misidentified as *E. polata* (Duponchel 1830). For the taxon *Larentia brullei* Lefebvre, 1836), currently considered a junior synonym of *E. punctipes*, the type locality was given as "pays des Eskimaux" (Lefebvre 1836: 399), which in Scoble (1999: 273) was interpreted to be "Lappland" (just one of four areas mentioned as origins of taxa newly described in the paper, the other being, besides "pays des Eskimaux", Greenland and Labrador). Munroe (1951: 290) believed, based on Lefebvre's plate, that *brullei* originated from Greenland, whereas more recently Troubridge (1997: 136) plausibly argued that *brullei* originated from "one of the islands of the Canadian Arctic Archipelago".

Operophtera bruceata **(Hulst, 1886) Figs 6D, 15V–Z**
FWL 9–14 mm (male), 1 mm (female); FWB 5.5–6.5 mm (male), 0.5 mm (female). Strong sexual dimorphism, female brachypterous. Male antenna simple, setae ('cilia') on lower (unscaled) surface long (as antennomere diameter, or more). FW broad, thinly covered with scales, apex markedly more rounded than in the two preceding species. Delicate wavy transverse fasciae variably developed, darkened central area sometimes present; unicolourous, dark greyish brown specimens occur. Female greyish to blackish brown.

Larva (Fig. 6D) stocky and fairly short compared to other Greenland geometrid larvae. Colour variable, green to dark blackish green with 2 latero-dorsal longitudinal narrow stripes; light and dark green larvae occurring among each other (OK; de Lesse 1951: 59–60: abdominal chaetotaxy).

BIOLOGY. Larva feeding in June-July on *Salix glauca* (Wolff 1964: 29), *Betula nana*, *Vaccinium uliginosum* and *Sorbus groenlandica* (OK), and also on herbs like *Rumex acetosa* (Johannesen 1990: 7). When young the larva lives between leaves in a loose spinning; larger larvae feed freely on the leaves. The adult males fly in mid-September to mid-November, with most specimens occurring in first half of October. Pupae from larvae found in three localities in Greenland in late June and July and kept outdoors in Denmark, emerged in second half of September. From two of these localities only females emerged (OK).

Specimens from heaths in the Island of Nanortalik (S) are smaller (FWL 9–12 mm) than specimens from elsewhere in Greenland (FWL 11–14 mm). The larva of this form feeds on herbs, especially *Rumex acetosa*, and is stated to be darker green than larvae of the main form. It is fully grown about 2 weeks later than the main form (which occurs more inland in the Island of Nanortalik where it feeds on *Salix glauca* and *Betula nana*) even though it flies 1–2 weeks earlier (Johannesen 1990: 6–7).

DISTRIBUTION. S, sSW. Locally very common; nSW (only Christianshåb and Eqi: 69°45'N).
Nearctic, Canada and northern USA (Wolff 1964: 28). Downes (1966: 1142) hypothised that *O. bruceata* may have been brought to Greenland from eastern North America ('Vinland') by Norse seafarers.

Eupithecia gelidata **Möschler, 1860 Figs 6E, 7A, 7C, 14N–P**
Syn.: *hyperboreata* Staudinger, 1861
FWL 8–11 mm, FWB 4–6 mm. Vertex (top of head) appearing grey, because most scales are dark or bicoloured whitish/dark, but with dark colouration intense and extensive; pure-white scales restricted to 1–2 posterior row(s). FW rather slender, triangular. Pattern varying from almost plain grey to pronouncedly contrast-rich; broad whitish outer cross-line (±distinctly divided by darker line) often prominent. Occasional particularly striking specimens have the basal area light, whereby the dark central area stands markedly out. In the male genitalia (Fig. 7A) the valvae resemble those of *E. pusillata* in having the lower distal corner produced, pointed; in the phallus there is, besides strong, melanized tooth- or plate-like cornuti, a sizable sclerotized plate densely set with small spines. Female bursa copulatrix (Fig. 7C) approximately pear-shaped, posterior sclerotized 'neck' (colliculum) not elongated anteriad; posterior row(s) of internal spines in bursa sac not markedly enlarged, and not unilaterally extending posteriad.

Larva (Fig. 6E) variable in colour: unicolourous light green or dark green, light brown or dark red-brown, with light or darker longitudinal lines on these backgrounds; sometimes ornamented with dark triangles along the dorsal part of the abdomen. In the field the larvae of the two *Eupithecia* species can be fairly easily distinguished by their different colouration and food plants. With faded alcohol preserved material devoid of food plant information, species identification is very difficult.

BIOLOGY. The larva feeds from mid-July especially on *Ledum groenlandicum*, but also on *Betula nana*, *Salix glauca* and *Vaccinium uliginosum* (OK), eating the upper parenchyma. It hibernates as a pupa (OK). Flight season from from early June to early August. *E. gelidata* is among the most abundant moths in the southern half of Greenland.

DISTRIBUTION. S, SW, NW (north to Upernavik: 72°45'N), NE (north to Hold with Hope: 73°40'N), SE (south to Saqqisikuik/Skjoldungen: 63°13'N). Widespread and common.

Holarctic. Throughout northern Europe and northern Asia. Widely distributed in North America (Mironov 2003: 105–106).

NOTE. Beside the characters mentioned above, *E. gelidata* can be separated from *E. pusillata* by its earlier flight time (specimens from June and first half of July are most likely *gelidata*), and by the distribution (specimens from outside S and SW are most likely *gelidata*).

Eupithecia pusillata **(Denis & Schiffermüller, 1775) Figs 6F, 7B, D, 14Q–R**
Syn.: *sobrinata* Hübner, 1817; *scoriata* Staudinger, 1857; *altenaria* auct.
Wolff 1964: *Eupithecia sobrinata* Hb.

FWL 10–12 mm, FWB 5–7 mm. Overall similar to *E. gelidata*, but on average larger and more broad-winged. Vertex appearing whitish because most scales are pure whitish or bicoloured whitish/dark, but with dark colouration weaker and less extensive than in *E.gelidata*. Wings more brownish and less contrast-rich than in *E. gelidata*; FW base never lighter than central area (itself often rather indistinct and never conspicuously darkened); in postdiscal area short black streaks on variable number of longitudinal veins sometimes conspicuous. Some specimens may be almost plain grey. In the male genitalia (Fig. 7B) the phallus contains no counterpart of *E. gelidata*'s spinose sclerotized plate. Female bursa copulatrix (Fig. 7D) more elongate than in *E. gelidata*, posterior sclerotization in 'neck' (colliculum) extended anteriad; posterior row(s) of internal spines in bursa sac particularly large and coarse, unilaterally extending posteriad.

Larva (Fig. 6F) green or light brown; mid-dorsal line broad, reddish brown, often broken up into discrete spots; anal shield pale with darker mid-dorsum. Staudinger (1857: 266) stated the larvae to be variable in colour and markings in Iceland.

BIOLOGY. In Iceland larvae feed exclusively on *Juniperus*; they have been found in June (Wolff 1971b: 75). Flight season in Greenland from late July to early September; moths can be seen swarming over *Juniperus* (Wolff 1964: 33).

DISTRIBUTION. S, SW (north to Kapisillit: 64°10'N), SE (only found at Saqqisikuik/Skjoldungen: 63°13'N).

Palaearctic. Iceland, Europe throughout Siberia (Mironov 2003: 142).

NOTES. In earlier publications, *e.g.*, Wolff (1964: 33) and Ferguson (in Hodges *et al.* 1983: 106) *E. pusillata* is considered a Holarctic species, but according to Bolte (1990) *pusillata* is in the Nearctic region replaced by two closely related species, *E. interruptofasciata* Packard, 1873 and *E. niphadophilata* (Dyar, 1904), both of which also feed on *Juniperus*, but differ from *pusillata* in the genitalia.

The Greenland population is ascribed to ssp. *scoriata* Staudinger, described from Iceland (Mironov 2003: 142); males of this taxon have the strong, apical (in the uneverted vesica) cornutus shorter than in nominotypical *pusillata* specimens from continental Europe.

NOCTUOIDEA

In the classification adopted here (in accordance with Nieukerken *et al.* 2011, following Zahiri *et al.* 2011) the Greenland noctuoids are placed in two separate families,

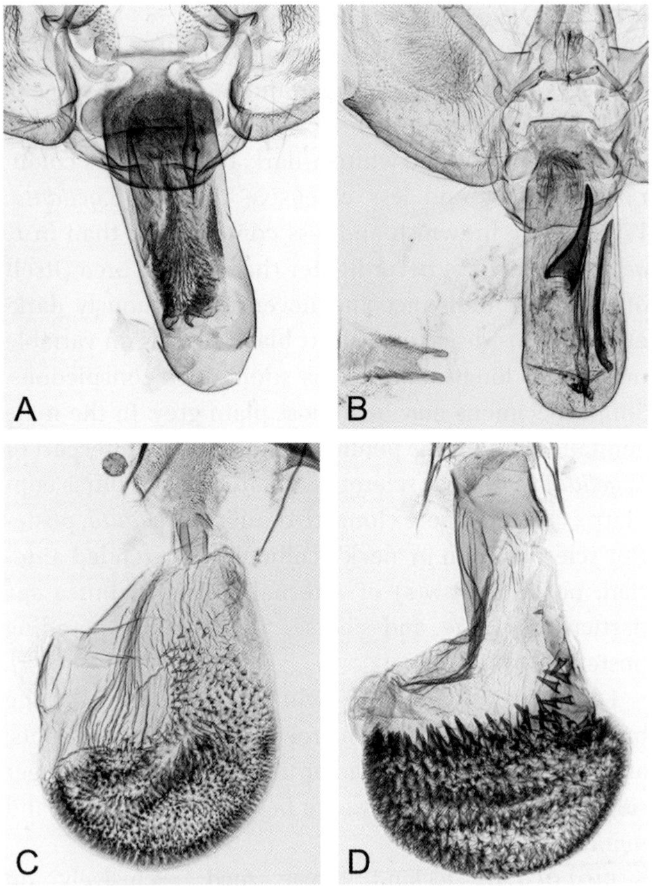

FIGURE. 7 Eupithecia *genitalia, photographed from slide preparations of KOH–macerated specimens. A,* E. gelidata *male phallus. B,* E. pusillata *male phallus. C,* E. gelidata *female bursa copulatrix. D,* E. pusillata *female bursa copulatrix.*

Erebidae and Noctuidae, the latter being the most species-rich lepidopteran family in Greenland. The two families are not separable by easily observable characters. All large and robust moths in the country belong here. The sequence of noctuid species follows Lafontaine & Schmidt (2010).

Key to species

ADULTS

1. Hindwing warm yellow with black margin.......... .... *Syngrapha borea*
– Hindwing differently coloured 2
2. Forewing in middle with light silvery or light golden γ-shaped mark 3
– Forewing in middle without light silvery or light golden γ-shaped mark 5
3. Median area of forewing blackish-brown, markedly contrasting to basal and apical areas; HW ciliae white ....*Syngrapha parilis*
– Median area of forewing only slightly contrasting to basal and apical areas; HW ciliae ± distinctly chequered light and dark 4
4. Forewing dark greyish brown with a bronze lustre; HW underside with ciliae less distinctly chequered light and dark, without distinct dark central spot ... . *Autographa gamma*
– Forewing dark grey without bronze lustre; HW underside with ciliae contrastingly chequered white and black, with distinct dark central spot........... .... *Syngrapaha u-aureum*
5. Antennae clearly < 0.5× FW length, in ♂ strongly bipectinate to apex, in ♀ simple and particularly short; ocelli absent; facies as in Figs 16A–C .... *Gynaephora groenlandica*
– Antennae > 0.5× FW length, if bipectinate (*Agrotis* ♂) only in basal 2/3 6
6. Tibiae, at least on mid and hind legs, with strong spines (stout scelerotized setae) projecting out from scale/hair vestiture (Fig. 8C); facies as in Figs 17E–S . ... 7 (Noctuinae–Noctuini)
– Tibiae without such spines 13
7. Fore tibia about as long as, or shorter than basal tarsal segment (Fig. 8C)................................8
– Fore tibia distinctly longer than basal tarsal segment... 11
8. Basal part of FW with broad anterior margin pale, markedly contrasting with ground colour elsewhere on wing...9
– FW without contrasting pale colouration along basal part of anterior margin 10
9. FW relatively broader, L/W ratio ≈ 2.2; FW ground colour warm reddish-brownish outer wing field without cuneiform spots (Figs 17P–Q) .... *Euxoa westermanni*
– FW relatively more narrow, L/W ratio ≥ 2.3; FW ground colour very variable, but not warm reddish-brownish; cuneiform spots often prominent in outer area (Figs 17M–O)................................... ....(some forms of) *Euxoa drewseni*
10. Larger species, FWL ≥ 18 mm; on FW prominent cuneiform streak extending outwards from reniform stigma; HW ground colour whitish in central area; male antennae bipectinate on basal 2/3............. ....*Agrotis ipsilon*
– Smaller species, FWL ≤ 18 mm; pattern very variable, FW without distinct cuneiform streak extending outwards from reniform stigma; HW ground colour

more uniformly greyish; male antennae weakly serrate on basal 2/3 .... *Euxoa drewseni*

11. Large robust species (FWL 19–25 mm) .... *Eurois occulta*

– Usually smaller species (FWL ≤ 19 mm) 12

12. FW ground colour light brownish grey, dark markings rather contrasting (Figs 17E–F) .... *Rhyacia quadrangula*

– FW ground colour dark brown, hence dark markings little contrasting (Figs 17K–L) .... *Spaelotis clandestina*

13. Compound eyes with interfacetal hairs prominent (Fig. 8B) 14

– Compound eyes with interfacetal hairs absent, or short and delicate, hardly detectable 16

14. FW ground colour light greyish-brown, conspicuous dark-filled reniform stigma markedly contrasting with surrounding wing area .... *Anarta trifolii*

– FW ground colour usually dark brownish or blackish, only rarely light greyish-brown; dark core in reniform stigma not contrasting with surrounding wing area 15

15. On average smaller species, both wing pairs usually little contrast-rich; light orbicular stigma (with dark core) usually most conspicuous element in FW pattern .... *Lasionycta leucocycla*

– On average larger and more contrast-rich species; in FW light reniform stigma (with dark core) and light-edged transverse lines usually prominent .... *Polia richardsoni*

16. Larger species, FWL almost always > 12 mm. Facies as Figs 16Q–Y. FW cilia not pronouncedly chequered 17

– Smaller species, FWL 10–12 mm. Facies as Figs 16K–P, FW cilia often pronouncedly chequered 18

17. Eye margins lined with markedly elongate hairs (Fig. 8A); FW pointed, ground colour shades of grey with little or no brown (Figs 16Q–T); male antennae with scale-free part of individual segments markedly convex, hence antenna appearing 'serrate' (Fig. 11D); female with anal papillae ± bilaterally flattened, smoothly rounded, long-haired (removal of hairs/ scales with a fine brush may be required for observation, Fig. 11F) .... *Mniotype adusta*

– Eye margins not lined with markedly elongate hairs; FW more rounded, ground colour usually appearing decidedly brown (Figs 16U–Y); male antennae with individual segments near-cylindrical, hence antenna appearing 'filiform' (Fig. 11A); female with anal papillae dorsoventrally flattened, short-haired (Fig. 11C) .. .. *Apamea zeta*

18. FW with area between antemedial and medial lines lighter than area between medial and postmedial lines; cilia chequering very distinct; FW and HW undersides both with dark band extending backwards from costa and enclosing central spots which therefore are not discrete; genitalia as in Figs 10A, C .... *Sympistis lapponica*

– FW with median field (between ante- and postmedial lines) usually darkened towards dorsum or in its entirety, area between antemedial and medial lines not lighter than area between medial and postmedial lines, moths sometimes almost uniformly dark blackish brown; cilia chequering often less distinct; wing undersides without dark bands from costa, usually with discrete and (particularly on FW) prominent central spots; genitalia as in Figs 10B, D .... *Sympistis zetterstedtii*

LARVAE

The single erebid (*Gynaephora*) is unique among Greenland Lepidoptera in having a long-haired larva (Fig. 9B). Larvae of noctuids can be distinguished as follows:

1. Abdomen with prolegs only on segments V–VI and terminal segment 2

– Abdomen with full complement of prolegs (on segments III–VI and terminal segment) 5

2. SV–group bisetose on abdominal segment I. Larva green with dark green mid-dorsal line bordered with white; broad pale addorsal line running between D1 and D2 setae .... *Autographa gamma*

– SV–group trisetose on 1. abdominal segment. Ornaments variable 3

3. Meso- and metathorax with a single SV seta .... *Syngrapha borea*

– Meso- and metathorax with 2 SV setae 4

4. Skin strongly spinulate, head with dark reticulate pattern, 'radula' with 7 ridges .... *Syngrapha parilis*

– Skin nearly smooth, head with very pale reticulate pattern, 'radula' with 10 ridges .... *Syngrapha u-aureum*

5. Mid-dorsal line divided by narrow dark longitudinal stripe, margins whitish; premental setae about as long or longer than 1. segment of labial palp; seta SD1 on abdominal segment IX strong 6

– Mid-dorsal line, if present, not divided; premental setae shorter than 1. segment of labial palp; seta SD1 on abdominal segment IX very slender ('hair-like') 7

6. Premental setae very long, about 2 × 1st segment of labial palp (Fig. 2E), black V-shaped pattern of dorsal region bordered by yellow. Larva on *Salix herbacea*. .... *Sympistis lapponica*
– Premental setae shorter, about as long as 1. segment of labial palp or shorter, black V-shaped pattern of dorsal region bordered by whitish grey. Larva on *Dryas*. .... *Sympistis zetterstedtii*
7. Spinneret (see Fig. 2B) tubular, long, ventral tooth of mandible absent, seta L2 of prothorax setaceous, ornaments absent on back except pinacula......... .... *Apamea zeta*
– Spinneret dorsoventrally flattened, ventral tooth of mandible present, seta L2 of prothorax hair-like, ornaments present on back, sometimes also pinacula 8
8. Spinneret rather long with smooth or shortly serrated upper lip and straight lower lip, ornaments usually without segmental series of cuneate spots on subdorsal line 9
– Spinneret short with fringed upper lip and notch in lower lip, ornaments with segmental series of cuneate spots above subdorsal line (except *Euxoa* and *Agrotis*) 12
9. Dark spots above subdorsal and spiracular lines, spiracular line broad with white dorsal and ventral margins and orange middle region .... *Anarta trifolii*
– Dark spots absent above subdorsal and spiracular lines, spiracular line narrow, usually indistinct 10
10. Head and T1 shield black, setae D1 and D2 on dark pinacula. Dorsal region with segmental series of diamond-shaped pattern; subdorsal line white, broader than mid-dorsal line; pleural and ventral zones blackish................. *Lasionycta leucocycla*
– Head and T1 shield brown, pinacula D1 and D2 absent, setal base spots pale. Dorsal diamond-shaped pattern present or absent; subdorsal line narrow, fragmented; ventral region of body pale.......... 11
11. Larva dark brown with noticeable large white base spots of setae D1 and D2: T1 shield strong, shining, subdorsal line with broad blackish margin .... *Polia richardsoni*
– Larva green or pale brown, setal base spots small, yellowish white; T1 shield weaker, like adjacent trink wall; middorsal and subdorsal lines very narrow and fragmented, margins weak .... *Mniotype adusta*
12. Epicranial sulcus shorter than height of facial triangle (see Fig. 2A), spinneret shorter than wide with strongly modified, fringed upper lip and bipartite lower lip. Abdominal prolegs short; crochets in semicircle, their number lower (no more than 15 on the proleg of III). T1 shield strongly sclerotized....... 13
– Epicranial sulcus about as long as, or longer than, height of facial triangle, spinneret length variable. Abdominal prolegs longer, crochets in longitudinal row, their number usually higher (more than 15 on III). T1 shield weaker 15
13. Head black, reticulate pattern often visible between setae L1–P1, fields pale, thoracic shield black, middorsal and subdorsal lines hardly visible (rare migrant in Greenland)................ *Agrotis ipsilon*
– Head with dark reticulate pattern, fields blackish and bands pale, thoracic shield brownish with narrow middorsal line, middorsal and subdorsal lines visible on body 14
14. Black frontal stripe of head between A2 and P1 of uniform width, stemmatal zone blackish brown. Thoracic shield blackish brown with white, narrow middorsal line and broader subdorsal line........... .... *Euxoa drewseni*

FIGURE 8 *Noctuid structural characters. A,* Mniotype adusta, *head showing naked compound eyes lined with markedly elongate hairs. B,* Polia richardsoni, *head showing prominent interfacetal hairs on compound eyes. C,* Agrotis ipsilon, *fore and middle legs with strong tibial spines (here also present on basal tarsal segment, particularly numerous on middle leg).*

– Head capsule brown, with distinctive pattern of darker spots .... *Euxoa westermanni*

15. Larva brown with whitish middorsal, subdorsal and spiracular lines, dorsal margin of subdorsal line with segmental series of rather small, dark spots, transverse stripe absent on 8. abdominal segment........ .... *Rhyacia quadrangula*

– Larvae darker blackish grey with noticeable wedge-shaped spots especially on abdominal segments VII–VIII above subdorsal line from MD1 to D2 setae, transverse stripe on 8. abdominal segment present caudad from D2 setae..............................16

16. Mandible without inner tooth. Subdorsal line whitish enlarged towards seta D2, dorsal region with segmental series with blackish arrow-like pattern, transverse stripe on segment VIII weak, spiracular line wavy, whitish .... *Spaelotis clandestina*

– Mandible with inner tooth. Subdorsal line white, narrow, dorsal region with segmental series of wedge-shaped spots above subdorsal line from MD1 to D2, transverse stripe broad on 8. segment, narrower also on other abdominal segments, spiracular line broad, white, with reddish central part.......... .... *Eurois occulta*

EREBIDAE–LYMANTRIINAE

Members of this group ('tussock moths') were until recently treated as a family of their own; in the arrangement adopted by Lafontaine & Schmidt 2010, Zahiri *et al.* 2011 and Nieukerken *et al.* 2011 they are a subfamily in the Erebidae. The pronounced hairiness of their larvae has counterparts in few other noctuid taxa, and none in the Greenland fauna.

Gynaephora groenlandica (Wocke, 1874) Figs 9A–B, 16A–C

FWL 15–21 mm, FWB 8–11 mm. Robust moth, unique in Greenland fauna by its antennae (in males strongly bipectinate) being unusually short relative to the forewing. Weakly patterned wings and long, stout female postabdomen similarly distinctive. Proboscis tiny. Only larger Greenland moth with reduced proboscis.

Larva (Fig. 9B) distinctive among Greenland caterpillars by dense trunk vestiture of long, brown secondary setae. Particularly dense and long tufts present dorsally on T2, T3 and AIX; a patch of paler setae dorsally on AI–III. Prolegs long. The early stages of *G. groenlandica* are described in details, figured and compared with those of the related *G. rossii* (Curtis 1835) by Morewood & Lange (1977). The larva hairs can cause skin irritation (*op. cit.* 132).

BIOLOGY. The larva prefers *Salix* (Wolff 1964: 27). Morewood & Lange (1997: 132–133) list 10 further host plants, of which only *Dryas integrifolia* were used regulary. A cocon in coll. ZMUC has several leaves of *Betula nana* spun to its surface. The eggs are laid in masses covered by hair from the abdomen of the female, either on the cocoon from where the female emerged, nearby in the vegetation or on the ground. Eggs laid on cocoons are very vulnerable to predation from birds. The larva stage lasts for many (up to 11, Ferguson 1978) years. After each hibernation the larvae are active from snow melting and a few weeks onwards, eating the young shoots and developing leaves and catkins of *Salix* etc. They then stop eating entering into diapause, often placing themselves exposed on rocks etc. Full grown larvae begin spinning cocoons very soon after becoming active in the spring. The cocoons are fastened to the surface in exposed sites with maximum insolation. The cocoon has a double layer which is thought to act as a "micro-greenhouse", and the temperature within it has been shown to be higher than the surrounding substrate temperature. Most adults emerge in June and early July. Males fly high, fast and erratically during day, whereas females apparently do not fly at all and may stay in the cocoon until found by a male (Ferguson 1978: 21; Kristiansen 1982: 146; Morewood & Lange 1997: 121–137).

There is an extensive literature dealing with the biology (ecology, physiology) of *G. groenlandica*; for further references see Morewood & Lange (1997).

DISTRIBUTION. SW, NW, N, NE. From the Nuuk area (64°N) to Tasiussaq (73°20'N) in west, around the north coast to Kap Neumayer (82°18'N) and down to Gåseland (70°15'N) in east. Larvae can be plentiful, especially in North Greenland, but adult moths are rarely seen. A record from SV: Ivigtut (61°13'N) requires confirmation.

Nearctic. Outside of Greenland only found in a few widely separated localities in the far northern Canadian Arctic (Ryan & Hergert 1977: 396 (map); Ferguson 1978: 21). A record from Ward Hunt Island (83°05'N) may represent the northernmost occurrence of a lepidopteran species (http://www.answers.com/topic/lepidoptera-1).

NOTE. All records of Arctiidae (larvae) from Greenland are based on misidentification of larvae of this moth, as already pointed out by Wocke (in Homeyer 1874: 409). This is also true for the record of *Pararctia lapponica* (Thunberg) by Danks (1981: 507).

NOCTUIDAE–PLUSIINAE

All Greenland members of this subfamily have a conspicuous, silvery or golden γ-like mark (sometimes fragmented)

FIGURE 9 *Noctuoids. A,* Gynaephora groenlandica, *newly emerged female (right) on her cocoon on which she has already deposited eggs, and male (left). B–N, larvae of* Gynaephora groenlandica *(B),* Autographa gamma *(C),* Syngrapha parilis *(D),* Sympistis lapponica *(E),* Sympistis zetterstedtii *(F–G; G shows little contrast-rich early instar feeding on* Dryas *flower),* Apamea zeta *(H),* Mniotype adusta *(I),* Polia richardsoni *(J),* Lasionycta leucocycla *(K),* Eurois occulta *(L),* Euxoa adumbrata *(M),* Agrotis ipsilon *(N).*
(PHOTOS: K. FISCHER (A), K. SILVONEN (B–F, H–N), J. BÖCHER (G))

near the FW centre. Plusiine larvae are easily recognized by lacking prolegs on abdominal segments III–IV, hence approaching the even more reduced complement in Geometridae.

***Autographa gamma* (Linnaeus, 1758) Figs 9C, 16D–E**
FWL 14–18 mm, FWB 8–10 mm. FW ground colour varying from lighter to darker greyish brown, bronzy lustre distinctive. Silvery γ- mark sometimes severed in outer part, most often on one wing only. Hindwing underside with indistinct central spots and cilia chequering. No spines on hind tibiae.

Larva (Fig. 9C) large, stocky. Head green, or black with reticulate pattern. Trunk green, with contrasting narrow sub-dorsal longitudinal line, skin spinulated, with spines coloured as skin.

BIOLOGY. Polyphageous on herbaceous plants. In Greenland a pupa has been found between spun leaves of *Gnaphalium norvegicum* at 26th June; the moth emerged 4th July (OK), hence it was almost certainly a descendant of a spring immigrant. Larval breeding in *Lactuca* (Iceberg lettuce) cultures is also recorded (P. Nielsen, pers. comm.). Adults are active both during day and night and visit flowers. Flight season late June to late August.

DISTRIBUTION. S, SW, SE. In West Greenland found between Tuapaat (60°05'N) and Sisimiut (66°66'N), and in East Greenland between Saqqisikuik/Skjoldungen (63°13'N) and Ravns Fjord (68°30'N).

Palaearctic. Throuthout the Palaearctic Region, but most common in the western part (Goater *et al.* 2003: 224). Iceland (Wolff 1971b: 110–111). There are no confirmed records of *A. gamma* from the Nearctic outside Greenland (Lafontaine & Poole 1991: 84; D. Lafontaine, *in litt.*).

NOTES. *A. gamma* is considered unable to hibernate in cooler climates (*e.g.*, Ahola & Silvonen 2005), and hence Greenland specimens must be European immigrants or their immediate descendents. While very small specimens often occur in Europe, they are so far not known from Greenland. The species is not observed every year, but may occasionally occur in large numbers as, *e.g.*, in 1950 and 1988.

Greenland *A. gamma* have the γ-shaped mark unusually variable; subdivision as well as shortening of its outer end (on one or both wings) is relatively more frequent than in north European samples.

Syngrapha u-aureum (Guenée, 1852) Figs 16F–G

FWL 12–16 mm, FWB 6–9 mm. Somewhat similar to *Autographa gamma*, but FW ground colour darker, markings more prominent, transverse fasciae more waved/zigzagged; γ-shaped mark silvery white to light golden, most often (80–90% of specimens) divided; absence of bronzy lustre distinctive. HW underside with distinct median spot(s) and cilia chequering. Hind tibia in front of terminal pair of 'spurs' with at least 1 'spine' (similar to the numerous ones on tarsi; this character may be useful for separating the species from smaller and worn *A. gamma*).

Larval head dark olive green. Trunk smooth, with 3 pairs of white lines on dorsum and yellow lateral line extending through the spiracles (Canada: Lafontaine & Poole 1991: 108–109).

BIOLOGY. Host plant *Vaccinium uliginosum* (Johannesen 1990: 8); pupae found under stones (OK). Adults fly in sunshine, visiting flowers of, *e.g.*, *Viscaria*, in dark nights attracted to UV light. Flight season early July to mid-August.

DISTRIBUTION. S, SW, NE, SE. In West Greenland found between Ilua Fjord: Kangikitsoq (60°20'N) and Sarqaq (70°N), and between Saqqisikuik/Skjoldungen (63°13'N) and Mestersvig (72°10'N) in East Greenland.

Nearctic. Outside Greenland only in eastern Canada (Manitoba, Quebec, Labrador, and New Foundland) and north-eastern USA (Maine, New Hampshire and New York) (Lafontaine & Poole 1991: 109–110, map).

Syngrapha borea (Aurivillius, 1890) Figs 16H–I

FWL 12–15 mm, FWB 7–9 mm. The HW colouration renders this species unmistakable among Greenland moths. FW ground colouration variable, γ-shaped mark sometimes divided; HW black marginal band variable in width.

Larval head capsule with conspicuous brown-spotted pattern. Trunk pale, with delicately but densely granulated skin

BIOLOGY. Hostplant unknown. Adults fly in sunshine during day and visit flowers of, *e.g.*, *Viscaria*. In dark nights they are attracted to UV light. Flight season from mid-July to mid-August.

DISTRIBUTION. S, SW, NW, SE. Found between Pamialluk (60°05'N) and Eqi (69°45'N), and at Saqqisikuik/Skjoldungen (63°13'N) on the southeast cost.

Nearctic. Alaska, Yukon, Northwest Territories southwards to British Columbia and Alberta in the Rocky Mountains (Lafontaine & Poole 1991: 111, map).

NOTE. Specimens from Greenland (nominotypical subspecies) differ from those of western North America by smaller size, paler grey forewing ground colour, and narrower dark marginal band on the hindwing (Lafontaine & Poole 1991: 112).

Syngrapha parilis (Hübner, 1809) Figs 9D, 16J

FWL 11–15 mm, FWB 6–9 mm. FW median area in dorsal part blackish-brown, strongly contrasting with adjacent areas. HW dark, with broad light band beyond mid-length; cilia not checkered. Specimens with almost unicolorous dark grey-brown wings have been named f. *nigrescens* Knaben, 1937.

Larva (Fig. 9D) with head yellowish, with dark reticulate pattern. Trunk spinulated; brown in early and green in later (penultimate and last) instars. Mid-dorsal line yellowish white, divided, enclosing green median stripe: addorsal and subdorsal lines delicate, spiracular line broader, all pale yellowish white (Europe).

BIOLOGY. In northwestern Europe larvae are known to feed on *Betula* and *Vaccinium* (and to accept *Polygonum* in captivity); in Ellesmere Island found on *Salix*. Adults are diurnal and have been caught in numbers visiting flowering heather in sunshine (Knaben 1937: 21). Flight season mid-June to early August.

DISTRIBUTION. SW, NW, N, NE. Northern half of Greenland from Kangerlussuaq/Sdr. Strømfjord (67°N) in west through Midsommersøerne (82°15'N) in north to Jameson Land (70°N) in east.

Holarctic. Throughout northern Canada from Ellesmere Island to Alaska, southwards in the Rocky Mountains to Alberta (Lafontaine & Poole 1991: 112–123, map). Northern

Scandinavia, Novaya Zemlya to eastern Siberia (Goater *et al.* 2003: 239). Absent from Iceland (*contra* Goater *et al. op cit.*).

NOCTUIDAE–ONCOCNEMIDINAE

A subfamily represented in Greenland by two congeneric species, which are among the least easily distinguishable moths in the country. They are small, robust noctuids immediately separable from the superficially somewhat similar hadenines in *Polia* and *Lasionycta* by having unhaired eyes (which, however, are 'lashed', *i.e.*, edged by long hairs, like *Mniotype* shown in Fig.8A) and little or no contrast between the marginal and central/inner HW areas.

Sympistis lapponica (Thunberg, 1791) Figs 2E, 9E, 10A, C, 16K–M

FWL 10–12 mm, FWB 5–6 mm. Patagia ('collar') scales of different colours arranged in overall regular succession: white – dark brown – white(/yellowish). FW with outer area of median field (between medial and postmedial lines) markedly darker than inner area. Extent of white-grey scales highly variable, in dark specimens only present as thin lines and dots, in light specimens most of basal and apical third white-grey. Underside basal and central areas with marked dark suffusion (on HW in particular), thereby contrasting strongly with white postmedial fasciae (in turn contrasting strongly with dark marginal areas); dark fasciae usually extending from costa along inner margins of white postmedial fasciae, central spots not discrete. Intensity of light-grey scale suffusion variable; dark specimens sometimes almost uniformly blackish (f. *tenebricosa* Möschler). In male genitalia (Fig. 10A) sclerotized phallus tube shorter than valve; vesica armature without very long band of close-set cornuti. Female with corpus bursae relatively markedly shorter (just above twice ductus length) than in *S. zetterstedtii*, appendix bursae (a rounded lobe originating close to the ductus/corpus transition) prominent (Fig. 10C).

Larva (Fig. 9E) stout, with broad, pale mid-dorsal line and darker dorso-lateral longitudinal lines. Coxal basis of all thoracic legs heavily sclerotised. Best distinguished from *S. zetterstedtii* by the contrast-rich alternating dark and white diagonal dorso-lateral trunk markings. Dark brown markings of uniform width, with narrow white or yellowish borders. Premental setae about twice as long as first segment of labial palp (Fig. 2E).

BIOLOGY. Larvae probably of this species have been found under *Dryas* plants (Nielsen 1910: 66). In Finland it primarily feeds on *Salix*, first instar larvae apparently accepting only pollen, later instars feeding on leaves. Other host plants include *Betula*, *Empetrum* and *Vaccinium* (references in Ahola & Silvonen 2005). Adults fly in sunshine and visit flowers. Flight season end June to mid-August.

DISTRIBUTION. S, SW, E. Between Julianehåb (60°43'N) and Tugtuligssuaq (75°40'N) in West Greenland and between Kap Dalton (69°24'N) and Jameson Land (70°40'N) in East Greenland. A record from Thule (Kristiansen 1974: 214) is based on a misidentified specimen of *S. zetterstedtii* (Kristiansen 1975: 252). Holarctic. Throughout northern Canada to Alaska, southwards to Labrador, Quebec and southern Northwest Territories (Lafontaine & Wood 1997: 750). In the Palearctic region in northern, subarctic ranges of Europe (Ronkay & Ronkay 1995: 41) and Asia, including Novaya Zemlya (Kononenko 2005: 140).

Sympistis zetterstedtii Staudinger, 1857 Figs 9F–G, 10B, D, 16N–P)

Wolff 1964: *Sympistis* (*Anarta*) *zetterstedtii* Staud.
Ronkay & Ronkay 1995: *Sympistis nigrita* (Boisduval, 1840).

FWL 10–12 mm, FWB 5–6 mm. Patagia ('collar') scales of different colours more mixed, these scales consistently narrow, with at most 2 apical teeth. FW with outer area of median field not darker than inner area, entire field usually darkened towards dorsum. Intensity of light-grey scale suffusion variable; dark specimens sometimes almost uniformly blackish. Underside basal and central areas with little dark suffusion (at least on FW), dark fasciae not developed; central spots appearing distinct on overall greyish or (particularly on FW) whitish background. In male genitalia (Fig. 10B) sclerotized phallus tube as long as, or longer than, valve; vesica armature including a very long band of close-set cornuti. Female with corpus bursae relatively markedly longer (>3× ductus length) than in *S. lapponica*, appendix bursae indistinct (Fig. 10D).

Larva strongly resembling that of *S. lapponica*, but dorso-lateral markings differing in shape: dark brown diagonal flecks broad between setae D1 and D2, tapering backwards; white borders of flecks broad. Premental setae only slightly longer than first segment pf labial palp.

BIOLOGY. The host plant in Greenland is *Dryas*, as it is in Ellesmere Island and N. Europe (Mackay 1972b; Ahola & Silvonen 2005); the larvae may appear very conspicuous while feeding on the flowers (J. Böcher, pers. comm., Fig. 9G). Flight season mid-June to end September. Adults visit flowers of *Dryas* and *Silene* in the evening, occasionally in high numbers (Knaben 1937: 24). According to Dreisig (1981: 41) they are mainly active in the early afternoon because they are dependent on high temperatures.

DISTRIBUTION. SW, NW, N, NE, E. In West Greenland between Grønnedal (61°20'N) and Upernavik (72°45'N), in North Greenland from Tugtuligssuaq (75°40'N) and Dundas (72°33'N) in west through Midsommer Elv (82°15'N) in northernmost Greenland to Ravns Fjord (68°30'N) in east. The species is rare in S Greenland, becoming more abundant towards northeast.
Holarctic. Throughout northern half of North America southwards to Washington in western USA (Lafontaine & Wood 1997: 750). In the Palaearctic Region northern and central Asia (Kononenko 2005: 141) to northern Scandinavia.

NOTES. Following Ronkay & Ronkay (1995: 43) the taxon *zetterstedtii* (type locality Lapland) has been considered a subspecies of *S. nigrita* (type locality Switzerland) in most newer literature. However, the DNA barcode (CO1) indicates a species-level distinction between the Alpine *S. nigrita* and the Holarctic *S. zetterstedti* (Mutanen *et al.* 2012); this view was subsequently followed, and supported by details in the genitalia of both sexes, by Lafontaine & Schmidt (2013a, b), and it is adopted here. The population in Greenland is referable to ssp. *kolthoffi* (Aurivilius, 1890). Troubridge (2008: 65) lists this taxon and *S. besla* (Skinner & Mengel) as synonyms of *S. nigrita* ssp. *zetterstedtii*, but that is a mistake, and *kolthoffi* is the oldest name for Nearctic populations of this species (Lafontaine & Schmidt 2010: 141).

NOCTUIDAE-NOCTUINAE: APAMEINI

The two Greenland species now included in this and the following tribe respectively can usually be easily distinguished, but both are very variable and a few forms may not be immediately referable to one or the other by superficial examination.

Apamea zeta (Treitschke, 1825) Figs 9H, 11A–C, 16U–Y

Syn.: *exulis* Lefebvre, 1836; *groenlandica* Duponchel, 1838; *downesi* Mikkola, 2009
Wolff 1964: *Apamea* (*Crymodes*) *exulis* Lef.
Wolff 1971a: *Apamea maillardi exulis* Lef.

FWL 14–19 mm, FWB 7–10 mm. Eye margins without lining of markedly elongate hairs. Antennae in both sexes simple, filiform (Fig. 11A), with ciliae about half antennal diameter. FW often appearing less pointed than in *M. adusta*. FW pattern highly variable, light grey to blackish brown, sometimes extensively suffused with yellowish or brownish scales; medial field in some specimens dark to black, in others base and distal part of the wing darkened and medial field light; suffusion of light scales along principal longitudinal veins may be striking. Moths from East Greenland usually have more yellow suffusion than the overall more brownish specimens from West Greenland. Male valve with apical part markedly expanded (Fig. 11B). Female with anal papillae dorsoventrally flattened, somewhat pointed, short-haired, but with stout, short apical setae; posterior margin of segment VII markedly concave (Fig. 11C).

Larva (Fig. 9H) with brown head. Trunk pale greyish white, pinaculae pale, glossy. T1 and anal shields yellowish grey, glossy as pinaculae; pinaculum of L-group setae on T1 particularly elaborate. Most setae conspicuously long.

BIOLOGY. Described in detail from Iceland (Staudinger 1857: 243–246): Host is *Poa alpina* and probably other grasses; larva first feeding inside lower stem or root, later

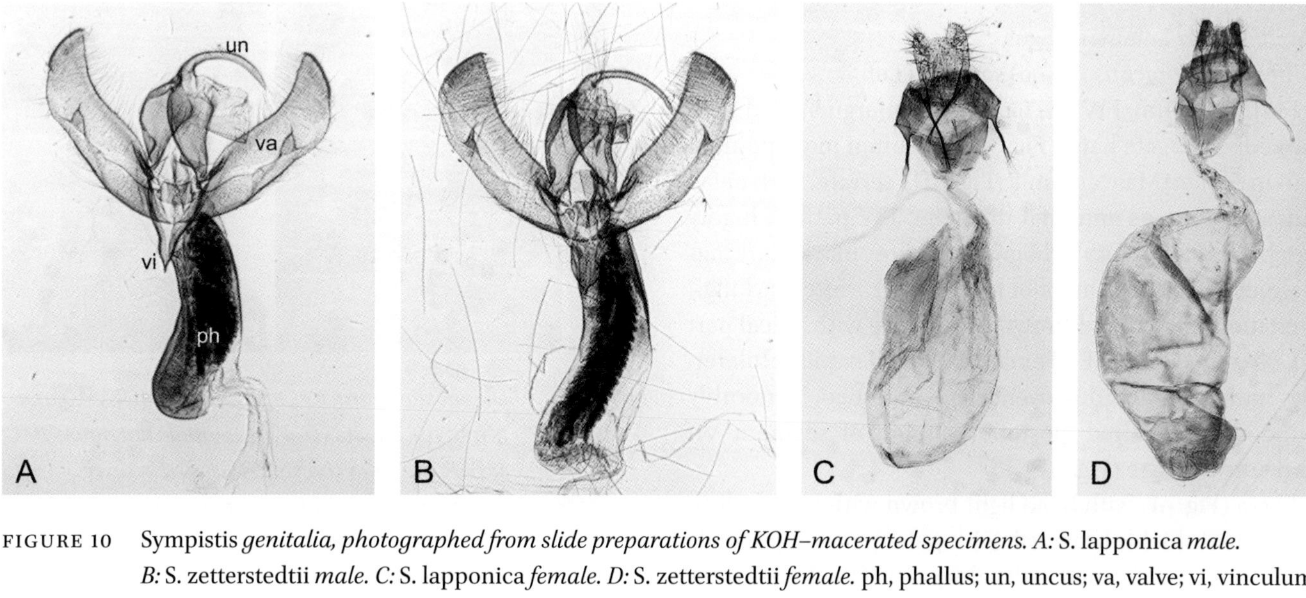

FIGURE 10 Sympistis *genitalia, photographed from slide preparations of KOH-macerated specimens. A:* S. lapponica *male. B:* S. zetterstedtii *male. C:* S. lapponica *female. D:* S. zetterstedtii *female.* ph, phallus; un, uncus; va, valve; vi, vinculum.

from outside. Larval growth rate variable. Staudinger was convinced that some individuals have a 2- or even 3-years life cycle. Flight season late June to early September. Böcher ([2001]: 146) gives *Empetrum nigrum* as host plant in Greenland. In southern Greenland *zeta* has been caught in large numbers in automatic light traps (P. Nielsen, pers. comm.). Knaben (1937: 20) reported a specimen found high up in the mountains in East Greenland.

DISTRIBUTION. Presumably **G.** Between Pamialluk (60°05'N) and Upernavik (72°45'N), at Nedre Midsommer Sø (82°15'N) (Mackay 1972b: 866), and between Lindenow Fjord (60°15'N) and Sabine Ø (74°40'N)].

Holarctic. Canada. Siberia, Fennoscandia, Svalbard, Iceland (Zilli *et al.* 2006: 154).

NOTE. This Greenland moth has had a confusing taxonomic history in recent decades (Wolff 1971b; Mikkola & Lafontaine 1986), and its status may not be definitely settled. A species-level separation of the taxon *groenlandica* Duponchel, 1838 from *exulis* (Zilli *et al.* 2006) was based on differences in the DNA 'barcode' which have subsequently proved problematical (Mikkola *et al.* 2009: 94), and it is not supported by morphological differences.

Mikkola *et al.* (2009: 95–96) stated that specimens from northern areas (of the Nearctic region) are particularly small and have slightly reduced, ellipsoid eyes, and they described a new subspecies *A. exulis downesi* from Ellesmere Island based on these characters. An examination of the variability of these traits in the extensive Greenland material available of the species remains to be made.

NOCTUIDAE–NOCTUINAE: XYLENINI

Mniotype adusta (Esper, 1788) Figs 8A, 9I, 11D–F, 16Q–T

Syn.: *sommeri* Lefebvre, 1836*

Wolff 1964: *Eumichtis* (*Crino*) *sommeri* Lef.

FWL 11–18 mm, FWB 6–10 mm. Eye margins lined with markedly elongate hairs (Fig. 8A). FW often more pointed than in *A. zeta*. Male antenna (Fig. 11D) 'serrate', with ciliae almost as long as antennal diameter. FW pattern highly variable, ranging from light grey with fine, black markings to almost black with more or less distinct white markings, often mottled with red-brown. Male valve with apical part not expanded (Fig. 10E). Female with anal papillae bilaterally rather than dorsoventrally flattened, smoothly rounded, long-haired; posterior margin of segment VII markedly concave (Fig. 11F).

Larva (Fig. 9I) with head light brown with lighter dots. T1 and anal shields brownish with paler pattern; pinacula small, oval, dark rimmed with pale centre (OK observations on preserved larva); trunk colour variable, with darker, more or less broken mid-dorsal and subdorsal lines; spiracular stripe pale, lined by dark ventral subdorsal zone. Young larvae may be green or brown; most turning brown when full-grown, but some remaining green (Wolff 1971b: 11).

BIOLOGY. Larva polyphageous (Staudinger 1857: 249); from Greenland recorded feeding on *Empetrum*, and also bred from *Lactuca* (Iceberg lettuce) cultures (P. Nielsen, pers. comm.). Johannesen (1990: 8) recorded a frozen larva found in mid-winter on the snow successfully completing development. Pupae may be found under stones. Flight season mid-June to late August, adults visit flowering *Silene acaulis* and are attracted to light. For further details on biology on Iceland see Staudinger (1857) and Wolff (1971b).

DISTRIBUTION. S, SW, SE. Between Pamialluk (60°05'N) and Sarqaq (70°N) and between Saqqisikuik/Skjoldungen (63°13'N) and Mestersvig (72°10'N).

NOTES. The status of the Labrador/Greenland/Iceland/Faroe *Mniotype* as either a separate species *M. sommeri* or a subspecies of the continental-Eurasian *M. adusta* has for long time been discussed, *e.g.*, by Wolff (1971b: 100–101), who concluded that the differences pointed to "a strong subspecific separation, but not to a specific distinction". The North American populations are currently considered to belong to *M. adusta* (D. Lafontaine, *in litt.*), and while Lafontaine & Schmidt (2010: 153) refer to *sommeri* as "a Greenland and Iceland form of this otherwise Eurasian species" they did not assign it subspecific status, but explicitly synonymized it with *adusta*. We here note that

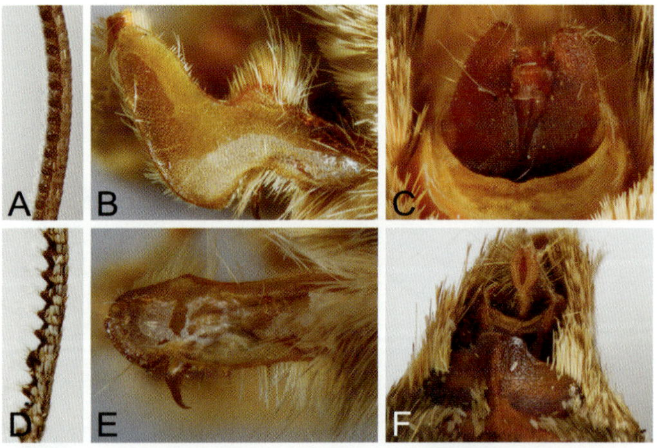

FIGURE 11 *Diagnostic characters of* Apamea zeta *(A–C) and* Mniotype adusta *(D–F). A, D, male antennae. B–C & E–F, genitalia characters as seen in dried specimens after ± extensive removal of hairs/scales: male valves, lateral (B & E) and female abdominal tips, ventral (C & F).*

the larvae are slightly different, that of *M. sommeri* having a darker reddish-brown dorsal trunk zone, and its first labial palp segment is shorter.

NOCTUIDAE–NOCTUINIAE: HADENINI

Distinguishable by long 'interfacetal' hairs on the compound eyes (Fig. 8B; the two native Greenland taxa (*Polia richardsoni, Lasionycta leucocycla*) otherwise have a superficial resemblance to the *Sympistis* species.

Anarta trifolii (Hufnagel, 1766) Fig. 16Å

Wolff 1964: not recorded.

FWL 14–17 mm, FWB 7–9 mm. Markedly differing from other Greenland hadenines: FW distinctive by light ground colour, large and conspicuously dark-filled reniform stigma, and whitish subterminal line produced into conspicuous 'W'.

Larval colour varying from green to dark brown; mid- and subdorsal lines weak, spiracular line broad, central areas forming continuous orange/yellowish stripe, margins white. Spiracles in black or (in some green specimens) green spots; dark spots may also be present above subdorsal line in anterior parts of trunk segments.

BIOLOGY. Larva (nocturnal) in northern Europe primarily feeding on *Atriplex* and *Chenopodium*, but also on many other plants (not *Trifolium*!). In northern Europe it has two generations per year. In Greenland only known from three adults caught in a light trap between 27.vii. and 13.viii.1983, and one from the same locality 22.-31.vii.2009.

DISTRIBUTION. SW. The Greenland specimens are all from Narsarsuaq (61°10'N) (S. Kjeldgaard, P. Nielsen).

Holarctic. Widely distributed throughout North America and Eurasia, except from the northernmost areas (Hacker 1998: 730–731). Frequently migratory.

NOTE. This is the first published record of *A. trifolii* from Greenland. While it cannot be excluded that the Narsarsuaq specimens were introduced as larvae, the recapture in 2009 points towards that the species is now established in the area.

Polia richardsoni (Curtis, 1835) Figs 8B, 9J, 17A–D

Wolff 1964: *Hadena* (*Anarta*) *richardsoni* Curt.
McCabe 1980: *Polia richardsoni squara* Smith, 1908.

*Aplecta richardsoni groenlandica** Heydemann, 1944 **syn. n.**

FWL 12–17 mm, FWB 6–8 mm. FW varying from rather uniform dark grey to black with whitish grey markings; rarely extensively suffused with yellowish scales. Reniform stigma usually prominent, whitish, with ± extensive dark core, sometimes united to orbicular stigma by white line. On HW broad dark marginal band usually markedly contrasting with light (whitish) central area; central area occasionally (particularly in females) more suffused with dark scales. See *Lasionycta leucocycla* below.

Larva (Fig. 9J) with head rounded, pale beige with weak reticulate markings. Longitudinal trunk lines fragmented, partly obliterated by black margins, spiracular stripe narrow with white markings. Dorsal zone yellowish brown, subdorsal zone blackish. Spots at base of dorsal setae MD1, D1 and D2 pure white and large, setae of SD1–V1 groups on dark brown pinacula. Nielsen (1910: 66) gave a description based on larvae found in NE.

BIOLOGY. Larvae have been found under *Dryas* plants (Nielsen 1910: 66). The host plants spectrum is known to be broad in Ellesmere Island and northern Europe: *Dryas* spp., *Oxyria* spp., *Saxifraga oppositifolia, Salix arctica, Silene acaulis, Phyllodoce coerulea, Loiseleuria procumbens* (Mackay 1972b; Ahola & Silvonen 2008); larva overwintering twice in captivity. Flight season early May to late August. Adults visit flowers of *Silene* and *Dryas* in the evening (Knaben 1937: 22).

DISTRIBUTION. S, SW, NW, N. From Ilua (59°55'N) to Kap Morris Jesup (83°39'N). Locally common in mid-latitude and northern parts of Greenland, becoming rarer towards south.

Holarctic. Canada and northern USA (McCabe 1980: 23). Throughout Siberia (including Novaja Zemlja) to Scandinavia (Kononenko 2005: 167).

NOTES. McCabe (1980: 21–239) reviewed the subspecies and synonyms of *P. richardsoni* in North America. Both ssp. *squara** (Smith, 1908) and ssp. *groenlandica** (Heydemann, 1944) were described from Greenland. An examination of the 217 Greenland specimens of *P. richardsoni* in the ZMUC shows that any geographical variation is blurred by the substantial individual variation. If the Greenland populations are considered to be subspecifically distinct from the nominate form (described from 69°48'N, 92°23'W in Canada) as suggested by McCabe, they should be referred to as ssp. *squara*.

Lasionycta leucocycla (Staudinger, 1857)* Figs 9K, 16Z–Ø

Wolff 1964: *Lasiestra* (*Anarta*) *leucocycla* Staud.

FWL 11–13 mm, FWB 6–7 mm. Somewhat similar to *Polia richardsoni*, but usually smaller and much less contrast-rich: FW with light transverse lines little distinctive; light orbicular stigma (usually with dark core) usually conspicuous, reniform stigma barely identifiable; HW central/inner area only occasionally pure white or yellowish, but still almost always distinctly delimited from dark margin. FW suffusion of white scales somewhat variable; rarely almost black specimens with only few white scales occur. Females distinctive among Greenland noctuids by the extensible abdominal tip ('oviscapt') being particularly elongate, strongly sclerotized and shining, unscaled, ± protruding beyond scaled segment VII. Differing from the more distantly related, but similarly sized *Sympistis* species by the 'hairy' eyes and usually quite marked contrast between the marginal and central/inner HW areas.

Larva (Fig. 9K) with head black, prothoracic shield with white dorsal and subdorsal lines, otherwise black. Trunk brown, middorsal line indistinct, subdorsal line continuous, white, rather broad; spiracular line undeveloped; pinacula and cuffs of prolegs present, dark brown.

BIOLOGY. Host plant in Greenland unknown, in Finland (ssp. *dovrensis* Wocke, 1864) reared from *Dryas*. The moth is diurnal in the northern part of its range, to the south also nocturnal (Dreisig 1981: 39; Lafontaine *et al.* 1986: 258). Larvae can be found throughout the summer (Mackay 1972b), and the species probably has a two-years life cycle. Flight season mid-July to mid-August.

DISTRIBUTION. SW, NW. Confined to the central part between Maniitsoq/Sukkertoppen (65°25'N) and Thule (77°35'N). Apparently a rare species.

Holarctic. Canada and northern USA (New Hampshire, Colorado, Oregon). Siberia, northern Fennoscandia, incl. mountains of central Norway (Lafontaine *et al.* 1986: 257, 277, map).

NOCTUIDAE–NOCTUINAE: NOCTUINI

Preserved specimens pertaining to this large tribe can be immediately recognized by the spinose tibiae, which among Greenland noctuids are paralleled only in (most) plusiines, to which the moths otherwise have no close resemblance. Living moths differ from most other noctuids by not resting with wings in roof-like position, but holding their wings flat over the abdomen, with forewings overlapping and leading edges near-parallel.

Euxoa adumbrata (Eversmann, 1842) Figs 9M, 17M–O

Syn.: *drewseni* (Staudinger, 1857)*

Wolff 1964: *Chorizagrotis* (*Euxoa*) *drewseni* Staud.

FWL 11–17 mm, FWB 4–6 mm. FW highly variable: ground colour ranging from greyish brown to almost unicolorous dark with or without reddish brown lustre; sometimes variegated, with whitish to light-brown areas along costa (thereby somewhat resembling *E. westermanni*) and between veins; stigmata 'filling' may or may not be markedly lighter than ground colour; markings ± distinct, wedge-shaped spots in subterminal area usually distinctive.

Larva (Fig. 9M) resembling that of *E. westermanni*, but darker, with blackish vertical stripe on head between setae A2 and P1.

BIOLOGY. Adults fly from middle of July to middle of August.

DISTRIBUTION. S, SW, NE. In west between Narsarsuaq (61°10'N) and Ritenbenk (69°45'N). In east between Mestersvig (72°10'N) and Moskusoksefjord (73°45'N).

Holarctic. In several subspecies from eastern Europe to Greenland. The population in Greenland belongs to spp. *drewseni* (Staudinger, 1857) (Lafontaine 1987: 30; Lafontaine & Schmidt 2010: 101, 160).

Euxoa westermanni (Staudinger, 1857)* Figs 17P, 17Q

FWL 14–17 mm, FWB 7–9 mm. More broad-winged than *E. drewseni* and FW pattern less variable. Ground colour shades of warm reddish brown, sometimes ± mottled with black; field between ante- and postmedian fasciae sometimes markedly darker than surrounding fields.

Larval head capsule brown with distinctive pattern of darker spots. Pinaculae large, brown and glossy. T1 shield heavily sclerotised, darkened latero-anteriorly than posteriorly; with narrow, dorsal white stripe, posteriorly with characteristic pattern of dark spots.

BIOLOGY. Behaviour of early stages unknown. Adults fly from beginning of July to end of August

DISTRIBUTION. S, SW, NE, SE. In west north to Qeqertarsuaq/Godhavn (69°05'N). In east found at Saqqisikuik/Skjoldungen (63°13'N) and Lollands Elv, Jameson Land (70°35'N). *E. westermanni* was formerly considered a rare species (Wolff 1964: 17–18), but collectings with automatical light traps in SW: Upernaviarsuk proved it to be common there (P. Nielsen, pers. comm.). Nearctic. Known from disjunct localities in Canada (Lafontaine 1987: 44–45).

Agrotis ipsilon (Hufnagel, 1766) Figs 9N, 17R–S

Wolff (1964): not recorded.

FWL 18–22 mm, FWB 7–9 mm. Male antennae bipectinate in basal 2/3, simple in female. FW ground colour sexually dimorphic, females being darker than males, sometimes nearly black; wedge-shaped black mark outside reniform signa and 2 similar marks on subterminal line distinctive. Hindwing shiningly white, with darker veins and ± extensive peripheral fuscous suffusion.

BIOLOGY. Likely not breeding in Greenland. Larvae known to be nocturnal, feeding on a wide range of herbaceous plants and shrubs and often building subterranean tunnels from which it forages. In Canada hibernating as pupa (Lafontaine 2004: 252).

DISTRIBUTION. SW. Only known from one male and one female caught in light traps at Upernaviarsuk (60°45'N) 8.viii.1983 and 22–29.vii.1985 (P. Nielsen).

Cosmopolitan.

NOTE. This is the first record of *A. ipsilon* from Greenland. This well known long distance migrant may have come from either Europe or North America.

***Rhyacia quadrangula* (Zetterstedt, 1839)* Figs 17E–F**
Wolff 1964: *Caradrina* (*Rhyacia*) *quadrangula* Zett.

FWL 15–18 mm, FWB 7–9 mm. FW ground colour varying from lighter to darker brownish grey (females tend to be darkest), never dark brown as in *Spaelotis clandestina*. Black markings ± distinct, bar between orbicular and reniform stigmata usually prominent. Transverse fasciae may be single or double; when double, 'filling' not or hardly lighter than ground colour.

Larva superficially similar to that of *Eurois occulta* (L.). Head brown, with darker brown reticulate pattern and submedian arcs. Trunk pale grey with light brown dorsal and dorsolateral patches; lateral line below spiracles pale grey; lower sides and venter dark grey; ground colour overlaid with black speckling, especially on back. Abdominal segments I–VIII with elongate, black spots anterior to seta D2. Spiracles pale brown with black rim (Lafontaine 1998: 58).

BIOLOGY. The larva reportedly causes extensive damage to pasture grasses in southern Greenland (Fibiger 1993: 30). Flight season is end June to end September (a single specimen is labelled as found in late May). Newly emerged moths can be fund in grass tussocks (Johannesen 1990: 8). In South Greenland they have been caught in numbers in light traps (P. Nielsen, pers. comm.).

DISTRIBUTION. SW, between Pamiagluk (60°05'N) and Uummannaq (70°40'N). A specimen in ZMUC labelled 'East Greenland', but without exact locality, is considered mislabelled.

Holarctic ('Nearctic-Icelandic). North-eastern part of Canada (Lafontaine 1998: 58). Iceland (Wolff 1971b: 87). Records from central Asia (Fibiger 1993: 30) are doubtful (Kononenko 2005: 183).

***Spaelotis clandestina* Harris, 1841 Figs 17K–L**
FWL 15–19 mm, FWB 7–9 mm. Dark FW ground colour sometimes with reddish tinge along fore margin; distinctness of blackish markings and light transverse fasciae variable, the latter may be almost invisible. Worn specimens may apear almost unicolorous.

Larva large (up to 40 mm) and conspicuous. Somewhat similar to *E. occulta*, readily distinguishable by trunk markings (see key). Anal shield sometimes with dark, dorsal circle (MA).

BIOLOGY. Hostplant in Greenland not known with certainty; Johannesen (1990: 8) reported it feeding on *Ledum groenlandicum*. In Canada the larva is a pest on a variety of trees, shrubs and herbaceous plants (Lafontaine 1998: 76). Adults fly from middle of July to end of August. Moths can be found resting under stones. In South Greenland *clandestina* has been caught in some numbers in light traps (P. Nielsen, pers. comm.).

DISTRIBUTION. S, SW, SE. Between Qaqortoq/ Julianehåb (60°35'N) and Sarqaq (70°N) in west, and Danells Fjord (60°50'N) in east.

Nearctic. Throughout Canada (apart from arctic parts) and the USA (Lafontaine 1998: 77).

NOTE. *S. clandestina* is closely related to the palaearctic *S. suecica* (Aurivillius, 1890) with which it has been considered conspecific (*e.g.*, Kononenko *et al.* 1989). Specific distinctness, indicated by subtle male genitalia differences, was inferred by Wolff (1964: 19) and more recently Lafontaine (1998: 76).

***Eurois occulta* (Linnaeus, 1758) Figs 9L, 17G–J**
FWL 18–25 mm, FWB 9–12 mm. This large and robust moth has a quite variable FW pattern, the ground colour ranging from light grey to almost black. FW markings usually distinct, but becoming rather obsolete in some dark forms.

Larva large (reaching ca 60 mm) and distinctive. Head capsule with dark spotted pattern. Trunk dark greyish-brown, with interrupted white lines along dorsum; spiracular line broad, white with elongate yellowish areas around spiracles, dorsally bordered by black spots.

BIOLOGY. The abundance in Greenland is markedly fluctuating. The larva seems to prefer *Betula nana*, but also feeds on *Salix, Vaccinium* and lower plants including grasses, especially during mass-occurence. In such situations nearly all low vegetation can be destroyed, partly by the caterpillars' defoliating, partly by the coverage of their excrements and dead bodies (OK). Pupation under stones or among moss. *E. occulta* can be a pest in cultivated areas, and there is some evidence (Iversen 1934) that its damage to the vegetation contributed significantly to the extinction of the Norse in Greenland. Flight season early July to early September.

DISTRIBUTION. S, SW, SE. North to Qasigiannguit/ Christianshåb (69°N) in west, and Skjoldungen (63°13'N) in east. Its abundance varies much between different years.

Holarctic. In North America from New Foundland to Alaska, southwards to New Mexico, but not in artic Canada (Lafontaine 1998: 85). Throughout the Palaearctic Region.

NOTE. The population in Greenland has been referred to ssp. *implicata* (Lefebvre, 1836) (Wolff 1964: 19; Lafontaine 1998: 85), which on average is smaller, with duller wing colouration and more indistinct markings than specimens from North America.

[***Graphiphora augur*** **(Fabricius, 1775)**
A specimen of this holarctic species was caught in a light trap at Upernaviarsuk in S, 8.viii.1983 (P. Nielsen). While the species is not known as a long-distance migrant, the adult does have a habit of hiding in crevices, often in buildings. Hence this isolated Greenland occurrence is considered to be more likely a result of accidental introduction.]

Acknowledgements

Above all we are indebted to K. Silvonen (Aalto, Finland) for placing the numerous live photographs of live Nordic/arctic caterpillars at our disposal. We are also indebted to several other colleagues for information: L. Aarvik (Zoologisk Museum, Oslo), R. Brown (Mississippi Entomological Museum, Mississippi), D.R. Davis (US National Museum of Natural History, Washington DC), M. Fibiger† (Sorø), J.-K. Jensen (Nólsoy, Faroe Islands), C. Gielis (Lexmond, The Netherlands), J. Kullberg (Zoological Museum, Helsinki), E. Laasonen (Helsinki), B. Landry (Muséum d'Histoire Naturelle, Genève), J. Grieshuber (Bad Griesbach), R. Jacobsen (Nordby, Samsø), D. Lafontaine and J.-F. Landry (Canadian National Collection of Insects, Arachnids and Nematodes, Ottawa), M. Nuss (Staatliche Naturhistorische Sammlungen Dresden), E. Ólafsson (Icelandic Institute of Natural History, Reykjavik), P. Nielsen (Copenhagen) and J. Viidalepp (Estonian Agricultural University, Tartu).

References

Aarvik, L. (2011): Familiy Tortricidae. *In*: Karsholt, O. & van Nieukerken, E.J. (eds): *Lepidoptera, Moths. – Fauna Europaea* version 2.4, http://www.faunaeur.org (visited 2.6.2013).

Adamski, D., Landry, J.-F., Nazari; V & Priest, R., 2014. Three new species of leaf-mining Gelechiidae (Lepidoptera) from Canada and northeastern United States. *Journal of the Lepidopterists' Society* 68: 101–123.

Agassiz, D.J.L. (1996): *Yponomeutidae (including Roeslerstammiidae)*. *In*: Emmet, A.M.: The Moths and Butterflies of Great Britain and Ireland 3, pp. 39–114, pls 9–1. – Colchester.

Ahola, M. & Silvonen, K. (2005–08): *Pohjoisen Euroopan yökkösten toukat. Larvae of Northern European Noctuidae* 1: 1–567 (2005); 2: 1–672 (208). – Vaasa.

Arenberger, E. (2005): *Pterophoridae* III. *In*: Gaedike, R. (ed): Microlepidoptera Palaearctica 12: 1–191. – Keltern.

Aurivillius, C. (1900): Lepidoptera och Coleoptera insamlada under professor A.G. Nathorst's arktiske expeditioner 1898 och 1899, under den svenska expeditionen till Beeren Eiland 1899 och under konservator G. Kolthoff's expedition till Grönland 1900. – *Öfversigt af Kongliga Vetenskaps-Akademiens Förhandlinger* 1900: 1135–1144.

Barca, E. (1937): Mikrolepidopteren aus Nordostgrönland. – *Skrifter om Svalbard og Ishavet* 71: 31–37, pl. 2–3.

Benander, P. (1950): Fjärilar, Lepidoptera: Småfjärilar, Microlepidoptera. Vecklarefjärilar, Tortricina. – *Svensk Insektfauna* 10(2): 1–173, pls. 1–9. Stockholm.

Błeszyński, S. (1965): Crambidae. *In*: Amsel, H.G., Gregor, F. & Reisser, H: *Microlepidoptera Palearctica*. 1(1): i–xlvii, 1–553; 1(2): pls 1–133. – Wien.

Bird, C.D., Hilchie, G.J., Kondla, N.G., Pike, E.M. & Sperling, F.A.H. (1995): *Alberta Butterflies*. 347 pp. – Edmonton, AB.

Boertmann, D. (2007): *Grønlands rødliste*. 152 pp. – Nuuk & Århus.

Boisduval, J.B.A.D. de, 1832–41. *Icones historiques des Lépidoptères.....d'Europe etc*. 1: 1–252, pls. 1–47; 2: 1–208, pls. 48–84. Paris. [not seen].

Bolte, K.B. (1990): Guide to the Geometridae of Canada (Lepidoptera). VI. Subfamily Larentiinae. I. Revision of the genus *Eupithecia*. – *Memoirs of the entomological Society of Canada* 151: 1–253.

Brown, J.W., Baixeras, J., Brown, J.R., Horak, M., Komai, F., Metzler, E.H., Razowski, J. & Tuck, K. (2005): Tortricidae (Lepidoptera). *World Catalogue of Insects* 5: 1–741. – Stenstrup.

Böcher, J. [2001]: *Insekter og andre smådyr – i Grønlands fjeld og ferskvand*. 302 pp. – Nuuk.

Coulson, S.J., Hodkinson, I.D., Webb, N.R. & Convey, P. (2003): A high-Arctic population of *Pyla fusca* (Lepidoptera, Pyralidae) in Svalbard? – *Polar Biology* 26: 283–285.

Danks, H.V. (1981): *Arctic arthropods. A review of systematics and ecology with particular reference to the North American fauna*. 608 pp. – Ottawa.

Downes, J.A. (1964): Arctic insects and their environment. – *The Canadian Entomologist* 96: 279–307, pl. 1–2.

Downes, J.A., (1966): The Lepidoptera of Greenland; some geographic considerations. – *The Canadian Entomologist* 98: 1135–1144.

Dreisig, H. (1981): Daily flight activity of moths in the continuous daylight of the arctic summer. – *Holarctic Ecology* 4: 36–42.

Emmet, A.M., Langmaid, J.R., Bland, K.P., Corley, M.F.V. & Razowski, J. (1996): Coleophoridae. *In*: Emmet, A.M. (ed.): *The Moths and Butterflies of Great Britain and Ireland* 3: 126–338, pls 1–8, 12–15. – Colchester.

Ferguson, D.C. (1978): Noctuoidea: Lymantriidae. *In*: Dominick, R.B. *et al.* (eds): *The Moths of America North of Mexico* 22(2): 1–110, i–x, pls A, 1–8. – London.

Ferris, C.D. (1982): Revision of the North American *Colias hecla* Lefèbvre (Pieridae: Coliadinae). – *Bulletin of the Allyn Museum* 71: 1–19.

Fibiger, M. (1993): Noctuinae II. – *Noctuidae Europaea* 2: 1–220. Sorø.

Gaedike, R. (2009): A new synonym of *Trichophaga tapetzella* (Linnaeus, 1758) (Lepidoptera: Tineidae). – *Entomologist's Gazette* 60: 85–87.

Goater, B. (1986): *British Pyralid Moths. A Guide to their identification*. Frontispiece + 175 pp. – Colchester.

Goater, B., Ronkay, L. & Fibiger, M. (2003): Catocalinae & Plusiinae. – *Noctuidae Europaea* 10: 1–452. Sorø.

Grieshuber, J., Lukhtanov, V.A. & Lvovsky, A.L. (2001): Die Lectotypen von *Colias hyperborea* Grum-Grshimailo, 1900, *Colias aquilionaris* Grum-Grshimailo, 1900, und *Colias hecla* var. *orientalis* Grum-Grshimailo, 1900 (Lepidoptera: Pieridae). – *Entomologische Zeitschrift, Stuttgart* 111: 168–172.

Hacker, H. (1998): Revision der Gattungen *Hadula* Staudinger, 1889 (=*Discestra* Hampson, 1905, =*Aglossestra* Hampson, 1905; =*Cardiestra* Boursin, 1963), *Anartomorpha* Alphéraky, 1892, *Trichanarta* Hampson, 1895, *Anarta* Ochsenheimer, 1816 und *Cardepia* Hampson, 1905 mit Beschreibung einer neuen Gattung *Hadumorpha* gen. n. (Lepidoptera, Noctuidae). – *Esperiana* 6: 577–843, pls 1–13.

Harper, M.W., Langmaid, J.R. & Emmet, A.M. (2002): Oecophoridae. – *In*: Emmet, A.M. & Langmaid, J. (eds): *The Moths and Butterflies of Great Britain and Ireland* 4(1): 43–177, pls 1–5.1–326.

Hasenfuss, I. (1960): Die Larvalsystematik der Zünsler (Pyralidae). – *Abhandlungen zur Larvalsystematik der Insekten* 6: 1–263. pp. – Berlin, Akademie-Verlag.

Henriksen, K.L. (1939): A revised index of the insects of Grønland, containing a supplement to the insect list in Kai L. Henriksen & Will. Lundbeck: Grønlands Landarthropoder (Medd. om Grønl. Bd. 22, 1917). – *Meddelelser om Grønland* 119(10): 1–112.

Henriksen, K.L. & Lundbeck, W. (1917): Landarthropoder (Insecta et Arachnida). – *Meddelelser om Grønland* 22: 481–821.

Henriksen, H.J. & Kreutzer, I. (1982): *Skandinaviens Dagsommerfugle i Naturen*. 215 pp. – Odense.

Hodges, R.W. (1974): Gelechioidea: Oecophoridae. *In:* Dominick, R.B. *et al.* (eds): *The Moths of America North of Mexico* 6(2): 1–142, pls 1–7, i–x. – London.

Hodges, R.W. *et al.* (eds) (1983): *Check List of the Lepidoptera of America North of Mexico including Greenland*. xxiv + 284 pp. – London.

Hodges, R.W. (1998): The Gelechioidea. Pp. 131–158. *In*: Kristensen, N.P. (ed.): Lepidoptera, Moths and Butterflies, vol. 1: Evolution, Systematics, and Biogeography. *Handbook of Zoology/Handbuch der Zoologie* 35(4): i–vi, 1–491 pp. – Berlin, New York.

Homeyer, A. von, (1874): Lepidopteren. – *Die zweite deutsche Nordpolarfahrt in den Jahren 1869 und 1870, unter Führung des Kapitän Karl Koldewey* 2: 407–410.

Huemer, P. & Karsholt, O. (2010): Gelechiidae II (Gelechiinae: Gnorimoschemini). *In*: Huemer, P., Karsholt, O. & Nuss, M. (eds): *Microlepidopetra of Europe* 6: 1–586. – Stenstrup.

ICZN (1961): *International code of zoological nomenclature adopted by the XV international congress of Zoology*. xviii + 176 pp. – London.

ICZN (1999): *International code of zoological nomenclature*. 4th ed. xxix + 306 pp. London.

Iversen, J. (1934): Moorgeologische Untersuchungen auf Grönland. Ein Beitrag zur Beleuchtung der Ursachen des Unterganges der mittelalterlichen Nordmännerkultur. – *Meddelelser fra Dansk Geologisk Forening* 8: 341–358.

Johannesen, P.O. (1990): Observationer af sommerfugle i Syd- og Sydvestgrønland (Lepidoptera). – *Entomologiske Meddelelser* 58: 1–8.

Kaila, L., Mutanen, M. & Nyman, T. (2011): Phylogeny of the megadiverse Gelechioidea (Lepidoptera): Adaptations and determinents of success. – *Molecular Phylogenetics and Evolution* 61: 801–809.

Kaitila, J.-P. (1996): Suomen jäytäjäkoiden (Gelechiidae) elintavat. – The biology of the species of Gelechiidae occuring in Finland (Lepidoptera: Gelechiidae). – *Baptria* 21: 81–105.

Karsholt, O. & Rutten, T. (2005): The genus *Bryotropha* Heinemann in the western Palaearctic (Lepidoptera: Gelechiidae). – *Tijdschrift voor Entomologie* 148: 77–207.

Knaben, N. (1937): Makrolepidopteren aus Nordostgrönland. – *Skrifter om Svalbard og Ishavet* 71: 3–30, pl. 1–2.

Kononenko, V. (2005): An annotated check list of the Noctuidae (s.l.) (Lepidoptera, Noctuoidea: Nolidae, Erebidae, Micronoctuidae, Noctuidae) of the Asian part of Russia and the Ural region. – *Noctuidae Sibiricae* 1: 1–242. Sorø.

Kononenko, V.S., Lafontaine, J.D. & Mikkola, K. (1989): An annotated Check List of Noctuid Moths (Lepidoptera, Noctuidae) of Beringia. – *Entomologicheskoe Obozrenie* 68: 549–567 [In Russian; English translation: *Entomological Review* 69: 117–138, 1990].

Koponen, S. (1981): *Coleophora glaucicolella* (Lepidoptera, Coleophoridae) new to the fauna of Greenland. – *Annales entomologica Fennica* 47: 28.

Kozlov, M.V., Kullberg, J. & Dubatolov, V.V. (2006): Lepidoptera of the Taymyr peninsula, northwestern Siberia. – *Entomologica Fennica* 17: 136–152.

Kristiansen, H.E. (1974): Sommerfuglefangst på Grønland. – *Lepidoptera* 2: 209–214.

Kristiansen, H.E. (1975): Rettelse. – *Lepidoptera* 2: 252.

Kristiansen, H.E. (1982): Knud Rasmussen Mindeekspedition 1979. – *Lepidoptera* 4: 137–147.

Kuznetsov, V.I. (1989): 21. Family Tortricidae (Olethreutidae, Cochylidae) – *In*: Medvedev, G.S. (ed.): *Keys to the Fauna of the European part of the USSR* 4(1): Tortricid Moths: 279–956.

Kyrki, J. (1989): Reassessment of the genus *Rhigognostis* Zeller, with descriptions of two new and notes on further seven Palaearctic species (Lepidoptera: Plutellidae). – *Entomologica Scandinavica* 19: 437–453.

Lack, D. (1934): Some Insects from the Scoresby Sound Region, East Greenland, with the Account of the Fauna of a Nunatak. – *The Annals and Magazine of Natural History* (Ser. 10) 14: 599–606.

Lafontaine, J.D. (1987): Noctuoidea, Noctuidae (part): Noctuinae, Noctuini (part – *Euxoa*). *In*: Dominick, R.B. *et al.* (eds): *The Moths of America North of Mexico* 27(2): 1–237. – Washington DC.

Lafontaine, J.D. (1998): Noctuoidea, Noctuidae (part): Noctuinae, Noctuini (part). *In*: Dominick, R.B. *et al.* (eds): *The Moths of America North of Mexico* 27(3): 1–348. – Washington DC.

Lafontaine, J.D. (2004): Noctuoidea, Noctuidae (part): Noctuinae, Agrotini. *In*: Hodges, R.W. *et al.* (eds): *The Moths of America North of Mexico* 27(1): 1–385. – Washington DC, London.

Lafontaine, J.D., Kononenko, V.S. & McCabe, T.L. (1986): A review of the *Lasionycta leucocycla* complex (Lepidoptera: Noctuidae) with descriptions of three new subspecies. – *The Canadian Entomologist* 118: 255–279.

Lafontaine, J.D. & Poole, W. (1991): Noctuoidea, Noctuidae (part), Plusiinae. *In*: Dominick, R.B. *et al.* (eds): *The Moths of America North of Mexico* 25(1): 1–182. – Washington DC.

Lafontaine, J.D. & Schmidt, B.C. (2010): Annotated check list of the Noctuoidea (Insecta, Lepidoptera) of North America north of Mexico. – *ZooKeys* 40: 1–239.

Lafontaine, J.D. & Schmidt, B.C. (2013a): Comments on differences in classification of the superfamily Noctuoidea (Insecta, Lepidoptera) between Eurasia and North America. – *ZooKeys* 64: 209–217.

Lafontaine, J.D. & Schmidt, B.C. (2013b): Additions and corrections to the check list of the Noctuoidea (Insecta, Lepidoptera) of North America north of Mexico. – *ZooKeys* 264: 227–236.

Lafontaine, J.D. & Wood, D.M. (1997): Butterflies and Moths (Lepidoptera) of the Yukon. *In:* Danks, H.V. & Downes, J.A. (eds): *Insects of the Yukon:* 723–785. – Biological Survey of Canada, Ottawa.

Landry, J.-F. (1991): Systematics of Nearctic Scythrididae (Lepidoptera: Gelechioidea): phylogeny and classification of supraspecific taxa, with a review of described species. – *Memoirs of the entomological Society of Canada* 160: 1–341.

[Langer, T.W.]: (1967): Mindre meddelelser. – *Lepidoptera* 1(5): 102–104.

Langer, T.W. (1968): Udforskningen af Grønlands sommerfuglefauna 1780–1890 – et historisk overblik. – *Naturens Verden* 1968: 56–64, 112–128, 161–178.

Layberry, R.A., Hall, P.W. & Lafontaine, J.D. (1998): *The Butterflies of Canada.* viii + 280 pp, 32 pls. – Toronto., Buffalo, London.

Lefebvre, M.A. (1836): Description de quelques Lépidoptères nocturnes hyperboréens. – *Annales de la Société Entomologique de France* 5: 389–401, pl. 10.

Lesse, H. de (1951): Expéditions Polaires Françaises (Missions Paul-Emilie Victor). Zoologie. – 4e note: Macrolepidoptera. – *Annales de la Société Entomologique de France* 118(1949): 51–78.

Lesse, H. de & Viette, P. (1949): Expéditions Polaires Françaises (Missions Paul-Emilie Victor). Campagne 1949 au Groenland. Zoologie. – Première Note: Microlepidoptera. – *Annales de la Société Entomologique de France* 115(1946): 81–92.

Lukhtanov, V. & Lukhtanov, A. (1994): Die Tagfalter Nordwestasiens (Lepidoptera, Diurna). *Herbipoliana* 3: 1–440. – Marktleuthen.

Mackay, M.R. & Downes, J.A. (1969): Arctic Olethreutinae (Lepidoptera: Tortricidae). – *The Canadian Entomologist* 101: 1048–1053.

Mackay, M.R. (1972a): Larval sketches of some Microlepidoptera, chiefly North American. – *Memoirs of the entomological Society of Canada* 88: 1–83.

Mackay, M.R. (1972b): The larvae of Canadian Arctic Noctuidae. – *The Canadian Entomologist* 104: 859–872.

Mackay, M.R. (1972c): The larva of *Nomophila noctuella* (Lepidoptera: Pyralidae). – *The Canadian Entomologist* 104: 1479–1481.

McCabe, T.L. (1980): A reclassification of the *Polia* complex for North America (Lepidoptera: Noctuidae). – *New York State Museum Bulletin* 432: 1–141.

Meyrick, E. [1928]: *A revised handbook of Lepidoptera.* vi + 914 pp. – London.

Meyrick, E. (1937): *Exotic Microlepidoptera* 5(5): 129–160. – Marlborough.

Mikkola, K. (1968): Transportation by Air of *Nomophila noctuella* Schiff. (Lep., Pyralidae) to Greenland from North America in 1954. – *Entomologiske Meddelelser* 36: 509–510.

Mikkola, K. & Lafontaine, J.D. (1986): A preliminary note on the taxonomy of the *Apamea zeta* complex, with the first report of *A. zeta* from Fennoscandia (Lepidoptera, Noctuidae). – *Notulae Entomologica* 66: 91–95.

Mikkola, K., Lafontaine, J.D. & Gill, J.D. (2009): Noctuoidea, Noctuidae (part): Xyleninae (Part): Apameini (Part – Apamea group of genera). *In*: Hodges, R.W. *et al.* (eds): *The Moths of America North of Mexico* 26(9): 1–192. Washington DC.

Miller, W.E. & Jalava, J. (2000): Boreal Olethreutini 2. (Tortricidae): Wing and genitalia illustrations, a new synonymy, and a new Holarctic addition. – *Journal of the Lepidopterists' Society* 54: 47–51.

Mironov, V. (2003): Larentiinae II (Perizomini and Eupitheciini). – *The Geometrid Moths of Europe* 4: 1–464.

M'Lachlan, R. (1878): Report on the Insecta (including Arachnida) collected by Capt. Feilden and Mr. Hart between the parallels of 78° and 83° north latitude, during the recent Arctic expedition. – *Journal of the Linnean Society of London* (Zoology) 14: 98–122. [not seen].

Morewood, W.D. & Lange, P. (1997): Immature stages of high arctic *Gynaephora* species (Lymantriidae) and notes on their biology at Alexandra Fiord, Ellesmere Island, Canada. – *Journal of the Research on Lepidoptera* 34 (1995): 119–141.

Munroe, E. (1951): The geographic variation of *Dasyuris polata* (Duponchel) in North America (Lepidoptera: Geometridae). – *The Canadian Entomologist* 83: 290–294.

Munroe, E. (1972): Pyraloidea, Pyralidae, comprising subfamilies Scopariinae, Nymphulinae. *In*: Dominick, R.B. *et al.* (eds): *The Moths of America North of Mexico* 13(1A): 1–134, pls 1–4,A–C, G–H. – London.

Munroe, E. (1973): A supposedly Cosmopolitan insect: The Celery Webworm and allies, genus *Nomophila* Hübner (Lepidoptera: Pyralidae: Pyraustinae). – *The Canadian Entomologist* 105: 177–216.

Munroe, E. & Solis, M.A. (1999): The Pyraloidea. Pp. 233–256. *In*: Kristensen, N.P. (ed.): Lepidoptera, Moths and Butterflies, vol. 1: Evolution, Systematics, and Biogeography. *Handbook of Zoology/Handbuch der Zoologie* 35(4): i–x, 1–491. – Berlin, New York.

Mutanen, M., Hausmann, A., Hebert, P.D.N., Landry, J.-F., Waard, J.R. de & Huemer, P. (2012): Allopatry as a Gordian Knot for Taxonomists: Patterns of DNA Barcode Divergence in Arctic-Alpine Lepidoptera. – *Plos ONE* 7(10): 1–9, e47214. doi:10.1371/journal.pone.0047214.

Neunzig, H.H. (1987): Pyralidae (Pyraloidea). *In*: Stehr, F. (ed.): *Immature Insects* 1: 462–494. – Dubuque: Kendall/Hunt.

Neunzig, H.H. (1990): Pyraloidea, Pyralidae (part). Phycitinae (part). *In*: Dominikc, R.B. *et al.* (eds): *The Moths of America North of Mexico* 15(3): 1–165. – Washington DC.

Neunzig, H.H. (2003): Pyraloidea, Pyralidae (part). Phycitinae (part). *In*: Dominick, R.B. *et al.* (eds): *The Moths of America North of Mexico* 15(5): 1–338. – Washington DC.

Nielsen, J.C. (1910): A catalogue of the insects of North-East Greenland with descriptions of some larvae. – *Meddelelser om Grønland* 43: 55–68, pl. 7–8.

Nieukerken, E.J. van, Kaila, L., Kitching, I.J., Kristensen, N.P., Lees, D.C., Minet, J., Mitter, C., Mutanen, M., Regier, J.C., Simonsen, T.J., Wahlberg, N., Yen, S.-H., Zahiri, R., Adamski, D., Baixeras, J., Bartsch, D., Bengtsson, B.Å., Brown, J.W., Bucheli, S.R., Davis, D.R., De Prins, J., De Prins, W., Epstein, M.E., Gentili-Poole, P., Gielis, C., Hättenschwiler, P., Hausmann, A., Holloway, J.D., Kallies, A., Karsholt, O., Kawahara, A.Y., Koster, J.C. (Sjaak), Kozlov, M.V., Lafontaine, J.D., Lamas, G., Landry, J.-F., Lee, S., Nuss, M., Park, K.-T., Penz, C., Rota, J., Schmidt, B.C., Schintlmeister, A., Sohn, J.-C., Solis, M.A., Tarmann, G.M., Warren, A.D., Weller, S., Yakovlev, R.V., Zolotuhin, V.V. & Zwick, A. (2011): Order Lepidoptera Linnaeus, 1758. *In*: Zhang, Z.-Q. (ed.): Animal Biodiversity: An outline of higher classification and survey of taxonomic richness. – *Zootaxa* 3148: 212–221.

Nuss, M. (2005): Scopariinae. *In*: Goater, B., Nuss, M. & Speidel, W.: Pyraloidea I (Crambidae: Acentropinae, Evergestinae, Heliothelinae, Schoenobiinae, Scopariinae). *In*: Huemer, P. & Karsholt, O. (eds): *Microlepidoptera of Europe* 4: 1–304. – Stenstrup.

Obraztsov, N.S. (1963): Some North American moths of the genus *Acleris* (Lepidoptera: Tortricidae). – *Proceedings of the United States National Museum* 114: 213–270, pls 1–18.

Packard, A.S. (1876): A monograph of the geometrid moths or Phalaenidae of the United States. – *Report of the United States Geological (and Geographical) Survey of the Territories* 10: i–iv, 1–607, pls 1–13. Washington DC.

Palm, E. (1986): Nordeuropas Pyralider – med særligt henblik på den danske fauna (Lepidoptera: Pyralidae). – *Danmarks Dyreliv* 3: 1–287. København.

Patočka, J. & Turčani, M. (2005): *Lepidoptera Pupae. Central European Species* 1–2, 863 pp. – Stenstrup, Apollo Books.

Pelham-Clinton, E.C. (1985): Tineidae. *In*: Heath, J. & Emmet, A.M. (eds): *The Moths and Butterflies of Great Britain and Ireland* 2: 152–207, pls 8–9. – Colchester.

Povolný, D. (1967): Genitalia of some Nearctic and Neotropic members of the tribe Gnorimoschemini (Lepidoptera, Gelechiidae). – *Acta entomologica Musei nationalis Pragae* 37: 51–127.

Povolný, D. (1998): New taxa and faunistic records of the tribe Gnorimoschemini from the Nearctic Region (Lepidoptera, Gelechiidae). – *Stapfia* 55: 327–347.

Povolný, D. (2002): *Iconographia tribus Gnorimoschemini (Lepidoptera, Gelechiidae) Regionis Palaearcticae*. 110 pp., 16 colour pls., 87 pls. – Bratislava.

Razowski, J. (1984): Tortricini. *In*: Amsel, H.G., Gregor, F., Reisser, H. & Roesler, R.-U. (eds): *Microlepidoptera Palaearctica* 6(1): i–xv, 1–376; 6(2): 101 pls. – Karlsruhe.

Robinson, G.S. & Nielsen, E.S. (1993): Tineid genera of Australia. – *Monographs of Australian Lepidoptera* 2: xv + 344 pp. Melbourne.

Robinson, G.S. & Sattler, K. (2001): *Plutella* in the Hawaiian Islands: Relatives and Host-Races of the Diamondback Moth (Lepidoptera: Plutellidae). – *Bishop Museum Occasional Papers* 67: 1–27.

Ronkay, G. & Ronkay, L. (1995): Cuculliinae II. – *Noctuidae Europaea* 7: 1–224. Sorø.

Rutten, T. & Karsholt, O. (2004): Review of the Nearctic species of *Bryotropha* (Lepidoptera, Gelechiidae). – *Zootaxa* 740: 1–42.

Ryan, J.K. & Hergert, C.R. (1977): Energy budget for *Gynaephora groenlandica* (Homeyer) and *G. rossii* (Curtis) (Lepidoptera: Lymantriidae) on Truelove Lowland. *In*: Bliss, L.C. (ed.): *Truelove Lowland, Devon Island, Canada: A high arctic ecosystem*. 714 pp. – University of Alberta Press.

Rydell, J., Roininen, H., Philip, K.W. & Karhu, A. (2001): Lepidoptera collected in the Canadian Arctic during the Tundra Northwest 99 expedition. – *Entomologica Fennica* 12: 131–138.

Scoble, M.J. (ed.) (1999): *Geometrid Moths of the World, a Catalogue (Lepidoptera, Geometridae)*. 1016 + 129 pp. – Collingwood & Stenstrup.

Scott, J.A. (1986): *The Butterflies of North America. A Natural History and Field Guide*. xiii + 583 pp., 64 pls. – Stanford.

Simonsen, T. J. & Wahlberg, N. (2010): The evolutionary history of *Boloria* (Lepidoptera: Nymphalidae): phylogeny, zoogeography and larval–foodplant relationships. – *Systematics and Biodiversity* 8: 513–529.

Shields, O. (1992): World distribution of the *Vanessa cardui* group (Nymphalidae). – *Journal of the Lepidopterists' Society* 46: 235–238.

Sinev, S. Yu (ed.) (2008): *Katalog Cheshuekrylykh (Lepidoptera) Rossii (Catalogue of the Lepidoptera of Russia)*. 424 pp. – St. Petersburg & Moscow.

Skinner, H. (1892): [Editorial note]. – *Entomological News* 3: 49, pl. 2.

Skinner, H. & Mengel, L.W. (1892): Greenland Lepidoptera. – *Proceedings of the Academy of natural Sciences of Philadelphia* 44: 156–159.

Staudinger, O. (1857): Reise nach Island zu entomologischen Zwecken untergenommen. – *Stettiner entomologische Zeitung* 18: 209–289.

Stehr, F. (ed.) (1987): *Immature Insects* 1, – Dubuque. Kendall/Hunt, 754 pp.

Strand, E. (1905) Coleoptera, Hymenoptera, Lepidoptera und Araneae. – *Report of the Second Norwegian Arctic Expedition in the "Fram" 1898–1902*, 1(3): 1–30.

Troubridge, J.T. (1997): Revision of the Nearctic species of the genus *Entephria* Hübner (Lepidoptera: Geometridae, Larentiinae). – *Entomologica scandinavica* 28: 121–139.

Troubridge, J.T. (2008): A generic realignment of the Oncocnemidini *sensu* Hodges (1983) (Lepidoptera: Noctuidae: Oncocnemidinae), with description of a new genus and 50 new species. – *Zootaxa* 1903: 1–95.

Vestergaard, K. (1956): Konsultationsarbejdet. *Årsberetning* 1953–1954: 6–15. – Statens Skadedyrslaboratorium, Springforbi, Danmark.

Walsingham, T. de Grey (1889): Steps towards a revision of Chamber's Index, with notes and descriptions of new species (cont.) – *Insect life* 2: 23–26, 51–54, 77–81, 116–120.

Wilterding, J.H. III & Balogh, G.J. (2002): Review of the North American gray *Pyla* Grote (Lepidoptera: Pyralidae: Phycitinae) with description of a new species from western United States. – *Proceedings of the entomological Society of Washington* 104: 485–504.

Wolff, N.L. (1964): The Lepidoptera of Greenland. – *Meddelelser om Grønland* 159(11): 1–74, pls 1–21.

Wolff, N.L. (1971a): Tillæg og rettelser til fortegnelsen over Grønlands sommerfugle. – *Entomologiske Meddelelser* 39: 71–79.

Wolff, N.L. (1971b): Lepidoptera. – *The Zoology of Iceland* 3(45): 1–193, pls 1–15.

Wolff, N.L. (1975): *Nomophila*-arterne (Lepidoptera, Pyralidae) fra det nordatlantiske område. – *Entomologiske Meddelelser* 43: 129–135.

Wood, J.H. (1892): Our rush-feeding Coleophorae. – *Entomologist's monthly Magazine* 28: 117–122, 169–176, pl. 1.

Zahiri, R., Kitching, I.J., Lafontaine, J.D., Mutanen, M., Kaila, L., Holloway, J. & Wahlberg, N. (2011): A new molecular phylogeny offers hope for a stable family level classification of the Noctuoidea (Lepidoptera). – *Zologica Scripta* 40: 158–173.

Zetterstedt, J.W. (1839): Lepidoptera. – *Insecta lapponica* 4: 868–1013. Lipsiae.

Zilli, A., Ronkay L. & Fibiger, M. (2006): Apameini. – *Noctuidae Europaea* 8: 1–323.

FIGURE 12 *A–B, Gracillariidae:* Phyllonorycter ledella. *C–D, Coleophoridae:* Coleophora glaucicolella (*C*), Coleophora alticolella (*D*). *E–F, Tineidae:* Monopis laevigella (*E, Danish specimen*), Tineola bisselliella (*F, Danish specimen*). *Plutellidae:* Plutella xylostella (*G–H*), Rhigognostis senilella (*I–J*). *K & N, Oecophoridae:* Endrosis sarcitrella (*K*), Hofmannophila pseudospretella (*N*). *Elachistidae-Depressariinae:* Agonopterix antennariella (*L–M*). *Xyloryctidae-Scythridinae:* Scythris noricella (*O–P*). *Gelechiidae:* Bryotropha similis (*Q*), Gnorimoschema valesiella (*R*), Gnoriomoschema vibei (*S–T*). **The specimens figured in this and the following figures are not shown at the same magnification. Unlesss otherwise stated the specimens illustrated in this and following figures are from Greenland. Magnifications: A–B, 4x; C–T: 2.5x.**

FIGURE 13 *Tortricidae:* Acleris caryosphena (*A–D*), Acleris arcticana (*E–J*), Phiaris inquietana (*K–L*), Argyroploce mengelana (*M–N*), Epinotia indecorana (*O–Q*). Magnification: 2.5x.

FIGURE 14 *A, Pterophoridae:* Stenoptilia mengeli. *B–E, Pyralidae:* Pyla fusca (*B–C*), Plodia interpunctella (*D, Danish specimen*), Ephestia elutella (*E, Danish specimen*). *H–M, Crambidae:* Tehama bonifatella (*F–G*), Gesneria centuriella (*H–J*), Udea torvalis (*K–L*), Nomophila nearctica (*M*). *N–R, Geometridae:* Eupithecia gelidata (*N–P*), Eupithecia pusillata (*Q–R*). Magnification 2x.

FIGURE 15 *A–B, Pieridae:* Colias hecla (*A male, B female*). *C–F, Lycaenidae:* Lycaena phlaeas (*C–D, D from continental N. America*), Agriades glandon (*E male, F female*). *G–O, Nymphalidae:* Vanessa atalanta (*G, Danish specimen*), Vanessa cardui (*H*), Boloria polaris (*I & K females K underside ; J male*), Boloria chariclea (*L & N–O females, O underside ; M male*). *P–Z, Geometridae:* Psychophora sabini (*P–R*), Entephria punctipes (*S–U*), Operophtera bruceata (*V–Y males, Z female*). Natural size.

FIGURE 16 *A–C, Erebidae-Lymantriinae:* Gynaephora groenlandica (*A female, BC males*). *D–Å, Noctuidae:* Autographa gamma (*D–E*), Syngrapha u-aureum (*F–G*), Syngrapha borea (*H–I*), Syngrapha parilis (*J*), Sympistis lapponica (*K–M*), Sympistis zetterstedtii (*N–P*), Mniotype adusta (*Q–T*), Apamea zeta (*U–Y*), Lasionycta leucocycla (*Z–Ø*), Anarta trifolii (*Å*). Natural size.

FIGURE 17 *Noctuidae:* Polia richardsoni (*A–D*), Rhyacia quadrangula (*E–F*), Eurois occulta (*G–J*), Spaelotis clandestina (*K–L*), Euxoa adumbrata (*M–O*), Euxoa westermanni (*P–Q*), Agrotis ipsilon (*R–S*, *Danish specimens*). Natural size.

CHAPTER 16

Siphonaptera (Fleas)

Katharina Dittmar and Michael W. Hastriter

Fleas are small, heavily sclerotized, wingless, ectoparasitic holometabolous insects, with sucking mouthparts. The body is laterally flattened with numerous backward projecting spines and bristles; the short antennae are found in grooves on the side of the head. The hind legs are strong with enlarged coxae, femora and tibiae, modified for jumping. Eggs are white, ovoid, non-adhesive. The 3 larval instars are apodous and eyeless. The pupae are enclosed in silken cocoons. Recent molecular studies support their monophyly, and suggest a close relationship to the Mecoptera (Whiting *et al.*, 2008; Whiting, 2002; Misof *et al.*, 2014).

Morphology: adults

The morphological diversity among flea families and species is extraordinary; a glossary published in 1971 by Rothschild and Traub provides the standard for interpretations of structures by systematists. The morphological features most important for identification appear in Figs 1 and 2.

The head consists of an anterior (frons) and a dorso-posterior occipital area usually separated by the antennal fossa and an interantennal suture (if present). The preantennal part of the head bears mouthparts and sometimes genal ctenidia (comb-like series of highly modified flattened setae). In many fleas an eye-spot is present on each side, which presumably represents a rudimentary eye (Taylor *et al.*, 2005). In some species the frons has a frontal tubercle. The mouthparts are sucking with 3 piercing stylets (epipharynx and 2 maxillary laciniae); maxillae, maxillary, and labial palps are well developed. The antenna is 3-segmented (scape, pedicel and flagellum, [='clavus' in siphonapterists' terminology]), lodged in the antennal fossa. The clavus is usually comprised of nine segments, some may be fused. Male antennae are usually longer than those of females, and may be vertically extendable over the head. The occiput occasionally has an occipital groove that is always more pronounced in males (Fig. 1A–B).

Prothorax with pronotum and prosternosome, mesothorax with mesonotum, mesepisternum, mesepimeron, separated by pleural rod and mesosternum; metathorax with metanotum, metepisternum, metasternum and metepimeron (Fig. 1A–B). The pronotum might bear pronotal ctenidia. The thoracic terga have one or more rows of setae. The mesepimeron is occasionally with setae. The jumping ability of fleas is embodied in the pleural arch, formed by fusion of the ventral end of the notal ridge and the transverse ridge of metanotal area (Fig. 1A–B, D).

The abdomen has 10 terga (tg I–X) and 8 sterna (st II–IX). The sensilial plate (sensilium, present in males and females) on tergum X has a variable number of sensory pits and bears an anal stylet (Fig. 1A). In males terga VIII and IX, and sterna VIII and IX are highly modified and part of the copulatory apparatus. Abdominal terga have one or more rows of setae, and fully developed ctenidia or spinelets are sometimes present. Tergum VII usually has antesensilial bristles (except Vermipsyllidae and some Lycopsyllidae). Sternum VIII of female modified with various lobes and sinuses.

The legs have well-developed coxae, tibiae with a variable number of bristles on the dorsal margin, and five tarsal segments. The hind coxae have an inner internal rod and outer internal ridge.

The genital apparatus of the male is composed of highly modified terminal segments (terga VIII and IX, and sterna VIII and IX) and a phallosome (aedeagus). Tergum IX (clasper) comprises dual internal apodemes, manubria and external basimere (lobes and processes) and usually an articulated telomere (a finger or movable process). The female have spermathecae (used for sperm storage; divided into a head 'bulga' and a tail 'hilla') and a bursa copulatrix. Extended from the bursa copulatrix are two ducts connecting to the spermathecae. Most fleas have only one spermatheca. In such cases, the second duct is vestigial, or blind, and known as ductus obturatus. (Fig. 2A–F).

Immature stages

Siphonaptera eggs are usually white, ovoid, and non adhesive. They can be found in the surroundings of the host's living space or occasionally in the fur of the host animal.

Flea larvae are small, legless, and sparsely covered with setae. They have a pair of blunt terminal appendages, and eyes are lacking. Usually, there are 3 larval instars, which

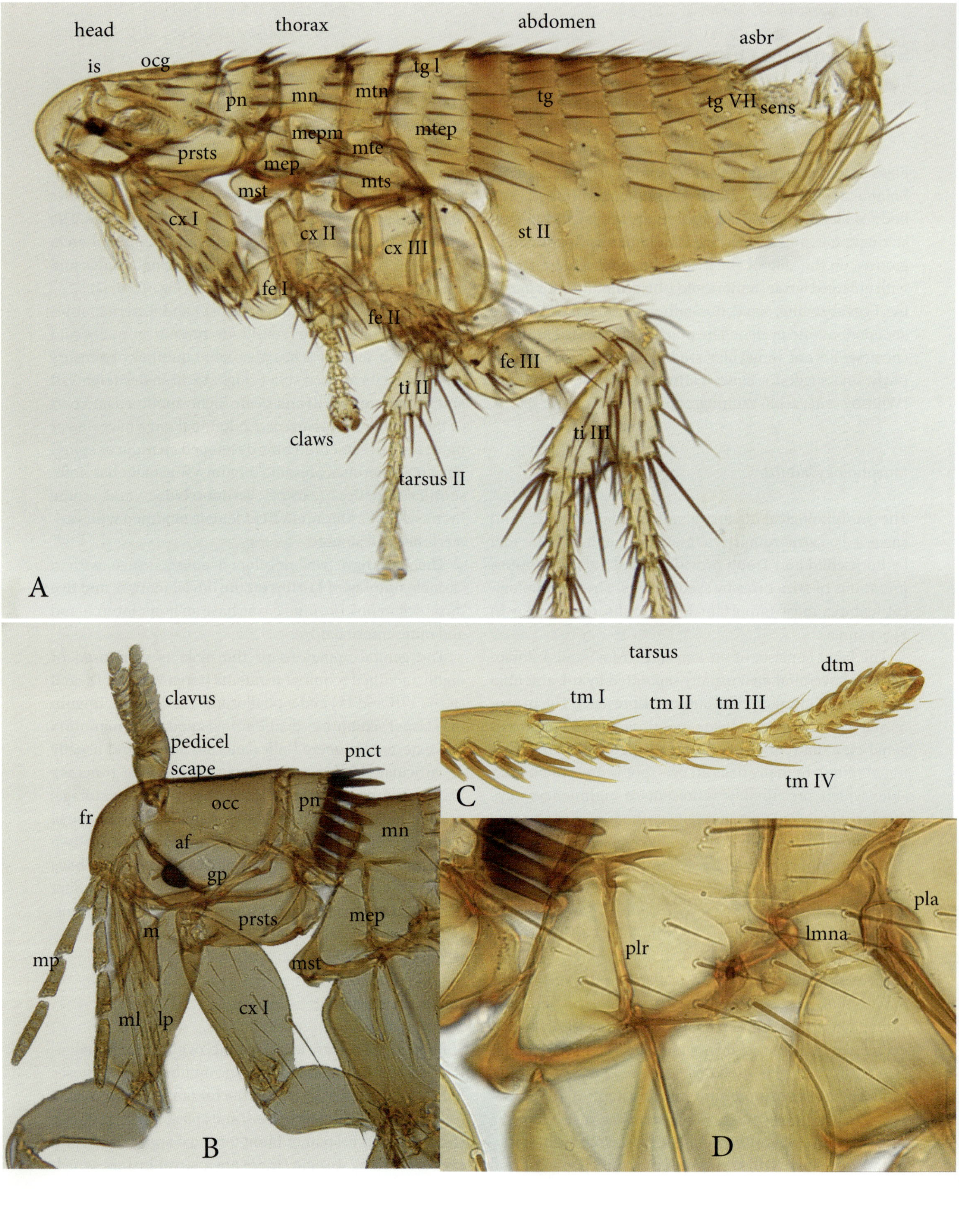
head
thorax
abdomen
asbr
is
ocg
pn
mn
mtn
tg I
tg
tg VII
sens
prsts
mepm
mte
mtep
mep
mst
mts
cx I
cx II
cx III
st II
fe I
fe II
fe III
ti II
ti III
claws
tarsus II
A
clavus
pedicel
scape
pnct
occ
pn
fr
mn
af
gp
prsts
mep
m
mp
mst
ml
lp
cx I
B
tarsus
dtm
tm I
tm II
tm III
tm IV
C
pla
lmna
plr
D

differ little, except in size, and in most cases an egg-burster (egg-tooth) is present in LI. Only the head and mouthparts are sclerotized. Flea larvae are seldom seen because they develop in the nesting material or in the surroundings of the host. They feed on organic material, including droplets of dried, digested blood from the adults. When fully developed, the larva spins a silken cocoon, into which debris is incorporated. Developmental diapause is possible in this stage (*e.g.* overwintering, waiting for hosts). For most flea larvae species identification is impossible, due to a lack of distinguishing morphology, as well as a lack of comparative studies.

Greenland fauna

Just 8 species of fleas have been documented from Greenland. They belong to the families Pulicidae: *Euhoplopsyllus glacialis glacialis* Taschenberg, 1880; *Pulex irritans* Linnaeus, 1758; Ceratophyllidae: *Ceratophyllus (Ceratophyllus) gallinae* Schrank, 1803; *Ceratophyllus (C.) vagabundus vagabundus* Boheman, 1866; *Ceratophyllus (Emmareus) garei* Rothschild, 1902; *Ceratophyllus (E.) borealis* Rothschild, 1907; *Megabothris (Amegabothris) groenlandicus* Wahlgren, 1903; and Vermipsyllidae: *Chaetopsylla (Chaetopsylla) globiceps* Taschenberg, 1880.

Simplified key to the Siphonaptera of Greenland (only applicable to known Greenland species. If specimen does not fit this key, please refer to broader literature).

1. Outer internal ridge of mid coxae absent. Sensilium with 14 pits each side. Frontal tubercle absent 2
– Outer internal ridge of mid coxae present. Sensilium with more than 14 pits each side. Frontal tubercle present 3
2. Clavus of antenna asymmetrical, last segments anteriorly fused. Pleural rod of mesothorax and pronotal ctenidium present .... *Euhoplopsyllus glacialis glacialis*
– Club of antenna symmetrical, elliptical shape. Pleural rod of mesothorax and pronotal ctenidium absent .... *Pulex irritans*
3. Pronotal ctenidium present. Antesensilial bristles present 4
– Pronotal ctenidium absent. Females with elongated body. Antesensilial bristles absent .... *Chaetopsylla (Chaetopsylla) globiceps*
4. Telomere adorned with darkly pigmented, stout setae on caudal margin. Terminal spiracle (spiracle VIII) greatly enlarged .... *Megabothris (Amegabothris) groenlandicus*
– Setae on caudal margin of telomere slender, not stout or darkly pigmented. Spiracle VIII not enlarged 5
5. Penis rods only slightly extending beyond apex of aedeagal apodeme (not coiled). Tooth (spur) on posteroventral margin of prosternosome (both sexes). Bulga of spermatheca wider in middle than ends (lemon shaped) ....*Ceratophyllus (Emmareus) garei* & *C. (E.) borealis*
– Penis rods extending far beyond apex of aedeagal apodeme (coiled like watch spring). Prosternosome without tooth (spur). Sides of bulga of spermatheca somewhat parallel (sausage shaped) 6
6. Apex of telomere acutely angled. Hilla distinctly shorter than bulga .... *Ceratophyllus (Ceratophyllus) v. vagabundus*
– Apex of telomere obtusely rounded. Hilla subequal in length to bulga .... *Ceratophyllus (Ceratophyllus) gallinae*

***Euhoplopsyllus glacialis glacialis* (Taschenberg, 1880)**
Figs. 3, 4
Henriksen 1939: *Hoplopsyllus glacialis* Tasch.

Clavus of antenna with ventral side distinctly segmented. Large eye-spot, frons more slanting in female

FIGURE 1 (LEFT) *Morphological details and terms. A, overview of an adult male flea* (Parapulex *sp.*); (*asbr*) antesensilial bristle; (*cx I*) coxa I; (*cx II*) coxa II; (*cx III*) coxa III; (*fe I*) femur I; (*fe II*) femur II; (*fe III*) femur III; (*is*) interantennal suture; (*mep*) mesepisternum; (*mepm*) mesepimeron; (*mg*) male genitalia; (*mn*) mesonotum; (*mst*) mesosternum; (*mte*) metepisternum; (*mtep*) metepimeron; (*mtn*) metanotum; (*mts*) metasternum; (*sens*) sensillium [pygidium]; (*ocg*) occipital groove; (*pn*) pronotum; (*prsts*) prosternosome; (*stII*) sternum II; (*ti II*) tibia II; (*ti III*) tibia III; (*tg I*) tergum I. *B, overview of head and first thoracic segment;* (*af*) antennal fossa; (*cx I*) coxa I; (*f*) frons; (*gp*) genal process; (*lp*) labial palpus; (*m*) maxilla; (*mep*) mesepisternum; (*ml*) maxillary laciniae; (*mn*) metanotum; (*mp*) maxillary palpus; (*mst*) mesosternum; (*occ*) occiput; (*pn*) pronotum; (*pnct*) pronotal ctenidium; (*pst*) prosternosome. *C, overview of tarsus;* (*dtm*) distotarsomere; (*tm I*) tarsomere I; (*tm II*) tarsomere II; (*tm III*) tarsomere III; (*tm IV*) tarsomere IV. *D, pleural arch and pleural rod;* (*lmna*) *lateral metanotal area;* (*pla*) pleural arch; (*plr*) pleural rod.

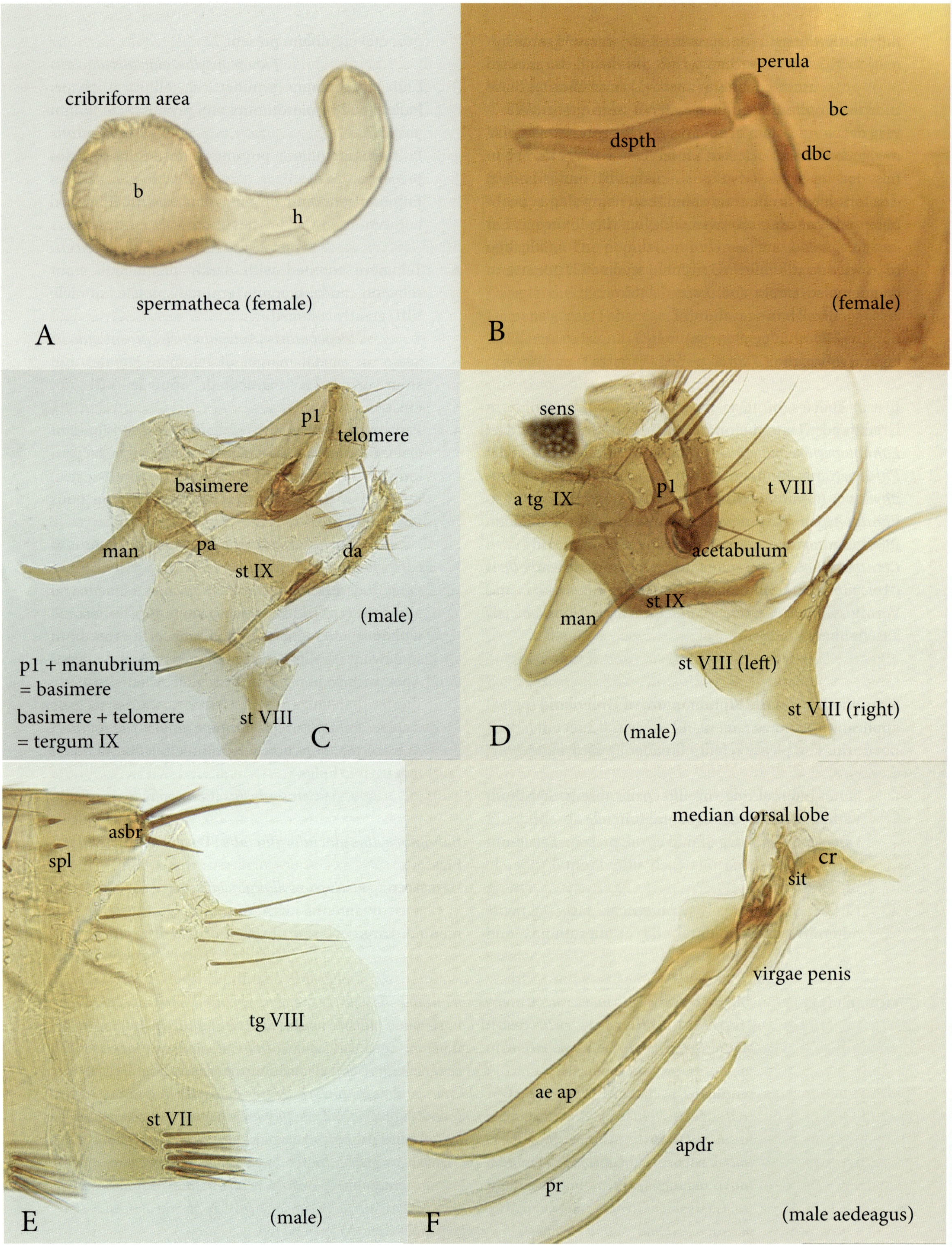
cribriform area
b
h
spermatheca (female)
A
perula
bc
dspth
dbc
(female)
B
p1
telomere
basimere
man
pa
da
st IX
(male)
p1 + manubrium
= basimere
basimere + telomere
= tergum IX
st VIII
C
sens
a tg IX
p1
t VIII
acetabulum
st IX
man
st VIII (left)
st VIII (right)
(male)
D
asbr
spl
tg VIII
st VII
E
(male)
median dorsal lobe
cr
sit
virgae penis
ae ap
apdr
pr
F
(male aedeagus)

than in male. Labial palpus reaching fore coxa, with terminal segment as long as first segment of middle tarsus. Pronotal ctenidium with 7–8 sharply pointed, slender spines on each side. Abdominal tergites with row of 12–15 setae, male sternum III to VI with 4 to 5 setae, female with 5 or 6. Hind tibia with 8 to 9 dorsal notches with stout bristles (including apical notch). Hind coxa with patch of 12–20 short spiniform bristles on inner surface. – *Euhoplopsyllus g. glacialis* is very similar to *Euhoplopsyllus g. lynx* Baker, 1904, and they could, in fact, be geographical variants of the same species. They are separable only by the larger size (2.5 to 2.75 mm) of *E. g. glacialis*. *Male* (Fig. 3): P1 (fixed process) of clasper divided into 2 lobes (L1 & L2). Quadrate dorsal lobe with 12 long setae. Narrow ventral lobe with characteristic short, stout, apical spiniform seta and 3 (2) subventral or ventral setae. Lateral portion of sternum VIII strongly rounded ventrally with longitudinal row of 5 long setae. Manubrium of clasper variable in form. P3 much longer than P1, narrow at beginning and broadening apically with some thin bristles. Distal arm of sternum IX club shaped, widening gradually towards apex. Apical margin of distal arm with spiniform setae, ventral margin with numerous small, pale bristles. *Female* (Fig. 4): Basal abdominal sternum with usual postmedian pair of long setae, and small ventral setae. Sternum VII with a row of 12–19 setae on both sides. Tergum VIII with a vertical row of 4–6 setae, occasionally with 1 or 2 adjacent setae. Tergum VIII with characteristic finger-like lobe. Bulga of spermatheca subspherical, much shorter than hilla, bulga and hilla clearly distinguishable.

BIOLOGY. Mainly reported from arctic hares (*Lepus arcticus*). Single reports from *Mustela erminea* (ermine) and *Alopex lagopus* (arctic fox) are known, however, these hosts can be regarded as accidental associations.

DISTRIBUTION. S, SW, NW, N, NE. Endemic to Greenland.

Pulex irritans (Linnaeus, 1758) Figs 5, 6

Large and conspicuous eye-spot. No genal and pronotal ctenidia, but genal process is present. Labial palpus up to 3/4 length of fore coxa. Anterior angle of oral margin sometimes bearing a single, small spinelet, possibly representing a vestigial genal ctenidium. Single ocular seta anterior to eye-spot, plus single seta near the base of maxilla. Occiput with one strong seta. Row of small spiniforms on mesal surface of the hind coxa, near apex. *Male* (Fig. 5): P1 (fixed process) of clasper tubular, very large and distal end bending downwards, covering P2 and P3 completely. Dorsal side of P1 with series of small marginal and submarginal setae. P2 and P3 about 3/4 the length of P1. Ventral margin of P3 with 8–9 small, pale setae. Manubrium long and slightly curved upward. *Female* (Fig. 6): Sternum VII with 11–13 setae and incrassation, resulting in small, slightly projecting dorsal lobe. Tergum VII with strong antesensilial bristle on small protuberance, flanked by 2 small lateral bristles. Spermatheca with subglobular bulga, hilla longer than bulga.

BIOLOGY. *Pulex irritans* seems to be not very host-specific, and has been reported from a variety of mammals and birds. However, it has been reported mainly from humans and dogs in Greenland.

DISTRIBUTION. (G?). Found in a number of towns and settlements. On the "Danmark Ekspedition" it was collected on hares.

Cosmopolitan.

NOTE: *P. irritans* has a worldwide distribution and shows considerable populational variation in some morphological characters. In the past, this has led to the description of several subspecies, which are now synonymized. It is debated whether *P. simulans* Baker is a junior synonym of *P. irritans*. Molecular studies (with limited sampling) suggest that they may belong to one large metapopulation (Dittmar and Whiting, 2003). *Pulex* spp. have been suspected of a role in the epidemiology of plague (Laudisoit *et al.*, 2007).

FIGURE 2 (LEFT) *Genitalia details and terms. A, female spermatheca;* (*b*) bulga of spermatheca [head]; (*h*) hilla of spermatheca [tail]. *B, duct of spermatheca and bursa copulatrix;* (*bc*) bursa copulatrix; (*dbc*) duct of bursa copulatrix; (*dspth*) duct of spermatheca. *C, basimere, telomere and sternum VIII;* (*da*) distal arm of sternum IX; (*pa*) proximal arm of sternum IX; (*man*) manubrium; (*p1*) fixed process; (*st VIII*) sternum VIII; (*st IX*) sternum IX. *D, basimere, telomere, sternum VIII, and tergum IX;* (*a t IX*) apodeme of tergum IX; (*man*) manubrium; (*p1*) fixed process; (*sens*) sensillium; (*st VIII*) sternum VIII; (*st IX*) sternum IX. *E, terminal segments, male;* (*asbr*) antesensilial bristles; (*spl*) spinelets; (*st VII*) sternum VII; (*tg VIII*) tergum VIII. *F, aedeagus;* (*ae ap*) aedeagal apodeme, (*apdr*) apodemal rod; (*app*) apical part of aedaegal apodeme; (*cr*) crochet; (*pr*) penis rods; (*v*) vesicle (of phallosome); (*vp*) virgae penis; (*sit*) sclerotized inner tube.

***Ceratophyllus* (*Ceratophyllus*) *gallinae* (Schrank, 1903) Figs 7, 8**

Large, well sclerotized eye-spot, frontal tubercle present. Pronotal ctenidia present, with 8–10 spines on each side. Genal process with apical portion pointed and sclerotized. Marginal row of mesonotum with one strong dorsal seta, thorax slightly compressed. Prosternosome with strongly sclerotized margins, but without spur. Distotarsomere III with 5 pairs of lateral plantar bristles, and one pair of preapical plantar bristles. *Male* (Fig. 7): Internal apodeme of tergum IX triangular, with triangular lobe, separated from manubrium. Telomere with 4 setae on caudal margin, setae 2 and 3 stronger than others, longer than in other species of the "gallinae" complex. Width of manubrium tapering towards apex. Penis rods and apodemal rods making at least 1 convolution. Proximal arm of sternum IX nearly straight, upper end anvil shaped. Distal arm of sternum IX half-moon shaped, with 4 weakly sclerotized setae. Sternum VIII with short base, and elongated body with no apical setae. Crochet finger like. *Female* (Fig. 8): Basal part of ductus spermatheca not strongly dilated or wrinkled, basal part of ductus obturatus strongly sclerotized, bulga of spermatheca cylindrical and broad, much longer than hilla. Ductus spermatheca often inserted at apex of bulga. Sternum VII with 2 rows of setae. Tergum VIII usually with 5 strong marginal setae.

BIOLOGY. Most common hosts are passerine birds, nesting in shrubs and other sites above ground. Secondarily, it has become a major flea of poultry (European hen flea), and is said to have been introduced to Greenland with settlers and their domestic poultry in the 18th century. In Greenland it has been found on *Plectrophenax nivalis* (snow bunting) and *Dicrostonyx torquatus* (arctic lemming).

DISTRIBUTION. NE.

Palaearctic; introduced to many parts of the world.

***Ceratophyllus* (*Ceratophyllus*) *vagabundus vagabundus* (Boheman, 1866) Figs 9, 10**

Well-developed eye-spot, heavily pigmented. Pronotal ctenidium with 14–15 spines on each side, no genal ctenidia, genal process short, and apically pointed. Weakly developed frontal tubercle. Two occipital setae, margin of proepimeron without spur. Metepimeron without setae. Because a dissection of *C.* (*C.*) *vagabundus vagabundus* was not available, we used *Ceratophyllus* (*C.*) *vagabundus insularis*, which is a very closely related subspecies. We felt a dissection could better portray pertinent morphological characters. *Male* (Fig. 9): Internal apodeme of tergum IX triangular, separated from manubrium. Telomere with 4–5 setae (larger gap between 2 and 3) on posterior margin, apical margin oblique. Telomere about 3 times as long as wide. Width of manubrium of clasper tapering towards apex. Penis rods and apodemal rods making at least 1 convolution. Proximal arm of sternum IX formed as double wave, apex hook shaped. Distal arm of sternum IX half-moon shaped, with 6 strong setae. Tergum VIII with rounded basal portion and usually 5 dorsal setae on body. Sternum VIII elongated, with 2 strong basal setae and 3 strong apical setae. *Female* (Fig. 10): Enlarged part of ductus spermathecae and bursa copulatrix sclerotized. Bursa copulatrix with characteristic perula. Hilla of spermatheca about one third of length of bulga. Bulga cylindrical, with globular end. Ductus obturatus (blind duct) not sclerotized.

BIOLOGY. Parasite on ground-nesting birds (*e.g. Larus* spp., *Pagophila eburnea*).

DISTRIBUTION. NE (only a single record from Holmsvig, 1934).

Holarctic.

***Ceratophyllus* (*Emmareus*) *borealis* (Rothschild, 1907)**

This species is sympatric with *C.* (*E.*) *garei* across its range and is so similar morphologically to *C.* (*E.*) *garei* that the description of *C.*(*E.*) *garei* will suffice to depict this species (see below). Possibly, *C.* (*E.*) *borealis* and *C.* (*E.*) *garei* represent the same species. This species is listed in Henriksen's 'Revised index' based on a record by Carpenter (1938).

***Ceratophyllus* (*Emmareus*) *garei* (Rothschild, 1902) Figs 11, 12**

Well developed eye-spot, heavily pigmented. Pronotal ctenidia usually with 12–13 spines on each side, no genal ctenidia genal process short, well developed and apically pointed. Weakly developed frontal tubercle. Two strong occipital setae. Margin of proepimeron with characteristic, small, sclerotized "tooth". Metepimeron with one row of setae. One antesensilial bristle. *Male* (Fig. 11): Internal apodeme of tergum IX triangular, separated from manubrium. Telomere with 1 strong apical seta, and 4 evenly spaced setae on caudal margin. Manubrium U-shaped. Penis rods and apodemal rods with at most one (usually less) convolution. Proximal arm of sternum IX formed as double-wave, with upper end anvil shaped. Distal arm of sternum IX half-moon shaped with 3–4 weakly sclerotized setae. Sternum VIII with short base, elongated body with very broad vexillium and finger-like process; apex with 2 stout long setae and one short seta. Crochet hook-shaped. *Female* (Fig. 12): Basal part of ductus spermathecae dilated and wrinkled, weakly sclerotized ductus obturatus, bulga about as long as hilla, insertion of the ductus spermatheca into bulga at an angle of at least 45° with the long axis of the bulga.

BIOLOGY. Hosts are ground-dwelling birds, nesting in moist sites near the ground. In areas where *C.* (*C.*) *gallinae* is absent, it may replace it as a serious pest of poultry. In spring, *C.* (*Emmareus*) *garei* may be found in high numbers on ground-nesting birds.

DISTRIBUTION. SW, NE (only recorded from Qeqertarssuaq/Godhavn and Ella Ø).

Holarctic, arctic-temperate.

Megabothris (*Amegabothris*) *groenlandicus* (Wahlgren, 1903) Figs 13, 14

Henriksen 1939: *Ceratophyllus groenlandicus* Wahlgr.

Pronotal ctenidium with 10 or 11 spines per side, genal ctenidium absent, genal process well developed, apically pointed. Medium sized, normally sclerotized eye-spot. Postantennal region with no more than 2 setae at the base of the clavus of antenna. Typically 3 ocular setae. Frontal tubercle absent. Spiracular fossa of tergum VIII enlarged, 3 antesensilial bristles, median the longest. *Male* (Fig. 13): Basimere elongated into lobe, internal apodeme triangular. Telomere twice as long as fixed process (P^1). 3 modified setae on caudal margin of telomere. 2 ventral on protuberance (one sharp, one blunt) widely separated from blunt spiniform at dorsocaudal apex. Proximal end of telomere very narrow (half the width of apex). Sternum VIII with forked base, club-shaped body, apex with 2 long, fine setae. Rows of 7–8 small setae along ventral margin. Crochet hook-shaped. Penis rod not coiled, apodemal rod of sternum IX extending to apex of aedeagal apodeme. *Female* (Fig. 14): Bursa copulatrix smooth sided. Large section of duct of bursa copulatrix and portion of ductus obturatus heavily sclerotized. Bulga of spermatheca barrel-shaped and slightly longer than hilla. Apex of hilla striated. Sternum VII with 12–18 short setae and 6–7 long ones, usually with distinct lobe, sometimes with a sinus.

BIOLOGY. Primarily associated with *Dicrostonyx* spp. (lemming). Preferred hosts *Dicrostonyx torquatus* and *D. hudsonius*, but also reported on *Lemmus sibiricus* and other microtines.

DISTRIBUTION. In Greenland only reported from NE.

Nearctic Region.

NOTE. This species was not represented in the collection of the ZMUC (Zoological Museum of Copenhagen); the account is based on Canadian material lent by the Canadian National Collection of Insects, Arachnids and Nematodes, Ottawa, Canada.

Chaetopsylla (*Chaetopsylla*) *globiceps* (Taschenberg 1880) Figs 15, 16

Henriksen 1939: *Vermipsylla globiceps* Tasch.

Large flea, with elongated appearance. Very characteristic, well developed frontal tubercle. Ctenidia absent, genal process pointed with apical part sclerotized. Large, well developed eye-spot, but lightly pigmented. Labial palpus 5-segmented, extending beyond fore coxa. One long and one short stout apicolateral spine on tibia I. At least one hair-like seta on segment IV of hind tarsus extending to or beyond apex of segment V. Five lateral plantar bristles on distotarsomere. Abdominal terga with narrow supraspiracular region and broader subspiracular region. Two rows of setae on abdominal terga, with anterior row reduced to one to 2 setae on posterior segments. Metepimeron with 2 rows of setae. Antesensilial bristles absent. *Male* (Fig. 15): Basimere broadly rounded, fringed with long setae. Acetabulum of telomere situated mesad at upper half of basimere, apex of telomere barely reaching margin of basimere. Manubrium tapering ventrad. Sclerotized inner tube narrow and elongate. Median dorsal lobe bulbous. Ventral portion of aedeagal pouch sclerotized, lateral lobes lacking. *Female* (Fig. 16): Spermatheca with small ovoid bulga. Hilla narrow and 2.5 X longer than bulga. Bursa copulatrix short, broad and lightly sclerotized. Apical margin of sternum VII with a deep, triangular sinus (sometimes more rounded).

BIOLOGY. *Chaetopsylla globiceps* has been reported mainly from carnivores. Known hosts in Greenland include *Alopex lagopus* (arctic fox) and *Mustela erminea* (ermine).

DISTRIBUTION. S, NE (Qaqortoq/Julianehåb, Eskimonæs etc.).

Holarctic.

References

Carpenter, G.D.H. (1938): Notes on insects collected in West Greenland by the Oxford University Greenland Expedition, 1936. With descriptions of a new species of *Angitia* (Hymenoptera: Ichneumonidae) by A. Roman, and of *Fannia* (Diptera, Anthomyidae) by J.E. Collin. – *The Annals and Magazine of Natural History* (11) 1: 529–553.

Cheetham, T.B. (1988): Male genitalia and the phylogeny of the Pulicoidea (Siphonaptera). – Theses Zoologicae 8; vii + 224 pp.

Dittmar, K. & Whiting, M.F. (2003): Genetic and phylogeographic structure of populations of *Pulex simulans* (Siphonaptera) in Peru inferred from two genes (CytB and CoII). – *Parasitology Research* 91: 55–59.

Henriksen, K.L. (1939): A revised index of the insects of Grønland. – *Meddelelser om Grønland* 119 (10): 112 pp.

Holland, G.P. (1985): The fleas of Canada, Alaska, and Greenland (Siphonaptera). – *Memoirs of the Entomological Society of Canada*, Ottawa 117/Supplement S130, 632 pp.

Hopkins, G.H.E. & Rothschild, M. (1953): *An illustrated catalogue of the Rothschild Collection of Fleas (Siphonaptera) 1: Tungidae, Pulicidae.* – British Museum of Natural History. London, 361 pp.

Hopkins, G.H.E. & Rothschild, M. (1956): *An illustrated catalogue of the Rothschild Collection of Fleas (Siphonaptera) 2: Coptopsyllidae, Vermipsyllidae, Stephanocircidae, Ischnopsyllidae, Hypsophthalmidae, and Xiphiopsyllidae.* – British Museum of Natural History. London, 445 pp.

Laudisoit, A., Leirs, H., Makundi, M., Van Dongen, S., Davis, S., Neerinckx, S., Deckers, J. & Libois, R. (2007): Plague and the human flea, Tanzania. – *Emerging Infectious Diseases* 13: 687–693.

Misof, B., Liu, S.-l., Meusemann, K. [99 authors] (2014): *Phylogenomics resolves the timing and pattern of insect evolution.* – Science 346 (6210): 763–767.

Traub, R., Rothschild, M. & Haddow, J. (1983): *The Rothschild Collection of fleas: The Ceratophyllidae. Key to genera and host relationships.* – British Museum of Natural History. London, 288 pp.

Whiting, M.F. (2002): Mecoptera is paraphyletic: multiple genes and phylogeny of Mecoptera and Siphonaptera. – *Zoologica Scripta* 31: 93–104.

Whiting, M.F., Whiting, A.S., Hastriter, M.W. & Dittmar, K. (2008): A molecular phylogeny of fleas (Insecta: Siphonaptera): origins and host associations. – *Cladistics* 24: 677–707.

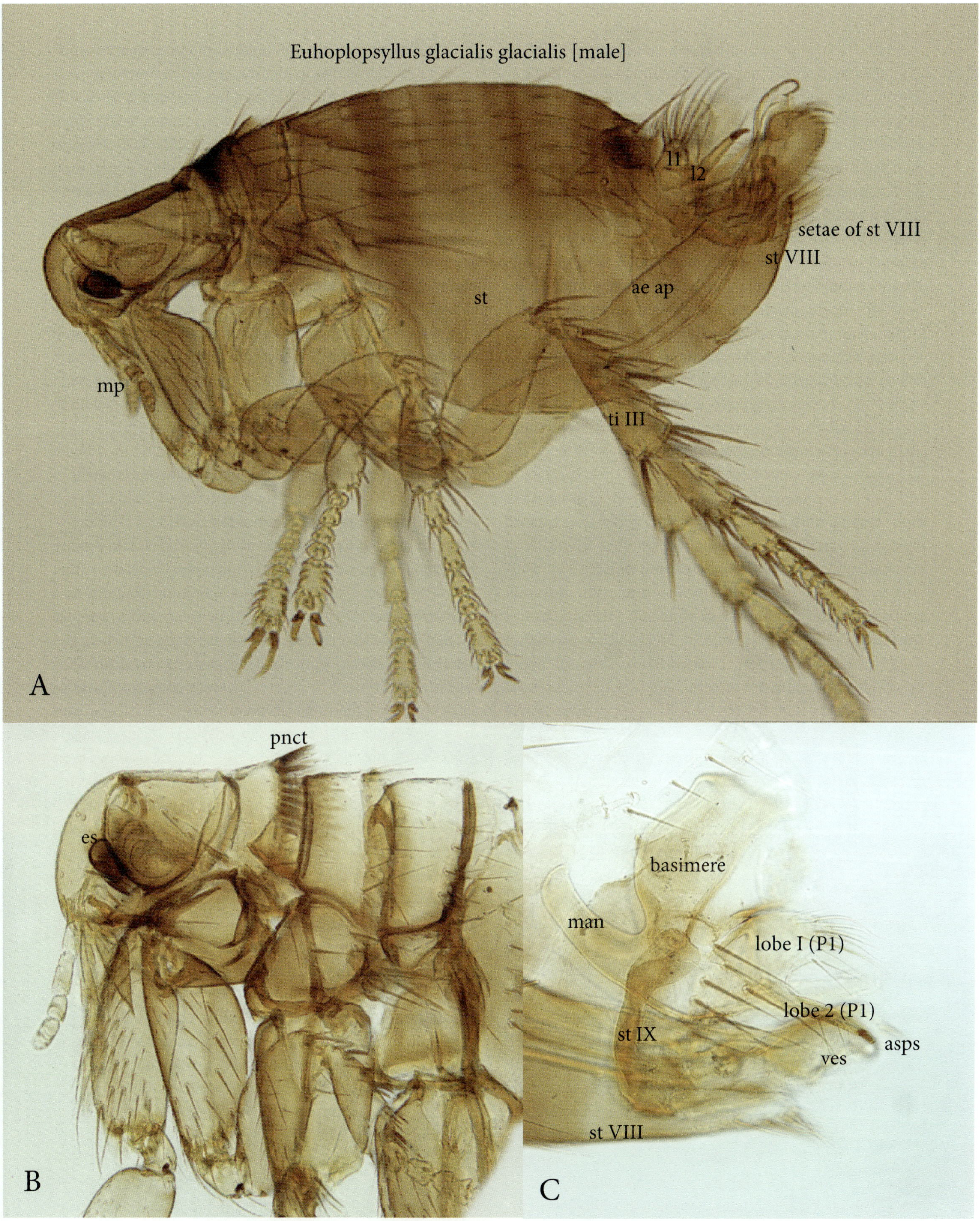

FIGURE 3 *Male* Euhoplopsyllus glacialis glacialis. A, *overview;* (*ae ap*) aedeagal apodeme*;* (*mp*) maxillary palpus*;* (*st*) sternum*;* (*st VIII*) sternum VIII*;* (*ti III*) tibia III. *B, head and pronotum;* (*es*) eye-spot*;* (*pnct*) pronotal ctenidium. *C, terminalia;* (*asps*) apical spiniform seta*;* (*man*) manubrium*;* (*st VIII*) sternum VIII*;* (*st IX*) sternum IX*;* (*ves*) vesicula.

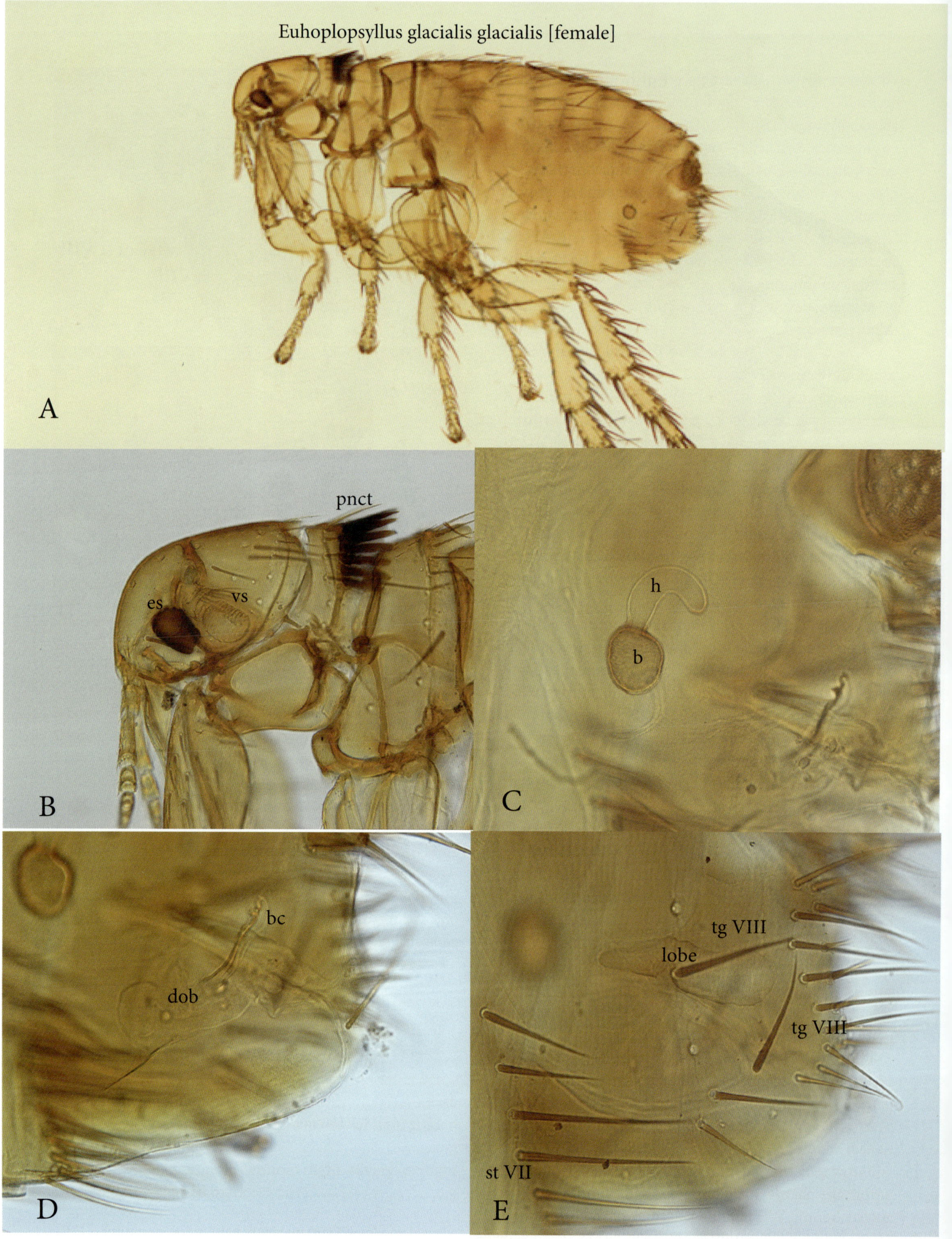

FIGURE 4 *Female* Euhoplopsyllus glacialis glacialis. *A, overview. B, head and anterior thorax; (es)* eye-spot*; (pnct)* pronotal ctenidium*; (vs)* ventral (segmented) side of antenna. *C–E, genitalia; (b)* bulga*; (bc)* bursa copulatrix*; (dob)* ductus obturatus*; (h)* hilla*; (st VII)* sternum*; (tg VIII)* tergum *VIII.*

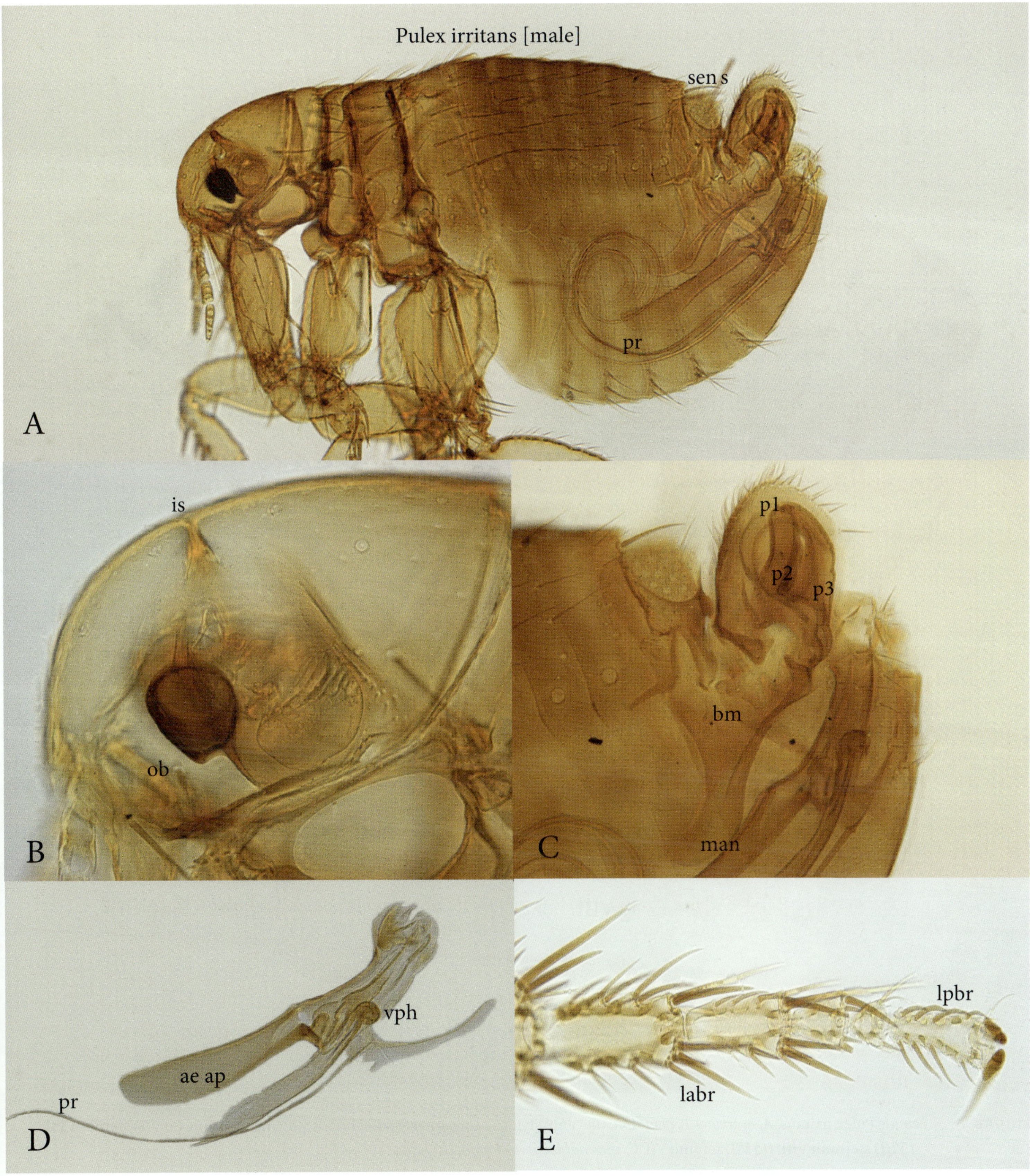

FIGURE 5 *Male* Pulex irritans. *A, overview.* (*pr*) penis rods*;* (*sen*) sensillium. *B, head;* (*is*) interantennal suture*;* (*ob*) ocular bristle. *C–D, terminalia;* (*ae ap*) aedeagal apodeme*;* (*pr*) penis rods*;* (*bm*) basimere*;* (*man*) manubrium*;* (*p1*) fixed process of clasper*;* (*p2*) movable process 1*;* (*p3*) movable process 2. (*vph*) vesicula of phallosome. *E, distotarsomere III;* (*labr*) lateral apical bristles*;* (*lpbr*) lateral plantar bristles.

FIGURE 6 *Female* Pulex irritans. *A, overview. B, genitalia;* (*as*) anal stylet*;* (*asbr*) antesensilial bristle*;* (*ic*) incrassation*;* (*st VII*) sternum VII*;* (*st VIII*) sternum VIII*;* (*tg VII*) tergum VII. *C, spermatheca;* (*b*) *bulga;* (*h*) *hilla.*

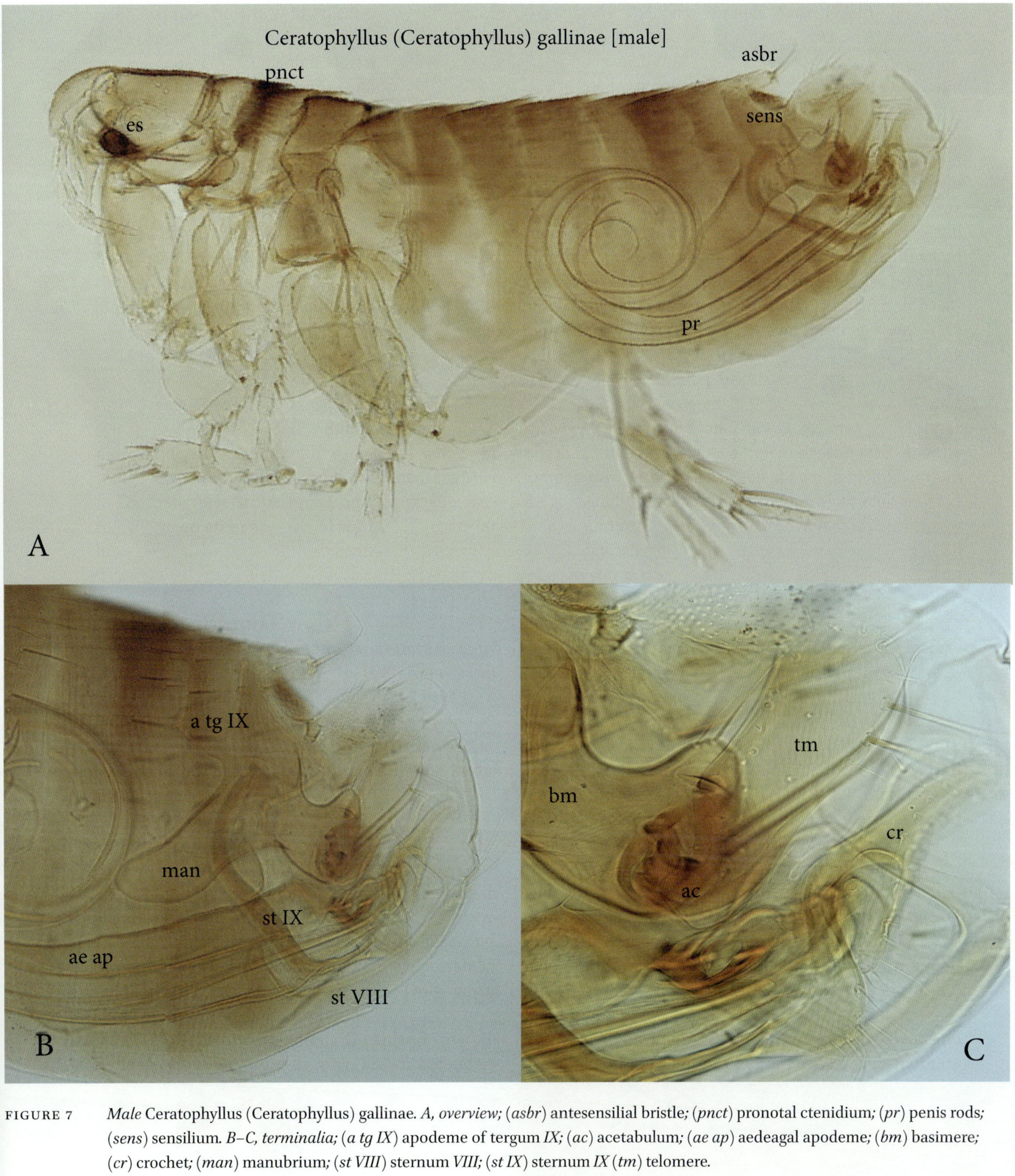

FIGURE 7 *Male* Ceratophyllus (Ceratophyllus) gallinae. *A, overview; (asbr)* antesensilial bristle*; (pnct)* pronotal ctenidium*; (pr)* penis rods*; (sens)* sensilium. *B–C, terminalia; (a tg IX)* apodeme of tergum *IX; (ac)* acetabulum*; (ae ap)* aedeagal apodeme*; (bm)* basimere*; (cr)* crochet*; (man)* manubrium*; (st VIII)* sternum *VIII; (st IX)* sternum *IX (tm)* telomere.

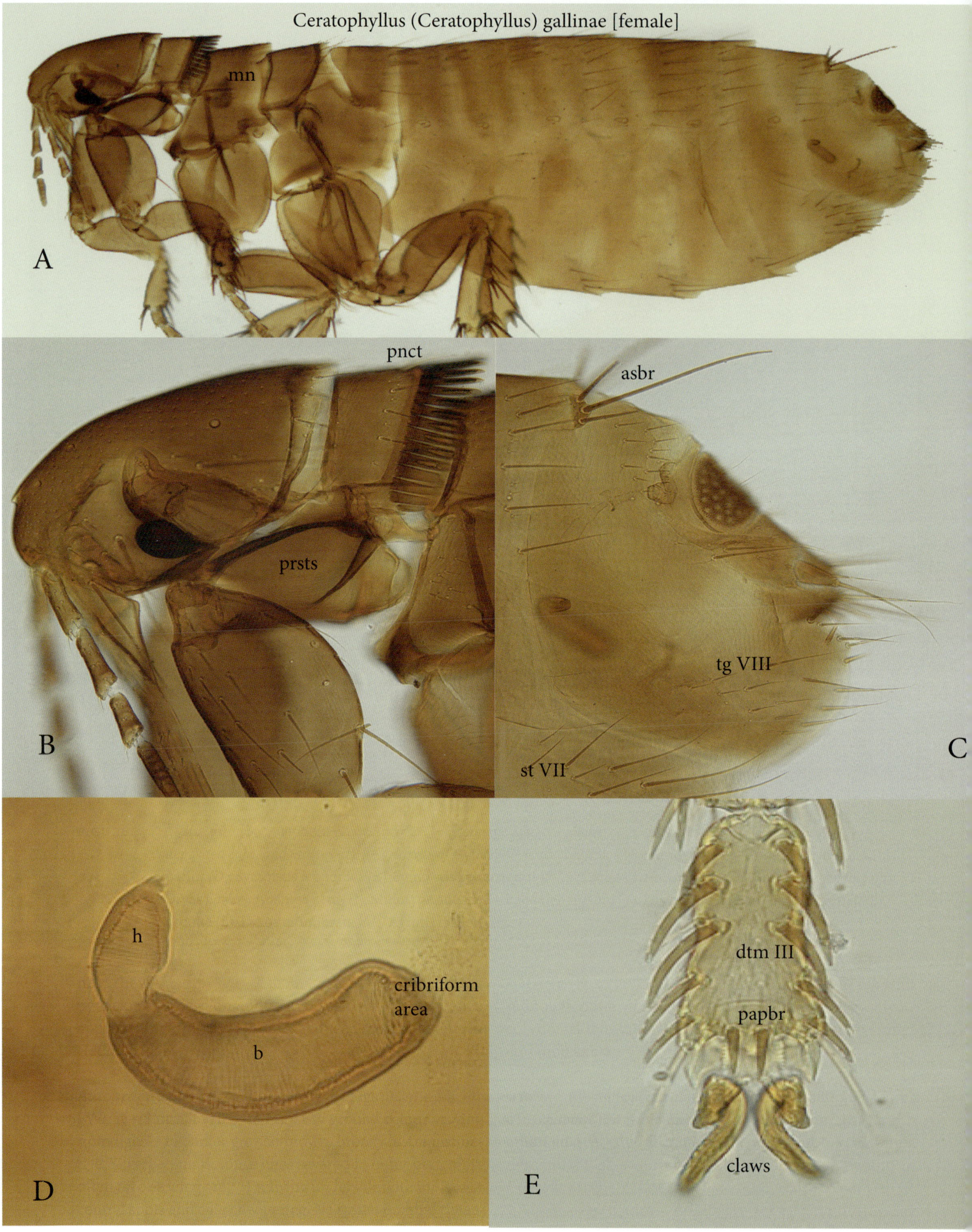

FIGURE 8 *Female* Ceratophyllus (Ceratophyllus) gallinae. *A, overview.* (*mn*) mesonotum. *B, head;* (*pnct*) pronotal ctenidium; (*prsts*) prosternosome *C, terminalia;* (*asbr*) antesensilial bristles; (*st VII*) sternum *VII;* (*tg VIII*) tergum *VIII. D, spermatheca;* (*b*) bulga; (*h*) hilla. *E, distotarsomere* (*papbr*) preapical plantar bristles.

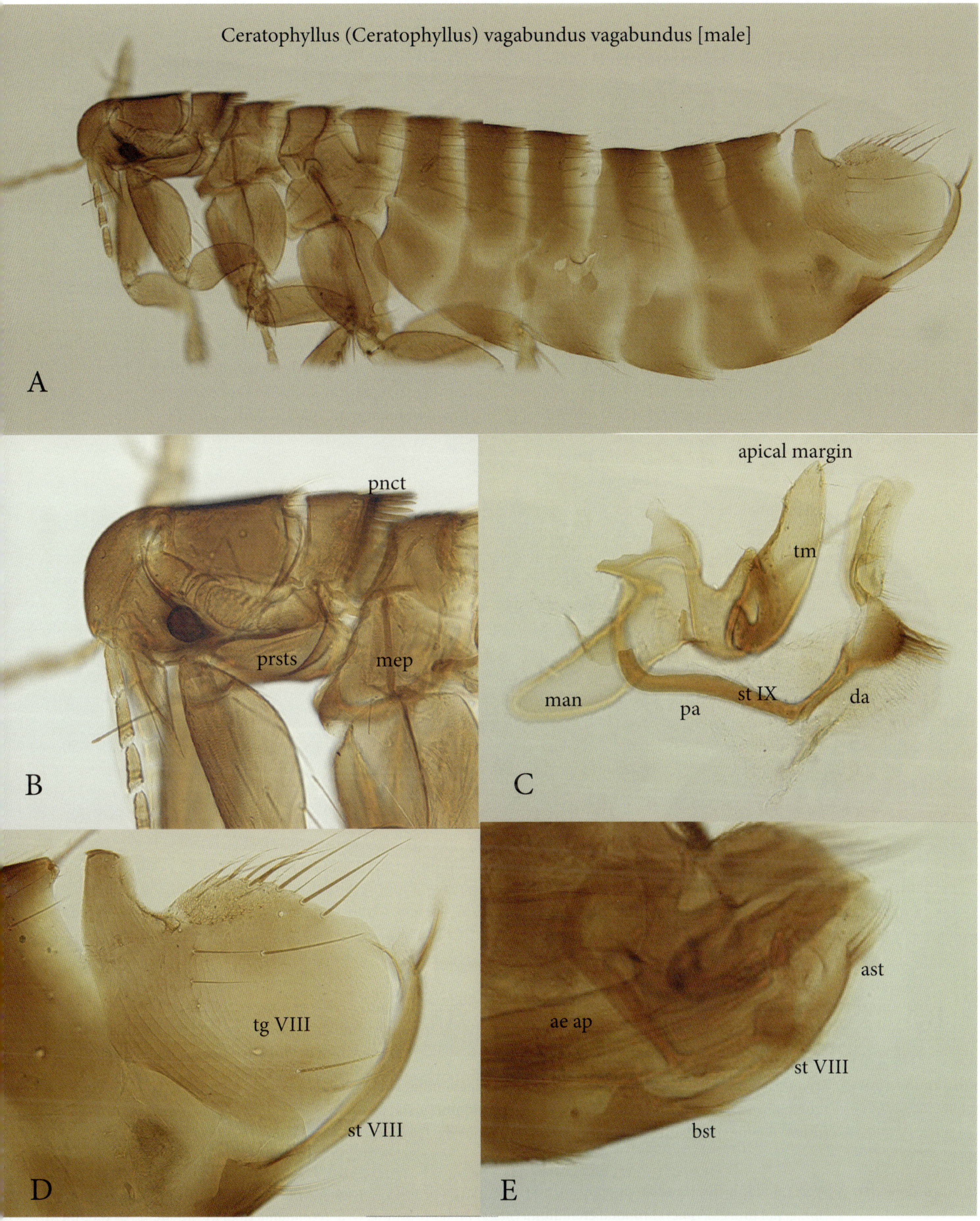

FIGURE 9 *Male* Ceratophyllus (Ceratophyllus) vagabundus vagabundus. *A, overview. B, head and pronotum ;* (*mep*) mesepisternum; (*pnct*) pronotal ctenidium; (*prsts*) prosternosome. *C–E, terminalia;* (*ae ap*) aedaegal apodeme; (*ast*) apical setae of sternum *VIII;* (*bst*) basal setae of sternum *VIII;* (*da*) distal arm; (*man*) manubrium; (*pa*) proximal arm; (*st VIII*) sternum *VIII;* (*st IX*) sternum *IX;* (*tg VIII*) tergum *VIII;* (*tm*) telomere.

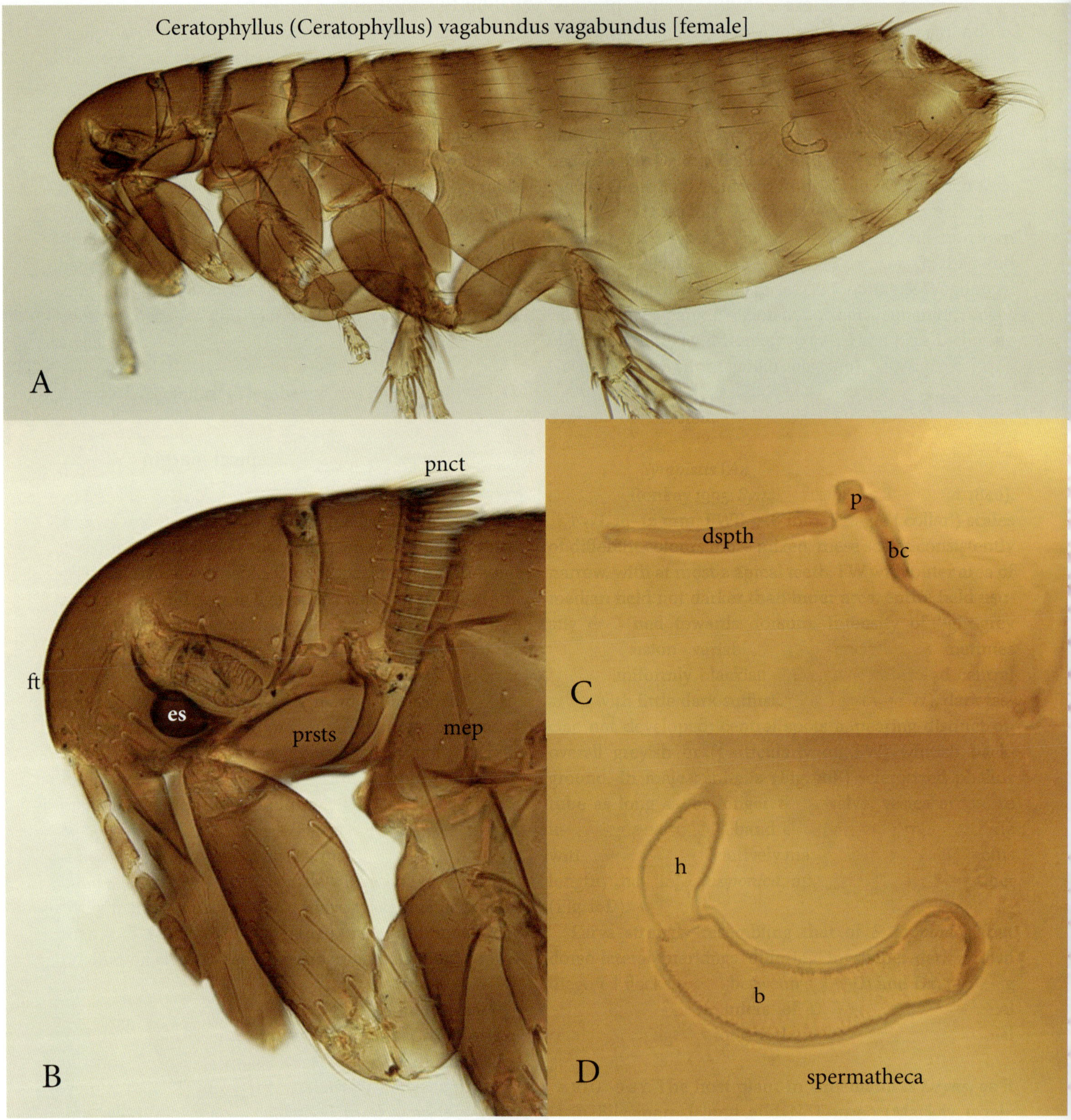

FIGURE 10 (ABOVE) *Female* Ceratophyllus (Ceratophyllus) vagabundus vagabundus. *A, overview. B, head and pronotum;* (*es*) eye-spot; (*ft*) frontal tubercle; (*mep*) mesepisternum; (*pnct*) pronotal ctenidium (*prsts*) prosternosome. *C, female genitalia;* (*bc*) bursa copulatrix; (*dspth*) ductus spermatheca; (*p*) perula of bursa copulatrix. *D, spermatheca;* (*b*) bulga; (*h*) hilla.

FIGURE 11 (RIGHT) *Male* Ceratophyllus (Emmareus) garei. *A, overview;* (*pr*) penis rods. *B, head and pronotum;* (*es*) *eye-spot;* (*ft*) frontal tubercle; (*gp*) genal process; (*pnct*) pronotal ctenidium; (*prsts*) prosternosome. *C–E, terminalia;* (*a tg IX*) apodeme of tergum *IX;* (*ae ap*) aedeagal apodeme; (*ar*) apodemal rod; (*as*) apical setae of sternum *VIII;* (*asbr*) *antesensilial bristle;* (*cr*) crochet; (*da*) distal arm; (*man*) manubrium; (*pa*) proximal arm; (*p1*) *fixed process of clasper;* (*sens*) *sensilium;* (*st VIII*) sternum *VIII;* (*st IX*) sternum *IX;* (*tm*) *telomere;* (*v*) vesicula of phallosome.

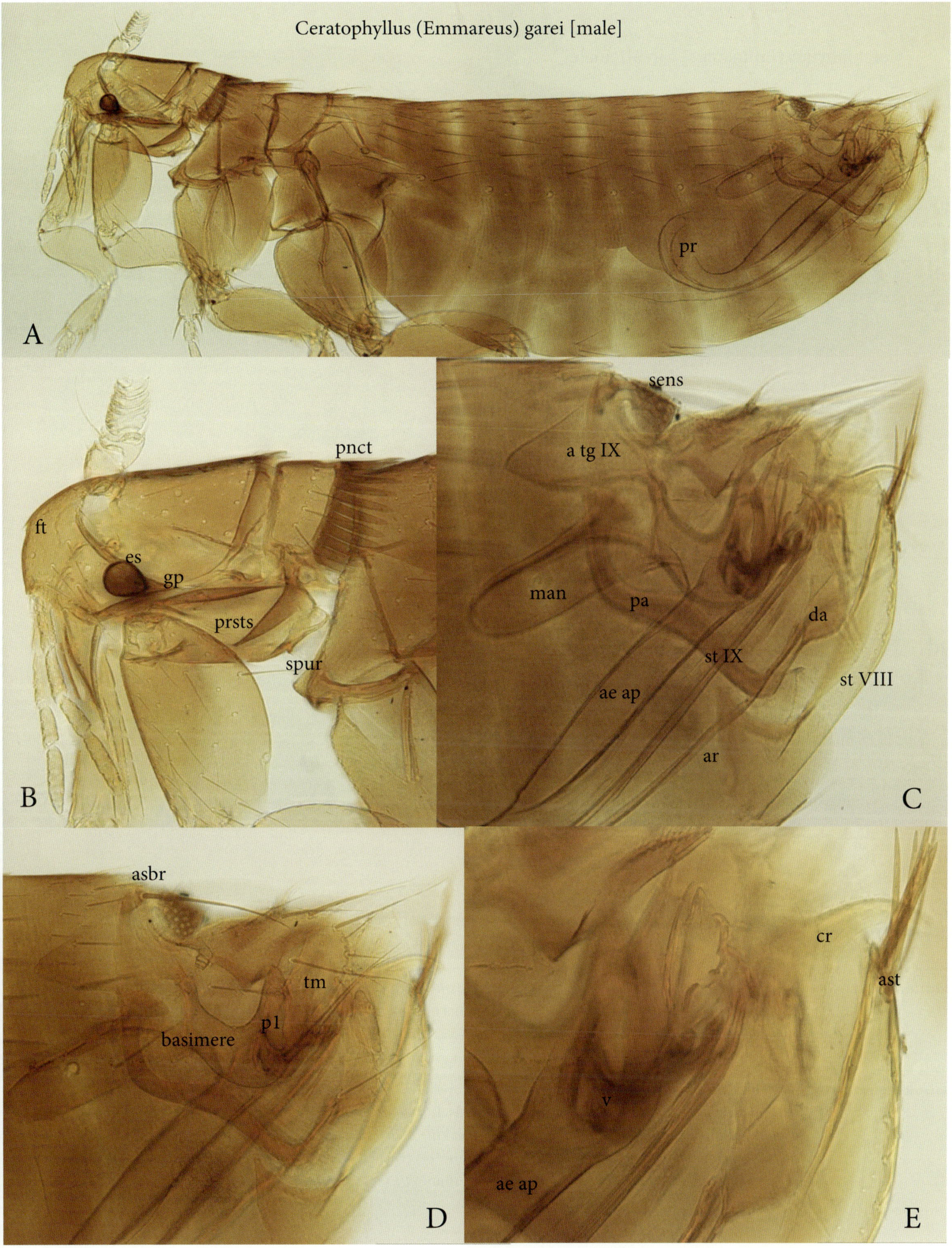
Ceratophyllus (Emmareus) garei [male]
A
pr
B
ft
es
gp
prsts
spur
pnct
C
sens
a tg IX
man
pa
da
st IX
st VIII
ae ap
ar
D
asbr
tm
p1
basimere
E
cr
ast
v
ae ap

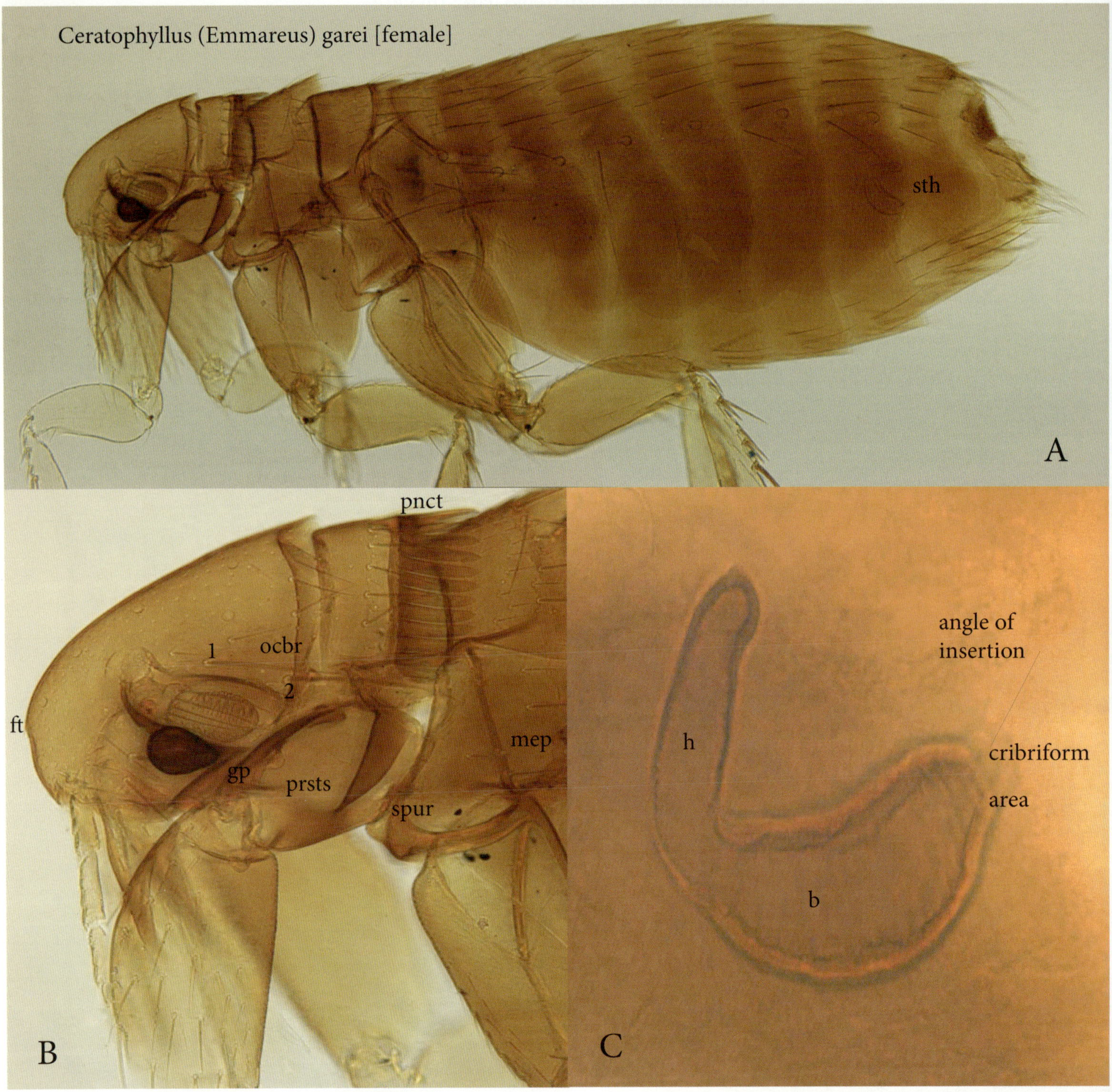

FIGURE 12 (ABOVE) *Female* Ceratophyllus (Emmareus) garei. *A, overview;* (*sth*) spermatheca. *B, head and pronotum;* (*ft*) frontal tubercle; (*gp*) genal process; (*mep*) mesepimeron, (*ocbr*) occipital bristles; (*pnct*) pronotal ctenidium; (*prsts*) prosternosome. *C, spermatheca;* (*b*) bulga; (*h*) hilla.

FIGURE 13 (RIGHT) *Male* Megabothris (Amegabothris) groenlandicus. *A, overview;* (*a tg IX*) apodeme of tergum *IX;* (*ae ap*) aedeagal apodeme; (*ar*) apodemal rods; (*man*) manubrium; (*pr*) penis rods. *B, head and pronotum;* (*es*) eye-spot; (*gp*) genal process; (*pnct*) pronotal cternidium; (*prsts*) prosternosome. *C, tergum VIII;* (*asbr*) antesensilial bristles; (*sens*) sensilium; (*tg VIII*) tergum *VIII. D–E, terminalia;* (*ar*) apodemal rod; (*bm*) basimere; (*cr*) crochet; (*man*) manubrium; (*p1*) fixed process of clasper; (*pr*) penis rod; (*sit*) sclerotized inner tube; (*st VIII*) sternum *VIII;* (*st IX*) sternum *IX;* (*tm*) telomere.

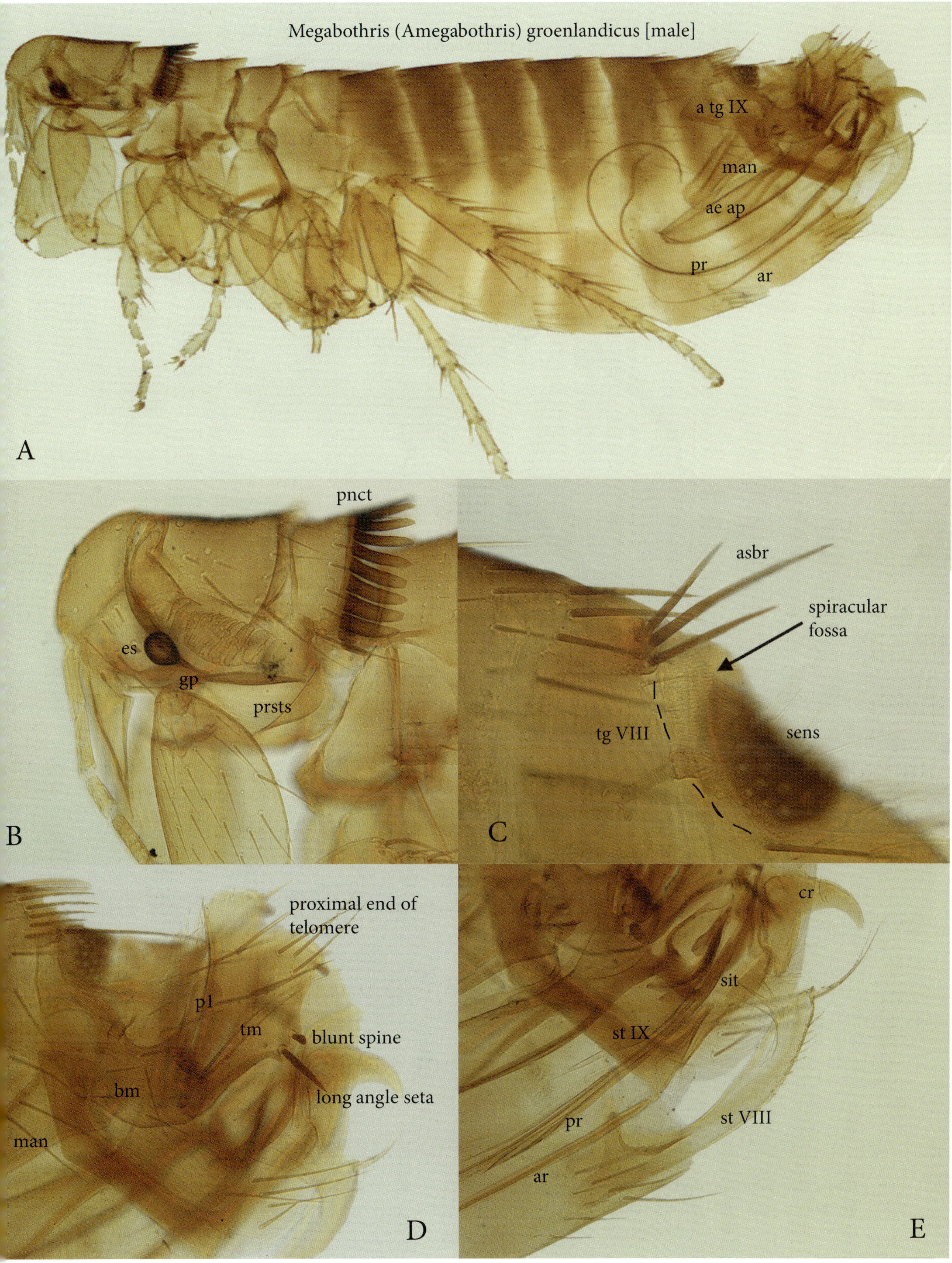
Megabothris (Amegabothris) groenlandicus [male]
a tg IX
man
ae ap
pr
ar
A
pnct
es
gp
prsts
B
asbr
spiracular fossa
tg VIII
sens
C
proximal end of telomere
p1
tm
blunt spine
bm
long angle seta
man
D
cr
sit
st IX
pr
st VIII
ar
E

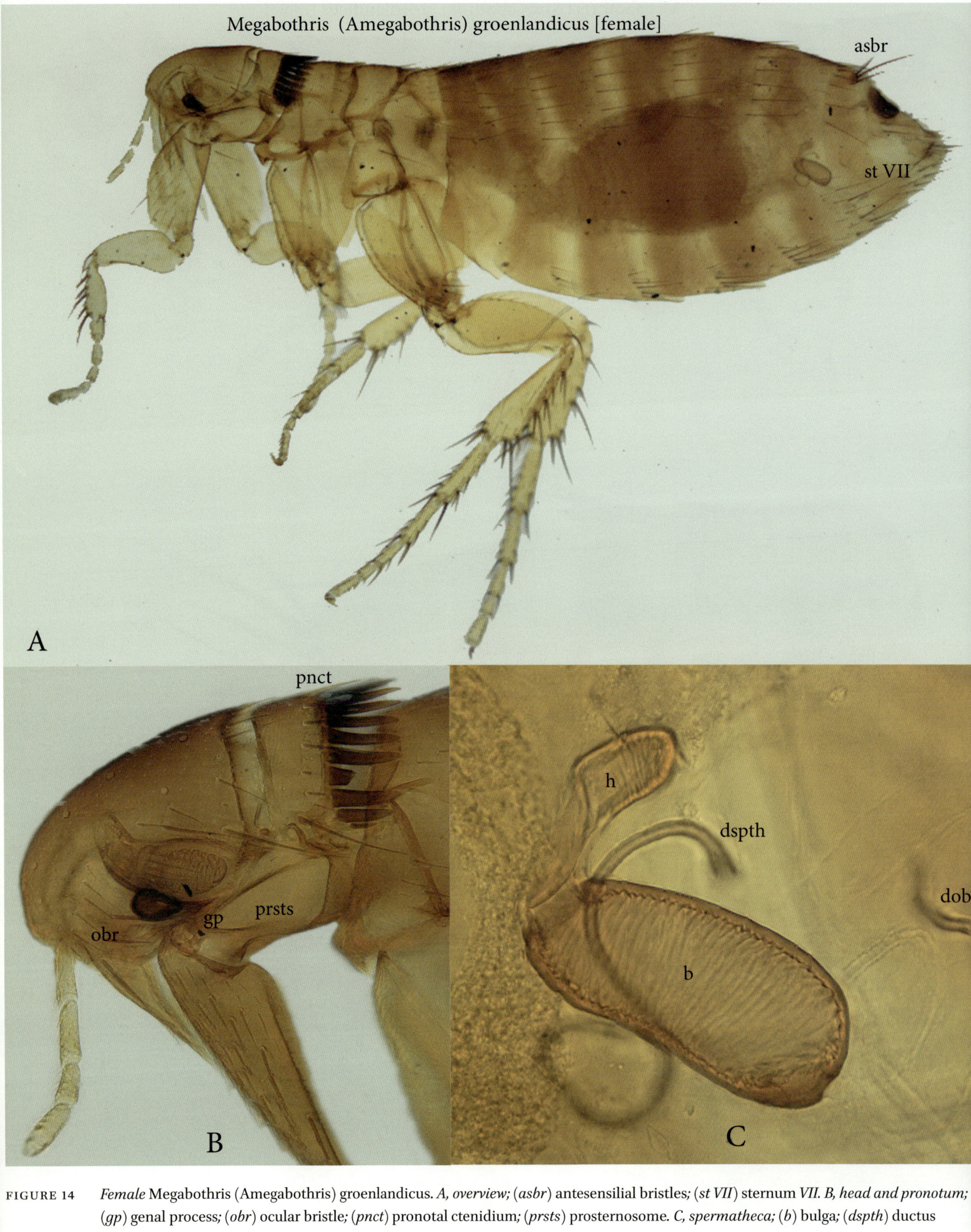

FIGURE 14 *Female* Megabothris (Amegabothris) groenlandicus. *A, overview;* (*asbr*) antesensilial bristles*;* (*st VII*) sternum *VII. B, head and pronotum;* (*gp*) genal process*;* (*obr*) ocular bristle*;* (*pnct*) pronotal ctenidium*;* (*prsts*) prosternosome. *C, spermatheca;* (*b*) bulga*;* (*dspth*) ductus spermatheca*;* (*dob*) ductus obturatus*;* (*h*) *hilla.*

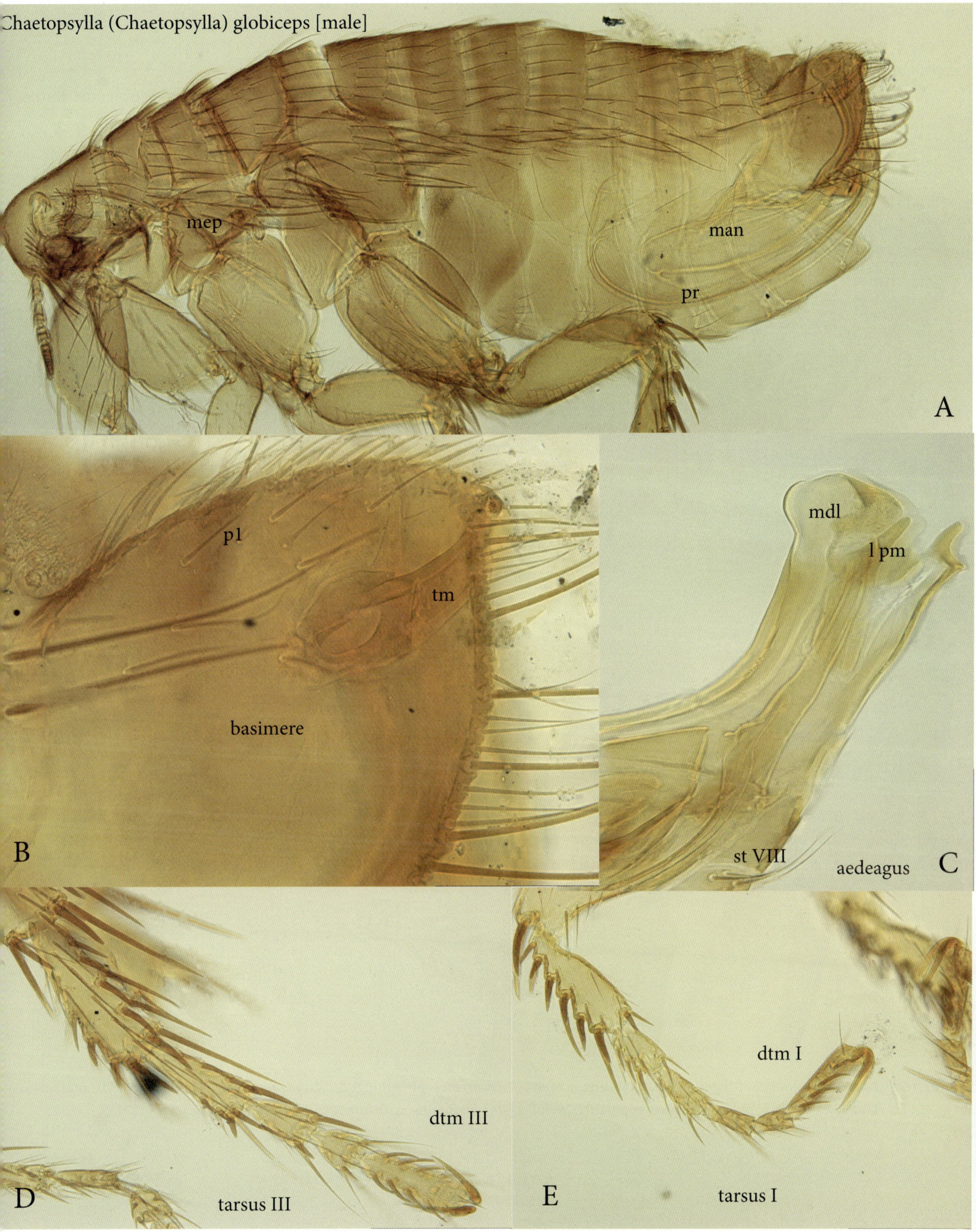

FIGURE 15 *Male* Chaetopsylla (Chaetopsylla) globiceps. *A, overview;* (*man*) manubrium; (*mep*) mesepimeron; (*pr*) penis rods. *B, tergum IX;* (*p1*) fixed process of clasper; (*tm*) telomere. *C, aedeagus;* (*l pm*) lateral portion of paramere; (*mdl*) median dorsal lobe; (*st VIII*) sternum *VIII. D, tarsus III;* (*dtm III*) distotarsomere *III. E, tarsus I;* (*dtm I*) distotarsomere *I.*

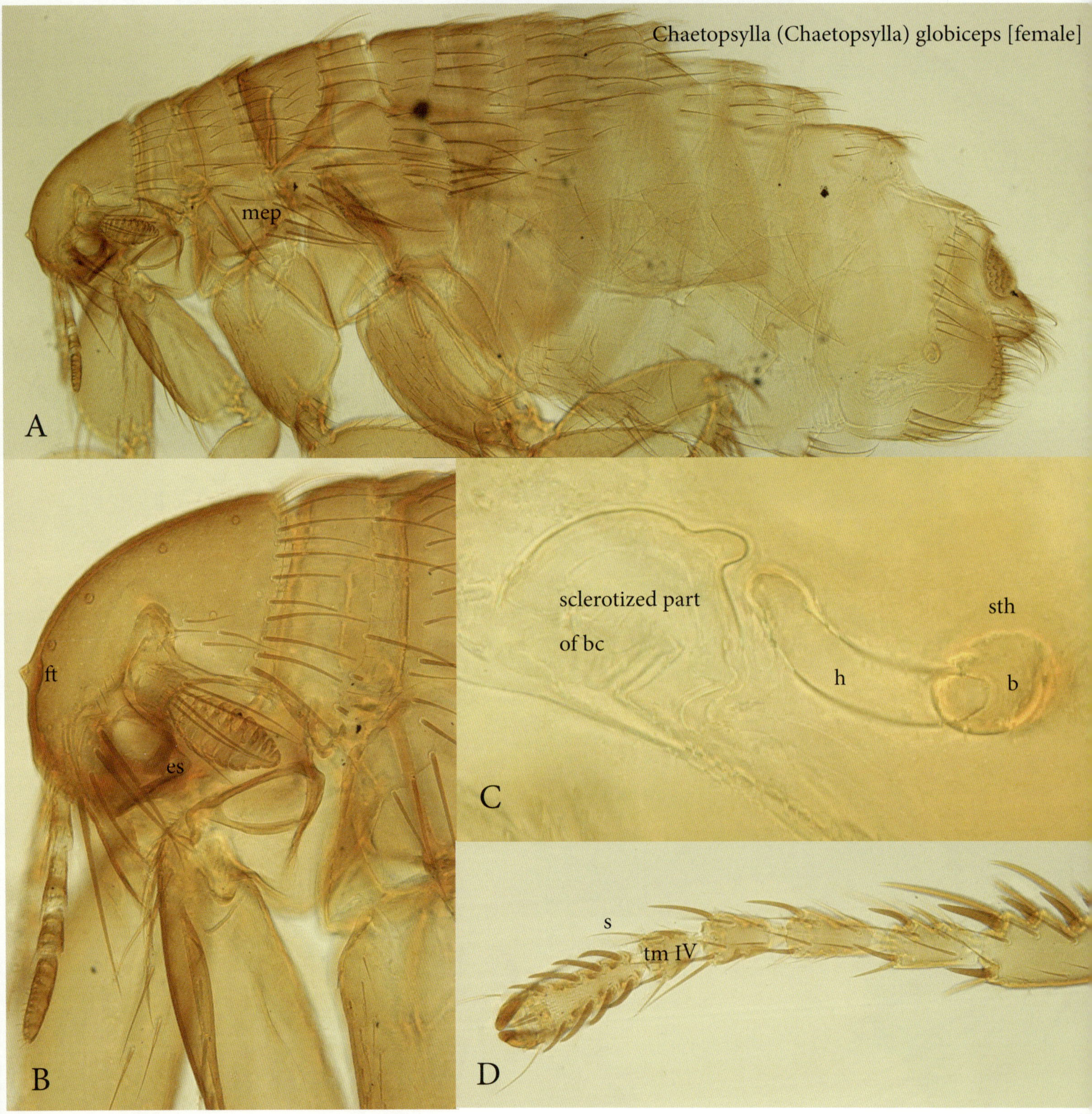

FIGURE 16 *Female* Chaetopsylla (Chaetopsylla) globiceps. *A, overview;* (*mep*) mesepimeron. *B, head and pronotum;* (*es*) eyespot; (*ft*) frontal tubercle. *C, spermatheca;* (*b*) bulga; (*bc*) bursa copulatrix; (*h*) hilla; (*sth*) spermatheca. *D, tarsus I;* (*s*) apicolateral seta; (*tm*).

CHAPTER 17

Diptera
(*Two-winged or 'true' flies*)

17.1. Generalities and family keys

Thomas Pape and Verner Michelsen

Diptera constitute one of the 'megadiverse' insect orders with about 160.000 species currently described. They occur abundantly in all non-marine ecosystems, but in low to high arctic environments Diptera become increasingly dominant to other insects, both in terms of species numbers and density of individuals. With 370+ species currently on record from Greenland, Diptera just outnumber all remaining true insects in the area. Adult Diptera tend to be aerial or frequenting exposed surfaces on the ground or in the vegetation, often visiting flowers for nectar or pollen. Diptera larvae may be aquatic, living secluded in moist organic detritus or humus-enriched soil, forming burrows or galls in fresh plant or fungal tissues, or living as animal parasites or parasitoids.

Diptera are readily seen as monophyletic and as adults easily told apart from other winged insects by the transformation of the hind wings into halteres and by the unique formation of soft, lobe-like labella derived from the labial palps.

Morphology: adults

Adult Diptera usually have prominent compound eyes, which may occupy the greater part of the head surface, especially in males. All three ocelli are retained in most families, but either the median or all three ocelli have disappeared in some groups. The great diversity of antennal structure in adult Diptera has always played a major taxonomic role. Families belonging to the 'lower' or 'nematoceran' grade of Diptera have generally an elongate and many-segmented antenna, while the antenna in Brachycera tends to be short and compact, and apart from the two first segments (scape and pedicel) consisting of only a few visible flagellomeres. The typical fly antenna of most 'higher' Brachycera have an enlarged, more or less hanging basal flagellomere (postpedicel) from which the remaining flagellum arises as a bristle-like formation known as the arista.

The mouth parts of adult Diptera are overall highly specialised and only able to take fluid meals. The major device for ingestion, and basic to all Diptera, are the paired, soft labella formed at the tip of the labium. They replace the labial palps of other insects. Adult flies can only ingest solid food of a fine texture that can be suspended in saliva, such as pollen grains. The saliva is distributed through a system of creases, which in the Brachycera take the form of regular canals (pseudotracheae), on the oral side of the labella. Food particles suspended in fluid are conveyed to the functional mouth opening at the labellar bases and sucked up through a canal formed by the labrum and tongue by the strongly musculated cibarial sucking pump situated at level of the clypeus. Functional mandibles have only been retained in relatively few families, and only among blood-feeding ('biting') females. When present the mandibles are slender, blade-like and apically finely serrated, and they have only retained the posterior articulation with the head capsule. In the absence of mandibles, some groups of Diptera have invented alternative stabbing devices from other parts of the mouth apparatus in order to attain predatory or blood-feeding habits.

The thorax of adult Diptera is characterized by the enormous development of the wing-bearing mesothorax, which indicates an unusual strength of the indirect flight muscles. The metathorax is on the contrary strikingly small as a consequence of the unique transformation of the hind wings into small, club-shaped appendages known as halteres. The fore wings are highly varied in shape and venation and provide a rich source of important and easily accessible taxonomic characters. The legs tend to be fairly weak and slender and have as a rule retained 5-articled tarsi. The chaetotaxy of the thorax and legs is of major taxonomic importance in most adult Diptera.

The abdomen varies greatly in adult Diptera. In the nematoceran grade Diptera and some non-cyclorrhaphan Brachycera, the abdominal segments (except I and VIII–X) are simple and uniform (except decreasing in size posteriorly) with tergites and sternites of roughly similar size. There is a strong tendency among most Brachycera that the abdominal tergites expand laterally at the expense of the sternites. The posterior abdominal segments VIII–X are sexually modified in all Diptera, but this segmental specialization becomes more pronounced in the higher, cyclorrhaphan Brachycera that typically have segments VI–X well set apart from the remaining abdomen. These modified segments form the so-called terminalia or postabdomen, that in females is variously adapted for ovi- or larviposition. Especially in the male sex, the terminalia provide in general the richest and most reliable source of characters for species identification.

Immature stages

Larval Diptera are consistently lacking thoracic legs, although abdominal prolegs and various types of creeping welts may help them move around. Larvae belonging to the nematoceran grade of Diptera have a distinct, sclerotised head capsule, rather complete mouthparts often equipped with a complex filter apparatus and normal mandibles, and have usually retained 10 visible abdominal segments. Larvae of the higher, cyclorrhaphan Brachycera are highly modified maggots: the head is partly invaginated and devoid of a head capsule and contains a complex, internal cephaloskeleton supporting the specialized mandibles operating vertically and near-parallel as mouthhooks. The morphological interpretation of the cephalic parts of a maggot is controversial: A median hook-like structure between the mouthhooks that may represent the labrum is often present in first instar larvae and functions as an egg-burster. The secondary, functional mouth is usually flanked by oral ridges equipped with curtains of cirri and may function as a filter-feeding apparatus. The antennae are represented by peg-like structures set in large sockets. The maxillae are reduced to sense organs dorsolaterally displaced from the mouth opening. A maggot has only eight visible abdominal segments (last three merged into an anal division).

The pupa of most families in the nematoceran and non-cyclorrhaphan brachyceran grades has antennae, legs and wings closely adpressed to the body (obtect pupa), more rarely are these appendages free (exarate pupa, *e.g.*, Cecidomyiidae). All members of the Cyclorrhapha have an exarate pupa, but they maintain the 3rd-instar larval cuticle as a protective, hardened case around the pupa resulting in the formation of a puparium. During eclosion from the puparium, the anterior part either splits open at random as in the 'aschizan grade' of Cyclorrhapha (*e.g.*, Phoridae and Syrphidae), or along preformed hatching lines characteristic of the Schizophora.

Greenland fauna

The composition of the small Diptera fauna of Greenland is unusual in several respects. Noteworthy is the absence of soldier flies, snipe flies, horse flies, robber flies, stiletto flies and bee flies, *i.e.*, all families belonging to the 'lower' Brachycera. Gall-building cecidomyiids (subfamily Cecidomyiinae) are practically absent, and some of the larger acalyptrate families (Tephritidae, Lauxaniidae, Chloropidae, Drosophilidae) appear to be entirely absent. Other groups are proportionally rich in species, notably Chironomidae among the nematoceran Diptera and Muscidae and Anthomyiidae among the Calyptratae. Parasitic Diptera are poorly represented in Greenland, represented only by some Tachinidae developing in lepidopteran caterpillars, two species of botflies attacking reindeer, and a single species of snail-killing Sciomyzidae. Partial or complete reduction of the wings has taken place independently in a number of dipteran groups, sometimes as a response to a harsh and cold climate, but no Greenland species show obvious signs of wing reduction.

Principal groups and family key

The traditional subordinal division of Diptera into 'Nematocera' (those with long, filiform antennae) and Brachycera (those with short, compact antennae) is practical but unacceptable in a phylogenetic context, because the 'Nematocera' is a paraphyletic assemblage with some families more closely related to the Brachycera than to other nematocerans. Still, it may seem convenient to maintain a nematoceran grade when referring to the non-brachyceran Diptera. Species of this grade include the well-known crane flies, gnats, mosquitoes, midges *etc.*, and are primarily characterized by the longish and slender antennae with many flagellomeres and usually 5-articled maxillary palps, but most species are also notably slender-bodied and with long, delicate legs.

The remaining Diptera, belonging to the monophyletic Brachycera, tend to be more robust and compactly built than nematoceran Diptera. A low number of antennal flagellomeres and maxillary palpomeres are characteristic of adult Brachycera. A noteworthy larval specialization is a shift in the movement of the larval mandibles (or mouthhooks) from horizontal to oblique (in 'lower' Brachycera) or subvertical (near-parallel) in cyclorrhaphous Brachycera.

The 'lower' Brachycera form a paraphyletic assemblage, while the 'higher' Brachycera constitute the monophyletic Eremoneura defined by several, relatively subtle characters such as only three larval instars and a one-segmented adult maxillary palp. The Eremoneura consist of the Empidoidea and the Cyclorrhapha. The Empidoidea (dance flies and long-legged flies) are defined by being predatory as adults, and by females having only one sclerotised spermatheca.

The Cyclorrhapha have highly derived larvae of the maggot-type and pupate within a puparium (see above). Most cyclorrhaphans belong to the Schizophora, the species of which use a large, sac- or bladder-like structure (the ptilinum) emerging at the tip of the frons to push open the puparium. The ptilinum has no other function and retracts soon after emergence, leaving the ptilinal suture. Another notable schizophoran apomorphy is the 360° rotation of the ninth and following segments of the male abdomen detectable from the coil of the genital duct around the hindgut and the partial rotation and asymmetry of the pregenital segments VII–VIII.

The Calyptratae (house flies, blow flies and their relatives) form a well-supported group within the Schizophora defined, *e.g.*, by having prestomal teeth derived from the labellar pseudotracheae near the functional mouth opening, a cleft on the antennal pedicel, and a lobe-like lower calypter (secondarily reduced in the Scathophagidae). The remaining, non-calyptrate Schizophora are probably not monophyletic, but conventionally referred to as the 'acalyptrate grade'. The acalyptrates comprise many families for which the phylogenetic relationships are still very uncertain.

Key to the families of Diptera of Greenland

ADULTS

1. Antennal flagellum thread-like, built of a series of more or less uniform, bead-like to cylindrical articles. Palp generally 3–5 articled. Cell cup open, i.e. vein CuA_2 ending in hind margin of the wing (Fig. 1A) .... 2 (NEMATOCERAN GRADE)
– Antennal flagellum consisting of a thick basal part (postpedicel) and a slender apical part differentiated into a style or arista built of 2–3 articles. Palp 1–2 articled. Cell cup closed well before wing margin (Fig. 1B) or reduced (Fig. 1C), i.e. CuA_2 joining A_1, or absent .... 15 (BRACHYCERA)
2. Scutum with a V-shaped transverse suture. Wing with two anal veins, A_1 and A_2 both reaching wing margin (Figs 1A, D); discal cell (d) usually present (Fig. 1D)..3
– Transverse suture of scutum absent or incomplete medially, never V-shaped. Wing with only one anal vein (A_1) reaching margin, the other (A_2) fading out before margin, or absent; discal cell absent (Figs 1E–K) 5
3. Ocelli present. A_2 short and retracted (Fig. 1D)....... .... Trichoceridae, 17.3
– Ocelli absent. A_2 more than half as long as A_1 (Fig. 1A)..4
4. Terminal (5th) article of palp longer than two preceding articles together. Head capsule forming a prominent rostrum terminating in a thorn-like nasus .... Tipulidae, 17.2
– Terminal article of palp shorter than two preceding articles together. Head capsule weakly rostrate, without a nasus .... Limoniidae, 17.2
5. Mouth parts (proboscis) very long and slender. Body, legs, and wing veins (Fig. 1E) more or less covered with appressed scales................ Culicidae, 17.4
– Mouth parts short. Body without scales, wing with or without narrow scales 6
6. Small, short-legged and hairy "moth flies". Wings broadly lanceolate (Fig. 1F), held roof-like over the abdomen ....Psychodidae, 17.11
– Without this combination of characters 7
7. Antennal flagellum built of 4–10 short and compact articles; antenna not much longer than length of head 8
– Antennal flagellum built of ten or more, usually elongate and slender articles; antenna usually more than twice as long as length of head 11
8. Ocelli present 9
– Ocelli absent 10
9. Palp 5-articled. Crossvein bm–cu present; pterostigma present (Fig. 1G) ......Bibionidae, 17.9
– Palp 1-articled. Crossvein bm–cu absent; pterostigma absent (Fig. 1H).....................Scatopsidae, 17.9
10. Wing broad, only about twice as long as broad; M with two branches, arising from a very short or no stem (Fig. 1I)......................... Simuliidae, 17.5
– Wing elongate, about three times as long as broad; M unbranched (often pale and difficult to see) (Fig. 1J).................. Chironomidae, in part, 17.7
11. Tiny, delicate flies. Wing venation more or less reduced, M_{1+2} only present in Lestremiinae (Fig. 1K). Legs without apical tibial spurs ....Cecidomyiidae, 17.12
– At least one apical spur present on fore and hind tibiae 12
12. Ocelli present. Antennal scape and pedicel of similar size...13
– Ocelli absent. Pedicel enlarged and globular, especially in males.....................................14
13. Eyes meeting above antennae as a narrow bridge. Wing with stem and fork of M about equal in length, the fork bell-shaped (Fig. 2A).........Sciaridae, 17.8
– Eyes separate above antennae. Wing with fork of M much longer than stem, lanceolate rather than bell-shaped (Fig. 2B) ....Mycetophilidae (incl. Bolitophilidae), 17.10
14. Vein M with two branches (Fig. 2C). Postnotum without a longitudinal groove .... Ceratopogonidae, 17.6

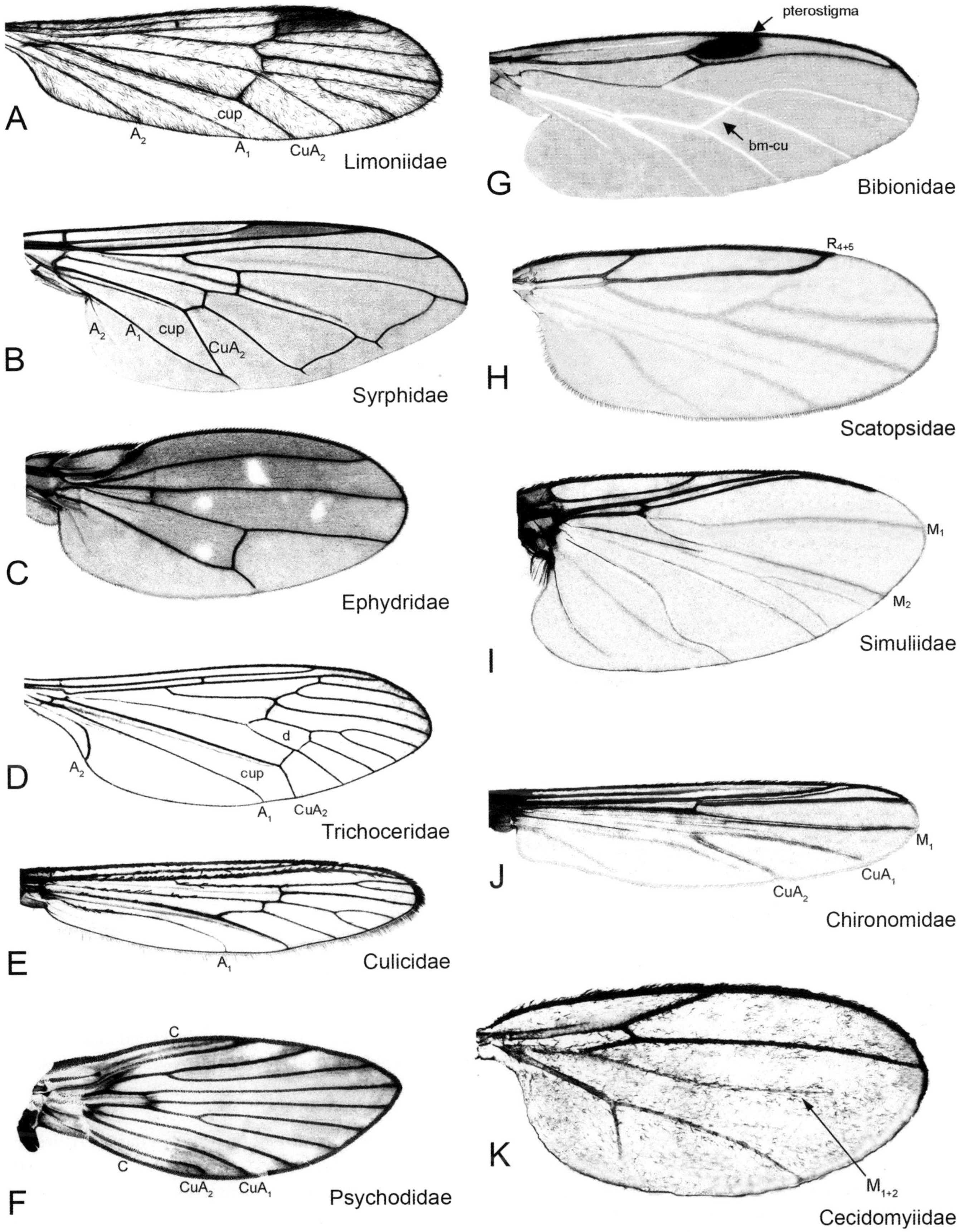

FIGURE 1 *Right wing, dorsal view, of Diptera from various families of nematocerans (A, D–K) and Brachycera (B–C).*

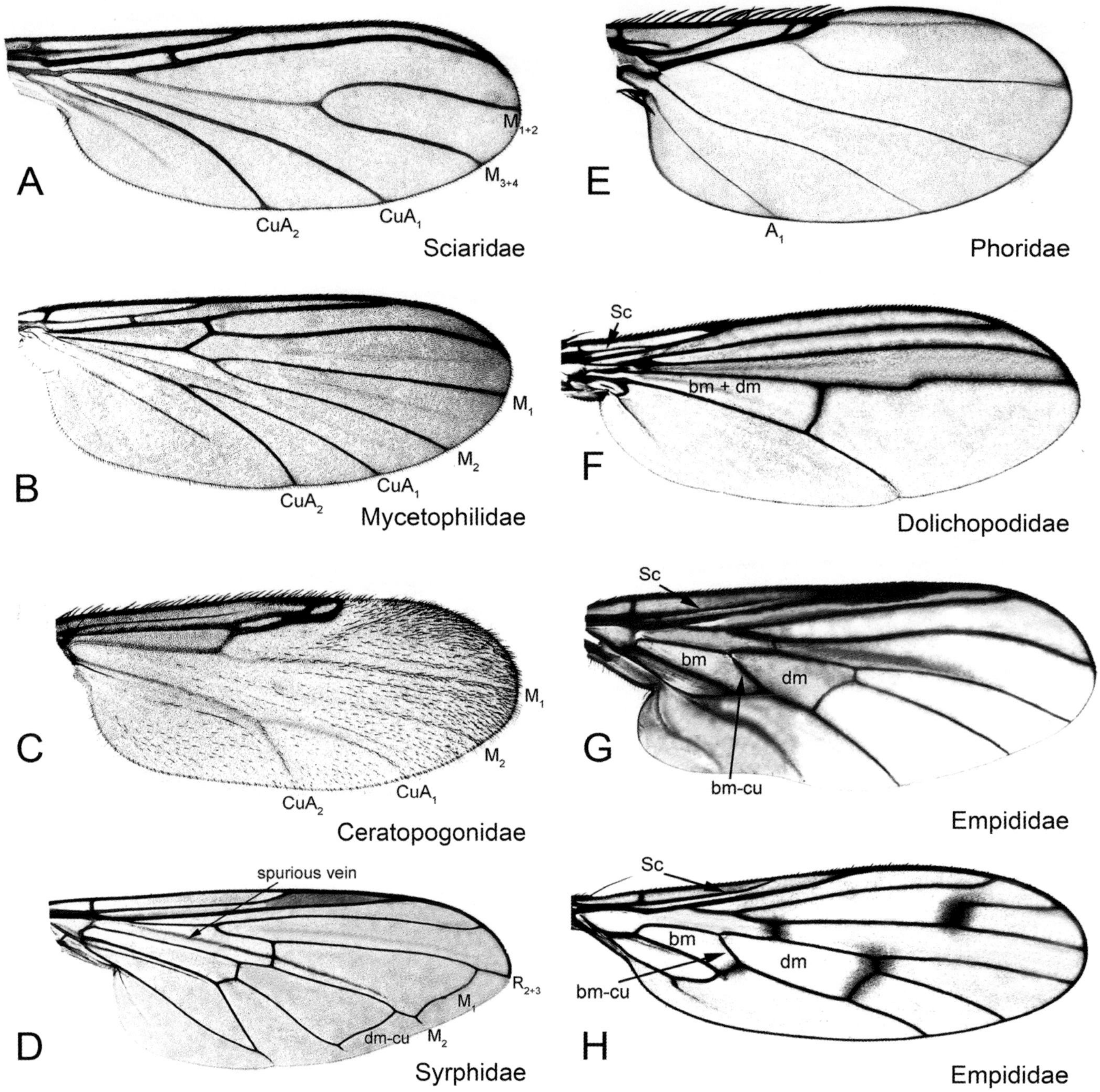

FIGURE 2 *Right wing, dorsal view, of non-schizophoran Diptera from various families of nematocerans (A–C) and Brachycera (D–H).*

– Vein M unbranched (Fig. 1J). Postnotum with a longitudinal groove......... Chironomidae, in part, 17.7

15. Ptilinal suture and lunule absent..................... ....16 (NON-SCHIZOPHORAN BRACHYCERA)

– Ptilinal suture and lunule present.................... ....19 (CYCLORRHAPHA SCHIZOPHORA)

16. Spurious vein visible between veins Rs and M; M_1 curved forward more or less in line with crossvein dm–cu and joining unbranched R_{4+5} in a crossvein-like manner (Fig. 2D)................ Syrphidae, 17.16

– Spurious vein absent; M_1 not curved forward in a crossvein-like manner............................... .17

17. Branches of vein R strongly thickened and crowded into anterior base; other veins weaker snd crossing the wing obliquely and more or less parallel to each other (Fig. 2E)..........................Phoridae, 17.15

– Wing venation without the above characteristics 18

18. Cells bm and dm confluent (crossvein bm–cu absent); Sc abruptly curved posteriorly and fused with R_1 (Fig. 2F).......................... Dolichopodidae, 17.14

– Cell bm separated from cell dm by crossvein bm–cu; Sc joining C or ending freely, never abruptly joining R_1 (Figs 2G, H)........................ Empididae, 17.13

19. Greater ampulla present as a knob-like swelling below wing base. Antennal pedicel (2nd article) with a dorsal seam. Lower calypter usually conspicuous, lobe-like except in Scathophagidae .... 20 (CALYPTRATAE)
– Greater ampulla absent. Antennal pedicel without a dorsal seam except in Ephydridae. Lower calypter always quite inconspicuous, linear .... 26 (ACALYPTRATE GRADE)
20. Large, hairy flies often resembling bees. Mouth parts reduced, non-functional. Head inflated, with small, sunken antennae.. Oestridae (incl. Hypodermatidae), 17.27
– Not with the above combination of characters. Mouth parts functional 21
21. Meron with a subvertical row of bristles. Vein M_1 angled forward distal to crossvein m–cu (Fig. 3A) 22
– Meron usually bare, rarely with some fine irregular setae. Vein M_1 nearly straight, only exceptionally with a forward curvature (Figs 3B–D) 23
22. Subscutellum strongly developed, prominently and evenly convex in profile............ Tachinidae 17.27
– Subscutellum absent............Calliphoridae, 17.27
23. Lower hind part of head with fine pale setae. Lower calypter not developed........Scathophagidae, 17.26
– Head throughout with black setae, but if exceptionally with some fine pale setae on lower hind part of head then lower calypter developed............. 24
24. Vein Sc in its distal half running straight towards wing margin (Fig. 3B). Hind tibia with a true dorsal seta near middle, aligned with the dorsal preapical seta ....Fanniidae, 17.26
– Vein Sc distally with a bend towards vein C (Figs 3C, D). Hind tibia usually with setae in anterodorsal and posterodorsal position, but never with a true dorsal seta near middle.................................. 25
25. Vein A_1 continuing to wing margin as a faint fold (Fig. 3C). Scutellum with fine hairs ventrally at tip, but if bare (*Fucellia*) then vein C with row of scattered spinules antero-ventrally, or wing darkly clouded on at least distal part ....Anthomyiidae, 17.26
– Vein A_1 fading out well before wing margin (Fig. 3D). Scutellum bare ventrally at tip... Muscidae, 17.26
26. Vein Sc complete, reaching vein C well separated from tip of vein R_1 (Figs 3E–G)...................... 27
– Vein Sc more or less reduced distally or inseparably from distal part of vein R_1 (Figs 3H–J) 32
27. Vein C without subcostal break (Fig. 3E).......... 28
– Vein C with subcostal break (Figs 3F, G)...........31
28. Posterior thoracic spiracle with one or more fine bristles on lower margin..............Sepsidae, 17.18
– Posterior thoracic spiracle without bristles on lower margin... 29
29. Postvertical bristles divergent. Wing membrane with a dark reticulate pattern ....Sciomyzidae 17.17
– Postvertical bristles convergent. Wing membrane clear 30
30. Small (wing length less than 3mm) flies covered in dense whitish grey dusting ....Chaemaemyiidae, 17.23
– Larger brownish black flies with a depressed body and strong legs [family of likely occurrence in Greenland .... Coelopidae]
31. Vein C with a row of scattered spine-like setae (Fig. 3F). Postvertical bristles strong and convergent .... Heleomyzidae, 17.21
– Vein C without a row of spine-like setae (Fig. 3G). Postvertical bristles divergent ..... Piophilidae, 17.19
32. Vein C without humeral break (Fig. 3H).......... 33
– Vein C with humeral break (Figs 3I, J)............. 34
33. Postvertical bristles divergent. Vein A_1 distinct (Fig. 3H) ....Agromyzidae, 17.20
– Postvertical bristles convergent or reduced. Vein A_1 reduced ....Canacidae, 17.24
34. Basal article of hind tarsus notably short and thick. Vein A_1 developed but fading out before wing margin (Fig. 3I) ....Sphaeroceridae, 17.22
– Hind tarsus not thus modified. Vein A_1 very reduced or absent (Fig. 3J) .... Ephydridae, 17.25

Key to the families of Diptera of Greenland

MATURE LARVAE

The key below has to a large extent been adapted from the larval family key in the Manual of Nearctic Diptera (McAlpine *et al.* 1981, 1987), and most illustrations are reproduced from this authoritative work. The figures are in most cases of Nearctic species not found in Greenland, but it is hoped that they will serve to exemplify the characters used in the key. The key is constructed for larvae in their last instar, but there is unfortunately no common feature, which will tell if a given larva is mature or not.

1. Larva small (ca. 1 mm). A small postcephalic segment present between head capsule and prothorax in addition to three thoracic and nine abdominal segments (Fig. 4A). Head capsule very small, lightly sclerotized, rounded or only slightly longer than broad and with paired cephalic bars laterally; mouthparts indistinctly differentiated 2

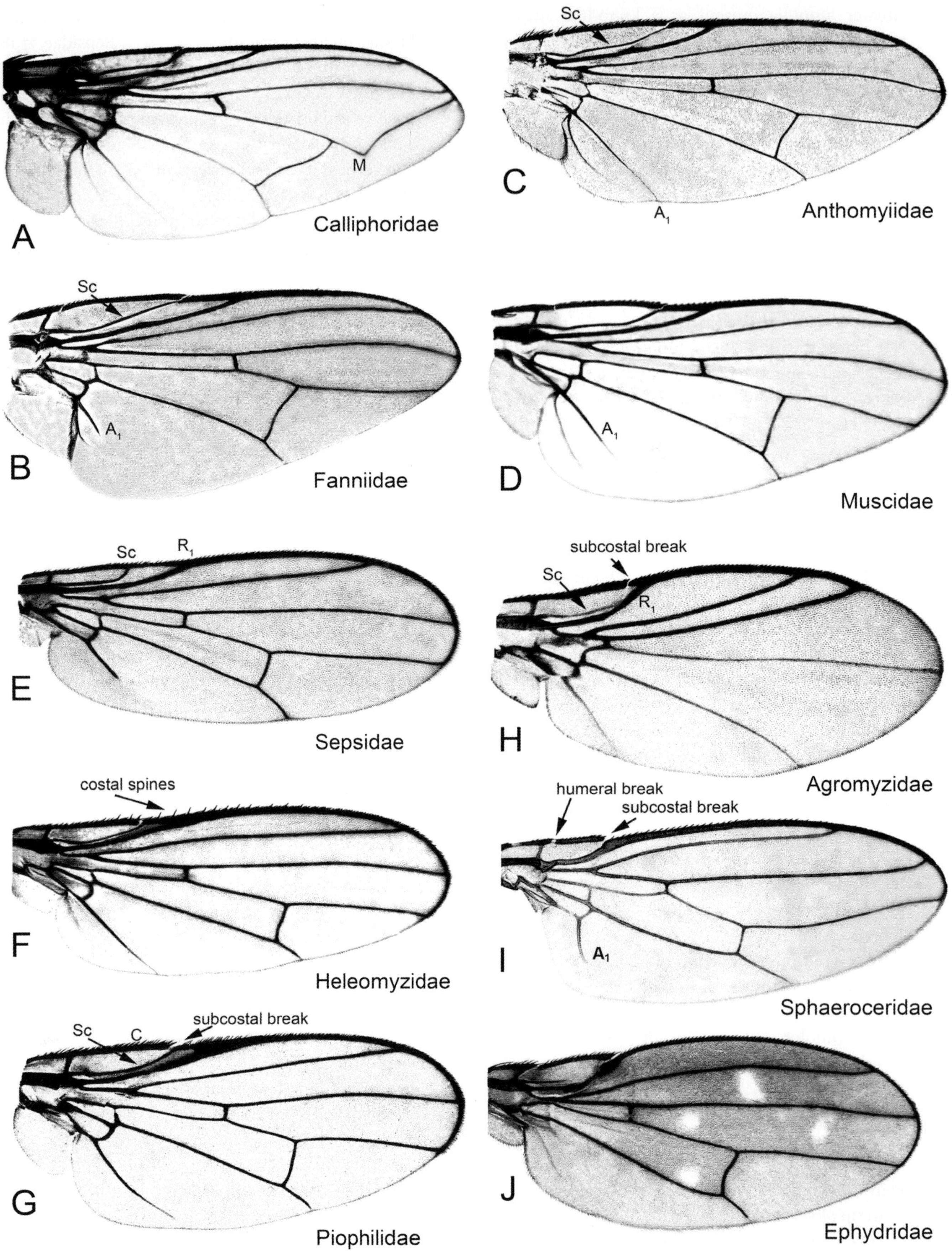

FIGURE 3 *A–I. Right wing, dorsal view, of schizophoran Diptera.*

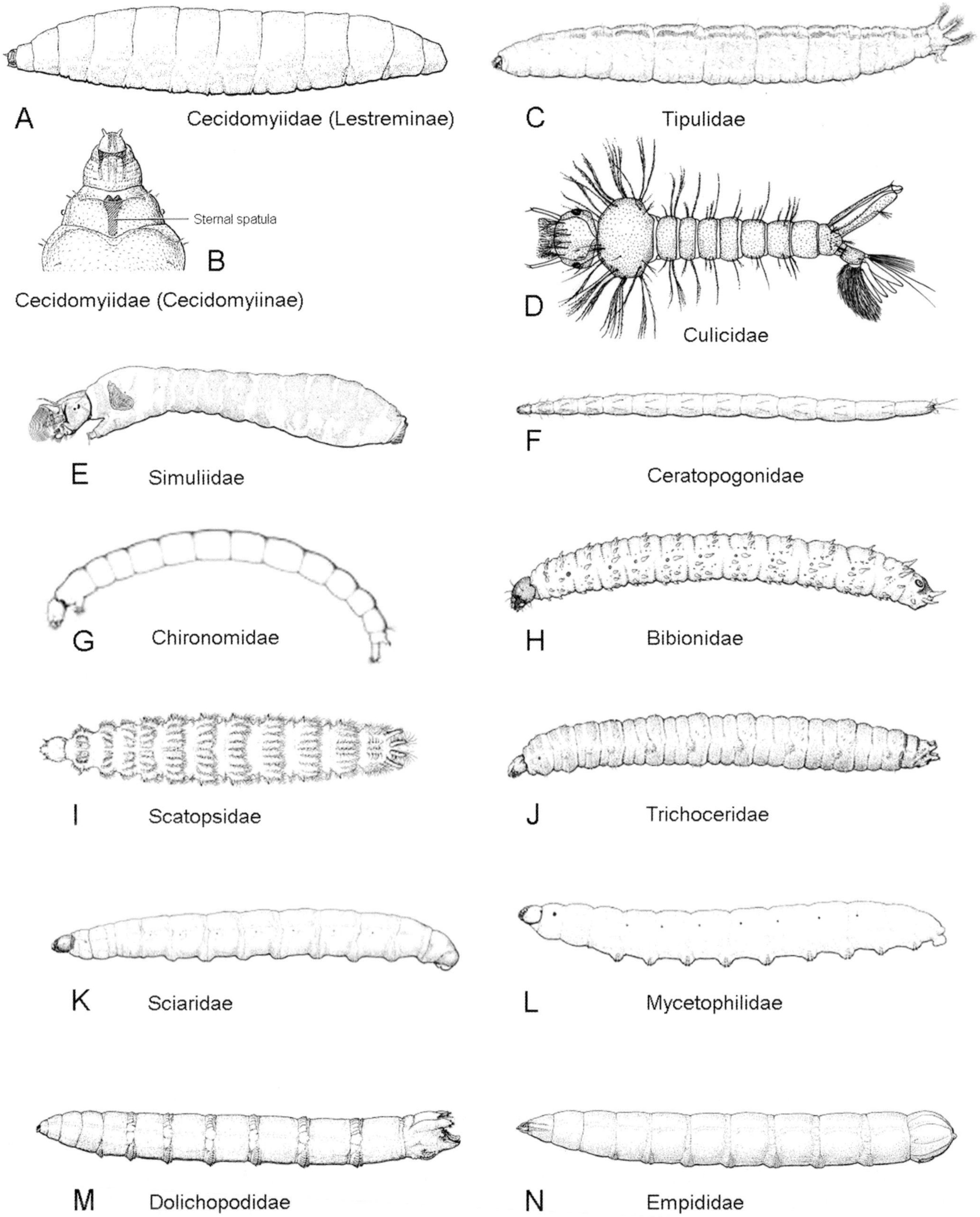

FIGURE 4 *A–N: Last instar larva of nematoceran Diptera. Illustrations from Manual of Nearctic Diptera vol. 1 (McAlpine* et al. *1981) except for D, which was kindly provided by Peter Skidmore.*

– Size and color various. Postcephalic segment not differentiated. Head capsule or internal cephaloskeleton with well-differentiated mouthparts; paired cephalic bars lacking................................3

2. Gall-building species, colour orangish or reddish, with a variously shaped sclerotised rod ventrally on first thoracic segment ('sternal spatula') (Fig. 4B)........................ Cecidomyiidae (Cecidomyiinae), 17.12

– Free-living in soil and leaf litter, colour whitish or greyish, no sclerotised rod ventrally on first thoracic segment........ Cecidomyiidae (Lestremiinae), 17.12

3. Head capsule usually complete and permanently exposed, more rarely partially retracted within thorax and incomplete as a result of excisions in the capsule from behind (Tipulidae). Mandibles moving against one another in a horizontal or oblique plane (oblique downward direction in Trichoceridae), usually with two or more apical teeth, rarely hook-like or sickle-shaped.....................................4

– Head capsule variously reduced posteriorly, always partially or almost fully retracted into the thorax or entirely replaced by an internal cephaloskeleton. Mandibles moving parallel to one another in a vertical plane, usually hook- or sickle-shaped.. 13 (BRACHYCERA)

4. Head capsule longitudinally excised dorsally to varying degrees, in extreme cases reduced to several slender rods; capsule capable of partial or complete retraction within thorax. Respiratory system with only posterior spiracles (metapneustic), and the spiracular disc carrying the posterior spiracles usually surrounded by lobe-like projections of variable length (Fig. 4C). Humid or wet earth.................................Tipulidae (incl. Limoniidae), 17.2

– Head capsule complete and incapable of retraction within thorax. Respiratory system not as above; abdomen terminally with or without lobe-like projections...5

5. Free-living in stagnant water. Thoracic segments indistinctly differentiated, forming a round or somewhat flattened compound segment that is wider than any of the abdominal segments. Posterior spiracles mounted on a common siphon (tube-like structure), set at an angle to the longitudinal axis (Fig. 4D).............................. Culicidae, 17.4

– Not free-living (i.e., attached to or living in some substrate). Thorax not particularly set off from head and abdomen. No common spiracular siphon at tip of abdomen (at most with each spiracle on a short process)...6

6. Larva aquatic or in wet soil. Respiratory system without spiracles (apneustic)............................ 7

– Larva not aquatic. Respiratory system with at least one pair of functional spiracles......................9

7. On surface structures in flowing water. Head capsule usually with a pair of conspicuous folding labral fans dorsolaterally. Prothorax with an unpaired proleg ventromedially. Abdomen slender anteriorly, more swollen posteriorly; terminal segment ending in a ring or circlet of numerous radiating rows of minute hooked spines (Fig. 4E)............. Simuliidae, 17.5

– Not in flowing water. Prothoracic or terminal prolegs, if present, paired even if only by a slight separation of the apical spines or crochets. Body cylindrical and with no particular swellings; terminal segment sometimes with a pair of prolegs, but never with a ring or circlet of hooked spines......................8

8. Head distinctly elongate, often pointed. Larva slender, smooth, with uniform bead-like segments; prolegs not developed; short or long setae present scattered on most segments (Fig. 4F). Larva in moist soil......................... Ceratopogonidae, 17.6

– Head usually rounded or spherical, sometimes slightly elongate. First thoracic segment with a pair of ventral prolegs (may be partly fused) (Fig. 4G). Body segments, except sometimes the terminal one, lacking prominent tubercles and setae. Larva aquatic or in wet soil..................... Chironomidae, 17.7

9. Respiratory system with 10 pairs of spiracles (holopneustic). Body rather robust, all segments usually with tubercles or spinous processes (Fig. 4H).. Bibionidae, 17.9

– Respiratory system never holopneustic and usually with at least the posterior pair of spiracles reduced. Only terminal abdominal segment sometimes having tuberculate or spinous processes, although other segments sometimes with broad tumid swellings associated with creeping welts 10

10. Segments extensively setose (Fig. 4I). Each spiracle on a short process. Mandibles moving in an oblique downward direction; labrum slender and somewhat laterally compressed, with a dense brush of short setae on its ventral apex and epipharyngeal surface. Terminal abdominal segment with a pair of dorsally sclerotized lobes or a broad sclerotized shelf below posteriorly directed spiracles and behind the anus. Posterior spiracles either sessile or each situated at the apex of a sclerotized tubular process. Posterior segment with four fleshy lobes (Fig. 4I). In excrement and decaying matter.........Scatopsidae, 17.9

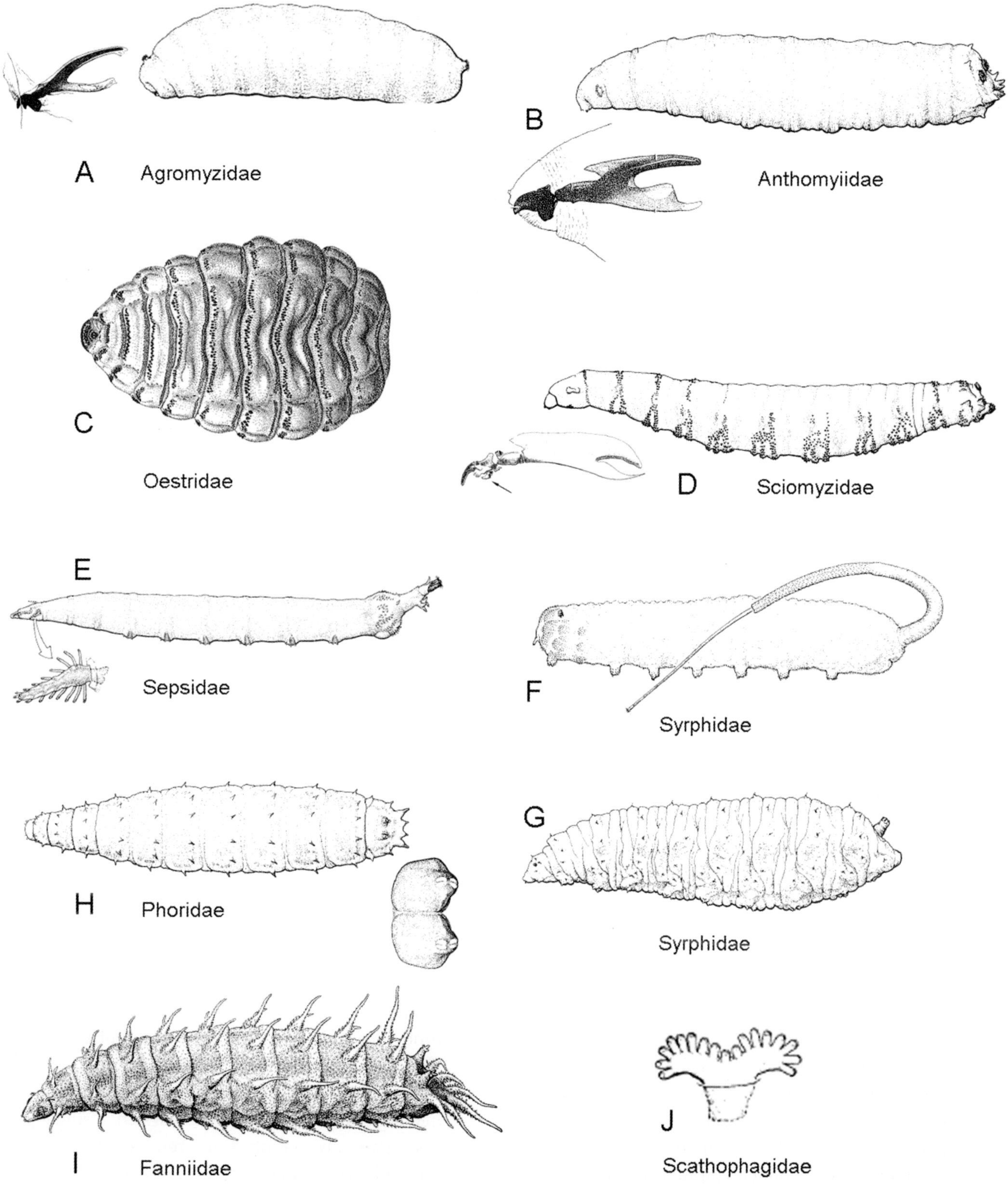

FIGURE 5 *A–J: Last instar larva of brachyceran Diptera. Illustrations from Manual of Nearctic Diptera vol. 2 (McAlpine* et al. *1987) except for C, which is reproduced from Grunin (1962, Fauna SSSR 19).*

– Segments not particularly setose. Posterior spiracles not at the end of a short process 11

11. Mandible and maxilla not flattened, without a serrate inner margin. With one pair of spiracles on the first thoracic segment, one pair at the terminal abdominal segment (amphipneustic). Body segments secondarily divided; posterior spiracles situated on apex of terminal abdominal segment and surrounded by four fleshy lobes (Fig. 4J) .... Trichoceridae, 17.3

– Mandible or maxilla or both flattened or lamellate, with a serrate inner margin. Respiratory system different (never with only anterior and posterior spiracles) 12

12. Head capsule shiny black. Abdominal creeping welts without sclerotized spicules (Fig. 4K) .... Sciaridae, 17.8

– Head capsule variously coloured. Abdominal creeping welts with sclerotized spicules (Fig. 4L) .... Mycetophilidae, 17.10

13. Sclerotized portions of head capsule always present and usually partially exposed externally. Labrum, mandibles and maxillae recognizable (although small) .... 14 (EMPIDOIDEA)

– 'Maggot head' entirely without external sclerotized parts; only a membranous pseudocephalic segment anterior to prothorax, normally bearing two pairs of papilla-like projections considered to be vestiges of antennae and maxillary palps; characteristically shaped cephaloskeleton completely within prothorax. Labrum and maxillae not clearly recognisable as such, mandibles present as a pair of near-parallel mouthhooks .... 15 (CYCLORRHAPHA)

14. Posterior projections of the internal head capsule (metacephalic rods) moderately expanded to spatulate apically. Terminal abdominal segment with four primary lobes surrounding posterior spiracles or evenly rounded; either six or seven abdominal creeping welts present (Fig. 4M) .... Dolichopodidae, 17.14

– Metacephalic rods slender throughout. Terminal abdominal segment either bearing a single median protuberance below posterior spiracles, or if more than one terminal lobe present, then respiratory system often without spiracles (apneustic) and seven or eight pairs of abdominal prolegs with apical hook-like crochets present (Fig. 4N) .... Empididae, 17.13

15. Larva mining in plants 16

– Larva not mining in plants 17

16. Anterior spiracles near one another on dorsal surface of prothorax. Mouthhook with longitudinal axis at oblique or right angles to remainder of cephaloskeleton, and usually bearing two or more pairs of equal-sized anteriorly directed teeth (Fig. 5A) .... Agromyzidae, 17.20

– Anterior spiracles situated laterally. Mouthhook with longitudinal axis largely in line with remainder of cephaloskeleton, and either bearing less than two pairs of teeth or bearing two or more pairs of teeth of unequal size (Fig. 5B) .... Anthomyiidae (phytophagous species), 17.26

17. Parasitic on reindeer (larva subdermally under the skin or in nasopharyngeal cavities). Strong spines encircling most segments (Fig. 5C) .... Oestridae, 17.27

– Not parasitic on reindeer. Spinulation usually not very pronounced 18

18. Insect parasitoids .... Tachinidae, 17.27

– Not insect parasitoids 19

19. Larva predator or parasitoid on freshwater snails. Cephaloskeleton with a ventral arch below base of anteriorly toothed or serrated mouthhooks (Fig. 5D) .... Sciomyzidae, 17.17

– Biology different. Cephaloskeleton lacking a ventral arch below mouthhooks 20

20. Spicules and microtrichiae extensively covering terminal abdominal segment only; anterior spiracle with elongate chamber with stalked pores projecting to either side (Fig. 5E) .... Sepsidae, 17.18

– Not this combination of characters 21

21. Posterior spiracles more or less distinctly elevated above plane of terminal segment or supported on a more or less raised or extended structure (siphon, protuberance, stalk or similar) 22

– Posterior spiracles situated level with (more rarely below) the plane of the terminal segment 30

22. Posterior spiracles placed on a short common process distinctly set off (i.e., 'chimney-like') from remaining body (Fig. 5F) OR on a long and retractile tube ('rat tail') (Fig. 5G); spiracular plates united along median margin .... Syrphidae, 17.16

– Posterior spiracles elevated above surface of abdominal segment on a pair of processes OR a single process gradually set off from remaining body and normally with spiracular plates well-separated 23

23. Posterior spiracles each with slits arranged in two pairs that are placed one behind the other (Fig. 5H) .... Phoridae, 17.15

– Posterior spiracles not so 24

24. Body with fleshy outgrowths laterally on all segments (Fig. 5I) .... Fanniidae, 17.26

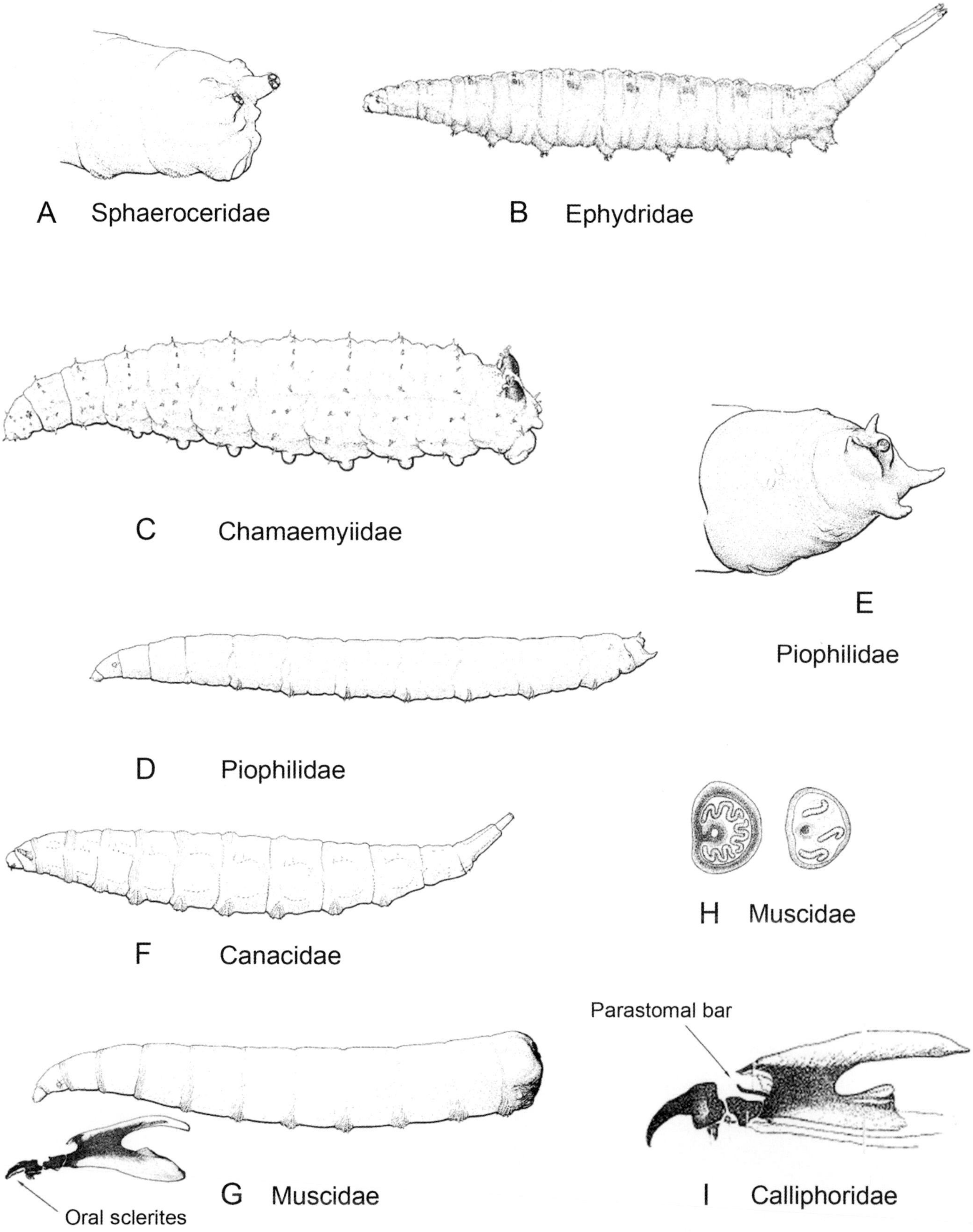

FIGURE 6 *A–J: Last instar larva of brachyceran Diptera. Illustrations from Manual of Nearctic Diptera vol. 2 (McAlpine* et al. *1987).*

– Body without fleshy outgrowths.................. 25

25. Anterior spiracle bicornuate, with buds arranged on two more or less distinct, diverging arms (Fig. 5J).... .... Scathophagidae, 17.26

– Anterior spiracle different, usually fan-shaped or indistinct 26

26. Anterior spiracle fan-shaped or tree-like. Posterior spiracles on two distinctly separated subcylindrical lobes two or three times longer than diameter of spiracular plate (Fig. 6A).........Sphaeroceridae, 17.22

– Anterior and posterior spiracles of different shape.. 27

27. Posterior spiracles at the apices of branches diverging from a common base, this 'terminal fork' apically sclerotized (Fig. 6B) .... Ephydridae, 17.25

– Posterior spiracles of different shape............. 28

28. Posterior spiracles on more or less elongate, heavily sclerotized spiracular processes (Fig. 6C). Predator on coccids ....Chamaemyiidae, 17.23

– Posterior spiracles of different shape. Not feeding on coccids... 29

29. Posterior spiracles situated on the median sloping surfaces of spiracular prominences, which appear capable of withdrawing or folding onto one another (Figs 6D, E). Larva in older, dry carcasses or in cheese, salted fish, ham etc., not in seaweed................. .... Piophilidae, 17.19

– Posterior spiracles situated at the apex of a respiratory tube, separated by a narrow furrow (Fig. 6F). Larva probably in seaweed........... .... Canacidae, 17.24

30. Cephaloskeleton without parastomal bars, with variously shaped accessory oral sclerites (Fig. 6G). Posterior spiracles either with serpentine spiracular slits or slits distinctly angled (Fig. 6H).............. .... Muscidae, 17.26

– Cephaloskeleton usually with parastomal bars, without accessory oral sclerites. Posterior spiracular slits different, usually straight...........................31

31. Mouthhooks with a single tip (without teeth) (Fig. 6I). Larva necrophagous in carrion or hematophagous in bird's nests........Calliphoridae, 17.27

– Mouthhooks with or without teeth (Fig. 5B). Larva in litter or fungi... ....Anthomyiidae (non-phytophagous species), 17.26

General literature on Diptera

McAlpine, J.F., Peterson, B.V., Shewell, G.E., Teskey, H.J., Vockeroth, J.R. & Wood, D.M. (eds) (1981, 1987): *Manual of Nearctic Diptera*, Vols 1 & 2. Research Branch, Agriculture Canada, Monographs 27 & 28.

McAlpine, J.F & Wood, D.M. (eds) (1989): *Manual of Nearctic Diptera*, Vol. 3. Research Branch, Agriculture Canada, Monograph 32.

Papp, L. & Darvas, B. (eds) (1997–2000): *Contributions to a Manual of Palaearctic Diptera*, Vols 1–4. Science Herald, Budapest.

17.2. Tipuloidea (Craneflies: Tipulidae and Limoniidae)

Pjotr Oosterbroek, Fenja Brodo, Vladimir Lantsov and Jaroslav Starý

Following Starý (1992) it became general usage to recognize four families within the superfamily Tipuloidea, viz., the Tipulidae, Limoniidae, Pediciidae and Cylindrotomidae; as far as known only the two first-mentioned are represented in Greenland. Some workers have retained the concept of a single family, Tipulidae (Byers 1992), and after the completion of the present chapter a combined morphological and molecular analysis has led Petersen *et al.* (2010) to advocate recognition of just two families (Pediciidae and Tipulidae), considering Limoniidae paraphyletic. Here we retain the Starý classification, adopted in our previous publication on Greenland craneflies (Oosterbroek *et al.* 2007) prompted by our involvement in the present project.

The Tipuloidea differ from other nematocerans by lacking ocelli, having a V-shaped mesonotal suture, and having two anal veins reaching the hind margin of the wing. Worldwide more than 15.000 species have been described.

The Greenland fauna of tipuloids comprises 15 species, including older and single records for three species.

The first records of Greenland Tipulidae were by Otto Fabricius, who in 1780 mentioned 4 species of *Tipula*. One of these, *Tipula rivosa*, most probably refers to *Tipula arctica* Curtis; the other three belonging to non-tipuloid families (Lundbeck 1901). The Tipuloidea recorded or described from Greenland since 1780 were enumerated by Henriksen & Lundbeck (1918) and Henriksen (1939), and have been reviewed in more detail by Mannheims & Theowald (1971). Since then it appears that only few new collecting data for Greenland Tipuloidea have been published (Cotton 1979, Messersmith 1982), both only mentioning *Tipula arctica*. Since our above-mentioned 2007 survey of Greenland craneflies, a detailed taxonomic inquiry into some arctic *Symplecta* (Starý & Brodo 2009) has led to reassessment of the status of its Greenland members.

Key to species

(for structural terms see family accounts and figure legends)

1. Antenna with more than 10 segments; flagellar segments (all segments beyond the second) rather uniform (Figs 1A, C, D). Upper part of head (vertex) without ocelli. 2 anal veins reaching the hind margin of the wing (Figs 1B, H). Mesonotum with V-shaped transverse suture .... Tipuloidea: Tipulidae and 'Limoniidae', 2

– Other combination of characters ....Non-tipuloid Diptera

2. Head with rostrum and palps well developed, rostrum sometimes with nasus (Fig. 1A). Larger species, body length 9 mm or (usually) more .... Tipulidae, 3

– Rostrum and palps short, rostrum without nasus (Fig. 1D). Smaller species, body length usually about 5 mm, only in *D. rhicnoptiloides* larger and in some specimens even slightly beyond 9 mm ....Limoniidae, 8

3. Flagellar segments (all antennal segments beyond the second) with verticils (several thin but long bristles at the basal part of the segment); flagellar segments rounded and basally thickened (Fig. 1C), or kidney-shaped (Fig. 2C), but not serrate 4

– Flagellar segments without verticils and serrate (Fig. 1A) .... *Prionocera*

4. Vein Rs short, arising more or less opposite crossvein Sc–R which connects vein Sc with vein R (Fig. 1H) .. .. *Nephrotoma*, 5

– Vein Rs longer, arising more toward the wing base with respect to crossvein Sc–R (Fig. 1B) .... *Tipula*, 6

5. Head, thorax and abdomen largely yellow with dark brown markings; lateral stripes on thoracic dorsum outcurved at front end, outcurved part dull or velvety; abdomen with narrow dorsal stripe and lateral spots (Figs 1E–G). Hind margin of male sternite 8 simple, in middle with rectangular appendage set with short strong setulae (Fig. 1I). Female cercus with pointed tip (Fig. 1J) ..... *Nephrotoma flavescens*

– Body largely light to dark brown with grey pruinosity, especially on thorax and abdomen; outcurved part of black dorsal thoracic stripes shiny; abdomen with broad dorsal and lateral stripes. Hind margin of male sternite 8 concave, its corners provided with incurved bristles, medially with an appendage of variable shape but without short strong setulae (Fig. 1L–N). Female cercus with blunt tip (Fig. 1K) .. .. *Nephrotoma lundbecki lundbecki*

6. Wing with conspicuous markings (Fig. 2A, 5). Dorsal thoracic stripes separated by (at least traces of) brown. Male antennal segments 4–11 kidney-shaped

(Fig. 2C). Male tergite 9 with a broad saucer-shaped sclerotized plate (Fig. 2B). Female terminalia as in Fig. 2D, cercus serrate, hypovalve without lateral flanges but with short, narrow, flexible tips .... *Tipula* (*Vestiplex*) *arctica*

– Wing with a light brown tinge and without markings except for pterostigma. Dorsal thoracic stripes separated by lighter grey. Male antennal segments simple (more or less as in Fig. 1D). Male tergite 9 emarginate distally, forming two lobes (Fig. 2E). Female with cercus not serrate and hypovalve broadly rounded distally and with lateral flanges (Figs 2H–I) .... *Tipula* (*Arctotipula*), 7

7. Nasus present. Two halves of male sternite 9 separated by a long narrowly U-shaped incision that ends near the hind margin of sternite 8 (Fig. 2F). Female cercus relatively short; lateral flanges of hypovalvae lying next to the dorsal margin of hypovalvae (Fig. 2H) .... *Tipula* (*Arctotipula*) *besselsoides*

– Nasus absent. Two halves of male sternite 9 separated by a V-shaped incision that ends at some distance from the hind margin of sternite 8 (Fig. 2G). Female cercus very short; lateral flanges of hypovalvae lying above the dorsal margin of hypovalvae (Fig. 2I) .... *Tipula* (*Arctotipula*) *besselsi besselsi*

8. Three R–veins reach the wing margin (in the section between veins Sc and M1+2 these are R1, R3+4 and R5; Figs 3A, B) .... *Dicranomyia*, 9

– Four R–veins reach the wing margin (in the section between veins Sc and M1 or Sc and M1+2 these are R1, R3, R4 and R5; Figs 3E–G, 4A, C, E, F) 10

9. Thorax yellow to orange brown, with a distinct darker coloured median stripe dorsally. Male terminalia: basal outgrowth of inner (ventral) gonostylus, so-called rostrum, simple, bearing spines at outer margin (Fig. 3C) .... *Dicranomyia* (*Dicranomyia*) *modesta*

– Thorax brown and pruinose, dorsal stripe hardly indicated. Male terminalia: basal outgrowth of inner gonostylus with two long spines at outer margin and series of bristles or hairs at inner margin (Fig. 3D) .. .. *Dicranomyia* (*Idiopyga*) *piscataquis*

10. Vein R2+3+4 as long or longer than vein R3 (Fig. 3E) .... *Rhabdomastix caudata*

– Vein R2+3+4 much shorter than vein R3 (Figs 3F, 4F) 11

11. Vein Rs with an appendix, so-called spur, at basal deflection (Fig. 3F). Head, legs and thoracic dorsum with short black spines, on thorax in two rows .... *Dactylolabis* (*Dactylolabis*) *rhicnoptiloides*

– Vein Rs without appendix (Figs 3G, 4A, C, E, F). Slender hairs (not spines) on head, legs and thorax ... 12

12. Cell r3 without crossvein (Figs 3G, 4A, B); wing membrane with numerous macrotrichia ..... *Ormosia*, 13

– Cell r3 with crossvein (Figs 4E, F); macrotrichia present on veins only ....*Symplecta*, 15

13. Discal cell absent (Fig. 3G). Male tergite 9 conspicuously enlarged (Figs 3H, I) .... *Ormosia* (*Ormosia*) *affinis*

– Discal cell present (Figs 4A, C). Male tergite 9 small (Figs 4B, D) 14

14. Discal cell small and distinctly shorter than outer part of vein CuA1 (Fig. 4A). Male terminalia: inner gonostylus simple (Fig. 4B) .... *Ormosia* (*Ormosia*) *fascipennis*

– Discal cell large and as long or longer than outer part of vein CuA1 (Fig. 4C). Male terminalia: inner gonostylus with strong short spines at excavated inner margin (Fig. 4D) ...... *Ormosia* (*Ormosia*) *subnubila*

15. Discal cell asymmetrical, with cell m1+2 considerably shorter than cell m3; crossvein Sc–R not close to origin of Rs, mostly placed rather far beyond, often nearly at half length of Rs; A2 strongly sinuous distally (Fig. 4E) .... *Symplecta* (*Symplecta*) *hybrida*

– Discal cell not as above, cell m1+2 and cell m3 of about equal length; crossvein Sc–R mostly placed close to origin of Rs; A2 weakly sinuous distally (Fig. 4F) .... *Symplecta* (*Symplecta*) *scotica*

TIPULIDAE

Tipulidae as conservatively circumscribed are medium to large sized, slender Diptera, and include some of the largest forms among the nematocerans (indeed the largest of all Greenland Diptera belongs here). Adults are distinguished by the following combination of characters: Elongate maxillary palps, the last segment usually being longer than the preceding segments together. Usually 13 antennal segments. Rostrum well developed and often with nasus. Compound eyes large, separated and bare. Ocelli absent. Thorax with a transverse V-shaped mesonotal suture. Wings elongate with 2 anal veins reaching wing margin. Vein Sc usually ending in R (not in C). No short crossvein proximal to tip of Sc. Cell m1 usually present. Wings sometimes reduced. Legs very long and slender. Tibiae without or with 1 or 2 apical spurs. Abdomen long and slender. Male terminalia (hypopygium) with a variety of characteristic features. Female terminalia ('ovipositor') include 2 pairs of elongate valves (cerci and hypovalvae).

Larvae elongate cylindrical with a tough cuticle ('leatherjackets'). Body consisting of 3 thoracic and 8 immediately

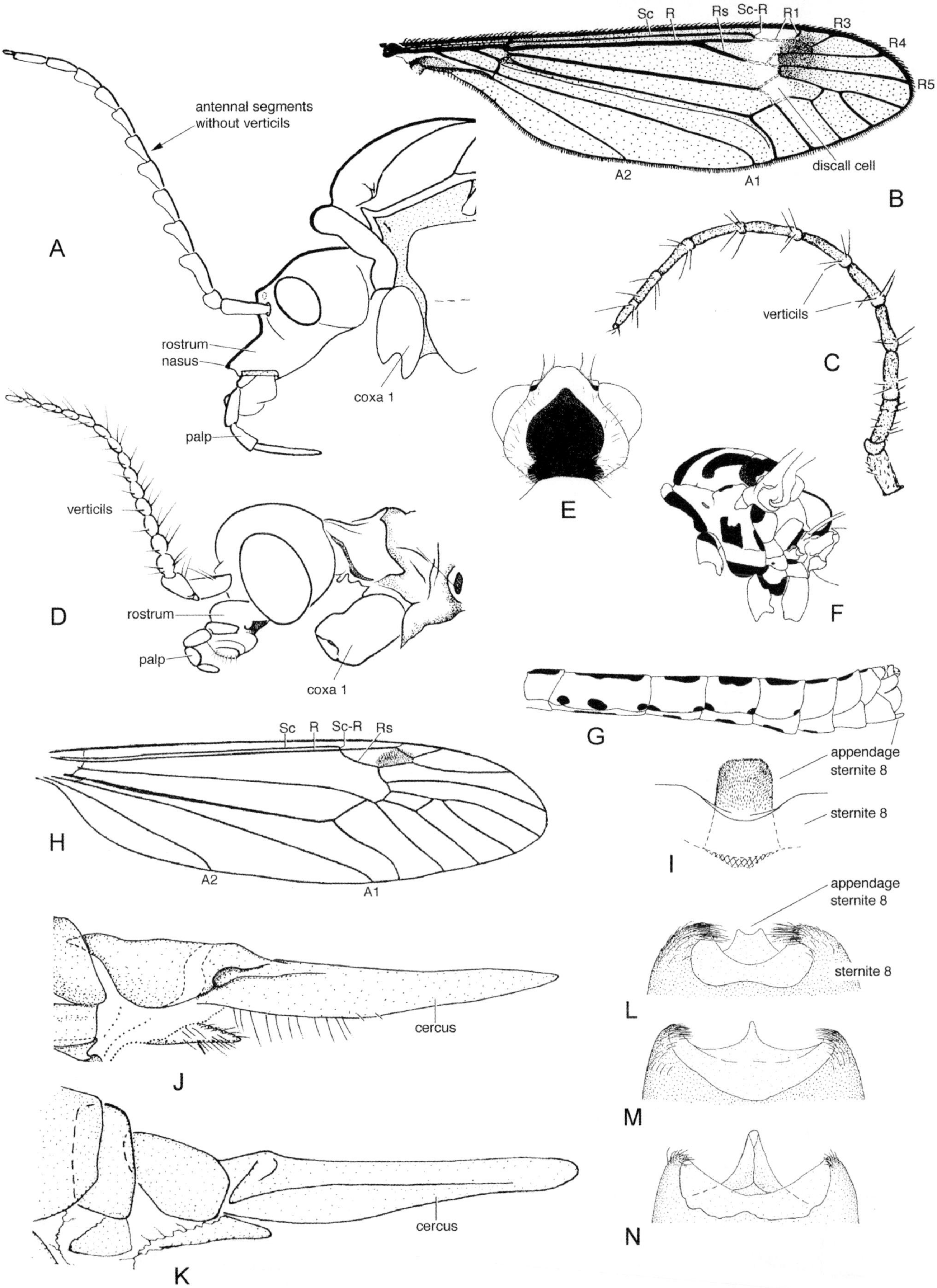

FIGURE 1 *A, antenna and head, male* Prionocera *sp.; B, wing,* Prionocera *sp.; C, antenna, male* Nephrotoma flavescens*; D, antenna and head, male* Erioconopa trivialis*; E, head,* Nephrotoma flavescens*, dorsal view; F, thorax* Nephrotoma flavescens*, lateral view; G, abdomen, male* Nephrotoma flavescens*, lateral view; H, wing,* Nephrotoma *sp.; I, sternite 8, male* Nephrotoma flavescens*, ventral view; J, cercus, female* Nephrotoma flavescens*, lateral view; K, cercus, female* Nephrotoma l. lundbecki*, lateral view; L–N, different sternite 8 shapes, male* Nephrotoma l. lundbecki. (*A, B: Brodo 1987; C, E–J: Oosterbroek 1978; D: Dienske 1987; K, L–N: Oosterbroek 1979*).

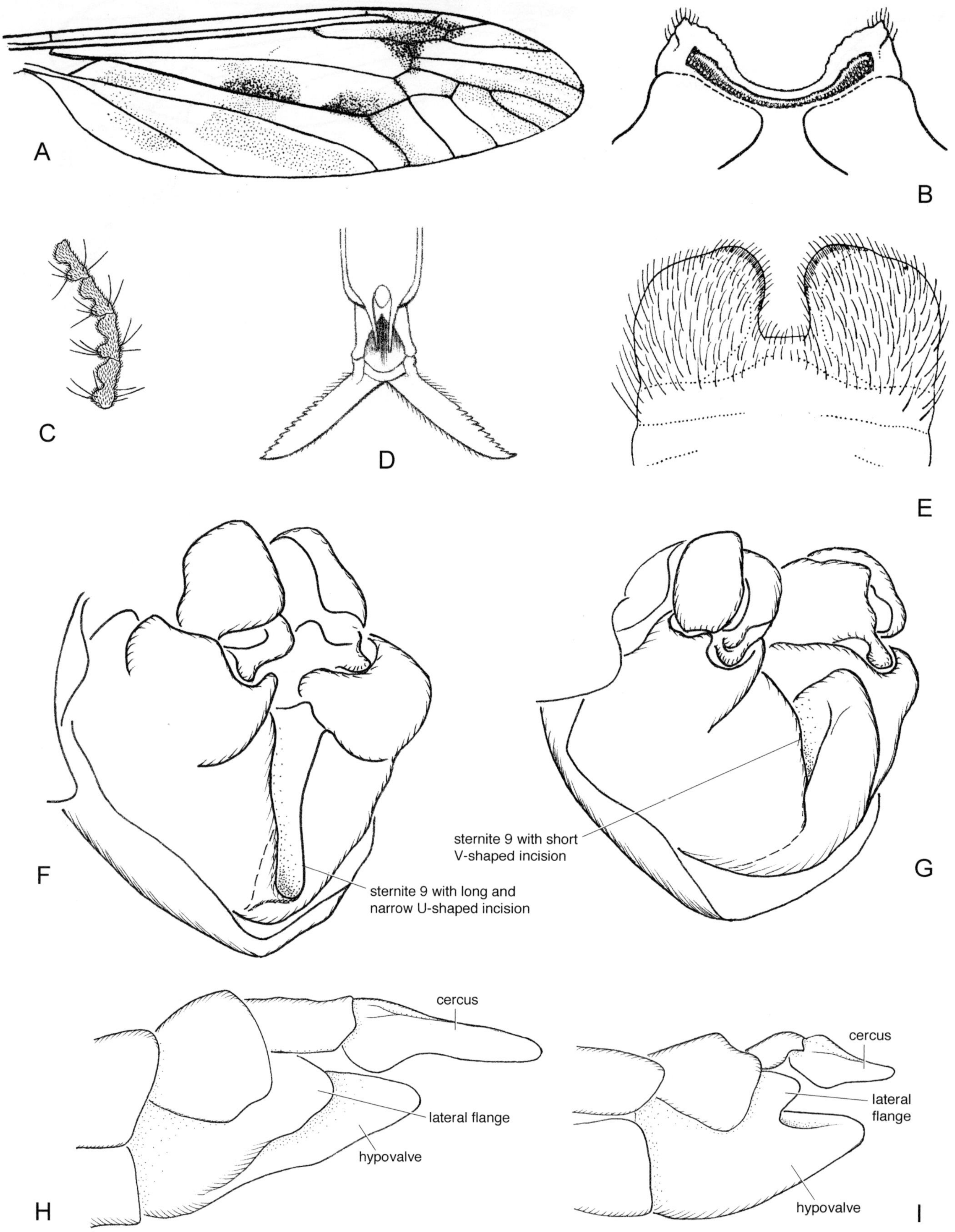

FIGURE 2 *A, wing,* Tipula (Vestiplex) arctica*; B, male tergite 9,* T. (V.) arctica*, dorsal view; C, antennal segments 4–7, male* T. (V.) arctica*; D, cerci, female* T. (V.) arctica*; E, male tergite 9,* Tipula (Arctotipula) b. besselsi*, dorsal view; F, hypopygium,* T. (A.) besselsoides*, oblique-rear view; G, hypopygium,* T. (A.) b. besselsi*, oblique-rear view; H, female terminalia,* T. (A.) besselsoides*, lateral view; G, female terminalia,* T. (A.) b. besselsi*, lateral view.* (*A: Alexander 1919; B, C: Savchenko 1964; D: Curtis 1835; E: Mannheims 1953; F–I: Oosterbroek* et al. *2007*).

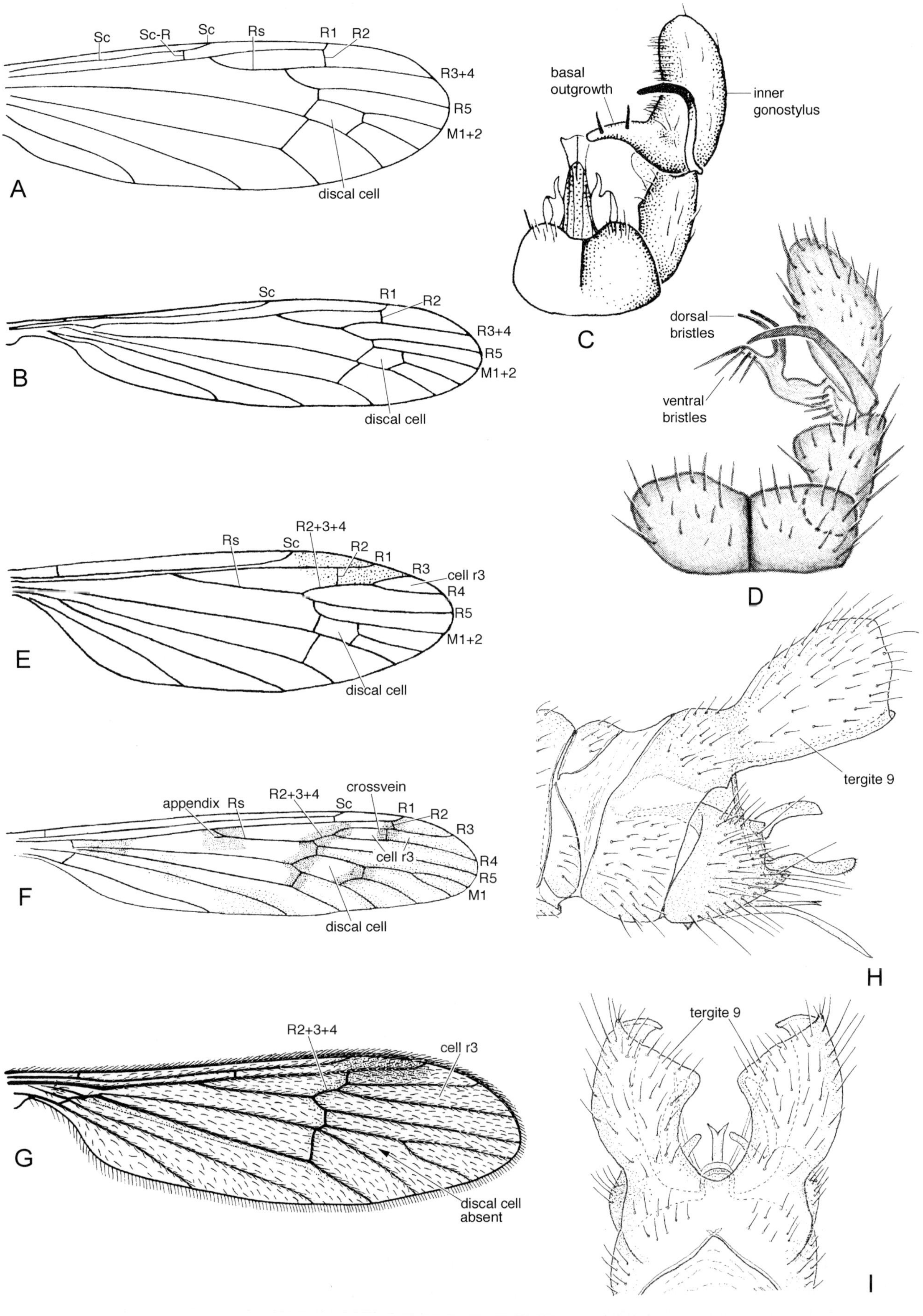

FIGURE 3 *A, wing,* Dicranomyia (Dicranomyia) modesta*; B, wing,* Dicranomyia (Idiopyga) piscataquis*; C, hypopygium,* D. (D.) modesta*, dorsal view; E, hypopygium,* D. (I.) piscataquis*, dorsal view; E, wing,* Rhabdomastix caudata*; F, wing,* Dactylolabis (Dactylolabis) rhicnoptiloides*; G, wing,* Ormosia (Ormosia) affinis*; H–I, hypopygium,* O. (O.) affinis *in lateral (H) and dorsal (I) view. (A: Dienske 1987; B: Lackschewitz 1964; C: Geiger 1986; D: Nielsen 1951; E: Mannheims & Theowald 1971; F: Alexander 1919; G: Alexander & Byers 1981; H & I: Tjeder 1970).*

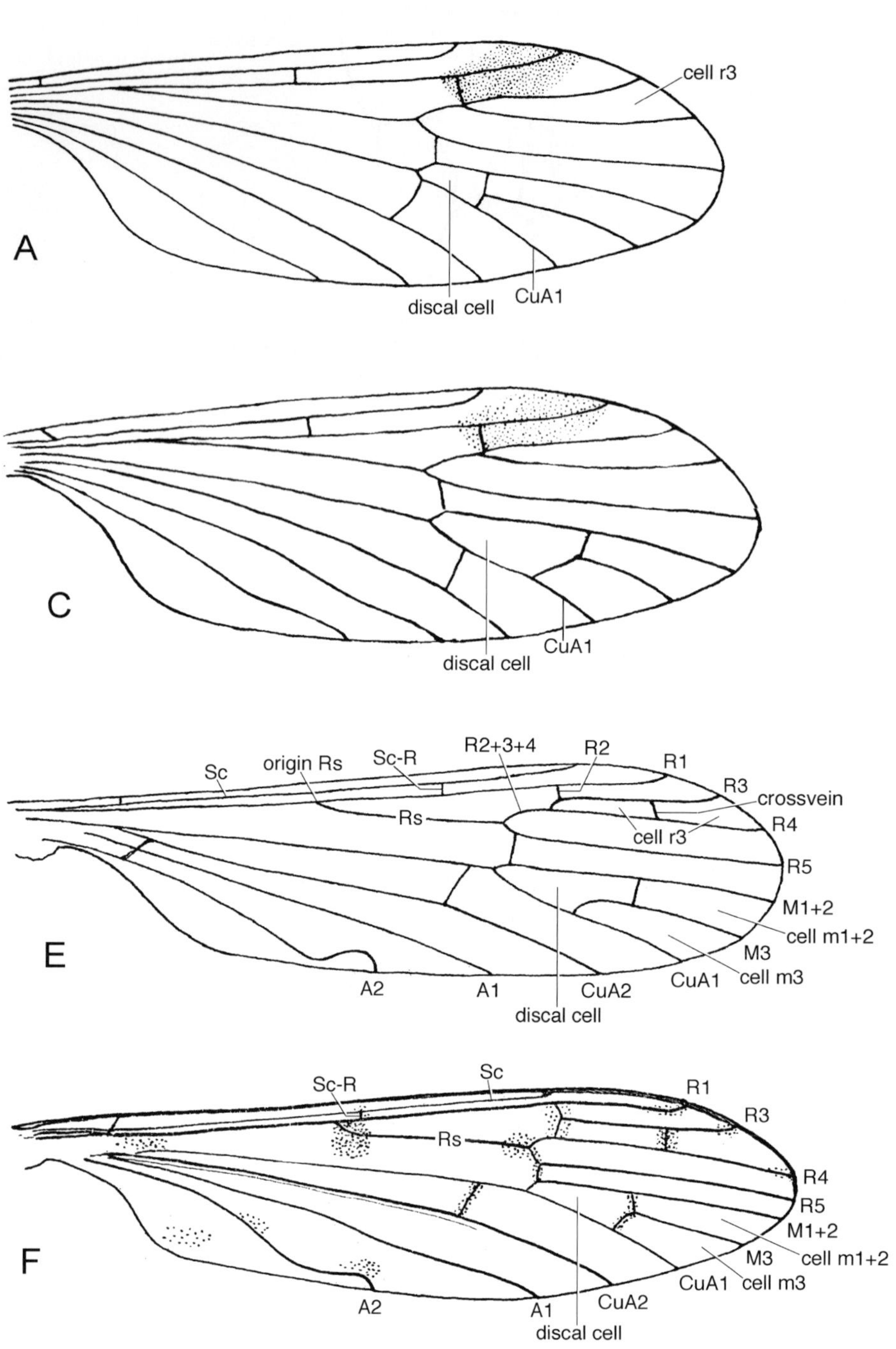

FIGURE 4 *A, wing,* Ormosia (Ormosia) fascipennis*; B, hypopygium,* O. (O.) fascipennis*, oblique dorsal view; C, wing,* O. (O.) subnubila*; D, hypopygium,* O. (O.) subnubila*, oblique dorsal view; E, wing,* Symplecta (Symplecta) hybrida*; F, wing,* S. (S.) scotica. *(A–D: Mannheims & Theowald 1971; E: Theowald 1971; F: Edwards 1938).*

identifiable abdominal segments. Head capsule massive, hemicephalous and largely enclosed in prothoracic segment. Tracheal system metapneustic. Truncate posterior end of larvae divided into a spiracular field with 2 spiracles and 6 lobes, and an anal field with anus and sometimes anal papillae or pushing organs; spiracular lobes sometimes partly reduced.

Pupae obtect and usually provided with short mesothoracic horns. Eyes, palpi, antennae, wings and legs clearly visible within sheaths. Abdominal segments, especially 4–7, usually with spines along posterior margin.

Larvae rarely aquatic, usually in semi-aquatic or terrestrial environments, such as along streams, lakes, ponds, or marshes, in moist to wet cushions of mosses or liverworts, wood or sodden logs in various stages of decay, leaf litter, mud, sandy or gravelly riversides, dry soils of pasturelands, lawns, steppes and even semi-deserts. Life-cycle consisting of a usually brief egg stage (up to two weeks), four larval instars, and a short pupal stage (also up to two weeks) before emergence of the short-lived adult. Summer as well as winter diapause of certain stages have been noted. Species under temperate conditions produce one or two generations a year, but depending on the species and the environmental conditions, the entire cycle may be as short as ten weeks or as long as six years. Larvae feed on a variety of material such as decaying plant and animal matter, mosses and algae. A few species, especially in *Tipula* (*Tipula*) and *Nephrotoma*, are destructive feeders on pasture grasses, seedlings and crops and may cause severe commercial losses. Some aquatic species in *Tipula* (*Arctotipula*) and *Prionocera* are apparently facultatively carnivorous. Adults do not appear to feed to any great extent, but can be observed taking liquids at flowers or wet substrates and surfaces.

Tipulidae are to be found from high latitudes (in north to 82°30'N) to equatorial forests and from sea-level to about 5200 m elevation. Usually they are associated with moist and temperate environments in forested or more open areas, such as swamps, marshes and meadows. Some species may be very abundant locally and of great ecological importance as food for a variety of animals. The family comprises about 4000 described species worldwide.

***Nephrotoma flavescens* (Linnaeus, 1758)**
Henriksen 1939: *Nephrotoma lineata* Scop.

Head Fig. 1E; male antenna Fig. 1C; thorax Fig. 1F; male abdomen Figs 1G, I; female postabdomen Fig. 1J

Separable (together with *N. lundbecki*) from all other Greenland tipulids by characters in the antenna (Fig. 1C) and wing venation (Fig. 1H).

BIOLOGY. The larva lives in the soil of grasslands, gardens, scrubs, etc.

DISTRIBUTION. SW. Only one record (Qasigiannguit/Christianshåb, Lundbeck 1898). It seems likely that the single Greenland record refers to an accidental transport.

Palaearctic, boreal-southern temperate (east to western part of Russia; not known from the Faroes and Iceland).

***Nephrotoma lundbecki lundbecki* (Nielsen, 1907)**
Male sternite Figs 1L-N; female postabdomen Fig. 1K

BIOLOGY. A widespread, common species, found for example in all the Canadian arctic ecosystems, extending from northern Ellesmere Island to the boreal forest (Brodo 1990), but primarily in cold, exposed and open habitats. On Ellesmere Island the larvae have been found in the soil of moist slopes, in troughs between tundra hummocks, and in wet mud at the edge of water bodies. Usually these larvae were not close to plant material, however, one larva was found in association with roots of *Polygonum viviparum*. Probably these larvae feed on micro-organisms or detritus in the soil (Brodo 2000). In the Altai and Amur regions the species was found in snowberry (*Symphoricarpos*) vegetations at altitudes between 2000 and 2100 m. (Savchenko *et al.* 1972) and tall herbaceous *Calamagrostis* meadows (Zinovjev & Savchenko 1962).

DISTRIBUTION. S, SW, NW (north to Inglefield Fjord), N (Peary Land, Centrum Sø), NE (from Jameson Land northward to Danmarks Havn). Not found in SE.

Circumpolar, high arctic-northern temperate. The subspecies *alexanderi* Oosterbroek and Tangelder, 1987, is known from Canada (Manitoba) and U.S.A. (Michigan).

***Prionocera* sp.**
Henriksen 1939: *Stygeropis parri* Kby.

A female tipulid recorded from Greenland by Ross (1826) as *Ctenophora parrii* is likely a *Prionocera*, but species identity cannot be ascertained. No additional material of the genus has subsequently been found.

BIOLOGY. The genus is a characteristic component of the wetlands in the taiga and forest-tundra regions of North America and the taiga and tundra regions in the Palaearctic. *Prionocera* species have truly aquatic larvae (Alexander 1919; Theowald 1967; Lantsov 1984; Lantsov & Chernov 1987; Brodo 1987, 1990, 2000). Larvae of *Prionocera* in the Arctic are phytophagous, feeding on mosses and algae (Lantsov 1984, Lantsov & Chernov 1987). Occasional and facultative necrophagy and carnivory (larvae of tipulids and tabanids) have been documented (Lantsov 1984, Brodo 1990).

DISTRIBUTION. SW. The single female specimen was recorded from "Hvalfiske Øer" which according to Mannheims and Theowald (1971) are small islands close to Qeqertarsuaq/Disko Island. The genus *Prionocera* is primarily boreal.

***Tipula* (*Arctotipula*) *besselsi besselsi* Osten Sacken, 1877**
Henriksen 1939: not recorded.

Male tergite 9 Fig. 2E; hypopygium Fig. 2G; female terminalia Fig. 2I.

BIOLOGY. According to observations of the subgenus *Arctotipula* in the arctic tundra of the Taimyr Peninsula

(Lantsov & Chernov 1987, Lantsov 1997), the larvae are predators on Chironomidae larvae, Plecoptera larvae and Tipulidae eggs as well. According to these and observations by us at the arctic tundra of the Taimyr Peninsula and the Canadian arctic (Nunavut), females oviposit in wet sand of river banks, and the larvae are truly aquatic, occurring in small rivers and wide stony rivulets among stones near the banks and in flooded stony splits of rivers. Larger larvae (3rd and 4th instar) tunnel along just below the surface in slowly moving to stagnant water bodies. They thereby leave networks of lines in the silt of creek beds, thus advertising their presence and making this taxon a particularly useful environmental indicator, for example of climatic changes (Brodo 2000).

DISTRIBUTION. NW (Polaris Bugt), N (Nedre Midsommersø).

Holarctic, high arctic-boreal (in Eurasia only found in eastern Siberia: Magadan). The subspecies *centrasiatica* Savchenko, 1961, is known from eastern Russia (Irkutsk province) and Mongolia.

***Tipula* (*Arctotipula*) *besselsoides* Alexander, 1919**
Henriksen 1939: not recorded.

Hypopygium Fig. 2F; female terminalia Fig. 2H.

BIOLOGY. Nothing is known, but the biology is likely to be similar to that of the other species in the subgenus *Arctotipula*.

DISTRIBUTION. N. Only 1 record, from Peary Land (Nedre Midsommersø, D.M. Wood & J.E.H. Martin, 28 June 1966, 1 female).

Nearctic, arctic (Canada: North West Territories).

***Tipula* (*Vestiplex*) *arctica* Curtis, 1835**
Habitus Fig. 5; wing Fig. 2A; male antennal segments 4–7 Fig. 2C; male tergite 9 Fig. 2B; female cerci Fig. 2D.

BIOLOGY. One of the most characteristic species of the circumpolar arctic insect fauna. The body size in both sexes of the species decreases towards the north, probably as one adaptation for saving energy (Hemmingsen & Jensen 1957, Brodo 2000). Both in the south and in the north of Greenland there are records at rather high altitudes (up to about 600–700 m; Hemmingsen & Jensen 1957). Species habitats are sides of lakes and marshy places (Ross 1826), stony tundra, wet lowland tundra with dry patches, coastal lowland. For oviposition, females prefer rather loose soil, which is not too dry, but also not wet; *T. arctica* belongs to the tipulids with deep-boring oviposition habits (Hemmingsen 1956). The larvae live in the soil, for example under *Dryas* clumps, and are probably saprophagous (detritophagous; Brodo 2000). Johansen's (1910) life-history observations at Danmarks Havn in NE Greenland, are worth repeating:

"The larvae of the large *Tipula arctica* hibernate down in the ground, and if we dig at places which are mouldy and not too dry, *e.g.* under *Cassiope* tufts, we easily find their large maggots. A little after the end of June they assume the pupa stage and begin to work up towards the surface, and here and there under stones etc. we may find circular holes leading down to a vertical passage, in which sits the active pupa with the head end upwards. It is thus well-protected against enemies, whilst at the same time having plenty of air. At the end of the month appear the imagines; it is the male especially, which is seen and only in flight; if it rests on the ground namely, its colour blends perfectly with that of the surroundings it prefers (bare, sunny, stony spots with patches of vegetation of *Cassiope*, grass etc.).....Towards the middle of August the summer already begins to wane. The large *Tipula arctica* imagines are no longer seen".

The duration of the life cycle in Greenland was estimated at two years (Nielsen 1910). Regarding *Tipula* (*Pterelachisus*) *carinifrons* Holmgren, 1883, with a life cycle as long as up to 6 years (arctic tundra west of Taimyr; Lantsov 1982) and the comparable big size of *T.* (*A.*) *arctica* specimens, the duration of the life cycle in northern Greenland may be estimated as at least 3 or 4 years, especially under less favourable climatic conditions.

Larvae and adults serve as an essential part of the food of certain birds (Curtis 1835, Hemmingsen & Jensen 1957, De Korte 1986, Brodo 1990).

DISTRIBUTION. S, SW, NW, N, NE. *Tipula arctica* is the most frequently collected crane fly from Greenland, being recorded already by Otto Fabricius in 1780 (as *Tipula rivosa*). Its Greenland distribution is discussed in detail by Hemmingsen & Jensen (1957), including its absence along the east coast south of 69°N. The species is found in all northern landscapes of Eurasia and North America, from the boreal forest in the south as far north as the high arctic (Lundstrøm 1915; Riedel 1919; Alexander 1922, 1924b;

FIGURE 5 Tipula (V.) arctica. (PHOTO: J. BÖCHER)

Lackschewitz 1936; Carpenter 1938; Oliver 1963; Savchenko 1964; Lantsov & Chernov 1987), in the USA also found in the southern Rocky Mountains of Colorado (Brodo 1990), in Greenland as far north as 82°30'N.

Holarctic, arctic-boreal (in Eurasia collected in entire northern Russia including Novaya Zemlya).

LIMONIIDAE

Small to medium-sized slender-bodied nematocerans distinguished by the following combination of characters: Last segment of maxillary palpus mostly short, about as long as the two preceding segments together. Antenna usually with 14 to 16 segments. Rostrum usually short (elongate in some extralimital taxa). Nasus always absent. Ocelli absent. Compound eyes bare. Thorax with transverse V-shaped mesonotal suture. Wings elongate with 2 anal veins reaching wing margin. Vein Sc ending in C; crossvein Sc–R usually present before tip of Sc. M1+2 sometimes absent. Wings sometimes reduced. Legs usually very long and slender. Tibiae with 0, 1 or 2 apical spurs. Abdomen usually long and slender. Male terminalia (hypopygium) with a variety of characteristic features. 2 pairs of gonostyli usually present, sometimes 1 or 3. Female terminalia ('ovipositor') variously modified, but in general including a pair of dorsal valves, the cerci, which are usually elongate, upturned and pointed, and a pair of ventral valves, called hypovalvae or hypogynial valves.

Larvae elongate, hemicephalic and metapneustic, rarely apneustic. Head capsule distinct, well sclerotized anteriorly, deeply incised ventrally and sometimes dorsolaterally, rigid parts reduced to 6 longitudinal bars in most Chioneinae and Limnophilinae. Posterior two-thirds or more of head capsule enclosed by and retractable within prothoracic segment. Abdominal segments smooth or with transverse rows of fine hairs, in several instances with creeping welts or fleshy projections. Terminal segment bearing posterior spiracles; spiracular disk usually surrounded by 5 or fewer lobe-like projections of variable length. Usually 4 anal lobes present.

Pupae obtect, elongate. Eyes prominent. Mesothoracic horns usually simple, ranging from short sessile to very elongate. Antennal sheaths long. Tarsal sheaths arranged side by side, not superimposed. Abdomen parallel-sided or almost so, usually more or less smooth except for welts. Anal segment usually with spines, prominent marginal or other abdominal spines especially in Hexatominae.

Larvae usually in various aquatic and semi-aquatic environments during most of their development, moving to margins or dryer places for pupation. Usually in more or less wet, organic soils or decaying vegetation in or along streams, lakes, swamps, marshes and woodlands. Other habitats include drier soils (*Dicranoptycha*, some species of *Cheilotrichia*, *Dicranomyia* and *Limonia*), intertidal zones or brackish water (some Limoniini), wet cliffs, piles or bridge piers supporting algal growths (some species of *Limonia*, *Orimarga*, *Elliptera*, *Dactylolabis*), mosses or liverworts (various Limoniinae), decaying wood, or sodden logs in streams, where larvae commonly feed on fungal mycelia (*Gnophomyia*, *Teucholabis*, *Lipsothrix*), sandy or gravelly borders of streams with moderate humus (many Eriopterinae), woody or fleshy fungi (*Metalimnobia*). Most larvae feed on decaying plant material, many Limoniinae feed on algae and the like; Limnophilinae are carnivorous.

Limoniidae as conservatively circumscribed, with some 10.500 described species worldwide, is a very large dipteran assemblage.

Ormosia (*Ormosia*) *affinis* (Lundbeck, 1898)

Wing Fig. 3G; hypopygium Figs 3H, I.

BIOLOGY. A semi-aquatic species (Young & Gelhaus 2000). For northwestern U.S.A. (New York, New England) the following biotopes have been listed: dryer woods above swamp, *Sphagnum* bog, arborvitae swamp, very old bogs, swales, in hellebore patches (Alexander 1924, 1925, 1927, 1936). Biotopes for far eastern Russia are open fir-larch woodland (Sakhalin) and mixed forests (Kuril Islands) (Savchenko & Krivolutskaya 1976).

DISTRIBUTION. S, sSW (north to Nuup Kangerlua/Godthåbsfjord: Kapisillit).

Holarctic, temperate (U.S.A.: Washington to Maine, south to California, Ohio and Pennsylvania; Canada: British Columbia, Ontario, Quebec, Nova Scotia and Newfoundland); Far eastern Russia (Sakhalin, Kuril Islands).

Ormosia (*Ormosia*) *fascipennis* (Zetterstedt, 1838)

Wing Fig. 4A; hypopygium Fig. 4B.

BIOLOGY. A species of boreal and mountainous biotopes (Savchenko & Parkhomenko 1980), found in polygonal tundra (Taimyr Peninsula; E.N. Savchenko, pers. comm.), dry slopes along river banks with dwarf birch shrub (Polar Ural), as well as small mires between hills or isolated mountains in far eastern Russia (Chukotka; Savchenko & Parkhomenko 1980), and brooks, small rivers, swamps, moist soil in Romania (Ujvarosi 2005). Greenland material was caught flying over willow scrub and in heathland at sea level.

DISTRIBUTION. G (Pituffik/Thule, Peary Land (Nedre Midsommersø), Gåseland, Mestersvig, Saqqisikuik/Skjoldungen (only record from SE)).

Circumpolar and holarctic, widely distributed. The species has a very spotty distribution in the Canadian Arctic, having been collected on Axel Heiberg Island and at Lake Hazen, Ellesmere Island (Oliver 1963), but not found some 300 km south on the Fosheim Peninsula (Brodo 2000). Furthermore, only females have been collected.

***Ormosia* (*Ormosia*) *subnubila* Alexander, 1920**
Henriksen 1939: not recorded.

Wing Fig. 4C; hypopygium Fig. 4D.

DISTRIBUTION. S (Eqaluit-Landet at Qaqortoq/Julianehaab, Ivittuut), SW (Sisimiut/Holsteinsborg, Qeqertarsuaq/Godhavn).

Nearctic (Alaska, Colorado).

***Rhabdomastix caudata* (Lundbeck, 1898)**
Henriksen 1939: *Gonempeda caudata* Lundb.

Wing Fig. 3E.

Only females are known. Collected in willow scrub (Carpenter 1938).

DISTRIBUTION. SW (Sermiligaarsuk, Kapisillit, Strømfjordshavn/Camp Lloyd, Marrait (Nuussuaq), NE (Hekla Havn)).

Nearctic (Baffin Land).

***Symplecta* (*Symplecta*) *hybrida* (Meigen, 1804)**
Henriksen 1939: *Helobia hybrida* Meig.

Wing Fig. 4E.

BIOLOGY. A very widespread species, found in a large variety of biotopes, such as near water reservoirs, flooded meadows, swamps, on wet rocks along shores, near hot springs, intertidal zones of small rivers in silt, sand and leaf fallings of *Alnus*, brooks in hilly districts, lowland boggy or brackish meadows, etc. (Wrage 1982, Lantsov & Chernov 1987, Reusch 1988, Podeniene 2001, Chandler & Crossley 2003). Oviposition and larval development take place in wet sand near water, in open terrain (Alexander 1920, for Iceland), as well as in deciduous forests (Beling 1879). The larvae make differently shaped excavations in the substrate, most probably feeding on a variety of organic matter (Alexander 1920, Savchenko 1982). In Western Europe adults occur during all vegetation periods but with a maximum abundance at the beginning and at the end of the season, demonstrating a bivoltine life cycle (Savchenko 1982).

DISTRIBUTION. NW. Only recorded once in Greenland (Pituffik/Thule, 1 male).

Circumpolar and holarctic, arctic-tropical, also alpine. Very widely distributed.

***Symplecta* (*Symplecta*) *scotica* (Edwards, 1938)**
Henriksen 1939: *Helobia hybrida* Meig *partim*.

Oosterbroek *et al.* (2007): *S. scotica, S. sheldoni* (Alexander, 1955) [misidentification]; *S. sunwapta* (Alexander, 1952).

Wing Fig. 4F.

BIOLOGY. Most Greenland records are from heaths. According to Alexander (1952) the species was not uncommon in wet boggy meadows at the Sunwapta Pass, Alberta, Canada (type locality of *S. sunwapta*). The specimens were swept from the tundra-like vegetation, including dwarf birch and willow, with grasses and sedges. Conditions were cold and bitter, even on sunny days, because of the glaciers of Mount Athabasca and the Columbia Icefield. In the typical tundra subzone of the Taimyr Peninsula, the species is usually abundant near water biotopes, presumably in shingle and oozy shoals of rivulets and small rivers. It occurs together with other species of Limoniidae, such as *Dicranota polaris, Arctoconopa obscuripes, Molophilus pleuralis, Molophilus propinquus* (Lantsov & Chernov 1987, as *Symplecta intermedia*). In temperate Europe known from brooks in hilly districts and lowland boggy meadows (Chandler & Crossley 2003).

DISTRIBUTION. S, SW (north to Qeqertarsuup Tunua/Disko Bugt: Saqqaq (Klitelven) and Nuussuaq (Marraat)), NW (Pituffik/Thule), NE (Sabine Ø).

Holarctic, arctic-temperate; recorded from Alaska to far east Russia.

NOTES. The material from Sabine Ø [Island] was recorded under the name *punctipennis* by Henriksen & Lundbeck (1918). The junior synonymy of *S. sunwapta* was established by Starý & Brodo (2009).

***Dactylolabis* (*Dactylolabis*) *rhicnoptiloides* (Alexander, 1919)**
Henriksen 1939: not recorded.

Wing Fig. 3F.

BIOLOGY. On the high arctic Devon Island, Y. Chernov found the species in 1991 (pers. comm.) at a variety of biotopes: plateau near slopes, polar desert, coastal sea lowland with dry crushed stone patches, stony tundra, wet lowland tundra with dry patches. On Ellesmere Island the larvae have been found under *Dryas* clumps (Brodo 2000) and larvae were also collected in pitfall traps and in the mud and silt beside water bodies, in ice-wedge depressions between tundra hummocks, and by clumps of *Carex* (Oosterbroek *et al.* 2007). On northern Ellesmere Island this species emerges slightly later than the other tipuloid species, not appearing before the first week of July and it is on the wing into the second week of August (Oosterbroek *et al.* 2007).

DISTRIBUTION. N (Kap København), NE (Holm Land).

Nearctic, arctic-boreal (from Alaska to Ellesmere Island, Quebec and Labrador). In the Canadian arctic, the species is found from the boreal forest tundra as far north as the high arctic (Brodo 1990).

***Dicranomyia* (*Dicranomyia*) *modesta* (Meigen, 1818)**
Wing Fig. 3A; hypopygium Fig. 3C.

BIOLOGY. One of the most common eurytopic species of the genus (Savchenko 1982, Salmela 2004), occurring in a large variety of wet biotopes, such as different kinds of forest, shrub, bog and meadow communities. In mountain regions it does not reach above the tree line (Tjeder 1958). In at least part of the range there are two or even more generations a year (Savchenko 1982, Salmela 2001). Females lay 130–160 eggs. The larvae are terrestrial to semi-aquatic, with habitats ranging from spring and brook environments to brackish water (Lindner 1959) or even halophilous conditions (Schmidt 1913).

DISTRIBUTION. S (recorded in Greenland only once, from Tunulliarfik/Eriks Fjord, Lundbeck 1898).

Circumpolar, Holarctic, boreal-southern temperate, widely distributed.

***Dicranomyia* (*Idiopyga*) *piscataquis* (Alexander, 1941)**
Henriksen 1939: not recorded.

Wing Fig. 3B; hypopygium Fig. 3D.

BIOLOGY. In Greenland it was collected on a river bank with grass and *Salix* vegetation (Nielsen 1951). In Juneau, Alaska, it was collected along a trail cutting through *Salix-Alnus-Tsuga* thickets. In Canada it was trapped in clear-cut logging areas, Carmanah Valley, Vancouver Island (Oosterbroek *et al.* 2007).

DISTRIBUTION. S, SW (north to Kangerlussuaq/Søndre Strømfjord airport).

Holarctic, temperate.

References

Alexander, C.P. (1919): The crane-flies collected by the Canadian arctic expedition, 1913–18. – *In*: Anderson, R.M., *Report of the Canadian arctic expedition 1913–18, Vol. III: Insects; Part C: Diptera*: 3c-30c.

Alexander, C.P. (1920): The crane-flies of New York. Part II. Biology and phylogeny. – *Memoirs, Cornell University Agricultural Experiment Station* 38: 691–1133.

Alexander, C.P. (1922): The crane-flies (Superfamily Tipuloidea, Order Diptera). – *Report of the scientific results of the Norwegian expedition to Novaya Zemlya 1921*, 5, Videnskapsselskapet, Kristiania, pp. 1–16.

Alexander, C.P. (1924): Report of the second Norwegian arctic expedition in the Fram 1898–1902. – *Norsk Entomologisk Tidsskrift* 1: 296–297.

Alexander, C.P. (1925): The crane-flies (Tipulidae) of New England: first supplementary list. – *Occasional Papers of the Boston Society of Natural History* 5: 169–174.

Alexander, C.P. (1927): The crane-flies (Tipulidae) of New England: second supplementary list. – *Occasional Papers of the Boston Society of Natural History* 5: 223–231.

Alexander, C.P. (1936): The crane-flies (Tipulidae) of New England: fourth supplementary list. – *Occasional Papers of the Boston Society of Natural History* 8: 273–292.

Alexander, C.P. (1941): New Nearctic crane-flies (Tipulidae, Diptera). Part XV. – *Canadian Entomologist* 73: 85–90.

Alexander, C.P. (1952): Undescribed species of crane-flies from the western United States and Canada (Dipt.: Tipulidae). Part XIII [concl.]. – *Entomological News* 63: 267–271.

Alexander, C.P. (1955): The crane flies of Alaska and the Canadian Northwest (Tipulidae, Diptera). The genus *Erioptera* Meigen. – *Miscellaneous Publications, Museum of Zoology, University of Michigan* 90: 1–33.

Alexander, C.P. & Byers, G.W. (1981): Tipulidae. – *In*: McAlpine, J.F. *et al.* (eds), *Manual of Nearctic Diptera*. Vol. 1. Biosystematic Research Centre, Ottawa, Ontario, Monograph 27: 153–190.

Beling, T. (1879): Zweiter Beitrag zur Naturgeschichte (Metamorphose) verschiedener Arten aus der Familie der Tipuliden. – *Verhandlungen der Zoologisch-Botanischen Gesellschaft in Wien* 28: 21–56.

Brodo, F. (1987): A revision of the genus *Prionocera* (Diptera: Tipulidae). – *Evolutionary Monographs* 8: 1–93.

Brodo, F. (1990): Crane flies (Diptera: Tipulidae) of the arctic islands. – *In*: Harrington, C.R. (ed.), *Canada's missing dimension*, Vol. 2. Canadian Museum of Nature, Ottawa: 471–484.

Brodo, F. (2000): The insects, mites, and spiders of Hot Weather Creek, Ellesmere Island, Nunavut. – *GSC Bulletin* 529: 145–173.

Byers, G.W. (1992): Crane flies – three families or one. – *Acta Zoologica Cracoviensia* 35: 37–41.

Carpenter, G.D.H. (1938): Notes on insects collected in West Greenland by the Oxford University Greenland expedition, 1936. – *Annals and Magazine of Natural History* 11: 529–553.

Chandler, P.J. & Crossley, R. (2003): *Symplecta chosenensis* (Alexander, 1940) (Diptera, Limoniidae) new to Britain, with comments on the status of *S. scotica* (Edwards, 1938). – *Dipterists Digest* 10: 49–54.

Cotton, M.J. (1979): Records of insects from north-east Greenland. – *Entomologist's Monthly Magazine* 113: 213–217.

Curtis, J. (1835): Descriptions, & c. of the insects brought home by commander James Clark Ross, R.N., F.R.S., & c. *In*: Ross, J., *Appendix to the narrative of a second voyage in search of a north-west passage and of a residence in the arctic regions during the years 1829, 1830, 1831, 1832, 1833*. London: lix–lxxx.

Dienske, J.W. (1987): An illustrated key to the genera and subgenera of the western palaearctic Limoniidae (Insecta,

Diptera) including a description of the external morphology. – *Stuttgarter Beiträge zur Naturkunde (A)* 409: 1–52.

Edwards, F.W. (1938): British short-palped craneflies. Taxonomy of adults. – *Transactions of the Society for British Entomology* 5: 1–168.

Fabricius, O. (1780): *Fauna Groenlandica*. J.G. Rothe, Hafniae et Lipsiae [=Copenhagen & Leipzig]: i–xvi, 1–452.

Geiger, W. (1986): Diptera Limoniidae. – *Insecta Helvetica Fauna* 8: 1–131.

Hemmingsen, A.M. (1956): Deep-boring ovipository instincts of some crane-fly species (Tipulidae) of the subgenera *Vestiplex* Bezzi and *Oreomyza* Pok. and some associated phenomena. – *Videnskabelige Meddelelser fra Dansk Naturhistorisk Forening* 118: 243–315.

Hemmingsen, A.M. & Jensen, B. (1957): The occurrence of *Tipula (Vestiplex) arctica* Curtis in Greenland and its decreasing body length with increasing latitude. – *Meddelelser om Grønland* 159: 3–20.

Henriksen, K.L. (1939): A revised index of the insects of Grønland containing a supplement to the insect list in Kai L. Henriksen and Will. Lundbeck: Grønlands Landarthropoder. – *Meddelelser om Grønland* 119: 1–112.

Henriksen, K.L. & Lundbeck, W. (1918): Landarthropoder (Insecta et Arachnida). – *Meddelelser om Grønland* 22: 481–823.

Johansen, F. (1910): General remarks on the life of insects and arachnids in North-East Greenland. *In*: Johansen, F. & Nielsen, I.C., The insects of the Danmark expedition. – *Meddelelser om Grønland* 43: 35–54.

Korte, J. de (1986): *Ecology of Long-tailed Skua,* Stercorarius longicaudatus *at Scoresby Sund, East Greenland. Part 5: Food and feeding habits*. Thesis University of Amsterdam: 1–30.

Lackschewitz, P. (1936): Das Genus *Tipula* in der Arktis und dem borealen Waldgebiet Eurasiens. – *Trudy Zoologicheskogo Instituta, Akademiya Nauk SSSR, Leningrad* 4: 245–312.

Lackschewitz, P. (1964): New and little-known palaearctic crane-flies of the family Limoniidae (Diptera, Tipuloidea). – *Entomologicheskoe Obozrenie* 43: 710–733 (in Russian).

Lantsov, V.I. (1982): Adaptive peculiarities of the life cycle of the arctic crane-fly *Tipula carinifrons* (Diptera, Tipulidae). – *Ekologiya* 1: 71–76 (in Russian) (English translation: 1982, *Soviet Journal of Ecology* 13: 67–71).

Lantsov, V.I. (1984): Ecology, morphology and taxonomy of arctic crane-flies of the genus *Prionocera* (Diptera: Tipulidae). – *Zoologicheskii Zhurnal* 63: 1196–1204 (in Russian).

Lantsov, V.I. (1997): Trophic relations of aquatic crane-flies (Diptera, Tipulidae) in the arctic. – *In*: *Diptera (Insecta) in Ecosystems, Zoological Institute, Russian Academy of Sciences, St. Petersburg*: 73–75 (in Russian).

Lantsov, V.I. & Chernov, Y.I. (1987): *Tipuloid crane-flies in the tundra zone*. Moscow: 1–175 (In Russian).

Lindner, E. (1959): Beiträge zur Kenntnis der Larven der Limoniidae (Diptera). – *Zeitschrift für Morphologie und Ökologie der Tiere* 48: 209–319.

Linnaeus, C. (1758): *Systema naturae per regna tria naturae, secundum classes, ordines, genera, species, cum caracteribus, differentiis, synonymis, locis*. Ed. 10. Salvii, Holmiae [=Stockholm], 1: i–iv, 1–824.

Lundbeck, W. (1898): Diptera groenlandica. – *Videnskabelige Meddelelser fra Dansk Naturhistorisk Forening i Kjøbenhavn* 1898: 236–314.

Lundbeck, W. (1901): Diptera groenlandica. – *Videnskabelige Meddelelser fra Dansk Naturhistorisk Forening i Kjøbenhavn* 1900: 281–316.

Lundström, C. (1915): Diptera-Nematocera aus den arctischen Gegenden Siberiens. – *Memoires de l'Academie Imperiale des Sciences de St.-Petersbourg* (8) 29(8): 1–33.

Mannheims, B. (1953). 15. Tipulidae. – *In*: Lindner, E. (ed.), *Die Fliegen der palaearktischen Region*, 3(5)1, Lief. 173: 113–136.

Mannheims, B. & Theowald, Br. (1971): Die Tipuliden von Grönland (Diptera, Tipulidae). – *Bonner Zoologische Beiträge* 22: 332–342.

Meigen, J.W. (1804): *Klassifikazion und Beschreibung der europäischen zweiflügeligen Insekten (Diptera Linn.)*. Braunschweig, 1: i–xxviii, 1–152; 2: i–vi, 153–314.

Meigen, J.W. (1818): *Systematische Beschreibung der bekannten europäischen zweiflügeligen Insekten*. Aachen, 1: i–xxxvi, 1–333.

Messersmith, D.H. (1982): A report on a collection of Diptera from Iceland and Greenland. – *Fauna Norvegia, Serie B* 29: 36–39.

Nielsen, I.C. (1907): The insects of East-Greenland. – *Meddelelser om Grønland* 29: 363–409.

Nielsen, I.C. (1910): A catalogue of the insects of north-east Greenland with description of some larvae. *In*: Johansen, F. & Nielsen, I.C., The insects of the Danmark expedition. – *Meddelelser om Grønland* 43: 55–68.

Nielsen, P. (1951): *Limonia (Dicranomyia) vibei* n.sp. from Grønland (Dipt. Tipul.). – *Entomologiske Meddelelser* 26: 185–186.

Oliver, D.R. (1963): Entomological studies in the Lake Hazen area, Ellesmere Island, including a list of species of Arachnida, Collembola and Insects. – *Arctic* 16: 175–180.

Oosterbroek, P. (1978): The western Palaearctic species of *Nephrotoma* Meigen, 1803 (Diptera, Tipulidae), part 1. – *Beaufortia* 27: 1–137.

Oosterbroek, P. (1979): The western Palaearctic species of *Nephrotoma* Meigen, 1803 (Diptera, Tipulidae), part 4, including a key to the species. – *Beaufortia* 29: 129–197.

Oosterbroek, P., Tangelder, I.R.M. (1987): A new subspecies, *Nephrotoma lundbecki alexanderi*, described from south of the maximum extent of the Laurentide Ice and new distribution records for other nearctic *Nephrotoma* (Diptera: Tipulidae). – *Canadian Entomologist* 119: 49–53.

Oosterbroek, P., Brodo, F., Lantsov, V. & Stay, S. (2007): The Tipulidae and Limoniidae of Greenland. – *Entomologiske Meddelelser* 75: 3–33.

Osten Sacken, C.R. (1877): Report on the Diptera brought home by Dr. Bessels from the arctic voyage of the Polaris in 1872. – *Proceedings of the Boston Society of Natural History* 19: 41–43.

Petersen, M.J., Bertone, M.A., Wiegmann, B. M. & Courtney, G.W. (2010): Phylogenetic synthesis of morphological and molecular data reveals new insights into the higher-level classification of Tipuloidea (Diptera). – *Systematic Entomology* 35: 526–545.

Podeniene, V. (2001): Notes on the larvae of *Rhabdomastix* (*Sacandaga*) *laeta* (Loew, 1873) (Diptera, Limoniidae). – *Acta Zoologica Lituanica* 11: 385–387.

Reusch, H. (1988): Faunistische und phänologische Untersuchungen über Stelzmücken des Niedersachsischen Tieflandes (Diptera: Limoniidae). – *Braunschweiger Naturkundliche Schriften* 3: 171–203.

Riedel, M.P. (1919): Resultats scientifiques de l'expedition des frères Kuznecov (Kouznetzov) a l'Oural Arctique en 1909, sous la direction de H. Backlund, 8. Nematocera polyneura. – *Zapiski Rossiiskoi Akademii Nauk, Leningrad* (8) 28(8): 1–10.

Ross, J.C. (1826): Appendix, Zoology. *In*: Ross, J.C., *Journal of a third voyage for the discovery of a North-West passage from the Atlantic to the Pacific performed in the years 1824–25, in his majesty's ships Hecla and Fury, under the orders of captain William Edward Parry, R.N., F.R.S., and commander of the expedition.* Murray, London: 111–115.

Salmela, J. (2001): Adult craneflies (Diptera: Nematocera) around springs in southern Finland. – *Entomologica Fennica* 12: 139–152.

Salmela, J. (2004): Semiaquatic flies (Diptera, Nematocera) of three mires in the southern boreal zone, Finland. – *Memoranda Societatis pro Fauna et Flora Fennica* 80: 1–10.

Savchenko, E.N. (1961): Crane-flies (Diptera, Tipulidae), Subfam. Tipulinae, Genus *Tipula* L., 1. – *Fauna USSR, Diptera*, 2(3) (N.S.) 79: 1–488 (in Russian).

Savchenko, E.N. (1964): Crane-flies (Diptera, Tipulidae), Subfam. Tipulinae, Genus *Tipula* L., 2. – *Fauna USSR, Diptera*, 2(4) (N.S.) 89: 1–503 (in Russian).

Savchenko, E.N. (1982): Limoniid-flies (subfamily Eriopterinae). – *Fauna Ukrainy* 14(3): 1–335 (in Ukrainian).

Savchenko, E.N. & Krivolutskaya, G.O. (1976): Limoniidae of the south Kuril Islands and south Sakhalin. *Akad. Nauk. ukr. SSR*, Kiev: 1–160 (In Russian).

Savchenko, E.N. & Parkhomenko, S.I. (1980): Preliminary data on the Limoniid-flies (Diptera, Limoniidae) fauna of Chukotka. – *Freshwater Fauna of the Far East*, Vladivostok: 95–109 (in Russian).

Savchenko, E.N., Violovich, N.A. & Narchuk, E.P. (1972): Review of the crane-flies (Diptera, Tipulidae) of the Altai mountains. – *Entomologicheskoe Obozrenie* 51: 74–95 (in Russian) (English translation: 1972, *Entomological Review* 51: 45–57).

Schmidt, R. (1913): Die Salzwasserfauna Westfalens. – *Jahresbericht des Westfälischen Provinzial-Vereins für Wissenschaft und Kunst* 1913: 29–94.

Starý, J. (1992): Phylogeny and classification of Tipulomorpha, with special emphasis on the family Limoniidae. – *Acta Zoologica Cracoviensia* 35: 11–36.

Starý, J. & Brodo, F. (2009): Arctic species of the subgenus *Symplecta sensu stricto* (Diptera: Limoniidae). – *Canadian Entomologist* 141: 1–30.

Theowald, Br. (1967): Familie Tipulidae (Diptera, Nematocera). Larven und Puppen. – *Bestimmungsbücher zur Bodenfauna Europas* 7: 1–100.

Tjeder, B. (1958): A synopsis of the Swedish Tipulidae, 1. Subfam. Limoniinae: tribe Limoniini. – *Opuscula Entomologica* 23: 133–169.

Ujvarosi, L. (2005): Limoniidae and Pediciidae (Insecta: Diptera) assemblages along mountainous streams: additions to assess the biodiversity in wet habitats in Carpathians, Romania. – *Acta Biol. Debr. Oecol. Hung*. 13: 233–248.

Wrage, H.-A. (1982): Ökologie der Stelzenmücken (Limoniidae) des Litorals und angrenzender Gebiete im Nordseekustenbereich (Diptera, Nematocera). – *Faunistisch-Ökologische Mitteilungen*, Supplement 3: 1–48.

Young, C.W. & Gelhaus, J.K. (2000): *Crane Flies of Pennsylvania: Preliminary checklist and database development with emphasis on aquatic species*. Report submitted to Pennsylvania Wild Resource Conservation Fund: i–ii, 1–256 (available from the authors).

Zetterstedt, J.W. (1838): *Sectio tertia. Diptera. Dipterologis Scandinaviae amicis et popularibus carissimus. Insecta Lapponica*. Lipsiae [=Leipzig]: 477–868.

Zinovjev, G.A. & Savchenko, E.N. (1962): On the fauna of crane-flies (Diptera, Tipulidae) of the Amur and Zea rivers interarea. – *Entomologicheskoe Obozrenie* 41: 554–571 (in Russian).

17.3. Trichoceridae (Winter gnats)

Christine Dahl and Ewa Krzemińska

Both sexes of Trichoceridae are non-biting, cranefly-like nematocerans with small rounded heads, inconspicuous mouthparts, long palps, long antennae, and hyaline wings with the second anal vein short, sharply bent to wing margin, the thorax without a complete dorsal, transversal suture as in Tipulidae, a slender abdomen, long legs and species specific setation (Fig. 1A). The male hypopygium has more or less complicated basal and distal parts of the forceps (see Figs 5A, G; 6A, G), the female a slightly bent ovipositor of variable length (see Figs 5D, K; 6D, J). The main wing venation is similar to that of 'limoniine' tipulids, but with the second anal vein short and sharply bent (Fig. 1B), except the genus *Diazosma* with a straighter second anal vein.

The eggs are oval, hyaline, easily desiccated, laid in decaying organic matter. The larvae (Fig. 2) are elongated with a chitinous head capsule with well developed mandibles and maxillae and two terminal tracheal openings. The pupae (Fig. 3) have movable abdomen; they stay in the same place, and the duration of this stage is short. Male adults are more short-lived and more susceptible to desiccation than females. Adults of boreal species can withstand temperatures down to -4°C for some time.

The family contains *c.* 160 species worldwide of which c. 20 are described on the basis of females only. The family is divided into two subfamilies Trichocerinae and Paracladurinae each with three genera (Dahl & Krzemińska 1997, Krzemińska 2005). Trichocerids are present in all continents, except for Africa; some species reach the Antarctic Islands, autochthonous (Alexander 1955) or introduced (Volonterio *et al.* 2013). Most species have been recorded from cool or cold climates.

Despite the abundance of adults in autumn, especially males in holarctic areas, the larvae and pupae are not easily found because of their patchy distribution. Larvae develop in decaying leaf-litter and vegetables, in South Georgia in birds' nests (Dahl 1970a). Both larvae and adults were found in rodent burrows in northern Finland in March, June and August (Dahl 1968).

Males of boreal species form huge swarms in late autumn when temperature is not below +6° C (Dahl 1970b). They may also be found on snow in early winter. Females can survive winter under snow in frost-free cavities or in caves. In Greenland only adults have been sampled, all during spring and summer.

All known Greenland taxa belong to the genus *Trichocera*, subfamily Trichocerinae; this genus contains most of the holarctic species (Dahl & Alexander 1976) and is divided into three subgenera (Starý 1998, Krzemińska 2002) based

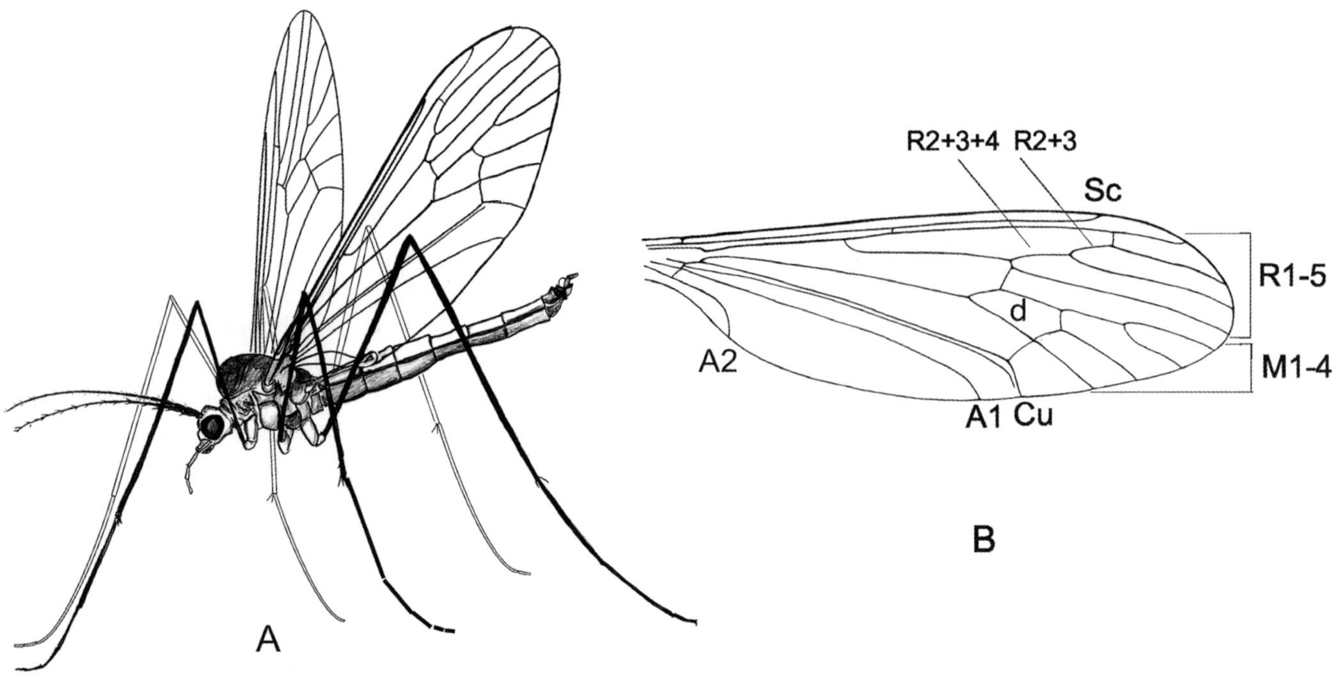

FIGURE 1 *A,* Trichocera saltator, *habitus; B,* Trichocera *wing. (A1 & A2, anal veins; Cu, cubital vein; d, discal cell; M, media; R, radius; Sc, subcosta).*

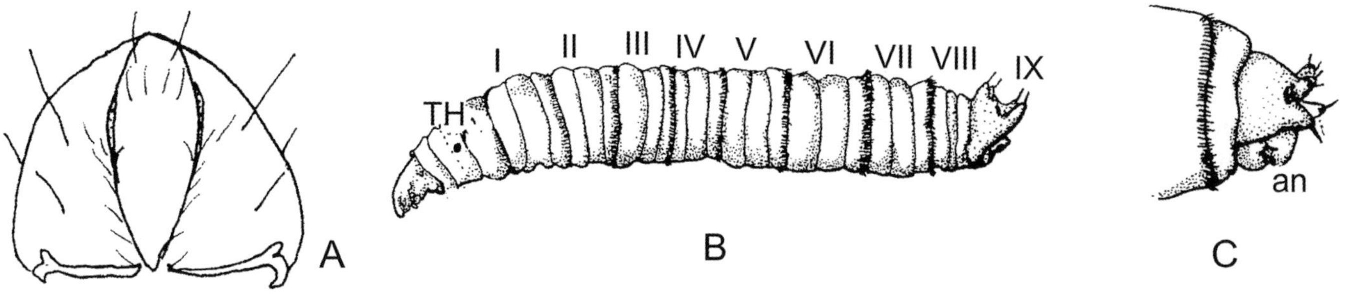

FIGURE 2 Trichocera *larva. A, head capsule; B, habitus. (an, anal shield; TH, thorax; Roman numerals indicate abdominal segments). (2B, C redrawn from Dahl & Krzemińska 1997).*

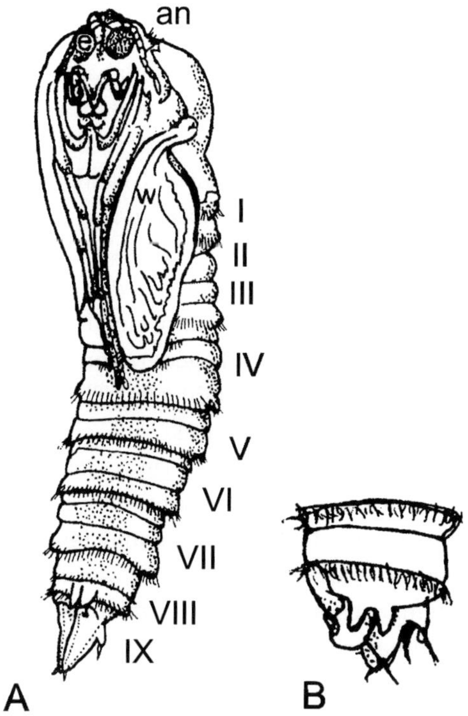

FIGURE 3 Trichocera *pupa. A, habitus of female; B, last segments of male. (an, antennal sheath; w, wing sheath; Roman numerals indicate abdominal segments). (3A, B redrawn from Dahl & Krzemińska 1997).*

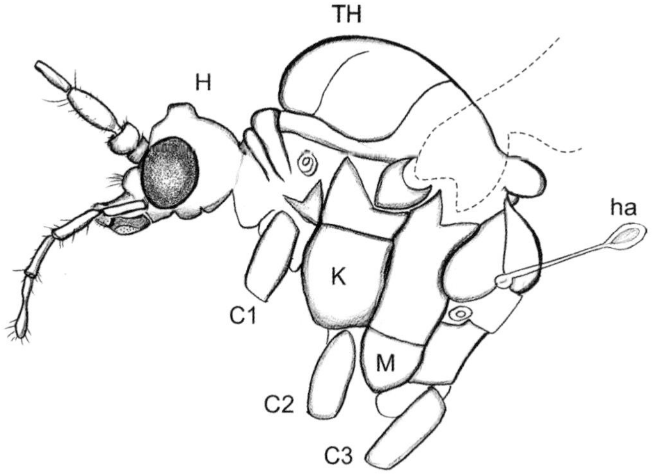

FIGURE 4 Trichocera *head (H) and thorax (TH). C, coxae; K, katepisternum; M, meron.*

on male characters. Adult members of the genus are characterized by having the katepisternum broad, twice the size of the meron (Fig. 4); first tarsomere longer than second; crossvein m-cu in outer anal angle of discal cell (Fig. 1B).

Key to Greenland *Trichocera* species

Terminology follows Dahl & Krzemińska (1997); wing venation according to Krzemińska *et al.* (2009).

MALES

1. Wings with several dark patches, distal part of forceps with a slight tubercle, bridge rounded, not fused, 1st flagellar segment not rounded (Figs 6A–C) .... *maculipennis*
– Wings without patches 2
2. Distal part of forceps with conspicuous armature (Figs 5A–C) .... *lutea*
– Distal part of forceps without conspicuous armature 3
3. Distal part of forceps with a pronounced tubercle, bridge slightly pointed, not fused, 1st flagellar segment short, rounded (Figs 5H–J) .... *borealis*
– Distal part of forceps without tubercle, bridge rounded, not fused, 1st–3rd flagellar segments very long (Figs 6G–I) .... *columbiana*

FEMALES

1. Wings with several dark patches, ovipositor with pointed tip, distinct sclerotized border of the setulose area, wide opening of the vaginal plate and

rounded supravaginal plate with several bristles (Figs 6D–F) .... *maculipennis*
– Wings without patches 2
2. Ovipositor very short and thick, sharply bent, 9th tergite markedly swollen, supravaginal plate rounded with two bristles (Figs 5D–G).................. *lutea*
– Ovipositor long, only moderately bent 3
3. Ovipositor moderately long with blunt tip, vaginal foramen moderately wide, vaginal plate with broad supravaginal plate with 2–6 bristles (Figs 5K–M) ... . *borealis*
– Ovipositor longer than last abdominal segments, stout with rounded tip, vaginal plate widely open with a slender supravaginal plate with 2 bristles (Figs 6J–L) .... *columbiana*

***Trichocera* (*Metatrichocera*) *lutea* (Becher, 1882)**
Figs 5A–F

Henriksen 1939: not recorded.

Both sexes: Wings clear; proportions between R2+3+4 and R2+3 slightly variable. Flagellar segments 1–3 short and swollen, the following somewhat longer; palpae long. *Female* (Figs 5D–F): Ovipositor short, moderately bent, 8th sternite with long, narrow hypogynal valves, 9th tergite swollen; vaginal plate with large foramen, supravaginal plate rounded with 2 bristles and apodeme of vaginal plate unforked. *Male* (Figs 5A–C): Basal part of forceps ventrally balloon-like, distal part with two basal outgrowths, distally with one large, flat outgrowth; bridge narrow, fused, apex acute; 9th sternite notched with lateral and caudal bristles, 9th tergite with two sets of bristles on rounded lateral parts; aedeagus with short parameres and straight basal apodeme.

Larva and pupa not known.

BIOLOGY. No special traits of habitat selection have been reported. The larva might be found in detritus-rich biotopes on or in the soil. In Scandinavia the adults have been found in mixed forests in late summer, and in Greenland *lutea* has been reported as a summer species.

DISTRIBUTION. S, SW (only few records), NE (several records).

Circumpolar, arctic-subarctic (also scarcely boreal and alpine). The species is abundant on Jan Mayen and Spitsbergen (Dahl 1957, 1973).

***Trichocera* (*Saltrichocera*) *borealis* Lackschewitz, 1934**
Figs 5H–M

Henriksen 1939: *Trichocera regelationis* L.

Both sexes: Wings greyish without discernible dark spots on wing venation. Antennae usually short, flagellar segments short, rounded, palpi shorter than flagellar segments 1–3 together (redescription: Dahl & Krzemińska 2008). *Female* (Figs 5K–M): Ovipositor long to only medium long, slightly bent and with a blunt end. The 8th sternite deeply notched with hypogynal valves elongated, slender. The vaginal plate with pointed supravaginal plate with 2–6 bristles, a medium-sized opening and a knob-like apodeme. *Male* (Figs 5H–J): Basal part of forceps not balloon-like, bridge divided, the pointed ends touching but not fused, distal part only with small knob at base; 9th sternite straight, narrow, with numerous bristles caudally, evenly spread out; aedeagus pear-shaped, parameres long and slightly bent; parameral apodeme more or less bent, lateral apodeme ear-shaped.

Larva and pupa (Figs 2, 3): Descriptions given in Dahl (1973) on material from Spitsbergen. Nine free body segments. Head capsule in dorsal view of IV instar larva broad, short with narrow frontoclypeus and weak setae at the edges of clypeites; mandibles with shovel-like terminal teeth and a well expressed dental process; 9th segment with caudal spiracles and 4 small lobes, last segment not balloon-like.

BIOLOGY. Larvae might be found in detritus-rich biotopes with numerous leaves of *Betula* and *Salix* on the ground. In Greenland the adults seem to occur throughout summer. The species has one yearly generation.

DISTRIBUTION. E, W (localities in Dahl 1967). The most common trichocerid species in Greenland.

NOTE. Two males found at Mestersvig (1953) fall into the most extreme type of *borealis*; they were by Mannheims identified as *T. regelationis*. In both males and females Dahl (1967) found a considerable intraspecific variability of some male and female characters; any correlation with geography/climate remains to be worked out.

***Trichocera* (*Saltrichocera*) *maculipennis* Meigen, 1818**
Figs 6A–F

Henriksen 1939: not recorded.

Both sexes with three conspicuous brown patches on their wings at Rs, R2+3+4, at the crossvein r-r, at the front of the discal cell and on the crossvein m-cu. In ethanol these patches can fade. It is one of the larger species, especially the female. Flagellar segment 1 swollen, 1/5 longer than segment 2 and the following ones. Palpi longer than flagellar segments 1–3 together. *Female* (Figs 7D–F): Ovipositor with a more or less pointed tip and a distinct, sclerotized border of the setulose area, shorter and more bent than in *T. borealis*; vaginal apodeme with large opening, rounded supravaginal plate with several bristles, the basal part form a very shallow fork or is flat. *Male* (Figs 6A–C): Basal part of forceps with pointed bridge, distal part stout with a pronounced tubercle, longer than basal part; adeagus pear shaped, parameres centrally broad with somewhat bent basal part and the lateral apodeme with a small lateral tip; 9th sternite with a single line of setae and a wave-like caudal border with pronounced edges, 9th tergite straight.

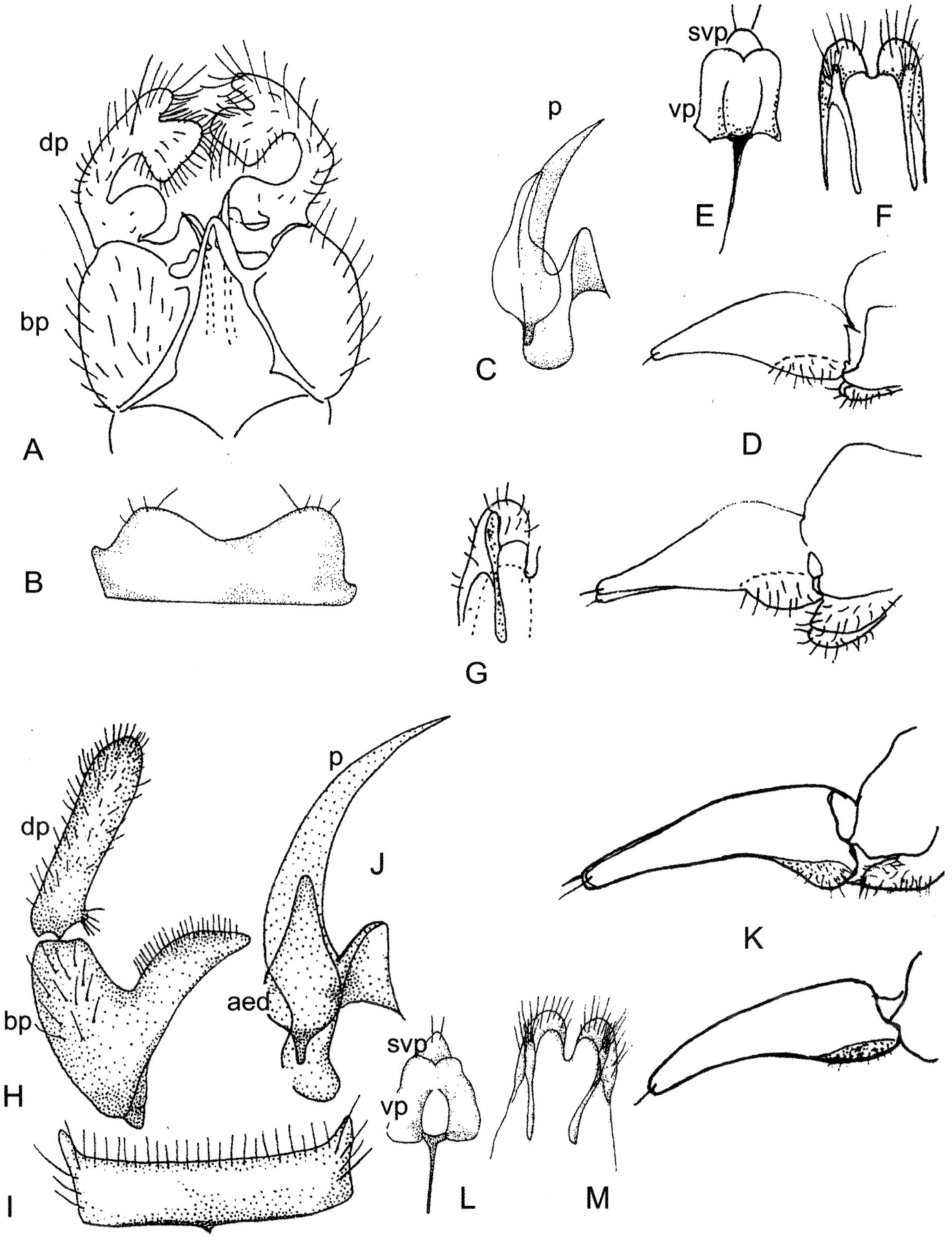

FIGURE 5 *A–G, genitalia details of* Trichocera lutea *male (A–C) and female (D–G). A, hypopygium; B, ninth sternite; C, aedeagus-paramere complex; D, ovipositor (of two specimens, showing individual variation); E, vaginal/supravaginal plates; F–G, ninth sternite. H–M, genitalia details of* Trichocera borealis *male (H–J) and female (K–M). H, hypopygium; I, ninth sternite; J, aedeagus-paramere complex; K, ovipositor (of two specimens, showing individual variation); L, vaginal/supravaginal plates; M, hypogynal valvae (aed, aedeagus; bp, basal part part of forceps; dp, distal part of forceps; hv, hypogynal valvae; p, paramere; svp, supravaginal plate; vp, vaginal plate). (A redrawn from Dahl 1957; B–C, H–J, L–M redrawn from Dahl 1967).*

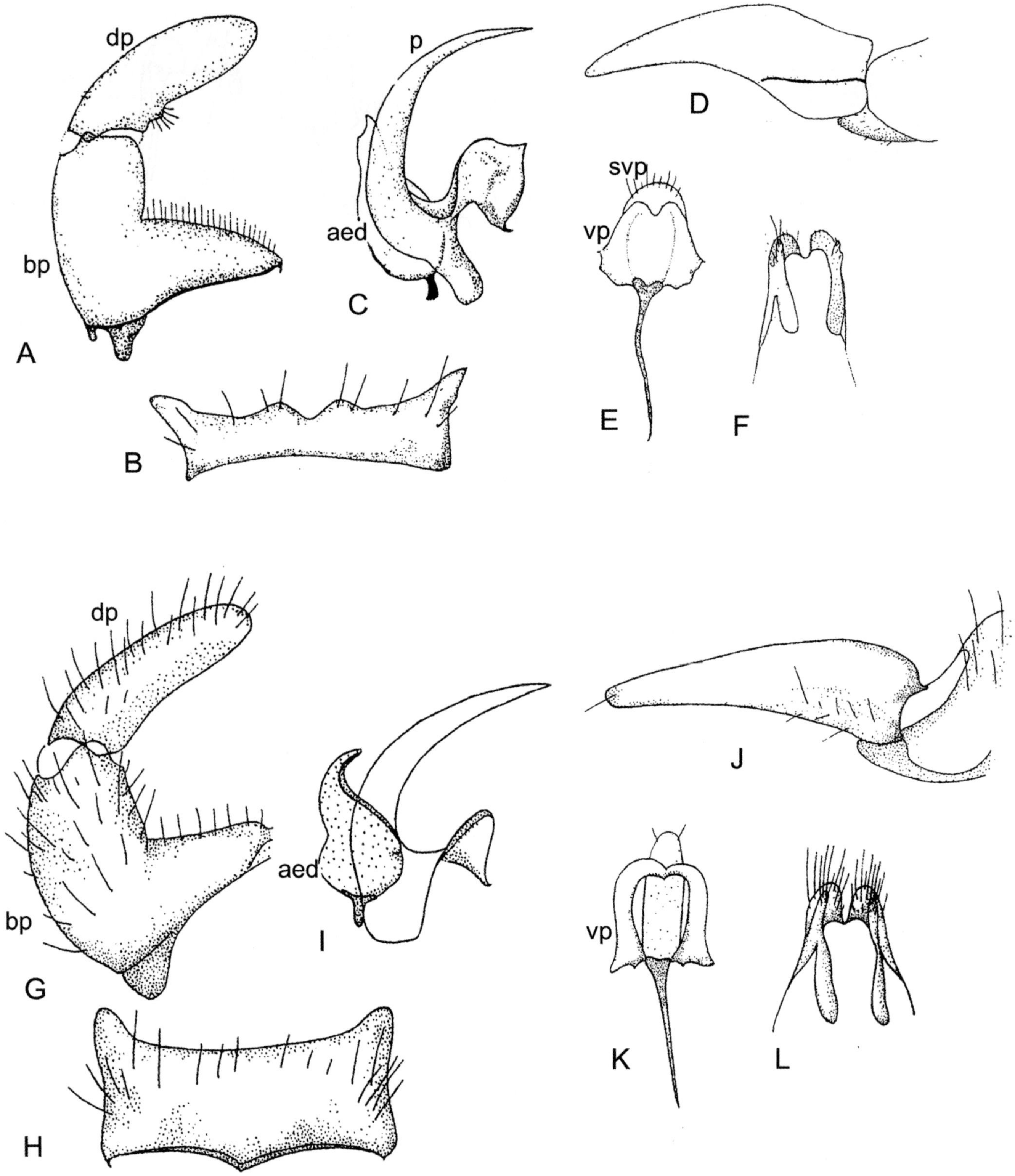

FIGURE 6 *A–F, genitalia details of* Trichocera maculipennis *male (A–C) and female (D–F). A, hypopygium; B, ninth sternite; C, aedeagus-paramere complex; D, ovipositor; E, vaginal/supravaginal plates; F, ninth sternite. G–L, genitalia details of* Trichocera columbiana *male (G–I) and female (J–L). G, hypopygium; H, ninth sternite; I, aedeagus-paramere complex; J, ovipositor; K, vaginal/supravaginal plates; L, hypogynal valvae (aed, aedeagus; bp, basal part of forceps; dp, distal part of forceps; hv, hypogynal valvae; p, paramere; svp, supravaginal plate; vp, vaginal plate). (A–D from Dahl 1966; G–L redrawn from Dahl 1967).*

Larva known from Britain by the description of Brindle (1962), characteristic by anal lobes devoid of any blackened surfaces.

BIOLOGY. The species is known from very different synanthropic habitats in non-arctic environments and has been found in Europe in caves, mines and on grass roofs.

In boreal areas *maculipennis* appears to have two generations, one in spring and one in autumn. In subarctic/arctic areas it seems to be a summer species (Dahl, 1970).

DISTRIBUTION. W (localities in Dahl, 1967), probably introduced to Greenland.

Palearctic species living in temperate to arctic regions; recently found introduced by man to maritime Antarctis (Volonterio *et al.* 2013).

Trichocera (*Saltrichocera*) *columbiana* Alexander, 1927
Figs 6G–L

Henriksen 1939: not recorded.

Large to medium species with hyaline wings and very long flagellar segments (segment 1–6), palpal segments 2–4 longer than 4–5, but shorter than flagellar segments 1–3 in both males and females. One tibial spur on foreleg, medium stout. The species resembles *T.* (*S.*) *major* Edwards. *Female* (Figs 6D–F): Length of ovipositor in Alaskan material variable (Dahl 1967) but with a more rounded tip than in *borealis* or in *Trichocera* (*M.*) *arctica* Lundström. Foramen wide, vaginal apodeme knob-like, supravaginal plate round with two bristles, different from *arctica*. *Male* (Figs 6G–I): 9th tergite straight with pointed edges, 9th sternite with central part somewhat bowl-like and groups of bristles laterally. 8th sternite visibly more sclerotized than the others, bridge stout with straight tip, not fused; distal part of forceps stout, only slightly longer than the basal part, without a distinct tubercle, but with a group of stout bristles in the same position; parameres broad in the central part, the basal apodeme of the same breadth, lateral apodeme ear-like, aedeagus pear-shaped.

Larva and pupa: Figs 48–51 in Dahl (1967) represent nearctic material; there are no previous records known from Greenland.

BIOLOGY. On Ellef Ringnes Island McAlpine (1965) found larvae and one half-moulted pupa together with adults in burrows of *Dicrostonyx groenlandicus*.

DISTRIBUTION. NE (Zackenberg Station, 2 males). New species to Greenland.

Nearctic, arctic-temperate? (Alaska, Canada, western U.S.A.; Pratt 2003).

[*Trichocera* (*T.*) cf. *japonica* Matsumura, 1915

DISTRIBUTION. NE (Zackenberg Station). Only one male is present in the material, a casual introduction? Pending the finding of more specimens, *T. japonica* is not yet considered a member of the native Greenland trichocerid fauna.

Nearctic (Canada, northern U.S.A.; Dahl 1967).]

References

Alexander, C.P. (1955): The crane-flies of the subantarctic islands of New Zealand (Diptera). *Records of the Dominion Museum* 2: 233–239.

Brindle, A. (1962): Taxonomic notes on the larvae of British Diptera. II. Trichoceridae and Anisopodidae. *The Entomologist* 95: 285–288.

Dahl, C. (1957): Die Gattung *Trichocera* in Spitzbergen, Bäreninsel und Jan Mayen (Dipt.). – *Opuscula Entomologica* 30: 227–237.

Dahl, C. (1967): Notes on arctic and subarctic Trichoceridae (Dipt. Nem.) from Canada, Alaska and Greenland. – *Opuscula Entomologica* 32: 49–78.

Dahl, C. (1968): Taxonomy and distribution of Trichoceridae (Dipt. Nemat.) from Finland and adjacent areas of the USSR. – *Opuscula Entomologica* 33: 365–370.

Dahl, C. (1970a): Diptera: Trichoceridae of South Georgia. – *Pacific Insect Monographs* 23: 271–273.

Dahl, C. (1970b): Distribution, phenology and adaptation to arctic environment in Trichoceridae (Dipt. Nemat.). – *Oikos* 21: 185–202.

Dahl, C. (1973): Notes on the arthropod fauna of Spitsbergen III. 14. Trichoceridae (Dipt.) of Spitsbergen. – *Annales Entomologici Fennici* 39: 49–59.

Dahl, C. & Alexander, C.P. (1976): A world catalogue of Trichoceridae Kertész, 1902, (Diptera). – *Entomologica scandinavica* 7: 7–18.

Dahl, C. & Krzemińska, E. (1997): Family Trichoceridae. E. *In*: Papp L. & Darvas, E. (eds). *Contibutions to a Manual of Palearctic Diptera* 2: 227–236. Science Herald, Budapest.

Dahl, C. & Krzemińska, E. (2008): Designation and description of lectotype *Trichocera* (*Saltrichocera*) *borealis* Lackschewitz, 1934 (Diptera, Trichoceridae). – *Norwegian Journal of Entomology* 55: 131–135.

Krzemińska, E. (2002): A new subgenus and two new species of the genus *Trichocera* Meigen, 1803 (Diptera: Trichoceridae). – *Annales Zoologici* (Warszawa) 52: 2001–2005.

Krzemińska, E. (2005): Subfamily Paracladurinae. III. Phylogenetic biogeography: two new genera and three new species described (Diptera, Trichoceridae). – *New Zealand Journal of Zoology* 32: 317–357.

Krzemińska, E., Krzemiński, W. & Dahl, C. (2009): Monograph of fossil Trichoceridae (Diptera). Over 180 million years of evolution. Institute of Systematics and Evolution of Animals, Polish Academy of Sciences, 171 pp.

McAlpine, J.F. (1965): Insects and related terrestrial invertebrates of Ellef Ringnes Island. – *Arctic*: 18: 73–103.

Pratt, H.D. (2003): The Winter Crane Flies of North America north of Mexico (Diptera: Trichoceridae). – *Proceedings of the Entomological Society of Washington* 105: 901–914.

Starý, J. (1998): New species of *Trichocera* Meigen, 1803, with a re-assessment of the subgenera *Trichocera* s.str. and *Metatrichocera* Dahl, 1966 (Diptera, Trichoceridae). – *Studia dipterologica* 5: 175–183.

Volonterio, O., Ponce de Leon, R., Convey, P. & Krzemińska, E. (2013): First record of Trichoceridae (Diptera) in the maritime Antarctic. *Polar Biology* 36: 1125–1131.

17.4. Culicidae (Mosquitoes)

Christine Dahl

Nematocerans which in both sexes have a long proboscis, in females containing long stylets enabling piercing and blood sucking. Abdomen slender, hyaline fore wings elongate (Figs 1A, D) and halteres knob like; legs and antennae long (Figs 1A–C), the latter in males with whorls of long setae (Fig. 1C).The body, legs and wings have a vestiture of white, brown or black scales (Figs 1E–F). Females with visible ovipositor, males with complex hypopygium (Fig. 3), which is turned 180° after eclosion. Eggs elliptic, laid on soil or water surfaces. Larvae (Figs 2A–C, 4) have four instars and are aquatic. Their mouthparts are adapted for scraping and/or biting and their foreheads bear long

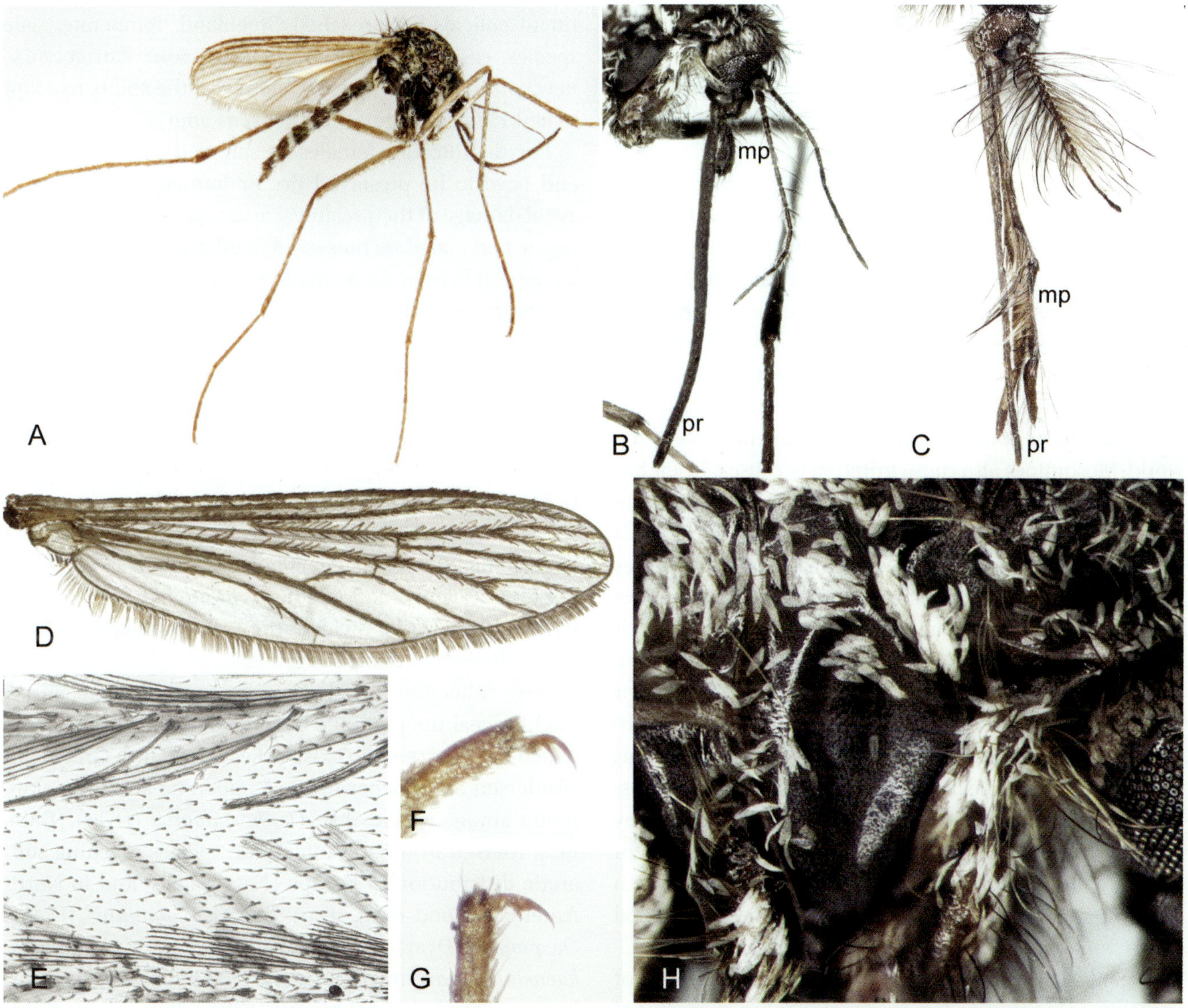

FIGURE 1 *A–F & H,* Ochlerotatus impiger. *A, ♀ habitus; B, ♀ head, lateral; C, ♂ head, lateral; D, wing; E, wing upperside scales along two consecutive longitudinal veins, underside scales appearing as shadows; F, apex of ♀ front leg showing pretarsal claw; H, ♀ posterior head and thorax, showing scale types present on lateral surface; note that the thoracic scale vestiture is somewhat incomplete (as will be the case in many collection specimens, since scales easily become rubbed off); G,* Ochlerotatus nigripes, *apex of ♀ front leg, showing pretarsal claw. mp, maxillary palp; pr, proboscis. (Specimens identified by B.V. Pedersen)*
(PHOTO: N.P. KRISTENSEN)

setae forming so-called mouth brushes which create water currents for particle uptake. Pupae (Fig. 2D) are aquatic, breathing through two thoracic spiracles, and their morphology permits considerable activity as swimming and diving also of the pharate adults. All life stages are dependent on atmospheric air.

The family includes more than 3500 species belonging to three subfamilies; the Greenland taxa pertain to the Culicinae-Aedini, reclassified by Reinert *et al.* (2004). For some common species the long list of synonyms (Knight & Stone 1977) indicates their variability of characters over geographical areas. For reliable identification of all life stages examination of many characters are necessary. Mosquitoes are found worldwide with exception of Iceland, the Antarctic and a few extreme, isolated Arctic, Antarctic and South Pacific islands. The highest species richness and character diversification occur in tropical and subtropical regions.

Different life cycle patterns exist. Under subarctic and arctic conditions two patterns prevail: uni-voltine or sometimes bi-voltine species with different phenologies with females or larvae surviving cold periods, or uni-voltine species with different female seasonal phenologies and with eggs surviving in dry and cold habitats.

Females can be carriers of viruses, bacteria or micro-organisms, some of which cause serious, even deadly diseases in man or domestic animals in different parts of the world. Mosquitoes also cause irritation because of their biting habits and their enormous numbers, especially in northern subarctic and arctic areas. Most warm-blooded animals serve as blood sources. Females and males of most species take their first meals from plant nectar, the males to achieve enough energy for swarming and survival, the females for finding male swarms and later to enable their search for a first blood meal. Most species need blood for maturing their egg batches. However, some can lay at least one batch without a blood meal, they are 'autogenous'. In arctic species this ability remains to be studied. The danger of disease transmission lies in the flexibility of female blood intake. They compensate for insufficient blood intake or interruption of a blood meal by biting several times and shifting between available vertebrate hosts. Only few species take blood meals from both birds and mammals (including humans).

Greenland is known for the overwhelming number of mosquitoes in summer, but despite this only two species of Culicidae have been reliably reported from Greenland. Their recent distribution in the country is fairly wide. Both species are circumpolar, arctic and uni-voltine. In south-west and west Greenland Nielsen and Nielsen (1966) carried out studies of the biology of mosquitoes with special emphasis on female feeding and male swarming behaviour. On Ellesmere Island Corbet (1966) and Corbet and Danks (1973, 1975) compared the biology of both species. At these high latitudes males swarm at daytime. The swarms attract females which react to male sounds and/or visual clues. Flight activity was observed at 1.7° C (Danks 1981). Females laid their eggs in soil depressions, in which next spring snow-melt pools occurred. Separate field studies in Greenland of the two species are desirable to reveal any differences in egg laying, phenology and biting behaviour. Such studies will be essential should virus transmission ever occur in the country. Future comparison of West and East coast populations with continental materials, based on molecular techniques, might elucidate from which continent culicids have reached Greenland. Temperate, agile species, easily spread by human activities, can colonize new continents very quickly and reveal the ability to adapt to new climatic regimes (Becker *et al.* 2010).

For identification females are sampled in great numbers and have to be preserved dry or immediately frozen to avoid damage of their scaling. Larvae (preferably instar IV), pupae and males are preserved in ethanol. Preparations of larvae and hypopygia are made on slides. For identification a microscope with at least 100x magnification is necessary.

Taxonomy of Greenland Culicidae

The two Greenland culicids both belong to a taxon *Ochlerotatus* Lynch-Arribalzaga, which has been alternately considered a separate genus and a subgenus in the very large genus *Aedes* Meigen. Following Reinert *et al.* (2004) the former view is adopted here, and *Ochlerotatus* is now itself subgenerically divided; the Greenland species both belong to the nominate subgenus.

New collections, carried out by mosquito specialists, might reveal the presence of some additional species for which habitats seem to exist in East and South Greenland (Andersen and Böcher 1994). A subarctic and temperate mountainous species like *Oc. hexodontus* (Dyar) (Dahl, 1974, Nielsen 1979), and temperate species reaching subarctic distribution in Europe (Natvig 1948) and in North America (Wood *et al.* 1979), like *Oc. communis* (DeG.), *Oc. pionips* (Dyar), *Oc. punctor* (Kirby) and *Culiseta alaskaensis* (Ludlow) might be found if sampling, especially of larvae, is carried out throughout the whole growing season. *Culiseta* adults would be recognizable by the presence of prominent setae just in front of the first spiracle, and the larvae by the position of the multibranched seta 1–S near the base of the siphon (not near/beyond mid-length as in *Ochlerotatus*, Fig. 4). For further information on culicids in NE America reference should be made to

Wood *et al.* (1979) and Thielman & Hunter (2007), for NW European taxa to Dahl (1997).

Identification of the Greenland species is difficult, since the differences are quite subtle and a number of allegedly diagnostic characters have proved to be too variable to be operational; partly denuded females are especially difficult to distinguish. Hence mixed Greenland mosquito material still might be kept under the name of the first recorded species ("*Aedes*" *nigripes*) in collections. The best species characters are found in males and in IVth instar larvae (Dahl 1997). Danks and Corbet (1973) have given keys to the earlier instars and pupae of the two species.

Explanation of morphological terms used in key and descriptions are given in Figures 1B–C; 2A–C; 3C.

Key to Greenland species of Culicidae

ADULTS

The two sexes are immediately separable by the long and strong setae whorl-like covering the antennal flagellum in males, whereas the females have few and short setae. Also, the palpi in males are about as long as the proboscis, in females tiny appendages at the proboscis base (Figs 1B–C).

1. Claws only medium long, markedly bent, subbasal tooth long, claw and tooth near-parallel (Fig. 1F). *Male genitalia* (Figs 3A–C): lobes of 'tergite IX' (anatomically IXth sternite) with few (<10) black spines. Gonocoxites with basal lobes somewhat angular in dorsal view (in permanent slide mounts the outline may change according to degree of compression). Phallosome sclerotization posteriorly weak, restricted to lateral apices (which may be ± concealed by the 'paraprocts') .... *Ochlerotatus impiger*
– Claws longer and more straight, subbasal tooth short, claw and tooth clearly diverging (Fig. 1G). *Male genitalia* (Figs 3D, E): lobes of IXth 'tergite' larger and with more numerous (>10) black spines. Gonocoxite with basal lobes in dorsal view more rounded. Claspettes only with short setae at base. Phallosome sclerotization posteriorly markedly strong/melanized around entire near-circular terminal indentation .... *Ochlerotatus nigripes*

LARVAE (4TH INSTAR)

1. Medium-sized (sometimes small, fragile) larvae. Head capsule broad (Fig. 4A). Anal papillae long, but < 1.5× length of anal segment; saddle open, one or

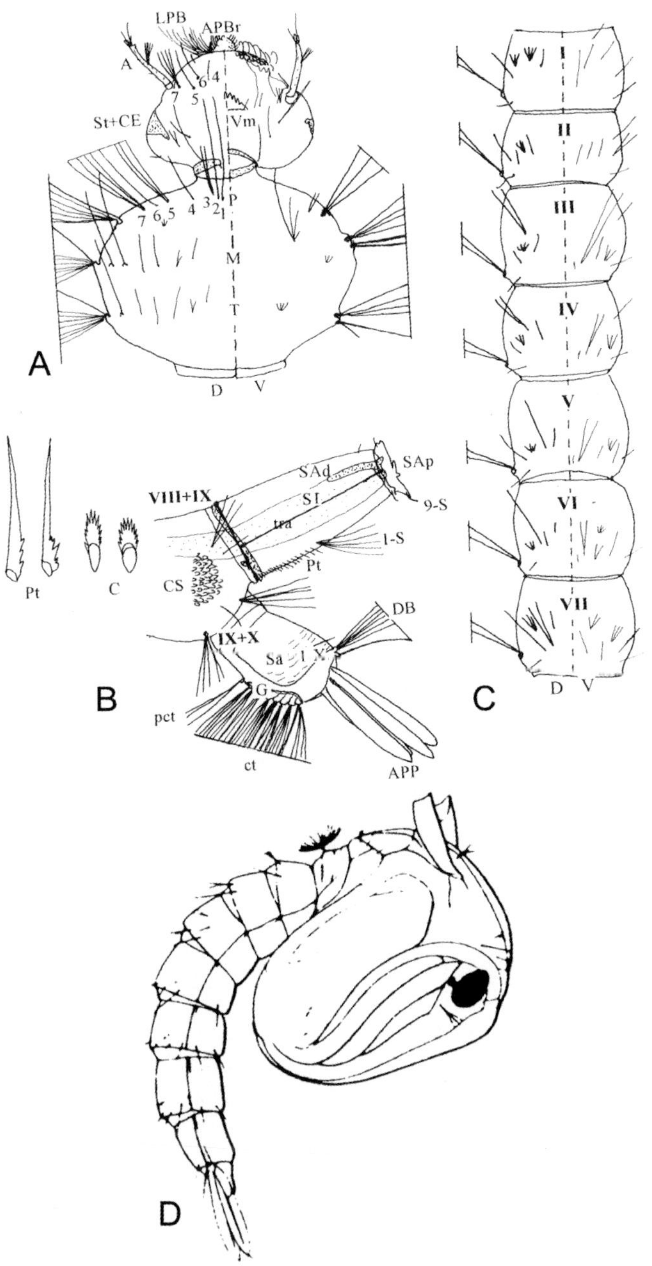

FIGURE 2 *Culicine mosquito, immature stages. A, larva, head and thorax, dorsal (D) left, ventral (V) right; B, larval terminal abdominal segments, lateral; C, larva, preabdomen, dorsal (D) left, ventral (V) right; D, pupa, lateral. Arabic numerals denote setae,* **emboldened** *Roman numerals denote abdominal segment numbers. A, antenna; APbr, anteromedian palatal brush; APP, anal papillae; C, comb scales; CS, comb scale patch; ct, cratal tufts; DB, dorsal brush; G, grid; LPB, lateral palatal brush (so-called 'mouth brush'); M, mesothorax; P, prothorax; pct, precratal tuft; Pt, pecten teeth; Sa, saddle; SAd, spiracular apodeme; SAp, siphonal plate (spiracular apparatus); SI, siphon; St+CE, stemma and (developing) compound eye; T, metathorax; tra, trachea; Vm, ventral mentum; l–S, 1–X & 9–S, setae. After Dahl (1997).*

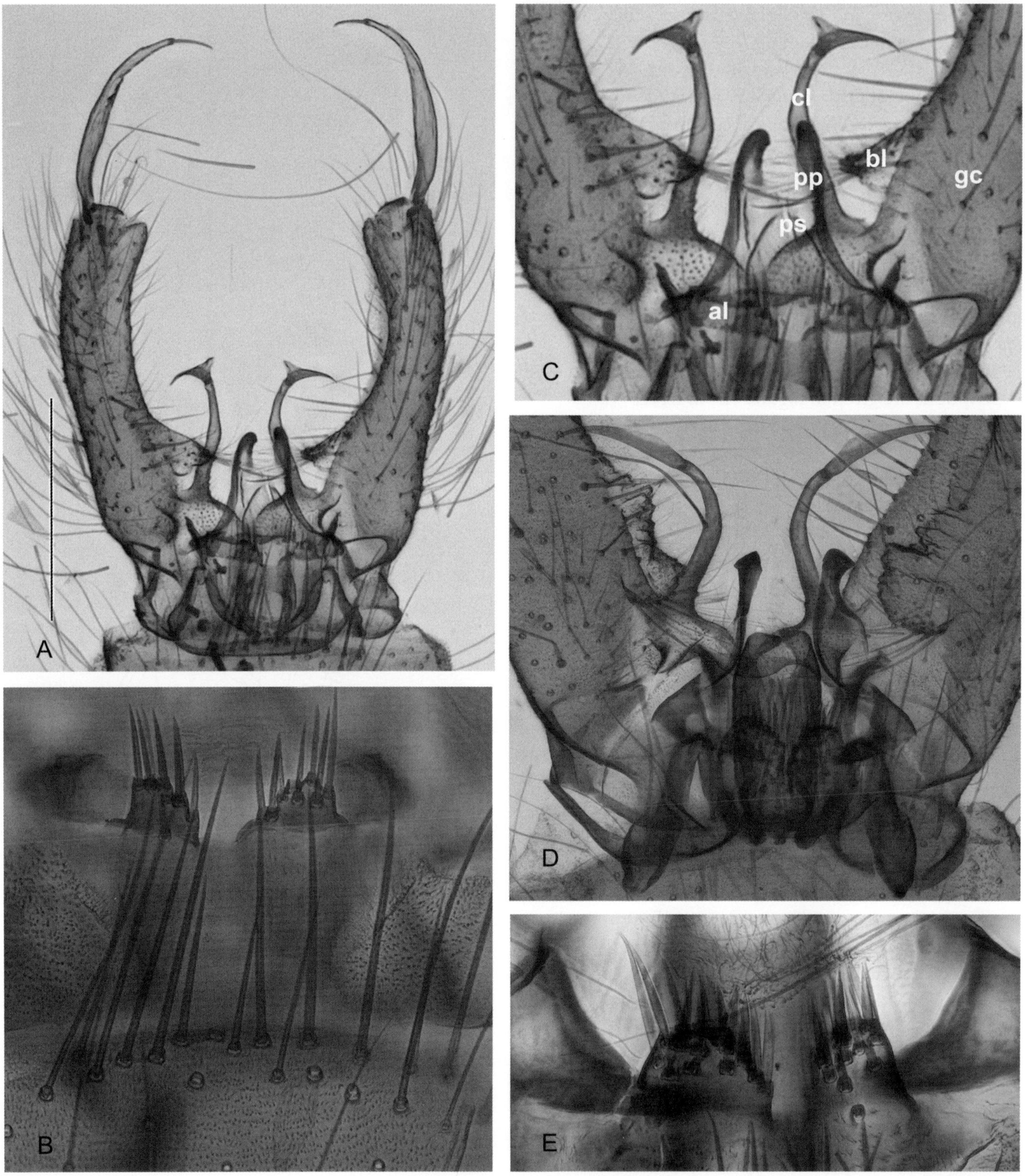

FIGURE 3 Ochlerotatus, *male genital segments, dorsal views, magnification in A lower than in other figures. A–C,* Oc. impiger. *A, overview; B, 'tergite IX lobes' and adjacent posterior margin of tergum VIII (another specimen than in A & C); C, central genital field. D–E,* Oc. nigripes. *D, central genital field; E, 'tergite IX lobes' (another specimen than in D), magnification as in B. al, [topographically] dorsoanterior lobe of segment IX; bl, basal gonocoxite lobe; cl, claspette; gc, gonocoxite; pp, paraproct; ps, phallosome. (B.V. Pedersen praep. & det.*
(PHOTO: N.P. KRISTENSEN)

two free branches of ventral brush (Fig. 4D). Siphon with one type of pecten teeth with a small, basal spine (Fig. 4B), in a closed row. Comb scales rounded with short median spine and small lateral spines, all in an irregular patch (Fig. 4C) .... *Ochlerotatus impiger*

– Larger larvae with round head capsule (Fig. 4E). Saddle closed; anal papillae much (>1.5×) longer than anal segment (Fig. 4H). Siphon long and stout; pecten teeth of two types, one without basal spine and irregularly positioned near the siphonal brush, the other with basal spine (Fig. 4F), positioned in a closed row. Comb scales large with an elongated spine, in two or three irregular rows (Fig. 4G) .... *Ochlerotatus nigripes*

Ochlerotatus impiger **(Walker, 1848) Figs 1A–F, 3A–C, 4A–D**

Henriksen 1939: not recorded.
Nielsen & Nielsen 1966: *Aedes nearcticus* Dyar, 1919.

Female with medium long, black setae on scutum and mesonotum. All other setae dark brown or yellow. Wings with few white scales on base of costa and radius. The number of white scales on costa may prove variable between populations; this requires closer studies. Legs with elongate white scales ventrally on coxa, femur and tibia. Tarsi I–III with some white scales basally, not arranged in discrete bands. Tarsal claws medium long, markedly bent, subbasal tooth long, claw and tooth near-parallel. Proboscis with dark and few scattered white scales. Vertex, occiput, postpronotum and lateral scutum with scattered yellowish (in newly hatched specimens more whitish) scales, dorsal scutum with brownish dark scales divided medially by a narrow, naked stripe and laterally with white scales. Postcoxal patch present. Meskatepisternum with several indistinct white patches and some scattered scales, mesepimeron with some setae and a whitish scale patch. Abdomen dorsally with white bands on dark scaled segments, ventrally white, sometimes with darker bands. Ovipositor small, round. *Male* (Figs 1C, 3A–C) with proboscis and palpi dark and the latter somewhat shorter than the proboscis. Vertex and occiput with scattered white scales and dark stout setae, coxae with long, dark brown or black setae. Scales and setae on scutum as in female, more sparse white scaling on meskatepisternum and mesepimeron. Legs with dark scales. Claws as in female. Abdomen with indistinct white bands, sometimes also ventrally. IX dorsal plate ('tergite', anatomically IX sternite) with two narrow lobes with few, stout black spines. Gonocoxites with more or less angular basal lobes with numerous long setae, the dorsal ones longest, not spine-like. Distal lobes weakly developed, stylus with long and slender tooth. Phallosome sclerotization posteriorly weak, restricted to lateral apices.

Larva, instar IV (Figs 4A–C) medium sized, fragile, with broad head capsule. Head setae 6 and 5 single, 7 double. Prothoracic setae 3 and 2 half the length of 1, all three single (variations of one seta can occur in setae 1 and 2). Setae 4 small, single; 5 double; 6 single; 7 with three branches. Saddle open, only one or two free branches of ventral brush, anal papillae long, slender. Siphon with pecten teeth of one type with a small basal spine and in a closed row. Comb scales rounded with a relatively short mid spine and very small lateral spines, all in an irregular patch.

HABITAT. Confined to arctic and subarctic tundra and low shrub localities. In subarctic areas it can occur at higher elevations at the edge of shrub vegetation as single species in flat pools or in valleys with other species (Dahl *et al.* 2004).

BIOLOGY. The flight activity of adults, hatching, egg laying, adult phenology and activity and the temperature dependence of larvae have been studied on Ellesmere Island by Corbet and Danks (1973, 1975). Comparative studies for this species from Greenland are needed.

DISTRIBUTION. N, NE, SE, S, sSW, nSW, NW.

Circumpolar, arctic and subarctic.

NOTES: Greenland material of *Oc. impiger* was for the first time convincingly separated from *Oc. nigripes* by Nielsen & Nielsen (1966) and recorded under the name of *nearcticus* Dyar; this name was used by Natvig (1945, 1948) for Scandinavian specimens, whereby he established the holarctic distribution of yet another *Ochlerotatus* species.

In Canada the species was reported as *impiger* (Wood *et al.* 1979), followed by Gutsevitch *et al.* (1974) for the Russian north coasts and for Scandinavia by Nielsen (1979).

Ochlerotatus nigripes **(Zetterstedt, 1838) Figs 1G, 3D, 3E, 4E–H, 5**

Henriksen 1939: *Aedes nigripes* Zett.

The species was described on material from Greenland by Zetterstedt (1838) and has been referred to as the most "famous nuisance species" of the circumpolar, high arctic areas. *Female* with very long black bristles on head, scutum, lateral thorax and coxae. All other bristles weaker, black especially on front leg coxa. Wings with dark scaling, only few white scales in a patch at the base of costa and occasionally on the radius and other wing veins. Legs with mixed dark and white scales, ventrally with white ones. First tarsomere on front leg with some whitish scales. Tarsal claws longer and more straight than in *Oc. impiger*, subbasal tooth short, claw and tooth clearly diverging.

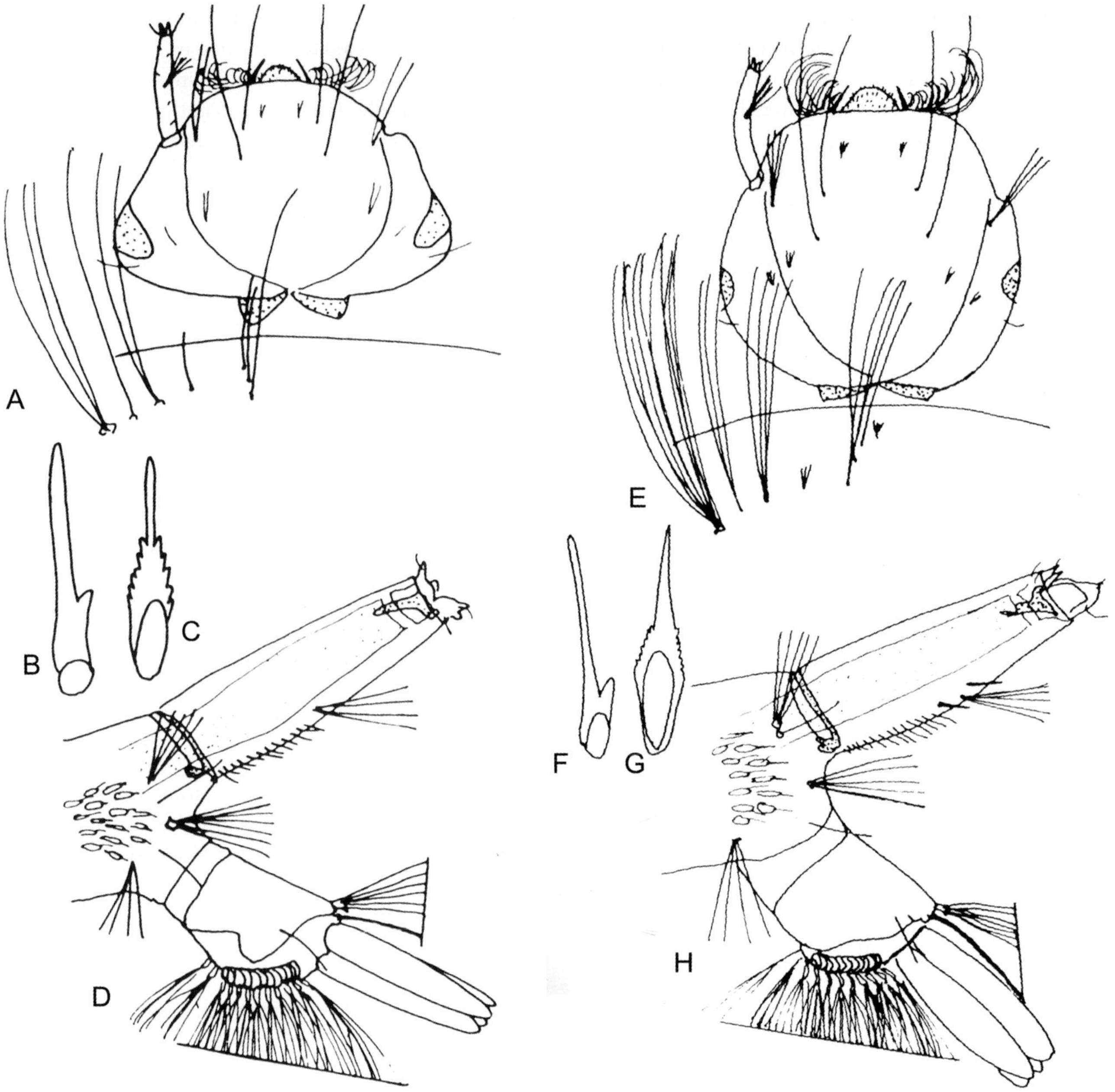

FIGURE 4 *Diagnostic details of Greenland* Ochlerotatus *larvae. A–D,* Oc. impiger. *A, head dorsal; B, pecten tooth; C, comb scale; D, last abdominal segments and siphon, lateral view. E–G,* Oc. nigripes. *E, head dorsal; F, pecten tooth; G, comb scale; H, last abdominal segments and siphon, lateral view. After Dahl (1997).*

Proboscis dark scaled, palpi stout, with scattered white scales. Vertex, postpronotum with few, scattered, whitish scales, lateral scutum with whitish patches. Dorsal scutum with dark scales and medially with a broad, naked stripe. Postcoxal patch present. Diffuse white scaling on legs variable. Meskatepisternum with one distinct patch covering its anterior and median parts, mesepimeron with narrow white scale patch. Abdomen dorsally with whitish bands, ventrally either whitish or with darker bands. Ovipositor small, round. *Male* with long stout black setae on all parts of the body. Scale colour as in females, but fewer scales present on metkatepisternum, mesepimeron and abdomen. Tarsal claws as in female. In genitalia (Figs 3D–E) IXth 'tergite' lobes larger than in *Oc. impiger*, with many stout black spines. Gonocoxite with rounded basal lobes with numerous, long setae, the dorsal ones longest. Distal lobes indistinct, stylus with medium long, stout tooth. Phallosome sclerotization posteriorly markedly strong/ melanized around entire near-circular terminal indentation.

Larva, instar IV (Figs 4D–F), large, with dark grey integument and rounded head capsule. Head setae 6 and 5 single, 7 with several branches. Prothoracic setae 1 single, 2 with two or three equally long and stout branches, 3 small, insignificant, 4 small single, 5 and 6 with three or four equally long branches, 7 with five or six branches. Saddle closed, without free branches of ventral brush, anal papillae very long, slender. Siphon long, stout, pecten teeth of two types, distal ones without lateral spine, three to four irregularly spaced near the siphonal brush. The other type of pecten teeth long with a basal spine and in a closed row. Comb scales large with one elongated spine, situated in two or three irregular rows. Instar III and younger larvae are much more robust than those of *Oc. impiger*.

HABITAT. Confined to treeless high arctic tundra areas and to subarctic regions north of or above the tree-line. It occurs together with *Oc. impiger* and rarely in the same subarctic tundra areas as *Oc. hexodontus* and other species with northern tundra distribution.

BIOLOGY. In permanent ponds at Lake Hazen Camp on Ellesmere Island *nigripes* was locally the most common species (Corbet and Danks 1973). Under subarctic conditions in Sweden *Oc. nigripes* larvae were found in pools at 400 m a.s.l. and rarely at higher altitudes in flat pools occupying them as only species (Dahl *et al.* 2004). The larvae are extremely shy.

FIGURE 5 *Female* Ochlerotatus nigripes *sucking nectar in flower of* Dryas integrifolia. (PHOTO: J. BÖCHER)

DISTRIBUTION. N, NE, SE, sSW, nSW, NW.

The holarctic distribution of this species is widely recognized (Becker *et al.* 2010).

NOTES: The record of *Aedes nigripes* as far south as from Poland (Skierska 1962) is based on a larva, most probably misidentified due to a picture in Martini (1931) purportedly showing a *nigripes* larva, which is incorrect. *Aedes triseriatus* (Say), recorded from Narsarsuaq (sSW) by Messershmith (1971), is an eastern North American species, which breeds in tree holes (Wood *et al.* 1979); the Greenland record almost certainly represents a misidentification.

Appendix: Questionably referred material

Six small female mosquito specimens, very denuded and bleached, are reportedly from SE: Angmassalik (15.8.1933 N. Tinbergen leg.). Their front claws are as in the genus *Aedes sensu* Reinert *et al.* (2004), and they could belong to the *Ae. cinereus*-complex. The specimens are so denuded, and were collected so far north of the known distribution of members of the latter, that the record is currently considered questionable, pending availability of fresh material from the same locality.

References

Andersen, S. & Böcher J. (1994): Skjoldungens insektfauna. – *Forskning i Grønland.* [*Tusaat*] 1–2: 56–60. København

Becker, N., Petric, D., Zgomba, M., Boase, C., Dahl, C., Madoon, M. & Kaiser, A. (2010): *Mosquitoes and their control* (2nd edn.). Springer, Berlin/Heidelberg, 577 pp.

Corbet, P.S. (1966): Diel Patterns of Mosquito Activity in a High Arctic Locality: Hazen Camp, Ellesmere Island, NWT. – *Canadian Entomologist* 98: 1238–1252.

Corbet, P.S. & Danks, H.V. (1973): Seasonal Emergence and Activity of Mosquitoes (*Diptera: Culicidae*) in a High Arctic Locality. – *Canadian Entomologist* 105: 837–872.

Corbet, P.S. & Danks, H.V. (1975): Egg-laying habits of mosquitoes in the high Arctic. – *Mosquito News* 35: 8–14.

Dahl, C. (1974): Circumpolar *Ochlerotatus* (*Ochlerotatus*) species in North Fennoscandia. – *Mosquito Systematics* 6: 57–73.

Dahl, C. (1997): Diptera Culicidae, Mosquitoes. *In*: A.N. Nilsson (ed.) *Aquatic Insects of North Europe – A taxonomic Handbook.* vol. 2: 165–186.

Dahl, C., Nielsen, L.T. & Petersson, E. (2004): Mosquito larvae (Diptera: Culicidae) in snow-melt pools in a Swedish Lapland area. *Journal of Vector Ecology* 29: 109–123.

Danks, S.V. (1981): *Arctic Arthropods*. Entomological Society of Canada, Ottawa, 680 pp.

Gutsevich, A.V., Monchadskii, A.S. & Shtakelberg, A.A. (1974): Mosquitoes Family Culicidae. *In*: *Fauna of the U.S.S.R. Diptera* 3:4. Academy of Sciences of the USSR Zoological Institute. New Series 100, 408 pp.

Knight, K.L. & Stone, A. (1977): *A Catalog of the Mosquitoes of the World*. The Thomas Say Foundation VI. The Geo.W. King Company Baltimore MA, 611 pp.

Martini, E. (1931): Culicidae. *In*: Lindner E. (ed.) *Die Fliegen der Palaearktischen Region*. Stuttgart, 398 pp.

Messershmith, D.H. (1971): Extension of the distributional range of *Ochlerotatus triseriatus* (Say) to Greenland. – *Mosquito Systematics Newsletter* 3: 1–7.

Natvig, L.R. (1945): Notes on *Culex alpinus* Linnaeus and *Ochlerotatus nigripes* (Zett.). *Norsk Entomologisk Tidskrift* 7: 99–106.

Natvig, L.R. (1948): Contribution to the knowledge of the Danish and Fennoscandian mosquitoes Culicini. *Norsk Entomologisk Tidskrift Supplement* 1, 567 pp.

Natvig, L.R. (1954): The type specimens of *Ochlerotatus nigripes* (Zett.). *Norsk Entomologisk Tidskrift* 9: 86–92.

Nielsen, E.T. & Nielsen H. (1966): Observations on Mosquitoes in Greenland. *Meddelelser om Grønland* 170(3): 1–27.

Nielsen, L.T. (1979): A Comperative Taxonomic and Distributional Study of Holarctic *Ochlerotatus* Mosquitoes. *National Geographic Society Research Reports* 20: 495–506.

Reinert, J.F., Harbach, R.E. & Kitching, I.E. (2004): Phylogeny and classification of Aedini (Diptera; Culicidae), based on morphological characters of all life stages. – *Zoological Journal of the Linnean Society* 142: 289–368.

Skierska, B. (1962): Discovery of the sub arctic mosquito species *Ochlerotatus* (*Ochlerotatus*) *nigripes* Zetterstedt in Poland. – *Bulletin of the Institute of Marine Medicine Gdansk* 13: 59–66.

Thielman, A.C. & Hunter, F.F. (2007): Photographic Key to the Adult Female Mosquitoes (Diptera: Culicidae) of Canada. – *Canadian Journal of Arthropod Identification* doi: 10.3752/cjai.2007.04 http://www.biology.ualberta.ca/bsc/ejournal/th_04/th_04.html.

Wood, D.M., Dang, P.T. & Ellis, R.A. (1979): The Mosquitoes of Canada Diptera: Culicidae. *In*: *The Insects and Arachnids of Canada*. Part 6. Publ. 1686. Hull, Quebec, 390 pp.

Zetterstedt, J.W. (1838): *Insecta Lapponica descripta*. Lipsiae, Voss, 868 pp.

17.5. Simuliidae (Black-flies)

Frank Jensen

Small flies with body length 2–5 mm and wing length 2.0–4.6 mm. The body is stout with broad wings, short antennae, short proboscis, long and slender palps. The head is low slung on the deeply arched thorax, giving the imago a humpbacked appearance. Colour dark brown to dark grey with conspicuous lighter to silvery-white spots on scutum, abdomen and legs (hence the Danish name in Greenland: "kamikfluer"). Larvae and pupae live in swiftly running water.

In adults (Fig. 1) the compound eyes are large, in males touching at frons, with large facets in upper part, small facets in lower part. Female eyes not meeting at frons, facets all of same size. In *Simulium vittatum* there are about 1500 facets in male and 1000 in female. No ocelli present. Antennae short, consisting of 11 segments. 1st and 2nd segment of same shape as the rest; hairs on antennae short and inconspicuous; males never with plumose antennae like in Chironomidae. Thorax laterally with a conspicuous, large bare pleural membrane (anepisternal cleft), larger than in most other dipterans. Wings clear, short and broad with well developed anal lobe. Only 2 crossveins present. Upper surface of wings with macrotrichia (*Prosimulium* spp. only have microtrichia) and spinules (Figs 2A–B). The whole wing covered with minute nipples perpendicular to the membrane (visible only in SEM) which are thought to render the wings non-wettable; in *S. vittatum* there are about 164 million nipples per wing. Legs relatively short. Tarsus of hind leg with two unusual structures: *pedisulcus* and *calcipala* (Figs 2C–E) which are very weak in *Prosimulium*, but well developed in *Simulium*. The abdomen is elongate with segmentation clearly identifiable, although females gorged in blood may have a nearly spherical abdomen. The sexual organs are very important for a precise identification of both sexes; due to their minuteness it is often necessary to make slide mounts of them.

The obtect pupa is very characteristic. Its colour changes with time from light cream to light brown. Shortly before adult eclosion the pupa seems black because the fully formed imago is visible through the skin. The size is 2.5–5.0 mm exclusive of gills. These are very different in shape and size and have great taxonomic value. Pupae are essentially encased in a cocoon spun of silk. The basically slipper-shaped cocoon is also very characteristic (in *Simulium* and *Prosimulium*). The pupa is held in place in the cocoon by hooks on abdomen. When adults emerge they blow air from their tracheae into the 'ecdysial space' between the pupal and adult cuticle until the pressure bursts the former; the adult is released in a small air bubble which rises fast to the water surface. A mass emergence looks like small raindrops breaking the water surface.

FIGURE 1 *Habitus of adult female* Simulium vittatum.

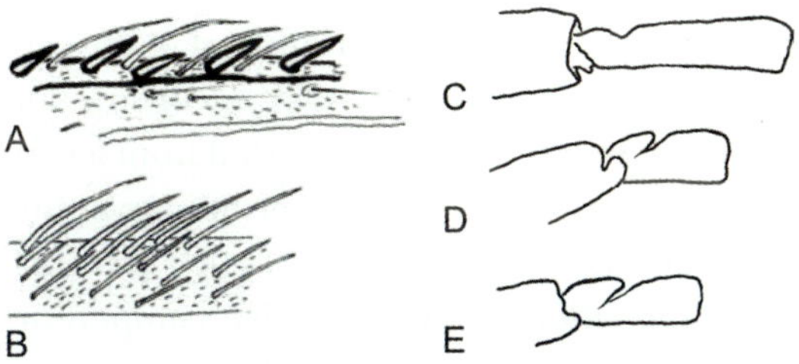

FIGURE 2 *Structural characters in adult Simuliidae. A–B, front edge of wing of* Simulium rostratum *(A) and* Prosimulium ursinum *(B), C–E, calcipala and pedisulcus of hindlegs of* Prosimulium ursinum *(C),* Simulium rostratum *(D) and* Simulium vittatum *(E).*

The larva is also very characteristic and, like the pupa, cannot be confused with any other stream-living taxa. The size of the full-grown larvae is 4–12 mm. The body is elongated and broadest near its posterior end that bears a circlet of hook-rows anchoring the larvae in the current. In the fore end a downwards directed proleg also has a small circlet of hook rows used in locomotion. The head bears large setal fans borne on a two-segmented 'stalk' arising from the basal corners of the labrum. There are 7 larval instars, 6 in *Simulium rostratum*. The 'last larval instar' is actually already a pharate pupa and is easily recognised as the pupal gill (referred to as the 'gill histoblast' in simuliid literature) is dark, and its configuration can be observed after dissection.

Black-fly larvae produce silk in the large U-shaped labial glands which extend into the trunk. The silk is very sticky when fresh and adheres firmly to almost all substrates. It is used to anchor the larvae in the current: a small pad of silk is placed on the substrate, and the hooks on the posterior circlet and the hook-circlet of the proleg are sunk into the pad. The larvae are able to move like a geometrid caterpillar by stretching out and making a silk pad onto which they then can anchor the proleg, then drawing the hind end up and securing the hold with the posterior hook-circlet. If conditions become unsuitable they can also, like a spider, let themselves downstream on a long silken thread until they find a better place. In times of drought the larvae may let themselves drift down the stream; this normally takes place at night (the light intensity is detected by the stemmata). When mature the larvae change the composition of silk and use it to manufacture a cocoon in which they stay during pupation.

Black-fly larvae constitute an important part of the ecosystem of most streams where they may occur in enormous numbers. Maximum numbers reported from a variety of biotopes are ranging from 160 to 1.200.000 larvae/m^2, and peak populations may be as high as 100 per cm^2. Larvae of *Simulium vittatum* are aggressive competitors for space and will always be evenly spaced.

The egg is asymmetrical, 0.15–0.54 mm, subtriangular ovoid; it is light cream and sticky when laid, with time yellow or light brown. Eggs are deposited either singly, in small groups, in strings or in flat layers. The eggs of *P. ursinum* can overwinter and survive desiccation, while those of *S. rostratum* are highly susceptible to dryness.

The preferred substrates for egg-laying are stones or plants etc. that break the water surface, so that the females have access to wetted or submerged areas. Females of both *S. vittatum* and *S. rostratum* enter the water to some depth to lay eggs. The number of eggs per female is in the range of 50–600. Gravid females often show communal laying behaviour and huge egg masses up to 1 cm thick may be found.

All larvae of Greenland species are filter feeders. They use their cephalic fans to catch particles that are licked off and swallowed. The cephalic fans consist of some rays that are spread out fanlike and held against the current. A row of microtrichia on the rays effectively enlarge their area, and probably most species cover rays and microtrichia in a mucosubstance enabling them to catch particles more effectively. In *S. vittatum* the mucosubstance traps bacteria. The particles filtered range in size from 0.5 mm to 0.091 μm. The larvae mainly live on filamentous algae, diatoms, bacteria and animal matter. When feeding in a weak current the larvae stand almost vertically from the substrate; in stronger currents they lie nearly parallel to it. They do not change their position when feeding, which means that they have to twist their body up to 180 degrees in order to present the concave side of their fans to the current.

Apart from filter feeding the larvae can also scrape algae etc. from the surface on which they sit, and the larger larvae may eat smaller larvae and other invertebrates.

If feeding at all, adult male black-flies feed only on flower nectar. Also females of most species suck nectar to obtain energy for flying. For most species a blood-meal from a bird or mammal is obligatory in order to be able to mature eggs. In Greenland and many other countries, especially in the Low Arctic and Subarctic, black-flies (females) are well known for attacking man and beast and may be a serious nuisance during summer when they arrive abundantly to get blood meals. When great numbers of black-flies are present, biting can be serious. Up to 6.000 females have been reported (not in Greenland) simultaneously attacking a single cow, which may be killed (in 15 minutes!) due to the toxic substances used to prevent blood from clotting the mouthparts.

Many predators directly or indirectly regulate black-fly numbers. Main predators are fish, mainly during drifting of the larvae, and when the pupae hatch. Likewise ovipositing females are eaten and in some rivers they contribute significantly to the diet of fish. Passerine birds take some adults but more important are ducks (especially *Anas platyrhynchos*) that may consume large numbers of larvae and pupae. In Iceland harlequin duck (*Histrionicus histrionicus*) is reported to have simuliids as its main food. The most important parasites are Mermithidae, Microsporidia and fungal parasites of which the trichomycete fungus *Harpella melusinae* has been found on *P. ursinum* and *S. rostratum*.

Being pests the black-flies have been extensively studied in many countries, and they were among the first groups to be investigated histologically due to the giant polytene chromosomes in the larval salivary glands. Nearly all the morphological species studied constitute groups of sibling species. These are morphologically very much alike, but by their chromosomes can be shown to be reproductively isolated. *Simulium vittatum*, which is very common in Greenland, is an example of such a morpho-species comprising several sibling species. Nearly the whole black-fly material collected in Greenland consists of mature females. This is unfortunate since the other developmental stages are far easier to work with taxonomically. Many nominal species are recorded from Greenland but this and preceding studies have shown that in reality there are only 3 species (morphospecies) known. This paucity of species may well be real, but some rare species might well have escaped discovery. In the material are some very small larvae of *Prosimulium* which do not look like *P. ursinum*.

TECHNICAL NOTES. Simuliid larvae and pupae are best collected by sampling stones and aquatic plants where water velocity is high. Both larvae and pupae prefer smooth surfaces. Most larvae will be on top of the stones and most pupae on the downstream or bottom sides. In large or deep streams a very useful collecting device is a length of flag-line with a 40 × 3 × ½ cm flat iron bent in a sharp angle to snare the stems of water plants or trailing vegetation. It is advised always to make a sample of pupae to be able to check the determination of larvae. Pupae are best left *in situ* in their cocoon on the plant, so scissors are essential. If the pupae are found on larger stones, they can be taken off with a pair of fine pointed forceps. Always pull the pupa off from the hind end to avoid tearing the cocoon that may be of great taxonomical value. Black-flies are easily collected on artificial substrates, and an anchored piece of plastic will become colonized within days. Adults are easily caught (the females will find you!). The males often swarm on the lee side of shrubs near the water, or can be found resting on the underside of leaves on the bank side vegetation. Ovipositing females are most often found at sundown, their preferred ovipositing sites being plants or stones etc. breaking the water surface. Spider webs near the bank are also a good place to look. However, obtaining adults for identification purposes from pupae will ensure positive association with the pupa. Often it is possible to find pupae that have the larval head cuticle trapped in the gills, so that positive association is ensured between larva, pupa and adult.

It is relatively easy to rear larvae: A cylinder glass with water flow from the bottom will produce enough current to enable the larvae to feed. They may be fed with algae or very finely ground goldfish food. When the larvae pupate, they are best left until the imago is formed within the cocoon. At this time the pupae turn black. They may then be removed and placed singly on a piece of damp filter paper, or better sphagnum moss, in a small vial, until the pupa hatches. Keep the adult alive for some hours before killing it. This allows the cuticle to harden and assume the adult colouration. Most species are very difficult to maintain in culture as adults, since they usually demand a blood meal, and they are reluctant to mate in captivity. However *S. vittatum* has been held in captivity for several generations.

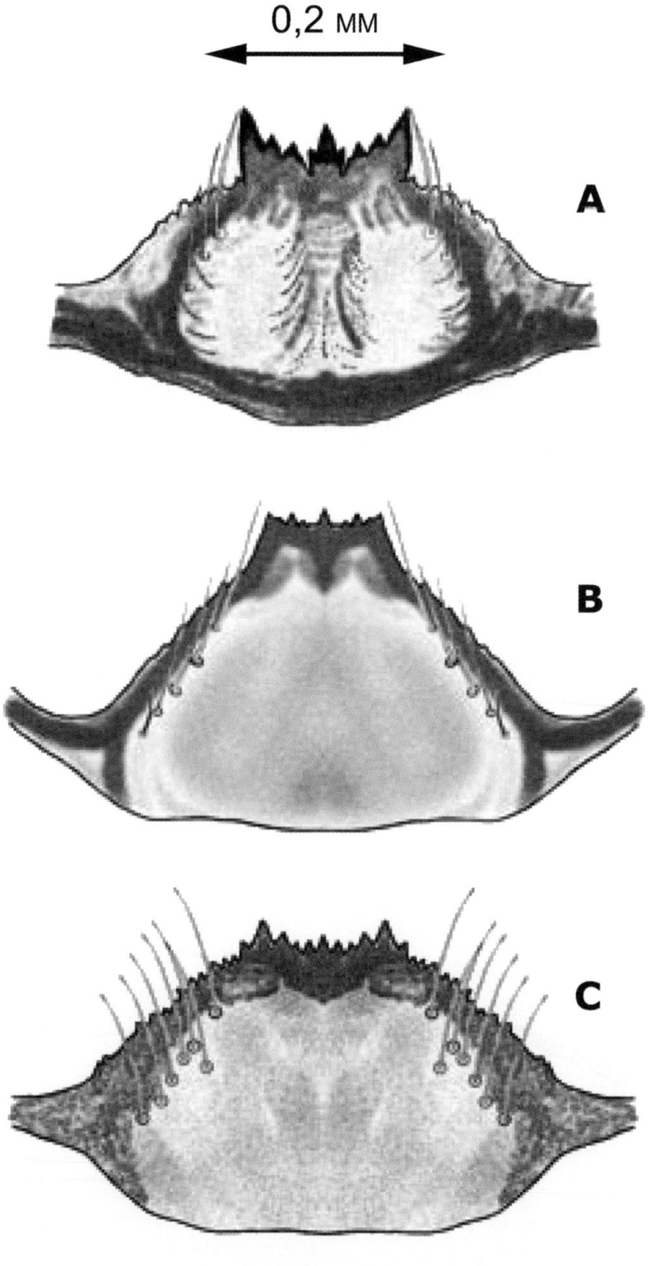

FIGURE 3 *Submentum of larvae drawn to the same scale of* Prosimulium ursinum (*A*), Simulium rostratum (*B*) *and* Simulium vittatum (*C*).

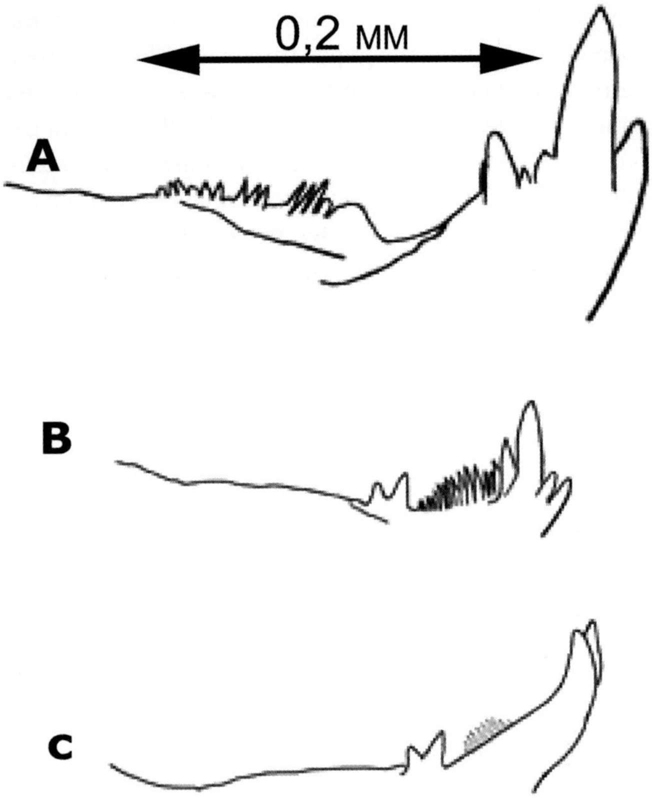

FIGURE 4 *Mandible of larvae drawn to same scale of* Prosimulium ursinum (*A*), Simulium rostratum (*B*) *and* Simulium vittatum (*C*).

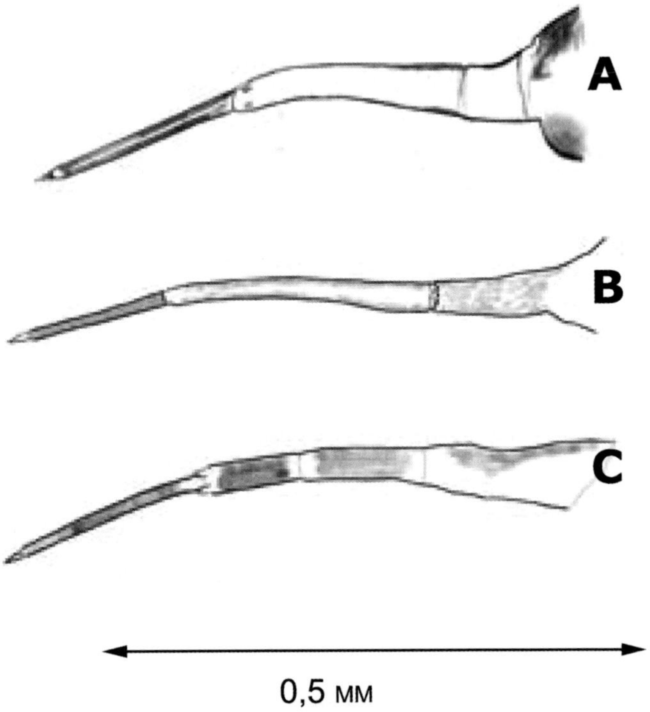

FIGURE 5 *Antennae of larvae drawn to same scale of* Prosimulium ursinum (*A*), Simulium rostratum (*B*) *and* Simulium vittatum (*C*).

Key to species

LARVAE

1. Submentum with large dark teeth, the middle tooth trifid (Fig. 3). Mandible with serrated secondary teeth (Fig. 4). Head capsule large in relation to body, the edges nearly touching back of apotome (anterior/dorsal part of head capsule, delimited by cleavage lines). Antennae with basal sections transparent and an abrupt change in diameter down to the thin uniformly dark and cylindrical distal section (Fig. 5). Anal gills simple and sausage-shaped (Fig. 6) .... *Prosimulium ursinum*
– Submentum with smaller teeth and the middle tooth never trifid. Antennae never with basal sections transparent 2
2. Anal gills simple. Head marks on apotome 'positive', *i.e.*, they are darker than the adjacent areas (although sometimes only weakly so); surrounding areas do not form a central H-shaped mark (Fig. 7) .... *Simulium* (*Psilozia*) *vittatum*
– Anal gills with secondary lobes (Fig. 6C). Head marks on apotome negative (*e.g.* Fig. 7B.), hence they are as light or lighter than the surrounding areas, which may form a central H-shaped mark. Pupal gill dissected from histoblast with 16 filaments .... *Simulium* (*s.str.*) *rostratum*

PUPAE

1. Last abdominal segment with two long and strong spines on dorsal side (Fig. 8A). Pupal cocoon without any definite shape and of loose weave; often embedded with sand grains and detritus and often only covering half of the pupa (although sometimes covering the whole). Pupal gill with 14 filaments arranged with 2 upper groups of filaments with a short common trunk. Branching of gill with 2 groups of 4 filaments and a group with 6 filaments (see Fig. 9A) .... *Prosimulium ursinum*
– Last abdominal segment without two long strong spines on dorsal side (at most with small blunt spines or tiny hooks or double or triple graphnel-like hooks). Pupal cocoon with definite shape and firm weave 2
2. Gill of pupa with 6 filaments (Fig. 9B) .... *Simulium* (*s.str.*) *rostratum*
– Gill of pupa with 16 filaments (Fig. 9C) .... *Simulium* (*Psilozia*) *vittatum*

ADULT MALES

1. In the wing radius 2 is forked (may be difficult to detect; observation from several angles is necessary).

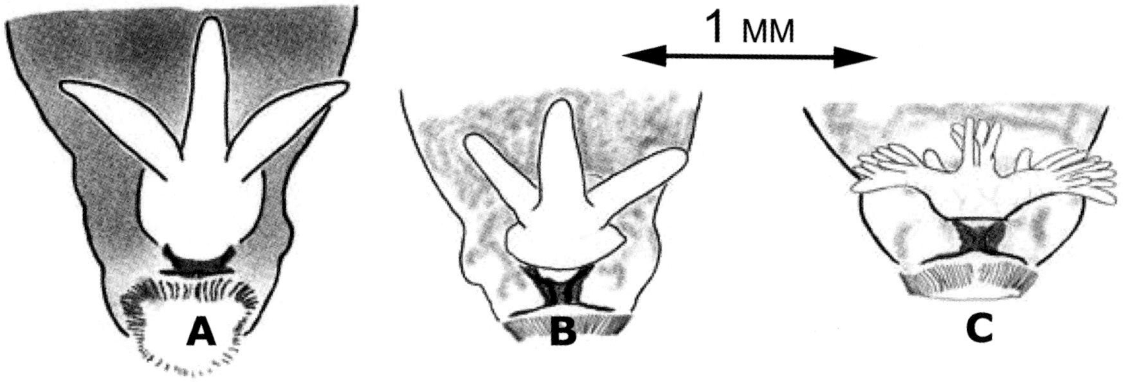

FIGURE 6 *Anal gills of larvae drawn to same scale of* Prosimulium ursinum (*A*), Simulium vittatum (*B*) *and* Simulium rostratum (*C*).

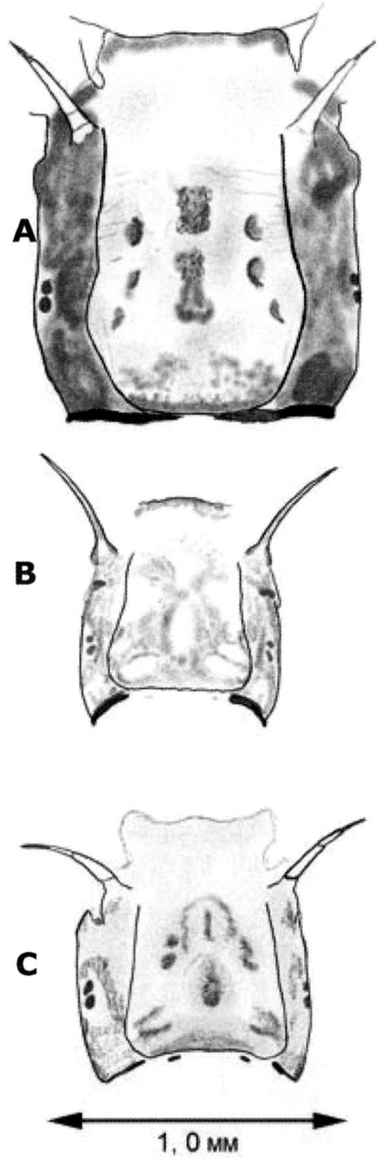

FIGURE 7 *Apotome of larvae drawn to same scale of* Prosimulium ursinum (*A*), Simulium rostratum (*B*) *and* Simulium vittatum (*C*).

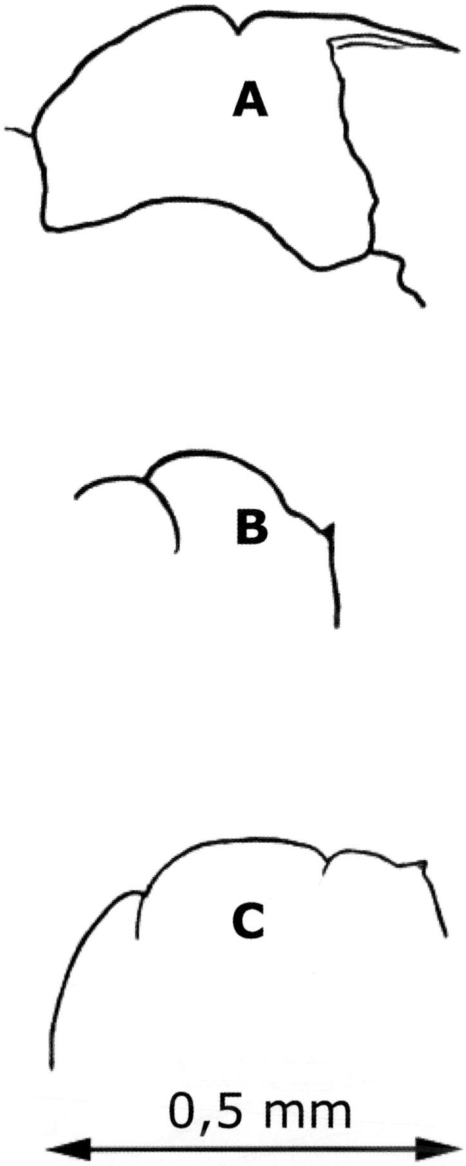

FIGURE 8 *Tip of pupal abdomen with anal hooks drawn to same scale of* Prosimulium ursinum (*A*), Simulium rostratum (*B*) *and* Simulium vittatum (*C*).

Three front veins with hairs of different length, but never with spines Fig. 2B. Hind leg basitarsus with no or a very small calcipala and second tarsal segment without pedisulcus at all .... *Prosimulium* (NB! *P. ursinum* males have never been found, but known congeners will key out here).

– Radius 2 without any branching but may be thicker distally than proximally. Three front veins with both spines and long and very short hairs (Fig. 2A) 2

2. Hind basitarsus with a small calcipala and second hind tarsal segment with a pedisulcus. Abdomen with grey markings. Tips of male styles armed with 3–4 large horns, styles small in relation to coxites (Fig. 10) .... *Simulium* (*Psilozia*) *vittatum*

– Hind basistarsus with large calcipala and and second hind tarsal segment with a deep pedisulcus. Dorsum with clear silvery white markings. Foretibia and abdomen with silvery white markings. Styles with 1 small terminal spine and is large in relation to coxites (Fig. 11) .... *Simulium* (*s.str.*) *rostratum*

ADULT FEMALES

1. In the wing radius 2 is forked. Dorsum without any pattern and clad in rough hairs. Legs long, thin and uniformly coloured (Fig. 12A), Hind basitarsus without a calcipala (Fig. 2C), second hind tarsal segment

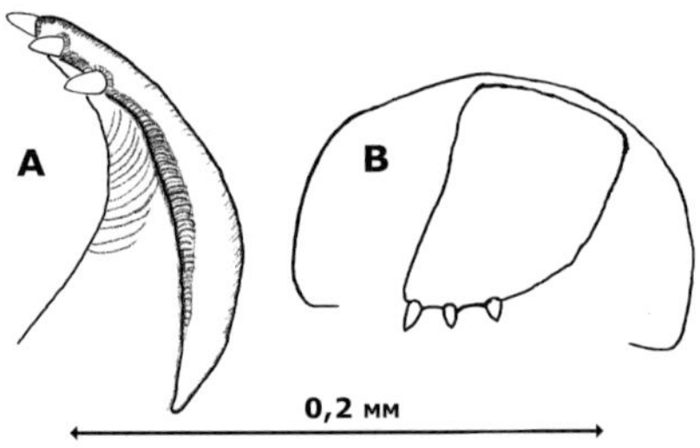

FIGURE 10 *Male stylus of* Simulium vittatum. *Tip of stylus seen from the side and above (A), stylus and coxite seen from above (B).*

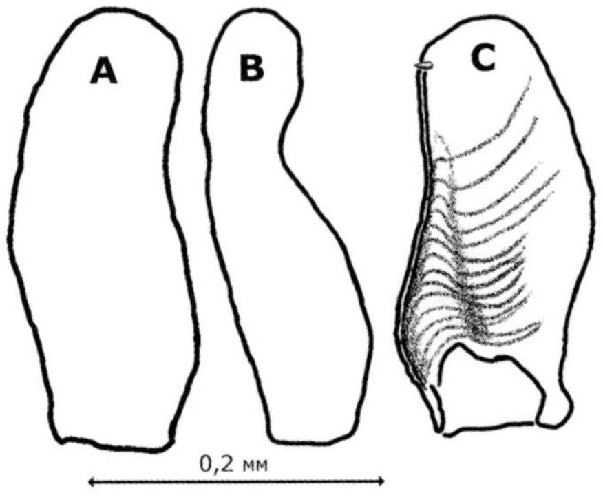

FIGURE 11 *Male stylus of* Simulium rostratum *seen from behind (A), from the outside (B) and from the inside (C).*

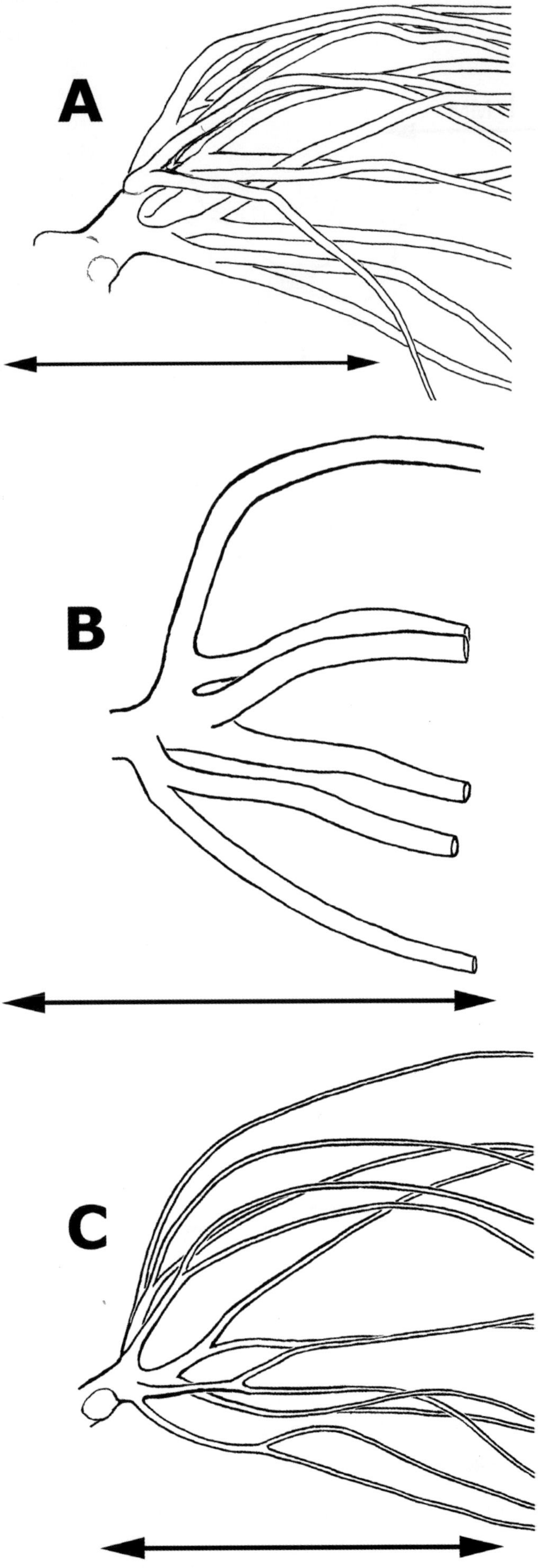

FIGURE 9 *Pupal gill (bars are 1 mm long) of* Prosimulium ursinum *(A),* Simulium rostratum *(B) and* Simulium vittatum *(C).*

without a very weakly developed pedisulcus (care must be taken when observing this as most species have very hairy tarsi and the hairs obscure the pedisulcus) .... *Prosimulium ursinum*

– Dorsum of female with pattern. Hind basitarsus with calcipala, second hind tarsal segment with a pedisulcus 2

2. Dorsum with 3 more or less clear longitudinal stripes and light mark on fore tibia .... *Simulium (Psilozia) vittatum*

– Dorsum without stripes, more or less dark grey not clearly 3-striped; fore tibia with silvery-white mark . ... *Simulium (s.str.) rostratum*

Prosimulium ursinum **(Edwards, 1935) Figs 2B, 2C, 3A, 4A, 5A, 6A, 7A, 9A, 12A, 13A**

Henriksen 1939: *Simulium (Prosimulium) ursinum* Edw.

Radius with 2 forks. Thorax covered in coarse hairs and groove above katepisternum incomplete and shallow. General colour brown. Pupa in shapeless cocoon, 14 branches in gill. Larva with large cephalic capsule with tridentate middle tooth of submentum, abdomen rather abruptly expanded at 5th segment (Fig. 13A).

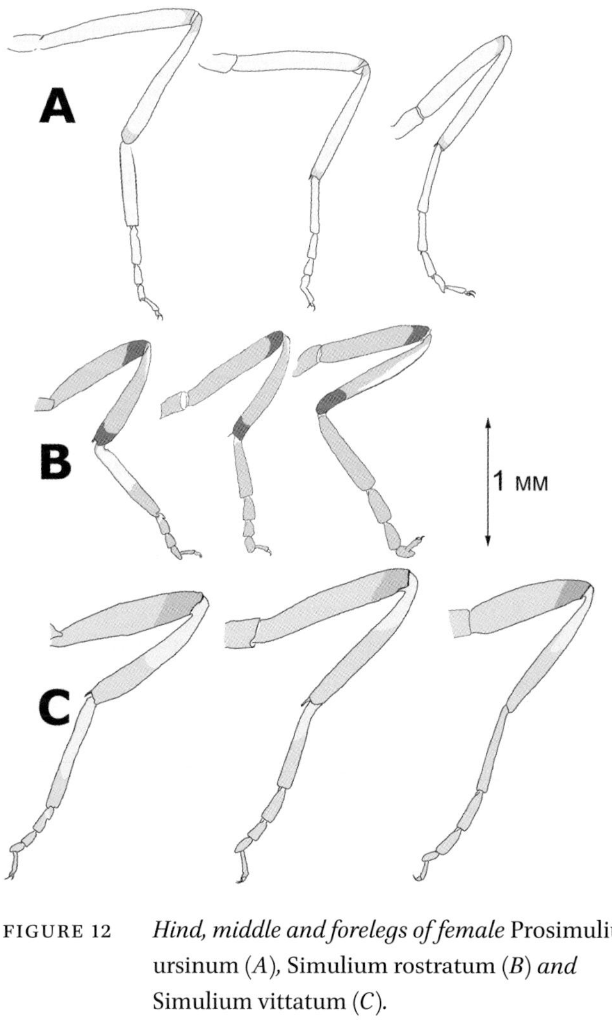

FIGURE 12 *Hind, middle and forelegs of female* Prosimulium ursinum *(A),* Simulium rostratum *(B) and* Simulium vittatum *(C).*

BIOLOGY. No males have ever been found and the species thus must reproduce parthenogenetically. The species does not possess biting mouthparts and does not need a bloodmeal to produce the *c.* 50 eggs, being obligate autogenous and allotriploid. The larvae prefer submerged vegetation over stones; they are able to live in meltstreams near glaciers and produce viable parthenogenetic eggs directly from the pupal stage. The pharate adult disintegrates within the cocoon and releases the eggs into the water. However, lower down the streams the same species do produce flying females.

DISTRIBUTION. S, SW (north to Qasigiannguit/Christianshåb), NE (Jameson Land: Lollandselv). A single female found in 2009 in Zackenberg at 74°28'N (Ilmonen and Várkonyi, 2011) is the most northern find in Greenland.

Holarctic?, arctic-alpine. The *Prosimulium* complex is found from northern Scandinavia over Iceland and Greenland to arctic parts of Canada, but not in Siberia. The most northern black-fly is from 74°30'N (Bjørnøya).

Simulium (Pzilopelmia) vittatum **Zetterstedt, 1838 (species complex) Figs 1, 2E, 3C, 4C, 5C, 6C, 7C, 8C, 9C, 10, 12C, 13C, 14**

The *S. vittatum* complex is known to comprise a number of cryptic species, but material from Greenland, which is the type locality, has not been examined with respect to chromosome complement. Females with 3-striped scutum. Male gonopods with styles very small compared with coxae (Fig. 10). Pupae with slipper-shaped cocoon and 16 branches in gill. Larvae with positive colouring of apotome.

BIOLOGY. *S. vittatum* does not need a blood meal for the first egg laying, which is based on reserves from the larval life. However, it needs blood for a second batch. The species varies much in size, and large females lay considerably more eggs than smaller females (counts vary between 618 and 312). The eggs are laid in a string. Alleged *S. vittatum* in Iceland sometimes does not attack humans, though they may occur in enormous numbers. In Greenland it is well documented that the species bites man. Possibly there are two sibling species in the N. Atlantic, or they are indeed conspecific but have different physiological requirements. The species mostly sucks blood on mammals, but 13 females in the Greenland material have been collected on birds (N. Tinbergen leg.). Flight ranges up to 104 km has been measured. The larvae can tolerate a wide spectrum of stream velocities, 0.08 m/s to 1.31 m/s. They are very tolerant of temperature and have been collected in waters from 0 to 33°C. Larvae of both *S. rostratum* and *S. vittatum* seem to prefer the outlet of streams from lakes, perhaps because of higher temperatures or richer food here. *S. vittatum* can survive a flood by burrowing into the substrate to a depth of 30 cm or by creeping into rock fissures. It is also able to survive for

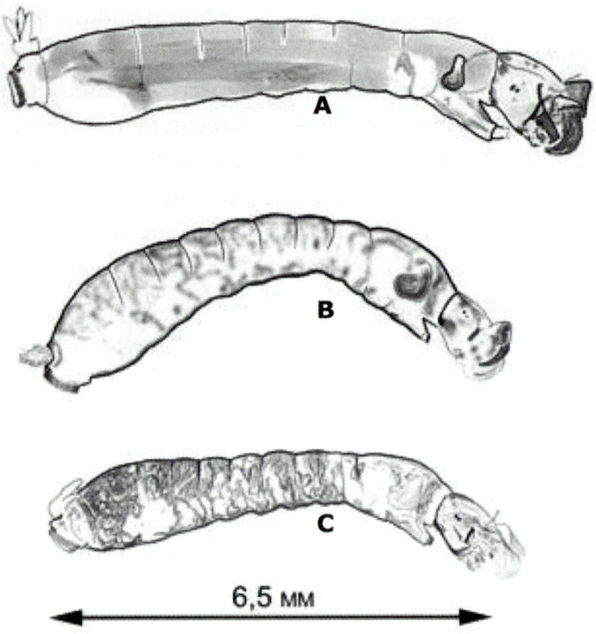

FIGURE 13 *Whole larva of* Prosimulium ursinum (*A*), Simulium rostratum (*B*) *and* Simulium vittatum (*C*).

FIGURE 14 Simulium vittatum *sucking blood on a finger.* (PHOTO: J. BÖCHER)

several days in stagnant water. The larvae are always found on stones, vegetation or other fixed objects, whereas the pupae mainly sit on vegetation and the underside of stones away from the main current. Most often there are several pupae together on suitable substrates.

DISTRIBUTION. S, SW (north to Uumannaq Fjord: Appat), NE (only Mestersvig), SE (only Ammassalik).

Holarctic-North Atlantic, arctic-boreal. The *S. vittatum* complex is found in USA and Canada, Greenland, Iceland and the Faroes but not yet recorded in Scandinavia.

***Simulium* (*Simulium*) *rostratum* (Lundström, 1911) Figs 2A, 2D, 3B, 4B, 5B, 6B, 7B, 8B, 9B, 11, 12B, 13B**

Henriksen 1939: *S. rostratum* belongs to the *S. verecundum* species complex and is isomorphic to *S. verecundum*.

Adult thorax with groove above katepisternum narrow and fully semicircular. Vein Rs undivided. Both sexes with silvery spots on fore tibiae. Pupae with 6–branched gill. Larva with negative colouring of apotome.

BIOLOGY. While the larva of *S. vittatum* prefers the gentler slopes, *S. rostratum* does not mind that the water is broken and boulder strewn. *S. rostratum* is obligately anautogenous, can produce up to 250 eggs and will not start its egg-laying until the light intensity is declining by 2% a minute. Not very much is known about its biting, but it has been recorded on humans and (ouside Greenland) on horses and around caribous and cattle.

DISTRIBUTION. S, SW (north to Uummannaq Fjord: Perlerfiup Kangerlua), SE (Saqqisikuik/Skjoldungen).

Holarctic?, low arctic-boreal (Europe, Greenland, Canada, USA.).

References

Adler, P.H., Currie, D.D. and Wood, D.M. (2004): *The Black Flies (Simuliidae) of North America*. Cornell University Press, Ithaca, NY.

Crosskey, R.W. (1990): *The Natural History of Blackflies*. – Wiley & sons, Chichester.

Ilmonen, J. & Várkonyi, G. (2011): First record of *Prosimulium ursinum* (Diptera: Simuliidae) in Northeast Greenland. – *Entomologiske Meddelelser* 79: 27–30.

17.6. Ceratopogonidae

Søren Achim Nielsen and Katrine Bøg

Ceratopogonids are small sized, 1–5 mm, slender to moderately robust nematocerous flies. They are closely related to the Chironomidae but the females have mouthparts that can bite and the wing venation is different. The wings are held horizontally at rest, one above the other, and are sometimes boldly patterned. Early stages are usually found in moist or aquatic habitats. The larvae are apneustic, often elongated, and some are strong swimmers (Wirth, 1952a). Ceratopogonidae occur in all continents except Antarctica. The family includes more than 5000 species in over 100 genera (Borkent and Wirth, 1997); its division into 4 subfamilies (Leptoconopinae, Forcipomyiinae, Dasyheleinae and Ceratopogoninae) is generally accepted (Downes and Wirth, 1981). The adult structure in the three last-mentioned subfamilies is fairly uniform and can be treated in a single general description; the Leptoconopinae stand somewhat apart, but no species of this subfamily are recorded from Greenland.

In adults the eyes are approximated or meeting at the midline, sometimes bare or with fine pubescence. Antenna with 13 flagellomeres showing marked sexual dimorphism since the male flagellomeres 1–8 usually bear whorls of long setae, giving a plumose appearance. Female proboscis in biting species about as long as the head, consisting of a rigid labrum, blade-like mandibles and maxillary processes (laciniae), and a hypopharynx. The mandibles are normally armed with strong teeth, maxillary blade with fine teeth. Non-biting females and all males have reduced armature. Maxillary palp with 5 segments of which 1 and 2 are sometimes incompletely separated, segment 3 has a sensory organ enclosed in a pit or on the surface.

The thorax is dorsally convex and extending anteriorly over the head (Fig. 1A). The wings usually have 2 radial cells (Fig. 1B), but the first, second or both may be reduced or absent in some genera. The legs are rather short and stout. Female claws may be short and equal, long and equal, long and more or less unequal with or without inner and outer teeth. Empodia usually rudimentary, well-developed only in most Forcipomyiinae. Male claws are usually small, simple and equal.

The female abdomen has 10 well identifiable segments and 1–3 spermathecae of different shape. In some genera, *e.g. Bezzia*, internal sclerotized gland rods are present.

In males the first 8 tergites and sternites are normal. Tergite 9 is usually enlarged covering the elements of the terminalia (Fig. 11G). The gonopods are two-segmented; aedeagus extends back from sternite 9 and is articulated with the gonocoxites, which are often triangular with sclerotized lateral arms or ventral plate. The parameres are free or more or less fused above the aedeagus, in some species asymmetric, *e.g.* in *Dasyhelea*. Segment 10 is weakly developed, located below tergite 9.

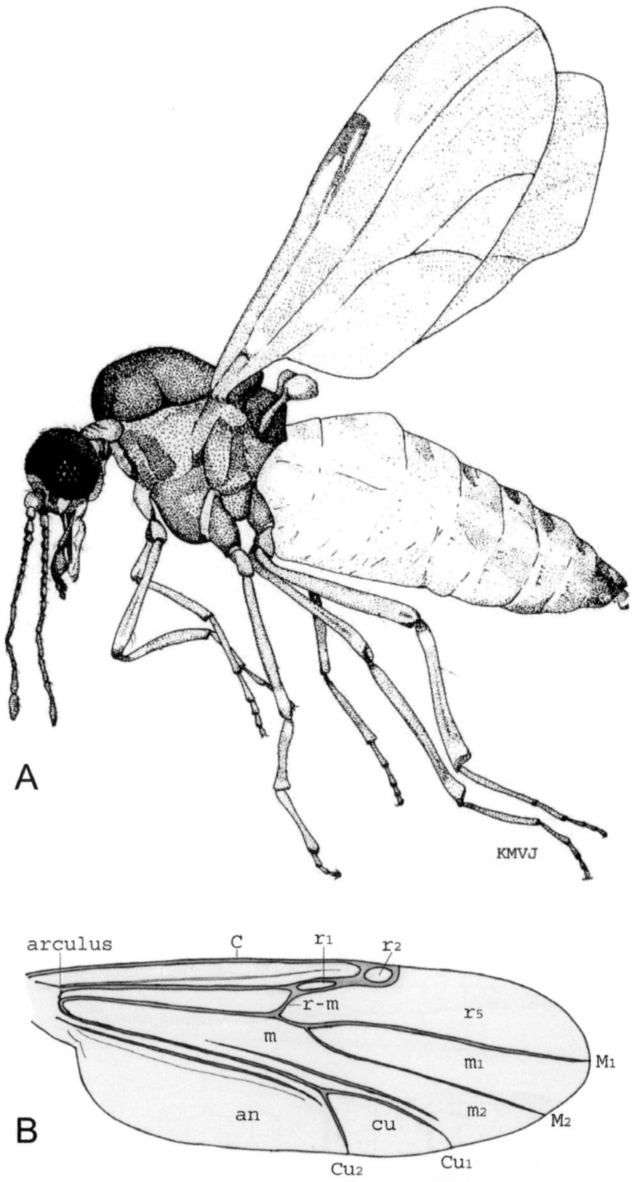

FIGURE 1 *Ceratopogonid* (Culicoides) *female habitus (A); ceratopogonid wing venation (B). an: anal cell; C: costal vein; cu: cubital cell;* Cu_1, Cu_2*: 1st and 2nd cubital vein; m: median cell;* M_1, M_2*: 1st, 2nd median veins;* r_1, r_2*: 1st, 2nd radial cells, the latter is alternatively referred to as cell* r_2*; r–m: cross-vein. (A drawn by K.-M. Vagn Jensen).*

The eggs of Ceratopogonidae are ovoid to elongate or horseshoe-shaped, in *Dasyhelea* encased in a gelatinous cover. The chorion is often sculptured and may bear numerous projections. In northern Europe the eggs normally hatch within 72 hours, but some species (such as *Culicoides grisescens*) may overwinter as eggs; normally the eggs do not withstand prolonged drought (Boorman, 1997).

The larvae vary in form among the genera. The head is usually sclerotized with a short collar between head and thorax. In *Forcipomyia, Atrichopogon* and some *Dasyhelea* the head is hypognathous, in Ceratopogoninae prognathous. The body may bear a complex systems of setae set on tubercles, and the prothorax bears a bilobed anterior proleg (*Forcipomyia, Atrichopogon*) (Figs 2A–D) (Boorman, 1997). In other genera the body is smooth and elongate, without prolegs (except the first instar of some *Culicoides*) or other appendages, and sometimes with minute setae. The head capsule is pigmented and often bears a pair of eye spots; the thoracic segments may also be pigmented (Boorman, 1997). Of particular importance in the identification of larvae is the form of the pharyngeal skeleton. The body consists of 3 distinct thoracic and 9 abdominal segments without external spiracles. The majority of the ceratopogonid larvae are unknown.

The pupae are relatively uniform in structure. They are compact and have a pair of prothoracic respiratory horns with numerous spiracles. The form of these horns and the spiracular arrangement are important diagnostic characters, as are the chaetotaxy of the abdominal segments and the operculum (Fig. 2E) (Blanton and Wirth, 1979). In most *Forcipomyia* the pupae are sedentary, in other genera they are partially buried in mud or dead leaves, or float freely in the water, and many Dasyhelinae and Ceratopogoninae pupae can move by twisting the abdomen (Boorman, 1997). The pupal stages may last from a couple of days to several weeks (Boorman, 1997).

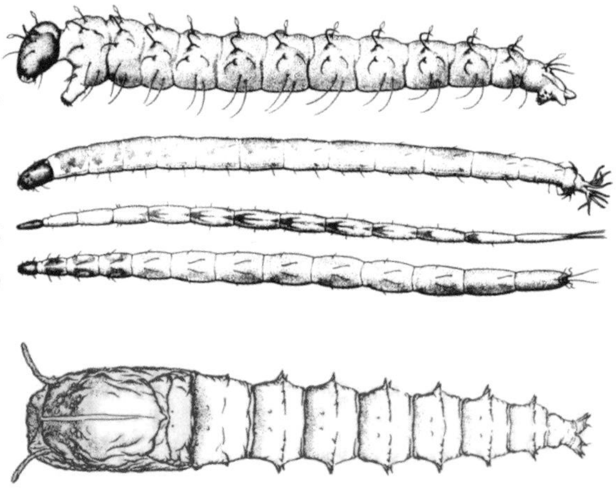

FIGURE 2 *Ceratopogonid immatures. A–D, larvae of* Forcipomyia *sp. (A);* Dasyhelea *sp. (B);* Bezzia *sp. (C);* Culicoides *sp. (D). E, pupa of* Dasyhelea *sp. A–D after Downes & Wirth (1981); E after Szadziewski* et al. *(1997).*

The larvae of Ceratopogonidae are found in a wide variety of habitats from damp earth and damp leaf litter to mud and marshy ground, ponds and rivers; in general they require some degree of moisture (Boorman, 1997). Some of the larvae belonging to Ceratopogoninae are able to swim with eel-like motion. There are 4 larval instars; the larval development may take from a couple of days to many months depending on the temperature. Larvae feed on micro-organisms and detritus or predate small nematodes or insect larvae (Boorman, 1997).

Most adult Ceratopogonidae are found near the larval habitats and the active dispersal in many species is largely restricted by the males' limited flight range. Ceratopogonids may be carried into the upper atmosphere and transported by the wind over long distances. The females usually require a meal of blood or other sources of protein for egg-development, but some species may be autogenous, at least for the first egg batch (Boorman, 1997). Many insectivorous genera capture their prey from swarms of Chironomidae or Ephemeroptera (Downes, 1978). In adult males the mouth parts are reduced; accordingly their only nourishment is sugar and water from nectar of flowers or from honeydew (Szadziewski *et al.*, 1997). In all *Dasyhelea* and some *Forcipomyia* both sexes feed exclusively on nectar. The females of *Culicoides* and some *Forcipomyia* suck blood from mammals and birds, while others feed on frogs (*Forcipomyia*). Some species of *Forcipomyia* are ectoparasitic on other insects. The females of all Ceratopogoninae (except *Culicoides*) prey on small insects.

The identification of Ceratopogonidae is based on structural features of the wings and legs and specific identification often involves examination of wing patterns, antennae and genitalia. Palp characters are also extremely useful. Many characters are difficult to see in pinned specimens and Ceratopogonidae are therefore usually slide-mounted in balsam or euparal for study.

Six ceratopogonid genera have been recorded from Greenland: *Forcipomyia* (*Lasiohelea*), *Dasyhelea*, *Brachypogon* (*Isohelea*), *Ceratopogon*, *Bezzia* and *Culicoides*. The Ceratopogonidae of Greenland were discussed by several authors, *e.g.* Andersen (1937) and last reviewed by Henriksen (1939). In the Greenland material from the Copenhagen Zoological Museum 9 species of Ceratopogonidae are registrated. It has only been possible to identify 2 of them to the species level, though all relevant keys to the Palaearctic ceratopogonid fauna have been consulted. According to A. Borkent (pers. comm.) the ceratopogonid fauna of northern Canada is not well understood and there are no keys. Consultation of several original (old) species descriptions led to no conclusive identifications.

Taxonomic characters and definitions

The terminology adopted here follows that of Downes and Wirth (1981). In both sexes the antennae have 15 segments: a short ring-like scape, an enlarged pedicel and a flagellum composed of 13 segments (flagellomeres). The *antennal ratio* of the female is the ratio of the sum of the lengths of the apical 5 segments of the flagellum (11–15) to the sum of the basal 8 (3–10) segments. It is essential that the calculations are based on summing the measurements of individual segments. *Sensilla* are thin-walled papillae arising from a pit and surrounded by a number of setulae. The distribution of the sensilla on the different flagellar segments is an important diagnostic character especially in the genus *Culicoides*.

The *wing length* is measured from arculus to wing tip. The *costal ratio* is the ratio of the costal length measured from the humeral cross-vein to the wing length. For vein/cell nomenclature see Fig. 1B.

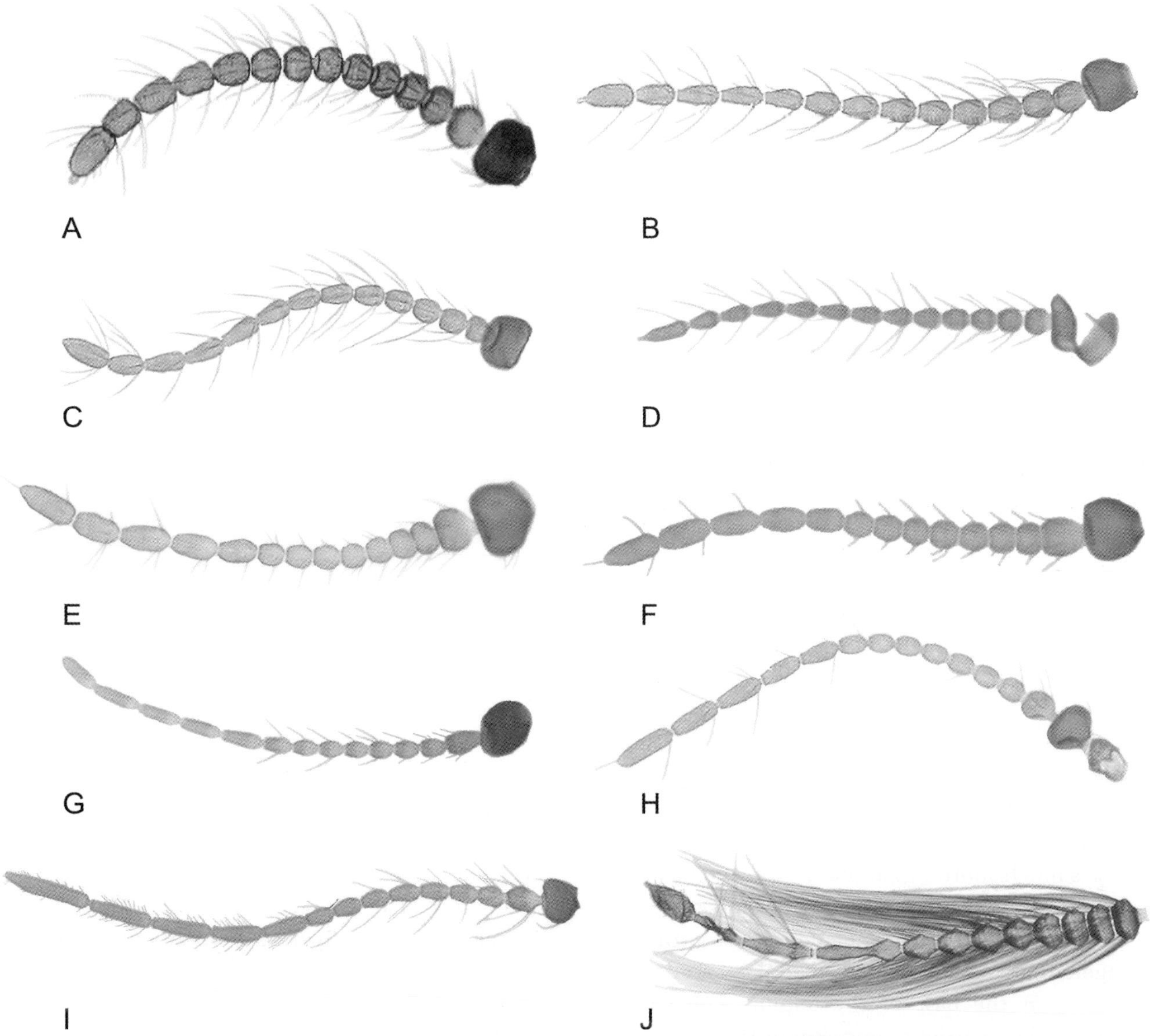

FIGURE 3 *A–I, female antennae of* Forcipomyia *sp. 1 (A);* Forcipomyia *sp. 2 (B);* Dasyhelea *sp. 1 (C);* Dasyhelea *sp. 2 (D);* Brachypogon (Isohelea) *sp. (E);* Ceratopogon abstrusus *(F);* Bezzia *sp. (G);* Culicoides *sp. 1 (H);* Culicoides sordidellus *(I). J, male antenna of* Dasyhelea *sp. 2.*

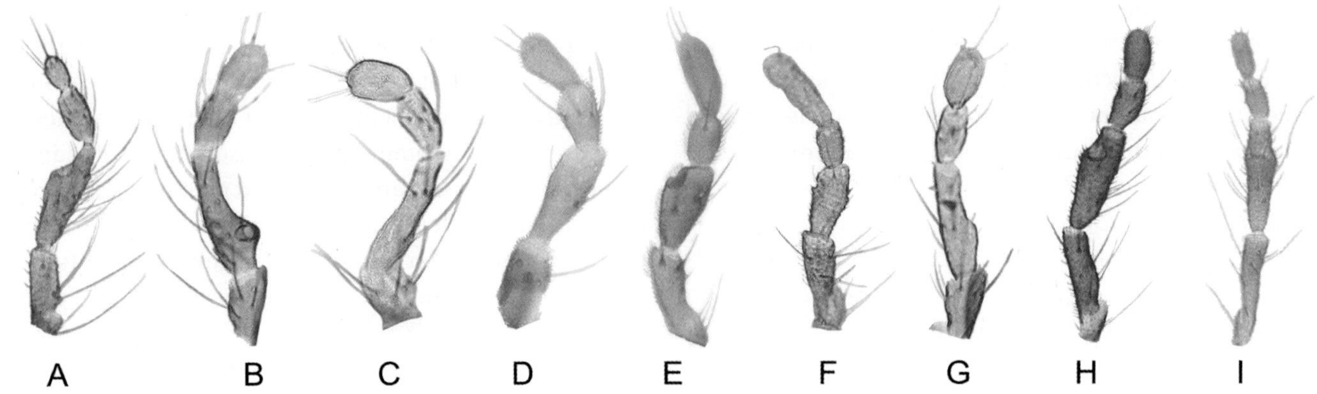

FIGURE 4 *Female palps (basal palpomere not consistently included) of* Forcipomyia *sp. 1 (A);* Forcipomyia *sp. 2 (B);* Dasyhelea *sp. 1 (C);* Dasyhelea *sp. 2 (D);* Brachypogon (Isohelea) *sp. (E);* Ceratopogon abstrusus *(F);* Bezzia *sp. (G);* Culicoides *sp. 1 (H);* Culicoides sordidellus *(I).*

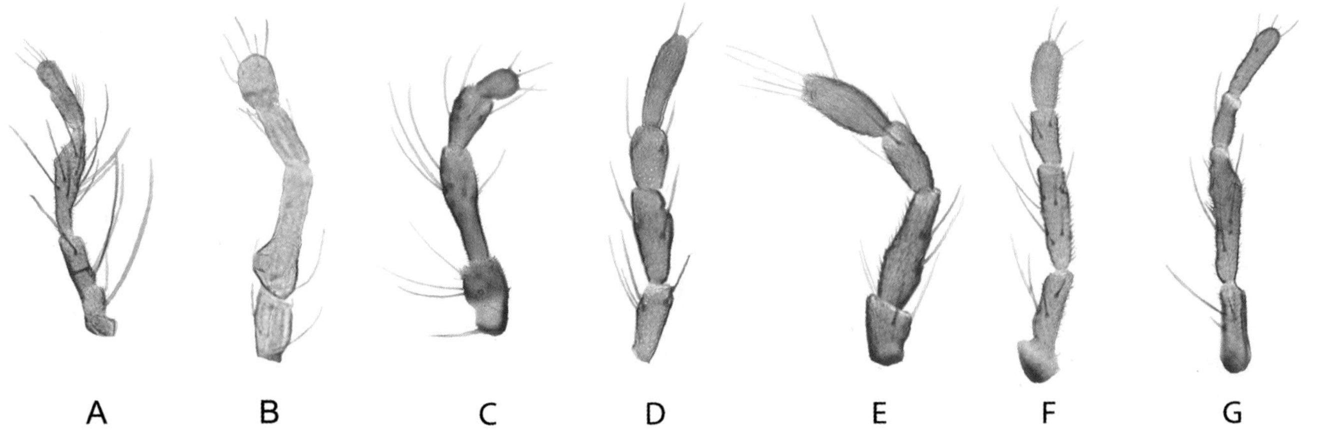

FIGURE 5 *Male palps (basal palpomere not consistently included) of* Forcipomyia *sp. 1 (A);* Forcipomyia *sp. 2 (B);* Dasyhelea *sp. 2 (C);* Brachypogon (Isohelea) *sp.(D);* Bezzia *sp. (E);* Culicoides *sp. 1 (F);* Culicoides sordidellus *(G).*

Characters of the *female spermathecae* (Fig. 10): the number and size of the spermathecae and the shape of the *neck* (a short sclerotized part of the ducts which join to a common duct).

The *male hypopygium* provides valuable characters for distinguishing the species (Fig. 11). Tergum 9 forms an expanded plate where the hind corners are frequently expanded as a pair of apicolateral processes. Sternum 9 is much shorter than the tergum and usually with a caudomedian excavation on its hind margin, where the aedeagus articulates at the lateral corners. The 2-segmented forceps-shaped genital appendages arise laterally at the base of the tergum. The basal segment (basistyle) carries 2 processes at the base: the ventral and dorsal roots. The parameres are usually a pair of sclerotized internal rodlike sclerites with knobbed base.

Key to genera and species of Ceratopogonidae recorded from Greenland

1. Empodia well-developed, claws strongly curved (Figs 8A, B, 9A, B). Wings usually with numerous macrotrichia (Figs 6A, B, 7A, B) ....2 (Forcipomyiinae)
– Empodia small or vestigial, claws more gently curved (Figs 8C–I, 9C–G); macrotrichia of wings usually less numerous or absent 3
2. Small species; length of wings about 1 mm; antennal ratio about 1; third palp segment slender, sensory pit round, shallow (Figs 4A, 5A); 1 spermatheca with a distinct neck (Fig. 10A); hypopygium as in Fig. 11A .. .. *Forcipomyia* sp. 1
– Larger species; length of wings about 1.3 mm; antennal ratio about 1.4; third palp segment swollen in the

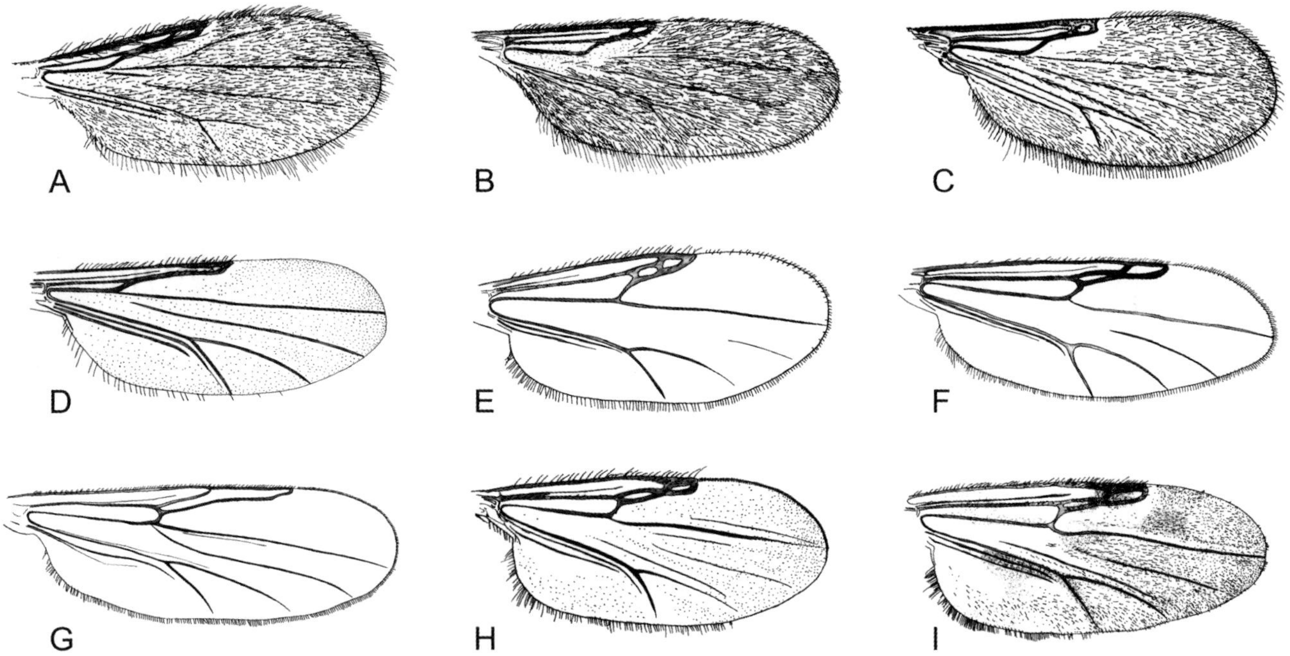

FIGURE 6 *Female wings of* Forcipomyia *sp. 1 (A);* Forcipomyia *sp. 2 (B);* Dasyhelea *sp. (C);* Dasyhelea *sp. 2 (D);* Brachypogon (Isohelea) *sp. (E);* Ceratopogon abstrusus *(F);* Bezzia *sp. (G);* Culicoides *sp. 1 (H);* Culicoides sordidellus *(I).*

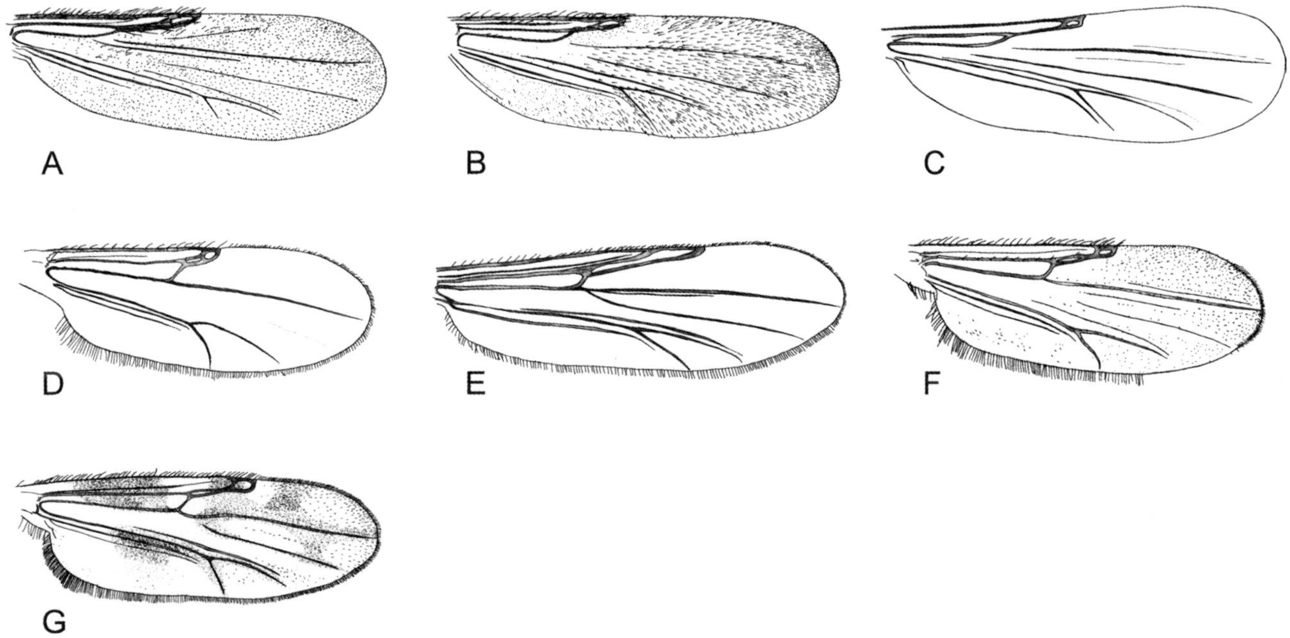

FIGURE 7 *Male wings of* Forcipomyia *sp. 1 (A);* Forcipomyia *sp. 2 (B);* Dasyhelea *sp. 2 (C);* Brachypogon (Isohelea) *sp. (D);* Bezzia *sp. (E);* Culicoides *sp. 1 (F);* Culicoides sordidellus *(G).*

proximal third (Figs 4B, 5B); 2 ovoid spermathecae (Fig. 10B) with short necks; hypopygium as in Fig. 11B ....*Forcipomyia* sp. 2

3. Flagellomeres sculptured. Cell r_1 nearly or completely closed; cell r_2 truncated, usually ending at or before middle of wing (Figs 6C, D, 7C). Eyes with short pubescence. Female and male claws small and equal (Figs 8C, D, 9C) .... 4 (Dasyheleinae: *Dasyhelea*)

– Flagellomeres not sculptured. Either cell r_1 or cell r_2 or both well-developed, ending beyond middle of the wing. Eyes usually bare. Female claws variable .. .. 5 (Ceratopogoninae)

FIGURE 8 *Apical female tarsomeres of* Forcipomyia *sp. 1 (A);* Forcipomyia *sp. 2 (B);* Dasyhelea *sp. 1 (C);* Dasyhelea *sp. 2 (D);* Brachypogon (Isohelea) *sp. (E);* Ceratopogon abstrusus *(F);* Bezzia *sp. 1 (G);* Culicoides *sp. 1 (H);* Culicoides sordidellus *(I).*

FIGURE 9 *Apical male tarsomeres of* Forcipomyia *sp. 1 (A);* Forcipomyia *sp. 2 (B);* Dasyhelea *sp. 2 (C);* Brachypogon (Isohelea) *sp. (D);* Bezzia *sp. (E);* Culicoides *sp. 1 (F);* Culicoides sordidellus *(G).*

4. Small species; wing length 0.96 mm; palp segment 3 narrowing towards apex, palp segments 4 and 5 broader than the 3rd (Fig. 4C) .... *Dasyhelea* sp.1
– Larger species; wing length 1.46 mm; palp segment 3 smoothly widened until distal third, palp segments 4 and 5 narrower than 3rd (Fig. 4D); hypopygium as in Fig. 11C............................... *Dasyhelea* sp. 2
5. Fork at junction of veins M_1 and M_2 distally to R–M; M_2 sometimes obsolete basally (Figs 6E, F, H, I, 7D, F, G) 6
– Fork at junction of veins M_1 and M_2 at or before R–M (medial fork sessile); M_2 complete at base; wing with one radial cell (Figs 6G, 7E);female abdomen with internal sclerotized gland rods; hypopygium as in Fig. 11E .... *Bezzia* sp.
6. Female and male claws small, equal, and simple (Figs 8H, I, 9F, G). Cell r_1 and cell r_2 both well-developed, similar in size; wings with abundant macrotrichia (Figs 6H, I, 7F, G); eyes bare........... .... 7 (*Culicoides*)
– Female claws large and equal with inner tooth; cell r_1 and cell r_2 both well-developed; macrotrichia of wings scanty or absent; eyes with pubescence 8
7. Medium-sized species; wing length 1.58 mm; eyes broadly separated; 3 spermathecae (Fig. 10H); hypopygium as in Fig. 11G ....*Culicoides* sp.1
– Large species; wing length 2.24 mm; eyes contiguous; two functional spermathecae and a rudimentary third (Fig. 10I); hypopygium as in Fig. 11F .... *Culicoides sordidellus*

8. Larger species; wing length 1.65 mm; costal ratio greater than 0.65; three spermathecae (Fig. 10F); eyes broadly separated; (male not seen) .... *Ceratopogon* sp.
– Smaller species; wing length 1.26 mm; costal ratio less than 0.65; two spermathecae (Fig. 10E); eyes contiguous; hypopygium as in Fig. 11D .... *Brachypogon* (*Isohelea* Kieffer)

FORCIPOMYIINAE

Forcipomyia Meigen

Henriksen (1939) reports one unidentified *Forcipomyia* species from East Greenland (Kap Alfred, Kempe Fjord), but neither of the species described below correspond to the description presented by Henriksen.

Forcipomyia sp. 1 Figs 3A, 4A, 5A, 6A, 7A, 8A, 9A, 10A, 11A

Female: Wing length 1.00 mm. Eyes contiguous, bare. Antenna: lengths of flagellar segments (µm) 35–26–26–26–26–26–23–26–32–38–38–36–67; antennal ratio 1.02; antennal segment 13 of characteristic shape (Fig. 3A). Palp segment 3 (Fig. 4A) slender, sensory pit round, shallow. Lengths of palp segments 3–5 (µm) 93–41–26. Tarsi (Fig. 8A) with well-developed empodium; claws strongly curved. Wings (Fig. 6A) with costa terminating near middle of wing (costal ratio 0.53; total length 0.43 mm). Microtrichia minute; macrotrichia weak but abundant. Cell r_1 closed and cell r_2 short but well developed. One ovoid, well-developed spermatheca (Fig. 10A) with a distinct neck (µm: 104). *Male*: Wing length 1.22 mm; wing as in Fig. 7A, similar to female wing but more slender. Also the palps (Fig. 5A) and tarsi (Fig. 9A) are very similar to those of the female. Hypopygium (Fig. 11A) with slender gonostyles; sternum 9 with a caudomedian excavation, ventral membrane not spiculate; tergum 9 rounded with 2 distinct bristles on prominences. Aedeagus with a basal arc. Basal arms slender and convergent, parameres long and twisted.

DISTRIBUTION. SW (Qeqertarsuaq/Godhavn, NE (Zackenberg, Mestersvig, Jameson Land: Coloradodal, Muslingeelvdal).

Forcipomyia sp. 2. Figs 3B, 4B, 5B, 6B, 7B, 8B, 9B, 10B, 11B

Female: Wing length 1.30 mm. Eyes contiguous, bare. Antenna (Fig. 3B): lengths of flagellar segments (µm) 64–44–46–49–49–49–49–49–52–52–55–52–70; antennal ratio 1.42; antennnal segment 13 of characteristic shape. Third palp segment (Fig. 4B) swollen in the proximal third where a round and moderately deep sensory pit is present. Lengths of palp segments 3–5 (µm) 113–43–41. Tarsi (Fig. 8B) with well-developed empodium; claws strongly curved. Wings (Fig. 6B) with costa terminating before the middle of the wing (costal ratio 0.47; total length 0.68 mm). Microtrichia minute; macrothrichia well-developed and numerous. Cell r_1 closed and cell r_2 short but well developed. Two ovoid well-developed spermatheca (Fig. 10B) with short necks (µm: 122–125). *Male*: Wing length 1.43 mm. Wing as in Fig. 7B; similar to female wing but more slender. Also the palps (Fig. 5B) and tarsi (Fig. 9B) are very similar to those of the female. The hypopygium (Fig. 11B) with slender gonostyles. Sternum 9 with a caudomedian excavation, ventral membrane not spiculate. Aedeagus with basal arc, distal portion long and slender with a simple rounded tip. Parameres long and filamentous.

DISTRIBUTION. S (Narsarsuaq), SW (Kangerlussuatsiaq/Evighedfjorden, Kangerlussuaq/Søndre Strømfjord, Qeqertarsuaq/Godhavn, Qeqertarsuup Tunua/Disko Bay: Qeqertasussuk).

DASYHELEINAE

Dasyhelea Kieffer

Henriksen (1939) reports *Dasyhelia scutellata* (Meigen) from West Greenland. However, the original description is only based on morphological characters of dried individuals and is inadequate.

Dasyhelea sp. 1. Figs 3C, 4C, 6C, 8C, 10C

Female: Wing length 0.96 mm. Eyes contiguous with short pubescence. Antenna (Fig. 3C): lengths of flagellar segments (µm) 35–29–32–32–35–36–35–38–46–46–46–44–58, antennal ratio 1.13, total length 0.51 mm; 1st flagellomere with sensilla coeloconica. Palp segment 3 (Fig. 4C) long and very slender, narrowing against apex. Palp segments 4 and 5 broader than 3. Sensory pits inconspicuous. Lengths of palp segments 3–5 (µm) 75–29–38. Tarsi (Fig. 8C) with small, equal and simple claws. Wings (Fig. 6C) with cell $r_1$completely closed, cell r_2 square-ended and ending before the middle (costal ratio 0.44). Microtrichia minute, macrothrichia abundant. One spermatheca (Fig. 10C) with oblique and relatively long neck (75 µm). *Male* unknown.

DISTRIBUTION. S (Narsarsuaq), SW (Qeqertarsuaq/Godhavn; Qeqertarsuup Tunua/Disko Bugt: Qeqertasussuk; Kangerlussuaq/Søndre Strømfjord).

Dasyhelea sp. 2. Figs 3D, 3J, 4D, 5C, 6D, 7C, 8D, 9C, 10D, 11C

Female: Wing length 1.46 mm. Eyes contiguous, with pubescence. Antenna (Fig. 3D): lengths of flagellar segments (µm) 64–38–44–49–52–55–55–58–61–64–58–52–96; antennal ratio 1.25, total length 0.74 mm. Palp segment 3 (Fig.4D) long and smoothly until the distal third, subsequently narrowing against apex. Sensory pits inconspicuous. Tarsi (Fig. 8D) with small, equal and simple claws. Wings (Fig. 6D): cell r_1 completely closed, cell r_2 abruptly closed and ending at the middle of the wing (costal ratio 0.50). Microtrichia minute. Macrothrichia sparse. One spermatheca (Fig. 10D) with crooked neck (93 µm). *Male*: Wing length 1.52 mm. Wings (Fig. 7C) similar to the female wings but more slender (costal ratio 0.51). Macrotrichia sparse. Palp segment 3 (Fig. 5C) slender, expanded against apex. Sensory pits inconspicuous. Antennal flagellomeres sculptured (Fig. 3J). Tarsi (Fig. 9C) with short, simple, equal claws. Hypopygium (Fig. 11C): short, stout dististyles with enlarged, knobbed tip. Sternum 9 without caudomedian excavation, tergum 9 gently rounded distally, with small apicolateral processes and a caudomedian protrusion. Aedeagus with basal arch angular and short, median part consists of two converging sclerotized arms. Parameres asymmetric.

DISTRIBUTION. S (Narsarsuaq), SE (Saqqisikuik/ Skjoldungen: Qoornoq)

CERATOPOGONINAE

***Brachypogon* (*Isohelea* Kieffer)**

This genus is not mentioned by Henriksen (1939) and it has not been possible to identify the species.

***Brachypogon* (*Isohelea*) sp. Figs 3E, 4E, 5D, 6E, 7D, 8E, 9D, 10E, 11D**

Female: Wing length 1.26 mm. Eyes contiguous with short pubescence. Antenna (Fig. 3E): lengths of flagellar segments (µm) 131–55–49–46–58–55–61–64–110–122–131–122–165; antennal ratio 0.80; first antennal flagellar segment with sensilla coeloconica, total length 1.17 mm. Palp segment 3 (Fig. 4E) short and compact, sensory pit round, shallow. Lengths of palp segments 3–5 (µm) 52–32–46. Katepisternum of thorax with one or more setae. Tarsi (Fig. 8E): claws large with a small basal tooth, equal. Wings (Fig. 6E) milky. Costa terminating beyond the middle of the wings (costal ratio 0.62); microtrichia minute, macrothrichia absent. Cell r_1 and cell r_2 short with adjacent veins thickened. 2 ovoid well-developed, unequal spermathecae (Fig. 10E) with short necks (µm: 74–55). *Male*: Wing length 1.22 mm. Wing (Fig. 7D) similar to the female wing but more slender. Palp segment 3 (Fig. 5D) more slender than that of the female, and tarsi (Fig. 9D) with short, simple, equal claws. Hypopygium (Fig. 11D) with slender gonostyles. Sternum 9 with broad, shallow, caudomedian excavation. Ventral membrane not spiculate. Tergum 9 with rounded margin and short apicolateral and caudomedian protrusions. Aedeagus long, triangular and tapering. Parameres knife-shaped and at basis connected by a sclerotized bridge.

DISTRIBUTION. SW (Qeqertarsuaq/Godhavn, Nuup Kangerlua/Godthåbsfjord: Kapisillit: Saqqaq), NE (Zackenberg, Mestersvig, Jameson Land: Coloradodal, Muslingeelvdal, Lollands Elv).

***Ceratopogon abstrusus* Borkent and Grogan, 1995. Figs 3F, 4F, 6F, 8F, 10F**

Female: Wing length 1.65 mm. Eyes broadly separated with short pubescence. Antenna (Fig. 3F): lengths of flagellar segments (µm) of 58–29–30–35–35–35–35–38–46–58–59–62–81 antennal ratio 0,96, total length 0.60 mm; first flagellomere with 3 sensilla coeloconica. 3rd palp segment (Fig. 4F) long and very slender, narrowing against apex; sensory pits inconspicuous. Lengths of palp segments 3–5 (µm) 73–41–84. Tarsi (Fig. 8F) with large, equal claws with inner tooth on forelegs, midlegs and hindlegs. Wings (Fig. 6F): cell r_1 and cell r_2 complete and costa terminating beyond the middle of the wings (costal ratio 0.72); very few macrothrichia at wing apex. 3 spermathecae (Fig. 10F) with short necks, µm: (72–72–39). *Male*s not found.

DISTRIBUTION. NE (Mestersvig, Jameson Land: Muslingeelvdal).

***Bezzia* Kieffer**

This genus is not mentioned in Henriksen (1939) and it has not been possible to identify the species.

***Bezzia* sp. Figs 3G, 4G, 5E, 6G, 7E, 8G, 9E, 10G, 11E**

Female: Wing length 2.07 mm. Costal ratio 0.71. Eyes narrowly separated, bare. Antennae (Fig. 3G): lengths of flagellar segments (µm) 93–48–46–44–46–46–49–54–80–81–83–90–87. Antennal ratio 1.01, total length 0.85 mm. Palp segment 3 (Fig. 4G) long, basal part broader, distal part slender. Lengths of palp segments 3–5 (µm) 104–58–55. Tarsi (Fig. 8G) with medium sized, equal and simple claws. Wings (Fig. 6G): cells r_1 and r_2 confluent (costal ratio 0.71), microtrichia minute, macrotrichia absent. Abdomen with internal sclerotized gland rods; two spermathecae (Fig. 10G), ovoid with long and slender necks (µm: 128–93). *Male*: Wing length 1.93 mm; costal ratio 0.66. Wings similar to the female wings but more slender (Fig. 7E). Palp segment 3 as in Fig. 5E, claws similar to female claws (Fig. 9E). Hypopygium (Fig. 11E) with short, stout dististyles shaped like duck-heads.

DISTRIBUTION. SW (Kangerlussuaq/Søndre Strømfjord, Nuup Kangerlua/Godthåbsfjord: Kapisillit: Saqqaq).

***Culicoides* Latreille**

Henriksen (1939) records two species of *Culicoides*: *C. minutissimus* Zett. and *C. sordidellus* Zett. *C. sordidellus* is only registrated from Greenland. The females of species 1 of the genus has some resemblance with the above mentioned *C. minutissimus* described by Søgaard Andersen (1937) and is also registrated in NE but there are important diagnostic differences, *e.g.* that the antennal sensilla of female *C.* sp. 1 are distributed on all flagllomeres whereas in *C. minutissimus* (Campbell and Pelham-Clinton, 1960; Kremer, 1965) only flagellomeres 3 to 10 bear sensilla Further, *C.* sp. 1 is much bigger than *C. minutissimus*. Therefore *C.* sp. 1 is new to Greenland but it has not been possible to identify to species by use of keys to the Palearctic fauna.

***Culicoides* sp. 1 Figs 3H, 4H, 5F, 6H, 7F, 8H, 9F, 10H, 11G**
Female: Wing length 1.58 mm. Eyes broadly separated, bare. Antennae (Fig. 3H): lengths of flagellar segments (µm) 55–38–38–39–41–39–44–61–64–75–78–96; antennal ratio 0.71, total antennal length: 0.69 mm. Distribution pattern of antennal sensilla: 3–15(n = 5; 1,5–0,8–1,0–0,8–1,1–0,88–1,0–0,8–0,9–0,9–1,0–0,8–0,1). Palp segment 3 (Fig. 4H) long and slender with one or two shallow sensory pits near apex. Lengths of palp segments 3–5 (µm) 90–46–44. Tarsi (Fig. 8H) with small, equal and simple claws. Wings (Fig. 6H) milky; radius, subcosta and radial veins brown pigmented. Costa terminating beyond the middle of the wings (costal ratio 0.61); microtrichia dense; macrothrichia abundant. Cell r_1 and cell r_2 complete, formed by more heavily sclerotized radial branches. Two spermathecae plus a more or less rudimentary third one (Fig. 10H) with short necks (µm: 72–70–43). *Male*: Wing length 1.64 mm. Wings (Fig. 7F) similar to the female wings but more slender. The 3rd palp segment (Fig. 5F) more slender than that of the female, and tarsi (Fig. 9F) with short, simple, equal claws. Diststyles of hypopygium (Fig. 11G) slender. Sternum 9 with dish-shaped caudomedian excavation. Ventral membrane not spiculate. Tergum 9 with large triangular apicolateral processes and deep caudomedian notch. Aedeagus broad and stout, basal arch extending to a third of total length; distal part with broad truncate tip. Parameres separate, tapering, slender distally.

DISTRIBUTION. NE (Zackenberg, Jameson Land: Muslingeelvdal).

***Culicoides sordidellus* (Zetterstedt, 1838) Figs 3I, 4I, 5G, 6I, 7G, 8I, 9G, 10I, 11F**
Female: Wing length 2.24 mm. Eyes contiguous, bare. Antennae (Fig. 3I): lengths of flagellar segments (µm) 73–54–55–57–58–54–55–55–91–93–104–128–171; antennal ratio 0.78, antennal total length 1.05 mm. Distribution pattern of antennal sensilla: 3–15 (n = 3; 3,3–0–0–0–0–0–0–0–1–1,7–1–2,3–1,7). Palp segment 3 (Fig. 4I) long and slender, gently swollen shortly before apex with more diffuse, shallow sensory pits. Tarsi (Fig. 8I) with small, equal and simple claws. Wings (Fig. 6I) brown pigmented with whitish spots and an hourglass-shaped spot in r5 as figured. Radius, subcosta and radial veins brown. Costa terminating beyond the middle of the wings (costal ratio 0.78); microtrichia dense; macrothrichia abundant. Cell r_1 and cell r_2 complete, formed by more heavily sclerotized radial branches. Two functional spermathecae (Fig. 10I) with short necks and a rudimentary third (µm: 118–104–33). *Male*: Wing length 1.97 mm. Wings (Fig. 7G) similar to the female wings but more slender and the brown pigmentation less conspicuous. Palp segment 3 (Fig. 5G) more slender than that of the female and not swollen near apex. Tarsi (Fig. 9G) with short, simple, equal claws. Diststyles of hypopygium (Fig. 11F) slender with enlarged, rounded tips. Sternum 9 with shallow caudomedian excavation. Ventral membrane not spiculate. Tergum 9 rounded distally, with small apicolateral processes and a caudomedian protrusion. Aedeagus broad and stout, basal arch extending to a third of total length; distal portion with broad truncate tip. Parameres separate, tapering and becoming slender and diverge distally.

DISTRIBUTION. S (Narsarsuaq, Upernaviarsuk), SW (Kangerlussuaq/Søndre Strømfjord), SE (Saqqisikuik/Skjoldingen: Qoornoq).

Acknowledgement

We thank Dr. Boy Overgaard Nielsen Aarhus, Denmark, for critical and helpful comments on the manuscript.

References

Andersen, F. Søgaard (1937): Über die Metamorphose der Ceratopogoniden und Chironomiden Nordost-Grönlands. – *Meddelelser om Grønland* 16(1): 1–95.

Blanton, F.S. and Wirth, W.W. (1979). *The sandflies (Culicoides) of Florida (Diptera: Ceratopogonidae)*. Arthropods of Florida and Neighbouring Land Areas. Vol 10. Florida Department of Agriculture and Consumer Services, Gainsville, Florida, 204 pp.

Boorman, J. (1997): Family Ceratopogonidae. *In*: Papp, L. & Darvas, B. (eds): *Contribution to a Manual of Palaearctic Diptera* 2: 349–368.

Borkent, A. & Wirth, W.W. (1997): World Species of Biting Midges (Diptera: Ceratopogonidae. – *Bulletin of the American Museum of Natural History* 233: 1–257.

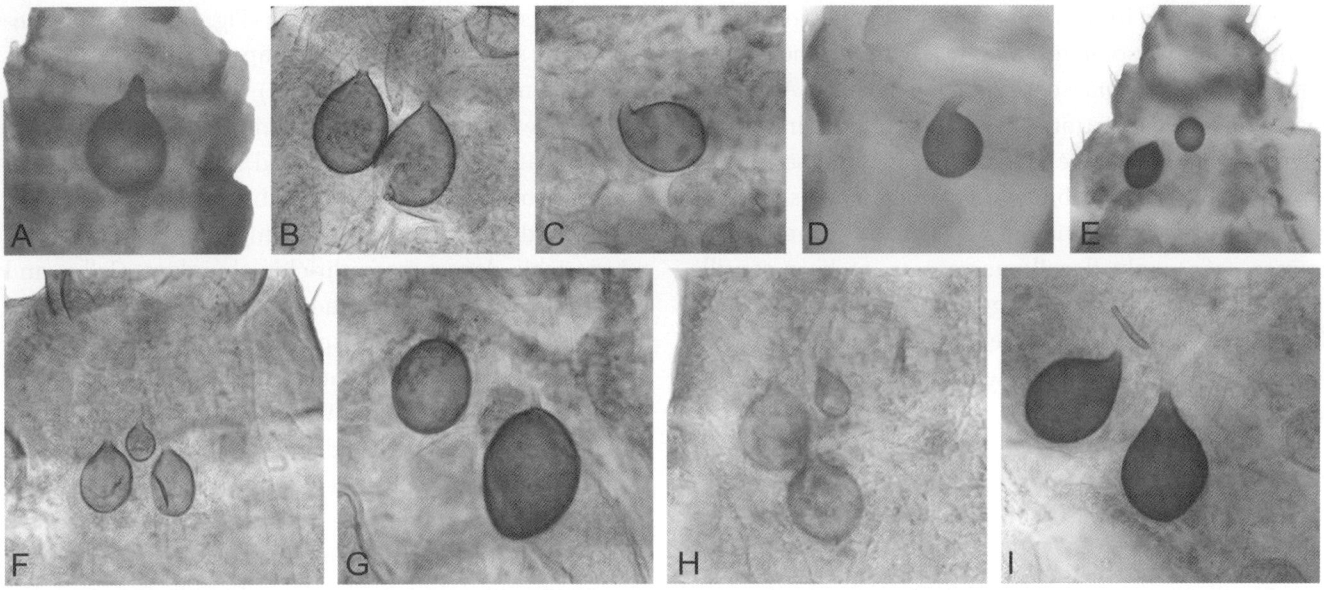

FIGURE 10 *Spermathecae of* Forcipomyia *sp. 1 (A);* Forcipomyia *sp. 2 (B);* Dasyhelea *sp. 1 (C);* Dasyhelea *sp. 2 (D);* Brachypogon (Isohelea) *sp. (E);* Ceratopogon abstrusus *(F);* Bezzia *sp. (G);* Culicoides *sp. 1 (H);* Culicoides sordidellus *(I).*

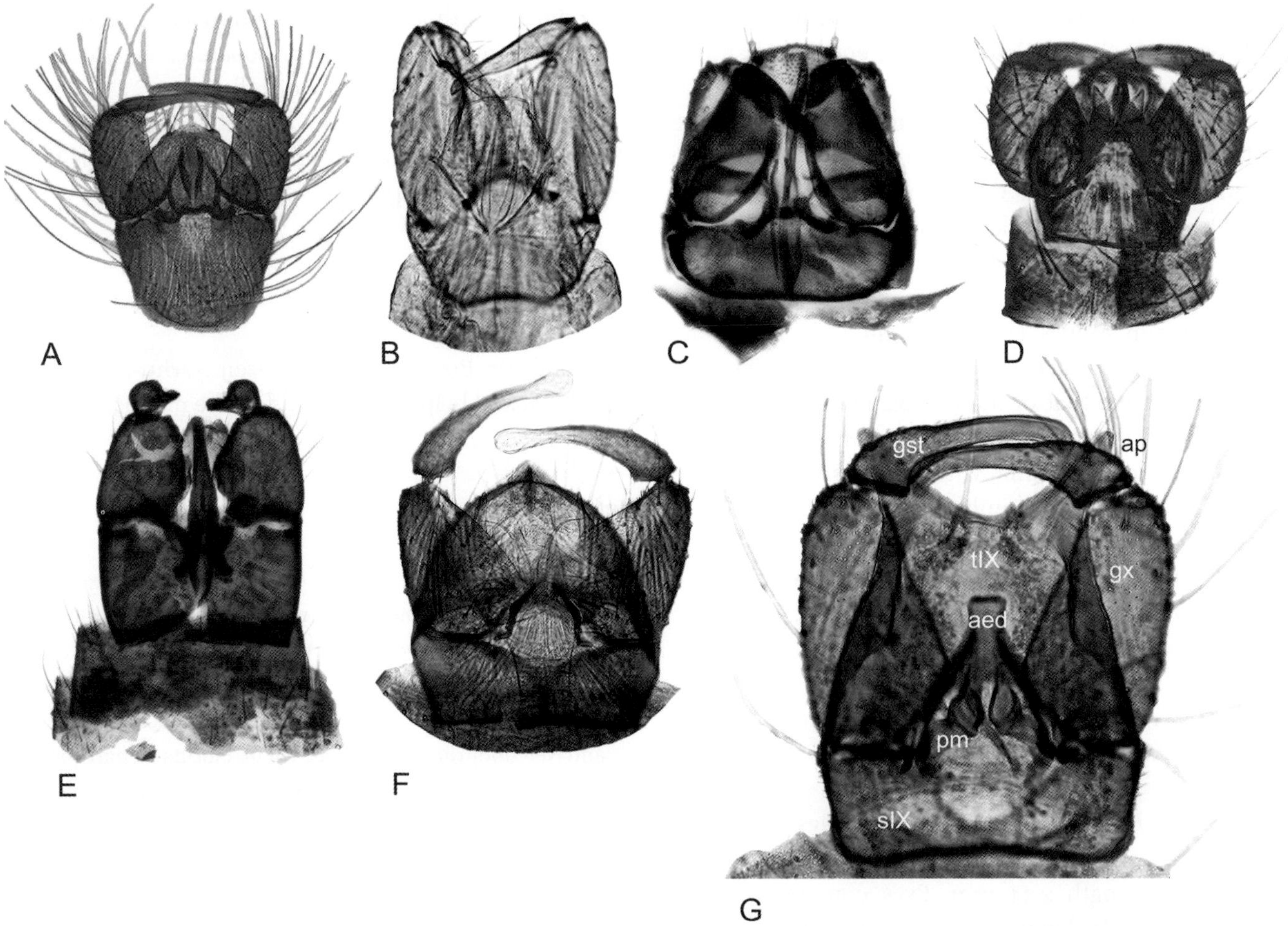

FIGURE 11 *Hypopygia of* Forcipomyia *sp. 1 (A);* Forcipomyia *sp. 2 (B);* Dasyhelea *sp. 2 (C);* Brachypogon (Isohelea) *sp. (D);* Bezzia *sp. (E);* Culicoides sordidellus *(F);* Culicoides *sp. 1 (G). aed: aedeagus; ap: apicolateral process of tergite IX; gst: gonostyles; gx: gonocoxite; pm: parameres; sIX: sternite IX; tIX: tergite IX.*

Campbell, J.A. & Pelham-Clinton, E.C. (1960): A taxonomic review of the British species of »*Culicoides*« Latreille (Diptera, Ceratopogonidae). – *Proceedings of the Royal Society of Edinburgh* 67(3): 181–302.

Downes, J.A. (1977): Evolution of feeding habits in Ceratopogonidae. – *Mosquito News* 37: 279–280.

Downes, J.A. (1978): Feeding and mating in the insectivorous Ceratopogonidae (Diptera). – *Memoirs of the entomological Society of Canada* 104: 1–62.

Downes, J.A. & Wirth, W.W. (1981): Ceratopogonidae. *In*: McAlpine, J.F., Peterson, B.V., Shewell, G.E., Teskey, H.J., Vockeroth, J.R., and Wood, D.M. (eds): *Manual of Nearctic Diptera* 1. Research Branch Agriculture Canada Monograph 27, pp 393–421.

Henriksen, K. (1939): A revised index of the insects of Grønland. – *Meddelelser om Grønland* 119(10): 1–112.

Henriksen, K. & Lundbeck, W. (1917): Grønlands Landarthropoder. – *Meddelelser om Grønland* 22: 484–821.

Kremer, M. (1965): Contribution à l'étude du genre *Culicoides* Latreille particulièrement en France. *Encyclopedie Entomologique* 39: 299.

Lundbeck, W. (1898): Diptera groenlandica. – *Videnskabelige Meddelelser fra Naturhistorisk Forening i Kjøbenhavn* 1898: 236–314.

Szadziewski, R., Krywinski, J. & Gilka, W. (1997): Diptera Ceratopogonidae, Biting Midges. *In*: Nilsson, A. (ed.) *Aquatic Insects of North Europe, A Taxonomic Handbook*, 2: 243–263.

Wirth, W.W. (1952). The Heleidae of California. – *University of California Publications in Entomology* 9: 95–266.

17.7. Chironomidae (Non-biting midges)

Claus Lindegaard

Chironomids are very abundant in both species and numbers of individuals and is one the most important secondary producer groups in freshwater ecosystems. Especially at high altitude and in northern areas their importance in freshwater is indisputable. This is still more striking in Greenland, where immigration of many groups has been hampered by the isolated geographic position of Greenland. The chironomids, however, are in the adult stage easily dispersed by wind, and the most abundant and common species are widely distributed over most of the Northern Hemisphere – implying that the larval ecosystems are available. Immigration to Greenland has mainly taken place from the northern part of Canada, where especially the sounds between Ellesmere Island and North Greenland are relatively small barriers for chironomids. This implicates that species able to live in the high arctic Ellesmere Island have been the main source from which the Greenlandic chironomids descend. From North Greenland they have spread south the west coast and east to Peary Land and south the east coast. These main immigration routes have restricted the number of species in the southern part of Greenland, where the diversity is much lower than at corresponding latitudes in North America and Eurasia. A second immigration route may be from Northern Fennoscandia via Svalbard. However, here the distances over open sea are much longer and this route may be less important. For the time being, the knowledge of the Greenland chironomid fauna is so modest that it is not possible to decide if a Palaearctic contribution has taken place, in spite of that a few species (*e.g. Sergentia coracina, Tanytarsus niger*) are known only from East Greenland.

Morphology

A short survey of terms in chironomid morphology is presented below. More terms are used in the keys and explained by referring to specific figures. A complete survey of terms is given by Sæther (1980) and Sæther *et al.* (2000).

ADULTS. Adult chironomids range in length from 1.5 to 25 mm. The thorax and abdomen may be uniformly coloured or scutum may be striped by separate colours and the abdominal tergites with a pattern of colours. The most obvious difference between sexes is that the female antennae are much less densely plumose.

The antenna consists of a large globular segment 2 and a flagellum of up to 14 "segments" termed flagellomeres (Fig. 5O). Segment 1 is usually suppressed. The flagellum normally carries whorls of long setae giving the antennae its characteristic plumose appearance. In Tanypodinae and Podonominae the penultimate and in the remaining subfamilies the last flagellomere is the longest. The eyes are very large and often protruding dorsally (Figs 12J, 44I–J). In a number of species they appear to be pubescent due to the presence of fine microtrichia between the facets (Figs 12J, 34G). The mouthparts are strongly reduced with normally only the maxillary palps well developed (Figs 12J, 19G).

The 3-segmented thorax bears the legs, wings and halters (Figs 1E, 4E, 32I). The antepronotum forms a narrow collar anterior to mesothorax, which is greatly enlarged and comprises dorsally the scutum, scutellum and postnotum. In most species postnotum possesses a median groove or keel. The thorax may bear numerous groups of setae which may be of great taxonomic value (Fig. 32I).

The legs of the chironomids are moderately long, especially the fore legs, and consist of a small coxa and trochanter, a long femur and tibia, and five tarsal segments (tarsomeres) of various length. Tarsomere 5 bears two claws. The distal end of the tibia on middle and hind leg often carry 1–2 spines that are of taxonomic importance (Figs 5H–N). In Chironominae a comb of additional smaller spines (the tibial comb) may be present. Pulvilli (pad-like processes beneath each claw) are well developed in some genera (Fig. 51F).

The wing margins and veins are usually fringed with macrotrichia, whereas the wing membrane may be bare, or invested to a varying degree with macrotrichia. The wing venation is moderate complex, but contains few crossveins. The crossvein MCu is present only in Podonominae, Tanypodinae, Diamesinae, and Prodiamesinae. Vein R_{2+3} is forked only in Tanypodinae and may in many species be faint or absent (Figs 5A–G).

The abdomen consists of 8 segments plus genitalia termed the hypopygium (Fig. 1E). The hypopygium demonstrates a considerable variation in structure, but is always comprised of a modified segment 9 and the major appendages called gonocoxite and gonostylus (Figs 4A–D). In the Chironominae the gonocoxite possesses two prominent lobes termed superior volsella (appendage 1) and inferior volsella (appendage 2). These lobes may be more or less reduced or absent in the remaining subfamilies. The Tanytarsini possess further a median volsella. Tergite 9 may end in a prominent anal point.

Eggs. The eggs are more or less elongated. They range from 0.15 mm to 0.6 mm in length and the ratio length/width vary from 1.5 in Telmatogetoninae to about 4.5 in some Tanypodinae. The eggs of most species show an inconspicuous porosity, although some species have a distinct micropyle. The colour of chorion often undergoes a marked change during egg development. The eggs of most species are laid in rows within two layers of gelatine. The clarity of egg arrangement varies among taxa and with the consistency of the gelatine. The typical structure is a line of eggs arranged like a series of horse-shoes alternating from left to right through the mass, which resembles a pseudo-spiral. More details on chironomid eggs and egg masses are given by Nolte (1993).

Immature stages

Larvae. Fully grown 4th instar larvae range in length from 2 mm to about 30 mm. The body consists of head, thorax and abdomen. The colour may be white, yellow, green, blue, violet, rose, red, or brown, but often changes drastically after preservation.

The head is eucephalic with a completely sclerotized capsule; it is not retractile into the thorax. The colour ranges from pale yellow, to brown or black. Great variation in the structure of the head is found throughout the family and is used taxonomically. One to three eye spots are placed laterally (Figs 1A–C, 2, 3A–B). The antennae consist of 4–8 segments and may be short (eventually reduced) or longer than the head capsule (Fig. 2). They are retractile into the head capsule only in Tanypodinae (Fig. 2A). In the other subfamilies the antenna are placed directly on the head or on shorter or longer pedestals (Fig. 2D). The mandibles have several teeth (non-predators) or only one sharp terminal tooth (predators). The mentum is well-developed with sclerotized teeth in most chironomids (Figs 2B–D) except in the subfamily Tanypodinae, where ligula has replaced the function of mentum (Fig. 2A).

The thorax consists of the first three segments behind the head. First segment usually bears a pair of parapods (prolegs, Figs 1A–C), which may be partly fused. Prior to pupation the thorax becomes swollen and clearly demarcated from the abdomen.

The abdomen consists of nine segments. The posterior segment bears a pair of parapods (prolegs), 4 anal tubules (anal papillae), and a pair of procercus with 2 lateral setae and a bunch of longer apical setae (Figs 1A–B). The appendages of the anal segment may be more or less reduced in terrestrial and semi-terrestrial species (Fig. 1C). In a few Chironomini genera (in Greenland only *Chironomus*) some species have 2 pairs of ventral tubuli on abdominal segment 8 and in addition a short pair of lateral tubuli may be present at abdominal segment 7 (Figs 1B, 59D–F). Body setae are usually not conspicuous, but a number of Tanypodinae tribes have a distinct lateral fringe of swim-setae, and Tanytarsini have a characteristic, but often not easily seen, bifurcate seta lateral on most abdominal segments. Some Orthocladiinae genera (*e.g. Cricotopus, Tvetenia*) may also possess characteristically body setae (Figs 23D, 58E).

Pupae. The pupae range in length from about 1.5 mm to 25 mm. When the adult ecdyses, the cephalothorax splits along a Y-shaped mid-dorsal suture.

The head shield region consists of eye, antenna, and mouthpart sheaths plus the frontal apotome. The frontal apotome is usually with frontal setae, which may be placed on cephalic tubercles (Figs 60H, 62H, 63E).

The thorax bears the legs, halters and wing sheaths. The latter may bear on their dorsal margin near the apex a tubercle (the 'nose', Fig. 49G), and on the outer surface bordering its apex a row of small tubercles or points (the 'pearl row', Figs 22G, 28H). Anteriorly on each side of the thorax is the 'thoracic horn', a breeding organ (Fig. 1D). In most Podonominae, Telmatogetoninae, and Tanypodinae it is a double-walled tube with a smaller or larger plastron plate placed (sub)apically (Figs 6D, 10E–F). In the remaining chironomids the thoracic horn is a closed, hollow filament, without any obvious internal structure. Except in the Chironomini it is an unbranched tube, externally smooth, toothed, scaled or bearing setae. In the Chironomini it is branched at least once, more usually it takes the form of a dense tuft. The thoracic horn may be absent especially in semi-terrestrial and terrestrial species. However, the thoracic horn may break off quite easily. Some Tanypodinae bear a row of conical tubercles, the thoracic comb, near the thoracic horn (Fig. 8G). The thorax usually has a number of taxonomically significant groups of setae (*e.g.* Figs 19F, 54H).

The abdomen consists of eight similar segments and one visible additional segment modified of three terminal segments into the anal lobes and genital sheaths (Fig. 1D). Tergites 2–8 often bear distinctive groupings of large spines, recurved hooks, or shagreen spinules (Figs 3D–M). The posterolateral corners on segment 8 are often armed with points or teeth and these may form a contiguous comb, which may surmount a cuticular projection or spur

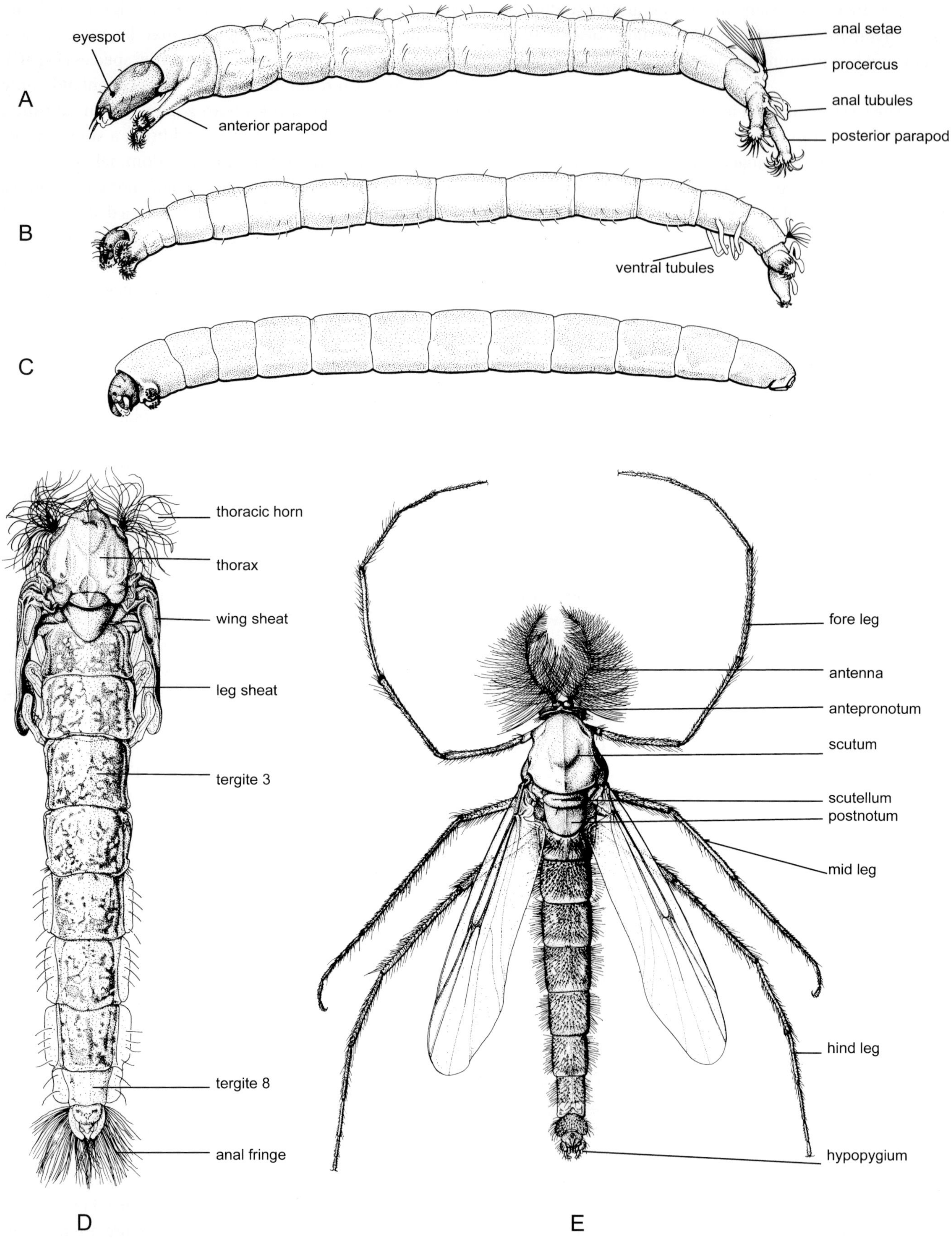

FIGURE 1 *Chironomidae larvae, pupa and adult male. A–C. Lateral view of larvae. A: Tanypodinae, B: Chironominae* (Chironomus *sp.*)*, C: Orthocladiinae* (*terrestrial genus*)*. C: Dorsal view of pupa* (Chironomus *sp.*)*, E: Dorsal view of adult male* (Chironomus *sp.*)*. A–C from Oliver* et al. (*1978*)*; D–E from Lindegaard* (*1997*)*.*

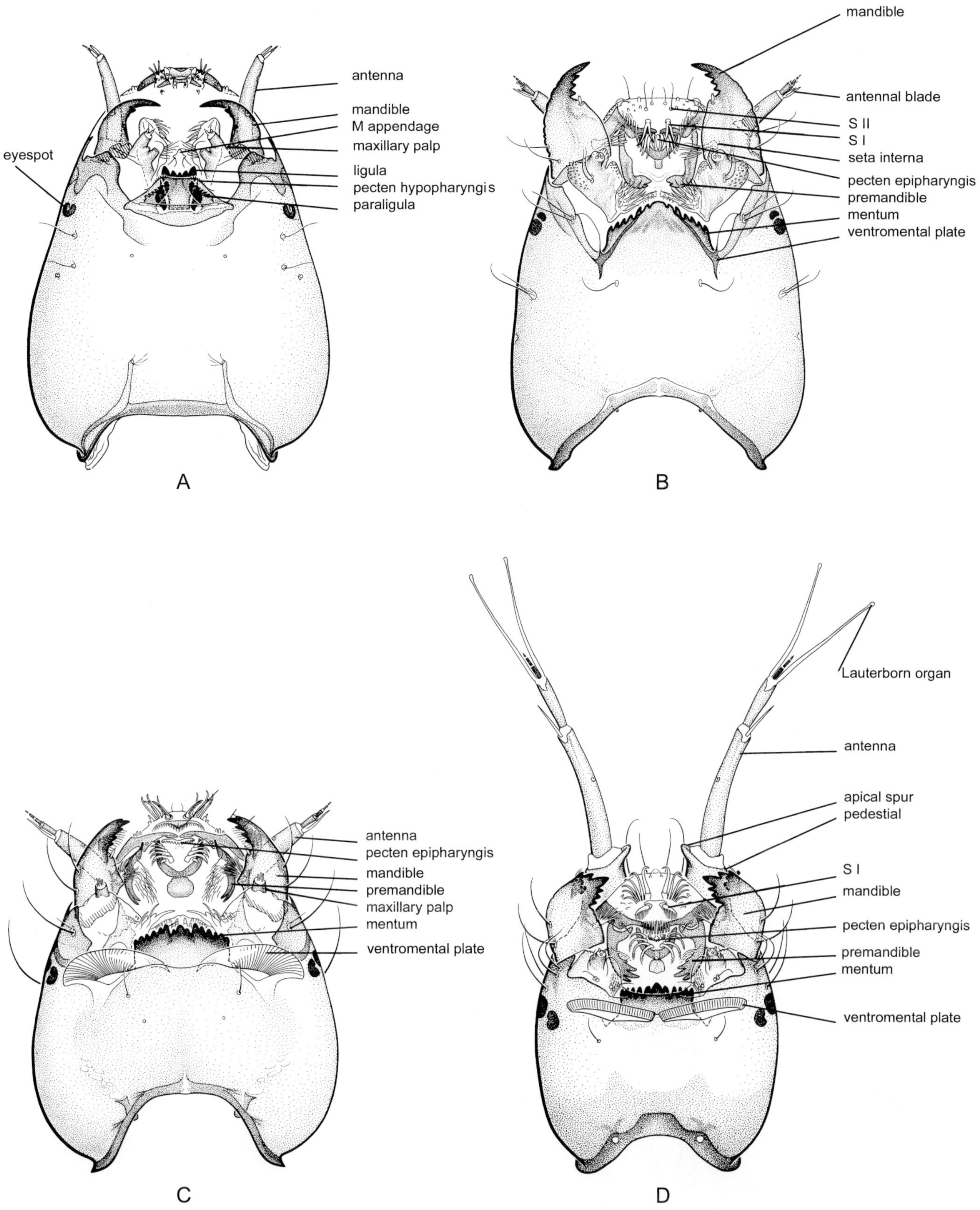

FIGURE 2 *Chironomidae larvae. A–D: Ventral view of larval heads. A: Tanypodinae, B: Orthocladiinae, C: Chironomini, D: Tanytarsini. A–D from Oliver* et al. *(1978).*

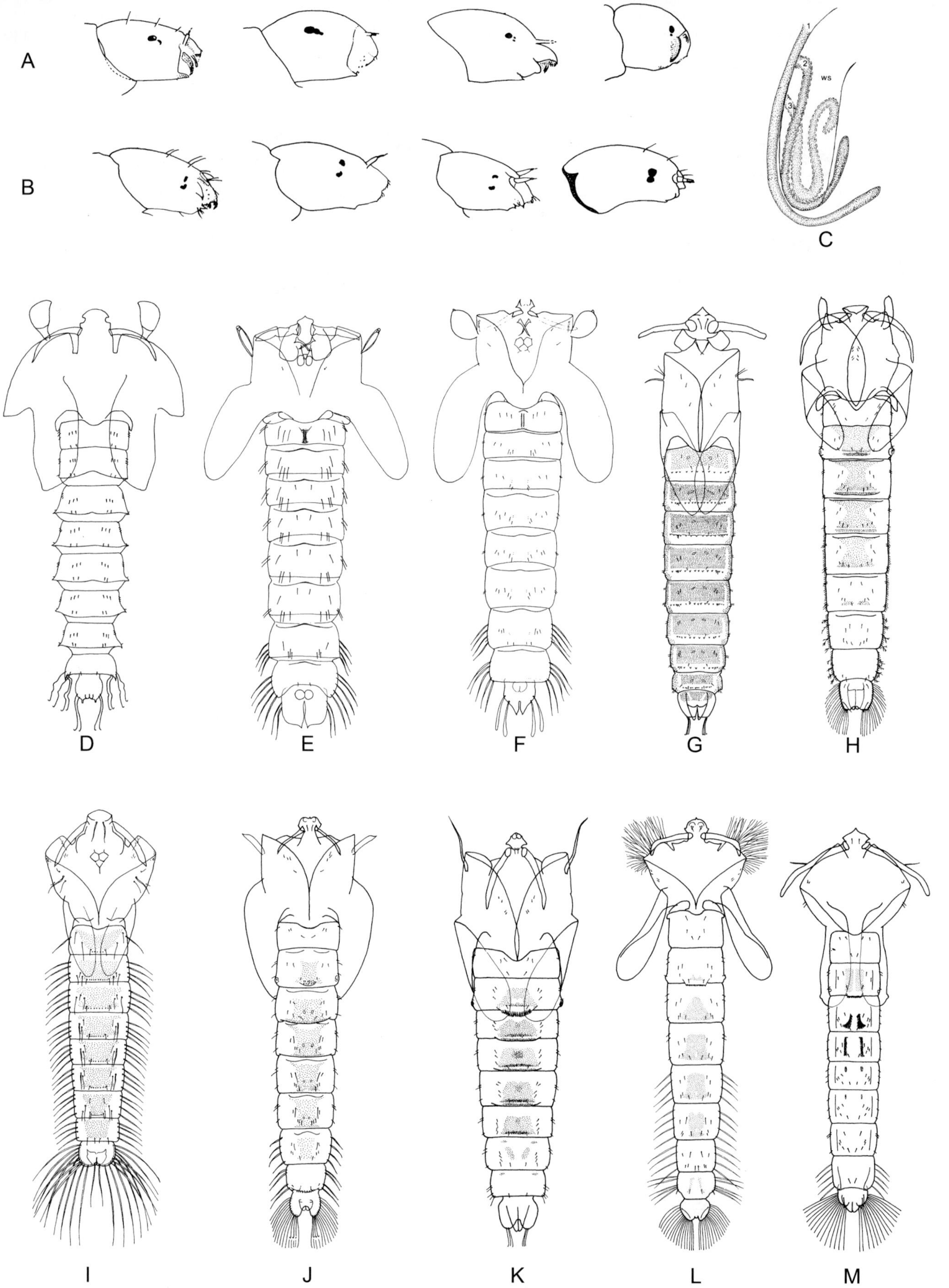

FIGURE 3 *Chironomidae larvae and pupae. A–B: Lateral view of larval head capsules showing the position of eyespots. A: Orthocladiinae, B: Chironomini. C: Lateral view of leg sheaths recurved under wing sheaths. D–M: Dorsal view of pupal exuvia. D: Podonominae* (Parochlus), *E: Tanypodinae* (Procladius), *F: Tanypodinae* (Ablabesmyia), *G: Diamesinae* (Diamesa), *H: Prodiamesinae* (Monodiamesa), *I: Orthocladiinae* (Corynoneura), *J: Orthocladiinae* (Heterotrissocladius), *K: Orthocladiinae* (Orthocladius), *L: Chironomini* (Dicrotendipes), *M: Tanytarsini* (Tanytarsus). *A–B from Lindegaard (1997); C from Coffmann (1978); D–M from Langton (1991).*

(Figs 59K–J, 69G, K–L). The caudal lobes often bear a fringe of long hairs, and/or one or more stout spine-like to hair-like setae (macrosetae, *e.g.* Figs 18D–E). These are frequently hooked apically. The genital plates are swellings beneath the anal segment which may extend well beyond the posterior margin of the anal lobes.

BIOLOGY

Life cycle. All species of Chironomidae have four larval instars. Length and width of head capsule almost doubles between instars. The larvae of most species build non-movable tubes or loose silk-web retreats, but Tanypodinae, Diamesinae and probably a few Orthocladiinae live freely borrowing in soft substrate or cling to vegetation and stones. Pupation takes place within the larval tube or retreat. In most chironomids the duration of the pupal stage is very short, normally a few days, but in arctic lakes pupae can survive winter and emergence takes place through cracks in the ice in early spring. Before emergence the pharate adult produces gases, which are held back within the pupal skin. A combined swimming activity and presence of gases allow the pupa to ascend very quickly to the water surface, where ecdysis takes place. During the process of ascending to surface and ecdysis the pupae are very vulnerable to predation by fish, and chironomid pupae are in fact the main food resource for stickleback and landlocked arctic char populations in Greenland.

Habitats. Many chironomid species are opportunists and may adapt to live in different biotopes and under different ecological conditions. These species are widely distributed and may be very abundant. They grow fast when temperature and food are optimal, but are able to slow down growth rate if the environmental factors are unfavourable. Common to arctic chironomids are their ability to tolerate low temperatures and exploit low quality food. Therefore, the life cycle may be extended to several years compared to warmer environments where the same species may have one or several generations per year. However, large differences in temperature regime occur in Greenlandic freshwater habitats. In South Greenland the summer period is relatively long and temperatures in shallow ponds may rice to 20°C or more. In such habitats growth is rapid and one or more generations a year are possible. Opposite to these warm habitats are the large and deep lakes and ponds in North Greenland, where temperature hardly exceeds 4°C. Development here may take several years. Thus, Butler (1982) found that *Chironomus* spp. in arctic ponds in Alaska took seven years from egg to adult. In North Greenland the thickness of ice cover may grow up to 2 m and many ponds and shallow parts of lakes may freeze to the bottom. Several chironomids rebuild during early winter their tubes into over wintering cocoons. Inside these the larvae are tightly folded; the larvae dehydrate which often result in a shrinking and their metabolism decreases drastically. In this way they can survive several months of freezing. Alternatively, the larvae penetrate deeper into the sediment to avoid freezing (Danks 1971).

Chironomidae is a very successful aquatic insect group and may be found in all types of freshwater biotopes. A fairly large part of the Orthocladiinae is semi-terrestrial living in moist habitats close to permanent water bodies (*e.g. Bryophaenocladius*, *Limnophyes*, *Metriocnemus* and *Smittia*); a few species live exclusively in marine environments (*e.g. Halocladius*).

Respiration. Chironomid larvae are apneustic, which means that the tracheal system is closed and all larvae depend on an oxygen transport from the surrounding water through the body surface to the blood. Most chironomids are not able to regulate there oxygen uptake, but depend fully on high oxygen contents in their habitats to secure the necessary gradient for maintaining a high diffusion rate of oxygen. However, most Chironomini posses haemoglobin in the blood and are therefore able to maintain a constant oxygen uptake with decreasing ambient oxygen concentration. Consequently, Chironominae may dominate the chironomid communities in biotopes with fluctuating oxygen concentrations. Members of the genus *Chironomus* are especially suitable to survive periods with low oxygen supply and are dominant inhabitants of organically polluted rivers and the profundal zone of eutrophic lakes. *Chironomus anthracinus* is characteristic of the profundal zone of natural eutrophic lakes in the temperate areas of the Northern Hemisphere and are known to survive several months under apparently anoxic conditions.

Trophic relations. In a large family as the Chironomidae all types of feeding behaviour and food preferences are represented, but the dominant types are algae grazers, detrital collectors and predators. The Tanypodinae larvae are generally carnivorous although the 1st and 2nd instar larvae feed also on algae. The mouthparts of Tanypodinae are adapted for predation on small invertebrates including chironomid larvae. The head capsule and mouthparts are not heavily sclerotized as in most other chironomid larvae, allowing of some elasticity for swallowing the prey whole. The mandibles are falciform and act like pincers to seize the prey and pull it into contact with a sclerotized toothed ligula. Muscles then cause the ligula to rotate thereby pushing the prey into the pharynx. The head capsule and mouthparts of the other subfamilies are primarily adapted for scraping algae or detritus off surfaces. The mentum is developed as a toothed plate and may function as a scraper; the mandibles move obliquely and are adapted for biting across the mental plate pushing the food into the mouth. The Diamesinae and many Orthocladiinae primarily feed on algae, while some Orthocladiinae and most Chironominae eat detrital material.

Members of Chironomidae are abundant in most freshwater habitats and in temperate areas larval densities of 50.000 individuals per m^2 may often be recorded. Under special conditions even higher densities have been reported. Although many chironomids are small, the high densities combined with relatively high turn over rates make them extremely important as secondary producers and food for higher trophic levels. Chironomids contribute from 18% to 96% (average 60%) of total benthic production in lakes and from 2% to 81% (average 22%) in streams Lindegaard (1989). The common feature for ecosystems where chironomid dominated was some sort of stress factor, resulting in very high primary production/allochthonous input of organic matter and often combined with special physical and/or chemical factors. Such factors may be eutrophication of lakes with nutrients or pollution of streams with sewage resulting in oxygen depletion. In stable ecosystems without any special factor the production of chironomids rarely exceeds 25% of total fauna. An exception is Greenlandic biotopes where the absence of most other invertebrate groups results in an overall dominance by chironomids. Thus, chironomids made up 96% of the benthic secondary production in the oligotrophic Lake 95 in the Narsaq area (Lindegaard 1989, Lindegaard & Mæhl 1993).

The typical succession of chironomid subfamilies/tribes along a river continuum found in temperate rivers (Lindegaard & Brodersen 1995) is in principle also to be found in Greenlandic running waters. However, in Greenland only the upper part of the river continuum is present and is dominated by Diamesinae and Orthocladiinae. Tanytarsini (*Micropsectra*) may be present, but Chironominae are almost absent except in outlet from lakes.

The Greenland fauna

The lakes in Greenland and their fauna are not well studied, but the chironomid communities follow the same pattern as in temperate lakes. The littoral zone possesses large and complex communities, whereas species richness and diversity decrease with depth. The factors determining chironomid assemblages in the profundal zone of regularly stratified lakes are an interaction primarily between the quality and quantity of food and the oxygen conditions. The available food originates exclusively from precipitating organic matter produced in the photic zone, except in relatively shallow oligotrophic lakes, which also may have a profundal production of phytobenthos. The chironomids characteristic of oligotrophic lakes are often free living and thus able to increase their range of foraging to seek scarce food at this end of the trophic spectrum. Food is more abundant in eutrophic lakes and the majority of chironomids live in fixed tubes from which they feed by browsing around and/or by filtration within the tube. The larval tubes allow respiration movements and facilitate compensation for the microstratification of oxygen occurring in these lakes in the sediment/water interface. Moreover these larvae can – due to the presence of haemoglobin – maintain a high oxygen uptake with decreasing ambient oxygen concentration.

The larger and deeper lakes in Greenland are all cold and poor in nutrient and are classified as oligotrophic. Ponds and shallower lakes in vegetated areas are warmer during summer and may receive a significant amount of nutrient from the surroundings. These waterbodies may be classified as mesotrophic or eutrophic. The glacial lakes immediately connected to the glaciers having an extremely low primary production and temperatures hardly exceeding 4°C are typically ultraoligotrophic. Characteristic chironomid communities are associated with these various types of lakes (Sæther 1979). In Greenland the chironomid fauna in glacial lakes is characterized by species belonging to *Pseudodiamesa*, *Oliveridia*, *Heterotrissocladius*, *Hydrobaenus* and to a lesser degree *Micropsectra*. The same species may be found in oligotrophic lakes in some distance from the glaciers, but the littoral communities in these lakes are much more diverse, characterized also by a number of Chironominae (*Paracladopelma* and *Polypedilum*). Further, in sheltered areas with large sedimentation it may be possible to find *Chironomus hyperboreus*. The shallow lakes and ponds in South and West Greenland are eutrophic, characterized by large populations of chironomids. The most prominent are *Ablabesmyia*, *Corynoneura*, *Psectrocladius*, *Chironomus*, *Dicrotendipes*, *Paratanytarsus* and *Tanytarsus*. In dry areas the evaporation is so high that some lakes accumulate different salts in the bottom water resulting in formation of a chemocline. These low-saline lakes are more or less permanently stratified and hypolimnion consequently without oxygen and the profundal zone below the chemocline without chironomids. However, the chironomids may be very abundant above the chemocline and often dominated by *Tanytarsus gracilentus*.

In spite of the Chironomidae being the absolutely dominant group in Greenlandic freshwaters, the knowledge of species richness and distribution is very restricted. A few collections from various expeditions 50–100 years ago (references in Henriksen 1939) are together with Søgaard

Andersen's research in Ella Ø, Northeast Greenland (Andersen 1937) our main sources for chironomid faunistics in Greenland. However, a number of species are described on specimens from these collections by various taxonomists at that time. Some of these early descriptions are subsequently revised, often by including specimens from other parts of the world. The nomenclature of Chironomidae has developed since the survey of Henriksen (1939) and the current status of his list is shown in Table 1. A large chironomid material from North Greenland was collected by the Canadian Peary Land expeditions during the 1960s. Unfortunately, this material has never been worked out, but various taxonomists have subsequently included material from it in revisions of some genera. In connection with ecological projects I have made collections of larvae and pupae in lakes and rivers in the Narsaq (1975) and Kangerlussuaq area (2003). Part of this material has been identified to genus or species level and is included in the current checklist (Table 1), which otherwise is based on a review of the existing literature. A large material of adult chironomids taken in window traps at the Zackenberg Station (NE Greenland) is deposited in the Zoological Institute, Århus University, but this material has not yet been identified.

The problems of taxonomy also greatly influence the quality of faunistic works, and the presented checklist from Greenland may be considered very preliminary. The total number of approximately 125 species recorded from Greenland (Table 1) is low and in spite that considerably more species may be expected in Greenland, the final number probably will be much lower than the 500–800 species recorded from the same latitudes in Fennoscandia and Canada. However, it is somewhat higher than the 80 species recorded in Iceland (Hfransdottir, 2005). The low number of species in Greenland may be explained by the main immigrations route from Ellesmere Island to North Greenland restricting the possible sources to be species tolerating the high arctic climate.

NOTES ON IDENTIFICATION. The family Chironomidae comprises several thousand species. They are divided into 11 subfamilies, but only 6 are recorded in Greenland. The identification of Chironomidae has been hampered by a very large amount of inadequate descriptions from the first half of the 19th century, but recently critical revisions have successively decreased the number of species (Ashe & Cranston 1990, Oliver *et al.* 1990).

Due to the combined efforts of many taxonomists, identification of larval, pupal and adult male chironomid

TABLE 1 *Chironomidae recorded from Greenland with references newer than Henriksen (1939).*

Name	References
Podonominae	
Parochlus kiefferi (Garrett, 1925)	Henriksen 1939, no. 64, 66, Brundin 1966, Oliver *et al.* 1990
Trichotanypus posticalis (Lundbeck, 1898)	Henriksen 1939, no. 65, Brundin 1966, Oliver *et al.* 1990
Tanypodinae	
Ablabesmyia pulchripennis Lundbeck, 1898	Henriksen 1939, no. 62, Roback 1965, 1985, Oliver *et al.* 1990
Arctopelopia melanosoma (Goethghebuer, 1933)	Henriksen 1939, no. 61, Oliver *et al.* 1990
**Procladius crassinervis* Zetterstedt, 1838	Henriksen 1939, no. 58, 59, Roback 1966
**Procladius culiciformis* Limnaeus, 1767	Roback 1971
Procladius paragretis Roback, 1971	Oliver *et al.* 1990
Psectrotanypus pictipennis Zetterstedt, 1838	Henriksen 1939, no. 63, Roback 1965, Oliver *et al.* 1990 Oliver & Dillon 1997
Diamesinae	
Diamesa aberrata Lundbeck, 1898	Henriksen 1939, no. 138, Hansen & Cook 1976, Oliver *et al.* 1990
Diamesa bertrami Edwards, 1935	Henriksen 1939, no. 137, Hansen & Cook 1976, Oliver *et al.* 1990
Diamesa bohemani Goethghebuer, 1932	Hansen & Cook 1976, Oliver *et al.* 1990
Diamesa chorea Lundbeck, 1898	Henriksen 1939, no. 140, Hansen & Cook 1976, Oliver *et al.* 1990
Diamesa geminata Kieffer, 1926	Henriksen 1939, no. 135, Hansen & Cook 1976, Sæther *et al.* 1984, Oliver *et al.* 1990

TABLE 1 *Chironomidae recorded from Greenland with references newer than Henriksen (1939)* (cont.)

Name	References
Diamesa lindrothi Goethghebuer, 1931	Henriksen 1939, no. 139, Hansen & Cook 1976, Oliver *et al.* 1990,
Diamesa simplex Kieffer, 1926	Hansen & Cook 1976, Sæther *et al.* 1984, Oliver *et al.* 1990
**Diamesa davisi* Edwards 1933	Thienemann 1941, Oliver *et al.* 1990 questionable record by Serra-Tosio 1972 and Hansen & Cook 1976
**Diamesa hyperborea* Holmgren, 1869	Henriksen 1939, no. 142, not in Greenland Oliver *et al.* 1990
Pseudodiamesa branickii (Nowicki, 1873)	Henriksen 1939, no. 134, ?Pagast 1947, Oliver 1976, Ashe & Cranston 1990, Oliver *et al.* 1990
"*Pseudokiefferiella*" undescribed species, Oliver 1986	New to Greenland
Prodiamesinae	
Monodiamesa bathyphila (Kieffer, 1918)	Oliver *et al.* 1990
Prodiamesa olivacea (Meigen, 1818)	Ashe & Cranston 1990
Orthocladiinae	
**Bryophaenocladius vernalis* (Goetghebuer, 1921)	Henriksen 1939, no. 116
**Camptocladius stercorarius* De Geer, 1781	Henriksen, 1939, no. 129, Strenzke 1940
Chaetocladius adsimilis Goetghebuer, 1933	Henriksen 1939, no. 99, Oliver *et al.* 1990
Corynoneura arctica Kieffer, 1923	Oliver & Dillon 1997, description Hirvenoja 1988
Corynoneura celeripes Winnertz, 1852	Oliver *et al.* 1990
Corynoneura scutellata Winnertz, 1846	Henriksen 1939, no. 133, Ashe & Cranston 1990, Oliver *et al.* 1990
Cricotopus (*Cricotopus*) *cumulatus* Hirvenoja, 1973	Oliver & Dillon 1988, Oliver *et al.* 1990
Cricotopus (*Cricotopus*) *gelidus* (Kieffer, 1922)	Oliver *et al.* 1990
**Cricotopus* (*Cricotopus*) *glacialis* misidentification = *laricomalis*	Henriksen, 1939, no. 91
Cricotopus (*Cricotopus*) *lestralis* (Edwards, 1924)	Hirvenoja 1973, Oliver & Dillon 1988, 1997, Oliver *et al.* 1990
Cricotopus (*Cricotopus*) *pilosellus* Brundin, 1956	Oliver & Dillon 1997, Oliver *et al.* 1990
Cricotopus (*Cricotopus*) *tibialis* (Meigen, 1804) (= *polaris* Kieffer, Oliver *et al.* 1990)	Henriksen 1939, no. 90, 93, 94, Hirvenoja 1973, Oliver & Dillon 1988, 1997, Oliver *et al.* 1990, Sæther *et al.* 1984
Cricotopus (*Isocladius*) *laricomalis* Edwards, 1932	Hirvenoja 1973, Oliver *et al.* 1990
Diplocladius cultriger Kieffer, 1908	Oliver *et al.* 1990
Eukiefferiella claripennis (Lundbeck, 1898)	Henriksen 1939, no. 113, Oliver 1970, Oliver *et al.* 1990
Halocladius variabilis (Stæger, 1839)	Henriksen 1939, no. 92, Hirvenoja 1973, Oliver *et al.* 1990
Heterotrissocladius changi Sæther, 1975	
Heterotrissocladius oliveri Sæther, 1975	Henriksen 1939, no. 106
Heterotrissocladius marcidus group	
Hydrobaenus conformis (Holmgren, 1869)	Sæther 1976, Oliver *et al.* 1990
Hydrobaenus fusistylus Goetghebuer, 1933	Henriksen 1939, no. 100, Sæther 1976, Oliver *et al.* 1990
Krenosmittia sp.	New to Greenland
Limnophyes anderseni Sæther, 1990	Sæther 1990
Limnophyes asquamatus Andersen, 1937 (= *hamiltoni* Sæther, 1969)	Cranston & Oliver 1988, Sæther 1990, Oliver *et al.* 1990
**Limnophyes atomarius* Zetterstedt, 1850	Henriksen 1939, no. 117, Oliver *et al.* 1990 nomen dubium (= *Corynoneura atomaria*), Ashe & Cranston 1990
Limnophyes brachytomus (Kieffer), 1922 (= *borealis* var. *squamatus* Andersen, 1937)	Henriksen 1939, no. 120, Cranston & Oliver 1988, Sæther 1990, Oliver *et al.* 1990
Limnophyes eltoni Edwards, 1922	Henriksen 1939, no. 119, Cranston & Oliver 1988, Sæther 1990, Oliver *et al.* 1990

Name	References
Limnophyes minimus (Meigen, 1818) (= *groenlandicus* Andersen, 1937)	Henriksen 1939, no. 121, Sæther 1990
Limnophyes pentaplastus (Kieffer, 1921) (= *prolongatus* (Kieffer, 1921)	Cranston & Oliver 1988, Sæther 1990, Oliver *et al.* 1990
Limnophyes pumilio (Holmgren, 1869) (= *globifer* Lundström, 1915)	Henriksen 1939, no. 118, 121, Cranston 1979, Sæther 1990, Oliver *et al.* 1990
Mesocricotopus sp. cf. *thienemanni* (Goetghebuer, 1940)	New to Greenland
Metriocnemus albolineatus (Meigen, 1818)	Henriksen 1939, no. 84, Oliver *et al.* 1990
Metriocnemus eurynotus (Holmgren, 1883)	Henriksen 1939, no. 85, Ashe & Cranston 1990, Oliver *et al.* 1990
Metriocnemus fuscipes (Meigen, 1818)	Henriksen 1939, no. 86, Pagast *et al.* 1941, Ashe & Cranston 1990, Oliver *et al.* 1990
Metriocnemus picipes (Meigen, 1818)	Sæther 1995, Ashe & Cranston 1990, Oliver *et al.* 1990
Metriocnemus polaris Kieffer, 1926 (=? *eurynotus* Holmgren)	Sæther *et al.* 1984, Oliver *et al.* 1990
Metriocnemus tristellus Edwards, 1929	Ashe & Cranston (1990), Oliver *et al.* 1990
Metriocnemus ursinus (Holmgren, 1869)	Henriksen 1939, no. 82, Oliver *et al.* 1990
Oliveridia tricornis (Oliver, 1976)	Oliver (1976), Sæther 1976, Oliver *et al.* 1990
Orthocladius (*Eudactylocladius*) *gelidorum* (Kieffer, 1923), (= *nanseni* Kieffer, 1926)	Sæther *et al.* 1984, Cranston 1998
Orthocladius (*Eudactylocladius*) *gelidus* Kieffer, 1922	Andersen 1937, Henriksen 1939, no. 110, Brundin 1956, Cranston 1984, Oliver *et al.* 1990
Orthocladius (*Eudactylocladius*) *priomixtus* (Holmgren, 1869)	Henriksen, 1939, no. 104, Sæther 2004
Orthocladius (*Eudactylocladius*) *subletteorum* Cranston 1998 (= *Orthocladius* (*Eudactylocladius*) *mixtus* (Holmgren, 1869), misidentifications)	Henriksen 1937, no. 108, Oliver *et al.* 1990, Oliver & Dillon 1997 Cranston 1998
Orthocladius (*Eudactylocladius*) sp. cf. *dubitatus* Johannsen, 1942	New to Greenland
Orthocladius (*Eudactylocladius*) sp. cf. *olivaceus* Kieffer, 1911	New to Greenland
Orthocladius (*Euorthocladius*) *difficilis* Lundbeck, 1898	Henriksen 1939, no. 107, Oliver 1970, Oliver *et al.* 1990, Soponis 1990
Orthocladius (*Euorthocladius*) *rivicola* Kieffer, 1911	Soponis 1990, Oliver *et al.* 1990
Orthocladius (*Euorthocladius*) *thienemanni* Kieffer, 1906	Soponis 1990, Oliver *et al.* 1990
Orthocladius (*Mesorthocladius*) *frigidus* (Zetterstedt, 1838)	Henriksen 1939, no. 109, Soponis 1987, Oliver *et al.* 1990
Orthocladius (*Mesorthocladius*) *rousellae* Soponis, 1990	Soponis 1990
Orthocladius (*Mesorthocladius*) *vaillanti* Langton & Cranston, 1991	New to Greenland
Orthocladius (*Orthocladius*) *charensis* Soponis, 1977	Soponis 1977, Oliver *et al.* 1990
Orthocladius (*Orthocladius*) *decoratus* (Holmgren, 1869)	Henriksen 1939, no. 114, Ashe & Cranston 1990, Oliver *et al.* 1990
Orthocladius (*Orthocladius*) *groenlandensis* Goetghebuer, 1933	Henriksen 1939, no. 102, Soponis 1977, Oliver *et al.* 1990
Orthocladius (*Orthocladius*) *knabeni* Goetghebuer, 1933	Henriksen 1939, no. 101, Soponis 1977, Oliver *et al.* 1990

TABLE 1 *Chironomidae recorded from Greenland with references newer than Henriksen (1939)* (cont.)

Name	References
Orthocladius (Orthocladius) knuthi Soponis, 1977	Soponis 1977, Oliver *et al.* 1990
Orthocladius (Orthocladius) lapponicus Goetghebuer, 1940	Soponis 1977, Oliver *et al.* 1990
**Orthocladius (Orthocladius) rubicundus* (Meigen, 1818)	Henriksen 1939, no. 115
Orthocladius (Pogonocladius) consobrinus (Holmgren, 1869)	Henriksen 1939, no. 111, Pinder & Cranston 1976 (description), Oliver *et al.* 1990
Orthocladius spitzbergensis Kieffer, 1919	Henriksen 1939, no. 103, Oliver *et al.* 1990
Paracladius quadrinodosus Hirvenoja, 1973	Oliver 1976, Oliver *et al.* 1990
Parakiefferiella nigra Brundin, 1949	Sæther *et al.* 2000
Parametriocnemus graminicola (Lundbeck, 1898)	Henriksen 1939, no. 105, Oliver 1970, Oliver *et al.* 1990
Parametriocnemus lundbeckii Johannsen, 1905	Henriksen 1939, no. 88, Oliver *et al.* 1990
Paraphaenocladius impensus Walker, 1856) (= *debilipennis* Lundbeck, 1898)	Henriksen 1939, no. 89, Sæther & Wang 1995, Oliver *et al.* 1990
Paraphaenocladius brevinervis (Holmgren, 1869) (= *Metriocnemus polaris* Kieffer, 1926)	Sæther 1995
Psectrocladius barbatimanus Kieffer, 1926 (=? *barbimanus*)	Sæther *et al.* 1984, Oliver *et al.* 1990
Psectrocladius barminanus (Edwards, 1929)	Henriksen 1939, no. 97, Oliver *et al.* 1990
Psectrocladius fennicus Storå, 1939	Langton, 1985, Oliver *et al.* 1990
Psectrocladius limbatellus (Holmgren, 1869)	Henriksen 1939, no. 96, Langton 1985, Ashe & Cranston 1990, Oliver *et al.* 1990
Psectrocladius octomaculatus group	New to Greenland
**Psectrocladius sordidellus* (Zetterstedt, 1838)	Henriksen 1939, no. 98, (not Nearctic , Oliver *et al.* 1990)
Pseudosmittia nanseni Kieffer, 1926	Sæther *et al.* 1984, Cranston & Oliver 1988, Oliver *et al.* 1990
Pseudosmittia oxoniana (Edwards, 1922)	Andersen 1937, Ashe & Cranston 1990, Oliver *et al.* 1990
Rheocricotopus (Psilocricotopus) chapmani Edwards, 1935	Henriksen 1939, no. 95, Sæther 1985, Oliver *et al.* 1990
Rheocricotopus (Rheocricotopus) sp.	New to Greenland
Smittia aterrima (Meigen, 1818)	Oliver & Dillon 1997, Oliver *et al.* 1990
Smittia extrema (Holmgren, 1869)	Henriksen 1939, no. 127, Ashe & Cranston 190, Oliver *et al.* 1990
**Smittia flexinervis* (Kieffer, 1911)	Henriksen 1939, no. 132
Smittia parva Lundbeck, 1898	Henriksen 1939, no. 131, Oliver *et al.* 1990
Smittia polymorpha Andersen, 1937	Henriksen 1939, no. 126, Oliver *et al.* 1990
Smittia subnudipennis Goetghebuer, 1933	Henriksen 1939, no. 125, Oliver *et al.* 1990
Smittia velutina Lundbeck, 1898	Henriksen 1939, no. 130, Oliver *et al.* 1990
Thienemanniella sp.	New to Greenland
Tokunagaia obriaini Hayes & Murray, 1988	Oliver & Dillon 1997
Tokunagaia sp. cf. *rectangularis* (Goetghebuer, 1940)	New to Greenland
Tvetenia sp. cf. *bavarica* (Goetghebuer, 1934)	New to Greenland
Orthocladiinae genus A	New to Greenland
Chironominae	
Chironomini	
Chironomus hyperboreus Stæger, 1845	Henriksen 1939, no. 68, Wülker & Butler 1983, Oliver *et al.* 1990
Chironomus riparius Meigen, 1804	Henriksen 1939, no. 70, Oliver *et al.* 1990

Name	References
Chironomus staegeri Lundbeck, 1898	Henriksen 1939, no. 69, Oliver *et al.* 1990
Dicrotendipes modestus (Say, 1823)	Epler 1987, 1988, Oliver *et al.* 1990
**Dicrotendipes nervosus* (Stæger, 1839)	Henriksen 1939, no. 71, Oliver *et al.* 1990
**Endochironomus oldenbergi* Goetghebuer, 1932	Henriksen 1939, no. 67, Grodhaus 1987, Oliver *et al.* 1990
Parachironomus sp.	
Paracladopelma nigritula group	Lindegaard *et al.* 1978
Phaenopsectra incompta (Zetterstedt, 1838)	Henriksen 1939, no. 87, Oliver *et al.* 1990
Polypedilum sp. cf. *laetum* (Meigen, 1818)	New to Greenland
Sergentia coracina (Zetterstedt, 1850)	Henriksen 1939, no. 72, Oliver & Dillon 1997, Wülker *et al.* 1998, Oliver *et al.* 1990
Stictochironomus unguiculatus (Malloch, 1934)	Oliver *et al.* 1990, Oliver & Dillon 1997
Tanytarsini	
Constempellina brevicosta (Edwards, 1937)	Oliver & Dillon 1997
Corynocera oliveri Lindeberg, 1970	Brodersen & Bennike 2003
Micropsectra brundini Säwedal, 1979	Säwedal 1979, Oliver *et al.* 1990
Micropsectra logani (Johannsen, 1928) (= *groenlandica* Andersen, 1937)	Oliver & Dillon 1997, Oliver *et al.* 1990, Stur & Ekrem 2006
**Micropsectra insignilobus* Kieffer, 1924	Henriksen 1939, no. 74, Oliver *et al.* 1990
Micropsectra junci Meigen, 1818	Henriksen 1939, no. 75, 80, Oliver *et al.* 1990
Micropsectra lindrothi Goetghebuer, 1931	Lindegaard *et al.* 1979, Oliver *et al.* 1990
Micropsectra polita (Malloch, 1915)	New to Greenland
Micropsectra recurvata Goetghebuer, 1928	Henriksen 1939, no. 70, Lindegaard *et al.* 1978, Oliver *et al.* 1990
Micropsectra tori Säwedal, 1981	Säwedal 1981, Oliver *et al.* 1990
Paratanytarsus laccophilus (Edwards, 1929)	Oliver & Dillon 1997
Paratanytarsus natvigi (Goetghebuer, 1923)	Henriksen 1939, no. 77, Reiss & Säwedal 1981, Oliver *et al.* 1990
Paratanytarsus setosimanus (Goetghebuer, 1933)	Henriksen 1939, no. 78, Reiss & Säwedal 1981, Ashe & Cranston 1990, Oliver *et al.* 1990
Paratanytarsus tenuis (Meigen, 1830)	Henriksen 1939, no. 81, Reiss & Säwedal 1981, Ashe & Cranston 1990, Oliver *et al.* 1990
Tanytarsus andersini Reiss et Fittkau, 1971	Reiss & Fittkau 1971, Oliver *et al.* 1990
Tanytarsus gracilentus (Holmgren, 1883)	Lindeberg 1968, Oliver & Dillon 1997, Brodersen *et al.* 2001
Tanytarsus gregarius Kieffer, 1909	New to Greenland
Tanytarsus niger Andersen, 1937	Henriksen 1939, no. 79, Oliver *et al.* 1990

* Indicates that the identity of the species and/or the record from Greenland is uncertain. Extensive lists of synonyms are given by Ashe & Cranston (1990) and Oliver *et al.* (1990).

genera from the Holarctic Region is now possible (Wiederholm 1983, 1986, 1989, Sæther *et al.* 2000). The adult female chironomids may be keyed to subfamily, tribe, and partly genus by Sæther (1977). A number of genera or species groups have been revised including both palaearctic and nearctic species and species identification may be possible. References are given in the notes to the genera in question. However, it must be stressed that identification to species level is extremely difficult and for many genera must await critical revisions. Further, in some genera a search for original descriptions (references in Ashe & Cranston 1990, Oliver *et al.* 1990) and comparisons with reference collections are necessary.

Key to subfamilies and Chironomine tribes for adult males

The following key is primarily based on wing nervation. However, a few species within most subfamilies have reduced wings without distinct nervation; none of these species are yet recorded from Greenland, but may be identified according to Wiederholm (1989).

1. Wing with crossvein M–Cu present (Figs 5A–C) ...2
– Wing with crossvein M–Cu absent (Figs 5D–G) 5
2. Wing vein R_{2+3} absent (Fig. 6G). Crossvein R–M situated close to the middle of wing. Gonostylus simple or with two large apical lobes, each with terminal to subterminal seta or megaseta (Figs 6F, 7F) ....Podonominae
– Wing vein R_{2+3} present (Fig. 5A) 3
3. Wing vein R_{2+3} forked (*e.g.* Figs 5B, 10H); if simple, wing membrane with setae ....Tanypodinae
– Wing vein R_{2+3} simple (Figs 5C, 12I); wing membrane usually bare, at most with a few setae on distal half 4
4. Wing with FCu usually proximal to crossvein M–Cu (Fig. 12I); if opposite or slightly distal, then wing shortened ....Diamesinae
– Wing with FCu opposite or distal to crossvein M–Cu (Figs 5C, 18F); wing not short ....Prodiamesinae
5. Antepronotal lobes widely separated. Postnotum without a median groove or keel. Antennal flagellum with 6 flagellomeres (Fig. 16E). Fore coxa enlarged; tarsomere 4 cordiform on all legs; tarsomere 5 sometimes trilobed. Marine species not yet recorded from Greenland .... [Telmatogetoninae]
– This combination of characters not present. Postnotum with a median keel or groove (Fig. 4E) except in the marine *Clunio* (not yet recorded from Greenland). Tarsomere 5 never trilobed 6
6. Gonostylus movable and usually folded inward toward the midline (Fig. 4B); tarsomere 1 always shorter than tibia on fore legs and hind tibial comb, if present, consisting of free spines (Fig. 5I) ....Orthocladiinae
– Gonostylus rigid, fused to the gonocoxite and directed backwards (Figs 4C, D); tarsomere 1 longer than tibia on fore leg. If gonostylus is slightly movable a really folding inwards is rare and hind tibial comb consists of fused spines (Figs 5J–N) .... Chironominae, 7
7. Wing membrane with macrotrichia, especially near apex; squama always without setae (Fig. 5G) .... Tanytarsini
– Wing membrane with or without macrotrichia; if macrosetae present then squama always with setae. Crossvein R–M oblique to long axis of the wing (Fig. 5F).. 8
8. Superior volsella transverse. Eyes widely separated; fore tibia with comb and long spur; hypopygium without anal point, but anal tergite with median, broadly conical, apically crenated, extension. This tribe is not yet recorded from Greenland .... [Pseudochironomini]
– Superior volsella longitudinal or oblique (Fig. 4C); fore tibia without comb, but eventually with spur (Figs 5K, L) .... Chironomini

Key to subfamilies and Chironominae tribes for 4th and 3rd instar larvae

This key is designed for 4th and 3rd instar larvae only. The different instars can be separated on the length and width of the head capsule that almost doubles between instars.

1. Antenna retractile into head capsule (Fig. 2A). Eye single, usually kidney-shaped (Fig. 2A). Prementum with ligula distinctively developed (Fig. 2A)......... ....Tanypodinae
– Antenna not retractile; placed directly on head capsule or on pedestals (Fig. 2B–D). Usually 2 (or 3) eyes on each side, eventually placed closely together (Figs 3A–B). Ligula normally weaker developed. Mentum nearly always strongly developed (Figs 2B–D) 2
2. Premandibles absent. Procercus 8–10 times as high as wide (Fig. 7B), antennal segment 3 annulated (Fig. 7C) .... Podonominae
– Premandibles present. Procercus not more than 4 times as high as wide, usually much less (*e.g.* Fig. 14B), or absent. Antennal segment 3 annulated or not.. 3
3. Antennal segment 3 annulated (Fig. 12C), except in *Protanypus* (not yet recorded form Greenland), in which head capsule has numerous closely-spaced short setae and postoccipital margin with a long, ventrolateral, posteriorly-directed projection on each side .... Diamesinae
– Antennal segment 3 never annulated (*e.g.* Fig. 21C). Head capsule without the above-mentioned characters 4
4. Ventral part of mentum expanded laterally to form large ventromental plates which are striated, but never beneath with hairs resembling a beard (Figs 2C, 2D). Eyes placed above each other, either

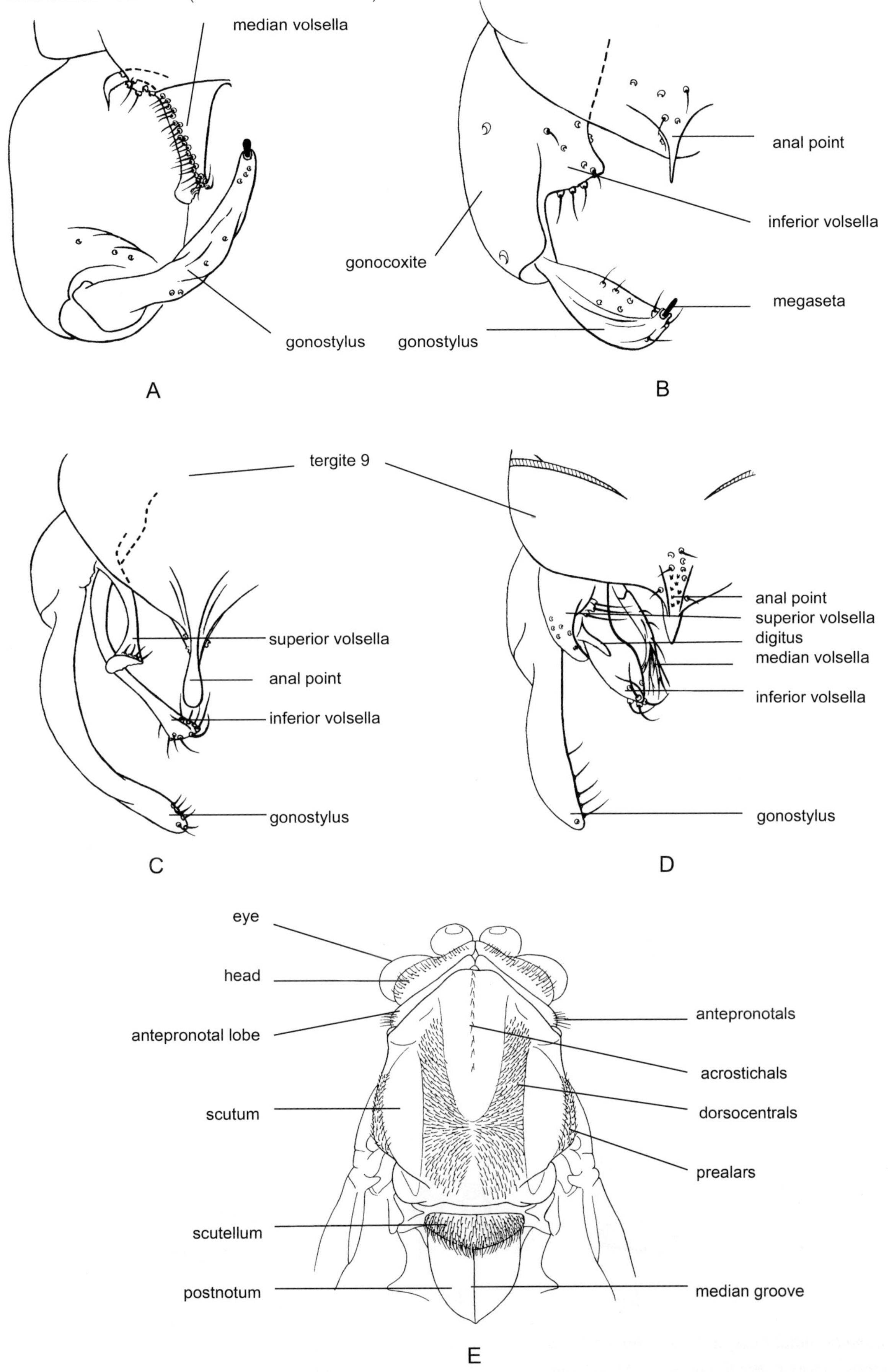

FIGURE 4 *Chironomidae adult males 1. A–D: Doral view of hypopygium. A: Tanypodinae, B: Orthocladiinae, C: Chironomini, D: Tanytarsini. E: Dorsal view of head and thorax of* Cricotopus pilosellus. *A–D from Pinder (1978); E from Brundin (1956).*

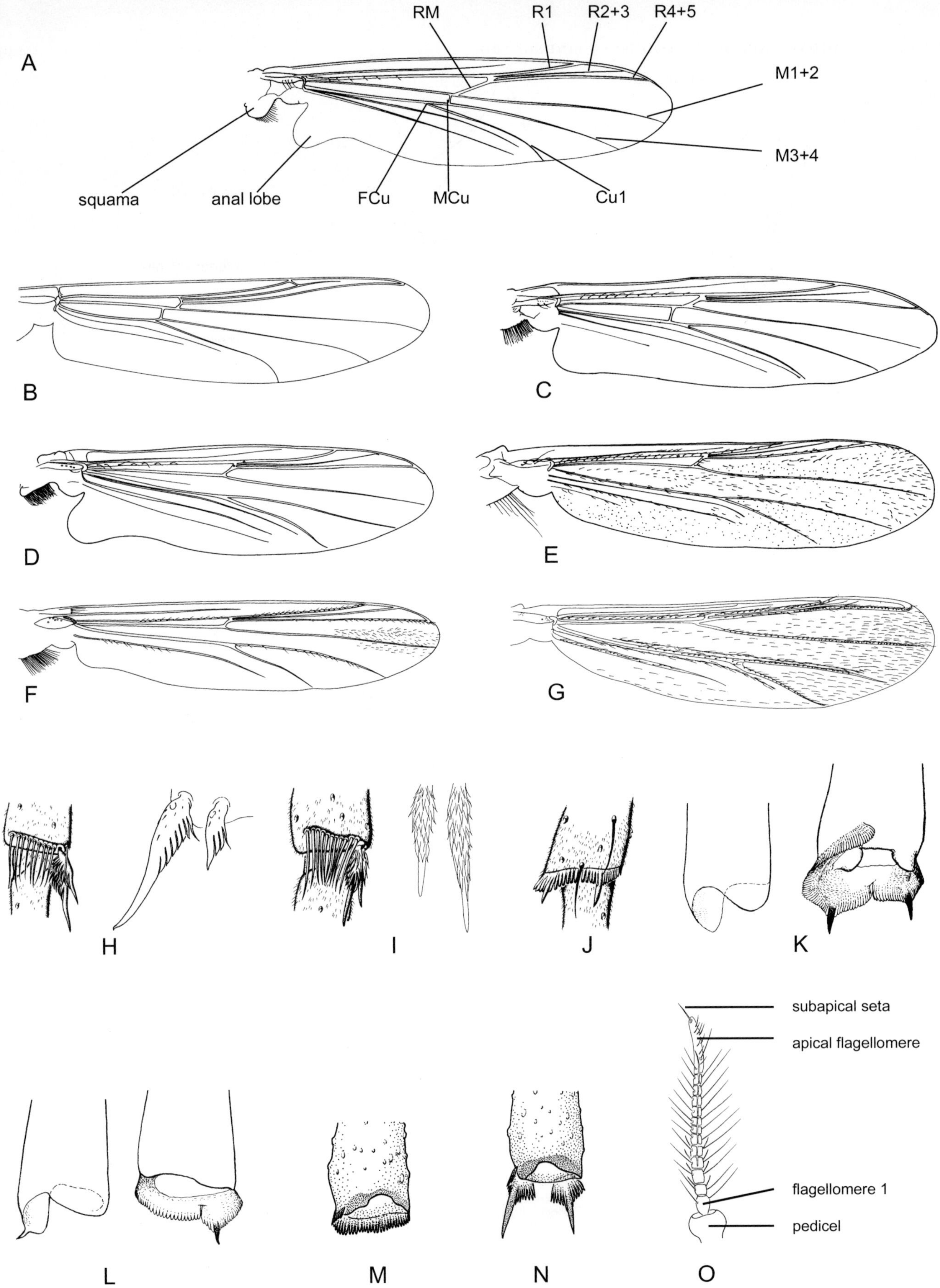

FIGURE 5 *Chironomidae adult males 2. A–G: Wings and wing venation. A: Tanypodinae* (Ablabesmyia), *B: Diamesinae* (Pseudodiamesa), *C: Prodiamesinae* (Monodiamesa), *D: Orthocladiinae* (Hydrobaenus), *E: Orthocladiinae* (Metriocnemus), *F: Chironomini* (Sergentia), *G: Tanytarsini* (Tanytarsus). *H–N: Tibial combs and spurs. H: Hind tibia of Tanypodinae* (Ablabesmyia), *I: Orthocladiinae* (Metriocnemus), *J: Chironominae. K–L: Fore and hind tibia of K:* Chironomus, *L:* Polypedilum. *M–N: Hind tibia of M:* Micropsectra, *N:* Tanytarsus. *O: Antenna. A–G from Wiederholm (1989); H–J from Pinder (1978), Murray & Fittkau (1989), Brundin (1956); K–L from Johannsen & Townes (1952); M–N from Sæther* et al. *(2000); O from Oliver & Dillon (1989).*

separated or closely together (Fig. 3B). Living larvae nearly always red ....Chironominae, 5

– Ventromental plates practically absent (Fig. 2B) or if present never with striation and often with a beard beneath (Fig. 18B). Eyes placed beside or obliquely above each other; often very closely together (Fig. 3A). Living larvae almost never red 7

5. Ventromental plates separated medially only by less than width of median mental tooth (Fig. 2 D). If ventromental plates are separated by the width of 3 median mental teeth, then antenna is placed on distinct pedestal bearing a prominent apical spur (species with this character are not yet recorded from Greenland) 6

– Ventromental plates widely separated medially (Fig. 2 C). Antenna shorter than half the length of head and placed directly on head or on pedestal shorter than wide (Fig. 2 C) ....Chironomini

6. Antennae often longer than half the length of head and always placed on distinct pedestals longer than wide (Fig. 2D) ....Tanytarsini

– Antennae shorter than half the length of head and placed on pedestals shorter than wide. This tribe is not yet recorded from Greenland .... [Pseudochironomini]

7. Ventromental plates strongly expanded, with beard beneath (Figs 17A, 18B). Antenna with 4 segments (Fig. 18C). Eyes often placed oblique above each other .... Prodiamesinae

– Ventromental plates smaller or if strongly expanded either beard absent or antenna not with 4 segments. Eyes beside each other and often close together; anterior eyespot often smaller than the posterior, resembling a comma (Fig. 3A) 8

8. Prementum with M appendage divided into many fine, often serrate, branches resembling a single brush (Fig. 16A). Premandible short and broad. Antenna short with 4 segments. Marine species not yet recorded from Greenland........................ .... [Telmatogetoninae]

– Prementum variably developed, but never with median brush. Premandible longer and slender. Antenna usually with 5–6 segments; when reduced segmentation indistinct .... Orthocladiinae

Key to subfamilies and Chironominae tribes for pupae

1. Thoracic horn with from 2 to more than 100 branches (Figs 1D, 64E) 2

– Thoracic horn absent or consisting of a single branch (*e.g.* Fig. 50H) 3

2. Thoracic horn with 2 short, plump, apically rounded branches. This tribe is not yet recorded from Greenland .... [Pseudochironomini]

– Thoracic horn with more than 2 slender, apically pointed branches (Figs 1D, 64E). If only 2 branches, then abdominal tergites 2–6 with distinct pair of rectangular fields of short points (*Lauterborniella*, not yet recorded from Greenland)................... .... Chironomini

3. Apparently only 8 abdominal segments, because segments 8 and 9 are developed into a subcircular terminal disc fringed with short hairs, but without lateral macrosetae (Fig. 16C). Marine species not yet recorded from Greenland .... [Telmatogetoninae]

– Abdomen with 9 segments. If anal lobes broadly rounded and fringed with short hairs, then never in combination with segment 8. Anal lobe normally with 2 lateral or 3–5 apical setae 4

4. Thoracic horn well developed with either a conspicuous terminal plastron (Figs 6D, 11E, F) or if plastron inconspicuous with a distinct reticulation on the surface (Figs 8G, H). If thoracic horn is small and without plastron, then anal lobes with 2 lamellar macrosetae near the midline of lateral margin (cf. Fig. 10F) 5

– Thoracic horn without plastron or distinct reticulation, or absent. Anal lobe never with 2 lamellar macrosetae 6

5. Lateral margin of anal lobes with 2 lamellar or stiff macrosetae (Figs 9F, 11G), with or without an additional setal fringe of fine hairs or spines ....Tanypodinae

– Anal lobes always terminally pointed or with terminal slender spurs. Anal lobes never with a fringe of setae, but always with more than 2 macrosetae, although these setae may be spine-like, minute hair-like or lamellar and waved (Figs 6E, 7D). Distal part of fore and mid leg sheaths practically straight...... ....Podonominae

6. Posterolateral margin of abdominal segment 8 always with at least one sclerotized spine, often with a comb of spines (anal comb, Fig. 70G). Anal lobes nearly always with a fringe of filamentous setae, but without macrosetae (Figs 68F, 73I, J). Tergites 3–4 usually with central patches of spines (Fig. 69G). Thoracic horn often with long hairs (Figs 68H–J) ... . Tanytarsini

– Posterolateral margin of abdominal segment 8 without sclerotized spine (except some *Zalutschia* spp. not yet recorded from Greenland) or comb of spines. Anal lobe nearly always with macrosetae and with or without a fringe of filamentous setae (Figs 3I–K). Thoracic horn if present without long hairs (*e.g.* Figs 43O–R) 7
7. Fore and mid leg sheaths with the tarsal sections extended straight along the midline between the wing sheaths, hind legs recurved under the wing sheaths (Fig. 12G) .... Diamesinae
– Generally all leg sheaths recurved under the wing sheaths (Fig. 3C) 8
8. Lateral fringe on anal lobes and large thoracic horn present (Figs 17D, E, 18C, E). Spines on abdominal tergites absent (except in male tergite 9 in *Odontomesa*, not yet recorded from Greenland). Macrosetae either absent and then 2 pairs of dorsal setae on anal lobes present, or 3–5 macrosetae on anal lobes and then 2 pairs of weak dorsal setae present, or 4–5 anal macrosetae present and then dorsal setae absent .... Prodiamesinae
– The above combination of characters absent........ .... Orthocladiinae

PODONOMINAE

The two Greenland species of Podonominae are easily recognized in all three stages. The larvae are characterized by the long procerci 4–8 times longer than broad, the absence of premandibles and the annulated 3rd antennal segment; the pupae by the combination of large thoracic horn with distinct plastron plate, straight distal part of the fore and mid leg sheaths and large wavy setae on abdominal segment 8 and 9; the adult males by the wing vein combination of crossvein MCu present, R_{2+3} absent, FCu distal to MCu and the special appearance of gonostylus (Figs 6F & 7F).

Key to the Greenland Podonominae

MALE ADULTS

1. Gonostylus with 2 large apical lobes (Fig. 6F)........ .... *Parochlus kiefferi*
– Gonostylus simple with a row of large megasetae along frontal margin (Fig. 7F) .... *Trichotanypus posticalis*

LARVAE

1. Procercus almost hyaline coloured (Fig. 6C); the trilobed median tooth of mentum of same length or larger than the first lateral tooth (Fig. 6A) .... *Parochlus kiefferi*
– Procercus hyaline anteriorly, blackish posteriorly from base to apex (Fig. 7B); the trilobed median tooth distinctly shorter than the first lateral tooth (Fig. 7A) .... *Trichotanypus posticalis*

PUPAE

1. Posterior angles of abdominal segments 3–8 protracted into large transverse processes (Fig. 6E) .... *Parochlus kiefferi*
– Posterior angles of abdominal segments 3–8 not protracted into large transverse processes (Fig. 7D) .... *Trichotanypus posticalis*

***Parochlus kiefferi* (Garrrett, 1925) Fig. 6**
P. tibialis (Stæger, 1845), preoccupied.
Henriksen 1939: *Tanypus tibialis* Stæger.

P. kiefferi is the only Holarctic *Parochlus* species. It was originally described by Stæger (1845) from Greenland, but the name *tibialis* was preoccupied. All instars are easily recognizable.

BIOLOGY. The larvae are strictly cold-stenothermal, living in cool mountain springs and streams, in North America also in large rivers. They are free living, feeding on diatoms and detritus.

DISTRIBUTION. S, nSW (widely distributed in the Narsaq and Kangerlussuaq/Søndre Strømfjord areas), SE (Lake Fjord; Edwards 1935).

Holarctic, south to the Alps in Europe and to New Mexico in America.

***Trichotanypus posticalis* (Lundbeck, 1898) Fig. 7**
Henriksen 1939: *Podonomus posticalis* Lundb.

Described from Greenland (Qeqertarsuup Tunua/Disko Bugt: Sydostbugt). All instars are easily recognizable.

BIOLOGY. The larvae live among mosses in both running and stagnant waters, even in small pools. They are eurythermal, tolerating temperatures up to 20°C.

DISTRIBUTION. S, nSW (common, but not abundant in the Kangerlussuaq/Søndre Strømfjord area).

Circumpolar, widely distributed in northern parts of Fennoscandia, Siberia and Canada.

TANYPODINAE

Four genera of Tanypodinae are known from Greenland. The larvae are characterized by their retractile antennae and well developed ligula; the pupae by the two stout lamellar macrosetae laterally on anal lobe; the adults by the combination of wing crossvein MCu present and the presence of a forked R_{2+3}.

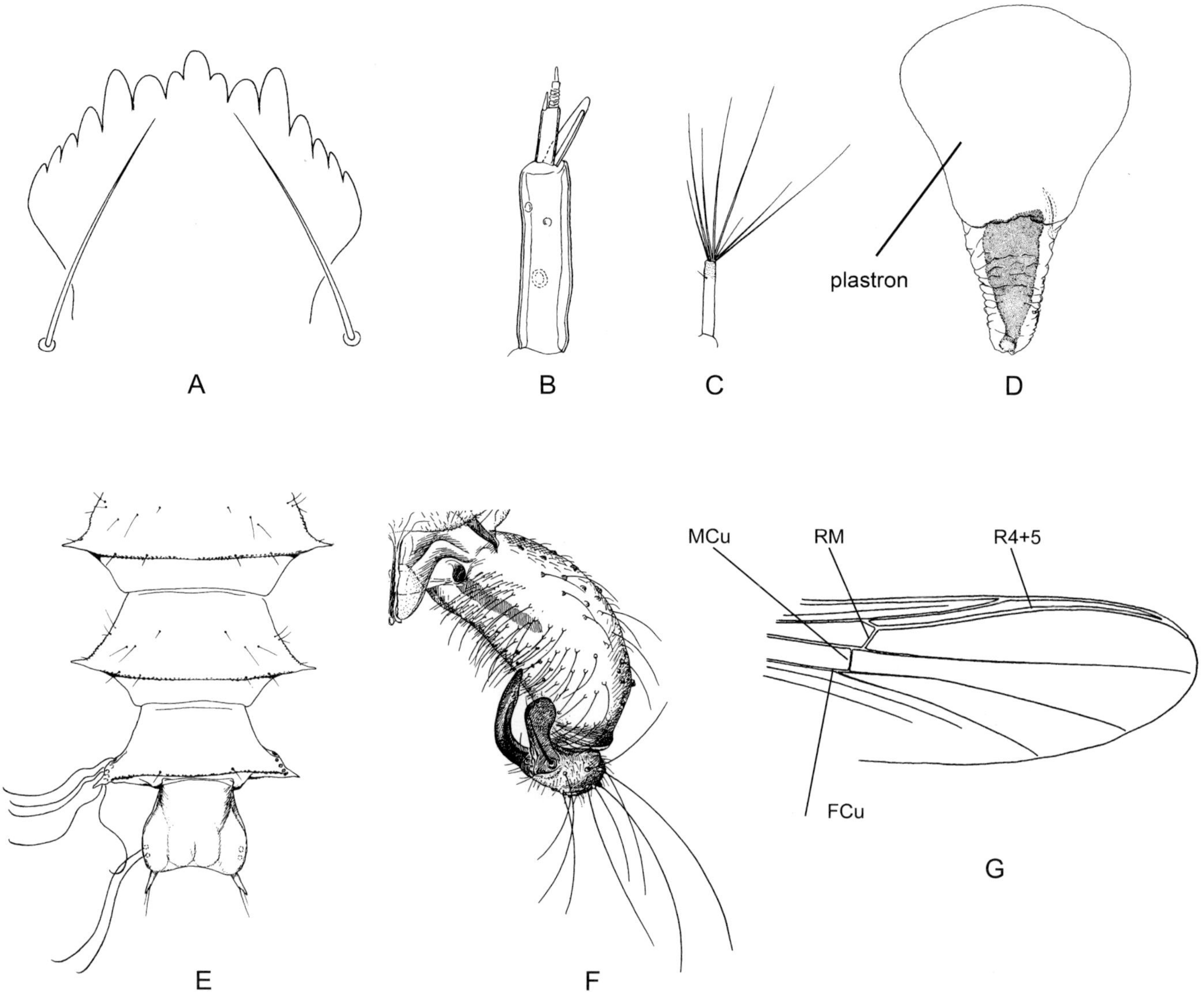

FIGURE 6 Parochlus kiefferi. *A–C: Larva. A: mentum, B: procercus with anal setae, C: antenna. D–E: Pupa. D: thoracic horn, E: tergites 7–9. F–G: Adult. F: dorsal view of male hypopygium, G: distal part of wing. A–G from Brundin (1983, 1986, 1989).*

Key to the Greenland Tanypodinae

MALE ADULTS

1. Wing crossvein M–Cu placed some distance before FCu (Fig. 10H) .... *Procladius*
– Wing crossvein M–Cu placed beyond FCu (Fig. 11H) 2
2. Costa produced beyond R_{4+5} by a distance at least as long as RM (Fig. 11H) .... *Psectrotanypus*
– Costa not or only slightly produced beyond R_{4+5} 3
3. Tibiae with 3–4 dark rings (Fig. 8J) .... *Ablabesmyia*
– Tibiae unicoloured .... *Arctopelopia*

LARVAE

1. Abdominal segments relatively broad, with fringe of swim-setae; head relatively short with cephalic index (length: width) 0.65–1.00 2
– Abdominal segments relatively slender, without fringe of swim-setae; head longer with cephalic index 0.40–0.67 3
2. Ligula with 4 teeth of equal length (Fig. 11A) .... *Psectrotanypus*
– Ligula with 5 teeth, the median tooth distinctly lower than lateral teeth (Fig. 10A) .... *Procladius*
3. Two dark claws on posterior parapod (Fig. 8E); basal segment of maxillary palp subdivided into 2

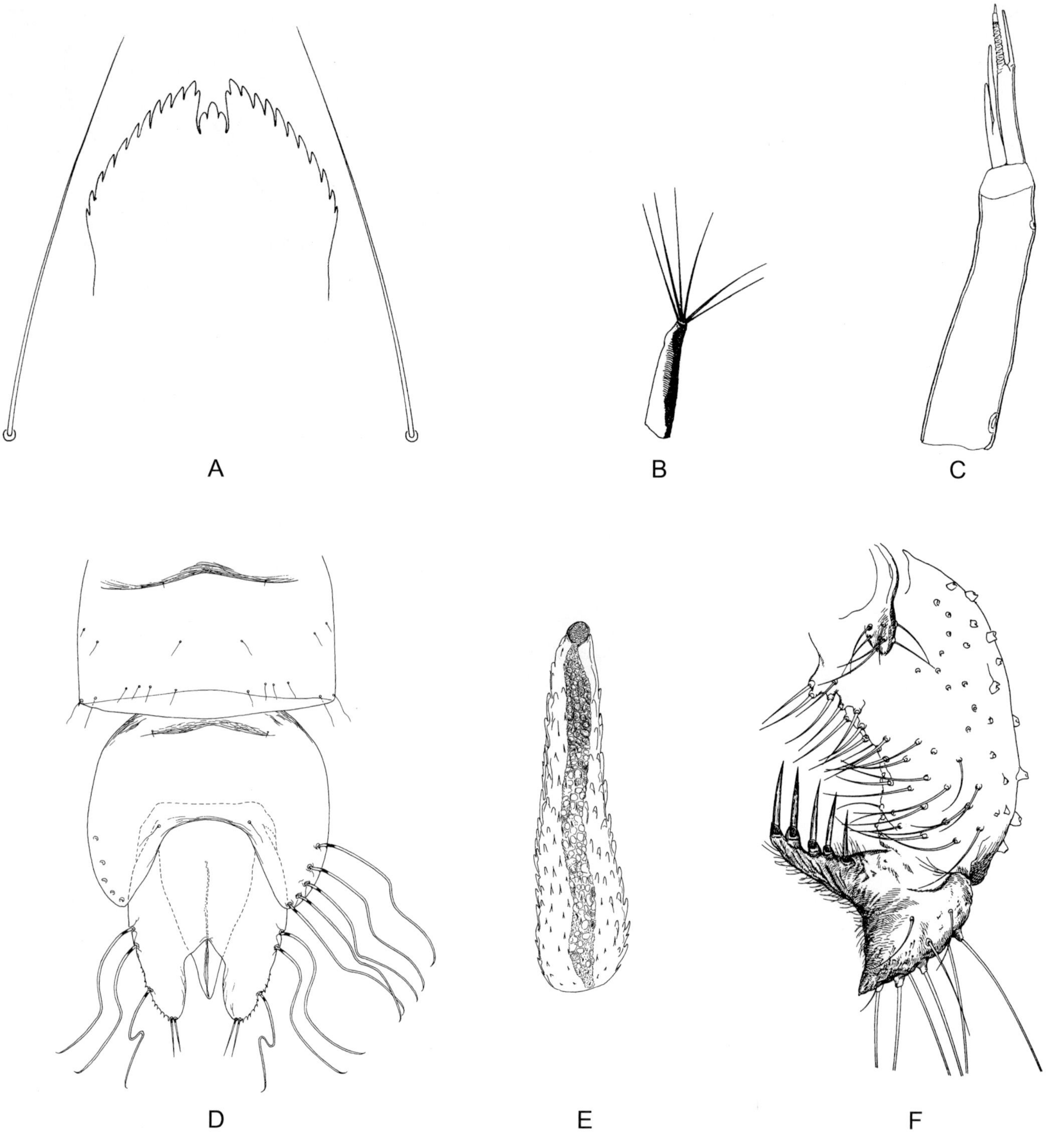

FIGURE 7 Trichotanypus posticalis. *A–C: Larva. A: mentum, B: procercus with anal setae, C: antenna. D–E: Pupa. D: thoracic horn, E: tergites 7–9. F: Adult, dorsal view of male hypopygium. A–F from Brundin (1983, 1986, 1989).*

segments (Figs 8C, 8D); mandible with large basal tooth (Fig. 8 B) .... *Ablabesmyia*

– All claws on posterior parapods unicoloured; basal segment of maxillary palp not subdivided (Fig. 9E); mandible with small basal tooth (Fig. 9B) .... *Arctopelopia*

PUPAE

1. Outer and inner border of anal lobe with a dense fringe of hair-like spinules (Fig. 11G) .... *Psectrotanypus*

– Anal lobe at most serrated (Fig. 10G) 2

2. Anal lobe with outer border rounded, the two lobes together forming a semi-circular plate. Outer border fringed with smaller or larger spines (Fig. 10G) ...... *Procladius*
– Anal lobe triangular. Outer border with only minute spines (Figs 8I, 9I) 2
3. Thoracic horn bulbous with a reticulate appearance; indistinct plastron enclosed on a hyaline protuberance; thoracic comb present (Figs 8G, H) .... *Ablabesmyia*
– Thoracic horn flattish with spines; plastron and thoracic comb absent (Fig. 9G) .... *Arctopelopia*

***Ablabesmyia pulchripennis* Lundbeck, 1898 Fig. 8**
Henriksen 1939: *Tanypus pulchripennis* Lundb.

Described from Greenland (Aasiaat/Egedesminde) and redescribed by Roback (1966, 1971). Larvae and pupae also described by Roback (1985).

BIOLOGY. The larvae live in stagnant waters and may be abundant in small lakes and ponds. They are carnivorous, feeding on small chironomid larvae. They can survive on algae (diatoms), but the growth rate is then very low. Very common in lakes and ponds. Adults swarm numerously above pools in the beginning of July (Henriksen & Lundbeck 1917).

DISTRIBUTION. S, SW.

Nearctic, widely distributed in North America as far south as Kansas.

***Arctopelopia melanosoma* (Goetghebuer, 1933) Fig. 9**
Henriksen 1939: *Tanypus malanosoma* Goet.

Described from NE Greenland (Loch Fyne) and recorded by Andersen from Ella Ø. Redescribed by Fittkau (1962).

BIOLOGY. The larvae live in oxygen rich lakes as free-living predators.

DISTRIBUTION. S, SW, NE. Widely distributed in the Narsaq and Kangerlussuaq/Søndre Strømfjord areas, but not abundant.

Holarctic, northern parts.

***Procladius* Skuse, 1889 Fig. 10**
The taxonomy of the genus *Procladius* needs revision and the species cannot be determined with certainty for the time being. Roback (1971) provided a key for the Nearctic adult males. The larvae .. live primarily in standing waters and are probably common in all ponds and lakes in Greenland. They burrow in the sediment, feeding primarily on other chironomid larvae. Many species can survive short anoxic periods.

****Procladius crassinervis* Zetterstedt, 1838 and **P. culiciformis* (Linnaeus 1767) Fig. 10**
Henriksen (1939) mentioned two species from Greenland: *P. choreus* (Meigen, 1804) and *P. crassinervis* Zetterstedt, 1838. The first was reported by Andersen (1937) from NE Greenland (Ella Ø). This species was later by Roback (1966) determined as *P. crassinervis,* which was also reported by Edwards (1931) from W Greenland. Roback (1971) synonymised *P. crassinervis* with *P. culiciformis*. However, Ashe & Cranston (1990) maintained both species as valid species. Nor Oliver *et al.* (1990) or Ashe & Cranston (1990) record any of these species from Greenland.

***Procladius paragretis* Roback, 1971 Fig. 10**
Henriksen 1939: not recorded.

Described on adults from Greenland (Kangerlussuaq/Søndre Strømfjord) and only known from this locality. It differs from *P. culiciformis* in the distystyle, which in *P. culiciformis* is narrow and curved, in *P. paragretis* large and triangular (Figs 10I–J).

***Psectrotanypus pictipennis* Zetterstedt, 1838 Fig. 11**
Henriksen 1939: *Tanypus pictipennis* Zett.

Described on material from Greenland (without locality given). It was redescribed by Roback (1966).

BIOLOGY. The larvae live in small ponds. The number of exuviae in each locality is low indicating a low abundance. The larvae of *Psectrotanypus* spp. are known for their ability to live in mud with a low content of oxygen. They are considered carnivorous.

DISTRIBUTION. S, SW. Edwards (1931) recorded it from Qussuk (Nuup Kangerlua/Godthåbsfjord). I have found it in both the Narsaq and the Kangerlussuaq/Søndre Strømfjord areas.

So far recorded only from Greenland.

DIAMESINAE
Two or three genera of Diamesinae are known from Greenland. The larvae are recognized by the annulated 3rd antennal segment as in Podonominae, but differs from these by lacking or having only small procerci at most 1.5 times longer than broad. The pupae are recognized by the fore and mid leg sheaths with the tarsal sections laying straight along the midline between wing sheaths as in Podonominae, but the Greenland species of Diamesinae can be separated from Podonominae by lacking or having only small thread formed thoracic horns. The adults are recognized on the presence of wing crossvein Mcu, a simple vein R_{2+3} and Fcu placed proximally to Mcu.

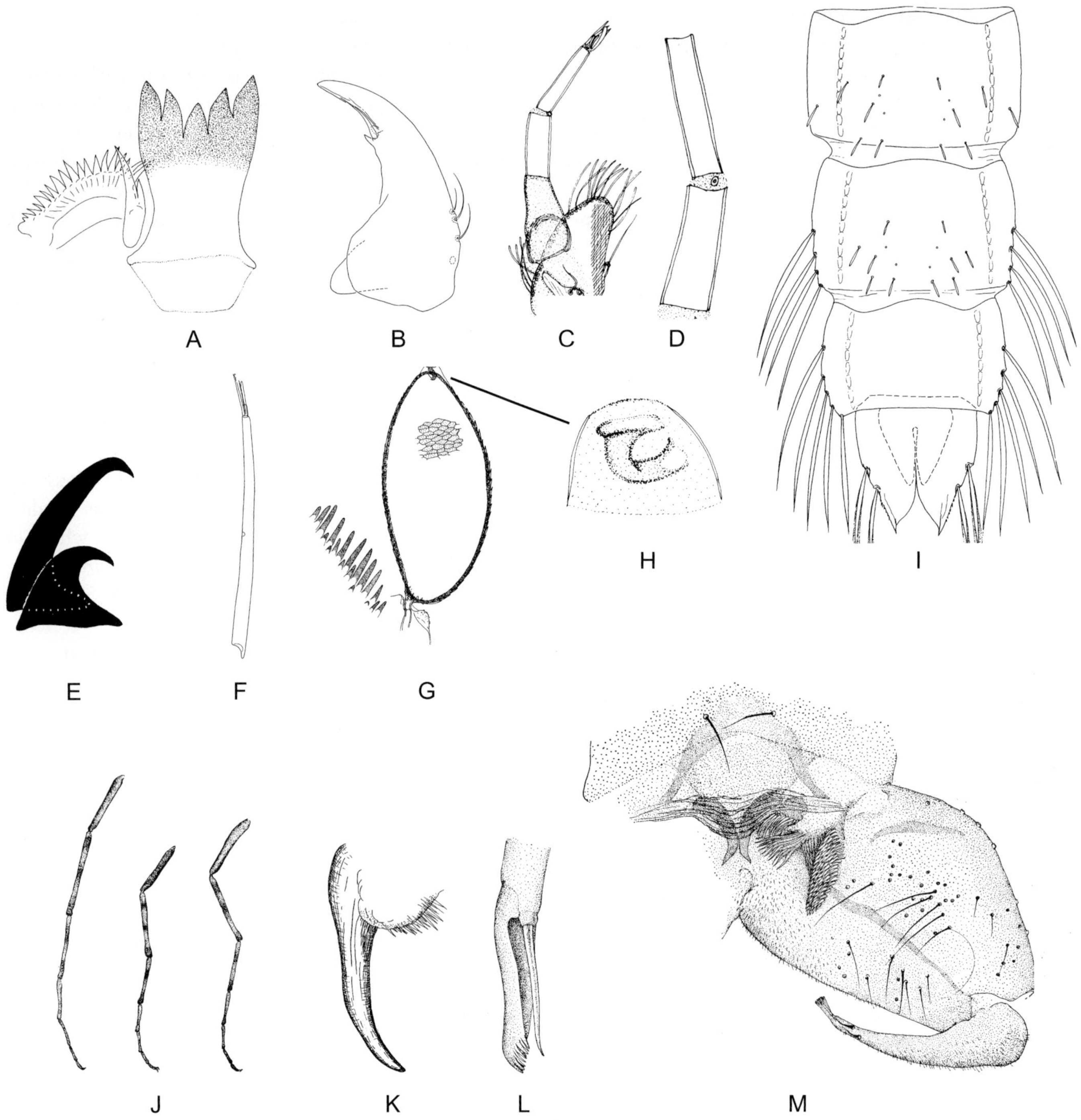

FIGURE 8 Ablabesmyia. *A–F: Larva. A: ligula, paraligula and pecten hypopharyngis, B: mandible, C: maxilla, D: basal segment of maxillary palp, E: dark claws of posterior parapod, F: antenna. G–I: Pupa. G: thoracic horn, H: plastron and horn sac neck of thoracic horn, I: tergites 6–8 and anal lobe. J–M: Adult male. J: fore- mid- and hind-leg, K: superior volsella, L: apical end of gonostylus, M: hypopygium. A–F, H and J–M:* A. pulchripennis *from Roback (1966, 1971, 1985); G–I:* Ablabesmyia *sp. from Fittkau & Murray (1986).*

Key to genera of Greenland Diamesinae

MALE ADULTS

1. Tarsomere 4 cordiform, shorter than tarsomere 5 (Fig. 12K) 2
– Tarsomere 4 cylindrical. Hypopygium as in Fig. 14 .. .. *Pseudodiamesa branickii*
2. Eye with microtrichia only between ommatidia near inner margin of eye .... *Pseudokiefferiella* ....

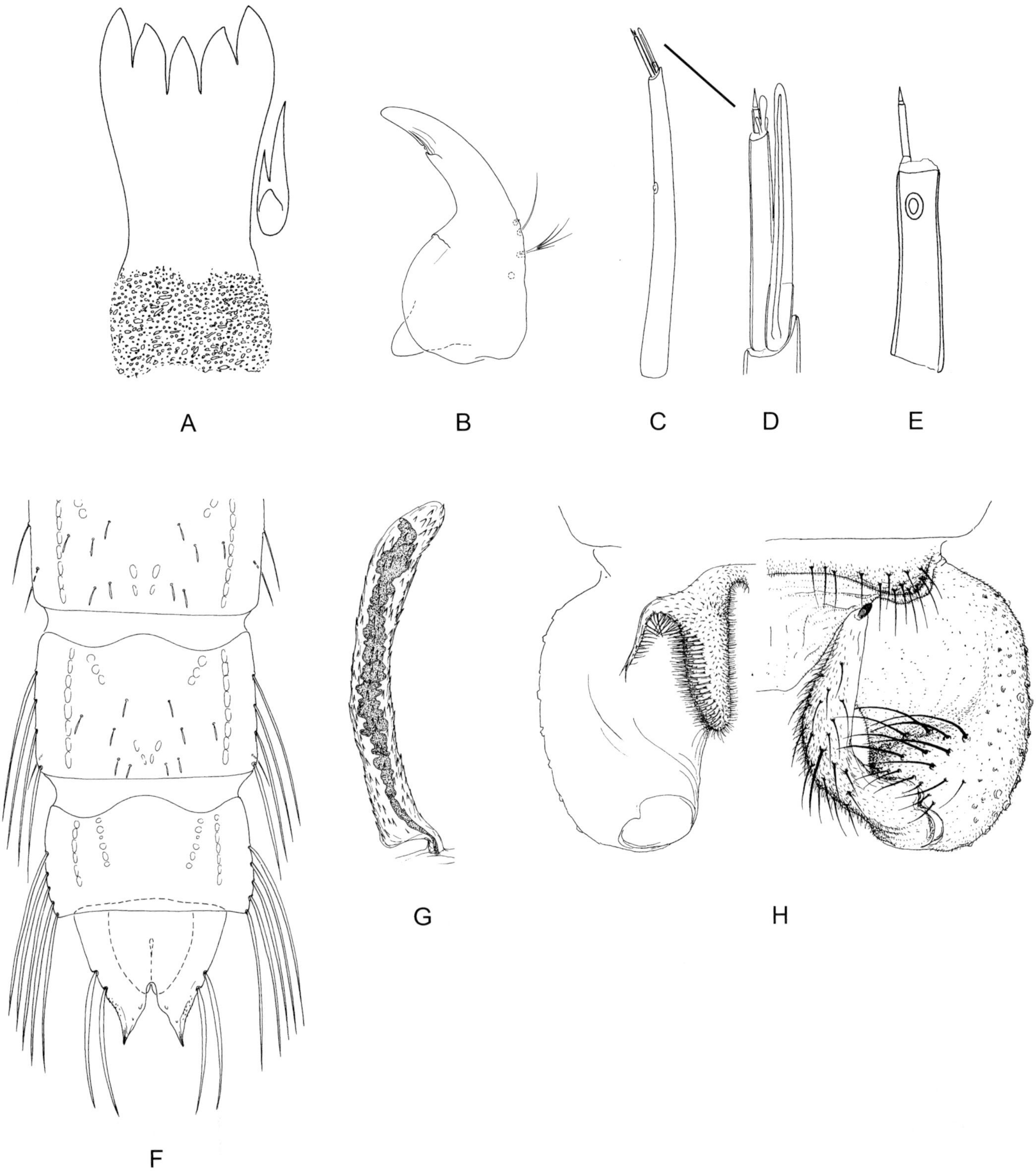

FIGURE 9 Arctopelopia. *A–E: Larva. A: ligula and paraligula, B: mandible, C: antenna, D: apical segments of antenna, E: maxillary palp. F–G: Pupa. F: tergites 7–8 and anal lobe, G: thoracic horn. H: Right dorsal and left ventral view of male hypopygium. A–E:* Arctopelopia *sp. from Fittkau & Roback (1983) and Fittkau & Murray (1986); F–H.* A. melanosoma *from Fittkau (1962).*

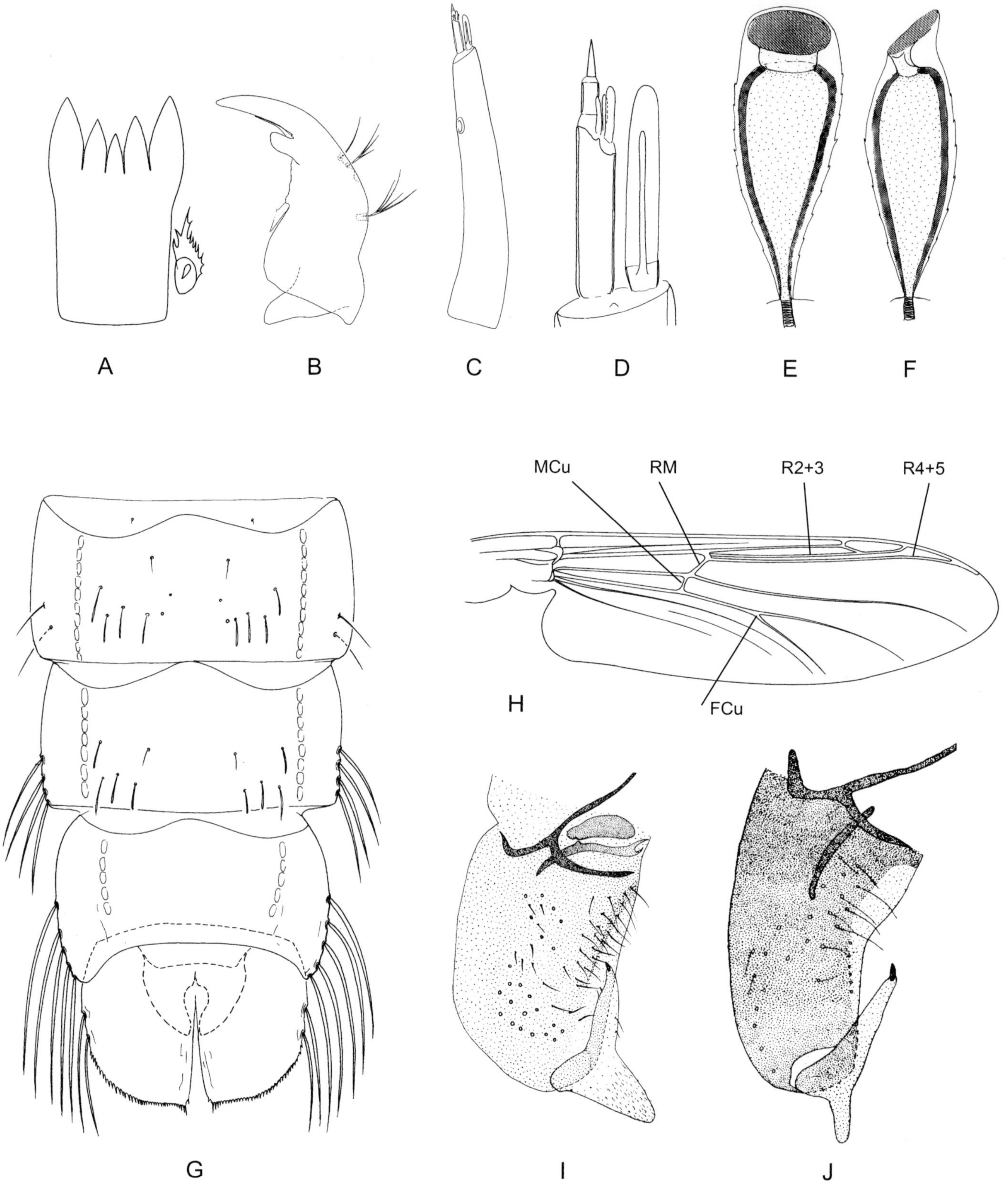

FIGURE 10 Procladius. *A–D: Larva. A: ligula and paraligula, B: mandible, C: antenna, D: apical segments of antenna. E–G: Pupa. E: thoracic horn, frontal view, F: thoracic horn, lateral view, G: tergites 6–8 and anal lobe. H–J: Adult. H: wing, I: hypopygium of* P. paragretis, *J: hypopygium of* P. culiciformis. *A–D:* Procladius *sp. from Fittkau & Roback (1983); E–F from Roback, 1980; G from Fittkau & Murray (1986), H from Murray & Fittkau (1889); I–J from Roback (1971).*

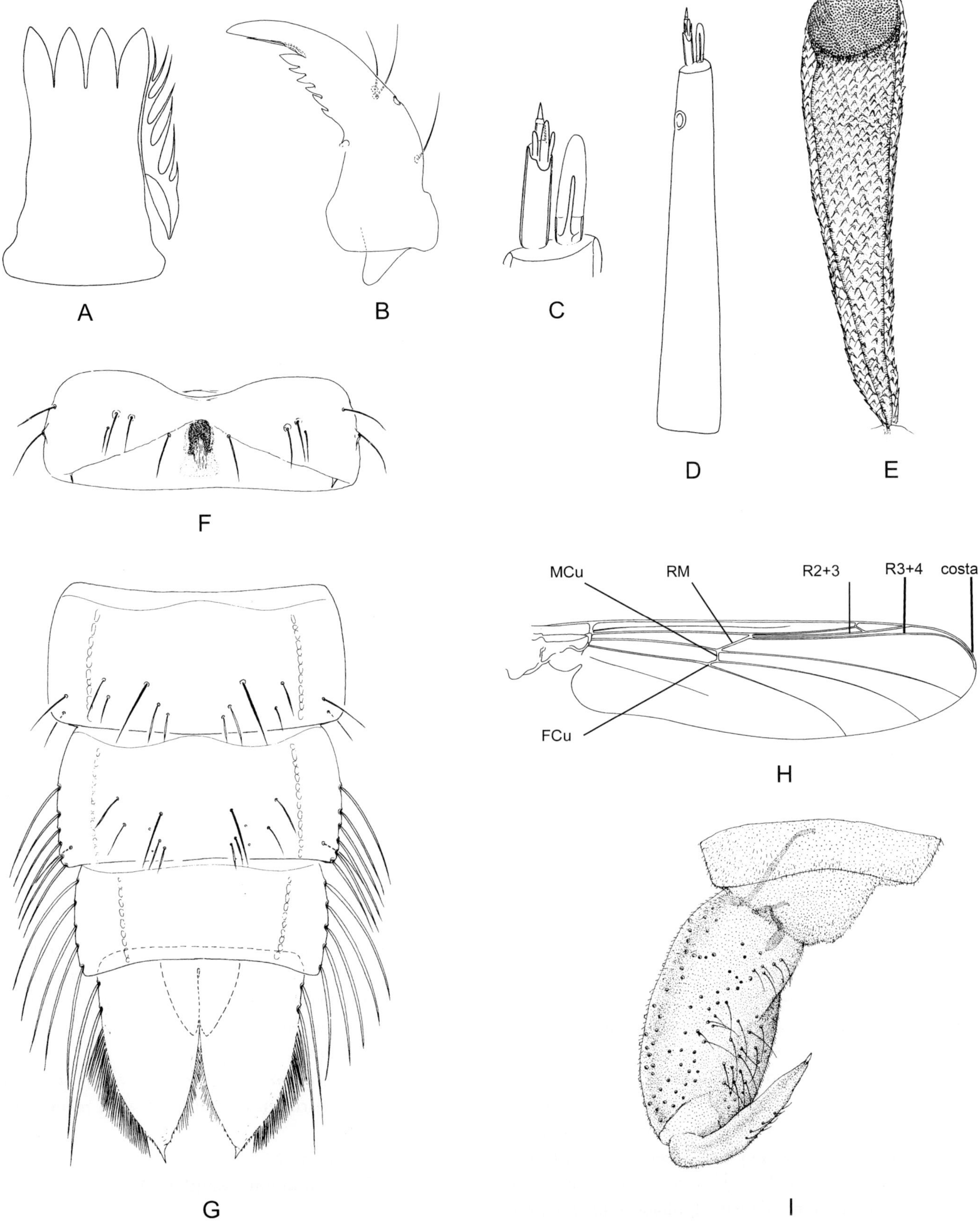

FIGURE 11 Psectrotanypus. *A–D: Larva. A: ligula and paraligula, B: mandible, C: apical segments of antenna, D: antenna. E–G: Pupa. E: thoracic horn, F: tergite 1, G: tergites 6–8 and anal lobe. H–I: adult. H: wing, I: male hypopygium. A–D:* Psectrotanypus *sp. from Fittkau & Roback (1983); E–G from Fittkau & Murray (1986); H from Murray & Fittkau (1989); I:* P. pictipennis *from Roback (1971).*

– Eye with microtrichia between all ommatidia (Fig. 12J) .... *Diamesa*

LARVAE

1. Median tooth of mentum triangular, separated from first lateral teeth by deep notches (Fig. 14A). Procercus about 1.5 as high as broad (Fig. 14B) .... *Pseudodiamesa*
– Median tooth of mentum rounded or bilobed, narrowly separated from the lateral teeth (Figs 12A, B, 15A) 2
2. Procercus absent or weakly developed (Fig. 12E); body setae pale and indistinct .... *Diamesa*
– Procercus higher than broad (Fig. 15B); body setae strong and dark .... *Pseudokiefferiella*

PUPAE

1. Strong often dark thorn-like spines on abdominal tergites and sternites (Figs 12F, H). Thoracic horn often long, slender and filiform (Fig. 12F) ....*Diamesa*
– Thorn-like spines on abdominal sternites absent. Thoracic horn absent or filiform 2
2. Segments 2–7 with spine-like process on posterolateral corner (Figs 15E, F); thoracic horn filiform ....*Pseudokiefferiella*
– Segments 2–7 without spine-like processes (Figs 14C, D); thoracic horn absent .... *Pseudodiamesa*

***Diamesa* Meigen, 1835 Figs 12, 13**

A large genus with about 90 Holarctic species. The nearctic species are revised by Hansen & Cook (1976) and the adult males may be identified following their key and descriptions. The association between larvae, pupae and adults are unknown for many species. Until now 7 species of *Diamesa* are recorded from Greenland, and some species have been mentioned from Greenland, but the identifications are questionable. *D. hyperborea* (Holmgren, 1869), recorded and identified on a female by Edwards (1935) and quoted by Henriksen (1939, no.142) as *D. ursus* (Kieffer, 1918) is not accepted by Oliver *et al.* (1990). The same applies to *D. septima* (Kieffer, 1922) quoted by Henriksen (1939, no.141). The identity of the taxon referred to by Henriksen (1939, no. 136) as *Syndiamesa pubitarsis* Zett. is uncertain. Earlier records are questionable unless they have been included in the review by Hansen & Cook (1976), who also outlined some of the problems as referred below. Probably more *Diamesa* species are to be found in Greenland.

The larvae are cold-adapted and mostly found in cold mountain streams, free-living, primarily on stones and rock surfaces. They feed on diatoms, other algae and detritus. Many species are specific for glacial streams. In the European lowland a few species are found in springs and cold streams. In northern areas some species may also be present in the littoral zone of lakes.

***Diamesa aberrata* Lundbeck, 1898**

Henriksen 1939: *Diamesa* (*Psilodiamesa*) *aberrata* Lundb.

Described from Greenland (Nuuk/Godthåb). Hansen & Cook (1976) confirmed the identity. Edwards (1935) recorded 4 specimens from NE Greenland (Jameson Land); one of these is *D. aberrata,* one is *D. chorea*, and two are *D. simplex* (Hansen & Cook 1976). Andersen (1937), Fig. 56) figured the hypopygium of *D. aberrata,* but this is in fact *D. simplex* (Hansen & Cook 1976).

DISTRIBUTION. sSW (Nuuk/Godthåb), NE (several localities).

Holarctic. Widely distributed in the Palaearctic region (Ashe & Cranston, 1990), also recorded from Canada and U.S.A. (Oliver *et al.* 1990).

***Diamesa bertrami* Edwards, 1935**

Described from NE Greenland (Kap Dalton), but the holotype specimen was not seen by Hansen & Cook (1976).

DISTRIBUTION. NE (only recorded from the Blosseville Kyst: Kap Dalton).

Palaearctic (including Bear Island), arctic-alpine.

***Diamesa bohemani* Goethghebuer, 1932**

Henriksen 1939: not recorded.

One specimen from S Greenland has been confirmed by Hansen & Cook (1976).

DISTRIBUTION. S (no locality specified, Hansen & Cook 1976), ?SE (Kangerlussuaq, Henriksen 1937, record not confirmed).

Holarctic, widely distributed in Europe and Canada.

***Diamesa chorea* Lundbeck, 1898**

Henriksen 1939: *Diamesa* (*Psilodiamesa*) *chorea* Lundb.

Described from SW Greenland and redescribed by Hansen & Cook (1976). One of Edwards' (1935) specimens from NE Greenland was also confirmed as *D. chorea* (Hansen & Cook 1976).

BIOLOGY. Males observed in dancing swarms at 300 m a.s. (Nuussuaq; Henriksen & Lundbeck 1917, record not confirmed).

DISTRIBUTION. SW, NE.

Nearctic (Rocky Mountains south to Wyoming).

***Diamesa davisi* Edwards 1933**

Henriksen 1939: not recorded.

Thienemann (1941) quoted *D. davisi* recorded from Greenland, but without locality specified. Neither Serra-Tosio (1972) nor Hansen & Cook (1976) believe this record

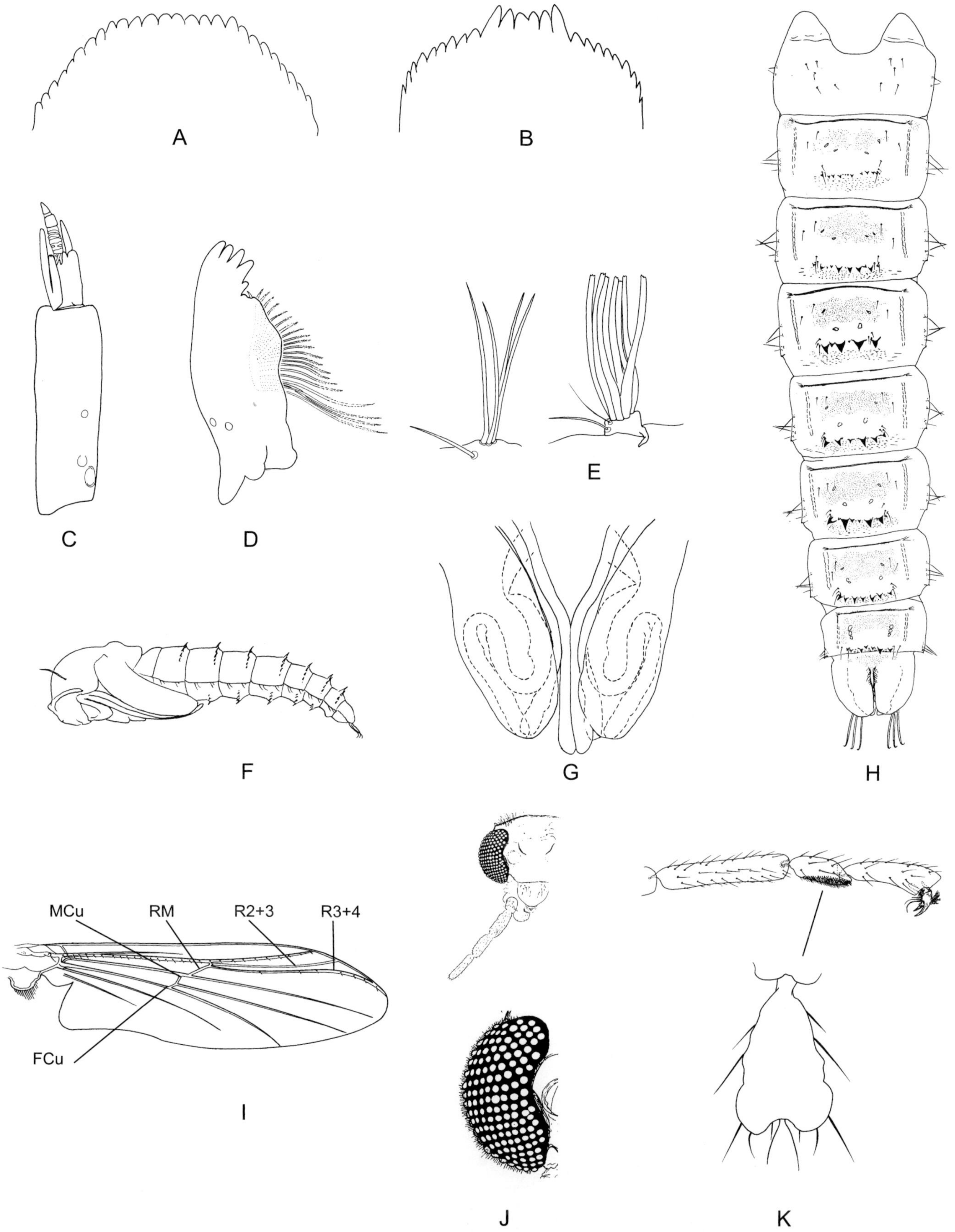

FIGURE 12 Diamesa 1. *A–E: Larva. A–B: mentum, C: antenna, D: mandible, E: procercus with anal setae. F–H: Pupa. F: lateral view of pupa, G: wing and leg sheaths, ventral view, H: tergites 1–9. I–K: Adult male. I: wing, J: head and eye seen from front; eye with microtrichia between ommatidia, K: lateral view of tarsomeres 3–5 with dorsal view of cordiform tarsomere 4. A–E from Oliver (1983); F from Serra-Tosio (1972); H–I from Oliver (1986, 1989); J–K from Hansen & Cook (1976).*

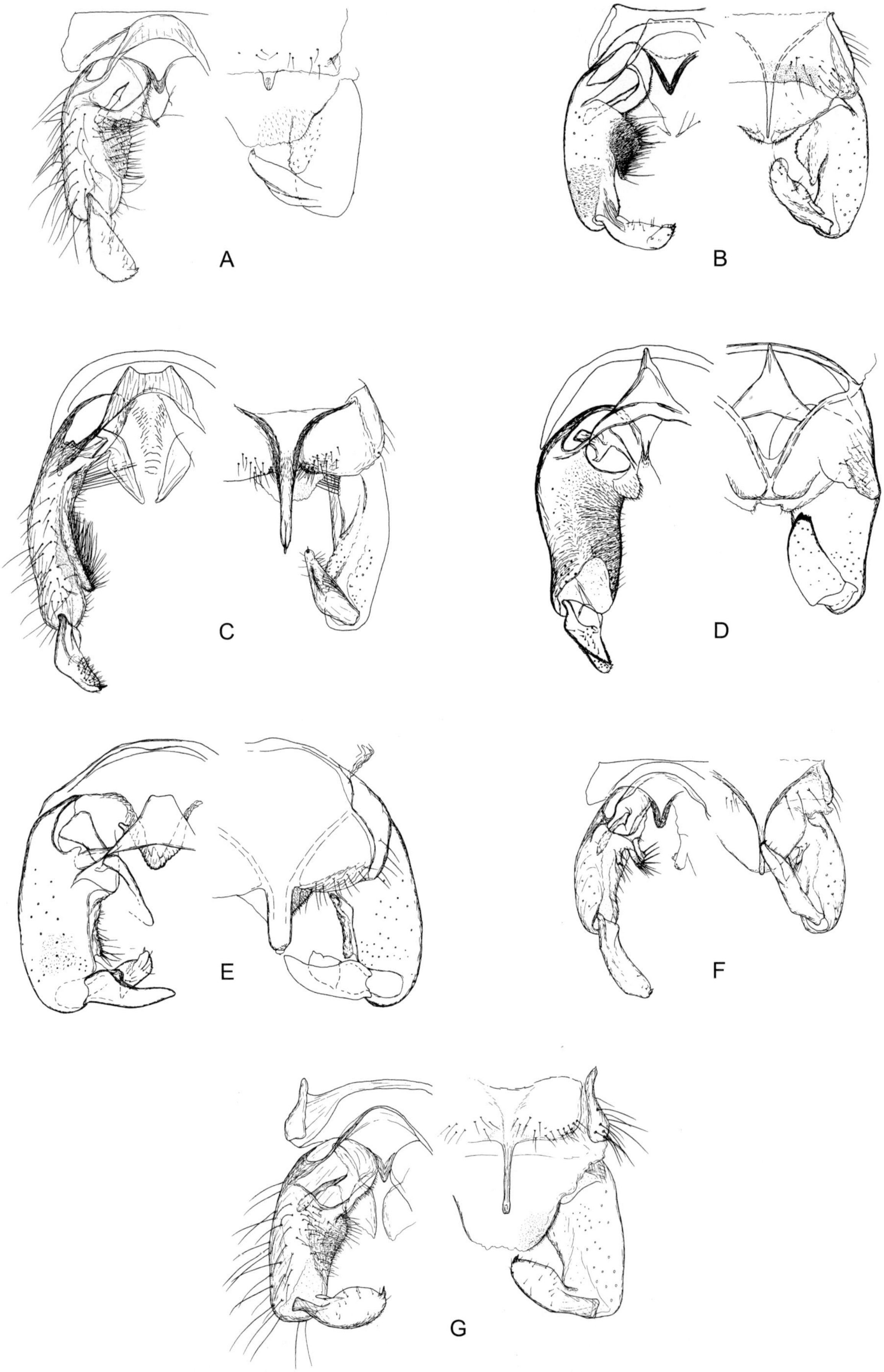

FIGURE 13 Diamesa 2. *A–G: Right dorsal and left ventral view of hypopygium of adult male. A:* D. aberrata, *B:* D. bertrami, *C:* D. bohemani, *D:* D. davisi, *E:* D. geminata, *F:* D. lindrothi, *G:* D. simplex. *A–G from Hansen & Cook (1976).*

is valid, but the species is maintained as occurring in Greenland by Oliver *et al.* (1990). In a revision of the *davisi* group none of the species are recorded from Greenland (Willassen 1985).

DISTRIBUTION. ? Greenland.

Nearctic (Canada, U.S.A.).

***Diamesa geminata* Kieffer, 1926**

Henriksen 1939: *Syndiamesa biappendiculata* Goetg.

Described on material from NW Greenland (Foulke Fjord: Reindeer Point). Hansen & Cook (1976) suggest *D. biappendiculata* (Goetghebuer, 1928) described on material from Jameson Land, NE Greenland, (Remy, 1928) to be a synonym of *D. geminata.* Also two males from Etah (Peary Land) are included in the redescription by Hansen & Cook (1976). Further, Sæther *et al.* (1984) report it from East Greenland (Liverpool Land).

DISTRIBUTION. NW, N, NE.

Holarctic (eastern Russia, Canada: Northwest Territories).

****Diamesa hyperborea* Holmgren, 1869**

Henriksen 1939: *Diamesa* (*Adiamesa*) *?ursus.*

Henriksen (1937, no. 142) quoted Edward's (1935) record of *D. ursus* (Kieffer, 1918), a synonym of *D. hyperborea* from SE Greenland (Lake Fjord). However, the record is a female and at present it is not possible to identify female species of *Diamesa*. The record is thus uncertain (Hansen & Cook 1976).

***Diamesa lindrothi* Goetghebuer, 1931**

Henriksen 1939: *Diamesa* (*Psilodiamesa*) *lindrothi* Goetg.

Recorded by Edwards (1935) from SE Greenland (Lake Fjord). This record is approved by Hansen & Cook (1976) and is so far the only Greenland record.

DISTRIBUTION. SE (only known from Lake Fjord).

Europe (widely distributed in Iceland, Norway, Sweden and also known from the Alps; Serra-Tosio 1972).

***Diamesa simplex* Kieffer, 1926**

Henriksen 1939: not recorded.

DISTRIBUTION. S, N, NE. According to Hansen & Cook (1976) recorded from S Greenland, Jameson Land (Edwards 1935, misidentified as *D. aberrata*), NE Greenland (Andersen 1937, misidentified as *D. aberrata*), Peary Land, Inglefield Land: Etah.

Nearctic, arctic-alpine (Alaska, northern Canada, Rocky Mountains south to Wyoming).

***Pseudodiamesa branickii* (Nowicki, 1873) Fig. 14**

Henriksen 1939: *Syndiamesa branickii* Now.

BIOLOGY. A character species of α-oligotrophic lakes (Sæther 1979). I have not recorded larvae or pupal exuviae from the glacial lakes in the Kangerlussuaq/Søndre Strømfjord or the Narsaq regions, where the other character species of α-oligotrophic lakes, *Oliveridia tricornis*, is common. Emergence takes place early in spring, while the lakes are still partly covered by ice.

DISTRIBUTION. N, NE (recorded by Oliver (1976) from Peary Land: Nedre Midsommersø, and by Goetghebuer (1933) from Wollaston Forland).

Holarctic, widely distributed (Ashe & Cranston 1990, Oliver *et al.* 1990).

"*Pseudokiefferiella*" undescribed species Fig. 15

Henriksen 1939: not recorded.

Oliver (1986) mentioned that un-associated pupal exuviae from the Nearctic region have most of the characters of *Pseudokiefferiella*, but lack the posterior spines on tergite 1 and that the posterior spines on tergites 2–8 are strong.

DISTRIBUTION. sSW. One exuvium with these characteristics is recorded from the Kangia Fjord region.

Nearctic, northern parts.

[Telmatogetoninae *Telmatogeton japanicus* Tokunaga, 1933 Fig. 16

BIOLOGY. A marine species living in the tidal zone on stones, often in connection with green algae (*Ulva*, *Enteromorpha*). The larvae build small tubes of fine sand attached to stones and feed by grazing the biofilm on stones and macro-algae.

DISTRIBUTION. Not yet recorded from Greenland. Originally a Pacific species, but now widely distributed along the coastal zone of the Holarctic Region, probably spread by intercontinental shipping. Recently recorded from Iceland (Hrafnsdottir 2005) and may likely be spread to Greenland.]

PRODIAMESINAE

Two species of Prodiamesinae are recorded from Greenland. The larvae are recognized by the large ventromental plates without striation but with beard beneath. No single character separates the pupae from all other subfamilies. The presence of large, simple thoracic horns, lateral fringe on anal lobe and the absence of posterolateral spur on abdominal segment 8 will separate the Prodiamesinae pupae from most other Greenland chironomids. The adults may be recognized on the presence of wing crossvein Mcu combined with a simple R_{2+3} and Fcu placed distally to Mcu.

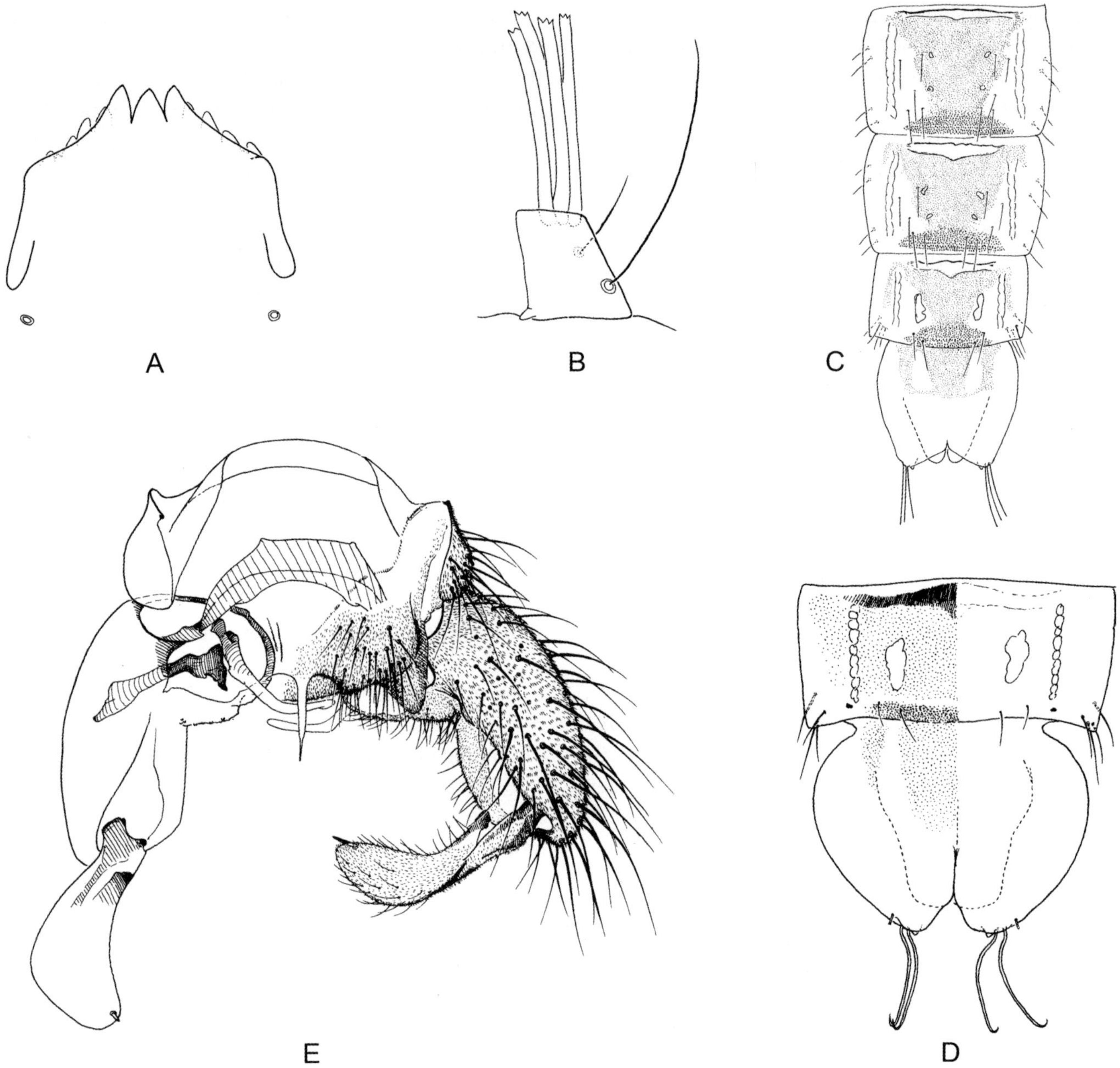

FIGURE 14 Pseudodiamesa. *A–B. Larva. A: mentum, B: procerci. C–D: Pupa. C: Tergites 7–8 and anal lobe of* P. artica. *D: anal lobe of* P. branickii. *E: Hypopygium of* P. branickii. *A–C from Oliver (1983, 1986); D–E: from Serra-Tosio (1976).*

Key to the Greenland Prodiamesinae

MALE ADULTS

1. Gonostylus bilobed (Fig. 18G) .... *Prodiamesa olivacea*
– Gonostylus simple (Fig. 17F) .... *Monodiamesa bathyphila*

LARVAE

1. Median mental tooth bilobed, smaller than the first trilobed lateral teeth; ventromental plate broad with a beard of strong black setae (Figs 18A, B) .... *Prodiamesa olivacea*
– Median mental tooth concave and broad; ventromental plate narrow with a weak beard beneath (Fig. 17A) .... *Monodiamesa bathyphila*

PUPAE

1. In addition to lateral swim setae, 4–5 recurved macro-setae distally on anal lobe (Fig. 18D, E) .... *Prodiamesa olivacea*

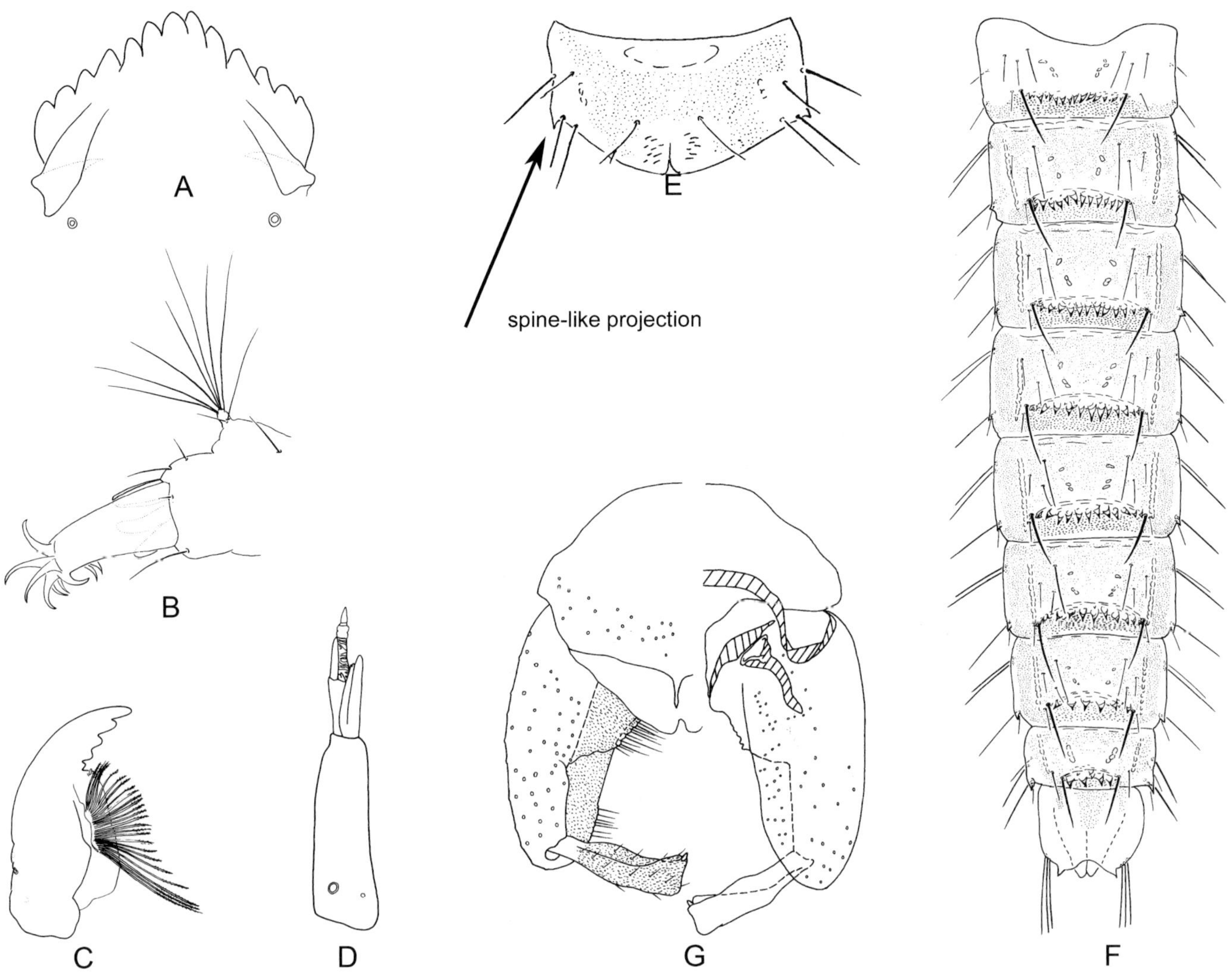

FIGURE 15 Pseudokiefferiella. *A–D: Larva. A: mentum, B: anal end, C: mandible, D: antenna. E–F: Pupa. E: tergite 8, F: tergites 1–9, G: Hypopygium of adult male. A–D from Oliver (1983); E–F from Oliver (1986); G from Oliver (1989).*

– Macro-setae on anal lobe absent (Fig. 17D) .... *Monodiamesa bathyphila*

Monodiamesa bathyphila **(Kieffer, 1918) Fig. 17**
Henriksen 1939: not recorded.

Mentioned from Greenland by both Ashe & Cranston (1990) and Oliver *et al.* (1990), but without any locality given. All stages are easily recognizable.

BIOLOGY. The larvae live in sandy sediments in oligotrophic to mesotrophic lakes and may also be found in sandy streams.

DISTRIBUTION. ? Greenland.

Holarctic, widespread.

Prodiamesa olivacea **(Meigen, 1818) Fig. 18**
Henriksen 1939: not recorded.

Mentioned from Greenland by Ashe & Cranston (1990), but without locality given. All stages are so characteristic, that they cannot be misidentified.

BIOLOGY. The larvae live in sandy sediments, primarily in running waters, and may be abundant in sediment enriched with organic matter. However, oxygen depletion is tolerated for only a few hours.

DISTRIBUTION. ? Greenland.

Holarctic, widely distributed. Very common in the European lowland, where the larvae occur in almost all streams and rivers and sparsely reported also from lakes.

ORTHOCLADIINAE

More than 20 genera with about 70 species of Orthocladiinae are recorded from Greenland. Probably

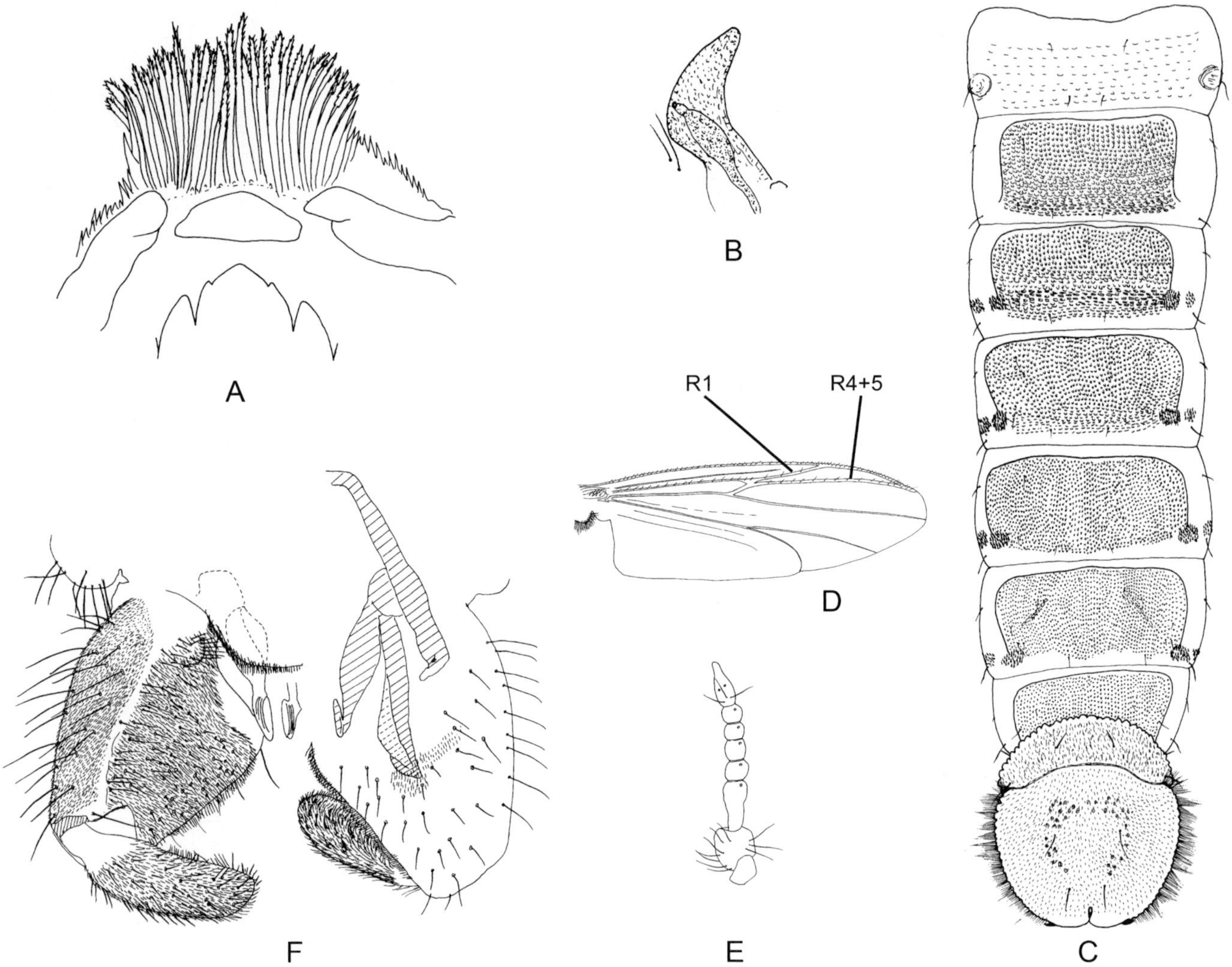

FIGURE 16 Telmatogeton. *A: Mentum of larva. B–C: Pupa. B: thoracic horn, C: tergites 1–9. D–F: Adult male. D: wing, E: antenna, F: left dorsal and right ventral view of hypopygium. A–F from Cranston (1983, 1986, 1989).*

both more species and more genera of this subfamily are likely to be found in Greenland. Therefore, the following keys are preliminary. Keys to all known Holarctic genera are provided by Wiederholm (1983) and Sæther *et al.* (2000).

Key to genera of Greenland Orthocladiinae

MALE ADULTS

1. Veins R_1 and R_{4+5} short, thick and fused with costa in thick clavus, ending before midpoint of wing (Figs 22H, 55G) 2
– Veins R_1 and R_{4+5} narrow, elongate separated from costa until apex beyond midpoint of wing 3
2. Fore trochanter with dorsal keel (Fig. 22I); transverse sternapodeme dilated, inverted V– or U– formed, with no trace of oral projection (Figs 22K–M) .... *Corynoneura*
– Fore trochanter without dorsal keel; transverse sternapodeme narrow, straight or convex, with slight to moderate oral projections (Fig. 55H) .... *Thienemanniella*
3. Eye hairy (cf. Fig. 12J) 4
– Eye bare or pubescent (*i.e.* microtrichia, if present, not extending beyond ommatidia) 11
4. Gonostylus bifurcate (Fig. 25G) .... *Diplocladius*
– Gonostylus simple or at most with subapical lobe 5
5. Dorsocentrals decumbent (Fig. 4E) 6
– Dorsocentrals erect (Fig. 32I) 9
6. Anal point present (Fig. 46G) .... *Paracladius*
– Anal point absent 7
7. Prealars extending anteriorly to level of median anepisternum; dorsocentrals not joint dorsally at midline of scutum (Fig. 27H).......... *Halocladius*

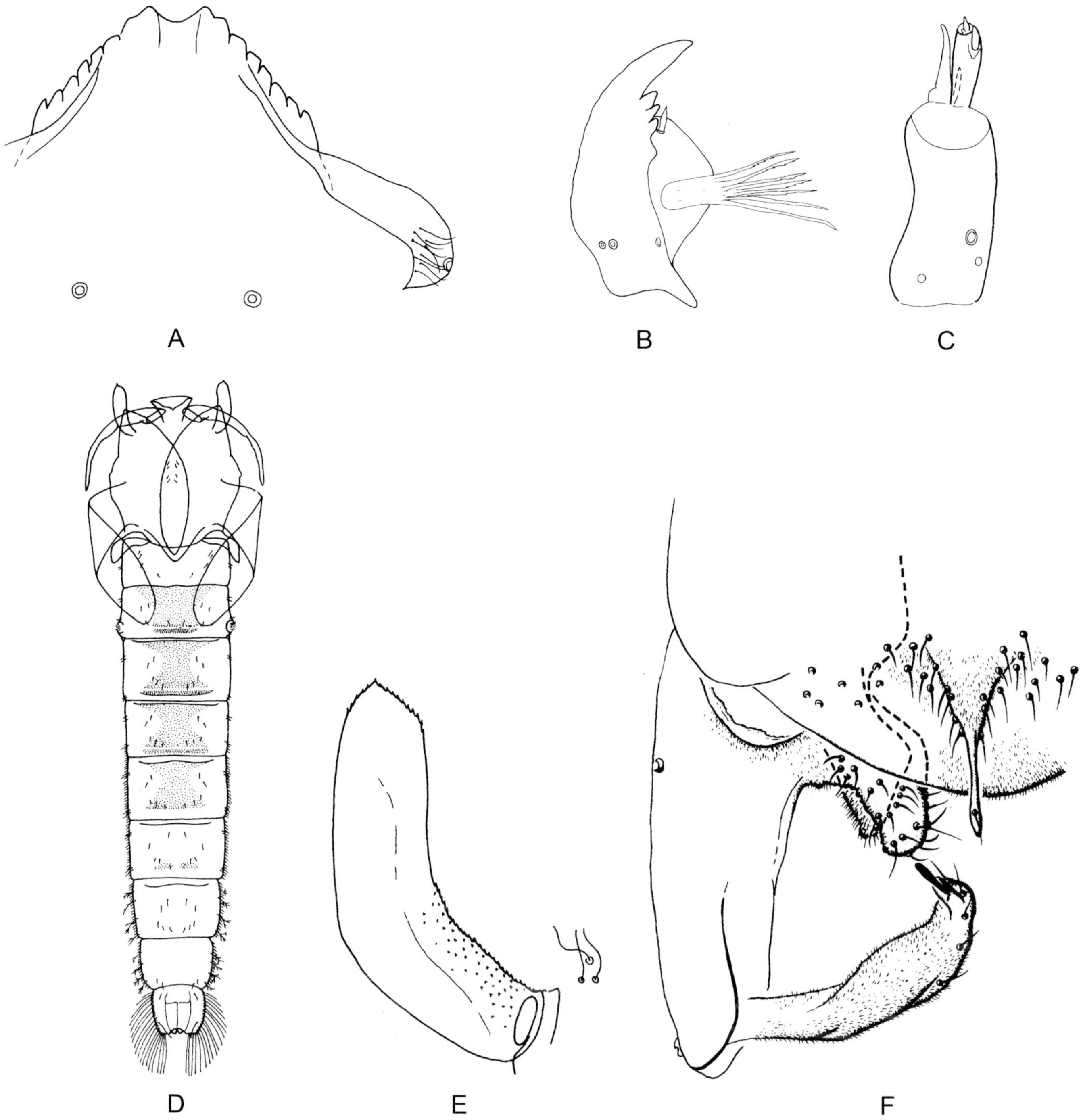

FIGURE 17 Monodiamesa. *A–C: Larva. A: mentum, B: mandible, C: antenna. D–E: Pupa. D: dorsal view of pupal excuvia, E: thoracic horn. F: Hypopygium of adult male. A–C from Sæther (1986); D–E from Langton (1991); F from Pinder (1978).*

– Prealars usually not extending anteriorly to level of median anepisternum (Fig. 27I); if extending, then dorsocentrals joined dorsally at midline of scutum . ... *Cricotopus*, 8

8. Superior volsella absent or more or less flattened, never distinctly produced. Inferior volsella often bilobed or double (Figs 24A–E) .... *Cricotopus* (*Cricotopus*)

– Superior volsella distinct, rounded or produced to form a hump. Inferior volsella always present and simple (Fig. 24F) .... *Cricotopus* (*Isocladius*)

9. Antenna with strong subapical seta (Fig. 54J) .... *Smittia*

– Antenna without strong subapical seta 10

10. Anal point very broad with scattered long setae (Fig. 34H) .... *Mesocricotopus*

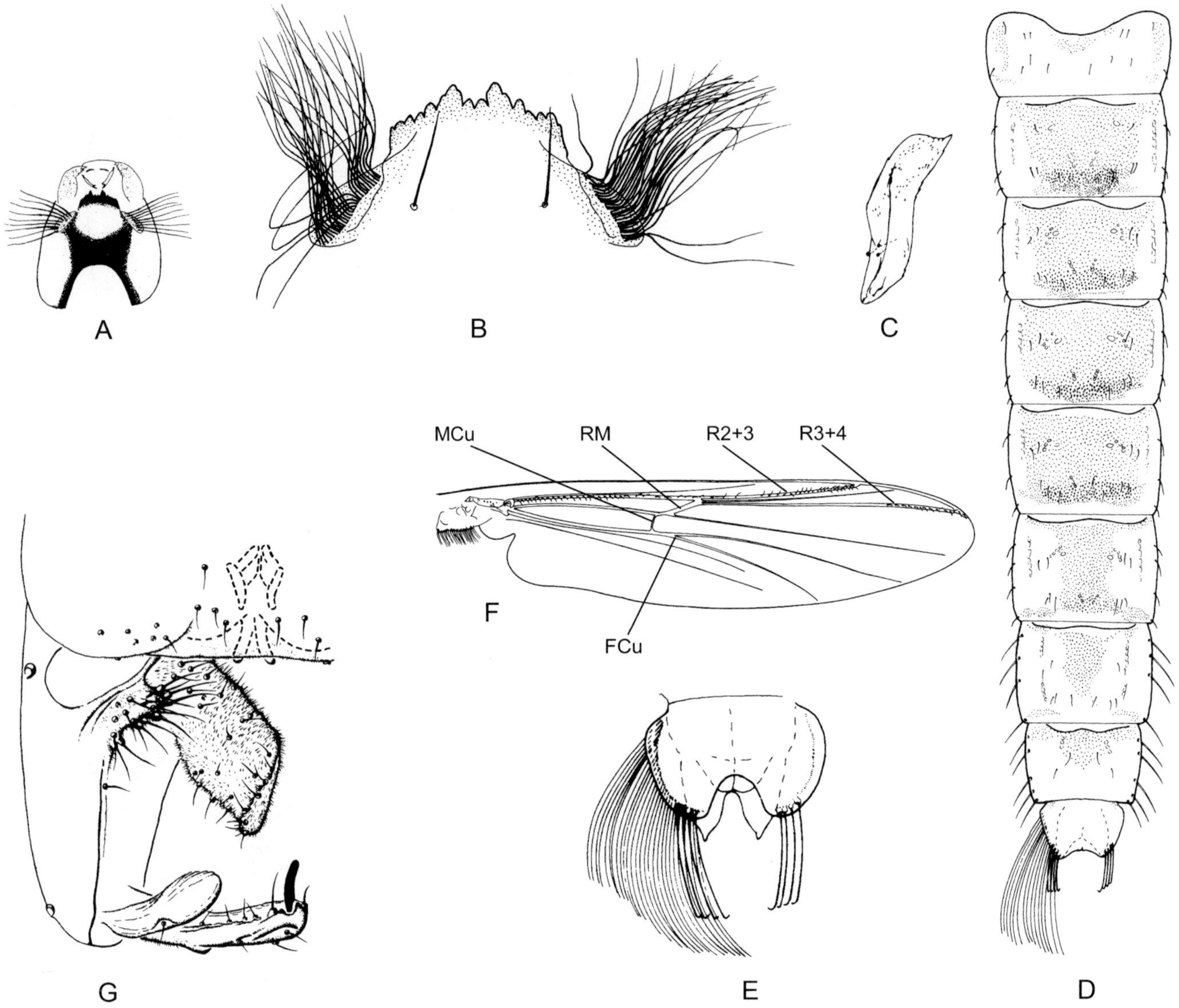

FIGURE 18 Prodiamesa olivacea. *A–B: Larva. A: ventral view of head, B: mentum. C–E. Pupa. C: thoracic horn, D: tergites 1–9, E: anal lobe, F–G: Adult male. F: wing, G: hypopygium. A–B from Moller Pillot (1984); C–F from Sæther (1986, 1989a); G from Pinder (1978).*

– Anal point narrow with few long setae (Figs 53G, H) .... *Rheocricotopus*

11. Wing membrane with setae at least on tip (Fig. 5E) 12

– Wing membrane bare (Fig. 5D) 15

12. R_{4+5} always and costa usually ending proximally to apex of M_{3+4} (Fig. 49K) .... *Paraphaenocladius*

– R_{4+5} usually and costa always ending opposite or distally to apex of M_{3+4} (Fig. 29C) 13

13. Costa not produced; apex of R_{4+5} rounded, with slight notch at apex of costa (Fig. 29C) .... *Heterotrissocladius*

– Costa produced (Fig. 48L); if only slightly produced, then apex of R_{4+5} meeting costa at an acute angle, without notch at apex of costa 14

14. Cu_1 more or less straight (Fig. 5E) .... *Metriocnemus*

– Cu_1 curved or sinuous (Fig. 48L) .... *Parametriocnemus*

15. Epimeron 2, posterior mesanepisternum 2 and usually preepisternum and dorsal antepronotum with setae. Often some humerals and/or prescutellars lanceolate (Fig. 32I) .... *Limnophyes*

– Epimeron 2, posterior mesanepisternum 2 and dorsal antepronotum always, preepisternum usually, bare. No lanceolate setae on scutum 16

16. Lateral spines on spurs of hind and mid tibia diverge from shaft of spur (Fig. 21G) .... *Chaetocladius*

– Lateral spines on spurs of hind and mid tibia pressed to shaft of spur (Fig. 5I) 17

17. Squama with at least 1 seta 18
– Squama bare 27
18. Pulvilli pad-like or pectinate, at least ½ length of claw (Fig. 51F); Cu_1 straight .... *Psectrocladius*
– Pulvilli absent, vestigial or small, never ½ length of claw.. 19
19. Hind tibia comb absent or vestigial, reduced to spiniform seta .... *Bryophaenocladius*
– Hind tibia comb normally developed (Fig. 5I) 20
20. Anal point absent 21
– Anal point present 22
21. Virga consisting of a dense cluster of spines (Fig. 56I). Eye with short wedge-shaped dorsomedial extension .... *Tokunagaia*
– Virga consisting of a small field of fine spinules. Eye rounded, without dorsomedial extension .... *Eukiefferiella*
22. Palp segment 3 with apical, finger-like projection (Fig. 19G) .... *Bryophaenocladius*
– Palp segment 3 without apical projection......... 23
23. Clypeus distinctly swollen, wider than pedicel width (Fig. 29A) 24
– Clypeus never distinctly swollen, always narrower than pedicel width (Fig. 29B) 25
24. Anal point broadened at apex; tergite 9 with a pair of strongly sclerotized, black projections (Fig. 37H); eye pubescent .... *Oliveridia tricornis*
– Anal point conical; tergite 9 without a pair of strongly sclerotized, black projections (Fig. 29D); eye bare .. .. *Heterotrissocladius subpilosus* group
25. Acrostichals absent 26
– Acrostichals present (cf. Figs 4E, 32I) 27
26. Superior volsella distinct, collar-like, hook-shaped or triangular (*e.g.* Fig. 44D, cf. Fig. 40 for explanation) . ... 32 (*Orthocladius*)
– Superior volsella scarcely defined, gently rounded or absent (Fig. 57J) .... *Tvetenia*
27. Acrostichals starting close to anterior margin of scutum, weak to strong, never hooked .... *Orthocladius*
– Acrostichals starting some distance from anterior margin of scutum, short, often hooked .... *Hydrobaenus*
28. Pulvilli absent or vestigial .... *Pseudosmittia*
– Pulvilli present and distinct (cf. Fig. 51F) 29
29. Costa not or only moderately extended beyond R_{4+5}. Anal point not extending beyond posterior margin of tergite 9 (Fig. 52H) .... *Pseudosmittia*
– Costa strongly extended beyond R_{4+5}. Anal point extending to or beyond posterior margin of tergite 9 (Fig. 20G) 30
30. 30. Anal point long, densely covered with coarse microtrichia (Fig. 20G) .... *Camptocladius*
– Anal point short and broad, bare or more or less covered with microtrichia (Figs 31G, 47J) 31
31. Scutum gently rounded, with microtrichial tuft (Fig. 47I) .... *Parakiefferiella*
– Scutum without microtrichial tuft ..... *Krenosmittia*
32. Hypopygium with well-developed dorsal part of inferior volsella (Fig. 40C), virga present (Fig. 40C) or absent 33
– Hypopygium without well-developed dorsal part of inferior volsella (Figs 38J–O), virga absent .... *Orthocladius* (*Eudactylocladius*)
33. Superior volsella well-developed, triangular to rounded (Fig. 44D)...... *Orthocladius* (*Orthocladius*)
– Superior volsella reduced or collar-like (Fig. 44A) 34
34. Anal point robust with rounded apex (Fig. 40) ... 35
– Anal point with pointed apex (Fig. 44E) 38
35. Eyes widely separated (Fig. 44J) 36
– Eyes extended dorsomedially (Fig. 44I) 37
36. Anal lobe of wing well developed, projecting (cf. Fig. 45H) .... *Orthocladius* (*Mesorthocladius*) *rousellae*
– Anal lobe of wing at most slightly producing (cf. Fig. 5E) .... *Orthocladius* (*Euorthocladius*)
37. Vein R_{4+5} bare; ventral part of inferior volsella not extended below dorsal part (Figs 41L, 41M) .... *Orthocladius* (*Mesorthocladius*)
– Vein R_{4+5} with setae or ventral part of inferior volsella prominently extended below dorsal part (Fig. 44E).. .... *Orthocladius* (*Orthocladius*)
38. Eyes widely separated (Fig. 44J) .... *Orthocladius* (*Pogonocladius*) *consobrinus*
– Eyes extended dorsomedially (Fig. 44I) .... *Orthocladius* (*Orthocladius*)

LARVAE

1. Anal end without procerci (Figs 19C, 20C, 52C); procercal setae present or absent; terrestrial or semiterrestrial species 2
– Procerci present with a variable number of procercal setae (Figs 31C, 49F, 50B); mostly aquatic species .. 6
2. Preanal and anal segments and posterior parapods bent at right angels to the axis of the rest of the body (Fig. 19C) 3
– Preanal and anal segments on same axis as rest of the body (Fig. 20C)................................ 4
3. Posterior parapods divided into two parts; anal tubules absent; one procercal seta present. Larvae are difficult to separate from *Bryophaenocladius* and are not yet with certainty recorded from Greenland .... [*Gymnometriocnemus*]

– Posterior parapods undivided; anal tubules present; no procercal setae present (Fig. 19C) .... *Bryophaenocladius*
4. S I and S II both bifid (Fig. 20D) 5
– S II never bifid (Fig. 54E) .... *Smittia*
5. Posterior parapods, procerci, procercal setae and anal claws absent (Fig. 20C) ....*Camptocladius*
– Posterior parapods and anal claws present (Fig. 52C) .... *Pseudosmittia*
6. Antenna at least half length of head (Figs 22A, 55A) 7
– Antenna shorter than half length of head 8
7. Antenna 4-segmented, longer than head (Figs 22A, 22E) ....*Corynoneura*
– Antenna 5-segmented, only slightly longer than half head-length (Figs 55A–B) ....*Thienemanniella*
8. One of the procercal setae as long as a quarter of body-length (Fig. 31C)................. *Krenosmittia*
– No procercal setae as long as a quarter of body-length 9
9. Preanal segment extending backwards over anal segment such that procercal setae are directed posteriorly (Fig. 49F) .... *Paraphaenocladius*
– Posterior segments normal 10
10. Ventromental plates conspicuous and extending laterally to outer teeth of flattened mentum (Fig. 21A); beard absent. S I never bifid 11
– Ventromental plates usually weak or vestigial; if large then beard present (*e.g.* Fig. 25A), or S I bifid, or both 15
11. Distinct to weak, paired, separate, sometimes pectinate labral lamellae present between or below S I setae (Fig. 21B) ....*Chaetocladius*
– Labral lamellae, if present, weak and simple, not paired, or at least fused at base 12
12. Antenna with 7 segments, segment 7 vestigial (Figs 28C–E) .... *Heterotrissocladius*
– Antenna with at most 6 segments, if 6 segments the sixth vestigial 13
13. S I with fewer than 10 branches, frequently with only 3–4; premandible with one apical tooth (Fig. 47B); first pair of mental teeth closely fused to median tooth (Fig. 47A) .... *Parakiefferiella*
– S I with at least 15 branches (Figs 30C, 37E); premandible with more than one apical tooth (Fig. 37C) 14
14. Mentum with a single wide median tooth (Fig. 37A) .... *Oliveridia*
– Mentum with paired median tooth (Fig. 30A) ....*Hydrobaenus*
15. Distinct beard present beneath ventromental plates (Figs 25A, 50A, D, 53A) 16
– Beard absent or very weak and only visible with difficulty 19
16. Mentum with four small median teeth; beard conspicuous with lateral setae larger than median setae (Fig. 25A)............................... *Diplocladius*
– Mentum with one or two median teeth; beard clearly visible with uniform setae (Figs 50A, 53A) 17
17. S I bifid (Fig. 53 B); procercus with an apical posteriorly directed spur ....18 (*Rheocricotopus*)
– S I palmate with 3–10 subeqaul lobes (Figs 50C, G); procercus with one or more sclerotized spurs at base (Fig. 50B) ....*Psectrocladius*
18. Body setae long and strong, the longest seta more than 100 µm ....*Rheocricotopus* (*Psilocricotopus*)
– Body setae short, up to 85 µm .... *Rheocricotopus* (*Rheocricotopus*)
19. Anal tubules absent; weak indistinct beard present; marine species ....*Halocladius*
– Anal tubules present; freshwater species 20
20. Median tooth of mentum broad, low, domeshaped and paler than lateral teeth (Fig. 46A); S I bifid (Fig. 46C); weak beard present, but difficult to see .. .. *Paracladius*
– Median tooth of mentum not paler than lateral teeth 21
21. S I simple or with very weak apical dentation; inner margin of mandibles with spines (Figs 26B, 56F) .. 22
– S I clearly bifid, palmate or pectinate; inner margin of mandibles with or without spines............. 23
22. S I and S II simple and strong; S III simple and weak (Fig. 56E); inner margin of mandible clearly serrated with long spines (Fig. 56F) .... *Tokunagaia*
– S I simple and weak; inner margin of mandible serrated with short spines (Fig. 26B) ......*Eukiefferiella*
23. S I bifid 24
– S I palmate or pectinate 31
24. Mentum with paired median teeth (Fig. 34A); antennal blade extends well beyond fifth antennal segment Fig. 34D) .... *Mesocricotopus*
– Mentum with single median tooth; antennal blade does not reach terminal antennal segment 25
25. Abdominal segments laterally with tufts of setae (Fig. 23 D). Some *Limnophyes* also have tufts of setae, but then S I weakly serrated or simple .... *Cricotopus*, 26
– Abdominal segments at most with simple setae. Some *Cricotopus* have simple body setae and may be impossible to distinguish from *Orthocladius* .... *Orthocladius*, 27
26. Pecten epipharyngis consist of 3 scales (Fig. 23E) ... . *Cricotopus* (*Cricotopus*)

– Pecten epipharyngis consist of a single conical scale (Fig. 23F) .... *Cricotopus* (*Isocladius*)
27. Seta interna on mandible present (Figs 38B, 39B, F, 43B, F) 28
– Seta interna on mandible absent (Figs 41B, F, J) .... *Orthocladius* (*Mesorthocladius*)
28. Antenna with 4 segments (Fig. 45C) .... *Orthocladius* (*Pogonocladius*) *consobrinus* p.
– Antenna with 5 segments 29
29. Head capsule brown 30
– Head capsule yellow ....*Orthocladius* (*Orthocladius*)
30. Head capsule reddish brown .... *Orthocladius* (*Eudactylocladius*)
– Head capsule light to dark brown, not reddish .... *Orthocladius* (*Euorthocladius*)
31. Abdominal segments laterally with tufts of setae (cf. Fig. 23D) 32
– Abdominal segments without tufts of setae 33
32. S1 simple with at most small serrations apically .... *Limnophyes*
– S1 divided in several irregular more or less strongly serrated branches (Fig. 58) .... Orthocladiinae genus A
33. Some setae of abdomen at least half the length of the segment bearing them; two lateral setae very close together (Fig. 57E) .... *Tvetenia*
– No abdominal setae as long as this 34
34. Premandible without a brush; labral lamellae absent or weak 35
– Premandible with a brush; labral lamellae distinct (Figs 21B, 35E; few *Chaetocladius* have indistinct labral lamellae) 37
35. Base of mentum distinctly and uniquely sclerotized; setae submenti well within the base of mentum (Figs 48A, D); S I distinctly serrated (Figs 48C, F); weak labral lamellae present ... *Parametriocnemus*
– Base of mentum not uniquely sclerotized; setae submenti placed at base of mentum (Fig. 47A); S I weaker serrated; labral lamellae absent 36
36. Ventromental plates distinct, extending to outer mental teeth; no tooth at base of mentum (Fig 47A) .... *Parakiefferiella*
– Ventromental plates absent or vestigial; distinct tooth at base of mentum (Figs 32A, 32D) .... *Limnophyes*
37. Labral lamellae consist of bell-shaped structures more or less deeply divided, plumosed and fused at base (Fig. 35E); procercus often with a distal posteriorly directed spur; procercal setae short and strong (Fig. 35C); antennae often reduced (Fig. 35F)....... .... *Metriocnemus*
– Labral lamellae if present separated at base and either pectinate, plumose or simple (Fig. 21B); procercus without spur; procercal setae of normal length; antennae shorter than mandible, but never reduced ....*Chaetocladius*

PUPAE

1. Anal lobe with a full fringe or partial fringe of setae and 3 macrosetae; fringe setae may be sparse or dense, short or long 2
– Anal lobes without a fringe, with or without 3 macrosetae; anal lobe may be reduced 14
2. Thoracic horn absent 3
– Thoracic horn present 4
3. Wing sheaths with pearl rows (Fig. 22G) ....*Corynoneura*
– Wing sheaths without pearl row (Fig. 55E) .... *Thienemanniella*
4. Most of the lateral setae on segment 8 lamellar, usually broadly lamellar (Fig. 50I) 5
– All lateral setae on segment 8 hair-like (*e.g.* Figs 38G, H) 13
5. Tergite 4 with distinct spine patches or rows in the median field (*e.g.* Fig. 50I) 6
– Tergite 4 with simple shagreen only, distinct spine patches or rows not present (Figs 28K, L) 8
6. Tergite 4 with at least 1 distinct median spine group and 1 or more posterior rows of spines 7
– Tergite 4 without median spine group, but with 1 or more posterior spine rows (Fig. 53F) .... *Rheocricotopus* (*Psilocricotopus*)
7. One median spine patch on tergites 4–6 or 2 median spines patches on tergites 3–6 or 4–6; spine patches distinct (Figs 50J–L).................. *Psectrocladius*
– One median spine patch on tergites 4–6 (eventually on tergites 3–6); spines in patches not much stronger than the shagreen spinules (Fig. 53E) .... *Rheocricotopus* (*Rheocricotopus*)
8. Pedes spurii B present on segment 2, sometimes vestigial (Figs 28J, K, L) 9
– Pedes spurii B absent on segment 2 (Fig. 28I) ....*Heterotrissocladius subpilosus* group
9. Pedes spurii B present on segment 3 (Fig. 34F) .... *Mesocricotopus*
– Pedes spurii B absent on segment 3 (*e.g.* Fig 30I) ..10
10. Fringe of setae on anal lobe most concentrated in the basal half of each lobe; setae of fringe short and often sparse (Figs 30E, F) 11
– Fringe of setae on anal lobe not concentrated in the basal half of each lobe; setae of fringe short or long 12

11. Shagreen on tergites 3–5 extensive (Fig. 30I); anal lobes without apical spines .... *Hydrobaenus*
– Shagreen on tergites 3–5 restricted to a narrow, central, longitudinal field, which is connected medially to a posterior band of spinules (Fig. 36G); anal lobes with minute apical spinules .... *Oliveridia*
12. Posterior margin of sternite 8 with special structures, a row of spines in the male and 2 strong chitinized, triangular lobes in the female (Fig. 28L) .... *Heterotrissocladius*
– Posterior margin of sternite 8 of both male and female without such structures .... *Hydrobaenus*
13. Wing sheaths with pearl row; pedes spurii B on segment 2 long tapering to a point (Figs 48I, K) .... *Heterotrissocladius*
– Wing sheaths without pearl row; pedes spurii B on segment 2 normal; anal lobe fringe consisting of a few short not true setae (Fig. 42D) .... *Orthocladius* (*Mesorthocladius*) *vaillanti*
14. Anal lobes with 3 terminal or lateral macrosetae (+1 median seta in *Tvetenia*).......................... 15
– Anal lobes with 0–2 or more than 3 macrosetae; anal lobe sometimes reduced.......................... 33
15. Each anal lobe with 3 strong terminal macrosetae and a median seta of nearly equal size. Tergites 3–5 with few recurved spines (hooklets) on the posterior margin (Fig. 57G) .... *Tvetenia*
– Anal lobe without median setae 16
16. Thoracic horn present 17
– Thoracic horn absent 30
17. Tergite 3–5 with a single row of recurved spines (hooklets) on the posterior margin (Fig. 26E) .... *Eukiefferiella*
– Tergite 3–5 without recurved spines (hooklets) on the posterior margin 18
18. Wing sheaths with pearl rows; pedes spurii B on segment 2 long and tapering (Figs 48I, K) .... *Parametriocnemus*
– Wing sheaths without pearl rows; pedes spurii B on segment 2 not long and tapering...................19
19. Anal lobes pointed at apex, points may be long or short (Figs 31D, E, 47G, H) 20
– Anal lobes rounded or truncate at apex, but accessory spines may be present at the apex 21
20. Anal lobes extended in long tapering points; posterior margins of tergites 2–7 with rows of spines (Figs 31D, E) .... *Krenosmittia*
– Anal lobes with relatively short tapering points; posterior margins of tergites 2–7 without rows of spines (Fig. 47H) .... *Parakiefferiella*
21. Tergites 2–8 with posterior rows of long needle-like spines (Fig. 32G) .... *Limnophyes*
– Tergites 2–8 without posterior rows of long needle-like spines 22
22. Anal lobes with cylindrical median sections and lateral flat sections; anal lobe macrosetae inserting at the apex of the cylindrical section (Fig. 25E) .... *Diplocladius*
– Anal lobes without this combination 23
23. Hooklets absent on posterior margin of segment 2, although minute anteriorly directed spinules may be present (Figs 38F, H) 24
– Hooklets present on posterior margin of segment 2 25
24. Tergites 2(3)–6 with a pair of median spine patches (Fig. 38E) .... *Orthocladius* (*Eudactylocladius*)
– Tergites 2(3)–6 without a pair of median spine patches (Fig. 21F); anal lobe megasetae spine-like .... *Chaetocladius*
25. Tergites 3–7 with a pair of median spine patches (Fig. 45G) .... *Orthocladius* (*Pogonocladius*) *consobrinus*
– Tergites 3–7 without a pair of median spine patches 26
26. Thoracic horn short and broad with numerous spinules (Fig. 46E) .... *Paracladius*
– Thoracic horn slender, if short and broad then without spinules 27
27. Thoracic horn filamentous, tapered from a swollen base, 600–900 µm long (Fig. 42E) .... *Orthocladius* (*Mesorthocladius*) *frigidus*
– Thoracic horn slender, not filamentous, 150–400 µm long (Figs 23H, I, 43O, R) 28
28. Hooklets on posterior margin of segment 2 usually in 2 distinct rows which extend along half or more of the segment (Fig. 23G); setae of anal lobe often short and straight; anal lobes without terminal spines .... *Cricotopus*, 29
– Hooklets on posterior margin of segment 2 usually in 3 or more irregular rows, often restricted to one third of the segment (Fig. 43M); setae of anal lobe straight or curved; anal lobes sometimes with terminal spines (Fig. 43N) ....*Orthocladius* (*Orthocladius*)
29. Shagreen on tergites usually undivided, if divided into anterior and posterior fields, then anterior field extending to anterolateral margin of tergite and all anal macrosetae subequal in size (Fig. 23K) .... *Cricotopus* (*Isocladius*)
– Shagreen on tergites usually distinctly divided into anterior and posterior fields, if fields not distinctly divided, then anterior field not extending to

anterolateral margin of tergite and medial anal macrosetae smaller than rest of the macrosetae (Fig. 23J) .... *Cricotopus* (*Cricotopus*)

30. Tergites 3–5 with a single row of moderately large recurved spines (hooklets) on the posterior margin (Figs 57F, 57G) .... *Tokunagaia*
– Tergites 3–5 without a single row of recurved spines 31
31. Tergites 2–8 with posterior rows of very long needle-like spines (Fig. 32G) .... *Limnophyes*
– Tergites 2–8 without posterior rows of very long needle-like spines 32
32. Posterior margin of tergite 2 without a group or row of hooklets; anal lobe macrosetae less (often much less) than the length of the anal lobe (Figs 35H, J) .. .. *Metriocnemus*
– Posterior margin of tergite 2 with a group or row of hooklets; anal lobe macrosetae short (about ¼ of the length of the anal lobe, Figs 27E, F); marine species .... *Halocladius*
33. Anal lobe (actually tergite 9) with 4 very fine macrosetae (Fig. 20F); terrestrial species ...*Camptocladius*
– Anal lobe with 0–2 macrosetae; aquatic or semi aquatic species 34
34. Anal lobe with at least 1 distinct macroseta 35
– Anal lobe without macrosetae; anal lobe may be reduced 37
35. Thoracic horn absent ....*Metriocnemus*
– Thoracic horn present, but may be very small 36
36. Pedes spurii B present on segment 2 and 3 (Fig. 42B); wing sheaths without pearl row .... *Orthocladius* (*Mesorthocladius*) *rousellae*
– Pedes spurii B present on segment 2, but absent on segment 3; wing sheaths with pearl row (Fig. 49G) .. .. *Paraphaenocladius*
37. Thoracic horn present, but may be very small; tergites not completely covered with evenly sized shagreen (Figs 39J, M) .... *Orthocladius* (*Euorthocladius*)
– Thoracic horn absent; tergites almost completely covered with evenly sized shagreen (Fig. 52G) 38
38. Conjunctives armed at least dorsally (Fig. 52G) .... *Pseudosmittia*
– Conjunctives generally unarmed, although extensive shagreen of tergites may extend onto conjunctives (Fig. 19D) 39
39. Precorneals and antepronotals thorn- or spike-like, if hair-like base about 3 µm (Fig. 19F) .... *Bryophaenocladius*
– Precorneals and antepronotals hair-like, narrow 1 µm or less (Figs 52E, 54H) 40
40. Antepronotal setae 0–1 (Fig. 54H) .... *Smittia*
– Antepronotal setae 2–3 (Fig. 52E) .....*Pseudosmittia*

***Bryophaenocladius vernalis* (Goetghebuer, 1921) Fig. 19**

Henriksen 1939: *Orthocladius* (*O.*) cf. *vernalis*.

B. cf. *vernalis* is recorded by Henriksen (1939, no. 116) from the Kangerlussuaq/Søndre Strømfjord region, but the identification is uncertain and it is not mentioned by Ashe & Cranston (1990) and Oliver *et al.* (1990). However, I have found larvae belonging to this genus in streams both in the Kangerlussuaq/Søndre Strømfjord region and in the Qeqertarsuaq/Disko area.

BIOLOGY. The larvae of *Bryophaenocladius* live mainly in terrestrial or semi-terrestrial environments, but also in submersed mosses in ponds and lakes.

DISTRIBUTION. nSW.

Central Europe.

***Camptocladius stercorarius* De Geer, 1781 Fig. 20**

Henriksen 1939: *Smittia stercoraria* De Geer.

Recorded from "North Greenland" (nSW) by Holmgren (1872), referred by Lundbeck (1898) and Henriksen (1939, no. 129). Also Strenzke (1940) mentioned it as found in Greenland. However, no new findings are published and it is not mentioned from Greenland by Ashe & Cranston (1990) and Oliver *et al.* (1990).

BIOLOGY. The larvae live in soil rich in organic matter and are common in dung from cows and horses. They feed on decomposing microbes.

DISTRIBUTION. ?N.

Holarctic (including Iceland), widely distributed.

***Chaetocladius* Kieffer, 1911 Fig. 21**

The genus *Chaetocladius* needs revision and there is at present no valid key for identification of species. I have found larvae, pupae and adults of at least 3 different *Chaetocladius* species in streams in the Kangerlussuaq/Søndre Strømfjord and the Narsaq area.

The *Chaetocladius* larvae cover a large range of habitats from pristine springs to waste water in percolating systems of sewage plants. They are partly truly aquatic or often semi-aquatic in wet mosses along borders of streams and ponds, feeding on the biofilm covering the substrate.

***Chaetocladius adsimilis* Goetghebuer, 1933**

Henriksen 1939: *Orthocladius* (*Chaetocladius*) *adsimilis* Goetg.

DISTRIBUTION. NE. Described on specimens from Geographical Society Ø, but not recorded later.

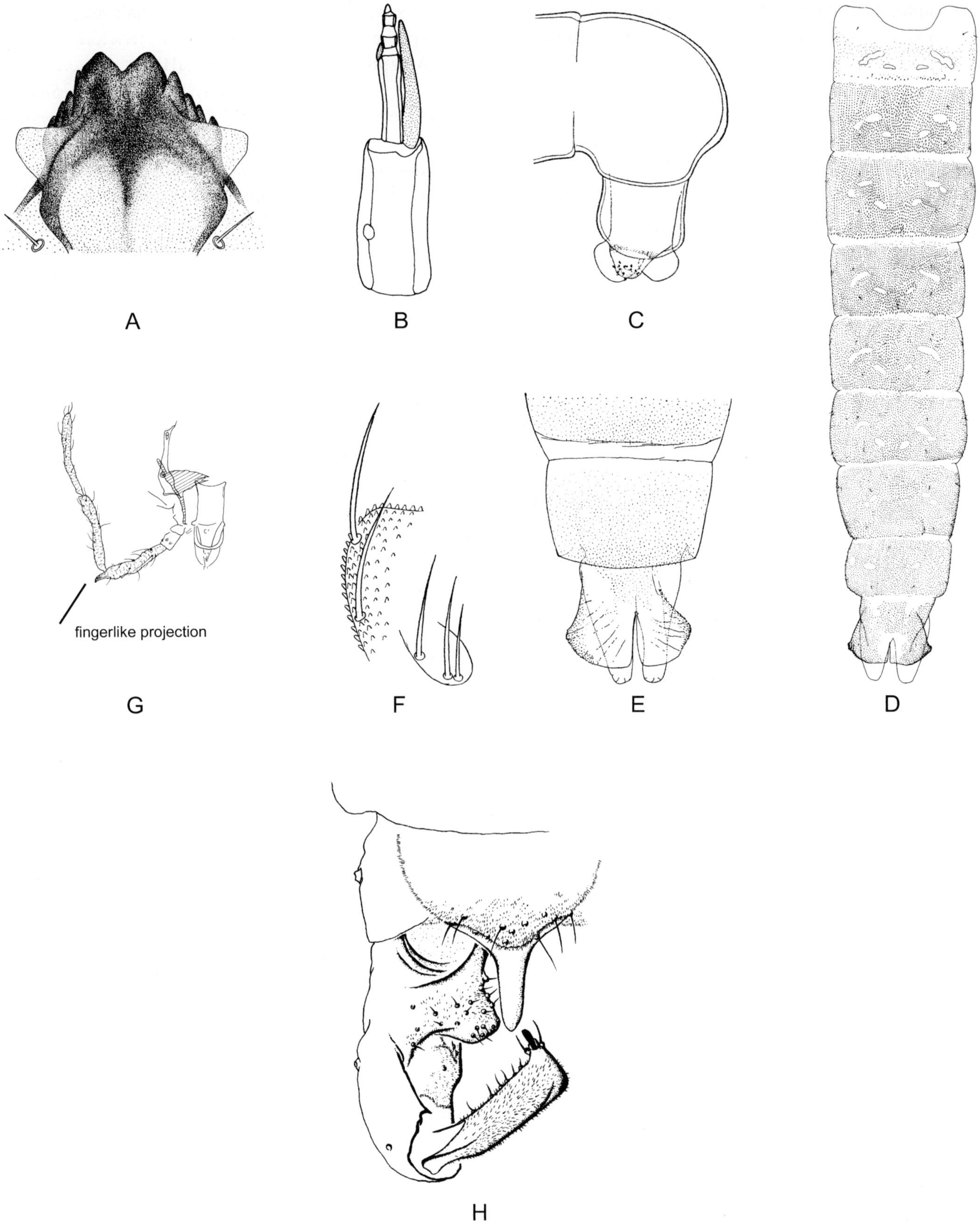

FIGURE 19 Bryophaenocladius. *A–C: Larva. A: mentum, B: antenna, C: anal end. D–F: Pupa. D: tergites 1–9. E: anal lobe, F: thorax area with antepronotals and precorneals. G–H: Adult male. G: palp showing fingerlike projection on segment 3, H: hypopygium of* B. vernalis. *A & E from Strenzke (1942); B from Cranston (1983); C from Thienemann (1944); D from Coffman* et al. *(1986); F from Strenzke (1951); G from Sæther (1973); H from Pinder (1978).*

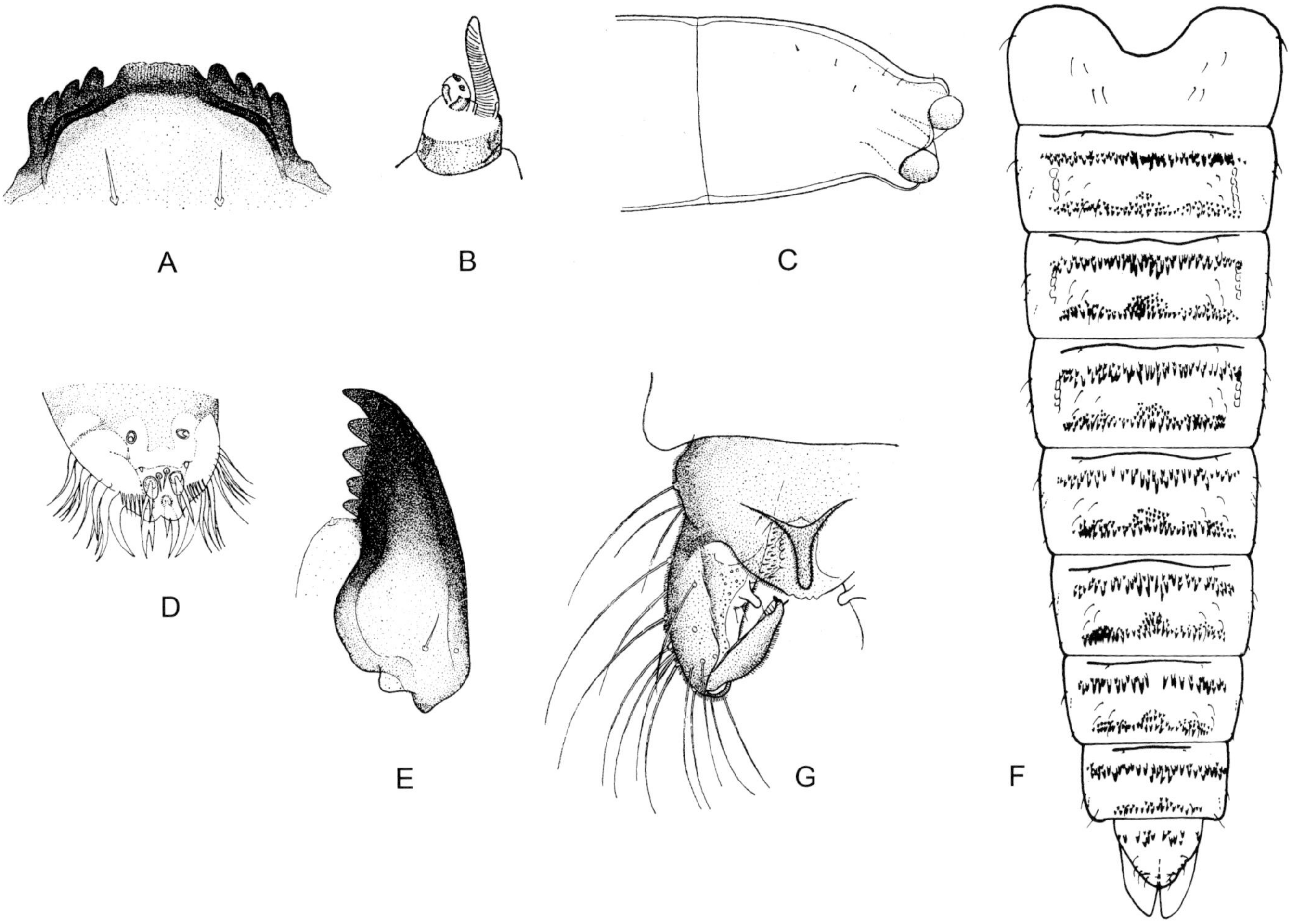

FIGURE 20 Camptocladius. *A–E: Larva. A: mentum, B: antenna, C: anal end, D: labrum, E: mandible. F: Tergites 1–9 of pupa. G: Hypopygium of adult male. A–E & G from Strenzke (1940); F from Coffmann* et al. *(1986).*

Corynoneura Winnertz, 1846 Fig. 22

The genus needs revision. Incomplete keys to adult males are given by Schlee (1968) and Hirvenoja (1988). Recently Sæther & Wang (2009) provided a key to all adult males of the genus. Larvae of *Corynoneura* are abundant in all ponds and smaller lakes in the vegetated areas in the Narsaq and the Kangerlussuaq/Søndre Strømfjord areas and are probably present all over Greenland. Some species have also been recognized in fast flowing mountain streams. They are free living, feeding on deposited material in surface sediments. The genus is Holarctic and widely distributed.

***Corynoneura arctica* Kieffer, 1923**

Henriksen 1939: not recorded.

Described on material from Novaya Zemlya and redescribed by Hirvenoja (1988).

DISTRIBUTION. ? Greenland (Oliver & Dillon 1997, without specified locality).

Nearctic, arctic, widely distributed (Oliver & Dillon 1997).

***Corynoneura celeripes* Winnertz, 1852**

Henriksen 1939: not recorded.

DISTRIBUTION. ? S, SW. Reported from Maniitsoq/Sukkertoppen (as *C. atra* Winnertz, 1852) and Qaqortoq/Julianehåb by Lundbeck (1898), but was not quoted by Henriksen (1939). Oliver *et al.* (1990) report *C. celeripes* from Greenland without specified locality.

Holarctic, widely distributed.

***Corynoneura scutellata* Winnertz, 1846**

Reported by Goetghebuer (1933) from Loch Fyne and by Andersen (1937) from Ella Ø, both NE. In the same paper Andersen synonymized *C. arctica*, *C. atra* and *C. celeripes* with *C. scutellata*. However, at present *arctica*, *celeripes* and *scutellata* are all valid species. *C. scutellata* is redescribed by Hirvenoja (1988). Further, *C. scutellata* is reported from Greenland by both Ashe & Cranston (1990) and Oliver *et al.* (1990).

DISTRIBUTION. ? Greenland (the records need confirmation).

Holarctic, widely distributed.

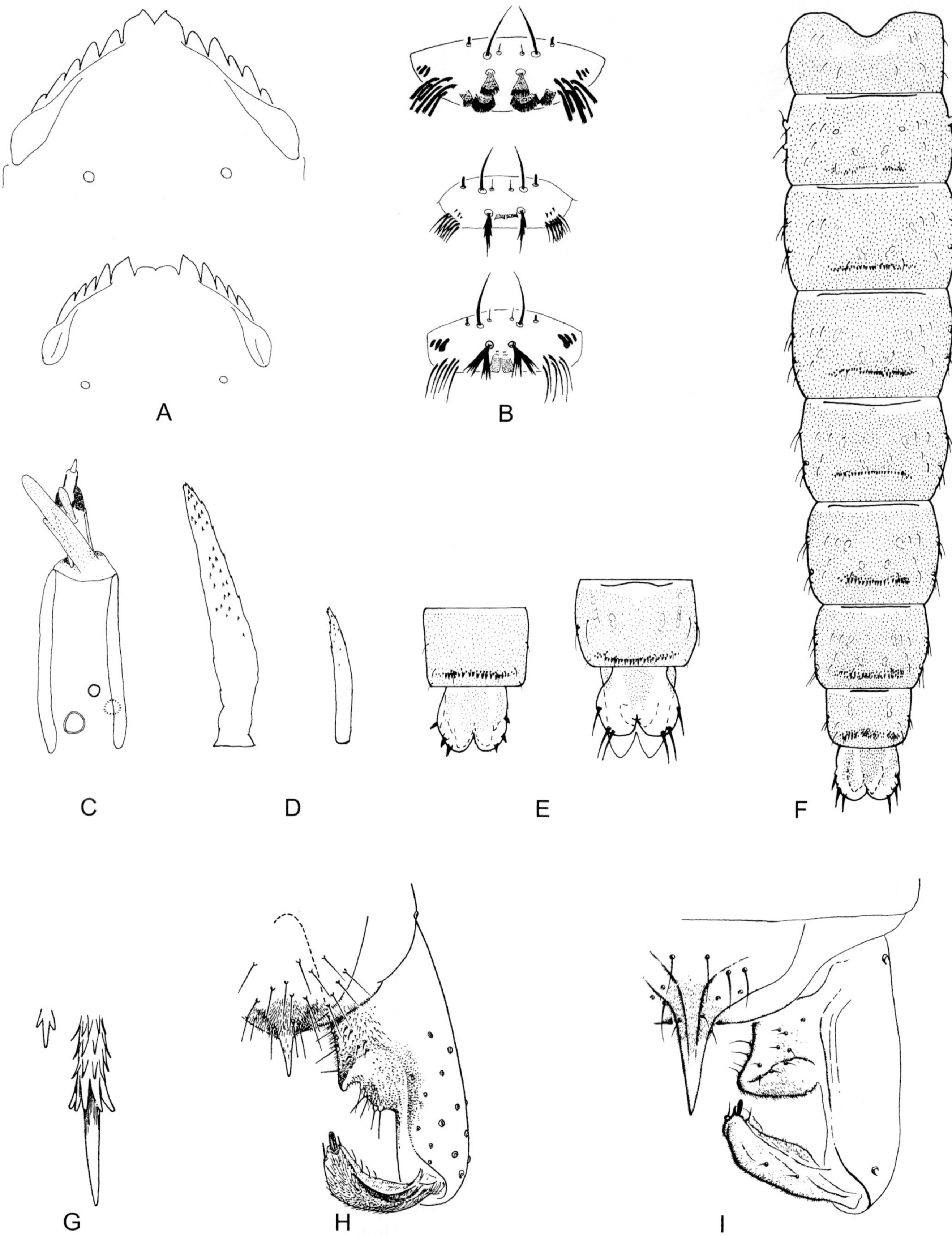

FIGURE 21 Chaetocladius *spp. A–C: Larva. A: mentum, B: labrum, C: antenna. D–E: Pupa. D: thoracic horn, E: anal lobe, F: tergites 1–9. G–I: Adult male. G: spurs at hind tibia, H: hypopygium of* C. gracilis *I: hypopygium of* C. piger. *A–C from Moller Pillot (1984); D–F from Coffman* et al. *(1986); G–H from Brundin (1956); I from Pinder (1978).*

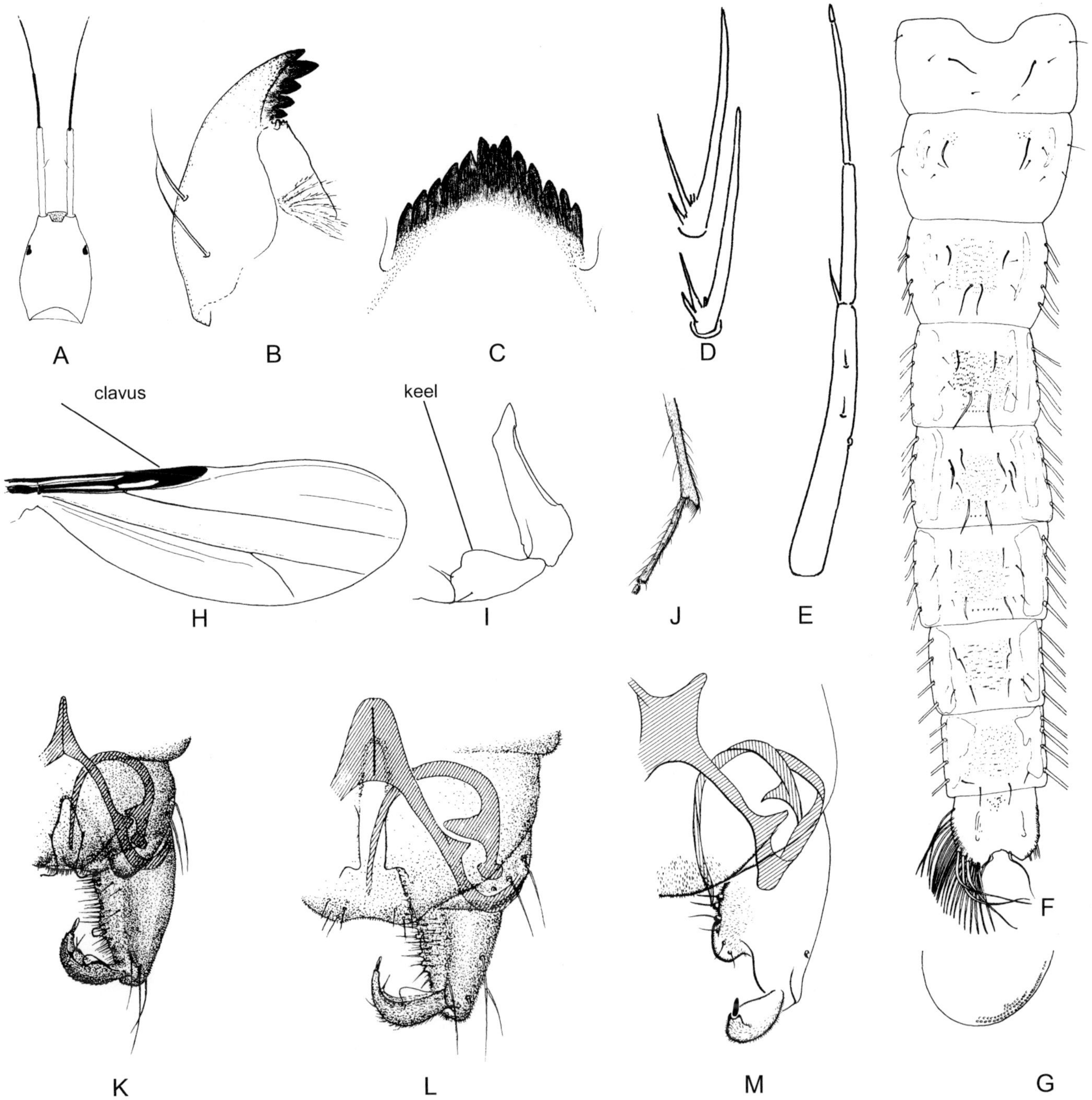

FIGURE 22 Corynoneura. *A–E: Larva. A: dorsal view of head, B: mandible, C: mentum, D: basiventral spine of posterior parapod, E: antenna. F–G: Pupa. F: tergites 1–9, G: pearl row on distal end of wing sheath. H–M: Adult male. H: wing, I: fore trochanter, J: fore tibia and tarsus 1, K: hypopygium of* C. scutellata, *L: hypopygium of* C. arctica, *M: hypopygium of* C. celeripes. *A from Moller Pillot (1984); B–E & H–K from Hirvenoja (1988); F–G from Coffman* et al. *(1986); J & M from Pinder (1978).*

Cricotopus van der Wulp, 1874 Figs 23–24

The genus *Cricotopus* is very large, probably with more than 100 species in the Holarctic area. The Palaearctic species were thoroughly revised by Hirvenoja (1973). He provides keys to adults, pupae and larvae. The arctic male adult species of the Nearctic region are reviewed by Oliver & Dillon (1988). The larvae are mainly found in ponds and lakes, but a few species are restricted to running waters. They live in small tubes made by detrital particles, feeding on algae and other micro-organisms.

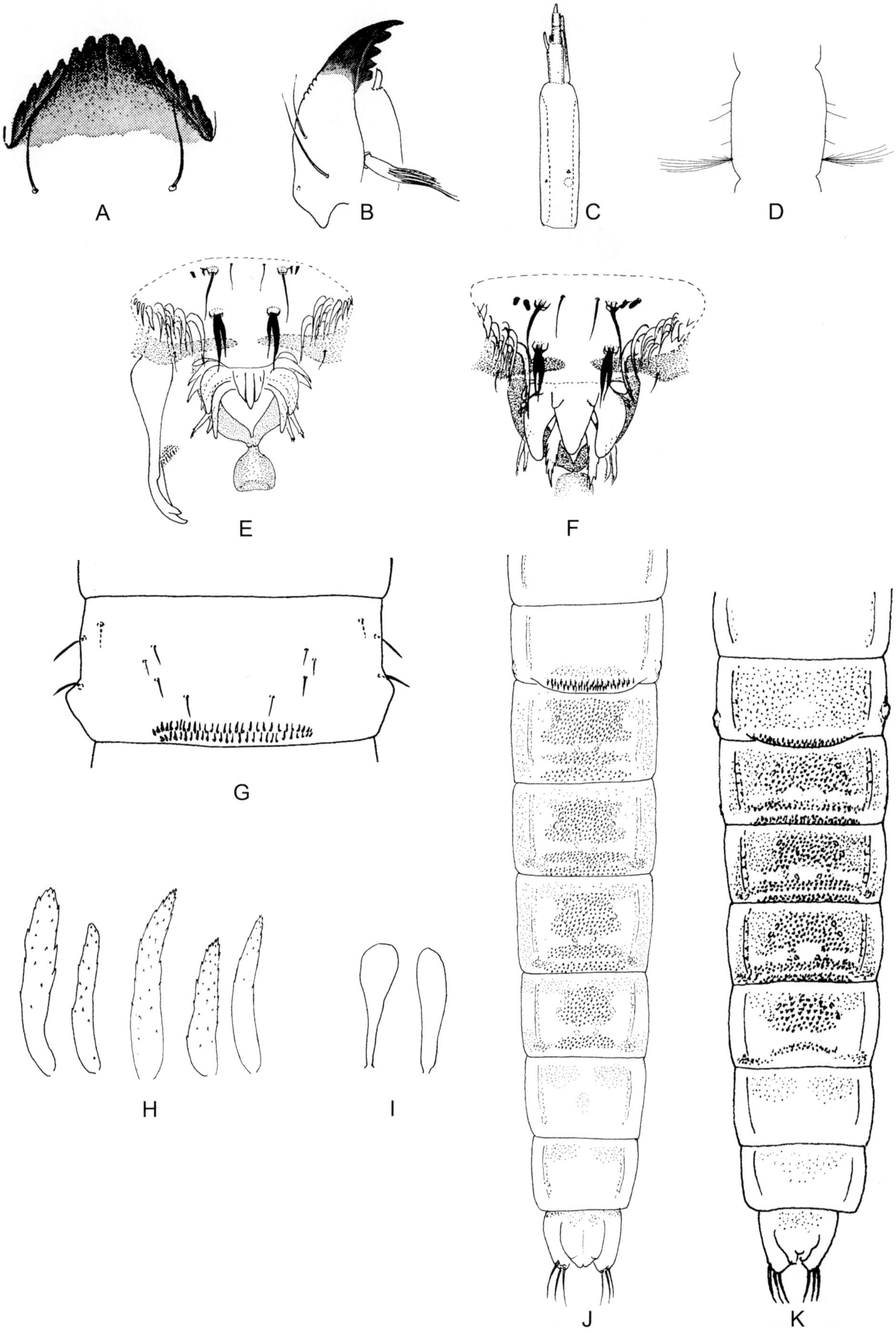

FIGURE 23 Cricotopus *1. A–F: Larva. A: mentum, B: mandible, C: antenna, D: dorsal view of abdominal segment. E–F: labrum and epipharyngeal region of C.* (Cricotopus) *and C.* (Isocladius). *G–K: Pupa. G: tergite 2 with posterior hooklets, H–I: thoracic horn of C.* (Cricotopus) *and C.* (Isocladius), *J–K: tergites 2–9 of* C. (Cricotopus) *and* C. (Isocladius). *A–C & E–K from Hirvenoja (1973); D from Moller Pillot (1984).*

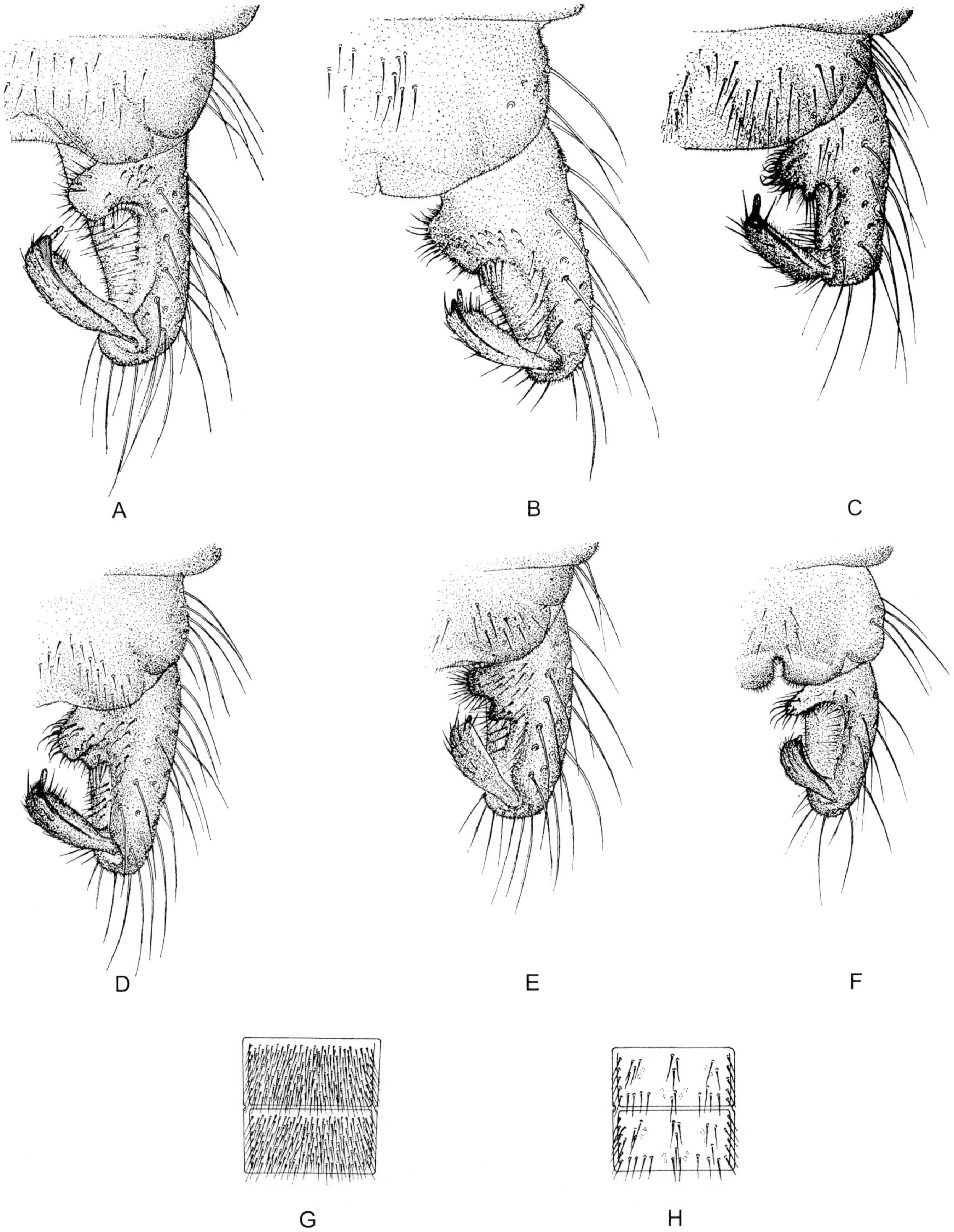

FIGURE 24 Cricotopus 2. *A–F: Dorsal view of hypopygium of A:* C. (Cricotopus) cumulatus, *B:* C. (C.) gelidus, *C:* C. (C.) lestralis, *D:* C. (C.) pilosellus, *E:* C. (C.) tibialis, *F:* C. (Isocladius) laricomalis. *G–H: Tergites 3–4 of G:* C. (C.) pilosellus *H:* C. (I.) laricomalis. *A–H from Hirvenoja (1973).*

***Cricotopus* (*Cricotopus*) *cumulatus* Hirvenoja, 1973**
Henriksen 1939: not recorded.

Described on adults from Peary Land (Nedre Midsommersø). Larva and pupa are unknown.

BIOLOGY. Oliver & Dillon (1988) suggested the larvae to live in large oligotrophic lakes.

DISTRIBUTION. N.

Holarctic, arctic (northern Sweden and northern Canada; Oliver *et al.* 1990).

***Cricotopus* (*Cricotopus*) *gelidus* (Kieffer, 1922)**
Henriksen 1939: not recorded.

Described on material from Novaya Zemlya. It is not included in the survey of the Nearctic arctic *Cricotopus* species given by Oliver & Dillon (1988). Larva and pupa are unknown.

DISTRIBUTION. ? Greenland (without specified locality; Oliver *et al.* 1990).

Holarctic (Norway; Schnell & Aagaard 1996. Canada: Northwest Territories; Oliver *et al.* 1990).

***Cricotopus* (*Cricotopus*) *lestralis* (Edwards, 1924)**
Henriksen 1939: not recorded.

Described from Spitsbergen and redescribed by Hirvenoja (1973) and Oliver & Dillon (1988). Larva and pupa are described by Oliver & Dillon (1988).

BIOLOGY. The larvae live in small vegetated ponds (Oliver & Dillon 1988).

DISTRIBUTION. N (Peary Land: Nedre Midsommersø).

Nearctic (northern Canada, widely distributed).

***Cricotopus* (*Cricotopus*) *pilosellus* Brundin, 1956**
Henriksen 1939: not recorded.

The larva is unknown.

DISTRIBUTION. ? Greenland. The species is not recorded from Greenland by Hirvenoja (1973) or by Oliver & Dillon (1988), but Oliver *et al.* (1990) and Oliver & Dillon (1997) included it as occurring in Greenland, although without specification of locality.

Holarctic, widely distributed in northern parts.

***Cricotopus* (*Cricotopus*) *tibialis* (Meigen, 1804)**
= *C. basalis* Stæger, 1845.

Henriksen 1939: *Cricotopus basalis* Stæger, *C. polaris* Kieff, *Trichocladius ?baciliger* Kieff.

BIOLOGY. The larvae are reported from nearly all kinds of freshwater, often associated with filamentous algae. They are eurythermal, found in habitats with large temperature ranges.

DISTRIBUTION. SW, N. Oliver & Dillon (1988) confirm records from Maniitsoq/Sukkertoppen and Peary Land: Nedre Midsommersø, but probably the species is widely distributed in Greenland.

Holarctic, arctic-temperate, widely distributed.

***Cricotopus* (*Isocladius*) *laricomalis* Edwards, 1932**
Henriksen 1939: *Cricotopus glacialis* Edw. [misidentification].

BIOLOGY. The larvae live in permanent ponds.

DISTRIBUTION. NE (only recorded from Ella Ø).

Holarctic, widely distributed in northern parts.

NOTE. Andersen (1937) reported *C. glacialis* Edwards, 1922 from Ella Ø (NE), but according to Hirvenoja (1973) and Oliver & Dillon (1988) it is in fact *C. laricomalis*.

***Diplocladius cultriger* Kieffer, 1908 Fig. 25**
Henriksen 1939: not recorded.

Easily recognizable in all three stages. Recorded by Oliver *et al.* (1990) from Greenland, but without specified locality given.

BIOLOGY. The larvae are primarily found in running waters and are common in most European lowland streams. Emergence takes place in early spring, often before snow melt.

DISTRIBUTION. nSW. I have found larvae and pupae in low numbers in a few streams in the Kangerlussuaq/Søndre Strømfjord and Qeqertarsuaq/Disko areas.

Holarctic, widely distributed in northern parts.

***Eukiefferiella claripennis* Lundbeck, 1898 Fig. 26**
Henriksen 1939: *Orthocladius* (*O.*) *claripennis* Lundb.

Described from Qaqortoq/Julianehåb (S) and redescribed by Oliver (1970).

BIOLOGY. The larvae are primarily found in running waters and are common in most European lowland streams. They are eurythermal, living on plants and stones and feeding on algae.

DISTRIBUTION. S, SW. I have found larvae and pupae in large numbers in streams in the Narsaq, Kangerlussuaq/Søndre Strømfjord and Qeqertarsuaq/Disko areas. Probably widespread in western Greenland, but not yet recorded from E Greenland.

Holarctic, common and widespread.

***Halocladius variabilis* (Stæger, 1839) Fig. 27**
Henriksen 1939: *Cricotopus vitripennis* Meigen.

Described on individuals from Denmark (possibly Greenland) without locality given.BIOLOGY. A marine species. The larvae live on stones and algae in brackish water along the shoreline, feeding on micro-algae and detritus.

DISTRIBUTION. S, SW. Lundbeck (1898) found the species widespread and often in great numbers along the coastline of western Greenland.

Europe, widespread along the coast.

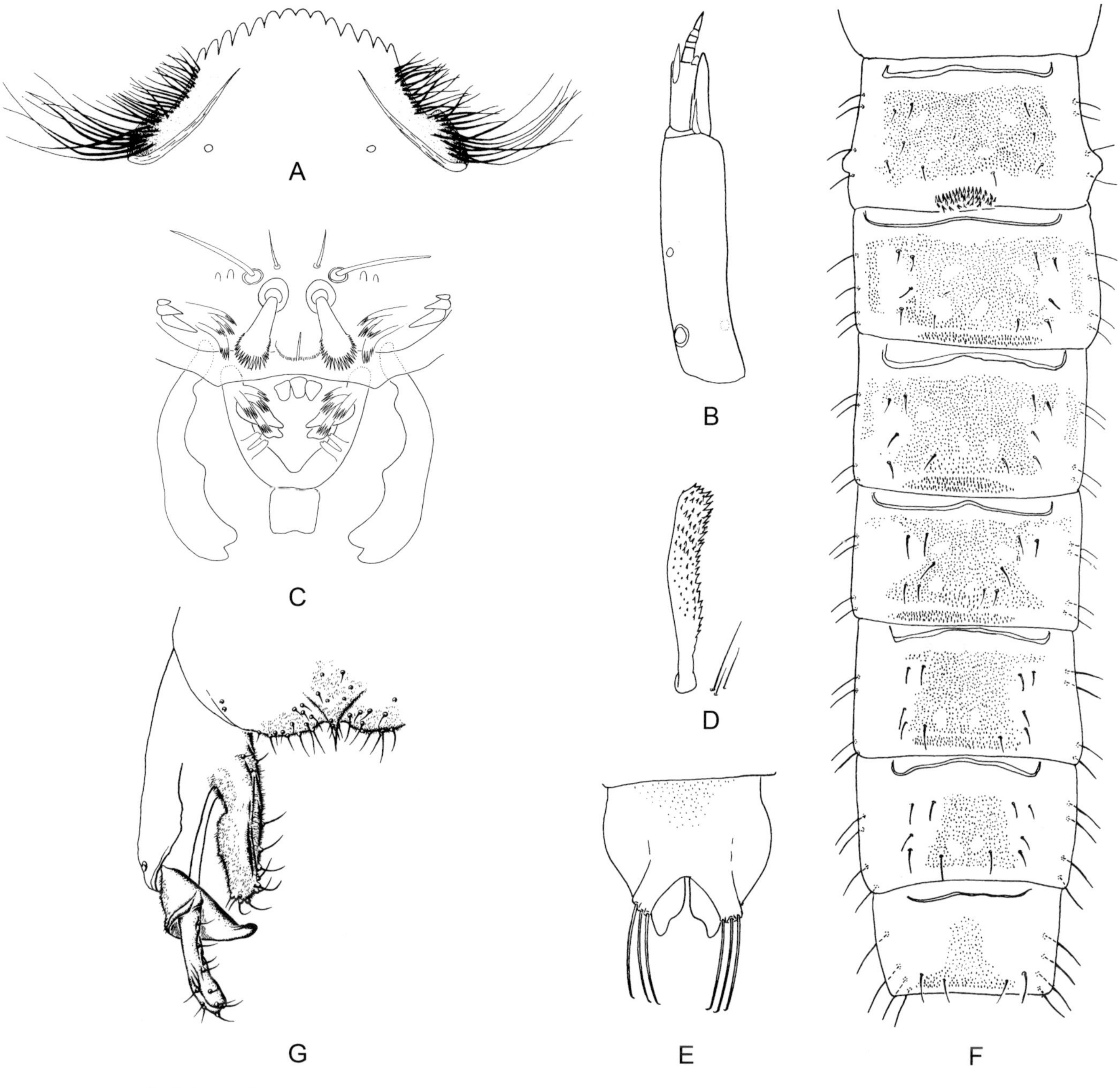

FIGURE 25 Diplocladius cultriger. *A–C: Larva. A: mentum, B: antenna, C: labrum and epipharyngeal region. D–F: Pupa. D: thoracic horn, E: anal lobe, F: tergites 2–8. G: Hypopygium of adult male. A from Moller Pillot (1984); B–C from Cranston* et al. *(1983); D–F from Coffmann* et al. *(1986); G from Pinder (1978).*

***Heterotrissocladius* Spärck, 1923 Figs 28, 29**

The genus was revised by Sæther (1975a). He described a number of new species from the Nearctic Region, some of which also occur in Greenland. However, the palaearctic and the nearctic species are very close and may be impossible to separate. The taxon recorded by Henriksen (1939, no. 87) as *Metriocnemus incomptus* Zett. belongs in this genus, but its identity remains uncertain.

Heterotrissocladius species are characteristic of oligotrophic lakes and the larvae are abundant in many Greenlandic lakes. Although they have maximum abundance in the lower littoral zone, they are able to maintain a substantial population down into the profundal zone. In such lakes the amount of food declines drastically with depths. Contrary to most chironomids, the larvae are sluggish animals moving very little. This indicates a low metabolism which in combination with an

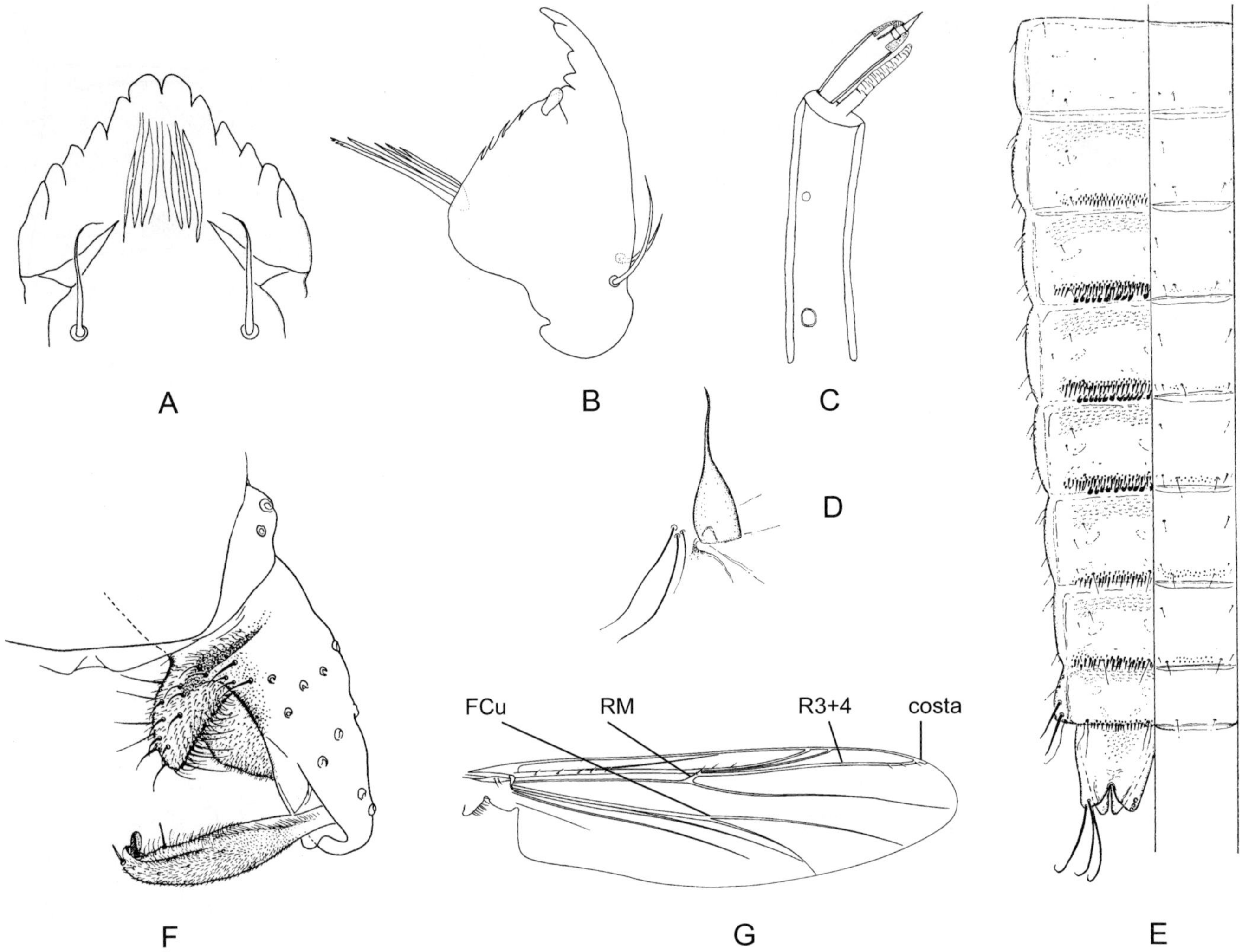

FIGURE 26 Eukiefferiella claripennis. *A–C: Larva. A: mentum, B: mandible, C: antenna. D–E: Pupa. D: thoracic horn, E: left: tergites 1–9, right: sternites 1–8. F–G: Adult male. F: hypopygium, G: wing. A–B from Cranston* et al. *(1983); C from Moller Pillot (1984); D–E from Lehmann (1972); F from Brundin (1956); G from Cranston* et al. *(1989).*

ability to utilize efficiently the energy-poor faecal pellets from zooplankton enable them to grow under these unfavourable conditions. A one-year life cycle was found in Lake 95 (Narsaq area, S Greenland), but two contemporary cohorts existed with emergence in midsummer and late autumn/early spring, respectively (Lindegaard & Mæhl 1993).

Heterotrissocladius changi **Sæther, 1975**
Henriksen 1939: not recorded.

BIOLOGY. The larvae are found in the lower littoral to upper profundal zone of oligotrophic lakes.

DISTRIBUTION. S. Reported from Lake 95 in the Narsaq area (Lindegaard *et al.* 1978, Lindegaard & Mæhl 1993).

Nearctic, widely distributed in northern parts.

Heterotrissocladius oliveri **Sæther, 1975**
Henriksen 1939: *Orthocladius (Heterotanytarsus) subpilosus* Kieff. [misidentification].

Belongs to the *subpilosus* group, where adults and pupae are easily recognized by the swollen clypeus (Fig. 29A).

BIOLOGY. A strongly cold-stenothermal species, found in ultra-oligotrophic lakes. Emergence often takes place through cracks in the ice.

DISTRIBUTION. S, SW, N, NE. Reported from Lake 95 in the Narsaq area (Lindegaard *et al.* 1978, Lindegaard & Mæhl 1993) and is common and abundant in the lakes of the Kangerlussuaq/Søndre Strømfjord area. Probably distributed all over Greenland.

Nearctic, widely distributed in northern parts.

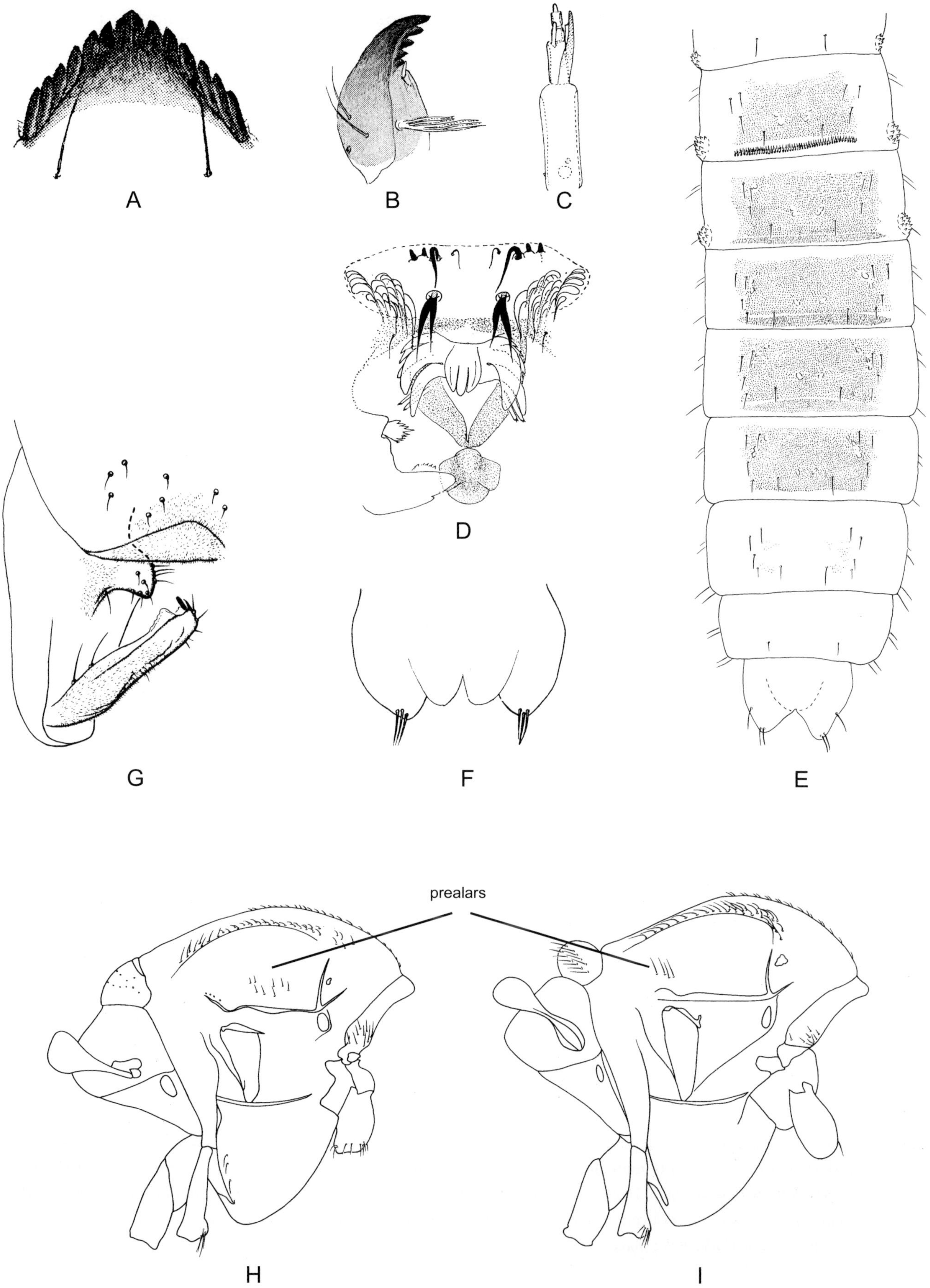

FIGURE 27 Halocladius variabilis. *A–D: Larva. A: mentum, B: mandible, C: antenna, D: labrum and epipharyngeal region. E–F: Pupa. E: tergites 1–9 of* Halocladius *sp., F: anal lobe. G: Hypopygium of adult male. H–I: Lateral view of thorax. H:* Halocladius, *I:* Cricotopus. *A–D & F from Hirvenoja (1973); E from Coffmann* et al. *(1986); G from Pinder (1978); H–I from Cranston* et al. *(1989).*

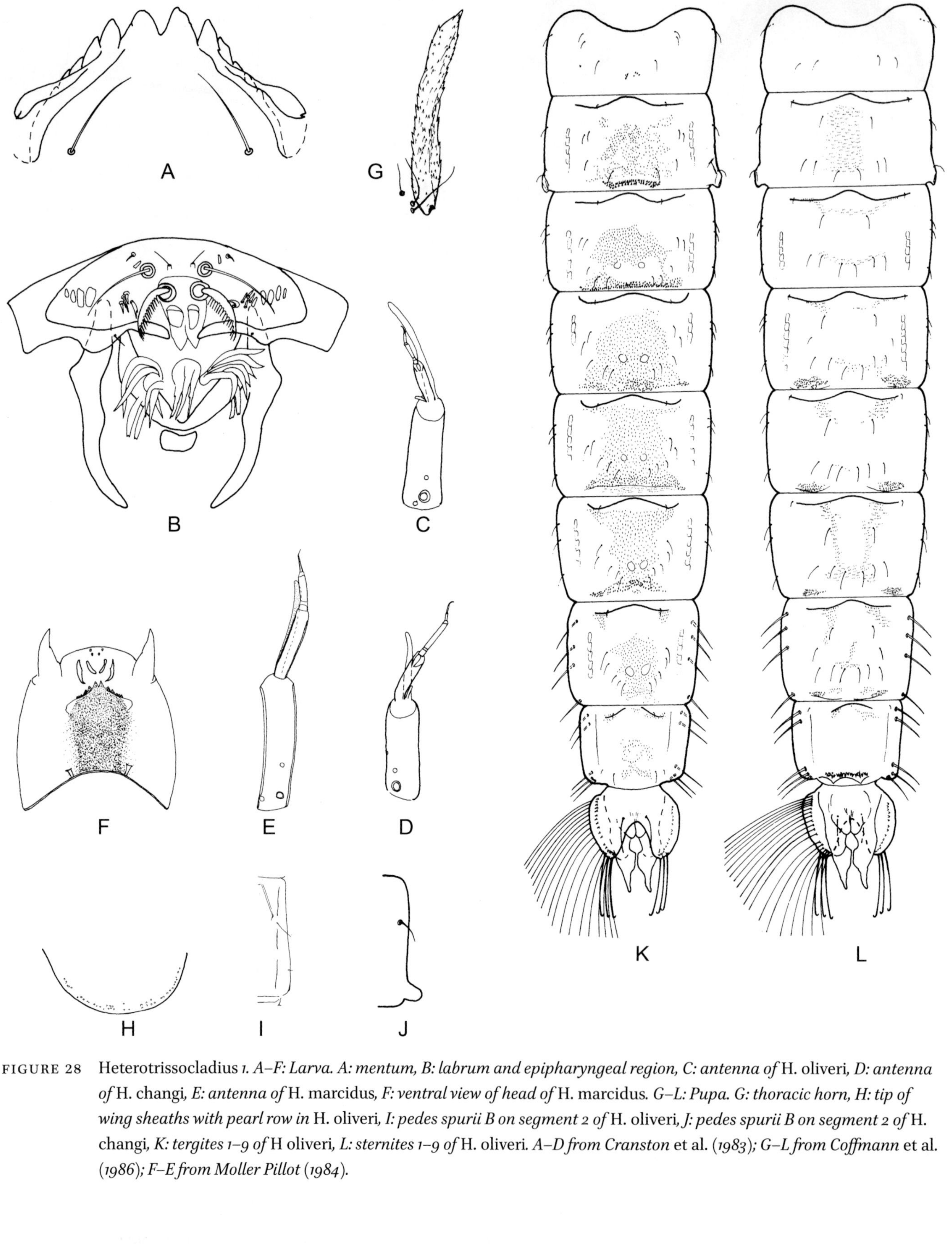

FIGURE 28 Heterotrissocladius *1. A–F: Larva. A: mentum, B: labrum and epipharyngeal region, C: antenna of* H. oliveri, *D: antenna of* H. changi, *E: antenna of* H. marcidus, *F: ventral view of head of* H. marcidus. *G–L: Pupa. G: thoracic horn, H: tip of wing sheaths with pearl row in* H. oliveri, *I: pedes spurii B on segment 2 of* H. oliveri, *J: pedes spurii B on segment 2 of* H. changi, *K: tergites 1–9 of* H oliveri, *L: sternites 1–9 of* H. oliveri. *A–D from Cranston* et al. *(1983); G–L from Coffmann* et al. *(1986); F–E from Moller Pillot (1984).*

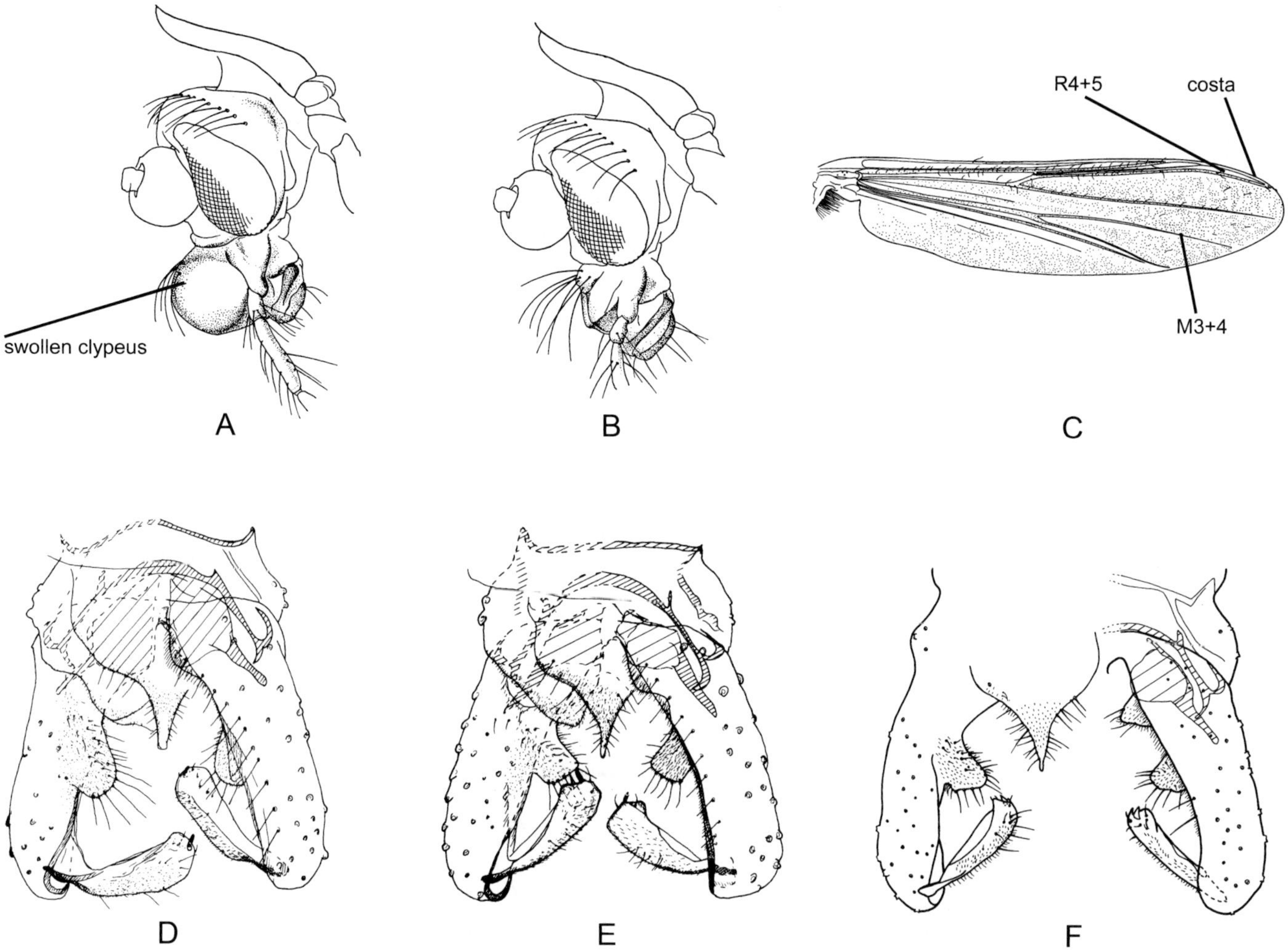

FIGURE 29 Heterotrissocladius 2. *A–F: Adult male. A: head of* H. subpilosus *group, B: head of* H. marcidus *group, C: wing of* H. marcidus *group, D: hypopygium of* H. oliveri, *E: hypopygium of* H. changi, *F: hypopygium of* H. marcidus. *A–B from Brundin (1956); C from Cranston* et al. *(1989); D–F from Sæther (1975a).*

NOTE. H. *subpilosus* was recorded from Liverpool Land (NE Greenland; Edwards (1935) and Peary Land (Nedre Midsommersø; Oliver, 1976), but according to Sæther (1975a) it is probably *H. oliveri.*

***Heterotrissocladius marcidus* group**

Larvae belonging to this group are easily recognized by the darkened gula (Fig. 28 F). They are common and abundant in lakes of the Kangerlussuaq/Søndre Strømfjord area (nSW).

***Hydrobaenus conformis* (Holmgren, 1869) Fig. 30**

Henriksen 1939: not recorded.

Described on one specimen from Svalbard. Redescribed by Sæther (1976), who included specimens from Peary Land in the material reviewed.

BIOLOGY. The larvae are characteristic of oligotrophic lakes.

DISTRIBUTION. N.

Circumpolar, arctic-subarctic.

***Hydrobaenus fusistylus* Goetghebuer, 1933 Fig. 30**

Henriksen 1939: *Orthocladius* (*Chaetocladius*) *fusistylus* Goetg.

Described on material from East Greenland. Redescribed by Sæther (1976), who included material from Peary Land (Nedre Midsommersø) in the material reviewed.

BIOLOGY. The larvae live in arctic and high alpine lakes. A very common species in the glacial lakes of the Kangerlussuaq/Søndre Strømfjord area.

DISTRIBUTION. SW, N, NE (widely distributed in Greenland).

Nearctic (northern Canada, Rocky Mountains).

***Krenosmittia* sp. Fig. 31**

Henriksen 1939: not recorded.

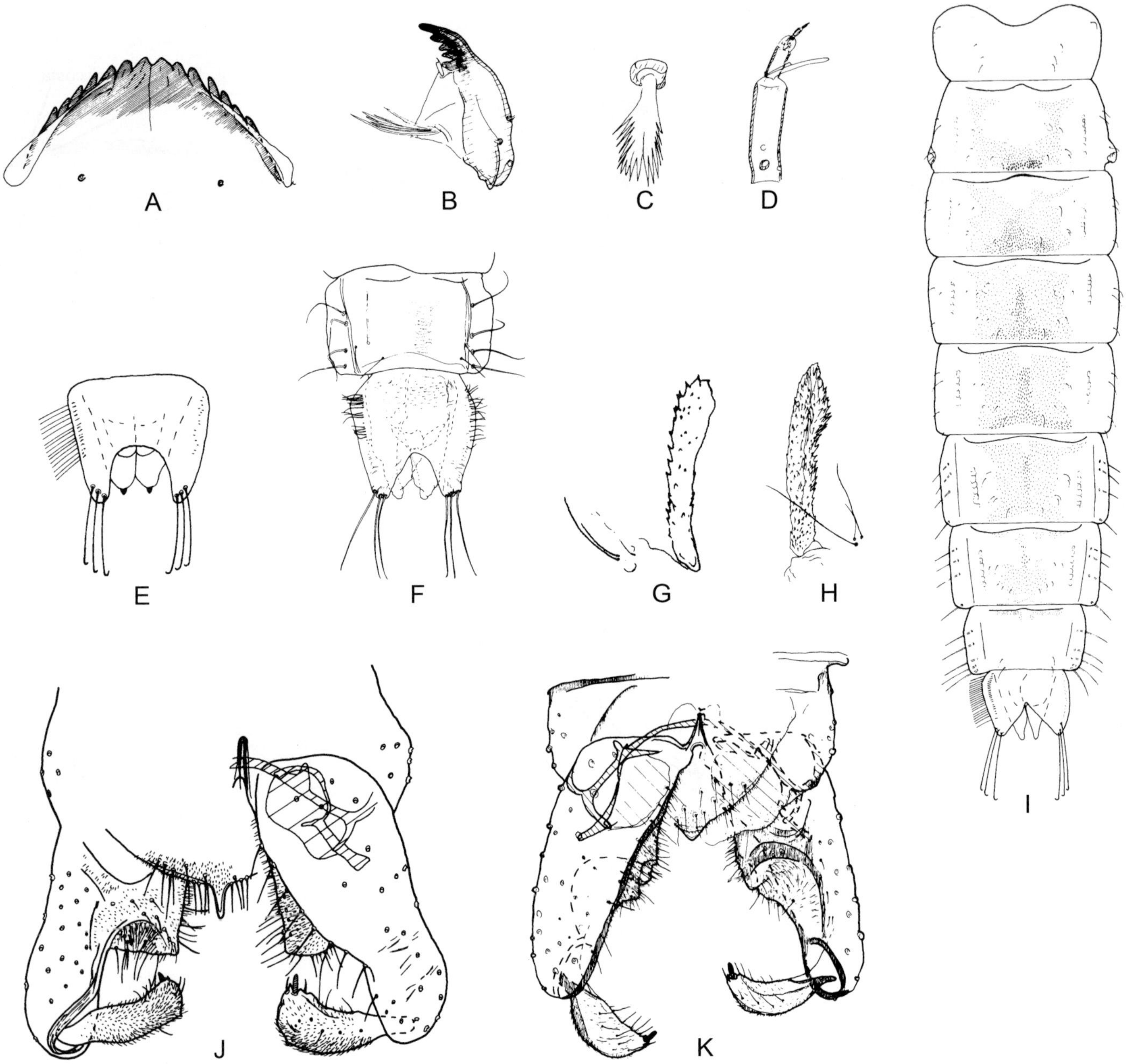

FIGURE 30 Hydrobaenus. *A–D: Larva of* H. fusistylus. *A: mentum, B: mandible, C: S I, D: antenna. E–I: Pupa. E: anal lobe of* H. fusistylus, *F: anal lobe of* H. conformis, *G: thoracic horn of* H. fusistylus, *H: thoracic horn of* H. conformis, *I: tergites 1–9 of* Hydrobaenus *sp. J–K: Hypopygium of adult male of J:* H. conformis, *K:* H. fusistylus. *A–D, F–H & K from Sæther (1976); E & I from Coffmann* et al. *(1986); K from Cranston* et al. *(1989).*

Keys to the Palaearctic adult males and pupae of *Krenosmittia* are given by Tuiskunen & Lindeberg (1986) and Langton (1991). However, some undescribed Nearctic species exist (Cranston *et al.* 1989).

A female pupae and a pupal exuvium of *Krenosmittia*, was found in S Greenland (Kvane Elv in the Narsaq area; Riget 1981) and in a lake in the Kangerlussuaq/Søndre Strømfjord region. According to Langton (1991) they are close to, but probably not conspecific with *K. camptophlebs* (Edwards, 1929) or *K. halvorseni* (Cranston & Sæther, 1986).

BIOLOGY. The cold-stenothermal larvae live primarily in running waters.

DISTRIBUTION. S, SW. I have found larvae of *Krenosmittia* sp. in Kvane Elv, Narsaq area and Robinson River, Kangerlussuaq/Søndre Strømfjord region (Friberg *et al.* 2001).

Holarctic genus.

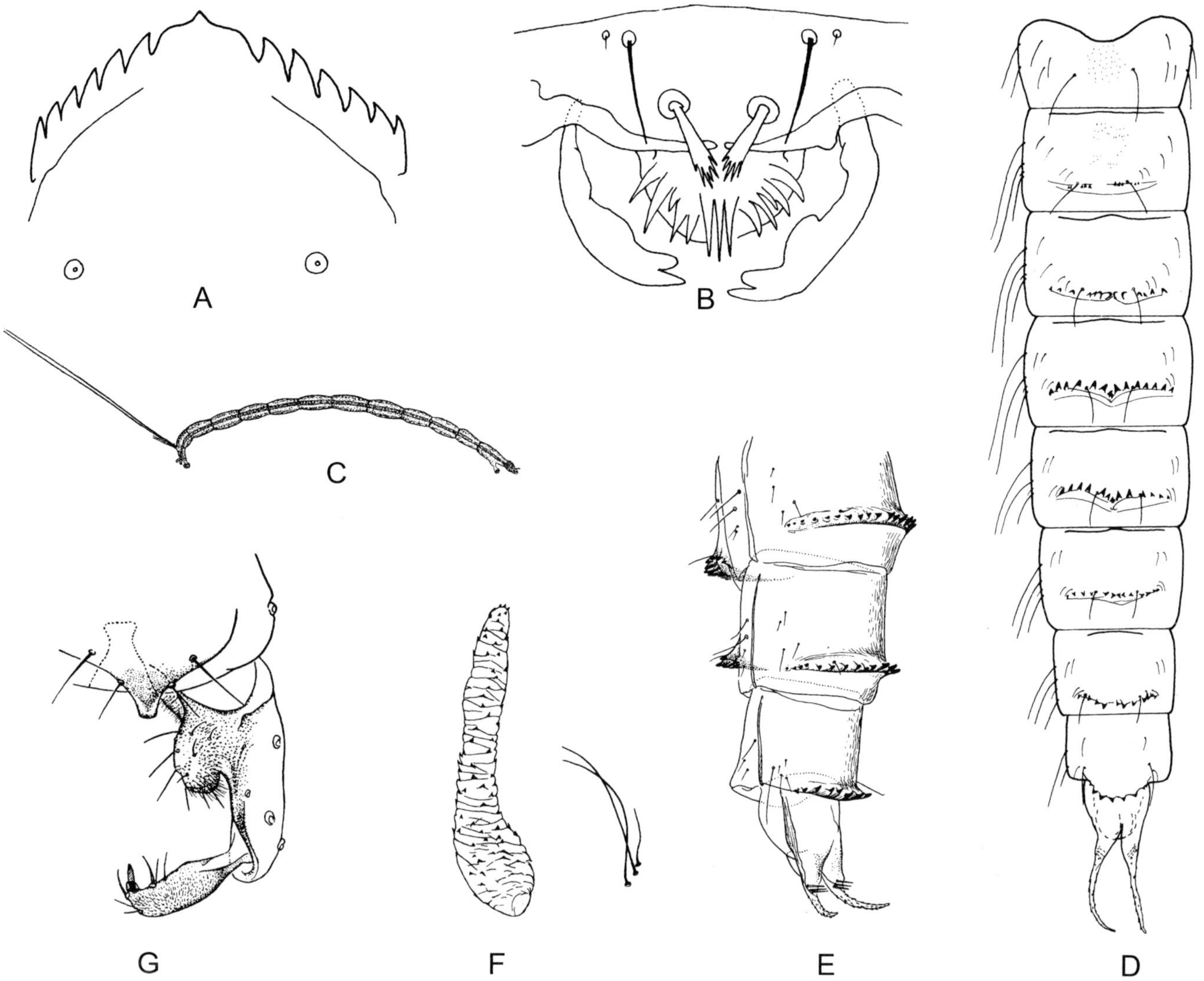

FIGURE 31 Krenosmittia. *A–C: Larva. A: mentum, B: labrum and epipharyngeal region, C: larval habitus. D–E: Pupa. D: Tergites 1–9 of* C. camptophlebs, *E: lateral view of abdominal segments 6–9 of* C. boreoalpina, *F: thoracic horn of* C. boreoalpina. *G: Hypopygium of* C. boreoalpina. *A–B from Cranston* et al. *(1983); C from Thienemann (1944); D & F from Coffmann* et al. *(1986); E & G from Brundin (1956).*

***Limnophyes* Eaton, 1875 Figs 32–33**

About 50 species of *Limnophyes* are described. Sæther (1990) made a comprehensive revision of the genus, and earlier identifications are no longer valid unless they are included in this review. Many species are found in arctic or subarctic regions and 7 species are so far known from Greenland. *L. atomarius* (Zetterstedt, 1850) recorded by Lundbeck (1898) and referred to by Henriksen (1939, no. 117) is according to Ashe & Cranston (1990) a *nomen dubium*, and the identity of the taxon recorded by Henriksen (1939, no. 123 as *Limnophyes* cf. *minima* Meig. remains similarly uncertain. The association between larvae, pupae and adults are unknown for many species. The larvae are found in aquatic, semi-terrestrial or terrestrial biotopes.

***Limnophyes anderseni* Sæther, 1990**

Henriksen 1939: *Limnophyes borealis* var. *squamatus* Søgaard Andersen.

Described on material collected by Andersen (1937) from NE Greenland (Ella Ø) and determined by him as *L. borealis* var. *squamatus*. Larva and pupa are unknown.

DISTRIBUTION. NE. The species is only known from the type locality.

***Limnophyes asquamatus* Andersen, 1937**

Henriksen 1939: *Limnophyes eltoni* var. *asquamatus* Søgaard Andersen.

Described on material from NE Greenland (Ella Ø) and redescribed by Sæther (1990). The larva is unknown.

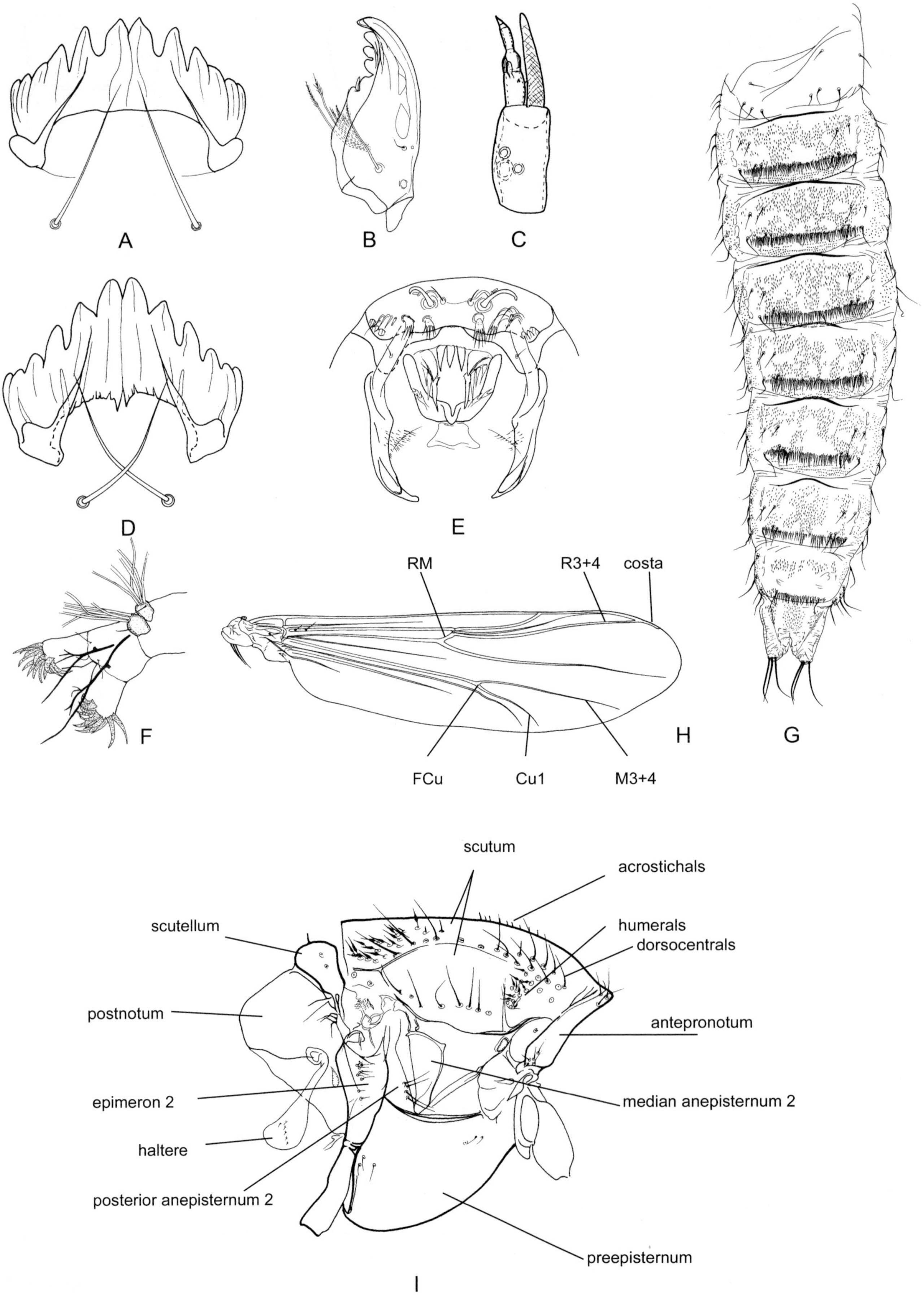

FIGURE 32 Limnophyes *1. A–F: Larva. A: mentum, B: mandible, C: antenna of* L. pentaplastus*, D: mentum of* L. minimus*, E: labrum and epipharyngeal region of* L. pentaplastus*, F: anal end of* L. pumilio*. G: Pupal tergites 1–9 of* L. pentaplastus*. H–I: Adult male. H: wing, I: lateral view of thorax of* L. eltoni*. A–E & G–I from Sæther (1990); F from Cranston (1979).*

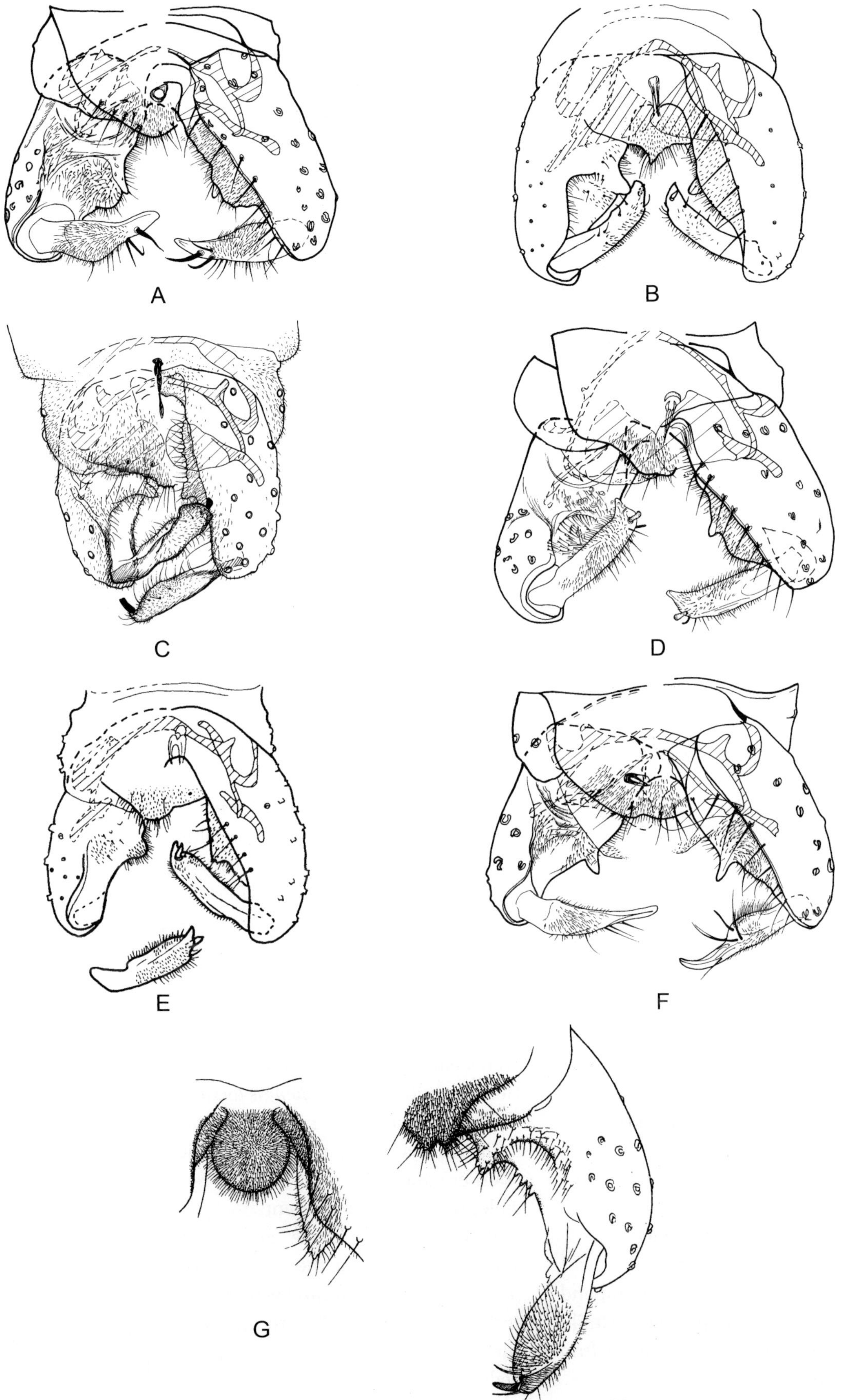

FIGURE 33 Limnophyes 2. *A–H: Hypopygium of adult males. A:* L. andersini, *B:* L. asquamatus, *C:* L. brachytomus, *D:* L. eltoni, *E:* L. minimus, *F:* L. pentaplastus, *G–H:* L. pumilio, *G: ventral view of globular median lobe. A–B & C–E from Sæther (1990); C from Sæther (1975b); G–H from Brundin (1947).*

BIOLOGY. The larvae are probably truly aquatic (Sæther, 1990).

DISTRIBUTION. NE.

Holarctic, apparently widely distributed.

***Limnophyes brachytomus* (Kieffer), 1922**

Henriksen 1939: *Limnophyes borealis* Goetg.

L. borealis Goetghebuer, 1933

Limnophyes borealis var. *squamatus* Andersen, 1937 (Henriksen 1939, no. 120).

L. brachytomus was redescribed by Sæther (1990) partly on material from Greenland. Larva and pupa are unknown.

DISTRIBUTION. SW (Amerloq), NE (Ella Ø, Kap Herschell).

Holarctic, including Bear Island and Spitsbergen.

***Limnophyes eltoni* Edwards, 1922**

Redescribed by Oliver (1962) and Sæther (1990). Larva and pupa are unknown.

DISTRIBUTION. N (only recorded from Peary Land: Nedre Midsommersø; Cranston & Oliver (1988).

Palaearctic, arctic (Bear Island, Spitsbergen; Sæther 1990).

***Limnophyes minimus* (Meigen, 1818)**

Henriksen 1939: *Limnophyes groenlandicus* Søgaard Andersen.

Redescribed by Sæther (1990) partly on material from NE Greenland (Ella Ø; Andersen 1937).

BIOLOGY. The larvae live in semi-aquatic biotopes as wet mosses, margins of springs and small streams.

DISTRIBUTION. NE.

Holarctic, possibly worldwide.

***Limnophyes pentaplastus* (Kieffer, 1921)**

L. prolongatus (Kieffer, 1921)

Henriksen 1939: not recorded.

BIOLOGY. The larvae are found in wet mosses and leaves in springs and seepings.

DISTRIBUTION. N. Only recorded from Peary Land: Nedre Midsommersø (Cranston & Oliver 1988). This record is accepted by Sæther (1990).

Holarctic, widely distributed.

***Limnophyes pumilio* (Holmgren, 1869) Fig. 33O**

Henriksen 1939: *Limnophyes globifer* Lundstr.

The male is easily recognized by the globular appendage (Fig. 33O). Only *L. torulus* Sæther, 1990 has a similar appendage. This species, however, is only known from Norway and may be a variation of *L. pumilio*.

BIOLOGY. The larvae are known from the littoral zone of lakes.

DISTRIBUTION. NE (Jameson Land; Edwards 1935, confirmed by Cranston 1979).

Holarctic, widely distributed in northern parts.

***Mesocricotopus* sp. cf. *thienemanni* (Goetghebuer, 1940) Fig. 34**

Henriksen 1939: not recorded.

I have found pupal exuviae of a *Mesocricotopus* species in glacial streams in the Kangerlussuaq/Søndre Strømfjord region. Only one species, *M. thienemanni*, has been described. The pupal exuviae fit with the description of this species, but according to Coffmann *et al.* (1986) at least one more undescribed species occurs in lotic environments in North America.

BIOLOGY. Found in littoral to profundal zones of oligotrophic and mesotrophic lakes as well as in running waters.

DISTRIBUTION. nSW (Kangerlussuaq/Søndre Strømfjord).

Holarctic, widely distributed in northern parts.

***Metriocnemus* van der Wulp, 1874 Figs 35, 36**

The genus *Metriocnemus* comprises about 40 species. It is partially revised by Sæther (1989b, 1995), who also provided a key to adult males. 7 species are recorded from Greenland. The association between larvae, pupae and adults are unknown for many species. The larvae are aquatic, semi-aquatic or terrestrial.

***Metriocnemus albolineatus* (Meigen, 1818)**

Henriksen 1939: *Metriocnemus atratulus* Zett.

Recorded by Lundbeck (1895) from Maniitsoq/Sukkertoppen and Sisimiut/Holsteinsborg. Oliver *et al.* (1990) maintain the record from Greenland, but without further comments.

DISTRIBUTION. SW.

Holarctic, widely distributed.

***Metriocnemus eurynotus* (Holmgren, 1883)**

Henriksen 1939: *Metriocnemus longitarsus* Goetg.

The variation within *M. eurynotus* is quite large and indicates that more than one species may be involved (Sæther 1989b). The larvae may be recognized by the dark violet rings on each segment.

BIOLOGY. The larvae are eury-thermal, free-living and feed on the on the surface biofilm in madicolous (hygropetric) habitats and wet mosses. Small streams and springs are preferred, but the species is also found in the surf zone of alpine lakes as well as in percolating filter beds of sewage plants.

DISTRIBUTION. SW, NE. Recorded by Goetghebuer (1933) from NE (Kap Herschell). Larvae are common in the

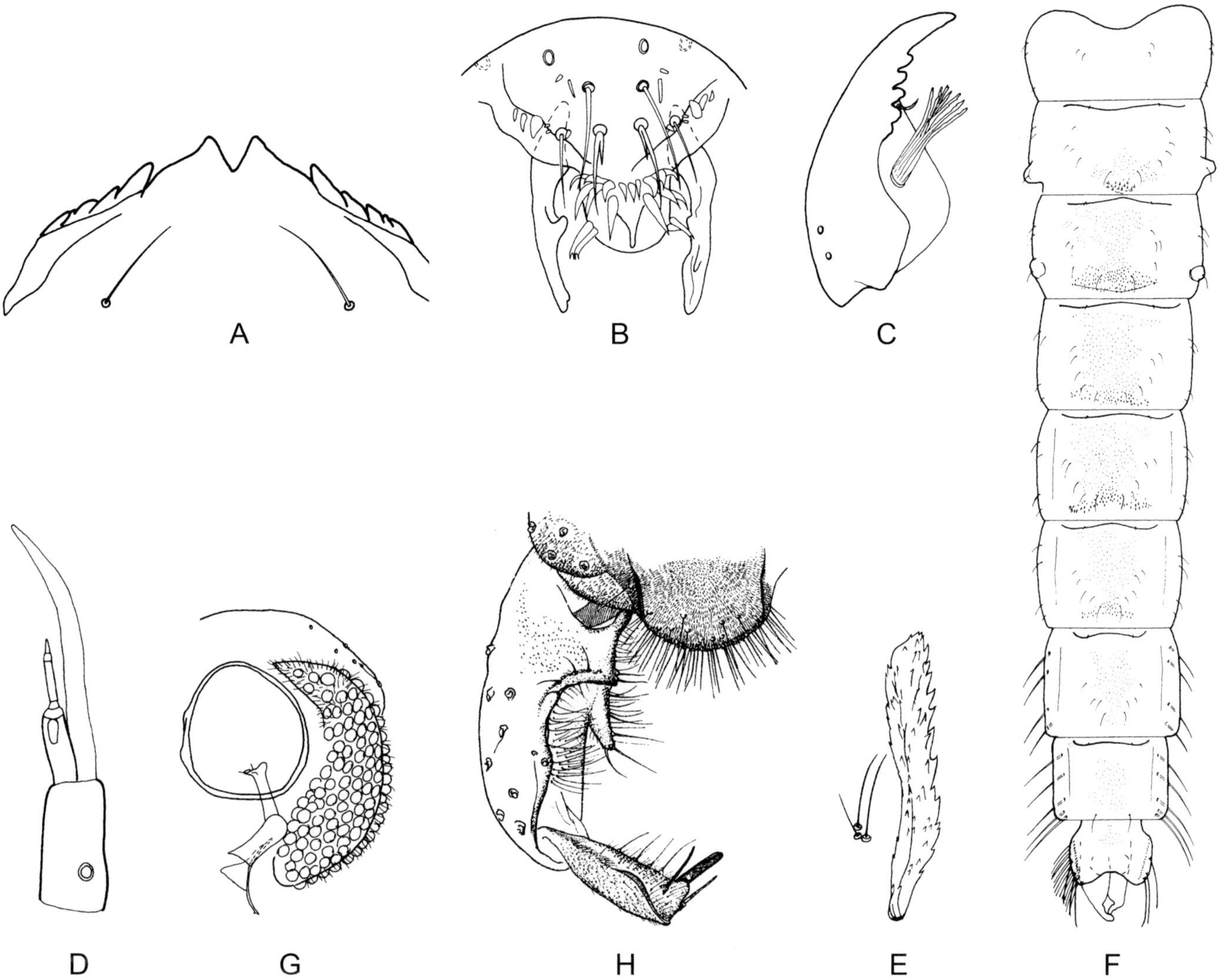

FIGURE 34 Mesocricotopus thienemanni. *A–D: Larva. A: mentum, B: labrum and epipharyngeal region, C: mandible, D: antenna. E–F: Pupa. E: thoracic horn, F: tergites 1–9. G–H: Adult male. G: frontal view of eye, H: hypopygium. A–D from Cranston* et al. *(1983); E–F from Coffmann* et al. *(1986); G from Cranston* et al. *(1989); H from Brundin (1947).*

Kangerlussuaq/Søndre Strømfjord area and the species is probably widely distributed in Greenland.

Holarctic, widely distributed and common.

***Metriocnemus fuscipes* (Meigen, 1818)**

BIOLOGY. The larvae are mainly terrestrial in humid soil with mosses, but may also be found in wet mosses in springs. They feed on the substrate biofilm.

DISTRIBUTION. S, SW, NE (Tyroler Fjord: Revet, Kap Herschell; Goetghebuer 1933).

Holarctic, widely distributed.

***Metriocnemus picipes* (Meigen, 1818)**

Henriksen 1939: not recorded.

Mentioned by Stæger (1845) from Greenland, but Lundbeck (1898) thought it might be *M. fuscipes*. The record from Greenland was maintained by both Ashe & Cranston (1990) and Oliver *et al.* (1990). Sæther (1995) redescribed *M. picipes* and stated that the Greenlandic records need reconfirmation.

***Metriocnemus polaris* Kieffer, 1926**

Henriksen 1939: not recorded.

Redescribed by Sæther *et al.* (1984) on material from Ellesmere Island and Greenland (Foulke Fjord: Reindeer Point). They stated that *M. polaris* may be a synonym of *M. eurynotus*, but it has not been confirmed in later revisions of the genus.

DISTRIBUTION. NW.

Canada (Ellesmere Island).

***Metriocnemus tristellus* Edwards, 1929**

Henriksen 1939: not recorded.

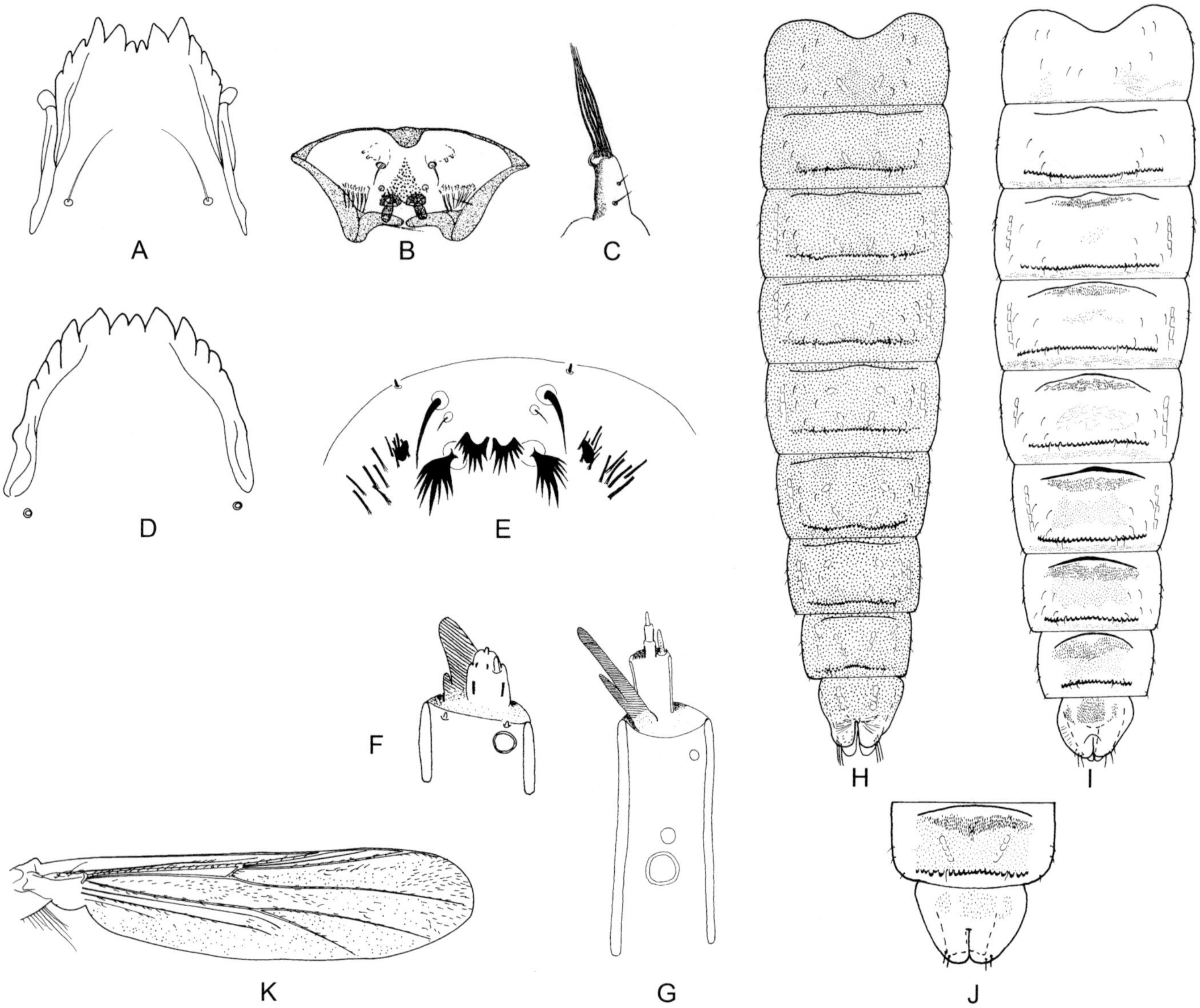

FIGURE 35 Metriocnemus *1. A–G: Larva. A: mentum, B: labrum, C: procercus of* M. fuscipes, *D: mentum, E: labrum of* M. eurynotus, *F: antenna of* M. fuscipes, *G: antenna of* M. eurynotus. *H–J: Pupa. H: tergites 1–9 of* M. fuscipes, *I: tergites 1–9 of* Metriocnemus *sp., J: tergite 8 and anal lobe of* M. eurynotus. *K: wing of* Metriocnemus *sp. A & D from Cranston* et al. *(1983); B & C from Pagast* et al. *(1941); E–G from Moller Pillot (1984); K from Cranston* et al. *(1989).*

Mentioned from Greenland by Ashe & Cranston (1990) and Oliver *et al.* (1990), but without further comments.

DISTRIBUTION. (Greenland).

Reported from England and Finland.

Metriocnemus ursinus **(Holmgren, 1869)**

Described on material from Spitsbergen and redescribed by Sæther (1989b), who did not include specimens from Greenland.

BIOLOGY. The larvae are free-living and feed on substrate biofilm in wet mosses in springs and the terrestrial surroundings.

DISTRIBUTION. nSW, NE (Henriksen 1939).

Holarctic, widely distributed in northern parts.

Oliveridia tricornis **(Oliver, 1976) Fig. 37**

Henriksen 1939: not recorded.

Described on material including males from Peary Land (Nedre Midsommersø).

BIOLOGY. A character species of ultra-oligotrophic lakes (Sæther 1979). I have found *O. tricornis* in glacial lakes in the Kangerlussuaq/Søndre Strømfjord and Narsaq areas. The larvae are strictly cold-stenothermal. The adults emerge as soon as the ice cover breaks up. They cluster to the rocks at

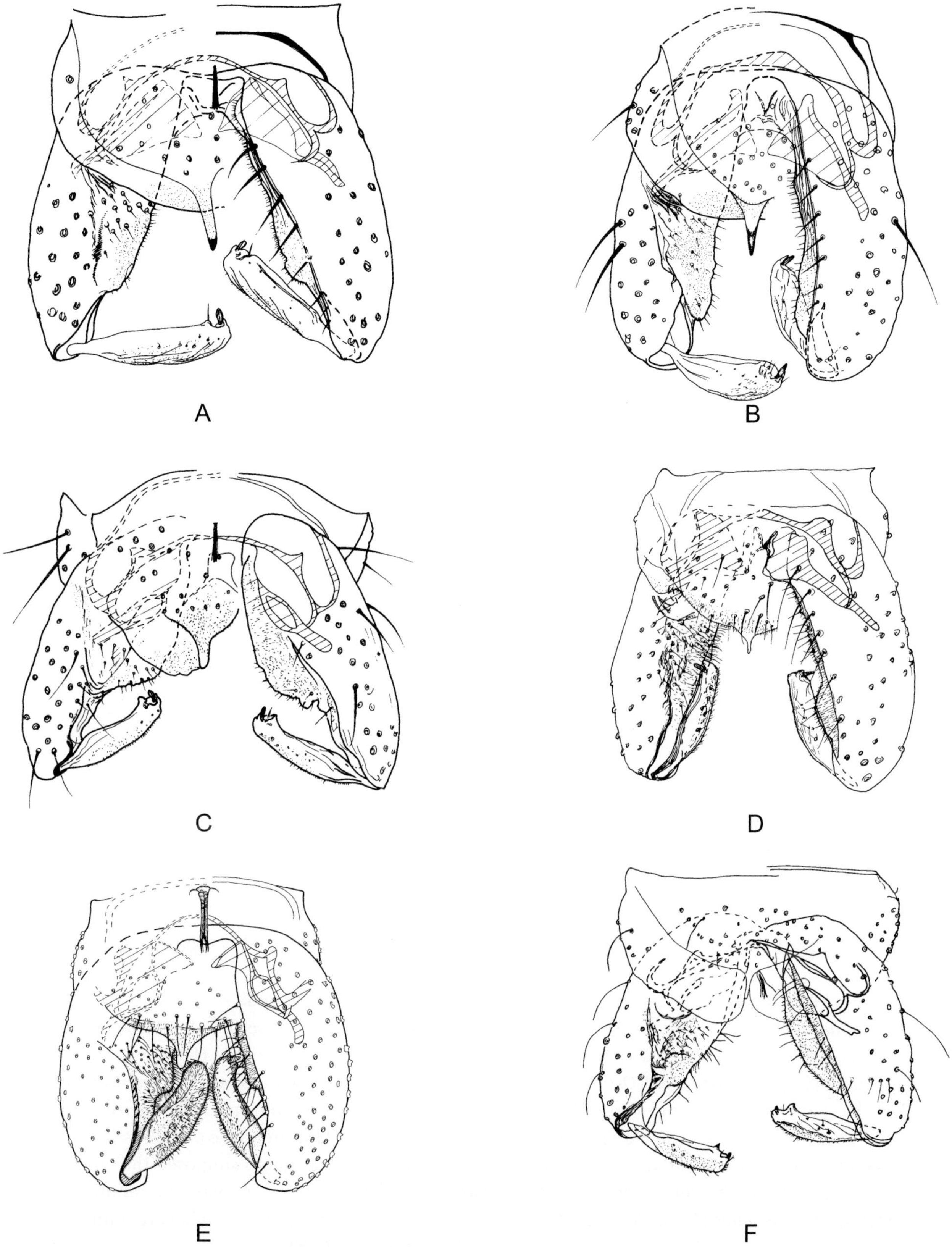

FIGURE 36 Metriocnemus 2. *Fig. A–F: Hypopygium of adult male. A:* M. albolineatus, *B:* M. fuscipes, *C:* M. eurynotus, *D:* M. picipes, *E:* M. polaris, *F:* M. ursinus. *A–C & F from Sæther (1989b); D from Sæther (1995); E from Sæther* et al. *(1984).*

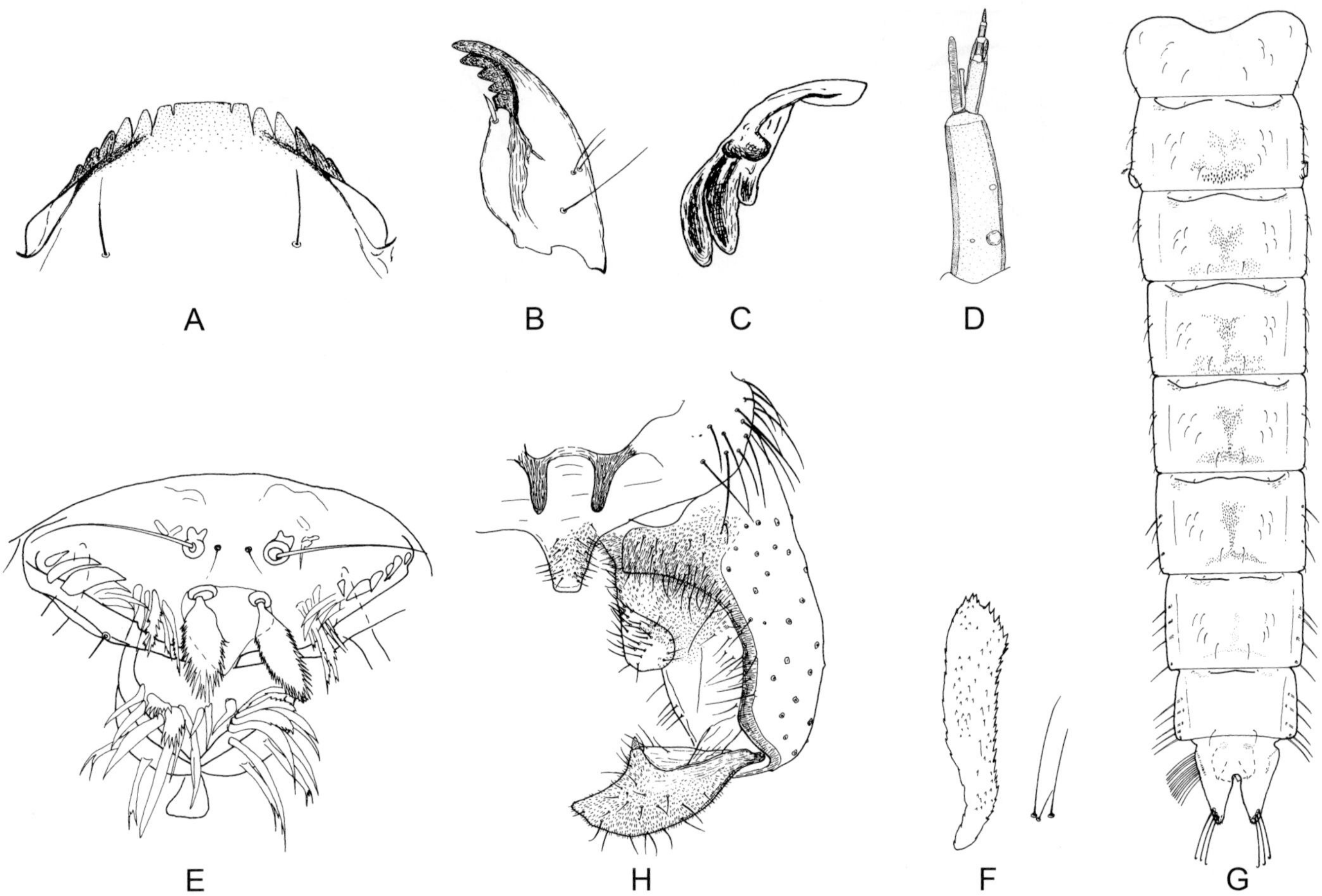

FIGURE 37 Oliveridia tricornis. *A–E: Larva. A: mentum, B: mandible, C: premandible, D: antenna, E: labrum and epipharyngeal region. F–G: Pupa. F: thoracic horn, G: tergites 1–9. H: Adult male hypopygium. A–D & H from Oliver (1976); E from Cranston* et al. *(1983); F–G from Coffmann* et al. *(1986).*

the margin of lakes, where males are very active in mating. The adults can fly but rarely form swarms (Oliver 1976).

DISTRIBUTION. S, SW, N.

Holarctic, arctic, widely distributed.

Orthocladius van der Wulp, 1874 Figs 38–45

The genus *Orthocladius* is divided into several subgenera: The nearctic *Orthocladius* (*Eudactylocladius*) has been revised by Cranston (1998) and Sæther (2004), *Orthocladius* (*Euorthocladius*) by Soponis (1990) and *Orthocladius* (*Orthocladius*) by Soponis (1977). The single species of *Orthocladius* (*Pogonocladius*) has been redescribed by Pinder & Cranston (1976). Recently, Sæther (2005) established a new subgenus *Orthocladius* (*Mesorthocladius*).

Orthocladius (_Eudactylocladius_) sp. cf. _dubitatus_ Johannsen, 1942

Henriksen 1939: not recorded.

Pupal exuviae from the outlet of Lake Ferguson in the Kangerlussuaq/Søndre Strømfjord area keyed out to *O.* (*E.*) *dubitatus*.

DISTRIBUTION. Nearctic, widely distributed in North America.

Orthocladius (_Eudactylocladius_) _gelidorum_ (Kieffer, 1923) Fig. 38

Henriksen 1939: not recorded.

O. (*E.*) *nanseni* was redescribed by Sæther *et al.* (1984) on material from NW Greenland (Foulke Fjord: Reindeer Point). Cranston (1998) proposed *O.* (*E.*) *nanseni* as a synonym of *O.* (*E.*) *gelidorum* and added specimens from Peary Land (Nedre Midsommersø).

BIOLOGY. Cranston (1998) supposed the larvae to live primarily in more or less permanent, summer flowing streams and rivers, but pupal exuviae are found in lakes.

DISTRIBUTION. N, NW.

Holarctic (widely distributed in northern Canada, also found in Norway).

Orthocladius (_Eudactylocladius_) _gelidus_ Kieffer, 1922 Fig. 38

Henriksen 1939: not recorded.

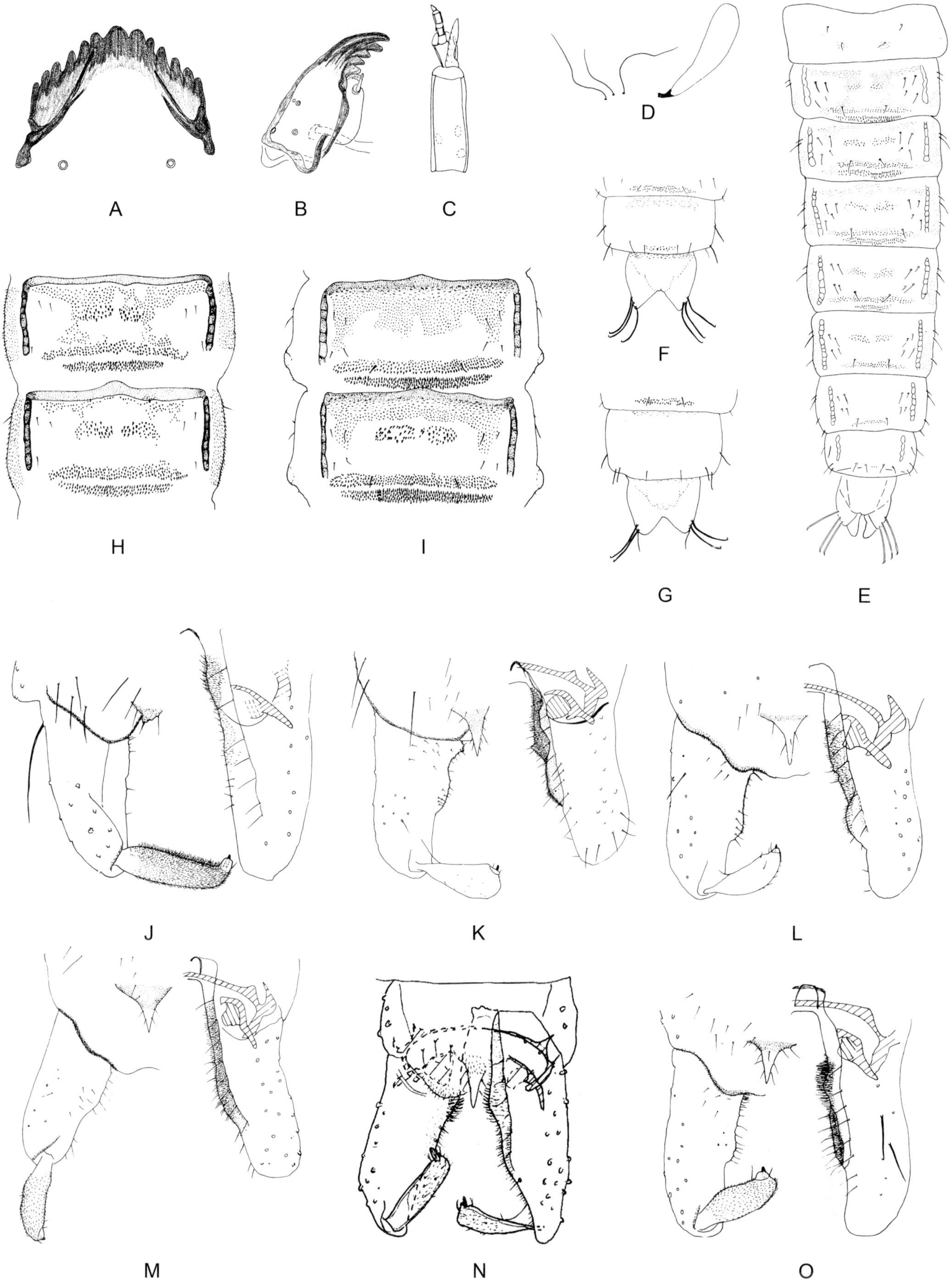

FIGURE 38 Orthocladius (Eudactylocladius). *A–C: Larva. A: mentum, B: mandible, C: antenna. D–I: Pupa. D: thoracic horn, E: tergites 1–9, F: anal lobe of* O. (E.) gelidus,*G: anal lobe of* O. (E.) olivaceus, *H: tergites 2–3 of* O. (E.) gelidus, *I: tergites 2–3 of* O. *(E.)* dubitatus. *J–O: Hypopygium of adult male of J:* O. (E.) *dubitatus, K:* O. (E.) gelidorum, *L:* O. (E.) gelidus, *M:* O. (E.) olivaceus, *N:* O. (E.) priomixtus, *O:* O. (E.) subletteorum. *A–C from Soponis (1977); D, F–M & O from Cranston (1998); E from Coffmann* et al. *(1986); N from Sæther (2004).*

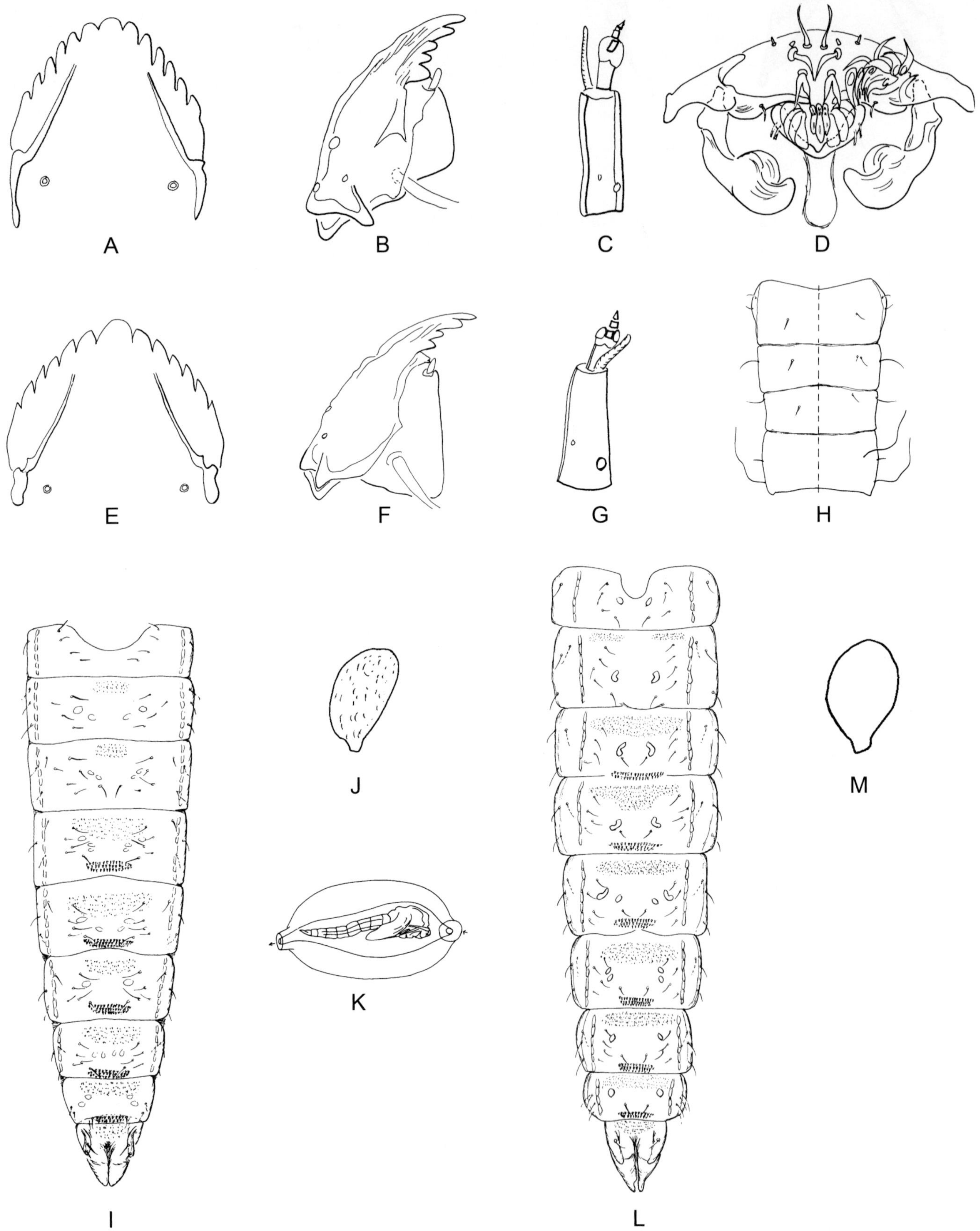

FIGURE 39 Orthocladius (Euorthocladius) *1. A–H: Larva. A & E: mentum, B & F: mandible, C & G: antenna, D: labrum and epipharyngeal region, H: setation of abdominal segments. I–M: Pupa. I & L: tergites 1–9, J & M: thoracic horn, K: pupa in gelatinous tube. A–H & I–J:* O. (E.) rivicola, *E–H & K–M:* O. (E.) thienemanni. *A–M from Soponis (1990).*

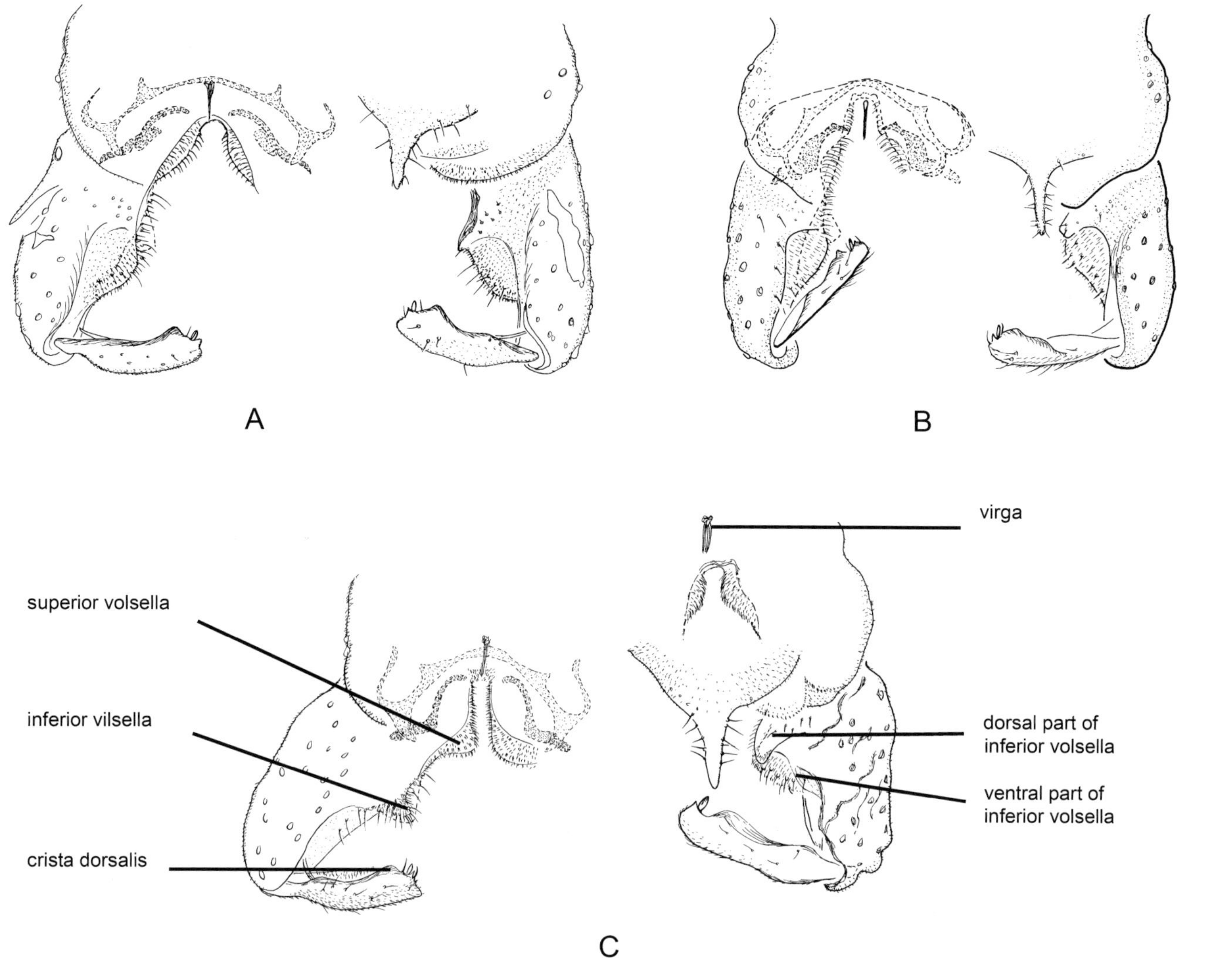

FIGURE 40 Orthocladius (Euorthocladius) 2. *Hypopygium of adult males; right: dorsal view, left: ventral view. A:* O. (E). difficilis, *B:* O. (E.) rivicola, *C:* O. (E.) thienemanni. *A–C from Soponis (1990).*

BIOLOGY. Cranston (1998) supposed the larvae to live primarily in more or less permanent, summer flowing streams and rivers, but pupal exuviae are found in lakes.

DISTRIBUTION. SW, NE (Ella Ø, Kap Alfred; Andersen 1937). These records were accepted by Cranston (1984) and Oliver *et al.* (1990). I have found the species in a pond in the Kangerlussuaq/Søndre Strømfjord area.

Holarctic, arctic (widely distributed, but not as far north as *O.* (*E.*) *gelidorum*).

***Orthocladius* (*Eudactylocladius*) sp. cf. *olivaceus* Kieffer, 1918**

Henriksen 1939: not recorded.

Pupal exuviae from lakes and ponds in the Kangerlussuaq/Søndre Strømfjord area keyed out to *O.* (*E.*) *olivaceus*.

DISTRIBUTION. SW.

Holarctic, widespread in Europe and western North America.

***Orthocladius* (*Eudactylocladius*) *priomixtus* Sæther, 2004 Fig. 38**

Henriksen 1939: *Orthocladius* (*Dactylocladius*) ? *microcomus* Kieff.

Described on material from Norway. *O.* (*E.*) *microcomus* Kieffer, 1922 recorded by Andersen (1937) is probably the same species. *O.* (*E.*) *priomixtus* is very similar to *O.* (*E.*) *subletteorum* and may be a form of this species (Sæther (2004).

DISTRIBUTION. NE (only recorded from Ella Ø).

Europe (only recorded from northern Fennoscandia; Sæther 2004).

***Orthocladius* (*Eudactylocladius*) *subletteorum* Cranston 1998 Fig. 38**

Henriksen 1939: *Orthocaldius* (*O.*) *mixtus* Holmgr.

Described on material including one specimen from Maniitsoq/Sukkertoppen. *O.* (*E.*) *subletteorum* includes nearctic specimens earlier misidentified as *O.* (*E.*) *mixtus* (Holmgren, 1869).

DISTRIBUTION. nSW.

Holarctic (only known from northern Canada and Iceland; Cranston 1998, Hfransdottir, 2005).

***Orthocladius* (*Euorthocladius*) *difficilis* Lundbeck, 1898 Fig. 40**

Henriksen 1939: *Orthocladius* (*O.*) *difficilis* Lundb.

Described on material from Qeqertarsuup Kangerlua/ Disko Bugt and redescribed by Oliver (1970). The larva and pupa are unknown.

DISTRIBUTION. nSW (only recorded from Qeqertarsuup Tunua/Disko Bugt: Kangersuneq).

Nearctic (northern Canada).

***Orthocladius* (*Euorthocladius*) *rivicola* Kieffer, 1911 Fig 39, 40**

Henriksen 1939: not recorded.

O. (*E.*) *rivicola* is redescribed by Soponis (1990).

BIOLOGY. The larvae live in stony streams, where they build tubes similar to those of *O.* (*E.*) *thienemanni.*

DISTRIBUTION. N (only reported from Peary Land: Nedre Midsommersø).

Holarctic, widely distributed.

***Orthocladius* (*Euorthocladius*) *thienemanni* Kieffer, 1906 Figs 39, 40**

Henriksen 1939: not recorded.

BIOLOGY. Larvae and pupae are found in fast flowing waters, where they live in gelatinous tubes encrusted with sand grains and diatoms. Prior to pupating the tube is enlarged and the pupae live in half-ellipsoid tubes. By undulating they cause water to flow through openings in both ends of the tube (Fig. 39 K).

DISTRIBUTION. S, nSW. I have found the species in both the Narsaq and the Kangerlussuaq/Søndre Strømfjord areas.

Holarctic, widely distributed.

***Orthocladius* (*Mesorthocladius*) *frigidus* (Zetterstedt, 1838) Figs 41, 42**

Henriksen 1939: *Orthocaldius* (*O.*) *frigidus* Zett.

O. (*M.*) *frigidus* has been moved from subgenus *Euorthocladius* to *Orthocladius* and back again several times, but according to Sæther (2005) it now belongs to a new subgenus: *Mesorthocladius*. Originally described on specimens from Greenland and redescribed by Soponis (1987).

BIOLOGY. The larvae live primarily in running waters, but are also found in the surf zone of northern and high altitude lakes. They are either free-living or live in tubes, often in algae and moss vegetation.

DISTRIBUTION. S, SW, NE (common and widely distributed).

Holarctic, widely distributed.

***Orthocladius* (*Mesorthocladius*) *rousellae* Soponis, 1990 Figs 41, 42**

Henriksen 1939: not recorded.

Described partly on material from Peary Land (Nedre Midsommersø). The pupa is easily recognized.

BIOLOGY. The larvae live in glacial lakes and streams.

DISTRIBUTION. S, SW (Narsaq and Kangerlussuaq/ Søndre Strømfjord areas), N (Peary Land). Probably widespread in Greenland.

Nearctic, arctic.

***Orthocladius* (*Mesorthocladius*) *vaillanti* Langton & Cranston, 1991 Figs 41, 42**

Henriksen 1939: not recorded.

The pupa is described by Langton & Cranston (1991) and Caldwell (1998) and is easily recognizable on the "fringed" anal lobe and the elongated thoracic horn, which is generally broken off leaving a small brown stalk.

BIOLOGY. The larvae are found in mountain streams.

DISTRIBUTION. S, SW (I have found pupal exuviae in Kvane Elv, Narsaq, and in the outlet from Lake Ferguson, Kangerlussuaq/Søndre Strømfjord).

Holarctic, but not often recorded.

***Orthocladius* (*Orthocladius*) *charensis* Soponis, 1977 Figs 43, 44**

Henriksen 1939: not recorded.

Described partly on material from Peary Land (Nedre Midsommersø).

BIOLOGY. The larvae live at rocky shores of arctic lakes.

DISTRIBUTION. N.

Nearctic, arctic (Canada).

***Orthocladius* (*Orthocladius*) *decoratus* (Holmgren, 1869) Fig. 44**

BIOLOGY. Not known, but probably living in the littoral zone of arctic lakes.

DISTRIBUTION. NE. Records by Remy (1928) from Kap Stewart and Goetghebuer (1933) from Vega Sund and Revet, Claveringfjorden. These records are accepted by Soponis (1977) and Oliver *et al.* (1990).

Holarctic, arctic (widely distributed).

***Orthocladius* (*Orthocladius*) *groenlandicus* Goetghebuer, 1933 Fig. 44**

Henriksen 1939: *Orthocladius* (*Dactylocladius*) *groenlandicus* Goetg.

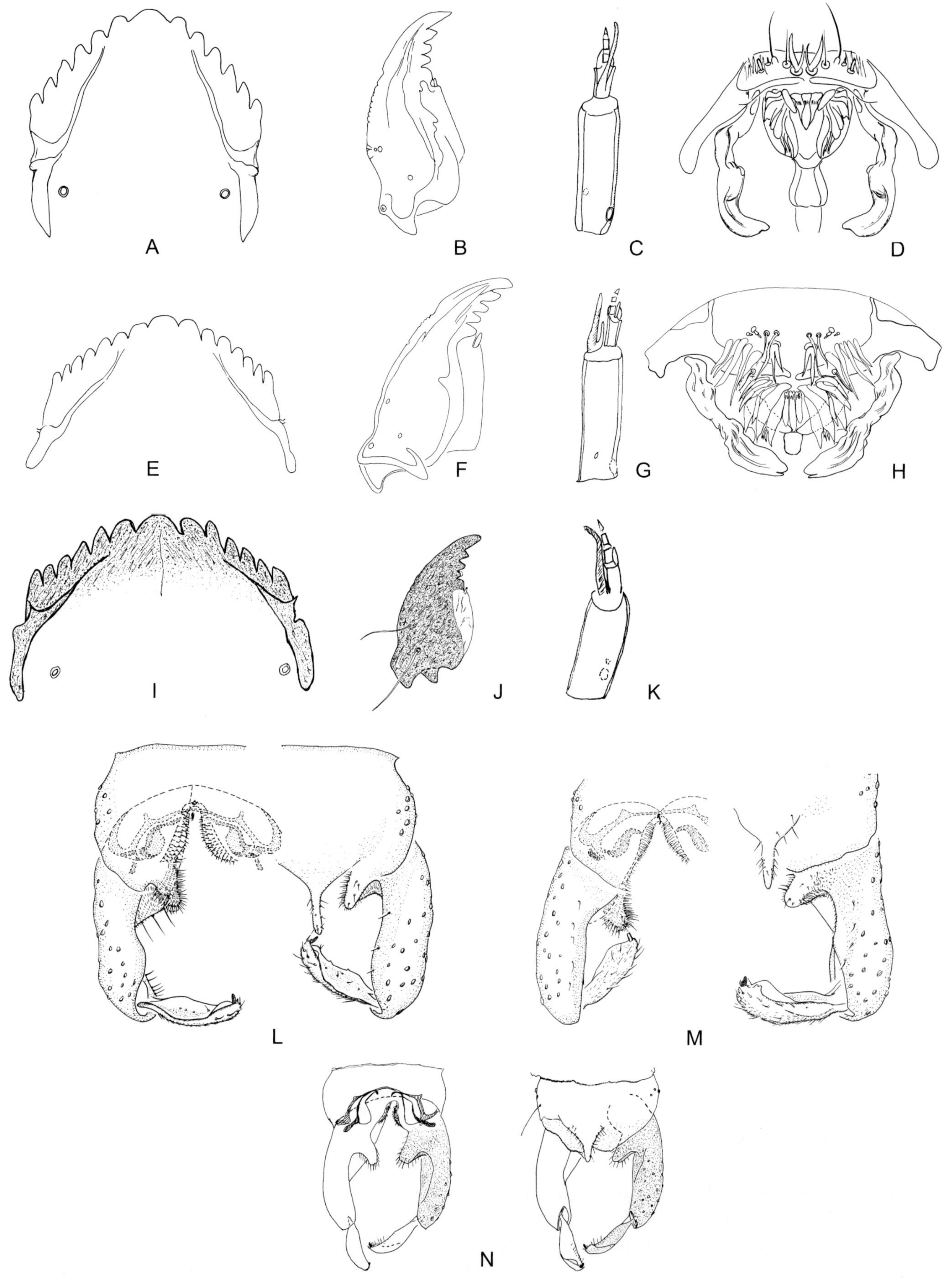

FIGURE 41 Orthocladius (Mesorthocladius) *1. A- K: larva. A, E & I: mentum, B, F & J: mandible, C, G & K: antenna, D & H: labrum and epipharyngeal region. A–D:* O. (M.) frigidus, *E–H:* O. (M.) rousellae, *I–K:* O. (M.) vaillanti. *L–M: Hypopygium of adult male; right: dorsal view, left: ventral view. L:* O. (M.) frigidus, *M:* O. (M.) rousellae, *N:* O. (M.) vaillanti. *A–D & M from Soponis (1987); E–H from Soponis (1990); I–K & N from Caldwell (1998).*

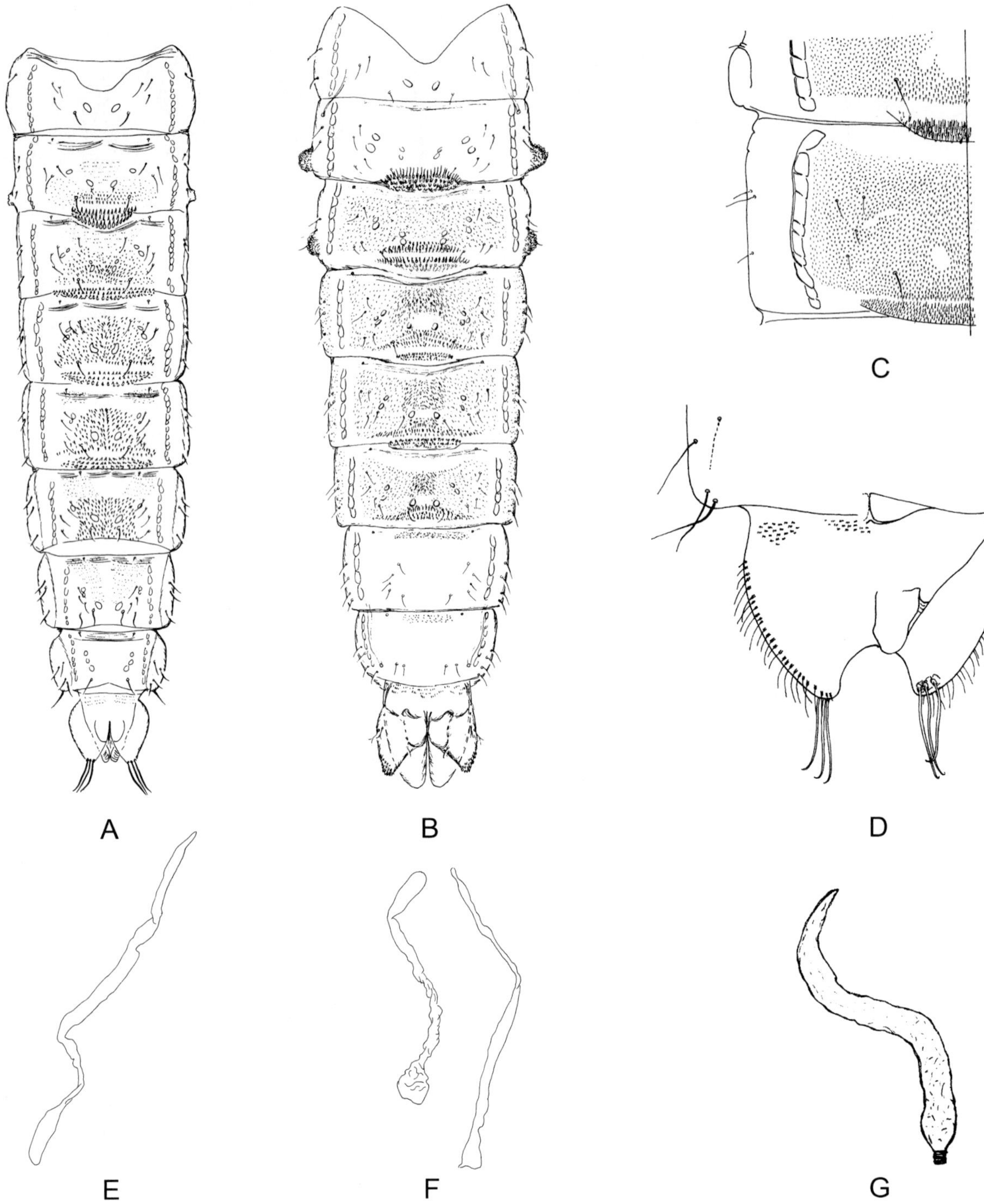

FIGURE 42 Orthocladius (Mesorthocladius) 2. *A–G: pupa. A–B: tergites 1–9, C: tergites 2–3, D: tergite 9, E–G: thoracic horn. A & E:* O. (M.) frigidus, *B & F:* O. (M.) rousellae, *C, D & G:* O. (M.) vaillanti *A & E from Soponis (1987); B & F from Soponis (1990); C–D from Langton & Cranston (1991); G from Caldwell (1998).*

Described on material from NE Greenland (Vega Sund). The description is revised by Soponis (1977). The type specimen is the only record, larva and pupa are unknown.

***Orthocladius* (*Orthocladius*) *knabeni* Goetghebuer, 1933**
Fig. 44
Henriksen 1939: *Orthocladius* (*Dactylocladius*) *knabeni* Goetg.

Described on material from NE Greenland. The description could not be revised by Soponis (1977) as the type material is lost, but she acknowledged it as a valid species based on the description given by Goetghebuer (1933). Larva and pupa unknown.

DISTRIBUTION. NE (Vega Sund, Hurry Inlet: Fame Øer).

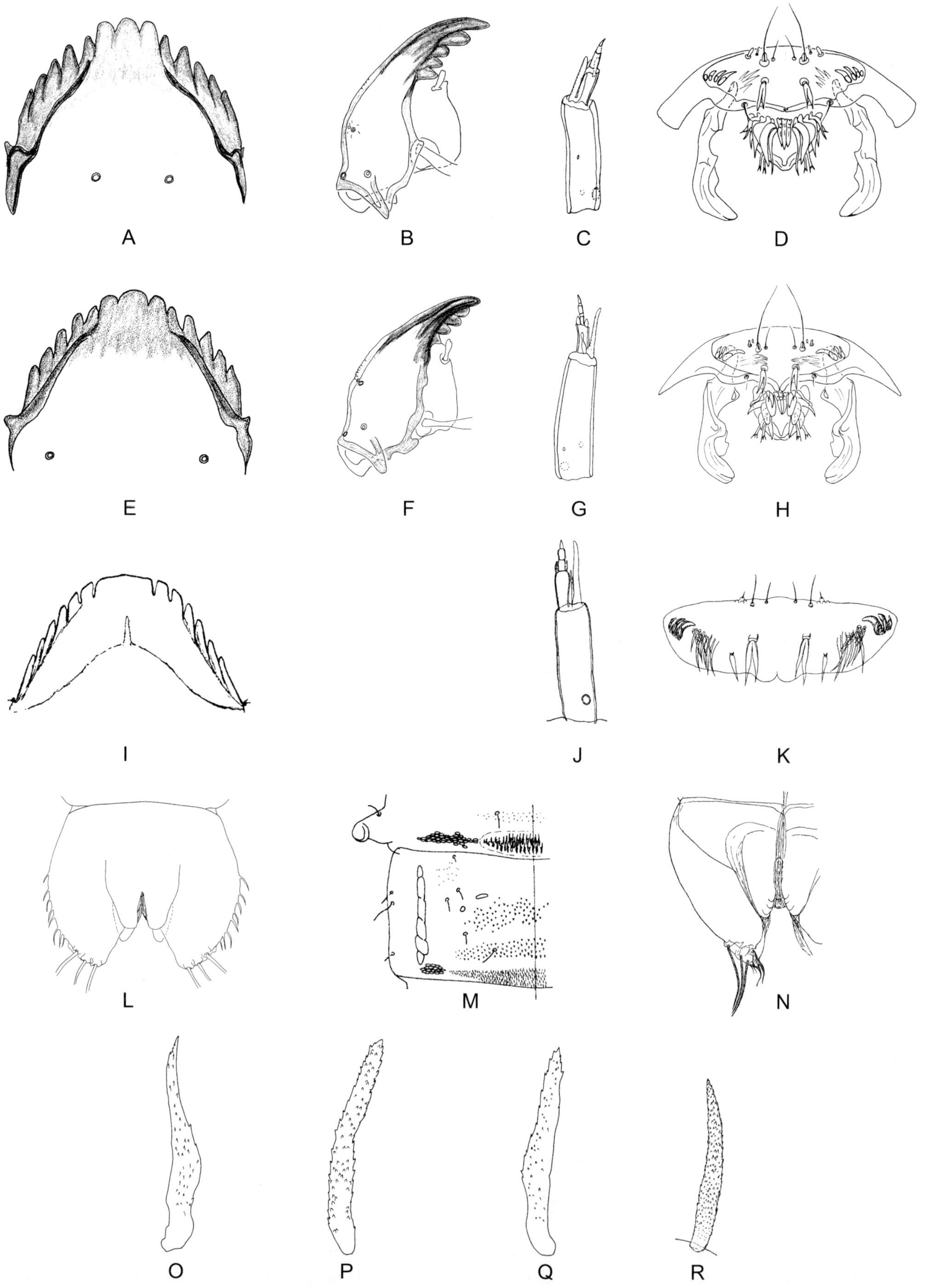

FIGURE 43 Orthocladius (Orthocladius) *1. A–K: Larva. A–D:* O. (O.) charensis, *E–H:* O. (O.) lapponicus, *I–K:* O. (O.) rubicundus. *1st column: mentum, 2nd column: mandible, 3rd column: antenna, 4th column: labrum and epipharyngeal region. L–R: pupa. L & N: tergite 9, M: tergites 2–3, O–R: thoracic horn. L & O:* O. (O.) charensis, *M, N & R:* O. (O.) rubicundus, *P:* O. (O.) knuthi, *Q:* O. (O.) lapponicus. *A–H, L & P–Q from Soponis (1977); I–K, N & R from Thienemann & Krüger (1937); M from Langton & Cranston (1991).*

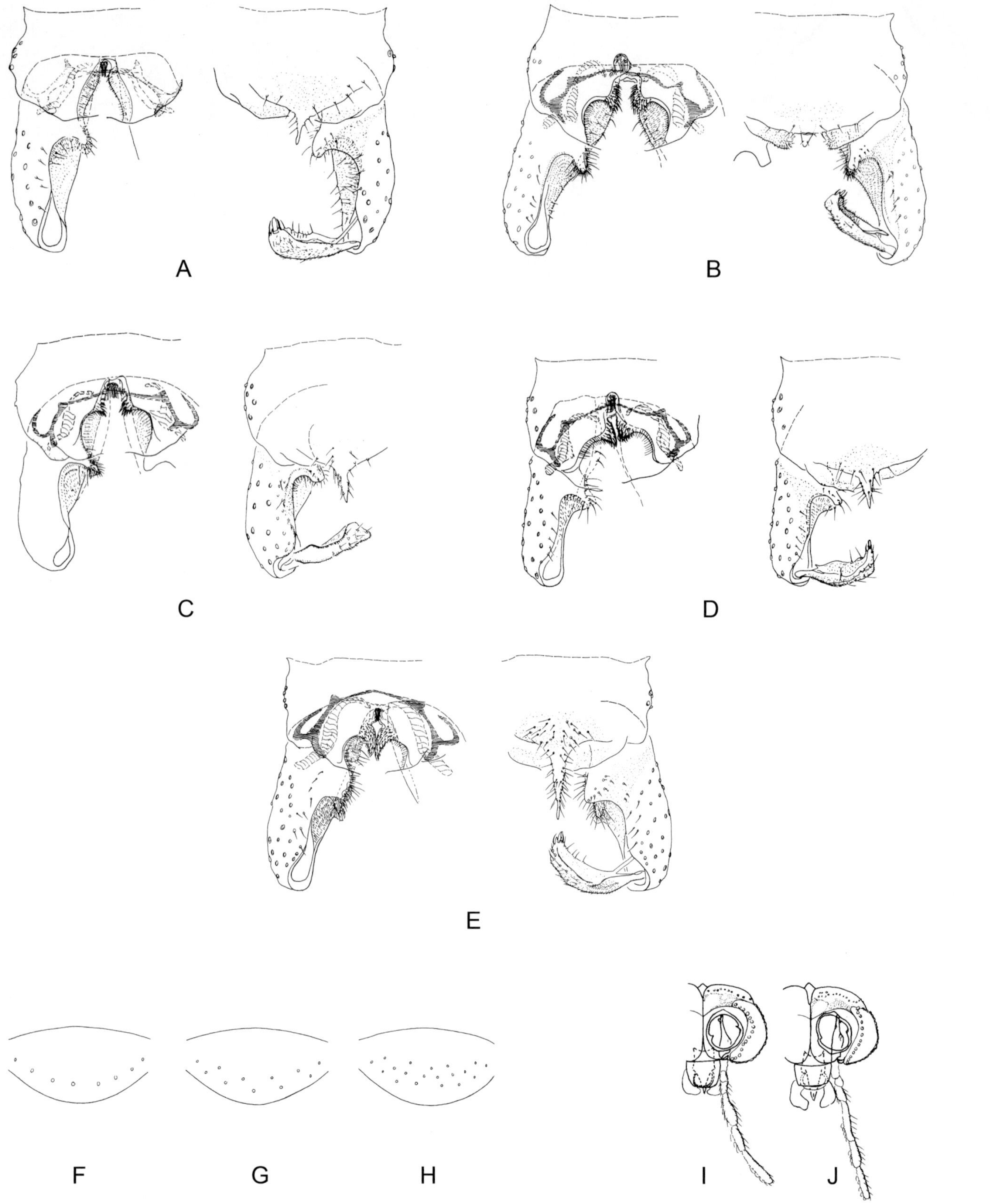

FIGURE 44 Orthocladius (Orthocladius) 2. *Hypopygium of adult males; right: dorsal view, left: ventral view. A:* O. (O.) charensis, *B:* O. (O.) decoratus, *C:* O. (O.) groenlandicus, *D:* O. (O.) knuthi, *E:* O. (O.) lapponicus. *F–H: Arrangement of scuttellar setae. F: uniserial, G: biserial, H: multiserial. I–J: Frontal view of head. I: eye expanded dorsomedially, J: eye not expanded dorsomedially. A–E from Soponis (1977); F–J from Soponis (1990).*

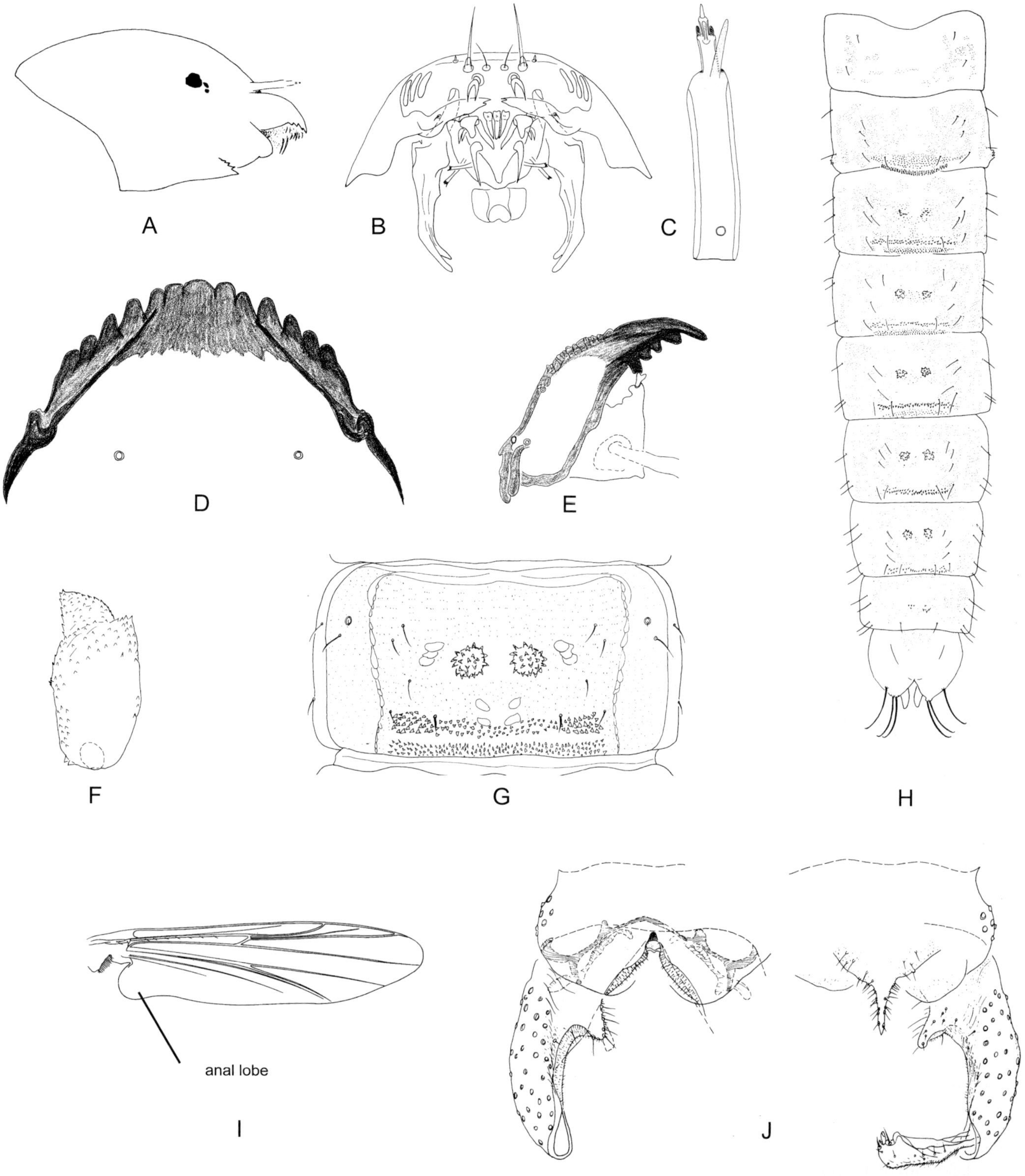

FIGURE 45 Orthocladius (Pogonocladius) consobrinus. *A–E: Larva. A: lateral view of head, B: labrum and epipharyngeal region, C: antenna, D: mentum, E: mandible. F–H: Pupa. F: thoracic horn, G: tergite 3, H: tergites 1–9. I–J: Adult male. I: wing, J: hypopygium. A & C from Moller Pillot (1984); B, D–E & G from Soponis (1977); F & H from Coffmann* et al. *(1986); I–J from Cranston* et al. *(1989).*

***Orthocladius* (*Orthocladius*) *knuthi* Soponis, 1977 Figs 43, 44**
Henriksen 1939: not recorded.

Described on material from Peary Land (Nedre Midsommersø). The larva is unknown.

BIOLOGY. Found in the littoral zone of arctic lakes.

DISTRIBUTION. N.

Holarctic (only known from Spitsbergen and a single locality in Canada, Northwest Territories).

***Orthocladius* (*Orthocladius*) *lapponicus* Goetghebuer, 1940 Figs 43, 44**
Henriksen 1939: not recorded.

BIOLOGY. The larvae have been found in the moss zone (4–15 m depth) of arctic Char Lake, Canada (Welch 1973).

DISTRIBUTION. N. Reported from Peary Land (Nedre Midsommersø; Soponis 1977).

Holarctic (only known from Sweden and Canada: Northwest Territories).

****Orthocladius* (*Orthocladius*) *rubicundus* (Meigen, 1818) Figs 43, 44**
Henriksen 1939: *Orthocladius* (*O.*) cf. *rubicunda* Meig.

Reported by Carpenter (1938) from the Kangerlussuaq/Søndre Strømfjord area. However, this record is not accepted by neither Soponis (1977) nor Oliver *et al.* (1990).

DISTRIBUTION. ?nSW.

Holarctic, widely distributed in Europe (Oliver & Dillon 1997).

***Orthocladius* (*Pogonocladius*) *consobrinus* (Holmgren, 1869) Fig. 45**
Henriksen 1939: *Orthocladius* (*O.*) cf. *consobrinus* Holmgr.

The adult of *O.* (*P.*) *consobrinus* is redescribed by Pinder & Cranston (1976).

BIOLOGY. The larvae live primarily in the littoral zone of lakes.

DISTRIBUTION. nSW (outlet from Moraine Lake, Qeqertarsuaq/Disko; Friberg *et al.* 2001), NE (Ella Ø; Andersen 1937).

Holarctic, widely distributed.

****Orthocladius spitzbergensis* Kieffer, 1919**
Henriksen 1939: *Orthocladius* (*Dactylocladius*) *spitzbergensis* Kieff.

Reported from NE Greenland (Kap Stewart) by Remy (1928). The description of *O. spitzbergensis* is inadequate and Ashe & Cranston (1990) regarded it a *nomen dubium*, whereas Oliver *et al.* (1990) maintained it as a valid species recorded from Greenland.

***Paracladius quadrinodosus* Hirvenoja, 1973 Fig. 46**
Henriksen 1939: not recorded.

BIOLOGY. The larvae are known from both lotic and lentic waters. They are often recorded from large cold lakes.

DISTRIBUTION. N. Reported from Peary Land (Nedre Midsommersø; Oliver 1976).

Holarctic, arctic.

***Parakiefferiella* Thienemann, 1936 Fig. 47**
About 30 species of *Parakiefferiella* are known from the Holarctic Region. *P. nigra* Brundin, 1949 is reported from Greenland by Sæther *et al.* (2000) without further details. I have found pupal exuviae of a *Parakiefferiella* species in Lake Ferguson, Kangerlussuaq/Søndre Strømfjord. According to Tuiskunen (1986) they belong to another species, because the thoracic horn at the distal end is covered with small points, which are absent in *P. nigra*.

The *Parakiefferiella* species are known from both lentic and lotic habitats.

The genus is known from a number of localities in the Holarctic Region.

***Parametriocnemus graminicola* (Lundbeck, 1898) Fig. 48**
Henriksen 1939: *Orthocladius* (*Heterotanytarsus*) *graminicola* Lundb.

Described on material from SW Greenland. Redescribed by Sæther (1969) and Oliver (1970).

BIOLOGY. A cold stenothermal species. The larvae are found in cold springs and fast flowing streams and rivers.

DISTRIBUTION. nSW (known from Qeqertarsuup Tunua/Disko Bugt: Kangersuneq and Amerloq: Utorqaat).

Nearctic, arctic (Canada).

***Parametriocnemus lundbeckii* Johannsen, 1905 Fig. 48**
Henriksen 1939: *Metriocnemus lundbecki* Johansen [*sic*].

P. lundbeckii is according to Oliver *et al.* (1990) a new name for *Chironomus nanus* described by Lundbeck (1898) on material from S Greenland (Ivittuut). Redescribed by Sæther (1969), who stated that it is very close to the European *P. stylatus* (Kieffer, 1924).

BIOLOGY. A relatively cold stenothermal species. The larvae are found in springs and fast flowing streams and rivers.

DISTRIBUTION. S (Tunulliarfik, Ivittuut).

Nearctic, widely distributed.

***Paraphaenocladius impensus* (Walker, 1856) Fig. 49**
Henriksen 1939: *Metriocnemus debilipennis* Lundb.

Reported from Greenland by Lundbeck (1898), who described it as *P. debilipennis* on specimens from S Greenland (Ivittuut). It is redescribed by Oliver (1970) and Sæther & Wang (1995).

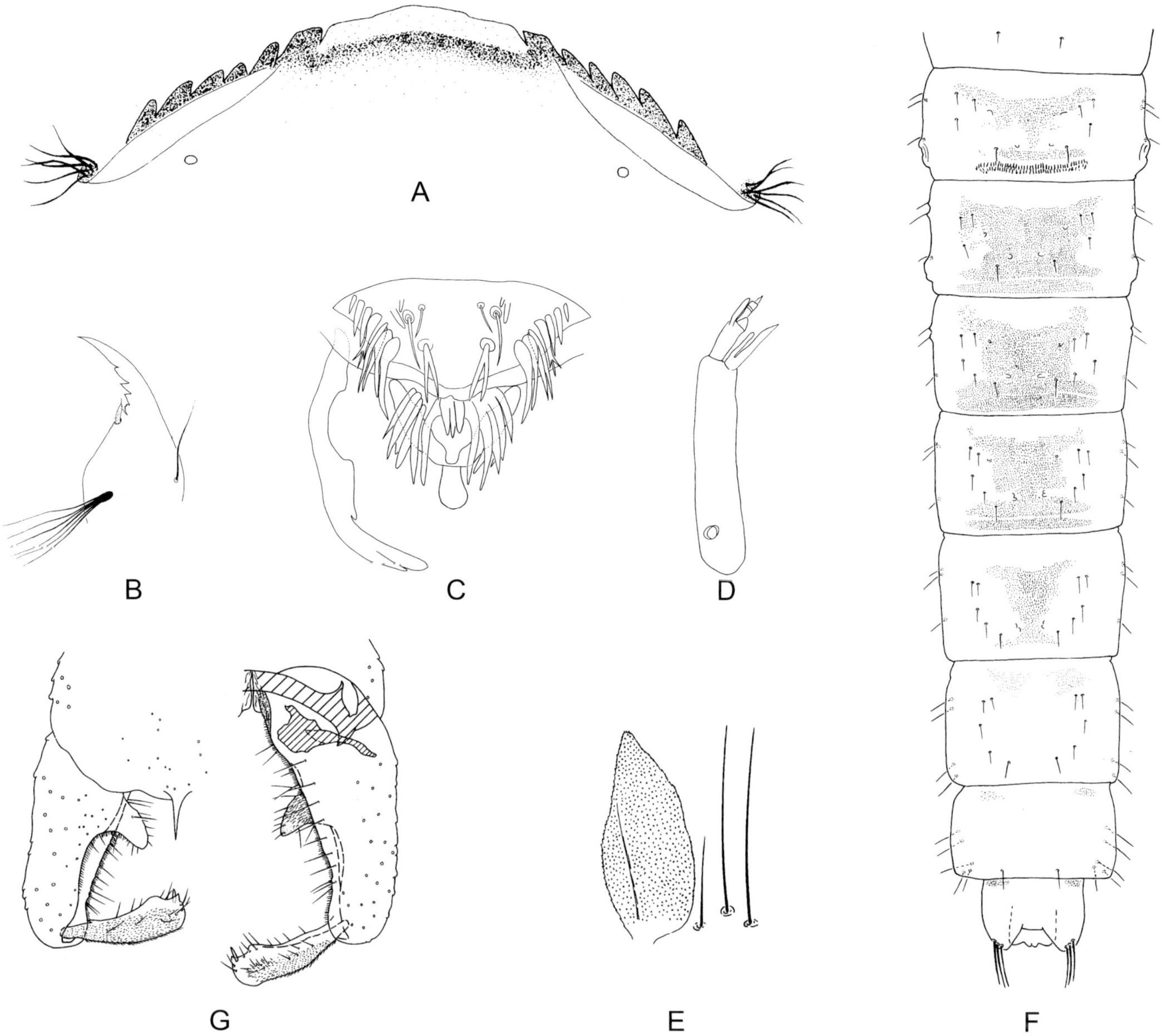

FIGURE 46 Paracladius. *A–D: Larva. A: mentum, B: mandible, C: labrum and epipharyngeal region, D: antenna. E–F: Pupa of* P. quadrinosus. *E: thoracic horn, F: tergites 1–9. G: Left dorsal and right ventral view of hypopygium of adult male of* P. quadrinosus. *A from Moller Pillot (1984); B–D from Cranston* et al. *(1983); E–F from Coffmann* et al. *(1986); G from Cranston* et al. *(1989).*

BIOLOGY. The larvae live in moist soil at the margin of lakes, springs, streams and in meadows. They are free-living, probably feeding on the substrate biofilm.

DISTRIBUTION. S (Tunilliarfik,), SW (Neria). Probably widespread at least in western Greenland.

Holarctic, widely distributed.

Paraphaenocladius brevinervis **(Holmgren, 1869) Fig. 49**
Henriksen 1939: not recorded.

P. brevinervis is redescribed by Sæther & Wang (1995) partly on material from Peary Land (Nedre Midsommersø). Larva and pupa are unknown.

BIOLOGY. The larvae live in wet soil covered with mosses. Adults are caught in traps over such biotopes (Sendstad *et al.* 1976).

DISTRIBUTION. N.

Holarctic, arctic (known from Spitsbergen and the Canadian Arctic Archipelago).

Psectrocladius **Kieffer 1906 Figs 50, 51**
So far only species belonging to the subgenus *Psectrocladius* (*Psectrocladius*) have been recorded from Greenland. It is a large subgenus with about 70 species in the Holarctic Region. The subgenus is in need of revision, and identification is

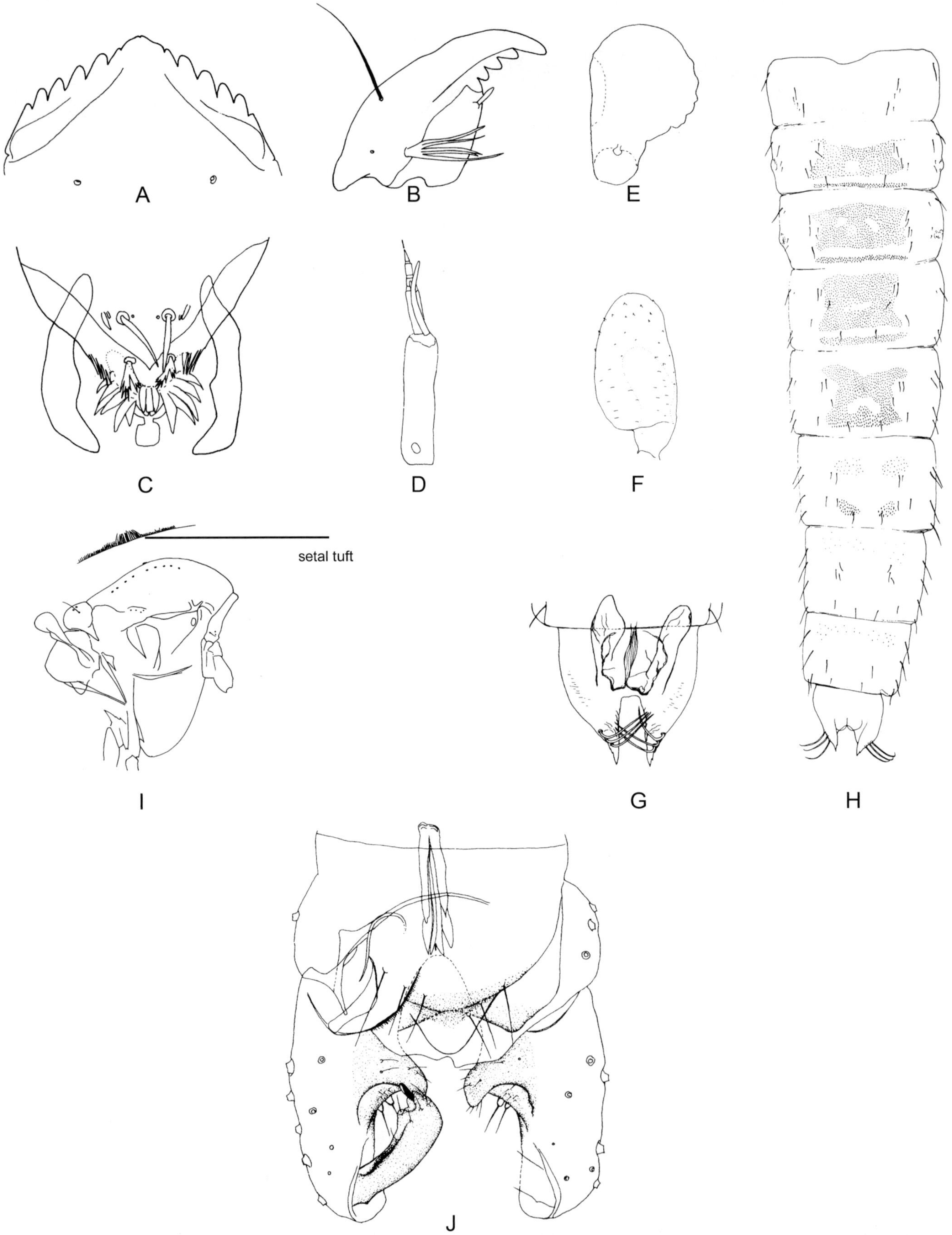

FIGURE 47 Parakiefferiella. *A–D: Larva. A: mentum, B: mandible, C: labrum and epipharyngeal region, D: antenna of* P. bathophila. *E–H: Pupa. E: thoracic horn of* P. nigra, *F: thoracic horn, G: anal lobe, H: tergites 1–9 of* P. bathophila. *I–J: Adult male: I: lateral view of thorax showing setal tuft, J: hypopygium of* P. nigra. *A–D from Cranston* et al. *(1983); E–G & J fra Tuiskunen (1986); H from Coffmann* et al. *(1986); I from Cranston* et al. *(1989).*

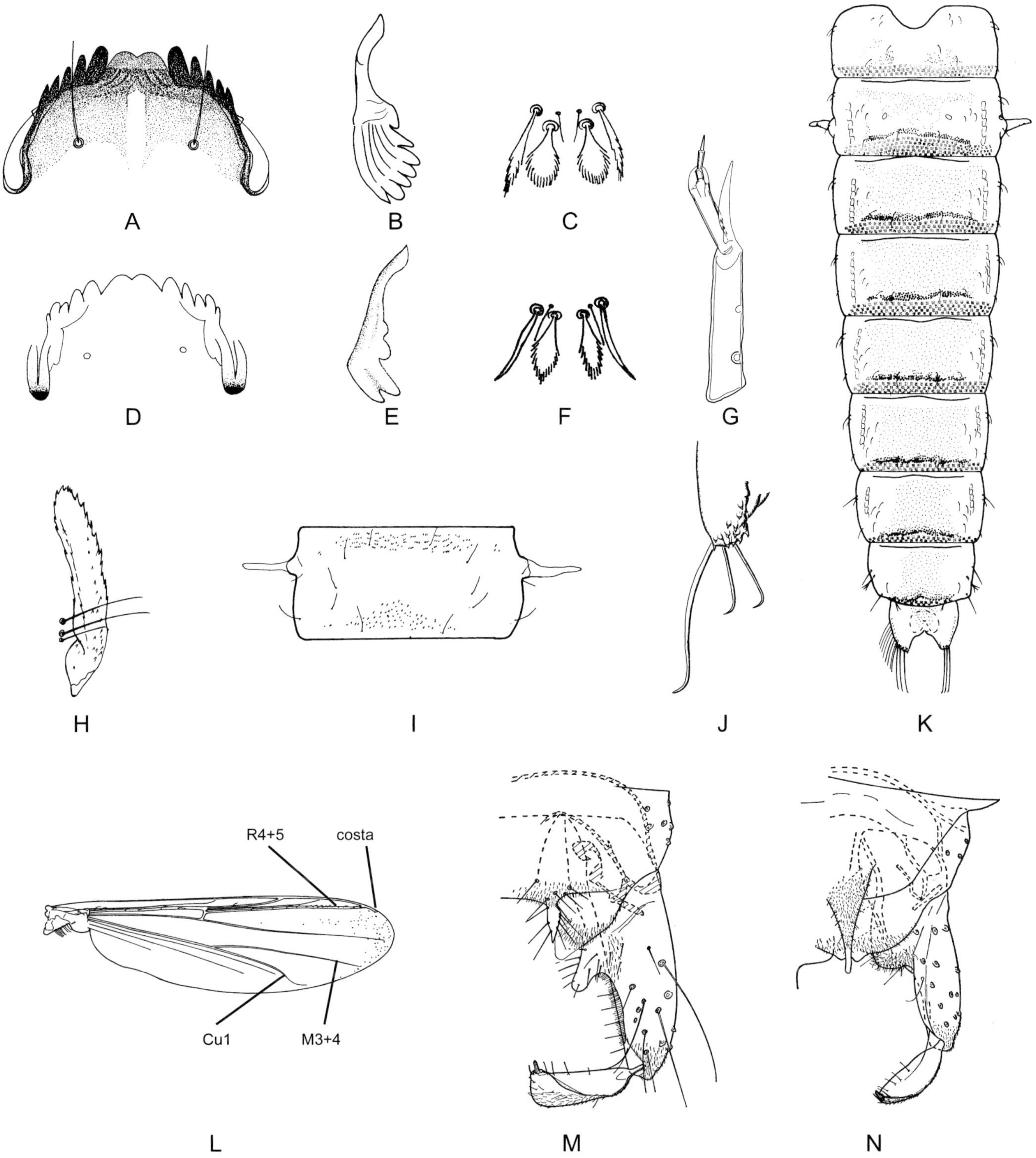

FIGURE 48 Parametriocnemus. *A–G: Larva. A: mentum, B: premandible C: S I & S II of* P. boreoalpinus/graminicola, *D: mentum, E: premandible, F: S I & S II, G: antenna, of* P. stylatus/lundbecki. *H–K: Pupa. H: thoracic horn, I: sternite 2, J: anal lobe of* P. boreoalpinus/graminicola. *K: tergites 1–9 of* P. lundbecki. *L–N: Adult male. L: wing of* P. graminicola, *M: hypopygium of* P. graminicola, *N: hypopygium of* P. lundbecki. *A from Gowin & Thienemann (1942); B: from Cranston* et al. *(1983); C & F from Kownacka &* Kownacki *(1967); D from Moller Pillot (1984); E: from Kownacki & Zosidze (1973); G, I & M–N from Sæther (1969); J from Furse* et al. *(1984); H & K from Coffmann* et al. *(1986); L from Cranston* et al. *(1989).*

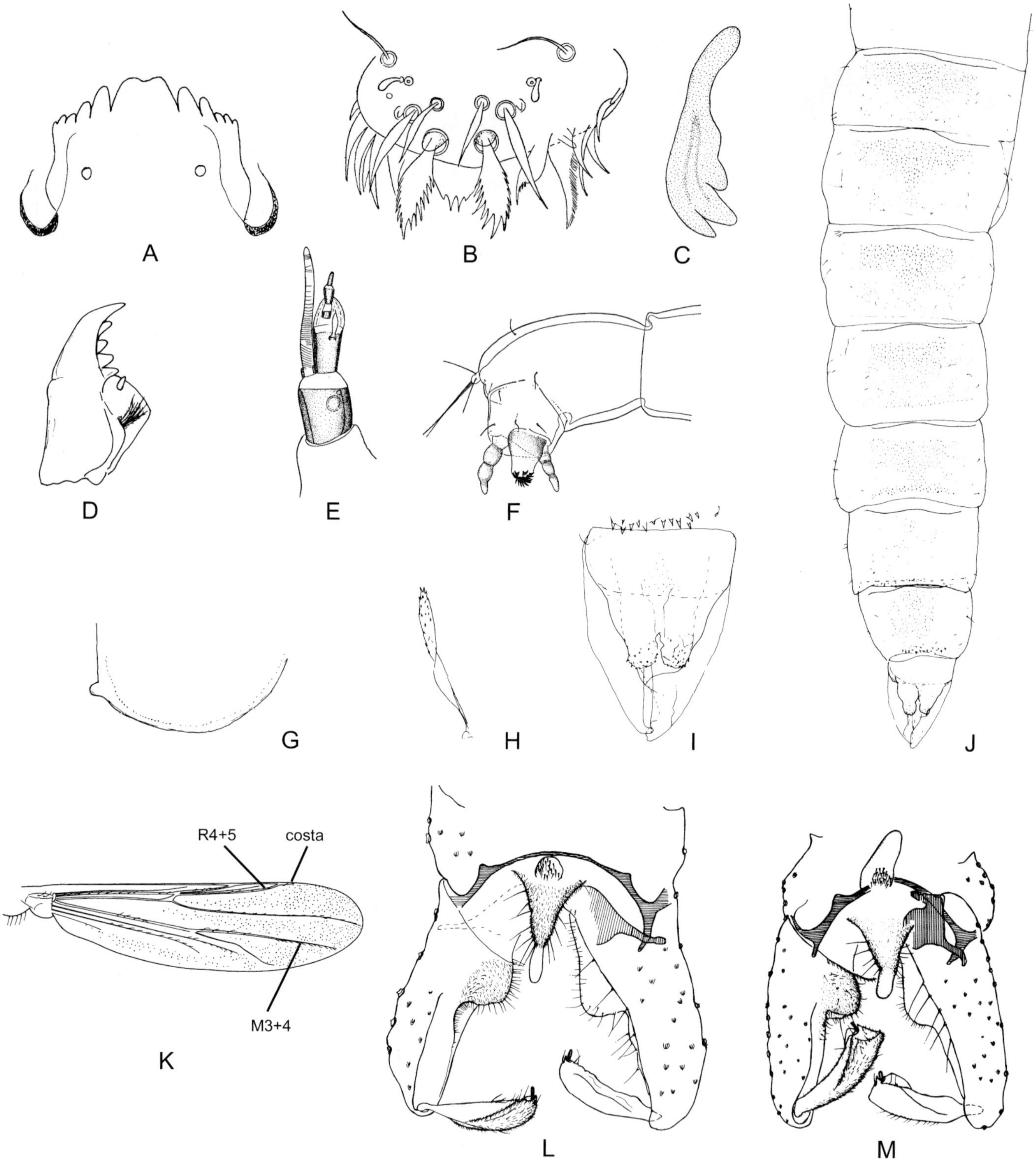

FIGURE 49 Paraphaenocladius. *A–F: Larva. A: mentum, B: labrum, C: premandible, D: mandible, E: antenna, F: anal end. G–J: Pupa. G: wing sheath with pearl row and nose, H: thoracic horn, I: anal lobe, J: tergites 1–9. K–M: Adult male. K: wing, L: hypopygium of* P. impensus, *M: hypopygium of* P. brevinervis. *A from Moller Pillot (1984); B–C & E–F from Thienemann & Strenzke (1941); D & G–M from Sæther & Wang (1995).*

uncertain. 4 species are reported from Greenland, but more may be expected. They all belong to the *limbatellus* group, in which the pupae may be recognized by the presence of double median spine patches on tergites 4–6. According to pupal exuviae at least one more species belonging to the *octomaculatus* group is common in western Greenland. This group may be recognized by the presence of double spine patches on tergites 3–6. The questionable species, *P. sordidellus,* belongs to the *sordidellus* group, in which the pupa may be recognized on the presence of frontal setae and single median point patches on tergites 4–6. Keys for larvae and pupae to the subgenera of *Psectrocladius* are provided by Cranston *et al.* (1983) and Coffmann *et al.* (1986).

The larvae are predominantly found in ponds and lakes, where they live in tubes of small sand grains and detritus attached to the substrate. They feed on detritus and biofilm.

***Psectrocladius barbatimanus* Kieffer, 1926**
= ? *P. barbimanus* Edwards, 1929.

Henriksen 1939: not recorded.

Described on material from Ellesmere Island. According to Sæther *et al.* (1984) the type material is damaged and a redescription is impossible. They suggested it to be a synonym of *P. barbimanus*. However, Oliver (1963) recognized the species from sampling in Ellesmere Island and Oliver *et al.* (1990) maintain *P. barbatimanus* as a valid species, also occurring in Greenland.

***Psectrocladius barbimanus* (Edwards, 1929) Fig. 50I**
Henriksen 1939: *P. armatus* Søgaard Andersen.

Recorded by Andersen (1937) from NE Greenland (Ella Ø) and described as *P. armatus*. This species is easily recognized especially as pupae by the anal lobe with 5 macrosetae (Fig. 50 I).

DISTRIBUTION. S, SW, NE. Very common in the lakes of the Kangerlussuaq/Søndre Strømfjord and Narsaq areas and probably common all over Greenland.

Holarctic, widely distributed.

***Psectrocladius fennicus* Storå, 1939**
Henriksen 1939: not recorded.

Some of the specimens reported from Greenland (Nuup Kangerlua/Godthåbsfjord and Liverpool Land) by Edwards (1931, 1935) and identified as *P. limbatellus* are according to Langton (1985) in fact *P. fennicus*.

DISTRIBUTION. sSW, NE.

Holarctic, widely distributed in northern parts.

***Psectrocladius limbatellus* (Holmgren, 1869)**
BIOLOGY. The larvae live primarily in lakes and ponds.

DISTRIBUTION. SW (Nuup Kangerlua/Godthåbsfjord: Qussuk, Kangerlussuaq/Søndre Strømfjord), NE (Liverpool Land, Ella Ø, Kempes Fjord). Probably common all over Greenland.

Holarctic, widely distributed.

***Psectrocladius octomaculatus* group**
DISTRIBUTION. S, SW. Pupal exuviae are common in ponds and lakes in the Kangerlussuaq/Søndre Strømfjord area.

Species belonging to the *P. octomaculatus* group are widely distributed in the Holarctic Region.

***Psectrocladius sordidellus* (Zetterstedt, 1838)**
Henriksen 1939: *Psectrocladius* ? *sordidellus* Zett.

P. sordidellus is recorded by Lundbeck (1898), but based on a heavily damaged female specimen (from Aalatsivik). Oliver *et al.* (1990) did not mention it and the record must be considered questionable. I have not found pupal exuviae of this type in S and SW Greenland.

DISTRIBUTION. ? nSW.

Palaearctic, widely distributed.

***Pseudosmittia nanseni* Kieffer, 1926 Fig. 52**
Henriksen 1939: *Smittia longitibia* Goetg.

P. nanseni is redescribed by Sæther *et al.* (1984) partly on material from Foulke Fjord (Reindeer Point) and again by Cranston & Oliver (1988) partly on material from Peary Land (Nedre Midsommersø).

BIOLOGY. Larva and pupa are unknown. Most *Pseudosmittia* species are terrestrial or semi-terrestrial; the larvae are known to live in humus-rich soil in meadows and heath lands or in wet mosses in springs and along margins of lakes and ponds.

DISTRIBUTION. N and NE (Myggbukta; Goetghebuer (1933), Jameson Land; Edwards (1935) Ella Ø; Andersen (1937) as *Smittia longitibia*).

Nearctic (northern Canada).

NOTE. Sæther & Ferrington (2003) synonymized *Smittia longitibia* Goetghebuer, 1933 with *P. nanseni*.

****Pseudosmittia oxoniana* (Edwards, 1922)**
Henriksen 1939: not recorded.

Reported by Edwards (1935) from Jameson Land and by Andersen (1937) from Ella Ø. Both records are questionable.

BIOLOGY. The larvae are purely aquatic.

DISTRIBUTION. ? NE.

Palaearctic, known from few places in the northern parts.

***Rheocricotopus* (*Psilocricotopus*) *chapmani* Edwards, 1935 Fig. 53**
Henriksen 1939: *Trichocladius chapmani* Edw.

Described on specimens from SE Greenland (Lake Fjord) and redescribed by Sæther (1985). Larva and pupa unknown.

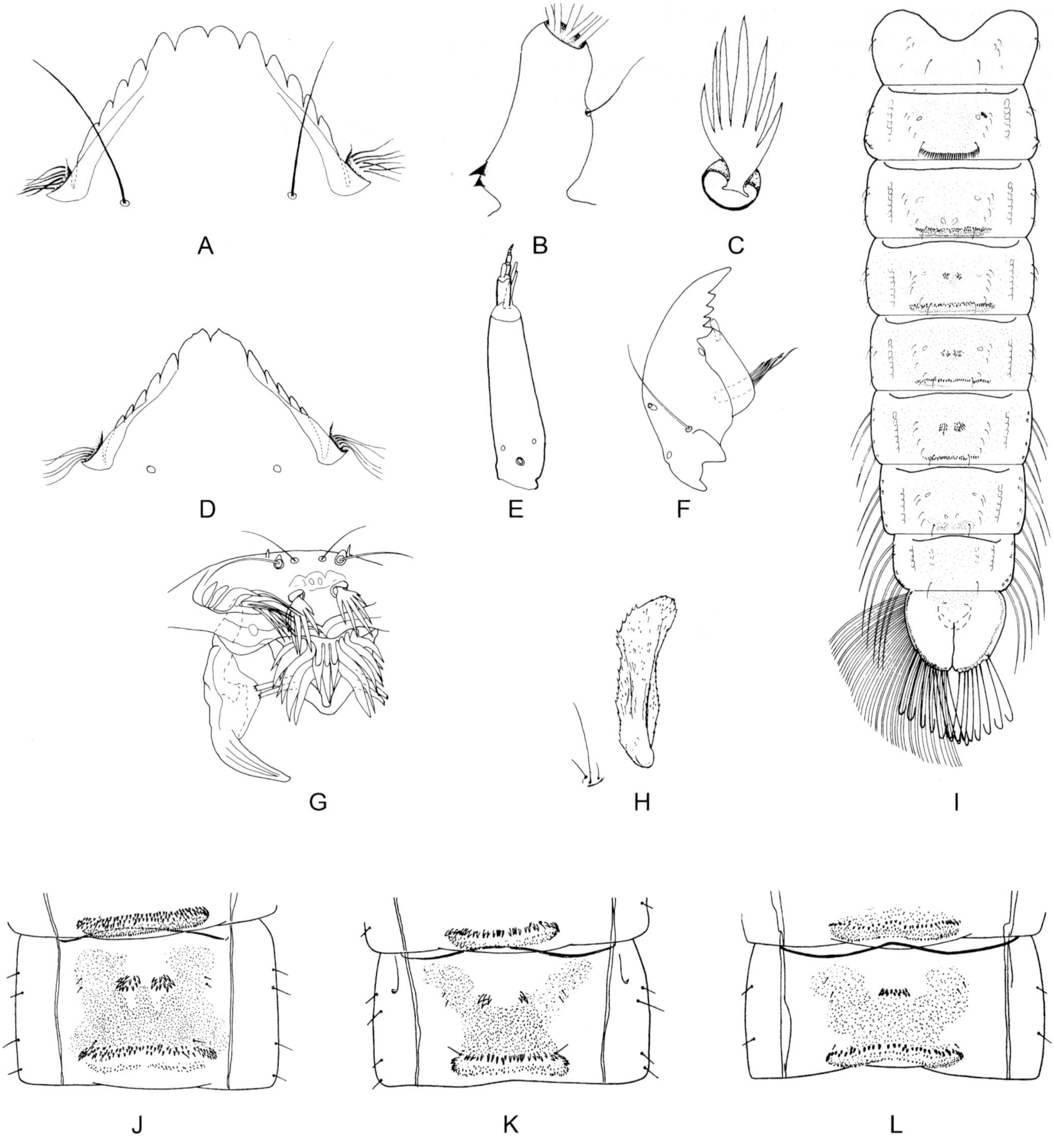

FIGURE 50 Psectrocladius *1. A–G: Larva. A: mentum, B: procercus, C: S II of* P. barbimanus. *D: mentum, E: antenna, F: mandible, G: labrum and epipharyngeal region of* P. sordidellus. *H–L: Pupa. H: thoracic horn, I: tergites 1–9 of* P. barbimanus, *J: tergite 3 of.* P. fennicus, *K: tergite 3 of.* P. limbatellus, *L: tergite 3 of.* P. sordidellus. *A–C from Moller Pillot (1984); D–G from Cranston* et al. *(1983); H–I from Coffmann* et al. *(1986); J–L from Brundin (1949).*

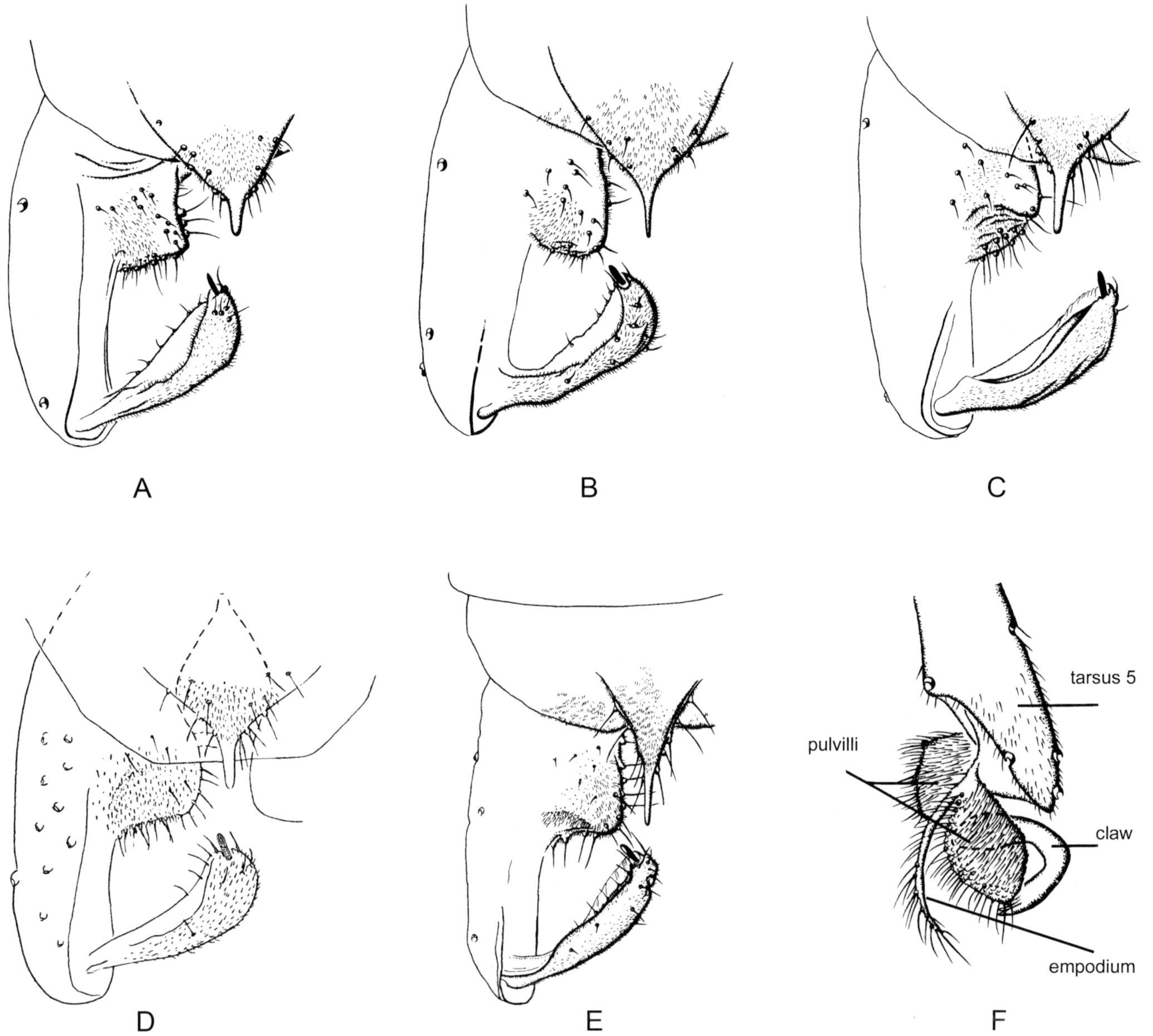

FIGURE 51 Psectrocladius 2. *A–E: Hypopygium of adult male. A:* P. barbimanus, *B:* P. fennicus, *C:* P. limbatellus, *D:* P. octomaculatus, *E:* P. sordidellus. *F: Lateral view of tip of tarsus showing pulvilli and empodium of* Psectrocladius *sp. A–C & E–F from Pinder (1978); D from Wülker (1956).*

DISTRIBUTION. SE.

Palaearctic (known from Norway and Finland).

***Rheocricotopus* (*Rheocricotopus*) sp. Fig. 53**

Henriksen 1939: not recorded.

BIOLOGY. The larvae live mainly in running water, but in northern areas and at high altitude they may be found in the surf zone of lakes.

DISTRIBUTION. S, SW. I have found larvae and pupae of a *R.* (*Rheocricotopus*) species in streams in the Qeqertarsuaq/Disko and Narsaq areas.

The subgenus is widely distributed in the Holarctic Region.

***Smittia* Holmgren, 1869 Fig. 54**

The genus *Smittia* is a large genus with about 40 species and in need of revision. Not less than 4 of 7 recorded species from Greenland are described on material from various localities in Greenland. Larvae and pupa of most species are unknown. The record of *S. flexinervis* by Remy (1928) is uncertain. *S. aterrima* is the only widespread species. One species was found hiding in the flowers of *Pedicularis lanata* (Philipp *et al.* 1996). The larvae are mainly terrestrial, living in humid soil rich in organic matter. Some species are found close to aquatic biotopes.

***Smittia aterrima* (Meigen, 1818)**

Henriksen 1939: not recorded.

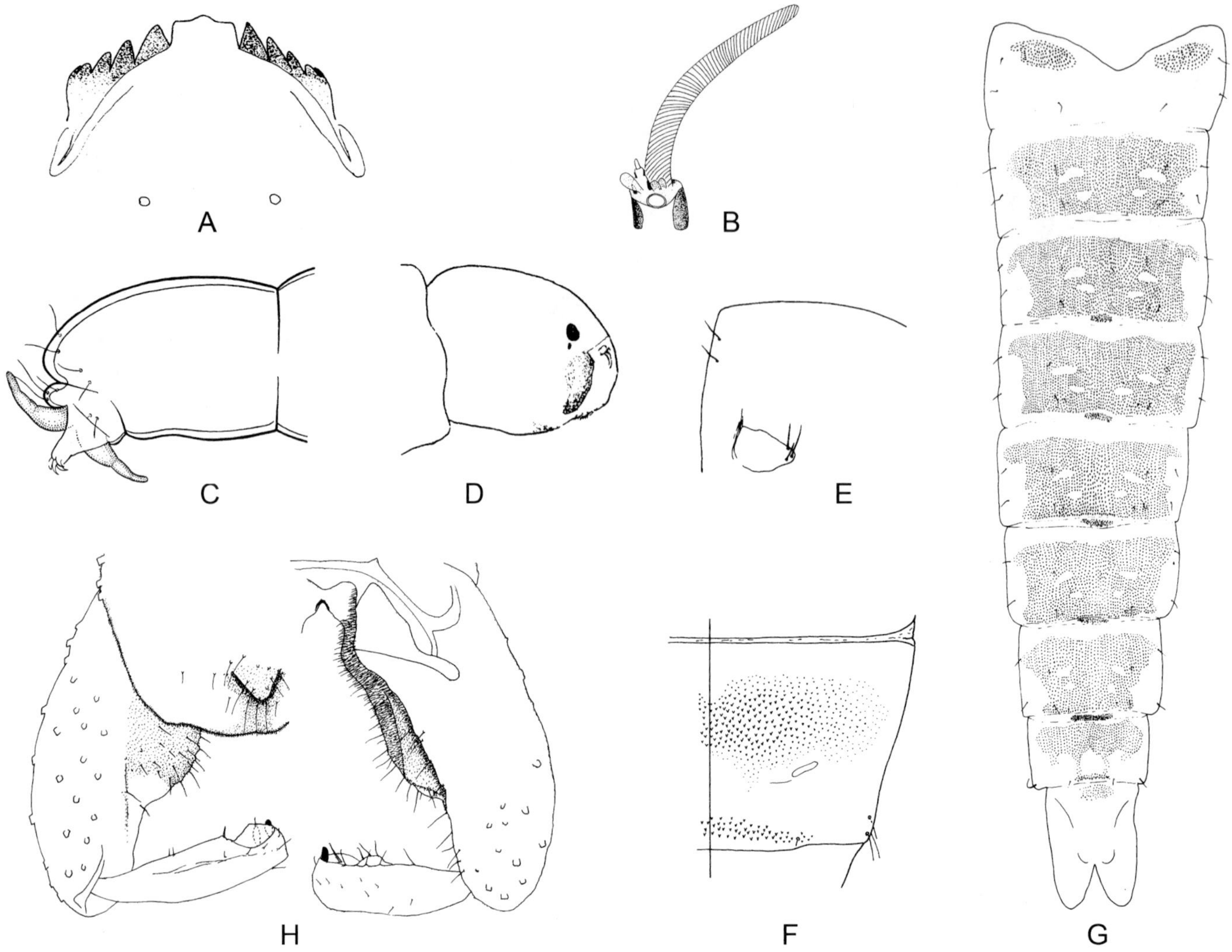

FIGURE 52 Pseudosmittia. *A–D: Larva. A: mentum, B: antenna, C: anal end, D: lateral view of head. E–G: Pupa. E: thorax area with antepronotals and precorneals, F: tergite 8, G: tergites 1–9 of* P. oxoniana. *H: Hypopygium of adult male of* P. nanseni. *A–B & D from Moller Pillot (1984); C from Strenzke (1951); E & G from Coffmann* et al. *(1986); F from Langton (1994); H from Cranston & Oliver (1988).*

Oliver *et al.* (1990) and Oliver & Dillon (1997) mentioned *S. aterrima* from Greenland, but without localities.

DISTRIBUTION. ? Greenland.

Holarctic, widespread.

Smittia extrema (Holmgren, 1869)

DISTRIBUTION. nSW (Qeqertarsuup Kangerlua/Disko Bugt; Lundbeck 1898), NE (widely distributed; Goetghebuer 1933).

Circumpolar, arctic.

***_Smittia flexinervis_ (Kieffer, 1911)**

Henriksen 1939: *Smittia* ? *longicosta* Edw.

Described from Bear Island and recorded by Remy (1928) from Kap Stewart (NE Greenland). This record is questionable and not approved by neither Ashe & Cranston (1990) nor Oliver *et al.* (1990).

DISTRIBUTION. ? NE.

Palaearctic, arctic (known from Spitsbergen and Bear Island).

Smittia parva Lundbeck, 1898

Described on females from S and SW Greenland. It is redescribed by Oliver (1970).

DISTRIBUTION. S, SW (Lundbeck 1898). Only females are known.

Only known from Greenland.

Smittia polymorpha Andersen, 1937

Described on material from NE Greenland (Ella Ø etc.)

DISTRIBUTION. NE. Seems to be widely distributed on Ella Ø and surroundings in Kong Oscar Fjord.

Only known from Greenland.

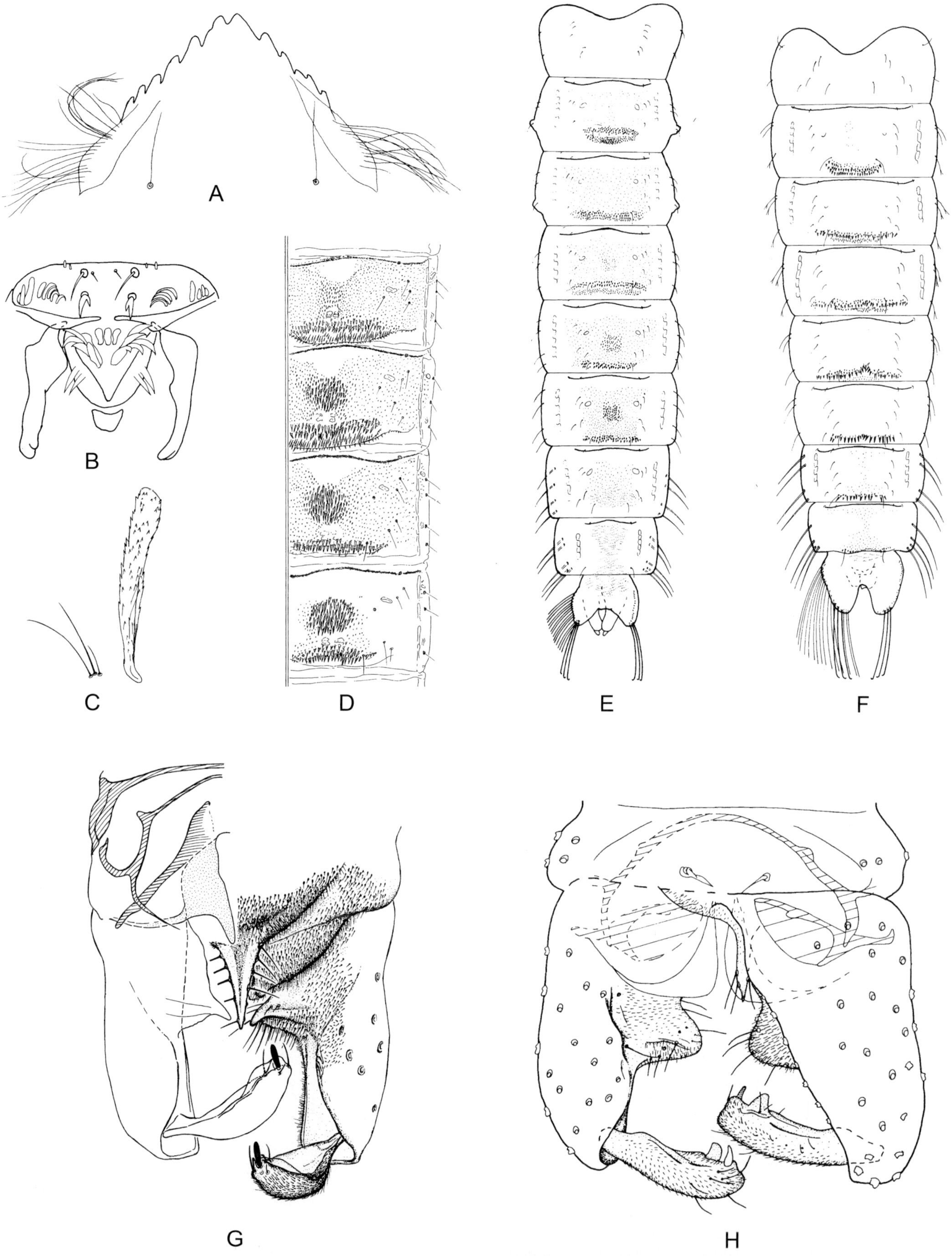

FIGURE 53 Rheocricotopus. *A–B: Larva. A: mentum, B: labrum and epipharyngeal region. C–F: Pupa. C: thoracic horn, D: tergites 3–6, E: tergites 1–9 of* R. (Rheocricotopus). *F: tergites 1–9 of* R. (Psilocricotopus). *G–H: Hypopygium of adult male. G:* R. (R.) effusus, *H:* R. (P.) chapmani. *A from Moller Pillot (1984); B from Cranston* et al. *(1983); C & E–F from Coffmann* et al. *(1986); D & G from Lehmann (1969); H from Sæther (1985).*

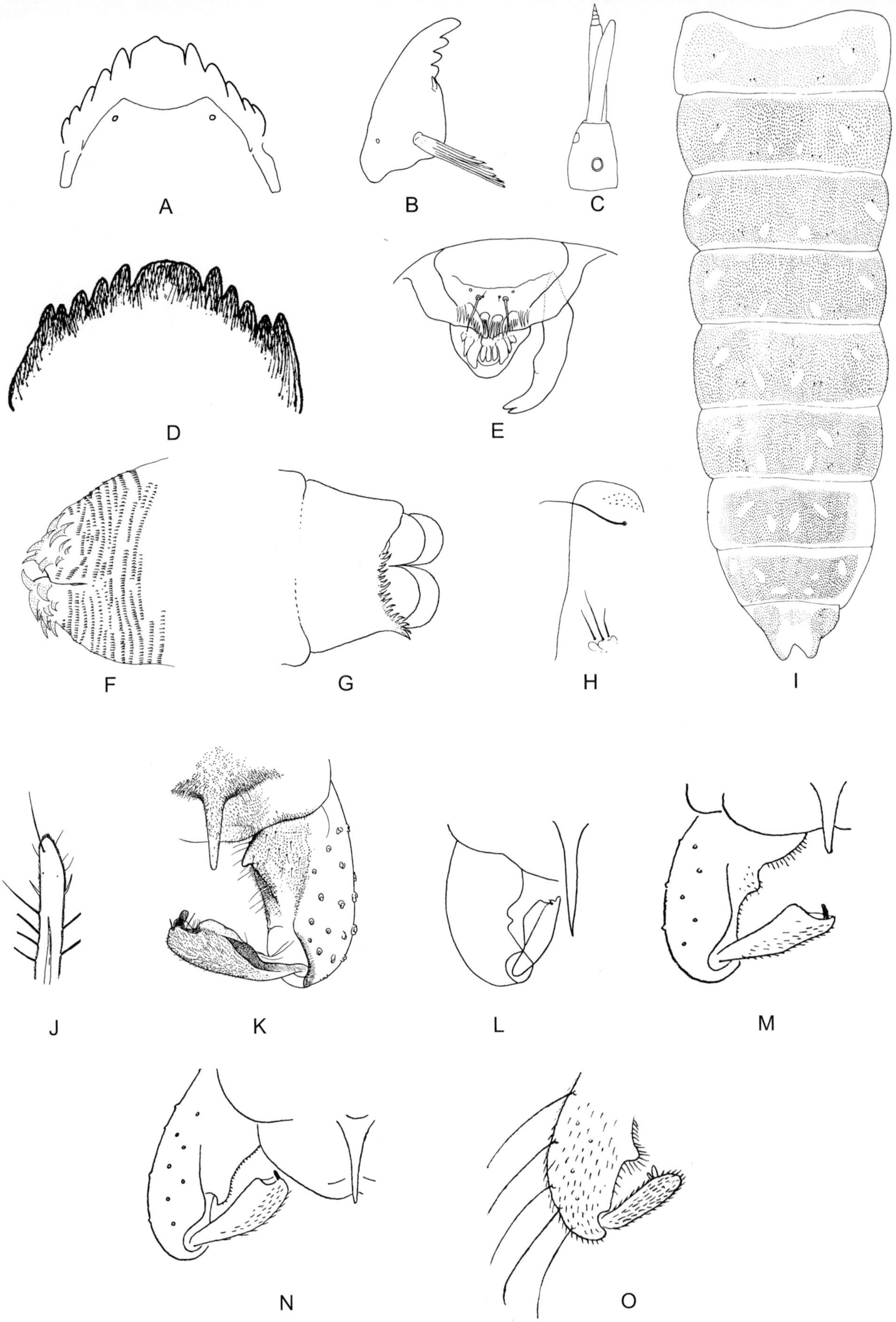

FIGURE 54 Smittia *1. A–G: Larva. A & D: mentum, B: mandible, C: antenna, E: labrum and epipharyngeal region, F: anterior parapods, G: anal end. H–I: Pupa. H: Thorax area with antepronotals and precorneals, I: tergites 1–9. J–O: Adult male. J: tip antenna, K–O: hypopygium. D, G and J–K of* S. aterrima*; L of* S. polymorpha*; M of* S. extrema*; N of* S. subnudipennis*; O of* S. flexinervis*; others of* Smittia *spp. A–C & E–F from Cranston* et al. *(1983); D & G from Johannsen (1937); H–I from Coffmann* et al. *(1986); J from Cranston* et al. *(1989); K from Brundin (1956); L from Andersen (1937); M–N from Goetghebuer (1933); O from Goetghebuer (1940–1950).*

***Smittia subnudipennis* Goetghebuer, 1933**

Described on specimens from NE Greenland by Goetghebuer (1933).

DISTRIBUTION. NE (Myggbukta; Goetghebuer 1933, Ella Ø and surroundings (Andersen 1937).

Only known from Greenland.

***Smittia velutina* Lundbeck, 1898**

Described on females from S and SW Greenland. Oliver (1970) redescribed Lundbeck's type material and stated that the same females are common at Hazen Camp, Ellesmere Island (NWT, Canada) and that they perhaps are the females of *S. extrema*.

DISTRIBUTION. S, SW (presumably widespread).

Nearctic (only known from Canada: Ellemere Island).

***Thienemanniella* Kieffer, 1911 Fig. 55**

Henriksen 1939: not recorded.

I found larvae belonging to *Thienemanniella* in large numbers in a stream on Qeqertarsuaq/Disko (Friberg *et al.* 2001). Hestenes & Sæther (2000) reviewed the Nearctic *Thienemanniella* species and provided preliminary keys.

The larvae are free-living and very mobile, feeding on algae and detritus. They are primarily found in running waters, but may also be abundant in the littoral zone of lakes, especially at high altitude and in northern areas.

***Tokunagaia* Sæther, 1973 Fig. 56**

The genus *Tokunagaia* was established from the *Eukiefferiella rectangularis* group by Sæther (1973) and revised by Halvorsen & Sæther (1987), who provided keys for adult males and pupae. However, the pupae are extremely similar and identification to species difficult (Langton 1991).

***Tokunagaia obriaini* Hayes & Murray, 1988 Fig. 56**

Henriksen 1939: not recorded.

T. obriaini is described on material from Canada: Northwest Territories. Oliver & Dillon (1997) referred it as occurring in Greenland.

BIOLOGY. The larvae are found in running waters.

DISTRIBUTION. S, SW. I have found larvae and pupae in a number of streams in the Narsaq and Kangerlussuaq/Søndre Strømfjord areas.

Nearctic, arctic (Canada).

***Tokunagaia* sp. cf. *rectangularis* (Goetghebuer, 1940) Fig 56**

Henriksen 1939: not recorded.

Pupal exuviae from a number of streams in the Kangia Fjord and Kangerlussuaq/Søndre Strømfjord areas keyed out as *T. rectangularis*. They are easily separated from *T. obriaini* by the longer posterior points on the abdominal tergites.

DISTRIBUTION. SW.

Palaearctic, northern Fennoscandia.

***Tvetenia* sp. cf. *bavarica* (Goetghebuer, 1934) Fig. 57**

Henriksen 1939: not recorded.

Immature stages easily recognizable – larvae by the long setae on abdominal segments; the pupae by the extra medial macrosetae.

BIOLOGY. The larvae build small tubes of detritus attached to plants, algae and mosses in running waters.

DISTRIBUTION. S. I have found larvae, pupae and adults of a species near *T. bavarica* in streams in the Narsaq area.

Palaearctic, widely distributed.

Orthocladiinae genus A Fig. 58

Henriksen 1939: not recorded

Most larvae of the *Cricotopus* genus are recognized by the combination of bifid S 1 on labrum and setal tufts on the lateral part of the abdominal segments. However, larvae with setal tufts and S 1 more or less strongly serrated have been found in lakes close to, but not directly connected to the ice field in the Kangerlussuaq/Søndre Strømfjord area (Hamerlik & Brodersen, in prep.). These larvae may belong to an unknown genus.

CHIRONOMINAE, TRIBE CHIRONOMINI

Key to genera of Greenland Chironomini

ADULT MALES

1. Antenna with 11 flagellomeres 2
– Antenna with 13 flagellomeres (Fig. 5O) 4
2. Inferior volsella never with long dense setae, though it may be with microtrichia; superior volsella pediform to rectangular with microtrichia (Fig. 62K) .... *Paracladopelma*
– Inferior volsella always well developed and with long setae (Figs 59L, 60P, Q) 3
3. Inferior volsella long, slender, strongly bowed dorsoventrally, enlarged apically with long setae to the tip only (Figs 60P, Q) .... *Dicrotendipes*
– Inferior volsella broad subcylindrical with long setae covering the distal half (Fig. 59L) .... *Chironomus*
4. Wing membrane with macrotrichia (Fig. 5F) 5
– Wing membrane without macrotrichia 6
5. Fore tarsus with beard, *i.e.* the longest setae are more than 4 times as long as the width of tarsal segment . ... *Phaenopsectra*

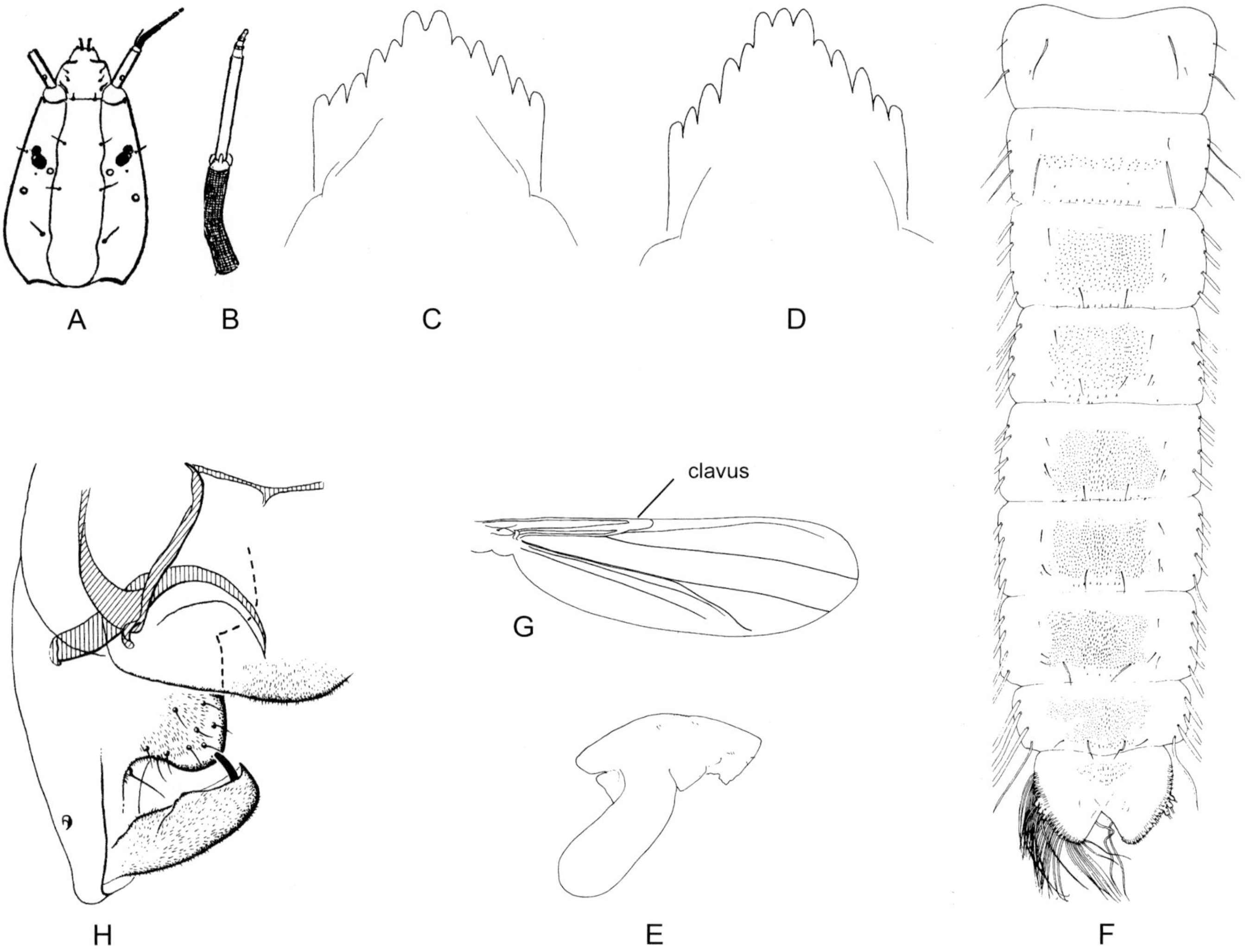

FIGURE 55 Thienemanniella. *A–D: Larva. A: dorsal view of head, B: segment 2–5 of antenna, C & D: mentum. E–F: Pupa. E: wing sheath without pearl row, F: tergites 1–9. G–H: Adult male. G: wing, H: hypopygium of* T. majuscule. *A–B from Thienemann (1944); C–D from Moller Pillot (1984); E–F from Coffmann* et al. *(1986); H from Pinder (1978).*

– Fore tarsus without beard .... *Sergentia*

6. Tergite 8 anteriorly tapered (Fig. 64G) .... *Polypedilum*

– Tergite 8 anteriorly not tapered 7

7. Both combs on hind tibia with spur (cf. Fig. 5K) .... *Endochironomus*

– Only one of the combs on hind tibia with spur (cf. Fig. 5L) .... *Stictochironomus*

LARVAE

(Larva of *Endochironomus oldenbergi* is unknown)

1. Antenna with 6 segments; Lauterborn organs alternate on segments 2 and 3 (Fig. 66E); four median teeth higher than the lateral teeth; the median teeth are often worn down (Figs 66A, D) .... *Stictochironomus*

– Antenna with 5 segments; Lauterborn organs opposite on segment 2 (*e.g.* Fig. 59H) 2

2. Maxillary palp elongate, protruding from the head like an extra pair of antennae (Fig. 62A). Mentum pale with the most lateral teeth brown (Figs 66A, D) .... *Paracladopelma*

– Maxillary palp of normal size; all teeth in mentum dark 3

3. Frontoclypeal apotome with a large mark distally (Figs 60C, G); ventromental plates narrower than mentum (Figs 60A, D) .... *Dicrotendipes*

– Frontoclypeal apotome without a mark distally 4

4. Mentum divided into three parts; the median ends of the ventromental plates are anteriorly produced and separate the 4 median teeth from the rest of mentum (Figs 61A, 63A, 65A) 5

– Mentum not divided 7

5. Frontal apotome separated from clypeus by a fine suture; pecten epipharyngis with extra toothlets (Fig. 61E) .... *Endochironomus*

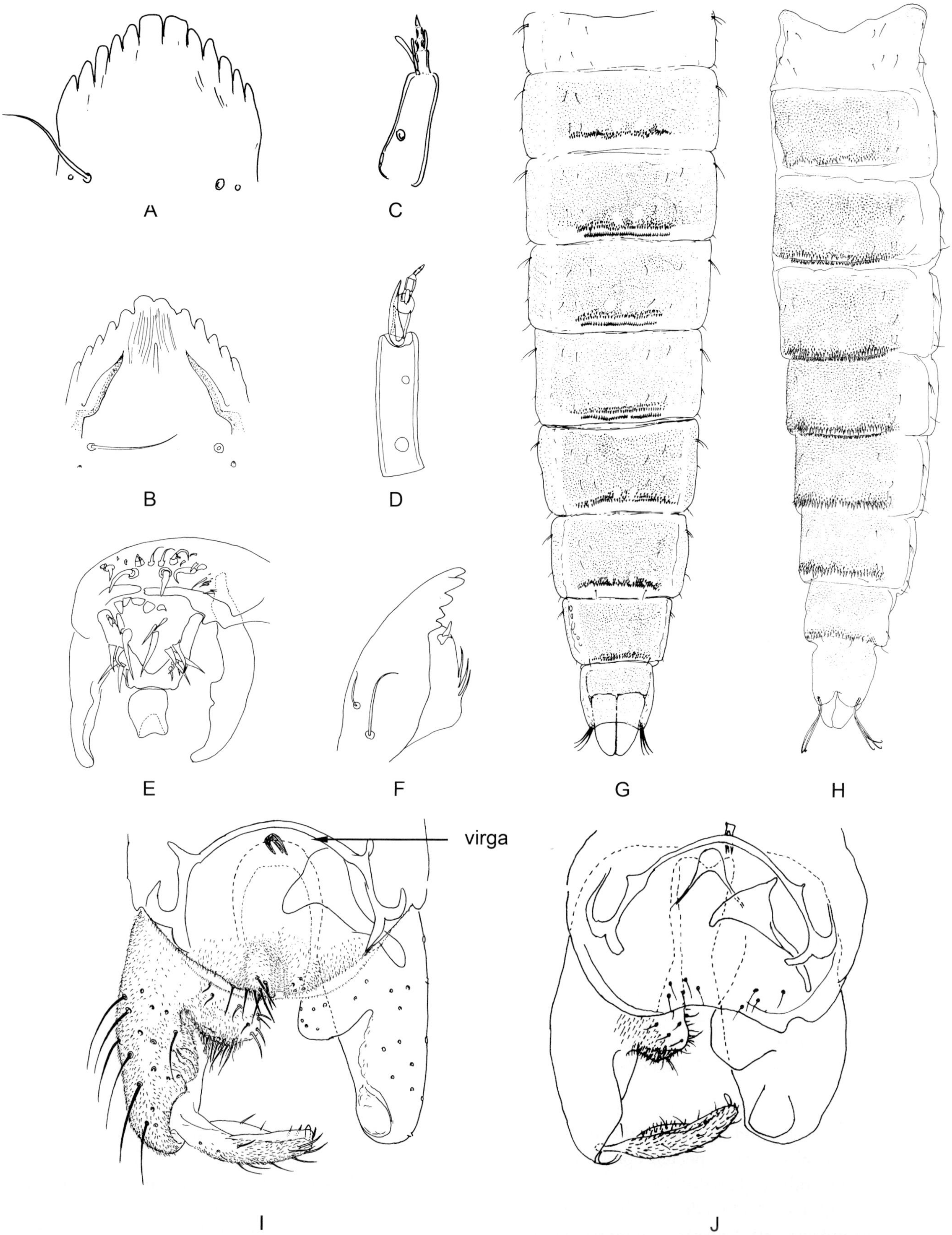

FIGURE 56 Tokunagaia. *A–F: Larva. A: mentum & C: antenna of T.* obriaini. *B: mentum, D: antenna, E: labrum and epipharyngeal region and F: mandible of.* T. rectangularis. *G–H: Pupa. F: tergites 1–9 of* T. obriaini *and G: tergites 1–9 of* T. rectangularis. *I–J: Hypopygium of adult male of I:* T. obriaini *and J:* T. rectangularis. *A, C, G & I from Hayes & Murray (1988); B–F, H & J from Halvorsen & Sæther (1987).*

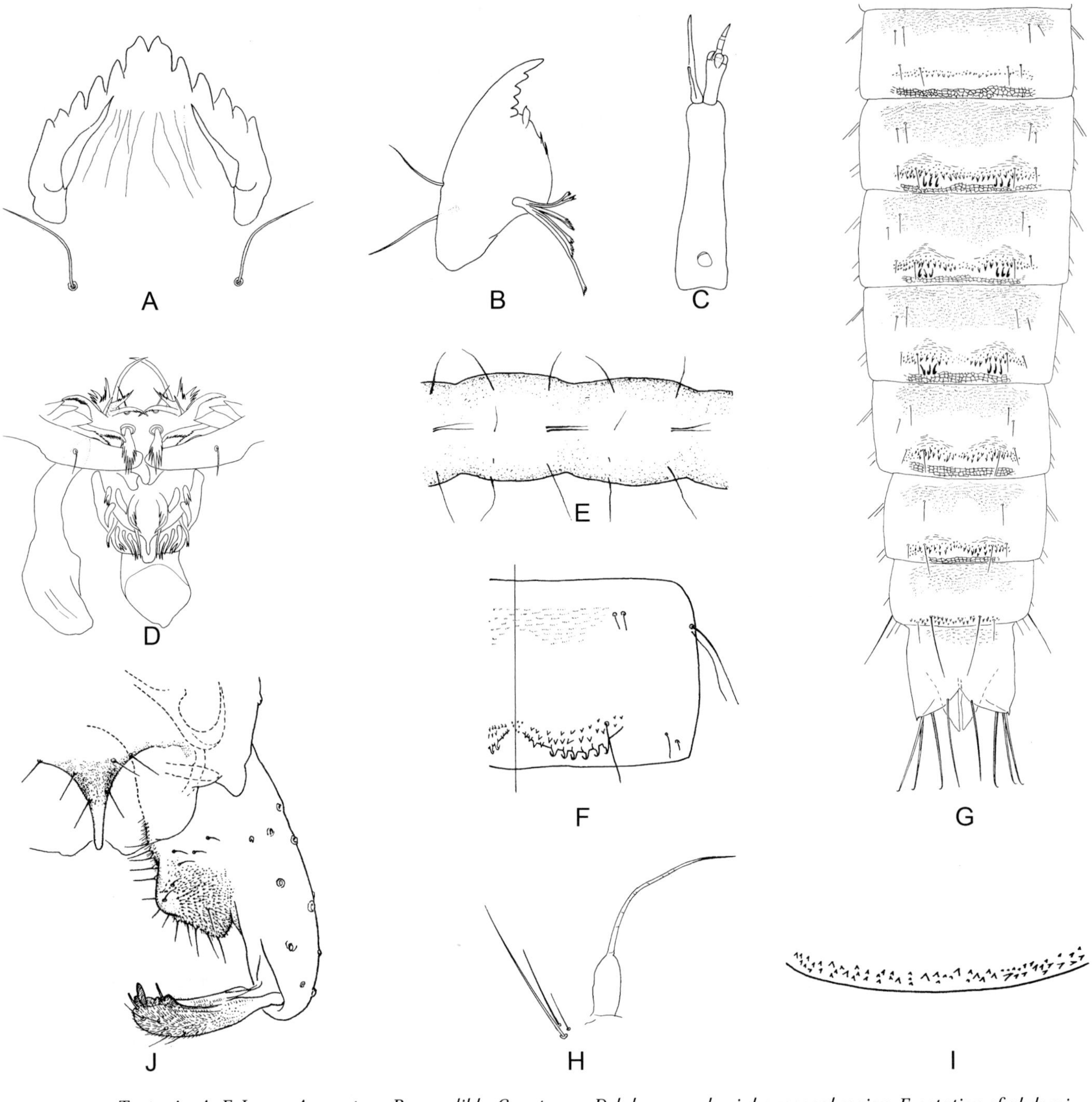

FIGURE 57 Tvetenia. *A–E: Larva. A: mentum, B: mandible, C: antenna, D: labrum and epipharyngeal region, E: setation of abdominal segments. F–I: Pupa. F: tergite 3, G: tergites 2–9, H: thoracic horn, I: pearl row at tip of wing sheaths. J: Hypopygium of adult male (near* T. calvescens *and* T. bavarica*). A–D from Cranston* et al. *(1983); E from Moller Pillot (1984); F & I from Langton (1991); G & H from Coffmann* et al. *(1986); J from Brundin (1956).*

– Frontal apotome and clypeus not divided, forming a frontoclypeal apotome; pecten epipharyngis without extra toothlets (Figs 63C, 65D) 6

6. The 2 central teeth of median teeth of mentum narrower and distinctly lower than outer pair of median teeth; mandible with 3 inner teeth (Figs 63A, B) .... *Phaenopsectra*

– The 2 central teeth of median teeth of mentum narrower and only slightly lower than outer pair of median teeth; mandible with 4 inner teeth (Figs 65A, B) .... *Sergentia*

7. Frontoclypeal apotome with convex anterior margin; median tooth of mentum trilobed (Fig. 59A), with or without two pairs of ventral tubuli on abdominal segment 8 (Figs 59D–F) .... *Chironomus*

– Frontoclypeal apotome with straight anterior margin; median tooth of mentum not trilobed; the

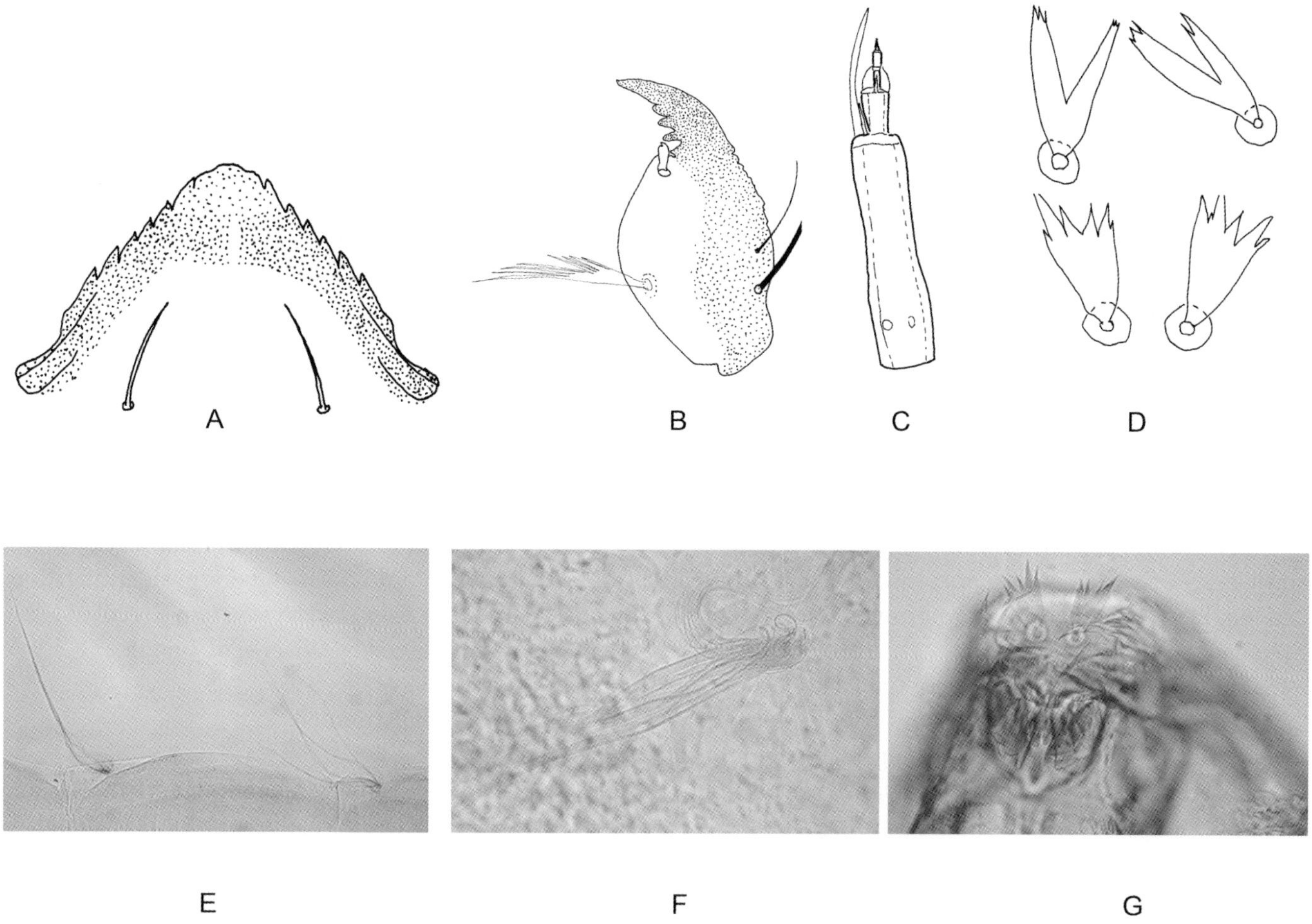

FIGURE 58 *Orthocladiinae genus A. A–G: Larva. A: mentum, B: mandible, C: antenna, D: variation in S 1 setae, E–F: lateral tufts of setae on abdominal segments, G: labrum and epipharyngeal region. E–G.*
(PHOTOS: L. HAMERLIK)

2 central median teeth of mentum higher than the outer median teeth (Fig. 64A); without ventral tubuli .... *Polypedilum*

PUPAE

(The pupa of *Endochironomus oldenbergi* is unknown)

1. Cephalic tubercle with circular field of spinules (Figs 63E, 65E) 2
– Cephalic tubercle without circular field of spinules or absent 3
2. Conjunctives 1–2 to 4–5 with dark lateral fleck (Fig. 63G) .... *Phaenopsectra*
– Conjunctives 1–2 to 4–5 without dark lateral fleck (Fig. 65G) .... *Sergentia*
3. Thoracic horn with at most 20 branches (Fig. 64E) . ... *Polypedilum*
– Thoracic horn a brush of 30 to several hundred fine branches (Fig. 1D) 4
4. Frontal seta long, robust and broad (Fig. 66G) .... *Stictochironomus*
– Frontal seta very short and slender (Fig. H) 5
5. Pedes spurii B absent on segment 2 (Figs 62I, J) .... *Paracladopelma*
– Pedes spurii B present on segment 2 (Figs 59K, 60O) 6
6. Conjunctives 3–4 and/or 4–5 with a short, slender L seta, only visible by careful examination; characteristic anal spur present, base with pale fenestrae (Fig. 59J) .... *Chironomus*
– Conjunctives without L setae; long slender S-shaped anal spur curved outwards and apically produced into a fine hair-like point; in Greenland specimens the anal spur most often double (Figs 60K, L) .... *Dicrotendipes*

***Chironomus* Meigen, 1803 Fig. 59**

About 100 species of *Chironomus* are described, but only incomplete keys are available and the genus needs revision. 3 species have been reported from Greenland, but chromosome analyses on larval material from Ellesmere Island (NWT, Canada) revealed 5 *Chironomus* species

(Wülker and Martin 2000). Therefore, more species are expected to occur in Greenland.

The larvae inhabit mainly lakes and ponds of all kinds, live in soft sediment and are known for their ability to survive low oxygen concentrations for weeks or months.

***Chironomus hyperboreus* Stæger, 1845**
Described by Stæger and redescribed by Andersen (1937) on material from N and NE Greenland. The larvae belong to the northern freshwater group of *Chironomus* larvae without ventral tubuli (Fig. 59 D), which otherwise are characteristic of the marine *C. salinarius* group. The chromosomes of these freshwater species are described by Wülker & Butler (1983).

BIOLOGY. The larvae are found in lakes and ponds.

DISTRIBUTION. S, nSW, N, NE. Very common in ponds and lakes in the Kangerlussuaq/Søndre Strømfjord area and on Ella Ø. Probably common over large parts of Greenland.

Nearctic, widespread in northern Canada (Oliver *et al.* 1990). However, chromosome analyses have revealed that several closely related species belong to the group of freshwater species in which the larvae are without tubuli. Earlier records from Canada should be revised.

***Chironomus riparius* Meigen, 1804**
The larva has 2 pairs of ventral tubuli, but no lateral tubuli (Fig. 59E).

BIOLOGY. Larvae of this type are common together with larvae of *C. hyperboreus* in lakes and ponds in the Narsaq and Kangerlussuaq/Søndre Strømfjord areas. It is known as character species of streams polluted with organic matter.

DISTRIBUTION. S, SW (widespread and common).

Holarctic, widely distributed.

***Chironomus staegeri* Lundbeck, 1898**
Originally described by Stæger (1845) on specimens from Greenland as a variant of *C. hyperboreus.* Lundbeck (1898) redescribed this variant as a new species. Sublette & Sublette (1971) redescribed the species (adult males, pupae and larvae) on material from various localities in U.S.A. However, their description may not be of the same species as the original *C. staegeri* from Greenland. Curry (1956) and Epler (2001) placed the larvae of *C. staegeri* in the *C. plumosus* group with lateral tubuli on abdominal segment 7 (Fig. 59 F). Larvae of this type are not yet recorded from Greenland.

DISTRIBUTION. nSW (Aulatsivik and Aasiaat/Egedesminde) according to Lundbeck (1898).

? Nearctic (Sublette & Sublette 1971).

***Dicrotendipes modestus* (Say, 1823) Fig. 60**
Henriksen 1939: not recorded.

Redescribed by Contreras-Lichtenberg (1986) and Epler (1987). The latter included adult males from Greenland in his redescription. Both authors provided keys to larvae, pupae and adult males of the genus *Dicrotendipes*.

BIOLOGY. The tube-dwelling larvae are mainly found in lakes and ponds where they live in the surface sediment feeding on algae, detritus and associated micro-organisms.

DISTRIBUTION. SW. Pupal exuviae are very common in the Kangerlussuaq/Søndre Strømfjord area.

Holarctic, widely distributed.

***Dicrotendipes nervosus* (Stæger, 1839) Fig. 60**
Henriksen 1939: *Chironomus brevitibialis* Zett.

Reported from Nuuk/Godthåb by Lundbeck (1898). This is the only record and may be a misidentification. Epler (1988) did not refer *D. nervosus* from Greenland, but the record is maintained by Oliver *et al.* (1990).

DISTRIBUTION. sSW.

Holarctic, widely distributed.

***Endochironomus oldenbergi* Goetghebuer, 1932 Fig. 61**
Reported by Goetghebuer (1933) from NE Greenland. It is a problematic species. Ashe & Cranston (1990) considered it a *nomen dubium*, and in his revision of the genus Grodhaus (1987) could only "describe" it from Goetghebuers original description. Andersen (1937) suggested that it might be *Sergentia coracina*. However, Oliver *et al.* (1990) maintained the record of *E. oldenbergi* from Greenland.

DISTRIBUTION. NE (only recorded from Loch Fyne).

Palaearctic (Germany).

***Paracladopelma nigritula* group, Fig. 62**
Jackson (1977) revised the genus and provided keys to larvae, pupae and adult males. Larvae and pupae of a *Paracladopelma* species belonging to the *nigritula* group were found commonly in lakes in the Narsaq and Kangerlussuaq/Søndre Strømfjord areas (Lindegaard *et al.* 1978). It was, however, not possible to identify the species.

BIOLOGY. The larvae live in sandy sediment of oligotrophic lakes. In the European lowland they are also common in sandy streams. Species of *Paracladopelma* are cold stenothermal, preferring oligotrophic biotopes.

DISTRIBUTION. S, SW.

The *P. nigritula* group is widely distributed in the Holarctic Region.

***Phaenopsectra incompta* (Zetterstedt, 1838) Fig. 63**
Henriksen 1939: *Metriocnemus incomptus* Zett.

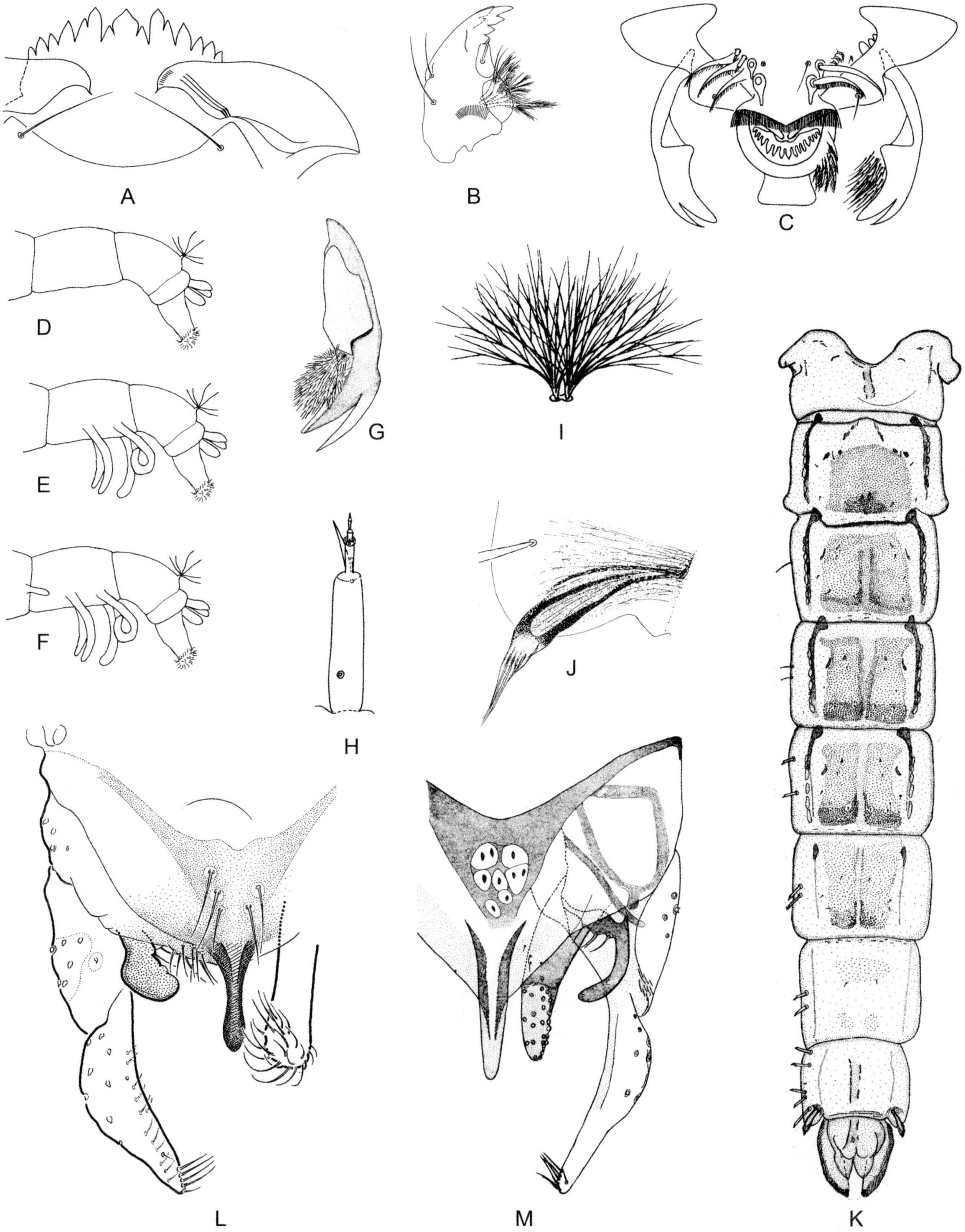

FIGURE 59 Chironomus. *A–H: Larva. A: mentum, B: mandible, C: labrum and epipharyngeal region, D–F: anal end of D:* C hyperboreus, *E:* C. riparius, *F:* C. staegeri, *G: premandible, H: antenna. I–K: Pupa. I: thoracic horn, J: anal spur, K: tergites 1–9. L–M: Hypopygium of adult male of L:* C. riparius, *M:* C. staegeri. *A–C, E & H–I from Pinder & Reiss (1983, 1986); G, K & M from Sublette & Sublette (1971); I from Lenz (1959); L from Strenzke (1959).*

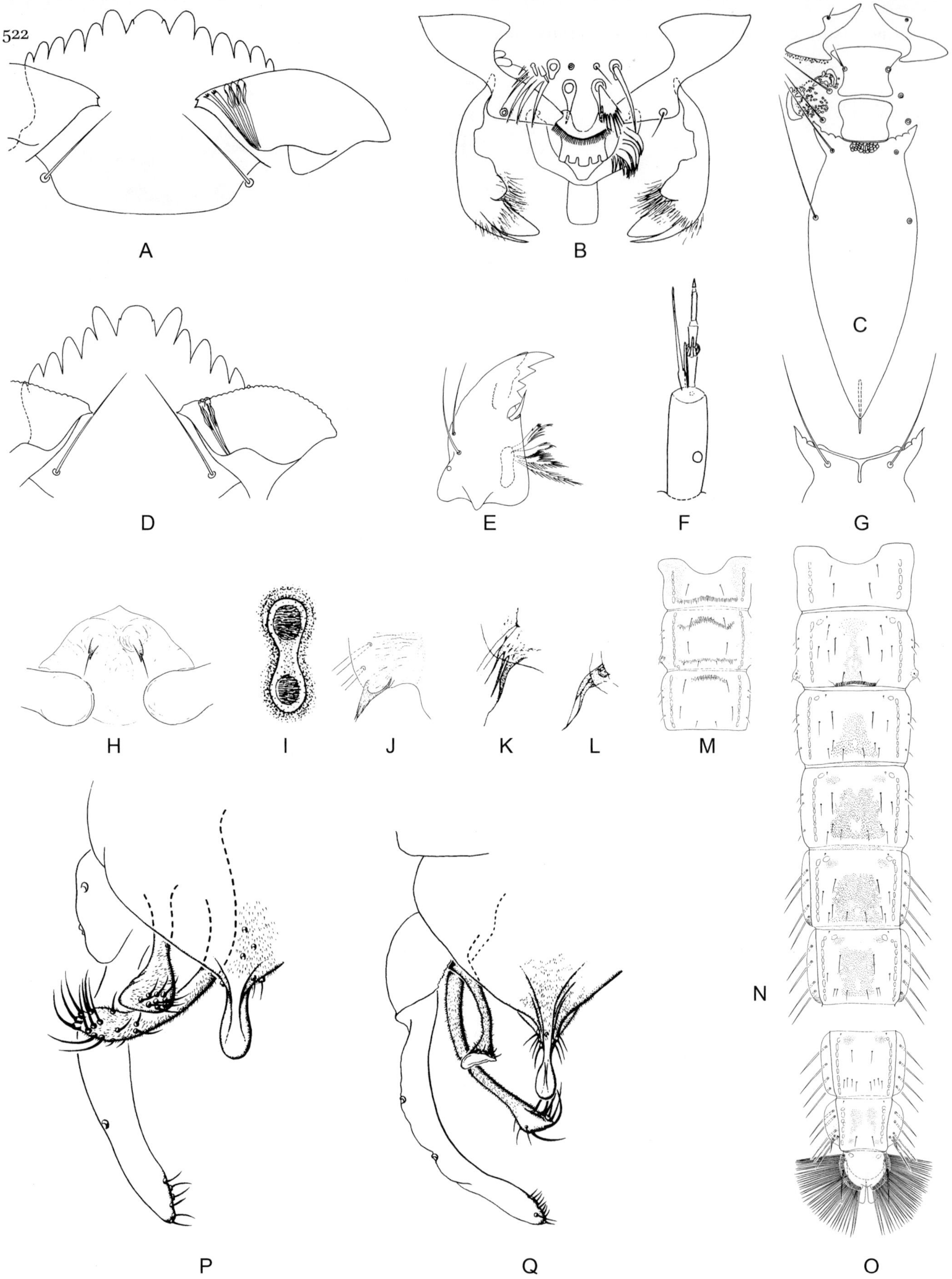

FIGURE 60 Dicrotendipes. *A- G: Larva. A: mentum, B: labrum and epipharyngeal region, C: dorsal sclerites of head of* D. modestus, *D: mentum, E: mandible, F: antenna, G: apical part of frontal apotome of* D. nervosus. *H–O: Pupa. H: frontal apotome, I: basal ring, J: anal spur of* L. nervosus, *K–L: variation in anal spur of* L. modestus, *M: sternites 1–3, N: tergites 1–6, O: tergites 7–9. P–Q: Hypopygium of adult male of P:* D. modestus, *Q:* D. nervosus. *A–G from Pinder & Reiss (1983); H–J & M–O from Pinder & Reiss (1986); K–L from Contreras-Lichtenberg (1986); P–Q from Pinder (1978).*

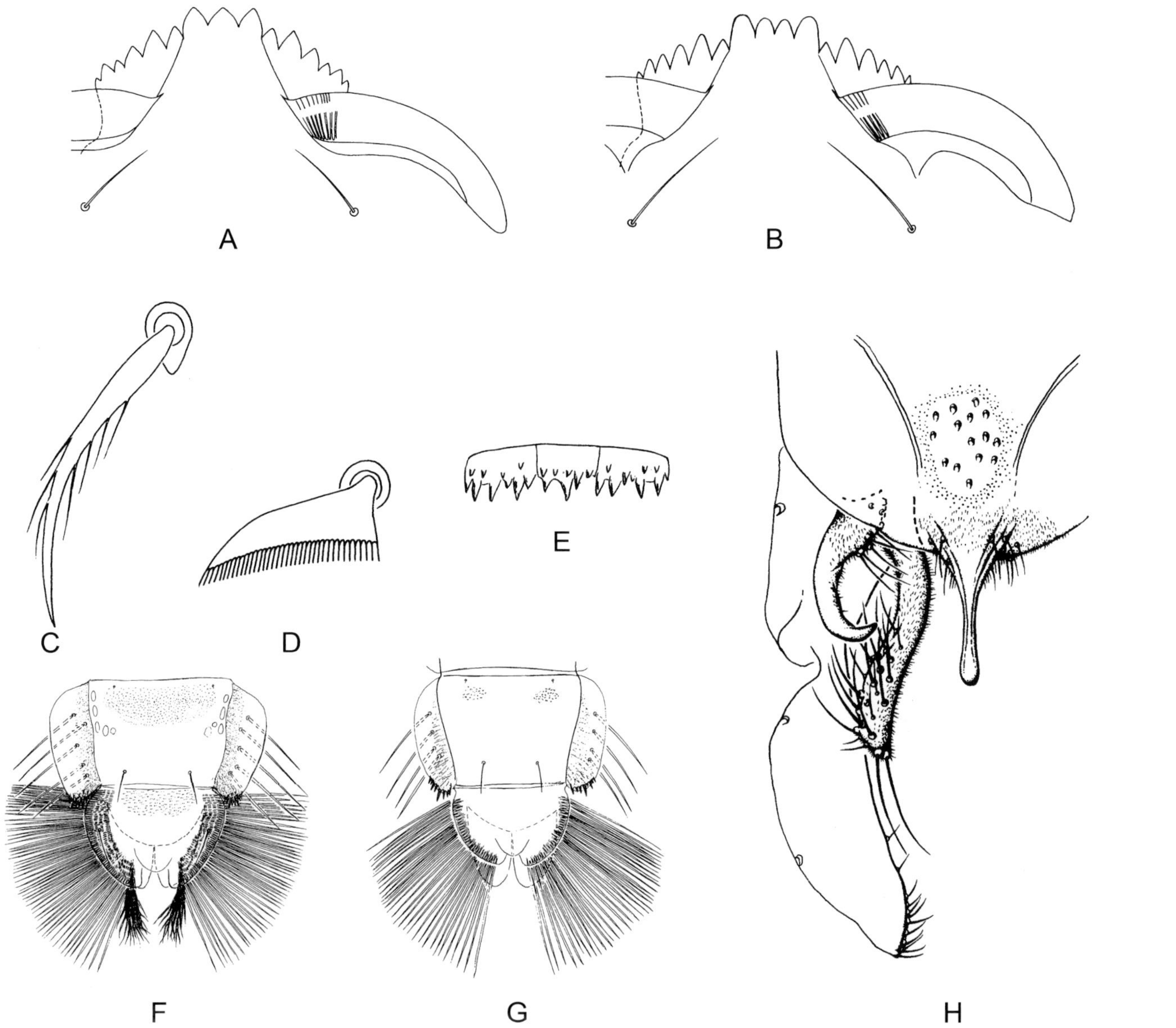

FIGURE 61 Endochironomus, Synendotendipes. *A–E: Larva. A: mentum of* E. tendens, *B: mentum of* S. impar, *C: S I, D: S II, E: pecten epipharyngis. F–G: Anal end of pupa. F:* E. albipennis, *G:* S. impar. *H: Hypopygium of adult male of* S. dispar. *A–D from Pinder & Reiss (1983); E from Lenz (1959); F–G from Pinder & Reiss (1986); H from Pinder (1978).*

P. incompta is a problematic record. Lundbeck (1898) identified one female as *Chironomus incomptus* and placed it as a species of *Heterotrissocladius*. Later Edwards (1929) argued that Zetterstedt's *incomptus* was a *Phaenopsectra*. Henriksen (1939, no. 87) once again placed it as a *Heterotrissocladius* (*Metriocnemus*) species. Ashe & Cranston (1990) listed it as a *nomen dubium* under both *Heterotrissocladius* and *Phaenopsectra*, whereas Oliver *et al.* (1990) maintained it as a valid *Phaenopsectra* species recorded from Greenland. It is therefore included in the keys. The genus *Phaenopsectra* is closely related to *Sergentia*.

***Polypedilum* sp. cf. *laetum* (Meigen, 1818) Fig. 64**
Henriksen 1939: not recorded.

I found larvae and pupae which according to Maschwitz & Cook (2000) key out to *P. laetum* in Lake Ferguson (Kangerlussuaq/Søndre Strømfjord).

BIOLOGY. The larvae live in sandy sediment in the littoral zone of lakes.

DISTRIBUTION. nSW.

Holarctic, widely distributed.

***Sergentia coracina* (Zetterstedt, 1850) Fig. 65**
Henriksen 1939: *Pentapedilum* (*Sergentia*) *coracina* Zett.

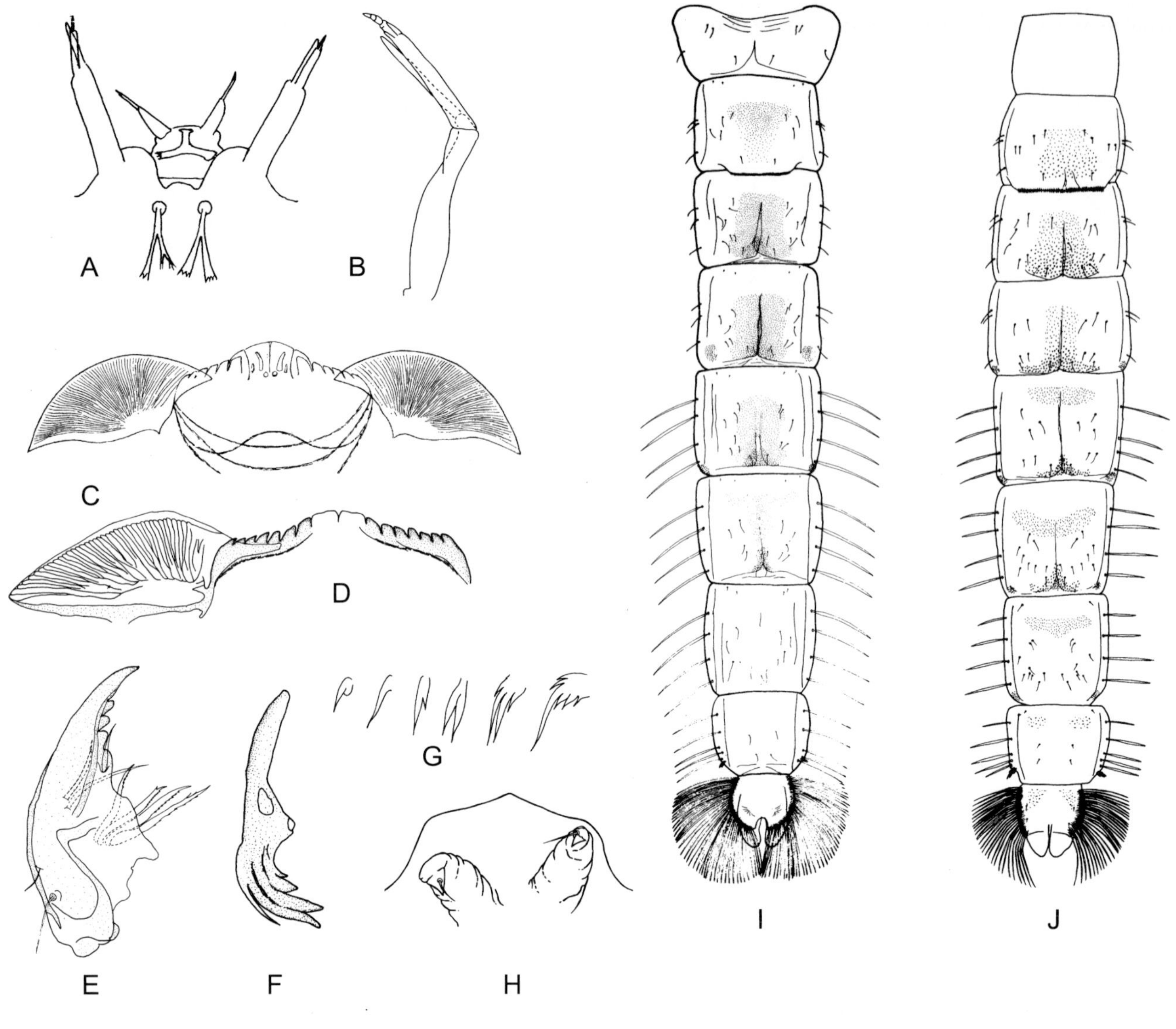

FIGURE 62 Paracladopelma. *A–G: Larva. A: dorsal view of anterior part of head showing the large divided S 3 setae, B: antenna, C–D: mentum, E: mandible, F: premandible. G–J: Pupa. G: variation in anal spur, H: frontal apotome, I–J: tergites 1–9. K–L: Hypopygium of adult male. A–C, H–I & K:* P. nigritula*, D–G, J & L:* P. winnelli*. A from Moller Pillot (1984); B–C from Lenz (1959); D–F, I–J & L from Jackson (1977); G from Langton (1991); H from Pinder & Reiss (1986); K from Pinder (1978).*

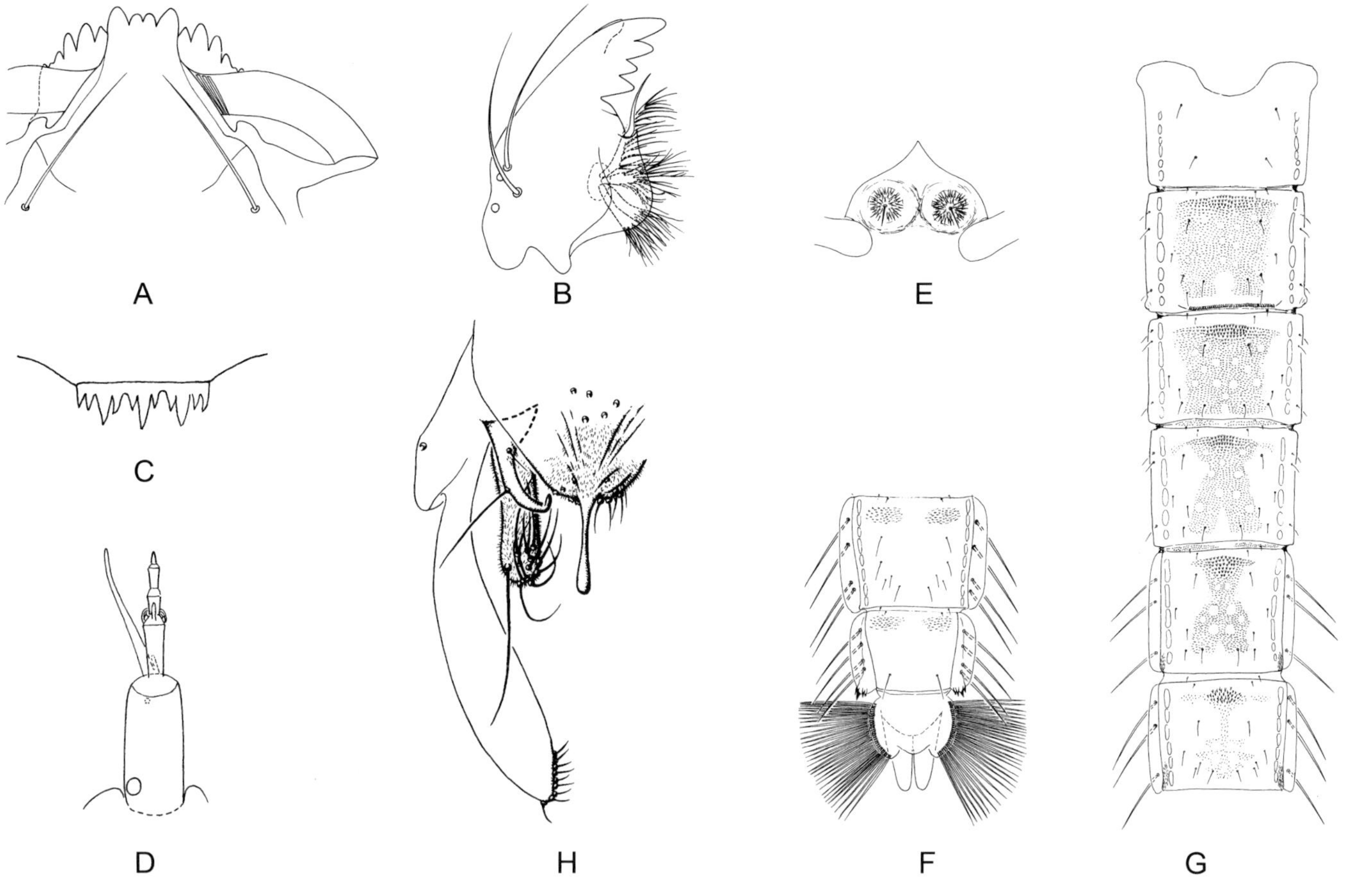

FIGURE 63 Phaenopsectra. *A–D: Larva. A: mentum, B: mandible, C: pecten epipharyngis, D: antenna. E–G: Pupa. E: frontal apotome and cephalic tubercle with circular fields of spinules, F: tergites 7–9, G: tergites 1–6. H: Hypopygium of adult male of* P. flavipes. *A–B & D from Pinder & Reiss (1983); C from Moller Pillot (1984); E–F from Pinder & Reiss (1986); H from Pinder (1978).*

BIOLOGY. A character species of oligotrophic-mesotrophic lakes (Brundin 1956, Sæther 1979). It is a cold-stenothermal northern species and in central Europe confined to the profundal of larger lakes (Wülker *et al.* 1998).

DISTRIBUTION. NE (only recorded from Ella Ø; Andersen 1937).

Holarctic, widely distributed.

***Stictochironomus unguiculatus* (Malloch, 1934) Fig. 66**

Henriksen 1939: not recorded.

Described from arctic Canada and reported from Greenland by Oliver *et al.* (1990) and Oliver & Dillon (1997) without further references. The larva is unknown.

BIOLOGY. *Stictochironomus* larvae live primarily in sandy sediment in the littoral zone of lakes and slow flowing rivers. In oligotrophic lakes they may be found in the profundal zone.

DISTRIBUTION. ? Greenland.

Nearctic, arctic (widely distributed).

CHIRONOMINAE, TRIBE TANYTARSINI

Key to the Greenland Tanytarsini

ADULT MALES

1. All palp segments short (Fig. 68I). Fore leg long compared to short mid and hind legs (Fig. 68H). Tibiae without combs, but with 2 thorn-like spurs distally .. .. *Corynocera*
– Without this combination of characters 2
2. Costa ending proximally to, or opposite apex of M_{3+4} .... *Constempellina*
– Costa ending distinctly distally to apex of M_{3+4} (Fig. 5G) 3
3. Combs of mid and hind tibia clearly separated (Fig. 5N) .... *Tanytarsus*
– Combs of mid and hind tibia fused or only narrowly separated (Fig. 5M) 4
4. Tibial combs with 2 spurs .... *Paratanytarsus*
– Tibial combs without spurs; occasionally a single spur may be present (Fig. 5M) .... *Micropsectra*

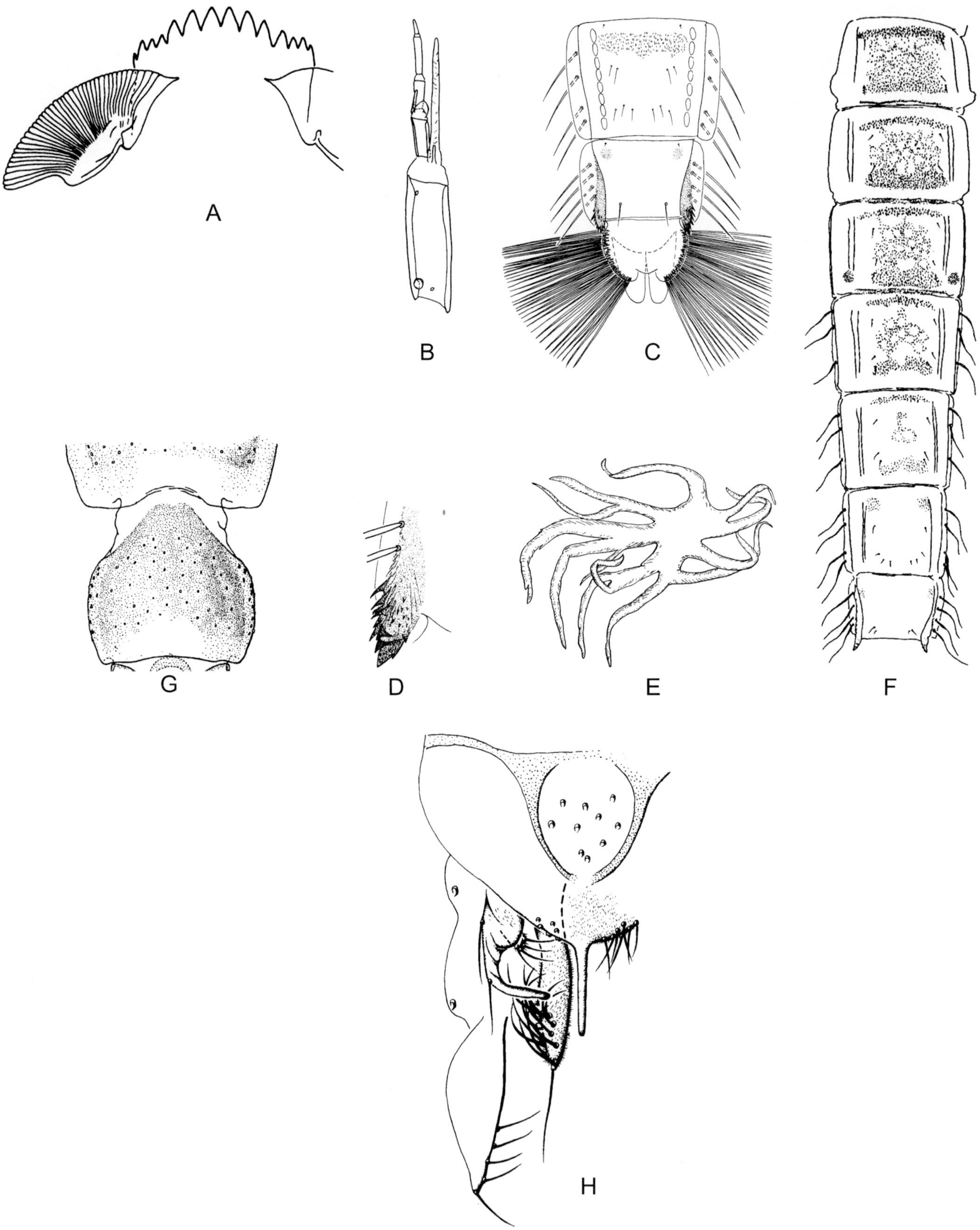

FIGURE 64 Polypedilum laetum. *A–B: Larva. A: mentum, B: antenna. C–F: Pupa. C: anal lobe, D: anal spur, E: thoracic horn, F: tergites 2–8. G–H: Adult male. G: tergite 8, H: hypopygium. A–B & F from Maschwitz & Cook (2000); C–D from Pinder & Reiss (1986); E from Lenz (1959); G from Sæther* et al. *(2000); H from Pinder (1978).*

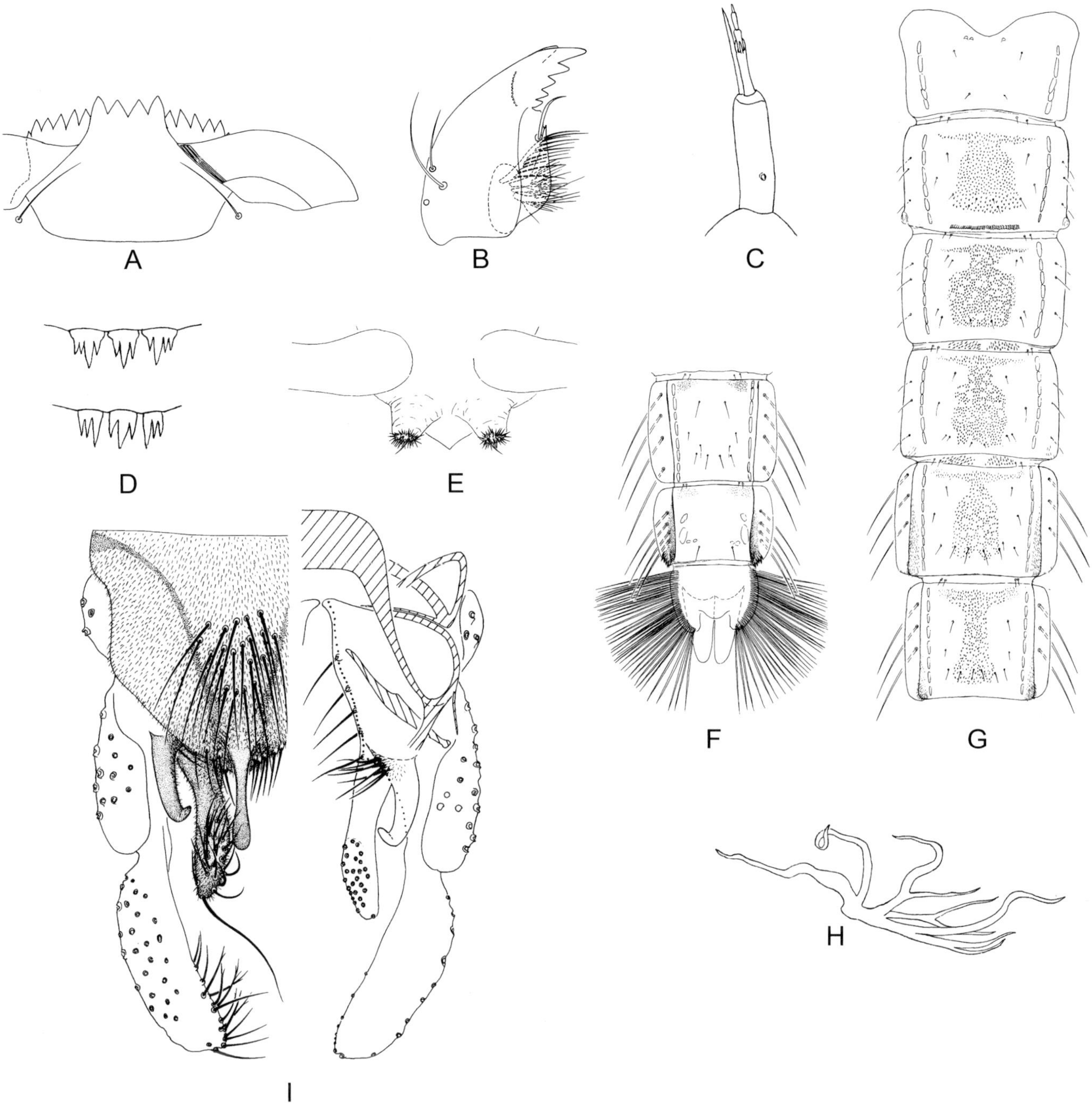

FIGURE 65 Sergentia coracina. *A–D: Larva. A: mentum, B: mandible, C: antenna, D: pecten epipharyngis. E–G: Pupa. E: frontal apotome and cephalic tubercle with circular fields of spinules, F: tergites 7–9, G: tergites 1–6, H: thoracic horn. I: Hypopygium of adult male. A–D from Pinder & Reiss (1983); E–G from Pinder & Reiss (1986); H from Lenz (1959); I from Cranston* et al. *(1989).*

LARVAE

1. Median part of mentum, consisting of 3 broad teeth, distinctly delineated from lateral teeth; mandible with characteristic arrangement of teeth and large ventral lobe (Figs 68A, B) .... *Corynocera*
– Median part of mentum not distinctly delineated from the lateral teeth; mandible of normal form (*e.g.* Figs 69A, E) 2
2. Ventromedial plates separated medially with at least the width of three median teeth (Fig. 67A) .... *Constempellina*
– Ventromental plates almost in contact medially (*e.g.* Fig. 68A) 3
3. Premandible with 3–5 teeth (Figs 73G, H); Lauterborn organs small, situated on long pedicels; antennal pedestal often without spur (Figs 73C, D) .... *Tanytarsus*

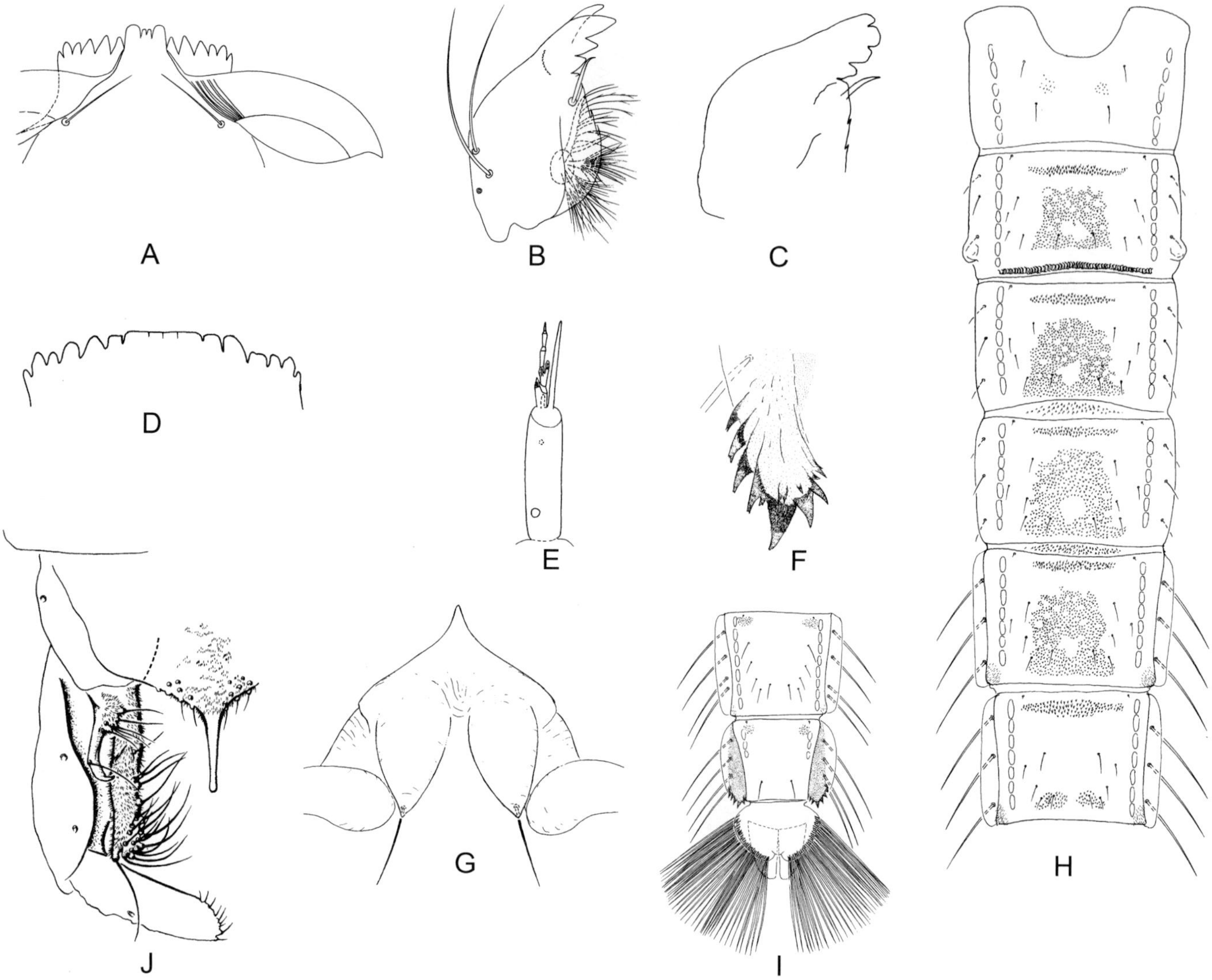

FIGURE 66 Stictochironomus. *A–E: Larva. A: mentum, B: mandible, C: worn mandible, D: worn mentum, E: antenna. F–H: Pupa. F: anal spur, G: frontal apotome, H: tergites 1–6, I: tergites 7–9. J: Hypopygium of adult male* (S. sticticus). *A–B & E from Pinder & Reiss (1983); F–H from Pinder & Reiss (1986); J from Pinder (1978).*

– Premandible bifid (Fig. 69D); Lauterborn organs small, situated on long pedicels or large, situated on short pedicels; antennal pedestal with or without spur (Figs 69B, C) 4

4. Lauterborn organs small, situated on long pedicels; antennal pedestal often with spur (Figs 69B, C) .... *Micropsectra*

– Lauterborn organs large, situated on relatively short pedicels; antennal pedestal without spur (Fig. 71D) ... . *Paratanytarsus*

PUPAE

1. Segment 8 with 2°6 strong lateral thorn-like spines (Fig. 67H) .... *Constempellina*

– Segment 8 without lateral thorn-like spines, but with distinct anal comb (*e.g.* Fig. 68F) 2

2. Anal lobe with setae confined to the posterior half (Fig. 68F) .... *Corynocera*

– Anal lobe with complete fringe of setae (Fig. 73I) 3

3. One or more of tergites 4–6 with a median spinule structure (Figs 71I–L) .... *Paratanytarsus laccophilus*

– Spinule groups on tergites always paired 4

4. Thoracic horn with fringe of long setae (Figs 69H–J); tergite 3 with pair of curved spine or spinule patches (Figs 69F–G) ....*Micropsectra*

– Thoracic horn bare or with small teeth or short setae (Figs 73M, N); tergite 3 usually with paired longitudinal straight spine or spinule patches (Figs 73I, J) ....*Tanytarsus*

Constempellina brevicosta (Edwards, 1937) Fig. 67
Henriksen 1939: not recorded.

Reported from Greenland by Oliver & Dillon (1997) without specified locality.

BIOLOGY. The larvae live in the sandy littoral and sublittoral of lakes where they construct transportable cases of small sand grains. They are resistant to freezing for several months (Grimås & Widerholm 1979).

DISTRIBUTION. ? Greenland.

Holarctic, widely distributed in northern part.

Corynocera oliveri Lindeberg, 1970 Fig. 68
Henriksen 1939: not recorded

Larval head capsules of *C. oliveri* have been recorded from surface sediments cores in W Greenland lakes (Brodersen & Anderson 2002, Brodersen & Bennike 2003). However, the identification of larval head capsules may be problematic and the record may turn out to be a species in the *Tanytarsus lugens* group.

BIOLOGY. The larvae live in soft sediments of shallow lakes and ponds. *Corynocera* larvae are considered cold stenothermal, but larvae of *C. ambigua* are also found in small summer warm ponds in temperate parts of Europe (Brodersen & Lindegaard 1999). Emergence takes place very early in spring, often while the lakes are partially covered with ice. The adults do not fly, but are whirling around on the water surface to mate.

DISTRIBUTION. nSW. Several lakes in the Kangerlussuaq/Søndre Strømfjord area (Brodersen & Anderson 2002).

Palaearctic (known only from northern Fennoscandia).

Micropsectra Kieffer, 1909 Figs 69, 70
Micropsectra is a large genus with more than 100 Holarctic species (Stur & Ekrem 2006). The genus needs revision and no complete key exists. 6 or 7 species of *Micropsectra* have been recorded from Greenland, but more may be expected. Most species are cold stenothermal and confined to running water or, in the European lowland, to large oligotrophic lakes. Some species are more eurythermal and have wider distribution. In high altitudes and northern areas *Micropsectra* species are common in all types of lakes. The larvae live in long tubes of detritus and feed on algae and other micro-organisms.

Micropsectra atrofasciata group
In drift samples from glacial lakes in the Kangerlussuaq/Søndre Strømfjord area female pupae of a *Micropsectra* species were very abundant. According to Torbjørn Ekrem (pers. com.) they belong to a parthenogenetic species close to or conspecific with *M. atrofasciata* (Kieffer, 1918).

BIOLOGY. Obligatory or facultative parthenogenesis is recorded in both temperate and arctic chironomids. It has been described from Ellesmere Island (Oliver & Danks 1972, Langton 1998), but even though it may be a common phenomenon in Greenlandic chironomids, it is probably not a special adaptation to arctic conditions.

Micropsectra brundini Säwedal, 1979
Henriksen 1939: not recorded.

Described on material from the Narsaq area (S Greenland). It is very close to *M. insignilobus* and earlier records of this species in Greenland and arctic Canada may be *M. brundini*, but see notes on *M. insignilobus*. Larvae and pupae are unknown.

DISTRIBUTION. S (only known from the type locality).

Micropsectra logani (Johannsen, 1928)
Henriksen 1939: *Micropsectra groenlandica* Søgaard Andersen. *M. groenlandica* was described by Andersen (1937) based on material from NE Greenland. Brundin (1949) supported its status as a valid species. However, Stur & Ekrem (2006) in a revision of the *M. atrofasciata* group established *M. groenlandica* as a synonym of *M. logani*.

BIOLOGY. Known from lakes and running waters in the northern parts of Europe, but is confined to cold springs and the profundal zone of oligotrophic lakes in central Europe.

DISTRIBUTION. S (Lindegaard *et al.* 1978 found it in lakes in the Narsaq area), Ella Ø (only record from NE).

Holarctic, widely distributed in northern parts.

Micropsectra insignilobus Kieffer, 1924
Henriksen 1939: not recorded.

Redescribed by Säwedal (1976). Earlier records of *M. insignilobus* from Greenland are probably *M. brundini* (Säwedal 1979) and *M. insignilobus* might be removed from the Greenlandic check list. However, according to Torbjørn Ekrem (pers. com.) *M. brundini* may be a synonym of *M. insignilobus*.

BIOLOGY. Known only from oligotrophic lakes.

DISTRIBUTION. N (Peary Land).

Holarctic (known from Fennoscandia and Canada: Ellesmere Island).

Micropsectra junci Meigen, 1818
Henriksen 1939: *Tanytarsus junci* Meig. & *Micropsectra gmundensis* Egg.

Redescribed by Säwedal (1976).

BIOLOGY. In Europe found in cold springs and small streams.

DISTRIBUTION. S (Qaqortoq/Julianehåb, Lundbeck 1898). Oliver *et al.* (1990) approved the record.

Europe, northern and central.

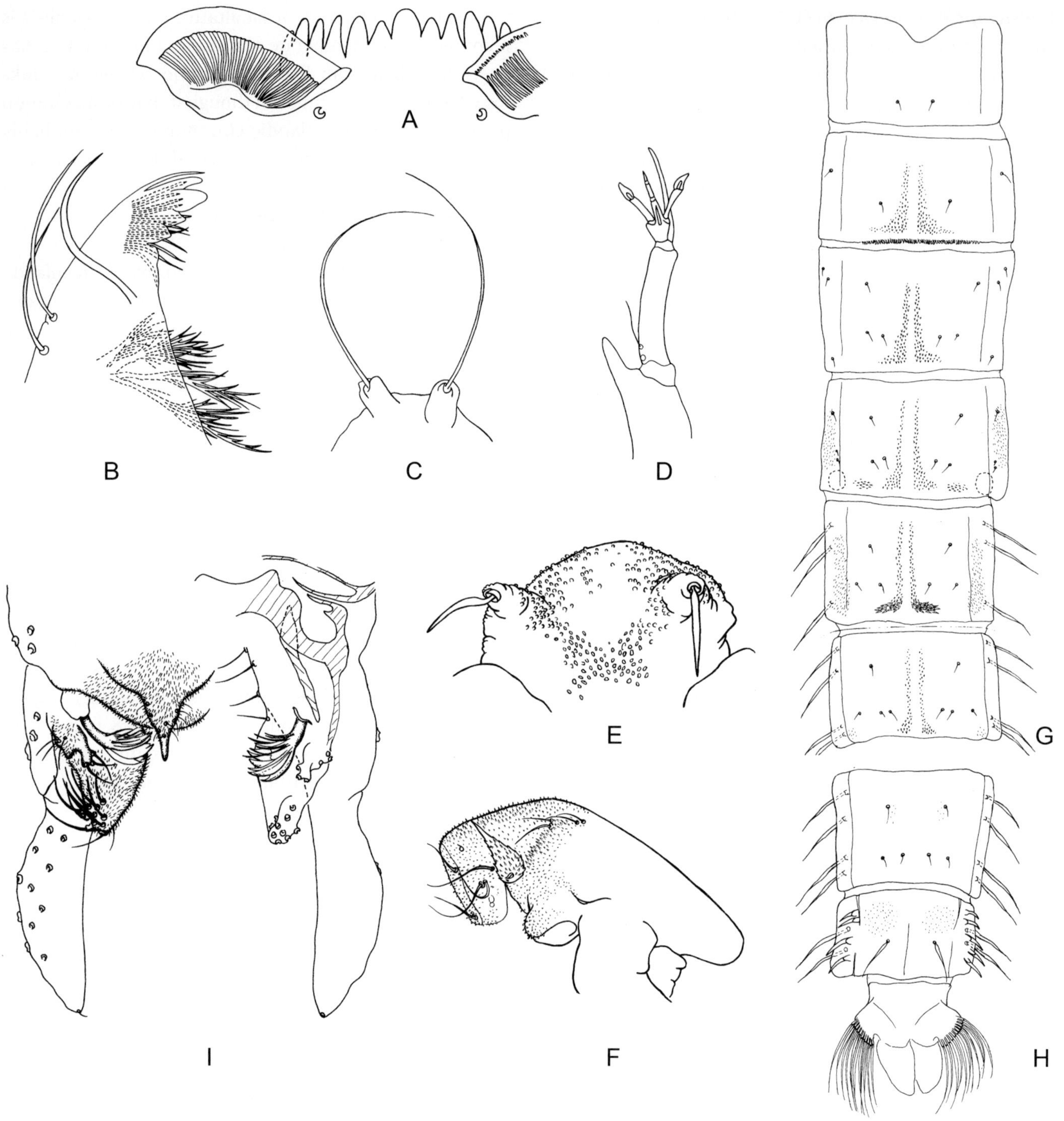

FIGURE 67 Constempellina brevicosta. *A–D: Larva. A: mentum, B: tip of mandible, C: dorsal head setae, D: antenna. E–H: Pupa. E: frontal apotome, F: thorax with thoracic horn, G: tergites 1–6, H: tergites 7–9. I: Hypopygium of adult male. A–D from Pinder & Reiss (1983); E–H from Pinder & Reiss (1986); I from Cranston* et al. *(1989).*

Micropsectra lindrothi **Goetghebuer, 1931**
Henriksen 1939: not recorded.

Described on material from SE Greenland. Redescribed by Säwedal (1976). The adult male is easily recognised on the leaf-shaped lamellar bristles at tip of the median volsella (Fig. 70 C).

BIOLOGY. Prefers cold springs and small streams in north-western Europe, but is also found in the littoral zone of lakes in Iceland.

DISTRIBUTION. S (recorded from the Narsaq area by Lindegaard *et al.* 1978), SE (Lake Fjord).

Europe, widely distributed.

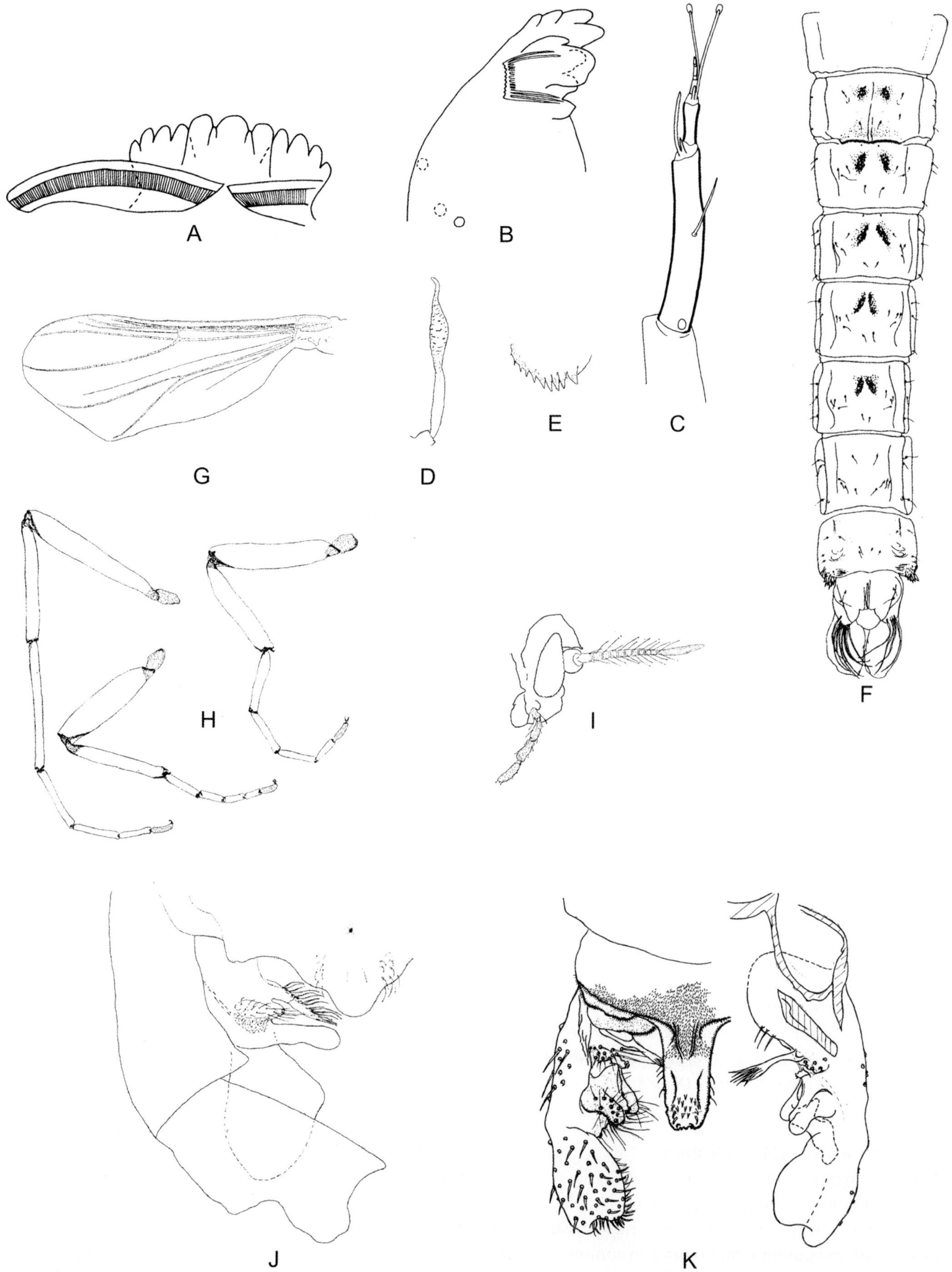

FIGURE 68 Corynocera oliveri. *A–C: Larva. A: mentum, B: mandible, C: antenna of* C. ambigua. *D–F: Pupa. D: thoracic horn, E: anal spur, F: tergites 1–9. G–K: Adult male. G: wing, H: fore, mid and hind leg, I: lateral view of head showing palp and antenna, J: hypopygium of* C. oliveri, *K: hypopygium of* C. ambigua. *A–C from Pinder & Reiss (1983); D–E & H–J from Lindeberg (1970); F from Sæther* et al. *(2004); K from Cranston* et al. *(1989).*

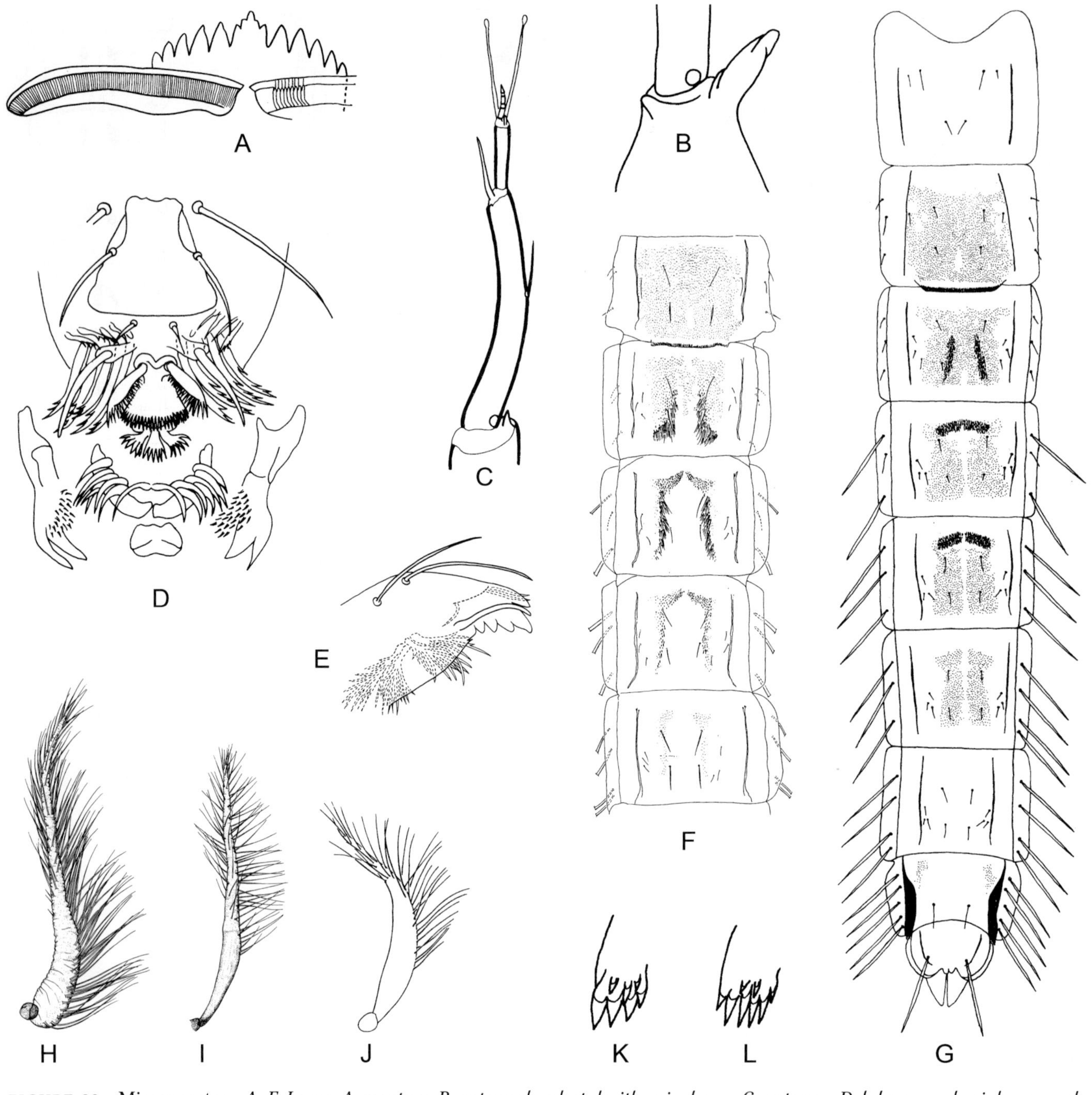

FIGURE 69 Micropsectra *1. A–E: Larva. A: mentum, B: antennal pedestal with apical spur, C: antenna, D: labrum and epipharyngeal region, E: tip of mandible. F–L: Pupa. F: tergites 2–6 of* M. polita, *G: tergites 1–9 of* M. junci, *H–I: variation in thoracic horn* (M. notescens, M. insignilobus, M. polita), *K–L: anal spur of* M. lindrothi *and* M. lindebergi. *A–E from Pinder & Reiss (1983); F & J from Oliver & Dillon (1994); G–I from Säwedal (1976); K–L from Langton (1991).*

***Micropsectra polita* (Malloch, 1915)**
Henriksen 1939: not recorded.

Redescribed by Oliver & Dillon (1994).

BIOLOGY. The larvae prefer low-order streams and springs with maximum temperatures in summer ranging between 10° and 18°C (Oliver & Dillon 1994).

DISTRIBUTION. nSW. Adult males were found at the outlet from Lake Ferguson, Kangerlussuaq/Søndre Strømfjord, during July 2002 (det. Torbjørn Ekrem).

Nearctic, widely distributed (Oliver *et al.* 1990, Oliver & Dillon 1994).

***Micropsectra recurvata* Goetghebuer, 1928**
Redescribed by Reiss (1974) and easily recognized as adult male on the very slender inferior volsella, which in the distal part is bent in medial direction (Fig. 70 E).

BIOLOGY. Considered a cold-stenothermal species living in cold springs and streams.

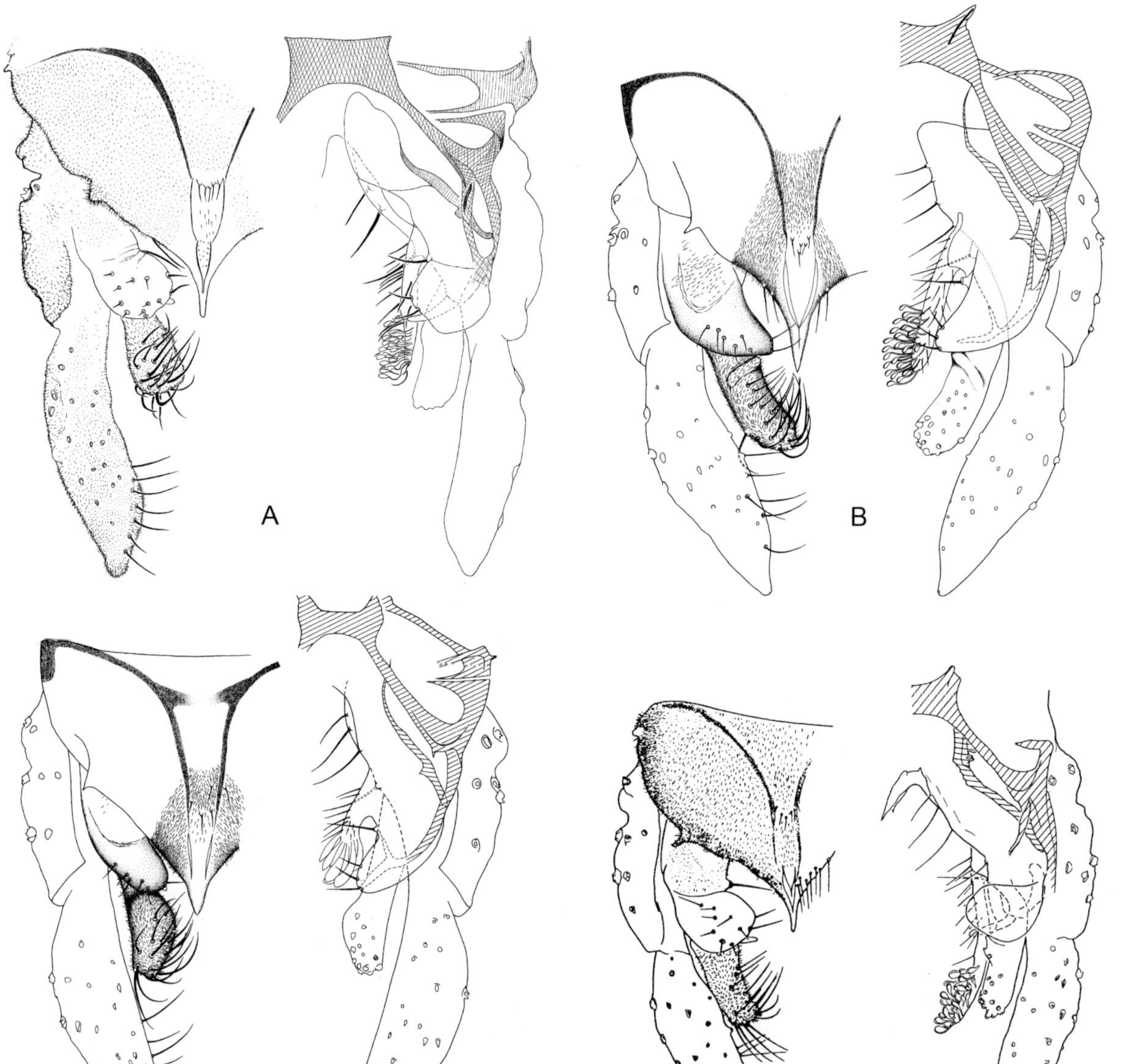

FIGURE 70 Micropsectra 2. *A–F: Hypopygium of adult male; left: dorsal view, right: ventral view; in A and F median volsella is inserted. A:* M. logani, *B:* M. junci, *C:* M. lindrothi, *D:* M. polita, *E:* M. recurvata, *F:* M. tori. *A from Stur & Ekrem (2006); B–C from Säwedal (1976); D from Oliver & Dillon (1994); E from Reiss (1974); F from Säwedal (1981).*

DISTRIBUTION. S (Narsaq area, Lindegaard *et al.* (1978), SE (Lake Fjord, Edwards 1935).

Holarctic (widely distributed in Europe and also reported from North America).

Micropsectra tori **Säwedal, 1981**
Henriksen 1939: not recorded.

Described on material from the Narsaq area, S Greenland. Only known from the type locality. Larva and pupa are unknown.

Paratanytarsus **Thienemann & Bause, 1913 Figs 71, 72**
A large genus with more than 30 Palaearctic species. The genus needs revision and no complete key exists. 4 species are recorded from Greenland, all included in keys given by Reiss & Säwedal (1981). They also give references for descriptions of larvae, pupae and adults of the Greenlandic species.

The larvae live in standing waters of all types where they build tubes of detritus and feed on algae and other micro-organisms.

Paratanytarsus laccophilus **(Edwards, 1929)**
Henriksen 1939: not recorded.

Reported from Greenland by Oliver & Dillon (1997), but without further references.

BIOLOGY. The larvae live in shallow water of lakes, ponds and pools.

DISTRIBUTION. S, nSW. (I have found pupal exuviae commonly in a number of lakes and ponds in the Narsaq and Kangerlussuaq/Søndre Strømfjord areas).

Holarctic (widely distributed in the Palaearctic region, also recorded from Canada).

Paratanytarsus natvigi **(Goetghebuer, 1933)**
Henriksen 1939: *Phoenopelma* sp. indet.

Described on material from NE Greenland (Loch Fyne). It is further recorded by Andersen (1937) from Ella Ø (NE Greenland) as "*Phoenopelma*" sp.

BIOLOGY. The larvae live in shallow waters of small lakes and ponds.

DISTRIBUTION. NE (only known from Ella Ø and Loch Fyne).

Holarctic, widely distributed in northern parts.

Paratanytarsus setosimanus **(Goetghebuer, 1933)**
Henriksen 1939: *Ditanytarsus setosimanus* Goetg.

Described on material from NE Greenland (Loch Fyne) and later recorded from Ella Ø (NE Greenland; Andersen (1937).

BIOLOGY. The larvae are found in shallow water in small lakes and ponds.

DISTRIBUTION. NE (only known from Loch Fyne and Ella Ø).

Holarctic (northern Scandinavia and Canada: Northwest Territories).

Paratanytarsus tenuis **(Meigen, 1830)**
Henriksen 1939: *Tanytarsus tenuis* Meig.

BIOLOGY. The larvae live in shallow water in the littoral zone of lakes.

DISTRIBUTION. S. Recorded by Lundbeck (1898) from Ivittuut. This record is maintained by both Ashe & Cranston (1990) and Oliver *et al.* (1990).

Europe, widely distributed.

Tanytarsus **van der Wulp, 1874 Figs 73, 74**
The genus *Tanytarsus* has more than 100 Palaearctic species. In spite of partial revisions by Reiss & Fittkau (1971), Ekrem *et al.* (2003) and Ekrem (2004) it is not possible to identify the species with certainty. 4 species are recorded from Greenland, and judging from pupal exuviae from the Kangerlussuaq/Søndre Strømfjord area at least two more species occur in SW Greenland. According to Torbjørn Ekrem (pers. com.) they are close to – but not conspecific – with *T. latiforceps* Edwards & Thienemann, 1941 and *T. aculeatus* Brundin, 1949. The larvae of most *Tanytarsus* species are eurythermal and live in standing waters of all types, where they may be very abundant. A few species are cold stenothermal and constrained to springs and streams. They build tubes of detritus and feed on algae and other micro-organisms.

Tanytarsus anderseni **Reiss & Fittkau, 1971**
Henriksen 1939: not recorded.

Described on material taken by Andersen (1937) on Ella Ø (NE Greenland) and partly used for his description of *T. niger*.

BIOLOGY. The larvae probably live in the same lakes on Ella Ø as *T. niger*.

DISTRIBUTION. NE (only recorded from Ella Ø).

Nearctic (Canada).

Tanytarsus gracilentus **(Holmgren, 1883)**
Henriksen 1939: not recorded.

BIOLOGY. The larvae live in all kinds of standing waters and may also be found in brackish water. They may be very numerous, *e.g.* in Lake Mývatn, Iceland, where the abundance in periods exceed 100,000 individuals per m^2 (Lindegaard & Jónasson 1979).

DISTRIBUTION. nSW (common in lakes in the Kangerlussuaq/Søndre Strømfjord area; Brodersen *et al.* 2001), N (Peary Land; Lindeberg 1968).

Holarctic, widely distributed in northern parts.

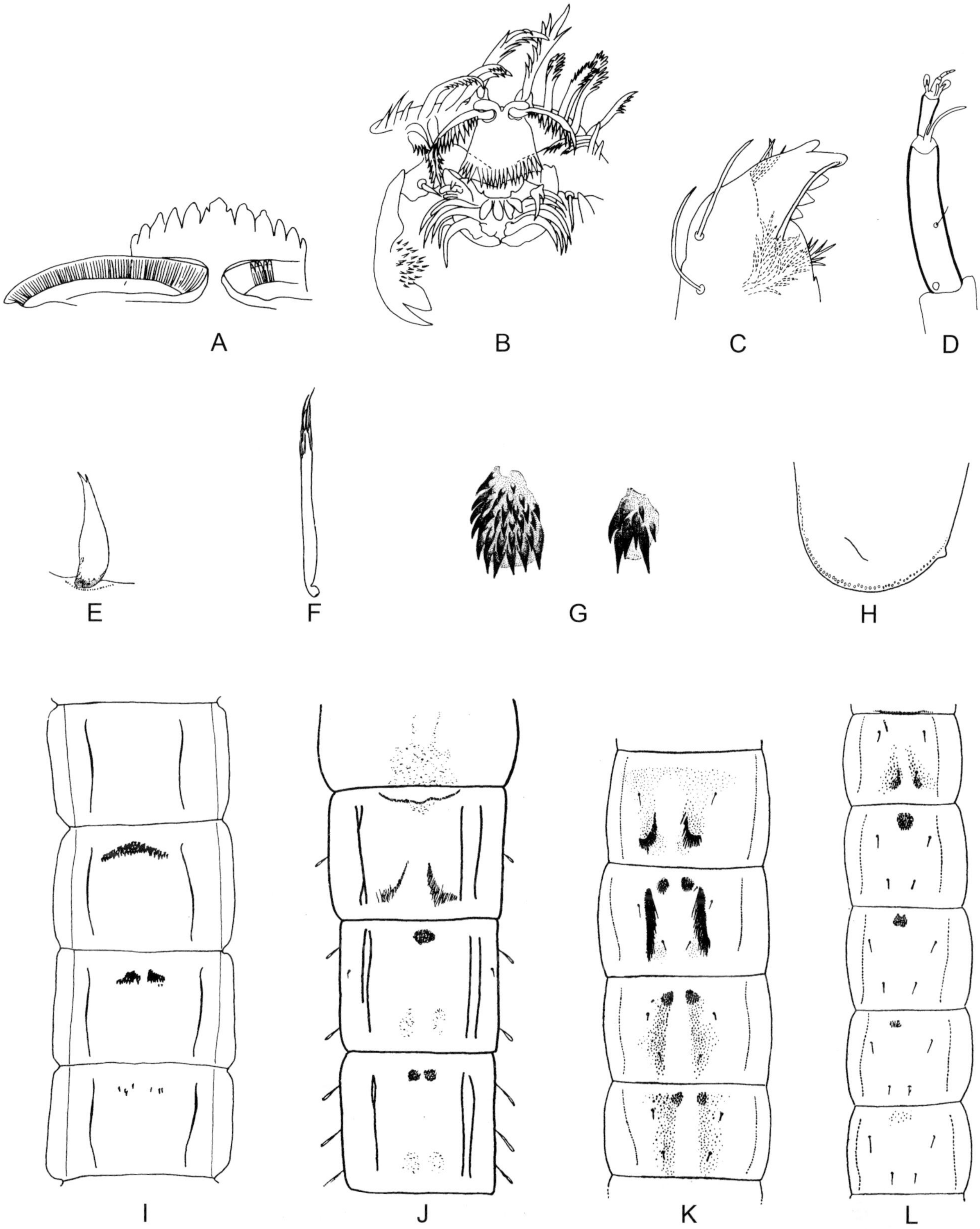

FIGURE 71 Paratanytarsus *1. A–D: Larva. A: mentum, B: labrum and hypopharyngeal region, C: mandible, D: antenna. E–L: Pupa. E: thoracic horn of* P. laccophilus, *F: thoracic horn of* P. natvigi, *G: anal spur of* P. setosimanus, *H: wing sheath with nose and pearl row. I–L: tergites 3–6 (7) of I:* P. laccophilus, *J:* P. natvigi, *K:* P. setosimanus, *L:* P. tenuis. *A–D from Pinder & Reiss (1983); E, G–H & K–L from Thienemann (1951); F & J from Andersen (1937); I from Reiss & Säwedal (1981).*

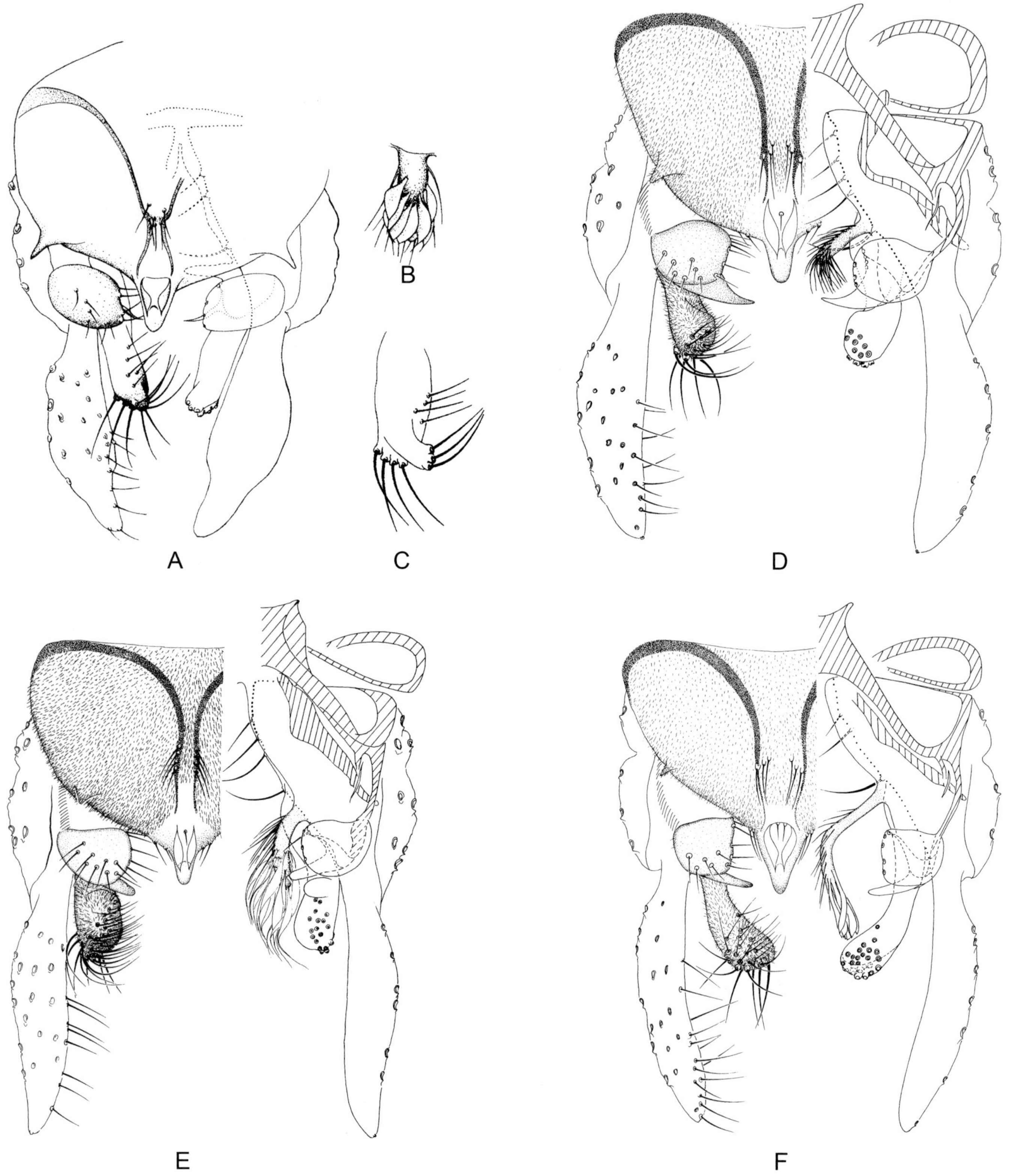

FIGURE 72 Paratanytarsus 2. *A–F: Hypopygium of adult male. A–C:* P. laccophilus, *B: median volsella, C: inferior volsella, D:* P. natvigi, *E:* P. setosimanus, *F:* P. tenuis. *A–C from Reiss (1968); D–F from Reiss & Säwedal (1981).*

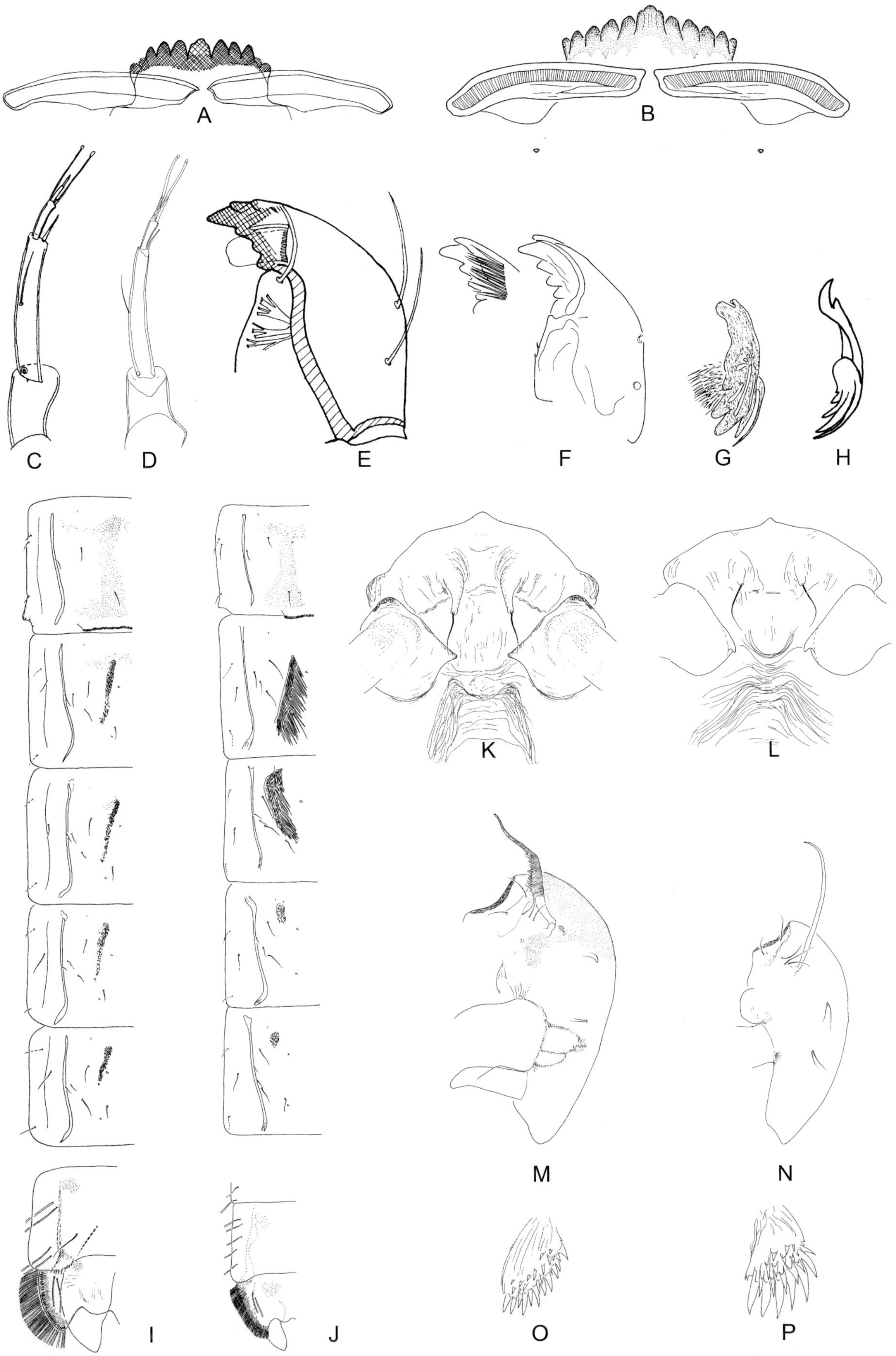

FIGURE 73 Tanytarsus 1. *A–H: Larva. A & B: mentum, C & D: antenna, E & F: mandible, G & H: premandible. I–P: Pupa. I & J: tergites 2–9, K & L: frontal apotome, M & N: lateral view of thorax, O & P: posterolateral comb of segment 8. A, C, E, G, I, K, M & O:* T. gracilentus, *B, D, F & H:* T. gregarius/inaequalis, *J, L, N & P:* T. gregarius. *A, C, E & G from Paasivirta (1972); B, D, F, H, J, L, N & P from Ekrem (2004); I, K, M & O from Ekrem* et al. *(2003).*

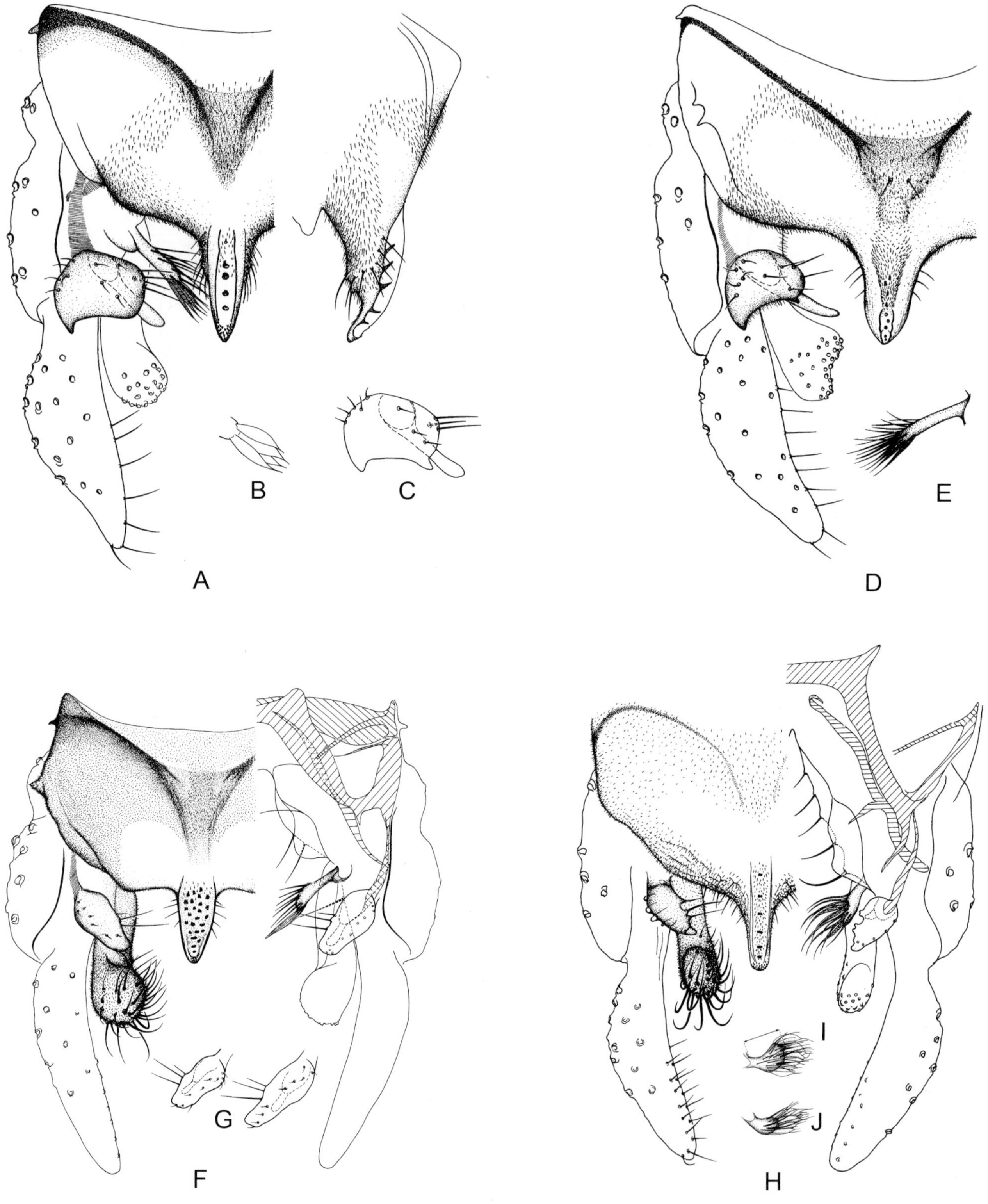

FIGURE 74 Tanytarsus 2. *A–C:* T. anderseni. *A: hypopygium, left dorsal view, right lateral view of anal point, B: median volsella, C: digitus and superior volsella. D–E:* T. niger. *D: dorsal view of hypopygium, E: median volsella. F–G:* T. gracilentus. *F: hypopygium, right dorsal view, left ventral view, G: superior volsella. H & J:* T. inaequalis. *H: hypopygium, right dorsal view, left ventral view, J: median volsella. I: median volsella of* T. gregarius. *A–J from Reiss & Fittkau (1971).*

***Tanytarsus gregarius* Kieffer, 1909**

Henriksen 1939: not recorded.

BIOLOGY. The larvae mainly live in the littoral zone of lakes, but also inhabit the profundal zone of oligotrophic lakes. A high abundance may be present in temperate lakes, but only small numbers are found in Greenland standing waters.

DISTRIBUTION. nSW (ponds and lakes in the Kangerlussuaq/Søndre Strømfjord area).

Holarctic, widely distributed in northern parts.

***Tanytarsus niger* Andersen, 1937**

Described on material from NE Greenland (Ella Ø) and redescribed by Brundin (1947) and Reiss & Fittkau (1971).

BIOLOGY. The larvae live in small (arctic) lakes and ponds. In the small lakes on Ella Ø the maximum abundance of larvae was 1900 individuals per m^2. The European populations of *T. niger* are univoltine with emergence in early spring.

DISTRIBUTION. NE (only recorded from Ella Ø).

Circumpolar, arctic, but also known (as a relict ?) from a number of places in temperate parts of Canada and Europe (including Denmark).

References

Andersen, F.S. (1937): Ueber die Metamorphose der Ceratopogoniden und Chironomiden Nordost-Grönlands. – *Meddelelser om Grønland* 116, 1: 1–95.

Ashe, P., & Cranston, P.S. (1990): Family Chironomidae. – *In*: Soós, A. & Papp, L. (eds) *Catalogue of Palaearctic Diptera*. Vol. 2 Psychodidae-Chironomidae, pp. 113–499. Budapest.

Brodersen, K.P. & Lindegaard, C. (1999): Mass occurrence and sporadic distribution of *Corynocera ambigua* Zetterstedt (Diptera, Chironomidae) in Danish lakes. Neo- and paleolimnological records. – *Journal of Paleolimnology* 22: 41–52.

Brodersen, K.P., Lindegaard, C. & Anderson, N J. (2001): Holocene temperature and environmental reconstruction from lake sediments in the Søndre Strømfjord region, southern West Greenland. – *Geology of Greenland Survey Bulletin* 189: 59–64.

Brodersen, K.P. & Anderson, N.J. (2002): Distribution of chironomids (Diptera) in low arctic West Greenland lakes: trophic conditions, temperature and environmental reconstruction. – *Freshwater Biology* 47: 1137–1157.

Brodersen, K.P. & Bennike, O. (2003): Interglacial Chironomidae (Diptera) from Thule, Northwest Greenland: matching modern analogues to fossil assemblages. – *Boreas* 32: 560–565.

Brodersen, K.P., Pedersen, O., Lindegaard, C. & Hamburger, K. (2004): Chironomids (Diptera) and oxy-regulatory capacity: An experimental approach to paleolimnological interpretation. – *Limnology and Oceanography* 49: 1549–1559.

Brundin, L. (1947): Zur Kenntnis der schwedischen Chironomiden. – *Arkiv för Zoologi* 39 A, 3: 1–95 + 25 Tafeln.

Brundin, L. (1949): Chironomiden und andere Bodentiere der südschwedischen Urgebirgsseen. Ein Beitrag zur Kenntnis der bodenfaunischen Charakterzüge schwedisher oligotropher Seen. – *Institute of Freshwater Research Drottningholm, Report* 30: 1–914.

Brundin, L. (1956): Zur Systematik der Orthocladiinae (Dipt. Chironomidae). – *Institute of Freshwater Research Drottningholm, Report* 37: 1–185.

Brundin, L. (1966): Transantarctic relationships and their significance, as evidenced by chironomid midges. With a monograph of the subfamilies Podonominae and Aphroteniinae and the Austral Heptagyiae. – *Kunglia Svenska Vetenskapakademiens Handlingar* 4. Serien, Band 11. Nr 1: 1–472.

Brundin, L. (1983): 4. The larvae of Podonominae (Diptera: Chironomidae) of the Holarctic region – Keys and diagnoses. – *Entomologica scandinavica Supplement* 19: 23–31.

Brundin, L. (1986): 4. The pupae of Podonominae (Diptera: Chironomidae) of the Holarctic region – Keys and diagnoses. – *Entomologica scandinavica Supplement* 28: 19–30.

Brundin, L. (1986): 4. The adult males of Podonominae (Diptera: Chironomidae) of the Holarctic region – Keys and diagnoses. – *Entomologica scandinavica Supplement* 34: 23–36.

Butler, M.G. (1982): A 7-year life cycle for *Chironomus* species in arctic Alaskan tundra ponds (Diptera: Chironomidae). – *Canadian Journal of Zoology* 60: 58–70.

Caldwell, B.A. (1998): Description of the adult male and larva of *Orthocladius* (*Orthocladius*) *vaillanti* (Diptera: Chironomidae). – *Journal of the Kansas Entomological Society* 71: 234–240.

Carpenter, G.D.H. (1938): Notes on insects collected in West Greenland by the Oxford University Greenland Expedition, 1936. – *Annals & Magazine of Natural History* (ser. 11) 1: 529–553.

Coffmann, W.P. (1978): Chironomidae. – *In*: Merrit, R.W. & Cummins, K.W. (eds) *An introduction to the aquatic insects of North America*, pp. 345–376. Kendal/Hunt Publ. Co., Dubuque, Iowa.

Coffmann, W.P., Cranston, P.S. Oliver, D.R. & Sæther, O.A. (1986): The pupae of Orthocladiinae (Diptera: Chironomidae) of the Holarctic region – Keys and diagnoses. – *Entomologica scandinavica Supplement* 28: 147–296.

Contreras-Lichtenberg, R. (1986): Revision der in der Westpaläarktis verbreiteten Arten des Genus *Dicrotendipes* Kieffer, 1913 (Diptera, Nematocera, Chironomidae). – *Annales Naturhistorisches Museum in Wien* 88/89 B: 663–726.

Cranston, P.S. (1979): The development stage of *Limnophyes globifer* (Lundström) (Diptera: Chironomidae). – *Hydrobiologia* 67: 19–25.

Cranston, P.S. (1982): A key to the larvae of the British Orthocladiinae (Chironomidae). – *Freshwater Biological Association Scientific Publications* 45: 1–152.

Cranston, P.S. (1983): 3. The larvae of Telmatogetoninae (Diptera: Chironomidae) of the Holarctic region – Keys and diagnoses. – *Entomologica scandinavica Supplement* 19: 17–22.

Cranston, P.S. (1984): The taxonomy and ecology of *Orthocladius* (*Eudactylocladius*) *fucimanus* (Kieffer), a hygropetric chironomid (Diptera). – *Journal of Natural History* 18: 873–895.

Cranston, P.S. (1986): 3. The pupae of Telmatogetoninae (Diptera: Chironomidae) of the Holarctic region – Keys and diagnoses. – *Entomologica scandinavica Supplement* 28: 13–17.

Cranston, P.S. (1989): 3. The adult males of Telmatogetoninae (Diptera: Chironomidae) of the Holarctic region – Keys and diagnoses. – *Entomologica scandinavica Supplement* 23: 17–21.

Cranston, P.S. (1998): Nearctic *Orthocladius* subgenus *Eudactylocladius* revised (Diptera: Chironomidae). – *Journal of the Kansas Entomological Society* 71: 272–295.

Cranston, P.S., Oliver, D.R. & Sæther, O.A. (1983): 7. The larvae of Orthocladiinae (Diptera: Chironomidae) of the Holarctic region – Keys and diagnoses. – *Entomologica scandinavica Supplement* 19: 149–291.

Cranston, P.S. & Sæther, O.A. (1986): Rheosmittia (Diptera: Chironomidae): a generic validation and revision of the western Palaearctic species. – *Journal of Natural History* 20: 31–51.

Cranston, P.S., Oliver, D.R. & Sæther, O.A. (1989a): 7. The adult males of Orthocladiinae (Diptera: Chironomidae) of the Holarctic region – Keys and diagnoses. – *Entomologica scandinavica Supplement* 34: 165–352.

Cranston, P.S. & Oliver, D.R. (1988): Additions and corrections to the Nearctic Orthocladiinae (Diptera: Chironomidae). – The *Canadian Entomologist* 120: 425–462.

Cranston, P.S., Dillon, M.E., Pinder, L.C.V. & Reiss, F. (1989b): 7. The adult male of Chironominae (Diptera: Chironomidae) of the Holarctic region – Keys and diagnoses. – *Entomologica scandinavica Supplement* 34: 353–502.

Curry, La Verne, L. (1956): Notes on the ecology of and taxonomy of the midges *Tendipes* (*Tendipes*) *staegeri* (Lundbeck) = *Chironomus staegeri* Lundbeck (Diptera). – *Entomological News* 67: 225–236.

Danks, H.V. (1971): Overwintering of some north temperate and arctic Chironomidae. II Chironomid biology. – The *Canadian Entomologist* 103: 1875–1910.

Edwards, F.W. (1924): Results of the Merton College Expedition to Spitsbergen, 1923 – No. 4. Diptera Nematocera. – *Annals & Magazine of Natural History* (ser. 9), 14: 162–174.

Edwards, F.W. (1929): British non-biting midges (Diptera, Chironomidae). – *Transactions of the Entomological Society of London* 77: 279–430.

Edwards, F.W. (1931): Oxford University Greenland expedition, 1928. – Diptera Nematocera. – *Annals & Magazine of Natural History Series* (ser. 10) 8: 617–618.

Edwards, F.W. (1935): Diptera Nematocera from East Greenland. – *Annals & Magazine of Natural History* (ser. 10) 15: 467–473.

Ekrem, T. (2004): Immature stages of European *Tanytarsus* species I. The *eminulus-*, *gregarius*, *lugens-* and *mendax* species groups (Diptera, Chironomidae). – *Deutches entomologische Zeitschrift* 51: 97–146.

Ekrem, T., Sublette, M.F. & Sublette, J.E. (2003): North American *Tanytarsus* I. Descriptions and keys to species in the *eminulus*, *gregarious*, *lugens* and *mendax* groups (Diptera: Chironomidae). – *Annals of the Entomological Society of America* 96: 265–328.

Epler, J.H. (1987): Revision of the Nearctic *Dicrotendipes* Kieffer, 1913 (Diptera: Chironomidae). – *Evolutionary Monographs*: 1–102 + 37 plates.

Epler, J.H. (1988): Biosystematics of the Genus *Dicrotendipes* Kieffer, 1913 (Diptera: Chironomidae: Chironominae) of the world. – *Memoirs of the American Entomological Society* 36: 1–214.

Epler, J.H. (2001): *Identification manual for the larval Chironomidae (Diptera) of North and South Carolina.* – North Carolina Department of Environment and Natural Resources, Raleigh, North Carolina. Available from http://www.sjrwmd.com/technicalreports/pdfs/SP/SJ2001-SP13.pdf.

Fittkau, E.J. (1962): Die Tanypodinae (Diptera: Chironomidae). Die Tribus Anatopyniini, Macropelopiini und Pentaneurini. – *Abhandlungen Larvalsystematik Insekten* 6: 1–453.

Fittkau, E.J. & Roback, S.S. (1983): The larvae of Tanypodinae (Diptera: Chironomidae) of the Holarctic region – Keys and diagnoses. – *Entomologica scandinavica Supplement* 19: 33–110.

Fittkau, E.J. & Murray D.A. (1986): The pupae of Tanypodinae (Diptera: Chironomidae) of the Holarctic region – Keys and diagnoses. – *Entomologica scandinavica Supplement* 28: 31–113.

Friberg, N., Milner, A.M., Svendsen, L.M., Lindegaard, C. & Larsen, S.E. (2001): Macroinvertebrate stream communities along regional and physico-chemical gradients in western Greenland. – *Freshwater Biology* 46: 1753–1764.

Furse, M.T., Armitage, P.D. & Wright, J.F. (1984): *Parametriocnemus boreoalpinus* Gowin (Diptera: Chironomidae) new to Britain, with notes on the separation of larvae and pupae of the British species. – *Entomologist's Gazette* 35: 265–270.

Goetghebuer, M. (1933): Chironomides du Groenland oriental, du Svalbard et de la Terre de Francois Joseph. – *Skrifter om Svalbard og Ishavet* 53: 19–31.

Goetghebuer, M. (1937–1938): Tendipedidae (Chironomidae) b) Subfamilie Tendipedinae A. Die Imagines. – *In*: Lindner, E. (ed.) *Die Fliegen der palaearktischen Region* 13 c. Tendipedidae-Tendipedinae, pp. 1–139. Stuttgart.

Goetghebuer, M. (1940–1950): Tendipedidae (Chironomidae) f) Subfamilie Orthocladiinae A. Die Imagines. – *In* Lindner, E. (ed.) *Die Fliegen der palaearktischen Region* 13 g. Tendipedidae-Orthocladiinae, pp. 1–208. Stuttgart.

Goetghebuer, M. (1955): Deux Chironomides nouveaux du Groenland (Dipt. Nemat.). Expeditions polaires françaises (Groenland): Collection A. Bauer du Musée Zoologique de Strassburg). – *Bulletin Association Philomathique d'Alsace et de Lorraine* 9: 167.

Gowin, F. & Thienemann A. (1942). Zwei neue Orthocladiinen-Arten aus Lunz (Niederdonau). – *Zoologischer Anzeiger* 140: 101–109.

Grimås, U. & Widerholm, T. (1979). Biometry and biology of *Constempellina brevicosta* (Chironomidae) in a subarctic lake. – *Holarctic Ecology* 2: 119–124.

Grodhaus, G. (1987): *Endochironomus* Kieffer, *Tribelos* Townes, *Synendotendipes*, n. gen., and *Endotribelos*, n. gen. (Diptera: Chironomidae) of the Nearctic Region. – *Journal of the Kansas Entomological Society* 60: 167–247.

Halvorsen, G.A. & Sæther, O.A. (1987): Redefinition and revision of the genus *Tokunagaia* Sæther, 1973 (Diptera: Chironomidae). – *Entomologica scandinavica Supplement* 29: 173–188.

Hansen, D.C. & Cook, E. (1976): The systematics and morphology of the Nearctic species of *Diamesa* Meigen, 1835 (Diptera: Chironomidae). – *Memoirs of the American Entomological Society* 30: 1–203.

Hayes, B.P. & Murray, D.A. 1988: *Tokunagaia obriaini* – a new species of Orthocladiinae (Diptera: Chironomidae) from Northwest Territories, arctic Canada. – *Entomologica scandinavica* 18: 361–366.

Henriksen, K.L. (1939). A revised index of the insects of Grønland. – *Meddelelser om Grønland* 119, 10: 1–112.

Henriksen, K.L. & Lundbeck, W. (1917): Grønlands Landarthropoder (Insecta et Arachnida Groenlandica). – *Meddelelser om Grønland* 22: 481–823.

Hestenes, T. & Sæther, O.A. (2000): Three new *Thienemanniella* Kieffer species with a review of the Nearctic species. – *In* Hoffrichter, O. (ed.) Late 20th Century Research on Chironomidae: an Anthology from the 13th International Symposium on Chironomidae, pp. 103–127. Aachen.

Hfransdottir, T. (2005): Diptera 2 (Chironomidae). – *The Zoology of Iceland* III, 48b: 1–169.

Hirvenoja, M. (1973): Revision der Gattung *Cricotopus* van der Wulp und ihrer Verwandten (Diptera, Chironomidae). – *Annales Zoologici Fennici* 10: 1–363.

Hirvenoja, M. (1988): *Corynoneura brundini* spec. nov. Ein Beitrag zur Systematik der Gattung *Corynoneura*. – *Spixiana Supplement* 14: 213–238.

Holmgren, A.E. (1872): Insekter från Nordgrønland, samlade af Prof. A. E. Nordensköld år 1870. – *Öfversigt af Kongliga VetenskapsAkademiens Förhandlingar.* 29 (6): 97–105.

Jackson, G.A. (1977): Nearctic and Palaearctic *Paracladopelma* Harnisch and Saetheria n. gen. (Diptera: Chironomidae). – *Journal of the Fisheries Research Board of Canada* 34: 1321–1359.

Johannsen, O.A. (1937): Aquatic Diptera. Part III. Chironomidae: Subfamilies Tanypodinae, Diamesinae and Orthocladiinae. – *Cornell University Agricultural Experiment Station Memoirs* 205: 1–84 + 18 plates.

Johannsen, O.A. & Townes H.K. (1952): Guide to the Insects of Connecticut. Part VI. The Diptera or true flies. – *State of Connecticut, State Geological and Natural History Survey Bulletin* 80: 1–147.

Kieffer, J.J. (1923): Nouvelle contribution à l'étude des Chironomides de la Nouvelle-Zemble. – *Report of the Scientific Results of the Norwegian Expedition to Nowaya Zemlya* 9: 3–11.

Kieffer, J.J. (1926): Chironomiden der 2. FRAM–Expedition (1898–1902). – *Norsk entomologisk Tidsskrift* 2: 78–89.

Kownacka, M. & Kownacki, A. (1967): *Parametriocnemus boreoalpinus* Gowin et Thienemann 1942 (Tendipedidae, Diptera) new species for the Tatra Mts. – *Acta Hydrobiologia, Krakow* 9: 187–191.

Kownacki, A. & Zodsidze, R. (1973): *Parametriocnemus stylatus adzhaicus* n. ssp. (Chironomidae, Diptera). – *Bulletin de l'Academie Polonaise des Sciences, Serie des sciences biologiques* Cl. II, 21: 127–130.

Langton, P.H. (1985): Review of type specimens of the *limbatellus* group, with a provisional key to the known females of *Psectrocladius* Kieffer (Diptera: Chironomidae). – *Entomologica scandinavica* 15: 477–485.

Langton, P.H. (1991): A key to pupal excuviae of West Palaearctic Chironomidae. – *P.H. Langton, Huntingdon, Cambridgeshire*, 386 pp.

Langton, P.H. (1998): *Micropsectra silvesterae* n. sp. and *Tanytarsus heliomesonyctios* n. sp. (Diptera, Chironomidae), two parthenogenetic species from Ellesmere Island, Arctic Canada. – *Journal of the Kansas Entomological Society* 71: 208–215.

Langton, P.H. & Cranston, P.S. (1991) Pupae in nomenclature and identification: West Palaearctic *Orthocladius* s. str. (Diptera: Chironomidae) revised. – *Systematic Entomology* 16: 239–252.

Lehmann, J. (1969): Die europäischen Arten der Gattung *Rheocricotopus* Thien. und Harn. und drei neue Artvertreter dieser Gattung aus der Orientalis (Diptera, Chironomidae). – *Archiv für Hydrobiologie* 66: 348–381.

Lehmann, J. (1972): Revision der europäischen Arten (Puppen ♂♂ und Imagines ♂♂) der Gattung *Eukiefferiella* Thienemann. – *Beiträge zur Entomologie* 22: 347–405.

Lenz, F. (1959): Die Metamorphose der Tendipedinae. – *In* Lindner, E. (ed.) *Die Fliegen der palaearktischen Region* 13 c. Tendipedidae-Tendipedinae, pp. 139–260. Stuttgart.

Lindeberg, B. (1968): Population differences in *Tanytarsus gracilentus* (Holmgr.) (Dipt., Chironomidae). – *Annales Zoologici Fennici* 5: 88–91.

Lindeberg, B. (1970): Tanytarsini (Diptera, Chironomidae) from northern Fennoscandia. – *Annales Zoologici Fennici* 7: 303–312.

Lindegaard, C. (1989): Secondary production of zoobenthos in freshwater ecosystems. A review with special reference to Chironomidae (Diptera). – *Acta Biologica Debrecina. Supplementum Oecologica Hungarica* 3: 231–240.

Lindegaard, C. (1997): Diptera Chironomidae, non-biting midges. – *In* Nilsson, A. (ed.) *Aquatic Insects of North Europe. A taxonomic Handbook, Vol. 2: Odonata-Diptera*, pp. 265–294. Apollo Books, Stenstrup.

Lindegaard, C. & Jónasson, P.M. (1979): Abundance, populations dynamics and production of zoobenthos in Lake Mývatn, Iceland. – *Oikos* 32: 202–227.

Lindegaard, C., Mæhl, P. & Nielsen, B.H. (1978): Zoobenthos of lakes located within and outside the Ilimausaq intrusion in South Greenland. – *Verhandlungen Internationale Vereinigung für Theoretische und Angewandte Limnologie* 20: 159–164.

Lindegaard, C. & Mæhl, P. (1993): Abundance, population dynamics and production of Chironomidae (Diptera) in an ultraoligotrophic lake in South Greenland. – *Netherlands Journal of Aquatic Ecology* 26: 576–581.

Lindegaard, C. & Brodersen, K.P. (1995): Distribution of Chironomidae (Diptera) in the river continuum. – *In*: Cranston, P.S. (ed.). *Chironomids from genes to ecosystem*, pp. 257–271. CSIRO Publication, Melbourne.

Lundbeck, W. (1898): Diptera groenlandica. – *Videnskabelige Meddelelser fra dansk naturhistorisk Forening* 5, 10: 236–314.

Maschwitz, D.E. & Cook, E.F. (2000): Revision of the Nearctic species of the genus *Polypedilum* Kieffer (Diptera: Chironomidae) in the subgenera *P.* (*Polypedilum*) Kieffer and *P.* (*Uresipedilum*) Oyewo and Sæther. – *Bulletin of the Ohio Biological Survey, New Series* 12, 3: 1–135.

Moller Pillot, H.K.M. (1984): De larven der Nederlandse Chironomidae (Diptera) (Orthocladiinae sensu lato). – *Nederlandse Faunistische Mededelingen* 1B: 1–175.

Murray D.A. & Fittkau, E.J. (1989): The adult males of Tanypodinae (Diptera: Chironomidae) of the Holarctic region – Keys and diagnoses. – *Entomologica scandinavica Supplement* 34: 37–123.

Nolte, U. (1993): Egg masses of Chironomidae (Diptera). – *Entomologica scandinavica Supplement* 43: 1–75.

Oliver, D.R. (1962): A review of the subfamily Orthocladiinae (Chironomidae, Diptera) of Bear Island. – *Astarte* 20: 1–19.

Oliver, D.R. (1963): Entomological studies in the Lake Hazen Area, Ellesmere Island, including lists of species of Arachnida, Collembola and Insecta. – *Arctic* 16: 175–180.

Oliver, D.R. (1970): Designation and description of lecotypes of the six Greenland Orthocladiinae (Dipt. Chironomidae) described by Lundbeck in 1898. – *Entomologica scandinavica* 1: 102–108.

Oliver, D.R. (1976): Chironomidae (Diptera) of Char Lake, Cornwallis Island, N. W. T., with description of two new species. – *The Canadian Entomologist* 108: 1053–1064.

Oliver, D.R. (1983): 7. The larvae of Diamesinae (Diptera: Chironomidae) of the Holarctic region – Keys and diagnoses. – *Entomologica scandinavica Supplement* 19: 115–138.

Oliver, D.R. (1986): 7. The pupae of Diamesinae (Diptera: Chironomidae) of the Holarctic region – Keys and diagnoses. – *Entomologica scandinavica Supplement* 28: 119–137.

Oliver, D.R. (1989): 7. The adult males of Diamesinae (Diptera: Chironomidae) of the Holarctic region – Keys and diagnoses. – *Entomologica scandinavica Supplement* 34: 129–154.

Oliver, D.R. & Danks, H.V. (1972): Sex ratios of some high arctic Chironomidae (Diptera). – *The Canadian Entomologist* 104: 1413–1417.

Oliver, D.R. & Dillon, M.E. (1988): Review of *Cricotopus* (Diptera: Chironomidae) of the Nearctic arctic zone with description of two new species. – *The Canadian Entomologist* 120: 463–496.

Oliver, D.R. & Dillon, M.E. (1989): The adult males of Chironomidae (Diptera) of the Holarctic region – Key to subfamilies. – *Entomologica scandinavica Supplement* 34: 11–15.

Oliver, D.R. & Dillon, M.E. (1994): Systematics of some species of *Micropsectra* (Diptera: Chironomidae) living in low-order streams in southern Ontario, Canada. – *The Canadian Entomologist* 126: 199–217.

Oliver D.R. & Dillon, M.E. (1997): Chironomids (Diptera: Chironomidae) of the Yukon arctic North Slope and Herschel Island. – *In*: Danks, H.V. & Downes, J.A. (eds) *Insects of the Yukon. Biological Survey of Canada* (*Terrestrial Arthropods*), pp. 615–635. Ottawa.

Oliver, D.R., McClaymont, D. & Poussel, M.E. (1978): A key to some larvae of Chironomidae (Diptera) from the Mackenzie and Porcupine River watersheds. – *Fisheries & Marine Service Technical Report* 791: 1–73.

Oliver, D.R., Dillon, M.E. & Cranston, P.S. (1990): A catalog of Nearctic Chironomidae. – *Research Branch Agriculture Canada, Publication* 1857/B. 89 pp.

Paasirvita, L. (1972): Taxonomy, ecology and swarming behaviour of *Tanytarsus gracilentus* Holmgr. (Diptera, Chironomidae) in Valassaaret, Gulf of Bothnia, Finland. – *Annales Zoologici Fenneci* 9: 255–264.

Pagast, F., Thienemann, A. & Krüger, F. (1941): Terrestrische Chironomiden. VIII. *Metriocnemus fuscipes* Mg. und *Metriocnemus terrester* n. sp. Pagast. – *Zoologischer Anzeiger* 133: 202–213.

Pagast, F. (1947): Systematik und Verbreitung der um die Gattung *Diamesa* groupierten Chironomiden. – *Archiv für Hydrobiologie* 41: 435–596.

Philipp, M., Woodell, S.R.J., Böcher, J. & Mattson, O. 1996. Reproductive biology of four species of *Pedicularis* (Scrophulariaceae) in West Greenland. – Arctic and Alpine Research 28, 4: 403–413.

Pinder, L.C.V. (1978): A key to adult males of British Chironomidae (Diptera) the non-biting midges. – Freshwater Biological Association Scientific Publication 37: 1–169 + 189 figs.

Pinder, L.C.V. & Cranston, P.S. (1976): Morphology of the male imagines of *Orthocladius* (*Pogonocladius*) *consobrinus* and *O. glabripennis* with observation of the taxonomic status of *O. glabripennis* (Diptera: Chironomidae). – *Entomologica scandinavica* 7: 19–23.

Pinder, L.C.V. & Reiss, F. (1983): 7. The larvae of Chironominae (Diptera: Chironomidae) of the Holarctic region – Keys and diagnoses. – *Entomologica scandinavica Supplement* 19: 293–435.

Pinder, L.C.V. & Reiss, F. (1986): 7. The pupae of Chironominae (Diptera: Chironomidae) of the Holarctic region – Keys and diagnoses. – *Entomologica scandinavica Supplement* 28: 299–456.

Reiss, F. (1974): Revision des Typen-Materials einiger Tanytarsini-Arten (Chironomidae, Diptera) aus dem Museum Brüssel. – *Entomologisk tidskrift* 95, Supplementum: 203–211.

Reiss, F. (1968): Ökologische und systematische Untersuchungen an Chironomiden (Diptera) des Bodensees (Teil 1). – *Archiv für Hydrobiologie* 64: 176–246.

Reiss, F. & Fittkau, E.J. (1971): Taxonomie und Ökologie europäisch verbreiteter *Tanytarsus*-Arten (Chironomidae, Diptera). – *Archiv für Hydrobiologie Supplement* 40: 75–200.

Reiss, F. & Säwedal, L. (1981): Keys to males and pupae of the Palaearctic (excl. Japan) *Paratanytarsus* Thienemann & Bause, 1913, n. comb., with descriptions of three new species (Diptera: Chironomidae). – *Entomologica scandinavica, Supplement* 15: 73–104.

Remy, P. (1928): Quelques Arthropodes récoltés dans la region du Scoresby-Sund (Groenland oriental) au cours de la croissière arctique du Pourquoi-pas? en 1926. – *Meddelelser om Grønland* 74: 85–95.

Riget, F. (1981): Ferskvandsbiologiske undersøgelser i Narssaq Elv, 1981. – *Grønlands Fiskeriundersøgelser*, Copenhagen, 48 pp.

Roback, S.S. (1966): Notes on Tanypodinae types in European museums (Dipt.: Chironomidae). – *Entomological News* 77: 113–132.

Roback, S.S. (1971): The adults of the subfamily Tanypodinae (+Pelopiinae) in North America (Diptera: Chironomidae). – *Monographs of the Academy of Natural Sciences of Philadelphia* 17:1–410.

Roback, S.S. (1980): The immature chironomids of the Eastern United States IV. Tanypodinae – Procladiini. – *Proceedings of the Academy of Natural Sciences of Philadelphia* 132:1–63.

Roback, S.S. (1985): The immature chironomids of the Eastern United States VI. Pentaneurini – genus *Ablabesmyia*. – *Proceedings of the Academy of Natural Sciences of Philadelphia* 137: 153–212.

Sæther, O.A. (1969): Some Nearctic Podonominae, Diamesinae, and Orthocladiinae (Diptera: Chironomidae). – *Bulletin of the Fisheries Research Board of Canada* 170: 1–154.

Sæther, O.A. (1973): Four species of *Bryophaenocladius* Thien., with notes on other Orthocladiinae (Diptera: Chironomidae). – *The Canadian Entomologist* 105: 51–60.

Sæther, O.A. (1975a): Nearctic and Palaearctic *Heterotrissocladius* (Diptera: Chironomidae). – *Bulletin of the Fisheries Research Board of Canada* 193: 1–67.

Sæther, O.A. (1975b): Twelve new species of *Limnophyes* Eaton, with keys to Nearctic males of the genus (Diptera: Chironomidae). – *Canadian Entomologist* 107: 1029–1056.

Sæther, O.A. (1976): Revision of *Hydrobaenus*, *Trissocladius*, *Zalutschia*, *Paratrissocladius*, and some related genera (Diptera: Chironomidae). – *Bulletin of the Fisheries Research Board of Canada* 195: 1–287.

Sæther, O.A. (1977): Female genitalia in Chironomidae and other Nematocera: morphology, phylogenies, keys. – *Bulletin of the Fisheries Research Board of Canada* 197: 1–211.

Sæther, O.A. (1979): Chironomid communities as water quality indicators. – *Holarctic Ecology* 2: 65–74.

Sæther, O.A. (1980): Glossary of chironomid morphology terminology (Diptera: Chironomidae). – *Entomologica scandinavica Supplement* 14: 1–51.

Sæther, O.A. (1983): 8. The larvae of Prodiamesinae (Diptera: Chironomidae) of the Holarctic region – Keys and diagnoses. – *Entomologica scandinavica Supplement* 19: 141–147.

Sæther, O.A. (1985): A review of the genus *Rheocricotopus* Thienemann & Harnisch, 1931, with the description of three new species. – *Spixiana Supplement* 11: 59–108.

Sæther, O.A. (1986): 8. The pupae of Prodiamesinae (Diptera: Chironomidae) of the Holarctic region – Keys and diagnoses. – *Entomologica scandinavica Supplement* 28: 139–145.

Sæther, O.A. (1989a): 8. The adult males of Prodiamesinae (Diptera: Chironomidae) of the Holarctic region – Keys and diagnoses. – *Entomologica scandinavica Supplement* 34: 155–163.

Sæther, O.A. (1989b): *Metriocnemus* van der Wulp: a new species and a revision of species described by Meigen, Zetterstedt, Stæger, Holmgren, Lundström and Strenzke (Diptera: Chironomidae). – *Entomologica scandinavica* 19: 393–430.

Sæther, O.A. (1990): A review of the genus *Limnophyes* Eaton from the Holarctic and Afrotropical regions (Diptera: Chironomidae, Orthocladiinae). – *Entomologica scandinavica Supplement* 35: 1–135.

Sæther, O.A. (1995): *Metriocnemus* van der Wulp: Seven new species, revision of species, and new records (Diptera: Chironomidae). – *Annales de Limnologie* 31: 35–64.

Sæther, O.A. (2004): Three new species of *Orthocladius* subgenus *Eudactylocladius* (Diptera: Chironomidae) from Norway. – *Zootaxa* 508: 1–12.

Sæther, O.A. (2005): A new subgenus and new species of *Orthocladius* van der Wulp, with a phylogenetic evaluation of the validity of the subgenera of the genus (Diptera: Chironomidae). – *Zootaxa* 974: 1–56.

Sæther, O.A., Sublette, J.E. & Willassen, E. (1984): Chironomidae (Diptera) from the 2nd Fram Expedition (1898–1902) to arctic North America described by J. J. Kieffer. – *Entomologica scandinavica* 15: 249–275.

Sæther, O.A. & Wang, X. (1995): Revision of the genus *Paraphaenocladius* Thienemann, 1924 of the world (Diptera: Chironomidae, Orthocladiinae). – *Entomologica scandinavica Supplement* 48: 1–69.

Sæther, O.A., Ashe, P. & Murray, D.A. (2000): A.6. Family Chironomidae. – *In*: Papp, L. & Darvas, B. (eds) *Contributions to a Manual of the Palaearctic Diptera (with special reference to flies of economic importance) Appendix*, pp. 113–334. Science Herald, Budapest.

Sæther, O.A. & Ferrington, L.C., Jr. (2003): Nomenclature notes on some orthoclads (Diptera: Chironomidae). – *Zootaxa* 322: 1–7.

Sæther, O.A. & Wang, X. (2009): *Corynoneura* Winnertz from East Asia, with a systematic review of the genus (Diptera: Chironomidae: Orthocladiinae). *Zootaxa* 2287: 1–44.

Sävedal, L. (1976): Revision of the *notescens*-group of the genus *Micropsectra* Kieffer, 1909 (Diptera: Chironomidae). – *Entomologica scandinavica* 7: 109–144.

Sävedal, L. (1979): *Micropsectra brundini* n. sp. (Diptera: Chironomidae) and geographical variation in the *Micropsectra insignilobus* aggregate. – *Entomologica scandinavica Supplement* 10: 133–138.

Säwedal, L. (1981): Description of *Micropsectra tori* n. sp. from Greenland, with notes on the *recurvata*-group (Diptera: Chironomidae). – *Entomologica scandinavica* 12: 27–30.

Schlee, D. (1968): Vergleichende Merkmalsanalyse zur Morphologie und Phylogenie der *Corynoneura*-Gruppe (Diptera, Chironomidae). Zugliech eine allgemeine Morphologie der Chironomiden-Imago. – *Stuttgarter Beiträge zur Naturkunde* 180: 1–150.

Schnell, Ø.A. & Aagaard, K. (1996): Chironomidae-Fjærmygg. – *In* Aagaard, K. & Dolmen, D. (eds) *Limnofauna Norvegica*, pp. 210–248. Trondheim.

Sendstad, E., Solem, J.O. & Aagaard, K. (1976): Studies of terrestrial chironomids (Diptera) from Spitsbergen. – *Norwegian Journal of Entomology* 24: 91–98.

Serra-Tosio, B. (1972): Ecologie et biogeography des Diamesini d'Europe (Diptera, Chironomidae). – *Travaux du Laboratoire d'Hydrobiologie et de Pisciculture de Grenoble* 63: 5–173.

Serra-Tosio, B. (1976): Chironomides des Alpes: Le genre *Pseudodiamesa* (Diptera, Chironomidae). – *Travaux Scientifiques du Parc National de la Vanoice* 7: 117–138.

Soponis, A.R. (1977): A revision of the Nearctic species of *Orthocladius* (*Orthocladius*) van der Wulp (Diptera: Chironomidae). – *Memoirs of the Entomological Society of Canada* 102: 1–187.

Soponis, A.R. (1987): Notes on *Orthocladius* (*Orthocladius*) *frigidus* (Zetterstedt) with a redescription of the species (Diptera: Chironomidae). – *Entomologica scandinavica Supplement* 29: 123–131.

Soponis, A.R. (1990): A revision of the Holarctic species of *Orthocladius* (*Euorthocladius*) (Diptera: Chironomidae). – *Spixiana Supplement* 13: 1–68.

Stæger, C. (1845): Grønlands antliater. – *Naturhistorisk Tidsskrift* (2) 1: 346–369.

Strenzke, K. (1940): Terrestrische Chironomiden. V. *Camptocladius stercorarius* De Geer. – *Zoologis her Anzeiger* 132: 115–123.

Strenzke, K. (1942): Terrestrische Chironomiden. XII. *Bryophaenocladius subvernalis* Edw. – *Zoologischer Anzeiger* 137: 10–18.

Strenzke, K. (1951): Systematik, Morphologie und Ökologie der terrestrischen Chironomiden. – *Archiv für Hydrobiologie Supplementum* 18: 207–414.

Strenzke, K. (1959): Revision der Gattung *Chironomus* Meig. I. Die Imagines von 15 norddeutschen Arten und Unterarten. – *Archiv für Hydrobiologie* 56: 1–42.

Stur, E. & Ekrem, T. (2006): A revision of West Palaearctic species of the *Micropsectra atrofasciata* species group (Diptera: Chironomidae). – *Zoological Journal of the Limnaean Society* 146: 165–225.

Sublette, J.E. & Sublette, M.F. (1971): Description of the immature stages and adults of the *Chironomus staegeri* group. – *Studies in Natural Sciences, Portales, New Mexico* 1, 1: 6–89.

Thienemann, A. (1941a): Lappländische Chironomiden und ihre Wohngewässer (Ergebnisse von Untersuchungen im Abiskogebiet in Swedisch-Lappland). – *Archiv für Hydrobiogie Supplementum* 17: 1–253.

Thienemann, A. (1944): Bestimmungstabellen für die jetzt bekannten Larven und Puppen der Orthocladiinen (Diptera Chironomidae). – *Archiv für Hydrobiogie* 39: 551–664.

Thienemann, A. (1951): *Tanytarsus*-Studien II. Die Subsectio *Paratanytarsus*. – *Archiv für Hydrobiologie, Supplement* 18: 595–632.

Thienemann, A. & Krüger, F. (1937): "*Orthocladius*" *abiskoensis* Edwards und *rubicundus* (Mg.), zwei "Puppen-Species" der Chironomiden. – *Zoologischer Anzeiger* 117: 257–267.

Thienemann, A. & Strenzke, K. (1941): Terrestrische Chironomiden. VII. Die Gattung *Paraphaenocladius* Th. – *Zoologischer Anzeiger* 133: 137–146.

Tuiskunen, J. (1986): The Fennoscandian species of *Parakiefferiella* Thienemann (Diptera, Chironomidae, Orthocladiinae). – *Annales Zoologici Fennici* 23: 175–196.

Tuiskunen, J. & Lindeberg, B. (1986): Chironomidae (Diptera) from Fennoscandia north of 68°N, with a description of ten new species and two new genera. – *Annales Zoologici Fennici* 23: 361–393.

Welch, H.E. (1973): Ecology of Chironomidae (Diptera) in a polar lake. – *Journal of the Fisheries Research Board of Canada* 33: 227–247.

Wiederholm, T. (ed.) (1983): Chironomidae (Diptera) of the Holarctic Region – Keys and diagnoses. Part 1 – Larvae. – *Entomologica scandinavica Supplement* 19: 1–457.

Wiederholm, T. (ed.) (1986): Chironomidae (Diptera) of the Holarctic Region – Keys and diagnoses. Part 2 – Pupae. – *Entomologica scandinavica Supplement* 28: 1–482.

Wiederholm, T. (ed.) (1989): The adult males of Chironomidae (Diptera) of the Holarctic Region – Keys and diagnoses. – *Entomologica scandinavica Supplement* 34: 1–532.

Willassen, E. (1985): A review of *Diamesa davisi* Edwards and the *davisi* group. – *Spixiana* 11: 109–137.

Wülker, W. (1956): Zur Kenntnis der Gattung *Psectrocladius* Kieff. (Dipt., Chironom.). – *Archiv für Hydrobiologie, Supplement* 24: 1–66.

Wülker, W. & Butler M. G. (1983): Karyosystematics and morphology of Northern *Chironomus* (Diptera: Chironomidae): Freshwater species with larvae of the *salinarius*-type. – *Entomologica scandinavica* 14: 121–136.

Wülker, W., Kiknaddze, I.I., Kerkis, I.E., Nevers, P. (1998): Chromosomes, morphology, ecology and distribution of *Sergentia baueri*, spec. nov., *S. prima* Proviz & Proviz, 1997 and *S. coracina* Zett., 1824. – *Spixiana* 22: 69–81.

Wülker, W. & Martin, J. (2000): Northernmost *Chironomus* karyotypes (Insecta, Diptera, Chironomidae). – *Spixiana* 23: 151–156.

Zetterstedt, J.W. (1838): Dipterologis Scandinaviae. Sect. 3: Diptera. – *In Insecta Lapponica . Sect. 3: Diptera*, pp 477–868. Lipsiae [= Leipzig].

Addendum to Chironomidae (Non-biting midges)

Claus Lindegaard

Since the manuscript was finished in 2010, a number of revisions of chironomid genera have been published and some corrections and additions are necessary. The identification of Holarctic Chironomidae larvae to genera has been revised and emended with a number of new genera (Andersen *et al.* 2013a), which to some extent makes the identification of pupae and adult males by Wiederholm (1986, 1989) outdated. However, it is in practise not a problem for the Greenlandic genera, except for the old genus *Pseudosmittia* (see below).

***Ablabesmyia pulchripennis* Lundbeck, 1898**
Keys to pupae and larvae of the Nearctic *Ablabesmyia*, including *A. pulchripennis*, are given by Sæther (2011).

***Procladius* Skuse, 1889**
Keys to females, pupae and larvae of *Procladius* living in Lake Winnepeg, Canada, are given by Sæther (2010). They include *P. paragretis* and *P. culiciformis*.

***Pseudodiamesa branickii* (Nowicki, 1873)**
Serra-Tosio (1976) redescribed adults and pupae of *P. branickii*. Larvae of the *P. branickii* group can be separated from the *P. nivosa* group by characters given in Ilyashuk *et al.* (2010). Ilyashuk & Ilyashuk (2011) mentioned that *P. branickii* from Eurasia consists of at least four sibling species based on chromosomal analyses.

[*Telmatogeton japanicus* Tokunaga, 1933
The specimens from Iceland are described by Sæther (2009) as a new species, T. *murrayi* Sæther 2009, very close to *T. japanicus*. Not yet recorded from Greenland.]

***Bryophaenocladius* Thienemann, 1934**
Wang *et al.* (2004) provided a key for the known Nearctic males of the genus.

***Corynoneura* Winnertz, 1846**
Keys to adults, pupae and larvae of the Nearctic *Corynoneura* are given by Fu & Sæther (2012). The keys include the three species previously mentioned from Greenland.

***Heterotrissocladius* Spärck, 1923**
A report on fish and invertebrates of three lakes, a river and a stream in Qorlortorsuaq, Nanortalik, S Greenland, includes the chironomids determined by Øyvind Schnell, Bergen (Bergersen 1993). They recorded *H. subpilosus* Kieffer, 1911 and *H. marcidus* (Walker, 1856) and suggested that these species belong to a special Palaearctic element in South Greenland in spite that the location is within the same region as Narsaq, where the Nearctic *H. oliveri* and *H. changi* occur. However, the differences between *H. subpilosus* and *H. oliveri* are very small, and in fact I was not myself convinced about the identity of the Greenlandic species. Therefore, specimens from the Narsaq area were checked by Ole Sæther, who recognized the Nearctic species, *H. oliveri* and *H. changi*. *H. subpilosus* does not occur in Iceland but is common in Svalbard and Jan Mayen, so it may still be possible that both species occur in Greenland. Further, Brodersen & Anderson (2002) reported four different *Heterotrissocladius*-types according to larval headcapsules in surface sediment samples from 47 lakes in the Kangerlussuaq/Søndre Strømfjord area. Detailed analyses of all stages are needed to reveal the actual number and identity of the *Heterotrissocladius* species in Greenland.

***Paracladius quadrinodosus* Hirvenoja, 1973**
Headcapsules of a *Paracladius* sp. were found in sediment cores in two lakes on Store Koldewey, NE Greenland (Schmidt *et al.* 2011). In both cases in layers older than 1000 years.

***Parakiefferiella nigra* Brundin, 1949**
Walker *et al.* (1992) described the larvae of *P. nigra* and stated it is common throughout the Arctic as well as subarctic Europe and North America, including a record from Peary Land. Schmidt *et al.* (2011) reported headcapsules of *P. nigra*-type from surface sediment in one lake on Store Koldewey, NE Greenland.

***Paraphaenocladius* cf. *nasthecus* Sæther & Wang 1995**
Bergersen (1993) found this species in a river in the Qorlortorsuaq area, Nanortalik, probably belonging to a species of *Paraphaenocladius* not recorded before in Greenland. *P. nasthecus* is known from northern North America and China. A key to the known species of *Paraphaenocladius* is given by Sæther & Wang (1995).

***Paratrichocladius* Santos Abreu 1918**
Bergersen (1993) recorded this genus in a river in the Qorlortorsuaq area, Nanortalik. Both larvae and pupae of this genus are difficult to separate from some *Cricotopus* and/or *Orthocladius* species. The adult males separate from *Cricotopus* in having erect dorsocentrals and from *Orthocladius* by lacking the anal point. This genus has not been recorded from Greenland before.

***Psectrocladius* Kieffer, 1906**
Sæther & Langton (2011) provided preliminary keys to males, females, pupae and larvae of the known species of the *P. limbatellus* group.

***Pseudosmittia* Edwards, 1932**
A comprehensive worldwide revision of the genus is made by Ferrington & Sæther (2011) resulting in splitting the old genus *Pseudosmittia* into three new genara: *Pseudosmittia* Edwards, 1932, *Allocladius* Kieffer, 1913 and *Hydrosmittia* Ferrington & Sæther, 2011.

They give revised descriptions and keys to all stages of the known species. In total 132 species are described, of which three are recorded from Greenland.

***Allocladius nanseni* Kieffer, 1926**
Records as for *Pseudosmittia nanseni*.

***Hydrosmittia oxonianus* (Edwards, 1922)**
Earlier records stated for *Pseudosmittia oxonianus* are still uncertain, but the presence in Greenland was confirmed by a male recorded from Kvane River, Narsaq, 1981 (Ferrington & Sæther 2011).

***Hydrosmittia virgo* (Strenzke, 1950)**
The record from Greenland is known from 11 larvae deposited in Natur-Museum Senckenberg, Germany, and labelled: Greenland, 1949, leg. Delasse, det. K. Strenzke, No. 10, without further details (Ferrington & Sæther 211). This species is new to Greenland.

***Smittia* Holmgren, 1869**
The genus is still in need of revision, but a number of European *Smittia* species are redescribed by Moller Pillot (2008), including *S. aterrima*, together with comprehensive notes on their ecology. Kevan (2007) recorded females of *S. velutina* as the most prominent visitors at flowers in Hazen Camp, Ellesmere Island (NWT, Canada), and suggested that they use the nectar to mature their eggs. This species may be common in flowers also in Greenland (cf. Phillip *et al.* 1996). Kevan (2007) further assumed the species to be parthenogenetic, as he never observed any male and as the females willingly deposited eggs.

Orthocladiinae genus A

The larva is described by Andersen *et al.* (2013b) under the name: Orthocladiinae genus "Greenland".

***Chironomus* Meigen, 1803**
Proulx *et al.* (2013) provided a key for a number of Canadian larval *Chironomus*, but also described *C. stœgeri* as belonging to the *C. plumosus*-type, even though this larval type has not been reported from Greenland.

***Sergentia coracina* (Zetterstedt, 1850)**
Headcapsules of *S. coracina*-type are reported from sediment cores in two lakes on Store Koldewey, NE Greenland (Schmidt *et al.* 2011). In both cases in layers older than 1000 years.

***Constempellina brevicosta* (Edwards, 1937)**
A redescription of adult male, pupa and larva of *C. brevicosta* was given by Zorina (2013). No further records from Greenland.

***Corynocera oliveri* Lindeberg, 1970**
Stur & Ekrem (2011) described the larvae of *Tanytarsus heliomesonyctios* Langton, 1998 and found a number of characters similar to those described for *C. oliveri* (*in* Brooks *et al.* 2007). They therefore suggested that headcapsules from sediment cores used by palaeolimnologists and identified as *C. oliveri*/*T. lugens*-type may be *T. heliomesonyctios*. This species is parthenogenetic and so far only known from Ellesmere Island, Canada, and Svalbard, Norway, but may, if the above assumption is correct, be widely distributed in the northern Holarctic.

***Micropsectra* Kieffer, 1909**
Anderson *et al.* (2013) provided a tentative key to adult males and pupae of known Nearctic *Micropsectra* species. They maintain *M. tori* and *M. insignilobus* as valid species, but point out that they are morphologically very similar.

***Micropsectra atrofasciata* group**
In paleoecological studies headcapsules from sediment cores are often identified as *M. radialis*-type. However,

M. radialis Goetghebuer, 1939 is not yet recorded from the Nearctic region. The larvae may be identified by the larval keys in Stur & Ekrem (2006, 2011).

References

Andersen, T., Cranston, P.S. & Epler, J.H. (eds) (2013a): The larvae of Chironomidae (Diptera) of the Holarctic Region – Keys and diagnoses. – *Insect Systematics & Evolution*, Supplement 66: 1–571.

Andersen, T., Sæther, O.A., Cranston, P.S. & Epler J.H. (2013b): The larvae of the Holarctic Orthocladiinae – Keys and diagnoses. – *Insects Systematics & Evolution*, Supplement 66: 189–386.

Anderson, A.M., Stur, E. & Ekrem, T. (2013): Molecular and morphological methods reveal cryptic species of Nearctic *Micropsectra* (Diptera: Chironomidae). – *Freshwater Science* 32: 892–921.

Bergersen, R. (1993): Freshwater fish and invertebrates of Qorlortorsuaq, Vatnahverfi, South Greenland. – *Tromura, Tromsø Museums Rapportserie* 73: 1–71.

Brodersen, K.P. & Anderson, N.J. (2002): Distribution of chironomids (Diptera) in low arctic West Greenland lakes: trophic conditions, temperature and environmental reconstruction. – *Freshwater Biology* 47: 1137–1157.

Brooks, S.J., Langdon, P.G. & Heiri, O. (2007): The identification and use of Palaearctic Chironomidae larvae in paleoecology. – QRA Technical Guide No. 10, Quaternary Research Association, London, 276 pp.

Ferrington, L.C. & Sæther, O.A. (2011): a revision of the genera *Pseudosmittia* Edwards, 1932, *Allocladius* Kieffer, 1913, and *Hydrosmittia* gen. n. (Diptera: Chironomidae, Orthocladiinae). – *Zootaxa* 2849: 1–314.

Fu, Y. & Sæther, O.A. (2012): *Corynoneura* Winnertz and *Thienemanniella* Kieffer from the Nearctic Region (Diptera: Chironomidae: Orthocladiinae). – *Zootaxa* 3536: 1–61.

Ilyashuk, B.P. & Ilyashuk, E.A. (2011): Moustached *Pseudodiamesa* is still in waiting for a modern cytogenetic approach and a taxonomic revision: a reply to Willassen. – *Chironomus Newsletter* 24: 27–28.

Ilyashuk, B.P., Ilyashuk, E.A., Makarchenko, E.A. & Heiri, O. (2010): Midges of the genus *Pseudodiamesa* Goetghebuer (Diptera, Chironomidae): current knowledge and palaeoecological perspective. – *Journal of Paleolimnology* 44: 667–676.

Kevan, P.G. (2007): Diaheliotaxis and ombrophobia in an anthophilous high arctic midge, *Smittia velutina* (Lundbeck, 1898) (Chironomidae). – *Chironomus Newsletter* 20: 29–31.

Langton, P.H. (1998): *Micropsectra silvesterae* n. sp. and *Tanytarsus heliomesonyctios* n. sp., (Diptera: Chironomidae), two parthenogenetic species from Ellesmere Island, Arctic Canada. – *Journal of the Kansas entomological Society* 71: 208–215.

Moller Pillot, H.K.M. (2008): Identification and ecology of the genus *Smittia* Holmgren in the Netherlands (Diptera: Chironomidae). – *Tijdschrift voor Entomologie* 151: 245–270.

Philipp, M., Woodell, S.R.J., Böcher, J. & Mattson, O. (1996): Reproductive biology of four species of *Pedicularis* (Scrophulariaceae) in West Greenland. – *Arctic and Alpine Research* 28: 403–413.

Proulx, I., Martin, J., Carew, M. & Hare, L. (2013): Using various lines of evidence to identify *Chironomus* species (Diptera: Chironomidae) in eastern Canadian lakes. – *Zootaxa* 3741: 401–458.

Sæther, O.A. (2009): *Telmatogeton murrayi* sp. n. from Iceland and *T. japonicus* Tokunaga from Madeira (Diptera: Chironomidae). – *Aquatic Insects* 31: 31–44.

Sæther, O.A. (2010): *Procladius* Skuse from Lake Winnepeg, Manitoba, Canada, with keys to some females and immature stages of the genus (Diptera: Chironomidae). – *Zootaxa* 2726: 34–58.

Sæther, O.A. (2011): Notes on Canadian *Ablabsmyia* Johansen, with keys to known Nearctic immatures of the genus (Diptera: Chironomidae). – *Zootaxa* 3069: 43–62.

Sæther, O.A. & Langton, P.H. (2011): New Nearctic species of the *Psectrocladius limbatellus* group (Diptera: Chironomidae). – *Aquatic Insects* 33: 133–163.

Sæther, O.A. & Wang, X. (1995): Revision of the genus *Paraphaenocladius* Thienemann, 1924 of the world (Diptera: Chironomidae, Orthocladiinae). – *Entomologica scandinavica Supplement* 48: 1–69.

Serra-Tosio, B. (1972): Ecologie et biogeography des Diamesini d'Europe (Diptera, Chironomidae). – *Travaux du Laboratoire d'Hydrobiologie et de Pisciculture de Grenoble* 63: 5–173.

Schmidt, S., Wagner, B., Heiri, O., Klug, M., Bennike, O. & Melles, M. (2011): Chironomids as indicators of the Holocene climatic and environmental history of two lakes in Northeast Greenland. – *Boreas* 40: 116–130.

Stur, E. & Ekrem, T. (2006): A revision of West Palaearctic species of the *Micropsectra atrofasciata* species group (Diptera: Chironomidae). – *Zoological Journal of the Linnean Society* 146: 165–225.

Stur, E. & Ekrem, T. (2011): Exploring unknown life stages of Arctic Tanytarsini (Diptera: Chironomidae) with DNA barcoding. – *Zootaxa* 2743: 27–39.

Wang, X., Liu, Z. & Epler, J.H. (2004): New species of *Bryophaenocladius* Thienemann from the Nearctic Region (Diptera: Chironomidae: Orthocladiinae). – *Zootaxa* 581: 1–10.

Walker, I.R., Oliver, D.R. & Dillon, M.E. (1992): The larva and habitat of *Parakiefferiella nigra* Brundin (Diptera: Chironomidae). – *Netherlands Journal of Aquatic Ecology* 26: 527–531.

Wiederholm, T. (ed.) (1986): Chironomidae (Diptera) of the Holarctic Region – Keys and diagnoses. Part 2 – Pupae. – *Entomologica scandinavica Supplement* 28: 1–482.

Wiederholm, T. (ed.) (1989): The adult males of Chironomidae (Diptera) of the Holarctic Region – Keys and diagnoses. – *Entomologica scandinavica Supplement* 34: 1–532.

Zorina, O.V. (2013): A systematic review of the genus *Constempellina* Kieffer (Diptera: Chironomidae) from the Far East, with description of a new species. – *Zootaxa* 3694: 201–212.

17.8. Sciaridae

Pekka Vilkamaa

Sciaridae are small (1–4 mm), dark coloured, slender nematocerous flies that are world-wide distributed. More than 1900 valid species are known, and the number is constantly rising. The best known sciarid fauna is that of the Palaearctic with about 800 species, but those of the Nearctic and especially the tropical regions and Southern Hemisphere are very poorly known.

Sciaridae are rather uniform in general appearance. The most easily observable family character is the distinctive wing venation (Fig. 1) and the narrow eye bridge connecting the left and right compound eye. The morphology of Sciaridae is most comprehensively treated by Menzel & Mohrig (2000), also by Menzel & Mohrig (1997) and Steffan (1966, 1981). The only existing key to the Palaearctic genera is provided by Menzel & Mohrig (2000). The male hypopygial structures offer the best (and in some genera the only) characters to differentiate between species, and in many cases they have also applicable generic characters. In most genera, the females lack any reliable diagnostic characters.

The Sciaridae are most numerous both in species and individuals in temperate and tropical forests. The subarctic or arctic areas have been studied only occasionally, usually by describing new species found. Of the arctic islands, the sciarid fauna of Spitsbergen is most actively studied. Holmgren (1869) described a number of species, further records were later given by Edwards (1922, 1923), and Lengersdorf (1930, 1935) recorded four species and described two species. Tuomikoski (1967) identified eight species from the material of a Finnish expedition to Spitsbergen and defined the taxonomic position of some of them. Four species were recorded by Edwards (1935) from Bear Island, 3 from Jan Mayen (Edwards 1923), and 3 species were recorded from the Pribilof Islands (Malloch 1923). Hippa & Vilkamaa (1994) described one species (*Camptochaeta quadriceps*) and recorded two other of that genus from Ellesmere Island.

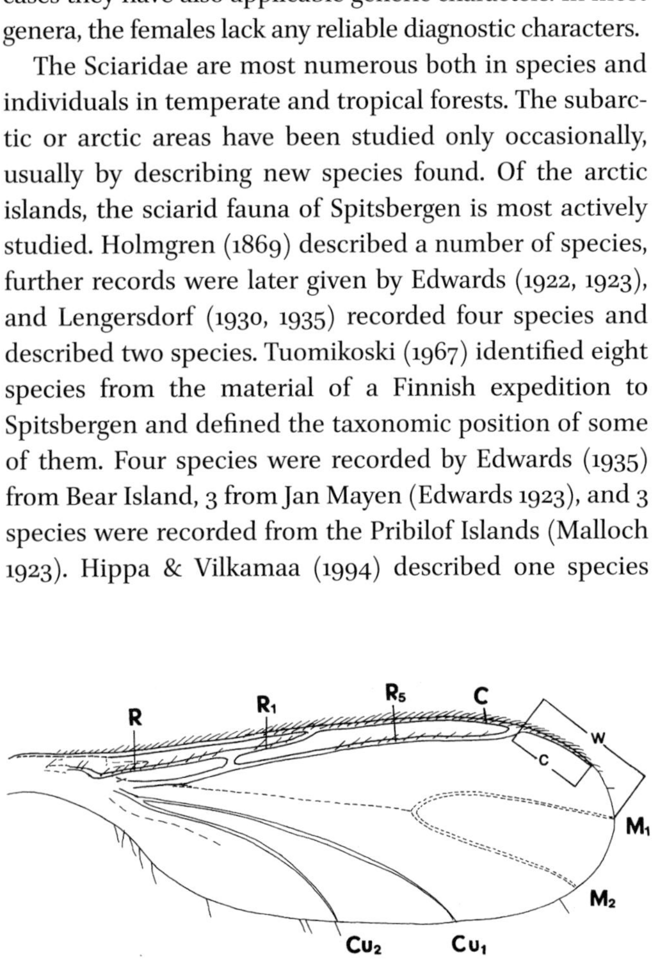

FIGURE 1 *Sciarid wing (after Vilkamaa & Hippa 2005).*

The records of Sciaridae from Greenland are old and based only on the data of a few studies. Holmgren (1872) described 1 species, *Sciara groenlandica* (=*Lycoriella* (*Hemineurina*) *modesta* Staeger, 1840). Rübsaamen (1898) described 8 species, the names of only 2 of which have survived as valid. Edwards (1935b) observed 2 species from a Greenland material. In the most comprehensive treatment of the Diptera of Greenland, Lundbeck (1898) recorded 16 species of Sciaridae, some of which as new to science. Lundbeck's material, housed in the Natural History Museum of Denmark (ZMUC), Copenhagen, is still unrevised (Vilkamaa, in prep.), except for the type material which was studied by Menzel & Mohrig (2000). Henriksen (1939) listed 22 sciarid species from Greenland, many indicated as being only tentatively identified.

A revision of type material has so far established the identity of 7 species of Sciaridae of Greenland. My preliminary survey of the large material in the collection in ZMUC has revealed a total of 15 species from the country, including two undescribed ones. This number of species seems to be much lower than that observed in the forested areas of Finnish Lapland (Salmela & Vilkamaa, unpubl.).

Key to the genera of Sciaridae in Greenland

1. Apex of front tibia with a comb-like row of setae ... 2
– Apex of front tibia with a patch of setae 3
2. Right and left spur of middle and hind tibia subequal in length .... *Bradysia*
– Right and left spur of middle and/or hind tibia very subequal in length .... *Scatopsciara*
3. Antennal flagellomeres very short, gonostylus without apical tooth or megasetae ...... *Schwenckfeldina*
– Antennal flagellomere normal, gonostylus with apical tooth and megasetae 4
4. Intercoxal area of hypopygium with a setose lobe or groups of setae, gonostylar megasetae without basal bodies 5
– Intercoxal area of hypopygium without a lobe or groups of setae, gonostylar megasetae with basal bodies .... *Camptochaeta*
5. Colour dark, gonostylus usually tumid and mesially impressed ....*Lycoriella* (*Hemineurina*)

– Colour pale, gonostylus slender and mesially not impressed .... *Lycoriella* (*Lycoriella*)

Checklist of identified Greenland Sciaridae

Species marked by an asterisk are listed by Henriksen (1939) with the species name in question.

Bradysia **Winnertz**
Bradysia forcipulata (Lundbeck, 1898)*
1 additional species recognized.

Camptochaeta **Hippa & Vilkamaa**
5 species recognized.

Lycoriella **Frey**
Lycoriella **(*Lycoriella*)** ***attenuata*** (Rübsaamen, 1898)*
Lycoriella **(*Hemineurina*)** ***cochleata*** (Rübsaamen, 1898)*
Lycoriella **(*Hemineurina*)** ***modesta*** (Staeger, 1840)
Lycoriella **(*Hemineurina*)** ***vitticollis*** (Holmgren, 1883)
2 additional species of *Lycoriella* (*Hemineurina*) recognized.

Scatopsciara **Frey**
Scatopsciara **(*Scatopsciara*)** ***morionella*** (Holmgren, 1883)

Schwenckfeldina **Frey**
Schwenckfeldina tridentata (Rübsaamen, 1898)

Addendum

The present account, written more than a decade ago, still represents the state-of-the-art as far as published information about Greenland sciarids is concerned. However, resumed and still uncompleted revisionary work on the fauna has revealed more species than recorded in the present chapter. Hence an expansion of the information presented here will expectedly be published within the next few years.

References

Edwards, F.W. (1922): Results of the Oxford University Expedition to Spitzbergen. 1921. No 14. Diptera Nematocera.– *The Annals and Magazine of Natural History* ser. 9 (10): 193–215.

Edwards, F.W. (1923): On the Nematocerous Diptera of Jan Mayen Island. – *The Annals and Magazine of Natural History* ser. 9 (11): 235–240.

Edwards, F.W. (1924): Results of the Merton College Expedition to Spitzbergen, 1923. No 4. Diptera Nematocera. – *The Annals and Magazine of Natural History* ser. 9(14): 162–174.

Edwards, F.W. (1935a): Diptera from Bear Island. – *The Annals and Magazine of Natural History* ser 15(10): 531–543.

Edwards, F.W. (1935b): Diptera Nematocera from East Greenland. – *The Annals and Magazine of Natural History* ser. 10(15): 467–473.

Hippa, H. & Vilkamaa, P. (1994): The genus *Camptochaeta* gen .n. (Diptera, Sciaridae). – *Acta Zoologica Fennica* 194: 1–85.

Holmgren, A.E. (1869): Bidrag till kännedomen om Beeren Eileands och Spetsbergens insekt-fauna. – *Kongliga Svenska Vetenskaps-Akademiens Handlingar* 8: 50–55.

Lengersdorf, F. (1930): III. Ordo: Diptera. Subordo: Orthorhapha. Gruppe: Nematocera, Oligoneura. 1. Fam. Sciaridae. *In*: Thor, S. (ed.). Beiträge zur Kenntniss der invertebraten Fauna von Svalbard. – *Skrifter om Svalbard og Ishavet* 27: 51–53.

Lengersdorf, F. (1935): *Neosciara delicata* nov. spec. – *Norsk Entomologisk Tidsskrift* 4: 75.

Malloch, J.R. (1923): Insects, arachnids and chilopods of the Pribilof Islands, Alaska. Diptera (Except Tipulidae, Rhyphidae and Calliphoridae). – *North American Fauna* 46: 170–227.

Menzel, F. & Mohrig, W. (1997): Family Sciaridae. *In*: Papp, P. & Darvas, B. (eds), *Contributions to a Manual of Palaearctic Diptera, Volume 2, Nematocera and Lower Brachycera*, pp. 51–69. Science Herald, Budapest.

Menzel, F. & Mohrig, W. (2000): Revision der paläarktischen Trauermücken (Diptera: Sciaridae). – *Studia dipterologica Supplement* 6 [1999]: 1–761

Rübsaamen, E.H. (1898): Grönländische Mycetophiliden, Sciariden, Cecidomyiden, Psylliden, Aphiden und Gallen. – *Bibliotheca Zoologica* 20(8): 103–119.

Steffan, W.A. (1966): A generic revision of the family Sciaridae (Diptera) of America north of Mexico. – *University of California Publications in Entomology* 44: 1–77.

Steffan, W.A. (1981): Sciaridae. *In*: McAlpine, J.F. *et al.* (eds), *Manual of Nearctic Diptera*. Vol. 1: 247–255. Research Branch Agriculture Canada. Monograph 27. Ottawa, Ontario.

Tuomikoski, R. (1967): Mycetophilidae and Sciaridae from Spitzbergen collected by Dr. J. Kaisila in 1965. – *Annales Entomologici Fennici* 33: 43–51.

Vilkamaa, P. & Hippa, H. (2005): Phylogeny of *Peyerimhoffia* Kieffer, with the revision of the species (Diptera: Sciaridae). – *Insect Systematics and Evolution* 35: 457–480.

17.9. Bibionidae & Scatopsidae

Jean-Paul Haenni

The two families treated here are not particularly closely related, but both belong to lineages that arose during the early evolution of the bibionomorph assemblage of nematoceran flies. Each of them is represented in Greenland by just a single species.

BIBIONIDAE

Small to large flies (length of body 4 to 15 mm), with robust setose body, swollen fore femora, fore tibiae armed with series of spines or strong apical spurs. Strong sexual dimorphism, males (Fig. 2) with large holoptic eyes, females with small, largely separated eyes. Antennae short with flagellomeres shortened, palpi 4-segmented. Wings with anterior veins generally darkened, pterostigma often well-marked, R_s unforked (in Bibioninae), posterior veins hyaline or slightly tinged. Larvae up to 25 mm long, subcylindrical, with a complete cephalic capsule, 12-segmented with characteristic rows of numerous processes and spines, holopneustic. Pupae elongate, slender. (For more details on morphology, see Skartveit, 1997.)

Larvae phytosaprophagous, feeding on decaying plant material, mainly in decomposing litter, sometimes in rotten wood. Larvae of some species may attack roots and stems of living plants and are occasionly considered as minor agricultural pests. Adults are diurnal, often slowly moving flies, that are more frequent and diverse in open biotopes, especially in spring. Male specimens often form large aerial swarms.

The Nearctic genera are keyed in Hardy (1981), and the Palaearctic genera in Skartveit (1997), both of which include references for identification of species of particular genera. Known larvae are keyed in Skartveit (2002). General European distribution may be found in Skartveit (2004).

The single Greenland species of Bibionidae is the Palaearctic *Dilophus femoratus* Meigen, 1804; it is here recorded for the first time.

***Dilophus femoratus* (Meigen, 1804) Figs 1, 2**

Members of the genus *Dilophus* are comparatively small-sized flies easily separable from other Bibionidae and Diptera by the presence of transverse rows of spines on the anterior tibiae (Fig. 1B) and on pronotum. This species may be easily recognized from most Palaearctic and Nearctic species by the milky white wings in both sexes, with anterior veins weakly tinged, posterior veins hardly visible, not contrasting at all with the membrane (Fig. 1A). Fore tibiae with 2 rows of spines, the median row transverse, oblique, consisting of 4 strong spines (Fig. 1B). Male genitalia very distinctive, with a deep square-shaped posterior incision on 9th sternite (Fig. 1C), and gonocoxites apically nearly right-angled, in the shape of a prehistoric axe (Fig. 1D). Female black in general colour, with fore coxae yellow-rufous and all femora at least partly yellow-rufous; head elongated in front of the eyes. (For more figures, see Haenni, 1982.)

BIOLOGY. A saprophagous species of which the larvae live communally in the upper layer of soil in various open, semi-open and forested biotopes, from sea-level to high mountains (up to 2000 m in central Europe). Larva and development described by Morris (1922). A common univoltine summer species, which may be occasionally very abundant.

DISTRIBUTION. S (only recorded from Narsarsuaq; 9♂♂ 5♀♀, labelled "*Dilophus?caurinus* McAtee", ZMUC.).

A widespread Palaearctic species, present in most parts of Europe including Iceland and Scandinavia, and in Northern Asia (Haenni, 1982; Skartveit, 2004).

NOTE. Several *Bibio* species have recently been proved to have a boreal Holarctic distribution (Fitzgerald & Skartveit, 1997) and further collections in suitable parts of Greenland may possibly produce a few other boreoalpine species of Bibionidae (*e.g. Dilophus borealis* Skartveit, *D. caurinus* McAtee, *Bibio brunnipes* (Fabricius), *B. lautaretensis* Villeneuve, *B. rufipes* (Zetterstedt), *B. siebkei* Mik).

SCATOPSIDAE

Minute to small (length of body 0.5 to 4.5 mm) but comparatively strongly built midges, generally dark brown to black, sometimes with fuscous or yellow markings. Body often somewhat elongated and parallel-sided. Eyes generally holoptic, forming a supraantennal bridge. Antennae comparatively short with shortened flagellomeres. Palpi one-segmented. Wing venation reduced, with posterior veins generally faint. Legs usually unmodified. Abdominal pleurites longitudinally ribbed. Larvae 3–7 mm long, 12-segmented, dorso-ventrally flattened, with a complete cephalic capsule, peripneustic, generally with a pair of prominent posterior spiracles. Known pupae obtect, enclosed in last instar larval skin. For more details on morphology, see Haenni (1997).

Known larvae are saprophagous, living in various media, mainly of vegetal, but also of animal origin. They have been found in all kinds of decaying vegetal material (rotten plants and fruits), in rotten wood, under decaying bark of

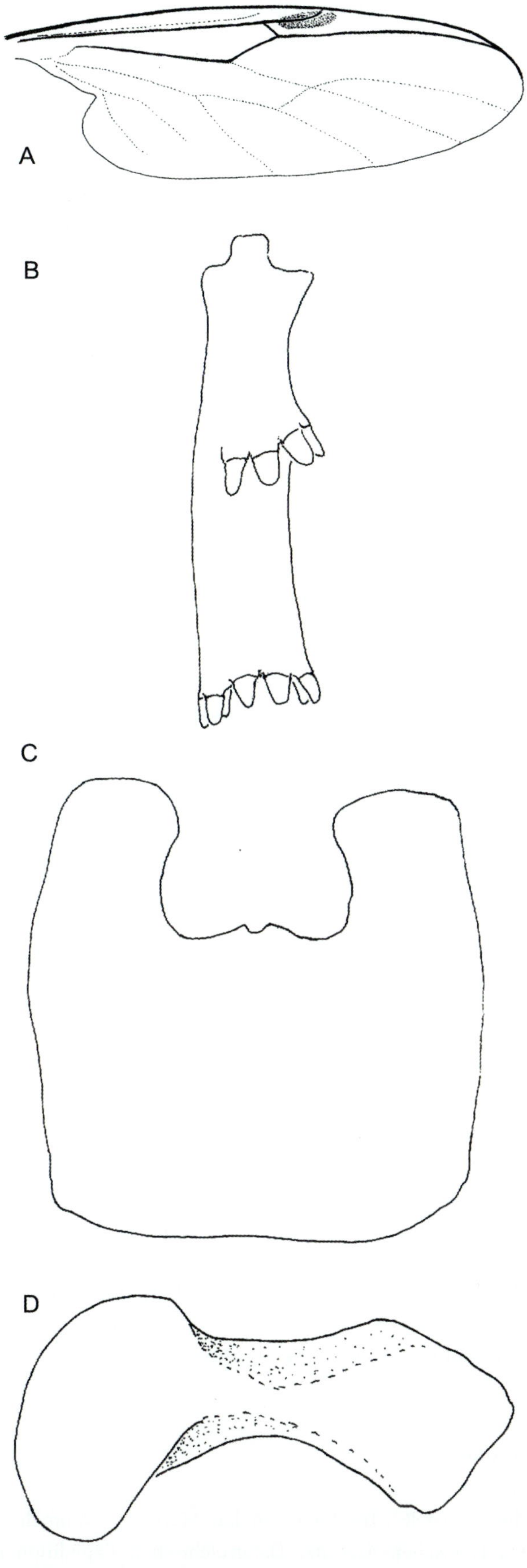

FIGURE 1 Dilophus femoratus (*Meig.*): *A, female wing; B, male fore tibia; C, male sternite 9, ventrally (diagrammatic); D, male gonocoxite.*

FIGURE 2 Dilophus femoratus (*Meig.*), *male habitus (not Greenland specimen).*

dead trees, in soil, leaf litter, dung, fungi, etc. For a review of immature stages and their media of development, see Haenni & Vaillant (1994). Adults are diurnal, inconspicuous midges that may be found in nearly all biotopes (including anthropogenic ones for a few species). They are most frequent and diverse in open, fresh and marshy areas.

The Nearctic genera are keyed in Cook (1981), and the Palaearctic genera in Haenni (1997), both of which include references for identification of species of particular genera. General European distribution may be found in Haenni (2004).

Scatopse Linnaeus, 1758

Species of genus *Scatopse* may be separated from other Scatopsidae by the presence of a stump vein directed anteriorly on M_1 (Fig. 3A), together with wing membrane and posterior veins hyaline, and first segment of hind tarsi of male spinose, shorter than second segment (Fig. 3B).

Scatopse notata (Linnaeus, 1758) Fig. 3

The largest species of Scatopsidae in the northern hemisphere (up to 4.5 mm long), shiny black, with a pair of conspicuous yellow spots at the posterior corners of the dorsum, additional yellow markings often present on the pleurae, wings and posterior veins hyaline (Figs 3A, C). It can be separated from other species of genus *Scatopse* (*S. alpestris* Cook and *S. lapponica* Duda, which may possibly occur in Greenland) by the venation of wing: second costal section much longer than first and third section in *S. notata* (about as long or hardly longer in the other species); distal section of R_{4+5} with 3–5 rows of setulae on the underside (1 row of setulae in the other species). The shape of male and female genitalia (Cook, 1956) is also distinctive.

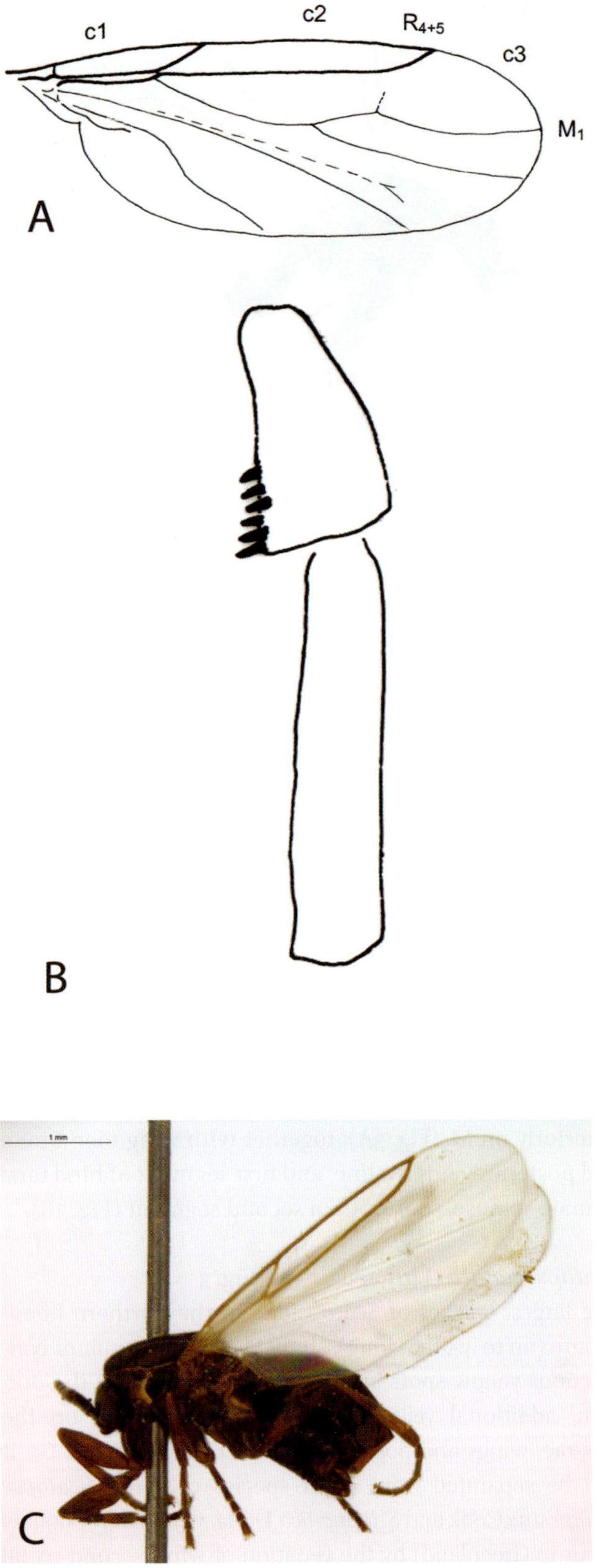

FIGURE 3 Scatopse notata (*L.*)*: A, wing* (*diagrammatic; c1, c2, c3: costal sections*)*; B, male left hind tarsus, segments 1 and 2* (*pilosity omitted*)*; C, male habitus*).

BIOLOGY. A polysaprophagous species developing in various decaying matters (Haenni & Vaillant, 1994), including compost and refuse heaps, human excrements and various dung. *Scatopse notata* is a strongly synanthropic species, but it occurs also under natural conditions in the Palaearctic region. Under favourable conditions the life-cycle may be achieved within 3–4 weeks and several generations may develop in a year.

DISTRIBUTION. S. Recorded as common at Qaqortoq/ Julianhåb and Igaliko by Henriksen & Lundbeck (1917) (28 specimens: Julianehaab, 25.VIII.[18]89, 4♂♂ 3♀♀, Lundbeck; same, 11.IX.[18]89, 4♂♂; same, 29.IV.1900, 1♀, G. Meldorf; Igaliko, 9.VI.1898, 1♂ 3♀♀, G. Meldorf; same, Juli 1898, 8♀♀, G. Meldorf [this is the material mentioned in Henriksen & Lundbeck (1917)]; Julianehaab, Eqaluit, 25. VIII.1951, 1♂ 1♀, Chr. Vibe; Upernaviarssuk, 28.VII-1. VIII.1987, 1♂ 1♀, P. Nielsen; all ZMUC).

Holarctic, including Iceland, Scandinavia, Siberia and Alaska; also caught on Jan Mayen Island at the end of the 19th century where it is the only Scatopsidae recorded so far (1♀ labelled: "5/88/278 Jan Mayen 1884", NHMW; this specimen had been wrongly identified and recorded as *Scatopse pulicaria* (Meigen) by Becher, 1886). This synanthropic species has become widespread all over the world by human activities and trade. This is most certainly also the case for Greenland.

NOTES. The presence of *S. notata* is most probably restricted to human settlements in Greenland, as it seems rather untrustworthy that it could develop under natural conditions. On the other hand, a few other boreal species of Scatopsidae (*e.g. Rhegmoclemina vaginata* (Lundstr.), *Scatopse alpestris* Cook, *S. lapponica* Duda, *Thripomorpha* spp.) may possibly occur in favourable areas of Greenland.

Acknowledgements

I am pleased to thank Thomas Pape and Jens J. Böcher, Natural History Museum of Denmark, who kindly arranged a loan of the material and helped with various advices, and Ruth Contreras-Lichtenberg, formerly at the Naturhistorisches Museum Wien, who made available the material of the Austrian expedition to Jan Mayen.

References

Becher, E. (1886): Insekten von Jan Mayen. Gesammelt von Dr. F. Fischer, Arzt der Österreichischen Expedition auf Jan Mayen. *In*: Kaiserliche Akademie der Wissenschaften (ed.), Die Internationale Polarforschung 1882–1883. Die

Österreichische Polarstation Jan Mayen ausgerüstet durch seine Excellenz Graf Hanns Wilczek, geleitet von k.k. Corvetten-Capitän Emil Edlen von Wohlgemuth. Beobachtungs-Ergebnisse. – *Österreichische Polarstation Jan Mayen (Zoologie)* 3(6): 59–66.

Cook, E.F. (1956): A contribution toward a monograph of Scatopsidae (Diptera). Part VI. The Genera *Scatopse* Geoffroy and *Holoplagia* Enderlein. – *Annals of the entomological Society of America* 49(6): 593–611.

Cook, E.F. (1981): Scatopsidae. *In*: McAlpine, J.F. *et al.* (eds.), *A Manual of Nearctic Diptera*. Volume 1: 313–319. Research Branch, Agriculture Canada, Monograph 27.

Fitzgerald, S. & Skartveit, J. (1997): Holarctic distributions in the genus *Bibio* (Diptera: Bibionidae) – *Entomologica scandinavica* 28: 103–119.

Haenni, J.-P. (1982): Révision des espèces européennes du groupe de *Dilophus febrilis* (L.), avec description d'une espèce nouvelle (Diptera, Bibionidae). – *Revue suisse de Zoologie* 89(2): 337–354.

Haenni, J.-P. (1997): Family Scatopsidae. *In*: Papp, L. & Darvas, B. (eds.), *Contributions to a Manual of Palaearctic Diptera. Volume 2. Nematocera and Lower Brachycera*; 255–272. Science Herald, Budapest.

Haenni, J.-P. & Vaillant, F. (1994): Description of dendrolimnobiontic larvae of Scatopsidae (Diptera) with a review of our knowledge of the preimaginal stages of the family. – *Mitteilungen der Schweizerischen entomologischen Gesellschaft* 67(1–2): 43–59.

Haenni, J.-P. (2004): Fauna Europaea: Insecta, Diptera, Scatopsidae. *In*: De Jong, H. (ed.), *Fauna Europaea: Diptera "Nematocera"*. Fauna Europaea version 1.2 (7 March 2005), http://www.faunaeur.org.

Hardy, D.E. (1981): Bibionidae. *In*: McAlpine, J.F. *et al.* (eds.) *A Manual of Nearctic Diptera*. Volume 1: 217–222. Research Branch, Agriculture Canada, Monograph 27.

Henriksen, K.L. (1939): A revised index of the Insects of Grønland. – *Meddelelser om Grønland* 119(10): 1–112.

Henriksen, K.L. & Lundbeck, W. (1917): Grønlands Landarthropoder (Insecta et Arachnida). – *Meddelelser om Grønland* 22: 481–821.

Morris, H.M. (1922): The larval and pupal stages of the Bibionidae. Part II. – *Bulletin of entomological Research* 13: 189–195.

Skartveit, J. (1997): Family Bibionidae. *In*: Papp, L. & Darvas, B. (eds.), *Contributions to a Manual of Palaearctic Diptera. Volume 2. Nematocera and Lower Brachycera*; 41–50. Science Herald, Budapest.

Skartveit, J. (2002): The larvae of European Bibioninae (Diptera, Bibionidae). – *Journal of Natural History* 36: 449–485.

Skartveit, J. (2004): Fauna Europaea: Insecta, Diptera, Bibionidae. *In*: De Jong, H. (ed.), *Fauna Europaea: Diptera "Nematocera"*. Fauna Europaea version 1.2 (7 March 2005), http://www.faunaeur.org.

17.10. Mycetophilidae, Bolitophilidae & Keroplatidae (Fungus gnats)

Geir Söli, Eirik Rindal and Øivind Gammelmo

Small to medium sized, slender to moderately robust gnats. Many species are about the size of mosquitoes, and most of them have simple antennae and humped thorax. A characteristic feature is their distinctly elongated coxae. Strong bristles are usually present on thorax and tibia. Also characteristic are the presence of 3, more rarely 2 ocelli and compound eyes that do not meet above antennal sockets. They have very small, or entirely missing, tarsal pads (puvilli), but normally well developed tibial spurs and the abdomen is narrowly inserted on thorax, leaving most of the mediotergite free. The terminology follows that of Söli (1997) (see also Figs 1, 3).

Larvae of many genera and numerous species are still unknown, but usually they are cylindrical and slender, with a strongly sclerotized head capsule. The pupa can be distinguished from other superfamilies by having the combination of distinctly visible spurs and leg sheaths side by side rather than overlapping.

Mycetophilids are most abundant in humid areas, especially moist woodland. During the day adults of many species congregate in moist dark places such as overhanging stream banks and cavities under tree roots. Some species visit flowers, among them those with elongated mouthparts (*e.g. Gnoriste*). Larvae of many species live in fleshy or woody fungi; both polyphagy and oligophagy have been reported. Most groups of fungi, including Myxomycetes, have been reported to be infested by fungus gnats. Larvae of several species have been found in dead wood, under bark, or in nests of birds and squirrels; most or all of the larvae are probably mycetophagous (for references see Söli *et al.* 2000). Larvae of some Keroplatidae spin webs and capture and feed on small arthropods.

In cool temperate regions, adults of several species hibernate under bark, in hollow plants and in caves etc. during periods of drought and cold.

The Greenland fauna is fairly well documented compared to that in the Nearctic region. Lundbeck (1898) reported 11 species and more species were later added by Henriksen & Lundbeck (1917), Edwards (1931, 1935), Carpenter (1937, 1938), and Henriksen (1939). Up to present, 17 species have been recorded from Greenland. Of these, one belongs to the family Bolitophilidae, the remaining to the family Mycetophilidae. The present review adds 18 species, one Keroplatidae and 17 Mycetophilidae; of which 8 have previously not been recorded from the Nearctic region. Four species previously reported from Greenland were not found in the studied material. The identification of some of these species, however, are questioned.

The Sciaroidea are generally considered a monophyletic group. There is, however, great divergence in the ranking of sub-groups by different authors (for a review, see Söli *et al.* 2000). A preliminary phylogeny of the Mycetophilidae is presented in Söli (1997).

Fungus gnats occur in all continental areas except Antarctica, from northern Greenland south to Tierra del Fuego, and on many oceanic islands. About 4.500 species (Sciaridae excl.) have been described, more than 1.450 of them from the Palaearctic, but numerous more species await description.

The present compilation was completed in 2006. Recent studies of Mycetophilidae in northern Fennoscandia, Russia and Svalbard, which also involves DNA barcoding, have provided new knowledge which unfortunately was not possible to include here.

Key to families and genera of Sciaroidea in Greenland

1. Both median and posterior fork nearly y-shaped and with distinct stem (Fig. 1A) 3
– Median fork y-shaped, posterior fork not so (Figs 1B, C) 2
2. Stem of median fork and R fused for a short distance. Wings clear; distinct posterior angle (Fig. 1C). Antennae very long, longer than wing .... (Keroplatidae) *Macrocera*
– Stem of median fork and R connected by distinct crossvein ta. Wing with distinct dark spot where R1 meets costa; posterior angle rounded (Fig. 1D). Antennae distinctly shorter than wing (Bolitophilidae) .... *Bolitophila*
3. Eye bridge present. Stem and fork of M subequal in length, fork distinctly bell-shaped .... Sciaridae (17.8)
– Eyes not meeting above antennae. Wings with stem of M much shorter than fork, fork lanceolate rather than bell-shaped .... Mycetophilidae, 4
4. Wing membrane either with microtrichia irregularly arranged or bare; macrotrichia present or absent. Ocelli variable in position, often far from eye margins .... Subfamily Sciophilinae, 5

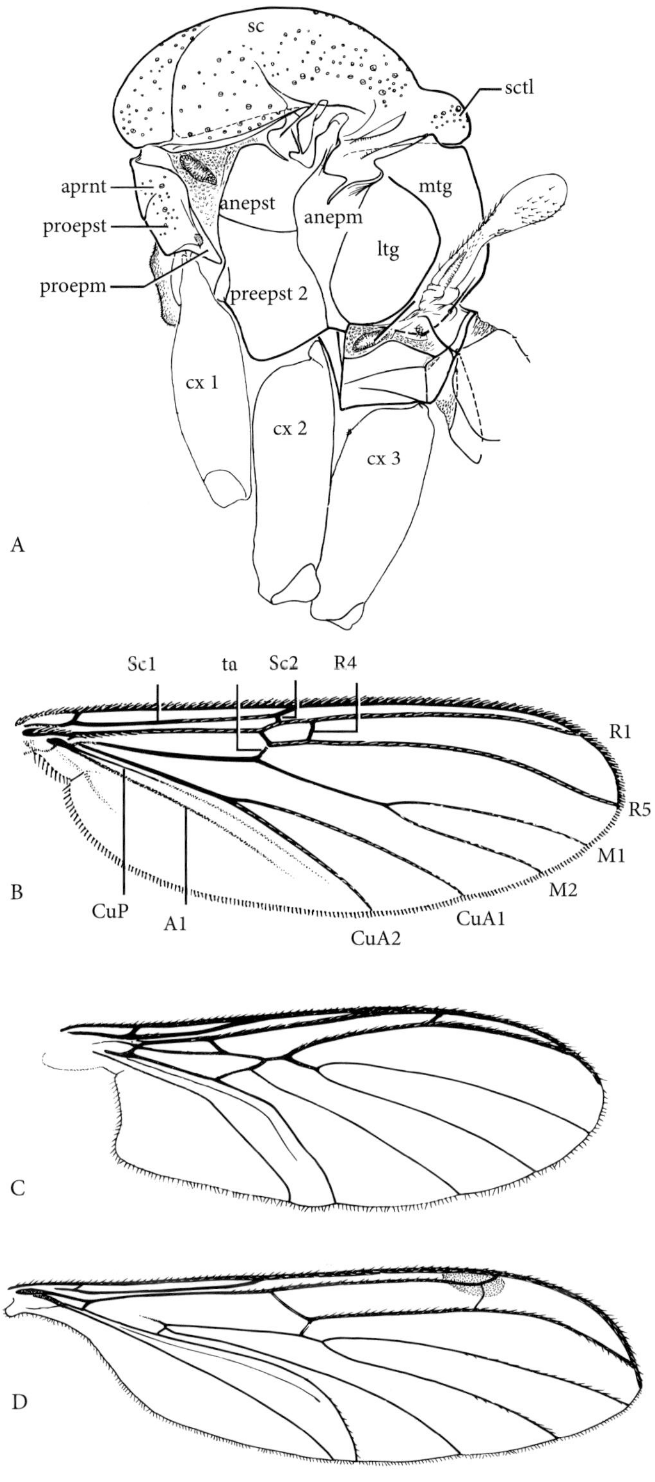

FIGURE 1
Thorax of (A) Boletina*; wings of (B)* Mycomya*, (C)* Macrocera *and (D)* Bolitophila*. A1, anal vein; anepm, anepimeron; anepst, anepisternum; aprnt, antepronotum; CuA1 and CuA2, anterior branches of cubitus; CuP, posterior branch of cubitus; cx, coxa; ltg, laterotergite; M1 and M2, branches of media; mtg, mediotergite; preepsst, preepisternum; proepm, proepimeron; proepst, proepisternum; R1, anterior branch of radius; R4 and R5, posterior branches of radius; Sc, subcosta; sc, scutum; sctl, scutellum; ta, anterior transversal.*

– Microtrichia always present and, especially near wing margin, arranged in more or less regular longitudinal lines; macrotrichia absent. Lateral ocelli close to eye margin .... Subfamily Mycetophilinae, 9

5. Fine tibial setae arranged in regular longitudinal rows .... *Mycomya*

– Tibial setae irregularly arranged.....................6

6. Wing membrane with many distinct macrotrichia in addition to microtrichia. Point of furcation of CuA beyond point of furcation of M............ *Sciophila*

– Wing membrane without macrotrichia; microtrichia present. Point of furcation of CuA before or opposite level of point of furcation of M.....................7

7. Sc1 ending free; apical spur of fore tibia, and longer spur of mid tibia, not longer than apical tibial diameter .... *Speolepta*

– Sc1 ending in C; apical spurs all well developed and distinctly longer than apical tibial diameter 8

8. Mouthparts forming a long slender proboscis several times as long as height of head...... *Gnoriste*

FIGURE 2 Mycetophila fungorum, *habitus (not Greenland specimen).*

– Mouthparts not elongated, shorter than height of head .... *Boletina*
9. Anepisternum bare or with short fine setae........10
– Anepisternum with strong bristles at least near upper margin.......................................15
10. Point of furcation of CuA beyond level of point of furcation of M...................................... 11
– Point of furcation of CuA before or opposite level of point of furcation of M 12
11. R5 nearly straight .... *Exechia*
– R5 curved towards costa at apex.........*Exechiopsis*
12. A1 strong, extending beyond point of furcation of CuA...13
– A1 short, not extending beyond point of furcation of CuA...14
13. Anepisternum with short setae............*Tarnania*
– Anepisternum bare ....*Rymosia*
14. Hind tibia with one or more short fine posterior bristles on apical third. Scutum with subappressed bristles on most of disc .... *Brevicornu*
– Hind tibia without posterior bristles. Scutum with distinct bristles on anterior half only.........*Allodia*
15. Anepimeron with small and large setae; all tibia with strong, well developed bristles..........*Mycetophila*
– Anepimeron bare or with small and weak setae; tibia with less developed bristles.........................16
16. Point of furcation of CuA well beyond point of furcation of M; Sc1 ending free .... *Phronia*
– Point of furcation of CuA before point of furcation of M; Sc1 usually ending in R .... *Trichonta*

FAMILY KEROPLATIDAE

A large and diverse family with more than 800 described species. Unlike most other species of Sciaroidea, being fungivorous or scavangers, the larvae of numerous Keroplatidae are predacious. The family has a worldwide distribution with most representatives in the tropics. *Macrocera* is the only genus recorded from Greenland.

***Macrocera* Meigen, 1803 Fig. 1C**

Species of *Macrocera* are found in all zoogeographic regions and easily recognized by their long antennae. Males can be identified on details in their genitalia, while females at present can not be satisfactorily identified. The Holarctic species have been treated by *e.g.* Chandler (1990) and Zaitzev (1994). Seemingly uncommon in Greenland and the genus has previously not been recorded here. Unfortunately, only females were present in the studied material.

DISTRIBUTION. S, SW (north to Nuup Kangerlua/ Godthåbsfjord), SE (only Skjoldungen).

FAMILY BOLITOPHILIDAE

The family includes about 60 species in one single genus, *Bolitophila*. The species are all slender, medium-sized, delicate gnats with rather long and narrow wings with a typical and quite constant venation. Except for a newly described oriental species (Ševčík & Papp 2004), the family is confined to the Holarctic region, including the Mediterranean part of North Africa.

***Bolitophila* Meigen, 1818**

The genus is divided into two subgenera, *Bolitophila* s.s. and *Cliopisa* Enderlein of which the species recorded from Greenland belongs to the former.

***Bolitophila cinerea* Meigen, 1818 Figs 1D, 3A**

The species is best identified on the male gonostylus ending in a strongly sclerotized, flattend rim.

BIOLOGY. Larvae have been bred from a wide variety of fungi, most commonly *Hypholoma* (Strophariaceae) and *Pholiota* (Cortinariaceae) (Hutson *et al.* 1980).

DISTRIBUTION. SW, N (Etah, Peary Land: Kap København). The species, which appears rare in

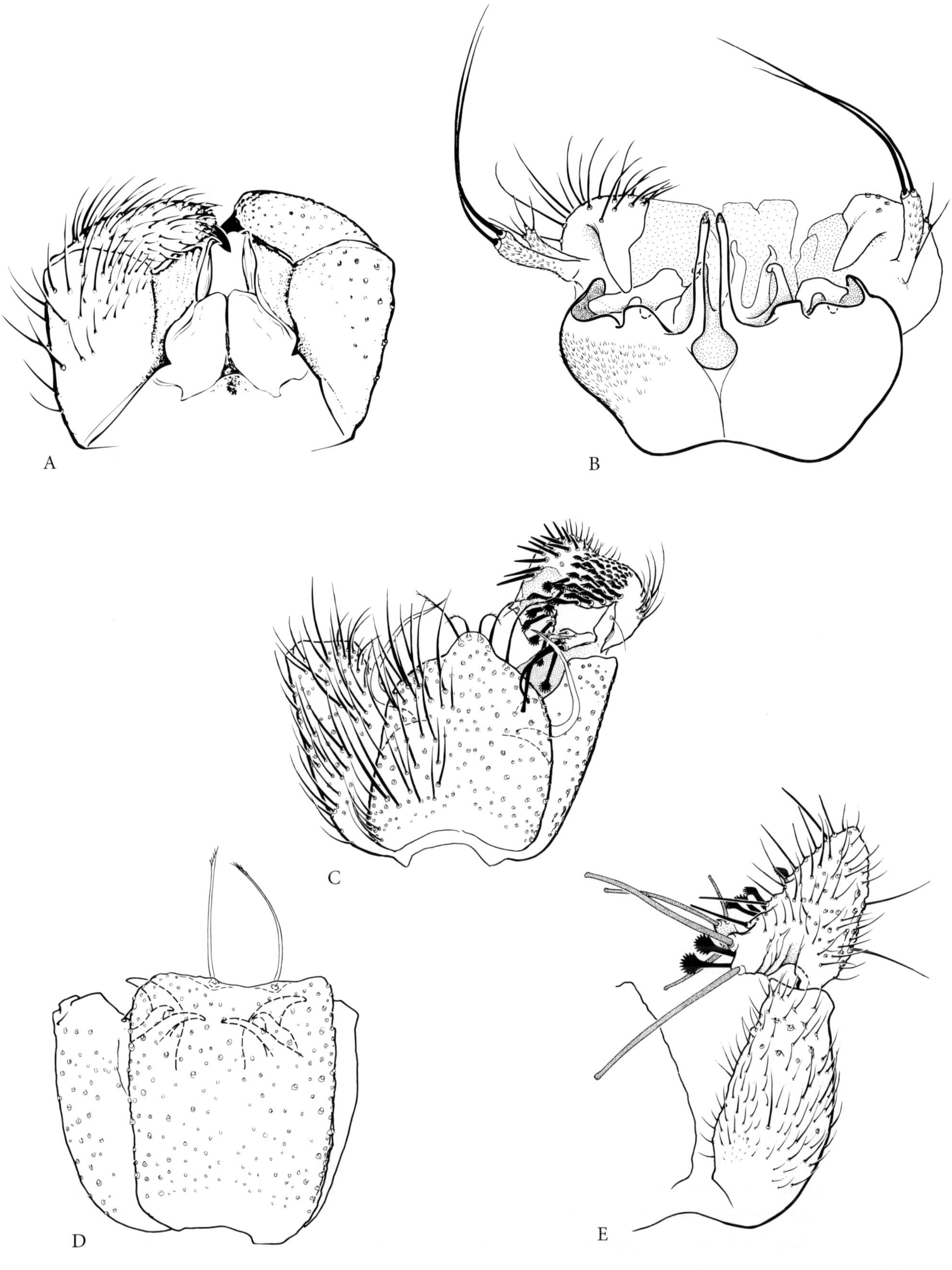

FIGURE 3 *Male genitalia of (A)* Bolitophila cinerea *Meigen, dorsal view, tergite 9 and proctiger removed; (B)* Mycomya islandica *Väisänen, ventral view; (C)* Sciophila festiva *Zaitzev, dorsal view; and (D–E) unidentified species in the* Sciophila lutea-*group, (D) gonocoxite and gonostylus, ventral view, (E) Tergite 9 and gonocoxites, dorsal view.*

Greenland, was recorded for the first time from Greenland (Etah) by Carpenter (1937).

Holarctic, arctic-temperate (common and widespread in Europe, but not recorded from Iceland and Svalbard.)

FAMILY MYCETOPHILIDAE

By far the largest family of Sciaroidea, with about 3,000 species in 150 genera. Adult mycetophilids are common in damp and shady places in shrubs and forests or along stream banks in open country, but several species may also be found in wetlands, heathland and in montane regions. Larvae of mycetophilids are mainly bound to the sporophores of fleshy Basidiomycetes (especially Agaricaceae, Amanitaceae, Boletaceae, and Russulaceae) inside which they feed on spores and hyphae. Some species can be bred from Polyporaceae (bracket fungi) and rarely from Ascomycetes and Myxomycetes. Some are polyphagous, while the majority is restricted to particular genera or families of fungi.

Mycomya Rondani, 1856

One of the largest genera of fungus gnats, with more than 180 Holarctic species. A thorough revision of these is found in Väisänen (1984).

Larvae of *Mycomya* spin slimy webs and have often been reported from rotten wood or under bark, probably associated with mycelium of wood-rotting fungi. Our knowledge about the biology of numerous species is still insufficient.

Mycomya islandica Väisänen, 1984 Fig. 3B

Henriksen 1939: *Mycomyia tenuis* Walker, 1856 (misidentifcation).

The species belongs to the *tenuis*-group in which the males have a very characteristic outlining of tergite 9, with distinct lateral appendages bearing two very long, highly conspicuous dark setae.

DISTRIBUTION. NE (Scoresby Sund: Gåseland, Lake Fjord; Jameson Land: Kap Stewart). Seems to be very rare in Greenland (Väisänen 1984).

NOTE. The two specimens from Scoresby Sund, a male and a female, were labelled "*Sciophila ?apicalis*", and published as so by Henriksen & Lundbeck (1917: 565). *S. apicalis* (Winnertz, 1863) is a junior synonym of *Mycomya tenuis* (Walker, 1856). The same specimens were later published as belonging to *M. tenuis* by Edwards (1935) (see also Henriksen 1939).

Holarctic, arctic. (Northernmost Europe, including Russia; Canada: Yukon, Manitoba).

Sciophila Meigen, 1818

Wing surface covered with both micro- and macrotrichia or macrotrichia only. At present most species may only be satisfactorily identified by characters in the highly complex male genitalia. Zaitzev (1982) revised the Holarctic species, but numerous species have been described later, rendering the systematics even more complicated. Hence, the genus is today in need of a thorough revision.

Sciophila festiva Zaitzev, 1982 Fig. 3C

Henriksen 1939: not recorded.

DISTRIBUTION. NE (Mestersvig). The species has previously not been recorded from Greenland.

Nearctic, arctic (Canada: NW Territories.)

Sciophila sp. (*lutea*-group) Figs 3D, E

The outlining of the genitalia indicates that this species belongs to the *lutea*-group. The distal, dorsomedian part of the gonocoxite has 3 distinct hooks, 1 large, pointing outwards, and 2 smaller, pointing inwards (Fig. 3D). The species seemingly can not be ascribed to any known species, and will be described elsewhere. A description of the species demands for a review of available Nearctic material and is beyond the scope of this study.

DISTRIBUTION. SW (Disko Bugt: Qeqertasussuk; Uummannaq Fjord: Tasiusaq).

Species of the *lutea*-group have been recorded from both Nearctic and Palaearctic regions.

Sciophila cf. *hirta* Meigen, 1818

COMMENT. Listed from West Greenland by Henriksen (1939), referring to Henriksen & Lundbeck (1917: 565): *Lasiosoma hirtum*. The studied specimen had its genitalia partly destroyed, and a conclusive identification could thus not be made.

DISTRIBUTION. N (only found in Peary Land: Kap København).

Holarctic, arctic-temperate.

Speolepta Edwards, 1925

In the Holarctic *Speolepta* all stages are regular inhabitants of caves, but adults may occasionally be found outside caves. The single male in the material probably belongs to an undescribed Nearctic species previously mentioned from North America (*e.g.* Laffoon 1965; Vockeroth 1981). Unfortunately, the specimen is in a very poor condition.

DISTRIBUTION. SW (only Disko Bugt: Qeqertasussuk: 68°35'N 51°05'W).

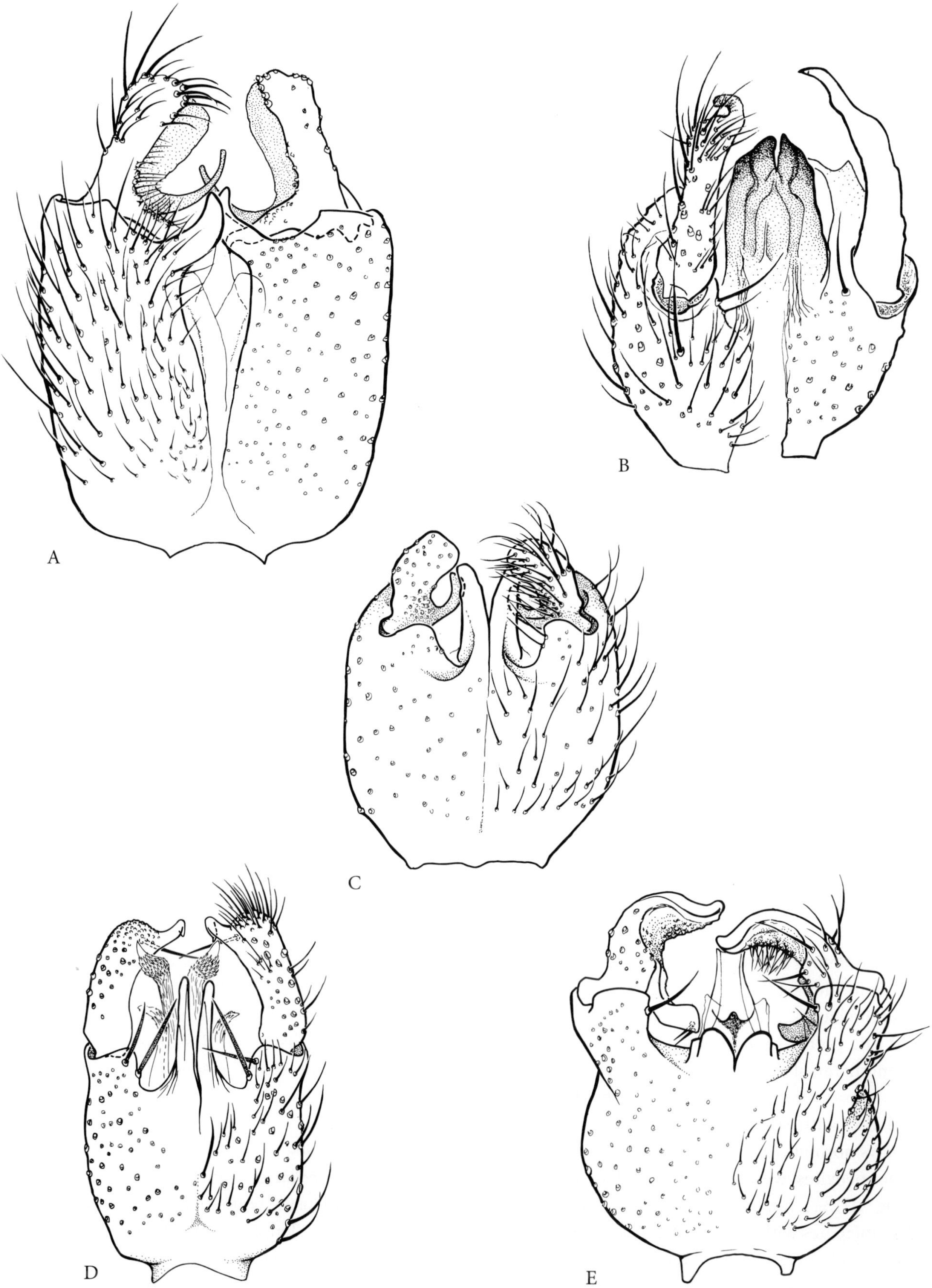

FIGURE 4 *Male genitalia, ventral view, of (A)* Gnoriste longirostris *Siebke, (B)* Boletina maculata *Holmgren, (C)* Boletina groenlandica *Staeger, (D)* Boletina digitata *Lundström, and (E)* Boletina arctica *Holmgren.*

***Gnoriste* Meigen, 1818**

All species have prolonged mouthparts, in most species reaching to the tip of the fore coxae. The biology is poorly known, but some species have been reported to visit flowers (Olesen & Warncke 1989; Okuyama *et al.* 2004).

***Gnoriste longirostris* Siebke, 1863 Fig. 4A**

Henriksen 1939: *Gnoriste groenlandica* Lundbeck, 1898

COMMENT. Lundbeck (1898) described *Gnoriste groenlandica* based on material from"Tunugdliarfik-fjord" and from "Tasersuak paa Nugsuak-Halvøen". The species has later been recorded from both East and West Greenland (Henriksen & Lundbeck 1917; Edwards 1935; Henriksen 1939).

BIOLOGY. Apparently most common in montane areas (Chandler 2004), but in Norway also recorded from lowland forest. Biology unknown, but the species may be associated with bryophytes (Falk & Chandler 2005).

DISTRIBUTION. S, SW (north to the Nuussuaq peninsula: Saqqaq Tasersua), NW (only Scoresby Sund: Lake Fjord).

Holarctic (Scotland, Norway; North. America: Connecticut).

***Boletina* Staeger, 1840**

A huge and very homogenous genus with more than 100 described species. Most species have been described from North America and Eurasia. Despite being among the most common mycetophilids in a wide variety of biotopes, a comprehensive revision of the genus is still wanting.

Breeding biology of most species is poorly known, but larvae of *Boletina* have been reported from rotting wood to mosses and liverworts (Hutson *et al.* 1980). Adults visit flowers (Okuyama *et al.* 2004).

Key to males of *Boletina* in Greenland

1. Laterotergite with setae .... *maculata*
– Laterotergite bare 2
2. Cu1 and Cu2 both with numerous setae along their entire length; tarsal claws distinctly club-shaped.... .... *groenlandica*
– Cu1 and Cu2 with setae on their apical half only; tarsal claws normal 3
3. Fused gonocoxites with two long median protrusions (Fig. 4D) .... *digitata*
– Fused gonocoxites with two, blunt and short median protrusions (Fig. 14) .... *arctica*

***Boletina maculata* Holmgren, 1870 Fig. 4B**

Henriksen 1939: not recorded.

DISTRIBUTION. SW (only Ilulissat/Jakobshavn). Not previously recorded from Greenland, nor from the Nearctic region.

Palaearctic, arctic-temperate. (Europe east to northern Russia).

***Boletina groenlandica* Staeger, 1845 Fig. 4C**

BIOLOGY. In montane tundra, usually above tree line (Falk & Chandler 2005).

DISTRIBUTION. S, SW (north to Qeqertarsuup Tunua/ Disko Bugt: Saqqaq), SE (Ammassalik, Saqqisikuik/ Skjoldungen).

Circumpolar, arctic-temperate.

***Boletina digitata* Lundström, 1914 Fig. 4D**

Henriksen 1939: not recorded.

DISTRIBUTION. SW, NW (north to Upernavik). Not previously recorded from Greenland, nor from the Nearctic region

Holarctic, arctic-subtropical (widespread in Europe east to central Russia, also recorded from the Near East).

***Boletina arctica* Holmgren, 1872 Fig. 4E**

Henriksen 1939: not recorded.

DISTRIBUTION. S, SW, NW (north to Pituffik/Thule).

Circumpolar, arctic-boreal (alpine).

Other species

***Boletina sciarina* Staeger, 1840**

This species was listed by Henriksen (1939) from W Greenland (Utorqait in Amerdloq), referring to Carpenter (1938: 546 "*B.* sp. cf. *sciarina*"). Drawings of the male genitalia may be found in *e.g.* Hutson *et al.* (1980), Bei-Bienko (1969), or Landrock (1927, 1940). Moreover, as numerous new species closely related to *B. sciarina* have been described during the last decades, it seems likely that the listing of *B. sciarina* is based on a misidentification, and that the species should be omitted from the list.

***Exechia* Winnertz, 1863**

Several species of *Exechia* are known to hibernate in caves, others can be found inside hollow stems of *Phragmites* and umbelliferous plants. Jensen (2005) found that *Exechia frigida* in Greenland has the flight period when the temperature is between 1,5°C and 9,5°C, with a peak at 5,0°C.

At present the species can only be identified with certainty on characters in the male genitalia.

Key to males of *Exechia* in Greenland

1. Gonostyli elongate, more than 3 times as long as wide 2
– Gonostyli rounded, less than 3 times as long as wide 4
2. Gonocoxite with two distinct median protrusions ventrally, each with a strong, blunt megaseta; no median (Fig. 5A) .... *lundstroemi*
– Gonocoxites without median protrusion 3
3. Gonostylus with a minute incision distally, only visible under strong magnification (Fig. 5C) .... *frigida*
– Gonostylus distally distinctly and deeply bilobed (Figs 5D, F) .... *spinuligera*
4. Gonocoxite with a broad v-shaped incision ventrally; dorsalmost part of gonostylus with a row of dark setae (Figs 5G, H) .... *micans*
– Gonocoxites with two rounded lobes ventrally; dorsalmost part of gonostylus without row of dark setae (Figs 5I, J) .... *nitidicollis*

***Exechia lundstroemi* Landrock, 1923 Fig. 5A**
Henriksen 1939: not recorded.

DISTRIBUTION. S, SW (only found in Narsarsuaq and Kangerlussuaq/Søndre Strømfjord). Not previously recorded outside the Palaearctic region.

Holarctic, arctic-temperate (widespread in Europe, also recorded from eastern Palaearctis).

***Exechia frigida* (Boheman, 1865) Figs 5B, C**
Henriksen 1939: *Exechia frigida* Holmgr.

DISTRIBUTION. G (a very common species, found almost everywhere in Greenland).

Circumpolar, arctic-temperate.

***Exechia spinuligera* Lundström, 1912 Figs 5D–F**
Henriksen 1939: not recorded.

DISTRIBUTION. S, SW (north to Qeqertarsuup Tunua/Disko Bugt: Saqqaq), NE (Scoresby Sund: Gåseland; Jameson Land: Lollands Elv). Not previously recorded outside the Palaearctic region.

Holarctic, arctic-temperate (very common in Europe).

***Exechia micans* Laštovka & Matile, 1974 Figs 5G, H**
Henriksen 1939: not recorded.

DISTRIBUTION. S, SW (north to Kangerlussuaq/Søndre Strømfjord). Not previously recorded outside the Palaearctic region.

Holarctic, arctic-temperate.

***Exechia nitidicollis* Lundström, 1913 Figs 5I–K**
DISTRIBUTION. S, SW (north to Kangerlussuaq/Søndre Strømfjord), NE (Jameson Land: Muslingeelvdal, Lollands Elv).

Holarctic, arctic-temperate.

Other species

***Exechia bicincta* (Staeger, 1840)**
This species is listed by Henriksen (1939) from W Greenland, referring to Henriksen & Lundbeck (1917: "*Exechia interrupta*"). Figures of the male genitalia may be found in *e.g.* Zaitzev (1999).

DISTRIBUTION. SW (only 1 specimen from Qeqertarsuup Tunua/Disko Bugt: Sudostbugt).

Palaearctic, arctic-temperate (in Europe common and widespread, but not recorded from the northernmost areas; also from eastern Palaearctis and the Near East).

***Exechia fusca* (Meigen, 1804)**
The species is listed by Henriksen (1939) from both East and West Greenland, referring to Henriksen & Lundbeck (1917: "*Exechia fungorum*"). According to Henriksen (1939) "Edwards proposes that the Greenland specimens determined to be this species really belongs to the next species (*frigida* Holmgr.)". Figures of the male genitalia may be found in *e.g.* Zaitzev (1999).

COMMENT. We find it likely that the listing of *E. fusca* is based on a misidentification of *E. frigida* as noted by Henriksen (1939).

DISTRIBUTION. Holarctic, arctic-subtropical (recorded from most of Europe, including Iceland, also from eastern Palaearctis, the Near East and northern Africa.)

***Exechiopsis* Tuomikoski, 1966**
The genus has more than 70 species and is presently divided into the subgenera *Exechiopsis* and *Xenexechia* Tuomikoski, 1966. The genus is closely related to *Exechia* (Rindal & Söli 2006) and is also known to contain species that hibernate in caves. The genus is not previously recorded from Greenland.

Unidentified species (female)

DISTRIBUTION. SW (Qeqertarsuup Tunua/Disko Bugt: Qeqertasussuk).

***Brevicornu* Marshall, 1896**
The genus is widely distributed, with more than 40 species recorded from the Palaearctic region. The larvae are often

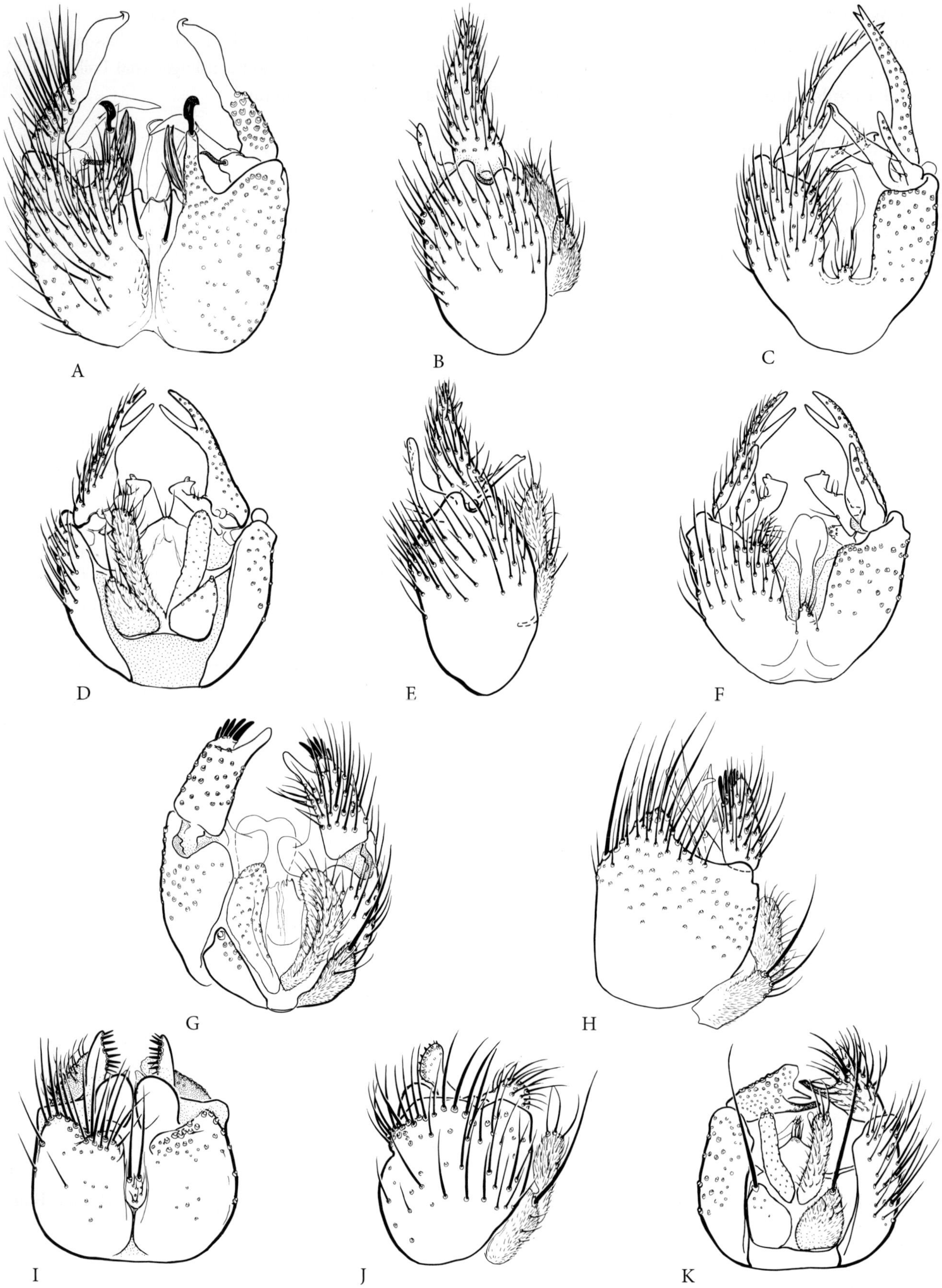

FIGURE 5 *Male genitalia of* (*A*) Exechia lundstroemi *Landrock, ventral view;* (*B–C*) Exechia frigida (*Boheman*), (*B*) *lateral, and* (*C*) *ventral view;* (*D–F*) Exechia spinuligera *Lundström,* (*D*) *dorsal,* (*E*) *lateral, and* (*F*) *ventral view;* (*G–H*) Exechia micans *Laštovka & Matile,* (*G*) *dorsal and* (*H*) *lateral view; and* (*I–K*) Exechia nitidicollis *Lundström,* (*I*) *dorsal,* (*J*) *lateral, and* (*K*) *ventral view.*

found on soft terrestrial fungi. The species are identified by characters in the male genitalia, but a comprehensive identification key does not exist.

***Brevicornu griseolum* (Zetterstedt, 1852) Fig. 6A**
Henriksen 1939: not recorded.

DISTRIBUTION. NE (Jameson Land: Langelands Elv). The species has previously not been recorded outside the Palaearctic region.

Palaearctic, arctic-temperate (in Europe east to central Russia; also recorded from the Near East).

***Brevicornu fuscipenne* (Staeger, 1840) Figs 6B–D**
Henriksen 1939: not recorded.

DISTRIBUTION. SW (Kangerlussuaq/Søndre Strømfjord; Disko Bugt: Qeqertasussuk), NE (Dove Bugt: Lindhard Ø; Jameson Land: Gurreholm Dal). Not previously recorded from Greenland.

Holarctic, arctic-temperate (widespread in Europe).

***Tarnania* Tuomikoski, 1966**
A small genus that has been reared from many species of terrestrial fungi in the families Tricholomataceae and Cortinariaceae. In England adults have been recorded from June to October in wooded fringes of heathland and woodland on acid soils (Falk & Chandler 2005).

***Tarnania tarnanii* (Dziedzicki, 1910) Figs 6E, F**
Henriksen 1939: not recorded.

DISTRIBUTION. S, SW (north to Qeqertarsuup Tunua/ Disko Bugt), SE (Saqqisikuik/Skjoldungen). Not previously recorded from Greenland.

Holarctic, arctic-temperate (widespread in Europe).

***Rymosia* Winnertz, 1863**
A relatively large genus with about 80 described species. One species has previously been recorded from Greenland, *R. lacki* (see below).

Only females were found in the material studied. Several specimens are present, and based on characters in their terminalia they must be ascribed to 2 species. Unfortunately, the genus has not been thoroughly revised, and it is thus not possible to identify species on female characters alone.

***Rymosia lacki* Edwards, 1935**
COMMENTS. The species was described from East Greenland: Constable Dal on Jameson Land. Drawings of male terminalia may be found in Edwards (1935).

DISTRIBUTION. NE. The species is only known from the type specimens.

***Mycetophila* Meigen, 1803**
This large genus with several species groups is found in all zoogeographical regions.

For a majority of species, a correct identification can only be done based on characters in the male genitalia. The genus is in need of thorough revision, and the delineation of several species is unsatisfactory. This holds in particular for the *ruficollis*-group to which two of the species recorded from Greenland belong, namely *M. ruficollis* and *M. evanida*. This problematic and complex group has been treated by Laštovka (1972). Awaiting further revisions and taking geographic variation into consideration, the number of species in the Greenlandic material must be regarded as tentative.

***Mycetophila fungorum* (De Geer, 1776) Figs 2, 6G, J**
DISTRIBUTION. S, SW (north to Qeqertarsuup Tunua/ Disko Bugt: Qegertasussuk), SE (Saqqisikuik/Skjoldungen).

Holarctic, arctic-temperate (recorded from most parts of Europe, but not from Iceland or Svalbard; also from eastern Palaearctis. The Nearctic distribution highly uncertain; the species is listed from the Nearctic by Laffoon (1965), but not so by Chandler (2004).

***Mycetophila evanida* Laštovka, 1972 Figs 6H, K**
Henriksen 1939: not recorded.

DISTRIBUTION. SW (only Nuup Kangerlua/Godthåbsfjord: Kapisillit). Not previously recorded outside the Palaearctic region.

Palaearctic, arctic-temperate (widespread in Europe).

***Mycetophila ruficollis* Meigen, 1818 Figs 6I, L**
Henriksen 1939: not recorded.

DISTRIBUTION. S (only Narsarsuaq). Not previously recorded outside the Palaearctic region.

Palaearctic, arctic-temperate (widespread in Europe).

***Phronia* Winnertz, 1863**
The genus contains more than 90 species worldwide. Adults are frequently collected along smaller streams. The larvae are flattened and slug-like. For most species the biology is poorly known, but in some cases larvae have been observed grazing on slime moulds and fungi growing on the surface of decaying logs. Jensen (2005) showed that *Phronia*, probably *P. egregia*, in Greenland has its flight period when the temperature is between 3,0°C and 5,0°C, with a peak at 4,0°C. (Jensen, however, misidentified *Phronia* as *Mycetophila*.)

In Greenland, species of *Phronia* are the only species in the subfamily Mycetophilinae with a short cubital fork. The Holarctic species of the genus have been revised by Gagné (1975).

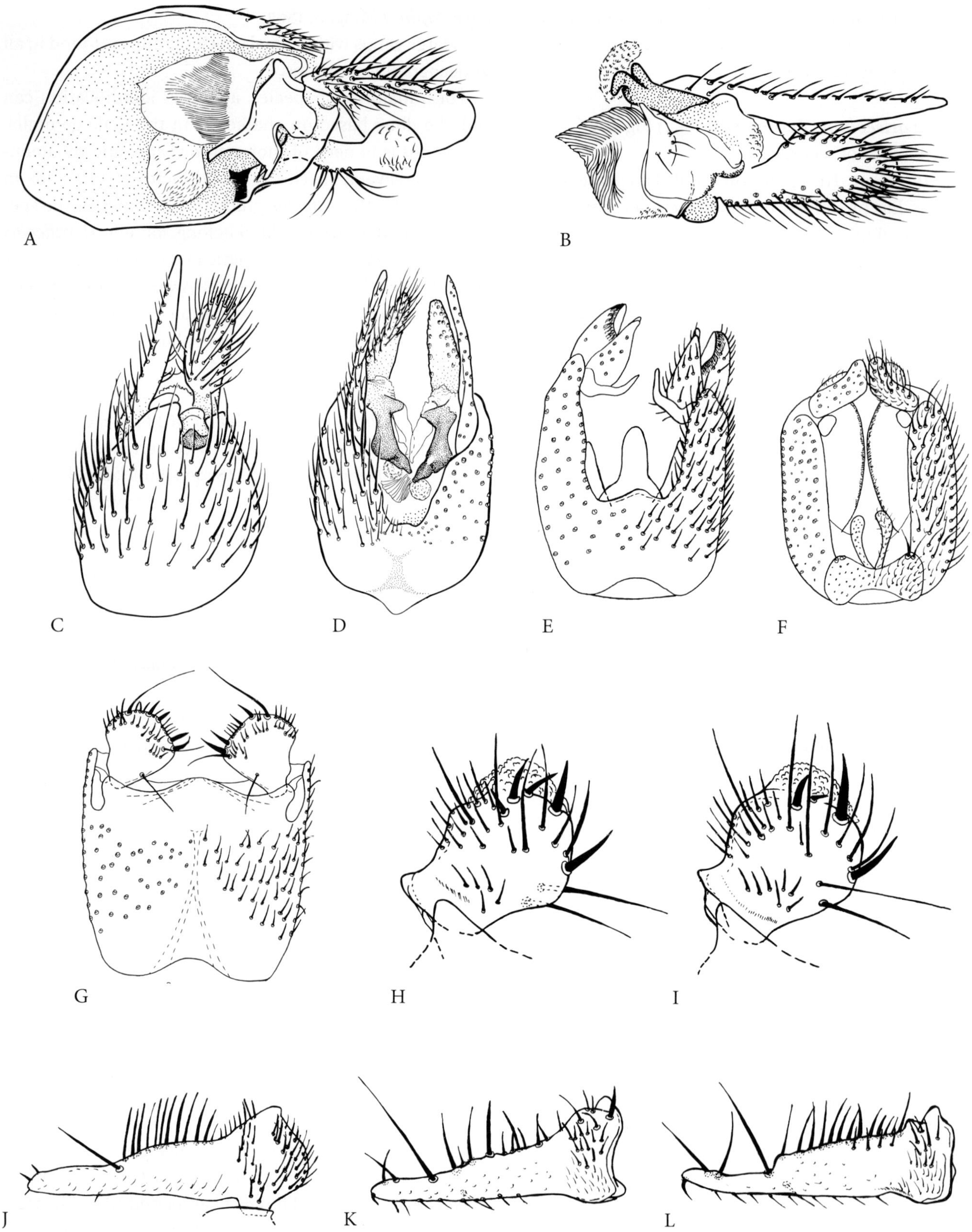

FIGURE 6 *Male genitalia of* (*A*) Brevicornu griseolum (*Zetterstedt*), *gonocoxite longitudinally divided, interior view;* (*B–D*) Brevicornu fuscipenne (*Staeger*), (*B*) *lateral, interior view of right gonostylus.* (*C*) *lateral and* (*D*) *ventral view of gonocoxites and gonostyli;* (*E–F*) Tarnania tarnanii (*Dziedzicki*), (*E*) *ventral and* (*F*) *dorsal view of gonocoxite and gonostyli;* (*G, J*) Mycetophila fungorum (*De Geer*), (*G*) *ventral view of gonocoxites and gonostyli,* (*J*) *gonostylus, lateral view;* (*H, K*) Mycetophila evanida *Laštovka,* (*H*) *ventral and* (*K*) *lateral view of gonostylus;* (*I, L*) Mycetophila ruficollis *Meigen,* (*I*) *ventral and* (*L*) *lateral view of gonostylus.*

Phronia braueri **Dziedzicki, 1889 Fig. 7A**
Henriksen 1939: not recorded.

Males may be identified on their large, curved gonostylus.

DISTRIBUTION. SW (only Nuup Kangerlua/Godthåbsfjord: Kapisillit). Not previously recorded from Greenland.

Circumpolar, arctic-temperate (widespread in Europe and North America north to Alaska. Also recorded from eastern Palaearctis and the Near East).

Phronia egregia **Dziedzicki, 1889 Fig. 7B**
Henriksen 1939: not recorded.

The median, ventral protrusion of the fused gonocoxites is a characteristic feature for the species, together with the pointed, more or less triangular gonostylus.

DISTRIBUTION. SW (Kangerlussuatsiaq/Evighedsfjorden, Kangerlussuaq/Søndre Strømfjord: Arnangarngup Kua), NE (Adolf S. Jensen Land: Påskedalen; Daniel Bruun Land; Lindhard Ø). Not previously reported from Greenland.

Circumpolar, arctic-temperate (widespread in Europe east to northern Russia; also recorded from eastern Palaearctis; several localities in northernmost North America, including Ellesmere Island and Baffin Island).

Phronia exigua **(Zetterstedt, 1852) Fig. 7C**
Males may be identified on their broad, triangular gonostyli.

DISTRIBUTION. S, SW, NW (north to Upernavik: Aappilattoq).

Circumpolar, arctic-subtropical (widespread in Europe, also eastern Palaearctis, the Near East, northern Africa, Madeira; scattered records from most parts of northern North America, including Alaska).

Trichonta **Winnertz, 1863**
A genus of about 70 species. The biology is poorly known, but larvae have been reared from various wood-encrusting fungi. The genus is closely related to *Phronia,* to which its members also have a general morphological resemblance. Species of *Trichonta* may, however, be identified on the longer cubital fork. Holarctic species of *Trichonta* have been revised by Gagné (1981).

Trichonta delicata **Gagné, 1981 Figs 7D, E**
Henriksen 1939: not recorded.

Males may be identified on the long and slender inner appendage of the gonostyli, pointing cephalad in resting position. The gonostylus also has a most characteristic inner appendage with a row of small, curved hooks.

DISTRIBUTION. S (only Narsarsuaq). Not previously recorded from Greenland.

Holarctic, arctic-temperate (seemingly with a rather restricted

DISTRIBUTION. Finland and Norway, USA: Montana, Canada: British Columbia, Ontario, Quebec).

Trichonta vulgaris **Loew, 1869 Figs 7F–H**
Henriksen 1939: not recorded.

Males with broad gonostyli; inner appendage small and rounded.

DISTRIBUTION. SW (only Kangerlussuatsiaq/Evighedsfjorden and Nuup Kangerlua/Godthåbsfjord: Kargiusap Nunaa).

Circumpolar, arctic-temperate (widespread in Europe east to northern Russia, also eastern Palaearctis and the Near East; scattered records from North America, including Alaska).

Other species

Lundbeck (1898) recorded a male of *Trichonta obesa* Winnertz, 1863 from West Greenland. Gagné (1981), however, included *T. obesa* as *incerta sedis* because the type specimen (a female) lacks abdomen. Gagné (1981) suggests that the species referred to by Lundbeck (1898) may have been *Trichonta venosa* (Staeger, 1840). Gagné's suggestion is based on the fact that Lundbeck in his identification relied on Mik (1880) who did not manage to separate *T. venosa* from *T. obtusa.*

Other genera recorded from Greenland

Allodia Winnertz, 1863
The genus contains more than 85 species with poorly known biology. Larvae of some species are associated with Ascomycetes, others are found in a variety of fungi. The genus is divided into two subgenera: *Allodia* s.s. and *Brachycampta.*

Allodia unicolor Lundbeck, 1898
Listed by Henriksen (1939) from SW (Nuussuaq peninsula: Tasersuaq), referring to Henriksen & Lundbeck (1917) as "*Brachycampta unicolor*".

COMMENT. The species was described based on one female from Tasersuaq (Lundbeck 1898); no more specimens of this species are known to the authors.

DISTRIBUTION. Only known from Greenland.

References

Bei-Bienko, G.Y. (ed.): (1988). *Keys to the insects of the European part of USSR.* Volume 5, Part I. Smithsonian Institution

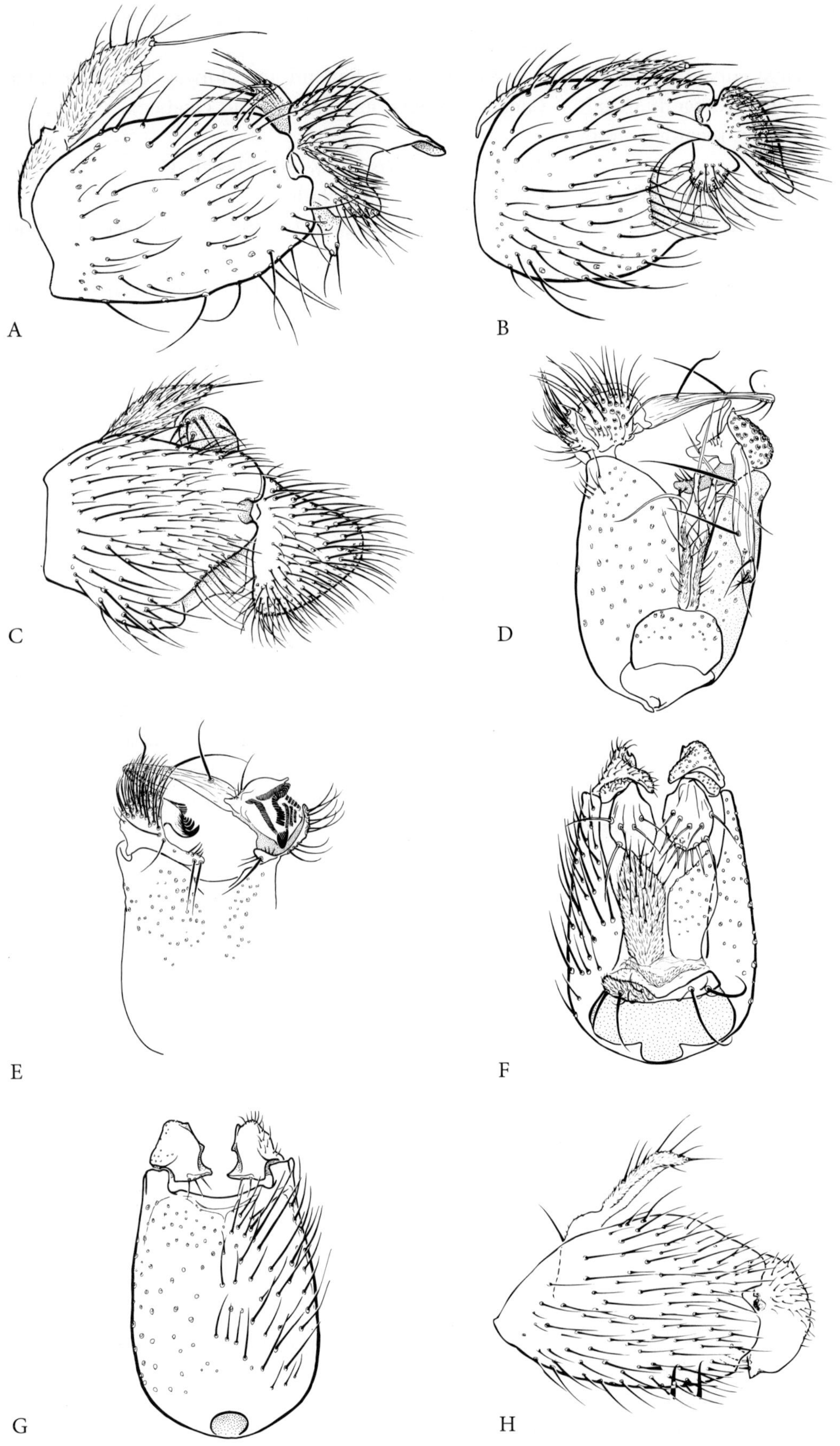

FIGURE 7 *Male genitalia of (A)* Phronia braueri *Dziedzicki, lateral view; (B)* Phronia egregia *Dziedzicki, lateral view; (C)* Phronia exigua *(Zetterstedt), lateral view; (D–E)* Trichonta delicate *Gagné, (D) semi-dorsal and (E) semi-ventral view; (F–H)* Trichonta vulgaris *Loew, (F) dorsal, (G) ventral and (H) lateral view.*

Libraries and The National Science Foundation. Washington D.C. 1234 pp. [Translation of the Russian edition.]

Carpenter, G.D.H. 1937. Notes on the insects collected in Northwest Greenland by the Oxford University Ellesmereland Expedition 1934–35. With description of a new Species of *Rhamphomyia* (Dipt.) by J.E. Collin. *Annals and Magazine of Natural History* Ser. 10, 20: 401–409.

Carpenter, G.D.H. 1938. Notes on the insects collected in West Greenland by the Oxford University Greenland Expedition 1936. With description of a new Species of *Angitia* (Hymenoptera, Ichneumonidae) by A. Roman, and of *Fannia* (Diptera, Anthomyidae) by J. E. Collin. *Annals and Magazine of Natural History* Ser. 11, 1: 529–553.

Chandler, J.P. 1990. Notes on *Macrocera* Meigen (Mycetophiloidea, Keroplatidae) including *M. nigropicea* new to Britain. *Dipterists Digest* 3: 27–31.

Chandler, J.P. 2004. Fauna Europaea: Mycetophilidae. *In*: de Jong, H. (ed.): Fauna Europaea: Diptera "Nematocera". Fauna Europaea version 1.1, http://www.faunaeur.org. (Accessed February 2007.)

Edwards, F.W. 1931. Oxford University Greenland Expedition, 1928. – Diptera, Nematocera. *Annals and Magazine of Natural History* Ser. 10, 8: 617–618.

Edwards, F.W. 1935. Diptera Nematocera from East Greenland. *Annals and Magazine of natural History* Ser. 10, 15: 467–473.

Falk, S.J. & Chandler, P. 2005. A review of the scarce and threatened flies of Great Britain, ISSN 1473–0154.

Gagné, R.J. 1975. A revision of the Nearctic species of the genus *Phronia* (Diptera: Mycetophilidae). *Transactions of the American Entomological Society* 101: 227–318.

Gagné, R.J. 1981. A monograph of *Trichonta* with a model for the distribution of Holarctic Mycetophilidae (Diptera). *Technical Bulletin, United States Department of Agriculture* 1638: 1–64.

Henriksen, K. 1939. A revised index of the insects of Grønland. *Meddelelser om Grønland* 119, 10: 1–112.

Henriksen, K. & W. Lundbeck. 1917. Grønlands Landarthropoder. *Meddelelser om Grønland* 22, 31: 481–821.

Hutson, A.M., Ackland, D.M. & Kidd, L.N. 1980. Mycetophilidae (Bolitophilinae, Ditomyiinae, Diadocidiinae, Keroplatinae, Sciophilinae and Manotinae). *Handbooks for the Identification of British Insects* 9 (3): 1–112.

Jensen, C.W. 2005. Undersøgelse over grønlandske svampemygs (Diptera: Mycetophilidae) flyveaktivitet i relation til klimatiske faktorer. Baseret på indsamlinger fra Zackenbergdalen, Nordøstgrønland. MSc-thesis, University of Roskilde, Institute of Biology and Chemistry; 58 pp. [in Danish].

Laffoon, J.L. 1965. Superfamily Mycetophiloidea, Family Mycetophilidae (Fungivoridae). *In*: Stone, A., Sabrosky, C.W., Wirth, W.W., Foote, R.H. & Coulson, J. (eds): *A catalogue of the Diptera of America north of Mexico.* United States Department of Agriculture, Agricultural Handbook 276: 196–229. Washington, D.C.

Landrock, K. 1927. Fungivoridae (Mycetophilidae). *In*: Lindner, E. (ed.): *Die Fliegen der Paläarktischen Region.* 2: 1–196 Schweizerbart, Stuttgart.

Landrock, K. 1940. Zweiflügler oder Diptera. VI: Pilzmücken oder Fungivoridae. *In*: Dahl, F. (ed.): *Die Tierwelt Deutschlands* 38: 1–166.

Lastovka, P. 1972. Holarctic species of *Mycetophila ruficollis*-group (Diptera, Mycetophilidae). *Acta Entomologica Bohemoslovaca* 69: 275–294.

Lundbeck, W. 1898. Diptera groenlandica. *Videnskabelige Meddelelser fra den Naturhistoriske Forening i Kjøbenhavn* 1898: 236–314.

Mik, J. 1880. Dipterologische Mittheilungen. *Verhandlungen der Zoologisch-Botanischen Gesellschaft in Wien* 30: 587–610, pl. XVII.

Olesen, J.M. & Warncke, E. 1989. Temporal changes in pollen flow and neighbourhood structure in a population of *Saxifraga hirculus* L. *Oecologia* 79: 205–211.

Okuyama, Y., Kato, M. & Murakami, N. 2004. Pollination by fungus gnats in four species of the genus *Mitella* (Saxifragaceae). *Botanical Journal of the Linnean Society* 144: 449–460.

Rindal, E. & Söli, G. 2006. Phylogeny of the subfamily Mycetophilinae (Diptera: Mycetophilidae). *Zootaxa* 1302: 43–59.

Ševčík, J. & Papp, L. 2004. Bolitophilidae (Diptera) from Taiwan: a family new to the Oriental region. *Acta Zoologica Academiae Scientiarum Hungaricae* 50 (1): 55–62.

Söli, G.E.E. 1997. On the morphology and phylogeny of Mycetophilidae, with a revision of *Coelosia* Winnertz (Diptera, Sciaroidea). *Entomologica scandinavica Supplement* 50: 1–140.

Söli, G.E.E., Vockeroth, J.R. & Matile, L. 2000. Families of Sciaroidea. *In*: Papp, L. & Darvas, L. (eds): *Contributions to a manual of Palaearctic Diptera. Appendix.* Pp. 49–93. Budapest, Science Herald.

Väisänen, R. 1984. A monograph of the genus *Mycomya* Rondani in the Holarctic region (Diptera, Mycetophilidae). *Acta Zoologica Fennica* 177: 1–346.

Vockeroth, J.R. 1981. Mycetophilidae, *In*: McAlpine *et al.* (eds): Manual of Nearctic Diptera. 1. Research Branch, Agriculture Canada, Ottawa. Monograph No. 27: 223–247.

Zaitzev, A.I. 1982. Fungus gnats of the genus *Sciophila* Meig. of the Holarctic. Akademia Nauk USSR. Moscow. 76 pp. [in Russian].

Zaitzev, A.I. 1994. *Fungus gnats of the fauna of Russia and adjacent regions. Part. 1. (Ditomyiidae, Bolitophilidae, Diadocidiidae, Keroplatidae, Mycetophilidae).* Moscow "Nauka", Moscow. 288 pp. [in Russian].

Zaitzev, A. 1999. Key to the insects of Russian Far East. Vol. VI. Diptera and Siphonaptera. Pt 1. Vladivostok. Dal'nauka. 665 pp. [in Russian].

17.11. Psychodidae

Bo W. Svensson

Small (1–5 mm) thickly haired nematocerans. Mainly grey or yellowish grey, but wings often with white, brown or black tufts of hairs, sometimes iridescent.

The family Psychodidae is one of the most taxonomically complex groups of nematocerans. It comprises about 500 species in Europe (Ježek & Barták 2000, Oosterbroek 2006), but some are probably still to be found and females of many species are undescribed. About 85 species are known from the Nordic countries.

Mandibles absent; adult flies non-piercing except for the bloodsucking mainly sub-tropical–tropical Phlebotominae (sand flies, on humans spreading the severe disease leishmaniasis) and some females in the subfamily Sycoraxinae. Ocelli absent; eyes reniform and eye bridge usually present, but in the largest and best known subfamily Psychodinae often incomplete. Antennae fairly long with (12) 13–16 segments (26–113 in Bruchomyiinae), flagellomeres barrel- or bottle-shaped. Maxillary palps long, usually four-segmented (outside Europe sometimes three- or five-segmented). Wing (Fig. 1A) hairy, usually more or less lancet-shaped, in a few taxa rounded, *e.g. Sycorax* and *Trichomyia* (both in Denmark and Sweden). Nine to ten longitudinal veins and R-fork and M-fork always present; crossveins absent or restricted to basal half of the wing. Scent organs are frequent in many males, i.e. 'cornicula' on head, and allurement organs on pro- and mesothorax (on the former sometimes referred to as 'patagia').

Moth flies have a characteristic short and erratic flight, and adults are mainly nocturnal. While resting wings are held roof-like or broadly V-shaped besides the abdomen. (Quate & Vockeroth 1981, Wagner 1998).

Larvae of Psychodinae inhabit a wide range of substrates, but mostly in moist or semiaquatic habitats, *e.g.* water-filled tree-holes, garbage, humus and leaf litter, excrements, not least of cattle (especially in *Psychoda* s.l.), lavatories and toilets, rotten vegetables, potatoes and fruit, mushrooms, and in sand and mud alongside water (Vaillant 1971).

Very short life cycles from 1–4 weeks have been reported in many Psychodini, which usually are bivoltine or polyvoltine in northern Europe (Wagner 1998).

Psychoda* (*Logima*) *satchelli **Quate, 1955 Fig. 1**
Antenna with flagellomeres bottle-shaped. One retinaculum. Labium with prominent finger-like apical process,

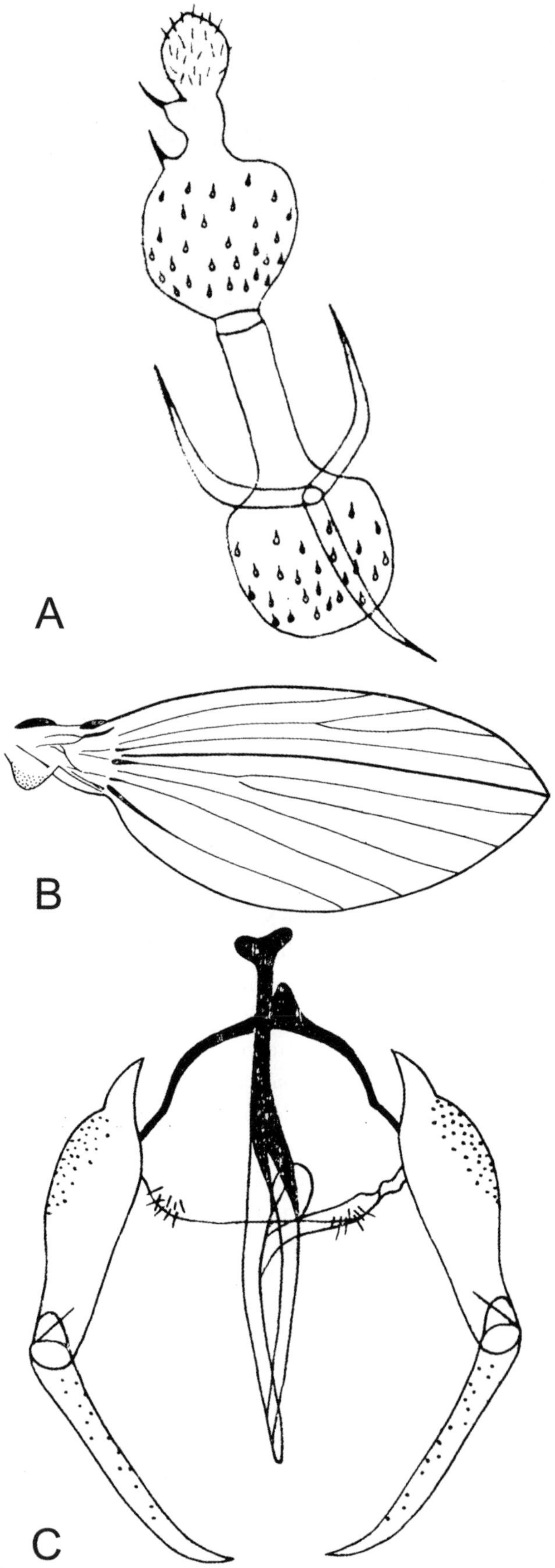

FIGURE 1 Psychoda satchelli. *A, apical flagellomeres (12–14); B, wing; C, coxopodites, harpagones and copulatory organ, dorsal view. After Ježek (1990).*

aedeagus asymmetric. Both wing forks complete. Antennal segment 14 of slightly various shape, usually not elongate but rather often more or less rounded. Swollen part between segments 13 and 14 with a strong lateral spine. Segment 13 with a subapical spine. Wing length 1.5–2.7 mm, usually 1.9–2.2 mm. Male genitalia as Fig. 1C.

BIOLOGY. Larval habitats are largely unknown, but adults have been found at ditches and banks of streams and rivers, in toilets, at silo pits and soil heaps and in barns, as well as in light traps and citronella bait traps.

DISTRIBUTION. S (Narsarsuaq, 22–31 July, 2009, Munk & Kjeldgaard, some tens of males and some females from this sample examined), the first record of this only psychodid species found in Greenland; nSW (Kangerlussuaq, July 2013, a small series obtained by Malaise trapping, T. Pape).

NOTES. According to Ježek (1984, 1990) the species belongs to the genus *Logima* Eaton, including 21 species in the world and five in Europe, of which three are very similar. Bravo *et al.* (2006), however, recently proposed to maintain the several genera split off from *Psychoda* Latreille (see Ježek 1984) in this genus, which is here adopted.

P. satchelli (wing length 1.9–2.7 mm) was described from California; it is common and widely distributed in North America and Canada, and is there one of the most abundant species of Psychodidae (Quate 1955). It is only recorded from four countries in Europe (Fauna Europaea 2010), but is certainely overlooked (Svensson, pers. obs.) and is widely distributed in Sweden (Svensson 2009) and Finland (Salmela 2008). The male genitalia of *satchelli* vary somewhat across the Holactic. Males from southernmost Greenland (Narsarsuaq) are intermediate, or rather, more in accordance with specimens from Sweden and Finland (Svensson, pers. obs.) and Central Europe (Ježek 1990), than with the original description by Quate (1955). The variation of the female genitalia seems insufficiently documented in the figures given by Quate (Ježek 1990 and original observations on many Nordic specimens). A careful taxonomic investigation of this species seems needed.

References

Ježek, J. (1984): Six new genera of the tribe Psychodini End. (Diptera, Psychodidae). – *Acta Faunistica Entomologica Musei Nationalis Pragae* 17: 133–154.

Ježek, J. (1990): Redescription of nine common Palaearctic and Holarctic species of Psychodini End. (Diptera: Psychodidae). – *Acta Entomologica Musei Nationalis Pragae* 43: 33–83.

Ježek, J. & Barták, M. (2000): Psychodidae. *In*: Barták, M. & Vaňhara, J. (eds.): Diptera in an industrially affected region (north-western Bohemia, Bílina and Duchcov environs). I. *Folia Facultatis Scientiarum Naturalium Universitatis Masarykianae Brunensis, Biologia* 104: 93–100.

Oosterbroek, P. (2006): *The European Families of the Diptera*. KNNV Publishing, Utrecht.

Quate, L.W. (1955): A revision of the Psychodidae (Diptera) in America north of Mexico. – *University of California Publications in Entomology* 10: 103–273.

Quate, L.W. & Vockeroth, J.R. (1981): Psychodidae. – *In*: McAlpine, J.F. *et al.* (eds): *Manual of Nearctic Diptera* 1: 293–300.

Salmela, J. (2008): Semiaquatic fly (Diptera, Nematocera) fauna of fens, springs, headwater streams and alpine wetlands in the northern boreal ecoregion, Finland. – *w-album* 6: 3–63.

Svensson, B.W. (2009): Fjärilsmyggfaunan i ett hagmarksområde och en ladugård i östra Blekinges skogsland. Med en översikt av familjen Psychodidae:s morfologi, systematik och utforskande, samt särskilt de svenska *Psychoda* s.l.-arternas biologi. [The moth fly fauna in a pasture- woodland area and in a barn in the eastern part of the province of Blekinge, SE Sweden. Insecta: Diptera, Psychodidae, with a review of the morphology, ecology and research history of the family Psychodidae, and especially the biology of the Swedish *Psychoda* s.l. species]. – *Entomologisk Tidskrift* 130: 185–208.

Vaillant, F. (1971): Psychodidae-Psychodinae. – *In*: Lindner, E. (ed.): *Die Fliegen der palaearktischen Region* 287: 44–45.

Wagner, R. (1998): Diptera Psychodidae. – *In*: Papp, P. & Darvas, B. (eds.): *Contributions to a Manual of Palaearctic Diptera* 2: 205–226.

17.12. Cecidomyiidae (Gall midges)

Mathias Jaschhof

Cecidomyiidae range among the largest families of the order Diptera, mainly due to the tremendous diversity, met on all taxonomic levels, of the youngest subfamily, the Cecidomyiinae. Judged from the fossil record, diversification of the mainly plant-feeding Cecidomyiinae started in the Palaeogene—some 100 million years after the gall midge clade had evolved—and this is certainly to be seen in co-evolution with flowering plants (Gagné 1989). Roughly 4/5 of the about 6.000 currently known gall midge species belong to this one subfamily (Gagné 2004, Evenhuis *et al.* 2007).

Cecidomyiids, or cecids, are delicate, fragile flies most of which have a body length in the range 1–4 mm. There is a pronounced sexual dimorphism. In both sexes the compound eyes are usually joined at the top of the head to form a bridge. The long antennae often have 12 or 14 flagellomeres, and those of most Porricondylinae and Cecidomyiinae bear thread-like sensoria called circumfila. The vestiture on the flagellum is in parts arranged in rows or whorls, and it is often longer than the width of the individual flagellomere, especially in males (Fig. 1A). Wing venation (Fig. 1B) is greatly reduced; the costa is continuous around the wing and usually has a break close to the wing apex, separating the strong anterior from the weaker posterior portions. The long legs lack tibial spurs, and in Porricondylinae and Cecidomyiinae the first tarsomere is much shorter than the second. The male abdomen is apically blunt due to the terminalia, which often are more or less dorsoflexed. The female abdomen is tapered and often modified to form a protrusible ovipositor, which is particularly long and elaborate when the eggs are placed in crevices of living plants (many Cecidomyiinae) or dead wood (many Lestremiinae and Porricondylinae). Even though cecids are thought by many to be rather uniform morphologically and by a first glance may appear comparatively 'simple', the structural diversity is considerable, notably in the antennae, wings and terminalia (see, *e.g.*, Gagné 1981).

The white, yellow or reddish larvae of cecids are usually less than 5 mm long, flat-cylindrical, tapered at both ends, and have a distinct body segmentation (Fig. 1C). The tiny head capsule and the absence of legs and other striking body appendages are characters further contributing to their 'wormlike' habitus. The mouthparts, modified to ingest a liquid diet, are greatly reduced. The integument bears sensory organs, *i.e.*, papillae and/or setae in fixed, group-specific patterns. Unique to the family, the third (rarely also the second) instar has on the ventral side of thorax an elongate, sclerotized dermal structure, called the sternal spatula, which may become variously reduced or lost in some subgroups.

The common name 'gall midge' actually applies only to the strongly derived gall-inducing forms among the subfamily Cecidomyiinae. The large majority of species in this subfamily produce galls in plants, but apart from plant feeders, Cecidomyiinae also contain fungus feeders, predators (on aphids, mites, and other invertebrates) and parasitoids (of aphids and psyllids). The Cecidomyiinae therefore exhibit the greatest biological diversity within the family. Members of the other three subfamilies, Catotrichinae (7 known species), Lestremiinae (wood midges, 600 species) and Porricondylinae (*c.* 600 species), retain the primitive mode of mycetophagy, with their larvae consuming fungal hyphae on decomposing biomass. Noteworthy, the oldest known fossil Cecidomyiidae, dating back to the Jurassic/Cretaceous boundary, was classified by Kovalev (1990) within the extant genus *Catotricha* Edwards.

The tiny size of cecids does not add to the attractiveness of these midges as objects of scientific research, as it calls for the application of specific study techniques and an extensive practical experience in order to develop a flair for them. While many Diptera systematists consider Cecidomyiidae the most demanding family of nematoceran Diptera, less than ten cecidomyiidologists are currently actively working on the group (including several continuing their work in retirement), trying to cope with a world fauna that to the larger extent is unknown in every respect. Taxonomic knowledge of tropical and southern hemisphere faunas is particularly poor, and the number of species actually occurring there is impossible to estimate. Even the relatively well-worked Nearctic and Palaearctic gall midge faunas offer research opportunities for more than the next generation of cecidomyiidologists. Such a situation might surprise, considering that there are many major crop pests among the phytophagous gall midges, and some zoophagous species are important enemies of various agricultural pests.

The reader is referred to Gagné (2004), Yukawa & Rohfritsch (2005), and Roskam (2005), in order to acquire condensed information on our current knowledge of the phylogeny, classification, biology and ecology of cecidomyiids. For advice on group-specific anatomy and study methods, including collecting, rearing and preparing specimens, Gagné (1989, 1994) should be consulted for plant-feeding forms and Jaschhof (1998) for fungus-feeding forms. For identification of genera occurring in Greenland, the key by Gagné (1981) is recommended. For a reliable species identification, an expert opinion is strongly recommended.

This account of Cecidomyiidae from Greenland is based on specimens, few in number and generally in poor condition, collected on various occasions in the second half of the 20th century and deposited in the Natural History Museum of Denmark. The material is probably a by-catch or residue rather than the result of specific collecting efforts and therefore not representative of the Greenland fauna. Specialised collections of gall midges have apparently not yet been attempted in any part of Greenland. Nonetheless, the small material studied yielded members of all the subfamilies with the exception of Catotrichinae.

Key to subfamilies

1. Five tarsal segments, with first segment longer than second. Ocelli usually present. Vein M1+2 usually present 2

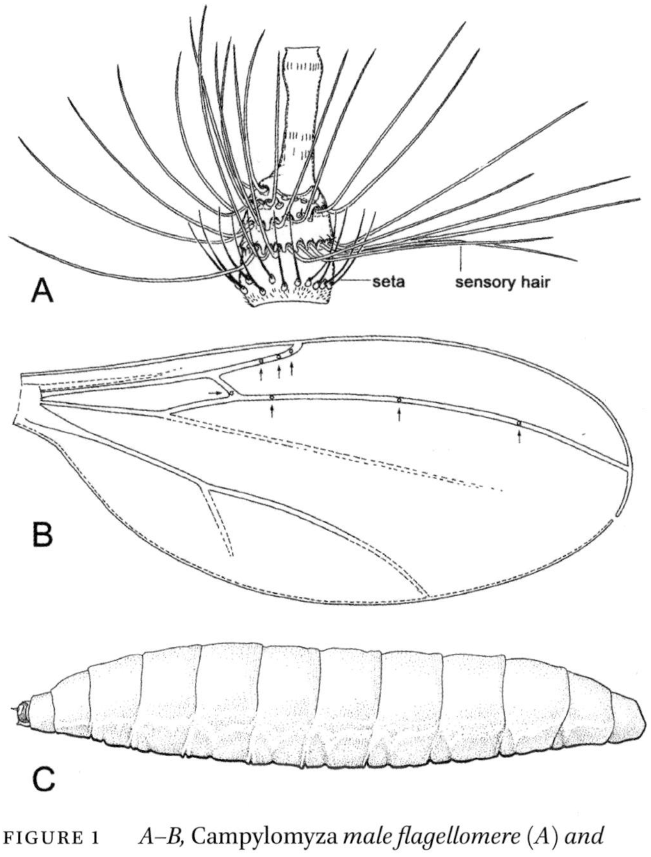

FIGURE 1 *A–B,* Campylomyza *male flagellomere (A) and wing (B), arrows indicate 'sensory buds'. C, larva of a lestremiine cecidomyiid. A–B from Jaschhof (1998), C from Gagné (1981).*

– Five tarsal segments, with first tarsal segment much shorter than second, or fewer than five tarsal segments. Ocelli absent. Vein M1+2 usually absent 3

2. Antennal flagellum with vestiture irregularly arranged. Necks of flagellomeres usually with setiform sensilla. Vein CuA1 basally joining the horizontal vein replacing M.. .... [Catotrichinae, not known from Greenland]

– Antennal flagellum with parts of vestiture arranged in rows and/or whorls. In males (Fig. 1A), crenulate whorls usually present. Necks of flagellomeres bare. Vein CuA1 basally free or joining CuA2 (Fig. 1B) .... Lestremiinae

3. Basal section of Rs as strong as other wing veins. Male gonocoxites united ventrobasally. Female cercus usually two-segmented......... Porricondylinae

– Basal section of Rs usually absent or weaker than other veins. Male gonocoxites free ventrobasally. Female cercus one-segmented ...... Cecidomyiinae

Lestremiinae

Of this subfamily, three specimens were identified as belonging to genus ***Campylomyza*** Meigen. Larvae of *Campylomyza* spp. are fungivorous and live in various kinds of humus-enriched habitats (Mamaev & Krivosheina 1965, Jaschhof 1998). *Campylomyza* is mainly Holarctic in distribution, with a few extralimital records needing confirmation, or referring to species which presumably were introduced by man into more southern latitudes (Jaschhof 1998, Jaschhof & Jaschhof 2003). As usual in the Lestremiinae, species of *Campylomyza* occupy extensive distribution areas, and several of the about 20 Holarctic species are known to penetrate far into northern latitudes, which in Scandinavia is even beyond 70°N (Jaschhof 1996 and unpubl.). Two of the three specimens here studied are males presumably belonging to an unnamed species of the ***Campylomyza flavipes*** group. Poor condition of the available specimens, along with the fact that species of the *flavipes* group are extremely difficult to distinguish from one another, are obstacles to describing the new species. The third specimen, a female, possibly belongs to ***Campylomyza alpina*** Siebke, 1863, a species fairly commonly found throughout the northern Palaearctic region (Jaschhof 1998). The only previously published record of lestremiines from Greenland (Lundbeck 1898) very probably also refers to a *Campylomyza* sp., even though Lundbeck listed the species under the name *Cecidomyia atra* Meigen, 1804, which was later shown to belong to subgenus *Xylopriona* Kieffer within genus *Monardia* Kieffer (Edwards 1938, Jaschhof 1998). *Xylopriona atra*

(Meigen) is widespread Holarctic in distribution, but absent in the extreme north (Jaschhof unpubl.), and Lundbeck's (1898) record for Greenland, also appearing in subsequent catalogues (Stone *et al.* 1965, Gagné 2004), should be considered as erroneous.

Porricondylinae, Cecidomyiinae

These two subfamilies are represented in the available study material by a few female specimens, but in a condition that does not allow an identification.

Addendum

Readers are recommended to consult the recent version of *A Catalog of the Cecidomyiidae of the World* (Gagné & Jaschhof 2014) for updates on major changes in the classification of the family, subordinated taxa added since 2004, and new taxonomic literature.

References

Edwards, F.W. (1938): On the British Lestremiinae, with notes on exotic species.—5. (Diptera, Cecidomyiidae). – *Proceedings of the Royal Entomological Society of London* 7: 199–210.

Evenhuis, N.L., Pape, T., Pont, A.C. & Thompson, F.C. (eds) (2007); Biosystematic Database of World Diptera, Version 9.5. URL: http://www.diptera.org/biosys.htm.

Gagné, R.J. (1981): Cecidomyiidae. Pp. 257–292. *In*: McAlpine, J.F., Peterson, B.V., Shewell, G.E., Teskey, H.J., Vockeroth, J.R. & Wood, D.M. (eds), *Manual of Nearctic Diptera*. Vol. 1. Research Branch, Agriculture Canada Monograph No. 27.

Gagné, R.J. (1989): *The Plant-feeding Gall Midges of North America*. Cornell University Press, Ithaca and London; xi + 356 pp.

Gagné, R.J. (1994): *The Gall Midges of the Neotropical Region*. Cornell University Press, Ithaca and London; xi + 352 pp.

Gagné, R.J. (2004): A Catalog of the Cecidomyiidae (Diptera) of the World. *Memoirs of the Entomological Society of Washington* 25: 1–408.

Gagné, R.J. & Jaschhof, M. (2014): *A catalog of the Cecidomyiidae (Diptera) of the world*. Third Edition. Digital version 2. URL: http://www.ars.usda.gov/SP2UserFiles/Place/12454900/Gagne_2014_World_Cecidomyiidae_Catalog_3rd_Edition.pdf.

Jaschhof, M. (1996): Zur Gallmücken-Fauna Lapplands mit Beschreibung neuer Arten aus den Gattungen *Aprionus*, *Peromyia* und *Neurolyga* (Cecidomyiidae, Lestremiinae). – *Studia dipterologica* 3: 338–355.

Jaschhof, M. (1998): Revision der "Lestremiinae" (Diptera, Cecidomyiidae) der Holarktis. – *Studia dipterologica Supplement* 4: 1–552.

Jaschhof, M. & Jaschhof, C. (2003): Wood midges of New Zealand (Cecidomyiidae, Lestremiinae). Part I: Introductory notes and tribes Lestremiini, Strobliellini, Campylomyzini and Pteridomyiini Jaschhof trib. nov. – *Studia dipterologica* 10: 97–132.

Kovalev, V. (1990): [Diptera.] Pp. 123–177. *In*: Rasnitsyn, A.P. (ed.), [*Late Meszoic Insects of Eastern Transbaikalia*.] Moscow, Nauka Press.

Lundbeck, W. (1898): Diptera groenlandica. – *Vidensk. Meddel. Naturhist. For. Kjøbenhavn* 1898 (ser. 5, 10): 236–314, 2 pls.

Mamaev, B.M. & Krivosheina, N.P. (1965): [*The Larvae of Gall Midges (Diptera, Cecidomyiidae)*.] Akademiya Nauk USSR, Moscow; 277 pp.

Roskam, H.C. (2005): Phylogeny of Gall Midges (Cecidomyiidae). Pp. 305–319. *In*: Raman, A., Schaefer, C.W. & Withers, T.M. (eds), *Biology, Ecology and Evolution of Gall-inducing Arthropods*. Vol. 1. Science Publishers, Enfield, New Hampshire.

Stone, A., Sabrosky, C.W., Wirth, W.W., Foote, R.H. & Coulson, J.R. (eds) (1965): *A Catalog of the Diptera of America North of Mexico*. Agricultural Research Service, U. S. Department of Agriculture. Washington, D. C.; iv + 1696 pp.

Yukawa, J. & Rohfritsch, O. (2005): Biology and Ecology of Gall-inducing Cecidomyiidae (Diptera). Pp. 273–304. *In*: Raman, A., Schaefer, C.W. & Withers, T.M. (eds), *Biology, Ecology and Evolution of Gall-inducing Arthropods*. Vol. 1. Science Publishers, Enfield, New Hampshire.

17.13. Empididae (Dance or Dagger flies)

Miroslav Barták

This large family (as currently circumscribed with ca 2.500 described species worldwide) is closely related to the Dolichopodidae with which they share an antennal design (with a prominent ± elongate conical/triangular, 'postpedicellus' followed by a stout 'stylus') that is unique among Greenland Brachycera. The two are perhaps most easily separated by the wing venation, which in all Greenland dolichopodids is so characteristic by the presence of just one single, and very prominent, cross-vein (extending between the two posterior longitudinal veins); also the Greenland dolichopodids, unlike the empidids, have a greenish metallic sheen.

Adult empidids are mostly predators. Just two genera (representing two different subfamilies) are recorded from Greenland. Of the five Greenland *Ramphomyia* species no less than three are currently unknown outside this country.

Empidid genera

1. Body dull greyish; proboscis short, only small labellum and palps projecting beyond head capsule (Fig. 1A)....................................*Clinocera*
– Body shining, blackish; piercing proboscis very long and prominent (Fig. 1B)...............*Rhamphomyia*

Clinocera stagnalis **(Haliday, 1833) Fig. 1A**
Henriksen 1939: *Hydrodromia stagnalis* Collin

Igaliko, Kapisillit, Qaqortoq, Kangersuneq, Bjørnesund, Fiskenæsfjorden, Tissaluk, Narsarsuaq, Kangamiut, Maniitsoq, Disko Fjord (Angujartutit, varm kilde [hot spring]).

DISTRIBUTION. S, SW (north to Qeqertarsuup Tunua/ Disko Bay). A single specimen was collected at a hot spring surrounded by snow at Kuanerssuit Sulluat in Kangerluk/ Disko Fjord, on April 29 (1969, J. Böcher).

Holarctic. Examined Greenland specimens agree with those from N. America in having the apex of the male clasping cercus rounded, hence contrasting with the nipple-like apex characterizing Palaearctic specimens, including those from Iceland (Sinclair 2008).

Key to Rhamphomyia *species*

1. Costal seta absent. Ac setae usually at least 4-serial . ...*nigrita*
– Costal seta present. Ac setae 2–3 serial.............2

FIGURE 1 *Habitus of* Clinocera stagnalis (*A*), Rhamphomyia nigrita *male* (*B*), R. nigrita *female* (*C*).

2. Dc almost uniserial (usually with at most 1–2 setae standing outside the row). 4–6 sc. Propleura with setae. The 1st sternum without setae. Anal vein complete. Female tibia 3 without anterior osmeterium ....*hoeli*
– Characters in other combinations..................3
3. Propleura bare. Anal vein incomplete. Male genitalia very long, almost as half of abdomen. Female: tibia 3 without anterior osmeterium, tibia 2 with irregularly long ventral setae, abdominal tergite 5 with long (at least half length of tergite) hind marginals, wing with anal lobe clouded.....................*filicauda*
– Propleura with setae. Anal vein complete. Male genitalia short. Female: tibia3 with anterior osmeterium, tibia 2 with two dense rows of evenly very short ventral setae, tergite 5 with very short hind marginals (at most 1/4 length of tergite), anal lobe pale..........4
4. Smaller species (wing less than 4.2 mm). dc uniserial in hind part. Usually 6–10 sc. Male: phallus with right-angled fold in middle. Female: femur 3 in apical third with several av setae which are at least as long as femur 3 is high ....*hirticula*
– Larger species (wing more than 4.3 mm). dc multiserial behind. Usually more than 12 sc. Male: phallus evenly curved. Female: femur 3 in apical third with av setae much shorter than femur 3 is high (with the possible exception of the terminal one) ....*hirtula*

Rhamphomyia (s. str.) nigrita Zetterstedt, 1838 Fig. 1B, C

Halter clear yellow, more than 6 sc setae, male genital lamella greatly enlarged, covering dorsum of abdomen, female hind tibia without anterior osmeterium, wing evenly brown. Male genitalia: Frey (1954–56) fig. 420.

DISTRIBUTION. G (common everywhere in Greenland). Probably circumpolar.

***Rhamphomyia* (*Pararhamphomyia*) *hoeli* Frey, 1949**

Females very similar to those of *R. hirticula*, but postpronotum with one long and several very short setae, all basitarsi short setose ventrally. In *R. hirticula*: postpronotum with several long setae, basitarsi with rather long spines ventrally. Male genitalia: Frey (1954–56) fig. 365.

DISTRIBUTION. NW (only Thule Air Base), NE (found in a restricted area around Kong Oscar Fjord).

Only known from Greenland.

***Rhamphomyia* (*Pararhamphomyia*) *filicauda* Lundbeck, 1918**

Male genitalia: Frey (1954–56) fig. 373.

DISTRIBUTION. NW (Thule Air Base), N (Peary Land: Kap København), NE (Dronning Louise Land, Lindhard Ø, Mestersvig, Jameson Land). A northern species, very common.

Only known from Greenland.

***Rhamphomyia* (*Pararhamphomyia*) *hirticula* Collin 1937**

Male genitalia: Frey (1954–56) fig. 224.

DISTRIBUTION. NW. Only collected at Thule Air Base. Seemingly very rare.

Only known from Greenland.

***Rhamphomyia hirtula* Zetterstedt, 1842**

Male genitalia: Collin (1961, p. 402) fig. 139.

DISTRIBUTION. S, SW, NW (north to Aappilattoq Island, about 60 km north of Upernavik), N (Peary Land: Kap København), NE (Ymers Ø, Jameson Land, Hekla Havn, Mestersvig, Blosseville Kyst: Søkongen Bugt, D'Aunay Bugt), SE (Kangilinnguit/Skjoldungen). Seems to be very common, possibly G.

Palaearctic (only known from Iceland and Scotland).

References

Collin, J.E. (1961): *British Flies, Empididae*, VI. 782 pp. Cambridge.

Frey, R. (1954–56): Empididae. *In*: Lindner E. (ed.): *Die Fliegen der palaearktischen Region* 4 (4): 400–639.

Sinclair, B.J. (2008): *The Systematics of new World* Clinocera *Meigen*. vii + 245 pp. NRC Research Press, Ottawa.

17.14. Dolichopodidae (Long-legged flies)

Marc Pollet

With over 7100 species, the family is among the most species rich in the Diptera. The majority features brilliant green to blue colours, but also mainly yellow, brown or black forms exist. Body size ranges from about 1 to 10 mm, legs are generally rather to very long, mouthparts protruding (less than in Empididae) and wing venation rather simple, with vein M1(+2) often showing a distinct bend. Males of many species show distinct ornaments on antennae, legs, wings and even body which seem to play a role during courtship (Fig. 4A, B).

Since most species are distinctly hygrophilous, the highest species richness and abundances are encountered in humid biotopes such as swamps, saltmarshes, carrs, humid forests, humid heathland and peatmoors, and on shallow and sparsely vegetated banks of water bodies. Other species prefer vertical surfaces for mating and foraging and are mainly found on tree trunks and fences. Many *Medetera* species seem to breed in trees, and biotopes with a very special dolichopodid fauna are rotholes and sapruns.

All species are considered predatory on soft-bodied invertebrates such as aphids, springtails, mites and enchytraeid worms. This holds true for the larval stages as well, except for the larvae of *Thrypticus* that live as stem miners. The larval development in most species takes place in the soils of the habitats of the adults. Larvae of coastal species are found in wet sandy soils of beaches, among decaying algae or even within or amid barnacles.

The dolichopodid fauna of Greenland is staggeringly poor as compared to other Palaearctic and Nearctic regions at the same latitude, except for Iceland. Apart from a dubious Colorado record of *Dolichopus groenlandicus*, both *D. dasyops*, *D. groenlandicus* and *D. humilis* can be termed Nearctic circumpolar species, whereas *D. plumipes* has a much wider distribution in the Holarctic. The fact that only *Dolichopus* is represented in Greenland is not surprising as this is the dominant genus in all northern regions in the Holarctic.

The Greenland distribution of *D. dasyops* and *D. humilis*, which seem to be restricted to the Thule/Pituffik region (NW), is striking as is the fact that *D. groenlandicus* was not encountered here. This kind of restricted distributions often suggests recent human introduction, which would imply human transport of suitable materials (soil, plant material) prior to 1940 (= first records of both species). As the Thule Airbase was first erected in 1951, a recent introduction of at least *D. dasyops* must be excluded. A more plausible explanation is natural dispersal from the American mainland via Ellesmere Island. It remains, however, enigmatic why both species did not disperse further south. It is possible that unfavourable conditions in the Melville Bugt area to a large extent have prohibited dispersal.

Abbreviations*:* ad: anterodorsal; ap: apical; av: anteroventral; pd: posterodorsal; pv: posteroventral; fore leg: leg I, mid leg: leg II, hind leg: leg III. Tarsomeres of legs are indicated as 1 to 5, starting from the basis (metatarsus) towards the apex (5th tarsomere). In the key to males special attention is drawn to the sex-linked features only present in male specimens, the so-called Male Secondary Sexual Characters (MSSC). In most cases, these are not only important for diagnostic purposes but are also phylogenetically relevant.

For a more detailed account of the species and pertinent references, consult Pollet & Böcher (2005).

Key to species

MALES

1. All postocular bristles black 2
– Lower postocular bristles yellow to white 3
2. Fourth and fifth tarsomere of leg I strongly flattened and black, 5th tarsomere about as deep as long (Fig. 4A). Antenna with scape yellow on ventral half (Fig. 4C) .... *Dolichopus dasyops*
– Leg I without flattened tarsomeres. Antenna entirely black (Fig. 4F) .... *Dolichopus groenlandicus*
3. Metatarsus II pennate laterally (Fig. 4B). Tibia II with a longitudinal narrow brown streak anterodorsally, whitish at apex. Femur III with one ad preapical bristle. Antenna mainly yellow with 1st flagellomere black on apicodorsal half (Fig. 4E)................. .... *Dolichopus plumipes*
- Leg II unmodified. Femur III with two ad preapical bristles. Antenna entirely black, with scape remarkably long, about 2× as long as wide (Fig. 4D)........ .... *Dolichopus humilis*

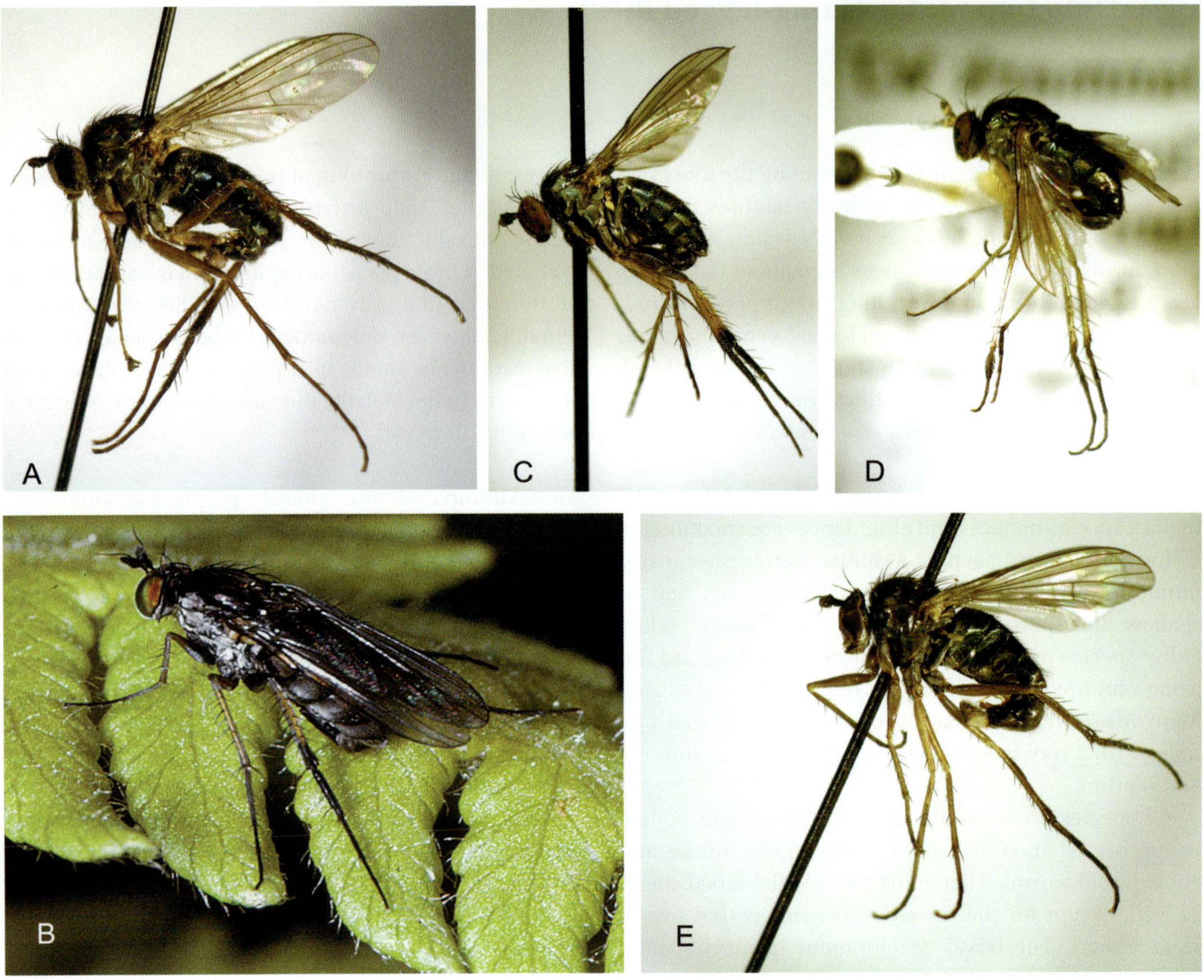

FIGURE 1 *General habitus (male). A,* Dolichopus dasyops; *B, C,* Dolichopus groenlandicus; *D,* Dolichopus plumipes; *E,* Dolichopus humilis.

FEMALES

1. All postocular bristles black 2
- Lower postocular bristles yellow to white 3

2. Antenna with scape yellow on ventral half. All femora entirely dark yellow .... *Dolichopus dasyops*
– Antenna entirely black. Femur III entirely, and femur I largely dark brown .... *Dolichopus groenlandicus*

3. Antenna mainly yellow with 1st flagellomere black on apicodorsal half. Femur III with 1 ad preapical bristle. Femora and coxa I entirely yellow .... *Dolichopus plumipes*
- Antenna entirely black, with scape longer than wide. Femur III with 2 ad preapical bristles. Coxa I entirely blackish brown, and femora I and III largely brown . ... *Dolichopus humilis*

***Dolichopus dasyops* Malloch, 1919 Figs 1A, 2A, 3A, 4A, 4C**

Largest *Dolichopus* species in Greenland with body and wing length over 5.5 mm (Fig. 1A). The face is shining whitish yellow (pale golden). Thorax shining green bronze, with greenish grey pleura. Wing transparent without costal stigma (Fig. 3A). Legs yellowish brown to brown Tarsus I with 4th and 5th tarsomere flattened laterally (Fig. 4A). Abdomen shining bluish green. Cercus obscure yellow, transparent, with broad apical black margin and strong curved bristles on apicodorsal inner margin (Fig. 2A).

DISTRIBUTION. NW. Collected exclusively in the Thule/Pituffik area.

Holarctic (not recorded in western Palaearctic), arctic.

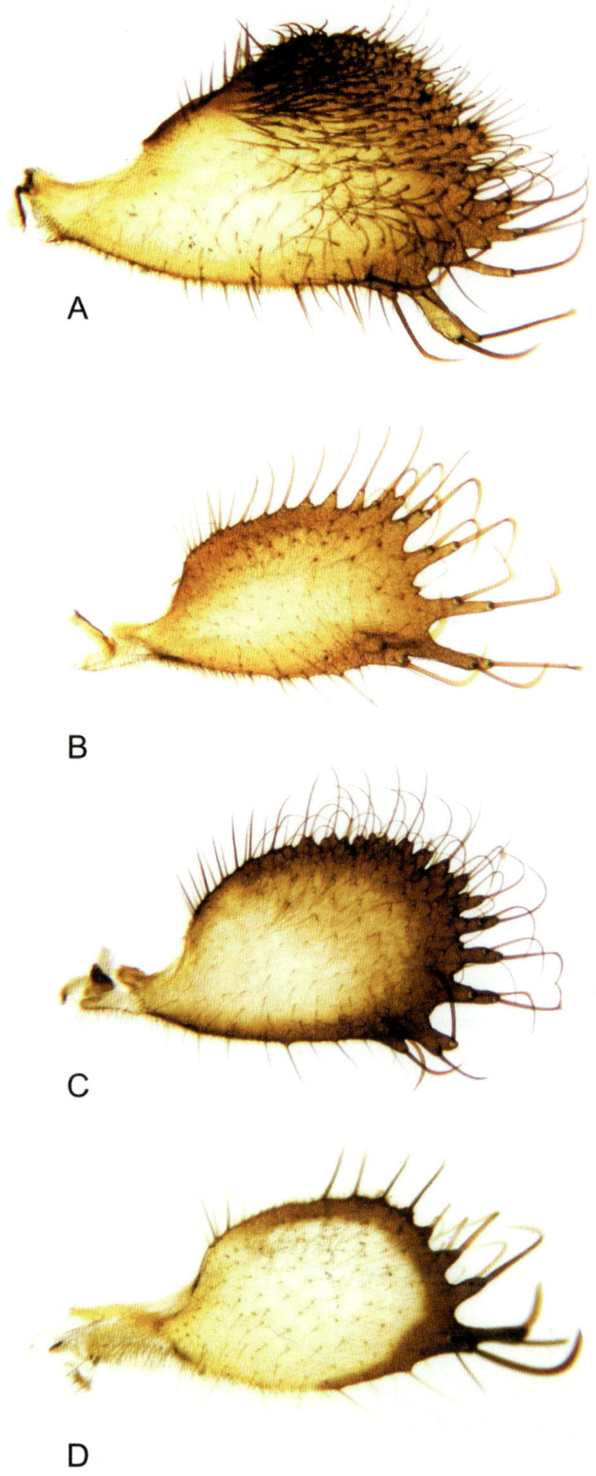

FIGURE 2 *Cercus (genital lamella). A,* Dolichopus dasyops; *B,* Dolichopus groenlandicus; *C,* Dolichopus humilis; *D,* Dolichopus plumipes.

Dolichopus groenlandicus Zetterstedt, 1843 Figs 1B, 1C, 2B, 3B, 4F

Body length *c.* 4.8 mm, wing length *c.* 4.7 mm (Fig. 1B) although smaller specimens (body and wing length about 4.0 mm) from northern SW have been seen. In the male the face is rusty brown, in the female greyish white with yellowish tinge. Thorax shining bronze on dorsum with pleura dusted bluish grey. Wing smoked with small costal stigma (MSSC, Fig. 3B). Legs, including coxae, largely dark brown with tibiae mainly yellow. Abdomen metallic bluish green, dusted bluish grey as seen from frontolateral angle. Hypopygium with rather small brown cercus (Fig. 2B).

BIOLOGY. The courtship behaviour of this species is frequently observed on the banks of fresh waters. By means of wing postures and wing movements the males try to contact the females.

DISTRIBUTION. S, SW, NW, SE, NE. Most common *Dolichopus* species in Greenland. More than 90% of the collected specimens originate from S and SW.

Nearctic, low arctic-northern temperate.

Dolichopus humilis Van Duzee, 1921 Figs 1E, 2C, 3C, 4D

Both body and wing length about 4.5 mm (Fig. 1C). Face shining silvery white. Thorax bronze green with pleura greenish grey dusted. Wing rather broad, 2.4× as long as wide, with strong costal stigma (MSSC); vein M1 with distinct bend (Fig. 3C). Legs yellowish brown to dark brown, with tibia and metatarsus of leg II mainly yellow. Abdomen metallic green, greyish dusted laterally. Cercus obscure yellowish white with blackish band along apical and ventral margin, with long curved marginal bristles (Fig. 2C).

DISTRIBUTION. NW. Possibly restricted to the Thule/Pituffik area, where it has been collected more abundantly than the sympatric *D. dasyops*.

Circumpolar, arctic.

Dolichopus plumipes (Scopoli, 1763) Figs 1D, 2D, 3D, 4B, 4E

Smallest *Dolichopus* species of Greenland, body length *c.* 4.4 mm, wing length *c.* 4.2 mm (Fig. 1 E).

Male with face pale golden yellow, in female face greyish white with weak yellow tinge. Thorax shining green bronze, with pleura greenish grey. Wing with hind margin concave at basis, without costal stigma (Fig. 3D). Legs, including coxa I mainly pale yellow. Tibia II narrow, elongate, pale yellow with flat bare dorsal face featuring a distinct dark brown stripe. Tarsus II entirely black, with metatarsus flattened, pennate laterally (Fig. 4B). Abdomen shining green, laterally with grey dusting. Cercus ovoid, transparent white to yellow with narrow black marginal band, with denticulate bristles at apex and curved ventral bristles (Fig. 2D).

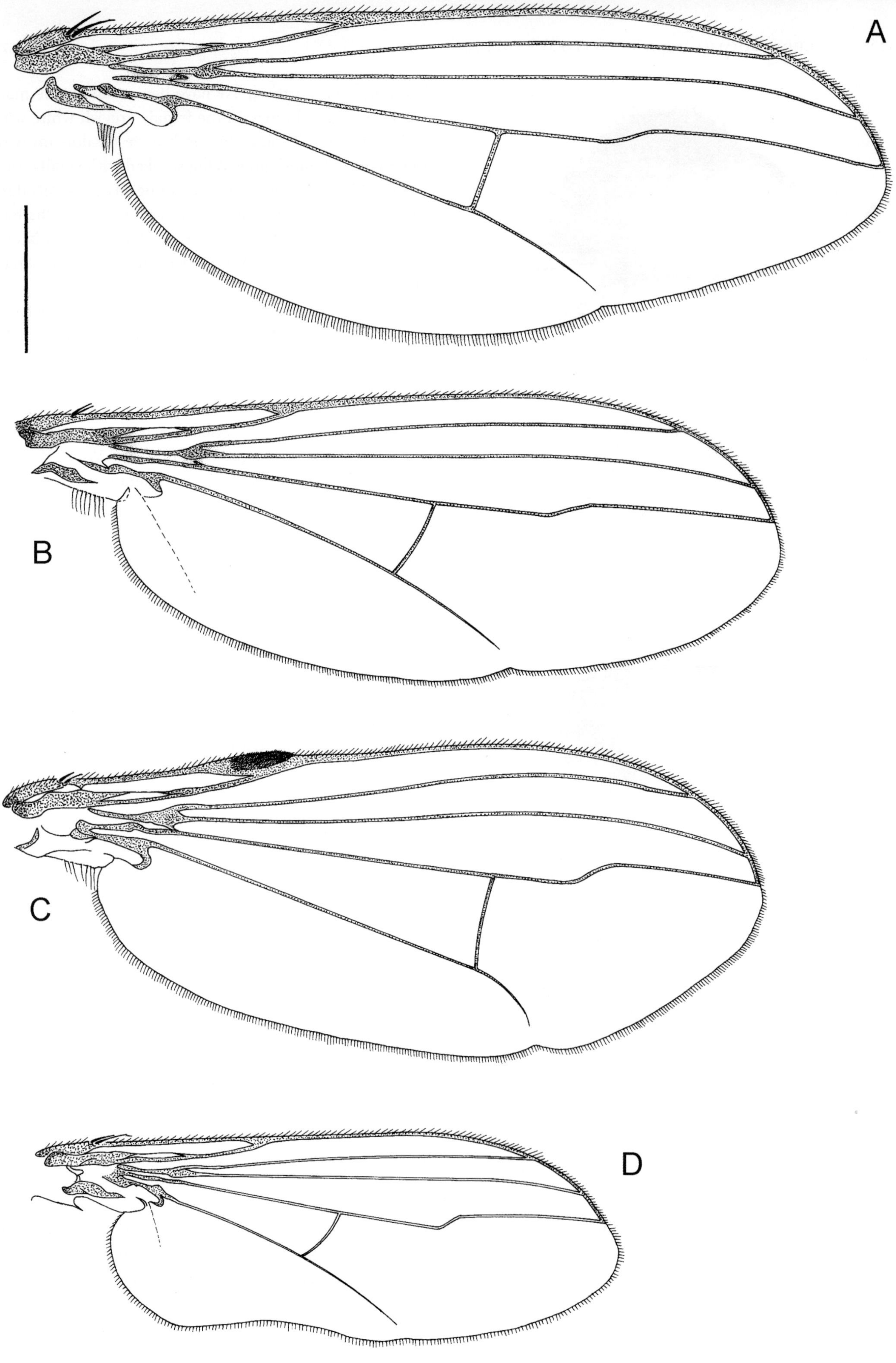

FIGURE 3 *Wing. A,* Dolichopus dasyops*; B,* Dolichopus groenlandicus*; C,* Dolichopus humilis*; D,* Dolichopus plumipes. *Scale = 1 mm.*

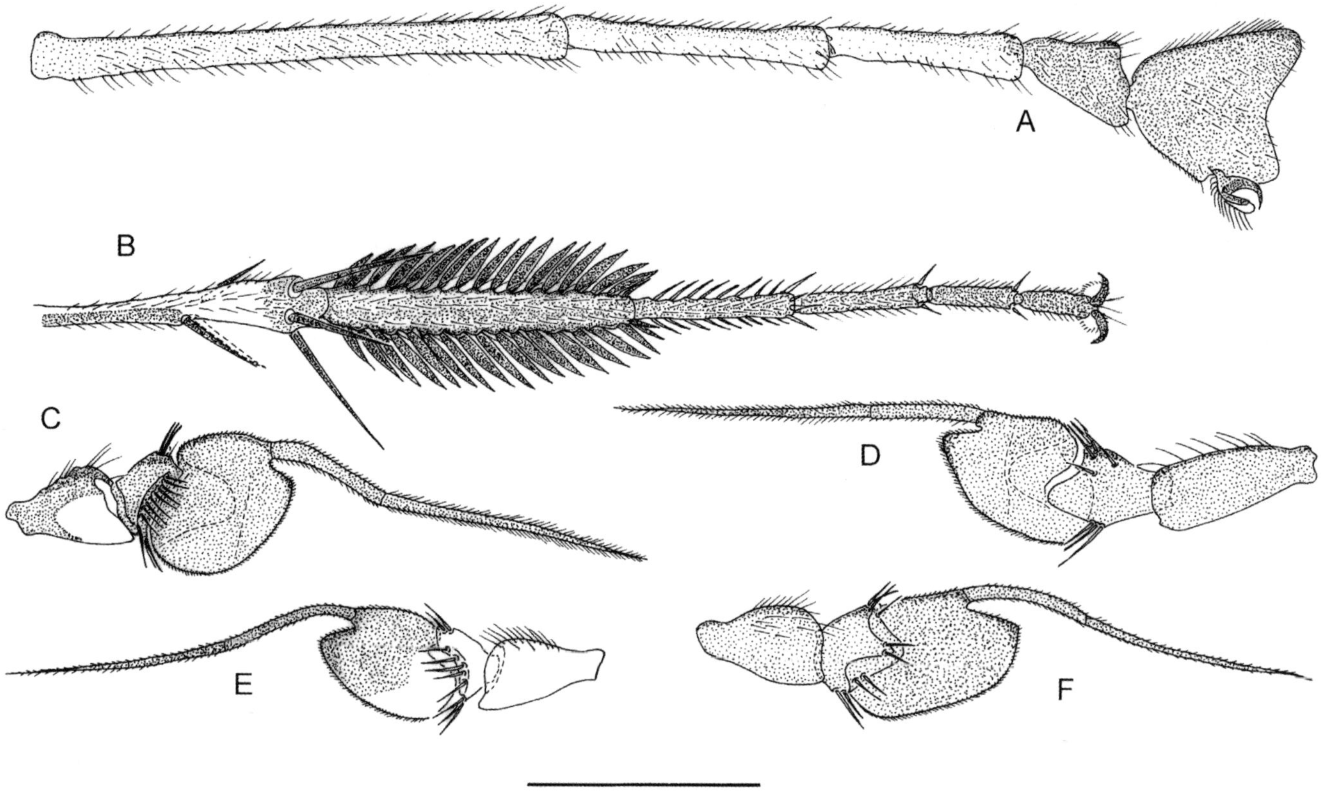

FIGURES 4 *Male secondary sexual sharacters. A, tarsus I of* Dolichopus dasyops; *B, tarsus II of* Dolichopus plumipes. *C–F, antennae of* Dolichopus dasyops (*C*); Dolichopus humilus (*D*); Dolichopus plumipes (*E*); Dolichopus groenlandicus (*F*). *Scale = 0.5 mm.*

BIOLOGY. *D. plumipes* is the second most common *Dolichopus* species in Greenland, representing 22% of the specimens examined. More than 90% of all specimens were collected in southernmost Greenland (S). In Europe this is an euryotopic species of grasslands and marshlands, which avoids oligotrophic biotopes (peatbogs and humid heathland) and dense forests.

DISTRIBUTION. S, sSW, NE.
Holarctic, low arctic-southern temperate.

Reference

Pollet, M. & Böcher, J. (2005): The Dolichopodidae (Diptera) of Greenland. – *Entomologiske Meddelelser* 73: 3–15.

17.15. Phoridae (Scuttle flies)

R. Henry L. Disney

The Phoridae comprise a large family of small flies that are thought to exhibit a greater variety of larval habits than any other family of insects. These include feeders on decaying organic matter ranging from rotting plants to human corpses, fruit feeders, fungivores, kleptoparasites feeding on the stores or intercepting the feed of social insects, specialised predators, parasitoids and parasites (Disney, 1994).

Most adults, including all the species recorded from Greenland, are readily recognised by their wing venation, characterised by a shortened costa and radial veins and by the remaining veins, behind the latter, being much thinner and lacking any cross veins beyond the basal quarter (Fig. 9A). Elsewhere, especially among species associated with ants and termites, flightless females are common.

The larvae (Fig. 12B) have the posterior spiracles closely approximated, each with four openings arranged in pairs, one behind the other; the anterior spiracles simple and unbranched. Many have numerous short processes on the body segments. The pupal respiratory horns (Fig. 13C) pierce the eclosion plates of the puparium (Fig. 12A). The larvae of parasitoid and aquatic species (the latter including some maritime and marine species) exhibit a wide range of adaptive departures from this basic type.

Apart from a tramp species, the larval habits of the species recorded from Greenland remain unknown. It will be especially interesting to discover the larval habits of the two sibling species, *M. clara* and *M. groenlandica*. The rough phenologies of the adults of the two species, as indicated by the numbers collected in each month over the years, suggest that *M. clara* has its peak in August (like *M. cirriventris*) but *M. groenlandica* has its peak in July. Another clear difference is that the ratios of the sexes obtained was 2.9 : 1 (males to females) for *M. clara* but 20 : 1 for *M. groenlandica*. The ratio of 6.8 : 1 for *M. cirriventris* was intermediate. These figures suggest that there is a significant difference between the behaviours of the two sibling species.

In the collected gravid females that have not begun to lay eggs the number of mature eggs was counted. As the number of eggs matured in a single ovarian cycle (the egg batch size) tends to vary with the size of the fly, the length of the hind femur (HF) was recorded. All of these species have eggs that are smooth and of simple standard form (as fig. 2.3(a) in Disney, 1994), so merely the lengths and breadths of mature eggs is noted for each species.

The phorid fauna of Greenland is clearly impoverished, with only six recorded species, all in the giant genus *Megaselia*. By contrast more than 300 species in 23 genera have been recorded from the British Isles (Disney, 1983, 1989). The five previously reported species from Greenland (Schmitz, 1956) are Holarctic in distribution, but *M. rufipes* is a tramp species that was almost certainly transported by man from Europe to the Nearctic Region. During the preparation of this chapter a new species was added (Disney, 2004c). The lack of any species of the genera *Phora* Latreille and *Triphleba* Rondani is surprising, as both include cold adapted species. Indeed, not only does *Triphleba* occur in Iceland (Prescher *et al.*, 2005), but some species of this genus are notorious for being active in extremely cold conditions (*e.g.* Mostovski & Disney, 2002; Soszynska & Durska, 2002). The list of Phoridae for Iceland is almost twice as long. Only two of the species, *M. clara* and *M. pleuralis*, are also found in Greenland. The Nearctic species of *Megaselia* are keyed by Borgmeier (1964, 1966). The full listing of the literature will be found in Borgmeier (1968).

The methods for best preserving the Phoridae are discussed by Disney (2001). This concluded that scuttle flies should be initially preserved in fluid, 70–80% ethanol being recommended, and be subsequently mounted on slides.

Key to species

1 With distinct hypopygium at tip of abdomen, ending in an anal tube comprising the cerci fused to the proctiger (Figs 1, 3B, 4A, 5B, 7B, 10B)...... Male, 2

– Tip of abdomen with the pair of cerci well separated (*e.g.* Fig. 11) ....Female, 7

2 Abdominal tergites and epandrium (the dorsal shield of the hypopygium) with conspicuous feathered bristles (Figs 10A, B) .... *Megaselia rufipes*

– Not so...3

3 Mesopleuron bare. Hypopygium as in Fig. 1......... .... *Megaselia arcticae*

– Mesopleuron with hairs and sometimes one or more bristles as well. Hypopygium otherwise............4

4 Haltere knob greyish brown and as dark as stem ...5

– Haltere knob straw yellow and even when a little dusky clearly contrasting with the dark stem. Hypopygium as in Figs 7B, 8 .... *Megaselia pleuralis*

5 Rear margin of abdominal sternum 6 with a pair of hair palisades on small ridges and hypopygium as in Fig. 3B...................... *Megaselia cirriventis*
– Without these hair palisades and hypopygium otherwise.. 6
6 Basitarsus of front leg asymmetric (Fig. 6A). Hairs below basal half of hind femur as in Fig. 5A. Hypopygium as in Fig. 5B............................ .... *Megaselia groenlandica*
– Basitarsus of front leg almost symmetrical. Hairs below basal half of hind femur weaker and less crowded. Hypopygium as in Fig. 4A................ .... *Megaselia clara*
7 Mesopleuron bare................................... 8
– Mesopleuron with hairs and sometimes a small bristle as well... 9
8 Haltere entirely greyish brown. Femora brown. Abdominal tergites 5–7 as in Fig. 2B................. .... *Megaselia arcticae*
– Haltere knob straw yellow. Femora largely straw yellow, but hind femur may be partly brown. Abdominal segments 6–10 as in Fig. 11. T5–T7 as in Fig. 9B... .... *Megaselia rufipes*
9 Haltere knob greyish brown and as dark as stem 10
– Haltere knob straw yellow and even when a little dusky clearly contrasting with dark stem. Abdominal tergite 6 to epiproct as in Fig. 7A .... *Megaselia pleuralis*
10. Mesopleuron with a differentiated bristle at rear margin of hair patch. Abdominal tergites 5–8 as in Fig. 3A .... *Megaselia cirriventris*
– Mesopleuron with hairs only. Abdominal tergites 5–8 otherwise (Figs 4B, 6B)......................... 11
11. Wings >2.0 mm long, usually >2.3 mm, and with a paler grey tinge and the thin veins 4–6 being a paler brownish grey. T5–T7 as in Fig. 4B.................... .... *Megaselia clara*
– Wings <2.3 mm long, usually <2.1 mm, and darker grey tinge and the thin veins 4–6 being brown. T5–T7 as in Fig. 6B................. *Megaselia groenlandica*

Megaselia arcticae **Disney, 2004c Figs 1, 2**

This is a member of a complex of species related to *M. giraudii* (Egger). The European species have been recently

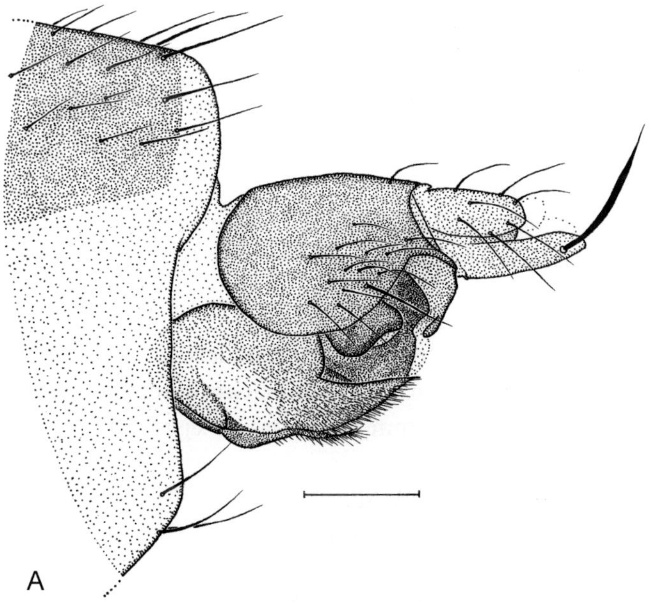

FIGURE 1 Megaselia arcticae *male, left face of hypopygium. Scale bar = 0.1 mm.*

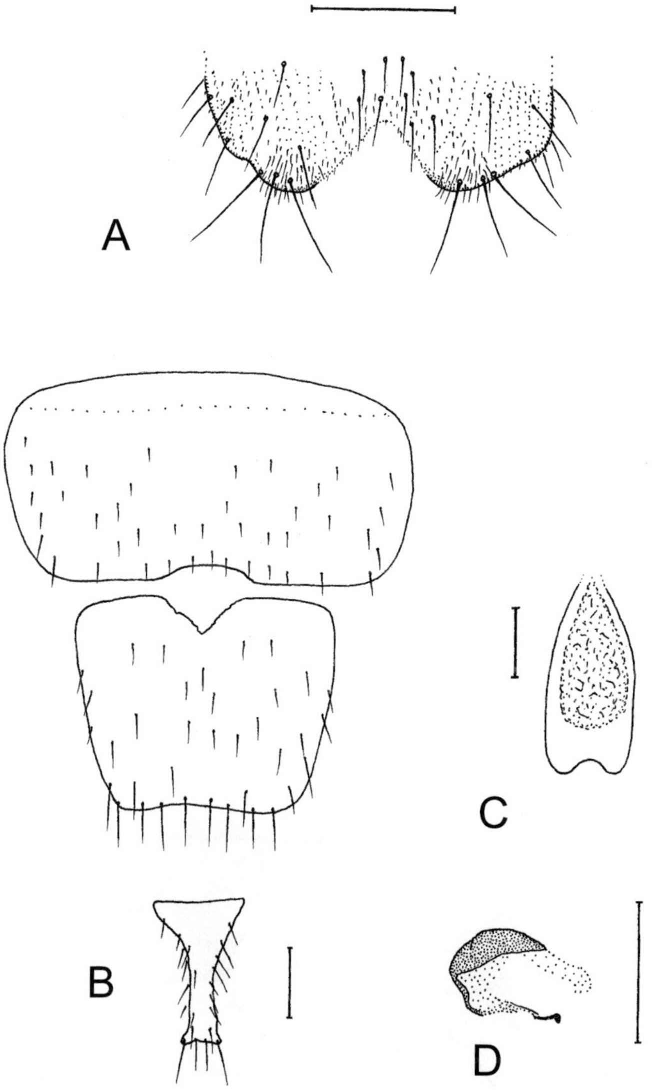

FIGURE 2 Megaselia arcticae *female, abdominal features. A, hind margin of sternum 8. B, tergites 5–7. C, Dufour's crop mechanism. D, furca (internalised sternite 9). Scale bars = 0.1 mm.*

revised (Buck & Disney, 2001), but the relevant section of the Nearctic fauna has yet to be revised. In the keys covering the relevant group of Nearctic species (Borgmeier, 1966) the males of *M. arcticae* will run to couplet 6 lead 1 of Group VI, to *M. orestes* Borgmeier. They are immediately distinguished by the strongly differentiated hairs at the tip of the proctiger, larger size (wing length exceeding 1.5 mm), darker palps with shorter bristles and costal section 2 clearly longer than 3. In the keys to the European species (Buck & Disney, 2001) the males of *M. arcticae* run to couplet 20, to *M. immodensior* Disney and *M. mimodensior* Buck. It is clearly a sibling species of these two, both of which are immediately distinguished by their pale straw yellow palps and haltere knobs and paler brown or brownish yellow legs. In *M. arcticae* the palps are light brown, the haltere knobs brown and the legs dark brown. The hypopygia of the three species are very similar. The females run to couplet 22, where neither lead applies. Its furca (vaginal sclerite) differs from that of any of the European species.

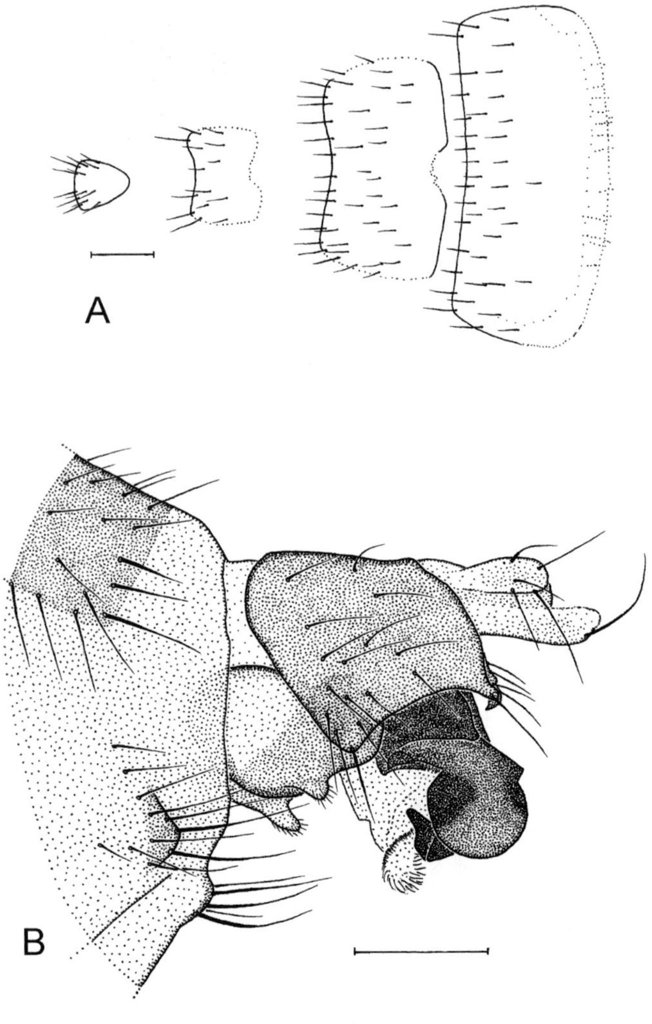

FIGURE 3 Megaselia cirriventris. *A, female, abdominal tergites 5–8. B7, male, left face of hypopygium. Scale bars = 0.1 mm.*

BIOLOGY. Ratio of males to females 2 : 1 (n = 12).

DISTRIBUTION. NE (only recorded from Zackenberg).

Megaselia cirriventris Schmitz, 1929 Fig. 3

When providing a redescription of this species Schmitz (1958: 482) erroneously synonymised *M. piliventris* Schmitz (1937: 119; a replacement name for *M. pilifera* Schmitz, 1936: 227) with it and then produced a hybrid description (Disney, 2004a). Critical features are shown in Fig. 3.

BIOLOGY. Caught June-September. Ratio of males to females is 6.8 : 1 (n = 422). Normally females of *Megaselia* possess a gland valve complex (Dufour's crop mechanism) in the neck of the crop that is associated with the dispatch of protein-rich meals to the crop instead of the stomach (Disney, 1987). In this species no DCM has been found, but the crop frequently had a suspension of rectangular fragments (typically measuring 0.003–0.01 × 0.01–0.027 mm). The eggs measure c. 0.32 × 0.13mm. Gravid females carried from 20 eggs (HF = 0.53 mm long) to 32 eggs (HF = 0.67 mm long).

DISTRIBUTION. NW, NE. Only known from Greenland.

Megaselia clara (Schmitz, 1921) Fig. 4

This is a sibling species of *M. groenlandica*. Indeed the females of these two species are almost indistinguishable. While *M. clara* tends to have paler wings with paler thin veins they are occasionally darker than usual. Likewise the wings of *M. groenlandica* are typically darker with darker thin veins, but occasional specimens occur with paler wings. The wings of *M. clara* are longer (measured from long bristle at base of costa to wing tip – see Fig. 17), but the ranges of variation just overlap:

Wing length (mm)	**1.6**	**1.7**	**1.8**	**1.9**	**2.0**	**2.1**	**2.2**	**2.3**	**2.4**	**2.5**	**2.6**	**2.7**
M. clara males	-	-	-	-	-	-	-	-	12	8	3	1
M. clara females	-	-	-	-	-	1	1	0	5	6	11	3
M. groenlandica males	-	1	3	7	2	2	2	1	-	-	-	-
M. groenlandica females	2	-	3	3	1	1	-	-	-	-	-	-

BIOLOGY. Ratio of males to females 2.9 : 1 (n = 134). Eggs measure 0.34–0.40× 0.15 mm. Gravid females carried 26 eggs (HF = 0.94 mm long) and 28 eggs (HF 1.02 mm long).

DISTRIBUTION. SW, SE.

Holarctic, arctic and alpine.

Megaselia groenlandica (Lundbeck, 1901) Figs 5, 6

A sibling species of *M. clara*. The females are almost indistinguishable (see under *M. clara* above).

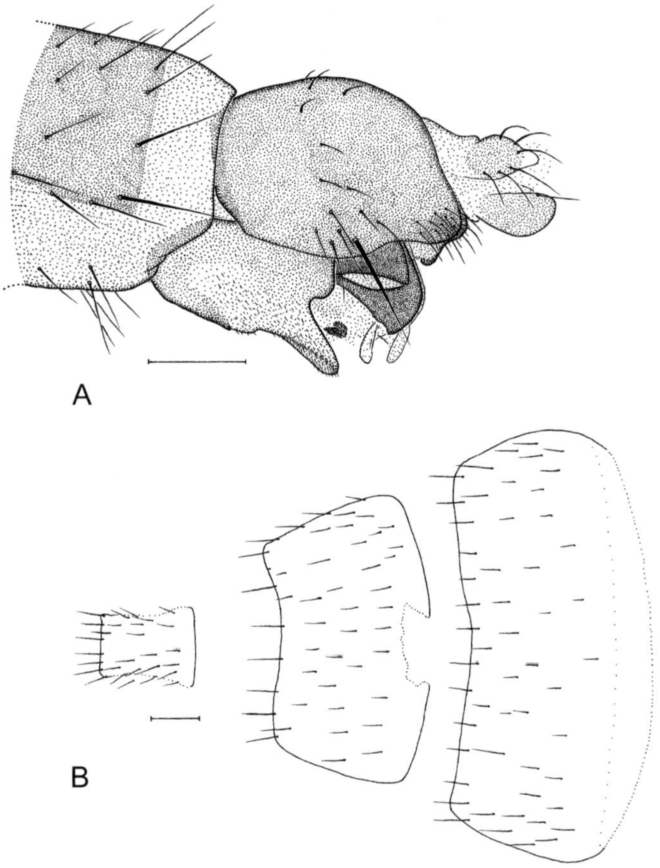

FIGURE 4 Megaselia clara. *A, male, left face of hypopygium. B, female, abdominal tergites 5–7. Scale bars = 0.1 mm.*

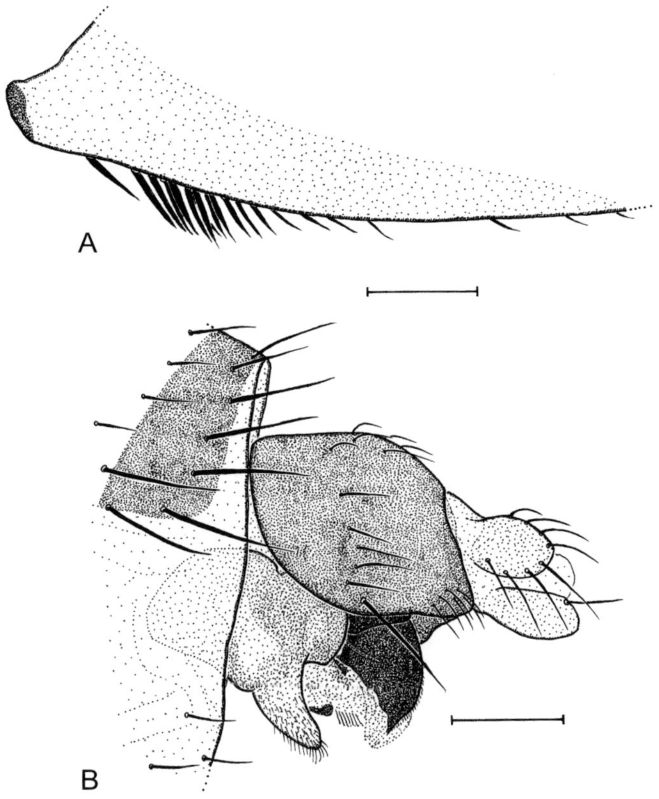

FIGURE 5 Megaselia groenlandica *male. A, hairs below base of hind femur. B, left face of hypopygium. Scale bars = 0.1 mm.*

BIOLOGY. Ratio of males to females 20 : 1 (n = 189). The eggs measure 0.42× 0.17 mm. Gravid females carried 22–24 eggs (HF ranges from 0.6 to 0.8 mm). Observed in flowers of *Dryas integrifolia* and *Ranunculus nivalis* (Philipp *et al.*, 1990). Brenner (2003) reported that in the Austrian Alps flight activity extended from mid morning to late afternoon.

DISTRIBUTION. SW, NW (north to Upernavik). Holarctic, arctic and alpine.

Megaselia pleuralis **(Wood, 1909) Figs 7, 8**

Schmitz (1957) reported this species from Greenland.

BIOLOGY. Found in a wide range of habitats and in Europe its phenology pattern is typically bivoltine, with a summer peak and an autumn peak (Disney *et al.*, 1981).

Larvae have been reared from pigeon dung (*Columba livia*) (Disney, 1994) and from a gall of *Dasineura urticae* (Cecidomyiidae) (Schmitz, 1957). In Europe adults have been recorded visiting flowers of *Angelica archangelica, A. sylvestris* (Umbelliferae), *Leontodon autumnalis* (Compositae) and *Hedera helix* (Araliaceae) (Disney, 1994). They are more readily attracted to yellow water (pan) traps than to white traps (Disney, 2004b).

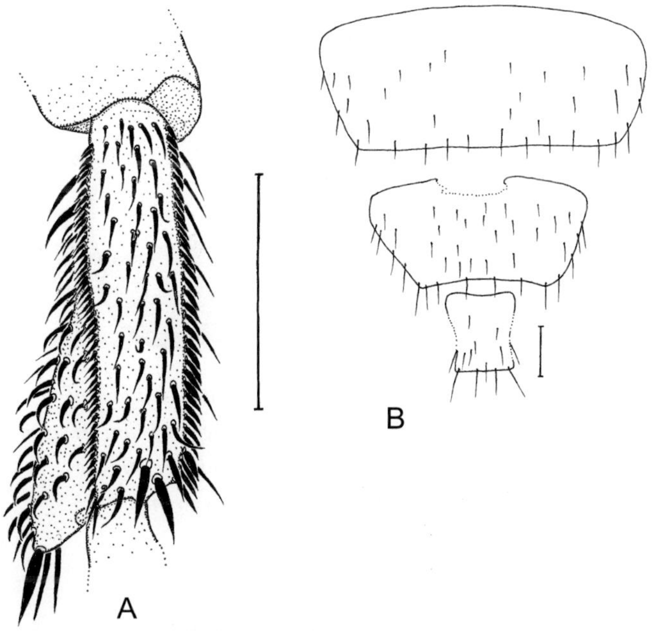

FIGURE 6 Megaselia groenlandica. *A, male, posterior face of front basitarsus. B, female, abdominal tergites 5–7. Scale bars = 0.1 mm.*

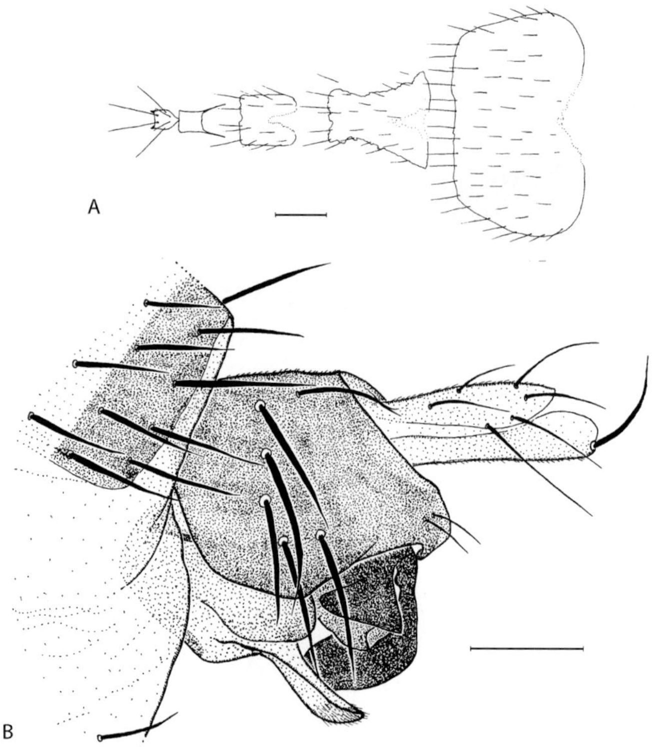

FIGURE 7 Megaselia pleuralis. *A, female, abdominal tergite 6 to epiproct. B, left face of hypopygium. Scale bars = 0.1 mm.*

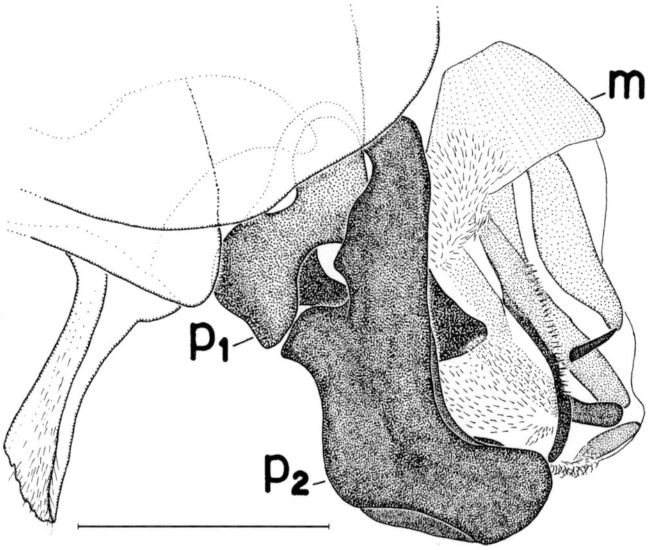

FIGURE 8 Megaselia pleuralis *male, detail of left face of hypopygium showing lobe of hypandrium (H) and the extruded penis complex embraced by the paraphyses (P1, P2). Scale bar = 0.1 mm.*

DISTRIBUTION. Schmitz gave no locality.

Originally Holarctic, temperate, but has been transported by man as far as Tasmania (Disney, 2003).

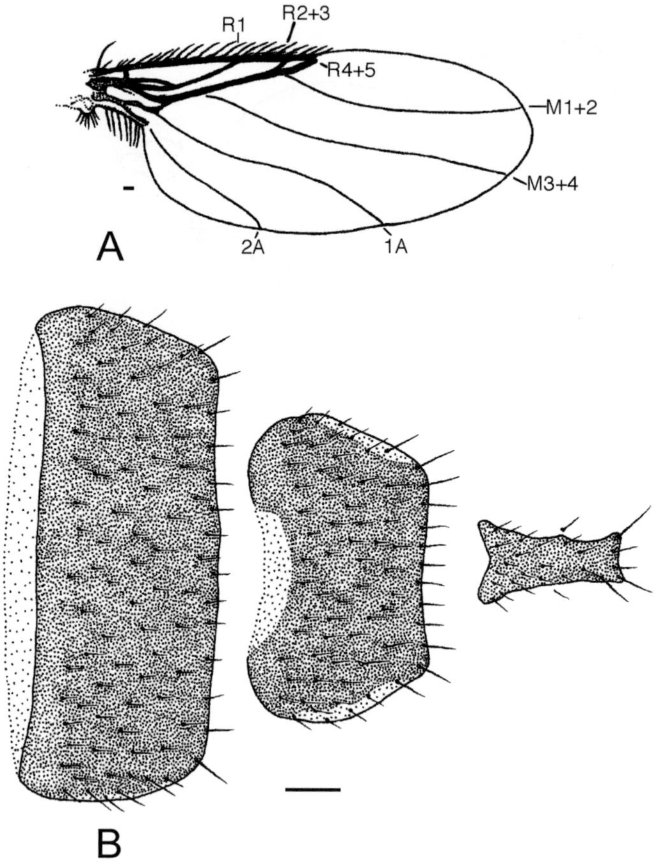

FIGURE 9 Megaselia rufipes. *A, right wing, with principal veins labelled. B, female abdominal tergites 5–7. Scale bars = 0.1 mm.*

Megaselia rufipes **(Meigen, 1804) Figs 9–13**

BIOLOGY. The eggs measure 0.5–0.6 × 0.20–0.25 mm. In England recorded batch sizes in gravid females are 24–38 eggs according to the size of the fly. The larva and puparium as Figs 12–13. The larvae exploit a wide range of decaying organic materials (Disney, 1994), especially those associated with human dwellings and including the organic material and micro-organisms in trickling filter sewage systems (Painter, 1980).

DISTRIBUTION. NW (one female from Hekla Havn, 1892).

Originally a western Palaearctic species that man has transported around the world (outside the tropics), even to Tasmania and New Zealand (Disney, 2003). Its status as a tramp species is reflected in the list of synonyms it has acquired.

References

Borgmeier, T. (1964): Revision of the North American Phorid flies. Part II. The species of the genus *Megaselia*, subgenus

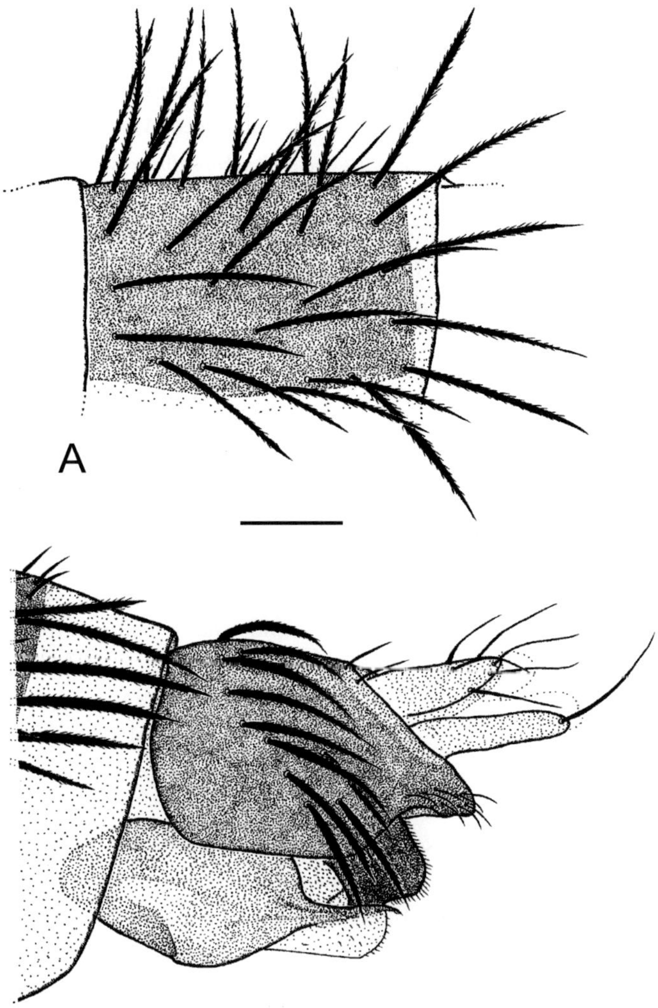

FIGURE 10 Megaselia rufipes *male. A, side view of abdominal tergite 4. B, left face of hypopygium. Scale bar = 0.1 mm.*

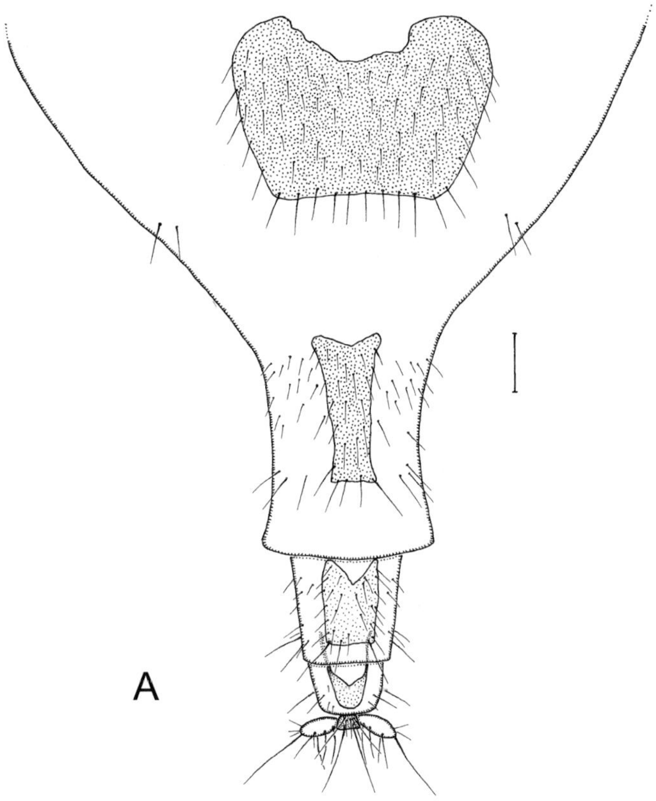

FIGURE 11 Megaselia rufipes *female, dorsal view of end of abdomen from segment 6 onwards. Scale bar = 0.1 mm.*

Aphiochaeta (Diptera, Phoridae). – *Studia Entomologica, Petropolis* 7, 257–416.

Borgmeier, T. (1966): Revision of the North American Phorid flies. Part III. The species of the genus *Megaselia*, subgenus *Megaselia* (Diptera, Phoridae). – *Studia Entomologica, Petropolis* **8**, 1–160 (1965).

Borgmeier, T. (1968): A catalogue of the Phoridae of the world (Diptera, Phoridae), – *Studia Entomologica, Petropolis* **11**, 1–367.

Brenner, S. (2003): Tageszeitliche Flugaktivität von Buckelfliegen (Insecta: Diptera: Phoridae) in der alpinen Stufe der Ötztaler Alpen (Österreich). – *Berichte des Naturwissenschaftlich-Medizinischen Vereins in Innsbruck* **90**, 207–218.

Buck, M. & Disney, R.H.L. (2001): *Megaselia giraudii* and *M. densior* species complexes of Europe, including ecological notes (Diptera, Phoridae). – *Beiträge zur Entomologie* **51**, 73–154.

Disney, R.H.L. (1983): Scuttle Flies – Diptera, Phoridae (except *Megaselia*). – *Handbooks for the Identification of British Insects* **10(6)**, 1–81, with 186 figs.

Disney, R.H.L. (1987): Observations on a peculiar mechanism in the crop of some Phoridae (Diptera) and its taxonomic value. – *Journal of Natural History* **21**, 275–280.

Disney, R.H.L. (1989): Scuttle Flies – Diptera Phoridae Genus *Megaselia*. – *Handbooks for the Identification of British Insects* **10(8)**, 1–155.

Disney, R.H.L. (1994): *Scuttle Flies: The Phoridae.* London: Chapman & Hall. xii + 467pp.

Disney, R.H.L. (2001): The preservation of small Diptera. – *Entomologist's Monthly Magazine* **137**, 155–159.

Disney, R.H.L. (2003): Tasmanian Phoridae (Diptera) and some additional Australasian species. – *Journal of Natural History* **37**, 505–639.

Disney, R.H.L. (2004a): Rescue of *Megaselia piliventris* (Schmitz) (Dipt., Phoridae) from synonymy. *Entomologist's Monthly Magazine* **140**, 27–29.

Disney, R.H.L. (2004b): Species preferences for white versus yellow water traps for scuttle flies (Dipt., Phoridae). – *Entomologist's Monthly Magazine* **140**, 31–35.

Disney, R.H.L. (2004c): A new species of *Megaselia* Rondani (Dipt., Phoridae) from Arctic Greenland. – *Entomologist's Monthly Magazine* 140, 309–311.

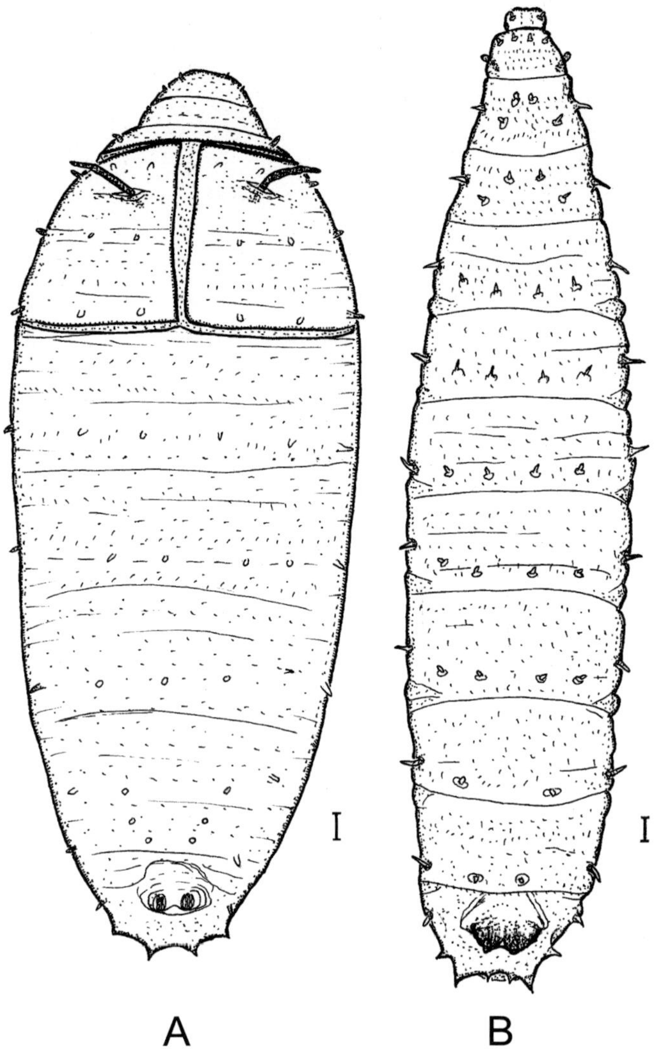

FIGURE 12 Megaselia rufipes *dorsal views of pre-adult stages. A, puparium, with eclosion plates beginning to detach. B, third instar larva. Scale bars = 0.1 mm.*

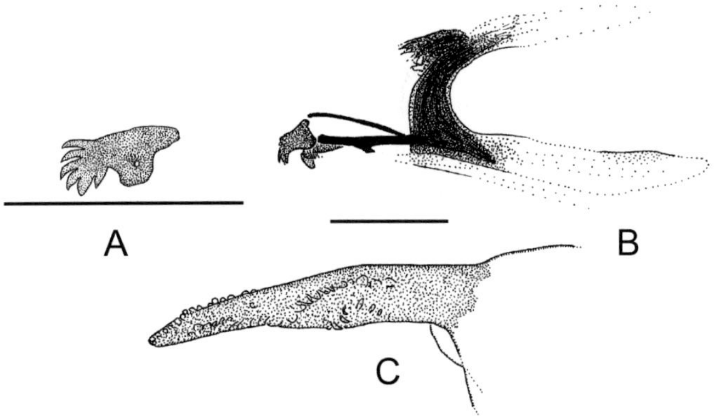

FIGURE 13 Megaselia rufipes *details of pre-adult stages. A, third instar larva, left mouth hook. B, third instar larva, left face of cephalopharyngeal skeleton and mouth hook. C, pupa, respiratory horn. Scale bars = 0.1 mm.*

Disney, R.H.L., Coulson, J.C. & Butterfield, J. (1981): A survey of the scuttle flies (Diptera: Phoridae) of upland habitats in Northern England. – *Naturalist, Hull* **106**, 53–66.

Lundbeck, W. (1901): Diptera groenlandica. *Videnskabelige Meddelelser fra Dansk naturhistorisk Forening i Kjøbenhavn* **1900**, 281–316.

Meigen, J.W. (1804): *Klassifikation und Beschreibung der europäschen zweiflügen Insekten.* Braunschweig.

Mostovski, M.B. & Disney, R.H.L. (2002): A remarkable new species of *Triphleba* Rondani (Diptera: Phoridae). – *Studia dipterologica* **8**, 557–562 (2001).

Painter, H.A. (1980): A survey of filter fly nuisances and their remedies. – *Water Research Centre Technical Report* **155**, 1–40.

Philipp, M., Böcher, J., Mattson, O. & Woodell, R.J. (1990): A quantitative approach to the sexual reproductive biology and population structure in some arctic flowering plants: *Dryas integrofolia, Silene acaulis* and *Ranunculus nivalis. Meddelelser om Grønland Bioscience* **34**, 60 pp.

Prescher, S., Bonet, J. & von Tschirnhaus, M. (2005): New records of scuttle flies (Diptera: Phoridae) from Iceland, with a survey of all simultaneously collcted Dipteran families. – *Entomologica Fennica* **16**, 237–245.

Schmitz, H. (1921): Neue europäische *Aphiochaeta*-Arten IV. (Phoridae, Dipt.). – *Entomologische Berichten de Nederlandsche Entomologische Vereeniging* **5**, 319–327.

Schmitz, H. (1929): Ergebnisse der Groenlandreise des "Pourquoi pas?" 1926. Eine neue *Megaselia*-Art (Untergattung *Aphiochaeta*) von Ost-Groenland. – *Natuurhistorisch Maandblad* **18**, 85–86.

Schmitz, H. (1936): Phoriden von der französisch-spanischen Grenze bei Hendaya. – *Tijdschrift voor Entomologie* **79**, 222–229.

Schmitz, H. (1937) Spanische Phoriden des Madrider Museums. – *Broteria* **33**, 119–127.

Schmitz, H. (1956): Phoridae. *In*: Lindner E (ed.), *Die Fliegen der palaearktischen Region* **4(33)** (Lieferung 187), 369–416.

Schmitz, H. (1957): Phoridae. *In*: Lindner E (ed.), *Die Fliegen der palaearktischen Region* **4(33)** (Lieferung 196), 417–464.

Schmitz, H. (1958): Phoridae. *In*: Lindner E (ed.), *Die Fliegen der palaearktischen Region* **4(33)** (Lieferung 202), 465–512.

Soszynska, A. & Durska, E. (2002): Cold-adapted species of *Triphleba* Rondani (Diptera: Phoridae). – *Annales Zoologici (Warszawa)* **52**, 279–283.

Wood, J.H. (1909): On the British species of *Phora*. Part II contd. – *Entomologist's Monthly Magazine* **45**, 24–29, 59–63, 113–120, 143–149, 191–195, 240–244.

17.16. Syrphidae (Hoverflies)

Jens Böcher, Pavel Láska, Libor Mazánek and Tore R. Nielsen

Adult syrphids are among the most popular dipteran, and indeed insect, groups due to their conspicuous colouration and behaviour. In many countries their distribution and biology have been studied in detail. Most species are diurnal and very active. They are strongly attracted to flowers and are important pollinators, both eating pollen and imbibing nectar. Their exposed feeding is probably responsible for their mimicry of aculeate Hymenoptera, and mimicry is more frequent and better developed in Syrphidae than in any other family of Diptera, except Conopidae. The common name ('hover flies') refers to their ability of hovering at one spot in the air – and suddenly of changing to a new position – while their wings are vibrating with extreme rapidity. The family is nearly world-wide in distribution. About 6,000 species and ca. 200 genera are recognized.

The body is usually shining, black or metallic green, very often with yellow or orange spots on head, thorax and especially abdomen; usually it has a covering of dense short hairs. The facial profile is highly variable and sometimes sexually dimorphic (Fig. 4). Antennae are (in Greenland species) borne on frontal prominences, the segments vary greatly in length and shape. The compound eyes are very large; in males the head is holoptic or nearly so. The distribution of short hairs ('pile') and bristles (if present) on thorax is taxonomically important. The wing-venation is very characteristic (*e.g.* Fig. 15C): the mid-cell (r5) and disc-cell (d) are closed by, respectively, M1+2 and M3+4, both distally bent anteriorly, together forming a 'false' margin parallel with the real margin. An extra longitudinal, so-called 'spurious vein,' is found between the veins R and M. A pterostigma is usually present, extending to apex of vein R1. The abdomen is extremely varied in shape, usually subovate, in some cases short and broad, more frequently elongate and petiolate.

The eggs are almost uniform in shape, elongate-ovoid, without processes. Their surface sculpturing is varied, providing important taxonomic characters in some groups (Chandler 1968, Kula 1993).

The larvae are distinguishable from those of other Diptera primarily by their posterior pair of spiracles (the 'spiracular plate') being elevated on a process ranging from a single short protuberance to a very long and retractile respiratory tube. The cuticle is tough and leathery, sometimes with a system of small spines; the segmentation is obscure due to transverse corrugation of the body. The dorsum of prothorax has longitudinal grooves.

Seven biological types of syrphid larvae may be recognized (Torp 1994), of which only two are represented in Greenland. In the *aphidovorous type* the body is much attenuated anteriorly and the posterior spiracular tubes very short. The body is frequently patterned with green and brown, and due to a ventral flattening and a highly changeable shape, the appearance is rather slug-like. Larvae of this type are predacious on insects, mainly aphids, and typical of the Syrphinae; the mouthparts are modified for gripping, piercing and tearing prey. In the *aquatic saprophagous type* ('rat-tail') the body terminates in a long, flexible respiratory process which may be extended to several times the length of the rest of the body. Mouthparts of the saprophages are modified for filter-feeding. This type is in Greenland represented by *Helophilus*. Terrestrial saprophagous and phytophagous larval types are thus not found in Greenland. Larval biology in Greenland has not been studied for any species.

Key to subfamilies

1. Humeri bare (Fig. 1A); head strongly convex posteriorly and closely adpressed to thorax so that humeri are partly or entirely hidden. Male: tergite 5 visible in dorsal view .... Syrphinae
– Humeri haired (Fig. 1B); head not strongly convex posteriorly, humeri clearly exposed. Male: tergite 5 not visible in dorsal view................Eristalinae

Key to genera of Syrphinae

1. Face and scutellum black, often with greenish lustre. Tergites 3 and 4 most often with yellow, grey or bluish spots. *Males*: tibia and/or fore tarsus flattened and/or broadenend, with special bristles or pile. If both fore tibia and basitarsus are cylindrical, then fore femur with a long curled bristle behind at tip. *Females*: fore tarsus often dorsoventrally flattened and slightly broadened ....*Platycheirus*
– Face and/or scutellum largely yellow 2
2. Anterior anepisternum pilose, at least posterodorsally; habitus as in Figs 13E, G–I .... *Parasyrphus*
– Anterior anepisternum bare 3

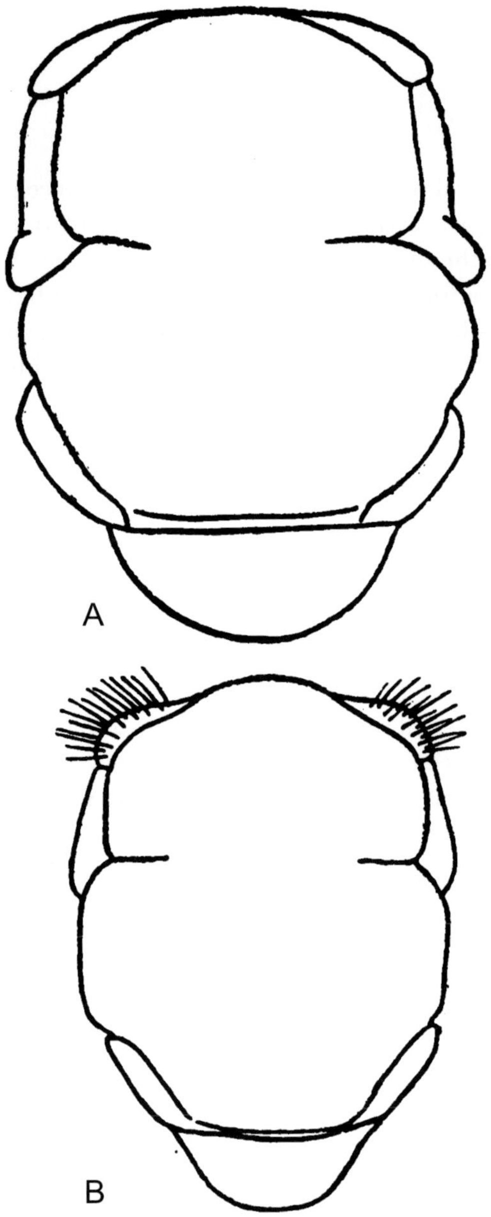

FIGURE 1 *Thorax in dorsal view of* Syrphus *(A) and* Cheilosia *(B).*

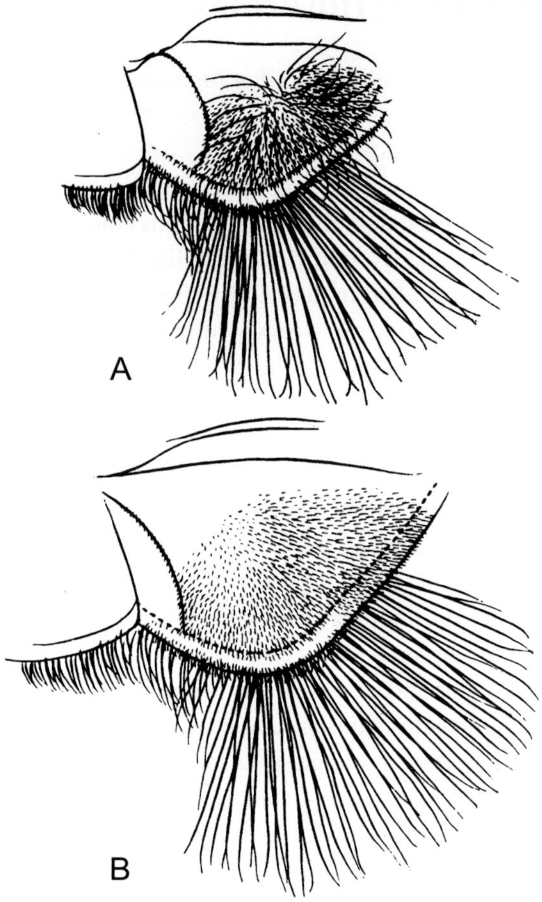

FIGURE 2 *Lower calypter of* Syrphus torvus *(A) and* Epistrophe grossulariae *(B).*

3. Thoracic dorsum: lateral margins with a sharply-defined yellow band, at least from humerus to suture, often extending to postalar knob .... *Sphaerophoria*
– Thoracic dorsum: lateral margins without sharply-defined band, black or at most with a dull yellow band, not sharply contrasting with median parts .. 4
4. Calypter with lower lobe pilose, especially on posteromedian portion (Fig. 2A). Eyes haired, in the female the hairs are short and less dense.......... .... *Syrphus torvus*
– Calypter with lower lobe bare (Fig. 2B)........... 5
5. Eyes densely haired. Face with a broad black median stripe from upper mouth-edge towards base of antennae. Antennae uniformly blackish. Tergites 3 and 4 with yellow, narrow lunulate spots, which do not reach the side margins. Sternite 2 with a blackish oval patch . Wing membrane behind marginal veins not undulated............... *Dasysyrphus pinastri*
– Eyes bare or with very short and scattered hairs. Wing membrane behind marginal veins rather broad and slightly undulated .... *Eupeodes*

Dasysyrphus pinastri **(De Geer, 1776) Fig. 13A**
Henriksen 1939: *Syrphus lunulatus* Meig.

BIOLOGY. In Europe in conifer forests and plantations, in montane and subarctic *Betula* forests; in Scandinavia occasionally in the northern tundra. The larva feeds on aphids, preferably on shrubs.

DISTRIBUTION. S, SW. Possibly quite rare (Tasermiut: Quinqa, Narsarsuaq, Kangerlussuaq/Søndre Strømfjord).

Palaearctic, low arctic-temperate (Iceland, northern and central Europe south to the Pyrenees, eastwards into Turkey and Russia through Siberia to Yakutia).

Eupeodes **Osten Sacken**
Margin of wing behind marginal veins broad with slightly undulated wing membrane; arista distant from the base

of basal flagellomere (more than 1/4 the length of the latter); eyes bare or only with minute sparse hairs; metasternum haired; anterior anepisternum bare. Greenland specimens represent usually very dark forms with greatly reduced yellow patterns on abdomen that respond to extreme low temperature during pupation (Dušek & Láska 1974). Relying on yellow colour patterns for identification of Greenland specimens can sometimes be misleading even at generic level (*e.g. Eupeodes/Parasyrphus groenlandicus*).

Key to species of Eupeodes (males and females)

1. Alula completely covered in microtrichia especially in males, in females basal part of alula sometimes with vague small bare area in the middle; tergite 5 with lateral margins at least partly yellow, sternite 5 of females with noticeable smaller dark spot than that on previous sternites; pale hairs on thorax off-white to pale yellow .... *punctifer*
– Basal part of alula with distinct bare area in the middle especially in females, in males bare area sometimes vague; tergite 5 completely black only with posterior margin narrowly brown to yellow, sternite 5 of females usually completely black as previous sternites; pale hairs on thorax almost dead-white 2
2. Scutellum pale pilose, sometimes only several black hairs intermingled on disk; face dorsally to facial tubercle yellow, median dark stripe indistinct .... *rufipunctatus*
– Scutellum black pilose on disc; face with distinct and broad dark median stripe that overlaps facial tubercle almost to antennal sockets .... *vockerothi*

***Eupeodes* (*Metasyrphus*) *punctifer* (Frey, 1934) Fig. 13B**

Face with distinct dark median stripe overlapping facial tubercle towards antennal sockets. Alula completely covered in microtrichia. Pale hairs on thorax off-white to pale yellow; scutellum black-haired. Tergite 5 with lateral margins at least partly yellow; sternite 5 of females with dark spot noticeably smaller than conterparts on previous sternites. Oral margin continuously brown-black; dust spots on frons of females present, rather small, triangular. Postero-basal long hairs on front femur almost all black. Tergites 3 and 4 usually with well developed yellow deflected spots.

BIOLOGY. Arctic and alpine communities and taiga. Flowers visited: *Dryas integrifolia* (Philipp *et al.* 1990), (outside Greenland:) *Dryas octopetala, Silene acaulis.*

FIGURE 3 Parasyrphus tarsatus (*female*) *in flower of* Angelica archangelica.

DISTRIBUTION. S, SW, NW, N, NE (not reported from SE).

Circumpolar, arctic.

***Eupeodes* (*Metasyrphus*) *rufipunctatus* (Curran, 1925) Fig. 13C**

Henriksen 1939: *Metasyrphus nigroventris* Fluke.

Face dorsally to facial tubercle yellow, median dark stripe indistinct. Basal part of alula with distinct large, bare area in the middle. Pale hairs on thorax almost whitish; scutellum pale, pilose, in some cases a few black hairs intermingled on disk. Tergite 5 usually completely black, with only posterior margin narrowly brown to yellow; sternites (including sternite 5) of females usually completely black. Oral margin continuously black; dust spots on frons of females reduced. Postero-basal long hairs on front femur almost all black. Yellow pattern on tergites usually more or less reduced.

BIOLOGY. Larva not described. Outside Greenland collected on *Salix* with both aphids and coccids present.

DISTRIBUTION. SW, NW, N, NE.

Holarctic, arctic-alpine (Canada, USA: alpine zone of British Columbia, Washington, California). In Palaearctis only Iceland.

***Eupeodes* (*Metasyrphus*) *vockerothi* (Fluke, 1952) Fig. 13D**

Henriksen 1939: not recorded.

Face with distinct and broad dark median stripe overlapping facial tubercle almost to antennal socket. Basal part of alula usually with distinct bare area in the middle. Pale hairs on thorax almost white; scutellum black pilose on disc. Tergite 5 completely black, with only posterior margin narrowly brown to yellow; all sternites of females

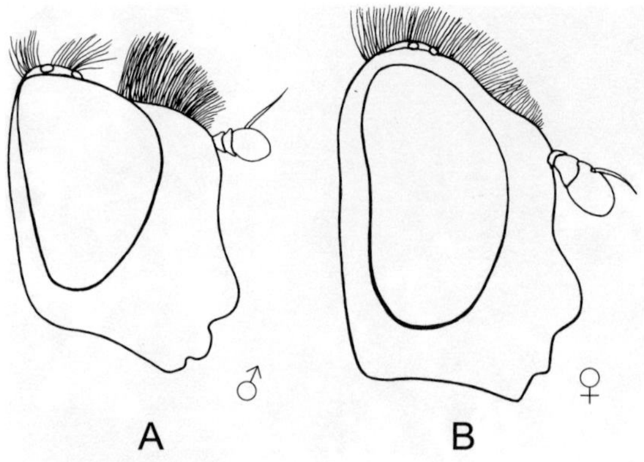

FIGURE 4 Parasyrphus groenlandicus. *Head in lateral view of male (A) and female (B).*

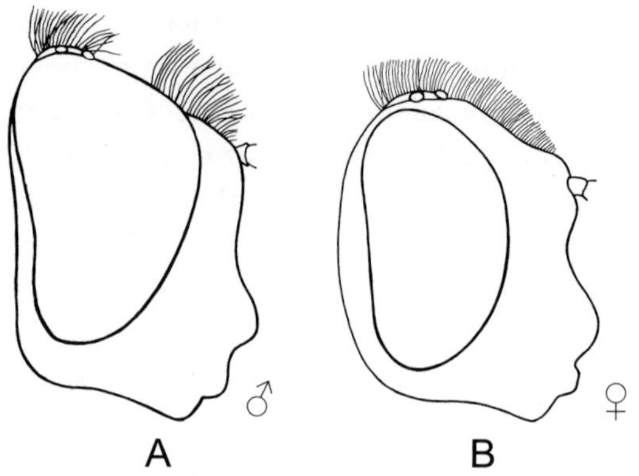

FIGURE 5 Parasyrphus tarsatus. *Head in lateral view of male (A) and female (B).*

usually completely black or almost so. Lateral border of oral cavity narrowly brown to yellow, paler than black oral margin; dust spots on frons of females distinct, usually nearly connected in the middle of frons. Postero-basal long hairs on front femur partly pale.

DISTRIBUTION. S, SW, NW, NE, SE.

Nearctic, northern (previously confused with similar species under *E. luniger*, distribution data need revision).

***Parasyrphus* Matsumura**

Eyes hairy, face without distinct black median stripe.

Key to species of Parasyrphus

1. Male 2
– Female 3
2. Frons large, very broad and swollen, with long and dense hairs. Eye angle about 130° (Fig. 4A). Tergites 2–4 with yellow-orange spots which may be small or even lacking . . . . *groenlandicus*
– Frons smaller, narrower and less swollen, with shorter and less dense hairs. Eye angle about 100° (Fig. 4B). Tergites 2–4 usually with yellow-orange spots, in arctic specimens the spots may be small or lacking . . . . *tarsatus*
3. Face more than half as wide as head (but variable) . . . . *groenlandicus*
– Face half as wide as head (variable) . . . . *tarsatus*

***Parasyrphus groenlandicus* (Nielsen, 1910) Figs 4, 13E**

Henriksen 1939: *Lasiopticus groenlandicus* Nielsen

BIOLOGY. Arctic and alpine communities and also recorded from taiga. Flowers visited (Norway): *Dryas octopetala, Rubus chamaemorus.*

DISTRIBUTION. G.

Circumpolar, arctic and alpine.

***Parasyrphus tarsatus* (Zetterstedt, 1838) Figs 3, 5, 13G–I**

Henriksen 1939: *Epistrophe tarsata* Zett.

BIOLOGY. Deciduous and conifer forest and taiga, subalpine *Betula/Pinus* forest, arctic/alpine dwarf *Betula/Salix* shrubs. Males hover at 1–2 m height and settle on bare ground, stones or rocks in the sun; both sexes often sun-bathe on foliage of *Betula* in the morning. Flowers visited: *Ranunculus nivalis, Dryas integrifolia, Silene acaulis* (Philipp *et al.* 1990), (Norway) white Apiaceae, *Ledum palustre, Papaver, Potentilla crantzii, Ranunculus, Rubus chamaemorus, Salix, Saxifraga aizoides, Taraxacum.*

DISTRIBUTION. S, SW, NW, SE.

Circumpolar, arctic-boreal, widely distributed.

***Platycheirus* Lepeletier & Serville**

Key to species of Platycheirus

1. Male 2
– Female 6
2. Fore tibia and tarsus slender throughout 3
– Fore tibia slightly to distinctly broadened towards apex, or basitarsus of fore leg distinctly broadened, or both 4
3. Apical half of posterior surface of fore femur with a regular row of rather long and stiff black setae. The seta close to femoral apex longer and with curved tip. Basitarsus of fore leg posteriorly short-haired . . . . *lundbecki*
– Apical half of posterior surface of fore femur with uniform fine hairs. Basitarsus of fore leg with about six long and weak posterior bristles, the last one or two bristles with curved apices . . . . *carinatus*

4. Fore tibia slender, only slightly broadened near apex and distinctly narrower than basitarsus. Basitarsus about 1.5 times as long as broad. Mid leg entirely black .... *groenlandicus*
– Fore tibia broadened from base to apex, at apex as broad as or broader than basitarsus. 5
5. Fore femur behind at base with a cluster of 2–3 long white bristly and flattened hairs, their tips wavy. Basitarsus of fore leg distinctly longer than wide and with a black dot posteriorly at base; its posterior margin slightly and evenly rounded, and therefore not parallel with the anterior margin. Second tarsal segment only slightly wider than long. Mid leg orange yellow, femur sometimes a little darkened at base and behind. Hind basitarsus somewhat swollen, a little thicker than hind tibia at apex .... *hyperboreus*
– Fore femur behind at base without a cluster of 2–3 long white bristly and flattened hairs. Basitarsus of fore leg as wide as long, subsquare. Second tarsal segment nearly 1.5 times as wide as long. Mid leg with femur and tibia black, except for the yellow knee region. Hind basitarsus slender, not thicker than hind tibia at apex..................... *varipes*
6. Abdomen shining black, without maculae on the tergites. Frons shining black, only very lightly dusted, without dust spots....................... *carinatus*
– Abdomen with greyish or silvery spots on tergites 2–5. Frons uniformly dusted, or with dust spots 7
7. Fore and mid legs orange yellow or reddish, the last tarsal segments may be somewhat darkened. Fore femur at base with a tuft of 2–3 long white hairs, their apices wavy. Hind femur often yellow at base. Face rather vertical, only slightly protruding .... *hyperboreus*
– Legs mainly black, the knees yellow. Fore femur at base without a tuft of bristly white hairs. Hind femur all black. Face more protruding 8
8. Frons covered with uniform light greyish or brownish dusting. 3rd antennomere roundish, not longer than wide. Tergites 3–4 with more or less distinct greyish spots, often best seen from behind .... *groenlandicus*
– Frons with isolated dust spots. 3rd antennomere elongated, longer than wide..................... 9
9. Thoracic scutum with short and some scattered longer hairs. The short hairs are scarcely as long as width of 3rd antennomere. Silvery spots on tergite 3 subtriangular, narrowest at median line of tergite and distinctly widening towards the lateral margins. Wing completely covered with microtrichia........ .... *varipes*
– Thoracic scutum uniformly covered with longer hairs, the hairs are as long as width of 3rd antennomere. Silvery spots on tergite 3 subrectangular, widening only slightly towards the lateral margins. Wing covered with microtrichia, except for a small bare area at base of cells br and bm .... *lundbecki*

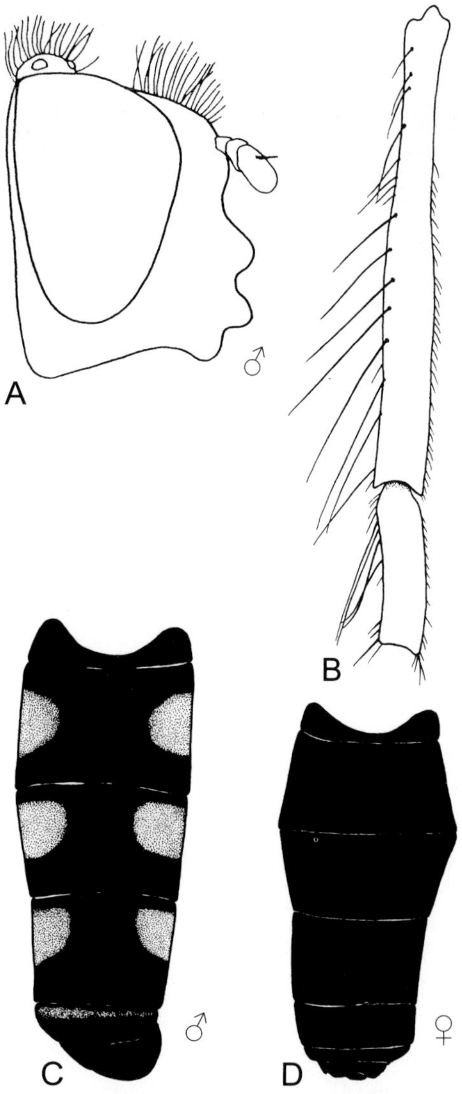

FIGURE 6 Platycheirus carinatus. *A, male head in lateral view. B, tibia and first tarsal segment. C–D, abdomens of male (C) and female (D).*

Platycheirus carinatus **(Curran, 1927) Figs 6, 13J,K**

BIOLOGY. Found in arctic and alpine communities and boreal barren mountain tops. Collected in flowers of *Dryas*.

DISTRIBUTION. N, NW, NE.

Circumpolar, arctic-alpine.

FIGURE 7 Platycheirus hyperboreus (*female*). (PHOTO: G. VÁRKONYI)

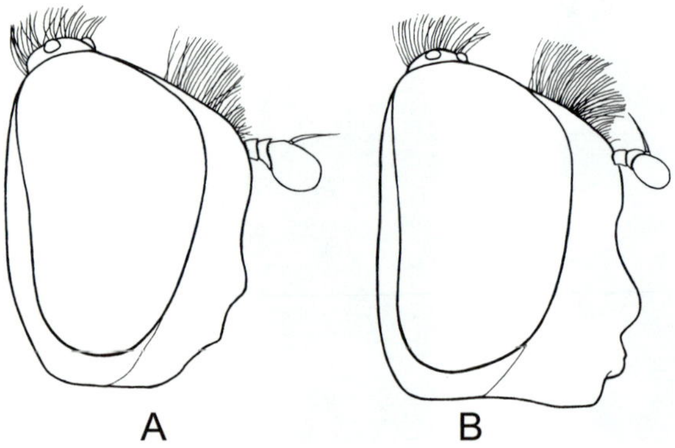

FIGURE 8 *Male heads in lateral view of* Platycheirus hyperboreus (*A*) *and* Platycheirus varipes (*B*).

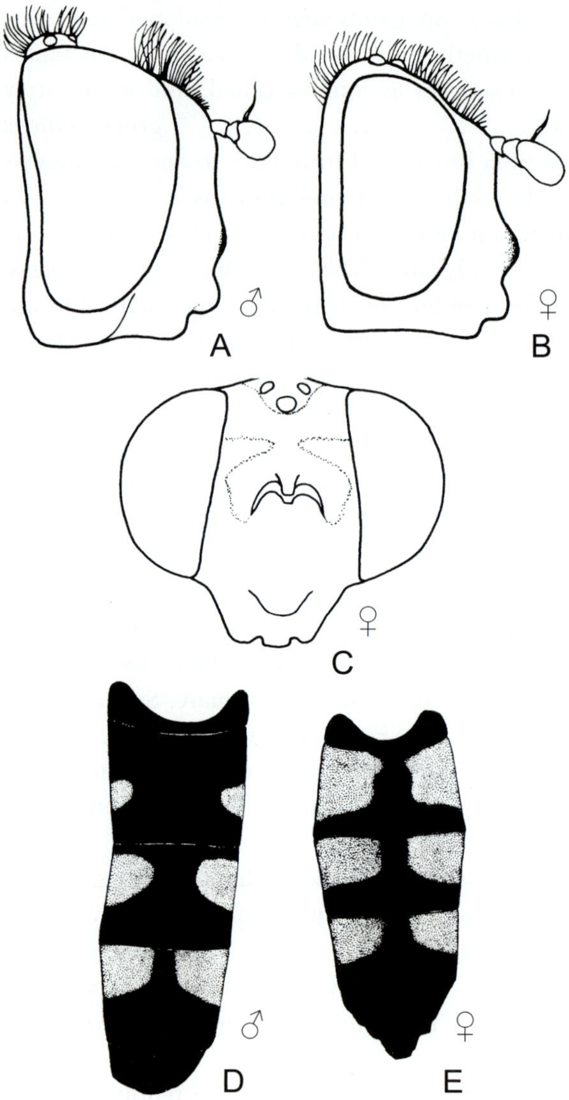

FIGURE 9 Platycheirus lundbecki. *A, head of male in lateral view. B, head of female in lateral view*). *C, head of female in frontal view. D–E, abdomens of male* (*D*) *and female* (*E*).

Platycheirus groenlandicus **Curran, 1927 Figs 10C, 13L**

BIOLOGY. Found in arctic and alpine *Betula/Salix* dwarf shrubs, sub-alpine *Betula* forest, and open areas in taiga. Flowers visited (Norway:) *Ranunculus, Salix, Sedum*.

DISTRIBUTION. NW, NE (Pituffik/Thule, Zackenberg, Ella Ø, Mestersvig).

Circumpolar, arctic-boreal and alpine.

Platycheirus hyperboreus **(Stæger, 1845) Figs 7, 8A, 10A, 14A–C**

Henriksen 1939: *Platychirus hyperboreus* Stæger.

BIOLOGY. Lives in wetland and marshes. Flowers visited (Norway:) *Carex, Ranunculus*, Apiaceae. The larva is aphidophagous.

DISTRIBUTION. S, SW. Very common.

Circumpolar, low arctic-boreal and alpine.

Platycheirus lundbecki **(Collin, 1931). Figs 9, 14D–F**

Henriksen 1939: *Melanostoma lundbecki* Collin.

BIOLOGY. Besides lakes in taiga and in arctic-alpine communities. Flowers visited: *Dryas integrifolia* (Philipp *et al.* 1990), (Norway:) *Salix*.

DISTRIBUTION. S, SW, NE. Common, but seems to be absent from N, NW and SE. Circumpolar, arctic-alpine.

Platycheirus varipes **Curran, 1923 Figs 8B, 10B, 14G,H**

Henriksen 1939: ? *Platychirus albimanus* Fabr.

BIOLOGY. Found in alpine and arctic communities, also *Betula* scrub. Flowers visited (Norway:) small, white Brassicaceae, *Ranunculus, Salix*.

DISTRIBUTION. S, SW, SE.

Circumpolar, low arctic-alpine.

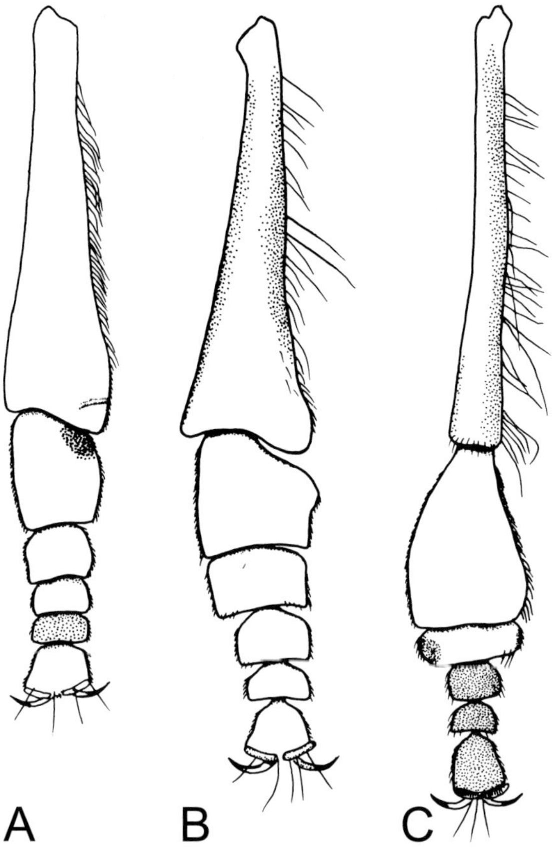

FIGURE 10 *Male left fore tibia and tarsus of* Platycheirus hyperboreus (*A*), Platycheirus varipes (*B*) *and* Platycheirus groenlandicus (*C*).

***Sphaerophoria* Lepeletier & Serville**

Male: abdomen long, distinctly longer than apex of wings. Hind femur with dense and strong setulae posteroventrally on apical three-quarters. Dorsal lobe of surstylus elongate, with broad triangular bare flattened area at base. Female: hind femur ventrally with a broad median stripe which is without black bristles.................. *scripta*

(See also the distinguishing characters of *S. abbreviata*).

***Sphaerophoria abbreviata* Zetterstedt, 1859 Figs 13K, 14I**

Published from Narsarsuaq, one female dated 16. VII 1976 (Haarto & Koponen 2003).

According to Haarto (pers. comm.) the female of S. *abbreviata* is distinguishable from that of *S. scripta* by the black frontal stripe broadening downwards, and by having the hairs on scutellum black (in *scripta* the frontal black stripe usually narrows downwards, and the hairs on scutellum are yellow, except for single black hairs along hind margin). In the Greenland material examined by us we found these differences too variable for a reliable identification.

BIOLOGY. The larva feeds on aphids.

DISTRIBUTION. S (Narsarsuaq).

Circumpolar, low arctic-temperate.

***Sphaerophoria scripta* (Linnaeus, 1758)**
Figs 14J–L, 15A

BIOLOGY. Open ground, grassland up to and including alpine and arctic communities, grassy clearings in dry woodland; heaths, garrigue and suburban gardens. Predominantly in coastal salt marshes at the northern edge of its range, further south distinctly anthropophilic, occurring with various crops and along hedges and roadside verges.

Larva described and figured by Goeldlin (1974); aphid-feeding on herbaceous plants, including various crop plants. Flowers visited (Norway): Apiaceae, *Achillea millefolium, Erigeron, Leontodon, Ranunculus.*

DISTRIBUTION. S, SW, NE, SE.

Circumpolar, low arctic-subtropical. A migratory species.

***Syrphus* Fabricius**

***Syrphus torvus* Osten Sacken, 1875 Figs 15B,C**

BIOLOGY. Conifer and deciduous forests, alpine and arctic dwarf-shrub heaths. Flowers visited: *Silene acaulis* (Philipp *et al.* 1990), (Norway:) Apiaceae, *Glaux maritima, Hieracium, Ranunculus, Salix, Sorbus, Taraxacum*. Larva described and figured by Dušek & Láska (1964); aphid feeding, occurs on trees, bushes and shrubs.

DISTRIBUTION. SW, NE, SE.

Circumpolar, arctic-temperate.

Eristalinae

Arista bare. Eyes bare, not quite meeting in the male. Thoracic dorsum with greyish or yellowish longish dust stripes, face with a bare black stripe. In Greenland only represented by two species of *Helophilus* in which the median stripe of the face is black and only the basal part of hind tibia is yellow. Fore tarsi black.

Key to species of Helophilus

1. Yellow spots on tergite 2 normally with a whitish dust spot in medial corners. Tergites 3–4 (in female also tergite 5) with strong greyish spots or lunulae. Hind part of thoracic dorsum partly with black hairs, disc of scutellum mainly black-haired. Profile of face less pointed (Fig. 12A) ....*groenlandicus*

– Yellow spots on tergite 2 without a whitish dust spot in medial corners. Tergites 3–4 black, sometimes with two more or less obvious oblique dust spots. Hind part of thoracic dorsum and disc of scutellum usually purely-yellow haired. Profile of face rather deep and pointed (Figs. 12B) ....*lapponicus*

FIGURE 11 Helophilus groenlandicus (*female*).

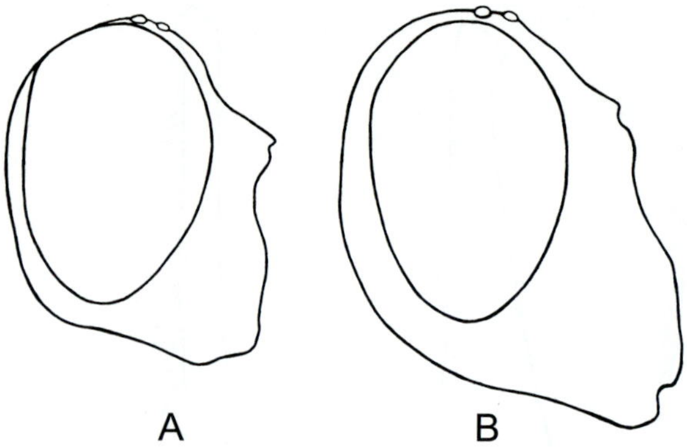

FIGURE 12 *Male heads in latereral view of* Helophilus groenlandicus (*A*) *and* Helophilus lapponicus (*B*).

FIGURE 13 *A,* Dasysyrphus pinastri (*female*)*; B,* Eupeodes punctifer*; C,* Eupeodes rufipunctatus (*female*)*; D,* Eupeodes vockerothi (*female*)*; E,* Parasyrphus groenlandicus (*female*)*; F,* Parasyrphus groenlandicus (*male*)*; G,* Parasyrphus tarsatus (*female*)*; H,* Parasyrphus tarsatus (*female*)*; I,* Parasyrphus tarsatus (*male*)*; J,* Platycheirus carinatus (*female*)*; K,* Sphaerophoria abbreviata (*female*)*; L,* Platycheirus groenlandicus (*female*).

FIGURE 14 *A,* Platycheirus hyperboreus (*female*); *B,* Platycheirus hyperboreus (*male*); *C,* Platycheirus hyperboreus (*male*); *D,* Platycheirus lundbecki (*female*); *E,* Platycheirus lundbecki (*male*); *F,* Platycheirus lundbecki (*male*); *G,* Platycheirus varipes (*female*); *H,* Platycheirus varipes (*male*); *I,* Sphaerophoria abbreviata (*female*); *J,* Sphaerophoria scripta (*female*); *K,* Sphaerophoria scripta (*female*); *L.* Sphaerophoria scripta (*male*).

***Helophilus groenlandicus* (Fabricius, 1780)**
Figs 11, 12A, 15D,E

BIOLOGY. In Greenland occurring close to wet areas like fens and bogs. Also recorded from taiga. Associated with wet decaying organic material, in particular accumulations of decaying vegetation in ponds and mud, where the larva lives in an oxygen-deficit medium. Flowers visited: *Dryas integrifolia* (Philipp *et al.* 1990), (Norway) *Matricaria, Stellaria.*

DISTRIBUTION. S, SW, NW, NE, SE.

Circumpolar, arctic-northern temperate, alpine.

***Helophilus lapponicus* Wahlberg, 1844**
Figs 12B, 15F,G

Henriksen 1939: *Helophilus borealis* Stæger

BIOLOGY. Like *H. groenlandicus* generally in humid areas like fens and bogs, in which the larva lives in wet decaying organic material, in particular accumulations of decaying vegetation in ponds and mud. Flowers visited: *Dryas integrifolia* (Philipp *et al.* 1990).

DISTRIBUTION. S, SW, NW, NE, SE.

Circumpolar, arctic and alpine.

FIGURE 15 *A,* Sphaerophoria scripta (*male*); *B,* Syrphus torvus (*female*); *C,* Syrphus torvus (*male*); *D,* Helophilus groenlandicus (*male*); *E,* Helophilus groenlandicus (*female*); *F,* Helophilus lapponicus (*male*); *G,* Helophilus lapponicus (*male*).

References

Collin, J.E. (1931): The Oxford University Expedition to Greenland, 1928. Diptera (Orthorrhapha Brachycera and Cyclorrhapha) from Greenland. – *The Annals and Magazine of Natural History Ser.*10: 67–91.

Curran, C.H. (1923): Two undescribed syrphid flies from New England. – *Occasional Papers from the Boston Society of Natural History* 5: 65–67.

Curran, C.H. (1927): Synopsis of males of the genus *Platycheirus* St. Fargeau and Serville with description of new Syrphinae (Diptera). – *American Museum Novitates* 247: 1–13.

Dušek, J., Láska, P. (1974): Influence of temperature during pupal development on the colour of Syrphid adults (Syrphidae, Diptera). (IIIrd Meeting of Czechoslovak dipterists). – *Folia Facultatis Scientiarum Naturalium Universitas Purkynianae Brunensis* 15, Biol. 43: 77–81.

Fabricius, O. (1780): *Fauna Groenlandica.* – Hafniae et Lipsiae: 1–452, spec.: 184–221.

Fluke, C.L. (1933): Revision of the *Syrphus* flies of America north of Mexico (Diptera, Syrphidae, *Syrphus* s. l.). Part I. – *Transactions of the Wisconsin Academy of Sciences, Arts and Letters* 28: 63–127, 3 pls.

Fluke, C.L. (1952): The *Metasyrphus* species of North America (Syrphidae). – *American Museum Novitates* 1690: 1–10.

Haarto, A. & Koponen, S. (2003): Syrphid flies from southern Greenland. – *Entomologiske Meddelelser* 71: 63.

Henriksen, K.L. (1939): A revised index of the insects of Grønland, containing a supplement to the insect list in Kai L. Henriksen & Will. Lundbeck: Grønlands Landarthropoder (Medd. om Grønl. Bd. 22. 1917). – *Meddelelser om Grønland* 119(10): 1–112

Nielsen, J.C. (1910): A catalogue of the insects of north-east Greenland with descriptions of some larvae. *In:* Johansen, F. and Nielsen, J.C. *The insects of the "Danmark" expedition . – Meddelelser om Grønland* 43: 55–68.

Nielsen, T.R. (1997): The Hoverfly genera *Anasimyia* Schiner, *Helophilus* Meigen, *Parhelophilus* Girschner and *Sericomyia* Meigen in Norway. – *Fauna Norvegica* Serie B 44: 107–122.

Philipp, M., Böcher, J., Mattson, O. & Woodell, R.J. (1990): A quantitative approach to the sexual reproductive biology and population structure in some arctic flowering plants: *Dryas integrifolia, Silene acaulis* and *Ranunculus nivalis.* – *Meddelelser om Grønland Bioscience* 34, 60 pp.

Speight, M.C.D. (2003): *Species account of European Syrphidae (Diptera): species of the Atlantic, Continental and Northern*

Regions. *In*: Speight, M.C.D., Castella, E., Obrdlik, P. & Ball, S. (eds.) Syrph the Net, the database of European Syrphidae 39: 1–209. Dublin.

Staeger, C. (1845): Grønlands antliater. *Naturhistorisk Tidsskrift* 1844–1845 2: 346–369.

Torp, E. (1994): *Danmarks Svirrefluer*. Danmarks Dyreliv 6. Apollo Books, Stenstrup, 490 pp.

Torp Pedersen, E. (1989): Book review: Catalogue of Palaearctic Diptera. Volume 8: Syrphidae (L.V. Peck). – *Entomologica Scandinavica* 20: 478.

Vockeroth, J.R. (1992): The Flower Flies of the Subfamily Syrphinae of Canada, Alaska, and Greenland. Diptera: Syrphidae. *In*: *The Insects and Arachnids of Canada* 18, 456 pp Agriculture Canada.

17.17. Sciomyzidae

William L. Murphy and Thorkild Munk†

Mostly cool-adapted, small to medium-sized, dull grayish-brown flies found in a wide variety of freshwater and semi-aquatic habitats that support populations of the aquatic, shoreline, and terrestrial species of snails, slugs, and fingernail clams on which the larvae feed. The family is characterized as follows: lower calypter absent; ptilinal suture clearly defined; no vibrissae; antennae three-segmented, with bare to plumose arista; postvertical orbital setae parallel to divergent; wing with costa unbroken and subcosta reaching the costa; and all tibiae with one dorsal pre-apical bristle. The puparia often have an upturned posterior end. Overwintering is usually as mature larva in the puparium.

The family Sciomyzidae currently includes five subfamilies, 63 genera and 543 valid species (Murphy 2014). Members of the family occur on all continents except Antarctica. A single species is known from Greenland (Narsarsuaq).

Knutson and Vala (2011) reviewed sciomyzid behavior, ecology, life cycles, morphology, and identification; they also provided keys to adults, larvae, and puparia and a checklist of world species. Murphy *et al.* (2012) analyzed key features of behavioral and phenological groups, behavior and development, population biology, bioindicators, ecosystem service provision and conservation, phylogenetics, molecular studies, evolutionary biology, and biocontrol.

FIGURE 1 *Typical adult* Dictya *species* (*female*).

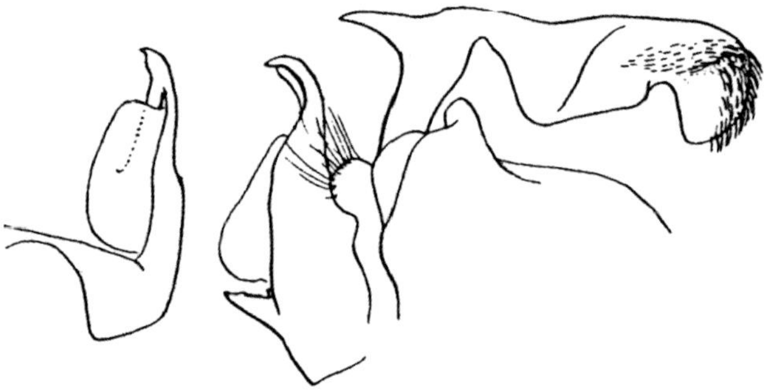

FIGURE 2 Dictya umbroides *male terminalia, anterior view* (*left*) *and lateral view* (*right*) (*both inverted*).

Dictya umbroides Curran, 1932. Figs 1–2

Species of *Dictya* are placed in the sciomyzine tribe Tetanocerini, characterized by lack of a proepisternal seta. They have a distinctive white face with a small, black spot in the center. Most setae and many setulae have a dark somatic base, some of which are confluent, presenting a mottled appearance. Some large, dull, brownish spots also are present. The wings are darkened along the front margin, becoming grayish on the posterior half, with small, translucent spots giving a mottled appearance (Fig. 1). *Dictya umbroides* is distinguished from congeners by the setose prosternum, surstylus with dorsal tip angulate and projecting well beyond the dorsal margin of the surstylus, ventral process of the hypandrium strongly bent forward and with an anterodorsal flange, and by the ventral process of the epandrium with a moderately small, narrow posterior lobe and a slightly projecting anterior lobe (Fig. 2). Female not described. The combination of pedicel matt on the dorsal side and parafrontal spots much smaller than 1/3 the diameter of the ocellar triangle identifies *Dictya umbroides* as a member of the "typical group" as characterized by Steyskal (1954). Length 4.1–5.0 mm, one of the smallest members of the genus.

BIOLOGY: The specimens first found in Greenland were collected in a Malaise trap situated in open scrub of about 2,5 m tall *Betula pubescens* on rather dry ground, but about 60 m away from a marshy area with ponds along a small stream. The site is about 1 km east of the easternmost end of the runway of the Narsarsuaq airport. Valley and Berg (1977) reared *D. umbroides* in the laboratory from adults collected in Ontario, Canada, and Oregon, U.S.A. The latter habitat was a small spring that flowed into a lake, with emergent vegetation composed mainly of grasses and sedges growing along its sides. Adults lived for as long as 137 days in captivity.

Females laid eggs as early as six days after emergence. Greatest number of eggs laid by a female was 943. Eggs produced larvae in three days. Duration of larval stages was as follows: first instar, 3–4 days; second instar, 2–9 days; third instar, 5–8 days. Each generation lasted 26–41 days.

DISTRIBUTION: S. Found only in Narsarsuaq, July 2004 and July 2010 (T. Munk).

Alaska east to Northwest Territories, south to British Columbia, east to Newfoundland; also New Mexico, Colorado, Michigan, and Ohio.

References

Knutson, L.V. & Vala, J.-C. (2011): *Biology of Snail-Killing Sciomyzidae Flies*. Cambridge, UK: Cambridge Univ. Press. xix + 506 pp.

Murphy, W.L. (2014): Rediscovery of *Dictya disjuncta* Orth (Diptera: Sciomyzidae) with descriptions of all life stages and life cycle. – *Proceedings of the Entomological Society of Washington* 116 (2): 172–187.

Murphy, W.L., Knutson, L.V., Chapman, E.G., Mc Donnell, R.J., Williams, C.D., Foote, B.A. & Vala, J.-C. (2012): Key aspects of the biology of snail-killing Sciomyzidae flies. – *Annual Review of Entomology* 57: 425–447.

Steyskal, G.C. (1954): The American species of the genus *Dictya* Meigen (Diptera, Sciomyzidae). – *Annals of the Entomological Society of America* 47: 511–539.

Vala, J.-C., Murphy, W.L., Knutson, L. & Rozkošný, R. (2012): A cornucopia for Sciomyzidae (Diptera). – *Studia dipterologica* 19: 67–137.

Valley, K.R. & Berg, C.O. (1977): Biology and immature stages of snail-killing Diptera of the genus *Dictya* (Sciomyzidae). – *Search Agriculture, Entomology* 18 (Ithaca) 7 (2): 1–44.

17.18. Sepsidae

Verner Michelsen

Sepsids are small to medium-sized acalyptrate flies characterized as being 'ant-like', a description actually applying only to members of the principal subfamily Sepsinae, with their spherical head, petiolate abdomen and dark, shining integument. A diagnostic feature of the entire family is the presence of black setae at the lower hind margin of the posterior thoracic spiracle (Fig. 1B, D). Sepsid larvae feed on a variety of decaying organic substances.

The family comprises two subfamilies and is represented in all zoogeographical regions. More than 300 species have been described, one in Orygmatinae, all remaining species in Sepsinae. Each subfamily is represented by a single species in Greenland. Principal reference and source for the present account: Pont & Meier (2002).

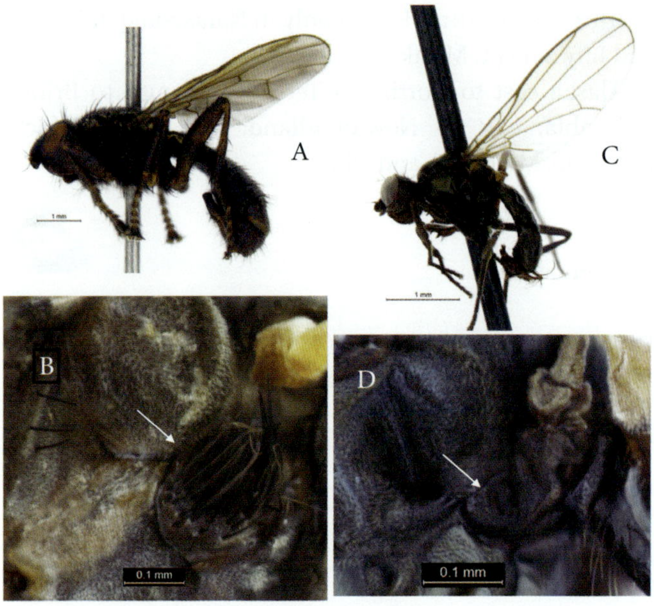

FIGURE 1 *Sepsidae.* Orygma luctuosum *male, habitus (A), and posterior thoracic spiracle (B);* Themira arctica *male, habitus (C), and posterior thoracic spiracle (D).*

Key to Greenland Sepsidae

1. Larger (wing length nearly 5 mm or more), head and body flattened, legs very robust. Head with small eyes and very broad genae .... *Orygma luctuosum* (Orygmatinae)

– Smaller (wing length less than 4 mm), rather delicate and ant-like. Head spherical in profile, with large eyes and narrow genae .... *Themira arctica* (Sepsinae)

***Orygma luctuosum* Meigen, 1830. Fig. 1A, B**

Henriksen 1939: not recorded.

First recorded from Greenland by Vibe (1950).

The stout, bristly orygmatine has little superficial similarity to sepsines.

BIOLOGY: Associated with wrack beds on seashores. Behaviour-wise distinctive by usually holding the wings close to the body while running on the larval food substrate, and relying on burrowing rather than flying as an escape strategy. Larvae feed on decaying kelps (*Laminaria, Fucus*).

DISTRIBUTION. SW.

Amphi-atlantic (E North America, NW Europe).

***Themira arctica* (Becker, 1915). Fig. 1C, D**

Henriksen 1939: not recorded.

First recorded from Greenland by Vibe (1950).

As in almost all sepsines males have the fore legs characteristically modified: femur ventrally with a peg-like and a pointed process, tibia ventrally excavated and with a blunt process. Males also have a broadly V-shaped abdominal sternite IV bearing long, slender tufts of setae curving over the abdomen.

BIOLOGY: Sepsine adults are mostly observed on/near the decomposing organic matter in which the larvae breed. They display characteristic waving movements of the partly-opened wings while running, as well as elaborate courtship/copulatory behaviours; the modified male forelegs enable a secure grasp of the female wing bases during the copulatory act. In Eurasian tundras *T. arctica* has been found on carcasses and offal. The larvae may feed on waterfowl dung, as do those of many congeners.

DISTRIBUTION. nSW.

Circumpolar, arctic-subarctic.

References

Pont, A.C. & Meier, R. (2002): The Sepsidae of Europe. – *Fauna Entomologica Scandinavica* 37: 1–221.

Vibe, C. (1950): Some insects new to Greenland. – *Entomologiske Meddelelser* 25: 419–420.

17.19. Piophilidae (Cheese skippers)

Andrey L. Ozerov

Small flies (ca. 3–6 mm), usually black (Fig. 2). Frons entirely setulose, similar in width in both sexes. Antenna short, arista bare. One pair of well-developed vibrissae. Scutum almost bare to densely hairly, usually shining. Wings longer than abdomen, hyaline. Costa broken only once near *Sc*. Subcosta complete, reaching *C*. Anal vein usually reaching wing margin at least as fold. Femora and tibiae simple. Abdomen shining, with short setulae. Segments 6–8 asymmetrical. Tergites 6 and 7 usually absent or present as fragments of tergite 6. Sternites 6–8 mostly dorsally situated. Male genitalia as a whole symmetrical, but hypandrium usually asymmertical. Ovipositor telescopic. Two spermathecae present, usually elongate, sclerotized.

Eggs 0.5–0.9 mm long, white, elongated, banana-shaped: one surface convex, the other slightly concave. Chorion with fine honeycomb-like reticulation and hatching line in the form of a weak fold.

Third instar larvae 5–10 mm long, slender, conico-cylindrical, and elongate (Fig. 1G). Cuticle white, shining and dense. Abdominal segment 7 with 1–2 pairs of protuberances laterodorsally. Abdominal segment 8 with a pair of same protuberances laterally near posterior margin, and a pair of ventral protuberances; there is a protuberance above each posterior spiracle. Anal plate small. Anterior spiracles with 2–16 lobes (Fig. 1D). Posterior spiracles sessile, spiracular plate small, with 3 radially situated slits. Accessory oral sclerite absent; pharyngeal sclerite with dorsal bridge; parastomal bar well-developed (Fig. 1C). Larvae breed mainly in rotting wood or carrion of vertebrate animals and in high-protein foodstuffs. The larva is capable of jumping: It first bends like a ring, fastening its mouthhooks to the tubercles of the posterior end of the body. Then the muscles contract, after which it quickly straightens out and thereby jumps.

Puparia brown to blackish, strongly sclerotized, subcylindrical, flattened dorsoventrally anteriorly or dorsally. Protuberances and spinules as in third instar larva.

The generic classification follows Ozerov (2004).

Key to species of Greenland Piophilidae

1. Scutellum with wart-like processes distal of each apical scutellar seta. Body with dense long hairs .... *Lasiopiophila pilosa*
– Scutellum without processes apically 2
2. *pprn* and presutural *ial* absent. Scutum delicately shagreened, with 3 rows of short setulae along *ac* (unpaired) and *dc* lines .... *Piophila casei*
– *pprn* and presutural *ial* present...................... 3
3. Frons entirely black................................. 4
– Frons yellow or reddish in lower part .... *Allopiophila vulgaris*
4. Notopleuron with greyish pollen above notopleural setae. Facial carina near lower margin with triangular opening .... *Allopiophila nigerrima*
– Notopleuron without greyish pollen above notopleural setae. Facial carina near lower margin without triangular opening.......... *Allopiophila arctica*

Allopiophila arctica (Holmgren, 1883). Fig. 1F
Similar to *Allopiophila nigerrima* (see below), but smaller (3.0–5.0 mm), differs well by stucture of male genitalia. Caudal end of larva as in Fig. 1F.

BIOLOGY. In Canada adults were recorded on the surface of a one year old grave, and some specimens were collected from dog faeces (McAlpine 1977).

DISTRIBUTION. NW (only recorded from Dundas, former Thule).

Circumpolar, arctic (Canada, Europe, Asia).

Allopiophila nigerrima (Lundbeck, 1901)
Henriksen 1939: *Piophila nigerrima* Lundb.

The largest of the Greenland Piophilidae (4.0–7.0 mm). Frons, thorax and abdomen entirely black. Legs black, only apex of femora and base of tibiae yellowish. Scutum uniformly bearing hairs. *pprn* and *ph* present. Anepisternum bare. Scutellum without tubercles. The larva is unknown. Biology unknown.

DISTRIBUTION. NW (Inglefield Land, Marshall Bugt; Hall Land, ca. 40 km south of Newman Bay).

Nearctic, high arctic (besides Greenland in northern Canada).

Allopiophila vulgaris (Fallén, 1820)
Henriksen 1939: *Piophila vulgaris* Fall.

Frons yellow in lower half and usually blackish in upper half. Face, gena and palpus yellow. Antenna yellow, but postpedicel sometimes darkened. Scutum uniformly bearing short hairs, *pprn* and *ph* present. Anepisternum bare. Scutellum without tubercles.

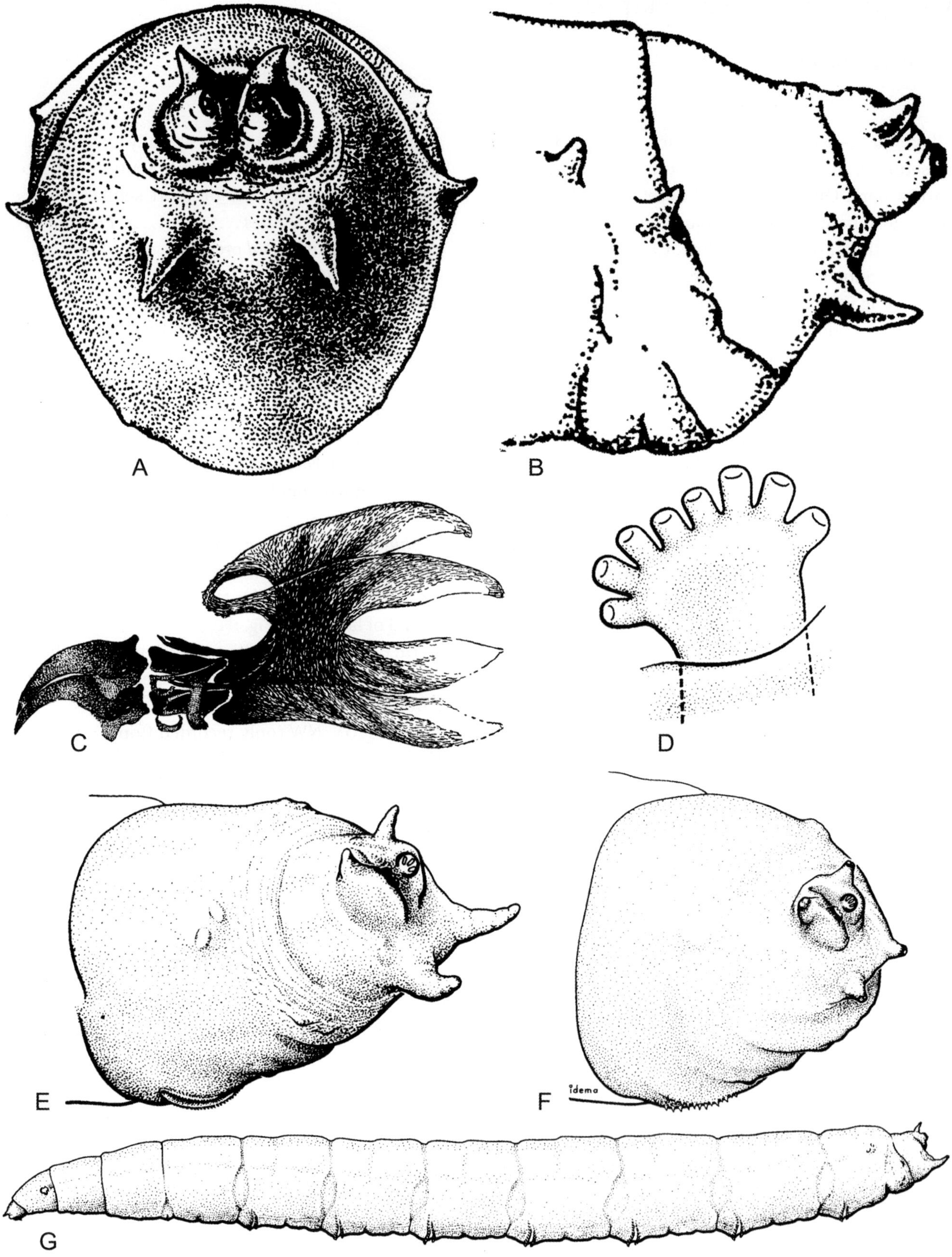

FIGURE 1 *Piophilidae, larval characters. A, caudal end of larva of* Piophila casei, *posterior view; B, same, lateral view; C, cephaloskeleton of larva of* Piophila casei, *lateral view; D, anterior spiracle of larva of* Lasiopiophila pilosa; *E, caudal end of larva of* Lasiopiophila pilosa, *posterolateral view; F, caudal end of larva of* Allopiophila arctica, *posterolateral view; G, larva of* Lasiopiophila pilosa, *lateral view. A–C after Zimin (1948); D–G after McAlpine (1977).*

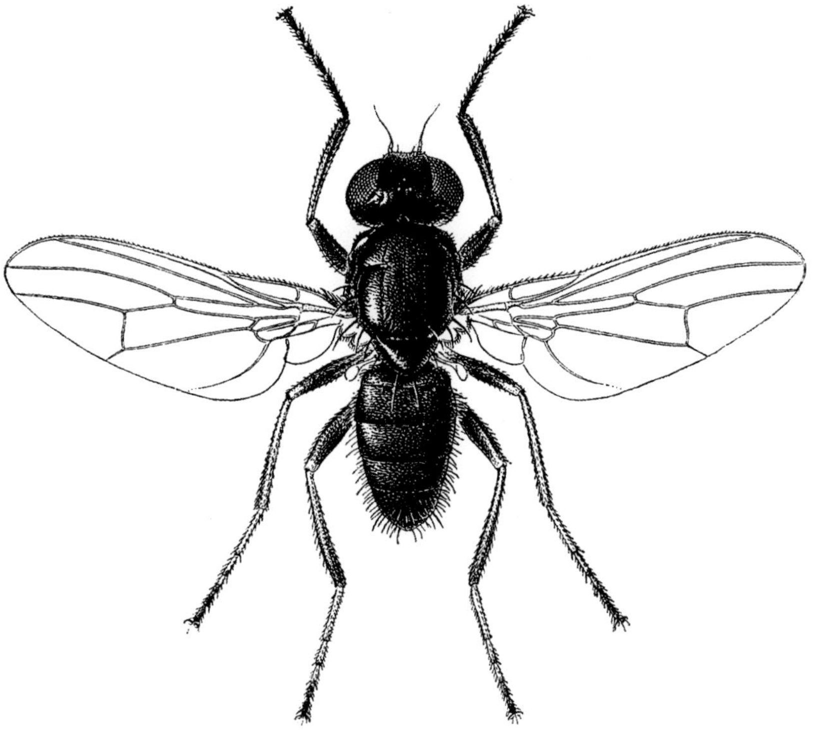

FIGURE 2
Piophila casei, *male habitus. After Stackelberg (1956).*

The larva described by Hennig (1943), similar to that of *A. arctica.*

BIOLOGY. Observed on dead bodies of animals and their bones, and on faeces. Most likely connected with carrion in Greenland.

DISTRIBUTION. G.

Holarctic, arctic-subtropical, widespread.

***Lasiopiophila pilosa* (Staeger, 1845). Figs 1D–E, G**

Henriksen 1939: *Piophila pilosa* Stæger

Frons black, but in lower third yellow. Face and parafacial yellow. Gena yellow in upper half under eye and black on the remaining surface. Scutum uniformly bearing relatively long hairs (slightly shorter than setae). *pprn* present, *ph* absent. Anepisternum with long hairs in posterior half. Scutellum with two short apical tubercles. Caudal end of larva as in Fig. 1E.

BIOLOGY. No data for Greenland. Possibly adults together with adults of *A. vulgaris* may be found on carcasses of mammals.

DISTRIBUTION. G, apart from N.

Circumpolar, arctic.

***Piophila casei* (Linnaeus, 1758) (Cheese skipper). Figs 1A–C, 2**

Frons in upper half blackish-yellow to black, in lower half reddish yellow; sometimes yellow in low third of frons only. Face, gena, palpus and antenna yellow. Scutum thinly shagreened, with 3 rows short hairs between lines *ac* and along each *dc* lines (Fig. 1H). *pprn* and *ph* absent. Anepisternum in posterior half bearing scattered hairs. Scutellum without tubercles.

Caudal end of larva as in Figs 1A–B.

BIOLOGY. Well-known as a serious pest in the food industry; the larva live in cheese, salted fish, ham etc, in old carcasses of various animals. No data for Greenland, but possibly only in places where fish are processed. Adults only collected in August.

DISTRIBUTION. nSW (only reported from Saqqaq).

Cosmopolitan.

References

Hennig, W. (1943): Piophilidae. *In*: Lindner, E. (ed.): *Die Fliegen der palaearktischen Region* 5 (1): 1–53.

Henriksen, K.L. (1939): A revised index of the insects of Grønland. – *Meddelelser om Grønland* 119 (10): 1–112.

McAlpine, J.F. (1977): A revised classification of the Piophilidae, including 'Neottiophilidae' and 'Thyreophoridae' (Diptera: Schizophora). *Memoirs of the Entomological Society of Canada* 103, I–VI+1–66 pp.

Ozerov, A.L. (2004): K klassifikatsii dvukrylykh semeystva Piophilidae (Diptera) [On classification of the family

Piophilidae (Diptera)]. *Zoologicheskiy zhurnal* 83: 1353–1360 [in Russian with English summary].
Stackelberg, A.A. (1956): Sinantropnye dvukrylye fauny SSSR. – *Opredeliteli po faune SSSR, izdavaemye Zoologicheskim institutom AN SSSR* 60, 164 pp [in Russian].
Zimin, L.S. (1948): Opredelitel' lichinok sinantropnykh mukh Tadzhikistana (po III stadii). – *Opredeliteli po faune SSSR, izdavaemye Zoologicheskim institutom AN SSSR* 28, 116 pp [in Russian].

17.20. Agromyzidae

Leif Lyneborg† and Jens Böcher

Agromyzid flies are small (only few millimeters long) dark greyish acalyptrate flies (Fig. 2A). Useful diagnostic features of an agromyzid are found on the head (Fig. 1A) and the wings (Figs 1B–G). The head has a pair of diverging setae behind the ocellar triangle (Fig. 1A: 7) and presence along the frontal eye margins of several pairs of strong setae, the lower ones of which (Fig. 1A: 9) are incurved, whereas the upper ones (Fig. 1A: 8) are outcurved. In the wing the subcostal cell (Figs 1B–C) offers important characters: Firstly it always shows a break of the costa, secondly the position of this break relative to the termination of subcosta separates the two subfamilies, Agromyzinae and Phytomyzinae. All Greenland agromyzids belong to the last mentioned subfamily.

All agromyzids are dependent on live plant tissue for their development. The females possess a rigid oviscapt by help of which they place the eggs in the tissue. The larvae may form leaf-mines (see Fig. 2B for an example) or they live hidden in tunnels in stems or flower heads. The hosts of the Greenland agromyzids are unknown. The larvae are maggots. Pupation takes place inside the host plant or in the ground below. Adults can be swept from vegetation.

Nine species in four genera are recorded from Greenland (Griffith 1966). The species can be separated with the following key, which is mainly based on external characters. In order to achieve more certain identifications, the male genitalia should be investigated. For wing characters see Figs 1D–G. The "sections" of costa denote the parts of costa between insertions of other veins (1st section from origin to insertion of sub-costa, 2nd section from this insertion to insertion of radius 2+3, 3rd section from this to insertion of radius 4+5, 4th section from this to insertion of media 1+2; consult Spencer 1976).

Key to the Greenland species of Agromyzidae

1. Costal vein not extending beyond vein R_{4+5} (Figs 1F, G) 2
– Costal vein reaches vein M_{1+2} (Figs 1D, E) 3
2. Both cross-veins present, positioned close to the wing-base (Fig. 1F) .... *Napomyza parvicella*
– Lower cross-vein absent (Fig. 1G) .... *Phytomyza*, 5
3. Wing tip at vein M_{1+2} (Fig. 1E) .... *Phytoliriomyza arctica*
– Wing tip between veins R_{4+5} and M_{1+2} (Fig. 1D) .... *Cerodontha*, 4
4. Male aedeagus as in Figs 4A, B. Female with somewhat enlarged third antennal segment (somewhat resembling that of male, Fig. 3A) .... *Cerodontha* (*Dizygomyza*) *lindrothi*
– Male aedeagus as in Fig. 4C, D. Female with small, rounded third antennal segment .... *Cerodontha* (*Dizygomyza*) *thulensis*
5. Rear upper orbital bristle very weak or absent, only 1 lower orbital bristle 6
– Both upper orbital bristles of about equal length... 7
6. Scutellum, sides of mesonotum, pleura, frons and first 2 antennal segments yellow. The relationship between lengths of costal sections 2 and 4 is 2.2–3.0. Hypophallus clearly sclerotised; distiphallus variable, but not forming more than about 5 coils (Fig. 5E) .... *Phytomyza ranunculi*
– Thorax and scutellum almost completely dark. First two antennal segments dark. The relationship between lengths of costal sections 2 and 4 is 1.6–2.0. Hypophallus virtually unsclerotised; distiphallus enormously long, forming about 10 coils (Fig. 5J) .... *Phytomyza vibeana*
7. Frons yellow. Femora marked with yellow. Acrostichals 2–3 rowed. Oviscapt very elongate, of length as rest of abdomen. Distiphallus small and remote from the paraphalli (Figs 5H, I) .... *Phytomyza varipes*
– Frons dark at least centrally. Femora dark except at their tips. Oviscapt not so elongate 8
8. Acrostichals in 2 rows. Aedagus as in Figs 5B, C. Ejaculatory bulb as in Fig. 5A .... *Phytomyza fuscula*
– Acrostichals in about 4 rows. Aedeagus as in Fig. 5D. Ejaculatory bulb minute .... *Phytomyza opacella*

Cerodontha (*Dizygomyza*) *lindrothi* Griffiths, 1964. Figs 3A, 4A–B

Externally very similar to the palaearctic *C.* (*D.*) *luctuosa* Meigen and C. (*D.*) *effusi* Kari, but can clearly be distinguished by the male aedeagus (Fig. 4A–B).

BIOLOGY. Unknown. The most likely larval food plant is *Carex* spp.

DISTRIBUTION. S, NE (only recorded from three places in the Qaqortoq/Julianehåb Bugt area and from Scoresby Sund: Hekla Havn on Danmark Ø).

Also known only from Iceland.

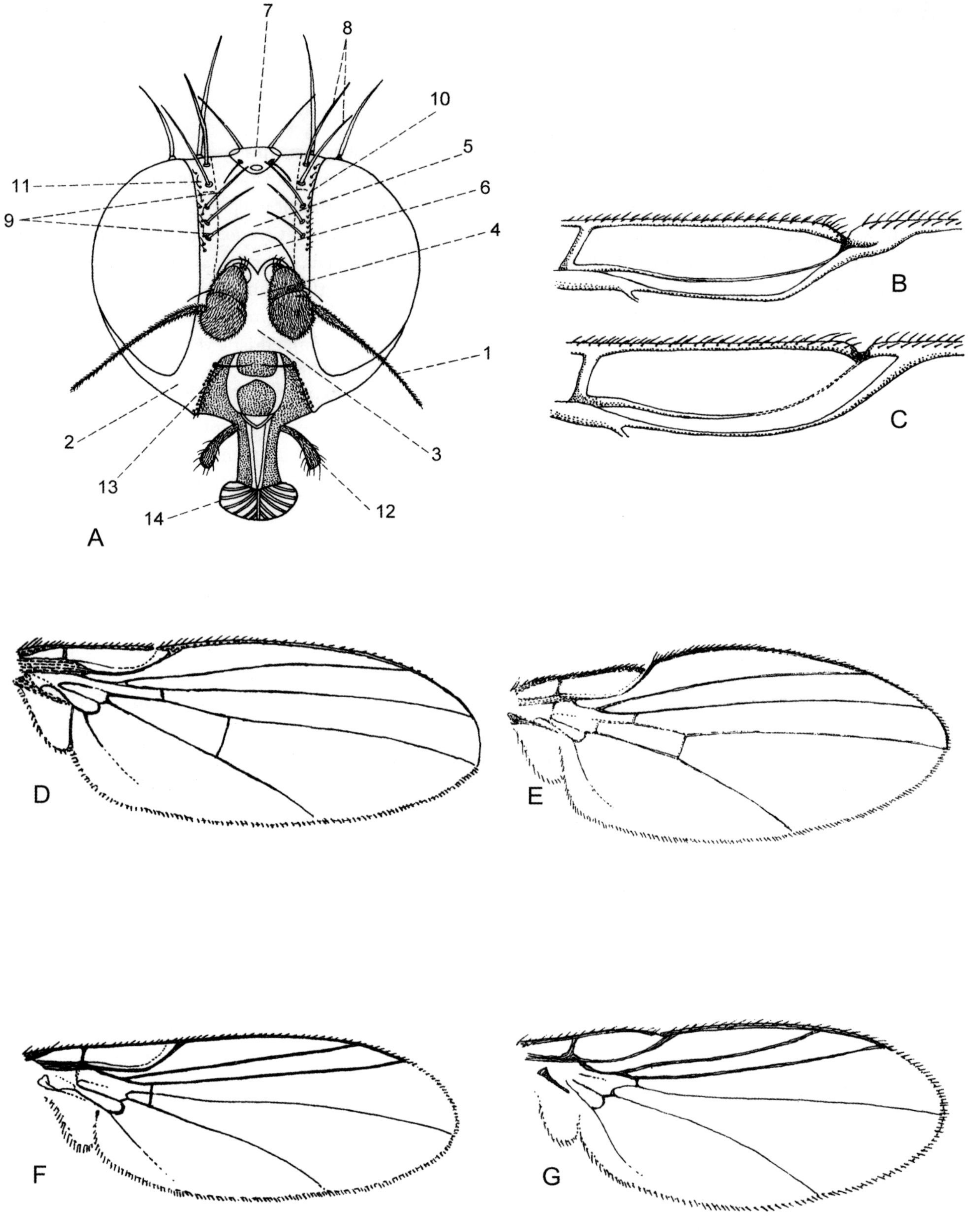

FIGURE 1 *A, head of an agromyzid fly, frontal view: 1, arista; 2, jowl; 3, face; 4, facial keel; 5, frons; 6, lunule; 7, ocellar triangle; 8, upper orbital bristles (ors); 9 lower orbital bristles (ori); 10, orbital setulae; 11, orbit; 12, maxillary palp; 13, vibrissa; 14 labellum of proboscis; B–C, anterior basal wing region of agromyzines (B) and phytomyzines (C); D–G, wings of* Cerodonta *(D),* Phytoliriomyza *(E),* Napomyza *(F),* Phytomyza *(G). After Spencer (1976).*

FIGURE 2 Phytomyza fuscula. *Habitus of adult (A); larval mine in grass* (Arrhenatherum) *leaf (B; after Spencer 1976).*

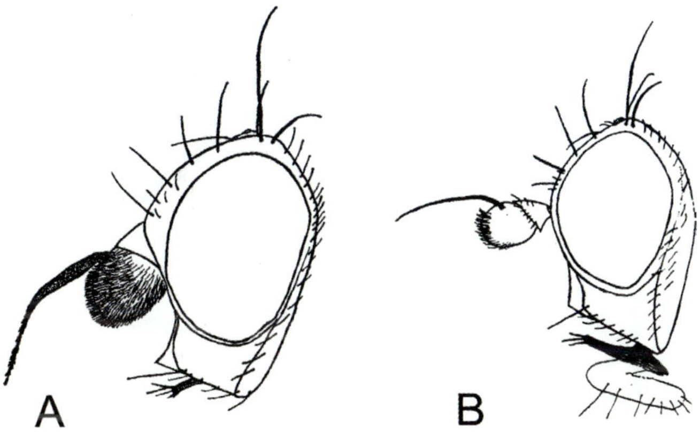

FIGURE 3 Cerodonta *heads, lateral view. Male* C. lindrothi *(A); female* C. thulensis *(B). After Griffiths (1966).*

***Cerodontha* (*Dizygomyza*) *thulensis* Griffiths, 1966. Figs 3B, 4C–D**
Female head and aedeagus, see Figs 3B, 4C–D.

BIOLOGY. Unknown. The most likely larval food plants are *Carex* spp.

DISTRIBUTION. NW. Only recorded from Pituffik/Thule.

Nearctic, arctic.

***Phytoliriomyza arctica* (Lundbeck, 1900). Fig. 4E**
Henriksen (1939): *Agromyza arctica* Lundb.

The colouration is variable and especially the colour of the frons ranges from yellow to dark brown. Aedeagus, see Fig. 4E.

BIOLOGY. The larval host plant in Greenland is unknown but is probably a species of Asteraceae.

DISTRIBUTION. S, SW (north to *c.* 70.5°N), NE (north to *c.* 73°N), SE. A common and widely distributed species.

Circumpolar and Holarctic, arctic-subtropical. Unusually wide distribution in several zoogeographical regions, including Brazil (Neotropical region).

***Napomyza parvicella* (Coquillet, 1902). Figs 4F–H**
Aedeagus and ejaculatory bulb, see Figs 4F–H.

BIOLOGY. Collected on *Papaver radicatum* in the Pituffik/Thule area 650 m a.s., but it is not proved that this plant is the host.

DISTRIBUTION. NW (hitherto only recorded from the Pituffik/Thule area).

High-arctic, also recorded from Alaska: St. Paul Island.

Phytomyza Fallén, 1810. A huge genus with hundreds of species. Five species have been recorded from Greenland, among them are some of the most common and widest distributed species in the family.

***Phytomyza fuscula* Zetterstedt, 1838. Figs 2, 5A–C**
Henriksen 1939: *Phytomyza nigritella* Zett.

Aedeagus and ejaculatory bulb, see Figs 5A–C.

BIOLOGY. The larva is leaf-miner on various Poaceae and in northern Scandinavia it has been a pest on cereals (Spencer 1976).

DISTRIBUTION. S, SW, NW, N (Jørgen Brønlund Fjord), NW (Scoresby Sund: Røde Ø, Hekla Havn, Mestersvig).

In Greenland the most common and widespread species of Agromyzidae.

Holarctic, arctic-temperate. Not known from Iceland and the Faroes and not reported from Russia.

***Phytomyza opacella* Hendel, 1935. Fig. 5D**
Henriksen 1939: *Phytomyza obscurella* Fall.

Aedeagus, see Fig. 5D.

BIOLOGY. Unknown, but the larva is probably a miner on Poaceae.

DISTRIBUTION. S, SW (north to Uummannaq).

Palaearctic, low arctic-temperate and alpine.

***Phytomyza ranunculi* (Schrank, 1803). Figs 5E–F**
Henriksen 1939: *Phytomyza maculipes* Zett.

Aedeagus, see Figs 5E–F.

BIOLOGY. The larva forms conspicuous white mines on leaves of *Ranunculus* spp. In Scandinavia there are several annual generations, and larvae can be found feeding even in winter, in plants covered by snow. Pupation takes place externally (Spencer 1976).

DISTRIBUTION. S (several localities), SE (Ammassalik).

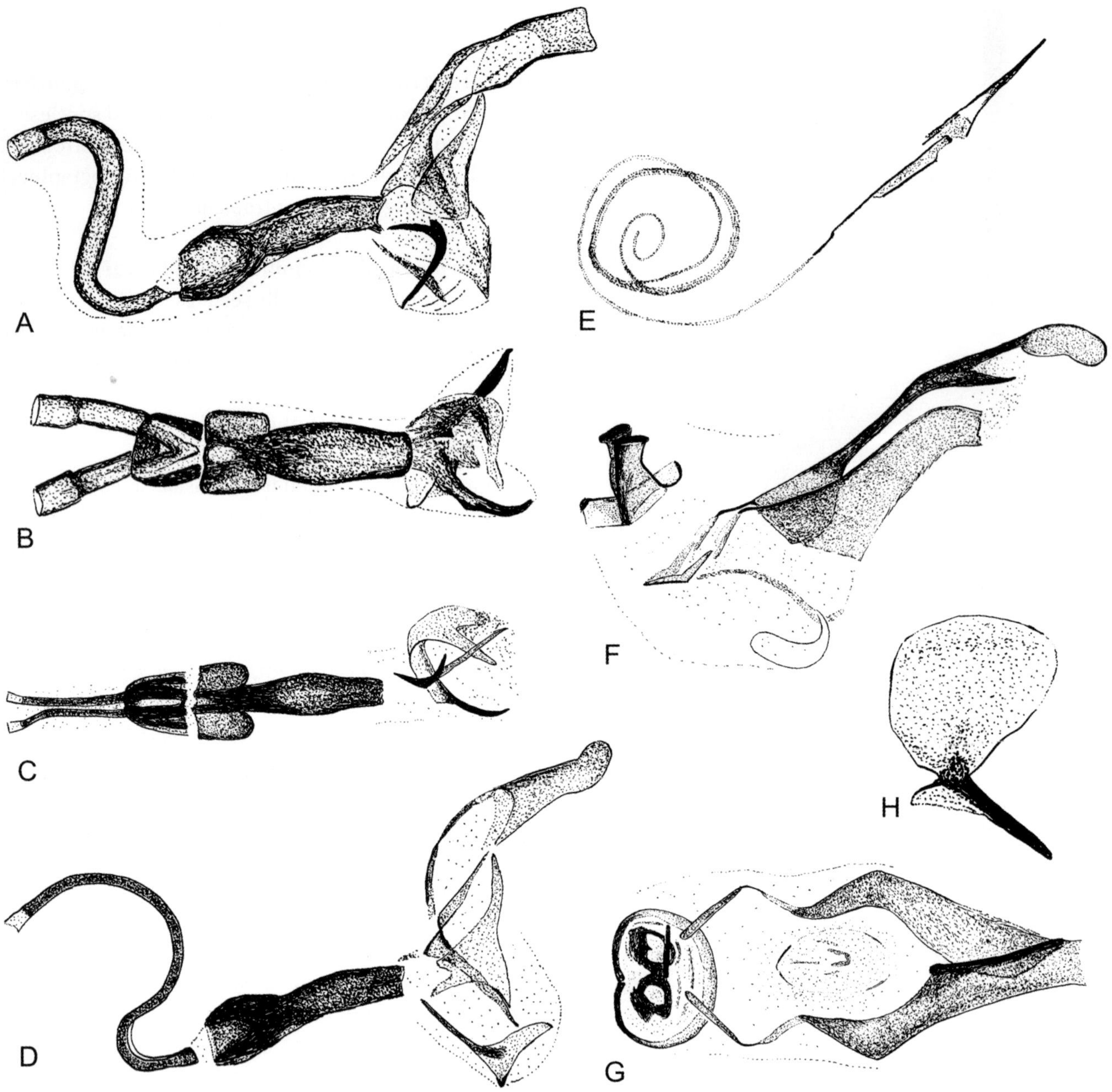

FIGURE 4 *Male genitalia of* Cerodonta, Phytoliriomyza *and* Napomyza. *A–B,* C. lindrothi, *aedeagus in lateral (A) and ventral (B) view; C–D,* C. thulensis, *aedeagus in ventral (C) and lateral (D) view; E,* P. arctica, *aedeagus in lateral view; F–H,* N. parvicella, *aedeagus in lateral (F) and ventral (G) view, ejaculatory bulb (H). After Griffiths (1966).*

Holarctic, low arctic-subtropical and alpine, widely distributed.

Phytomyza varipes **Macquart, 1835. Figs 5G–I**
The pale third antennal segment in both sexes and the conspicuously long oviscapt make this a distinctive species. For male genitalia see Figs 5G (ejaculatory bulb) and 5H–I (aedeagus, distiphallus, paraphalli).

BIOLOGY. The larva feeds on the seeds in fruits of *Rhinanthus* spp. The oval, shining black puparium is found inside the fruits. In Iceland the imago has been bred from *Rhinanthus minor*.

DISTRIBUTION. S (only recorded from Qassiarsuk and Qanisaartut).

Holarctic, low arctic-temperate (Europe, Canada: Nova Scotia).

Phytomyza vibeana **Griffiths, 1966. Figs 5J–K**
Aedeagus and ejaculatory bulb, see Figs 5J and K, respectively.

DISTRIBUTION. NW, NE (only recorded from Upernavik and Scoresby Sund: Hekla Havn on Danmark Ø).

Nearctic, low arctic.

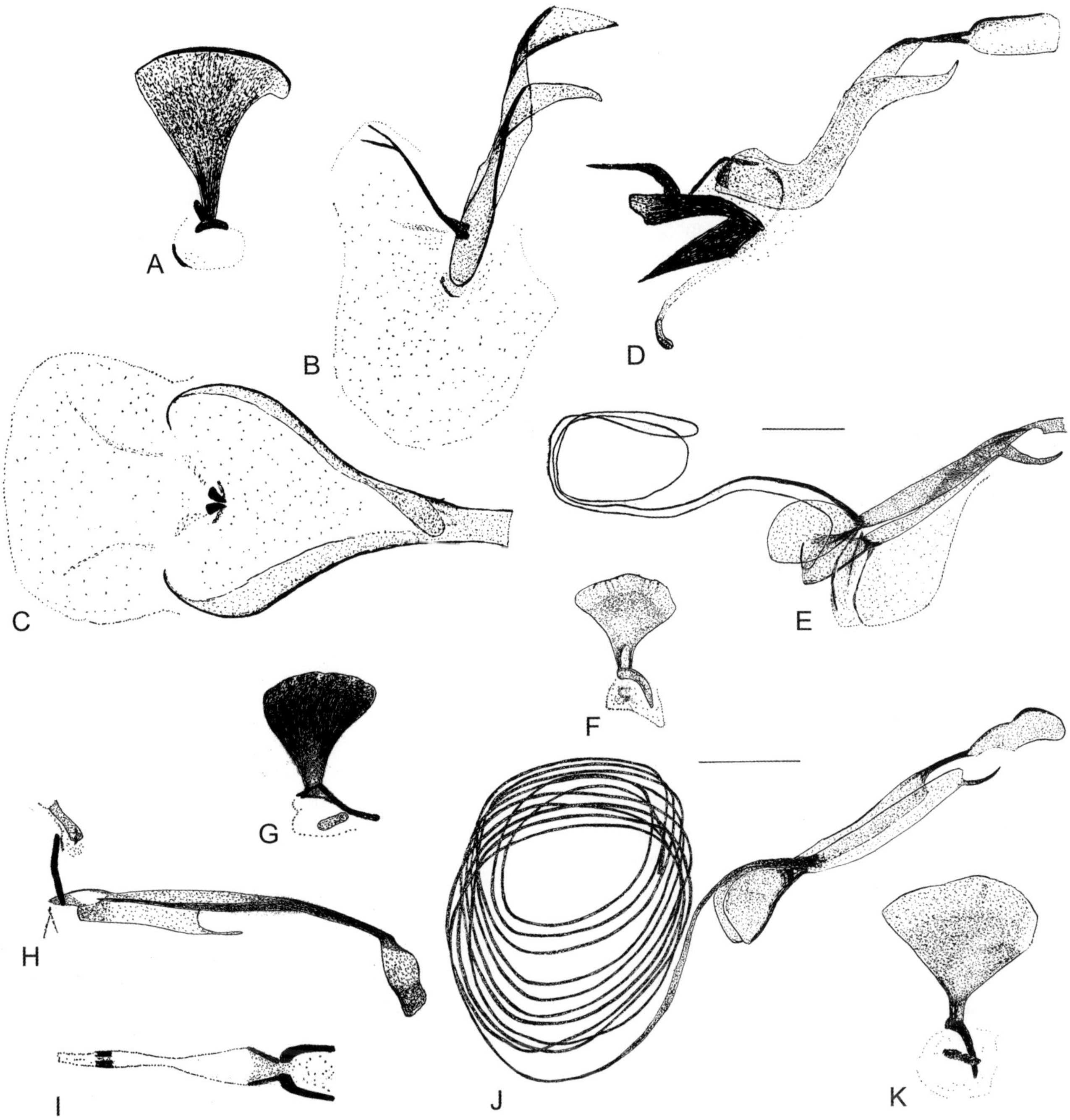

FIGURE 5 *Male genitalia of* Phytomyza. *A–C,* P. fuscula, *ejaculatory bulb (A), aedeagus in lateral (B) and ventral (C) view; D,* P. opacella, *aedeagus in lateral view; E–F,* P. ranunculi, *aedeagus in lateral view (E), ejaculatory bulb (F); G–I,* P.varipes, *ejaculatory bulb (G) aedeagus in lateral view (H), distiphallus and paraphalli in ventral view (I); J–K,* P. vibeana, *aedeagus in lateral view (J), ejaculatory bulb (K). A–D, J–K after Griffiths (1966); E–F and G–I after Griffiths (1964).*

References

Griffiths, G.C.D. (1964): The agromyzid fauna of Iceland and the Faroes, with appendices on the *Phytomyza milii* and *robustella* groups (Diptera, Agromyzidae). – *Entomologiske Meddelelser* 32: 393–450.

Griffiths, G.C.D. (1966): The Agromyzidae (Diptera) of Greenland. – *Meddelelser om Grønland* 170 (4): 32 pp.

Henriksen, K.L. (1939): A revised index of the insects of Grønland. – *Meddelelser om Grønland* 119 j(110): 1–101.

Spencer, K.A. (1976): *The Agromyzidae (Diptera) of Fennoscandia and Denmark*. Fauna Entomologica Scandinavica 5, 1–2. Scandinavian Science Press Ltd., Klampenborg, Denmark, 606 pp.

17.21. Heleomyzidae

David Horsfield

This is a family of over 65 genera and 22 tribes in the world and is cosmopolitan, though mainly a group of the temperate zones (Gorodkov 1984, McAlpine 1985). Five species belonging to four genera have been identified from Greenland in this study. The Greenland Heleomyzidae consists of medium sized flies (c. 3–7 mm), pruinose and coloured yellow, brown or grey. The family is distinguished from other acalypterate Diptera by the following combination of characters: Oral vibrissae present, convergent postocellar bristles, costal spines present (not in the non-Greenland *Borboropsis* Czerny); also characterised by one or two pairs of orbital bristles, orbital plates present, tibiae with dorsal preapical bristles (except for some specimens of the non-Greenland *Borboropsis* Czerny and *Oldenbergiella* Czerny), costal break near union of subcosta with costa, subcosta entirely separate and ocellar bristles behind anterior ocellus (Gill 1962, Gorodkov 1984).

Papp (1998) treats the family as consisting of three subfamilies (Suilliinae, Heteromyzinae and Heleomyzinae). All four genera found in Greenland belong to the subfamily Heleomyzinae and are Holarctic in distribution. Keys to the subfamilies and Palaearctic genera are provided by Papp (1998) while keys to the Nearctic genera are provided by Gill and Peterson (1987) and to the Nearctic species by Gill (1962).

Morphological details of eggs, larvae and puparia of the family are given by Hennig (1952), Papp (1998), Ferrar (1987) and Gill and Peterson (1987). Larvae are of the typical Cyclorrhaphous maggot; they are cylindrical, elongate, tapering anteriorly and bluntly rounded to truncate posteriorly. There are up to 5 pairs of tubercles on the last segment caudally. Larvae are amphipneustic with the anterior spiracles usually fan-shaped and with 6–12 round lobes while the posterior spiracles are on cylindrical projections. The cephalopharyngeal skeleton has a long parastomal bar. Ventral pharyngeal ridges, thought to be characteristic of saprophagous larvae, are noted as present by Ferrar (1987), though they were not found by Skidmore (1966) or Rotheray (2012). The puparium is barrel-shaped and narrowed anteriorly.

Larvae of Heleomyzinae are presumed to be saprophagous, breeding in various kinds of decaying organic matter, including carrion, mammal faeces, vegetable matter, bird nests and fungi (Buxton 1960, Ferrar 1987, Garnett and Foote 1967, Hennig 1952 and Skidmore 1962, 1966). Some species are associated with man, especially in the Arctic and in alpine zones, and under these severe climatic conditions in early spring are capable of breeding in large numbers in toilets, refuse and dead bodies (Gorodkov 1984, 1989).

Adults are found in moist and shady places and in the vicinity of larval food such as in mammal's burrows, bird's nests and in caves (Ferrar 1987, Hackman 1963, Skidmore 1962).

General distribution data in the species accounts are based on Gill (1965) and Gorodkov (1984, 1989).

Key to the genera and species of Greenland Heleomyzidae

1. Prosternal bristles absent 2
– Prosternal bristles present 3
2. Mid tibia with several bristles (anterodorsal and posterodorsal) on the shaft. One orbital bristle. Male genitalia: Fig. 1................. *Oecothea specus*
– Mid tibia lacking bristles on the shaft. Two orbital bristles with the anterior as long as posterior. Male genitalia: Fig. 2 .... *Neoleria prominens*
3. Prosternum with one pair of bristles; rarely two pairs. Anterior orbital bristle shorter than posterior. Anepimeron hairy on lower front part. Anepisternum extensively hairy. Male genitalia: Fig. 3.............. .... *Scoliocentra fraterna*
– Prosternum with two or more (usually 4–8) pairs of bristles. Anterior orbital bristle as long as posterior. Anepimeron bare. Anepisternum bare but may have bristles on posterior margin .... *Heleomyza*, 4
4. Anepisternum rarely with a single small bristle at middle of posterior margin. Male with long, curved process of sternite 10 bent backwards gradually throughout length (Fig. 4)....... *Heleomyza serrata*
– Anepisternum with 1–3 bristles at middle of posterior margin. Male with apical portion of process of sternite 10 expanded to at least twice as broad as the base (Fig. 5) .... *Heleomyza borealis*

***Oecothea specus* (Aldrich, 1897). Fig. 1.**
Body and wing length *c.* 5.5–6.0 mm. The general body colouration is dark grey with head largely yellow with some grey. The wings are hyaline with yellow veins and infuscated cross-veins. The legs are yellow-brown. This is the only species with bristles on the shaft of the mid-tibia (2–3 strong anterodorsal bristles and 1–2 strong

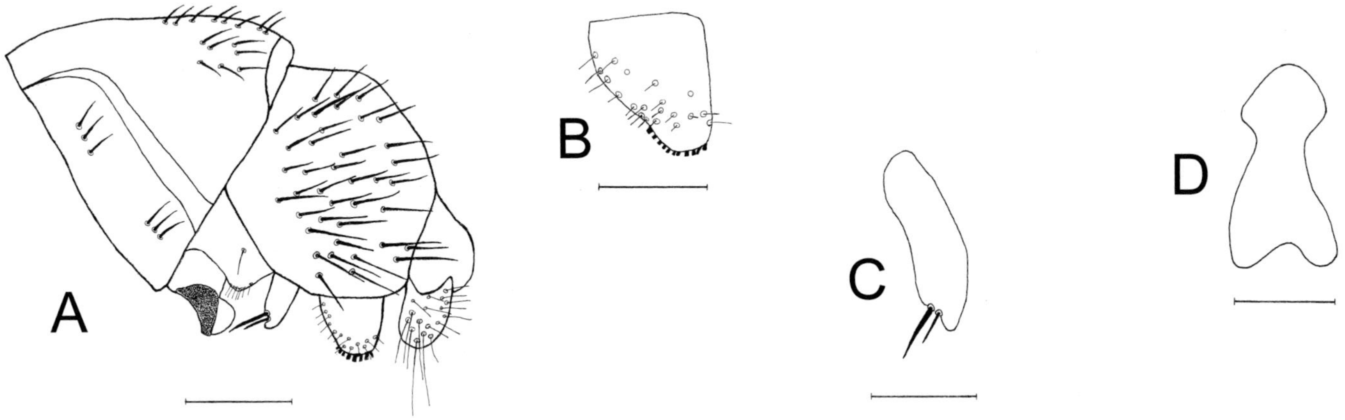

FIGURE 1 *Oecothea* specus, *male genitalia. A, apex of male abdomen (left lateral view), B, surstylus (lateral, left side); C, paramere (lateral, left side); D, outline of metaphallic plate. Scale bars: A, D, 0.2 mm; B–C, 0.1 mm.*

posterodorsal bristles), with a single orbital bristle and with the first flagellomere angled at the junction of the dorsal and anterior surfaces. Like *Neoleria prominens* there are no prosternal bristles. The dorsocentral bristles are characteristic, with the first 3 bristles about equal in length and the 4th larger.

Male genitalia (Fig. 1). The surstylus (gonocoxite of Gorodkov 1959) has a wide apical end with several rows of short, stout, peg-like bristles (Fig. 1B). The apical end of the metaphallic plate is expanded into a heart-shaped disc (Fig. 1D).

The last larval instar is described by Lobanov (1970).

BIOLOGY. Hackman (1963) records adults in burrows of small mammals and Gill (1962) cites records from birds' nests and caves. Gorodkov (1989) notes that the species occurs in burrows and that it is not rare in houses and on windows. The species is recorded from houses in Iceland (Nielsen, Ringdahl and Tuxen 1954).

In Greenland adults have been taken in July and August. Séguy (1934) gives records of larva from the stem of a rotten cabbage, from pieces of rabbit skins and from burrows of various mammals. Skidmore (1962) gives breeding records from a bird's nest and "potting soil". Hackman (1963) reared the species from eggs laid by adults in rodent excrement and cheese placed in vole burrows.

DISTRIBUTION. N, NE (only found in Heilprin Land, Ella Ø, and Zackenberg).

Holarctic, Neotropical, Oriental, Australian, high arctic-southern temperate.

NOTE. Species identification of *Oecothea* follows Gorodkov (1959, 1989) rather than Gill (1962). Gill (1962) describes two closely related species *Oe. fenestralis* and *Oe. specus* (Aldrich) differing only in certain characters of the male genitalia. This same variation is described by Gorodkov (1959) as occurring within *Oe. fenestralis*.

Neoleria prominens **(Becker, 1897). Fig. 2.**
Henriksen 1939: *Leria septentrionalis* Collin.

Body length *c.* 4–5 mm; wing length *c.* 4–5.5 mm. The general body colouration is dark ash-grey and the head is grey with part yellow. The wings are hyaline with yellow veins. The legs are yellow-brown. Like *O. specus* there are no prosternal bristles while there are 2 orbital bristles, of equal length, rather than the 1 of *O. specus*. There are 4 dorsocentral bristles, 3 strongly developed; this is the only species in which the second bristle (first post-sutural) is

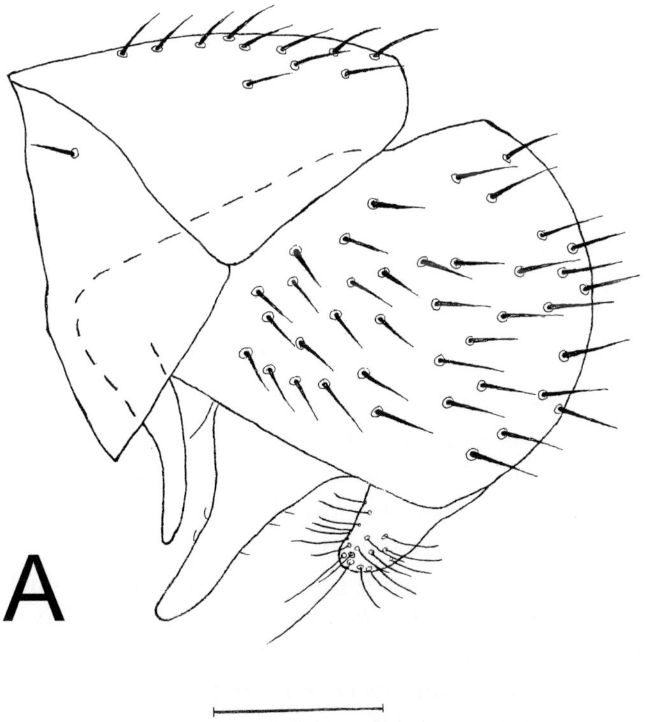

FIGURE 2 *Neoleria prominens, apex of male abdomen (left lateral view). Scale bar 0.2 mm.*

weak and little bigger than surrounding hairs on the dorsum.

Male genitalia (Fig. 2). The surstylus is roughly funnel-shaped and comes to a point.

BIOLOGY. Gorodkov (1989) notes that the species occurs in tundra and forest-tundra. Nielsen, Ringdahl and Tuxen (1954) record the species (as *Neoleria septentrionalis* Collin) in Iceland from home-field and meadow, on flowers and also on carrion.

In Greenland adults have been taken in February and between May and August. According to Gorodkov (1989) this species is a common synanthrope with the larvae occurring in latrines, refuse and dead bodies; it occurs only rarely in undisturbed nature. Collin (1923) records adults from the nest of a glaucous gull (*Larus hyperboreus*). Gill (1962) records adults collected from garbage and Weber (1954) collected adults from the carcase of a dog in arctic Alaska.

DISTRIBUTION. S, SW, NW, NE, SE (not recorded from N).

Holarctic (Circumpolar), high arctic-northern temperate.

***Scoliocentra fraterna* Loew, 1863. Fig. 3**

Body length *c.* 4.5 - 7 mm; wing length *c.* 4.5 - 8 mm. The mainly bluish-grey thorax contrasts with the yellow or reddish-brown abdomen while the head is mainly yellow with some grey. The wings are hyaline with a faint yellow-brown tinge and yellow veins. The legs are yellow.

This species is easily recognisable as the only species with 1 pair (rarely 2) of prosternal bristles and with hairs on the anepisternum and anepimeron. The dorsocentrals are characteristic with the anterior 2 about 2/3 the length of the posterior 2. There are 2 orbital bristles with the anterior bristle about 2/3 the length of the posterior orbital.

Male genitalia (Fig. 3). The anterior surface of the surstylus is hollowed out so the two sides are roughly at right angles. There are hairs on the outer face of the surstylus apically along the whole length and grouped in the middle on the anterior side. Small spines are visible through the surstylus on the inner face, occurring densely grouped apically and anteriorly while extending basally, scattered on the posterior half (Fig. 3B). The arm of the process of sternite 10 is covered in short, stout spines (Fig. 3C).

BIOLOGY. Adults of the genus have been found in small mammal burrows (Hackman 1963) and in caves (Skidmore 1962, Garnett and Foote 1967). *Scoliocentra villosa* has been bred from dead chickens by Skidmore (1966).

DISTRIBUTION. S, SW.

Nearctic, Palaearctic (only Far East of Russia), low arctic-southern temperate.

***Heleomyza serrata* (Linnaeus, 1758). Fig. 4**

Body length *c.* 4–6 mm; wing length *c.* 4.5–6 mm. The slate-grey thorax contrasts with the yellow abdomen and the head is grey and partly yellow. The wings are hyaline with yellow veins and the legs are yellow.

This species and *Heleomyza borealis* are the only species in which there are numerous (4–8) pairs of prosternal bristles and in which there may be bristles on the posterior margin of the anepisternum. In *H. serrata* the anepisternum is usually completely bare, though sometimes there is a single small bristle. There are 4 dorsocentral bristles with the anterior 2 only slightly shorter than the posterior 2. The orbital bristles are 2 in number, both strong and about equal in length.

Male genitalia (Fig. 4). The length of the process of sternite 10 is greater than the length of the epandrium. The apex of the epandrium is not extended. The process of sternite 10 is curved, being bent backwards gradually throughout its length and has fine hairs on the outer face (Fig. 4B) and on the inner face there are strong, short, spines grouped centrally and with finer spines basally (Fig. 4C).

The last larval instar is described by Lobanov (1970).

BIOLOGY. Gorodkov (1962) states that the major part of the area of the species lies in the forest zone and that the species does not penetrate far into the tundra. According to Gorodkov (1962) it congregates in caves and adults are also recorded from small mammal burrows (Hackman 1963). Nielsen, Ringdahl and Tuxen (1954) state that in Iceland the species is often found in houses and also on flowers of *Archangelica* and *Salix*.

According to Gorodkov (1962) *H. serrata* is a facultative synanthropic species. Gill (1962) cites rearing records from birds' nests, dung, "hen manure" and fungi, and also noted that he had collected it (presumably as adults) in traps baited with excrement. Garnett and Foote (1967) cite records of turnips and dung as larval food.

DISTRIBUTION. S (only known from Upernaviarsuk).

Holarctic, low arctic-southern temperate.

***Heleomyza borealis* (Boheman, 1865). Fig. 5**

Heleomyza czernyi (Collart, 1933)

The body length is variable *c.* 2.5–5 mm; wing length *c.* 3–5 mm. The general body colouration is greyish brown and the head is grey and partly yellow. The wings are hyaline with yellow veins. The legs have grey femur and tibia, and tarsi are yellow.

Like *H. serrata* this species has numerous (usually 4–6) pairs of prosternal bristles. This is the only species in which there is always at least one strong bristle, often accompanied by 1–2 weaker bristles, at the middle of the

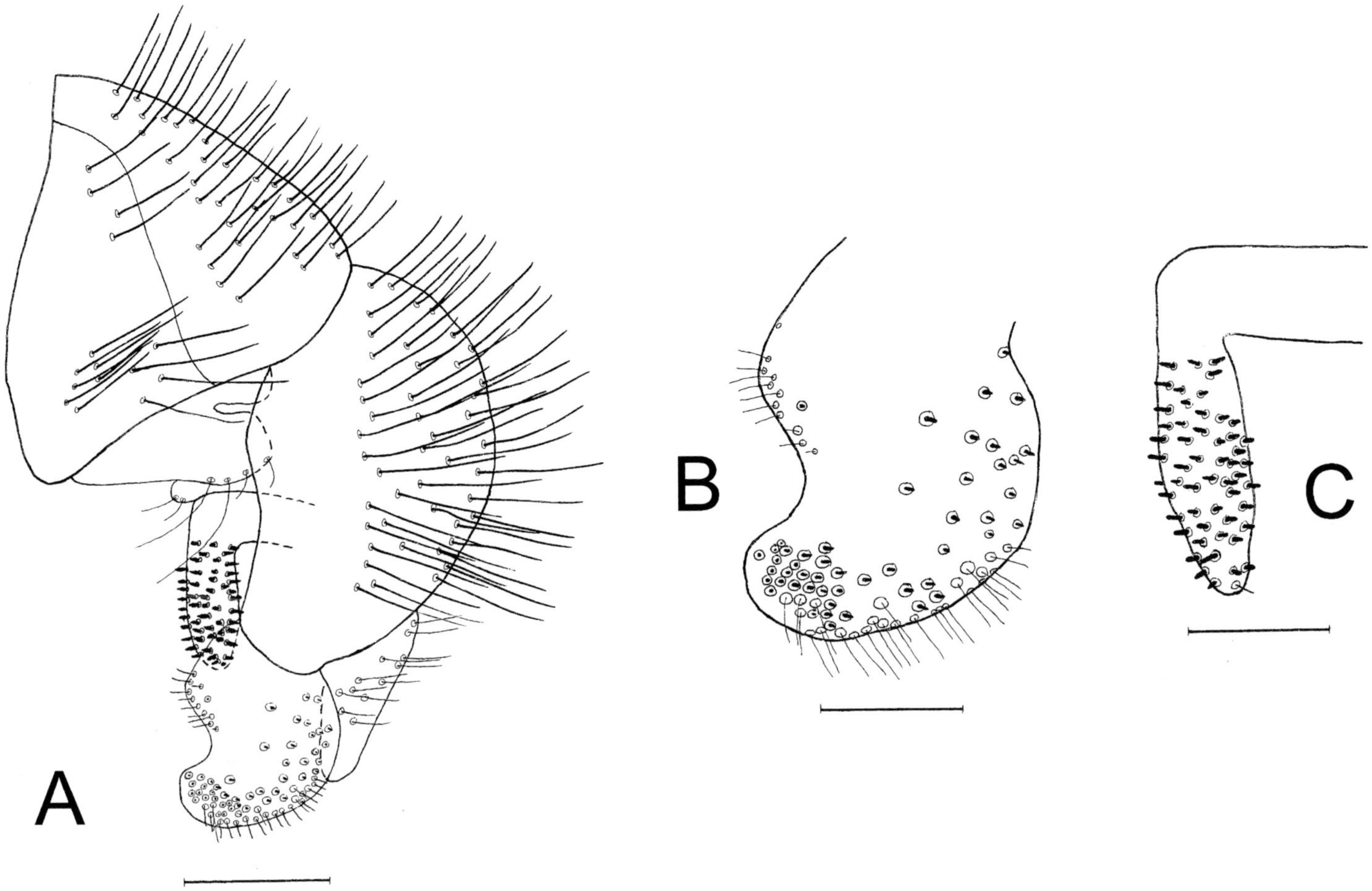

FIGURE 3 Scoliocentra fraterna, *male genitalia. A, apex of male abdomen (left lateral view); B, surstylus (lateral, left side, outer face, showing hairs on outer face and small spines on inner face); C, process of sternite 10 (lateral, left side, outer face). Scale bars: A, 0.2 mm; B–C, 0.1 mm.*

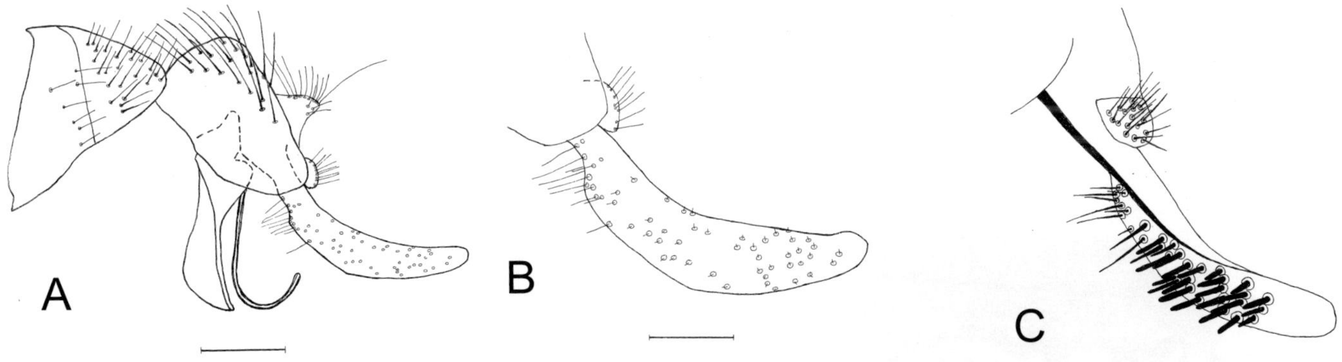

FIGURE 4 Heleomyza serrata, *male genitalia. A, apex of male abdomen (left lateral view); B, process of sternite 10 (lateral, left side, outer face); C, process of sternite 10 (sub-lateral, right side, inner face). Scale bars: A, 0.2 mm; B–C, 0.1 mm.*

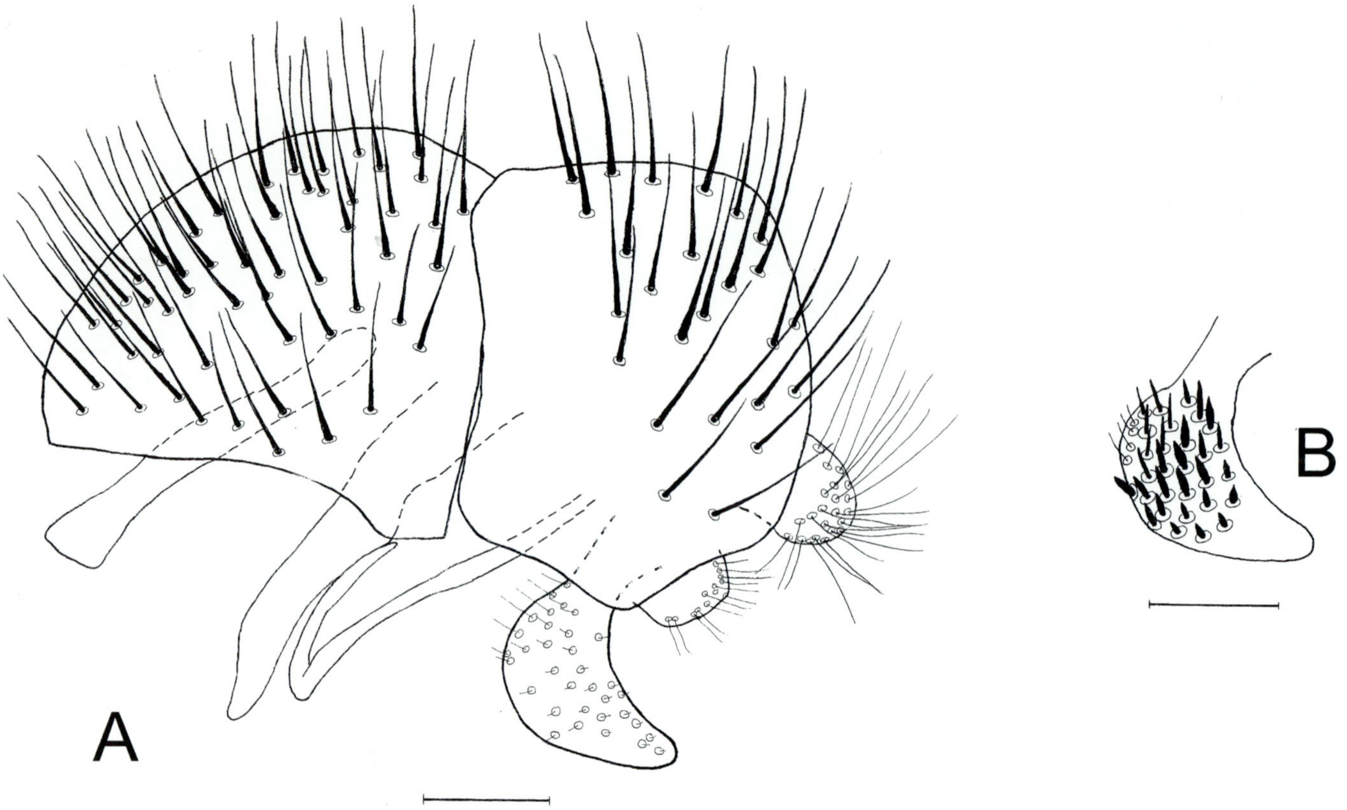

FIGURE 5 Heleomyza borealis, *male genitalia. A, apex of male abdomen (left lateral view): B, process of sternite 10 (lateral, right side, inner face). Scale bars 0.1 mm.*

posterior margin of the anepisternum. The dorsocentral and orbital bristles are similar to *H. serrata*.

Male genitalia (Fig. 5). The length of the process of sternite 10 is much less than the length of the epandrium. The whole of the central portion of the inner face of the process of sternite 10 is covered by strong, short spines (Fig. 5B).

BIOLOGY. *Heleomyza* species are usually synanthropic in tundra areas and sometimes accumulate in large numbers in caves (Gorodkov 1989). The species was found in a cave by Collart (1933). Gorodkov (1989) states that *Heleomyza* species often occur in refuse, in latrines, on dead bodies of animals, etc. Rotheray (2012) has rearing records of *H. borealis* from a dead sheep on St. Kilda, Scotland, and describes the puparium. According to Lyneborg (1968) in the Faroes the species is often taken in houses; larvae have been found in hay and a male specimen was reared from a puparium found in a nest of eider.

FIGURE 6 Heleomyza borealis, *habitus (courtesy R. Thomason).*

DISTRIBUTION. S, SW, NE, SE.

Holarctic, high arctic-southern temperate.

Notes on Henriksen's checklist

Henriksen (1939) gives the names of three species of Heleomyzidae for Greenland. These are: *Leria inscripta* Meigen, *Leria modesta* Meigen and *Leria septentrionis* Collin. The names in current use are respectively *Neoleria inscripta* (Meigen), *Heleomyza modesta* (Meigen) and *Neoleria prominens* (Becker) (Gill 1965, Gorodkov 1984).

N. inscripta differs from *N. prominens* in the yellow colouration of the postpronotal lobe and/or part of the scutellum and proepisternum while the dorsocentral bristles are all clearly stronger than the surrounding hairs (especially comparing the first post-suturals). The male genitalia of *N. inscripta* are similar in structure to *N. prominens* but are smaller (Collin 1943, Gill 1962). Examination of the specimens upon which previous records were based would be required to authentic these earlier records.

According to Gorodkov (1962) information in the literature on *H. modesta* is, with few exceptions, not to be relied upon. This is because this is one of three species (including *H. borealis*) that differ only in the structure of the process of sternite 10, a character mostly not utilised by earlier workers. Gorodkov (1962) predicted the occurrence of *H. borealis* in Iceland and Greenland, from where *H. modesta* had been reported. Confirmation of *H. modesta* in Greenland is required.

References

Buxton, P.A. (1960): British Diptera associated with fungi. III. Flies of all families reared from about 150 species of fungi. – *Entomologist's monthly Magazine* 96: 61–94.

Collart, A. (1933): Description d'un Hélomyzida troglophile nouveau de Belgique (Diptera). – *Bulletin & Annales de la Société entomologique de Belgique* 73: 402–405.

Collin, J.E. (1923): Diptera (Orthorrhapha Brachycera and Cyclorrhapha) from Spitsbergen and Bear Island. Results of the Oxford University Expedition to Spitsbergen, 1921. – No. 22. – *Annals and Magazine of natural History* 11 (9): 116–123.

Collin, J.E. (1943): The British species of Helomyzidae (Diptera). – *Entomologist's monthly Magazine* 79: 234–251.

Ferrar, P. (1987): *A guide to the breeding habits and immature stages of Diptera Cyclorrhapha.* Entomonograph 8. E.J. Brill and Scandinavian Science Press, Leiden and Copenhagen.

Garnett, W.B. and Foote, B.A. (1967): Biology and immature stages of *Pseudoleria crassata* (Diptera: Heleomyzidae). – *Annals of the Entomological Society of America* 60: 126–134.

Gill, G.D. (1962): The heleomyzid flies of America north of Mexico (Diptera: Heleomyzidae). – *Proceedings of the United States National Museum* 113: 495–603.

Gill, G.D. (1965): Family Heleomyzidae. *In* Stone, A. (ed.) *A catalog of the Diptera of America north of Mexico*, Agricultural Handbook 276, pp. 808–816. U.S. Department of Agriculture, Agricultural Research Service.

Gill, G.D. and Peterson, B.V. (1987): Chapter 89. Heleomyzidae. *In* McAlpine, J.F. (ed.) *Manual of Nearctic Diptera*, 2, pp. 973–980. Research Branch, Agriculture Canada, Ottawa.

Gorodkov, K.B. (1959): Revision of the Palaearctic species of the genus *Oecothea* Hal. (Diptera, Helomyzidae). – *Entomological Review, Washington* 38: 812–828.

Gorodkov, K.B. (1962): Revision of the Palaearctic species of the genus *Leria* R.-D. (Diptera, Helomyzidae). – *Entomological Review, Washington* 41: 398–416.

Gorodkov, K.B. (1984): Family Heleomyzidae. *In* Soós, Á. and Papp, L. (eds.) *Catalogue of Palaearctic Diptera*, 10, pp. 15–45.

Gorodkov, K.B. (1989): Family Helomyzidae (Heleomyzidae). *In* Bei-Bienko, G.Y. (ed.) *Keys to the Insects of the European Part of the USSR Volume V Part II Diptera and Siphonoptera*, pp. 510–537. E.J. Brill, Leiden.

Hackman, W. (1963): Studies on the dipterous fauna in burrows of voles (*Microtus, Clethrionomys*) in Finland. – *Acta Zoologica Fennica* 102: 1–64.

Hennig, W. (1952): *Die Larvenformen der Dipteren, 3*. Akademie-Verlag, Berlin.

Henriksen, K.L. (1939): A revised index of the insects of Grønland. – *Meddelelser om Grønland* 119 (10): 1–101.

Lobanov, A.M. (1970): On morphology of last instar larvae of Helomyzidae (Diptera). – *Zoologicheskii Zhurnal* 49 (11): 1671–1675. [In Russian, English summary.]

Lyneborg, L. (1968): Diptera Brachycera et Cyclorrhapha. *In* Jensen, A.S., Lundbeck, W. and Mortensen, T. (eds.) *Zoology of the Faroes*, XLII a, pp. 1–79, Copenhagen.

McAlpine, D.K. (1985): The Australian Genera of Heleomyzidae (Diptera: Schizophora) and a Reclassification of the Family into Tribes. – *Records of the Australian Museum* 36: 203–251.

Nielsen, P., Ringdahl, O. and Tuxen, S.L. (1954): Diptera 1 (exclusive of Ceratopogonidae and Chironomidae). *In* Friðriksson, A., Guðmundsson, F., Hannesson, P., Jensen, A.D., Spårck, R., Saemundsson, B., Tuxen, S.L. and Täning Vedel, A. (eds.) *The Zoology of Iceland*, III (48 a), pp. 1–189. Ejnar Munksgaard, Copenhagen and Reykjavik.

Papp, L. (1998): Families of Heleomyzoidea. *In* Papp, L. and Darvas, B. (eds.) *Contributions to a Manual of Palaearctic Diptera*, 3, pp. 425–455. Science Herald, Budapest.

Rotheray, G.E. (2012): Morphology of the puparium and breeding sites of eight species of Heleomyzidae (Diptera). – *Journal of Natural History* 46: 2075– 2102.

Ryan, J.K. (1977): Appendix 7. Invertebrates of Truelove Lowland. – *In*: Bliss, L.C. (ed.) *Truelove Lowland, Devon Island, Canada: A High Arctic Ecosystem*, pp. 699–703. University of Alberta Press, Edmonton.

Séguy, E. (1934): *Diptères (Brachycères) (Muscidae Acalypterae et Scatophagidae)*. Fauna de France, 28. Lechevalier, Paris.

Skidmore, P. (1962): Notes on the Helomyzidae of Lancashire and Cheshire, including records from other parts of north west England. – *Entomologist* 95: 193–198; 226–236.

Skidmore, P. (1967): The biology of *Scoliocentra villosa* (Mg.) (Dipt., Heleomyzidae). – *Entomologist's monthly Magazine* 102: 94–98.

Weber, N.A. (1954): Arctic Alaskan Diptera. *Proceedings of the Entomological Society of Washington* 56 (2): 86–91.

17.22. Sphaeroceridae

Jindřich Roháček

Very small to small (0.7–5.5 mm) relatively robust flies (Fig. 1), easily recognized from other acalyptrate families by the shortened and dilated hind basitarsus. Usually brown to black, more rarely with yellow to red markings. Chaetotaxy variable, setae reduced in Sphaerocerinae but rich and long in Limosininae and most Copromyzinae. Gena rather high (eye relatively small), frons with distinct interfrontalia. Wing: C with 2 breaks, Sc reduced, venation relatively complete in Sphaerocerinae (Fig. 2A) and Copromyzinae (Fig. 2B), reduced in Limosininae (Figs 2C–E). For more detail, and also for immatures, see Roháček (1998).

Both larvae and adults of all known species are saprophagous. The larvae develop in various organic matter apparently feeding as microsaprophages on microorganisms destructing rotting plants, dung, carrion or fungi. Adults occupy habitats with breeding media of larvae and can be found on diverse decayed plant remnants (grass, moss, compost, wood debris, leaf litter, mud, litoral debris), fungi and matter of animal origin (dung, manure, carrion) in open as well as woodland, preferably in damp, habitats.

Indentifications: The Palaearctic genera of Sphaeroceridae are keyed in Roháček (1998) which also includes references to identification aids to particular genera. Nomenclature follows Roháček *et al.* (2001), Papp (2008) and Marshall *et al.* (2011); general distribution and further references can also be found in the this catalog. More precise distributional data from Europe are given in Roháček (2010).

FIGURE 1 Sphaerocera curvipes, *female. Body length 4 mm.* (PHOTO: J. ROHÁČEK)

Six species of Sphaeroceridae are recorded here (all for the first time) from Greenland, one belonging to the subfamily Copromyzinae, one to Sphaerocerinae and four to Limosininae. Of these only one (*Norrbomia fumipennis*) is a native circumboreal species, all the others are cosmopolitan, subcosmopolitan or widespread holarctic species with high degree of synanthropy, which were introduced to most parts of the world (probably also to Greenland) by human agency.

Key to the identification of species recorded from Greenland

1. C reaching to vein M; cells bm and cup closed (Figs 2A, B) 2
– C reaching only to or somewhat beyond R_{4+5}; M shortened, not reaching wing margin; cell bm coalesced with cell dm; cell cup open (Figs 2C–E)... .... 3 (subfamily Limosininae)
2. Setae and hairs on mesonotum reduced to short spines or blunt minute spinulae on warts; scutellum without apical sc setae and basal sc reduced or lost; hind tibia without dorsal preapical seta; CuA1 reachig wing margin (Fig. 2A) [subfamily Sphaerocerinae]. Scutellum with only 2 lateral teeth; mesonotal setae reduced but not on warts; femora swollen (posterior very robust and curved); fore tarsus uniformly dark brown; hind basitarsus with ventral excavation and a small spine (Fig. 1) .... *Sphaerocera curvipes*
– Mesonotum with hairs and long setae, scutellum with at least 2 pairs of sc setae; hind tibia with a long dorsal preapical seta; CuA1 not reaching wing margin (Fig. 2B) [subfamily Copromyzinae]. Hind tibia with ventroapical spur and a long anteroventral seta at 2/3 or 3/4; postocular setae in single regular row; genal seta short; mid tibia with only 1 (preapical) anterodorsal seta; only 1 long humeral seta and apical section of M subequal or longer than that between cross-veins; wing membrane distinctly brownish .... *Norrbomia fumipennis*
3. Disc of scutellum densely setulose (Fig. 3E); vein A1 distinctly angular; C with 2nd sector longer than 3rd (Fig. 2E); discal cell angular posteriorly; only 2 sternopleural setae and body uniformly dark greyish brown .... *Coproica vagans*
– Disc of scutellum bare; vein A1 sinuate (Figs 2C, D) 4
4. Vein R_{4+5} upcurved, ending farther from wing apex than projection of M would; C strongly produced beyond apex of R_{4+5} (Fig. 2D); discal cell with shortened section between cross-veins (shorter than length of posterior cross-vein, see Fig. 2D).......... .... *Pullimosina heteroneura*
– Vein R_{4+5} straight, reaching margin near apex of wing; C ending at or very slightly beyond apex of R_{4+5} (Fig. 2C) .... 5 (*Spelobia*)
5. Male head with face yellow (female face dark brown); haltere more or less yellow; mid basitarsus short, with thick ventral setulae; male 4th sternum unarmed; female spermathecae with smooth surface.......................... *Spelobia luteilabris*
– Head of both sexes with face blackish brown; haltere brown with darker knob; mid basitarsus longer, with fine ventral setulae (Fig. 3D); male 4th sternum with a small posteromedial comb of short spinulae (Fig. 3B); female spermathecae with transversely grooved surface .... *Spelobia pseudosetaria*

COPROMYZINAE

***Norrbomia fumipennis* (Stenhammar, 1855)**

This species of *Norrbomia* is characterized by darkened wings, entirely microtomentose frons and 3 dc setae. It is most similar to *N. yukonensis* Marshall & Norrbom,1993 (known only from Canada: Yukon) which differs from *N. fumipennis* mainly by details of the male and female genitalia (see Marshall & Norrbom 1993: Figs 32–34, 53–55, 66–72).

BIOLOGY: The species is known to develop in burrows of rodents of the genera *Lemmus* and *Microtus* (Hackman 1965; Marshall & Norrbom 1993).

DISTRIBUTION. S (first record).

A circumboreal species, widespread in northern areas of Canada, USA, Norway, Sweden, Finland and Russia, also known from Iceland and Svalbard (Roháček *et al.* 2001).

SPHAEROCERINAE

***Sphaerocera curvipes* Latreille, 1805. Figs 1, 2A**

The largest palaearctic species of Sphaerocerinae (3.5–4.3 mm), easily recognized from its strongly swollen femora (Fig. 1), dark fore tarsi and a short ventral spine on incrassate and ventrally emarginate hind basitarsus (Kim 1968: fig. 7).

BIOLOGY: A very common coprophagous species, secondarily also occurring on other decaying organic matter (carrion, fungi, plant refuse), particularly abundant on manure of domestic animals.

DISTRIBUTION. S (first record).

Cosmopolitan, northernmost records from Alaska, Faroe Is., Iceland (see Roháček *et al.* 2001).

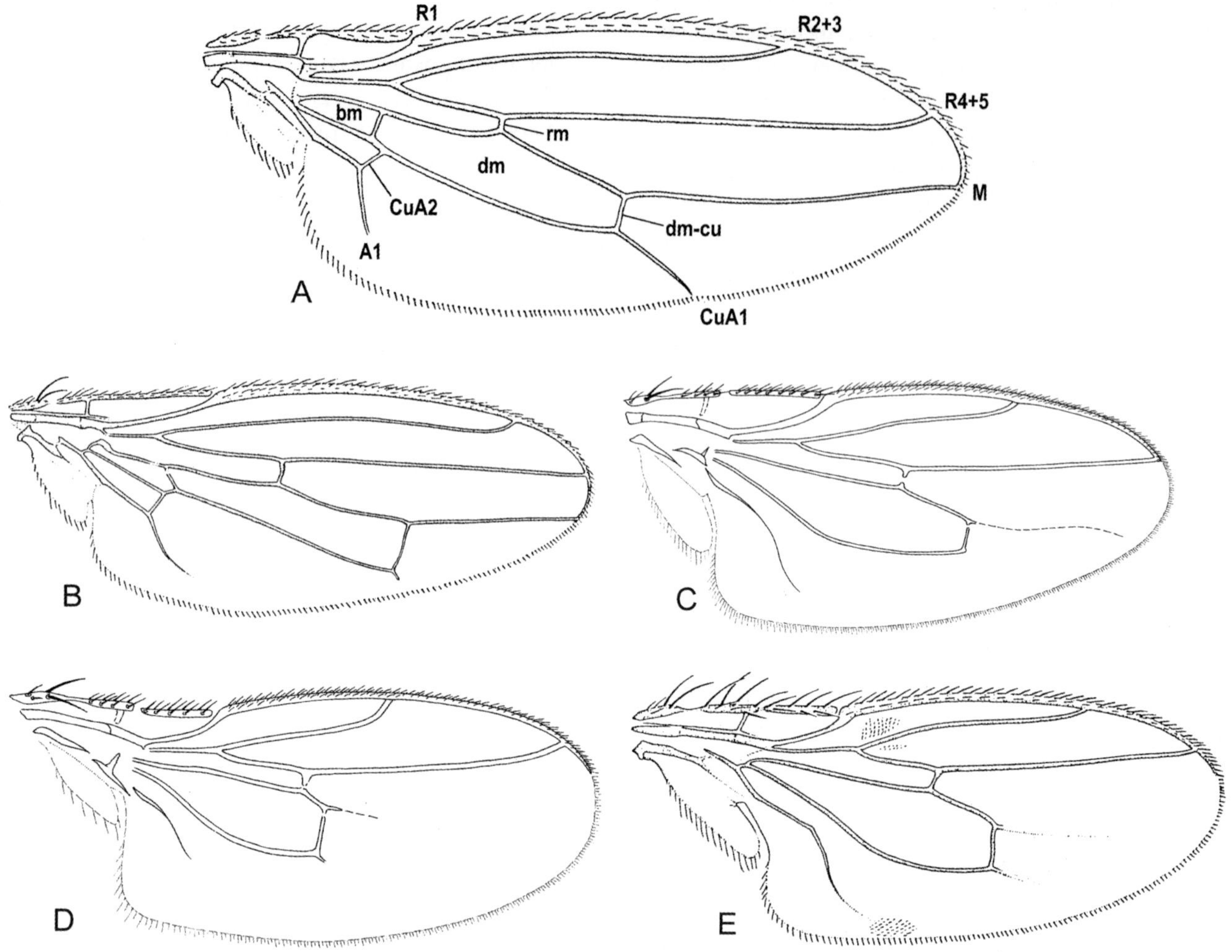

FIGURE 2 *Wings of Sphaeroceridae. A,* Sphaerocera curvipes*; B,* Copromyza *sp.; C,* Spelobia luteilabris*; D,* Pullimosina heteroneura*; E,* Coproica *sp.*

LIMOSININAE

Spelobia luteilabris **(Rondani, 1880) Fig. 2C**

A *Spelobia* species characterized by dichroic face (yellow in male, brown in female) and usually yellow halteres but mainly by structures of the male and female genitalia (see Roháček 1982–1985: Figs 612–620).

BIOLOGY: A polysaprophagous species forming both synanthropic and wild populations. It seems to be more frequent within the human environment where it occurs mainly in henhouses and poultry farms, rabbit-hutches, cellars, urinals, refuse or compost heaps etc. (Roháček 1982–1985).

DISTRIBUTION. S (first record).

Widespread in the Holarctic Region and also introduced to New Zealand; northernmost records are from Alaska, Iceland, Faroe Is. (Roháček *et al.* 2001).

Spelobia pseudosetaria **(Duda, 1918) Figs 3A–D**

A *Spelobia* species with dark face and halteres and with somewhat elongate mid basitarsus, which can be separated from closely allied species with paired dorsal setae in distal half of on mid tibia (Fig. 3C) only by characters in the male and female genitalia and, particularly, by small posteromedial row of spines on male 4th sternum (Fig. 3B), see also Roháček (1982–198: Figs 575–590).

BIOLOGY: Another polysaprophagous species, more or less confined to human settlements living in decayed vegetation in compost and refuse heaps, on rabbit dung, in cellars and occurring (adults) also indoors. Wild populations can be mainly found in caves and burrows of mammals (Roháček 1982–1985).

DISTRIBUTION. S, SW (first record).

Widespread in the Holarctic Region but also recorded from New Zealand (obviously introduced); northernmost records are from Iceland (Roháček *et al.* 2001).

Pullimosina **(*Pullimosina*) *heteroneura*** **(Haliday, 1836) Fig. 2D**

A small species having shortened discal (dm) cell, with section on M between anterior (r-m) and posterior (dm-cu)

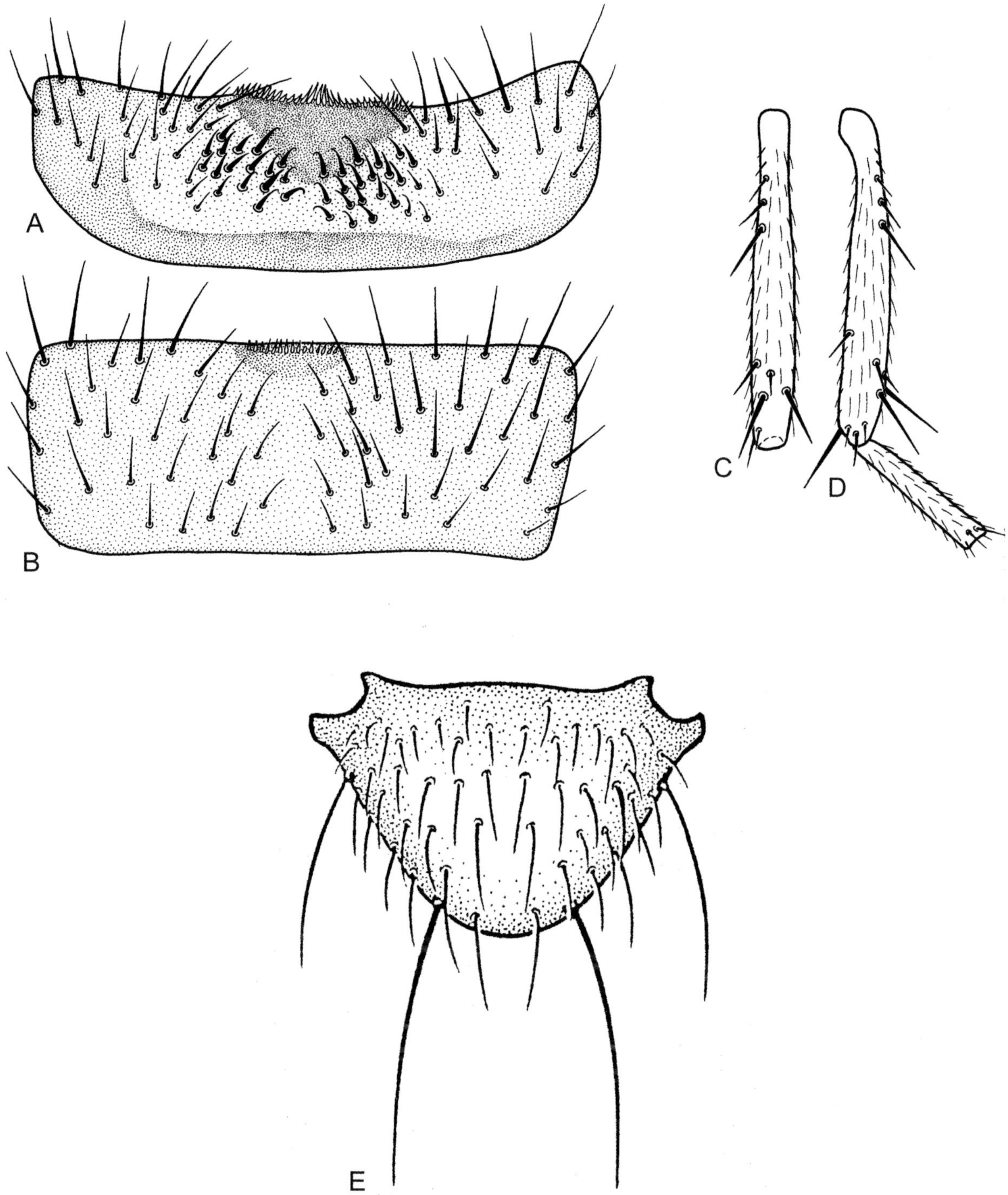

FIGURE 3 *Characters of Limosininae. A,* Spelobia pseudosetaria, *male 5th sternum, B, male 4th sternum, C, mid tibia dorsally; D, mid tibia and basitarsus anteriorly; E,* Coproica *sp., scutellum dorsally.*

cross-veins distinctly shorter than posterior cross-vein (Fig. 2D). For male and female genitalia see Roháček (1982–198: Figs 728–738). Only *P.* (*P.*) *zayensis* Marshall, 1986 has similar venation (but with discal cell yet more shortened); however, this species is termophilous and occurs only in southern areas of the Holarctic Region; its genitalia are figured in Marshall (1986: Figs 45–48, 85–87).

BIOLOGY: A polysaprophagous species developing in various decayed matter, often within human settlements, including dung and refuse heaps (Roháček 1982–1985).

DISTRIBUTION. S (first record).

Subcosmopolitan, widespread in the Holarctic Region but known (probably introduced) also from Afrotropical, Australasian and Neotropical Regions; the northernmost records are from Iceland (Roháček *et al.* 2001).

Coproica vagans **(Haliday, 1833)**
A dark *Coproica* species; C with second sector longer than third (Fig. 2E), discal (dm) cell angulate, female cerci with long sinuate hair-like setae, male gonostylus short (low), attenuated anteriorly, with long setae ventrally (Pitkin 1988: Fig. 449).

BIOLOGY: A predominantly coprophagous species, most common on dung heaps and manure, but also occuring on other decayed material (carrion, rotting plants), particularly in refuse heaps and similar habitats with accumulated organic matter.

DISTRIBUTION. S (first record).

A cosmopolitan species (see Roháček *et al.* 2001). Northernmost records are from Iceland, northern Finland and northwestern Russia.

References

Kim, K.C. (1968): Revision of *Sphaerocera*, with description of a new genus Afromyia (Diptera: Sphaeroceridae). – *Annals of the Entomological Society of America* 61(2): 296–312.

Hackman, W. (1965): On the genus *Copromyza* Fall. (Diptera, Sphaeroceridae), with special reference to the Finnish species. – *Notulae Entomologicae* 45: 33–46.

Marshall, S.A. (1986): A revision of the Nearctic species of the genus *Pullimosina* (Diptera, Sphaeroceridae). – *Canadian Journal of Zoology* 64: 522–536.

Marshall, S.A. & Norrbom, A.L. (1993): A revision of the New World species of *Norrbomia* (Diptera: Sphaeroceridae), including all American species previously placed in *Borborillus*. – *Insecta Mundi* 6(3–4) (1992): 151–181.

Marshall, S.A., Roháček, J., Dong, H. & Buck, M. (2011): The state of Sphaeroceridae (Diptera: Acalyptratae): a world catalog update covering the years 2000–2010, with new generic synonymy, new combinations, and new distributions. – *Acta Entomologica Musei Nationalis Pragae* 51 (1): 217–298.

Papp, L. (2008): New genera of the Old World Limosininae (Diptera, Sphaeroceridae). – *Acta Zoologica Academiae Scientiarum Hungaricae* 54, suppl. 2: 47–209.

Pitkin, B.R. (1988): Lesser dung flies. Diptera: Sphaeroceridae. *Handbooks for the Identification of British Insects*, Vol. 10, pt. 5e, 175 pp., Royal Entomological Society of London, London.

Roháček, J. (1982–1985): A monograph and re-classification of the previous genus *Limosina* Macquart (Diptera, Sphaeroceridae) of Europe. Parts I–IV. – *Beiträge zur Entomologie, Berlin* 32: 195–282 (Part I, 1982), 33: 3–195 (Part II, 1983)), 33: 203–255 (Part III, 1983), 35: 101–179 (Part IV, 1985).

Roháček, J. (1998): 3.43. Family Sphaeroceridae. *In*: Papp, L. & Darvas, B. (eds): *Contributions to a Manual of Palaearctic Diptera*. Vol. 3., Higher Brachycera. 880 pp. (p. 463–496), Science Herald, Budapest.

Roháček, J. (2010): Sphaeroceridae. *In*: Pape, T. (ed.): *Fauna Europaea, Diptera Brachycera*. Fauna Europaea, version 2.3, http://www.faunaeur.org.

Roháček J., Marshall S.A., Norrbom A.L., Buck M., Quiros D.I. & Smith I. (2001): World catalog of Sphaeroceridae (Roháček J., ed.), 414 pp., Slezské zemské muzeum, Opava.

17.23. Chamaemyiidae

Thomas Pape

Small flies with uniformly hyaline wings and dense, silvery grey pruinosity (Fig. 1). Legs blackish with orange knees. Abdominal tergites 4–5 with or without a pair of small blackish or olive brown spots.

Head slightly broader than thorax. Eye shortly pubescent (appearing bare). Frons 0.5× head width and equibroad in male and female, with sides more or less parallel, and with a very few, scattered setulae; 2 reclinate fronto-orbital bristles, a pair of strong proclinate ocellar bristles; smaller convergent postocellar bristles; well developed inner and outer vertical bristles. Lunule broadly exposed, bare. Face receding, shallowly concave and without antennal grooves. Vibrissa absent; gena with few subvibrissal and genal setae. Antenna short and porrect; first flagellomere oval, less than twice as long as wide; pedicel about as long as broad, without a dorsal seam, with a conspicuous dorsal bristle and median bristles; arista very shortly pubescent. Proboscis short; palpus small. Thorax with one postpronotal, one presutural supra-alar, two notopleural, one postsutural supra-alar, and two postalar bristles present; one presutural and two postsutural dorsocentral bristles present; weak prescutellar acrostichal bristle present; acrostichal setulae in two rows. Scutellum with two pairs of marginal bristles. Wing moderately broad, with well-developed anal lobe and alula; costa unbroken, extending to media; subcosta complete and free; first radial vein approaching subcosta closely near apex of the latter but then departing from it and ending in costa at some distance from subcosta; basal medial cell and discal medial cell separated by a crossvein; posterior cubital cell present.

Only one species of Chamaemyiidae is known from Greenland: *Chamaemyia geniculata* (Zetterstedt, 1838), which is recorded here from Greenland for the first time.

Chamaemyia geniculata **(Zetterstedt, 1838). Fig. 1.**
This species may be recognized from other acalyptrate flies in Greenland by its overall silvery pruinosity, the receding head profile and the orange knees.

BIOLOGY. Unknown, but as the larvae of most species of this family are free-living predators of adelgids, aphids, and scale insects (McAlpine 1987), the present species should be expected to share this life habit.

DISTRIBUTION: S (only recorded from Narsarsuaq).

The species is widespread in the Palaearctic Region and known from Alaska and British Columbia in the Nearctic Region (Pape & Thompson 2010, Gaimari 2012).

FIGURE 1 Chamaemyia geniculata (*Zetterstedt, 1838), male, habitus.*

References

Gaimari, S.J. (2012) Fauna Europaea: Chamaemyiidae. *In*: Pape, T. & Beuk, P. (eds) Fauna Europaea: Diptera. *Fauna Europaea* version 2.5. http://www.faunaeur.org [accessed 20 May 2013].

McAlpine, J.F. (1987) Chapter 88. Chamaemyiidae. *In*: McAlpine, J.F.; Peterson, B.V.; Shewell, G.E.; Teskey, H.J.; Vockeroth, J.R. & Wood, D.M. (eds) *Manual of Nearctic Diptera*. Vol. 2. Agriculture Canada Monograph, 28. pp. 965–971.

Pape, T. & Thompson, F.C. (eds) (2010). *Systema Dipterorum*, version 1.0. http://www.diptera.org [accessed 20 May 2013].

17.24. Canacidae

Wayne N. Mathis

Minute to small flies, body length 0.91–5.00 mm. *Head:* Postocellar (=postvertical) setae usually developed; posterior fronto-orbital seta lateroclinate; vibrissa weakly differentiated. Arista dorsal. Face sometimes characterized by a shiny protuberance just above vibrissal socket (*Tethina* Haliday in Curtis) or nearby; face strongly depressed and short or with medial carina (*Tethina*) or distinctly convex. Gena bare, except for ventral or nearly ventral row of setae (peristomal setae), or even with a few dorsoclinate, strong setae (Canacinae). *Thorax*: Precoxal bridge developed. Prescutellar acrostichal setae developed; presutural dorsocentral setae differentiated; anepisternum with 2–3 developed posterior setae, also bearing a dorsoclinate seta at posterodorsal corner; 1 katepisternal seta; proepisternal seta developed. Wing generally hyaline, bearing fine, dense microtrichia; subcosta usually weakened apically, close to vein R_1; vein A_1 (A_1+CuA_2) short; vein A_2 long, present as a fold. *Abdomen*: Pregenital sclerites of male short and fused; male tergite 6 fused with sternite 8, forming a usually symmetrical, pregenital sclerite; male sternite 7 lacking; postgonites firmly connected laterally to base of phallapodeme or medially to it, distinctly anterior to basiphallus; hypandrium forming a sheath or phallic mantle around the postgonite and basiphallus; epandrium bearing 1–2 pairs of surstyli ventrally, sometimes anterior surstylus lacking; posterior surstylus partially articulated or fused with epandrium; mediobasal corner of surstylus connected to broad interparameral sclerite; cercus very short; postabdomen of female more-or-less retracted telescopically; 2 sclerotized spermathecae, variable in shape with a narrower cylindrical extension into spermathecal duct; cercus subcylindrical to tapered distally, sometimes bearing stout to pointed, spine-like setulae.

The Canacidae include 313 species that are ranked into 27 genera and 6 subfamilies (Munari and Mathis 2010 and updates).

Species of the subfamily Tethininae are mostly halophiles/thalassophiles, occurring primarily in coastal marine biotopes. Adults of thalassophilous species are commonly found in coastal marine areas (Karl 1930), including the intertidal zone, wrack heaps (usually brown algae that are most abundant along temperate seashores bathed by cold currents), salt marshes, dune vegetation, lagoon-littoriparian zones and on salty soils or bare sand. Unlike most species of Canacidae, those of the subfamily Pelomyiinae are not restricted to littoral habitats. These inland habitats are sometimes saline.

The immature stages are incompletely known. Ferrar (1987) provided some observations on the puparia of *Tethina grisea* (Fallén). Gorczytza (1988) reported on the spatial and seasonal distribution of some European species from a study using color traps on the Frisian Islands of Mellum and Memmert. In nature, an abundance of individuals and a paucity of species usually characterize sites where canacids occur.

Key to subfamilies, genera and species of Canacidae occurring in Greenland

1. Gena bare; peristoma bearing row of strong, dorsally curved setae; frons with 2 rows of interfrontal setulae, often seta-like; cell bm and discal cell distinct; crossvein bm present; prescutellar acrostichal setae present; scutum bearing numerous rows of coarse setae arising from punctures; postpronotum with 3 or more setae, the ventral seta dorsoclinate............... [Tethininae, *Tethina*; this near-cosmopolitan genus comprises more than half of the described tethinids; so far no representatives have been found in Greenland, but their occurrence in the country would not be surprising.]

– Gena with scattered setulae; peristoma bearing row of weak setulae; frons bare or with a few weak and scattered setulae; cell bm and discal cell confluent, crossvein bm absent; prescutellar acrostichal setae absent; scutum smooth, at most with a few scattered setulae; postpronotum with 1 or 2 setae, lacking a ventral seta that is dorsoclinate (Pelomyinae)....... ....*Pelomyiella mallochi* (Sturtevant)

Pelomyiella mallochi **(Sturtevant, 1923). Figs 1–2**

Henriksen 1939: not recorded

Distinguished from congeners by the following combination of characters: Body largely olivaceous grayish microtomentose. 1.3–2.4 mm.

Head higher than long; frons bearing some setulae in addition to larger setae; fronto-orbital and orbital setae usually with similar orientation, mostly reclinate or lateroclinate; fronto-orbital setae 2, although anterior seta much shorter than posterior seta, sometimes difficult to

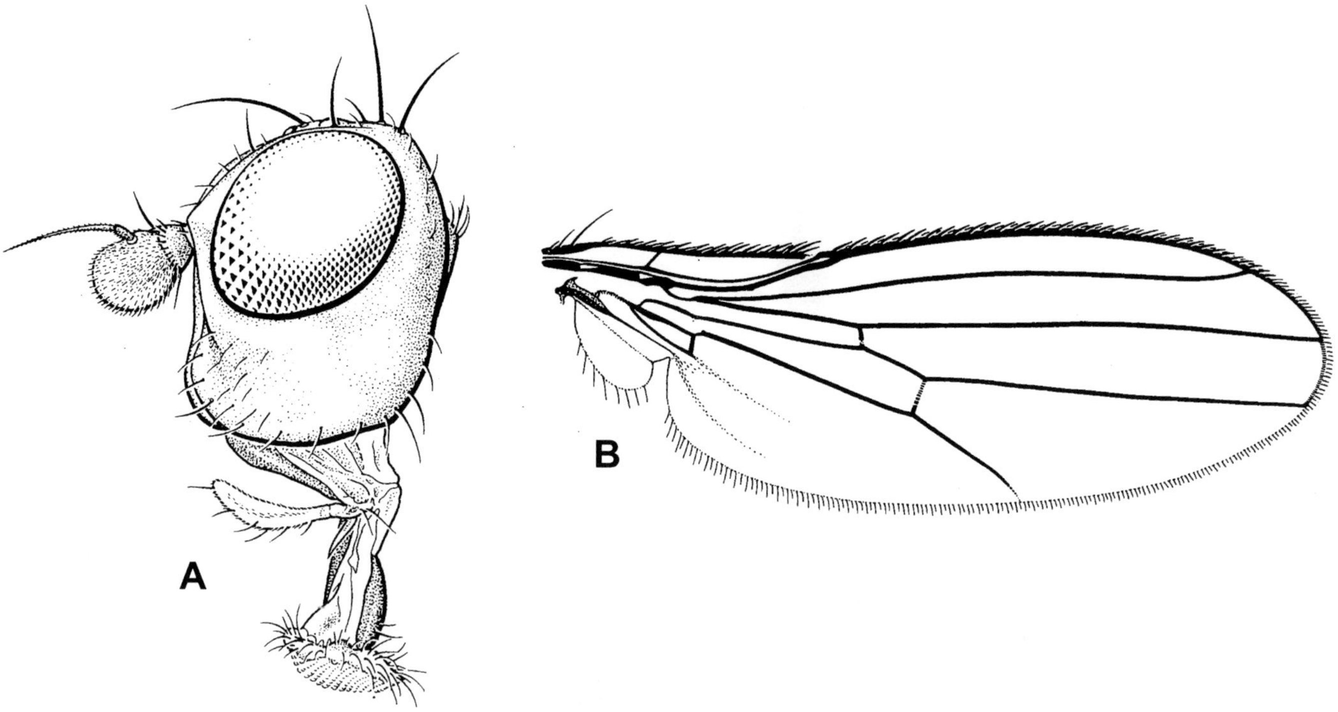

FIGURE 1 Pelomyiella mallochi; *A, head, lateral view; B, right wing, dorsal view. (From Manual of Nearctic Diptera).*

see; paravertical setae more or less convergent. Antenna largely black although 1st flagellomere often with some yellowish coloration apically and ventrally. Face vertical, not produced; face and peristoma microtomentose, without shiny stripes. Eye round, appearing bare, setulae very sparse or lacking. Gena high, more than half eye height; gena bearing few to many scattered, pale setulae between eye and ventral row of setulae. Palpus and proboscis usually normally developed; clypeus small, if exposed not protruding anteriorly beyond oral margin. Mesoscutum with numerous rows of coarse setulae arising from punctures; scutellar disc bare; postpronotum with 3 or more setae, ventral seta curved upward; acrostichal setulae lacking; prescutellar acrostichal setae present. Wing with costa not spinose; vein A1+CuA2 short, much shorter than discal cell; wing usually shorter, about twice as long as wide (less often 2.5 to 3 times); cell bm and discal cell distinct. Legs, including coxae, mostly gray; tibiae usually gray to yellows; mid and hind tibiae evenly setulose, lacking anterodorsal or posterodorsal setae. Abdominal tergites wider than long; tergite 6 well differentiated from short syntergosternite 7+8, the latter forming a dorsal pregenital sclerite. Male terminalia with epandrium extended ventrad as a ventral lobe; anterior surstylarlike lobe broadly fused with epandrium and situated ventrad of epandrium; surstylus greatly reduced, bearing some setulae, articulated with sternite 10, situated about level with cerci along posterior margin of epandrium; phallus long, sinuous, bearing dorsal pubescence.

The larva is unknown.

BIOLOGY. Unlike most Pelomyinae, this species occurs inland as well as along maritime coasts (Roháček 1992).

DISTRIBUTION. SW (Kangerlussuaq/Søndre Strømfjord), NW (Pituffik/Thule).

Holarctic/circumboreal, low arctic-subtropical.

NOTE. In the Nearctic region this species occurs sympatrically with *P. maritima* along the coast of Texas and with *P. melanderi* in western North America. The status of *P. mallochi*, *P. maritima*, and *P. melanderi* remains unresolved, as the primary types are females, and we have no males of *P. maritima* from the type locality. Morphologically, the issue is further complicated by few or no differences in structures of the male terminalia between *P. mallochi*, *P. melanderi*, and *P. maritima*.

References

Ferrar, P. (1987): A guide to the breeding habits and immature stages of Diptera Cyclorrhapha. – *Entomonograph* 8 (part 1: text): 1–478, (part 2: figures): 479–907. Leiden-Copenhagen: E.J. Brill/Scandinavian Science Press.

Gorczytza, H. (1988): Die Tethiniden der Nordseeinseln Mellum und Memmert (Diptera: Tethinidae). – *Drosera* 1988: 303–310.

FIGURE 2 Pelomyiella mallochi *habitus.*

Henriksen, K.L. (1939): A revised index of the insects of Grønland. – *Meddelelser om Grønland* 119: 1–101.

Karl, O. (1930): Thalassobionte und thalassophile Diptera Brachycera. – *Die Tierwelt der Nord- und Ostsee* 19(2): 33–84.

Munari, L. & Mathis, W.N. (2010): World catalog of the family Canacidae (including Tethinidae) (Diptera), with keys to the supraspecific taxa. – *Zootaxa* 2471: 1–84.

Roháček, J. (1992): Tethinidae (Diptera) of Czechoslovakia: a faunistic survey. – *Časopis Slezského Zemského Muzea, Opava* (A) 41: 127–131.

Sturtevant, A.H. (1923): New species and notes on synonymy and distribution of Muscidae Acalypteratae (Diptera). – *American Museum Novitates* 76: 1–12.

17.25. Ephydridae (Shore flies)

Tadeusz Zatwarnicki and Wayne N. Mathis

Very small to medium-sized flies of 1–6 mm. Body usually dark coloured, dull to shining. Head very variable in shape; arista distinctly pectinate dorsally, plumose or bare; subcranial cavity may be small to very large (Figs 1A–B, D–E, 2B, F, 3A, G, 4G). Wings hyaline to extensively darkened or patterned, costa with humeral and subcostal breaks, subcostal vein incomplete, cell Cu*p* open (Figs 1C, I, 2A, G, 3B, F, 4A–B). Adults are often found on or near the soil in humid or wet localities where they mate, oviposit or feed on a wide variety of matters like algae, honeydew, carrion, etc. Otherwisc, adults may be found at some distance from aquatic or semiaquatic habitats, *e.g.* on tree trunks and in vegetation. Some species have been recorded as pests in crops and greenhouses.

Most species have aquatic or semi-aquatic larvae that may have developed structural or physiological adaptations to specific habitats. Limnic larvae may be found from fresh water to very saline biotopes where they live of algae, bacteria, yeasts or even decaying animal matter and excrements. Some larvae are phytophagous, mining in leaves or stems. The shape is varying, from dorsoventrally flattened to cylindrical (Figs 1F, 4H). The integument is variously wrinkled, tuberculated, and covered with numerous setae or spinulae. Posterior abdominal segment is rather tapered, forming a cap-like disc or ending in lateral processes or a more or less elongated, sometimes branched or retractile respiratory tube bearing posterior spiracles and sometimes accessary filaments.

Eight shore-fly species were recorded from Greenland in previous faunistic papers; one species, *Gymnoclasiopa matanuska*, is here recorded for the first time from the country.

Key to the species

1. Face not produced anteroventrally, forming keel medially; orbital setae reduced, proclinate and/or ventroclinate (Figs 1A, B, D) 2
– Facial prominence large, gaping; orbital setae large and lateroclinate (Figs 1E, 2B, F, 3A, G, 4G).........3
2. Presutural or sutural dorsocentral seta inconspicuous or absent............*Gymnoclasiopa matanuska*
– Presutural or sutural dorsocentral seta present, conspicuous (Fig. 1D) .... *Philygria vittipennis*
3. Midface and lower facial margin without setae (Fig. 1E)...............................*Parydra pusilla*
– Midface and lower facial margin setose, often with long setae (Figs 2B, F, 3A, G, 4G)....................4
4. Frons with only 1 orbital seta present (Figs 2B, F)...5
– Two fronto-orbital setae present (Figs 3A, G, 4G)...6
5. Wing with costal vein extended to vein M, with pattern of many white and dark spots (Fig. 2A) .... *Limnellia stenhammari*
– Wing with costal vein extended to or just beyond apex of vein R_{3+4}; spots on wings white (Fig. 2G).... .... *Scatophila cribrata*
6. Presutural dorsocentral seta absent (Fig. 4G); gena with a distinct seta among fine setulae; wings with white spots (Figs 4A, B)............................7
– Presutural dorsocentral present (Figs 3A, G); gena without a distinct seta; wings without distinct markings (Figs 3B, F) 8
7. Shape of wing spot 3 usually round or at least widest in middle, and veins in vicinity of the spot are straight (Fig. 4B); costal section II (the straight distance between veins R_1 and R_{2+3}) longer than III (distance between veins R_{2+3} and R_{4+5}) (Fig. 4B); male tergites 3–4 equal in length; female sternite 7 divided medially into 2 sclerites (Fig. 4L) ....*Scatella stagnalis*
– Shape of wing spot 3 usually narrowest in middle, i.e. shaped like body of a violin, and veins in vicinity of spot divergent from it (Fig. 4A); costal section II longer than III; male tergite 4 markedly longer than 3; female sternite 7 a single sclerite (Fig. 4K) ....*Scatella tenuicosta*
8. Posterior orbital seta closer to inner vertical seta than to anterior orbital seta (Fig. 3G); gena at least 1/3 times as high as eye; frontal vitta shiny; lateral scutellar setae nearly as strong as subapical scutellar setae (Fig. 3G)........... *Thinoscatella quadrisetosa*
– Posterior orbital seta closer to anterior orbital seta than to inner vertical seta (Fig. 3A); gena at most 1/4 times as high as eye; frontal vitta dull, at most subshiny; lateral scutellar setae much less than half length of subapical scutellar setae (Fig. 3A)......... ....*Lamproscatella brunnipennis*

Gymnoclasiopa matanuska **Mathis and Zatwarnicki, 2012. Figs 1A–B**

Henriksen 1939: not recorded.

2.1–3.2 mm. Dorsum generally microtomentose grey, microtomentum on abdomen sparse and fine; frons densely

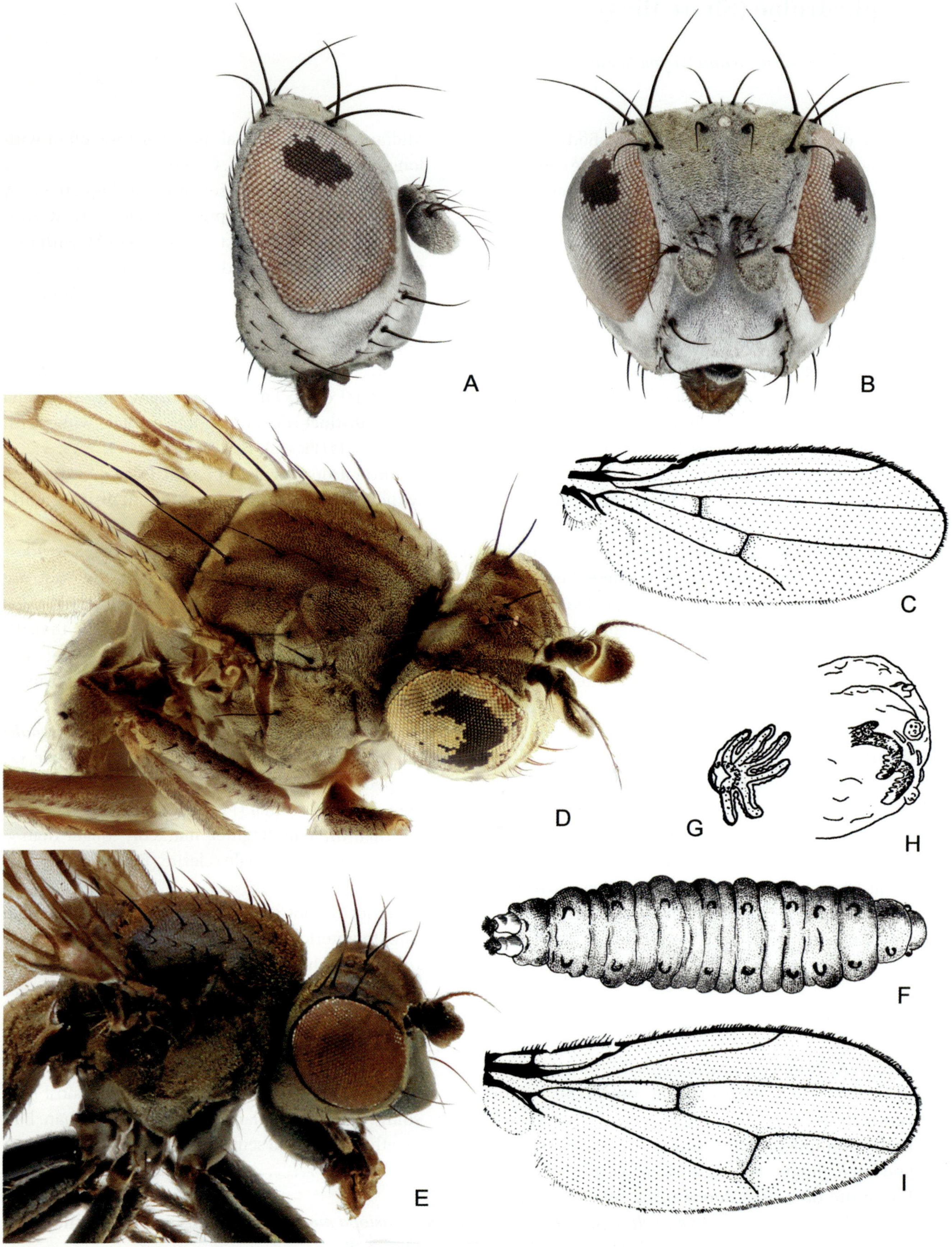

FIGURE 1 *Ephydridae. A–B,* Gymnoclasiopa matanuska. *A, head, lateral view; B, same, anterior view. C–D,* Philygria vittipennis. *C, wing; D, head and thorax, oblique view. E–I,* Parydra pusilla. *E, head and thorax, oblique view; F, larva, dorsal view; G, anterior spiracle of larva, lateral view; H, anterior section of larva with mouthhooks, oblique view; I, wing.* (*C, I from Papp 1975; F–H from Nielsen* et al. *1954*).

microtomentose, grey to blackish grey (Fig. 1A); antenna black; basal flagellomere covered with silvery grey microtomentum; arista bearing 5 dorsal rays; generally silvery white colored; gena silvery white, relatively high, higher than height of basal flagellomere; maxillary palpus black (Fig. 1B). Thorax and abdomen generally densely microtomentose, grey to tannish grey, dull. Wing hyaline; costal vein ratio 0.38–0.52; M vein ratio 0.50–0.67. Coxae, femora and tibia concolorous densely microtomentose grey to blackish dorsally, ventrally with some reddish coloration; fore- and hindtarsi mostly black; midtarsus with tarsomeres 1 and 2 somewhat yellowish to reddish, especially ventrally, apical tarsomeres black.

The larva is undescribed and has not been found in Greenland.

BIOLOGY. In Alaska and Mongolia found in limnic biotopes, primarily muddy to sandy shores.

DISTRIBUTION. SW (Kangerlussuaq/Søndrestrøm Air Base, 18.VI.1952, 1♀; W. J. Brown, Canadian National Collection).

Holarctic, subarctic-boreal.

Philygria vittipennis **(Zetterstedt, 1838). Figs 1C–D**

Henriksen 1939: *Hydrina* ? *vitipennis* Zett.

1.7–2.0 mm. Mostly grey. Fronto-orbital setae greatly reduced, inconspicuous, as 3–5, proclinate, hair-like setulae; fronto-orbital plate dark brown (Fig. 1D); antenna dark brown, lighter ventrally; arista appearing essentially bare, length of dorsal rays less than basal aristal width; face and parafacial greyish yellow; palpus yellow. Mesonotum with dark stripes along acrostichal and dorsocentral setal tracks (Fig. 1D); posterior notopleural seta inserted at much higher level than anterior seta; wing hyaline except for darkened crossvein r-m and dm-cu (Fig. 1C). Abdomen microtomentose on basal tergites, but with tergites 4–5 (♂) or 4–6 (♀) becoming shiny black and sparsely microtomentose.

The larva is undescribed and has not been found in Greenland.

BIOLOGY. In Scandinavia the species has distinct xerophilous tendencies (Dahl 1959). The ecological distribution consequently includes dry biotopes (*e.g.* dune heaths).

DISTRIBUTION. S (Qassiarsuk, Narsarsuaq), SW (Kangerlussuaq/Søndre Strømfjord, Qeqertarsuaq/Godhavn, Uummannaq Fjord: Ikerasak), NE (Scoresby Sund: Gåseland).

Palaearctic, subarctic-subtropical.

Parydra pusilla **(Meigen, 1830). Figs 1E–I**

1.5–1.9 mm. Two well-developed, lateroclinate, fronto-orbital setae (Fig. 1E); face in lateral view straight to concave or convex, usually densely microtomentose, silvery white to grey; anterior oral margin in lateral view extended beyond antennal bases; facial carina near antennal bases not extended beyond level of oral margin; facial series of setae with only dorsal 1–2 setae developed, inserted near parafacial; oral margin lacking ventroclinate setae; clypeus exposed and visible. Wing (Fig. 1I) hyaline to faintly infumate; costal vein extended to vein M; vein R_{2+3} relatively short; crossvein r-m located distinctly beyond level of junction of vein R_1 with costa; crossveins r-m and dm-cu with darker spot over them and faintly white areas laterad of these darker, crossvein spots.

Larval mouthhooks (Fig. 1H) sickle-shaped, without row of fine teeth apicoventrally. Anterior spiracle with 3–7 branches (Fig. 1G).

BIOLOGY. In Scandinavia only found in limnic biotopes, primarily muddy shores (Dahl 1959).

Adults may often be seen walking on the water's surface for short distances. They appear to graze the mud surface where they probably obtain algae as food.

DISTRIBUTION. sSW (only found at Nuup Kangerlua/Godthåbsfjord: Kapisillit).

Palaearctic, subarctic-subtropical.

Lamproscatella brunnipennis **(Malloch, 1920). Figs 3A–E**

1.8–2.6 mm. Brownish and greyish-blue, sometimes heavily microtomentose. Mesofrons generally dull; face only moderately protruding; golden brown, sometimes greyish or olivaceous. Posterior fronto-orbital bristle placed approximately in the middle between the anterior fronto-orbital bristles (Fig. 3A). Wing relatively clear, with no maculation (Fig. 3B). Basal pair of lateral scutellar setae not more than half length of apical pair (Fig. 3A). Shape of tergite 5 of male subtrapezoidal, posterior margin broadly rounded to subtruncate; shape of surstylus digitiform, straight; epandrium narrowing gradually to merger with surstyli; gonite produced ventrally, rounded, with submucronate lateral projection in anterior view (Figs 3D–E).

The larva is undescribed and has not been found in Greenland.

BIOLOGY. The habitat is maritime seashores, which can be partially submerged at high tide, usually with vegetation of grasses and sedges.

DISTRIBUTION. S (only Narsaq), NW (Pituffik/Thule Air Base), N (Inglefield Land).

Holarctic, subarctic-boreal.

Limnellia stenhammari **(Zetterstedt, 1846). Figs 2A–E**

1.5–2.3 mm; one well-developed, lateroclinate fronto-orbital seta; setulose portion of face mostly unicolorous, brown to greyish brown, occasionally with faintly coloured cinereous band along ventral margin (Fig. 2B); gena short.

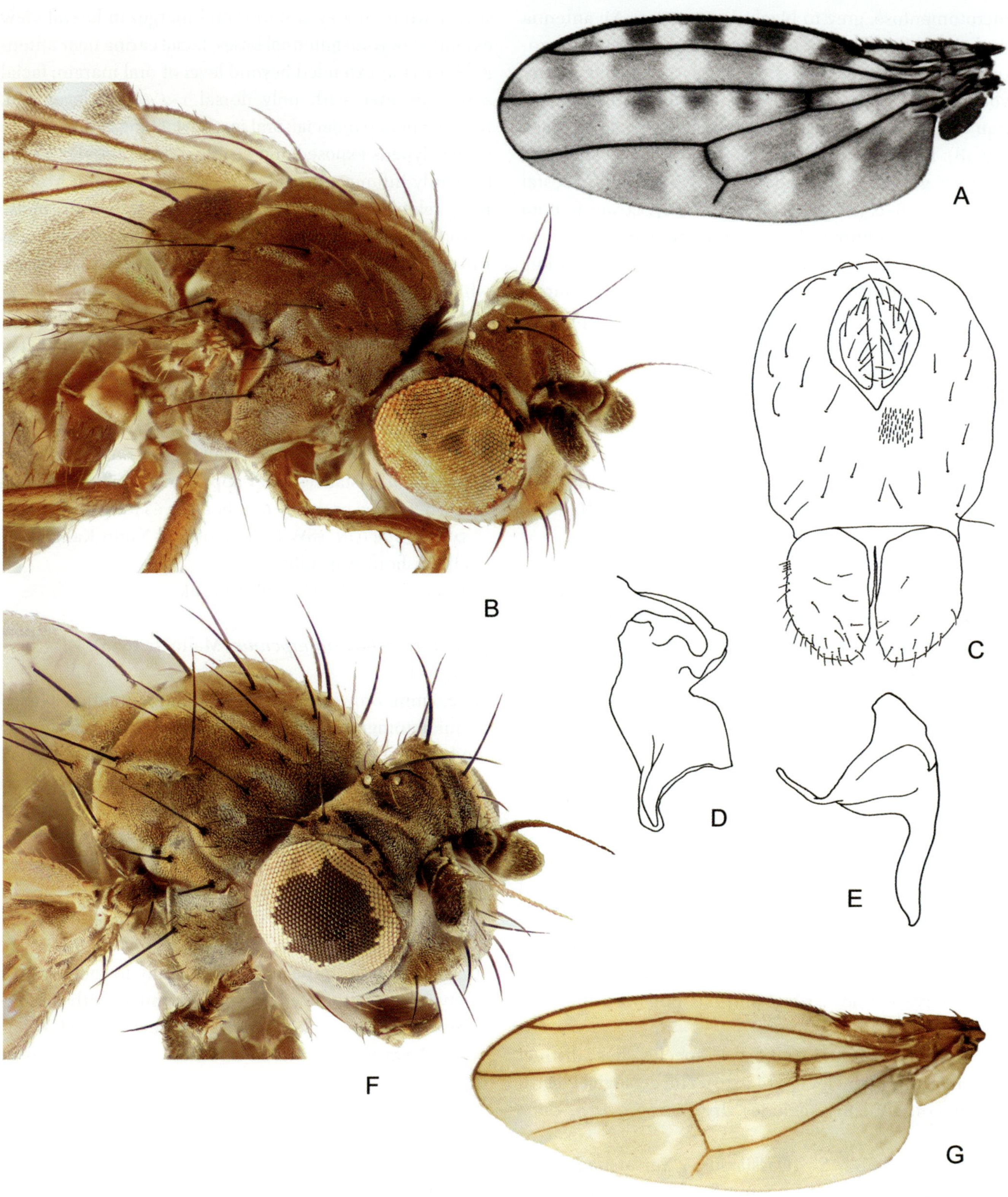

FIGURE 2 *Ephydridae. A–E,* Limnellia stenhammari. *A, wing; B, head and thorax, oblique view; C, epandrium and surstyli, posterior view D, aedeagus and phallapodeme, lateral view; E, gonite, lateral view; F–G,* Scatophila cribrata. *F, head and thorax, oblique view; G, wing; (A, C–E from Mathis 1978).*

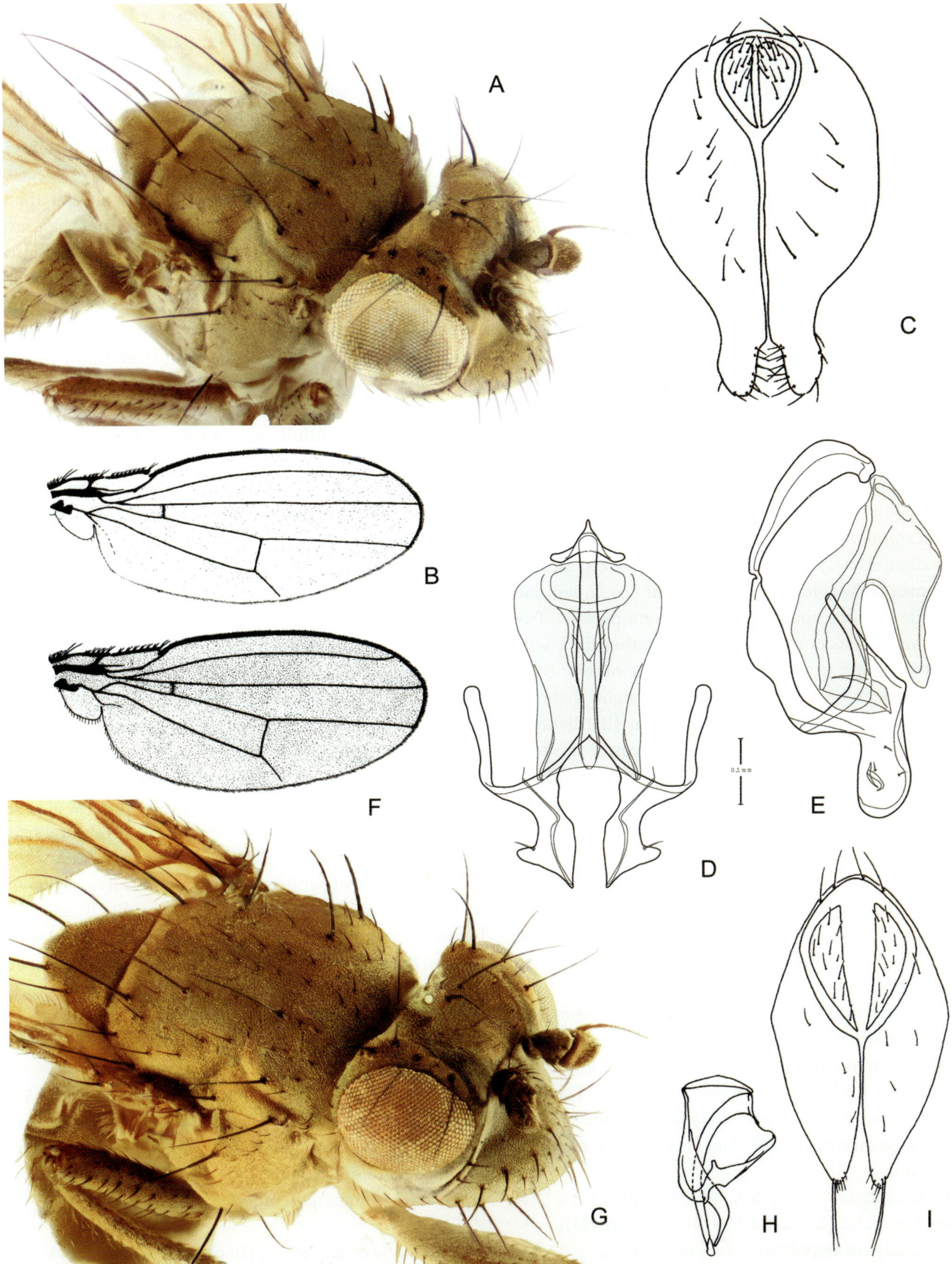

FIGURE 3 *Ephydridae. A–E,* Lamproscatella brunnipennis. *A, head and thorax, oblique view; B, wing; C, epandrium, posterior view; D, male internal terminalia, posterior view; E, same, lateral view. F–I,* Thinoscatella quadrisetosa. *F, wing; G, head and thorax, oblique view; H, male internal terminalia, lateral view; I, male epandrium, posterior view.* (*E from Ólafsson 1991; B, G–I, from Mathis 1979*).

Thorax mostly brownish black to black with cinereous vittate and guttate markings; acrostichal setulae in two rows, extended to scutellar base; dorsocentral setae 2 (0+2) (Fig. 2B). Wing maculate (Fig. 2A), generally brown with numerous white spots and markings; cell r_{4+5} with three darker spots closely appressed to vein R_{4+5}; costal vein extended to vein M. Legs generally unicolourous, black; halter brownish yellow to black. Abdomen black; posterior tergites becoming smooth and shiny; male terminalia as in Figs 2 C–E.

The larva undescribed and has not been found in Greenland.

BIOLOGY. In Scandinavia this species exhibits weak xerophilous tendencies, preferring drier meadow and dune heath biotopes (Dahl 1959).

DISTRIBUTION. S (Qassiarsuk, Upernaviarsuk), SW (Nuup Kangerlua/Godthåbsfjord: Kapisillit; Kangerlussuaq/Søndre Strømfjord).

Holarctic, widely distributed boreal-subtropical.

Scatella stagnalis **(Fallén, 1813). Figs 4B, D, F, H–J, L**

Approximately 2.5 mm. Body unicolourous brown, sometimes with golden dust. Two well-developed, lateroclinate, fronto-orbital setae; genal seta present. Two postsutural dorsocentral setae. Wings rather dark with five spots, spot 5 usually weaker than the others; costa rather thick; costal section II (the straight distance between veins R_1 and R_{2+3}) longer than III (distance between veins R_{2+3} and R_{4+5}) (Fig. 4B); costal spinulae closely spaced. Posteroventral setae on forefemur about as long as width of femur. Male tergite 3 nearly equals tergite 4 in length. Male genitalia with tip of gonite somewhat rounded below the hook (Fig. 5I); epandrium as in Fig. 4F. Female sternite 7 divided medially into 2 sclerites (Fig. 4L).

Larval mouthhooks with row of fine teeth apicoventrally (Fig. 4J). Anterior spiracle with 3 or more tapered branches (Fig. 4I).

BIOLOGY. Found in a variety of biotopes, limnic as well as maritime beaches and banks, floating algae and aquatic plants, muddy ground and wet rocks, and in any other humid place. Adults are even found in dry areas. The species utilizes a variety of particulate food matter, ranging from blue-green algae to true algae (in particular diatoms) and detritus.

DISTRIBUTION. G [S, SW, NW (Pituffik/Thule), N (Nyboe Land), NE (Gåseland, Ella Ø), SE (Ørsted Ø)].

Essentially cosmopolitan, subarctic-tropical.

Scatella tenuicosta **(Collin, 1930). Figs 4A, C, E, G, K**

Variable size, usually 1.5–2.0 mm. Unicolourous brown, sometimes rusty. Two well-developed, lateroclinate, fronto-orbital setae; genal seta present. Two postsutural dorsocentral setae, no postsutural acrostichal setae. Wings variably clear (dark in larger specimens), with five spots, spot 3 most often formed like the body of a violin, with the veins bent in vicinity of the spot; costal vein extended to vein M; costa slender; costal section II shorter than III (Fig. 4A); costal spinulae closely spaced. Posteroventral setae on forefemur slightly shorter than width of femur. Male tergite 3 obviously shorter than tergite 4. Male genitalia with tip of gonite more or less angled below the hook (Fig. 4D); epandrium as in Fig. 4E. Female sternite 7 a single plate (Fig. 4K).

BIOLOGY. Found in abundance both in limnic and maritime localities under a variety of conditions. Tuxen (1944) reported the larvae from a hot spring having a temperature of *c.* 48°C. Larval food consists mainly of blue-green algae. Adults rasp algae off the stones with their specially adapted scraperlike pseudotracheae. Larvae move freely among the algae not far below water surface, tearing off pieces of algal filaments with the mouth hooks.

DISTRIBUTION. G (almost the same Greenland distribution as in *S. stagnalis*).

Holarctic, boreal-subtropical.

Scatophila cribrata **(Stenhammar, 1844). Figs 2F–G**

1.6–2.2 mm. One well-developed, lateroclinate, fronto-orbital seta. Face grey but with a brown stripe extended along and from interfoveal carina to just above facial setae at oral margin (Fig. 2F). Mesonotum dark brown except for short, grey, longitudinal stripes; posterior apex of scutellum with grey spot; 3 dorsocentral setae, presutural seta well developed; halter black (Fig. 2F). Wing generally infumate (Fig. 2G) with white spots (2 spots in cell r_{2+3}; 3 spots in cell r_{4+5}; an angulate, linear spot in discal cell just before crossvein dm-cu; 2 spots in cell m, more basal spot narrow, irregularly C-shaped; cell cua_1 with a single spot at same level as spot in discal cell); costal vein short, extended only to vein R_{4+5}.

The larva is undescribed and has not been found in Greenland.

DISTRIBUTION. S (Itilleq, Qaqortoq/Julianehåb, Qasiarssuk, Narsarsuaq, Ivittuut), SW (Nuup Kangerlua/Godthåbsfjord: Qooqqut, Kapisillit; Kangerlussuaq/Søndre Strømfjord, Ameralik, Qeqertarsuaq/Godhavn).

Holarctic, subarctic-temperate.

Thinoscatella quadrisetosa **(Becker, 1896). Figs 3F–H**

1.8–2.8 mm (averaging 2.45 mm); mostly olivaceous brown to dark brown, but with some faintly grey areas, particularly on head; 2 well-developed, lateroclinate, fronto-orbital setae (Fig. 3G); protruding face bearing 2–3

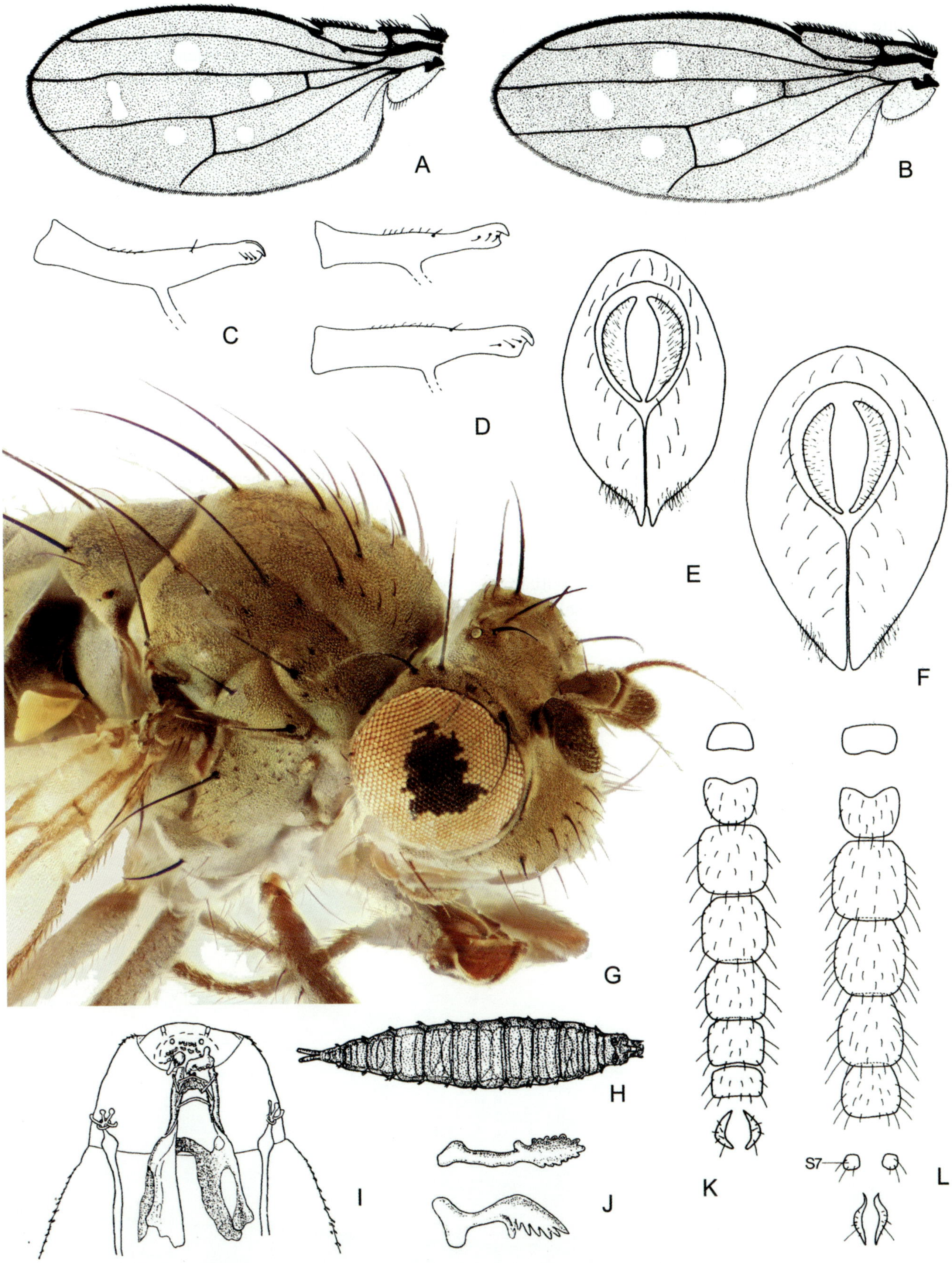

FIGURE 4 *Ephydridae. A, D, E, G, K.* Scatella tenuicosta, *A, wing; D, gonite, lateral view; E, epandrium, posterior view; G, head and thorax, oblique view; K, female abdominal sternites. B, C, F, H–J, L.* Scatella stagnalis. *B, wing; C, gonite, lateral view (two forms); F, epandrium, posterior view; H, larva, posterior view I, anterior section of larva with mouthhooks, ventral view; J, mouthhook in dorsal and lateral view, K–L, female abdominal sternites; (A–F, K–L from Ólafsson 1991; C from Sen 1931; I–J from Tuxen 1936).*

dorsoclinate facial setae (ventral seta longest) and numerous facial setulae; genal seta lacking. Mesofrons brown, shining, parafrons deep blue, orbits brown, face greyish-brown, sometimes more greyish. Eye relatively small; anterior portion of gena and parafacial unicolorous, distinctly lighter in colour than face, greyish white, becoming darker posteriorly. Mesonotum (Fig. 3G) microtomentose to subshiny, mostly unicolorous, olivaceous brown to greyish brown, more microtomentose toward anterior margin; three dorsocentral setae. Wing (Fig. 3F) uniformly infuscate, unspotted; veins moderately thick; costal spinulae closely spaced; costal vein extended to vein M. Abdomen mostly unicolorous, subshiny, olivaceous brown. Structures of male terminalia as in Figs 3H–I.

The larva is undescribed and has not been found in Greenland.

BIOLOGY. Predominantly a seashore species, found in abundance in upper part of the tidal zone. The habitat is usually flats of coarse sand covered with a layer of blue-green algae.

DISTRIBUTION. S (Itilleq), SW (Nuuk/Godthåb, Qeqertarsuaq/Disko), NW (Pituffik/Thule Air Base).

Circumboreal, subarctic-temperate.

References

Dahl, R.G. (1959): Studies on Scandinavian Ephydridae (Diptera Brachycera). – *Opuscula Entomologica* (Suppl. 15), 224 p.

Henriksen, K.L. (1939): A revised index of the insects of Grønland. – *Meddelelser om Grønland* 119(10): 1–112.

Mathis, W.N. (1978): A revision of the Nearctic species of *Limnellia* Malloch (Diptera: Ephydridae). – *Proceedings of the Biological Society of Washington* 91(1): 250–293.

Mathis, W.N. (1979): Studies of Ephydrinae (Diptera: Ephydridae), II: Phylogeny, classification, and zoogeography of Nearctic *Lamproscatella* Hendel. – *Smithsonian Contributions to Zoology* 295: 41+iv pp.

Mathis, W.N. & Zatwarnicki, T. (2012): A revision of the New World species of *Gymnoclasiopa* Hendel (Diptera: Ephydridae). – *ZooKeys* 248: 1–69.

Nielsen, P., Ringdahl, O. & Tuxen, S.L. (1954): Diptera 1 (exclusive of Ceratopogonidae and Chironomidae). – *Zoology of Iceland* 3, pt. 48a, Copenhagen, Reykjavik, 189 pp.

Ólafsson, E. (1991): Taxonomic revision of western Palaearctic species of the genera *Scatella* R.-D. and *Lamproscatella* Hendel, and studies on their phylogenetic position within the subfamily Ephydrinae (Diptera: Ephydridae). – *Entomologica Scandinavica*, Suppl. 37: 1–100.

Papp, L. (1975): Vízilegyek – Ephydridae. – *In*: *Fauna Hungariae* 15/6, Budapest, 128 pp.

Sen, P. (1931): Preliminary observations on the early stages of *Scatella stagnalis* Fall. (Ephydridae, Diptera). – *Journal of the Marine Biological Association* 16: 847–851.

Tuxen, S.L. (1936): Die Arten der Gattung *Scatella* (*Ephydridae*) in heissen Quellen. – *Opuscula Entomologica* 1: 105–111.

Tuxen, S.L. (1944): The hot springs, their animal communieties and their zoogeographical significance. – *In: Zoology of Island* 1, part II, 206 pp.

17.26. Anthomyiidae, Fanniidae, Muscidae and Scathophagidae (The Muscidae family group)

Verner Michelsen

The group of calyptrate Diptera conventionally treated as the Muscoidea consists of 4 well defined families all occurring in Greenland. Anthomyiidae by 32 species and Muscidae by 37 species are among the most important families of Diptera in Greenland in terms of species diversity and overall abundance. Many species play an important role as pollinators of arctic flowering plants. The small, mainly boreal to high arctic family Scathophagidae represented by 6 species in Greenland is less diverse, but includes a few abundant species frequenting dung and seashore beds of brown algae. Most Fanniidae occur in forested areas and only one of the 3 species on record from Greenland is possibly indigeneous, the others certainly introduced by man.

The following combination of characters should enable separation of adult muscoid flies from other families of Diptera occurring in Greenland: (1) antenna with a longitudinal seam on pedicel and a basidorsal arista on postpedicel (3rd antennomere); (2) meron normally bare, never with a vertical row of setae; (3) distal section of vein M straight, exceptionally angled forward. Adult muscoids can be identified to family by using the key to the Diptera families found in Greenland (17.1).

The arrangement is alphabetical by family, genus and species. For each family is given a key to the adults of both sexes. A few additional species have been recorded in the literature from time to time, but are certainly based on misidentifications (Michelsen 2006) and thus excluded from the present treatment. For each species the full scientific name is followed by a page reference to the recent catalogue of the Muscoidea of Greenland by Michelsen (2006) which gives an overview of the fauna with new species records and an exhaustive review of earlier records with details on synonymy and misidentifications. Therefore, only the most important synonyms used in the literature on Greenland Diptera are specified in the present account.

ANTHOMYIIDAE

The Anthomyiidae have *c.* 2000 described species, but many more await description. The family is most diverse in the cooler, boreal and elevated zones of the Northern Hemisphere, and the family is well represented in Greenland by 32 species in 12 genera. Apart from 3 exclusively Nearctic species, the remaining 29 species have Holarctic, often circumpolar distributions. Most species have a southern distribution in Greenland, but 4 species occur primarily in the high arctic parts.

Anthomyiid flies are overall very similar to Muscidae, but are generally recognized by having some tiny, hair-like setulae beneath the tip of scutellum, vein A_1 extended to the wing margin as a fine fold or shadow, and a slender, often parallel-sided and flattened male abdomen.

Most species are strongly sexually dimorphic and many species have males that occasionally participate in aerial swarming. Many species are strongly anthophilous, feeding on nectar, less frequently also on pollen. Anthomyiids undoubtedly play a significant role as pollinators under arctic conditions. Larval diets vary considerably. Several groups have saprophagous larvae developing in all kinds of decomposing organic debris, while the larvae of others groups are more or less host specialized, feeding on living tissues of vascular plants or fungi. Factual knowledge and phylogenetic inference suggest that the larvae of Greenland Anthomyiidae are saprophagous in 17 species, phytophagous in 12 species and fungivorous in 3 species.

Key to species (males, females)

1. Males (abdomen caudally obtuse, sternite V enlarged, deeply incised at hind margin) 1
– Females (abdomen caudally tapering, usually exposing tip of oviscapt with paired cerci) 32
2. Head dichoptic, female-like, having eyes separated by a broad frons 3
– Head holoptic, having eyes narrowly separated by largely suppressed frons 9
3. Wing at least apically with a darkly clouded area. Thoracic spiracles enlarged, whitish coloured 4
– Wing without demarcated darker areas. Thoracic spiracles unremarkable with respect to size and colour..5
4. Gena in profile about as deep as eye. Frons provided with very fine setae. Abdomen caudally thick and truncated............................... *Fucellia vibei*

FIGURE 1 *Muscoid flies, males:* Pegoplata tundrica, *Anthomyiidae* (*A*); Spilogona arctica, *Muscidae* (*B*); Scathophaga furcata, *Scathophagidae* (*C*).
(PHOTOS: J. BÖCHER)

– Gena in profile less deep than eye. Frons provided with unremarkable setae. Abdomen not caudally enlarged .... *Fucellia pictipennis*

5. Vein C distally with av-row of widely spaced, short spines. Scutellum bare beneath tip 6

– Vein C bare or finely setulose ventrally. Scutellum with tiny setulae beneath tip 8

6. Palpus and tibiae reddish brown to yellow. Hind femur ventro-basally with a rounded, swelling bearing clusters of setulae and spinules.............
.... *Fucellia tergina*

– Palpus black, tibiae dark reddish brown to black. Hind femur different................................7

7. Anterior declivity of thorax with some blunt setulae of moderate length. Hind femur ventro-basally without swelling but with small cluster of setulae...
.... *Fucellia ariciiformis*

– Anterior declivity of thorax with some very short, spinose setulae. Hind femur ventro-basally with an inner laminate projection beset with spinules at tip and an outer cluster of setulae....................
.... *Fucellia fucorum*

8. Antennal postpedicel enlarged, arista notably short and basally thickened. Scutellum between apical pair of scutellar setae with setulae finer than setulae at lateral margins............. *Myopina crassipalpis*
– Antennal postpedicel and arista normal in size and proportions. Scutellum at apical margin with setulae of same size as setulae at lateral margins ....*Botanophila moriens*
9. Notopleuron with some setulae in addition to 2 notopleural setae.................................10
– Notopleuron bare apart from 2 notopleural setae... 12
10. Prementum grey-dusted. Notopleural setulae standing between 2 notopleural setae..................... ....*Paregle audacula*
– Prementum shiny black. Notopleural setulae standing around posterior notopleural seta............. 11
11. Arista pubescent. Setulae on anterior declivity of thorax coarse and blunt. Mid tibia with av-seta...... ....*Zaphne nearctica*
– Arista plumose. Setulae on anterior declivity of thorax unremarkable. Mid tibia without av-seta .... *Zaphne divisa*
12. Vein C bare on v-surface at least distal to insertion of vein Sc..13
– Vein C throughout finely setulose on v-surface 27
13. Vein C finely setulose basal to insertion of vein Sc... 14
– Vein C bare on v-surface throughout..............17
14. Genital sclerite large, ventrally densely beset with coarse, blunt setulae...............................15
– Genital sclerite normal in respect to size and vestiture 16
15. Pregenital sclerite shiny black. Mid tibia with 1(–2) ad-seta............................. *Zaphne frontata*
– Pregenital sclerite extensively grey-dusted. Mid tibia with 2(–3) ad-setae................. *Zaphne tundrica*
16. Sternite V with apically hooked lobes .... *Zaphne brunneifrons*
– Sternite V with simple lobes .... *Zaphne occidentalis*
17. Hind tibia with 2 pd-setae. Prementum non-tomentose, shiny black 18
– Hind tibia with 3–4 pd-setae. Prementum more or less matt, tomentose 19
18. Prealar seta coarse and blunt. Scutellum between apical pair of scutellar setae with setulae finer than setulae at lateral margins. Sternite V with short, simple lobes...................... *Pegoplata tundrica*
– Prealar seta normal. Scutellum between apical pair of scutellar setae either bare or with setulae of same size as those at lateral margins. Sternite V with each lobe bearing a thin, leaf-like process .... *Paradelia arctica*
19. Fore tarsomere 1 with ventro-basal setula notably coarse. Caudal segments of abdomen black subshining through sparse dusting .... *Lasiomma picipes*
– Fore tarsomere 1 with ventro-basal setula unremarkable. Caudal segments of abdomen matt by greyish dusting as the remaining abdomen..............20
20. Face and vibrissal angle projected far beyond fronto-parafacial angle. Palpus and haustellum long and slender .... *Botanophila rubrigena*
– Face and vibrissal angle level with or only reaching slightly beyond fronto-parafacial angle. Palpus and haustellum not long and slender 21
21. Longest aristal pubescence more than 2× as long as greatest aristal diameter. Prealar seta same length as notopleural setae............... *Botanophila profuga*
– Longest aristal pubescence less than 1.5× as long as greatest aristal diameter. Prealar seta much shorter than notopleural setae............................ 22
22. Mid tarsomere 1 with a row of long d-setulae..... 23
– Mid tarsomere 1 without long d-setulae........... 24
23. Hind tibia with 1–4 short pv-setae in basal half. Sternite V with each lobe bearing 2 straight, close-set, apically truncated setulae at inner apical margin................................ *Delia fabricii*
– Hind tibia with extensive row of short pv-setae. Sternite V without modified setulae at apex .... *Delia groenlandica*
24. Hind tibia with a complete comb of fine, erect pv-setae 25
– Hind tibia with a short, irregular row of fine pv-setae 26
25. Fore tibia at apex with coarse, truncated pv-seta. Hind femur with row of pv-setae confined to distal half or third .... *Delia platura*
– Fore tibia at apex with slender, tapering pv-seta. Hind femur with entire row of short, erect pv-setae. .... *Delia setigera*
26. Sternites III–IV expanded, at midline completely devoid of setulae, while at lateral margins with some notably long and strong setae. Sternite V unremarkable .... *Delia echinata*
– Sternites III–IV unremarkable. Sternite V with each lobe bearing a dense comb of setulae along inner margin......................... *Botanophila betarum*
27. Hind tibia with 1–3 short p-setae. Robust, dark-legged species with long prealar seta................ .... *Eutrichota tunicata*
– Hind tibia without p-setae......................... 28

28. Hind femur with pv-setae shorter than half depth of femur 29
– Hind femur with 1–4 pv-setae at least fully as long as depth of femur 30
29. Hind femur with pv-row of short, notably coarse and erect setulae. Sternite V with setulose lobes large and strongly projected ventrad........ *Pegomya notabilis*
– Hind femur with pv-row of short and fine setulae. Sternite V with setulose lobes small and weakly projected ventrad................*Pegomya circumpolaris*
30. Lobes of sternite V apically bare and turned ventrad ....*Pegomya flavifrons*
– Lobes of sternite V apically setulose................31
31. Body covered in light ash-grey dusting. Antennal postpedicel short, less than 1.5× as long as wide; arista very short-pubescent, strongly thickened basally. Genal setulae in 2–3 irregular rows .... *Pegomya icterica*
– Body covered in brownish grey dusting. Antennal postpedicel c. 2.0× as long as wide; arista pubescent, normally thickened basally. Genal setulae arranged in single row......................... *Pegomya zonata*
32. Lower calypter much shorter than upper calypter, linear to slightly convex along hind margin....... 33
– Lower calypter distinctly lobe-like, ranging from half to full length of upper calypter 40
33. Very small (wing length less than 3.5 mm). Antennal postpedicel in profile barely longer than wide; arista short, strongly thickened basally. Grey-dusted prementum thick, apically attenuate; labella small .... *Egle groenlandica*
– Larger (wing length more than 4.0 mm)........... 34
34. Wing at least apically with a dark-clouded area. Thoracic spiracles enlarged, white-coloured...... 35
– Wing without dark-clouded areas. Thoracic spiracles unremarkable 36
35. Gena in profile about as deep as an eye. Setae on frons notably short and fine ....*Fucellia vibei*
– Gena in profile less deep than an eye. Setae on frons of normally strong ....*Fucellia pictipennis*
36. Vein C distally with av-row of widely spaced, short spines. Scutellum bare beneath tip................ 37
– Vein C bare or finely setulose ventrally. Scutellum with tiny setulae beneath tip 39
37. Palpus and tibiae reddish brown to yellow.......... ....*Fucellia tergina*
– Palpus black, tibiae dark reddish brown to black 38
38. Anterior declivity of thorax with some blunt setulae of moderate length. Mid femur with a series of av-setae in basal half...............*Fucellia ariciiformis*
– Anterior declivity of thorax with some very short, spinose setulae. Mid femur without av-setae in basal half ....*Fucellia fucorum*
39. Scutellum between apical pair of scutellar setae with setulae finer than setulae at lateral margins. Body and legs blackish subshining through thin cover of brown dusting Cerci with subapical incision .... *Myopina crassipalpis*
– Scutellum between apical pair of scutellar setae with setulae of same size as those at lateral margins. Body and legs covered in dense greyish dusting. Cerci not incised apically.................*Botanophila moriens*
40. Notopleuron with some setulae in addition to 2 notopleural setae...................................41
– Notopleuron bare apart from 2 notopleural setae... 43
41. Prementum tomentose, matt. Notopleural setulae standing between 2 notopleural setae ... *Paregle audacula*
– Prementum non-tomentose, shiny black. Notopleural setulae standing around posterior notopleural seta 42
42. Arista pubescent. Setulae on anterior declivity of thorax coarse and blunt. Mid tibia with av-seta...... ....*Zaphne nearctica*
– Arista plumose. Setulae on anterior declivity of thorax unremarkable. Mid tibia without av-seta ....*Zaphne divisa*
43. Vein C bare on v-surface at least distal to insertion of vein Sc 44
– Vein C throughout finely setulose on v-surface ... 58
44. Vein C finely setulose on v-surface basal to insertion of vein Sc 45
– Vein C bare on v-surface throughout.............. 48
45. Mid and hind tibiae yellow 46
– Mid and hind tibiae blackish 47
46. Frontal stripe uniformly brownish black, occasionally dirty yellowish on lower part; orbital setulae absent or in a single row outside lower orbital setae .... *Zaphne brunneifrons*
– Frontal stripe blackish on upper 3/4 and orange-red on lower 1/4; orbital setulae in 2 rows outside lower orbital setae .... *Zaphne occidentalis*
47. Mid tibia with 1(–2) ad-seta (1 strong)............... ....*Zaphne frontata*
– Mid tibia with 2(–3) ad-setae (2 strong)............. .... *Zaphne tundrica*
48. Hind tibia with 2 pd-setae. Prementum shiny black. 49
– Hind tibia with 3–5 pd-setae. Prementum more or less covered in grey dusting....................... 50
49. Palpus apically strongly dilated. Prealar seta blunt. Scutellum between apical pair of scutellar setae with

setulae finer than setulae at lateral margins.......... .... *Pegoplata tundrica*
– Palpus unremarkable. Prealar seta normal, tapering. Scutellum at apical margin, between apical scutellar setae, either bare or with a few setulae of same size as those at lateral margins ....*Paradelia arctica*
50. Face and vibrissal angle projected far beyond fronto-parafacial angle. Palpus and haustellum extended and slender....................*Botanophila rubrigena*
– Face and vibrissal angle level with or only reaching slightly beyond fronto-parafacial angle. Palpus and haustellum short....................................51
51. Longest aristal pubescence more than 2× as long as greatest aristal diameter. Prealar seta same length as notopleural setae................*Botanophila profuga*
– Longest aristal pubescence less than 1.5× as long as greatest aristal diameter. Prealar seta much shorter than notopleural setae............................. 52
52. Upper occiput, beneath postocular setae, practically bare.. 53
– Upper occiput extensively setulose 56
53. Lower posterior katepisternal seta nearly as long and strong as upper posterior seta. Distal oviscapt (epiproct and cerci) with stout, recurved spines..... .... *Delia fabricii*
– Lower posterior katepisternal seta much shorter and weaker than upper posterior seta. Distal oviscapt only with setulae.................................. 54
54. Fore tarsomeres 2–4 expanded .... *Delia groenlandica*
– Fore tarsus slender................................ 55
55. Body tomentum grey, in places with a brown tinge. Mid and hind tibiae dark-coloured. Hind femur with av-setae largely confined to distal half. Mid-dorsal dark markings on abdominal tergites III and IV forwardly expanded.....................*Delia platura*
– Body tomentum whitish grey with a faint yellowish brown tinge. Mid and hind tibiae yellowish in basal parts. Hind femur with row of av-setae extended almost to base. Mid-dorsal dark markings on abdominal tergites III and IV narrow to obsolete........... .... *Delia setigera*
56. Presutural acrostichal setae absent. Cerci clothed in fine pubescence, matt ....*Botanophila betarum*
– Presutural acrostichal setae 1–3 pairs. Cerci extensively smooth and shiny 57
57. Presutural acrostichal setae 3 subequal pairs. Cerci slender, cylindrical................*Lasiomma picipes*
– Presutural acrostichal setae 1 strong pair preceded by a much finer pair. Cerci somewhat depressed, widest at base .... *Delia echinata*
58. Hind tibia in basal half with a short p-seta. Robust, dark-legged species with orange-red frontal stripe; interfrontal setae absent; prealar seta very long ....*Eutrichota tunicata*
– Hind tibia without p-seta.......................... 59
59. Oviscapt short and thick, sternites 6 and 7 with dense fields of spiny setulae 60
– Oviscapt long and slender, without fields of modified setulae 61
60. Arista strongly thickened in basal third, minutely pubescent. Genal setulae in two rows. Light bluish grey dusted species with anterior parts of head yellow to orange-yellow.............*Pegomya icterica*
– Arista slightly thickened basally, short-pubescent. Genal setulae in one row. Brownish grey dusted species with anterior parts of head orange-red.......... .... *Pegomya flavifrons*
61. Frontal stripe with a pair of setae. Presutural rows of acrostichal setae wider apart than their distance to rows of dorsocentral setae. Mid and hind tarsi slightly thickened, distal tarsomeres ventrally densely clothed in hair-like sensilla .... *Pegomya circumpolaris*
– Frontal stripe bare, at the most with a pair of fine setulae. Presutural rows of acrostichal setae closer together than their distance to rows of dorsocentral setae. Mid and hind tarsi unremarkable 62
62. Postgenal setulae all dark. Cerci in profile longish, parallel-sided...................... *Pegomya notabilis*
– Postgenal setulae mainly dark, but intermingled with some pale ones. Cerci in profile short, apicoventrally rounded .... *Pegomya zonata*

Botanophila **Lioy, 1864**

A large, morphologically diverse genus with hundreds of described species after the inclusion of the species earlier assigned to *Pegohylemyia* Schnabl & Dziedzicki. The genus is primarily defined by male genitalic characters. It is remarkably rich in species in northern and elevated parts of North America and Eurasia, yet only 4 species with circumpolar distributions are known from Greenland.

The adults (except possibly *B. moriens*) are anthophilous, feeding on nectar. The immature stages are not known for the Greenland species, but the larvae are expectedly phytophagous or phyto-saprophagous. Apparently, the genus avoids the high arctic parts of Greenland.

Botanophila betarum **(Lintner, 1883) Fig. 2A**

Michelsen 2006: 107. Earlier recorded from Greenland as '*moesta* Holmgren,' a name actually referring to a species of *Lasiomma*.

FIGURE 2 *Habitus of Anthomyiidae:* Botanophila betarum, *male (A);* Delia fabricii, *male (B) and female (C);* Eutrichota tunicata, *male (D);* Fucellia pictipennis, *male (E);* Fucellia ariciiformis, *female (F);* Myopina crassipalpis, *male (G);* Paregle audacula, *female (H);* Pegomya notabilis, *male (I).*

DISTRIBUTION. S, SW, NW, SE (north to Upernavik and Ammassalik).

Circumpolar, low arctic–subarctic.

Botanophila moriens **(Zetterstedt, 1845)**
Michelsen 2006: 107.

DISTRIBUTION. SW (Qeqertarsuaq/Godhavn), NE (Mestersvig, Jameson Land).

Holarctic, low arctic–alpine (Norway, Sweden, Scotland, Italy and Canada).

Botanophila profuga **(Stein, 1916)**
Michelsen 2006: 108. Records of the common European '*fugax* Meigen' from Greenland in the early literature often refer to the present species.

DISTRIBUTION. S, SW, NE, SE (common in southern parts, north to Uummannaq Fjord and Jameson Land).

Circumpolar, arctic–boreal and alpine.

***Botanophila rubrigena* (Schnabl, 1915)**

Michelsen 2006: 108. Also recorded from Greenland as '*sobrina* Collin' and '*nuoljensis* Ringdahl', both synonyms.

DISTRIBUTION. S, SW, SE (locally frequent in the southern parts, reaching north to Qeqertarsuaq/Disko and Ammassalik).

Circumpolar, low arctic–subarctic and alpine.

***Delia* Robineau-Desvoidy, 1830**

A large, widespread genus with hundreds of morphologically relatively uniform species. 5 species are known from Greenland.

The adult flies are frequently encountered in low vegetation or on open ground, but also visit flowers. The larvae of all species are expectedly phytophagous, living as leaf-miners in various dicots, or as borers in shoots or floral parts of Poaceae (grasses, cereals) and various dicot families. The genus seems to avoid the high arctic parts of Greenland and includes 3 Holarctic and 1 (*D. groenlandica*) Nearctic species. Two species (*D. echinata, D. platura*) are remarkably widespread, tolerating subtropical to low arctic climatic conditions.

***Delia echinata* (Séguy, 1923)**

Michelsen 2006: 108. The name '*florilega* Zetterstedt' has been misapplied for the present species in the early literature on Greenland Diptera.

BIOLOGY. In Greenland (SW, Grædefjord) reared from larva forming a leaf-mine on *Cerastium alpinum* (O. Karsholt *leg.*). Elsewhere reported as a leaf-miner and stem-borer of various Caryophyllaceae and Chenopodiaceae.

DISTRIBUTION. S, SW, NE, SE (mainly southern, north to Uummannaq Fjord and Germania Land).

Circumpolar, arctic–subtropical and alpine, very widespread.

NOTE. Females are difficult to separate from those of *D. platura*; in addition to the character given in the key it differs by having a short row of a-setae on fore femur.

***Delia fabricii* (Holmgren, 1872) Figs 2B, C**

Michelsen 2006: 108.

BIOLOGY. Available knowledge is mainly from Norway and Russia, reviewed by Griffiths (1992: 1312 ff.). Overwinters as eggs deposited in late summer in small batches in the soil near the host plant represented by various grasses, *e.g.*, species of *Poa* and *Agropyron*. The larva hatches in spring and feeds on shoots of its host grass, causing "dead-heart" injuries. Mature larvae pupate in the soil in early summer. The adults appear a couple of weeks later.

DISTRIBUTION. S, SW, NE, SE. A common grassland species north to Uummannaq Fjord and Zackenberg.

Circumpolar, arctic–boreal and alpine.

***Delia groenlandica* Griffiths, 1993**

Michelsen 2006: 108.

DISTRIBUTION. S (Narsarsuaq), SW (Qeqertarsuaq/Godhavn). Apparently a very local species.

Nearctic, low arctic–subarctic and alpine (Western Cordilleras).

***Delia platura* (Meigen, 1826)**

Michelsen 2006: 108. Earlier recorded from Greenland by the synonym '*cilicrura* Rondani'.

BIOLOGY. Larvae ubiquistic, mainly phytophagous but even reported to feed on insect eggs and other animal material. Most larvae apparently live as "seed maggots" attacking the young shoots of a variety of dicots, including cruciferous and leguminous crops.

DISTRIBUTION. S, SW, NE, SE (north to Uummannaq Fjord and Ella Ø).

Cosmopolitan, low arctic–subtropical and alpine. Known from all biogeographic regions except Antarctica.

***Delia setigera* (Stein, 1920)**

Michelsen 2006: 109.

DISTRIBUTION. S (Narsarsuaq), SW (Itileq, Kuussuaq, Uummannak Fjord: Alfred Wegener Halvø). Only 6 specimens collected in Greenland.

Circumpolar, low arctic–temperate and alpine.

***Egle* Robineau-Desvoidy, 1830**

A genus with nearly 30 described species largely confined to boreal, subarctic and montane parts of the Holarctic Region. The group that contains some of the smallest known anthomyiid species, is defined by details of the male and female terminalia and – more obviously – by its biology: All species are univoltine and among the first anthomyiid flies to appear in spring after overwintering in the pupal stage, but they are also the first to disappear by end of spring. The adults feed on both pollen and nectar from their larval host plants. The larvae are known to feed on seed capsules of Salicaceae (*Salix, Populus*) and Tamaricaceae (*Myricaria*).

A single species of *Egle* is known from Greenland, but only from female specimens. With a wing length of less than 3.5mm it is significantly smaller than other anthomyiid species found in Greenland. The very short antenna and almost linear lower calypter also help to identify the species.

Egle groenlandica Michelsen, 2010

Michelsen 2006: 109, as *Egle* sp. The species has been misidentified as '*parva* Robineau-Desvoidy', '*atomaria* Zetterstedt' and '*minuta* Meigen' in the literature on Greenland Diptera.

BIOLOGY. Most species of *Egle*, and surely also the one found in Greenland, stand in a close, mutually beneficial relation to species of *Salix*. The adult flies, feeding on pollen from male catkins and nectar from female catkins, tend to carry massive amounts of pollen grains ventrally on their body and legs and thus must play a significant role as pollinators of their dioecious host plants.

DISTRIBUTION. S, SW, NW, NE, SE (north to Upernavik Laksefjord and Zackenberg).

Occurrence outside Greenland not documented so far.

NOTES. Male unknown; female easily identified by the characters of the key.

Griffiths (2003) suggested an affiliation of the only species found in Greenland with his *Egle steini* superspecies, but he was unable to associate it with any of the 7 North American species recognized in the male sex. Michelsen (2010) further showed that it differed from the females of all *Egle* spp. found in Europe and accordingly described it as a new species without knowing the male sex. Based on the fact that *Egle groenlandica* is now known by more than 100 specimens from several localities in Greenland, but exclusively by females, Michelsen (*op.cit.*) suggested that it could be parthenogenetic, at least in Greenland. If so, this is the first known case of parthenogenesis among muscoid Diptera!

Eutrichota Kowarz, 1893

The genus, with about 70 known species, is strikingly more diverse in North America than in Eurasia; a few Oriental and Neotropical species are also known. Very little biological information is available. The adults hardly visit flowers. The larvae of some species are known to live burrows and nests of rodents, where they expectedly feed on debris. Only one, high arctic species is known from Greenland.

Eutrichota tunicata (Zetterstedt, 1846) Fig. 2D

Michelsen 2006: 109. Also recorded from Greenland as '*analis* Schnabl', a synonym.

DISTRIBUTION. N (Peary Land: Jørgen Brønlund Fjord), NE (Zackenberg, Lindhard Ø, Sabine Ø, Clavering Sund, Myggbukta, Mestersvig), SE (Hekla Havn, Manby Halvø).

Circumpolar, mainly high arctic.

Fucellia Robineau-Desvoidy, 1841

A genus of seaweed flies with nearly 30 known species and mainly Holarctic in distribution, though a few introduced or even endemic species are known from other regions. The species are easily recognized by the feminized male head, reduced lower calypter and av-series of scattered spinules on vein C.

Most species of *Fucellia*, including 3 species found in Greenland (*F. ariciiformis*, *F. fucorum*, *F. tergina*), are strongly halophilous and largely confined to sea shores with beds of decaying brown algae in which the saprophagous larvae develop. A smaller group of *Fucellia*, characterized by patterns of dark clouding on the wing wings and enlarged, whitish thoracic spiracles, seems to prefer silty or sandy ground, including dune areas, at edges of fresh or brackish waters. This group is represented in Greenland by 2 species (*F. pictipennis*, *F. vibei*).

Fucellia ariciiformis (Holmgren, 1872) Fig. 2F

Michelsen 2006: 109.

DISTRIBUTION. S, SW, NW, NE. The most abundant and widespread species of *Fucellia* in Greenland, but not found north of Etah (NW) and Hochstetter Forland (NE).

Holarctic, arctic, coastal (Russian Far East, Alaska, Canada).

Fucellia fucorum (Fallén, 1819)

Michelsen 2006: 109.

DISTRIBUTION. S, SW (north to Qeqertarsuaq/Godhavn).

Circumpolar, low arctic–temperate.

Fucellia pictipennis Becker, 1907 Fig. 2E

Michelsen 2006: 109.

The darkly clouded wings and large, whitish thoracic spiracles readily separate the species from all other species except *F. vibei*.

DISTRIBUTION. NW (Etah), N (Peary Land: Jørgen Brønlund Fjord, Danmarks Fjord), NE (Zackenberg; Clavering Ø, Ella Ø, Mestersvig), SE (Hekla Havn).

Holarctic, high arctic, littoral (Russian Far East, Alaska, Canada).

Fucellia tergina (Zetterstedt, 1845)

Michelsen 2006: 109. The synonymous name '*intermedia* Lundbeck' has been widely used for the present species in the literature on Greenland Diptera.

DISTRIBUTION. S, SW (north to Uummannaq Fjord: Ukkusissat Fjord), SE (Saqqisikuik/Skjoldungen).

Low arctic–subtropical, wide-ranging sea-shore species known from most biogeographical regions.

Fucellia vibei Collin, 1951
Michelsen 2006: 110; first record from Greenland.

DISTRIBUTION. N. Only 1 specimen recorded from Greenland, Peary Land (Jørgen Brønlund Fjord, south coast, on clayey substrate in tidal zone at stream outlet; Danish Pearyland Expedition 1947–50, leg. P. Johnsen).

Nearctic, high arctic (previously known only from Canada: Northwest Territories (Huckett 1965: 27). The type area, Ellesmere Island, lies only 25km from NW Greenland.

Lasiomma Stein, 1916
A poorly defined genus with more than 35 known species, many occurring in the northern Holarctic area; 1 species in Greenland.

Lasiomma picipes (Meigen, 1826)
Michelsen 2006: 110. Recorded from Greenland under the synonymous names '*octoguttata* Zetterstedt' and '*moesta* Holmgren'.

A small, inconspicuous species reminding of a *Delia*, but differs in the male sex by the shiny black postabdomen (segments VII–IX).

BIOLOGY. The larvae are known to feed on debris, preferably in nests of birds.

DISTRIBUTION. S, SW, NW, NE, SE. Common in the southern parts, north to Tasiusaq (north of Upernavik) and Jameson Land.

Circumpolar, low arctic–temperate and alpine, widespread.

Myopina Robineau-Desvoidy, 1830
A small genus consisting of stocky, terrestrial species with notably small eyes and correspondingly broad frons and genae in both sexes. Only 4 known species confined to the Holarctic Region; 3 species have been found solely in arctic areas, 1 species in Greenland.

Myopina crassipalpis Ringdahl, 1937 Fig. 2G
Michelsen 2006: 110; first record from Greenland.

BIOLOGY: Like other species of *Myopina* probably a littoral species best found on sand and silt at lakes and streams. Larvae unknown, but saprophagous in related genera.

DISTRIBUTION. NE (Mestersvig, Godfred Hansen Ø, Zackenberg, Hochstetter Forland: Langelv, Jameson Land: Langelands Elv, Colorado Dal).

Holarctic, mainly high arctic (Sweden: Jämtland; Alaska; Canada: Yukon, Northwest Territories).

Paradelia Ringdahl, 1933
Less than 20 described species largely confined to northern and elevated parts of the Holarctic Region. Larval habits are barely known, but they are believed to live in organic debris, *e.g.* in nests of birds or mammals and in rotten fungi; 1 species in Greenland.

Paradelia arctica (Ringdahl, 1933)
Michelsen 2006: 110. Earlier treated as a variety or subspecies of *Paradelia intersecta* (Meigen), but Griffiths (1987) presented evidence for treating it as a valid species.

DISTRIBUTION. G, but not common.

Probably circumpolar, arctic (Norway, Sweden, Alaska, Canada).

Paregle Schnabl & Dziedzicki, 1911
A genus with 10 described species largely confined to northern and elevated parts of the Holarctic Region; 1 species in Greenland.

Paregle audacula (Harris, 1780) Fig. 2H
Michelsen 2006: 110. The name '*radicum* Linnaeus' was used earlier for the present species, but is today correctly assigned to the notorious cabbage-root maggot (*Delia radicum*).

BIOLOGY. The adults are strongly anthophilous, perhaps showing a preference for flowering Compositae. The larvae develop mainly in carnivore and human faeces.

DISTRIBUTION. S, SW, NW, SE. Common locally in the southern parts; north to Upernavik and Scoresby Sund.

Circumpolar, arctic–subtropical, very widespread, somewhat anthropochorous.

Pegomya Robineau-Desvoidy, 1830
A large genus with several hundred species found mainly in the Holarctic Region, but even the Old and New World tropics have smaller numbers of species. An important genus in Greenland represented by 5 species, of which 2 are larval leaf-miners of dicots and 3 feed on bolete mushrooms. The Greenland species of *Pegomya* may be distinguished from other anthomyiids by the combination of a body and leg colour being to varying extent yellow to reddish brown, vein C with v-setulae, and hind tibia with only 2 pd-setae.

Pegomya circumpolaris Ackland & Griffiths _in_ Griffiths, 1983
Michelsen 2006: 111. The present species has in Greenland been confused with other mushroom-feeding *Pegomya* species. See under *P. notabilis*.

BIOLOGY. Unknown, but the larvae surely feed on bolete mushrooms as is known for closely related species.

DISTRIBUTION. S, SW, SE (north to Uummannak Fjord and Saqqisikuik/Skjoldungen).

Circumpolar, low arctic–boreal.

***Pegomya flavifrons* (Walker, 1849)**
Michelsen 2006: 111. Also recorded from Greenland by the synonym '*albimargo* Pandellé'.

BIOLOGY. A larval leaf-miner on a wide range of Caryophyllacae. The host plant(s) in Greenland is unknown.

DISTRIBUTION. S, SW, SE (north to Qeqertarsuaq/Disko).

Circumpolar, low arctic–subtropical, remarkably widespread.

***Pegomya icterica* (Holmgren, 1872)**
Michelsen 2006: 111. The species, originally described from Greenland, has been incorrectly synonymized with *P. haemorrhoum* (Zetterstedt).

BIOLOGY. Outside Greenland known as a larval leaf-miner on *Oxyria digyna* (Polygonaceae), but it very likely even attacks species of *Rumex*.

DISTRIBUTION. S, SW, NW, NE, SE. Very widespread, reaching north to Tasiusaq (north of Upernavik) and Sabine Island.

Holarctic, arctic and alpine. Also known from Iceland, Norway, Sweden, Alaska, Canada: Western Cordilleras (south to Oregon).

***Pegomya notabilis* (Zetterstedt, 1846) Fig. 2I**
Michelsen 2006: 111.

BIOLOGY. Larvae fungivorous, feeding on species of *Boletus* and *Leccinum* (Boletaceae). The female inserts her eggs in clusters well beneath the upper cap surface of the host bolete. Fungal host(s) in Greenland unknown.

DISTRIBUTION. S, SW, SE (north to Kangerlussuaq/Søndre Strømfjord and Saqqisikuik/Skjoldungen).

Circumpolar, low arctic–temperate.

NOTE. Students of Greenland Diptera had earlier difficulties with the taxonomy of the 3 species of *Pegomya* associated with bolete mushrooms: *P. notabilis, P. zonata* and *P. circumpolaris*. Any of these species may be hiding under records of '*rufipes* Fallén, 1825' (preoccupied) and the misapplied names '*conformis* Fallén', '*hyoscyami* Panzer', '*tenera* Zetterstedt' and '*pilosa* Stein'.

***Pegomya zonata* (Zetterstedt, 1838)**
Michelsen 2006: 111. The present species has in Greenland been confused with other mushroom-feeding *Pegomya* species. See under *P. notabilis*.

BIOLOGY. Similar to that of *P. notabilis*, except that the female deposits her eggs one by one in the spore layer of bolete mushrooms. Fungal host(s) in Greenland unknown.

DISTRIBUTION. S, SW (north to Uummannaq Fjord).

Circumpolar, low arctic–temperate.

***Pegoplata* Schnabl & Dziedzicki, 1911**
Less than 30 described species, mainly Holarctic; 1 species is known from Greenland. Earlier assigned to *Nupedia* Karl, now regarded as a synonym of *Pegoplata*. The larvae are saprophagous, some of them preferably living in bovine droppings.

***Pegoplata tundrica* (Schnabl, 1915) Fig. 1A**
Michelsen 2006: 111. Also recorded from Greenland under the synonym '*pseudodissecta* Ringdahl'.

DISTRIBUTION. S, SW (north to Kangerlussuaq/Søndre Strømfjord).

Circumpolar, low arctic–subarctic and alpine (Western Cordilleras).

***Zaphne* Robineau-Desvoidy, 1830**
More than 50 species found almost exclusively in northern and elevated parts of the Holarctic region; with totally 6 species this is an important genus in Greenland. Species of *Zaphne* were earlier assigned to *Acroptena* Pokorny, a synonym, or accommodated in the valid genus *Hydrophoria* Robineau-Desvoidy. Robust, mostly blackish and abundantly setose anthomyiids with a shiny black prementum. Most species are found among low vegetation in marshy places on muddy or silty ground at water edges. The adults possess enlarged prestomal teeth and occationally feed on chironomids and other weak insects. The immatures are poorly known but expectedly the larvae live as scavengers or predators in moist, humus rich soils.

***Zaphne brunneifrons* (Zetterstedt, 1838)**
Michelsen 2006: 111.

DISTRIBUTION. S, SW, SE (north to Qeqertarsuaq/Disko and Scoresby Sund). Widespread in southern Greenland.

Circumpolar, low arctic–boreal.

NOTE. Greenland records of '*brunneifrons* Zetterstedt' before Griffiths (1998) include either or both of the very similar species *Z. brunneifrons* and *Z. occidentalis* (*q.v.*).

***Zaphne divisa* (Meigen, 1826)**
Michelsen 2006: 112.

DISTRIBUTION. S, SW, SE (north to Uummannaq and Zackenberg).

Circumpolar, arctic–temperate.

***Zaphne frontata* (Zetterstedt, 1838) Fig. 3A**
Michelsen 2006: 112.

DISTRIBUTION. S, SW, NW, NE, SE. Widespread and common, north to Kap Seddon (NW) and Ella Ø and Geographical Society Ø (NE).

Circumpolar, arctic and alpine.

NOTE. Records of '*frontata* Zetterstedt' in the early literature on Greenland Diptera also include the very similar *Z. tundrica* (*q.v.*).

Zaphne nearctica **(Huckett, 1965)**
Michelsen 2006: 112.

DISTRIBUTION. SW (Kangerlussuaq/Søndre Strømfjord, Ilulissat/Jakobshavn, Saqqaq).

Holarctic, low arctic–subarctic. (Siberia, Alaska, northern Canada).

Zaphne occidentalis **(Malloch, 1920)**
Michelsen 2006: 112.

DISTRIBUTION. SW, NE, SE (found between Kangerlussuaq/Søndre Strømfjord and Qeqertarsuaq/Disko and in East Greenland north to Zackenberg).

Nearctic, arctic–temperate; disjunct in North America (widespread in western U.S.A. and southwestern Canada, also across low arctic Alaska and Canada to Greenland).

NOTE. The species has been confused with *Z. brunneifrons* (*q.v.*) both taxonomically and nomenclaturally.

Zaphne tundrica **(Schnabl *in* Schnabl & Dziedzicki, 1911)**
Michelsen 2006: 112.

DISTRIBUTION. SW, NE, SE. Widespread north to Ukkusissat Fjord and Zackenberg. Apparently absent in S.

Circumpolar, arctic.

NOTE. Confused with *Z. frontata* (Zetterstedt) in the early literature. Subsequently recorded from Greenland as '*verticina* Zetterstedt, 1838,' but that name was shown by Griffiths (1998) to refer to a closely related species occurring in Fennoscandia and Canada only.

FANNIIDAE

A small family with *c.* 370 described species, but worldwide in distribution. Most species are conventionally referred to the heterogenous genus *Fannia*, of which 3 species have been found in Greenland.

The adults resemble both Anthomyiidae and Muscidae but are distinguished by the distally straight subcostal vein, very short anal vein and strictly dorsal submedian seta on hind tibia; male abdomen long-ovoid as in many Muscidae but strongly depressed as in most Anthomyiidae. Fanniid larvae are very characteristic, dorsoventrally flattened with a leathery integument drawn out into 3 pairs of plumate processes on each segment, *i.e.* strongly deviating from the streamlined maggot shape typical of the other muscoid families.

Adult fanniids are strongly sexually dimorphic. Males are pronouncedly aerial, often engaged in hovering flights, singly or in small swarms. Both sexes frequent flowers in order to obtain nectar. The larvae are saprophagous, feeding on many kinds of decomposing organic debris.

Key to species (males, females)

1. Males: eyes narrowly separated by frons 2
– Females: eyes widely separated by frons 4
2. Abdomen to some extent translucent yellow, at least latero-basally. Mid coxa without spine-like setae .... *Fannia canicularis*
– Ground colour of abdomen wholly black. Mid coxa anteriorly with 1–3 spine-like, apically recurved setae 3
3. Mid coxa anteriorly with 3 spine-like setae. Mid tibia with a shiny ventral swelling. Hind tibia with ad- and av-setae inconspicuous .... *Fannia scalaris*
– Mid coxa antero-apically with 1 spine-like seta. Mid tibia simple. Hind tibia with comb-like rows of long, fine ad- and av-setae .... *Fannia bifimbriata*
4. Row of small setulae on upper parafacial extends below level of aristal base............................ .... *Fannia canicularis*
– Row of setulae on upper parafacial ends above level of aristal base 5
5. Larger, wing length more than 5.0 mm. Fore tibia with 1 ad-seta at distal quarter. Hind tibia with 3–4 av-setae............................... *Fannia scalaris*
– Smaller, wing length less than 5.0 mm. Fore tibia without ad-seta. Hind tibia with 5–8 av-setae........ .... *Fannia bifimbriata*

Fannia **Robineau-Desvoidy, 1830**

Among the 3 species of *Fannia* known from Greenland only *F. bifimbriata* Collin may be indigenous to Greenland – or at least able to maintain viable populations outside human settlements. The occurrence of the other two species, the 'lesser housefly' (*F. canicularis*) and the 'latrine fly' (*F. scalaris*) in Greenland is clearly a result of chance introduction. Both species are very widespread, but increasingly synanthropic to the north and evidently unable to live independent of human settlements under arctic climates.

Fannia bifimbriata **Collin *in* Carpenter, 1938 Fig. 3B**
Michelsen 2006: 112.

DISTRIBUTION. S (Nunap Isua /Kap Farvel area; Tasiusaq), SW (Kangerlussuatsiaq /Evighedsfjord area, Itileq, Amerloq). North to the Sisimiut /Holsteinsborg

FIGURE 3 *Habitus of Anthomyiidae (A), Fanniidae (B) and Muscidae (C–I):* Zaphne frontata, *male (A);* Fannia bifimbriata, *male (B);* Drymeia groenlandica, *male (C);* Drymeia segnis, *male (D);* Graphomya ungava, *female, dorsal (E) and lateral (F);* Hydrotaea anxia, *male (G);* Lispe uliginosa, *female (H);* Lophosceles frenatus, *male (I).*

area. The only fanniid species that is possibly indigenous in Greenland.

Nearctic, subarctic–boreal and montane, widespread.

NOTE. Records of '*armata* Meigen' and '*manicata* Meigen' from Greenland are probably all misidentifications of the present species, that was originally described from Greenland.

Fannia canicularis **(Linnaeus, 1761) (Lesser Housefly)**
Michelsen 2006: 113.

BIOLOGY. The larvae live in a wide variety of decaying organic substrates. It is frequently found breeding in animal manure in confined livestock facilities.

DISTRIBUTION. S, SW, NE. Certainly introduced by man and an obligate synanthropic species in Greenland.

Cosmopolitan, anthropochorous. Natural range reaching from the cool temperate to subtropical zones of the Northern Hemisphere.

Fannia scalaris **(Fabricius, 1794) (Latrine Fly)**
Michelsen 2006: 113; first record from Greenland.

BIOLOGY. Known as the 'latrine fly' because of its occurrence in human privies, but the larvae live in a wide variety of decaying organic substrates, including domestic garbage. Reared from puparia found in compost (Narsarsuaq). Introduced by man.

DISTRIBUTION. S (Upernaviarsuk, Qaqortoq / Julianehåb, Narsarsuaq).

Holarctic, widespread boreal–subtropical, somewhat anthropochorous.

MUSCIDAE

A large family with *c.* 5.240 described species distributed worldwide. In terms of both species diversity and general abundancy it is the most important family of Diptera Brachycera in Greenland. A total of 37 species in 10 genera are known from Greenland; among them are 27 Holarctic and 10 Nearctic species. The high arctic parts of Greenland are inhabitated by no less than 22 species belonging to the genera *Spilogona* (19), *Drymeia* (2) and *Phaonia* (1).

The adults are most easily confused with anthomyiids, but the scutellum is without fine setulae beneath tip, vein A_1 does not extend to the wing margin, and the male abdomen is more or less ovoid and subcylindrical.

Most species are strongly sexually dimorphic, yet the habit of aerial swarming has apparently only been retained in *Hydrotaea*, *Drymeia* and some allied genera. The adults of all Greenland muscids, except *Lispe uliginosa*, probably visit flowers more or less regularly in order to obtain nectar and/or pollen. Accordingly, muscid flies constitute no doubt the most significant family of pollinating insects in the high arctic parts of Greenland. Species of *Lispe*, *Limnophora* and *Spilogona* all possess enlarged prestomal teeth and may to varying degree hunt and feed on soft-bodied arthropods such as adult or immature chironomids and simuliids. Most of the Greenland species of Muscidae expectedly have larvae that hatch from the egg in the II or III (final) instar. Such larvae are at least partly predatory, hunting smaller dipterous larvae, oligochetes, etc. in semi-dry to aquatic situations. Some larvae live freely in humus-rich soil, others prefer more specialized habitats such as cushions of wet moss or green algae, carrion, dung, rotten fungi, burrows and nests of mammals and birds.

Key to species (males, females)

1. Anepimeron with a cluster of setulae 2
– Anepimeron bare 4
2. Apical section of vein M bent sharply forward. Prosternum setulose .... *Musca domestica*
– Vein M straight. Prosternum bare 3
3. Palpus yellow, apically strongly expanded. Parafacial setulose .... *Lispe uliginosa*
– Palpus black, slender. Parafacial bare .... *Drymeia segnis*
4. Node of Rs, and often base of vein R_{4+5}, with one or more d- and v-setulae 5
– Node of Rs and vein R_{4+5} bare 8
5. Prealar seta present. Posterior notopleural seta surrounded by some setulae 6
– Prealar seta absent. Notopleuron bare apart from 2 notopleural setae 7
6. Arista plumose. Apical section of vein M bent smoothly forward. Calypteres whitish............... .... *Graphomya ungava*
– Arista pubescent. Vein M straight. Calypteres yellowish .... *Mydaea palpalis*
7. Vibrissal angle barely projected beyond fronto-parafacial angle. Sternite I with a few setulae. Frons broad in both sexes............................ .... *Limnophora rotundata*
– Vibrissal angle strongly projected beyond fronto-parafacial angle. Sternite I bare. Frons narrow in male, broad in female..... *Limnophora groenlandica*
8. Mid femur with preapical a-seta.................... 9
– Mid femur without a-seta near apex 15
9. Notopleuron with some setulae in addition to 2 notopleural setae.................................. 10
– Notopleuron bare except for 2 notopleural setae 13
10. Fore tibia without p-seta(e) 11
– Fore tibia with one or more p-seta(e) 12
11. Fore tibia without pv-seta(e). Male: fore femur subapically with 2 ventral cuticular teeth; hind tibia simple. Female: mid tibia without row of pd-setae .. .. *Hydrotaea anxia*
– Fore tibia with 3–4 pv-setae on distal half. Male: fore femur simple; hind tibia ventro-apically forming a compressed, apically rounded process. Female: mid tibia with a complete row of pd-setae.......... .... *Drymeia groenlandica*
12. Fore tibia distally with a dense clothing of longish, semi-erect v-setulae. Mid tibia with 2 or more v- and pv-setae .... *Phaonia pallidisquama*
– Fore tibia without longish v-setulae. Mid tibia without v- and pv-setae............... *Phaonia bidentata*

13. Hind tibia with pd-setae inconspicuous or absent. Prealar seta absent. Male: abdomen somewhat compressed, segments IV–V appearing deeper than wide. Female: haustellum notably thick, polished black on prementum.. *Spilogona arctica*
– Hind tibia with a prominent submedian pd-seta. Prealar seta(e) present............................14
14. Single prealar seta more than half as long as posterior notopleural seta. Mid tibia with 1–2 pv-setae. Male: hind femur strongly bowed, preapically with a brush of setae born on a v-swelling .... *Lophosceles frenatus*
– Duplicated prealar setae notably short and spinose, less than half as long as posterior notopleural seta. Mid tibia without pv-seta. Male: hind femur weakly bowed, preapically without v-swelling, but with loose av- and pv-groups of setae.. *Lophosceles minimus*
15. Vibrissal angle projected far beyond fronto-facial angle. Proboscis long, haustellum more or less slender, cylindrical 16
– Vibrissal angle behind, in line with or slightly projected (*S. malaisei*) beyond fronto-facial angle. Proboscis short, haustellum more or less conical.... 23
16. Mid tibia with 1–5 short setae on av- to pv-surfaces. Hind femur without pv-setae.......................17
– Mid tibia without setae ventrally. Hind femur in basal part with some longish pv-setae (except short and unremarkable in *S. sanctipauli*!).........18
17. Thorax darkly subshining through thin dark brown dusting. Abdomen dorsally brownish black without indication of usual paired markings. Calypteres and halter yellow. Female: hypoproct with fine, recurrent spines............................ *Spilogona tundrae*
– Thorax matt, covered in dense brownish black and ash grey dusting. Abdomen dorsally with large, paired, partly confluent dark markings. Calypteres white, light brownish at margins; knob of yellow halter usually darkened. Female: hypoproct setulose ... . *Spilogona dorsata*
18. Larger, wing length 5–7mm. Subdorsal break in vertical row of anepisternal setae occupied by setula(e).. 19
– Smaller, wing length rarely exceeding 5mm. Subdorsal break in vertical row of anepisternal setae bare.. 20
19. Scutellum with some setulae on lateral margins below level of lateral setae. Hind femur with longish pv-setae. Tarsomere 1 of hind leg without v-seta near base. Female: hypoproct setulose.. *Spilogona almqvistii*
– Scutellum bare on lateral margins. Hind femur with pv-setae shorter than depth of femur.. Tarsomere 1 of hind leg with short v-seta near base. Female: hypoproct with fine, recurrent spines.. *Spilogona tendipes*
20. Dusting on dorsum of thorax extensively matt brownish black, only slightly greyish on postpronotal lobes; knob of halter dark in both sexes 21
– Dusting on dorsum of thorax light greyish on at least anterior mesonotum and postpronotal lobes; knob of halter dark in male, yellow in female........... 22
21. Lower part of face gradually produced forward. Male parafrontalia matt brownish black, separated by distinct frontal vitta. Calypteres small, dirty yellowish brown. Female: hypoproct with fine, recurrent spines............................ *Spilogona arcticola*
– Lower part of face rather abruptly produced forward. Male parafrontalia silvery grey, more or less contiguous. Calypteres of normal size, whitish. Female: hypoproct setulose....... *Spilogona extensa*
22. Prementum wholly shiny black. Pleural parts of thorax covered in greyish dusting. Basal pv-setae on hind femur shorter than femoral depth. Male terminalia enlarged.................. *Spilogona sanctipauli*
– Prementum with some grey dusting mid-ventrally, rarely more extensively. Upper pleural parts of thorax covered in brownish black dusting. Basal pv-setae on hind femur fully as long as femoral depth. Male terminalia normal....... *Spilogona megastoma*
23. Dorsocentrals behind suture 3 evenly spaced setae 24
– Dorsocentrals behind suture 4 setae, if exceptionally 3 then setae unevenly spaced..................... 26
24. Hind femur without pv-setae in basal half. Male: frons throughout broader than length of antennal postpedicel. Female: hypoproct with recurrent spines............................... *Spilogona alpica*
– Hind femur with some pv-setae in basal half. Male frons on upper part narrower than length of postpedicel. Female: hypoproct setulose 25
25. Spinules in basal third of vein C notably long and strong. Mid tibia with 1(–2) ad-seta .... *Spilogona tornensis*
– Spinules of vein C normal sized. Mid tibia without ad-seta............................ *Spilogona baltica*
26. Mid tibia with 1–3 av- to pv-setae (sometimes short and inconspicuous!) 27
– Mid tibia without setae on ventral surfaces....... 30
27. Hind femur with pv-setae in basal third.......... 28

– Hind femur without pv-setae 29
28. Larger, wing length reaching 6.0 mm; robust species with coarse setae on body and legs. Notopleuron with some setulae in addition to 2 notopleural setae. Male abdomen ovoid, somewhat depressed, dorsally grey with large, trapezoid dark markings on tergites III–IV. Female: parafrontalia narrow, less than 1/3 as wide as frontal vitta .... *Spilogona trigonifera*
– Smaller, wing length reaching 5.0 mm; setae of body and legs fine. Notopleuron bare apart from usual 2 notopleural setae. Male abdomen conical, uniformly brownish black. Female: parafrontalia broad, at least 1/2 as wide as frontal vitta .... *Spilogona novaesibiriae*
29. Halter ochre yellow. Male: abdomen with paired dark markings almost confluent mid-dorsally, but well demarcated by greyish dusting laterally......... .... *Spilogona zaitzevi*
– Halter brownish black. Male: abdomen uniformly brownish black.............. *Spilogona melanosoma*
30. Prementum shiny, practically without any dusting 31
– Prementum dull, covered extensively in grey dusting 33
31. Small species, wing length in general less than 4.5 mm. Vibrissal angle reaching slightly beyond fronto-parafacial angle. Male: mid femur without ventral spinose seta(e). Female: abdomen darkly subshining through thin brown dusting; hypoproct setulose .... *Spilogona malaisei*
– Medium sized species, wing length more than 4.5 mm. Vibrissal angle in line with fronto-parafacial angle. Male: mid femur with one or several ventral spinose seta(e). Female: abdomen dull greyish with paired dark markings on tergites III–IV; hypoproct with recurrent spinules 32
32. Subdorsal break in vertical row of anepisternal setae occupied by setula. Male: eyes densely haired; mid femur in basal half with a pv-row of spinose setae. Female: eyes shortly but distinctly haired; mid femur in basal half with a row of simple pv-setae.......... ....*Spilogona puberula*
– Subdorsal break in vertical row of anepisternal setae bare. Male: eyes bare; mid femur near base with single spinose v-seta. Female: eyes bare; mid femur with v-seta near base sometimes absent.................. .... *Spilogona monacantha*
33. Subdorsal break in vertical row of anepisternal setae occupied by setula. Very small, wing length 3.0–4.5 mm. Hind femur without pv-setae in basal half. Male: broadly subovate abdomen dorsally silvery grey tomentose with contrasting paired black markings on tergites III–IV......*Spilogona deflorata*
– Subdorsal break in vertical row of anepisternal setae bare 34
34. Males 35
– Females... 39
35. Halter blackish, contrasting with purely white calypteres. Abdomen bluish grey tomentose, trapezoid dark marks on tergites III and IV narrowly separated dorsally, laterally extended as hind marginal bands . ...*Spilogona micans*
– Halter yellow to light brownish.................... 36
36. Abdomen broadly subovate, somewhat depressed, trapezoid dark markings on tergites III and IV widely separated dorsally. Sternite I setulose ....*Spilogona opaca*
– Abdomen slender, subconical. Sternite I bare..... 37
37. Medium-sized, wing length exceeding 5mm. Vein Sc basally with 1–3 v-setulae... *Spilogona trianguligera*
– Small species, wing length less than 4.5mm. Vein Sc bare.. 37
38. Hind femur without distinct pv-setae in basal part. Vein C spinules in basal part of wing short and inconspicuous................... *Spilogona denudata*
– Hind femur with some longish and fine pv-setae in basal half. Vein C spinules quite long and strong in basal part of wing ....*Spilogona tornensis*
39. Hypoproct setulose 40
– Hypoproct with recurrent spines...................41
40. Larger, wing length more than 6 mm. Sternite I with a few setulae .... *Spilogona opaca*
– Smaller, wing length barely exceeding 4 mm. Sternite I bare ....*Spilogona tornensis*
41. Larger, wing length more than 6 mm. Vein Sc basally normally with 1–3 v-setulae. Mid tibia with 2–3 ad-setae .... *Spilogona trianguligera*
– Smaller, wing length less than 6 mm. Vein Sc bare. Mid tibia with 0–1 short ad-seta 42
42. Hind femur with some longish and fine pv-setae in basal half ....*Spilogona micans*
– Hind femur without pv-setae in basal half........... .... *Spilogona denudata*

Drymeia Meigen, 1826

A Holarctic genus with nearly 60 described species showing greatest species diversity in the northern and elevated parts of Asia. The 2 species occurring in Greenland have earlier been assigned to *Pogonomyia* Rondani, *Eupogonomyia* Malloch and *Pogonomyioides* Malloch, genera now regarded as synonyms of *Drymeia*.

Adult *Drymeia* frequent flowers regularly in order to obtain nectar. The predatory larvae supposedly live freely in humus rich soils.

***Drymeia groenlandica* (Lundbeck, 1901) Fig. 3C**
Michelsen 2006: 113.

DISTRIBUTION. nSW, NW, NE, SE. South to Kangerlussuaq/Søndre Strømfjord and Scoresby Sund.

Nearctic, high arctic–subarctic (Canada).

***Drymeia segnis* (Holmgren, 1883) Fig. 3D**
Michelsen 2006: 113.

DISTRIBUTION. NW, N, NE, SE (arctic species ranging from Pituffik/Thule to Scoresby Sund).

Circumpolar, high arctic (Novaya Zemlya, Alaska, Canada).

***Graphomya* Robineau-Desvoidy, 1830**
The taxonomy of the Eurasian and North American species of *Graphomya* is notoriously difficult and unsettled. Earlier, a single variable species, *G. maculata* (Scopoli) was recognized in North America, but Arntfield (1975) split the Nearctic component of *Graphomya* into 9 species, none of them belonging to true *G. maculata* described from Europe. One Nearctic species occurs in Greenland. The emended spelling '*Graphomyia*' was used until recently.

Adult *Graphomya* are strongly anthophilous, often seen nectar-feeding on umbelliferous flowera. Thhe predatory larvae are aquatic or semiaquatic and reported feeding on rat-tailed larvae of eristaline Syrphidae.

***Graphomya ungava* Arndtfield, 1975 Figs 3E, F**
Michelsen 2006: 113. Also recorded from Greenland as '*maculata* Scopoli' and '*picta* Zetterstedt', both misidentifications of European species.

DISTRIBUTION. S, SW, SE (north to Qeqertarsuaq/Disko and Saqqisikuik/Skjoldungen).

Nearctic, low arctic–subarctic (Alaska, Canada).

***Hydrotaea* Robineau-Desvoidy, 1830**
A genus with more than 80 recognized species, most of them found in the temperate to subtropical parts of the Northern Hemisphere.

***Hydrotaea anxia* (Zetterstedt, 1838) Fig. 3G**
Michelsen 2006: 114. Also recorded from Greenland under the synonym '*bispinosa* Zetterstedt'.

BIOLOGY. Belongs to the "*dentipes* group" of *Hydrotaea* in which the larvae have retained 3 instars and live as facultative predators on other fly larvae in putrifying substances such as garbage, dung and carrion.

DISTRIBUTION. S, SW (north to Qeqertarsuaq/Godhavn).

Circumpolar, low arctic–subarctic.

***Limnophora* Robineau-Desvoidy, 1830**
A large and poorly known, almost cosmopolitan genus with some 350 recognized species and most diverse in the Old World tropics. Only 2 species occur in Greenland, of which one has earlier been assigned to the synonymous genus or subgenus *Pseudolimnophora* Strobl. Species of *Limnophora* resemble species of *Spilogona*, but are readily separated by the setulose prosternum and presence of d- and v-setulae on the node of wing vein Rs.

Biologically, adult and larval *Limnophora* agree closely with *Spilogona*, except that most species show a stronger preference for running waters for larval development. Recent observations (Werner & Pont 2006) show that some species deposit their eggs singly among aquatic algae or mosses on stones or other firm items in the splash zone of fast-running streams, situations rich in simuliid larvae, the preferred prey of many *Limnophora* larvae.

***Limnophora groenlandica* Malloch, 1920**
Michelsen 2006: 114. Also recorded from Greenland under the synonym '*sinuata* Collin'.

DISTRIBUTION. SW, NE, SE (ranging from Qeqertarsuatsiaat/Fiskenæsset north to Uummannaq Fjord and from Saqqisikuik/Skjoldungen north to Dove Bugt).

Holarctic species with an atlantic, arctic–boreal distribution in Europe: Iceland, Faroe Islands, northern Fennoscandia.

***Limnophora rotundata* (Collin, 1930)**
Michelsen 2006: 114. Also recorded from Greenland as '*triangula* Fallén' and '*nigripes* Robineau-Desvoidy', both misidentifications of European species.

DISTRIBUTION. S (Nunap Isua/Kap Farvel area; Tasiusaq), SW (Allumersat/Bjørnesund, Qeqertarsuatsiaat/Fiskenæsfjorden, Kangerlussuatsiaq/Evighedsfjord), SE (Saqqisikuik/Skjoldungen area).

Holarctic, low arctic–boreal, apparently widespread.

***Lispe* Latreille, 1796**
A large, cosmopolitan genus with more than 150 recognized species. One species is known from Greenland.

Adult *Lispe* do not visit flowers, but have strong predatory habits, favouring soft-bodied adult and larval Diptera as prey. The agile flies are often seen running on moist surfaces of silt or mud alongside fresh or brackish waters, hunting for prey or engaged in specialized courtship behaviour.

***Lispe uliginosa* Fallén, 1825 Fig. 3H**
Michelsen 2006: 115.

DISTRIBUTION. SW (from Nuup Kangerlua/ Godthåbsfjord north to Uummannaq Fjord).

Widespread, northern Holarctic species.

***Lophosceles* Ringdahl, 1922**
A small genus with less than 10 species, confined to boreal and elevated parts of the Northern Hemisphere. Two species occur in Greenland. The adults of a few species have been observed visiting umbelliferous flowers in places with lush vegetation, but other biological information is not available.

***Lophosceles frenatus* (Holmgren, 1872) Fig. 3I**
Michelsen 2006: 115.

DISTRIBUTION. S, SW, SE (north to Saqqaq and Saqqisikuik/Skjoldungen).

Holarctic, low arctic–subarctic and alpine (northern and central Europe, Alaska, Canada).

***Lophosceles minimus* (Malloch, 1919)**
Michelsen 2006: 115.

DISTRIBUTION. S, SW, NW, NE, SE (north to Upernavik and Fligely Fjord at Kuhn Ø).

Holarctic, low arctic–subarctic (Norway, Sweden, Alaska, Canada).

***Musca* Linnaeus, 1758**
A genus with nearly 70 recognized species, most of them found in the Old World tropics.

***Musca domestica* Linnaeus, 1758 (House Fly)**
Michelsen 2006: 116.

BIOLOGY. Larval development in cold climates is dependent on man-made habitats like heaps of fermenting dung or garbage.

DISTRIBUTION. SW (Ivittuut 1938). Surely accidentally introduced; it is unlikely that this termophilous species is able to establish viable populations anywhere in the arctic zone.

Cosmopolitan, synanthropic.

***Mydaea* Robineau-Desvoidy, 1830**
A genus with *c.* 50 recognized species found primarily in the temperate forest zones of the Northern Hemisphere. A single species is known from Greenland.

The adults are seen on foliage and larval substrates rather than flowers. The larvae live as predators in more or less specialized decomposing substrates such as bovine droppings, rotten fungi or mammalian borrows.

***Mydaea palpalis* Stein, 1916**
Michelsen 2006: 117; first record from Greenland.

DISTRIBUTION. SW (Kangerluarsunguaq/Grædefjord, Kangerlussuaq/Søndre Strømfjord: Arnangarnup Kua). Evidently a local and rare species.

Holarctic, low arctic–subarctic and alpine. (Alaska, Canada, Iceland, Fennoscandia, Russia, Czech Republic).

***Phaonia* Robineau-Desvoidy, 1830**
A widespread genus with hundreds of recognized species. Notably diverse in the Holarctic forest zones. Only 2 species of *Phaonia* occur in Greenland, both mesophilous and strongly anthophilous, feeding on both pollen and nectar. The larvae of related species in Europe are predatory, living freely in humus rich soil.

***Phaonia bidentata* Ringdahl, 1933 Fig. 4A**
Michelsen 2006: 117. Also recorded from Greenland as '*serva* Meigen', '*subfuscinervis* Zetterstedt' and '*consobrina* Zetterstedt', all misidentifications of European species.

DISTRIBUTION. nSW, NW, N, NE, SE (south to Kangerlussuaq/Søndre Strømfjord and Scoresby Sund).

Nearctic, arctic (Canada).

***Phaonia pallidisquama* (Zetterstedt, 1849)**
Michelsen 2006: 117. Also recorded from Greenland as '*plumbea* Meigen' and '*morio* Zetterstedt', both misidentifications.

BIOLOGY. In a Greenland study (Philipp *et al.* 1990) it was observed that the species frequently visits flowers of *Dryas integrifolia* to feed and to bask. Nine times as many females as males were caught in flowers, presumably because of the greater need for nutrition in the egg-producing females. These often spend more than 10 minutes in each flower, dividing the time between feeding on pollen and nectar and basking. The species was also frequently seen visiting flowers of *Potentilla vahliana*.

DISTRIBUTION. S, SW (north to Uummannaq Fjord: Ingia Fjord).

Holarctic, low arctic–subarctic (Fennoscandia, Canada).

***Spilogona* Schnabl, 1911**
A genus with some 270 known species, of which the great majority occur in the northern and elevated parts of the Northern Hemisphere. A very important genus in Greenland with 24 species, of which 18 are Holarctic and 6 Nearctic in distribution. A large proportion of the species have their core distribution in the northern, high arctic parts of Greenland.

Spilogona adults are to varying degrees anthophilous, feeding on nectar. They all possess enlarged prestomal teeth and are thus potentially capable of supplying this diet by hunting adults and immatures of soft-bodied insects like simuliids and chironomids. The immature stages are generally poorly known, but the larvae are certainly free living, obligative predators on dipterous larvae and other organisms, probably among algae and on wet surfaces along fresh or brackish waters.

NOTES. Species of the present genus were earlier treated in *Limnophora*. The true diversity of the Greenland fauna of *Spilogona* was utterly underestimated until the appearance of the outstanding revision by Collin (1930) incorporating characters of the male terminalia. Before Collin most of the smaller species of *Spilogona* found in Greenland were misidentified as '*triangulifera* Zetterstedt', a species not found in Greenland until now. Because *Spilogona* is a diverse and taxonomically difficult group in Greenland descriptive notes are supplied for all species to supplement the characters given in the species key.

Spilogona almqvistii (Holmgren, 1880)
Michelsen 2006: 117.

Large (wing length 5.1–7.0mm), abundantly but finely setose. Prementum shiny black. Calypteres dirty white, lower calypter enlarged; knob of halter dark. Abdomen in both sexes with large, almost confluent, paired dark markings on abdomen. Female hypoproct setulose.

DISTRIBUTION. nSW, NW, N, NE, SE. Common in northern Greenland, reaching south to Kangerlussuaq/Søndre Strømfjord and Scoresby Sund.

Circumpolar, arctic.

Spilogona alpica (Zetterstedt, 1845)
Michelsen 2006: 117.

Very small (wing length 2.8–4.2mm), velvety brownish black. Prementum thinly pruinose. Calypteres white, lower calypter enlarged, halter black. Male frontal vitta as wide as distance between outer margins of posterior ocelli. Paired dark abdominal markings small and indistinct in male, absent in female. Female hypoproct spinulose.

DISTRIBUTION. S, SW, NE, SE (north to Uummannaq and Scoresby Sund).

Holarctic, low arctic–subarctic, in Europe alpine.

Spilogona arctica (Zetterstedt, 1838) Figs 1B, 4B, C
Michelsen 2006: 117.

Small to medium sized (wing length 3.1–5.7mm). Proboscis short; male prementum slender, shiny except for some thin tomentum mid-ventrally; female prementum tickened, entirely polished black. Mid femur with preapical anterior seta. Female hypoproct setulose.

DISTRIBUTION. S, SW, SE. Evidently the most abundant species of muscoid fly in the southern parts of Greenland, but not found north of Sigguup Nunaa/Svartenhuk and Ammassalik.

Probably circumpolar, low arctic–boreal. Palaearctic distribution poorly known due to confusion with *S. contractifrons* (Zetterstedt).

Spilogona arcticola Huckett, 1965
Michelsen 2006: 118; first record from Greenland.

Small (wing length 4.0–4.9mm), velvety brownish black. Both sexes with no indication of paired black markings on abdomen. Prementum shiny black except tomentose along a mid-ventral stripe.

DISTRIBUTION. N (Peary Land: Kap København), NE (Zackenberg).

Nearctic, high arctic (Canada: Northwest Territories).

Spilogona baltica (Ringdahl, 1918)
Michelsen 2006: 118.

Small (wing length 3.5–4.8mm), resembling *S. arctica* but prementum extensively tomentose in both sexes, dorsocentrals always 2 + 3, mid femur without preapical anterior seta and male abdomen less deep.

DISTRIBUTION. S, SW, SE (north to Qeqertarsuaq/Godhavn and Saqqisikuik/Skjoldungen).

Circumpolar, low arctic–temperate, mainly coastal; widespread.

Spilogona deflorata (Holmgren, 1872)
Michelsen 2006: 118.

Small (wing length 3.0–4.5mm) with white calypteres and dirty brownish halter. Male brownish black, except dorsum of broadly ovoid abdomen silvery tomentose with a pair of large, trapezoid black markings on tergite III and a pair of smaller triangular markings on tergite IV. Female resembling *S. baltica* and *S. arctica* but differs from the former by a tomentose prementum and 2 + 4 dorsocentrals, from the latter by absence of preapical anterior seta on mid femur and from both by having recurrent spinules on hypoproct.

DISTRIBUTION. G. Known from all parts of Greenland but scarce southernmost.

Nearctic, arctic (Alaska, Canada).

Spilogona denudata (Holmgren, 1869)
Michelsen 2006: 118.

Small (wing length 3.4–5.0), dark grayish, with weakly projected lower face and and grey tomentose prementum.

Paired dark abdominal markings poorly indicated in both sexes. Very similar to *S. tornensis*, but male without longish pv-setae in basal half of hind femur and female on hypoproct with some strong, recurrent spines.

DISTRIBUTION. SW, NW, N, NE, SE (south to Kangerlussuaq/Søndre Strømfjord and Scoresby Sund).

Circumpolar, arctic.

Spilogona dorsata **(Zetterstedt, 1845)**

Michelsen 2006: 119. Also recorded from Greenland under the synonym '*hyperborea* Boheman'.

Medium sized (wing length 4.0–6.2mm), blackish species easily identified by the characters given in the key.

DISTRIBUTION. SW, NW, N, NE, SE. Abundant and widespread, although more scarce to absent in southermost parts.

Circumpolar, arctic.

Spilogona extensa **(Malloch, 1920)**

Michelsen 2006: 119.

Small (wing length 3.8–4.9mm), velvety brownish black. Male with large paired black markings faintly indicated on abdomen, same markings barely visible in female. Prementum entirely shiny black.

DISTRIBUTION. NW, N, NE, SE (ranging south to Nuussuaq and Scoresby Sund).

Nearctic, mainly high arctic (Canada).

Spilogona malaisei **(Ringdahl, 1920)**

Michelsen 2006: 120.

Very small (wing length 3.2–4.5), resembling *S. deflorata*, but prementum shiny black and slender in both sexes; knob of halter yellow. Broadly ovoid male abdomen with black markings standing on light grey rather than silvery background. Female abdomen black, unmarked, with a black subshine throught thin dark tomentum; hypoproct setulose.

DISTRIBUTION. S, SW, NW, NE, SE. Widespread except in northernmost parts.

Holarctic, high arctic–subarctic (Fennoscandia, Alaska, Canada).

Spilogona megastoma **(Boheman, 1866)**

Michelsen 2006: 120.

Small (wing length 3.2–5.0mm), in possessing a projecting face, a ventrally bare mid tibia and a setulose female hypoproct rather resembling *S. sanctipauli* and *S. extensa*, but distinguished by the characters given in the key.

DISTRIBUTION. nSW, NW, NE, SE (ranging south to Qeqertarsuaq/Godhavn and Scoresby Sund area, but hitherto not found in N).

Circumpolar, arctic.

Spilogona melanosoma **(Huckett, 1932)**

Michelsen 2006: 120.

Medium sized (wing length 4.5–6.0mm). Prementum tomentose. Male: head and body entirely velvety brownish black without markings on abdomen; calypteres pale yellowish; head of halter black. Female: head and body covered in light brownish grey tomentum; paired dark markings slightly visible on abdomen; head of halter yellow; hypoproct with recurrent spinules.

DISTRIBUTION. NW, N, NE (ranging from Pituffik/Thule to Zackenberg).

Nearctic, mainly high arctic (Alaska, Canada).

Spilogona micans **(Ringdahl, 1918)**

Michelsen 2006: 120.

Small (wing length 3.9–5.4mm). Proboscis short, prementum tomentose. Hind femur basally with setae in postero-ventral position. Head and body in male bluish-grey tomentose, abdomen ovoid, on tergites III–IV with dark brown triangular markings extending laterally as hind-marginal bands. Female greyish tomentose, except brown on subsquarish discal area of mesonotum, upper part of mesopleuron and paired, triangular markings on abdomen; hypoproct with recurrent spines.

DISTRIBUTION. nSW, NW, NE, SE (reaching south to Kangerlussuaq/Søndre Strømfjord and Scoresby Sund, not found in N).

Holarctic, high arctic–subarctic (Fennoscandia, Austria, Alaska, Canada).

Spilogona monacantha **(Collin, 1930)**

Michelsen 2006: 120. Greenland records of '*semiglobosa* Zetterstedt' are misidentifications of the present species.

Medium sized (wing length 4.6–6.4mm). Prementum shiny black. Male light bluish grey tomentose on head and body, abdomen long-ovoid with large, trapezoid black markings on tergites III–IV. Female grey-tomentose, patterned with brown stripes on mesonotum and brown paired markings on abdomen; hypoproct with recurrent spinules.

DISTRIBUTION. nSW, NW, NE, SE (south to Kangerlussuaq/Søndre Strømfjord and Scoresby Sund, not found in N).

Nearctic, arctic (Alaska, Canada).

Spilogona novaesibiriae **(Frey, 1915)**

Michelsen 2006: 120, as *S. obsoleta* (Malloch, 1920), synonym used up to the present.

Small species (wing length 3.3–5.0mm). Proboscis short, haustellum rather thick in female, prementum with very thin tomentum, thus shiny black. Calypteres small, yellowish in male, whitish in female; halter brown to brownish black in male, brownish yellow in female. Male thorax and abdomen uniformly brownish black. Female thorax light brownish grey with brown striping pattern indicated on mesonotum; abdomen grey with paired dark grey markings faintly indicated on tergites III–IV; hypoproct setulose.

DISTRIBUTION. NW, N, NE, SE (ranging from Pituffik/Thule to Scoresby Sund).

Holarctic, mainly high arctic (East Siberia: Novosibirskiye Islands, Alaska, Canada).

Spilogona opaca **(Schnabl, 1915)**

Michelsen 2006: 120. Also recorded from Greenland under the synonym '*freyi* Ringdahl'.

Medium to large (wing length 5.0–6.5mm). Resembling *S. trigonifera* but prementum more densely tomentose, notopleuron without accessory setulae, mid tibia without setae ventrally and trapezoid abdominal markings are more remote.

DISTRIBUTION. SW, NE (between Allumersat/Bjørnesund and Qeqertarsuaq/Godhavn; only recorded from Ella Ø and Jameson Land in NE).

Circumpolar, low arctic–subarctic.

Spilogona puberula **(Ringdahl, 1918)**

Michelsen 2006: 121; first record from Greenland.

Easily identified by the key characters. Medium sized (wing length 5.3–6.1mm).

DISTRIBUTION. S (Kangilinnguit/Grønnedal: 400–600 m), SW (Qeqertarsuaq/Godhavn: Østerlien), NW (Upernavik), NE (Zackenberg).

Holarctic (Norway, Sweden, Alaska), high arctic–subarctic.

Spilogona sanctipauli **(Malloch, 1921) Fig. 4D**

Michelsen 2006: 121.

Small to medium-sized (wing length 3.3–5.2mm), in possessing a projecting face, a ventrally bare mid tibia and a setulose female hypoproct rather resembling *S. megastoma* and *S. extensa*, but distinguished by the characters given in the key.

BIOLOGY. In a Greenland study (Philipp *et al.* 1990) it was observed that the species was responsible for about half the total insect visits to flowers of *Dryas integrifolia*, where it fed on both pollen and nectar. About 90% of the specimens were males that possibly used the flowers as posts from which to set off and couple with females in flight. Flowers of *Potentilla vahliana* and *Ranunculus nivalis* were used in the same way.

DISTRIBUTION. nSW, NW, N, NE, SE. Widespread and common in the northern parts, but apparently absent south of Kangerlussuaq/Søndre Strømfjord and Scoresby Sund.

Circumpolar, arctic.

Spilogona tendipes **(Malloch, 1920)**

Recorded for the first time from Greenland by Sorokina & Michelsen (2014). The identification is somewhat tentative, as based on 3 females only found recently at Zackenberg (NE Greenland).

A large species (wing length exceeding 6mm) with a projecting lower face, hind femur with pv-setae, sternite I with setulae and female hypoproct with recurrent spines.

DISTRIBUTION. NE (Zackenberg).

Circumpolar, arctic (Russian Far East: Wrangel Island, Alaska, Canada).

Spilogona tornensis **(Ringdahl, 1926)**

Michelsen 2006: 121.

Very small to small species (wing length 2.8–4.2mm), dark grayish, with weakly projected lower face and and grey tomentose prementum. Paired dark abdominal markings poorly indicated in both sexes. Very similar to *S. denudata*, but some longish pv-setae present in basal half of hind femur and female hypoproct setulose.

DISTRIBUTION. N, NE, SE (ranging from Peary Land to Scoresby Sund).

Holarctic, mainly high arctic (Fennoscandia, Alaska, Canada).

Spilogona trianguligera **(Zetterstedt, 1838)**

Recorded for the first time from Greenland by Sorokina & Michelsen (2014).

Medium-sized species. The only male specimen available from Greenland matches perfectly the descriptions of the European *Spilogona trianguligera* and the synonymous North American *S. setinervis* (Huckett, 1932).

DISTRIBUTION. NE (Jameson Land).

Holarctic species known from Alaska, northern Canada, Great Britain, Faroe Islands, Norway and Finland.

Spilogona trigonifera **(Zetterstedt, 1838)**

Michelsen 2006: 121.

Medium sized (wing length 4.3–6.0mm). Proboscis short, haustellum rather thick in female, prementum with very thin tomentum, thus shiny black. Calypteres large, lower calypter strongly projecting, pale yellowish; halter yellow. Male thorax and abdomen uniformly brownish black. Female thorax and abdomen somewhat shiny, mainly olive-grey tomentose; abdominal tergites III–IV with distinct darker triangular markings; hypoproct setulose.

DISTRIBUTION. S, SW, NE, SE (north to Qeqertarsuaq/ Godhavn and Scoresby Sund: Jameson Land).

Nearctic, low arctic–subarctic (Alaska, Canada).

Spilogona tundrae **(Schnabl, 1915)**

Michelsen 2006: 121. Also recorded from Greenland under the synonym '*macropyga* Frey.'

Medium sized (wing length 4.3–6.2mm), blackish species easily identified by the characters given in the key.

DISTRIBUTION. NW, N, NE, SE (ranging from Pituffik/ Thule to Scoresby Sund).

Circumpolar, mainly high arctic.

Spilogona zaitzevi **(Schnabl *in* Becker & al., 1915)**

Michelsen 2006: 121. Recorded from Greenland under the synonym '*latilamina* Collin'.

Medium sized (wing length 4.2–5.9mm). Resembling *S. melanosoma* but differs in male by yellow halter and distinct lateral demarcation of large, almost confluent black markings on abdomen. The female is more darkly grey tomentose on head and body and entirely bare on lateral margins of scutellum.

DISTRIBUTION. N, NE (south to Scoreby Sund).

Eastern Palaearctic and Nearctic, mainly high arctic.

SCATHOPHAGIDAE

A small family with *c.* 425 described species worldwide. Most species are found in the northern temperate to arctic and elevated parts of the Northern Hemisphere. Scathophagids are by 6 species in 3 genera surprisingly poorly represented in Greenland, considering that the group is rich in species living beyond the tree limit elsewhere in the circumpolar zone. The Greenland species consist of 1 exclusively Nearctic, 4 Holarctic and 1 species (*Scathophaga litorea*) confined to the western Palaearctic area. Half the species occur primarily in the northern, high arctic parts of Greenland. An account of the Greenland species with distributions maps is given by Pape (1987).

The absence of setae on frontal stripe, golden setulae on ventral head and linear lower calypter serve to separate scathophagids from other muscoid Diptera. Scathophagids are medium to large, often strongly setose and/or densely "hairy" flies. Sexual dimorphy is pronounced in some genera but never affects the head which is broad-fronted in both sexes.

Scathophagids are meso- or hygrophilous, but particularly diverse in wetland habitats. Adult scathophagids have enlarged prestomal teeth and occasionally feed on smaller Diptera; some species also frequent flowers. The family is known as "dung flies", but actually most species have phytophagous or carnivorous larvae, and only a few species of *Scathophaga* use cowpats and other kinds of dung regularly as 'hunting ground' for both adults and larvae.

Key to species (males, females)

1. Katepisternum with 3 setae. Aristomere II strikingly long, meeting aristomere III at an obtuse angle...... .... *Gonarcticus arcticus*
– Katepisternum with 1 seta. Arista unremarkable 2
2. Prealar seta short and spine-like. Fore tibia ventrally with 2–3 irregular rows of short, black spinules .... *Microprosopa haemorrhoidalis*
– Prealar seta unremarkable. Fore tibia without ventral spinules 3
3. Legs mainly orange yellow. Anepimeron bare, exceptionally with a few pale setulae...................... 4
– Legs mainly or entirely black. Anepimeron with a variable number of mostly black setulae standing in a cluster 5
4. Prosternum bare. Mid and hind femora at least distally with some black setae in addition to the clothing of setulae .... *Scathophaga furcata*
– Prosternum with some golden setulae. Mid and hind femora without black setae .... *Scathophaga apicalis*
5. Palpus black. Parafrontalia strikingly broad and abundantly setose, about same width as frontal stripe .... *Scathophaga nigripalpis*
– Palpus yellow. Parafrontalia much narrower than frontal stripe .... *Scathophaga litorea*

Gonarcticus **Becker, 1894**

A small genus with 2 Palaearctic and 2 Nearctic species.

Gonarcticus arcticus **(Becker, 1907) Fig. 4E**

Michelsen 2006: 122.

DISTRIBUTION. NW (Pituffik/Thule, Marshall Bugt), N (Peary Land: Kap København, Jørgen Brønlund Fjord), NE (Ella Ø, Mestersvig, Jameson Land), SE (Hekla Havn).

Nearctic, mainly high arctic (Alaska, Canada).

Microprosopa **Becker, 1894**

A genus with *c.* 4 Holarctic, 6 Palaearctic and 3 Nearctic species.

Microprosopa haemorrhoidalis **(Meigen, 1826) Fig. 4F**

Michelsen 2006: 122.

DISTRIBUTION. S, SW, SE (north to Qeqertarsuaq/ Disko and Saqqisikuik/Skjoldungen.

Circumpolar, low arctic–subarctic and alpine.

FIGURE 4 *Habitus of Muscidae (A–D) and Scathophagidae (E–I):* Phaonia bidentata, *male (A);* Spilogona arctica, *male (B) and female (C);* Spilogona sanctipauli, *male (D);* Gonarcticus arcticus, *male (E);* Microprosopa haemorrhoidalis, *male (F);* Scathophaga apicalis, *male (G);* Scathophaga furcata, *male (H);* Scathophaga litorea, *male (I).*

Scathophaga Meigen, 1803
An important genus with *c.* 60 species, most of them confined to the Holarctic Region, but a few species are endemic to the Oriental, Afrotropical and Neotropical regions. 4 species are known from Greenland. They were earlier assigned to *Scopeuma* Meigen or *Scatophaga* Meigen, both synonyms of *Scathophaga*.

Scathophaga apicalis Curtis _in_ Ross, 1835 Fig. 4G
Michelsen 2006: 122. Recorded from Greenland by the synonymous name '*lanata* Lundbeck.'

DISTRIBUTION. SW, NW, N, NE (south to Nuuk/Godthåb and Ammassalik).

Circumpolar, mainly high arctic.

Scathophaga furcata **(Say, 1823) Figs 1C, 4H**
Michelsen 2006: 122. Also recorded from Greenland by the synonymous names '*squalida* Meigen' and '*fuscinervis* Zetterstedt.'

DISTRIBUTION. S, SW, NW, NE, SE. Very common in the southern parts, reaching north to Upernavik and Zackenberg.

Circumpolar, low arctic–temperate and alpine, widespread.

Scathophaga litorea **(Fallén, 1819) Fig. 4I**
Michelsen 2006: 122. Also recorded from Greenland under the synonymous names '*impudica* Reiche' and '*nigripes* Holmgren.'

BIOLOGY. Mainly found on sheltered coasts where the predatory larvae develop in algal wreck-beds.

DISTRIBUTION. S, SW, NE, SE. Very common in southern parts, reaching a little north of Upernavik and to Jameson Land.

Western Palaearctic, low arctic–temperate, widespread, confined to coastal environments.

Scathophaga nigripalpis **Becker, 1907**
Michelsen 2006: 123.

DISTRIBUTION. NE (Zackenberg, Hochstetter Forland, Mestersvig), SE (Hekla Havn). Apparently mainly confined to NE from Zackenberg to Scoresby Sund.

Circumpolar, mainly high arctic.

References

Arntfield, P.W. (1975): A revision of *Graphomya* Robineau-Desvoidy (Diptera: Muscidae) from North America. – *Canadian Entomologist* 107: 257–302.

Collin, J.E. (1930): A revision of the Greenland species of the anthomyid genus *Limnophora* sens. lat. (Diptera), with figures of the male genitalia of these and many other Palaearctic species. – *Transactions of the Entomological Society of London* 78: 255–281, pls. 15–27.

Griffiths, G.C.D. (1982–2003): Anthomyiidae. – *Flies of the Nearctic Region* 8(2), 1–15: 2635 pp. Schweizerbart'sche Verlagsbuchhandlung, Stuttgart.

Huckett, H.C. (1965): The Muscidae of Northern Canada, Alaska and Greenland (Diptera). – *Memoirs of the Entomological Society of Canada* 42: 1–369.

Michelsen, V. (2006): Annotated catalogue of the Anthomyiidae, Fanniidae, Muscidae and Scathophagidae (Diptera: Muscoidea) of Greenland. – *Steenstrupia* 29: 105–126.

Michelsen, V. (2009): Revision of the willow catkin flies, genus *Egle* Robineau-Desvoidy (Diptera: Anthomyiidae), in Europe and neighbouring areas. – *Zootaxa* 2043: 1–76.

Pape, T. (1984): Grønlands møgfluer (Diptera: Scathophagidae) [The dung-flies of Greenland]. – *Entomologiske Meddelelser* 51: 44–48.

Philipp, M., Böcher, J., Mattsson, O. & Woodell, S.R.J. (1990): A quantitative approach to the sexual reproductive biology and population structure in some arctic flowering plants: *Dryas integrifolia*, *Silene acaulis* and *Ranunculus nivalis*. – *Meddelelser om Grønland, Bioscience* 34: 1–60.

Sorokina, V. & Michelsen, V. (2014): Contributions to the taxonomy and faunistics of some arctic species of *Spilogona* Schnabl (Diptera: Muscidae). – *Zootaxa* 3814: 512–520.

Werner, D. & Pont, A.C. (2006): The feeding and reproductive behaviour of the Limnophorini (Diptera: Muscidae). – *Studia dipterologica, Supplement* 14: 79–114.

17.27. Calliphoridae, Oestridae, Tachinidae

Thomas Pape

The blow flies (Calliphoridae), bot flies (Oestridae) and tachinid flies (Tachinidae) are members of the Tachinidae family-group or superfamily Oestroidea. These are typical calyptrate flies with well-developed lower calypteres, a bulbous swelling below wing base (greater ampulla), a complete seam on the antennal pedicel, and scutum with a more or less complete transverse suture. The species are in general more bristly than species of the remaining calyptrate families, although the vestiture of the bot flies is composed of dense, hair-like setae. Blow flies, bot flies and tachinid flies can be separated from other calyptrates in Greenland by the presence of a distal bend of wing vein M and a row of setae on the meron below the metathoracic spiracle (transformed into a patch of hair-like setae in the bot flies). Biologically the species are associated with carrion and faeces or develop as parasites in vertebrates or parasitoids in invertebrates.

The Calliphoridae contain about 1000 species worldwide, the Oestridae is a small family with about 150 species, while the Tachinidae is a very large family with about 8.000 species described so far and many more to be discovered in the warm temperate and tropical countries of the world.

CALLIPHORIDAE (BLOW FLIES)

Metallic blue, violet, bluish or greenish calyptrate flies, rarely almost black. Larvae are typical maggots living in decomposing meat or in birds' nests. Rognes (1991) and Whitworth (2006) provide modern keys to the identification of Fennoscandian and most North American species.

FIGURE 1 *Calliphoridae. A.* Calliphora uralensis ♀. *B.* Cynomya mortuorum ♂. *C.* Protocalliphora tundrae ♀. *D.* Protophormia atriceps ♀. *E.* Protophormia terraenovae ♂.

Key to species

1. Head with at least genal groove orangish and with some occipital setae white. Lower calypter with black setae on dorsal surface; stem vein bare on the dorsal surface 2
– Head entirely black, occipital setae black. Lower calypter bare on dorsal surface; stem vein with setae on the dorsal surface 3
2. Head exclusive of antennae mainly orange: anterior part of fronto-orbital plate, all of parafacial plate, facial plate, anterior half of gena, and palps bright orange; dark parts confined to posteriormost part of gena, postgena and occiput. Abdomen shining metallic blue or blue-green, occasionally even olive or violet; without any 'dusting' .... *Cynomya mortuorum*
– Head mainly dark: all of gena, postgena and occiput black; only facial plate, genal groove, and palps orange (gena may be light brown or have a streak of orange). Abdomen metallic blue with thin, whitish 'dusting' (microtomentum)........................... .... *Calliphora uralensis*
3. Lower calypter dark brown; upper calypter with black hair-like setae on the up-turned surface (wings in resting position). Abdomen without 'dusting' (microtomentum)................................... 4
– Lower calypter white, inclusive the hair-like setae on the rim; upper calypter bare on the up-turned surface (at most with white hair-like setae laterally). Abdomen with thin but conspicuous microtomentum ('dusting') .... *Protocalliphora tundrae*
4. Vibrissal angle drawn forwards and distinctly anterior to a vertical line through the antennal insertion. Lower calypter tongue-shaped with median margin diverging from scutellum. Thoracic spiracles very large .... *Protophormia atriceps*
– Vibrissal angle slightly anterior to a vertical line through the antennal insertion. Lower calypter broad posteriorly with median margin following scutellum for some distance. Thoracic spiracles of normal size........... *Protophormia terraenovae*

Calliphora uralensis **Villeneuve, 1922 Fig. 1A**

Henriksen 1939: *Calliphora uralensis* Villen.

NOTE: Hall's (1965) record of *C. terraenovae* Macquart from Greenland appears to be an error for *C. uralensis* (Rognes 1991), and it was repeated by Whitworth (2006).

BIOLOGY. Breeding in smaller as well as larger vertebrate carcases.

DISTRIBUTION. S, SW (north to 71°15'N; Rognes 1991). SE (Saqqisikuik/Skjoldungen, Ammassalik).

Palaearctic, temperate to arctic.

Cynomya mortuorum **(Linnaeus, 1761) Fig. 1B**

NOTE: The very similar *Cynomya cadaverina* Robineau-Desvoidy co-exists with *C. mortuorum* in the Russian Far East and Alaska and replaces *C. mortuorum* entirely in the 'gap' between Alaska and Greenland. Earlier records of *C. cadaverina* from Greenland (Hough 1898, Sailer & Lienk 1951, Hall 1965, Danks 1981) are misidentifications of *C. mortuorum* (Rognes 1991).

BIOLOGY. Breeding in small and medium-sized vertebrate carcases like dead rodents and birds.

DISTRIBUTION. S, SW, SE, NE, NW (north to Qaanaaq and Ella Ø, Zackenberg).

Holarctic (Palaearctis, Alaska), temperate to arctic.

Protocalliphora tundrae **Sabrosky, Bennet & Whitworth, 1989 Fig. 1C**

Henriksen 1939: *Protocalliphora azurea* Fall.

NOTE: Although some species of the genus *Protocalliphora* have strikingly narrow distributions, a high arctic New World endemic would *a priori* seem unlikely. Sabrosky *et al.* (1989) mention that *P. tundrae* is very similar to the West Palaearctic *P. nuortevai* Grunin and that these could even be vicariant subspecies.

BIOLOGY: Breeding in birds' nests where the larvae suck blood of the nestlings. Bred from the nests of snow bunting in Peary Land (Sabrosky *et al.* 1989).

DISTRIBUTION. SW (Kangerlussuaq/Søndre Strømfjord, Qeqertarsuaq/Disko), N (Peary Land), NE (Scoresby Sund: Hekla Havn).

Nearctic, arctic.

Protophormia atriceps **(Zetterstedt, 1845) Fig. 1D**

Henriksen 1939: *Phormia atriceps* Zett.

BIOLOGY: Breeding in vertebrate carcases of all sizes, from rodents to walruses (Rognes 1983), with development time extending to 1–2 years under harsh conditions (Danks 1981). The mature larvae form their puparia in the southfacing parts of the fur in order to take most advantage of the sun's heat (McAlpine 1965, Rognes 1983). Overwinters as adults as well as immatures.

DISTRIBUTION. SW, NW, N, NE, SE (south to Qeqertarsuaq/Godhavn and Saqqisikuik/Skjoldungen).

Circumpolar, arctic.

Protophormia terraenovae **(Robineau-Desvoidy, 1830) Fig. 1E**

Henriksen 1939: *Phormia groenlandica* Zett.

BIOLOGY. Breeding in vertebrate carcases of all sizes. Occasionally involved in sheep myiasis, and reported as attacking and killing living reindeer in Alaska (Hall 1948). The larvae form puparia in the fur with a preference for the south-facing edge along the ground, or they migrate up to 0.5 m away from the carcase in the topsoil (pers. obs.). Overwinters primarily as adults but even as immatures. The species may build up into nuisance numbers around refuse dumps.

DISTRIBUTION. G (apparently throughout Greenland, although the occurrence along the east coast is poorly documented).

Holarctic, arctic-temperate.

Unconfirmed species and misidentifications

Henriksen (1939:84) lists *Lucilia* sp. from West Greenland (81°30'N) based on Henriksen & Lundbeck (1917:626). This is most likely a misidentification, possibly even of *Cynomya mortuorum*, which sometimes has a green metallic hue. I have not been able to track the original material upon which the record was based.

Henriksen (1939:84) lists *Acrophaga alpina* (Zetterstedt) from East Greenland (74°35'N) based on Henriksen & Lundbeck (1917:626). This could be a reference to *C. stelviana* (Brauer & Bergenstamm), as this was the old concept behind the name until revised by Rognes (1986). *Calliphora stelviana* has a Holarctic boreal and subalpine distribution, but no material is known from Greenland. The modern interpretation of *Acrophaga alpina* of the older authors is *Calliphora genarum* (Zetterstedt), which likewise has a Holarctic distribution that excludes Greenland (Rognes 1991). I have not been able to track the original material upon which Henriksen & Lundbeck's record of *Acrophaga alpina* was based, but it is unlikely that *C. genarum* and *C. stelviana* would be overlooked as these differ from *C. uralensis* by their white rather than infuscated or brown lower calypteres. As Henriksen & Lundbeck (1917) mention the possibility that their *Acrophaga alpina* "is identical with *Cynonyia* (sic) *mortuorum*", I would consider a misidentification as most likely.

Messersmith (1982:39) records one specimen of *Calliphora vomitoria* (Linnaeus) from Narsarsuaq, but he states that this is "probably same as *vicina* Robineau-Desvoidy and *erythrocephala* Meigen". I have not been able to track the original material upon which the record was based, but the failure of the author to distinguish between *vicina* and *vomitoria* suggests that a confusion with *C. uralensis* is quite likely. Rognes (1991) states that the range of *C. vicina* apparently does not include Greenland, although Messersmith's record is not explicitly discussed. Rognes further states for *C. uralensis* that "this species is surprisingly often confused with *vicina* in collections" (p. 84).

Richard & Harmston (1972) record *Pollenia rudis* (Fabricius), listed under family Muscidae, as collected from a south-facing slope in the Maniitsoq/Sukkertoppen area in SW Greenland. No material has been recovered and the record is here considered as most likely a misidentification. Species of *Pollenia* are obligate parasitoids of earthworms (Szpila 2003), and although the few Greenland earthworms possibly could act as hosts, it is unlikely that a native Greenland species of *Pollenia* would have escaped notice. Intercepting a few specimens introduced either as adults or bred from puparia in imported soil cannot be ruled out, but such a scenario is not considered very likely under the circumstances outlined by Richard & Harmston (1972).

OESTRIDAE (BOT FLIES)

Large, robust calyptrates with dense, hair-like clothing setae and transverse colour-bands that gives the species a bumblebee-like appearance (Nilssen *et al.* 2000). Mouthparts reduced and non-functional. Larvae grow into large, robust maggots, with strong spines on most segments.

Larvae of all bot flies are obligate parasites of mammals, and the larvae are nourished by the hemolymph and other exudates provoked by their presence either in boils under the skin or attached to the mucous membranes in the naso-pharyngeal cavities or in the alimentary canal.

The two species occurring in Greenland are most reliably separated by the wing venation (Fig. 2B,D), but they differ even in the general habitus and colour of the abdomen, with *Cephenemyia trompe* being slightly more robust and with stouter legs and a larger head, and the abdomen is usually predominantly black with some scattered orange or yellowish hair-like setae, while *Hypoderma tarandi* is more slender and has the abdomen clothed extensively in reddish setae (Fig. 2A,C). Larvae are readily separated by their location in the host. Both species are obligate parasites of the reindeer, with larvae of *C. trompe* lodging in the nasal or frontal sinuses, those of *H. tarandi* sitting in warbles under the skin. Bot flies rather than mosquitoes and blackflies are the major cause of insect harassment of reindeer (Hagemoen *et al.* 2002). Although ovi- or larvipositing bot flies inflict no pain to their hosts, the reindeer are psychologically stressed by flying bot flies and display various behaviours from annoyance to panic escape reactions. Both bot fly species (as well as other nuisance flies) may cause a reindeer to shake or move its head swiftly or shake its body violently, while stomping the legs is a stereotypic reaction against warble flies, and sudden nose dropping, rubbing of the nose in the ground, and violent sneezing is elicited specifically against approaching nasal bot flies. Reindeer restlessly running for hours, resting on snow patches or standing on windy ridges or hilltops are seeking shelter from bot flies – and neither from aggressive mosquitoes nor from excessive summer heat (Hagemoen *et al.* 2002).

FIGURE 2 *Oestridae. A–B.* Cephenemyia trompe ♂*, habitus (A) and wing (B) [Norwegian Lappland, courtesy of Arne Nilssen]. C–D.* Hypoderma tarandi ♂*, habitus (C) and wing (D) [Norwegian Lappland, courtesy of Arne Nilssen].*

Key to species

1. Pedicel setose and tomentose. First flagellomere much larger than pedicel. Facial plate without a keel and narrowing ventrally. Wing vein M with a sharp, right-angled bend and a small extension from angle. Abdomen robust, outline almost circular, setosity mostly black with scattered orange or yellow setae on tergites 4–5. Female with short, non-telescopic ovipositor...................... *Cephenemyia trompe*
– Pedicel bare and shining black. First flagellomere at most as large as pedicel. Facial plate with a high keel separating the antennae completely, plate very broad ventrally. Wing vein M with an obtuse, not particularly sharp bend and no extension. Abdomen oval in outline, setosity straw yellow on tergite 1, orange or reddish on tergites 3–5. Female with long, telescopic ovipositor............*Hypoderma tarandi*

***Cephenemyia trompe* (Modeer, 1786) (Nasal bot fly of the reindeer) Fig. 2A, B**

Henriksen 1939: not recorded.

NOTE: Museum specimens (adults) only seen from Kangerlussuaq/Søndre Strømfjord (dating from 1981). *Cephenemyia trompe* was introduced to Greenland (Itivnera, Godthåbsfjorden), when domestic reindeer were brought in from Norway in 1952, resulting in infestation of the local reindeer population (Vibe 1981, Lassen & Aastrup 1981).

BIOLOGY. Female flies inject newly hatched larvae towards the muzzle of the reindeer, aiming at a small area just above the upper lip (Anderson 1989, Anderson & Nilssen 1990). The larvae will seek down into the mouth or up into the nostrils, and they creep along the mucus membranes up into the pharyngeal and nasal cavities. Development is about 10 months, with most of larval growth taking place in spring, and mature larvae leave the host and pupate in the ground (Grunin 1966). Adult males aggregate on hilltops for mating (Downes 1985), and the female produces on average close to a thousand eggs (Nilssen 1997).

DISTRIBUTION. SW.

Circumpolar, low arctic-boreal.

***Hypoderma tarandi* (Linnaeus, 1758) (Dermal or warble bot fly of the reindeer) Fig. 2B, C**

Henriksen 1939: not recorded.

NOTE: The distribution probably matches that of its reindeer host, but particularly for the reindeer population of SW Greenland, the introduction of reindeer from Norway in 1952 brought dermal bot flies to a population that at that time was considered to have no bot fly parasites (Vibe 1981, Lassen & Aastrup 1981).

BIOLOGY. Female flies glue series of eggs onto single hairs, each egg gripping the hair with a small clasper. In warm weather female flies will land directly on the reindeer to oviposit, while in cooler weather the female may walk up to the reindeer, turn around and continue by backing until the ovipositor gets into contact with the fur of the host. The eggs hatch in about 4–5 days, and the first instar larvae bury into the skin and migrate subcutaneously for 3–4 months towards the dorsum of the host, where they settle, forming a boil. The larva maintains a small pore in the boil for respiration. Development is about 10 months, with most of larval growth taking place in spring. Mature larvae leave the host by backing out of the boil, fall to the ground and pupate in the topsoil (Grunin 1965). Mating takes place in rocky areas along rivers and streams or in rocky areas in drying river beds (Anderson *et al.* 1994), and the female produces on average about 600 eggs (Nilssen 1997).

DISTRIBUTION. SW (from Allumersat/Bjørnesund north to Kangerlussuaq/Søndre Strømfjord and Itilleq), NW (Pituffik/Thule Air Base: Baker Lake).

Circumpolar, low arctic-boreal.

TACHINIDAE (TACHINID FLIES)

The adults are typical members of the Tachinidae family-group but generally more bristly. Recognized by their strongly convex subscutellum that bulges below the scutellum. Larvae are internal parasitoids of insects.

Key to species

1. Pedicel elongate and longer than first flagellomere; the latter is about as long as wide in lateral view. Hind coxa with setae on posterior margin 2
– Pedicel shorter than first flagellomere. Hind coxa bare on posterior margin 3
2. Wing base orange. Parafacial ventrally without strong bristles among the setae...................... .... *Tachina ampliforceps*
– Wing base not particularly coloured, at most light brownish. Parafacial ventrally with 2–3 strong bristles among the setae .... *Peleteria aenea*
3. First and second aristomeres both 3–4× as long as broad. Wing vein R_{4+5} setose almost to crossvein r–m. Wing cell R_{4+5} with long petiole...................... .... *Periscepsia stylata*
– First and second aristomeres both less than 2× as long as broad. Wing vein R_{4+5} setose only at base. Wing cell R_{4+5} without petiole.......................4
4. Eye densely setose .... *Trafoia arctica*
– Eye bare .... *Exorista thula*

Exorista thula Wood, 2002 Fig. 3A

Henriksen 1939: *Tachina fasciata* Fall.

BIOLOGY. Bred from *Gynaephora groenlandica* in Heilprin Land. Known as a parasitoid of *Gynaephora* spp. in arctic Canada (Moorewood & Wood 2002).

DISTRIBUTION. N (Heilprin Land), NE (Germania Land, Lindhard Ø, Kong Oscar Fjord: Nathorst Land, Ella Ø, Scoresby Sund: Gåseland, Hekla Havn, Jameson Land).

Nearctic (arctic Canada: Yukon, Nunavut, Northwest Territories).

Peleteria aenea (Stæger, 1849) Fig. 3B

Henriksen 1939: *Peleteria aenea* Zett.

BIOLOGY. Bred from *Gynaephora groenlandica* (Henriksen & Lundbeck 1917). In SE Greenland, females have been observed occurring singly at altitudes of 300–400 m visiting thyme flowers. Males were found in the lowland 'taking stations' on small sandy patches, fighting off other males (Andersen 1993).

DISTRIBUTION. S (Pamialluk, Narsarsuaq), SW (north to Kangerlussuaq/Søndre Strømfjord), NE (widespread, north to Dronning Louise Land and Bessel Fjord), SE (Saqqisikuik/Skjoldungen).

Nearctic (arctic Canada: Nunavut).

Periscepsia stylata (Brauer & Bergenstamm, 1891) Fig. 3C

Henriksen 1939: *Petina stylata* B. & B.

BIOLOGY. Bred from *Gynaephora groenlandica* (Henriksen & Lundbeck 1917).

DISTRIBUTION. SW (Kangerlussuaq/Søndre Strømfjord, Qeqertarsuaq/Disko), NW (Upernavik), N (Jørgen Brønlund Fjord, Heilprin Land), NE (Germania Land, Mestersvig, Scoresby Sund: Gåseland, Hekla Havn, Stormkap), SE (Ammassalik, Saqqisikuik/Skjoldungen). Apparently absent in S.

Holarctic (northeastern Palaearctis, arctic Canada), arctic-boreal.

Tachina ampliforceps (Rowe, 1931) Fig. 3D

Henriksen 1939: not recorded.

Messersmith (1982) *Nowickia hispida.*

Böcher (2001) *Tachina hispida.*

BIOLOGY. Bred from *Polia purpurissata* (Lepidoptera: Noctuidae) in North America (Arnaud 1978, as *Nowickia hispida*).

FIGURE 3 *Tachinidae. A.* Exorista thula ♀. *B.* Peleteria aenea ♂. *C.* Periscepsia stylata ♀. *D.* Tachina ampliforceps ♂. *E.* Trafoia arctica ♀.

DISTRIBUTION. S (south to Narsarsuaq), SW (north to Kangerlussuaq/Søndre Strømfjord).

Nearctic, temperate to arctic.

***Trafoia arctica* (Sack, 1923) Fig. 3E**

Henriksen 1939: *Lypha* (*Tenuirostra*) *arctica* Sack.

BIOLOGY. Unknown. The European species *Trafoia gemina* has been bred from a geometrid moth (Tschorsnig & Herting 1994).

DISTRIBUTION. N (Hall Land, Peary Land), NE (Bessel Fjord, Ile de France).

Listed from Amerloq: Utoqqat (SW) and Jackson Ø (NE) by Henriksen (1939), but no material was recovered in the collections of the Zoological Museum, Copenhagen (ZMUC).

Circumpolar, arctic.

Unconfirmed species and misidentifications

Richard & Harmston (1972) record *Chetogena claripennis* (Macquart, 1848) [under *Phorocera claripennis* Macq.] as collected from the Maniitsoq/Sukkertoppen area. No material has been recovered, and the record is here considered a misidentification, possibly of *Peleteria aenea*. *Chetogena claripennis* is widely distributed in North America and its northern range spans from Alaska to Ontario (O'Hara & Wood 2004), but there are no occurrences in easternmost Canada nor from north of the tree line (J.E. O'Hara, pers. comm.).

References

Andersen, S. (1993): Collecting in Greenland. – *Tachinid Times* 6: 2.

Anderson, J.R. (1989): Use of deer models to study larviposition by wild nasopharyngeal bot flies (Diptera: Oestridae). – *Journal of Medical Entomology* 26: 234–236.

Anderson, J.R. & Nilssen, A.C. (1990): The method by which *Cephenemyia trompe* (Modeer) larvae invade reindeer (*Rangifer tarandus*). – *Rangifer Special Issue* 3: 291–297.

Anderson, J.R., Nilssen, A.C. & Folstad, I. (1994): Mating behavior and thermoregulation of the reindeer warble fly, *Hypoderma tarandi* L. (Diptera: Oestridae). – *Journal of Insect Behavior* 7: 679–706.

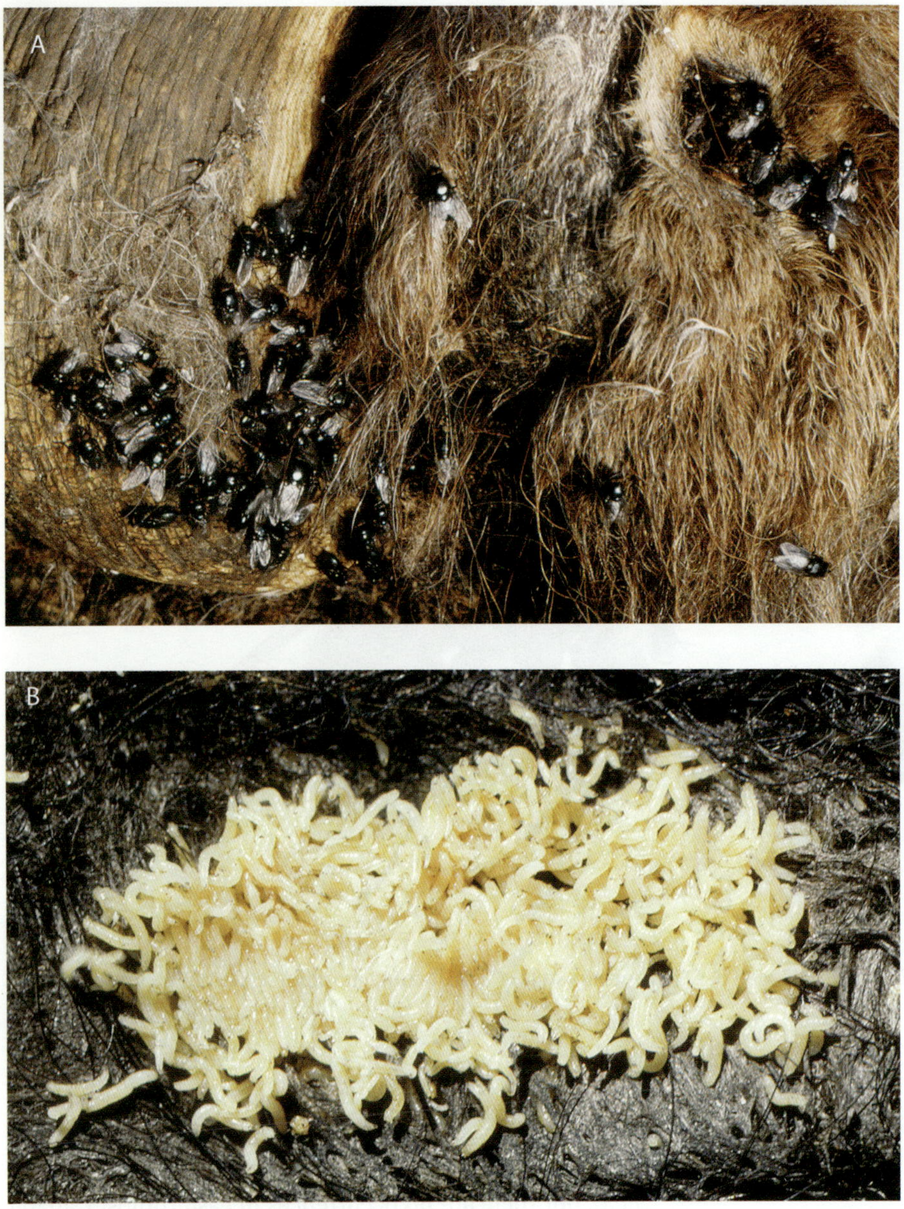

FIGURE 4
Egg-laying Protophormia terraenovae *(A) and larvae of same species (B) on carcass of musk ox.*
(PHOTO: J. BÖCHER)

Arnaud, P.H. jr. (1978): A host parasite catalog of North American Tachinidae (Diptera). – *United States Department of Agriculture Miscellaneous Publication* 1319: 1–860.

Böcher, J. (2001): *Insekter og andre smådyr – i Grønlands fjeld og ferskvand.* – Forlaget Atuagkat; 302 pp. [In Danish.]

Danks, H.V. (1981): *Arctic arthropods. A review of systematics and ecology with particular reference to the North American fauna.* – Ottawa; 608 pp.

Downes, C.M., Smith, S.M., Theberge, J.B. & Dewar, H.J. (1985): Hilltop aggregation sites and behavior of male *Cephenemyia trompe* (Diptera: Oestridae). – *Canadian Entomologist* 117: 321–326.

Grunin, K.J. (1965): Hypodermatidae. *Die Fliegen der Paläarktischen Region* (ed. by E. Lindner) 8: 1–154, 41 pls., addendum 155–160. (Pp. 1–40 published 31.xii.1964; pp. 155–160 published 30.ix.1969.)

Grunin, K.J. (1966): Oestridae. *Die Fliegen der Paläarktischen Region* (ed. by E. Lindner) 8: 1–96, 25 pls.

Hagemoen, R.I. & Reimers, E. (2002): Reindeer summer activity pattern in relation to weather and insect harassment. – *Journal of Animal Ecology* 71: 883–892.

Hall, D.G. (1948): *The blowflies of North America.* Thomas Say Foundation, Lafayette, Indiana; 6 + 477 pp., 51 pls.

Hall, D.G. (1965): Family Calliphoridae. *In*: Stone, A., Sabrosky, C.W., Wirth, W.W., Foote, R.H. & Coulson, J.R. (eds), *A catalog of the Diptera of America north of Mexico*, pp. 922–932. – Agricultural Handbook 276.

Henriksen, K.L. (1939): A revised index of the insects of Grønland. – *Meddelelser om Grønland* 119 (10): 1–101.

Henriksen, K.L. & Lundbeck, W. (1917): Grønlands landarthropoder (Insecta et Arachnida Groenlandica). – *Meddelelser om Grønland* 22: 481–823.

FIGURE 5
"Disruptive colouring" of Tachina *ampliforceps.*
(PHOTO: J. BÖCHER)

Hough, G. de N. (1898): Two new American species of *Cynomyia,*—a study in chaetotaxy. – *Entomological News* 9: 105–111.

Lassen, P. & Aastrup, P. (1981): Studies on the semi-domestic reindeer population (*Rangifer tarandus tarandus* L.) at Itivnera, West Greenland. – *Danske Vildtundersøgelser* 35: 1–36. [In Danish with English summary.]

McAlpine, J.F. (1965): Insects and related terrestrial invertebrates of Ellef Ringnes Island. – *Arctic* 18: 73–103.

Messersmith, D.H. (1982): A report of a collection of Diptera from Iceland and Greenland. – *Fauna Norvegica*, Ser. B 29: 36–39.

Moorewood, W.D. & Wood, D.M. (2002): Host utilization by *Exorista thula* Wood (sp. nov.) and *Chetogena gelida* (Coquillett) (Diptera: Tachinidae), parasitoids of arctic *Gynaephora* species (Lepidoptera: Lymantriidae). – *Polar Biology* 25: 575–582.

Nilssen, A.C. (1997): Flight capacity of the reindeer warble fly, *Hypoderma tarandi* (L.), and the reindeer nose bot fly, *Cephenemyia trompe* (Modeer) (Diptera: Oestridae). – *Ecological Entomology* 22: 294–304.

Nilssen, A.C., Anderson, J.R. & Bergersen, R. (2000): The Reindeer Oestrids *Hypoderma tarandi* and *Cephenemyia trompe* (Diptera: Oestridae): Batesian mimics of bumblebees (Hymenoptera: Apidae: *Bombus* spp.)? – *Journal of Insect Behavior* 13: 307–320.

O'Hara, J.E. & Wood, D.M. (2004): Catalogue of the Tachinidae (Diptera) of America north of Mexico. – *Memoirs of Entomology, International* 18: 1–410.

Richard, P.W. & Harmston, F.C. (1972): Diptera and other arthropods of the Sukkertoppen Tasersiaq area, southwest Greenland. – *Arctic* 24: 107–114.

Rognes, K. (1983): Additions to the Norwegian fauna of calypterate Diptera (Tachinidae, Calliphoridae, Muscidae). – *Fauna norvegica* Ser. B. 30: 88–93.

Rognes, K. (1986): The identity of *Sarcophaga alpina* Zetterstedt, 1838 and the interpretation of *Acrophaga* Brauer & Bergenstamm, 1891 (Diptera: Calliphoridae). – *Entomologica scandinavica* 16: 351–354.

Rognes, K. (1991): Blowflies (Diptera, Calliphoridae) of Fennoscandia and Denmark. – *Fauna Entomologica Scandinavica* 24: 1–272.

Sabrosky, C.W., Bennett, G.F. & Whitworth, T.L. (1989): *Bird blow flies* (Protocalliphora) *in North America (Diptera: Calliphoridae).* – Smithsonian Institution Press, Washington, DC and London; 312 pp.

Sailer, R.I. & Lienk, S.E. (1951): Blow flies in Alaska. – *Canadian Entomologist* 83: 208–211.

Szpila, K. (2003): Actual impact of the blowfly-parasitoids (Diptera: Calliphoridae) on earthworms (Oligochaeta: Lumbricidae) of cultivated fields, against the literature data. – *Wiadomości Entomologiczne* 22: 211–220.

Tschorsnig, H.-P. & Herting, B. (1994) Die Raupenfliegen (Diptera: Tachinidae) Mitteleuropas: Bestimmungstabellen und Angaben zur Verbreitung und Ökologie der einzelnen Arten. – *Stuttgarter Beiträge zur Naturkunde Serie A (Biologie)* 506: 1–170.

Vibe, C. (1981): Landpattedyr. *In*: Nørrevang, A. & Lundø, J. (eds), *Grønlands natur*, pp. 477–499. – Copenhagen, Politikens Forlag. [In Danish.]

Whitworth, T. (2006): Keys to the genera and species of blow flies (Diptera: Calliphoridae) of America North of Mexico. – *Proceedings of the Entomological Society of Washington* 108: 689–725.

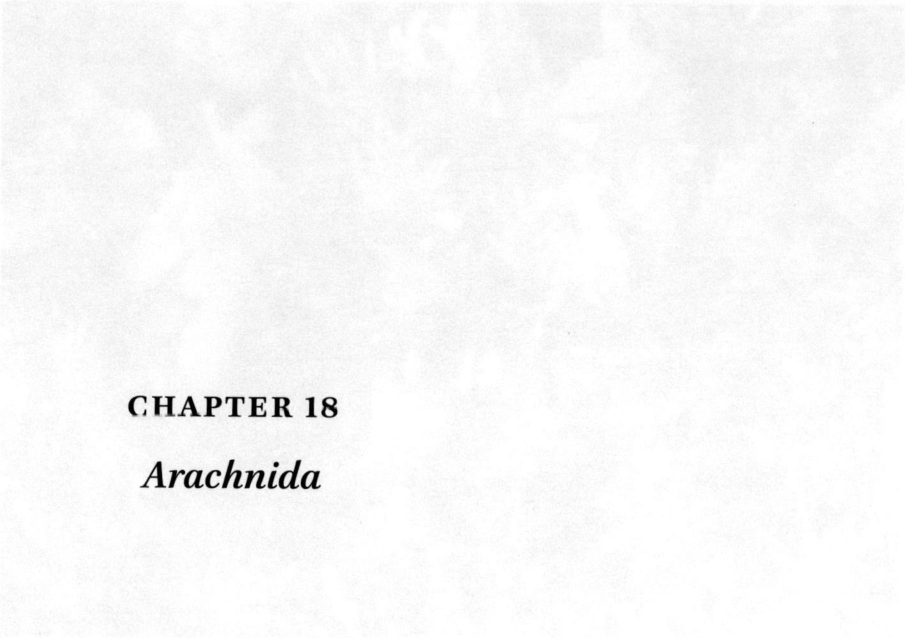

CHAPTER 18

Arachnida

18.1. Araneae (Spiders)

Yuri M. Marusik

Spiders comprise one of the largest orders of animals with more than 45,000 described species, both extant (living) and extinct (fossil). Only five orders of insects have more species (Coleoptera, Diptera, Lepidoptera, Hymenoptera, Hemiptera). Spiders are distributed worldwide, except for Antarctica, and occur in all types of terrestrial biotopes; some are even aquatic.

Morphology: adults

The body is divided into 2 parts, *prosoma* (=cephalothorax) and *opistosoma* (=abdomen). Compared to the opistosoma, the prosoma is strongly sclerotized. It is encapsulated in 2 main sclerites, the dorsal *carapace* (Fig. 1A) and the ventral *sternum* (Fig. 1B). The carapace is pear- or egg-shaped and may be divided into the cephalic (or head) part and the thoracic part (Fig. 1A). The former part bears 8 eyes arranged in 2 or 3 rows. Space between anterior median eyes and margin of carapace is termed clypeus (Figs 1C–D). The cephalic and thoracic parts are separated by anterior radial stripes or a shallow grove. Prosoma bears 6 pairs of appendages (Figs 1A–B): chelicera, pedipalps and four pairs of walking legs. Coxae of pedipalps have anterior outgrowths which forms maxillae. An unpaired appendage between maxillae is called labium (Fig. 1B). The pedipalp (or palp) consists of 6 segments: coxa, trochanter, femur, patella, tibia and tarsus. In adult males the palps are strongly modified and serve as copulatory organs (Figs 1C, E. 11A, C etc). The walking legs comprise 7 segments (Fig. 1A). The tarsus is split into metatarsus (or pretarsus) and tarsus. Tarsus bears 2–3 claws. In most species legs and palps bear strong macrosetae, called spines. In arachnology legs are marked by numbers I–IV.

Prosoma and legs may be uniformly coloured, although in several cases they have a distinct pattern (Araneidae, Lycosidae, Thomisidae, many Theridiidae). Abdomen is of oval shape, and except in *Ceratinella* and *Pelecopsis* the Greenlandic species have no abdominal sclerites. Members of the genera mentioned have a dorsal scutum (sclerotized shields) (Fig. 1A). Abdominal appendages are represented by three pairs of spinnerets. They may form a compact group (Figs 1L–N) or be stretched into one transversal row (Fig. 1R). In dictynids a transversal plate in front of spinnerets is termed cribellum (Fig. 1N). Ventrally abdomen is clearly separated into two parts by an epigastral furrow lying in anterior fourth of abdomen (Fig. 1B). Adult females have a distinct sclerotization above the epigastral field called *epigyne* (Figs 1B,F–G). It serves as a copulatory organ. In most Greenland species the tracheal opening is placed just in front of the spinnerets, but *Hahnia* is an exception by having the opening between spinnerets and the epigastral furrow (Fig. 1R). In the majority of Greenland species the abdomen is uniformly coloured, while in Araneidae, Lycosidae, Tetragnathidae, Theridiidae, Thomisidae the abdomen often shows a species specific pattern.

The most complicated structures in spiders are their copulatory organs: male palp and female epigyne. These organs are most diagnostic (species specific).

Male palp. In many species (Agelenidae, Gnaphosidae, Hahniidae, many Linyphiidae, Salticidae, Thomisidae) tibia bears one or more apophyses (Figs 1C, 1E, 26C) which may be placed laterally, ventrally or dorsally. Tarsus of the male palp is modified into a cymbium (spoon like structure) which bears a bulbus (or bulb), a set of sclerites used for storing sperm and for copulation (Figs 1C, 1E). In Araneidae, Tetragnathidae and Theridiidae cymbium has an outgrowth called paracymbium (Figs 1C–E) near the base of cymbium, while in Theridiidae the paracymbium is placed in the upper part of cymbium. It is indistinct in *Theridion* and *Thymoites*. Although bulbus in all Greenland species is rather complicated, several species have (at first look) a simple-shaped bulbus, for example *Hahnia*. The bulb contains several outgrowths (apophyses) (Figs 1E, 3A,C,E,G, 16A). All species have a more or less complicated/distinct embolus, a sclerite used for inserting sperm. Other parts of bulbal apophyses are called median (or tegular) (Figs 3A, 16C,E) apophysis, conductor, supra-tegular apophysis, lamella-characteristica, etc. Bulbus may have one or more apophyses in addition to embolus.

Epigyne. The epigyne is an unpaired structure placed above the median part of the epigastral furrow in females. It may be flat without fovea or outgrowths (*Hahnia*, *Praestigia*, *Tetragnatha*), or very complicated and large (*Araneus*, *Improphantes*, *Lepthyphantes*) bearing fovea (*Gnaphosa*, *Neriene*, *Porrhomma*) or one (*Araneus*, *Oreonetides*, etc) or more outgrowths (*Bathyphantes*). Several species may have both fovea and outgrowth (*Gnaphosa*, *Pardosa*). A flexible epigynal outgrowth is called scape. The inflexible part can be formed by the base of the epigyne (*Improphantes*, *Oreonetides*) or from an outgrowth of the fovea (septum).

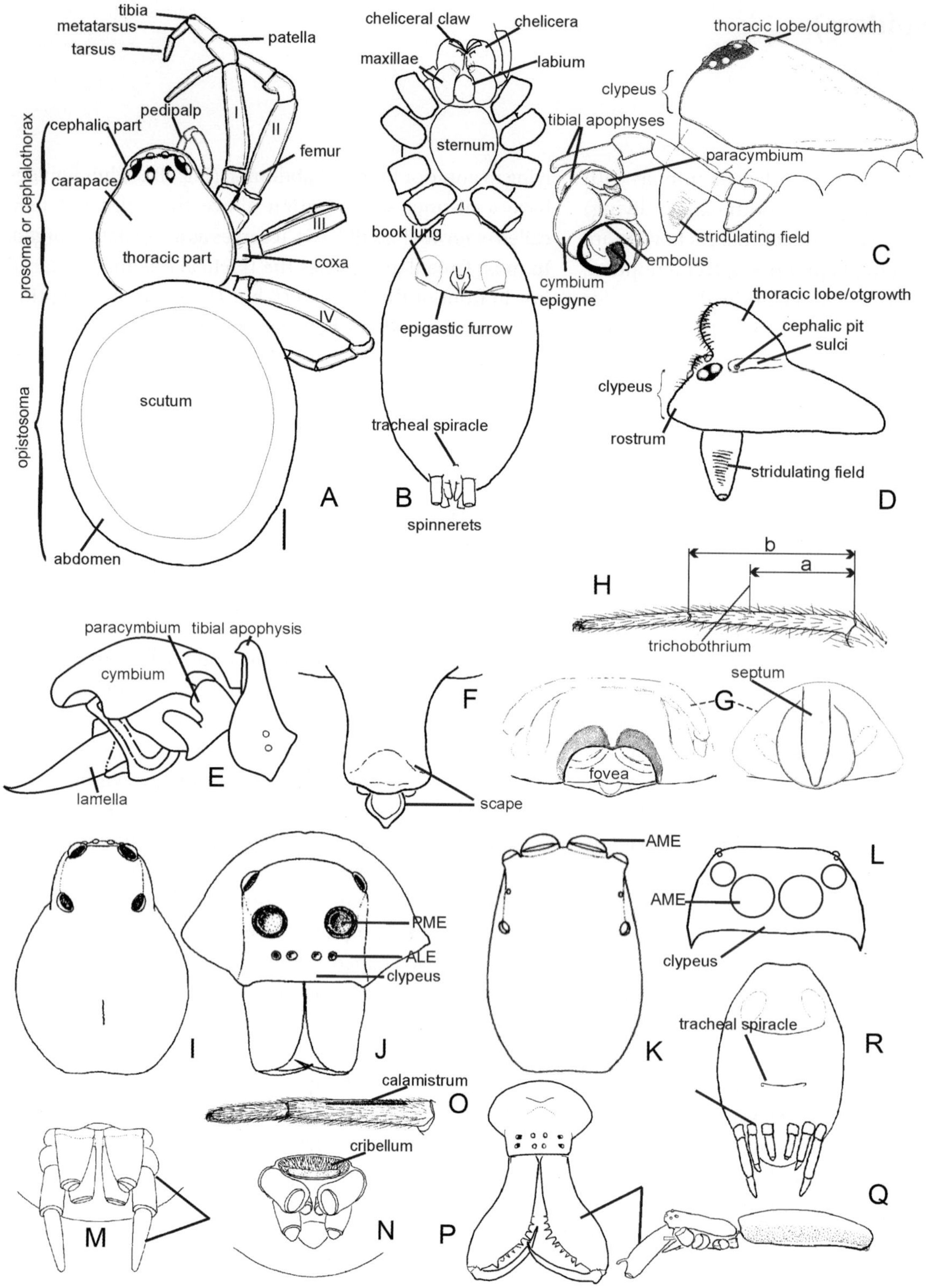

FIGURE 1 *Morphology of spiders. A,* Ceratinella ornatula (♀)*; B,* Gnaphosa lapponum (♀)*; C,* Tarsiphantes latithorax (♂)*; D,* Scotinotylus alpinus (♂)*; E,* Hilaira herniosa *♂ palp retrolateral; F,* Oreonetides vaginatus, *epigyne ventral; G, epigyne of* Neriene (*left*) *and* Hybaucheniudium (*right*)*; H, linyphiid leg; I, J, lycosid carapace above and in front; K, L, salticid carapace above and in front; M, N, spinnerets of* Tegenaria domestica*; N, cribellum of dictynids; O, calamistrum of dictynids; P, Q,* Tetragnatha *sp. in front and lateral; R, abdomen of* Hahnia glacialis, *ventral.*

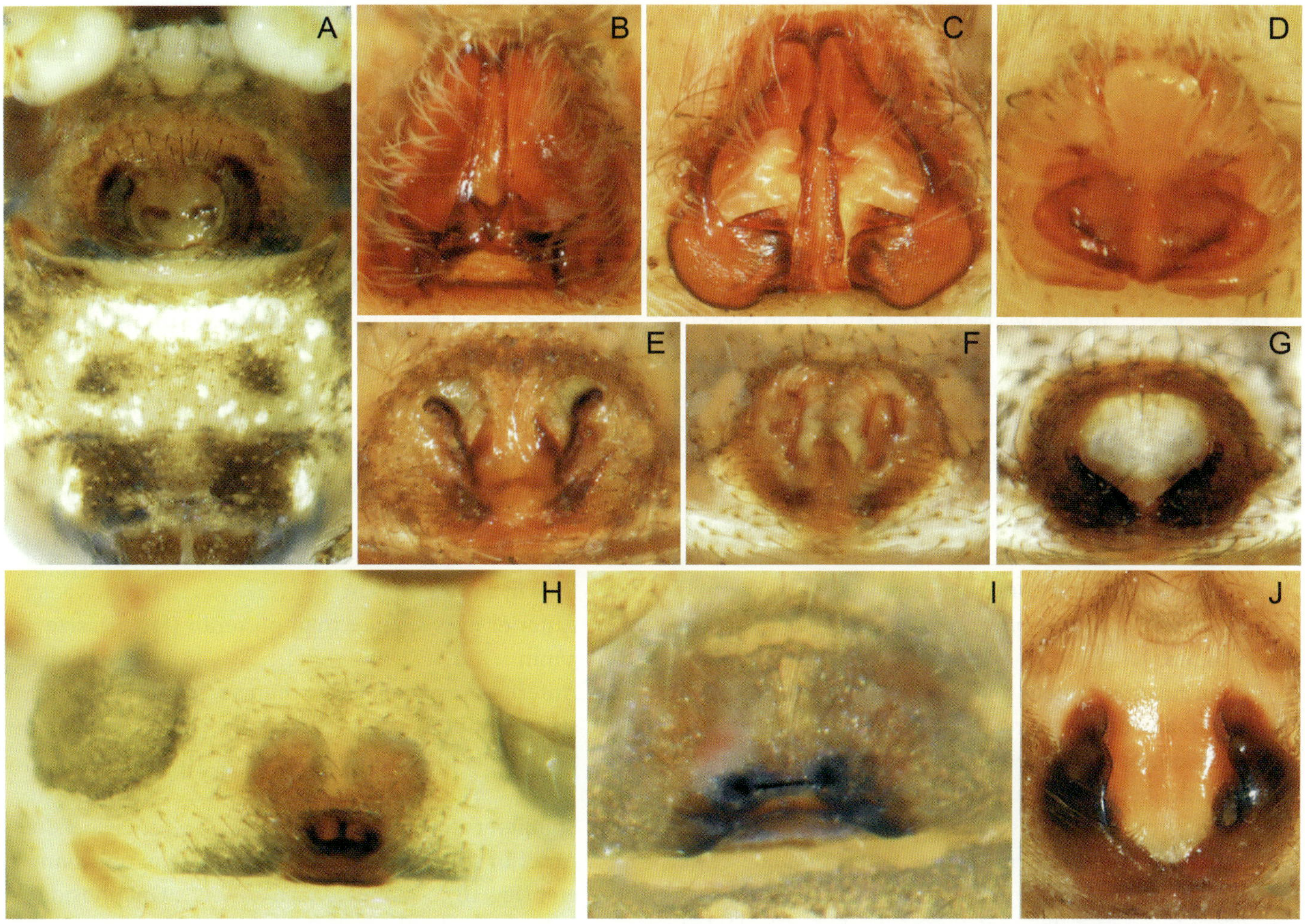

FIGURE 2 *Epigynes of* Ohlertidion lundbecki (*A*), Pardosa furcifera (*B*), P. groenlandica (*C*), P. hyperborea (*D*), Xysticus deichmanni (*E–F*), X. durus (*G*), Thymoites oleatus (*H*), Enoplognatha intrepida (*I*) *and* Haplodrassus signifier (*J*).

Biology and life cycle

All spiders are predators, feeding on insects and other arthropods, including terrestrial crustaceans and their own kin. For catching prey they use different strategies. Some make webs of different construction, others discover prey by walking. Certain species use *e.g.* flowers as sites for ambush, waiting for their victims. Most spiders are general hunters and can feed on any type of prey, although a few are specialised on specific groups. Rather many spiders eat mainly ants or termites, and several species feed chiefly or exclusively on other spiders. All Greenland species seem to be generalists.

The life cycle of spiders includes egg, first instar (which stays in the egg-cocoon), several instars, adult. The cycle may be short, some months, or very long, lasting for several years. Long cycles occur in northern or large-sized spiders.

All species of spiders display some kind of maternal care. In the simplest case, females make egg-cocoons, fill them with eggs and place them in a hidden place (Araneidae, Hahniidae, Tetragnathidae), while most species care about their brood for a longer time. In several groups (Gnaphosidae, Thomisidae, Philodromidae, Salticidae, Linyphiidae, etc) females guard their own egg-cocoons. Gnaphosids and some other spiders enclose themselves in sealed retreats where they also place their cocoons. Female wolf spiders (Lycosidae) carry their egg cocoons attached to spinnerets beneath the abdomen. After hatching spiderlings, the mothers carry them on their abdomen until they become too heavy. For several species it has been shown that mothers can feed their offsprings by catching food for them, or by regurgitating food.

Spiders spend the winter in different stages, although some always as eggs, juveniles, or subadults (last juvenile stage, ready to moult at spring). Certain species overwinter in special retreats, others just in hidden places. Rather many spiders living in the north especially linyphiids, can remain active throughout the winter and even hunt while temperature is below 0°C.

Identification

Most spiders can be identified only as adults. This mature stage may be recognised in males by the strongly swollen and sclerotized palps bearing several discrete sclerites (Figs 1C, 12S,U), in females by the presence of an epigyne (sclerotized plate often bearing outgrowths, placed between book-lung opercula just over the epigastric furrow; Fig. 1B). Unfortunately, most spiders living in Greenland are small, and it is rather difficult to observe the diagnostic characters in palp and epigyne. This requires some experience and knowledge of terminology. Because of these difficulties, the author has tried to avoid using copulatory organs in keys and diagnoses and, when possible, to use only somatic characters that are easy to see.

Identification sources for Greenland species are spread among many publications. Most of the Holarctic species can be identified by Roberts (1987, 1995, 1998). For Nearctic species the best source are Paquin & Dupérré (2003) and revisions from the series "The insects and arachnids of Canada" (Dondale & Redner, 1978; 1990; Dondale *et al.*, 2003). Arctic species may be identified thanks to Leech (1966). Many good illustrations can be found in Holm's and Brændegård's publications. Agnarsson's treatment of Icelandic spiders (1996) includes all Holarctic and several arctic species.

Keys are given to families and to species in families with more than 2 Greenland representatives. Species in families with 2 species (Dictynidae, Gnaphosidae, Thomisidae) can be distinguished according to notes under each species.

Diagnostic illustrations are given for all Araneidae, Linyphiidae, Lycosidae, Dictynidae. For other families represented by several species (Gnaphosidae, Thomisidae, Theridiidae) figures for some species or sexes are not given in cases when they have peculiar characters, or can be identified by exclusion. Figures used in this work are partly original and partly taken from other sources (Holm's papers, Dondale and co-authors, Paquin & Dupérré, 2003). In the latter case they have been modified.

Greenland fauna

With 75 species recorded from Greenland the spiders constitute an unusually high proportion of the terrestrial arhropod fauna of the country. They are also a comparatively well studied element of this fauna, due particularly to the work of Brændegård and Holm.

All spider taxa found in Greenland are here presented in alphabetic order: families, genera within families and species within genera. The reader is also referred to the checklist by Larsen & Scharff (2003) and Marusik *et al.* (2006). Synonyms are given only when other species names have been used in the latest preceding treatments of the Greenland Araneae (Brændegård 1946, Holm 1967, Larsen & Scharff 2003).

Key to families

(see Table 1 for tabular survey of characters)

1. Eyes in three rows: 4–2–2 (Figs 1J–L) 2
– Eyes in 2 rows (Figs 1A, P) 3
2. Anterior eyes smaller than posterior (Fig. 1J), size > 5 mm .... Lycosidae
– Anterior eyes much larger than posterior (Figs 1K, L), size 5–7 mm .... Salticidae
3. Spinnerets placed in one row (Fig. 1R), size < 3 mm. .... Hahniidae
– Spinnerets forms compact group 4
4. Anterior spinnerets spaced by more than one diameter (Fig. 1B), > 5 mm.................. Gnaphosidae
– Spinnerets closely spaced........................... 5
5. Posterior spinnerets long, with 2 joints equal in length (Fig. 1M)........................... Agelenidae
– Posterior spinnerets not longer than anterior....... 6
6. Legs laterigrade (I–IV stretching sideway, Figs 18A, F, 26), tarsi with 2 claw 7
– Legs prograde (I–II stretching forward, III–IV backward), tarsi with 3 claws 8
7. Abdomen pentagonal-roundish, as long as wide (Fig 26) .... Thomisidae
– Abdomen longer than wide (Fig. 18F) .... Philodromidae
8. Body elongate (Fig. 1Q), abdomen and chelicera more than 2 times longer than wide, chelicera with more than 5 teeth (Fig. 1P) .... Tetragnathidae
– Abdomen globular, width > ½ length 9
9. Abdomen with cribellum (Fig. 1N), and tarsi of leg IV with calamistrum (Fig. 1O), < 4 mm, legs without spines, abdomen with pattern........... Dictynidae
– Cribellum and calamistrum absent 10
10. Clypeus smaller than AME, ♀ abdomen globular (Figs 6A, B), with distinct pattern, tibia I in ♂ with numerous lateral spines, > 4 mm .... Araneidae

TABLE 1 *Key to families (Ag, Agelenidae; Ar, Araneidae, Di, Dictynidae; Ha, Hahnidae, Li, Linyphiidae; Ly, Lycosidae; Ph, Philodromidae; Sa, Salticidae; Te, Tetragnathidae; Te, Theridiidae, To, Thomisidae)*

	Ag	Ar	Di	Gn	Ha	Li	Ly	Ph	Sa	Te	Te	To
Eyes in three/two rows	2	2	2	2	2	2	3	2	3	2	2	2
Spinnerets in one row					yes							
Anterior spinnerets spaced				yes								
Posterior spinnerets long with two joints of equal size	yes											
Cribellum and calamistrum present			yes									
Lateral eyes in group		yes	yes		yes	yes				+–	yes	
Legs stretching aside								yes				yes
Chelicera with stridulating field						yes						
Abdomen length at least twice longer than wide										yes		
Ventral serrate hairs on tarsus IV											yes	
Tarsal claws	3	3	3	2	3	3	3	2	2	3	3	2

– Clypeus higher than AME (Figs 1C, D), body without pattern, abdomen globular (size < 4 mm) or not, ♂ tibia without numerous spines 11

11. Tarsus IV without serrate hairs, chelicerae with stridulating field (Figs 1C, D), paracymbium placed basally (Figs 1C, E) and hook-shaped, palpal tibia never extended retrolaterally (Figs 1E, F, 14G), epigyne with median plate (Figs 12A, B, G, H, 15B), visible from caudal view.................... Linyphiidae

– Tarsus IV with serrate hairs, chelicerae without stridulating field, paracymbium when visible placed apically finger-shaped palpal tibia extended retrolaterally (Fig. 21A), without apophyses, epigyne without median plate.................... Theridiidae

AGELENIDAE (FUNNELWEB- OR COBWEB SPIDERS)

A relatively small family with about 1150 species distributed worldwide. Some of them are litter dwellers, others live in the herb and bush strata. There are several synantropic species, one of which has been introduced to Greenland.

Tegenaria domestica (Clerck, 1757)

Easily distinguishable, even juveniles, thanks to long posterior spinnerets and three pairs of round light spots on sternum.

BIOLOGY. Exclusively synantropic; at least in temperate areas never found outside buildings.

DISTRIBUTION. A single juvenile specimen was found in a house at Jørgen Brønlund Fjord, Peary Land (Brændegård 1960, but missing in the check list by Larsen & Scharff 2003). The species is to be expected from several Greenland towns and settlements.

Cosmopolitan.

ARANEIDAE (ORBWEB SPIDERS)

The third largest spider family with more than 3000 species, distributed worldwide. Most araneids are tree, bush or grass dwellers, building orb webs. Some species make special retreats from silk and plant material where they stay during daytime and mount the web only if prey drops into it. Some species have no retreat and stay beside or in the hub of their web. Adults occur in warm periods only. All Greenland species seem to hibernate in their egg-cocoons, and juveniles appear in spring. The Greenland Araneidae are rather easily distinguished by their colour pattern, even juveniles, and by the shape of copulatory organs. Occurrence of an additional species, *Araniella*

FIGURE 3 *Web of araneid spider (possibly* Larinioides *sp.) in scrub of* Salix glauca*.* (PHOTO: J. BÖCHER)

displicata (Hentz, 1847), is very probable, as it reaches 55°N along the east coast of Canada.

Key to species

1. Abdomen glossy and oval, tegulum with spine-like outgrowth, median apophysis with one outgrowth, epigyne without scape, 3–4.3 mm (Figs 6B, 11C, D) ....*Hypsosinga groenlandica*
– Abdomen round or slightly triangular and broader near the front end, epigyne with scape, tegulum without outgrowth, more than 4.5 mm.............2
2. Epigyne with large triangular scape, median apophsyis transverse, with 3 outgrowths (Figs 4A, B), abdominal pattern with white dots.................................. ....*Araneus groenlandicola*
– Epigyne with short scape, median apophysis with 2 outgrowths, stretching outside the bulb (Figs 4E–H), abdomen with distinct folium.......................3
3. Scape finger-like, longer than wide, median apophysis longer than wide, its upper arm twice as thin as basal (Figs 4G, H) ....*Larinioides cornutus*
– Scape wide, if elongate it is wider at tip than at base, median apophysis as long as wide, both arms subequal in width (Figs 4E, F)....................... ....*Larinioides patagiatus*

***Araneus groenlandicola* (Strand, 1906). Figs 4A–B**
Holm 1967: *Araneus quadratus* Cl.
Larsen & Scharff 2003: *Araneus quadratus* Clerck, 1757.
Both abdominal colour (greenish to orange) and pattern (cross formed by white spots) and copulatory organs are diagnostic. The scape is large with a wide base, and median apophysis have three outgrowths. The scape is often broken after copulation, and can look different from illustration.
BIOLOGY. Orb webs are built among grass, often near water or in moist meadows. Web vertical, placed less than 1 m above ground. Retreat always present.
DISTRIBUTION. S, SW (north to 64°08'N).
Nearctic (from Alberta south to Minnesota).

***Hypsosinga groenlandica* Simon, 1889. Figs 4C, D, 6B**
Holm 1967: *Singa* (*Hypsosinga*) *groenlandica* Simon.
Colour pattern, oval shaped abdomen and copulatory organs (lack of scape, median apophysis with one outgrowth, spine-like extension of tegulum) are diagnostic.
BIOLOGY. The webs seems to be placed close to the ground among grass. Retreats absent.
DISTRIBUTION. S, SW (north to Qeqertarsuaq/Disko), SE (north to 63°14'N).
Nearctic, low arctic-temperate, alpine.

***Larinioides cornutus* (Clerck, 1757). Figs 4G, H, 6A**
Holm 1967: *Araneus cornutus* Cl.
Similar to *L. patagiatus*, but can easily be separated from it by lighter colouration and copulatory organs (small finger-like scape, bifid median apophysis that is longer than wide).
BIOLOGY. Orb web vertical in grass vegetation, always placed near water and less than 1 m above ground. Retreat always present.
DISTRIBUTION. S, sSW.
Circumpolar, low arctic-temperate.

***Larinioides patagiatus* (Clerck, 1757). Figs 4E–F, 5**
Holm 1967: *Araneus ocellatus* Cl.

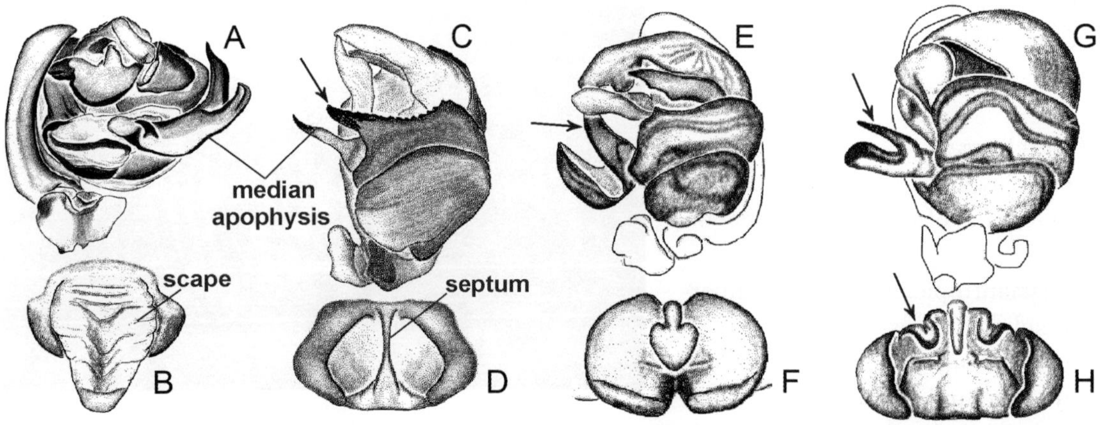

FIGURE 4 *Male palp (A, C, E, G) and epigyne (B, D, F, H) of araneids.* Araneus groenlandicola *(A, B),* Hypsosinga groenlandica *(C, D),* Larinioides patagiatus *(E, F) and* L. cornutus *(G, H).*

FIGURE 5 Larinioides patagiatus *with bag of* Simulium vittatum.
(PHOTO: J. BÖCHER)

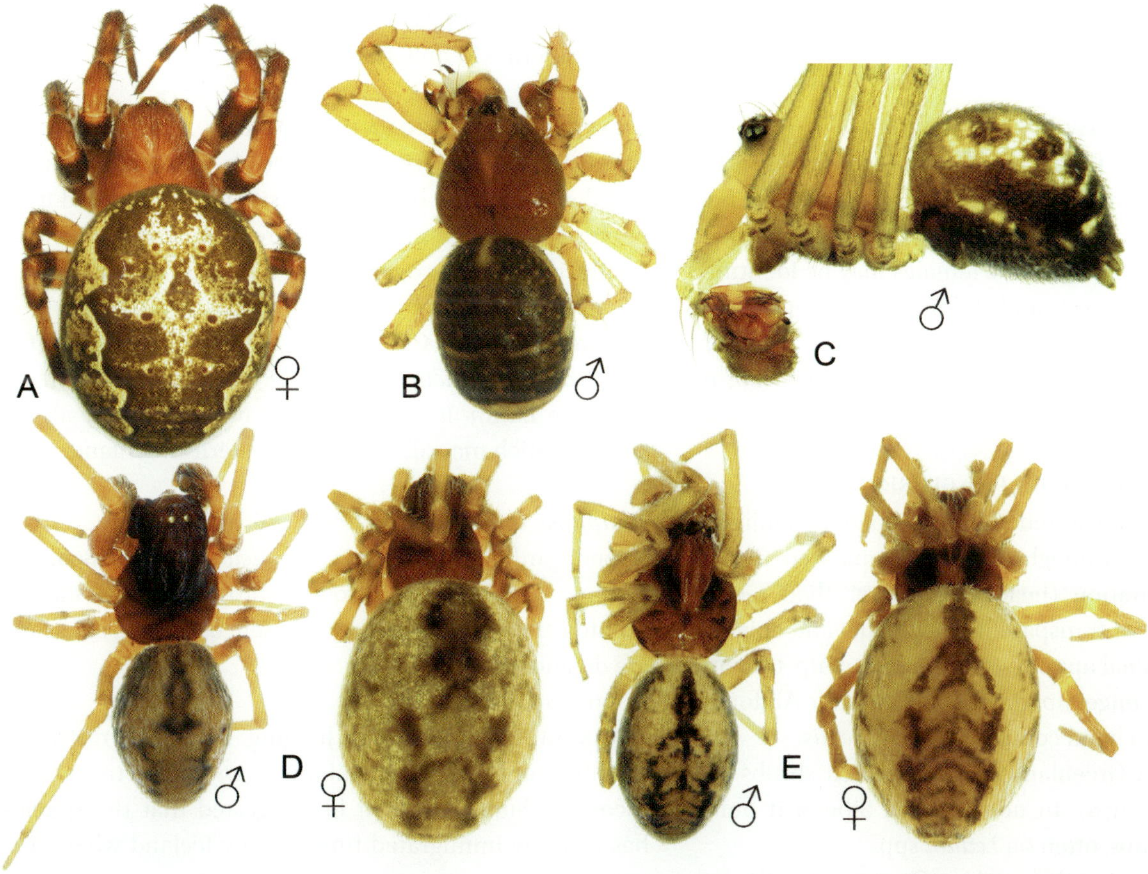

FIGURE 6 *Habitus of* Larinioides cornutus (*A*), Hypsosinga groenlandica (*B*), Lepthyphantes turbatrix (*C*), Dictyna major (*D*) *and* Emblyna borealis (*E*).

FIGURE 7 *Dictynid spider in web on* Artemisia borealis. (PHOTO: J. BÖCHER)

Similar to *L. cornutus*, but can be separated from it by darker colouration and features of copulatory organs (wider scape, median apophysis as long as wide).

BIOLOGY. Vertical orb web built among grass, bushes or trees, placed 0.5 to 2 m above ground. Retreats always present.

DISTRIBUTION. sSW.

Holarctic, low arctic-temperate.

DICTYNIDAE (MESHWEB SPIDERS)

A relatively small family with less than 600 species, distributed worldwide. Dictynids are recognised by having cribellum and calamistrum. Males are easily identified by the shape of the palp, while females may be identified by the colour of carapace and abdomen, the shape of epigyne openings and of vulva, or sometimes by transparent receptacula. Most species make small mesh-like webs on vegetation or among stones (Fig. 7).

Dictyna major Menge, 1869. Figs 6D, 8B–D

In general appearance and shape of epigyne similar to *Emblyna borealis* although it can be separated from it by darker colouration (head area darker than rest of carapace), shape of transparent receptacula and epigynal openings, long tibial apophysis of the male palp, thin and longer embolus, longer upper arm of conductor. Abdomen has wider median band compared to *E. borealis*.

BIOLOGY. In S Greenland found on birch shrubs and trees (Koponen, 1978). In northeastern Siberia it builds webs on low shrubs, often on *Ledum* spp.

DISTRIBUTION. S, SW (north to Qeqertarsuaq/Disko), SE.

Holarctic, low arctic-boreal.

Emblyna borealis (O. P.-Cambridge, 1877). Figs 6E, 8A, E, F

Brændegård 1946: *Dictyna borealis* Cambridge, 1877.

Holm, 1967: *Dictyna borealis* O.P. -Cambridge.

Similar in general appearance and shape of epigyne to *Dictyna major*, but can be separated from it by lighter colouration (head is lighter than other parts of carapace), shape of transparent receptacula and epigynal openings, short tibial apophysis of the male palp, thick and shorter embolus, shorter upper arm of conductor.

BIOLOGY. In Peary Land it was collected under stones in dry localities such as south facing slopes and heaths with *Dryas*, *Cassiope*, *Carex* and *Vaccinium*, up to 300 m a.s. One egg cocoon contained 14 eggs. Brændegård suggests that this species has triennial life cycle (Brændegård 1946, 1960). Specimens from Ellesmere Island (Leech 1966) are most frequently found on dry, south facing sheltered slopes. It seems that all stages except adult can overwinter. In captivity it feeds on small Diptera like Chironomidae and Ceratopogonidae, whereas Cyclorraphous dipterans were refused. The species seems to use ballooning for dispersal (Brændegård 1938).

DISTRIBUTION. G.

NE Siberia - Nearctis, arctic.

GNAPHOSIDAE (GROUND SPIDERS)

The Gnaphosidae with over 2100 species is distributed worldwide, but the highest diversity is found in arid temperate areas. Most gnaphosids are ground dwellers. The family is easily recognised by the anterior spinnerets, which are spaced about one diameter.

Gnaphosa orites Chamberlin, 1922 (probably the same as *Micaria constricta* Emerton, 1894), that is present on adjacent Baffin Island, is a circumpolar, arctic-alpine species which may also occur in western Greenland.

Gnaphosa lapponum (L. Koch, 1866). Fig. 9A–B

Brændegård 1967: *Gnaphosa islandica* Sørensen, 1898.

Dark colouration (brown, brown-black), sharply pointed tibial apophysis and presence of epigynal scape are diagnostic.

BIOLOGY. Unknown.

DISTRIBUTION. NE. – Judging from the restricted distribution in East Greenland from 70°30' to 70°45'N (Scoresby Sund) it might be suggested that the species has recently immigrated from nearby Iceland where it is common.

West Palaearctic (Europe-western Siberia), low arctic-temperate.

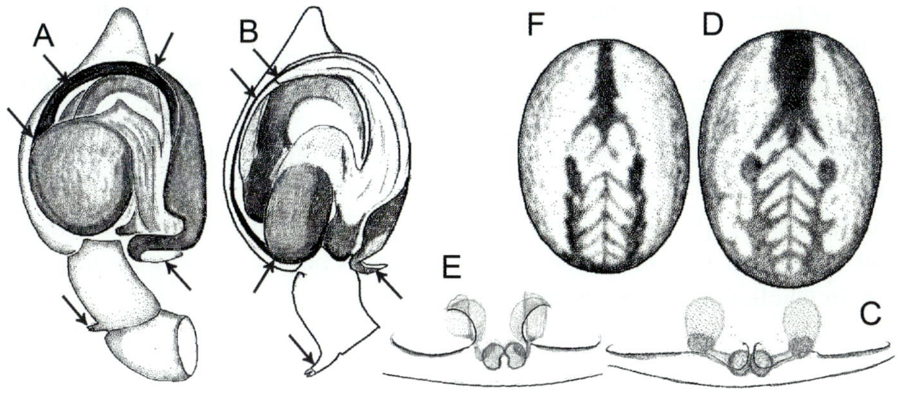

FIGURE 8 Dictyna major (*B–D*) *and* Emblyna borealis (*A, E, F*). *A, B, ♂ palp; C, E, epigyne; D, F, ♀ abdomen, dorsal. Arrows show diagnostic details.*

Haplodrassus signifer **(C.L. Koch, 1839). Figs 2J, 9A, C, 10**
Brændegård 1946: *Drassodes signifer* C.L. Koch, 1839.

Light colouration (yellow to yellow brown), rounded tip of tibial apophysis and epigyne with dark strongly sclerotized sides and without scape are diagnostic.

BIOLOGY. The species inhabits dry and warm localities without dense vegetation. According to Holm (1967) collected in *Empetrum* heath, herb field, snow beds with *Salix herbacea,* and in *Fucus* on the sea shore.

DISTRIBUTION. S, SW (north to Qeqertarsuaq/ Disko), SE.

Circumpolar, low arctic-temperate.

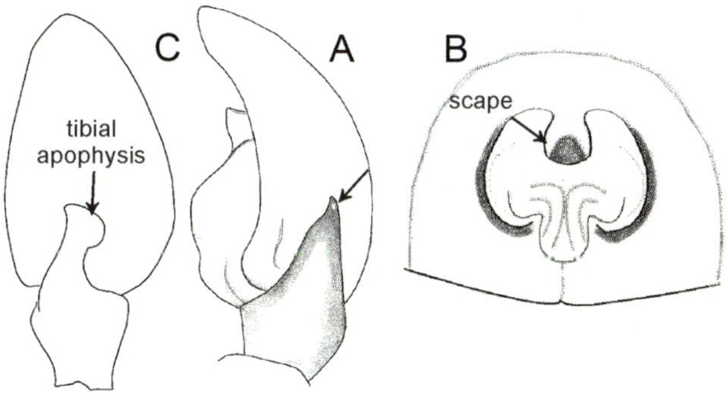

FIGURE 9 *Male palp of* Haplodrassus signifer (*A*) *and epigyne of* Gnaphosa lapponum (*B*).

HAHNIIDAE (COMB-TAILED OR LESSER COBWEB SPIDERS)

A small family with less than 250 species, but distributed worldwide. All Holarctic hahniids are ground dwellers. Members of this family can easily be distinguished by their spinnerets arranged in one row, and the wide spiracle placed between spinnerets and epigastric furrow.

Hahnia glacialis **Sørensen, 1898. Fig. 1R**
Marusik et al. 2006

BIOLOGY. The Species occurs in relatively warm places, *e.g.* beneath stones in lichen heaths.

DISTRIBUTION. S, SW (north to 69°46'N), SE.

NE Siberia - Nearctis, arctic-boreal.

FIGURE 10 Haplodrassus signifer *guarding egg cocoon in a retreat below a stone.* (PHOTO: J. BÖCHER)

LINYPHIIDAE (DWARF OR MONEY SPIDERS)

The second largest spider family with about 4500 species distributed worldwide. The highest diversity is found in the boreal (northern temperate) zone, but even in the Arctic it is a dominating family, where linyphiids occur in all terrestrial biotopes, although most inhabit litter. Linyphiids may hibernate in all stages and are often active during winter; even the copulation sometimes takes place in winter. It seems that some species may have more than one generation per year.

Linyphiids may only be confused with Theridiidae, especially the genus *Robertus*. Linyphiids have always a stridulation field on chelicera (sometimes difficult to observe) and have no serrated hairs on tarsus IV.

48 species have been found in Greenland, but it is very likely that several more may be found.

Because there are so many species of this family in Greenland, and most of them are less than 3 mm, it is rather difficult to observe and identify them. Besides shape of copulatory organs and of male carapace, there are several additional characters to be used such as 1) number of tibial spines on different legs, 2) presence or absence of trichobothrium on metatarsus IV, 3) position of trichobothrium on metatarsus I (Fig. 1H), 4) presence and size of stridulation field in chelicera, etc. All these characters are difficult for observation, and they are not used here, although indicated in the tables.

Key to Linyphiidae species and groups (see Table 8 for additional characters)

1. Abdomen with scutum 2
– Abdomen without scutum........................... 3
2. Carapace in ♂ unmodified, without raised parts and pits.......................... *Ceratinella ornatula*
– Head raised with sulci and pits, thorax with fine pits (Fig 11P)...................... ♂ of *Pelecopsis parallela*
3. Males 4
– Females... 6
4. Carapace modified: with raised or extended cephalic or thoracic parts, or strong spines, or thick or dense hairs 5
– Carapace unmodified......................... Table 3
5. Lateral side of carapace and chelicera with spines (Fig. 12M).................................... *Erigone*
– Carapace without spines..................... Table 2
6. Epigyne raised over rest of abdomen and extended behind epigastric furrow at least on half of its length (Figs 12D, 14M–P, 14V)........................ Table 4
– Epigyne not raised over rest of abdomen or if raised it is no extended behind epigastric furrow......... 7
7. Epigyne with one or more outgrowth or raised over other parts of abdomen (*e.g.*, Figs 11A, C, E, G, I, K, M, O, R, T, V, X, Z, 14H)........................... Table 5
– Epigyne without outgrowths 8
8. Epigyne with distinct fovea (depression) (Figs 14R, T) 9
– Epigyne without distinct fovea 10
9. Fovea wider than long (Fig. 14R), abdomen with pattern............................... *Neriene peltata*
– Fovea as long as wide (Fig. 14T), abdomen without pattern........................ *Porrhomma convexum*
10. Dorsal side of epigyne with distinct structures formed by separate sclerites .... Table 6
– Dorsal side of epigyne flat, without any furrow...... .... Table 7

Agyneta jacksoni **(Brændegård, 1937). Figs 14A, B, E**

Holm 1967: *Meioneta rurestris* (C.L. Koch).

Larsen & Scharff 2003: *Meioneta jacksoni* Brændegård, 1937.

Females rather similar to other *Agyneta* species, but can be separated by little difference between anterior and posterior median eyes, and by lighter colouration. Males of the two species can easily be separated by the shape of lamella characteristica.

BIOLOGY. According to Holm (1967) occuring sparsely in bogs and heaths. In Canada found in arctic tundra south to treeline and in alpine tundra; often under stones or in bogs (Saaristo & Koponen, 1998).

DISTRIBUTION. S, SW, SE.

Eastern Nearctic, low arctic and alpine.

Agyneta nigripes **(Simon, 1884). Figs 14C–E**

Brændegård 1946: *Micryphantes nigripes jacksoni* Brændegård, 1937.

Holm 1967: *Meioneta nigripes* (Simon).

Larsen & Scharff 2003: *Meioneta nigripes* (Simon, 1884).

Females very similar to those of *A. jacksoni*, but have darker colouration of legs, and prominent posterior median eyes larger than anterior one (1.5–2 times). Males can easily be separated from other Greenland species by the shape of lamella characteristica.

BIOLOGY. Seems to be fairly eurytopic in the Arctic. According to Brændegård (1946) it inhabits pronouncedly arid areas in East Greenland and was collected beneath sparse vegetation on a sandy shore in northern Canada (Saaristo & Koponen 1998).

DISTRIBUTION. SW, N, NE, SE. In some areas its range overlaps that of *A. jacksoni.*

Circumpolar, arctic and alpine; southernmost records in the Alps.

Bathyphantes eumenis **(L. Koch, 1879). Figs 14F, J**

Bathyphantes eumenoides Holm, 1967.

By the shape of the epigyne easily distinguished from other Greenland Linyphiidae.

BIOLOGY. Unknown for the Greenlandic population. In Siberia and Yukon Territory it occurs among stones in a wide range of biotopes with moist litter.

DISTRIBUTION. SW (north to Qeqertarsuaq/Disko).

Circumpolar, arctic-boreal.

Bolephthyphantes index **(Thorell, 1856). Figs 14G–I**

Brændegård 1946: *Bolyphantes index* Thorell, 1856.

Shape of copulatory organs (strong and blunt macroseta on palpal patella, bifid dorsal plate of epigyne),

TABLE 2 *Key to ♂ of Linyphiidae with modified carapace*

1) Rostrum present (r) or not. 2) Type of carapace outgrowth: ocular area (o), postocular (p), thoracic (t). 3) Presence of sulci (s) and cephalic pits (p). 4) Modified (short and dense, or long) hairs present (h). 5) Size of tibial apophysis: comparable to tibia or longer (l), shorter (s), massive (m) or absent (a); number if more than 1. 6) Tip of tibial apophysis: blunt (B), sharp (S), number if more than 1.

	1	2	3	4	5	6	Special characters
Bolephthyphantes index				h	a		Strong seta of palpal patella, abdominal pattern
Diplocephalus barbiger	r	o	sp	h	l	B	
Dismodicus decemoculatus	r	o	sp		s	S–B	
Hilaira vexatrix		t		h	sm		Metatarsal spines
Hilaira herniosa		t		h	l	BS	Metatarsal spines
Hybauchenidium gibbosum		t	sp	h	sm	BS	
Mecynargus borealis	r	o		h	a		
Mecynargus paetulus		o			s	B	
Metopobactrus prominulus		o			s	S	
Oreoneta frigida		p		h	l	B	Metatarsal spines
Pelecopsis parallela		o	sp		l	S	Abdominal scutum
Pocadicnemis americana		o			s	~S	
Praestigia groenlandica		o		h	l	S	
Scotinotylus alpinus	r	p	sp	h	s2	S2	
Scotinotylus evansi		o		h	s	S	
Scotinotylus sacer	r	p	sp	h	s	S	
Silometopoides pampia		o	p		l	B	
Typhochrestus latithorax		p		h	l2	BS2	
Walckenaeria c. brevicula		o			l	LB	
Walckenaeria castanea		o	sp		l	LS	
Walckenaeria clavicornis		o			l2	SB2	
Walckenaeria karpinskii		o			l2	BB2	

TABLE 3 *Key to ♂ of Linyphiidae with unmodified carapace*

1) Tibial apophysis present (p), L – large (>or = tibia). 2) Tibial apophysis: sharp (s), blunt (b). 3) Paracymbium complicated: with pockets, or spines (c) or simple (s). 4) Paracymbium higher than tibia (h) or smaller (s). 5) Abdominal pattern present (p) or absent (a). 6) Pre-epigastric field modified (p). 7) Abdomen color: dark (d), light (l) intermediate (i).

	1	2	3	4	5	6	7	Special characters
Agyneta jacksoni			c	h	a		ld	PME subequal to other eyes
Agyneta nigripes			c	h	a		d	PME much larger than other eyes
Bathyphantes eumenis			s	h	p-a		d	
Collinsia holmgreni	p	s	c	s	a		d	swollen palpal tibia
Collinsia spetsbergensis	p	s	s	s	a		d	
Collinsia thulensis			s	s	a		dl	

TABLE 3 *Key to ♂ of Linyphiidae with unmodified carapace* (cont.)

	1	2	3	4	5	6	7	Special characters
Diplocentria rectangulata	p	s	s	s	a		l	
Improphantes complicatus			c	h	a		l	
Islandiana princeps	p	b	c	h	a		l	paracymbium with 3 strong setae
Lepthyphantes turbatrix			c	h	p		d	cymbium near base with triangle outgrowth, abdominal pattern
Mecynargus morulus	p	b	s		a	P	d	
Mecynargus sphagnicola			c	s	a		d	
Neriene peltata			s	s	p		i	
Oreonetides vaginatus			c	h	a		l	
Porrhomma convexum			s	h	a		l	
Sciastes extremus	pL	s	s	s	a		l	
Semljicola obtusus	p	b	s	s	a	P	ld	
Sisicus apertus			s	s	a			very short carapace
Tiso aestivus	pL	b	s	h-s	a		d	long palpal patella, and tibial apophysis
Typhochrestus pygmaeus	p	b	s	s	a		?l	
Wabasso quaestio	p	b	s	s	a		i	Metatarsus I with spines

TABLE 4 *Key to ♀ of Linyphiidae with extended epigyne*
1) Scape present (p) or absent (a). 2) Abdominal pattern present (p). 3) Coloration dark (d) or light (l).

	1	2	3	Special characters
Hilaira herniosa	a		d	scape as wide as long
Improphantes complicatus	p		l	scape as wide as long
Lepthyphantes turbatrix	p	p	d	scape triangular
Oreonetides vaginatus	p		l	scape longer than wide

TABLE 5 *Key to ♀ of Linyphiidae with epigynal outgrowth or epigyne with raised parts*
1) Epigyne with scape (s) or septum (e), number if more than 1. 2) Epigyne divided by furrow into two parts (f). 3) Epigyne with raised (r) parts (extending over the plate of epigyne). 4) Abdominal pattern present (p) or absent (a). 5) Coloration of abdomen: dark (d), light (l).

	1	2	3	4	5	Special characters
Bathyphantes simillimus	S2		r	p-a	d	2 scapes, upper one finger-like
Bolephthyphantes index	S2		r	p	l	bifid (bilobed) and broad upper scape-plate
Diplocentria rectangulata				a	l	small outgrowth with socket
Diplocephalus barbiger		f		a	d	posterior parts with club
Hilaira vexatrix			r	a	o	raised, opening visible in caudal view

	1	2	3	4	5	Special characters
Hybauchenidium gibbosum	E		r	a	d	epigyne divided by vertical septum
Scotinotylus alpinus	S		r	a	l	scape as long as wide
Scotinotylus evansi	S		r	a	l	scape as long as wide
Scotinotylus sacer	S		r	a	l	scape longer than wide
Silometopoides pampia			r	a	d	central transverse part partly raised
Walckenaeria clavicornis		f		a	d	epigynal lobe with bulge
Walckenaeria karpinskii		f		a	d	epigynal lobes with almost parallel sides

TABLE 6 *Key to ♀ of Linyphiidae with epigynes lacking outgrowth, but with distinct furrows*
1) Septum present (p). 2) Epigyne with "filled fovea" (central part) well separated from sides and up (p), "fovea" partly subdivided (s). 3) Shape of "fovea": M – M-shaped (or V-shaped), R – round, H –heart-shaped, T – trapezoidal, L – lense-shaped; P – pear-shaped; A – transverse subdivided. 4) Median plate visible in undissected epigyne (p). 5) Coloration of abdomen: dark (d), light (l).

	1	2	3	4	5	Special characters
Agyneta jacksoni	p	~ps	M		ld	PME ~ other eyes
Agyneta nigripes	p	~ps	M		d	PME > other eyes
Collinsia holmgreni		p	R		d	
Collinsia spetsbergensis					d	
Collinsia thulensis		p	H	~p	dl	
Dismodicus decemoculatus		ps	~R		d	
Erigone spp.					d	Surface with strong or fine transversal furrows, edge strongly chitinized, chelicera with denticles on promarginal surface
Hilaira vexatrix					o	
Mecynargus borealis		p	T		d	
Mecynargus morulus		ps	M		d	
Mecynargus sphagnicola		p	L		d	
Metopobactrus prominulus					ld	
Oreoneta frigida		p?s	A	p	~	Triangle outgrowth, median plate with 2 pits
Pelecopsis parallela				p	d	3 time wider than high
Pocadicnemis americana				p	d	
Sciastes extremus					l	
Semljicola obtusus		p	Δ		ld	
Silometopoides pampia					d	
Sisicus apertus	p	ps	H		l	Carapace broad (Fig. 18Y)
Tiso aestivus				p	d	Semioval
Typhochrestus latithorax		ps	M		d	
Typhochrestus pygmaeus		p	P		?l	
Walckenaeria c. brevicula		ps	A	p	d	Transversal
Walckenaeria castanea		ps	A	p	d	Transversal

TABLE 7 *Key for ♀ of Linyphiidae with simple epigynes (lacking outgrowth or furrows)*
1) Shape of dorsal plate of epigyne triangle (T). 2) Distance between receptacula larger or smaller than diameter (Ø) of receptacula. 3) Coloration of abdomen: dark (d), light (l) or intermediate (i).

	1	2	3	Special characters
Diplocentria rectangulata	T	<Ø	l	
Hilaira vexatrix			i	
Islandiana princeps	T	>Ø	l	median plate triangle; V-shaped pattern in epigyne
Mecynargus paetulus			d	
Metopobactrus prominulus		>Ø	ld	V-shaped pattern in epigyne
Praestigia groenlandica			d	ocular area with small horn-cornicule

TABLE 8 *Additional characters for separation of Linyphiidae*
1) Total length, mm. 2) Position of trichobothrium on metatarsus I ♂/♀*. 3) Dorsal tibial spines on legs I–IV. 4) Presence of trichobothrium on metatarsus IV.

	1	2	3	4
Agyneta jacksoni	1.6–2.2	0.22	2222	+
Agyneta nigripes	1.6–2.2	0.21–0.27	2222	+
Bathyphantes simillimus	2.1–2.6	0.27	2222	–
Bolephthyphantes index	2.5–3.4	0.15–0.20	2222	–
Ceratinella ornatula	1.8–2.2	0.40	1111	+
Collinsia thulensis	1.7–2.2	0.39–0.50	2221	–
Collinsia holmgreni	1.9–2.5	0.56–0.63	2221	–
Collinsia spetsbergensis	1.8–2.4	0.58–0.62	2221	–
Diplocentria rectangulata	1.2–1.45	0.35–0.44	2221	–
Diplocephalus barbiger	1.7–1.8	0.5	2211	–
Dismodicus decemoculatus	1.7–2.0	0.8–0.95	1111/0000	+
Erigone arctica arctica	2.6–3.2	0.4–0.5	2221	
Erigone arctica soerenseni	2.6–3.0	0.52	2221	–
Erigone psychrophila	2.0–3.1	0.4–0.5	2221	–
Erigone tirolensis	2.1–2.8	0.4–0.5	2221	–
Erigone whymperi	2.3–3.5	0.47–0.51	2221	–
Hilaira herniosa	4.0–4.6	0.64–0.65	2222	+
Hilira vexatrix	2.7–3.5	0.79–0.85	2222	+
Hybauchenidium gibbosum	2.5–2.8	0.67–0.7	2211	+
Improphantes complicatus	1.7–2.2	0.18–0.22	2222	–
Islandiana princeps	1.5–1.9	0.51	2221	–
Lepthyphantes turbatrix	2.0–2.4	0.22–0.23	2222	–
Mecynargus borealis	1.2–1.6	0.41–0.52	2221/0	–
Mecynargus morulus	1.6–2.0	0.68–0.77	2222/0	–
Mecynargus paetulus	1.6–2.1	0.45–0.5	2221/0	–
Mecynargus sphagnicola	1.4–1.8	0.44	2221	–
Metopobactrus prominulus	1.4–2.0	0.8–0.5	1111	+
Neriene peltata	2.2–3.7	0.12–0.20	2222	–
Oreoneta frigida	2.7–4.0	0.6–0.7	2222	+

	1	2	3	4
Oreonetides vaginatus	3–3.8	0.3–0.39	2222	–
Pelecopsis parallela	1.3–2.0	0.5–0.59	1111/0000	–
Pocadicnemis americana	1.7–2.2	0.8–0.95	1111	+
Porrhomma convexum	2.2–3.2	0.4–0.49	2222	–
Praestigia groenlandica	2.1–2.4	0.82/0.88	2211	+
Scotinotylus alpinus	2.0–2.7	0.43–0.5	2221	–
Scotinotylus evansi	1.75–1.23	0.45–0.55	2221	–
Scotinotylus extremus	1.8–2.25	0.5/0.51	2222	+
Scotinotylus sacer	1.6–1.8	0.56/0.4	2221	–
Semljicola obtusus	1.4–1.5	0.56/0.57	2221	–
Silometopoides pampia	1.5–2.0	0.8	1111	–
Sisicus apertus	1.2–1.4	0.35–0.5	2222	–
Tiso aestivus	1.5–2.2	0.5–0.59	1111	–
Tarsiphantes latithorax	1.45–1.9	0.5	2222	+
Typhochrestus pygmaeus	1.7–1.9	No data	2221	+
Wabasso quaestio	1.3–1.65	0.45/0.4	2221	–
Walckenaeria castanea	2.5–3.9	Ca 0.44	2211	+
Walckenaeria c. brevicula	2.0–2.5	0.5/0.5	2211	+
Walckenaeria clavicornis	2.0–2.8	Ca 0.5	2211	+
Walckenaeria karpinskii	2.0–3.0	Ca 0.51	2211	+

* position of trichobothrium on metatarsus I is expressed as decimal fraction a/b in Fig. 10H.

abdominal pattern, and strong hairs on cephalic part of the male carapace are diagnostic. Cephalic hairs somewhat similar to those in *Thymoites oleatus* (Theridiidae), but ocular area without transverse furrow.

BIOLOGY. Unknown in Greenland. In eastern Siberia it occurs in open places, *e.g.* among grass in small meadows close to birch coppice or ericaceous heaths.

DISTRIBUTION: SW. Only one Greenland record (65°50'N).

Palaearctic, low arctic-boreal, northernmost record in Polar Ural (about 67°N).

Ceratinella ornatula **(Crosby & Bishop, 1925). Fig. 1A**

Marusik *et al.* 2006: *Ceratinella ornatula* Crosby & Bishop, 1925.

Abdominal scutum in both sexes and lack of cephalic lobes in male are diagnostic.

BIOLOGY. Unknown.

DISTRIBUTION. nSW (only known from Qeqertarsuaq/Disko, where 2 specimens were receently found).

Nearctic, boreal.

Collinsia holmgreni **(Thorell, 1871). Figs 13C, D**

Brændegård 1940: *Coryphaeolana holmgreni* Thorell, 1871.

Separated from other Greenland species by the shape of copulatory organs.

BIOLOGY. In western Greenland most typically in snow beds, but also found on moist, moss-covered ground in heaths and bordering brooks (Holm 1967). In Chukotka (northeastern extremity of Asia) it also inhabits places with moist ground, although in North Cisokhotia restricted to *Chosenia-Populus* forests along rivers. Here a rather common and abundant species.

DISTRIBUTION. S, SW, NW, NE, SE.

Circumpolar, arctic and alpine.

Collinsia spetsbergensis **(Thorell, 1871). Figs 14A, B**

Brændegård 1946: *Microerigone spitsbergensis* Thorell, 1871.

Recognized by long transparent tapering lamella in male palp, epigyne with long transverse furrow and lack of median furrow.

BIOLOGY. Appears to be very common in some Greenland areas. In Scandinavia exclusively in lichen vegetation above 1400 m a.s. In Ellesmere Island specimens were found in river deltas with fine, muddy sand covered with dense *Equisetum* and some *Salix*. Adults may be found throughout the season, but do not seem to overwinter (Leech 1966).

DISTRIBUTION. nSW (Qeqertarsuaq/Disko only), N, NE, SE (north of 65°35'N).

Circumpolar, arctic (southernmost in Iceland).

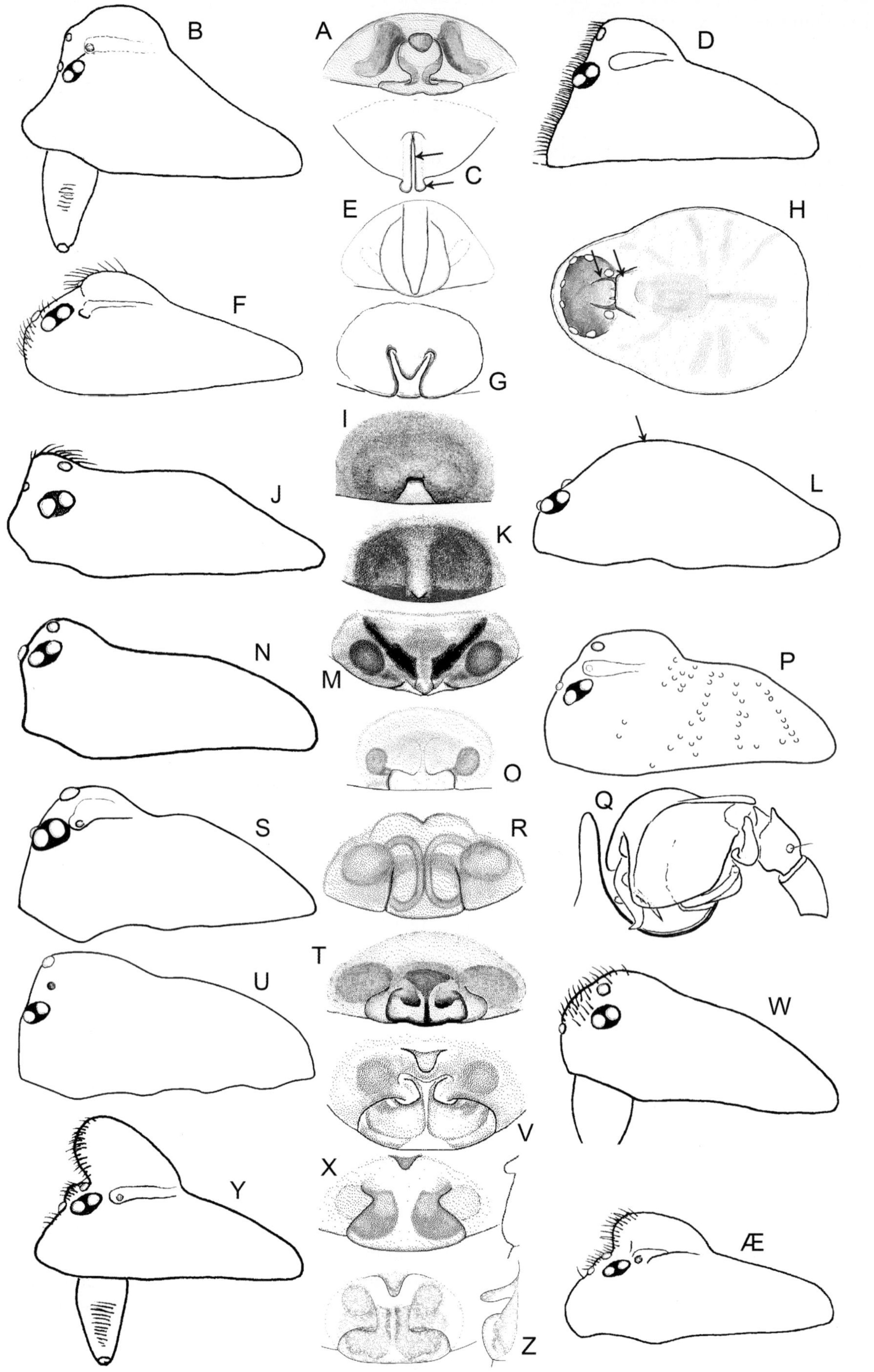

FIGURE 11 *Distinguishing characters in Linyphiidae species with modified male carapace. Q: ♂ palp; C, E, G, I, K, M, O, R, T, V, X, Z: epigyne, ventral; remaining figures are carapace, dorsal. A, B,* Dismodicus decemoculatus*; C, D,* Diplocephalus barbiger*; E, F,* Hybauchenidium gibbosum*; G, H,* Tarsiphantes latithorax*; I, J,* Mecynargus borealis*; K, L,* M. paetulus*; M, N,* Metopobactrus prominulus*; O, P,* Pelecopsis parallela*; Q–S,* Pocadicnemis americana*; T, U,* Silometopoides pampia*; V, W,* Scotinotylus evansi*; X, Y,* S. alpinus *and Z, Æ,* S. sacer.

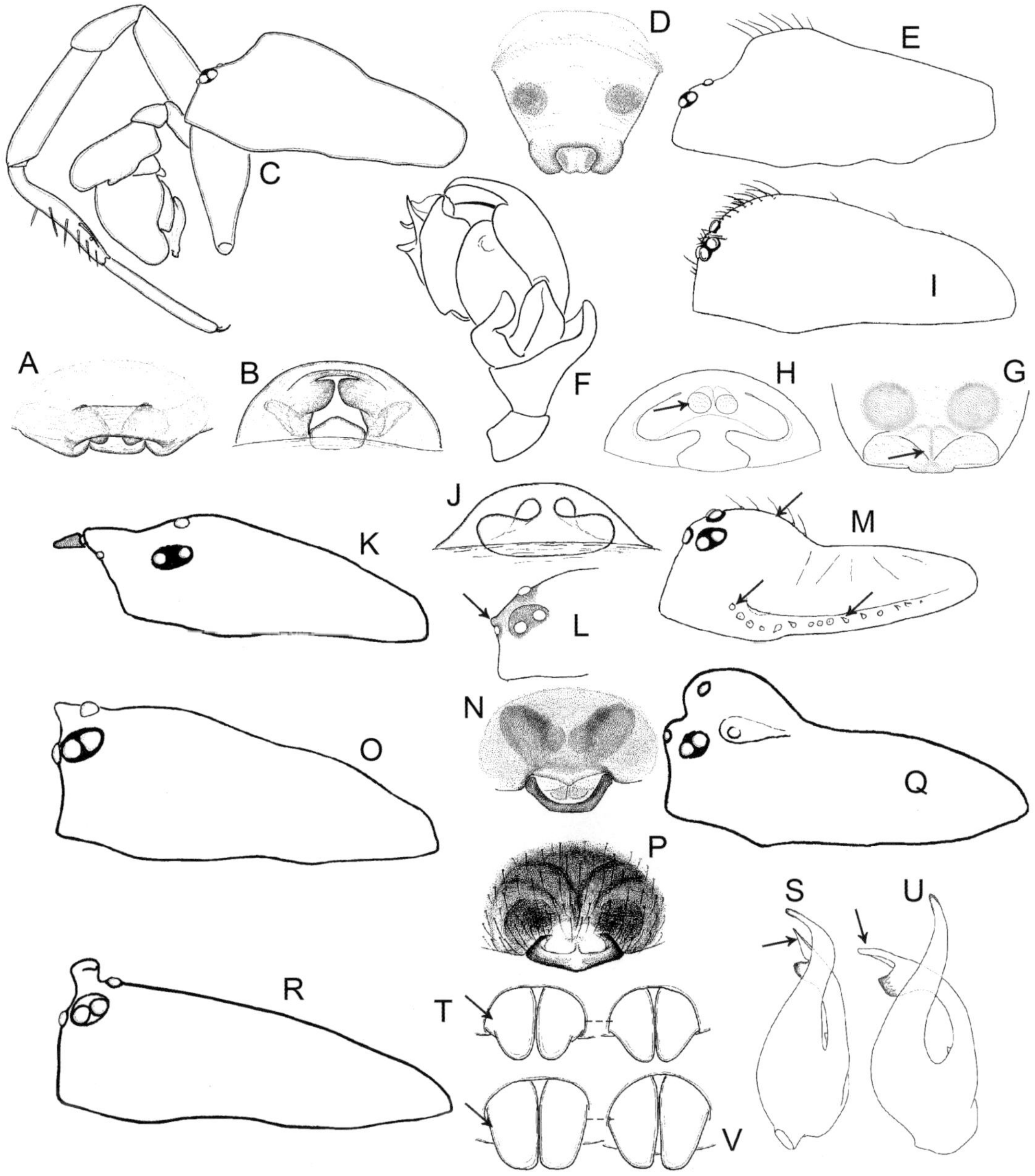

FIGURE 12 *Distinguishing characters in Linyphiidae species with modified ♂ carapace. F, S, U, ♂ palp; A, D, G, N, P, T, V, epigyne ventral; B, H, J, epigyne caudal; C, E, I, K, M, O, Q, R, ♂ carapace; L, ♀ carapace. A–C,* Hilaira vexatrix*; D, E,* H. herniosa*; F–I,* Oreoneta frigida*; J–L,* Praestigia groenlandica*; M,* Erigone *sp.; N, O,* Walckenaeria cuspidata brevicula*; P, Q,* W. castanea*; R, type characteristic of both* W. clavicornis *and* W. karpinskii*; S, T,* W. clavicornis*; U, V,* W. karpinskii.

***Collinsia thulensis* (Jackson, 1934). Figs 13E, F**

Brændegård 1946: *Coryphaeolana thulensis* Jackson, 1934.

Resembles *C. holmgreni*, but can easily be separated by shape of epigyne and more swollen palpal tibia.

BIOLOGY. Collected under stones in North and East Greenland (Brændegård 1960). In Ellesmere Island common on gravely parts of river deltas with scanty vegetation of *Dryas integrifolia* and *Salix arctica*. It is mostly active during the early season, where breeding season is restricted to the end of June. Adults do not seem to overwinter. In captivity it was fed with mites and collemboles, but refused flies (Leech 1966).

DISTRIBUTION. NW, N, NE (in Greenland only north of 70°50'N).

Nearctic and Greenland-Spitsbergen, arctic. Only Alaskan records south of 70°N.

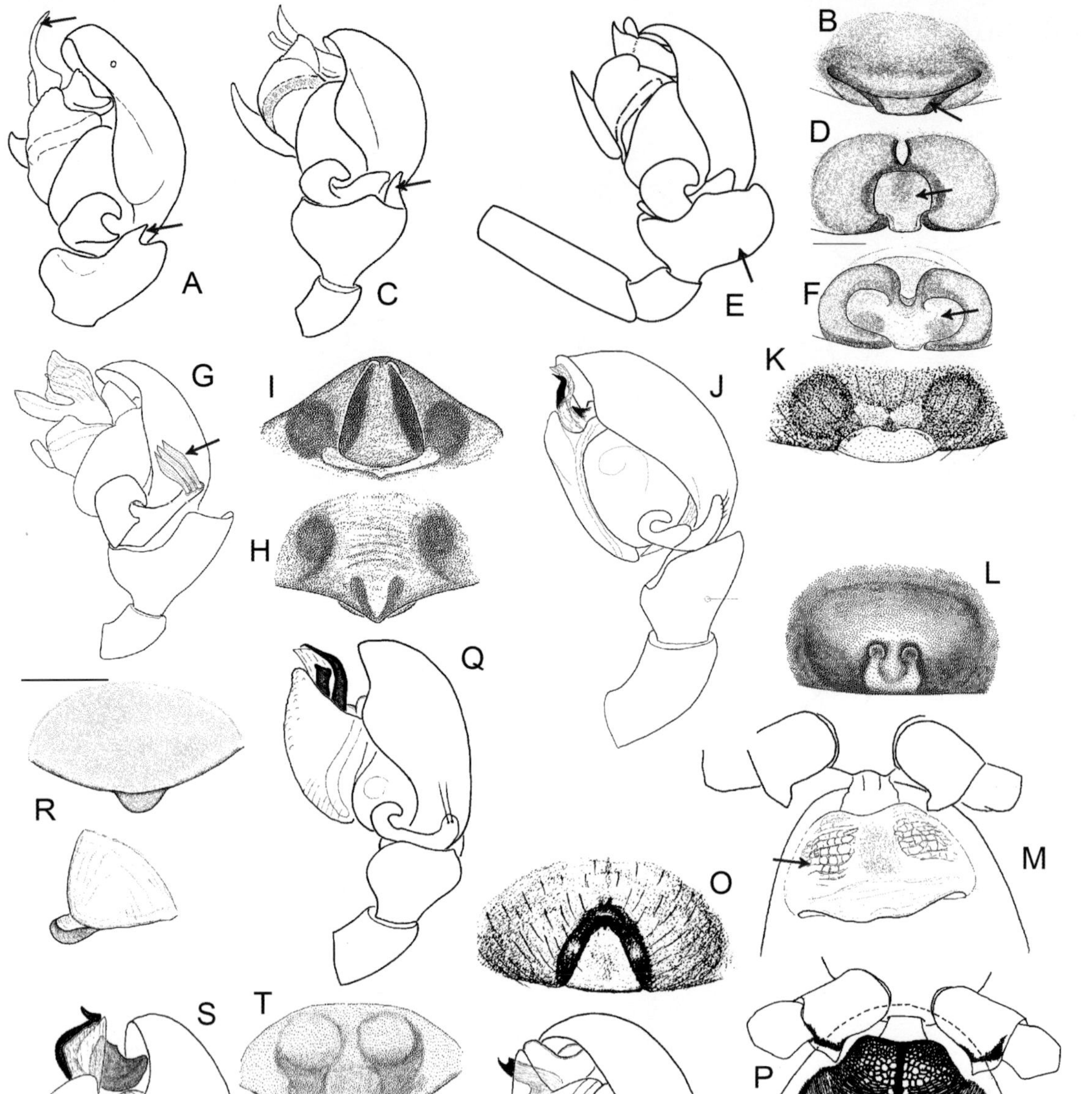

FIGURE 13 *Distinguishing characters in Linyphiidae with unmodified ♂ carapace. A, C, E, G, J, N, Q, S, U, Y, Æ, ♂ palp; B, D, F, H, I, K, L, O, R, T, V–X, Z, Ø, epigyne ventral; M, P: ♂ abdomen, ventral. A, B,* Collinsia spetsbergensis*; C, D,* C. holmgreni*; E, F,* C. thulensis*; G–I,* Islandiana princeps*; J, K,* Mecynargus sphagnicola*; L, M,* M. morulus*; N, P,* Semljicola obtusus*; Q, R,* Wabasso quaestio*; S, T,* Diplocentria rectangulata*; U–X,* Typhochrestus pygmaeus*; Y, Z,* Tiso aestivus*; Æ, Ø,* Sciastes extremus.

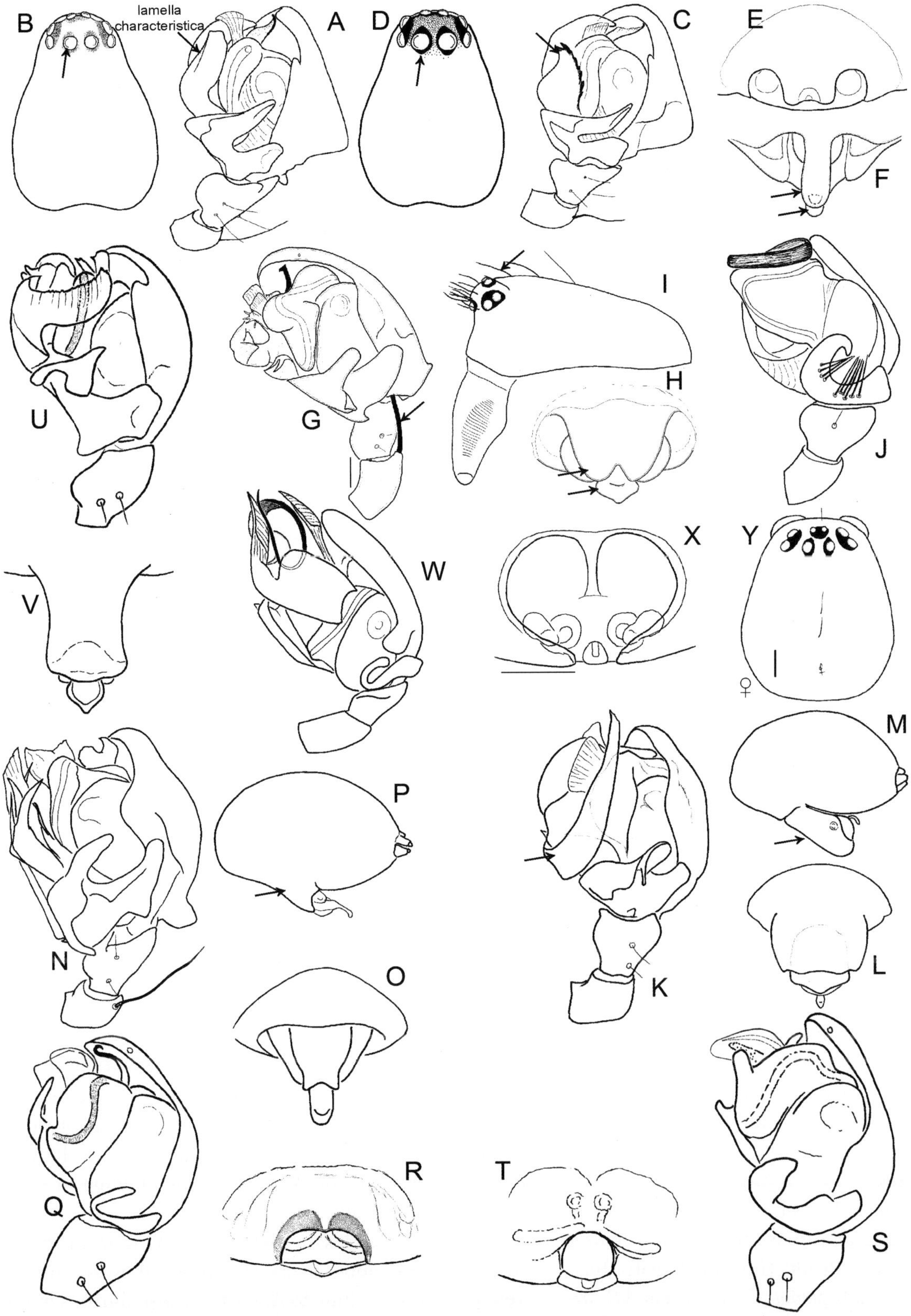

FIGURE 14 *Distinguishing characters in linyphiids. A, C, G, J, N, Q, S, U, W, ♂ palp; E, F, H, L, O, R, T, V, X, epigyne ventral; B, D, Y, ♀ carapace, dorsal; I, ♂ carapace and P, M: ♀ abdomen, lateral. A, B,* Agyneta jacksoni*; C, D,* A. nigripes*; E, type characteristic of both* A. jacksoni *and* A. nigripes*; G–I,* Bolephthyphantes index*; F, J,* Bathyphantes eumenis*; K–M,* Improphantes complicatus*; N–P,* Lepthyphantes turbatrix*; Q, R,* Neriene peltata*; S, T,* Porrhomma convexum*; U, V,* Oreonetides vaginatus*; W, X:* Sisicus apertus.

***Diplocentria rectangulata* (Emerton, 1914). Figs 14S, T**
Marusik *et al.* 2006: *Diplocentria rectangulata* (Emerton, 1915).

Epigyne is similar to that in *Wabasso quaestio* and may be distinguished by lack of pit on the tip of epigynal plate. Male palp has a characteristic, small spine-like lateral tibial apophysis.

BIOLOGY. Litter dweller. In continental North America and Europe adults occur throughout the season.

DISTRIBUTION. sSW. Only one specimen (a female) is known from a single Greenland locality (Nuup Kangerlua/Godthåbsfjord: Kapisillit).

Circumpolar, boreal.

***Diplocephalus barbiger* (Roewer, 1955). Figs 11C, D**
Holm 1967: *Diplocephalus barbatus* (L. Koch).

Males can be distinguished by the shape of carapace, females by peculiar outgrowths of epigyne.

BIOLOGY. In Ellesmere Island it was found on gravely sections of river deltas with scattered surface vegetation. According to Leech (1966) the webs are built in cracks in the ground, rarely on the surface. The highest activity was observed around June 20.

DISTRIBUTION. NW. In Greenland only known from a single locality (Etah).

Arctic, but absent from N. Europe; most of the records are north of 70°N.

***Dismodicus decemoculatus* (Emerton, 1882). Figs 11A, B**
Holm 1967: *Dismodicus bifrons decemoculatus* (Emerton).

Larsen & Scharff 2003: *Dismodicus variegatus* Jackson, 1938.

Strongly raised cephalic part of carapace and the shape of epigyne are characteristic.

BIOLOGY. In willow thickets, where it often occurs on the leaves and in the litter (Holm 1967).

DISTRIBUTION. SW (north to Qeqertarsuaq/Disko).

Nearctic, temperate. – It is likely that *D. decemoculatus* is conspecific with the Palaearctic *D. bifrons* (Blackwall, 1841). In that case the range is circumpolar.

***Erigone* Audouin, 1862**
Males of *Erigone* can be fairly easily separated by the shape of palpal femur, patella and tibia shown in the figures. Females are much more difficult. Three forms with thickened edge of epigyne (2 subspecies of *E. arctica* and *E. tirolensis*) may be separated by means of inner structure. Although males of *E. whymperi* and *E. psychrophila* are rather different in shape of patellar apophyses and base of femur, their females are almost indistinguishable. Ranges of these species are partly overlapping, and therefore their identification can not be definitive if males are lacking.

Key to Erigone species

1. Male 2
– Female.. 5
2. Dorsal tibial apophysis has no lamella, or if it has it is unbroken (Figs 15A, D)........................ *arctica*
– Dorsal tibial apophysis lamella-shaped, lamella broken (Figs 15F, I, L) 3
3. Palpal patella with long, gradually tapering turned apophysis as long as patella............ *psychrophila*
– Patellar apophysis not turned, shorter than patella (Figs 15I, L) 4
4. Palpal femur strongly turned near the base, palpal apophysis almost 2 times shorter than palp, upper half of femur with 3–4 spines (including apical) (Fig. 15L).................................. *whymperi*
– Palpal femur straight, patellar apophysis almost as long as patella, upper half of femur with 6 spines (Fig. 15I) .... *tirolensis*
5. Epigyne weakly sclerotised, covered with numerous fine wrinkles/furrows (Figs 15G, H, M, N) 6
– Epigyne strongly sclerotised at least near edge, surface with few furrows (Figs 15B, C, E)...............7
6. Receptacula converging-parallel, upper edge placed over receptacula (Figs 15G, H) .... *psychrophila*
– Receptacula diverging, upper edge placed just over receptacula (Figs 15M, N)................. *whymperi*
7. Receptacula touching or almost touching each other, median plate rectangular (Figs 15B, C, E)..... *arctica*
– Receptacula spaced by more than one receptaculum diameter, median plate triangular (Figs 15J, K)....... .... *tirolensis*

***Erigone arctica arctica* (White, 1852). Figs 15A, C**
Brændegård 1940: *Erigone sibirica orientalis* Brændegård, 1940.

Elongate, not broken dorsal tibial apophysis and closely spaced receptacula in epigyne are diagnostic. The subspecies can be distinguished by the shape of copulatory organs. *E. a. arctica* has longer tibial apophysis, relatively shorter receptacula and larger epigyne opening in comparison to median plate.

BIOLOGY. Collected by Berlese sampling from moors and meadows along rivers (Brændegård 1940). Adults were present from June to August and also during winter. The species appears able to overwinter in different stages, and the life cycle is probably extended over several years (Brændegård 1946).

DISTRIBUTION. N, NE, sSE. In western Greenland replaced with *E. a. soerenseni*.

Northeastern Siberia – Nearctic, arctic.

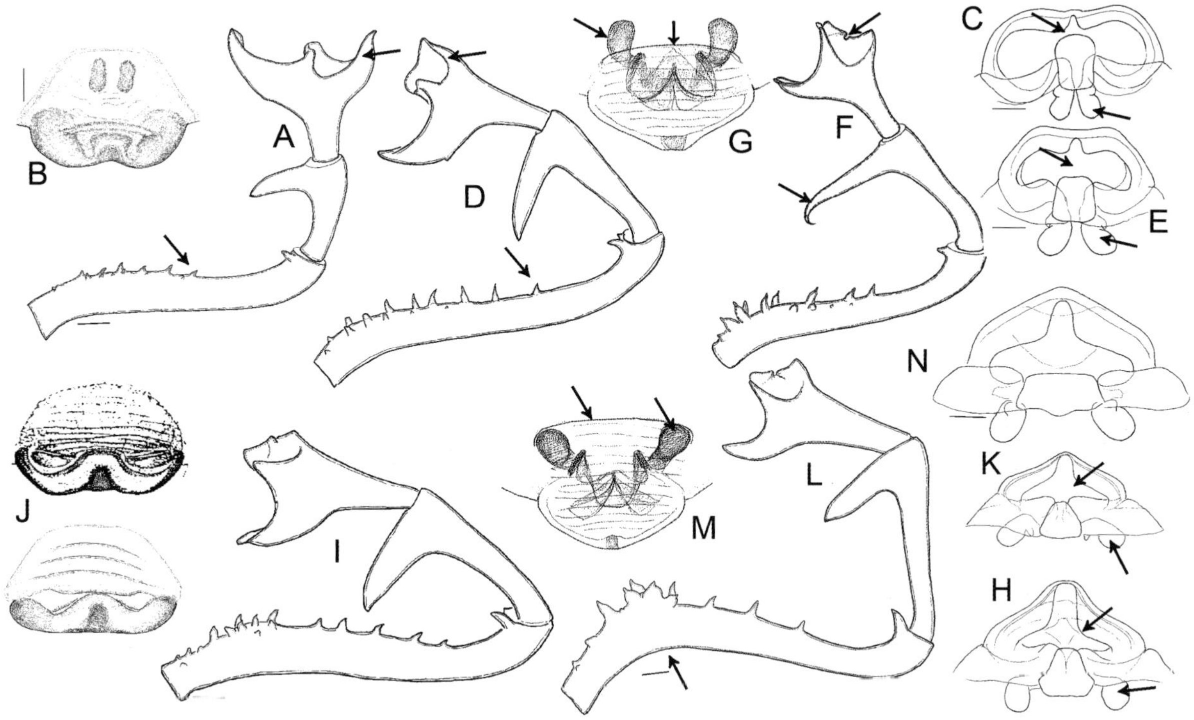

FIGURE 15 *Distinguishing characters in* Erigone *spp. A, D, F, I, L, ♂ palp; B, G, J, M, epigyne ventral; C, E, H, K, N, epigyne caudal. A, C,* E. arctica arctica*; B, D–E,* E. a. soerenseni*; F–H,* E. psychrophila*; I–K,* E. tirolensis*; L–N,* E. whymperi.

Erigone arctica soerenseni **Holm, 1956. Figs 15B, D, E**
Marusik *et al.* 2006

Elongate, not broken dorsal tibial apophysis and closely spaced receptacula in epigyne are diagnostic for the species. The subspecies can be distinguished by the shape of copulatory organs. *E. a. soerenseni* has shorter dorsal tibial apophysis, relatively larger receptacula and smaller epigyne opening in comparison to median plate.

BIOLOGY. Reported under stones on *Salix herbacea* covered soil near the seashore (Holm 1967).

DISTRIBUTION. S, sSW. The subspecies seems trestricted to southern West Greenland.

NOTE. In North America so far only the nominative subspecies of *E. arctica* is known. It is thus interesting that a separate, possibly endemic subspecies (*E. a. soerenseni*) is found in adjacent western Greenland, and that *E. a. arctica* occurs in East Greenland.

Erigone psychrophila **Thorell, 1871. Figs 15F–H**
Marusik *et al.* 2006

Males may be separated from Greenland congeners by long, tapering and curved patellar apophysis. Females are similar to *E. whymperi* and almost indistinguishable. With some uncertainty females of *E. psychrophila* may be recognised by the position of upper edge over receptacula, converging receptacula and longer furrow of fovea.

BIOLOGY. On Ellesmere Island (Leech 1966) restricted to vegetated, marshy areas at the edges of ponds and quiet streams and to water-saturated vegetated slopes. The species, including adults, overwinters in vegetation and can often be found moving under water in slush snow during spring melt. According to Holm (1958) the species prefers dry places in Spitsbergen. In upper Kolyma (northeast Siberia) it is very abundant on pebbly banks of small creeks.

DISTRIBUTION. NW, N, NE, SE.

Circumpolar, arctic and alpine. Southernmost records in Scotland.

Erigone tirolensis **L. Koch, 1872. Figs 15I–K**
Marusik *et al.* 2006

Males of this species can be confused with *E. whymperi*, but may be separated by the shape of palpal dorsal apophysis and the number of ventral femoral spines in palp. *E. tirolensis* has almost straight femur, while *E. whymperi* has strongly turned femur near the base. Females can be distinguished from the congeners by the shape of vulva and the width of sclerotized plate, which is thick (broad) in *E. tirolensis* and thin in *E. whymperi.*

BIOLOGY. Unknown.

DISTRIBUTION. In Greenland only known from the northern half of SE (65°30'N to 70°50'N).

Circumpolar, arctic and alpine (Scotland and the Alps).

***Erigone whymperi* O. P.-Cambridge, 1877. Figs 15L–N**
Marusik *et al.* 2006: *Erigone whymperi* O. Pickard-Cambridge, 1877.

Males similar to *E. tirolensis*, but may be separated by the shape of dorsal tibial apophysis, strongly turned palpal femur, and fever ventral femoral spines. Females very similar to *E. psychrophila* and only distinguished with some uncertainty by diverging receptacula, upper edge of epigyne placed over receptacula, and shorter furrow of fovea.

BIOLOGY. In moist, open localities like mires and moss cover adjoining brooks and tarns as well as on heaths with low *Salix* vegetation. Numerous specimens have also been collected by sifting *Fucus* wrack on the seashore (Holm 1967).

DISTRIBUTION. S, SW (north to 70°32'N), SE (north to 70°50'N).

General range unclear, but most probably restricted to Greenland and eastern Nearctic.

[*Erigone atra* Blackwall, 1833

Listed in Holm's (1967) table, in the column corresponding to West Greenland species. It was never mentioned in the text, and has not been found among the Greenland museum material. It has accordingly been excluded from the list.]

[*Erigone penessa* Thorell, 1878

Described from Greenland and since then never found. It may be a synonym of *E. whymperi* or other erigonids. The lack of types does not allow to synonymise it.]

***Hilaira herniosa* (Thorell, 1875). Figs 1E, 12D–E**
Marusik *et al.* 2006

From other Greenland *Hilaira* it may be separated by the shape of male carapace, presence of long, dark tapering apophysis in male palp, and extended epigyne.

BIOLOGY. In Siberia it occurs in different habitats with moist litter (*Sphagnum* bogs, litter under *Salix*). Adults remain active even at -6°C, and a 50% mortality is found at -25°C.

DISTRIBUTION. SW (62° to 63°13'N).

Circumpolar, low arctic-boreal and alpine.

***Hilaira vexatrix* (O. P.-Cambridge, 1877). Figs 12A–C**
Brændegård 1946: *Hilaira curvitarsis* Sørensen, 1898.

From other *Hilaira* species distinguished by shape of male carapace, palp without long tapering outgrowth, and the epigyne which is not extended.

BIOLOGY. According to Holm (1967) the species prefers moist places with a fairly luxuriant field layer and is never found under stones. It was collected by sifting moss and by pitfall trapping in dwarf-schrub heaths and willow scrub. Brændegård (1946, 1960) describes highly differing preferences: arid conditions, most frequently under stones. On Ellesmere Island (Leech 1966) it occurs at ponds and small streams with a dense vegetation and many rocks, under which it crawls to overwinter, also in 1–2 cm deep cracks in the ground. The species is active on the ground as soon as the soil temperature is above freezing, even though the air temperature is well below zero. There are different peaks of activity: In June courting and mating take place and the females lay eggs, and a second peak is found in late July until the end of August and most probably refers to the second generation. Egg-cocoons contain 8–11 eggs. A great number of immatures die by cannibalism.

DISTRIBUTION. nSW, NW, N, NE.

Eastern Siberia-Nearctic, arctic. In Siberia east of Lena River delta (71°35'N) and north of the Arctic Circle in Eastern Yakutia and Chukotka.

***Hybauchenidium gibbosum* (Sørensen, 1898). Figs 13E, F**
Holm 1967: *Hybocoptus gibbosus* (Sørensen).

Males may be distinguished from all linyphiids by the shape of carapace, and females by the peculiar epigyne.

BIOLOGY. Unknown in Greenland. Holm (1967) collected a single female by sifting seaweed. In Canada reported in moss. A common species in Northeastern Siberia, where it occurs within the forest belt in various biotopes with meso- or hygrophytic vegetation. Adults appear about mid August.

DISTRIBUTION. S, sSW (northernmost 64°30'N).

Northeastern Siberia-Nearctis, boreal.

***Improphantes complicatus* (Emerton, 1882). Figs 14K–M**
Brændegård 1946: *Lephthyphantes audax* Sørensen, 1898.

Holm 1967: *Leptyphantes complicatus* (Emerton).

Distinguished by the shape of the male palp and a very large epigyne.

BIOLOGY. Among moss and litter on heaths and in willow thickets. It occurs both in very wet places like *Sphagnum* bogs, and under stones in dry wind-eroded lichen heaths.

DISTRIBUTION. S, SW (north to Disko), southern SE.

Circumpolar, arctic-temperate and alpine.

***Islandiana princeps* Brændegård, 1932. Figs 13G–I**
Marusik *et al.* 2006

Male palp, female epigyne and pale colouration distinguish this species.

BIOLOGY. Holm (1967) collected the species under stones on dry ground with scattered vegetation, and by pitfall trapping in a *Harrimanella –Phyllodoce* heath at a snow bed.

DISTRIBUTION. S, SW (north to Qeqertarsuaq/Disko), NE (north to 74ºN).

Nearctic and Iceland, arctic-boreal.

Lepthyphantes turbatrix **(O. P.-Cambridge, 1877). Figs 14N–P, 5B, 22A**

Marusik *et al.* (2006): *Leptyphantes turbatrix* (O. Pickard-Cambridge, 1877).

Male palp and female epigyne are characteristic.

BIOLOGY. Collected chiefly under stones in different types of heaths and in snow beds with *Salix herbacea* (Holm 1967).

DISTRIBUTION. S, SW (north to 70°32'N), SE (north to 60°25'N).

Nearctic, low arctic-boreal.

Mecynargus borealis **Jackson, 1930. Figs 11I, J**

Brændegård 1946: *Typhochraestus borealis* Jackson.

Holm 1967: *Conigerella borealis* (Jackson).

Males easily separated from other linyphiids by small size and conical elevation of ocular area. Shape of epigyne and small size allow identification of females.

BIOLOGY. Most specimens were collected by sifting mosses and lichens in different types of heaths (Holm 1967).

DISTRIBUTION. SW (north to Disko), SE, southern half of NE.

Holarctic, arctic.

Mecynargus morulus **(O. P.-Cambridge, 1873). Figs 13L, M**

Brændegård 1946: *Rhaebothorax morulus* Cambridge, 1873.

Holm 1967: *Rhaebothorax morulus* (O. P.-Cambridge).

Males may be distinguished from congeners by special reticulation of genital area and by shape of the palp. Epigyne simple but very characteristic.

BIOLOGY. In Greenland common in *Hylocomium* moss in heaths (Holm 1967).

DISTRIBUTION. S, SW (north to 67ºN), SE.

Palaearctic, east to Polar Ural, arctic-boreal and alpine.

Mecynargus paetulus **(O. P.-Cambridge, 1875). Figs 1K, L**

Holm 1967: *Rhaebothorax paetulus* (O. P.-Cambridge).

Males may be distinguished from congeners and from the similar *Semljicola obtusus* by concave thoracic part. Shape of epigyne is species specific.

BIOLOGY. Collected under stones on moist ground or among *Sphagnum*.

DISTRIBUTION. S, SW (north to Qeqertarsuaq/Disko).

Circumpolar, arctic-boreal.

Mecynargus sphagnicola **(Holm, 1939). Figs 13J, K**

Holm 1967: *Rhaebothorax sphagnicola* Holm, 1939.

Both males and females may easily be separated from congeners by the shape of copulatory organs.

BIOLOGY. The rarest species of *Mecynargus* in Greenland, where it is found in moist localities, frequently among *Sphagnum* in bogs (Holm 1967).

DISTRIBUTION. S, SW (north to Qeqeretarsuaq/Disko).

Circumpolar, arctic-boreal and alpine. In Europe restricted to Scandinavia.

Metopobactrus prominulus **(O. P.-Cambridge, 1872). Figs 11M, N**

Marusik *et al.* 2006: *Metopobactrus prominulus* (O. P.-Cambridge, 1872).

Males have slightly raised cephalic part. Epigyne is rather variable in general appearance, but differ from all other species.

BIOLOGY. Found in both dry and rather moist heaths and among litter in *Salix* thickets (Holm 1967).

DISTRIBUTION. SW (known only from Qeqertarsuaq/Godhavn and Kangerlussuaq/Søndre Strømfjord).

Circumpolar, low arctic-southern temperate.

Neriene peltata **(Wider, 1834). Figs 14Q, R**

Holm 1967: *Linyphia peltata* Wider & Reuss.

Both sexes may be identified by specific colouration and shape of copulatory organs.

BIOLOGY. In Great Britain found in a wide range of biotopes including grassland, heathland, marshes, scrub, woodland and gardens where it occurs on the ground in litter, in vegetation and in low undergrowth (Harvey *et al.* 2002).

DISTRIBUTION. nSW. Only one female collected in Søndre Isortoq (65°20'N). Because members of this genus are not found in the Arctic, and this species occurs in Fennoscandia only south of 65ºN, the record in Greenland is probably caused by mislabeling or occasional introduction (Holm 1967).

W. Palaearctic, temperate (Europe and Caucasus south to Turkey and Armenia).

Oreoneta frigida **(Thorell, 1872). Figs 12F–I**

Brændegård 1946: *Hilaira frigida* Thorell, 1872.

Holm 1967: *Hilaira frigida* (Thorell).

Larsen & Scharff 2003: *Hilaira frigida* (Thorell, 1872).

Males are separated from other large linyphiids by convex thoracic part and shape of palpal tibial apophysis. Females may be recognised by wide epigyne without extensions like in *Hilaira herniosa*, or the clear fovea, like in *Hilaira vexatrix*.

BIOLOGY. According to Holm (1967) found under stones in snow beds with *Salix herbacea*, in moist dwarf-shrub heaths and in meadows. It is often abundant among mosses in mires and also occurs in the litter layer of willow scrub. One of the most abundant spiders in Greenland.

DISTRIBUTION. S, SW, NW (north to 73°45'N), NE (north to 72°50'N), SE.

Amphiatlantic, arctic-boreal. Seems to be restricted to Greenland and Norway.

Oreonetides vaginatus **(Thorell, 1872). Figs 14U, V**
Brændegård 1946: *Oronetides vaginatus* (Thorell, 1872).

Holm 1967: *Oreonetides vaginata* (Thorell).

Large size, light colouration, extended epigyne and shape of the palp are characteristic.

BIOLOGY. Colleted under stones and by pitfall trapping in snow beds with *Salix herbacea,* under stones and among moss in moist heaths, and among litter in willow scrub (Holm 1967).

DISTRIBUTION. S, SW (north to Qeqertarsuaq/Disko).

Circumpolar, low arctic-boreal and alpine.

Pelecopsis parallela **(Wider, 1834). Figs 11O, P**
Marusik *et al.* 2006

Males may be distinguished by small pits arranged in radial rows, raised cephalic part and small abdominal scutum. Females are recognised by the shape of epigyne and cephalic punctuation.

BIOLOGY. In Britain this species is reported from a variety of biotopes, including moss, detritus in woods, under stones and especially in both calcareous and acid grasslands.

DISTRIBUTION. nSW (found in one locality only, Akulliit/Mellemfjord on Qeqertarsuaq/Disko: 69°42'N, 54°35'W).

Palaearctic, arctic-boreal: from Europe northeast to Chukotka Peninsula.

Pocadicnemis americana **Millidge, 1976. Figs 11Q–S**
Holm 1967: *Pocadicenis pumila* (Blackwall).

Larsen & Scharff 2003: *Pocadicnemis pumila* (Blackwall, 1841).

It was earlier thought that the Greenland population belonged to the Palaearctic *P. pumila* (Blackwall, 1841).

Male palp and epigyne are very characteristic in this species. Epigyne is rather variable, but due to longitudinal ducts it can not be misidentified.

BIOLOGY. Collected in pitfall traps in bogs, *Empetrum* heaths and herb fields, also by sifting moss and litter in *Salix* thickets – always in small numbers (Holm 1967).

DISTRIBUTION. S, SW (north to Qeqertarsuaq/Disko), southern SE.

Nearctic, low arctic-boreal.

Porrhomma convexum **(Westring, 1851). Figs 14S, T**
Marusik *et al.* 2006

Copulatory organs are characteristic.

BIOLOGY. One couple was collected under stones at a brook (Koponen 1982).

DISTRIBUTION. S, sSW.

Nearctic-W. Palaearctic, temperate.

Praestigia groenlandica **Holm, 1967. Figs 12J–L**
Marusik *et al.* 2006

Males may be recognised by the peculiar shape of carapace with cornicule and cap, and females by very small horn (cornicule) in ocular area. Epigyne has no special features (fovea or furrows), but may be recognised by lifting the epigyne and observe the posterior side.

BIOLOGY. In Greenland collected in various, but always moist localities: *Carex-* and *Carex-Eriophorum* bogs, heaths with mosses and low *Salix glauca*, snow beds with *Salix herbacea*, and litter in willow scrubs (Holm 1967).

DISTRIBUTION. SW (only three known localities from 65°17'N north to Qeqertarsuaq/Disko).

Subendemic of Greenland: SW Greenland, Northern Labrador and Ellesmere, arctic.

Sciastes extremus **Holm, 1967. Figs 13Æ, Ø**
Marusik *et al.* 2006

Light colouration and shape of copulatory organs are characteristic.

BIOLOGY. Prefers moderately moist localities and occurs in luxuriant *Empetrum-Vaccinium uliginosum* heaths with low *Salix*, among litter in willow scrubs, in *Cassiope tetragona-Phyllodoce* heaths and snow beds with *Sibbaldia procumbens* and *Salix herbacea*.

DISTRIBUTION. S, SW (north to Qeqertarsuaq/Disko).

The species is only known from Greenland.

Scotinotylus alpinus **(Banks, 1896). Figs 1D, 11X, Y**
Brændegård 1946: *Coryphaeolana lapidicola* (Sørensen).

Holm 1967: *Cochlembolus alpinus* (Banks).

Males distinguished by the high outgrowth (lobe) of cephalic part. Females very similar to *S. sacer*, but separated by the relative length of epigynal scape and shape of median plate.

BIOLOGY. Found under stones in very dry, wind-eroded heaths as well as on moss covered soil of snow beds (Holm 1967).

DISTRIBUTION. S, SW (north to Qeqertarsuaq/Disko), SE.

Nearctic and Siberia east of Yenisei River, arctic-boreal.

Scotinotylus evansi **(O. P. –Cambridge, 1894). Figs 11V, W**
Brændegård 1946: *Caledonia evansi* Cambridge, 1894.

Shape of carapace, male palpal tibia and epigynal plate are specific.

BIOLOGY. Generally found under stones.

DISTRIBUTION. SE (north of 65°35'N) and southernmost NE (north to 71°23'N).

Palaearctic (only Europe west of Ural), arctic and alpine.

Scotinotylus sacer **(Crosby, 1929). Figs 11Z, Æ**
Holm, 1967: *Cochlembolus sacer* Crosby.

Shape of male carapace and relative length of epigynal scape eparate this species from other congeners.

BIOLOGY. Rarely found under stones, but collected in following biotopes: litter in a willow thicket, herb cover on a slope, *Empetrum-Vaccinium uliginosum* heath and *Carex* bog (Holm 1967).

DISTRIBUTION. nSW (only Qeqertarsuaq/Godhavn).

Siberia and Nearctis, arctic-boreal.

Semljicola obtusus **(Emerton, 1914). Figs 13N–P**
Holm 1967: *Latithorax obtusus* (Emerton).

Males recognised by special reticulation over the genital area and palp. The epigyne is similar to that in *Typhochrestus pygmaeus.*

BIOLOGY. Occurs among mosses in different rather moist biotopes: luxuriant dwarf shrub heaths, at brooks or in litter in *Salix* thickets; also found abundantly in a snow bed dominated by *Salix herbacea* (Holm 1967). In Hudson Bay islands it inhabits stony and mossy tundras (Koponen 1992). In Quebec it was found in *Sphagnum* bogs and heaths, and in Alaska in spruce forests (Saaristo & Eskov 1996).

DISTRIBUTION. S, SW (north to Qeqertarsuaq/Disko).

Nearctic, low arctic and alpine.

Silometopoides pampia **(Chamberlin, 1948). Figs 11T, U**
Marusik *et al.* 2006

Distinguished by male carapace with raised cephalic part and a pair of pits, long tibial apophysis and shape of epigyne.

BIOLOGY. On Ellesmere Island it occurs on south- and southwest facing densely vegetated slopes which are permanently water-saturated (Leech 1966).

DISTRIBUTION. NW (only found at Pittufik/Thule Airport).

Chukotka and Nearctic, arctic (only known west of Chukotka).

Sisicus apertus **(Holm, 1939). Figs 14W–Y**
Marusik *et al.* 2006

Male palp and shape of epigyne are very characteristic.

BIOLOGY. In Fennoscandia it occurs especially in *Sphagnum* bogs. Also reported among moss (especially *Hylocomium*) in *Betula* forests.

DISTRIBUTION. SW.

Siberia + Nearctis, boreal.

Tarsiphantes latithorax **(Strand, 1905). Figs 1C, 11G, H**
Marusik *et al.* 2006: *Tarsiphantes latithorax* Strand, 1905, stat.rev.

Shape of carapace, tibial apophysis and epigyne distinguish this species.

BIOLOGY. On Ellesmere Island distributed throughout soggy, vegetated areas and near pond edges. Has been collected under water in slush snow at spring melt and on wet, south facing slopes and depressions with sedges or mosses (Leech 1966).

DISTRIBUTION: NW (only known from Pittufik/Thule Airbase, but a wide occurrence in northern Greenland is very probable).

Holarctic (not in Europe), arctic (southernmost in upper Kolyma, ca 62°N).

Tiso aestivus **(L. Koch, 1872). Figs 13Y, Z**
Marusik *et al.* 2006

Lack of cephalic pits allows separating males from the other species with long tibial apophysis, namely *Silometopoides pampia.* Females can de distinguished by the shape of epigyne.

BIOLOGY. Inhabits rather moist places like snow beds with *Sibbaldia-Salix herbacea* as well as herb fields in dry localities and lichen heaths on mountain slopes (Holm 1967).

DISTRIBUTION. S, SW (north to Qeqertarsuaq/Disko), SE (north to 70°40'N).

Circumpolar, arctic-boreal.

Typhochrestus pygmaeus **(Sørensen, 1898). Figs 13U–X**
Marusik *et al.* 2006

Male palp and epigyne distinguish this species.

BIOLOGY Found in a *Empetrum-Vaccinium uliginosum* heath with low *Salix glauca* near the seashore (Holm 1967). Seems to be very rare in Greenland.

DISTRIBUTION. S, SW, NW (north to 74°44'N).

Nearctic, arctic.

Wabasso quaestio **(Chamberlin, 1948). Figs 13Q, R**
Holm 1967: *Diplocentria replicata* Holm, 1967.

Male palp and epigyne distinguish this species.

BIOLOGY. Both in very moist localities as *Sphagnum* bogs and in dry lichen heaths. May also inhabit litter in dense *Salix glauca* thickets. (Holm 1967).

DISTRIBUTION. SW (north to Qeqertarsuaq/Disko).

Eastern Nearctic, boreal.

Walckenaeria castanea **(Emerton, 1882). Figs 12P, Q**
Marusik *et al.* 2006

Male carapace, shape of tibial apophysis and epigyne distinguish this species.

BIOLOGY. Two females were found under stones by a small river (Koponen, 1982).

DISTRIBUTION. S (only found once at Narsaq).

Nearctic north of 40°N, temperate.

Walckenaeria clavicornis (Emerton, 1882). Figs 12R–T

Holm 1967: *Cornicularia clavicornis* Emerton.

Very similar to *W. karpinskii.* Males may be separated by the shape of dorsolateral tibial apophysis, which is sharply pointed in *W. clavicornis* and blunt in *W. karpinskii.* Females may be distinguished by the epigyne, but chiefly by looking at series: *W. clavicornis* has epigynal plates with strong lateral bulge at the base, while in *W. karpinskii* there are no bulges and sides of plates are almost parallel.

BIOLOGY. Mosses in bogs and luxuriant heaths (Holm 1967).

DISTRIBUTION. nSW, NW, southern NE.

Nearctic and northeastern Siberia, arctic and alpine, but exact range is not clear.

Walckenaeria cuspidata brevicula (Crosby & Bishop, 1931). Figs 12N, O

Holm 1967: *Cornicularia cuspidata* (Blackwall).

Males can be recognised by small horn (cornicule) in eye area and females by the shape of epigyne.

BIOLOGY. In the Qeqertarsuaq/Disko area all specimens were found in herb field vegetation on slopes (Holm 1967).

DISTRIBUTION. Only known from 2 localities in SW (Kangerluarsuk Tulleq and Qeqertarsuaq/Godhavn).

Nearctic, low arctic.

Walckenaeria karpinskii (O. P.-Cambridge, 1873). Figs 12R, U, V

Holm 1967: *Cornicularia karpinskii* (O.P. -Cambridge).

Very similar to *W. clavicornis.* Males may be separated by the shape of dorsolateral tibial apophysis, which is sharply pointed in *W. clavicornis* and blunt in *W. karpinskii.* Females may be distinguished by epigyne, but chiefly by looking at a series: *W. clavicornis* has epigynal plates with strong lateral bulge at the base, while in *W. karpinskii* there are no bulges and sides of plates are almost parallel.

BIOLOGY. Has been found among moss and under stones in rather moist dwarf-shrub heaths and by sifting *Sphagnum* in a mire (Holm 1967). On Ellesmere Island it lives in cracks in the ground and goes out only if the humidity is over 90% (Leech 1966).

DISTRIBUTION. S, southernmost SW and SE. The range does not seem to overlap with that of the sibling species, *W. clavicornis.* – Although a southern species in Greenland, it occurs in the High Arctic on Ellesmere Island (81°49'N).

Circumpolar, arctic-boreal and alpine.

LYCOSIDAE (WOLF SPIDERS)

The fourth largest spider family with 2400+ known species and worldwide distribution. Most are wandering spiders, although in temperate areas many are burrowing. Some wolf spiders, chiefly in subtropical and tropical areas, are web builders. All boreal and arctic species are wandering.

Lycosidae have four small eyes in 1 anterior row and 2 pairs of much larger eyes behind. Body shape is very uniform in wolf spiders, although some are more robustly built than others.

Lycosids may easily be recognised by their fast running away when approached. Also the habit of females to carry their whitish egg-cocoons attached to spinnerets is distinguishing. However, the females of *Alepecosa* and *Arctosa* dig a small hole in the soil and stay there with the cocoon until the youngs hatch from the eggs. The Greenland species may generally be recognised in the field by their colour pattern. However, the most important taxonomic characters are the shape of the copulatory organs.

Key to species and genera according to somatic characters

1. Carapace without pattern, abdomen with light heart mark, body stumpy, chelicera with 3 teeth on the inner row .... *Arctosa*
– Carapace with median and submarginal light bands, abdomen with light or dark heart mark 2
2. Carapace with broken submarginal band 3
– Submarginal band unbroken 6
3. Abdomen with light variegated pattern, chelicera with 2 retromarginal teeth .... *Alopecosa*
– Abdomen without variegated pattern, chelicera with 3 retromarginal teeth 4
4. Median band on carapace poorly visible, on groove only .... *Pardosa groenlandica*
– Median band of carapace clear 5
5. Total length 8–11 mm, carapace > 3.8 mm .... *Pardosa albomaculata*
– Total length 6–8 mm, carapace < 3.5 mm .... *Pardosa glacialis*
6. Abdomen with light median band, carapace < 3 mm .... *Pardosa hyperborea*
– Abdomen without light median band, carapace < 3 mm .... *Pardosa furcifera*

Key to species by copulatory organs

1. Males 2
– Females 9

2. Embolus unhidden by tegulum and tegular apophysis (Figs 16A, C, E, G, I) 3
– Embolus hidden by tegulum and tegular apophysis (Figs 16K, M, O) 7
3. Tegular apophysis stretching horizontally (width > height) (Figs 16C, E, G, I) 4
– Tegular apophysis stretching vertically (width = or < height) 6
4. Tegular apophysis gradually tapering (Fig. 16C) .... *Pardosa albomaculata*
– Tegular apophysis not gradually tapering (Fig. 16E, I) 5
5. Tegular apophysis with kind of saddle, terminal apophysis hooked (Fig. 16E) .... *Pardosa glacialis*
– Tegular apophysis without saddle, terminal apophysis finger-like (Fig. 16I) .... *Pardosa groenlandica*
6. Tegular apophysis triangular (Fig. 16G) .... *Pardosa furcifera*
– Tegular apophysis about rhomboidal (Fig. 16A) .... *Pardosa hyperborea*
7. Cymbium without claws ...... *Alopecosa exasperans*
– Cymbium with 2 claws, abdomen with white lanceolate spot 8
8. Palea with distinct spine (Fig. 16K) .... *Arctosa insignita*
– Palea without spine (Fig. 16M) .... *Arctosa alpigena*
9. Epigynal plate (septum) wider than high 10
– Epigynal septum higher than wide 12
10. Septum fused with upper part of epigyne by stem (Fig. 16P) .... *Alopecosa exasperans*
– Septum (plate) not connected with the upper part of epigyne 11
11. Epigynal plate transversal, rectangular, width/height = 3/1 (Fig. 16N) .... *Arctosa alpigena*
– Epigynal plate trapeziodal-triangular, width/height = 2/1 (Fig. 16L) .... *Arctosa insignita*
12. Fovea distinct 13
– Fovea absent 15
13. Fovea bottle- (vase-) shaped, with distinct margins (Fig. 16D) .... *Pardosa albomaculata*
– Fovea not bottle-shaped 14
14. Fovea with distinct margins, about circular (or as inverted heart) (Fig. 16J) ...... *Pardosa groenlandica*
– Fovea about triangular, with distinct margins at the base only (Figs 16H) .... *Pardosa furcifera*
15. Septum as high as wide (Figs 16B) .... *Pardosa hyperborea*
– Septum higher than wide, spade-shaped (Figs 16F) .... *Pardosa glacialis*

***Alopecosa exasperans* (O.P. -Cambridge, 1877). Figs 16O–Q, 18D.**
Holm 1967: *Tarentula exasperans* (Cambr.).

This single representative of the genus is easily recognised by the shape of copulatory organs and variegated pattern.

BIOLOGY. In Peary Land collected on a mountain slope and in *Dryas* vegetation (Brændegård 1960). On Ellesmere Island (Leech 1966) it is the most heliophilic species among the spider species found. Collected mainly on dry south- and southwest facing slopes near clumps of *Dryas integrifolia. A. exasperans* was never found together with *Pardosa glacialis.* Peak of activity was observed around July 2, and first adults appeared June 20. The largest number of males were found between June 20 and July 10, later only single specimens. The first egg cocoons observed on July 6. Cocoons contained about 70 eggs and were laid in small digs about 2 cm deep. There are 7 or 8 instars and the whole life cycle may last 6 or 7 years. In experiments the species prefered small Diptera, though Collembola can be eaten if caught. Blow flies (*Phormia* and *Protocalliphora* spp.) were refused even when the wings were cut off, perhaps because *A. exasperans* has very small chelicera for a spider of its size.

DISTRIBUTION. NW, N. The species has been reported from Qeqertarsuaq/Disko and Nuup Kangerlua/Godthåbsfjord, but in both cases probably based on mislabelling.

Nearctic, high arctic.

Key to Greenland 'Arctosa'

1. Males 2
– Females 3
2. Palea with spine .... *insignita*
– Palea without spine .... *alpigena*
3. Epigynal plate transversal, rectangular, width/height = 3/1 .... *alpigena*
– Epigynal plate trapeziodal-triangular, width/height = 2/1 .... *insignita*

***Arctosa alpigena* (Doleschall, 1852). Figs 16M, N**
Marusik *et al.* 2006: *"Arctosa" alpigena* (Doleschall, 1852).

Very similar to *A. insignita.* Both species can be recognised from other Greenland lycosids by the peculiar lanceolate white band on abdomen. In *A. alpigena* the length of the band is 1/3 of abdomen, in contrast to almost ½ in *A. insignita.* The species may also be distinguished by the shape of copulatory organs: epigynal plate transversal and 3 times wider than high in *A. alpigena*, triangular in *A. insignita*; palea (retroapical part of bulbus) without

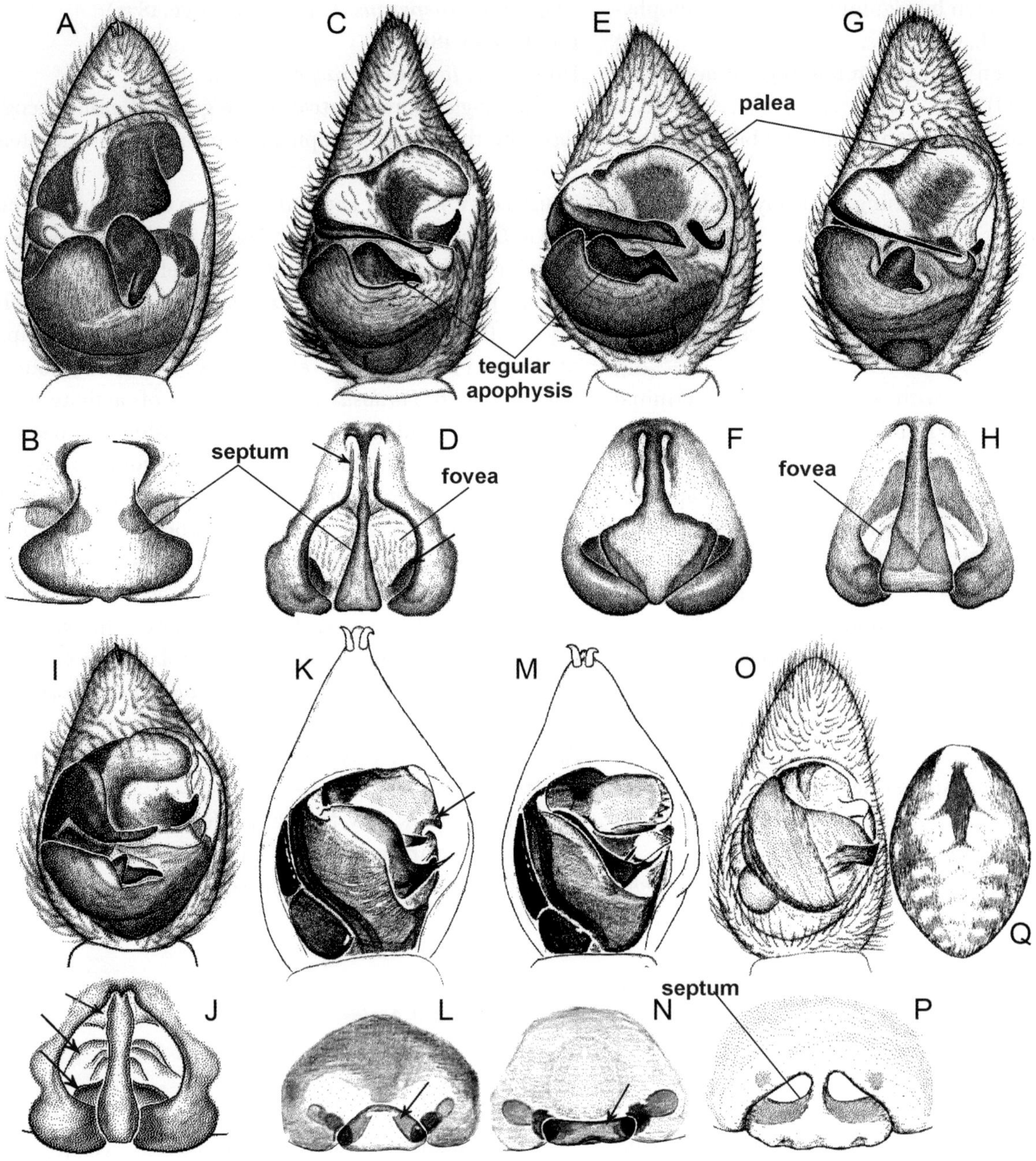

FIGURE 16 *Distinguishing characters in wolf spiders. A, C, E, G, I, K, M, O, ♂ palp; B, D, F, H, J, L, N, P, epigyne ventral; Q: ♀ abdomen, dorsal. A, B,* Pardosa hyperborea*; C, D,* Pardosa albomaculata*; E, F,* P. glacialis*; G, H,* P. furcifera*; I, J,* P. groenlandica*; K, L,* Arctosa insignita*; M–N,* A. alpigena*; O–Q,* Alopecosa exasperans.

triangular outgrowth in *A. alpigena*. The two (sibling?) species also have separate ranges in Greenland, with *A. alpigena* in East Greenland and *A. insigita* in West Greenland (however, with a small overlap of the two species in S).

BIOLOGY. At Kangerlussuaq (Blosseville Kyst) collected under stones at about 200 m a.s.l. (Brændegård 1946). In Chukotka it inhabits differernt biotopes from ericaceous and *Betula* heaths to open and dry arctic steppe.

DISTRIBUTION. S, SE (north to Hekla Havn, 70°30'N). A record from Kangerlussuaq/Søndre Strømfjord may be caused by a misidentification.

Circumpolar, low arctic-boreal and alpine. One of the most widespread Greenland lycosids.

Arctosa insignita **(Thorell, 1872). Figs 16K, L, 18C**

Marusik *et al.* 2006: *"Arctosa" insignita* (Thorell, 1872).

Distinguishing characters: see *A. alpigena*.

FIGURE 17 Pardosa groenlandica, *female with egg cocoon.* (PHOTO: J. BÖCHER)

BIOLOGY. Most common in bogs and moist heaths, but may also occur in dry localities like *Empetrum*- and lichen heaths on mountain slopes (Holm 1967).

DISTRIBUTION. S, SW (north to Qeqertarsuaq/Disko). Ranges of the two *Arctosa* species are slightly overlapping in S.

Nearctic, low arctic and alpine. Occurs also in Chukotka Peninsula.

Pardosa albomaculata **Emerton, 1885. Figs 16C, D**

Similar to *P. groenlandica* from which it may be distinguished by longer embolus, sharply pointed bifid conductor, and flask-shape of epigynal fovea (neck-less in *P. groenlandicola*).

BIOLOGY. Only 2 specimens reported at elevations 145 and 560 m. It inhabits stony beaches and stone piles, sometimes at the foot of glaciers (Dondale & Redner 1990)

DISTRIBUTION. S. It is likely that this species was formerly misidentified.

Nearctic, arctic-alpine.

Pardosa furcifera **(Thorell, 1875). Figs 2B, 16G, H, 18G**

Tegular apophysis as wide as high and epigyne with long septum but without distinct margins of the fovea are characteristic.

BIOLOGY. On rather moist ground and in *Salix herbacea* snow beds as well as in the moss layer of heaths adjoining brooks. Also in herb fields on slopes and in *Salix glauca* thickets (Holm 1967).

DISTRIBUTION. S, SW.

Nearctic, low arctic and alpine.

Pardosa glacialis **(Thorell, 1872). Figs 16E, F**

Marusik *et al.* 2006

Shape of epigynal septum, terminal apophysis and long transversal tegular apophysis are characteristic.

BIOLOGY. In heaths as well as in bogs. As a rule it is not found together with *P. groenlandica* or *P. furcifera* (Holm 1967). On Ellesmere Island *P. glacialis* occurs everywhere except in windy places. It overwinters in soil cracks and under stones. The highest activity of males was observed at the end of June. In females two activity peaks were observed: around June 24 and at the end of August. Egg cocoons contain about 50 eggs. There are 7 instars, and each instar lasts one year, so the whole life cycle takes about 6 years. In an experiment, *P. glacialis* fed most readily on Chironomidae, and when hungry also on smaller cyclorrhaphous flies. Blow flies and other large flies were rejected. The species is highlycannibalistic, but has never been observed eating other spider species. *P. glacialis* is a prey for buntings (*Emberiza*) and knot (*Calidris canutus*). At least 1% of specimens investigated was infected by *Hexamermis* (Nematoda), although the rate of parasitism may be higher (Leech 1966)

FIGURE 18 *Habitus of* Thanatus arcticus (*A,F*), Pardosa groenlandica (*B,H*), P. furcifera (*G*), Arctosa insignita (*C*), Alopecosa exasperans (*D*) *and* Pardosa hyperborea (*E*).

DISTRIBUTION. G (*P. glacialis* is the spider species with largest Greenland range).

Nearctic, arctic. Also known from easternmost Chukotka.

Pardosa groenlandica **(Thorell, 1872). Figs 2C, 16I, J, 17, 18E**
Marusik *et al.* 2006

Recognised by long epigynal septum and well developed margins of roundish fovea, undivided conductor and thick finger-like terminal apophysis.

BIOLOGY. The typical biotope in Greenland is herb fields on mountain slopes as well as *Salix herbacea* and *Sibbaldia procumbens* communities of snow beds (Holm 1967). It is also found under stones in heaths and at the seashore. According to Hammel (2005), *P. groenlandica* is the lycosid species found at highest altitude on Lyngsmarksfjeldet at Qeqertarsuaq/Godhavn, namely 800 m.a.s., 200 m above the highest reaching *P. glacialis*. The life cycle seems to last several years (Brændegård 1946) and is possibly highly flexible. Hammel (2005) demonstrated a likely two-year cycle at Qeqertarsuaq/Godhavn.

DISTRIBUTION. S, SW (north to Qeqertarsuaq/Disko), NE, SE.

Nearctic, low arctic and alpine.

Pardosa hyperborea **(Thorell, 1872). Figs 16A, B**
Brændegård 1946: *Pardosa saltuaria*, L. Koch 1870.

Easily recognized by body pattern and colour, which is lighter than in other lycosids.

Wide septum, lack of fovea, and tegular apophysis higher than wide are characteristic.

BIOLOGY. The biotope seems to be fairly luxuriant, south facing heath vegetation.

DISTRIBUTION. S, SW (north to Qeqertarsuaq/Disko), SE (north to 70°50'N).

Holarctic, low arctic-boreal (absent in northeastern Siberia).

PHILODROMIDAE (RUNNING CRAB SPIDERS)

A relatively small family with slightly over 500 species distributed almost worldwide, although absent in Australia and New Zealand. All philodromids are wandering spiders. They hunt not only on the ground like wolf spiders but also on vegetation (grass, bushes, trees, rocks).

Thanatus arcticus **Thorell, 1872. Figs 18A, F, 19**
Marusik *et al.* 2006

Body shape and eye pattern can be recognised even in juveniles.

BIOLOGY. In West Greenland adult specimens were collected in *Empetrum*- *Cassiope*- and lichen heaths, while immature specimens also occurred in many other places. Some females were found under stones, guarding their egg cocoons, which are large, white and flat, attached to the under surface of the stone (Holm 1967). In northeastern Siberia found in almost all biotopes of forest belt and mountain tundra.

FIGURE 19 Thanatus arcticus *in ambush on flower of* Dryas integrifolia.
(PHOTO: J. BÖCHER)

DISTRIBUTION. S, SW (north to Qeqertarsuup Tunua/Disko Bugt).

Circumpolar, low arctic-temperate.

SALTICIDAE (JUMPING SPIDERS)

The largest spider family with over 5600 species. Most are found in tropical and subtropical areas, very few in the Arctic. Jumping spiders can be recognised by very large anterior median eyes and small posterior eyes.

Salticus scenicus **(Clerck, 1757). Figs 1K, L**
Marusik *et al.* 2006

Most probably this species was observed by Otto Fabricius in Paamiut/Frederikshåb (Fabricius 1780). Since then it was mentioned among Greenland spiders only once (Sørensen 1898, with reference to Fabricius). In spite of missing material in museums, there is no doubt that *Salticus* sp. was really observed. It is, however, not certain whether it was *S. scenicus* or *S. zebraneus* (C. L. Koch, 1837) which is also partly synantropical (occurs on buildings, at least in Denmark). Like *Tegenaria domestica* it was probably introduced to Greenland from Europe.

TETRAGNATHIDAE (LONG-JAWED ORBWEB SPIDERS)

A large family of spiders with about thousand species distributed worldwide. The majority are orbweavers, although some species have lost the ability to spin webs and hunt other spiders or litter dwelling arthropods.

***Tetragnatha extensa* (Linnaeus, 1758). Figs 1P, Q, 20**
Marusik *et al.* 2006

Recognised even as juveniles due to stick-like shape, long chelicera with numerous teeth, and silvery pigmentation on abdomen.

BIOLOGY. Holm (1967) caught it on bushes of *Alnus crispa*. The species is always restricted to biotopes close to open fresh water The fairly weak looking orbweb has a hole in the center and is placed horizontally. The food is mainly midges and mosquitoes.

DISTRIBUTION. S, SW (north to Qeqertarsuaq/Disko), southernmost SE.

Widespread in northern hemisphere, ranging from low arctic into the tropics.

THERIDIIDAE (COMB-FOOTED SPIDERS)

Fifth largest spider family with 2300+ known species distributed worldwide. Although originally web builders, many groups have lost the ability to spin webs and have turned to cleptoparasitism or myrmecophagy.

Greenlandic theridiids, especially *Robertus*, can be confused with Linyphiidae or Araneidae (*Hypsosinga*), but definitely recognised by a comb of serrated setae on tarsus IV (difficult to observe) and lack of a stridulating field on chelicera. Theridiid males have apical paracymbium (hidden in *Theridion* and *Thymoites*) and extended retrolateral margin of palpal tibia.

Key to species

1. Abdomen and carapace without pattern............ .... *Robertus fuscus*
– Abdomen with pattern 2

FIGURE 20 Tetragnatha extensa *in web on* Poa glauca. (PHOTO: J. BÖCHER)

2. Abdomen globular with folium pattern (Fig. 22B), male with modified carapace (Fig. 20D), epigyne with 2 closely spaced openings (Figs 2H, 20C)....... .... *Thymoites oleatus*
– Abdomen globular or elongate, male carapace not modified..3
3. Abdomen grayish with folium, legs without rings, carapace without pattern, male chelicera with large tooth, epigyne without fovea (has no margins), 2.9–5.0 mm (Figs 21E, F, 23A) ....*Enoplognatha intrepida*
– Abdomen with dark coloured median band (Figs 23C, 25B), carapace dark with darker head, legs with rings, male chelicera without strong tooth, epigyne with fovea (lateral margins distinct), 1.5–2.8 mm (Figs 21G, 22C, 23B)............................. .... *Ohlertidion lundbecki*

***Enoplognatha intrepida* (Sørensen, 1898). Figs 2I, 21E, F, 23–24**
Marusik *et al.* 2006

Separated from other theridiids by presence of strong tooth on male chelicera, strongly chitinized rim on male abdomen (anterior part), lack of pattern on carapace and lack of rings on legs; epigyne without distinct fovea and openings.

BIOLOGY. In Greenland mostly found under stones in dry heaths. Egg cocoons spherical, 3–4.5 mm in diameter, containing 24–64 eggs. Cocoons attached beneath stones and guarded by the females (Holm 1967).

DISTRIBUTION. S, SW (north to Qeqertarsuaq/Disko), SE (north to 63°14'N).

Nearctic, arctic-boreal and alpine.

***Robertus fuscus* (Emerton, 1894). Figs 21A, B**
Marusik *et al.* 2006

Recognised by uniformly coloured body and nonglobular abdomen. Copulatory organs are also specific. *Robertus* species may be confused with linyphiids.

BIOLOGY. Prefers humid places and was in Greenland collected in a *Sphagnum* mire (Koponen 1982).

DISTRIBUTION. S, SW

Nearctic, boreal.

***Ohlertidion lundbecki* (Sørensen, 1898). Figs 2A, 21G, 22C**
Marusik *et al.* (2006): *Theridion ohlerti lundbecki* (Sørensen, 1898).

Easily recognised, even juveniles, by globular abdomen with typical pattern. From the similarly shaped *Thymoites oleatus* separated by a distinct median band, dark carapace, unmodified male carapace, and distinct lateral margins of fovea. Copulatory organs are also very characteristic.

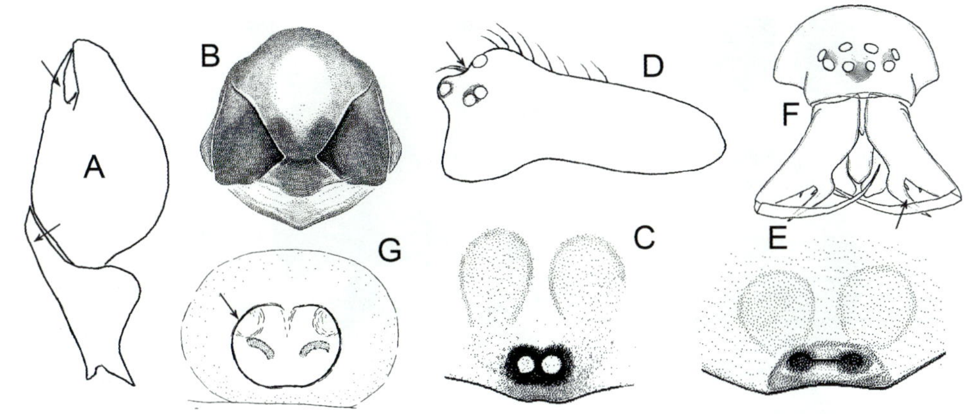

FIGURE 21 *Distinguishing characters in Theridiidae. A, ♂ palp; B, C, E, G, epigyne ventral; D, F, ♂ carapace. A, B,* Robertus fuscus*; C, D,* Thymoites oleatus*; E, F,* Enoplognatha intrepida*; G,* Ohlertidion lundbecki.

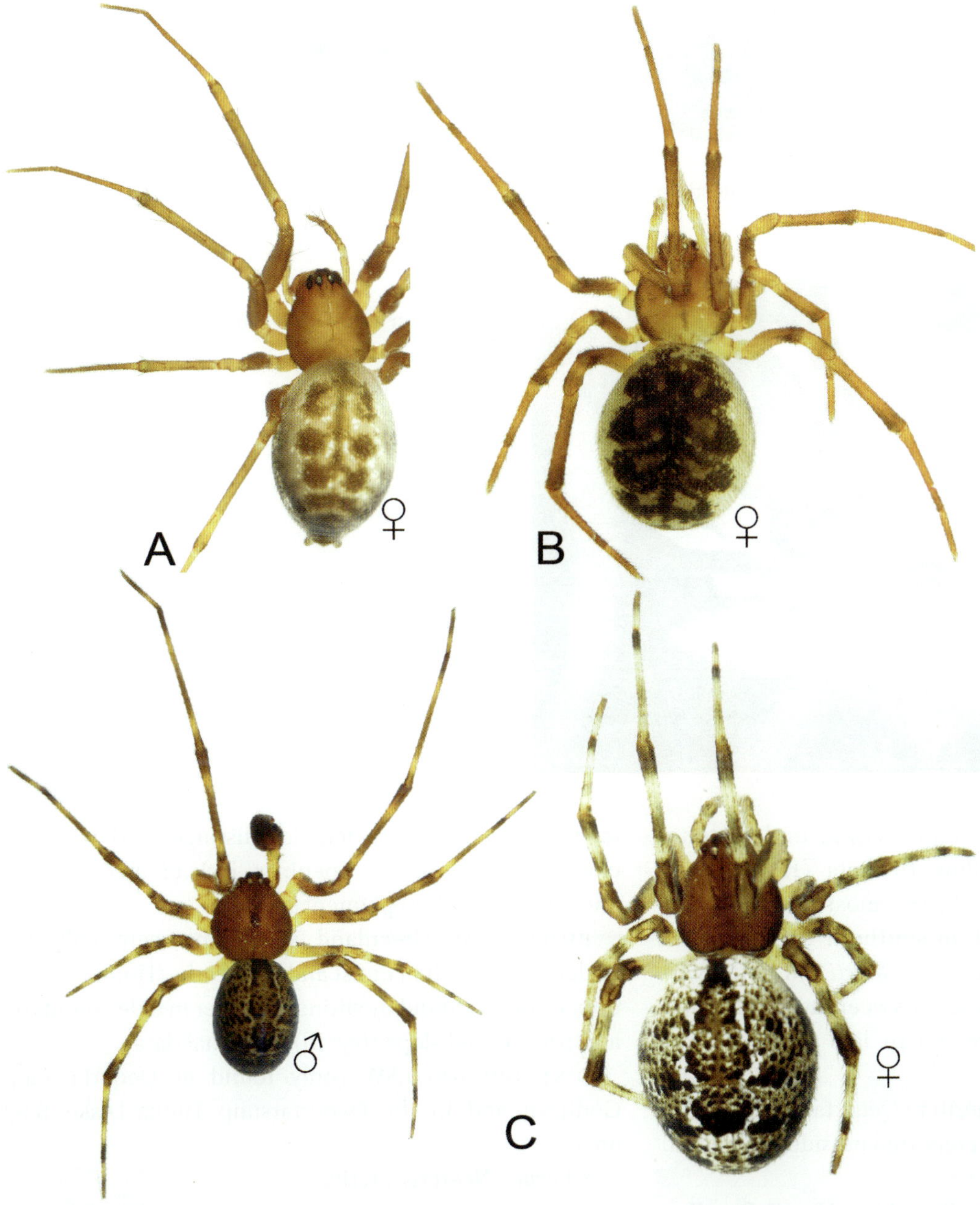

FIGURE 22 *Habitus of* Lepthyphantes turbatrix (*A*), Thymoites oleatus (*B*), Ohlertidion lundbecki (*C*).

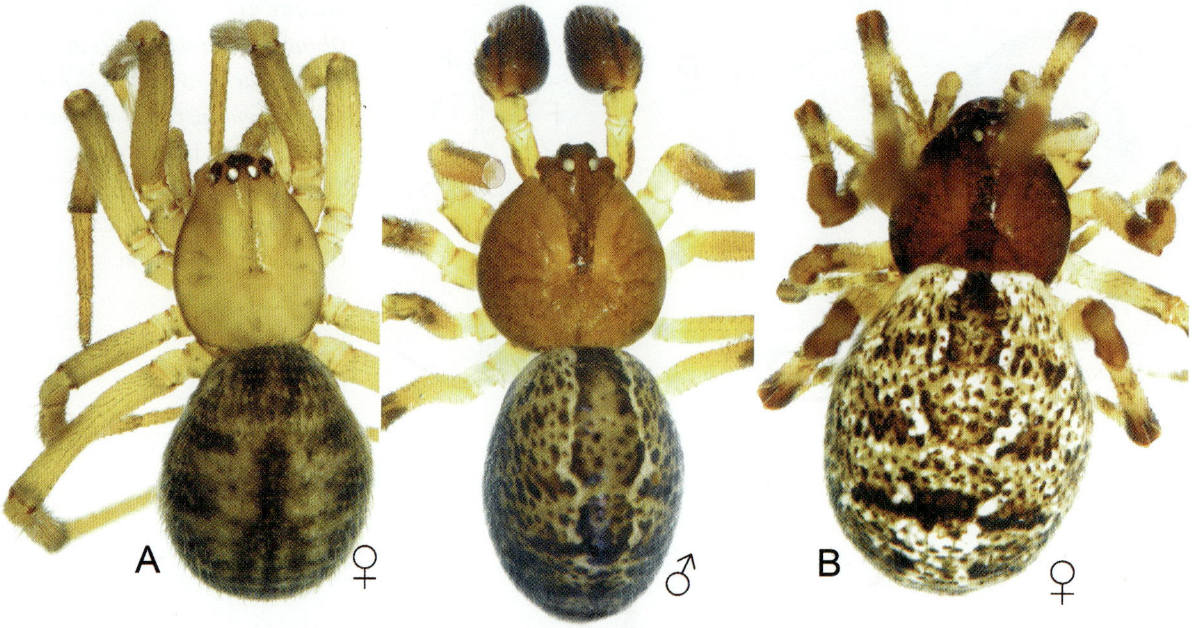

FIGURE 23 *Habitus of* Enoplognatha intrepida (*A*), Ohlertidion lundbecki (*B*).

FIGURE 24
Ohlertidion lundbecki *on flower of* Chamerion latifolium.
(PHOTO: J. BÖCHER)

BIOLOGY. Reported on *Alnus crispa*, under stones and among litter in willow thickets (Holm 1967). According to Koponen (1978) the most abundant spider species of birch shrub in southern Greenland. In upper Kolyma a close relative, *O. ohlerti* (Thorell, 1870), was collected in large number by sweeping shrub thickets (willow, birch, ericaceous) in late summer (YM, personal data).

DISTRIBUTION. S, SW (north to Qeqertarsuaq/Disko).

This species has not been reported outside Greenland.

Thymoites oleatus **(L. Koch, 1879). Figs 2H, 21C, D, 22B**
Holm 1967: *Theridion oleatum* L. Koch.

Globular abdomen and the pattern are specific. Males may be recognised by modified head with raised cephalic lobe and dense hairs; females distinguished by small closely spaced epigyne openings, lack of median band and white (guanine) pigment spots.

BIOLOGY. All Greenland specimens were collected under stones both in moist and dry heaths (Holm 1967). In upper Kolyma found by sifting leaf litter in alder stands on a north exposed slope (650 m) (personal data).

DISTRIBUTION. nSW (only found at Qeqertarsuaq/Godhavn and in the Qeqertarsuup Tunua/Disko Bugt area).

Siberia + Nearctis, arctic.

THOMISIDAE (CRAB SPIDERS)

A large spider family with about 2170 species known from all over the world. Northern species easily recognised by

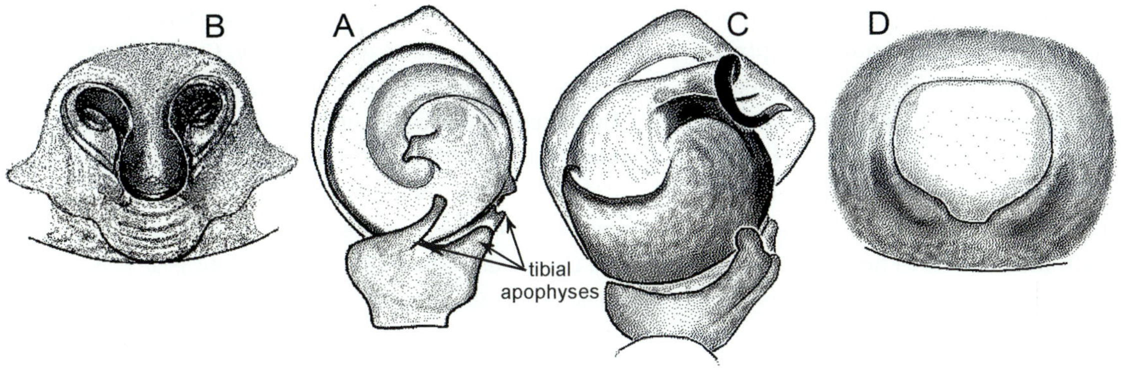

FIGURE 25 *Distinguishing characters in* Xysticus *spp. A, C, ♂ palp; B, D, epigyne ventral. A, B,* X. deichmanni, *C, D,* X. durus.

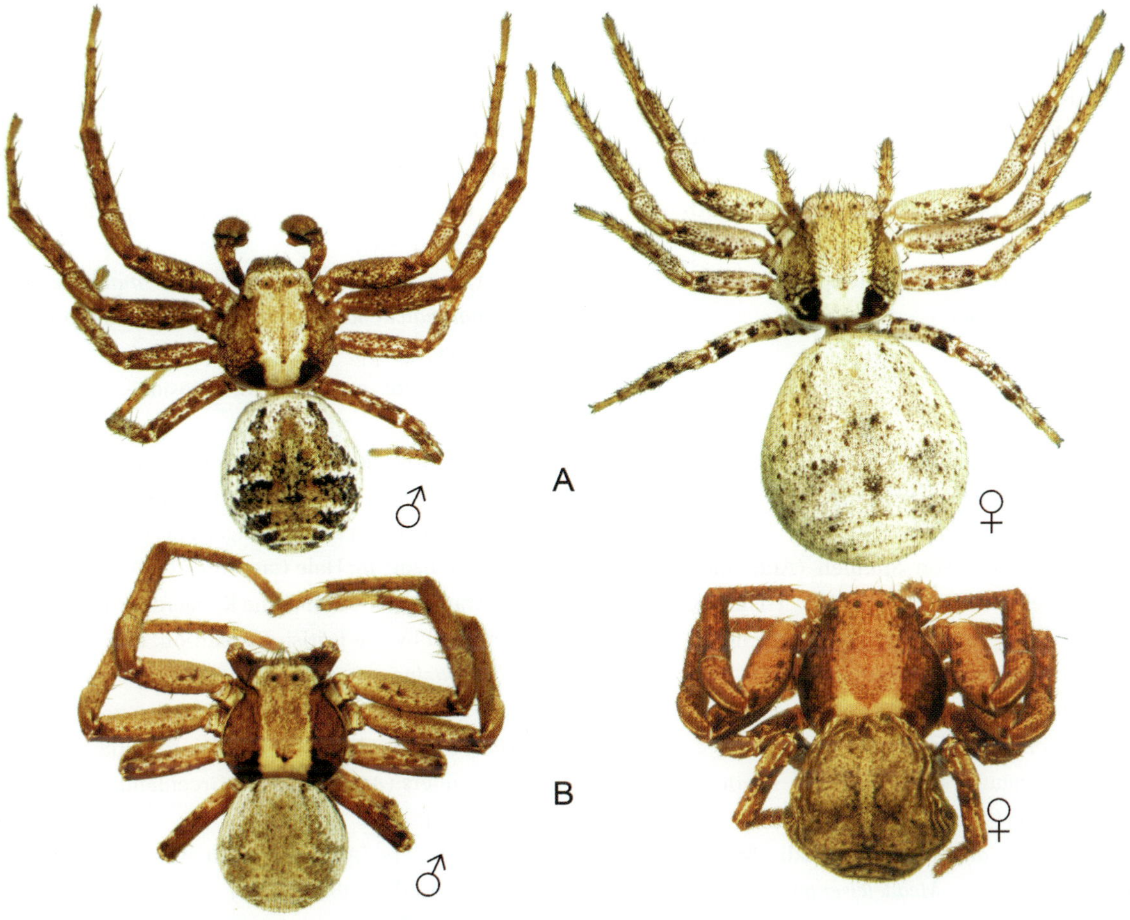

FIGURE 26 *Habitus of* Xysticus deichmanni (*A*), X. durus (*B*).

the body shape. Thomisidae inhabit different strata (litter, herbs and bushes, tree trunks and canopy). Some are wandering spiders, others sit-and-wait predators. Many have cryptic colouration. The occurrence of a third third crab spider species, *Misumena vatia* Clerck, 1758 in southern Greenland, is probable, because this species is found north of the polar circle in Siberia and Alaska and is present in coastal Labrador.

Xysticus deichmanni Sørensen, 1898.
Figs 2E, F, 25A, B, 26A
Brændegård 1946: *Xysticus labradorensis* Keyserling.

Recognised by colour pattern, presence of strong tegular ridge, thick spiralled (screwed) embolus, and lack of any epigynal outgrowths.

BIOLOGY. Brændegård (1940) collected the species among stones in house ruins and under stones in grass vegetation. On Ellesmere Island (Leech 1966) most specimens were taken on the ground in dry and warm places, occasionally observed in flowers of *Dryas*. Two peaks of male activity: around June 24 and at the end of August, whereas females were most active in July. Both sexes may hibernate, and it seems that fertilised females overwinter to lay eggs in the following season. A single cocoon contained 28 eggs. The main food seems to be small Diptera, especially Chironomidae and Culicidae.

DISTRIBUTION. NE.

Nearctic, arctic-boreal.

***Xysticus durus* (Sørensen, 1898). Figs 2G, 25C, D, 26B**
Marusik *et al.* 2006

Recognised by colour pattern, presence of small tegular apophysis in the centre of the bulbus, thin coiled embolus, and presence of median septum in female epigyne.

BIOLOGY. Holm (1967) collected the species by pitfall trapping in *Empetrum*-, *Vaccinium*- and *Polygonum*-heaths as well as in herb slopes and a *Carex* bog. Koponen (1978) found numerous juveniles on birch shrubs in South Greenland.

DISTRIBUTION. S, SW (north to Qeqertarsuaq/Disko), SE.

Nearctic, low arctic and alpine.

References

Agnarsson, I. (1996): *Íslenskar könguær*. – Fjölrit Náttúfræđistofnunar, Reykjavík, 175 pp.

Breuss, W. (2004): Eine Ausbeute von Spinnen (Arachnida, Araneae) von Franz-Josef-Land (Russland). *In*: Thaler, K. (ed.), Diversität und Biologie von Webspinnen, Skorpionen under anderen Spinnentieren. – *Denisia* 12: 535–545.

Brændegård, J. (1937): Spiders (Araneina) from southeast Greenland. – *Meddelelser om Grønland* 108 (4): 1–14.

Brændegård, J. (1938): Aeronautic spiders in the Arctic. – *Meddelelser om Grønland* 119 (5): 1–9.

Brændegård, J. (1939): *Arctosa alpigena* Doleschall 1852 og *Arctosa omdohmoys* Thorell 1872. – *Meddelelser om Grønland* 108 (7): 1–12.

Brændegård, J. (1940): Spiders (Araneina) from northeast Greenland. – *Meddelelser om Grønland* 125 (8): 1–29.

Brændegård, J. (1946): The spiders (Araneina) of East Greenland. A faunistic and zoogeographical investigation. Zoology of East Greenland 14. – *Meddelelser om Grønland* 121 (14): 1–128.

Brændegård, J. (1960): The spiders (Araneida) of Peary Land, North Greenland. – *Meddelelser om Grønland* 149(6): 1–24.

Buckle, D. J., Carroll, D., Crawford, R.L. & Roth, V.D. (2001): Linyphiidae and Pimoidae of America north of Mexico: checklist, synonymy, and literature. – *Fabreries*, Suppl. 10: 89–191.

Dondale, C.D. & Redner, J.H. (1978): *The insects and arachnids of Canada, Part 5. The crab spiders of Canada and Alaska, Araneae: Philodromidae and Thomisidae*. – Research Branch, Agriculture Canada, Publ. 1663: 1–255.

Dondale, C.D. & Redner, J.H. (1990): *The insects and arachnids of Canada. Part 17. The wolf spiders, nurseryweb spiders, and lynx spiders of Canada and Alaska (Araneae: Lycosidae, Pisauridae, and Oxyopidae)*.– Agriculture Canada Publ. 1856: 1–383.

Dondale, C.D., Redner, J.H., Marusik, Yu.M. (1997): Spiders (Araneae) of the Yukon. *Insects of the Yukon* (*ed.* Danks, H.V.): 73–113. Ottawa.

Dondale, C.D., Redner, J.H., Paquin P. & Levi, H.W. (2003): The insects and arachnids of Canada. Part 23. *The orb-weaving spiders of Canada and Alaska (Araneae: Uloboridae, Tetragnathidae, Araneidae, Theridiosomatidae)*. NRC Research Press, Ottawa, 371 pp.

Eskov, K.Yu. (1994): Catalogue of the linyphiid spiders of northern Asia (Arachnida, Araneae, Linyphiidae). Sofia-Moscow: PENSOFT Publ. 144 pp.

Eskov, K.Yu., Marusik, Yu.M. (1992): On the Siberio-Nearctic spider genus *Silometopoides* (Araneida: Linyphiidae). – *Reichenbachia* 29: 97–103.

Fabricius, O. (1780): *Fauna Grønlandica, systematice sistens animalia Grønlandica occidentalis hactenus indagata etc.* Halfniæ et Lipsiæ, 452 pp.

Gertsch, W.J. (1934): Notes on American Lycosidae. – *American Museum Novitates* 693: 1–25.

Hammel, J.U. (2005): *Ökologie und Phylogenie grönländischer Wolfspinnen (Lycosidae, Araneae)*. Biologischen Institut, Abteilung Zoologie, Universität Stuttgart. 123 pp.

Holm, Å. (1939): Araneae. *In*: Hale Carpenter, G. D., & Holm, E.: Insecta and Araneae collected in the Kangerdlugsuak Region of East Greenland by the British Expedition, 1935–36. – *The Annals and Magazine of natural History* (11) 3: 72–80.

Holm, Å. (1956): Notes on Arctic spiders of the genera *Erigone* Aud. and *Hilaira* Sim. – *Arkiv för Zooogi* 9: 453–468.

Holm, Å.(1958): Spiders (Araneae) from Greenland. – *Arkiv för Zoologi* 11: 525–534.

Holm, Å. (1967): Spiders (Araneae) from west Greenland. – *Meddelelser om Grønland* 184 (1): 1–99.

Jackson, A.R. (1934): Notes on Arctic spiders obtained in 1933. – *The Annals and Magazine of natural History* (10) 14: 611–620.

Jackson, A.R. (1938): Notes on Arctic spiders obtained in 1933–1936. – *Proceedings of the zoological Society of London* (B) 107: 543–551.

Koponen, S. (1978): Notes on herbivorous insects of the birch in southern Greenland. – *Reports from the Kevo Subarctic Research Station* 14: 13–17.

Koponen, S. (1982): Spiders (Araneae) from Narssaq, southern Greenland. – *Entomologiske Meddelelser* 49: 117–119.

Koponen, S. (1992): Spider fauna (Araneae) of the low arctic Belcher Islands, Hudson Bay. – *Arctic* 45: 358–362.

Larsen, S. & Scharff, N. (2003): The spiders of Greenland – a checklist (Aracnida: Araneae). – *Entomologiske Meddelelser* 71: 53–61.

Leech, R.E. (1966): The spiders (Araneida) of Hazen Camp 81°49'N, 71°18'W. – *Quaestiones entomologicae* 2: 143–212.

Marusik, Yu.M., Böcher, J. & Koponen, S. (2006): The collection of Greenland spiders (Aranei) kept in the Zoological Museum, University of Copenhagen. – *Arthropoda Selecta* 14 (1): 59–80.

Marusik, Y.M., Eskov, K.Yu., Koponen, S. & Vinokurov, N.N. (1993): A check-list of the spiders (Aranei) of Yakutia, Siberia. – *Arthropoda Selecta* 2: 63–79.

Marusik, Yu.M, Logunov, D.V. & Koponen, S. (2000): *Spiders of Tuva, South Siberia.* IBPN FEB RAS. 252 pp

Marusik, Yu.M. & Koponen, S. (2002): *Vermontia,* a linyphiid genus new to Palaearctic and new records of spiders from northeast Asia (Araneae). – *Bulletin of the British arachnological Society* 12: 149–165.

Millidge, A.F. (1976): Re-examination of the erigonine spiders "*Micrargus herbigradus*" and "*Pocadicnemis pumila*" (Araneae: Linyphiidae). – *Bulletin of the British arachnological Society* 3: 145–145.

Ovtsharenko, V.I., Platnick, N.I. & Song, D.X. (1992): A review of the North Asian ground spiders of the genus *Gnaphosa* (Araneae, Gnaphosidae). – *Bulletin of the American Museum of Natural History* 212: 1–88.

Paquin, P. & Dupérré, N. (2003): Guide d'identification des araignées de Québec.- *Fabreries*, Suppl. 11: 1–251.

Palmgren, P. (1975): Die Spinnenfauna Finnlands und Ostfennoskandiens. VI. Linyphiidae 1. – *Fauna Fennica* 28: 1–102.

Roberts, M.J. (1987): *The spiders of Great Britain and Ireland 2. Linyphiidae.* – Harley Books, Colchester. 204 pp.

Roberts, M.J. (1995): *Collins Field Guide: Spiders of Britain & Northern Europe.* – Harper Collins, London. 383 pp.

Roberts, M.J. (1998): *Spinnengids.* – Tirion, Baarn, Netherlands. 397 pp.

Saaristo, M.I. & Eskov, K.Yu (1996): Taxonomy and zoogeography of the hypoarctic erigonine spider genus *Semljicola* (Araneida, Linyphiidae). – *Acta Zooogici Fennica* 201: 47–69.

Saaristo, M.I. & Koponen, S. (1998): A review of northern Canadian spiders of the genus *Agyneta* (Araneae, Linyphiidae), with descriptions of two new species. – *Canadian Journal of Zoology* 76: 566–583.

Saaristo, M.I. & Marusik Yu.M. (2003 (2004)): Revision of the Holarctic spider genus *Oreoneta* Kulczyński, 1894 (Arachnida; Araneae; Linyphiidae). – *Arthropoda Selecta* 12 (3–4): 185–220.

18.2. Opiliones (Harvestmen)

Jens Böcher

The body outline is near-elliptical, with prosoma broadly joined to abdomen without constriction. Centre of the prosomal carapace bearing a tubercle with an eye on each side. Openings of a pair of scent glands are located on the anterior margin of carapace. The chelicerae are small, slender and triangulate, the last two segments forming a pair of pincers. The pedipalps are fairly leglike, similar to those in spiders. In most species the legs are extremely long and slender, many times exceeding the length of the body. The tarsus is always multisegmented and highly flexible. The coxae of first and second pairs of legs, like those of the pedipalps, bear a process which projects anteriorly. The second pair of legs is longest and during walking often held forwards like antennae. As an escape-mechnism many species are capable of shedding one or more legs (autotomy).

The abdomen is visibly segmented. The sterna of the first two abdominal segments are fused to form a plate, operculum, which appears to be situated between the last two pairs of coxae. The operculum bears the two genital openings.

In contrast to other arachnids, harvestmen are mostly omnivorous. Food is seized by the pedipalps and passed to the chelicerae, which hold and crush it. The coxal processes pass food into the mouth. Food is thus not limited to liquids but includes small particles. Harvestmen are not equipped with venom or ability to produce silk

Respiration by means of tracheae, spiracles being located on each side of the second sternum. In long-legged forms secondary spiracles are present on tibiae.

The female reproductive system includes a long, tubular ovipositor, originating from the mid-ventral part of abdomen; counterparts are absent in other arachnids except certain mites. While males may fight other males, mating is not preceded by any courtship. The male extends a long, tubular penis which usually passes between the female's chelicerae before entering her genital orifice. Shortly after mating the female seeks a humid location for depositing the eggs.

FIGURE 1 Mitopus morio, *habitus.*

Most species are found in humid surroundings and most are nocturnal.

In Greenland only one species:

Mitopus morio **(Fabricius, 1799) Fig. 1**
BIOLOGY. The species is very common in most plant communities, even in fairly dry localities, but most abundant on luxuriant herb slopes. The species is not nocturnal but can be seen in activity even in sunshine. In Greenland everything points to one yearly generation and hibernation in the egg-stage.

DISTRIBUTION. S, SW, SE (north to Sisimiut/Holsteinsborg and Ammasalik).

Circumpolar, subarctic-subtropical.

Reference

Meinertz, T. (1973): The occurrence of *Mitopus morio* (Fabr.) Phalangidae, Opiliones in Greenland and in other northern regions. – *Meddelelser om Grønland* 191 (7): 16 pp.

18.3. *Acari* (*Mites and Ticks*)

18.3.1. Generalities and key to higher taxa

Olga Makarova

Mites (Acari or Acarina) constitute an extremely diverse group of mainly small arachnids (Lindquist *et al.* 2009). They are known among Devonian fossils (Dubinin 1962, Norton *et al.* 1988, Kethley *et al.* 1989) which are about 400 million years old. Thus mites are among the oldest known terrestrial arthropods together with collembolans, myriapods, other arachnids, *etc*. The number of described mite species exceeds 54 500 (Zhang 2011), but the number of known species is considered only a small part (about 5%) of the existing species diversity (Evans 1992, Walter *et al.* 1996). Their adult size varies from 0.08 mm (gall mites, follicle mites) to *c.* 30 mm (engorged ixodid mites), but most representatives are 0.40–0.80 mm in length.

Mites have very different diets, life styles and modes of fertilization, reproduction, and development (see Walter & Proctor 2013); pertinent information is briefly given below in the treatment of the separate suborders. The group is really ubiquitous inhabiting all soil types, the surface of glaciers, subterranean waters, plants, deep seas, different freshwater bodies. Numerous species dwell in temporary biotopes: carrion, droppings, mushrooms, sea debris, stored products, animal nests. Unlike other Arachnida, which are free-living, a vast number of mites display various symbiotic associations with other animals (different invertebrates, mammals, birds), being commensals, phoretic travellers, paraphages, parasites, *etc.*

Morphology

Unlike most arachnids, mites lack a distinct segmentation. In general their body consists of the anterior *gnathosoma* and the posterior *idiosoma* (Figs 1E, F). The gnathosomal tubes (or the gnathosomal base) is formed by the remnants of body segments I–II and the coxae of the *palps* (*pedipalps*), *pp* (Fig. 1F). Two pairs of appendages are attached here: the plesiomorphically 3-segmented *chelicerae, ch,* of very different structure (Figs 1M, 2F–H, M), and 2–5 segmented palps, which often combine the sensorial, holding and grasping functions (Figs 1F, K, L, 2N). In spite of a superficial resemblance to a 'head', the gnathosoma does not contain the brain and eyes (if present), which are located in/on the idiosoma. The underside of the gnathosomal base, the *subcapitulum, sbc,* is generally produced into a beak-like *hypostome, hyp,* which in addition to the adoral setae and the unpaired labrum often bears the paired (blade-like, dentate, horn-like, *etc.*) *rutella, ru* (Fig. 1L), or *corniculi, co* (Fig. 1B), presumably of setal origin. The idiosoma may be subdivided into (1) the anterior part, the *propodosoma* (bearing legs I, II; covered by the prodorsal shield or *prodorsum*), and the posterior part, the *hysterosoma*, by the *sejugal furrow, sjg* (Figs 1J, 2A, M), or (2) the *podosoma* (bearing legs I–IV) and the *opisthosoma*. In the latter case the tagmal division is rarely shown by the *postpedal furrow* and usually recognizable only in the sclerotization pattern (Figs 1C, D). Often the anterior part of idiosoma dorsally bears 1–2 pairs of *trichobothria, trb* (Figs 1J, 2A–D), *i.e.* the specialized sensillae arising from conical depressions, that are termed *bothridia, bt* (Fig. 1N). Ventrally, the distinct *sternum* (sternal shield, *stsh*) may be found between the leg coxae (Figs 1A, E) or the coxae are fused with the body wall, forming a plate-like *coxisternum* or paired *epimera, ep* (Fig. 1O). The respiratory system, if present, opens by paired spiracular orifices, or *stigmata, st,* whose number and location serve as important taxonomical features on the order-suborder level. The stigma may be surrounded by the *spiracular plate, sp* (Fig. 1E) or associated with the *peritreme, prt,* a groove of different structure and length (Fig. 1A). The acarine legs originally comprise 7 podomeres: *coxa, trochanter, femur, genu, tibia, tarsus* and *pretarsus*, but both a secondary division and fusion of articles often occur. The pretarsus commonly terminates in paired *claws, cl* (Figs 1I, 2J), and/or an unpaired median *empodial claw* (*empodium*), *ecl,* of different form (Figs 2J–L). Besides, a sucker-like lobate *pulvillus, plv,* is usually present (Figs 1I, 2O). All these terminal structures together with the pretarsus form the *ambulacrum, amb* (Fig. 1G). Leg I may be lacking an ambulacrum (Fig. 1H), being often a sensory rather than ambulatory appendage.

Systematics

Walter *et al.* (1996) suggested the following combination of diagnostic features for the Acari:

- hexapod prelarva (lost in Parasitiformes and many derived Acariformes);
- hexapod larval stage;
- three octopod nymphal stages (variously abbreviated in derived taxa);

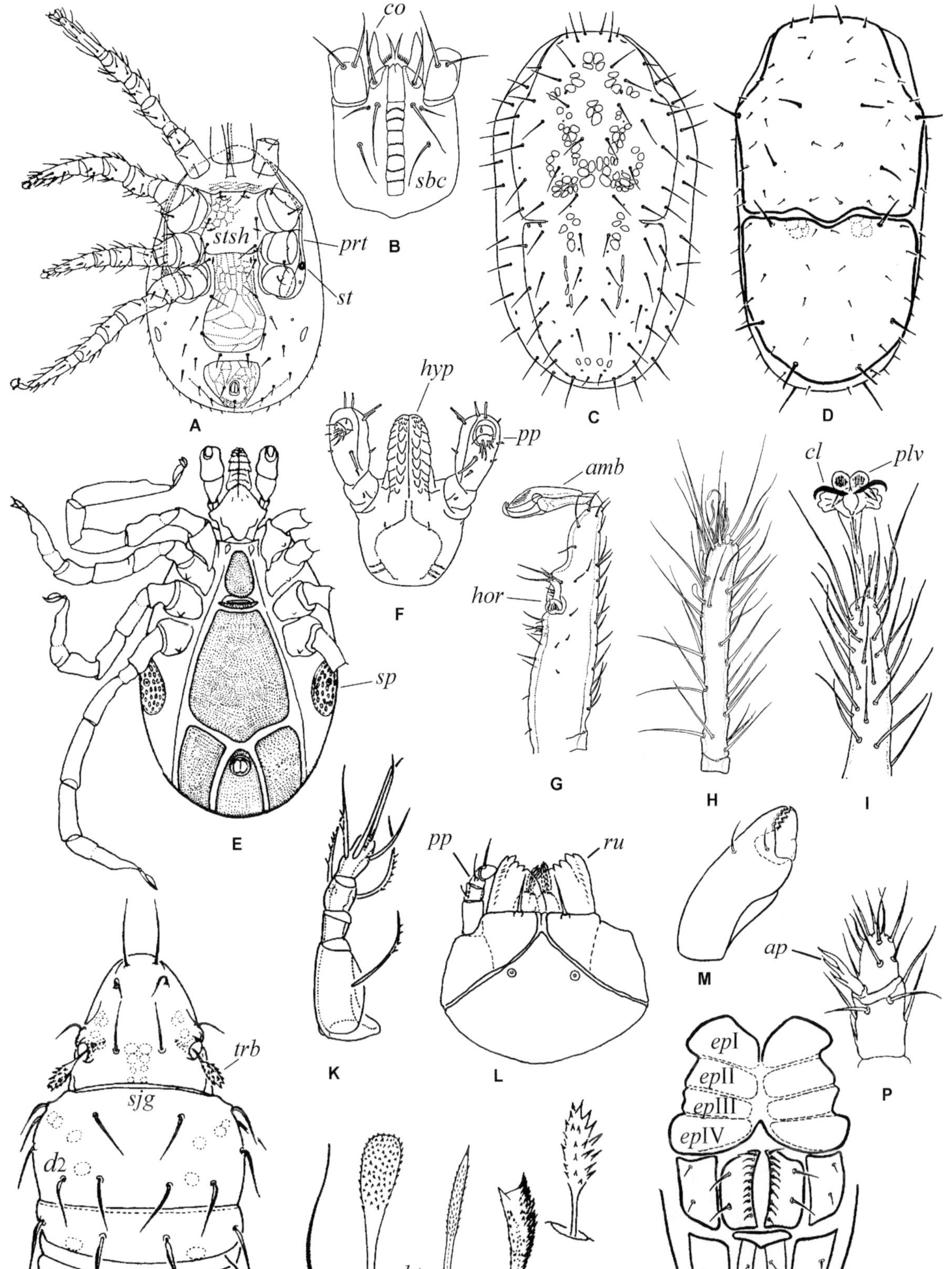

FIGURE 1 *A, mesostigmatic mite* Hypoaspis *sp., ventral aspect; B, mesostigmatic mite* Arctoseius *sp., subcapitulum, C, dorsal aspect; D, mesostigmatic mite* Cornigamasus *sp., dorsal aspect; E, ixodid mite, ventral aspect, schematic, F, gnathosoma ventrally, G, tarsus I; H, mesostigmatic mite* Arctacarus *sp., tarsus I; I, mesostigmatic mite* Veigaia *sp., tarsus I distally; J, oribatid mite* Liochthonius *sp., dorsal aspect; K, oribatid mite* Suctobelba *sp., palp; L, oribatid mite* Camisia *sp., gnathosoma ventrally; M, oribatid mite* Diapterobates *sp., chelicera; N, diversity of oribatid trichobothria; O, oribatid mite, ventral aspect, schematic; P, mesostigmatic mite* Parasitus *sp., tibia and tarsus of palp. A, after Bregetova 1956; G, F, after Filippova 1977; J, after Moritz 1976; P, after Lange 1955.*

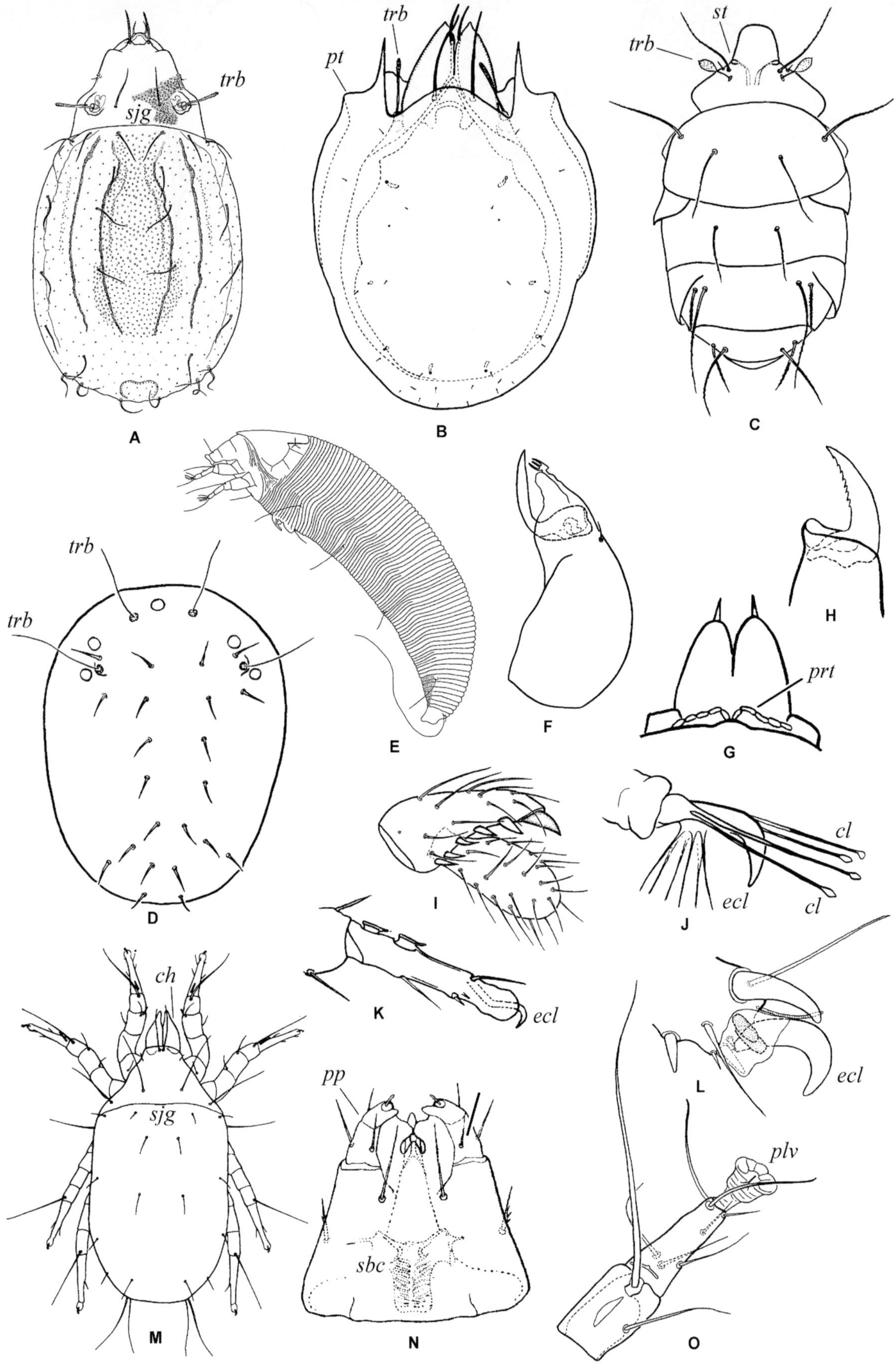

FIGURE 2 *A, oribatid mite* Platynothrus *sp., dorsal aspect; B, oribatid mite* Achipteria *sp., dorsal aspect; C, prostigmatic mite* Bakerdania *sp., dorsal aspect; D, prostigmatic mite* Cyta *sp., dorsal aspect; E, prostigmatic mite* Neoleipothrix *sp.; lateral aspect; F–H, chelicerae of prostigmatic mites, F,* Penthaleus *sp., G,* Stigmaeus *sp., H,* Podothrombium *sp.; I, prostigmatic mite* Erytraeus *sp., palpal thumb-claw complex; J, prostigmatic mite* Panonychus *sp., ambulacrum IV; K–N, members of astigmatic mite family Acaridae, K, tarsus IV of male, L, tarsus I distally, M, dorsal aspect, N, subcapitulum; O, astigmatic mite* Dermatophagoides *sp., tarsus I. E, after Chetverikov 2006; K, M, N, after Zakhvatkin 1941.*

- gnathosoma delimited by a circumcapitular suture;
- palpcoxal endites fused medially forming a hypostome;
- hypostome with rutella or corniculi (lost in many derived Acariformes);
- loss of external evidence of opisthosomal segmentation, *i.e.* without tergites and sternites;
- ingestion of particulate food (lost in many derived taxa).

Most authors recognize two or three major lineages (differently ranked) of mites, *viz.* the **Parasitiformes** (Anactinotrichida including Opilioacarida) and **Acariformes** (Actinotrichida). Members of Parasitiformes are lacking the optically active actinochitin in their setae. In contrast, the representatives of Acariformes possess setae containing actinochitin. It has even been suggested that the mites as conventionally delimited are diphyletic, in as much as the Anactinotrichida and Actinotrichida allegedly have their closest relatives among different non-acarine arachnids (Zakhvatkin, 1952, van der Hammen 1989). Mite monophyly was supported in a detailed analysis by Lindquist (1984), but continues to be challenged by subsequent work; see the recent accounts by Danlop & Alberti (2007), Shultz (2007) and references therein. So, as the explicit evidence for diphyletic origins of Acari, a specific pattern of gnathosomal musculature may support the unit [Anactinotrichida + Ricinulei], whereas the association [Actinotrichida + Solifugae] is supported by sperm and testis morphology (Danlop & Alberti 2007).

Here we follow the system mainly developed by Norton *et al.* 1993, but for pragmatic reasons the traditional concept of Prostigmata (Kethley 1982, 1990; Evans 1992) is retained. The same applies to the adoption of the widespread treatment of Oribatida and Astigmata as equally ranking taxa, rather than treating the latter as subordinate in the former (as given in Norton & Behan-Pelletier 2009, O'Connor 2009; see remarks in chapter 18.3.7).

The existence of two vernacular names ("mites" and "ticks") does not reflect a proper taxonomical subdivision. "Ticks" is only a name for part of the Parasitiformes, namely the haematophagous ixodid mites (suborder Ixodida), which parasitize different vertebrates, including man and domestic animals. In Greenland Ixodida is represented by a single species (*Ixodes uriae* White, 1852) whereas the mites number at least 289 recorded species belonging to the orders Parasitiformes (suborder Mesostigmata – 63 species) and Acariformes (suborder Oribatida – 110 species, suborder Astigmata – 17 species, suborder Prostigmata s.l. – 99 species. Water mites (8 species, all members of Prostigmata) are described and keyed in chapter 18.3.5).

Greenland/arctic fauna

The investigation of Greenland mites dates back to O. Fabricius (1780) and P. Kramer (1897). A special investigation of gall mites (Eriophyoidea) was carried out by Sophie Rostrup (1900) and several aspects of the moss mite (Oribatida) natural history as well as their faunistics and taxonomy were studied by the outstanding Danish acarologist Marie Hammer (1937, 1944, 1946, 1952, 1953, 1954, *etc.*). Some Greenland mites were found as subfossils (Graversen 1931, Haarløv 1967). The first and also the last total lists of the Greenland acarofauna were published by Trägårdh (1904, 1912), Henriksen & Lundbeck (1917), and Jørgensen, later Hammer, (1934). After their publications only a few records were added, mainly as descriptions of new species (Haarløv 1942, 1967; Strenzke *et al.* 1955; Sellnick 1960; Chant & Hansell 1971; Hunter & Husband 1973; Karg 1984, 1985; Southcott 1992), though several important expeditions were organized to the key areas of Greenland. Nonetheless, the present treatment of their collections has increased the number of known Greenland terrestrial mites from about 160 (some records are doubtful) to 290 species. The fauna of mites associated with Greenland vertebrates, however, remains almost unknown; its study must await future efforts.

Arctic landscapes harbour hundreds of mite species (Danks 1981, Hodkinson *et al.* 2013, Coulson *et al.* 2014) and soil mites are particularly diverse. Most of them have wide geographical ranges (transholarctic polyzonal or arcto-boreal species, cosmopolitans, *etc.*), but many occur mainly in the tundra zone and polar deserts. Some arctic species have ranges extending southward into the northern taiga, where they usually prefer cold habitats (snow beds, bogs, karst funnels, *etc.*). A considerable number of mites, which previously have been considered as arctic species, were later found in southern Siberian mountains, being actually arctomontane in their distribution (Makarova 2013). Among true arctic mites, some species inhabit the Low Arctic (southern tundra and typical tundra subzones in the Russian tradition), but a few species populate mainly the High Arctic (arctic tundra subzone and polar desert zone). Many arctic mite species are circumpolar or semicircumpolar. In the last case the gap is often found just in the Greenland sector. The markedly different proportions of specialized cryobiont species (arctic and arctomontane) in Greenland Oribatida (4%) and Mesostigmata (45%) compared to the Arctic as a whole (18% and 31% respectively; Hodkinson *et al.* 2013) were suggested to be related to differences in their geological age and evolutionary rate (Makarova & Böcher 2009).

In general, the diversity of soil mites in the Holarctic decreases gradually to the north (Danks 1990, Behan-Pelletier

1999, Maraun *et al.* 2007), but unevenly in different suborders. The suborder Prostigmata undergoes minimal taxonomical reduction northward (Makarova 2002b). The soil in one tundra biotope is often populated by 30–50 free living mite species, and a single core of upper soil layer (5x5x5 cm) can harbour 15–30 mite species. Even the polar desert communities in the vicinity of glaciers are inhabited by a dozen of mite taxa comprising mycophagous, algophagous, omnivorous and predatory species. In polar deserts, in a single biome of the Earth, the diversity of arachnids (mainly mites) exceeds that of insects (Makarova 2002a, Chernov 2004).

The density of soil mite populations in the arctic soils usually varies within the limits 40–2500 specimens/dm^2 (references in Makarova 2002b). In the southern tundra, the mite biomass may be comparable to the total biomass of vertebrate animals (Krivolutsky 1976). The strong prevalence of oribatid mites in the taiga zone diminishes in the tundra zone, and in arctic polar deserts they are to a great extent replaced by members of Prostigmata, as also in the true arid deserts (references in Makarova 2002b).

Within one arctic locality, the soil mite communities of different biotopes vary, sometimes markedly (Thomas & McLean 1988, Byzova *et al.* 1995, Makarova 2002a, Gwiazdowicz & Coulson 2011, *etc*). Only a few ubiquitous species inhabit either *Dryas* heaths or patchy tundra sites or sedge-moss bogs. Besides, rather particular groupings of mites exist in bumble-bee nests, tidal zone biotopes, turf of polar fox hills, lemming burrows and nests, debris of sea bird colonies and the enriched soil surrounding them. Numerous specialized mite species live on vertebrates themselves and often one host species is home to several species of mites (Gill & Strandtmann 1977).

In general, the diversity of animal taxa in Greenland is relatively poor due to recent colonization following the almost total glaciation (Downes 1988). The age of the Greenland biota probably amounts to only a few thousand years (Bennike *et al.* 2000). But mites are represented by the most of the families characteristic of arctic landscapes and by almost three hundreds of species belonging to all main ecological groups. Thus, the formation of acarocenoses in young Greenland landscapes obviously is a relatively rapid process. This is evidently caused by the tiny size of these omnipresent animals possessing an enormous capacity of dispersal (Makarova & Böcher 2009). In Greenland the ratio between Palaearctic and Nearctic mesostigmatan species is 4:1 (Makarova & Böcher 2009) as in other soil-dwelling arthropod taxa (Böcher 1988, Babenko & Fjellberg 2006). On the contrary, the Greenland Oribatida display a greater Nearctic impact (Makarova & Böcher 2009). So far there is no adequate explanation of such strong differences between these small-sized soil-inhabiting groups.

Methodology: sampling and examination

Many methods have been developed for sampling mites of different life styles (various pitfalls, extracting procedures, hunting devises). For the identification of mites, commonly very small animals, the employment of various types of microscopy is necessary. The techniques of collecting, fixation, bleaching and preparations of mites have been treated in a number of manuals (Krantz 1978, Evans 1992, Walter & Proctor 2001, Walter & Krantz 2009, *etc.*).

The following keys are valid for the Greenland fauna only and are necessarily preliminary, because only few special investigations on Greenland mites have been conducted, and the parasitic groups are still almost unknown in Greenland. Morphological sketches, serving mainly as explanations of features used in the given keys are presented in the introduction to each suborder.

Key to orders and suborders of subclass Acari*

1. Sternum and free leg coxae usually present (Figs 1A, E); idiosoma without sejugal furrow between legs II and legs III; with a pair of stigmata laterally to coxae II–IV (Fig. 1A) or a pair of spiracular plates provided with numerous aeropyles posterior to coxae IV (Fig. 1E); palp often with subterminal claw remnant, *apotele, ap* (Fig. 1P).........2 (order Parasitiformes)
– Without free sternum and leg coxae but with coxisternum or epimera (Fig. 1O); sejugal furrow usually present (Figs 1J, 2A, M); stigmata, if present, never situated laterally to legs II–IV; palp without apotele. .... 3 (order Acariformes)
2. Subcapitulum with well developed hypostome armed with rows of retrorse denticles (Fig. 1F); no elongated peritremes associated with stigmata, often spiracular plates developed (Fig. 1E); tarsus I with elaborate sensory Haller's organ, *hor* (Fig. 1G).. .... suborder Ixodida (=Metastigmata), 18.3.3
– Hypostome lacking rows of retrorse denticles, but usually with corniculi (Fig. 1B); stigmata associated with elongate peritremes (Fig. 1A); sensory elements of tarsus I not forming elaborated structure (Figs 1H, I)...... suborder Mesostigmata (=Gamasida), 18.3.2
3. Palps small and comprising 1 or 2 free podomeres (Fig. 2N); chelicerae markedly chelate (as on Fig. 1M); ambulacra of legs comprising an empodial claw with an associated membranous pulvillus (Figs 2K, L) or a sucker-like pulvillus situated on elongate pretarsus (Fig. 2O); adult idiosoma usually

* Mainly based on Evans & Till (1979) with modification.

weakly sclerotized; stigmata and tracheae never present; idiosoma without trichobothria (Fig. 2M) and never covered dorsally by overlapping sclerites; body never vermiform.............................. .... suborder Astigmata (=Acaridida), 18.3.7

– Palps conspicuous and usually with 3 to 5 free podomeres (Figs 1K, L); if small with fewer podomeres, then idiosoma either vermiform (Fig. 2E) or with overlapping sclerites dorsally in female (Fig. 2C); chelicerae chelate or modified (Figs 2F–H); leg ambulacra of different structure; idiosomal sclerotization varying; respiratory system usually present; prodorsum often with trichobothria (Figs 1J, 2A–D) 4

4. Subcapitulum with conspicuous rutella (Fig. 1L); chelicerae typically chelate-dentate (Fig. 1M); one pair of prodorsal trichobothria (fusiform, comb-like, clavate, *etc.*; rarely piliform) almost invariably present (Figs 1J, N, 2A, B); palps simple, tibia never with distal claw; tracheal system, if present, with openings associated with legs or prodorsal bothridia; often heavily sclerotized in adult stage, sometimes with ridges (Fig. 2A) or wing-like expansions of idiosoma, *pteromorphs, pt* (Fig. 2B), which is never vermiform .... suborder Oribatida (=Cryptostigmata), 18.3.6

– Subcapitulum only rarely with rutella; chelicerae rarely chelate-dentate, usually variously modified into piercing stylets (Fig. 2G) or hook-like organs (Figs 2F, 2H); prodorsal trichobothria, when present, often setiform (Fig. 2D); palps various, often with tibia and tarsus forming "thumb-claw" complex (Fig. 2I); tracheae, if present, opening by paired stigmata situated between or on cheliceral bases (Fig. 2G) or on dorsal surface of prodorsum (Fig. 2C) and often associated with peritremes; often weakly sclerotized mites, never with ridges or wing-like expansions of idiosoma, which may be vermiform (Fig. 2E) or covered dorsally by a number of discrete, usually overlapping sclerites (Fig. 2C)............... .... suborder Prostigmata (=Actinedida), 18.3.4

Acknowledgments

This treatment of all Greenland mites could not have been written without the assistance of numerous colleagues: A. Babenko, A. Baker, B. Bayartogtokh, L. Beck, V. Behan-Pelletier, A. Bhattacharyya, J. Bloszik, J. Böcher, A. Bochkov, O. Butenko, D. Chant, Yu. Chernov, F. Chetverikov, M. Colloff, S. Ermilov, N. Filippova, A. Fjellberg, S. Golovatch, D. Gwiazdowicz, B. Halliday, W. Karg, A. Khaustov, P. Klimov, G. Krantz, N.P. Kristensen, E. Lindquist, Yu. Lopatina, S. Mahunka, P. Mašan, N. Matveyeva, V. Michelsen, S. Mironov, F. Monson, M. Moritz, R. Norton Z. Olszanowski, M. Potapov, E. Sidorchuk, T. Solhøy, U. Shtanchaeva, M. Stanyukovich, J. Stary, L. Subias, A. Tanasevitch, G. Weigmann, S. Woas. They all provided me with information, necessary publications or critique of early manuscript drafts. The *Carlsberg Foundation*, the *Russian Fond of Basic Research* and the Programs "*Life Origin and Biosphere Formation*" and "*Living Nature*" of the Russian Academy of Sciences supported me during certain aspects of this work. Tables of photos and figures have been prepared for publication with the invaluable participation of K. Makarov. I am very grateful to all.

References

Babenko, A.B. & Fjellberg, A. (2006): Collembola Septentrionale. A Catalogue of Springtails of the Arctic Regions. – KMK Scientific Press Ltd., Moscow. 190 pp.

Behan-Pelletier, V.M. (1999): Oribatid mite fauna of northern ecosystems: products of evolutionary adaptations or physiological constraints. – *In*: Mitchell, R., Horn, D.J., Needham, G.R. & Welbourn, W.C. (eds.). Acarology IX. Proceedings of IX Inernational Congress of Acarology. Columbus, Ohio, 1996. V. 2: 87–93. Ohio Biological Survey, Columbus.

Bennike, O., Björck, S., Böcher, J. & Walker, I.R. (2000): The Quaternary arthropod fauna of Greenland: a review with new data. – *Bulletin of the Geological Society of Denmark* 47: 111–134.

Böcher, J. (1988): The Coleoptera of Greenland. – *Meddelelser om Grønland, Bioscience* 26: 1–100.

Bregetova, N.G. (1956): Gamasid Mites (Gamasoidea). *Identification Keys of the USSR Fauna.* V. 61. – Academy of Sciences of the USSR, Moscow-Leningrad. 248 pp. [In Russian].

Byzova, J.B., Uvarov, A.V. & Petrova, A.D. (1995): Seasonal changes in communities of soil invertebrates in tundra ecosystems of Hornsund, Spitsbergen. – *Polish Polar Research* 16(3–4): 245–266.

Chant, D.A. & Hansell, R.I.C. (1971): The genus *Amblyseius* (Acarina: Phytoseiidae) in Canada and Alaska. – *Canadian Journal of Zoology* 49(5): 705–758.

Chernov, Yu.I. (2004): The animal world of the polar desert on the Devon Island Plateau, Canadian Arctic Archipelago. – *Entomological Review* 84 (Suppl. 1): S15–S24.

Chetverikov, Ph.E. (2006): Eriophyid mites of the subfamily Phyllocoptinae (Acari, Eriophyidae) from sedges (Cyperaceae). – *Acarina* 14 (1): 57–68.

Coulson, S.J., Convey, P., Aakra, K., Aarvik, L., Ávila-Jiménez, M.L., Babenko, A., Biersma, E., Boström, S., Brittain, J.E., Carlsson, A., Christoffersen, K.S., De Smet, W.H., Ekrem, T., Fjellberg, A., Füereder, L., Gustafsson, D., Gwiazdowicz, D.J., Hansen, L.O., Holmstrup, M., Hullé, M., Kaczmarek, L., Kolicka, M., Kuklin, V.,

Lakka, H.-K., Lebedeva, N., Makarova, O., Maraldo, K., Melekhina, E., Ødegaard, F., Pilskog, H.E., Simon, J.C., Sohlenius, B., Solhøy, T., Søli, G., Stur, E., Tanasevitch, A., Taskaeva, A., Velle, G., Zawierucha, K. & Zmudczynska-Skarbek, K. (2014): The terrestrial and freshwater invertebrate biodiversity of the archipelagoes of the Barents Sea; Svalbard, Franz Josef Land and Novaya Zemlya. – *Soil Biology & Biochemistry* 68: 440–470.

Danks, H.V. (1981): Arctic Arthropods. – Tyrell Press Limited, Ottawa. 608 pp.

Danks, H.V. (1990): Arctic Insects: instructive diversity. – *In*: Harington, C.R. (ed.). Canada's Missing Dimension: Science and History in the Canadian Arctic Islands. V. II: 444–470. Canadian Museum of Nature, Ottawa.

Downes, J.A. (1988): The post-glacial colonization of the North Atlantic Islands. – *Memoirs of the Entomological Society of Canada* 144: 55–92.

Dubinin, V.B. (1962): The superclass Chelicerata. – *In*: Rodendorf, B.B. (ed.). Osnovy paleontologii [Fundamentals of Paleontology]: 375–530. Academy of Sciences of the USSR, Moscow [In Russian].

Dunlop, J.A. & Alberti, G. (2007): The affinities of mites and ticks: a review. – *Journal of Zoological Systematics and Evolutionary Research* 46(1): 1–18.

Evans, G.O. (1992): Principles of Acarology. – CAB International, Wallingford. 563 pp.

Evans, G.O. & Till, W.M. (1979): Mesostigmatic mites of Britain and Ireland (Chelicerata: Acari: Parasitiformes). An introduction to their external morphology and classification. – *Transactions of the Zoological Society of London* 35: 139–270.

Fabricius, O. (1780): Fauna Groenlandica. – Hafniae et Lipsiae, Copenhagen and Leipzig. 211 pp.

Filippova, N.A (1977): Ixodid Ticks of the Subfamily Ixodinae. – *Fauna of the USSR. Arachnida* N.S. V. 4(4): 1–396. "Nauka" Publishing House, Leningrad [In Russian].

Gill, D. & Strandtmann, R.W. (1977): Ectoparasites of the collared lemming (*Dicrostonyx torquatus*) on Bathurst Island, N.W.T., Canada. – *Journal of Medical Entomology* 14(1): 101–106.

Graversen, C.B. (1931). Notizen über grönländische Oribatiden. – *Meddelelser om Grønland* 91(2): 1–18.

Gwiazdowicz, D.J. & Coulson, S.J. (2011): High Arctic gamasid mites (Acari, Mesostigmata): community composition on Spitsbergen, Svalbard. – *Polar Research* 30(8311): 1–7.

Haarløv, N. (1942): A morphologic-systematic-ecological investigation of Acarina. – *Meddelelser om Grønland* 128(1): 1–71.

Haarløv, N. (1967): Arthropoda (Acarina, Diptera) from subfossil layers in West Greenland. – *Meddelelser om Grønland* 184(3): 1–17.

Hammer, M. (1937): A quantitative and qualitative investigation of the microfauna communities of the soil at Angmagssalik and in Mikis Fjord. – *Meddelelser om Grønland* 108(2): 1–53.

Hammer, M. (1944): Studies on the oribatids and collemboles of Greenland. – *Meddelelser om Grønland* 141(3): 1–210.

Hammer, M. (1946): The zoology of east Greenland. Oribatids. – *Meddelelser om Grønland* 122(1): 1–39.

Hammer, M. (1952): The oribatid and Collembola fauna in some soil samples from Søndre Strømfjord. – *Entomologiske Meddelelser* 26(5): 404–414.

Hammer, M. (1953): Collemboles and oribatids from the Thule District (North West Greenland) and Ellesmere Island (Canada) collected by J.C. Troelsen and Chr. Vibe. – *Meddelelser om Grønland* 136(5): 1–16.

Hammer, M. (1954): Collemboles and oribatids from Peary Land (North Greenland). – *Meddelelser om Grønland* 127(5): 1–28.

Henriksen, K.L. & Lundbeck, W. (1917): Grønlands Landarthropoder (Insecta et Arachnida). – *Meddelelser om Grønland* 22: 481–821.

Hodkinson, I.D., Babenko, A., Behan-Pelletier, V., Böcher, J., Boxshall, G., Brodo, F., Coulson, S.J., De Smet, W., Dozsa-Farkas, K., Elias, S., Fjellberg, A., Fochetti, R., Foottit, R., Hessen, D., Hobaek, A., Holmstrup, M., Koponen, S., Liston, A., Makarova, O., Marusik, Yu.M., Michelsen, V., Mikkola, K., Mustonen, T., Pont, A., Renaud, A., Rueda, L.M., Savage, J., Smith, H., Samchyshyna, L., Velle, G., Viehberg, F., Vikberg, V., Wall, D.H., Weider, L.J., Wetterich, S., Yu, Q., & Zinovjev, A. (2013): Chapter 7. Terrestrial and Freshwater Invertebrates. – *In*: Meltofte, H. (ed.). Arctic Biodiversity Assessment. Status and Trends in Arctic Biodiversity: 195–223. Conservation of Arctic Flora and Fauna, Akureyri.

Hunter, P.E. & Husband, R.W. (1973): *Pneumolaelaps* (Acarina) mites from North America and Greenland. – *Florida Entomologist* 56(2): 77–91.

Jørgensen, M. (1934): Ein revidiertes Verzeichnis über grönländische Milben. – *Zoologischer Anzeiger* 107(1–2): 40–47.

Karg, W. (1984): Zur Kenntnis der Untergattung *Pneumolaelaps* Berlese, 1925 der Raubmilbengattung *Hypoaspis* Canestrini, 1884. – *Deutsche Entomologische Zeitschrift, Neue Folge* 31(1–3): 35–39.

Karg, W. (1985): Die mit Arten der Gattung *Bombus* Latreille vergesellschafter Raubmilben der Gattung *Parasitus* Latreille, 1795 (Acarina, Parasitiformes). – *Zoologischer Jarbücher, Syst.* 112: 525–535.

Kethley, J. (1982): Acariformes. Prostigmata. – *In*: Parker, S.P. (ed.). Synopsis and Classification of Living Organisms 2: 117–145. McGraw Hill, New York.

Kethley, J. (1990): Acarina: Prostigmata (Actinedida). – *In*: Dindal, D.E. (ed.). Soil Biology Guide: 667–756. J. Wiley & Sons Publ., New York.

Kethley, J.B., Norton, R.A., Bonamo, P.M. & Shear, W.A. (1989). A terrestrial alicorhagiid mite (Acari: Acariformes) from the Devonian of New York. – *Micropaleontology* 35: 367–373.

Kramer, P. (1897): Grönländische Milben. Zoologische Ergebnisse der von der Gesellschaft für Erdkunde zu Berlin unter Leitung Dr. von Drygalski aus gesandten Grönlandexpedition nach Dr. Vanhoffen's Sammlung bearbeitet, VI. – *Bibliographia Zoologica* 20: 77–83.

Krantz, G.W. (1978): A Manual of Acarology. 2nd edition. – Oregon State University Bookstores, Corvallis. 509 pp.

Krivolutsky, D.A. (1976): Role of oribatid mites in biogeocoenoses. – *Zoologicheskii Zhurnal, Moscow* 55(2): 226–236 [In Russian].

Lange, A.B. (1955): Order Parasitiformes (Reuter) Zachv. 1. Suborder Mesostigmata Can. General review. – *In*: Bregetova, N.G. (ed.) Mites of Rodents of the USSR Fauna. *Identification Keys of the USSR Fauna.* V. 59: 217–243. Academy of Sciences of the USSR, Moscow-Leningrad [In Russian].

Lindquist, E.E. (1984): Current theories on the evolution of major groups of Acari and on their relationships with other groups of Arachnida, with consequent implications for their classification. – *In*: Griffiths, D.A. & Bowman, C.E. (eds.). Acarology VI, 1: 28–62. J.Wiley & Sons Publishing, New York.

Lindquist, E.E., Krantz, G.W. & Walter, D.E. (2009): Chapter 8. Classification. – *In*: Krantz, G.W. & Walter, D.E. (eds.). A Manual of Acarology. 3rd Edition: 97–103. Texas Tech. University Press, Lubbock.

Makarova, O.L. (2002a): Acarocenoses (Acariformes, Parasitiformes) in polar deserts: 1. Mite assemblages of the Severnaya Zemlya Archipelago: structure of fauna and abundance. – *Entomological Review* 82(7): 839–856.

Makarova, O.L. (2002b): Acarocenoses (Acariformes, Parasitiformes) in polar deserts: 2. Cenotic relations, structure of communities, and the proportion of suborders. – *Entomological Review* 82(7): 857–875.

Makarova, O.L. (2013): Gamasid mites (Parasitiformes, Mesostigmata) of the European Arctic and their distribution patterns. – *Entomological Review* 93(1): 113–133.

Makarova, O. & Böcher, J. (2009): Diversity and geographical ranges of Greenland mites (Acari: Oribatida and Mesostigmata). – *In*: Golovatch, S.I., Makarova, O.L., Babenko, A.B. & Penev, L.D. (eds.). Species and Communities in Extreme Environments: 165–186. Pensoft Publishers & KMK Scientific Press, Sofia-Moscow.

Maraun, M., Schatz, H. & Scheu, S. (2007): Awesome or ordinary? Global dibersity pattern of oribatid mites. – *Ecography* 30: 209–216.

Moritz, M. (1976): Revision der europäischen Gattungen und Arten der Familie Brachychthoniidae (Acari, Oribatei). Teil 2. – Mitteilungen aus dem Zoologischen Museum in Berlin 52: 227– 319.

Norton, R.A. & Behan-Pelletier, V.M. (2009): Chapter 15. Suborder Oribatida. – *In*: Krantz, G.W. & Walter, D.E. (eds.). A Manual of Acarology. 3rd Edition: 430–564. Texas Tech Univ. Press, Lubbock.

Norton, R.A., Bonamo, P.M., Grierson, J.D. & Shear, W.A. (1988): Oribatid mite fossils from a terrestrial Devonian deposit near Gilboa, New York State. – *Journal of Paleontology* 62(2): 259–269.

Norton, R.A., Kethley, J.B., Johnston, D.E. & O'Connor, B.M. (1993): Phylogenetic perspectives on genetic systems and reproductive modes of mites. – *In*: Wrench, D.L. & Ebbert, M.A. (eds.). Evolution and Diversity of Sex Ratio in Insects and Mites: 8–99. Chapman & Hall, New York.

O'Connor, B.M. (2009): Chapter 16. Cohort Astigmatina. – *In*: Krantz, G.W. & Walter, D.E. (eds.). A Manual of Acarology. 3rd Edition: 565–658. Texas Tech Univ. Press, Lubbock.

Rostrup, S. (1900): Grønlandske Phytoptider. – *Videnskabelige Meddelelser fra den naturhistorisk Forening i København*: 241–249.

Sellnick, M. (1960): *Zercon hammerae*, nov. spec., eine neue Milbenart aus Ost-Grönland (Acarina, Zerconinae). – *Entomologiske Meddelelser* 29: 216–220.

Shultz, J.W. (2007): A phylogenetic analysis of the arachnid orders based on morphological characters. – *Zoological Journal of the Linnean Society* 150: 221–255.

Southcott, R.V. (1992): Revision of the larvae of *Leptus* Latreille (Acarina: Erythraeidae) of Europe and North America, with descriptions of post-larval instars. – *Zoological Journal of the Linnean Society* 105: 1–153.

Strenzke, K. von, de Lesse, H. & Denis, J. (1955): Microfaune du sol de l'Eqe, Groenland; vol. 1, arachnides. Hermann, Paris. [Expéditions polaires francaises, VII; pub. as Actualités scientifiques et industrielles, 1232.] 84 pp.

Thomas, R.H. & MacLean, S.E. (1988): Community structure in soil Acari along a latitudinal transect of Tundra sites in Northern Alaska. – *Pedobiologia* 31: 113–138.

Trägårdh, I. (1904): Monographie der arktischen Acariden. – *Fauna Arctica* 4: 1–78.

Trägårdh, I. (1912): Acari. Danmarks-Ekspeditionen til Grönlands Nordøstkyst 1906–1908, Bd. III, No. 14. – *Meddelelser om Grønland* 43: 417–426.

van der Hammen, L. (1989): An Introduction to Comparative Arachnology. – SPB Academic Publishing, The Hague. 576 pp.

Walter, D.E. & Krantz, G.W. (2009): Chapter 7. Collection, Rearing, and Preparing Specimens. – *In*: Krantz, G.W. & Walter, D.E. (eds.). A Manual of Acarology. 3rd Edition: 83–96. Texas Tech Univ. Press, Lubbock.

Walter, D.E., Krantz, G. & Lindquist, E.E. (1996): Acari. The Mites. Version 13, December 1996. – http://tolweb.org/Acari/2554/1996.12.13. *In*: The Tree of Life Web Project, http://tolweb.org/.

Walter, D.E. & Proctor, H.C. (2013): Mites: Ecology, Evolution & Behaviour. Life at a Microscale. 2nd Edition – Springer, Dordrecht – Heidelberg – New York – London. 494 pp.

Walter, D.E. & Proctor, H.C. (2001): Mites in Soil. An Interactive Key to Mites & Other Soil Microarthropods. – http://www.publish.csiro.au/web_cds_demo/mites/UserGuide.htm.

Zakhvatkin, A.A. (1941): Tyroglyphoid mites (Tyroglyphoidea). – *Fauna of the USSR. Arachnida.* V. 6(1): 1–474. Academy of the Sciences, Moscow-Leningrad [In Russian].

Zakhvatkin, A.A. (1952): The division of the Acarina into orders and their position in the system of the Chelicerata. – *In*: *Parazitologicheskii Sbornik* 14: 5–46. Zoologicheskii Institut AN SSSR, Leningrad [In Russian].

Zhang, Z.-Q. (2011): Phylum Arthropoda von Siebold, 1848. – *In*: Zhang, Z.-Q. (ed.). Animal Biodiversity: An Outline of Higher-Level Classification and Survey of Taxonomic Richness. – *Zootaxa* (special volume) 3148: 99–103.

18.3.2. Mesostigmata (Gamasida, Gamasid mites)

Olga Makarova

A large cosmopolitan group of medium-sized mites. The suborder comprises about 109 families and 11 400 described species (Beaulieu et al. 2011). Free-living mesostigmatans inhabit soil, litter, mosses, plants, and nests of different animals, being mainly predators of other arthropods, nematodes, enchytraeids, *etc.* A few taxa are considered mycetophagous, detritophagous, facultatively polleniphagous, *etc.* Most gamasids inhabiting enriched temporal habitats have established phoretic associations with other animals, mainly insects (Figs 17I, 21D). Besides that, among Mesostigmata there are numerous well adapted parasites (internal and external) of insects and vertebrates with different level of specificity. Until now only 20 species of Mesostigmata were recorded in Greenland (Trägårdh 1904, 1931; Henriksen & Lundbeck 1917; Jørgensen 1934; Haarløv 1942; Sellnick 1960; Hunter & Husband 1973; Karg 1984, 1985). During the preparation of this book the number was increased to 63, but new finds are possible, especially in southern Greenland.

Morphology

Body length varies within the limits of 200–2500 µm. Two tagmata are discernible: the *idiosoma*, IS, and *gnathosoma*, GS (Figs 2C). Idiosoma is without sejugal furrow (Figs 1A–E) and consists of *podosoma* (anterior body part with legs I–IV) and *opisthosoma* (posterior body part). Sometimes a slight constriction between podosoma and opisthosoma is noticeable. Idiosoma may be pale with hardly visible sclerites, but usually it is covered partially or entirely by a number of shields (red-brown, amber, orange, yellowish). The interscutal membrane may be striated or granulated and is normally whitish-yellow. Seen from above, podosoma and opisthosoma are covered by the *podonotal* and *opisthonotal shields* respectively (Fig. 1B, *psh* and *osh*), or an entire *holodorsal shield* is present (Fig. 1C, *hsh*), which sometimes may be possessed by lateral incisions (Figs 6F, 8A). In some members of the cohort Uropodina, the holodorsal shield is surrounded by an entire or subdivided *marginal shield* (Fig. 1D, E, *msh*). Rarely the opisthonotal shield is divided in several sclerites (Fig 1A). The dorsal chaetome may be oligotrichous (44 pairs of setae), hypotrichous (sometimes less than 20 pairs of setae) or hypertrichous. The dorsal chaetotaxy (number and disposition of setae) is of great taxonomical importance. Several nomenclature systems have been proposed for a designation of dorsal setae and poroids (gland pores and slit-organs). The ventral sclerotization varies significantly. Near the posterior margin of the *gnathosoma,* GS, the *tritosternum* is present (Figs 1G, 2C, *trt*) that consists of the rectangular or trapezoid *base* and 1–2 *laciniae*. In females normally the paired *presternal shields, prsh,* the entire *sternal shield, stsh* (with 3 pairs of setae), and the *genital aperture* are located in the intercoxal region (Fig. 2C). *Metasternal shields, mstsh,* with a pair of setae are often developed behind the sternal shield. The female genital orifice is covered by the *genital shield, gnsh,* the anterior margin of which may be rounded (Fig. 16G), attenuated into a spike (Fig. 4A) or endowed with a hyaline flap (Fig. 14A). Beneath the genital shield, *vaginal sclerite(s)* (Fig. 6J, *vg*) or the intricate *endogynium* (Figs 5F, M) are sometimes developed. Rarely the small *adgenital shields* are formed on each side of genital shield (Figs 15A, D, *adg*). The *anus, an,* is subterminal ventral (Fig. 2C). Females often have a separate *anal shield, ansh,* bearing 3 *perianal setae* (1 *postanal seta, pa,* and paired *adanal setae, ad*) surrounding small anal valves. Rarely a pair of *euanal setae* is present on the anal valves. The area of aciculae, termed the *cribrum, crb,* is situated posteriad the postanal seta. The anal shield may be united with the *ventral shield, vnsh,* into a *ventrianal shield* bearing ventral and perianal setae (Fig. 10I). The genital shield may also be fused with the ventral shield into an *opisthogenital shield* (Fig. 13A) or a large *hologastric shield* (merged genital, ventral and anal shields) is present (Fig. 13L). *Stigmata, stg,* are situated laterad coxae II–IV and associated with *peritremes, prt,* of different length flanked by the *peritrematal shields, prtsh* (Fig. 2C). The last are often partially or fully fused with *exopodal shields, exsh,* and a dorsal shield. Behind the peritrematal shields 1–2 pairs of *metapodal shields, mpdsh,* are usually present in Dermanyssina (Figs 2C, 11A). The ventral sclerotization of males differs markedly from that of females. In males the genital orifice covered by the *genital valve, gnv,* is located either near the anterior margin of the *sternitigenital shield* (Fig. 2D) or within a shield at the level of coxae II–III (in this case the genital aperture is protected by 2 small sclerites, the anterior of which often has a pair of *eugenital setae*; Fig. 2E). Behind the sternitigenital shield often a large ventrianal shield is developed (Fig. 9K), or an entire *holoventral shield* may be formed. Rarely the ventral and anal shields are free (Fig. 8I).

The well visible, tubular *gnathosoma* is movably articulated to the idiosoma (Fig. 2C). The anterior margin of the *gnathosomal base, gnb,* dorsally forms the *tectum, tec* (Fig. 1F), which usually has one, several or numerous projections (Fig. 1J) and rarely is flap-like, smooth (Fig. 13B). The ventral surface of the gnathosoma, or *subcapitulum, scp,* is anteriorly extended into the *hypostome, hyp,* which bears 3 pairs of *hypostomal setae* (*C*1–3) and horn-like *corniculi, co* (Fig. 1G). Another pair of setae *C*4 (*palpcoxal setae*) is inserted in the middle of the subcapitulum. Between these setae the *deutosternum, dt,* or subcapitular groove, is located. It is provided with several rows of small denticles. The *palps, pp,* consist of *trochanter* (*tr*), *femur* (*fm*), *genu* (*gn*), *tibia* (*tib*) and *tarsus* (*tar,* Fig. 1F), of which the last is endowed with a 2–4–tined claw-like movable structure (Fig. 1H) or *apotele, ap* (modified *ambulacrum*). The *chelicera, ch,* normally comprises 3 joints. Two distal joints are often dentate and form the proper *chela*. The *fixed digit, d.f.,* has a single seta located dorsally at the base of chela and usually a setiform sensory structure, *pilus dentilis* (*p.d.*) is present ventrally (Fig. 1L). The arthrodial membrane at the base of the *movable digit, d.m.*, is often produced into a coronet of setulae (Fig. 4K) or 1–2 brush-like processes (Fig. 12F). In males the chelicera may be adapted for sperm transfer. In the cohort Dermanyssina the male movable digit is provided with a *spermatodactyl, smd* (free or partially free process, Fig. 1I). In males of the cohort Parasitina the *spermatotreme, smt* (hole-like or slit-like opening) is developed on the movable digit (Fig. 1L).

The *legs* comprise 7 free podomeres (Fig. 2C): *coxa* (*cx*), *trochanter* (*tr*), *femur* (*fm*), *genu* (*gn*), *tibia* (*tib*), *tarsus* (*tar*), and *ambulacrum* (*amb*). The femora and tarsi II–IV are secondarily divided by a *peripodomeric suture* (consisting of slit-organs) into *basifemur* and *telofemur* or into *basitarsus* and *telotarsus*, respectively. The ambulacrum (Fig. 1K) usually comprises the *pretarsus* (*pt*), 2 *claws* (*cl*), and the *pulvillus* (*plv*), which may be sucker-like or consisting of some roundish or pointed lobes. In most females belonging to the cohort Dermanyssina the paired openings (*solenostomes*) of the sperm access system are located in the arthrodial cuticle of *acetabula* III or IV, or on the legs III. They are connected with paired ducts, which are united in a median *sacculus* (Fig. 2A). In some taxa paired sclerotic *spermathecae* are present (Fig. 2B). The strongly sclerotized part of the spermatheca is subdivided into *atrium* and *calyx* opening into the thin-walled *vesicle*. In males leg II and/or IV are often armoured with apophyses or spurs (hypertrophied setae or cuticular excrescences, Figs 6C, 12B) which help to hold the female during copulation.

Life cycle

Mesostigmatans pass through 5 stages: egg, larva (sometimes suppressed), protonymph, deutonymph, and adult. Oviparity is common, but ovoviviparity and larviparity are also known in several taxa. The larva is hexapod. The octopod nymphs lack a genital orifice. The protonymph always has insignificant peritremes. A heteromorphic phoretic deutonymph is present in the ontogenesis of some Uropodina. The reproduction may be sexual or parthenogenetic (arrhenotokous or thelytokous). The sperm transfer is by spermatophores during copulation, and commonly the male chelicerae are involved.

Identification and information sources

The identification keys are based mainly on adult (often female) features. In some cases deutonymphal characteristics are used. The given length corresponds to the idiosomal length (without gnathosoma) with some exceptions (*Arctoseius*), when the length of a dorsal shield is used. Characters used in the keys are adjusted to the Greenland fauna composition. The comprehensive work by Evans & Till (1979) was broadly used in the compiling of keys for higher taxa. Rank of higher taxa follows Norton et al. (1993), though some significant changes were suggested recently (Lindquist et al. 2009). Numerous distributional and biological data were obtained from the important publications by Bregetova et al. (1977), Behan (1978), Evans (1992), Karg (1993), Walter & Proctor (2013).

Key to cohorts

ADULTS

1. Hypostomal (*C*1–3) and palpcoxal (*C*4) setae forming more or less a longitudinal row of setae on either side of deutosternum (Fig. 16C); tritosternum with base wider than long and closed to coxae I or with narrow base partially or entirely covered by the greatly enlarged coxae I (Fig. 17D); genital shield in female usually entirely located within intercoxal region (Figs 16G, 17B); genital orifice of male between coxae II–IV (Fig. 17C); pedofossae (depressions into which legs II–IV can be withdrawn for protection) often present; idiosoma with entire (Fig. 17A) or fragmented (Fig. 16A) marginal shield; male chelicera lacking spermatodactyl (Fig. 16J) .... Uropodina

– Posterior hypostomal setae (*C*2–3) usually forming more or less transverse row (Figs 1G, 3C, 10H);

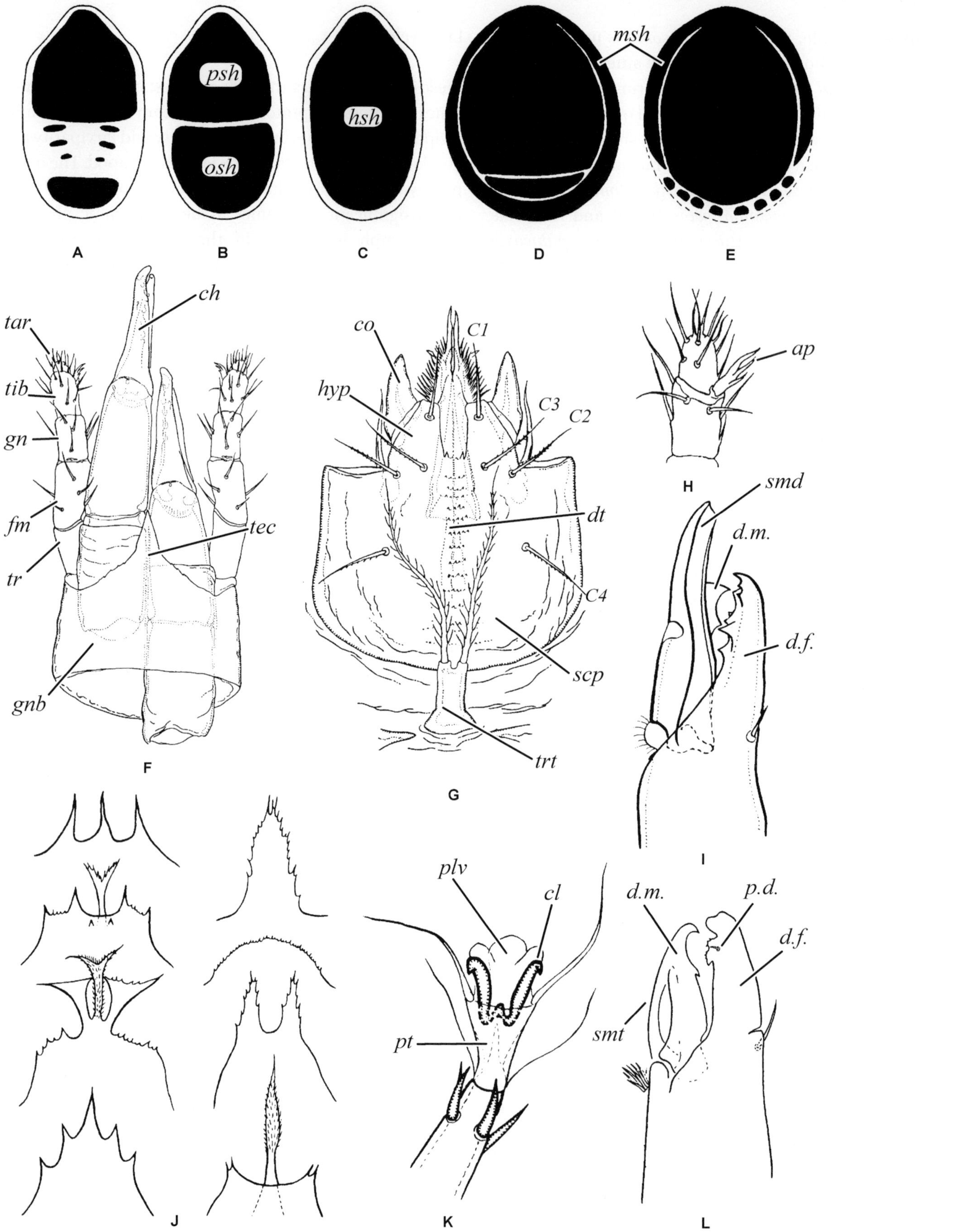

FIGURE 1 *A–E, adult dorsal sclerotization of mesostigmatic mites, variants; F,* Euryparasitus *sp., female, gnathosoma, dorsal aspect, G, subcapitulum; H,* Parasitus *sp., female, tibia and tarsus of palp; I,* Arctoseius *sp., male, chela, K, female, ambulacrum II; J, gnathotectum of mesostigmatic mites, variants; L,* Vulgarogamasus *sp., male, chela. F–H, after Lange 1955.*

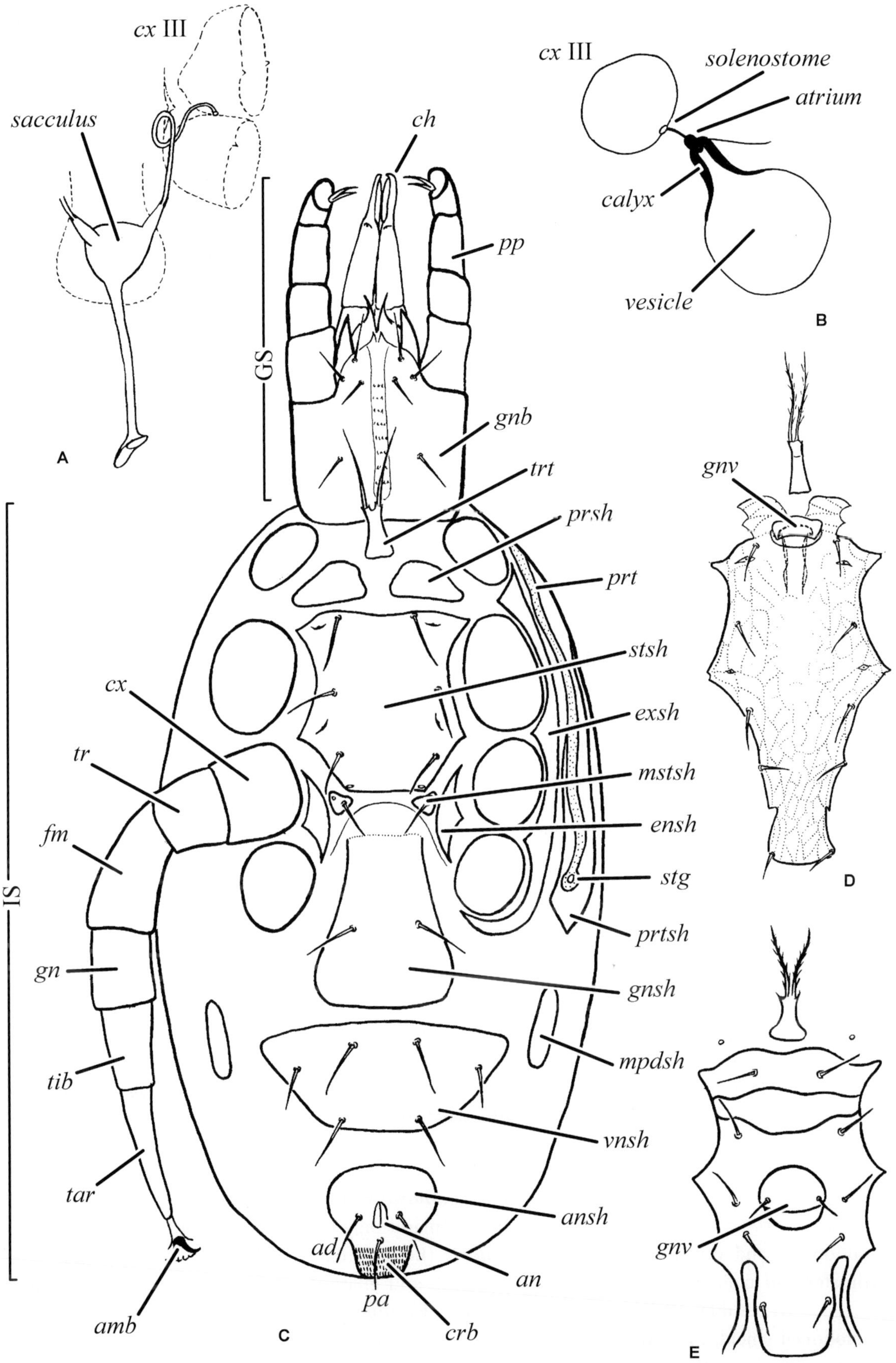

FIGURE 2 *A,* Arctoseius babenkoi, *female, spermatheca; B,* Neoseiulus *sp., female, spermatheca; C, female ventral sclerotization of gamasid mites, schematic; D,* Arctoseius *sp., male, sternitigenital shield; E,* Zercon *sp., male, sternitigenital shield. A, after Makarova 2000.*

tritosternum commonly biramous and widely removed from coxae I (Figs 7L, 8C), rarely reduced to tiny sclerite (Fig. 5C) or absent; no pedofossae; idiosoma never with marginal shield; male chelicera with (Figs 8F, 13H) or without spermatodactyl, or with spermatotreme (Figs 1L, 4K) 2

2. Idiosoma broadly triangular in outline and covered by 2 subequal shields whose lateral margins are serrated and possessed by setae (Figs 15E, F); opisthonotal shield usually with 4 large fossa-like structures near its posterior margin; peritreme short, not extending beyond the anterior margin of coxa III and often inverted comma-shaped (Figs 15D, G); genital orifice of male between coxae II and III (Fig. 15C); movable digit of male chelicera without spermatodactyl; leg II of male unarmed; palp apotele 2-tined .... Zerconina, p. 730

– Outline of idiosoma ovate or pyriform (Figs 6A, 7K); margins of dorsal shield(s) commonly not serrated; opisthonotal fossa-like structures absent; peritreme of different length, often long (Figs 9J, 11A); male chelicera with spermatodactyl (Fig. 8E) or spermatotreme (Fig. 1L) 3

3. Metasternal shields in female large, flanking triangular genital shield (Figs 3B, 4A); genital shield with well sclerotized anterior spike; movable digit of male chelicera without spermatodactyl, but with spermatotreme (Fig. 4K) .... Parasitina, p. 710

– Without large metasternal shields in female; anterior margin of female genital shield more often rounded, usually membranous (Figs 8C, 11A); male chelicera with spermatodactyl (Figs 9H, 13H) .... Dermanyssina, p. 716

PARASITINA

Contains one family, Parasitidae. Mites of this family mostly predators, free-living in humus, litter, mosses (subfamily Pergamasinae) or in temporal habitats (tidal debris, animal nests, droppings, carrion, *etc.*) and their deutonymphs disperse on insects (most of Parasitinae).

PARASITIDAE

Key to genera

ADULTS

1. Female with podonotal and opisthonotal shields (Figs 3A, 4C); tritosternum in male well developed or absent .... subfamily Parasitinae, 2

– Female with holodorsal shield; male with base of biramous tritosternum strongly reduced and covered by lamina of genital orifice (Fig. 5C) .... subfamily Pergamasinae, 4

2. Opisthogastric shield in both sexes (or opisthogastral region in deutonymphs) hypertrichous, typically with more than 20 pairs of setae (Fig. 3H); male with normally developed tritosternum; associated with *Bombus* .... *Parasitellus*

– Opisthogastric shield in adults with not more than 15 pairs of setae (Fig. 4D); male with or without tritosternum 3

3. Corniculi in adults and deutonymph long, slender, parallel and extending beyond anterior margin of palp trochanter (Fig. 3C); anterior margin of opisthonotal shield strongly concave (Figs 3A, D); male without tritosternum .... *Cornigamasus*

– Corniculi short (Fig. 4F), not extending to anterior margin of palp trochanter; anterior margin of opisthonotal shield usually straight (Figs 4C, E); male with normally developed tritosternum.............. .... *Vulgarogamasus*

4. Gnathotectum with 5 projections (Fig. 6B); both sexes with opisthonotal region of holodorsal shield polytrichous (Fig. 6A), female with more than 30 pairs of setae, male with more than 45 pairs of setae ....*Pergamasus*

– Gnathotectum trident (Fig. 5G); both sexes with opisthonotal region of holodorsal shield (with or without transverse suture in the male) oligotrichous (Fig. 5H), with a maximum of 25 pairs of setae in female and 35 pairs in male......................... 5

5. In female 1 pair of rather big, long presternal shields (Fig. 5K); in male movable digit of chelicera with 1 tooth (Fig. 5I).......................... *Paragamasus*

– In female between 2 triangular presternal shields 1, several, or numerous small sclerites developed (Fig. 5A); in male movable digit of chelicera with 2 teeth (Fig. 5D).......................... *Lysigamasus*

PARASITINAE

Members of this subfamily are common in Greenland, even in the far north. They occur in litter and mosses and are especially common in bumblebee nests and in the tidal zone of the sea.

***Cornigamasus* Evans, 1979. Fig. 18A**

In Greenland only one species recorded, ***Cornigamasus lunaris*** (Berlese, 1882). Adults and deutonymphs with corniculi and gnathotectum of very characteristic form (Figs 3C, E), anterior margin of opisthonotal shield undulate (Figs 3A, D). Phoresy of deutonymphs on small dipterans very common. Usually development in organically enriched sites. Female 800–1100 μm, male 710–750 μm. Palaearctic (Europe including Iceland and Novaya Zemlya,

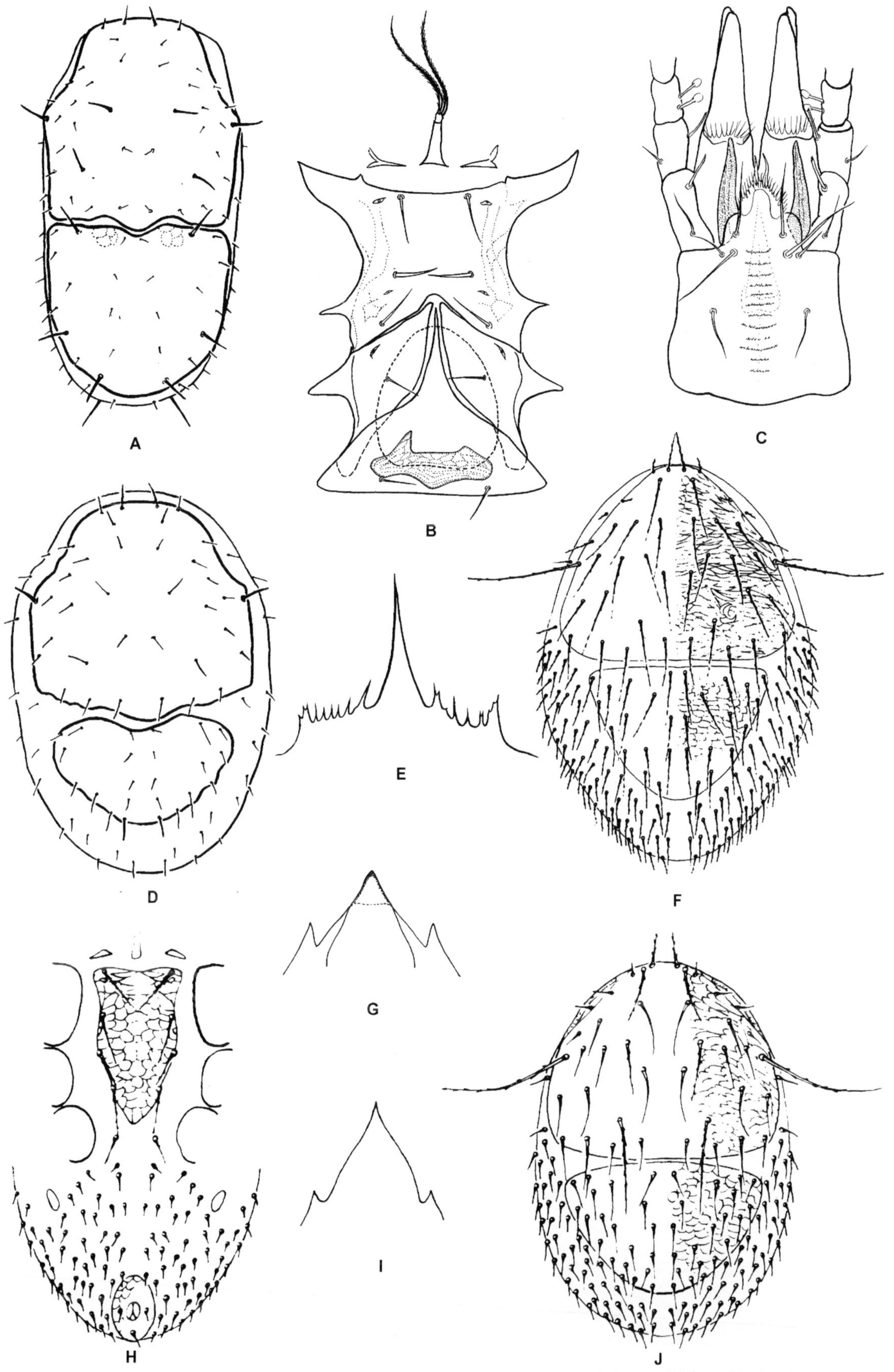

FIGURE 3 *A,* Cornigamasus lunaris, *female, dorsal aspect, B, sternitigenital region, C, gnathosoma ventally, D, deutonymph, dorsal aspect, E, gnathotectum; F,* Parasitellus papei, *deutonymph, dorsal aspect, G, gnathotectum; H,* Parasitellus arcticus, *deutonymph, ventral aspect, I, gnathotectum, J, dorsal aspect. F–J, after Karg 1985.*

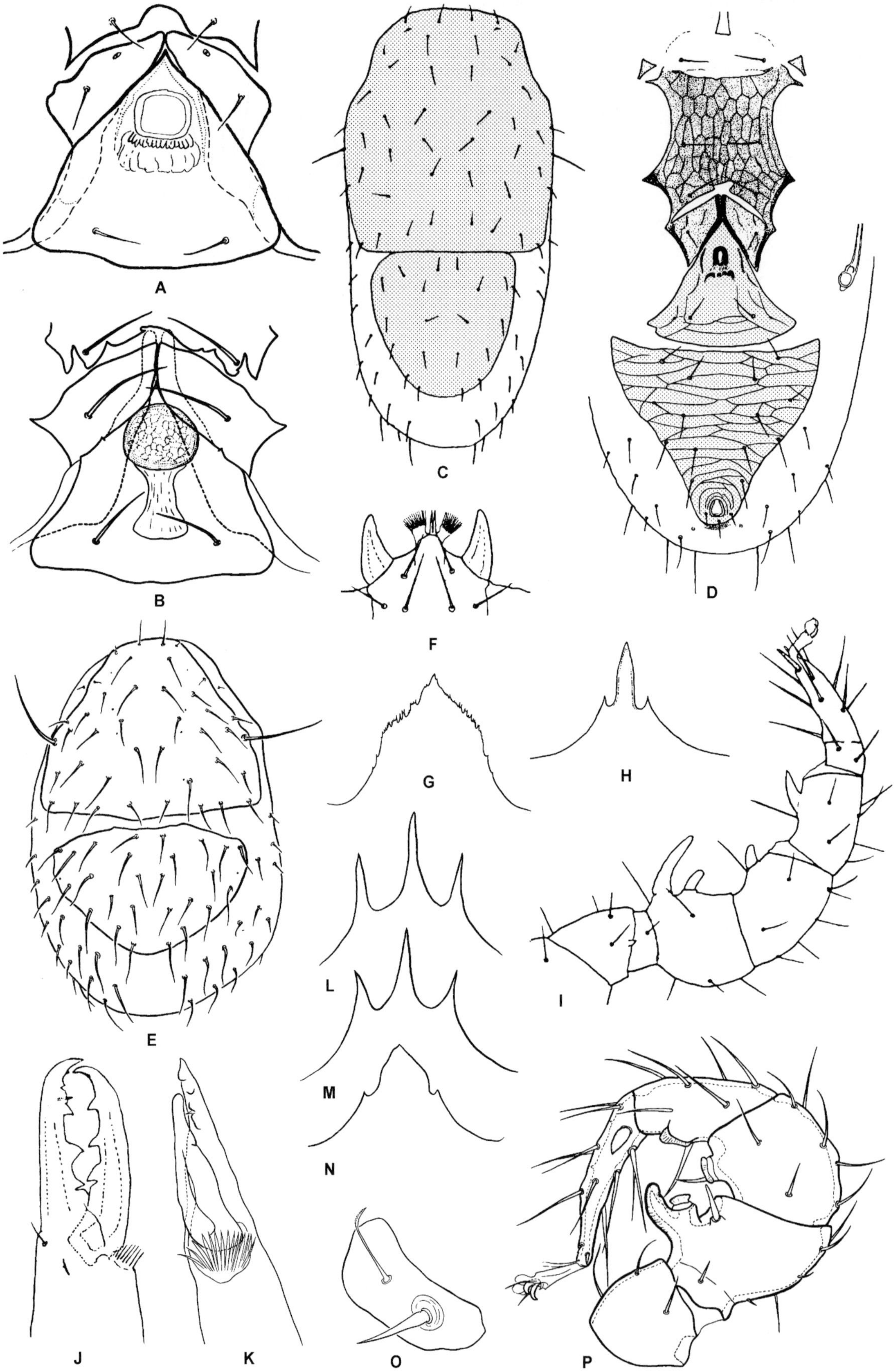

FIGURE 4 *A,* Vulgarogamasus immanis, *female, genital region, G, male, gnathotectum; B,* Vulgarogamasus *sp., female, genital region, M, gnathotectum, E, deutonymph, dorsal aspect, L, gnathotectum, K, male, chela, N, gnathotectum, O, palp trochanter, P, leg II, coxa removed; C,* Vulgarogamasus halophilus, *female, dorsal aspect, D, ventral aspect, F, hypostome, H, gnathotectum, J, chela, I, male, leg II, coxa removed. C, D. F, H–J, after Sellnick 1957.*

Siberia including Taymyr, northern Africa); southeastern Asia.

***Parasitellus* Willmann, 1939. Fig. 18B**

Mites belonging to this genus are associated with bumblebees exclusively. Most common in northern areas of Greenland. Adults infrequent, exist short time in *Bombus*-nests. On insects, only deutonymphs occur.

Key to species

DEUTONYMPHS

1. Opisthonotal shield with 14–15 pairs of strong, long, sparsely barbed setae (Fig. 3F); middle point of gnathotectum robust, stronger than lateral points, obtuse or rounded (Fig. 3G). 940–1190 μm. Circumpolar.............................. *papei* (Karg, 1985)
– Opisthonotal shield with 16–22 pairs of different sized, usually smooth setae, anterior setae twice larger than posterior ones (Fig. 3J); middle point of gnathotectum sharp, attenuated (Fig. 3I). 850–1030 μm. Circumpolar............... *arcticus* (Karg, 1985)

***Vulgarogamasus* Tichomirov, 1969. Figs 18C, D**

Species of this genus are especially common in tidal zone of sea coasts and in mammal nests.

Key to species

ADULTS

1. Enormous mites, more than 2000 μm; gnathotectum with numerous teeth sometimes grouping in clusters (Fig. 4G); in female endogynium with roundish and folded structures (Fig. 4A); sea shore dwellers. Female 2130–2700 μm, male 2040–2350 μm. Northern Europe (including Iceland, Spitsbergen)........ .... *immanis* (Berlese, 1903)= *Parasitus fucicola* Trägårdh, 1904, types tested (Makarova 2014)
– Smaller mites, length less than 1500 μm; gnathotectum trident (Figs 4H, L, M) 2
2. Length of idiosoma not more than 750 μm; in female sternal shield longer than wide, endogynium with large chitinous lump (Fig. 4D); in male gnathotectum like female gnathotectum (Fig. 4H), leg II as in Fig. 4I, palp trochanter and tarsus II without apophysis or spur; on sea coasts. Female 600–750 μm, male 460 μm. Northern Europe................. .... *halophilus* (Sellnick, 1957)
– Length of idiosoma not less than 1000 μm; in female sternal shield wider than long, endogynium with weakly sclerotized circle connected with the “neck” (Fig. 4B); in male middle point of gnathotectum very broad (Fig. 4N), palp trochanter with basal seta as a spur on a small apophysis (Fig. 4O), tarsus II with finger-shaped apophysis (Fig. 4P); in deutonymph (most common stage in soil) gnathotectum with 3 rather thin projections (Fig. 4L), podonotal shield with 13 pairs of setae (Fig. 4E). Female 1280 μm, male 1075 μm, deutonymph 990–1080 μm. Greenland, East Eropean Arctic............................... ....sp. near *oudemansi* (Berlese, 1903)

PERGAMASINAE

Representatives of this subfamily occur mainly in southern areas of Greenland. Free-living in soil, polyphagous predators of small invertebrates. Deutonymphs are never phoretic.

***Lysigamasus* Karg, 1971. Fig. 18G**

In Greenland hitherto known 1 species, ***Lysigamasus runciger*** (Berlese, 1903). In female no toothed structures in endogynium (Figs 5A, B). In male holodorsal shield subdivided by transverse suture, apophyses on leg II large (Fig. 5E). Female 720–860 μm, male 725–840 μm. Palaearctic (including Iceland).

***Paragamasus* Hull, 1918. Figs 18H, I**

Key to species

ADULTS

1. In female endogynium with 2 longitudinal rows of large teeth (Fig. 5F); in male holodorsal shield subdivided by transverse suture (Fig. 5H), apophyses on leg II very strong, especially on femur (rectangular in some aspect), trochanter II also possessed by apophysis (Fig. 5J). Female 980–1120 μm, male 1000–1200 μm. Western Europe (including Iceland) .... *P. (Paragamasus) robustus* (Oudemans, 1902)
– In female endogynium with 2 crown-shaped structures (Fig. 5M); in male holodorsal shield without transverse suture, apophyses on leg II rather small (Fig. 5L). Female 820–970 μm, male 650–750 μm. Western Europe (including Iceland) .... *P. (Aclerogamasus) integer* (Bhattacharyya, 1963)

***Pergamasus* Berlese, 1904. Figs 18E, F**

In Greenland only one species recorded, ***Pergamasus (Pergamasus) crassipes*** (Linnaeus, 1758). This large predatory mite inhabits only southernmost areas of Greenland. Holodorsal shield polytrichous (Fig. 6A). Gnathotectum with 5 projections(Fig. 6B). In female endogynium with forked outgrowth (Fig. 6D), its form varies (Fig. 6E). In male apophyses

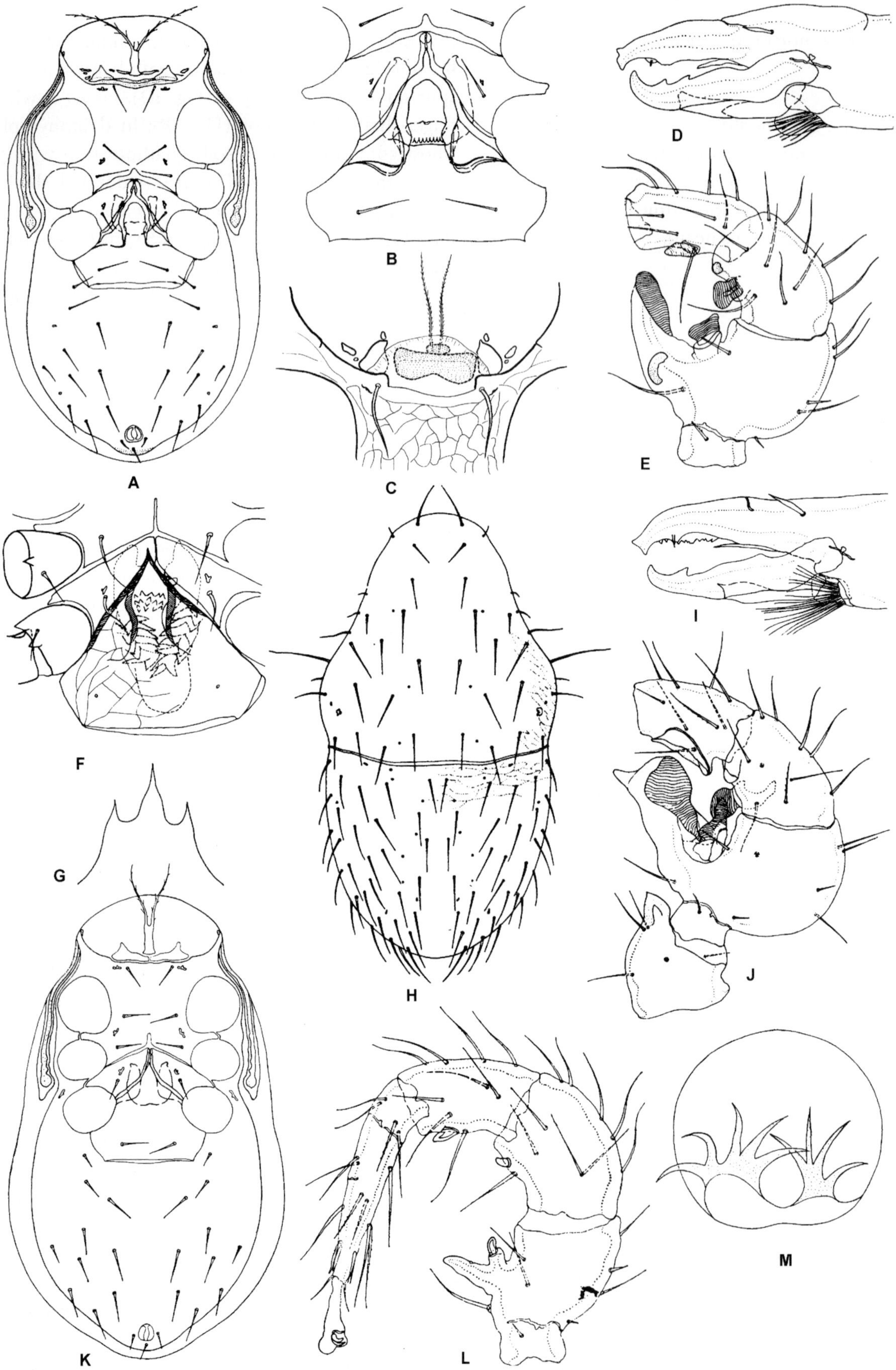

FIGURE 5 *A,* Lysigamasus runciger, *female, ventral aspect, B, genital region, C, male, presternal region, D, chela, E, femur, genu, and tibia of leg II; F,* Paragamasus robustus, *female, genital region, H, male, dorsal aspect, I, chela, J, leg II, coxa and tarsus removed; G,* Paragamasus integer, *female, gnathotectum, K, ventral aspect, M, endogynium, L, male, leg II, coxa and trochanter removed. A, B, D, E, G–M, after Bhattacharyya 1963 F, after Karg 1993.*

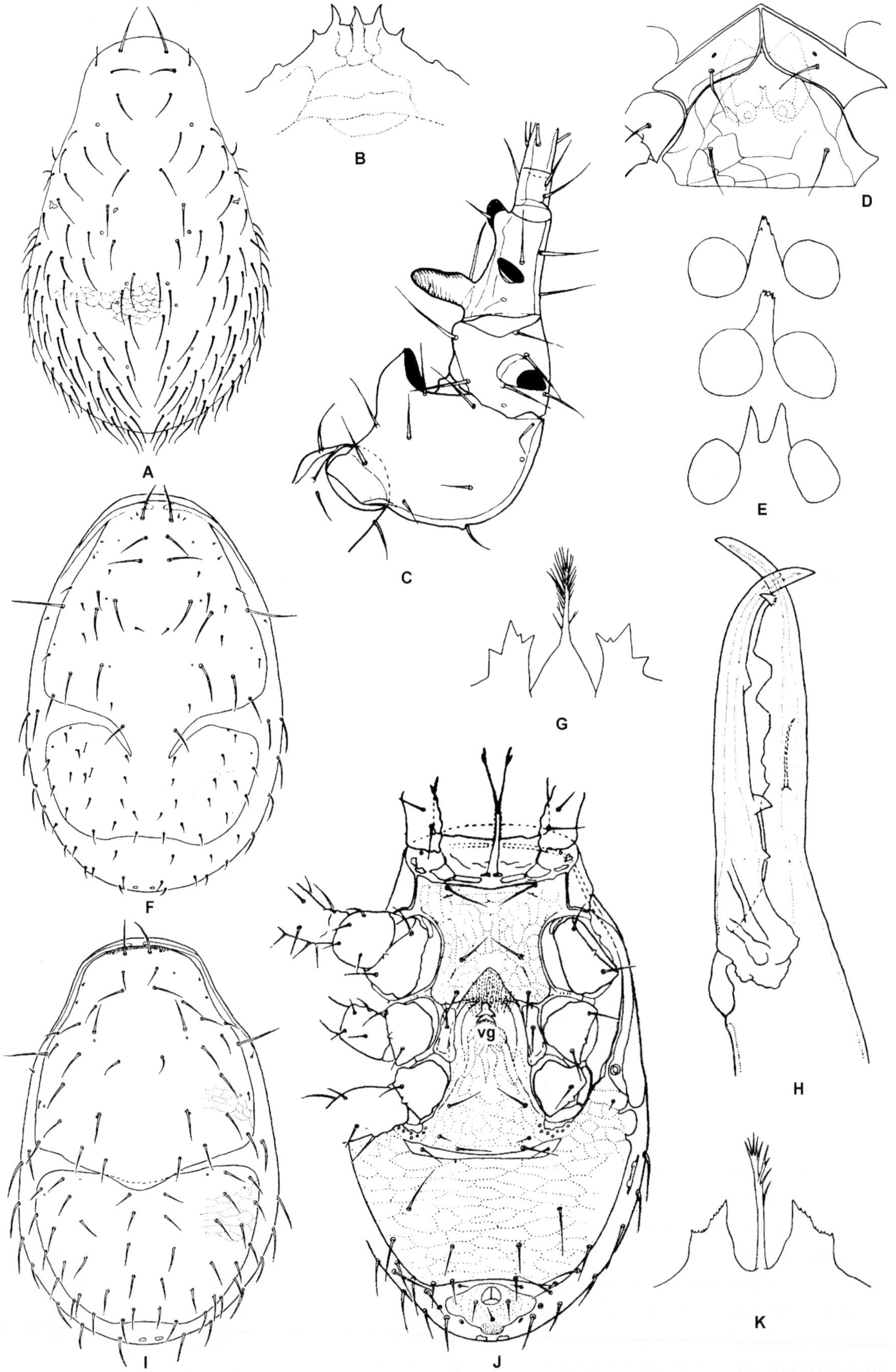

FIGURE 6 *A,* Pergamasus crassipes, *male, dorsal aspect, B, gnathotectum, C, femur, genu, tibia of leg II, D, female, genital region, E, endogynium, variants of form; F,* Veigaia cerva, *female, dorsal aspect, G, gnathotectum, H, chela; I,* Veigaia kochi, *dorsal aspect, J, ventral aspect, K, gnathotectum. A, B, after Bhattacharyya 1963; C–E, after Karg 1993; F–K, after Bregetova et al. 1977.*

on leg II prominent, tibia II with 3 apophyses (Fig. 6C). Female 1360–1500 µm, male 620–1000 µm. Holarctic, South America, Australia, New Zealand.

DERMANYSSINA

Key to families

ADULTS

1. Tritostrenum reduced to small sclerite or absent (Fig. 14L), never with paired laciniae; idiosomal and palpal chaetotaxy markedly reduced (Figs 14K, M), setae often in the form of microsetae; peritreme greatly reduced or absent; parasites of respiratory tract of birds .... Rhinonyssidae, p. 730
– Biramous tritosternum present (Figs 7C, L); extreme hypotrichy of idiosoma rare; trochanter, femur and genu of palp usually with 2, 5, and 6 setae respectively (Fig. 1F); free-living species, nest inhabitants or ectoparasites 2
2. Palp apotele 3- or 4-tined and with closely associated hyaline scale-like appendage (Fig. 7D); in female vaginal sclerite present (Fig. 6J); Greenland species develop without males........... Veigaiidae, p. 725
– Palp apotele without associated hyaline scale-like appendage, 2- or 3-tined (Figs 1H, 13K); in female vaginal sclerite usually absent (excepting some Macrochelidae) 3
3. Two dorsal shields (Fig. 12A), or in female posterior dorsal shield separated on 5 unequal parts (Fig. 12I); palp apotele 3-tined .... Halolaelapidae, p. 728
– Dorsal shield entire, at most with lateral incisions (Fig. 8A); palp apotele 2- or 3-tined 4
4. Holodorsal shield markedly hypotrichous, with not more than 20 pairs of setae (Figs 11J, L); inside female idiosoma in the region of coxae III sclerotized parts of spermatheca visible (Figs 2B, 11N–Q)... .... Phytoseiidae, p. 727
– Holodorsal shield with at least 27 pairs of setae, if less (female of *Thinoseius*), then sternal shield fragmented (Fig. 12P) and palp genu with 5 setae...... 5
5. Dorsal shield markedly hypertrichous; in female genital shield and sometimes sternal and anal shields with accessory setae (Fig. 14C); associated with bird and mammal nests...................... .... Haemogamasidae, p. 730
– Dorsal shield commonly oligotrichous, additional setae when present (Laelapidae) restricted to the region of *J* and *Z* series of setae (Figs 13D, F, *Jx*, *Zx*) 6
6. Both sexes with 1 or more of coxae with retrorse non-setigerous spurs (Fig. 14J), excluding anterior spine of coxa II; ectoparasites of mammals .... Hirstionyssidae, p. 730
– Both sexes without non-setigerous retrose spurs on coxae (excluding anterior spine of coxae II)....... 7
7. Tarsus II in both sexes with 1 or 2 distal setae developed into stout spurs (Fig. 12M); palp apotele 3-tined; female with trapezoid opisthogenital shield (Fig. 12L) or (if the former fused with anal shield) with large hologastric shield; peritrematal and exopodal shields fused, flanking opisthogenital shield; male with entire holoventral shield and legs II armoured .... Pachylaelapidae, p. 729
– Tarsus II in both sexes without massive spurs but sometimes with enlarged setae; palp apotele 2- or 3-tined.. 8
8. Leg I without claws; arthrodial membrane at base of movable digit of chelicera produced into 2 long, brush-like processes (Fig. 12F); peritreme forming U-shaped loop near stigma (Fig. 12K); genital shield of female with associated lateral sclerites; palp apotele 3-tined .... Macrochelidae, p. 729
– Leg I usually with claws; arthrodial membrane at base of movable digit of chelicera without brush-like processes; peritreme not forming U-shaped loop near stigma; genital shield of female without associated lateral sclerites (Figs 8C, 12P, 14A); palp apotele 2- or 3-tined..................................... 9
9. Palp genu with 5 setae; adults display sharp sexual dimorphism; female without sclerotized sternal shields (Fig. 12P), at most 1 pair of sternal setae on sclerotized cuticle; holodorsal shield in female reduced, with a maximum of 17 pair of setae; in male well developed sternitigenital and anal shields present, holodorsal shield normally with 30 pairs of setae .... Eviphididae (*Thinoseius*), p. 729
– Palp genu with 6 setae; female with sternal shield well developed, bearing 3 (rarely 4) pairs of setae (Figs 9J, 10K, 13L); holodorsal shield in both sexes normal, with 30 and more pairs of setae; in male large ventrianal shield (Fig. 9K) or entire holoventral shield present, rarely ventral shield free (Fig. 8I).... 10
10. In female genital shield relatively small, slightly projecting behind coxae IV, lacking setae (Figs 10B, I) or only with a single pair of setae (in this case metasternal seta associated with small sclerite, Fig. 11A); in male 2 large shields on ventral surface (sternitigenital and ventrianal, Fig. 8H), seldom ventral shield free in some *Arctoseius* (Fig. 8I); lateral margins of dorsal shield in both sexes with or without incisions (Figs 7F, G); gnathotectum with 2–3

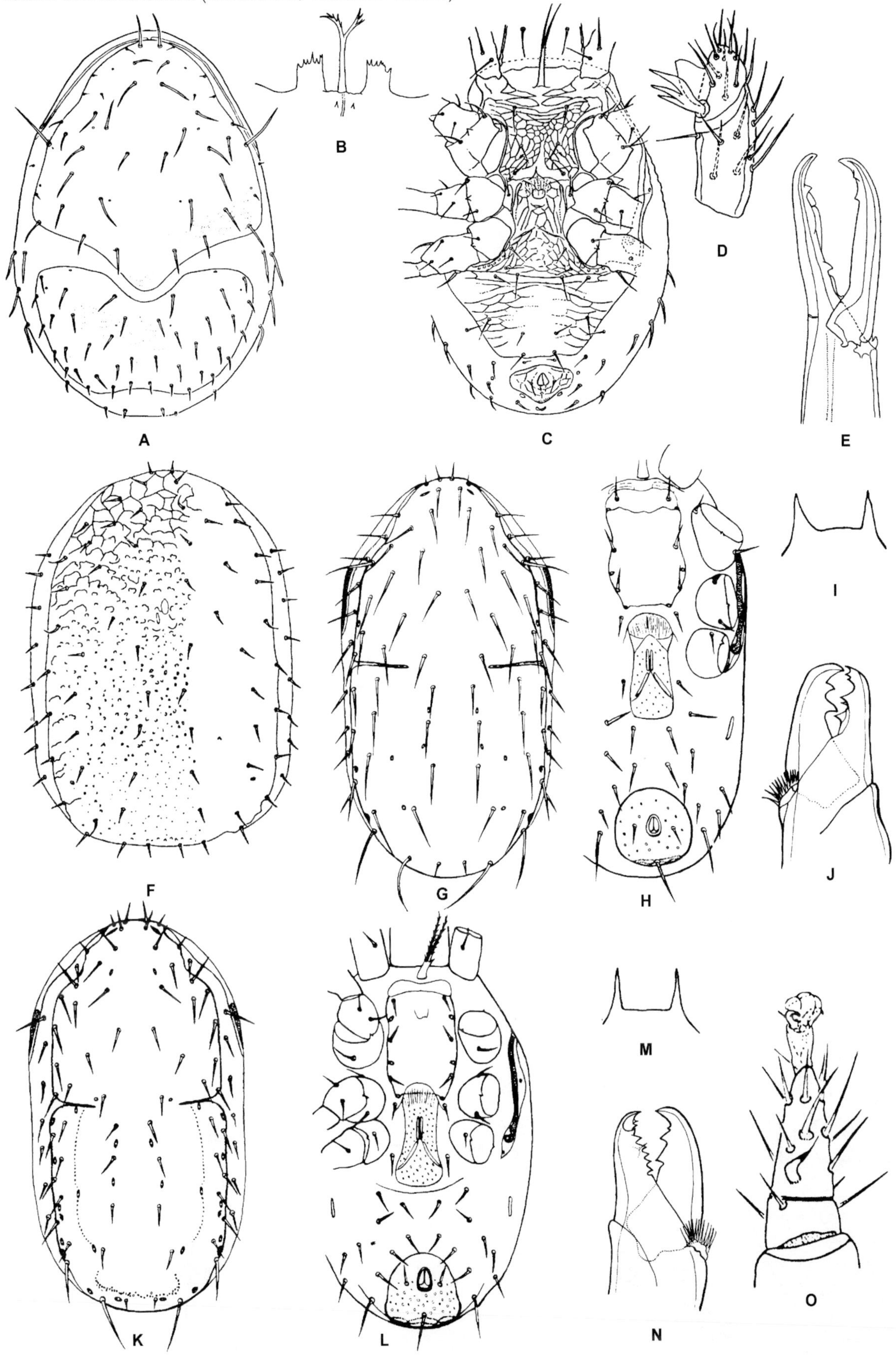

FIGURE 7 *A,* Veigaia nemorensis, *female, dorsal aspect, B, gnathotectum, C, ventral aspect, D, tibia and tarsus of palp, E, chela; F,* Arctoseius minutus, *female, dorsal aspect; G,* Arctoseius semiscissus, *female, dorsal aspect, H, ventral aspect, I, gnathotectum, J, chela; K,* Arctoseius cetratus, *female, dorsal aspect, L, ventral aspect, M, gnathotectum, N, chela, O, male, tarsus II. A–E, after Bregetova et al. 1977; F, after Gwiazdowicz 2007; G–N, after Karg 1993; O, after Lindquist 1964.*

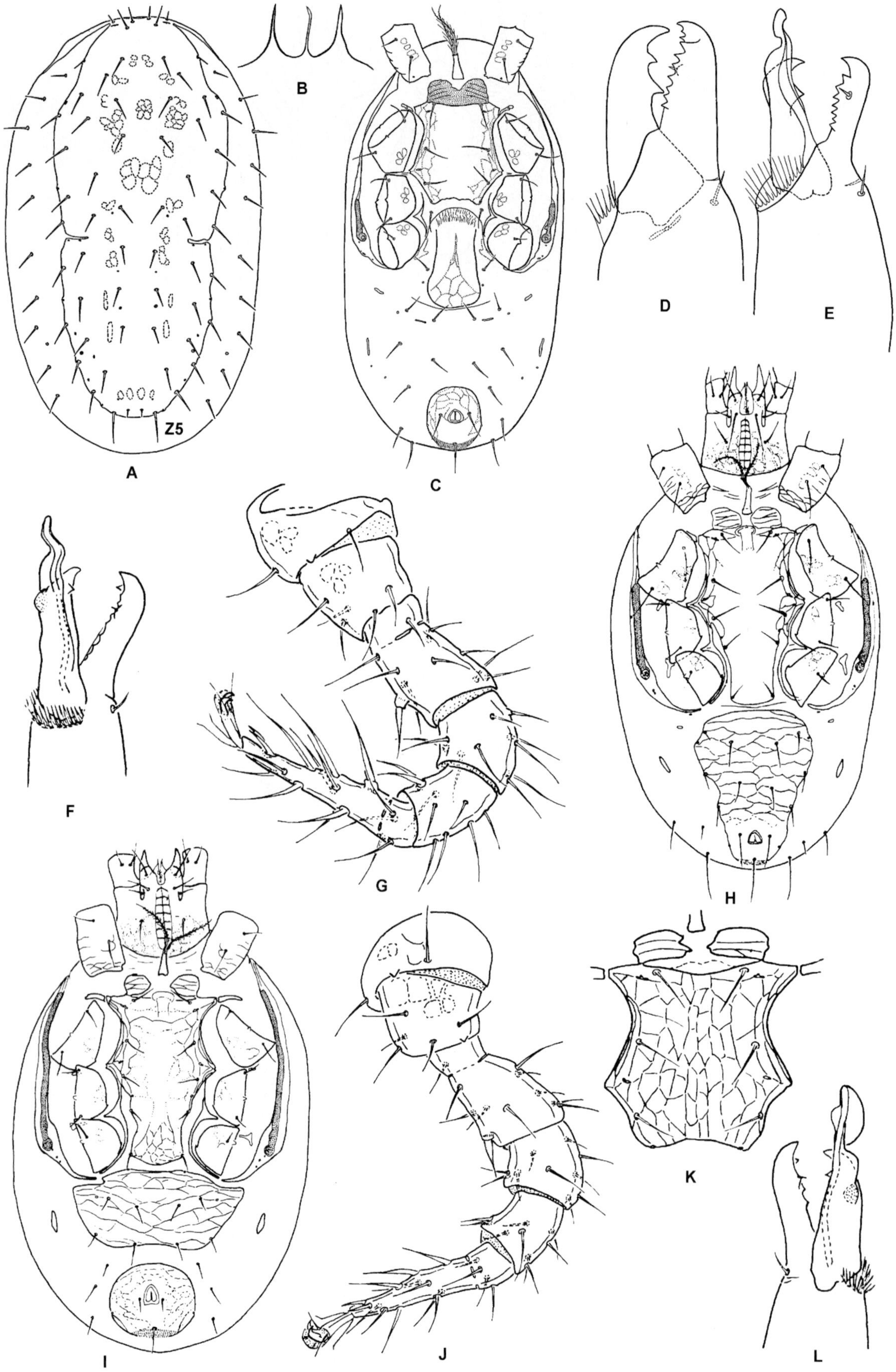

FIGURE 8 *A,* Arctoseius haarlovi, *female, dorsal aspect, B, gnathotectum, C, ventral aspect, D, chela, E, male, chela; F,* Arctoseius multidentatus, *male, chela, G, leg II, H, ventral aspect; I,* Arctoseius weberi, *male, ventral aspect, J, leg II, L, chela, K, female, sternal region. A–E, after Lindquist & Makarova 2011; F–L, after Lindquist 1961.*

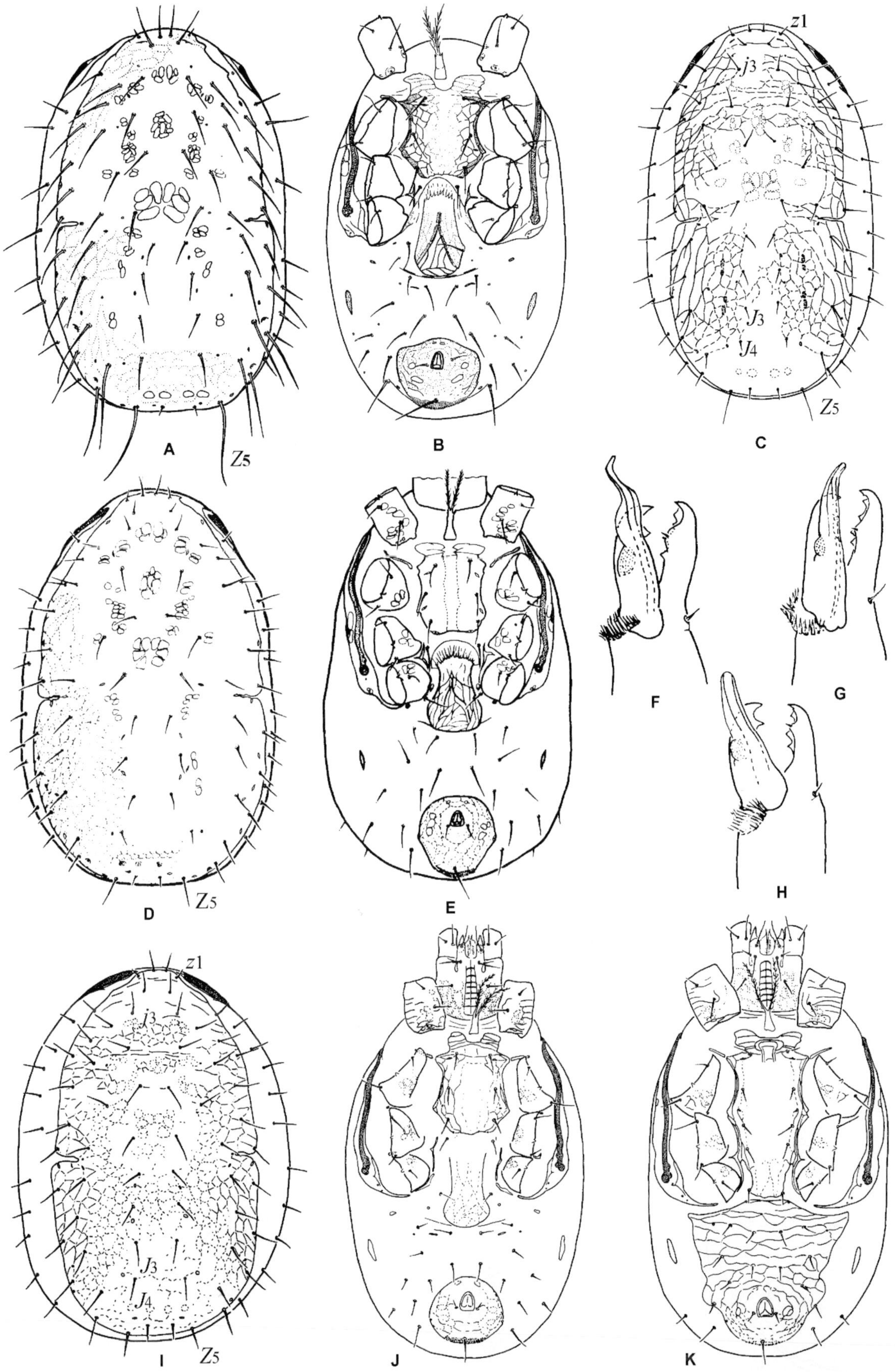

FIGURE 9 *A,* Arctoseius tschernovi, *female, dorsal aspect, B, ventral aspect; C,* Arctoseius idiodactilus, *female, dorsal aspect, F, male, chela; D,* Arctoseius tajmyricus, *female, dorsal aspect, E, ventral aspect; G,* Arctoseius ornatus, *male, chela, I, female, dorsal aspect; H,* Arctoseius minor, *male, chela, K, ventral aspect, J, female, ventral aspect. A, B, after Makarova 2000a; C, F–K, after Lindquist 1961; D, E, after Petrova & Makarova 1991.*

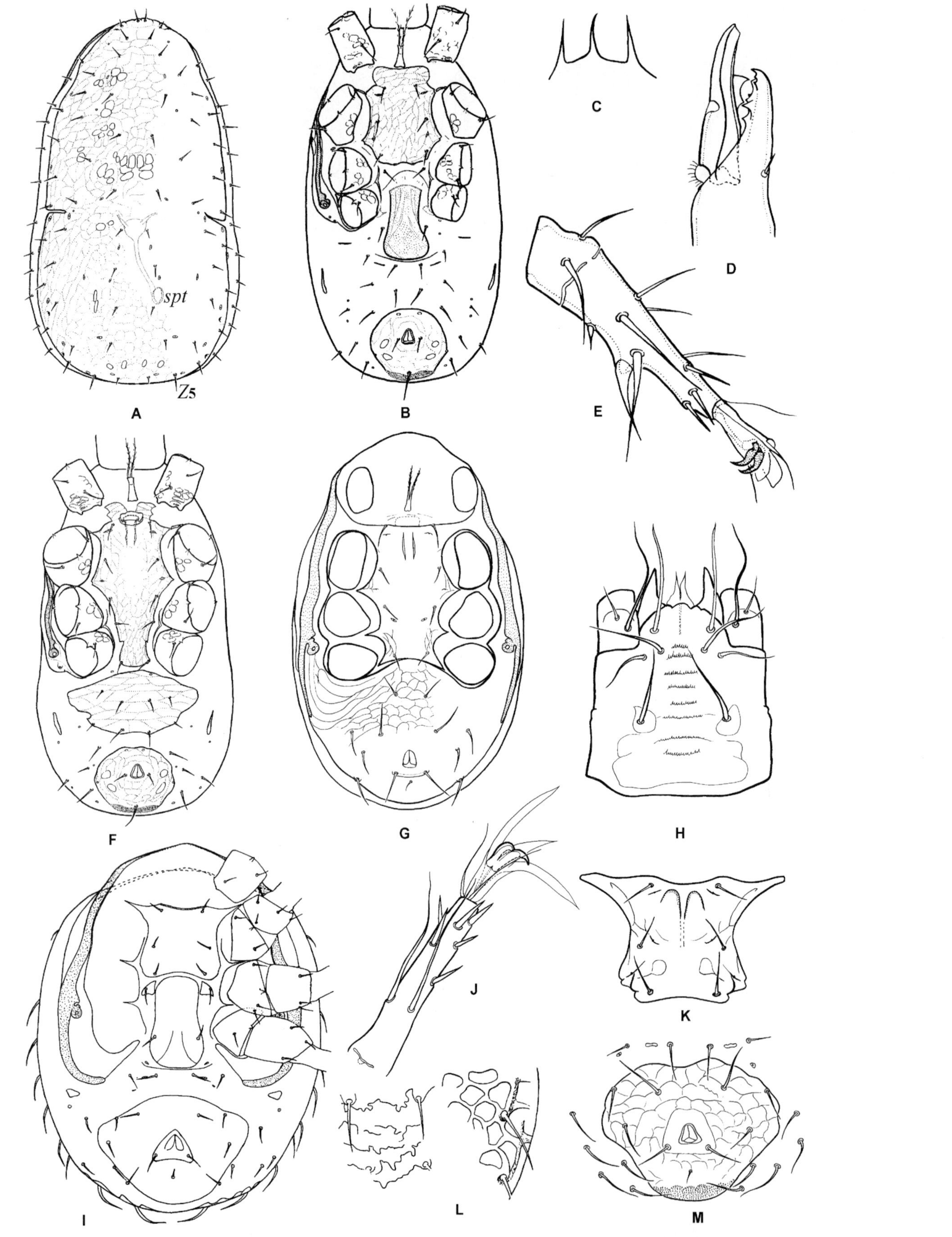

FIGURE 10 *A,* Arctoseius babenkoi, *female, dorsal aspect, B, ventral aspect, C, gnathotectum, D, male, chela, E, tarsus II, F, ventral aspect; G,* Cheiroseius necorniger, *male, ventral aspect, H, female, subcapitulum, J, telotarsus and ambulacrum II, K, sternal shield, M, ventrianal region; I,* Cheiroseius groenlandicus, *female, ventral aspect, L, dorsal shield ornamentation. A–F, after Makarova 2000; I, L, after Haarløv 1942.*

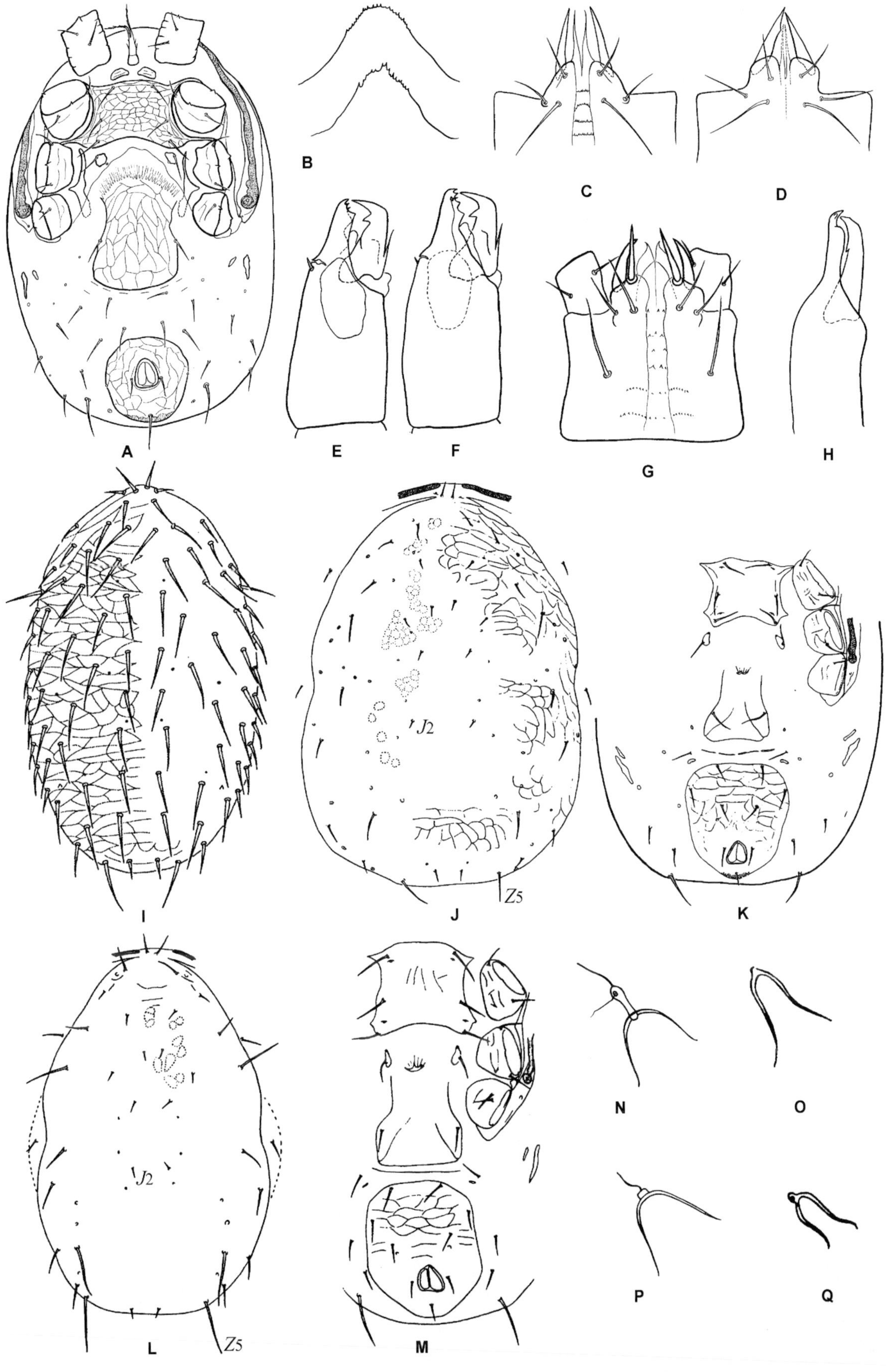

FIGURE 11 *A,* Proctolaelaps parvanalis, *female, ventral aspect, B, gnathotectum, variants, C, hypostome, E, chela, paraxial aspect, F, chela, antaxial aspect; D,* Proctolaelaps *sp. 1 near* bombophila, *female, hypostome, H, chela; G,* Proctolaelaps pygmaeus, *female, subcapitulum, I, dorsal aspect; J,* Neoseiulus apcutic, *female, dorsal aspect, K, ventral aspect, O, inner part of spermatheca; L,* Neoseiulus tobon, *female, dorsal aspect, M, ventral aspect, Q, inner part of spermatheca; N,* Neoseiulus ellesmerei, *female, inner part of spermatheca; P,* Neoseiulus *cf.* inak, *female, inner part of spermatheca. G, I, after Evans 1958; J–M, O, Q, after Chant 2007.*

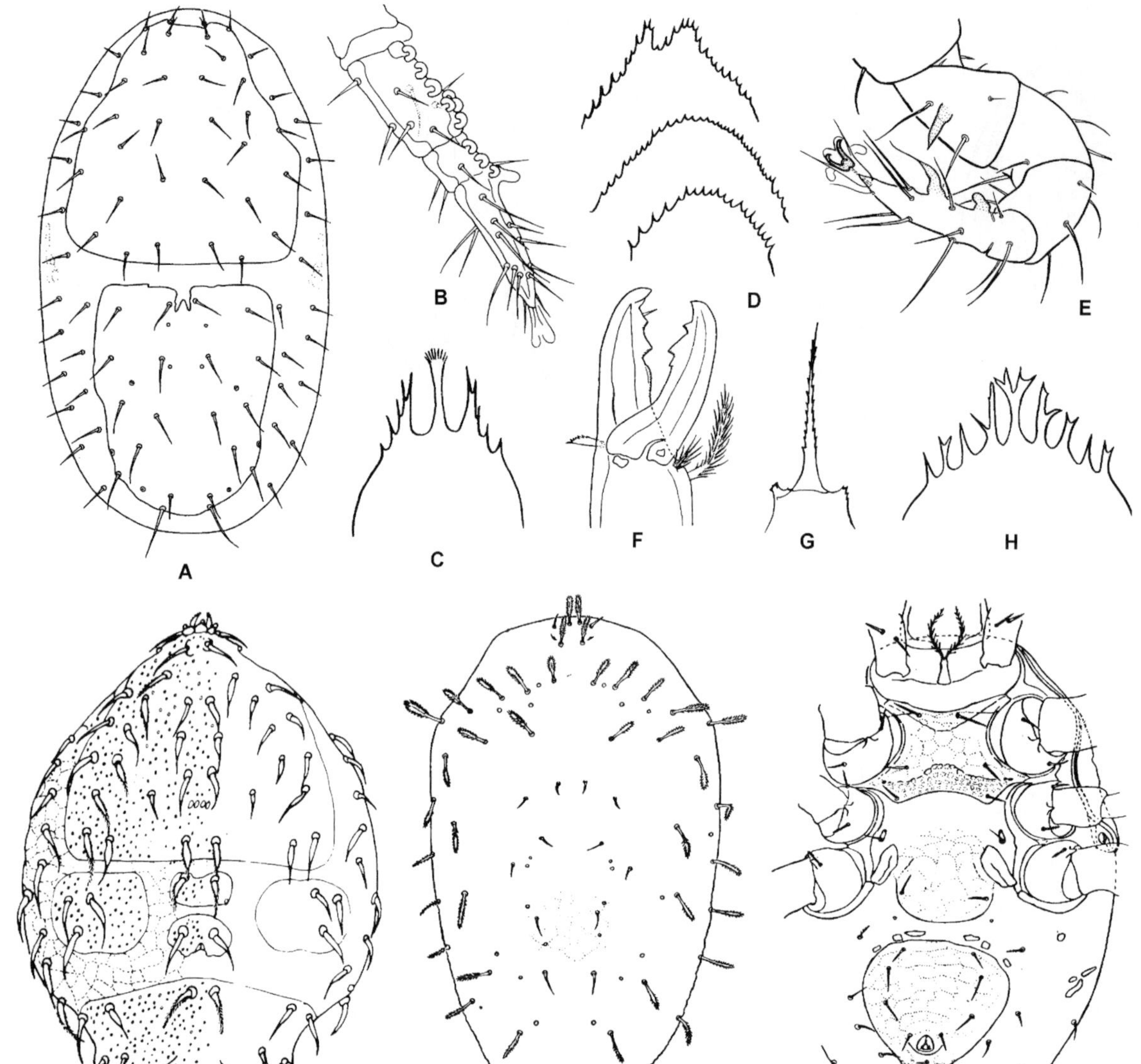

FIGURE 12 *A,* Halolaelaps coxalis, *female, dorsal aspect, B, male, tibia and tarsus IV, C, deutonymph, gnathotectum; D,* Halolaelaps gerlachi, *female, gnathotectum, variants of form, E, male, leg II, coxa and trochanter removed; F,* Macrocheles montanus, *female, chela, J, dorsal aspect, K, ventral aspect; G,* Saprosecans baloghi, *female, gnathotectum, I, dorsal aspect; H,* Thinoseius spinosus, *deutonymph, gnathotectum, N, chela, O, female, gnathotectum, P, ventral aspect. A, B, after Blaszak & Ehrnsberger 1993; F, J, K, after Bregetova et al. 1977; G, I, after Karg 1993.*

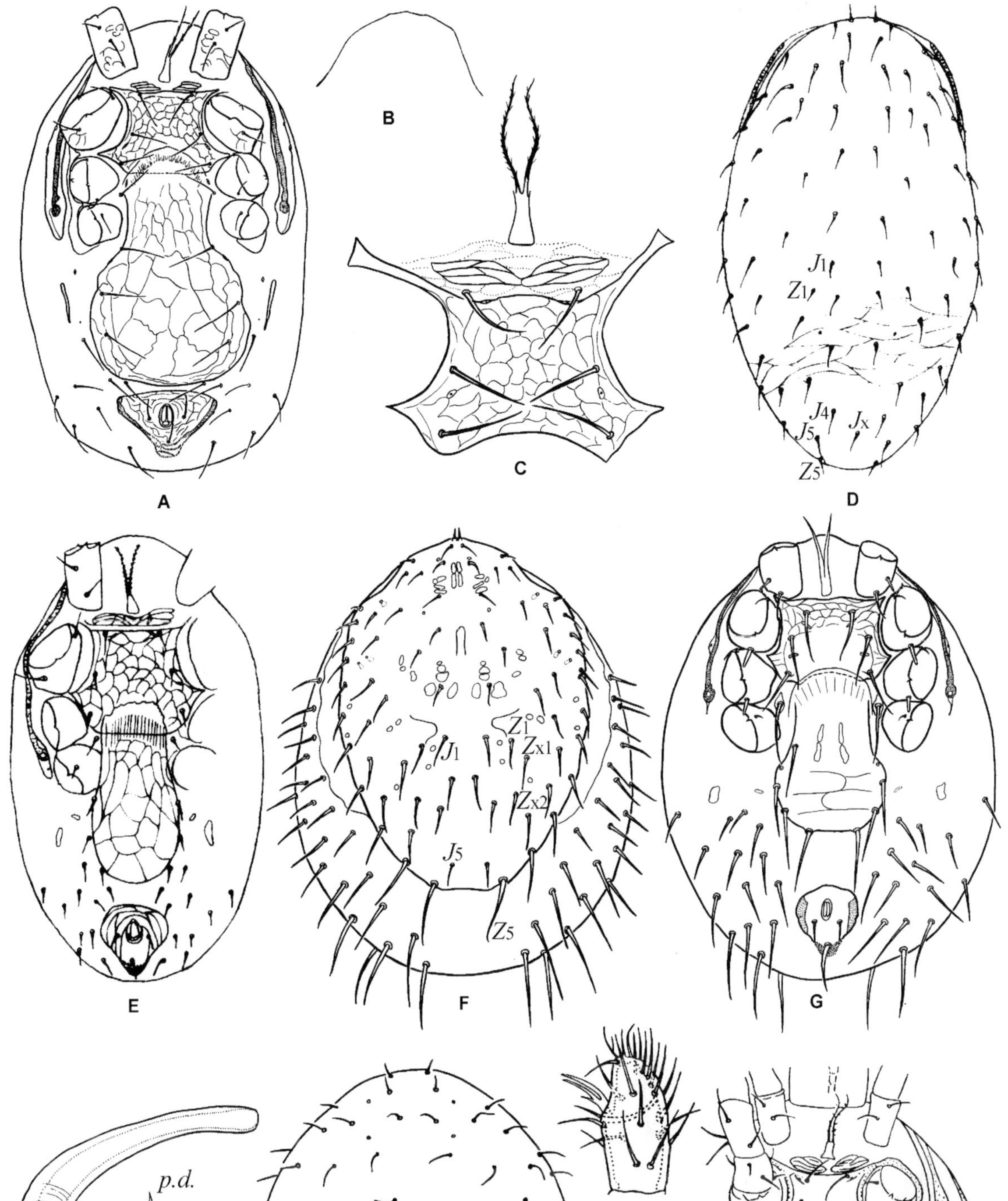

FIGURE 13 *A,* Gymnolaelaps missouriensis, *female, ventral aspect, B, gnathotectum; C,* Hypoaspis *cf.* heterosetosus, *female, sternal region; D,* Hypoaspis lubricoides, *female, dorsal aspect, E, ventral aspect; F,* Laelaps semitectus, *female, dorsal aspect, G, ventral aspect, I, chela, H, male, chela; J,* Ololaelaps sellnicki, *male, dorsal aspect, K, tibia and tarsus of palp, L, female, ventral aspect. D, E, after Karg 1993; J–L, after Bregetova et al. 1977.*

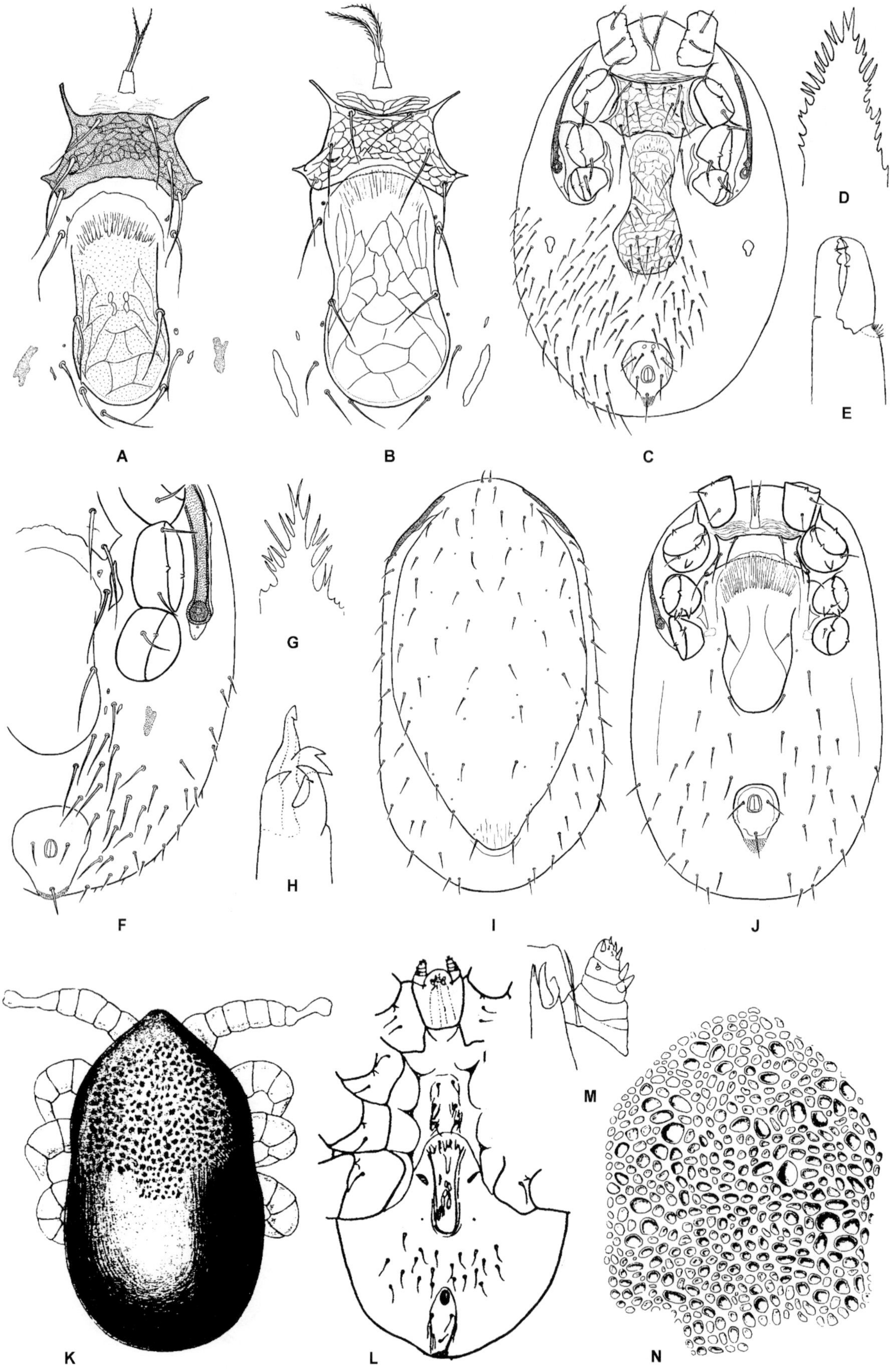

FIGURE 14 *A,* Pneumolaelaps groenlandicus, *female, sternitigenital region, F, ventral aspect; B,* Pneumolaelaps patae, *female, sternitigenital region; C,* Haemogamasus ambulans, *female, ventral aspect, D, gnathotectum, E, chela, G, male, gnathotectum, H, chela; I,* Hirstionyssus isabellinus, *female, dorsal aspect, J, ventral aspect; K,* Rhinonyssus levinseni, *female, dorsal aspect, L, ventral aspect, M, chela and palp, N, dorsal sclerotization. D, E, G, H, after Bregetova 1949; K, M, N, after Trägårdh 1904; L, after Butenko 1984.*

projections (Figs 7M, 8B) or rounded with smooth or serrated margin (Fig. 11B) .... Ascidae, p. 725

– In female genital shield large, distinctly projecting behind coxae IV (Figs 13E, 14F), or fused with ventral shield in opisthogenital shield with 1–4 pairs of setae (Fig. 13A), or entire hologastric shield (merged genital, ventral and anal shields) presents with 5–7 pairs of setae (Fig. 13L); in male entire holoventral shield; lateral margins of dorsal shield in both sexes without incisions; gnathotectum membranous, rounded, with smooth or serrated margin..................... .... Laelapidae, p. 729

VEGAIIDAE (FIG. 18J)

In Greenland the family is represented by one genus, ***Veigaia*** Oudemans, 1905. Active predators in soil (main prey is collembolans); Greenland species develop without males.

Key to species

FEMALES

1. Dorsal shield united but with deep lateral incisions directed posteriorly (Fig. 6F); chela enormous, 4–5 times longer than corniculi (Fig. 6H); gnathotectum as in Fig. 6G. 810–960 µm. Holarctic, Neotropic...... ....*cerva* (Kramer, 1878)

– Two dorsal shields (Figs 6I, 7A); chela of normal size (Fig. 7E)...2

2. Larger mites (length more than 1000 µm); anterior margin of opisthonotal shield almost straight (Fig. 6I); posterior margin of ventral shield with 2 pairs of setae (Fig. 6J); gnathotectum as in Fig. 6K. 1330–1450 µm. Holarctic (including Alaska, Novaya Zemlya, Taymyr) ....*kochi* (Trägårdh, 1901)

– Smaller mites (length less than 1000 µm); anterior margin of opisthonotal shield medially concave (Fig. 7A); posterior margin of ventral shield with 1 pair of setae (Fig. 7C); gnathotectum as in Fig. 7B. 650–940 µm. Holarctic ....*nemorensis* (C.L. Koch, 1839)

ASCIDAE

Ecologically diverse family. Greenland species composition comprises unspecialized soil predators (*Arctoseius*), carnivorous dwellers of over-wetted sites (*Cheiroseius*) and residents of animal (mammals, birds, insects) nests and shelters (some *Proctolaelaps*), where they feed on nematodes, arthropods and micromycetes. Females of some species disperse on dipterans (Fig. 17I).

Key to genera

ADULTS

1. Anterior hypostomal setae, 1 seta on palp trochanter (Fig. 10H), and 2 setae on tarsi II–IV long, whip-like (Fig. 10J); legs II–IV with median lobes of pulvilli slender, acute; peritreme extending far behind stigma (Fig. 10I); adanal setae inserted posteriad hind margin of anus (Fig. 10M); female with ventoanal shield bearing 3–4 pairs of ventral setae in addition to 3 perianal setae; in male ventrianal and peritrematal shields merged (Fig. 10G).......... .... *Cheiroseius*

– Anterior hypostomal setae, setae on palp trochanter and tarsi II–IV never whip-like; legs II–IV with median lobes of pulvilli rounded (Figs 1K, 7O); peritreme not extending behind stigma (Figs 10B, 11A); adanal setae inserted anteriad hind margin of anus; female with anal shield bearing only 3 perianal setae (Fig. 7L); in male ventrianal (or ventral) and peritrematal shields separated (Figs 8H, 10F) 2

2. Gnathotectum biramous or triramous (Figs 7I, 8B); body dorsum often not entirely covered by dorsal shield (Figs 8A, 9I), which is typically with pair of lateral incisions (except *Arctoseius minutus*); female with genital shield devoid of setae (Fig. 7H) . ... *Arctoseius*

– Gnathotectum usually multidentate (Fig. 11B) or smooth; dorsal shield covers body dorsum entirely, without incisions (Fig. 11I); female with genital shield bearing a pair of setae (Fig. 11A).............. .... *Proctolaelaps*

Arctoseius Thor, 1930. Figs 19A–C

The key comprises 12 species including *Arctoseius ornatus* Evans, 1955 and *A. tschernovi* Makarova, 2000 (common circumpolar species) not yet recorded in Greenland. Finds of other species of *Arctoseius* are also very probable. Reliable identification requires investigation of males. Given total length refers to length of dorsal shield.

*Key to species**

ADULTS

1. Dorsal shield without lateral incision, its reticulation unclear (Fig. 7F); gnathotectum triramous; rather small mites. Female 270–325 µm, male 250–260 µm. Holarctic (including Low Arctic)..................... .... *minutus* (Halbert, 1915)

* Adopted from Lindquist 1961, Makarova 2000b.

– Dorsal shield with lateral incision (Fig. 7G), usually smooth or distinctly reticulate; gnathotectum biramous (Fig. 7M) or triramous (Fig. 8B) 2

2. Gnathotectum biramous (Figs 7I, M); incisions of dorsal shield very deep, their length about 1/3 of dorsal shield width (Figs 7G, K); tibia IV with 6 setae; peritremes do not extend anteriad beyond mid-level of coxae II (Fig. 7L); in female sternal shield weakly sclerotized, nearly rectangular with rounded angles 3

– Gnathotectum triramous (Figs 8B, 10C); incisions of dorsal shield shorter (Figs 8A, 9I); tibia IV with 7 setae; peritremes of different length 4

3. Dorsal setae relatively long, most of them extend to base of posterior seta in series (Fig. 7G); fixed digit of chela with 3–4 teeth (Fig. 7J); in male setae on tarsus II uniform, acicular. Female 355–420 µm, male 270–310 µm. Holarctic (including High Arctic), Australian .... *semiscissus* (Berlese, 1892)

– Dorsal setae shorter, most of them do not extend to base of posterior seta in series (Fig. 7K); fixed digit of chela with 7–11 small teeth (Fig. 7N); in male telotarsus II bears ventrally seta with dilated base and filiform apex (Fig. 7O). Female 270–360 µm, male 250–280 µm. Holarctic (including High Arctic), Australian .... *cetratus* (Sellnick, 1940)

4. Peritremes very short, not extending anteriad beyond mid-level of coxae II (Figs 8C, H); dorsal shield rather narrow, not covering body fully (Fig. 8A); caudal setae *Z5* positioned close together, distance between them usually 4–6 times smaller than width of dorsal shield; perianal setae differ in length only slightly; fixed digit of chela with numerous teeth (Figs 8D, F); in male femora and tarsi II bear modified setae (Fig. 8G), spermatodactyl bent (Figs 8E, F) 5

– Peritremes well developed, extending anteriad beyond the level of anterior margin of coxae II (Figs 8I, 9E); dorsal shield wider, almost completely covering body (Figs 9D, 10A); caudal setae *Z5* usually separated by a greater distance; postanal seta twice as long as adanal ones 6

5. Peritremes extremely short, extending anteriad only to mid-level of coxae III (Fig. 8C); male chela as in Fig. 8E; smaller mites. Female 368–468 µm, male 332–376 µm. Circumpolar, arctic (including High Arctic)... .... *haarlovi* Lindquist et Makarova, 2011= *Arctoseius laterincisus* Thor, 1930 sensu Haarløv, 1942

– Peritremes longer, extending anteriad to mid-level of coxae II (Fig. 9O); male chela as in Fig. 8F, leg II as in Fig. 8G; larger mites. Female 566–677 µm, male 488–550 µm. Circumpolar, arctic .... *multidentatus* Evans, 1955

6. Presternal plates well sclerotized, clearly separated from sternal (or sternitigenital) shield, with distinct linear ornamentation (Fig. 8K); peritremes extend anteriad at most to the level of posterior margin of coxae I (Fig. 8I); fixed digit of chela with 5–8 teeth in median row; in male ventral shield free, spermatodactyl asymmetrical spoon-shaped (Fig. 8L), ventral surface of femur II distally bears specialized seta with dilated base (Fig. 8J). Female 462–561 µm, male 416–466 µm. Circumpolar, arctic-subarctic .... *weberi* Evans, 1955

– Presternal plates membranous, fused with sternal (or sternitigenital) shield (Figs 9E, 10B); peritremes longer; fixed digit of chela with only 2–3 teeth in median row; in male spermatodactyl of another form (Figs 9F, H) 7

7. Larger mites (length of dorsal shield in female no less than 640 µm, in male no less than 480 µm) with long legs (tarsi III usually 4.5–5.5 times as long as wide); in male ventral shield free...................8

– Smaller mites (length of dorsal shield in female less than 600 µm, in male less than 450 µm) with shorter legs; in male ventral shield fused with anal shield (Fig. 9K) or free (Fig. 10F) 9

8. Dorsal setae in general long (Fig. 9A), longest caudal setae *Z5* 104–148 µm; in female sternal shield well sclerotized, reticulate, with distinct anterior and posterior angles (Fig. 9B). Female 640–700 µm, male 548–600 µm. Circumpolar, high arctic .... *tschernovi* Makarova, 2000

– Setae of dorsal shield moderately long (Fig. 9D), longest caudal *Z5* 42–76 µm, in female sternal shield weakly and non-uniformly sclerotized, smooth, with vague outlines (Fig. 9E). Female 624–760 µm, male 488–600 µm. Circumpolar, arctic..................... .... *tajmyricus* Petrova et Makarova, 2000

9. Dorsal setae in general longer (Figs 9I, C), longest caudal setae *Z5* 30–60 µm; in male spermatodactyl rather narrow apically (Figs 9F, G) 10

– Dorsal setae short (Fig. 10A), longest caudal setae *Z5* 20–27 µm; in male spermatodactyl rather broad and truncate apically (Figs 9H, 10D) 11

10. Dorsocentral setae *J3* and *J4* very short (18–21 µm), their length only 1/2–2/3 (rarely 3/4) that of setae *j3* (Fig. 9C); peritremes shorter, never extending anteriad to bases of setae *z1*; in male spermatodactyl sinuous, abruptly narrowing distally beyond the apex of movable digit (Fig. 9F). Female 392–524

µm, male 324–403 µm. Circumpolar, arctic, alpine in North America...... *idiodactilus* Lindquist, 1961

– Dorsocentral setae *J3* and *J4* almost as long as most setae of dorsal shield; their length (26–39 µm) no less than 3/4 that of setae *j3* (Fig. 9I); peritremes very long, wide, in most cases extending anteriad to bases of setae *z1*; in male spermatodactyl straight, gradually narrowing toward apex (Fig. 9G). Female 458–592 µm, male 376–444 µm. Circumpolar, arctic-subarctic.......................... *ornatus* Evans, 1955

11. In female inner part of spermatheca usually well visible (Figs 2A, 10A, *spt*); sternal shield with complete reticulate ornamentation, almost as long as wide (Fig. 10B); in male ventral shield free (Fig. 10F), ventral surface of tarsus II bears 2 specialized setae (large and small) with dilated bases (Fig. 10E). Female 400–496 µm, male 344–400 µm. Circumpolar, high arctic................ *babenkoi* Makarova, 2000

– In female inner part of spermatheca not visible; sternal shield centrally without reticulate ornamentation, clearly longer than wide (Fig. 9J); in male ventral and anal shields merged (Fig. 9K), tarsus II without specialized setae. Female 352–439 µm, male 296–360 µm. Circumpolar, arctic.. *minor* Lindquist, 1961

***Cheiroseius* Berlese, 1916. Fig. 18K**
Key to species

FEMALES

1. Anterior part of sternal shield with a pair of semicircular depressions (Fig. 10K); anterior margin of ventrianal shield commonly sinuate, medially slightly concave (Fig. 10M). In females from Europe leg I (580–780 µm) distinctly longer than idiosoma (550–745 µm). In Greenland females leg I (650–680 µm) shorter or equal to idiosoma (660–710 µm). Male (Fig. 10G) idiosomal length 440–470 µm. Palaearctic, Ethiopian .... *necorniger* (Oudemans, 1903) aberr.

– Anterior part of sternal shield without particular ornamentation; anterior margin of ventrianal shield medially usually convex (Fig. 10I). Female 620 µm. NE Greenland. Northern Nearctic (including Ellesmere Island, Devon Island).....*groenlandicus* Haarløv, 1942

***Proctolaelaps* Berlese, 1923. Fig. 18L**
Key to species

FEMALES

1. Well sclerotized mites; body shields distinct, light brown; sternal shield wide, almost twice broader than long (Fig. 11A); corniculi narrow, straight, almost parallel (Fig. 11C); gnathotectum with dentate margin (Fig. 11B); movable digit of chela hook-shaped (Figs 11E, F). Female 460–500 µm, male 340–360 µm. Circumpolar, arctic (including High Arctic)......................... *parvanalis* Thor, 1930

– Poorly sclerotized mites; body shields whitish......2

2. Anterior hypostomal setae thickened (Fig. 11G); corniculi of normal form, horn-like; gnathotectum dentate; fixed digit of chela with a row of numerous teeth; dorsal setae rather long, most of them extend to insertion of posterior seta in series (Fig. 11I). Female 340–410 µm; male 290–380 µm. Cosmopolitan .... *pygmaeus* (Müller, 1860)

– Anterior hypostomal setae needle-shaped (Fig. 11D); corniculi narrow 3

3. Fixed digit of chela with a few (1–2) teeth (Fig. 11H); corniculi convergent (Fig. 11D). Female 430 µm. Only known from N Greenland (? in *Bombus*-nest) ....*Proctolaelaps* sp. 1 near *bombophila* (Westerboer, 1963)

– Fixed digit of chela with a row of numerous teeth; corniculi near parallel. Female 470 µm. Only known from East Greenland ....*Proctolaelaps* sp. 2

Both *Proctolaelaps* sp. 1 near *bombophila* and *Proctolaelaps* sp. 2 represented in the ZMUC-collection only by a single poorly preserved specimen. Records of other members of *Proctolaelaps* are probable.

PHYTOSEIIDAE

For reliable identification of species belonging to this family, the examination of some delicate structures of female insemination apparatus is necessary. Paired sclerotic parts of spermatheca are visible inside body near coxae III (Fig. 2B). They consist of atrium (input division) and calyx (main division), whose form and size vary strongly and serve as important diagnostic features of species. At least 7 phytoseiid species live in Greenland, but only members of *Neoseiulus* are common. Numerous phytoseiid species inhabit plants where they prey mainly on phytophagous mites. Some species live in mosses and litter. Many phytoseiids may consume pollen.

Key to genera

ADULTS

1. Dorsocentral setae *J2* present (Figs 11J, L) .... *Neoseiulus*

– Dorsocentral setae *J2* absent.........*Proprioseiopsis*

Neoseiulus **Hughes, 1948. Fig. 19D**
Key to species

FEMALES

1. Dorsal setae more or less homogenous, short or very short, even the longest caudal setae *Z5* not longer than 42 µm (Fig. 11J); dorsal shield distinctly reticulate. 2
– Dorsal setae notably differ in length, some of them rather long, the longest caudal setae *Z5* not shorter than 60 µm (Fig. 11L); dorsal shield faintly reticulate 3
2. Atrium of spermatheca conspicuous, calyx cup-shaped (Fig. 11N); caudal dorsal setae *Z5* 39–40 µm; rather narrow mite, idiosomal length/width ratio 1.8. 440–470 µm. Circumpolar, arctic (including Ellesmere Island). . . . . cf. *ellesmerei* (Chant & Hansell, 1971)
– Atrium of spermatheca not prominent, calyx bell-shaped (Fig. 11O); caudal dorsal setae *Z5* 31–41 µm; rather broad mite, idiosomal length/width ratio 1.4 (Fig. 11J). 460 µm. Circumpolar, arctic (including Ellesmere Island, Devon Island) . . . . *apkutik* (Chant & Hansell, 1971)
3. Atrium of spermatheca swollen and compact, calyx broad cup-shaped, its width equal to or more than length (Fig. 11P); caudal dorsal setae *Z5* 88–90 µm. 500 µm. Only known from SW Greenland. Nearctic, arctic-alpine . . . . cf. *inak* (Chant & Hansell, 1971)
– Atrium of spermatheca not so distinct, calyx flask-shaped, rather narrow, its length almost twice the width (Fig. 11Q); caudal dorsal setae *Z5* 60 µm (Fig. 11L). 410 µm. Nearctic, arctic (including Devon Island) . . . . *tobon* (Chant & Hansell, 1971)

NOTE. At least 1 additional species, ***Neoseiulus*** sp. near ***tobon*** (Chant & Hansell, 1971) with most dorsal setae longer, inhabits western and northern Greenland (known also from arctic Russia). Finds of other representatives of *Neoseiulus* are very probable in southern parts of Greenland.

Proprioseiopsis **Muma, 1961**
In Greenland mites of this genus are evidently infrequent. Two specimens (male and female) of ***Proprioseiopsis*** sp. near ***okanagensis*** (Chant, 1957) are recorded from the Kap Farvel area (southernmost Greenland). Another species, ***Proprioseiopsis*** sp. (1 male), originated from East Greenland.

HALOLAELAPIDAE
Members of this family commonly occur in marine littoral habitats and accumulations of decaying organic matter, where they prey on other invertebrates. Deutonymphs use amphipods, dipterans and beetles for dispersal.

Key to genera

ADULTS

1. Female with pair of posterolateral protuberances bearing setae (Fig. 12I); opisthonotal shield divided into 5 sclerites of different size; interscutal membrane reticulate, its ornamentation formed by small tubercles; dorsal setae barbed, dilated basally; gnathotectum with very long central projection (Fig. 12G) . . . . *Saprosecans*
– Female without posterolateral protuberances; opisthonotal shield of both sexes entire, its anterior margin medially with or without incision (Figs 12A, 17G); interscutal membrane smooth, granulate or striate, not reticulate; dorsal setae needle-shaped smooth or brush-like; gnathotectum of different form (Figs 12C, D) . . . . *Halolaelaps*

Halolaelaps **s. l. Berlese et Trouessart, 1889. Fig. 19H**
Key to species

ADULTS, DEUTONYMPHS

1. Gnathotectum with 3 groups of teeth (Fig. 12C); in female, male and deutonymph medial incision on anterior margin of opisthonotal shield distinctive, wide, forked (Fig. 12A); in male tibia and tarsus IV ventrally with garland of tubercles (Fig. 12B); smaller mites. Female 380–440 µm, male 350–405 µm, deutonymph 310 µm. Northern Europe. . . . . *H.* (*Saprogamasellus*) cf. *coxalis* Willmann, 1957
   NOTE. Only immature specimens have hitherto been found in the Greenland material.
– Gnathotectum as a whole triangular or rounded, its margin multidentate (Fig. 12D); medial incision on anterior margin of opisthonotal shield absent (in male and deutonymph) or very small, hardly visible (in female, Fig. 17G); male tarsus IV with 2 basally dilated setae, much more remarkable equipment developed on leg II (Fig. 12E); larger mites. Female 640–770 µm, male 560–620 µm, deutonymph 520–550 µm. Northern Europe, arctic (including Svalbard) . . . . *H.* (*Halogamasellus*) *gerlachi* Hirschmann, 1966

Saprosecans **Karg, 1964. Figs 19E, F**
Represented by the distinctive species ***Saprosecans baloghi*** Karg, 1964 with characteristic body form and integument ornamentation (Fig. 12I); male unknown;

deutonymph phoretic on small dipterans. Female 415–430 μm, deutonymph 250–300 μm. SW Greenland. Holarctic (including Canadian Arctic Archipelago, Taymyr, New Siberian Islands).

NOTE. It is very likely that "*Zercon lundbecki* Trägårdh, 1904" (description based on phoretic deutonymph ex *Anthomyia* sp., Diptera, from Greenland, Figs 17E, H) also belongs to Halolaelapidae, possibly to the genus *Saprolaelaps* Leitner, 1946 (Fig. 19G). If that is the case, in adults genu III is with 8 setae, coxa II with a distinctive anterior spine; male with holoventral shield.

PACHYLAELAPIDAE (FIG. 19J)

Free-living predatory mites. Only a single specimen belonging to this family is available from SW Greenland, a female of a ***Pachylaelaps*** sp. near ***denticulatus*** Hirschmann & Krauss, 1965. Legs II robust, tarsus II with 2 characteristic strong spines (Fig. 12M), sclerotic parts of spermatheca in the form of long wide tubes originating from coxae IV and directed postero-medially (Fig. 12L). Female 880 μm. Only known from SW Greenland.

MACROCHELIDAE (FIG. 19I)

Unspecialized predators, common in organically enriched microbiotopes as well as in forest litter. Single recorded member of this family, ***Macrocheles montanus*** (Willmann, 1951), inhabits southern Greenland. Female with remarkable uneven reticulation of sternal shield (Fig. 12K), sharply heterogeneous setae of dorsal shield and frontal dorsal setae drawn together (Fig. 12J). Female 1070–1210 μm. SW Greenland, Palaearctic (Europe, West Siberia, Taymyr).

EVIPHIDIDAE (FIG. 20F)

This genus has not been mentioned from the Arctic before. But littoral ***Thinoseius spinosus*** (Willmann, 1939) seems to be regular in both West- and East Greenland north to 78° N. The deutonymph is known to be phoretic on small dipterans (Fig. 21D). Robust chelae of deutonymph possessed with large teeth (Fig. 12N) provide the clinging to insects. Other stages of development with graceful chelicerae probably feed upon nematodes. Female (Figs 12P, O) 750–830 μm, male 600 μm, deutonymph (Figs 12H, N) 540–570 μm. Europe (including Iceland, islands of Barents Sea), Far East (Kuril Islands).

LAELAPIDAE

The family contains numerous free-living predatory forms as well as species associated with insects, birds and mammals being inhabitants of nests or scavengers and parasites.

Key to genera

ADULTS

1. Female with hologastric shield and metasternal setae situated on sternal shield (Fig. 13L); both sexes with holodorsal shield arched and completely covering dorsum of the mite (Fig. 13J); palp apotele with 3 unequal tines (Fig. 13K) .... *Ololaelaps*
– Female with anal shield free and metasternal setae outside the sternal shield (Figs 13E, G); in both sexes holodorsal shield less convex; palp apotele with 2 or 3 tines 2
2. In female pilus dentilis on fixed digit notably enlarged, long (Fig. 13I); in both sexes idiosomal setae robust, straight, long and thick (Figs 13F, G); genu IV with 10 setae; chelicerae of male never chelate-dentate, with fixed digit reduced and movable digit partially or entirely fused with elongate grooved spermatodactyl (Fig. 13H)..................... *Laelaps*
– In female pilus dentilis on fixed digit short, setiform (Fig. 14E); in both sexes idiosomal setae smaller, if long never rigid (Figs 13D, 14F); genu IV with 9 setae; chelicera of male chelate-dentate, spermatodactyl free 3
3. Apotele with 3 subequal tines; both sexes with discrete well-sclerotized presternal shields (Fig. 13A); female with opisthogenital shield bearing 3 pairs of setae .... *Gymnolaelaps*
– Apotele with 2 tines; female with genital shield bearing 1 pair of setae (Figs 13E, 14B) 4
4. Opisthogastic and lateral unsclerotized cuticle hypertrichous (Fig. 14F); peritreme often broad; genu IV with 2 ventral setae; dwellers of *Bombus*-nests .... *Pneumolaelaps*
– Opisthogastic and lateral unsclerotized cuticle oligotrichous (Fig. 13E); peritreme never broad; genu IV usually with 1 ventral seta .... *Hypoaspis*

Gymnolaelaps **Berlese, 1903. Fig. 19L**

One species found in SW Greenland, ***Gymnolaelaps missouriensis*** (Ewing, 1909) (= *Laelaspis austriacus* Sellnick, 1935, synonymy by Hennessey & Farrier, 1988). In adults, integument characteristically vividly yellow; female with enlarged opisthogenital shield bearing 3 pairs of setae; anal shield broad triangular; opisthogenital and anal shields are close to each other (Fig. 13A); gnathotectum with smooth margin (Fig. 13B). Female 485–530 μm. Holarctic, Hawaii Islands.

Hypoaspis **Canestrini, 1884 s. l. Fig. 19K**
Key to species

FEMALES

1. Posterior margin of sternal shield nearly straight (Fig. 13E); dorsal setae rather short, never extended to insertion of posterior seta in series (Fig. 13D); opisthonotal region of dorsal shield with 1 unpaired medial seta (*J*x); in general smaller mite. Female 480–580 µm, male 440 µm. Europe... *lubricoides* Karg, 1971
– Posterior margin of sternal shield nearly concave (Fig. 13C); dorsal setae rather long, many of them extend to insertion of posterior seta in series; opisthonotal region of dorsal shield with 2–3 unpaired medial setae; in general larger mite. Female 540–625 µm. Europe .... cf. *heterosetosus* Bernhard in Hirschmann, 1969

Laelaps **C.L. Koch, 1836. Fig. 20C**
Represented by a single species, ***Laelaps semitectus*** L. Koch, 1879, permanent parasite of lemmings (mainly *Dicrostonyx* spp.). Idiosoma possessed by long thick setae facilitating the attachment to lemming fir (Figs 13F, G). Female 790–940 µm, male 700–750 µm. Palaearctic, arctic.

Ololaelaps **Berlese, 1904. Fig. 20B**
One member found in Greenland, the eurytopic ***Ololaelaps sellnicki*** Bregetova & Koroleva, 1964 (Figs 13J, L). Adults well sclerotized, with notably convex dorsal shield. Female 640–790 µm, male 520–600 µm. SW Greenland. Holarctic, arctic-boreal.

Pneumolaelaps **Berlese, 1920. Fig. 20A**
Species of this genus are obligatorily associated with bumblebees. The development of mites takes place in *Bombus*-nests; females disperse on adult insects.

Key to species

FEMALES

1. Dorsal setae decidedly heterogeneous, differing in length up to 5 times; genital shield rather narrow (Fig. 14A), distance between genital setae (105 µm) almost equal to their length (90 µm); presternal and metapodal shields unclear; smaller mite. 630–670 µm. In Greenland found on *Bombus polaris* and *B. hyperboreus*. Nearctic (Devon Island)...................................... *groenlandicus* (Trägårdh, 1904) = *Hypoaspis* (*Pneumolaelaps*) *arctos* Karg, 1984, types tested (Makarova, 2014).
– Difference in length of dorsal setae not so marked (less than 3 times); genital shield wider, distance between genital setae twice as long as their length (Fig. 14B); presternal and metapodal shields distinct; larger mite. 700–800 µm. Found in N Greenland on *Bombus polaris* and *B. hyperboreus*. Nearctic (Ellesmere Island)... *patae* Hunter & Husband, 1973

HAEMOGAMASIDAE (FIG. 20D)
In Greenland represented by 1 genus, *Haemogamasus* (and possibly by 2 species) noteworthy by the hypertrichy of idiosoma (Fig. 14C). Mites of this genus are facultative or obligate hematophagous ectoparasites of mammals and birds. The common Greenland species ***Haemogamasus ambulans*** (Thorell, 1872) inhabits litter, mosses and mammal nest, being facultative parasite. Female (Figs 14D, E) 910–1160 µm, male (Figs 14G, H) 700–740 µm. Holarctic (including High Arctic), Oriental (Korea).

HIRSTIONYSSIDAE (FIG. 20E)
Mites of this family are obligatory hematophagous ectoparasites of small mammals. In Greenland has been recorded 1 host-unspecific species, ***Hirstionyssus isabellinus*** (Oudemans, 1913) (Figs 14I, J), associated here with the arctic lemming, *Dicrostonyx groenlandicus*. Female 560–760 µm, male 510 µm. Holarctic (including High Arctic).

RHINONYSSIDAE
Mites of this family are endoparasites of birds inhabiting their respiratory tracts. Idiosoma weakly sclerotized, hypotrichous; peritremes strongly reduced or absent. From Greenland only ***Rhinonyssus levinseni*** (Trägårdh, 1904) is recorded ex nasal cavity of *Somateria mollissima* (Figs 14K–M). Several authors considered *R. levinseni* a synonym of *Rhinonyssus rhinolethrum* (Trouessart, 1895), but this is still under discussion. *Rhinonyssus rhinolethrum* is known to be parasitic on different ducks, geese, and swans in several regions (Holarctic, Oriental, Ethiopian). *Rhinonyssus levinseni* inhabits the nasal cavity of *Somateria* spp. and *Polysticta stelleri* (after: Butenko 1984). Female 1090–1430 µm, male 880–990 µm. Holarctic.

ZERCONINA
In Greenland represented by 1 family, Zerconidae. Free-living mites in litter, humus, mosses, turf in woods and meadows. The species are rather slowly mowing and known to be oligophagous predators (nematodophagy is common).

ZERCONIDAE

Key to genera

ADULTS

1. Anterior margin of ventrianal shield with 1 pair of setae (Fig. 15A); peritrematal shield with 3 setae (1 large ciliated, others short and smooth), its postero-lateral corner tapered posteriorly; all dorsal setae homogeneous, densely ciliated (Fig. 15B) .... *Parazercon*
– Anterior margin of ventrianal shield with 2 pairs of setae (Figs 15D, G); peritrematal shield with 2 setae (one large and ciliated, the other short and often smooth), truncate posteriorly; dorsal setae usually heterogeneous (Figs 15F, H); if more or less similar, then smooth or barbed distally (Fig. 15E) 2
2. Most setae of opisthonotal shield densely ciliated, inserted on tubercles, some setae smooth and short (Fig. 15H); opisthonothal shield ornamented by small pits; adgenital shields not developed; peritrematal shield close to podonotal shield (Fig. 15G) . ... *Mixozercon*
– Opisthonotal setae smooth or scarcely barbed (Figs 15E, F); at least anterior part of opisthonothal shield reticulated; adgenital shields well developed, each with 2–4 pores (Fig. 15D); peritrematal shield posteriorly removed from podonotal shield (Fig. 15C)....................................... *Zercon*

Mixozercon Halašková, 1963. Fig. 20H

Records of the single member, ***Mixozercon sellnicki*** (Schweizer, 1948), are confined to southern Greenland. It is a distinctive species with remarkable opisthonotal chaetome and ornamentation (Fig. 15H). Males unknown. Female 340–370 µm. Europe.

Parazercon Trägårdh, 1931. Fig. 20G

One species recorded in Greenland, ***Parazercon radiatus*** (Berlese, 1914), with characteristic dorsal chaetome (Fig. 15B). Populations usually without males. Female 345–365 µm. Holarctic.

Zercon C.L. Koch, 1836. Figs 20I–K

Key to species

FEMALES

1. Dorsal setae rather homogenous, barbered distally, the length of most of them differ less than 2 times (Fig. 15E). Female 520–590 µm, male 420–490 µm. The most common representative of the genus in Greenland. Only known from Greenland and Iceland .... *hammerae* Sellnick, 1960 = *Z. thulium* Athias-Henriot, 1980 types tested (Makarova, 2014) = Z. *triangularis* C.L. Koch, 1836 sensu Jørgensen, 1934; material studied by M. Hammer [Jørgensen] tested.

   NOTE. Originally described on protonymph and larva.
– Dorsal setae strongly heterogeneous, their lengths differ 3–5 times (Figs 15F, I).......................... 2
2. "Pore" *Po*3 on opisthonotal shield situated between insertions of setae *Z*3 and *Z*4 (Fig. 15F). Female 420–480 µm, male 340–360 µm. Europe (including Svalbard), arcto-boreal............................... .... *zelawaiensis* Sellnick, 1944 = *Z. curiosus* Trägårdh, 1910 sensu Jørgensen 1934; material studied by M. Hammer [Jørgensen] tested.
– "Pore" *Po*3 on opisthonotal shield situated between insertions of setae *I*4 and *Z*4 (Figs 15I, 17F). Female 460–510 µm, male 380–400 µm. Palaearctic, arctic-alpine (including Svalbard), ?Alaska .... *solenites* Haarløv, 1942 = *Z. curiosus* Trägårdh, 1910 sensu Haarløv 1942

   NOTE. *Zercon solenites* was described based on the deutonymph as a unique species possessed by long peritremes (Haarløv 1942). But this feature is a standard character for deutonymphs of the genus *Zercon*. Female mistakenly identified by Haarløv (1942) as *Zercon curiosus* does not correspond to this species (material studied by N. Haarløv tested). Similarity of both forms as well as their records from dry biotopes in one locality (Mørkefjord, NE Greenland), where other *Zercon* species were not found, point out their conspecificity. Recently all development stages of *Z. solenites* were described on the Svalbard's material (Gwiazdowicz et al. 2011).

UROPODINA

The cohort comprises mainly heat-loving mites. Along with litter residents there are numerous species associated with transient substrata (manure, compost, carrion, tidal debris, animal nests). The Greenland records are limited to the south-western area.

Key to families

ADULTS

1. Coxae I enlarged, flattened and capable of concealing gnathosoma and tritosternum (Fig. 17D); base of tritosternum not extremely broad; pedofossae often well developed .... Uropodidae

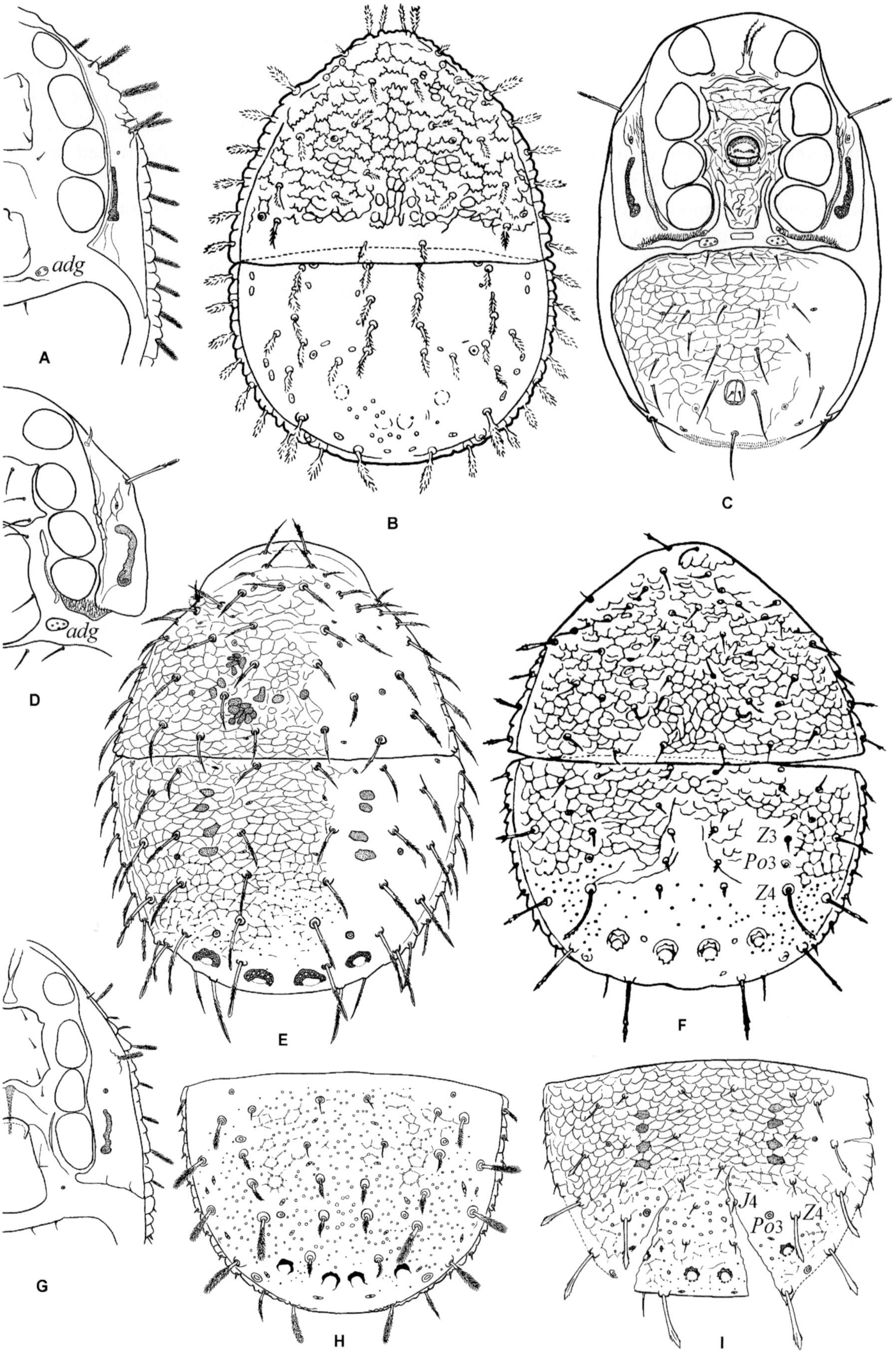

FIGURE 15 *A,* Parazercon radiatus, *female, peritrematal region, B, dorsal aspect; C,* Zercon hammerae, *male, ventral aspect, D, female, peritrematal region, E, dorsal aspect; F,* Zercon zelawaiensis, *female, dorsal aspect; G,* Mixozercon sellnicki, *female, peritrematal region, H, opisthonotal shield; I,* Zercon solenites, *female [from N. Haarløv's slide, Mørkefjord, NE Greenland], opisthonotal shield. B, F, after Halašková 1969.*

– Coxae I not enlarged and flattened and not capable of concealing gnathosoma and tritosternum (Fig 16C); base of tritosternum very broad and close to coxae I; pedofossae shallow or lacking .... Trachytidae

TRACHYTIDAE

Represented by 3 species belonging to the genus *Trachytes*, found only in SW Greenland. Idiosoma subtriangular in outline with a strongly developed vertex (guitar-shaped); dark yellow or brownish, well sclerotized mites. Slowly moving free-living *Trachytes* species inhabit forest-litter, mosses, turf of meadows; suspected to be mycetophagous. Greenland representatives develop without males.

Trachytes Michael, 1894. Figs 20L, 21A

Key to species

FEMALES

1. Interscutal membrane anteriad ventrianal shield medially with a pair of short thin setae (Fig. 16B); smaller mites. Female 500–600 µm. SW Greenland. Europe, arctic-alpine................................ .... *hirschmanni* Hutu, 1973
– Interscutal membrane anteriad ventrianal shield medially without setae (Figs 16E, G); bigger mites, idiosomal length not less than 600 µm.............. 2
2. Dorsal shield with angular crests and some setae inserted on small roundish plates (Fig. 16F); genital shield strongly dilated posteriorly (Fig. 16G). Female 600–700 µm. Holarctic .... *aegrota* (C.L. Koch, 1841)
– Dorsal shield without angular crests and roundish plates (Fig. 16D); genital shield slightly dilated posteriorly (Fig. 16E). Female 610–700 µm. SW Greenland. Palaearctic, boreal (Northern Europe, Siberia)....... .... *edleri* Hutu, 1983

UROPODIDAE (BOX MITES)

Extremely biologically diverse family, whose species inhabit litter, mosses, turf, different decaying organic substrata, wood beetle galleries, ant hills, nests and burrows of different animals. Uropodids feed on other small invertebrates, fungi, detritus, vascular plants etc. They disperse by well sclerotized deutonymphs phoretic on arthropods and vertebrates. During phoresy deutonymphs attach to the host by the caudal pedicle secreted by a particular hind gut gland.

Key to genera

ADULTS

1. Posterior region of dorsal shield with a transverse row of 4 strongly pilose or serrated setae; the area posterior to dorsal shield occupied by a weakly sclerotized shield bearing 4 stout foliate setae covered with minute spines (Fig. 17A)...... *Dinychus*
– Posterior region of dorsal shield without transverse row of pilose or serrated setae; the area posterior to dorsal shield at most with scutella bearing separate setae (Fig. 16I)........... *Uropoda* (*Phaulodinychus*)

Dinychus Kramer, 1882. Fig. 21B

Represented by 1 broadly distributed, free-living, morphologically distinct species, ***Dinychus perforatus*** Kramer, 1882 (Figs 17A–C). Found only in SW Greenland, not common. Female 780–940 µm, male 720–830 µm. Palaearctic (including Jan Mayen, southwestern Taymyr).

Uropoda Latreille, 1806. Fig. 21C

Until now recorded once in soil of wet beach meadow, SW Greenland. Only 2 males of ***Uropoda*** (***Phaulodinychus***) sp. found (Figs 16I, J) which could not be identified. Length of idiosoma 630–660 µm, legs with remarkable setae (Fig. 16H).

References

Behan, V.M. (1978): Diversity, distribution and feeding habits of North American arctic soil Acari. – Ph. D. Thesis, Dept. Ent., Macdonald College of McGill University. 422 pp.

Beaulieu, F., Dowling, A.P.G., Klompen, H., de Moraes, G.J. & Walter, D.E. (2011): Superorder Parasitiformes Reuter, 1909. – *In*: Zhang, Z.-Q. (Ed.). Animal Biodiversity: An Outline of Higher-Level Classification and Survey of Taxonomic Richness. *Zootaxa* (special volume) 3148: 123–128.

Bhattacharyya, S.K. (1963): A revision of the British mites of the genus *Pergamasus* Berlese s.lat. (Acari: Mesostigmata). – *Bulletin of the British Museum (Natural History), Zoology*: 11(2): 131–242.

Blaszak, R. & Ehrnsberger, R. (1993): Beiträge zur Kenntnis von *Halolaelaps* (*Saprogamasellus*) Götz, 1952 (Acari: Gamasida: Halolaelapidae). – Genus 4(3): 143–267.

Bregetova, N.G. (1949): Data on the fauna of mites of the family Haemogamasidae of the USSR. – Parasitologicheskii Sbornik, ZIN AN USSR 11: 161–185.

Bregetova, N.G., Wainstein, B.A., Kadite, B.A., Koroleva, E.V., Petrova, A.D., Tichomirov, S.I. & Shcherbak, G.I. (1977): Key to soil-inhabiting mites. Mesostigmata. – Akad. Nauk SSSR, Leningrad. 718 pp. [In Russian].

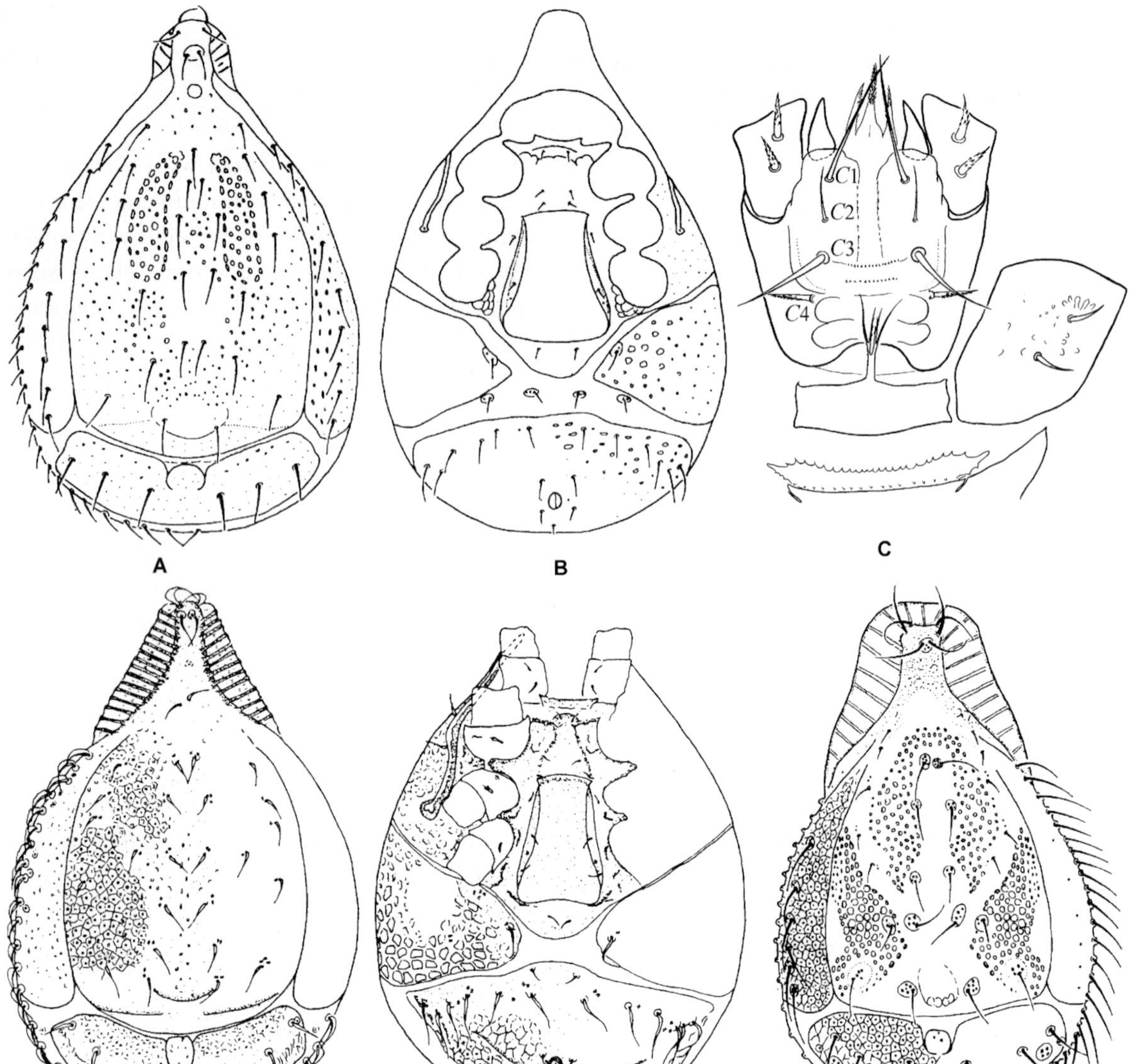

FIGURE 16 *A,* Trachytes hirschmanni, *female, dorsal aspect, B, ventral aspect; C,* Trachytes edleri, *female, subcapitulum and tritosternum, D, dorsal aspect, E, ventral aspect; F,* Trachytes aegrota, *female, dorsal aspect, G, ventral aspect; H,* Uropoda (Phaulodinychus) *sp., male, leg II, I, posterior part of dorsum, J, chelicera without basal segment. A, B, F, G, after Mašan 2001; D, E, after Hutu 1983 modified.*

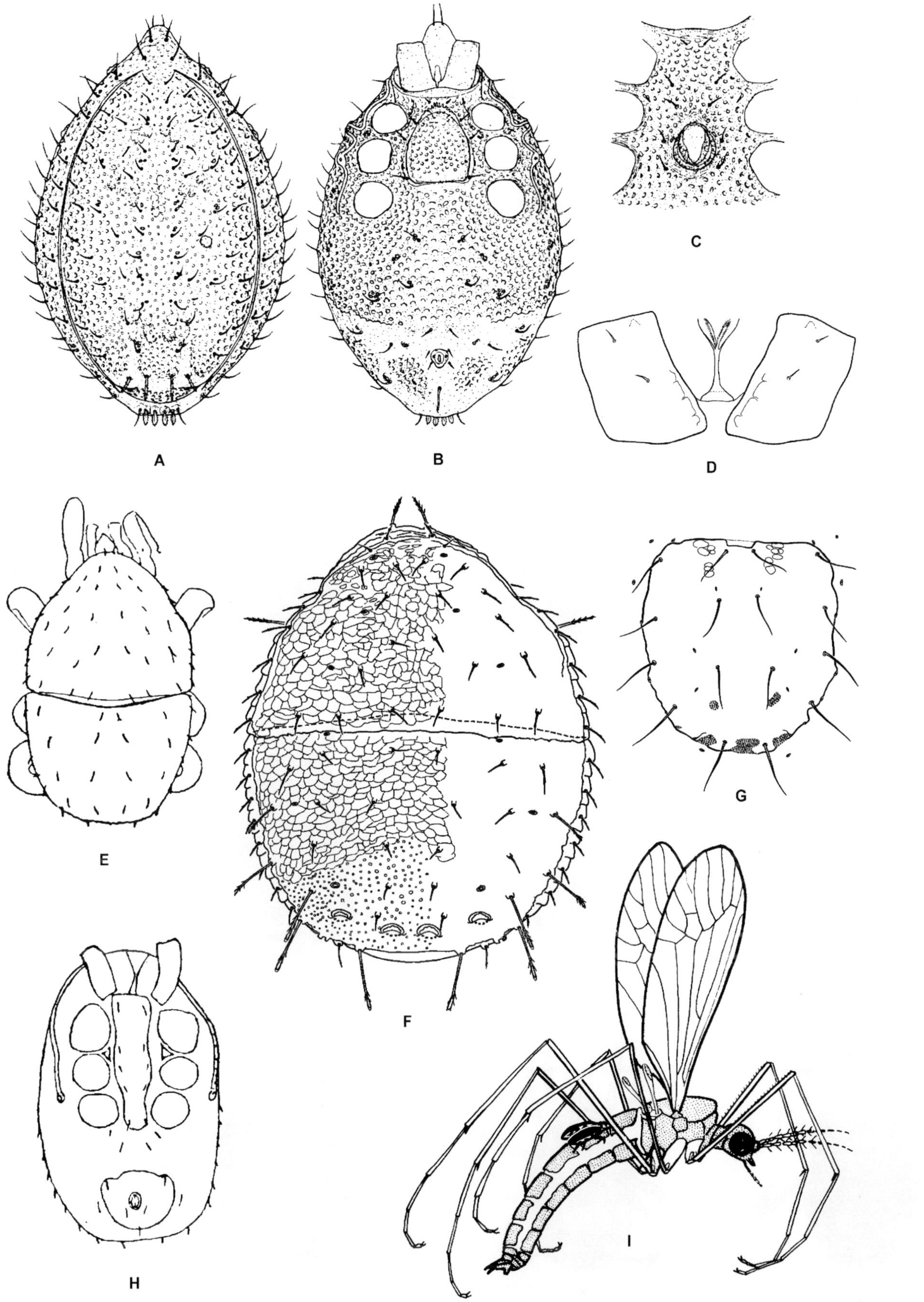

FIGURE 17 *A,* Dinychus perforatus, *female, dorsal aspect, B, ventral aspect, C, male, genital region, D, tritosternum and coxae I; E,* (?) Saprolaelaps *sp.* [= Zercon lundbecki *Trägårdh, 1904*], *deutonymph, dorsal aspect, H, ventral aspect; F,* Zercon solenites, *female, dorsal aspect; G,* Halolaelaps gerlachi, *female, opisthonotal shield; I,* Trichocera arctica (*Trichoceridae*) *carrying females of* Arctoseius tajmyricus. *A–C, after Trägårdh 1943; E, H, after Trägårdh 1904; F, after Gwiazdowicz et al. 2011; I, after Makarova 1995.*

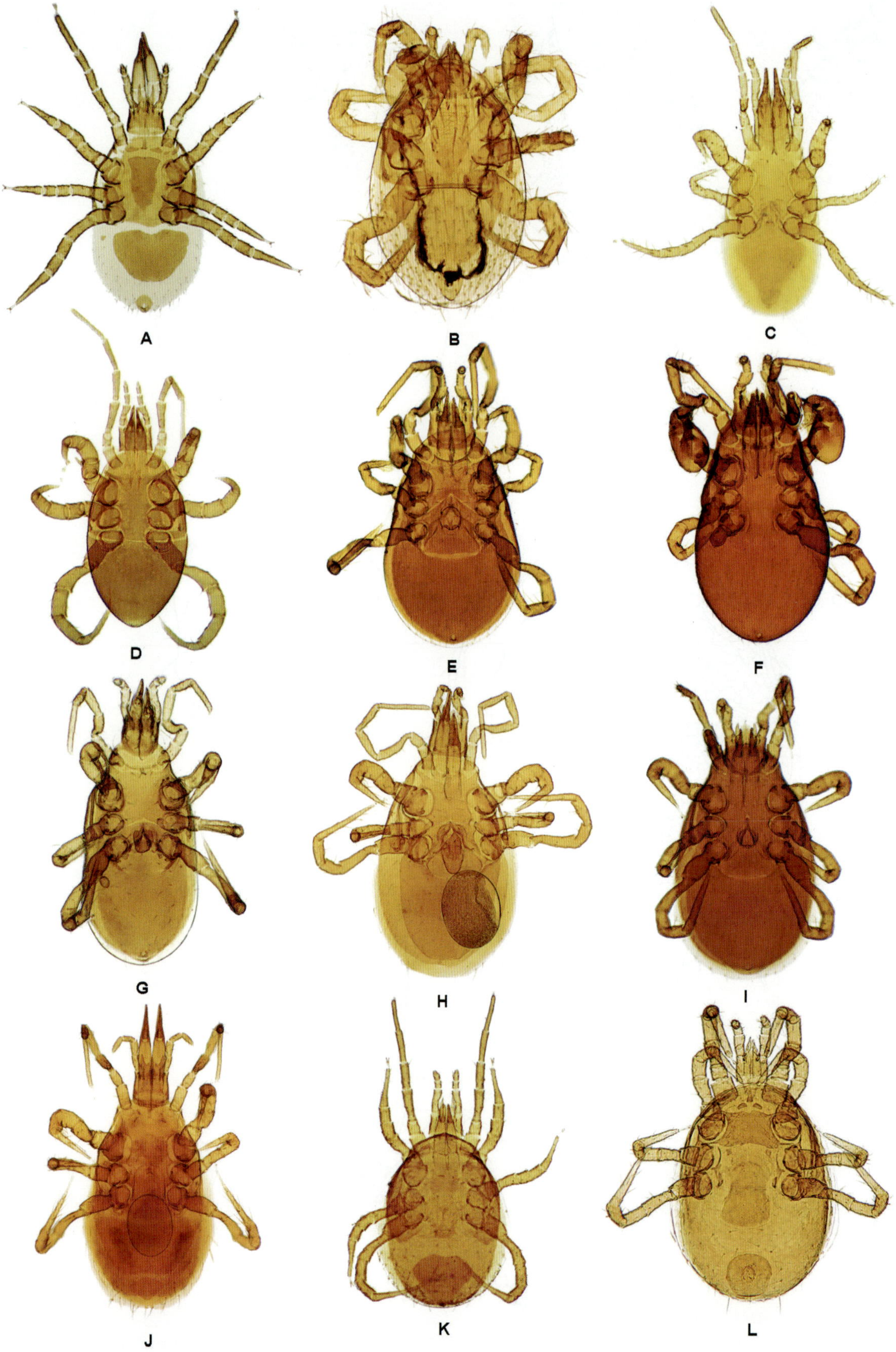

FIGURE 18 *Habitus of mesostigmatic mites. A,* Cornigamasus lunaris, *deutonymph; B,* Parasitellus papei, *deutonymph; C,* Vulgarogamasus halophilus, *female; D,* Vulgarogamasus immanis, *male; E,* Pergamasus crassipes, *female, F, male; G,* Lysigamasus runciger, *female; H,* Paragamasus robustus, *female; I,* Paragamasus integer, *female; J,* Veigaia kochi, *female; K,* Cheiroseius necorniger, *female; L,* Proctolaelaps parvanalis, *female.*

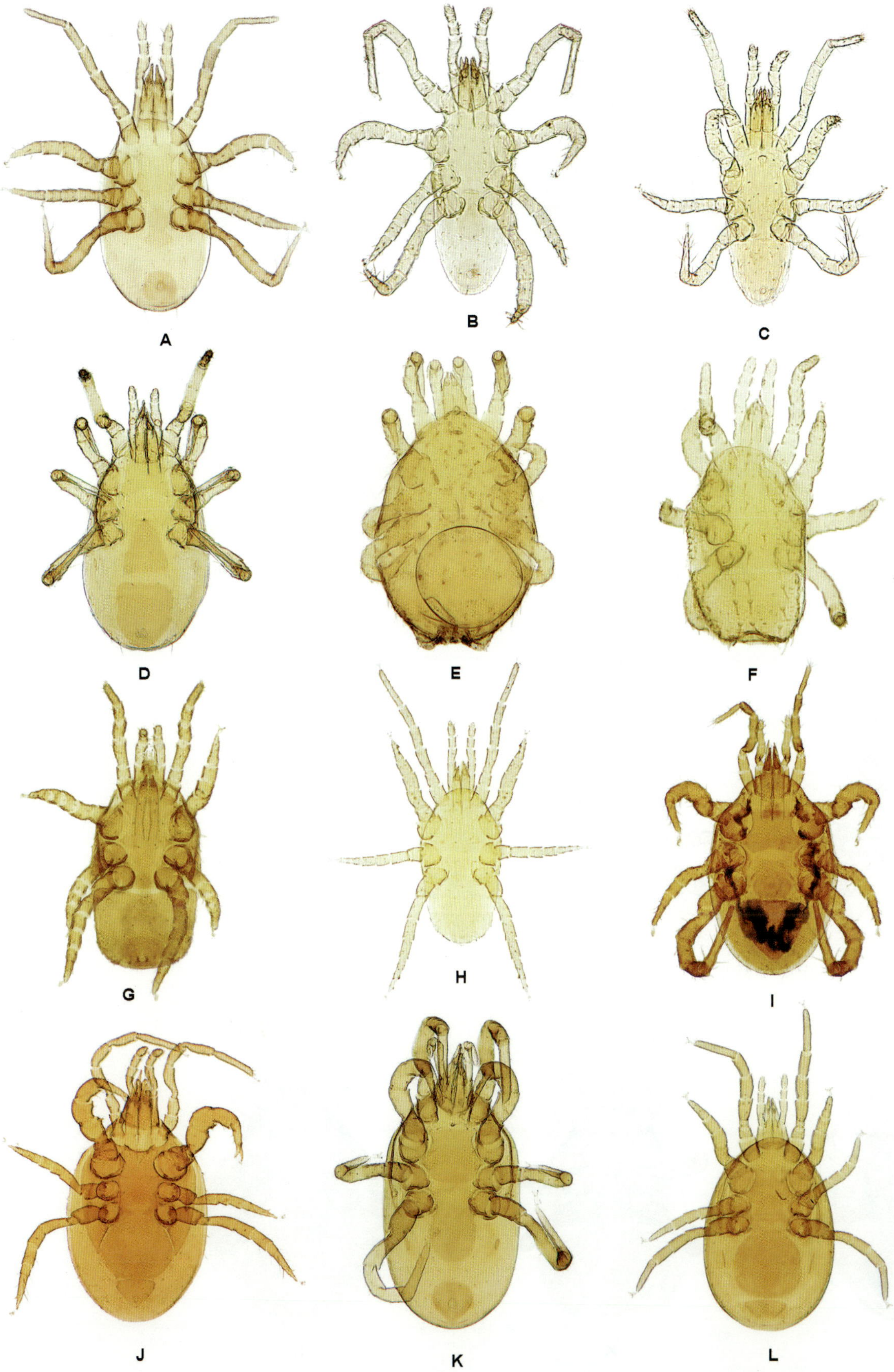

FIGURE 19 *Habitus of mesostigmatic mites. A,* Arctoseius weberi, *female; B,* Arctoseius haarlovi, *female, C, male; D,* Neoseiulus apkutik, *female; E,* Saprosecans baloghi, *female, F, deutonymph; G,* Saprolaelaps punctulatus *Leitner, 1946, deutonymph; H,* Halolaelaps gerlachi, *female; I,* Macrocheles montanus, *female; J,* Pachylaelaps *sp., female; K,* Hypoaspis lubricoides, *female; L,* Gymnolaelaps missouriensis, *female.*

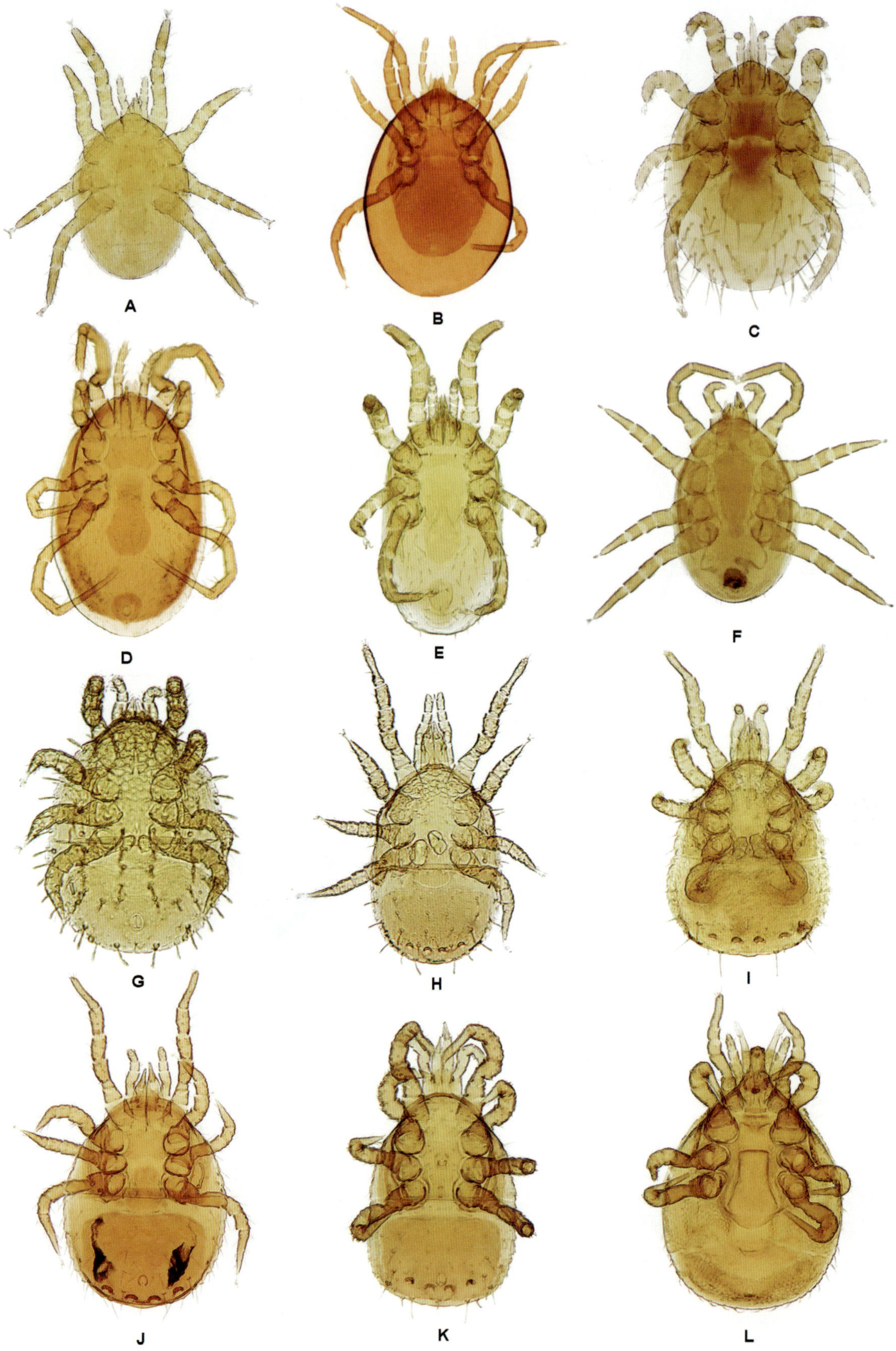

FIGURE 20 *Habitus of mesostigmatic mites. A,* Pneumolaelaps groenlandicus, *female; B,* Ololaelaps sellnicki, *female; C,* Laelaps semitectus, *female; D,* Haemogamasus ambulans, *female; E,* Hirstionyssus isabellinus, *female; F,* Thinoseius spinosus, *deutonymph; G,* Parazercon radiatus, *female; H,* Mixozercon sellnicki, *female; I,* Zercon zelawaiensis, *female; J,* Zercon hammerae, *female, K, male; L,* Trachytes edleri, *female.*

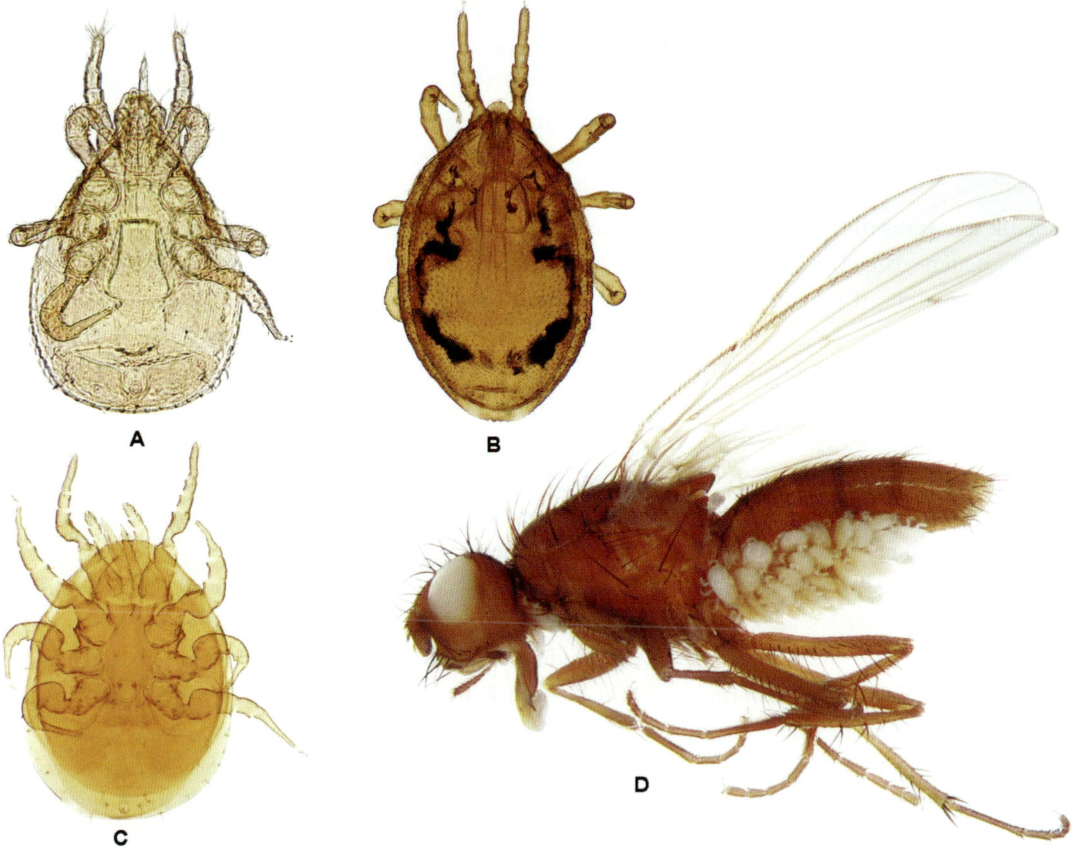

FIGURE 21 *A–C, habitus of mesostigmatic mites: A,* Trachytes hirschmanni, *female; B,* Dinychus perforatus, *female; C,* Uropoda (Phaulodinychus) *sp., male. D,* Fucelia ariciiformis *(Anthomyidae) carrying deutonymphs of* Thinoseius spinosus.

Butenko, O.M. (1984): Mites of the Family Rhinonyssidae from Non-Passerine Birds of the USSR. – Publishing House of the Moscow State University, Moscow. 188 pp. [In Russian].

Chant, D.A. (2007): Species of phytoseiid mites from northern tundra vegetation (Acari: Phytoseiidae). – *International Journal of Acarology* 33(3): 199–221.

Coulson, S.J. & Refseth, D. (2004): Chapter 3. The terrestrial and freshwater invertebrate fauna of Svalbard (and Jan Mayen). – *In*: Prestud, P. *et al.* (eds). *Skrifter* 201. A Catalogue of the Terrestrial and Marine Animals of Svalbard: 75–122. Norwegian Polar Institute, Tromsø.

Evans, G.O. (1958): A revision of the British Aceosejinae (Acarina: Mesostigmata). – *Proceedings of the Zoological Society of London* 131(2): 177–229.

Evans, G.O. (1992): Principles of Acarology. – CAB International, Cambridge. 563 pp.

Evans, G.O. & Till, W.M. (1979): Mesostigmatic mites of Britain and Ireland (Chelicerata: Acari: Parasitiformes). An introduction to their external morphology and classification. – *Transactions of the Zoological Society of London* 35: 139–270.

Gwiazdowicz, D.J. (2007): Ascid Mites (Acari, Mesostigmata) from Selected Forest Ecosystems and Microhabitats in Poland. – *Wydawnictwo Akademii Rolniczej im.* A. Cieszkowskiego, Poznan. 248 pp.

Gwiazdowicz, D.J., Teodorowicz, E. & Coulson, S.J. (2011): Redescription of *Zercon solenites* Haarløv, 1942 (Acari: Mesostigmata: Zerconidae) with a key to the Svalbard species of the genus *Zercon*. – *International Journal of Acarology* 37(Suppl. 1): 135–148.

Haarløv, N. (1942): A morphologic-systematic-ecological investigation of Acarina. – *Meddelelser om Grønland* 128(1): 1–71.

Halašková, V. (1969): Zerconidae of Czechoslovakia (Acari: Mesostigmata). – *Acta Universitatis Carolinae Biologica* 3–4: 175–352.

Hennessey, M.K. & Farrier, M.H. (1988): Systematic revision of thirty species of free-living, soil-inhabiting gamasine mites (Acari: Mesostigmata) of North America. – *Technical Bulletin of North Carolina State University* 285: 1–123.

Henriksen, K.L. & Lundbeck, W. (1917): Grønlands Landarthropoder (Insecta et Arachnida). – *Meddelelser om Grønland* 22: 481–821.

Hunter, P.E. & Husband, R.W. (1973): *Pneumolaelaps* (Acarina) mites from North America and Greenland. – *The Florida Entomologist* 56(2): 77–91.

Hunter, P.E. & Husband, R.W. (1973): *Pneumolaelaps* (Acarina) mites from North America and Greenland. – *The Florida Entomologist* 56(2): 77–91.

Hutu, M. (1983): Gangsystematik der Parasitiformes. Teil 428. Teilgänge, Stadien von 6 neuen *Trachytes*-Arten aus Rumänien und Sweden (Uropodini, Uropodinae). – *Acarologie* 30: 51–66.

Jørgensen, M. (1934): Ein revidiertes Verzeichnis Über grönländische Milben. – *Zoologischer Anzeiger* 107(1–2): 40–47.

Karg, W. (1984): Zur Kenntnis der Untergattung *Pneumolaelaps* Berlese, 1925 der Raubmilbengattung *Hypoaspis* Canestrini, 1884. – *Deutsche Entomologische Zeitschrift, Neue Folge* 31(1–3): 35–39.

Karg, W. (1985): Die mit Arten der Gattung *Bombus* Latreille vergesellschafter Raubmilben der Gattung *Parasitus* Latreille, 1795 (Acarina, Parasitiformes). – *Zoologischer Jarbücher, Syst.* 112: 525–535.

Karg, W. (1993): Acari (Acarina), Milben. Parasitiformes (Anactinochaeta). Cohors Gamasina Leach. Raubmilben. (2. Aufl.). – *Die Tierwelt Deutschlands* 59: 1–523.

Lange, A.B. (1955): Order Parasitiformes (Reuter) Zachv. 1. Suborder Mesostigmata Can. General review. – *In*: Bregetova, N.G. (ed.) Mites of Rodents of the USSR Fauna. *Identification Keys of the USSR Fauna.* V. 59: 217–243. Academy of Sciences of the USSR, Moscow-Leningrad [In Russian].

Lindquist, E.E. (1964): A revision of mites of the subfamily Blattisocinae (Acarina: Blattisocidae) in America north of Mexico. Ph. D. Thesis, Department of Zoology, University of California, Barkely. 416 pp.

Lindquist, E.E. (1961): Taxonomic and biological studies of the mites of the genus *Arctoseius* Thor from Barrow, Alaska (Acarina: Aceosejidae). – *Hilgardia* 30(11): 301–350.

Lindquist, E.E., Krantz, G.W. & Walter, D.E. (2009): Chapter 12, Order Mesostigmata. – *In*: Krantz, G.W. & Walter, D.E. (Eds.). A Manual of Acarology. 3rd Edition: 124–232. Texas Tech. University Press, Lubbock.

Lindquist, L.L. & Makarova, O.L. (2011): Two new circumpolar mite species of the genus *Arctoseius* (Parasitiformes, Mesostigmata, Ascidae). – *Entomological Review* 91(8): 1054–1072.

Makarova, O.L. (1995): On the association of gamasid mites of the genus *Arctoseius* (Mesostigmata, Ascidae) with winter crane flies (Diptera, Trichoceridae) at the north of Taymyr Peninsula. – *Arctic Insect News* 6 2–4.

Makarova, O.L. (2000a): To a study of mites of the genus *Arctoseius* (Parasitiformes, Ascidae) from the Far North: 1. Morphometric analysis of the taxonomic features of Arctic species and description of *Arctoseius tschernovi* sp. n. and *A. miranalis* sp. n. – *Entomological Review* 80(Suppl. 1): 112–130.

Makarova, O.L. (2000b): To a study of mites of the genus *Arctoseius* (Parasitiformes, Ascidae) from the Far North: 2. Description of *A. productus* sp. n. and *A. babenkoi* sp. n. and key to the High Arctic species. – *Entomological Review* 80(Suppl. 1): 131–142.

Makarova, O.L. (2014): The free-living mite fauna (Acari) of Greenland. – *Zoologicheskii Zhurnal, Moscow* 93(12): 1404–1419 [In Russian].

Mašan, P. (2001): Roztoce kohorty Uropodina (Acarina, Mesostigmata) Slovenska. – *Annotationes Zoologicae et Botanicae* 223: 1–320.

Norton, R.A., Kethley, J.B., Johnston, D.E. & O'Connor, B.M. (1993): Phylogenetic perspectives on genetic systems and reproductive modes of mites. – *In*: Wrench, D.L. & Ebbert, M.A. (eds.). Evolution and diversity of sex ratio in insects and mites: 8–99. Chapman & Hall, New York.

Petrova, A.D. & Makarova, O.L. (1991): *Arctoseius tajmyricus* – a new species of gamasid mites (Mesostigmata, Aceosejidae) phoretic on trichocerid flies (Diptera). – *Entomological Review* 70(5): 139–142.

Sellnick, M. (1940): Die Milbenfauna Islands. – *Göteborgs Kungl. Vetenskaps- och Vitterhets-Samhälles Handlingar. B* 6(14): 1–129.

Sellnick, M. (1957): Zwei neue Milbenarten aus Küstengrundwasser von Simrishamn. – *Kungliga Fysiografiska Sällskapets i Lund Förhandlingar* 27(2): 11–23.

Sellnick, M. (1960): *Zercon hammerae*, nov. spec., eine neue Milbenart aus Ost-Grönland (Acarina, Zerconinae). – *Entomologiske Meddelelser* 29: 216–220.

Trägårdh, I. (1904): Monographie der arktischen Acariden. – *Fauna Arctica* 4(1): 1–78.

Trägårdh, I. (1931): Terrestrial Acarina. – *Zoology of the Faroes* 2(49): 1–69.

Trägårdh, I. (1943): Zur Kenntnis der Prodinychidae (Acarina). – *Arkiv för Zoologi, Ser. A* 34(21): 1–29.

Walter, D.E. & Proctor, H.C. (2013): Mites: Ecology, Evolution & Behaviour. Life at a Microscale. 2nd Edition. Springer, Dordrecht – Heidelberg – New York – London. 494 pp.

18.3.3. Ixodida (Metastigmata, Ticks)

Olga Makarova

The Ixodida, or ticks, comprise obligate bloodsuckers of vertebrates. This rather small suborder (about 800 known species) is represented in Greenland by only *Ixodes uriae* White, 1852 [=*Ixodes putus* (Pickard-Cambridge, 1876)], the largest Greenland mite. The engorged female reaches 0.9–1.0 cm in length. The life cycle includes egg, larva, nymph, and adult.

The most remarkable morphological features of ticks, present in all active stages, are the highly modified *hypostome* and the *Haller's organ*. The hypostome, anterior part of *subcapitulum* (Fig. 1F, *sbc*), is an elongated holdfast structure armed with rows of retrorse denticles (Figs 1F, 2B, *hyp*) which anchors the tick to its host during bloodsucking. Haller's organ constitutes the complex sensory field located in the dorsal depression of tarsus I (Fig. 1G, *hor*) and comprising contact and olfactory receptor sensillae.

IXODIDAE

Ixodes uriae White, 1852. Figs 1–3.

Two tagmata are distinguishable: the *idiosoma* and *gnathosoma* (Fig. 1A, IS and GS respectively). The larval tick is hexapod (Figs 1A, C), lacks spiracular plates and possesses a small podonotal shield or *scutum* (Fig. 1A, *sc*). The scutum is 0.3–0.5 mm in length, bearing 4 pairs of setae, punctulate, with reticulate ornamentation (Fig. 1B). The *anus* opens caudoventrally (Fig. 1C, *an*). Nymph and adults are octopod and possess *spiracular plates* behind coxae IV (Figs 1D, H, I, *sp*). Nymphal scutum 0.6–0.8 mm in length; body setae more numerous than in larvae (Fig. 1D); genital opening (between coxae I–II) absent. Unengorged adults 2.8–3.3 mm. Female with scutum longer than wide, broadest anteriorly, punctate, bearing a few very small (20 µm) setae (Fig. 1H); *cervical furrows* (*cf*) reach posterior margin of scutum; elastic dorsal cuticle (or *alloscutum, al*) with numerous long (130 µm) flattened setae; *genital opening* is an arcuate slit (Fig. 1J, *go*); base of gnathosoma dorsally with transverse *porose areas* (Fig. 2A, *pa*); hypostome broadly rounded apically, dentition 2/2 (Fig. 2B); *palps* widely separated and inwardly curved (Fig. 2A, *pp*). Male dorsal shield (or *conscutum, csc*) with *scapulae* (anterior angles, *scp*) and cervical furrows prominent, strongly punctuate (Fig. 2G); setae of conscutum small (20–40 µm), rare; most ventral idiosomal setae thin and short, only some posterior setae spine-like; posterior margin of body with fringe of 5 tufts of long, flattened setae adjacent to each other (Fig. 2F); palps horn-like, upturned, tapering at tips (Fig. 2C); hypostome rudimentary, bilobed apically, with few vestigial denticles (Fig. 2D). All active stages have legs comprising 7 free podomeres (Figs 1H, J, 2F, G): *coxa* (*cx*), *trochanter* (*tr*), *femur* (*fm*), *genu* (*gn*), *tibia* (*tib*), *tarsus* (*tar*), and *ambulacrum* (*amb*). The tarsi II–IV are secondarily divided by a *peripodomeric suture* (consisting of slit-organs) into *basitarsus* and *telotarsus*, respectively. The ambulacrum (Figs 1G, 2E) usually comprises the *pretarsus* (*pt*), 2 *claws* (*cl*), and the membranous *pulvillus* (*plv*). In male, all tarsi have two apical denticles (Fig. 2G).

BIOLOGY. The species is reported from 52 bird species. Its main hosts are auks, tube-nosed sea birds, cormorants, sea gulls, penguins. In the North Atlantic, ticks are most common on guillemots (*Uria aalge, Uria lomvia*), black guillemot (*Cepphus grylle*), kittiwake (*Rissa tridactyla*), razorbill (*Alca torda*), and puffin (*Fratercula arctica*) (Eveleigh & Threlfall 1974, Filippova 1977, Mehl & Traavik 1983, Barton *et al.* 1996, *etc.*). On other birds (gulls, cormorants etc.) ticks were most often recorded when these were found together with the main hosts in common rookeries (Violovich 1962, Barton *et al.* 1996). The Greenland records are associated with kittiwake (*R. tridactyla*), Iceland gull (*Larus glaucoides*) and guillemot (*U. lomvia*) (Henriksen & Lundbeck 1917, Bequaert 1946).

Larvae, nymphs and females have been recorded from their hosts, while males are found only in and around nests in turf and rocky crevices. Because of the rudimentary character of the hypostome, males cannot attach themselves to a host; thus the copulation takes place in the breeding sites of hosts. Like other bird ticks, females and immatures feed often on the head and the upper region of the neck, where they cannot be removed by the bird's beak (Arthur 1963, Balashov 1998). The nestlings may have the majority of ticks attached to the body (Mehl & Traavik 1983). The body weight of tick females increases 20–30 times during feeding (Balashov 1998) (Figs 3A, B). All stages are able to develop at temperatures between +1°C and +20°C (Karpovich 1973). The life cycle is completed in 2–7 years depending on the climate type, weather conditions and microclimatic regime of the bird colony (Violovich 1962, Karpovich 1973, Eveleigh & Threlfall 1974, Barton *et al.* 1996). The feeding occurs only in first half of summer in agreement with the host's breeding cycle, as the birds leave their breeding places in late July – early

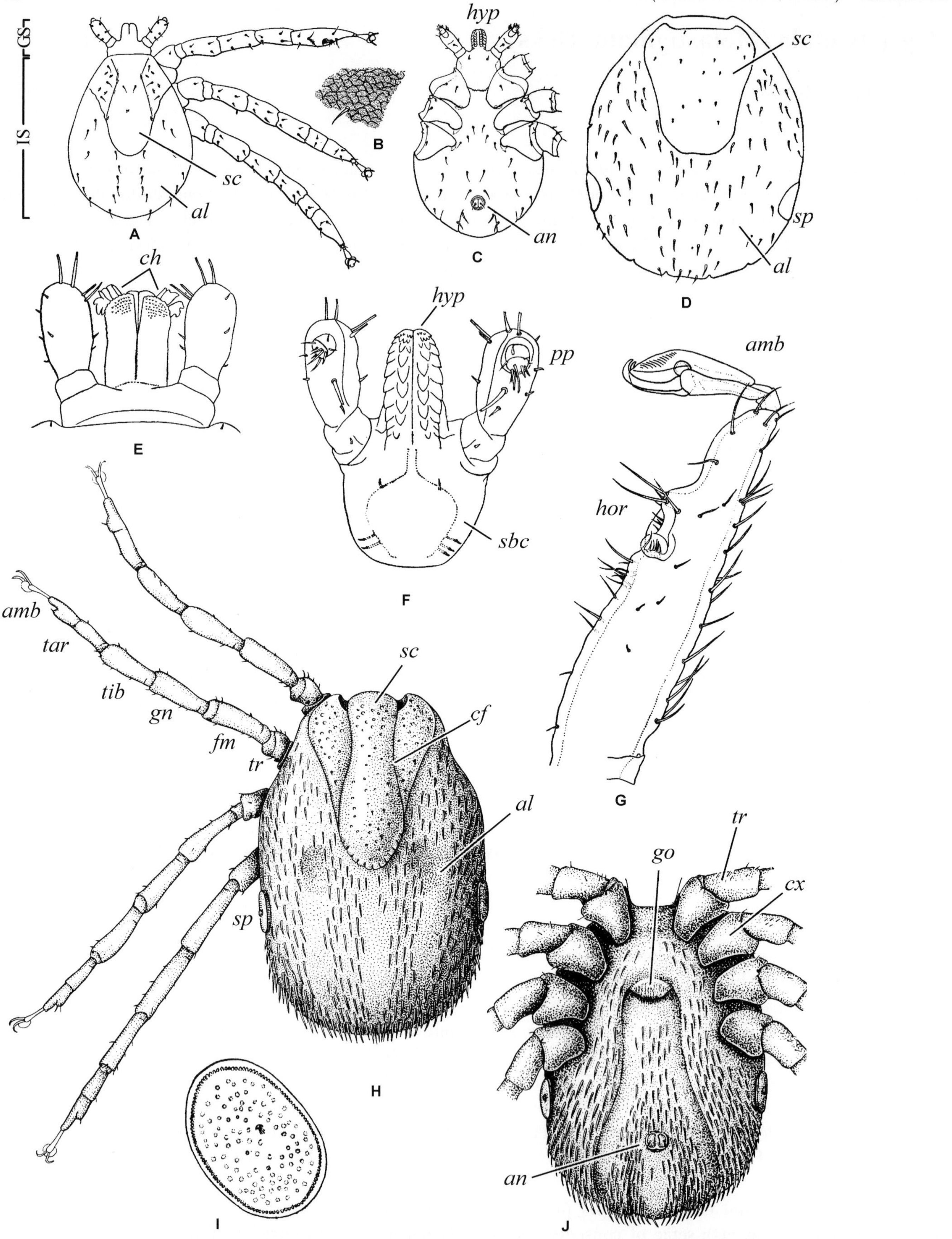

FIGURE 1 Ixodes uriae*: A, larva dorsally; B, larva, fragment of scutum; C, larva ventrally; D, nymph, idiosoma dorsally; E, nymph, gnathosoma dorsally; F, nymph, gnathosoma ventrally; G, female, tarsus I; H, female, idiosoma with legs dorsally; I, nymph, spiracular plate; J, female, idiosoma ventrally. All after Filippova 1977.*

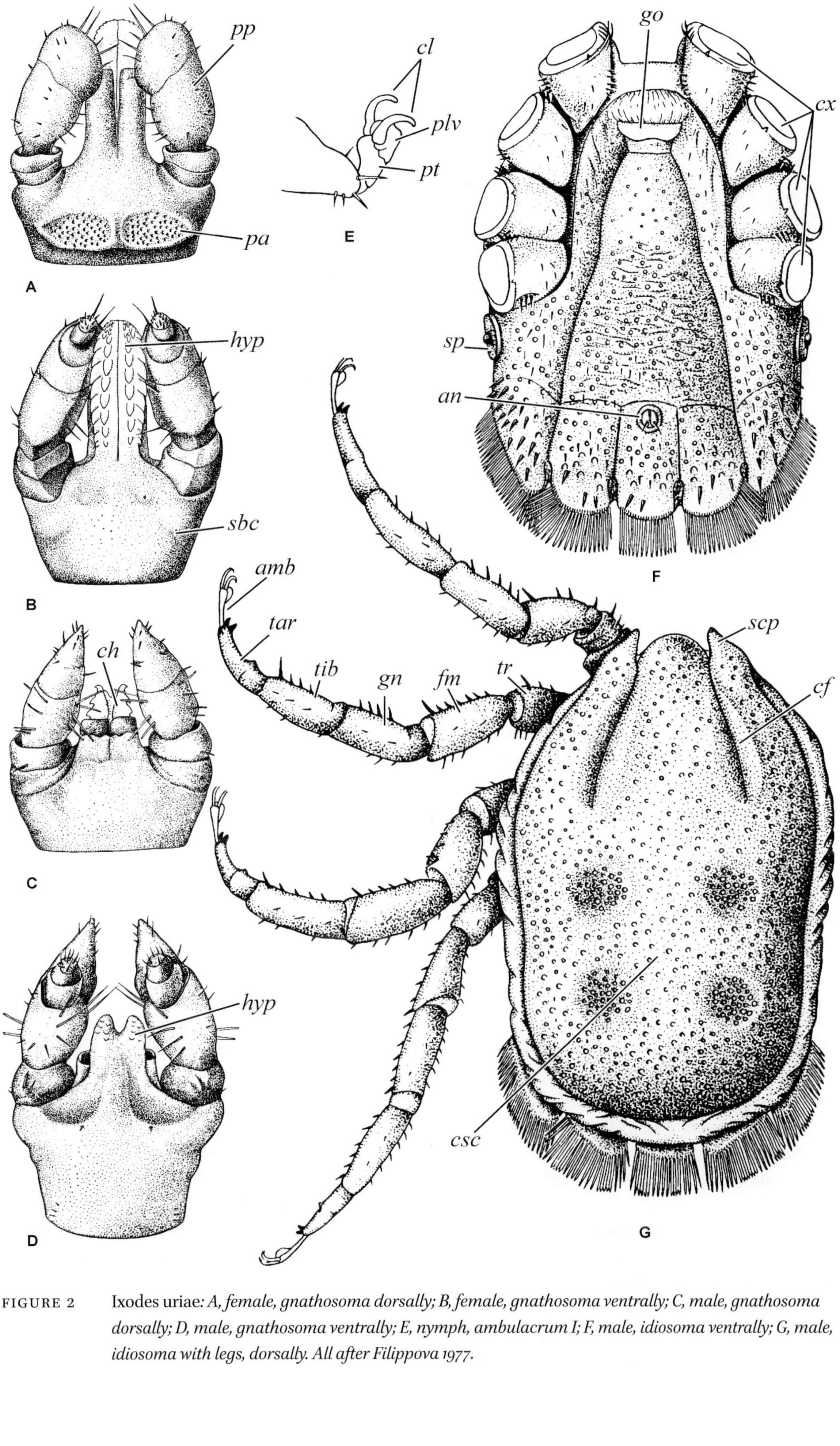

FIGURE 2 Ixodes uriae*: A, female, gnathosoma dorsally; B, female, gnathosoma ventrally; C, male, gnathosoma dorsally; D, male, gnathosoma ventrally; E, nymph, ambulacrum I; F, male, idiosoma ventrally; G, male, idiosoma with legs, dorsally. All after Filippova 1977.*

FIGURE 3 Ixodes uriae, *female: A, dorsal aspect, B, ventral aspect.*

August and return to land only next spring (Karpovich 1973). All adults and most nymphs spend the winter in an unfed state (Karpovich 1973, Mehl & Traavik 1983). Mass death of engorged ticks has been observed in a cold winter; thus only eggs and unfed ticks are considered well adapted to hibernation (Karpovich 1973). The supercooling point of *I. uriae* eggs (-28.7°C) is lowest among the development stages (Lee & Baust 1987).

Bloodsucking of *I. uriae* is directly harmful to birds. Heavily infested hosts, particularly nestlings, may be killed (Arthur 1963). In bird colonies hungry ticks can attack man and their bite is painful (Karpovich 1973, Mehl & Traavik 1983). After feeding on man, even fully engorged ticks die. In the Northern Hemisphere *I. uriae* is carrier of several neurotropic viruses, and its infection rate is often high (see references in Mehl & Traavik 1983). Experimental infections of several bird species with tick virus produced pathogenic conditions, sometimes with lethal outcome (Lvov *et al.* 1979).

DISTRIBUTION. *Ixodes uriae* is very widely distributed, circumpolar and bipolar, but recorded only from marine birds and their breeding places. The ticks inhabit coasts and islands of all oceans excluding tropical areas (Filippova 1977).

References

Arthur, D.R. (1963): British Ticks. – Butterworths, London. 213 pp.

Balashov, Yu.S. (1998): Ixodid Ticks – Parasites and Vectors of Diseases. – "Nauka" Publ. House, Saint Petersburg. 287 pp. [In Russian].

Barton, T.R., Harris, M.P., Wanless, S. & Elston, D.A. (1996): The activity periods and life-cycle of the tick *Ixodes uriae* (Acari: Ixodidae) in relation to host breeding strategies. – *Parasitology* 112(6): 571–580.

Bequaert, J.C. (1946): The ticks, or Ixodoidea, of the northeastern United States and Eastern Canada. – *Entomologica Americana* 25(3–4): 121–232.

Eveleigh, E.S. & Threlfall, W. (1974): The biology of *Ixodes uriae* in Newfoundland. – *Acarologia* 16(4): 621–635.

Filippova, N.A. (1977): Ixodid Ticks of Subfamily Ixodinae. – "Nauka" Publ. House, Saint Petersburg. 396 pp. [In Russian].

Henriksen, K.L. & Lundbeck, W. (1917): Landarthropoder (Insecta et Arachnida). Conspectus Faunae Groenlandicae, Pars secunda. – *Meddelelser om Grønland* 22: 481–823.

Karpovich, V.N. (1973): The life cycle of *Ceratixodes putus* (Pick.-Camb.) in Murmansk condition. – *Parazitologiya*, Leningrad 7: 128–134 [In Russian].

Lee, R.E. & Baust, J.G. (1987): Cold-hardiness in the Antarctic tick, *Ixodes uriae*. -- *Physiological Zoology* 6(4): 499–506.

Lvov, D.K., Gromashevski, V.L., Skvortsova, T.M *et al.* (1979): Arboviruses of high latitudes in the USSR. – *In*: Kurstak, E. (ed.). Arctic and Tropical Arboviruses: 21–37. Academic Press, New York.

Mehl, R. & Traavik, T. (1983): The tick *Ixodes uriae* (Acari, Ixodides) in seabird colonies in Norway. – *Fauna norvegica*. Ser. B 30: 94–107.

Violovich, N.A. (1962): About ecology of *Ceratixodes putus* Pick.-Cambr., 1878 (Ixodidae) on islands of Far East. – *Voprosy ekologii* 8: 20–21. Kiev University Publ. House, Kiev [In Russian].

18.3.4. Prostigmata (Actinedida)

Olga Makarova

The Prostigmata is a large and extremely diverse taxon morphologically and ecologically. It comprises terrestrial, aquatic and marine, parasitic and free living (predatory, saprophagous, phytophagous, microphytophagous) mites. Not long ago, the Prostigmata as conventionally circumscribed were interpreted as a paraphyletic group (Norton *et al.* 1993), and some families were transferred to other cohorts and later even to other suborders (Lindquist *et al.* 2009). However, for convenience we here follow the traditional understanding (Kethley 1982, Evans 1992). The suborder comprises about 160 families and more than 25 900 known species including 61 families and about 7 100 species of aquatic forms (Walter *et al.* 2011, Zhang *et al.* 2011). Ranks of higher taxa in the key below mainly follow Norton *et al.* (1993).

At present 90 species of 24 families of terrestrial prostigmatans are known in Greenland, but some unregistered parasitic families (Myobiidae, Psorergatidae, Syringophilidae and probably others) may undoubtedly be found on appropriate hosts.

The given key is extremely preliminary because no targeted investigations of the Greenland fauna have been made, and the parasitic groups are almost unknown. Some cases of recent recombination and synonymy could have been overlooked.

Morphology

Two main body tagmata can be recognized, the *idiosoma* and *gnathosoma*. The idiosoma, IS (Fig. 1B), varying in length from 80 μm to 16 mm, is usually ovoid, weakly sclerotized, with or without sclerites. Soft integument may be striated or granulated. Typically two body regions are recognizable, usually demarcated by the *dorsosejugal furrow, dsf*: *propodosoma*, PPS (possessing two anterior pairs of legs) and *hysterosoma*, HS. The dorsal surface of propodosoma, *prodorsum* (*pd*), often bears 3–6 pairs of setae, 0–2 pairs of *trichobothria, trb* (specialized tactile setae with enlarged, often recessed alveoli), and 0–2 pairs of *eyes, ey*; sometimes 1 unpaired eye is present near the anterior prodorsal margin. Frontal protuberance (*naso, ns*) may be present on the prodorsum endowed with a pair of setae or nude. Anterior extremity of the prodorsum is sometimes tapered into the *epirostrum, epr* (Fig. 3A). In trombidiid mites, a well sclerotized linear *crista metopica, cm* (Figs 6L, 8A) is present in the prodorsal region of active postlarval stages. In this case prodorsal trichobothria usually are situated on expansions (*sensillar plates*) of the crista metopica. One pair of cup-shaped sensory structures (*urnulae, ur*) is found in nymphs and adults of Balaustinae, Erythraeidae (Fig. 7C). Hysterosoma may be holotrichous with 16 pairs of hysteronotal setae, hypotrichous with a reduced setal pattern, or highly hypertrichous with a thick pelage of homonomous setae. *Hysteronotal sclerites, hns,* may be developed (Figs 1B, 9A, 12A) or not (Figs 2A, I, 6B, D), or proper *hysteronotal shield* may be present (Figs 8J, L). Rarely 1 pair of opisthodorsal trichobothria is present in the caudal region (Fig. 6D). Leg coxae are fused with the ventral body wall forming plate-like coxisternal elements or *epimeres, ep* (Fig. 1C). The genital orifice is situated behind IV legs, usually longitudinal (rarely transverse), covered by *genital valves, gv,* bearing genital setae. The *ovipositor,* eugenital setae, and 2–3 pairs of *genital papillae, gp* (osmoregulatory organs), are often present beneath the genital valves (Fig. 1C). These may be flanked by *aggenital setae, ag*. Larvae of species, whose nymphs and adults are equipped with genital papillae, have a pair of *urstigmata* (papilla-like structure, Claparède's organ) between coxae I and II (Figs 7L, 8E). *Anus, an,* is usually ventral subterminal (Figs 7L, 10J) or terminal (Figs 2J, 8M), rarely dorsal (Figs 3C, D), closed by anal valves, or open. In some lineages a coalescences of the genital and anal regions takes place, often accompanied by development of a retractile *aedeagus* in males. Fertilization occurs either by copulation or indirectly by deposition of spermatophores.

The gnathosoma, GS, is extremely diverse: cryptic inconspicuous (Figs 8E, 10B, 12J), with a snout-like elongate base (Figs 5I, 14D–G), or with raptorial hypertrophied palps (Fig. 9C), *etc*. The cheliceral bases may be free and separated (Figs 3C, E), contiguous, adnate (Figs 6A, B), or merged into a *stylophore* (Figs 10I). Sometimes the stylophore is united with the gnathosomal base forming a *rostrum* (Fig. 9C). The dorsal surface of rostrum is covered by the *rostral shield, rs*. The chelicerae are primarily chelate-dentate (Figs 3K, 5L), but the *fixed digit* is often reduced, while the *movable digit, d.m.*, is usually edentate and styliform (Figs 1B, 3E), rod-like (Fig. 10I), or sickle-shaped (Fig. 7N). The anterior part of *subcapitulum, sc,* is sometimes produced into a beak-like *hypostome, hyp,* bearing adoral setae (Figs 1A, 5I). A pair of strongly sclerotized blade-like toothed setae, *rutella, ru,* are in some Endeostigmata located at the antero-lateral margins of subcapitulum (Figs 1A, 2G). The tracheal system, when present, usually opens by paired *stigmata, st,* situated between the cheliceral bases (Figs 1A, 10I), on rostrum (Fig. 9C), or on the anterior part of the prodorsum (Fig. 12A). Stigmata are often associated with *peritremes, prt,* formed dorsally at the base of the gnathosoma, on the prodorsum, or on the cheliceral bases (Figs 1B, 10I). The *palp, pp,* numbers 1–5 free segments (Figs 1A, B, 10I). Palpal *tarsus, tar,* and *tibia, tib,* may be merged into a *tibiotarsus.* Palpal *femur, fm,* may be subdivided into *basifemur* and *telofemur.* Palptibia often terminates by a hypertrophied claw-like seta (*thumb-claw process, tc*), displacing the palptarsus from a terminal to a subterminal (ventral) positon (Figs 1B, 7B, F, G).

The *legs* consist originally of a full podomeric set (*trochanter, femur, genu, tibia, tarsus,* and *ambulacrum, amb*; Fig. 1B). Femur is often subdivided into the articulating *basifemur, bfm,* and *telofemur, tfm.* Different kinds of leg reduction may be found, especially in parasitic lineages of prostigmatans. The ambulacrum typically bears 2 lateral *claws, cl,* and a median *empodium, em* (claw-like, pad-like, rayed, *etc.*) (Fig. 1D), but either claws or an empodium may be lost. Besides normal setae some other types of cuticular structures are present on the legs – trichobothria, *solenidia, eupathidia, etc.* Trichobothria, if present, are developed on tibia and tarsi (Fig. 5H). Solenidia (erect protrusions of the cuticle lacking in alveoli) may be present on different leg segments. Eupathidia are modified setae with a hollow core, enlarged alveolus and truncate tip. Number, size, form and disposition of the cuticular structures mentioned are of a great taxonomical importance. Sometimes the leg setal equipment is hypertrichous.

Life cycle

The primary life-history pattern includes 7 stages – egg, prelarva, hexapod larva, three octopod nymphs, and adult (female or male). The nonfeeding prelarva is rarely mobile. In most cohorts juveniles and adults are homeomorphic, but in the cohort Parasitengona the normally parasitic larva is heteromorphic with respect to the nymphs and adults, and the protonymph and tritonymph are calyptostases (nonfeeding and nonwalking forms characterized by the regression of mouthparts and appendages and developing within the cuticle of the preceding stages). Larvae may be regressive by having not fully developed mouthparts or legs in contrast to nymphs and adults. The number of nymphal stages may be 3, 2, 1, sometimes (for example in Tarsonemina) there are no nymphs at all, only egg, larva, and adult. Females of some parasitic species are larviparous. Unlike immatures the adults possess functional genitalia. Primitively, mating is by spermatophores, but a direct insemination has evolved in Raphignathina and Tarsonemina accompanied by an intromittent aedeagus formarion.

Identification

The identification keys are based mainly on adult (sometimes female) features. In special cases larval features are used. The given length normally corresponds to the idiosomal length, but in some families the total length (gnathosoma + idiosoma) is used.

Key to families*

1. With 2 pairs of legs in all stages; legs without true claws, with highly branched empodium; idiosoma greatly elongated, annulate, vermiform, or fusiform (Figs 11A, H, 17G); plant inhabitants (cohort Eupodina, in part)....................... Eriophyidae
– With 3 or 4 pairs of legs; legs with or without true claws, with or without empodium; idiosoma of different form, rarely vermiform or fusiform.......... 2
2. Female with a pair of anterolateral prodorsal stigmata (Fig. 12A) and associated tracheae, and commonly with a pair of capitate trichobothria (both generally absent in males); peritremes absent; hysterosoma with 4–5 dorsal sclerites (Figs 12A, J), often overlapping (may be fused in male; Fig. 12K); cheliceral bases and subcapitulum integrated in small gnathosomatic capsule, Fig. 12J (cohort Tarsonemina) 3
– Female and male with stigmata between or on the cheliceral bases (Figs 1A, 10I), or stigmata absent; capitate prodorsal trichobothria may be present; peritremes often present in stigmate forms; hysterosoma dorsally without such sclerites; gnathosoma of different structure.............................. 5
3. Female with leg IV attenuate, 3-segmented (femur and genu fused), without claws or empodium (Fig. 12J); male with leg IV 3–4-segmented, possessing a single sessile claw (Fig. 12K) .... Tarsonemidae
– Leg IV of female with separate femur and genu, often with claws and membranous empodium; leg IV of male 5-segmented................................. 4

* Mainly based on Krantz 1978, Kethley 1990, Evans 1992 with reduction and modification.

4. Hysteronotal sclerites large, covering prodorsum completely, anterior margin striate, forming a roof over the gnathosoma (Figs 12E, G); distance between insertions of legs II and III equal to the distance between insertions of legs III and IV (Fig. 17J) .... Scutacaridae
– Hysteronotal sclerites not covering entire prodorsum (Fig. 12A); distance between insertions of legs II and III at least twice the distance between insertions of legs III and IV (Figs 12C, 17H, I) .... Pygmephoridae
5. Palps with a distinct tibial thumb-claw process in all stages that often displaces the diminished tarsus ventrally (Figs 7B, F, G, P), or with palpal tarsus not modified and palpal tibia with 1–3 distal claws (Fig. 6F); if larvae parasitic on arthropods, then palp genu with only 1 seta or nude; terrestrial forms, free-living or parasitic.................................... 6
– Palps simple or chelate, rarely fang-like; larvae of aquatic species parasitic on insects may have a palpal thumb-claw process, in this case palp genu with 2 setae; terrestrial, aquatic and marine species; free-living or parasitic.................................... 16
6. Adults and nymphs strongly hypertrichous, idiosomal setae not arranged in rows, often forming a thick pelage (Figs 16A, D, G, J); with 1 or 2 pairs of prodorsal trichobothria usually inserted in a median crista metopica (Figs 6L, 7C, H, J); leg femora divided into basifemur and telofemur; larvae heteromorphic, parasitic on arthropods; adults and nymphs free-living predators (cohort Parasitengona, subcohort Trombidia).................................... 11
– Body setae of both adults and nymphs relatively few, generally arranged in transverse rows; prodorsal trichobothria usually absent (if present, then not inserted in a crista); leg femora entire or divided; larvae homeomorphic, in habit similar to nymphs and adults.. 7
7. Chelicerae freely articulated at base; movable digit of chela strongly hooked (Fig. 6F); palpal tibia with 1–3 claws; palpal tarsus long and prominent, usually longer than palpal tibia and extending well beyond palpal claw(s); long radiating legs (Fig. 15G) densely covered with setae (cohort Anystina).. Anystidae
– Cheliceral bases contiguous, fused or partly fused; movable digit of chela stilettiform, aciculate, or whip-like; palpal tibia with only 1 claw (Fig. 10I); palpal tarsus of different size; legs never radiating (cohort Raphignathina, in part)..................... 8
8. Cheliceral bases forming stylophore which is fused with subcapitulum (Figs 9C, D, F, G); predators or parasites... 9
– Cheliceral bases closely contiguous or fused, but usually not integrated with other gnathosomal elements (Fig. 10I); phytophages or predators......... 10
9. Peritremes situated in the middle part of rostral shield or bordering this shield (Figs 9C, D); palps 5-segmented, palpal femur and genu separated; predators and vertebrate parasites ...... Cheyletidae
– Peritremes situated at base of gnathosoma posteriorly (Fig. 9F); palps 2-segmented, palpal femur and genu fused (Fig. 9G); parasitic in/on skin of birds .... Harpirhynchidae
10. Cheliceral stilets greatly elongated, whip-like, strongly recurved basally within stylophore (Fig. 10I); integument without proper shields (Fig. 10H); female genital aperture transverse, separated from anal valves; peritremes well developed on anterior margin of prodorsum; living on plants, phytophagous.. Tetranychidae
– Cheliceral bases contiguous or fused but not completely (as on Fig. 1B); movable digits short, stilettiform or sickle-shaped; dorsum entirely or partly covered by shields (Figs 8J, L, M, 9A); female genital and anal orifices covered by pair of common valves; peritremes absent; free-living in litter and on plants .... Stigmaeidae
11. Nymphs and adults (4 pairs of legs present); free-living 12
– Larvae (3 pairs of legs present); parasites of arthropods 14
12. Genital papillae absent; chelicerae retractile into idiosoma, with movable digit long, straight; 2 pairs of prodorsal trichobothria inserted into crista metopica (Figs 6L, 7C, E, H); 1 or 2 pairs of eyes...... .... Erythraeidae
– Genital papillae present (Fig. 8B); chelicerae not retractile into idiosoma, with movable digit rather short, curved, hook-like or sickle-shaped (Fig. 7N); 1 pair of prodorsal trichobothria inserted into crista metopica (Figs 7J, 8A); 2 pairs of eyes 13
13. Sensillar plate in middle of crista metopica (Fig. 7J); trichobothria inserted anteriad eyes; eyes situated on tall pedicle; palpal tibia terminates by one claw (Fig. 7P); spiniform setae on palpal tibia prominent, but not arranged in rows.......... Podothrombiidae
– Sensillar plate in the posterior part of crista metopica (Fig. 8A); trichobothria inserted posteriad eyes; eyes situated on rather low base; palpal tibia terminates by claw and 2 robust claw-like spines (Fig. 8C);

spiniform setae on palpal tibia arranged in 1 or 2 rows .... Microtrombidiidae

14. Urstigma (large papilla between I and II coxae) and anal opening absent; coxal fields I and II usually separated (Fig. 16E); 2 pairs of prodorsal trichobothria (Figs 7A, D, I) .... Erythraeidae
– Urstigma and anal opening present (Figs 7L, 8E); coxal fields I and II contiguous; single pair of prodorsal trichobothria (Figs 7K, 8D)...................... 15

15. Genu II and III each with 3 or 4 branched setae (Fig. 7M); opisthoventral setae numerous (Fig. 7L); ambulacrum III symmetrical...... Podothrombiidae
– Genu II and III each with 2 branched setae (Fig. 8F); opisthosoma ventrally with few setae (Fig. 8E); ambulacrum III symmetrical (Fig. 8H) or asymmetrical (Fig. 8G) .... Microtrombidiidae

16. Aquatic forms (freshwater or marine), free-living or parasites of invertebrates developing in water (larvae, if parasitic on insects, with palp genu having 2 setae) 17
– Terrestrial forms, free-living or parasitic (mainly on vertebrates) 18

17. Palps usually with only 3 or 4 segments; idiosoma typically with 4 dorsal and 4 ventral sclerites; genital papillae minute or absent; mainly marine species, in Greenland besides marine forms 3 freshwater species recorded (cohort Eupodina, in part) (see chapter 18.3.5) .... Halacaridae
– Palps usually with 5 free segments; idiosomal sclerotization various, but not as above; genital papillae usually present; principally freshwater species (cohort Parasitengona, in part) (see chapter 18.3.5).. .... subcohort Hydrachnidia

18. Prodorsal trichobothria present, differing in form (hair-like, clavate, *etc.*) from other prodorsal setae, usually inserted in distinctive bothridia (Figs 2A, I, 3G, 6C, D); leg femora usually subdivided.......... 19
– Prodorsal trichobothria absent; prodorsal setae similar in size and form or absent (Figs 9H, J, 10A); leg femora entire; living on vertebrates (cohort Raphignathina, in part).................................. 30

19. Tarsus I with one or more recumbent solenidia typically inserted in recessed area of cuticle (Figs 3J, 4E, G, H) (cohort Eupodina, superfamily Eupodoidea).... 20
– Tarsus I with solenidia erect or bent (never recumbent), arising from small circular depressions, sometimes absent....................................... 23

20. With large opposed cheliceral digits (Figs 3H, K, N, 4A, C, F); each recumbent solenidion of tarsus I produced proximally behind its insertation (*rhagidial organ*, Figs 3J, 4E, G).............. Rhagidiidae
– With small and sometimes distorted cheliceral digits; at least fixed digit reduced (Figs 3E, F); recumbent solenidia of tarsus I usually not produced proximally to insertion.. 21

21. Adult rather well sclerotized; dorsum of idiosoma ornamented in pentagonal pattern with distinctive "V"- or "Y"-shaped furrow (Fig. 3A).................. .... Penthalodidae
– Soft-bodied mites; idiosoma usually without reticulate ornamentation (Figs 3C, D) 22

22. Anal opening ventral or terminal (Fig. 2J); femur IV often enlarged (Figs 13D, E); body usually oval (Figs 2I, K).............................. Eupodidae
– Anal opening dorsal or terminal (Figs 3C, D); femur IV normally developed (Fig. 3C); body usually semispherical (Fig. 13H)..................... Penthaleidae

23. Subcapitulum with rutella (Fig. 2G); chelicerae free, chelate (Fig. 2E); 2–3 pairs of genital papillae present (cohort Endeostigmata) 24
– Subcapitulum without rutella (Figs 4N, 5I); chelicerae free (Figs 4O, 5E) or adnate (Fig. 6A), chelate or with fixed digit reduced; genital papillae present or absent (cohort Eupodina, in part)................. 26

24. Tarsi I–IV each with 2 claws and empodium (Fig. 2F); with 1 pair of prodorsal trichobothria (Fig. 2H); adults with 3 pairs of genital papillae .... Terpnacaridae
– Claws absent on all tarsi, claw-like empodia always present; with 1 or 2 pairs of prodorsal trichobothria; adults with 2 or 3 pairs of genital papillae 25

25. With 1 pair of prodorsal trichobothria (Figs 2A–C); adults often with 2 or 3 pairs of genital papillae .... Alicorhagiidae
– With 2 pairs of prodorsal trichobothria (Fig. 2D); adults with 3 pairs of genital papillae .... Nanorchestidae

26. Subcapitulum anteriorly elongated and snout-like (Figs 4N, 5I); chelicerae separate, chelate (Figs 4O, 5D, E, L) or movable digit hook-like, fixed digit reduced; 2 pairs of prodorsal trichobothria (Figs 4M, 5A); with trichobothria on various leg segments (cohort Eupodina, superfamily Bdelloidea) 27
– Subcapitulum not alongated; cheliceral bases fused or contiguous along their medial edges (Figs 6A, B); movable digit short, needle-like, fixed cheliceral digit reduced; 1 pair of prodorsal trichobothria (Figs 6A–D); leg trichobothria absent (cohort Eupodina, superfamily Tydeoidea) 28

27. Palpi strong, grasping, with terminal fang (Fig. 14J); with 2 or 3 pairs of genital papillae.................. .... Cunaxidae

– Palpi long, antenniform, often elbowed, typically with 1–2 long apical setae (Figs 5G, K); with 3 pairs of genital papillae .... Bdellidae
28. With *ereynetal organ, er,* situated in distal portion of tibia I, consisting of a solenidion recessed into inverted sac-like structure with narrow duct opening near seta (Fig. 6E); often with pair of opisthodorsal trichobothria (Fig. 6D) in addition to prodorsal trichobothria and with distal tarsal concavity on tarsi I–II (Fig. 6E); free-living or animal parasites.... .... Ereynetidae
– Without ereynetal organ; opisthosomal trichobothria and tarsal concavities absent (Figs 6A–C); mainly free living.. 29
29. Median prodorsal setae (*ro*) inserted either between or posterior to trichobothria (Fig. 6C); 6 pairs of medial hysteronotal setae; idiosoma narrowly oval; anus terminal .... Iolinidae
– Median prodorsal setae inserted anterior to trichobothria (Figs 6A, B); 5 pairs of medial hysteronotal setae; idiosoma oval or broadly oval; anus ventral...................................... Tydeidae
30. Idiosomal setae absent or represented by 1 or 2 pairs of alveoli; idiosoma elongate, annulate, worm-like (Fig. 9E); inhabitants of hair follicles, surface glands and ducts of vertebrates .... Demodicidae
– Idiosoma with more than 2 pairs of distinctly developed setae (Figs 9H–K, 10A, B).................... 31
31. Legs I strongly shortened; genu I with distinctly developed clasping organ (Figs 10A, B); parasites of small mammals .... Myobiidae
– Legs I–IV similarly developed; genu I without clasping organ (Figs 9I, K).............................. 32
32. Femora I–IV each with strong ventral spur (Fig. 9K); empodia pad-like, sometimes bilobed; tiny (100–150 μm) circular mites (Fig. 9J); parasites in skin of mammals .... Psorergatidae
– Femora I–IV without spurs (Fig. 9I); empodia each with double row of tenent hairs; elongate mites (Fig. 9H) living in bird quills........ Syringophilidae

ENDEOSTIGMATA (= PACHYGNATHINA)

ALICORHAGIIDAE

Small, soft-bodied mites, inhabiting moss, soil and litter. In Greenland only representatives of genus *Alicorhagia* found. Prodorsum with 1 pair of trichobothria, hysterosoma with regular rows of ciliated setae (Figs 2A, C); chelicerae chelate-dentate providing consumption on fungal hyphae, spores, and small soft-bodied invertebrates (Walter 1988, 2001). Among their preys nematodes, collembolans, and another mites. *Alicorhagia fragilis* spins silken thread from its mouth opening. Females use this thread for weaving a platform for egg deposition, and immatures apply it for cocoon construction (Walter 1988). Development without males (probably thelytoky; Norton *et al.* 1993).

***Alicorhagia* Berlese, 1910. Fig. 13A**

Key to species

1. Dorsal hysterosomal setae clavate, ciliated (Fig. 2A); 3 pairs of genital papillae. 330–370 μm. Circumpolar (arctic-subarctic)................ *clavipilus* Thor, 1931
– Dorsal hysterosomal setae tapering distally, plumose (Figs 2B, C); 2 pairs of genital papillae..............2
2. Most hysteronotal setae rather short (10–15 μm); central prodorsal setae shorter than distance between them (Fig. 2B). 220–260 μm. Holarctic (including High Arctic) .... *fragilis* Berlese, 1910
– Most hysteronotal setae of middle length (24–32 μm); central prodorsal setae longer than distance between them (Fig. 2C). 280–310 μm. ?Circumpolar (Queen Elizabeth Islands, Severnaya Zemlya Archipelago) .... *Alicorhagia* sp.

NANORCHESTIDAE (FIG. 13B)

Greenland members of this family belong to the genus *Nanorchestes*. Small mites, red in life, with roundish idiosoma lacking in sclerotization and covered by numerous branchy setae (Figs 2D, E) providing plastron formation (Rounsevell & Greenslade 1988); legs IV saltatory. Modified stiletto-like labrum obviously enables piercing of algal cells, the main diet of different *Nanorchestes* species inhabiting the tidal zone, as well as mosses, lichens and litter of inland biotopes. In North Greenland ***Nanorchestes cf. collinus*** Hirst, 1918 *sensu* Strandtmann 1982 recorded, but finds of other species possible. Female 210–250 μm, male 230 μm. Holarctic (including High Arctic).

TERPNACARIDAE (FIG. 13C)

In Greenland represented by 1 widespread as a whole and common in the Arctic species, ***Terpnacarus bouvieri*** Grandjean, 1939, with body of characteristic colour (gray-lilac) and structure (Figs 2F–H). These particulate-feeding mites consume fungal hyphae and spores (Walter 1988). In arctic areas they generally prefer rather dry, well drained habitats. Males unknown (Norton *et al.* 1993). Female 295–400 μm. Holarctic (including High Arctic).

EUPODINA

EUPODIDAE

Mites of this family occur everywhere in Greenland and populate mosses, litter, turf, *etc.* Weakly sclerotized, often with naso bearing anterior pair of dorsal idiosomal setae (Fig. 2P) and femora of legs IV enlarged, providing jumping (Figs 13D, E); some species red in life. Most Greenland representatives belong to genera known as mycophages and microeuryphages (see references in Makarova 2002).

Key to genera

1. Naso absent (Fig. 2I); anus terminal (Fig. 2J); 3 pairs of pseudoanal setae (*ps*1–3) .... *Claveupodes*
– Naso usually present (Figs 2K, N–P); anus ventrocaudal; 2 or 3 pairs of pseudoanal setae............2
2. Anterior prodorsal setae (*v*1) inserted posteriorly to naso, clavate, capitate or threadlike (Figs 2K, N, O).. .... *Cocceupodes*
– Anterior prodorsal setae inserted on naso, setiform (Fig. 2P)..................................... *Eupodes*

***Claveupodes* Strandtmann & Prasse, 1976**

An unnamed species of this rare monotypic genus (one specimen) found in shrub heath litter in SW Geenland. *Claveupodes* sp. differs from its congeners mainly by thread-like trichobothria (Fig. 2I) and structure of rhagidial organ on tarsus II. Female 208 μm.

***Cocceupodes* Thor, 1934. Fig. 13G**

*Key to species**

1. Anterior prodorsal setae (*v*1) threadlike (Fig. 2N); adults with 6 pairs of genital setae (Fig. 2L); hysteronotal setae short, barely half as long as distance to following seta. 200–250 μm. Holarctic (including Severnaya Zemlya Archipelago)..................... .... *paradoxus* Weiss-Fogh, 1948**
– Anterior prodorsal setae capitate or clavate (Figs 2K, O) ; adults with 4 or 6 pairs of genital setae; hysteronotal setae of different length....................2
2. Internal vertical setae capitate (Fig. 2O); adults with 6 pairs of genital setae; hysteronotal setae moderately short (in average half as long as distance to following seta). 280–350 μm. ? Europe ....cf. *stellatus* Strandtmann et Prasse, 1977
– Internal vertical setae clavate (Fig. 2K); adults with 4 or 6 pairs of genital setae; hysteronotal setae of different length 3
3. Adults with 6 pairs of genital setae. 450 μm. SW Greenland ....*Cocceupodes* sp.
– Adults with 4 pairs of genital setae (Fig. 2M) 4
4. Hysteronotal setae longer (most of them 32–44 μm), as long as distance to following seta or nearly so (Fig. 2K). 290–430 μm. Holarctic (including High Arctic) .... *breweri* Strandtmann, 1971
– Hysteronotal setae shorter (20–25 μm), about half as long as distance to following seta. 120–330 μm. Holarctic (including High Arctic) ....*mollicellus* (C.L. Koch, 1838)
= *Eupodes clavifrons* R. Canestrini, 1886
= *Cocceupodes curviclava* Thor, 1934

* Adopted from Strandtmann & Prasse 1976.

** Recently (Jesionowska 2010b) was included in the genus *Filieupodes* Jesionowska, 2010.

***Eupodes* C.L. Koch, 1835. Figs 13D–F**

Members of this genus are very common in Greenland and especially numerous in High Arctic landscapes. Soft-bodied, long-legged mites with naso bearing anterior prodorsal setae (Fig. 2P) and femur IV often greatly swollen providing jumping (Figs 13D, E). Their diet includes algae, micromycetes, small invertebrates. At least 5 species are recorded. Old finds of *Eupodes lapponicus* Trägårdh, 1910, *E. variegatus* C.L. Koch, 1838, and *E. viridis* Oudemans, 1906 (Jørgensen 1934, Haarløv 1942) need verification. Poor knowledge of *Eupodes* taxonomy and bad condition of preserved material prevent sure identifications.

PENTHALODIDAE (FIG. 13I)

Morphologically characteristic mites with coloured (dark yellow, greenish-yellow) ornamented body, Y-shaped idiosomal furrow and beak-like tri-lobed anterior extension of prodorsum (*epirostrum, epr*) partly covering dorsum of gnathosoma (Figs 3A, B). The single Greenland species, ***Penthalodes cf. ovalis*** (Dugès, 1834) (= *Penthaleus arcticus* Trägårdh, 1900), prefers bare ground patches in frost-boil tundra where it probably consumes green algae. 250–425 μm. Northern Holarctic (including High Arctic). Recent revision of this genus (Jesionowska 2010a) argues reinvestigation of Greenland specimens.

PENTHALEIDAE

Mites with remarkable dorsal position of anal opening (Figs 3C, D). Their body soft, semispherical, green or dark-blue; gnathosoma and legs often red-orange. The species penetrate epidermal cells of plants, sucking out their content. At least 3 species belonging to genus *Penthaleus* inhabit Greenland, but only one of them was named – the obviously thelytokous ***Penthaleus major*** (Dugès, 1834) (winter grain mite). Being phytophagous on numerous, mainly herbaceous plants species inclusive crops, *P. major* is an economically important mite species. Mass occurrence of this species in SW Greenland pastures and hayfields is reported by Nielsen (1984). Two annual generations of *P. major* were observed in Iceland (Hallas & Gudleifsson 2005); males are unknown. This mite is known from different natural vegetation zones of several continents, but some finds are suspected to be based on misidentifications (Qin & Halliday 1996).

Penthaleus **C.L. Koch, 1835. Fig. 13H**

Key to species

1. Chelicera with movable digit long (75–85 μm) and slender, sickle-like (Fig. 3E); dorsal idiosomal setae sparse (21 pairs), of middle size (35–68 μm) (Fig. 3D). 730–960 μm. Cosmopolitan (including central Taymyr)........................*major* (Dugès, 1834)
   = *Penthaleus haematopus* C.L. Koch, 1835
   = *Penthaleus insulanus* Thorell, 1872
– Chelicera with movable digit short (34–41 μm) and tapered, not sickle-like (Fig. 3F); dorsal idiosomal setae in different number, of middle size or rather short..2
2. Dorsal idiosomal setae more sparse (not more than 35 pairs), rigid, rather short (20–48 μm) (Fig. 3C). 620–740 μm. ?Greenland............*Penthaleus* sp. 1
– Dorsal idiosomal setae more numerous (more than 40 pairs), thin, of middle size (40–72 μm). Chelicera as on Fig. 3F. 660–940 μm. ?Greenland...*Penthaleus* sp. 2

RHAGIDIIDAE

Soft-bodied, rapidly moving, predatory mites, dwelling in litter, mosses, under stones, and regularly being found in caves. Thelytokous parthenogenesis has been proved for some species and is suspected for others. Rhagidiids spin silk for moulting, huntering and egg protection (Zacharda 1980). Members of this family often have robust chelate chelicerae, by means of which they prey on other small arthropods. The special T-like (in profile) or banana-shaped sensillae (*"rhagidial setae"*) are developed on the legs; on tarsi I and II these are arranged in special *"rhagidial organs"* of important taxonomic significance (Figs 4E, G). Rhagidial organ I is accompanied by *"stellate seta", sts* (famulus, specialized star-like seta). The structure of this organ (considered to be chemosensory in function) is specific at the generic level (Zacharda 1980). In Greenland rhagidiids are poorly known, but at least ten species belonging to seven genera are recorded.

Key to genera*

1. Chelicera with 1 seta (Figs 4F, J)....................2
– Chelicera with 2 setae (Figs 3H, K, N)..............4
2. Hypostome with only 1 pair of nude subterminal setae (Fig. 3L); stellate seta relatively large, with unusually long lateral branches (Fig. 3J)...*Hammenia*
– Hypostome with 2 pairs of nude subterminal setae (Fig. 4L); stellate seta small, with short lateral branches (Figs 4E, G, H)............................3
3. Rhagidial organ I with parallel rhagidial setae, stellate seta with long ciliated stem (Fig. 4G); cheliceral seta inserted distinctly proximally to joint of movable digit (Fig. 4F)............................*Shibaia*
– Rhagidial organ I with rhagidial setae arranged tandem, stellate seta with very short stem (Fig. 4H); cheliceral seta inserted distally to joint of movable digit (Fig. 4J).......................................*Thoria*
4. Distal cheliceral seta inserted in proximally open laterodorsal pit (Fig. 3H); trichobothria distinctly clavate (Fig. 3G)......................... *Coccorhagidia*
– Distal cheliceral seta inserted dorsally (Figs 3K, N, 4A); trichobothria clavate but very narrow (Fig. 3M) or filiform..5
5. Fixed digit dilated, with strikingly flattened inner margin (Fig. 4C) or with prebasal thorn (Fig. 4A)..*Robustocheles*
– Fixed digit slender, without flattened inner margin or prebasal thorn (Figs 3K, N)......................6
6. Dorsum of chelicera convex with deep saddle-shaped depression (Fig. 3N); cheliceral digits long, rather slender, strongly arched; both cheliceral setae inserted distally to joint of movable digit..*Poecilophysis*
– Dorsum of chelicera convex, but with shallow depression (Fig. 3K); cheliceral shears relatively short, almost straight; proximal cheliceral seta inserted on level of joint of movable digit or proximally to this joint..........................*Rhagidia*

* Adopted from Zacharda 1980.

Coccorhagidia **Thor, 1934. Fig. 14A**

This small genus is represented in Greenland by one species, ***Coccorhagidia pittardi*** Strandtmann, 1971. Easily recognizable mite with characteristic clavate trichobothria (Fig. 3G) and short chela (Fig. 3H). Female 500–690 μm. Circumpolar (including High Arctic), alpine in High Tatras.

Hammenia **Zacharda, 1980. Fig. 14C**

Monotypic genus represented by ***Hammenia macrostella*** Zacharda, 1980 found only in SW Greenland. This relatively small rhagidiid can be easily distinguished by clavate trichobothria (Fig. 3I) and gigantic stellate seta (Fig. 3J). Female 225–290 μm. Europe, southern Taymyr.

Poecilophysis **Cambridge, 1876. Fig. 13K**

In Greenland 1 species is found, ***Poecilophysis (Soprocheles) saxonica*** (Willmann, 1934) (=*Rhagidia shibai* Strandtmann, 1971). Trichobothria slightly clavate (Fig. 3M), fixed digit apically with 5 cusps (Fig. 3N). Female 480–530 μm. Holarctic (including Alaska and Severnaya Zemlya Archipelago).

***Rhagidia* Thorell, 1872. Fig. 13L**

In Greenland 2 species recorded. Large ***Rhagidia* (*Rhagidia*) *gelida*** Thorell, 1872 (Fig. 3K) is very abundant and widespread in the Arctic (Zacharda 1980). Female 1730 µm, male 1050–1500 µm. Arctic, circumpolar. An old find of *Rhagidia terricola* (C.L. Koch, 1835) (Haarløv 1942) needs verification.

***Robustocheles* Zacharda, 1980. Fig. 13J**

Key to species

1. Solenidion on tibia I distinctly dorsodistal (Fig. 4B); inner margin of fixed digit with large prebasal thorn pointed distally and 3 blunt apical cusps (Fig. 4A). Thelytokous. Female 520–610 µm. Europe, Alaska ... . *R.* (s. str.) *mucronata* (Willmann, 1936)
– Solenidion on tibia I dorsal, medioproximal (Fig. 4D); inner margin of fixed digit without prebasal thorn, but with 4 blunt apical cusps (Fig. 4C). Female 1100–1300 µm. Holarctic (including High Arctic) .... *R.* (*Lewia*) *hilli* (Strandtmann, 1971)

One more species of *Robustocheles* inhabits SW Greenland; its diagnostic is complicated.

***Shibaia* Zacharda, 1980**

This characteristic genus is represented in Greenland by ***Shibaia longisensilla*** (Shiba, 1969). Rather small mites with graceful chelicera (Fig. 4F) and stellate seta of very particular form (Fig. 4G). Development without males (Zacharda 1980). Female 420–560 µm. Holarctic (including High Arctic).

***Thoria* Zacharda, 1980. Fig. 14B**

Key to species

1. Thrichobothria not reaching posterior margin of prodorsum (Fig. 4I); epimeral formula 3–1–4–3; terminal palpal segment with 10 ciliated setae. Female 560–670 µm. Holarctic (including High Arctic), Hawaii .... *brevisensilla* Zacharda, 1980 = *Rhagidia uniseta* Thor, 1909 sensu Strandtmann 1971
– Trichobothria overlapping posterior margin of prodorsum (Fig. 4K); epimeral formula 3–1–4–5 to 3–1–4–8; terminal palpal segment with 12–13 ciliated setae. Female 1050 µm. Europe .... *uniseta* s. l. Thor, 1909

BDELLIDAE (*SNOUT MITES*)

Mites belonging to this family live almost everywhere, but particularly in open localities, especially in the tidal zone of different beaches. They are active, fast-running mites, predaceous on small dipterans, collembolans, soft-bodied mites, *etc.* Large (up to 3–4 mm) species with elongated hypostome elucidating their common appellation. The body is usually brightly coloured, often dark red with dark blue-black blotches. Bdellids can produce silken threads which serve for the protection of their eggs or themselves during molting (cocoon construction). These threads are used also in the attachment of preys to the substrate. At least ten species of six genera are found in Greenland. Total length including gnathosoma given.

Key to genera*

1. Venter of hypostome with 5–6 pairs of conspicuous setae (Fig. 5I) 2
– Venter of hypostome with 2 pairs of conspicuous setae (Fig. 4N) 5
2. Ttrichobothrium present on tibia II (Fig. 5H); prodorsum with 3 or 4 pairs of setae (including 2 pairs of trichobothria); palpal tibiotarsus cylindrical and often as long or longer than palpal basifemur (Fig. 5K); end palpal setae shorter than palpal femur 3
– Tibia II without trichobothrium; prodorsum with 4 pairs of setae (including 2 pairs of trichobothria; Figs 5A–C); palpal tibiotarsus truncate and considerably shorter than palpal basifemur (Fig. 5G); end palpal setae long (as long or longer than palpal femur). .... *Bdella*
3. Each chelicera with more than 2 setae, usually with 7–20 (Fig. 5E)............................ *Neomolgus*
– Each chelicera with 2 setae (Fig. 5L) 4
4. Prodorsum with 3 pairs of setae (including 2 pairs of trichobothria); posterior bothridium goblet-shaped (Fig. 5M); posterior trichobothrium usually minute .... *Thoribdella*
– Prodorsum with 4 pairs of setae (including 2 pairs of trichobothria); posterior bothridium simple, posterior trichobothrium not reduced in size............. .... *Octobdellodes*
5. Chelicera of normal length; chela massive, movable digit sickle-shaped (Fig. 4O); unpaired median eye present between anterior prodorsal trichobothria (Fig. 4M)....................................... *Cyta*
– Chelicera elongated; chela digits reduced, needle-like (Fig. 5D); without unpaired median eye .... *Biscirus*

---

* Adopted from Atyeo 1960.

Bdella **Latreille, 1795. Fig. 14D**

Key to species

1. Prodorsal apodemes ("sclerites") rather narrow, long (Fig. 5A); skin striae finely broken; palpal basifemur with 13 or more setae, palpal tibiotarsus with 7 setae. 700–1820 µm. Holarctic (mainly maritime littoral) .. .. *longicornis* (Linnaeus, 1758)
– Prodorsal apodemes broad, S-shaped (Figs 5B, C); skin striae weakly broken; palpal basifemur with 8–17 setae, palpal tibiotarsus with 6–7 setae 2
2. Prodorsal striae formed by scarcely broken or almost unbroken lines (Fig. 5B); palpal basifemur with 8–11 setae, palpal tibiotarsus with 6 setae. 710–1170 µm. Holarctic (including Iceland, ?Svalbard, ?Severnaya Zemlya Archipelago)......... *muscorum* Ewing, 1909 ? = *Bdella semiscutata* Thor, 1930
– Prodorsal striae formed by separate relatively short lines (Fig. 5C); palpal basifemur with 15–17 setae, palpal tibiotarsus with 7 setae. 1500–1680 µm. SW, S Greenland .... *Bdella* sp.

Biscirus **Thor, 1927. Fig. 14I**

One specimen of one species found in SW Greenland, ***Biscirus silvaticus*** (Kramer, 1881). Its chelicera elongated; chela digits reduced, needle-like (Fig. 5D). 1100–1880 µm. Holarctic (including Iceland).

Cyta **von Heyden, 1826. Fig. 14H**

Only one species, ***Cyta latirostris*** (Hermann, 1804), known from Greenland. It can be easily distinguished from other Greenland bdellids by thickened chelicerae with massive chelae (Fig. 4O) and unpaired median eye between anterior prodorsal trichobothria (Fig. 4M). It is a common bdellid species in Berlese extractions, prefering dry conditions. 660–1100 µm. Cosmopolitan (including Ellesmere Island, Spitsbergen, Severnya Zemlya Archipelago, Taymyr).

Neomolgus **Oudemans, 1937. Fig. 14E**

Key to species

1. Palpal genu with 5–8 setae, twice longer than palpal telofemur (Fig. 5F). 2060–2345 µm. Circumpolar, arctic-subarctic, littoral .... *littoralis* (Linnaeus, 1758)
– Palpal genu with 4 setae, equals in length telofemur (Fig. 5J). 1700–1860 µm. Northern Europe (including Iceland) .... cf. *clypeatus* Thor, 1930

Octobdellodes **Atyeo, 1960. Fig. 14G**

Identification of a single, damaged specimen of ***Octobdellodes*** sp., from SW Greenland was not possible.

Thoribdella **Grandjean, 1938. Fig. 14F**

Key to species

1. Posterior trichobothrium leaf-shaped (Fig. 5M); larger. 1200–1440 µm. Common in NW, SW and S Greenland .... *Thoribdella* sp. 1 aff. *lapidaria* (Kramer, 1881)
– Posterior trichobothrium setiform (Fig. 5N); smaller. 1020–1060 µm. SW Greenland......................... .... *Thoribdella* sp. 2

CUNAXIDAE (FIG. 14J)

Middle sized (400–600 µm), red or reddish mites, extremely rare in the Arctic, where they occur in dry meadow-like communities. Cunaxids feed on small arthropods in soil, litter, mosses, and on plants. Their palps are usually raptorial, possessed by spines for grasping preys. In SW Greenland one species found, ***Cunaxoides*** sp. (a single specimen).

TYDEIDAE (FIGS 14L, 15A, B)

The Tydeidae are small (100–500 µm), poorly sclerotized, relatively fast moving mites. A large and both morphologically and biologically extremely diverse family. It contains predators, plant feeders, pollen feeders, scavengers, microphytophagous; but most tydeids are unspecialized feeders (Krantz 1978, Krantz & Lindquist 1979). The colour of phytophages (*Lorryia*, many species of *Tydeus*) usually green, yellow-green, or amber, but can vary depending of gut content. The grazing on fungi is common, even for obligate plant-inhabiting members. Tydeids are recorded on plants, in mosses, algal mats, litter, humus, stored products, bird nests, *etc.* Some species are phoretic on insects or symbiotic in their nests. Certain tydeids can cause dermatoses by biting men and domestic animals. Free-living tydeids in Greenland are weakly sclerotized, small, striate forms. The family is poorly known in Greenland, but its members are ordinary in different communities. Old finds (Haarløv 1942) of *Tydeus croceus* (Linnaeus, 1758) and *Triophtydeus pinicolus* (Oudemans, 1929)* need verification. At least 2 genera recorded. The key is highly preliminary and fits only the available Greenland collection.

Key to genera

1. Femur III with 2 setae; medial striae just before dorsomedial setae *d1* longitudinal (Fig. 6A) .... *Lorryia* Oudemans, 1925

* Recently *Triophtydeus* Thor, 1932 was included in the tydeoid family Meyerellidae (Andre & Fain 2000).

– Femur III with 1 seta; medial striae just before dorsomedial setae *d1* transversal (Fig. 6B)................. .... *Tydeus* C.L. Koch, 1835

IOLINIDAE (FIGS 15C–F)

Small (100–450 μm), weakly sclerotized mites. The family comprises free-living and entomophilous species associated with different orders of insects. Greenland representatives are very small (150–250 μm), narrowly oval, colourless in slides. They all belong to the subfamily Tydaeolinae inhabiting soil, litter, mosses, and rarely higher plants. These mites are considered as mycetophages. The available Greenland material is scarce and badly preserved, but at least 4 genera may be recognized. The key is highly preliminary and fits only the available Greenland collection.

Key to genera

1. Trichobothria capitate or clavate (Fig. 6C).........2
– Trichobothria setiform or hair-like.................3
2. Tarsus I with 12 setae (Fig. 15F) ....*Coccotydaeolus* Baker, 1965
– Tarsus I with 11 setae (Fig. 15C) .... *Tydaeolus* Berlese, 1910 = *Coccotydeus* Thor, 1931
3. Two pairs of genital setae (Fig. 15D)................. ....*Microtydeus* Thor, 1931
– Three to four pairs of genital setae (Fig. 15E)........ ....*Paratriophtydeus* Baker, 1965

EREYNETIDAE (FIG. 14K)

Numerous representatives of this family inhabit nasal tracts of different vertebrates (amphibians, birds, mammals), certain ereynetids even parasitize terrestrial slugs and aquatic beetles. Free-living species are known to be phoretic on sciarids (Diptera) and staphylinids (Coleoptera). Most soil inhabiting ereynetids belong to *Ereynetes* Berlese, 1883. Its members are infrequently registered at high latitudes, where their microphytophagy (on algae, fungi, lichens) may be observed (Goddard 1982). Signs of omnivory are also noted. One unnamed soil species, ***Ereynetes*** **sp.**, has been found in material collected by the "3-year East Greenland Expedition" (1931–1934). This species is easily distinguished by the presence of a pair of opisthodorsal trichobothria (Fig. 6D) and distal tarsal concavities on tarsi I–II (Fig. 6E); *ereynetal organ, er*, situated in distal portion of tibia I, consisting of a solenidion recessed into inverted sac-like structure with narrow duct opening near seta. Female 240–245 μm. E Greenland.

ERIOPHYIDAE (*GALL MITES*) (FIG. 17G)

Tiny (from 80 μm, usually 250–200 μm), 4-legged, strictly phytophagous mites. They cause rust, gall formation (Fig. 18), bud and leaf modifications, leaf discolouration, *etc.* Most species are host specific on the species or generic level. The density of eriophyids is sometimes high and harmful for the plants, and in addition some mites transfer deleterious plant viruses. They are passively dispersed by air currents and insects, but have no special morphological or behavioral adaptations to phoresy (Lindquist 1998). One old article was devoted to Greenland eriophyids (Rostrup 1900). The information below originates from this publication, but modern generic key and species names are given with the use of Amrine & Stasny (1994). Six of 11 Greenland eriophyid species develop on different species of willow (*Salix*), 1 species on *Betula* spp., 1 species on *Pedicularis* spp., whereas 3 mite species inhabiting 3 other plant genera use one plant species alone.

Key to genera

1. Body vermiform, *annuli* (pseudosegmental rings) subequal dorso-ventrally (Figs 11A, D); prodorsal shield typically lacking a frontal lobe (Fig. 11C), or with a slight projection over gnathosoma base, or frontal lobe narrow, basally flexible (Fig. 11B) (subfamily Eriophyinae) 2
– Body usually fusiform; opisthosoma typically divided into broad and heavy dorsal annuli and narrow ventral annuli (Figs 11H–J); prodorsal shield usually with a broad-based and rigid frontal lobe over gnathosoma (Figs 11E–G) (subfamily Phyllocoptinae) 4
2. Frontal lobe on prodorsal shield present, rectangular in shape with rounded apical protuberances (Fig. 11B)................................... *Stenacis*
– Distinct frontal lobe on prodorsal shield absent (Figs 11C, D) 3
3. Tibia I seta present (Fig. 11D) .... *Aceria*
– Tibia I seta absent........................... *Acalitus*
4. *Scapular setae* (*sc*) ahead of posterior shield margin (Fig. 11E), directed forward, or upwards and centrad ....*Phyllocoptes*
– Scapular setae on or very near posterior shield margin (Fig. 11G), directed to the rear, usually divergent.....5
5. Frontal lobe broad and rounded, not acuminate ....*Aculus*
– Frontal lobe more acuminate, frequently ending in a sharp central point (Fig. 11G) .... *Aculops*

Stenacis **Keifer, 1970**

In Greenland represented by 1 species, ***Stenacis triradiatus*** (Nalepa, 1892) (Figs 11A, B), which is responsible for bud deformation of different willow species (*Salix*). Holarctic.

***Acalitus* Keifer, 1965**
At present 1 Greenland species recorded, ***Acalitus rudis*** (Canestrini, 1891), responsible for erineum formation and bud deformation of different species of birch (*Betula*). Holarctic.

***Aceria* Keifer, 1944**
At least 2 species in Greenland.
Aceria rhodiolae (Canestrini, 1892) (Fig. 11C) forms galls on inflorescence and leaves of *Sedum rosea*. Europe.
Aceria saxifragae (Rostrup, 1900) (Fig. 11D) vagrant, cause deformation of *Saxifraga oppositifolia*. Europe.

***Phyllocoptes* Nalepa, 1887**
At least 3 species inhabit Greenland plants.
Phyllocoptes phytoptoides Nalepa, 1892 (Fig. 11E) forms galls on willows (*Salix* spp.). Europe.
Phyllocoptes groenlandicus Rostrup, 1900 (Fig. 11H), described from buds of *Salix herbacea*, Greenland.
Phyllocoptes empetri Rostrup, 1900 (Figs 11F, I) responsible for gall formation in crowberry (*Empetrum nigrum*), Greenland.

***Aculops* Keifer, 1966**
So far 1 species recorded, ***Aculops pedicularis*** (Nalepa, 1892) (Fig. 11G), causing erineums and margin curling of leaves of different louseworts (*Pedicularis* spp). Europe.

***Aculus* Keifer, 1959. Figs 11J, 18**
At least 2 species live in Greenland.
Aculus parvus (Nalepa, 1892) (Fig. 11J) vagrant on *Salix* spp., causing leaf discoloration. Europe.
Aculus tetanothrix (Nalepa, 1889) forms galls and erineums as well as causes margin curving of leaves of *Salix* spp. Europe, Ellesmere Island.

One more species was mentioned from Greenland (ex *Salix herbacea*) as ***Cecidophyes*** sp. (Rostrup 1900). Its modern status is unclear.

ANYSTINA

ANYSTIDAE (WHIRLIGIG MITES) (FIG. 15G)
The family comprises rather large (500–2000 µm), orange or red, rapidly moving mites with long, radiating, hairy legs. Naso with a pair of trichobothria. Anystids are common on plants where they prey on phytophagous insects and other mites. Some species are used in biological pest control. One species is recorded in S Greenland, ***Anystis baccarum*** (Linnaeus, 1758), with characteristic palpal tibia bearing 3 strong spines distally (Fig. 6F). Probably thelytokous. Young female 1700–1800 µm. ?Cosmopolitan.

PARASITENGONA
The cohort comprises terrestrial and aquatic mites developing with distinct metamorphosis. Their heteromorphic larvae parasitize vertebrates or arthropods. Protonymphal and tritonymphal stages are transformed into calyptostases hidden under the integument of larva and deutonymph, respectively. Active postlarval stages, *viz.* deutonymph and adults, are mainly free-living predators.

TROMBIDIA (*VELVET MITES*)
The subcohort includes predominantly terrestrial species. Deutonymph and adult are characterized by the remarkable hypertrichy and the absence of idiosomal cuticular glands. All active stages with palpal thumb-claw process (Figs 7B, F, G, P), 1–2 pairs of prodorsal trichobothria and 1–2 pairs of eyes. Numerous genera are known only in larval or postlarval stages, explaining why some mites really have/had *two* names. Regarding the Greenland fauna this is the case with *Erythraeus-Bochartia, Abrolophus-Hauptmannia, Microtrombidium-Ettmuelleria*. This situation results in the existence of non-congruent larval and imaginal keys.

ERYTHRAEIDAE
Mites belonging to this family are red or reddish-brown, with 1–2 pairs of lateral eyes and 2 pairs of prodorsal trichobothria. Postlarval stages with chelicerae greatly elongated, retractile; genital papillae absent; trichobothria situated on anterodorsal crista metopica. Larvae without anal opening and urstigmata. Deutonymphs and adults prey upon small arthropods living in the soil, on plants, *etc.*, but the larvae parasitize insects, arachnids (including other mites), and myriapods; they very rarely are predaceous or polyphagous.

Key to genera

ADULTS

1. Prodorsum with 2 pairs of eyes (Fig. 7E) .... *Erythraeus*
– Prodorsum with 1 pair of eyes (Figs 7C, H) 2
2. Prodorsum with a pair of sensory pits ("*urnulae*") near eyes (Fig. 7C)....................... *Balaustium*
– Prodorsum without urnulae 3
3. Eyes anterior to middle of crista metopica (Fig. 7H). .... *Leptus*
– Eyes behind middle of crista metopica (Fig. 6L)..... .... *Abrolophus*

Key to genera

LARVAE

1. Prodorsum with 2 pairs of eyes (Fig. 7D); palpal thumb-claw process bifurcate (Fig. 7F) .... *Bochartia*
– Prodorsum with 1 pair of eyes (Figs 7A, I); palpal thumb-claw process entire (Fig. 7B)................2
2. Dorsal scutum distinctly triangular (Fig. 7I); non bothridial dorsal setae thick, pigmeneted, setulose.. .... *Leptus*
– Dorsal scutum quadrangular, pentagonal or roundly triangular (Fig. 7A); non bothridial dorsal setae weaker, not pigmented, ciliate.........*Hauptmannia*

Abrolophus **Berlese, 1891 for adults. Figs 15I, J**
(*Hauptmannia* Oudemans, 1910 for larvae)
That adults of genus *Abrolophus* are probably congeneric with larval genus *Hauptmannia* has been mentioned several times, and some proved cases are known, but no formal synonymization of both names has hitherto been proposed (Haitlinger 2003). Relatively small velvet mites with elongate body (Fig. 6G). Crista metopica with 2 distinct expansions surrounded by a scerotized area (Fig. 6L). Two species of *Abrolophus* (as *Erythraeus*) recorded in Greenland (Trägårdh 1904), but these data should be tested.

Key to species

ADULTS

1. Palpal tibial claw with small tooth on inner side (Fig. 6J); tarsi I–IV almost equal in length (Figs 6H, I). 1440 µm. Greenland......*unidentatus* Trägårdh, 1904
– Palpal tibial claw without tooth on inner side (Fig. 6K); tarsus I almost twice as long as tarsi II–IV. 860–1260 µm. ?Europe .... cf. *miniatus* (Hermann, 1804)

Hauptmannia **Oudemans, 1910 for larvae. Fig. 15K**
(*Abrolophus* Berlese, 1891 for adults)
Larvae of *Hauptmannia* parasitize insects, often different Thysanoptera. They have been caught on plants, in litter, and mosses. There is data about predatory habits of larvae in some species (see Walter & Proctor 2013). The Greenland larva of ***Hauptmannia*** sp. (one specimen, Fig. 7A) belongs to a group of species characterized by comb-like seta on palpal tarsus (Fig. 7B). 310–360 µm. Southern Taymyr.

Balaustium **(von Heyden, 1826) sensu Berlese 1946. Fig. 15H**
A genus with a confused taxonomic history, never thoroughly reviewed, comprising mites with a single pair of anterolateral eyes in adults (Fig. 7C) and one pair of sensory cup-shaped structures near eyes, urnulae (Halliday 2001). Mites of this genus occur in litter, on herbs and trees, in the littoral and even on building walls. Adults are known as predators (Hemiptera, Thysanoptera), feeders of green leaves and pollen, biters of humans (Grandjean 1946; Newell 1963, *etc.*). Larvae of *Balaustium* spp. are free-living predators and pollen eaters. In some species thelytoky is suspected. Old find (Haarløv 1942) of *Balaustium rubripes* (Berlese & Trouessart, 1889) needs verification. Only adults of ***Balaustium*** sp. (Fig. 15H) are found in SW Greenland. Adults 1260–1400 µm.

Erythraeus **Latreille, 1806 for adults. Figs 16A, B**
(?= ***Bochartia*** Oudemans, 1910 for larvae).
The synonymy of *Erythraeus* and *Bochartia* is very probable, but it is not proved (Southcott 1961, B. Halliday pers. comm. 2007). Still no formal synonymization of both names has been proposed. Adult mites of *Erythraeus* prey in litter. One adult species has hitherto been recorded in Greenland, ***Erythraeus phalangioides*** (De Geer, 1897). This morphologically variable mite has very long legs (especially legs IV) in all active stages of development. Adults are dark red, crista metopica of characteristic form and setal equipment (Fig. 7E), palpal tibia with 2–5 conical spines ventrally (Fig. 7G). Greenland specimens, without spines on palpal genu, belong to ***Erythraeus phalangioides*** var. ***gracilipes*** (Kramer, 1897). In pitfalls adults can be attacked by *Leptus*-larvae (Fig. 16C). Adults 2000–3700 µm. S, SW, NW, NE, SE Greenland. Northern Europe (including Iceland). Larvae of *Bochartia* (Fig. 16B) habitually resemble adults of *Erythraeus* (Fig. 16A) and are known as parasites of different insects, especially bugs and aphids. Members of *Bochartia* have never been mentioned in Greenland. Only two larvae of ***Bochartia*** sp. (Figs 7D, F, 16B) are available from pitfall material. Unfed larva 530–560 µm. S, SE Greenland.

Leptus **Latreille, 1796. Figs 16D, E**
Rather common mites in southern Greenland. Adults are brightly red with dense pelage of short pigmented setae (Fig. 7H). Larvae are also red in life, with long dark setae (Figs 7I, 16E), well known as extremely active parasites of different arthropods. In Greenland, larvae of ***Leptus kalaallus*** Southcott, 1992 were regularly found on harvestmen (*Mitopus morio*) attaching to their legs (Southcott 1992). In pitfall material their feeding on spiders (Fig. 16F) and other mites (Fig. 16C) was recorded. These cases may be artefacts, but such diet seems to be appropriate for larvae. Unengorged larva 380–530 µm. W, SW, S Greenland.

NOTE. An old description of larval *Leptus groenlandicus* (Trägårdh, 1904) from Greenland was not satisfactory, the type specimen is lost, and now this taxon is considered to be the *species inquirenda* (Southcott 1992). Modern intrageneric taxonomy of the genus *Leptus* is mainly based on larvae, but earlier authors (Trägårdh 1904, Jørgensen 1934) named adult members of this genus from Greenland as *Leptus vertex* (Kramer, 1886). Adults from the ZMUC-collection were not identical with *Leptus vertex* (Kramer, 1886) sensu Schweizer 1951. Length 1980–2950 µm.

PODOTHROMBIIDAE (FIGS 16G, H)

Members of this family have 1 pair of prodorsal trichobothria and 2 pairs of eyes (Figs 7J, K); with chelicerae rather short, not retractile, movable digits blade-like (Figs 7N, O). Deutonymph and adults red or dark-red, with palptibia possessing comb of spines dorsally and additional spine cluster ventrally (Fig. 7P). They can be found in litter, mosses or under stones, preying on small invertebrates. Larvae with anal opening and urstigmata (Figs 7K, L) parasitize aphids (Hemiptera). One species possibly inhabits Greenland. In earlier publications (Trägårdh 1904, Jørgensen 1934) two names were used, *Podothrombium bicolor* (Hermann, 1804) and *Trombidium bicolor* var. *curtipalpe* Thor, 1900, which were later considered as separate species. Both species are recorded in Europe, including Iceland and Svalbard. Available Greenland specimens from ZMUC more probably belong to ***Podothrombium strandi*** Berlese, 1910 since dorsal setae are dense, long (75–110 µm), distinctly ciliated; tarsus I shorter than tibia I. They have crista metopica expanded anteriorly, anterior margin of prodorsum well sclerotized, concave (Fig. 7J). Alive adults are bicoloured, with bright red appendages and red-brown idiosoma. Numerous larvae from the same locations (Figs 7K, L) also differ from the larva of *Podothrombium svalbardense* Oudemans, 1930 found on Svalbard together with adults of *P. bicolor* and *P. curtipalpe* (Thor 1930). Larvae of *P. svalbardense* were suspected to parasitize aphids (Thor 1930). Greenland *Podothrombium* larvae are recorded on Psillidae (Fig. 16I) and are possibly conspecific with *P. strandi*. Adults of *P. strandi* 1860–2900 µm, unengorged larva of *P.* (?) *strandi* 460–640 µm. Widespread in Greenland, northern Europe, Kuril Islands, and northern Labrador.

MICROTROMBIDIIDAE

Mites of this family have 1 pair of prodorsal trichobothria; chelicerae not retractile, rather short, their movable digit sickle-like; eyes doubled (Figs 8A, D). Deutonymphs and adults bright red or orange, with heavy pelage of setae (Fig. 16J). Palpal tibia dorso-medially with 1–2 distinct rows of spines, terminating with claw and 2 strong spines (Fig. 8C). Crista metopica anteriorly fused with a strongly sclerotized area (*vertex*) bearing several to over a dozen setae of different form than idiosomal setae (Fig. 8A). The species are free-living in plant litter, mosses, *etc.* The larvae parasitize insects, mainly Diptera, exceptionally other mites. Larval dorsum (Figs 8D, I) usually with 2 large shields: *scutum, sct* (anterior) and *scutellum, scl* (posterior). Adults found in Greenland belong to the genus *Microtrombidium*, whereas the larvae represent 2 genera.

Key to genera

LARVAE

1. Empodium present on tarsus III only (Fig. 8H); scutum not bended ventrally (Fig. 8I); ratio width/length of scutellum 1.5–1.7 .... *Parafeiderium*
– Empodium present on all tarsi; anterolateral parts of scutum bended ventrally (Fig. 8D); ratio width/length of scutellum about 3.0........... *Ettmuelleria*

Microtrombidium Haller, 1882 for adults. Fig. 16J

In Greenland one “adult species” found, ***Microtrombidium sucidum*** (L. Koch, 1879). In adults, the idiosoma relatively wide; palpal tibia as in Fig. 8C; anterior margin of prodorsum with characteristic dense row of setae (Fig. 8A); crista metopica with weakly sclerotized posterior projection; body setae densely ciliated. Adults 1180–2400 µm. Holarctic, arctic-alpine (including Barents Sea islands, southern Taymyr, northern Labrador).

Ettmuelleria Oudemans, 1911 for larvae. Fig. 16K

A microtrombidiid larva from Sarek area, Sweden, which was considered as conspecific to an adult of *Microtrombidium sucidum* by Trägårdh (1910), became type species for the separate genus *Ettmuelleria* Oudemans, 1911. But now ***Ettmuelleria sucida (Trägårdh, 1910) sensu*** Oudemans 1911, is regarded as corresponding to another adult species, not *M. sucidum* (Southcott 1994, Gabryś 1999, Makol & Gabryś 2002). In pitfall material from S Greenland were found 2 larvae which fit with Oudemans's description by a characteristic dorsal sclerotization (Fig. 8D), scutum bended ventrally, all tarsi with empodium, ambulacrum III asymmetrical (Fig. 8G), coxal setae ciliate (Fig. 8E), etc. Unengorged larva 370–420 µm. ?Holarctic.

Parafeiderium Vercammen-Grandjean & Cochrane, 1974. Fig. 16L

This small Holarctic genus (2 described species) is known only from larvae parasitizing Ceratopogonidae (Diptera) of different genera (Baker 1999). A single Greenland specimen of ***Parafeiderium*** sp. originated from pitfall, where it might have detached from some insect. Scutum and scutellum of characteristic form (Fig. 8I); tarsus I with distinct dorsal tubercle bearing eupathidial seta; empodium present on tarsus III only (Fig. 8H); ambulacrum III symmetrical; coxal setae bifid. Unengorged larva 265 µm. NE Greenland.

RAPHIGNATHINA

STIGMAEIDAE (FIGS 17A–D)

A large cosmopolitan family comprising predators and phytophages which are found in litter, in mosses, on trees, *etc.* Some predatory species can regulate abundance of economically important phytophagous arthropods and are used in agricultural biocontrol system. Several members of *Stigmaeus* and *Eustigmaeus* are recorded on different sandflies (subfamily Phlebotominae, Psychodidae), considered either parasitic (Fan & Zhang 2005) or accidentally phoretic (Gerson 1972, Evans 1992).

Key to genera

1. Prodorsum with 4 pairs of setae; hysterosoma with 1 extensive dorsal shield bearing 6 pairs of setae (Figs 8J, L) .... *Eustigmaeus* = *Ledermuelleria* Oudemans, 1923
– Prodorsum with 3 pairs of setae; hysterosoma dorsally covered by 1–2 unpaired shields along midline and at least 3 pairs of small paired sclerites (Figs 8M, 9A)... *Stigmaeus*

***Eustigmaeus* Berlese, 1910. Figs 17C, D**

Reddish, globular mites living on plant leaves, in ground litter, soils, *etc.*, but especially common in mosses. The mites suck out the cell content of moss leaves, and mosses alone constitute suitable food for completing their life cycle (Gerson 1972). Usually they feed on various moss species, but reproduce only by feeding on a few of them. Almost the whole development is possible under a water film, but not oviposition (Gerson 1972). Besides the sandfly-association, there are finds of *Eustigmaes* mites on tipulid flies, chironomid midges and ceratopogonids, but these are considered as fortuitous associations (Krantz & Lindquist 1979, Lindquist 1998). Mites of this genus are common only in the Low Arctic, mainly in overwetted biotopes.

Key to species

1. Dorsal shields smooth (Fig. 8J); dorsal setae strong, rather long (Fig. 8K), most of them 72–88 µm; idiosoma diamond-shaped viewed from above. Female 340–360 µm, male 270–300 µm. NE Greenland .... *Eustigmaeus* sp. ? *Stigmaeus pulchellus* Thor, 1930, Spitsbergen
– Dorsal shields gently foveolate (Fig. 8L); dorsal setae shorter (30–56 µm); idiosoma roundish viewed from above. Female 280–330 µm, male 220–230 µm. East Greenland, Nearctic (Ontario, Quebec, Ellesmere Island)........................... *arcticus* Wood, 1972

NOTE. One more species of this genus was recorded in Greenland long ago (Jørgensen 1934), ***Eustigmaeus rhodomela*** (C.L. Koch, 1841) (= ***Raphignathus patrius*** Berlese, 1885), but this find needs confirmation.

***Stigmaeus* C.L. Koch, 1836. Figs 17A, B**

The genus is rather common in arctic soils northwards to polar deserts, but numerous only in some wet, southern tundra communities. The species prey on small arthropods, frequently collembolans, whose pigment are often visible inside the mites. Hitherto 2 species recorded, but finds of other species are very probable in southern Greenland.

Key to species

1. Idiosoma small, narrowly oval (Fig. 8M); dorsal shields smooth; dorsal setae differ extremely in length (from 20 to 180 µm), smaller setae dulled and shortly barbed, longer setae hair-like and glabrous. Female 260–280 µm. SW Greenland................. .... *Stigmaeus* sp. aff. *longipilis* G. Canestrini, 1889
– Idiosoma large, broadly oval (Fig. 9A); dorsal shields reticulate; dorsal setae vary in length less than twice (65–115 µm), scarcely barbed, differing in form (some of them apically pointed, other truncate, with 1–2 tiny barbs on or near tip, Fig. 9B). Female 500–680 µm. NE Greenland, Alaska, Severnaya Zemlya Archipelago .... *parmatus* Summers, 1962
One more species of this genus was recorded in Greenland long ago (Jørgensen 1934), ***Stigmaeus siculus*** (Berlese, 1885), but this find needs confirmation.

CHEYLETIDAE

Among Cheyletidae are free-living forms populating litter, turf, tree bark, stored products, as well as species associated with insects and bird or mammal nests. Most cheyletids are predators on other small arthropods. Even species living in the fur of mammals and in the bird feathering (Fig. 9D) nevertheless often prey on their ectoparasites. Only a few cheyletid species are real parasites on their host. Hitherto only 1 species, ***Cheyletus eruditus*** (Schrank, 1781), has been recorded in Greenland, from a "glumiflores meadow" in NE (Haarløv 1942). This eurytopic mite (Fig. 9C) is known as an effective regulator of laboratory populations of granary mites and house dust mites. Development without males. Female 650–710 µm. Cosmopolitan.

DEMODICIDAE

Diminutive (usually 90–200 µm), annulate, often worm-like mites (Fig. 9E). They are commensals and parasitic in skin of man and other mammals causing different skin deseases which may be aggravated by staphylococcus infection. Their occurrence in human population can reach 70–100%, and grows with the age of a person (Sengbush & Hauswirth 1986). Demodicids are known to inhabit hair follicles and skin glands, where they feed on subcutaneous secretions and interstitial fluid, only occasionally piercing walls of epithelial cells and sucking their content. Sometimes mites penetrate epidermis and can be recorded in subdermal tissues, digestive tract and even in the circulatory system. As a rule, demodicids display host specificity on species level and typically 2 mite species inhabit 1 host species. Faces of people are populated also by 2 species: ***Demodex folliculorum*** (Simon, 1843) and ***Demodex brevis*** Akbulatova, 1966. *D. folliculorum* lives mainly on eyelids and forehead, *D. brevis* inhabits the nose and chin.

HARPIRHYNCHIDAE

Mites of this family are parasitic in/on the cuticle of bird skin. Hitherto unknown from Greenland, but common snipe (*Gallinago gallinago*), recorded in Greenland, is known as a host of *Harpirhynchoides capellae* (Fritsch, 1954), subfamily Harpirhynchinae (Figs 9F, G). In the ZMUC-collection some specimens of ***Harpirhynchoides*** sp. were revealed. Tarsi III–IV lack claws and empodia, terminating instead in series of long whip-like setae. The species was extracted from the skin of snow bunting (*Plectrophenax nivalis*) during the Danish Peary Land Expedition 1948–1949. ?Circumpolar.

MYOBIIDAE

Ectoparasites of different small mammals. The remarkably modified legs I (Figs 10A, B) provide attachment to hairs of their hosts. Myobiids feed on lymph and blood. They are rather common on rodents, but until now not found in Greenland. It is, however, very probable that they will be found on lemming (*Dicrostonyx groenlandicus*). In Canadian High Arctic (Bathurst Island) this rodent provides sustenance for the mite *Radfordia* (*Microtimyobia*) *arctica* Fain & Lukoschus, 1977 (= *Radfordia macdonaldi* Gill & Strandtmann, 1977) (Gill & Strandtmann 1977). Total length: female 400–450 μm, male 330–350 μm.

PSORERGATIDAE

Very small (130–200 μm) roundish mites living in the surface skin layers of mammals (Figs 9J, K). They cause dermatitis, mange or follicular infections of their hosts, being quite common on different animals. Domestic sheep and cattle are populated by appropriate species of *Psorergates*. Hitherto unknown in Greenland.

SYRINGOPHILIDAE (*QUILL MITES*)

Members of this family (Figs 9H, I) inhabit feather quills of birds where they feed on fluids by piercing quill walls. Different mite genera usually parasitize certain orders of birds. Species of Syringophilidae are highly host specific and occupy only particular type of feathers. One bird species can host up to 4 quill mite species (Kethley 1971). Syringophilids are common on birds, though so far not recorded from Greenland. So, *Chenophila branta* Kethley, 1970 is known ex Canada goos (*Branta bernicla*), *Philoxanthornea clarki* Kivganov, 1995 – ex mew gull (*Larus canus*), *Niglarobia helleri* (Oudemans, 1904) – ex ruddy turnstone (*Arenaria interpres*), *Creagonycha totana* (Oudemans, 1904) – ex sanderling (*Calidris alba*), *Syringophilopsis turdus* – ex fieldfare (*Turdus pilaris*). All these birds breed in Greenland.

TETRANYCHIDAE (*SPIDER MITES*)

Tetranychids are found everywhere, but just a few species are recorded from the Arctic, where only representatives of *Bryobia* are common. Most tetranychids are oligophagous or polyphagous phytophages. Long whiplike cheliceral stilettos allow them to feed by piercing parenchymal cells of plants. Some spider mites disperse on web threads like spiders, ballooning by means of air currents.

Key to genera

1. Propodosoma with 4 pairs of dorsal setae (altogether 16 pairs of dorsal setae), 2 pairs of them situated on remarkable frontal protuberances (Fig. 10C); true claws strong, empodium with tenent hairs (Fig. 10E); integument rather thick, sometimes folded; dorsal setae short and broadly spatulate, barbed .... *Bryobia*
– Propodosoma with 3 pairs of dorsal setae without frontal protuberances (altogether 12–13 pairs of dorsal setae, Fig. 10H); true claws weak, empodium without tenent hairs, but splintered (Fig. 10F); integument soft, densely striated; dorsal setae not flattened, usually setiform, finely or coarsely barbed 2
2. One pair of para-anal setae (Fig. 10J); empodium split distally, usually into 3 pairs of hairs (Fig. 10G) .... *Tetranychus*
– Two pairs of para-anal setae; empodium split basally, with proximoventral hairs (Fig. 10F)..... *Panonychus*

Bryobia **C.L. Koch, 1836. Fig. 17E**

These mites are particularly common on grasses and herbaceous plants. In Greenland the genus is represented by the highly variable species ***Bryobia praetiosa*** C.L. Koch, 1836 (Figs 10C–E, 17E) [? = *Bryobia cristata* (Dugès, 1834)]. Extremely polyphagous, eurytopic. Development without males. In central Europe several generations are possible in one growing season; overwintering in any stage. In the subarctic conditions of northern Iceland, *Bryobia cristata* was discovered in activity shortly after snow melting. It has 1 annual generation and does not produce aestivated eggs (Hallas & Gudleifsson 2004). Female 700–800 µm. Cosmopolitan.

Panonychus **Yokoyama, 1929**

Three females of ***Panonychus*** sp. (Fig. 10F) have been found in East Greenland. Identification was not possible.

Tetranychus **Dufour, 1832. Fig. 17F**

Two badly preserved females of ***Tetranychus*** sp. (Figs 10G, H, J, 17F) found in NE Greenland have not been identified.

TARSONEMINA (=HETEROSTIGMATA)

PYGMEPHORIDAE (FIG. 17H, I)

Small (200–300 µm), whitish, yellowish or light-brown mites, free-living in litter, mosses, and accumulations of organic matter, or associated with insect or mammal nests. Some species are phoretic on these animals themselves. They are piercing-sucking mycetophages with different feeding preferences (Krantz & Lindquist 1979, Walter & Proctor 2013). Widely in arctic tundra soils as well as in turf of lemming colonies live representatives of ***Bakerdania*** Sasa, 1961 (Figs 12A, C, 17H, I). These mites possess 2 pairs of prodorsal setae apart from trichobothria (Fig. 12A); dorsal seta on femur I usually unciform (Fig. 12B). At least 4 species inhabit Greenland.

SCUTACARIDAE (FIGS 17J, K)

Small (commonly 150–300 µm), often roundish, yellowish or brownish mites, found in litter, humus, composts, nests of insects, birds, and mammals. Free-living or animal-associated, but without features of parasitism or parasitoidism (Kaliszewski *et al.* 1995). Most species appear to be fungivorous. *Scutacarus acarorum* (Goeze, 1780), associated with holarctic *Bombus* spp. (Figs 12G–I) feeds exclusively on the mould fungus *Histioplasma capsulatum* growing in bumblebee nests (Schousboe 1986). Many species (only females) phoretic on insects, and often a special phoretomorph female is formed for dispersion. Most Scutacaridae reproduce by thelytokous parthenogenesis. Members of the family are regularly found in tundra landscapes, but are not common in the High Arctic. Until now the representatives of 2 genera are recorded in Greenland, ***Pygmodispus*** (1 species found) and ***Imparipes*** (2 species). The other genera known from the Arctic (*Scutacarus* and *Diversipes*) are also included in the generic key.

Key to genera

1. Leg IV 4-segmented (tibia and tarsus fused into tibiotarsus), without ambulacrum (Fig. 12H) ....*Scutacarus* Gros, 1845
– Leg IV 5-segmented (tibia and tarsus IV separate), with or without ambulacrum (Fig. 12F) 2
2. Tibia IV with 4 simple setae; inner surface of trochanter IV usually with spur-like projection (Fig. 12F)....................*Pygmodispus* Paoli, 1911
– Tibia IV with 3 simple setae; trochanter IV without spur-like projection 3
3. Genu and tibia IV free (movably connected); alveolar channel of setae *c2* usually weakly sclerotized ... .*Diversipes* Berlese, 1903
– Genu and tibia IV immovably fused; alveolar channel of setae *c2* usually strongly sclerotized (Fig. 12D) ....*Imparipes* Berlese, 1903

TARSONEMIDAE (FIG. 17L)

The family comprise numerous small mites (120–300 µm), free-living or associated with insects. Tarsonemids feed on different micromycetes, vascular plants, epiphytic algae and lichens, rarely predating on other mites. Entomophilous species parasitize mainly adult insects or prey on their eggs, or may be fungivorous in their nests (Lindquist 1986, Kaliszewski *et al.* 1995). These diminutive mites are uncommon in the Arctic, where only few species are recorded. On Ellesmere Island (close to Greenland) the circumpolar *Steneotarsonemus arcticus* Lindquist, 1986 is registered (Figs 12J, K). It inhabits grasses (*Poa glauca, Alopecurus alpinus*). Female 210–276 µm, male 146–166 µm. Several Greenland specimens in the ZMUC (Fig. 17L) are not identified, but belong to another genus. They have been found in SW Greenland.

References

Amrine, J.W. Jr. & Stasny, T.A. (1994): Catalog of the Eriophyoidea (Acarina: Prostigmata) of the World. – Indira Publishing House, West Bloomfield, Michigan, USA. 798 pp.

André, H.M. & Fain, A. (2000): Phylogeny, ontogeny and adaptive radiation in the superfamily Tydeoidea (Acari: Actinedida), with a reappraisal of morphological characters. – *Zoological Journal of the Linnean Society* 130 (3): 405–448.

Atyeo, W.T. (1960): A revision of the mite family Bdellidae in North and Central America (Acarina, Prostigmata). – *University Kansas Science Bulletin* 40 (8): 354–499.

Baker, A.S. (1999): Two new species of larval mites (Acari: Trombidioidea: Microtrombidiidae and Johnstonianidae) parasitising *Culicoides impunctatus*, the highland midge

(Insecta: Ceratopogonidae), in Scotlad. – *Systematic Parasitology* 44: 37–47.

Bochkov, A.V. (1995): Mites of the subgenus *Microtimyobia* (Acariformes, Myobiidae, *Radfordia*) found in the Russia and surrounding countries. – *Parazitologiya* 29: 480–492 [In Russian].

Bochkov, A.V. (2002): Classification and phylogeny of mites of the superfamily Cheyletoidea (Acari: Prostigmata) – Entomological Review 81: 488–513.

Bochkov, A.V. & Apanaskevich, D.A. (2001): Two new species of the family Syringophilidae (Acari: Cheyletoidea) from passeriform birds collected in Altai. – *Folia Parasitologica* 48: 321–325.

Bochkov, A.V., Mironov, S.V. & Gorgol, V.T. (1994): Ornithocheyletia phylloscopi sp.n. (Acariformes: Cheyletidae), new species of parasitic mites from the willow warbler. – *Acarina* 1-2: 73–80.

Davydova, M.S. (1978): Peculiarities of the gamasid mites distribution in landscapes and some microbiotopes. – *In*: Cherepanov, A.I. (ed.). Chlenistonogiye Sibiri: 23–29. "Nauka" Publishing House, Novosibirsk [In Russian].

Dubinin, V.B. (1955): Suborder Trombidiformes Reuter. III. Superfamily Demodicoidea Banks. – *In*: Bregetova, N.G. (ed.) Mites of Rodents of the USSR Fauna. *Identification Keys of the USSR Fauna*. V. 59: 141–152. Academy of Sciences of the USSR, Moscow-Leningrad [In Russian].

Evans, G.O. (1992): Principles of Acarology. – CAB International, Wallingford. 563 pp.

Fan, Q.-H. & Zhang, Z.-H. (2005): Raphignathoidea (Acari: Prostigmata). – *Fauna of New Zealand* 52: 400 pp.

Gabryś, G. (1999): The World Genera of Microtrombidiidae (Acari, Actinedida, Trombidioidea). – *Monographs of the Upper Silesian Museum* 2: 1–361.

Gerson, U. (1972): Mites of the genus *Ledermuelleria* (Prostigmata: Stigmaeidae) associated with mosses in Canada. – Acarologia 13 (2): 319–343.

Gill, D. & Strandtmann, R.W. (1977): Ectoparasites of the Collared Lemming (*Dicrostonyx torquatus*) on Bathurst Island, N.W.T., Canada. – *Journal of Medical Entomology* 14 (1): 101–106.

Goddard, D.G. (1982): Feeding biology of free-living Acari at Signy Island, South Orkney Islands. – *British Antarctic Survey Bulletin* 51: 290–293.

Grandjean, F. (1946): Observations sur les Acariens (9 série). – *Bulletin du Muséum National d'Histoire Naturelle, Second Series* 18: 337–344.

Haarløv, N. (1942): A morphologic-systematic-ecological investigation of Acarina. – *Meddelelser om Grønland* 128 (1): 1–71.

Haitlinger, R. (2003): *Hauptmannia bohdani* n. sp. from Poland (Acari: Prostigmata: Erythraeidae). – *Genus* 14 (4): 603–607.

Hallas, T.E. & Gudleifsson, B.E. (2004): Phenology of *Bryobia cristata* (Acari, Prostigmata) in hayfields in northern Iceland. – *Exprimental and Applied Acarology* 33: 103–107.

Hallas, T.E. & Gudleifsson, B.E. (2005): Life cycles of *Penthaleus major* (Dugés) (Acari, Prostigmata) in hayfields in northern Iceland. – *Icelandic Agricultural Sciences* 18: 39–44.

Halliday, R.B. (2001): Systematic and biology of the Australian species of *Balaustium* von Heyden (Acari: Erythraeidae). – *Australian Journal of Entomology* 40: 326–330.

Jesionowska, K. (2010a): A morphological study of the genus *Penthalodes* (Acari, Prostigmata, Eupodoidea, Penthalodidae) with description of a new species. – *Zootaxa* 2672: 29–49.

Jesionowska, K. (2010b): Cocceupodidae, a new family of eupodoid mites, with description of a new genus and two new species from Poland. Part I. (Acari: Prostigmata: Eupodoidea). – *Genus* 21(4): 637–658.

Kaliszewski, M., Athias-Binche, F. & Lindquist, E.E. (1995): Parasitism and parasitoidism in Tarsonemina (Acari: Heterostigmata) and evolutionary considerations. – *Advances in Parasitology* 35: 335–367.

Kethley, J. (1971): Population regulation in quill mites (Acarina: Syringophilidae). – *Ecology* 52: 1113–1118.

Kethley, J.B. (1982): Prostigmata. – *In*: Parker, S.D. (ed). Synopsis and Classification of Living Organisms. V. 2: 117–145. McGraw-Hill, New York.

Kethley, J. (1990): Acarina: Prostigmata (Actinedida). – *In*: Dindal, D.E. (ed). Soil Biology Guide: 667–756. J.Wiley & Sons Publ., New York.

Khaustov, A.A. (2008): Mites of the Family Scutacaridae of Eastern Palaearctic. – "Akademperiodyka" Publishing House, Kiev. 291 pp.

Khaustov, A.A. & Makarova, O.L. (2005): Two new mite species of the genus *Bakerdania* (Acari, Heterostigmata, Pygmephoridae) from polar deserts of the Severnaya Zemlya Archipelago. – *Entomological Review* 85(Suppl. 1): 141–146.

Krantz, G.W. (1978): A Manual of Acarology. 2nd edition. – Oregon State University Bookstores, Corvallis. 509 pp.

Krantz, G.W. & Lindquist, E.E. (1979): Evolution of phytophagous mites (Acari). – *Annual Review of Entomology* 24: 121–158.

Lindquist, E.E. (1986): The world genera of Tarsonemidae (Acari: Heterostigmata): a morphological, phylogenetic, and systematic revision, with a reclassification of family-group taxa in Heterostigmata. – *Memoirs of Entomological Society of Canada* 136: 1–517.

Lindquist, E.E. (1998): Evolution of phytophagy in trombidiform mites. – *Experimental and Applied Acarology* 22: 81–100.

Lindquist, E.E., Krantz, G.W. & Walter, D.E. (2009): Chapter 8, Classification. – *In*: Krantz, G.W. & Walter, D.E. (eds.). A Manual of Acarology. 3rd Edition: 97–103. Texas Tech. University Press, Lubbock.

Makarova, O.L. (2002): Acarocenoses (Acariformes, Parasitiformes) in polar deserts: 2. Cenotic relations, structure of communities, and the proportion of suborders. – *Entomological Review* 82(7): 857–875.

Makol, J. & Gabriś, G. (2002): A redescription of *Sucidothrombium sucidum* (L. Koch, 1879) (Acari: Actinotrichida: Microtrombidiidae) with characteristics of all active instars. – Annales Zoologici 52(3): 195–206.

Newell, I.M. (1963): Feeding habits in the genus *Balaustium* (Acarina, Erythraeidae), with special reference to attacks on man. – *Journal of Parasitology* 49: 498–502.

Nielsen, P. (1984): Entomologoske undersögelser I og omkring fåreholderområdet I Sydgrönland. – *Arbejdsgruppen vedrörende miljø og fåreavl. Royal Veterinary and Agricultural University, Copenhagen* 2: 1–28.

Norton, R.A., Kethley, J.B., Johnston, D.E. & O'Connor, B.M. (1993): Phylogenetic perspectives on genetic systems and reproductive modes of mites. – *In*: Wrench, D.L. & Ebbert, M.A. (eds.). Evolution and Diversity of Sex Ratio in Insects and Mites: 8–99. Chapman & Hall, New York.

Qin, T.K. & Halliday, R.B. (1996): The Australian species of *Chromotydaeus* Berlese and *Penthaleus* C.L. Koch (Acarina: Penthaleidae). – *Journal of Natural History* 30: 1833–1848.

Rodendorf, B.B. (1940): The identification key of predatory and pyemotid mites (Cheyletidae and Pediculoididae). – Uchonyye Zapiski MGU 42: 69–98.

Rostrup, S. (1900): Grønlandske Phytoptider. – *Videnskabelige Meddelelser fra den naturhistorisk Forening i København*: 241–249.

Rounsevell, D.E. & Greenslade, P. (1988): Cuticle structure and habitat in the Nanorchestidae (Acari: Prostigmata). – *Hydrobiologia* 165: 209–212.

Schousboe, C. (1986): On the biology of *Scutacarus acarorum* Goeze (Acarina: Trombidiformes). – *Acarologia* 27(2): 151–158.

Schweizer, J. (1951): Die Landmilben des schweizerischen Nationalparkes. 2. Teil: Trombidiformes Reuter 1909. – *Ergebnisse der wissenschaftlichen Untersuchung des schweizerischen Nationalparks* 3, N.F.(23): 49–172.

Sengbush, H.C. & Hauswirth, J.W. (1986): Prevalence of hair follicle mites, *Demodex folliculorum* and *D. brevis* (Acari: Demodicidae), in a selected human population in western New York, USA. – *Journal of Medical Entomology* 23: 384–388.

Southcott, R.V. (1961): Studies on the systematics and biology of the Erythraeoidea (Acarina), with a critical revision of the genera and subfamilies. – *Australian Journal of Zoology* 9(3): 367–610.

Southcott, R.V. (1992): Revision of the larvae of *Leptus* Latreille (Acarina: Erythraeidae) of Europe and North America, with descriptions of post-larval instars. – *Zoological Journal of the Linnean Society* 105: 1–153.

Southcott, R.V. (1994): Revision of the larvae of the Microtrombidiinae (Acarina: Microtrombidiidae), with notes on life histories. – *Zoologica, Stuttgart* 144: 1–155.

Strandtmann, R.W. (1982): Notes on *Nanorchestes* III. Four species from the Arctic Tundra (Acari: Endeostigmata: Nanorchestidae) – *Pacific Insects* 24: 69–77.

Strandtmann, R.W. & Prasse, J. (1976): Prostigmatic mites from the experimental farm in Etzdorf/Saalkreis, G.D.R. – *Abhandlungen und Berichte des Naturkundemuseums Görlitz* 50(2): 1–36.

Trägårdh, I. (1904): Monographie der arktischen Acariden. – *Fauna Arctica* 4: 1–78.

Thor, S. (1930): Beiträge zur Kenntnis der Invertebraten Fauna von Svalbard. – *Skrifter om Svalbard og Ishavet* 27: 1–155.

Walter, D.E. (1988): Predation and mycophagy by Endeostigmatid mites (Acariformes: Prostigmata). – *Experimental and Applied Acarology* 4: 159–166.

Walter, D.E. (2001): Endemism and cryptogenesis in 'segmented' mites: A review of Australian Alicorhagiidae, Terpnacaridae, Oehserchestidae and Grandjeanicidae (Acari, Sarcoptiformes). – *Australian Journal of Entomology* 40: 207–218.

Walter, D.E., Bolton, S., Uusitalo, M. & Zhang, Z.-Q. (2011): Suborder Endeostigmata Reuter, 1909. – In: Zhang, Z.-Q. (ed.) Animal Biodiversity: An Outline of Higher-Level Classification and Survey of Taxonomic Richness. *Zootaxa* (special volume) 3148: 139–140.

Walter, D.E. & Proctor, H.C. (2013): Mites: Ecology, Evolution & Behaviour. Life at a Microscale. 2nd Edition. – Springer, Dordrecht – Heidelberg – New York – London. 494 pp.

Zacharda, M. (1980): Soil mites of the family Rhagidiidae (Actinedida: Eupodoidea). Morphology, systematics, ecology. – *Acta Universitatis Carolinae Biologica* 1978: 489–785.

Zhang, Z.-Q., Fan, Q.-H., Pesic, V., Smit, H., Bochkov, A.V., Khaustov, A.A., Baker, A., Wohltmann, A., Wen, T., Amrine, J.W., Beron, P., Lin, J., Gabrys, G. & Husband, R. (2011): Order Trombidiformes Reuter, 1909. – *In*: Zhang, Z.-Q. (ed.) Animal Biodiversity: An Outline of Higher-Level Classification and Survey of Taxonomic Richness. *Zootaxa* (special volume) 3148: 129–138.

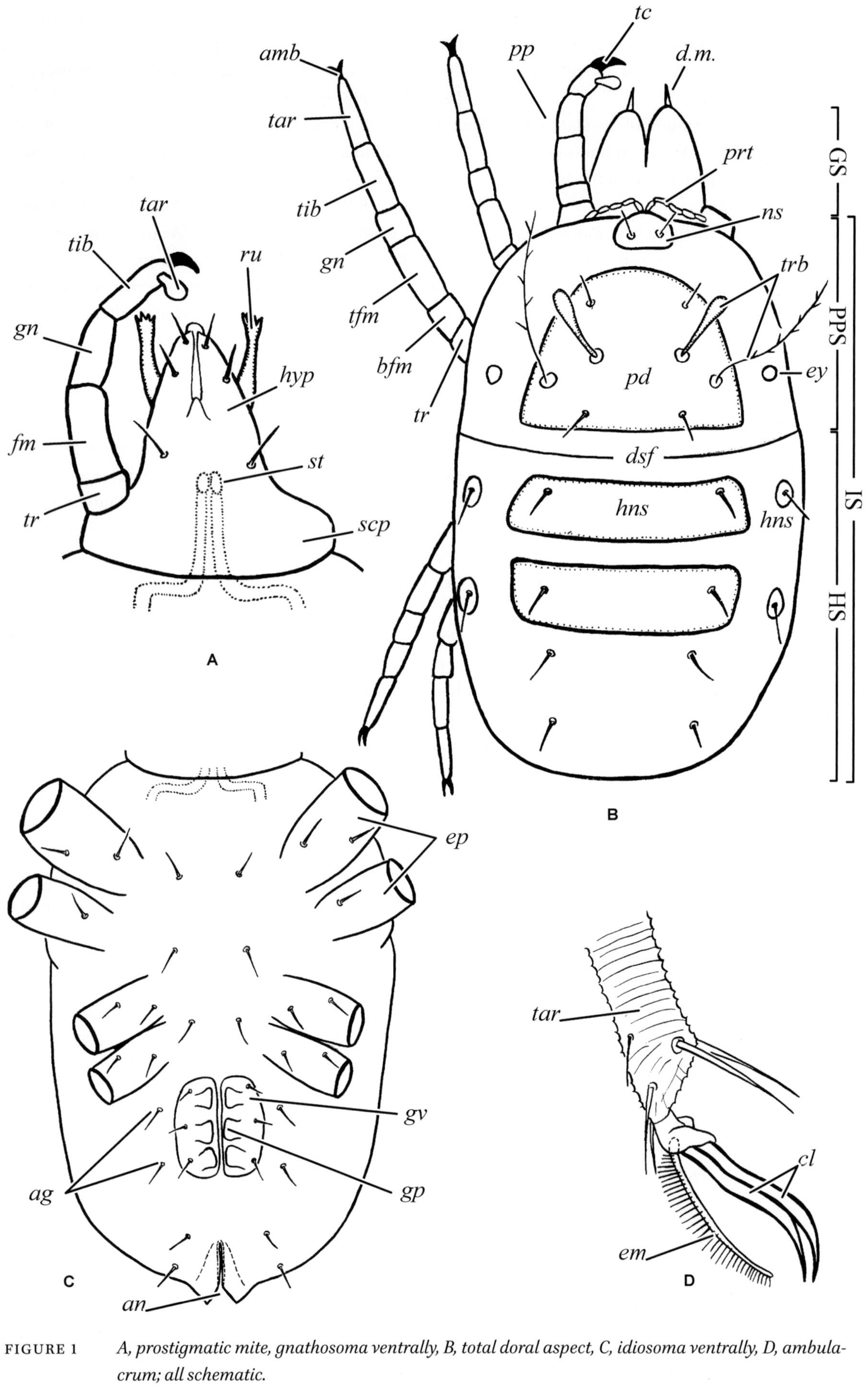

FIGURE 1 *A, prostigmatic mite, gnathosoma ventrally, B, total doral aspect, C, idiosoma ventrally, D, ambulacrum; all schematic.*

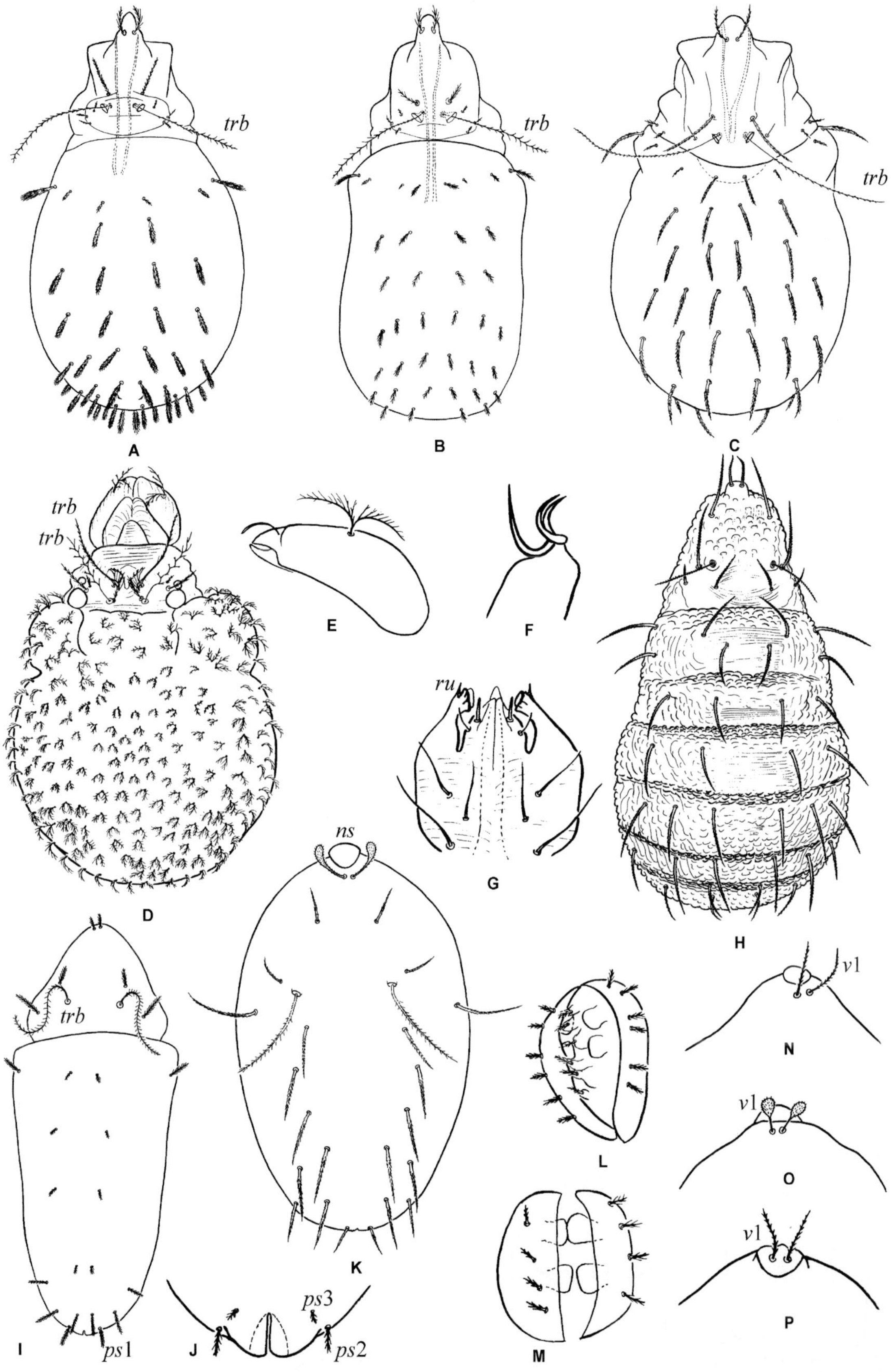

FIGURE 2 *A,* Alicorhagia clavipilus, *dorsal aspect; B,* Alicorhagia fragilis, *dorsal aspect; C,* Alicorhagia *sp., dorsal aspect; D,* Nanorchestes *cf.* collinus, *dorsal aspect, E, chelicerae; F,* Terpnacarus bouvieri, *ambulacrum I, G, subcapitulum, H, dorsal aspect; I,* Claveupodes *sp., dorsal aspect, J, anal region; K,* Cocceupodes breweri, *dorsal aspect, M, genital region; L,* Cocceupodes paradoxus, *genital region, N, idiosomal vertex; O,* Cocceupodes *cf.* stellatus, *idiosomal vertex; P,* Eupodes *sp., idiosomal vertex. All adults.*

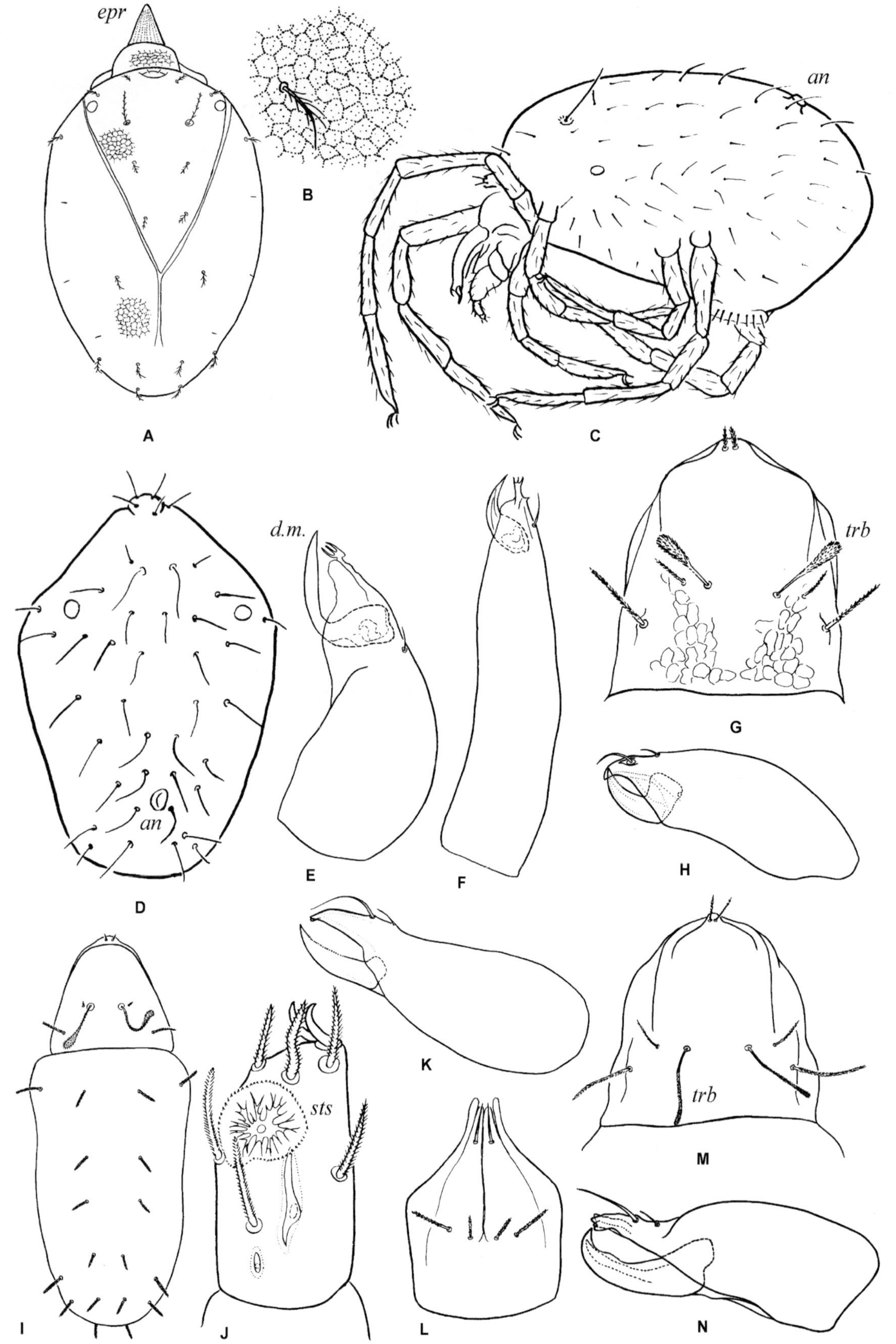

FIGURE 3 *A,* Penthalodes *cf.* ovalis, *dorsal aspect, B, dorsal ornamentation; C,* Penthaleus *sp. 1, lateral aspect; D,* Penthaleus major, *dorsal aspect, E, chelicera; F,* Penthaleus *sp. 2, chelicera; G,* Coccorhagidia pittardi, *prodorsum, H, chelicera; I,* Hammenia macrostella, *dorsal aspect, J, tarsus I, L, subcapitulum; K,* Rhagidia gelida, *chelicera; M,* Poecilophysis saxonica, *prodorsum, N, chelicera. All adults.*

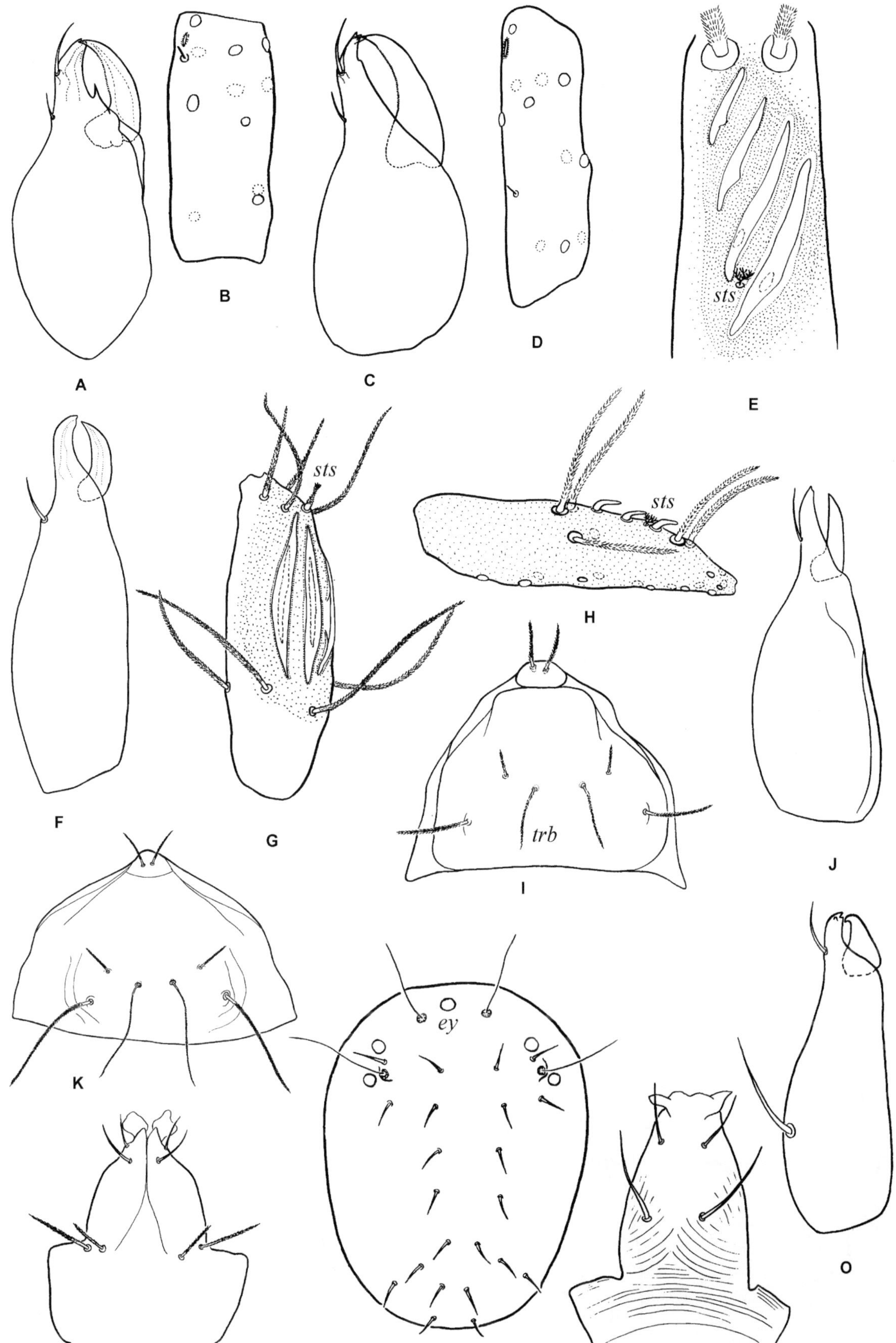

FIGURE 4 *A,* Robustocheles mucronata, *chelicera, B, tibia I; C,* Robustocheles hilli, *chelicera, D, tibia I, E, tarsus I distally; F,* Shibaia longisensilla, *chelicerae, G, tarsus I; H,* Thoria brevisensilla, *tarsus I, I, prodorsum, J, chelicera; K,* Thoria uniseta, *prodorsum, L, subcapitulum; M,* Cyta latirostris, *dorsal aspect, N, subcapitulum, O, chelicera. All adults.*

FIGURE 5 *A,* Bdella longicornis, *prodorsum; B,* Bdella muscorum, *prodorsal apodemes; C,* Bdella *sp., prodorsal apodemes, G, palp; D,* Biscirus silvaticus, *chelicera; E,* Neomolgus littoralis, *chelicera, F, palpal telofemur and genu, H, tibia II; I,* Neomolgus clypeatus, *subcapitulum, J, palpal telofemur and genu; K,* Thoribdella *sp. 1 aff.* lapidaria, *palp, L, chelicera, M, posterior prodorsal bothridium; N,* Thoribdella *sp. 2, posterior prodorsal bothridium. All adults.*

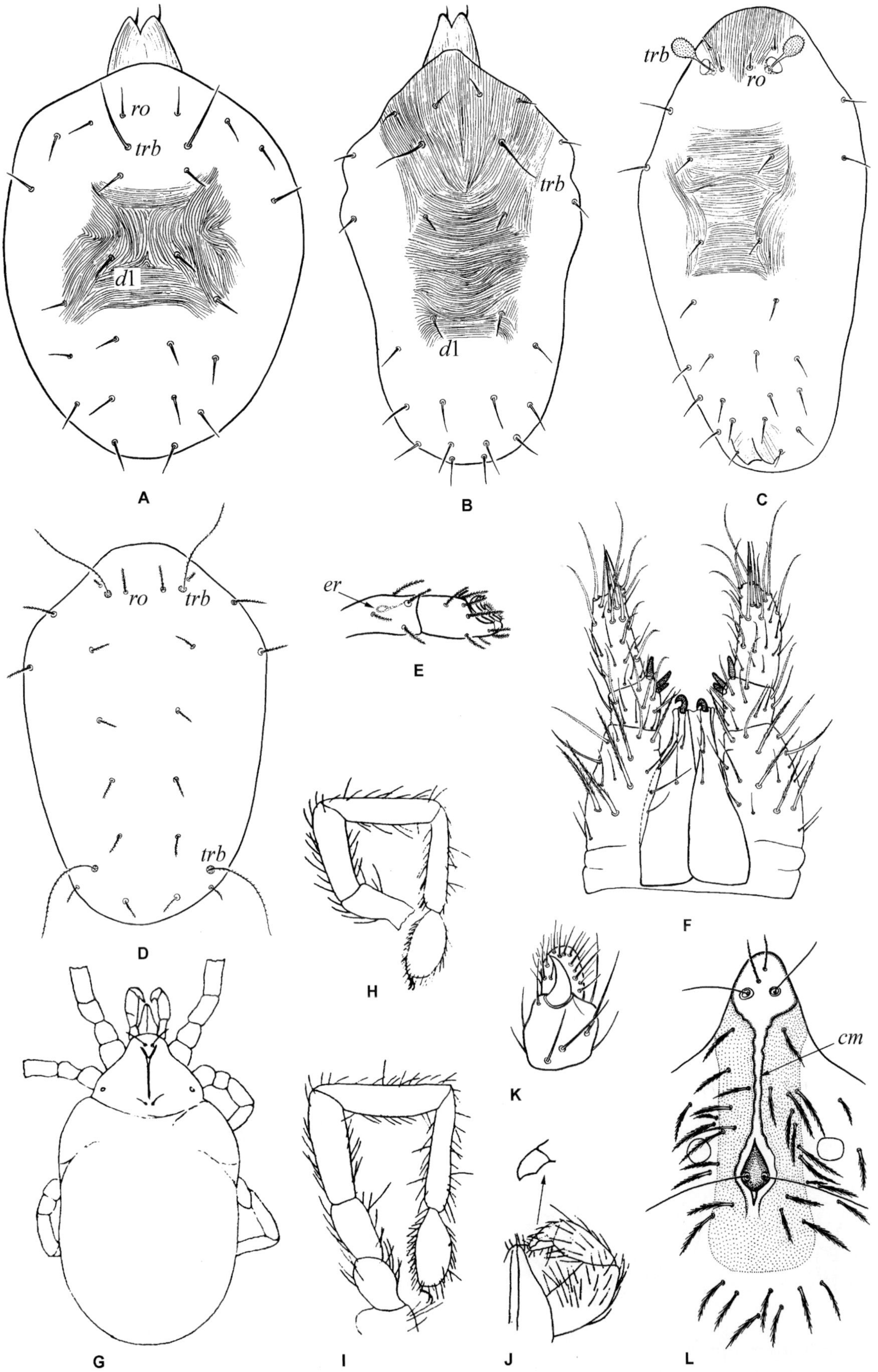

FIGURE 6 *A,* Lorryia *sp. dorsal aspect; B,* Tydaeus *sp., dorsal aspect; C,* Tydaeolus *sp., dorsal aspect; D,* Ereynetes *sp., dorsal aspect, E, tibia and tarsus I; F,* Anystis baccarum, *gnathosoma dorsally; G,* Abrolophus unidentatus, *dorsal aspect, H, leg I, I, leg II, J, gnathosoma; K,* Abrolophus *cf.* miniatus, *palpal tibia and tarsus, L, prodorsum. All adults. G–J, after Trägårdh 1904.*

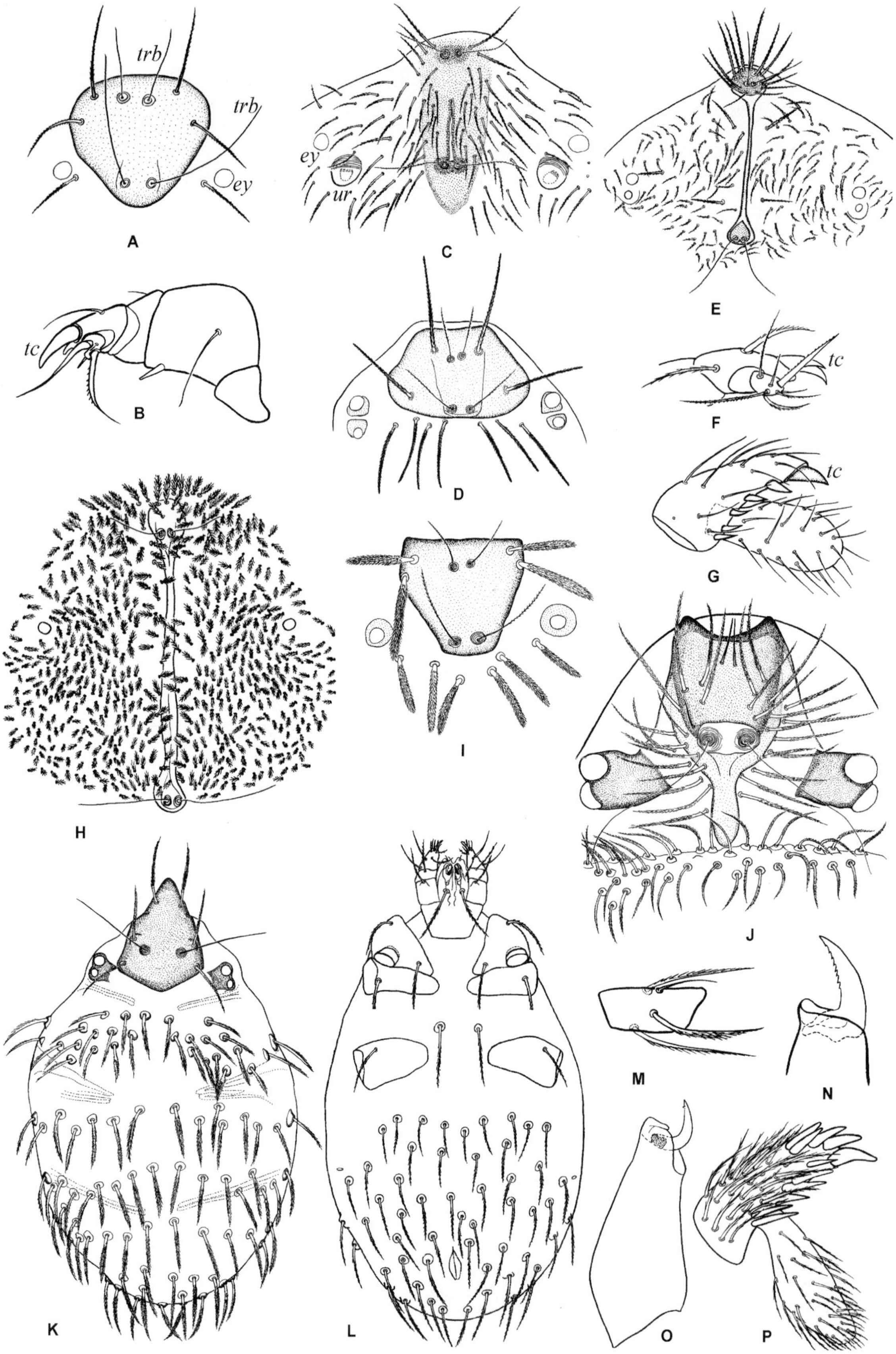

FIGURE 7 *A,* Hauptmannia *sp., larva, prodorsum, B, palp; C,* Balaustium *sp., adult, prodorsum; D,* Bochartia *sp., larva, prodorsum, F, palpal tibia and tarsus; E,* Erythraeus falangioides, *adult, prodorsum, G, palpal tibia and tarsus; H,* Leptus *sp., adult, prodorsum; I,* Leptus kalaallus, *larva, prodorsum; J,* Podothrombium strandi, *adult, prodorsum, N, chelicera, P, palpal tibia and tarsus; K,* Podothrombium (*?*) strandi, *larva, dorsal aspect, L, ventral aspect, M, genu II, O, chelicera.*

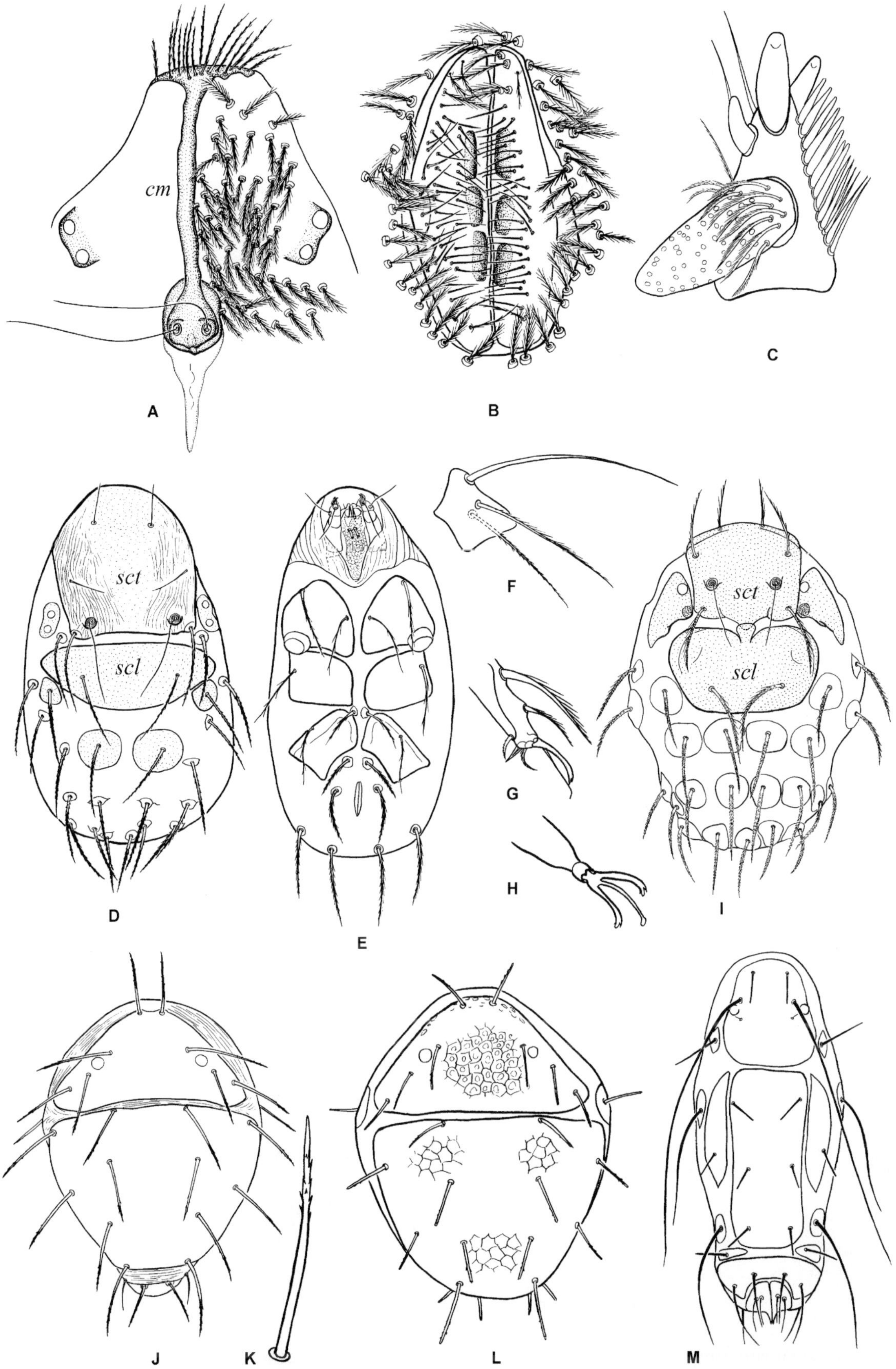

FIGURE 8 *A,* Microtrombidium sucidum, *adult, prodorsum, B, genital region, C, palpal tibia and tarsus; D,* Ettmuelleria (?) sucida, *larva, dorsal aspect, E, ventral aspect, F, genu III, G, ambulacrum III; H,* Parafeiderium *sp., larva, ambulacrum III, I, dorsal aspect; J,* Eustigmaeus *sp., female, dorsal aspect, K, dorsal seta; L,* Eustigmaeus arcticus, *female, dorsal aspect; M,* Stigmaeus *sp., female.*

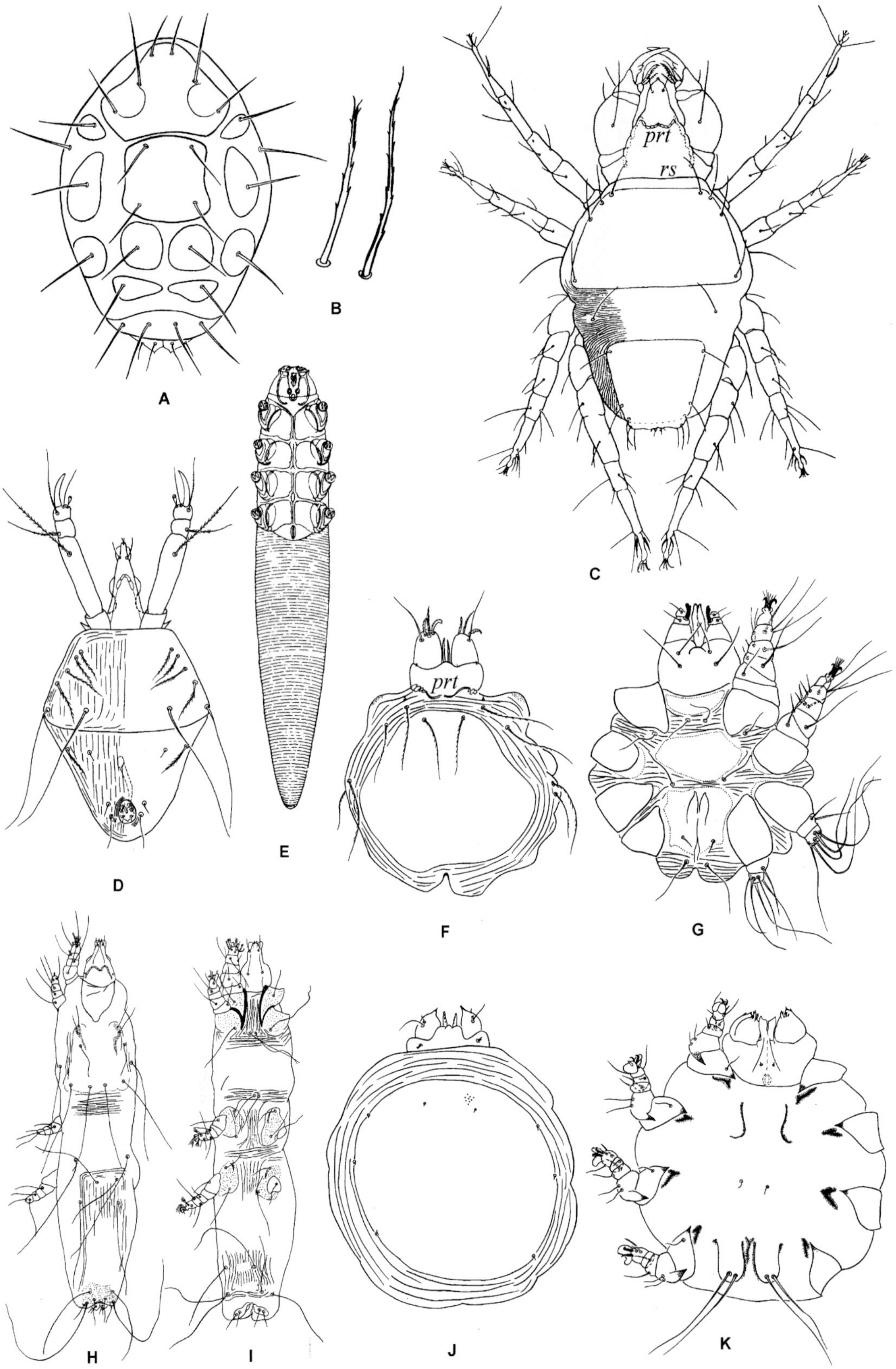

FIGURE 9 *A,* Stigmaeus parmatus, *dorsal aspect, B, dorsal setae; C,* Cheyletus eruditus, *dorsal aspect; D, member of* Ornithocheyletia, *dorsal aspect; E,* Demodex folliculorum, *ventral aspect; F,* Harpirhynchoides capellae, *dorsal aspect, G, ventral aspect; H, member of* Syringophilopsis, *dorsal aspect, I, ventral aspect; J, member of* Psorergates, *dorsal aspect, K, ventral aspect. All adults. C, after Rodendorf 1940; D, after Bochkov* et al. *1994; E, after Dubinin 1955; F, G, J, K, after Bochkov 2002; H, I, after Bochkov & Apanaskevich 2001.*

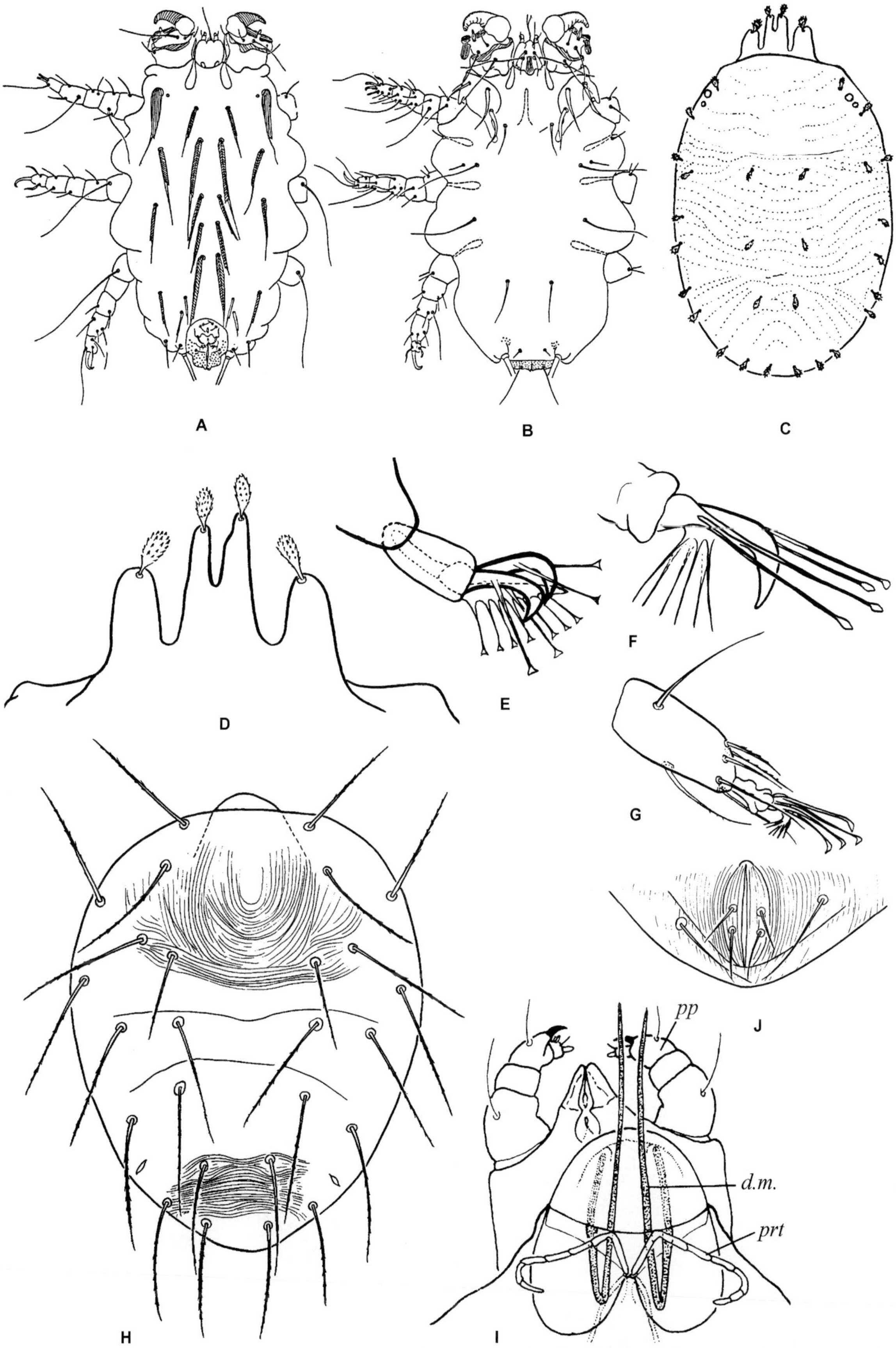

FIGURE 10 *A, member of* Radfordia, *dorsal aspect, B, ventral aspect; C,* Bryobia praetiosa, *dorsal aspect, D, frontal region of idiosoma, E, ambulacrum IV; F,* Panonychus *sp., ambulacrum IV; G,* Tetranychus *sp., ambulacrum IV, H, dorsal aspect, J, anal region; I, member of* Tetranychus, *gnathosoma dorsally; all adults. A, B, after Bochkov 1995; I, after Krantz 1978.*

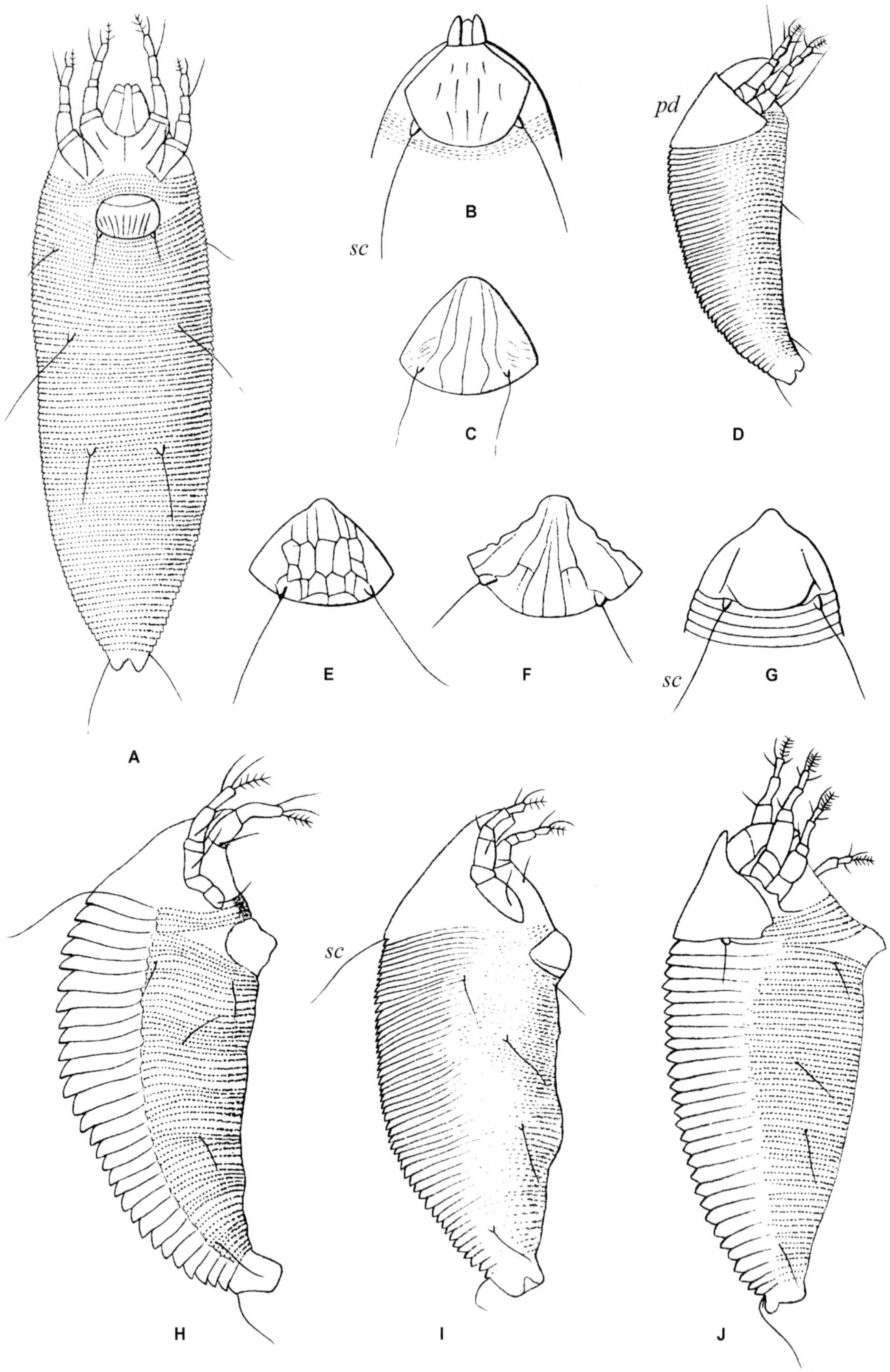

FIGURE 11 *A,* Stenacis triradiatus, *ventral aspect, B, prodorsal region; C,* Aceria rhodiolae, *prodorsum; D,* Aceria saxifragae, *lateral aspect; E,* Phyllocoptes phytoptoides, *prodorsum; F,* Phyllocoptes empetri, *prodorsum, I, lateral aspect; G,* Aculops pedicularis, *prodorsum; H,* Phyllocoptes groenlandicus, *lateral aspect; J,* Aculus parvus, *lateral aspect. All adults. After Rostrup 1900.*

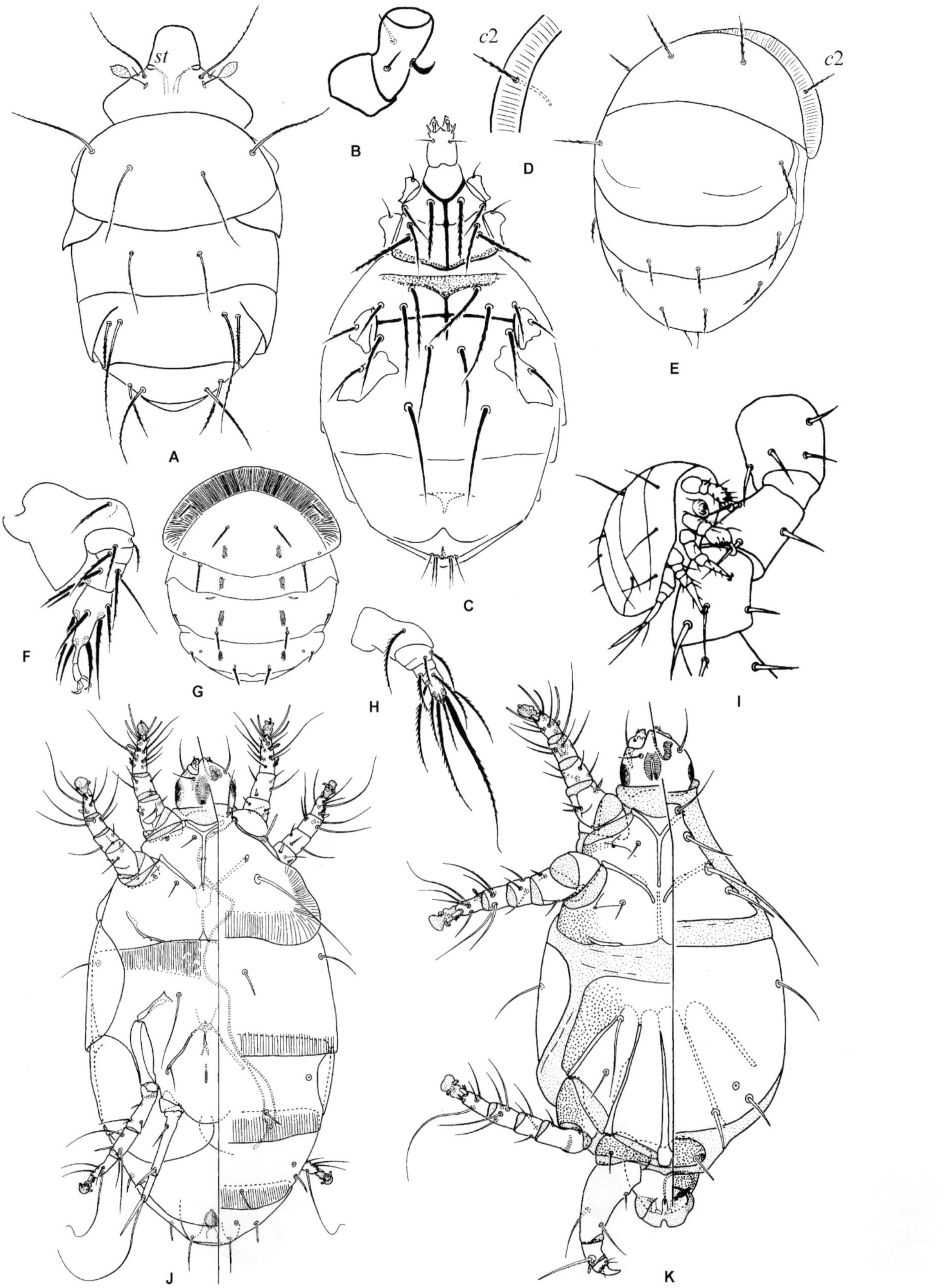

FIGURE 12 *A,* Bakerdania *sp., female, dorsal aspect, B, femur I; C,* Bakerdania arctobia, *female, ventral aspect; D,* Imparipes *sp., female, hysteronotal seta* c2 *and its alveolar channel, E, dorsal aspect; F, member of* Pygmodispus, *female, leg IV; G,* Scutacarus acarorum, *female, dorsal aspect, H, leg IV, I, hyperphoresy:* S. acarorum *on the leg of gamasid mite* Parasitellus fucorum *phoretic on* Bombus *sp.; J,* Steneotarsonemus arcticus, *female, K, male. C, after Khaustov & Makarova 2005; F–H, after Khaustov 2008; I, after Davydova 1978; J, K, after Lindquist 1986.*

FIGURE 13 *Habitus of prostigmatic mites. A,* Alicorhagia clavipilus*; B,* Nanorchestes *cf.* collinus*; C,* Terpnacarus bouvieri*; D,* Eupodes *sp. 1; E,* Eupodes *sp. 2; F,* Eupodes *sp. 3; G,* Cocceupodes breweri*; H,* Penthaleus *sp.; I,* Penthalodes *cf.* ovalis*; J,* Robustocheles hilli*; K,* Poecilophysis saxonica*; L,* Rhagidia gelida. *All females.*

FIGURE 14 *Habitus of prostigmatic mites. A,* Coccorhagidia pittardi*; B,* Thoria brevisensilla*; C,* Hammenia macrostella*; D,* Bdella longicornis*; E,* Neomolgus littoralis*; F,* Thoribdella *sp. 1; G,* Octobdellodes *sp.; H,* Cyta latirostris*; I,* Biscirus silvaticus*; J,* Cunaxoides *sp.; K,* Ereynetes *sp.; L,* Tydeus *sp. All females.*

FIGURE 15 *Habitus of prostigmatic mites. A,* Lorryia *sp. 1, female; B,* Lorryia *sp. 2, female; C,* Tydaeolus *sp., female; D,* Microtydeus *sp., female; E,* Paratriophtydeus *sp., female; F,* Coccotydaeolus *sp., female; G,* Anystis baccarum, *female; H,* Balaustium *sp., female; I,* Abrolophus *sp., female, J, male; K,* Hauptmannia *sp., larva.*

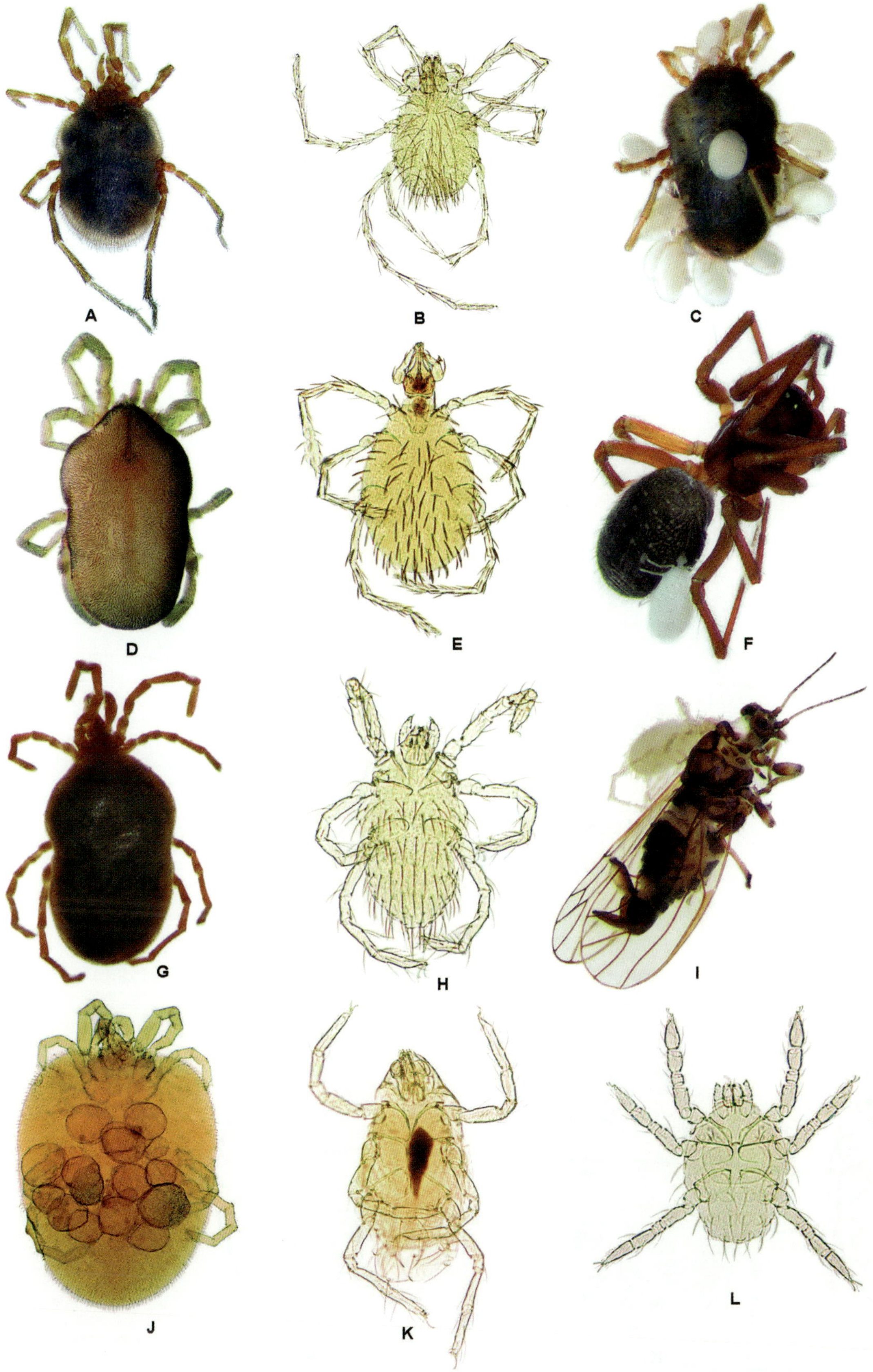

FIGURE 16 *Habitus of prostigmatic mites. A,* Erythraeus phalangioides, *adult; B,* Bochartia *sp., larva; C,* Erythraeus phalangioides, *adult, parasitized by larvae of* Leptus *sp.; D,* Leptus *sp., adult; E,* Leptus kalaallus, *larva; F, spider (Lyniphiidae) parasitized by the larva of* Leptus kalaallus*; G,* Podothrombium strandi, *adult; H,* Podothrombium *sp., larva; I, Psillidae parasitized by the larva of* Podothrombium *sp.; J,* Microtrombidium sucidum, *female; K,* Ettmuelleria sucida, *larva; L,* Parafeiderium *sp., larva.*

FIGURE 17 *Habitus of prostigmatic mites. A,* Stigmaeus *sp.; B,* Stigmaeus parmatus*; C,* Eustigmaeus arcticus*; D,* Eustigmaeus *sp.; E,* Bryobia praetiosa*; F,* Tetranychus *sp.; G, Eriophyidae gen. sp.; H,* Bakerdania *sp. 1; I,* Bakerdania *sp. 2; J,* Imparipes *sp. 1; K,* Imparipes *sp. 2; L, Tarsonemidae gen. sp. A–C, E–L, females, D, male.*

FIGURE 18 Salix glauca *with leaves heavily affected by gall mites,* Aculus *sp., – a very common phenomenon in southern Greenland.*

18.3.5. Hydrachnidia and Halacaridae (Water mites)

Reinhard Gerecke

In several groups of mites some particular species, or representatives of single genera, have adapted to submerged life, and two large clades include exclusively species bound to aquatic habitats. Halacarids originally evolved adaptations to life in marine biotopes and later invaded from there, via streams and ground water, to all types of fresh waters, including high montane springs, lakes and bogs. However, early ancestors of hydrachnids adapted to aquatic life probably in temporary flooded swamps and helocrene springs which gave rise to marked phylogentical radiations in fresh water biotopes. Only few species became adapted to seawater, restricted to algal mats near the coast.

In keys below the following abbreviations are adopted: Gl – glandular organ; L – length; W – width; I L, II L, III L, IV L – legs I–IV respectively (their segments numbering beginning from the base). Figure 1 contains some other morphological explanation.

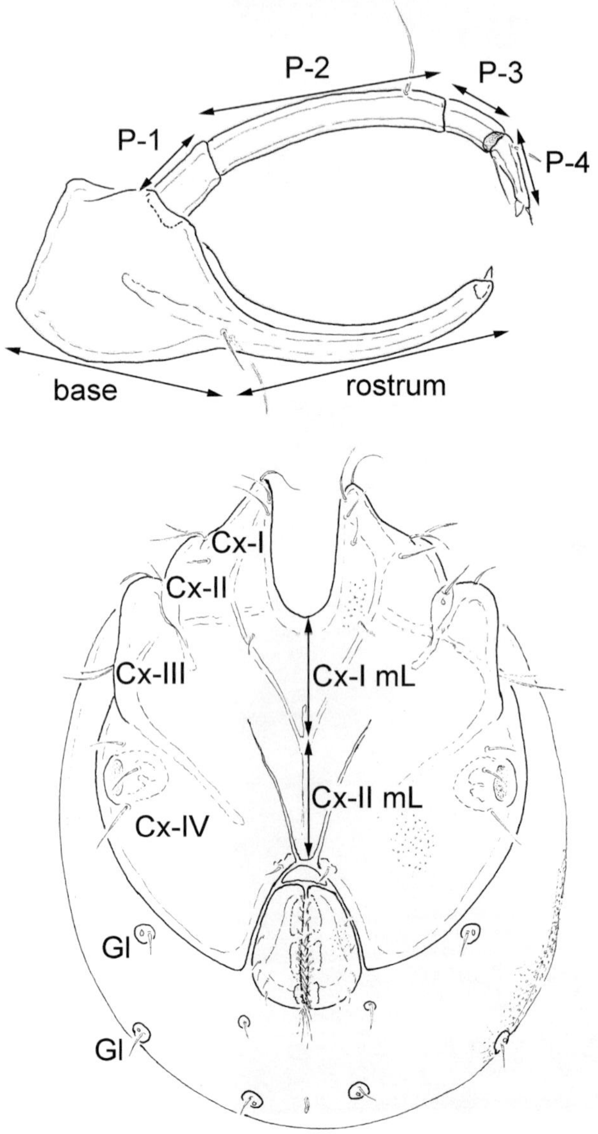

FIGURE 1 *Water mites, terminology: Upper part: Gnathosoma and palp; lower part: ventral idiosoma. Abbreviations: P-1-4 = segments of palp, Cx-1-4 = coxae 1–4 (in this species fused into a coxal shield); Gl = glandular organ; mL = median length of appropriate coxae.*

Key to stages

1. Three pairs of legs. . . . . larvae
– Four pairs of legs. In resting stages (proto- and tritonymphs of Hydrachnidia), legs completely absent, but exuviae of appendages of the preceding stage may be attached to the outer surface or structures, and eyes of the next stage may be seen through the integument. . . . . postlarval stages

Key to species

LARVAE

1. Idiosoma with 3 pairs of ventral setae only (anal area without setae) and at most 7 pairs of dorsal setae; palp with 4, legs with 5 movable segments; not parasitic . . . . Halacaridae
[Species not keyed out; Halacarid larvae differ only gradually from nymphs and adults and may be recognized on the basis of characters given for adults].
– Idiosoma with more than 4 pairs of ventral setae (at least 2 pairs in the anal area, Fig. 3D) and more than 7 pairs of dorsal setae. Number of segments of palps and legs variable, but in all species recorded from Greenland palp with 5, legs with 6 movable segments. Mostly parasitic on arthropods . . . . 2 (Hydrachnidia)
2. Coxal plates 2 and 3 separated from each other by continuous suture lines. Anal plate small and roundish (Figs 3D, 7E) 3

– Coxal plates 2 and 3 on each side medially fused, suture line visible in the lateral part of plates only. Anal plate larger, often subtriangular (Figs 5C, 6H)............4
3. Dorsal plate longitudinally striated, bearing 4 pairs of setae (Fig. 3E) ....*Sperchon* [In the study area only *S. brevirostris*, larval characters of this species insufficiently known].
– Dorsal plate reticulated, bearing 5 pairs of setae (Fig. 7F)....................................*Arrenurus* [In the study area only *A. membranator*, larva of this species undescribed].
4. No setae inserting at posterior edge of coxal plates 3. Only 1 pair of long setae at posterior idiosoma margin (Fig. 6H).. ..*Pionacercus* [In the study area only *P. norvegicus*].
– 2 pairs of setae inserting at posterior edge of coxal plates 3. 3 pairs of very long setae inserting at posterior idiosoma margin (Figs 5C, E) WW5
5. Small-sized (coxal field L < 300 µm, anal plate diameter < 30 µm, dorsal shield L < 200 µm). Posterior margin of coxae 3 truncate, partly concave and forming right mediocaudal angles. Anal plate not embraced by medial margins of coxae 3 (Fig. 5E). Dorsal shield slender (L/W > 2.0).................... .... *Lebertia groenlandica*
– Larger in size (coxal field L > 300 µm, anal plate diameter > 30 µm, dorsal shield L > 200 µm). Posterior margin of coxae 3 rounded, medially embracing the anal plate (Fig. 5C). Dorsal shield wider (L/W < 2.0)................................ *Lebertia porosa*

Key to species

POSTLARVAL STAGES

1. Legs 1 and 2 directed anteriorly, legs 3 and 4 posteriorly (Fig. 2A, 8A). Minute species (adult idiosomal length 200–600 µm) generally with little colouration (for an exception, see *Porolohmanella violacea*). Without idiosomal glandular organs. Palp (Figs 2C, F) with 4 movable segments. Legs (Figs 2B, G) without specialized (long, hair-like) swimming setae........ .... 2 (Halacaridae)
– Legs in a circular arrangement or restricted to the anteriormost part of the idiosoma. Generally larger in dimensions (adult idiosomal length 450–1.500 µm). Colouration often bright and conspicuous. With numerous pairs of regularly arranged idiosomal glandular organs (*e.g.*, Fig. 6B, 8C). Palp with 5 movable segments (*e.g.*, 3 B). Legs in many taxa with swimming setae (*e.g.*, 4E) .... 4 (Hydrachnidia)
2. Idiosoma enlarged (Fig. 2A). Palps and first legs stout and with various strong setae (Figs 2B, C). Claws of first leg forming an umbrella-like pecten............ .... *Soldanellonyx chappuisi*
– Idiosoma narrowed. Palps and first legs slender, bearing only fine setae (Figs 2E, G, I). Claws of all legs simple, sickle-shaped 3
3. Reddish in colour. Gnathosomal rostrum strongly projecting. Palps very long and slender (Fig. 2I). I–L–3 shorter and thicker, I–L–5 without a basal swelling (Fig. 2H)...........*Porolohmanella violacea*
– Without apparent colouration. Gnathosomal rostrum more stout. Palps shorter and thicker (Figs 2F, G). I–L–3 relatively long and slender, I–L–5 with a basal swelling (Fig. 2E)...... *Halacarellus subcrispus*
4. Adult idiosoma completely covered by a porose sclerotisation (Figs 7A, B — porosity visible also in the still unsclerotized integument of freshly hatched specimens). Gonopore area flanked by tongue-shaped fields bearing numerous minute osmoregulatory organs ('acetabula'). Palp robust and subchelate (P–4 with a ventrodistal projecting opposing the pointed, claw-like terminal segment). Legs bearing swimming setae. Males with the posterior idiosoma area and hind legs variously modified; the species known from Greenland with a long caudal petiole. Deutonymphs with soft integument, but unmistakable due to the characters of palp and genital field .... *Arrenurus membranator*
– Idiosoma dorsally and ventrocaudally with soft integument. Gonopore area with 3 (in adults, Fig. 8E) or 2 (in deutonymphs, Fig. 8F) osmoregulatory organs located on small plates in the soft integument between gonopore and genital flaps. Palp more slender, P–4 various in shape, but without ventrodistal extension. Legs with or without swimming setae. Sexual dimorphism may be expressed in the hind legs, but not the posterior idiosoma margin 5
5. Coxae fused into a unique shield (Fig. 4A). Coxal shield closely embracing the genital field and characterized by a Y-shaped suture line, basally formed by the medial margins of coxae 2 and distally on each side by the fused margins of coxae 1 and 2. Palp with P–3 medially bearing 5 (in subgenera not recorded from Greenland also 6 or 7) long setae, 3 of them arranged at the distal margin, further 2 dorsally (Figs 4B,D 6
– Coxae arranged in 3 or 4 groups, with coxae III+IV separated by membranous integument both medially

from each other and anteriorly from coxae I+II (Figs 3A, 7A). Palp setation different 7

6. Legs with swimming setae (Fig. 4E). Palp segment P–3 stout, segment P–4 distally narrowed, bearing all dorsal hairs concentrated near the distal segment edge(Fig. 4D) .... *Lebertia porosa*

– Legs without swimming setae (Fig. 4C). Palp segment P–3 elongated, segment P–4 nearly equal in height from the base to the tip, at least one dorsal hair in the centre of the segment (Fig. 4B) .... *Lebertia groenlandica*

7. Legs with swimming setae (Fig. 6D). Integument finely lineated. Genital field of adults with a pair of sclerotized plates bearing 3 osmoregulatory pores each (Fig. 6A). In deutonymphs medially fused platelets each bearing 2 acetabula (Fig. 6G). Strong sexual dimorphism in the shape of genital field and hind leg (Figs 6D, F). Palp stout, a ventral projection on P–4, but not on P–2 (Fig. 6C)..................... .... *Pionacercus norvegicus*

– Legs without swimming setae. Integument papillate (Figs 8C–D. Genital field of adults with movable

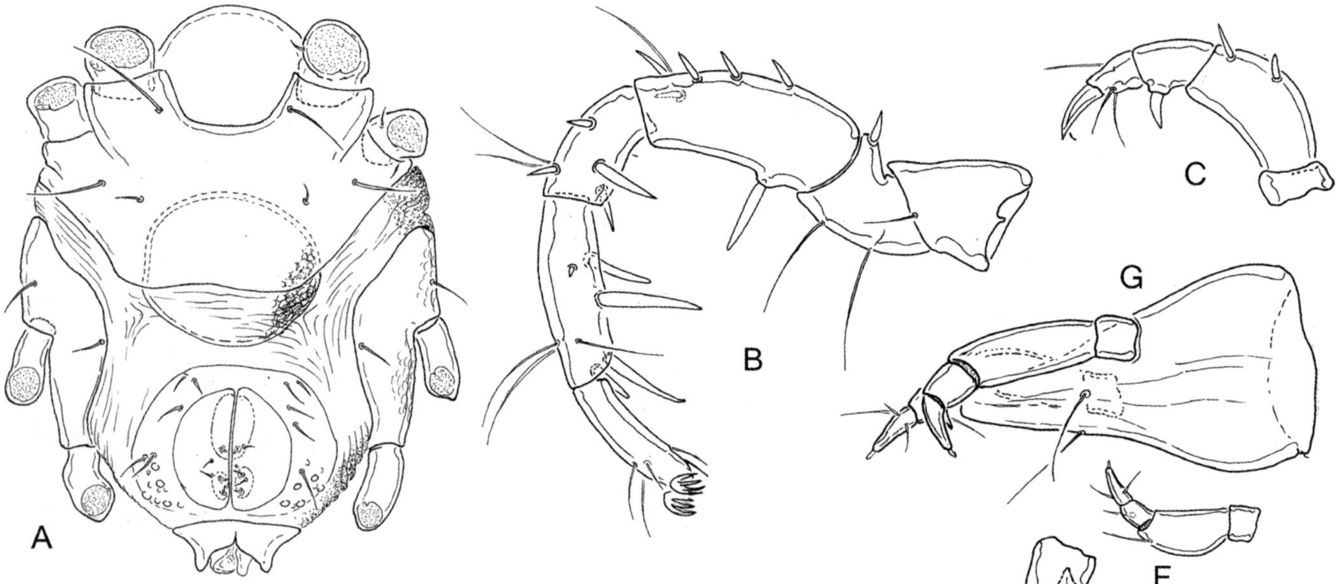

FIGURE 2 *Halacarid mites. A–C,* Soldanellonyx chappuisi, *female (Greenland, Tornarssuk). A, ventral view (legs omitted); B, first leg; C, palp; D–G,* Halacarellus subcrispus *(after Bartsch, 1978). D, female ventral view (legs omitted); E, female first leg; F, deutonymph palp; G, female gnathosoma. H–I,* Porolohmanella violacea, *female (Germany); H, first leg; I, gnathosoma.*

flaps covering 3 pairs of osmoregulatory pores which lie in the soft integument flanking the gonopore (Fig. 3A), in deutonymphs 2 pairs of acetabula embraced by a pair of sclerotized half-rings (Fig. 3C). Sexual dimorphism restricted to minute details of the genital field. Palp more slender, a ventral projection on P–2, but not on P–4 (Fig.3B)................. .... *Sperchon brevirostris*

HALACARIDAE

Halacarids differ from true freshwater mites in minute dimensions also in the adult stage (total length generally less than 500 µm), commonly pale colouration, characteristic arrangement of leg insertions (2 anterior pairs of legs directed anteriorly, 2 posterior pairs directed posteriorly, the two groups separate by a large gap) and the palp consisting of 4 segments only. The development of Halacaridae is "hemimetabolic", with stages morphologically grading into each other, and lacking particular heteromorphic parasitic or pupal stages. Consequently, it is possible to a large extent to classify immature specimens based on adult characters, keeping in mind that the genital field is not developed and dimensions as well as numbers of setae are smaller than in adults.

Halacarellus subcrispus **Bartsch, 1978. Fig. 2D–G.**

Adult idiosomal length in males 410–460 µm, in females 400–480 µm. Sclerites without colour, smooth or with a very fine porosity. Legs slender and with fine, mostly hair-like setae. A few peg-like setae on ventral surface of I–L–4–6, and II–IV–L–5. Genital field with 3 pairs of acetabula covered by genital sclerites. Anal pore in terminal position. Leg claws simple, those of I–L with a few fine denticles on inner margin. Gnathosomal rostrum triangular, equally narrowed. Unmistakable among Greenland freshwater halacarids by the combination of slender idiosoma, first legs with simple claws, and gnathosomal rostrum and palps not particularly elongated.

BIOLOGY. Found in mosses, algae and detritic sediments, mostly in the macrophyte belt of standing fresh waters. Halotolerant, character species of mesopsammal habitats in the intertidal seashore from where it obviously invaded the freshwater habitats in Greenland.

DISTRIBUTION. W, springs near the seashore (Bartsch 1978).

Palaearctic, arctic. Marine beaches near Tromsö, Norway, and on Svalbard (interstitial).

Porolohmanella violacea **Kramer, 1879. Fig. 2H–I.**

Adult idiosomal length 300–550 µm. Sclerites pink. All legs similar in shape, slender and with fine, hair-like setae. Genital sclerites each with 2 acetabula. Anal plate on ventral side. Leg claws simple. Gnathosomal rostrum extremely elongated and narrow. Palp slender, second segment particularly elongated, all segments bearing fine setae only. Unmistakable among Greenland halacarids by the elongated gnathosomal rostrum and palps, reddish colour, first legs slender and with simple claws.

BIOLOGY. Found in mosses, algae and detritic sediments, mostly in the macrophyte belt of standing fresh waters. Halotolerant. Only females are known.

DISTRIBUTION. SW (Nuuk/Godthaab, Eqitsoq, Taserârsuk nov Sinnifik at Qeqertarsuaq/Disko).

Holarctic.

Soldanellonyx chappuisi **Walter, 1917. Fig. 2A–C, 8A.**

Adult idiosomal length 350–450 µm (the female from Greenland 390 µm). Sclerites pale, body brownish, often with a white dorsal stripe. First legs enlarged and thickened, with strong, blunt setae on segments 3–5. Claws with multiplied tips arranged in umbrella shape. Gnathosomal rostrum short. Palp robust, with a strong peg-like seta arranged on a ventral socklet in distal half of third segment.

BIOLOGY. Living in ground water and deeper parts of lakes and springs. In springs all developmental stages are found the year round. Male not yet described.

DISTRIBUTION. S (hitherto only found on Toornaarsuk, Kap Farvel area).

Holarctic.

HYDRACHNIDIA

Hydrachnids (true freshwater mites) are characterized by generally relatively large dimensions (>500 µm) and often bright colours (freqently red or orange, also green, blue or dark brown to black), leg insertions arranged radially (with the exception of Hydrovolzioidea, not recorded from the Greenland) and the palp typically consisting of 5 segments.

SPERCHONTIDAE

Sperchon brevirostris **Koenike, 1895. Fig. 3A–C, 8 B–D.**

Sperchon brevirostris lineatus Thor, 1899: Lettevall 1961 (syn. of *S. brevirostris*: K. Viets 1956)

Adult idiosomal length about 1000 µm (males), 1000–1500 µm (females). Body colouration red, often brownish. Integument papillate, glandular plates enlarged. First coxal plates medially fused by a sclerotized bridge. Anal pore surrounded by a sclerotized ring. Gnathosomal rostrum short. Palp robust, with a prominent ventrodistal extension on P–2, P–4 long and slender, ventral peg-like setae small, not on sockets.

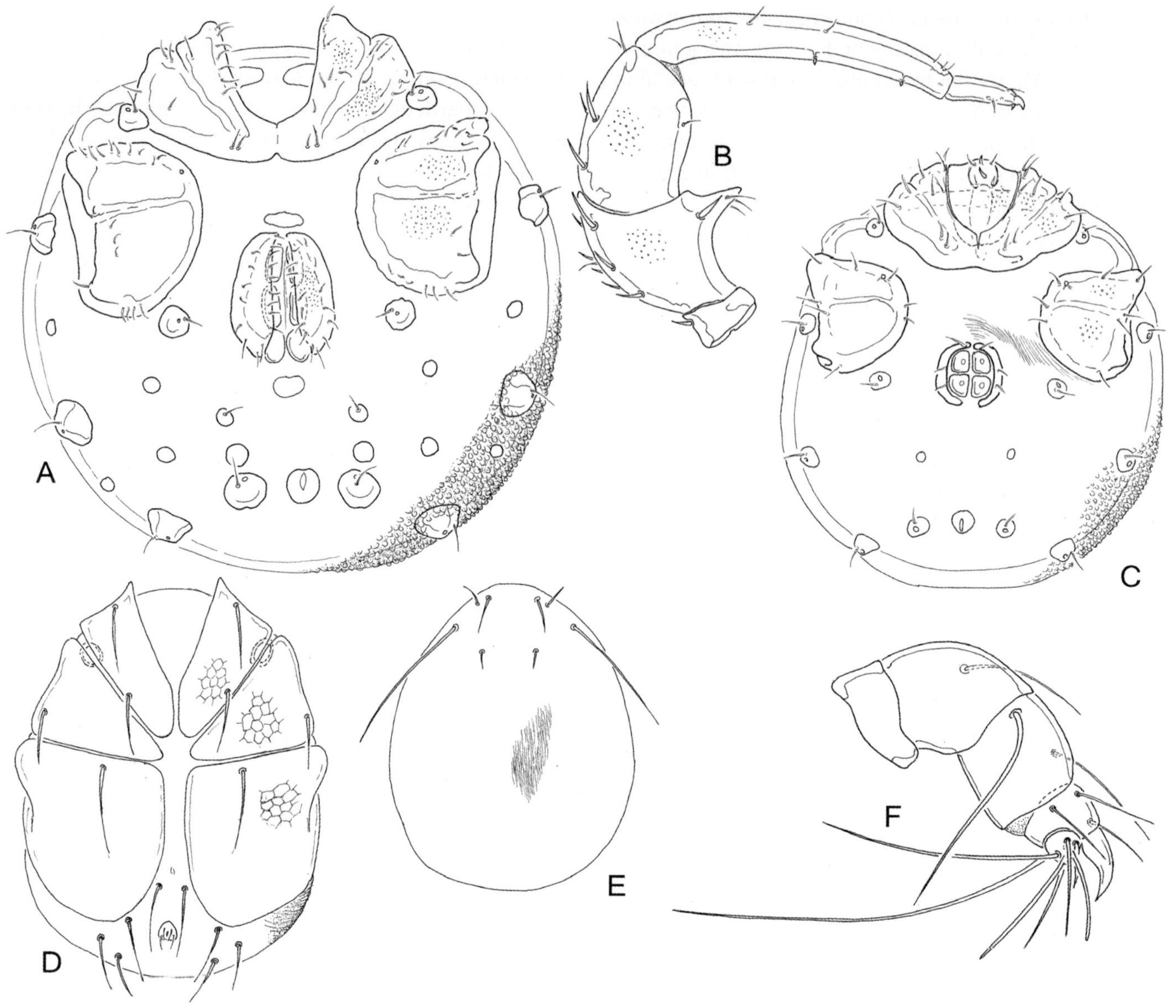

FIGURE 3 *Hydrachnidia-Sperchontidae,* Sperchon. *A–C,* S. brevirostris *(Montenegro, A–B from Di Sabatino et al. 2010). A, male ventral view (legs omitted); B, male palp; C, deutonymph ventral view (legs omitted); D–F,* S. mutilus *larva (after Martin 2003). D, ventral view; E, dorsal shield; F, palp.*

Deutonymph similar to adult, but differing in minor dimensions and lower numbers of setae, and in the provisional genital field formed by 2 pairs of acetabula embraced by paired lateral half-ring sclerites.

Larva only known from unsufficient descriptions. Information in the key based on *S. mutilus* which regarding adult morphology is similar. Distinction from larvae of other species present in Greenland possible on the basis of characters diagnostic for the genus.

BIOLOGY. Upper reaches of mountain streams of Central Europe, in Northern Europe also littoral in lakes, but avoiding both spring areas and higher order streams.

DISTRIBUTION. NE. In Greenland only one record (Loch Fyne, Hold With Hope; Lettevall 1962).

Western Palaearctic.

LEBERTIIDAE

***Lebertia* (*Pilolebertia*) *porosa* (Thor, 1900). 4D–E; 5 A–C.**

Lebertia insignis porosa Thor, 1900: Lettevall 1961.

Adult idiosomal length 800–1500 µm. Colour reddish brown with a white dorsomedial line, sclerotized parts reddish or greenish. Integument smooth, with a fine lineation on the ventral side. II–L, III–L and IV–L with high numbers of swimming setae as typical for the subgenus *Pilolebertia*. Differing from other species of this clade in the arrangement of setae and setal pores of the palp: P–3 mediodistal setae aequidistant, P–4 with one of the two ventral pores in the proximal half of the segment, distoventral pore distanced from the distal segment edge.

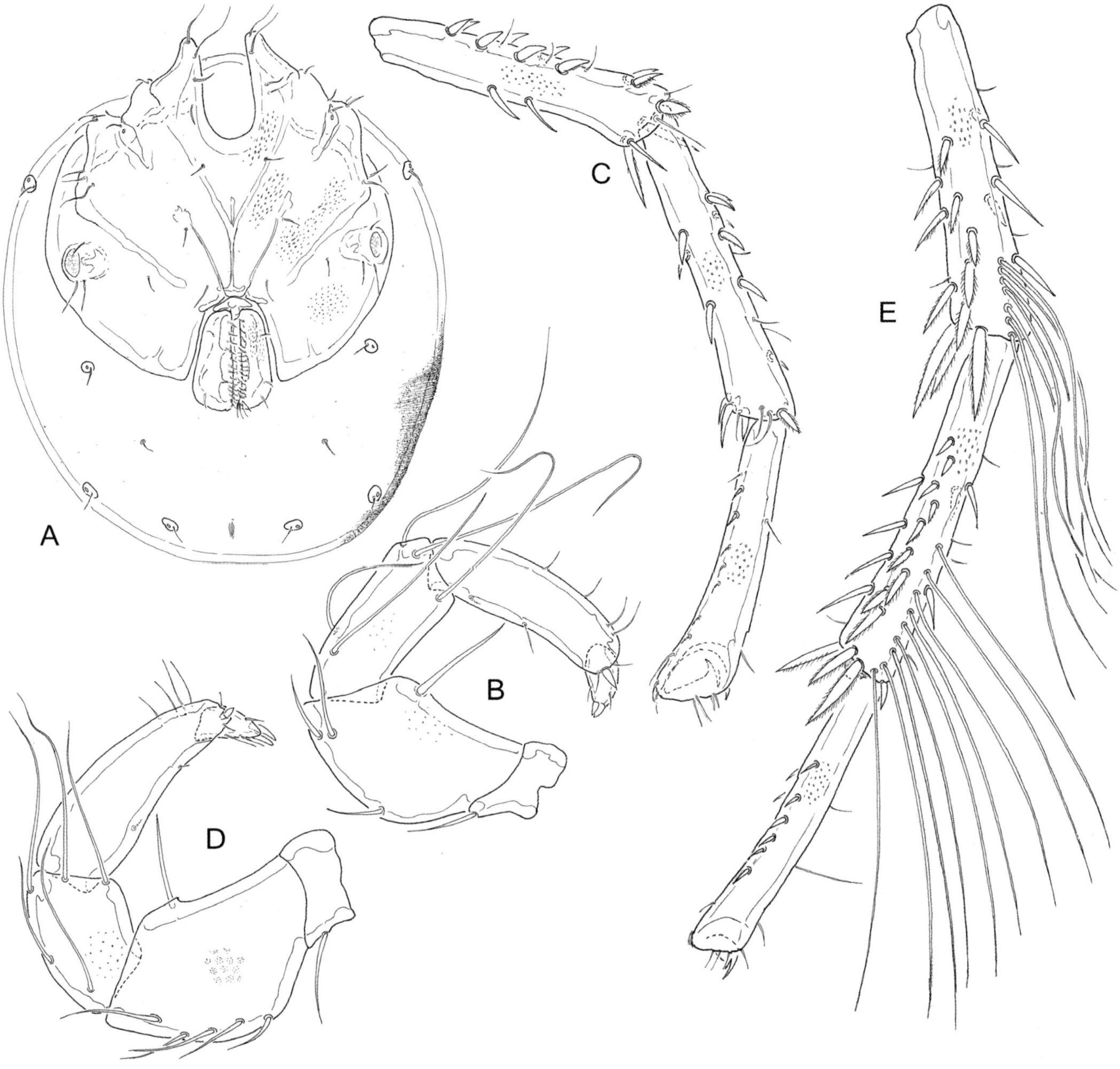

FIGURE 4 *Hydrachnidia-Lebertiidae,* Lebertia. *A–C,* L. groenlandica, *female (Greenland, Bjoernesund). A, ventral view; B, palp; C, hind leg, segments 4–6. D–E,* L. porosa, *female (Greenland, Søstersø). D, palp; E, hind leg.*

Deutonymph similar to adult, but different in minor dimensions and setae numbers, and in the provisional genital field formed by 2 pairs of acetabula surrounded by a sclerite ring. In comparison with the other *Lebertia* species present in Greenland easily recognized by the presence of swimming setae. – Larva differing from that of other species recorded from Greenland as described in the key. Measurements of specimens from Greenland (Zackenberg, n = 4, in µm): Idiosoma L 356–370 (540 in an engorged, squeezed specimen); coxal field L 330–336; dorsal shield L/W 324–348/200–210; anal plate L/W 38–56/38–47; coxae 1/2+3 median L 100–110/132–135; gnathosoma L 100–112. These larvae were not taken together with adults, but can be attributed to the species in agreement with the description given by Wainstein (1980).

BIOLOGY. Preferably in the macrophyte belt of larger lakes and in moss vegetation of stream pools.

DISTRIBUTION. N (Inglefield Land), NE (Germania Land, Zackenberg).

Holarctic, arctic (due to the high number of synonyms, many records need confirmation).

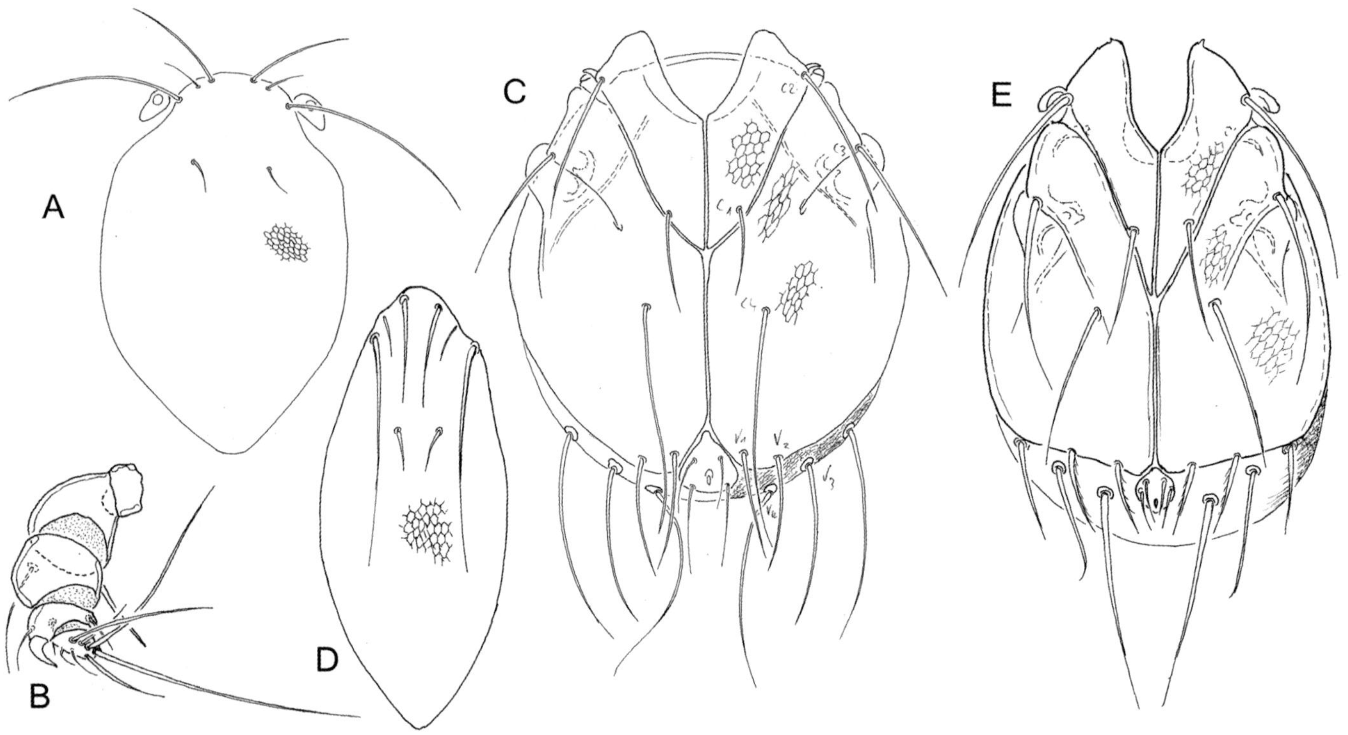

FIGURE 5 *Hydrachnidia-Lebertiidae,* Lebertia, *larvae. A–C,* L. porosa *(Greenland, Zackenberg). A, dorsal shield, B, palp, C, ventral view (legs omitted). D–E,* L. groenlandica *(Greenland, Godhavn). D, dorsal shield, E, ventral view (legs omitted).*

NOTE: Due to former taxonomists' frequent use of individually variable characters (*e.g.,* body colour, relative extension of coxae, artifacts in skin structure), a high number of synonyms has been published for this species. *Lebertia fabricii* (Thorell, 1972), first published as *Hygrobates fabricii* for a specimen recorded by Fabricius (1780) under the name of *Acarus aquaticus,* has not been described sufficiently to allow species identification; consequently, it is regarded a *species dubia* by K. Viets (1956). In view of the high abundance of *L. porosa* in the study area, records of *L. fabricii* most probably have to be referred to this species, as is the case with a specimen from ZMUC ("23/608. Danmark-Exped. *Leerbertia fabricii*") which is identical with *L. porosa.* A definitive synonymization of *L. porosa* with *L. fabricii* is impossible due to lack of type material of the latter.

***Lebertia* (*Lebertia*) *groenlandica* Lettevall, 1962. 4A–C; 5 D–E, 8E–G.**

Adult idiosomal length of males 1300–1400 µm, of females 1300–1600 µm. Colour reddish brown, with a white dorsomedial line. Integument with regular porosity and a fine striation, with rather long lines on the ventral side, but dorsally with only short, irregular lines or completely smooth. Medial suture Cx–1 (175–240 µm) distinctly longer than that of Cx–2 (110–162; ratio 1.2–1.6). Genital field L in males 190–200, in females 230–250 µm. Legs without swimming setae as typical for most species of the subgenus *Lebertia* s.str. Palp with segments P–2 and P–3 nearly equal in length (in males 118–133 and 128–139 µm respectively, in females 127–132 and 135–138 µm), P–3 with 5 long setae, 1 at the ventrodistal edge, 2 placed close to each other at the dorsodistal edge, and further 2 close to each other proximally at the dorsal margin. P–4 longer (in males 135–143, in females 143–153 µm), with maximum height in the distal part of segment and distally slightly bowed ventrad, with relatively long dorsal hair-like setae, a strong mediodistal peg-like seta and 2 ventral peg-like setae, their insertions dividing the ventral segment margin in sectors 1:3:3).

Deutonymph similar to adult, but different in minor dimensions and setae numbers, and in the provisional genital field formed by 2 pairs of acetabula surrounded by a sclerotized ring. In comparison with the other *Lebertia* species present in Greenland easily recognized by the absence of swimming setae. – Larva differing from that of other species recorded from Greenland as described in the key. In the following, a first partial description of the larvae included into the type series is given. These larvae were taken together with adults. In view of the generally poor fauna of Greenland and the dominant representation of postlarval *L. groenlandica* at

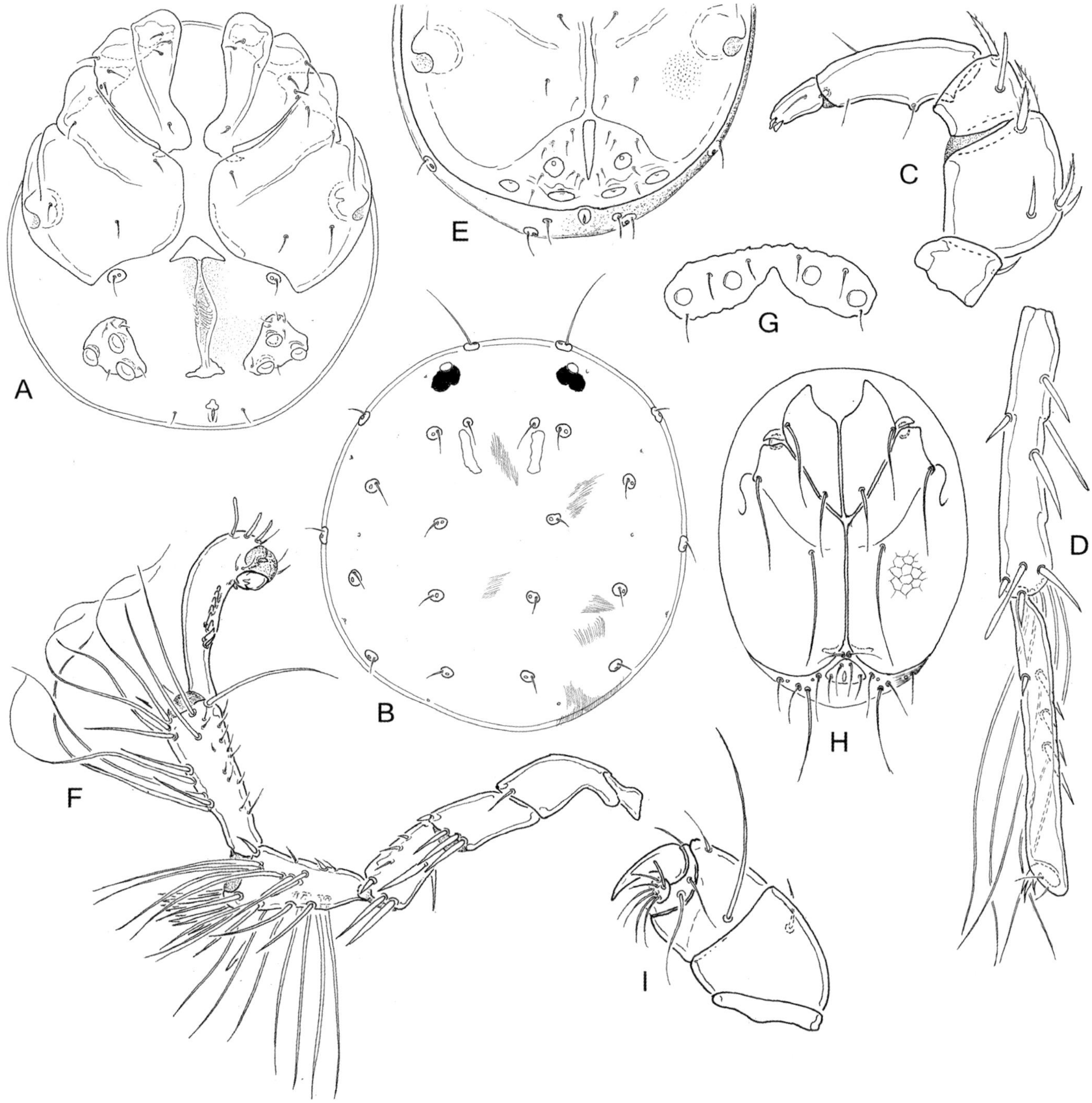

FIGURE 6 *Hydrachnidia-Pionidae,* Pionacercus. *A–G,* P. norvergicus. *A–D, female (Greenland, Hangoq). A, ventral view; B, dorsal view; C, palp; D, hind leg, terminal segments. E–F, male (The Netherlands, Utrecht, Rosweide, 02.09.1984 Smit). E, genital field; F, hind leg; G, deutonymph, provisional genital field (Greenland, Hangoq). H–I,* P. colymicus, *larva (after Tuzovskij 1982). H, ventral view; I, palp.*

the type locality, they can be attributed to this species with certainty.

As they are all slide mounted in undissected state and idiosoma characters are highly diagnostic, gnathsoma and idiosoma appendages are not taken in consideration. Measurements (slides 11, 18–21, n = 5, in µm): Idiosoma L 280–310; coxal field L 244–252; dorsal shield L/W 230/103; anal plate L/W 19–22/18–20; coxae 1/2+3 median L 83–94/99–105: gnathosoma L 98–105.

BIOLOGY. Collected from springs and streams with different types of substrate.

DISTRIBUTION. S, SW (endemic to Greenland). Spring areas near Qeqertarsuaq/Godhavn; river between Qorlortorsuup Ima and Qorlortorsuup Tasia (60°46'N,

45°13'W), here the only species of water mite and quite common (Bergersen 1995).

At the present state of knowledge it is impossible to judge the zoogeographical state of this species.

NOTES: In view of the scanty documentation of the North American lebertiid fauna it is possible that populations in Greenland belong to a Nearctic sister taxon of *Lebertia maglioi*. However, also the evolution *in situ* of an endemic species originating from an ancestor in common with *L. maglioi* cannot be excluded. In this group of species, a comparative study of continental and insular populations, including larval morphology, host preference and molecular evolution, would be of particular zoogeographical interest.

The late Carl Bader planned to describe a new taxon "*L. groenlandicamima*" from material collected in South Greenland (documented by notes preserved in the Natural History Museum Basel, NHMB, and as *nomen nudum* in BERGERSEN 1995). However, apart from slightly smaller dimensions, the specimens in question (preserved at NHMB) are in good agreement with the type series of *L. groenlandica*.

Lettevall (1962) gave a detailed description of this species, but his interpretation of the integument structure was not correct. Consequently, he compared this species only with the phylogenetically rather distant *L. convergens* Thor, 1911, *L. glabra* Thor, 1897 and *L. zschokkei* (Koenike, 1902), all characterized by strong sculpturation or lineation of the integument surface. A fine, locally obliterated integument striation is found in many species currently attributed to the subgenus *Lebertia* s.str.. *Lebertia groenlandica* is most similar to *L. maglioi*, a species recorded also from northern latitudes, including the Faroe Islands (Gerecke submitted) in

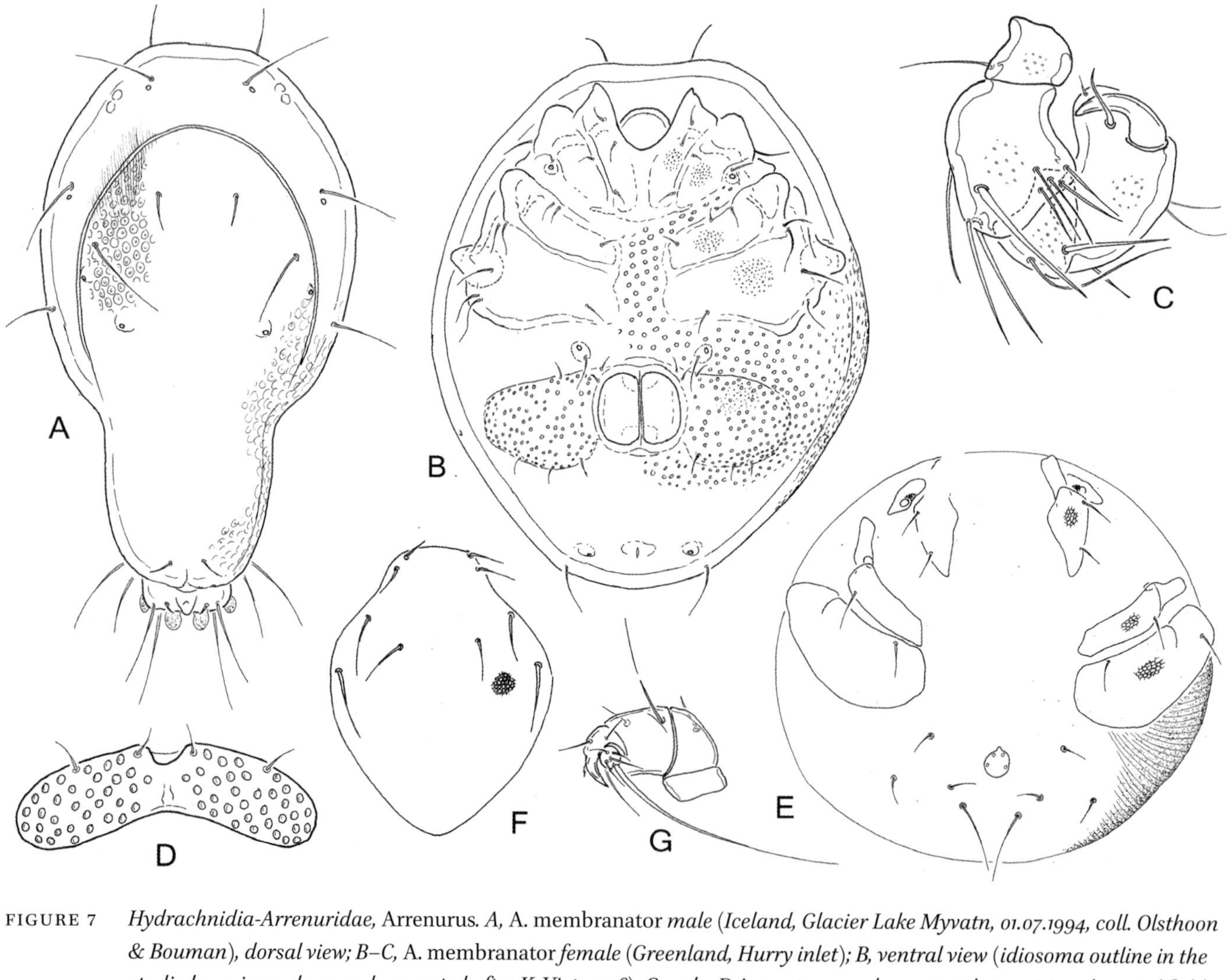

FIGURE 7 *Hydrachnidia-Arrenuridae,* Arrenurus. *A,* A. membranator *male (Iceland, Glacier Lake Myvatn, 01.07.1994, coll. Olsthoon & Bouman), dorsal view; B–C,* A. membranator *female (Greenland, Hurry inlet); B, ventral view (idiosoma outline in the studied specimen damaged, corrected after K. Viets 1936); C, palp. D* Arrenurus *sp. deutonymph, provisional genital field. E–G,* A. *cf.* membranator *larva (Greenland, Zackenberg). E, ventral view of an engorged specimen; F, dorsal shield; G, palp.*

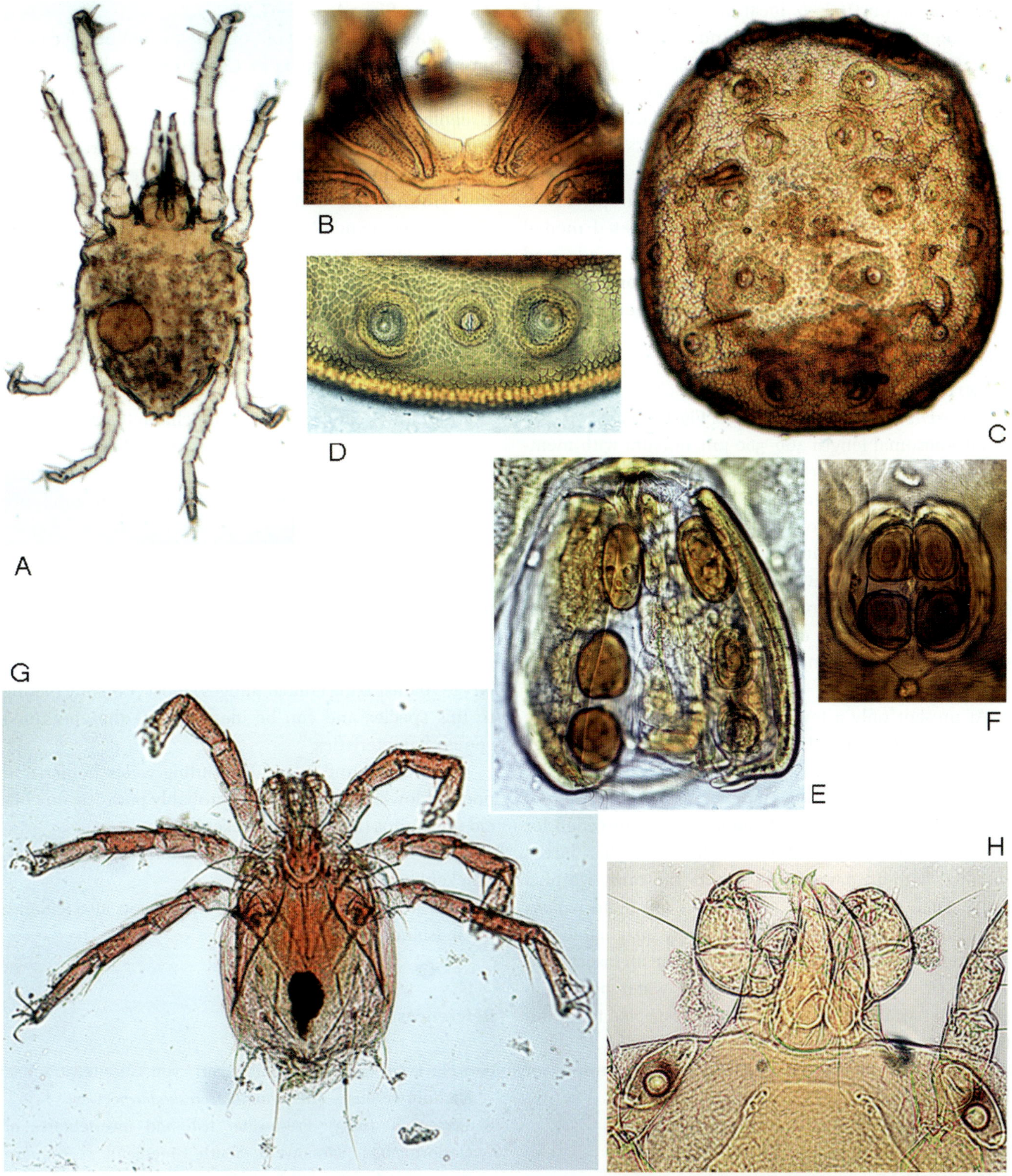

FIGURE 8 *Water mites. A,* Soldanellonyx chappuisi, *female, ventral view; B–E,* Sperchon brevirostris *male; B, medially fused Coxae I+II; C, dorsum. D, excretory pore area; E–H,* Lebertia groenlandica*; E, female, genital field, one genital flap opened; F, deutonymph, provisory genital organ; G, larva, ventral vie; H,* Arrenurus *cf.* membranator *larva, frontal area and gnathosoma.*

complete absence of swimming setae, low number of ventral setae on IV–L–6, location of setae on P–3, and the rather characteristic shape and setal equipment of P–4. *Lebertia maglioi* differs from *L. groenlandica* in minor dimensions (*e.g.* length genital field < 200, P–2 < 110, P–3 < 100, P–4 < 120 µm) and a relatively shorter medial suture of Cx–1 (mL ratio Cx–1/Cx–2 0.9–1.2). *Lebertia harenarum* Láska, 1953, a still little known species of the *L. maglio* – group from Central Europe, differs from both compared species in the elongated medial suture of Cx–1 (length ratio Cx–1/Cx–2+3 > 2.0) and from *L. groenlandica* also in minor dimensions (Gerecke submitted). No species of this group has been described so far in the larval stage.

PIONIDAE

Pionacercus norvegicus **Thor, 1898. Fig. 6A–G.**
Adult idiosomal length 460–500 µm, dorsum with membranous integument bearing a pair of small rod-shaped anterior sclerites. Genital field with 3 pairs of large acetabula, in males membranous but included in the coxal sclerotization, in females with acetabula located on triangular plates. Palp stout, P–4 with a characteristic ventral extension in the proximal part of the segment and an additional minor denticle in the distal part. Legs with swimming setae. Male IV–L strongly modified for spermatophore transfer (Fig. 6F).

At present only a few deutonymphs and females are known from Greenland. A detailed description of larvae and adults was given by Tuzovskij (2008, larvae and adults, 2013, adults only) on the base of a population from Jaroslawl province, Russia. Lundblad (1927) recorded for another species, *P. leuckarti* Piersig, 1894 that larvae moulted without parasitic feeding to the protonymphal stage. Such a lifestyle is improbable for *P. norvegicus* because in case of a water mite, the only imaginable way of colonizing a remote island is phoresy on an insect host.

BIOLOGY. Found in small standing water bodies, in southern part of the distribution area restricted to mountain lakes.

DISTRIBUTION. S, SW (collected at a number of localities).

Palaearctic: central and northern Europe.

ARRENURIDAE

Arrenurus is the most species-rich species Palaearctic water mite genus, and several species have been found colonizing also remote islands. Most important characters for species discrimination are found in the variously modified male cauda, while identification of females is often difficult.

Arrenurus (Megaluracarus) membranator **Thor, 1901. Figs 7A–C, E–G, 8H.**
Adult idiosomal length of both males (including cauda) and females about 1300 µm. Colouration yellowish with brown marks. Differing from all other water mites in Greenland by the complete idiosoma sclerotization with large porosity, the dorsal shield separated from remaining body sclerite by a narrow membranous furrow (round and continuous in females, horseshoe-shaped and caudally interrupted in males), the high number of acetabula on tongue-shape areas flanking the genital field, and the subchelate palp. Swimming setae are present on legs II–IV. Males of *A. membranator* differ from other species of the subgenus *Megaluracarus* (characterized by the long, projecting cauda) by the presence of 2 pairs of hyaline extensions at the posterior margin of cauda. Females are characterized by palp segment P–2 bearing only a few scattered medial setae, combined with acetabular fields rather large, slightly extending posteriad the level of the caudal gonopore edge.

Deutonymphs are rather different from adults (with membranous integument, not showing sexual dimorphism) and cannot be recognized at species level. However, as *A. membranator* is the only species of the genus known from Greenland, both deutonymphs and larvae (parasitizing culicid midges) most probably belong to this species and can be identified by the combined genus characteristics.

BIOLOGY. Found in small standing water bodies and pools of lowland rivers. Larvae probably parasitic on culicid mosquitos (as found at Zackenberg).

DISTRIBUTION. SW (Qeqertarsuaq/Godhavn), NE (Zackenberg, Hurry Inlet).

Palaearctic (central and northern Europe, also Iceland, Faroe Islands), widespread.

References

Bartsch, I. (1978): Halacaridae (Acari) von Gezeitenstränden Nordnorwegens. – *Mikrofauna Meeresboden* 70: 1–23.

Bergersen, R. (1993): Freshwater fish and invertebrates of Qorlortorsuaq, Vatnahverfi, South Greenland. – *Tromura Naturvitenskap* (Tromsø) 73: 1–71.

Bergersen, R. (1995): Is Greenland a zoogeographical unit of its own? – *Journal of Biogeography* 22: 1–6.

Di Sabatino, A.; Gerecke, R.; Gledhill, T. & Smit, H. (2010): Hydrachnidia, Hydrophantoidea and Lebertioidea. – *In*: Gerecke, R., ed: Süßwasserfauna von Mitteleuropa 7/2–2: Chelicerata: Acari II. Springer 1–234.

Fabricius, O. (1780): *Fauna Groenlandica*, systematice sistens animalia Groenlandiae occidentalis hactenus indigenata, quod nomen specificum, triviale, vernaculumque; synonyma auctorum plurium etc.... Hafniae et Lipsiae, J.G. Rothe: 16 + 452 pp., 1 table.

Gerecke, R. submitted. Revisional studies on the European species of the water mite genus *Lebertia* Neuman, 1880 (Acari: Hydrachnidia: Lebertiidae). – *Abhandlungen der Senckenbergischen Naturforschenden Gesellschaft Frankfurt* 566: 1–144.

Lettevall, U. (1962): On the Hydracarina of Greenland with a description of *Lebertia* (*Pseudolebertia*) *groenlandica* n. sp. – *Meddelelser om Grønland* 170(1): 1–40.

Lundblad, O. (1927): Die Hydracarinen Schwedens. I. Beitrag zur Systematik, Embryologie, Ökologie und Verbreitungsgeschichte der schwedischen Arten. – *Zoologisk Bidrag* 11: 181–540.

Lundblad, O. (1930): Hydracarina. – *Zoology of the Faroes*, Copenhagen 2 (48): 1–65.

Martin, P. (2003): Larval morphology of spring-living water mites (Hydrachnidia, Acari) from the Alps. – *Annales de Limnologie – International Journal of Limnology* 39(4): 363–393.

Smith, I.M. & Oliver, D.R. (1986): Review of parasitic associations of larval water mites (Acari:Parasitengona:Hydrachn ida) with insect hosts. – *Canadian Entomologist* 118: 407–472.

Smith, B.P. (1998): Loss of larval parasitism in parasitengonine mites. – *Experimental and Applied Acarology* 22(4): 187–199.

Tuzovskij, P.V. 1982. Novyj vid vodjanogo kleshcha roda *Pionacercus* Piersig, 1894 (Pionidae, Acariformes). – *Biologicheskie Nauki* (Zool.) 8: 29–34.

Tuzovskij, P.V. (2008): K sistematike vodjanogo kleshcha *Pionacercus norvegicus* (Acariformes, Pionidae). – *Zoologicheskie Jurnal* 87(2): 156–166.

Tuzovskij, P.V. (2013): Water mites of the genus Pionacerus Piersig, 1894 (Acari: Hydrachnidia, Pionidae) in Russia. – *Amurian zoological journal* V (3): 256–278.

Viets, K. (1925): Beiträge zur Kenntnis der Hydracarinen aus Quellen Mitteleuropas. – *Zoologische Jahrbücher, Abteilung für Systematik, Ökologie und Geographie der Tiere* 50(4–6): 451–596.

Viets, K. (1936): Wassermilben oder Hydracarina (Hydrachnellae und Halacaridae). – *In*: Dahl, F. (ed.): *Tierwelt Deutschlands*, Jena (G. Fischer), 31: 10+1–288, 32: 289–574.

Viets, K. (1956): *Die Milben des Süßwassers und des Meeres. Hydrachnellae et Halacaridae* (*Acari*). II. und III. Teil: Katalog und Nomenklator. – Jena (G. Fischer): 1–870.

Viets, K.O. (1987): Die Milben des Süßwassers (Hydrachnellae und Halacaridae [part.], Acari). II.: Katalog. – *Sonderbände des Naturwissenschaftlichen Vereins Hamburg* 8: 1–1012.

Wainstein, B.A. (1980): Opredelitel lichinok vodjanych kleshchej. – *Leningrad, Nauka*: 1–238.

18.3.6. Oribatida (=Cryptostigmata, Beetle mites)

Olga Makarova and Valerie Behan-Pelletier

The acarine suborder Oribatida, excluding the Astigmata, are much more homogenous ecologically than the other large mite suborders found in soil, the Prostigmata and Mesostigmata. Most oribatids are free living in litter, soil, suspended soil, and in arboreal or petrophilous lichens and mosses. Some oribatids inhabit the marine or freshwater littoral. Representatives of several genera can permanently exist in aquatic habitats. In soil mite communities, oribatids predominate in most biomes and biotopes excluding the driest and coldest ones. Their density often amounts to several hundred thousand specimens per square meter (Norton 1990) and their biomass in the Low Arctic (up to 12 kg per hectare) is comparable to the total biomass of vertebrates (Krivolutsky 1976). The taxonomic diversity of oribatids in one biotope can be strikingly high (up to a hundred species) in spite of a significant overlap in diet among species (Maraun *et al.* 2003). In the Low Arctic one biotope can be populated by 20–30 oribatid mite species. Communities of the High Arctic are usually impoverished and in the most severe conditions number only 1–7 species (McAlpine 1965; Makarova 2002).

Oribatids are mainly saprophagous and/or microphytophagous, and most species appear to be highly opportunistic in their food choice (Behan & Hill 1978, Norton 1994, Schneider *et al* 2004). Luxton (1972) following Schuster (1956) recognized the following trophic groups: the *macrophytophages, microphytophages* and *panphytophages*. Macrophytophages feed on higher plant material, usually modified by some decomposition. Among them the xylophages, phyllophages, lichenophages etc. can be identified. Microphytophages bite and swallow the fragments of fungal hyphae, spores, yeasts, algae, and some mites are considered bacteriophages. Panphytophages use both these diets. Numerous oribatids readily eat live nematodes or are scavengers on dead arthropods. Concerning feeding habits, Walter and Proctor (2013) designated most oribatids into two feeding groups: the *comminuting microbivore-detritovores* (clear majority) and *piercing-sucking microbivores*. So, representatives of Suctobelbidae with attenuated chelicerae (Fig. 1G, H) are believed to be feeding on fluids and do not produce gut boluses. The main role of oribatids in the litter decomposition lies in the fragmentation of organic matter and vectoring of microbial propagules (Seastedt 1984; Maraun & Scheu 2000).

Morphology

Most species have the *idiosoma* 300–700 µm in length but in general body size varies from 150 µm to 1.5 mm. Colour may be whitish, yellow, black (rarely orange and reddish), but more often various shades of brown predominate. Deepness of colour usually depends on the degree of cuticle sclerotization and melanization. In most families, there is an external secreted layer of nonchitinous nature called the *cerotegument*. It may cover all or parts of the integument and look like granulation, reticulation, a filamentous stratum etc. Body form varies greatly (ovoid, hemispherical, cylindrical etc.). Sexual dimorphism is usually inconspicuous or absent, rarely prominent. Usually 2 tagmata are discernible (Figs 1E, 2A): *proterosoma,* PTS (comprising *gnathosoma, gn,* and *propodosoma* bearing 2 anterior pairs of legs) and *hysterosoma,* HS (bearing 2 posterior pairs of legs). These two units are separated by the *sejugal groove* (Fig. 2B, *sjg*). The dorsal and dorsolateral region of proterosoma is called *prodorsum* (Fig. 1E, *prd*); the large dorsal shield of the hysterosoma is named *notogaster* (*ng*). In ventral aspect, the gnathosoma is usually protected in an anteroventral cavity known as the *camerostome* (Fig. 2A), the *coxisternal region* corresponds in general to the area of leg insertations, and the *anogenital region* (with appropriate orifices) is located behind the last pair of legs.

The prodorsum, anteriorly often terminates by the smooth or dentated *rostrum* (*rm*), which forms the roof and sides of the camerostome (Figs 1E, 2A). In some taxa, a frontal protuberance, the *naso,* may be developed (Figs 10I, J). The prodorsal surface may be smooth or have a range of different structures (tubercles, folds, ridges, pits etc.). The most common are *lamellae* and *costulae*. Lamellae are usually ribbonlike, relatively broad blades (Figs 9K, 15D, 16G), with a free edge (tectum) and a free distal end or *cusp* (Fig. 1E, *csp*). They may be separate, partly fused or connected by a *translamella* (*tl*). In some oribatids there are additional paired ridges on the prodorsum which are situated anteriad and laterad of lamellae (Fig. 2C), the *prolamellae* (*prl*) and *sublamellae* (*sl*), respectively. Costulae are low, narrow longitudinal thickenings of the cuticle without free tecta (Figs 9B, 11D, E, I). The prodorsum bears 5–6 pairs of setae (Figs 1E): the *rostral setae* (*ro*), the *lamellar setae* (*le*), the *interlamellar setae* (*in*), the *trichobothria* (*trb*) and 1–2 pairs of *exobothridial setae* (*ex*). The trichobothrium (or sensillus) is a highly modified sensory seta of rheotactile function. It usually consists of stalk and head. The form of trichobothria varies sharply (clavate, fusiform, pectinate, lanceolate, setiform etc.) almost never resembling the form of other prodorsal setae; it is rarely absent. Trichobothria are located posterolaterally on the prodorsum being inserted in caplike *bothridia* (*bt*) which may have a funnel-shaped rim. A lamellate *tutorium* (*tu*), usually with free cusp, may be developed laterad and ventrad of each lamella above the first leg insertation (Figs 1E, 2B, 15J). In some taxa, a *genal notch* (*gen*) is present on the margin of the camerostome (Figs 2B, 15J). In some families two paired apodemes insert internally from the posterior edge of the prodorsum, the dorsophragmata (Fig. 1E, *dph*) and the pleurophragmata (Fig. 1E, *pph*), and cheliceral muscles insert on these apodemes.

The *notogaster* of some more primitive families is subdivided by transverse scissures into 2 to 4 sclerites (Figs 3A, 4G) and a separation of the *pleural* (Figs 3J, 4B, *pl*) and *suprapleural plates* (Fig. 5B) may also occur, but most oribatids have an entire notogaster. Notogastral setae may be of very different size and form. The most complete set of notogastral setae numbers 16 pairs of setae, but this number often is strongly reduced (down to 4 pairs). Anterolateral winglike or auriculate projections known as *pteromorphs* (*pt*) are present in oribatids of certain families (Figs 1E, 14E, 18D, F). They may bend ventrally, covering the bases of the legs (Figs 16G, J). Pteromorphs may be movable or immovable. Movable pteromorphs possess a *hinge* (Figs 1E, 17J, *hi*) or the line of desclerotization (well seen in lateral aspect) near their base which, with associated muscles, enables their contraction to the sides of the body. The anterior part of the notogaster may bear different paired sclerotized structures: sharp horn-like or triangular projections known as *spinae adnatae* (Fig. 8B, *s. a.*), longitudinal crests or *cristae* (Fig. 12H, *cr*), nonmovable angular *humeral projections* (Fig. 14B, *hp*) etc. Sometimes the anterior edge of notogaster forms a flat projection overhanging the prodorsum. This structure is called the *anterior notogastral tectum* (*ant*) and may be even (Fig. 1E) or sinuate (Figs 18B, C). The posterior notogastral edge may be curved ventrally forming the *posterior notogastral tectum*, entire (Fig. 2D, *pnt*) or with overlapping lobes (Fig. 2E). The dorsal part of the sejugal groove, or the *dorsosejugal scissure*, may be concealed by the anterior notogastral tectum in some groups (Figs 1E, 18A, B). In some oribatids, a light spot or *lenticulus* (Figs 1E, 13F, 14A, *lt*) is present on the notogaster behind the dorsosejugal scissure: it is involved in photoreception. An *octotaxic system* (8 glandular elements) of either *porose areas* or *saccules* is present on the notogaster of most higher (brachypyline) oribatids. Porose areas look like the light spots, being places of concentrated cuticular pores. Usually 4 pairs of porose areas are present (Fig. 1E, *Aa*, *A1–3*). Saccules (*Sa*, *S1–3*), which are homologous to porose areas, are bag-shaped invagination with porose walls opening to the surface through an inconspicuous pore. They are located in the same positions as porose areas (Fig. 18G) but appear darker than the surrounding integument. In most oribatids large, paired *opisthonotal* (or *latero-abdominal*) *glands* open mediolaterally on the notogaster by circular or slit-like pores (Figs 1E, 6J, *gla*).

Ventral surface. The coxisternal region (or epimeral region) comprises 4 plate-like coxisternal elements called *epimeres* (Fig. 2G, *ep*I–IV). Epimeres are demarcated by *apodemes* (Fig. 2A) which constitute the sclerotized lamellate invaginations for muscle attachments. For example, *epimere* II is outlined anteriorly by the *apodeme* II (Fig. 2A, *ap*II) and posteriorly by the *sejugal apodeme* (*apsj*). In Damaeidae, the sejugal groove ventrolaterally is delineated by tubercle-shaped protuberances, called *parastigmatic enantiophyses* (Fig. 8B, *Sa, Sp*). The *genital plates* (Fig. 2A, *gp*) cover the genital orifice through which the female *ovipositor* (Fig. 2F) or male *spermatopositor* can be extruded; 3 pairs of *genital papillae* (Fig. 2F, *gpp*) are present in most adults. The *aggenital plates* (Fig. 2G, *agp*) are located laterad of genital plates. The anal opening is enclosed by a pair of *anal plates* (*anp*) adjoining to *adanal plates* (*adp*), often a small unpaired *preanal plate* (*prp*) is developed between the genital and anal orifices. Recognition of the so called 'lower oribatids' and 'higher oribatids' is primarily based on the structure of the anogenital region. In 'lower oribatids' the adanal and aggenital plates are relatively narrow; and if sclerotized, the aggenital region is never fully fused to coxisternum (*macropyline type*, Fig. 2G). In more derived ('higher') oribatid mite groups, the aggenital and adanal plates are expanded and fused forming the *ventral plate* (*vp*), in which the relatively smaller anal and genital plates are inserted; the ventral plate is also fused with the coxisternum (*brachypyline type*, Fig. 2A). Rarely a small unpaired *postanal plate* (*pop*) may be developed (Fig. 9I). In higher oribatids, tecta-like structures are associated with the bases of legs I and II, *pedotecta* I and II (Fig. 2A, *pd* I, II), which protect basal podomeres. A *discidium* is often present between acetabula III–IV (Fig. 2A, *dis*). A paired *circumpedal carina* may curve anteromedially from the edge of the ventral plate, running behind leg IV (Fig. 2A, *cpc*). The circumpedal carina may merge with the discidium and in some cases the anterior end projects as an inconspicuous, knife-like *custodium* (*cus*).

The *gnathosoma* lies in a camerostome and comprises the *subcapitulum* (*sc*) possessing a toothed or membranous *rutella* (Figs 1A, C, *ru*), paired *palps* consisting of 2–5 palpomeres (Fig. 1B, *pp*) and paired *chelicerae* (Figs 1F–I). The subcapitulum in most oribatids is subdivided by the labiogenal articulation into a posterior unpaired *mentum* (Fig. 1B, *mn*) and anterior paired *genae* (*gn*) (*stenarthric* or *diarthric*). In the most primitive families this articulation is absent (*anarthric*) (Fig. 1A). The chelicerae are normally compact and chelate with *movable digit* (*d. m.*) and *fixed digit* (*d. f.*) generally dentate (Fig. 1F). A few families (Suctobelbidae, Phenopelopidae etc.) have modified chelicerae with the distal part strongly narrowed, and the chela small (*peloptoid* chelicerae, Fig. 1I), or toothless or bearing inconspicuous denticles (*attenuate-edentate, suctorial* chelicerae, Figs 1G, H). Some Brachypylines have a small saccule (axillary saccule of the subcapitulum) opening at the adaxial base of the palp.

The *legs* normally have 6 free podomeres (Fig. 1K): *trochanter* (*tr*), *femur* (*fm*), *genu* (*gu*), *tibia* (*tib*), *tarsus* (*tar*), and *ambulacrum* (*amb*). The ambulacrum usually comprises 1 or 3 claws (respectively monodactylous or tridactylous leg) and a basilar piece. The leg setation is widely used in taxonomical research.

Life cycle

The typical set of life stages comprises egg, prelarva, larva, proto-, deuto-, tritonymph (Figs . 24I–L), and adult. Some hydrophilous taxa display ovoviviparity. The larva is the first mobile feeding stage. The development is anamorphic with body segments added terminally from the larval instar to the deutonymph. Larvae are hexapod and lacking genital papillae. Octopod nymphs I–III generally have 1, 2 or 3 pairs of genital papillae respectively. In higher oribatids, the adults are more strongly sclerotized and often have habitat and feeding preferences different from the immatures, as in holometabolous insects. Sperm transfer is usually indirect by stalked spermatophores. Copulation *per se* is unknown.

Greenland fauna

The suborder Oribatida without Astigmata includes about 170 families and more than 10 000 known species (Subías 2004, Schatz *et al.* 2011), but the discovery rate of new species, especially in the tropics, is very high. First records of Greenland oribatids date from Kramer (1897). By the beginning of the 20th century only 12 species were known from Greenland (Trägårdh 1904, Henriksen & Lundbeck 1917). But now the moss mites are the best known mite taxon in Greenland because of the careful long-term investigations by Marie Hammer. She first published a comprehensive check-list of the Greenland oribatids (Jørgensen [later Hammer] 1934) and then a monographic study on Greenland Oribatida (1944) concerning 57 species. Some of these were first found in Greenland by Niels Haarløv (1942). Later Hammer (1946, 1952, 1952a, 1953, 1954, 1960) and Strenzke *et al* (1955) added 30 species to the list including several species new to science. Subfossil oribatids were initially investigated by Graversen (1931) and Haarløv (1967). After the recent revision we enumerate 29 families and 110 Greenland species of Oribatida (Makarova & Böcher 2009), which are keyed below. Characters used in the keys (based on adult features) are mainly relevant for the Greenland fauna. Some previous keys were intensively used (Ghilarov 1975; Moritz 1976; Balogh & Mahunka 1983; Behan-Pelletier 1986, 1993, 1994; Colloff 1993; Olszanowski 1996; Weigmann 2006; Norton & Behan-Pelletier 2009; Bayartogtokh 2010 etc.).

Key to families[1]

1. Genua of all legs noticeably shorter than tibia, functioning as a "knee" (Fig. 2K); legs articulate with body in deep acetabula, as a "ball and socket" joint; venter of brachypyline type with coxisternal, aggenital and adanal regions fused into a single unit that carries their respective setae (Figs 2A, 9D)................8
– Genua of all legs similar to tibia in size and shape, not reduced to knee-like segment (Fig. 2J); leg articulation not in deep acetabula, at most in shallow depressions; venter of macropyline type: paired aggenital and adanal plates often distinguishable (Fig. 2G), coxisternal region often divided by sejugal articulation and usually not fused to aggenital region (Figs 4J, 5D) 2
2. Body ptychoid: legs can be fully withdrawn as a group and prodorsum rotated ventrally such that it touches the notogaster and ventral plate so that legs are fully covered, and the animal resembles a closed box (Figs 19G, H); body commonly laterally compressed or circular in cross section...........Phthiracaridae
– Body not ptychoid, commonly dorsoventrally compressed in cross section.........................3
3. Notogaster divided by 1 or 2 distinct transverse scissures (Figs 3A, I)....................................4
– Notogaster without transverse scissures (Figs 5H, 6F, 7B)...5

1 Mainly based on Norton & Behan-Pelletier (2009) with modification.

4. Notogaster with single transverse scissure (Fig. 3A); medium size (550–700 μm), yellow, orange or reddish mites ....Hypochthoniidae
– Notogaster with 2 transverse scissures (Figs 3B, D); small (120–350 μm), usually whitish or yellow mites .... Brachychthoniidae
5. Prodorsal bothridium and trichobothrium lost without trace (Figs 7C, D); subcapitulum without labiogenal articulation (Fig. 1A) .... Malaconothridae
– Prodorsal bothridium developed (exceptions: *Mucronothrus*, Fig. 6J; some *Trhypochthoniellus*, Fig. 7B); trichobothrium usually of different shape than other prodorsal setae (Figs 5H, K, 6E); subcapitulum with labiogenal articulation (Fig. 1B) 6
6. Aggenital setae (2 pairs) present; notogaster usually flat or slightly concave dorsally, often with more or less parallel sides in dorsal aspect (Figs 5K, 6C, G); body usually covered with adherent mineral or organic debris ....Crotoniidae[2]
– Aggenital setae absent; notogaster usually more or less convex dorsally and convex laterally (Figs 5H, 7A, B); body usually without perceptible adherent debris 7
7. Preanal plate present; 3–5 pairs of setae on epimere II; opisthonotal glands not prominent (Fig. 5H); prodorsum covered with distinct areoles or reticulation; brownish or orange mites ....Nothridae
– Preanal plate absent; 1 pair of setae on epimere II; opisthonotal glands clearly evident (Figs 6J, 7B); body integument hazel coriaceous.................. ....Trhypochthoniidae
8. Notogaster without octotaxic system of porose areas or saccules; pteromorphs absent; epimeres II, III and IV usually rectangular in shape (Fig. 7H); apodemes II and sejugal apodeme transverse (Fig. 9D); discidium and circumpedal carina absent 9
– Notogaster usually with octotaxic system of porose areas or saccules (Figs 15A, 18G); pteromorphs present or absent; epimeres II, III and IV never rectangular in shape; apodemes II and sejugal apodeme angled towards genital plate (Fig. 14C); discidium and circumpedal carina present....................21
9. Prodorsum without true lamellae, with (Figs 11F, 13F) or without costulae (Figs 7I, 8G)...................10
– Prodorsum with true lamellae (Figs 9K, 10F, 15H).... 17
10. Notogaster clearly sculptured, with 16 pairs of conspicuous setae (Figs 7F, G); large (about 750–1000 μm) dark species with strongly convex notogaster... .... Hermanniidae
– Notogaster without conspicuous sculpturing, commonly with fewer than 16 pairs of setae........... 11
11. Notogastral setae (other than *p* series) arranged in 2 conspicuous longitudinal rows (Figs 8A, B); well developed parastigmatic enantiophyses present (Fig. 8B, *Sa*, *Sp*).......................... Damaeidae
– Notogastral setae not arranged in 2 longitudinal rows; parastigmatic enantiophyses absent........12
12. Seta *d* (dorsal seta) inserted on proximal fifth of femora I–III, proximal to other femoral setae (Fig. 13E); 13, or 15 to 17 pairs of notogastral setae (secondary *h* setae present; Fig. 13F); trichobothrium reduced or absent (frequently broken); lenticulus present, circular, strongly convex (Fig. 13F) .... Hydrozetidae
– Seta *d* not most proximal seta on femora I–III; 10 to 15 pairs of notogastral setae (only 3 usual pairs of *h* setae); trichobothrium normal, reduced or absent; lenticulus, if present, not strongly convex.........13
13. Integument dark, but weakly sclerotized and easily deformed (Figs 13I–K); notogaster strongly overhanging ventral plate; trichobothrium very small or absent.......................... Ameronothridae
– Integument well sclerotized, not easily deformed; notogaster not strongly overhanging ventral plate; trichobothrium normally developed..............14
14. Anal plates with 3 to 9 pairs of setae (Fig. 9I); porose areas distoventrally on tibiae I to IV and proximoventrally on tarsi I to IV (Fig. 9J).........Eremaeidae
– Anal plates with 2 pairs of setae; tibiae and tarsi without porose areas...............................15
15. Gnathosoma modified: chelicerae attenuate-edentate (Figs 1G, H), rutella filmy (Fig. 1C); distal eupathidia on palptarsus fused basally, appearing like forked seta (Fig. 1D); prodorsum with pair of rounded fields (Figs 12I, N); very small mites (usually less than 300 μm) .... Suctobelbidae
– Gnathosoma normal: chelicerae with dentate chelae, rutella normally developed; prodorsum without pair of rounded fields.............................16
16. Costulae, if present, shorter than half length of prodorsum (Figs 11A, F, 12B); cristae, if present, not longer than 1/3 of notogaster (Fig. 11J)......Oppiidae
– Costulae longer than half of prodorsum (Fig. 12H); cristae 1/3 of notogaster or longer ... Quadroppiidae
17. Humeral region with 2 pairs of closely adjacent setae *c* (*c1* and *c2*) (Fig. 9F)..................... Liacaridae

2 We follow the arguments of Colloff & Cameron (2009) in the use of Crotoniidae for the family Camisiidae. Recent molecular and morphological studies show that Crotoniidae (*sensu stricto*) evolved within Camisiidae, which is then paraphyletic. Crotoniidae is the older family-group name and must be used for the merged set of genera.

– Humeral region with 0 or 1 pair of setae *c* (Figs 9K, 10C)..18
18. Integument smooth and shiny; lamellae converging or parallel, if parallel, not positioned close to lateral margin of prodorsum (Figs 9K, 10A)...............19
– Integument distinctly sculptured with coarse ridges, tubercles, or areoles (Figs 10C, F); lamellae usually broad, almost parallel, close to lateral margin of prodorsum 20
19. Genital plates large (Fig. 9L); distance between genital and anal plates commonly less than length of anal plates; usually smaller species (less than 350 µm); notogaster glabrous or with 10 pairs of notogastral setae. Length: 250–600 µm...Astegistidae
– Genital plates medium in size; distance between genital and anal plates commonly greater than length of anal plates; usually larger species; 8 pairs of notogastral setae present or vestigial (Fig. 10A). Length: 400–1100 µm....................Peloppiidae
20. Genal tooth present; circumpedal carina partially developed; tarsi and tibiae articulate normally; anal plate triangular in shape (Fig. 10F); interlamellar setae positioned in interbothridial region (Fig. 10G). Length: 225–300 µm..Tectocepheidae
– Genal tooth absent; circumpedal carina absent; tarsi functionally fused to tibiae; anal plate rectangular in shape (Fig. 10C); interlamellar setae usually positioned well anterior of bothridia or on lamellae. Length: 350–600 µm.................Carabodidae
21. Chelicerae peloptoid (Fig. 1I); well-developed, blocky, birefringent cerotegument covering most of body................................Phenopelopidae
– Chelicerae normal, chelate-dentate (Fig. 1F); cerotegument mostly inconspicuous or absent........ 22
22. Prodorsal lamellae broad and long (covering most of prodorsum) and partly fused medially (Figs 18F, G); pteromorphs immovable, with knifelike anterior projections ....Achipteriidae
– Prodorsal lamellae usually narrow (Figs 14D, 16E), rarely touching medially (but may be joined by translamella, Figs 14A, 15C); pteromorphs, if present, without knifelike anterior projections 23
23. Pedotectum I at most weakly developed as small lamina (Fig. 2C) or absent; prodorsum without genal notch; axillary saccule of the subcapitulum absent...24
– Pedotectum I well-developed as large lamina (Figs 15F, J); prodorsum with genal notch; axillary saccule present.................................... 27
24. Posterior notogastral tectum present (Fig. 2D); octotaxic system reduced to 1 or 2 pairs of saccules (Fig. 14A, *Sa* and *S3*); lenticulus often prominent ....Scutoverticidae
– Posterior notogastral tectum absent; octotaxic system present as 4 pairs of porose areas or saccules (Fig. 14D, F); lenticulus absent..................... 25
25. Pteromorphs well developed, with hinge (Fig. 14E); insertions of adductor muscles for pteromorphs evident on notogaster.....................Haplozetidae
– Pteromorphs, if present, without hinge (Fig. 14B); pteromorph adductor muscles absent 26
26. Prodorsum with prolamellae often present, sublamellae present (Fig. 2C); sternal furrow often present, extending from genital plates anteriorly on coxisternum (Fig. 14C); pteromorphs or humeral projections usually present (Fig. 14B); distal eupathidia on palptarsus in same plane; 4 pairs of genital setae..........................Scheloribatidae
– Prodorsum without prolamellae and sublamellae; sternal furrow absent; pteromorphs or humeral projections absent (Fig. 14D); distal eupathidia on palptarsus not in same plane; 4 or 5 pairs of genital setae...................................Oribatulidae
27. Notogaster without posterior tectum; 10 to 15 pairs of notogastral setae (Figs 15A, H).........Ceratozetidae
– Notogaster with posterior tectum (Fig. 2E); 10 pairs of notogastral setae , can be vestigial (Fig. 14G)... 28
28. Lamellar setae inserted on prodorsal surface (Fig. 14J); lamellar cusps absent (Figs 14G, I) ....Chamobatidae
– Lamellar setae inserted on lamellae or on lamellar cusps (Figs 16L, 17F); lamellar cusps usually present, occasionally absent .... Punctoribatidae

HYPOCHTHONIIDAE (FIG. 20A)

In Greenland only 1 species recorded, ***Hypochthonius rufulus*** C.L. Koch, 1836. Body of distinctive pentagonal shape in dorsal aspect, flat, orange or reddish colour; notogaster with single transverse scissure (Fig. 3A). 650–700 µm. Semicosmopolitan: Holarctic (including Ellesmere Island, Iceland, southern Taymyr), southeastern China, Seychelles.

BRACHYCHTHONIIDAE

Key to genera

1. Seta *d2* on anterior notogastral shield in extreme lateral position forming longitudinal row with setae *cp* and *e2* (Figs 3C, D); 2–4 pairs of suprapleural plates present (Fig. 5B).............................2

– Seta *d2* situated mediad setae *cp* and *e2* (Figs 3I, 4G); suprapleural plates absent (Fig. 4B) or only 1 pair present (Fig. 4H) 4
2. Four pairs of suprapleural plates present; insertions of lateral setae on anterior notogastral sclerite separated by sculptural line (Figs 3B, C); larger than 250 μm, wide and flat, vivid yellow mites (Fig. 20B) .... *Eobrachychthonius*
– Two or three pairs of suprapleural plates present; insertions of lateral setae on anterior notogastral sclerite not separated by sculptural line (Figs 3D, 4K); smaller mites (less than 250 μm); body with heavy ornamentation, pale yellow or yellow (Fig. 20C).. 3
3. All 3 adanal setae thin, simple, setiform; 2 pairs of suprapleural plates................. *Brachychthonius*
– Two adanal setae (*ad2* and *ad3*) dilated as blade (Figs 4J, 5B); 3 pairs of suprapleural plates .... *Sellnickochthonius*
4. All 3 adanal setae thin, simple, setiform (Fig. 5F).... .... *Verachthonius*
– Adanal seta *ad2* hypetrophic, dilated as blade (Figs 3H, 4H) 5
5. Notogastral shields with heavy ornamentation (Fig. 5C); genital plate with 6 setae (Fig. 5D)........ .... *Synchthonius*
– Notogastral shields without conspicuous ornamentation (Figs 3G, 4C); genital plate with 7 setae (Fig. 4F)... 6
6. Adanal plates fused posteriorly so that anal plates surrounded by U-shaped plate .... *Liochthonius*
– Adanal plates not fused posteriorly, distinctly separated........................... *Neoliochthonius*

Brachychthonius **Berlese, 1910**

In Greenland only 1 species recorded, ***Brachychthonius berlesei*** Willmann, 1928 (Fig. 3D). 200–210 μm. Holarctic (including Ellesmere Island, Iceland).

Eobrachychthonius **Jacot, 1936. Fig. 20B**

Key to species

1. Aggenital plate whole, not subdivided, with seta (Figs 3B, E). 270–340 μm. Holarctic (including Svalbard, Taymyr) .... *latior* (Berlese, 1910)
– Aggenital plate subdivided into 2 unequal parts (Figs 3C, F), aggenital seta situated on smaller posterior part. 250–340 μm. Semicosmopolitan: Holarctic (including Iceland, Svalbard), southeastern China, Neotropical, Subantarctic............................ .... *oudemansi* Hammen, 1952

Liochthonius **Hammen, 1959. Fig. 20F**

Key to species

1. Head of trichobothrium fusiform, apically rounded or acuminate, with sparse distinct barbs (Figs 3H, J) 2
– Head of trichobothrium apically bifurcate, with dense minute barbs (Figs 4B, D, F) 3
2. Notogastral setae comparatively long, tips of most dorsocental setae reach insertion of next most posterior setae (Figs 3G, H); interlamellar setae longer than distance between their bases. 160–220 μm. Holarctic (including Svalbard, Taymyr) .... *brevis* (Michael, 1888)
– Notogastral setae comparatively short, their tips usually not reaching insertion of next most posterior setae (Figs 3I, J); interlamellar setae shorter than distance between their bases. 150–180 μm. Semicosmopolitan: Palaearctic (including Yamal Peninsula, Taymyr), Philippines, India, Australia .... *simplex* (Forsslund, 1942)
3. Prodorsum with transverse ridge anteriad lamellar setae (Fig. 4B); similar structure developed posteriad setae *fi* on posterior notogastral sclerite; longitudinal rows of interbothridial sigillae parallel (Fig. 4A). 200–240 μm. Holarctic (including Ellesmere Island, Devon Island, Iceland, Svalbard, Taymyr), southeastern China .... *sellnicki* (Thor, 1930)
– Prodorsum and posterior notogastral sclerite without transverse ridge................................. 4
4. Distance between setae *c1* on anterior notogastral sclerite shorter or at least equal to distance between setae *e1* on second notogastral sclerite (Fig. 4C); longitudinal rows of interbothridial sigillae convergent; head of trichobothrium long (length:width ratio 3:1, Fig. 4D). 200–230 μm. Holarctic (including Iceland, High Arctic)....... .... *muscorum* Forsslund, 1964
– Distance between setae *c1* longer than the distance between setae *e1* (Fig. 4E); longitudinal rows of interbothridial sigillae parallel; head of trichobothrium shorter (length:width ratio 2.5:1, Fig. 4F). 180–200 μm. Holarctic (including Ellesmere Island, Iceland, Svalbard)................. .... *lapponicus* (Trägårdh, 1910)

Neoliochthonius **Lee, 1982**

In Greenland only 1 species recorded, ***Neoliochthonius piluliferus*** (Forsslund, 1942). Very small, nearly white mite with head of trichobothrium characteristically spherical (Figs 4G, H). 120–170 μm. Holarctic (including Svalbard).

Sellnickochthonius **Krivolutsky, 1964. Fig. 20C**

Key to species

1. Notogastral setae dilated as blade or phylliform (Figs 4I, J). 150–170 µm. Holarctic (including Taymyr) .... *zelawaiensis* (Sellnick, 1928)
– Notogastral setae thin, simple, smooth (Figs 4K, 5A) 2
2. Ornamental fields in the middle part of anterior notogastral shield not fused (2 groups with 3 pairs of fields each) (Figs 5A, B). 160 µm. Semicosmopolitan (Holarctic, New Zealand, Subantarctic) .... *suecicus* (Forsslund, 1942)
– Ornamental fields in the middle part of anterior notogastral shield fused, forming 2 large (anterior and posterior) fields (Fig. 4K). 170–200 µm. Holarctic (including Ellesmere Island, Iceland), Neotropical .. .. *immaculatus* (Forsslund, 1942)

Synchthonius **Hammen, 1952. Fig. 20E**

In Greenland only 1 species recorded, ***Synchthonius crenulatus*** (Jacot, 1938) (Figs 5C, D). 200–210 µm. Holarctic (including Iceland).

Verachthonius **Moritz, 1976. Fig. 20D**

In Greenland only 1 species recorded, ***Verachthonius montanus*** (Hammer, 1952) (Figs 5E, F). 220–240 µm. Nearctic (Rocky Mountains), Neotropical (Andes).

PHTHIRACARIDAE

Key to genera

1. Inner margin of ano-adanal plate with compact group of 4 rather long setae (Fig. 5G); body surface foveate; trichobothrium narrow, sickle-shaped, smooth or distally ciliate; notogastral setae needle-shaped or bacilliform, smooth or barbed .... *Atropacarus*
– Inner margin of ano-adanal plate with 2 setae; body surface smooth, punctate; trichobothrium short, either fusiform or rounded distally, usually smooth; notogastral setae thin, long, acuminate, smooth..... .... *Phthiracarus*

Atropacarus **Ewing, 1917. Fig. 20G**

In Greenland only 1 species recorded, ***Atropacarus (Atropacarus) striculus*** (C.L. Koch, 1836). Body pale brown; notogastral setae rigid, bacilliform, distally ciliate (Fig. 5G); body size varies greatly. Length of prodorsum 170–290 µm, length of notogaster 320–590 µm. SW, S Greenland. Holarctic (including Iceland, Svalbard).

Phthiracarus **Perty, 1839. Fig. 20H**

Greenland mites belonging to this genus (not less than 2 species) need re-examination. Hammer (1944, 1952) recorded 2 species, *Phthiracarus piger* (Scopoli, 1763) and *Phthiracarus borealis* (Trägårdh, 1910), probably following Willmann (1931). Both names have been mentioned many times and the records argue for a holarctic distribution in both cases. Recently the former species was considered "species inquirenda" until it is re-described from neotypic material (Niedbała 1992). The same author (Niedbała 1992, 2002) considered *Hoploderma borealis* Trägårdh, 1910 to be unsatisfactorily described for placement in keys, although belonging to the genus *Phthiracarus*.

NOTHRIDAE (FIG. 20I)

In Greenland only 1 species recorded, ***Nothrus borussicus*** Sellnick, 1928. Large orange or rufous mite; body dorsally roughly foveate; notogastral setae robust, bacilliform, smooth, caudal setae longest (Fig. 5H). 950–990 µm. Holarctic (including Iceland).

CROTONIIDAE

Key to genera

1. Anogenital region posteriorly bordered by transverse line formed by posterior margin of notogaster (Fig. 5J); aggenital setae on unsclerotized cuticle; 3 pairs of anal setae; body usually covered with adherent soil particles............................. *Camisia*
– Anogenital region posteriorly bordered by angular line formed by lateral margins of notogaster (Fig. 6H); aggenital setae on medial margin of aggenital plate; 2 pairs of anal setae; body usually more "clean"... 2
2. Notogaster without longitudinal crests (Figs 6C–E) . ... *Heminothrus*
– Notogaster with 2 pairs of longitudinal crests (Figs 6F, G, I) 3
3. Trichobothrium longer than other prodorsal setae (Fig. 6F); 2 pairs of setae inserted between medial crests of notogaster .... *Neonothrus*
– Trichobothrium not longer than other prodorsal setae (Figs 6G, I); no setae inserted between medial crests of notogaster .... *Platynothrus*

Camisia **Heyden, 1826. Fig. 20J**

Key to species

1. Legs monodactylous; notogastral setae distinctly phylliform (Fig. 5I); 14 pairs of genital setae. 860–1140 µm. Holarctic (including Iceland, Svalbard), south-eastern China .... *lapponica* (Trägårdh, 1910)

– Legs tridactylous; notogastral setae simple, ciliate or spiniform; 9 or 18–20 pairs of genital setae.......2
2. Posterior margin of notogaster rounded (Fig. 5K); lamellar setae situated on small tubercles; tips of these tubercles and often tips of lamellar setae not reaching rostrum; 18–20 pairs of genital setae; body integument foveate. 710–830 μm. Holarctic (Alaska, Yukon, Northern Europe including Iceland, Svalbard), Chile ....*foveolata* Hammer, 1955
– Posterior margin of notogaster angular, with broad ledge or paired projections bearing caudal setae (Figs 6A, B); lamellar setae inserted on distinct apophyses, tips of these apophyses extending anteriad of rostrum; 9 pairs of genital setae 3
3. Caudal projections of notogaster large, funnel-shaped (Fig. 6A); notogaster laterally smooth; dorsocentral ridges weak, without connection. 930–1100 μm. Holarctic (including Iceland, Svalbard, Taymyr) ....*biverrucata* (C.L. Koch, 1839)
– Caudal lobes of notogaster comparatively small (Fig. 6B); notogaster laterally strongly indented with short transverse ridges; well developed dorsocentral longitudinal ridges connected by transverse ridge. 800–1000 μm. Holarctic (including Queen Elizabeth Islands, Iceland, Svalbard), India, Central America.. .... *horrida* (Hermann, 1804)

Heminothrus **Berlese, 1913. Figs 20K, 24I**
Key to species
1. Two to three pairs of caudal notogastral setae on large apophyses (Fig. 6C); notogastral setae heterogeneous, dorsocentral setae rather short, marginal setae 2–3 times longer; smaller mites. 660–720 μm. Holarctic, Taiwan...................... .... *longisetosus* Willmann, 1925
– Caudal setae inserted on body margin, at most on small tubercles (Fig. 6D); notogastral setae less different in length......................................2
2. Trichobothrium, interlamellar and lamellar setae nearly of same length (Fig. 6D); trichobothrium rather long, bacilliform; caudal setae very long, some of them as long as half of notogaster length; central part of notogaster punctate. 950–1000 μm. Holarctic (including Iceland), southeastern China, India...... .... *thori* (Berlese, 1904) = *Capillonothrus thori* in Norton & Ermilov (2014).
– Interlamellar seta twice longer than trichobothrium and lamellar seta (Fig. 6E); trichobothrium short, clavate; caudal setae shorter, not longer than 1/3 of notogaster length; central part of notogaster clearly reticulate. 1000–1080 μm. Palaearctic (including Iceland, Svalbard), California........ ....*capillatus* (Berlese, 1914) = *Platynothus capillatus septentrionalis* Sellnick, 1944

Neonothrus **Forsslund, 1955. Fig. 20L**
In the genus only 1 species, ***Neonothrus humicolus*** Forsslund, 1955. Middle-sized yellowish-brown mite; medial notogastral crests stronger than lateral ones; trichobothrium very long, setiform, gently barbed (Fig. 6F). 600–650 μm. Holarctic.

Platynothrus **Berlese, 1913. Fig. 21A**
Key to species
1. Notogastral setae short, tips of most dorsocentral setae not reaching insertion of next most posterior seta (Fig. 6G); 11 pairs of genital setae. 720–860 μm. Northern Holarctic (including Iceland, Svalbard) ... . *punctatus* (L. Koch, 1879)
– Notogastral setae longer, tips of most dorsocentral setae extending posteriad insertion of next most posterior seta (Fig. 6I); usually 13–15 pairs of genital setae. 925–940 μm. Semicosmopolitan (including Iceland, Svalbard)..........*peltifer* (C.L. Koch, 1839)

TRHYPOCHTHONIIDAE
In Greenland the family is represented by 4 species belonging to 3 genera and characterized by very prominent opisthonotal glands and coriaceous hazel integument.

Key to genera
1. Legs monodactylous; anal plates with 2 setae; rostral setae closely adjacent; bothridium reduced, on disc-like elevation; trichobothrium setiform (Fig. 6J)..... ....*Mucronothrus*
– Legs tridactylous; anal plate with a single seta; rostral setae well separated; bothridium well developed or reduced; trichobothrium clavate (Fig. 7A), fusiform (Fig. 19F) or absent (Fig. 7B) 2
2. Adanal plate with 3 setae; notogastral setae barbed (Fig. 7A) or smooth; bothridium well developed, trichobothrium barbed, clavate or fusiform; 6–11 pairs of genital setae...............*Trhypochthonius*
– Adanal plate with 2 setae; notogastral setae smooth, setiform (Fig. 7B); bothridium and trichobothrium well developed, small or fully reduced (Fig. 19F); trichobothrium, if present, smooth, fusiform; 10–18 pairs of genital setae............ *Trhypochthoniellus*

Mucronothrus **Trägårdh, 1931. Fig. 24C**
Mites of this genus populate mainly freshwater habitats such as spring, meltwaters, seepages etc. In Greenland the genus is represented by ***Mucronothrus nasalis*** (Willmann, 1929) (Fig. 6J) with characteristic rostral mucro bearing rostral setae and reticulated notogaster. 580–800 µm. S, SW Greenland. Cosmopolitan (including Iceland).

Trhypochthonius **Berlese, 1904. Fig. 21B**
In Greenland one species recorded, ***Trhypochthonius tectorum*** (Berlese, 1896), with notogastral setae of different length, barbed, thickened distally (Fig. 7A). 580–780 µm. Semicosmopolitan: Holarctic (including Iceland), Oriental, Neotropical.

Trhypochthoniellus **Willmann, 1928. Fig. 24B**
Mites of this genus are very hydrophilous, often dwelling in wet moss carpets.

Key to species

1. Notogastral setae differ in length not more than 2 times (Fig. 19E); interlamellar seta shorter, 100–130 µm. 440–580 µm. Cosmopolitan: Holarctic (including Ellesmere Island, Iceland), St. Helena Island, Australia, Neotropical..... *longisetus* (Berlese, 1904)
= *Trhypochthoniellus setosus* Willmann, 1928
NOTE. Very variable species. Trichobotrium can be well developed (forma *longiseta*), small or fully reduced (forma *setosa*) (Fig. 19F). These differences occur in one population and even in one specimen asymmetrically (Weigmann 2006). In northern Nearctic, only *Trhypochthoniellus longisetus* ssp. *canadensis* Hammer, 1952 is found (Behan-Pelletier 1997, 1999; Walter *et al.* 2013), but the single Greenland record (Hammer 1960) was not identified to subspecies.

– Notogastral setae differ in length more than 3 times (Fig. 7B); interlamellar seta longer, 145–195 µm. 530–570 µm. S Greenland (Kap Farvel area)............................. sp. near *brevisetus* Kuriki, 2005

MALACONOTHRIDAE

Key to genera

1. Tarsi tridactylous, elongate; hysterosoma dilated posteriorly (Figs 7D, E)............*Trimalaconothrus*

– Tarsi monodactylous, short; hysterosoma cylindrical (Fig. 7C) .... *Malaconothrus*

Malaconothrus **Berlese, 1904. Fig. 21D**
In Greenland the genus is represented by ***Malaconothrus mollisetosus*** Hammer, 1952 (Fig. 7C), the possible junior synonym of *M. gracilis* Hammen, 1952, which in turn is a possible junior synonym of *Malaconothrus monodactylus* (Michael, 1888) (Subías, 2004). Middle-sized, hazel, leathery species with very short legs possessing large claws; outwardly it resembles some acarid mites. 410–450 µm. Holarctic (including Iceland), Neotropical.

Trimalaconothrus **Berlese, 1916. Fig. 21C**

Key to species

1. Distance between medial notogastral setae *c1* and *d1* the same as the distance between setae *d1* and *e1* (Fig. 7D); prodorsum and notogaster without longitudinal ridges; notogaster posteriorly without angular cavities. 420–570 µm. SW, S, SE Greenland. Holarctic (Europe, Alaska), Oriental ....*foveolatus* Willmann, 1931

– Distance between medial notogastral setae *c1* and *d1* much shorter than distance between setae *d1* and *e1* (Fig. 7E); prodorsum and notogaster medially with longitudinal ridges; notogaster posteriorly with angular cavities. 290–340 µm. Cosmopolitan ....*maior* (Berlese, 1910)
= *Malaconothrus novus* Sellnick, 1921

HERMANNIIDAE (FIG. 21E)
In Greenland represented by the genus ***Hermannia*** Nicolet, 1855.

Key to species

1. Notogaster with distinct reticulate pattern (Fig. 7F); notogastral setae bacilliform. 770–880 µm. Northern Holarctic (including Ellesmere Island, Iceland, Svalbard) .... *reticulata* Thorell, 1871

– Notogaster without reticulate ornamentation; notogastral setae dilated.................................2

2. Notogaster ornamented by separate tubercles posteriorly arranged in longitudinal rows; notogastral setae long, auriculate (Figs 7G, H). 900–1050 µm. Europe (including Iceland) ....*nodosa* Michael, 1888
= *Hermannia scabra* (L. Koch, 1879) sensu Michael 1898; Sellnick 1928; Willmann 1931, 1952 (Woas 1978).

– Notogaster ornamented by light round foveae (Fig. 7I); notogastral setae wide, nearly leaf-shaped. 880–1220 µm. Holarctic (including Queen Elizabeth Islands)......................*subglabra* Berlese, 1910?
= *Hermannia pulchella* Willmann, 1952 (Forsslund 1957), but see Woas (1981).

DAMAEIDAE

Represented in Greenland by at least 9 species. However, this family is poorly known, old descriptions do not meet modern requirements, especially with regard to leg chaetotaxy. Status of some species as well as some records need clarification. Thus keys are preliminary, adopted to Greenland records, but include all available information.

Key to genera

1. Prodorsum laterally between insertions of legs I and II with large angular apophysis (propodolateral apophysis) (Fig. 8A); posterior part of prodorsum behind bothridia with 2 pairs of tubercles .... *Damaeus*
– Prodorsum laterally without apophysis, rounded (Fig. 8B); posterior part of prodorsum behind bothridia with 0–1 pair of tubercles (Figs 8C, H)........... .... *Epidamaeus*

Damaeus **C.L. Koch, 1835. Fig. 21F**

In Greenland represented by 1 remarkable species, ***Damaeus*** sp. near ***gracilipes*** (Kulczynski, 1902) (Fig. 8A). 820–1080 μm. S Greenland, ?Canada.

NOTE. *Belba gracilipes* Kulczynski, 1902 is mentioned only in old publications devoted to Greenland oribatids (Jørgensen 1934, as *Belba clavipes* (Hermann, 1804); Hammer 1944). A certain *Damaeus* sp. near *gracilipes* Kulczynski, 1902 was later recorded in Québec, Canada (Smith 1978, cited in Marshall *et al.* 1987). The European species *Damaeus gracilipes* was recently redescribed based on Spanish material (Pérez-Íñigo 1997). Presently it is considered the junior synonym of *Damaeus auritus* C.L. Koch, 1835 (Subías 2004). During preparation of this volume mites resembling *Damaeus gracilipes* were found in pitfall material from Kap Farvel (southernmost Greenland), but they distinctly differ from the European form. Greenland records of "*Damaeus gracilipes*" need verification. Both these mites have needle-shaped, straight trichobothrium; posterior part of prodorsum behind bothridia with 2 pairs of tubercles; legs very long and thin, segments without remarkable distal thickening. But the Greenland mites differ by presence of a pair of tubercules beneath anterior margin of notogaster (Fig. 8A), shorter trichobothrium, larger spinae adnatae, different leg chaetotaxy etc.

Epidamaeus **Bulanova-Zachvatkina, 1957. Figs 21G, H**

Key to species

1. Trichobothrium distally with expanded elongate barbed head, 130 μm in length (Fig. 8B); interlamellar setae short; prodorsum posteriorly with 1 pair of tubercles; leg IV almost twice length of body. 720–800 μm. SW Greenland. British Isles, Kuril Islands...................... *tenuipes* (Michael, 1885)
NOTE. The Greenland record (Hammer 1944) needs confirmation.
– Trichobothrium distally not swollen, rarely slightly broadened (Fig. 8E)................................2
2. Trichobothrium and interlamellar seta thick and straight, tapered but not attenuate, heavily barbed (Figs 8C, E); notogastral setae strong; prodorsum posteriorly with 1 pair of tubercles 3
– Trichobothrium, interlamellar and notogastral setae not as above; prodorsum posteriorly with 0–1 pair of tubercles..4
3. Notogastral setae knifelike or sickle-shaped, smooth (Fig. 8C); spinae adnatae straight triangular. 730–800 μm. Northern Holarctic (NE Asia, Primor'ye, Alaska, Canada)........*fortispinosus* Hammer, 1967
? = *Oribata trägårdhi* Graversen, 1931, subfossil, Greenland (Fig. 8F)
– Notogastal setae not dilated, heavily barbed (Fig. 8E); spinae adnatae curved hornlike. 700–790 μm. Holarctic (Northern Europe, West Beringia, northern Nearctic) .... *longitarsalis* (Hammer, 1952)
4. Prodorsum posteriorly with 1 pair of tubercles; trichobothrium long, flagellate (Figs 8D, G); smaller mites (415–460 μm)................................5
– Prodorsum posteriorly without tubercles; trichobothrium setiform (Fig. 8H) or with undulating tip (Fig. 8I); larger mites (570–825 μm) 6
5. Interlamellar setae long, flagellate (Fig. 8D), their length (97 μm) exceeds half length of trichobothrium (160 μm); spinae adnatae straight, directed anteriorly. 415–435 μm. Northern Nearctic (Alaska, Yukon)... ....*floccosus* Behan-Pelletier & Norton, 1985
– Interlamellar setae shorter (75 μm), setiform (Fig. 8G), not longer than half length of trichobothrium (170 μm); spinae adnatae curved, hornlike. 430–460 μm. W Greenland.....*trigonalis* (Strenzke, 1955)
6. Larger mites (body length more than 730 μm); interlamellar setae (Fig. 8H) 2.5 times shorter than trichobothrium (70 and 180 μm respectively). 740–825 μm. W Greenland........................... .... *inornatus* (Strenzke, 1955)
– Smaller mites, body length less than 730 μm; interlamellar setae 3.5 and more times shorter than trichobothrium 7
7. Notogastral setae brown, hard, erect (Fig. 8I); interlamellar setae 55 μm long, trichobothrium 194 μm.

570–720 µm. SW Greenland. Northern Nearctic (Alaska, Canada) .... *koyukon* Behan-Pelletier & Norton, 1985
= *Belba tatrica* (Kulczynski, 1902) sensu Hammer 1952
– Notogastral setae light, soft, slightly curved (Fig. 9A); interlamellar setae very short. 680–720. NW Greenland, Taymyr .... *groenlandicus* (Hammer, 1953)

EREMAEIDAE

Key to genera

1. Notogaster with 11 pairs of setae (Fig. 9B); postanal plate absent (Fig. 9D) .... *Eremaeus*
– Notogaster with 10 pairs of setae; postanal plate present (Fig. 9I) .... *Eueremaeus*

***Eremaeus* C.L. Koch, 1835**

In Greenland only 1 species known, ***Eremaeus translamellatus*** Hammer, 1952, which may be easily recognized because of well developed translamellar ridges (Figs 9B–D). 520–750 µm. Holarctic.

***Eueremaeus* Mihelčič, 1963. Fig. 21K**

The Greenland material needs re-examination. At least 2 species found.

Key to species

1. Lamellar costulae short, straight, well separated (Fig. 9G); interlamellar setae shorter than distance between their bases; prodorsum posteriorly without tubercles. 530 µm. Northern Canada, ?Europe (including Iceland)................. *Eueremaeus* sp. 1
= *Eremaeus oblongus* C.L. Koch, 1836 sensu Hammer 1952
NOTE. Recently *Eremaeus oblongus* C.L. Koch, 1836 was synonymized with redescribed *Eremaeus hepaticus* C.L. Koch, 1836 (Beck & Woas 1991), but figures represented in Hammer (1952) do not agree with this description. Hammer mentioned differences between Canadian specimens and mites from Greenland, Iceland and continental Europe concerning body size and/or leg ornamentation. Weigmann (2006) considered *Eueremaeus oblongus* (C.L. Koch, 1835) and *Eremaeus hepaticus* C.L. Koch, 1835 separate species.
– Lamellar costulae long, arcuate, diverging posteriorly (Fig. 9H); interlamellar setae longer than distance between their bases; prodorsum posteriorly with 2–3 rows of tubercles of different size. 590–600 µm. SW Greenland ....
Eueremaeus sp. 2 near *michaeli* Behan-Pelletier, 1993
NOTE. Distribution of *Eueremaeus michaeli*: Canada (British Columbia).

LIACARIDAE (FIG. 24D)

In Greenland only 1 species known, ***Procorynetes globifer*** (Kramer, 1897) [= *Liacarus* (*Procorynetes*) *globifer* in Subías 2004], with very remarkable pattern of prodorsal sclerotization formed by distally fused lamellae with apical trident (Fig. 9E); trichobothrium distally globular, with short barbs (Fig. 9F); 6 pairs of genital setae. 450–660 µm. Greenland, Northern Europe (including Faroe Islands).

ASTEGISTIDAE (FIG. 21J)

In Greenland only 1 species recorded, ***Cultroribula dentata*** Willmann, 1950, with prominent structure of rostrum and lamellae (Figs 9K, L). 320–350 µm. SW Greenland. Holarctic.

PELOPPIIDAE (FIGS 21I, 24J)

In Greenland 1 genus recorded, ***Ceratoppia*** Berlese, 1908.

Key to species

1. Apical rostral tooth much larger than other rostral teeth (Figs 10A, D). 600–1100 µm. Holarctic (including Iceland, Svalbard), northern Oriental .... *bipilis* (Hermann, 1804)
– All rostral teeth of same size (Figs 10B, E). 700–1000 µm. Holarctic (including Svalbard).................. ....*sphaerica* (L. Koch, 1879)
= *Ceratoppia hoeli* Thor, 1930 (Hammer 1952a)

CARABODIDAE (FIG. 21L)

In Greenland only 1 species recorded, ***Carabodes labyrinthicus*** (Michael, 1879). The heavy body ornamentation and lamellar structure allow easy recognition (Fig. 10C). 500–520 µm. SW, S, SE Greenland. Holarctic (including Iceland, Svalbard).

TECTOCEPHEIDAE (FIGS 24A, K)

In Greenland only 1 genus, ***Tectocepheus*** Berlese, 1896, represented by 2 cosmopolitan species. Their identity is the subject of discussion (Laumann *et al.* 2007).

Key to species

1. Lamellar cusps attenuate distally, lamellar setae inserted on their tips; translamella narrow (Fig. 10F). 290–350 µm. Cosmopolitan (including Ellesmere Island, Devon Island, Iceland, Svalbard) .... *velatus* (Michael, 1880)
– Lamellar cusps wide and rounded distally, lamellar setae inserted anterolaterally; translamella wide

(Fig. 10G). 350–360 µm. Cosmopolitan (including Svalbard) .... *sarekensis* Trägårdh, 1910

OPPIIDAE

Given keys are highly preliminary. The generic system of this family is not stable, *e.g.*, Weigmann (2006) considers *Moritzoppia* and *Rhinoppia* subgenera of *Oppiella*, whereas these genera retain their status in Subías (2004).

Key to genera

1. Prodorsum mainly smooth, without distinct costulae, but pair of small, backwardly directed interbothridial crests present (Figs 11A, C); 10 pairs of notogastral setae; trichobothrium fusiform or lanceolate, unilaterally ciliate with long ciliae (Figs 11B, C); 6 pairs of genital setae .... *Rhinoppia*
– Distinct interbothridial costulae present; 4–5 pairs of genital setae 2
2. Dorsosejugal scissure straight (Figs 11J, 12C, E) or slightly arched (Fig. 11H), never extending into posterior part of prodorsum; genital plates with 5 pairs of setae (Fig. 11I) .... *Oppiella*
– Dorsosejugal scissure strongly convex, parabolic or semicircular, extending into posterior part of prodorsum (Figs 10H, 11D, 12B); genital plates with 4–5 pairs of setae 3
3. Rostral setae close to each other, inserted on a small naso (Figs 10I, J); trichobothrium long, distally dilated, smooth or finely barbed (Fig. 10H); 5 pairs of genital setae .... *Dissorhina*
– Rostrum smooth, rostral setae not close to each other (Figs 11D, F); trichobothrium shorter, distally globular or fusiform, with short barbs; usually 4 pairs of genital setae (Figs 11G, 12A) .... *Moritzoppia*

Dissorhina **Hull, 1916. Fig. 22B**

In Greenland represented by 1 characteristic species, ***Dissorhina ornata ornata*** (Oudemans, 1900) (Figs 10H–J). 245–350 µm. Holarctic (including Devon Island, Iceland, Svalbard), Seychelles.

Rhinoppia **Balogh, 1983. Fig. 22A**

Key to species

1. Interlamellar seta short (Fig. 11A); anterior margin of notogaster medially truncate, ciliae of trichobothrium more or less uniform (Fig. 11B). 285–360 µm. Semicosmopolitan (Palaearctic, New Zealand, Hawaii Islands)............... *obsoleta* (Paoli, 1908)
– Interlamellar seta rather long (Fig. 11C); anterior margin of notogaster medially slightly convex with pair of small projections; trichobothrium with ciliae of different length. 320 µm. Holarctic (including Iceland), Senegal .... *subpectinata* (Oudemans, 1901) sensu Woas 1986

Moritzoppia **Subías & Rodriguez, 1987. Fig. 22C**

Key to species

1. Trichobothrium with almost spherical head (Figs 11E, F). 280–320 µm. Northern Holarctic (including Ellesmere Island, Severnaya Zemlya Archipelago) ..... *clavigera* (Hammer, 1952) aberr.

   NOTE. Colloff and Seyd (1991) redescribed *M. clavigera* based on syntypes from Coppermine, Canada. Greenland specimens have longer dorsal setae (*in* 40–44 µm instead of 20 µm, notogastral seta *c2* 44–48 µm instead of 31 µm). Recently (Subías & Arillo 2001) *M. clavigera* was considered a subspecies of *Moritzoppia unicarinata* (Paoli, 1908), but more careful investigations are necessary.
– Trichobothrium fusiform distally (Figs 11D, 12B)..... 2
2. Anterior margin of notogaster with 2 pairs of very small projections (medial projections broad, roundish, poorly visible, lateral projections triangular) (Fig. 12B); trichobothrium clearly unilaterally barbed. 280–320 µm. Palaearctic .... *neerlandica* (Oudemans, 1900) sensu Weigmann 2006 ? = *Moritzoppia translamellata* (Willmann, 1923). Holarctic (including Ellesmere Island, Svalbard, Iceland)

   NOTE. Hammer (1944) noted that Greenland and Iceland specimens of *Oppia translamellata* Willmann, 1923 are lacking the distinct translamella. Later Hammer (1952) illustrated *Oppia translamellata* from Canada (Fig. 12G).
– Anterior margin of notogaster without projections, strongly convex; barbs of trichobothrium hardly visible (Fig. 11D). 260–320 µm. Northern Holarctic (including High Arctic) .... *similis* (Gordeeva & Grishina, 1991) ? = *Oppia translamellata* Willmann, 1923 sensu Hammer 1952 part.
? = *Oppia unicarinata* Paoli, 1908 sensu Strenzke *et al.* 1955

Oppiella **Jacot, 1937. Fig. 22D**

Key to species

1. Head of trichobothrium clavate, smooth (Fig. 11H) or with several small distal denticles; costulae short, interbothridial surface without zig-zag ridges.

320–330 μm. Holarctic (including Iceland, Svalbard), Australia................*splendens* (C.L. Koch, 1841)
– Trichobothrium of another form or unilaterally ciliate (Figs 11J, 12C, E); costulae long or zig-zag ridges present between bothridia 2
2. Trichobothrium fusiform, attenuate distally with a few (usually 5–7) very long ciliae (Fig. 11J); rostrum smooth; anterior margin of notogaster with wide rectangular ledge. 220–360 μm. Cosmopolitan (including Queen Elizabeth Islands, Iceland, Svalbard) ....*nova* (Oudemans, 1902)
= *Oppia neerlandica* Oudemans, 1900 sensu Willmann 1931
= *Oppia washburni* Hammer, 1952
– Trichobothrium dilated distally, with long ciliae (Figs 12C, E); rostrum with 2 incisions (Fig. 12D) or small sharp projection (Fig. 12F)....................3
3. Rostrum with 2 incisions, rostral setae inserted behind them (Figs 12C, D). 250–270 μm. Holarctic (including Ellesmere Island, Iceland) .... *maritima* (Willmann, 1929)
– Rostrum apically with small sharp projection (Figs 12E, F). 230–240 μm. Holarctic (Western Europe, northern Nearctic).......*acuminata* (Strenzke, 1951)

QUADROPPIIDAE (FIG. 22E)

In Greenland only 1 species recorded, ***Quadroppia quadricarinata*** (Michael, 1885), easily distinguished by its prominent prodorsal and notogastral sclerotization (Fig. 12H); sternal field between epimeres III and IV narrowing anteriad (Fig. 12K). 190–230 μm. Semicosmopolitan (Holarctic, southeastern China, Madagascar).

SUCTOBELBIDAE

Key to genera

1. Anterior margin of notogaster with distinct tubercles (Figs 12I, J) or angular projections (Fig. 13A); rostral setae sharply curved, with long ciliae (Figs 12L, N)..........................*Suctobelbella*
– Anterior margin of notogaster smooth (Fig. 13C); rostral setae almost straight, with short barbs (Fig. 13D) .... *Suctobelba*

***Suctobelba* Paoli, 1908**

In Greenland only 1 species represented, ***Suctobelba trigona*** (Michael, 1888), with trichobothrium smooth (Fig. 13C), rostrum with only 1 pair of lateral incisions (Fig. 13D), prodorsal ornamentation posteriad rostral setae formed by transverse parallel lines. 215–280 μm. Palaearctic (including Iceland, Faroe Islands).

***Suctobelbella* Jacot, 1937. Fig. 22F**

Key to species

1. Rostrum anteriorly convex, with 5–6 pairs of lateral teeth (Figs 12L, M); trichobothrium fusiform, distinctly unilaterally ciliate (Fig. 12I); larger mites, 250–310 μm. Palaearctic, Neotropical................ .... *subtrigona* (Oudemans, 1900)
– Rostrum anteriorly concave, with 2–3 pairs of lateral teeth behind apical lobes (Figs 12N, 13B); trichobothrium distally dilated, with rounded (Fig. 12N) or acuminate tip (Fig. 13A); smaller mites, 195–220 μm. 2
2. Trichobothrium distally clavate, with rounded tip, unilaterally weakly barbed (Fig. 12N); apical lobe of rostrum truncate or rounded; anterior margin of notogaster with small tubercles (Fig. 12J). 195 μm. W Greenland............. *delessei* (Strenzke, 1955)
– Trichobothrium distally broadly fusiform, with acuminate tip, smooth or scarcely barbed (Fig. 13A); apical lobe of rostrum bidentate (Fig. 13B); anrerior margin of notogaster with large tubercles (Fig. 13A). 195–220 μm. Holarctic (including Ellesmere Island, Iceland), Neotropical ....*acutidens* (Forsslund, 1941)

HYDROZETIDAE (FIG. 22G)

In Greenland represented by 1 species, ***Hydrozetes confervae*** (Schrank, 1781), noted by its remarkable morphological variability. Hydrophilous mite with relatively long legs, converging but not fused lamellae, short clavate trichobothrium (often broken) and prominent lenticulus (Figs 13E, F). 450–560 μm. Holarctic.

AMERONOTHRIDAE (FIG. 22H)

In Greenland only 1 genus found, ***Ameronothrus*** Berlese, 1896, represented by 3 species.

Key to species

1. Notogaster convex medially, with broad flat peripheral band (Fig. 13J); covered with microtubercles of equal size arranged irregularly on convex part and radiatingly on flattened band; tarsi tridactylous; terrestrial and littoral species. 540–580 μm. Holarctic (including Iceland), Antilles......................... .... *maculatus* (Michael, 1882)
– Notogaster slightly convex, with numerous mostly longitudinal zig-zag ridges (Figs 13I, K); tarsi tridactylous or monodactylous; littoral semiaquatic species 2

2. Tarsi monodactylous (Fig. 13G). 740–880 μm (Fig. 13K). Northern Holarctic (including Iceland, Svalbard) .... *nigrofemoratus* (L. Koch, 1879)
– Tarsi tridactylous (Fig. 13H). 660–800 μm (Fig. 13I). Northern Holarctic (including Iceland, Svalbard) .... *lineatus* (Thorell, 1871)

SCUTOVERTICIDAE (FIG. 24E)

Only 1 species, ***Scutovertex minutus*** (C.L. Koch 1835), was mentioned for Greenland (Shtanchaeva & Netuzhilin 2003), but this record needs verification. *Scutovertex minutus* is known from the Palaearctic (Subías 2004) including Svalbard (Bayartogtokh *et al.* 2011) and Iceland (Gjelstrup & Solhøy 1994). In the collection of ZMUC, Copenhagen, a single specimen of ***Scutovertex*** sp. (Fig. 14A) 575 μm in length was found that originated from Saqqaq, nSW Greenland (70°N).

SCHELORIBATIDAE (FIG. 22I)

In Greenland 1 species recorded, ***Liebstadia similis*** (Michael, 1888), with prodorsum and notogaster fused (no dorsosejugal scissure); pteromorphs like small triangular projections (Fig. 14B); sternal furrow extending from genital plates anteriorly on coxisternum (Fig. 14C); lamellae close to lateral margins of prodorsum; tarsi monodactylous. 500–600 μm. Holarctic (including Iceland, Svalbard), India, Australia, New Zealand.

ORIBATULIDAE (FIG. 22J)

In Greenland only 1 species found, ***Oribatula tibialis*** (Nicolet, 1855), with lamellae widened and angular anteriorly (Fig. 14D); lamellar setae inserted on their anterolateral angles; pteromorphs nearly absent (as very small humeral projections); tarsi homotridactylous. 410–540 μm. Holarctic (including Iceland, Svalbard), India.

HAPLOZETIDAE (FIG. 22K)

In Greenland represented by 1 genus, ***Peloribates*** Berlese, 1908.

Key to species

1. Notogastral setae rather long (only slightly shorter than prodorsal setae), clearly barbed (Fig. 14E); anterior margin of notogaster distinctly convex. 480–520 μm. Holarctic................. *pilosus* Hammer, 1952
– Notogastral setae short (half length of prodorsal setae), thin, barely barbed (Fig. 14F); anterior margin of notogaster almost straight. 450–500 μm. Northern Holarctic (Yamal Peninsula, Taymyr, Alaska, Canada) .... *canadensis* Hammer, 1952

CHAMOBATIDAE (FIG. 22L)

In Greenland represented by 1 genus, ***Chamobates*** Hull, 1916, with body strongly convex, notogastral setae tiny or vestigial, rostrum possessing teeth or/and incisions.

Key to species

1. Rostrum medially round, with 2 rather small lateral teeth (Figs 14G, H). 340–435 μm. Holarctic (including Iceland, Svalbard, Faroe Islands), Seychelles .... *cuspidatus* (Michael, 1884)
– Rostrum medially trident with projections separated by deep incisions (Figs 14I, J). Palaearctic (including Svalbard). 420–465 μm .... *birulai* (Kulczynski, 1902) sensu Trägårdh 1904

NOTE. *Notaspis birulai* Kulczynski, 1902 was initially described from Svalbard. Its rostrum was characterized and pictured as acute anteriad. Trägårdh (1904) considered this species only as the variety of *Oribata cuspidata* Michael, 1884 and gave the figures (19, 20) displaying the trident rostrum. Later authors treated *Chamobates birulai* (Kulczynski, 1902) as the separate species present in Europe and northern Siberia (*e.g.* Beck & Woas 1991, Krivolutsky *et al.* 1995, Weigmann 2006, Bayartogtokh *et al.* 2011). Both species can coexist in one forest (Beck & Woas 1991).

CERATOZETIDAE

Key to genera

1. Lamellae short, without cusps (Figs 15D, K); notogaster with 10 pairs of setae 2
– Lamellae with cusps (Figs 15A, H); notogaster with 10–15 pairs of setae..................................3
2. Tutorium narrow, without cusp or at most with very short, blunt cusp (Fig. 15E); translamella present or absent; dorsophragmata separated (Fig. 15D); notogastral setae rather long .... *Edwardzetes*
– Tutorial cusp broad, lamelliform (Fig. 15J); translamella absent; dorsophragmata fused (Fig. 15K); notogastral setae of medium size............. *Iugoribates*
3. Tutorial cusp broad, lamelliform, dentate, sinuate or undulating distally (Figs 16C, F, J); bothridium cup-shaped, without distinct lobes (Fig. 16I).............4
– Tutorial cusp tapering to point, not broad, lamelliform (Figs 15F, I, 16B); bothridium with well developed lobes (Figs 15A, G, H)..........................6
4. Notogastral setae *dm* present (notogaster with 13 pairs of setae) (Figs 15B, C) .... *Diapterobates*

– Notogastral setae *dm* absent (notogaster with 10–12 pairs of setae) (Figs 16E, H) 5
5. Translamella present (Figs 16G, H); lamellae broad; lamellar cusps well developed, with dens (Figs 16F, I) .... *Trichoribates*
– Translamella absent (Fig. 16E); lamellae narrow; lamellar cusps short (about 11 μm), narrow (tapering to width of lamellar seta), without dens (Fig. 16C)... .... *Svalbardia*
6. Lenticulus absent (Figs 14K, 15A); notogaster with 10–11 pairs of short and thin or diminutive setae; tarsi mono- or tridactylous .... *Ceratozetes*
– Lenticulus present (Figs 15G, H, 16A); notogaster with 10 or more, often 14–15 pairs of rather long, barbed setae; tarsi tridactylous 7
7. Translamella absent; lamellae large, broad, converging; lamellar cusps broad, with dens (Fig. 16D); 10 pairs of notogastral setae (rather long, barbed)...... .... *Oromurcia*
– Translamella present or absent; 14–15 pairs of notogastral setae (Figs 15G, H, 16A) 8
8. Lamellae strongly converging, reaching to level of insertion of rostral setae (Fig. 15G); tutorial cusp reaching almost to, or anteriad, insertion of rostral setae (Fig. 15F); translamella present .... *Fuscozetes*
– Lamellae not or slightly converging, more often not reaching anteriad level of insertion of rostral setae (Figs 15H, 16A); tutorial cusp very short, not reaching insertion of rostral setae (Figs 15I, 16B); translamella present or absent.................................... 9
9. Notogaster with 15 pairs of setae (seta *c1* present) (Fig. 15H) .... *Ghilarovizetes*
– Notogaster with 14 pairs of setae (seta *c1* absent) (Fig. 16A) .... *Melanozetes*

Ceratozetes **Berlese, 1908. Fig. 23A**

Key to species

1. Tarsi monodactylous; translamella absent (Fig. 14K); lamellar cusps usually with lateral dens; rostrum with pair of lateral teeth; trichobothrium slightly claviform, bilaterally barbed. 350–430 μm. Holarctic (including Iceland), Neotropical .... *thienemanni* Willmann, 1943
– Tarsi tridactylous; translamella present (Fig. 15A); lamellar cusp without distinct lateral dens; rostrum rounded or undulating; trichobothrium clavate, rounded distally, with heavily barbed head and smooth stalk. 440–530 μm. Greenland .... sp. near *spitsbergensis* Thor, 1934 = "*Sphaerozetes piriformis* ? (Nic.) = *Oribata piriformis* var. *setiger* Trägårdh, 1910" sensu Jørgensen 1934; materialstudied by M. Hammer [Jørgensen] revised ? = *Sphaerozetes piriformis* (Nicolet, 1855) sensu Thor 1930

NOTE. Differs from *Ceratozetes spitsbergensis* by presence of seta *v'* on genua I and II, form of lamellar cusp, tutorium, genal tooth, trichobotrium, dorsophragmata etc.

Diapterobates **Grandjean, 1936. Figs 23B, 24L**

Key to species

1. Anterior porose areas (*Aa*) small, roundish (Fig. 15B); pleurophragma with heavily sclerotized, fimbriate, posterior extensions; translamella, if present, long, narrow; seta *c2* noticeably longer than other notogastral setae of medium size; smaller mites. 400–480 μm. Holarctic (mainly boreal), Nepal................ .... *variabilis* Hammer, 1955
– Anterior porose areas rather large, elongated (Fig. 15C); pleurophragma without heavily sclerotized, fimbriate, posterior extensions; translamella broad; all notogastal setae of equal size, rather long, rigid; larger mites. 580–720 μm. Northern Holarctic (including High Arctic) .... *notatus* (Thorell, 1871)

Edwardzetes **Berlese, 1914. Fig. 23C**

In Greenland represented by 1 species, ***Edwardzetes edwardsi*** (Nicolet, 1855), with very characteristic structure of prodorsum (Fig. 15E), notogaster bearing 10 pairs of rather long barbed setae, all porose areas roundish (Fig. 15D). 700–860 μm. SW, S, SE Greenland. Europe, mainly boreal (including Iceland, Svalbard), Caucasus.

Fuscozetes **Sellnick, 1928. Fig. 24F**

In Greenland represented by 1 species, ***Fuscozetes sellnicki*** Hammer, 1952, with very characteristic structure of prodorsum (Fig. 15F), 14 pairs of barbed notogastral setae, notogastral porose areas placed on small tubercles (Fig. 15G). 420–510 μm. Northern Holarctic (including High Arctic).

Ghilarovizetes **Shaldybina, 1969. Fig. 23E**

In Greenland represented by 1 species, ***Ghilarovizetes longisetosus*** (Hammer, 1952), with notogaster bearing 15 pairs of rather long (80–97 μm) barbed setae (Figs 15H, I). 590–730 μm. Northern Holarctic (Magadan Region, Chukotka, Alaska, Canada).

***Iugoribates* Sellnick, 1944. Fig. 23D**

In Greenland represented by ***Iugoribates gracilis*** Sellnick, 1944 with prominent tutorium and pedotectum I (Figs 15J, K). 570–680 µm. Circumpolar (including High Arctic), not found in southern Greenland.

***Melanozetes* Hull, 1916**

In Greenland only 1 species recorded, ***Melanozetes meridianus*** Sellnick, 1928, with notogaster bearing 14 pairs of medium-sized (60–70 µm), stout, barbed setae (Figs 16A, B). 600–660 µm. Holarctic (including High Arctic).

***Oromurcia* Thor, 1930. Fig. 23F**

In Greenland only 1 species recorded, ***Oromurcia bicuspidata*** Thor, 1930. Lamellae of characteristic form with fully parallel medial margins and prominent lateral dens anteriorly (Fig. 16D); lamellar seta inserted in medial angle of cusp; rostrum with 2 teeth; 10 pairs of smooth middle size notogastral setae; all porose areas roundish. 570–650 µm. SW, S, SE Greenland. Northern Europe (including Iceland, Svalbard), Yamal Peninsula.

***Svalbardia* Thor, 1930. Fig. 23G**

A monotypic arctic genus containing the single species ***Svalbardia paludicola*** Thor, 1930 with characteristic form of lamellae (Figs 16C, E). The species prefers moist and wet biotopes. 520–620 µm. Circumpolar (including Devon Island, Svalbard), Transbaikalia.

***Trichoribates* Berlese, 1910. Fig. 23H**

Key to species

1. Translamella broad (Fig. 16G); notogaster with 3 pairs of saccules; notogastral setae rather short (30–40 µm) and thin. 530–620 µm. Northern Holarctic (West Beringia, Canada including Queen Elizabeth Islands) .... *polaris* Hammer, 1953
   NOTE. *Trichoribates polaris* Hammer, 1953 was accidentally specified for Greenland (Behan-Pelletier 1985), but its occurrence there is very probable.
– Translamella narrow, arched posteriorly (Figs 16I, J); notogaster with 4 pairs of porose areas, porose area *A1* divided (Figs 16H, K); notogastral setae large (55–90 µm), barbed. 550–600 µm. N Europe (including Iceland) .... *biarea* Gjelstrup & Solhøy, 1994
   NOTE. A third species, *Trichoribates trimaculatus* (C.L. Koch, 1835), was erroneously mentioned for Greenland in the monograph of M. Hammer (1944). All her slides kept in the ZMUC with label "*T. t.*" contain specimens belonging to *Diapterobates notatus* (also Ceratozetidae).

PUNCTORIBATIDAE (= MYCOBATIDAE)

Key to genera

1. Anterior tectum of notogaster with convex, rounded medial process and pair of thickened bands bordering medial process laterally (Fig. 16L); pteromorph with at most partial hinge......... *Zachvatkinibates*
– Anterior tectum of notogaster evenly rounded, without medial process and thickened bands (Figs 17I–K); pteromorph with complete hinge......... *Mycobates*

***Mycobates* Hull, 1916. Fig. 23J**

Key to species

1. Tarsi monodactylous; porose areas unclear (Fig. 17A); trichobothrium short, head longer than stalk (Fig. 17B). 405–455 µm. Northern Holarctic (eastern Canada, Northern Europe)........................... .... *exigualis* Behan-Pelletier, 1994
– Tarsi tridactylous; porose areas clearly evident (Figs 17D, K) 2
2. Dorsophragmata fused (Fig. 17D); area bordered by translamella and lamellar cusps distinctly U-shaped (Fig. 17E); integument over notogastral porose areas with scored structure (Fig. 19A). 600–670 µm. Holarctic (NE Asia, Japan, Nearctic)................. .... *incurvatus* Hammer, 1952
– Dorsophragmata separated (Figs 17J, K); area bordered by translamella and lamellar cusps rectangular (Figs 17F–H); integument over notogastral porose areas never scored 3
3. Integument over notogastral porose areas smooth; trichobothrium short, head longer than stalk (Fig. 17C). 490–580 µm. Holarctic (including Iceland, Svalbard) .... *sarekensis* (Trägårdh, 1910)
– Integument over notogastral porose areas tuberculate (Figs 19B–D); trichobothrium long, head shorter than stalk (Fig. 17H)................................4
4. Interlamellar seta thin, shorter than 80 µm, not or just reaching translamella (Figs 17F, I); integument of ventral and anal plates microtuberculate; integument of notogastral porose areas composed of uniform tubercles close together but distinct (Fig. 19B). 440–570 µm. Northern Nearctic (Alaska, Canada)... .... *dryas* Behan-Pelletier, 1994
– Interlamellar seta thick, longer than 100 µm, reaching anteriad translamella (Fig. 17H), or flagellate, reaching anteriad lamellar cusps (Fig. 17K); integument of notogastral porose areas of different structure (Figs 19C, D)...................................5
5. Integument with loosely spaced small tubercles over notogastral porose areas (Fig. 19C); interlamellar and

notogastral setae flagellate (Fig. 17K). 480–520 µm. Northern Nearctic (Alaska, Canada) .... *beringianus* Behan-Pelletier, 1994
– Integument with tightly spaced, partly fused tubercles over notogastral porose areas (Fig. 19D); interlamellar and notogastral setae tapered to acuminate (Figs 17H, J). 530–630 µm. Northern Holarctic (NE Asia, Alaska, Canada) .... *conitus* Hammer, 1952

***Zachvatkinibates* Schaldybina, 1973. Fig. 23I**

In Greenland represented by ***Zachvatkinibates*** cf. ***shaldybinae*** Behan-Pelletier & Eamer, 2005. *Zachvatkinibates shaldybinae* (Fig. 16L) known from eastern Canada (Ontario, Québec, Nova Scotia). Greenland specimens in general smaller (total length 432–483 µm instead of 470–528 µm), with longer posterior notogastral setae (32–40 µm, in Canadian specimens about 25 µm), anal and adanal setae (32–36 µm instead of 16–18 µm). Both Canadian and Greenland forms live in marine littoral biotopes. 430–480 µm. SW Greenland, ?eastern Canada.

PHENOPELOPIDAE

Key to genera

1. Chelicera peloptoid (elongated, abruptly narrowed distally, with small chela, Fig. 1I); rostrum narrow (Figs 18B, C); lamellae inconspicuous; pleurophragmata elongated, extending well into opisthosoma (Figs 18A, B); lenticulus well circumscribed (Fig. 18D) .... *Eupelops*
– Chelicera normally developed; rostrum broad, evenly rounded; lamellae broad conspicuous, with large cusps (Fig. 18H); pleurophragmata normally developed, not elongated; lenticulus often indistinct .... *Propelops*

***Eupelops* Ewing, 1917. Figs 23K, L**

The Greenland *Eupelops* species need careful re-examination. Three species are recorded in southern Greenland. The key mainly illustrates their differences.

Key to species

1. Notogastral setae *h3*, *lp* and porose area *A1* close together, forming a characteristic complex (Fig. 18A); notogastral setae relatively short, heterogeneous (two posterolateral pairs tiny), smooth or shortly barbed, posterior notogastral setae not longer than 40 µm; trichobothrium clavate; anterior tectum of notogaster medially concave. 450–550 µm. Greenland .... *Eupelops* sp. ? = *Pelops bilobus* Sellnick, 1928 in Hammer 1944

NOTE. The identification (as "*Pelops ?bilobus*") was doubtful already for Hammer (1944). *Eupelops* sp. differs from *Eupelops curtipilus* (Berlese, 1916) [=*Pelops bilobus* Sellnick, 1928 (synonymy in Weigmann 2006)] by longer notogastral setae (longest setae 28–36 µm instead 10–15 µm in *E. curtipilus*) and wider space between lamellar cusps.

– Notogastral setae *h3*, *lp* and porose area *A1* not close together (Figs 18B, D); at least some posterior notogastral setae longer than 65 µm...................... 2

2. Notogastral setae more or less uniform — stout, bacilliform or poorly arched, barbed, apically truncate or rounded (Fig. 18E); trichobothrium clavate; area between lamellar cusps broad, at most twice longer then wider (Fig. 18C); anterior tectum of notogaster almost straight medially (Fig. 18D). 630–720 µm. Northern Holarctic (including Iceland) .... *septentrionalis* (Trägårdh, 1910)
– Notogastral setae heterogeneous, 1–2 posterior pairs of setae longer, spatulate, clearly barbed (Fig. 18B); trichobothrium fusiform; area between lamellar cusps narrow, at least three times longer than wider; anterior tectum of notogaster with 2–3 projections (medial projection may be developed or not). 540–680 µm. Holarctic (including Iceland) .... *plicatus* (C.L. Koch, 1836) ? = *Eupelops uraceus* (C.L. Koch, 1836) in Hammer 1944

NOTE. Modern description of *Eupelops uraceus* does not exist. *Eupelops plicatus* is the most common member of the genus in Greenland.

***Propelops* Jacot, 1937**

In Greenland only 1 species recorded, ***Propelops groenlandicus*** (Sellnick, 1944), with conspicuous long and broad lamellae reaching anteriad rostrum, connected by very short translamella (Fig. 18H). 440–470 µm. West Beringia, northern Nearctic.

ACHIPTERIIDAE

Key to genera

1. Notogaster with porose areas......... *Parachipteria*
– Notogaster with saccules (Fig. 18G)....... *Achipteria*

***Achipteria* Berlese, 1885. Fig. 24H**

In Greenland 1 species recorded, ***Achipteria coleoptrata*** (Linnaeus, 1758), with diminutive notogastral setae (Fig. 18G). 540–650 µm. SW Greenland. Holarctic, India, St. Helena Island.

***Parachipteria* Hammen, 1952. Fig. 24G**
In Greenland 1 species recorded, ***Parachipteria punctata*** (Nicolet, 1855), with notogastral setae of middle size (Fig. 18F). 550–580 µm. SW Greenland. Holarctic (including Svalbard), St. Helena Island.

References

Balogh, J. & Mahunka, S. (1983): The Soil Mites of the World. Volume 1. Primitive Oribatid Mites of the Palaearctic Region. – Akademai Kiado, Budapest & Elsevier, Amsterdam. 372 pp.

Bayartogtokh, B. (2010): Oribatid Mites of Mongolia (Acari: Oribatida). – KMK Scientific Press Ltd., Moscow. 371 pp. [In Russian].

Bayartogtokh, B., Schatz, H. & Ekrem, T. (2011): Distribution and diversity of the soil mites of Svalbard, with redescriptions of three known species (Acari: Oribatida). – *International Journal of Acarology* 37(6): 467–484.

Beck, L. & Woas, S. (1991): Die Oribatiden-Arten (Acari) eines südwestdeutschen Buchenwaldes I. – *Carolinea* 49: 37–82.

Behan, V.M. (1978): *Camisia labradorica* (Acari: Oribatei: Camisiidae): a new soil mite species from the Canadian subarctic. – *Canadian Entomologist* 110: 547–550.

Behan, V.M. & Hill, S.B. (1978): Feeding habits and spore dispersal of oribatid mites in the North American Arctic. – *Revue d'Ecologie et Biologie du Sol* 15: 497–516.

Behan-Pelletier, V.M. (1984): *Ceratozetes* (Acari: Ceratozetidae) of Canada and Alaska. – *Canadian Entomologist* 116: 1449–1517.

Behan-Pelletier, V.M. (1985): Ceratozetidae of the western North American arctic. – *Canadian Entomologist* 117: 1287–1366.

Behan-Pelletier, V.M. (1986): Ceratozetidae (Acari: Oribatei) of the western North American subarctic. – *Canadian Entomologist* 118: 991–1057.

Behan-Pelletier, V.M. (1993): Eremaeidae (Acari: Oribatida) of North America. – *Memoirs of the Entomological Society of Canada* 168: 1–193.

Behan-Pelletier, V.M. (1994): *Mycobates* (Acari: Oribatida: Mycobatidae) of America north of Mexico. – *Canadian Entomologist* 126: 1301–1361.

Behan-Pelletier, V.M. (1997): Oribatid mites (Acari: Oribatida) of the Yukon. – *In*: Danks, H.V. & Downes, J.A. (eds.). Insects of Yukon: 115–149. *Biological Survey of Canada (Terrestrial Arthropods)*, Ottawa.

Behan-Pelletier, V.M. (1999): Oribatid mite fauna of northern ecosystems: products of evolutionary adaptations or physiological constraints. – *In*: Mitchell, R., Horn, D.J., Needham, G.R. & Welbourn, W.C. (eds.). Acarology IX. Proceedings of IX Inernational Congress of Acarology. Columbus, Ohio, 1996. V. 2: 87–93. Ohio Biological Survey, Columbus.

Behan-Pelletier, V.M. & Eamer, B. (2005): *Zachvatkinibates* (Acari: Oribatida: Mycobatidae) of North America, with descriptions of sexually dimorphic species. – *Canadian Entomologist* 137: 631–647.

Behan-Pelletier, V.M. & Norton, R.A. (1983): *Epidamaeus* (Acari: Damaeidae) of arctic western North America and extreme northeast U.S.S.R. – *Canadian Entomologist* 115: 1253–1289.

Behan-Pelletier, V.M. & Norton, R.A. (1985): *Epidamaeus* (Acari: Damaeidae) of subarctic western North America and extreme Northeastern U.S.S.R. – *Canadian Entomologist* 117: 277–319.

Colloff, M.J. (1993): A taxonomic revision of the oribatid mite genus *Camisia* (Acari: Oribatida). – *Journal of Natural History* 27: 1325–1408.

Colloff, M.J. & Cameron, S.L. (2009): A revision of the oribatid mite genus *Austronothrus* Hammer (Acari: Oribatida): sexual dimorphism and a re-evaluation of the phylogenetic relationships of the family Crotoniidae. - *Invertebrate systematics* 23(2): 87–110.

Colloff, M.J. & Seyd, E.L. (1991): A new species of *Moritzoppia* from montane sites in the British Isles, with a redescription of *M. clavigera* (Hammer, 1952) (Acari: Oribatida: Oppiidae). – *Journal of Natural History* 25: 1067–1074.

Coulson, S.J. (2007): The terrestrial and freshwater invertebrate fauna of the High Arctic archipelago of Svalbard. – *Zootaxa* 1448: 41–58.

Forsslund, K.-H. (1957): Notizen űber Oribatei (Acari). I. *Arkiv för Zoologi* 10: 583–593.

Ghilarov, M.S. (1975, ed.): A Key to Soil-inhabiting Mites (Sarcoptiformes). – Nauka, Moscow. 491 pp. [In Russian].

Gjelstrup, P. & Solhøy, T. (1994): Oribatid mites (Acari). – *The Zoology of Iceland III*(57e): 1–75.

Gordeyeva, E.V. & Grishina L.G. (1991): New species of oribatid mites of the family Oppiidae (Sarcoptiformes, Oribatei) from Siberia. – *Zoologicheskii Zhurnal, Moscow* 70(6): 39–49 [In Russian].

Graversen, C.B. (1931): Notizen űber grönländische Oribatiden. – *Meddelelser om Grønland* 91(2): 1–18.

Haarløv, N. (1942): A morphologic-systematic-ecological investigation of Acarina. – *Meddelelser om Grønland* 128(1): 1–71.

Haarløv, N. (1967): Arthropoda (Acarina, Diptera) from subfossil layers in West Greenland. – *Meddelelser om Grønland* 184(3): 1–17.

Hammer, M. (1937): A quantitative and qualitative investigation of the microfauna communities of the soil at Angmagssalik and in Mikis Fjord. – *Meddelelser om Grønland* 108(2): 1–53.

Hammer, M. (1944): Studies on the oribatids and collemboles of Greenland. – *Meddelelser om Grønland* 141(3): 1–210.

Hammer, M. (1946): The zoology of east Greenland. Oribatids. – *Meddelelser om Grønland* 122(1): 1–39.

Hammer, M. (1952): Investigations on the microfauna of northern Canada. Part I: Oribatidae. – *Acta Arctica* 4: 1–108.

Hammer, M. (1952): The oribatid and Collembola fauna in some soil samples from Søndre Strømfjord. – *Entomologiske Meddelelser* 26(5): 404–414.

Hammer, M. (1953): Collemboles and oribatids from the Thule District (North West Greenland) and Ellesmere Island (Canada) collected by J.C. Troelsen and Chr. Vibe. – *Meddelelser om Grønland* 136(5): 1–16.

Hammer, M. (1954): Collemboles and oribatids from Peary Land (North Greenland). – *Meddelelser om Grønland* 127(5): 1–28.

Hammer, M. (1955): Alaskan oribatids. – *Acta Arctica* 7: 1–36.

Hammer, M. (1960): A few more oribatids from Greenland and Lapland. – *Entomologiske Meddelelser* 29: 360–363.

Henriksen, K.L. & Lundbeck, W. (1917): Grønlands Landarthropoder (Insecta et Arachnida). – *Meddelelser om Grønland* 22: 481–821.

Jørgensen, M. (1934): Ein revidiertes Verzeichnis über grönländische Milben. – *Zoologischer Anzeiger* 107(1–2): 40–47.

Kramer, P. (1897): Grönländische Milben. – *Bibliotheca Zoologica* 20: 77–83.

Krivolutsky, D.A. (1976): Role of oribatid mites in biogeocoenoses. – *Zoologicheskii Zhurnal, Moscow* 55(2): 226–236 [In Russian].

Krivolutsky, D.A., Lebren, P., Kunst, M. *et al.* (1995): The Oribatid Mites. – Nauka, Moscow. 224 pp. [In Russian].

Laumann, M., Norton, R.A., Weigmann, G., Scheu, S., Maraun, M. & Heethoff, M. (2007): Speciation in the parthenogenetic oribatid mite genus *Tectocepheus* (Acari, Oribatida) as indicated by molecular phylogeny. – *Pedobiologia* 51(2): 111–122.

Luxton, M. (1972): Studies on the oribatid mites of a Danish woodland soil. I. Nutritional biology. – *Pedobiologia* 12: 434–463.

Luxton, M. (1989): Michael's british Damaeids (Acari: Cryptostigmata). – *Journal of Natural Hystory* 23: 1367–1372.

Makarova, O.L. (2002): Acarocenoses (Acariformes, Parasitiformes) in polar deserts: 1. Mite assemblages of the Severnaya Zemlya Archipelado: structure of fauna and abundance. – *Entomological Review* 82(7): 839–856.

Makarova, O.L. (2014): The free-living mite fauna (Acari) of Greenland. – *Zoologicheskii Zhurnal*, Moscow 93 (12): 1235–1251 [In Russian].

Makarova, O.L. & Böcher, J., (2009): Diversity and geographical ranges of Greenland mites (Acari: Oribatida and Mesostigmata). – *In*: Golovatch, S.I., Makarova, O.L., Babenko, A.B. & Penev, L.D. (eds.). Species and Communities in Extreme Environments: 165–186. Pensoft Publishers & KMK Scientific Press, Sofia–Moscow.

Maraun, M., Martens, H., Migge, S., Theenhaus, A. & Scheu, S. (2003): Adding to 'the enigma of soil animal diversity': fungal feeders and saprophagous soil invertebrates prefer similar food substrates. – *European Journal of Soil Biology* 39: 85–95.

Maraun, M. & Scheu, S. (2000): The structure of oribatid mite communities (Acari, Oribatida): patterns, mechanisms and implications for future research. – *Ecography* 23(3): 374–383.

Marshall, V.G., Reeves, R.M. & Norton, R.A. (1987): Catalogue of the Oribatida (Acari) of continental Unites States and Canada. – *Memoirs of the Entomological Society of Canada* 139: 1–418.

McAlpine, J.F. (1965): Insect and related terrestrial invertebrates of Ellef Ringnes Island. – *Arctic* 18: 73–103.

Michael, A.D. (1888): British Oribatidae. V. 1, 2. – Ray Society, London. 657 pp.

Moritz, M. (1970): Revision von *Suctobelba trigona* (Michael, 1888), ein Beitrag zur Kenntnis der europäischen Arten der Gattung *Suctobelba* Paoli, 1908 sensu Jacot, 1937 (Acari, Oribatei, Suctobelbidae). – *Mitteilungen aus dem Zoologischen Museum in Berlin* 46: 135–166.

Moritz, M. (1971): Beiträge zur Kenntnis der Oribatiden (Acari) Europas. 3. *Suctobelbella alloenasuta* n. sp. und Suctobelba messneri n. sp. sowie die bisher aus der DDR bekannten Arten der *nasalis-subtrigona*-Gruppe (Suctobelbidae). – *Mitteilungen aus dem Zoologischen Museum in Berlin* 47: 85–98.

Moritz, M. (1974): Beiträge zur Kenntnis der Oribatiden (Acari) Europas. 5. *Suctobelbella difissa* n. sp. und *Suctobelbella acutidens* (Forsslund, 1941) (Suctobelbidae). – *Abhandlungen und Berichten des Naturkundemuseums Görlitz* 48(3): 1–8.

Moritz, M. (1976): Revision der europäischen Gattungen und Arten der Familie Brachychthoniidae (Acari, Oribatei). Teil 1, 2. – *Mitteilungen aus dem Zoolodischen Museum in Berlin* 52: 27–136, 227–319.

Niedbała, W. (1992): Phthiracaroidea (Acari, Oribatida). Systematic Studies. – PWN, Warsawa & Elsevier, Amsterdam-Oxford-New York-Tokyo. 612 pp.

Niedbała, W. (2002): Ptyctimous mites (Acari, Oribatida) of the Nearctic Region. – *Monographs of the Upper Silesian Museum* 4. 261 pp.

Norton, R.A. (1990): Acarina: Oribatida. – *In*: Dindal, D.E. (ed.). Soil Biology Guide: 779–803. J. Wiley & Sons Publ., New York.

Norton, R.A. (1994): Evolutionary aspects of oribatid mite life histories and consequences for the origin of the Astigmata. – *In*: Houck, M.A. (ed.). Mites: Ecological and Evolutionary Studies of Life-History Patterns: 99–135. Chapman & Hall, New York.

Norton, R. A. & Behan-Pelletier, V. M. (2009): Suborder Oribatida. Chapter 15. – *In*: Krantz, G.W. & Walter, D.E. (eds.). A Manual of Acarology: 430–564. Texas Tech Univ. Press, Lubbock.

Norton, R.A., Behan-Pelletier, V.M. & Wang, H. (1996): The aquatic oribatid mite genus *Mucronothrus* in Canada and the western U.S.A. (Acari, Trhipochthoniidae). – *Canadian Journal of Zoology* 74: 926–949.

Norton, R.A. & Ermilov, S.G. (2014): Catalogue and historical overview of juvenile instars of oribatid mites (Acari: Oribatida). – *Zootaxa* 3833(1): 1–132.

Olszanowski, Z. (1996): A monograph of the Nothridae and Camisiidae of Poland (Acari: Oribatida: Crotonoidea). – *Genus*, Suppliment: 1–201.

Paoli, G. (1908): Monografia del genere *Dameosoma* Berl. e generi affini. – *Redia* 5: 31–91.

Pérez-Íñigo, C. (1997): Acari. Oribatei, Gymnonota I. – Fauna Iberica 9: 374 pp.

Schatz, H., Behan-Pelletier, V.M., OConnor, B.M. & Norton, R.A. (2011): Suborder Oribatida van der Hammen, 1968. – In: Zhang, Z.-Q. (ed.) Animal Biodiversity: An Outline of Higher-Level Classification and Survey of Taxonomic Richness. *Zootaxa* (special volume) 3148: 141–148.

Schneider, K., Renker, K., Scheu, S. & Maraun, M. (2005): Feeding biology of oribatid mites: a minireview. – *In*: Weigmann, G., Alberti, G., Wohltmann, A. & Ragusa, S. (eds.). Acarine Biodiversity in the Natural and Human Sphere. Proceedings of the V Symposium of the European Association of Acarologists (Berlin 2004). *Phytophaga* (Palermo) 14(2004): 247–256.

Schuster, R. (1956): Der Anteil der Oribatiden an den Zersetzungsvorgängen im Boden. – *Zoologischer Jarbücher*, Abteilung Systematische, Ökologische und Geographie den Tiere 45: 1–33.

Seastedt, T.R. (1984): The role of microarthropods in decompositions and mineralization processes. – *Annual Review of Entomology* 29: 25–46.

Shtanchaeva, U.Ya. & Netuzhilin, I.A. (2003): A review of the world fauna of oribatid mites from the family Scutoverticidae (Acari, Oribatida) with description of new species. – *Entomological Review* 83, Suppliment 1: S46–S66.

Strenzke, K. von, de Lesse, H. & Denis, J. (1955): Microfaune du sol de l'Eqe, Groenland; vol. 1, arachnides. Hermann, Paris. [Expéditions polaires francaises, VII; pub. as Actualités scientifiques et industrielles, 1232.] 84 pp.

Subías, L.S. (2004): Listado sistemático, sinonimico y biogeográfico de los ácaros oribátidos (Acariformes, Oribatida) del mundo (1758–2002). – *Graellsia* 60 (numero extraordinario): 3–305.

Subías, L.S. & Arillo, A. (2001): Acari. Oribatei, Gymnonota II. – *Fauna Iberica* 15: 289 pp.

Trägårdh, I. (1904): Monographie der arktischen Acariden. – *Fauna Arctica* 4: 1–78.

Walter, D.E., Latonas, S. & Byers, K. (2013): Almanac of Alberta Oribatida. Part 1. Ver. 2.3. The Royal Alberta Museum, Edmonton, AB. http://www.royalalbertamuseum.ca/natural/insects/research/research.htm.

Walter, D.E. & Proctor, H.C. (2013): Mites: Ecology, Evolution & Behaviour. Life at a Microscale. 2nd Edition – Springer, Dordrecht – Heidelberg – New York – London. 494 pp.

Weigmann, G. (2006): Hornmilben (Oribatida). Die Tierwelt Deutschlands, 76. Teil. Keltern, Goecke & Evers. 520 pp.

Willmann, C. (1931): Moosmilben oder Oribatiden (Oribatei). – In: Dahl, F. (ed.). Die Tierwelt Deutschlands, 22. Teil. Spinnentiere oder Arachnoidea: 79–200. V. G. Fisher, Jena.

Woas, S. (1978): Die Arten der Gattung *Hermannia* Nicolet 1855 (Acari, Oribatei). 1. – *Beiträge zur naturkundlichen Forschung in Südwestdeutschland* 37: 113–141.

Woas, S. (1981): Die Arten der Gattung *Hermannia* Nicolet 1855 (Acari, Oribatei). 2. – *Andrias* 1: 89–100.

Woas, S. (1986): Beitrag zur Revision der Oppioidea sensu Balogh, 1972 (Acari, Oribatei). – *Andrias* 5: 21–224.

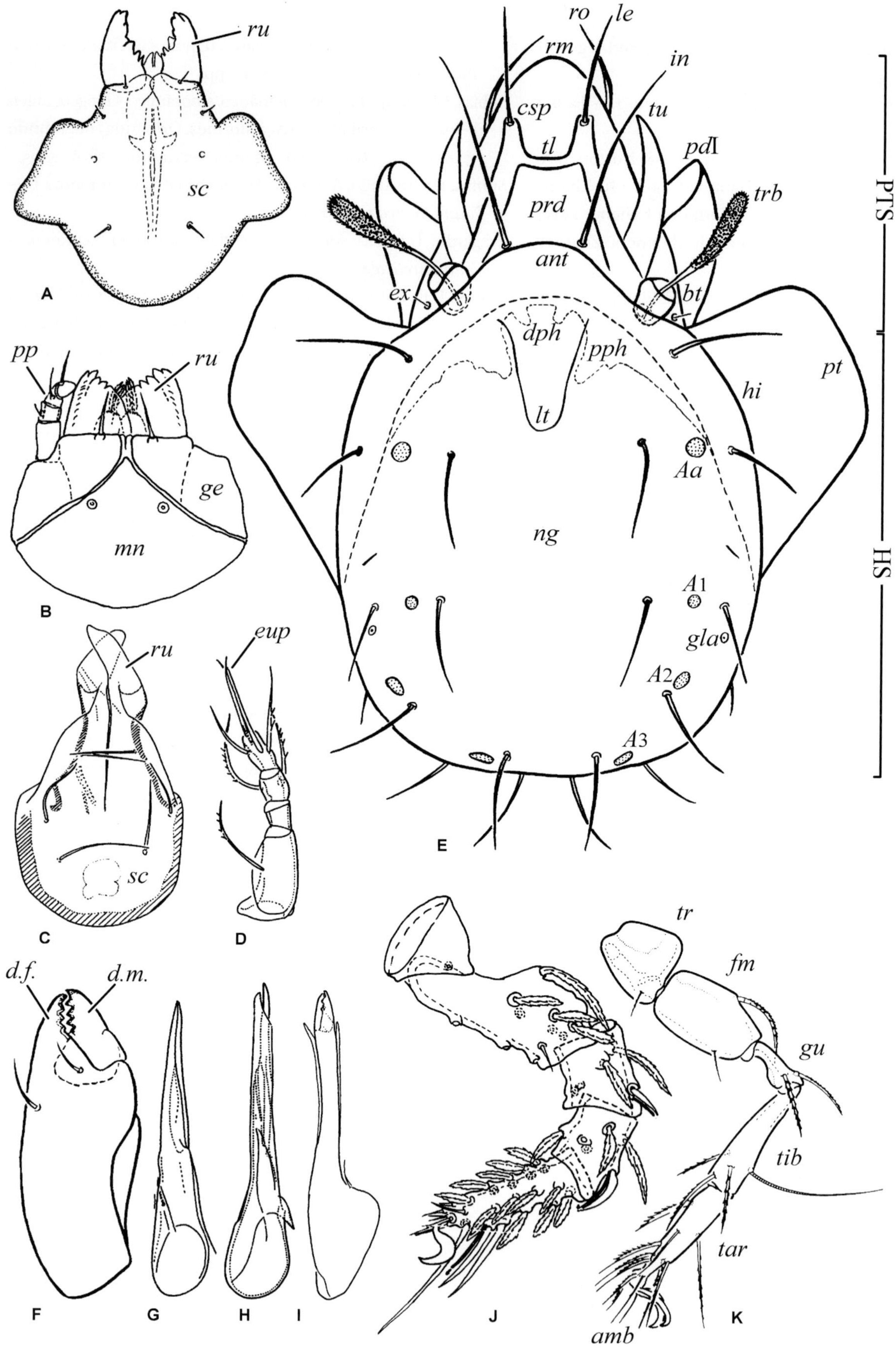

FIGURE 1 *A,* Trimalaconothrus *sp., subcapitulum; B,* Camisia lapponica*, gnathosoma, ventral aspect, J, leg I; C,* Suctobelba trigona*, subcapitulum, D, palp, G, chelicera; E, brachypyline oribatid mite, dorsal aspect, schematic; F,* Diapterobates *sp., chelicera; H,* Suctobelbella subtrigona*, chelicera; I,* Eupelops *sp., chelicerae; K,* Trichoribates polaris*, leg IV. B, J, after Behan 1978; C, D, G, H, after Woas 1986; K, after Behan-Pelletier 1985.*

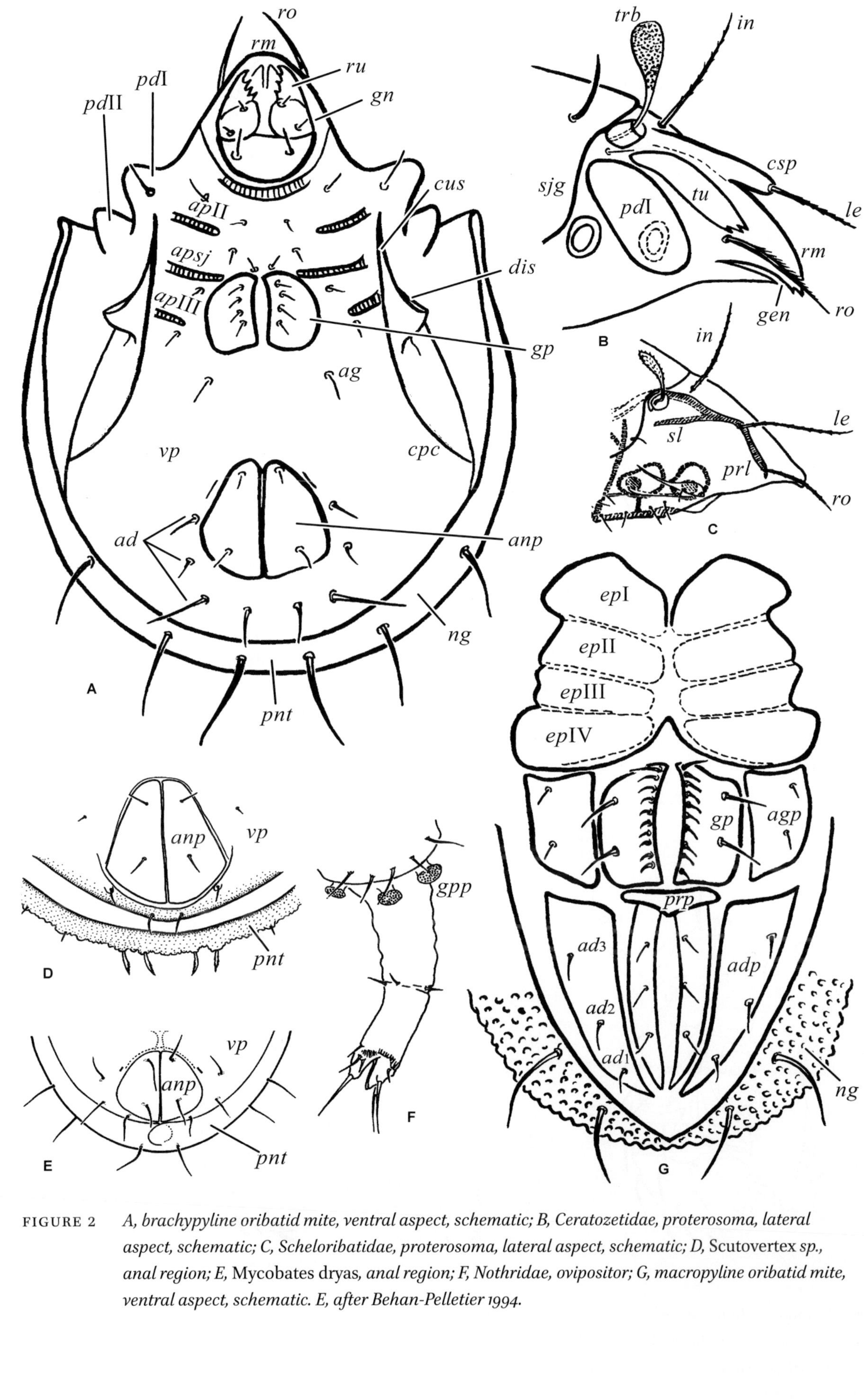

FIGURE 2 *A, brachypyline oribatid mite, ventral aspect, schematic; B, Ceratozetidae, proterosoma, lateral aspect, schematic; C, Scheloribatidae, proterosoma, lateral aspect, schematic; D,* Scutovertex *sp., anal region; E,* Mycobates dryas*, anal region; F, Nothridae, ovipositor; G, macropyline oribatid mite, ventral aspect, schematic. E, after Behan-Pelletier 1994.*

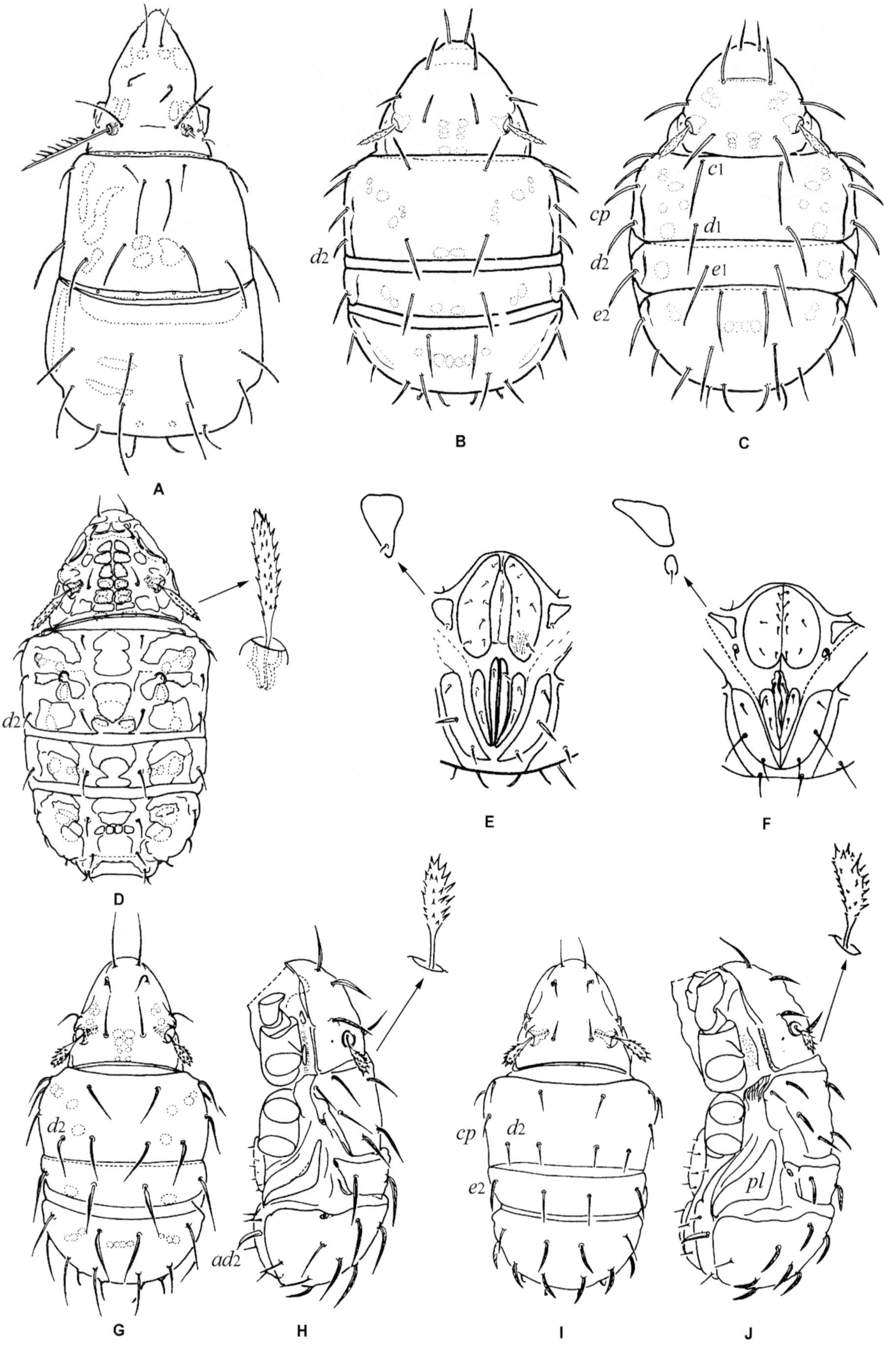

FIGURE 3 *A,* Hypochthonius rufulus, *dorsal aspect; B,* Eobrachychthonius latior, *dorsal aspect, E, anogenital region; C,* Eobrachychthonius oudemansi, *dorsal aspect, F, anogenital region; D,* Brachychthonius berlesei, *dorsal aspect; G,* Liochthonius brevis, *dorsal aspect, H, lateral aspect; I,* Liochthonius simplex, *dorsal aspect, J, lateral aspect. A–C, E, F, after Weigmann 2006; D, G–J, after Moritz 1976.*

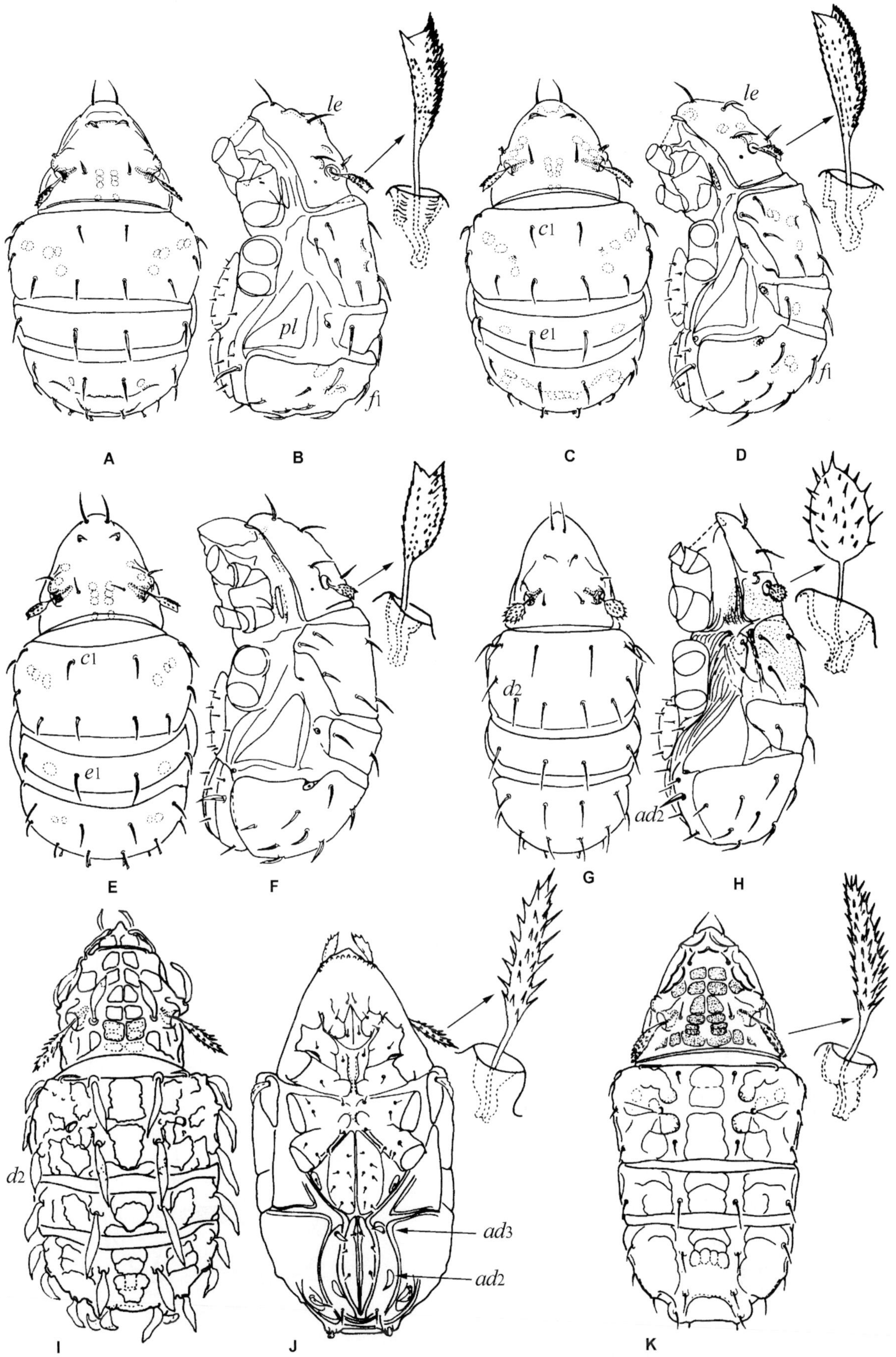

FIGURE 4 *A,* Liochthonius sellnicki, *dorsal aspect, B, lateral aspect; C,* Liochthonius muscorum, *dorsal aspect, D, lateral aspect; E,* Liochthonius lapponicus, *dorsal aspect, F, lateral aspect; G,* Neoliochthonius piluliferus, *dorsal aspect, H, lateral aspect; I,* Sellnickochthonius zelawaiensis, *dorsal aspect, J, ventral aspect; K,* Sellnickochthonius immaculatus, *dorsal aspect. All after Moritz 1976.*

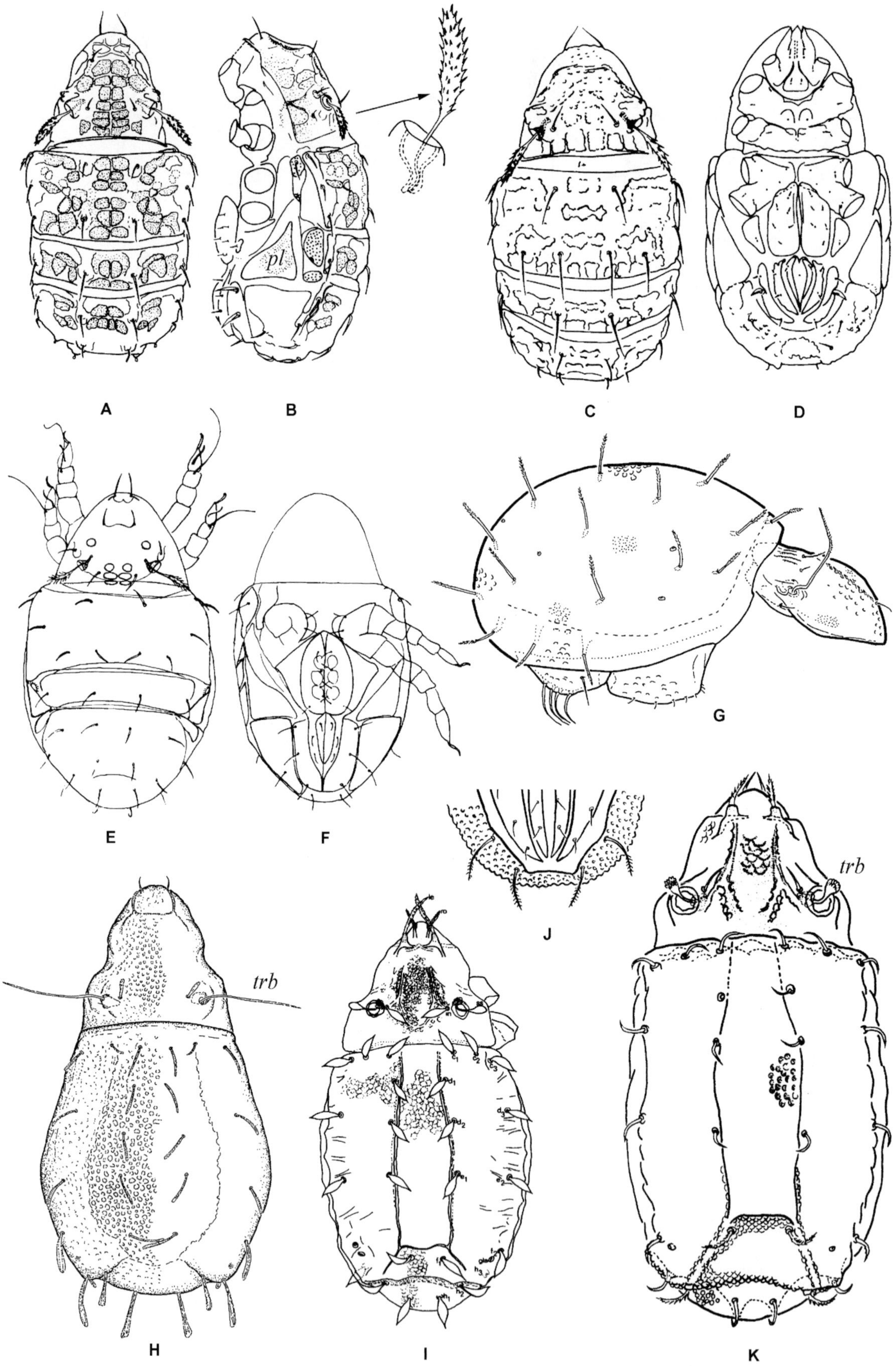

FIGURE 5 *A,* Sellnickochthonius suecicus, *dorsal aspect, B, lateral aspect; C,* Synchthonius crenulatus, *dorsal aspect, D, ventral aspect; E,* Verachthonius montanus, *dorsal aspect, F, ventral aspect; G,* Atropacarus striculus, *lateral aspect; H,* Nothrus borussicus, *dorsal aspect; I,* Camisia lapponica, *dorsal aspect; J,* Camisia *sp., ventrocaudal region, schematic; K,* Camisia foveolata, *dorsal aspect. A–D, after Moritz 1976; E, F, after Hammer 1952; G, after Weigmann 2006; H, after Olszanowski 1996; I, after Behan 1978; K, after Colloff 1993.*

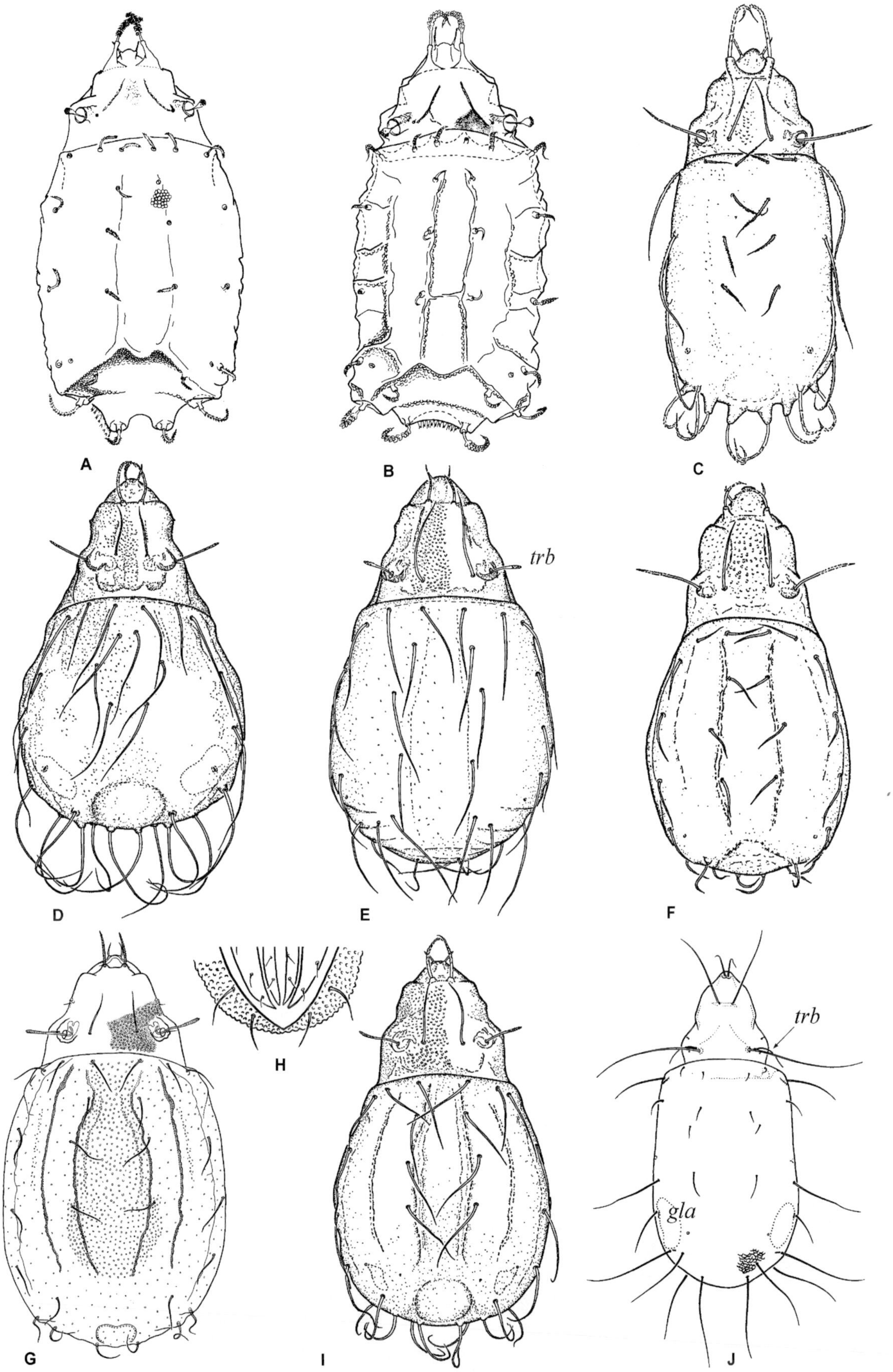

FIGURE 6 *A,* Camisia biverrucata, *dorsal aspect; B,* Camisia horrida, *dorsal aspect; C,* Heminothrus longisetosus, *dorsal aspect; D,* Heminothrus thori, *dorsal aspect; E,* Heminothrus capillatus, *dorsal aspect; F,* Neonothrus humicolus, *dorsal aspect; G,* Platynothrus punctatus, *dorsal aspect; H,* Heminothrus *sp., ventrocaudal region, schematic; I,* Platynothrus peltifer, *dorsal aspect; J,* Mucronothrus nasalis, *dorsal aspect. A, B, after Colloff 1993; C–F, I, after Olszanowski 1996; J, after Norton et al. 1996.*

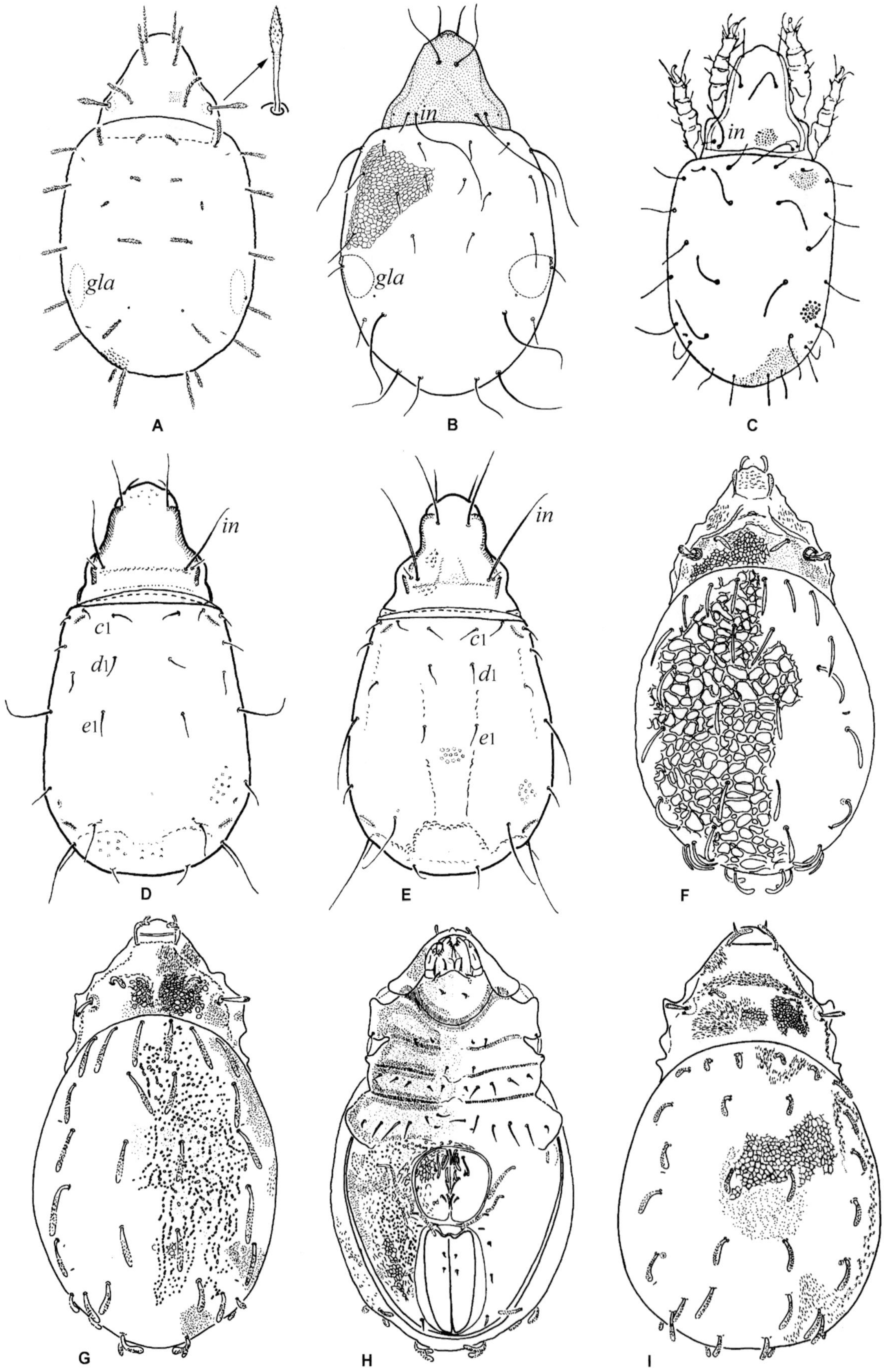

FIGURE 7 *A,* Trhypochthonius tectorum, *dorsal aspect; B,* Trhypochthoniellus *sp. near* brevisetus, *dorsal aspect; C,* Malaconothrus mollisetosus, *dorsal aspect; D,* Trimalaconothrus foveolatus, *dorsal aspect; E,* Trimalaconothrus maior, *dorsal aspect; F,* Hermannia reticulata, *dorsal aspect; G,* Hermannia nodosa, *dorsal aspect, H, ventral aspect; I,* Hermannia subglabra, *dorsal aspect. A, D, E, after Weigmann 2006; C, after Hammer 1952; F, G, H, after Woas 1978; I, after Woas 1981.*

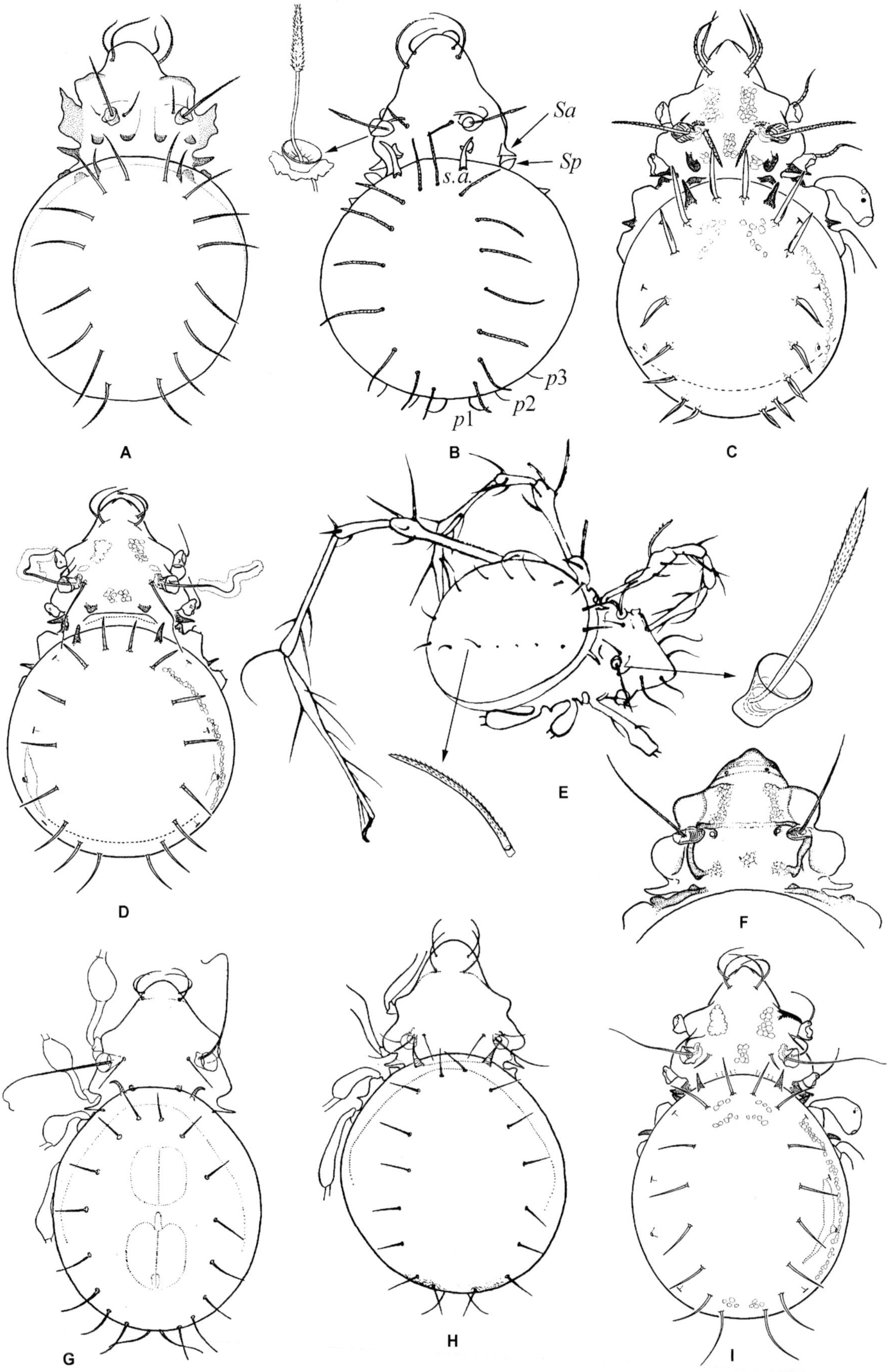

FIGURE 8 *A,* Damaeus *sp. near* gracilipes, *dorsal aspect; B,* Epidamaeus tenuipes, *dorsal aspect; C,* Epidamaeus fortispinosus, *dorsal aspect; D,* Epidamaeus floccosus, *dorsal aspect; E,* Epidamaeus longitarsalis, *lateral aspect; F,* "Oribata trägårdhi"*, prodorsum, dorsal aspect; G,* Epidamaeus trigonalis, *dorsal aspect; H,* Epidamaeus inornatus, *dorsal aspect; I,* Epidamaeus koyukon, *dorsal aspect. B, after Luxton 1989; C, after Behan-Pelletier & Norton 1983; D, I, after Behan-Pelletier & Norton 1985; E, after Hammer 1952; F, after Graversen 1931; G, H, after Strenzke et al. 1955.*

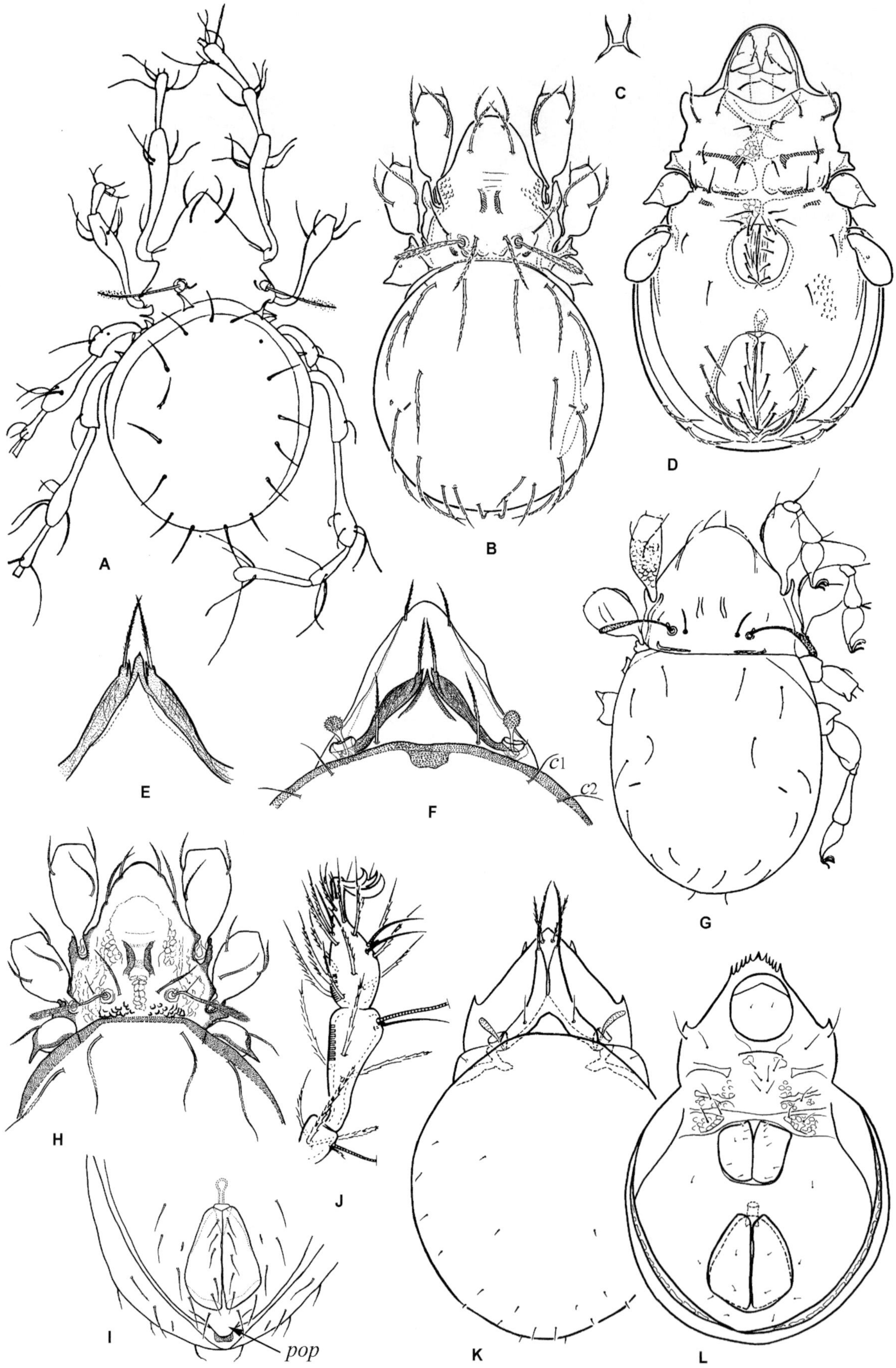

FIGURE 9 *A,* Epidamaeus groenlandicus, *dorsal aspect; B,* Eremaeus translamellatus, *dorsal aspect, C, costulae, D, ventral aspect; E,* Procorynetes globifer, *lamellae, F, partial dorsum; G,* Eueremaeus *sp. 1, dorsal aspect; H,* Eueremaeus *sp. 2 near* michaeli, *partial dorsum, I, anal region; J,* Eueremaus *sp., tibia and tarsus I; K,* Cultroribula dentata, *dorsal aspect, L, ventral aspect. A, after Hammer 1953; B, D, J, after Behan-Pelletier 1993; C, G, after Hammer 1952.*

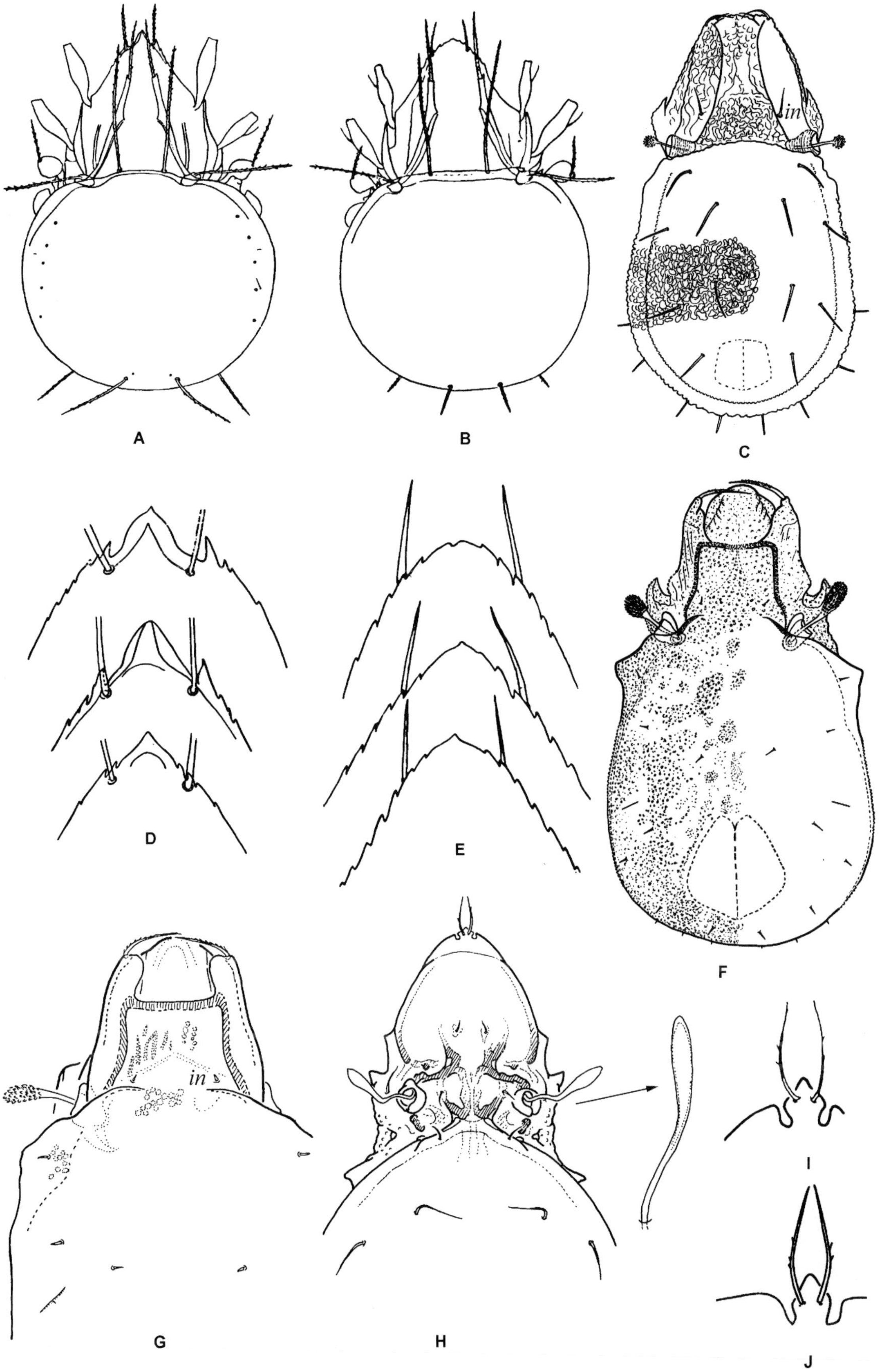

FIGURE 10 *A,* Ceratoppia bipilis, *dorsal aspect, D, variation of rostrum; B,* Ceratoppia sphaerica, *dorsal aspect, E, variation of rostrum; C,* Carabodes labyrinthicus, *dorsal aspect; F,* Tectocepheus velatus, *dorsal aspect; G,* Tectocepheus sarekensis, *partial dorsum; H,* Dissorhina ornata, *partial dorsum, I, J, rostrum. A, B, D, E, after Hammer 1955; G–I, after Weigmann 2006.*

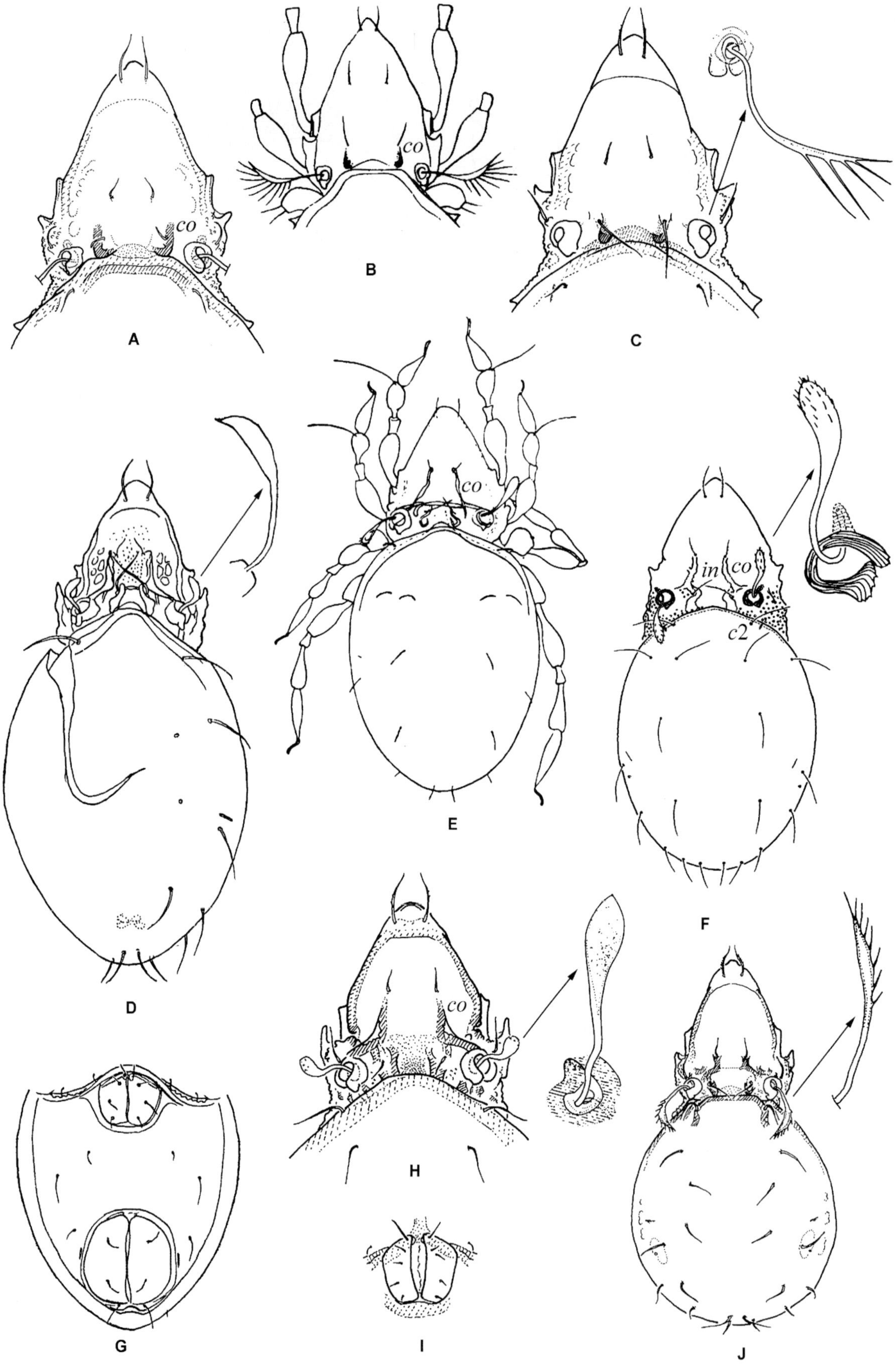

FIGURE 11 *A, B,* Rhinoppia obsoleta, *partial dorsum; C,* Rhinoppia subpectinata, *partial dorsum; D,* Moritzoppia similis, *dorsal aspect, G, anogenital region; E, F,* Moritzoppia clavigera, *dorsal aspect; H,* Oppiella splendens, *partial dorsum, I, genital plates; J,* Oppiella nova, *dorsal aspect. A, C, H–J, after Weigmann 2006; trichobothria in C, H, after Woas 1986; B, after Paoli 1908; D, G, after Gordeeva & Grishina 1991; E, after Hammer 1952; F, after Colloff & Seyd 1991.*

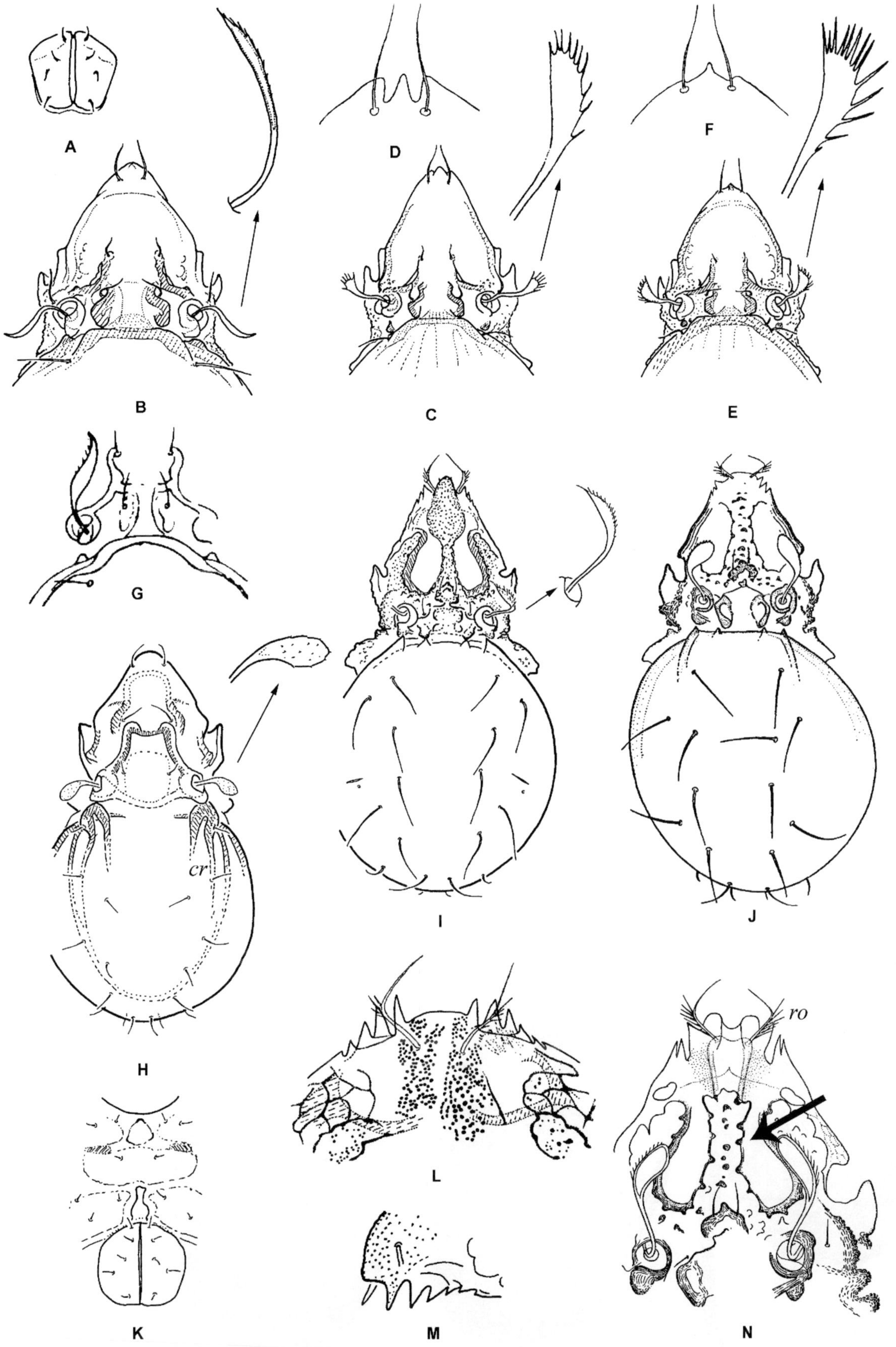

FIGURE 12 *A,* Moritzoppia neerlandica, *genital plates, B, partial dorsum; C,* Oppiella maritima, *partial dorsum, D, rostrum; E,* Oppiella acuminata, *partial dorsum, F, rostrum; G,* Moritzoppia translamellata *(Willmann, 1923) sensu Hammer 1952, partial dorsum; H,* Quadroppia quadricarinata, *dorsal aspect, K, epimeral-genital region; I,* Suctobelbella subtrigona, *dorsal aspect, L, rostrum, dorsal aspect, M, rostrum, lateral aspect; J,* Suctobelbella delessei, *dorsal aspect, N, prodorsum (arrow marks framed round field). A–C, E, H, K, after Weigmann 2006; G, after Hammer 1952; I, M, after Moritz 1971; L, after Woas 1986, modified; J, N, after Strenzke et al. 1955.*

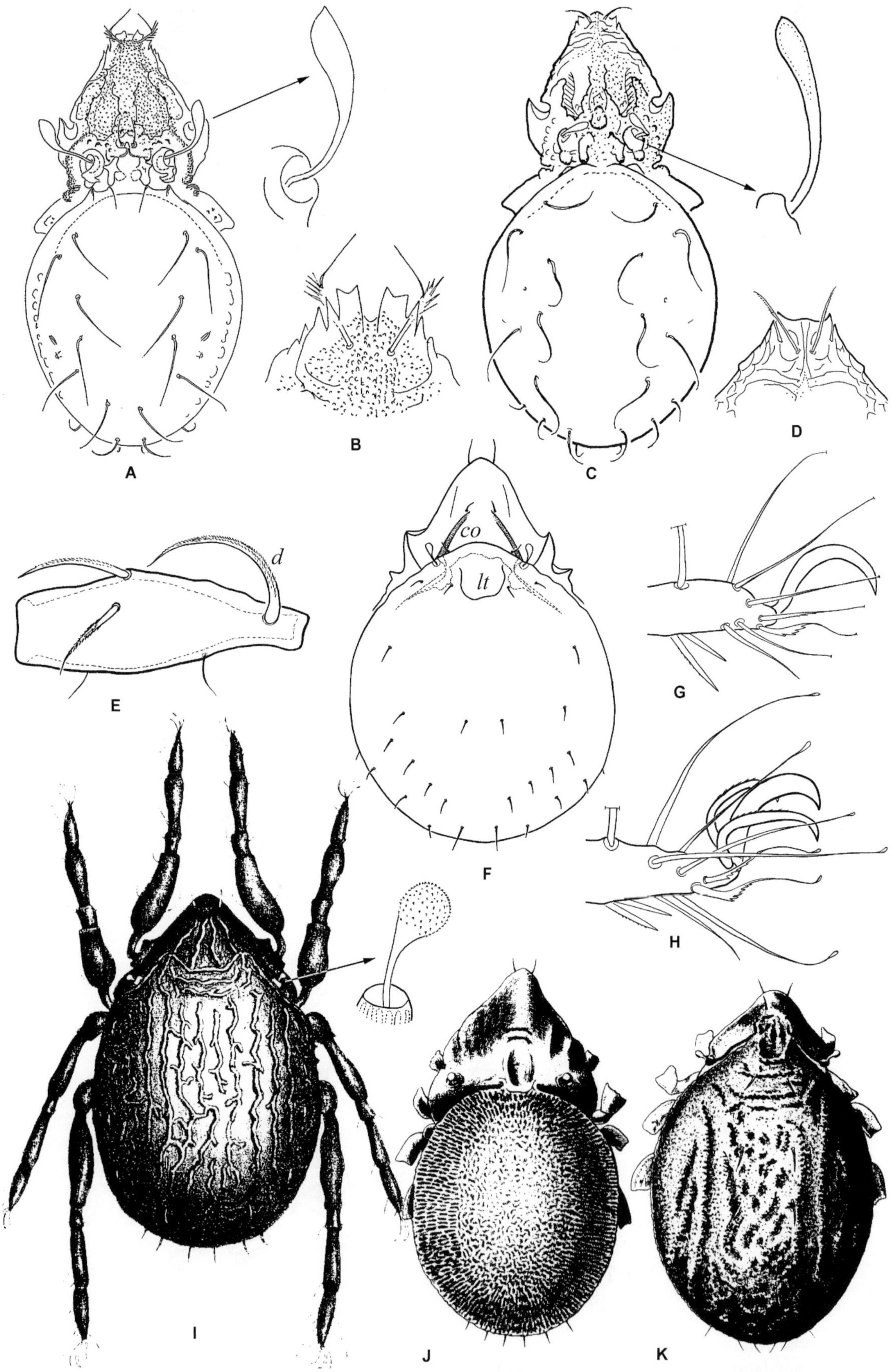

FIGURE 13 *A,* Suctobelbella acutidens, *dorsal aspect, B, rostrum, dorsal aspect; C,* Suctobelba trigona, *dorsal aspect, D, rostrum, dorsal aspect; E,* Hydrozetes confervae, *femur III, F, dorsal aspect; G,* Ameronothrus nigrofemoratus, *ambulacrum IV, K, dorsal aspect; H,* Ameronothrus lineatus, *ambulacrum IV, I, dorsal aspect; J,* Ameronothrus maculatus, *dorsal aspect. A, B, after Moritz 1974; C, D, after Moritz 1970; I, after Michael 1888; J, K, after Trägårdh 1904.*

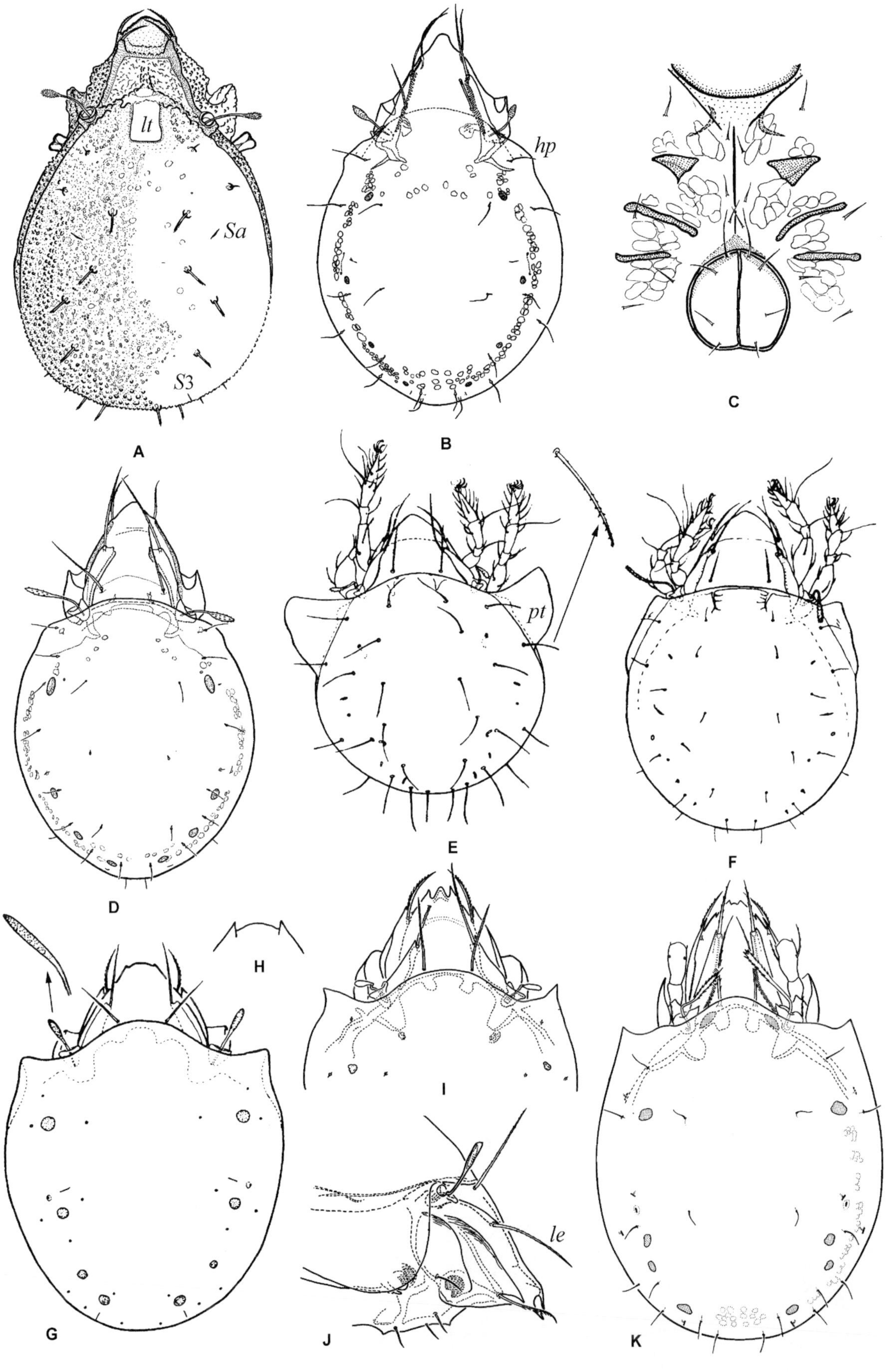

FIGURE 14 *A,* Scutovertex *sp., dorsal aspect; B,* Liebstadia similis, *dorsal aspect, C, coxisternal region; D,* Oribatula tibialis, *dorsal aspect; E,* Peloribates pilosus, *dorsal aspect; F,* Peloribates canadensis, *dosal aspect; G,* Chamobates cuspidatus, *dorsal aspect, H, rostrum; I,* Chamobates birulai, *proterosoma, dorsal aspect, J, proterosoma, lateral aspect; K,* Ceratozetes thienemanni, *dorsal aspect. B, D, G, H, after Bayartogtokh 2010; E, F, after Hammer 1952; I, J, after Beck & Woas 1991; K, after Behan-Pelletier 1984.*

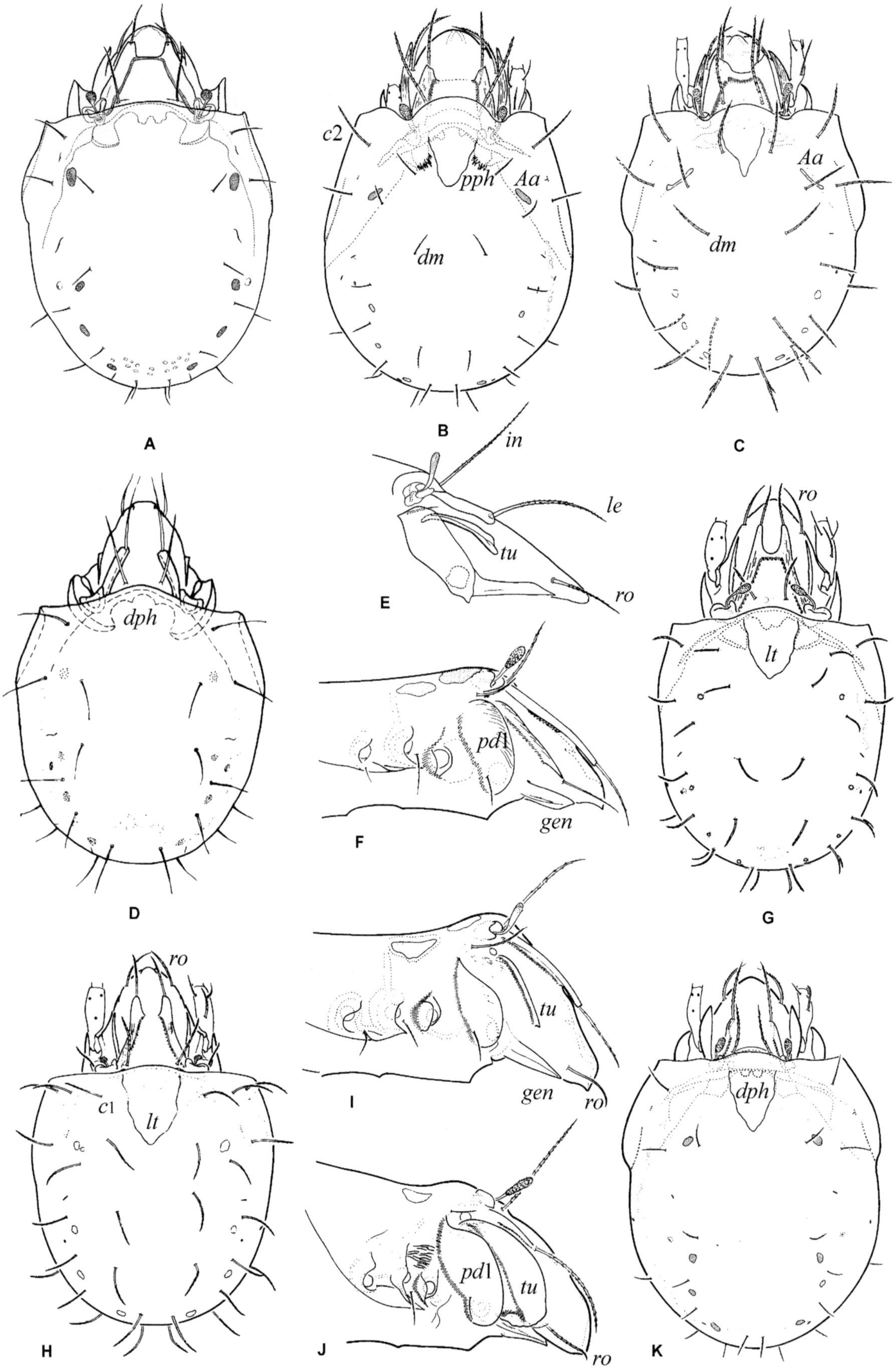

FIGURE 15 *A,* Ceratozetes *sp. near* spitsbergensis, *dorsal aspect; B,* Diapterobates variabilis, *dorsal aspect; C,* Diapterobates notatus, *dorsal aspect; D,* Edwardzetes edwardsi, *dorsal aspect, E, prodorsum, lateral aspect; F,* Fuscozetes sellnicki, *prodorsum and podosoma, lateral aspect, G, dorsal aspect; H,* Ghilarovizetes longisetosus, *dorsal aspect, I, prodorsum and podosoma, lateral aspect; J,* Iugoribates gracilis, *prodorsum and podosoma, lateral aspect, K, dorsal aspect. B, C, F–K, after Behan-Pelletier 1985; D, after Ghilarov 1975.*

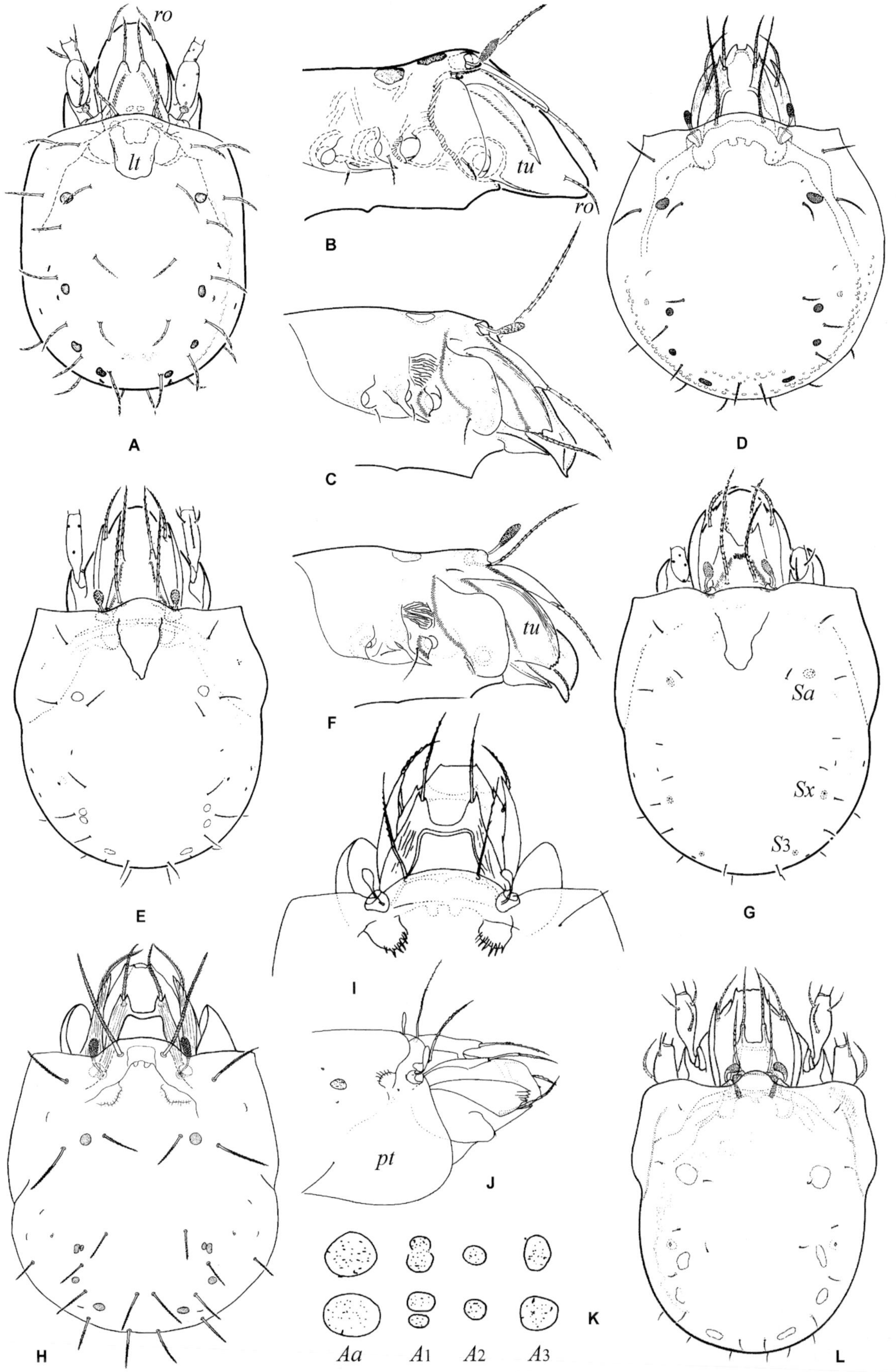

FIGURE 16 *A,* Melanozetes meridianus, *dorsal aspect, B, prodorsum and podosoma, lateral aspect; C,* Svalbardia paludicola, *prodorsum and podosoma, lateral aspect, E, dorsal aspect; D,* Oromurcia bicuspidata, *dorsal aspect; F,* Trichoribates polaris, *prodorsum and podosoma, lateral aspect, G, dorsal aspect; H,* Trichoribates biarea, *dorsal aspect, I, prodorsum and podosoma, dorsal aspect, J, prodorsum and podosoma, lateral aspect, K, notogastral porose areas; L,* Zachvatkinibates shaldybinae, *dorsal aspect. A, B, after Behan-Pelletier 1986; C, E–G, after Behan-Pelletier 1985; I–K, after Gjelstrup & Solhøy 1994, modified; L, after Behan-Pelletier & Eamer 2005.*

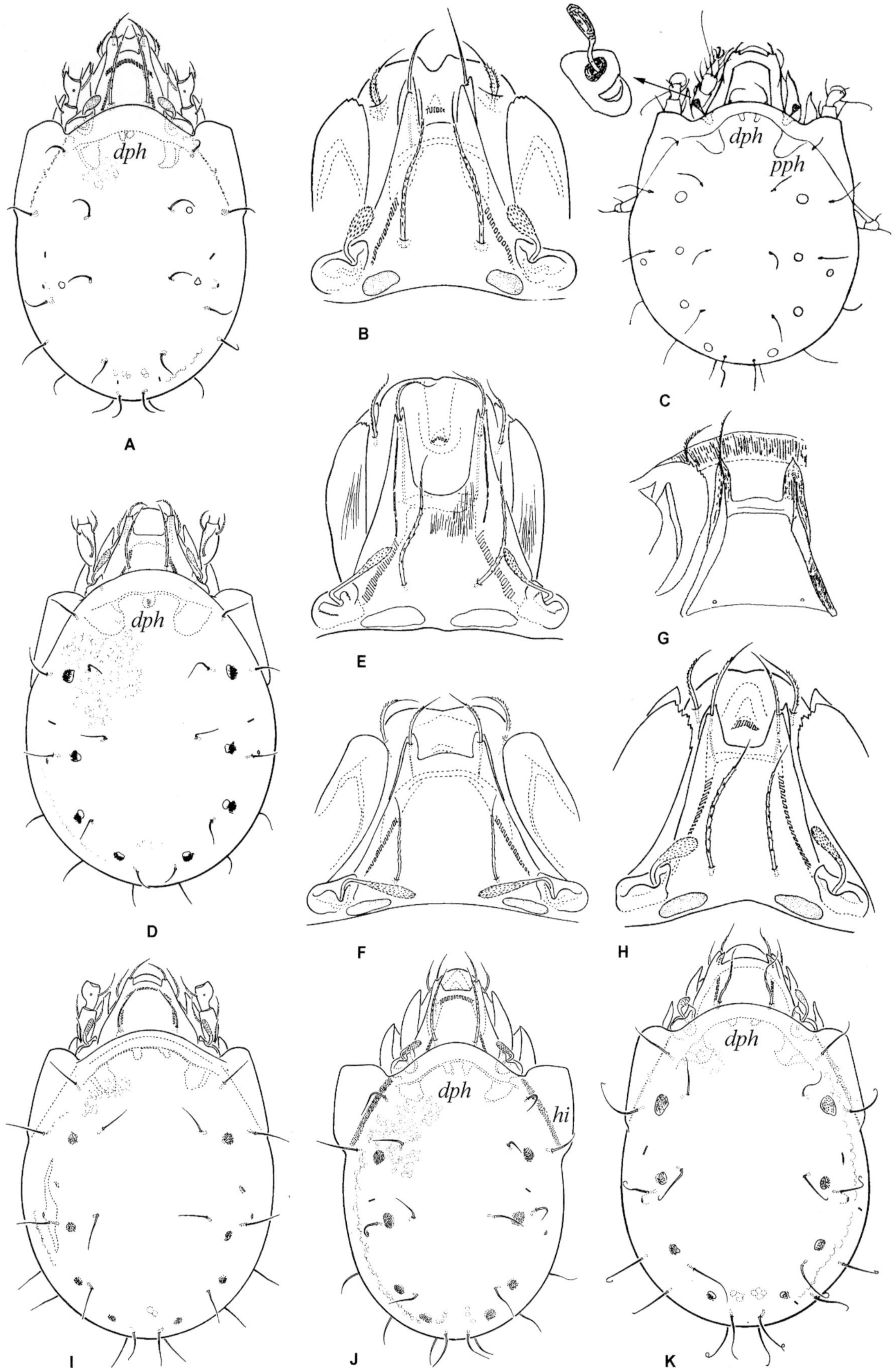

FIGURE 17 *A,* Mycobates exigualis, *dorsal aspect, B, prodorsum; C,* Mycobates sarekensis, *dorsal aspect, G, prodorsum; D,* Mycobates incurvatus, *dorsal aspect, E, prodorsum; F,* Mycobates dryas, *prodorsum, I, dorsal aspect; H,* Mycobates conitus, *prodorsum, J, dorsal aspect; K,* Mycobates beringianus, *dorsal aspect. A, B, D–F, H–K, after Behan-Pelletier 1994; C, G, after Hammer 1952.*

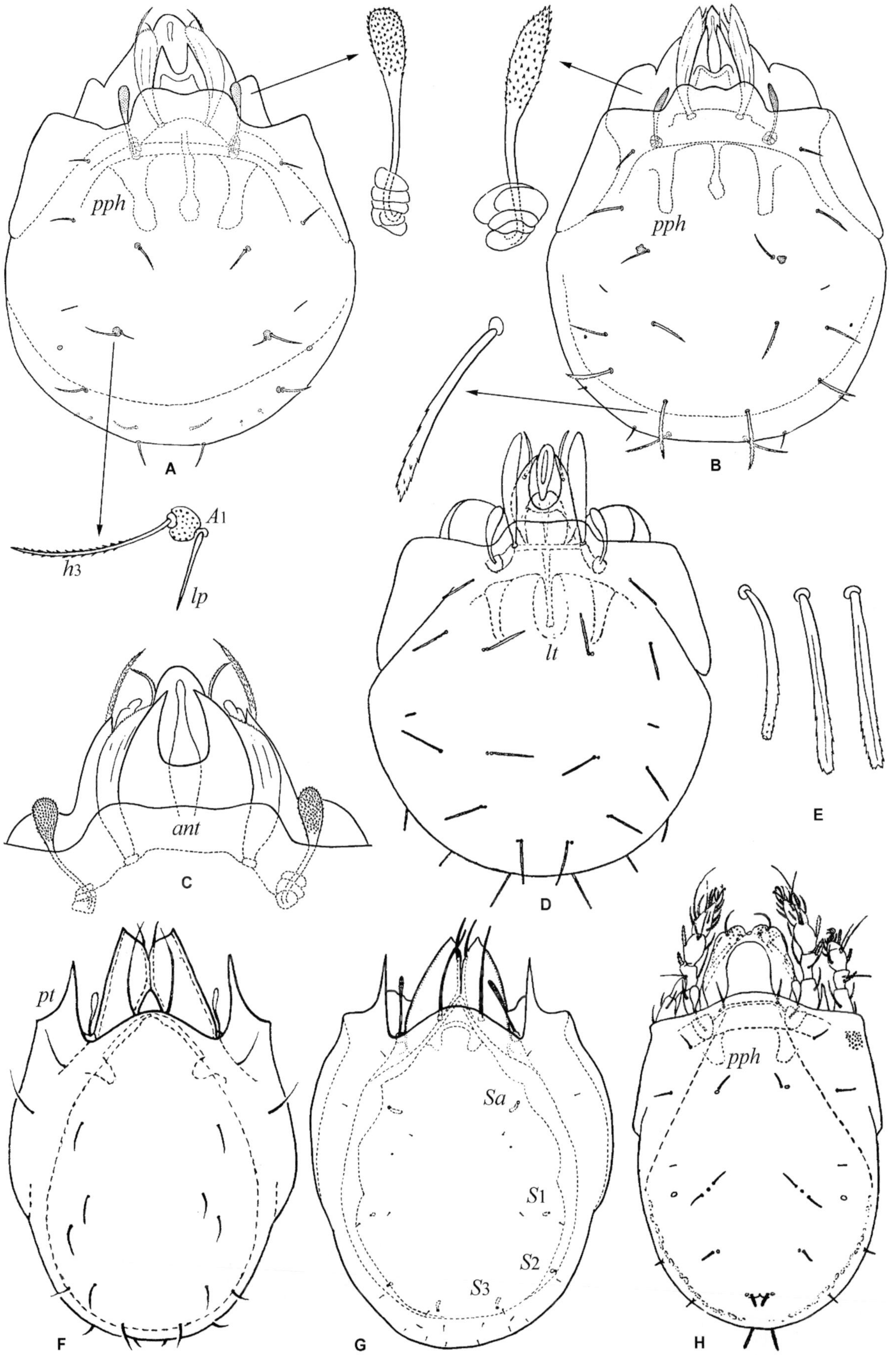

FIGURE 18 *A,* Eupelops *sp., dorsal aspect; B,* Eupelops plicatus, *dorsal aspect; C,* Eupelops septentrionalis, *prodorsum, D, dorsal aspect, E, notogastral setae; F,* Parachipteria punctata, *dorsal aspect; G,* Achipteria coleoptrata, *dorsal aspect; H,* Propelops groenlandicus, *dorsal aspect. D, H, after Hammer 1952; F, after Ghilarov 1975.*

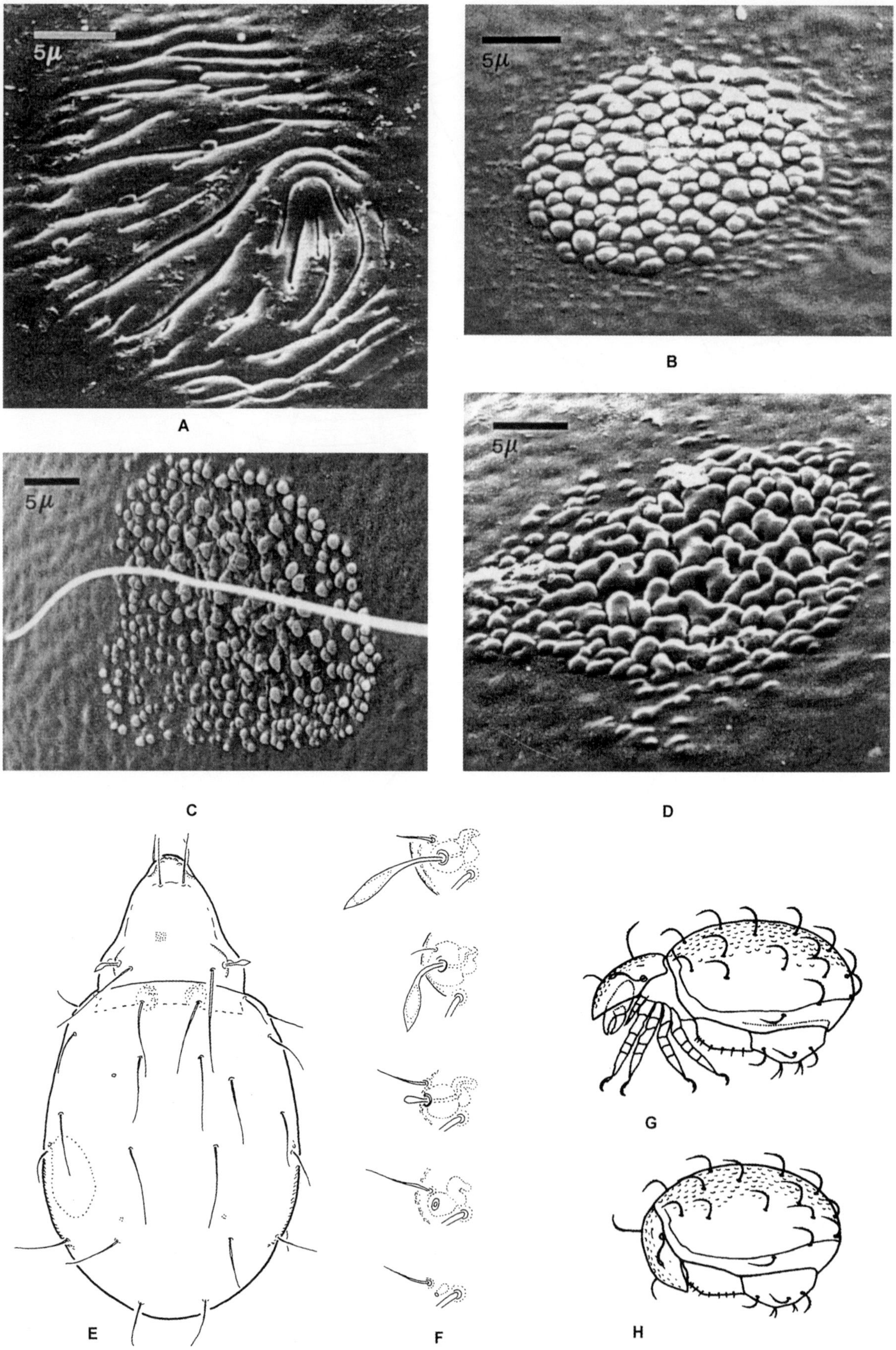

FIGURE 19 *A,* Mycobates incurvatus, *notogastral porose area* Aa; *B,* Mycobates dryas, *notogastral porose area* Aa; *C,* Mycobates beringianus, *notogastral porose area* A1; *D,* Mycobates conitus, *notogastral porose area* Aa; *E,* Trhypochthoniellus longisetus, *dorsal aspect, F, reduction levels of trichobothrium and bothridium; G, H, Phthiracaridae gen. sp., lateral aspect. A–D, after Behan-Pelletier 1994; E, F, after Weigmann 2006; G, H, after Krivolutsky et al. 1995.*

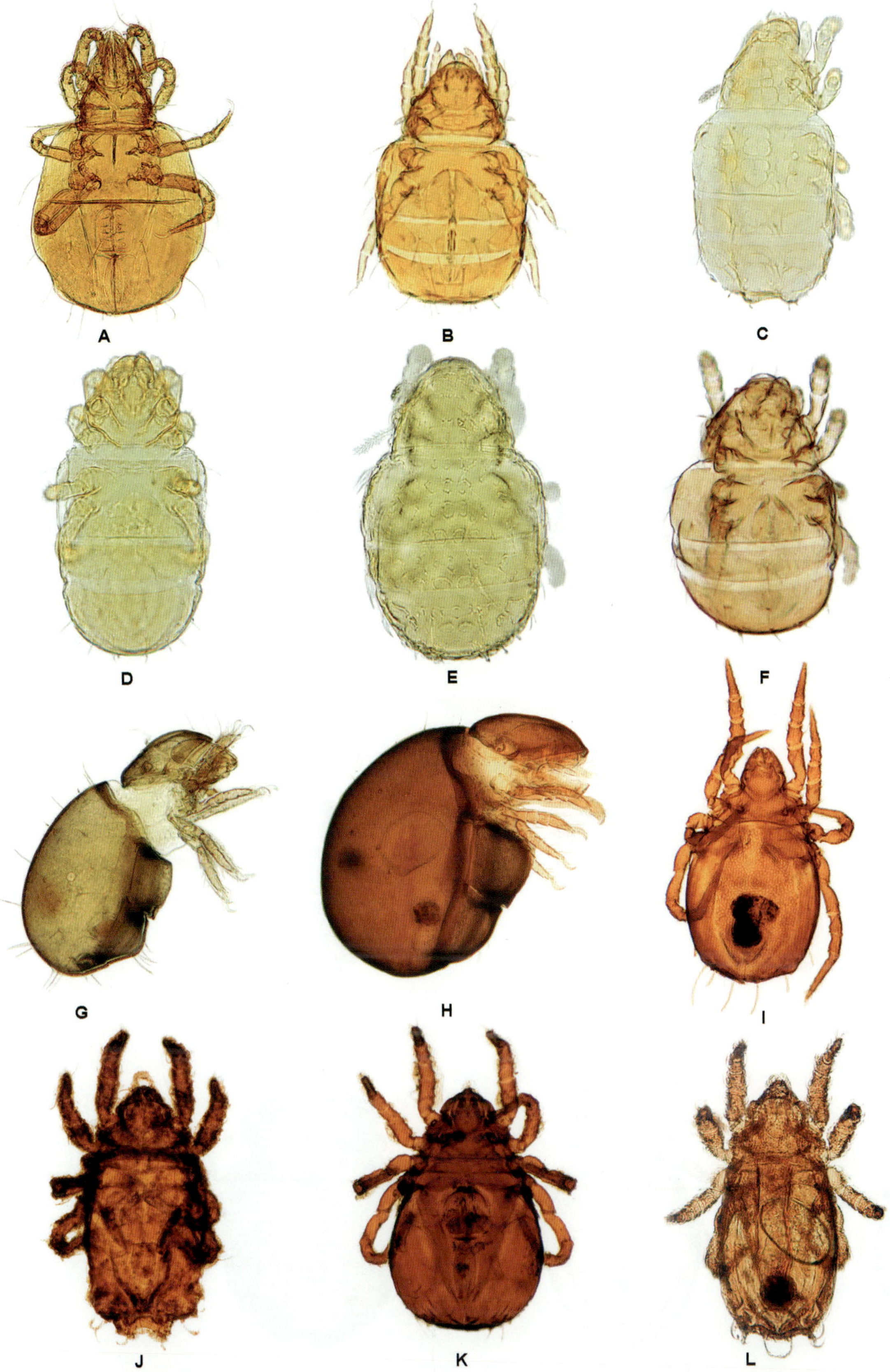

FIGURE 20 *Habitus of oribatid mites. A,* Hypochthonius rufulus*; B,* Eobrachychthonius latior*; C,* Sellnickochthonius suecicus*; D,* Verachthonius montanus*; E,* Synchthonius crenulatus*; F,* Liochthonius muscorum*; G,* Atropacarus striculus*; H,* Phthiracarus *sp.; I,* Nothrus borussicus*; J,* Camisia biverrucata*; K,* Heminothrus thori*; L,* Neonothrus humicolus. *All adults.*

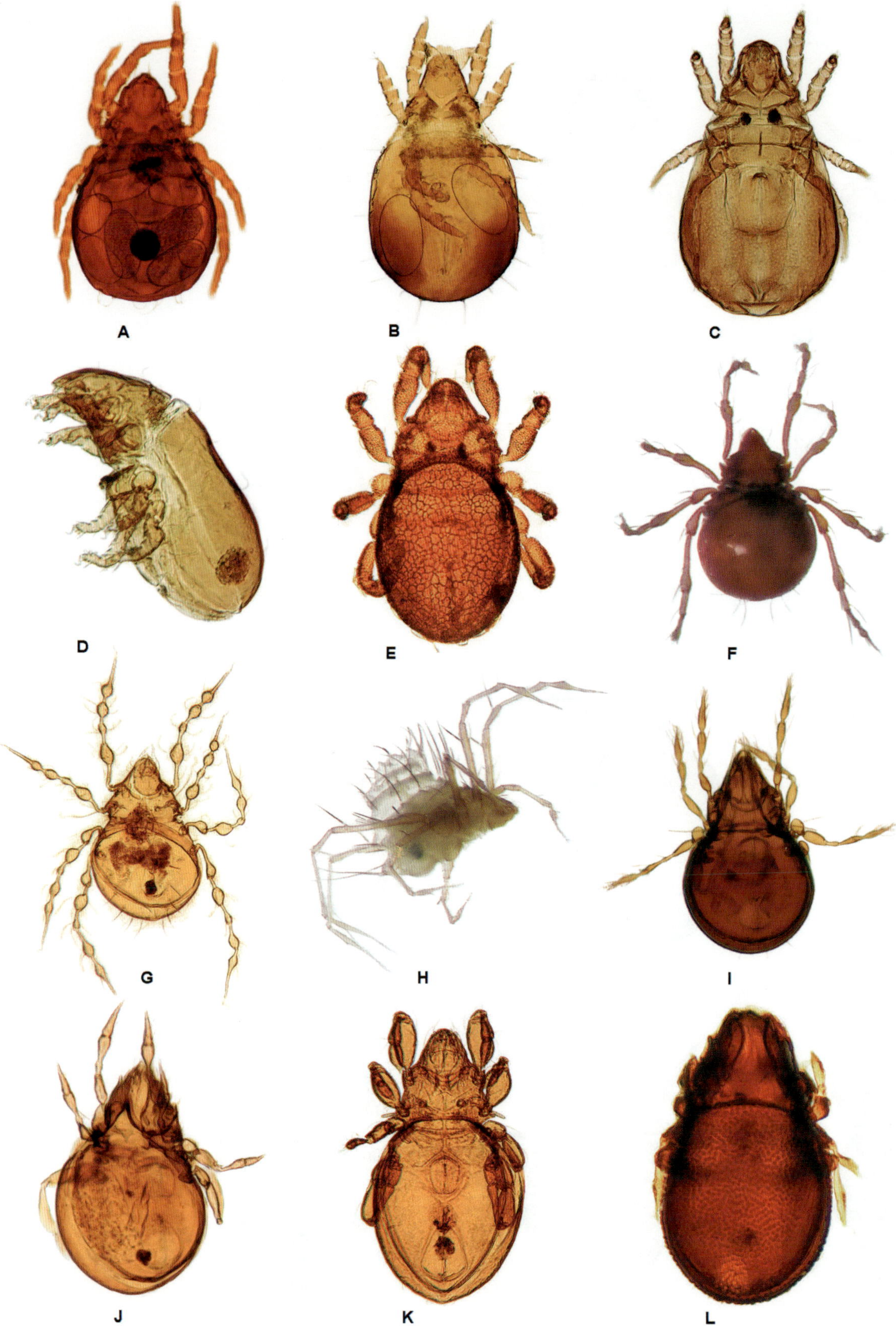

FIGURE 21 *Habitus of oribatid mites. A,* Platynothrus peltifer*; B,* Trhypochthonius tectorum*; C,* Trimalaconothrus maior*; D,* Malaconothrus mollisetosus*; E,* Hermannia reticulata*; F,* Damaeus *sp.; G,* Epidamaeus floccosus*; H, Damaeidae gen. sp.; I,* Ceratoppia sphaerica*; J,* Cultroribula dentata*; K,* Eueremaeus *sp.; L,* Carabodes labyrinthicus*. A–G, I–L, adult; H, tritonymph.*

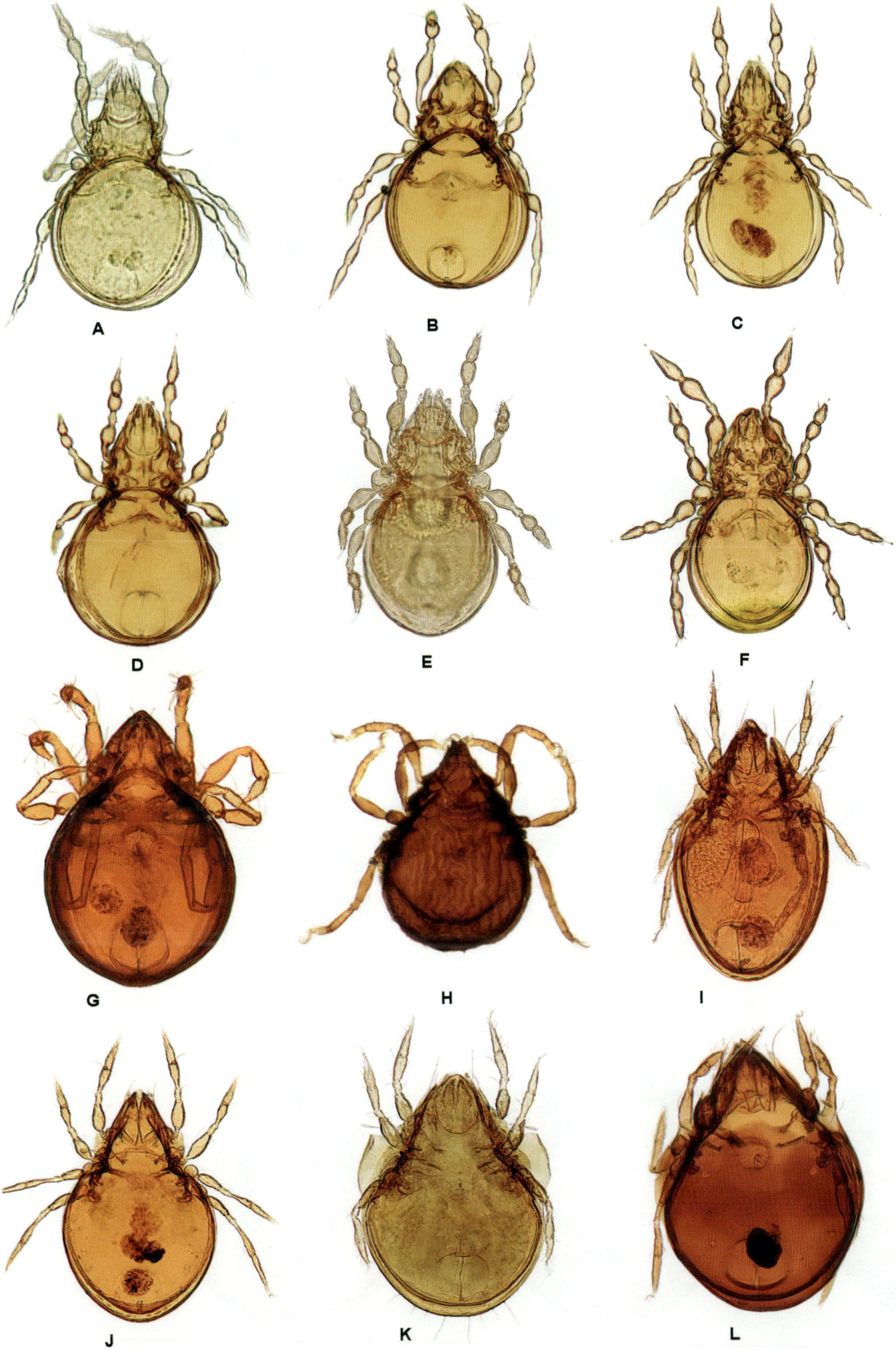

FIGURE 22 *Habitus of oribatid mites. A,* Rhinoppia obsoleta*; B,* Dissorhina ornata*; C,* Moritzoppia similis*; D,* Oppiella nova*; E,* Quadroppia quadricarinata*; F,* Suctobelbella acutidens*; G,* Hydrozetes confervae*; H,* Ameronothrus lineatus*; I,* Liebstadia similis*; J,* Oribatula tibialis*; K,* Peloribates pilosus*; L,* Chamobates cuspidatus*. All adults.*

FIGURE 23 *Habitus of oribatid mites. A,* Ceratozetes thienemanni*; B,* Diapterobates notatus*; C,* Edwardzetes edwardsi*; D,* Iugoribates gracilis*; E,* Ghilarovizetes longisetosus*; F,* Oromurcia bicuspidata*; G,* Svalbardia paludicola*; H,* Trichoribates biarea*; I,* Zachvatkinibates *cf.* shaldybinae*; J,* Mycobates sarekensis*; K,* Eupelops plicatus*; L,* Eupelops *sp. All adults.*

FIGURE 24 *Habitus of oribatid mites. A,* Tectocepheus velatus*; B,* Trhypochthoniellus *sp.; C,* Mucronothrus nasalis*; D,* Procorynetes globifer*; E,* Scutovertex *sp.; F,* Fuscozetes sellnicki*; G,* Parachipteria punctata*; H,* Achipteria coleoptrata*; I,* Heminothrus capillatus*; J,* Ceratoppia bipilis*; K,* Tectocepheus velatus*; L,* Diapterobates notatus*. A–H, adult; I–L, tritonymph.*

18.3.7. Astigmata (Acaridida)

Olga Makarova

In the recent 'A Manual of Acarology' unlike the previous ideas this extremely diverse mite group was included in the suborder Oribatida as the cohort Astigmatina within the supercohort Desmonomatides (Norton & Behan-Pelletier 2009, O'Connor 2009). This taxonomical innovation was supported by serious morphological (O'Connor 1984, Norton 1998) and biochemical (Sakata & Norton 2001) arguments. But the spermatological (Liana & Witaliński 2005) and molecular marker data (Domes *et al.* 2007) yielded very different results, that is why we keep here the traditional understanding of the Astigmata as the separate sarcoptiform suborder.

The diversity of Astigmata can be estimated to 77 families containing about 6 000 described species (Schatz *et al.* 2011). The suborder comprises free-living mites as well as highly specialized entomophilous forms and species intimately associated with vertebrates. Among Astigmata there are mycetophagous, saprophagous, microbivorous, predatory, and parasitic species. These usually slowly moving mites are rather rare in zonal arctic soils, but regularly inhabit the turf of lemming and polar fox hills as well as the litter of meadows on slopes. They may be very numerous in lemming burrows, bumblebee nests, bird colonies, and human settlement sites (compost, manure of domestic animals, garbage, stored products, house dust, *etc.*).

Morphology

Most astigmatids are weakly sclerotized, commonly lacking shields (Figs 5A, D, H, J–L). The body is divided into two regions, the *gnathosoma*, GS, and the *idiosoma*, IS (Fig. 1A). The idiosoma is usually subdivided into the *propodosoma* (PPS) and the *hysterosoma* (HS) by the *sejugal furrow* (*sf*). The idiosomal chaetome is oligotrichous. The number, size and disposition of setae are of great taxonomical importance. The dorsal surface of propodosoma, or *prodorsum*, usually has two pairs of vertical setae (*vi, ve*), two pairs of scapular setae (*si, se*), and a pair of small supracoxal setae (Fig. 1A). Prodorsal trichobothria, stigmata, and peritremes are absent. Respiration is mostly integumental. Openings of paired *hysterosomal glands* (*gla*) are normally visible between dorsolateral setae *d2* and *e2* (Figs 1A, 2L, 3A, E). In females, the large primary genital aperture, or *oviporus* (*op*), is located ventrally between the legs. It may be longitudinal (Figs 1B, 3L) or transversal (Figs 4D, F, G, J), covered by valve(s), or slit-like. The female insemination opening, or *gonopore* (*gp*), is situated terminally and leads to the *bursa copulatrix*, *bc* (Figs 1A, 3A, C, 4A). The male genital apparatus usually has a sclerotized *aedeagus* (*aed*), a tube- or gutter-like structure used for inoculation (Figs 1E, 3H). Commonly two pairs of knob-like or finger-shaped *genital papillae* (*gpp*) flank the genital orifice in both sexes (Figs 1B, D). Several pairs of setae (adanal, *ad*, and pseudanal, *ps*, ones) are inserted in the anal region (Fig. 1B). A pair of roundish *paranal suckers* (*pas*) in male (Figs 1E , 5E, K) provides fixation on the female hysterosoma during copulation. The gnathosoma is usually small with diminutive two-segmented *palps*, *pp* (Figs 1B, C). The *chelicerae*, *ch*, are chelate-dentate with *fixed* and *movable digits* (*d.f.* and *d.m.*; Figs. 1A, G), sometimes dentate with a reduced movable digit (Histiostomatidae). The *legs* consist of 6 podomeres: *trochanter* (*tr*), *femur* (*fm*), *genu* (*gn*), *tibia* (tib), *tarsus* (*tar*), and *ambulacrum* (*amb*) (Fig. 1B). The *coxae* are completely fused with the ventral idiosoma. The base of ambulacrum, pretarsus (*pt*), is without true claws, but often with an *empodial claw* (*ecl*), supported by two sclerotized *condylophores*, *cd* (Figs 1F, 2K). There are true setae, *solenidia*, *sd* (erect protrusions of the cuticle lacking alveoli), and *famuli*, *fa* (small, generally conical chaemoreceptors) on legs (Figs 1F, 3F). Setae and solenidia occur on different podomeres, whereas the famulus is present only on tarsus I and rarely II. The male often has a pair of small *copulatory suckers*, *cs*, on tarsi IV (modified setae; Figs 1H, 3G). The heteromorphic deutonymph (*hypopus*) is commonly well sclerotized, lacking mouthparts, with the gnathosoma transformed into the *palposoma*, *pls* (Figs 1I, 2J, M, 3J). Entomophilous hypopi, being phoretic on insects, possess a posteroventral *sucker plate*, *sp*, which contains several pairs of different suckers (Figs 1I, 2D, M, 3J). Hypopi phoretic on mammals usually have posteroventral ridged claspers, which hold a host's hair.

Life cycle

In females the oviporus serves only for egg-laying. The female insemination is internal via the terminal bursa copulatrix. The life cycle of mites belonging to Astigmata includes a maximum of six stages: egg, larva, protonymph, heteromorphic deutonymph (hypopus), tritonymph, and adults. The hypopus is generally highly resistant to harsh environmental conditions and usually non-feeding (Figs 5C, G, I). It

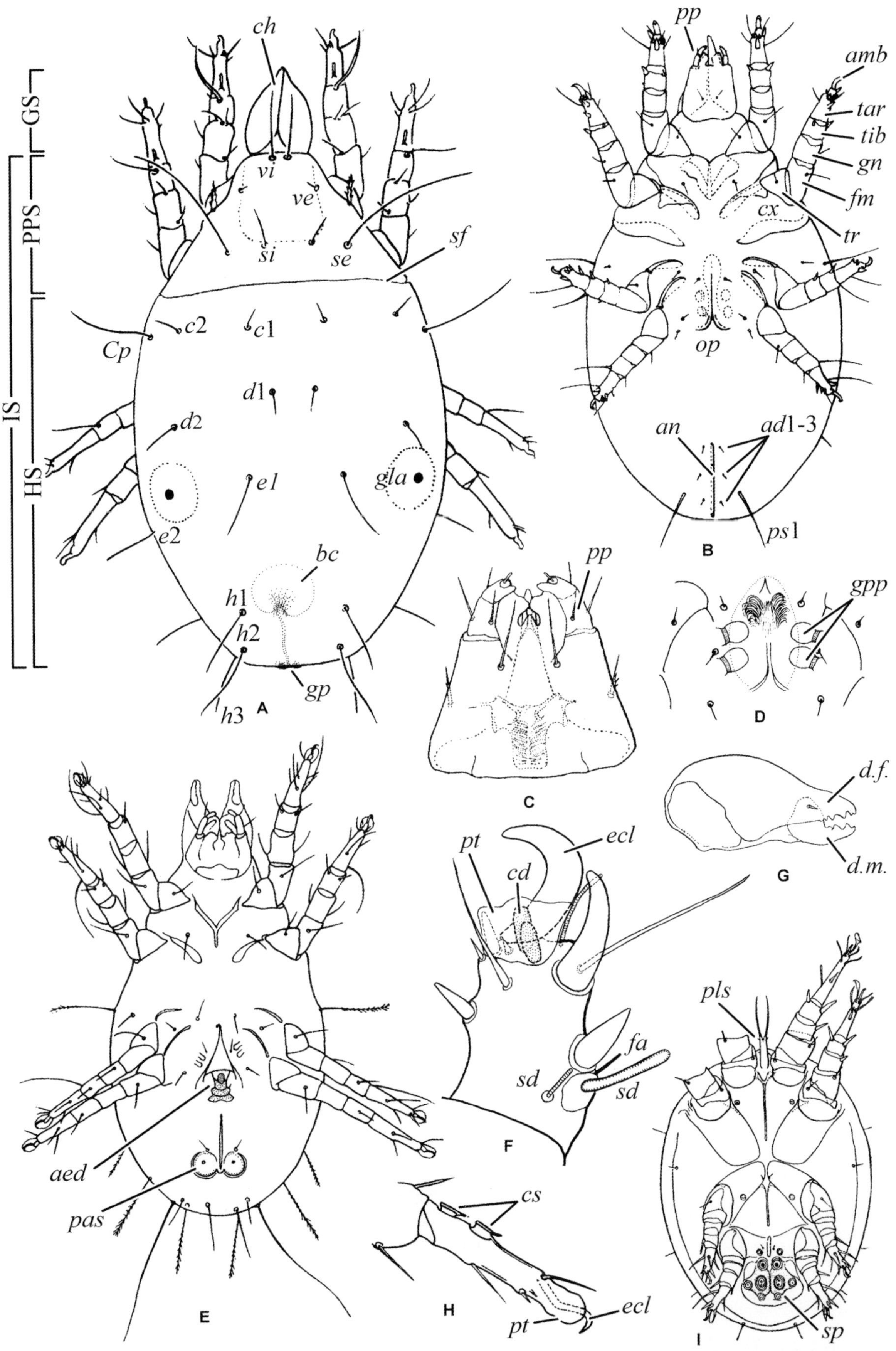

FIGURE 1 *Acaridae div., schematic. A, female, dorsal aspect; B, female, ventral aspect; C, female, gnathosoma ventrally; D, female, genital region; E, male, ventral aspect; F, female, tarsus I; G, female, chelicera; H, male, tarsus IV; I, hypopus, ventral aspect. A–C, E, G–I, after Zakhvatkin 1941.*

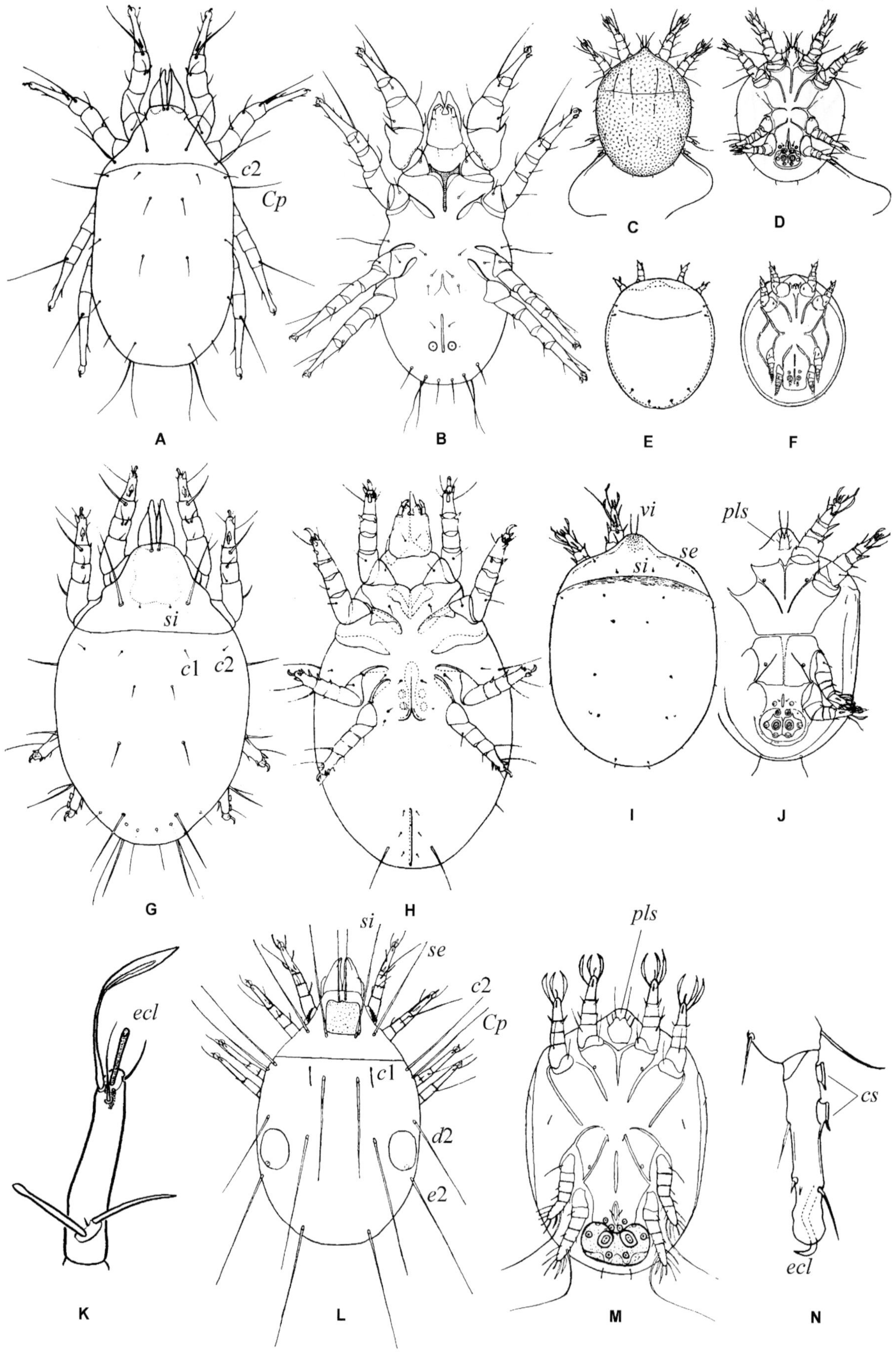

FIGURE 2 *A,* Acarus siro, *female, dorsal aspect, B, male, ventral aspect, C, hypopus (phoretic), dorsal aspect, D, ventral aspect; E,* Acarus *sp.* (siro-*complex*), *resting hypopus, dorsal aspect, F, ventral aspect; G,* Rhizoglyphus robini, *male, dorsal aspect, H, female, ventral aspect, I, hypopus, dorsal aspect, J, ventral aspect; K,* Rhizoglyphus *sp., hypopus, tarsus I; L,* Kuzinia laevis, *female, dorsal aspect, M, hypopus, ventral aspect, N, male, tarsus IV. A–J, L–N, after Zakhvatkin 1941.*

is often a dispersing stage, phoretic on insects or mammals. In some species, the hypopal stage may be lacking in favorable situations. Two types of hypopi exist – the phoretic (Figs 1I, 2M, 3J) and the quiescent ones (Figs 2F, 3M). Members of the parasitic family Hypoderatidae have aphagous adults, but feeding hypopus develop in the subcutaneous tissues of birds, consuming nutrients through the integument. Mites of some genera (including many *Tyrophagus*) do not form hypopi. Members of the group Psoroptidia (three superfamilies; mainly parasites of birds and mammals) do not have a heteromorphic deutonymphal stage at all.

Identification

The Greenland astigmatic mites are poorly known. We give here very preliminary identification keys for free-living species recorded in the soil, houses, or on insects, and only a short review of available data concerning parasitic forms. Characters used in the keys are relevant only to the Greenland fauna composition.

FREE-LIVING ASTIGMATA[1]

Key to families

ADULTS

1. Hysterosoma with two pairs of ventral sclerotized rings, *rg* (Fig. 3L); chelicerae dentate, movable digit vestigial .... Histiostomatidae
– Without ventral sclerotized rings; chelicerae chelate-dentate with well developed movable digit (Fig. 1G) 2
2. Tibia I with one seta and one solenidion (Fig. 4C); vertical seta (*vi, ve*) absent; idiosomal cuticle striated ....Pyroglyphidae, partly (*Dermatophagoides*)
– Tibia I usually with two setae and one solenidion (Figs 3D, F); setae *vi* commonly present (Figs 3A, E); body integument usually smooth without striae 3
3. Sejugal furrow present (Figs 2A, G); tarsi usually with a pair of short sclerotized condylophores supporting a well-developed empodial claw in a fleshy pretarsus (Fig. 3D); males usually with a pair of paranal suckers (Fig. 1E) and two suckers on tarsus IV (Fig. 3G)....... .... Acaridae
– Sejugal furrow absent (Fig. 4A); tarsal condylophores long, poorly visible; empodial claws reduced (Fig. 4B); males usually without paranal and tarsal suckers .... Glycyphagidae (*Glycyphagus*)

HYPOPI

1. Tarsus I with one solenidion, tibia I with two solenidia (Fig. 3K); legs III and IV generally directed anteriorly (Fig. 3J) .... Histiostomatidae
– Tarsus I with two or three solenidia, tibia I with one solenidion (Figs 3D, F); legs III and IV generally directed posteriorly (Figs 1I, 2D, J, M) 2
2. Posteroventral sucker plate usually with anterior and median suckers well-developed and setae *ps1* and *ps2* conoidal (Figs 1I, 2J, M) .... Acaridae
– Posteroventral sucker plate with anterior and median suckers vestigial or absent and setae *ps1* and *ps2* in form of ribbed claspers, small vestigial conoids, or absent; sometimes without phoretic adaptations, staying within the protonymphal skin (inert hypopus; Fig. 3M) .... Glycyphagidae (*Glycyphagus*)

ACARIDAE

Saprophagous, mycetophagous, and phytophagous species which are most common in stored food. Hypopi are usually entomophilous with distinct sucker plate. In favorable conditions, the populations can grow very quickly, causing enteritis and different allergic reactions in humans.

Key to genera

ADULTS

1. External vertical setae (*ve*), internal scapular setae (*si*), hysteronotal setae *c1* and *c2* are microsetae (Fig. 2G) or absent (Fig. 3A); empodial claws very large, robust, especially on legs I–II (Fig. 3D) 2
– External vertical setae, internal scapular setae, hysteronotal setae *c1* and *c2* well developed (Fig. 3E); empodial claws not outstanding (Figs 2N, 3F, G) ... 3
2. Setae *si, c1* and *c2* absent (Fig. 3A); female anal opening flanked by at most 2 pairs of setae (Fig. 3B); male with hysteronotal shield .... *Schwiebea*
– Setae *si, c1* and *c2* present (Fig. 2G); female anal opening flanked by more than 2 pairs of setae (Fig. 2H); male without hysteronotal shield .... *Rhizoglyphus*
3. Humeral seta (*Cp*) much longer than dorsolateral seta *c2* (Fig. 2A); two solenidia on genu I differ in length more than three times .... *Acarus*
– Humeral seta and dorsolateral seta *c2* large, almost equal in size (Figs 2L, 3E) 4

1 Initially based on Philips 1990, with modification.

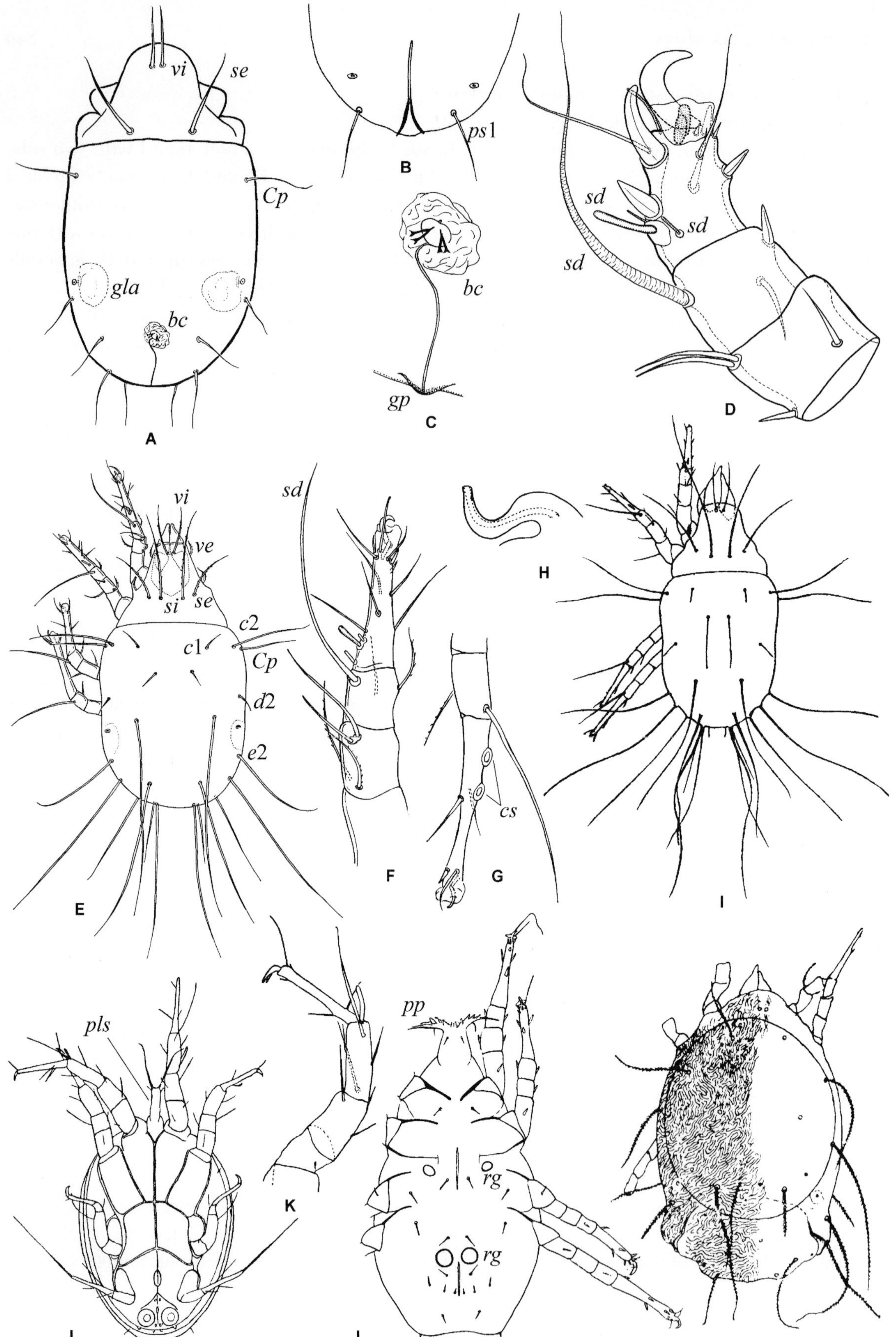

FIGURE 3 *A,* Schwiebea *cf.* danielopoli, *female, doral aspect, B, anal region, C, bursa copulatrix; D,* Schwiebea *sp., genu, tibia, and tarsus I; E,* Tyrophagus similis, *female, dorsal aspect, F, genu, tibia, and tarsus I, G, male, tarsus IV; H,* Tyrophagus putrescentiae, *male, aedeagus, I, female, dorsal aspect; J,* Myianoetus digitiferus, *hypopus, ventral aspect, K, leg I; L,* Anoetus *sp., female, ventral aspect; M,* Glycyphagus domesticus, *hypopus, dorsal aspect. E–G, after Hughes 1961, with modification; H, I, M, after Zakhvatkin 1941; J, K, after Trägårdh 1904.*

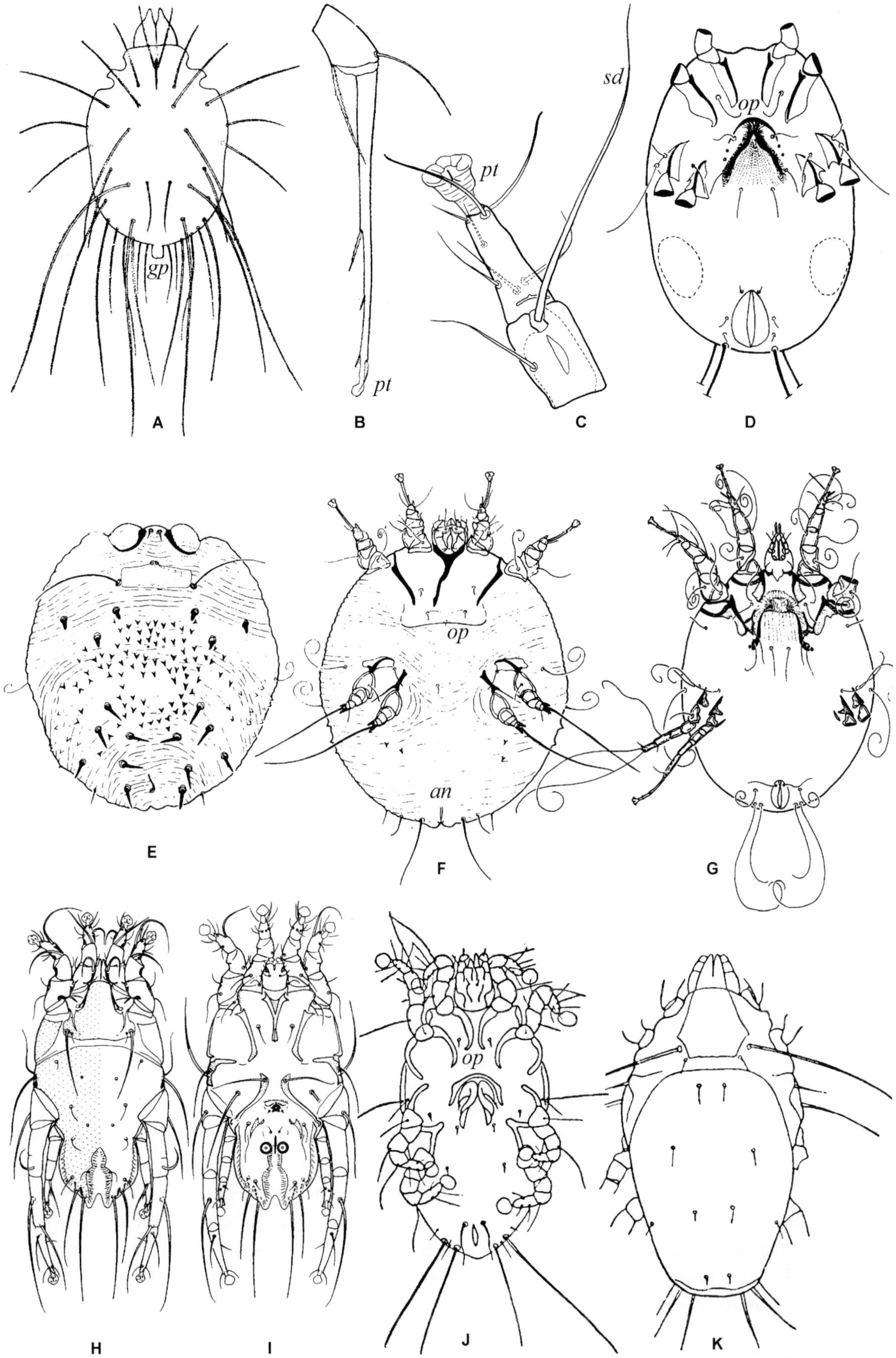

FIGURE 4 *A,* Glycyphagus domesticus, *female, dorsal aspect, B, tibia and tarsus IV; C,* Dermatophagoides pteronyssinus, *female, tibia and tarsus I, D, ventral aspect; E,* Sarcoptes scabiei, *female, dorsal aspect, F, ventral aspect; G,* Psoroptes equi, *female, ventral aspect; H,* Tectingrassia calidris *Vasjukova & Mironov, 1986* (*ex* Calidris alpina), *male, dorsal aspect, I, ventral aspect; J,* Bychovskiata charadrii, *female, ventral aspect, K, dorsal aspect. A, B, after Zakhvatkin 1941; D–G, after Krantz 1978; H, I, after Vasyukova & Mironov 1991; J, K, after Dubinin 1956.*

4. Dorsolateral seta *d2* not shorter or slightly shorter than setae *Cp*, *c2*, *e2* (Fig. 2L); tibiae and tarsi I–II with longitudinal keel; in bumblebee nests .... *Kuzinia*
– Dorsolateral seta *d2* much shorter than setae *Cp*, *c2*, *e2* (Fig. 3E); tibiae and tarsi I–II without longitudinal keel .... *Tyrophagus*

Key to genera

HYPOPI

1. Basal seta of genua I and II pectinate; scapular setae (*si*, *se*) usually long, reaching posterior edge of propodosomal shield (Fig. 2C) .... *Acarus*
– Basal seta of genua I and II smooth; scapular setae (*si, se*) usually very short (Fig. 2I) 2
2. Palposoma shorter than its basal width (Fig. 2M); associated with bumblebees ....*Kuzinia*
– Palposoma much longer than its basal width (Fig. 2J) 3
3. Vertical setae *vi* usually very small, about equal to scapular setae *si* .... *Schwiebea*
– Vertical setae *vi* usually much longer than scapular setae *si* (Fig. 2I) .... *Rhizoglyphus*

***Acarus* Linnaeus, 1758. Fig. 5B**

In Greenland one species is found, ***Acarus siro*** Linnaeus, 1758 (Figs 2A–D), recorded in old cheese, "Zwei Jahre alt im Lande" (W. Lundbeck), Ritenbenk, West Greenland (Trägårdh 1904), and in a lemming hill turf in material collected by the Danish Peary Land Expedition (1948–1949). *Acarus siro* is known as a dominant species in granaries, but found also in other commercial food deposits, meal, house dust, hay, compost heaps, bird and mammal nests. Griffiths (1970) showed that *Acarus siro* auct. is a species complex. Two species of this complex form zoochorous hypopi (phoretic on insects, birds, mammals) and one species forms the immobile hypopus with the smooth integument, reduced legs and short dorsal setae (Figs 2E, F). The reexamination of the old Greenland material of adults confirmed their belonging to *Acarus siro* s. str. Phoretic hypopi of this species have strongly foveate dorsal surface and long dorsal (especially prodorsal) setae (Fig. 2C). Female 350–670 μm, male 320–430 μm, hypopus 200–260 μm. Cosmopolitan.

***Kuzinia* Zachvatkin, 1941. Fig. 5C**

In Greenland hypopi of ***Kuzinia laevis*** (Dujardin, 1849) (=*Tyroglyphus fucorum* Oudemans, 1902; = *Tyroglyphus lundbecki* Trägårdh, 1904) are very common everywhere on *Bombus* spp., often numerous (Fig. 2M). Feeding stages (Figs 2L, N) inhabit only bumblebee nests. Female 520–700 μm, male 480–600 μm, hypopus 300–390 μm (Greenland hypopi 242–288 μm). Palaearctic.

***Rhizoglyphus* Claparède, 1869. Fig. 5I**

Two hypopi of ***Rhizoglyphus*** sp. (Fig. 5I) were extracted from the soil sample during the 3-year East Greenland Expedition (1931–1934). Members of this genus are rarely found in the Low Arctic, usually in grassy biotopes. The common *Rhizoglyphus robini* Claparède, 1869 (Figs 2G–K) probably inhabits human stored products (especially vegetables and fruits). This morphologically varying, hydrophilous species can be a serious pest in granaries and vegetable stores. Males may be homeomorphic or heteromorphic in comparison with a female. Hypopi are foretic on different dipterans and small beetles (Zakhvatkin 1941). Female 680–930 μm, male 380–780 μm, hypopus 250–370 μm. Cosmopolitan.

***Schwiebea* Oudemans, 1923. Figs 5D, E**

These are rather common mites in the Low Arctic. Usually present in small numbers in the soil, but sometimes numerous in plant debris accumulations, lemming nests *etc*. In temperate regions, the most members of *Schwiebea* occur in and under bark of rotting wood. Some species seem to be completely aquatic in habits. Mites of this genus produce hypopi facultatively. At least 3 species are found in Greenland – ***Schwiebea (Schwiebea)*** cf. ***danielopoli*** Fain, 1982 (Figs 3A–C), ***Schwiebea (Sch.)*** sp. 1 (Fig. 3D) and ***Schwiebea (Sch.)*** sp. 2. They are easily recognizable by their prominent robust claws of legs I–II, strong dorsal spurs on tarsi I–II (Fig. 3D) and substantially reduced idiosomal chaetome (Figs 3A, B). *Schwiebea danielopoli* Fain, 1982 is known from Europe and Taymyr Peninsula. Female 555–670 μm, male 570 μm.

***Tyrophagus* Oudemans, 1923. Fig. 5A**

Until now only ***Tyrophagus similis*** Volgin, 1949 (=*T. dimidiatus* (Hermann, 1804) sensu Hughes 1961) (Figs 3E–G) was found in Greenland. This characteristic species inhabits mainly litter and turf, particularly common in pastures, may sometimes be collected on plants, in mammal and *Bombus* nests. Sometimes *T. similis* was considered as a pest of crops. Female 400–610 μm, male 400–410 μm. Cosmopolitan.

Usually species of *Tyrophagus* feed on fungi, detritus and living plants, but they may also ingest nematodes (Walter *et al.* 1986). Many species do not form heteromorphic deutonymphs. Probably another member of this genus, *Tyrophagus putrescentiae* Schrank, 1781, may be found in Greenland in the future. This ubiquitous species (Figs 3H, I) inhabits both natural and anthropogenic sites.

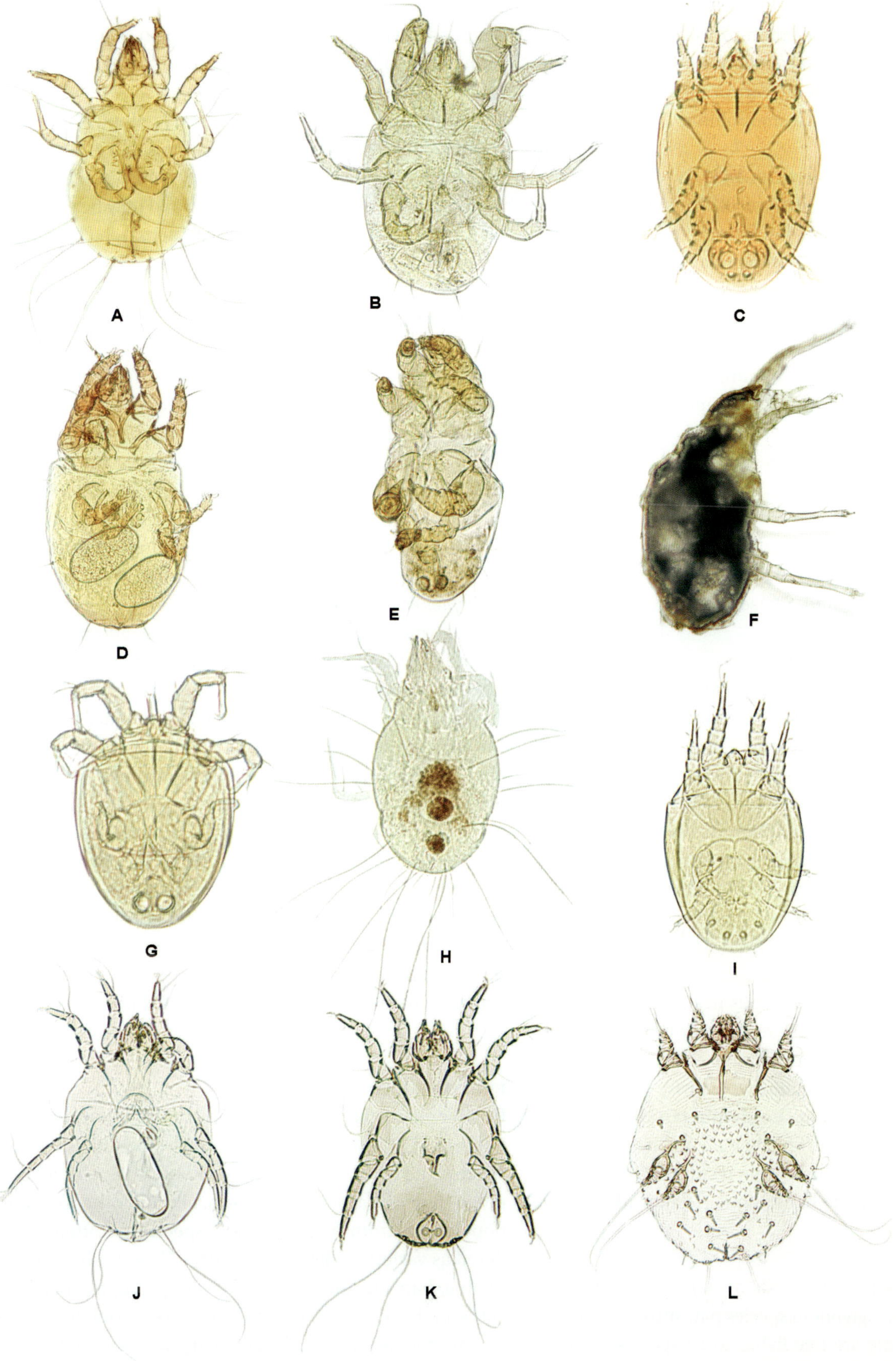

FIGURE 5 Habitus of astigmatic mites. *A,* Tyrophagus similis, *female; B,* Acarus siro, *male; C,* Kuzinia laevis, *hypopus; D,* Schwiebea *sp., female, E, male; F,* Myianoetus *sp., female; G,* Myianoetus digitiferus, *hypopus; H,* Glycyphagus domesticus, *female; I,* Rhizoglyphus *sp., hypopus; J,* Dermatophagoides pteronyssinus, *female, K, male; L,* Sarcoptes scabiei, *female.*

It is well known as a pest in laboratory cultures of insects and was recorded as a causal agent of human enteric and urogenital acariasis (Baker & Warton 1952). Female 320–415 μm, male 280–350 μm. Cosmopolitan.

HISTIOSTOMATIDAE (=ANOETIDAE) (FIGS 5F, G)

Small or middle sized, whitish mites. The palps and chelicerae are highly modified for filtering microorganisms (Fig. 3L). Hysterosoma of nonhypopal stages with two pairs of ventral *sclerotized rings* (*rg*) suspected to be organs of an osmoregulation (Witaliński *et al.* 2002). Most species prefer wet, sometimes aqueous, enriched environments and may be found in mammal nests, plant debris accumulations, excrements, *etc.* Hypopi are common on arthropods, particularly beetles and dipterans. In Greenland one species was recorded, ***Myianoetus digitiferus*** (Trägårdh, 1904). Its hypopi (Figs 3J, K) were collected on *Hydrotaea anxia* (Zetterstedt, 1838), Muscidae (Orpissuit, West Greenland) (Trägårdh 1904) and in samples of the lemming hill turf in Peary Land. Hypopus 210–250 μm. Greenland, Iceland, Surtsey.

GLYCYPHAGIDAE (FIG. 5H)

Members of this family inhabit litter and turf, but are particularly common in mammal nests and anthropogenic accumulations of organic matter (stored products, hay, composts, *etc.*). They form an important component of the acarofauna of house dust (Hughes 1976). Some species are endofollicular parasites. Phoretic hypopy disperse mainly on mammals. Inert hypopi (with reduced legs and fastening structures) remain within the protonymphal skin (Fig. 3M) and are dispersed passively. In Greenland represented by 1 species, ***Glycyphagus domesticus*** (De Geer, 1778) (Figs 4A, B). It is known mainly as a synantropic species, especially numerous in hay and composts (Zakhvatkin 1941), but single records from bird nesting places, grassy turf on meadows and other natural substrates are also known (Hughes 1961, 1976). Female 350–750 μm, male 270–400 μm. Cosmopolitan. It has been found on Franz Josef Land, Spitsbergen and in Greenland, but very probably these finds were results of sample contamination in human habitations (Zakhvatkin 1941, Hughes 1961).

PYROGLYPHIDAE (PART) (FIGS 5J, K)

Although many pyroglyphoid species parasitize vertebrates (mainly birds) some are free-living. Most species occur in mammal and bird nests, stored products, human habitations (beddings, furniture, dust of floors), feeding on skin scales, feather and hair matter, and probably also on fungi and pollen (Wharton 1976). Pyroglyphids have not yet been recorded in Greenland. Two cosmopolitan species populate human dwellings throughout the world, *Dermatophagoides farinae* (Hughes, 1961) and *Dermatophagoides pteronyssinus* (Trouessart, 1897) (Figs 4C, D), causing atopic dermatitis, rhinitis, conjunctivitis, bronchial asthma, and other allergic diseases. Abundance of mites in house dust usually increases in periods without heating and may be especially high in traditional village buildings (up to 1000 specimens per gram of dust).

PARASITIC ASTIGMATA

SARCOPTIDAE (FIG. 5L)

Members of this family live on or in the skin of mammals including man. The itch-mite, ***Sarcoptes scabiei*** (De Geer, 1758) (Figs 4E, F), is a highly variable species comprising more than 10 varieties according to the host species. They are parasitic mainly on domestic animals, but known also from wild hosts. Sometimes the heavy infestation results in death of hosts, suggesting a fairly recent parasite-host association. It was suggested (Fain 1978), that *Sarcoptes scabiei* arose from another sarcoptid genus parasitizing monkeys, and that domestic animals rather recently got this mite from man. Their systematic status (species, subspecies, varieties, strains) is a subject of discussion, because the morphological differences between varieties are small. Some forms originating from different hosts display reproductive isolation. *Sarcoptes scabiei scabiei* (=*Sarcoptes scabiei* var. *hominis*) was rather common on people in West Greenland in the beginning of the 20th century, causing scabies or sarcoptic mange. In unsanitary, overpopulated conditions the transmission of scabies can be intensive. After infestation on human skin erythematous patches appear which cause severe itching, and scratched places may be subjected to secondary infections. Female 270–410 μm, male 170–210 μm. Cosmopolitan.

PSOROPTIDAE

In Greenland one species was recorded, ***Psoroptes equi*** (Hering, 1838) (=*Psoroptes longirostris* Mégnin, 1880). This species, the common scab mite (Fig. 4G), has a broad host range (horses, cattle, sheep, goats, elk, and others) and its populations from particular host species are often considered as varieties (Baker *et al.* 1956). They produce severe itching, hair loss and pruritis. Severe psoroptic mange in cattle and sheep may be fatal (Roberts & Meleney 1970). Female 760–860 μm, male 580–680 μm. Cosmopolitan.

Undoubtedly, another psoroptid species lives in the ears of Greenland domestic cats and dogs, *Otodectes cynotis* (Hering, 1838). Ear mites may cause strong irritation in the external auditory canal.

FEATHER MITES

Feather mites belong to three superfamilies comprising over 2700 species (Schatz *et al.* 2011) inhabiting feathers and skin of birds throughout the world. Some of them feed on feather pith or skin but others consume feather oils and material trapped in it (scurf, pollen, fungi, *etc.*) without causing structural damage (Proctor 2003). One bird species may be populated by several mite species which live only on feathers of special type or even on particular parts of feathers. Usually the preferred habitat is well reflected in mite morphology, especially in the males (Figs 4H, I). Feather mite phylogeny often parallels host phylogeny, however the 'host-jumping' cases are found in several mite lineages (Proctor 2003).

Greenland feather mites should be studied more carefully. Until now we have only data from an old investigation (Trägårdh 1904). Here is only given a list of these finds with some comments.

PTEROLICHIDAE

Pterolichus (*Pseudalloptes*) *bimucronatus* (Trouessart, 1884) was recorded on the rock ptarmigan, *Lagopus mutus,* in West Greenland (Trägårdh 1904). This find is highly doubtful. It is very probable that mites developing on *Lagopus mutus* belong to an undescribed species of the genus *Tetraolichus* Atyeo & Gaud, 1992 (S. Mironov, pers. comm. 2007).

AVENZOARIIDAE

Bychovskiata charadrii (Canestrini, 1878) (=*Pterolichus charadrii*) was found on the ringed plover, *Charadrius hiaticula,* in West Greenland (Trägårdh 1904). This mite species (Figs 4J, K) also inhabits other sandpipers of the genera *Charadrius, Actitis, Philomachus, Gallinago,* and *Tringa* (Dubinin 1956, Vasyukova & Mironov 1991). Female 380–400 µm, male 320–330 µm. Cosmopolitan.

Zachvatkinia puffini (Buchholz, 1869) (=*Pteronyssus puffini*) was mentioned on the Arctic tern, *Sterna paradisaea,* and Sabine's gull, *Xema sabini,* in West Greenland (Trägårdh 1904). But species of the *puffini*-group are associated with birds of the order Procellariiformes (Mironov 1989). Accordingly, reinvestigations are necessary. The mite found on *Sterna paradisaea* very probably belongs to *Zachvatkinia sternae* (Canestrini et Fanzago, 1876) (S. Mironov, pers. comm. 2007).

ANALGIDAE

Analges (Analges) longispinosus Tyrrell, 1882 was found on the snow bunting, *Plectrophenax nivalis,* in West Greenland (Trägårdh 1904). Holarctic.

ALLOPTIDAE

Alloptes (*Alloptes*) *crassipes* (Canestrini, 1878) was mentioned from the black guillemot, *Cepphus grylle,* in West Greenland (Trägårdh 1904). However, this mite species is common on sandpipers of the genus *Calidris* (Dubinin 1951). Reinvestigation is needed. Obviously mites obtained from *Cepphus grylle* belong to *Alloptes* (*Alloptes*) *cepphusi* Dubinin, 1951, common on this host (S. Mironov, pers. comm. 2007).

Alloptes (*Conuralloptes*) *bisetatus* (Haller, 1882) was recorded on the purple sandpiper, *Calidris maritima,* in West Greenland (Trägårdh 1904). But this species is considered to be associated only with different terns (Dubinin 1951). Reinvestigation is necessary. Most probably the mites found belong to *Alloptes* (*Conuralloptes*) *eroliae* Dubinin, 1951 known just from *Calidris maritima* (S. Mironov, pers. comm. 2007).

References

Baker, E.W., Evans, T.M., Gould, D.J., Hull, W.B. & Keegan, H.L. (1956): A Manual of Parasitic Mites of Medical or Economic Importance. – National Pest Control Association, New York. 170 pp.

Baker, E.W. & Wharton, G.W. (1952): An Introduction to Acarology. – Macmillan, New York. 465 pp.

Domes, K., Althammer, M., Norton, R.A., Scheu, S. & Maraun, M. (2007): The phylogenetic relationship between Astigmata and Oribatida (Acari) as indicated by molecular markers. – *Experimenatal & Applied Acarology* 42: 159–171.

Dubinin, W.B. (1951): Feather Mites (Analgesoidea). Part I. Introductuion to their Study. – *Fauna SSSR, Paukoobraznye* 6(5). Zoological Institute, Acad. Nauk SSSR, N.S. No 43. 364 pp. [In Russian].

Dubinin, W.B. (1956): Feather Mites (Analgesoidea). Part III. Family Pterolichidae. – *Fauna SSSR, Paukoobraznye* 6(7). Zoological Institute, Acad. Nauk SSSR, N.S. No 63. 814 pp. [In Russian].

Fain, M.D. (1978): Epidemiological problems of scabies. – *International Journal of Dermatology* 17(1): 20–30.

Griffiths, D.A. (1970): A further systematic study of the genus *Acarus* L., 1758 (Acaridae, Acarina), with a key to species. – *Bulletin of the British Museum of Natural History, Zoology* 19(2): 85–118.

Henriksen, K.L. & Lundbeck, W. (1917): Landarthropoder (Insecta et Arachnida). Conspectus Faunae Groenlandicae, Pars secunda. – *Meddelelser om Grønland* 22: 481–823.

Hughes, A.M. (1961): Terrestrial Acarina III. Acaridiae. – *The Zoology of Iceland* III(57c): 12 pp.

Hughes, A.M. (1976): The Mites of Stored Food and Houses. – Ministry of Agriculture, Fisheries and Food Technical Bulletin No 9. HMSO, London. 400 pp.

Krantz, G.W. (1978): A Manual of Acarology. 2nd edition. – Oregon State University Bookstores, Corvallis. 509 pp.

Liana, M. & Witaliński, W. (2005): Sperm structure and phylogeny of Astigmata. – *Journal of Morphology* 265(3): 318–324.

Mironov, S.V. (1989): New subgenus and three new species of feather mite genus *Zachvatkinia* from Procellariiformes birds. – *Parasitologia, St. Petersburg* 23(4): 309–318.

Norton, R.A. (1998): Morphological evidence for the evolutionary origin of Astigmata (Acari: Acariformes). – *Experimental & Applied Acarology* 22: 559–594.

Norton, R.A. & Behan-Pelletier, V.M. (2009): Chapter 15. Suborder Oribatida. – *In*: Krantz, G.W. & Walter, D.E. (eds.) A Manual of Acarology: 430–564. Texas Tech Univ. Press, Lubbock.

O'Connor, B.M. (1982): Astigmata. – In: Parker, S.P. (ed.). Synopsis and Classification of Living Organisms 2: 146–169. McGraw-Hill, New York.

O'Connor, B.M. (1984): Phylogenetic relationships among higher taxa in the Acariformes, with particular reference to the Astigmata. – *In*: Griffiths, D.A. & Bowman, C.E. (eds.) Acarology VI, 1: 19–62. Ellis Horwood, Chichester.

O'Connor, B.M. (2009): Chapter 16. Cohort Astigmatina. – In: Krantz, G.W. & Walter, D.E. (eds.) A Manual of Acarology: 565–658. Texas Tech Univ. Press, Lubbock.

Philips, I.R. (1990): Acarina: Astigmata (Acaridida). – In: Dindal, D.L. (ed.). Soil Biology Guide: 757–778. John Wiley, New York.

Proctor, H.C. (2003): Feather mites (Acari: Astigmata): ecology, behavior, and evolution. – *Annual Review of Entomology* 48: 185–209.

Roberts, I.H & Meleney, W.P. (1970): Ear-scab mites, *Psoroptes cuniculi* (Acarina: Psoroptidae), in captive mule deer. – *Journal of Parasitology* 56: 1039–1040.

Sakata, T. & Norton, R.A. (2001): Opisthonotal gland chemistry of early derivative oribatid mites (Acari) and its relevance to systematic relationships of Astigmata. – *International Journal of Acarology* 27: 281–292.

Schatz, H., Behan-Pelletier, V.M., O'Connor, B.M. & Norton, R.A. (2011): Suborder Oribatida van der Hammen, 1968. – *In*: Zhang, Z.-Q. (ed.) Animal Biodiversity: An Outline of Higher-Level Classification and Survey of Taxonomic Richness. *Zootaxa* (special volume) 3148: 141–148.

Trägårdh, I. (1904): Monographie der arktischen Acariden. – *Fauna Arctica* 4: 1–78.

Vasyukova, T.T. & Mironov, S.V. (1991): Feather Mites from Anseriformes and Charadriiformes of Yakutia (Systematics). – "Nauka" Publ. House, Novosibirsk. 201 pp. [In Russian].

Walter, D.E., Hudgens, R.A. & Freckman, D.W. (1986): Consumption of nematodes by fungivorous mites, *Tyrophagus* spp. (Acarina, Astigmata, Acaridae). – *Oecologia (Berlin)* 70(3): 357–361.

Wharton, G.W. (1976): House dust mites. – *Journal of Medical Entomology* 12: 577–621.

Witaliński, W., Liana, M. & Alberti, G. (2002): Fine structure and probable function of ring organs in the mite *Histiostoma feroniarum* (Acari: Actinotrichida: Acaridida: Histiostomatidae). – *Journal of Morphology* 253: 255–263.

Zakhvatkin, A.A. (1941): Tyroglyphoid Mites (Tyroglyphoidea). – *Fauna SSSR, Paukoobraznye* 6(1). Zoological Institute, Acad. Nauk SSSR, N.S. No 28. 475 pp. [In Russian].

CHAPTER 19

Chilopoda (Centipedes)

Jens Böcher

The chilopod head bears a single pair of simple, filiform antennae and in most cases ocelli. Three pairs of forward-directed mouth appendages are situated on the ventral side of the head: mandibles, 1st maxillae, 2nd maxillae. A pair of 'poison-claws' (maxillipeds) formed by the modified first trunk legs cover the mouthparts from below; each of these contains a poison gland. Centipedes are carnivorous, preying mostly on small arthropods.

In mature members of the lineage (Lithobiomorpha) to which the Greenland representatives belong the poison-claw segment is followed by 15 leg-bearing segments. Between the last of these and the terminal telson are two small, limbless segments, the pregenital and genital segment; the gonopores are located on the latter. The fairly strong legs are laterally directed. Respiration is by tracheae opening through lateral spiracles. Sperm transmission is indirect. Following a courtship the male typically constructs a small silken web on which a spermatophore is placed; it is subsequently taken up by the female and inserted into her gonopore. The newly eclosed juveniles have 7 pairs of legs behind the poison-claws; additional leg-bearing segments are added with the following moults.

FIGURE 1 Lamyctes emarginatus.

Only one species is free-living in Greenland. However, a single specimen of the ubiquitous *Lithobius forficatus* (Linnaeus, 1758) has once been taken in a building in Ivittuut, an obvious human introduction.

HENICOPIDAE

***Lamyctes emarginatus* (Newport, 1844) Fig. 1**

Böcher & Enghoff 1984: *Lamyctes fulvicornis* Meinert, 1868

About 10 mm. The hindmost pair of legs are long and antenna-like.

BIOLOGY. Unlike most other centipedes this species prefers fairly dry and exposed biotopes with low vegetation.

The species is parthenogenetic in most of its range; males are only known from the Canary Islands, the Azores and in Australia. *L. emarginatus* is one of the terrestrial arthropods which might have been introduced by the Norse people. However, it is globally so widespread, with a number of isolated ocurrences, that it may as well be native to Greenland.

DISTRIBUTION. S (only found in Narsarsuaq and Kangillinguit/Grønnedal).

Holarctic (except Siberia), subarctic-subtropical, but also found in the Australian region.

Reference

Böcher, J. & Enghoff, H. (1984): A centipede in Greenland: *Lamyctes fulvicornis* Meinert, 1868 (Chilopoda, Lithobiomorpha, Henicopidae). – *Entomologiske Meddelelser* 52: 49–50.

CHAPTER 20

Introduced orders

Jens Böcher

A few insect species introduced into Greenland belong to orders with no native representatives in the country. Some of them are probably introduced repeatedly without being able to establish themselves in Greenland, even indoors. All three species are mentioned in Henriksen (1939).

Dermaptera (Earwigs)

The common European ***Forficula auricularia*** Linnaeus, 1758 has been found once. Like most earwigs it is likely near-omnivorous (feeding on plant as well as animal material), and has maternal care.

Orthoptera (Grasshoppers and Crickets)

One species has been found once, ***Diestrammena asynamora*** (Adelung, 1902) (Henriksen 1939: *Tachycines asynamorus*) (Rhaphidophoridae). An Asian species which is common in European greenhouses. It is a wingless predator, feeding on other insects.

Dictyoptera (Mantises, Cockroaches, Termites)

A single species of cockroach, ***Blatta orientalis*** Linnaeus, 1758 has been recorded once. The flattened body shape, with overlapping wings, very long antennae and conspicuous cerci are characteristic features. A harmless nocturnal omnivore.

Reference

Henriksen, K.L. (1939): A revised index of the insects of Grønland. – *Meddelelser om Grønland* 119(10): 1–112.

Glossary

A glossary of many of the terms (in singular and plural where likely relevant) applied in this work to the treatment of hexapods and spiders. The mite terms are explained/illustrated in the respective introductory chapters. Many hexapod terms are used across all/several orders, others have more restricted use, or are used in different senses in different taxa (and/or by different authors). Uppercase three-or-four-letters abbreviations of higher taxa, sometimes followed by abbreviations of lower-level taxa, are given where considered useful.

A(nalis)	posterior principal longitudinal vein(s) in insect wings
ac setae	DIPT, acrostichal setae
acrostichals	DIPT, aligned bristles close to midline on mesonotum
aedeagal	pertaining to aedeagus
aedeagus	male intromittent organ, or part thereof
alula	DIPT, term used in different senses for basal lobes of the posterior wing margins; in DIPT-syrph it is the sizable, most distal lobe
AME	ARAN, anterior median eyes
anepisternum	upper part of the episternum, which may be demarcated by a sulcus/furrow/slit from the lower part (katepisternum)
antenn|a/-ae	most anterior appendage pair on the head of insects and myriapods
antennomere	segments in antenna
anus	hind gut opening
apex	tip, end
apical	at or near apex
apneustic	a tracheal system without open spiracles
apodeme	infolding from cuticle-bearing body wall; may come in many forms (plate-like to rod-like/tubular) and usually accommodating muscle insertions/origins
apotome	anteromedian part of head capsule of immature hexapod, delimited by cleavage lines
areola	HYM, median area of propodeum
areolet	HYM-Ich, second submarginal cell in FW (small cell in wing center)
arist|a/-ae	slender, bristle-like or plumose part of antenna beyond basal flagellomere in cyclorrhaphan flies
aristomeres	the individual segments in the arista
basimere	SIPH, main part of paramere
basitarsus	most proximal tarsomere
bothridia	rounded depressions from which trichobothria arise
bristle	a large, stout seta
bulbus	ARAN, organ on male palp, used to storage sperm, and for copulation
bulga	SIPH, part of flea spermatheca
C, c(osta)	most anterior principal longitudinal vein in insect wings
calamistrum	ARAN, a comb or brush-like series of curved hairs on metatarsus IV used for spinning silk from cribellum
calcar	HYM, variably modified apcical tibial spur
calcipala	DIPT-Sim, apical process on first tarsomere of hind leg
calypter	DIPT, posterior lobe(s) of wing base
carapace	a shield, for instance the fused terga of the archnid prosoma
card|o/-ines	most basal piece of hexapod maxilla
carina	a keel

cauda	HEM-Aphid, variously modified terminal abdominal segment
cephalic lobe/ outgrowth	ARAN, an outgrowth of carapace emerging from ocular area, or carrying at least one pair of eyes.
cephalic part	ARAN, the anterior (head) portion of carapace
cercus/-ci	HEX, the segmental appendage of the posteriormost abdominal segment
chela	pincer
chelicere	ARACH, the most anterior appendage-pair
chorion	the egg shell
cibarium	the 'pre-oral' part of the anterior food tract, i.e., the cavity in front of the morphological mouth, enclosed by the mouth parts
circumfila	DIPT-Cec, specialized composite chemoreceptive sensilla on the antennal flagellum
claspette	DIPT, inner appendage borne on basal part of male gonopod
clava	HYM, terminal ± widened part of antennal flagellum, may comprise more segments
clavate	club-shaped
clavus	HEM-Het, posterior/inner basal part of corium of hemelytron of, delimited by a furrow; SIPH, antennal flagellum
claw	small sclerotized and curved appendage borne on pretarsus, usually paired
***-clinate**	bent towards *
clypeus	HEX, anteromedian area of head capsule, apically bearing labrum and variably delimited from more dorsal/proximal frons; ARAN, the area between the anterior row of eyes and the edge of carapace
cordiform	heart-shaped
corium	HEM-Het, inner firm/pigmented part of hemelytron
cornicul|a/ae	DIPT-Psych, process on adult cranium bearing scent organ
cornut|us/-i	sclerotized spine-like process on evaginable inner sac (endotheca) in male phallic organ
cox|a/-ae	most basal free segment of the leg
cribellum	ARAN, a sieve-like plate in front of spinnerets used for producing fine silk.
crossvein	transverse vein between principal longitudinal veins in insect wing
ctenidium	SIPH, comb-like series of stout bristles
Cu(bitus)	principal longitudinal vein in insect wings, located just in front of anal vein(s)
cuneus	HEM-Het, distal area sometimes demarcated along front margin of corium
cymbium	ARAN, the broadened-hollowed terminal part of the male palp, to which the bulbus is attached.
dc setae	DIPT, dorsocentral setae
dens	COLL, the paired segments of the furca, arising from the manubrium, and each bearing the mucro
distiphallus	distal section of male phallus
dorsocentrals bristles/setae	DIPT, sublateral seta row on mesoscutum, laterad from acrostichals
dorsum	the upper side of a tagma or segment
elytr|on/-a	firm/pigmented fore wing, which in repose covers and protects the abdomen and hind wing (when present)
empodium	unpaired, variably shaped lobe below the paired claws of the pretarsus
epandrium	dorsal sclerotization on abdominal segment IX
epigastric furrow	ARAN, a transverse .groove across the anterior ventral part of the abdomen
epigyne	ARAN, the sclerotized region of the adult female spiders covering the internal genitalia and located between book lungs and anterior of epigastric furrow
epimer|on/-a	HEX, part of thoracic pleuron behind pleural ridge
epiproct	terminal part of abdomen, located (morphologically) above the anus
epistern|um/-a	part of thoracic pleuron in front of pleural ridge

epomia	HYM-Ich, oblique ridge crossing transverse furrow on pronotal side
exarate	pupal type with antennal, mouthpart, wing and leg sheaths free from general body surface, hence permitting considerable mobility for the pharate adult
face	anterior part of head capsule, may be morphologically composite
facial orbit	HYM, lateral margins of lower face bordering compound eyes
fasci|a/ae	transverse band on an insect wing
fem|ur/-ora	third free segment of the leg
flagell|um/-a	the (usually long) distal part of the antenna of true insects beyond the second (pedicellus)
flagellomeres	the segments in the antennal flagellum
forceps	tongs, sometimes used for male gonopods
fovea	ARAN, a depression (peat) of epigyne bearing insemination openings
frons	anteromedian part of head
fronto-orbital bristles	DIPT-Cyclor, row of bristles on anterior head surface, adjacent to eye margin
fronto-orbital plate	DIPT-Cyclor, median area on adult head above parafacial area
funicle	funiculus
funiculus	HYM, subterminal segments in antennal flagellum
furca	COLL, the posterior abdominal appendages with a common basal shaft (manubrium) bearing the paired segments dens and mucro
FW	fore wing
gale|a/-ae	inner lobe borne on maxillary stipes, laterad from lacinia
gaster	HYM, part of abdomen in apocritans; variably applied to entire metasoma or just the portion behind the petiole
gena|-ae	lateral part of head capsule below/behind the compound eye
gloss|a/-ae	inner lobe borne on prelabium, mediad from paraglossa
glymma	HYM-Ich, lateral concavity on metasomal segment 1 between its base and the spiracle
gonite	DIPT, paramere
gonocox|a/-ae	ventral appendage on abdominal segment IX in male insects, usually considered the basal part of the segmental appendage
gonocoxite	proximal segment of two-segmented ventral appendage on abdominal segment IX in male insects, usually considered to be the proximal part of the segmental appendage
gonoforceps	HYM, male gonopods flanking phallic organ
gonopod	one-or two-segmented ventral appendage on abdominal segment IX in male insects, usually considered to be the segmental appendage
gonostyl|us/-i	distal segment of two-segmented ventral appendage on abdominal segment IX in male insects, usually considered to be the distal part of the segmental appendage
grapnel	DIPT-Sim, cuticular hook named after its resemblance to devices for anchoring ships
halophile	salt-loving
halter/-ers	DIPT, reduced and modified hind wings
hemelytr|on/-a	specialized fore wing type in HEM-Het, with firm/pigmented inner part and membranous distal part
hemicephalous (/-cephalan/ -cephalic)	DIPT, larvae with a partially retained, externally visible head capsule
hilla	SIPH, part of spermatheca
holoptic	insect head with compound eyes enlarged and contiguous in mid-line
humer|us/-i	DIPT, pro-postnotal lobe
HW	hind wing
hypandrium	DIPT, formation ascribed to male abdominal venter IX

hypoclypeal depression HYM-Brac, space formed by emargination of clypeus and concave labrum
hypophallus sclerites in male intromittent organ
hypopharynx tongue-like median lobe forming the 'floor' of the preoral cavity, mainly formed by medioventral parts of mandibular and maxillary segments; DIPT, usually used for the 'lonchus', which also comprises a labial component
hypopygium term used in various senses for formations of abdominal apex
hypovalvae processes from female abdominal venter VIII

idiobiont a parasitoid larva which prevents its host from continuing development, paralyzing it permanently
inferior appendage TRIC, male gonopod
infuscate dark ('smoky') blackish-brown

katepisternum the lower part of the episternum, which may be demarcated by a sulcus/furrow/slit from the upper part (the anepisternum)
koinobiont a parasitoid larva which allows its host to continue development for some time

labellum/-lae DIPT, the more or less pronouncedly bilobed soft-walled apical part of adult labium (palp homologue), bearing 'pseudotracheae'
labrum anteromedian lobe (of debatable morphological origin) from head capsule, ± extensively covering paired mouth appendages, forming functional 'upper lip'.
labium medially fused posterior mouth appendage pair in hexapods (behind maxillae), forming functional 'lower lip'
lacinia/-niae most medial inner lobe borne on maxillary stipes, mediad from galea
laterigrade ARAN, crab-like orientation with legs directed outward from the body
laterotergite DIPT, lateral part of longitudinally divided tergum
latimeres HYM, male gonopods flanking phallic organs
Lauterborn organs DIPT-Chir, composite sentory organ on second antennal segment in larva
ligula composite median process of prelabium, including glossa and paraglossae
linea calva HYM, non-microtrichiated line in wing

macrotrichi|um/-a seta
malar space HYM, shortest distance between compound eye and mandibular base
mandible anterior mouth appendage pair in chilopods and hexapods; usually strong biting jaws, but variously modified (sometimes stylet-like), reduced or lost in some subordinate lineages
manubri|um/-a COLL, basal shaft of furca; SIPH, handle-like extension of male clasper body
maxilla/-ae HEX, second pair of mouth appendages, located beween mandibles and labium; CHIL, the homologous appendage pair is '1st maxilla', while the labium homologue is '2nd maxilla'
maxillary palp segmented appendage borne on the outer apex of maxillary stipes
maxilliped first appendage pair behind the 'true mouth appendages' in some arthropod groups; the chilopod 'poison claw' is a maxilliped
M(media) principal longitudinal vein in insect wings between R/Rs and Cu systems
median apophysis ARAN, an apophysis rising from the center of the bulbus
median plate ARAN, a sclerite often hidden, at least partly, in epigastric furrow
mediotergite DIPT, median part of longitudinally divided tergum
mentum distalmost part of postlabium (where the latter is subdivided); in DIPT-Chir scale-like
meron/-ra the posterior portion of an insect coxa which is divided by a longitudinal costal ridge; in higher DIPT mesomeron incorporated into the lateral body wall
mes- pertaining to the mesothorax
mesoseries arrangement of proleg crotchets in a row on the median side of the planta
mesosoma HYM, thorax + propodeum in apocritans

mesothorax	second thoracic segment in hexapods
met-	pertaining to the metathorax
metapneustic	a tracheal system with only posterior spiracles open
metasoma	HYM, abdomen behind 'wasp-waist' in apocritans
metathorax	third (hindmost) thoracic segment in hexapods
microtrichium/ia	small, non-socketed, purely cuticular 'hair'; the term is sometimes restricted to such hairs of which more than one is formed per underlying epidermal cell
mola, molar	grinding surface of mandible
moult	the transition from one instar to another in arthropods; involves the separation of epidermis and cuticle, the formation of a new cuticle, and (with a few notable exceptions such as the late larval instars of higher flies) the shedding of the old one
mucro	COLL, the small terminal segment of the furca, borne on the dens
nasale	in COL larvae anteromedian process (of variable morphological composition) from head capsule
nasus	DIPT-Tipul, apical process on rostrum
notaul|us/-i	HYM, line(s) on mesoscutum delimiting a median mesocutal lobe from lateral mesoscutal lobes
notopleural	DIPT, one or two strong setae found posterior to the postpronotal (or humeral) lobe
notum	the tergum of a thoracic segment
obtect	pupal type with antennal, mouthpart, wing and leg sheath glued by a secretion to general body surface, hence little mobile
occipital	situated on occiput
occiput	upper posterior surface of head capsule
ocellus/-celli	simple eyes (at most three, one median and two paramedian) borne on anteromedian part of head capsule in many hexapods (not paraneopteran nymphs and endopterygote larvae)
operculum	a lid
opisthosoma	ARACH, legless posterior main part of body behind prosoma/cephalothorax
oral sclerite	DIPT, part of the maggot cephaloskeleton
orbital bristles/setae	DIPT-Cyclor = fronto-orbital or more dorsally positioned setae/bristles
osmeterium	scent-producing area
ovipositor	device on the insect female postabdomen formed by paired appendages on segment VIII and IX and serving the placement of the eggs (in Hymenoptera-Apocrita also, and in the Apocrita-Aculeata only, serving injection of venom), alternatively often used for oviscapt; in Opiliones used for a functionally analogous process adjacent to the gonopore
oviscapt	a posteriorly narrowed female postabdomen, the segments of which are 'telescoped' into each other; the design permits oviposition in narrow crevices
palea	ARAN, the upper-retrolateral part of bulbus in Lycosidae, just above the tip of embolus
palp/-i	HEX, distal/external appendage of maxilla and labium, serial homologue of main part of locomotory limb; HEM-Psyl, female genital appendages; ARAN, the second appendage of prosoma, its coxa forms spider's 'maxillae', and the male is palp strongly modified and used for copulation.
palp|us/-i	palp
PAO	COLL, postantennal organ (sense organ situated behind antennal base)
paracymbium	ARAN, an appendage of the cymbium, located in basal (Araneidae, Linyphiidae, Tetragnathidae) or apical (Theridiidae) part. Usually it is rigid, but can be flexible (Linyphiidae)
parafacial	DIPT-Cyclor, area between ptilinal fissure and compound eye
paraglossa	outer lobe borne on prelabium, laterad from glossa
parameres	processes flanking male phallic organ, may have different morphological origins

parapods	DIPT-Chir, prolegs on larval prothorax and terminal abdominal segments
parastomal bar	DIPT, part of the maggot cephaloskeleton
patagium/-ia	LEP, a protuberance on the prothoracic dorsum; DIPT-Psych, scent organ
patella	ARACH, leg segment between femur and tibia (in ACAR = genu)
pecten	a comb
pedes spurii	prolegs
pedicel	pedicellus
pedicell|us/-i	second segment of insect antenna
pedipalp	the second pair of appendages of chelicerates, also referred to as palp
pedisulcus	DIPT-Sim, notch in hind margin of second tarsomere of hind leg
peritreme	sclerotization surrounding a spiracle
petiole	a stalk; HYM, constricted abdominal segment(s) immediately behind propodeum in apocritans
phallosome	part of phallus
phallus	male intromittent organ
pharynx	anterior part of food tract, following immediately behind cibarium
planta	terminal surface of the proleg of a larval proleg
pleural ridge/sulcus	internal strengthening ridge (sometimes accompanied by an external groove) on thoracic pleural sclerotizations; in wing-bearing segments running between pleurocoxal articulation and pleural wing process
pleural wing process	a process from the upper margin of the pleural sclerotization of a winged segment; it articulates with a sclerite ('2nd axillary') in the wing base
pleurite	a ± discrete sclerotization in the pleural area
pleuron/-ra	lateral area between tergum and sternum in a segment; somctimes used specifically for a sclerotization in this area
plumose	feather-like
PME	ARAN, posterior median eyes
postlabium	basal main part of hexapod labium, its surface sclerotization (if present) may be subdivided into a basal submentum and a distal mentum
postnotum	posterior part of a notal plate, often ± separated from preceding part (and morphologically belonging to following segment)
postocellar	behind ocelli
postpronotal lobe	DIPT, lateral lobe of adult's pro-postnotum, closely associated with mesonotum
precoxal bridge	sclerotized connection between a thoracic sternum and pleuron in front of the leg base
prelabium	distal main part of hexapod labium, bears the labial palp, glossa and paraglossa
premandible	DIPT, paired processes on labrum
prepectus	HYM, small sclerite situated between pronotum and mesopleuron
prescutum	the anteriormost part of a pterothoracic notum
pretarsus	the distal segment leg of a hexapod leg
pro-	in compound words often = pertaining to prothorax
procercus	DIPT-Chir, setose tubercle in front of larval anus
proctiger	abdominal segment X
prograde	ARAN, leg orientation with the anterior pair directed forward and the posterior pairs backwards
proleg	± leg-like appendages (of debatable morphological nature, but at least in some cases likely serial homologues of thoracic legs) on abdominal segments of some endopterygote larvae (and occasionally pupae)
propectus	HYM, functional unit (comprising fused cervical sclerites and propleron plus prosternum and forelegs) connecting head with bulk of thorax
propneustic	a tracheal system with only anterior spiracles open
propodeum	HYM, first abdominal segment in apocritans, closely connected to metathorax and situated in front of 'wasp-waist'
prothorax	the anterior thoracic segments

pteropleuron	the pleuron of a winged segment
pterostigma	subapical sclerotized and/or pigmented marking on front margin of wing (sometimes just FW)
pterothorax	the wing-bearing segments of a winged insect
ptilinal fissure	DIPT-Cyclor, inverted-U shaped scar/groove on adult head demarcating evagination site of ptilinum at adult emergence
pulvill|us/-i	small softwalled lobe(s) on the pretarsus below the claws; unlikely homologous in all arthropods
pulvinus	PSO-Pht, a modified hyaline labrum assisting mandibles in anchoring the louse to the host integument
puparium	the specialized cocoon of a cyclorrhaphan fly, formed by the sclerotized cuticle of the penultimate larval instar
R, r(adius)	one of the principal (and often, together with the subcostal, the strongest) longitudinal veins of an insect wing, situated between Sc and M
Radius (or radial) sector	posterior main branch of R, was perhaps an independent vein in ancestral pterygotes
Radula	LEP, in some Noct groups arrangements of small transverse crests/rugosities located laterally on the dorsal hypopharyngeal surface
reniform	kidney-shaped
retinaculum	COLL, modified limbs on III clasping furca in repose; LEP, device on forewing for clasping hindwing frenulum (wing-coupling); DIPT- Psych: strong seta of cercopod in the male genitalia
retrolateral	ARAN, there are two lateral sides, one is fore- (or pro-) lateral, another is back- (or retro-) lateral; all linyphiid palps are illustrated from the retrolateral side in this work
rhinaria	HEM-Aph, sensory organs on antennae of
rostrum	HEX, an anteroventrally produced head capsule; HEM, the sucking tube comprising mandibular and maxillary stylets enclosed in a labial sheath; ARAN, an outgrowth of clypeus or ocular area directed forward
Rs	Radius sector
saddle	DIPT-Cul, dorsal sclerotization on terminal abdominal segment in larvae
saprophagous	living from decomposing organic matter
Sc, sc	subcosta
scale	a broad, flattened and usually un-innervated seta
scape	HEX = scapus; ARAN, an unpaired finger, tongue, or lip like outgrowth of the upper part of epigyne, free on the other side
scapus	HEX, the basal segment of the antenna
scutellar/s	setae/bristles on scutellum
scutellum	the hindmost part of a pterothoracic notum; sometimes (e.g., HEM-Cic &-Het, COL) used specifically for the ± triangular portion visible between the fore wing bases in the resting insect
scutum	INS, the part of a pterothoracic notum in front of the scutellum (and behind the prescutum, when the latter is demarcated); ARAN, a sclerotized plate on abdomen, rarely present in Greenland species
septum	ARAN, a longitudinal partition dividing epigynal fovea in 2 parts.
seta/-ae	hair formed by one ('trichogen') epidermal cell, and arising from a flexible socket formed by another ('tormogen') epidermal cell; often innervated, being associated with one (typical for mechanoreceptors) or more (typical for e.g., chemoreceptors) sensory neuron(s)
setose	covered with setae
setosity	seta vestiture
setula	small seta
setulose	covered with small setae
siphon	a tube; DIPT, respiratory tube in aquatic larvae
siphuncul|us/-i	HEM-Aph, paired dorsal tubular outgrowths on an abdominal segment

speculum	HYM, non-microtrichiated field in wing
spermatheca	sac-like, sometimes sclerotized, organ in female genital apparatus serving storage of sperm after copulation
spine	strong macroseta or pointed cuticular outgrowth; ARAN, also sclerotized plate on abdomen (present in 2 Greenland species)
spinula	a small spine
spiracle	opening of tracheal system, often equipped with closing and/or filtering devices
spur	sizable (passively movable) spine-like outgrowth on a leg; it is multicellular and may itself bear hairs, scales and sensilla
squama	a scale; DIPT-Chir, proximal wing lobe
stemm|a/-ata	simple lateral eye type of endopterygote larva
stemmaticum	triangular area on head capsule bearing ocelli
sternal spatula	DIPT-Cec, a ventral sclerite on the thorax of mature (rarely also 2nd instar) larvae
sternite	sternum (in many authors' usage)
stern|um/-a	sclerotized ventral plate of a segment
stigm|a/-ata	spiracle; HEX, sometimes used for wing markings (in wing membrane, or in scale vesititure)
stigmal vein	in taxa with rediced venation a short vein extending posteriorly from marginal vein
stip|es/-ites	subbasal piece of maxilla; it bears the galea and lacinia on the inner side and the maxillary palp on the outer apex
stridualting field	ARAN, lateral side of chelicere with numerous fine parallel ridges
style	stylus
styl|us/-i	distal segment on two-segmented gonopod in pterygote insects
Subcosta	principal longitudinal vein in insect wing, located just behind C; often (together with R), the strongest
submentum	HEX, the proximal surface sclerotization of a subdivided postlabium
sulc|us/-i	strengthening ridge on a sclerite formed by cuticular invagination; may or may not be accompanied by an external groove
superior appendage	TRIC, passively movable paired appendages on abdominal segment X in males
suture	ambiguous term used for a variety of linear/curvilinear formations on cuticle, often = sulcus; alternatively used for unsclerotized lines, particularly ecdysial lines
tagm|a/-ata	functional unit of an arthropod body formed by a series of adjacent body segments
tars|us/-i	the distal leg segment; usually subdivided into socalled tarsal segments or tarsomeres
tegular apophysis	ARAN, an outgrowth of tegulum (part of bulbus)
tegulum	ARAN, part of bulbus, heavily sclerotized
thoracic lobe/ outgrowth	ARAN, an outgrowth of carapace emerging behind the ocular area, and carrying no eyes
telomere	SIPH, movable process on clasper
telson	posteriormost segment of an arthropod, often indistinguishable in post-embryonic stages
tenent hairs	COLL, apically widened setae near apex of tibiotarsus
tergite	tergum (in some authors' usage)
terg|um/-a	sclerotized dorsal plate of a segment
thalassophile	preferring habitats adjacent to the sea
thyridia	HYM-Ichn, sculptured surface areas in impressions (= gastrocoeli) anterolaterlly on metasomal tergum 2; TRICH, wing cell
tibi|a/-ae	leg segment between femur and tarsus
tibial apophysis	ARAN, an outgrowth of male palpal tibia; it can be ventral, lateral or dorsal
tomentose	pubescense of short hairs
tongue	mouthparts modified for suction of fluid nutrients, e.g., the lepidopteran proboscis
trichobothri|um/-a	ARACH, a long fine hair, rising almost vertically from a roundish socket
trochanter	small leg segment between coxa and femur

unguiculus	COLL, lower part/appendage of claw
unguis	claw
vaginal	structures associated with female genital opening
valve	term used for a variety of genital appendages, such as ovipositor components in HEM and HYM, or the male gonopod in LEP
vein	± linear/curvilinear, usually hollow, sclerotized strengthening device in insect wing; its lumen contains blood, and often a nerve and a trachea
venation	configuration of wing veins
venter	ventral side of a tagma or segment
ventral tube	COLL, a mid-ventral outgrowth on first abdominal segment
vermiform	'worm-like', simple and elongate
vertex	upper/dorsal part of head capsule
verticils	slender, strong bristles on basal part of a flagellomere
vibrissae/ vibrissal setae	DIPT, setae located on vibrissal angle
vibrissal angle	DIPT, lower anterolateral corners of head capsule
virga	DIPT-Chir, internal spine group in male phallus
volsell|a/-ae	median appendage of gonopod, used particularly in HYM and DIPT-Chir
Ø	ARAN, diameter

Taxonomic Index

Valid names of families, genera and species occurring in Greenland are included with the page of their main treatment. Names of subspecies are indexed only when the taxon is treated in a separate entry.